TEACHER'S EDITION

HOLT

MIDDLE SCHOOL Math

Course 1

INDIANA EDITION

Jennie M. Bennett
David J. Chard
Audrey Jackson
Jim Milgram
Janet K. Scheer
Bert K. Waits

HOLT, RINEHART AND WINSTON
A Harcourt Education Company
Orlando • Austin • New York • San Diego • Toronto • London

INDIANA EDITION

INDIANA

Explanation of Correlation

*The following document is a correlation of **HOLT MIDDLE SCHOOL MATH, Course 1** to the Indiana Academic Standards Mathematics, Grade 6. The format for this correlation follows the same basic format established by the standards, modified to accommodate the addition of page references. The correlation provides a cross-reference between the skills in the standards and representative page numbers where those skills are taught or assessed.*

The references contained in this correlation reflect Holt, Rinehart and Winston's interpretation of the Math objectives outlined in the Indiana curriculum.

KEY TO REFERENCES

Prefix	Explanation
SE	*Student's Edition*

Printed in the United States of America

ISBN 0-03-070989-X

123 048 05 04 03

INDIANA EDITION

INDIANA

Detailed Correlation of Holt Middle School Math, *Course 1* to the *Indiana Academic Standards Mathematics, Grade 6*

Standard 1

Number Sense

Students compare and order positive and negative integers, decimals, fractions, and mixed numbers. They find multiples and factors.

6.1.1	Understand and apply the basic concept of negative numbers (e.g., on a number line, in counting, in temperature, in "owing").	**SE**	450–453
6.1.2	Interpret the absolute value of a number as the distance from zero on a number line, and find the absolute value of real numbers.	**SE**	450–453
6.1.3	Compare and represent on a number line positive and negative integers, fractions, decimals (to hundredths), and mixed numbers.	**SE**	5–6, 93, 94, 167–170, 178–180, 450–452, 454–457
6.1.4	Convert between any two representations of numbers (fractions, decimals, and percents) without the use of a calculator.	**SE**	167–170, 172–173, 182–183, 184–185, 418–421, 422–425
6.1.5	Recognize decimal equivalents for commonly used fractions without the use of a calculator.	**SE**	167–169, 422–425
6.1.6	Use models to represent ratios.	**SE**	396–397, 398
6.1.7	Find the least common multiple and the greatest common factor of whole numbers. Use them to solve problems with fractions (e.g., to find a common denominator to add two fractions or to find the reduced form for a fraction).	**SE**	156–157, 158–159, 160–163, 201, 212–215, 216–219, 232–235, 242–245, 246–249

INDIANA EDITION

INDIANA

Standard 2

Computation

Students solve problems involving addition, subtraction, multiplication, and division of integers. They solve problems involving fractions, decimals, ratios, proportions, and percentages.

6.2.1	Add and subtract positive and negative integers.	**SE**	465–468, 469–472
6.2.2	Multiply and divide positive and negative integers.	**SE**	473–475, 476–479
6.2.3	Multiply and divide decimals.	**SE**	106, 108–109, 110–111, 114–115, 116–117, 118, 119, 120, 121, 122, 123, 124–126, 127–130, 427
6.2.4	Explain how to multiply and divide positive fractions and perform the calculations.	**SE**	192–193, 194–195, 210–215, 216–219, 220–225, 226
6.2.5	Solve problems involving addition, subtraction, multiplication, and division of positive fractions and explain why a particular operation was used for a given situation.	**SE**	188–189, 190–191, 226–229, 252–253, 256–261
6.2.6	Interpret and use ratios to show the relative sizes of two quantities. Use the notations: a/b, a to b, a:b.	**SE**	392–395
6.2.7	Understand proportions and use them to solve problems.	**SE**	396–401, 402–404, 405–408, 409–411, 412–415, 426–427
6.2.8	Calculate given percentages of quantities and solve problems involving discounts at sales, interest earned, and tips.	**SE**	424, 426–427, 428–429, 430–431, 432–439
6.2.9	Use estimation to decide whether answers are reasonable in decimal problems.	**SE**	96, 97, 98, 99, 102–105, 121, 122
6.2.10	Use mental arithmetic to add or subtract simple fractions and decimals.	**SE**	102–105, 236–239

Standard 3

Algebra and Functions

Students write verbal expressions and sentences as algebraic expressions and equations. They evaluate algebraic expressions, solve simple linear equations, and graph and interpret their results. They investigate geometric relations.

6.3.1	Write and solve one-step linear equations and inequalities in one variable and check the answers.	**SE**	49–50, 51–53, 58–59, 61, 62–65, 66–68, 69–72, 73–77, 92, 134–137, 256–259, 482–485
6.3.2	Write and use formulas with up to three variables to solve problems.	**SE**	134–137, 500–503, 504–507, 508–510, 516–519, 530–533, 534–537, 538–541, 679
6.3.3	Interpret and evaluate mathematical expressions that use grouping symbols such as parentheses.	**SE**	20–23
6.3.4	Use parentheses to indicate which operation to perform first when writing expressions containing more than two terms and different operations.	**SE**	24–27
6.3.5	Use variables in expressions describing geometric quantities.	**SE**	500–503, 504–507, 508–510, 514–519, 530–533, 534–537, 538–541
6.3.6	Apply the correct order of operations and the properties of real numbers (e.g., identity, inverse, commutative, associative, and distributive properties) to evaluate numerical expressions. Justify each step in the process.	**SE**	20–23, 24–27, 48–49, 50–51
6.3.7	Identify and graph ordered pairs in the four quadrants of the coordinate plane.	**SE**	294–295, 458–461
6.3.8	Solve problems involving linear functions with integer values. Write the equation and graph the resulting ordered pairs of integers on a grid.	**SE**	602–603, 604, 605, 606, 607
6.3.9	Investigate how a change in one variable relates to a change in a second variable.	**SE**	598–601

INDIANA EDITION

INDIANA

Standard 4

Geometry

Students identify, describe, and classify the properties of plane and solid geometric shapes and the relationships between them.

6.4.1	Identify and draw vertical, adjacent, complementary, and supplementary angles and describe these angle relationships.	**SE** 322–323, 332–335, 336–337, 338–339, 378–379
6.4.2	Use the properties of complementary, supplementary, and vertical angles to solve problems involving an unknown angle. Justify solutions.	**SE** 332–335
6.4.3	Draw quadrilaterals and triangles from given information about them.	**SE** 344–347, 348–351
6.4.4	Understand that the sum of the interior angles of any triangle is 180° and that the sum of the interior angles of any quadrilateral is 360°. Use this information to solve problems.	**SE** 344–346, 352–355, 353, 354
6.4.5	Identify and draw two-dimensional shapes that are similar.	**SE** 362–364
6.4.6	Draw the translation (slide) and reflection (flip) of shapes.	**SE** 365–368, 610–611, 612, 613–615, 616–617, 618–619, 620–621, 622–623
6.4.7	Visualize and draw two-dimensional views of three-dimensional objects made from rectangular solids.	**SE** 524–525, 527, 528–529

Standard 5

Measurement

Students deepen their understanding of the measurement of plane and solid shapes and use this understanding to solve problems. They calculate with temperature and money, and choose appropriate units of measure in other areas.

6.5.1	Select and apply appropriate standard units and tools to measure length, area, volume, weight, time, temperature, and the size of angles.	**SE** 106, 107, 108, 109, 110, 111, 326–329, 330, 331, 500–503, 504–507, 508–510, 530–533, 534–537, 538–541, 542, 543, 673, 674
6.5.2	Understand and use larger units for measuring length by comparing miles to yards and kilometers to meters.	**SE** 675
6.5.3	Understand and use larger units for measuring area by comparing acres and square miles to square yards and square kilometers to square meters.	**SE** 677
6.5.4	Understand the concept of the constant p as the ratio of the circumference to the diameter of a circle. Develop and use the formulas for the circumference and area of a circle.	**SE** 514–519
6.5.5	Know common estimates of p (3.14, 22/7) and use these values to estimate and calculate the circumference and the area of circles. Compare with actual measurements.	**SE** 514–519
6.5.6	Understand the concept of significant figures and round answers to an appropriate number of significant figures.	**SE** 138–139
6.5.7	Construct a cube and rectangular box from two-dimensional patterns and use these patterns to compute the surface area of these objects.	**SE** 528–529, 530–533
6.5.8	Use strategies to find the surface area and volume of right prisms and cylinders using appropriate units.	**SE** 530–533, 534–537, 538–541
6.5.9	Use a formula to convert temperatures between Celsius and Fahrenheit.	**SE** 675, 679
6.5.10	Add, subtract, multiply, and divide with money in decimal notation.	**SE** 105, 113, 125, 126, 129, 134–137

INDIANA EDITION

INDIANA

Standard 6

Data Analysis and Probability

Students compute and analyze statistical measures for data sets. They determine theoretical and experimental probabilities and use them to make predictions about events.

6.6.1	Organize and display single-variable data in appropriate graphs and stem-and-leaf plots, and explain which types of graphs are appropriate for various data sets.	**SE**	272–273, 284–289, 305–309
6.6.2	Make frequency tables for numerical data, grouping the data in different ways to investigate how different groupings describe the data. Understand and find relative and cumulative frequency for a data set. Use histograms of the data and of the relative	**SE**	290–293, 297–300
6.6.3	Compare the mean, median, and mode for a set of data and explain which measure is most appropriate in a given context.	**SE**	275–277, 278–281, 310
6.6.4	Show all possible outcomes for compound . events in an organized way and find the theoretical probability of each outcome	**SE**	570–571, 572–573, 574–579
6.6.5	Use data to estimate the probability of future events.	**SE**	554–555, 556–557, 558–559, 560–561, 562–563, 564–565, 566–567, 580–581, 582–583, 587
6.6.6	Understand and represent probabilities as ratios, measures of relative frequency, decimals between 0 and 1, and percentages between 0 and 100 and verify that the probabilities computed are reasonable.	**SE**	554–557

INDIANA EDITION

INDIANA

Standard 7

Problem Solving

Students make decisions about how to approach problems and communicate their ideas.

6.7.1	Analyze problems by identifying relationships, telling relevant from irrelevant information, identifying missing information, sequencing and prioritizing information, and observing patterns.	**SE**	31–33, 34–35, 57, 152–153, 154–155, 283, 301–302, 303–304, 356–357, 358–359, 369, 371, 486–487, 570–571, 572–573, 581, 583
6.7.2	Make and justify mathematical conjectures based on a general description of a mathematical question or problem.	**SE**	131–132
6.7.3	Decide when and how to break a problem into simpler parts.	**SE**	508–510
6.7.4	Apply strategies and results from simpler problems to solve more complex problems.	**SE**	xxvi, 12–13, 14–15, 508–510
6.7.5	Express solutions clearly and logically by using the appropriate mathematical terms and notation. Support solutions with evidence in both verbal and symbolic work.	**SE**	7, 11, 15, 23, 27, 30, 33, 51, 55, 61, 65, 68, 72, 75, 95, 99, 105, 109, 117, 126, 130, 133, 137, 155, 159, 163, 170, 175, 181, 185, 191, 195, 197, 215, 219, 225, 229, 235, 239, 245, 249, 255, 259, 274, 281, 287, 293, 296, 300, 304, 325, 329, 335, 339, 347, 351, 355, 359, 364, 368, 372, 375, 395, 401, 404, 408, 411, 415, 425, 429, 435, 437, 453, 457, 461, 468, 472, 475, 479, 485, 486–487, 503, 507, 510, 513, 519, 527, 537, 541, 557, 561, 567, 573, 577, 583, 601, 607, 612, 615, 619, 623
6.7.6	Recognize the relative advantages of exact and approximate solutions to problems and give answers to a specified degree of accuracy.	**SE**	417
6.7.7	Select and apply appropriate methods for estimating results of rational–number computations.	**SE**	8–9, 10–11, 236–239
6.7.8	Use graphing to estimate solutions and check the estimates with analytic approaches.	**SE**	604–607
6.7.9	Make precise calculations and check the validity of the results in the context of the problem.	**SE**	569, 609
6.7.10	Decide whether a solution is reasonable in the context of the original situation.	**SE**	569, 609
6.7.11	Note the method of finding the solution and show a conceptual understanding of the method by solving similar problems.	**SE**	28–29, 30–31, 369–370, 371, 373–374, 508–510

Assessment and Higher-Order Thinking Skills

What Are Higher-Order Thinking Skills?

Higher-order thinking skills, sometimes called critical thinking skills, are not a new phenomenon on the education scene. In 1956, Benjamin Bloom published a book that listed a taxonomy of education objectives in the form of a pyramid similar to the one in the following illustration:

Bloom's Taxonomy of Educational Objectives

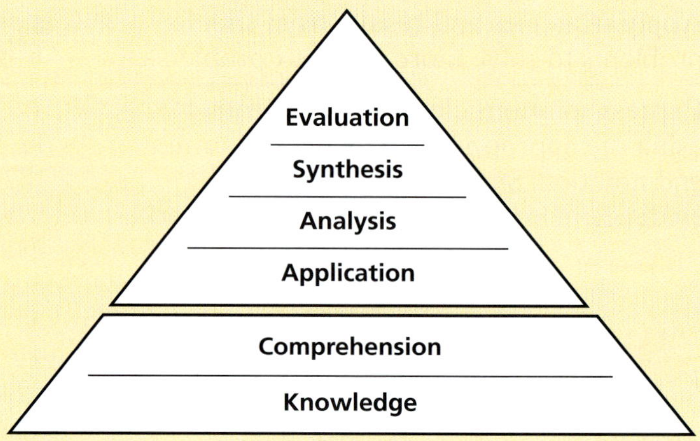

- **Knowledge** is the simplest level of education objectives and is not considered a higher-order thinking skill. It requires the learner to remember information without having to fully understand it. Tasks that students perform to demonstrate knowledge are recalling, identifying, recognizing, citing, labeling, listing, reciting, and stating.

  EXAMPLES

  1. *What is the formula for the area of a trapezoid?*
  2. *What quadrant is the point (2, −6) located in?*
  3. *What is the reciprocal of* $\frac{2}{3}$*?*

INDIANA EDITION

INDIANA

- **Comprehension** is not considered a higher-order thinking skill either. Learners demonstrate comprehension when they paraphrase, describe, summarize, illustrate, restate, or translate. Information isn't useful unless it's understood. Students can show they've understood by restating the information in their own words or by giving an example of the concept.

  EXAMPLES

  1. *Explain the difference between the points (4, 5) and (5, 4).*
  2. *Interpret the information in the graph below.*
  3. *Give an example of an irrational number.*

Many teachers tend to focus the most on knowledge and comprehension—and the tasks performed at these levels are important because they provide a solid foundation for the more complex tasks at the higher levels of Bloom's pyramid.

However, offering students the opportunity to perform at still higher cognitive levels provides them with more meaningful contexts in which to use the information and skills they have acquired, thus allowing them to more easily retain what they have learned.

When teachers incorporate **application, analysis, synthesis,** and **evaluation** as objectives, they allow students to utilize **higher-order thinking skills.**

- **Application** involves solving, transforming, determining, demonstrating, and preparing. Information becomes useful when students apply it to new situations—predicting outcomes, estimating answers—this is application.

  EXAMPLES

  1. *Organize the forms of pollution from most damaging to least damaging.*
  2. *Using the scale of 1 inch equals 200 miles, determine the point-to-point distance between Boston and Atlanta.*
  3. *Put the information below into a bar graph.*

INDIANA EDITION

INDIANA

- **Analysis** includes classifying, comparing, making associations, verifying, seeing cause-and-effect relationships, and determining sequences, patterns, and consequences. You can think of analysis as taking something apart in order to better understand it. Students must be able to think in categories in order to analyze.

  EXAMPLES

  1. *What math skills do you use when reading a circle graph?*
  2. *Use the function table to write a rule for* y *in terms of* x.
  3. *How can you use the LCM of 3 and 5 to find the sum of* $\frac{1}{3}$ *and* $\frac{1}{5}$*?*

- **Synthesis** requires generalizing, predicting, imagining, creating, making inferences, hypothesizing, making decisions, and drawing conclusions. Students create something which is new to them when they use synthesis. It's important to remember, though, that students can't create until they have the skills and information they have received in the comprehension through analysis levels.

  EXAMPLES

  1. *Make a scale drawing of your classroom.*
  2. *Write a word problem that can be represented by the equation* $3 + y = 5$*.*
  3. *Poll your classmates about their favorite breakfast food and display your results in an appropriate graph.*

- **Evaluation** involves assessing, persuading, determining value, judging, validating, and solving problems. Evaluation is based on all the other levels. When students evaluate, they make judgments, but not judgments based on personal taste. These judgments must be based on criteria. It is important for students to evaluate because they learn to consider different points of view and to know how to validate their judgments.

  EXAMPLES

  1. *Which of the following describes the correct way to round* $-3\frac{7}{8}$*?*
  2. *Based on the ratios of protein to serving size and fat to serving size, which muffin do you think is healthier? Explain.*
  3. *Do you think the statistics given in the article are accurate? Why or why not?*

INDIANA EDITION

INDIANA

Why is it Important for Students to Work with Higher-Order Thinking Skills?

For one thing, if students can determine the levels of questions that will appear on their tests, they will be able to study using appropriate strategies. Bloom's leveling of questions provides a useful structure in which to categorize test questions, since tests will characteristically ask questions within particular levels.

Also, thinking is a skill that can be taught. When you have students practice answering questions at all the levels of Bloom's taxonomy, you are helping to scaffold their learning. Information just becomes trivia unless that information is understood well enough to build more complicated concepts or generalizations. When students can comprehend—not just recall—the information, it becomes useful for future problem solving or creative thought. Think of information as a building material—like a board. It could be used to build something, but it is just useless litter unless you understand how to make use of it.

Below are some question stems you—or your students—could use to create questions for each of the levels of higher-order thinking:

Application

1. Make a diagram to show ________.

2. Use (a formula, manipulatives, mental math, a problem solving strategy, etc.) to find _________.

3. (Find, determine, calculate, compute, etc.) ________.

4. Explain how the (principle, theorem, concept) is evident in ________.

5. In what way is _________ a ______?

Analysis

1. Which (strategies, operations, etc.) would you use to solve this problem?

2. Find a pattern in ____________.

3. What other (properties, rules, definitions) are similar? Explain.

4. Compare and contrast ____________.

5. How does the value of ________ affect the value of ________?

INDIANA

INDIANA EDITION

Synthesis

1. Write a problem that can be solved by __________ .
2. Use information in (your science book, a newspaper article, etc.) to write a problem.
3. Create a new way to classify ______.
4. Design your own _______to show ______.
5. Create a new way to __________.

Evaluation

1. Is (an answer, an estimate, etc.) reasonable? Explain.
2. Do you have enough information to solve this problem?
3. Which ________ best represents _________?
4. Which solution method (is most efficient, is most accurate, gives the most information, etc.)?
5. What is the important information in this problem?

Short Response

Short-response questions are questions that require a short written answer as well as the solution to a mathematical problem. You will probably be given blank space in your answer booklet to answer these types of questions. Do not write outside of this space. You should plan to spend about five minutes answering each short-response question.

Read the question carefully, and use the information from the problem in your solution. Be sure that you answer the question completely. Show your work. Write your answer in the answer blank, if one is provided.

Some short-response questions may ask you to make a graph. A blank grid may be provided for you, or you may have to draw your own. You may lose points if you do not title your graph (when appropriate), label the axes, and use appropriate scales, so be sure you do these things.

Short-response questions are scored differently than other types of questions. They are usually given 0, 1, or 2 points. If an answer shows no understanding of the mathematics involved in the problem, or if a question is left unanswered, that response will receive 0 points. If a response shows an understanding of the procedure, but the final answer is incorrect, then that response will receive 1 point. A response will also receive 1 point if the answer is correct but does not clearly show understanding of the procedure. A response will receive 2 points if the answer is correct and demonstrates a complete understanding of the mathematics involved.

Because of this scoring system, you should always try to answer short-response questions even if you are unsure of the correct answer. If part of your answer is correct, you could receive partial credit.

In short-response questions, you will usually not be penalized for spelling or grammatical mistakes or for leaving off the units in your answer.

INDIANA

INDIANA EDITION

Sample Question

For triangle ABC, $m\angle A = 48°$ and $m\angle B = 112°$. Show the steps necessary to find the measure of angle C.

Sample Responses

All triangles have 180°. $\begin{array}{r} 112 \\ +48 \\ \hline 160 \end{array}$ So C must be $180 - 160$. $\begin{array}{r} 180 \\ -160 \\ \hline 20 \end{array}$ $C = 20$	*This response will receive 2 points. The student demonstrates that he or she understands how to find m∠C, and the final answer is correct.*
$\begin{array}{r} 112 \\ +48 \\ \hline 150 \end{array}$ $\begin{array}{r} 180 \\ -150 \\ \hline 30 \end{array}$ C is 30°.	*This response will receive 1 point. The student shows an understanding of how to find m∠C, but he or she made an error in calculation, making the final answer incorrect.*
20°	*This response will also receive 1 point. The final answer is correct, but because the student showed no work, it is not clear that he or she understands the procedure or the reasoning.*
$A = 48$ $B = 112$	*This response will receive 0 points. There is no indication that the student understands any aspect of the problem.*

TEACHER'S EDITION

HOLT

MIDDLE SCHOOL Math

Course 1

INDIANA EDITION

Jennie M. Bennett
David J. Chard
Audrey Jackson
Jim Milgram
Janet K. Scheer
Bert K. Waits

HOLT, RINEHART AND WINSTON
A Harcourt Education Company
Austin • Orlando • Chicago • New York • Toronto • London • San Diego

STAFF CREDITS

Editorial

Lila Nissen, *Vice President*
Robin Blakely, *Associate Director*
Joseph Achacoso, *Assistant Managing Editor*
Threasa Boyar, *Editor*

Student Edition
April Warn, *Senior Editor*
Katie Seawell, *Editor*

Teacher's Edition
Kelli Flanagan, *Senior Editor*
Ronald Fowler, *Associate Editor*
Monica Robinson, *Associate Editor*

Ancillaries
Mary Fraser, *Executive Editor*
Higinio Dominguez, *Associate Editor*

Technology Resources
John Kerwin, *Executive Editor*
Robyn Setzen, *Senior Editor*
Patricia Platt, *Senior Technology Editor*
Manda Reid, *Technology Editor*

Copyediting
Denise Nowotny, *Copyediting Supervisor*
Patrick Ricci, *Copyeditor*

Support
Jill Lawson, *Senior Administrative Assistant*
Benny Carmona, III, *Editorial Coordinator*

Design

Book Design
Marc Cooper, *Design Director*
Tim Hovde, *Senior Designer*
Lisa Woods, *Designer*
Teresa Carrera-Paprota, *Designer*
Bruce Albrecht, *Design Associate*
Ruth Limon, *Design Associate*
Holly Whittaker, *Senior Traffic Coordinator*

Teacher's Edition
José Garza, *Designer*
Charlie Taliaferro, *Design Associate*

Cover Design
Pronk & Associates

Image Acquisition
Curtis Riker, *Director*
Tim Taylor, *Photo Research Supervisor*
Stephanie Friedman, *Photo Researcher*
Elaine Tate, *Art Buyer Supervisor*
Sam Dudgeon, *Senior Staff Photographer*
Victoria Smith, *Staff Photographer*
Lauren Eischen, *Photo Specialist*

New Media Design
Ed Blake, *Design Director*

Media Design
Dick Metzger, *Design Director*
Chris Smith, *Senior Designer*

Graphic Services
Kristen Darby, *Director*
Eric Rupprath, *Ancillary Designer*
Linda Wilbourn, *Image Designer*

Prepress and Manufacturing

Mimi Stockdell, *Senior Production Manager*
Susan Mussey, *Production Supervisor*
Rose Degollado, *Senior Production Coordinator*
Sara Downs, *Production Coordinator*
Jevara Jackson, *Senior Manufacturing Coordinator*
Ivania Lee, *Inventory Analyst*
Wilonda Ieans, *Manufacturing Coordinator*

Printed in the United States of America

ISBN 0-03-070989-X

1 2 3 4 5 6 7 8 9 048 10 09 08 07 06 05 04 03

Holt Middle School Math Indiana Edition Course 1

Teacher's Edition Contents

(peony), Don Voelker/Painet Inc.
(stamp), United States Postal Service
(cardinal), Kenneth Roberson/Index Stock Imagery/PictureQuest

Welcome to Holt Middle School Math Indiana Edition!

Have you ever wondered if anyone really uses the math that you learn in school? As you work through *Holt Middle School Math,* you will find that they certainly do. Not only that, but people are using math almost right in your backyard.

We've found the places where math is being used or is visible in Indiana. Residents of Indiana use math to solve problems and answer questions every day. You can do the same!

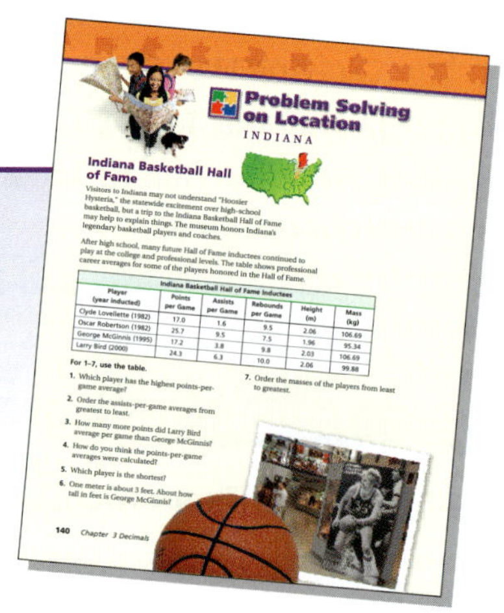

In Course 1, you will visit the **Indiana Basketball Hall of Fame**, in New Castle, Indiana. This museum honors Indiana's legendary basketball players and coaches. You will analyze data about some of the professional players who have been inducted into the Hall of Fame. You will use decimals to describe their points per game, rebounds per game, and other statistics.

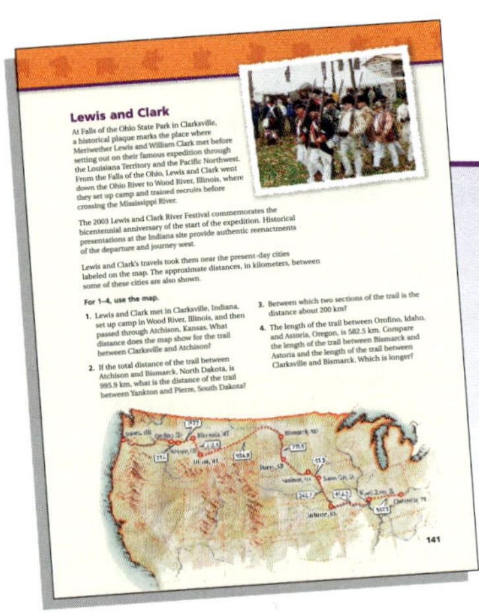

Then you will travel to **Falls of the Ohio State Park**, in Clarksville, Indiana, where a historical plaque marks the place where Meriwether Lewis and William Clark met before setting out on their famous expedition. From the Falls of the Ohio, Lewis and Clark traveled to Wood River, Illinois, to set up camp and train recruits before crossing the Mississippi River. Calculate the distances they traveled along the rest of their great journey west.

In Course 2, take a ride along the **Hoosier Hills Bicycle Route**. The route is about 27 miles long and runs from Liberty Park to Versailles State Park. Both motor vehicles and bicycles share this route. What are the chances that the weather will affect your ride? Use probability concepts to answer this and other questions.

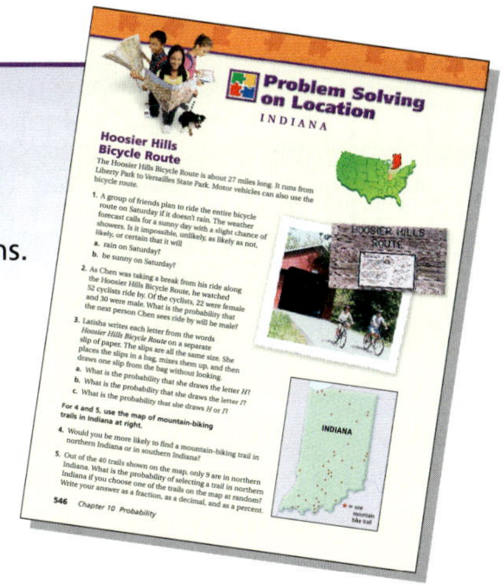
Problem Solving on Location
INDIANA
Hoosier Hills Bicycle Route

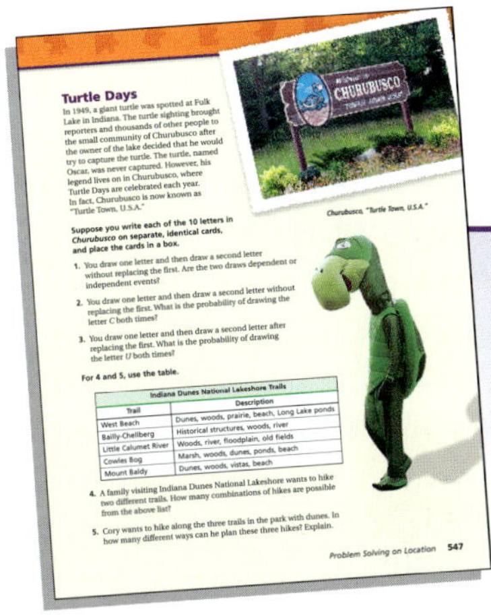
Turtle Days

Your next stop? **Indiana Dunes National Lakeshore**. You'll learn about the different trails there. You might want to hike all of the trails with dunes. How many different ways could you do this? Counting principles will help you to answer these types of questions.

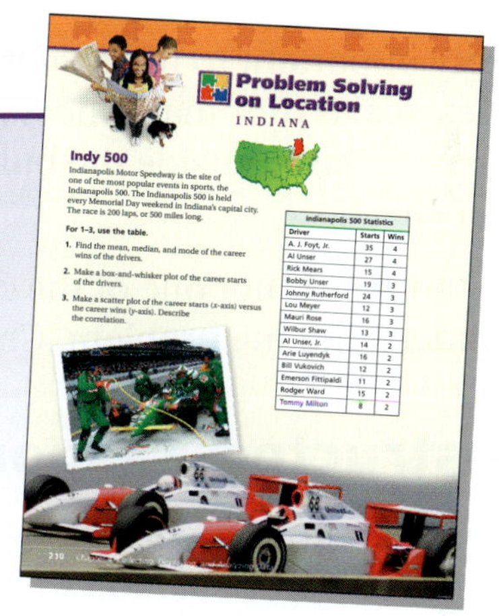
Problem Solving on Location
INDIANA
Indy 500

In Course 3, attend the most famous event in Indiana—the **Indianapolis 500!** Indianapolis Motor Speedway, in Indianapolis, is the site of this race, which takes place every Memorial Day weekend. The race is 200 laps, or 500 miles long. Analyze data about several Indy 500 drivers. How many times have they raced, and how many times have they won? You'll use measures of central tendency to describe this data, and then you'll make different data displays.

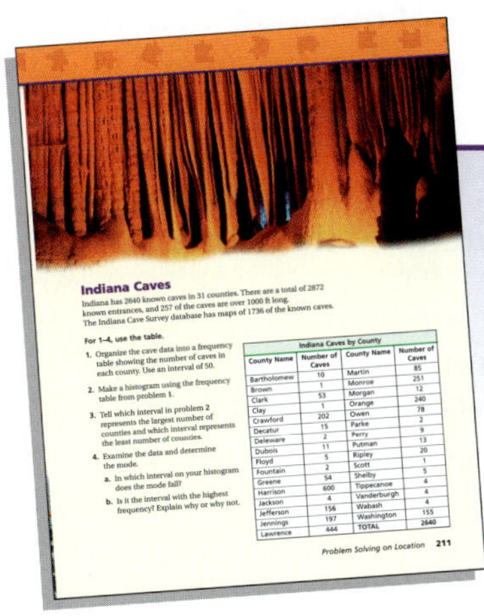
Indiana Caves

Finally, examine part of the **Indiana Cave Survey database.** This database has maps of 1,736 of the 2,640 known caves in Indiana. These caves are located in 31 counties. There are a total of 2,872 known entrances, and 257 of the caves are over 1,000 feet long. You'll analyze and display data about the caves and the counties in which they are located.

As you can see, there is plenty of math to be found in your home state. The next time you're out and about in Indiana, see what kinds of math you can find!

Indiana Statewide Testing for Educational Progress (ISTEP+) for the Middle Grades

The ISTEP+ is divided into different parts called sessions. Each session contains different types of questions. To do your best on the ISTEP+, you should learn how to answer each type of question.

The Grade 6 ISTEP+ mathematics tests includes two types of questions: multiple choice and constructed response.

Multiple Choice

Read each question carefully and work the problem. Choose your answer from among the answer choices given, and fill in the corresponding bubble on your answer sheet.

It is very important to fill in your answer sheet correctly. When shading in bubbles, make your marks heavy and dark. Fill in the bubbles completely, but do not shade outside the bubbles. Do not make any stray marks on your answer sheet.

If your answer is not one of the choices, read the question again. Be sure that you understand the problem. Check your work for possible errors.

Constructed Response

Constructed response questions are questions that require a short answer or essay as well as the solution to a mathematical problem. You will be given blank space in your answer booklet to answer these types of questions. There will be space for your calculations as well as a blank for your final answer. Do not write outside of this space.

Use the information from the problem in your solution. Be sure that you answer the question completely. Show your work. Be sure to write your answer in the answer blank.

Some constructed response questions may ask you to make a graph. A blank grid will be provided for you. You will lose points if you do not title your graph (when appropriate), label the axes, and use appropriate scales, so be sure you include all of these.

Constructed response questions can receive partial credit, so you should always try to answer these questions even if you are unsure of the correct answer. If part of your answer is correct, you will receive partial credit.

The Grade 8 ISTEP+ contains a third type of question—gridded response.

Gridded Response

Some questions require you to place your answer in a special grid on your answer sheet. This type of question is called "gridded response." Work the problem and find an answer. Then write your answer in the grid provided. There is often more than one correct way to write your answer in the response grid.

Sample Gridded-Response Question

1 A bowl of fruit contains three oranges, four apples, and three bananas. If Amy chooses one piece of fruit at random, what is the probability that she will choose an apple?

Sample Correct Answers

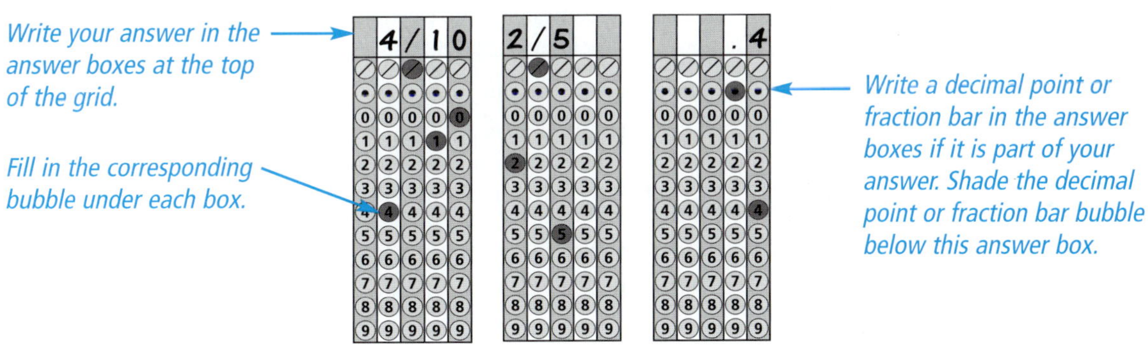

Notice that you may write your answers as either fractions or decimals. However, you may **not** write mixed numbers, such as $13\frac{1}{4}$, in a response grid. If you tried to fill in $13\frac{1}{4}$, it would be read as $\frac{131}{4}$ and would be counted as wrong. If your answer is a mixed number, write it as an improper fraction $\left(\frac{53}{4}\right)$ or as a decimal (13.25) before filling in your grid.

Practice, Practice, Practice

On the following pages, you will find a practice standardized test. This test has been designed to resemble the ISTEP+ and the types of questions it contains. Use this test to practice answering these questions, as well as to review some of the math that you will be tested on. The more comfortable and familiar you can become with the ISTEP+, the better your chances of success!

ISTEP+ Test Practice

Note: Test items are correlated to the Indiana Mathematics Standards for Grade 6. The practice test was designed to be representative of these standards, but it should not be considered a comprehensive assessment of their content.

Item Correlations and Text References

Session A

1. Standards 6.2.3, 6.5.10

See Lessons 3-3 and 3-6.

2. Standard 6.6.6

See Lessons 6-4 and 11-3.

3. Standard 6.2.7

See Lessons 8-1 and 8-2.

4. Standard 6.1.4

See Lesson 4-4.

ISTEP+ Test Practice

Session A

Directions: Choose the best response for each question.

1 Sam gathered 12 buckets of clams, Kristi gathered 9 buckets, and Mario gathered 6 buckets. They sold the clams for \$3.50 a bucket. How much money did they earn? **D**

A \$21.50 **C** \$42

B \$27 **D** \$94.50

2 The bar graph shows the number of crayons of each color in a bucket.

CRAYONS IN BUCKET

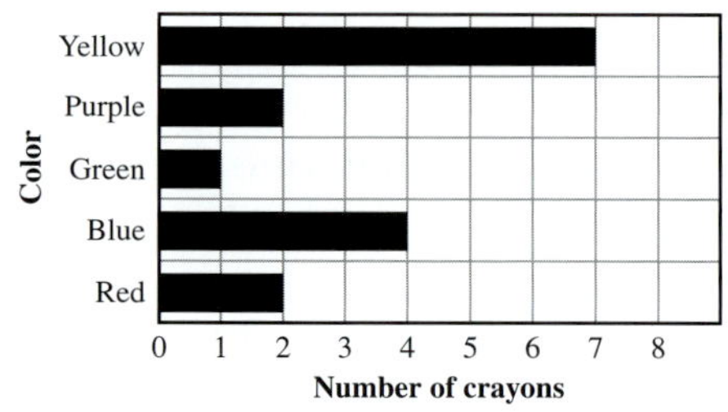

If you select a crayon at random, what is the probability that you will choose a purple crayon? **B**

A $\frac{1}{4}$ **C** $\frac{1}{16}$

B $\frac{1}{8}$ **D** $\frac{7}{8}$

3 Devon can type 40 words in one minute. At this rate, how long will it take her to type 900 words? **D**

A 20.5 minutes **C** 22 minutes

B 21 minutes **D** 22.5 minutes

4 Which decimal is equivalent to $\frac{3}{4}$? **A**

A 0.75 **C** 7.5

B 1.75 **D** 75.0

5 What is the greatest common factor of 46 and 69? **B**

A 3 **C** 46

B 23 **D** 138

6 Of the 100 members in the school marching band, 79 also play in the school concert band. Which of the following represents the ratio of students who play in both bands to students who play in the school marching band? **B**

A 100:79 **C** 100:21

B 79:100 **D** 21:100

7 What type of transformation is shown below? **B**

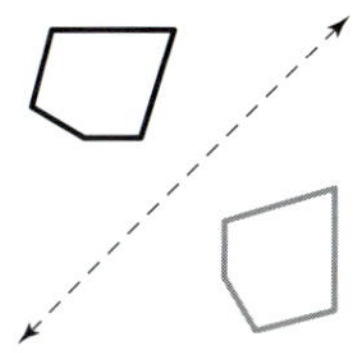

A rotation **C** translation

B reflection **D** dilation

8 What is the top view of this arrangement of cubes? **D**

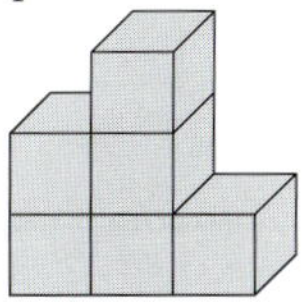

A  **C** 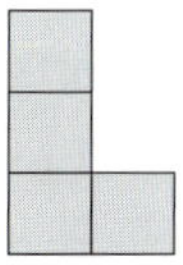

B  **D** 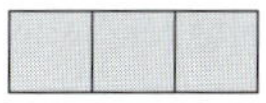

Go On

Item Correlations and Text References

5. Standard 6.1.7
See Lesson 4-3.

6. Standard 6.2.6
See Lesson 8-1.

7. Standard 6.4.6
See Lesson 7-10.

8. Standard 6.4.7
See Hands-On Lab 10B.

Item Correlations and Text References

9. Standard 6.3.1
See Lesson 2-2.

10. Standard 6.3.1
See Lessons 2-2 and 2-4.

11. Standard 6.2.8
See Lessons 8-9 and 8-10.

12. Standard 6.1.3
See Lesson 4-4.

13. Standards 6.2.3, 6.5.10
See Lessons 2-2, 3-3, and 3-6.

14. Standard 6.2.3
See Lesson 3-6.

ISTEP+ Test Practice

9 Becky has \$100 in her bank account. She withdraws d dollars, leaving a balance of \$63. Which equation describes this situation? **B**

A $d - 100 = 63$ **C** $d - 63 = 100$

B $100 - d = 63$ **D** $63 - d = 100$

10 A coat costs \$65 more than a dress. The price of the coat is \$113. How much does the dress cost? **B**

A \$24 **C** \$65

B \$48 **D** \$178

11 A video game is regularly priced at \$40. When Benny bought the game, it was on sale for 30% off the regular price. What was the cost of the discounted video game? **C**

A \$12 **C** \$28

B \$24 **D** \$36

12 Which number BEST represents the location of point C on the number line below? **A**

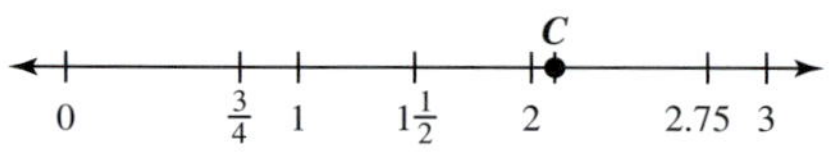

A 2.1 **C** 2.6

B $2\frac{1}{2}$ **D** $2\frac{4}{5}$

13 The sixth-grade class held a car wash. The students charged \$3.25 to wash a car and \$5.75 to wash a van. If they washed 42 cars and 13 vans, how much money did they raise? **D**

A \$74.75 **C** \$178.75

B \$136.50 **D** \$211.25

14 Keith has 3 bookcases that are each 1.6 meters wide. He wants to place them side by side along his bedroom wall. What is the width that the bookcases will measure when they are side by side? **B**

A 3.2 meters **C** 5.6 meters

B 4.8 meters **D** 6.4 meters

15 Two angles are complementary. What is the sum of their angle measures? **A**

A 90° **C** 180°

B 100° **D** 360°

16 Which number is less than $8\frac{3}{8}$? **A**

A 8.3 **C** 8.4

B $8\frac{5}{7}$ **D** $8\frac{2}{3}$

17 Which polygon appears to be similar to the one below? **C**

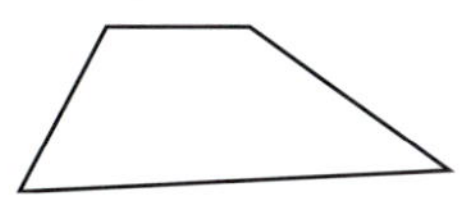

A

C

B

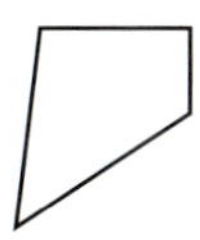

D

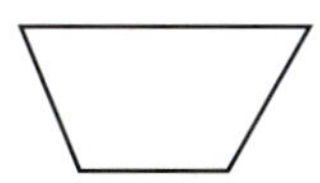

18 Your muscles make up about 40% of your overall mass. What is the muscle mass of a person whose total body mass is 60 kilograms? **B**

A 20 kilograms **C** 33 kilograms

B 24 kilograms **D** 100 kilograms

19 The mean amount of rainfall for three consecutive days was 2 inches. On the first two days, there were 5 inches of rain and 1 inch of rain. How much rain fell on the third day? **A**

A 0 inches **C** 4 inches

B 1 inch **D** 6 inches

Go On

Item Correlations and Text References

15. Standard 6.4.1

See Lesson 7-3.

16. Standard 6.1.3

See Lessons 4-6 and 4-7.

17. Standard 6.4.5

See Lesson 8-4.

18. Standard 6.2.8

See Lesson 8-9.

19. Standard 6.6.3

See Lesson 6-2.

Item Correlations and Text References

20. Standards 6.3.7, 6.3.8
See Lessons 6-6 and 9-3.

21. Standard 6.1.7
See Lesson 4-5.

22. Standard 6.3.3
See Lesson 1-4.

23. Standard 6.4.1
See Lesson 7-3.

24. Standard 6.2.6
See Lessons 8-1 and 8-2.

ISTEP+ Test Practice

20 Which point is on the line graphed below? **A**

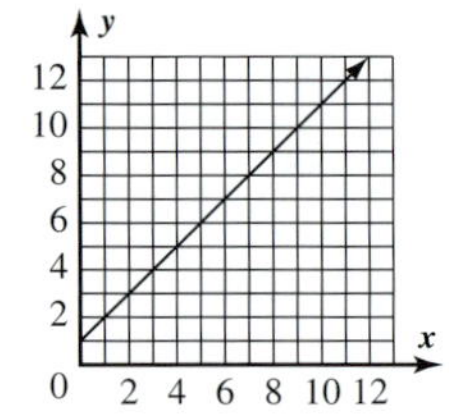

A (1, 2)
B (3, 2)
C (4, 4)
D (4, 3)

21 Which fraction is equivalent to $\frac{20}{24}$? **B**

A $\frac{6}{5}$
B $\frac{5}{6}$
C $\frac{10}{11}$
D $\frac{3}{5}$

22 What is the value of $(3 + 5) \times 4 - 4$? **A**

A 0
B 12
C 19
D 28

23 Which angle could be supplementary to an angle that measures 142°? **B**

A
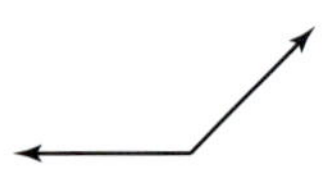

C

B
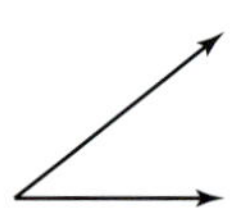

D
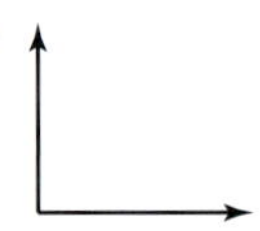

24 In 1903, Wilbur Wright flew the *Kitty Hawk* about 826 feet at an average speed of 14 feet per second. How long did the flight last? **C**

A 4 seconds
B 14 seconds
C 59 seconds
D 77 seconds

25 What is the measure of angle A? **C**

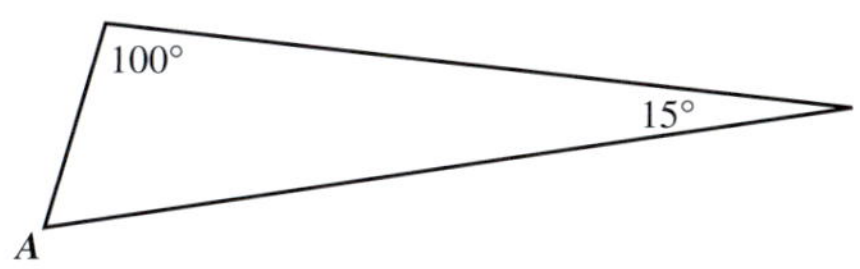

A 180° **C** 65°

B 115° **D** 15°

26 Which point is in Quadrant IV? **B**

A (1, 3) **C** ($^{-}6$, $^{-}6$)

B (2, $^{-}5$) **D** ($^{-}3$, 4)

27 What is the next figure in this pattern? **A**

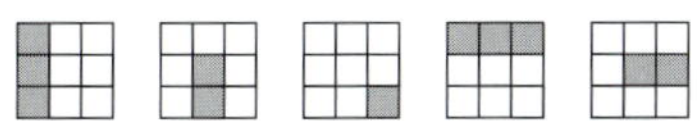

A 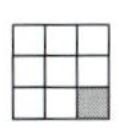

C 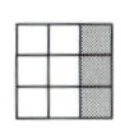

B 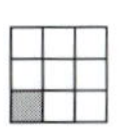

D 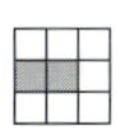

28 Simon bought b bags of ice that cost \$0.99 each. The total cost can be represented by the expression $0.99b$. What is the total cost if $b = 4$? **A**

A \$3.96 **C** \$4.95

B \$4.04 **D** \$5.94

29 For which value of x will the expression $^{-}3 \cdot 5 \cdot x$ be positive? **D**

A $x = 1$ **C** $x = \frac{1}{2}$

B $x = 0$ **D** $x = {}^{-}6$

30 What is the absolute value of $^{-}29$? **C**

A $^{-}29$ **C** 29

B 0 **D** 58

Go On

Item Correlations and Text References

25. Standard 6.4.4
See Lesson 7-5.

26. Standard 6.3.7
See Lesson 9-3.

27. Standard 6.7.1
See Lesson 7-8.

28. Standards 6.2.3, 6.3.1, 6.5.10
See Lessons 2-1 and 3-6.

29. Standard 6.2.2
See Lesson 9-6.

30. Standard 6.1.2
See Lesson 9-1.

Item Correlations and Text References

31. Standard 6.6.6
See Lesson 11-1.

32. Standard 6.3.7
See Lessons 6-6 and 9-3.

33. Standard 6.4.1
See Lesson 7-3.

ISTEP+ Test Practice

31 Sari tossed a coin ten times and recorded her results in this table.

COIN-TOSSING EXPERIMENT

Heads	Tails
𝍸 II	III

What is Sari's experimental probability of tossing heads? **C**

A $\frac{3}{10}$

B $\frac{1}{2}$

C $\frac{7}{10}$

D $\frac{10}{10}$

32 Which coordinates BEST represent the location of point *P*? **D**

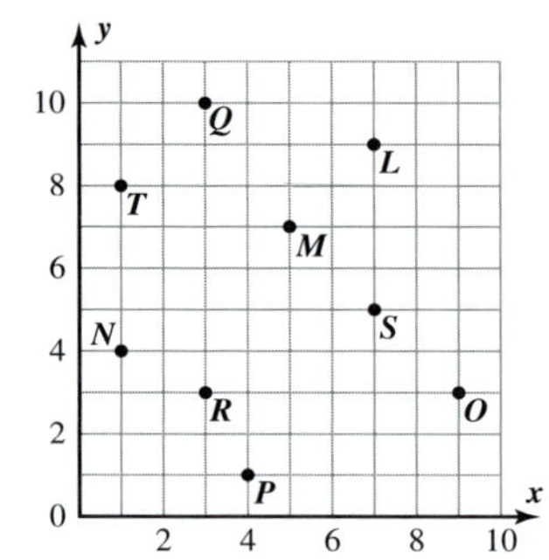

A (1, 4) **C** (4, 0)

B (0, 4) **D** (4, 1)

33 Which angle is adjacent to angle *ADE*? **B**

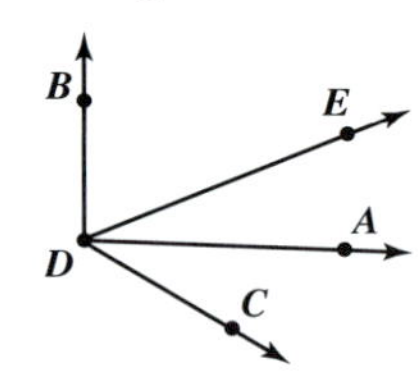

A angle *CDE* **C** angle *BDC*

B angle *BDE* **D** angle *ADB*

34 To display his math class's test scores, Mr. Fuentes made a line plot.

MATH TEST SCORES

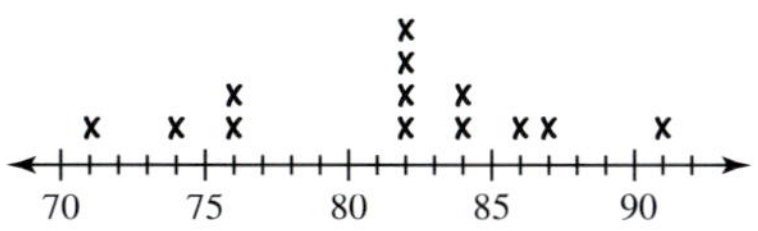

What is the MODE of the test scores? **C**

A 71 **C** 82

B 76 **D** 91

35 Which point is BEST represented by the coordinates (4, 6)? **C**

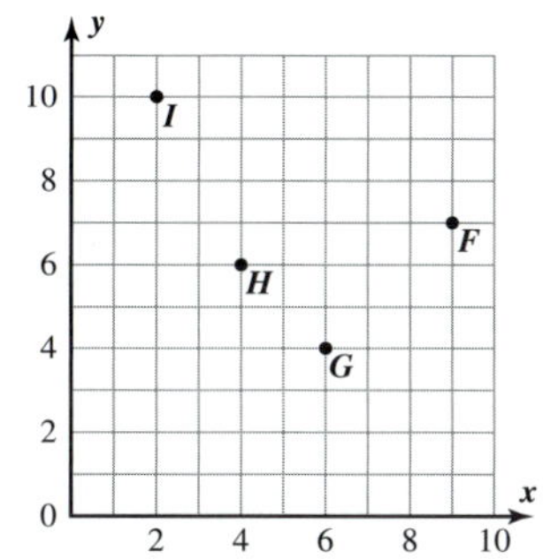

A *F* **C** *H*

B *G* **D** *I*

36 What is the measure of angle *Y*? **C**

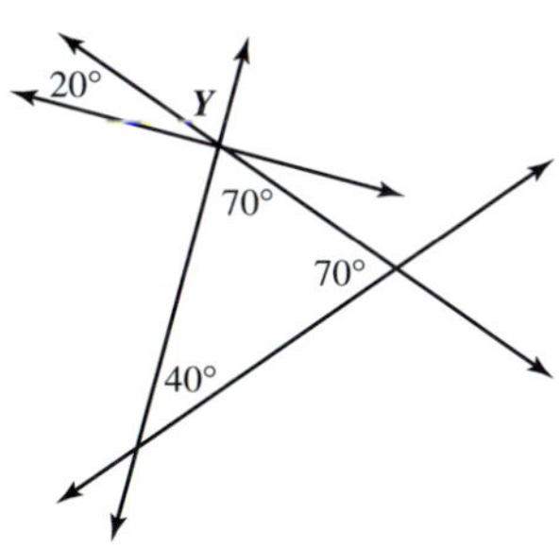

A 20° **C** 70°

B 40° **D** 90°

Go On

Item Correlations and Text References

34. Standard 6.6.3

See Lesson 6-2 and Skills Bank p. 672.

35. Standard 6.3.7

See Lessons 6-6 and 9-3.

36. Standards 6.4.1, 6.4.2

See Lesson 7-3.

Item Correlations and Text References

37. Standard 6.2.8
See Lesson 8-9.

38. Standard 6.5.8
See Lesson 10-8.

39. Standards 6.2.4, 6.2.5
See Lesson 5-3.

40. Standard 6.3.1
See Lessons 2-2 and 2-6.

41. Standard 6.6.3
See Lesson 6-2.

ISTEP+ Test Practice

37 A tour group left Indiana and traveled 120 miles by bus before stopping for a break. This distance was 60% of the total distance to their next stop. What was the total distance to their next stop? **D**

A 72 miles **C** 192 miles
B 180 miles **D** 200 miles

38 What is the volume of a cube whose side length measures 6 inches? **B**

Volume of cube $= s^3$

A 18 cubic inches **C** 36 cubic inches
B 216 cubic inches **D** 729 cubic inches

39 A bag of cat food holds 160 ounces. The bag's label states that one serving is $1\frac{1}{2}$ ounces. How many full servings are in one bag? **B**

A 80 **C** 120
B 106 **D** 145

40 During December, 21,828 passengers flew on a certain airline. This was four times the number of passengers who flew on the same airline in October. How many passengers flew on the airline in October? **A**

A 5,457 **C** 78,866
B 5,500 **D** 87,312

41 For his science project, Ben planted seeds at two different temperatures and recorded the number of days that it took each seed to sprout. His results are given in the table.

Seed	1	2	3	4	5	6	7	8
Temperature (°C)	25	25	25	25	30	30	30	30
Time to Sprout (days)	8	8	5	6	5	4	7	6

What is the MEAN (average) number of days it took the seeds planted at 25°C to sprout? **D**

A 5.5 **C** 6.125
B 6.33 **D** 6.75

42 How many degrees difference is there between 42°C and $^{-}7$°C? **C**

A 35°C   **C** 49°C

B 48°C   **D** 50°C

43 At a health club, the swimming pool contains 86,000 liters of water. The maintenance staff uses 20 liters of chlorine to purify the water. How many liters of water can be purified by one liter of chlorine? **B**

A 43   **C** 43,000

B 4,300   **D** 430,000

44 What is the least common multiple of 34 and 51? **C**

A 17   **C** 102

B 51   **D** 204

45 Butterflies fly on warm, sunny days to get their body temperatures up to 80°F. What is the approximate Celsius equivalent of 80°F? **B**

A 12°C   **C** 62°C

B 27°C   **D** 76°C

46 Charlie has a piece of wood that is $8\frac{1}{2}$ feet long. For a project, he needs $\frac{3}{4}$-foot segments. How many segments of this length can he cut? **C**

A 6   **C** 11

B $6\frac{3}{8}$ **D** $11\frac{1}{3}$

47 What is the surface area of this cylinder, to the nearest hundredth? **C**

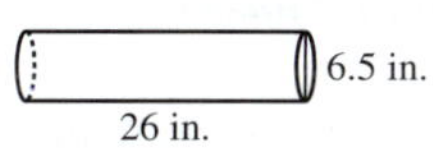

Area of circle = πr^2
Circumference of circle = $2\pi r$
Area of rectangle = lw
Use π = 3.14

A 301.67 square inches   **C** 596.99 square inches

B 563.83 square inches   **D** 1,098.50 square inches

Go On

Item Correlations and Text References

42. Standard 6.2.1
See Lesson 9-5.

43. Standard 6.2.7
See Lesson 8-2.

44. Standard 6.1.7
See Lesson 5-5.

45. Standard 6.5.9
See Skills Bank p. 675.

46. Standards 6.2.4, 6.2.5
See Lessons 3-9 and 5-3.

47. Standard 6.5.8
See Lesson 10-7.

Item Correlations and Text References

48. Standard 6.6.3
See Lesson 6-2.

49. Standards 6.4.1, 6.4.2
See Lesson 7-3.

Session B

1. Standards 6.1.7, 6.7.5
See Lesson 4-5.

2. Standards 6.1.4, 6.7.5
See Lessons 8-7 and 8-8.

3. Standard 6.7.5
See Lessons 4-9 and 8-9.

4. Standards 6.1.4, 6.2.5, 6.7.5
See Lessons 8-7 and 8-8.

ISTEP+ Test Practice

48 Kaylie surveyed nine of her friends about how many people lived in their households. Her results are shown in the table below.

NUMBER OF PEOPLE LIVING IN A HOUSEHOLD								
3	4	3	6	2	7	10	4	4

What is the MEDIAN of Kaylie's data set? **B**

A 2 **C** 5

B 4 **D** 8

49 What is the measure of angle *ABD*? **D**

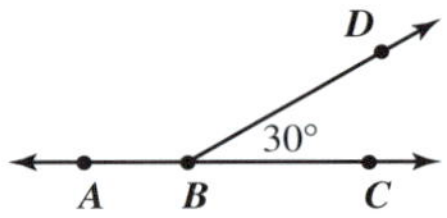

A 60° **C** 100°

B 70° **D** 150°

Session B

Directions: Write your response to each question. Show all of your work.

Dexter and Tom made a pizza. Now do numbers 1 through 4 about their pizza.

1 Dexter cut the pizza into 12 equal pieces, and ate 4 of the pieces. What fraction of the pizza did Dexter eat? Write your answer in simplest form.

Draw a picture to explain your answer.

2 Tom ate half of the pizza. What percent of the pizza did Tom eat?

Show your work or explain your answer in words.

3 How many pieces of pizza did Dexter and Tom eat in all?

Use words to explain your solution.

4 What percent of the pizza is left?

You MUST show your work.

Answers

1. A complete and correct response includes valid mathematical work that leads to a correct answer of $\frac{1}{3}$ and a drawing similar to the one below.

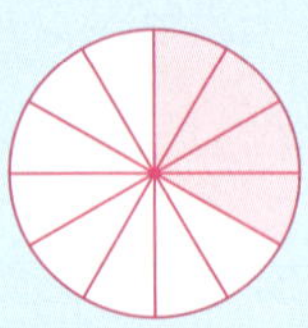

2. A complete and correct response includes valid mathematical work or a logical explanation that leads to a correct answer of 50%.

3. A complete and correct response includes valid mathematical work that leads to a correct answer of 10 pieces.

4. A complete and correct response includes valid mathematical work that leads to a correct answer of 16.67%.

Poinsettias are the most popular potted plants in the United States. Now do numbers 5 and 6 about poinsettias.

5 One year, the U.S. Department of Agriculture reported that nearly 60 million poinsettia plants were purchased in the United States. During the same year, poinsettia sales totaled \$22 million. Suppose each poinsettia plant cost the same amount. Write an equation that could be used to determine the price (p) of each plant.

6 Now use your equation to find the price of each poinsettia plant.

Show your work.

7 A quilt has 15 stripes. The first stripe is white, the second stripe is blue, and the third stripe is pink. If this pattern continues, how many times does the pattern appear in the quilt?

Show your work or explain your answer in words.

8 Find the value of $3 \times 9 + 5 - 2$. Write the steps and properties you used.

You MUST show your work.

9 Jo wants an average of 90 on her 8 science tests. What is the LEAST number of total points that she will need for all 8 tests to achieve her goal?

You MUST show your work.

10 At the town park, there is a circular jogging path all along the outer edge of the park. The path is 9.4 kilometers.

Draw a diagram to represent this situation.

How far is the path from the center of the park? Use 3.14 as an approximation for π and round your answer to the nearest tenth of a kilometer.

You MUST show your work.

11 A penny, a nickel, and a dime are all tossed at the same time. Draw a tree diagram to show all possible outcomes of this experiment. How many possible outcomes are there?

You MUST show your work.

Stop

Item Correlations and Text References

5. Standards 6.3.1, 6.7.5
See Lesson 2-2.

6. Standards 6.3.1, 6.5.10, 6.7.5
See Lesson 2-6.

7. Standards 6.7.1, 6.7.5
See Lesson 7-8.

8. Standards 6.3.6, 6.7.5
See Lesson 1-4.

9. Standards 6.6.3, 6.7.5
See Lesson 6-2.

10. Standards 6.5.4, 6.5.5, 6.7.5
See Lesson 10-5.

11. Standards 6.6.4, 6.7.5
See Lesson 11-4.

5. A complete and correct response includes an equation equivalent to $60p = 22$.

6. A complete and correct response includes valid mathematical work that leads to a correct answer of about \$0.37.

7. A complete and correct response includes valid mathematical work or a logical explanation that leads to a correct answer of 5.

8. A complete and correct response includes valid mathematical work that leads to a correct answer of 30.

9. A complete and correct response includes valid mathematical work that leads to a correct answer of 720 points.

10. A complete and correct response includes a diagram similar to the one below and valid mathematical work that leads to a correct answer of 1.5 km.

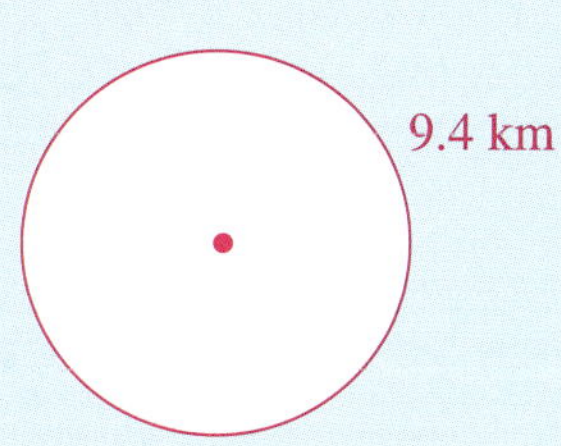

11. A complete and correct response includes a tree diagram similar to the one below and a correct answer of 8 possible outcomes.

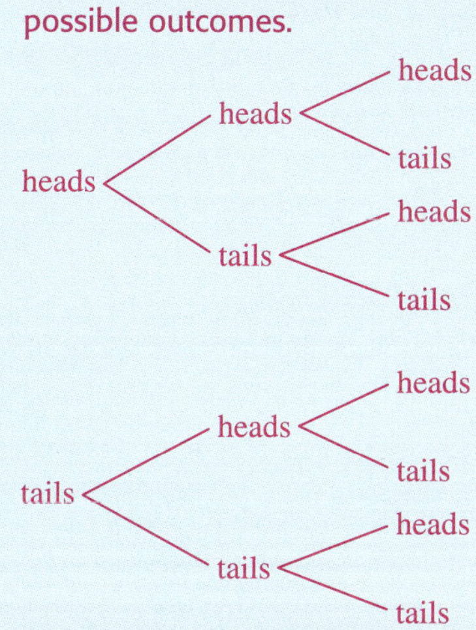

Reviewers

Thomas J. Altonjy
Assistant Principal
Robert R. Lazar Middle School
Montville, NJ

Jane Bash, M.A.
Math Education
Eisenhower Middle School
San Antonio, TX

Charlie Bialowas
District Math Coordinator
Anaheim Union High School District
Anaheim, CA

Lynn Bodet
Math Teacher
Eisenhower Middle School
San Antonio, TX

Louis D' Angelo, Jr.
Math Teacher
Archmere Academy
Claymont, DE

Troy Deckebach
Math Teacher
Tredyffrin-Easttown Middle School
Berwyn, PA

Mary Gorman
Math Teacher
Sarasota, FL

Brian Griffith
Supervisor of Mathematics, K–12
Mechanicsburg Area School District
Mechanicsburg, PA

Ruth Harbin-Miles
District Math Coordinator
Instructional Resource Center
Olathe, KS

Kim Hayden
Math Teacher
Milford Jr. High School
Milford, OH

Susan Howe
Math Teacher
Linie Kiln Middle School
Fulton, MD

Paula Jenniges
Austin, TX

Ronald J. Labrocca
District Mathematics Coordinator
Manhasset Public Schools
Manhasset, NY

Victor R. Lopez
Math Teacher
Washington School
Union City, NJ

George Maguschak
Math Teacher/Building Chairperson
Wilkes-Barre Area
Wilkes-Barre, PA

Dianne McIntire
Math Teacher
Garfield School
Kearny, NJ

Kenneth McIntire
Math Teacher
Lincoln School
Kearny, NJ

Francisco Pacheco
Math Teacher
IS 125
Bronx, NY

Vivian Perry
Edwards, IL

Vicki Perryman Petty
Math Teacher
Central Middle School
Murfreesboro, TN

Jennifer Sawyer
Math Teacher
Shawboro, NC

Russell Sayler
Math Teacher
Longfellow Middle School
Wauwatosa, WI

Raymond Scacalossi
Math Chairperson
Hauppauge Schools
Hauppauge, NY

Richard Seavey
Math Teacher–Retired
Metcalf Jr. High
Eagan, MN

Sherry Shaffer
Math Teacher
Honeoye Central School
Honeoye Falls, NY

Gail M. Sigmund
Math Teacher
Charles A. Mooney Preparatory School
Cleveland, OH

Jonathan Simmons
Math Teacher
Manor Middle School
Killeen, TX

Jeffrey L. Slagel
Math Department Chair
South Eastern Middle School
Fawn Grove, PA

Karen Smith, Ph.D.
Math Teacher
East Middle School
Braintree, MA

Bonnie Thompson
Math Teacher
Tower Heights Middle School
Dayton, OH

Mary Thoreen
Mathematics Subject Area Leader
Wilson Middle School
Tampa, FL

Paul Turney
Math Teacher
Ladue School District
St. Louis, MO

Audrey Jackson
Principal
Claymont Elementary School
Ballwin, MO

Classroom and Learning Environment

"Effective mathematics teaching requires understanding what students know and need to learn and then challenging and supporting them to learn it well."

National Council of Teachers of Mathematics 2000

The fundamental goal of *Holt Middle School Math* is to provide teachers with the necessary tools and understanding of school mathematics to ensure student success at all levels.

Differentiating Instruction

Holt Middle School Math enables teachers to easily differentiate instruction. By implementing the process of scaffolding, the program provides continuous support for challenging work. The section planner in *Holt Middle School Math* assists the teacher in planning and pacing for all students at all levels, while the lesson plans provided with the program help the teacher determine which resources to use to differentiate instruction. The exercises ensure that students have ample opportunities, with guidance, to master the skills taught in the lessons and to then apply these skills with critical thinking. Each lesson includes multiple examples and opportunities to reach all learners through extensions, journal activities, the use of manipulatives, and home connections.

Fostering Successful Instructional Strategies

Holt Middle School Math promotes successful learning by supporting numerous teaching strategies, including direct instruction and cooperative learning. The Hands-On and Technology Labs are ideal for cooperative learning in heterogeneous groups, and the Explorations are designed for discovery learning. The Focus on Problem Solving and Think and Discuss features in each chapter are intended to stimulate student interaction. The exercise sets at the end of each lesson are tied to specific examples to encourage students to direct their own learning and to foster parental help on assignments. Thus, *Holt Middle School Math* can be used to accommodate various styles of the teacher as well as the students.

Creating a Community of Learners

The way we think about our classrooms might be different today, but our goal is the same—mathematics success for all. This program strives to assist teachers in creating the positive environment necessary to build a community of competent and confident learners in a mathematics class. Imagine a classroom where diversity in learning is the norm and the teacher responds to the learners' needs with flexible strategies, open dialogue, and ongoing assessment. Students will learn best when learning opportunities are natural and when connections can be easily made. *Holt Middle School Math* aids teachers in maximizing the capacity of each student every day.

RESEARCH

Glickman, C. 2002.
Leadership for learning.
Alexander, VA: Association for Supervision and Curriculum Development.

National Council of Teachers of Mathematics. 2000.
Principles and standards for school mathematics.
Reston, VA: National Council of Teachers of Mathematics.

Senge, P. 1994.
The fifth discipline fieldbook.
New York, NY: Doubleday.

Tomlinson, C. 1995.
How to differentiate instruction in mixed-ability classrooms.
Alexander, VA: Association for Supervision and Curriculum Development.

Wiggins, G., and J. McTighe. 1998.
Understanding by design.
Alexander, VA: Association for Supervision and Curriculum Development.

David J. Chard, Ph.D.
Assistant Professor and Director of Graduate Studies in Special Education
University of Oregon
Eugene, OR

Accessibility for All Learners

"Students exhibit different talents, abilities, achievements, needs, and interests in mathematics. Nevertheless, all students must have access to the highest-quality mathematics instructional programs."

National Council of Teachers of Mathematics 2000

One of the primary goals of *Holt Middle School Math* is to provide teachers with a resource for teaching students new skills and strategies important for developing their comprehension of mathematics.

Coherent Pedagogical Approach

This program was designed with instructional features that represent a coherent pedagogical approach to mathematics instruction. Each lesson begins with carefully wrought examples of all of the skills, concepts, and strategies addressed. Additionally, the program's examples and counter-examples assist students in understanding the distinct boundaries that exist within each concept and the context in which particular skills and strategies are useful (Bransford, Brown, and Cocking 2000).

Procedural Fluency Development

A second goal is to develop procedural fluency in specific mathematical skills (Mathematics Learning Study Committee 2001). For many students with cognitive disabilities, insufficient practice hampers their ability to develop this fluency. *Holt Middle School Math* provides students with ample opportunities to practice specific computation and problem-solving procedures. Once fluent, students will then have ready access to these tools for use in more sophisticated mathematics.

Sufficient Scaffolding

Key to any instructional program is sufficient scaffolding to support student learning (Vygotsky 1962). In a typical middle school math classroom, some students will require substantial assistance in developing strategies for solving problems. Still others will already have the knowledge necessary to solve problems with little support. In this program, the instructional framework builds the background knowledge essential for ensuring that students are able to solve increasingly complex problems. Scaffolding is utilized in a number of ways throughout the program, from graduated difficulty of new content and applications to frequent opportunities for review, substantive reteaching lessons, and additional examples for extended instruction.

RESEARCH

Bransford, J. D., A. L. Brown, and R. R. Cocking, eds. 2000.
How people learn: Brain, mind, experience, and school.
Washington, DC: National Research Council.

Gersten, R., D. J. Chard, and S. Baker. 2002.
A meta-analysis of research on mathematics instruction for students with learning disabilities.
Eugene, OR: Eugene Research Institute.

Mathematics Learning Study Committee. 2001.
Adding it up: Helping children learn mathematics.
Washington, DC: National Academy Press.

National Council of Teachers of Mathematics. 2000.
Principles and standards for school mathematics.
Reston, VA: National Council of Teachers of Mathematics.

Vygotsky, L. 1962.
Thought and Language.
Cambridge, MA: MIT Press.

Jim Milgram, Ph.D.
Professor of Mathematics
Stanford University
Stanford, CA

RESEARCH

Morris, Anne K., and Vladimir M. Stoutsky. 1998.
Understanding of logical necessity: Developmental antecedents and cognitive consequences.
Child Development 69 (3): 721–41.

Mathematics Learning Study Committee. 2001.
Adding it up: Helping children learn mathematics.
Washington, DC: National Academy Press.

Schmidt, William, Richard Houang, and Leland Cogan. 2002.
A coherent curriculum: The case of mathematics.
American Educator 26 (2): 1–17.

Wu, H. H. 2001.
How to prepare students for algebra.
American Educator 25 (2): 10–17.

Transition to Advanced Mathematics

"...throughout the grades from pre-K through 8 all students can and should ... understand mathematical ideas, compute fluently, solve problems, and engage in logical reasoning."

Mathematics Learning Study Committee 2001

Middle school mathematics instruction occurs at a critical time in students' development and must address the needs specific to this period of learning.

From Foundation Skills to Advanced Topics

When students enter the middle grades, they must prepare for the transition to more advanced mathematical topics such as algebra and geometry while enhancing their basic arithmetic knowledge. It is crucial that they develop abstract reasoning and symbolic manipulation skills. In *Holt Middle School Math* these areas are carefully developed using methods aligned with standard best practices. The program addresses national and state standards while recognizing that some mathematical topics require more sophisticated instruction. All instructional materials, including the vocabulary lists, examples, and reference materials, reflect accurate mathematics. The integrity of the math represented in the program is strictly maintained so that the instructional design contributes positively to students' understanding of the discipline.

Instructional Sequencing

While the introduction of advanced topics requires that students broaden their understanding of mathematical ideas, it also reflects the hierarchical and sequential nature of mathematics as a discipline (Schmidt, Houang, and Cogan 2002). Students need to see the relationships between the math they are learning and real-world scenarios. To foster the development of these connections, the sequence of instruction within each grade and across this program accounts for the elements of mathematics that should be taught first in order to prepare students for later insights. The presentation of mathematical concepts in this program is aligned with that of the most successful international programs.

Enhancing the Role of the Teacher

Middle school mathematics instruction must be supported by materials that assist teachers in helping students successfully learn and do more complex mathematics. Care has been taken to ensure that each instructional lesson develops enough background information so that teachers can demonstrate to their students how the concepts they learn today will tie in to their later mathematical education. Teachers can use this foundation material as a resource when relaying information to their classes. In this way, *Holt Middle School Math* is an asset for teachers as well as for their students.

Jennie M. Bennett, Ed.D.
Instructional Mathematics Supervisor
Houston Independent School District
Houston, TX

RESEARCH

Artzt, Alice F., and Shirel Yaloz-Femia. 1999.
Mathematical reasoning during small-group problem solving. In Developing mathematical reasoning in grades K–12.
Reston, VA: National Council of Teachers of Mathematics.

Jensen, Eric. 1998.
Teaching with the brain in mind.
Alexandria: VA: Association for Supervision and Curriculum Development.

Krulik, Stephen, and Jesse A. Rudnick. 1999.
Innovation tasks to improve critical- and creative-thinking skills. In Developing mathematical reasoning in grades K–12.
Reston, VA: National Council of Teachers of Mathematics.

Levine, Mel. 2002.
A Mind at a Time.
New York, NY: Simon & Schuster.

Schell, Vicki J. 1981.
Learning partners: Reading and mathematics.
Paper presented at 14th annual meeting of the Missouri State Council of the International Reading Association. 13-17 May, at Columbia, MO.

Sullivan, Peter, and David Clarke. 1991.
Catering to all abilities through 'good' questions.
Arithmetic Teacher 39 (2): 14–18.

Strategic Problem Solving

"The single best way to grow a better brain is through challenging problem solving."

Eric Jensen, 1998

Unlike simple numeric computation problems, word problems present unique challenges to some students. One of the goals of this program is to teach students strategies to comprehend and solve word problems.

Development of Critical Thinking Skills

Holt Middle School Math ensures that students have the necessary tools to approach word problems strategically by teaching problem solving as a planned step-by-step process (Levine 2002). Using Polya's method for solving problems (understand the problem, create a plan, carry out the plan, and look back) activates students' critical thinking skills and engages students in making decisions and thinking logically. This program provides students with in-school systemic problem-solving experiences in which these critical skills are developed. Students generate different strategies for solving a problem, select the most feasible one, and arrive at a reasonable solution. These problem-solving skills help students understand how to approach real-world problems strategically both inside and outside the classroom.

Reading Connections

Reading comprehension is necessary to all subjects at all levels, including mathematics (Schell 1981). When students read word problems, they must synthesize or integrate their ideas and determine the operations to use. Reading is therefore a pivotal partner in problem solving; it sets the stage for understanding the problem itself. In this program students are asked to state the details of a problem, identify the necessary information, restate the problem in their own words, and demonstrate knowledge of mathematical vocabulary.

Asking Good Questions

Good questions engage the student in a more active role in learning. When the teacher asks such questions, learning is student-centered rather than teacher-centered. Good questions stimulate and activate communication between the teacher and students and allow students to respond in their own way when solving math problems. These questions can guide students at all levels to experience success with problem solving. Another goal of *Holt Middle School Math* is to assist teachers by suggesting good questions through features such as Focus on Problem Solving and Reaching All Learners.

Janet K. Scheer, Ph.D.
Executive Director
Create A Vision[TM]
Foster City, CA

Concrete Understanding

"When students gain access to mathematical representations and the ideas they represent, they have a set of tools that significantly expand their capacity to think mathematically."

National Council of Teachers of Mathematics 2000

H*olt Middle School Math* makes use of mathematical modeling and provides many options for the use of manipulatives to enhance student understanding of abstract concepts.

Manipulatives for Concrete Understanding

Educational research demonstrates the effectiveness of hands-on learning in supplementing understanding of mathematical ideas for some students (Threadgill-Sowder and Juilfs 1980). This is especially important in the middle grades, when students are exposed to increasingly abstract concepts. While some middle school students are ready to embrace these abstract topics, others still need the concrete foundation that manipulatives can supply. This program utilizes algebra tiles, pattern blocks, and two-color counters in Hands-On Labs to model topics such as fraction operations and grouping of terms in algebraic expressions. Additionally, the Reaching All Learners features provide teachers with concrete methods for presenting selected topics.

Bridging: Concrete to Symbolic Understanding

Most theories of developmental learning support the use of physical tools to establish a foundation for abstract thought. For this approach to be successful, students must make connections between the manipulatives with which they are working and the abstract mathematical concepts the materials represent; in this way, concrete action is transferred to symbolic understanding (Bohan and Shawaker 1994). It is essential that sufficient context and introduction to a manipulative lesson be provided so that students are able to form connections to symbolic representations in a guided manner (Stein and Bovalino 2001). The modeling activities within this program are intentionally placed after foundation skills have been developed and before symbolic computation is emphasized.

Opportunities to Expand Knowledge

Another advantage of manipulative lessons rooted in foundational skills is that students are given the opportunity to discover new mathematical concepts. The discovery-based knowledge that is developed in this program's Hands-On Labs is solidified by lessons that formalize the mathematical rules and symbolic representations of the concept. In this way, concrete understanding facilitates application of the learned mathematical concepts.

RESEARCH

Bohan, Harry J., and Peggy Bohan Shawaker. 1994.
Using manipulatives effectively: A drive down rounding road.
Arithmetic Teacher 41 (5): 246–48.

National Council of Teachers of Mathematics. 2000.
Principles and standards for school mathematics.
Reston, VA: National Council of Teachers of Mathematics.

Stein, Mary Kay, and Jane W. Bovalino. 2001.
Manipulatives: One piece of the puzzle.
Mathematics Teaching in the Middle School 6 (6): 356–59.

Threadgill-Sowder, Judith, and Patricia Juilfs. 1980.
Manipulative versus symbolic approaches to teaching logical connectives in junior high school: An aptitude x treatment interaction study.
Journal for Research in Mathematics Education 11 (5): 367–74.

Bert K. Waits, Ph.D.
Professor Emeritus of Mathematics
The Ohio State University
Columbus, OH

RESEARCH

Graham, A. T., and M.O.J. Thomas. 2000.
Building a versatile understanding of algebraic variables with a graphic calculator.
Educational Studies in Mathematics 41 (3): 265–82.

Hollar, Jeannie C., and Karen Norwood. 1999.
The effects of a graphing-approach intermediate algebra curriculum on students' understanding of function.
Journal for Research in Mathematics Education 30 (2): 220–26.

National Commission on Mathematics and Science Teaching for the 21st Century. 2000.
Before it's too late:
The Glenn Commission report.

National Council of Teachers of Mathematics. 2000.
Principles and standards for school mathematics.
Reston, VA: National Council of Teachers of Mathematics.

Technology to Enhance Learning

"Technology is essential in teaching and learning mathematics; it influences the mathematics that is taught and enhances students' learning."

National Council of Teachers of Mathematics 2000

A wide array of technological tools is available for use in mathematics classrooms, including graphing calculators, spreadsheet programs, and geometry software. *Holt Middle School Math* makes use of these tools to reinforce student learning.

Integrated Use of Technology

Research has demonstrated that technology, when used appropriately, can improve students' mathematical understanding and problem-solving skills (Hollar and Norwood 1999). Similarly, technological tools can help teachers challenge students to use and understand mathematics in real-world scenarios. Through the use of integrated technology labs, this program gives students a solid foundation for understanding how to use technology appropriately to learn mathematics. The program puts the NCTM Technology Principle into action in every chapter.

Balanced Curriculum

Holt Middle School Math acknowledges that students must utilize all available tools in the mathematics-learning process. Traditional paper-and-pencil skills are emphasized throughout the program but are supplemented by technology components. Proficiency in mental math computation is stressed as well. Thus, technology is presented not as an end in itself but rather as a means for understanding and application. Current research supports this use of technology. The many technology labs in the program are devoted to the pedagogical use of spreadsheets and dynamic geometry software. Students also use graphing calculators, which are powerful, portable computers with built-in software for graphing, data analysis, and statistics.

Professional Development

This program helps teachers achieve one of the Glenn Commission's major goals in its report to the nation on the crisis in mathematics and science education today. The report states that in order to "inform efforts to promote higher student achievement, teachers should actively work to improve [their] knowledge and skills to incorporate educational technology into [their] learning and teaching" (National Commission on Mathematics and Science Teaching 2000).

Problem Solving Handbook

CHAPTER 1 Number Toolbox

Assessment

go.hrw.com State Test Preparation Online . . . . . KEYWORD: MR4 TestPrep

Interdisciplinary LINKS

Student Help

internet connect go.hrw.com
Homework Help Online
6, 10, 14, 22, 26, 29, 32
KEYWORD: MR4 HWHelp

Algebra *Indicates algebra included in lesson development*

Introduction to Algebra

CHAPTER 2

Assessment

Interdisciplinary LINKS

Student Help

internet connect

Homework Help Online

50, 54, 60, 64, 67, 71, 74

KEYWORD: MR4 HWHelp

CHAPTER 3

Decimals

Assessment

go.hrw.com State Test Preparation Online .... KEYWORD: MR4 TestPrep

Interdisciplinary LINKS

Student Help

internet connect go.hrw.com

Homework Help Online

94, 98, 104, 108, 116, 122, 125, 129, 132, 136

KEYWORD: MR4 HWHelp

Algebra *Indicates algebra included in lesson development*

Number Theory and Fractions

CHAPTER 4

Interdisciplinary LINKS

Student Help

internet connect

Homework Help Online

154, 158, 162, 169, 174, 180, 184, 190, 194

KEYWORD: MR4 HWHelp

CHAPTER 5 Fraction Operations

Assessment

Interdisciplinary LINKS

Student Help

internet connect go.hrw.com
Homework Help Online
214, 218, 224, 228, 234, 238, 244, 248, 254, 258
KEYWORD: MR4 HWHelp

Algebra *Indicates algebra included in lesson development*

Collect and Display Data

CHAPTER 6

Assessment

Interdisciplinary LINKS

Student Help

internet connect go.hrw.com

Homework Help Online

273, 276, 280, 286, 292, 295, 299, 303, 306

KEYWORD: MR4 HWHelp

CHAPTER 7

Plane Geometry

Assessment

go.hrw.com State Test Preparation Online . . . Keyword: MR4 TestPrep

Interdisciplinary LINKS

Student Help

internet connect go.hrw.com

Homework Help Online

324, 328, 334, 338, 346, 350, 354, 358, 363, 367, 371, 374

KEYWORD: MR4 HWHelp

Algebra *Indicates algebra included in lesson development*

Ratio, Proportion, and Percent

CHAPTER 8

CHAPTER 9

Integers

Assessment

Interdisciplinary LINKS

Student Help

internet connect go.hrw.com

Homework Help Online

452, 456, 460, 467, 471, 474, 478, 484

KEYWORD: MR4 HWHelp

Algebra *Indicates algebra included in lesson development*

Perimeter, Area, and Volume

CHAPTER 10

Assessment

Interdisciplinary LINKS

Student Help

internet connect

Homework Help Online

502, 506, 512, 518, 526, 532, 536, 540

KEYWORD: MR4 HWHelp

CHAPTER 11

Probability

Interdisciplinary LINKS

Student Help

internet connect go.hrw.com

Homework Help Online

556, 560, 566, 572, 576, 582

KEYWORD: MR4 HWHelp

Assessment

State Test Preparation Online . . . . . KEYWORD: MR4 TestPrep

Algebra *Indicates algebra included in lesson development*

Functions and Coordinate Geometry

CHAPTER 12

Math Skills for **Life Science**

Math Skill	Where Taught
Balance equations (change on one side requires an equivalent change on the other).	Lessons 2-1, 2-2, 2-3, 2-4, 2-5, 2-6, 2-7, 3-10, 5-4, 5-10, 8-2, 8-9, 8-10, 9-8
Combinations	Hands-On Lab 11B
Convert between standard and metric units.	Hands-On Lab 3C
Convert between temperature units.	Lessons 2-4, 12-2
Convert between units in one dimension, two dimensions, or three dimensions.	Skills Bank p. 682
Display and read data in bar graphs, line graphs, and circle graphs.	Lessons 6-1, 6-2, 6-3, 6-4, 6-5, 6-6; Chapter 6 Extension; Hands-On Lab 8B
Find angle measures.	Lesson 7-2
Find areas.	Lesson 10-2
Find average (mean).	Lesson 6-2
Find probability.	Lessons 11-1, 11-2, 11-3, 11-4, 11-5, 11-6
Find surface area.	Lesson 10-5
Find surface area to volume ratios.	Skills Bank p. 677
Find volume and perimeter.	Lessons 10-1, 10-6, 10-7
Graph on a coordinate plane.	Lesson 6-6
Operate with exponents.	Lessons 1-3, 3-5; Chapter 9 Extension
Operate with fractions.	Lessons 4-8, 4-9, 4-10, 5-1, 5-2, 5-3, 5-4, 5-7, 5-8, 5-9
Operate with large whole numbers.	Lessons 1-2, 1-3, 1-4, 1-5, 1-6, 1-7
Operate with percents.	Lessons 8-7, 8-8; Chapter 8 Extension
Operate with powers of ten.	Lessons 1-3, 3-5
Operate with units of rate and convert to compatible units.	Lesson 8-1; Skills Bank p. 676
Read metersticks and inch rulers.	Lessons 3-4, 4-2; Hands-On Labs 3C, 4C
Round decimal and whole numbers.	Lessons 1-2, 3-2
Set up and solve proportions.	Lessons 8-1, 8-2; Hands-On Lab 8B
Simplify fractions.	Lessons 4-1, 4-7
Solve equations.	Lessons 2-1, 2-2, 2-3, 2-4, 2-5, 2-6, 2-7, 3-10, 5-4, 5-10, 8-2, 8-9, 8-10, 9-8
Solve literal equations (rewrite formulas).	Skills Bank p. 678
Translate words to equations or algebraic expressions.	Lessons 2-1, 2-2
Understand nonlinear relationships, such as half-life.	Hands-On Lab 12A; Skills Bank p. 680
Understand exponential growth behavior.	Lesson 1-3; Hands-On Lab 12A; Skills Bank p. 679
Use and operate with scientific notation.	Lesson 3-5
Use appropriate one-, two-, and three-dimensional measurement units.	Lessons 3-4, 4-2, 5-5, 8-4, 10-1, 10-2, 10-3, 10-4, 10-5, 10-6, 10-7, 10-8, 10-9; Hands-On Lab 4C; Skills Bank p. 676
Use scale models.	Lessons 8-4, 8-5, 8-6; Hands-On Lab 10C
Write equivalent percents, fractions, and decimals.	Lessons 4-1, 4-2, 8-8
Write equivalent ratios and ratios in simplest form.	Lesson 8-1

Math Skills for Earth Science

Math Skill	Where Taught
Convert between standard and metric units.	Hands-On Lab 3C; Skills Bank p. 675
Convert between temperature units.	Lessons 2-4, 12-2; Skills Bank p. 675
Display and read data in bar graphs, line graphs, and circle graphs.	Lessons 6-1, 6-2, 6-3, 6-4, 6-5, 6-6; Chapter 6 Extension; Hands-On Lab 8B
Find angle measures.	Lesson 7-2
Find areas.	Lesson 10-2
Find average (mean).	Lesson 6-2
Find surface area.	Lesson 10-5
Find volume.	Lessons 10-6, 10-7
Graph on a coordinate plane.	Lesson 6-6
Operate with exponents.	Lessons 1-3, 3-5; Chapter 9 Extension
Operate with fractions.	Lessons 4-8, 4-9, 4-10, 5-1, 5-2, 5-3, 5-4, 5-7, 5-8, 5-9
Operate with large whole numbers.	Lessons 1-2, 1-3, 1-4, 1-5, 1-6, 1-7
Operate with percents.	Lessons 8-7, 8-8; Chapter 8 Extension
Operate with powers of ten.	Lessons 1-3, 3-5; Skills Bank p. 666
Operate with units of rate and convert to compatible units.	Lesson 8-1
Read metersticks and inch rulers.	Lessons 3-4, 4-2; Hands-On Labs 3C, 4C
Round decimal and whole numbers.	Lessons 1-2, 3-2; Skills Bank p. 663
Set up and solve proprotions.	Lessons 8-1, 8-2, Hands-On Lab 8B
Simplify fractions.	Lessons 4-1, 4-7
Solve equations.	Lessons 2-1, 2-2, 2-3, 2-4, 2-5, 2-6, 2-7, 3-10, 5-4, 5-10, 8-2, 8-9, 8-10, 9-8
Solve literal equations (rewrite formulas).	Skills Bank p. 678
Translate words to equations or algebraic expressions.	Lessons 2-1, 2-2
Understand nonlinear relationships, such as Richter scale and half-life.	Hands-On Lab 12A; Skills Bank p. 680
Use and operate with scientific notation.	Lesson 3-5
Use appropriate one-, two-, and three-dimensional measurement units.	Lessons 3-4, 4-2, 5-5, 8-4, 10-1, 10-2, 10-3, 10-4, 10-5, 10-6, 10-7, 10-8, 10-9; Hands-On Lab 4C
Use scale models.	Lessons 8-4, 8-5, 8-6; Hands-On Lab 10C
Write equivalent percents, fractions, and decimals.	Lessons 4-1, 4-2, 8-8
Write equivalent ratios and ratios in simplest form.	Lesson 8-1

Math Skills for **Physical Science**

Math Skill	Where Taught
Balance equations (change on one side requires an equivalent change on the other).	Lessons 2-1, 2-2, 2-3, 2-4, 2-5, 2-6, 2-7, 3-10, 5-4, 5-10, 8-2, 8-9, 8-10, 9-8
Convert between standard and metric units.	Hands-On Lab 3C
Convert between temperature units.	Lessons 2-4, 12-2
Convert between units in one dimension, two dimensions, or three dimensions.	Skills Bank p. 681
Display and read data in bar graphs, line graphs, and circle graphs.	Lessons 6-1, 6-2, 6-3, 6-4, 6-5, 6-6; Chapter 6 Extension; Hands-On Lab 8B
Find angle measures.	Lesson 7-2
Find areas.	Lesson 10-2
Find average (mean).	Lesson 6-2
Graph on a coordinate plane.	Lesson 6-6
Operate with exponents.	Lessons 1-3, 3-5; Chapter 9 Extension
Operate with fractions.	Lessons 4-8, 4-9, 4-10, 5-1, 5-2, 5-3, 5-4, 5-7, 5-8, 5-9
Operate with integers.	Lessons 9-4, 9-6, 9-7, 9-8; Hands-On Labs 9A, 9B; Chapter 9 Extension
Operate with large whole numbers.	Lessons 1-2, 1-3, 1-4, 1-5, 1-6, 1-7
Operate with percents.	Lessons 8-7, 8-8; Chapter 8 Extension
Operate with powers of ten.	Lessons 1-3, 3-5
Operate with units of rate and convert to compatible units.	Lesson 8-1
Read metersticks and inch rulers.	Lessons 3-4, 4-2; Hands-On Labs 3C, 4C
Round decimal and whole numbers.	Lessons 1-2, 3-2
Set up and solve proportions.	Lessons 8-1, 8-2; Hands-On Lab 8B
Simplify fractions.	Lessons 4-1, 4-7
Solve equations.	Lessons 2-1, 2-2, 2-3, 2-4, 2-5, 2-6, 2-7, 3-10, 5-4, 5-10, 8-2, 8-9, 8-10, 9-8
Solve literal equations (rewrite formulas).	Skills Bank p. 678
Translate words to equations or algebraic expressions.	Lessons 2-1, 2-2
Understand inverse relationships.	Lessons 12-1, 12-2; Hands-On Lab 12A
Understand linear relationships.	Lessons 12-1, 12-2; Hands-On Lab 12A
Understand nonlinear relationships such as acceleration, pH scale, and half-life.	Hands-On Lab 12A; Skills Bank p. 680
Understand parallel and perpendicular line relationships.	Lesson 7-4; Hands-On Lab 7B
Use and operate with scientific notation.	Lesson 3-5
Use and understand binary numbers.	Chapter 1 Extension
Use appropriate one-, two-, and three-dimensional measurement units.	Lessons 3-4, 4-2, 5-5, 8-4, 10-1, 10-2, 10-3, 10-4, 10-5, 10-6, 10-7, 10-8, 10-9; Hands-On Lab 4C
Use scale models.	Lessons 8-4, 8-5, 8-6; Hands-On Lab 10C
Write equivalent percents, fractions, and decimals.	Lessons 4-1, 4-2, 8-8
Write equivalent ratios and ratios in simplest form.	Lesson 8-1
Find volume.	Lessons 10-6, 10-7

NCTM Standards For Grades 6–8

Number and Operations

Understand numbers, ways of representing numbers, relationships among numbers, and number systems

COURSE 1	COURSE 2	COURSE 3
Lessons 1-1, 1-3, 3-1, 3-5, 4-1, 4-2, 4-3, 4-4, 4-5, 4-6, 4-7, 5-5, 8-1, 8-2, 8-7, 8-8, 8-9, 8-10, 9-1, 9-2; Hands-On Labs 3A, 3B, 4A, 4B, 8A; Chapters 1, 4, 8, 9 Extensions	Lessons 2-1, 2-2, 2-4, 2-5, 2-6, 2-7, 2-8, 2-9, 3-1, 3-7, 3-8, 3-9, 3-10, 5-1, 5-2, 6-1, 6-3, 6-5; Hands-On Labs 2B, 8C; Chapter 3 Extension	Lessons 2-6, 2-9, 3-1, 3-10, 7-1, 8-1, 8-2, 8-3, 8-4, 8-5, 8-6, 8-7; Hands-On Lab 7A; Technology Labs 3A, 8B; Chapters 3, 8 Extensions

Understand meanings of operations and how they relate to one another

COURSE 1	COURSE 2	COURSE 3
Lessons 1-4, 1-5, 2-4, 2-5, 2-6, 2-7, 3-9, 3-10, 5-1, 5-2, 5-3, 5-4, 5-7, 5-8, 5-9, 5-10, 9-8; Hands-On Labs 3B, 3D, 5A, 5B, 5C, 5D, 9A, 9B, 9C; Technology Lab 1A	Lessons 2-3, 2-7, 2-8, 2-11, 2-12, 3-3, 3-4, 3-5, 4-2, 4-3, 4-4, 4-5, 4-6, 4-7, 4-8, 4-10, 4-11, 4-12, 6-1, 6-4, 8-2, 8-7, 8-8; Hands-On Labs 2B, 3A, 3B, 4A, 4B, 4C, 6A, 8B; Technology Labs 2A, 2, 4; Chapter 3 Extension	Lessons 1-1, 1-2, 1-3, 1-4, 1-5, 1-6, 2-1, 2-2, 2-3, 2-7, 3-2, 3-4, 3-5, 3-6, 3-7, 3-8, 3-9, 10-1, 10-2, 10-4; Hands-On Labs 2A, 3B, 10A; Technology Lab 12B; Chapter 3 Extension

Compute fluently and make reasonable estimates

COURSE 1	COURSE 2	COURSE 3
Lessons 1-2, 1-3, 1-4, 1-6, 3-2, 3-3, 3-6, 3-7, 3-8, 4-8, 4-9, 5-6, 6-2, 8-4, 8-5, 8-6, 8-7, 8-8, 8-9, 9-4, 9-5, 9-6, 9-7, 11-4; Hands-On Lab 11B	Lessons 3-3, 3-4, 3-5, 4-1, 4-2, 4-3, 4-4, 4-5, 4-7, 4-8, 4-9, 4-10, 4-11, 5-2, 5-3, 5-6, 5-7, 6-2, 6-3, 6-5, 6-6; Hands-On Lab 6A; Chapter 10 Extension	Lessons 1-6, 2-1, 2-2, 2-3, 2-6, 2-7, 2-8, 3-2, 3-3, 3-4, 3-5, 3-9, 4-3, 7-2, 7-3, 8-1, 8-2, 8-3, 8-4, 8-5, 8-6, 8-7, 9-6; Hands-On Lab 3B; Chapter 8 Extension

Algebra

Understand patterns, relations, and functions

COURSE 1	COURSE 2	COURSE 3
Lessons 1-7, 6-6, 12-1, 12-2; Hands-On Lab 12A	Lessons 1-7, 1-8, 12-1, 12-2, 12-3, 12-4, 12-5, 12-6, 12-7; Hands-On Labs 1C, 5A, 7F, 10A, 12A; Technology Lab 12; Chapters 7, 8, 11 Math-Ables	Lessons 1-7, 1-8, 1-9, 2-8, 11-1, 11-2, 11-5, 11-7, 12-1, 12-2, 12-3, 12-4, 12-5, 12-6, 12-7, 12-8; Hands-On Labs 9B, 12A; Technology Lab 1A

Represent and analyze mathematical situations and structures using algebraic symbols

COURSE 1	COURSE 2	COURSE 3
Lessons 2-1, 2-2, 2-3, 2-4, 2-5, 2-6, 2-7, 3-10, 5-4, 5-10, 6-6, 9-3, 9-8; Hands-On Labs 3B, 3D, 5A, 5B, 5C, 5D, 9A, 9B, 9C; Chapter 2 Extension	Lessons 2-7, 2-10, 2-11, 2-12, 3-6, 4-6, 4-12, 6-4, 11-1, 11-2, 11-3, 11-4, 11-5, 11-6, 11-7, 12-5, 12-8; Hands-On Labs 2C, 3C, 11A; Chapters 11, 12 Extensions	Lessons 1-1, 1-2, 1-6, 2-4, 2-5, 3-7, 7-4, 10-4, 10-5, 10-6, 11-3, 11-4; Hands-On Lab 2A; Technology Lab 11A

Use mathematical models to represent and understand quantitative relationships

COURSE 1	COURSE 2	COURSE 3
Lessons 2-4, 2-5, 2-6, 2-7, 3-10, 5-4, 5-10, 9-8, 12-1, 12-2; Hands-On Lab 12A	Lessons 1-9, 12-1, 12-3, 12-4, 12-6; Hands-On Labs 1C, 2C, 3A, 3B, 3C, 4A, 4B, 4C, 5A, 6A, 11A, 12A; Technology Labs 1B, 1C, 1, 3, 5, 9, 12	Lessons 4-2, 4-5, 10-1, 10-2, 10-3, 11-6; Hands-On Labs 6A, 7A, 7B, 7C, 8A, 10A; Chapter 11 Extension

Analyze change in various contexts

COURSE 1	COURSE 2	COURSE 3
Lessons 12-3, 12-4, 12-5, 12-6; Hands-On Lab 12A	Lessons 2-7, 6-5, 6-6, 12-3, 12-5, 12-7; Hands-On Lab 6A	Lessons 5-5, 5-7, 6-6, 6-7, 6-8, 6-9, 7-4, 7-6, 7-9, 8-4, 11-2; Hands-On Labs 5C, 7C

Geometry

Analyze characteristics and properties of two- and three-dimensional geometric shapes and develop mathematical arguments about geometric relationships

COURSE 1	COURSE 2	COURSE 3
Lessons 7-1, 7-2, 7-3, 7-4, 7-5, 7-6, 7-7, 7-8, 7-9, 8-4, 10-1, 10-2, 10-3, 10-4, 10-5, 10-6, 10-7, 10-8, 10-9; Hands-On Labs 7C, 10A, 10B, 10C; Technology Labs 7, 10	Lessons 5-5, 7-1, 7-2, 7-3, 7-4, 7-5, 7-6, 7-7, 7-8, 7-9, 8-8, 9-1, 9-5; Hands-On Labs 5A, 7A, 8B	Lessons 5-1, 5-2, 5-3, 5-4, 5-6, 6-3; Hands-On Labs 5A, 6A, 6B

Specify locations and describe spatial relationships using coordinate geometry and other representational systems

COURSE 1	COURSE 2	COURSE 3
Lessons 6-6, 8-6, 9-3, 12-3, 12-4, 12-5, 12-6; Chapter 9 Math-Ables	Lessons 3-2, 7-10, 12-5; Hands-On Labs 5A, 7A, 7B, 7C, 7D, 7E; Technology Lab 5	Lessons 1-7, 1-8, 5-5, 6-1, 6-2, 6-3, 6-4, 11-2, 11-3; Hands-On Labs 5C, 7C; Chapter 11 Math-Ables

Apply transformations and use symmetry to analyze mathematical situations

COURSE 1	COURSE 2	COURSE 3
Lessons 7-10, 7-11, 7-12; Hands-On Lab 7C	Lessons 7-10, 7-11; Hands-On Lab 7F; Technology Lab 7	Lessons 5-7, 5-8, 5-9, 7-5; Hands-On Labs 5C, 7C; Chapter 6 Extension

Use visualization, spatial reasoning, and geometric modeling to solve problems

COURSE 1	COURSE 2	COURSE 3
Lessons 7-1, 7-5, 7-6, 7-7, 7-8, 7-10, 7-11, 7-12, 8-4, 8-5, 10-1, 10-2, 10-3, 10-4, 10-5, 10-6, 10-7, 10-8, 10-9, 12-3, 12-4, 12-5, 12-6; Hands-On Labs 7A, 7B, 10B, 10C; Chapter 7 Extension	Lessons 3-3, 3-4, 4-5, 9-1, 9-2, 9-3, 9-4, 9-5; Hands-On Labs 2C, 3A, 3B, 3C, 4A, 4B, 4C, 5A, 6A, 7A, 7B, 7C, 7E, 9A, 9B, 11A, 12A; Technology Labs 5, 6, 7, 8, 9; Chapter 9 Extension; Chapters 4, 8, 9 Math-Ables	Lessons 5-1, 5-2, 5-3, 5-4, 5-5, 5-6, 5-7, 5-8, 5-9, 6-1, 6-2, 6-3, 6-4, 6-5, 6-6, 6-7, 6-8, 6-9, 6-10, 7-5, 7-8, 7-9; Hands-On Labs 5A, 5B, 5C, 6A, 6B, 7A, 7B, 7C, 8A; Chapter 7 Extension; Chapter 6 Math-Ables

Measurement

Understand measurable attributes of objects and the units, systems, and processes of measurement

COURSE 1	COURSE 2	COURSE 3
Lessons 3-4, 7-2, 8-3, 8-6, 10-1, 10-2, 10-7, 10-8, 10-9; Hands-On Labs 3C, 4C	Lessons 5-4, 5-5, 8-1, 8-2, 9-5; Hands-On Labs 5A, 8A	Lessons 6-5, 7-3, 7-7, 7-8, 7-9; Hands-On Lab 7C

Apply appropriate techniques, tools, and formulas to determine measurements

COURSE 1	COURSE 2	COURSE 3
Lessons 8-5, 8-6, 10-1, 10-2, 10-3, 10-4, 10-5, 10-7, 10-9; Hands-On Labs 3C, 10A; Chapter 3 Extension	Lessons 5-4, 5-5, 5-6, 5-7, 8-2, 8-3, 8-4, 8-5, 8-6, 8-8, 9-2, 9-3, 9-4, 9-5; Hands-On Labs 8B, 9B; Technology Labs 8, 9; Chapters 8, 9 Extensions	Lessons 6-1, 6-2, 6-4, 6-6, 6-7, 6-8, 6-9, 6-10, 7-2, 7-6, 7-7, 7-8, 7-9; Hands-On Labs 7B, 7C; Technology Lab 8B

Data Analysis and Probability

Formulate questions that can be addressed with data and collect, organize, and display relevant data to answer them

COURSE 1	COURSE 2	COURSE 3
Lessons 6-1, 6-4, 6-5, 6-7, 6-8, 6-9; Hands-On Lab 8B; Chapter 6 Extension	Lessons 1-1, 1-3, 1-4, 1-5, 1-6, 1-7, 1-8, 1-9; Hands-On Labs 1C, 7D; Technology Labs 1A, 1B	Lessons 4-1, 4-2, 4-3, 4-5, 4-7; Hands-On Labs 4A, 4B

Select and use appropriate statistical methods to analyze data

COURSE 1	COURSE 2	COURSE 3
Lessons 6-2, 6-3, 6-4, 6-5, 6-7, 6-8, 6-9; Hands-On Lab 6A; Chapter 6 Extension	Lessons 1-1, 1-2, 1-3, 1-4, 1-5, 1-6, 1-7, 1-8, 1-9; Technology Lab 1B	Lessons 4-2, 4-3, 4-4, 4-5, 4-6; Hands-On Labs 4A, 4B; Chapter 4 Extension

Develop and evaluate inferences and predictions that are based on data

COURSE 1	COURSE 2	COURSE 3
Lessons 6-7, 6-8, 11-6; Hands-On Labs 8B, 11A	Lessons 1-2, 1-8, 1-9, 10-1, 10-2, 10-4, 12-3; Hands-On Lab 1C	Lessons 4-7, 9-1, 9-2, 9-3, 9-4, 9-5, 9-7, 9-8, 11-7; Technology Lab 9A

Understand and apply basic concepts of probability

COURSE 1	COURSE 2	COURSE 3
Lessons 11-1, 11-2, 11-3, 11-4, 11-5, 11-6; Hands-On Lab 11B; Chapter 11 Extension	Lessons 10-1, 10-2, 10-3, 10-4, 10-5, 10-6, 10-7; Hands-On Lab 10A; Chapter 10 Extension	Lessons 9-1, 9-2, 9-3, 9-4, 9-5, 9-7, 9-8; Technology Lab 9A

Problem Solving

Build new mathematical knowledge through problem solving

Appears throughout each course in features such as the following:

Problem Solving on Location, Math-Ables, and *Performance Assessment* at the end of each chapter
Problem Solving Applications within each chapter
For example:

COURSE 1
Pages 13, 36–37, 78–79, 86, 128, 140–141, 200, 206, 227, 262, 312, 318

COURSE 2
Pages 48–49, 83, 120, 182, 248–249, 294, 404, 458–459, 502, 592–593, 600, 644

COURSE 3
Pages 48–50, 75, 108, 128, 150–151, 162–164, 218, 338, 351, 434–436, 442, 536, 578–580

Solve problems that arise in mathematics and in other contexts

Appears throughout each course in features such as the following:

Problem Solving on Location and *Math-Ables* at the end of each chapter
Interdisciplinary exercise sets in each chapter and *Problem Solving Applications* within each chapter
For example:

Pages 11, 15, 23, 70, 161, 198–199, 260–261, 285, 310–311, 382, 440, 490

Pages 50, 153, 180–181, 250, 292–293, 399, 402–403, 493, 502, 548, 594, 642–643

Pages 22, 67, 100–102, 134, 226, 264, 268–270, 328, 330–332, 413, 448, 501, 528–530, 575, 620

Apply and adapt a variety of appropriate strategies to solve problems

Appears throughout each course in features such as the following:

Problem Solving Skill lessons, *Focus on Problem Solving,* and *Problem Solving Applications* within each chapter
Problem Solving Handbook at the front of the book
For example:

Pages xviii–xxix, 28–29, 31–32, 52–53, 113, 165, 187, 353, 406, 455

Pages xviii–xxix, 19, 161, 192–195, 219, 230–233, 277, 387, 485, 565, 608–611

Pages xviii–xxix, 8–12, 25, 33, 92–95, 141, 145, 249, 350–354, 361, 415, 421, 456, 498, 513, 607

Monitor and reflect on the process of mathematical problem solving

Appears throughout each course in features such as the following:

Focus on Problem Solving within each chapter and *Problem Solving Applications* at the end of each chapter
Problem Solving Handbook at the front of the book
For example:

Pages xviii–xxix, 36–37, 78–79, 140–141, 198–199, 231, 283, 343, 438–439, 521

Pages xviii–xxix, 48–49, 248–249, 402–403, 443, 500–501, 529, 573, 592–593

Pages xviii–xxix, 48–49, 56, 83, 100–101, 170, 195, 210–211, 276, 299, 388–389, 396, 415, 461, 486–487, 494, 561, 632–633, 640

Reasoning and Proof

Recognize reasoning and proof as fundamental aspects of mathematics

Appears throughout each course in features such as the following:

Think and Discuss in every lesson and lab
Hands-On Labs and *Technology Labs* in every chapter
Math-Ables at the end of each chapter
Additional examples include:

COURSE 1	COURSE 2	COURSE 3
Lessons 1-6, 1-7, 2-2, 3-9, 6-8, 7-6, 7-7, 7-8, 8-2, 8-3, 10-4, 11-4; Chapter 4 Extension	Lessons 2-1, 7-7, 7-9, 10-6, 10-7, 12-2, 12-3, 12-8	Lessons 5-1, 5-2, 5-3, 5-4, 5-5, 12-1, 12-2, 12-3

Make and investigate mathematical conjectures

COURSE 1	COURSE 2	COURSE 3
Lessons 1-7, 7-8; Hands-On Labs 4C, 7B, 8A, 10A, 11B, 12A; Technology Labs 1, 7, 10, 12; Chapters 1, 9 Extensions	Lessons 7-3, 7-5, 10-1, 10-2, 10-4, 10-5; Hands-On Labs 3A, 3B, 4C, 5A, 7D, 12A; Technology Labs 2, 7, 8, 10; Chapters 8, 9, 10 Extensions	Lessons 2-8, 4-7, 6-6, 6-7, 6-8, 6-9, 9-3, 9-7, 9-8, 11-7, 12-1, 12-2, 12-3; Hands-On Labs 3B, 6A, 7B, 12A; Technology Lab 9A

Develop and evaluate mathematical arguments and proofs

Appears throughout each course in features such as the following:

Think and Discuss in every lesson and lab
Hands-On Labs and *Technology Labs* in every chapter
Write About It in the exercise sets
Additional examples include:

COURSE 1	COURSE 2	COURSE 3
Lessons 2-4, 2-5, 2-6, 2-7, 3-10, 4-1, 5-4, 5-10, 7-5, 7-6, 9-8, 11-4; Chapters 4, 9 Extensions	Lessons 2-2, 2-9, 2-10, 3-3, 3-4, 5-5, 7-8, 8-4, 8-5, 8-6, 9-2, 12-6, 12-7	Lessons 4-6, 5-1, 5-2, 5-3,5-6, 6-1, 6-2, 6-3, 6-4, 6-6, 6-7, 6-8, 6-9, 6-10, 7-6, 7-7; Chapter 7 Extension

Select and use various types of reasoning and methods of proof

Appears throughout each course in features such as the following:

Think and Discuss in every lesson and lab
Focus on Problem Solving and *Problem Solving Applications* within each chapter
Performance Assessment at the end of each chapter
Problem Solving Handbook at the front of the book
Choose a Strategy in the exercise sets
Additional examples include:

COURSE 1	COURSE 2	COURSE 3
Lessons 1-7, 2-4, 2-5, 2-6, 2-7, 3-10, 4-8, 4-9, 5-1, 5-4, 5-7, 5-10, 7-6, 7-8, 8-2, 8-3, 9-8, 11-1, 11-2, 11-3; Hands-On Labs 4B, 8A, 10A, 10B, 10C	Lessons 2-10, 7-8, 9-4, 12-6; Hands-On Labs 2C, 7C, 10A, 11A; Technology Labs 5, 7, 8, 9; Chapters 11, 12 Extensions; Chapters 1, 5, 10, 12 Math-Ables	Lessons 2-8, 4-7, 5-1, 5-2, 5-3, 5-4, 6-6, 6-7, 6-8, 6-9, 9-3, 9-7, 9-8, 11-7, 12-1, 12-2, 12-3; Hands-On Labs 3B, 4A, 5C, 6A, 7B, 12A; Technology Lab 9A

Communication

Organize and consolidate mathematical thinking through communication

Appears throughout each course in features such as the following:

Think and Discuss in every lesson and lab
Write About It in the exercise sets
For example:

COURSE 1	COURSE 2	COURSE 3
Pages 5, 7, 9, 13, 21, 63, 67, 70, 72, 93, 97, 99, 103, 133, 197, 225, 300, 329, 401	7, 38, 67, 99, 117, 165, 177, 195, 239, 247, 319, 347, 391, 399, 479, 567, 615, 631	Pages 5, 11, 30, 41, 56, 67, 75, 89, 94, 103, 144, 175, 187, 205, 246, 271, 301, 409, 470, 485, 506, 599, 626

Communicate mathematical thinking coherently and clearly to peers, teachers, and others

Appears throughout each course in features such as the following:

Focus on Problem Solving within each chapter and *Problem Solving Applications* at the end of each chapter
Problem Solving Handbook at the front of the book
For example:

COURSE 1	COURSE 2	COURSE 3
Pages 27, 55, 65, 95, 99, 123, 126, 185, 213, 215, 227, 247, 273, 274, 285, 295	Pages 17, 63, 95, 107, 159, 173, 209, 233, 315, 331, 365, 433, 515, 571, 585, 635	Pages 15, 27, 35, 44, 61, 73, 97, 103, 120, 165, 191, 194, 282, 293, 355, 404, 433, 481, 506, 531, 566, 581, 623

Analyze and evaluate the mathematical thinking and strategies of others

Appears throughout each course in features such as the following:

Think and Discuss in every lesson and lab
What's the Error? and *What's the Question?* in the exercise sets
For example:

COURSE 1	COURSE 2	COURSE 3
Pages 77, 117, 155, 249, 281, 304, 323, 333, 345, 370, 393, 410, 425, 533, 607	Pages 27, 44–47, 85, 99, 113, 133, 177, 205, 247, 315, 385, 399, 543, 619, 639	Pages 20, 42, 51, 63, 77, 97, 116, 125, 143, 213, 233, 284, 349, 403, 450, 457, 475, 505, 592, 612, 616

Use the language of mathematics to express mathematical ideas precisely

Appears throughout each course in features such as the following:

Think and Discuss in every lesson and lab
Write About It and *Write a Problem* in the exercise sets
For example:

COURSE 1	COURSE 2	COURSE 3
Pages 15, 61, 99, 161, 163, 168, 173, 223, 237, 245, 259, 364, 421, 519, 573, 619	Pages 7, 13, 27, 81, 153, 169, 199, 217, 271, 311, 385, 417, 427, 490, 533, 589, 619, 635	Pages 7, 27, 47, 61, 82, 97, 103, 120, 165, 203, 227, 258, 271, 289, 345, 379, 423, 481, 518, 594, 631

Connections

Recognize and use connections among mathematical ideas

COURSE 1	COURSE 2	COURSE 3
Lessons 3-4, 5-9, 8-3, 8-4, 8-5, 8-6, 12-3, 12-4, 12-5, 12-6; Hands-On Labs 3C, 4C; Chapter 4 Extension	Lessons 2-8, 5-1, 5-2, 5-4, 5-5, 5-6, 6-1, 7-3, 7-4, 7-5, 7-6, 7-7, 7-9, 9-5, 12-1, 12-5, 12-6; Hands-On Lab 5A; Technology Lab 5; Chapter 12 Extension	Lessons 5-5, 6-1, 6-2, 6-3, 6-4, 7-6, 7-7, 7-8, 7-9; Hands-On Labs 7A, 7B, 7C, 8A; Chapter 7 Extension

Understand how mathematical ideas interconnect and build on one another to produce a coherent whole

Course 1	Course 2	Course 3
Lessons 3-4, 5-9, 8-3, 8-4, 8-5, 8-6, 12-3, 12-4, 12-5, 12-6; Hands-On Labs 3C, 4C; Chapter 4 Extension	Lessons 3-2, 3-7, 3-8, 3-9, 6-1, 6-5, 7-10, 7-11, 8-3, 8-4, 8-5, 8-6, 8-8, 9-2, 9-3, 9-4, 12-3, 12-4; Technology Lab 7; Chapters 8, 9 Extensions	Lessons 1-6, 5-5, 6-1, 6-2, 6-3, 6-4, 7-6, 7-7, 7-8, 7-9, 10-1, 10-2, 12-4, 12-5, 12-6, 12-7; Hands-On Labs 3B, 7A, 7B, 7C, 8A; Technology Lab 12B; Chapters 6, 7, 11 Extensions

Recognize and apply mathematics in contexts outside of mathematics

Appears throughout each course in features such as the following:

Problem Solving on Location at the end of each chapter
Interdisciplinary exercise sets within each chapter
Chapter openers at the beginning of each chapter
Application problems in the exercise sets
Additional examples include:

Course 1	Course 2	Course 3
Lessons 1-1, 2-7, 3-1, 3-5, 4-6, 4-8, 6-1, 6-4, 6-6, 6-7, 7-8, 7-12, 8-6, 8-9, 9-1, 10-1, 10-2, 10-9, 11-1, 11-6, 12-6; Chapter 8 Extension	Lessons 1-1, 1-2, 1-3, 1-4, 1-5, 1-6, 1-7, 1-8, 1-9, 2-7, 2-10, 5-4, 5-7, 6-2, 6-5, 6-6, 10-1, 10-2, 10-4, 10-5, 12-8; Hands-On Labs 7F, 8A; Chapter 9 Math-Ables	Lessons 1-1, 1-3, 1-4, 2-2, 2-3, 2-4, 2-5, 2-8, 3-2, 3-3, 3-6, 3-7, 6-3, 6-4, 6-6, 6-9, 8-2, 11-1, 11-6; Hands-On Lab 4A

Representation

Create and use representations to organize, record, and communicate mathematical ideas

COURSE 1	COURSE 2	COURSE 3
Lessons 1-1, 1-3, 2-2, 3-1, 3-5, 4-4, 4-6, 6-1, 6-4, 6-5, 6-7, 6-8, 6-9, 8-5, 12-1, 12-2; Hands-On Lab 4A; Technology Labs 6A, 6; Chapters 1, 6 Extensions	Lessons 1-3, 1-4, 1-5, 1-6, 1-7, 1-8, 2-3, 2-7, 3-1, 3-2, 3-7, 3-8, 3-10; Hands-On Lab 8B; Technology Labs 1A, 1B, 2, 6, 9; Chapter 9 Math-Ables	Lessons 1-7, 1-8, 1-9, 4-1, 4-2, 4-5, 4-6, 4-7, 5-5, 6-1, 6-2, 6-3, 6-4, 6-5, 7-6, 7-7, 7-8, 11-1, 11-2, 11-3, 11-4, 12-4, 12-5, 12-6, 12-7; Hands-On Labs 2A, 3B, 4A, 5C, 6A, 6B, 7A, 8A, 7B, 10A, 12A; Technology Labs 1A, 4B, 9A, 12B

Select, apply, and translate among mathematical representations to solve problems

Course 1	Course 2	Course 3
Lessons 4-4, 4-7, 6-1, 6-4, 6-5, 6-7, 6-8, 6-9, 8-7, 8-8, 12-1, 12-2; Hands-On Lab 4A; Chapter 6 Extension	Lessons 1-9, 2-1, 2-2, 2-7, 3-7, 3-8, 3-9, 3-10, 5-4, 6-1, 8-1, 11-4, 12-3; Hands-On Labs 6A, 8A; Technology Labs 1B, 2A, 2; Chapter 1 Math-Ables	Lessons 1-9, 2-9, 7-1, 7-2, 7-3, 8-1, 8-2, 8-3, 8-4, 8-5, 8-6, 8-7; Hands-On Labs 2A, 3B, 4A, 5C, 6A, 6B, 7A, 7B, 8A, 10A, 12A; Technology Labs 1A, 9A, 12B

Use representations to model and interpret physical, social, and mathematical phenomena

Course 1	Course 2	Course 3
Lessons 2-3, 2-4, 2-5, 2-6, 2-7, 3-10, 5-4, 5-10, 6-1, 6-4, 6-5, 6-7, 6-8, 6-9, 8-6, 9-8, 12-2; Hands-On Labs 11A, 8B; Technology Labs 6A, 6, 11; Chapters 2, 6 Extensions	Lessons 3-4, 4-5, 5-7; Hands-On Labs 2C, 3A, 3B, 3C, 4A, 4B, 4C, 5A, 6A, 11A, 12A; Technology Labs 5, 6, 7, 8, 9, 12; Chapters 9, 10, 12 Math-Ables	Lessons 1-8, 1-9, 4-1, 4-2, 4-5, 4-6, 4-7, 7-7, 7-8, 7-9, 9-3, 11-1, 11-2, 11-3, 11-4, 11-5, 11-6, 11-7, 12-4, 12-5, 12-6, 12-7, 12-8; Hands-On Labs 2A, 3B, 4A, 5C, 6A, 6B, 7A, 7B, 7C, 8A, 10A, 12A; Technology Labs 11A, 12B

Stepping into the Future

All the Ways You Teach

Being ready for the future means making sure no student is left behind. Throughout the *Student Edition, Teacher Edition, Premier Online Edition*, and program resources for ***Holt Middle School Math,*** you'll find the help you need to teach, assist, and assess every ability level in your class without spending hours of extra time assigning special assignments and grading extra worksheets.

All the Ways They Learn

The perfect middle school math curriculum for the future ensures that students can master math concepts found in state tests, they are able to learn at their own speed and style, and that they are able to manipulate information to use it in everyday terms with everyday technology.

We're With Them Every Step of the Way

Meeting State Standards and Assessment

This program reflects the increased expectation of what students should "know and be able to do" by introducing algebra early and offering test preparation at the lesson and chapter levels and online through **go.hrw.com**.

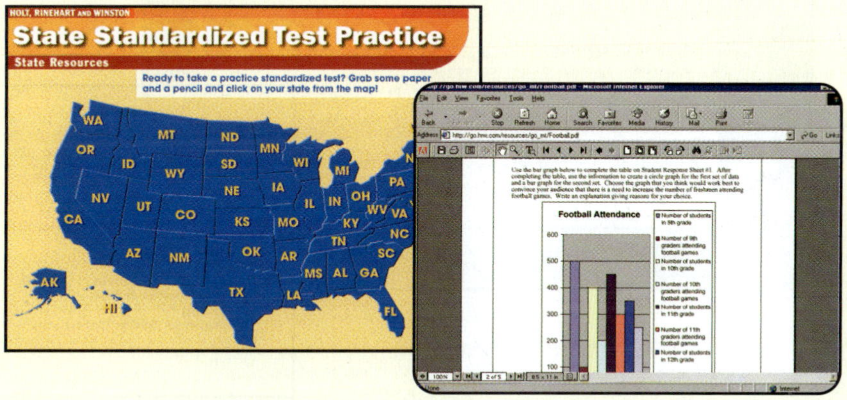

Differentiating Instruction

By offering early intervention and examples that allow students to get help early and practice at their own level of learning, this program helps you reach all learners, including students who need extra help and advanced learners.

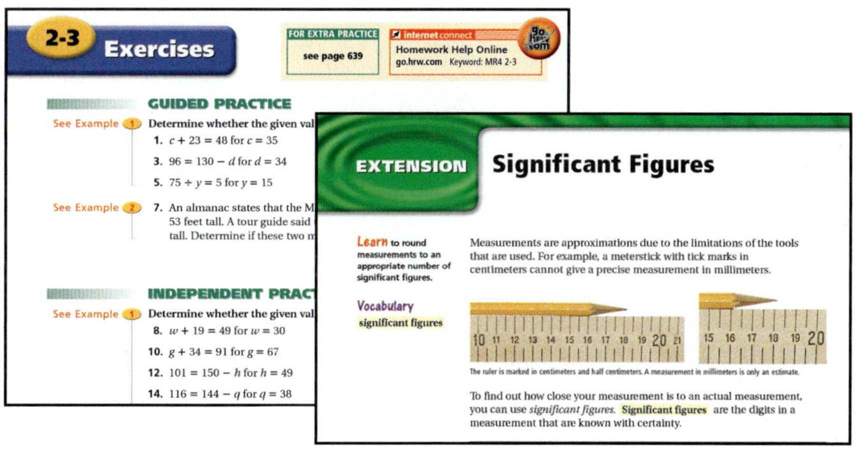

Integrating Technology

This program offers you state-of-the-art technology integrated with your curriculum saving you time and increasing your efficiency with the new test and practice generator on the ***One-Stop Planner® CD-ROM,*** the ***Are You Ready? Intervention CD-ROM,*** and the ***Holt Middle School Math Premier Online Edition.***

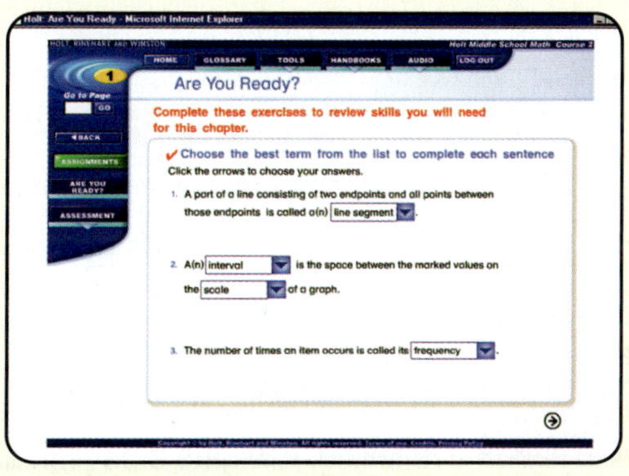

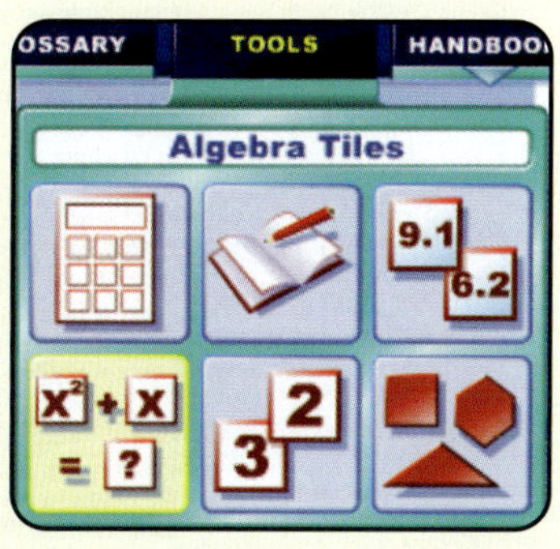

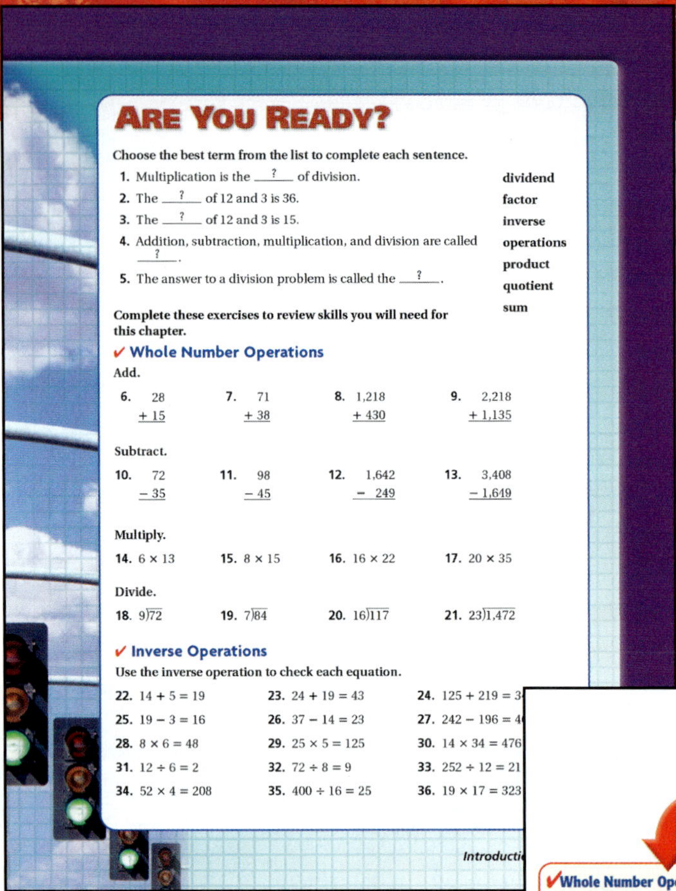

ARE YOU READY?

Choose the best term from the list to complete each sentence.

1. Multiplication is the __?__ of division.
2. The __?__ of 12 and 3 is 36.
3. The __?__ of 12 and 3 is 15.
4. Addition, subtraction, multiplication, and division are called __?__.
5. The answer to a division problem is called the __?__.

dividend
factor
inverse
operations
product
quotient
sum

Complete these exercises to review skills you will need for this chapter.

✔ Whole Number Operations

Add.

6. 28 + 15 7. 71 + 38 8. 1,218 + 430 9. 2,218 + 1,135

Subtract.

10. 72 − 35 11. 98 − 45 12. 1,642 − 249 13. 3,408 − 1,649

Multiply.

14. 6 × 13 15. 8 × 15 16. 16 × 22 17. 20 × 35

Divide.

18. 9)72 19. 7)84 20. 16)117 21. 23)1,472

✔ Inverse Operations

Use the inverse operation to check each equation.

22. 14 + 5 = 19 23. 24 + 19 = 43 24. 125 + 219 = 3
25. 19 − 3 = 16 26. 37 − 14 = 23 27. 242 − 196 = 4
28. 8 × 6 = 48 29. 25 × 5 = 125 30. 14 × 34 = 476
31. 12 ÷ 6 = 2 32. 72 ÷ 8 = 9 33. 252 ÷ 12 = 21
34. 52 × 4 = 208 35. 400 ÷ 16 = 25 36. 19 × 17 = 323

Introducti

Assessing Prior Knowledge

INTERVENTION

Diagnose and Prescribe

Evaluate your students' performance on this page to determine whether intervention is necessary or whether enrichment is appropriate. Options that provide instruction, practice, and a check are listed below.

Resources for Are You Ready?

- ***Are You Ready? Intervention and Enrichment***
- **Recording Sheet for Are You Ready?** *Chapter 2 Resource Book* . p. 00
- ***Are You Ready? Intervention CD-ROM***

Diagnose and Prescribe

Holt Middle School Math includes intervention strategies that diagnose students' difficulties with mathematics while providing intervention resources that will bring success to every learner.

Intervention Tools

Each chapter begins with **Are You Ready?** assessing students' knowledge of prerequisite skills, and helping you assign either more help or enrichment depending on where students stand.

Intervention

Intervention Components

Are You Ready? Intervention and Enrichment

This workbook provides additional help for students who have difficulty with a particular math concept through direct instruction, conceptual models, and scaffolded practice. Enrichment masters for every lesson enhance critical-thinking skills, as well as extend lesson objectives. Also available: ***Are You Ready? Intervention CD-ROM***.

Readiness Activities

This helpful resource contains activities for every chapter and helps students review prerequisite skills needed to complete each lesson.

Math: Reading and Writing in the Content Area

These activities provide strategies for students to help organize their thinking, navigate a page in a math book, and master vocabulary.

Meeting State Standards

Assessment that Gets Results

Throughout each chapter, students have access to a series of assessment resources, including **Chapter Review, Chapter Test, Performance Assessment,** and **Standardized Test Prep.**

Show What You Know

Students are asked to select work from the recently completed chapter—including section reviews, homework assignments, etc.—and create a portfolio from those pieces.

Short Response

Students are asked to execute a series of tasks, including creating their own problems with explanations of how they arrived at their conclusions and collecting magazine or newspaper articles that mention mathematical concepts.

Extended Problem Solving

Students are asked to solve a set of problems using a graph, table, etc. Students are able to use the strategy of their choice to complete the series.

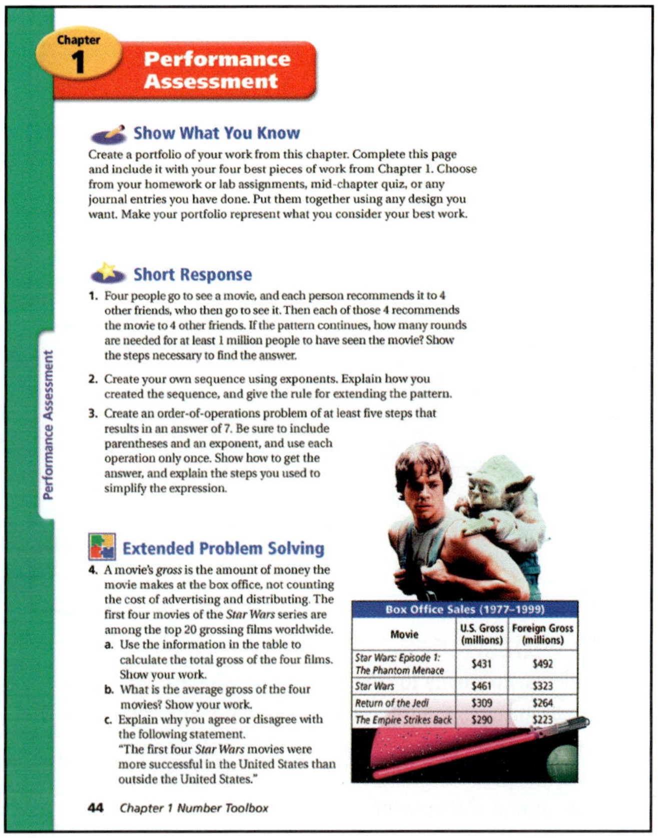

Chapter 1 **Performance Assessment**

Show What You Know

Create a portfolio of your work from this chapter. Complete this page and include it with your four best pieces of work from Chapter 1. Choose from your homework or lab assignments, mid-chapter quiz, or any journal entries you have done. Put them together using any design you want. Make your portfolio represent what you consider your best work.

Short Response

1. Four people go to see a movie, and each person recommends it to 4 other friends, who then go to see it. Then each of those 4 recommends the movie to 4 other friends. If the pattern continues, how many rounds are needed for at least 1 million people to have seen the movie? Show the steps necessary to find the answer.
2. Create your own sequence using exponents. Explain how you created the sequence, and give the rule for extending the pattern.
3. Create an order-of-operations problem of at least five steps that results in an answer of 7. Be sure to include parentheses and an exponent, and use each operation only once. Show how to get the answer, and explain the steps you used to simplify the expression.

Extended Problem Solving

4. A movie's *gross* is the amount of money the movie makes at the box office, not counting the cost of advertising and distributing. The first four movies of the *Star Wars* series are among the top 20 grossing films worldwide.
   a. Use the information in the table to calculate the total gross of the four films. Show your work.
   b. What is the average gross of the four movies? Show your work.
   c. Explain why you agree or disagree with the following statement. "The first four *Star Wars* movies were more successful in the United States than outside the United States."

Box Office Sales (1977–1999)		
Movie	U.S. Gross (millions)	Foreign Gross (millions)
Star Wars: Episode 1: The Phantom Menace	$431	$492
Star Wars	$461	$323
Return of the Jedi	$309	$264
The Empire Strikes Back	$290	$223

Performance Assessment

44 *Chapter 1 Number Toolbox*

Assessment Resources

- Inventory Assessment
- Section Quizzes
- Chapter Tests
- Performance Assessment
- Cumulative Tests
- End-of-year Test

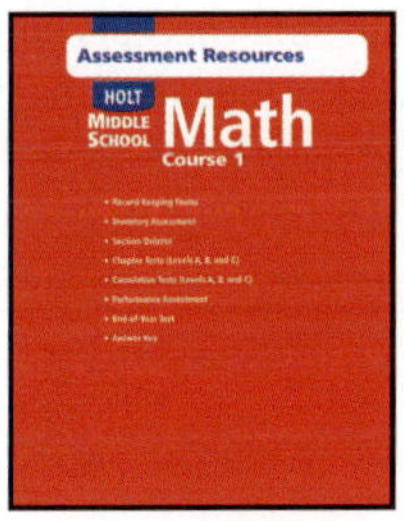

Test Prep Toolkit

Countdown to Testing Transparencies

Multiple choice and critical-thinking questions are featured on these transparencies, preparing students for state assessment by building problem-solving skills.

Standardized Test Prep Workbook

This resource includes a two-page test for every chapter as well as two state-specific tests, a diagnostic test and **Test-Taking Tips**.

Standardized Test Prep CD-ROM

This convenient assessment tool provides an easy way to create practice worksheets and tests that correlate to your state assessment.

Standardized Test Prep Video

This video gives visual demonstrations of math problems correlated to the book and to state standards. Hints and suggestions guide students toward solutions, while timed practice gives students experience solving multiple choice and short answer problems.

Assessment

Focus on Problem Solving

To support problem-solving skills needed in state testing, this feature gives students a real-world scenario with the steps, examples, and practice problems needed to fully master the concept.

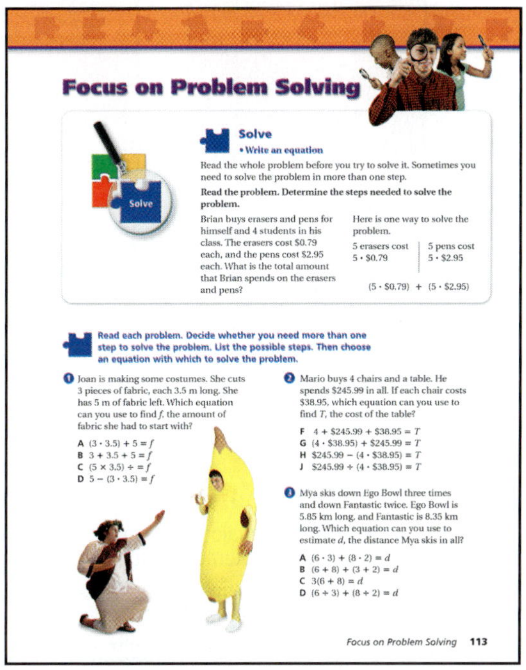

Focus on Problem Solving

Solve
• Write an equation

Read the whole problem before you try to solve it. Sometimes you need to solve the problem in more than one step.

Read the problem. Determine the steps needed to solve the problem.

Brian buys erasers and pens for himself and 4 students in his class. The erasers cost $0.79 each, and the pens cost $2.95 each. What is the total amount that Brian spends on the erasers and pens?

Here is one way to solve the problem.

5 erasers cost 5 · $0.79 | 5 pens cost 5 · $2.95

(5 · $0.79) + (5 · $2.95)

Read each problem. Decide whether you need more than one step to solve the problem. List the possible steps. Then choose an equation with which to solve the problem.

1. Joan is making some costumes. She cuts 3 pieces of fabric, each 3.5 m long. She has 5 m of fabric left. Which equation can you use to find f, the amount of fabric she had to start with?
   A $(3 \cdot 3.5) + 5 = f$
   B $3 + 3.5 + 5 = f$
   C $(5 \times 3.5) \div = f$
   D $5 - (3 \cdot 3.5) = f$
2. Mario buys 4 chairs and a table. He spends $245.99 in all. If each chair costs $38.95, which equation can you use to find T, the cost of the table?
   F $4 + \$245.99 + \$38.95 = T$
   G $(4 \cdot \$38.95) + \$245.99 = T$
   H $\$245.99 - (4 \cdot \$38.95) = T$
   J $\$245.99 \div (4 \cdot \$38.95) = T$
3. Mya skis down Ego Bowl three times and down Fantastic twice. Ego Bowl is 5.85 km long, and Fantastic is 8.35 km long. Which equation can you use to estimate d, the distance Mya skis in all?
   A $(6 \cdot 3) + (8 \cdot 2) = d$
   B $(6 + 8) + (3 + 2) = d$
   C $3(6 + 8) = d$
   D $(6 \div 3) + (8 \div 2) = d$

Focus on Problem Solving 113

Assessment

For Students

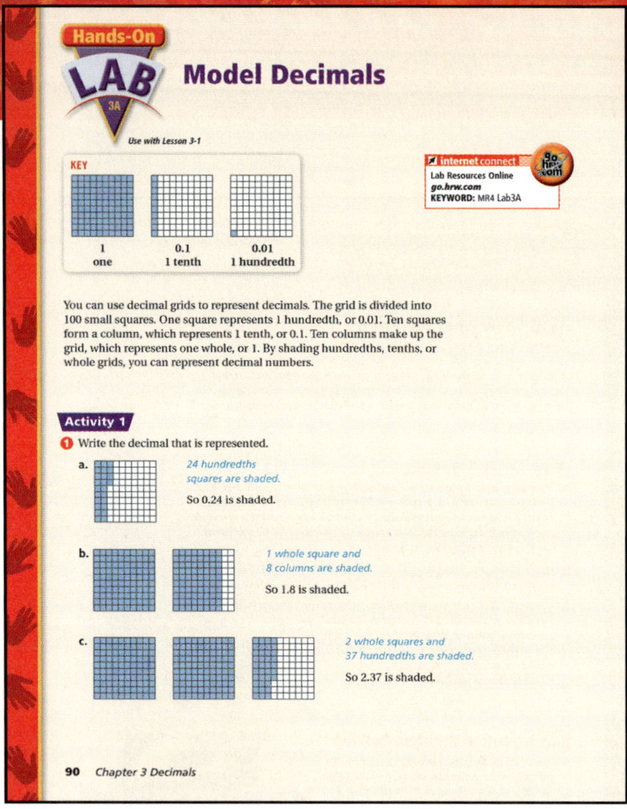

Hands-On LAB 3A

Model Decimals

Use with Lesson 3-1

internet connect
Lab Resources Online
go.hrw.com
KEYWORD: MR4 Lab3A

KEY

1 one	0.1 1 tenth	0.01 1 hundredth

You can use decimal grids to represent decimals. The grid is divided into 100 small squares. One square represents 1 hundredth, or 0.01. Ten squares form a column, which represents 1 tenth, or 0.1. Ten columns make up the grid, which represents one whole, or 1. By shading hundredths, tenths, or whole grids, you can represent decimal numbers.

Activity 1

1 Write the decimal that is represented.

a. *24 hundredths squares are shaded.* So 0.24 is shaded.

b. *1 whole square and 8 columns are shaded.* So 1.8 is shaded.

c. *2 whole squares and 37 hundredths are shaded.* So 2.37 is shaded.

90 *Chapter 3 Decimals*

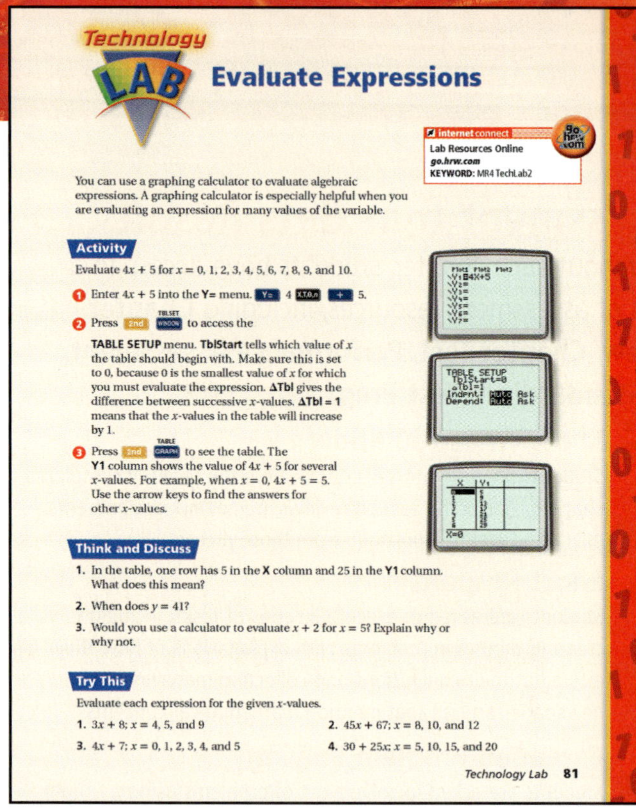

Technology LAB

Evaluate Expressions

internet connect
Lab Resources Online
go.hrw.com
KEYWORD: MR4 TechLab2

You can use a graphing calculator to evaluate algebraic expressions. A graphing calculator is especially helpful when you are evaluating an expression for many values of the variable.

Activity

Evaluate $4x + 5$ for $x = 0, 1, 2, 3, 4, 5, 6, 7, 8, 9,$ and 10.

1 Enter $4x + 5$ into the **Y=** menu. Y= 4 X,T,θ,n + 5.

2 Press 2nd WINDOW to access the **TABLE SETUP** menu. **TblStart** tells which value of x the table should begin with. Make sure this is set to 0, because 0 is the smallest value of x for which you must evaluate the expression. **ΔTbl** gives the difference between successive x-values. **ΔTbl = 1** means that the x-values in the table will increase by 1.

3 Press 2nd GRAPH to see the table. The **Y1** column shows the value of $4x + 5$ for several x-values. For example, when $x = 0$, $4x + 5 = 5$. Use the arrow keys to find the answers for other x-values.

Think and Discuss

1. In the table, one row has 5 in the **X** column and 25 in the **Y1** column. What does this mean?
2. When does $y = 41$?
3. Would you use a calculator to evaluate $x + 2$ for $x = 5$? Explain why or why not.

Try This

Evaluate each expression for the given x-values.

1. $3x + 8$; $x = 4, 5,$ and 9
2. $45x + 67$; $x = 8, 10,$ and 12
3. $4x + 7$; $x = 0, 1, 2, 3, 4,$ and 5
4. $30 + 25x$; $x = 5, 10, 15,$ and 20

Technology Lab 81

Kinesthetic Models

Give students an activity to step up learning. Models, algebra tiles, graphs, pictures that represent real-world math, and other visual representations are all incorporated into **Hands-On Lab** giving the reinforcement needed for rigorous content. Lab Resources also available online.

Visual Appeal

A graphing calculator can give learners the support they need to illustrate concepts and help retention of skills. **Technology Lab** walks students through a set of problems using their calculators. Lab Resources also available online.

Reaching All Learners

Alternate Openers: Exploration Transparencies

This notebook of transparencies provides an alternate way to teach each lesson.

Interactive Problem Solving

Blackline masters enable students to work problems for the lesson with step-by-step prompts.

Success for English Language Learners

These masters present the same concepts as the student lesson using fewer words and more visuals. Also includes teacher support with suggested activities and teaching tips.

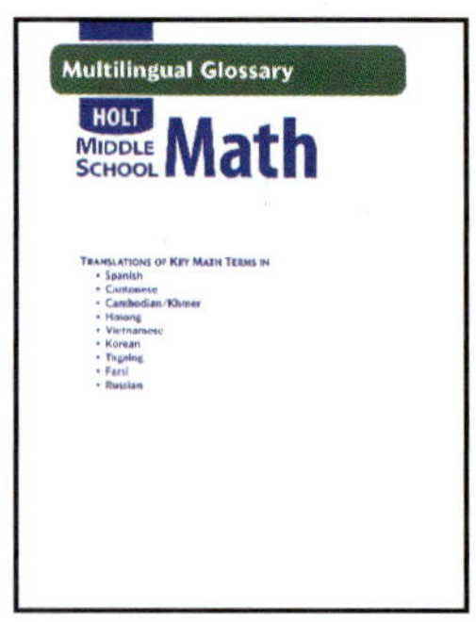

Multilingual Glossary

This glossary contains translations of key mathematical terms in Spanish, Cantonese, Cambodian/Khmer, Hmong, Vietnamese, Korean, Tagalog, Farsi, and Russian.

SPANISH STUDENT EDITIONS AVAILABLE!

Spanish Resources

- Libro de Trabajo: Guía Interactiva de Estudio (Spanish Interactive Study Guide Workbook)
- Libro de Trabajo: Tarea y Práctica (Spanish Homework and Practice Workbook)
- Activades de Apoyo Familia (Family Involvement Activities)

Internet Connect:

Students can link to homework help online directly related to chapter content with this in-text feature.

internet connect
Homework Help Online
go.hrw.com Keyword: MR4 2-7

Differentiating Instruction

Spark Interest

Show your students that math is needed in the real world.

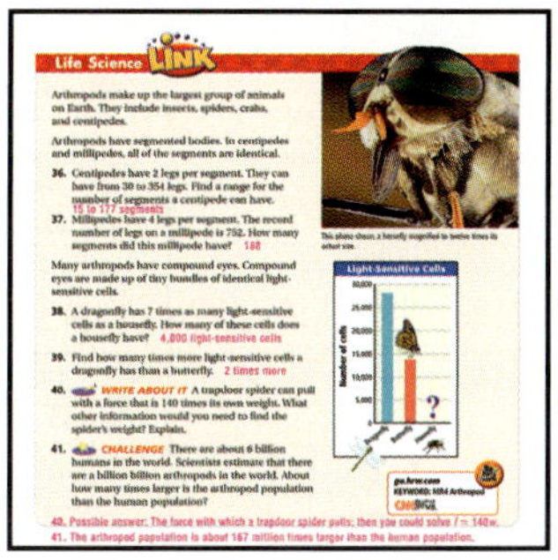

Interdisciplinary Link and Internet Activities

Interdisciplinary Links are featured throughout the program, generating interest and showing students how important math is in our everyday lives. Internet activities online at **go.hrw.com**.

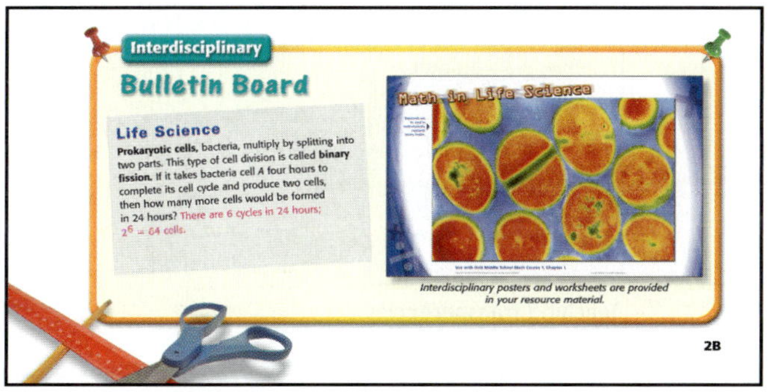

Interdisciplinary Bulletin Board

This feature gives you suggestions about how to update your class visually to complement the section you're about to teach.

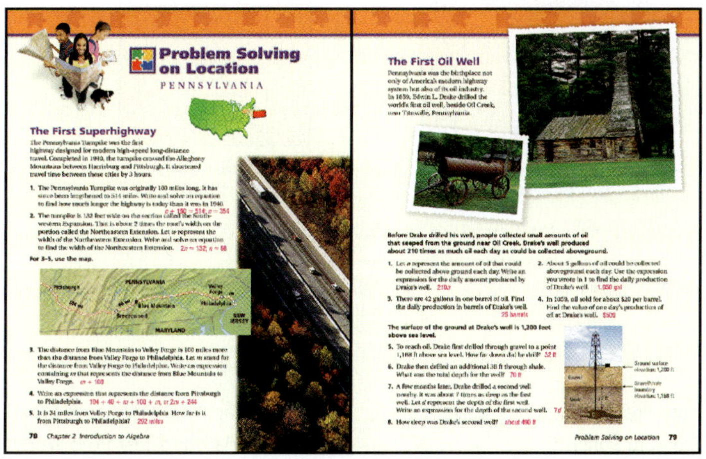

Problem Solving on Location

This feature takes students on a journey to places across the nation. students become familiar with geography and also learn a math lesson specific to that location, helping them sharpen their critical-thinking skills needed for state testing.

Teacher to Teacher

What is Differentiated Instruction?

Differentiated instruction answers the question of how to teach every student in a classroom—regardless of ability level or background—in a way that is equal, effective, and efficient.

Differentiated instruction is more and more becoming the alternative to a "one-size-fits-all classroom." Instead of only offering core information and basic instruction of skills, a curriculum centered around differentiated instruction offers information tailored to a student's readiness, interest, and profile (the way they learn).

How Do I Reach All of My Students?

Holt Middle School Math ensures differentiated instruction by offering content, resources, and technology that speak to all types of learners simultaneously. A visual learner, an auditory learner, a kinesthetic learner, and an English-language learner can all benefit from features and resources such as:

- **Hands-On Lab**
- **Technology Lab**
- **Career Link**
- **Interdisciplinary Links**
- **Internet Activities**
- **Interdisciplinary Bulletin Board**
- **Problem Solving on Location**

What better way to ensure understanding than through a program that helps you do all of this without having to maintain a separate agenda for each student? Differentiated instruction gives each student what he or she needs while they are learning with everyone else in the class. State and national requirements are also taught throughout the program. The differentiated instruction curriculum in ***Holt Middle School Math*** takes care of the teacher as well, making everything streamlined and easy to manage.

Background information on Differentiated Instruction was researched from Carol Ann Tomlinson's book *How to Differentiate Instruction in Mixed-Ability Classrooms* (Alexandria, VA: Association for Supervision and Curriculum Development, 1995). Tomlinson is Professor of Educational Leadership, Foundations, and Policy at the Curry School of Education, UVA. ASCD (Assoc. for Supervision and Curriculum Development) publishes much of her work.

Holt offers a comprehensive and systematic training program to complement ***Holt Middle School Math*** providing high-quality and accessible professional learning opportunities designed to relate to the unique needs of the educator. For more information on professional development services provided by Holt, email us at **holtinfo@hrw.com.**

HOLT Professional Development

Program Resources

Chapter Resources Booklet

These comprehensive books for each of the twelve chapters include all of the items needed to extend and reinforce the students' and teacher's books. Blackline masters, transparencies, and more are found all in one place making it convenient to plan your lesson.

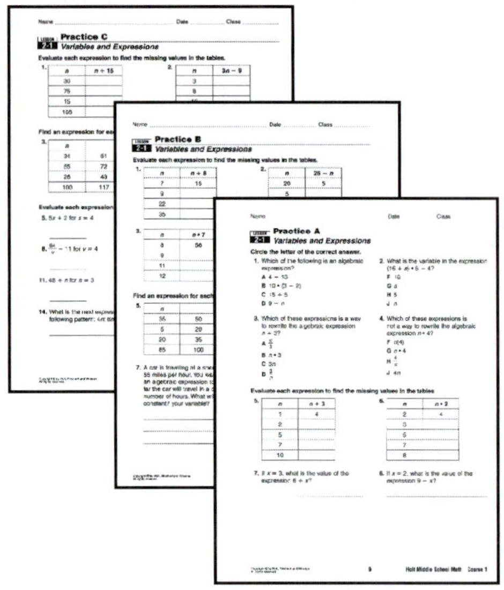

- Practice A, B, and C
- Reteach Masters
- Challenge Masters
- Problem Solving Masters
- Puzzles, Twisters, & Teasers
- and More!

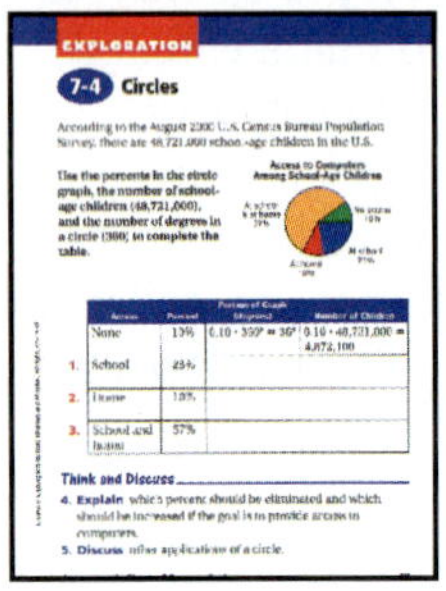

Exploration Transparency

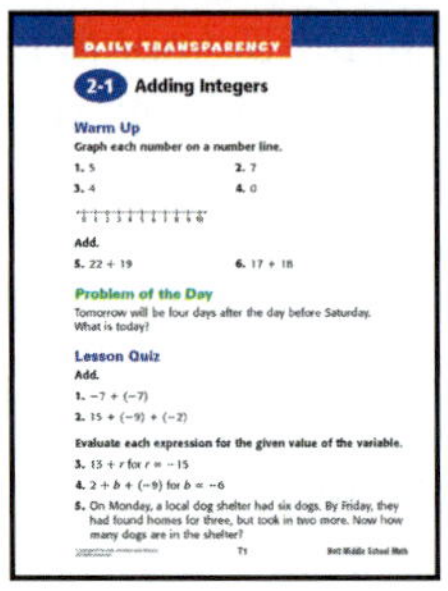

Daily Transparency

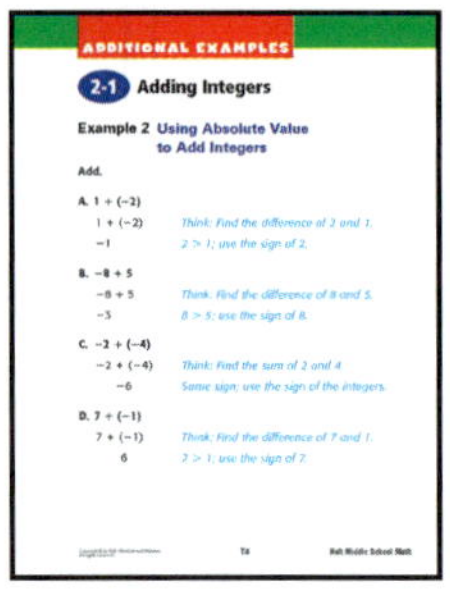

Additional Examples Transparency

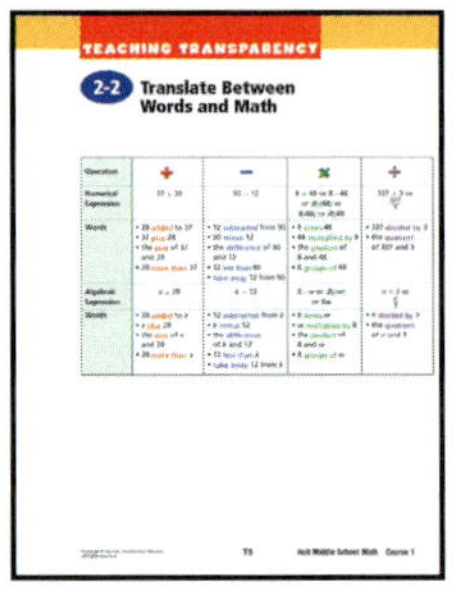

Teaching Transparencies

Assessment and Everyday Teaching Resources

- Assessment Resources
  - Inventory Assessment
  - Section Quizzes
  - Chapter Tests
  - Performance Assessment
  - Cumulative tests
  - End-of-year Test
- Test Prep Tool Kit
- Lesson Plans
- Solution Key
- Answer Transparencies

Homework and Practice Workbook

Interactive Study Guide Workbook

Standardized Test Prep Workbook

Reaching All Learners

- Alternate Openers: Explorations Transparencies
- Are You Ready? Intervention and Enrichment
- Consumer and Career Math
- Family Involvement Activities
- ACTIVIDADES DE APOYO FAMILIAR (Family Involvement Activities)
- Hands-On Lab Activities
- Interdisciplinary Posters and Worksheets
- Interactive Problem Solving
- Math: Reading and Writing in the Content Area
- Multilingual Glossary
- Technology Lab Activities
- Success for English Language Learners
- Readiness Activities
- Interactive Study Guide Workbook
- Standardized Test Prep Workbook
- Homework and Practice Workbook
- Libro de Trabajo: Guía Interactiva de Estud (Spanish Interactive Study Guide Workbook
- Libro de Trabajo: Tarea y Práctica (Spanish Homework and Practice Workboo

Technology that Engages and Expands Learning

Holt Middle School Math **offers an array of technology products that promote mathematics teaching and learning.**

One-Stop Planner® CD-ROM with New! Test and Practice Generator

This convenient tool for planning and managing lessons contains all the print-based teaching resources plus customizable lesson plans. You'll also be able to create tests and quizzes that correlate to your state assessment with the new test generator. All of these resources are accessible with the click of the mouse!

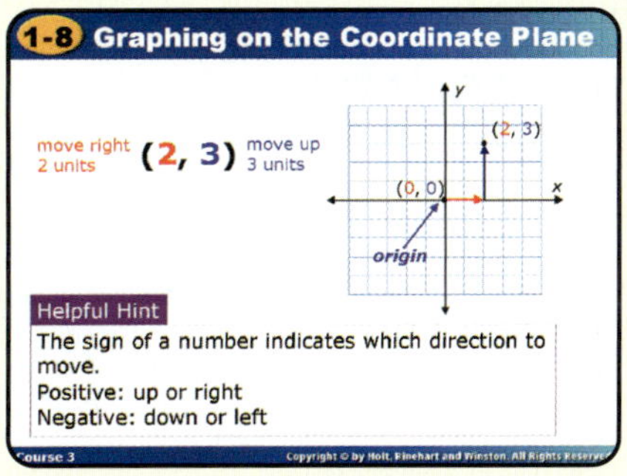

Lesson Presentations CD-ROM

This resource contains colorful, animated electronic lesson presentations—a convenient alternative to the blackboard or overhead projector! This program can also be used by individual students as a tutorial.

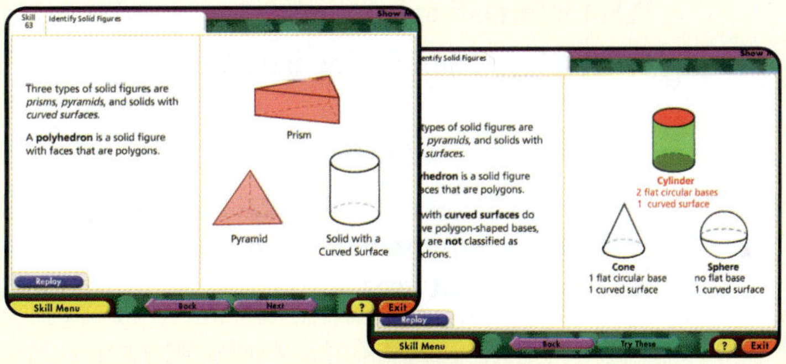

Are You Ready? Intervention CD-ROM

This CD-ROM provides an easy method of evaluating students' knowledge and administering additional help with prerequisite skills necessary for success. Students can work independently with computer-guided instruction and practice and can have their skills checked by computer-administered testing.

Electronic Textbooks Lighten the Load

Textbooks from **Holt Online Learning** are portable, expandable, interactive, and yet weigh nothing at all. You'll find interactive exercises and feedback, homework help, presentation materials and much more.

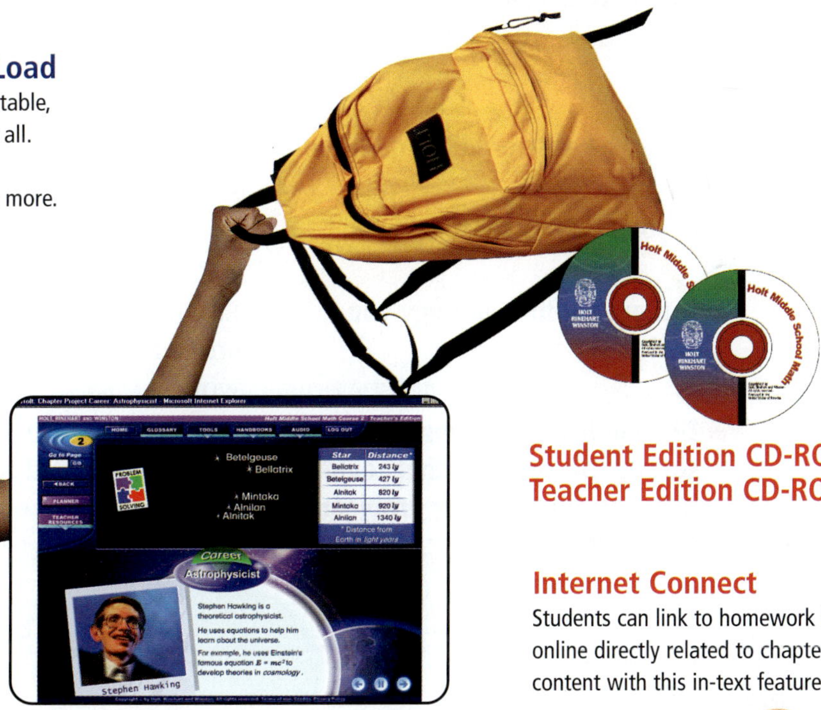

Student Edition CD-ROM
Teacher Edition CD-ROM

Internet Connect

Students can link to homework help online directly related to chapter content with this in-text feature.

internet connect
Homework Help Online
go.hrw.com Keyword: MR4 2-7

Premier Online Edition

The *Premier Online Edition* features content and assessment correlated to state standards.

Problem Solving Handbook

The Problem Solving Plan

In order to be a good problem solver, you first need a good problem-solving plan. The plan used in this book is detailed below.

UNDERSTAND the Problem

What are you asked to find?	Restate the question in your own words.
What information is given?	Identify the facts in the problem.
What information do you need?	Determine which facts are needed to answer the question.
Is all the information given?	Determine whether all the facts are given.
Is there any information given that you will not use?	Determine which facts, if any, are unnecessary to solve the problem.

Make a PLAN

Have you ever solved a similar problem?	Think about other problems like this that you successfully solved.
What strategy or strategies can you use?	Determine a strategy that you can use and how you will use it.

SOLVE

Follow your plan.	Show the steps in your solution. Write your answer as a complete sentence.

LOOK BACK

Have you answered the question?	Be sure that you answered the question that is being asked.
Is your answer reasonable?	Your answer should make sense in the context of the problem.
Is there another strategy you could use?	Solving the problem using another strategy is a good way to check your work.
Did you learn anything while solving this problem that could help you solve similar problems in the future?	Try to remember the problems you have solved and the strategies you used to solve them.

Using the Problem Solving Plan

During summer vacation, Nicholas will visit first his cousin and then his grandmother. He will be gone for 5 weeks and 2 days, and he will spend 9 more days with his cousin than with his grandmother. How long will he stay with each family member?

UNDERSTAND the Problem

Identify the important information.

- Nicholas's visits will total 5 weeks and 2 days.
- He will spend 9 more days with his cousin than with his grandmother.

The answer will be how long he will stay with each family member.

Make a PLAN

You can draw a diagram to show how long Nicholas will stay. Use boxes for the length of each stay. The length of each box will represent the length of each stay.

SOLVE

Think: There are 7 days in a week, so 5 weeks and 2 days is 37 days in all. Your diagram might look like this:

Cousin	? days	9 days	= 37 days
Grandmother	? days		

Cousin	14 days	9 days	37 − 9 = 28	*Subtract 9 days from the total days.*
Grandmother	14 days		28 ÷ 2 = 14	*Divide this number by 2 for the 2 places he will visit.*

So Nicholas will stay with his cousin for 23 days and with his grandmother for 14 days.

LOOK BACK

Twenty-three days is 9 days longer than 14 days. The total of the two stays is 23 + 14, or 37 days, which is the same as 5 weeks and 2 days. This solution fits the description of Nicholas's trip given in the problem.

Draw a Diagram

When problems involve objects, distances, or places, you can **draw a diagram** to make the problem easier to understand. You will often be able to use your diagram to solve the problem.

Problem Solving Strategies

Draw a Diagram	Make a Table
Make a Model	Solve a Simpler Problem
Guess and Test	Use Logical Reasoning
Work Backward	Use a Venn Diagram
Find a Pattern	Make an Organized List

All city blocks in Sunnydale are the same size. Tina starts her paper route at the corner of two streets. She travels 8 blocks south, 13 blocks west, 8 blocks north, and 6 blocks east. How far is she from her starting point when she finishes her route?

Understand the Problem

Identify the important information.

- Each block is the same size.
- You are given Tina's route.

The answer will be the distance from her starting point.

Make a Plan

Use the information in the problem to **draw a diagram** showing Tina's route. Label her starting and ending points.

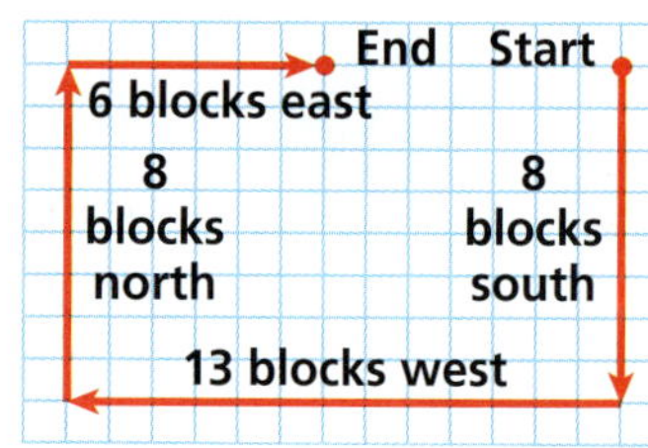

Solve

The diagram shows that at the end of Tina's route she is 13 − 6 blocks from her starting point.

$13 - 6 = 7$

When Tina finishes, she is 7 blocks from her starting point.

Look Back

Be sure that you have drawn your diagram correctly. Does it match the information given in the problem?

PRACTICE

1. Laurence drives a carpool to school every Monday. He starts at his house and travels 4 miles south to pick up two children. Then he drives 9 miles west to pick up two more children, and then he drives 4 miles north to pick up one more child. Finally, he drives 5 miles east to get to the school. How far does he have to travel to get back home? **4 mi**

2. The roots of a tree reach 12 feet into the ground. A kitten is stuck 5 feet from the top of the tree. From the treetop to the root bottom, the tree measures 32 feet. How far above the ground is the kitten? **15 ft**

Make a Model

If a problem involves objects, you can sometimes **make a model** using those objects or similar objects to act out the problem. This can help you understand the problem and find the solution.

Problem Solving Strategies

Draw a Diagram	Make a Table
Make a Model	Solve a Simpler Problem
Guess and Test	Use Logical Reasoning
Work Backward	Use a Venn Diagram
Find a Pattern	Make an Organized List

Alice has three pieces of ribbon. Their lengths are 7 inches, 10 inches, and 12 inches. Alice does not have a ruler or scissors. How can she use these ribbons to measure a length of 15 inches?

Understand the Problem

Identify the important information.

- The ribbons are 7 inches, 10 inches, and 12 inches long.

The answer will show how to use the ribbons to measure 15 inches.

Make a Plan

Measure and cut three ribbons or strips of paper to **make a model.** One ribbon should be 7 inches long, one should be 10 inches long, and one should be 12 inches long. Try different combinations of the ribbons to form new lengths.

Solve

When you put any two ribbons together end to end, you can form lengths of 17, 19, and 22 inches. All of these are too long.

Try placing the 10-inch ribbon and the 12-inch ribbon end to end to make 22 inches. Now place the 7-inch ribbon above them. The remaining length that is **not** underneath the 7-inch ribbon will measure 15 inches.

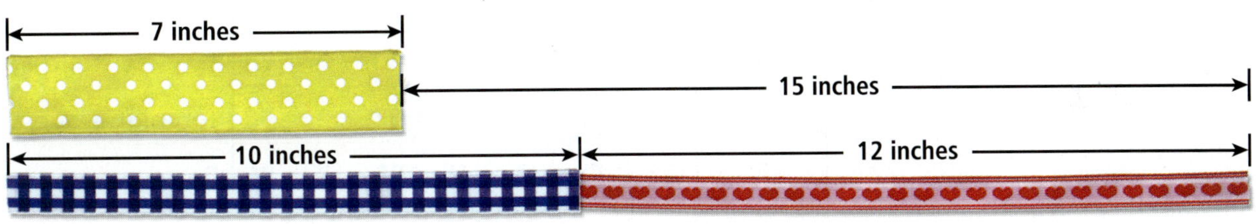

Look Back

Use another strategy. Without using ribbon, you could have **guessed** different ways to add or subtract 7, 10, and 12. Then you could have **tested** to see if any of these gave an answer of 15:

$10 + 12 - 7 = 15$

PRACTICE

Possible answers: Make 5 inches by folding the 10-inch piece in half; make 9 inches by placing the 12-inch piece and the 7-inch piece together and placing the 10-inch piece above them.

1. Find other lengths that you can measure with the three pieces of ribbon.
2. Andy stacks four cubes, one on top of the other, and paints the outside of the stack (not the bottom). How many faces of the cubes are painted?

17

Guess and Test

If you do not know how to solve a problem, you can always make a **guess**. Then **test** your guess using the information in the problem. Use the result to make a better guess. Repeat until you find the correct answer.

Problem Solving Strategies

Draw a Diagram	Make a Table
Make a Model	Solve a Simpler Problem
Guess and Test	Use Logical Reasoning
Work Backward	Use a Venn Diagram
Find a Pattern	Make an Organized List

There were 25 problems on a test. For each correct answer, 4 points were given. For each incorrect answer, 1 point was subtracted. Tania answered all 25 problems. Her score was 85. How many correct and incorrect answers did she have?

Understand the Problem

Identify the important information.

- There were 25 problems on the test.
- A correct answer received 4 points, and an incorrect answer lost 1 point.
- Tania answered all of the problems and her score was 85.

The answer will be the number of problems that Tania got correct and incorrect.

Make a Plan

Start with a **guess** for the number of correct answers. Then **test** to see whether the total score is 85.

Solve

Make a first guess of 20 correct answers.

Correct	Incorrect	Score	Result
20	5	$(20 \times 4) - (5 \times 1) = 80 - 5 = 75$	Too low—guess higher
23	2	$(23 \times 4) - (2 \times 1) = 92 - 2 = 90$	Too high—guess lower
22	3	$(22 \times 4) - (3 \times 1) = 88 - 3 = 85$	Correct ✓

Tania had 22 correct answers and 3 incorrect answers.

Look Back

Notice that the guesses made while solving this problem were not just "wild" guesses. Guessing and testing in an organized way will often lead you to the correct answer.

PRACTICE

1. The sum of Joe's age and his younger brother's age is 38. The difference between their ages is 8. How old are Joe and his brother? Joe is 23; his brother is 15.

2. 12 books; 5 cost $0.50.

2. Amy bought some used books for $4.95. She paid $0.50 each for some books and $0.35 each for the others. She bought fewer than 8 books at each price. How many books did Amy buy? How many cost $0.50?

Work Backward

Some problems give you a sequence of information and ask you to find something that happened at the beginning. To solve a problem like this, you may want to start at the end of the problem and **work backward.**

Problem Solving Strategies

Draw a Diagram	Make a Table
Make a Model	Solve a Simpler Problem
Guess and Test	Use Logical Reasoning
Work Backward	Use a Venn Diagram
Find a Pattern	Make an Organized List

Jaclyn and her twin sister, Bailey, received money for their birthday. They used half of their money to buy a video game. Then they spent half of the money they had left on a pizza. Finally, they spent half of the remaining money to rent a movie. At the end of the day, they had $4.50. How much money did they have to start out with?

Understand the Problem

Identify the important information.

- The girls ended with $4.50.
- They spent half of their money at each of three stops.

The answer will be the amount of money they started with.

Make a Plan

Start with the amount you know the girls have left, $4.50, and **work backward** through the information given in the problem.

Solve

Jaclyn and Bailey had $4.50 at the end of the day.

They had twice that amount before renting a movie. $2 \times \$4.50 = \$9$

They had twice that amount before buying a pizza. $2 \times \$9 = \$18$

They had twice that amount before buying a video game. $2 \times \$18 = \$36$

The girls started with $36.

Look Back

Using the starting amount of $36, work from the beginning of the problem. Find the amount they spent at each location and see whether they are left with $4.50.

Start:	$36
Video game:	$\$36 \div 2 = \$18$
Pizza:	$\$18 \div 2 = \$9$
Movie rental:	$\$9 \div 2 = \$4.50$ ✓

PRACTICE

1. The Lauber family has 4 children. Chris is 5 years younger than his brother Mark. Justin is half as old as his brother Chris. Mary, who is 10, is 3 years younger than Justin. How old is Mark? 31

2. If you divide a mystery number by 4, add 8, and multiply by 3, you get 42. What is the mystery number? 24

Find a Pattern

In some problems, there is a relationship between different pieces of information. Examine this relationship and try to **find a pattern.** You can then use this pattern to find more information and the solution to the problem.

Problem Solving Strategies

Draw a Diagram	Make a Table
Make a Model	Solve a Simpler Problem
Guess and Test	Use Logical Reasoning
Work Backward	Use a Venn Diagram
Find a Pattern	Make an Organized List

Students are using the pattern at right to build stairways for a model house. How many blocks are needed to build a stairway with seven steps?

Understand the Problem The answer will be the total number of blocks in a stairway with seven steps.

Make a Plan Try to **find a pattern** between the number of steps and the number of blocks needed.

Notice that the first step is made of one block. The second step is made of two blocks, the third step is made of three blocks, and the fourth step is made of four blocks.

Step	Number of Blocks in Step	Total Number of Blocks in Stairway
2	2	1 + 2 = 3
3	3	1 + 2 + 3 = 6
4	4	1 + 2 + 3 + 4 = 10

To find the total number of blocks, add the number of blocks in the first step, the second step, the third step, and so on.

Solve The seventh step will be made of seven blocks. The total number of blocks will be 1 + 2 + 3 + 4 + 5 + 6 + 7 = 28.

Look Back Use another strategy. You can **draw a diagram** of a stairway with 7 steps. Count the number of blocks in your diagram. There are 28 blocks.

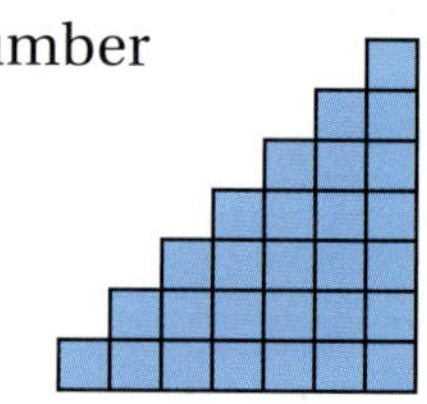

PRACTICE

1. A cereal company adds baseball cards to the 3rd box, the 6th box, the 11th box, the 18th box, and so on of each case of cereal. In a case of 40 boxes, how many boxes will have baseball cards? **6 boxes**

2. Describe the pattern and find the missing numbers.

1; 4; 16; 64; 256; ▒; ▒; 16,384 **1,024; 4,096**

Make a Table

When you are given a lot of information in a problem, it may be helpful to organize that information. One way to organize information is to **make a table.**

Problem Solving Strategies

Draw a Diagram	**Make a Table**
Make a Model	Solve a Simpler Problem
Guess and Test	Use Logical Reasoning
Work Backward	Use a Venn Diagram
Find a Pattern	Make an Organized List

Mrs. Melo's students scored the following on their math test: 90, 80, 77, 78, 91, 92, 73, 62, 83, 79, 72, 85, 93, 84, 75, 68, 82, 94, 98, and 82. An A is given for 90 to 100 points, a B for 80 to 89 points, a C for 70 to 79 points, a D for 60 to 69 points, and an F for less than 60 points. Find the number of students who scored each letter grade.

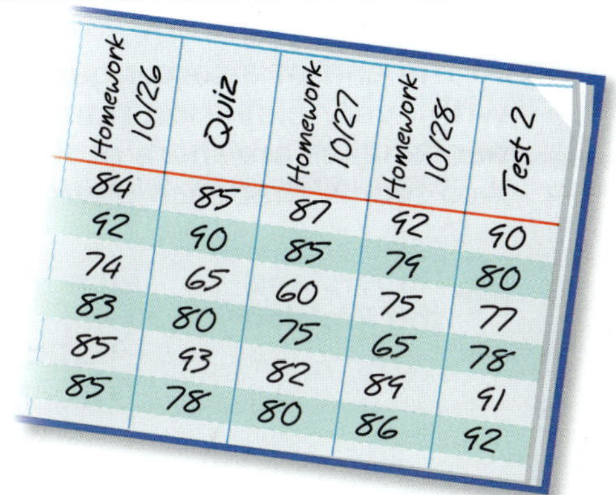

Understand the Problem

Identify the important information.

- You have been given the list of scores and the letter grades that go with each score.

The answer will be the number of each letter grade.

Make a Plan

Make a table to organize the scores. Use the information in the problem to set up your table. Make one row for each letter grade.

Solve

Read through the list of scores. As you read each score, make a tally in the appropriate place in your table. There are 20 test scores, so be sure you have 20 tallies in all.

Letter Grade	Number
A (90–100)	卌 I
B (80–89)	卌 I
C (70–79)	卌 I
D (60–69)	II
F (below 60)	

Mrs. Melo gave out six A's, six B's, six C's, two D's, and no F's.

Look Back

Use another strategy. Another way you could solve this problem is to **make an organized list.** Order the scores from least to greatest, and count how many scores are in each range.

62, 68, 72, 73, 75, 77, 78, 79, 80, 82, 82, 83, 84, 85, 90, 91, 92, 93, 94, 98

D (62, 68) C (72–79) B (80–85) A (90–98)

PRACTICE

1. The debate club has 6 members. Each member will debate each of the other members exactly once. How many total debates will there be? **15**

2. At the library, there are three story-telling sessions. Each one lasts 45 minutes, with 30 minutes between sessions. If the first session begins at 10:00 A.M., what time does the last session end? **1:15 P.M.**

Solve a Simpler Problem

Sometimes a problem contains large numbers or requires many steps. Try to **solve a simpler problem** that is similar. Solve the simpler problem first, and then try the same steps to solve the original problem.

Problem Solving Strategies

Draw a Diagram	Make a Table
Make a Model	**Solve a Simpler Problem**
Guess and Test	Use Logical Reasoning
Work Backward	Use a Venn Diagram
Find a Pattern	Make an Organized List

At the end of a soccer game, each player shakes hands with every player on the opposing team. How many handshakes are there at the end of a game between two teams that each have 20 players?

Understand the Problem

Identify the important information.

- There are 20 players on each team.
- Each player will shake hands with every player on the opposing team.

The answer will be the total number of handshakes exchanged.

Make a Plan

Solve a simpler problem. For example, suppose each team had just one player. Then there would only be one handshake between the two players. Expand the number of players to two and then three.

Solve

When there is 1 player, there is $1 \times 1 = 1$ handshake. For 2 players, there are $2 \times 2 = 4$ handshakes. And for 3 players, there are $3 \times 3 = 9$ handshakes.

If each team has 20 players, there will be $20 \times 20 = 400$ handshakes.

Players Per Team	Diagram	Handshakes
1	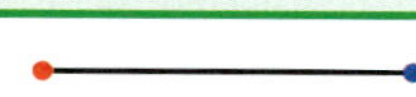	1
2	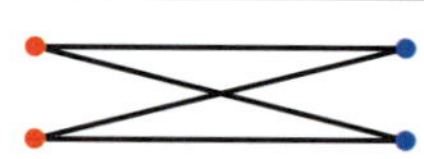	4
3	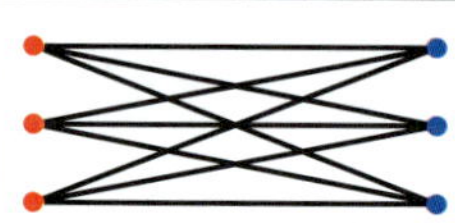	9

Look Back

If the pattern is correct, for 4 players there will be 16 handshakes and for 5 players there will be 25 handshakes. Complete the next two rows of the table to check these answers.

PRACTICE

1. Martha has 5 pairs of pants and 4 blouses that she can wear to school. How many different outfits can she make? **20**

2. What is the smallest 5-digit number that can be divided by 50 with a remainder of 17? **10,017**

Use Logical Reasoning

Sometimes a problem may provide clues and facts that you must use to answer a question. You can **use logical reasoning** to solve this kind of problem.

Problem Solving Strategies

Draw a Diagram	Make a Table
Make a Model	Solve a Simpler Problem
Guess and Test	**Use Logical Reasoning**
Work Backward	Use a Venn Diagram
Find a Pattern	Make an Organized List

Kevin, Ellie, and Jillian play three different sports. One person plays soccer, one likes to run track, and the other swims. Ellie is the sister of the swimmer. Kevin once went shopping with the swimmer and the track runner. Match each student with his or her sport.

Understand the Problem

Identify the important information.

- There are three people, and each person plays a different sport.
- Ellie is the sister of the swimmer.
- Kevin once went shopping with the swimmer and the track runner.

The answer will tell which student plays each sport.

Make a Plan

Start with clues given in the problem, and **use logical reasoning** to find the answer.

Solve

Make a table with a column for each sport and a row for each person. Work with the clues one at a time. Write "yes" in a box if the clue applies to that person. Write "no" if the clue does not apply.

	Soccer	Track	Swim
Kevin		no	no
Ellie			no
Jillian			

- Ellie is the sister of the swimmer, so she is not the swimmer.
- Kevin went shopping with the swimmer and the track runner. He is not the swimmer or the track runner.

So Kevin must be the soccer player, and Jillian must be the swimmer. This leaves Ellie as the track runner.

Look Back

Compare your answer to the clues in the problem. Make sure none of your conclusions conflict with the clues.

PRACTICE

Karin–plain cheese, Brent–pepperoni, Lola–ham-pineapple

1. Karin, Brent, and Lola each ordered a different slice of pizza: pepperoni, plain cheese, and ham-pineapple. Karin is allergic to pepperoni. Lola likes more than one topping. Which kind of pizza did each person order?

2. Leo, Jamal, and Kara are in fourth, fifth, and sixth grades. Kara is not in fourth grade. The sixth-grader is in chorus with Kara and has the same lunch time as Leo. Match the students with their grades.

Leo–4th, Kara–5th, Jamal–6th

Use a Venn Diagram

You can **use a Venn diagram** to display relationships among sets in a problem. Use ovals, circles, or other shapes to represent individual sets.

Problem Solving Strategies

Draw a Diagram	Make a Table
Make a Model	Solve a Simpler Problem
Guess and Test	Use Logical Reasoning
Work Backward	**Use a Venn Diagram**
Find a Pattern	Make an Organized List

Robert is taking a survey to see what kinds of pets students have. He found that 70 students have dogs, 45 have goldfish, and 60 have birds. Some students have two kinds of pets: 17 students have dogs and fish, 22 students have dogs and birds, and 15 students have birds and goldfish. Five students have all three kinds of pets. How many students in the survey have only birds?

Understand the Problem

List the important information.

- You know that 70 students have dogs, 45 have goldfish, and 60 have birds.

The answer will be the number of students who have only birds.

Make a Plan

Use a Venn diagram to show the sets of students who have dogs, students who have goldfish, and students who have birds.

Solve

Draw and label three overlapping circles. Work from the inside out. Write "5" in the area where all three circles overlap. This represents the number of students who have a dog, a goldfish, and a bird.

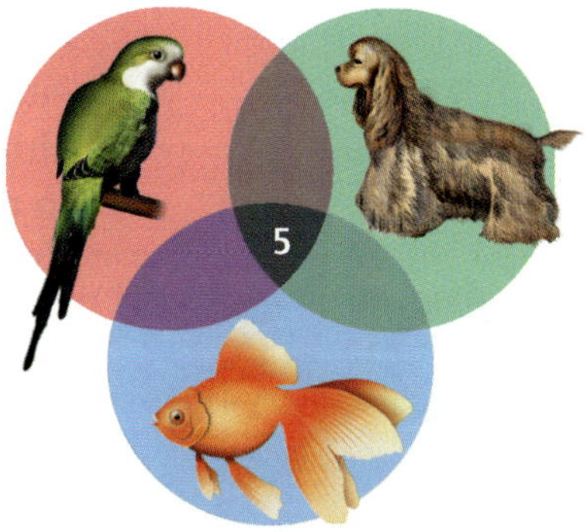

Use the information in the problem to fill in other sections of the diagram. You know that 60 students have birds, so the numbers within the bird circle will add to 60.

So 18 students have only pet birds.

Look Back

When your Venn diagram is complete, check it carefully against the information in the problem. Make sure your diagram agrees with the facts given.

PRACTICE

1. How many students have only dogs? **26**
2. How many students have only goldfish? **8**

Make an Organized List

In some problems, you will need to find how many different ways something can happen. It is often helpful to **make an organized list.** This will help you count the outcomes and be sure that you have included all of them.

Problem Solving Strategies

Draw a Diagram	Make a Table
Make a Model	Solve a Simpler Problem
Guess and Test	Use Logical Reasoning
Work Backward	Use a Venn Diagram
Find a Pattern	**Make an Organized List**

In a game at an amusement park, players throw 3 darts at a target to score points and win prizes. If each dart lands within the target area, how many different total scores are possible?

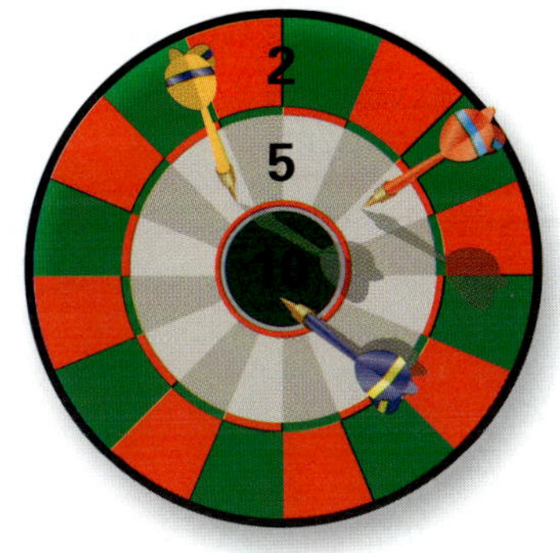

Understand the Problem

Identify the important information.

- A player throws three darts at the target.

The answer will be the number of different scores a player could earn.

Make a Plan

Make an organized list to determine all possible outcomes and score totals. List the value of each dart and the point total for all three darts.

Solve

You can organize your list by the number of darts that land in the center. All three darts could hit the center circle. Or, two darts could hit the center circle and the third could hit a different circle. One dart could hit the center circle, or no darts could hit the center circle.

3 Darts Hit Center	2 Darts Hit Center	1 Dart Hits Center	0 Darts Hit Center
10 + 10 + 10 = 30	10 + 10 + 5 = 25	10 + 5 + 5 = 20	5 + 5 + 5 = 15
	10 + 10 + 2 = 22	10 + 5 + 2 = 17	5 + 5 + 2 = 12
		10 + 2 + 2 = 14	5 + 2 + 2 = 9
			2 + 2 + 2 = 6

Count the different outcomes. There are 10 possible scores.

Look Back

You could have listed outcomes in random order, but because your list is organized, you can be sure that you have not missed any possibilities. Check to be sure that every score is different.

PRACTICE

1. A restaurant has three different kinds of pancakes: cinnamon, blueberry, and apple. If you order one of each kind, how many different ways can the three pancakes be stacked? 6

2. How many ways can you make change for a quarter using dimes, nickels, and pennies? 12

Chapter 1

Number Toolbox

Section 1A

Whole Numbers and Exponents

Lesson 1-1
Comparing and Ordering Whole Numbers

Lesson 1-2
Estimating with Whole Numbers

Lesson 1-3
Exponents

Section 1B

Using Whole Numbers

Technology Lab 1A
Explore the Order of Operations

Lesson 1-4
Order of Operations

Lesson 1-5
Mental Math

Lesson 1-6
Choose the Method of Computation

Lesson 1-7
Find a Pattern

Extension
Binary Numbers

Pacing Guide for 45-Minute Classes

Chapter 1

DAY 1	DAY 2	DAY 3	DAY 4	DAY 5
Lesson 1-1	Lesson 1-2	Lesson 1-3	Mid-Chapter Quiz Technology Lab 1A	Lesson 1-4
DAY 6	**DAY 7**	**DAY 8**	**DAY 9**	**DAY 10**
Lesson 1-5	Lesson 1-6	Lesson 1-7	Extension	Chapter 1 Review
DAY 11				
Chapter 1 Assessment				

Pacing Guide for 90-Minute Classes

Chapter 1

DAY 1	DAY 2	DAY 3	DAY 4	DAY 5
Lesson 1-1 Lesson 1-2	Lesson 1-3 Technology Lab 1A	Mid-Chapter Quiz Lesson 1-4 Lesson 1-5	Lesson 1-6 Lesson 1-7	Extension Chapter 1 Review
DAY 6				
Chapter 1 Assessment Lesson 2-1				

Across the Series

HARCOURT GRADE 5

- Compare and order whole numbers.
- Estimate and operate with whole numbers.
- Use order of operations, and use number properties to compute mentally.
- Choose an appropriate method of computation, and justify your choice.
- Recognize, describe, and extend patterns in sequences.

COURSE 1

- **Compare and order whole numbers.**
- **Estimate with whole numbers.**
- **Represent numbers by using exponents.**
- **Use the order of operations.**
- **Use number properties to compute mentally.**
- **Choose an appropriate method of computation, and justify your choice.**
- **Recognize, describe, and extend patterns in sequences.**

COURSE 2

- Represent numbers by using exponents.
- Express numbers in scientific notation and standard notation.
- Use the order of operations to evaluate numerical and algebraic expressions including exponents.

Across the Curriculum

LANGUAGE ARTS LINK

Math: Reading and Writing in the Content Area . . . . pp. 1–7

Focus on Problem Solving

Solve . . . . SE p. 17

Journal . . . . TE pp. 7, 11, 15, 23, 27

Write About It . . . . SE, last page of each lesson

SOCIAL STUDIES LINK

Geography . . . . SE p. 7

Social Studies . . . . SE pp. 5, 11, 23

SCIENCE LINK

Life Science . . . . SE pp. 15, 27

Earth Science . . . . SE p. 27

Astronomy . . . . SE pp. 28, 29

TE = *Teacher's Edition* **SE** = *Student Edition*

Interdisciplinary

Bulletin Board

Life Science

Prokaryotic cells, bacteria, multiply by splitting into two parts. This type of cell division is called **binary fission.** If it takes bacteria cell *A* four hours to complete its cell cycle and produce two cells, then how many more cells would be formed in 24 hours? There are 6 cycles in 24 hours; $2^6 = 64$ cells.

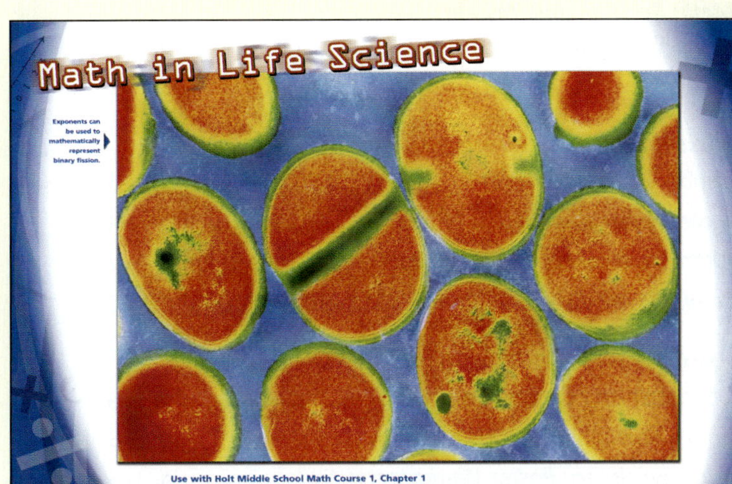

Interdisciplinary posters and worksheets are provided in your resource material.

Chapter 1

Resource Options

Chapter 1 Resource Book

Student Resources

Practice (Levels A, B, C) . . . . pp. 8–10, 17–19, 27–29, 37–39, 46–48, 56–58, 65–67
Reteach . . . . pp. 11, 20–21, 30, 40, 49–50, 59, 68
Challenge . . . . pp. 12, 22, 31, 41, 51, 60, 69
Problem Solving . . . . pp. 13, 23, 32, 42 52, 61, 70
Puzzles, Twisters & Teasers . . . . . pp. 14, 24, 33, 43, 53, 62, 71
Recording Sheets . . . . pp. 3, 7, 16, 26, 36, 45, 55, 64, 74, 77–79
Chapter Review . . . . pp. 72–73

Teacher and Parent Resources

Chapter Planning and Pacing Guide . . . . p. 4
Section Planning Guides . . . . pp. 5, 34
Parent Letter . . . . pp. 1–2
Teaching Tools . . . . pp. 77–81
Teacher Support for Chapter Project . . . . p. 75
Transparencies . . . . pp. T1–T25
- Daily Transparencies
- Additional Examples Transparencies
- Teaching Transparencies

Reaching All Learners

English Language Learners

Success for English Language Learners . . . . pp. 1–14
Math: Reading and Writing in the Content Area . . . . pp. 1–7
Spanish Homework and Practice . . . . pp. 1–7
Spanish Interactive Study Guide . . . . pp. 1–7
Spanish Family Involvement Activities . . . . pp. 1–8
Multilingual Glossary

Individual Needs

Are You Ready? Intervention and Enrichment . . . pp. 21–28, 145–148, 225–228, 405–406
Alternate Openers: Explorations . . . . pp. 1–7
Family Involvement Activities . . . . pp. 1–8
Interactive Problem Solving . . . . pp. 1–7
Interactive Study Guide . . . . pp. 1–7
Readiness Activities . . . . pp. 1–2
Math: Reading and Writing in the Content Area . . . . pp. 1–7
Challenge . . . . CRB pp. 12, 22, 31, 41, 51, 60, 69

Hands-On

Hands-On Lab Activities . . . . pp. 1–5
Technology Lab Activities . . . . pp. 1–4
Alternate Openers: Explorations . . . . pp. 1–7
Family Involvement Activities . . . . pp. 1–8

Applications and Connections

Consumer and Career Math . . . . pp. 1–4
Interdisciplinary Posters . . . . Poster 1, TE p. 2B
Interdisciplinary Poster Worksheets . . . . pp. 1–3

Transparencies

Alternate Openers: Explorations . . . . pp. 1–7
Exercise Answers Transparencies
Chapter 1 Resource Book . . . . pp. T1–T25
- Daily Transparencies
- Additional Examples Transparencies
- Teaching Transparencies

Technology

Teacher Resources

***Lesson Presentations* CD-ROM** . . . . Chapter 1
***Test and Practice Generator* CD-ROM** . . . . Chapter 1
***One-Stop Planner* CD-ROM** . . . . Chapter 1

Student Resources

***Are You Ready? Intervention* CD-ROM**
Skills 4, 3, 34, 54

internet connect go.hrw.com

Homework Help Online	**KEYWORD:** MR4 HWHelp1
Math Tools Online	**KEYWORD:** MR4 Tools
Glossary Online	**KEYWORD:** MR4 Glossary
Chapter Project Online	**KEYWORD:** MR4 PSProject1
Chapter Opener Online	**KEYWORD:** MR4 Ch1
CNN Student News	**KEYWORD:** MR4 CNN1

SE = *Student Edition* **TE** = *Teacher's Edition* **AR** = *Assessment Resources* **CRB** = *Chapter Resource Book* **MK** = *Manipulatives Kit*

Chapter 1

Assessment Options

Assessing Prior Knowledge

Determine whether students have the required prerequisite concepts and skills.

Are You Ready? . . . . SE p. 3
Inventory Test . . . . AR pp. 1–4

Test Preparation

Provide review and practice for chapter and standardized tests.

Standardized Test Prep . . . . SE p. 45
Spiral Review with Test Prep . . . . . SE, last page of each lesson
Study Guide and Review . . . . SE pp. 40–42
Test Prep Tool Kit

Technology

***Test and Practice Generator* CD-ROM**

internet connect

State-Specific Test Practice Online KEYWORD: MR4 TestPrep

go.hrw.com

Performance Assessment

Assess students' understanding of chapter concepts and combined problem-solving skills.

Performance Assessment . . . . SE p. 44
Includes scoring rubric in TE
Performance Assessment . . . . AR p. 104
Performance Assessment Teacher Support . . . . AR p. 103

Portfolio

Portfolio opportunities appear throughout the Student and Teacher's Editions.

Suggested work samples:

Problem Solving Project . . . . TE p. 2
Performance Assessment . . . . SE p. 44
Portfolio Guide . . . . AR p. xxxv
Journal . . . . TE pp. 7, 11, 15, 23, 27
Write About It . . . . SE, last page of each lesson

Daily Assessment

Obtain daily feedback on students' understanding of concepts.

Spiral Review and Test Prep . . . . . . SE, last page of each lesson

Also Available on Transparency in Chapter 1 Resource Book

Warm Up . . . . TE, first page of each lesson
Problem of the Day . . . . TE pp. 4, 8, 12, 20, 24, 28
Lesson Quiz . . . . TE, last page of each lesson

Student Self-Assessment

Have students evaluate their own work.

Group Project Evaluation . . . . AR p. xxxii
Individual Group Member Evaluation . . . . AR p. xxxiii
Portfolio Guide . . . . AR p. xxxv
Journal . . . . TE pp. 7, 11, 15, 23, 27

Formal Assessment

Assess students' mastery of concepts and skills.

Section Quizzes . . . . AR pp. 5, 6
Mid-Chapter Quiz . . . . SE p. 16
Chapter Test . . . . SE p. 43
Chapter Tests (Levels A, B, C) . . . . AR pp. 31–36
Cumulative Tests (Levels A, B, C) . . . . AR pp. 127–138
Standardized Test Prep
Cumulative Assessment . . . . SE p. 45
End-of-Year Test . . . . AR pp. 271–274

Technology

***Test and Practice Generator* CD-ROM**

Make tests electronically. This software includes:

- Dynamic practice for Chapter 1
- Customizable tests
- Multiple-choice items for each objective
- Free-response items for each objective
- Teacher management system

SE = *Student Edition* **TE** = *Teacher's Edition* **AR** = *Assessment Resources* **CRB** = *Chapter Resource Book* **MK** = *Manipulatives Kit*

Chapter 1

Chapter 1 Tests

Three levels (A,B,C) of tests are available for each chapter in the *Assessment Resources.*

Level A

Chapter 1 Chapter Test
Form A

Write < or >.

1. 7,489 [>] 7,305
2. 43,653 [<] 46,320
3. Write in order from least to greatest: 437; 513; 248.
   248; 437; 513
4. Write in order from greatest to least: 825; 854; 837.
   854; 837; 825

Round to the largest place value to estimate.

5. 456 + 328
   500 + 300 = 800
6. 3,184 + 1,594
   3,000 + 2,000 = 5,000
7. 639 − 163
   600 − 200 = 400
8. 7,153 − 3,635
   7,000 − 4,000 = 3,000

Round to the place value indicated to estimate the sum or difference.

9. 27,503 − 2,635; thousands
   28,000 − 3,000 = 25,000
10. 3,450 − 1,325; hundreds
    3,500 − 1,300 = 2,200

Write in exponential form.

11. 5 × 5 × 5 × 5
    5^4
12. 8 × 8 × 8 × 8 × 8
    8^5

Write as repeated multiplication.

13. 7^3
    7 × 7 × 7
14. 3^4
    3 × 3 × 3 × 3

Find each value.

15. 4^3
    64
16. 3^3
    27
17. 5^1
    5

Compare using <, >, or = .

18. 8^2 [>] 52^1
19. 4^3 [<] 75
20. 10^2 [=] 100

Chapter 1 Chapter Test
Form A, continued

Simplify each expression.

21. 24 − 15 ÷ 3
    19
22. 36 ÷ 6 × 3
    18
23. 6 − 2^2 + 4
    6
24. 3^2 + 25 ÷ 5
    14
25. 30 ÷ (12 − 7)
    6
26. 4^2 − (5 + 3)
    8

Use mental math to solve.

27. 2 + 9 + 8 + 6
    25
28. 7 + 9 + 11
    27
29. 4 × 9 × 5
    180
30. 2 × 8 × 5
    80
31. 6 × 25
    150

Identify a pattern. Replace ? with missing terms.

32. 13, 23, 34, 46, ?, 73, 88, ?
    59, 104
33. 6, 12, 24, 48, 96, ?, ?
    192, 384
34. 85, 73, 61, 49, ?, ?
    37, 25

Solve.

35. Mammoth Cave in Kentucky covers 52,419 acres. Bryce Cave in Utah covers 35,835 acres. How many more acres does Mammoth Cave cover?
    16,584 acres
36. A bat beats its wings 1,200 times per minute. How many beats will its wings make in 1 hour?
    72,000 beats
37. The Colorado River is 1,450 miles long. The Rio Grande is 1,900 miles long. How much longer is the Rio Grande than the Colorado River?
    450 miles
38. Mark mowed 84 lawns over the summer. He charged $18 per lawn. How much money did he make during the summer?
    $1,512

Level B

Chapter 1 Chapter Test
Form B

Write < or >.

1. 37,589 [>] 37,409
2. 436,532 [<] 438,503
3. Write in order from least to greatest: 3,290; 3,966; 3,078.
   3,078; 3,290; 3,966
4. Write in order from greatest to least: 8,254; 8,549; 8,375.
   8,549; 8,375; 8,254

Round to the largest place value to estimate.

5. 3,620 + 4,485
   4,000 + 4,000 = 8,000
6. 13,248 + 17,509
   10,000 + 20,000 = 30,000
7. 2,626 − 1,693
   3,000 − 2,000 = 1,000
8. 6,558 − 3,249
   7,000 − 3,000 = 4,000

Round to the place value indicated to estimate the sum or difference.

9. 6,658 − 5,250; thousands
   7,000 − 5,000 = 2,000
10. 51,728 + 23,250; thousands
    52,000 + 23,000 = 75,000

Write in exponential form.

11. 4 × 4 × 4 × 4 × 4 × 4
    4^6
12. 6 × 6 × 6 × 6
    6^4

Write as repeated multiplication.

13. 3^3
    3 × 3 × 3
14. 5^4
    5 × 5 × 5 × 5

Find each value.

15. 4^5
    1,024
16. 7^1
    7
17. 3^6
    729

Compare using, <, >, or =.

18. 2^3 [>] 1^4
19. 10,000 [<] 10^5
20. 2^4 [=] 4^2

Chapter 1 Chapter Test
Form B, continued

Simplify each expression.

21. 25 − 15 ÷ 3
    20
22. 17 + 36 ÷ 6 × 3 − 4
    31
23. 57 − 3^3 + 18
    48
24. 13 + 2^4 − (15 + 8)
    6
25. 15 + 30 ÷ (25 − 19) − 17
    3
26. 4^2 + 72 ÷ 9 − 18
    6

Use mental math to solve.

27. 28 + 9 + 32 + 7
    76
28. 7 + 29 + 11 + 23
    70
29. 2 × 8 × 7 × 5
    560
30. 7 × 35
    245
31. 42 × 6
    252

Identify a pattern. Replace ? with missing terms.

32. 111, 93, 75, ?, 39, ?
    57, 21
33. 5, 8, 14, 23, 35, ?, ?, ?
    50, 68, 89
34. 47, 50, 45, 48, 43, 46, ?, ?
    41, 44

Solve.

35. In 1966, 103,224 acres of land in Florida were used to grow grapefruit. Thirty years later, 144,416 acres were used. What was the increase in acreage?
    41,192 acres
36. A lion sleeps about 15 hours each day. How many hours does a lion sleep in one year?
    5,475 hours
37. The first people to climb Mount Everest started from their base camp at 5,486 meters and climbed to the summit at 8,848 meters. How far did they climb?
    3,362 meters
38. The school theater has 36 rows with 25 seats in each row. How many people can sit in the theater?
    900 people

Level C

Chapter 1 Chapter Test
Form C

Write < or >.

1. 37,489 [<] 37,498
2. 436,532 [>] 436,523
3. Write in order from least to greatest: 1,437; 1,473; 1,374.
   1,374; 1,437; 1,473
4. Write in order from greatest to least: 8,254; 8,245; 8,425.
   8,425; 8,254; 8,245

Round to the largest place value to estimate.

5. 4,362 + 3,548
   4,000 + 4,000 = 8,000
6. 27,349 + 15,409
   30,000 + 20,000 = 50,000
7. 18,693 − 11,285
   20,000 − 10,000 = 10,000
8. 62,375 − 28,117
   60,000 − 30,000 = 30,000

Round to the place value indicated to estimate the sum or difference.

9. 142,375 − 28,117; ten thousands
   140,000 − 30,000 = 110,000
10. 25,369 + 12,609; hundreds
    25,400 + 12,600 = 38,000

Write in exponential form.

11. 7 × 7 × 7 × 7 × 7 × 7 × 7
    7^7
12. 3 × 3 × 3 × 3 × 3
    3^5

Write as repeated multiplication.

13. 4^3
    4 × 4 × 4
14. 5^5
    5 × 5 × 5 × 5 × 5

Find each value.

15. 6^4 = 1,296
16. 2^9 = 512
17. 5^7 = 78,125

Compare using <, >, or =.

18. 7^3 [>] 4^4
19. 5^3 [<] 2^7
20. 8^2 [=] 4^3

Chapter 1 Chapter Test
Form C, continued

Simplify each expression.

21. 4 + 8 ÷ 2 − 3
    5
22. 7 + 4 × 3 + 16 ÷ 2
    27
23. (7 + 4) × 3 + 16 ÷ 2
    41
24. 7 × 3^2 + 38 − 5
    96
25. 4^2 − 16 ÷ 4 + (18 − 5)
    25
26. (5^2 + 8 ÷ 2) × (45 ÷ 5 − 7)
    58

Use mental math to solve.

27. 34 + 8 + 26 + 42
    110
28. 15 + 23 + 17 + 45
    100
29. 4 × 6 × 3 × 5
    360
30. 6 × 43
    258
31. 25 × 13
    325

Identify a pattern. Replace ? with missing terms.

32. 15, 17, 21, 27, ?, 45, 57, ?
    35, 71
33. 36, 40, 20, 24, 12, 16, ?, ?, ?
    8, 12, 6
34. ?, 2^2, 9, 4^2, ?, ?
    1, 25, 6^2

Solve.

35. The Park Row Building in New York is 386 feet high. The Empire State Building is 1,472 feet high. How much taller is the Empire State Building?
    1,086 feet
36. An adult giant panda eats about 850 pounds of bamboo shoots a month. How many pounds would a panda eat in one year?
    10,200 pounds
37. Tim played a video game 4 times. His scores were 56, 123, 78 and 95. How many points did he have after 4 games?
    352 points
38. The school gym has 6 sections of bleachers. Each section seats 125 people. How many seats are in the gym?
    750 seats

Test and Practice Generator CD-ROM

Create and customize multiple versions of the same tests with corresponding answers for any chosen chapter objectives.

Chapter 1 State and Standardized Test Preparation

Test Taking Skill Builder and Standardized Test Practice are provided for each chapter in the *Test Prep Tool Kit.*

TEST TAKING SKILL BUILDER

Test Taking Strategy **Multiple Choice: Find a Pattern**

Chapter 1

You can sometimes find the answer to a multiple choice question by using the answer choices to work backwards.

Example 1 What is the missing number in the sequence?

2 8 18 ___ 50 72

A 28 **B** 32 **C** 40 **D** 46

Solution: Look for a pattern to find how the numbers given in the sequence were found. Look for something in common with the numbers.

$1 \cdot 2 = 2$

$4 \cdot 2 = 8$

$9 \cdot 2 = 18$ Each number in the sequence is a multiple of 2.

$25 \cdot 2 = 50$

$36 \cdot 2 = 72$

Notice that the factor multiplied by 2 is a perfect square.

$1^2 = 1$, $2^2 = 4$, $3^2 = 9$, $5^2 = 25$, and $6^2 = 36$.

So, $4^2 = 16$, and $16 \times 2 = 32$. The missing number in the sequence is 32. The correct answer is choice B.

Answer choices to multiple choice questions usually contain distracters. Distracters are values that are arrived at by making a common misjudgment or a simple error in a calculation. You need to double check your calculation and reread the question statement in order to avoid choosing one of the given distracters as an answer choice.

Example 2 Use the pattern to find the units digit of 3^{10}.

$3^1 = \underline{3}$ $3^2 = \underline{9}$ $3^3 = 2\underline{7}$ $3^4 = 8\underline{1}$ $3^5 = 24\underline{3}$

F 3 **G** 9 **H** 7 **I** 1

Solution: Once you notice that the sequence of numbers is 3, 9, 7, 1 you can continue the pattern to find the units digit of 3^{10}. 3^6 has a $\underline{9}$ in the units digit. 3^7 has a $\underline{7}$ in the units digit. 3^8 has a $\underline{1}$ in the units digit. 3^9 has a $\underline{3}$ in the units digit. 3^{10} has a $\underline{9}$ in the units digit.

The correct answer is choice G.

Notice that the other answer choices contain the distracters 3, 7, and 1. If you had made a simple error in your reasoning, you might have selected the wrong answer choice.

Test Taking Strategy

Chapter 1, continued

Exercises Possible answers are given.

Multiple Choice Which value is the missing number in the sequence?

46 42 38 ___ 30 26

A 36 **B** 35 **C** 34 **D** 33

1. Explain how you can work backwards to determine the missing value.

   I can substitute each given answer choice into the pattern and see if that value is part of the sequence.

2. What is the pattern and which value completes the sequence?

   Each value is four less than the previous value. The missing number is 34, Choice C.

Multiple Choice Use the following pattern to find the units digit of 4^{12}.

$4^1 = \underline{4}$ $4^2 = 1\underline{6}$ $4^3 = 6\underline{4}$ $4^4 = 25\underline{6}$ $4^5 = 1{,}02\underline{4}$

F 2 **G** 4 **H** 6 **I** 8

3. Which answer choices are NOT considered to be distracters? Why?

   Choice F and Choice I; The pattern only includes the numbers 4 and 6 and the choices given in F and I cannot be arrived at by a simple mistake or misjudgment.

4. The correct answer is 6, Choice H. Explain why Choice G is a distracter.

   Choice G is a distracter because you could possible make a minor error in calculating the 12 value of the sequence, resulting in the answer of 4.

STANDARDIZED TEST PRACTICE

Standardized Test Practice

Chapter 1

Select the best answer for Questions 1–6.

1. Which is a true statement?
   A 34,567 > 35,678
   B 2,345 > 3,254
   C 123,984 > 122,894
   D 813,432 < 812,423

2. Davey's Food Service has provided the food and supplies for a local road race the last few years. Use the table to estimate the number of cups the road race will use in 4 years.

Davey's Food Service	
Item	**Number Supplied per Year**
Sports Drink	75 gallons
Cups	534
Energy Bars	185

   F 500 cups
   G 1,500 cups
   H 2,000 cups
   I 2400 cups

3. Five men each buy 12 plastic bats for $2 each. How much did they spend?
   A $10
   B $22
   C $120
   D $240

4. Everyday Wylie picks up aluminum cans. On day one he picked up two cans. Each day after that he picked up twice as many cans as the day before. Which expression would you use to find the number of cans he picked up on the sixth day?
   F 1^6
   G 2^6
   H 2^4
   I 6^2

5. There are 4 people in the Marker family. Each person takes about an 8-minute shower. The family wants to conserve water due to a drought. They know that their showerhead uses about 3 gallons of water per minute. How many gallons of water does the family use to take showers each day?
   A 24 gallons
   B 48 gallons
   C 96 gallons
   D 112 gallons

6. There are 6 teachers and 244 students traveling to the art museum for a field trip. One bus will hold 40 people. How many buses are needed altogether?
   F 4 buses
   G 5 buses
   H 6 buses
   I 7 buses

Standardized Test Practice

Chapter 1, continued

Gridded Response
Solve the problems. Use the answer sheet to write and grid-in your answer.

7. Nicholas sold a baseball card for $52. Its value had increased $2 each of the six years he owned it. How much did Nicholas originally pay for the card?

8. What is the area, per 1,000 square kilometers, of the continent with the smallest area?

Size of the Continents	
Continents	**Size (sq km x 1,000)**
Africa	30,065
Asia	44,579
Antarctica	13,209
Australia	7,687
Europe	9,938
North America	24,256
South America	17,819

9. Laura's family took a family vacation. The first day they drove 420 miles; which was one third of the distance of the entire trip. The next day the car had a flat tire, so they were only able to drive 60 miles. How many more miles does Laura's family have to drive?

Short Response
Solve the problems. Use the answer sheet to write your answers.

10. There are 15 players on a soccer team. Each player receives a jersey, shorts, and socks. A jersey costs $15, shorts cost $12, and socks cost $3. Write and evaluate an expression for the total cost of the team's outfit.

11. Harriet evaluated the following problem on a quiz and the teacher marked it as incorrect. Explain in words what Harriet did wrong. What is the correct answer?

    $3 + 4 \times 6 + 8 = 7 \times 14 = 98$

Extended Response

12. In a process called *mitosis*, a cell divides to form two new cells. Then those new cells divide again to form new cells and those new cells divide yet again to form more new cells, etc.
    a. Start with one cell and write how many cells there will be after the first division, the second division, and the third division. Make a table to show the pattern.
    b. Explain the pattern that you notice.
    c. If there are 56 cells during a stage of mitosis, how many cells will there be during the next stage? Explain in words how you determined your answer.

Sheet

Ⓗ Ⓘ See Lesson 1-
● Ⓓ See Lesson 1-
Ⓗ ● See Lesson 1-

9. 780

See Lesson 1-6.

10. $15(15 + 12 + 3) = 15(30) = 450$; The total cost of the team's unifor[…] $450.

11. Harriet added first. She should have multiplied the 4 and 6 first and t[…] added. (See Lesso[…])
    $3 + 4 \times 6 + 8 = 3 + 24 + 8 = 35$
    The correct answer is 35.

(See Lesso[…])

Extended Response
Write your answers for Problem 12 on the back of this paper.
See Lesson 1-7.

internet connect go.hrw.com

State-Specific Test Practice Online

KEYWORD: MR4 TestPrep

Test Prep Tool Kit

- Standardized Test Prep Workbook
- Countdown to Testing transparencies
- State Test Prep CD-ROM
- Standardized Test Prep Video

Customized answer sheets give students realistic practice for actual standardized tests.

Chapter 1

Number Toolbox

Why Learn This?

Remind students that mathematics and arithmetic are used in many jobs. Have students look at the table. Point out that a veterinary technician must know an animal's weight when determining how much food the animal requires.

Using Data

To begin the study of this chapter, have students:

- Order the animals from heaviest to lightest. elephant, hippopotamus, giraffe, buffalo, zebra
- Estimate the amount of food (in pounds) that a zebra eats in one month. about 900 pounds
- Find the difference in the amount of food eaten in one day by an elephant and a giraffe. 585 pounds

Chapter 1

Number Toolbox

TABLE OF CONTENTS

African Plant-Eating Animals

Animal	Weight (lb)	Daily Food Intake (lb)
Buffalo	1,500	45
Elephant	11,000	660
Giraffe	2,500	75
Hippopotamus	5,500	90
Zebra	950	30

Career ***Veterinary Technician***

Do you like caring for animals? Veterinary technicians perform many of the same tasks for veterinarians as nurses do for doctors. Veterinary technicians also do research that can help animals. To care for animals, technicians must know what the animals need to eat and how they behave with other types of animals. Large plant-eating animals, many of which live in Africa, need to eat specific kinds of grasses and trees. The table above shows the approximate weight of some animals and the approximate amount of food the animals eat each day.

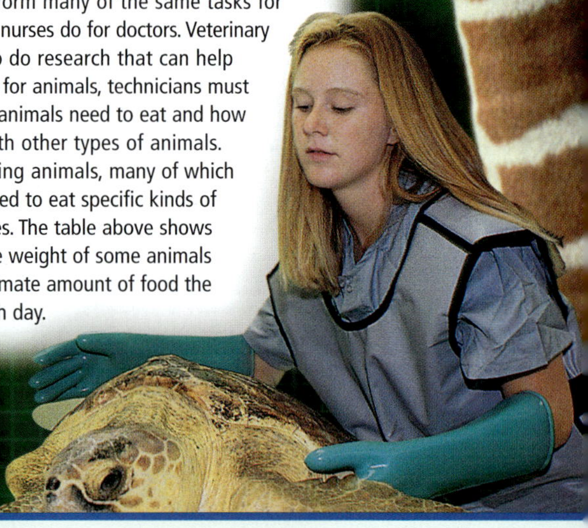

internet connect

Chapter Opener Online
go.hrw.com
KEYWORD: MR4 Ch1

Problem Solving Project

Life Science Connection

Purpose: To use estimates, calculations, and number lines to solve problems

Materials: Daily Food Intake worksheet

internet connect

Chapter Project Online: ***go.hrw.com***
KEYWORD: MR4 PSProject1

Understand, Plan, Solve, and Look Back

Have students:

- ✓ Complete the Daily Food Intake worksheet to discover the relationship between herbivore weight and amount of food required per day.
- ✓ Research the number of acres, hectares, or square miles required to sustain a family of three giraffes in the wild.
- ✓ Research hydroponics. Some of the food for herbivores in wild animal parks is grown hydroponically. Why?
- ✓ Check students' work.

ARE YOU READY?

Choose the best term from the list to complete each sentence.

place value
estimate
product
expanded
standard
period

1. The answer in a multiplication problem is called the __?__. **product**
2. 5,000 + 400 + 70 + 5 is a number written in __?__ form. **expanded**
3. A(n) __?__ tells about how many. **estimate**
4. The number 70,562 is written in __?__ form. **standard**
5. Ten thousands is the __?__ of the 4 in 42,801. **place value**

Complete these exercises to review skills you will need for this chapter.

✔ Compare Whole Numbers

Compare. Write <, >, or =.

6. 245 ■ 219 **>**
7. 5,320 ■ 5,128 **>**
8. 64 ■ 67 **<**
9. 784 ■ 792 **<**

✔ Round Whole Numbers

Round each number to the nearest hundred.

10. 567 **600**
11. 827 **800**
12. 1,642 **1,600**
13. 12,852 **12,900**
14. 1,237 **1,200**
15. 135 **100**
16. 15,561 **15,600**
17. 452,801 **452,800**

Round each number to the nearest thousand.

18. 4,709 **5,000**
19. 3,399 **3,000**
20. 9,825 **10,000**
21. 26,419 **26,000**
22. 12,434 **12,000**
23. 4,561 **5,000**
24. 11,784 **12,000**
25. 468,201 **468,000**

✔ Whole Number Operations

Add, subtract, multiply, or divide.

26. 18 × 22 **396**
27. 135 ÷ 3 **45**
28. 247 + 96 **343**
29. 358 − 29 **329**

✔ Evaluate Whole Number Expressions

Evaluate each expression.

30. 3 × 4 × 2 **24**
31. 20 + 100 − 40 **80**
32. 5 × 20 ÷ 4 **25**
33. 6 × 12 × 5 **360**

Assessing Prior Knowledge

INTERVENTION

Diagnose and Prescribe

Evaluate your students' performance on this page to determine whether intervention is necessary or whether enrichment is appropriate. Options that provide instruction, practice, and a check are listed below.

Resources for Are You Ready?

- ***Are You Ready? Intervention and Enrichment***
- **Recording Sheet for Are You Ready?** *Chapter 1 Resource Book . . . . . . p. 3*

Are You Ready? Intervention **CD-ROM**

ARE YOU READY?

Were students successful with Are You Ready?

NO INTERVENE

YES ENRICH

✔ Compare Whole Numbers 6–9
Intervention Practice, Skill 4
CD-ROM Intervention Activities, Skill 4

✔ Round Whole Numbers 10–25
Intervention Practice, Skill 3
CD-ROM Intervention Activities, Skill 3

✔ Whole Number Operations 26–29
Intervention Practice, Skill 34
CD-ROM Intervention Activities, Skill 34

✔ Evaluate Whole Number Expressions 30–33
Intervention Practice, Skill 54
CD-ROM Intervention Activities, Skill 54

Are You Ready? Enrichment, pp. 407–408

Section 1A

Whole Numbers and Exponents

One-Minute Section Planner

Lesson	Materials	Resources
Lesson 1-1 Comparing and Ordering Whole Numbers **NCTM:** Numbers and Operations, Communication, Connections, Representation **NAEP:** Number Properties 1i ☑ SAT-9 ☑ SAT-10 ☑ ITBS ☑ CTBS ☐ MAT ☐ CAT	**Optional** Teaching Transparencies T2–T3 *(CRB)*	• *Chapter 1 Resource Book,* pp. 6–14 • Daily Transparency T1, CRB • Additional Examples Transparencies T4–T5, CRB • *Alternate Openers: Explorations,* p. 1
Lesson 1-2 Estimating with Whole Numbers **NCTM:** Numbers and Operations, Communication **NAEP:** Number Properties 2b ☑ SAT-9 ☑ SAT-10 ☑ ITBS ☑ CTBS ☑ MAT ☐ CAT	**Optional** Recording Sheet for Reaching All Learners *(CRB, p. 77)*	• *Chapter 1 Resource Book,* pp. 15–24 • Daily Transparency T6, CRB • Additional Examples Transparencies T7–T8, CRB • *Alternate Openers: Explorations,* p. 2
Lesson 1-3 Exponents **NCTM:** Numbers and Operations, Problem Solving, Communication, Representation **NAEP:** Algebra 3b ☑ SAT-9 ☑ SAT-10 ☑ ITBS ☑ CTBS ☑ MAT ☑ CAT	**Optional** Teaching Transparency T10 *(CRB)*	• *Chapter 1 Resource Book,* pp. 25–33 • Daily Transparency T9, CRB • Additional Examples Transparencies T11–T12, CRB • *Alternate Openers: Explorations,* p. 3
Section 1A Assessment		• Mid-Chapter Quiz, SE p. 16 • Section 1A Quiz, AR p. 5 • *Test and Practice Generator* CD-ROM

SAT = *Stanford Achievement Tests* **ITBS** = *Iowa Test of Basic Skills* **CTBS** = *Comprehensive Test of Basic Skills/Terra Nova*
MAT = *Metropolitan Achievement Tests* **CAT** = *California Achievement Test*

NCTM = Complete standards can be found on pages T27–T33. **NAEP** = Complete standards can be found on pages A31–A35.

SE = *Student Edition* **TE** = *Teacher's Edition* **AR** = *Assessment Resources* **CRB** = *Chapter Resource Book* **MK** = *Manipulatives Kit*

Section Overview

Ordering Whole Numbers — *Lesson 1-1*

Why? Comparing and ordering numbers is the beginning of developing number sense.

1,234
1,254

1,234 < 1,254

To order numbers, you can compare and order numbers by using **place value**.

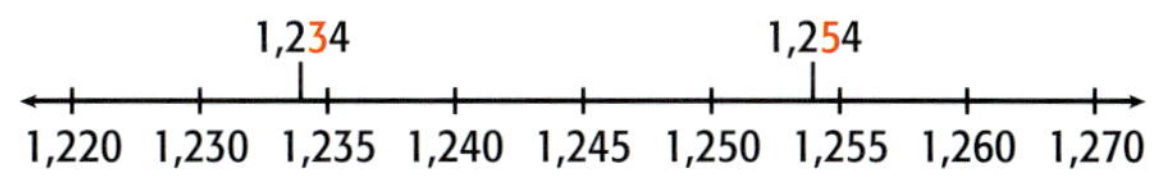

Numbers on a number line are ordered from least to greatest from left to right.

Estimating Whole Numbers — *Lesson 1-2*

Why? Estimating helps you check your answers or determine whether the result of an operation is reasonable.

When rounding, look at the digit to the **right** of the place to which you are rounding.

- If that digit is 5 or greater, round up.
- If that digit is less than 5, round down.

Estimate 235 × 829 to the nearest hundred.

200 × 800 = 1,600

Estimate 87 ÷ 28 to the nearest ten.

90 ÷ 30 = 3

Compatible numbers are numbers close to the numbers in the problem that you can calculate mentally.

Estimate 3,256 + 6,930 using compatible numbers.

3,000 + **7,000** ≈ 10,000

Representing Numbers Using Exponents — *Lesson 1-3*

Why? Exponents provide a shorthand method of representing numbers.

An **exponent** tells how many times the **base** is used as a factor.

The exponent is 5.

$3^5 = 3 \times 3 \times 3 \times 3 \times 3$

The base is 3.

3^5 is read as "three to the fifth power."

1-1 Organizer

Pacing: Traditional 1 day
Block $\frac{1}{2}$ day

Objective: Students compare and order whole numbers using place value or a number line.

Indiana Academic Standards
6.7.11

Warm Up

Compare. Use <, >, or =.

1. 8 ▢ 9 < **2.** 27 ▢ 14 >
3. 56 ▢ 23 > **4.** 10 ▢ 15 <
5. 11 ▢ 12 < **6.** 37 ▢ 16 >

Problem of the Day

Subtract your age from your age multiplied by 100. Divide the result by 11, and then divide the quotient by 9. What number do you get?
The answer will be the student's age.

Available on Daily Transparency in CRB

Math Humor

A line was injured and started to feel numb. Later, the situation got worse, and the line felt even more numb. What kind of line was it?

A "number" line

1-1 Comparing and Ordering Whole Numbers

Learn to compare and order whole numbers using place value or a number line.

The midyear world population in 1990 was 5,283,755,345 people. The world population by midyear 2010 is projected to be 6,823,634,553 people.

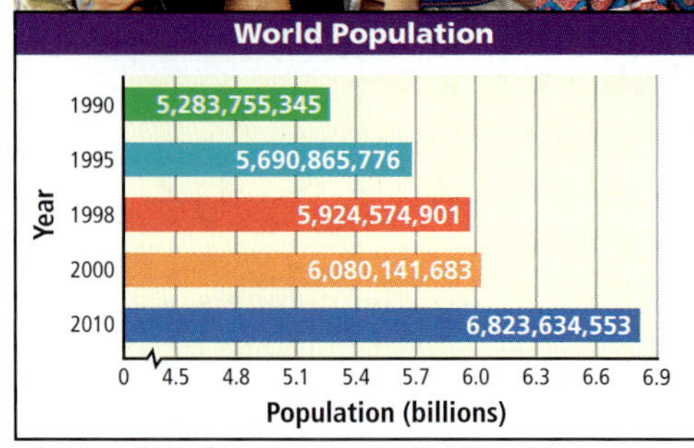

Source: U.S. Bureau of the Census, International Data Base, 2000

You can use place value to read and understand large numbers. In the place value chart below, 3 has a value of 3 millions, 3 ten thousands, or 3 ones, depending on its position in the number.

Place Value

Hundreds	Tens	Ones	Hundreds	Tens	Ones	Hundreds	Tens	Ones	Hundreds	Tens	Ones
		6 ,	8	2	3 ,	6	3	4 ,	5	5	3
Billions			Millions			Thousands			Ones		

Standard form: 6,823,634,553

Expanded form: 6,000,000,000 + 800,000,000 + 20,000,000 + 3,000,000 + 600,000 + 30,000 + 4,000 + 500 + 50 + 3

Word form: six billion, eight hundred twenty-three million, six hundred thirty-four thousand, five hundred fifty-three

1 Introduce

Alternate Opener

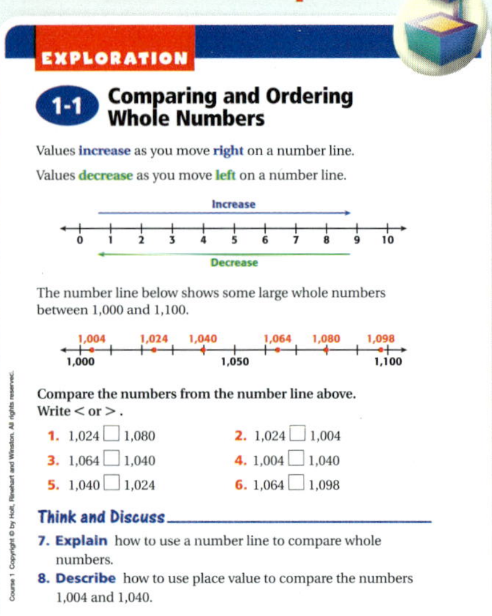
EXPLORATION

1-1 Comparing and Ordering Whole Numbers

Values **increase** as you move **right** on a number line.
Values **decrease** as you move **left** on a number line.

Increase
0 1 2 3 4 5 6 7 8 9 10
Decrease

The number line below shows some large whole numbers between 1,000 and 1,100.

1,004 1,024 1,040 1,064 1,080 1,098
1,000 1,050 1,100

Compare the numbers from the number line above. Write < or >.

1. 1,024 ▢ 1,080 2. 1,024 ▢ 1,004
3. 1,064 ▢ 1,040 4. 1,004 ▢ 1,040
5. 1,040 ▢ 1,024 6. 1,064 ▢ 1,098

Think and Discuss

7. Explain how to use a number line to compare whole numbers.
8. Describe how to use place value to compare the numbers 1,004 and 1,040.

Motivate

Ask students to estimate the number of sixth-grade, seventh-grade, and eighth-grade students in your school. (You may wish to get exact numbers from the school office.) Then have students decide which grade has the most students and which has the fewest students.

Exploration worksheet and answers on Chapter 1 Resource Book pp. 7 and 82

2 Teach

Lesson Presentation

Guided Instruction

In this lesson, students learn to compare and order whole numbers using place value or a number line. Show students how to use place value to compare two numbers with equal numbers of digits. Then, have students use a number line to order three numbers with different numbers of digits. Use number lines and place value in both examples so students can see that both methods work for comparing and ordering.

Teaching Tip Extend the examples by using numbers that the students generate, e.g., populations of nearby towns, number of boys and girls in the school, and so on.

EXAMPLE 1 Using Place Value to Compare Whole Numbers

Belgium's 2001 population was 10,258,762 people. The Czech Republic's 2001 population was 10,264,212 people. Which country had more people?

Belgium:	1 0, 2 [5] 8, 7 6 2	*Start at the left and compare digits in the same place value position. Look for the first place where the values are different.*
Czech Republic:	1 0, 2 [6] 4, 2 1 2	

50 thousand is less than 60 thousand.
10,258,762 is less than 10,264,212.

The Czech Republic had more people.

To order numbers, you can compare them using place value and then write them in order from least to greatest. You can also graph the numbers on a number line. As you read the numbers from left to right, they will be ordered from least to greatest.

EXAMPLE 2 Using a Number Line to Order Whole Numbers

Remember!

< means "is less than."
3 < 5   120 < 504

> means "is greater than."
17 > 9   212 > 83

Order the numbers from least to greatest.
923; 835; 1,266

Graph the following numbers on a number line:
The number 923 is between 900 and 1,000.
The number 835 is between 800 and 900.
The number 1,266 is between 1,200 and 1,300.

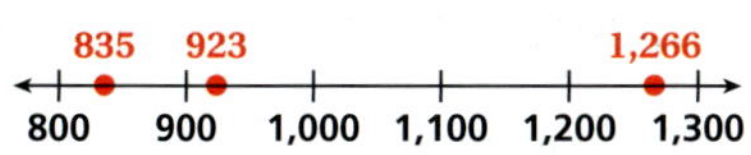

The numbers are ordered when you read the number line from left to right.

The numbers in order from least to greatest are 835, 923, and 1,266.

Think and Discuss

1. **Give** the place value of the digit 3 in each of the following numbers: 2,307,912; 2,370,912; 2,703,912.
2. **Read** each of the following numbers: 937,052; 3,012,480; 8,135,712,004.
3. **Look** at the bar graph at the beginning of the lesson. In which years was the population between 5,500,000,000 and 6,500,000,000?

COMMON ERROR ALERT

Remind students to start at the leftmost place value, not the right, when they are comparing and ordering numbers. Tell students that the order in which they compare digits when comparing numbers is the same order in which they read digits when reading numbers.

Additional Examples

Example 1

Belize's 2000 population was 249,183 people. Iceland's 2000 population was 276,365 people. Which country had more people?
Iceland had more people.

Example 2

Order the numbers from least to greatest: 675; 1,044; 497
The numbers in order from least to greatest are 497, 675, and 1,044.

Example 2 note: Point out that when whole numbers with different numbers of digits are being compared, the number with more digits is the greater of the two numbers.

3 Close

Reaching All Learners
Through Number Sense

To help students with place value comparisons, have them associate place value with denominations of currency. For example, 392 would be represented as 3 hundred-dollar bills, 9 ten-dollar bills, and 2 one-dollar bills.

Summarize

Demonstrate ordering numbers using both place value and number lines. Give students two sets of numbers; have students use place value to order one of the sets and a number line to order the other set.

Answers to Think and Discuss

1. hundred thousand; hundred thousand; thousand
2. nine hundred thirty-seven thousand, fifty-two; three million, twelve thousand, four hundred eighty; eight billion, one hundred thirty-five million, seven hundred twelve thousand, four
3. 1995, 1998, and 2000

1-1 PRACTICE & ASSESS

Students may want to refer back to the lesson examples.

Assignment Guide

If you finished Example 1 assign:
Core 1–2, 6–8, 15–20, 32–42
Enriched 1–2, 6–8, 15–18, 29–30, 32–42

If you finished Example 2 assign:
Core 1–14, 15–27 odd, 28, 32–42
Enriched 1, 3, 6–8, 15–42

Notes

1-1 Exercises

FOR EXTRA PRACTICE
see page 636

internet connect
Homework Help Online
go.hrw.com Keyword: MR4 1-1

GUIDED PRACTICE

See Example 1

1. Mount McKinley, in Alaska, is 20,320 feet tall. Mount Aconcagua, in Argentina, is 22,834 feet tall. Which mountain is taller? **Mount Aconcagua is taller.**
2. The area of the Caribbean Sea is 971,400 square miles. The area of the Mediterranean Sea is 969,100 square miles. Which sea is smaller in area? **The Mediterranean Sea is smaller.**

See Example 2

Order the numbers from least to greatest.

3. 726; 349; 642 **349; 642; 726**
4. 513; 915; 103 **103; 513; 915**
5. 497; 1,264; 809 **497; 809; 1,264**

INDEPENDENT PRACTICE

See Example 1

6. The attendance in 1999 at a theme park was 17,459,000 people. The attendance in 1999 at a water park was 15,200,000 people. Which park had the higher attendance? **the theme park with 17,459,000 in attendance**
7. According to the table, which river is longer, the Missouri or the Mississippi? **The Mississippi River is longer.**
8. A New York City driving range reported 413,497 golf balls were hit by customers last year. A Philadelphia range reported customers hit 408,959 golf balls. Which range had more golf balls hit? **New York City**

River Length (mi)	
Mississippi	2,340
Missouri	2,315
Ohio	618
Red	1,290
Rio Grande	1,900

See Example 2

Order the numbers from least to greatest.

9. 367; 597; 279 **279; 367; 597**
10. 619; 126; 480 **126; 480; 619**
11. 946; 705; 810 **705; 810; 946**
12. 423; 1,046; 805 **423; 805; 1,046**
13. 1,523; 2,913; 111 **111; 1,523; 2,913**
14. 1,764; 1,359; 666 **666; 1,359; 1,764**

PRACTICE AND PROBLEM SOLVING

Compare. Write < or >.

15. 46,495 ▢ 46,594 **<**
16. 162,648 ▢ 126,498 **>**
17. 3,654 ▢ 3,645 **>**
18. 512,105 ▢ 512,099 **>**
19. 29,448 ▢ 29,488 **<**
20. 913,203 ▢ 913,600 **<**

Order the numbers from greatest to least.

21. 591; 924; 341 **924; 591; 341**
22. 601; 533; 823; 149 **823; 601; 533; 149**
23. 291; 911; 439; 747 **911; 747; 439; 291**
24. 2,649; 3,461; 1,947 **3,461; 2,649; 1,947**
25. 5,349; 5,389; 5,480 **5,480; 5,389; 5,349**
26. 7,467; 7,239; 7,498 **7,498; 7,467; 7,239**

Math Background

The decimal system was developed by a Hindu mathematician before the year 1000. The system did not reach Europe until the thirteenth century. It took hundreds of years after that for the decimal system to be broadly adopted.

Comparing and ordering whole numbers is simple with the decimal system. Because only the 10 digits 0, 1, 2, 3, 4, 5, 6, 7, 8, and 9 are used, and because a digit's place indicates its value, you can read and compare any two numbers from left to right.

RETEACH 1-1

LESSON 1-1 **Reteach**
Comparing and Ordering Whole Numbers

You can use place value to compare or order whole numbers.
Use < or > to compare the numbers.

289,865 ▢ 289,765

Thousands			Ones			Compare the digits from left to right.
H	T	O	H	T	O	
2	8	9	**8**	6	5	First Number
2	8	9	**7**	6	5	Second Number

8 > 7
So, 289,865 > 289,765

Write < or > to compare the numbers.

1.

Thousands			Ones		
H	T	O	H	T	O
		3	5	4	7
		3	5	3	2

3,547 **>** 3,532

2.

Thousands			Ones		
H	T	O	H	T	O
		9	5	3	6
		9	6	3	5

9,536 **<** 9,635

Write the numbers in order from least to greatest.
976; 859; 924

Ones		
H	T	O
9	7	6
8	5	9
9	2	4

Compare the numbers in pairs.
976 > 859, 976 > 924, and 859 < 924.
So the numbers from least to greatest are 859; 924; 976.

Write the numbers in order from least to greatest.

3.

Ones		
9	5	4
9	4	5
9	6	9

945; 954; 969

4.

Ones		
3	4	3
3	3	4
4	3	4

334; 343; 434

5.

Ones		
8	9	4
8	9	2
9	6	5

892; 894; 965

PRACTICE 1-1

LESSON 1-1 **Practice B**
Comparing and Ordering Whole Numbers

Write < or > to compare the numbers.

1. 69 **<** 96
2. 117 **>** 107
3. 958 **<** 9,124
4. 3,567 **>** 3,561
5. 18,443 **>** 1,844
6. 64,209 **<** 64,290

Write the numbers from least to greatest.

7. 58; 166; 85 **58; 85; 166**
8. 115; 151; 111 **111; 115; 151**
9. 269; 29; 96 **29; 96; 269**
10. 308; 3,800; 3,080 **308; 3,080; 3,800**
11. 1,864; 824; 1,648 **824; 1,648; 1,864**
12. 4,663; 4,336; 43,666 **4,336; 4,663; 43,666**

Write the numbers from greatest to least.

13. 35; 53; 13 **53; 35; 13**
14. 807; 800; 708 **807; 800; 708**
15. 249; 392; 248 **392; 249; 248**
16. 555; 600; 535 **600; 555; 535**
17. 7,320; 6,000; 6,305 **7,320; 6,305; 6,000**
18. 999; 9,559; 5,995 **9,559; 5,995; 999**
19. Delaware and Rhode Island are the two smallest states. Delaware covers 1,955 square miles, and Rhode Island covers 1,045 square miles. What is the smallest state in the United States? **Rhode Island**
20. Vermont and Wyoming have the smallest populations in the United States. The population of Vermont is 608,827. The population of Wyoming is 493,782. Which state has the smallest population? **Wyoming**

History LINK

Before calculators and computers were invented, the abacus was used to solve math problems. On an abacus, rods represent place values, and beads that slide on the rods are used as counters.

go.hrw.com KEYWORD: MR4 Abacus

CNN Student News

27. ***GEOGRAPHY*** The three biggest states in the continental United States are California, 159,869 square miles; Montana, 147,047 square miles; and Texas, 267,277 square miles. Write the states in order from smallest area to largest area. **Montana, California, Texas**

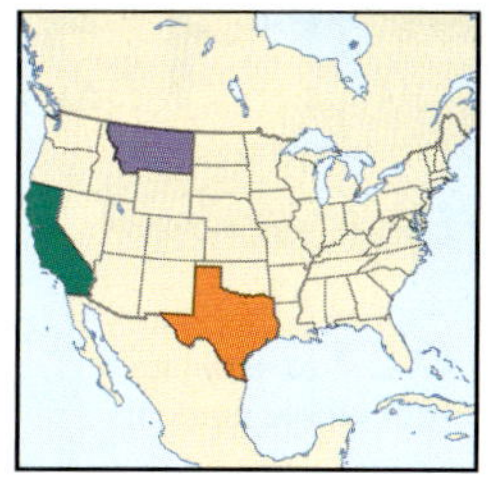

28. The two drawings show another way to represent numbers. The rod on the far left of each drawing represents the hundred thousands place. The number of beads on a rod tells the value for that place. Which drawing represents the greater number? **Drawing B**

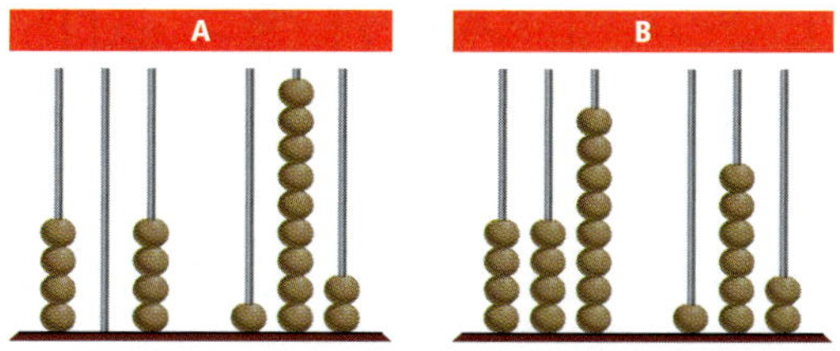

29. ***WHAT'S THE ERROR?*** A student said 19,465,405 is greater than 19,465,425. Explain the error. Write the statement correctly.

30. ***WRITE ABOUT IT*** Explain how you would compare 19,465,146 and 19,460,146.

31. ***CHALLENGE*** In Roman numerals, letters represent numbers. For example, I = 1, V = 5, X = 10, L = 50, and C = 100. Letters in Roman numerals are written next to each other; this is how the value of the number is shown. To read the numbers below, add the values of all of the letters. What numbers do the following represent?

a. CLX **160** **b.** LVI **56** **c.** CIII **103**

Spiral Review

Write the value of the red digit in each number. (Previous course)

32. 649,809 **800** **33.** 349,239 **300,000** **34.** 27,463 **20,000** **35.** 16,239 **6,000**

Write each number in word form. (Previous course)

36. 1,645 **37.** 24,498 **38.** 306,927 **39.** 4,605,926

Write each number in standard form. (Previous course)

40. two hundred thirty-four thousand, six hundred seventy-nine **234,679**

41. fifteen million, nine hundred three thousand, one hundred eight **15,903,108**

42. **TEST PREP** Which number has a 4 in the thousands place? (Previous course) **B**

A 10,400 **B** 14,307 **C** 3,742,619 **D** 8,405,361

Answers

29. Possible answer: The student should compare the tens place and note that $0 < 2$. Correctly written, it is $19{,}465{,}405 < 19{,}465{,}425$.

30. Possible answer: Compare each place value starting with ten millions. When a place value is less in one of the numbers, that number is less than the other number.

36. one thousand, six hundred forty-five

37. twenty-four thousand, four hundred ninety-eight

38. three hundred six thousand, nine hundred twenty-seven

39. four million, six hundred five thousand, nine hundred twenty-six

Journal

Have students write about when they would use place value or a number line to order numbers and why.

Test Prep Doctor

For Exercise 42, students will need to remember the order of the place-value positions. They may sketch a place-value table on scratch paper to do so. Students who answered **A, C,** or **D** have confused the position of the thousands place.

CHALLENGE 1-1

LESSON 1-1 **Challenge**
Ancient Calculators

Long before place-value charts were invented, people used a tool called an abacus to show and read numbers. The Chinese began using the *Suan Pan* abacus, shown below, about 800 years ago.

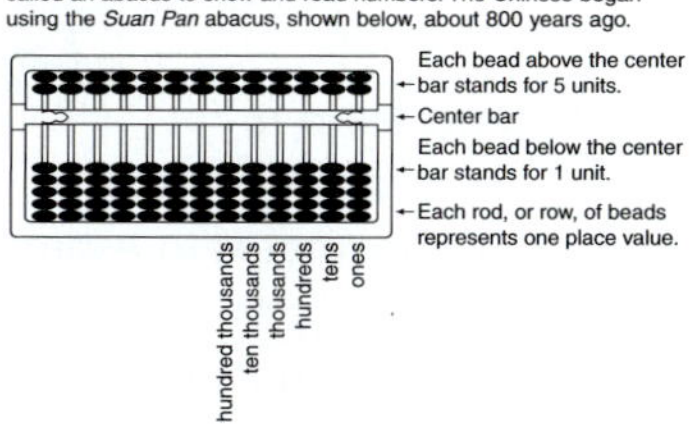

To use an abacus to show and read a number, move the beads to the center bar for each place value and add. For example:

2 8 4 0 5 = 28,405

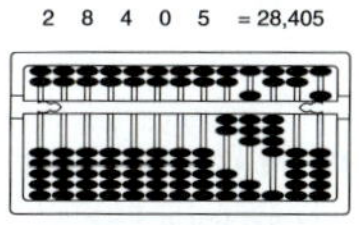

6 3 1 7 0 9 = 631,709

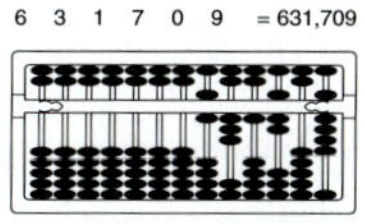

Write the number shown on each abacus. Then write < or > to compare the numbers.

1.

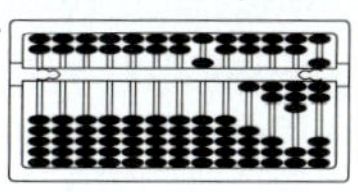

2.

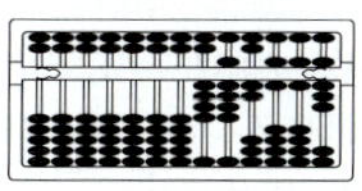

501, 237 [>] 492,668

PROBLEM SOLVING 1-1

LESSON 1-1 **Problem Solving**
Comparing and Ordering Whole Numbers

Use the tables below to answer each question.

Most Populated Countries	
Brazil	174,468,575
China	1,273,111,290
India	1,029,991,145
Indonesia	228,437,870
United States	278,058,881

Largest Countries (square mi)	
Brazil	3,265,059
Canada	3,849,646
China	3,705,408
Russia	6,592,812
United States	3,539,224

1. Which country has the greatest population in the world? How many people live there?
China; 1,273,111,290

2. How many countries in the world have more than one billion people? What are those countries?
2; China and India

3. What is the largest country in the world?
Russia

4. Which country's area is closest to 4,000,000 square miles?
Canada

5. Which country has a population less than two hundred million?
Brazil

6. Which countries have populations greater than the United States?
China and India

7. What is the error in the following statement? Canada is larger than the United States, but smaller than China.
Canada is larger than China.

8. Based on population and size, which country do you think is more crowded, Brazil or the United States? Explain.
U.S.; They are about the same size, but U.S. has more people.

9. Write the countries in order by population from greatest to least.
China, India, United States, Indonesia, Brazil

10. Write the countries in order by size from smallest to largest.
Brazil, United States, China, Canada, Russia

Lesson Quiz

Compare. Write < or >.

1. 47,328 ▢ 47,238 $>$

2. 933,826 ▢ 933,520 $>$

3. Order the numbers from least to greatest: 726, 847, 221.
221, 726, 847

4. The area of Panama is 78,200 square kilometers, and the area of Lithuania is 65,200 square kilometers. Which country is smaller in area? **Lithuania**

Available on Daily Transparency in CRB

1-2 Organizer

Pacing: Traditional 1 day
Block $\frac{1}{2}$ day

Objective: Students estimate with whole numbers.

Indiana Academic Standards
6.7.7, 6.7.11

Warm Up

Find each sum.

1. 3,214 + 5,490 **8,704**
2. 9,225 + 8,652 **17,877**
3. 3,210 + 1,200 **4410**
4. 8,774 + 2,156 **10,930**

Problem of the Day

Continue the number pattern below. Explain the pattern you found.

3, 6, 10, 15, ____, ____

21, 28; One possible pattern is to increase the difference between consecutive terms by one more than the difference between preceding consecutive terms.

Available on Daily Transparency in CRB

Math Fact

The estimated distance that light will travel in a year is 5,880,000,000,000 miles, or 1 light-year.

1-2 Estimating with Whole Numbers

Learn to estimate with whole numbers.

Vocabulary
compatible number
underestimate
overestimate

Sometimes in math you do not need an exact answer. Instead, you can use an estimate. Estimates are close to the exact answer but are usually easier and faster to find.

When estimating, you can round the numbers in the problem to *compatible numbers*. **Compatible numbers** are close to the numbers in the problem, and they can help you do math mentally.

EXAMPLE 1 Estimating a Sum or Difference by Rounding

Estimate each sum or difference by rounding to the place value indicated.

Remember!

When rounding, look at the digit to the right of the place to which you are rounding.

- If that digit is 5 or greater, round up.
- If that digit is less than 5, round down.

A 5,439 + 7,516; thousands

5,000	*Round 5,439 down.*
+ 8,000	*Round 7,516 up.*
13,000	

The sum is about 13,000.

B 62,167 − 47,511; ten thousands

60,000	*Round 62,167 down.*
− 50,000	*Round 47,511 up.*
10,000	

The difference is about 10,000.

An estimate that is less than the exact answer is an **underestimate**.

An estimate that is greater than the exact answer is an **overestimate**.

1 Introduce

Alternate Opener

EXPLORATION

1-2 Estimating with Whole Numbers

Harvard Middle School is collecting aluminum cans to recycle for a fund-raiser. The number of cans that each grade collected is shown in the bar graph.

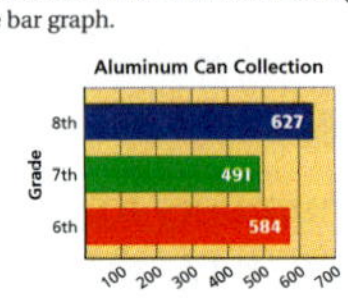

1. Estimate the total number of cans that Harvard Middle School collected.

The principal announced that nearly 2,000 cans were collected. The newspaper reported that over 1,700 cans were collected.

2. Is either of these reports correct?
3. Why are the reports different?

Think and Discuss

4. **Describe** how you reached your estimate.
5. **Identify** some words that indicate whether an amount is an estimate or approximation.

Motivate

Have students plan a party for 12 people. They'll need to estimate the number of plates and napkins and the amount of food and drink to buy. Tell them that sometimes you do not need to use exact numbers. Planning a party is one situation where an estimate is sufficient.

Exploration worksheet and answers on Chapter 1 Resource Book pp. 16 and 84

2 Teach

Lesson Presentation

Guided Instruction

In this lesson, students learn to estimate with whole numbers. Show students how to round numbers to estimate sums, differences, and products and to use compatible numbers to estimate quotients. Point out that the compatible numbers used to estimate a quotient are not necessarily found by rounding.

Teaching Tip Show students how to find compatible numbers for estimating quotients. Instead of always rounding numbers, they may want to choose a pair of numbers that are easy to divide mentally. For example, use 420 ÷ 70 to estimate 435 ÷ 75.

EXAMPLE 2 Estimating a Product by Rounding

Remember!

The area of a rectangle is found by multiplying the length by the width.

$A = \ell \times w$

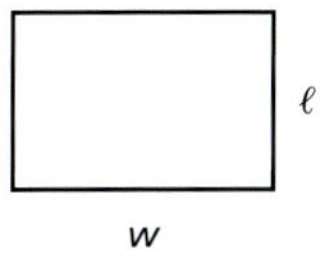

The sixth-grade class is preparing to paint a mural on one wall of the school. First the students need to paint the entire area of the wall white. The wall is a rectangle 9 feet tall and 27 feet wide. One quart of paint will cover an area of 100 square feet. How many quarts of white paint should the students buy?

First find the area of the wall in square feet.

$9 \times 27 \rightarrow 9 \times 30$ — *Overestimate the area of the wall.*

$9 \times 30 = 270$ — *The actual area is **less than** 270 square feet.*

If one quart of paint will cover 100 square feet, then 2 quarts will cover 200 square feet. Three quarts of paint will cover 300 square feet.

The students should buy three quarts of paint.

EXAMPLE 3 Estimating a Quotient Using Compatible Numbers

Mrs. Byrd will drive 120 miles to take Becca to the state fair. She can drive 65 mi/h. About how long will the trip take?

To find how long the trip will be, divide the miles Mrs. Byrd has to travel by how many miles per hour she can drive.

miles ÷ miles per hour

$120 \div 65 \rightarrow 120 \div 60$ — *120 and 60 are compatible numbers.* ***Underestimate** the speed.*

$120 \div 60 = 2$ — *Because she **underestimated** the speed, the actual time will be **less than** 2 hours.*

It will take Mrs. Byrd about two hours to reach the state fair.

Think and Discuss

1. **Suppose** you are buying items for a party and you have \$50. Would it be better to overestimate or underestimate the cost of the items?
2. **Describe** situations in which you might want to estimate.

Additional Examples

Example 1

Estimate each sum or difference by rounding to the place value indicated.

A. 12,345 + 62,167; ten thousands
The sum is about 70,000.

B. 4,983 − 2,447; thousands
The difference is about 3,000.

Example 2

The sixth-grade class wants to paint a wall. The wall is a rectangle 9 feet tall and 18 feet wide. One quart of paint will cover 100 ft^2. How many quarts of paint should the students buy?
They should buy 2 quarts of paint.

Example 3

Mr. Dehmel will drive 243 miles to the fair at 65 mi/h. About how long will his trip take? The actual time will be less than 4 hours.

Example 3 note: If students attempt to round 65 up to 70, remind them to compatible numbers when estimating quotients.

3 Close

Reaching All Learners
Through Number Sense

Have students work in pairs to rewrite each expression so that an estimate can be given. For the benefit of the class, have one pair explain what they did to find an estimate in each expression. (Use the recording sheet provided on Chapter 1 Resource Book p. 77.)

1. 29,502 + 24,098 **30,000 + 20,000**
2. 94,142 − 47,071 **90,000 − 50,000**
3. 23 × 2,802 **20 × 3,000**
4. 178,932 ÷ 6 **180,000 ÷ 6**

Summarize

Emphasize that estimates should be easy enough to obtain mentally. Remind students to round to the highest place value for sums, differences, and products. Encourage them to be mathematically creative when looking for compatible numbers to estimate quotients.

Answers to Think and Discuss

1. Possible answer: Overestimate; that way you will be sure to have enough money to buy everything you need.
2. Possible answers: when figuring the amount of time needed to complete a project; when saving money to buy something; when deciding how much paper is needed to wrap a gift

1-2 PRACTICE & ASSESS

Students may want to refer back to the lesson examples.

Assignment Guide

If you finished Example 1 assign:
Core 1–2, 5–8, 11–16, 20, 21, 26–36
Enriched 11–18, 20–36

If you finished Example 2 assign:
Core 1–3, 5–9, 11–17 odd, 20–22, 26–36
Enriched 11–36

If you finished Example 3 assign:
Core 1–14, 19–22, 26–36
Enriched 9–36

Notes

1-2 Exercises

FOR EXTRA PRACTICE
see page 636

Internet connect
Homework Help Online
go.hrw.com Keyword: MR4 1-2

GUIDED PRACTICE

See Example 1 **Estimate each sum or difference by rounding to the place value indicated.**

1. 4,689 + 2,469; thousands 7,000 **2.** 50,498 − 35,798; ten thousands 10,000

See Example 2 **3.** The graph shows the number of bottles of water used in three bicycle races last year. If the same number of riders enter the races each year, estimate the number of bottles that will be needed for races held in May over the next five years. 1,500 bottles of water

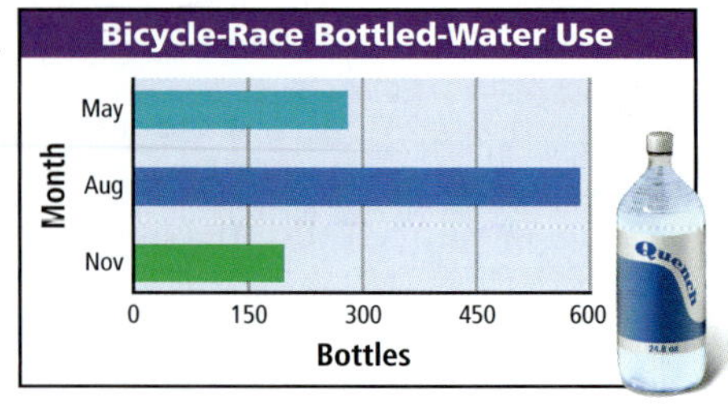

See Example 3 **4.** If a local business provided half the bottled water needed for the August bicycle race, about how many bottles did the company provide? 300 bottles

INDEPENDENT PRACTICE

See Example 1 **Estimate each sum or difference by rounding to the place value indicated.**

5. 6,570 + 3,609; thousands 11,000 **6.** 49,821 − 11,567; ten thousands 40,000

7. 3,912 + 1,269; thousands 5,000 **8.** 37,097 − 20,364; ten thousands 20,000

See Example 2 **9.** The recreation center has provided softballs every year to the city league. Use the table to estimate the number of softballs the league will use in 5 years. 150 softballs

See Example 3 **10.** The recreation center has a girls' golf team with 8 members. About how many golf balls will be available to each girl on the team? 40 golf balls

Recreation Center Balls Supplied

Sport	Number of Balls
Basketball	21
Golf	324
Softball	28
Table tennis	95

PRACTICE AND PROBLEM SOLVING

Estimate each sum or difference by rounding to the greatest place value.

11. 152 + 269 500 **12.** 797 − 234 600

13. 6,152 − 3,195 3,000 **14.** 9,179 + 2,206 11,000

15. 82,465 − 38,421 40,000 **16.** 38,347 + 17,039 60,000

17. 639,069 + 283,136 900,000 **18.** 777,060 − 410,364 400,000

Math Background

Another way to estimate a sum or product is to round both terms down to obtain one estimate and to round both terms up to obtain a second estimate. Together, the two estimates form a range within which the actual answer must lie. This is true only when adding or multiplying. It is not true when subtracting or dividing.

For example, 457 × 341 can be estimated by the products 400 × 300 = 120,000 and 500 × 400 = 200,000. You then can conclude that the actual product lies between 120,000 and 200,000.

RETEACH 1-2

LESSON 1-2 **Reteach**
Estimating with Whole Numbers

In mathematics, you can find an estimate when an exact answer is not needed. An estimate is close to the exact answer.

You can use rounding to estimate sums and differences.

A. Estimate the sum by rounding to the hundreds.
3,478 → 3,500
+ 7,136 → + 7,100
10,600

B. Estimate the difference by rounding to the thousands.
23,848 → 24,000
− 16,132 → − 16,000
8,000

Estimate each sum or difference by rounding to the place value indicated.

1. hundreds
789 → 800
+ 453 → + 500
1,300

2. thousands
4,987 → 5,000
− 2,348 → − 2,000
3,000

3. tens
456 → 460
+ 875 → + 880
1,340

4. tens
876 → 880
− 432 → − 430
450

5. hundreds
6,898 → 6,900
+ 2,671 → + 2,700
9,600

6. thousands
1,857 → 2,000
+ 3,598 → + 4,000
6,000

7. hundreds
8,813 → 8,800
− 2,384 → − 2,400
6,400

8. thousands
9,128 → 9,000
− 4,716 → − 5,000
4,000

PRACTICE 1-2

LESSON 1-2 **Practice B**
Estimating with Whole Numbers

Estimate each sum or difference. Possible answers:

1. 67 + 14 80 **2.** 583 − 329 300 **3.** 94 − 36 50

4. 2,856 + 2,207 5,000 **5.** 276 + 316 600 **6.** 6,020 − 3,688 2,000

7. 34,465 + 19,002 50,000 **8.** 78,135 − 19,431 60,000 **9.** 216,135 + 165,800 400,000

Estimate each product or quotient.

10. 59 ÷ 6 10 **11.** 51 • 8 400 **12.** 83 ÷ 4 21

13. 9 • 27 270 **14.** 49 ÷ 6 8 **15.** 53 • 8 400

16. 147 ÷ 5 30 **17.** 118 ÷ 6 20 **18.** 79 • 5 400

19. Sailfish are the fastest fish in the world. They can swim 68 miles an hour. About how far can a sailfish swim in 3 hours?
about 210 miles

20. At a height of 3,281 feet, Angel Falls in Venezuela is the tallest waterfall in the world. Niagara Falls in the United States is only 190 feet tall. About how much taller is Angel Falls?
about 3,000 feet taller

21. Ali, a gardener, is preparing to fertilize a lawn. The lawn is 30 yards by 25 yards. One bag of fertilizer will cover an area of 100 square yards. How many bags of fertilizer does Ali need to buy?
9 bags

Social Studies LINK

Use the bar graph for Exercises 19–25.

19. On one summer day there were 2,824 sailboats on Lake Erie. Estimate the number of square miles available to each boat. **10 square miles**

20. If the areas of all the Great Lakes are rounded to the nearest thousand, which two of the lakes would be the closest in area? **Erie and Ontario**

21. About how much larger is Lake Huron than Lake Ontario? **40,000 square miles**

22. The Great Lakes are called "great" because of the huge amount of fresh water they contain. Estimate the total area of all the Great Lakes combined. **280,000 square miles**

23. ***WHAT'S THE QUESTION?*** Lake Erie is about 50,000 square miles smaller. What is the question?

24. ***WRITE ABOUT IT*** Explain how you would estimate the areas of Lake Huron and Lake Michigan to compare their sizes.

25. ***CHALLENGE*** Estimate the average area of the Great Lakes. **60,000 square miles**

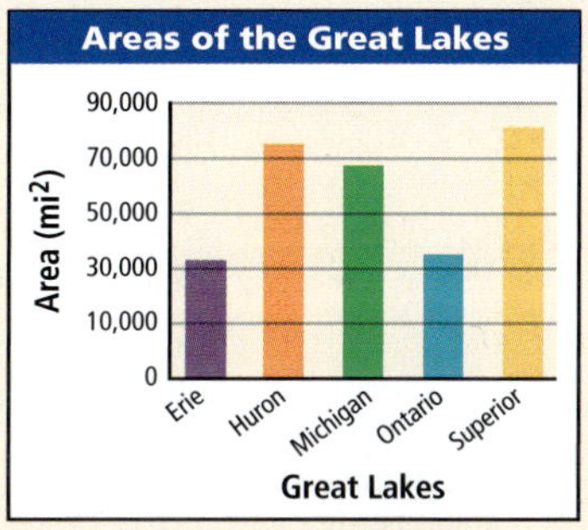

Area includes the water surface and drainage basin within the United States and Canada.

Spiral Review

Find each product or quotient. (Previous course)

26. 148 ÷ 4 **37** **27.** 523 × 46 **24,058** **28.** 1,054 ÷ 31 **34**

29. 223 × 16 **3,568** **30.** 522 ÷ 18 **29** **31.** 1,107 ÷ 27 **41**

Write each number in standard form. (Lesson 1-1)

32. 3,000 + 200 + 70 + 3 **3,273** **33.** 10,000 + 500 + 20 + 1 **10,521**

34. 500,000 + 60,000 + 300 + 3 **560,303** **35.** 70,000 + 7 **70,007**

36. **TEST PREP** Which of the following numbers is the standard form of eight hundred twenty-three thousand seven? (Lesson 1-1) **C**

A 800,237 **B** 823,700 **C** 823,007 **D** 8,237,000

Interdisciplinary LINK

Social Studies

Exercises 19–25 involve the areas of the Great Lakes. Physical features of the United States and Canada, including the Great Lakes, are studied in middle-school social studies programs, such as Holt, Rinehart & Winston's *People, Places, and Change.*

Answers

23. How does the size of Lake Superior compare with the size of Lake Erie?

24. Possible answer: Round each area to the nearest thousand and then subtract.

Journal

Have students write about two situations in which estimations would be helpful.

Test Prep Doctor

For Exercise 36, students need to translate from the written form of a number to its standard form. Students who chose **D** have confused the millions place with the hundred-thousands place. Students who chose **A** or **B** have used the correct digits in the correct order, but have placed some of them in the wrong place-value positions.

CHALLENGE 1-2

LESSON 1-2 Challenge
A Shopping Spree!

You have just won a $2,000 shopping spree at Electronics City! Use estimation and the store's advertisement below to make two different shopping lists of what you can buy without going over your spending limit.

LIST 1 LIST 2

Estimated Cost ________ Estimated Cost ________

Actual Cost ________ Actual Cost ________

Answers will vary depending on students' chosen items. Check for correct estimation. All lists should total less than $2,000.

PROBLEM SOLVING 1-2

LESSON 1-2 Problem Solving
Estimating with Whole Numbers

Use the table below to answer each question.

Facts About the World's Oceans

Ocean	Area (square mi)	Greatest Depth (ft)
Arctic	5,108,132	18,456
Atlantic	33,424,006	30,246
Indian	28,351,484	24,460
Pacific	64,185,629	35,837

1. If the depths of all the oceans were rounded to the nearest ten thousand, which two oceans would have the same depth? **Arctic and Indian**

2. In 1960, scientists observed sea creatures living as far down as thirty thousand feet. In which ocean(s) could these creatures live? **Pacific and Atlantic**

3. Which ocean covers about thirty-five million square miles? **Atlantic**

4. Which ocean's depth is closest to twenty thousand feet? **Arctic**

5. If you wanted to compare the depths of the Pacific Ocean and the Atlantic Ocean, which place value would you use to estimate? **thousands**

6. The oceans cover about three-fourths of Earth's surface. Estimate the total area of all the oceans combined by rounding to the nearest million? **about 130 million sq. mi**

7. There are 5,280 feet in a mile. About how many miles deep is the deepest point in the Pacific Ocean? **about 7 miles**

8. Rounding to the greatest place value, about how much larger is the Indian Ocean than the Arctic Ocean? **about 25 million sq. mi**

9. The Atlantic Ocean is about 40 times larger than the world's largest island, Greenland. Use this information to estimate the area of Greenland. **about 800,000 sq. mi**

10. About how much larger would the Pacific Ocean have to be to have more area than the other three oceans combined? **about 2 million sq. mi**

Lesson Quiz

Estimate each sum or difference by rounding to the place value indicated.

1. 7,420 + 3,527; thousands **11,000**

2. 47,821 + 19,925; ten thousands **70,000**

3. 8,254 − 5,703; thousands **2000**

4. 66,845 − 24,782; ten thousands **50,000**

5. One quart of paint covers an area of 100 square feet. How many quarts are needed to paint a wall 8 feet tall and 19 feet wide? **2**

Available on Daily Transparency in CRB

1-3 Organizer

Pacing: Traditional 1 day
Block $\frac{1}{2}$ day

Objective: Students represent numbers by using exponents.

Indiana Academic Standards
6.7.1, 6.7.2, 6.7.5, 6.7.11

Warm Up

Multiply.

1. 3 × 3 × 3 — 27
2. 4 × 4 × 4 — 64
3. 2 × 2 × 2 × 2 — 16
4. 5 × 5 × 5 × 5 — 625

Problem of the Day

Replace the letters *a*, *b*, and *c* with the numbers 3, 4, and 5 to make a true statement.

$2^a + 2^a = b^c$ $\quad 2^5 + 2^5 = 4^3$

Available on Daily Transparency in CRB

Math Humor

In the math Olympics, the contestant from Havana calculated the largest result. Everyone else was squarin' while he was *Cuban*.

1-3 Exponents

Learn to represent numbers by using exponents.

Vocabulary
exponent
base
exponential form

The most recent eruption of Mount Vesuvius took place in 1944.

Since 1906, the height of Mount Vesuvius in Italy has increased by 7^3 feet. How many feet is this?

The number 7^3 is written with an exponent. An **exponent** tells how many times a number called the **base** is used as a factor.

Base → 7^3 ← Exponent $= 7 \times 7 \times 7 = 343$

So the height of Mount Vesuvius has increased by 343 ft.

A number is in **exponential form** when it is written with a base and an exponent.

Exponential Form	Read	Multiply	Value
10^1	"10 to the 1st power"	10	10
10^2	"10 squared," or "10 to the 2nd power"	10 × 10	100
10^3	"10 cubed," or "10 to the 3rd power"	10 × 10 × 10	1,000
10^4	"10 to the 4th power"	10 × 10 × 10 × 10	10,000

EXAMPLE 1 Writing Numbers in Exponential Form

Write each expression in exponential form.

A 4 × 4 × 4

4^3 *4 is a factor 3 times.*

B 9 × 9 × 9 × 9 × 9

9^5 *9 is a factor 5 times.*

EXAMPLE 2 Finding the Value of Numbers in Exponential Form

Find each value.

A 3^7

$3^7 = 3 \times 3 \times 3 \times 3 \times 3 \times 3 \times 3$
$= 2{,}187$

B 6^4

$6^4 = 6 \times 6 \times 6 \times 6$
$= 1{,}296$

1 Introduce

Alternate Opener

EXPLORATION

1-3 Exponents

1. Maria won the grand prize on a game show. She will be given $2 the first month, $4 the second month, $8 the third month, and so on as her payment is doubled each month for one year.
   a. Complete the table.
   b. How much will Maria receive in the fifth month?
   c. How much will Maria receive in the eighth month?
   d. Use a calculator to determine how much Maria will receive in the last month of the year.

Month	Amount ($)
1	2
2	4
3	8
4	
5	
6	
7	
8	

Think and Discuss

2. **Describe** the pattern in the table.
3. **Explain** how the values in the table compare with the values 2, 2^2, 2^3, 2^4, and so on.

Motivate

Review with students the term *repeated addition,* and have them give examples, e.g., 5 + 5 + 5 + 5. Discuss with students an easier way to write a repeated addition expression, such as writing 5 + 5 + 5 + 5 as 4 × 5. Have students give an example of a repeated multiplication expression, e.g., 7 × 7 × 7. Explain that an easier way to write this repeated multiplication is 7^3.

Exploration worksheet and answers on Chapter 1 Resource Book pp. 26 and 86

2 Teach

Lesson Presentation

Guided Instruction

In this lesson, students learn to represent numbers by using exponents. First, teach students to write repeated multiplication expressions in exponential form using the Teaching Transparency T10. Then, teach students to find the values of numbers in exponential form. Show students how to apply exponents to a problem-solving situation. Remind students that numbers that are multiplied are called factors.

Teaching Tip Have students practice correctly pronouncing the exponential terms, such as "four to the third power."

EXAMPLE 3 PROBLEM SOLVING APPLICATION

PROBLEM SOLVING

In case Dana's school closes, a phone tree is used to contact each student's family. The secretary calls 3 families. Then each family calls 3 other families, and so on. How many families will be notified during the 6th round of calls?

1 Understand the Problem

The **answer** will be the number of families called in the 6th round.

List the **important information:**

- The secretary calls 3 families.
- Each family calls 3 families.

2 Make a Plan

You can draw a diagram to see how many calls are in each round.

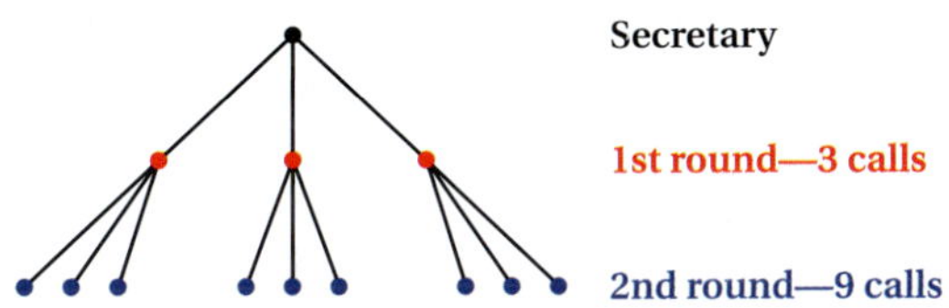

3 Solve

Notice that in each round, the number of calls is a power of 3.

1st round: 3 calls $= 3 = 3^1$

2nd round: 9 calls $= 3 \times 3 = 3^2$

So during the 6th round there will be 3^6 calls.

$3^6 = 3 \times 3 \times 3 \times 3 \times 3 \times 3 = 729$

During the 6th round of calls, 729 families will be notified.

4 Look Back

Drawing a diagram helps you visualize the pattern, but the numbers become too large for a diagram after the third round of calls. Solving this problem by using exponents can be easier and faster.

Think and Discuss

1. **Read** each number: 4^8, 12^3, 3^2.
2. **Give** the value of each number: 7^1, 13^2, 3^3.

COMMON ERROR ALERT

Students may multiply a base by its exponent instead of using the base as a factor the number of times indicated by the exponent. Remind them that in 4 to the 3rd power ($4 \times 4 \times 4$), 4 is used as a factor 3 times.

Additional Examples

Example 1

Write each expression in exponential form.

A. $5 \times 5 \times 5 \times 5$ 5^4

B. $3 \times 3 \times 3 \times 3 \times 3$ 3^5

Example 2

Find each value.

A. 2^6 64

B. 4^5 1,024

Example 3

A phone tree is used to contact families at Paul's school. The secretary calls 4 families. Then each family calls 4 other families, and so on. How many families will be notified during the fourth round of calls?

256 families will be notified.

Example 1 note: When writing expressions in exponential form, make sure that students count the number of times the factor appears, not the number of multiplication signs.

3 Close

Reaching All Learners

Through Number Sense

Have students write the repeated multiplication expression for each exponential term before computing its value until they become comfortable with exponential notations.

Summarize

Briefly review definitions of the new vocabulary in the lesson: *exponent*, *base*, and *exponential form*. Correctly pronounce for the class exponential terms and have them repeat your pronunciation.

Answers to Think and Discuss

1. four to the eighth power; twelve cubed, or twelve to the third power; three squared, or three to the second power
2. 7; 169; 27

1-3 PRACTICE & ASSESS

Students may want to refer back to the lesson examples.

Assignment Guide

If you finished Example 1 assign:
Core 1–4, 10–15, 29–36, 44–47, 54–64
Enriched 29–47, 54–64

If you finished Example 2 assign:
Core 1–8, 10–23, 41–47, 54–64
Enriched 20–23, 25–47, 54–64

If you finished Example 3 assign:
Core 1–24, 41–51, 54–64
Enriched 20–64

Answers

25. $16 \times 16 \times 16$
26. 22×22
27. $31 \times 31 \times 31 \times 31 \times 31 \times 31$
28. $46 \times 46 \times 46 \times 46 \times 46$
29. 4
30. $1 \times 1 \times 1 \times 1 \times 1 \times 1 \times 1 \times 1 \times 1$
31. $17 \times 17 \times 17 \times 17 \times 17 \times 17$
32. $8 \times 8 \times 8 \times 8 \times 8$

1-3 Exercises

FOR EXTRA PRACTICE see page 636

internet connect
Homework Help Online
go.hrw.com Keyword: MR4 1-3

GUIDED PRACTICE

See Example 1
Write each expression in exponential form.

1. $8 \times 8 \times 8$ 8^3 **2.** 7×7 7^2
3. $4 \times 4 \times 4 \times 4$ 4^4 **4.** $5 \times 5 \times 5 \times 5 \times 5$ 5^5

See Example 2
Find each value.

5. 4^2 16 **6.** 3^3 27 **7.** 5^4 625 **8.** 8^2 64

See Example 3
9. At Russell's school, one person will contact 4 people and each of those people will contact 4 other people, and so on. How many people will be contacted in the fifth round? 1,024 people

INDEPENDENT PRACTICE

See Example 1
Write each expression in exponential form.

10. $2 \times 2 \times 2 \times 2 \times 2 \times 2$ 2^6 **11.** $9 \times 9 \times 9 \times 9$ 9^4
12. $1 \times 1 \times 1$ 1^3 **13.** $6 \times 6 \times 6 \times 6 \times 6$ 6^5
14. $7 \times 7 \times 7 \times 7 \times 7 \times 7 \times 7$ 7^7 **15.** 3×3 3^2

See Example 2
Find each value.

16. 2^4 16 **17.** 3^5 243 **18.** 6^2 36 **19.** 9^2 81
20. 8^3 512 **21.** 1^4 1 **22.** 16^2 256 **23.** 10^8 100,000,000

See Example 3
24. To save money for a video game, you put one dollar in an envelope. Each day for 5 days you double the number of dollars in the envelope from the day before. How much have you saved after 5 days? 32 dollars

PRACTICE AND PROBLEM SOLVING

Write each expression as repeated multiplication.

25. 16^3 **26.** 22^2 **27.** 31^6 **28.** 46^5
29. 4^1 **30.** 1^9 **31.** 17^6 **32.** 8^5

Find each value.

33. 10^6 1,000,000 **34.** 73^1 73 **35.** 9^4 6,561 **36.** 80^2 6,400
37. 19^2 361 **38.** 2^9 512 **39.** 57^1 57 **40.** 5^3 125

Compare. Write $<$, $>$, or $=$.

41. 6^1 ■ 5^1 $>$ **42.** 9^2 ■ 20^1 $>$ **43.** 10^1 ■ $1{,}000{,}000^1$ $<$
44. 2^2 ■ 3^2 $<$ **45.** 5^3 ■ 11^2 $>$ **46.** 10^7 ■ 10^8 $<$

Math Background

The use of the terms *squared* and *cubed* is directly related to the measurements of area and volume. The area of a square with sides 5 units long is found by multiplying 5×5, or 5^2, or five squared. The volume of a cube with sides 5 units long is found by multiplying $5 \times 5 \times 5$, or 5^3, or five cubed.

RETEACH 1-3

CHAPTER 1-3 **Reteach**
Exponents

You can write a number in exponential form to show repeated multiplication. A number written in exponential form has a base and an exponent. An exponent tells you how many times a number, called the base, is used as a factor.

8^4 ← exponent
↑
base

Write the expression in exponential form.
$6 \cdot 6 \cdot 6$
6 is used as a factor 3 times.
$6 \cdot 6 \cdot 6 = 6^3$

Write each expression in exponential form.

1. $8 \cdot 8 \cdot 8 \cdot 8 \cdot 8$ 8^5 **2.** $3 \cdot 3$ 3^2 **3.** $5 \cdot 5 \cdot 5 \cdot 5$ 5^4 **4.** $7 \cdot 7 \cdot 7$ 7^3

You can find the value of expressions in exponential form.
Find the value.
2^5

Step 1: Write the expression as repeated multiplication.
$2^5 = 2 \cdot 2 \cdot 2 \cdot 2 \cdot 2$

Step 2: Multiply.
$2 \cdot 2 \cdot 2 \cdot 2 \cdot 2 = 32$

$2^5 = 32$

Find each value.

5. 12^3 1,728 **6.** 6^5 7,776 **7.** 10^4 10,000 **8.** 4^6 4,096

PRACTICE 1-3

LESSON 1-3 **Practice B**
Exponents

Write each expression in exponential form.

1. $9 \cdot 9$ 9^2 **2.** $7 \cdot 7 \cdot 7$ 7^3 **3.** $1 \cdot 1 \cdot 1 \cdot 1 \cdot 1$ 1^5
4. $5 \cdot 5 \cdot 5 \cdot 5$ 5^4 **5.** $2 \cdot 2 \cdot 2 \cdot 2 \cdot 2 \cdot 2$ 2^6 **6.** $10 \cdot 10 \cdot 10 \cdot 10$ 10^4

Find each value.

7. 6^2 36 **8.** 5^3 125 **9.** 10^3 1,000 **10.** 7^2 49
11. 2^5 32 **12.** 3^4 81 **13.** 25^1 25 **14.** 16^0 1

Compare using $<$, $>$, or $=$.

15. $8^0 < 7^1$ **16.** $10^2 < 11^2$ **17.** $8^2 = 4^3$
18. $3^4 > 5^2$ **19.** $2^5 < 9^2$ **20.** $6^2 > 3^3$

21. What whole number equals 25 when it is squared and 125 when it is cubed?
5

22. Use exponents to write the number 81 three different ways.
81^1; 9^2; 3^4

Life Science LINK

You are able to grow because your body produces new cells. New cells are made when old cells divide. Single-celled bodies, like bacteria, divide by *binary fission*, which means "splitting into two parts."

47. In science lab, Carol has a dish containing 4^5 cells. How many cells are represented by this number? **1,024 cells**

48. A certain colony of bacteria triples in length every 15 minutes. Its length is now 1 mm. How long will it be in 1 hour? (*Hint:* There are four cycles of 15 minutes in 1 hour.) **81 mm**

Use the bar graph for Exercises 49–53.

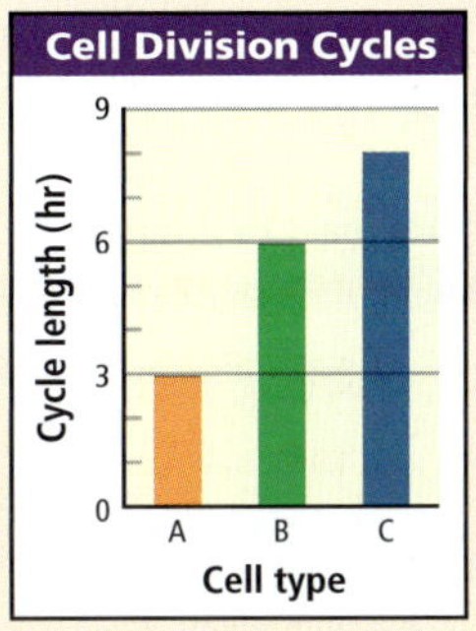

49. Determine how many times cell type A will divide in a 24-hour period. If you begin with one type A cell, how many cells will be produced in 24 hours? **8; 2^8, or 256**

50. Determine how many times cell type B will divide in a 24-hour period. If you begin with one type B cell, how many cells will be produced in 24 hours? **4; 2^4, or 16**

51. Determine how many times cell type C will divide in a 24-hour period. If you begin with one type C cell, how many cells will be produced in 24 hours? **3; 2^3, or 8**

52. *WRITE ABOUT IT* Explain how to find the number of type A cells produced in 48 hours.

53. *CHALLENGE* How many hours will it take one C cell to divide into at least 100 C cells? **56 hours**

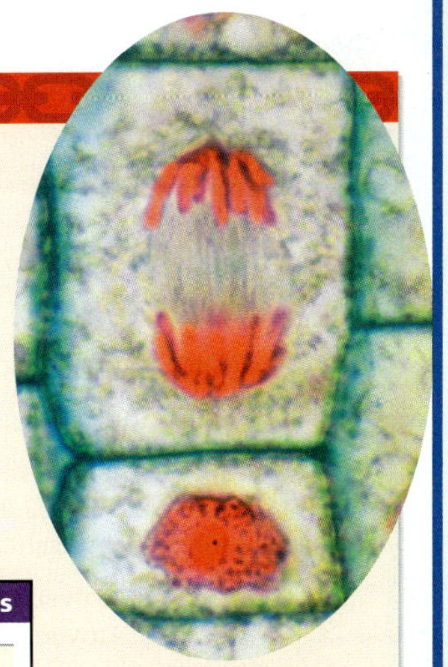

This plant cell shows the anaphase stage of mitosis. Mitosis is the process of nuclear division in complex cells called eukaryotes.

go.hrw.com
KEYWORD: MR4 Cell
CNN Student News

Spiral Review

Write each number in expanded form. (Lesson 1-1)

54. 269 **200 + 60 + 9**

55. 1,354 **1,000 + 300 + 50 + 4**

56. 32,498 **30,000 + 2,000 + 400 + 90 + 8**

57. 416,798 **400,000 + 10,000 + 6,000 + 700 + 90 + 8**

Round each number to the given place value. (Lesson 1-2)

58. 131; hundreds **100**

59. 796; tens **800**

60. 2,369; thousands **2,000**

61. 16,497; ten thousands **20,000**

62. 319,020; ten thousands **320,000**

63. 1,649,045; millions **2,000,000**

64. **TEST PREP** Which number is the closest estimate for 817 + 259? (Lesson 1-2) **D**

A 10,000 **B** 2,000 **C** 800 **D** 1,100

Interdisciplinary LINK

Life Science

Exercises 47–53 pertain to a type of cell division called binary fission. Binary fission is studied in middle-school life science programs such as *Holt Science & Technology*.

Answers

52. Divide 48 hours by the length of the cell-division cycle, 3 hours; 48 ÷ 3 = 16. Raise 2 to this power; 2^{16} = 65,536.

Journal

Have students write about how they would explain the use of exponents to a student who was absent from class.

Test Prep Doctor

For Exercise 64, students should round each addend to estimate the sum. Students who answered **A** incorrectly rounded 259 down to 200 instead of up to 300. Students who answered **B** incorrectly rounded both addends, possibly rounding both up to 1,000. Students who answered **C** disregarded the second addend.

CHALLENGE 1-3

LESSON 1-3 **Challenge**
Exponent Riddle

What is the greatest number that can be written with two digits?

Find the value of each expression below. Then in the box at the bottom of the page, write each expression's letter in the blank above its value. When you have found all the values, you will have solved the riddle.

E	3^3	27
H	5^2	25
I	2^4	16
N	34^0	1
O	9^2	81
P	4^3	64
R	6^2	36
T	7^2	49
W	10^2	100

__ __ __ __ __ __ __ __ __
1 16 1 27   49 81   49 25 27

__ __ __ __ __ __ __ __ __ __
1 16 1 49 25   64 81 100 27 36

PROBLEM SOLVING 1-3

LESSON 1-3 **Problem Solving**
Exponents

1. The Sun is the center of our solar system. The Sun is the star closest to our planet. The surface temperature of the Sun is close to 10,000°F. Write 10,000 using exponents.
   10^4
2. Patty Berg has won 4^2 major women's titles in golf. Write 4^2 in standard form.
   16
3. William has 3^3 baseball cards and 4^3 football cards. Write the number of baseball cards and footballs cards that William has.
   27 baseball cards and 64 football cards
4. Michelle recorded the number of miles she ran each day last year. She used the following expression to represent the total number of miles: 3 • 3 • 3 • 3 • 3 • 3 • 3. Write this expression using exponents. How many miles did Michelle run last year?
   3^7; 2,187 miles
5. In Tyrone's science class he is studying cells. Cell A divides every 30 minutes. If Tyrone starts with two cells, how many cells will he have in 3 hours?
   128 cells
6. Tanisha's soccer team has a phone tree in case a soccer game is postponed or cancelled. The coach calls 2 families. Then each family calls 2 other families. How many families will be notified during the 4th round of calls?
   16 families
7. The Akashi-Kaiko Bridge is the longest suspension bridge in the world. It is located in Kobe-Naruto, Japan and was completed in 1998. It is about 3^8 feet long. Write the approximate length of the Akashi-Kaiko Bridge in standard form.
   6,561 feet
8. The Strahov Stadium is the largest sports stadium in the world. It is located in Prague, Czech Republic. Its capacity is about 12^5. Write the capacity of the Strahov Stadium in standard form.
   248,832

Lesson Quiz

Write each expression in exponential form.

1. 12 × 12 × 12 **12^3**

2. 9 × 9 × 9 × 9 × 9 × 9 × 9 **9^7**

Find each value.

3. 20^2 **400**

4. 6^4 **1,296**

5. In a phone tree, each of 3 people will call 3 people, and then each of those will call 3 more. If there are 5 levels of the tree, how many people will be called? **243**

Available on Daily Transparency in CRB

Chapter 1 Mid-Chapter Quiz

Purpose: *To assess students' mastery of concepts and skills in Lessons 1-1 through 1-3*

Assessment Resources

Section 1A Quiz
Assessment Resources .... p. 5

***Test and Practice Generator* CD-ROM**

Additional mid-chapter assessment items in both multiple-choice and free-response format may be generated for any objective in Lessons 1-1 through 1-3.

Chapter 1 Mid-Chapter Quiz

Mid-Chapter Quiz

LESSON 1-1 (pp. 4–7)

Solve.

1. Which number is greater, 12,563,284 or 12,587,802? **12,587,802**
2. Which number is greater, 783,100,570 or 780,223,104? **783,100,570**
3. In 1998, there were 67,011,180 U.S. households with cable TV. In 1999, there were 67,592,000 U.S. households with cable TV. In which year did more U.S. households have cable TV? **1999**
4. In 2001, a university sold 1,981,299 tickets to its football games. In 2000, the same university sold 1,881,702 tickets. During which year were more tickets sold? **2001**

Order the numbers from least to greatest.

5. 1,052; 1,803; 1,231 **1,052; 1,231; 1,803**
6. 683; 542; 631 **542; 631; 683**
7. 2,305; 2,524; 3,012 **2,305; 2,524; 3,012**
8. 4,302; 5,019; 3,825 **3,825; 4,302; 5,019**
9. 4,344; 3,344; 3,444 **3,344; 3,444; 4,344**
10. 10,463; 14,063; 10,643 **10,463; 10,643; 14,063**

LESSON 1-2 (pp. 8–11)

Estimate each sum or difference by rounding to the place value indicated.

11. 61,582 + 13,281; ten thousands **70,000**
12. 86,125 − 55,713; ten thousands **30,000**
13. 7,903 + 2,654; thousands **11,000**
14. 34,633 − 32,087; thousands **3,000**
15. 1,896,345 + 3,567,194; hundred thousands **5,500,000**
16. 56,129,482 − 37,103,758; ten millions **20,000,000**
17. Marcus wants to make a stone walkway in his garden. The rectangular walkway will be 3 feet wide and 18 feet long. Each 2-foot by 3-foot stone covers an area of 6 square feet. How many stones will Marcus need? **9 stones**
18. Jenna's sixth-grade class is taking a bus to the zoo. The zoo is 156 miles from the school. If the bus travels an average of 55 mi/h, about how long will it take the class to get to the zoo? **about 3 hours**

LESSON 1-3 (pp. 12–15)

Write each expression in exponential form.

19. $7 \times 7 \times 7$ **7^3**
20. $5 \times 5 \times 5 \times 5$ **5^4**
21. $3 \times 3 \times 3 \times 3 \times 3 \times 3$ **3^6**
22. $10 \times 10 \times 10 \times 10$ **10^4**
23. $1 \times 1 \times 1 \times 1 \times 1$ **1^5**
24. $4 \times 4 \times 4 \times 4$ **4^4**

Find each value.

25. 3^3 **27**
26. 2^4 **16**
27. 6^2 **36**
28. 8^3 **512**
29. 1^8 **1**
30. 4^2 **16**
31. 5^4 **625**
32. 9^1 **9**

Notes

Focus on Problem Solving

Solve

• Choose the operation: addition or subtraction

Read the whole problem before you try to solve it. Determine what action is taking place in the problem. Then decide whether you need to add or subtract in order to solve the problem.

If you need to combine or put numbers together, you need to add. If you need to take away or compare numbers, you need to subtract.

Action	Operation	Picture
Combining Putting together	Add	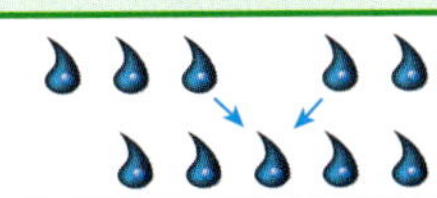
Removing Taking away	Subtract	
Comparing Finding the difference	Subtract	

Read each problem. Determine the action in each problem. Choose an operation in order to solve the problem. Then solve.

Most hurricanes that occur over the Atlantic Ocean, the Caribbean Sea, or the Gulf of Mexico occur between June and November. Since 1886, a hurricane has occurred in every month except April.

Number of Out-of-Season Hurricanes Since 1886

Month	Number
Jan	1
Feb	1
Mar	1
May	14
Dec	10

Use the table for problems 1 and 2.

1. How many out-of-season hurricanes have occurred in all?
2. How many more hurricanes have occurred in May than in December?
3. There were 14 named storms during the 2000 hurricane season. Eight of these became hurricanes, and three others became major hurricanes. How many of the named storms were not hurricanes or major hurricanes?

Focus on Problem Solving

Purpose: *To focus on choosing an operation: addition or subtraction*

Problem Solving Resources

Interactive Problem Solving . . . pp. 1–7

Math: Reading and Writing in the Content Area . . . . pp. 1–7

Problem Solving Process

This page focuses on the third step of the problem-solving process: **Solve**

Discuss

Have students discuss which actions take place in each problem and which operations the actions indicate.

1. combining; addition

2. finding the difference; subtraction

3. taking away; subtraction

Answers

1. $1 + 1 + 1 + 14 + 10 = 27$

2. $14 - 10 = 4$

3. $14 - (8 + 3) = 3$

Section 1B

Using Whole Numbers

One-Minute Section Planner

Lesson	Materials	Resources
Technology Lab 1A Explore the Order of Operations **NCTM:** Number and Operations **NAEP:** Number Properties 3f ☑ SAT-9 ☑ SAT-10 ☑ ITBS ☑ CTBS ☑ MAT ☑ CAT	**Required** Scientific calculators	• *Technology Lab Activities,* pp. 2–3
Lesson 1-4 Order of Operations **NCTM:** Number and Operations, Communication **NAEP:** Algebra 3b ☑ SAT-9 ☑ SAT-10 ☑ ITBS ☑ CTBS ☑ MAT ☑ CAT	**Optional** Teaching Transparency T14 *(CRB)*	• *Chapter 1 Resource Book,* pp. 35–43 • Daily Transparency T13, CRB • Additional Examples Transparencies T15–T16, CRB • *Alternate Openers: Explorations,* p. 4
Lesson 1-5 Mental Math **NCTM:** Number and Operations, Communication **NAEP:** Number Properties 5e ☐ SAT-9 ☑ SAT-10 ☑ ITBS ☑ CTBS ☐ MAT ☑ CAT	**Optional** Teaching Transparency T18 *(CRB)*	• *Chapter 1 Resource Book,* pp. 44–53 • Daily Transparency T17, CRB • Additional Examples Transparencies T19–T20, CRB • *Alternate Openers: Explorations,* p. 5
Lesson 1-6 Choose the Method of Computation **NCTM:** Number and Operations, Problem Solving, Reasoning and Proof, Communication **NAEP:** Number Properties 2b ☐ SAT-9 ☐ SAT-10 ☐ ITBS ☐ CTBS ☐ MAT ☐ CAT	**Optional** Recording Sheet for Reaching All Learners *(CRB, p. 78)*	• *Chapter 1 Resource Book,* pp. 54–62 • Daily Transparency T21, CRB • Additional Examples Transparency T22, CRB • *Alternate Openers: Explorations,* p. 6
Lesson 1-7 Find a Pattern **NCTM:** Algebra, Problem Solving, Reasoning and Proof, Communication **NAEP:** Algebra 1b ☑ SAT-9 ☑ SAT-10 ☑ ITBS ☑ CTBS ☑ MAT ☑ CAT	**Optional** Recording Sheet for Reaching All Learners *(CRB, p. 79)*	• *Chapter 1 Resource Book,* pp. 63–71 • Daily Transparency T23, CRB • Additional Examples Transparency T24, CRB • *Alternate Openers: Explorations,* p. 7
Extension Binary Numbers **NCTM:** Number and Operations, Representation **NAEP:** Number Properties 5f ☐ SAT-9 ☐ SAT-10 ☐ ITBS ☑ CTBS ☐ MAT ☐ CAT		• Additional Examples Transparency T25, CRB
Section 1B Assessment		• Section 1B Quiz, AR p. 6 • *Test and Practice Generator* CD-ROM

SAT = *Stanford Achievement Tests* **ITBS** = *Iowa Test of Basic Skills* **CTBS** = *Comprehensive Test of Basic Skills/Terra Nova*
MAT = *Metropolitan Achievement Tests* **CAT** = *California Achievement Test*

NCTM = Complete standards can be found on pages T27–T33. **NAEP** = Complete standards can be found on pages A31–A35.

SE = *Student Edition* **TE** = *Teacher's Edition* **AR** = *Assessment Resources* **CRB** = *Chapter Resource Book* **MK** = *Manipulatives Kit*

Section Overview

Using the Order of Operations — *Lesson 1-4*

Why? The order of operations ensures that everyone gets the same answer.

Order of Operations

1. Parentheses
2. Exponents
3. Multiply/Divide
4. Add/Subtract

Evaluate $9 \div (1 + 2) \times 4^2 - 5$.

$9 \div (1 + 2) \times 4^2 - 5$	*Perform operations within parentheses.*
$9 \div 3 \times 4^2 - 5$	*Find the values of numbers with exponents.*
$9 \div 3 \times 16 - 5$	*Divide.*
$3 \times 16 - 5$	*Multiply.*
$48 - 5$	*Subtract.*
43	

Using Number Properties — *Lesson 1-5*

Why? Number properties can help you perform calculations mentally.

Commutative Property

Addition: $2 + 3 = 3 + 2$
Multiplication: $2 \times 3 = 3 \times 2$

Associative Property

Addition: $(3 + 5) + 4 = 3 + (5 + 4)$
Multiplication: $(3 \times 5) \times 4 = 3 \times (5 \times 4)$

Distributive Property

$4 \times (8 + 2) = (4 \times 8) + (4 \times 2)$
and
$4 \times (8 - 2) = (4 \times 8) - (4 \times 2)$

Use number properties to evaluate mentally.

$14 + 22 + 16 + 28 = 14 + 16 + 22 + 28$ — ***Commutative Property***
$= (14 + 16) + (22 + 28)$ — ***Associative Property***
$= 30 + 50$
$= 80$

$7 \times 64 = 7 \times (60 + 4)$
$= (7 \times 60) + (7 \times 4)$ — ***Distributive Property***
$= 420 + 28$
$= 448$

Extending Patterns in Sequences — *Lessons 1-6, 1-7*

Why? By recognizing and extending number patterns, you can make predictions and solve problems involving function relationships.

A **sequence** is an ordered set of numbers.

×2 ×2 ×2 ×2

1, 2, 4, 8, 16, …

Each number in a sequence is called a **term.**

Multiply by 2 to get the next term.
The next three terms are as follows:

$16 \times 2 = \mathbf{32}$
$36 \times 2 = \mathbf{64}$
$64 \times 2 = \mathbf{128}$

Note: Some students will find different rules or patterns for the same sequence. There will always be many rules that will give rise to a given sequence.

1A Explore the Order of Operations

Pacing: Traditional 1 day
Block $\frac{1}{2}$ day

Objective: To use a scientific or graphing calculator to explore order of operations

Materials: Scientific or graphing calculator

Lab Resources

Technology Lab Activities ...... pp. 2–3

Using the Pages

This technology activity shows students how to evaluate expressions using the order of operations, which can be done on any graphing calculator. Specific keystrokes may vary, depending on the make and model of the graphing calculator used. The keystrokes given are for a TI-83 model. For keystrokes to other models, visit go.hrw.com.

The Think and Discuss problems can be used to assess students' understanding of the technology activity. The purpose of the Try This problems is to reinforce the correct order of operations by having students evaluate expressions both with pencil and paper and on a calculator.

Assessment

Suppose you used the following keystrokes:

2 [+] 3 [×] 4 [ENTER]

1. What would a nonscientific calculator display? **20**
2. What would a scientific or graphing calculator display? **14**

Technology LAB 1A: Explore the Order of Operations

Use with Lesson 1-4

internet connect
Lab Resources Online
go.hrw.com
KEYWORD: MR4 Lab1A

Look at the expression $3 + 2 \cdot 8$. To evaluate this expression, decide whether to add first or multiply first. Knowing the correct *order of operations* is important. Without this knowledge, you could get an incorrect result.

Activity 1

Evaluate $3 + 2 \cdot 8$ two different ways.

Add first, and then multiply by 8.	$3 + 2 = 5$ $5 \cdot 8 = 40$
Multiply first, and then add 3.	$2 \cdot 8 = 16$ $16 + 3 = 19$

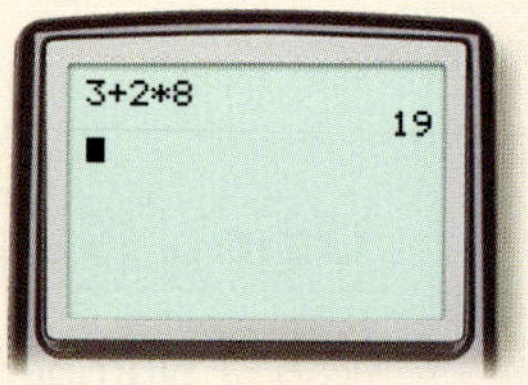

Now evaluate $3 + 2 \cdot 8$ using a graphing or scientific calculator.

The result, 19, shows that this calculator multiplied first, even though addition came first in the expression.

If there are no parentheses, then multiplication and division are done before addition or subtraction. If the addition is to be done first, parentheses *must* be used.

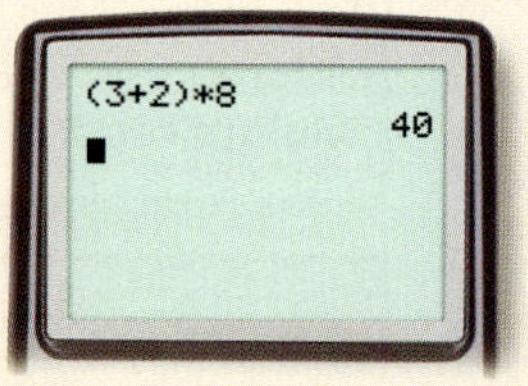

When you evaluate $(3 + 2) \cdot 8$ on a calculator, the result is 40. Because of the parentheses, the calculator adds before multiplying.

Graphing and scientific calculators follow a logical system called the algebraic order of operations. The order of operations tells you to multiply and divide before you add or subtract.

Think and Discuss

Division; multiplication and division must be done before addition or subtraction.

1. In $4 + 15 \div 5$, which operation do you perform first? How do you know?
2. Tell the order in which you would perform the operations in the expression $8 \div 2 + 6 \cdot 3 - 4$. **division, multiplication, addition, subtraction**

Try This

Evaluate each expression with pencil and paper. Check your answer with a calculator.

1. $4 \cdot 12 - 7$ **41**
2. $15 \div 3 + 10$ **15**
3. $4 + 2 \cdot 6$ **16**
4. $10 - 4 \div 2$ **8**

Activity 2

What should you do if the same operation appears twice in an expression? Use a calculator to decide which subtraction is done first in the expression 7 − 3 − 2.

If 7 − 3 is done first, the value of the expression is 4 − 2 = 2.

If 3 − 2 is done first, the value of the expression is 7 − 1 = 6.

On the calculator, the value of 7 − 3 − 2 is 2. The subtraction on the left, 7 − 3, is done first.

Addition and subtraction (or multiplication and division) are done from left to right.

Think and Discuss

1. In 15 + 5 + 4, does it matter which operation you perform first? Explain. **No; the result is 24 whether you add 15 + 5 first or 5 + 4 first.**
2. Does it matter which operation you perform first in 15 − 5 + 4? Explain. **Yes; if you subtract first, the result is 14, but if you add first, the result is 6. 15−5 should be performed first since addition and subtraction are done from left to right.**

Try This

Evaluate each expression. Check your answer with a calculator.

1. 8 − 6 − 1 **1**
2. 20 ÷ 5 ÷ 2 **2**
3. 3 · 6 · 2 **36**
4. 19 + 6 + 5 **30**

Activity 3

Without parentheses, the expression 8 + 2 · 10 − 3 equals 25. Insert parentheses to make the value of the expression 22.

What happens if you add first?	What happens if you subtract first?
(8 + 2) · 10 − 3	8 + 2 · (10 − 3)
10 · 10 − 3	8 + 2 · 7
100 − 3	8 + 14
97	22

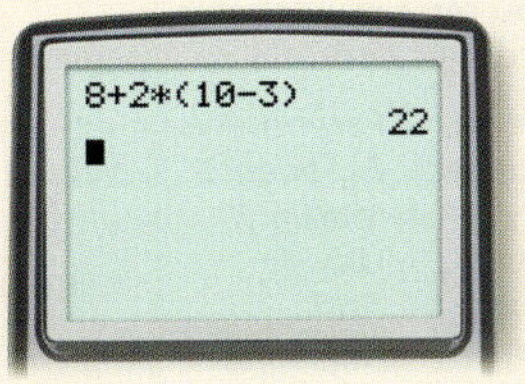

For the expression to equal 22, the subtraction must be done first.

Think and Discuss

1. To evaluate 13 + 5 · 255 on a calculator, you type 13 + 5 and then press the · key. But before you can type in the 255, the display changes to 18!
   a. Does this calculator follow the correct order of operations? Why? **No; the calculator evaluated 13 + 5 first.**
   b. How could you use this calculator to evaluate 13 + 5 · 255? **Possible answer: Multiply 5 · 255 first, and then add 13 to the result.**

Try This

Insert parentheses to make the value of each expression 12.

1. 56 − 40 + 4 **56 − (40 + 4)**
2. 3 − 1 · 10 − 4 **(3 − 1) · (10 − 4)**
3. 18 ÷ 2 + 1 + 6 **18 ÷ (2 + 1) + 6**

1-4 Organizer

Pacing: Traditional 1 day
Block $\frac{1}{2}$ day

Objective: Students use the order of operations.

Indiana Academic Standards
6.3.3, 6.3.6, 6.7.11

Warm Up

Perform the operations in order from left to right.

1. $8 + 4 - 2$ **10**
2. $9 \times 3 + 1$ **28**
3. $7 - 3 + 5$ **9**
4. $20 \div 4 + 6$ **11**

Problem of the Day

0 1 2 3 4 5 6 7 8 9 = 1

Put the appropriate plus or minus signs between the numbers so that the total equals 1.

$0 + 1 - 23 + 45 + 67 - 89 = 1$

Available on Daily Transparency in CRB

Math Humor

The forgetful student never made it through medical school because he couldn't remember the order of operations.

1-4 Order of Operations

Learn to use the order of operations.

Vocabulary
numerical expression
evaluate
order of operations

A **numerical expression** is a mathematical phrase that includes only numbers and operation symbols.

Numerical Expressions	$4 + 8 \div 2 \times 6$	$371 - 203 + 2$	$5{,}006 \times 19$

When you **evaluate** a numerical expression, you find its value.

Erika and Jamie each evaluated $3 + 4 \times 6$. Their work is shown below. Whose answer is correct?

Helpful Hint

The first letters of these words can help you remember the order of operations.

Please	*Parentheses*
Excuse	*Exponents*
My	*Multiply/*
Dear	*Divide*
Aunt	*Add/*
Sally	*Subtract*

When an expression has more than one operation, you must know which operation to do first. To make sure that everyone gets the same answer, we use the **order of operations**.

ORDER OF OPERATIONS

1. Perform operations in **parentheses**.
2. Find the values of numbers with **exponents**.
3. **Multiply** or **divide** from left to right as ordered in the problem.
4. **Add** or **subtract** from left to right as ordered in the problem.

$3 + 4 \times 6$	*There are no parentheses or exponents. Perform the multiplication first.*
$3 + 24$	*Add.*
27	*Erika has the correct answer.*

1 Introduce

Alternate Opener

EXPLORATION

1-4 Order of Operations

Calculators are programmed to perform operations in a certain order. Each keystroke sequence below results in 17.

3 [×] 5 [ENTER] [+] 2 [ENTER]   2 [+] 3 [×] 5 [ENTER]

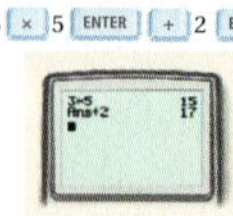

For each keystroke sequence, determine the order of operations the calculator follows.

1. [(] 2 [+] 3 [)] [×] 5 [ENTER]
2. 2 [^] 3 [−] 1 [×] 4 [ENTER]
3. 2 [^] [(] 3 [−] 1 [)] [×] 4 [ENTER]

Write the keystroke sequence for each expression.

4. $5 - 2^2$
5. $(2 - 3)^3 + 2$

Think and Discuss

6. **Explain** why there needs to be a rule for the order of operations.

Motivate

Have students evaluate $7 + 3 \times 4$. Tell them the expression represents 7 red marbles and 3 groups of 4 blue marbles. Then ask for the total number of marbles.

Exploration worksheet and answers on Chapter 1 Resource Book pp. 36 and 88

2 Teach

Lesson Presentation

Guided Instruction

In this lesson, students learn to use the order of operations. Show students how to evaluate numerical expressions containing more than one operation. Then have students apply the order of operations to a real-world (consumer) context. You may use Teaching Transparency T14 to remind students of the correct order of operations.

EXAMPLE 1 Using the Order of Operations

Evaluate each expression.

A $9 + 12 \times 2$

$9 + 12 \times 2$	*There are no parentheses or exponents.*
$9 + 24$	*Multiply.*
33	*Add.*

B $4 \times 3^2 + 8 - 16$

$4 \times 3^2 + 8 - 16$	*There are no parentheses.*
$4 \times 9 + 8 - 16$	*Find the values of numbers with exponents.*
$36 + 8 - 16$	*Multiply.*
$44 - 16$	*Add.*
28	*Subtract.*

C $8 \div (1 + 3) \times 5^2 - 2$

$8 \div (1 + 3) \times 5^2 - 2$	
$8 \div 4 \times 5^2 - 2$	*Perform operations within parentheses.*
$8 \div 4 \times 25 - 2$	*Find the values of numbers with exponents.*
$2 \times 25 - 2$	*Divide.*
$50 - 2$	*Multiply.*
48	*Subtract.*

EXAMPLE 2 *Consumer Application*

Regina bought 5 carved wooden beads for $3 each and 8 glass beads for $2 each. Evaluate the following expression to find the amount Regina spent for beads.

$5 \times 3 + 8 \times 2$

$5 \times 3 + 8 \times 2$

$15 + 16$

31

Regina spent $31 for beads.

Think and Discuss

1. Explain why $6 + 7 \times 10 = 76$ but $(6 + 7) \times 10 = 130$.

2. Tell how you can add parentheses to the numerical expression $2^2 + 5 \times 3$ so that 27 is the correct answer.

COMMON ERROR ALERT

Students using "**P**lease **E**xcuse **M**y **D**ear **A**unt **S**ally" to remember the order of operations may think that all multiplication is done before division and all addition is done before subtraction. Stress that this is not the case, as is explained in the chart on the first page of the lesson.

Additional Examples

Example 1

Evaluate each expression.

A. $15 - 10 \div 2$
10

B. $8 + 14 \div 2 + 6 \times 3^3$
177

C. $3 \times (4 + 7) + 2^2$
37

Example 2

Mr. Kellett bought 6 used CDs for $4 each and 5 used CDs for $3 each. Evaluate the following expression to find the amount Mr. Kellett spent on CDs. $6 \times 4 + 5 \times 3$

Mr. Kellett spent $39 on CDs.

3 Close

Reaching All Learners

Through Grouping Strategies

Have students work in groups to try to stump their classmates. Each group writes several numerical expressions, with no more than three operations, for which the order of operations must be used. For each expression, the group gives its actual value and an incorrect value that can be found by not following the order of operations. Groups then switch papers and choose the correct value for each expression.

Sample expression:

$4 + (6 + 8) \times 4$

60 or 72

(correct answer: 60)

Summarize

Review definitions of new vocabulary in the lesson: *numerical expression, evaluate,* and *order of operations*. Discuss how the terms relate to each other. Emphasize that multiplication and division have the same priority and that the same can be said of addition and subtraction.

Answers to Think and Discuss

1. Possible answer: Using the order of operations, $6 + 7 \times 10 = 6 + 70 = 76$, because you multiply 7×10 before adding. However, $(6 + 7) \times 10 = 13 \times 10 = 130$ because you perform the operation in parentheses before multiplying by 10.
2. $(2^2 + 5) \times 3$

1-4 PRACTICE & ASSESS

Students may want to refer back to the lesson examples.

Assignment Guide

If you finished Example 1 assign:
Core 1–6, 8–17, 20–27, 39–49
Enriched 19–49

If you finished Example 2 assign:
Core 1–19, 34–36, 39–49
Enriched 17–49

Notes

Answers

32. $2 \times (8 + 5) - 3 = 23$

33. $9^2 - 2 \times (15 + 16) - 8 = 11$

Math Background

Without the order of operations, one person could conclude $2 + 3 \times 4 = 14$ and another could conclude $2 + 3 \times 4 = 20$, and neither would be incorrect! A fundamental idea in arithmetic is that a numerical expression has a unique value. A simple explanation to why the correct answer is 14 and not 20 is based on the fact that $3 \times 4 = 4 + 4 + 4$. Thus the expression $2 + 3 \times 4$ can be written as $2 + 4 + 4 + 4$, which equals 14.

The purpose of the order of operations is to guarantee that every numerical expression has a unique value.

1-4 Exercises

FOR EXTRA PRACTICE see page 637

internet connect
Homework Help Online
go.hrw.com Keyword: MR4 1-4

GUIDED PRACTICE

See Example 1

Evaluate each expression.

1. $36 - 18 \div 6$ **33**

2. $7 + 24 \div 6 \times 2$ **15**

3. $11 + 2^3 \times 5$ **51**

4. $62 - 4 \times (15 \div 5)$ **50**

5. $5 \times (28 \div 7) - 4^2$ **4**

6. $5 + 3^2 \times 6 - (10 - 9)$ **58**

See Example 2

7. Coach Milner fed the team after the game by buying 24 Big Burger Deals for \$4 each and 7 Super Big Burger Deals for \$6 each. Evaluate the expression for the cost of the food: $24 \times 4 + 7 \times 6$. **\$138**

INDEPENDENT PRACTICE

See Example 1

Evaluate each expression.

8. $9 + 27 \div 3$ **18**

9. $2 \times 7 - 32 \div 8$ **10**

10. $45 \div (3 + 6) \times 3$ **15**

11. $100 \div 5^2 + 7 \times 3$ **25**

12. $4^2 + 48 \div (10 - 4)$ **24**

13. $6 \times 2^2 + 28 - 5$ **47**

14. $6^2 - 12 \div 3 + (15 - 7)$ **40**

15. $21 \div (3 + 4) \times 9 - 2^3$ **19**

16. $5 + 3 \times 2 + 12 \div 4$ **14**

17. $(3^2 + 6 \div 2) \times (36 \div 6 - 4)$ **24**

See Example 2

18. The nature park has a pride of 5 adult lions and 3 cubs. The adults eat 8 lb of meat each day and the cubs eat 4 lb. Use the expression to find the amount of meat consumed each day by the lions: $5 \times 8 + 3 \times 4$. **52 lb**

19. Angie read 4 books that were each 150 pages long and 2 books that were each 325 pages long. Evaluate the expression $4 \times 150 + 2 \times 325$ to find the total number of pages Angie read. **1,250 pages**

PRACTICE AND PROBLEM SOLVING

Evaluate each expression.

20. $12 + 3 \times 4$ **24**

21. $25 - 21 \div 3$ **18**

22. $60 \div (10 + 2) \times 4^2 - 23$ **57**

23. $10 \times (28 - 23) + 7^2 - 37$ **62**

24. $72 \div 9 - 2 \times 4$ **0**

25. $12 + (1 + 7^2) \div 5$ **22**

26. $(15 - 6)^2 - 34 \div 2$ **64**

27. $(2 \times 4)^2 - 3 \times (5 + 3)$ **40**

Add parentheses so that each equation is correct.

28. $2^3 + 6 - 5 \times 4 = 12$
$2^3 + (6 - 5) \times 4 = 12$

29. $7 + 2 \times 6 - 4 - 3 = 53$
$(7 + 2) \times 6 - (4 - 3) = 53$

30. $3^2 + 6 + 3 \times 3 = 36$
$3^2 + (6 + 3) \times 3 = 36$

31. $5^2 - 10 + 5 + 4^2 = 36$
$5^2 - 10 + (5 + 4^2) = 36$

32. $2 \times 8 + 5 - 3 = 23$

33. $9^2 - 2 \times 15 + 16 - 8 = 11$

RETEACH 1-4

LESSON 1-4 **Reteach**
Order of Operations

A mathematical phrase that includes only numbers and operations is called a numerical expression.

$9 + 8 \cdot 3 \div 6$ is a numerical expression.

To evaluate a numerical expression, you find its value.

You can use the order of operations to evaluate a numerical expression.

Order of Operations
1. Do all operations within parentheses.
2. Find the values of the numbers with exponents.
3. Multiply and divide in order from left to right.
4. Add or subtract in order from left to right.

Evaluate the expression.

$60 \div (7 + 3) + 7$	Do all operations within parentheses.
$60 \div 10 + 7$	Multiply and divide in order from left to right.
$6 + 7$	Add and subtract in order from left to right.
13	

Evaluate each expression.

1. $7 \cdot (12 + 8) - 6$
$7 \cdot$ 20 $- 6$
140 $- 6$
134

2. $10 \cdot (12 + 34) + 3$
$10 \cdot$ 46 $+ 3$
400 $+ 3$
463

3. $10 + (6 \cdot 5) - 7$
$10 +$ 30 $- 7$
40 $- 7$
33

4. $2^3 + (10 - 4)$ 14

5. $7 + 3 \cdot (8 + 5)$ 46

6. $36 \div 4 + 11 \cdot 8$ 97

7. $5^2 - (2 \cdot 8) + 9$ 18

8. $3 \cdot (12 \div 4) - 2^2$ 5

9. $(3^3 + 10) - 2$ 35

PRACTICE 1-4

LESSON 1-4 **Practice B**
Order of Operations

Evaluate each expression.

1. $10 + 6 \cdot 2$ 22

2. $(15 + 39) \div 6$ 9

3. $(20 - 15) \cdot 2 + 1$ 11

4. $(4^2 + 6) \div 11$ 2

5. $9 + (7 - 1) \cdot 2$ 21

6. $(2 \cdot 4) + 8 - (5 \cdot 3)$ 1

7. $5 + 18 \div 3^2 - 1$ 6

8. $8 + 5 \cdot 10 - 12$ 46

9. $14 + (50 - 7^2) \cdot 3$ 17

Insert parentheses so that each equation is correct.

10. $7 + 9 \cdot 3 - 1 = 25$ $(3 - 1)$

11. $2^3 - 7 \cdot 4 = 4$ $(2^3 - 7)$

12. $5 + 6 \cdot 9 \div 3 = 23$ $(9 \div 3)$

13. $12 \div 3 \cdot 2 = 2$ $(3 \cdot 2)$

14. $8 + 3 \cdot 6 - 4 - 1 = 13$ $(6 - 4)$

15. $4 \cdot 3^2 + 1 = 40$ $(3^2 + 1)$

16. $9 \cdot 0 + 5 - 3 = 42$ $(0 + 5)$

17. $15 \cdot 3^2 - 2^3 = 15$ $(3^2 - 2^3)$

18. $14 \div 2 + 5 \cdot 5 = 10$ $(2 + 5)$

19. Tyler walked 2 miles a day for the first week of his exercise plan. Then he walked 3 miles a day for the next 9 days. How many miles did Tyler walk in all?

41 miles

20. Paulo's father bought 8 pizzas and 12 bottles of juice for the class party. Each pizza cost \$9 and each bottle of juice cost \$2. Paulo's father paid with a \$100-bill. How much change did he get back?

\$4

Social Studies LINK

Archaeologists study cultures of the past by uncovering items from ancient cities. An archaeologist has chosen a site in Mexico for her team's next dig. She divides the location into rectangular plots and labels each plot so that uncovered items can be identified by the plot in which they were found.

Archaeologists uncovered pieces of pottery at the La Ventilla site in Mexico.

34. To prepare for the dig, the archaeologist must order a cover for the plot where the team is currently digging. Evaluate the expression $3 \times (2^2 + 6)$ to find the area of each plot in square meters. **30 m^2**

35. In the first week, the archaeology team digs down 2 meters and removes a certain amount of dirt. Evaluate the expression $3 \times (2^2 + 6) \times 2$ to find the volume of the dirt removed from the plot in the first week. **60 m^3**

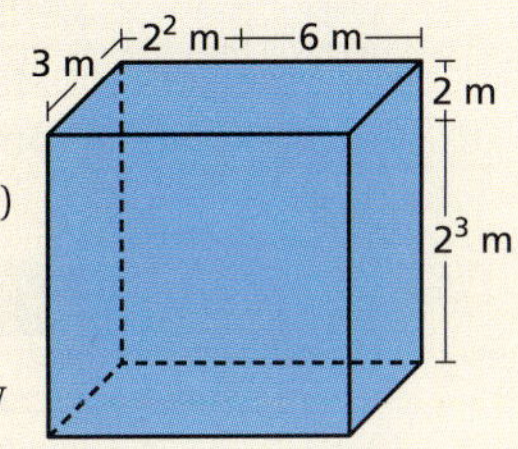

36. Over the next two weeks, the archaeology team digs down an additional 2^3 meters. Evaluate the expression $3 \times (2^2 + 6) \times (2 + 2^3)$ to find the total volume of dirt removed from the plot after 3 weeks. **300 m^3**

37. ***WRITE ABOUT IT*** Explain why the archaeologist must follow the order of operations to determine the area of each plot.

38. ***CHALLENGE*** Write an expression for the volume of dirt that would be removed if the archaeologist's team were to dig down an additional 3^2 meters after the first three weeks. **$3 \times (2^2 + 6) \times (2 + 2^3 + 3^2)$**

Spiral Review

Order the numbers from least to greatest. (Lesson 1-1)

39. 8,452; 8,732; 8,245 **8,245; 8,452; 8,732**

40. 984; 1,010; 991 **984; 991; 1,010**

41. 12,681; 12,751; 11,901 **11,901; 12,681; 12,751**

Estimate each sum or difference by rounding to the place value indicated. (Lesson 1-2)

42. 2,488 + 1,934; thousands **4,000**

43. 83,057 − 29,475; ten thousands **50,000**

44. 9,346 + 12,745; thousands **22,000**

Find each value. (Lesson 1-3)

45. 11^2 **121** **46.** 5^3 **125** **47.** 9^1 **9** **48.** 2^5 **32**

49. **TEST PREP** Which of the following is the value of the expression $3^4 + 9^1$? (Lesson 1-3) **C**

A 21 **B** 81 **C** 90 **D** 82

Interdisciplinary LINK

Social Studies

Exercises 34–38 focus on archaeological digs. The early histories of many regions are studied in middle-school social studies programs, such as Holt, Rinehart & Winston's *People, Places, and Change.* These histories are often based on remains and ruins uncovered during archaeological digs.

Answers

37. If she does not follow the order of operations, she will buy the wrong size cover.

Journal

Have students write about a situation outside of solving math problems in which the order in which things are done is important.

Test Prep Doctor

For Exercise 49, students need to know how to find the value of a number in exponential form and the value of a number with 1 as its exponent. Students who answered **A** multiplied 3 × 4 and then added 9. Students who answered **B** or **D** correctly found the value of 3^4 but used incorrect values for 9^1.

CHALLENGE 1-4

LESSON 1-4 **Challenge**
Crack the Expression Code

Each of these symbols stands for a different operation symbol:

♥ ♠ ♦ ♣

Each of these animals stands for a different whole number 1–4:

Use the equations below to find what each symbol and animal represents in the expression code.

OPERATIONS

1. ♥ = +
2. ♠ = −
3. ♦ = ×
4. ♣ = ÷

NUMBERS

5. = 4
6. = 2
7. = 1
8. = 3

PROBLEM SOLVING 1-4

LESSON 1-4 **Problem Solving**
Order of Operations

Evaluate each expression to complete the table.

Mammals with the Longest Tails

Mammal	Expression	Tail Length
Asian elephant	$2 + 3^2 \cdot 7 - (10 - 4)$	1. 59
Leopard	$5 \cdot 6 + 5^2$	2. 55
African elephant	$6 \cdot (72 \div 8) - 3$	3. 51
African buffalo	$51 + 6^2 \div 9 - 12$	4. 43
Giraffe	$4^3 - 3 \cdot 7$	5. 43
Red kangaroo	$11 + 48 \div 6 \cdot 4$	6. 43

Choose the letter for the best answer.

7. Adam and his two brothers went to the zoo. Each ticket to enter the zoo costs $7. Adam bought two bags of peanuts for $4 each, and one of his brothers bought a lion poster for $12. Which expression shows how much money they spent at the zoo in all?

A $7 + 4 + 12$
B $7 \cdot 3 + 4 + 12$
(C) $7 \cdot 3 + 4 \cdot 2 + 12$
D $(7 \cdot 3) + (4 \cdot 12)$

8. An elephant eats about 500 pounds of grass and leaves every day. There are 2 Africa elephants and 3 Asian elephants living in the City Zoo. How many pounds of grass and leaves do the zookeepers need to order each week to feed all the elephants?

F 2,500 pounds
(G) 17,500 pounds
H 3,000 pounds
J 21,000 pounds

9. The average giraffe is 18 feet tall. Which of these expressions shows the height of a giraffe?

A $4^2 - 2$
B $3 \cdot 12 \div 4 + 2$
(C) $3^3 \div 9 \cdot 6$
D $20 \div 5 + 5 - 6$

10. Some kangaroos can cover 30 feet in a single jump! If a kangaroo could jump like that 150 times in a row, how much farther would it need to go to cover a mile? (1 mile = 5,280 feet)

(F) 780 feet **H** 176 feet
G 26 feet **J** 5,100 feet

Lesson Quiz

Evaluate each expression.

1. $15 + 4 \times 2$ **23**
2. $(12 - 5)^2 - 10$ **39**
3. $3 + 9 \times 2 - 5$ **16**
4. $4^3 - 30 \div 2$ **49**
5. Chaz bought 4 football cards for $2 each and 8 baseball cards for $3 each. Evaluate the expression to find the amount Chaz spent on cards: $4 \times 2 + 8 \times 3$ **$32**

Available on Daily Transparency in CRB

1-5 Organizer

Pacing: Traditional 1 day
Block $\frac{1}{2}$ day

Objective: Students use number properties to compute mentally.

Indiana Academic Standards
6.3.3, 6.3.4, 6.3.6, 6.7.1, 6.7.11

Warm Up

Find each sum or product.

1. 17 + 15 **32** **2.** 29 + 39 **68**
3. 8(24) **192** **4.** 7(12) **84**
5. 3(91) **273** **6.** 6(15) **90**

Problem of the Day

Determine the secret number from the following clues:

- The number is a multiple of 5.
- It is divisible by 3.
- It is less than 200.
- Its tens digit equals the sum of its other two digits. **165**

Available on Daily Transparency in CRB

Math Humor

I can do mental math with fractions and decimals, but I have trouble computing with wholes in my head.

1-5 Mental Math

Learn to use number properties to compute mentally.

Vocabulary
Commutative Property
Associative Property
Distributive Property

Mental math means "doing math in your head." Shakuntala Devi is extremely good at mental math. When she was asked to multiply 7,686,369,774,870 by 2,465,099,745,779, she took only 28 seconds to multiply the numbers mentally and gave the correct answer of 18,947,668,177,995,426,462,773,730!

Most people cannot do calculations like that mentally. But you can learn to solve some problems very quickly in your head.

Many mental math strategies use number properties that you already know.

COMMUTATIVE PROPERTY (Ordering)	
Words	**Numbers**
You can add or multiply numbers in any order.	$18 + 9 = 9 + 18$ $15 \times 2 = 2 \times 15$

ASSOCIATIVE PROPERTY (Grouping)	
Words	**Numbers**
When you are only adding or only multiplying, you can group any of the numbers together.	$(17 + 2) + 9 = 17 + (2 + 9)$ $(12 \times 2) \times 4 = 12 \times (2 \times 4)$

EXAMPLE 1 **Using Properties to Add and Multiply Whole Numbers**

A Evaluate **12 + 4 + 18 + 46.**

$12 + 4 + 18 + 46$	*Look for sums that are multiples of 10.*
$12 + 18 + 4 + 46$	*Use the Commutative Property.*
$(12 + 18) + (4 + 46)$	*Use the Associative Property to make*
$30 + 50$	*groups of compatible numbers.*
80	*Use mental math to add.*

1 Introduce

Alternate Opener

EXPLORATION

1-5 Mental Math

1. Choose one expression from each pair to evaluate using mental math.
   a. 25 · 24
      (25 · 4) · 6
   b. 5 · 22
      (5 · 20) + (5 · 2)
   c. 13 + 44 + 27
      44 + (13 + 27)
   d. 41 + 32 + 9 + 18
      (41 + 9) + (32 + 18)
2. What makes the expressions you chose easier to evaluate?
3. What makes the expressions you did not choose more difficult to evaluate?

Think and Discuss

4. **Discuss** the mental math strategies you used.
5. **Compare** the first expression with the second expression in each pair. How are they alike? How are they different?

Motivate

Define *commute* (to travel back and forth), *associate* (to connect or combine), and *distribute* (to scatter or spread out). Connect the general meanings of these words to the meanings of the properties being introduced in this lesson.

Exploration worksheet and answers on Chapter 1 Resource Book pp. 45 and 90

2 Teach

Lesson Presentation

Guided Instruction

In this lesson, students learn to use number properties to compute mentally. Show students how to use the Commutative, Associative, and Distributive Properties to find sums and products mentally. (Properties are displayed on Teaching Transparency T18 in the Chapter 1 Resource Book.) You may introduce other mental math strategies to students. For example, when adding two numbers, you can add to one and subtract from the other to make them easier to compute mentally: $57 + 69 = (57 + 3) + (69 - 3) = 60 + 66 = 126$.

B Evaluate 5 × 12 × 2.

5 × 12 × 2	*Look for products that are multiples of 10.*
12 × 5 × 2	*Use the Commutative Property.*
12 × (5 × 2)	*Use the Associative Property to group compatible numbers.*
12 × 10	
120	*Use mental math to multiply.*

DISTRIBUTIVE PROPERTY	
Words	**Numbers**
When you multiply a number times a sum, you can	
• find the sum first and then multiply, or	6 × (10 + 4) = 6 × 14 = 84
• multiply by each number in the sum and then add.	6 × (10 + 4) = (6 × 10) + (6 × 4) = 60 + 24 = 84

When you multiply two numbers, you can "break apart" one of the numbers into a sum and then use the Distributive Property.

EXAMPLE 2 Using the Distributive Property to Multiply

Use the Distributive Property to find each product.

> **Helpful Hint**
> Break the greater factor into a sum that contains a multiple of 10 and a one-digit number. You can add and multiply these numbers mentally.

A 4 × 23

4 × 23 = 4 × (20 + 3)	*"Break apart" 23 into 20 + 3.*
= (4 × 20) + (4 × 3)	*Use the Distributive Property.*
= 80 + 12	*Use mental math to multiply.*
= 92	*Use mental math to add.*

B 8 × 74

8 × 74 = 8 × (70 + 4)	*"Break apart" 74 into 70 + 4.*
= (8 × 70) + (8 × 4)	*Use the Distributive Property.*
= 560 + 32	*Use mental math to multiply.*
= 592	*Use mental math to add.*

Think and Discuss

1. **Give examples** of the Commutative Property and the Associative Property.
2. **Name** some situations in which you might use mental math.

COMMON ERROR ALERT

When using the Distributive Property, some students may multiply only one addend. Stress that all addends inside the parentheses must be multiplied by the number outside the parentheses.

Additional Examples

Example 1

Evaluate.

A. 17 + 5 + 3 + 15 40

B. 4 × 13 × 5 260

Example 2

Use the Distributive Property to find each product.

A. 6 × 35 210

B. 9 × 87 783

3 Close

Reaching All Learners
Through Critical Thinking

Have students roll number cubes and generate numbers to be used in mentally applying the Distributive Property. Example: Three rolls of the number cube yield 6, 4, and 3. Use the numbers to write the expression 6 × 43 and mentally apply the Distributive Property.

6 × 43 = (6 × 40) + (6 × 3)
= 240 + 18
= 258

Summarize

Review the definitions of the new vocabulary in the lesson: *Commutative Property, Associative Property,* and *Distributive Property*. Discuss how the terms relate to and are different from each other.

Place emphasis on the Distributive Property, the property least familiar to the students. Remind students to "break apart" one number and to distribute the multiplication over the addends.

Answers to Think and Discuss

1. Possible answers:
   Commutative: 2 × 17 = 17 × 2
   Associative: (4 × 9) × 10 = 4 × (9 × 10)
2. Possible answers: calculating movie expenses, planning expenses for theme park visit

1-5 PRACTICE & ASSESS

Students may want to refer back to the lesson examples.

Assignment Guide

If you finished Example 1 assign:
Core 1–4, 13–16, 25–30, 53–59
Enriched 1–4, 13–16, 25, 26, 45, 53–59

If you finished Example 2 assign:
Core 1–24, 37–49 odd, 53–59
Enriched 17–59

Notes

Math Background

The multiplication of the term outside the parentheses gets distributed over all of the terms being added inside the parentheses. If there are more than two terms being added inside the parentheses, then all the addends get multiplied by the term outside the parentheses. For example, 4 × (2 + 3 + 5) equals (4 × 2) + (4 × 3) + (4 × 5).

1-5 Exercises

FOR EXTRA PRACTICE see page 637

internet connect
Homework Help Online
go.hrw.com Keyword: MR4 1-5

GUIDED PRACTICE

See Example 1 Evaluate.

1. 13 + 9 + 7 + 11 **40**
2. 19 + 18 + 11 + 32 **80**
3. 5 × 14 × 4 **280**
4. 4 × 16 × 5 **320**

See Example 2 Use the Distributive Property to find each product.

5. 5 × 24 **120** **6.** 8 × 52 **416** **7.** 4 × 39 **156** **8.** 6 × 14 **84**
9. 3 × 33 **99** **10.** 2 × 78 **156** **11.** 9 × 12 **108** **12.** 2 × 87 **174**

INDEPENDENT PRACTICE

See Example 1 Evaluate.

13. 15 + 17 + 3 + 5 **40**
14. 14 + 7 + 16 + 13 **50**
15. 5 × 25 × 2 **250**
16. 2 × 32 × 10 **640**

See Example 2 Use the Distributive Property to find each product.

17. 3 × 36 **108** **18.** 4 × 42 **168** **19.** 6 × 71 **426** **20.** 2 × 94 **188**
21. 5 × 25 **125** **22.** 6 × 62 **372** **23.** 7 × 21 **147** **24.** 8 × 41 **328**

PRACTICE AND PROBLEM SOLVING

Use mental math to find each sum or product.

25. 8 + 13 + 7 + 12 **40**
26. 2 × 25 × 4 **200**
27. 5 × 8 × 12 **480**
28. 5 + 98 + 95 **198**
29. 11 + 75 + 25 **111**
30. 8 × 11 × 5 **440**

Multiply using the Distributive Property.

31. 9 × 17 **153** **32.** 4 × 27 **108** **33.** 11 × 18 **198**
34. 7 × 51 **357** **35.** 2 × 28 **56** **36.** 9 × 42 **378**
37. 5 × 55 **275** **38.** 3 × 78 **234** **39.** 4 × 85 **340**
40. 6 × 36 **216** **41.** 8 × 24 **192** **42.** 11 × 51 **561**

43. ***BUSINESS*** Janice wants to order disks for her computer. She needs to find the total cost, including shipping and handling. If Janice orders 7 disks, what will her total cost be? **$175**

Description	Number	Unit Cost with Tax	Price
Computer Disk	7	$24.00	
		Shipping & Handling	$7.00
		Total	

RETEACH 1-5

LESSON 1-5 **Reteach**
Mental Math

Commutative Property
Changing the order of addends does not change the sum.
21 + 13 = 13 + 21

Changing the order of factors does not change the product.
5 • 7 = 7 • 5

Associative Property
Changing the grouping of addends does not change the sum.
(3 + 8) + 4 = 3 + (8 + 4)

Changing the grouping of factors does not change the product.
2 • (7 • 4) = (2 • 7) • 4

Distributive Property
When you multiply a number by a sum, you can
- Find the sum and then multiply. 3 • (8 + 4) = 3 • 12 = 36
  or
- Multiply the number by each addend and then find the sum.
  3 • (8 + 4) = (3 • 8) + (3 • 4) = 24 + 12 = 36

Identify the property shown.

1. 3 • (2 • 6) = (3 • 2) • 6 **associative**
2. 7 + 18 = 18 + 7 **commutative**
3. 4 • (8 + 5) = 4 • 13 **distributive**
4. 11 • 8 = 8 • 11 **commutative**
5. 3 • (8 + 4) = (3 • 8) + (3 • 4) **distributive**
6. (3 + 8) + 4 = 3 +(8 + 4) **associative**

Identify the property shown and the missing number in each equation.

7. 9 + 16 = y + 9 **commutative; $y = 16$**
8. 4 • (3 • 2) = (4 • n) • 2 **associative; $n = 3$**
9. 3 • (11 + 4) = 3 • a **distributive; $a = 15$**
10. 6 • (9 + 14) = b • 23 **distributive; $b = 6$**

PRACTICE 1-5

LESSON 1-5 **Practice B**
Mental Math

Find each sum or product.

1. 17 + 4 • 5 **37**
2. 25 • 3 • 4 **300**
3. 28 + 39 + 11 + 22 **100**
4. 12 + 7 + 8 + 13 **40**
5. 10 + 3 • 2 **16**
6. 9 • 8 • 5 **360**
7. 97 + 4 + 3 + 26 **130**
8. 2 • 6 • 5 **60**
9. 28 + 2 • 6 **40**

Use the Distributive Property to find each product.

10. 4 • 16 **64**
11. 8 • 31 **248**
12. 3 · 62 **186**
13. 2 • 46 **92**
14. 5 • 29 **145**
15. 7 • 22 **154**
16. 9 • 21 **189**
17. 6 • 15 **90**
18. 8 • 44 **352**
19. 4 • 29 **116**
20. 7 • 31 **217**
21. 5 • 57 **285**

22. Each ticket to a play costs $27. How much will it cost to buy 4 tickets? Which property did you use to solve this problem with mental math?
$108; Distributive Property

23. Mr. Stanley bought two cases of pencils. Each case has 20 boxes. In each box there is 10 pencils. Use mental math to find how many pencils Mr. Stanley bought.
400 pencils

24. When you consider that cows eat grass and the water needed to grow the grass that cows eat, it takes 65 gallons of water to produce one serving of milk! Use mental math to find how many gallons of water are needed to produce 5 servings of milk.
325 gallons

Poison-dart frogs are members of the family Dendrobatidae, which includes about 170 species. Many are brightly colored, with yellow, red, green, orange, blue, or black markings.

44. ***LIFE SCIENCE*** Poison-dart frogs can breed underwater, and the females lay from 4 to 30 eggs. What would be the total number of eggs if four female poison-dart frogs each laid 27 eggs? **108 eggs**

45. Paul is writing a story for the school newspaper about the landscaping done by his class. The students planted 15 vines, 12 hedges, 8 trees, and 35 flowering plants. How many plants were used in the project? **70 plants**

46. Rickie wants to buy 3 garden hoses at the home center clearance sale. How much will they cost? **$48**

Home Center Clearance Sale	
Table lamp	$15
Garden hose	$16
Ceiling fan	$52

47. The boys in Josh's family are saving money to buy 4 ceiling fans at the home center sale. How much will they need to save? **$208**

48. ***EARTH SCIENCE*** The temperature on Sunday was 58°F. The temperature is predicted to rise 4°F on Monday, then rise 2°F more on Tuesday, and then rise another 6°F by Saturday. What is the predicted temperature on Saturday? **70°F**

49. ***WHAT'S THE ERROR?*** A student wrote 5 + 24 + 25 + 6 = 5 + 25 + 24 + 6 by the Associative Property. What error did the student make? **The student used the Commutative Property instead of the Associative Property.**

50. ***WRITE A PROBLEM*** Give a problem that you could simplify using the Commutative and Associative Properties. Then, show the steps to solve the problem and label the Commutative and Associative Properties.

51. ***WRITE ABOUT IT*** Why can you simplify 5(50 + 3) using the Distributive Property? Why can't you simplify 5(50) + 3 using the Distributive Property?

52. ***CHALLENGE*** Explain how you could find the product of $5^2 \times 112$ using the Distributive Property. Evaluate the expression. **Possible answer: Break apart 112 as 100 + 10 + 2 and multiply by $5^2 = 25$. So, $25 \times 112 = (25 \times 100) + (25 \times 10) + (25 \times 2) = 2{,}500 + 250 + 50 = 2{,}800$.**

Spiral Review

Estimate each sum or difference to the place value indicated. (Lesson 1-2)

53. 12,876 + 17,986; thousands **31,000**

54. 72,876 + 15,987; ten thousands **90,000**

Evaluate each expression. (Lesson 1-3)

55. 3^4 **81**

56. 2^5 **32**

57. 4^3 **64**

58. 11^0 **1**

59. **TEST PREP** What is the correct value of $3^2 + (9 \div 3 - 2)$? (Lesson 1-4) **B**

A 7 B 10 C 18 D 4

CHALLENGE 1-5

LESSON 1-5 **Challenge**

Magic Squares

When you add the numbers in each row, each column, and each diagonal of a magic square, you get the same number—the magic sum.

In the magic square at the right, for example, the magic sum is 30.

9	7	14
15	10	5
6	13	11

Use mental math to complete each magic square and find the magic sum.

1.

7	1	10
9	6	3
2	11	5

Magic sum: **18**

2.

11	4	9
6	8	10
7	12	5

Magic sum: **24**

3.

1	15	14	4
12	6	7	9
8	10	11	5
13	3	2	16

Magic sum: **34**

4. Use mental math to create your own magic square using the numbers 1–9.

2	7	6
9	5	1
4	3	8

Magic sum: **Possible answer: 15**

PROBLEM SOLVING 1-5

LESSON 1-5 **Problem Solving**

Mental Math

The bar graph below shows the average amounts of water used during some daily activities. Use the bar graph and mental math to answer the questions.

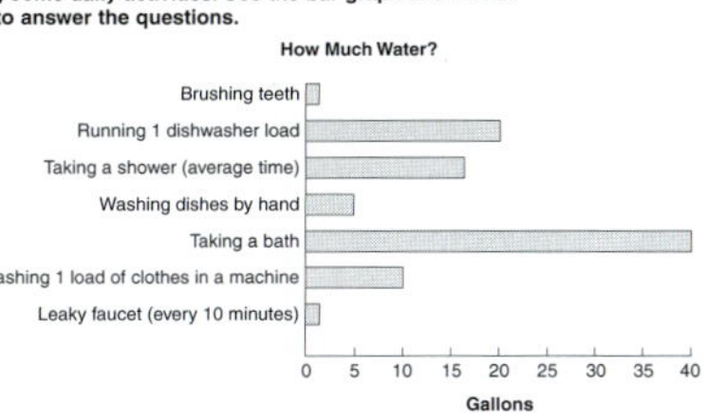

1. Most people brush their teeth three times a day. How much water do they use for this activity every week? **42 gallons**

2. How much water is wasted in a day by a leaky faucet? **288 gallons**

3. How much less water is used when you take a shower instead of a bath? **23 gallons**

4. How much water is used to wash four loads of laundry? **40 gallons**

5. Kenya used 24 gallons of water doing three of the activities listed in the table once. Which activities did she do? **taking a shower, brushing teeth, washing dishes by hand**

6. If you wash two loads of dishes by hand instead of using a dishwasher, how much water do you save? **30 gallons**

7. The average American uses 124 gallons of water a day. Name a combination of activities listed in the table that would equal that daily total. **Possible answer: taking a bath, washing 4 loads of laundry, brushing teeth two times, washing 2 dishwasher loads**

COMMON ERROR ALERT

When using the Distributive Property, students may randomly break apart one of the factors. For example, students may break apart 12 into 6 + 6 instead of 10 + 2. Caution them that the whole point of using the Distributive Property is to make the numbers easy to compute mentally.

Answers

50. Possible answer: 17 + (9 + 3); Associative: (17 + 9) + 3; Commutative 17 + 3 + 9

51. Possible answer: 5(50 + 3) is multiplying a number times a sum. The 5 can be multiplied by each of the numbers in the parentheses or times the sum, 53. 5(50) + 3 does not multiply a number times a sum. You must follow the order of operations and multiply 5(50) before adding 3.

Journal

Have students write about how the ability to multiply very large numbers mentally could help them in daily activities.

Test Prep Doctor

For Exercise 59, students need to remember the proper order of operations. Students who answered **A** may have followed the correct order, but they multiplied 3 × 2 to find 3^2. Students who answered **C** did the addition and subtraction before the division. Students who answered **D** performed the operations from left to right.

Lesson Quiz

Evaluate.

1. 18 + 24 + 2 + 6 **50**

2. 10 × 5 × 3 **150**

3. 13 + 42 + 7 + 8 **70**

Use the Distributive Property to find each product.

4. 8 × 12 **96** **5.** 6 × 15 **90**

6. Angie wants to buy 3 new video games. How much will she need to save if each game costs $27? **$81**

Available on Daily Transparency in CRB

1-6 Organizer

Pacing: Traditional 1 day
Block $\frac{1}{2}$ day

Objective: Students choose an appropriate method of computation and justify their choice.

Indiana Academic Standards
6.7.1, 6.7.3, 6.7.9, 6.7.11

Warm Up

Use mental math to find each fraction of 80.

1. $\frac{1}{2}$ 40 **2.** $\frac{1}{10}$ 8
3. $\frac{1}{4}$ 20 **4.** $\frac{1}{5}$ 16

Problem of the Day

About 23% of 600 firefighters are not on duty on any particular day. Steven estimates that about 200 firefighters have the day off. Is his estimate too high or too low? Explain.
too high; 25% of 600 is 150

Available on Daily Transparency in CRB

Math Fact

Historically, people who studied mathematics also studied astronomy. For example, Johannes Kepler and Sir Isaac Newton made contributions in both fields.

1-6 Choose the Method of Computation

Problem Solving Skill

Learn to choose an appropriate method of computation and justify your choice.

Earth has one moon. Scientists have determined that other planets in our solar system have as many as 39 moons. Mercury and Venus have no moons at all.

EXAMPLE 1 ***Astronomy Application***

How many known moons are in our solar system?

It might be hard to keep track of all of these numbers if you tried to add mentally. But the numbers themselves are small. You can use paper and pencil.

Planet	Moons
Mercury	0
Venus	0
Earth	1
Mars	2
Jupiter	39
Saturn	30
Uranus	21
Neptune	8
Pluto	1

Source: NASA, 2002

$$\begin{array}{r} 1 \\ 2 \\ 39 \\ 30 \\ 21 \\ 8 \\ +\ 1 \\ \hline 102 \end{array}$$

There are 102 known moons in our solar system.

EXAMPLE 2 ***Astronomy Application***

The average temperature on Earth is 59°F. The average temperature on Venus is 867°F. How much hotter is Venus's average temperature?

Venus temperature − Earth temperature
867 − 59

These numbers are small, and 59 is close to a multiple of 10. You can use mental math.

$(867 + 1) - (59 + 1)$ *Think: Add 1 to 59 to make 60. Add 1 to 867 to compensate.*
$868 - 60$
808

The average temperature on Venus is 808°F hotter than the average temperature on Earth.

1 Introduce

Alternate Opener

EXPLORATION

1-6 Choose the Method of Computation

Decide whether you would use mental math, pencil and paper, or a calculator to solve each problem. Then solve.

1. Susan makes $9.50 per hour. She worked 7 hours on Monday, 8 hours on Tuesday, 5 hours on Wednesday, and 10 hours on Friday. What is the total amount that Susan earned for the week?
2. Carlos is saving his money to buy a new bike. He earns $45 each week doing yard work, and the bike costs $189. How many weeks will he have to work to have enough money to buy the bike?
3. At a basketball game, 9,980 tickets were sold at $22 each. Find the total amount of money from ticket sales.
4. Rina counted the following numbers of books on each shelf in the storeroom: 24, 47, 26, 53, and 39. Find the total number of books.
5. A group of 12 people wants to rent a room at a pizza restaurant for a party. The room costs $75 to rent. Will $6 from each person be enough to cover the rent?

Think and Discuss

6. **Discuss** the method of computation you chose for each problem.
7. **Explain** how you decide whether to use pencil and paper or a calculator when you choose not to use mental math.

Motivate

Discuss other situations in which the degree of difficulty influences the choice that has to be made. Example: Should you walk, ride a bike, or take a car? Should you open a bag of snacks, make a sandwich, or cook a meal?

Exploration worksheet and answers on Chapter 1 Resource Book pp. 55 and 92

2 Teach

Lesson Presentation

Guided Instruction

In this lesson, students learn to choose an appropriate method of computation and to justify their choice. First, present students with three different methods of computation: paper and pencil, mental math, and using a calculator. Then provide some loose guidelines for making the choice: size of the numbers, the type of calculations, the number of calculations, etc.

EXAMPLE 3 ***Astronomy Application***

Every day, about 120 tons of cosmic dust—debris from outer space—enter Earth's atmosphere. How many tons of cosmic dust enter Earth's atmosphere each year?

tons per day × days per year *Think: There are 365 days in a year.*

120 × 365

These numbers are not compatible, so mental math is not a good choice.

You could use paper and pencil. But finding a product of 3-digit numbers requires several steps. Using a calculator will probably be faster.

Carefully enter the numbers on a calculator. Record the product.

120 × 365 = 43,800

Each year, about 43,800 tons of cosmic dust enter Earth's atmosphere.

Additional Examples

Example 1

Find the sum:

4 + 3 + 2 + 10 + 8 + 2 + 5 + 1

35

Example 2

Find the difference: 4,562 − 397.

4,165

Example 3

Find the quotient: 9,288 ÷ 24. 387

Think and Discuss

1. **Give an example** of a situation in which you would use mental math to solve a problem. When would you use paper and pencil?
2. **Tell** how you could use mental math in Example 2 if the problem were 867 + 59.

1-6 Exercises

FOR EXTRA PRACTICE
see page 637

internet connect
Homework Help Online
go.hrw.com Keyword: MR4 1-6

GUIDED PRACTICE

See Example 1

1. What is the total number of astronauts who have space flight experience? **364**

U.S.	Germany	France	Canada	Japan	Italy	Russia
244	9	8	7	5	3	88

See Example 2

2. In the 2000 Summer Olympic Games, 929 medals were given. The U.S. team brought home the most medals, 97. How many medals were not won by the U.S. team? **832**

See Example 3

3. A factory produces 126 golf balls per minute. How many golf balls can be produced in 515 minutes? **64,890**

1-6 PRACTICE & ASSESS

Assignment Guide

If you finished Example 1 assign:
Core 1, 4, 7–8, 21–27
Enriched 1, 4, 7–8, 18, 21–27

If you finished Example 2 assign:
Core 1–2, 4–5, 7–12, 21–27
Enriched 1–2, 4–5, 7, 8, 17–20, 21–27

If you finished Example 3 assign:
Core 1–17, 21–27
Enriched 1–6, 7–17 odd, 18–27

3 Close

Reaching All Learners

Through Grouping Strategies

Have students work in groups to build their confidence in the areas of problem solving and mental math skills. Provide the groups with the "Additional Examples." (A recording sheet is provided in the Chapter 1 Resource Book p. 78.) Have them first understand the problem and be able to restate the problem in their own words to another group member. Then the operations should be identified and the students should arrive at a method of computation. Consider grouping more proficient students with students who require more modeling.

Summarize

Remind the students that the choice of a method of computation depends on their understanding of the problem, the numbers involved in the problem, and the calculations needed to solve the problem. The most important factor in selecting a method of calculation is the proficiency of the individual student with each method.

Answers to Think and Discuss

1. Possible answers:
   mental math: finding the combined weight of 2 packages when the weight of one is close to a multiple of 10
   paper and pencil: finding the area of a room
2. Possible answer: Add 1 to 59 to make 60. Subtract 1 from 867 to compensate. 867 + 59 = (867 − 1) + (59 + 1) = 866 + 60 = 926

Lesson Quiz

Evaluate the expression and state the method of computation you used.

1. 17 + 6 + 24 + 35 + 3 + 5
   90; mental math
2. 63 × 197 12,411; paper and pencil
3. It takes Jupiter approximately 4,344 days to complete one revolution around the Sun. It takes Earth 365 days to revolve around the Sun. How many more days does it take Jupiter to revolve around the Sun than Earth?
   3,979 days

Available on Daily Transparency in CRB

7–16, 19. Complete answers on p. A1

INDEPENDENT PRACTICE

See Example 1

4. A carnival has a coin-toss game. The highest score is a total of all the squares on the board. What is that score? 72

6	9	5
10	20	8
3	7	4

See Example 2

5. It takes Mars 687 days to complete one revolution around the Sun. It takes Venus only 225 days to revolve around the Sun. How many more days does it take Mars to revolve around the Sun than Venus? 462

See Example 3

6. If each store in a chain of 108 furniture stores sells 135 sofas a year, what is the total number of sofas sold? 14,580

PRACTICE AND PROBLEM SOLVING

Evaluate the expression, and state the method of computation you used.

7. 5 + 24 + 7 + 1 + 64 + 2 + 8
   111
8. 16 + 2 + 4 + 13 + 5 + 1 + 14
   55
9. 828 × 623
   515,844
10. 742 − 167
    575
11. 41 + 169
    210
12. 499 − 201
    298
13. 57 × 198
    11,286
14. 338 + 12
    350
15. 3,813 × 117
    446,121
16. 337 − 124
    213
17. A satellite travels 985,200 miles per year. How many miles will it travel if it stays in space for 12 years? 11,822,400
18. ***WHAT'S THE QUESTION?*** An astronaut has spent the following minutes training in a tank that simulates weightlessness: 2, 15, 5, 40, 10, and 55. The answer is 127. What is the question?
    How many total minutes has she spent in weightlessness?
19. ***WRITE ABOUT IT*** Explain how you can decide whether to use pencil and paper, mental math, or a calculator to solve a subtraction problem.
20. ***CHALLENGE*** A list of possible astronauts was narrowed down by two committees. The first committee selected 93 people to complete a written form. The second selected 31 of those people to come to an interview. If 837 were not asked to complete a form, how many were on the original list? 930 astronauts

Spiral Review

Write each expression in exponential form. (Lesson 1-3)

21. 4 × 4 × 4 × 4 4^4
22. 2 × 2 × 2 × 2 × 2 2^5
23. 10 × 10 × 10 10^3

Evaluate each expression. (Lesson 1-4)

24. 4 × 14 + 12 ÷ 2 62
25. $16 \div 4^2 + 15 - 2$ 14
26. $5 + 2^2 (12 \div 3)$ 21

27. **TEST PREP** Which expression does **not** have the same value as 7 × (34 + 23)? (Lesson 1-5) C

A 7 × 57 **B** (7 × 34) + (7 × 23) **C** 7 × 34 + 23 **D** (7 × 50) + (7 × 7)

RETEACH 1-6

LESSON 1-6 **Reteach**
Choose the Method of Computation

Paper and pencil, mental math, and a calculator are three computation methods for solving problems.

- If there are many small numbers, use paper and pencil.
- If the numbers are small and easy, use mental math.
- If the numbers are large, use a calculator.

Before you solve a problem decide which computation method is the best.

Choose a computation method. Then solve.

At a book fair, 76 books were sold on the first day and 82 books were sold on the second day. How many books were sold during the two days?

Number of books sold on the first day + Number of books sold on the second day

76 + 82

The numbers are small and 82 is close to a multiple of 10. You can use mental math.

(76 + 2) + (82 − 2) = 78 + 80 = 158

During the two days, 158 books were sold.

Choose a computation method. Then solve.

1. Of the 248 books on display, 46 were nonfiction books. How many books were not nonfiction books?
   mental math, 202 books
2. Lisa bought 2 biographies for $5.37 each, a novel for $7.95, and a bookmark for $1.19. How much money did Lisa spend?
   paper and pencil, $19.88
3. Over two days, 234 students visited the book fair in groups of 18. How many groups visited the fair?
   calculator, 13 groups

PRACTICE 1-6

LESSON 1-6 **Practice B**
Choose the Method of Computation

1. Athletes from 197 countries competed at the 1996 Summer Olympic Games held in Atlanta, Georgia. That is 25 more countries that competed at the 1992 games held in Barcelona, Spain. How many different countries competed in Barcelona?
   Athletes from 172 countries competed in Barcelona.
2. At the 1996 Summer Olympic Games held in Atlanta, Georgia, 10,310 athletes competed. At the 1992 Summer Olympic Games held in Barcelona, Spain, 9,364 athletes competed. How many more athletes competed in Atlanta than in Barcelona?
   946 more athletes competed in Atlanta.
3. The marathon race is one of the oldest events in the Summer Olympic Games. Marathon competitors run a total of 26 miles 385 yards. There are 5,280 feet in a mile and 3 feet in a yard. How many yards long is the entire marathon race?
   The marathon is 46,145 yards long.
4. The world record for the fastest men's marathon race is 2 hours, 5 minutes, 42 seconds. The world record for the fastest women's marathon race is 2 hours, 20 minutes, 43 seconds. How much faster is the men's record marathon time?
   It is 15 minutes, 1 second faster.
5. The men's outdoor world record in the high jump is 2.45 meters or 8 feet 0.5 inches. The women's outdoor world record in the high jump is 2.09 meters or 6 feet 10.25 inches. How much higher is the men's high jump record? Write the answer in meters and feet.
   0.36 meter or 1 foot 2.25 inches
6. The men's world record in the 400-meter relay is 37.40 seconds, held by the U.S. If each of the four runners each ran 100 meters in the same time, how long did each runner run?
   9.35 seconds
7. Athletes from 13 nations competed in the first modern Olympics in 1896. Today, athletes from nearly 200 nations compete in the Summer Olympics. About how many more nations participate in the Olympics today than in 1896?
   about 187 nations

CHALLENGE 1-6

LESSON 1-6 **Challenge**
Finger Math

Chisenbop is an ancient method of computation using your fingers. One of the best-known forms of Chisenbop is used for basic multiplication computations. It works only when all the factors are greater than 5. Follow these steps to use this form of the Chisenbop method of computation. The product of 6 • 7 is shown as an example.

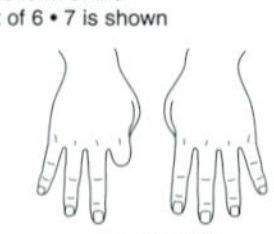

Step 1 Subtract 5 from the first factor. Turn down that number of fingers on your left hand.

Step 2 Subtract 5 from the second factor. Turn down that number of fingers on your right hand.

Step 3 Multiply the total number of turned-down fingers on both hands by 10.

Step 4 Find the product of the numbers of fingers that are **not** turned down on each hand.

Step 5 Add the two products from Step 3 and Step 4.

3 • 10 = 30
both hands
4 • 3 = 12
left hand **right hand**
30 + 12 = 42
So, 6 • 7 = 42.

Use the Chisenbop method to find each product.

1. 7 • 8 = 56
2. 6 • 9 = 54
3. 8 • 6 = 48
4. 8 • 9 = 72
5. 6 • 6 = 36
6. 9 • 7 = 63
7. 7 • 7 = 49
8. 7 • 9 = 63
9. 8 • 8 = 64
10. 6 • 8 = 48
11. 9 • 9 = 81
12. 9 • 8 = 72
13. When would you choose to use the Chisenbop method of computation? When would you choose not to use that method? Explain.
    Possible answer: I would use this method when both factors are greater than 5 and less than 11. I would not use this method when at least one of the factors is less than or equal to 5.

PROBLEM SOLVING 1-6

LESSON 1-6 **Problem Solving**
Choose the Method of Computation

Use the table below to answer the questions 1–6. For each question, write what method of computation you should use to solve it. Then write the solution.

Bones in the Human Body

Body Part	Number of Bones
Head	28
Throat	1
Spine	26
Chest	25
Shoulders	4
Arms	6
Hands	54
Legs	10
Feet	52

1. How many bones are in an average person's arms and hands altogether?
   mental math; 60 bones
2. How many more bones are in an average person's head than chest?
   mental math; 3 bones
3. Which part of the body has twice as many bones as the spine?
   mental math; feet
4. How many bones are in the body altogether?
   paper and pencil; 206 bones
5. A newborn baby has 350 bones. How many more bones does a newborn baby have than an adult?
   paper and pencil; 144 bones
6. How many bones are in each of an average person's feet, hands, legs, and arms?
   paper and pencil; feet: 26 bones; hands: 27 bones; legs: 5 bones; arms: 3 bones
7. The body's longest bones—thighbones and shinbones—are in the legs. The average thighbone is about 20 inches long, and the average shinbone is about 17 inches long. What is the total length of those four bones?
   paper and pencil; 74 inches
8. The body has 650 muscles. Seventeen of those muscles used to smile and 42 muscles are used to frown. How many more muscles are used to frown thanto smile?
   mental math; 25 muscles

1-7 Find a Pattern

Problem Solving Strategy

Learn to find patterns and to recognize, describe, and extend patterns in sequences.

Vocabulary
- perfect square
- sequence
- term

Whole numbers raised to the second power are called **perfect squares**. This is because they can be represented by objects arranged in the shape of a square.

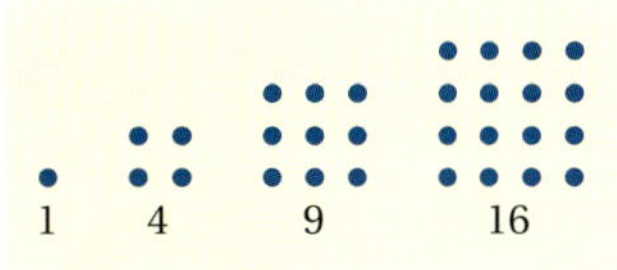

The perfect squares can be written as a sequence. A **sequence** is an ordered set of numbers. Each number in the sequence is called a **term**. In a sequence, there is often a pattern between one term and the next.

1	4	9	16	← Sequence
1st term	2nd term	3rd term	4th term	← Term
+3	+5	+7		

Helpful Hint

Look for a relationship between the 1st term and the 2nd term. Check if this relationship works between the 2nd term and the 3rd term, and so on.

You can use this pattern to find the fifth and sixth terms in the sequence. To get the fifth term, add 9. To get the sixth term, add 11.

1, 4, 9, 16, ■, ■, . . .

16 + 9 = **25**   25 + 11 = **36**

So the next two perfect squares are 25 and 36.

EXAMPLE 1 Extending Sequences with Addition and Subtraction

Identify a pattern in each sequence and name the next three terms.

A 3, 15, 27, 39, ■, ■, ■, . . .

3   15   27   39   ■   ■   ■
+12 +12 +12 +12 +12 +12

A pattern is to add 12 to each term to get the next term.

39 + 12 = 51   51 + 12 = 63   63 + 12 = 75

So 51, 63, and 75 will be the next three terms.

B 4, 15, 8, 19, 12, 23, 16, ■, ■, ■, . . .

4   15   8   19   12   23   16   ■   ■   ■
+11 −7 +11 −7 +11 −7 +11 −7 +11

A pattern is to add 11 to one term and subtract 7 from the next.

16 + 11 = 27   27 − 7 = 20   20 + 11 = 31

So 27, 20, and 31 will be the next three terms.

1-7 Organizer

Pacing: Traditional 1 day
Block $\frac{1}{2}$ day

Objective: Students find, recognize, describe, and extend patterns in sequences.

Indiana Academic Standards
6.7.1, 6.7.2, 6.7.11

Warm Up

Determine what could come next.

1. 3, 4, 5, 6, ____   7
2. 10, 9, 8, 7, 6, ____   5
3. 1, 3, 5, 7, ____   9
4. 2, 4, 6, 8, ____   10
5. 5, 10, 15, 20, ____   25

Problem of the Day available on Daily Transparency in CRB

Additional Example

Example 1

Identify a pattern in the sequence and name the next three terms.

A. 48, 42, 36, 30, ■, ■, ■, . . .
One pattern is to subtract 6 from each term. 24, 18, 12

B. 24, 34, 31, 41, 38, 48, ■, ■, ■, . . .
One pattern is to add 10 to one term and subtract 3 from the next. 45, 55, 52

1 Introduce

Alternate Opener

EXPLORATION

1-7 Find a Pattern

1. Examine the sequence of figures below and look for a pattern.

   a. Sketch the next two figures in your pattern. Count the number of line segments it takes to draw each.
   b. Copy and complete the table for your pattern.

Figure Number	1	2	3	4	5	6	7	8	9	10
Number of Line Segments	3	5	7							

Find the next three numbers in each sequence.

2. 1, 3, 5, 7, ____, ____, ____, . . .
3. 96, 84, 72, ____, ____, ____, . . .
4. 1, 3, 6, 10, 15, ____, ____, ____, . . .

Think and Discuss

5. **Describe** the pattern you noticed in the sequence of triangles.
6. **Explain** how you found the next three numbers in numbers 2–4.

Motivate

Give examples of sequences in math.
Example: 1, 2, 3, 4, . . .
2, 4, 6, 8, . . .
5, 10, 15, 20, . . .

Discuss with students what makes something a pattern. (A pattern is a sequence, together with some explicit rule that determines the next term in the sequence.)

Exploration worksheet and answers on Chapter 1 Resource Book pp. 64 and 94

2 Teach

Lesson Presentation

Guided Instruction

In this lesson, students learn to use the strategy "find a pattern" to recognize, describe, and extend sequences. First point out a pattern in the sequence of perfect squares. Then have students extend sequences with addition and subtraction. Then have them complete sequences with multiplication and division. Next, point out that some patterns are themselves a sequence and that a pattern could involve more then one operation.

Additional Example

Example 2

Identify a pattern in the sequence and name the missing terms.

A. 2, 6, 18, ■, 162, ■, . . .
One pattern is to multiply each term by 3. 54, 486

B. 12, 6, 24, 12, 48, ■, 96, 48, ■, 96, . . .
One pattern is to divide one term by 2 and multiply the next by 4. 24, 192

EXAMPLE 2 Completing Sequences with Multiplication and Division

Identify a pattern in each sequence and name the missing terms.

A 256, 128, 64, ■, 16, ■, . . .

256 → 128 → 64 → ■ → 16 → ■
(÷2, ÷2, ÷2, ÷2, ÷2)

A pattern is to divide each term by 2 to get the next term.

$64 \div 2 = 32$ $\quad$ $16 \div 2 = 8$
So 32 and 8 are the missing terms.

B 1, 6, 2, 12, 4, ■, 8, 48, ■, 96, . . .

1 → 6 → 2 → 12 → 4 → ■ → 8 → 48 → ■ → 96
(×6, ÷3, ×6, ÷3, ×6, ÷3, ×6, ÷3, ×6)

A pattern is to multiply one term by 6 and divide the next by 3.

$4 \times 6 = 24$ $\quad$ $48 \div 3 = 16$
So 24 and 16 are the missing terms.

Think and Discuss

1. **Tell** how you could check whether the next two perfect squares in the sequence 1, 4, 9, 16, . . . are 25 and 36.
2. **Explain** how to find the next term in the sequence 8, 4, 2, ■, ...

1-7 PRACTICE & ASSESS

Assignment Guide

If you finished Example 1 assign:
Core 1–4, 9–12, 18–19, 25–32
Enriched 1–4, 9–12, 21, 23, 25–32

If you finished Example 2 assign:
Core 1–8, 9–15 odd, 17–21, 25–32
Enriched 1–15 odd, 17–32

Answers

1–8. See p. A1.

1-7 Exercises

FOR EXTRA PRACTICE see page 637

internet connect
Homework Help Online
go.hrw.com Keyword: MR4 1-7

GUIDED PRACTICE

See Example 1 **Identify a pattern in each sequence and name the next three terms.**

1. 12, 24, 36, 48, ■, ■, ■, . . .
2. 105, 90, 75, 60, 45, ■, ■, ■, . . .
3. 7, 18, 16, 27, 25, ■, ■, ■, . . .
4. 44, 38, 42, 36, 40, ■, ■, ■, . . .

See Example 2 **Identify a pattern in each sequence and name the missing terms.**

5. 2, 6, ■, 54, 162, ■, . . .
6. 80, 8, 40, ■, 20, 2, ■, 1, . . .
7. 1, 6, 3, ■, 9, 54, ■, 162, . . .
8. 1,024, 256, ■, 16, ■, 1, . . .

Teach

Reaching All Learners

Through Number Sense

Triangular numbers are similar to perfect squares, except that they are represented by triangular instead of square arrays. Have students find the next three triangular numbers in the sequence by drawing. (A recording sheet is provided on Chapter 1 Resource Book p. 79.) Then have the students write the number sequence generated by the drawings to discover a number pattern that will continue the sequence.

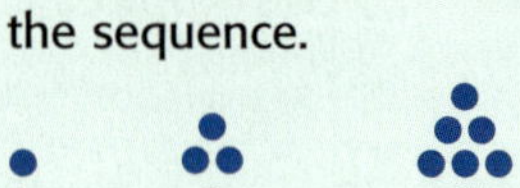

1 3 6

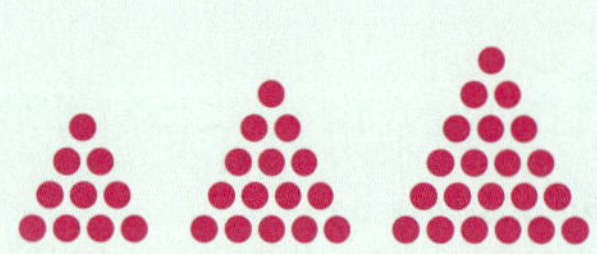

10 15 21

3 Close

Summarize

Give brief definitions of the new vocabulary in the lesson: *perfect square, sequence*, and *term.* Remind students that patterns can include the four operations or combinations of the operations. Also, a pattern that fits a sequence should not be considered the only one that could generate that sequence.

Answers to Think and Discuss

1. Possible answers: Draw the next two square arrays in the sequence, a 5 × 5 array and a 6 × 6 array; find the values of the next two numbers raised to the second power, 5^2 and 6^2.
2. Possible answer: First find a relationship between the given terms. One pattern is $8 \div 2 = 4$ and $4 \div 2 = 2$; divide each term by 2 to get the next term. $2 \div 2 = 1$, so 1 is the next term.

INDEPENDENT PRACTICE

See Example 1

Identify a pattern in each sequence and name the next three terms.

9. 9, 19, 30, 42, 55, ■, ■, ■, . . .
10. 95, 94, 92, 89, ■, ■, ■, . . .
11. 50, 55, 47, 52, 44, ■, ■, ■, . . .
12. 5, 3, 6, 4, 7, ■, ■, ■, . . .

See Example 2

Identify a pattern in each sequence and name the missing terms.

13. 1, 2, 6, ■, 120, ■, 5,040, . . .
14. 600, 300, ■, 30, 6, ■, . . .
15. 400, 100, ■, 50, 100, ■, 50, . . .
16. 120, 60, 180, ■, 270, ■, . . .

PRACTICE AND PROBLEM SOLVING

Use the pattern to write the first five terms of the sequence.

17. Start with 1; multiply by 3. **1, 3, 9, 27, 81**
18. Start with 12; add 12. **12, 24, 36, 48, 60**
19. Start with 100; subtract 7. **100, 93, 86, 79, 72**
20. Start with 2; square each term. **2, 4, 16, 256, 65,536**
21. The temperature was 45°F on Monday, 48°F on Tuesday, and 51°F on Wednesday. If the pattern continues, what temperature will it be on Friday? **57°F**
22. ***CHOOSE A STRATEGY*** The * shows where a piece is missing from the pattern. What piece is missing? **C**

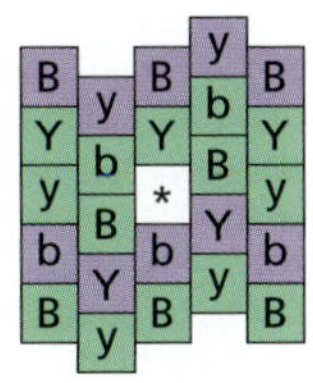

A y **B** B **C** y **D** Y

23. ***WRITE ABOUT IT*** How can you know whether a number is a perfect square? **The number is a perfect square if it can be shown to be a whole number raised to the second power.**
24. ***CHALLENGE*** Find the missing terms in the following sequence: ■, 2^3, 27, 4^3, 125, ■, 343, . . . **1, 6^3**

Spiral Review

Estimate each sum or difference by rounding to the nearest thousands place. (Lesson 1-2)

25. 5,237 − 1,586 **3,000**
26. 915,178 + 451,836 **1,367,000**
27. 39,187 − 24,999 **14,000**

Find each value. (Lesson 1-3)

28. 8^5 **32,768**
29. 5^3 **125**
30. 3^8 **6,561**
31. 4^4 **256**

32. **TEST PREP** Which is the correct value of $7^2 - (4 \times 7 - (3 - 2)) \div 3$? (Lesson 1-5) **C**

A 7 **B** 105 **C** 40 **D** 41

Lesson Quiz

Identify a pattern in each sequence and name the next three terms.

1. 12, 24, 36, 48, ■, ■, ■, ... **60, 72, 84**
2. 75, 71, 67, 63, ■, ■, ■, ... **59, 55, 51**

Identify a pattern in each sequence and name the missing terms.

3. 1000, 500, ■,125, ... **250**
4. 100, 50, 200, ■, 400, ■, ... **100, 200**

Available on Daily Transparency in CRB

Answers

9–16. See p. A1.

RETEACH 1-7

LESSON 1-7 **Reteach**
Find a Pattern

Find the next three numbers in the pattern.

8, 12, 16, 20, 24, □, □, □, ...

Step 1: Look at pairs of numbers to find the pattern.

8, 12, 16, 20, 24, □, □, □, ...

8 + 4 = 12   12 + 4 = 16   16 + 4 = 20

The pattern is to add 4.

Step 2: Use the pattern to name the next three numbers.

24 + 4 = 28   28 + 4 = 32   32 + 4 = 36

The next three numbers are 28, 32, and 36.

Find the next three numbers in each pattern.

1. 5, 8, 6, 9, 7 □, □, □, ... **10, 8, 11**
2. 90, 80, 70, 60, □, □, □, ... **50, 40, 30**
3. 2, 8, 4, 16, □, □, □, ... **8, 32, 16**
4. 10, 14, 18, 22, □, □, □, ... **26, 30, 34**
5. 13, 21, 29, 37, □, □, □, ... **45, 53, 61**
6. 24, 12, 16, 8, □, □, □, ... **12, 6, 10**
7. 14, 12,10, 8, □, □, □, ... **6, 4, 2**
8. 1, 7, 13, 19, □, □, □, ... **25, 31, 37**
9. 1, 3, 6, 10, □, □, □, ... **15, 21, 28**
10. 40, 38, 36, □, □, □, ... **34, 32, 30**
11. 54, 45, 36, □, □, □, ... **27, 18, 9**
12. 10, 25, 40, □, □, □, ... **55, 70, 85**
13. 36, 29, 22, □, □, □, ... **15, 8, 1**
14. 18, 36, 72, □, □, □, ... **144, 288, 576**

PRACTICE 1-7

LESSON 1-7 **Practice B**
Find a Pattern

Identify a pattern in each sequence, and name the next three terms.

1. 4, 8, 16, 32, □, □, □, ... **multiply by 2; 64, 128, 256**
2. 100, 95, 90, 85, □, □, □, ... **subtract 5; 80; 75; 70**
3. 8, 20, 32, 44, □, □, □, ... **add 12; 56; 68; 80**
4. 6, 12, 18, 24, □, □, □, ... **add 6; 30; 36; 42**
5. 9, 18, 27, 36, □, □, □, ... **add 9; 45, 54, 63**
6. 3, 6, 12, 24, □, □, □, ... **multiply by 2; 48, 96, 192**
7. 5, 10, 20, 40, □, □, □, ... **multiply by 2; 80, 160, 320**
8. 100, 125, 150, 175, □, □, □, ... **add 25; 200, 225, 250**
9. 20, 18, 16, 14, 12, □, □, □, ... **subtract 2; 10, 8, 6**
10. 4, 8, 12, 16, 20, □, □, □, ... **add 4; 24, 28, 32**

Identify a pattern in each sequence, and name the missing terms.

11. 300, 250, □, □, 100, □, 0, ... **subtract 50; 200; 150; 50**
12. 1, 15, □, 43, 57, □, 85, 99, ... **add 14; 29; 71**
13. 7, □, 21, 28, □, □, □, 56, ... **add 7; 14; 35; 42; 49**
14. 9, □, 13, □, □, □, 21, 23, ... **add 2; 11; 15; 17; 19**
15. □, 17, 15, □, 11, □, 7, □, 3, ... **subtract 2; 19; 13; 9; 5**
16. 100, 85, □, 55, □, 25, 10, ... **subtract 15; 70; 40**
17. A forest ranger in Australia took measurements of a eucalyptus tree for the past 3 weeks. The tree was 12 inches tall the first week, 19 inches the second week, and 26 inches the third week. If this growth pattern continues, how tall will the tree be next week?
33 inches
18. Maria puts the same amount of money in her savings account each month. She had $450 in the account in April, $600 in May, and $750 in June. If she continues her savings pattern, how much money will she have in the account in July?
$900

CHALLENGE 1-7

LESSON 1-7 **Challenge**
Picture Patterns

Draw the next three figures in each pattern.

1.
2.
3.
4.
5.
6.
7.

PROBLEM SOLVING 1-7

LESSON 1-7 **Problem Solving**
Find a Pattern

1. A giant bamboo plant was 5 inches tall on Monday, 23 inches tall on Tuesday, 41 inches tall on Wednesday, and 59 inches tall on Thursday. Describe the pattern. If the pattern continues, how tall will the giant bamboo plant be on Friday, Saturday, and Sunday?
Each day the giant bamboo plant grew 18 inches. The giant bamboo plant will be 77 inches tall on Friday, 95 inches tall on Saturday, and 113 inches tall on Sunday.
2. A scientist was studying a cell. After the second hour there were two cells. After the third hour there were four cells. After the fourth hour there were eight cells. Describe the pattern. If the pattern continues, how many cells will there be after the fifth, sixth, and seventh hour?
The number of cells doubled every hour. After the fifth hour there will be 16 cells. After the sixth hour there will be 32 cells. After the seventh hour there will be 64 cells.
3. The first place prize for a sweepstakes is $8,000. The third place prize is $2,000. The fourth place prize is $1,000. The fifth place prize is $500. Describe the pattern. What is the second place prize, and if the pattern continues, what are the sixth and seventh place prizes?
The prize money for each higher place in the sweepstakes doubles in value. The second place prize is $4,000. The sixth place prize is $250. The seventh place prize is $125.
4. The temperature was 59°F at 3:00 A.M., 62°F at 5:00 A.M., and 65°F at 7:00 A.M. Describe the pattern. If the pattern continues, what will the temperature be at 9:00 A.M., 11:00 A.M., and 1:00 P.M.?
Every two hours, the temperature increased by 3°F. The temperature will be 68°F at 9:00 A.M., 71°F at 11:00 A.M., and 74°F at 1:00 P.M.

CHAPTER 1 Extension

Pacing: Traditional 1 day
Block $\frac{1}{2}$ day

Objective: Students investigate the binary number system.

Using the Pages

In Lesson 1-1, students learned to compare whole numbers in the base-10 system. In this extension, students will learn to identify and compare binary numbers in base-2.

EXTENSION Binary Numbers

Learn to investigate the binary number system.

Vocabulary
base-10 system
binary number system

Our number system is called the **base-10 system** because each place value is 10 times greater than the place value to the right. Base-10 numbers contain the digits 0 through 9.

10^4	10^3	10^2	10^1	10^0
Ten thousands	Thousands	Hundreds	Tens	Ones
2	5	6	0	1

$$25{,}601 = 20{,}000 + 5{,}000 + 600 + 0 + 1$$
$$= (2 \times 10{,}000) + (5 \times 1{,}000) + (6 \times 100) + (0 \times 10) + (1 \times 1)$$

In the **binary number system**, each place value is 2 times greater than the place value to the right. Binary numbers contain only the digits 0 and 1.

2^4	2^3	2^2	2^1	2^0
Sixteens	Eights	Fours	Twos	Ones
1	0	1	1	1

$$10111 = (1 \times 16) + (0 \times 8) + (1 \times 4) + (1 \times 2) + (1 \times 1)$$
$$= 16 + 0 + 4 + 2 + 1$$
$$= 23$$

10111 in the binary system is equal to 23 in the base-10 system.

EXAMPLE 1 Converting Binary Numbers to Base 10

Find the base-10 value for each binary number.

A 1111

$$1111 = (1 \times 8) + (1 \times 4) + (1 \times 2) + (1 \times 1)$$
$$= 8 + 4 + 2 + 1$$
$$= 15$$

B 11001

$$11001 = (1 \times 16) + (1 \times 8) + (0 \times 4) + (0 \times 2) + (1 \times 1)$$
$$= 16 + 8 + 0 + 0 + 1$$
$$= 25$$

C 10101

$$10101 = (1 \times 16) + (0 \times 8) + (1 \times 4) + (0 \times 2) + (1 \times 1)$$
$$= 16 + 0 + 4 + 0 + 1$$
$$= 21$$

1 Introduce

Motivate

Modern computers do not work with decimal numbers. The basic unit of information on a computer is a *binary digit*, or *bit*. Bits can only be either a 0 or a 1. For this reason, computers use base-2, or binary, numbers. Information received by a computer is translated into and processed as 0's and 1's.

2 Teach

Lesson Presentation

Guided Instruction

In this extension, students investigate the binary number system. Show the students that base-10 numbers have place values that are 10 times greater than the place value to the right. Then show the students that in the binary system each place value is two times greater than the place value to the right. Next, show how to convert between the two systems.

To write a base-10 number as a binary number, "break apart" the base-10 number as a sum of powers of 2. Start with the highest power of 2 that is not more than the base-10 number.

EXAMPLE 2 **Converting Base-10 Numbers to Binary**

Find the binary number for each base-10 number.

A 9

$$9 = 8 + 0 + 0 + 1$$
$$= (1 \times 8) + (0 \times 4) + (0 \times 2) + (1 \times 1)$$
$$= 1001$$

B 13

$$13 = 8 + 4 + 0 + 1$$
$$= (1 \times 8) + (1 \times 4) + (0 \times 2) + (1 \times 1)$$
$$= 1101$$

C 27

$$27 = 16 + 8 + 0 + 2 + 1$$
$$= (1 \times 16) + (1 \times 8) + (0 \times 4) + (1 \times 2) + (1 \times 1)$$
$$= 11011$$

EXTENSION Exercises

Find the base-10 value for each binary number.

1. 101 5 **2.** 100 4 **3.** 111 7 **4.** 1000 8
5. 1011 11 **6.** 11111 31 **7.** 10 2 **8.** 10001 17

Write the expanded form of each binary number.

9. 11010 **10.** 10111 **11.** 11110 **12.** 11101
13. 10100 **14.** 10000 **15.** 10010 **16.** 11100

Find the binary number for each base-10 number.

17. 12 1100 **18.** 6 110 **19.** 18 10010 **20.** 10 1010
21. 1 1 **22.** 14 1110 **23.** 22 10110 **24.** 19 10011

Compare. Write <, >, or =. The number on the left is a base-10 number, and the number on the right is a binary number.

25. 20 ▒ 111 > **26.** 24 ▒ 11000 =
27. 3 ▒ 101 < **28.** 15 ▒ 1001 >

Additional Examples

Example 1

Find the base-10 value for each binary value.

A. 1101 13
B. 10010 18
C. 11111 31

Example 2

Find the binary number for each base-10 number.

A. 7 111
B. 14 1110
C. 21 10101

Answers

9. $(1 \times 16) + (1 \times 8) + (0 \times 4) + (1 \times 2) + (0 \times 1)$
10. $(1 \times 16) + (0 \times 8) + (1 \times 4) + (1 \times 2) + (1 \times 1)$
11. $(1 \times 16) + (1 \times 8) + (1 \times 4) + (1 \times 2) + (0 \times 1)$
12. $(1 \times 16) + (1 \times 8) + (1 \times 4) + (0 \times 2) + (1 \times 1)$
13. $(1 \times 16) + (0 \times 8) + (1 \times 4) + (0 \times 2) + (0 \times 1)$
14. $(1 \times 16) + (0 \times 8) + (0 \times 4) + (0 \times 2) + (0 \times 1)$
15. $(1 \times 16) + (0 \times 8) + (0 \times 4) + (1 \times 2) + (0 \times 1)$
16. $(1 \times 16) + (1 \times 8) + (1 \times 4) + (0 \times 2) + (0 \times 1)$

Teaching Tip To convert a base-10 number to a binary number, find a sum of powers of 2 that equals the number. For example, 27 = (▒ × 16) + (▒ × 8) + (▒ × 4) + (▒ × 2) + (▒ × 1), where each blank must be filled in by a 0 or a 1. The blanks should be filled in from left to right. 11011

3 Close

Summarize

Explain how to convert from binary to base-10 and from base-10 to binary using the sum of the powers of 2 as the vehicle for the conversions.

To convert from binary to base-10, multiply each binary digit by its corresponding place value and find the sum of the results.

To convert from base-10 to binary, "break apart" the base-10 number as a sum of powers of 2. Start with the highest power of 2 that is not more than the base-10 number.

Problem Solving on Location

Minnesota

Purpose: *To provide additional practice for problem-solving skills in Chapter 1*

Wind Energy

- After problem 2, have students consider the following problem: What is the total number of turbines at the three locations? 296
- After problem 3, have students consider the following problem: The total number of turbines at both of the Lake Benton locations is about 6 times the number at Agassiz Beach. About how many turbines are there at Agassiz Beach? about 50 What skills and operations were needed to solve the problems? estimation, addition, and division

Extension Have students research wind energy. Challenge them to find what portion of their state's energy is provided by wind and other sources of energy. Check students' work.

Problem Solving on Location

MINNESOTA

Wind Energy

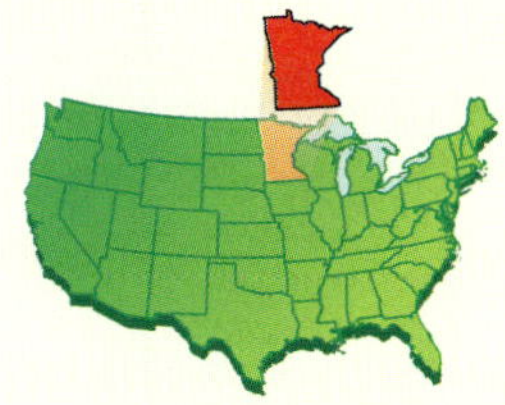

Wind energy has been used for centuries to pump water, grind grain, and power sailing vessels. Today, many communities use windmills, called wind turbines, to generate electricity. Several important wind turbine projects are located in Minnesota.

For 1–7, use the table.

Wind Turbines in Minnesota	
Location	**Turbines**
Lakota Ridge	15
Lake Benton I	143
Lake Benton II	138

1. Order the locations from the greatest number of turbines to the least. Lake Benton I, Lake Benton II, Lakota Ridge
2. How many more turbines are there at Lake Benton I than at Lake Benton II? 5
3. There are about 50 times as many wind turbines in California as there are at the three sites in Minnesota combined. Estimate the number of turbines that are in California. about 15,000

Electricity is measured in kilowatt-hours (kWh). The turbines at Lakota Ridge generate 30 million kWh per year. The turbines at Lake Benton II generate 355 million kWh per year.

4. About how many kilowatt-hours are generated each day at Lake Benton II? about 1 million kWh
5. Estimate the number of kilowatt-hours generated each year by each turbine at Lake Benton II. about 3 million kWh
6. About how many kilowatt-hours are generated each day at Lakota Ridge? about 100,000 kWh
7. How many kilowatt-hours are generated each year by each turbine at Lakota Ridge? 2 million kWh

Fishing

With over 4,500 lakes and rivers and a great variety of fish, Minnesota is a popular fishing and ice fishing spot. Fishing contests have become so popular that rules now limit the number of fishing contests allowed on a lake relative to the size of the lake. The number of fish and the size of the fish in many Minnesota lakes and rivers have decreased. For this reason there are now also rules for the number of fish and the size of fish that anglers are allowed to keep. Many anglers now practice catch-and-release fishing.

Fishing Contests with More Than 50 Boats or 100 People		
Lake Size (acres)	Maximum Number of Fishing Contests	Maximum Number of Contest Days
Less than 2,000	0	0
2,000–4,999	1	6
5,000–14,999	2	8
15,000–55,000	3	10
Greater than 55,000	No limit	No limit

For 1–3, use the table.

1. How many fewer contest days are there for lakes that are 2,500 acres than for lakes that are 25,000 acres? **2 fewer days**
2. How many more fishing contests are allowed on lakes that are 20,000 acres than on lakes that are 2,500 acres? **2 more contests**
3. Leech Lake is the third largest lake entirely within the boundaries of Minnesota. It is 111,527 acres. Is there a limit on the number of fishing contests allowed on this lake? Explain. **There is no limit because the lake is greater than 55,000 acres.**
4. The deepest area of Leech Lake is in Walker Bay, where the depth is 150 ft. If you are fishing in an area that is 35 feet deep, how much deeper is Walker Bay than where you are fishing? **115 ft**
5. Give a possible water temperature for Lake of the Woods if the water temperature in that lake is between 32°F and 39°F. **Possible answer: 35°F, or any other temperature between 32°F and 39°F**

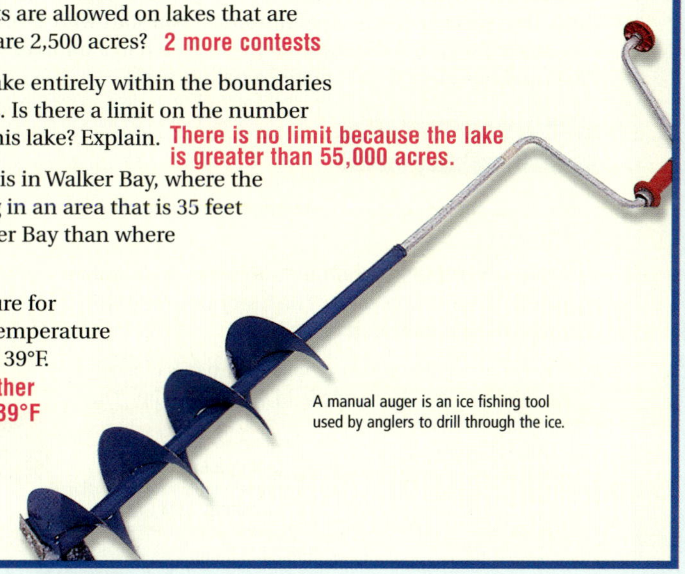

A manual auger is an ice fishing tool used by anglers to drill through the ice.

Fishing

- After problem 1, discuss the following: Cass Lake has an area of 15,596 acres. How many contest days are allowed at Cass Lake? 10 days
- After problem 4, have students consider the following problem: The muskellunge is a popular fish with anglers at Leech Lake. By law, a muskellunge shorter than 40 inches long must be thrown back into the lake. Must a 45-inch muskellunge be thrown back? Explain. Possible answer: No; the 45-inch fish exceeds the minimum length by 5 inches.

Extension Have students research the five largest lakes in their state. Have them present the data in a table and write three questions that could be solved using the data. Check students' work.

MATH-ABLES

Game Resources

Puzzles, Twisters & Teasers

Chapter 1 Resource Book

Palindromes

Purpose: *To apply the skill of addition to create palindrome numbers*

Discuss: Ask students to explain how the trick works. Ask them questions such as the following: How would you create a palindrome starting with the number 157? How many times must you repeat until the sum is a palindrome?
Possible answer: Begin with a number. Reverse its digits and add the resulting number to the original number. Repeat the process until the sum is a palindrome. It takes three steps:

157 + 751 = 908

908 + 809 = 1,717

1,717 + 7,171 = 8,888

Extend: First have students use the trick to create a palindrome by starting with the number 6. Then have them start with 8. Have students note the number of steps until a palindrome is formed. Challenge students to explain why starting with a number from 5 through 9 always yields a palindrome in two steps. Possible answer: 6 + 6 = 12; 12 + 21 = 33. 8 + 8 = 16; 16 + 61 = 77. Starting with a number from 5 through 9 yields a sum whose tens place is 1 and whose ones place is between 0 and 8. Interchanging the digits and adding means adding the same two numbers in different orders. There is no regrouping, so the sum will always be a palindrome.

Spin-a-Million

Purpose: *To enhance the study of place value by creating numbers from numerals on a spinner*

Discuss: Ask students what strategy they can use to help them win. Place smaller numerals in the smaller place values. Reserve the higher place values for larger numerals.

Extend: Given the numbers 2, 4, 3, 7, and 9, what is the largest number that could be formed in this game? the least number? 97,432; 23,479

MATH-ABLES

Palindromes

A *palindrome* is a word, phrase, or number that reads the same forward and backward.

Examples:

race car Madam, I'm Adam. 3710173

You can turn almost any number into a palindrome with this trick.

Think of any number. 283
Now add that number in reverse. + 382
665

Use the sum to repeat the previous step and keep repeating until the final sum is a palindrome.

665
+ 566
1,231

1,231
+ 1,321
2,552

It took only three steps to create a palindrome by starting with the number 283. What happens if you start with the number 196? Do you think you will ever create a palindrome if you start with 196? One man who started with 196 did these steps until he had a number with 70,928 digits and he still had not created a palindrome!

Spin-a-Million

The object of this game is to create the number closest to 1,000,000.

Taking turns, spin the pointer and write the number on your place-value chart. The number cannot be moved once it has been placed.

After six turns, the player whose number is closest to one million wins the round and scores a point. The first player to get five points wins the game.

internet connect
go.hrw.com
Go to ***go.hrw.com*** for a spinner and place value chart.
KEYWORD: MR4 Game1

Technology LAB

Find a Pattern in Sequences

internet connect
Lab Resources Online
go.hrw.com
KEYWORD: MR4 TechLab1

The numbers 4, 7, 10, 13, 16, 19, … form a sequence. To continue the sequence, identify a pattern. Here is a possible pattern:

4, 4 + 3 = 7, 7 + 3 = 10, 10 + 3 = 13,…

Activity

Use a spreadsheet to generate the first seven terms of the sequence above.

To start with 4, type **4** in cell A1.

To add 3 to the value in cell A1, type **=A1 + 3** in cell B1.

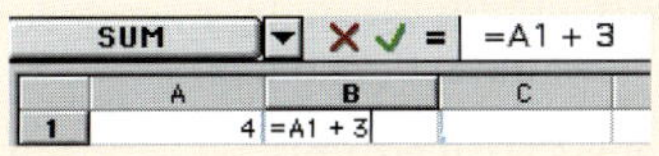

Press ENTER.

To continue the sequence, click the square in the lower right corner of cell B1, hold down the mouse button, and drag the cursor across through cell G1.

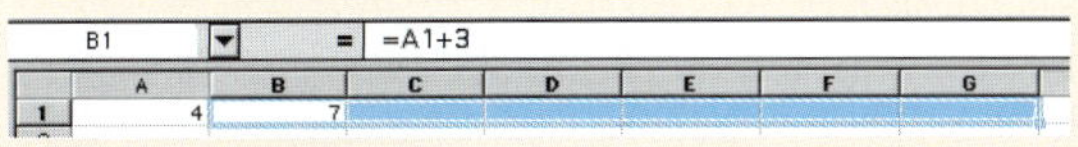

When you release the mouse button, A1 through G1 will list the first seven terms of the sequence.

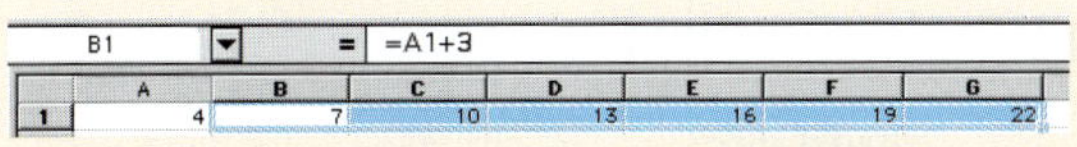

Think and Discuss

1. How do you use a sequence's pattern when you use your spreadsheet to generate the terms? **Possible answer: Enter the formula for the pattern into one cell. Then copy the formula into as many other cells as necessary.**

Try This

Identify a pattern in each sequence. Then use a spreadsheet to generate the first 12 terms.

1. 9, 14, 19, 24, 29, 34, …
2. 7, 13, 19, 25, 31, 37, …

Answers

Try This

1. 39, 44, 49, 54, 59, 64
2. 43, 49, 55, 61, 67, 73

Technology LAB

Find a Pattern in Sequences

Objective: To use a spreadsheet to perform repeated operations that generate a sequence

Materials: Spreadsheet software

Lab Resources

Technology Lab Activities . . . . p. 4

Using the Page

This technology activity shows students how to perform the same operation repeatedly, which can be done using a spreadsheet. Specific instructions may vary, depending on the spreadsheet program used. The instructions given are for Microsoft Excel. For instructions to other software, visit go.hrw.com.

The Think and Discuss problem can be used to assess students' understanding of the technology activity. Although Try This problems 1 and 2 can be done without a spreadsheet, they are meant to help students become familiar with using a spreadsheet to generate sequences.

Assessment

Suppose you entered 6 in cell A1, then entered =A1 + 4 in cell B1, and then pressed **ENTER**.

1. What number would appear in cell B1? 10
2. How do you use the spreadsheet to generate the next 4 terms? Click the fill handle (box in lower right corner of the cell) in cell B1, hold down, and drag across through cell F1.

Chapter 1 Study Guide and Review

Purpose: *To help students review and practice concepts and skills presented in Chapter 1*

Assessment Resources

Chapter Review

Chapter 1 Resource Book . . . pp. 72–73

***Test and Practice Generator* CD-ROM**

Additional review items in both multiple-choice and free-response format may be generated for any objective in Chapter 1.

Answers

1. sequence, term
2. base, exponent
3. order of operations
4. evaluate
5. 8,731; 8,735; 8,737; 8,740
6. 53,337; 53,341; 53,452; 53,456
7. 8,791; 81,790; 87,091; 87,901
8. 2,651; 22,561; 25,615; 26,551
9. 91,363; 93,613; 96,361; 96,631
10. 10,101; 10,110; 11,010; 11,110

Chapter 1 Study Guide and Review

Vocabulary

Term	Page	Term	Page
Associative Property	24	numerical expression	20
base	12	order of operations	20
Commutative Property	24	overestimate	8
compatible number	8	perfect square	31
Distributive Property	25	sequence	31
evaluate	20	term	31
exponent	12	underestimate	8
exponential form	12		

Complete the sentences below with vocabulary words from the list above. Words may be used more than once.

1. An ordered set of numbers is called a(n) __?__. Each number in a sequence is called a(n) __?__.
2. In the expression 8^5, 8 is the __?__, and 5 is the __?__.
3. The __?__ is a set of rules used to evaluate an expression that contains more than one operation.
4. When you __?__ a numerical expression, you find its value.

1-1 Comparing and Ordering Whole Numbers (pp. 4–7)

EXAMPLE

■ **Order the numbers from least to greatest.**

4,913; 4,931; 4,391

4,913 4,913 < 4,931
4,931

4,931 4,391 < 4,931
4,391

4,913 4,391 < 4,913
4,391

4,391 < 4,913 < 4,931

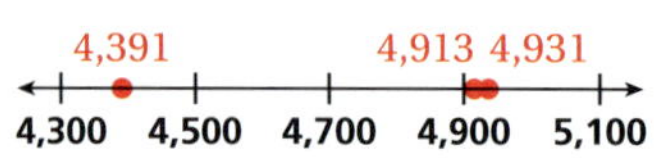

EXERCISES

Order the numbers from least to greatest.

5. 8,731; 8,737; 8,735; 8,740
6. 53,341; 53,337; 53,456; 53,452
7. 87,091; 8,791; 87,901; 81,790
8. 26,551; 25,615; 2,651; 22,561
9. 96,361; 96,631; 93,613; 91,363
10. 10,101; 11,010; 10,110; 11,110

1-2 Estimating with Whole Numbers (pp. 8–11)

EXAMPLES

- Estimate the sum 837 + 710 by rounding to the hundreds place.
  $800 + 700 = 1{,}500$
  The sum is about 1,500.
- Estimate the quotient of 148 and 31.
  $150 \div 30 = 5$
  The quotient is about 5.

EXERCISES

Estimate by rounding to the place value indicated.

11. 4,671 − 3,954; thousands

12. 3,123 + 2,987; thousands

13. 53,465 − 27,465; ten thousands

14. Ralph has 38 photo album sheets with 22 baseball cards in each sheet. About how many baseball cards does he have?

1-3 Exponents (pp. 12–15)

EXAMPLES

- **Write 6 × 6 in exponential form.**
  6^2 *6 is a factor 2 times.*

Find each value.

- 5^2
  $5^2 = 5 \times 5$
  $= 25$
- 6^3
  $6^3 = 6 \times 6 \times 6$
  $= 216$

EXERCISES

Write each expression in exponential form.

15. $5 \times 5 \times 5$	**16.** $3 \times 3 \times 3 \times 3$
17. $7 \times 7 \times 7 \times 7 \times 7$	**18.** 8×8
19. $4 \times 4 \times 4 \times 4$	**20.** $1 \times 1 \times 1$

Find each value.

21. 4^4	**22.** 2^4
23. 3^3	**24.** 1^5
25. 5^3	**26.** 10^2

1-4 Order of Operations (pp. 20–23)

EXAMPLE

- **Evaluate $8 \div (7 - 5) \times 2^2 - 2 + 9$.**

$8 \div (7 - 5) \times 2^2 - 2 + 9$

$8 \div 2 \times 2^2 - 2 + 9$ *Subtract in parentheses.*

$8 \div 2 \times 4 - 2 + 9$ *Simplify the exponent.*

$4 \times 4 - 2 + 9$ *Divide.*

$16 - 2 + 9$ *Multiply.*

$14 + 9$ *Subtract.*

23 *Add.*

EXERCISES

Evaluate each expression.

27. $9 \times 8 - 13$

28. $21 \div 3 + 4$

29. $6 + 4 \times 5$

30. $19 - 12 \div 6$

31. $30 \div 2 - 5 \times 2$

32. $(7 + 3) \div 2 \times 3^2$

33. $8 \times (7 + 5) \div 4^2 + 9 \div 3$

34. $3^2 \times 5 \div (10 \times 3 \div 2)$

Study Guide and Review

Answers

11. 1,000
12. 6,000
13. 20,000
14. 800
15. 5^3
16. 3^4
17. 7^5
18. 8^2
19. 4^4
20. 1^3
21. 256
22. 16
23. 27
24. 1
25. 125
26. 100
27. 59
28. 11
29. 26
30. 17
31. 5
32. 45
33. 9
34. 3

Answers

35. 30

36. 520

37. 80

38. 1,080

39. 40

40. 320

41. 100

42. 130

43. 168

44. 135

45. 204

46. 152

47. 216

48. 165

49. 52

50. 423

51. 62°

52. 186 bars

53. 17 and 28

54. 13 and 16

55. 40 and 120

56. 32 and 64

1-5 Mental Math (pp. 24–27)

EXAMPLES

Find each sum or product.

- $4 + 13 + 6 + 7$
  $4 + 6 + 13 + 7$
  $(4 + 6) + (13 + 7)$
  $10 + 20$
  30

- $5 \times 9 \times 6$
  $5 \times 6 \times 9$
  $(5 \times 6) \times 9$
  30×9
  270

- Use the Distributive Property to find the product.

  3×16
  $3 \times 16 = 3 \times (10 + 6)$
  $= (3 \times 10) + (3 \times 6)$
  $= 30 + 18$
  $= 48$

EXERCISES

Find each sum or product.

35. $9 + 5 + 1 + 15$ **36.** $8 \times 13 \times 5$
37. $31 + 16 + 19 + 14$ **38.** $6 \times 12 \times 15$
39. $17 + 12 + 8 + 3$ **40.** $16 \times 5 \times 4$
41. $11 + 23 + 27 + 39$ **42.** $13 \times 5 \times 2$

Use the Distributive Property to find each product.

43. 7×24 **44.** 9×15
45. 6×34 **46.** 8×19
47. 8×27 **48.** 5×33
49. 4×13 **50.** 9×47

1-6 Choose the Method of Computation (pp. 28–30)

EXAMPLE

- The average annual rainfall in Washington, D.C., is 39 inches. How much rain does Washington, D.C., average in 8 years?

  You may not be able to quickly multiply the numbers in your head, but the numbers are not so big that you must use a calculator. Use pencil and paper to find the answer.
  $39 \times 8 = 312$ inches

EXERCISES

51. The average high temperature for Washington, D.C., in January is 42°F. The record high temperature for Washington, D.C., is 104°F. How much hotter is the record temperature than the average high temperature in January?

52. There are 6 members on Lynn's chess team. If Lynn wants to give each member 31 mini chocolate bars—one for each day of the month—how many chocolate bars will she need?

1-7 Find a Pattern (pp. 31–33)

EXAMPLES

- Find the next two terms in the sequence.

  1, 3, 4, 7, ■, ■, . . .
  $1 + 3 = 4$ $4 + 7 = 11$
  $3 + 4 = 7$ $7 + 11 = 18$
  The next two numbers are 11 and 18.

EXERCISES

Find the next two terms in each sequence.

53. 1, 5, 6, 11, ■, ■, . . .
54. 1, 4, 7, 10, ■, ■, . . .
55. 1, 3, 4, 12, 13, 39, ■, ■, . . .
56. 2, 4, 8, 16, ■, ■, . . .

Chapter Test

Chapter 1

Solve.

1. Which number is greater, 16,880,953 or 16,221,773? **16,880,953**
2. Which number is greater, 22,481,093 or 23,662,840? **23,662,840**

Order the numbers from least to greatest.

3. 801; 798; 921 **798; 801; 921**
4. 4,835; 7,505; 4,310 **4,310; 4,835; 7,505**

Estimate each sum or difference by rounding to the place value indicated.

5. 8,743 + 3,198; thousands **12,000**
6. 62,524 − 17,831; ten thousands **40,000**

Estimate.

7. Kaitlin's family is planning a trip from Washington, D.C., to New York City. New York City is 227 miles from Washington, D.C., and the family can drive an average of 55 mi/h. About how long will the trip take? **about 4 hours**

Write each expression in exponential form.

8. $4 \times 4 \times 4 \times 4 \times 4$ **4^5**
9. $10 \times 10 \times 10$ **10^3**
10. $6 \times 6 \times 6 \times 6$ **6^4**

Find each value.

11. 2^3 **8**
12. 5^2 **25**
13. 4^4 **256**
14. 11^2 **121**

Evaluate each expression.

15. $12 + 8 \div 2$ **16**
16. $3^2 \times 5 + 10 - 7$ **48**
17. $12 + (28 - 15) + 4 \times 2$ **33**

Find each sum or product.

18. $15 + 23 + 47 + 5$ **90**
19. $5 \times 48 \times 2$ **480**
20. $44 + 18 + 12 + 6$ **80**

Use the Distributive Property to find the product.

21. 3×32 **96**
22. 52×6 **312**
23. 24×5 **120**
24. 81×6 **486**
25. At 5:00 A.M., the temperature was 41°F. By noon, the temperature was 69°F. By how many degrees did the temperature increase? **28°F**

Identify a pattern in each sequence and name the missing terms.

26. 8, 22, 36, 50, ■, ■, ■, . . . **add 14; 64, 78, 92**
27. 2, 3, 5, 8, 12, ■, ■, ■, . . . **add consecutive whole numbers beginning with 1; 17, 23, 30**

Chapter Test

Chapter Test

Chapter 1

Purpose: *To assess students' mastery of concepts and skills in Chapter 1*

Assessment Resources

Chapter 1 Tests (Levels A, B, C)
Assessment Resources . . . . . pp. 31–36

***Test and Practice Generator* CD-ROM**

Additional assessment items in both multiple-choice and free-response format may be generated for any objective in Chapter 1.

Chapter 1 Performance Assessment

***Purpose:** To assess students' understanding of concepts in Chapter 1 and combined problem-solving skills*

Assessment Resources

Performance Assessment
Assessment Resources . . . . p. 104

Performance Assessment Teacher Support
Assessment Resources . . . . p. 103

Answers

1–3. See p. A1.

4. See Level 3 work sample below.

Scoring Rubric for Problem Solving Item 4

Level 3

Accomplishes the purposes of the task.

Student gives clear explanations, shows understanding of mathematical ideas and processes, and computes accurately.

Level 2

Purposes of the task not fully achieved.

Student demonstrates satisfactory but limited understanding of the mathematical ideas and processes.

Level 1

Purposes of the task not accomplished.

Student shows little evidence of understanding the mathematical ideas and processes and makes computational and/or procedural errors.

Chapter 1 Performance Assessment

Show What You Know

Create a portfolio of your work from this chapter. Complete this page and include it with your four best pieces of work from Chapter 1. Choose from your homework or lab assignments, mid-chapter quiz, or any journal entries you have done. Put them together using any design you want. Make your portfolio represent what you consider your best work.

Short Response

1. Four people go to see a movie, and each person recommends it to 4 other friends, who then go to see it. Then each of those 4 recommends the movie to 4 other friends. If the pattern continues, how many rounds are needed for at least 1 million people to have seen the movie? Show the steps necessary to find the answer.
2. Create your own sequence using exponents. Explain how you created the sequence, and give the rule for extending the pattern.
3. Create an order-of-operations problem of at least five steps that results in an answer of 7. Be sure to include parentheses and an exponent, and use each operation only once. Show how to get the answer, and explain the steps you used to simplify the expression.

Extended Problem Solving

4. A movie's *gross* is the amount of money the movie makes at the box office, not counting the cost of advertising and distributing. The first four movies of the *Star Wars* series are among the top 20 grossing films worldwide.
   a. Use the information in the table to calculate the total gross of the four films. Show your work.
   b. What is the average gross of the four movies? Show your work.
   c. Explain why you agree or disagree with the following statement. "The first four *Star Wars* movies were more successful in the United States than outside the United States."

Box Office Sales (1977–1999)

Movie	U.S. Gross (millions)	Foreign Gross (millions)
Star Wars: Episode 1: The Phantom Menace	\$431	\$492
Star Wars	\$461	\$323
Return of the Jedi	\$309	\$264
The Empire Strikes Back	\$290	\$223

Student Work Samples for Item 4

Level 3

a. \$431 + \$492 + \$461 + \$323 + \$309 + \$264 + \$290 + \$223 = \$2,793

b. $\frac{\$2,793}{4}$ = \$698.25

c. I disagree with the statement because Star Wars: Episode I grossed more outside the U.S.

The student correctly found the total gross and average gross of the movies. The student correctly answered part c in a sentence.

Level 2

a. \$431
492
461
323
309
264
290
223
\$2,793

b. 349.125 / 8)2793.000 = \$349.13

c. I disagree because Episode I grossed less in the United States.

The student correctly found the total gross, but divided by 8 to find the average gross. The student correctly answered part c in a sentence.

Level 1

a. Episode I \$431 + \$492 = \$923

b. Episode I \$923 ÷ 2 = \$461.50

c. US \$431
Foreign \$309
I agree because it grossed more in the US.

The student found the total gross and average gross of one movie. The student did not compare the U.S. and foreign gross of each movie.

internet connect
State-Specific Test Practice Online
go.hrw.com Keyword: MR4 TestPrep

Standardized Test Prep

Chapter 1

Cumulative Assessment, Chapter 1

1. Which number is the greatest? B
   (A) 6,568,217 (C) 6,701,953
   (B) 6,739,549 (D) 6,589,211

2. What is five billion, two hundred fifty-two million, six hundred thousand, three hundred eleven in standard form? H
   (F) 5,252,603,011 (H) 5,252,600,311
   (G) 52,526,311 (J) 5,252,060,311

TIP! TEST TAKING TIP!
An exponent tells how many times a number called the base is used as a factor.

3. What is the value of 4^3? C
   (A) 12 (C) 64
   (B) 16 (D) 81

4. What is $5 \times 5 \times 5$ written in exponential form? J
   (F) 125 (H) 3^5
   (G) $100 + 20 + 5$ (J) 5^3

5. What is the value of $7 \times 3 + 2$? A
   (A) 23 (C) 35
   (B) 12 (D) 42

6. What is the value of $8^2 - (12 + 3) \times 2$? G
   (F) 98 (H) 110
   (G) 34 (J) 58

7. $6 \times 3 \times 4 = 3 \times 6 \times 4$ is an example of which property? B
   (A) Associative (C) Distributive
   (B) Commutative (D) Exponential

8. The bar graph shows the number of miles Jan biked each day last week. How many total miles did she bike? H

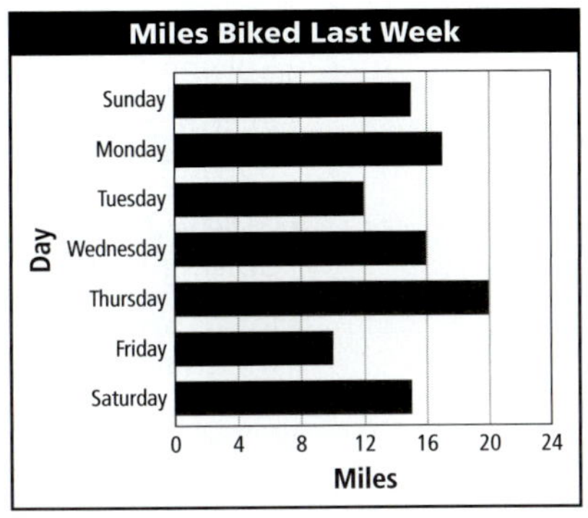

   (F) 15 miles
   (G) 17 miles
   (H) 105 miles
   (J) 110 miles

9. ***SHORT RESPONSE*** Find a pattern in the sequence 3, 6, 12, 24, 48, ... Use your pattern to find the next two terms.

10. ***SHORT RESPONSE*** Explain what 5^3 means. What is the value of 5^3?

Standardized Test Prep

Standardized Test Prep

Chapter 1

Purpose: *To provide review and practice for Chapter 1 and standardized tests*

Assessment Resources

Cumulative Tests (Levels A, B, C)
Assessment Resources . . . . pp. 127–138

State-Specific Test Practice Online
KEYWORD: MR4 TestPrep

Test Prep Doctor

Expand on the test-taking tip given for item 3 by reminding students that 4 is the base and 3 is the exponent. So $4^3 = 4 \times 4 \times 4$.

Point out to students that in item 4, answers **F** and **G** can be eliminated because they are not in exponential form. Answer **H** can be eliminated because the base is 5. The correct answer must be 5^3.

Answers

9. Multiply each term by 2 to find the next term; 96, 192

10. 5^3 means to use 5 as a factor 3 times; 125

Chapter 2

Introduction to Algebra

Section 2A

Understanding Variables and Expressions

Lesson 2-1
Variables and Expressions

Lesson 2-2
Translate Between Words and Math

Section 2B

Equations

Lesson 2-3
Equations and Their Solutions

Lesson 2-4
Solving Addition Equations

Lesson 2-5
Solving Subtraction Equations

Lesson 2-6
Solving Multiplication Equations

Lesson 2-7
Solving Division Equations

Extension
Inequalities

Pacing Guide for 45-Minute Classes

Chapter 2

DAY 12	DAY 13	DAY 14	DAY 15	DAY 16
Lesson 2-1	Lesson 2-2	Mid-Chapter Quiz Lesson 2-3	Lesson 2-4	Lesson 2-5
DAY 17	**DAY 18**	**DAY 19**	**DAY 20**	**DAY 21**
Lesson 2-6	Lesson 2-7	Extension	Chapter 2 Review	Chapter 2 Assessment

Pacing Guide for 90-Minute Classes

Chapter 2

DAY 6	DAY 7	DAY 8	DAY 9	DAY 10
Chapter 1 Assessment Lesson 2-1	Lesson 2-2 Lesson 2-3	Mid-Chapter Quiz Lesson 2-4 Lesson 2-5	Lesson 2-6 Lesson 2-7	Extension Chapter 2 Review
DAY 11				
Chapter 2 Assessment Hands-On Lab 3A				

Across the Series

HARCOURT GRADE 5

- Write and evaluate numerical and algebraic expressions involving whole numbers.
- Solve one-step whole number equations involving addition or multiplication.

COURSE 1

- **Identify and evaluate expressions.**
- **Translate between words and math.**
- **Determine whether a number is a solution of an equation.**
- **Solve one-step whole-number equations.**
- **Solve and graph one-step whole-number inequalities.**

COURSE 2

- Write and evaluate algebraic expressions that include exponents.
- Determine whether a number is a solution of an equation.
- Solve one-step equations.
- Solve and graph one-step inequalities.

Across the Curriculum

LANGUAGE ARTS LINK

Math: Reading and Writing in the Content Area . . . . pp. 8–14

Focus on Problem Solving
Understand the Problem . . . . SE p. 57

Journal . . . . TE pp. 51, 55, 61, 65, 72

Write About It . . . . SE, last page of each lesson

SOCIAL STUDIES LINK

Social Studies . . . . SE pp. 52, 63, 68

History . . . . SE p. 65

Geography . . . . SE pp. 68, 77

SCIENCE LINK

Science . . . . SE p. 55

Life Science . . . . SE pp. 59, 72

Physical Science . . . . SE pp. 65, 74

Earth Science . . . . SE p. 75

TE = *Teacher's Edition* **SE** = *Student Edition*

Interdisciplinary

Bulletin Board

Social Studies

You can get an idea about population density from this satellite photo of Earth. The population density of the United States is 74 people per square mile, which is 40 more people per square mile than in Argentina. Write and solve an equation to find the population density of Argentina.

$74 = p + 40$; $p = 34$ people/mi^2

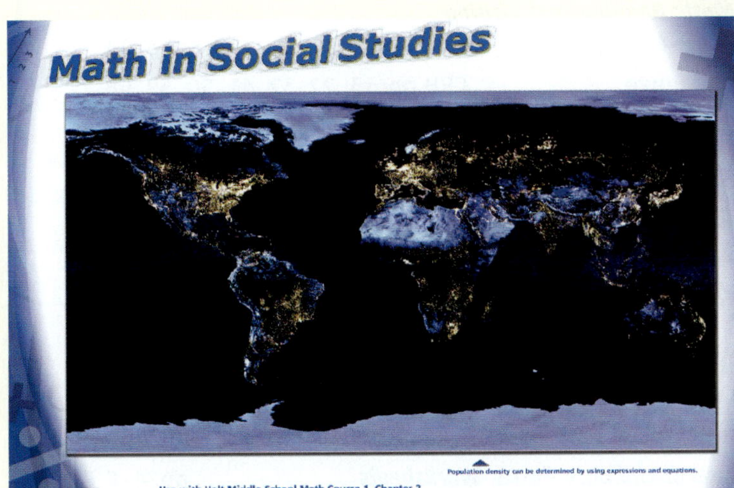

Interdisciplinary posters and worksheets are provided in your resource material.

Chapter 2

Resource Options

Chapter 4 Resource Book

Student Resources

Practice (Levels A, B, C) .... pp. 9–11, 18–20, 28–30, 37–39, 46–48, 55–57, 64–66
Reteach .... pp. 12, 21, 31, 40, 49, 58, 67
Challenge .... pp. 13, 22, 32, 41, 50, 59, 68
Problem Solving .... pp. 14, 23, 33, 42, 51, 60, 69
Puzzles, Twisters & Teasers ..... pp. 15, 24, 34, 43, 52, 61, 70
Recording Sheets .... pp. 3–4, 8, 17, 27, 36, 45, 54, 63, 73, 76, 78
Chapter Review .... pp. 71–72

Teacher and Parent Resources

Chapter Planning and Pacing Guide .... p. 5
Section Planning Guides .... pp. 6, 25
Parent Letter .... pp. 1–2
Teaching Tools .... pp. 76–78
Teacher Support for Chapter Project .... p. 74
Transparencies .... pp. T1–T26
- Daily Transparencies
- Additional Examples Transparencies
- Teaching Transparencies

Reaching All Learners

English Language Learners

Success for English Language Learners .... pp. 15–28
Math: Reading and Writing in the Content Area .... pp. 8–14
Spanish Homework and Practice .... pp. 8–14
Spanish Interactive Study Guide .... pp. 8–14
Spanish Family Involvement Activities .... pp. 9–16
Multilingual Glossary

Individual Needs

Are You Ready? Intervention and Enrichment pp. 145–148, 237–240, 407–408
Alternate Openers: Explorations .... pp. 8–14
Family Involvement Activities .... pp. 9–16
Interactive Problem Solving .... pp. 8–14
Interactive Study Guide .... pp. 8–14
Readiness Activities .... pp. 3–4
Math: Reading and Writing in the Content Area .... pp. 8–14
Challenge .... CRB pp. 13, 22, 32, 41, 50, 59, 68

Hands-On

Hands-On Lab Activities .... pp. 6–9
Technology Lab Activities .... pp. 5–9
Alternate Openers: Explorations .... pp. 8–14
Family Involvement Activities .... pp. 9–16

Applications and Connections

Consumer and Career Math .... ppp. 5–8
Interdisciplinary Posters .... Poster 2, TE p. 46B
Interdisciplinary Poster Worksheets .... pp. 4–6

Transparencies

Alternate Openers: Explorations .... pp. 8–14
Exercise Answers Transparencies
Chapter 1 Resource Book .... pp. T1–T26
- Daily Transparencies
- Additional Examples Transparencies
- Teaching Transparencies

Technology

Teacher Resources

***Lesson Presentations* CD-ROM** .... Chapter 2
***Test and Practice Generator* CD-ROM** .... Chapter 2
***One-Stop Planner* CD-ROM** .... Chapter 2

Student Resources

***Are You Ready? Intervention* CD-ROM**
Skills 34, 57

internet connect

go.hrw.com

Homework Help Online	**KEYWORD:** MR4 HWHelp2
Math Tools Online	**KEYWORD:** MR4 Tools
Glossary Online	**KEYWORD:** MR4 Glossary
Chapter Project Online	**KEYWORD:** MR4 PSProject2
Chapter Opener Online	**KEYWORD:** MR4 Ch2
CNN Student News™	**KEYWORD:** MR4 CNN2

SE = *Student Edition* **TE** = *Teacher's Edition* **AR** = *Assessment Resources* **CRB** = *Chapter Resource Book* **MK** = *Manipulatives Kit*

Chapter 2

Assessment Options

Assessing Prior Knowledge

Determine whether students have the required prerequisite concepts and skills.

Are You Ready? . . . . SE p. 47
Inventory Test . . . . AR pp. 1–4

Test Preparation

Provide review and practice for chapter and standardized tests.

Standardized Test Prep . . . . SE p. 87
Spiral Review with Test Prep . . . . . SE, last page of each lesson
Study Guide and Review . . . . SE pp. 82–84
Test Prep Tool Kit

Technology

***Test and Practice Generator* CD-ROM**

internet connect

State-Specific Test Practice Online KEYWORD: MR4 TestPrep

go.hrw.com

Performance Assessment

Assess students' understanding of chapter concepts and combined problem-solving skills.

Performance Assessment . . . . SE p. 86
Includes scoring rubric in TE
Performance Assessment . . . . AR p. 106
Performance Assessment Teacher Support . . . . AR p. 105

Portfolio

Portfolio opportunities appear throughout the Student and Teacher's Editions.

Suggested work samples:

Problem Solving Project . . . . TE p. 46
Performance Assessment . . . . SE p. 86
Portfolio Guide . . . . AR p. xxxv
Journal . . . . TE pp. 51, 55, 61, 65, 72
Write About It . . . . SE, last page of each lesson

Daily Assessment

Obtain daily feedback on students' understanding of concepts.

Spiral Review and Test Prep . . . . . . SE, last page of each lesson

Also Available on Transparency in Chapter 2 Resource Book

Warm Up . . . . TE, first page of each lesson
Problem of the Day . . . . TE, first page of each lesson
Lesson Quiz . . . . TE, last page of each lesson

Student Self-Assessment

Have students evaluate their own work.

Group Project Evaluation . . . . AR p. xxxii
Individual Group Member Evaluation . . . . AR p. xxxiii
Portfolio Guide . . . . AR p. xxxv
Journal . . . . TE pp. 51, 55, 61, 65, 72

Formal Assessment

Assess students' mastery of concepts and skills.

Section Quizzes . . . . AR pp. 7–8
Mid-Chapter Quiz . . . . SE p. 56
Chapter Test . . . . SE p. 85
Chapter Tests (Levels A, B, C) . . . . AR pp. 37–42
Cumulative Tests (Levels A, B, C) . . . . AR pp. 139–150
Standardized Test Prep
Cumulative Assessment . . . . SE p. 87
End-of-Year Test . . . . AR pp. 271–274

Technology

***Test and Practice Generator* CD-ROM**

Make tests electronically. This software includes:

- Dynamic practice for Chapter 2
- Customizable tests
- Multiple-choice items for each objective
- Free-response items for each objective
- Teacher management system

SE = *Student Edition* **TE** = *Teacher's Edition* **AR** = *Assessment Resources* **CRB** = *Chapter Resource Book* **MK** = *Manipulatives Kit*

Chapter 2

Chapter 2 Tests

Three levels (A,B,C) of tests are available for each chapter in the *Assessment Resources.*

Level A

Chapter 2 Chapter Test
Form A

Evaluate each expression to find the missing values in the tables.

1.

w	$w + 5$
3	8
4	9
5	10

2.

x	$\frac{12}{x}$
2	6
4	3
6	2

3.

a	$3a$
2	6
5	15
8	24

Find an expression for each table.

4.

w	$w - 4$
20	16
18	14
16	12

5.

y	$4y$
7	28
8	32
9	36

6.

z	$z + 8$
10	18
12	20
14	22

7. A peach has 85 calories. It has w calories more than an orange. Write an expression for the number of calories in an orange.
$85 - w$

8. A teacher has 24 students in her class. She wants to divide them into 4 equal groups. Write an expression to represent the number of students in each group.
$24 \div 4$

Write each phrase as a numerical or algebraic expression.

9. w plus 5 $w + 5$
10. take away x from 8 $8 - x$
11. the product of 10 and 7 $10 \cdot 7$
12. w divided by 5 $\frac{w}{5}$

Chapter 2 Chapter Test
Form A, continued

Write two phrases for each expression.

13. $f + 12$
Possible answer: twelve more than f; 12 added to f
14. $w \div 8$
Possible answer: w divided by 8; the quotient of w and 8
15. $110 - x$
Possible answer: take away x from 110; x less than 110
16. $5w$
Possible answer: the product of 5 and w; five times w

Determine whether the given value of the variable is a solution.

17. $7w = 21$ for $w = 3$
yes
18. $32 = 18 + y$ for $y = 11$
no
19. $12 + w = 30$ for $w = 18$
yes
20. $\frac{32}{q} = 16$ for $q = 16$
no

Solve each equation.

21. $6 + x = 12$ $x = 6$
22. $v + 8 = 26$ $v = 18$
23. $34 + m = 110$ $m = 76$
24. $18 = j - 10$ $j = 28$
25. $f - 9 = 10$ $f = 19$
26. $h - 12 = 18$ $h = 30$
27. $6h = 36$ $h = 6$
28. $16 = 4r$ $r = 4$
29. $7w = 49$ $w = 7$
30. $\frac{w}{4} = 7$ $w = 28$
31. $\frac{y}{5} = 6$ $y = 30$
32. $\frac{z}{11} = 7$ $z = 77$
33. Lina bought x rolls of film at $5.00 per roll and spent $25. How many rolls of film did Lina buy?
5 rolls
34. You plan to buy a DVD player 10 weeks from now. If the DVD player costs $260, how much money must you save each week in order to buy it?
$26.00

Level B

Chapter 2 Chapter Test
Form B

Evaluate each expression to find the missing values in the tables

1.

w	$w + 12$
3	15
4	16
5	17

2.

x	$\frac{x}{5}$
5	1
10	2
15	3

3.

a	$8a + 1$
3	25
5	41
8	65

Find an expression for each table.

4.

w	$w - 12$
20	8
18	6
16	4

5.

y	$8y$
7	56
8	64
9	72

6.

z	$z + 21$
10	31
20	41
40	61

7. The temperature at noon on an August day in Houston was 89°F. After a thunderstorm, it dropped 5°F. Write an expression for the new temperature.
$89 - 5$

8. Company A has 50 more employees than Company B. Let n be the number of employees in Company B. Write an expression for the number of employees in Company A.
$n + 50$

Write each phrase as a numerical or algebraic expression.

9. 32 more than w
$w + 32$
10. the product of 12 and 6
$12 \cdot 6$
11. 84 less than z
$z - 84$
12. the quotient of r and 140
$\frac{r}{140}$

Chapter 2 Chapter Test
Form B, continued

Write two phrases for each expression.

13. $f - 12$
Possible answer: twelve less than f; 12 subtracted from f
14. $w \div 15$
Possible answer: w divided by 15; the quotient of w and 15
15. $110 + x$
Possible answer: 110 plus x; x more than 110
16. $15w$
Possible answer: the product of 15 and w; fifteen times w

Determine whether the given value of the variable is a solution.

17. $12w = 144$ for $w = 12$
yes
18. $142 = 128 + y$ for $y = 14$
yes
19. $184 + w = 230$ for $w = 414$
no
20. $\frac{w}{64} = 23$ for $w = 16$
no

Solve each equation.

21. $86 + x = 112$ $x = 26$
22. $v + 108 = 126$ $v = 18$
23. $134 + m = 165$ $m = 31$
24. $108 = j - 10$ $j = 118$
25. $f - 49 = 25$ $f = 74$
26. $h - 52 = 120$ $h = 172$
27. $16h = 128$ $h = 8$
28. $84 = 4r$ $r = 21$
29. $17w = 187$ $w = 11$
30. $\frac{w}{14} = 7$ $w = 98$
31. $\frac{y}{35} = 9$ $y = 315$
32. $\frac{z}{25} = 12$ $z = 300$
33. Lina had x rolls of film developed for the yearbook staff at $8.00 per roll and spent $120. How many rolls of film did Lina have developed?
15 rolls
34. You plan to buy a DVD player 15 weeks from now. If the DVD player costs $180, how much money must you save each week in order to buy it?
$12.00

Level C

Chapter 2 Chapter Test
Form C

Evaluate each expression to find the missing values in the tables.

1.

w	$2w + 3$
3	9
4	11
5	13

2.

x	$\frac{12}{x} + 1$
2	7
4	4
6	3

3.

a	$3a - 4$
12	32
15	41
19	53

Find an expression for each table.

4.

w	$2w - 15$
8	1
12	9
16	17

5.

y	$9y$
7	63
12	108
17	153

6.

z	$z + 21$
10	31
31	52
51	72

7. 30 CDs weighing a total of d ounces are packed in a box that weighs b ounces. Find the weight of 5 packed boxes.
$5d + 5b$

8. The diameter of a silo is 4 feet less than the height. Let h represent the height. Write an expression for the diameter.
$h - 4$

Write each phrase as a numerical or algebraic expression.

9. 17 more than 51
$51 + 17$
10. x subtracted from 8, plus 12
$8 - x + 12$
11. the sum of $12x$ and 4
$12x + 4$
12. w divided by 5, subtracted from 9
$9 - \frac{w}{5}$

Chapter 2 Chapter Test
Form C, continued

Write two phrases for each expression.

13. $21 + y$
Possible answer: 21 plus y; y more than 21
14. $x - y$
Possible answer: the difference of x and y; y less than x
15. $110 \div 2x$
Possible answer: 110 divided by $2x$; the quotient of 110 and $2x$
16. $5w + 3$
Possible answer: the product of 5 and w, plus 3; five times w, plus 3

Determine whether the given value of the variable is a solution.

17. $17w = 221$ for $w = 3$
no
18. $132 = 18 + y$ for $y = 114$
yes
19. $w - 123 = 230$ for $w = 107$
no
20. $\frac{24}{w} = 8$ for $w = 3$
yes

Solve each equation.

21. $112 + x = 162$ $x = 50$
22. $v + 150 = 326$ $v = 176$
23. $234 + m = 243$ $m = 9$
24. $j - 26 = 126$ $j = 152$
25. $f + 94 = 184$ $f = 90$
26. $h - 436 = 23$ $h = 459$
27. $16h = 192$ $h = 12$
28. $112 = 4r$ $r = 28$
29. $71w = 213$ $w = 3$
30. $\frac{w}{17} = 7$ $w = 119$
31. $\frac{y}{12} = 15$ $y = 180$
32. $\frac{z}{16} = 8$ $z = 128$
33. A plumber needs 3 times as much solid pipe as perforated pipe to lay a drain line 200 feet long. How much solid pipe is needed?
150 feet
34. The length of a garden is 2 times the width. The perimeter of the garden is 144 feet. What is the length of the garden?
48 feet

Test and Practice Generator CD-ROM

Create and customize multiple versions of the same tests with corresponding answers for any chosen chapter objectives.

Chapter 2 State and Standardized Test Preparation

Test Taking Skill Builder and Standardized Test Practice are provided for each chapter in the *Test Prep Tool Kit.*

Test Taking Skill Builder

Test Taking Strategy **Extended Response**

Chapter 2

Extended Response questions are scored using a scoring rubric. You are expected to show all your work in detail. Use complete sentences to explain your thought process. Below is a scoring rubric that is used to score the answers to the following exercise.

Scoring Rubric

- **3-points:** Student writes and correctly solves an equation. Student defines the variable. Student answers the question in complete sentences.
- **2-points:** Student writes and correctly solves an equation but does not show all steps or does not give final answer in a complete sentences.
- **1-point:** Student gives correct answer but does not write and solve an equation, or student writes and solves incorrect equation.
- **0-points:** Student gives incorrect solution, or just gives a numerical answer with no equation or explanation.

Example Lisa has a collection of mystery books. She cleans her bookshelf and gives Cindy 18 books. She now has 23 books on her shelf. Write an equation to determine how many books Lisa originally had on her shelf, and then solve the equation. Show your work.

The following is an example of a 3-point response.

Let b = the original number of books on Lisa's shelf.

$b - 18 = 23$

$b - 18 + 18 = 23 + 18$ Write and solve an equation.

$b = 41$

Lisa originally had 41 books on her shelf.

The following is an example of a 2-point response.

$b - 18 = 23$

$b - 18 + 18 = 23 + 18$ Write and solve an equation.

$b = 41$

This is only a 2-point response because the student didn't define the variable or write the answer in a complete sentence.

Test Taking Strategy

Chapter 2, continued

Exercises

Use the scoring rubric for each question.

Scoring Rubric

- **3 points:** Student writes and correctly solves an equation. Student defines the variable. Student answers questions in complete a sentence and gives clear explanation of reasoning.
- **2 points:** Student writes and correctly solves an equation but does not show all steps or does not give the final answer in a complete sentence.
- **1 point:** Student gives correct answer but does not write and solve an equation, or student writes and solves incorrect equation.
- **0 points:** Student gives incorrect solution, or just gives a numerical answer with no equation or explanation.

1. Jeremy collects baseball cards. After selling 21 cards to his friend Mike, he has 105 cards remaining in his collection. Write an equation to determine how many cards Jeremy originally had in his collection, and then solve the equation. Show your work.

   2-point response:

   Let c = the original number of cards in the collection.

   $c - 21 = 105$

   $c = 126$

   Explain why the student received only 2 points for this answer.

   Possible answer: The student failed to show each step of the work. The student also failed to provide reasoning or write the answer in a complete sentence.

2. To receive all the points available on an answer to an extended response question, what must your answer always include?

   A a diagram or a sketch
   B formulas
   (C) an explanation or reasoning process
   D a visual data display

Standardized Test Practice

Standardized Test Practice

Chapter 2

Select the best answer for Questions 1–6.

1. Mario's truck holds 26 gallons of gas. The tank had 8 gallons in it, and Mario added 18 gallons when he fueled at the gas station. Which equation best represents this situation?

   A $8 + 18 = 26$
   B $18 - 8 = 26$
   C $18(8) = 26$
   D $\frac{26}{8} = 18$

2. Which expression is represented in the table?

n	?
18	3
24	4
30	5

   F $n - 15$
   G $6n$
   H $\frac{n}{6}$
   I $n + 15$

3. A farmer picks 87 pumpkins and puts them in a wagon to sell at the side of a road. If at the end of the day, he has 13 pumpkins left, how many pumpkins did he sell?

   A 64 pumpkins
   B 74 pumpkins
   C 78 pumpkins
   D 100 pumpkins

4. A football play starts on the 18 yard line and ends on the 42 yard line. How many yards were gained in the play?

   F 24 yards
   G 60 yards
   H 34 yards
   I 756 yards

5. There are usually 4 tiger cubs in a litter. If a tiger has 16 cubs in her lifetime, how many litters did she probably have?

   A 2 litters
   B 4 litters
   C 12 litters
   D 64 litters

6. During the month of November, Kendra saved d dollars. During the month of December, she saved \$4.00 more than the amount that she saved in November. Which expression represents the amount of money Kendra saved in December?

   F $d + \$4.00$
   G $d - \$4.00$
   H $\$4.00 \times d$
   I $\frac{\$4.00}{d}(25)$

Standardized Test Practice

Chapter 2, continued

Gridded Response

Solve the problems. Use the answer sheet to write and grid-in your answer.

7. During a class activity you and your partner measure your heights in inches. Your total combined height is 114 inches. If you are 56 inches tall, how many inches tall is your partner?
8. An entire cake has 340 calories. If you cut the cake into 10 pieces, how many calories does each piece have?
9. You have \$50 in the bank and each week you add \$20 to your account. The expression $50 + 20x$ represents the amount of money you will save after x weeks. How many dollars will you have in 8 weeks?
10. Maria baked cookies for her daughter's class. There are 18 students in her daughter's class plus a teacher and an educational aide. If each person gets 3 cookies, how many cookies did Maria bake?

Short Response

Solve the problems. Use the answer sheet to write your answers.

11. Your math partner says that in order to solve $5x = 105$ you subtract 5 from each side. Do you agree? Explain your reasoning.
12. Write an expression for the phrase twice the sum of a number, x, and fifteen. Explain in words how to evaluate the expression if $x = -6$, then evaluate the expression.
13. Christian wants to buy some oranges for 50 cents each. He has \$2. How many oranges can he buy? Write and solve an equation.

Extended Response

14. Use the figure.

   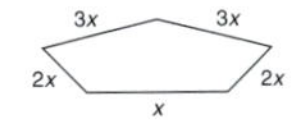

   a. Write an expression for the perimeter of the pentagon.
   b. If $x = 3$ in., what is the measure of each side? Show two different ways to calculate the perimeter.
   c. If the perimeter is 132 inches, what is x? What is the measure of each side? Show your work.

Sheet

H I See Lesson 2
C D See Lesson 2
H I See Lesson 2

10. 60

See Lesson 2

11. No; The equation says 5 times a number is 105. To find the number you have to divide both sides of the equation by 5.

12. $2(x + 15)$; Substitute the value -6 into the expression for x. Then use order of operations to evaluate the expression. $2(-6 + 15) = 18$ (See Lesson

13. $0.5x = 2$; $x = 4$; Christian can buy 4 oranges. (See Lesson

Extended Response (See Lesson

Customized answer sheets give students realistic practice for actual standardized tests.

internet connect

State-Specific Test Practice Online

KEYWORD: MR4 TestPrep

Test Prep Tool Kit

- Standardized Test Prep Workbook
- Countdown to Testing transparencies
- State Test Prep CD-ROM
- Standardized Test Prep Video

Chapter 2

Introduction to Algebra

Why Learn This?

Tell students that variables are used to take the place of quantities that change. For example, the number of cars heading north changes from one time period to the next. A traffic engineer might use a variable to represent the number of cars heading north and then use that variable in an equation to determine the timing of stoplights.

Using Data

To begin the study of this chapter, have students:

- Determine the number of cars heading east between noon and 2 P.M. 61
- Find the total number of cars passing through the intersection between 6 A.M. and 8 A.M. 255
- Offer suggestions as to why 114 cars head north during the morning rush hour, but only 11 head north during the evening rush hour. Possible answer: The intersection is located south of a city or other cluster of businesses.

Chapter 2

Introduction to Algebra

TABLE OF CONTENTS

internet connect
Chapter Opener Online
go.hrw.com
KEYWORD: MR4 Ch2

Number of Cars Traveling in Each Direction

	North	South	East	West
6–8 A.M.	114	36	48	57
8–10 A.M.	97	52	57	52
10 A.M.–noon	35	24	65	56
noon–2 P.M.	23	109	61	56
2–4 P.M.	18	138	70	72
4–6 P.M.	11	54	47	40

Career ***Traffic Engineer***

Have you ever wondered why traffic moves quickly through one intersection but slowly through another? Traffic engineers program stoplights so that vehicles can move smoothly through intersections. There are many variables at a traffic intersection—the number of vehicles that pass, the time of day, and the direction in which each vehicle travels are examples. Traffic engineers use this information to control the timing of stoplights. The table lists traffic movement through a given intersection during a given weekday.

Problem Solving Project

Social Studies Connection

Purpose: To use basic algebraic thinking to solve problems

Materials: Vehicle Travel worksheet

internet connect
Chapter Project Online: ***go.hrw.com***
KEYWORD: MR4 PSProject2

Understand, Plan, Solve, and Look Back

Have students:

- ✔ Complete the Vehicle Travel worksheet to use simple equations to help them estimate amounts of traffic flow.
- ✔ Make a table listing the variables that would influence the pattern of stoplights at a corner. How do these variables help determine how long the red and green lights should be on?
- ✔ Research an intersection near school or home and observe the traffic flow. What do they notice about the number of cars and trucks that enter the intersection at various times?
- ✔ Check students' work.

ARE YOU READY?

Choose the best term from the list to complete each sentence.

1. Multiplication is the ___?___ of division. inverse	dividend
2. The ___?___ of 12 and 3 is 36. product	factor
3. The ___?___ of 12 and 3 is 15. sum	inverse
4. Addition, subtraction, multiplication, and division are called ___?___. operations	operations
5. The answer to a division problem is called the ___?___. quotient	product
	quotient
	sum

Complete these exercises to review skills you will need for this chapter.

✔ Whole Number Operations

Add.

6. 28 + 15 = **43**
7. 71 + 38 = **109**
8. 1,218 + 430 = **1,648**
9. 2,218 + 1,135 = **3,353**

Subtract.

10. 72 − 35 = **37**
11. 98 − 45 = **53**
12. 1,642 − 249 = **1,393**
13. 3,408 − 1,649 = **1,759**

Multiply.

14. 6 × 13 **78**
15. 8 × 15 **120**
16. 16 × 22 **352**
17. 20 × 35 **700**

Divide.

18. 9)72 **8**
19. 7)84 **12**
20. 16)112 **7**
21. 23)1,472 **64**

✔ Inverse Operations

Use the inverse operation to check each equation. Possible answers are given.

22. 14 + 5 = 19 — **19 − 5 = 14**
23. 24 + 19 = 43 — **43 − 19 = 24**
24. 125 + 219 = 344 — **344 − 219 = 125**
25. 19 − 3 = 16 — **16 + 3 = 19**
26. 37 − 14 = 23 — **23 + 14 = 37**
27. 242 − 196 = 46 — **46 + 196 = 242**
28. 8 × 6 = 48 — **48 ÷ 6 = 8**
29. 25 × 5 = 125 — **125 ÷ 5 = 25**
30. 14 × 34 = 476 — **476 ÷ 34 = 14**
31. 12 ÷ 6 = 2 — **2 · 6 = 12**
32. 72 ÷ 8 = 9 — **9 · 8 = 72**
33. 252 ÷ 12 = 21 — **21 · 12 = 252**
34. 52 × 4 = 208 — **208 ÷ 4 = 52**
35. 400 ÷ 16 = 25 — **25 · 16 = 400**
36. 19 × 17 = 323 — **323 ÷ 17 = 19**

Assessing Prior Knowledge

INTERVENTION

Diagnose and Prescribe

Evaluate your students' performance on this page to determine whether intervention is necessary or whether enrichment is appropriate. Options that provide instruction, practice, and a check are listed below.

Resources for Are You Ready?

- ***Are You Ready? Intervention and Enrichment***
- **Recording Sheet for Are You Ready?** *Chapter 2 Resource Book. . . pp. 3–4*
- ***Are You Ready? Intervention* CD-ROM**

ARE YOU READY?

Were students successful with Are You Ready?

NO — INTERVENE

✔ **Whole Number Operations 6–21**
Intervention Practice, Skill 34
CD-ROM Intervention Activities, Skill 34

✔ **Inverse Operations . . .22–36**
Intervention Practice, Skill 57
CD-ROM Intervention Activities, Skill 57

YES — ENRICH

Are You Ready? Enrichment, pp. 409–410

Section 2A

Understanding Variables and Expressions

One-Minute Section Planner

Lesson	Materials	Resources
Lesson 2-1 Variables and Expressions **NCTM:** Algebra, Communication **NAEP:** Algebra 3b ☑ SAT-9 ☑ SAT-10 ☐ ITBS ☑ CTBS ☑ MAT ☑ CAT		• *Chapter 2 Resource Book,* pp. 7–15 • Daily Transparency T1, CRB • Additional Examples Transparencies T2–T3, CRB • *Alternate Openers: Explorations,* p. 8
Lesson 2-2 Translate Between Words and Math **NCTM:** Algebra, Problem Solving, Reasoning and Proof, Communication, Representation **NAEP:** Algebra 3a ☑ SAT-9 ☑ SAT-10 ☑ ITBS ☑ CTBS ☑ MAT ☑ CAT	**Optional** Teaching Transparency T5 *(CRB)*	• *Chapter 2 Resource Book,* pp. 16–24 • Daily Transparency T4, CRB • Additional Examples Transparencies T6–T7, CRB • *Alternate Openers: Explorations,* p. 9
Section 2A Assessment		• Mid-Chapter Quiz, SE p. 56 • Section 2A Quiz, AR p. 7 • *Test and Practice Generator* CD-ROM

SAT = *Stanford Achievement Tests* **ITBS** = *Iowa Test of Basic Skills* **CTBS** = *Comprehensive Test of Basic Skills/Terra Nova*
MAT = *Metropolitan Achievement Tests* **CAT** = *California Achievement Test*

NCTM = Complete standards can be found on pages T27–T33. **NAEP** = Complete standards can be found on pages A31–A35.

SE = *Student Edition* **TE** = *Teacher's Edition* **AR** = *Assessment Resources* **CRB** = *Chapter Resource Book* **MK** = *Manipulatives Kit*

Section Overview

Evaluating Algebraic Expressions — Lesson 2-1

Why? Evaluating algebraic expressions allows you to give meaning to symbolic representations, such as formulas and equations.

To **evaluate** an algebraic expression, substitute a number for the variable and then find the value.

An **algebraic expression** contains one or more variables and may contain operation symbols.

Evaluate the **algebraic expression** $3x + 5$ for $x = 2$.

$3x + 5$
$3 \cdot 2 + 5$
$6 + 5$
11

The algebraic expression $3x + 5$ has the value 11 when $x = 2$.

Translating Between Words and Algebra — Lesson 2-2

Why? Translating between words and algebra is used to solve real-world problems involving mathematics.

Operation	Action	Words	Expression	
+	Put together or combine.	• 5 added to 3 • 4 plus x • the sum of 7 and 2 • n more than 9	• $5 + 3$ • $4 + x$ • $7 + 2$ • $9 + n$	Numerical expression Algebraic expression Numerical expression Algebraic expression
−	Find how much more or how much less.	• 4 subtracted from b • 12 minus 5 • the difference of 7 and 3 • 2 less than y • take away 1 from 9	• $b - 4$ • $12 - 5$ • $7 - 3$ • $y - 2$ • $9 - 1$	Algebraic expression Numerical expression Numerical expression Algebraic expression Numerical expression
×	Put together groups of equal parts.	• t times 7 • 6 multiplied by s • the product of 4 and 3	• $t \times 7$, or $7t$ • $6 \times s$, or $6s$ • 4×3, or $(4)(3)$	Algebraic expression Algebraic expression Numerical expression
÷	Separate into equal groups.	• h divided by 2 • the quotient of 15 and 5	• $h \div 2$, or $\frac{h}{2}$ • $15 \div 5$, or $\frac{15}{5}$	Algebraic expression Numerical expression

2-1 Organizer

Pacing: Traditional 1 day
Block $\frac{1}{2}$ day

Objective: Students identify and evaluate expressions.

Indiana Academic Standards
6.3.6, 6.3.9, 6.7.1, 6.7.11

Warm Up

Simplify.

1. $4 + 7 \times 3 - 1$ 24

2. $87 - 15 \div 5$ 84

3. $6(9 + 2) + 7$ 73

4. $35 \div 7 \times 5$ 25

Problem of the Day

How can the digits 1 through 5 be arranged in the boxes to make the greatest product?

▢▢▢
× ▢▢

431
× 52

Available on Daily Transparency in CRB

Math Humor

Overheard in math class:
Teacher: What is 7*Q* plus 3*Q*?
Student: 10*Q*
Teacher: You're welcome!

2-1 Variables and Expressions

Learn to identify and evaluate expressions.

Vocabulary
variable
constant
algebraic expression

Inflation is the rise in prices that occurs over time. For example, you would have paid about \$7 in the year 2000 for something that cost only \$1 in 1950.

With this information, you can convert prices in 1950 to their equivalent prices in 2000.

1950	2000
\$1	\$7
\$2	\$14
\$3	\$21
\$*p*	\$*p* × 7

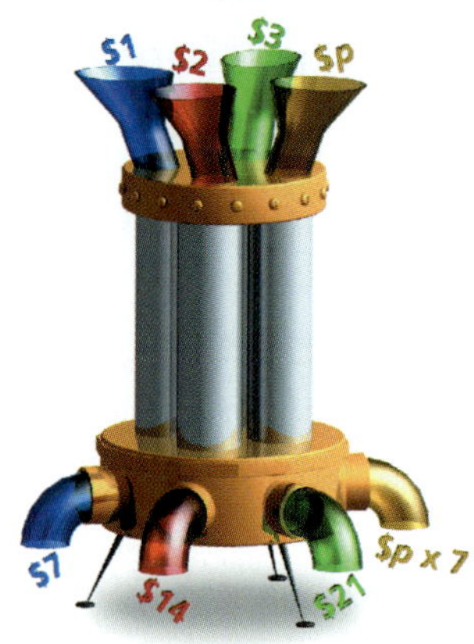

A **variable** is a letter or symbol that represents a quantity that can change. In the table above, *p* is a variable that stands for any price in 1950. A **constant** is a quantity that does not change. For example, the price of something in 2000 is always 7 times the price in 1950.

An **algebraic expression** contains one or more variables and may contain operation symbols. So $p \times 7$ is an algebraic expression.

Algebraic Expressions	NOT Algebraic Expressions
$150 + y$	$85 \div 5$
$35 \times w + z$	$10 + 3 \times 5$

To evaluate an algebraic expression, substitute a number for the variable and then find the value.

EXAMPLE 1 Evaluating Algebraic Expressions

Evaluate each expression to find the missing values in the tables.

A

w	$w \div 11$
55	5
66	▢
77	▢

Substitute for w in w ÷ 11.

$w = 55$; $55 \div 11 = 5$
$w = 66$; $66 \div 11 = 6$
$w = 77$; $77 \div 11 = 7$

The missing values are 6 and 7.

1 Introduce

Alternate Opener

EXPLORATION

2-1 Variables and Expressions

1. Look at the sequence of connected squares. The first square is made with 1 **black** segment and 3 red segments.

1 + 3 1 + 3 + 3 1 + 3 + 3 + 3

a. Sketch the next two squares.

b. To complete the table for the connected squares, count the number of segments it takes to draw each square.

Number of Connected Squares	1	2	3	4	5	10	20	100
Number of Segments	4	7						

c. How can you find the number of segments if you know the number of squares?

Think and Discuss

2. **Explain** the reasoning you used to find the number of segments in one hundred connected squares.

3. **Explain** the reasoning you could use to find the number of segments in one thousand connected squares.

Motivate

List these expressions on the board: $4 + 6$, 4×3, and $75 \div 5$. Remind students that expressions do not contain an equal sign or an answer. Now list $x + 6$, $y \times 3$, and $z \div 5$, and ask how these expressions are different. Explain that $x + 6$, $y \times 3$, and $z \div 5$ are *algebraic expressions.*

Exploration worksheet and answers on Chapter 2 Resource Book pp. 8 and 81

2 Teach

Lesson Presentation

Guided Instruction

In this lesson, students learn to identify and evaluate expressions. Discuss the definitions of *variable*, *constant*, and *algebraic expression*. First, have students provide examples of algebraic and non-algebraic expressions. Then teach students to evaluate algebraic expressions by substituting given values for the variable. Next teach students to find missing expressions that will generate the sequence for a table.

Evaluate each expression to find the missing values in the tables.

B

n	$4 \times n + 6$
1	10
2	■
3	■

Substitute for n in 4 × n + 6.
Use the order of operations.
$n = 1; 4 \times 1 + 6 = 10$
$n = 2; 4 \times 2 + 6 = 14$
$n = 3; 4 \times 3 + 6 = 18$

The missing values are 14 and 18.

Writing Math

When you are multiplying a number times a variable, the number is written first. Write "3x" and not "x3." Read 3x as "three x."

Multiplication and division expressions can be written without using the symbols × and ÷.

Instead of . . .	You can write . . .
$x \times 3$	$x \cdot 3 \quad x(3) \quad 3x$
$35 \div y$	$\frac{35}{y}$

EXAMPLE 2 Finding an Expression

Find an expression for each table.

A

x	■
6	48
7	56
8	64

$6 \cdot 8 = 48$
$7 \cdot 8 = 56$
$8 \cdot 8 = 64$

An expression is $x \cdot 8$, or $8x$.

B

x	■
12	1
24	2
36	3

$12 \div 12 = 1$
$24 \div 12 = 2$
$36 \div 12 = 3$

An expression is $\frac{x}{12}$.

Think and Discuss

1. **Name** a quantity that is a variable and a quantity that is a constant.
2. **Tell** whether each expression is an algebraic expression.
   a. $54 \div 2$ **b.** $45 + x$ **c.** $16y$ **d.** $24 \div 12$

Additional Examples

Example 1

Evaluate each expression to find the missing values in the tables.

A.

y	$5 \times y$
16	80
27	■ 135
35	■ 175

B.

z	$z \div 5 + 4$
20	8
45	■ 13
60	■ 16

Example 2

Find an expression for each table.

A.

n	■ $n \div 13$
39	3
52	4
65	5

B.

p	■ $7 \cdot p$
4	28
6	42
8	56

Example 2 note: Point out that finding an expression is similar to finding a rule for a pattern. Each row individually will satisfy several possible expressions; you must find an expression that works for all rows in the table.

3 Close

Reaching All Learners
Through Grouping Strategies

Have students work in groups to practice evaluating algebraic expressions. Have each group member write an algebraic expression. After the other group members take turns choosing numbers to substitute in the expression, have all the members evaluate for the chosen value. This process should continue until all members' expressions have been evaluated for several different values.

Summarize

Discuss brief definitions of the new vocabulary in the lesson: *variable*, *constant*, and *algebraic expression*. Remind students that multiplication and division can be represented in different ways within an algebraic expression and that the order of operations must be followed.

Answers to Think and Discuss

1. Possible answer: Your age is a variable quantity; the number of days in a week is a constant quantity.
2. **a.** no **b.** yes **c.** yes **d.** no

2-1 PRACTICE & ASSESS

Students may want to refer back to the lesson examples.

Assignment Guide

If you finished Example 1 assign:
Core 1–2, 5–6, 11–17 odd, 22–30
Enriched 11–19, 22–30

If you finished Example 2 assign:
Core 1–10, 11–17 odd, 18, 22–30
Enriched 7–30

Notes

Math Background

François Viète (1540–1603) was a lawyer in France who devoted his spare time to mathematics. In his book *In Artem*, he introduced the idea of representing unknown quantities using vowels and constants using consonants. He also used our present symbols + and – but had no symbol for equality. To write "equals" he would use the Latin word *aequatur*.

Viète is sometimes called the Father of Algebra.

2-1 Exercises

FOR EXTRA PRACTICE see page 638

internet connect
Homework Help Online
go.hrw.com Keyword: MR4 2-1

GUIDED PRACTICE

See Example 1 Evaluate each expression to find the missing values in the tables.

1.

n	$n + 7$
38	45
49	■ 56
58	■ 65

2.

x	$12x$
8	96
9	■ 108
10	■ 120

See Example 2 Find an expression for each table.

3. $x - 5$

x	■
50	45
45	40
40	35

4. $w + 9$

w	■
23	32
33	42
43	52

INDEPENDENT PRACTICE

See Example 1 Evaluate each expression to find the missing values in the tables.

5.

x	$4x$
50	200
100	■ 400
150	■ 600

6.

n	$2n - 2$
1	■ 0
6	■ 10
7	■ 12

See Example 2 Find an expression for each table.

7. $x \div 8$

x	■
0	0
72	9
88	11

8. $n + 25$

n	■
15	40
25	50
35	60

9. $x - 2$

x	■
8	6
4	2
2	0

10. $n \times 10$

n	■
50	500
75	750
100	1,000

RETEACH 2-1

LESSON 2-1 **Reteach**
Variables and Expressions

A variable is a letter or a symbol that stands for a number that can change. A constant is an amount that does not change.

A mathematical phrase that contains at least one variable is an algebraic expression. In the algebraic expression $x + 5$, x is a variable and 5 is a constant.

When you evaluate an algebraic expression, substitute a number for the variable and then find the value.

To evaluate the algebraic expression $m - 8$ for $m = 12$, first replace the variable m in the expression with 12.
$m - 8$
$12 - 8$
Then find the value of the expression.
$12 - 8 = 4$
The value of $m - 8$ is 4 when $m = 12$.

Evaluate each expression for the given value of the variable.

1. $x + 5$, for $x = 6$ — 11
2. $3p$, for $p = 5$ — 15
3. $z \div 4$, for $z = 24$ — 6
4. $w - 7$, for $w = 15$ — 8

To find an expression for a table of values, look for patterns.

x	■
3	12
4	16
5	20

Think: $3 \cdot ? = 12$ — $3 \cdot 4 = 12$
Think: $4 \cdot ? = 16$ — $4 \cdot 4 = 16$
Think: $5 \cdot ? = 20$ — $5 \cdot 4 = 20$

$x(4)$ is an expression for the table.

Find an expression for each table.

5. $x + 7$

x	■
3	10
5	12
7	14

6. $b \div 5$

b	■
25	5
15	3
10	2

7. $g \cdot 6$

g	■
2	12
6	36
8	48

8. $y - 2$

y	■
9	7
10	8
14	12

PRACTICE 2-1

LESSON 2-1 **Practice B**
Variables and Expressions

Evaluate each expression to find the missing values in the tables.

1.

n	$n + 8$
7	15
9	17
22	30
35	43

2.

n	$25 - n$
20	5
5	20
18	7
9	16

3.

n	$n \cdot 7$
8	56
9	63
11	77
12	84

4.

n	$24 \div n$
2	12
6	4
4	6
8	3

Find an expression for each table.

5.

n	$n + 15$
35	50
5	20
20	35
85	100

6.

n	$n \cdot 2$
13	26
9	18
30	60
25	50

7. A car is traveling at a speed of 55 miles per hour. You want to write an algebraic expression to show how far the car will travel in a certain number of hours. What will be your constant? your variable?
55 will be the constant, and the number of hours will be the variable.

8. Shawn evaluated the algebraic expression $x \div 4$ for $x = 12$ and gave an answer of 8. What was his error? What is the correct answer?
He used subtraction instead of division. The correct answer is 3.

PRACTICE AND PROBLEM SOLVING

Evaluate each expression for the given value of the variable.

11. $3h + 2$ for $h = 10$ **32**

12. $2x$ for $x = 15$ **30**

13. $4p - 3$ for $p = 20$ **77**

14. $\frac{c}{7}$ for $c = 56$ **8**

15. $3x + 17$ for $x = 13$ **56**

16. $5p$ for $p = 12$ **60**

17. The zloty is the currency in Poland. In 2002, 1 U.S. dollar was worth 4 zlotys. How many zlotys were equivalent to 8 U.S. dollars? **32 zlotys**

18. Use the graph to complete the table.

Cups of Water	Number of Lemons
8	▒ **24**
12	▒ **36**
w	▒ **$3w$**

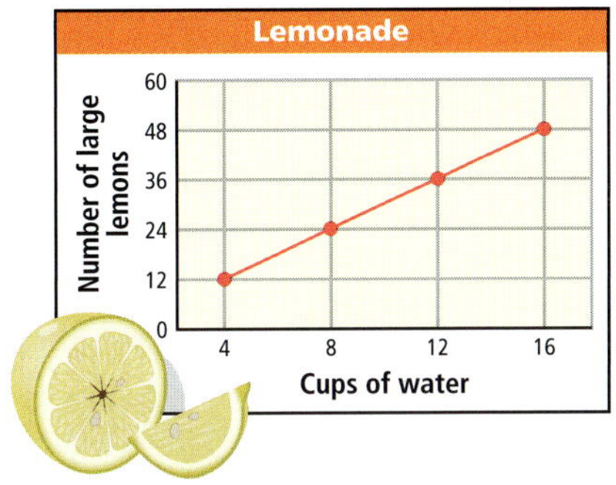

19. ***WHAT'S THE ERROR?*** A student evaluated the expression $x \div 2$ for $x = 14$ and gave an answer of 28. What did the student do wrong? **The student multiplied 14 by 2 instead of dividing 14 by 2.**

20. ***WRITE ABOUT IT*** A friend asks you to think of a number, double it, and then add 5. Write an algebraic expression to describe your friend's directions, and make a table of possible values.

21. ***CHALLENGE*** Using the algebraic expression $3n - 5$, what is the smallest whole-number value for n that will give you a result greater than 100? **36**

Money LINK

Why does a Polish coin show Australia and a kangaroo? This coin honors Pawel Edmund Strzelecki, a Polish explorer who explored and mapped much of Australia.

go.hrw.com KEYWORD: MR4 Money

CNN Student News

Spiral Review

Order the numbers from least to greatest. (Lesson 1-1)

22. 798; 648; 923 **648; 798; 923**

23. 1,298; 876; 972 **876; 972; 1,298**

24. 1,498; 2,163; 1,036 **1,036; 1,498; 2,163**

Estimate each sum or difference by rounding to the place value indicated. (Lesson 1-2)

25. 17,281 + 23,008; thousands **40,000**

26. 412,243 − 124,539; hundred thousands **300,000**

Write each expression in exponential form. (Lesson 1-3)

27. $3 \times 3 \times 3$ **3^3**

28. $5 \times 5 \times 5 \times 5 \times 5 \times 5$ **5^6**

29. $10 \times 10 \times 10 \times 10$ **10^4**

30. **TEST PREP** Which is the missing term in the sequence 3, 7, 15, 31, 63, ▒, 255, . . . ? (Lesson 1-7) **D**

A 67 B 95 C 103 D 127

Answers

20.

x	$2x + 5$
0	5
1	7
2	9
3	11

Journal

Have students write about situations, people, or things that are evaluated.

Test Prep Doctor

For Exercise 30, students need to remember that to find a rule for a sequence, they must find a similar relationship between all adjacent terms in the sequence, not just between two adjacent terms. Students who answered **A** assumed that because 3 + 4 = 7, the rule must be to add 4 to each term. Students who answered **B** assumed that because 31 + 32 = 63, the rule must be to add 32 to each term. The correct answer, **D**, applies the pattern of multiplying by 2 and then adding 1 or of adding double the quantity previously added.

CHALLENGE 2-1

LESSON 2-1 Challenge
Express Trains

Use the expression written on the side of each train's engine to find the missing values for the cars it pulls. Then choose your own value for the variable to fill in the last caboose on each train.

	Engine	Car 1	Car 2	Car 3	Caboose
1.	$n \div 7$	6 $n = 42$	**8** $n = 56$	**4** $n = 28$	**5** **$n = 35$**
2.	$2x + 5$	11 $x = 3$	**21** $x = 8$	**25** $x = 10$	**15** **$x = 5$**
3.	$c \div 12$	4 $c = 48$	**2** $c = 24$	**5** $c = 60$	**8** **$c = 96$**
4.	$5p - 9$	31 $p = 8$	**11** $p = 4$	**46** $p = 11$	**16** **$p = 5$**
5.	$7m + 2m$	45 $m = 5$	**18** $m = 2$	**81** $m = 9$	**127** **$m = 3$**

Possible answers are given on each caboose. Accept all answers that correctly match the chosen variable and the train's expression.

PROBLEM SOLVING 2-1

LESSON 2-1 Problem Solving
Variables and Expressions

Write the correct answer.

1. To cook 4 cups of rice, you use 8 cups of water. To cook 10 cups of rice, you use 20 cups of water. Write an expression to show how many cups of water you should use if you want to cook c cups of rice. How many cups of water should you use to cook 5 cups of rice?

   $2c$; 10 cups of water

2. Sue earns the same amount of money for each hour that she tutors students in math. In 3 hours, she earns \$27. In 8 hours, she earns \$72. Write an expression to show how much money Sue earns working h hours. At this rate, how much money will Sue earn if she works 12 hours?

   $9h$; \$108

3. Bees are one of the fastest insects on Earth. They can fly 22 miles in 2 hours, and 55 miles in 5 hours. Write an expression to show how many miles a bee can fly in h hours. If a bee flies 4 hours at this speed, how many miles will it travel?

   $11h$; 44 miles

4. A friend asks you to think of a number, triple it, and then subtract 2. Write an algebraic expression using the variable x to describe your friend's directions. Then find the value of the expression if the number you think of is 5.

   $3x - 2$; 13

Circle the letter of the correct answer.

5. The rupee is the currency in Pakistan. In 2002, 1 United States dollar was worth 60 rupees! How many rupees were equivalent to 3 United States dollars?
   A 3
   B 2
   (C) 180
   D 60

6. The peso is the currency in Mexico. In 2002, 1 United States dollar was worth 9 pesos. How many pesos were equivalent to 5 United States dollars?
   (F) 45
   G 4
   H 14
   J 54

Lesson Quiz

1. Evaluate the expression to find the missing values in the table.

x	$x - 5$
10	▒ **5**
7	▒ **2**
5	▒ **0**

2. Find an expression for the table.

x	▒ **$7x$**
1	7
3	21
5	35

3. Evaluate $8x$ for $x = 5$. **40**

4. Evaluate $4x - 1$ for $x = 12$. **47**

Available on Daily Transparency in CRB

2-2 Organizer

Pacing: Traditional 1 day
Block $\frac{1}{2}$ day

Objective: Students translate between words and math.

Indiana Academic Standards
6.7.1, 6.7.5, 6.7.11

Warm Up

Evaluate each expression for $x = 9$.

1. $7 + x$ 16 **2.** $4x$ 36
3. $2x + 1$ 19 **4.** $\frac{36}{x}$ 4

Problem of the Day

Draw a square around the numbers of four adjacent days on the calendar for this month. Add all the numbers in the square and subtract four times the first number. What number do you get? 16

Available on Daily Transparency in CRB

Math Fact

The word *algebra* is derived from the word *al-jabr*, which appeared in the title of al-Khwârizmî's treatise on algebra. Al-Khwârizmî lived during the late eighth and early ninth centuries.

2-2 Translate Between Words and Math

Problem Solving Skill

Learn to translate between words and math.

The earth's core is divided into two parts. The inner core is solid and dense, with a radius of 1,228 km. Let c stand for the thickness in kilometers of the liquid outer core. What is the total radius of the earth's core?

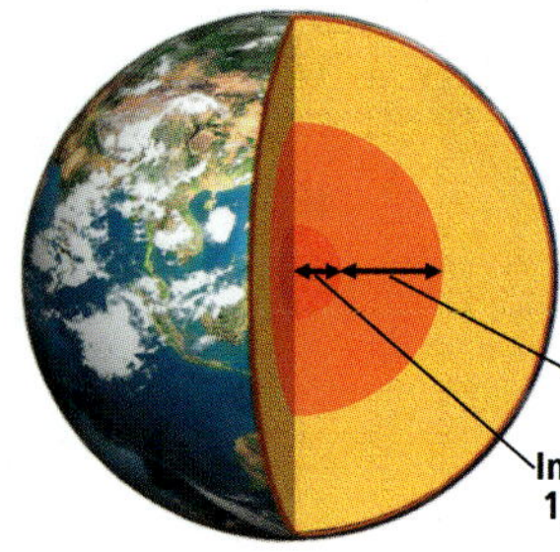

In word problems, you may need to identify the action to translate words to math.

Action	Put together or combine	Find how much more or less	Put together groups of equal parts	Separate into equal groups
Operation	Add	Subtract	Multiply	Divide

To solve this problem, you need to *put together* the measurements of the inner core and the outer core. To put things together, add.

$$1{,}228 + c$$

The total radius of the earth's core is $1{,}228 + c$ km.

EXAMPLE 1 ***Social Studies Applications***

A **The Nile River is the world's longest river. Let n stand for the length in miles of the Nile. The Amazon River is 4,000 miles long. Write an expression to show how much longer the Nile is than the Amazon.**

To *find how much longer,* subtract the length of the Amazon from the length of the Nile.

$$n \quad - \quad 4{,}000$$

The Nile is $n - 4{,}000$ miles longer than the Amazon.

B **Let s represent the number of senators that each of the 50 states has in the U.S. Senate. Write an expression for the total number of senators.**

To *put together 50 equal groups of s,* multiply 50 times s.

$$50s$$

There are $50s$ senators in the U.S. Senate.

1 Introduce

Alternate Opener

EXPLORATION

2-2 Translate Between Words and Math

Drawing pictures and using formulas can help you translate between words and math.

A basketball court is 50 ft wide by 94 ft long. What is its area?

Formula for area

A = length × width
$A = 94 \times 50$
$A = 4{,}700$

Area = ? 50 ft 94 ft

The area is 4,700 ft^2.

In some word problems, word order may be confusing. For example, the following problems can be translated in at least two different ways.

Rewrite each problem to make it clearer.

	Word Problem	Possible Translations	Better Word Problem
1.	Write the expression "4 times x plus 6."	$4x + 6$ or $4(x + 6)$	
2.	Translate "the square root of n minus 3."	$\sqrt{n} - 3$ or $\sqrt{n - 3}$	

Think and Discuss

3. Explain what you did to rewrite numbers 1 and 2 to make them easier to translate into math.

Motivate

Discuss the word *translate* with students and have students give examples of things that can be translated (e.g., English to Spanish). Similarly, you can translate mathematical situations expressed as words into symbols, and vice versa.

Exploration worksheet and answers on Chapter 2 Resource Book pp. 17 and 83

2 Teach

Lesson Presentation

Guided Instruction

In this lesson, students learn to translate between words and mathematical expressions. First, describe the actions that signal each operation. (You may use Teaching Transparency T5 in the Chapter 2 Resource Book.) Then teach students to translate words into algebraic expressions. Next, teach students to write numerical and algebraic expressions with words.

Teaching Tip Teach students to write all variables in a lowercase, cursive style. This eliminates confusion between x and $\times$, as well as between t and $+$.

There are several different ways to write math expressions with words.

Operation	+	−	×	÷
Numerical Expression	37 + 28	90 − 12	8 × 48 or 8 · 48 or (8)(48) or 8(48) or (8)48	327 ÷ 3 or $\frac{327}{3}$
Words	• 28 added to 37 • 37 plus 28 • the sum of 37 and 28 • 28 more than 37	• 12 subtracted from 90 • 90 minus 12 • the difference of 90 and 12 • 12 less than 90 • take away 12 from 90	• 8 times 48 • 48 multiplied by 8 • the product of 8 and 48 • 8 groups of 48	• 327 divided by 3 • the quotient of 327 and 3
Algebraic Expression	$x + 28$	$k - 12$	$8 \cdot w$ or $(8)(w)$ or $8w$	$n \div 3$ or $\frac{n}{3}$
Words	• 28 added to *x* • *x* plus 28 • the sum of *x* and 28 • 28 more than *x*	• 12 subtracted from *k* • *k* minus 12 • the difference of *k* and 12 • 12 less than *k* • take away 12 from *k*	• 8 times *w* • *w* multiplied by 8 • the product of 8 and *w* • 8 groups of *w*	• *n* divided by 3 • the quotient of *n* and 3

EXAMPLE 2 Translating Words into Math

Write each phrase as a numerical or algebraic expression.

A 287 plus 932

287 + 932

B *b* divided by 14

$b \div 14$ or $\frac{b}{14}$

EXAMPLE 3 Translating Math into Words

Write two phrases for each expression.

A $a - 45$

- *a* minus 45
- take away 45 from *a*

B (34)(7)

- the product of 34 and 7
- 34 multiplied by 7

Think and Discuss

1. **Tell** how to write each of the following phrases as a numerical or algebraic expression: 75 less than 1,023; the product of 125 and *z*.
2. **Give** two examples of "$a \div 17$" expressed with words.

Additional Examples

Example 1

A. Lake Superior is the largest lake in North America. Let *a* stand for the area in square miles of Lake Superior. Lake Erie has an area of 9,910 square miles. Write an expression to show how much larger Lake Superior is than Lake Erie.

$a - 9{,}910$

B. Let *p* represent the number of colored pencils in a box. If there are 26 boxes on the shelf, write an algebraic expression to represent the total number of pencils on the shelf.

$26 \cdot p$, or $26p$

Example 2

Write each phrase as a numerical or algebraic expression.

A. 987 minus 12 — 987 − 12

B. *x* times 45 — $45 \cdot x$, or $45x$

Example 3

Write two phrases for each expression.

A. $\frac{16}{b}$ — 16 divided by *b*; the quotient of 16 and b

B. (75)(32) — 75 times 32; the product of 75 and 32

3 Close

Reaching All Learners

Through Home Connection

Have students record real-world math situations they experience at home, using both words and mathematical symbols. Possible answer: Mom works out for the same length of time each day. How long does she work out in a week? 7*t*, where *t* represents the length of time she works out each day.

Summarize

Direct the students' attention to the chart of operations, expressions, and words. Review each numerical and algebraic expression while connecting to the words that describe each expression. Have the students duplicate the chart for their own reference.

Answers to Think and Discuss

1. 1,023 − 75; 125*z*; $125 \cdot z$; (125)(*z*)
2. *a* divided by 17; the quotient of *a* and 17

2-2 PRACTICE & ASSESS

Students may want to refer back to the lesson examples.

Assignment Guide

If you finished Example 1 assign:
Core 1, 8–9, 30–31, 36–42
Enriched 1, 8–9, 30–32, 36–42

If you finished Example 2 assign:
Core 1–3, 8–13, 30–31, 36–42
Enriched 10–15, 30–42

If you finished Example 3 assign:
Core 1–23, 30–31, 36–42
Enriched 10–42

Answers

16. 65 added to h; 65 more than h

17. take away 19 from 243; 243 minus 19

18. the quotient of 125 and n; 125 divided by n

19. 75 multiplied by 342; the product of 342 and 75

20. the quotient of d and 27; d divided by 27

21. the product of 45 and 23; 45 times 23

22. the sum of 629 and c; c more than 629

23. the difference of 228 and b; b less than 228

Math Background

We translate words into algebraic expressions using a consistent, universally understood system. This system has evolved over thousands of years. Archaeological records indicate that Babylonian mathematics had developed prose-based algebra by 2000 B.C.

The adoption of symbols to represent operations was also part of this evolution. The symbols + and − can be traced to Johann Widman (1498); the symbol · can be traced to Gottfried Leibniz (1698); and the symbol ÷ can be traced to Johann Heinrich Rahn (1659).

2-2 Exercises

FOR EXTRA PRACTICE see page 638

internet connect
Homework Help Online
go.hrw.com Keyword: MR4 2-2

GUIDED PRACTICE

See Example 1

1. The Big Island of Hawaii is the largest Hawaiian island, with an area of 4,028 mi^2. The next biggest island is Maui. Let m represent the area of Maui. Write an expression for the difference between the two areas. **$4{,}028 - m$**

See Example 2

Write each phrase as a numerical or algebraic expression.

2. 279 minus 125 **$279 - 125$**

3. the product of 15 and x **$15x$**

See Example 3

Write two phrases for each expression.

4. $r + 87$ The sum of r and 87; r plus 87

5. 345×196 the product of 345 and 196; 345 times 196

6. $476 \div 28$ the quotient of 476 and 28; 476 divided by 28

7. $d - 5$ the difference of d and 5; five less than d

INDEPENDENT PRACTICE

See Example 1

8. California has 21 more seats in the U.S. Congress than Texas has. If t represents the number of seats Texas has, write an expression for the number of seats California has. **$t + 21$**

9. Let x represent the number of television show episodes that are taped in a season. Write an expression for the number of episodes that are taped in 5 seasons. **$5x$**

See Example 2

Write each phrase as a numerical or algebraic expression.

10. 25 less than k **$k - 25$**

11. the quotient of 325 and 25 **$325 \div 25$**

12. 34 times w **$34w$**

13. 675 added to 137 **$137 + 675$**

14. the sum of 135 and p **$135 + p$**

15. take away 14 from j **$j - 14$**

See Example 3

Write two phrases for each expression.

16. $h + 65$

17. $243 - 19$

18. $125 \div n$

19. $342(75)$

20. $\frac{d}{27}$

21. $45 \cdot 23$

22. $629 + c$

23. $228 - b$

PRACTICE AND PROBLEM SOLVING

Translate each phrase into a numerical or algebraic expression.

24. 13 less than z **$z - 13$**

25. 15 divided by d **$15 \div d$**

26. 874 times 23 **$874(23)$**

27. m multiplied by 67 **$67m$**

28. the sum of 35, 74, and 21 **$35 + 74 + 21$**

29. 319 less than 678 **$678 - 319$**

RETEACH 2-2

LESSON 2-2 **Reteach**
Translate Between Words and Math

There are key words that tell you which operations to use for mathematical expressions.

Addition (combine)	Subtraction (less)	Multiplication (put together groups of equal parts)	Division (separate into equal groups)
add	minus	product	quotient
plus	difference	times	divide
sum	subtract	multiply	
total	less than		
increased by	decreased by		
more than	take away		

You can use key words to help you translate between word phrases and mathematics phrases.

A. 3 plus 5 $3 + 5$ **B.** 3 times x $3x$ **C.** 5 less than p $p - 5$ **D.** h divided by 6 $h \div 6$

Write each phrase as a numerical or algebraic expression.

1. 4 less than 8 $8 - 4$

2. q divided by 3 $q \div 3$

3. f minus 6 $f - 6$

4. d multiplied by 9 $d \cdot 9$

You can use key words to write word phrases for mathematical phrases.

A. $7k$
- the product of 7 and k
- 7 times k

B. $5 - 2$
- 5 minus 2
- 2 less than 5

Write a phrase for each expression.

5. $z \div 4$ z divided by 4

6. $5 \cdot 6$ 5 times 6

7. $m - 6$ 6 less than m

8. $s + 3$ s plus 3

PRACTICE 2-2

LESSON 2-2 **Practice B**
Translate Between Words and Math

Write an expression.

1. Terry's essay has 9 more pages than Stacey's essay. If s represents the number of pages in Stacey's essay, write an expression for the number of pages in Terry's essay.
$s + 9$

2. Let z represent the number of students in a class. Write an expression for the number of students in 3 equal groups.
$\frac{z}{3}$

Write each phrase as a numerical or algebraic expression.

3. 24 multiplied by 3 $24 \cdot 3$

4. n multiplied by 14 $n \cdot 14$

5. w added to 64 $64 + w$

6. the difference of 58 and 6 $58 - 6$

7. m subtracted from 100 $100 - m$

8. the sum of 180 and 25 $180 + 25$

9. the product of 35 and x $35x$

10. the quotient of 63 and 9 $63 \div 9$

11. 28 divided by p $28 \div p$

Write two phrases for each expression. Possible answers are given.

12. $n + 91$ n plus 91; 91 more than n

13. $35 \div r$ 35 divided by r; the quotient of 35 and r

14. $20 - s$ 20 minus s; s less than 20

15. Charles is 3 years older than Paul. If y represents Paul's age, what expression represents Charles's age?
$y + 3$

16. Maya bought some pizzas for \$12 each. If p represents the number of pizzas she bought, what expression shows the total amount she spent?
$12p$

Science LINK

The graph shows the number of U.S. space exploration missions from 1956 to 2000.

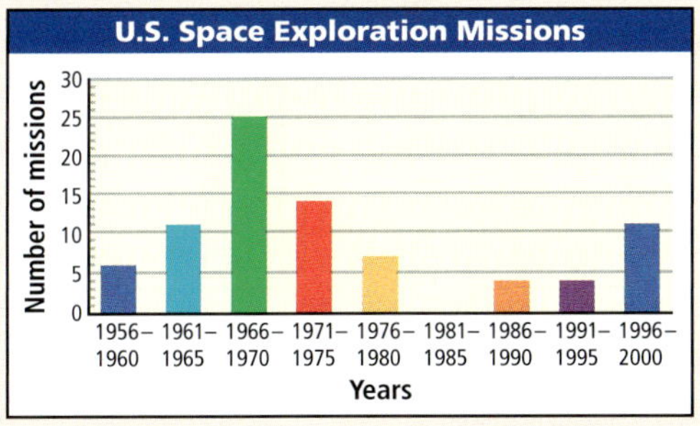

30. Between 1966 and 1970, the Soviet Union had m fewer space missions than the United States. Write an algebraic expression for this situation. **25 − *m***

31. Let d represent the number of dollars that the United States spent on space missions from 1986 to 1990. Write an expression for the cost per mission. **$d \div 4$**

32. ***WRITE A PROBLEM*** Use the data in the graph to write a word problem that can be answered with a numerical or algebraic expression.

33. ***WHAT'S THE QUESTION?*** The answer from the graph is $6 + 11 + 25 + 14 + 7 + 4 + 4 + 11$. What is the question?
What is the total number of U.S. space missions from 1956 to 2000?

34. ***WRITE ABOUT IT*** Let p stand for the number of missions between 1996 and 2000 that had people aboard. What operation would you use to write an expression for the number of missions without people? Why? Use the action in the problem to explain your answer.

35. ***CHALLENGE*** Write an expression for the following: two more than the number of missions from 1971 to 1975, minus the number of missions from 1986 to 1990. Then evaluate the expression. **(14 + 2) − 4; 12**

Spiral Review

Compare. Write <, >, or =. (Lesson 1-1)

36. 1,256,589 ■ 1,265,598 **<**

37. 2,568,987,254 ■ 2,568,987,254 **=**

Find each value. (Lesson 1-3)

38. 5^4 **625**

39. 2^5 **32**

40. 7^2 **49**

41. 11^3 **1,331**

42. TEST PREP $5^2 + (7 + 2) - 8 \cdot 3 = ?$ (Lesson 1-4) **A**

A 10 **B** 18 **C** 33 **D** 78

COMMON ERROR ALERT

When translating between phrases and algebraic subtraction expressions, students may reverse the numbers. For example, they may answer $13 - z$ instead of $z - 13$ for Exercise 24. Caution students to read the phrases carefully, especially those involving subtraction and division, where order is important.

Answers

32. Possible answer: What is the difference between the total number of missions t and missions from 1966 to 1970?

34. Possible answer: subtraction; the total missions minus the manned missions yields the number of unmanned missions.

Journal

In language arts, the word *expression* means a phrase or a sentence. Have students write about how an algebraic expression is like a phrase.

Test Prep Doctor

For Exercise 42, remind students how to find the value of a number in exponential form and to use the rules for order of operations. Students who answered **C** multiplied $5 \cdot 2$ instead of finding 5^2, and they performed operations from left to right instead of using the correct order. Students who answered **D** found the correct value for 5^2, but they performed the operations from left to right instead of following the rules for order of operations.

Lesson Quiz

1. Let x represent the number of minutes Kristen works out in one week. Write an expression for the number of minutes she works out in 4 weeks. **4*x***

Write each phrase as a numerical or algebraic expression.

2. 7 less than x **$x - 7$**

3. The product of 12 and w **12*w***

Write a phrase for each expression.

4. $17 + x$ ***x* more than 17 or *x* added to 17**

5. $n \div 12$ ***n* divided by 12 or the quotient of *n* and 12**

Available on Daily Transparency in CRB

CHALLENGE 2-2

LESSON 2-2 **Challenge**
Animal State

Follow the steps below in the exact order they are given. Do not skip ahead!

STEP 1 Pick a whole number 0–5.

STEP 2 Multiply the number by 3.

STEP 3 Square that product.

STEP 4 Add the digits in your result until you only have 1 digit. For example, 64 = 6 + 4 = 10 = 1 + 0 = 1.

STEP 5 If your sum is less than 5, add 5. If it is greater than 5, subtract 4.

STEP 6 Multiply your new sum or difference by 2.

STEP 7 Subtract 6 from that product.

STEP 8 Assign your new difference a letter in the alphabet starting with 1 = A, 2 = B, 3 = C, and so on.

STEP 9 Pick a state in the United States that begins with your letter.

STEP 10 Now look at the second letter in the name of your chosen state. Choose an animal that begins with that letter.

STEP 11 Share the state and animal you chose with a classmate. How do your choices compare? How do the numbers you chose in Step 1 compare?

All students should end with the state Delaware. Most students will choose an elephant, but accept any animal that begins with the letter E.

PROBLEM SOLVING 2-2

LESSON 2-2 **Problem Solving**
Translate Between Words and Math

Write the correct answer.

1. Holly bought 10 comic books. She gave a few of them to Kyle. Let c represent the number of comic books she gave to Kyle. Write an expression for the number of comic books Holly has left.
10 − *c*

2. Last week, Peter worked 40 hours for \$15 an hour. Write a numerical expression for the total amount Peter earned last week. Write an algebraic expression to show how much Peter earns in h hours at that rate.
40 • 15; 15*h*

3. The temperature dropped 5°F, and then it went up 3°F. Let t represent the beginning temperature. Write an expression to show the ending temperature.
***t* − 5 + 3**

4. Teri baked 48 cookies and divided them evenly into bags. Let n represent the number of cookies Teri put in each bag. Write an expression for the number of bags she filled.
48 ÷ *n*

5. Marisa purchased canned soft drinks for a family reunion. She purchased 1 case of 24 cans and several packages containing 6 cans each. If p represents the number of 6-can packages she purchased, what expression represents the total number of cans Marisa purchased for the reunion?
24 + 6*p*

6. Becky has 53 addresses listed in her e-mail address book. She forwarded a copy of an article to all of those people, except 5. Write an expression to show how many people she sent the article to.
53 − 5

Circle the letter of the correct answer.

7. Mei bought several CDs for \$12 each. Which of the following expressions could you use to find the total amount she spent on the CDs?
A $12 + x$
B $12 - x$
(C) $12x$
D $12 \div x$

8. Tony bought 2 packs of 50 plates and 1 pack of 30 plates. Which of the following expressions could you use to find the total number of plates Tony bought?
F 2 + 50 + 30
(G) (2 • 50) + 30
H (2 • 30) + 50
J 2(30 + 50)

Chapter 2

Mid-Chapter Quiz

Purpose: *To assess students' mastery of concepts and skills in Lessons 2-1 and 2-2*

Assessment Resources

Section 2A Quiz

Assessment Resources . . . . p. 7

***Test and Practice Generator* CD-ROM**

Additional mid-chapter assessment items in both multiple-choice and free-response format may be generated for any objective in Lessons 2-1 through 2-2.

Answers

Possible answers:

17. the sum of *n* and 19; 19 more than *n*

18. 12 times 13; the product of 12 and 13

19. *x* subtracted from 72; *x* less than 72

20. *t* divided by 12; the quotient of *t* and 12

21. *s* multiplied by 15; 15 times *s*

22. 27 divided by 9; the quotient of 27 and 9

23. *z* subtracted from 43; *z* less than 43

24. 93 divided by *k*; the quotient of 93 and *k*

25. the sum of 20 and *d*; *d* more than 20

Chapter 2

Mid-Chapter Quiz

Mid-Chapter Quiz

LESSON 2-1 (pp. 48–51)

Evaluate each expression to find the missing values in the tables.

1.

y	$23 + y$
17	40
27	50
37	60

2.

w	$w \times 3 + 10$
4	22
5	25
6	28

3.

x	$x \div 8$
40	5
48	6
56	7

Find an expression for each table.

4.

t	$t \times 3$
3	9
4	12
5	15

5.

y	$y \div 9$
36	4
45	5
54	6

6.

n	$n - 8$
10	2
15	7
20	12

LESSON 2-2 (pp. 52–55)

7. The small and large intestines are part of the digestive system. The small intestine is longer than the large intestine. Let n represent the length in feet of the small intestine. The large intestine is 5 feet long. Write an expression to show how much longer the small intestine is than the large intestine. $n - 5$

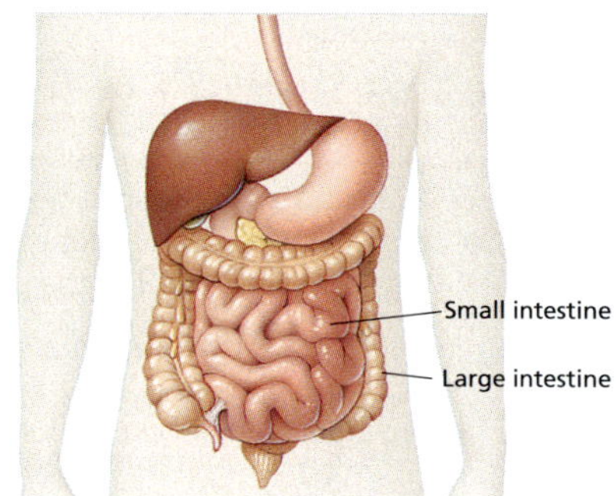

8. Let h represent the number of times your heart beats in 1 minute. Write an expression for the total number of times it beats in 1 hour. (*Hint:* 1 hour = 60 minutes) $60h$

Write each phrase as a numerical or algebraic expression.

9. 719 plus 210 $719 + 210$

10. t multiplied by 7 $7t$

11. the sum of n and 51 $n + 51$

12. 14 subtracted from p $p - 14$

13. the quotient of 510 and n $510 \div n$

14. the product of 52 and z $52z$

15. q divided by 3 $q \div 3$

16. the difference of 25 and g $25 - g$

Write two phrases for each expression.

17. $n + 19$

18. $12 \cdot 13$

19. $72 - x$

20. $\frac{t}{12}$

21. $15s$

22. $27 \div 9$

23. $43 - z$

24. $93 \div k$

25. $20 + d$

Focus on Problem Solving

Understand the Problem

• Identify too much or too little information

Problems often give too much or too little information. You must decide whether you have enough information to work the problem.

Read the problem and identify the facts that are given. Can you use any of these facts to arrive at an answer? Are there facts in the problem that are not necessary to find the answer? These questions can help you determine whether you have too much or too little information.

If you cannot solve the problem with the information given, decide what information you need. Then read the problem again to be sure you haven't missed the information in the problem.

Copy each problem. Circle the important facts. Underline any facts that you do not need to answer the question. If there is not enough information, list the additional information you need.

1. The reticulated python is one of the longest snakes in the world. One was found in Indonesia in 1912 that was 33 feet long. At birth, a reticulated python is 2 feet long. Suppose an adult python is 29 feet long. Let f represent the number of feet the python grew since birth. What is the value of f?

2. The largest flying flag in the world is 7,410 square feet and weighs 180 pounds. There are a total of 13 horizontal stripes on it. Let h represent the height of each stripe. What is the value of h?

3. The elevation of Mt. McKinley is 20,320 ft. People who climb Mt. McKinley are flown to a base camp located at 7,200 ft. From there, they begin a climb that may last 20 days or longer. Let d represent the distance from the base camp to the summit of Mt. McKinley. What is the value of d?

4. Let c represent the cost of a particular computer in 1981. Six years later, in 1987, the price of the computer had increased to $3,600. What is the value of c?

Answers

1. $f = 27$ ft
2. not enough information given
3. 13,120 ft
4. not enough information given

Focus on Problem Solving

Purpose: *To focus on making a plan*

Problem Solving Resources

Interactive Problem Solving . . pp. 8–14

Math: Reading and Writing in the Content Area . . . . pp. 8–14

Problem Solving Process

This page focuses on the first step of the problem-solving process: **Understand the Problem**

Discuss

Have students identify the information needed to solve each problem. Ask them whether all the information needed is given in the problem, and to identify any missing or extra information.

Possible answers:

1. You need to find the difference in length, so you need the length at birth and the adult length, and both are given. Everything else is extra information.
2. You need the height of the flag, which is missing. You would divide the height of the flag by the number of stripes, 13, to find the height of each stripe. If the width of the flag were given, you could find the height using the area. But, as the problem is written, the area and the weight of the flag are extra information.
3. You need to subtract the base-camp height from the peak height, and both are given. The length of the climb is extra information.
4. You need the cost of the computer in 1981, which is not given. All of the information is extra.

Section 2B

Understanding Variables and Expressions

One-Minute Section Planner

Lesson	Materials	Resources
Lesson 2-3 Equations and Their Solutions **NCTM:** Algebra, Communication, Representation **NAEP:** Algebra 4b ☐ SAT-9 ☐ SAT-10 ☑ ITBS ☐ CTBS ☑ MAT ☑ CAT	**Optional** Balance Scale Teaching Transparency T9 *(CRB)* Recording Sheet for Reaching All Learners *(CRB p. 76)*	• *Chapter 2 Resource Book,* pp. 26–34 • Daily Transparency T8, CRB • Additional Examples Transparencies T10–T11, CRB • *Alternate Openers: Explorations,* p. 10
Lesson 2-4 Solving Addition Equations **NCTM:** Number and Operations, Algebra, Reasoning and Proof, Communication, Representation **NAEP:** Algebra 4a ☑ SAT-9 ☑ SAT-10 ☑ ITBS ☑ CTBS ☑ MAT ☑ CAT	**Optional** Teaching Transparency T13 *(CRB)* Algebra tiles or counters *(MK, CRB p. 77)*	• *Chapter 2 Resource Book,* pp. 35–43 • Daily Transparency T12, CRB • Additional Examples Transparencies T14–T15, CRB • *Alternate Openers: Explorations,* p. 11
Lesson 2-5 Solving Subtraction Equations **NCTM:** Number and Operations, Algebra, Reasoning and Proof, Communication, Representation **NAEP:** Algebra 4a ☑ SAT-9 ☑ SAT-10 ☑ ITBS ☑ CTBS ☑ MAT ☑ CAT	**Optional** Recording Sheet for Reaching All Learners *(CRB p. 78)*	• *Chapter 2 Resource Book,* pp. 44–52 • Daily Transparency T16, CRB • Additional Examples Transparencies T17–T18, CRB • *Alternate Openers: Explorations,* p. 12
Lesson 2-6 Solving Multiplication Equations **NCTM:** Number and Operations, Algebra, Problem Solving, Reasoning and Proof, Communication, Representation **NAEP:** Algebra 4a ☑ SAT-9 ☑ SAT-10 ☑ ITBS ☑ CTBS ☑ MAT ☑ CAT		• *Chapter 2 Resource Book,* pp. 53–61 • Daily Transparency T19, CRB • Additional Examples Transparencies T20–T21, CRB • *Alternate Openers: Explorations,* p. 13
Lesson 2-7 Solving Division Equations **NCTM:** Number and Operations, Algebra, Reasoning and Proof, Communication, Connections, Representation **NAEP:** Algebra 4a ☑ SAT-9 ☑ SAT-10 ☑ ITBS ☑ CTBS ☑ MAT ☑ CAT	**Optional** Set of cards numbered 1–9 *(CRB p. 79)*	• *Chapter 2 Resource Book,* pp. 62–70 • Daily Transparency T22, CRB • Additional Examples Transparencies T23–T24, CRB • *Alternate Openers: Explorations,* p. 14
Extension Inequalities **NCTM:** Algebra, Representation **NAEP:** Algebra 4a ☐ SAT-9 ☐ SAT-10 ☐ ITBS ☑ CTBS ☐ MAT ☐ CAT		• Additional Examples Transparency T25, CRB
Section 2B Assessment		• Section 2B Quiz, AR p. 8 • *Test and Practice Generator* CD-ROM

SAT = *Stanford Achievement Tests* **ITBS** = *Iowa Test of Basic Skills* **CTBS** = *Comprehensive Test of Basic Skills/Terra Nova*
MAT = *Metropolitan Achievement Tests* **CAT** = *California Achievement Test*

NCTM = Complete standards can be found on pages T27–T33. **NAEP** = Complete standards can be found on pages A31–A35.

SE = *Student Edition* **TE** = *Teacher's Edition* **AR** = *Assessment Resources* **CRB** = *Chapter Resource Book* **MK** = *Manipulatives Kit*

Section Overview

Determining Solutions of Equations — *Lesson 2-3*

Why? Because equations are used to represent mathematical relationships in real situations, students can strengthen their problem-solving skills by learning to recognize and identify solutions to equations.

Situation: The Ferris wheel ride costs 3 tokens. After riding the Ferris wheel, Bailey had 5 tokens remaining. How many tokens did Bailey have before riding the Ferris wheel?

$$t - 3 = 5$$

$t = 9$ ***is not*** a solution because $9 - 3 = 5$ ***is not*** true.

$t = 8$ ***is*** a solution because $8 - 3 = 5$ ***is*** true.

Solving One-Step Equations — *Lessons 2-4 through 2-7*

Why? Many students can figure out the answers to problems without solving one-step equations. However, they will need to use the concepts learned at this level to solve equations involving fractions and decimals later in this course.

Equation	Operation	Inverse Operation	Isolating the Variable
$a + 9 = 17$	Addition	Subtraction	$a + 9 = 17$ $\underline{-9}\ \ \underline{-9}$ $a = 8$
$y - 11 = 25$	Subtraction	Addition	$y - 11 = 25$ $\underline{+11}\ \ \underline{+11}$ $y = 36$
$7b = 21$	Multiplication	Division	$7b = 21$ $\frac{7b}{7} = \frac{21}{7}$ $b = 3$
$\frac{x}{3} = 12$	Division	Multiplication	$\frac{x}{3} = 12$ $\frac{x}{3}(3) = 12(3)$ $x = 36$

2-3 Organizer

Pacing: Traditional 1 day
Block $\frac{1}{2}$ day

Objective: Students determine whether a number is a solution of an equation.

Indiana Academic Standards
6.3.1, 6.7.1, 6.7.9, 6.7.11

Warm Up

Evaluate each expression for $x = 8$.

1. $3x + 5$ **29** **2.** $x + 8$ **16**
3. $2x - 7$ **9** **4.** $8x \div 4$ **16**
5. $7x - 1$ **55** **6.** $x - 3$ **5**

Problem of the Day

Complete the magic square so that every row, column, and diagonal add up to the same total.

9	2	10
8	7	6
4	12	5

Available on Daily Transparency in CRB

Math Humor

What do "$x = 4$" and salt water have in common? Both can be called solutions.

2-3 Equations and Their Solutions

Learn to determine whether a number is a solution of an equation.

Vocabulary
equation
solution

An **equation** is a mathematical statement that two quantities are equal. You can think of a correct equation as a balanced scale.

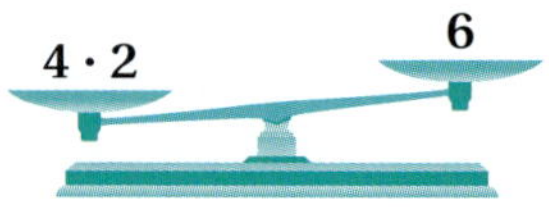

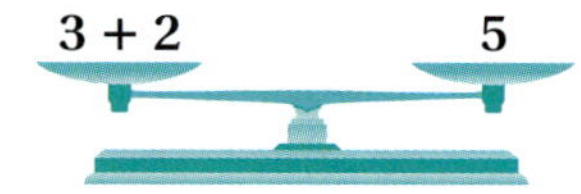

Equations may contain variables. If a value for a variable makes an equation true, that value is a **solution** of the equation.

Reading Math
The symbol ≠ means "is not equal to."

$s + 15 = 27$

$s = 12$

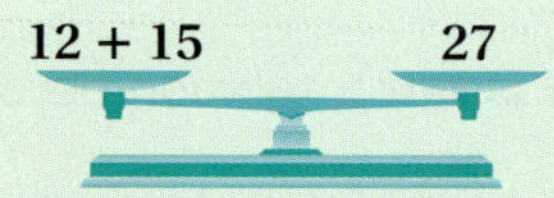

$s = 12$ is a solution because $12 + 15 = 27$.

$s = 10$

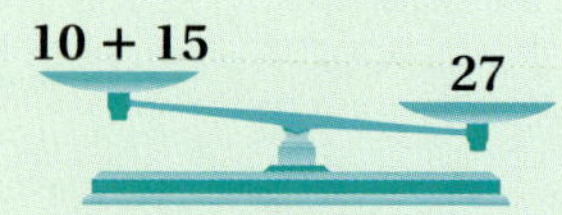

$s = 10$ is not a solution because $10 + 15 \neq 27$.

EXAMPLE 1 Determining Solutions of Equations

Determine whether the given value of each variable is a solution.

A $a + 23 = 82$ for $a = 61$

$a + 23 = 82$
$61 + 23 \stackrel{?}{=} 82$ *Substitute 61 for a.*
$84 \stackrel{?}{=} 82$ *Add.*

Since $84 \neq 82$, 61 is not a solution to $a + 23 = 82$.

1 Introduce

Alternate Opener

EXPLORATION

2-3 Equations and Their Solutions

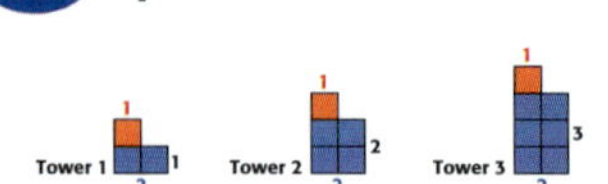

In the sequence of towers, the base of each tower is always 2 squares. Each tower has 1 red square on top. The heights of the towers vary. If we call the height of each tower *h*, we can represent this pattern with the following expression:

$2h + 1$ (Base of each tower: 2; Height: h; Red square on top: 1)

1. Use the pattern in the sequence of towers to draw a tower with 11 squares. Which tower number is it in the sequence?
2. Use the pattern to solve the equation $2h + 1 = 21$.
3. Look at the sequence of blue and white grids and draw a picture of the grid that has 10 blue squares.

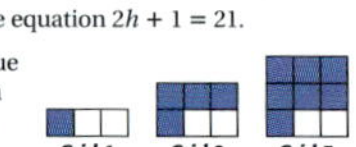

   a. Where in the sequence does this grid occur?
   b. Write an equation for the problem in 3a.

Think and Discuss

4. **Discuss** what is meant by "a solution to an equation."

Motivate

Have students experiment with a balance scale, placing different combinations of items on each side of the scale until it is balanced. Lead students to understand that when the scale is balanced, the sum of the weights of the items on the left equals the sum of the weights of the items on the right.

Exploration worksheet and answers on Chapter 2 Resource Book pp. 27 and 85

2 Teach

Lesson Presentation

Guided Instruction

In this lesson, students learn to determine whether a number is a solution of an equation. First teach students to determine whether given values for variables are solutions of equations. Then apply the concept to real-world problems.

Teaching Tip Always refer to an equation as a balanced mathematical statement and emphasize that only the correct solution will maintain the balance.

Determine whether the given value of each variable is a solution.

B $60 \div c = 6$ for $c = 10$

$60 \div c = 6$

$60 \div 10 \stackrel{?}{=} 6$ *Substitute 10 for c.*

$6 \stackrel{?}{=} 6$ *Divide.*

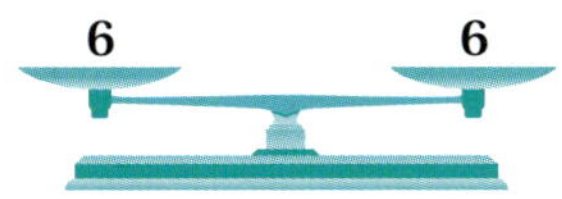

Because $6 = 6$, 10 is a solution to $60 \div c = 6$.

You can use equations to check whether measurements given in different units are equal.

For example, there are 12 inches in one foot. If you have a measurement in feet, multiply by 12 to find the measurement in inches: $12 \cdot \text{feet} = \text{inches}$, or $12f = i$.

If you have one measurement in feet and another in inches, check whether the two numbers make the equation $12f = i$ true.

EXAMPLE 2 *Life Science Application*

One science book states that a manatee can grow to be 13 feet long. According to another book, a manatee may grow to 156 inches. Determine if these two measurements are equal.

$12f = i$

$12 \cdot 13 \stackrel{?}{=} 156$ *Substitute.*

$156 \stackrel{?}{=} 156$ *Multiply.*

Because $156 = 156$, 13 feet is equal to 156 inches.

Think and Discuss

1. **Tell** which of the following is the solution to $y \div 2 = 9$: $y = 14$, $y = 16$, or $y = 18$. How do you know?
2. **Give an example** of an equation with a solution of 15.

COMMON ERROR ALERT

The equations in this lesson are relatively straightforward. Because of this, students may be careless, substituting the incorrect value for the variable. Caution students to double-check their work for accuracy.

Additional Examples

Example 1

Determine whether the given value of the variable is a solution.

A. $b - 447 = 1{,}203$ for $b = 1{,}650$

1,650 is a solution to $b - 447 = 1{,}203$.

B. $27x = 1{,}485$ for $x = 54$

54 is not a solution to $27x = 1{,}485$.

Example 2

Paulo says that the park is 19 yards long. Jamie says that the park is 664 inches long. Determine if these two measurements are equal.

19 yards is not equal to 664 inches.

3 Close

Reaching All Learners

Through Critical Thinking

Display a set of scales. (You may use Teaching Transparency T9 in the Chapter 2 Resource Book.) For each of the following equations, write the left side over the left scale and the right side over the right scale, and ask what the value of the variable must be for the scales to remain balanced. (A recording sheet is provided on Chapter 2 Resource Book page 76.)

1. $x + 3 = 7$ $x = 4$
2. $t - 5 = 3$ $t = 8$
3. $4y = 28$ $y = 7$
4. $\frac{w}{2} = 6$ $w = 12$

Summarize

Give brief definitions of the new vocabulary in the lesson: *equation* and *solution*. Have a student explain to the class how equations are like balanced scales. Ask the students how they know when they have found the right answer.

Answers to Think and Discuss

1. $y = 18$ because $18 \div 2 = 9$.
2. Possible answer: $x \div 3 = 5$

2-3 PRACTICE & ASSESS

Students may want to refer back to the lesson examples.

Assignment Guide

If you finished Example 1 assign:
Core 1–6, 8–19, 46–54
Enriched 21–34, 41–44, 46–54

If you finished Example 2 assign:
Core 1–20, 33–41 odd, 46–54
Enriched 21–54

Notes

2-3 Exercises

FOR EXTRA PRACTICE
see page 639

internet connect
Homework Help Online
go.hrw.com Keyword: MR4 2-3

GUIDED PRACTICE

See Example 1

Determine whether the given value of each variable is a solution.

1. $c + 23 = 48$ for $c = 35$ no
2. $z + 31 = 73$ for $z = 42$ yes
3. $96 = 130 - d$ for $d = 34$ yes
4. $85 = 194 - a$ for $a = 105$ no
5. $75 \div y = 5$ for $y = 15$ yes
6. $78 \div n = 13$ for $n = 5$ no

See Example 2

7. An almanac states that the Minnehaha Waterfall in Minnesota is 53 feet tall. A tour guide said the Minnehaha Waterfall is 636 inches tall. Determine if these two measurements are equal.
   53 feet is equal to 636 inches.

INDEPENDENT PRACTICE

See Example 1

Determine whether the given value of the variable is a solution.

8. $w + 19 = 49$ for $w = 30$ yes
9. $d + 27 = 81$ for $d = 44$ no
10. $g + 34 = 91$ for $g = 67$ no
11. $k + 16 = 55$ for $k = 39$ yes
12. $101 = 150 - h$ for $h = 49$ yes
13. $89 = 111 - m$ for $m = 32$ no
14. $116 = 144 - q$ for $q = 38$ no
15. $92 = 120 - t$ for $t = 28$ yes
16. $80 \div b = 20$ for $b = 4$ yes
17. $91 \div x = 7$ for $x = 12$ no
18. $55 \div j = 5$ for $j = 10$ no
19. $49 \div r = 7$ for $r = 7$ yes

See Example 2

20. Kent earns $6 per hour at his after-school job. One week, he worked 12 hours and received a paycheck for $66. Determine if Kent was paid the correct amount of money. (*Hint:* $6 • hours = total pay)
    No, he was not paid correctly.

PRACTICE AND PROBLEM SOLVING

Determine whether the given value of the variable is a solution.

21. $93 = 48 + u$ for $u = 35$ no
22. $112 = 14 \times f$ for $f = 8$ yes
23. $13 = m \div 8$ for $m = 104$ yes
24. $79 = z - 23$ for $z = 112$ no
25. $64 = l - 34$ for $l = 98$ yes
26. $105 = p \times 7$ for $p = 14$ no
27. $94 \div s = 26$ for $s = 3$ no
28. $v + 79 = 167$ for $v = 88$ yes
29. $m + 36 = 54$ for $m = 18$ yes
30. $x - 35 = 96$ for $x = 112$ no
31. $12y = 84$ for $y = 7$ yes
32. $7x = 56$ for $x = 8$ yes
33. $3x = 150$ for $x = 65$ no
34. $20k = 115$ for $k = 9$ no

Math Background

Much of the work students will do in their study of algebra will focus on equations. Because this chapter is an introduction to algebra, it is appropriate to highlight for students how equations relate to their earlier work in arithmetic.

Students realize that $2 + 2 = 4$ is a true statement and that $2 + 2 = 5$ is a false statement. Because of this, students may think that all algebraic equations have exactly one solution. Let them know that this is not the case, as they will discover through further work in algebra.

RETEACH 2-3

LESSON 2-3 **Reteach**
Equations and Their Solutions

An equation is a mathematical sentence that says that two quantities are equal.

Some equations contain variables. A solution for an equation is a value for a variable that makes the statement true.

You can write related facts using addition and subtraction.
$7 + 6 = 13$ $13 - 6 = 7$

You can write related facts using multiplication and division.
$3 \cdot 4 = 12$ $12 \div 4 = 3$

You can use related facts to find solutions for equations. If the related fact matches the value for the variable, then that value is a solution.

A. $x + 5 = 9$, when $x = 3$
Think: $9 - 5 = x$
$x = 4$
$3 \neq 4$
So $x = 3$ is not a solution of $x + 5 = 9$.

B. $x - 7 = 5$, when $x = 12$
Think: $5 + 7 = x$
$x = 12$
$12 = 12$
So $x = 12$ is a solution of $x - 7 = 5$.

C. $2x = 14$, when $x = 9$
Think: $14 \div 2 = x$
$x = 7$
$9 \neq 7$
So $x = 9$ is not a solution for $2x = 14$.

D. $x \div 5 = 3$, when $x = 15$
Think: $3 \cdot 5 = x$
$x = 15$
$15 = 15$
So $x = 15$ is a solution for $x \div 5 = 3$.

Use related facts to determine whether the given value is a solution for each equation.

1. $x + 6 = 14$, when $x = 8$ yes
2. $s \div 4 = 5$, when $s = 24$ no
3. $g - 3 = 7$, when $g = 11$ no
4. $3a = 18$, when $a = 6$ yes
5. $26 = y - 9$, when $y = 35$ yes
6. $b \cdot 5 = 20$, when $b = 3$ no
7. $15 = v \div 3$, when $v = 45$ yes
8. $11 = p + 6$, when $p = 5$ yes
9. $6k = 78$, when $k = 12$ no

PRACTICE 2-3

LESSON 2-3 **Practice B**
Equations and Their Solutions

Determine whether the given value of the variable is a solution.

1. $9 + x = 21$ for $x = 11$ No
2. $n - 12 = 5$ for $n = 17$ Yes
3. $25 \cdot r = 75$ for $r = 3$ Yes
4. $72 \div q = 8$ for $q = 9$ Yes
5. $28 + c = 43$ for $c = 15$ Yes
6. $u \div 11 = 10$ for $u = 111$ No
7. $\frac{k}{8} = 4$ for $k = 24$ No
8. $16x = 48$ for $x = 3$ Yes
9. $73 - f = 29$ for $f = 54$ No
10. $67 - j = 25$ for $j = 42$ Yes
11. $39 \div v = 13$ for $v = 3$ Yes
12. $88 + d = 100$ for $d = 2$ No
13. $14p = 20$ for $p = 5$ No
14. $6w = 30$ for $w = 5$ Yes
15. $7 + x = 70$ for $x = 10$ No
16. $6 \cdot n = 174$ for $n = 29$ Yes

Replace each ? with a number that makes the equation correct.

17. $5 + 1 = 2 +$? 4
18. $10 -$? $= 12 - 7$ 5
19. ? $\cdot 3 = 2 \cdot 9$ 6
20. $28 \div 4 = 14 \div$? 2
21. ? $+ 8 = 6 + 3$ 1
22. $12 \cdot 0 =$? $\cdot 15$ 0
23. Carla had $15. After she bought lunch, she had $8 left. Write an equation using the variable x to model this situation. What does your variable represent?
    $15 - x = 8$; x = the amount she spent on lunch
24. Seventy-two people signed up for the soccer league. After the players were evenly divided into teams, there were 6 teams in the league. Write an equation to model this situation using the variable x.
    $72 \div x = 6$

Replace each ▒ with a number that makes the equation correct.

35. 4 + 1 = ▒ + 2 **3**

36. 2 + ▒ = 6 + 2 **6**

37. ▒ − 5 = 9 − 2 **12**

38. 5(4) = 10(▒) **2**

39. 3 + 6 = ▒ − 4 **13**

40. 12 ÷ 4 = 9 ÷ ▒ **3**

41. Rebecca has 17 one-dollar bills. Courtney has 350 nickels. Do the two girls have the same amount of money? (*Hint*: First find how many nickels are in a dollar.)
17 ≠ 350 ÷ 20; No, they do not have the same amount of money.

42. ***CHOOSE A STRATEGY*** What should replace the question mark to keep the scale balanced? **C**

43. ***WRITE A PROBLEM*** Write an equation using a variable and information from the graph.

44. ***WRITE ABOUT IT*** Explain how to determine if a value is a solution to an equation.

45. ***CHALLENGE*** Is $n = 4$ a solution for $n^2 + 79 = 88$? Explain.
no; The equation is not balanced when $n = 4$.

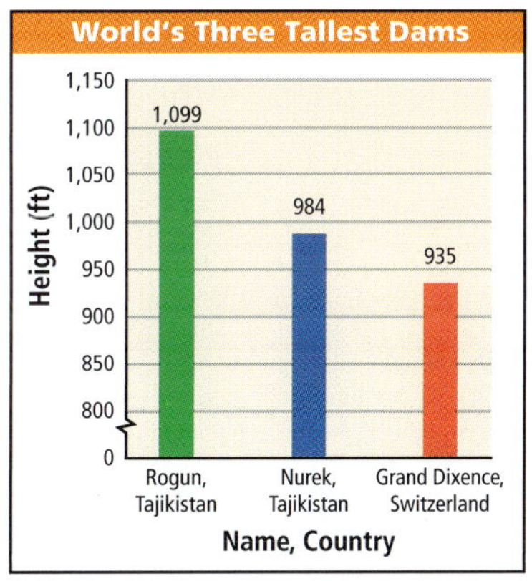

Spiral Review

Find each value. (Lesson 1-3)

46. 5^3 **125** **47.** 3^4 **81** **48.** 2^6 **64** **49.** 6^3 **216**

Evaluate each expression. (Lesson 1-4)

50. 50 − 2 × 10 **30**

51. $16 \div (6 + 2) + 2^2$ **6**

52. $20 \div (19 - 9) \times 3^2 - 5$ **13**

53. (8 + 7) ÷ 3 × (19 − 4) **75**

54. **TEST PREP** Which of the following phrases can be represented by $79x$? (Lesson 2-2) **C**

A 79 divided by x

B The difference of 79 and x

C The product of 79 and x

D The sum of 79 and x

COMMON ERROR ALERT

Some students may be confused about how to approach Exercises 35–40 because there are operations shown on each side of the equal sign. Advise students to perform the operation on the side of the equation that only has numbers, and think of the gray box on the other side of the equation as a variable.

Answers

43. Possible answer: $935 + h = 1{,}099$; How much higher is the Rogun Dam than the Grand Dixence Dam?

44. Substitute the value for the variable and see if both sides are equal.

Journal

Have students write about other ways scales are used in everyday life.

Test Prep Doctor

For Exercise 54, students will need to remember two things: translating numerical expressions into word phrases and the different notations for multiplication. Students who answered **A, B,** or **D** have confused the multiplication notation shown as being one of the other operations.

CHALLENGE 2-3

LESSON 2-3 **Challenge**
Keep It Balanced

Study the scales below. Then circle the solution below each scale that will keep it balanced.

1.

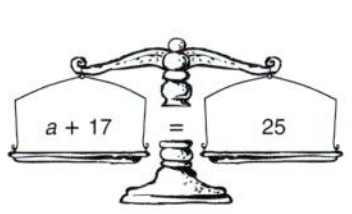

(circled) $a = 8$ $a = 9$

2.

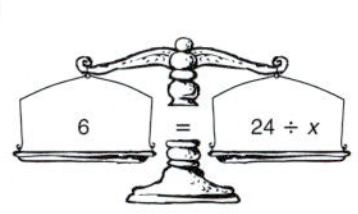

$x = 3$ (circled) $x = 4$

3.

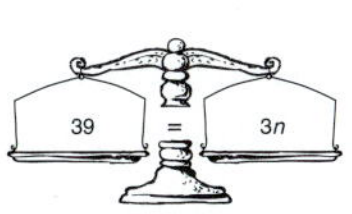

$n = 12$ (circled) $n = 13$

4.

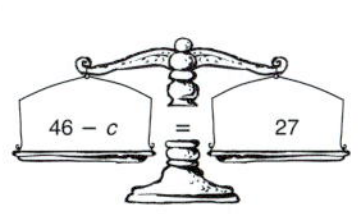

(circled) $c = 19$ $c = 29$

5.

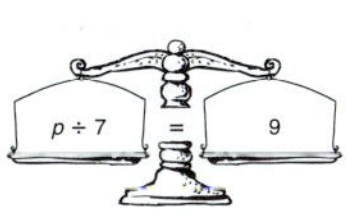

$p = 49$ (circled) $p = 63$

6.

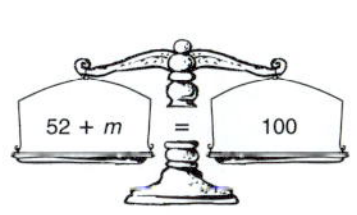

(circled) $m = 48$ $m = 58$

PROBLEM SOLVING 2-3

LESSON 2-3 **Problem Solving**
Equations and Their Solutions

Use the table to write and solve an equation to answer each question. Then use your answers to complete the table.

1. A hippopotamus can stay underwater 3 times as long as a sea otter can. How long can a sea otter stay underwater?
$3x = 15$; $x = 5$;
5 minutes

2. A seal can stay underwater 10 minutes longer than a muskrat can. How long can a muskrat stay underwater?
$x + 10 = 22$; $x = 12$;
12 minutes

3. A sperm whale can stay underwater 7 times longer than a sea cow can. How long can a sperm whale stay underwater?
$x \div 7 = 16$; $x = 112$;
112 minutes

How Many Minutes Can Mammals Stay Underwater?	
Hippopotamus	15
Human	**1**
Muskrat	**12**
Platypus	10
Polar bear	**2**
Sea cow	16
Sea otter	**5**
Seal	22
Sperm whale	**112**

Circle the letter of the correct answer.

4. The difference between the time a platypus and a polar bear can stay underwater is 8 minutes. How long can a polar bear stay underwater?
A 1 minute
(B) 2 minutes
C 3 minutes
D 5 minutes

5. When you divide the amount of time any of the animals in the table can stay underwater by itself, the answer is always the amount of time the average human can stay underwater. How long can the average human stay underwater?
F 6 minutes
G 4 minutes
H 2 minutes
(J) 1 minute

Lesson Quiz

Determine whether the given value of the variable is a solution.

1. $85 = 13x$ for $x = 5$ **no**

2. $w + 38 = 210$ for $w = 172$ **yes**

3. $8y = 88$ for $y = 11$ **yes**

4. $16 = w \div 6$ for $w = 98$ **no**

5. The local pizza shop charged Kylee $172 for 21 medium pizzas. The price of a medium pizza is $8. Determine whether Kylee paid the correct amount of money. (*Hint:* $8 · pizzas = total cost) **no**

Available on Daily Transparency in CRB

2-4 Organizer

Pacing: Traditional 1 day
Block $\frac{1}{2}$ day

Objective: Students solve whole number addition equations.

Indiana Academic Standards
6.3.1, 6.5.9, 6.7.1, 6.7.9, 6.7.10, 6.7.11

Warm Up

Determine whether each value is a solution.

1. $86 + x = 102$ for $x = 16$ **yes**
2. $18 + x = 26$ for $x = 4$ **no**
3. $x + 46 = 214$ for $x = 168$ **yes**
4. $9 + x = 35$ for $x = 26$ **yes**

Problem of the Day

After Renee used 40 m of string for her kite and gave 5 m to her sister for her wagon, she had 8 m of string left. How much string did she have to start with? **53 m**

Available on Daily Transparency in CRB

Math Fact

The symbol for equality, =, was first used in Robert Recorde's book *The Whetstone of Witte,* published in 1557.

2-4 Solving Addition Equations

Learn to solve whole-number addition equations.

Some surfers recommend that the length of a beginner's surfboard be 14 inches more than the surfer's height. If a surfboard is 82 inches, how tall should the surfer be to ride it?

The height of the surfer *combined* with 14 inches equals 82 inches. To combine amounts, you need to add.

Let h stand for the surfer's height. You can use the equation $h + 14 = 82$.

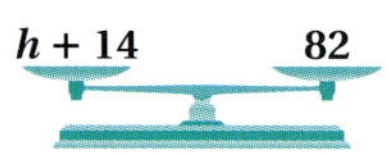

The equation h + 14 = 82 can be represented as a balanced scale.

To find the value of h, you need h by itself on one side of a balanced scale.

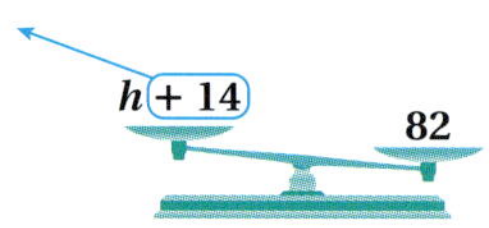

To get h by itself, first take away 14 from the left side of the scale. Now the scale is unbalanced.

To rebalance the scale, take away 14 from the other side.

A surfer using an 82-inch surfboard should be 68 inches tall.

Taking away 14 from both sides of the scale is the same as subtracting 14 from both sides of the equation.

$$\begin{array}{rcr} h + 14 & = & 82 \\ -\ 14 & & -\ 14 \\ \hline h \quad & = & 68 \end{array}$$

Subtraction is the inverse, or opposite, of addition. If an equation contains addition, solve it by subtracting from both sides to "undo" the addition.

1 Introduce

Alternate Opener

EXPLORATION

2-4 Solving Addition Equations

How much change from a dollar do you get when you buy something that costs 51 cents?

This problem can also be expressed as what number plus 51 is 100?

$n + 51 = 100$
$49 + 51 = 100$
$n = 49$ The change is 49¢.

Find the value of n in each equation.

1. $4 + n = 100$ $n =$ ______
2. $n + 45 = 100$ $n =$ ______
3. $19 + n = 100$ $n =$ ______
4. $n + 65 = 100$ $n =$ ______
5. $100 = 41 + n$ $n =$ ______

Think and Discuss

6. **Discuss** your strategies for solving the equations.
7. **Explain** how you can mentally find the solution to $n + 125 = 500$.

Motivate

Review the terms *constant*, *variable*, and *equation*. Have students give examples of addition equations that include a constant being added to a variable.

Exploration worksheet and answers on Chapter 2 Resource Book pp. 36 and 87

2 Teach

Lesson Presentation

Guided Instruction

In this lesson, students learn to solve whole-number addition problems. First teach students to use a pictorial model of a balanced scale to solve an addition equation. You may use Teaching Transparency T13 in the Chapter 2 Resource Book. Then teach students to solve addition equations by using subtraction, the inverse of addition. Next, have students check their answers.

Teaching Tip Have students draw balanced scale models when working through the exercises to give them a visual model to reinforce the concept.

EXAMPLE 1 Solving Addition Equations

Solve each equation. Check your answers.

A $x + 62 = 93$

$$\begin{aligned} x + 62 &= 93 \\ -62 \quad & \; -62 \\ x &= 31 \end{aligned}$$

62 is added to x.
Subtract 62 from both sides to undo the addition.

Check $x + 62 = 93$

$31 + 62 \stackrel{?}{=} 93$ *Substitute 31 for x in the equation.*

$93 \stackrel{?}{=} 93$ ✔ *31 is the solution.*

B $81 = 17 + y$

$$\begin{aligned} 81 &= 17 + y \\ -17 \quad & \; -17 \\ 64 &= y \end{aligned}$$

17 is added to y.
Subtract 17 from both sides to undo the addition.

Check $81 = 17 + y$

$81 \stackrel{?}{=} 17 + 64$ *Substitute 64 for y in the equation.*

$81 \stackrel{?}{=} 81$ ✔ *64 is the solution.*

EXAMPLE 2 *Social Studies Application*

Dyersberg, Newton, and St. Thomas are located along Ventura Highway, as shown on the map. Find the distance d between Newton and Dyersberg.

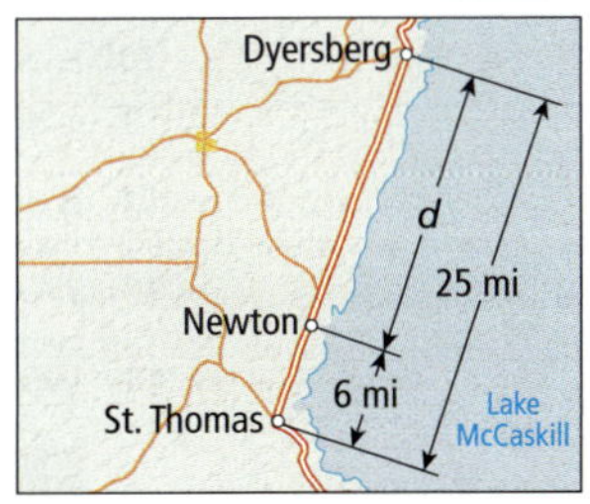

distance between Dyersberg and St. Thomas	=	distance between Newton and St. Thomas	+	distance between Newton and Dyersberg
25	=	6	+	d

$$\begin{aligned} 25 &= 6 + d \\ -6 \quad & \; -6 \\ 19 &= d \end{aligned}$$

6 is added to d.
Subtract 6 from both sides to undo the addition.

The distance between Newton and Dyersberg is 19 miles.

Think and Discuss

1. **Tell** whether the solution of $c + 4 = 21$ will be less than 21 or greater than 21. Explain.
2. **Describe** how you could check your answer in Example 2.

COMMON ERROR ALERT

Some students may add the amount subtracted from one side of the equation to the other side. For example: $h + 14 = 82$; $h + 14 - 14 = 82 + 14$; $h = 96$. Stress that to keep the equation "balanced" the same amount must be taken away from each side.

Additional Examples

Example 1

Solve each equation. Check your answers.

A. $x + 87 = 152$
$x = 65$

B. $72 = 18 + y$
$54 = y$

Example 2

Johnstown, Cooperstown, and Springfield are located in that order in a straight line along a highway. It is 12 miles from Johnstown to Cooperstown and 95 miles from Johnstown to Springfield. Find the distance d between Cooperstown and Springfield.

It is 83 miles from Cooperstown to Springfield.

3 Close

Summarize

Walk students through the process of how to solve addition equations involving variables and how to check solutions. Emphasize that equations are balanced mathematical statements and that the balance must be maintained while finding the solution. Have a volunteer demonstrate how to properly write the steps for finding the solution and checking the answer.

Reaching All Learners
Through Concrete Manipulatives

Students may have difficulty grasping the concept of using subtraction to solve addition equations. Have students use manipulatives such as algebra tiles or counters provided in the Manipulatives Kit to solve addition equations such as the following:

1. $x + 7 = 12$ 5
2. $9 + z = 20$ 11
3. $8 = 5 + y$ 3
4. $11 = 4 + y$ 7

Answers to Think and Discuss

1. Less than 21; you subtract 4 from both sides of the equation, which will make the answer 4 less than 21.
2. Substitute 19 for d in the equation $25 = 6 + d$. Because $6 + 19 = 25$, 19 is the correct solution.

2-4 PRACTICE & ASSESS

Students may want to refer back to the lesson examples.

Assignment Guide

If you finished Example 1 assign:
Core 1–6, 8–16, 36–43
Enriched 1–2, 18–30, 36–43

If you finished Example 2 assign:
Core 1–17, 19–31 odd, 32, 36–43
Enriched 14–43

Notes

2-4 Exercises

FOR EXTRA PRACTICE see page 639

internet connect
Homework Help Online
go.hrw.com Keyword: MR4 2-4

GUIDED PRACTICE

See Example 1

Solve each equation. Check your answers.

1. $x + 54 = 90$ $x = 36$
2. $49 = 12 + y$ $y = 37$
3. $n + 27 = 46$ $n = 19$
4. $22 + t = 91$ $t = 69$
5. $31 = p + 13$ $p = 18$
6. $c + 38 = 54$ $c = 16$

See Example 2

7. Lou, Michael, and Georgette live on Mulberry Street, as shown on the map. Lou lives 10 blocks from Georgette. Georgette lives 4 blocks from Michael. How many blocks does Michael live from Lou? 6 blocks

INDEPENDENT PRACTICE

See Example 1

Solve each equation. Check your answers.

8. $x + 19 = 24$ $x = 5$
9. $10 = r + 3$ $r = 7$
10. $s + 11 = 50$ $s = 39$
11. $b + 17 = 42$ $b = 25$
12. $12 + m = 28$ $m = 16$
13. $z + 68 = 77$ $z = 9$
14. $72 = n + 51$ $n = 21$
15. $g + 28 = 44$ $g = 16$
16. $27 = 15 + y$ $y = 12$

See Example 2

17. What is the length of a killer whale? 6 meters

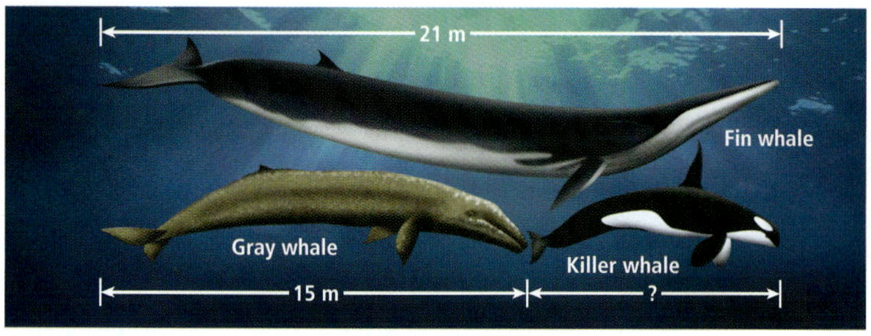

PRACTICE AND PROBLEM SOLVING

Solve each equation.

18. $x + 12 = 16$ $x = 4$
19. $n + 32 = 39$ $n = 7$
20. $23 + q = 34$ $q = 11$
21. $52 + y = 71$ $y = 19$
22. $73 = c + 35$ $c = 38$
23. $93 = h + 15$ $h = 78$
24. $125 = n + 85$ $n = 40$
25. $87 = b + 18$ $b = 69$
26. $12 + y = 50$ $y = 38$
27. $t + 17 = 43$ $t = 26$
28. $k + 9 = 56$ $k = 47$
29. $25 + m = 47$ $m = 22$

Math Background

In his *Elements, Book I*, Euclid listed five axioms that he called "common notions."

1. Things which are equal to the same thing are also equal to one another.
2. If equals be added to equals, the wholes are equal.
3. If equals be subtracted from equals, the remainders are equal.
4. Things which coincide with one another are equal to one another.
5. The whole is greater than the part.

RETEACH 2-4

LESSON 2-4 **Reteach**
Solving Addition Equations

To solve an equation, you need to get the variable alone on one side of the equal sign.

You can use tiles to help you solve addition equations.

Subtraction undoes addition, so you can use subtraction to solve addition equations.

One positive tile and one negative tile is called a zero pair because together they have a value of zero.

To solve $x + 3 = 5$, first use tiles to model the equation.

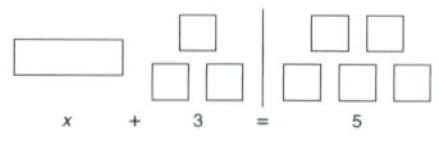

Next, add enough subtraction tiles to get the variable alone. Then add the same number of subtraction tiles to the other side of the equal sign.

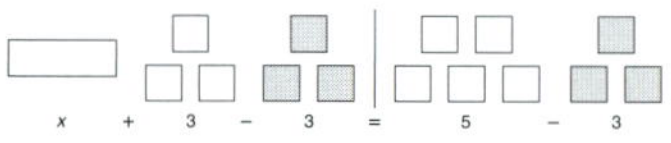

Then remove the greatest possible number of zero pairs from each side of the equal sign.

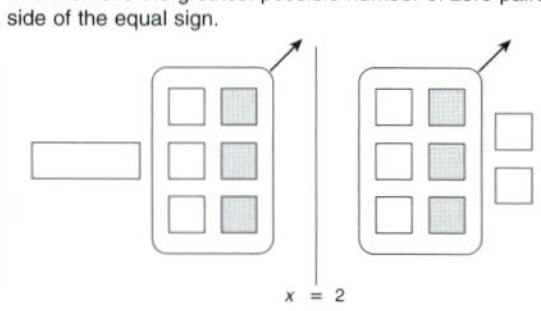

Check: $x + 3 = 5$
$2 + 3 = 5$
$5 = 5$

The remaining tiles represent the solution.
$x = 2$

Use tiles to solve each equation. Then check each answer.

1. $x + 8 = 11$ $x = 3$
2. $x + 4 = 9$ $x = 5$
3. $x + 7 = 13$ $x = 6$
4. $x + 1 = 3$ $x = 2$

PRACTICE 2-4

LESSON 2-4 **Practice B**
Solving Addition Equations

Solve each equation. Check your answers.

1. $s + 3 = 23$ $s = 20; 20 + 3 = 23$
2. $v + 10 = 49$ $v = 39; 39 + 10 = 49$
3. $q + 9 = 16$ $q = 7; 7 + 9 = 16$
4. $81 + m = 90$ $m = 9; 81 + 9 = 90$
5. $38 + x = 44$ $x = 6; 38 + 6 = 44$
6. $28 + n = 65$ $n = 37; 28 + 37 = 65$
7. $t + 31 = 50$ $t = 19; 19 + 31 = 50$
8. $25 + p = 39$ $p = 14; 25 + 14 = 39$
9. $19 + v = 24$ $v = 5; 19 + 5 = 24$

Solve each equation.

10. $m + 8 = 17$ $m = 9$
11. $r + 14 = 20$ $r = 6$
12. $25 + x = 32$ $x = 7$
13. $47 + p = 55$ $p = 8$
14. $19 + d = 27$ $d = 8$
15. $13 + n = 26$ $n = 13$
16. $q + 12 = 19$ $q = 7$
17. $34 + f = 43$ $f = 9$
18. $52 + w = 68$ $w = 16$

19. Kenya bought 28 beads, and Nancy bought 25 beads. It takes 35 beads to make a necklace. Write and solve two addition equations to find how many more beads they each need to make a necklace.
Kenya: $28 + b = 35$; $b = 7$ beads; Nancy: $25 + b = 35$; $b = 10$ beads

20. During a sales trip, Mr. Jones drove 15 miles east from Brownsville to Carlton. Then he drove several more miles east from Carlton to Sun City. The distance from Brownsville to Sun City is 35 miles. Write and solve an addition equation to find how many miles it is from Carlton to Sun City.
$15 + m = 35$; $m = 20$ miles

Science LINK

Popular items like the ball and the mood rings above are made of heat-sensitive materials. Changes in temperature, such as the warmth of your body heat, cause these materials to change color.

30. ***PHYSICAL SCIENCE*** Temperature can be measured in degrees Fahrenheit, degrees Celsius, or kelvins. To convert from degrees Celsius to kelvins, add 273 to the Celsius temperature. Complete the table.

	Kelvins (K)	°C + 273 = K	Celsius (°C)
Water Freezes	273	°C + 273 = 273	0
Room Temperature	296	°C + 273 = 296	23
Body Temperature	310	°C + 273 = 310	37
Water Boils	373	°C + 273 = 373	100

31. Alyssa has a temperature that is 2°C higher than normal body temperature. If Alyssa's temperature is 39°C, what is normal body temperature? **37°C**

32. ***HISTORY*** In 1520, the explorer Ferdinand Magellan tried to measure the depth of the ocean. He weighted a 370 m rope and lowered it into the ocean. This rope was not long enough to reach the ocean floor. Suppose the depth at this location was 1,250 m. How much longer would Magellan's rope have to have been to reach the ocean floor? **880 m**

33. ***WRITE A PROBLEM*** Use data from your science book to write a problem that can be solved using an addition equation. Solve your problem.

34. ***WRITE ABOUT IT*** Why are addition and subtraction called inverse operations? **They are inverse because one undoes the other.**

35. ***CHALLENGE*** In the magic square at right, each row, column, and diagonal has the same sum. Find the values of x, y, and z. **$x = 43$; $y = 73$; $z = 13$**

7	61	x
y	37	1
31	z	67

Spiral Review

Estimate each sum or difference by rounding to the place value indicated. (Lesson 1-2)

36. 6,832 + 2,078; thousands **9,000**

37. 52,854 − 25,318; ten thousands **20,000**

38. 49,135 − 12,798; thousands **36,000**

39. 78,497 + 19,980; ten thousands **100,000**

Evaluate each expression. (Lesson 1-4)

40. $3^3 - (15 - 8) + 4 \times 5$ **40**

41. $17 \times (5 - 3) + 2^4 \div 8$ **36**

42. $81 - 4 \times 3 + 18 \div (6 + 3)$ **71**

43. **TEST PREP** Which equation below is an example of the Associative Property of Addition? (Lesson 1-5) **C**

A $(7 \times 2) + 8 = 16 + 6$

B $2 \times 3 \times 5 = 2 \times 5 \times 3$

C $(5 + 7) + 9 = 5 + (7 + 9)$

D $4 \times (2 + 3) = (4 \times 2) + (4 \times 3)$

Answer

33. Possible answer: 37 mL of glycerin was added to a solution to bring the total volume to 93 mL. What was the original volume of the solution? $37 + x = 93$, $x = 56$ The original volume was 56 mL.

Journal

Have students write about why taking your shoes off before you stand on a scale is similar to solving an equation like $x + 2 = 95$.

Test Prep Doctor

For Exercise 43, students need to know the Associative, Commutative, and Distributive Properties. Students who answered **B** confused the Commutative Property with the Associative Property. Students who answered **D** confused the Distributive Property with the Associative Property. Answer **A** is not an example of any property.

CHALLENGE 2-4

LESSON 2-4 **Challenge**

The Temperature's Rising

Each pair of thermometers shows a beginning temperature on the left and an ending temperature on the right. Write and solve an addition equation to find the change in temperature shown on each pair of thermometers.

1.

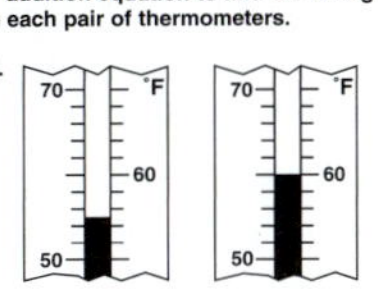

$55 + x = 60$; $x = 5$°F

2.

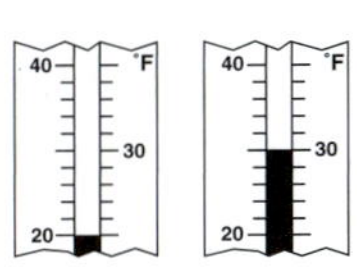

$20 + x = 30$; $x = 10$°F

3.

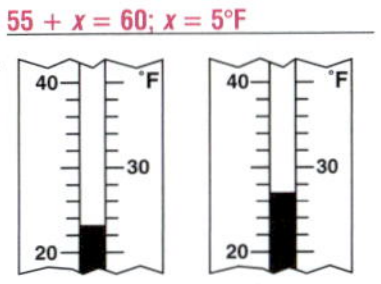

$23 + x = 27$; $x = 4$°F

4.

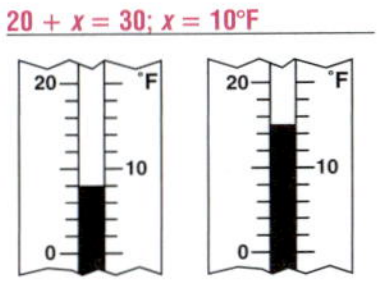

$8 + x = 15$; $x = 7$°F

5.

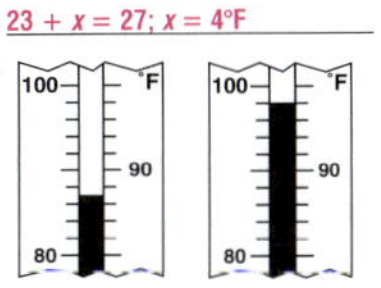

$87 + x = 98$; $x = 11$°F

6.

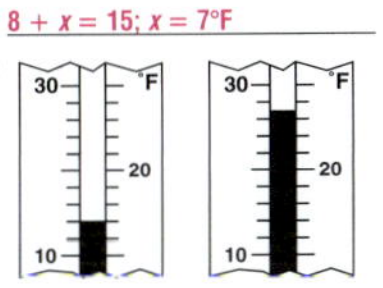

$14 + x = 27$; $x = 13$°F

PROBLEM SOLVING 2-4

LESSON 2-4 **Problem Solving**

Solving Addition Equations

Use the bar graph and addition equations to answer the questions.

Most-Populated States

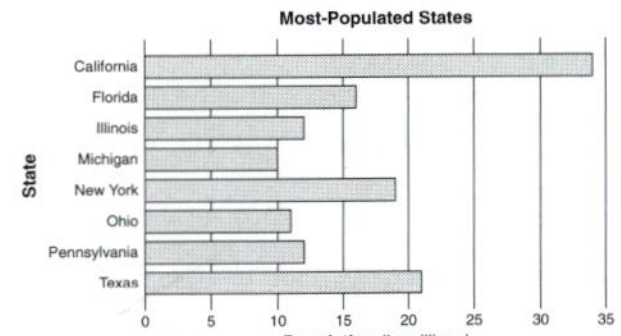

1. How many more people live in California than in New York?
$19 + x = 34$; $x = 15$;
15 million people

2. How many more people live in Ohio than in Michigan?
$10 + x = 11$; $x = 1$;
1 million people

3. How many more people live in Florida than in Illinois?
$12 + x = 16$; $x = 4$;
4 million people

4. How many more people live in Texas than in Pennsylvania?
$12 + x = 21$; $x = 9$;
9 million people

5. Which two states' populations are used in the equation $12 + x = 12$? What is the value of x?
Illinois and Pennsylvania; $x = 0$

6. What is the value of x in the equation $x + x + 2 = 34$? Which two states' populations are used in this equation?
$x = 16$; Florida and California

Circle the letter of the correct answer.

7. In 2000, the total population of the United States was 281 million. How many of those people did not live in one of the states listed in the table?
A 416 million
B 73 million
(C) 146 million
D 292 million

8. The combined population of Ohio and one other state is the same as the population of Texas. What is that state?
F California
G Florida
(H) Michigan
J Pennsylvania

Lesson Quiz

Solve each equation.

1. $x + 15 = 72$ $x = 57$
2. $81 = x + 24$ $x = 57$
3. $x + 22 = 67$ $x = 45$
4. $93 = x + 14$ $x = 79$
5. Kaitlin is 2 inches taller than Reba. Reba is 54 inches tall. How tall is Kaitlin? **56 inches**

Available on Daily Transparency in CRB

2-5 Organizer

Pacing: Traditional 1 day
Block $\frac{1}{2}$ day

Objective: Students solve whole-number subtraction equations.

Indiana Academic Standards
6.3.1, 6.7.1, 6.7.9, 6.7.10, 6.7.11

Warm Up

Solve each equation.

1. $x + 7 = 22$ $\quad x = 15$
2. $18 + x = 105$ $\quad x = 87$
3. $16 = x + 9$ $\quad x = 7$
4. $23 = x + 4$ $\quad x = 19$

Problem of the Day

Bruce has 25 CDs remaining after giving 14 to John, 17 to Mary, and 25 to Sue. How many CDs did he begin with? 81

Available on Daily Transparency in CRB

Math Fact

In Euclid's *Elements, Book I,* one of Euclid's axioms was, "If equals be added to equals, the wholes are equal."

2-5 Solving Subtraction Equations

Learn to solve whole-number subtraction equations.

When John F. Kennedy became president of the United States, he was 43 years old, which was 8 years younger than Abraham Lincoln was when Lincoln became president. How old was Lincoln when he became president?

Kennedy was President from 1961 to 1963.

Lincoln was President from 1861 to 1865.

Let a represent Abraham Lincoln's age.

Abraham Lincoln's age	−	8	=	John F. Kennedy's age
a	−	8	=	43

Remember that addition and subtraction are inverse operations. When an equation contains subtraction, use addition to "undo" the subtraction. Remember to add to both sides of the equation.

$$\begin{aligned} a - 8 &= 43 \\ +8 \quad & \quad +8 \\ a \quad &= 51 \end{aligned}$$

Abraham Lincoln was 51 years old when he became president.

EXAMPLE 1 Solving Subtraction Equations

A Solve $p - 2 = 5$. Check your answer.

$$\begin{aligned} p - 2 &= 5 && \text{2 is subtracted from } p. \\ +2 \quad & \quad +2 && \text{Add 2 to both sides to undo} \\ p \quad &= 7 && \text{the subtraction.} \end{aligned}$$

Check $p - 2 = 5$

$7 - 2 \stackrel{?}{=} 5$ *Substitute 7 for p in the equation.*

$5 \stackrel{?}{=} 5$ ✔ *7 is the solution.*

1 Introduce

Alternate Opener

EXPLORATION

2-5 Solving Subtraction Equations

After spending \$11, Jane has \$3 left in her purse. How much did she have initially?

Initial amount	\$11 spent	Amount left
n	−11	3

$n - 11 = 3$
$14 - 11 = 3$
$n = 14$ She had \$14 initially.

Find the value of n in each equation.

1. $n - 25 = 75$ $\quad n =$ ______
2. $n - 4 = 19$ $\quad n =$ ______
3. $n - 7 = 35$ $\quad n =$ ______
4. $n - 14 = 21$ $\quad n =$ ______
5. $n - 20 = 83$ $\quad n =$ ______

Think and Discuss

6. **Describe** your strategies for solving the subtraction equations.
7. **Explain** how you can find the solution to $n - 125 = 375$.

Motivate

Review with students how to solve addition equations. Have students give examples of equations that have a constant subtracted from a variable. Ask them to predict how to solve a subtraction equation.

Exploration worksheet and answers on Chapter 2 Resource Book pp. 45 and 89

2 Teach

Lesson Presentation

Guided Instruction

In this lesson, students learn to solve whole number subtraction equations. First, review the fact that addition and subtraction are inverse operations. Then teach students to use addition to solve subtraction equations, and check their solutions.

B Solve $40 = x - 11$. Check your answer.

$40 = x - 11$ — *11 is subtracted from x.*

$\underline{+11} \quad \underline{+11}$ — *Add 11 to both sides to undo the subtraction.*

$51 = x$

Check $40 = x - 11$

$40 \stackrel{?}{=} 51 - 11$ — *Substitute 51 for x in the equation.*

$40 \stackrel{?}{=} 40$ ✔ — *51 is the solution.*

C Solve $x - 56 = 19$. Check your answer.

$x - 56 = 19$ — *56 is subtracted from x.*

$\underline{+56} \quad \underline{+56}$ — *Add 56 to both sides to undo the subtraction.*

$x = 75$

Check $x - 56 = 19$

$75 - 56 \stackrel{?}{=} 19$ — *Substitute 75 for x in the equation.*

$19 \stackrel{?}{=} 19$ ✔ — *75 is the solution.*

Think and Discuss

1. **Tell** whether the solution of $b - 14 = 9$ will be less than 9 or greater than 9. Explain.
2. **Explain** how you know what number to add to both sides of an equation containing subtraction.

COMMON ERROR ALERT

Students just learned how to solve addition equations in the previous lesson. Because of this, they may try to subtract from each side of the subtraction equation to solve. Remind them to use the inverse operation, addition.

Additional Examples

Example 1

Solve each equation. Check your answer.

A. $y - 23 = 39$ $y = 62$

B. $78 = s - 15$ $93 = s$

C. $z - 3 = 12$ $z = 15$

2-5 Exercises

FOR EXTRA PRACTICE see page 639

internet connect — Homework Help Online go.hrw.com Keyword: MR4 2-5

GUIDED PRACTICE

See Example 1

Solve each equation. Check your answers.

1. $p - 8 = 9$ $p = 17$
2. $3 = x - 16$ $x = 19$
3. $a - 13 = 18$ $a = 31$
4. $15 = y - 7$ $y = 22$
5. $n - 24 = 9$ $n = 33$
6. $39 = d - 2$ $d = 41$

INDEPENDENT PRACTICE

See Example 1

Solve each equation. Check your answers.

7. $y - 18 = 7$ $y = 25$
8. $8 = n - 5$ $n = 13$
9. $a - 34 = 4$ $a = 38$
10. $c - 21 = 45$ $c = 66$
11. $a - 40 = 57$ $a = 97$
12. $31 = x - 14$ $x = 45$
13. $28 = p - 5$ $p = 33$
14. $z - 42 = 7$ $z = 49$
15. $s - 19 = 12$ $s = 31$

2-5 PRACTICE & ASSESS

Students may want to refer back to the lesson examples.

Assignment Guide

If you finished Example 1 assign:

Core 1–15, 17–31 odd, 32, 36–43

Enriched 12–43

3 Close

Reaching All Learners

Through Hands-On Experience

Have students work with subtraction equations written on a balanced set of scales. (You may use the recording sheet provided on page 78 in the Chapter 2 Resource Book.) Reinforce the concept of using the inverse operation to isolate the variable while maintaining the balance of the scales.

1. $x - 3 = 15$ $x = 18$
2. $y - 9 = 2$ $y = 11$
3. $7 = n - 6$ $n = 13$
4. $10 = z - 10$ $z = 20$

Summarize

Briefly describe how to solve subtraction equations involving variables and how to check solutions. Have a student volunteer to provide the next step in the solution of a subtraction equation. Have another student give an explanation for the step provided by the previous student.

Answers to Think and Discuss

1. Greater than 9; you add 14 to both sides of the equation to undo the subtraction, which means the solution will be 14 more than 9.
2. Add the number being subtracted from the variable to both sides of the equation to keep the equation balanced.

Lesson Quiz

Solve each equation.

1. $x - 9 = 21$ $x = 30$
2. $14 = x - 3$ $x = 17$
3. $x - 7 = 11$ $x = 18$
4. $16 = x - 14$ $x = 30$
5. $x - 9 = 11$ $x = 20$
6. Susan is taller than James. The difference in their height is 4 inches. James is 62 inches tall. How tall is Susan? 66 inches

Available on Daily Transparency in CRB

Answers

33–34. See p. A1.

PRACTICE AND PROBLEM SOLVING

Solve each equation.

16. $r - 57 = 7$ $r = 64$ **17.** $11 = x - 25$ $x = 36$ **18.** $8 = y - 96$ $y = 104$

19. $a - 6 = 15$ $a = 21$ **20.** $q - 14 = 22$ $q = 36$ **21.** $f - 12 = 2$ $f = 14$

22. $18 = j - 19$ $j = 37$ **23.** $109 = r - 45$ $r = 154$ **24.** $d - 8 = 29$ $d = 37$

25. $g - 71 = 72$ $g = 143$ **26.** $p - 13 = 111$ $p = 124$ **27.** $13 = m - 5$ $m = 18$

28. $20 = n - 4$ $n = 24$ **29.** $45 = k - 19$ $k = 64$ **30.** $t - 60 = 121$ $t = 181$

$r - 14{,}162 = 248$; $r = 14{,}410$ feet

31. *GEOGRAPHY* Mt. Rainier, in Washington, has a higher elevation than Mt. Shasta. The difference between their elevations is 248 feet. What is the elevation of Mt. Rainier? Write an equation and solve.

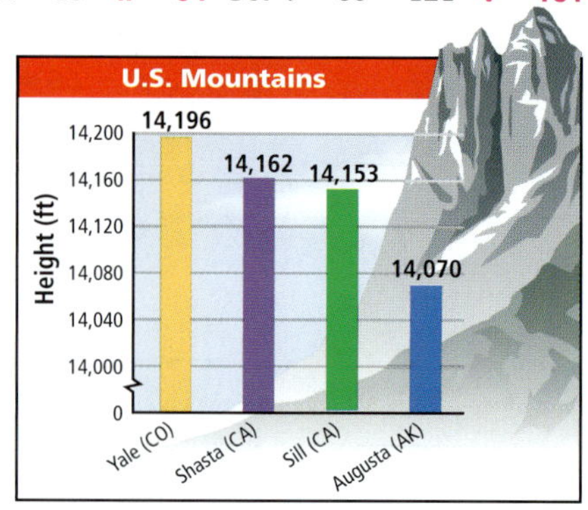

32. *SOCIAL STUDIES* In 2000, the population of Seoul, South Korea, was 16 million less than the population of Tokyo, Japan. The population of Seoul was 10 million. Solve the equation $10 = t - 16$ to find the population of Tokyo. 26 million

33. *WRITE ABOUT IT* Suppose $n - 15$ is a whole number. What do you know about the value of n? Explain.

34. *WHAT'S THE ERROR?* Look at the student paper at right. What did the student do wrong? What is the correct answer?

35. *CHALLENGE* Write "the difference between n and 16 is 5" as an algebraic equation. Then find the solution. $n - 16 = 5$; $n = 21$

Spiral Review

Find each value. (Lesson 1-3)

36. 5^3 125 **37.** 9^2 81 **38.** 3^4 81 **39.** 12^2 144

Label each as a numerical expression, an algebraic expression, or an equation. (Lesson 1-4, 2-1)

40. $75 \div 3$ numerical expression **41.** $x + 18 = 54$ equation **42.** $16 \times 12 + (8 - y)$ algebraic expression

43. TEST PREP For which equation is $b = 8$ a solution? (Lesson 2-3) D

A $13 - b = 8$ **B** $8 + b = 21$ **C** $b - 13 = 21$ **D** $b + 13 = 21$

RETEACH 2-5

LESSON 2-5 **Reteach**
Solving Subtraction Equations

To solve an equation, you need to get the variable alone on one side of the equal sign.

You can use tiles to help you solve subtraction equations.

Addition undoes subtraction, so you can use addition to solve subtraction equations.

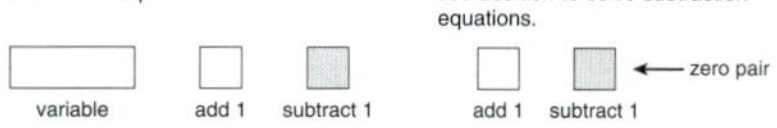

One positive tile and one negative tile is called a zero pair because together they have a value of zero.

To solve $x - 4 = 2$, first use tiles to model the equation.

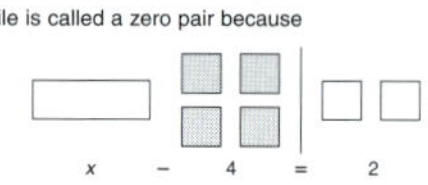

Next, add enough addition tiles to get the variable alone. Then add the same number of addition tiles to the other side of the equal sign.

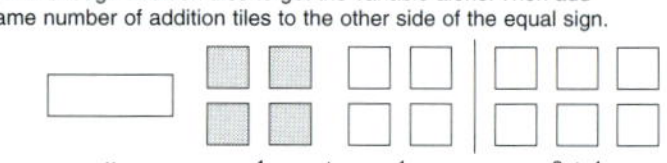

Then remove the greatest possible number of zero pairs from each side of the equal sign.

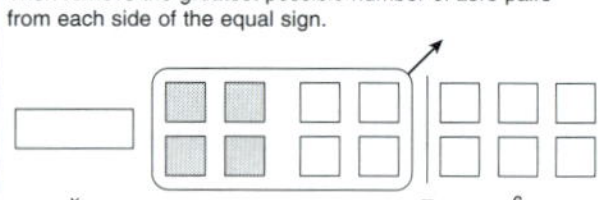

Check: $x - 4 = 2$
$6 - 4 = 2$
$2 = 2$

The remaining tiles represent the solution. $x = 6$

Use tiles to solve each equation. Then check each answer.

1. $x - 5 = 3$ $x = 8$
2. $x - 2 = 5$ $x = 7$
3. $x - 6 = 4$ $x = 10$
4. $x - 8 = 1$ $x = 9$
5. $x - 3 = 9$ $x = 12$
6. $x - 1 = 4$ $x = 5$
7. $x - 2 = 4$ $x = 6$
8. $x - 7 = 3$ $x = 10$

PRACTICE 2-5

LESSON 2-5 **Practice B**
Solving Subtraction Equations

Solve each equation. Check your answers.

1. $s - 8 = 12$ $s = 20$; $20 - 8 = 12$
2. $v - 11 = 7$ $v = 18$; $18 - 11 = 7$
3. $9 = q - 5$ $q = 14$; $9 = 14 - 5$
4. $m - 21 = 5$ $m = 26$; $26 - 21 = 5$
5. $34 = x - 12$ $x = 46$; $34 = 46 - 12$
6. $n - 45 = 45$ $n = 90$; $90 - 45 = 45$
7. $t - 19 = 9$ $t = 28$; $28 - 19 = 9$
8. $p - 6 = 27$ $p = 33$; $33 - 6 = 27$
9. $15 = v - 68$ $v = 83$; $15 = 83 - 68$

Solve each equation.

10. $7 = m - 5$ $m = 12$
11. $r - 10 = 22$ $r = 32$
12. $16 = x - 4$ $x = 20$
13. $40 = p - 11$ $p = 51$
14. $28 = d - 6$ $d = 34$
15. $n - 9 = 42$ $n = 51$
16. $q - 85 = 8$ $q = 93$
17. $f - 13 = 18$ $f = 31$
18. $47 = w - 38$ $w = 85$
19. Ted took 17 pictures at the aquarium. He now has 7 pictures left on the roll. Write and solve a subtraction equation to find out how many photos Ted had when he went to the aquarium. $x - 17 = 7$; $x = 24$ photos
20. Ted bought a dolphin poster for $12. He now has $5. Write and solve a subtraction equation to find out how much money Ted took to the aquarium. $x - 12 = 5$; $x = \$17$

CHALLENGE 2-5

LESSON 2-5 **Challenge**
The Price Is Right

Each of the grocery items on this page has a different price—$1, $2, $3, $4, or $5. Use logic and the subtraction equations below to figure out the price of each item. Then write the correct price on each item's price tag.

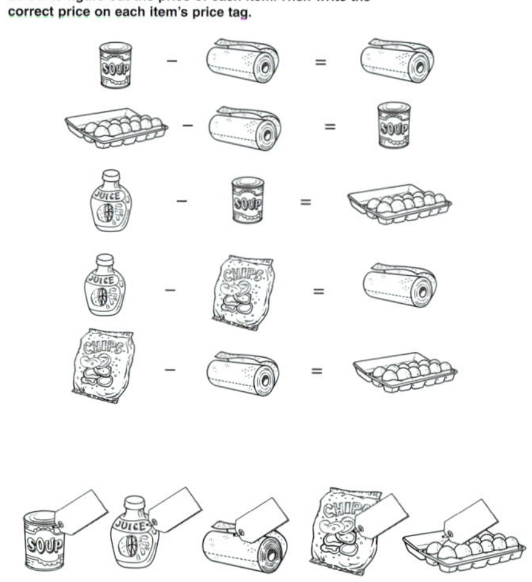

$2 $5 $1 $4 $3

PROBLEM SOLVING 2-5

LESSON 2-5 **Problem Solving**
Solving Subtraction Equations

Write and solve subtraction equations to answer the questions.

1. Dr. Felix Hoffman invented aspirin in 1899. That was 29 years before Alexander Fleming invented penicillin. When was penicillin invented?
$x - 29 = 1899$;
$x = 1928$; in 1928
2. Kimberly was born on February 2. That is 10 days earlier than Kent's birthday. When is Kent's birthday?
$x - 10 = 2$;
$x = 12$; February 12
3. Kansas and North Dakota are the top wheat-producing states. In 2000, North Dakota produced 314 million bushels of wheat, which was 34 million bushels less than Kansas produced. How much wheat did Kansas farmers grow in 2000?
$x - 34 = 314$;
$x = 348$; 348 million bushels
4. Scientists assign every element an atomic number, which is the number of protons in the nucleus of that element. The atomic number of silver is 47, which is 32 less than the atomic number of gold. How many protons are in the nucleus of gold?
$x - 32 = 47$;
$x = 79$; 79 protons
5. The spine-tailed swift and the frigate bird are the two fastest birds on earth. A frigate bird can fly 95 miles per hour, which is 11 miles per hour slower than a spine-tailed swift. How fast can a spine-tailed swift fly?
$x - 11 = 95$;
$x = 106$; 106 mph
6. The Green Bay Packers and the Kansas City Chiefs played in the first Super Bowl in 1967. The Chiefs lost by 25 points, with a final score of 10. How many points did the Packers score in the first Super Bowl?
$x - 25 = 10$;
$x = 35$; 35 points

Circle the letter of the correct answer.

7. The Rocky Mountains extend 3,750 miles across North America. That is 750 miles shorter than the Andes Mountains in South America. How long are the Andes Mountains?
A 3,000 miles
B 5 miles
C 180 miles
(D) 4,500 miles
8. When the United States took its first census in 1790, only 4 million people lived here. That was 277 million fewer people than the population in 2000. What was the population of the United States in 2000?
(F) 281 million
G 273 million
H 69 million
J 1,108 million

2-6 Solving Multiplication Equations

Learn to solve whole-number multiplication equations.

Armadillos are always born in groups of 4. If you count 32 babies, what is the number of mother armadillos?

To put together equal groups of 4, multiply. Let m represent the number of mother armadillos. There will be m equal groups of 4.

You can use the equation $4m = 32$.

Division is the inverse of multiplication. To solve an equation that contains multiplication, use division to "undo" the multiplication.

Remember!
$4m$ means "$4 \times m$."

$$4m = 32$$
$$\frac{4m}{4} = \frac{32}{4}$$
$$m = 8$$

There are 8 mother armadillos.

EXAMPLE 1 Solving Multiplication Equations

Solve each equation. Check your answers.

A $3x = 12$

$3x = 12$ — *x is multiplied by 3.*

$\frac{3x}{3} = \frac{12}{3}$ — *Divide both sides by 3 to undo the multiplication.*

$x = 4$

Check $3x = 12$

$3(4) \stackrel{?}{=} 12$ — *Substitute 4 for x in the equation.*

$12 \stackrel{?}{=} 12$ ✔ — *4 is the solution.*

B $8 = 4w$

$8 = 4w$ — *w is multiplied by 4.*

$\frac{8}{4} = \frac{4w}{4}$ — *Divide both sides by 4 to undo the multiplication.*

$2 = w$

Check $8 = 4w$

$8 \stackrel{?}{=} 4(2)$ — *Substitute 2 for w in the equation.*

$8 \stackrel{?}{=} 8$ ✔ — *2 is the solution.*

2-6 Organizer

Pacing: Traditional 1 day
Block $\frac{1}{2}$ day

Objective: Students solve whole-number multiplication equations.

Indiana Academic Standards
6.3.1, 6.7.1, 6.7.9, 6.7.10, 6.7.11

Warm Up

Divide.

1. $\frac{72}{12}$ 6
2. $\frac{65}{5}$ 13
3. $\frac{60}{6}$ 10
4. $\frac{40}{10}$ 4
5. $\frac{130}{5}$ 26
6. $\frac{91}{7}$ 13

Problem of the Day

Katie's little brother is building a tower with blocks. He adds 6 blocks and then removes 4 blocks. If he adds 8 more blocks, the tower will have twice as many blocks as when he started. How many blocks did he start with? 10 blocks

Available on Daily Transparency in CRB

Math Fact !!

The symbol × for multiplication was developed in England around 1600. It was not used widely in elementary arithmetic until after 1850. It is not used in algebra because of its resemblance to the letter *x*.

1 Introduce

Alternate Opener

EXPLORATION

2-6 Solving Multiplication Equations

Bill bought 7 tickets to a basketball game for $21. How much did each ticket cost?

t t t t t t t = $21

$7 \cdot t = 21$
$7 \cdot 3 = 21$
$t = 3$ Each ticket cost $3.

Find the value of n in each equation.

1. $4 \cdot n = 36$ $n =$ ______
2. $100 \cdot n = 500$ $n =$ ______
3. $50 = 5 \cdot n$ $n =$ ______
4. $24 \cdot n = 48$ $n =$ ______
5. $10 \cdot n = 240$ $n =$ ______

Think and Discuss

6. **Discuss** your strategies for solving the equations.
7. **Explain** how you can find the solution to $4 \cdot n = 200$.

Motivate

Review with students how to solve addition and subtraction equations. Have students give examples of equations that have a constant multiplied by a variable. Ask them to predict how to solve a multiplication equation.

Exploration worksheet and answers on Chapter 2 Resource Book pp. 54 and 91

2 Teach

Lesson Presentation

Guided Instruction

In this lesson, students learn to solve whole-number multiplication equations. First teach students to use division to solve multiplication equations. Then have students check their solutions.

Teaching Tip To demonstrate that division is the inverse of multiplication, have a student list the number sentences in the fact family for 3, 2, and 6.
$3 \times 2 = 6$; $2 \times 3 = 6$; $6 \div 2 = 3$; $6 \div 3 = 2$

COMMON ERROR ALERT

Caution students to be sure they are dividing in the correct order when solving multiplication equations. For example, to solve $4m = 32$, they should divide 32 by 4 as shown in the lesson opener, not 4 by 32.

Additional Examples

Example 1

Solve each equation. Check your answers.

A. $5p = 75$ $p = 15$

B. $16 = 8r$ $2 = r$

Example 2

The area of a rectangle is 56 square inches. Its length is 8 inches. What is its width? 7 inches

EXAMPLE 2 PROBLEM SOLVING APPLICATION

The area of a rectangle is 36 square inches. Its length is 9 inches. What is its width?

1 Understand the Problem

The **answer** will be the width of the rectangle in inches.

List the **important information:**

- The area of the rectangle is 36 square inches.
- The length of the rectangle is 9 inches.

Draw a diagram to represent this information.

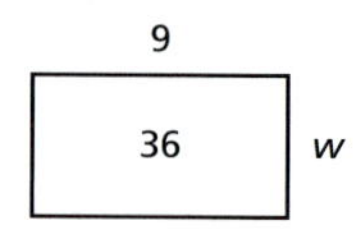

2 Make a Plan

You can write and solve an equation using the formula for area. To find the area of a rectangle, multiply its length by its width.

$$A = \ell w$$
$$36 = 9w$$

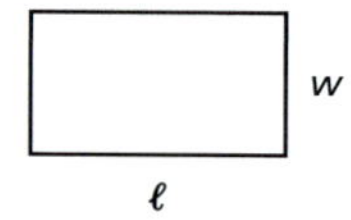

3 Solve

$36 = 9w$ *w is multiplied by 9.*

$\frac{36}{9} = \frac{9w}{9}$ *Divide both sides by 9 to undo the multiplication.*

$4 = w$

So the width of the rectangle is 4 inches.

4 Look Back

Arrange 36 identical squares in a rectangle. The length is 9, so line up the squares in rows of 9. You can make 4 rows of 9, so the width of the rectangle is 4.

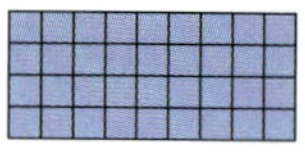

Think and Discuss

1. **Tell** what number you would use to divide both sides of the equation $15x = 60$.
2. **Tell** whether the solution of $10c = 90$ will be less than 90 or greater than 90. Explain.

3 Close

Summarize

Discuss with students how to solve multiplication equations involving variables and how to check solutions. Emphasize that the coefficient of the variable is the same number chosen to divide both sides of the equation by. Have a volunteer demonstrate the steps in solving and checking a multiplication equation.

Reaching All Learners

Through Number Sense

Have students use the fraction bar to indicate division of both sides of the equation. This provides an easy visual check to compare the coefficient of the variable and the number chosen to divide both sides of the equation by.

Answers to Think and Discuss

1. 15; because x is multiplied by 15
2. Possible answer: Less than 90; you divide both sides of the equation by 10, so the solution will be 9.

2-6 Exercises

FOR EXTRA PRACTICE see page 639

internet connect
Homework Help Online
go.hrw.com Keyword: MR4 2-6

GUIDED PRACTICE

See Example 1 Solve each equation. Check your answers.

1. $7x = 21$ $x = 3$
2. $27 = 3w$ $w = 9$
3. $90 = 10a$ $a = 9$
4. $56 = 7b$ $b = 8$
5. $3c = 33$ $c = 11$
6. $12 = 2n$ $n = 6$

See Example 2

7. The area of a rectangular deck is 675 square feet. The deck's width is 15 feet. What is its length? 45 feet

INDEPENDENT PRACTICE

See Example 1 Solve each equation. Check your answers.

8. $12p = 36$ $p = 3$
9. $52 = 13a$ $a = 4$
10. $64 = 8n$ $n = 8$
11. $20 = 5x$ $x = 4$
12. $6r = 30$ $r = 5$
13. $77 = 11t$ $t = 7$
14. $14s = 98$ $s = 7$
15. $12m = 132$ $m = 11$
16. $9z = 135$ $z = 15$

See Example 2

17. Colorado is almost a perfect rectangle. Its border from east to west is 387 mi, and its area is 104,247 mi^2. Estimate the length of Colorado's border from north to south.
$387w = 104{,}247$; about 250–350 miles

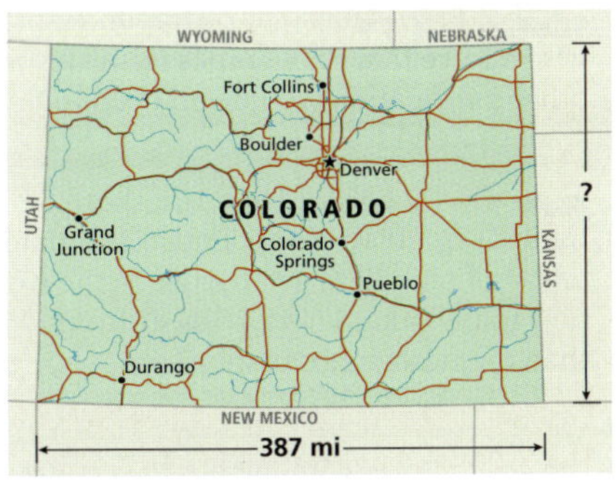

PRACTICE AND PROBLEM SOLVING

Solve each equation.

18. $5y = 35$ $y = 7$
19. $18 = 2y$ $y = 9$
20. $54 = 9y$ $y = 6$
21. $15y = 120$ $y = 8$
22. $4y = 0$ $y = 0$
23. $22y = 440$ $y = 20$
24. $3y = 63$ $y = 21$
25. $z - 6 = 34$ $z = 40$
26. $6y = 114$ $y = 19$
27. $161 = 7y$ $y = 23$
28. $135 = 3y$ $y = 45$
29. $y - 15 = 3$ $y = 18$
30. $81 = 9y$ $y = 9$
31. $4 + y = 12$ $y = 8$
32. $7y = 21$ $y = 3$
33. $a + 12 = 26$ $a = 14$
34. $10x = 120$ $x = 12$
35. $36 = 12x$ $x = 3$

2-6 PRACTICE & ASSESS

Students may want to refer back to the lesson examples.

Assignment Guide

If you finished Example 1 assign:
Core 1–6, 8–16, 19–33 odd, 42–48
Enriched 18–38, 40–48

If you finished Example 2 assign:
Core 1–17, 19–33 odd, 36–39, 42–48
Enriched 18–48

Notes

RETEACH 2-6

LESSON 2-6 **Reteach**
Solving Multiplication Equations

You can use tiles to help you solve multiplication equations.

Division undoes multiplication, so you can use division to solve multiplication equations.

To solve $3x = 12$, first use tiles to model the equation.

Next, divide each side of the equal sign into 3 equal groups

The number of tiles in one group represents the solution.

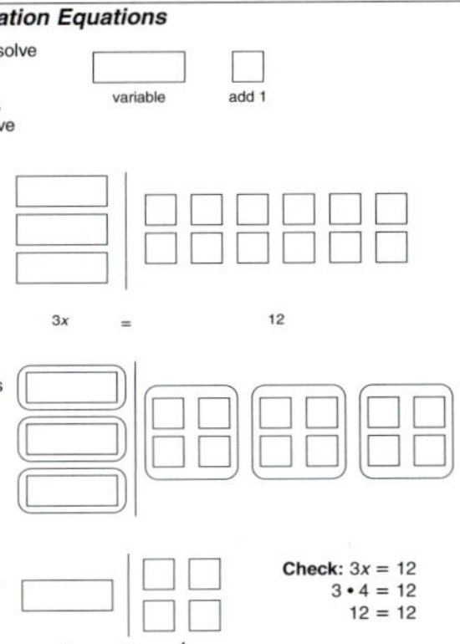

$x = 4$

Check: $3x = 12$
$3 \cdot 4 = 12$
$12 = 12$

Use tiles to solve each equation. Then check each answer.

1. $5x = 15$ $x = 3$
2. $2x = 6$ $x = 3$
3. $4x = 16$ $x = 4$
4. $8x = 24$ $x = 3$
5. $3x = 18$ $x = 6$
6. $6x = 12$ $x = 2$
7. $7x = 21$ $x = 3$
8. $9x = 9$ $x = 1$
9. $4x = 24$ $x = 6$
10. $3x = 9$ $x = 3$
11. $8x = 16$ $x = 2$
12. $5x = 25$ $x = 5$

PRACTICE 2-6

LESSON 2-6 **Practice B**
Solving Multiplication Equations

Solve each equation. Check your answers.

1. $8s = 72$ $s = 9; 8 \cdot 9 = 72$
2. $4v = 28$ $v = 7; 4 \cdot 7 = 28$
3. $27 = 9q$ $q = 3; 27 = 9 \cdot 3$
4. $12m = 60$ $m = 5; 12 \cdot 5 = 60$
5. $48 = 6x$ $x = 8; 48 = 6 \cdot 8$
6. $7n = 63$ $n = 9; 7 \cdot 9 = 63$
7. $10t = 130$ $t = 13; 10 \cdot 13 = 130$
8. $15p = 450$ $p = 30; 15 \cdot 30 = 450$
9. $84 = 6v$ $v = 14; 84 = 6 \cdot 14$

Solve each equation.

10. $49 = 7m$ $m = 7$
11. $20r = 80$ $r = 4$
12. $64 = 8x$ $x = 8$
13. $36 = 4p$ $p = 9$
14. $147 = 7d$ $d = 21$
15. $11n = 110$ $n = 10$
16. $12q = 144$ $q = 12$
17. $25f = 125$ $f = 5$
18. $128 = 16w$ $w = 8$
19. A hot-air balloon flew at 10 miles per hour. Using the variable h, write and solve a multiplication equation to find how many hours the balloon traveled if it covered a distance of 70 miles.
$10h = 70; h = 7$ hours
20. A passenger helicopter can travel 300 miles in the same time it takes a hot-air balloon to travel 20 miles. Using the variable s, write and solve a multiplication equation to find how many times faster the helicopter can travel than the hot air balloon.
$20s = 300; s = 15$ times faster

Math Background

The equations in this lesson are of the form $ax = b$ $(a \neq 0)$ where $\frac{b}{a}$ is a whole number. Multiplication equations in general do not require that $\frac{b}{a}$ be a whole number. To solve all equations of the form $ax = b$ $(a \neq 0)$, rational numbers are needed.

Interdisciplinary LINK

Life Science

Exercises 36–41 involve information about arthropods. Arthropods are studied in middle school life science programs, such as *Holt Science & Technology.*

Journal

Have students write about the following situation. Four friends want to share the cost of a sandwich. Why is dividing the cost of the sandwich by four the same as solving the equation $4n = 12$?

Test Prep Doctor

For Exercise 48, students need to remember to subtract to solve an addition equation. Students who answered **A** added 53 to 82 instead of subtracting 53 from 82. Students who answered **B** probably did the same thing but did the math incorrectly. Students who answered **D** may have understood the concept, but they did the math incorrectly.

Lesson Quiz

Solve each equation.

1. $10y = 300$ $y = 30$
2. $2y = 82$ $y = 41$
3. $63 = 9y$ $y = 7$
4. $78 = 13x$ $x = 6$
5. The area of a board game is 468 square inches. Its width is 18 inches. What is the length? 26 inches

Available on Daily Transparency in CRB

Life Science LINK

This photo shows a horsefly magnified to twelve times its actual size.

Arthropods make up the largest group of animals on Earth. They include insects, spiders, crabs, and centipedes.

Arthropods have segmented bodies. In centipedes and millipedes, all of the segments are identical.

36. Centipedes have 2 legs per segment. They can have from 30 to 354 legs. Find a range for the number of segments a centipede can have. 15 to 177 segments

37. Millipedes have 4 legs per segment. The record number of legs on a millipede is 752. How many segments did this millipede have? 188

Many arthropods have compound eyes. Compound eyes are made up of tiny bundles of identical light-sensitive cells.

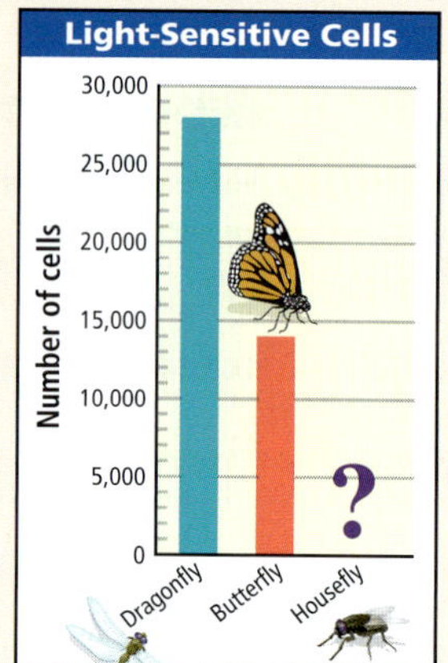

38. A dragonfly has 7 times as many light-sensitive cells as a housefly. How many of these cells does a housefly have? 4,000 light-sensitive cells

39. Find how many times more light-sensitive cells a dragonfly has than a butterfly. 2 times more

40. ***WRITE ABOUT IT*** A trapdoor spider can pull with a force that is 140 times its own weight. What other information would you need to find the spider's weight? Explain.

41. ***CHALLENGE*** There are about 6 billion humans in the world. Scientists estimate that there are a billion billion arthropods in the world. About how many times larger is the arthropod population than the human population?

go.hrw.com
KEYWORD: MR4 Arthropod
CNN Student News

40. Possible answer: The force with which a trapdoor spider pulls; then you could solve $f = 140w$.
41. The arthropod population is about 167 million times larger than the human population.

Spiral Review

Compare. Write <, >, or =. (Lesson 1-1)

42. 56,902 ▒ 56,817 >
43. 14,562 ▒ 14,581 <
44. 1,240,518 ▒ 1,208,959 >

Evaluate each expression. (Lesson 1-4)

45. $7 + 3 \times 4 - 2$ 17
46. $81 \div 3 - (5 + 6) \times 2$ 5
47. $3^1 \times 7 + (18 - 5) \times 4$ 73

48. **TEST PREP** Which value is a solution to the equation $n + 53 = 82$? (Lesson 2-3) C

A 135 **B** 125 **C** 29 **D** 27

CHALLENGE 2-6

LESSON 2-6 **Challenge**
Regulation Sizes

Write a multiplication equation for the area of each regulation field or court. Then solve the equations to find the missing measurements. Remember: Area = length • width, or $A = l \cdot w$.

1.
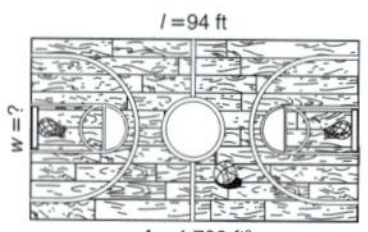

4,700 = 94w; w = 50

What is the width of a regulation basketball court?

50 feet

2.
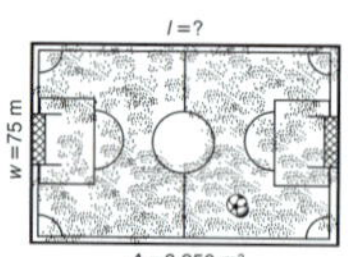

8,250 = 75l; l = 110

What is the length of a regulation soccer field?

110 meters

3.
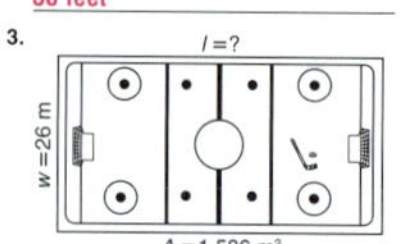

1,586 = 26l; l = 61

What is the length of a regulation ice hockey rink?

61 meters

4.
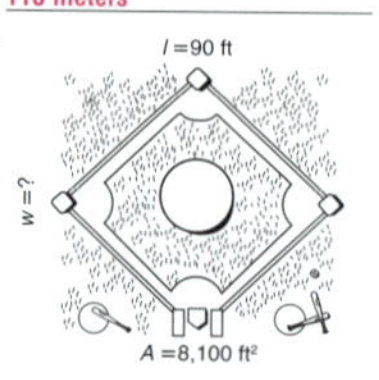

8,100 = 90w; w = 90

What is the length of a regulation baseball diamond?

90 feet

PROBLEM SOLVING 2-6

LESSON 2-6 **Problem Solving**
Solving Multiplication Equations

Write and solve a multiplication equation to answer each question.

1. In 1975, a person earning minimum wage made $80 for a 40 hour work week. What was the minimum wage per hour in 1975?
40x = 80; x = 2;
$2 per hour

2. If an ostrich could maintain its maximum speed for 5 hours, it could run 225 miles. How fast can an ostrich run?
5x = 225; x = 45;
45 miles per hour

3. About 2,000,000 people live in Paris, the capital of France. That is 80 times larger than the population of Paris, Texas. How many people live in Paris, Texas?
80x = 2,000,000; x = 25,000;
25,000 people

4. The average person in China goes to the movies 12 times a year. That is 3 times more than the average American goes to the movies. How many times a year does the average American go to the movies?
3x = 12; x = 4;
4 times a year

5. Recycling just 1 ton of paper saves 17 trees! If a city recycled enough paper to save 136 trees, how many tons of paper did it recycle?
17x = 136; x = 8;
8 tons

6. Seaweed found along the coast of California, called giant kelp, grows up to 18 inches a day. If a giant kelp plant has grown 162 inches at this rate, how many days has it been growing?
18x = 162; x = 9;
9 days

Circle the letter of the correct answer.

7. The distance between Atlanta, Georgia, and Denver, Colorado, is 1,398 miles. That is twice the distance between Atlanta and Detroit, Michigan. How many miles would you have to drive to get from Atlanta to Detroit?
A 2,796 miles
B 349.5 miles
Ⓒ 699 miles
D 1,400 miles

8. Jupiter has 2 times more moons than Neptune has, and 8 times more moons than Mars has. Jupiter has 16 moons. How many moons do Neptune and Mars each have?
Ⓕ 8 moons, 2 moons
G 2 moons, 8 moons
H 128 moons, 32 moons
J 32 moons, 128 moons

2-7 Solving Division Equations

Learn to solve whole-number division equations.

Japanese pearl divers go as deep as 165 feet underwater in search of pearls. At this depth, the pressure on a diver is much greater than at the water's surface. Water pressure can be described using equations containing division.

Multiplication is the inverse of division. When an equation contains division, use multiplication to "undo" the division.

EXAMPLE 1 Solving Division Equations

Solve each equation. Check your answers.

A $\frac{y}{5} = 4$

$\frac{y}{5} = 4$	*y is divided by 5.*
$5 \cdot \frac{y}{5} = 5 \cdot 4$	*Multiply both sides by 5 to undo the division.*
$y = 20$	

Check

$\frac{y}{5} = 4$	
$\frac{20}{5} \stackrel{?}{=} 4$	*Substitute 20 for y in the equation.*
$4 \stackrel{?}{=} 4$ ✔	*20 is the solution.*

B $12 = \frac{z}{4}$

$12 = \frac{z}{4}$	*z is divided by 4.*
$4 \cdot 12 = 4 \cdot \frac{z}{4}$	*Multiply both sides by 4 to undo the division.*
$48 = z$	

Check

$12 = \frac{z}{4}$	
$12 \stackrel{?}{=} \frac{48}{4}$	*Substitute 48 for z in the equation.*
$12 \stackrel{?}{=} 12$ ✔	*48 is the solution.*

2-7 Organizer

Pacing: Traditional 1 day
Block $\frac{1}{2}$ day

Objective: Students solve whole-number division equations.

Indiana Academic Standards
6.3.1, 6.7.1, 6.7.9, 6.7.10, 6.7.11

Warm Up

Solve.

1. $5x = 20$ $x = 4$
2. $7n = 84$ $n = 12$
3. $18 = y - 4$ $y = 22$
4. $21 = n + 3$ $n = 18$

Problem of the Day

If 11 is 4 less than 3 times a number, what is the number? 5

Available on Daily Transparency in CRB

Math Humor

When he saw the expression "x over eight" in the equation, the student thought the solution was for x to take some stomach medicine.

1 Introduce

Alternate Opener

EXPLORATION

2-7 Solving Division Equations

Four friends decided to share the cost of a gift for their dance teacher. After dividing the cost by 4, each friend's share is $25. What was the cost of the gift?

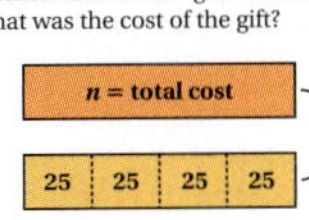

$\frac{n}{4} = 25 \longrightarrow \frac{100}{4} = 25$

$n = 100$

The gift cost $100.

Find the value of n in each equation.

1. $\frac{n}{2} = 50$ $n =$ ______
2. $\frac{n}{10} = 2$ $n =$ ______
3. $20 = \frac{n}{3}$ $n =$ ______
4. $\frac{n}{7} = 5$ $n =$ ______

Think and Discuss

5. **Discuss** your strategies for solving the equations.
6. **Explain** how you can find the solution to $\frac{n}{2} = 26$.

Motivate

Review with students how to solve multiplication equations. Have students give examples of equations that have a variable divided by a constant. Ask them to predict how to solve a division equation.

Exploration worksheet and answers on Chapter 2 Resource Book pp. 63 and 93

2 Teach

Lesson Presentation

Guided Instruction

In this lesson, students learn to solve whole number division equations. First explain that multiplication is the inverse of division, so multiplication is used to solve division equations. Then teach students to solve division equations. Next work through the science application with them.

COMMON ERROR ALERT

Students may try to divide each side of the division equation to solve. Remind them to use the inverse operation, multiplication, to solve.

Additional Examples

Example 1

Solve each equation. Check your answers.

A. $\frac{x}{7} = 5$
$x = 35$

B. $13 = \frac{p}{6}$
$78 = p$

Example 2

At Elk Meadows Park an aspen tree is one-third the height of a pine tree.

height of aspen $= \frac{\text{height of pine}}{3}$

The aspen tree is 14 feet tall. How tall is the pine tree?

The pine tree is 42 feet tall.

EXAMPLE 2 ***Physical Science Application***

Pressure is the amount of force exerted on an area. Pressure can be measured in pounds per square inch, or psi.

The pressure at the surface of the water is half the pressure at 30 ft underwater.

pressure at surface $= \frac{\textbf{pressure at 30 ft underwater}}{\textbf{2}}$

The pressure at the surface is 15 psi. What is the water pressure at 30 ft underwater?

Let p represent the pressure at 30 ft underwater.

$15 = \frac{p}{2}$ *Substitute 15 for pressure at the surface. p is divided by 2.*

$2 \cdot 15 = 2 \cdot \frac{p}{2}$ *Multiply both sides by 2 to undo the division.*

$30 = p$

The water pressure at 30 ft underwater is 30 psi.

Think and Discuss

1. Tell whether the solution of $\frac{c}{10} = 70$ will be less than 70 or greater than 70. Explain.

2. Describe how you would check your answer to Example 2.

3. Explain why $13 \cdot \frac{x}{13} = x$.

2-7 PRACTICE & ASSESS

Assignment Guide

If you finished Example 1 assign:
Core 1–3, 5–7, 9–17, 23–28
Enriched 1–3, 5–7, 9–15, 20–21, 23–28

If you finished Example 2 assign:
Core 1–19, 23–28
Enriched 4–28

2-7 Exercises

FOR EXTRA PRACTICE see page 639

internet connect
Homework Help Online
go.hrw.com Keyword: MR4 2-7

GUIDED PRACTICE

See Example 1 Solve each equation. Check your answers.

1. $\frac{y}{4} = 3$ $y = 12$ **2.** $14 = \frac{z}{2}$ $z = 28$ **3.** $\frac{r}{9} = 7$ $r = 63$

See Example 2 **4.** Irene mowed the lawn and planted flowers. The amount of time she spent mowing the lawn was one-third the amount of time it took her to plant flowers. It took her 30 minutes to mow the lawn. Let p represent the amount of time she spent planting flowers. Find the amount of time Irene spent planting flowers. 90 min

Reaching All Learners

Through Grouping Strategies

Give each group of students a set of nine cards numbered 1–9, provided on page 79 in the Chapter 2 Resource Book. On signal, each group picks a card at random and passes it to another group. The number on this card becomes the divisor in a division equation. On another signal, each group passes another card to the same group as before, and this card becomes the quotient. Groups solve the equations of the form (x ÷ first card) = (second card) and try to see which group finishes first.

3 Close

Summarize

Give brief descriptions of how to solve division equations involving variables and how to check solutions. Have a student volunteer to provide the next step in the solution of a division equation. Have another student give an explanation of the step provided by the previous student.

Answers to Think and Discuss

1. Greater than 70; you multiply both sides of the equation by 10 to undo the division, which means the solution will be the product of 70 and 10.
2. Substitute 30 for p in the equation. $15 = 30 \div 2$, so 30 is the correct solution.
3. Possible answer: You are multiplying by 13 and dividing by 13, so the operations undo each other.

INDEPENDENT PRACTICE

See Example 1

Solve each equation. Check your answers.

5. $\frac{d}{3} = 12$ $d = 36$ 6. $\frac{c}{2} = 13$ $c = 26$ 7. $7 = \frac{m}{7}$ $m = 49$

See Example 2

8. The area of Danielle's garden is one-twelfth the area of her entire yard. The area of the garden is 10 square feet. Let y represent the area of the yard. Find the area of the yard. $y = 120$ square feet

PRACTICE AND PROBLEM SOLVING

Find the value of c in each equation.

9. $\frac{c}{12} = 8$ $c = 96$ 10. $4 = \frac{c}{9}$ $c = 36$ 11. $\frac{c}{15} = 11$ $c = 165$
12. $14 = \frac{c}{5}$ $c = 70$ 13. $\frac{c}{4} = 12$ $c = 48$ 14. $\frac{c}{4} = 15$ $c = 60$
15. $30 = \frac{c}{6}$ $c = 180$ 16. $49 = \frac{c}{3}$ $c = 147$ 17. $\frac{c}{24} = 18$ $c = 432$

18. The Empire State Building is 381 m tall. At the Grand Canyon's widest point, 76 Empire State Buildings would fit end to end. Write and solve an equation to find the width of the Grand Canyon at this point. $\frac{w}{381} = 76$; $w = 28{,}956$ m

19. ***EARTH SCIENCE*** You can find the distance of a thunderstorm in kilometers by counting the number of seconds between the lightning flash and the thunder and then dividing this number by 3. If a storm is 5 km away, how many seconds will you count between the lightning flash and the thunder? 15 seconds

20. ***WRITE A PROBLEM*** Write a problem about money that can be solved with a division equation.

20. Possible answer: Each of 15 people gave \$5 to a fund. What was the total given? $\frac{w}{15} = 5$

21. ***WRITE ABOUT IT*** Use a numerical example to explain how multiplication and division undo each other. Possible answer: $9 \cdot 8 = 72$; $72 \div 8 = 9$

22. ***CHALLENGE*** Let m represent the amount of money Janine had on Monday. She spent half of it on Tuesday and half of what was left on Wednesday. On Thursday she had \$2. How much money did she have on Monday? \$8

Spiral Review

Identify a pattern in each sequence and name the next two terms. (Lesson 1-7)

23. 2, 6, 18, 54, ▇, ▇, ... 162, 486
24. 2, 5, 9, 14, 20, 27, ▇, ▇, ... 35, 44

Evaluate each expression for the given value of the variable. (Lesson 2-1)

25. $2y + 6$ for $y = 4$ 14 26. $\frac{z}{5}$ for $z = 40$ 8 27. $7r - 3$ for $r = 18$ 123

28. **TEST PREP** Which is an algebraic expression for the product of y and 4? (Lesson 2-2) A

A $4y$ B $4 + y$ C $\frac{y}{4}$ D $y - 4$

Lesson Quiz

Solve each equation. Check your answers.

1. $\frac{x}{10} = 7$ $x = 70$
2. $8 = \frac{x}{4}$ $x = 32$
3. $\frac{x}{9} = 11$ $x = 99$
4. $\frac{x}{15} = 7$ $x = 105$

5. The area of Sherry's flower garden is one-fourth the area of her vegetable garden. The area of the flower garden is 17 square feet. Let x represent the area of her vegetable garden. Find the area of her vegetable garden. 68 square feet

Available on Daily Transparency in CRB

RETEACH 2-7

LESSON 2-7 **Reteach**
Solving Division Equations

You can use multiplication and division to write related number facts.

$3 \cdot 4 = 12$ $12 \div 4 = 3$

Division and multiplication are inverse operations. They undo each other. So you can use multiplication to solve division equations.

To solve $\frac{x}{2} = 3$, think of a related number fact.

If $\frac{x}{2} = 3$, then $3 \cdot 2 = x$.

$3 \cdot 2 = x$
$x = 6$

Check: $\frac{x}{2} = 3$
$\frac{6}{2} = 3$ *substitute*
$3 = 3$

$x = 6$ is a solution for $\frac{x}{2} = 3$.

Use a related number fact to solve each equation. Then check each answer.

1. $\frac{x}{2} = 4$ $x = 8$ 2. $\frac{x}{8} = 2$ $x = 16$ 3. $\frac{x}{3} = 5$ $x = 15$ 4. $\frac{x}{4} = 3$ $x = 12$
5. $\frac{x}{5} = 1$ $x = 5$ 6. $\frac{x}{2} = 4$ $x = 8$ 7. $\frac{x}{6} = 3$ $x = 18$ 8. $\frac{x}{9} = 3$ $x = 27$
9. $\frac{x}{8} = 4$ $x = 32$ 10. $\frac{x}{2} = 9$ $x = 18$ 11. $\frac{x}{4} = 4$ $x = 16$ 12. $\frac{x}{7} = 3$ $x = 21$
13. $\frac{x}{5} = 4$ $x = 20$ 14. $\frac{x}{6} = 2$ $x = 12$ 15. $\frac{x}{9} = 4$ $x = 36$ 16. $\frac{x}{4} = 6$ $x = 24$

PRACTICE 2-7

LESSON 2-7 **Practice B**
Solving Division Equations

Solve each equation. Check your answers.

1. $\frac{s}{6} = 7$ $s = 42$; $\frac{42}{6} = 7$ 2. $\frac{v}{5} = 9$ $v = 45$; $\frac{45}{5} = 9$ 3. $12 = \frac{q}{7}$ $q = 84$; $12 = \frac{84}{7}$
4. $\frac{m}{2} = 16$ $m = 32$; $\frac{32}{2} = 16$ 5. $26 = \frac{x}{3}$ $x = 78$; $26 = \frac{78}{3}$ 6. $\frac{n}{8} = 4$ $n = 32$; $\frac{32}{8} = 4$
7. $\frac{t}{11} = 11$ $t = 121$; $\frac{121}{11} = 11$ 8. $\frac{p}{7} = 10$ $p = 70$; $\frac{70}{7} = 10$ 9. $7 = \frac{v}{8}$ $v = 56$; $7 = \frac{56}{8}$

Solve each equation.

10. $10 = \frac{m}{9}$ $m = 90$ 11. $\frac{r}{5} = 8$ $r = 40$ 12. $11 = \frac{x}{7}$ $x = 77$
13. $9 = \frac{p}{12}$ $p = 108$ 14. $15 = \frac{d}{5}$ $d = 75$ 15. $\frac{n}{4} = 28$ $n = 112$
16. $\frac{q}{2} = 134$ $q = 268$ 17. $\frac{u}{16} = 1$ $u = 16$ 18. $2 = \frac{w}{25}$ $w = 50$

19. All the seats in the theater are divided into 6 groups. There are 35 seats in each group. Using the variable s, write and solve a division equation to find how many seats there are in the theater. $\frac{s}{6} = 35$; $s = 210$ seats

20. There are 16 ounces in one pound. A box of nails weighs 4 pounds. Using the variable w, write and solve a division equation to find how many ounces the box weighs. $\frac{w}{16} = 4$; $w = 64$ ounces

CHALLENGE 2-7

LESSON 2-7 **Challenge**
What Does Algebra Mean?

About 1,200 years ago, Arab people invented the branch of mathematics called *algebra*. In fact, the word *algebra* comes from the Arabic word *al-jabr*. What does that word mean?

Solve each division equation below. Then in the box at the bottom of the page, write the variable in the blank above its value. When you have solved all the equations you will have found the answer to the question.

1. $\frac{s}{4} = 6$ $s = 24$ 8. $\frac{k}{9} = 5$ $k = 45$
2. $\frac{b}{3} = 5$ $b = 15$ 9. $6 = \frac{u}{2}$ $u = 12$
3. $9 = \frac{p}{4}$ $p = 36$ 10. $3 = \frac{n}{7}$ $n = 21$
4. $i \div 2 = 7$ $i = 14$ 11. $o \div 8 = 4$ $o = 32$
5. $8 = \frac{a}{6}$ $a = 48$ 12. $\frac{r}{7} = 5$ $r = 35$
6. $11 = f \div 2$ $f = 22$ 13. $t \div 9 = 3$ $t = 27$
7. $\frac{h}{7} = 6$ $h = 42$ 14. $8 = e \div 1$ $e = 8$

__ __ __ __ __ __ __ __ __ __
27 42 8 35 8 12 21 14 32 21

__ __ __ __ __ __ __ __
32 22 15 35 32 45 8 21

__ __ __ __ __
36 48 35 27 24

Answer: the reunion of broken parts

PROBLEM SOLVING 2-7

LESSON 2-7 **Problem Solving**
Solving Division Equations

Use the table to write and solve a division equation to answer each question.

Bakersville Sports League

Sport	Number of Teams	Players on Each Team
Baseball	7	20
Soccer	11	15
Football	8	24
Volleyball	12	9
Lacrosse	6	17
Basketball	10	10
Tennis	18	6

1. How many total people signed up to play soccer in Bakersville this year? $\frac{x}{11} = 15$; $x = 165$; 165 people
2. How many people signed up to play lacrosse this year? $\frac{x}{6} = 17$; $x = 102$; 102 people
3. What was the total number of people who signed up to play baseball this year? $\frac{x}{7} = 20$; $x = 140$; 140 people
4. Which two sports in the league have the same number of people signed up to play this year? How many people are signed up to play each of those sports? volleyball and tennis; 108 people

Circle the letter of the correct answer.

5. Which sport has a higher total number of players, football or tennis? How many more players?
   A football; 10 players
   B tennis; 144 players
   (C) football; 84 players
   D tennis; 18 players

6. Only one sport this year has the same number of players on each team as its number of teams. Which sport is that?
   (F) basketball
   G football
   H soccer
   J tennis

CHAPTER 2

Extension

Pacing: Traditional 1 day
Block $\frac{1}{2}$ day

Objective: Students solve and graph whole-number inequalities.

Using the Pages

In Lessons 2-3 through 2-7, students solved equations using addition, subtraction, multiplication, and division. In this extension, students will graph the solutions of inequalities on number lines. Students will also solve inequalities using addition, subtraction, multiplication, and division of whole numbers.

EXTENSION Inequalities

Learn to solve and graph whole-number inequalities.

Vocabulary
inequality

Reading Math
< means "is less than."
> means "is greater than."
≤ means "is less than or equal to."
≥ means "is greater than or equal to."

An **inequality** is a statement that two quantities are not equal.

$15 > 3$ $12 \leq 29$ $41 \geq 18$ $17 < 90$

An inequality may contain a variable, as in the inequality $x > 3$. Values of the variable that make the inequality true are solutions of the inequality.

x	$x > 3$	Solution?
0	$0 \overset{?}{>} 3$	No; 0 is **not** greater than 3, so 0 is not a solution.
3	$3 \overset{?}{>} 3$	No; 3 is **not** greater than 3, so 3 is not a solution.
4	$4 \overset{?}{>} 3$	Yes; 4 is greater than 3, so 4 is a solution.
12	$12 \overset{?}{>} 3$	Yes; 12 is greater than 3, so 12 is a solution.

This table shows that an inequality may have more than one solution. You can use a number line to show all of the solutions.

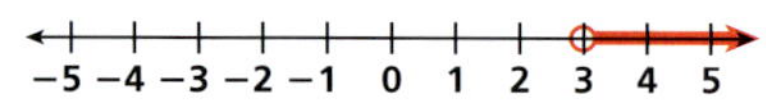

EXAMPLE 1 **Graphing Inequalities**

Graph the solutions to $w \leq 4$ on a number line.

The closed circle on the point 4 shows that 4 is a solution.

You can solve inequalities in the same way that you solved equations.

EXAMPLE 2 **Solving and Graphing Inequalities**

Solve each inequality. Graph the solutions on a number line.

A $y + 7 < 9$

$y + 7 < 9$ *7 is added to y.*
$\quad - 7 \quad - 7$ *Subtract 7 from both sides to undo the addition.*
$y < 2$

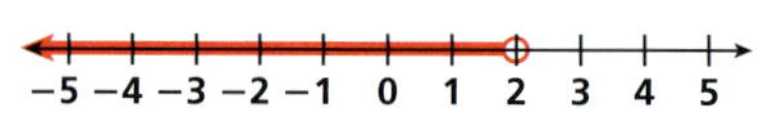

The open circle on the point 2 shows that 2 is not a solution.

1 Introduce

Motivate

When comparing two numbers a and b, exactly one of the following statements is true: $a < b$, $a = b$, or $a > b$.

2 Teach

Lesson Presentation

Guided Instruction

In this extension, students learn to solve and graph whole-number inequalities. Teach the students to graph inequalities. Then teach students to solve inequalities in a manner similar to how they solved equations. Have students graph their solutions on a number line.

Solve each inequality. Graph the solutions on a number line.

B $2m \geq 12$

$2m \geq 12$ — *m is multiplied by 2.*

$\frac{2m}{2} \geq \frac{12}{2}$ — *Divide both sides by 2 to undo the multiplication.*

$m \geq 6$

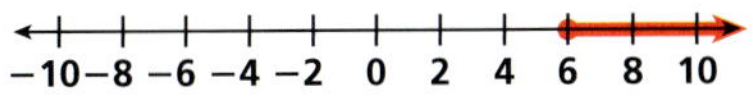

The closed circle on the point 6 shows that 6 is a solution.

EXTENSION Exercises

Graph the solutions to each inequality on a number line.

1. $w \leq 0$
2. $x > 5$
3. $z \geq 9$
4. $7 < t$
5. $m > 2$
6. $4 \geq q$
7. $a \leq 8$
8. $6 > x$
9. $y < 3$

Solve each inequality.

10. $3t \leq 27$ $t \leq 9$
11. $y - 5 \geq 0$ $y \geq 5$
12. $x + 4 < 10$ $x < 6$
13. $2c > 2$ $c > 1$
14. $\frac{d}{6} \geq 1$ $d \geq 6$
15. $r + 9 \leq 23$ $r \leq 14$
16. $15n < 75$ $n < 5$
17. $4 + r \leq 7$ $r \leq 3$
18. $f - 11 > 16$ $f > 27$
19. $2k < 8$ $k < 4$

Write an inequality for each sentence. Then graph your inequality.

20. c is less than or equal to two.
21. p is greater than 11.
22. At some lakes, people who fish must throw back any trout that is less than 10 inches long. Write an inequality that represents the lengths of trout that may be kept. $t \geq 10$
23. **GEOGRAPHY** Mt. McKinley is the highest point in the United States, with an altitude of 20,320 ft. Let a be the altitude of any other U.S. location. Write an inequality relating a to Mt. McKinley's altitude. $a < 20{,}320$
24. **WHAT'S THE ERROR?** A student graphed $x > 1$ as shown. What did the student do wrong? Draw the correct graph. The student drew a closed circle on 1 when it should have been open.

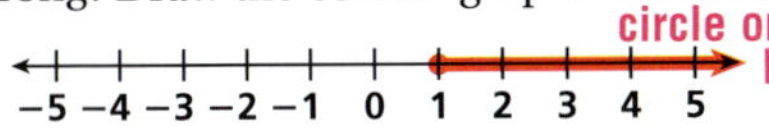

3 Close

Summarize

Review by graphing the following: $x > 3$, $x = 3$, and $x \leq 3$. Then have students volunteer ways that solving inequalities is the same as solving equations and ways that solving inequalities is different from solving equations.

Teaching Tip

For inequalities such as $7 < t$, have students rewrite the inequality placing the t to the left and reversing the inequality symbol. The inequality now reads $t > 7$. Having the variable come first makes the inequality easier to understand and to graph.

COMMON ERROR ALERT

In Exercise 4, some students may graph numbers that are less than 7 because the inequality reads "seven is less than t." Remind them that the inequality sign always "points to" the lesser of two numbers, which means that the values of t are greater than 7.

Additional Examples

Example 1

Graph the solutions to $x \geq 1$ on a number line.

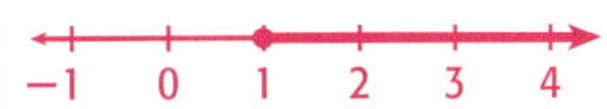

Example 2

Solve each inequality. Graph the solutions on a number line.

A. $z - 2 > 1$

$z > 3$

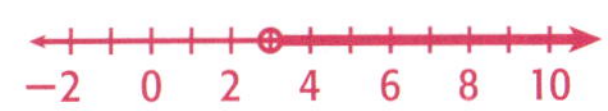

B. $\frac{x}{2} \leq 1$

$x \leq 2$

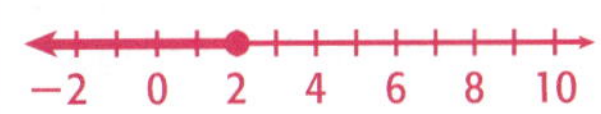

Answers

1. 
2. 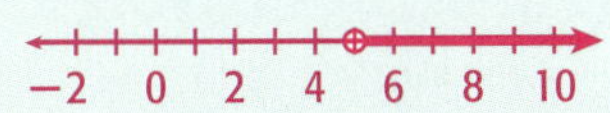
3. 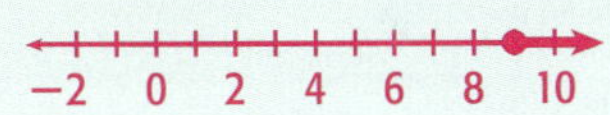
4. 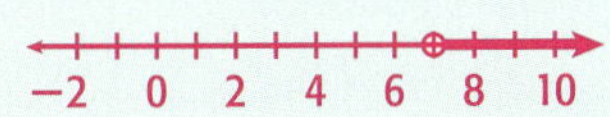
5. 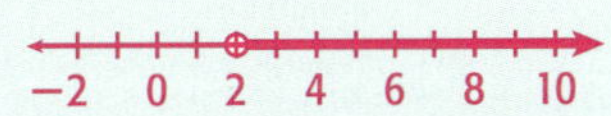
6. 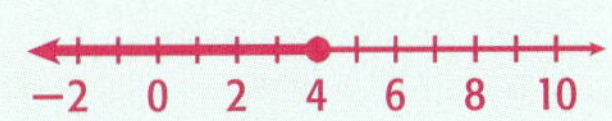
7. 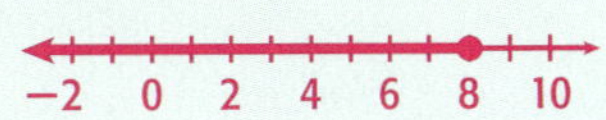
8. 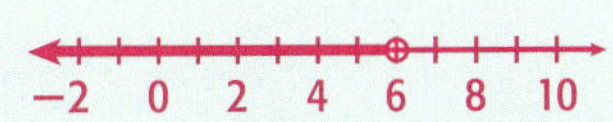
9. 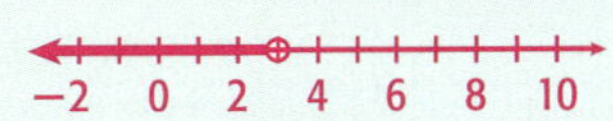

20. $c \leq 2$; 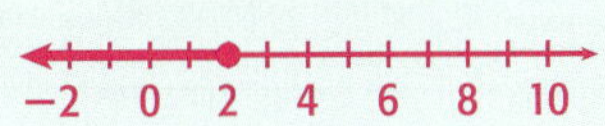
21. $p > 11$; 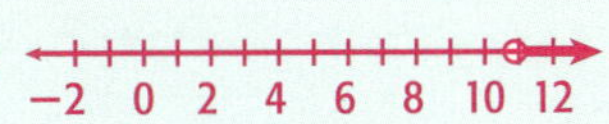

Problem Solving on Location

Pennsylvania

Purpose: *To provide additional practice for problem-solving skills in Chapters 1–2*

The First Superhighway

- After problem 1, ask students to write and solve a different equation to solve the problem. $514 - 160 = n$; n represents the amount by which the highway was lengthened.
- After problem 4, discuss with students how to determine which terms could be combined. Like terms can be combined. Add the constants $(104 + 40 + 100)$, and then add the m terms $(m + m)$. How can you evaluate the expression if you are given a value for m? Multiply the value of m by 2, and then add 244 to the product.

Extension After students have found all the distances, have them consider the scale of the diagram. Have them redraw the diagram so that the scale is more accurate.
Check students' work.

Problem Solving on Location

PENNSYLVANIA

The First Superhighway

The Pennsylvania Turnpike was the first highway designed for modern high-speed long-distance travel. Completed in 1940, the turnpike crossed the Allegheny Mountains between Harrisburg and Pittsburgh. It shortened travel time between these cities by 3 hours.

1. The Pennsylvania Turnpike was originally 160 miles long. It has since been lengthened to 514 miles. Write and solve an equation to find how much longer the highway is today than it was in 1940. $n + 160 = 514$; $n = 354$
2. The turnpike is 132 feet wide on the section called the Southwestern Expansion. That is about 2 times the road's width on the portion called the Northeastern Extension. Let w represent the width of the Northeastern Extension. Write and solve an equation to find the width of the Northeastern Extension. $2n = 132$; $n = 66$

For 3–5, use the map.

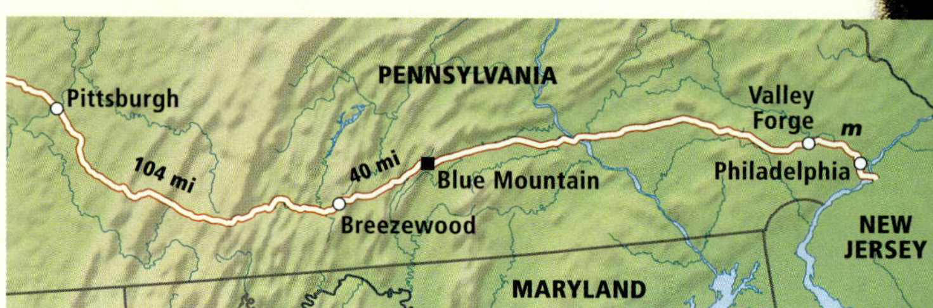

3. The distance from Blue Mountain to Valley Forge is 100 miles more than the distance from Valley Forge to Philadelphia. Let m stand for the distance from Valley Forge to Philadelphia. Write an expression containing m that represents the distance from Blue Mountain to Valley Forge. $m + 100$
4. Write an expression that represents the distance from Pittsburgh to Philadelphia. $104 + 40 + m + 100 + m$, or $2m + 244$
5. It is 24 miles from Valley Forge to Philadelphia. How far is it from Pittsburgh to Philadelphia? 292 miles

The First Oil Well

Pennsylvania was the birthplace not only of America's modern highway system but also of its oil industry. In 1859, Edwin L. Drake drilled the world's first oil well, beside Oil Creek, near Titusville, Pennsylvania.

Before Drake drilled his well, people collected small amounts of oil that seeped from the ground near Oil Creek. Drake's well produced about 210 times as much oil each day as could be collected aboveground.

1. Let a represent the amount of oil that could be collected above ground each day. Write an expression for the daily amount produced by Drake's well. **$210a$**
2. About 5 gallons of oil could be collected aboveground each day. Use the expression you wrote in **1** to find the daily production of Drake's well. **1,050 gal**
3. There are 42 gallons in one barrel of oil. Find the daily production in barrels of Drake's well. **25 barrels**
4. In 1859, oil sold for about $20 per barrel. Find the value of one day's production of oil at Drake's well. **$500**

The surface of the ground at Drake's well is 1,200 feet above sea level.

5. To reach oil, Drake first drilled through gravel to a point 1,168 ft above sea level. How far down did he drill? **32 ft**
6. Drake then drilled an additional 38 ft through shale. What was the total depth for the well? **70 ft**
7. A few months later, Drake drilled a second well nearby. It was about 7 times as deep as the first well. Let d represent the depth of the first well. Write an expression for the depth of the second well. **$7d$**
8. How deep was Drake's second well? **about 490 ft**

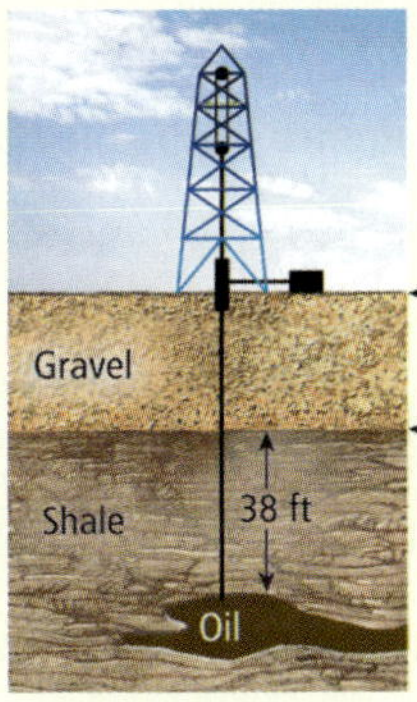

The First Oil Well

- After problem 3, ask the following question: How many gallons of oil are in 42 barrels? 1,764 gallons
- After problem 6, ask students to find the elevation at the bottom of the well. Have them explain their thinking. 1,130 feet above sea level; the total depth of the well is 70 ft, and 1,200 − 70 = 1,130.
- After problem 7, discuss why the variable d was chosen to represent the depth. When assigning a variable, it is often useful to use the first letter of the quantity being represented.

Extension Encourage students to research the story of Edwin L. Drake and the first oil well and to present it to their classmates through posters, models, etc.

MATH-ABLES

Game Resources

Puzzles, Twisters & Teasers
Chapter 2 Resource Book

Math Magic

Purpose: *To apply the problem-solving skill of writing and evaluating expressions to a magic trick*

Discuss: Ask students to explain how the charts were made. Possible answer: You can express any number using the expression $a + 3b + 9c$, with a, b, and c being 0, 1, or 2. For example, to express 7, let $a = 1$, $b = 2$, and $c = 0$. So 7 appears in the first chart once, in the second chart twice, and not in the third chart. How does algebra make this trick easier to understand? Writing the expression $a + 3b + 9c$ makes it easy to see why the trick works.

Extend: Challenge students to explore why the charts contain only the numbers from 1 to 26.
Using the expression $a + 3b + 9c$ and the values 0, 1, and 2 for the variables, the least value for the expression is $0 + 3(0) + 9(0) = 0$, and the greatest is $2 + 3(2) + 9(2) = 26$. Zero is not in the charts because it appears zero times in each of the three charts.

a	b	c	$a + 3b + 9c$
1	0	0	$1 + 3(0) + 9(0) = 1$
2	0	0	$2 + 3(0) + 9(0) = 2$
0	1	0	$0 + 3(1) + 9(0) = 3$
1	1	0	$1 + 3(1) + 9(0) = 4$
2	1	0	$2 + 3(1) + 9(0) = 5$
0	2	0	$0 + 3(2) + 9(0) = 6$
1	2	0	$1 + 3(2) + 9(0) = 7$
2	2	0	$2 + 3(2) + 9(0) = 8$
0	0	1	$0 + 3(0) + 9(1) = 9$
1	0	1	$1 + 3(0) + 9(1) = 10$
2	0	1	$2 + 3(0) + 9(1) = 11$
0	1	1	$0 + 3(1) + 9(1) = 12$
1	1	1	$1 + 3(1) + 9(1) = 13$
2	1	1	$2 + 3(1) + 9(1) = 14$
0	2	1	$0 + 3(2) + 9(1) = 15$
1	2	1	$1 + 3(2) + 9(1) = 16$
2	2	1	$2 + 3(2) + 9(1) = 17$
0	0	2	$0 + 3(0) + 9(2) = 18$
1	0	2	$1 + 3(0) + 9(2) = 19$
2	0	2	$2 + 3(0) + 9(2) = 20$
0	1	2	$0 + 3(1) + 9(2) = 21$
1	1	2	$1 + 3(1) + 9(2) = 22$
2	1	2	$2 + 3(1) + 9(2) = 23$
0	2	2	$0 + 3(2) + 9(2) = 24$
1	2	2	$1 + 3(2) + 9(2) = 25$
2	2	2	$2 + 3(2) + 9(2) = 26$

MATH-ABLES

Math Magic

Guess what your friends are thinking with this math magic trick.

Copy the following number charts.

1	10	19
2, 2	11, 11	20, 20
4	13	22
5, 5	14, 14	23, 23
7	16	25
8, 8	17, 17	26, 26

3	12	21
4	13	22
5	14	23
6, 6	15, 15	24, 24
7, 7	16, 16	25, 25
8, 8	17, 17	26, 26

9	15	21, 21
10	16	22, 22
11	17	23, 23
12	18, 18	24, 24
13	19, 19	25, 25
14	20, 20	26, 26

Step 1: Ask a friend to think of a number from 1 to 26.
Example: Your friend thinks of 26.

Step 2: Show your friend the first chart and ask how many times the chosen number appears. Remember the answer.
Your friend says the chosen number appears twice on the first chart. 2

Step 3: Show the second chart and ask the same question. Multiply the answer by 3. Add your result to the answer from step 2. Remember this answer.
Your friend says the chosen number appears twice. $3 \cdot 2 = 6$
The answer from step 2 is 2. $6 + 2 = 8$

Step 4: Show the third chart and ask the same question. Multiply the answer by 9. Add your result to the answer from step 3. The answer is your friend's number.
Your friend says the chosen number appears twice. $9 \cdot 2 = 18$
The answer from step 3 is 8. $18 + 8 = 26$ ← *Your friend's number*

How does it work?

Your friend's number will be the following:

(answer from step 2) + (3 · answer from step 3) + (9 · answer from step 4)

This is an expression with three variables: $a + 3b + 9c$. A number will be on a particular chart 0, 1, or 2 times, so a, b, and c will always be 0, 1, or 2. With these values, you can write expressions for each number from 1 to 26.

a	b	c	$a + 3b + 9c$
1	0	0	$1 + 3(0) + 9(0) = \mathbf{1}$
2	0	0	$2 + 3(0) + 9(0) = \mathbf{2}$
0	1	0	$0 + 3(1) + 9(0) = \mathbf{3}$

Can you complete the table for 4–26?

Technology LAB

Evaluate Expressions

internet connect
Lab Resources Online
go.hrw.com
KEYWORD: MR4 TechLab2

You can use a graphing calculator to evaluate algebraic expressions. A graphing calculator is especially helpful when you are evaluating an expression for many values of the variable.

Activity

Evaluate $4x + 5$ for $x = 0, 1, 2, 3, 4, 5, 6, 7, 8, 9$, and 10.

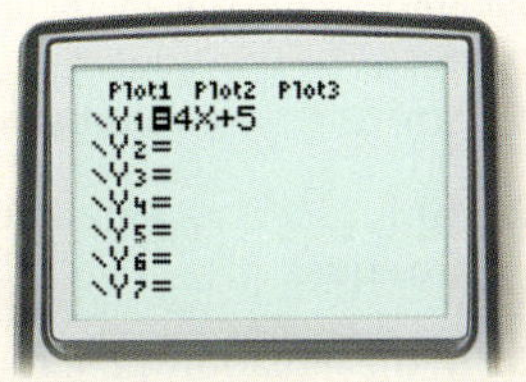

1. Enter $4x + 5$ into the **Y=** menu. [Y=] 4 [X,T,θ,n] [+] 5.

2. Press [2nd] [WINDOW] (TBLSET) to access the **TABLE SETUP** menu. **TblStart** tells which value of x the table should begin with. Make sure this is set to 0, because 0 is the smallest value of x for which you must evaluate the expression. **ΔTbl** gives the difference between successive x-values. **ΔTbl = 1** means that the x-values in the table will increase by 1.

3. Press [2nd] [GRAPH] (TABLE) to see the table. The **Y1** column shows the value of $4x + 5$ for several x-values. For example, when $x = 0$, $4x + 5 = 5$. Use the arrow keys to find the answers for other x-values.

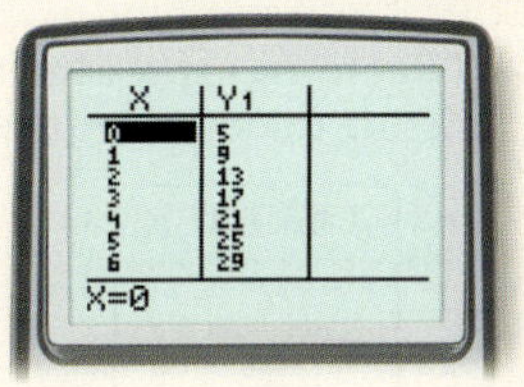

Think and Discuss

1. In the table, one row has 5 in the **X** column and 25 in the **Y1** column. What does this mean? When $x = 5$, $4x + 5 = 25$.
2. When does $y = 41$? when $x = 9$
3. Would you use a calculator to evaluate $x + 2$ for $x = 5$? Explain why or why not. No, because you are only evaluating the expression for one value and it can be easily and quickly done by using basic addition facts.

Try This

Evaluate each expression for the given x-values.

1. $3x + 8$; $x = 4, 5$, and 9
2. $45x + 67$; $x = 8, 10$, and 12
3. $4x + 7$; $x = 0, 1, 2, 3, 4$, and 5
4. $30 + 25x$; $x = 5, 10, 15$, and 20

Answers

Try This

1. 20; 23; 35
2. 427; 517; 607
3. 7; 11; 15; 19; 23; 27
4. 155; 280; 405; 530

Technology LAB

Evaluate Expressions

Objective: To use a graphing calculator to evaluate expressions

Materials: Graphing calculator

Lab Resources

Technology Lab Activities . . . . . pp. 8–9

Using the Page

This technology activity shows students how to use a graphing calculator to input an expression and a range of values and to have the calculator evaluate the expression for each value. Specific keystrokes may vary, depending on the make and model of the graphing calculator used. The keystrokes given are for a TI-83 model. For keystrokes to other models, visit go.hrw.com.

Think and Discuss problem 3 is meant to make students aware that a calculator is not always the fastest way to evaluate an expression. Although Try This problems 1–4 can be done without a graphing calculator, they are meant to help students become familiar with using a graphing calculator to evaluate expressions.

Assessment

1. How would you evaluate the expression $3x - 5 - x$ for $x = 3, 5$, and 7? Enter 3 [X,T,θ,n] [+] 5 [−] [X,T,θ,n] into the **Y=** menu. Then go into the TABLE SETUP menu and enter 3 for TblStart and 2 for ΔTbl. Press 2nd GRAPH and read the y-values for each x-value.
2. How could you simplify the expression before entering it into the calculator? Combine like terms to get $y = 2x - 5$.

Chapter 2 Study Guide and Review

Purpose: To help students review and practice concepts and skills presented in Chapter 2

Assessment Resources

Chapter Review

Chapter 2 Resource Book . . . pp. 71–72

***Test and Practice Generator* CD-ROM**

Additional review assessment items in both multiple-choice and free-response format may be generated for any objective in Chapter 2.

Answers

1. algebraic expression
2. equation
3. variable
4. constant
5. 7; 6
6. 6; 10
7. $p \times 6$
8. $s \div 2$

Chapter 2 Study Guide and Review

Study Guide and Review

Vocabulary

Complete the sentences below with vocabulary words from the list above. Words may be used more than once.

1. A(n) __?__ contains one or more variables.
2. A(n) __?__ is a mathematical statement that says two quantities are equal.
3. In the equation $12 + t = 22$, t is a __?__.
4. A(n) __?__ is a quantity that does not change.

2-1 Variables and Expressions (pp. 48–51)

EXAMPLE

- **Evaluate the expression to find the missing values in the table.**

n	$3n + 4$
1	7
2	■
3	■

$n = 1$ $3 \times 1 + 4 = 7$
$n = 2$ $3 \times 2 + 4 = 10$
$n = 3$ $3 \times 3 + 4 = 13$

The missing values are 10 and 13.

- **Find an expression for the table.**

x	■
4	16
5	20
6	24

$4 \cdot 4 = 16$
$5 \cdot 4 = 20$
$6 \cdot 4 = 24$

An expression is $x \cdot 4$, or $4x$.

EXERCISES

Evaluate each expression to find the missing values in the tables.

5.

y	$y \div 7$
56	8
49	■
42	■

6.

k	$k \times 4 - 6$
2	2
3	■
4	■

Find an expression for each table.

7.

p	■
9	54
10	60
11	66

8.

s	■
18	9
36	18
48	24

2-2 Translate Between Words and Math (pp. 52–55)

EXAMPLES

Write each phrase as a numerical or algebraic expression.

- 617 minus 191
  617 − 191
- *d* multiplied by 5
  $5d$ or $5 \cdot d$ or $(5)(d)$

Write two phrases for each expression.

- $a \div 5$
  - *a* divided by 5
  - the quotient of *a* and 5
- 67 + 19
  - the sum of 67 and 19
  - 19 more than 67

EXERCISES

Write each phrase as a numerical or algebraic expression.

9. 15 plus *b*
10. the product of 6 and 5
11. 9 times *t*
12. the quotient of *g* and 9

Write two phrases for each expression.

13. $4z$
14. $54 \div 6$
15. $3 - y$
16. $y - 3$
17. $15 + x$
18. $\frac{m}{20}$
19. 5,100 + 64

2-3 Equations and Their Solutions (pp. 58–61)

EXAMPLE

- **Determine whether the given value of the variable is a solution.**

  $f + 14 = 50$ for $f = 34$

  $34 + 14 \stackrel{?}{=} 50$ *Substitute 34 for f.*

  $48 \neq 50$ *Add.*

  34 is not a solution.

EXERCISES

Determine whether the given value of each variable is a solution.

20. $28 + n = 39$ for $n = 11$
21. $12t = 74$ for $t = 6$
22. $y - 53 = 27$ for $y = 80$
23. $96 \div w = 32$ for $w = 3$
24. $15x = 90$ for $x = 8$
25. $x - 61 = 17$ for $x = 75$

2-4 Solving Addition Equations (pp. 62–65)

EXAMPLE

- **Solve the equation $x + 18 = 31$.**

  $x + 18 = 31$ *18 is added to x.*

  $\underline{-18} \quad \underline{-18}$ *Subtract 18 from both sides to undo the addition.*

  $x = 13$

EXERCISES

Solve each equation.

26. $4 + x = 10$
27. $n + 10 = 24$
28. $c + 71 = 100$
29. $y + 16 = 22$
30. $44 = p + 17$
31. $94 + w = 103$
32. $23 + b = 34$
33. $56 = n + 12$
34. $39 = 23 + p$
35. $d + 28 = 85$

Study Guide and Review

Answers

9. $15 + b$
10. 6×5
11. $9t$
12. $g \div 9$
13. the product of 4 and *z*; 4 times *z*
14. 54 divided by 6; the quotient of 54 and 6
15. 3 minus *y*; the difference of 3 and *y*
16. *y* minus 3; the difference of *y* and 3
17. 15 plus *x*; the sum of 15 and *x*
18. *m* divided by 20; the quotient of *m* and 20
19. the sum of 5,100 and 64; 64 added to 5,100
20. yes
21. no
22. yes
23. yes
24. no
25. no
26. $x = 6$
27. $n = 14$
28. $c = 29$
29. $y = 6$
30. $p = 27$
31. $w = 9$
32. $b = 11$
33. $n = 44$
34. $p = 16$
35. $d = 57$

Answers

36. $k = 45$
37. $d = 9$
38. $p = 63$
39. $n = 67$
40. $r = 14$
41. $w = 144$
42. $h = 60$
43. $p = 167$
44. $v = 8$
45. $y = 9$
46. $c = 7$
47. $n = 2$
48. $s = 8$
49. $t = 10$
50. $a = 8$
51. $y = 8$
52. $r = 42$
53. $t = 15$
54. $y = 18$
55. $n = 72$
56. $z = 52$
57. $b = 100$
58. $n = 77$
59. $p = 90$

2-5 Solving Subtraction Equations (pp. 66–68)

EXAMPLE

■ Solve the equation.

$$\begin{aligned} c - 7 &= 16 \\ +7 \quad & \quad +7 \\ c &= 23 \end{aligned}$$

7 is subtracted from c.
Add 7 to each side to undo the subtraction.

EXERCISES

Solve each equation.

36. $28 = k - 17$
37. $d - 8 = 1$
38. $p - 55 = 8$
39. $n - 31 = 36$
40. $3 = r - 11$
41. $97 = w - 47$
42. $12 = h - 48$
43. $9 = p - 158$

2-6 Solving Multiplication Equations (pp. 69–72)

EXAMPLE

■ Solve the equation.

$$6x = 36$$
$$\frac{6x}{6} = \frac{36}{6}$$
$$x = 6$$

x is multiplied by 6.
Divide both sides by 6 to undo the multiplication.

EXERCISES

Solve each equation.

44. $5v = 40$
45. $27 = 3y$
46. $12c = 84$
47. $18n = 36$
48. $72 = 9s$
49. $11t = 110$
50. $7a = 56$
51. $8y = 64$

2-7 Solving Division Equations (pp. 73–75)

EXAMPLE

■ Solve the equation.

$$\frac{k}{4} = 8$$
$$4 \cdot \frac{k}{4} = 4 \cdot 8$$
$$k = 32$$

k is divided by 4.
Multiply both sides by 4 to undo the division.

EXERCISES

Solve each equation.

52. $\frac{r}{7} = 6$
53. $\frac{t}{5} = 3$
54. $6 = \frac{y}{3}$
55. $12 = \frac{n}{6}$
56. $\frac{z}{13} = 4$
57. $20 = \frac{b}{5}$
58. $\frac{n}{11} = 7$
59. $10 = \frac{p}{9}$

"PEANUTS" reprinted by permission of United Feature Syndicate, Inc.

Chapter Test

Chapter 2

Evaluate each expression to find the missing values in the tables.

1.

a	$a + 18$
10	28
12	30
14	32

2.

y	$y \div 6$
18	3
30	5
42	7

3.

n	$n \div 5 + 7$
10	9
20	11
30	13

Find an expression for each table.

4.

s	$s \div 12$
36	3
48	4
60	5

5.

t	$t \times 5$
6	30
7	35
8	40

6.

b	$b - 21$
100	79
75	54
50	29

7. There are more reptile species than amphibian species. Let n represent the number of living reptile species. There are 3,100 living species of amphibians. Write an expression to show how many more reptile species there are than amphibian species. $n - 3{,}100$

Write each phrase as a numerical or algebraic expression.

8. 26 more than n $n + 26$

9. g multiplied by 4 $4g$

10. the quotient of 180 and 15 $180 \div 15$

11. the difference of 100 and 17 $100 - 17$

Write two phrases for each expression. Possible answers given.

12. (14)(16) 14 times 16; the product of 14 and 16

13. $n \div 8$ n divided by 8; the quotient of n and 8

14. $p + 11$ p plus 11; the sum of p and 11

Determine whether the given value of the variable is a solution.

15. $5d = 70$ for $d = 12$ no

16. $29 = 76 - n$ for $n = 46$ no

17. $108 \div a = 12$ for $a = 9$ yes

18. $15 + m = 27$ for $m = 12$ yes

Solve each equation. Check your answers.

19. $a + 7 = 25$ $a = 18$

20. $121 = 11d$ $d = 11$

21. $3 = t - 8$ $t = 11$

22. $6 = \frac{k}{9}$ $k = 54$

23. Air typically has about 4,000 bacteria per cubic meter. If your bedroom is 30 cubic meters, about how many bacteria would you expect there to be in the air in your bedroom? about 120,000 bacteria

Chapter Test

Chapter 2

Purpose: *To assess students' mastery of concepts and skills in Chapter 2*

Assessment Resources

Chapter 2 Tests (Levels A, B, C)
Assessment Resources . . . . . pp. 37–42

***Test and Practice Generator* CD-ROM**

Additional assessment items in both multiple-choice and free-response format may be generated for any objective in Chapter 2.

Notes

Chapter 2 Performance Assessment

Purpose: *To assess students' understanding of concepts in Chapter 2 and combined problem-solving skills*

Assessment Resources

Performance Assessment
Assessment Resources . . . . p. 106

Performance Assessment Teacher Support
Assessment Resources . . . . p. 105

Answers

1–3. See p. A1.

4. See Level 3 work sample below.

Scoring Rubric for Problem Solving Item 4

Level 3

Accomplishes the purposes of the task.

Student gives clear explanations, shows understanding of mathematical ideas and processes, and computes accurately.

Level 2

Purposes of the task not fully achieved.

Student demonstrates satisfactory but limited understanding of the mathematical ideas and processes.

Level 1

Purposes of the task not accomplished.

Student shows little evidence of understanding the mathematical ideas and processes and makes computational and/or procedural errors.

Chapter 2 Performance Assessment

Performance Assessment

Show What You Know

Create a portfolio of your work from this chapter. Complete this page and include it with your four best pieces of work from Chapter 2. Choose from your homework or lab assignments, mid-chapter quiz, or any journal entries you have done. Put them together using any design you want. Make your portfolio represent what you consider your best work.

Short Response

1. Write the phrase "the quotient of n and 6" as an algebraic expression. Then evaluate the expression for $n = 72$.

2. The Morgans went on a car trip. They averaged 60 miles per hour. If h represents the number of hours they drove, write an expression for the distance the Morgans traveled in h hours.

3. Determine whether $x = 13$ is a solution for the equation $3x = 36$. If it is not a solution, solve the equation to find the value of x.

Extended Problem Solving

4. Three cities are located along the same highway. The distance between Artsville and Charlestown is 35 miles. Artsville is 18 miles from Burgston.

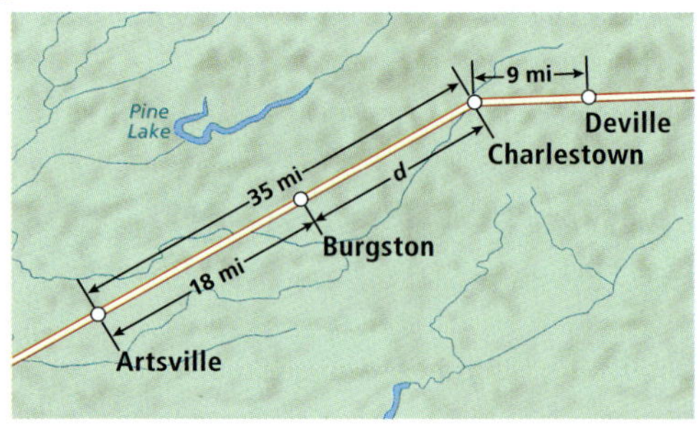

a. Let d represent the distance between Burgston and Charlestown. Write an equation that could be used to find the value of d.

b. To find the distance between Burgston and Charlestown, solve the equation you wrote in part **a.** Show your work.

c. An almanac states that Charlestown is 29,920 yards from Burgston. Is this the same as the distance you found in part **b**? Explain your answer in words. (*Hint:* There are 1,760 yards in a mile.)

Student Work Samples for Item 4

Level 3

a. 35 = d + 18
b. 35 = d + 18
−18   −18
17 = d
The distance between Burgston and Charlestown is 17 miles.
c. 17 × 1,760 = 29,920
The distances are the same because 17 groups of 1,760 yards is 29,920 yards.

The student wrote a correct equation and correctly solved for *d*. The student explained the answer to part c in words.

Level 2

a. 35 = 18 + d
b. d = 17
17 miles between Burgston and Charlestown
c. 29,920 ÷ 1,760 = 17

The student gave correct answers, but did not show all the work or explain part c in words.

Level 1

a. 35 = 18 + d
b. 35 = 18 + d
+18   + 18
53 = d
c. No

The student wrote a correct equation, but solved for *x* incorrectly. The student answered part c incorrectly and did not give any explanation.

internet connect
State-Specific Test Practice Online
go.hrw.com Keyword: MR4 TestPrep

Standardized Test Prep

Chapter 2

Cumulative Assessment, Chapters 1–2

1. Which is a solution to the equation $8a = 48$? C
   Ⓐ 8 Ⓑ 7 Ⓒ 6 Ⓓ 5

2. What is the value of 9^1? H
   Ⓕ 0 Ⓖ 1 Ⓗ 9 Ⓙ 19

3. Find the missing value in the table. A

n	$6 \times n - 7$
7	35
8	■

   Ⓐ 41 Ⓑ 36 Ⓒ 6 Ⓓ 55

4. Which means "the quotient of t and 6"? G
   Ⓕ $6 \div t$ Ⓖ $\frac{t}{6}$ Ⓗ $6t$ Ⓙ $6 + t$

5. What is the value of $3^3 - (15 \div 3) \times 2$? D
   Ⓐ 8 Ⓑ 44 Ⓒ 1 Ⓓ 17

6. Which number is greatest? J
   Ⓕ 12,301,542 Ⓖ 12,381,536 Ⓗ 12,311,518 Ⓙ 12,385,501

7. Nicole is 15 years old. She is 3 years younger than her sister Jackie. Solve the equation $j - 3 = 15$ to find Jackie's age. A
   Ⓐ 18 Ⓑ 17 Ⓒ 12 Ⓓ 5

TEST TAKING TIP!
Substitute the given value in each equation. See which value gives a correct equation.

8. Which of the following has a solution of 7? J
   Ⓕ $p + 14 = 20$ Ⓖ $7p = 42$ Ⓗ $\frac{p}{4} = 7$ Ⓙ $p - 2 = 5$

9. Michael plans to enter a biking and running competition. The total race course is 19 miles long. The map shows the course and the distance Michael will run. How many miles will Michael have to bike? B

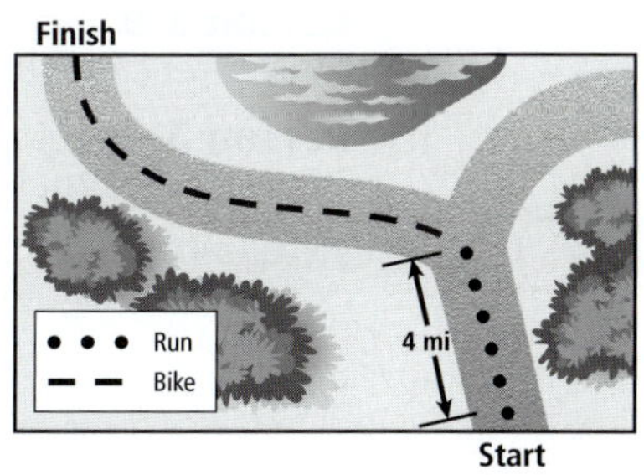

   Ⓐ 4 miles Ⓑ 15 miles Ⓒ 19 miles Ⓓ 24 miles

10. ***SHORT RESPONSE*** Show the steps to solve the equation $x + 5 = 11$. Explain why the only solution is $x = 6$.

11. ***SHORT RESPONSE*** A student said the solution of the equation $\frac{x}{7} = 14$ is $x = 2$. Explain the student's error and show the steps to solve the equation correctly.

Standardized Test Prep

Chapter 2

Purpose: *To provide review and practice for Chapters 1–2 and standardized tests*

Assessment Resources

Cumulative Tests (Levels A, B, C)
Assessment Resources. . . . pp. 139–150

State-Specific Test Practice Online
KEYWORD: MR4 TestPrep

Test Prep Doctor

Expand on the test-taking tip given for item 8 by pointing out to students that they could have solved each equation to see whether they would have obtained an answer of 7.

Point out to students that in item 7, another way to find the correct answer is to substitute the answer choices into the equation.

Answers

10. $x + 5 = 11$
$\underline{-5 = -5};$
$x = 6$

Six is the only number that, when added to five, gives a sum of eleven.

11. The student divided the right side of the equation by 7. The student should have multiplied both sides by 7.

$\frac{x}{7} = 14$

$7 \cdot \frac{x}{7} = 14 \cdot 7$

$x = 98$

Chapter 3

Decimals

Section 3A

Understanding Decimals

Hands-On Lab 3A
Model Decimals

Lesson 3-1
Representing, Comparing, and Ordering Decimals

Lesson 3-2
Estimating Decimals

Hands-On Lab 3B
Explore Decimal Addition and Subtraction

Lesson 3-3
Adding and Subtracting Decimals

Lesson 3-4
Decimals and Metric Measurement

Hands-On Lab 3C
Estimate Measurements

Section 3B

Multiplying and Dividing Decimals

Lesson 3-5
Scientific Notation

Hands-On Lab 3D
Explore Decimal Multiplication and Division

Lesson 3-6
Multiplying Decimals

Lesson 3-7
Dividing Decimals by Whole Numbers

Lesson 3-8
Dividing by Decimals

Lesson 3-9
Interpret the Quotient

Lesson 3-10
Solving Decimal Equations

Extension
Significant Figures

Pacing Guide for 45-Minute Classes

Chapter 3

DAY 22	DAY 23	DAY 24	DAY 25	DAY 26
Hands-On Lab 3A	Lesson 3-1	Lesson 3-2	Hands-On Lab 3B	Lesson 3-3
DAY 27	**DAY 28**	**DAY 29**	**DAY 30**	**DAY 31**
Lesson 3-4	Hands-On Lab 3C	Mid-Chapter Quiz Lesson 3-5	Hands-On Lab 3D	Lesson 3-6
DAY 32	**DAY 33**	**DAY 34**	**DAY 35**	**DAY 36**
Lesson 3-7	Lesson 3-8	Lesson 3-9	Lesson 3-10	Extension
DAY 37	**DAY 38**			
Chapter 3 Review	Chapter 3 Assessment			

Pacing Guide for 90-Minute Classes

Chapter 3

DAY 11	DAY 12	DAY 13	DAY 14	DAY 15
Chapter 2 Assessment Hands-On Lab 3A	Lesson 3-1 Lesson 3-2	Hands-On Lab 3B Lesson 3-3	Lesson 3-4 Hands-On Lab 3C	Mid-Chapter Quiz Lesson 3-5 Hands-On Lab 3D
DAY 16	**DAY 17**	**DAY 18**	**DAY 19**	**DAY 20**
Lesson 3-6 Lesson 3-7	Lesson 3-8 Lesson 3-9	Lesson 3-10 Extension	Chapter 3 Review Lesson 4-1	Chapter 3 Assessment Lesson 4-2

Across the Series

HARCOURT GRADE 5

- Read, write, and round decimals.
- Identify and write equivalent decimals.
- Estimate and find sums, differences, products, and quotients of decimals.
- Measure and convert length in metric units.

COURSE 1

- **Write, compare, and order decimals.**
- **Estimate and find decimal sums, differences, products, and quotients.**
- **Convert metric measurements.**
- **Write large numbers in scientific notation.**
- **Solve problems by interpreting the quotient.**
- **Solve equations involving decimals.**
- **Round measurements to an appropriate number of significant figures.**

COURSE 2

- Write and order decimals.
- Estimate, add, subtract, multiply, and divide with decimals, and solve decimal equations.
- Convert measurements within the metric measurement system, and determine acceptable levels of accuracy in measurements.

Across the Curriculum

LANGUAGE ARTS LINK

Math: Reading and Writing in the Content Area . . . . pp. 15–24
Focus on Problem Solving
Solve . . . . SE p. 113
Journal . . . . TE pp. 95, 99, 105, 109, 117, 123, 130, 137
Write About It . . . . SE pp. 95, 99, 105, 109, 117, 126, 130, 133, 137

SOCIAL STUDIES LINK

Social Studies . . . . SE pp. 117, 132
History . . . . SE p. 130
Geography . . . . SE p. 109

SCIENCE LINK

Astronomy . . . . SE p. 95
Life Science . . . . SE p. 117
Physical Science . . . . SE p. 123
Earth Science . . . . SE pp. 93, 117, 130
Health . . . . SE pp. 96, 99
Science . . . . SE p. 120

TE = *Teacher's Edition* **SE** = *Student Edition*

Interdisciplinary Bulletin Board

Physical Science

The density of a substance D is found by dividing its mass m by its volume V, $D = \frac{m}{V}$. Find the density of the nugget shown to determine whether it is real gold or fool's gold (iron pyrite).

$D = \frac{96.6}{5.0} = 19.32$; the nugget is real gold.

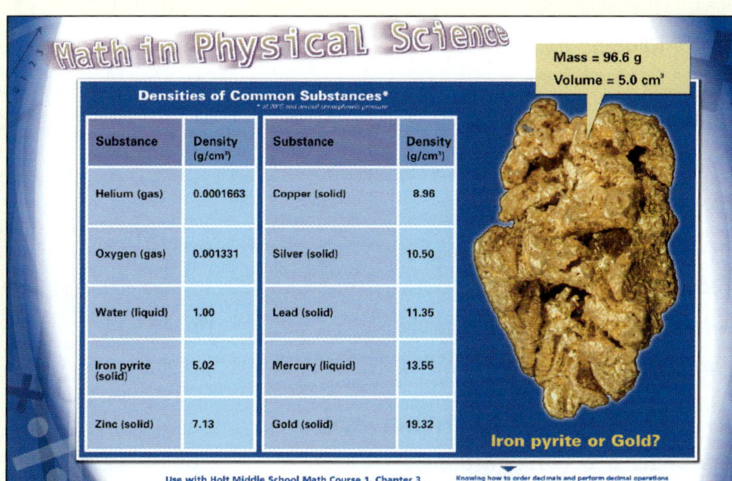

Interdisciplinary posters and worksheets are provided in your resource material.

Chapter 3

Resource Options

Chapter 3 Resource Book

Student Resources

Practice (Levels A, B, C). . . . . pp. 10–12, 20–22, 29–31, 38–40, 48–50, 57–59, 66–68, 75–77, 84–86, 93–95

Reteach. . . . . pp. 13–14, 23, 32, 41, 51, 60, 69, 78, 87, 96

Challenge . . . . pp. 15, 24, 33, 42, 52, 61, 70, 79, 88, 97

Problem Solving. . . . . pp. 16, 25, 34, 43, 53, 62, 71, 80, 89, 98

Puzzles, Twisters & Teasers . . . . pp. 17, 26, 35, 44, 54, 63, 72, 81, 90, 99

Recording Sheets . . . . . . pp. 3, 7–9, 19, 28, 37, 47, 56, 65, 74, 83, 92, 103, 106

Chapter Review . . . . pp. 100–102

Teacher and Parent Resources

Chapter Planning and Pacing Guide. . . . . p. 4

Section Planning Guides . . . . pp. 5, 45

Parent Letter . . . . pp. 1–2

Teaching Tools . . . . pp. 106–111

Teacher Support for Chapter Project . . . . p. 104

Transparencies . . . . pp. T1–T42

- Daily Transparencies
- Additional Examples Transparencies
- Teaching Transparencies

Reaching All Learners

English Language Learners

Success for English Language Learners . . . . pp. 29–48

Math: Reading and Writing in the Content Area . . . . pp. 15–24

Spanish Homework and Practice . . . . pp. 15–24

Spanish Interactive Study Guide . . . . pp. 15–24

Spanish Family Involvement Activities . . . . pp. 17–24

Multilingual Glossary

Individual Needs

Are You Ready? Intervention and Enrichment. . . pp. 13–16, 57–60, 145–148, 241–248, 409–410

Alternate Openers: Explorations . . . . pp. 15–24

Family Involvement Activities . . . . pp. 17–24

Interactive Problem Solving. . . . . pp. 15–24

Interactive Study Guide . . . . pp. 15–24

Readiness Activities . . . . pp. 5–6

Math: Reading and Writing in the Content Area . . . . pp. 15–24

Challenge. . . . . CRB pp. 15, 24, 33, 42, 52, 61, 70, 79, 88, 97

Hands-On

Hands-On Lab Activities. pp. 10–19

Technology Lab Activities . . . . p. 10

Alternate Openers: Explorations . . . . pp. 15–24

Family Involvement Activities . . . . pp. 17–24

Applications and Connections

Consumer and Career Math. . . . . pp. 9–12

Interdisciplinary Posters . . . . Poster 3, TE p. 88B

Interdisciplinary Poster Worksheets . . . . pp. 7–9

Transparencies

Alternate Openers: Explorations. . . . . pp. 15–24

Exercise Answers Transparencies

Chapter 3 Resource Book . . . . pp. T1–T42

- Daily Transparencies
- Additional Examples Transparencies
- Teaching Transparencies

Technology

Teacher Resources

***Lesson Presentations* CD-ROM**. . . . . Chapter 3

***Test and Practice Generator* CD-ROM** . . . . Chapter 3

***One-Stop Planner* CD-ROM** . . . . Chapter 3

Student Resources

***Are You Ready? Intervention* CD-ROM**
Skills 1, 12, 34, 58, 59

internet connect go.hrw.com

Homework Help Online	**KEYWORD:** MR4 HWHelp3
Math Tools Online	**KEYWORD:** MR4 Tools
Glossary Online	**KEYWORD:** MR4 Glossary
Chapter Project Online	**KEYWORD:** MR4 PSProject3
Chapter Opener Online	**KEYWORD:** MR4 Ch3
CNN Student News™	**KEYWORD:** MR4 CNN3

SE = *Student Edition* **TE** = *Teacher's Edition* **AR** = *Assessment Resources* **CRB** = *Chapter Resource Book* **MK** = *Manipulatives Kit*

Chapter 3

Assessment Options

Assessing Prior Knowledge

Determine whether students have the required prerequisite concepts and skills.

Are You Ready? . . . . SE p. 89
Inventory Test . . . . AR pp. 1–4

Test Preparation

Provide review and practice for chapter and standardized tests.

Standardized Test Prep . . . . SE p. 149
Spiral Review with Test Prep . . . . . SE, last page of each lesson
Study Guide and Review . . . . SE pp. 144–146
Test Prep Tool Kit

Technology

***Test and Practice Generator* CD-ROM**

internet connect

State-Specific Test Practice Online KEYWORD: MR4 TestPrep

go.hrw.com

Performance Assessment

Assess students' understanding of chapter concepts and combined problem-solving skills.

Performance Assessment . . . . SE p. 148
Includes scoring rubric in TE
Performance Assessment . . . . AR p. 108
Performance Assessment Teacher Support . . . . . AR p. 107

Portfolio

Portfolio opportunities appear throughout the Student and Teacher's Editions.

Suggested work samples:

Problem Solving Project . . . . TE p. 88
Performance Assessment . . . . SE p. 148
Portfolio Guide . . . . AR p. xxxv
Journal . . . . . TE pp. 95, 99, 105, 109, 117, 123, 130, 137
Write About It . . . . . SE pp. 95, 99, 105, 109, 117, 126, 130, 133, 137

Daily Assessment

Obtain daily feedback on students' understanding of concepts.

Spiral Review and Test Prep . . . . . . SE, last page of each lesson

Also Available on Transparency in Chapter 3 Resource Book

Warm Up . . . . TE, first page of each lesson
Problem of the Day . . . . . . TE, first page of each lesson
Lesson Quiz . . . . TE, last page of each lesson

Student Self-Assessment

Have students evaluate their own work.

Group Project Evaluation . . . . AR p. xxxii
Individual Group Member Evaluation . . . . . AR p. xxxiii
Portfolio Guide . . . . AR p. xxxv
Journal . . . . . TE pp. 95, 99, 105, 109, 117, 123, 130, 137

Formal Assessment

Assess students' mastery of concepts and skills.

Section Quizzes . . . . AR pp. 9–10
Mid-Chapter Quiz . . . . SE p. 112
Chapter Test . . . . SE p. 147
Chapter Tests (Levels A, B, C) . . . . . . AR pp. 43–48
Cumulative Tests (Levels A, B, C) . . . . . AR pp. 151–162
Standardized Test Prep
Cumulative Assessment . . . . SE p. 149
End-of-Year Test . . . . AR pp. 271–274

Technology

***Test and Practice Generator* CD-ROM**

Make tests electronically. This software includes:

- Dynamic practice for Chapter 3
- Customizable tests
- Multiple-choice items for each objective
- Free-response items for each objective
- Teacher management system

SE = *Student Edition* **TE** = *Teacher's Edition* **AR** = *Assessment Resources* **CRB** = *Chapter Resource Book* **MK** = *Manipulatives Kit*

Chapter 3

Chapter 3 Tests

Three levels (A,B,C) of tests are available for each chapter in the *Assessment Resources.*

Level A

Chapter 3 Chapter Test — *Form A*

Write each in standard form, expanded form, and words.

1. 3.1 — 3 + 0.1, three and one tenth
2. five and eight hundredths — 5.08, 5 + 0.08

Order the decimals from least to greatest.

3. 3.6, 3.2, 3.1 — 3.1, 3.2, 3.6
4. 5.87, 5.25, 5.46 — 5.25, 5.46, 5.87

Estimate. Round to the indicated place value.

5. 6.13 + 7.65; tenths — 13.8
6. 2.581 − 2.035, hundredths — 0.54

Estimate each product or quotient.

7. 4.1×9.2 — 36
8. $16.4 \div 8$ — 2

Estimate a range for each sum.

9. 9.6 + 3.1 + 5.8 — 17 to 19
10. 10.4 + 3.3 + 9.9 — 22 to 24

Find each sum or difference.

11. 10.2 + 7.0 — 17.2
12. 14.5 − 6.0 — 8.5

Evaluate 1.7 + x for each value of x.

13. $x = 1.2$ — 2.9
14. $x = 0.2$ — 1.9

Multiply or divide.

15. 3.86×100 — 386
16. $82.3 \div 10$ — 8.23

Use the abbreviation for the most appropriate metric unit.

17. The length of a raisin is about 12 mm.
18. The distance from New York to Orlando is approximately 1,720 km.

Convert each measure.

19. 32.6 cm = 326 mm
20. 70 kL = 70,000 L

Chapter 3 Chapter Test — *Form A, continued*

Write each number in scientific notation.

21. 5,000 — 5×10^3
22. 47,000 — 4.7×10^4

Write each number in standard form.

23. 8.3×10^3 — 8,300
24. 8.5×10^4 — 85,000

Find each product.

25. 5.4×0.5 — 2.7
26. 8.5×1.2 — 10.2

Evaluate $11x$ for each value of x.

27. $x = 2.1$ — 23.1
28. $x = 0.2$ — 2.2

Find each quotient.

29. $18 \div 3$ — 6
30. $3.2 \div 2$ — 1.6

Evaluate the expression $6.8 \div x$ for each value of x.

31. $x = 4$ — 1.7
32. $x = 0.2$ — 34

Find each quotient.

33. $6.5 \div 1.3$ — 5
34. $10.5 \div 2.5$ — 4.2

Solve each equation.

35. $y - 1.5 = 7.5$ — $y = 9$
36. $6j = 18$ — $j = 3$
37. $\frac{f}{7} = 3$ — $f = 21$
38. $18.5 + h = 28.5$ — $h = 10$
39. At $1.50 per dozen how many whole dozens of eggs can be bought for $6.00? — 4 dozen
40. Dennis bought 1,000 board feet of white pine for $100. What did he pay per board foot? — $0.10

Level B

Chapter 3 Chapter Test — *Form B*

Write each in standard form, expanded form, and words.

1. 6.024 — 6 + 0.02 + 0.004, six and twenty-four thousandths
2. four and seven thousandths — 4.007, 4 + 0.007

Order the decimals from least to greatest.

3. 13.6, 13.2, 13.62 — 13.2, 13.6, 13.62
4. 3.87, 3.2, 3.45 — 3.2, 3.45, 3.87

Estimate. Round to the indicated place value.

5. 36.134 + 7.65; tenths — 43.8
6. 2.5864 − 2.0356; hundredths — 0.55

Estimate each product or quotient.

7. $71.825 \div 8.01$ — 9
8. 120.4×2.985 — 360

Estimate a range for the sum.

9. 9.65 + 30.1 + 5.835 — 44 to 46

Estimate a range for the sum.

10. 10.435 + 30.4 + 89.0 — 129 to 130

Find each sum or difference.

11. 11.54 + 17.01 — 28.55
12. 41.8 − 6.7 — 35.1

Evaluate 3.79 + x for each value of x.

13. $x = 2.54$ — 6.33
14. $x = 0.354$ — 4.144

Multiply or divide.

15. $4.12 \times 1{,}000$ — 4120
16. $827.5 \div 10^5$ — 0.008275

Use the abbreviation for the most appropriate metric unit.

17. A bathtub holds approximately 106 L.
18. The distance of a long distance race is 6.1 km.

Convert each measure.

19. 0.97 cm = 9.7 mm
20. 7,000 L = 7 kL

Chapter 3 Chapter Test — *Form B, continued*

Write each number in scientific notation.

21. 62,000 — 6.2×10^4
22. 2,357,000 — 2.357×10^6

Write each number in standard form.

23. 7.421×10^6 — 7,421,000
24. 4.85×10^4 — 48,500

Find each product.

25. 1.72×0.3 — 0.516
26. 8.4×0.003 — 0.0252

Evaluate $23x$ for each value of x.

27. $x = 2.55$ — 58.65
28. $x = 3.612$ — 83.076

Find each quotient.

29. $19.5 \div 6$ — 3.25
30. $8.88 \div 3$ — 2.96

Evaluate the expression $7.2 \div x$ for each value of x.

31. $x = 5$ — 1.44
32. $x = 0.06$ — 120

Find each quotient.

33. $7.82 \div 3.4$ — 2.3
34. $17.5 \div 0.28$ — 62.5

Solve each equation.

35. $y - 5.4 = 7.5$ — $y = 12.9$
36. $6.6j = 26.4$ — $j = 4$
37. $\frac{f}{11} = 3.4$ — $f = 37.4$
38. $23.6 - h = 18.1$ — $h = 5.5$
39. David bought 1,000 feet of aluminum striping for $220. What did he pay per foot? — $0.22
40. At $1.25 per dozen how many whole dozens of eggs can be bought for $6.00? — 4 dozen

Level C

Chapter 3 Chapter Test — *Form C*

Write each in standard form, expanded form, and words.

1. 16.78 — 16 + 0.7 + 0.08, sixteen and 78 hundredths
2. fourteen and seven ten-thousandths — 14.0007, 14 + 0.0007

Order the decimals from least to greatest.

3. 113.06, 113.026, 113.620 — 113.026, 113.06, 113.620
4. 1,115.187, 1,115.129, 1,115.1187 — 1,115.1187, 1,115.129, 1,115.187

Estimate. Round to the indicated place value.

5. 136.534 + 117.85; tenths — 254.4
6. 122.2448 − 12.123, hundredths — 110.12

Estimate each product or quotient.

7. 41.53×19.2 — 800
8. $172.124 \div 8.01$ — 21.5

Estimate a range for the sum.

9. 92.1612 + 30.15 + 55.835 — 177 to 178

Estimate a range for the sum.

10. 192.124 + 130.8545 + 289.134 — 611 to 612

Find each sum or difference.

11. 424.5 − 166.7 — 257.8
12. 1,268.32 + 1,253.54 — 2,521.86

Evaluate 113.79 + x for each value of x.

13. $x = 25.36$ — 139.15
14. $x = -110.74$ — 3.05

Multiply or divide.

15. 19.123×10^2 — 1,912.3
16. $11{,}827.5 \div 10^6$ — 0.0118275

Use the abbreviation for the most appropriate metric unit.

17. The length of a paperclip — cm
18. The capacity of a large juice can — L

Convert each measure.

19. 0.12324 km = 123,240 mm
20. 72.254 cm = 0.72254 m

Chapter 3 Chapter Test — *Form C, continued*

Write each number in scientific notation.

21. 662,100 — 6.621×10^5
22. 2,357,000,000 — 2.357×10^9

Write each number in standard form.

23. 5.4211×10^{11} — 542,110,000,000
24. 4.8542×10^{-2} — 0.048542

Find each product.

25. 9.35×0.9 — 8.415
26. 65.4×0.17 — 11.118

Evaluate $48x$ for each value of x.

27. $x = 2.1261$ — 102.0528
28. $x = -14.02$ — −672.96

Find each quotient.

29. $24.0552 \div 12$ — 2.0046
30. $18.888 \div 4$ — 4.722

Evaluate the expression $9.2 \div x$ for each value of x.

31. $x = 0.5$ — 18.4
32. $x = -20$ — −0.46

Find each quotient.

33. $115.338 \div 2.82$ — 40.9
34. $100.6 \div 0.625$ — 160.96

Solve each equation and check your answer.

35. $y - 52.35 = 117.5$ — $y = 169.85$
36. $3.74j = 102.7378$ — $j = 27.47$
37. $\frac{f}{11.18} = 5.23$ — $f = 58.4714$
38. $112.58 - h = 118.12$ — $h = -5.54$
39. If a stalactite took 1,000 years to grow to a length of 4.52 inches, how much did it grow per year? — 0.00452 in.
40. Water weighs approximately 62.5 pounds per cubic foot. If a bathtub contains about 32 ft^3 of water, how much does the water in the bathtub weigh? — 2,000 pounds

Test and Practice Generator CD-ROM

Create and customize multiple versions of the same tests with corresponding answers for any chosen chapter objectives.

Chapter 3 State and Standardized Test Preparation

Test Taking Skill Builder and Standardized Test Practice are provided for each chapter in the *Test Prep Tool Kit.*

TEST TAKING SKILL BUILDER

Test Taking Strategy **Multiple Choice—Eliminate Answers**
Chapter 3

Eliminating answer choices that you know are incorrect or unreasonable is an excellent test taking strategy. Use mental math and estimation techniques to help you decide which answer choices to eliminate.

Example 1 Multiple Choice Tiffany, Lynn, and Rebeckah opened a lemonade stand to earn extra money. They earned $24.87. How much money will each girl receive if they split their earnings evenly?

A $6.99   B $8.29   C $8.33   D $9.00

Solution:
Use mental math to estimate the answer:
$24.87 \approx 24$
$\frac{24}{3} = 8$ The amount they receive should be slightly more than $8.
Eliminate Choice A since $6.99 is too small.
Eliminate Choice D since $9.00 is too large.
Choose between Choices B and C.
$\frac{24.87}{3} = 8.29$
Each girl will receive $8.29.
The correct answer is Choice B.

Example 2 James went clothes shopping at a department store. He chose 3 shirts, priced at $18.49, $27.89, and $21.39. He also picked out two pairs of jeans, priced at $28.59 and $32.29. Which is the closest estimate of the total cost of his purchases, before tax?

F $120   G $130   H $135   I $140

Solution:
Use mental math to estimate the answer:
Add the dollar amounts:
$18 + $27 + $21 + $28 + $32 = $126
Eliminate Choice F because it is too small.
Eliminate Choice I because it is too large.
Choose between Choices G and H.
The correct answer is Choice G.

Test Taking Strategy
Chapter 3, continued

Exercises Possible answers are given.

Identify two choices that you can eliminate immediately and explain your reasoning. Then solve the problem.

1. When Mrs. Johnson found money in the laundry, she would divide it evenly between her 4 children. This week she found $35.64. How much will each child receive?
   A $6.73   B $8.21   C $8.91   D $9.32
   Choice 1: Choice A is too small.
   Choice 2: Choice D is too large.
   The correct calculation is $\frac{35.64}{4} = 8.91$. The correct answer is Choice C.
2. Sally went to the movies on Saturday. She paid $6.75 for her ticket. She bought a soda for $2.39 and a bag of popcorn for $3.65. After the movie, she went to dinner and paid a total of $9.24. Which is the closest estimate to how much Sally spent?
   F $9   G $22   H $25   I $27
   Choice 1: Eliminate Choice F since it is unreasonable.
   Choice 2: Eliminate Choice I since $27 is too large.
   Choice G is the correct answer.
3. Jason and Samantha are planning a ski trip. They recorded the snowfall from the local newspaper. The first day it snowed 5.7 inches. The next four days it snowed 6.2 in., 5.4 in., 7.7 in., and 8.2 in. How much snow fell in total?
   A 29 in.   B 30 in.   C 33 in.   D 77 in.
   Choice 1: Eliminate Choice A since 29 in. is too small.
   Choice 2: Eliminate Choice D since 77 in. is unreasonable. The total snowfall must be less than 35 in. because if you round up the values you get, $6 + 7 + 6 + 8 + 9 = 36$. Choice C is the correct answer.

STANDARDIZED TEST PRACTICE

Standardized Test Practice
Chapter 3

Select the best answer for Questions 1–7.

1. Martins' Media Shop is selling three DVD's for $63.75. What is the price of each DVD?
   A $63.75
   B $60.75
   C $21.25
   D $19.12
2. You are stacking 110-pound boxes on a freight elevator. A sign on the elevator says, "Do not exceed 1,200 pounds." What is the maximum number of boxes you can stack on the elevator?
   F 8 boxes
   G 9 boxes
   H 10 boxes
   I 11 boxes
3. Mr. Nye wants to purchase 25 pairs of headphones for the computer lab. If the headphones are on sale for $4.99, about how much will it cost to buy the headphones?
   A $30   C $125
   B $50   D $200
4. Sidney spent $13.12 on bottled water. If bottled water cost $0.82 per gallon, how many gallons of water did Sidney buy?
   F 4 gallons
   G 10 gallons
   H 12 gallons
   I 16 gallons
5. The school relay team competed at the district meet. The runners times are shown below. What was the total time (in seconds) for the relay team?

Relay Team Results	
Runner	Time (seconds)
Marcus	20.2
Jose	19.3
Roberto	18.7
Steven	16.2

   A 18.6 seconds
   B 20.2 seconds
   C 37.2 seconds
   D 74.4 seconds
6. Stella works at a veterinarian's office 15 hours per week. If she earns $7.25 per hour, how much will she earn in one week?
   F $2.10
   G $7.75
   H $22.25
   I $108.75
7. What is the length of *a*, in meters?

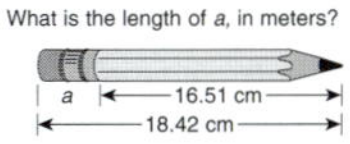

   A 0.3493 m
   B 0.304 m
   C 0.0211 m
   D 0.0191 m

Standardized Test Practice
Chapter 3, continued

Gridded Response
Solve the problems. Use the answer sheet to write and grid-in your answer.

8. When estimating the distance from the earth to the sun, Jenny needs to write the number 149,600,000 km in scientific notation. What exponent should Jenny use on base 10?
9. Mary went to lunch with $8.50. She bought an apple for $0.65, a bottle of water for $0.75, a turkey sandwich for $3.75, and a yogurt for $0.80. How much change should Mary receive?
10. As a lab team, you and your three partners measured the same piece of string. The measurements were 3.011 in., 3.01 in., 3.02 in., and 3.012 in. Which measurement was the smallest?

Short Response
Solve the problems. Use the answer sheet to write your answers.

11. Jill wrote a check for $21.18 and then had $115.62 left in her checking account. Write an equation to determine how much she had in her account before she wrote the check, then solve the equation.
12. Mrs. Henderson's math class ordered pizza for lunch so that they could stay in and study for their unit exam. They ordered 8 pizzas and six 2-liters of soda. The total cost, including tip, was $76.25. If 25 people chipped in the same amount of money, how much did each person pay? Show your work.

Extended Response

13. The longest side of the triangle is 21 units long.

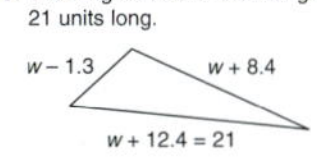

   a. Explain in words how to determine the lengths of the other two sides of the triangle.
   b. Find the unknown lengths.
   c. What is the perimeter of the triangle? Show your work.

Sheet
See Lesson 3-
See Lesson 3-
See Lesson 3-
10. 3.01
See Lesson 3-1.
Write your answers in the space provided.
11. $x - 21.18 = 115.62$; $x = 115.62 + 21.18$; $x = 136.80$; She had $136 in her account.
(See Lesson
12. 25)76.25; Each person chips in $3.05.
3.05
75
125
125
(See Lesso

Customized answer sheets give students realistic practice for actual standardized tests.

internet connect go.hrw.com
State-Specific Test Practice Online
KEYWORD: MR4 TestPrep

Test Prep Tool Kit

- Standardized Test Prep Workbook
- Countdown to Testing transparencies
- State Test Prep CD-ROM
- Standardized Test Prep Video

Chapter 3

Decimals

Why Learn This?

Tell students that decimals are essential to describe quantities accurately. For example, without decimals, the winning Men's High Jump heights from 1900 and 1928 would appear to be equal. The more decimal places used, the more accurate the measurement.

Using Data

To begin the study of this chapter, have students:

- Identify the years from the table in which the Women's 100 Meters was held. 1928, 1952, 1988, 2000
- Identify the winning distance of the Men's discus in 1988. 68.81 m
- Order the times in the Women's 100 Meters from fastest to slowest. 10.54 s, 10.75 s, 11.5 s, 12.2 s

Chapter 3

Decimals

TABLE OF CONTENTS

internet connect
Chapter Opener Online
go.hrw.com
KEYWORD: MR4 Ch3

Winning Olympic Performances

Year	Women's 100 Meters (s)	Women's Discus (m)	Men's 100 Meters (s)	Men's Discus (m)
1900	–	36.04	12.0	–
1928	12.2	39.62	10.8	47.32
1952	11.5	51.4	10.4	55.02
1988	10.54	72.3	9.92	68.81
2000	10.75	68.4	9.87	69.29

Career *Sports Historian*

Are people breaking records by running faster and jumping farther and higher? Records are kept for both professional and amateur sports. Many schools keep records of their individual athletes' and teams' performances. Keeping track of sports records is the job of sports historians. One of the most complete records is that of the Olympic games. The table shows the changes in the last century of the winning performances in some men's and women's Olympic sports.

PROBLEM SOLVING

Problem Solving Project

History Connection

Purpose: To solve problems using decimals

Materials: Olympic History worksheet

internet connect
Chapter Project Online: ***go.hrw.com***
KEYWORD: MR4 PSProject3

Understand, Plan, Solve, and Look Back

Have students:

- ✓ Complete the Olympic History worksheet to learn about some changes in Olympic results in the last century.
- ✓ Make a chart comparing the women's and men's times in the 100 meters. Graph the results.
- ✓ Discover why there aren't any women's results in these events for 1900. Have students find out if there are any Olympic events today that men compete in but women do not.
- ✓ Check students' work

ARE YOU READY?

Choose the best term from the list to complete each sentence.

1. In the metric system, the base unit for measuring length is the ___?___, and the base unit for measuring volume is the ___?___. meter; liter
2. In the expression 72 ÷ 9, 72 is the ___?___, and 9 is the ___?___. dividend; divisor
3. The answer to a subtraction expression is the ___?___. difference
4. A(n) ___?___ is a mathematical statement that says two quantities are equal. equation

difference
dividend
divisor
equation
liter
meter
quotient

Complete these exercises to review skills you will need for this chapter.

✔ Place Value of Whole Numbers

Identify the place value of each underlined digit.

5. 15<u>2</u> ten
6. <u>7</u>,903 thousand
7. <u>1</u>45,072 hundred thousand
8. 4,8<u>9</u>3,025 ten thousand
9. 1<u>3</u>,796,020 million
10. 1<u>4</u>5,683,032 ten million

✔ Add and Subtract Whole Numbers

Find each sum or difference.

11. \$425 − \$75 \$350
12. 532 + 145 677
13. 160 − 82 78

✔ Multiply and Divide Whole Numbers

Find each product or quotient.

14. \$320 × 5 \$1,600
15. 125 ÷ 5 25
16. 54 × 3 162

✔ Exponents

Find each value.

17. 10^3 1,000
18. 3^6 729
19. 10^5 100,000
20. 4^5 1,024
21. 8^3 512
22. 2^7 128

✔ Solve Whole Number Equations

Solve each equation.

23. $y + 382 = 743$ $y = 361$
24. $n - 150 = 322$ $n = 472$
25. $9x = 108$ $x = 12$

Assessing Prior Knowledge

INTERVENTION

Diagnose and Prescribe

Evaluate your students' performance on this page to determine whether intervention is necessary or whether enrichment is appropriate. Options that provide instruction, practice, and a check are listed below.

Resources for Are You Ready?

- ***Are You Ready? Intervention and Enrichment***
- **Recording Sheet for Are You Ready?** *Chapter 3 Resource Book . . . . . . p. 3*

Are You Ready? Intervention **CD-ROM**

ARE YOU READY?

Were students successful with Are You Ready?

NO — INTERVENE

✔ **Place Value of Whole Numbers 5–10**
Intervention Practice, Skill 1
CD-ROM Intervention Activities, Skill 1

✔ **Add and Subtract Whole Numbers 11–13**
Intervention Practice, Skill 34
CD-ROM Intervention Activities, Skill 34

✔ **Multiply and Divide Whole Numbers 14–16**
Intervention Practice, Skill 34
CD-ROM Intervention Activities, Skill 34

✔ **Exponents 17–22**
Intervention Practice, Skill 12
CD-ROM Intervention Activities, Skill 12

✔ **Solve Whole Number Equations 23–25**
Intervention Practice, Skills 58, 59
CD-ROM Intervention Activities, Skills 58, 59

YES — ENRICH

Are You Ready? Enrichment, pp. 411–412

Section 3A

Understanding Decimals

One-Minute Section Planner

Lesson	Materials	Resources
Hands-On Lab 3A Model Decimals **NCTM:** Number and Operations **NAEP:** Number Properties 1b ☑ SAT-9 ☑ SAT-10 ☐ ITBS ☐ CTBS ☑ MAT ☑ CAT	**Required** Decimal grids *(CRB p. 108)*	• *Hands-On Lab Activities,* p. 10
Lesson 3-1 Representing, Comparing, and Ordering Decimals **NCTM:** Number and Operations, Communication, Connections, Representation **NAEP:** Number Properties 1a ☑ SAT-9 ☑ SAT-10 ☑ ITBS ☐ CTBS ☑ MAT ☑ CAT	**Optional** Teaching Transparencies T2–T3 *(CRB)* Base-10 blocks *(MK)*	• *Chapter 3 Resource Book,* pp. 6–17 • Daily Transparency T1, CRB • Additional Examples Transparencies T4–T5, CRB • *Alternate Openers: Explorations,* p. 15
Lesson 3-2 Estimating Decimals **NCTM:** Number and Operations, Communication **NAEP:** Number Properties 2b ☑ SAT-9 ☑ SAT-10 ☐ ITBS ☐ CTBS ☑ MAT ☐ CAT		• *Chapter 3 Resource Book,* pp. 18–26 • Daily Transparency T6, CRB • Additional Examples Transparencies T7–T9, CRB • *Alternate Openers: Explorations,* p. 16
Hands-On Lab 3B Explore Decimal Addition and Subtraction **NCTM:** Number and Operations, Algebra **NAEP:** Number Properties 3f ☑ SAT-9 ☑ SAT-10 ☐ ITBS ☐ CTBS ☑ MAT ☑ CAT	**Required** Decimal grids *(CRB p. 108)*	• *Hands-On Lab Activities,* pp. 11–12
Lesson 3-3 Adding and Subtracting Decimals **NCTM:** Number and Operations, Communication **NAEP:** Number Properties 3a ☑ SAT-9 ☑ SAT-10 ☑ ITBS ☑ CTBS ☑ MAT ☑ CAT	**Optional** Newspaper advertisements	• *Chapter 3 Resource Book,* pp. 27–35 • Daily Transparency T10, CRB • Additional Examples Transparencies T11–T13, CRB • *Alternate Openers: Explorations,* p. 17
Lesson 3-4 Decimals and Metric Measurement **NCTM:** Measurement, Communication, Connections **NAEP:** Measurement 2b ☑ SAT-9 ☑ SAT-10 ☑ ITBS ☑ CTBS ☑ MAT ☑ CAT	**Optional** Teaching Transparencies T15–T16 *(CRB)* Items labeled with metric measurements	• *Chapter 3 Resource Book,* pp. 36–44 • Daily Transparency T14, CRB • Additional Examples Transparencies T17–T18, CRB • *Alternate Openers: Explorations,* p. 18
Hands-On Lab 3C Estimate Measurements **NCTM:** Measurement, Connections **NAEP:** Measurement 1c ☐ SAT-9 ☑ SAT-10 ☐ ITBS ☐ CTBS ☑ MAT ☐ CAT	**Required** One-quart and one-liter containers 1-kg weight Pound scale	• *Hands-On Lab Activities,* pp. 13–14
Section 3A Assessment		• Mid-Chapter Quiz, SE p. 112 • Section 3A Quiz, AR p. 9 • *Test and Practice Generator* CD-ROM

SAT = *Stanford Achievement Tests* **ITBS** = *Iowa Test of Basic Skills* **CTBS** = *Comprehensive Test of Basic Skills/Terra Nova*
MAT = *Metropolitan Achievement Tests* **CAT** = *California Achievement Test*

NCTM = Complete standards can be found on pages T27–T33. **NAEP** = Complete standards can be found on pages A31–A35.

SE = *Student Edition* **TE** = *Teacher's Edition* **AR** = *Assessment Resources* **CRB** = *Chapter Resource Book* **MK** = *Manipulatives Kit*

Section Overview

Comparing and Ordering Decimals — *Lesson 3-1*

Why? Ordered lists of decimals are frequently found in sports statistics, such as batting averages in baseball.

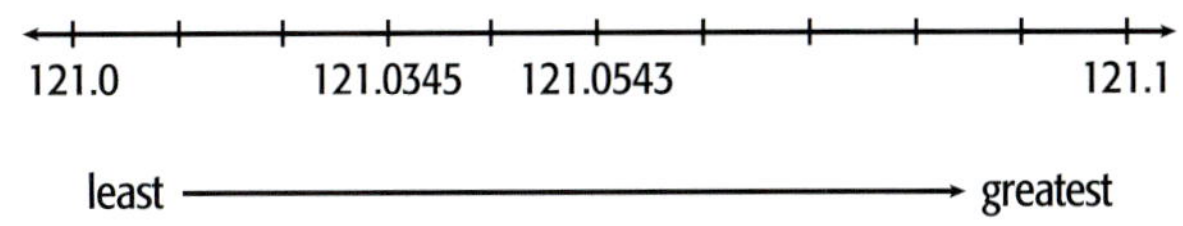

121.0**3**45 < 121.0**5**43

> To order numbers, you can compare them using **place value.**

> Numbers are **ordered** from least to greatest on the number line from left to right.

Estimating Decimals — *Lesson 3-2*

Why? You can use estimation to determine whether results of decimal operations are reasonable.

Estimate 3.**5**6 + 8.**3**1 to the nearest whole number.

4 + 8 = 12   The sum is about 12.

Estimate 9.7 ÷ 3.5.

9 ÷ **3** = 3   The quotient is about 3.

> When rounding, look at the digit to the **right of the place to which you are rounding**.
> - If that digit is 5 or greater, round up.
> - If that digit is less than 5, round down.

> **Compatible numbers** are close to the numbers in the problem, and they can help you do math mentally.

Adding and Subtracting Decimals — *Lesson 3-3*

Why? Using a checkbook requires adding and subtracting decimals.

Add 5 + 10.25 + 3.5.

```
   5.00
  10.25
+  3.50
  -----
  28.75
```

> Use zeros to write an equivalent number to the same number of decimal places as the other numbers.

Subtract 3.57 from 9.

```
  9.00
 −3.57
 -----
  5.43
```

> Align the decimal points.

Using the Metric System — *Lesson 3-4*

Why? The metric system is used throughout the world. To convert units within the metric system, multiply and divide by powers of ten.

> To **multiply** by the *n*th power of ten, move the decimal point *n* places right.

> To **divide** by the *n*th power of ten, move the decimal point *n* places left.

	Unit	Abbreviation	Approximate Comparison
Length	**Kilo**meter	km	Length of 10 football fields
	Meter	m	Width of a door
	Centimeter	cm	Width of your little finger
	Millimeter	mm	Thickness of a dime
Mass	**Kilo**gram	kg	Mass of a textbook
	Gram	g	Mass of a small paperclip
Capacity	Liter	L	Filled bottle of sparkling water
	Milliliter	mL	Half-filled eyedropper

Hands-On LAB

3A Model Decimals

Pacing: Traditional 1 day
Block $\frac{1}{2}$ day

Objective: To use decimal grids to represent decimals

Materials: Decimal grids

Lab Resources

Hands-On Lab Activities . . . . p. 10

Using the Pages

Discuss with students what each small square, row, and column represents. Discuss what number is represented by the entire grid.

Represent each decimal on a decimal grid.

1. 0.02

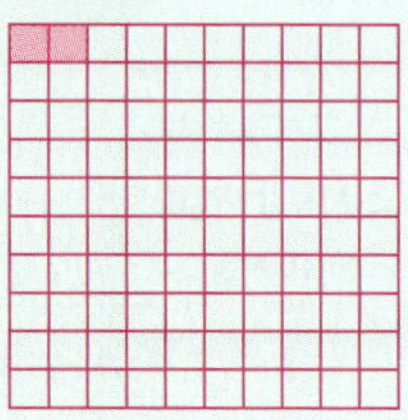

2. 3.4

Hands-On LAB 3A

Model Decimals

Use with Lesson 3-1

internet connect
Lab Resources Online
go.hrw.com
KEYWORD: MR4 Lab3A

KEY

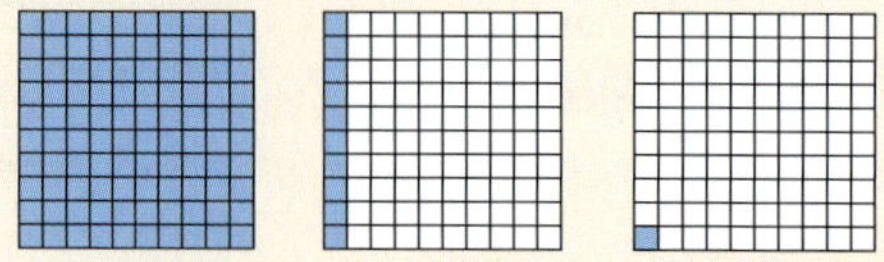

1	0.1	0.01
one	1 tenth	1 hundredth

You can use decimal grids to represent decimals. The grid is divided into 100 small squares. One square represents 1 hundredth, or 0.01. Ten squares form a column, which represents 1 tenth, or 0.1. Ten columns make up the grid, which represents one whole, or 1. By shading hundredths, tenths, or whole grids, you can represent decimal numbers.

Activity 1

1 Write the decimal that is represented.

a.

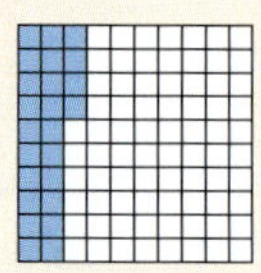

24 hundredths squares are shaded.

So 0.24 is shaded.

b.

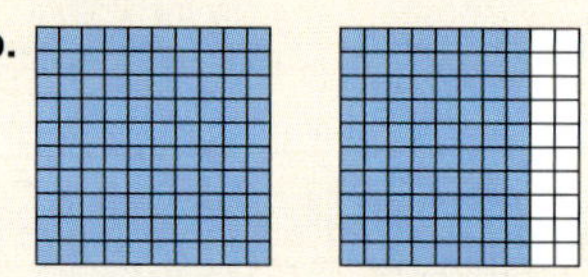

1 whole grid and 8 columns are shaded.

So 1.8 is shaded.

c.

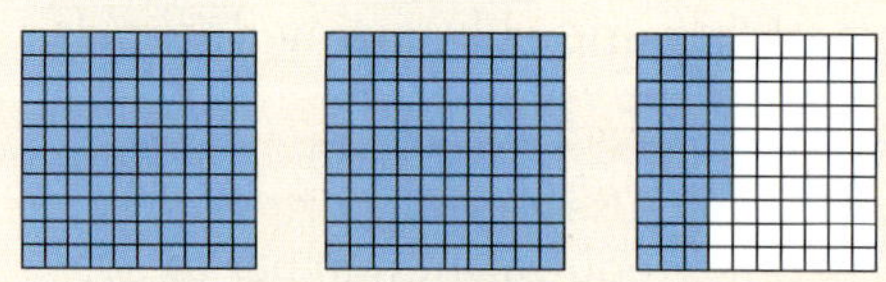

2 whole grids and 37 hundredths are shaded.

So 2.37 is shaded.

Think and Discuss

1. Tell how decimal grids show that 0.30 = 0.3. **Possible answer: Shading 30 small squares is the same as shading 3 columns of 10.**

Try This

Write the decimal that is represented.

1. 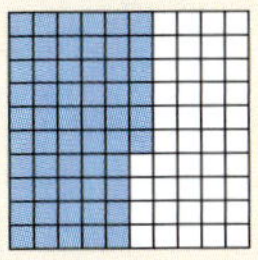 **0.56**

2. 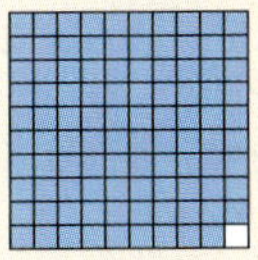 **0.99**

3. 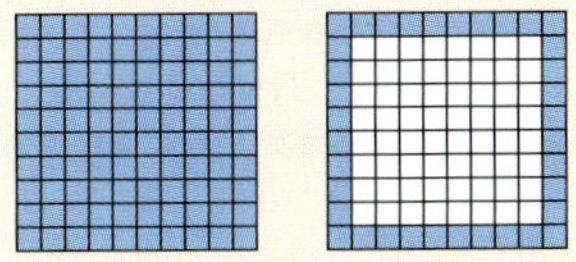 **1.36**

Activity 2

1 Use a decimal grid to represent each decimal.

a. 0.42

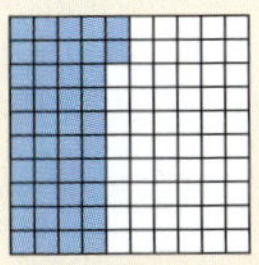

Shade 42 hundredths squares.

b. 1.88

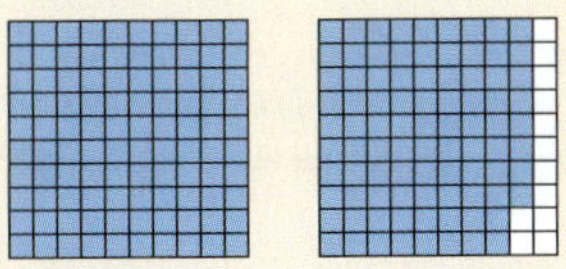

Shade 1 whole grid, 8 columns, and 8 small squares.

c. 2.75

Shade 2 whole grids, 7 columns, and 5 small squares.

Think and Discuss

1. Explain how to represent 0.46 by shading only 10 sections on the grid. (*Hint:* A section is a grid, column, or small square.)
Possible answer: Shade 4 columns, plus 6 small squares.

Try This

Use a decimal grid to represent each decimal.

1. 1.02 **2.** 0.04 **3.** 0.4 **4.** 2.14 **5.** 0.53

Answers

Activity 2

Try This

1. 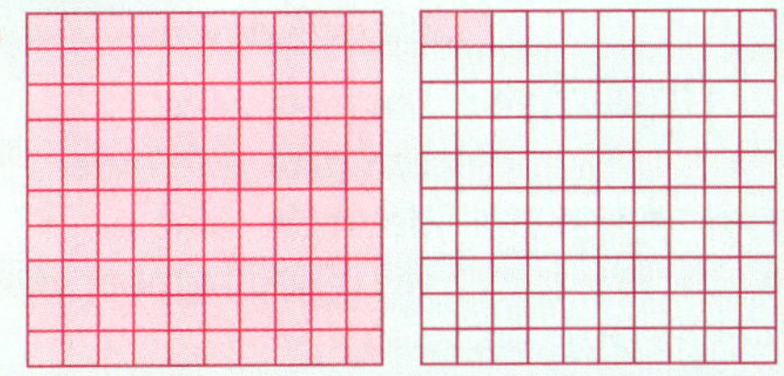

2. 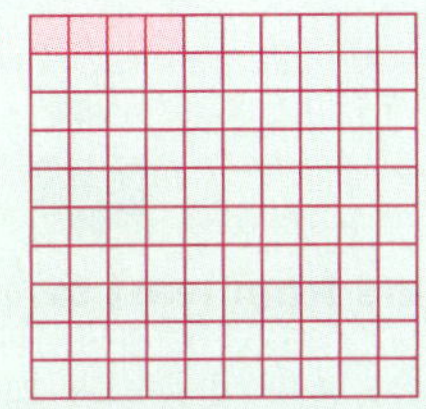

3. 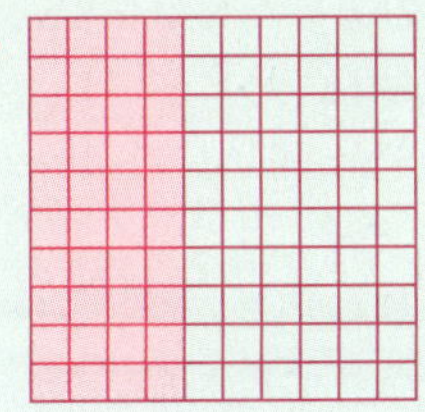

4. 

5. 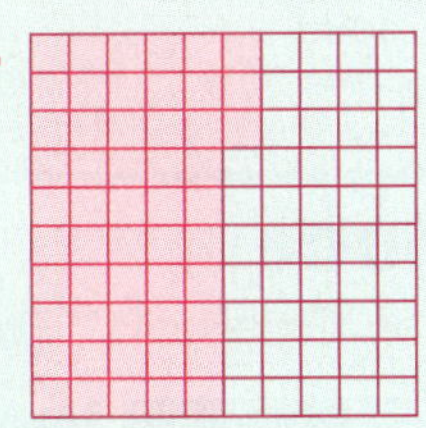

3-1 Organizer

Pacing: Traditional 1 day
Block $\frac{1}{2}$ day

Objective: Students write, compare, and order decimals using place value and number lines.

Indiana Academic Standards
6.1.3, 6.7.11

Warm Up

Order the numbers from least to greatest.

1. 242, 156, 224, 165
   156, 165, 224, 242
2. 941, 148, 914, 814, 721
   148, 721, 814, 914, 941
3. 345, 376, 354, 397
   345, 354, 376, 397

Problem of the Day

Lupe is taller than Reba and shorter than Miguel. Tory is shorter than Lupe but taller than Reba. List the four brothers and sisters in order from tallest to shortest. Miguel, Lupe, Tory, Reba

Available on Daily Transparency in CRB

Math Humor

Did you hear about the decimal that found studying very difficult? It never seemed to get the point.

3-1 Representing, Comparing, and Ordering Decimals

Learn to write, compare, and order decimals using place value and number lines.

The smaller the apparent magnitude of a star, the brighter the star appears when viewed from Earth. The magnitudes of some stars are listed in the table as decimal numbers.

Apparent Magnitudes of Stars	
Star	**Magnitude**
Procyon	0.38
Proxima Centauri	11.0
Wolf 359	13.5
Vega	0.03

Decimal numbers represent combinations of whole numbers and numbers between whole numbers.

Place value can help you understand and write decimal numbers.

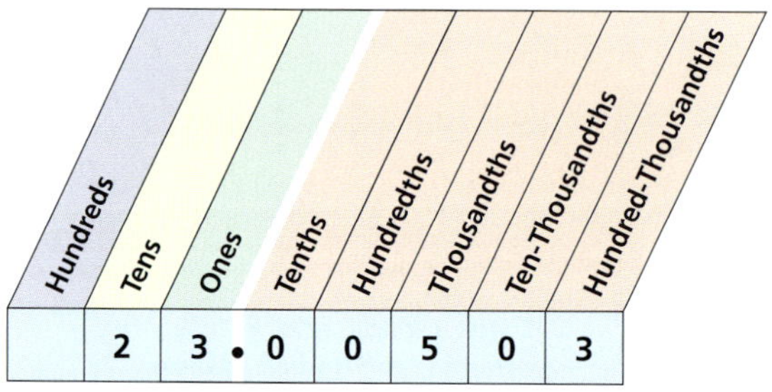

EXAMPLE 1 **Reading and Writing Decimals**

Reading Math
Read the decimal point as "and."

Write each decimal in standard form, expanded form, and words.

A 1.05

Expanded form: 1 + 0.05
Word form: one *and* five hundredths

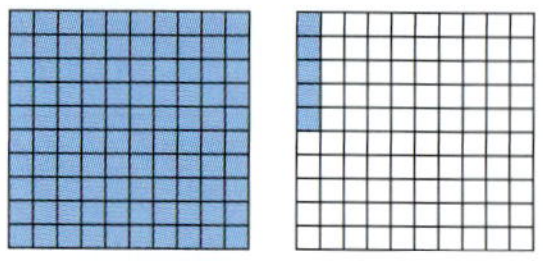

B 0.05 + 0.001 + 0.0007

Standard form: 0.0517
Word form: five hundred seventeen ten-thousandths

C sixteen and nine hundredths

Standard form: 16.09
Expanded form: 10 + 6 + 0.09

You can use place value to compare decimal numbers.

1 Introduce

Alternate Opener

EXPLORATION

3-1 Representing, Comparing, and Ordering Decimals

To model a decimal,

- color a 10-by-10-square grid for each whole in the decimal,
- color one 10-by-1-square strip for each tenth in the decimal, and color a small square for each hundredth in the decimal.

For example, the graph paper models the decimal 1.62.

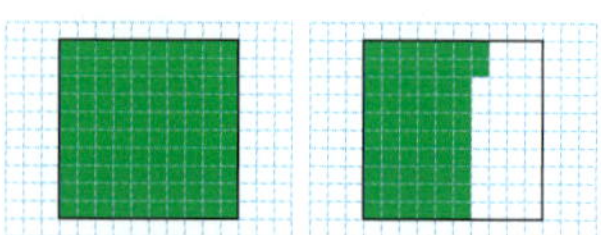

1. Draw a model for each decimal.
   a. 1.25 b. 2.13 c. 1.70 d. 1.7
2. Compare the models for **1c** and **1d.** What do you notice about these two decimals?
3. Order the decimals in numbers **1a–1d** from least to greatest. Explain your reasoning.

Think and Discuss

4. **Explain** why 0.4 = 0.40.
5. **Explain** why 0.5 is greater than 0.10 even though 10 is greater than 5.

Motivate

Review the terms *standard form, expanded form,* and *word form*. Have students give examples of whole numbers expressed in standard, expanded, and word form (e.g., 3,025; 3,000 + 20 + 5; three thousand, twenty-five). Explain that decimals can also be expressed in standard, expanded, and word form.

Exploration worksheet and answers on Chapter 3 Resource Book pp. 7–9 and 112

2 Teach

Lesson Presentation

Guided Instruction

In this lesson, students learn to write, compare, and order decimals using place value and number lines. First point out the place-value chart (Teaching Transparency T2 and T3 in the CRB). Emphasize the names of the places to the right of the decimal point. Then teach how to properly name and write decimals in standard, word, and expanded form. Teach students to compare two decimals by lining up the decimal points. Next, demonstrate how to order three decimals by using a number line.

Teaching Tip

Some students may benefit from using graph paper to align and compare decimal numbers.

EXAMPLE 2 ***Earth Science Application***

Rigel and Betelgeuse are two stars in the constellation Orion. The apparent magnitude of Rigel is 0.12. The apparent magnitude of Betelgeuse is 0.50. Which star has the smaller magnitude? Which star appears brighter?

0.12	*Line up the decimal points. Start from the left and compare the digits.*
0.50	*Look for the first place where the digits are different.*

1 is less than 5.
$0.12 < 0.50$

Rigel has a smaller apparent magnitude than Betelgeuse. The star with the smaller magnitude appears brighter. When seen from Earth, Rigel appears brighter than Betelgeuse.

EXAMPLE 3 **Comparing and Ordering Decimals**

Helpful Hint

Writing zeros at the end of a decimal does not change the value of the decimal.

0.3 = 0.30 = 0.300

Order the decimals from least to greatest.
14.35, 14.3, 14.05

14.35 14.30	$14.30 < 14.35$	*Compare two of the numbers at a time. Write 14.3 as "14.30."*
14.35 14.05	$14.05 < 14.35$	*Start at the left and compare the digits.*
14.30 14.05	$14.05 < 14.30$	*Look for the first place where the digits are different.*

Graph the numbers on a number line.

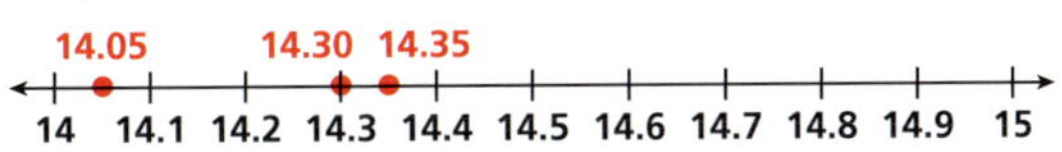

The numbers are ordered when you read the number line from left to right. The numbers in order from least to greatest are 14.05, 14.3, and 14.35.

Think and Discuss

1. **Explain** why 0.5 is greater than 0.29 even though 29 is greater than 5.
2. **Name** the decimal with the least value.
   0.29, 2.09, 2.009, 0.029
3. **Name** three numbers between 1.5 and 1.6.

COMMON ERROR ALERT

Some students try to align the digits farthest to the right, as with whole numbers, resulting in incorrect comparisons if the decimals have different numbers of decimal places. Many students will benefit from using graph paper to align their decimal numbers.

Additional Examples

Example 1

Write each in standard form, expanded form, and words.

A. 1.07
1 + 0.07; one *and* seven hundredths

B. 0.03 + 0.006 + 0.0009
0.0369; three hundred sixty-nine ten-thousandths

C. fourteen and eight hundredths
14.08; 10 + 4 + 0.08

Example 2

The star Wolf 359 has an apparent magnitude of 13.5. Suppose another star has an apparent magnitude of 13.05. Which star has the smaller magnitude?
The star with 13.05 has the smaller apparent magnitude.

Example 3

Order the decimals from least to greatest.

16.67, 16.6, 16.07
The numbers in order are 16.07, 16.6, 16.67.

3 Close

Reaching All Learners
Through Concrete Manipulatives

Have students work with base-10 blocks, provided in the Manipulatives Kit, to reinforce their understanding of place value in decimal numbers. Base-10 blocks provide visual, tactile, auditory, and kinesthetic learning modalities to support all learners. (Paper versions of base-10 blocks are provided on pages 109 and 110 of the Chapter 3 Resource Book.)

Summarize

Have students name the place-value positions to the right of the decimal point and give examples of decimals in standard, expanded, and word form. Relate the tenths and the hundredths places to dimes and pennies to help in recalling place-value names.

Answers to Think and Discuss

1. Possible answer: because five tenths is the same as 50 hundredths and 50 is greater than 29
2. 0.029
3. Possible answers: 1.51, 1.53, 1.57

3-1 PRACTICE & ASSESS

Students may want to refer back to the lesson examples.

Assignment Guide

If you finished Example 1 assign:
Core 1–4, 9–12, 17, 24–26, 37–42
Enriched 1–3, 9, 17–19, 24–27, 37–42

If you finished Example 2 assign:
Core 1–5, 9–13, 17–25 odd, 30–33, 37–42
Enriched 17–27, 30–42

If you finished Example 3 assign:
Core 1–16, 17–27 odd, 30–33, 37–42
Enriched 7–15 odd, 17–42

Answers

9. 7 + 0.08 + 0.009 + 0.0003; seven and eight hundred ninety-three ten-thousandths

3-1 Exercises

FOR EXTRA PRACTICE
see page 640

internet connect
Homework Help Online
go.hrw.com Keyword: MR4 3-1

GUIDED PRACTICE

See Example 1 **Write each decimal in standard form, expanded form, and words.**

1. 1.98 1 + 0.9 + 0.08; one and ninety-eight hundredths

2. ten and forty-one thousandths 10.041; 10 + 0.04 + 0.001

3. 0.07 + 0.006 + 0.0005 0.0765; seven hundred sixty-five ten-thousandths

4. 0.0472 0.04 + 0.007 + 0.0002; four hundred seventy-two ten-thousandths

See Example 2 **5.** Osmium and iridium are precious metals. The density of osmium is 22.58 g/cm^3, and the density of iridium is 22.56 g/cm^3. Which metal is denser? Osmium

See Example 3 **Order the decimals from least to greatest.**

6. 9.5, 9.35, 9.65 9.35, 9.5, 9.65

7. 4.18, 4.1, 4.09 4.09; 4.1; 4.18

8. 12.39, 12.09, 12.92 12.09, 12.39, 12.92

INDEPENDENT PRACTICE

See Example 1 **Write each decimal in standard form, expanded form, and words.**

9. 7.0893

10. 12 + 0.2 + 0.005 12.205; twelve and two hundred five thousandths

11. seven and fifteen hundredths 7.15; 7 + 0.1 + 0.05

12. 3 + 0.1 + 0.006 3.106; three and one hundred six thousandths

See Example 2 **13.** Two meteorites landed in Mexico. The one found in Bacuberito weighed 24.3 tons, and the one found in Chupaderos weighed 26.7 tons. Which meteorite weighed more? the Chupaderos meteorite

See Example 3 **Order the decimals from least to greatest.**

14. 15.25, 15.2, 15.5 15.2, 15.25, 15.5

15. 1.56, 1.62, 1.5 1.5, 1.56, 1.62

16. 6.7, 6.07, 6.23 6.07, 6.23, 6.7

PRACTICE AND PROBLEM SOLVING

Write each number in words.

17. 9.007 nine and seven thousandths

18. 5 + 0.08 + 0.004 five and eighty-four thousandths

19. 10.022 ten and twenty-two thousandths

Compare. Write <, >, or =.

20. 8.04 ■ 8.403 <

21. 0.907 ■ 0.6801 >

22. 1.246 ■ 1.29 <

23. one and fifty-two ten-thousandths ■ 1.0052 =

Write the value of the red digit in each number.

24. 3.026 six thousandths

25. 17.53703 three hundredths

26. 0.000598 five ten-thousandths

27. 425.1055 one tenth

Order the numbers from greatest to least.

28. 32.525, 32.5254, 31.6257

29. 0.34, 1.43, 4.034, 1.043, 1.424 4.034, 1.43, 1.424, 1.043, 0.34

28. 32.5254, 32.525, 31.6257

Math Background

The place-value system allows for the naming of whole numbers that are greater than any previously assigned value and for the naming of decimal numbers that are smaller than any previously assigned value.

RETEACH 3-1

LESSON 3-1 **Reteach**
Representing, Comparing, and Ordering Decimals

You can use place value to write decimals in standard form, expanded form, and word form.

To write 2.14 in expanded form, write the decimal as an addition expression using the place value of each digit.

2.14 can be written as 2 + 0.1 + 0.04.

Ones	Tenths	Hundredths	Thousandths	Ten Thousandths
2.	1	4		

When you write a decimal in word form, the number before the decimal point tells you how many wholes there are. The decimal point stands for the word "and."

Notice that the place value names to the right of the decimal begin with tenths, hundredths, and then thousandths. The "ths" ending indicates a decimal.

2.14 can also be written as *two and fourteen hundredths.*

1. How would you read a number with 4 decimal places?

The decimal should end with ten thousandths

Write each decimal in standard form, expanded form, and word form.

2.

Ones	Tenths	Hundredths	Thousandths	Ten Thousandths
5.	6	9	8	

5 + 0.6 + 0.09 + 0.008; five and six hundred ninety-eight thousandths

3.

Ones	Tenths	Hundredths	Thousandths	Ten Thousandths
0.	0	9	4	

0 + 0.09 + 0.004; ninety-four thousandths

4. 7 + 0.8

7.8; seven and eight tenths

5. twelve-hundredths

0 + 0.1 + 0.02; 0.12

PRACTICE 3-1

LESSON 3-1 **Practice B**
Representing, Comparing, and Ordering Decimals

Write each in standard form, expanded form, and words.

1. 2.07 2 + 0.07; two and seven hundredths
2. 5 + 0.007 5.007; five and seven thousandths
3. four and six tenths 4.6; 4 + 0.6
4. sixteen and five tenths 16.5; 10 + 6 + 0.5
5. 9 + 0.6 + 0.08 9.68; nine and sixty-eight hundredths
6. 1.037 1 + 0.03 + 0.007; one and thirty-seven thousandths
7. 2 + 0.1 + 0.003 2.103; two and one hundred three thousandths
8. eighteen hundredths 0.18; 0.1 + 0.08
9. 6.11 6 + 0.1 + 0.01; six and eleven hundredths

Order the decimals from least to greatest.

10. 3.578, 3.758, 3.875

3.578; 3.758; 3.875

11. 0.0943, 0.9403, 0.9043

0.0943; 0.9043; 0.9403

12. 12.97, 12.957, 12.75

12.75; 12.957; 12.97

13. 1.09, 1.901, 1.9, 1.19

1.09; 1.19, 1.9; 1.901

14. Your seventh and eighth ribs are two of the longest bones in your body. The average seventh rib is nine and forty-five hundredths inches long, and the average eighth rib is 9.06 inches long. Which bone is longer?

the seventh rib

15. The average female human heart weighs nine and three tenths ounces, while the average male heart weighs eleven and one tenth ounces. Which human heart weighs less, the male or the female?

the female heart

16. The state has $42.3 million for a new theater. The theater that an architect designed would cost $42.25 million. Can the theater be built for the amount the state can pay?

yes

17. Lyn traveled 79.47 miles on Saturday, 54.28 miles on Sunday, 65.5 miles on Monday, and 98.43 miles on Tuesday. Which day did she travel the greatest number of miles?

Tuesday

Astronomy LINK

Proxima Centauri, the closest star to Earth other than the Sun, was discovered in 1913. It would take about 115,000 years for a spaceship traveling from Earth at 25,000 mi/h to reach Proxima Centauri.

Use the table for Exercises 30–36.

30. Order the stars Sirius, Luyten 726-8, and Lalande 21185 from closest to farthest from Earth. Lalande 21185, Luyten 726-8, Sirius

31. Which star in the table is farthest from Earth? Ross 154

32. How far in light-years is Ross 154 from Earth? Write the answer in words and expanded form. Nine and forty-five hundredths, $9 + 0.4 + 0.05$

33. List the stars that are less than 5 light-years from Earth. Alpha Centauri, Proxima Centauri

34. ***WHAT'S THE ERROR?*** A student wrote the distance of Proxima Centauri from Earth as "four hundred and twenty-two hundredths." Explain the error. Write the correct answer.

35. ***WRITE ABOUT IT*** Which star is closer to Earth, Sirius or Lalande 21185? Explain how you can compare the distances of these stars. Then answer the question.

Distance of Stars from Earth

Star	Distance (light-years)
Alpha Centauri	4.35
Barnard's Star	5.98
Lalande 21185	8.22
Luyten 726-8	8.43
Proxima Centauri	4.22
Ross 154	9.45
Sirius	8.65

36. ***CHALLENGE*** Wolf 359 is located 7.75 light-years from Earth. If the stars in the table were listed in order from closest to farthest from Earth, between which two stars would Wolf 359 be located? Barnard's Star and Lalande 21185

Spiral Review

Compare. Write <, >, or =. (Lesson 1-1)

37. 4,897,204 ▒ 4,895,190 $>$ **38.** 133,099,588 ▒ 133,099,600 $<$

Write each expression in exponential form. (Lesson 1-3)

39. $3 \times 3 \times 3 \times 3 \times 3$ 3^5 **40.** $10 \times 10 \times 10 \times 10$ 10^4 **41.** $13 \times 13 \times 13$ 13^3

42. **TEST PREP** For which of the following equations is 8 a solution? (Lesson 2-6) B

A $7n = 63$ **B** $24 + p = 32$ **C** $t - 5 = 13$ **D** $\frac{a}{7} = 8$

Interdisciplinary LINK

Astronomy

Exercises 30–36 involve using data about the distances of well-known stars from Earth. Astronomy is studied in middle school earth science programs such as *Holt Science & Technology*.

Answers

34. Possible answer: Writing *hundred* after *four* was the error. The 4 is in the ones position. Correct answer: four and twenty-two hundredths.

35. Possible answer: Align the decimals and compare tenths; Lalande 21185

Journal

Have students write about a situation that does not involve money in which decimal numbers are used.

Test Prep Doctor

For Exercise 42, students need to remember how to find the solution of an equation with a variable. Students who answered **A** probably had the right idea but substituted 9 instead of 8 for *n*. Students who answered **C** probably also had the right idea, but they added instead of subtracting. Students who answered **D** probably assumed that 8 was the solution because it stands alone on one side of the equal sign.

CHALLENGE 3-1

LESSON 3-1 **Challenge**
Place Your Values

Complete the tables below to show different numbers that can be written with the same digits. Do not use the same digit more than once for each place value. Possible answers are given.

1. Use the digits 1, 3, 5, 7, and 9 to write four 5-digit numbers of increasing value.

Hundreds	Tens	Ones	Tenths	Hundredths	Thousandths	Ten-Thousandths
		1	3	5	7	9
		3	1	9	5	7
		5	7	3	9	1
		7	9	1	3	5

2. Use the digits 0, 2, 4, 6, 7, and 8 to write four 6-digit numbers of decreasing value.

Hundreds	Tens	Ones	Tenths	Hundredths	Thousandths	Ten-Thousandths
	8	7	6	4	0	2
	6	8	0	2	7	4
	4	0	8	7	2	6
	2	6	4	0	8	7

3. Use the digits 0, 1, 2, 3, 4, 5, and 6 to write four 7-digit numbers of increasing value.

Hundreds	Tens	Ones	Tenths	Hundredths	Thousandths	Ten-Thousandths
1	0	2	3	4	5	6
2	1	0	4	3	6	5
3	5	6	2	0	1	4
4	3	5	6	2	0	1

PROBLEM SOLVING 3-1

LESSON 3-1 **Problem Solving**
Representing, Comparing, and Ordering Decimals

Use the table to answer the questions.

Largest Marine Mammals

Mammal	Length (ft)	Weight (T)
Blue whale	110.0	127.95
Fin whale	82.0	44.29
Gray whale	46.0	32.18
Humpback whale	49.2	26.08
Right whale	57.4	39.37
Sperm whale	59.0	35.43

1. What is the heaviest marine mammal on Earth? the blue whale
2. Which mammal in the table has the shortest length? a gray whale
3. Which mammal in the table is longer than a humpback whale, but shorter than a sperm whale? a right whale
4. Which mammal weighs more than a right whale, but less than a blue whale? a fin whale
5. Which mammal weighs more, a right whale or a gray whale? a right whale
6. Which mammal measures forty-nine and two tenths feet long? a humpback whale
7. Which mammal weighs thirty-five and forty-three hundredths tons? a sperm whale

Circle the letter of the correct answer.

8. Which of the following lists shows mammals in order from the least weight to the greatest weight?
   - **A** sperm whale, right whale, fin whale, gray whale
   - **B** fin whale, sperm whale, gray whale, blue whale
   - **C** fin whale, right whale, sperm whale, gray whale
   - **(D)** gray whale, sperm whale, right whale, fin whale
9. Which of the following lists shows mammals in order from the greatest length to the least length?
   - **(F)** sperm whale, right whale, humpback whale, gray whale
   - **G** gray whale, humpback whale, right whale, sperm whale
   - **H** right whale, sperm whale, gray whale, humpback whale
   - **J** humpback whale, gray whale, sperm whale, right whale

Lesson Quiz

Write each in standard form, expanded form, and words.

1. 8.0342 $8 + 0.03 + 0.004 + 0.0002$; eight and three hundred forty-two ten-thousandths
2. $18 + 0.3 + 0.006$ 18.306; eighteen and three hundred six thousandths
3. eight and twelve hundredths 8.12; $8 + 0.1 + 0.02$
4. It takes Pluto 246.7 years to orbit the Sun, and it takes Neptune 164.8 years. Which planet takes longer to orbit the Sun? Pluto
5. Order the decimals from least to greatest: 16.35, 16.3, 16.5. 16.3, 16.35, 16.5

Available on Daily Transparency in CRB

3-2 Organizer

Pacing: Traditional 1 day
Block $\frac{1}{2}$ day

Objective: Students estimate decimal sums, differences, products, and quotients.

Indiana Academic Standards
6.2.9, 6.7.7, 6.7.11

Warm Up

Order the decimals from least to greatest.

1. 18.74, 18.7, 18.47
   18.47, 18.7, 18.74
2. 9.06, 9.66, 9.6, 9.076
   9.06, 9.076, 9.6, 9.66

Write each in words.

3. 3.072 three and seventy-two thousandths
4. 6.1258 six and one thousand two hundred fifty-eight ten-thousandths

Problem of the Day

Calculate your age in months.
Possible answer: 11 yr 8 mo = 140 mo

Available on Daily Transparency in CRB

Math Humor

The car owner wanted to know about how much it would cost to repair his headlights and bumper. I guess he was asking for a front-end estimate.

3-2 Estimating Decimals

Learn to estimate decimal sums, differences, products, and quotients.

Vocabulary
clustering
front-end estimation

Beth's health class is learning about fitness and nutrition. The table shows the approximate number of calories burned by someone who weighs 90 pounds.

Activity (45 min)	Calories Burned (App.)
Cycling	198.45
Playing ice hockey	210.6
Rowing	324
Water skiing	194.4

When numbers are about the same value, you can use *clustering* to estimate. **Clustering** means rounding the numbers to the same value.

EXAMPLE 1 *Health Application*

Beth wants to cycle, play ice hockey, and water ski. If Beth weighs 90 pounds and spends 45 minutes doing each activity, *about* how many calories will she burn in all?

198.45	→	200	*The addends cluster around 200.*
210.6	→	200	*To estimate the total number of calories,*
+ 194.4	→	+ 200	*round each addend to 200.*
		600	*Add.*

Beth burns about 600 calories.

EXAMPLE 2 Rounding Decimals to Estimate Sums and Differences

Remember!
When rounding, look at the digit to the right of the place to which you are rounding.
- If it is *5 or greater,* round *up.*
- If it is *less than 5,* round *down.*

Estimate by rounding to the indicated place value.

A 3.92 + 6.48; ones

3.92 + 6.48 *Round to the nearest whole number.*
4 + 6 = 10 *The sum is about 10.*

B 8.6355 − 5.039; hundredths

8.6355	8.64	*Round to the hundredths.*
− 5.039	− 5.04	*Align the decimals.*
	3.60	*Subtract.*

1 Introduce

Alternate Opener

EXPLORATION

3-2 Estimating Decimals

For each problem, estimate a solution. Then compute with a calculator to see how close your estimated solutions are to the actual solutions.

		Estimate	Actual
1.	Seven people want to share the cost of a \$33.75 boat rental. How much will each person pay?		
2.	You purchase items that cost the following amounts: \$4.95, \$1.29, \$6.67, \$4.19, and \$10.39. What is the total cost?		
3.	How much change is there from \$200.00 for an item that costs \$157.98?		
4.	Ron's gas tank holds 21 gallons and is empty. If gas costs \$1.499 per gallon, how much will it cost to fill Ron's tank?		

Think and Discuss

5. **Discuss** the estimation strategies you used.
6. **Describe** a situation in which all you need is an estimated solution and a situation in which you must calculate an exact solution.

Motivate

Have students brainstorm about real-world situations in which they may need to estimate decimal numbers (e.g., shopping, tipping, measurements). Then have students volunteer some estimation strategies to use in these situations.

Exploration worksheet and answers on Chapter 3 Resource Book pp. 19 and 115

2 Teach

Lesson Presentation

Guided Instruction

In this lesson, students learn to estimate decimal sums, differences, products, and quotients. First teach students to use clustering to estimate sums of numbers around the same value, and have them round decimals to estimate sums and differences. Then review compatible numbers and have students estimate products and quotients. Finally, teach front-end estimation and have students use the method to estimate sums.

EXAMPLE 3 Using Compatible Numbers to Estimate Products and Quotients

Remember!

Compatible numbers are close to the numbers that are in the problem and are helpful when you are solving the problem mentally.

Estimate each product or quotient.

A 26.76×2.93

$25 \times 3 = 75$ *25 and 3 are compatible.*

So 26.76×2.93 is about 75.

B $42.64 \div 16.51$

$45 \div 15 = 3$ *45 and 15 are compatible.*

So $42.64 \div 16.51$ is about 3.

You can also use *front-end estimation* to estimate with decimals. **Front-end estimation** means to use only the whole-number part of the decimal.

EXAMPLE 4 Using Front-End Estimation

Estimate a range for the sum.

$9.99 + 22.89 + 8.3$

Use front-end estimation.

9.99	→	9	*Add the whole numbers only.*
22.89	→	22	*The whole-number values of the decimals*
+ 8.30	→	+ 8	*are less than the actual numbers, so the*
	at least	39	*answer is an underestimate.*

The exact answer of $9.99 + 22.89 + 8.3$ is 39 or greater.

You can estimate a range for the sum by adjusting the decimal part of the numbers. Round the decimals to 0, 0.5, or 1.

0.99	→	1.00	*Add the decimal part of the numbers.*
0.89	→	1.00	*Add the whole-number estimate and the*
+ 0.30	→	+ 0.50	*adjusted estimate.*
		2.50	*The adjusted decimals are greater than the*
39.00 + 2.50 = 41.50			*actual decimals, so 41.50 is an overestimate.*

The estimated range for the sum is from 39.00 to 41.50.

Think and Discuss

1. **Tell** what number the following decimals cluster around: 34.5, 36.78, and 35.234.
2. **Determine** whether a front-end estimation without adjustment is always an overestimation or an underestimation.

COMMON ERROR ALERT

Check that students are using front-end estimation correctly. Some students may round the numbers to the nearest whole number; they should simply truncate the numbers, dropping the decimal portion.

Additional Examples

Example 1

Nancy wants to cycle, ice skate, and water ski for 30 minutes each. About how many calories will she burn in all? (Cycling = 165.5 cal, ice skating = 177.5 cal, and water skiing = 171.5 cal)

Nancy burns about 510 calories.

Example 2

Estimate by rounding to the indicated place value.

A. $7.13 + 4.68$; ones 12

B. $9.705 - 0.2683$; tenths 9.4

Example 3

Estimate each product or quotient.

A. 33.83×1.98 70

B. $72.77 \div 26.14$ 3

Example 4

Estimate a range for the sum.

$7.86 + 36.97 + 5.40$

The estimated range for the sum is from 48.00 to 50.50.

3 Close

Reaching All Learners

Through Curriculum Integration

Health Have students research approximate calories burned for other activities. Students can use this information to write and solve application problems similar to the one in Example 1.

Summarize

Review brief explanations for each type of estimation: *clustering*, *rounding*, *compatible numbers*, and *front-end estimation*. Have students work in pairs to solve estimation problems using the four techniques from the lesson. Then they can exchange papers and check each other's work.

Answers to Think and Discuss

1. Possible answer: 35
2. It is always an underestimation.

3-2 PRACTICE & ASSESS

Students may want to refer back to the lesson examples.

Assignment Guide

If you finished Example 1 assign:
Core 1, 11, 12, 27, 32, 36–41
Enriched 1, 11, 12, 27, 33, 36–41

If you finished Example 2 assign:
Core 1–5, 11–16, 27, 36–41
Enriched 11–16, 22, 24, 27, 32, 33, 35–41

If you finished Example 3 assign:
Core 1–8, 11–19, 27–30, 36–41
Enriched 12–33, 35–41

If you finished Example 4 assign:
Core 1–21, 27–31, 33, 36–41
Enriched 11–41

Notes

Math Background

Because of truncation, an estimate gained by front-end estimation will always be an underestimate. A quick way to obtain a guaranteed overestimate is to add the number of addends in the sum to the underestimate. This process works because you are adding 1 for each addend. Once you have the underestimate and the overestimate, you also have a range in which the true sum must lie.

3-2 Exercises

FOR EXTRA PRACTICE
see page 640

internet connect
Homework Help Online
go.hrw.com Keyword: MR4 3-2

GUIDED PRACTICE

See Example 1

1. Elba runs every Monday, Wednesday, and Friday. Last week she ran 3.62 miles on Monday, 3.8 miles on Wednesday, and 4.3 miles on Friday. About how many miles did she run last week? **about 12 miles**

See Example 2

Estimate by rounding to the indicated place value.

2. 2.746 − 0.866; tenths **1.8**
3. 6.735 + 4.9528; ones **12**
4. 10.8071 + 5.392; hundredths **16.20**
5. 5.982 − 0.4832; tenths **5.5**

See Example 3

Estimate each product or quotient.

6. 38.92 ÷ 4.06 **10**
7. 14.51 × 7.89 **120**
8. 22.47 ÷ 3.22 **7**

See Example 4

Estimate a range for each sum.

9. 7.8 + 31.39 + 6.95 **from 44 to 46.5**
10. 14.27 + 5.4 + 21.86 **from 40 to 42**

INDEPENDENT PRACTICE

See Example 1

11. Before Mike's trip, the odometer in his car read 146.8 miles. He drove 167.5 miles to a friend's house and 153.9 miles to the beach. About how many miles did the odometer read when he arrived at the beach? **about 450 miles**

12. The rainfall in July, August, and September was 16.76 cm, 13.97 cm, and 15.24 cm, respectively. About how many total centimeters of rain fell during those three months? **about 45 cm**

See Example 2

Estimate by rounding to the indicated place value.

13. 2.0993 + 1.256; tenths **3.4**
14. 7.504 − 2.3792; hundredths **5.12**
15. 0.6271 + 4.53027; thousandths **5.157**
16. 13.274 − 8.5590; tenths **4.7**

See Example 3

Estimate each product or quotient.

17. 9.64 × 1.769 **20**
18. 11.509 ÷ 4.258 **3**
19. 19.03 ÷ 2.705 **6**

See Example 4

Estimate a range for each sum.

20. 17.563 + 4.5 + 2.31 **from 23 to 25**
21. 1.620 + 10.8 + 3.71 **from 14 to 17**

PRACTICE AND PROBLEM SOLVING

Estimate by rounding to the nearest whole number.

22. 8.456 + 7.903 **16**
23. 12.43 × 3.72 **48**
24. 1,576.2 − 150.50 **1,425**

25. Estimate the quotient of 67.25 and 3.83. **17**

26. Estimate $79.45 divided by 17. **5**

RETEACH 3-2

LESSON 3-2 **Reteach**
Estimating Decimals

You can use rounding to estimate. Round to the indicated place value. Then add or subtract.

A. 3.478 + 7.136; tenths
3.478 — 7 ≥ 5, so round up — 3.5
7.136 — 3 < 5, so round down — +7.1
10.6
3.478 + 7.136 is about 10.6.

B. 12.848 − 6.124; hundredths
12.848 — 8 ≥ 5, so round up — 12.85
6.124 — 4 < 5, so round down — −6.12
6.73
12.848 − 6.124 is about 6.73.

Estimate by rounding to the indicated place value.

1. 1.04 + 9.37; tenths
1.04 rounds to **1.0**
9.37 rounds to **9.4**
estimate **10.4**

2. 2.17 + 3.56; tenths
2.17 rounds to **2.2**
3.56 rounds to **3.6**
estimate **5.8**

3. 6.753 − 4.245; hundredths
6.753 rounds to **6.75**
4.255 rounds to **4.26**
estimate **2.49**

You can use compatible numbers to estimate. Pick numbers that are close to the actual numbers that are easy to multiply or divide. Then multiply or divide.

A. 4.6 • 3.2
5 and 3 are compatible numbers.
5 • 3 = 15, so 4.6 • 3.2 is about 15.

B. 48.3 ÷ 13.2
48 and 12 are compatible numbers.
48 ÷ 12 = 4, so 48.3 ÷ 13.2 is about 4.

Use compatible numbers to estimate each product or quotient.

4. 9.4 • 5.6 **54**
5. 7.25 • 10.84 **77**
6. 84.8 ÷ 3.9 **21**
7. 21.9 ÷ 3.1 **7**
8. 8.3 • 7.6 **64**
9. 55.7 ÷ 6.9 **8**
10. 5.57 ÷ 2.7 **2**
11. 6.729 • 9.8 **70**

Possible answers are given.

PRACTICE 3-2

LESSON 3-2 **Practice B**
Estimating Decimals

Estimate. Round to the indicated place value.

1. 7.462 + 1.809; tenths **9.3**
2. 15.3614 − 2.0573; hundredths **13.30**
3. 56.4059 − 4.837; ones **51**
4. 0.60871 + 1.2103; hundredths **1.82**

Estimate each product or quotient. Possible answers are given.

5. 42.1 ÷ 5.97 **7**
6. 11.8 • 6.125 **72**
7. 63.78 ÷ 8.204 **8**
8. 7.539 • 3.0642 **24**
9. 80.794 ÷ 8.61 **9**
10. 19.801 • 2.78 **60**

Estimate a range for the sum. Possible answers are given.

11. 6.8 + 4.3 + 5.6 **from 15 to 17.5**
12. 12.63 + 9.86 + 20.30 **from 41 to 43.5**

13. Two sixth-grade classes are collecting money to buy a present for one of their teachers. One class collected $24.68 and the other class collected $30.25. About how much money did they collect in all? The gift they want to buy costs $69.75. About how much more money do they need?
about $55.00; about $15.00

14. On the highway, Anita drove an average speed of 60.2 miles per hour. At that speed, about how far can she travel in three and a half hours? At that same speed, about how many hours will it take Anita to drive 400 miles?
about 240 miles; about 7 hours

Use the table for Exercises 27–31.

Long-Distance Costs for Callers in the United States	
Country	Cost per Minute (¢)
Venezuela	22
Russia	9.9
Japan	7.9
United States	3.7

27. ***MONEY*** Round each cost in the table to the nearest cent. Write your answer using a dollar sign and decimal point. **$0.22, $0.10, $0.08, $0.04**

28. About how much does it cost to phone someone in Russia and talk for 8 minutes? **about 80 cents**

29. About how much more does it cost to make a 12-minute call to Japan than to make an 18-minute call within the United States? **(12 × 8) − (18 × 4) = 24, or about 24 cents**

30. Will the cost of a 30-minute call to someone within the United States be greater or less than $1.20? Explain.

31. Kim is in New York. She calls her grandmother in Venezuela and speaks for 20 minutes, then calls a friend in Japan and talks for 15 minutes, and finally calls her mother in San Francisco and talks for 30 minutes. Estimate the total cost of all her calls. **Possible answer: (20 × $0.20) + (15 × $0.08) + (30 × $0.04) = $4 + $1.20 + $1.20, or about $6.40**

32. ***HEALTH*** The recommended daily allowance (RDA) for iron is 15 mg/day for teenage girls. Julie eats a hamburger that contains 3.88 mg of iron. About how many more milligrams of iron does she need to meet the RDA? (Round to the nearest whole number.) **about 11 mg**

33. ***WRITE A PROBLEM*** Write a problem with three decimal numbers that have a total sum between 30 and 32.5. **Possible answer: 8.56 + 12.36 + 10.74**

34. ***WRITE ABOUT IT*** How do you adjust a front-end estimation? Why is this done?

35. ***CHALLENGE*** Place a decimal point in each number so that the sum of the numbers is between 124 and 127.
1059 + 725 + 815 + 1263 **105.9 + 7.25 + 0.815 + 12.63**

Spiral Review

Estimate each sum or difference by rounding to the place value indicated. (Lesson 1-2)

36. 6,319 + 13,804; thousands **20,000**

37. 25,680 − 18,502; ten thousands **10,000**

Evaluate each expression. (Lesson 1-4)

38. 15 + 18 ÷ 6 **18**

39. $4^2 + 19 \times 2 - 30$ **24**

40. $(26 - 14) \times 2^3 - 14 \div 2$ **89**

41. **TEST PREP** Which of the following is an example of the Distributive Property? (Lesson 1-5) **C**

A (8 + 2) + 6 = 8 + (2 + 6)

B 6 × 3 × 7 = 7 × 6 × 3

C 3 × (5 + 4) = (3 × 5) + (3 × 4)

D 6 × 1 = 1 × 6

Answers

30. Round 3.7 cents to 4 cents. 4 × 30 = $1.20. $1.20 is an overestimate, so it will cost less than $1.20.

34. You round the decimal part of the addends to either 1.00, 0.5, or 0, then add the estimated adjustment to the whole number estimate. It gives a range for the sum.

Journal

Have students write about how they can use estimation while shopping.

Test Prep Doctor

For Exercise 41, students need to have an understanding of the properties they have learned up to this point. Students who answered **A** have confused the Associative Property with the Distributive Property. Students who answered **B** or **D** have confused the Commutative Property with the Distributive Property.

CHALLENGE 3-2

LESSON 3-2 **Challenge**
Out to Lunch

Use the restaurant bills below to estimate the total cost of each meal. Then estimate the amount each person should pay to split each check evenly. Possible answers are given.

Number of People: 2

Quantity	Item	Price
2	Large Soda	$0.75 each
1	Cheeseburger	$4.55
1	BLT	$3.25
Estimated Total:		about $10.00
Estimated Cost Per Person:		about $5.00

Number of People: 4

Quantity	Item	Price
2	Slice of Pie	$1.45 each
1	Brownie	$1.25
3	Hot Tea	$0.60 each
Estimated Total:		about $6.00
Estimated Cost Per Person:		about $1.50

Number of People: 4

Quantity	Item	Price
2	Bowl of Soup	$2.89 each
1	Chicken Sandwich	$4.95
1	Chef Salad	$3.25
4	Coffee	$0.50 each
Estimated Total:		about $16.00
Estimated Cost Per Person:		about $4.00

Number of People: 5

Quantity	Item	Price
1	Pizza	$14.95
5	House Salad	$2.85 each
4	Large Soda	$0.75 each
1	Small Soda	$0.55
Estimated Total:		about $35.00
Estimated Cost Per Person:		about $7.00

PROBLEM SOLVING 3-2

LESSON 3-2 **Problem Solving**
Estimating Decimals

Write the correct answer. Possible answers are given.

1. Men in Iceland have the highest average life expectancy in the world—76.8 years. The average life expectancy for a man in the United States is 73.1 years. About how much higher is a man's average life expectancy in Iceland? Round your answer to the nearest whole year.
about 4 years

2. The average life expectancy for a woman in the United States is 79.1 years. Women in Japan have the highest average life expectancy—3.4 years higher than the United States. Estimate the average life expectancy of women in Japan. Round your answer to the nearest whole year.
about 82 years

3. There are about 1.6093 kilometers in one mile. There are 26.2 miles in a marathon race. About how many kilometers are there in a marathon race? Round your answer to the nearest tenths.
about 41.9 kilometers

4. At top speed, a hornet can fly 13.39 miles per hour. About how many hours would it take a hornet to fly 65 miles? Round your answer to the nearest whole number.
about 5 hours

5. The average male human brain weighs 49.7 ounces. The average female human brain weighs 44.6 ounces. What is the difference in their weights? Round your answer to the nearest whole ounce.
about 5 ounces

6. About 9.27 million people visit the Great Smokey Mountains National Park each year, and about 4.54 million people visit the Grand Canyon National Park. About how many visitors do the two parks receive each year in all? Round your answer to the nearest tenth.
about 13.8 million

Circle the letter of the correct answer.

7. Lydia earned $9.75 per hour as a lifeguard last summer. She worked 25 hours a week. About how much did she earn in 8 weeks?
A about $250.00
(B) about $2,000.00
C about $2,500.00
D about $200.00

8. Brent mixed 4.5 gallons of blue paint with 1.7 gallons of white paint and 2.4 gallons of red paint to make a light purple paint. About how many gallons of purple paint did he make?
(F) about 9 gallons
G about 8 gallons
H about 10 gallons
J about 7 gallons

Lesson Quiz

Estimate by rounding to the indicated place value.

1. 3.07442 + 1.352; tenths **4.5**

2. 7.305 − 4.12689; nearest whole number **3**

Estimate each product or quotient.

3. 6.75 × 1.82 **14**

4. 10.5 ÷ 3.42 **3**

5. The snowfall in December, January, and February was 18.26 cm, 29.36 cm, and 32.87 cm, respectively. About how many total centimeters of snow fell during the three months? **80**

Available on Daily Transparency in CRB

Hands-On LAB 3B Explore Decimal Addition and Subtraction

Pacing: Traditional 1 day
Block $\frac{1}{2}$ day

Objective: To use decimal grids to model addition and subtraction of decimals

Materials: Decimal grids

Lab Resources

Hands-On Lab Activities .... pp. 11–12

Using the Pages

Remind students that one small square represents 0.01, one row or column represents 0.1, and the entire grid represents 1.

Model each problem on a decimal grid. Then solve.

1. 0.2 + 0.33 **0.53**

2. 2.45 − 1.08 **1.37**

Explore Decimal Addition and Subtraction

Hands-On LAB 3B

Use with Lesson 3-3

internet connect
Lab Resources Online
go.hrw.com
KEYWORD: MR4 Lab3B

KEY

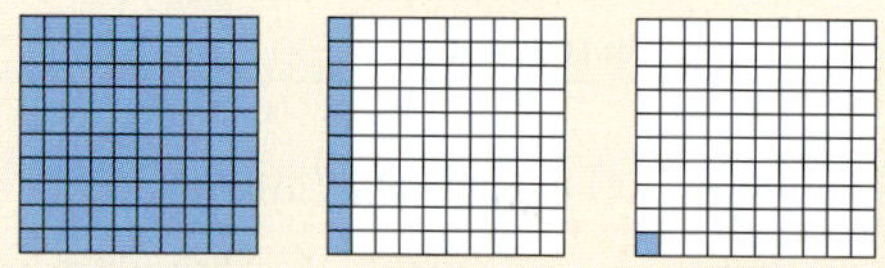

1	0.1	0.01
one	1 tenth	1 hundredth

You can model addition and subtraction of decimals with decimal grids.

Activity 1

1 Use decimal grids to find each sum.

a. 0.24 + 0.32

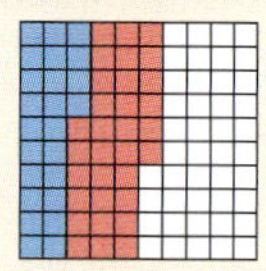

To represent 0.24, shade 24 squares.

To represent 0.32, shade 32 squares in another color.

There are 56 shaded squares representing 0.56.

0.24 + 0.32 = 0.56

b. 1.56 + 0.4

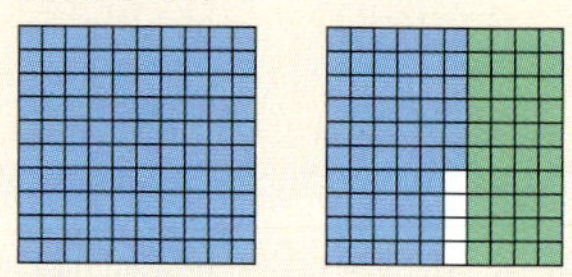

To represent 1.56, shade a whole grid and 56 squares of another.

To represent 0.4, shade 4 columns in another color.

One whole grid and 96 squares are shaded.

1.56 + 0.4 = 1.96

c. 0.75 + 0.68

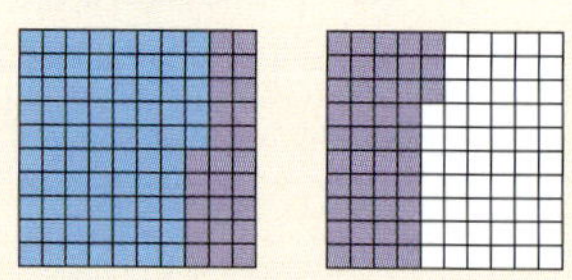

To represent 0.75, shade 75 squares.

To represent 0.68, shade 68 squares in another color. You will need to use another grid.

One whole grid and 43 squares are shaded.

0.75 + 0.68 = 1.43

Think and Discuss

1. How would you shade a decimal grid to represent 0.2 + 0.18?
Possible answer: To represent 0.2, shade 2 columns, and then shade 1 column plus 8 small squares in another color to represent 0.18.

Try This

Use decimal grids to find each sum.

1. 0.2 + 0.6 **0.8**	**2.** 1.07 + 0.03 **1.10**	**3.** 1.62 + 0.08 **1.70**
4. 0.45 + 0.29 **0.74**	**5.** 0.88 + 0.12 **1.00**	**6.** 1.29 + 0.67 **1.96**
7. 0.07 + 0.41 **0.48**	**8.** 0.51 + 0.51 **1.02**	**9.** 1.01 + 0.23 **1.24**

Activity 2

1 Use a decimal grid to find each difference.

a. 0.6 − 0.38

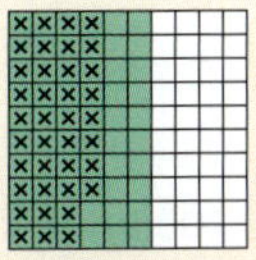

To represent 0.6, shade 6 columns.

Subtract 0.38 by removing 38 squares.

There are 22 remaining squares.

0.6 − 0.38 = 0.22

b. 1.22 − 0.41

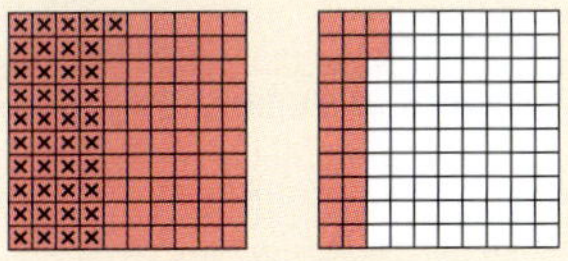

To represent 1.22, shade an entire decimal grid and 22 squares of another.

Subtract 0.41 by removing 41 squares.

There are 81 remaining squares.

1.22 − 0.41 = 0.81

Think and Discuss

1. How would you shade a decimal grid to represent 1.3 − 0.6?
Possible answer: Shade 1 complete grid plus 3 columns of another to represent 1.3. Then remove 6 columns: 3 from the second grid and 3 from the first. This leaves 7 columns representing 0.7.

Try This

Use decimal grids to find each difference.

1. 0.9 − 0.3 **0.6**	**2.** 1.2 − 0.98 **0.22**	**3.** 0.6 − 0.41 **0.19**
4. 1.6 − 0.07 **1.53**	**5.** 0.35 − 0.03 **0.32**	**6.** 2.12 − 0.23 **1.89**
7. 2.0− 0.86 **1.14**	**8.** 0.78 − 0.76 **0.02**	**9.** 1.06 − 0.55 **0.51**

Answers

Activity 1

Try This

1.

2.

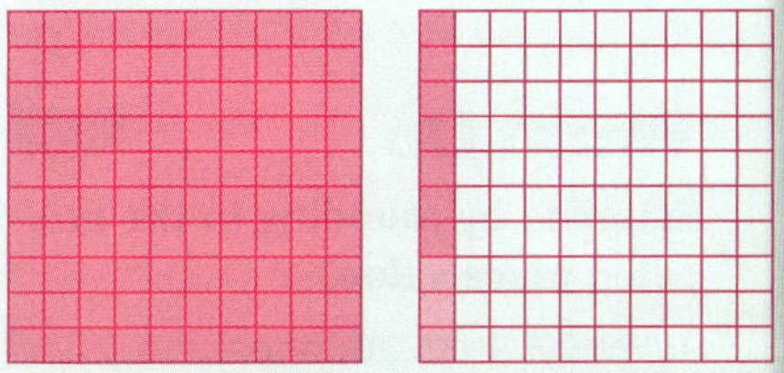

3.

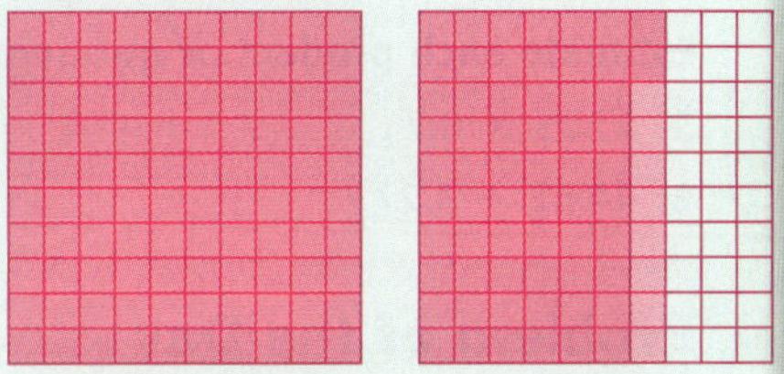

4.

5.

6.

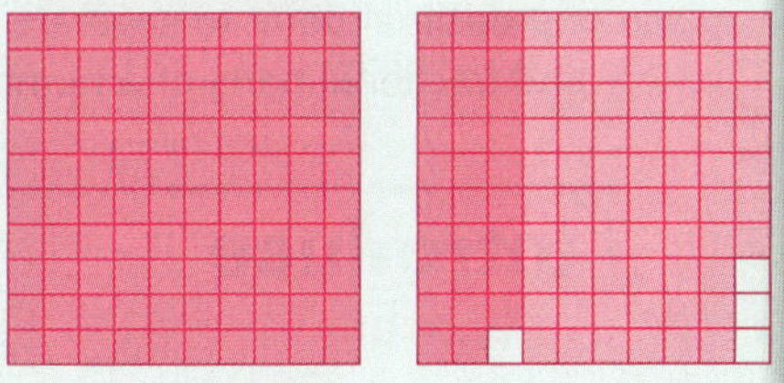

7.

8.

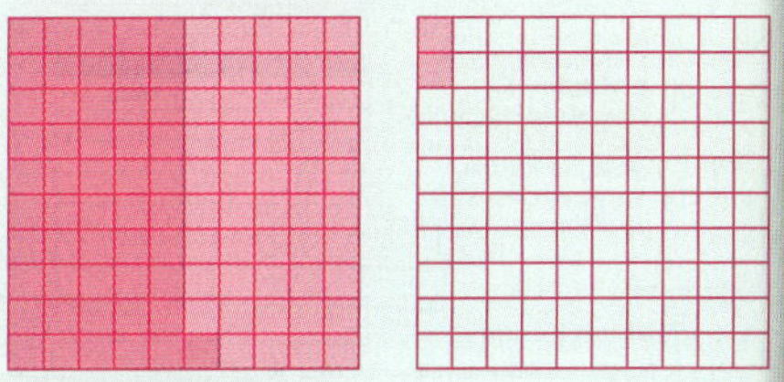

Answers

Activity 2

Try This

1–9. Complete answers on p. A1

9.

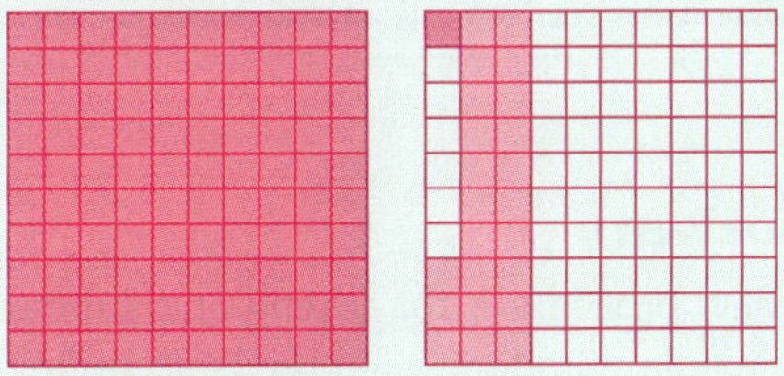

3-3 Organizer

Pacing: Traditional 1 day
Block $\frac{1}{2}$ day

Objective: Students add and subtract decimals.

Indiana Academic Standards
6.2.9, 6.2.10, 6.5.10, 6.7.7, 6.7.9, 6.7.10, 6.7.11

Warm Up

Estimate by rounding to the indicated place value.

1. 70.27 + 15.36; ones 85

2. 84.37 − 21.82; tenths 62.6

Estimate each product or quotient.

3. 27.25 × 8.7 270

4. 44.52 ÷ 3.27 15

Problem of the Day

Find a three-digit number that rounds to 440 and includes a digit that is the quotient of 24 and 3. Is there more than one possible answer? Explain your thinking.
438; no; the numbers that round to 440 are 435–444, 24 divided by 3 is 8, and 438 is the only number with 8 as a digit.

Available on Daily Transparency in CRB

Math Fact !!

In some countries a comma is used in place of a decimal point to separate whole and fractional parts of a number.

3-3 Adding and Subtracting Decimals

Learn to add and subtract decimals.

American gymnast Elise Ray won the 2000 U.S. Championships in the all-around, uneven bars, and floor-exercise events.

Elise Ray's Scores	
Event	**Points**
Floor exercise	9.8
Balance beam	9.7
Vault	9.425
Uneven bars	9.85

Elise Ray competed in the 2000 Summer Olympics in Sydney, Australia.

To find the total number of points, you can add all of the scores.

EXAMPLE 1 *Sports Application*

A **What was Elise Ray's total all-around score in the 2000 U.S. Championships?**

Find the sum of 9.8, 9.7, 9.425, and 9.85.

9.8 + 9.7 + 9.425 + 9.85 *Estimate by rounding to the nearest whole number.*
↓ ↓ ↓ ↓
10 + 10 + 9 + 10 = 39 *The total is about 39 points.*

Add.

```
  9.800     Align the decimal points.
  9.700
  9.425     Use zeros as placeholders.
+ 9.850
 38.775     Add. Then place the decimal point.
```

Since 38.775 is close to the estimate of 39, the answer is reasonable. Elise Ray's total all-around score was 38.775 points.

B **How many more points did Elise need on the uneven bars to have a perfect score of 10?**

Find the difference between 10 and 9.85.

```
 10.00    Align the decimal points.
− 9.85    Use zeros as placeholders.
  0.15    Subtract. Then place the decimal point.
```

Elise needed another 0.15 points to have a perfect score.

Helpful Hint

Estimating before you add or subtract will help you check whether your answer is reasonable.

1 Introduce

Alternate Opener

EXPLORATION

3-3 Adding and Subtracting Decimals

You have 19¢. How much more do you need to have $1.00? From 19¢ to 20¢, add 1¢. From 20¢ to $1.00, add 80¢. The answer is 1¢ + 80¢, or 81¢.

1. Draw arrows to connect each pair of amounts that would give you a sum of $1.00.

Amount 1	Amount 2
$0.19	$0.63
$0.25	$0.55
$0.76	$0.93
$0.07	$0.75
$0.65	$0.24
$0.37	$0.35
$0.45	$0.81

2. Compute the change from $10.00 on a purchase of each amount. From $1.25 to $2.00, add $0.75. From $2.00 to $10.00, add $8.00. So $10.00 − $1.25 = $8.75.

Amount	Change from $10.00
$1.25	$8.75
$2.76	
$3.07	
$4.65	
$5.37	
$6.45	
$7.59	

Think and Discuss

3. Name ten different pairs of numbers that each have a sum of $1.00.

4. Describe how you can use the strategy of "adding up" to find 200 − 176.25.

Motivate

Review procedures for adding and subtracting whole numbers: align digits according to place value (ones, tens, hundreds, etc.), perform the operation, and regroup when needed. Explain that the same steps are followed when adding and subtracting decimals.

Exploration worksheet and answers on Chapter 3 Resource Book pp. 28 and 117

2 Teach

Lesson Presentation

Guided Instruction

In this lesson, students learn to add and subtract decimals. First teach them to add four decimal addends, aligning decimal points and annexing zeros as placeholders. Then teach students to subtract a decimal from a whole number, aligning decimal points and using zeros as placeholders (that is, using zeros to write an equivalent number with the required number of decimal places). Next practice mental math strategies for adding and subtracting decimals. Finally, have students evaluate expressions for given variable values.

EXAMPLE 2 **Using Mental Math to Add and Subtract Decimals**

Find each sum or difference.

A $1.6 + 0.4$

$1.6 + 0.4$ *Think: 0.6 + 0.4 = 1*

$1.6 + 0.4 = 2$

B $3 - 0.8$

$3 - 0.8$ *Think: What number added to 0.8 is 1? 0.8 + 0.2 = 1 So 1 − 0.8 = 0.2.*

$3 - 0.8 = 2.2$

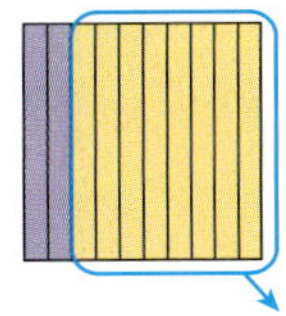

EXAMPLE 3 **Evaluating Decimal Expressions**

Remember!

You can place any number of zeros at the end of a decimal number without changing its value.

Evaluate $7.52 - s$ for each value of s.

A $s = 2.9$

$7.52 - s$	
$7.52 - 2.9$	*Substitute 2.9 for s.*
7.52	*Align the decimal points.*
− 2.90	*Use a zero as a placeholder.*
4.62	*Subtract.*
	Place the decimal point.

B $s = 4.5367$

$7.52 - s$	
$7.52 - 4.5367$	*Substitute 4.5367 for s.*
7.5200	*Align the decimal points.*
− 4.5367	*Use zeros as placeholders.*
2.9833	*Subtract.*
	Place the decimal point.

Think and Discuss

1. **Show** how you would write 2.678 + 124.5 to find the sum.
2. **Tell** why it is a good idea to estimate the answer before you add and subtract.
3. **Explain** how you can use mental math to find how many more points Elise Ray would have needed to have scored a perfect 10 on the floor exercise.

COMMON ERROR ALERT

Students may try to align decimal numbers to the right instead of on the decimal point when adding or subtracting in vertical format. Explain that the place-value positions in each number must line up.

Additional Examples

Example 1

A. What was Elise Ray's total for the events other than the floor exercise? Elise's total for the events was 28.975.

B. How many more points did Elise need on the vault to have a perfect score? **0.575**

Example 2

Find each sum or difference.

A. 1.8 + 0.2 **2.0**   **B.** 4 − 0.7 **3.3**

Example 3

Evaluate 6.73 − *x* for each value of *x*.

A. $x = 3.8$ **2.93**   **B.** $x = 2.9765$ **3.7535**

3 Close

Reaching All Learners
Through Home Connection

Have students look through newspaper ads, at home or in the library, for items that, when combined, total less than $50. Have them subtract their totals from $50 to find out how much change they would receive. Possible answer: I found shoes for $12.95, a CD for $13.98, and a skateboard for $19.49. The total cost is $46.42. The amount of change I would get from $50.00 is $3.58.

Summarize

Have students solve first a decimal addition problem and then a decimal subtraction problem (e.g., 8.34 + 3.7 + 5.029 and 12 − 7.62). Have them explain the steps they would follow to solve each. Emphasize the importance of aligning the decimals before solving the problem and the use of an estimate to check for reasonableness.

Answers to Think and Discuss

1. Write the problem vertically while aligning the decimal points and place values. Use two zeros to the right of 124.5 as placeholders.
2. Estimating before you add or subtract allows you to check if your answer is reasonable.
3. 0.8 + 0.2 = 1, so 9.8 + 0.2 = 10. She would have needed 0.2 more of a point.

3-3 PRACTICE & ASSESS

Students may want to refer back to the lesson examples.

Assignment Guide

If you finished Example 1 assign:
Core 1, 2, 11, 12, 25–27, 37, 39, 43–50
Enriched 2, 12, 25–27, 38–41, 43–50

If you finished Example 2 assign:
Core 1–6, 11–20, 25–27, 37, 39, 43–50
Enriched 2, 6, 12, 15–20, 25–29, 36–50

If you finished Example 3 assign:
Core 1–24, 30–31, 37, 39, 43–50
Enriched 2, 6, 10–12, 18–20, 22, 24–50

Notes

Math Background

Place values to the right of a decimal point continue the same pattern as the pattern to the left of the decimal point. Specifically, the value of each place to the right of the decimal point in any number is 10 times as great as the value of the next place to its right. And the value of the ones place is 10 times as great as the value of the tenths place. Because of this, algorithms for addition and subtraction of whole numbers can be extended to decimal numbers.

3-3 Exercises

FOR EXTRA PRACTICE see page 640

GUIDED PRACTICE

See Example 1 Use the table for Exercises 1 and 2.

Rea's Triathlon Training	
Sport	**Distance (mi)**
Cycling	14.25
Running	4.35
Swimming	1.6

1. How many miles in all is Rea's triathlon training? **20.2 miles**

2. a. How many miles did Rea run and swim in all? **5.95 miles**

b. How much farther did Rea cycle than swim? **12.65 miles**

See Example 2 Find each sum or difference.

3. $2.7 + 0.3$ **3** **4.** $6 - 0.4$ **5.6** **5.** $5.2 + 2.8$ **8** **6.** $8.9 - 4$ **4.9**

See Example 3 Evaluate $5.35 - m$ for each value of m.

7. $m = 2.37$ **2.98** **8.** $m = 1.8$ **3.55** **9.** $m = 4.7612$ **0.5888** **10.** $m = 0.402$ **4.948**

INDEPENDENT PRACTICE

See Example 1

11. During a diving competition, Phil performed two reverse dives and two dives from a handstand position. He received the following scores: 8.765, 9.45, 9.875, and 8.025. What was Phil's total score? **36.115**

12. Brad works after school at a local grocery store. How much did he earn in all for the month of October? **$567.38**

Brad's Earnings for October				
Week	1	2	3	4
Earnings	$123.48	$165.18	$137.80	$140.92

See Example 2 Find each sum or difference.

13. $7.2 + 1.8$ **9** **14.** $8.5 - 7$ **1.5** **15.** $3.3 + 0.7$ **4** **16.** $15.9 + 2.1$ **18**

17. $7 - 0.6$ **6.4** **18.** $7.55 - 3.25$ **4.3** **19.** $21.4 + 3.6$ **25** **20.** $5 - 2.7$ **2.3**

See Example 3 Evaluate $9.67 - x$ for each value of x.

21. $x = 1.52$ **8.15** **22.** $x = 3.8$ **5.87** **23.** $x = 7.21$ **2.46** **24.** $x = 0.635$ **9.035**

PRACTICE AND PROBLEM SOLVING

Add or subtract.

25. $5.62 + 4.19$ **9.81** **26.** $10.508 - 6.73$ **3.778** **27.** $13.009 + 12.83$ **25.839**

28. Find the sum of 0.0679 and 3.75. **3.8179** **29.** Subtract 3.0042 from 7.435. **4.4308**

RETEACH 3-3

LESSON 3-3 **Reteach**
Adding and Subtracting Decimals

You can use a place value chart to help you add and subtract decimals.

Add 1.4 and 0.9.

	Tens	Ones	Tenths	Hundredths	Thousandths
		1 .	4		
+		0 .	9		

So, 1.4 + 0.9 = 2.3.

Subtract 2.4 from 3.1.

	Tens	Ones	Tenths	Hundredths	Thousandths
		3 .	1		
−		2 .	4		

So, 3.1 − 2.4 = 0.7.

Find each sum or difference.

1.

	Tens	Ones	Tenths	Hundredths	Thousandths
		2 .	6		
+		1 .	1	5	

3.75

2.

	Tens	Ones	Tenths	Hundredths	Thousandths
		2 .	5	3	
−		1 .	7		

0.83

3. 4.3 + 1.4

	Tens	Ones	Tenths	Hundredths	Thousandths
		4 .	3		
+		1 .	4		

5.7

4. 14.4 − 3.8

	Tens	Ones	Tenths	Hundredths	Thousandths
	1	4 .	4		
−		3 .	8		

10.6

5. 7.3 + 8.5

	Tens	Ones	Tenths	Hundredths	Thousandths
		7 .	3		
+		8 .	5		

15.8

6. 12.34 − 6.9

	Tens	Ones	Tenths	Hundredths	Thousandths
	1	2 .	3	4	
−		6 .	9		

5.44

7. 6.3 − 2.5 3.8 **8.** 20.65 + 13.24 33.89 **9.** 8.9 − 1.95 6.95 **10.** 3.42 + 5.25 8.67

PRACTICE 3-3

LESSON 3-3 **Practice B**
Adding and Subtracting Decimals

Find each sum or difference.

1. 8.9 + 2.4 11.3 **2.** 12.7 − 9.6 3.1 **3.** 18.35 − 4.16 14.19

4. 7.21 + 11.6 18.81 **5.** 0.975 + 3.8 4.775 **6.** 20.66 − 9.1 11.56

7. Tiffany's job requires a lot of driving. How many miles did she travel during the month of February? 720.78

Miles Tiffany Traveled				
Week	1	2	3	4
Miles	210.05	195.18	150.25	165.30

8. Shelly babysits after school and on the weekends. How much did she earn in all for the month of April? $722.00

Shelly's Earnings for April				
Week	1	2	3	4
Earnings	$120.50	$180.75	$205.25	$215.50

Evaluate 5.6 − *a* for each value of *a*.

9. $a = 3.7$ 1.9 **10.** $a = 0.5$ 5.1 **11.** $a = 2.8$ 2.8

12. $a = 1.42$ 4.18 **13.** $a = 0.16$ 5.44 **14.** $a = 3.75$ 1.85

15. Allen bought a box of envelopes for $2.79 and a pack of paper for $4.50. He paid with a $10 bill. How much change should be receive? $2.71

16. From a bolt of cloth measuring 25.60 yards, Tina cut a 6.8-yard piece and an 11.9-yard piece. How much material is left on the bolt? 6.9 yards

Career LINK

The risk of death in a house fire can be reduced by up to 50% if the home has a working smoke alarm.

Evaluate each expression.

30. $8.09 - a$ for $a = 4.5$ **3.59**

31. $7.03 + 33.8 + n$ for $n = 12.006$ **52.836**

32. $b + (5.68 - 3.007)$ for $b = 6.134$ **8.807**

33. $(2 \times 14) - a + 1.438$ for $a = 0.062$ **29.376**

34. $5^2 - w$ for $w = 3.5$ **21.5**

35. $100 - p$ for $p = 15.034$ **84.966**

36. ***CAREERS*** A fire helmet must be sturdy enough to protect the firefighter's head from dangerous objects and extremely hot temperatures while still being as lightweight as possible. One fire helmet weighs 1.616 kg, and another fire helmet weighs 1.403 kg. What is the difference in weights? **0.213 kg**

37. ***MONEY*** Logan wants to buy a new bike that costs $135.00. He started with $14.83 in his savings account. Last week, he deposited $15.35 into his account. Today, he deposited $32.40. How much more money does he need to buy the bike? **$72.42**

38. ***SPORTS*** With a time of 60.35 seconds, Martina Moracova broke Jennifer Thompson's world record time in the women's 100-meter medley. How much faster was Thompson than Moracova when, in the next heat, she reclaimed the record with a time of 59.30 seconds? **1.05 seconds**

39. ***SPORTS*** The highest career batting average ever achieved by a professional baseball player is 0.366. Bill Bergen finished with a career 0.170 average. How much lower is Bergen's career average than the highest career average? **0.196**

40. ***WHAT'S THE QUESTION?*** A cup of rice contains 0.8 mg of iron, and a cup of lima beans contains 4.4 mg of iron. If the answer is 6 mg, what is the question? **How many milligrams of iron are in 2 cups of rice and 1 cup of lima beans?**

41. ***WRITE ABOUT IT*** Why is it important to align the decimal points before adding or subtracting decimal numbers? **Aligning the decimal points ensures that you add or subtract the digits in the same place value.**

42. ***CHALLENGE*** Evaluate $(5.7 + a) \times (9.75 - b)$ for $a = 2.3$ and $b = 7.25$. **$(5.7 + 2.3) \times (9.75 - 7.25)$; 8×2.5; 20**

Spiral Review

Find each value. (Lesson 1-3)

43. 7^3 **343** **44.** 4^3 **64** **45.** 10^5 **100,000** **46.** 12^2 **144** **47.** $10^2 + 2^3$ **108**

Find an expression for each table. (Lesson 2-1)

48.

x	2	4	9
■	7	17	42

$5x - 3$

49.

m	3	5	8
■	14	22	34

$4m + 2$

50. **TEST PREP** Which algebraic expression is the same as "the quotient of n divided by 6"? (Lesson 2-2) **B**

A $6n$ **B** $\frac{n}{6}$ **C** $n - 6$ **D** $6 + n$

CHALLENGE 3-3

LESSON 3-3 **Challenge**

A Penny Saved Is a Penny Earned

Next to each bank, describe three different coin combinations that equal the amount of money it holds. For each combination, use at least one quarter, one dime, one nickel, and one penny.

Possible combinations are given.

1. 

2 quarters, 3 dimes, 1 nickel, 4 pennies

2 quarters, 2 dimes, 2 nickels, 9 pennies

1 quarter, 5 dimes, 2 nickels, 4 pennies

2. 

4 quarters, 2 dimes, 1 nickel, 3 pennies

2 quarters, 4 dimes, 5 nickels, 13 pennies

3 quarters, 4 dimes, 2 nickels, 3 pennies

3. 

1 quarter, 3 dimes, 1 nickel, 5 pennies

1 quarter, 1 dime, 4 nickels, 10 pennies

1 quarter, 2 dimes, 3 nickels, 5 pennies

4. 

8 quarters, 2 dimes, 1 nickel, 5 pennies

4 quarters, 10 dimes, 4 nickels, 10 pennies

6 quarters, 3 dimes, 4 nickels, 30 pennies

PROBLEM SOLVING 3-3

LESSON 3-3 **Problem Solving**

Adding and Subtracting Decimals

Use the table to answer the questions.

Busiest Ports in the United States

Port	Imports Per Year (millions of tons)	Exports Per Year (millions of tons)
South Louisiana, LA	30.6	57.42
Houston, TX	75.12	33.43
New York, NY & NJ	53.52	8.03
New Orleans, LA	26.38	21.73
Corpus Christi, TX	52.6	7.64

1. How many more tons of imports than exports does the Port of New Orleans handle each year? **4.65 million tons**

2. How many tons of imports and exports are shipped through the port of Houston, Texas, each year in all? **108.55 million tons**

3. Which port handles more exports than imports each year? How many more tons of exports does it handle? **South Louisiana; 26.82 million tons**

4. Which two ports import the most goods each year? How many tons of imports do they ship altogether? **Houston and NY & NJ; 128.64 million tons**

5. Which port ships 0.39 more tons of exports each year than the port at Corpus Christi, Texas? **the Port of NY & NJ**

6. What is the difference between the imports and exports shipped in and out of Corpus Christi's port each year? **44.96 million tons**

Circle the letter of the correct answer.

7. What is the total amount of imports shipped into the nation's 5 busiest ports each year?
   (A) 238.22 million tons
   B 366.47 million tons
   C 128.25 million tons
   D 109.97 million tons

8. What is the total amount of exports shipped out of the nation's 5 busiest ports each year?
   F 366.47 million tons
   (G) 128.25 million tons
   H 109.97 million tons
   J 238.22 million tons

Journal

Have students write about a situation in which knowing an estimate is useful for making a decision.

Test Prep Doctor

For Exercise 50, students need to be familiar with mathematical terms as well as symbolic representations. Students who answered **C** or **D** have confused *quotient* with *difference* or *sum*. Students who answered **A** have either confused *quotient* with *product* or incorrectly assumed that $6n$ means "6 divided by n."

Lesson Quiz

Find each sum or difference.

1. $8.3 + 2.7$ **11**

2. $9.7 - 4$ **5.7**

3. $22.6 + 8.4$ **31**

4. Evaluate $12.76 - x$ for $x = 8.41$. **4.35**

5. During an ice-skating competition, Dawn received the following scores: 4.8, 5.2, 5.4. What was Dawn's total score? **15.4**

Available on Daily Transparency in CRB

3-4 Organizer

Pacing: Traditional 1 day
Block $\frac{1}{2}$ day

Objective: Students multiply and divide decimals by powers of ten and convert metric measurements.

Indiana Academic Standards
6.2.3, 6.5.1, 6.5.2, 6.7.5

Warm Up

Multiply.

1. 8.4 × 10 — 84
2. 8.72 × 10 — 87.2
3. 0.42 × 10 — 4.2
4. 732 × 100 — 73,200
5. 9.2 × 100 — 920

Problem of the Day

A nurse must administer a 4 mL dose of a drug daily to a patient. If there is one liter of this drug on hand, will it last the patient 150 days? If so, how much will remain? If not, how much more will be needed? yes; 0.4L will remain.

Available on Daily Transparency in CRB

Math Humor

Which metric measurement can you use to measure how bad something smells? A scent-i-meter

3-4 Decimals and Metric Measurement

Learn to multiply and divide decimals by powers of ten and to convert metric measurements.

You know that in a place-value chart each place value is ten times greater than the place value to its right.

The number of zeros in the power of ten, or the exponent in the power of ten, tells you how many places to move the decimal point.

EXAMPLE 1 Multiplying and Dividing by Powers of Ten

Multiply or divide.

A 4,325 × 1,000

4,325.000 — *There are 3 zeros in 1,000.*
To multiply, move the decimal point 3 places right.
= 4,325,000 — *Write 3 placeholder zeros.*

B 4,325 ÷ 1,000

4,325. — *There are 3 zeros in 1,000.*
To divide, move the decimal point 3 places left.
= 4.325

C $79.95 \div 10^4$

0079.95 — *The power of 10 is 4.*
Move the decimal point 4 places left.
= 0.007995 — *Write placeholder zeros.*

Reading Math
Prefixes:
kilo: thousand
centi: hundredth
milli: thousandth

The metric system uses powers of ten. The base unit for length is the *meter,* for mass the *gram,* and for capacity the *liter.* Prefixes are used to describe units that are greater or smaller than the base unit.

	Unit	Abbreviation	Approximate Comparison
Length	**Kilo**meter	km	Length of 10 football fields
	Meter	m	Width of a door
	Centimeter	cm	Width of your little finger
	Millimeter	mm	Thickness of a dime
Mass	**Kilo**gram	kg	Mass of a textbook
	Gram	g	Mass of a small paperclip
Capacity	Liter	L	Filled bottle of sparkling water
	Milliliter	mL	Half-filled eyedropper

1 Introduce

Alternate Opener

EXPLORATION

3-4 Decimals and Metric Measurement

You can use patterns to multiply or divide by powers of ten. The key is in the number of zeros in the power of ten.

1. The first row in the table below contains patterns for multiplication and division by powers of ten. Use these patterns to complete the table.

÷ 1,000	÷ 100	÷ 10	Number	× 10	× 100	× 1,000
0.007	0.07	0.7	7	70	700	7,000
			8.4			
			0.44			
			95.7			

Find the value of each multiplication.

2. 1.25 × 10
3. 1.25 × 100
4. 1.25 × 1,000
5. 4.2 × 10
6. 4.2 × 100
7. 4.2 × 1,000
8. 0.375 × 10
9. 0.375 × 100
10. 0.375 × 1,000

Find the value of each division.

11. 125 ÷ 10
12. 125 ÷ 100
13. 125 ÷ 1,000
14. 42.5 ÷ 10
15. 42.5 ÷ 100
16. 42.5 ÷ 1,000

Think and Discuss

17. **Discuss** the patterns you discovered in multiplying and dividing by powers of ten.
18. **Explain** how you know that 4.125 × 1,000 = 4,125.

Motivate

Review measurement units with students by having them list all the units they can think of for measuring length, weight/mass, and capacity. Explain that customary units, such as inches, pounds, and gallons, are not as widely used around the world as the metric units of meters, liters, and grams.

Exploration worksheet and answers on Chapter 3 Resource Book pp. 37 and 119

2 Teach

Lesson Presentation

Guided Instruction

In this lesson, students learn to multiply and divide decimals by powers of ten and to convert metric measurements. First teach students that multiplying or dividing by powers of ten will result in moving the decimal point to the right for multiplication and to the left for division. Explore the table of metric measures (Teaching Transparency T15, CRB). Have students practice identifying appropriate metric units. Next teach students to convert between metric measures (Teaching Transparency T16, CRB).

EXAMPLE 2 Choosing Appropriate Units

Use the abbreviation for the most appropriate metric unit.

A **A pencil is about 15 __?__ long.**

Think: A pencil is the length of about 15 little-finger widths.

A pencil is about 15 cm long.

B **The mass of an average man is about 75 __?__.**

Think: A man has a mass of about 75 textbooks.

The mass of a man is about 75 kg.

C **A pail holds about 20 __?__.**

Think: A pail could hold about 20 bottles of water.

A pail holds about 20 L.

In the metric system, each unit of measure is ten times greater than the unit to its right in a place-value chart.

1,000	100	10	1	0.1	0.01	0.001
Thousands	Hundreds	Tens	Ones	Tenths	Hundredths	Thousandths
Kilo-	Hecto-	Deca-	Base unit	Deci-	Centi-	Milli-

To convert units within the metric system, multiply or divide by powers of ten.

EXAMPLE 3 Converting Within the Metric System

Helpful Hint

To convert to smaller units, *multiply*.
To convert to larger units, *divide*.

Convert each measure.

A **The mass of a backpack is about 6,500 g. 6,500 g = __?__ kg**

6,500 g = (6,500 ÷ 1,000) kg — *1 kg = 1,000 g, so divide by 1,000.*

6,500 g = 6.5 kg — *Move the decimal point 3 places left.*

B **A glass holds about 0.3 L of milk. 0.3 L = __?__ mL**

0.3 L = (0.3 × 1,000) mL — *1 L = 1,000 mL, so multiply by 1,000.*

0.3 L = 300 mL — *Move the decimal point 3 places right.*

Think and Discuss

1. **Determine** whether any measure in meters can be converted to millimeters.
2. **Tell** how you know whether to multiply or divide when converting units of measure.

COMMON ERROR ALERT

Students may try to convert between different base units. For example, they may say that 1 **centi**meter equals 10 **milli**liters. The table on the previous page will help remind them of why this is not correct.

Additional Examples

Example 1

Multiply or divide.

A. 7,126 × 1,000 7,126,000

B. 7,126 ÷ 1,000 7.126

C. $46.34 \div 10^4$ 0.004634

Example 2

Use the abbreviation for the most appropriate metric unit.

A. A soda can is about 12 ▢ tall. cm

B. The mass of a pen is about 5 ▢. g

C. A sip of water is about 3 ▢. mL

Example 3

Convert each measure.

A. The height of a door is about 2 m. 2 m = ▢ km 0.002

B. The width of a door is about 0.85 m. 0.85 m = ▢ cm 85

3 Close

Reaching All Learners

Through Home Connection

Have students find 5–10 items at home that are labeled with metric measurements (e.g., food and drink items). Have them make a table showing each item, its metric measurement, and a conversion of each measurement to one equivalent metric measurement.

Summarize

Name the metric units covered in the lesson, and explain how to convert units within the metric system. Have a few students suggest items that could be measured, and have the class agree on the appropriate metric unit of measure.

Answers to Think and Discuss

1. Yes, multiply a measure in meters by 1,000 to get the equivalent measure in millimeters.
2. To change from larger units to smaller units, multiply. To change from smaller units to larger units, divide.

3-4 PRACTICE & ASSESS

Students may want to refer back to the lesson examples.

Assignment Guide

If you finished Example 1 assign:
Core 1–3, 12–17, 28–30, 42–50
Enriched 3, 12–16 even, 27–31, 37–39, 42–50

If you finished Example 2 assign:
Core 1–7, 12–22, 42–50
Enriched 3, 7, 12–20 even, 28–32, 38–50

If you finished Example 3 assign:
Core 1–26, 33, 35, 42–50
Enriched 2–26 even, 28–50

Notes

Math Background

The metric system today provides a uniform way for people to measure, and thus compare and communicate. Before the adoption of the metric system, there were 400 different ways in France alone to measure the area of land.

In 1791, the Paris Academy of Science proposed the metric system of measurement for use in France. By 1840, the metric system had become the only legal system in France and was on its way to being adopted in many countries around the world.

3-4 Exercises

FOR EXTRA PRACTICE see page 640

internet connect
Homework Help Online
go.hrw.com Keyword: MR4 3-4

GUIDED PRACTICE

See Example 1 **Multiply or divide.**

1. 5,937 × 100 **593,700**
2. $719.25 \div 10^3$ **0.71925**
3. 6.0912×10^5 **609,120**

See Example 2 **Use the abbreviation for the most appropriate metric unit.**

4. A piece of paper is about 28 __?__ long. **cm**
5. A carton of orange juice holds about 1 __?__. **L**
6. A pencil weighs about 5 __?__. **g**
7. A small glass holds about 250 __?__. **mL**

See Example 3 **Convert each measure.**

8. The mass of a cat is about 7 kg. 7 kg = __?__ g **7,000**
9. A race is about 5,000 m long. 5,000 m = __?__ km **5**
10. A container holds about 0.5 L of solution. 0.5 L = __?__ mL **500**
11. A picture frame is about 18 cm long. 18 cm = __?__ m **0.18**

INDEPENDENT PRACTICE

See Example 1 **Multiply or divide.**

12. 278 × 1,000 **278,000**
13. 15.09×10^3 **15,090**
14. 810.381 ÷ 100 **8.10381**
15. 74.1×10^4 **741,000**
16. $381.8 \div 10^5$ **0.003818**
17. 42,516 ÷ 10,000 **4.2516**

See Example 2 **Use the abbreviation for the most appropriate metric unit.**

18. A television screen is about 68 __?__ wide. **cm**
19. A boy is about 1.5 __?__ tall. **m**
20. A box of cereal weighs about 540 __?__. **g**
21. A spoon holds about 2 __?__. **mL**
22. A quarter is about 2.5 __?__ wide. **cm**

See Example 3 **Convert each measure.**

23. A sneaker is about 25 cm long. 25 cm = __?__ m **0.25**
24. A container holds about 2 L of milk. 2 L = __?__ mL **2,000**
25. A pen is about 0.18 m long. 0.18 m = __?__ cm **18**
26. A dumbbell weighs about 4.5 kg. 4.5 kg = __?__ g **4,500**
27. Jeremy walks about 10 km. 10 km = __?__ m **10,000**

RETEACH 3-4

LESSON 3-4 **Reteach**
Decimals and Metric Measurement

There are patterns in powers of ten.

$10^1 = 10$ $10^2 = 100$ $10^3 = 1{,}000$

The exponent equals the number of zeros in the power of ten.

You can use these patterns to multiply and divide by powers of ten. The number of zeros or the exponent of the power of ten tells you how many places to move the decimal point.

If you are dividing by a power of ten, move the decimal point left.
7,345 ÷ 100 = 73.45
two zeros, two places
73.45

If you are multiplying by a power of ten, move the decimal point right.
$23.4 \cdot 10^3 = 23{,}400$
exponent 3, three places
23,400

Multiply or divide.

1. $4.25 \cdot 10^4$ **42,500**
2. $1{,}347.8 \div 10^2$ **13.478**
3. 9.4 ÷ 1,000 **0.0094**
4. 18.05 · 100 **1,805**

The metric system uses powers of ten.

Each place value is 10 times as large as the place value to the right.						
1,000	100	10	1	0.1	0.01	0.001
thousands	hundreds	tens	ones	tenths	hundredths	thousandths
kilo	hecto	deka	meters	deci	centi	milli

2 m = ___ cm The centimeter unit is 2 places to the right of the meter.
2 m = 200 cm Move the decimal point 2 places to the right.

Use the chart to convert each measure.

5. 3.4 km = **3,400,000** mm
6. 7 dm = **0.007** hm
7. 4.32 dam = **43.2** m
8. 34.8 cm = **3.48** dm

PRACTICE 3-4

LESSON 3-4 **Practice B**
Decimals and Metric Measurement

Multiply or divide.

1. 7.5 • 100 **750**
2. 24.68 ÷ 10 **2.468**
3. 1.479 • 1,000 **1,479**
4. $316.2 \div 10^3$ **0.3162**
5. $0.69 \cdot 10^4$ **6,900**
6. $5.403 \div 10^2$ **0.05403**

Use the abbreviation for the most appropriate metric unit.

7. A large thermos holds about 1.5 **L**.
8. A computer screen is about 30.75 **cm** wide.
9. A beetle weighs about 0.68 **g**.
10. The distance from Dallas, Texas, to Denver, Colorado is 1,260 **km**.

Convert each measure.

11. 50 cm = **500** mm
12. 3.6 L = **3,600** mL
13. 6.5 kg = **6,500** g
14. 0.9 km = **900** m
15. 1.42 m = **142** cm
16. 12.85 mL = **0.01285** L
17. An official hockey puck can weigh no more than 170 grams. What is the puck's maximum weight in kilograms? **0.17 kilograms**
18. An official hockey puck is 2.54 centimeters thick. What is the official thickness of a hockey puck in millimeters? **25.4 millimeters**
19. An official hockey goal is 46.45 meters tall. What is the height of a hockey goal in centimeters? **4,645 centimeters**
20. Hockey pucks can be hit at speeds of up to 190 kilometers per hour! How many meters per hour is that? **190,000 meters per hour**

PRACTICE AND PROBLEM SOLVING

Find the value of each expression.

28. $2.39 \times 10{,}000$ **23,900**

29. $60.87 \div 10^3$ **0.06087**

30. 0.0863×10^3 **86.3**

31. $11.6 - 42 \div 10^2$ **11.18**

32. $(2.3 + 5.67) \times 10^5$ **797,000**

Convert each measure.

33. 6,000 cm = __?__ km **0.06**

34. 0.75 L = __?__ mL **750**

35. 7.54 kg = __?__ g **7,540**

36. 17.89 m = __?__ mm **17,890**

37. Table A is 91.4 centimeters long. Table B is 18.3 decimeters long. Which table is longer? How much longer? **Table B; 91.6 cm longer**

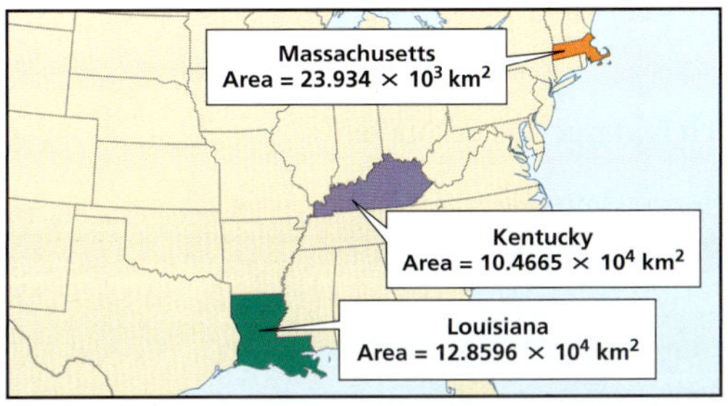

Use the map for Exercises 38 and 39.

38. ***GEOGRAPHY*** Which state has the greater area, Kentucky or Massachusetts? How much greater? **Kentucky, 80,731 km^2**

39. ***WHAT'S THE ERROR?*** A student wrote the following equation to find the area of Louisiana: $12.8596 \times 10^4 = 0.00128596$ km^2. Describe the error. Then write the correct answer.

40. ***WRITE ABOUT IT*** When you multiply or divide by a power of ten, how do you know how many places and in which direction to move the decimal point?

41. ***CHALLENGE*** About 3.281 feet equal 1 meter. About how many feet are equal to 2.5 meters? **7.5 ft**

Spiral Review

Solve each equation. (Lesson 2-6)

42. $b + 53 = 95$ **$b = 42$**

43. $10a = 340$ **$a = 34$**

44. $n - 24 = 188$ **$n = 212$**

45. $w + 20 = 95$ **$w = 75$**

46. $\frac{c}{5} = 60$ **$c = 300$**

47. $73 + p = 82$ **$p = 9$**

48. $r \times 10^2 = 300$ **$r = 3$**

49. $t \div 10^2 = 800$ **$t = 80{,}000$**

50. **TEST PREP** Which of the following is the standard form of "four and five hundred six thousandths?" (Lesson 3-1) **C**

A 4.0506 B 4.560 C 4.506 D 0.456

CHALLENGE 3-4

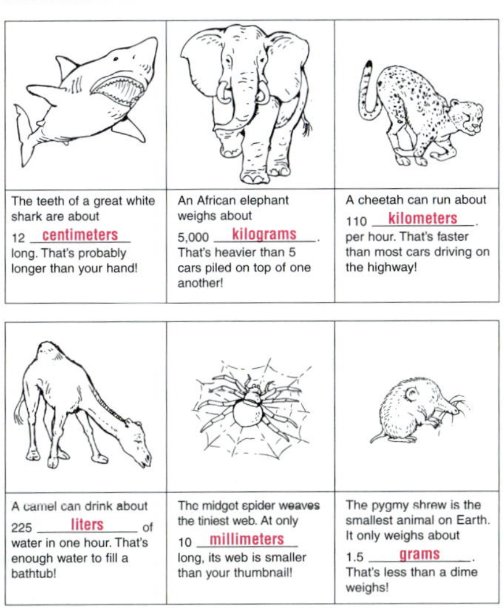

LESSON 3-4 **Challenge**
Metric Animals

Write the most appropriate metric unit for each of the measurements below.

The teeth of a great white shark are about 12 **centimeters** long. That's probably longer than your hand!

An African elephant weighs about 5,000 **kilograms**. That's heavier than 5 cars piled on top of one another!

A cheetah can run about 110 **kilometers** per hour. That's faster than most cars driving on the highway!

A camel can drink about 225 **liters** of water in one hour. That's enough water to fill a bathtub!

The midget spider weaves the tiniest web. At only 10 **millimeters** long, its web is smaller than your thumbnail!

The pygmy shrew is the smallest animal on Earth. It only weighs about 1.5 **grams**. That's less than a dime weighs!

PROBLEM SOLVING 3-4

LESSON 3-4 **Problem Solving**
Decimals and Metric Measurement

Write the correct answer.

1. The St. Gotthard Tunnel in Switzerland is the world's longest tunnel. It is 16.3 kilometers long. What is the tunnel's length in meters?
   16,300 meters
2. Ostriches are the world's heaviest birds. On average, they weigh 156,500 grams. How many kilograms does the average ostrich weigh?
   156.5 kilograms
3. The huge flower of the titam arum plant of Sumatra only lives for one day. During that time it grows 75 millimeters. What is the flower's height in centimeters?
   7.5 centimeters
4. The average male elephant drinks about 120,000 milliliters of water each day. How many liters of water do most male elephants drink each day?
   120 liters

Circle the letter of the correct answer.

5. If you want to measure the length of a pencil, which metric unit would be best to use?
   A meter
   B kilometer
   (C) centimeter
   D inch
6. If you want to measure how much a refrigerator weighs, which metric unit would be best to use?
   (F) kilogram
   G liter
   H pound
   J gram
7. Tim measured how much water his eyedropper can hold. Which of the following could be the correct measurement?
   A 2 liters
   (B) 2 milliliters
   C 2 grams
   D 2 kilograms
8. Which of the following best describes the distance from New York City to Boston, Massachusetts?
   F 330 meters
   G 330 centimeters
   (H) 330 kilometers
   J 330 millimeters

COMMON ERROR ALERT

Remind students to follow the order of operations rule in Exercises 28–32.

Answers

39. The decimal point was moved in the wrong direction; $12.8596 \times 10^4 = 128{,}596$

40. Possible answer: When multiplying, the decimal point is moved to the right. When dividing, the decimal point is moved to the left. The number of zeros in the power of 10, or the exponent in the power of 10, tells how many places to move the decimal point.

Journal

Have students write whether they think the metric system should be used in the United States.

Test Prep Doctor

For Exercise 50, all of the answer choices are shown in standard form, so students must demonstrate an understanding of decimal place value to select the correct answer. Students who answered **A** or **B** understand that the *and* after the 4 is used to separate the whole number from the decimal part of the number. Students who answered **D** have the correct number of decimal places, but they did not use zero as a placeholder and did not represent 4 as a whole number.

Lesson Quiz

Multiply or divide.

1. 16.04×10^2 **1,604**

2. $715.731 \div 100$ **7.15731**

Use the abbreviation for the most appropriate metric unit.

3. The length of a soccer field is about 100 __?__. **m**

4. The width of a computer screen is about 23 __?__. **cm**

Convert each measure.

5. The width of film for slides is 35 mm. 35 mm = __?__ m. **0.035**

6. The weight of an adult is about 70 kg. 70 kg = __?__ g. **70,000**

Available on Daily Transparency in CRB

Hands-On LAB 3C Estimate Measurement

Pacing: Traditional 1 day
Block $\frac{1}{2}$ day

Objective: To use rulers, containers, and scales to estimate measurements

Materials: Centimeter ruler, inch ruler, 1-liter container, 1-quart container, pound scale, 1-kilogram weight

Lab Resources

Hands-On Lab Activities . . . . . pp. 13–14

Using the Pages

Discuss with students the conversion equations that can be written by looking at the pictures in each activity.
1 in. ≈ 2.54 cm; 1 qt ≈ 1 L; 1 kg ≈ 2.2 lb

Complete each conversion.

1. 3.5 lb ≈ ▢ kg **1.59**

2. 5 qt ≈ ▢ mL **5,000**

3. 2 ft 5 in. ≈ ▢ cm **73.66**

Answers

Try This

1. Answers will vary. A 55-inch-tall child is about 1.4 meters tall.

Hands-On LAB 3C Estimate Measurements

Use with Lesson 3-4

Scientists usually give measurements in metric units, but most people in the United States still use the customary measurement system. To convert from one measurement system to the other, you can estimate units that are approximately equal.

Activity 1

You can see that 1 inch is about 2.5 centimeters.

1 An American football field is 100 yards long. How many meters are in 100 yards?

100 yd = (100 × 3) ft = 300 ft	
300 ft = (300 × 12) in. = 3,600 in.	*There are 12 inches in a foot.*
3,600 in. ≈ (3,600 × 2.5) cm ≈ 9,000 cm	*There are about 2.5 centimeters in 1 inch.*
9,000 cm = (9,000 ÷ 100) m ≈ 90 m	*There are 100 centimeters in 1 meter.*

There are about 90 meters in 100 yards.

Since 90 meters is close to 100 meters, a meter is about a yard.

2 There are about 5,000 feet in 1 mile. How many miles are in 1 kilometer?

1 mi ≈ 5,000 ft	
5,000 ft ≈ (5,000 ÷ 3) yd ≈ 1,666 yd	*There are 3 feet in 1 yard.*
1,666 yd ≈ 1,666 m	*One yard is about 1 meter.*
1,666 m ≈ (1,666 ÷ 1,000) km ≈ 1.6 km	*There are 1,000 m in 1 km.*

There are about 1.6 kilometers in 1 mile.

Think and Discuss

1. How could you find the number of inches in 1 meter? **Possible answer: 1 m = 100 cm; 100 cm ≈ (100 ÷ 2.5) in. ≈ 40 in. So, 1 m ≈ 40 in.**

Try This

1. Estimate your height in meters.

Teacher to Teacher

Jennifer Sawyer
Shawboro, North Carolina

Here is a way to get students interested in converting measurements through an interactive activity. Divide the class into groups of 2–4. Provide each group with a long strip of paper about 1 inch wide. Each group will mark a random length on the strip of paper and call the length "one unit." Then through folding, equally spaced units can be marked off to fill the strip. Any excess can be trimmed off the end. Students can measure three or four objects in the classroom using the ruler that they created and then also measure the objects in centimeters and inches, recording the measurements in a table. Meet as a group to discuss conversion factors.

Activity 2

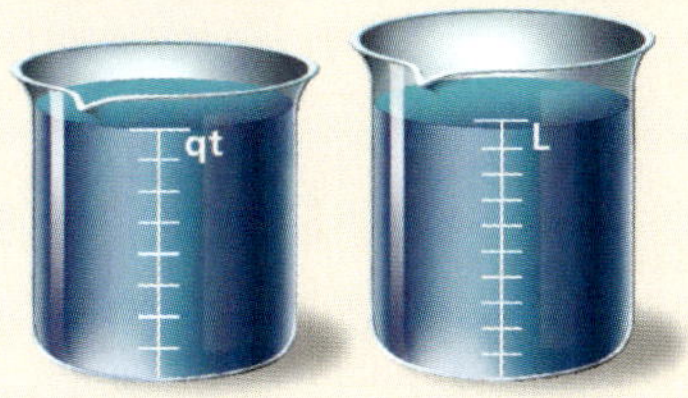

Fill a one-quart container with water. Pour the water into a one-liter container. A quart is a little less than a liter.

1 There are 4 quarts in a gallon. How many liters are in 1 gallon?

1 gal = 4 qt
4 qt ≈ 4 L *A quart is a little less than a liter.*

There are a little less than 4 liters in 1 gallon.

Think and Discuss

1. How could you find how many milliliters are in 1 gallon?

Try This

1. Estimate the volume in milliliters of a 20-gallon aquarium. **about 80,000 mL**

Activity 3

Place a 1-kilogram mass on a scale that gives weights in pounds. One kilogram weighs about 2.2 pounds.

1 There are 16 ounces in 1 pound. How many ounces are in 1 kilogram?

16 oz = 1 lb
1 kg ≈ 2.2 lb *One kilogram weighs about 2.2 pounds.*
2.2 lb ≈ (2.2 × 16) oz ≈ 35.2 oz *There are 16 ounces in 1 pound.*

There are about 35.2 ounces in 1 kilogram.

Think and Discuss

1. How could you find the number of grams in 1 ounce?

Try This

1. The mass of a medium-sized apple is about 100 grams. Estimate the number of apples that would weigh one pound. **4 or 5 apples**
2. The average weight of a newborn baby is 7 lb. Estimate this in kilograms. **about 3.15 kilograms**

Answers

Activity 2

Think and Discuss

1. Possible answer:
 1 gallon = 4 qt
 4 qt ≈ 4 L
 4L = 4,000 mL
 So 1 gallon ≈ 4,000 mL.

Activity 3

Think and Discuss

1. Possible answer:
 1 oz = (1 ÷ 16) lb = 0.0625 lb
 0.0625 lb ≈ (0.0625 ÷ 2.2) kg ≈ 0.028 kg
 0.028 kg ≈ (0.028 × 1,000) g ≈ 28 g

Chapter 3 Mid-Chapter Quiz

Purpose: *To assess students' mastery of concepts and skills in Lessons 3-1 through 3-4*

Assessment Resources

Section 3A Quiz
Assessment Resources .... p. 9

***Test and Practice Generator* CD-ROM**

Additional mid-chapter assessment items in both multiple-choice and free-response format may be generated for any objective in Lessons 3-1 through 3-4.

Chapter 3 Mid-Chapter Quiz

Mid-Chapter Quiz

LESSON 3-1 (pp. 92–95)

Write each decimal in standard form, expanded form, and words.

1. 4.012 4 + 0.01 + 0.002; four and twelve thousandths
2. ten and fifty-four thousandths 10 + 0.05 + 0.004; 10.054
3. On Monday Jamie ran 3.54 miles. On Wednesday he ran 3.6 miles. On which day did he run farther? Wednesday

Order the decimals from least to greatest.

4. 3.406, 30.08, 3.6 3.406, 3.6, 30.08
5. 10.10, 10.01, 101.1, 10.001 10.001, 10.01, 10.10, 101.1
6. 16.782, 16.59, 16.79 16.59, 16.782, 16.79

LESSON 3-2 (pp. 96–99)

7. Matt drove 106.8 miles on Monday, 98.3 miles on Tuesday, and 103.5 miles on Wednesday. About how many miles did he drive in all? about 300 miles

Estimate.

8. 8.345 − 0.6051; round to the hundredths 7.74
9. 16.492 − 2.613; round to the tenths 13.9
10. 18.79 × 4.68 100
11. 71.378 ÷ 8.13 9
12. 52.055 × 7.18 350

LESSON 3-3 (pp. 102–105)

13. Greg's scores at four gymnastic meets were 9.65, 8.758, 9.884, and 9.500. What was his total score for all four meets? 37.792

Find each sum or difference.

14. 0.47 + 0.03 0.50
15. 8 − 0.6 7.4
16. 2.2 + 1.8 4

Evaluate 8.67 − s for each value of s.

17. $s = 3.4$ 5.27
18. $s = 2.0871$ 6.5829
19. $s = 7.205$ 1.465

LESSON 3-4 (pp. 106–109)

Multiply or divide.

20. 516 × 10,000 5,160,000
21. 16.82×10^3 16,820
22. $521.7 \div 10^5$ 0.005217

Use the abbreviation for the most appropriate metric unit.

23. A fork is about 16 __?__ long. cm

Convert each measure.

24. 5,320 m = __?__ km 5.320
25. 1.6 L = __?__ mL 1,600

Focus on Problem Solving

Solve

• **Write an equation**

Read the whole problem before you try to solve it. Sometimes you need to solve the problem in more than one step.

Read the problem. Determine the steps needed to solve the problem.

Brian buys erasers and pens for himself and 4 students in his class. The erasers cost \$0.79 each, and the pens cost \$2.95 each. What is the total amount that Brian spends on the erasers and pens?

Here is one way to solve the problem.

5 erasers cost	5 pens cost
5 • \$0.79	5 • \$2.95

(5 • \$0.79) + (5 • \$2.95)

Read each problem. Decide whether you need more than one step to solve the problem. List the possible steps. Then choose an equation with which to solve the problem.

1. Joan is making some costumes. She cuts 3 pieces of fabric, each 3.5 m long. She has 5 m of fabric left. Which equation can you use to find f, the amount of fabric she had to start with? A

   A $(3 \cdot 3.5) + 5 = f$
   B $3 + 3.5 + 5 = f$
   C $(5 \times 3.5) \div 3 = f$
   D $5 - (3 \cdot 3.5) = f$

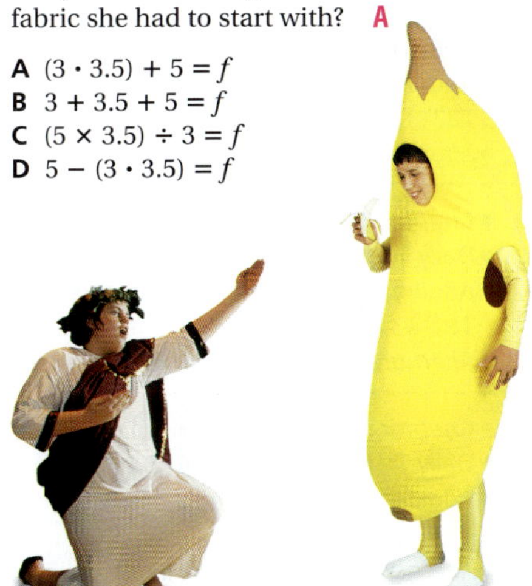

2. Mario buys 4 chairs and a table. He spends \$245.99 in all. If each chair costs \$38.95, which equation can you use to find T, the cost of the table? H

   F $4 + \$245.99 + \$38.95 = T$
   G $(4 \cdot \$38.95) + \$245.99 = T$
   H $\$245.99 - (4 \cdot \$38.95) = T$
   J $\$245.99 \div (4 \cdot \$38.95) = T$

3. Mya skis down Ego Bowl three times and down Fantastic twice. Ego Bowl is 5.85 km long, and Fantastic is 8.35 km long. Which equation can you use to estimate d, the distance Mya skis in all? A

   A $(6 \cdot 3) + (8 \cdot 2) = d$
   B $(6 + 8) + (3 + 2) = d$
   C $3(6 + 8) = d$
   D $(6 \div 3) + (8 \div 2) = d$

Answers

1. A; 15.5 m
2. H; \$90.19
3. A; 34 km

Focus on Problem Solving

Purpose: *To focus on writing an equation*

Problem Solving Resources

Interactive Problem Solving. pp. 15–24
Math: Reading and Writing in the Content Area . . . . pp. 15–24

Problem Solving Process

This page focuses on the third step of the problem-solving process:
Solve

Discuss

Have students discuss the steps required to solve each problem.
Possible answers:

1. Find the total amount of fabric used. Then add the amount left over to find the amount she had to start.
2. Find the total cost of the four chairs. Then subtract the cost from the total amount spent to find the amount spent on the table.
3. Estimate the total distance skied on each trail. Then add the distances together to estimate the total distance Mya skied.

Section 3B

Multiplying and Dividing Decimals

One-Minute Section Planner

Lesson	Materials	Resources
Lesson 3-5 Scientific Notation **NCTM:** Number and Operations, Communication, Connections, Representation **NAEP:** Number Properties 1f ☐ SAT-9 ☐ SAT-10 ☐ ITBS ☐ CTBS ☐ MAT ☐ CAT		• *Chapter 3 Resource Book,* pp. 46–54 • Daily Transparency T19, CRB • Additional Examples Transparencies T20–T21, CRB • *Alternate Openers: Explorations,* p. 19
Hands-On Lab 3D Explore Decimal Multiplication and Division **NCTM:** Number and Operations, Algebra **NAEP:** Number Properties 3f ☐ SAT-9 ☑ SAT-10 ☐ ITBS ☐ CTBS ☑ MAT ☑ CAT	**Required** Decimal grids *(CRB, p. 108)*	• *Hands-On Lab Activities,* pp. 15–16
Lesson 3-6 Multiplying Decimals **NCTM:** Number and Operations, Communication **NAEP:** Number Properties 3a ☑ SAT-9 ☑ SAT-10 ☑ ITBS ☑ CTBS ☑ MAT ☑ CAT	**Optional** Calculators Recording Sheet for Reaching All Learners *(CRB, p. 106)*	• *Chapter 3 Resource Book,* pp. 55–63 • Daily Transparency T22, CRB • Additional Examples Transparencies T23–T25, CRB • *Alternate Openers: Explorations,* p. 20
Lesson 3-7 Dividing Decimals by Whole Numbers **NCTM:** Number and Operations, Communication **NAEP:** Number Properties 3a ☑ SAT-9 ☑ SAT-10 ☑ ITBS ☑ CTBS ☑ MAT ☑ CAT	**Optional** Graph paper *(CRB, p. 107)*	• *Chapter 3 Resource Book,* pp. 64–72 • Daily Transparency T26, CRB • Additional Examples Transparencies T27–T29, CRB • *Alternate Openers: Explorations,* p. 21
Lesson 3-8 Dividing by Decimals **NCTM:** Number and Operations, Problem Solving, Communication **NAEP:** Number Properties 3a ☑ SAT-9 ☑ SAT-10 ☑ ITBS ☑ CTBS ☑ MAT ☑ CAT	**Optional** Calculators Index cards labeled with digits 0–9 *(CRB, p. 111)*	• *Chapter 3 Resource Book,* pp. 73–81 • Daily Transparency T30, CRB • Additional Examples Transparencies T31–T32, CRB • *Alternate Openers: Explorations,* p. 22
Lesson 3-9 Interpret the Quotient **NCTM:** Number and Operations, Problem Solving, Reasoning and Proof, Communication **NAEP:** Number Properties 5d ☑ SAT-9 ☑ SAT-10 ☑ ITBS ☑ CTBS ☑ MAT ☑ CAT		• *Chapter 3 Resource Book,* pp. 82–90 • Daily Transparency T33, CRB • Additional Examples Transparencies T34–T35, CRB • *Alternate Openers: Explorations,* p. 23
Lesson 3-10 Solving Decimal Equations **NCTM:** Number and Operations, Algebra, Reasoning and Proof, Communication, Representation **NAEP:** Algebra 4a ☐ SAT-9 ☐ SAT-10 ☐ ITBS ☐ CTBS ☑ MAT ☑ CAT		• *Chapter 3 Resource Book,* pp. 91–99 • Daily Transparency T36, CRB • Additional Examples Transparencies T37–T39, CRB • *Alternate Openers: Explorations,* p. 24
Extension Significant Figures **NCTM:** Measurement **NAEP:** Measurement 2e ☐ SAT-9 ☐ SAT-10 ☐ ITBS ☐ CTBS ☐ MAT ☐ CAT	**Optional** Teaching Transparency T40 *(CRB)*	• Additional Examples Transparencies T41–T42, CRB
Section 3B Assessment		• Section 3B Quiz, AR p. 10 • *Test and Practice Generator* CD-ROM

SAT = *Stanford Achievement Tests* **ITBS** = *Iowa Test of Basic Skills* **CTBS** = *Comprehensive Test of Basic Skills/Terra Nova*
MAT = *Metropolitan Achievement Tests* **CAT** = *California Achievement Test*

NCTM = Complete standards can be found on pages T27–T33. **NAEP** = Complete standards can be found on pages A31–A35.

SE = *Student Edition* **TE** = *Teacher's Edition* **AR** = *Assessment Resources* **CRB** = *Chapter Resource Book* **MK** = *Manipulatives Kit*

Section Overview

Writing Numbers in Scientific Notation — *Lesson 3-5*

Why? Scientific notation is used to express very large or very small numbers.

A number written in scientific notation has two parts that are multiplied.

$$1.2345 \times 10^4$$

The **first part** is a number that is greater than 1 and less than 10.

The **second part** is a power of ten.

Multiplying and Dividing with Decimals — *Lessons 3-6, 3-7, and 3-8*

Why? Multiplying and dividing decimals is used to convert currency.

Multiply the digits.
Then place the decimal point.

7.13	2 decimal places
× 0.2	+1 decimal place
1.426	3 decimal places

Place the decimal point by adding the number of decimal places in factors.

$0.36\overline{)11.2} \rightarrow 36\overline{)1120.00}$

```
      31.11
36)1120.00
  -108
     40
    -36
      40
     -36
       40
      -36
        4
```

Align the decimal point in the quotient.

Make the divisor a whole number by multiplying the divisor and dividend by the same power of ten.
$0.36 \times 10^2 = 36$
$11.2 \times 10^2 = 1120.00$

When a repeating pattern occurs, show three dots or draw a bar over the repeating part of the quotient.

$$31.111\ldots = 31.\overline{1}$$

Interpreting the Quotient — *Lesson 3-9*

Why? When you solve a division problem that has a remainder, you need to decide what the remainder represents.

When the question asks	→	You should
How many whole groups can be made when you divide?	→	Drop the decimal part of the quotient.
How many whole groups are needed to put all items from the dividend into a group?	→	Round the quotient up to the next highest whole number.
What is the exact number when you divide?	→	Use the entire quotient as the answer.

3-5 Organizer

Pacing: Traditional 1 day
Block $\frac{1}{2}$ day

Objective: Students write large numbers in scientific notation.

Indiana Academic Standards
6.2.3, 6.7.5, 6.7.11

Warm Up

Multiply.

1. 724×10^2 72,400
2. 837×10 8,370
3. 632.9×100 63,290
4. $18{,}256 \times 10$ 182,560
5. $10 \times 10 \times 10$ 1,000

Problem of the Day

A rope ladder is hanging from the back of a yacht. At 10:00 A.M., the water reaches the third step on the ladder. If every 2 hours the water rises the height of 2 steps, at what step would the water level be at 5:00 P.M.? Explain. the third, because the boat will rise with the water

Available on Daily Transparency in CRB

Math Fact

For very large and very small numbers, *mega-* indicates 10^6, *giga-* indicates 10^9, *micro-* indicates 10^{-6}, *nano-* indicates 10^{-9}, and *pico-* indicates 10^{-12}.

3-5 Scientific Notation

Learn to write large numbers in scientific notation.

Vocabulary
scientific notation

Pointillism is an art technique in which many small dots are placed close together to form a picture. The famous painting *A Sunday in the Park,* by Georges Seurat, is an example of pointillism. It is made of approximately 3,456,000 dots.

The dots in the painting are as small as 1/16 inch. It took Seurat about two years to complete the painting.

Scientific notation is a shorthand method for writing large numbers like 3,456,000.

Remember!
The number of zeros in the power of ten, or the exponent in the power of ten, tells you how many places to move the decimal point.

To write the number 3,456,000 in scientific notation, do the following:

3,456,000	*Move the decimal point left to form a number that is greater than 1 and less than 10.*
3,456,000	*Multiply that number by a power of ten.*
3.456×10^6	*The power of 10 is 6, because the decimal point is moved 6 places left.*

The number of dots written in scientific notation is 3.456×10^6.

A number written in scientific notation has two parts that are multiplied.

$$3.456 \times 10^6$$

The first part is a number that is greater than 1 and less than 10.
The second part is a power of 10.

EXAMPLE 1 Writing Numbers in Scientific Notation

Write each number in scientific notation.

A 700,000

700,000 *Move the decimal point 5 places left. The power of 10 is 5.*

$700{,}000 = 7 \times 10^5$

B 8,296,000

8,296,000 *Move the decimal point 6 places left. The power of 10 is 6.*

$8{,}296{,}000 = 8.296 \times 10^6$

1 Introduce

Alternate Opener

EXPLORATION

3-5 Scientific Notation

You can use exponents to represent powers of 10.

$10^1 = 10$
$10^2 = 10 \times 10 = 100$
$10^3 = 10 \times 10 \times 10 = 1{,}000$
$10^4 = 10 \times 10 \times 10 \times 10 = 10{,}000$

Find each product.

1. 6.25×10^1 2. 6.25×10^2 3. 6.25×10^3 4. 6.25×10^4

In scientific notation, numbers are written as the product of a power of 10 and a number starting with the ones place.

Number	Scientific Notation
6,250	6.25×10^3
62.50	6.25×10^1
625	6.25×10^2

Write each number in scientific notation by filling in the exponent for the power of 10.

5. $42.5 = 4.25 \times 10^{\square}$ 6. $425 = 4.25 \times 10^{\square}$
7. $4250 = 4.25 \times 10^{\square}$ 8. $42500 = 4.25 \times 10^{\square}$

Think and Discuss

9. **Describe** the process for writing numbers in scientific notation.
10. **Explain** how you know that $6.25 \times 10^6 = 6{,}250{,}000$.

Motivate

Write this number on the board: 5,880,000,000,000. Ask if anyone can read the number (5 trillion, 880 billion). Let the students know that this is approximately the number of miles light travels in one year (1 light-year). Tell students that they can write large numbers like these in a form called *scientific notation*.

Exploration worksheet and answers on Chapter 3 Resource Book pp. 47 and 121

2 Teach

Lesson Presentation

Guided Instruction

In this lesson, students learn to write large numbers in scientific notation. First, teach students to express a number that is written in standard form in scientific notation. Then teach students to express a number that is written in scientific notation in standard form.

Teaching Tip Be sure students understand that the exponent indicates the number of times the decimal point must be moved and not the number of zeros to be added.

Write each number in scientific notation.

C 58,000

58,000 — *Move the decimal 4 places left. The power of 10 is 4.*

$58{,}000 = 5.8 \times 10^4$

You can write a large number written in scientific notation in standard form. Look at the power of 10 and move the decimal point that number of places to the right.

EXAMPLE 2 Writing Numbers in Standard Form

Write each number in standard form.

A 8.753×10^2

8.753×10^2 — *The power of 10 is 2.*

8.753 — *Move the decimal point 2 places right.*

$8.753 \times 10^2 = 875.3$

B 3.2×10^7

3.2×10^7 — *The power of 10 is 7.*

3.2000000 — *Move the decimal point 7 places right. Use zeros as placeholders.*

$3.2 \times 10^7 = 32{,}000{,}000$

C 2.001×10^1

2.001×10^1 — *The power of 10 is 1.*

2.001 — *Move the decimal point 1 place right.*

$2.001 \times 10^1 = 20.01$

Think and Discuss

1. **Explain** how you can check whether a number is written correctly in scientific notation.
2. **Tell** why 782.5×10^8 is not correctly written in scientific notation.
3. **Tell** the advantages of writing a number in scientific notation over writing it in standard form. Explain any disadvantages.

COMMON ERROR ALERT

Some students may multiply the power of ten by a number between 10 and 100 when trying to write a number in scientific notation. Have them make sure that they use a number between 1 and 10.

Additional Examples

Example 1

Write each number in scientific notation.

A. 6,000,000 — 6×10^6

B. 411,000 — 4.11×10^5

C. 79,000,000 — 7.9×10^7

Example 2

Write each number in standard form.

A. 6.2174×10^3 — 6,217.4

B. 9.5×10^8 — 950,000,000

C. 4.83×10^5 — 483,000

3 Close

Reaching All Learners

Through Critical Thinking

Help students understand that writing a number in scientific notation is a two-part process. The first and most critical part is to create a number greater than 1 and less than 10 (i.e., a number that has only one digit to the left of the decimal point). Practice part one several times before proceeding to the next part. The second part is to count the number of places you have to move the decimal point to create the number in part one. The number of moves becomes the exponent of 10 in scientific notation.

Summarize

Remind students of what scientific notation is and give an example of a large number written in both standard form and scientific notation. Emphasize that a number in scientific notation is composed of a number between 1 and 10 multiplied by powers of ten.

Answers to Think and Discuss

1. Possible answer: Convert the number written in scientific notation back into standard form.
2. The part before the decimal point is not between 1 and 10. The correct way to write this number is 7.825×10^{10}.
3. Possible answer: Space and time can be saved by writing very large numbers in scientific notation. Space and time is wasted when scientific notation is used to write a smaller number such as in the tens or hundreds.

3-5 PRACTICE & ASSESS

3-5 Exercises

FOR EXTRA PRACTICE
see page 641

internet connect
Homework Help Online
go.hrw.com Keyword: MR4 3-5

Students may want to refer back to the lesson examples.

GUIDED PRACTICE

See Example 1 Write each number in scientific notation.

1. 62,000 6.2×10^4
2. 500,000 5.0×10^5
3. 6,913,000 6.913×10^6
4. 130,000 1.3×10^5
5. 7,015,000 7.015×10^6
6. 20,000 2.0×10^4

See Example 2 Write each number in standard form.

7. 6.793×10^6 6,793,000
8. 1.4×10^4 14,000
9. 3.82×10^5 382,000
10. 9.401×10^7 94,010,000
11. 3.3×10^3 3,300
12. 1.885×10^4 18,850

INDEPENDENT PRACTICE

See Example 1 Write each number in scientific notation.

13. 90,000 9.0×10^4
14. 186,000 1.86×10^5
15. 1,607,000 1.607×10^6
16. 240,000 2.4×10^5
17. 6,000,000 6.0×10^6
18. 16,900,000 1.69×10^7
19. 1,800 1.8×10^3
20. 12,865,000 1.2865×10^7
21. 50,400,000 5.04×10^7

See Example 2 Write each number in standard form.

22. 3.211×10^5 321,100
23. 1.63×10^6 1,630,000
24. 7.7×10^3 7,700
25. 2.14×10^4 21,400
26. 4.03×10^6 4,030,000
27. 8.1164×10^8 811,640,000
28. 6.33×10^5 633,000
29. 9.106×10^7 91,060,000
30. 5.5×10^2 550

PRACTICE AND PROBLEM SOLVING

Write each number in standard form.

31. 7.21×10^3 7,210
32. 1.234×10^5 123,400
33. 7.200×10^2 720
34. 2.08×10^5 208,000
35. 6.954×10^3 6,954
36. 5.43×10^1 54.3

Write each number in scientific notation.

37. 112,050 1.1205×10^5
38. 150,000 1.5×10^5
39. 4,562 4.562×10^3
40. 1,000 1×10^3
41. 65,342 6.5342×10^4
42. 95 9.5×10^1

Write each measurement using scientific notation.

43. 4 km = __?__ m 4×10^3
44. 3.78 km = __?__ cm 3.78×10^5
45. 18 L = __?__ mL 1.8×10^4
46. 75 kg = __?__ mg 7.5×10^7
47. 19.5 kg = __?__ g 1.95×10^4
48. 2 L = __?__ mL 2×10^3

Assignment Guide

If you finished Example 1 assign:
Core 1–6, 13–21, 51–52, 58–63
Enriched 37–42, 49–63

If you finished Example 2 assign:
Core 1–36, 40–42, 51–52, 58–63
Enriched 16–63

Notes

Math Background

Various methods have been used over the past 30 years to display numbers written in scientific notation on a calculator. For example, to express a number such as 4.35×10^{16}, most calculators use a notation such as 4.35 E16. The letter *E* signals exponent, with the base of 10 being assumed. Naturally, calculator users not familiar with scientific notation could become confused by such a display.

RETEACH 3-5

LESSON 3-5 **Reteach**
Scientific Notation

Scientific notation expresses a large number as the product of a number between one and ten and a power of ten.

To write 3,400 in scientific notation, move the decimal point to the left until the number falls between 1 and 10.

3,400 $1 < 3 < 10$, so move the decimal point 3 places to the left.

$3{,}400 = 3.4 \cdot 10^3$ (3 2 1) The number of times you move the decimal point left is the power of ten.

Express each number in scientific notation.

1. 175,000 $1.75 \cdot 10^5$
2. 298 $2.98 \cdot 10^2$
3. 5,764 $5.764 \cdot 10^3$
4. 83 $8.3 \cdot 10^1$
5. 40,300 $4.03 \cdot 10^4$
6. 2,000,000 $2 \cdot 10^6$
7. 51,010 $5.101 \cdot 10^4$
8. 190,025 $1.90025 \cdot 10^5$

You can express numbers written in scientific notation in standard form.

The power of ten tells you how many places to move the decimal point to the right.

$3.2 \cdot 10^4 = 32{,}000$ To write $3.2 \cdot 10^4$ in standard form, move the decimal point 4 places to the right.

Write each number in standard form.

9. $5.62 \cdot 10^3$ 5,620
10. $7.238 \cdot 10^2$ 723.8
11. $9.9 \cdot 10^5$ 990,000
12. $6.53 \cdot 10^1$ 65.3
13. $5.36 \cdot 10^4$ 53,600
14. $2.4 \cdot 10^2$ 240
15. $4.35 \cdot 10^3$ 4,350
16. $8 \cdot 10^5$ 800,000
17. $1 \cdot 10^4$ 10,000
18. $2.03 \cdot 10^3$ 2030
19. $1.12 \cdot 10^2$ 112
20. $3.002 \cdot 10^6$ 3,002,000

PRACTICE 3-5

LESSON 3-5 **Practice B**
Scientific Notation

Write each number in scientific notation.

1. 16,700 $1.67 \cdot 10^4$
2. 4,680 $4.68 \cdot 10^3$
3. 320,500 $3.205 \cdot 10^5$
4. 7,590,000 $7.59 \cdot 10^6$
5. 58,340,000 $5.834 \cdot 10^7$
6. 108,690,000 $1.0869 \cdot 10^8$

Write each number in standard form.

7. $3.25 \cdot 10^4$ 32,500
8. $7.08 \cdot 10^6$ 7,080,000
9. $1.209 \cdot 10^7$ 12,090,000
10. $6.8 \cdot 10^8$ 680,000,000
11. $0.51 \cdot 10^5$ 51,000
12. $0.006 \cdot 10^3$ 6

Identify the answer choice that is *not* equal to the given number.

13. 356,000
    A 300,000 + 56,000
    B $3.56 \cdot 10^5$
    (C) $3.56 \cdot 10^4$
14. $1.28 \cdot 10^6$
    (A) 100,000 + 28,000
    B 1,280,000
    C $12.8 \cdot 10^5$
15. 1,659,000
    A 1,600,000 + 59,000
    B $1.659 \cdot 10^6$
    (C) $16.59 \cdot 10^6$
16. $0.074 \cdot 10^3$
    A 70.0 + 4.0
    (B) $7.4 \cdot 10^5$
    C $7.4 \cdot 10^1$
17. In 2000, the population of Pennsylvania was 12,281,054. Round this figure to the nearest hundred thousand. Then write that number in scientific notation.
    12,300,000; $1.23 \cdot 10^7$
18. In 2000, the population of North Carolina was about $8.05 \cdot 10^6$, and the population of South Carolina was about $4.01 \cdot 10^6$. Write the combined populations of these two states in standard form.
    12,060,000

Earth Science LINK

This F/A-18 Hornet makes a vapor cloud by flying at Mach 0.98, just under the speed of sound.

go.hrw.com
KEYWORD: MR4 SOUND
CNN Student News

49. ***EARTH SCIENCE*** The speed of light is about 300,000 km/s. The speed of sound in air that has a temperature of 20°C is 1,125 ft/s. Write both of these values in scientific notation. 3.0×10^5 km/s; 1.125×10^3 ft/s

50. ***LIFE SCIENCE*** Genes carry the codes used for making proteins that are necessary for life. No one knows yet how many human genes there are. Estimates range from 3.8×10^4 to 1.2×10^5. Write a number in standard form that is within this range. Range 38,000 to 120,000. Possible answer: 100,000

Use the pictograph for Exercises 51 and 52.

51. Write the capacity of Rungnado Stadium in scientific notation. $150,000 = 1.5 \times 10^5$

52. Estimate the capacity of the largest stadium. Write the estimate in scientific notation. Possible answer: 2.4×10^5

World's Largest Stadiums

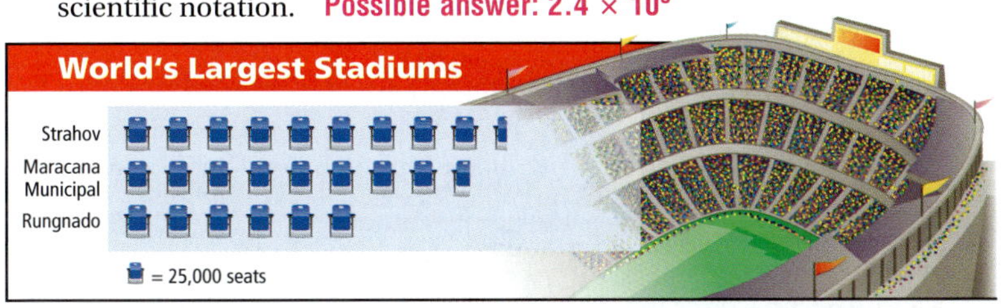

53. ***TECHNOLOGY*** In the year 2000, there were about 579 million computers in the world. Write this number in standard form and scientific notation. 579,000,000; 5.79×10^8

54. ***SOCIAL STUDIES*** The Library of Congress, in Washington, D.C., is the largest library in the world. It was founded in 1800 and has 24,616,867 books. Round the number of books to the nearest hundred thousand, and write that number in scientific notation. 24,600,000; 2.46×10^7

55. ***WHAT'S THE ERROR?*** A student said the number 56,320,000 written in scientific notation is 56.32×10^6. Describe the error. Then write the correct answer.

56. ***WRITE ABOUT IT*** How does writing numbers in scientific notation make it easier to compare and order the numbers?

57. ***CHALLENGE*** What is 5.32 written in scientific notation? 5.32×10^0

Spiral Review

Identify a pattern in each sequence, and name the next two terms. (Lesson 1-7)

58. 3, 6, 12, 24, 48, ▒, ▒, … Multiply by 2; 96, 192.

59. 1, 3, 8, 24, 29, 87, 92, ▒, ▒, … Multiply by 3, then add 5; 276, 281.

Solve each equation. (Lesson 2-5)

60. $a - 23 = 18$ $a = 41$

61. $y - 7 = 45$ $y = 52$

62. $x + 16 = 71$ $x = 55$

63. **TEST PREP** Which is the product of $30.62 \times 10{,}000$? (Lesson 3-4) D

A 0.003062 **B** 0.3062 **C** 30,620 **D** 306,200

Answers

55. Scientific notation uses a decimal number that is at least 1 but not more than 10. In scientific notation, the number would be written as 5.632×10^7

56. The powers of 10 can be compared. The number with the lowest power of 10 is the lesser number. If they have the same power of 10, the decimal numbers can be compared.

Journal

Have students write about an imaginary visit to an amusement park. Have them use scientific notation to list their observations in the park (e.g., the number of people who visit the park in one year, the time in seconds a roller-coaster ride lasts, etc.).

Test Prep Doctor

For Exercise 63, encourage students to eliminate wrong answer choices before choosing the correct answer. **A** and **B** can be eliminated right away because the product of two factors greater than 1 is greater than both factors. Students who answered **C** multiplied by 1,000 instead of by 10,000.

CHALLENGE 3-5

LESSON 3-5 **Challenge**
The Solar System

Write each planet's average distance from the Sun in standard form. Then use the distances to label the planets in our solar system.

	Planet	Average Distance From the Sun (mi)	
		Scientific Notation	Standard Form
1.	Earth	$9.29 \cdot 10^7$	92,900,000
2.	Jupiter	$4.836 \cdot 10^8$	483,600,000
3.	Mars	$1.416 \cdot 10^8$	141,600,000
4.	Mercury	$3.6 \cdot 10^7$	36,000,000
5.	Neptune	$2.794 \cdot 10^9$	2,794,000,000
6.	Pluto	$3.675 \cdot 10^9$	3,675,000,000
7.	Saturn	$8.87 \cdot 10^8$	887,000,000
8.	Uranus	$1.784 \cdot 10^9$	1,784,000,000
9.	Venus	$6.72 \cdot 10^7$	67,200,000

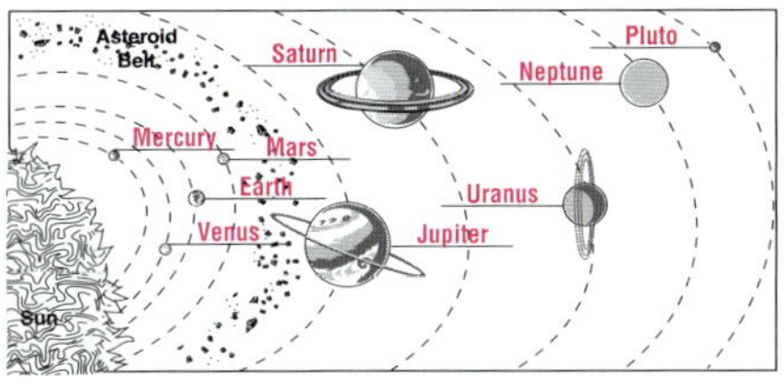

PROBLEM SOLVING 3-5

LESSON 3-5 **Problem Solving**
Scientific Notation

Write the correct answer.

1. The closest comet to approach Earth was called Lexell. On July 1, 1770, Lexell was observed about 874,200 miles from Earth's surface. Write this distance in scientific notation.
$8.742 \cdot 10^5$

2. Scientists estimate that it would take $1.4 \cdot 10^{10}$ years for light from the edge of our universe to reach Earth. How many years is that written in standard form?
14,000,000,000 years

3. In the United States, about 229,000,000 people speak English. About 18,000,000 people speak English in Canada. Write in scientific notation the total number of English speaking people in the United States and Canada.
$2.47 \cdot 10^8$ people

4. South Africa is the top gold-producing country in the world. Each year it produces $4.688 \cdot 10^8$ tons of gold! Written in standard form, how many tons of gold does South African produce each year?
468,800,000 tons

5. In 1872 Yellowstone became the first national park in the United States. Today, about $3.012 \cdot 10^6$ people visit Yellowstone National Park each year. What is that figure written in standard form?
3,012,000 people

6. Iowa and Illinois are the top corn-producing states. In 2000, farmers in Iowa grew 1,740,000 bushels of corn, and farmers in Illinois grew 1,669,000 bushels. Write their combined corn production in scientific notation.
$3.409 \cdot 10^6$ bushels

Circle the letter of the correct answer.

7. The temperature at the core of the Sun reaches 27,720,000°F. What is this temperature written in scientific notation?
A $2.7 \cdot 10^7$
B $2.72 \cdot 10^7$
C $2.772 \cdot 10^6$
(D) $2.772 \cdot 10^7$

8. Your body is constantly producing red blood cells—about $1.73 \cdot 10^{11}$ cells a day. How many blood cells is that written in standard form?
F 173,000,000 cells
G 17,300,000,000 cells
(H) 173,000,000,000 cells
J 1,730,000,000,000 cells

Lesson Quiz

Write each number in scientific notation.

1. 6,300 6.3×10^3

2. 70,400,000 7.04×10^7

Write each number in standard form.

3. 7.241×10^4 72,410

4. 8.2137×10^7 82,137,000

5. A Wall Street report indicated that a fast-moving stock had sold 3,295,000 shares. Write this number in scientific notation. 3.295×10^6

Available on Daily Transparency in CRB

Hands-On LAB 3D Explore Decimal Multiplication and Division

Pacing: Traditional 1 day
Block $\frac{1}{2}$ day

Objective: To use decimal grids to model multiplication and division of decimals

Materials: Decimal grids

Lab Resources

Hands-On Lab Activities. . . . . pp. 15–16

Using the Pages

Remind students that each small square represents 0.01, each row or column represents 0.1, and an entire grid represents 1.

Use decimal grids to find each product or quotient.

1. 2 · 0.35 **0.70**

2. 1.2 ÷ 2 **0.60**

Hands-On LAB 3D Explore Decimal Multiplication and Division

Use with Lessons 3-6 and 3-7

internet connect
Lab Resources Online
go.hrw.com
KEYWORD: MR4 Lab3D

KEY

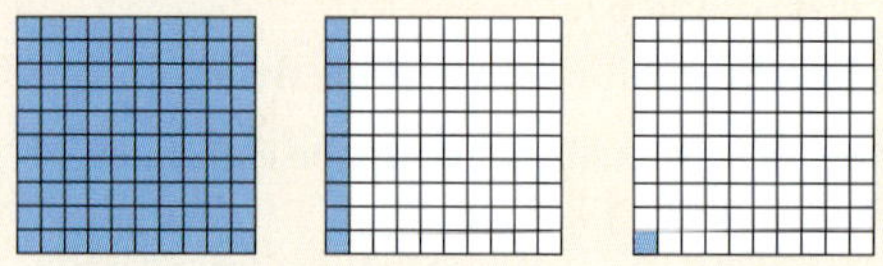

1	0.1	0.01
one	1 tenth	1 hundredth

You can use decimal grids to model multiplication and division of decimals.

Activity 1

1 Use decimal grids to find each product.

a. 3 · 0.32

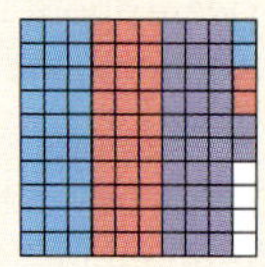

3 · 0.32 = 0.96

To represent 3 · 0.32, shade 32 small squares three times.

Use a different color to shade a different group of 32 small squares each time.

There are 96 shaded squares.

b. 0.3 · 0.5

0.3 · 0.5 = 0.15

To represent 0.3, shade 3 columns.

To represent 0.5, shade 5 rows in another color.

There are 15 squares in the area where the shading overlaps.

Think and Discuss

1. How is multiplying a decimal by a decimal different from multiplying a decimal by a whole number?
2. Why can you shade 5 rows to represent 0.5?

Try This

Use decimal grids to find each product.

1. 3 · 0.14 **0.42** **2.** 5 · 0.18 **0.90** **3.** 0.7 · 0.5 **0.35** **4.** 0.6 · 0.4 **0.24**

Answers

Activity 1

Think and Discuss

Possible answers:

1. When multiplying a decimal by a whole number *n*, model the decimal *n* times on the grid(s). The total number of squares is the product. When multiplying a decimal by a decimal, model one decimal with rows and the other with columns. The number of overlapping squares is the product.
2. Shading 5 rows or shading 5 columns both represent 0.5 because both 5 rows and 5 columns contain 50 squares. 50 squares is 0.50, the same as 0.5.

Try This

1.

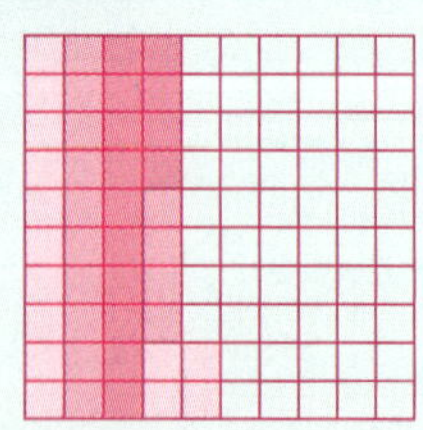

2.

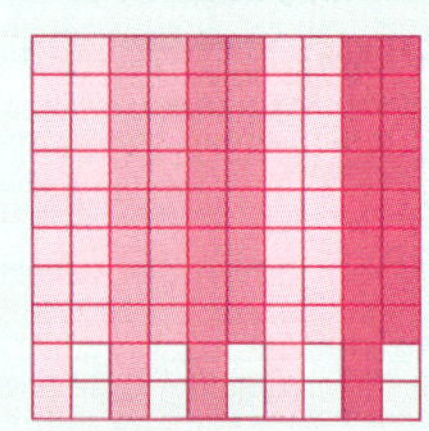

3.

4.

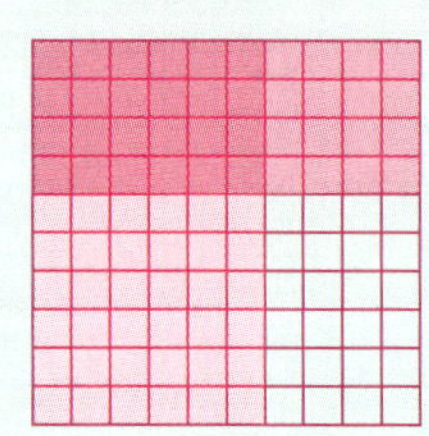

Activity 2

1. Use decimal grids to find each quotient.

a. 3.66 ÷ 3

Shade 3 grids and 66 small squares of a fourth grid to represent 3.66.

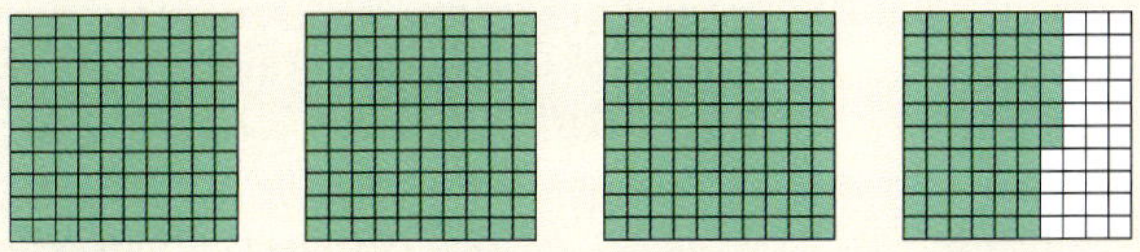

Divide the shaded wholes into 3 equal groups. Use scissors to divide the 66 hundredths into 3 equal groups.

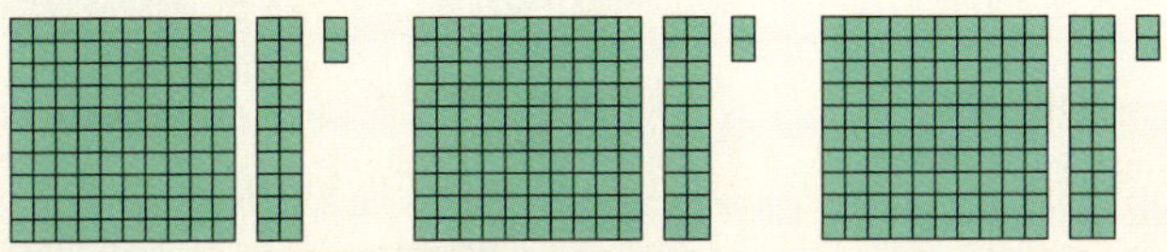

One whole grid and 22 small squares are in each group.

3.66 ÷ 3 = 1.22

b. 3.6 ÷ 1.2

Shade 3 grids and 6 columns of a fourth grid to represent 3.6. Cut apart the 6 tenths.

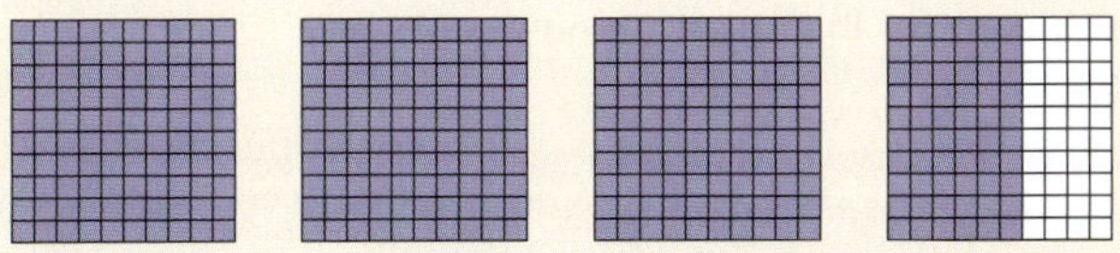

Divide the grids and tenths into equal groups of 1.2.

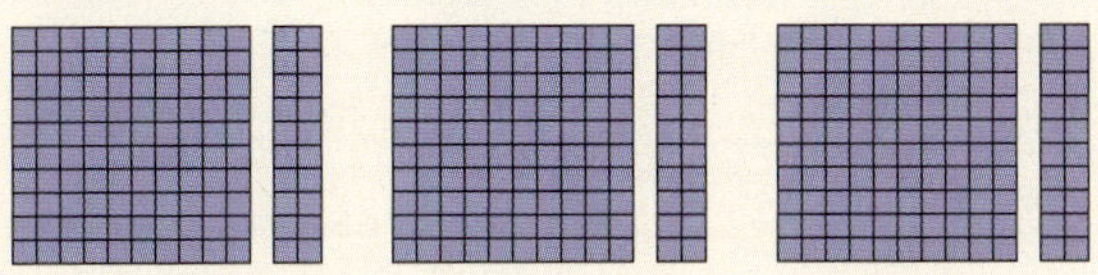

There are 3 equal groups of 1.2.

3.6 ÷ 1.2 = 3

Think and Discuss

1. Find 36 ÷ 12. How does this problem and its quotient compare to 3.6 ÷ 1.2?

Try This

Use decimal grids to find each quotient.

1. 4.04 ÷ 4 **1.01** **2.** 3.25 ÷ 5 **0.65** **3.** 7.8 ÷ 1.3 **6** **4.** 5.6 ÷ 0.8 **7**

Answers

Activity 2

Think and Discuss

1. 3; 36 ÷ 12 and 3.6 ÷ 1.2 have the same quotient; for 3.6 ÷ 1.2, you break up 3 whole grids and 6 columns into 3 groups of 1 grid and 2 columns each; for 36 ÷ 12, you break up 36 grids into 3 groups of 12 grids each.

Try This

1–4. Complete answers on p. A2

3-6 Organizer

Pacing: Traditional 1 day
Block $\frac{1}{2}$ day

Objective: Students multiply decimals by whole numbers and by decimals.

Indiana Academic Standards
6.2.3, 6.2.9, 6.7.7, 6.7.9, 6.7.10

Warm Up

Multiply.

1. 87 × 320 27,840
2. 943 × 800 754,400
3. 3,806 × 10 38,060
4. 1,207 × 100 120,700
5. 72 × 196 14,112
6. 120 × 523 62,760

Problem of the Day

Carmen and Rita sold homemade oatmeal cookies. After 8 days, they had sales totaling \$70. Each day, their sales were \$0.50 higher than the previous day. What were their sales on the first day? \$7.00

Available on Daily Transparency in CRB

Math Humor

Why is the computer-chip industry like multiplying two decimals between zero and one? It's always making a smaller product.

3-6 Multiplying Decimals

Learn to multiply decimals by whole numbers and by decimals.

Because the Moon has less mass than Earth, it has a smaller gravitational effect. An object that weighs 1 pound on Earth weighs only 0.17 pound on the Moon. How much does a 3-pound flag weigh on the Moon?

To find the weight of a 3-pound flag on the Moon, you can add decimals or multiply a decimal by a whole number.

Gravity on Earth is about six times the gravity on the surface of the Moon.

EXAMPLE 1 *Science Application*

Something that weighs 1 lb on Earth weighs 0.17 lb on the Moon. How much does a 3 lb flag weigh on the Moon?

3 × 0.17

You can think of multiplication by a whole number as a repeated addition.

$$\begin{array}{r} 0.17 \\ 0.17 \\ +\,0.17 \\ \hline 0.51 \end{array}$$

You can also multiply as you would with whole numbers.

Place the decimal point by adding the number of decimal places in the numbers multiplied.

0.17	*2 decimal places*
× 3	*+ 0 decimal places*
0.51	*2 decimal places*

A 3 lb flag on Earth weighs 0.51 lb on the Moon.

EXAMPLE 2 Multiplying a Decimal by a Decimal

Find each product.

A 0.2 × 0.6

Multiply. Then place the decimal point.

0.2	*1 decimal place*
× 0.6	*+ 1 decimal place*
0.12	*2 decimal places*

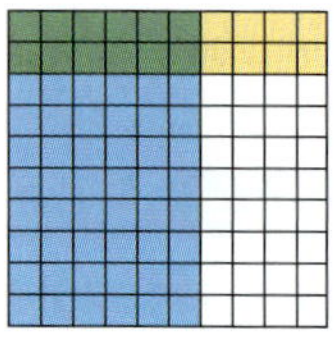

1 Introduce

Alternate Opener

EXPLORATION

3-6 Multiplying Decimals

When you multiply decimals, you can use estimation to help you determine the position of the decimal point in the product.

Estimate each product to the nearest whole number. Then use a calculator to see how reasonable your estimate is.

		Estimate	Actual
1.	4.235 × 16.9		
2.	0.78 × 568		
3.	56.1 × 23		
4.	15.6 × 2.15		

Estimate each product. Use this estimation to decide where to place a decimal point in the answer. Check with your calculator.

5. 70.5 × 4.4 = 3 1 0 2
6. 0.75 × 692 = 5 1 9 0
7. 56 × 3.125 = 1 7 5 0
8. 45.6 × 2.15 = 9 8 0 4
9. 4.17 × 1.2 = 5 0 0 4
10. 125.2 × 7.4 = 9 2 6 4 8

Think and Discuss

11. **Discuss** your strategies for estimating in numbers 1–4.
12. **Explain** how you know where to place the decimal point in a product.

Motivate

Have students give examples of how to use the multiplication algorithm to multiply 3-digit numbers by 2-digit numbers and 4-digit numbers by 1-digit numbers. Explain that the algorithm for multiplying decimals is the same; however, students must make sure that the decimal point is placed in the product correctly.

Exploration worksheet and answers on Chapter 3 Resource Book pp. 56 and 123

2 Teach

Lesson Presentation

Guided Instruction

In this lesson, students learn to multiply decimals by whole numbers and by decimals. First show students how repeated addition of a decimal is related to the multiplication of that decimal by a whole number. Then teach students to multiply a decimal by a decimal. Next have them evaluate multiplication expressions for given values of *x*.

Teaching Tip In working with positive numbers, make sure students understand that multiplication by a number less than one gives a product less than the other factor (e.g., 0.8 × 15 = 12 and 12 < 15).

Find each product.

B 0.05×0.9

$0.05 \times 1 = 0.05$ *Estimate the product. 0.9 is close to 1.*

Multiply. Then place the decimal point.

0.05	*2 decimal places*
× 0.9	*+ 1 decimal place*
0.045	*3 decimal places; use a placeholder zero.*

0.045 is close to the estimate of 0.05. The answer is reasonable.

C 3.25×4.8

$3 \times 5 = 15$ *Estimate the product. Round each factor to the nearest whole number.*

Multiply. Then place the decimal point.

3.25	
× 4.8	
2600	*2 decimal places*
13000	*+ 1 decimal place*
15.600	*3 decimal places*

15.600 is close to the estimate of 15. The answer is reasonable.

EXAMPLE 3 Evaluating Decimal Expressions

Remember!

These notations all mean multiply 3 times *x*.

$3 \cdot x \quad 3x \quad 3(x)$

Evaluate $3x$ for each value of x.

A $x = 4.047$

$3x = 3(4.047)$ *Substitute 4.047 for x.*

4.047	*3 decimal places*
× 3	*+ 0 decimal places*
12.141	*3 decimal places*

B $x = 2.95$

$3x = 3(2.95)$ *Substitute 2.95 for x.*

2.95	*2 decimal places*
× 3	*+ 0 decimal places*
8.85	*2 decimal places*

Think and Discuss

1. **Tell** how many decimal places are in the product of 235.2 and 0.24.
2. **Tell** which is greater, 4×0.6 or 4×0.006.
3. **Describe** how the products of 0.3×0.5 and 3×5 are similar. How are they different?

Additional Examples

Example 1

Something that weighs 1 lb on Earth weighs 0.17 lb on the Moon. How much would a 4 lb dumbbell weigh on the Moon? **0.68 lb**

Example 2

Find each product.

A. 0.3×0.4 **0.12** **B.** 0.07×0.8 **0.056**

C. 1.34×2.5 **3.350**

Example 3

Evaluate $5x$ for each value of x.

A. $x = 3.062$ **15.310** **B.** $x = 4.79$ **23.95**

3 Close

Reaching All Learners
Through Number Sense

Have students identify the expressions in the list that have a product of 0.48. You may want to use the recording sheet on page 106 of the Chapter 3 Resource Book.

0.2 × 2.4 ✔	0.3 × 1.6 ✔
0.2 × 0.24	3 × 0.16 ✔
2 × 2.4	3 × 1.6
2 × 0.24 ✔	0.3 × 0.16
0.4 × 1.2 ✔	0.6 × 0.8 ✔
0.4 × 0.12	6 × 0.8
4 × 1.2	8 × 0.06 ✔
4 × 0.12 ✔	8 × 0.6

Summarize

Outline the steps to follow when multiplying decimals. Emphasize the fact that the algorithm for multiplying decimals is the same as multiplying whole numbers except that the placement of the decimal in the product is determined by the total number of digits behind the decimal points in both factors.

Answers to Think and Discuss

1. 3 decimal places (to the thousandths place)
2. 4×0.6 is greater than 4×0.006.
3. Possible answer: Both have the number 15 in their product; the product of 3×5 is 15, and the product of 0.3×0.5 is 0.15.

3-6 PRACTICE & ASSESS

Students may want to refer back to the lesson examples.

Assignment Guide

If you finished Example 1 assign:
Core 1–2, 10–11, 38–39, 45–50
Enriched 2, 10, 38–41, 45–50

If you finished Example 2 assign:
Core 1–6, 10–19, 38–39, 45–50
Enriched 17–19, 26–33, 38–50

If you finished Example 3 assign:
Core 1–25, 38–39, 45–50
Enriched 18–50

Notes

Math Background

The shifting of the decimal point when multiplying decimals can be justified by considering the following example:

$$5.42 \times 1.6$$

$$\frac{542}{100} \times \frac{16}{10} = \frac{8{,}672}{1{,}000}$$

$$= 8.672$$

3-6 Exercises

FOR EXTRA PRACTICE see page 641

internet connect
Homework Help Online
go.hrw.com Keyword: MR4 3-6

GUIDED PRACTICE

See Example 1

1. Each can of cat food costs \$0.28. How much will 6 cans of cat food cost? **\$1.68**

2. Jorge buys 8 baseballs for \$9.29 each. How much does he spend in all? **\$74.32**

See Example 2 Find each product.

3. $\begin{array}{r} 0.6 \\ \times\ 0.4 \\ \hline \end{array}$ **0.24**
4. $\begin{array}{r} 0.008 \\ \times\ 0.5 \\ \hline \end{array}$ **0.0040**
5. $\begin{array}{r} 3.0 \\ \times\ 0.07 \\ \hline \end{array}$ **0.21**
6. $\begin{array}{r} 0.12 \\ \times\ 0.6 \\ \hline \end{array}$ **0.072**

See Example 3 Evaluate $5x$ for each value of x.

7. $x = 3.304$ **16.52**
8. $x = 4.58$ **22.90**
9. $x = 7.126$ **35.63**

INDEPENDENT PRACTICE

See Example 1

10. Gwenyth walks her dog each morning. If she walks 0.37 kilometers each morning, how many kilometers will she have walked in 7 days? **2.59 km**

11. Apples are on sale for \$0.49 per pound. What is the price for 4 pounds of apples? **\$1.96**

See Example 2 Find each product.

12. $\begin{array}{r} 0.9 \\ \times\ 0.03 \\ \hline \end{array}$ **0.027**
13. $\begin{array}{r} 4.5 \\ \times\ 0.5 \\ \hline \end{array}$ **2.25**
14. $\begin{array}{r} 0.31 \\ \times\ 0.7 \\ \hline \end{array}$ **0.217**
15. $\begin{array}{r} 1.6 \\ \times\ 0.08 \\ \hline \end{array}$ **0.128**

16. 0.007×0.06 **0.00042**
17. 0.04×3.0 **0.12**
18. 2.0×0.006 **0.012**
19. 0.005×0.003 **0.000015**

See Example 3 Evaluate $7x$ for each value of x.

20. $x = 1.903$ **13.321**
21. $x = 2.461$ **17.227**
22. $x = 3.72$ **26.04**
23. $x = 0.164$ **1.148**
24. $x = 5.89$ **41.23**
25. $x = 0.3702$ **2.5914**

PRACTICE AND PROBLEM SOLVING

Multiply.

26. 0.3×0.03 **0.009**
27. 1.4×0.21 **0.294**
28. 0.06×1.02 **0.0612**
29. 12.6×2.1 **26.46**
30. 3.04×0.6 **1.824**
31. 0.66×2.52 **1.6632**
32. $0.2 \times 0.94 \times 1.3$ **0.2444**
33. $1.54 \times 3.05 \times 2.6$ **12.2122**

Evaluate.

34. $6n$ for $n = 6.23$ **37.38**
35. $5t + 0.462$ for $t = 3.04$ **15.662**
36. $8^2 - 2b$ for $b = 0.95$ **62.1**
37. $4^3 + 5c$ for $c = 1.9$ **73.5**

RETEACH 3-6

LESSON 3-6 **Reteach**
Multiplying Decimals

You can use a model to help you multiply a decimal by a whole number.

Find the product of 0.12 and 4, using a 10 by 10 grid.

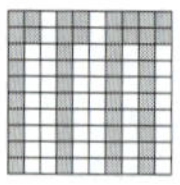

Shade 4 groups of 12 squares. Count the number of shaded squares. Since you have shaded 48 of the 100 squares, 0.12 • 4 = 0.48.

Find each product.

1. 0.23 • 3 **0.69**
2. 0.41 • 2 **0.82**
3. 0.011 • 5 **0.055**
4. 0.32 • 2 **0.64**
5. 0.15 • 3 **0.45**
6. 0.42 • 2 **0.84**
7. 0.04 • 8 **0.32**
8. 0.22 • 4 **0.88**

You can also use a model to help you multiply a decimal by a decimal.

Find the product of 0.4 and 0.6.
0.4 • 0.6 = 0.24

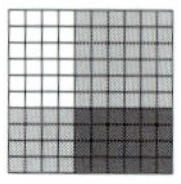

Find each product.

9. 0.2 • 0.8 **0.16**
10. 0.7 • 0.9 **0.63**
11. 0.5 • 0.5 **0.25**
12. 0.3 • 0.6 **0.18**
13. 0.5 • 0.2 **0.1**
14. 0.4 • 0.4 **0.16**
15. 0.1 • 0.9 **0.09**
16. 0.4 • 0.7 **0.28**

PRACTICE 3-6

LESSON 3-6 **Practice B**
Multiplying Decimals

Find each product.

1. $\begin{array}{r} 0.7 \\ \times\ 0.3 \\ \hline \end{array}$ **0.21**
2. $\begin{array}{r} 0.05 \\ \times\ 0.4 \\ \hline \end{array}$ **0.02**
3. $\begin{array}{r} 8.0 \\ \times\ 0.02 \\ \hline \end{array}$ **0.16**
4. $\begin{array}{r} 3.5 \\ \times\ 0.2 \\ \hline \end{array}$ **0.7**
5. $\begin{array}{r} 12.1 \\ \times\ 0.01 \\ \hline \end{array}$ **0.121**
6. $\begin{array}{r} 9.0 \\ \times\ 0.9 \\ \hline \end{array}$ **8.1**
7. 0.04 • 0.58 **0.0232**
8. 2.15 • 1.5 **3.225**
9. 1.73 • 0.8 **1.384**
10. 6.017 • 2.0 **12.034**
11. 3.96 • 0.4 **1.584**
12. 0.7 • 0.009 **0.0063**

Evaluate $8x$ for each value of x.

13. $x = 0.5$ **4**
14. $x = 2.3$ **18.4**
15. $x = 0.74$ **5.92**
16. $x = 3.12$ **24.96**
17. $x = 0.587$ **4.696**
18. $x = 14.08$ **112.64**

19. The average mail carrier walks 4.8 kilometers in a workday. How far do most mail carriers walk in a 6-day week? There are 27 working days in July, so how far will a mail carrier walk in July? **28.8 kilometers; 129.6 kilometers**

20. A deli charges \$3.45 for a pound of turkey. If Tim wants to purchase 2.4 pounds, how much will it cost? **\$8.28**

Physical Science LINK

Saturn is the second-largest planet in the solar system. Saturn is covered by thick clouds, and it is thought that there is no Earth-like surface beneath the clouds. Saturn's density is very low. Suppose you weigh 180 pounds on Earth. If you were able to stand on top of Saturn's clouds, you would weigh only 165 pounds. To find the weight of an object on another planet, multiply its weight on Earth by the gravitational pull listed in the table.

go.hrw.com
KEYWORD: MR4 Saturn
CNN Student News

Gravitational Pull of Planets (Compared with Earth)	
Planet	**Gravitational Pull**
Mercury	0.38
Venus	0.91
Mars	0.38
Jupiter	2.54
Saturn	0.93
Neptune	1.2

38. Christopher found a rock that weighs 5 pounds on Earth. How much would the rock weigh on Saturn? **4.65 pounds**

39. On which two planets would the weight of an object be the same? **Mercury and Mars**

40. A cat weighs 14 pounds on Earth. Would the cat weigh more or less on Venus? Explain. **Less; The gravitational pull on Venus is less than on Earth.**

41. An object weighs 9 pounds on Earth. How much would this object weigh on Mars? **3.42 pounds**

42. ***WRITE A PROBLEM*** Use the data in the table to write a word problem that can be answered by evaluating an expression with multiplication. Solve your problem.

43. ***WHAT'S THE ERROR?*** A student said that his new baby brother, who weighs 10 pounds, would weigh 120 pounds on Neptune. What is the error? Write the correct answer.

44. ***CHALLENGE*** An object weighs between 2.79 lb and 5.58 lb on Saturn. Give a range for the object's weight on Earth. **Possible answer: 2.79 ÷ 0.93 = 3, and 5.58 ÷ 0.93 = 6; the object must weigh between 3 lb and 6 lb.**

Galileo Galilei was the first person to look at Saturn through a telescope. He thought there were groups of stars on each side of the planet, but it was later determined that he had seen Saturn's rings.

43. Possible answer: The student forgot to place the decimal point in the answer. It should be 12 lb.

Spiral Review

Find a pattern in each sequence. (Lesson 1-7)

45. 1, 7, 3, 21, 17, 119, 115, ... **Multiply by 7, subtract 4.**

46. 1, 3^2, 25, 49, 9^2, 121, 169, ... **consecutive odd numbers squared**

Write each phrase as a numerical or algebraic expression. (Lesson 2-2)

47. the sum of 163 and 24 **163 + 24**

48. the product of 15 and *c* **15*c***

49. *y* divided by 8 **$y \div 8$ or $\frac{y}{8}$**

50. **TEST PREP** What is the sum of 5.6004 + 3.458? (Lesson 3-3) **A**

A 9.0584 **B** 5.9462 **C** 9.584 **D** 8.6462

Interdisciplinary LINK

Physical Science

Exercises 38–44 involve using information about the gravitational pull on various planets. The force of gravity is studied in middle school physical science programs such as *Holt Science & Technology*.

Answers

42. Possible answer: If an object weighs 100 pounds on Earth, how much will that object weigh on Jupiter? 2.54 × 100 = 254 pounds

Journal

Have students write about situations in a grocery store in which decimal numbers are multiplied by whole numbers or decimal numbers.

Test Prep Doctor

For Exercise 50, students need to remember to align the addends on the decimal points. Students who answered **B** aligned the digits on the right instead of at the decimal points. Students who answered **C** aligned the addends correctly but dropped the zero after the decimal point in the sum. Students who answered **D** aligned the addends correctly, but they put a zero as a placeholder in the second addend after the decimal point instead of at the end of the number.

CHALLENGE 3-6

LESSON 3-6 **Challenge**
Decimal Growth

Use the growth rate for each plant below to find how much it will grow in 1 week.

Eucalyptus Tree	Bristlecone Pine Tree	Trumpet Tree
Growth Rate: 2.5 cm per day	Growth Rate: 0.009 mm per day	Growth Rate: 0.28 in. per day
17.5 cm	0.063 mm	1.96 in

Use the growth rate for each plant below to find how much it will grow in 0.25 day.

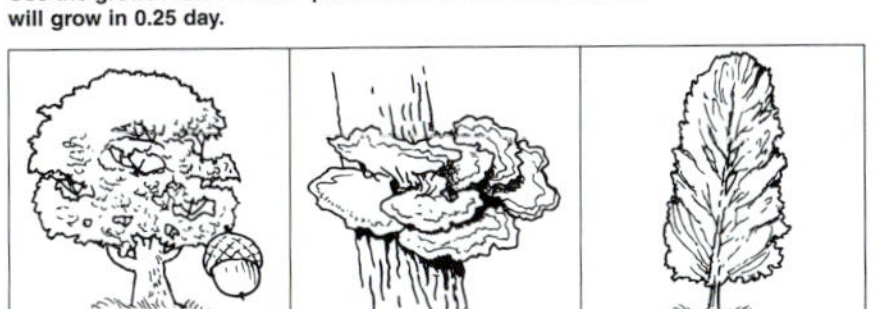

Oak Tree	Lichens	Poplar Tree
Growth Rate: 1.4 mm per day	Growth Rate: 0.0025 mm per day	Growth Rate: 0.118 in. per day
0.35 mm	0.000625 mm	0.0295 in.

PROBLEM SOLVING 3-6

LESSON 3-6 **Problem Solving**
Multiplying Decimals

Use the table to answer the questions.

United States Minimum Wage	
Year	**Hourly Rate**
1940	$0.30
1950	$0.75
1960	$1.00
1970	$1.60
1980	$3.10
1990	$3.80
2000	$5.15

1. At the minimum wage, how much did a person earn for a 40-hour workweek in 1950? **$30.00**
2. At the minimum wage, how much did a person earn for working 25 hours in 1970? **$40.00**
3. If you had a minimum-wage job in 1990, and worked 15 hours a week, how much would you have earned each week? **$57.00**
4. About how many times higher was the minimum wage in 1960 than in 1940? **about 3 times**
5. Ted's grandfather had a minimum-wage job at his neighborhood bakery in 1940. He worked 40 hours a week for the entire year. How much did Ted's grandfather earn at the bakery in 1940? **$624.00**
6. Companies pay minimum wage workers time-and-a-half for any hours worked over 40 per week. This means that the worker is paid their hourly wage plus half of that wage. If a person in 1960 worked 42 hours one week, how much did he or she earn? **$43.00**

Circle the letter for the correct answer.

7. Having one dollar in 1960 is equivalent to having $5.82 today. If you worked 40 hours a week in 1960 at minimum wage, how much would your weekly earnings be worth today?
   - A $40.00
   - B $5.82
   - **(C) $232.80**
   - D $2,328.00
8. In 2000, Cindy had a part-time job at a florist, where she earned minimum wage. She worked 18 hours each week for the whole year. How much did she earn from this job in 2000?
   - F $927.00
   - **(G) $4,820.40**
   - H $10,712.00
   - J $2,142.40

Lesson Quiz

Find each product.

1. 0.8 × 0.07 **0.056**
2. 0.006 × 0.07 **0.00042**

Evaluate 8*x* for each value of *x*.

3. *x* = 2.705 **21.64**
4. *x* = 0.804 **6.432**
5. "Pick your own" peaches sell for $0.95 per pound. You picked 92 pounds of peaches. How much were you charged? **$87.40**

Available on Daily Transparency in CRB

3-7 Organizer

Pacing: Traditional 1 day
Block $\frac{1}{2}$ day

Objective: Students divide decimals by whole numbers.

Indiana Academic Standards
6.2.3, 6.5.10, 6.7.9, 6.7.10

Warm Up

Divide.

1. 56,000 ÷ 8 **7,000**
2. 5,219 ÷ 17 **307**
3. 9,180 ÷ 12 **765**

Problem of the Day

In his pocket, Bill has $0.77 made up of 10 coins. What are the coins?
1 quarter, 3 dimes, 4 nickels, and 2 pennies

Available on Daily Transparency in CRB

Math Humor

Why is the student holding his math book to his ears and dancing? The teacher told him the book contained some good algorithms.

3-7 Dividing Decimals by Whole Numbers

Learn to divide decimals by whole numbers.

Ethan and two of his friends are going to share equally the cost of making a sculpture for the art fair.

To find how much each person should pay for the materials, you will need to divide a decimal by a whole number.

EXAMPLE 1 Dividing a Decimal by a Whole Number

Find each quotient.

A 0.75 ÷ 5

$$\begin{array}{r} 0.15 \\ 5\overline{)0.75} \\ -\ 5 \\ 25 \\ -\ 25 \\ 0 \end{array}$$

Place a decimal point in the quotient directly above the decimal point in the dividend.
Divide as you would with whole numbers.

B 2.52 ÷ 3

$$\begin{array}{r} 0.84 \\ 3\overline{)2.52} \\ -\ 2\ 4 \\ 12 \\ -\ 12 \\ 0 \end{array}$$

Place a decimal point in the quotient directly above the decimal point in the dividend.
Divide as you would with whole numbers.

EXAMPLE 2 Evaluating Decimal Expressions

Evaluate 0.435 ÷ x for each given value of x.

A $x = 3$

0.435 ÷ x
0.435 ÷ 3 *Substitute 3 for x.*

$$\begin{array}{r} 0.145 \\ 3\overline{)0.435} \\ -\ 3 \\ 13 \\ -\ 12 \\ 15 \\ -\ 15 \\ 0 \end{array}$$

Divide as you would with whole numbers.

B $x = 15$

0.435 ÷ x
0.435 ÷ 15 *Substitute 15 for x.*

$$\begin{array}{r} 0.029 \\ 15\overline{)0.435} \\ -\ 0 \\ 43 \\ -\ 30 \\ 135 \\ -135 \\ 0 \end{array}$$

Sometimes you need to use a zero as a placeholder.
15 > 4, so place a zero in the quotient and divide 15 into 43.

1 Introduce

Alternate Opener

EXPLORATION

3-7 Dividing Decimals by Whole Numbers

1. Four friends go on a vacation together. They decide to share all expenses evenly. Estimate the per person cost of each item, and then compute the actual cost with a calculator.

Item	Total Cost	Estimated Cost per Person	Actual Cost per Person
Cab fare	$50.00		
Pizza	$13.92		
Movie rental	$10.00		
Dinner	$76.20		
Boat ride	$35.96		

Estimate each quotient. Use this estimation to decide where to place a decimal point in the answer. Check with your calculator.

2. 125.2 ÷ 25 = 5 0 0 8
3. 40 ÷ 16 = 2 5
4. 7.5 ÷ 5 = 1 5
5. 75 ÷ 12 = 6 2 5

Think and Discuss

6. **Discuss** your strategies for estimating in number 1.
7. **Explain** how you know where to place the decimal point in a quotient.

Motivate

Have students give examples of how to use the division algorithm to divide 3-digit and 4-digit numbers by 1-digit and 2-digit numbers. Explain that the algorithm for dividing decimals is the same; however, students must make sure that the decimal point is placed in the quotient properly.

Exploration worksheet and answers on Chapter 3 Resource Book pp. 65 and 125

2 Teach

Lesson Presentation

Guided Instruction

In this lesson, students learn to divide decimals by whole numbers. First teach them to divide a decimal by a whole number using the division algorithm, and have them check the answer. Then teach them to evaluate division expressions for given whole number values of x. Finally, explain how to apply division in a consumer application.

EXAMPLE 3 Consumer Application

Ethan and two of his friends are making a papier-mâché sculpture using balloons, strips of paper, and paint. The materials cost $11.61. If they share the cost equally, how much should each person pay?

$11.61 should be divided into three equal groups.
Divide $11.61 by 3.

```
   $3.87
3)$11.61     Place a decimal point in the quotient directly
  − 9        above the decimal point in the dividend.
    2 6
  − 2 4      Divide as you would with whole numbers.
      21
    − 21
       0
```

Check

3.87 × 3 = 11.61

Each person should pay $3.87.

Remember!

divisor)dividend with *quotient* above

Multiplication can "undo" division. To check your answer to a division problem, multiply the divisor by the quotient.

Think and Discuss

1. **Tell** how you know where to place the decimal point in the quotient.
2. **Explain** why you can use multiplication to check your answer to a division problem.

Additional Examples

Example 1

Find each quotient.

A. 0.84 ÷ 3 — 0.28
B. 3.56 ÷ 4 — 0.89

Example 2

Evaluate 0.936 ÷ *x* for each given value of *x*.

A. $x = 9$ — 0.104
B. $x = 18$ — 0.052

Example 3

Jodi and three of her friends are making a tile design. The materials cost $10.12. If they share the cost equally, how much should each person pay? Each person should pay $2.53.

3-7 Exercises

FOR EXTRA PRACTICE see page 641

internet connect — Homework Help Online go.hrw.com Keyword: MR4 3-7

GUIDED PRACTICE

See Example 1 Find each quotient.

1. 1.38 ÷ 6 — 0.23
2. 0.96 ÷ 8 — 0.12
3. 1.75 ÷ 5 — 0.35
4. 0.72 ÷ 4 — 0.18

See Example 2 Evaluate 0.312 ÷ *x* for each given value of *x*.

5. $x = 4$ — 0.078
6. $x = 6$ — 0.052
7. $x = 3$ — 0.104
8. $x = 12$ — 0.026

See Example 3
9. Mr. Richards purchased 8 T-shirts for the volleyball team. The total cost of the T-shirts was $70.56. How much did each shirt cost? $8.82

3-7 PRACTICE & ASSESS

Assignment Guide

If you finished Example 1 assign:
Core 1–4, 10–13, 34–39
Enriched 19–24, 32–39

If you finished Example 2 assign:
Core 1–8, 10–17, 27, 30, 34–39
Enriched 14–17, 19–28, 31–39

If you finished Example 3 assign:
Core 1–18, 29–31, 34–39
Enriched 14–39

Reaching All Learners

Through Visual Organizers

If students are having difficulty keeping a division problem organized, encourage them to use graph paper (provided on page 107 of the Chapter 3 Resource Book) for setting up and working their division problems. Have them write each digit in a separate square to maintain the alignment of columns and rows.

3 Close

Summarize

Explain that the steps to follow when dividing a decimal by a whole number are the same steps involved in division of whole numbers except for the placement of the decimal in the quotient. Remind and encourage the students to check their solutions.

Answers to Think and Discuss

1. Place the decimal point in the quotient directly above the decimal point in the dividend.
2. Possible answer: Multiplication can "undo" division. The product of the divisor and the quotient should be the dividend.

Lesson Quiz

Find each quotient.

1. 3.12 ÷ 8 **0.39**

2. 5.68 ÷ 8 **0.71**

Evaluate the expression 1.25 ÷ *x* for the given value of *x*.

3. $x = 5$ **0.25**

4. $x = 25$ **0.05**

5. The tennis team is having 3 tennis rackets restrung. The total cost is $54.75. What is the average cost per racket? **$18.25**

Available on Daily Transparency in CRB

INDEPENDENT PRACTICE

See Example 1 — Find each quotient.

10. 0.91 ÷ 7 **0.13** **11.** 1.32 ÷ 6 **0.22** **12.** 4.68 ÷ 9 **0.52** **13.** 0.81 ÷ 3 **0.27**

See Example 2 — Evaluate 0.684 ÷ x for each given value of x.

14. $x = 3$ **0.228** **15.** $x = 4$ **0.171** **16.** $x = 18$ **0.038** **17.** $x = 9$ **0.076**

See Example 3 — **18.** Charles, Kate, and Kim eat lunch in a restaurant. The bill is $27.12. If they share the bill equally, how much will each person pay? **$9.04**

PRACTICE AND PROBLEM SOLVING

Divide.

19. 3.6 ÷ 4 **0.9** **20.** 15.35 ÷ 5 **3.07** **21.** 12.8592 ÷ 6 **2.1432** **22.** 0.729 ÷ 3 **0.243**

Find the value of each expression.

23. (0.49 + 0.0045) ÷ 5 **0.0989** **24.** (4.9 − 3.125) ÷ 5 **0.355**

Evaluate the expression $x \div 4$ for each value of x.

25. $x = 0.504$ **0.126** **26.** $x = 0.944$ **0.236** **27.** $x = 57.484$ **14.371** **28.** $x = 1.648$ **0.412**

29. Dance lessons cost $198.75 for 15 classes. What is the fee for one class? **$13.25**

30. At the grocery store, a 6 lb bag of oranges costs $2.04. Is this more or less expensive than the price shown at the farmers' market?

30. More expensive. 6 lb at farmers' market cost $1.80.

31. ***CHOOSE A STRATEGY*** Sarah had $1.19 in coins. Jeff asked her for change for a dollar, but she did not have the correct change. What coins did she have?
Possible answer: 3 quarters, 4 dimes, and 4 pennies

32. ***WRITE ABOUT IT*** When do you use a placeholder zero in the quotient?

32. when the divisor is greater than the portion of the dividend being divided into

33. ***CHALLENGE*** Evaluate the expression $x \div 2$ for the following values of x: 520, 52, and 5.2. Try to predict the value of the same expression for $x = 0.52$.
260, 26, 2.6; 0.26

Spiral Review

Order the numbers from least to greatest. (Lesson 1-1)

34. 3,673,809; 3,708,211; 3,671,935
3,671,935; 3,673,809; 3,708,211

35. 2,004,801; 225,971; 298,500,004
225,971; 2,004,801; 298,500,004

Compare. Write <, >, or =. (Lesson 3-1)

36. 7.0893 □ 7.0798 **>** **37.** 0.0312 □ 0.211 **<** **38.** 0.9571 □ 1.308 **<**

39. **TEST PREP** What is 56,930 expressed in scientific notation? (Lesson 3-5) **C**

A 0.5693×10^5 **B** 56.930×10^3 **C** 5.693×10^4 **D** 5.6930×10^3

RETEACH 3-7

LESSON 3-7 **Reteach**
Dividing Decimals by Whole Numbers

You can use decimal grids to help you divide decimals by whole numbers.

To divide 0.35 by 7, first shade in a decimal grid to show thirty-five hundredths.

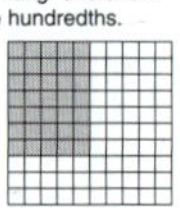

0.35 ÷ 7 means "divide 0.35 into 7 equal groups." Show this on the decimal grid.

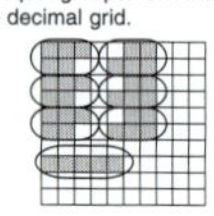

The number of units in each group is the quotient.

So, 0.35 ÷ 7 = 0.05.

Use decimal grids to find each quotient.

1. 0.24 ÷ 4

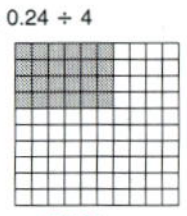

0.06

2. 0.48 ÷ 12

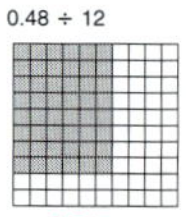

0.04

3. 0.50 ÷ 10

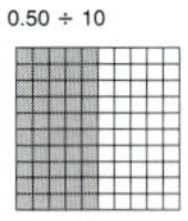

0.05

4. 0.98 ÷ 7

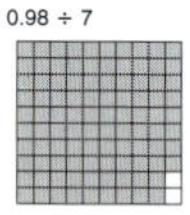

0.14

5. 0.6 ÷ 5 **0.12** **6.** 0.78 ÷ 6 **0.13** **7.** 0.99 ÷ 11 **0.09** **8.** 0.32 ÷ 4 **0.08**

PRACTICE 3-7

LESSON 3-7 **Practice B**
Dividing Decimals by Whole Numbers

Find each quotient.

1. 0.81 ÷ 9 **0.09** **2.** 1.84 ÷ 4 **0.46** **3.** 7.2 ÷ 6 **1.2**

4. 13.6 ÷ 8 **1.7** **5.** 4.55 ÷ 5 **0.91** **6.** 29.6 ÷ 8 **3.7**

7. 15.57 ÷ 9 **1.73** **8.** 0.144 ÷ 12 **0.012** **9.** 97.5 ÷ 3 **32.5**

10. 0.0025 ÷ 5 **0.0005** **11.** 2.84 ÷ 8 **0.355** **12.** 18.9 ÷ 3 **6.3**

Evaluate the expression 2.094 ÷ *x* for the given value of *x*.

13. $x = 2$ **1.047** **14.** $x = 4$ **0.5235** **15.** $x = 12$ **0.1745**

16. $x = 20$ **0.1047** **17.** $x = 15$ **0.1396** **18.** $x = 30$ **0.0698**

19. There are three grizzly bears in the city zoo. Yogi weighs 400.5 pounds, Winnie weighs 560.35 pounds, and Nyla weighs 618.29 pounds. What is the average weight of the three bears? **526.38 pounds**

20. The bill for dinner came to $75.48. The four friends decided to leave a $15.00 tip. If they shared the bill equally, how much will they each pay? **$22.62**

CHALLENGE 3-7

LESSON 3-7 **Challenge**
Get the Best Deal

Grocery stores often sell items in different quantities, package sizes, and unit prices. A unit price is the price for one unit of an item. To get the best deal, you should buy each item with the lowest unit price. Find each unit price and determine the best deal.

	1 for $0.69	6 for $2.70	12 for $4.80
	Unit price (per pound) **$0.69**	Unit price (per pound) **$0.45**	Unit price (per pound) **$0.40**

Best deal: **12 for $4.80**

	1 pound for $0.75	2 pounds for $1.70	5 pounds for $4.05
	Unit price (per pound) **$0.75**	Unit price (per pound) **$0.85**	Unit price (per pound) **$0.81**

Best deal: **1 pound for $0.75**

	6-ounce box for $1.98	12-ounce box for $3.72	16-ounce box for $5.28
	Unit price (per ounce) **$0.33**	Unit price (per ounce) **$0.31**	Unit price (per ounce) **$0.33**

Best deal: **12-ounce box for $3.72**

	6-pack for $1.08	12-pack for $2.64	24-pack for $4.08
	Unit price (per can) **$0.18**	Unit price (per can) **$0.22**	Unit price (per can) **$0.17**

Best deal: **24-pack for $4.08**

PROBLEM SOLVING 3-7

LESSON 3-7 **Problem Solving**
Dividing Decimals by Whole Numbers

Write the correct answer.

1. Four friends had lunch together. The total bill for lunch came to $33.40, including tip. If they shared the bill equally, how much did they each pay? **$8.35**

2. There are 7.2 milligrams of iron in a dozen eggs. Because there are 12 eggs in a dozen, how many milligrams of iron are in 1 egg? **0.6 milligrams**

3. Kyle bought a sheet of lumber 8.7 feet long to build fence rails. He cut the strip into 3 equal pieces. How long is each piece? **2.9 feet**

4. An albatross has a wingspan greater than the length of a car—3.7 meters! Wingspan is the length from the tip of one wing to the tip of the other wing. What is the length of each albatross wing (assuming wing goes from center of body)? **1.85 meters**

5. The City Zoo feeds its three giant pandas 181.5 pounds of bamboo shoots every day. Each panda is fed the same amount of bamboo. How many pounds of bamboo does each panda eat every day? **60.5 pounds**

6. Emma bought 22.5 yards of cloth to make curtains for two windows in her apartment. She used the same amount of cloth on each window. How much cloth did she use to make each set of curtains? **11.25 yards**

Circle the letter of the correct answer.

7. Aerobics classes cost $153.86 for 14 sessions. What is the fee for one session?
(A) $10.99
B $1.99
C about $25.00
D about $20.00

8. An entire apple pie has 36.8 grams of saturated fat. If the pie is cut into 8 slices, how many grams of saturated fat are in each slice?
F 4.1 grams
G 0.46 grams
(H) 4.6 grams
J 4.11 grams

3-8 Dividing by Decimals

Learn to divide whole numbers and decimals by decimals.

Julie and her family traveled to the Grand Canyon. They stopped to refill their gas tank with 13.4 gallons of gasoline after they had driven 368.5 miles.

To find the miles that they drove per gallon, you will need to divide a decimal by a decimal.

EXAMPLE 1 Dividing a Decimal by a Decimal

Find each quotient.

A $3.6 \div 1.2$

$1.2\overline{)3.6}$ — *Multiply the divisor and dividend by the same power of ten.*

$$\begin{array}{r} 3 \\ 12\overline{)36} \\ -\,36 \\ \hline 0 \end{array}$$

There is one decimal place in the divisor. Multiply by 10^1, or 10.

Think: $1.2 \times 10 = 12$ $3.6 \times 10 = 36$

Divide.

B $41.6 \div 0.39$

$0.39\overline{)41.6}$ — *Make the divisor a whole number by multiplying the divisor and dividend by 10^2, or 100.*

Think: $0.39 \times 100 = 39$ $41.6 \times 100 = 4{,}160$

$$\begin{array}{r} 106.66 \\ 39\overline{)4160.00} \\ -\,39 \\ \hline 26 \\ -\,0 \\ \hline 260 \\ -\,234 \\ \hline 26\,0 \\ -\,23\,4 \\ \hline 2\,60 \\ -\,2\,34 \\ \hline 26 \end{array}$$

Place the decimal point in the quotient. Divide.

When there is a remainder, place a zero after the decimal point in the dividend and continue to divide.

$106.66\ldots = 106.\overline{6}$ — *When a repeating pattern occurs, show three dots or draw a bar over the repeating part of the quotient.*

Helpful Hint

Multiplying the divisor and the dividend by the same number does not change the quotient.

$42 \div 6 = 7$
$\times 10 \downarrow \quad \times 10 \downarrow$
$420 \div 60 = 7$

$42 \div 6 = 7$
$\times 100 \downarrow \quad \times 100 \downarrow$
$4{,}200 \div 600 = 7$

3-8 Organizer

Pacing: Traditional 1 day
Block $\frac{1}{2}$ day

Objective: Students divide whole numbers and decimals by decimals.

Indiana Academic Standards
6.2.3, 6.5.10, 6.7.1, 6.7.9

Warm Up

Divide.

1. 4.8 ÷ 2 **2.4**
2. 16.1 ÷ 7 **2.3**
3. 0.36 ÷ 3 **0.12**
4. 25.28 ÷ 4 **6.32**
5. 6.25 ÷ 5 **1.25**

Problem of the Day

In the following magic square, 3.375 is the product of the numbers in every row, column, and diagonal. Fill in the missing numbers.

4.5	0.25	**3**
1	**1.5**	**2.25**
0.75	**9**	0.5

Available on Daily Transparency in CRB

Math Humor

What kinds of decimals do the most talking? Repeating decimals

1 Introduce

Alternate Opener

EXPLORATION

3-8 Dividing by Decimals

A CD store carries the packages of recordable CDs listed in the table. Use a calculator to find the cost of 1 CD and the purchasing power of $1.00 for each package.

	Item	Cost	Cost of 1 CD	Purchasing Power of $1.00
1.	Single CD	$1.19		
2.	5-pack	$4.95	4.95 ÷ 5 = 0.99	5 ÷ 4.95 = 1.01
3.	10-pack	$8.95		
4.	20-pack	$16.95		
5.	50-pack	$35.95		

The third column above shows the cost of one CD for different packages. This is called the ***unit cost.*** The fourth column shows how many CDs one dollar can buy. For example, if you buy the 5-pack at $4.95, one dollar buys 1.01 CD (a little more than one CD). This is called the ***purchasing power*** of $1.00.

6. Which package gives you the highest unit cost?
7. Which package gives you the lowest unit cost?
8. Which package gives you the greatest purchasing power per dollar? the least purchasing power per dollar?

Think and Discuss

9. **Describe** the relationship between unit cost and purchasing power by looking at the numbers in the third and fourth columns above.

Motivate

Solve the following problem for the class: 760 ÷ 8 **(95)**. Ask a student to use a calculator to solve 7,600 ÷ 80 **(95)**. Explain that multiplying the dividend and divisor by the same number, 10, did not change the solution. This fact is a key component of this lesson.

Exploration worksheet and answers on Chapter 3 Resource Book pp. 74 and 127

2 Teach

Lesson Presentation

Guided Instruction

In this lesson, students learn to divide whole numbers and decimals by decimals. First teach students to make the divisor a whole number when dividing a decimal by a decimal. Then teach them the notation used to signify repeating decimals. Finally, teach students to solve an application problem involving average gas mileage.

Teaching Tip Remind students that both the divisor *and* the dividend must be multiplied by the same power of ten to get the correct quotient.

COMMON ERROR ALERT

Some students may position the decimal for the quotient before multiplying the divisor and dividend by a power of ten. Remind them that the decimal is positioned only after multiplication by a power of ten has occurred.

Additional Examples

Example 1

Find each quotient.

A. 5.2 ÷ 1.3
4

B. 61.3 ÷ 0.36
$170.2\overline{7}$

Example 2

After driving 216.3 mi, the Yorks filled up with 10.5 gal of gas. On average, how many miles did they drive per gallon of gas?
They averaged 20.6 miles per gallon.

EXAMPLE 2 PROBLEM SOLVING APPLICATION

PROBLEM SOLVING

After driving 368.5 miles, Julie and her family refilled the tank of their car with 13.4 gallons of gasoline. On average, how many miles did they drive per gallon of gas?

1 Understand the Problem

The **answer** will be the average number of miles per gallon.

List the **important information:**

- They drove 368.5 miles.
- They used 13.4 gallons of gas.

2 Make a Plan

Solve a simpler problem by replacing the decimals in the problem with whole numbers.

If they drove 10 miles using 2 gallons of gas, they averaged 5 miles per gallon. You need to divide miles by gallons to solve the problem.

3 Solve

13.4)368.5 — *Multiply the divisor and the dividend by 10. Think: 13.4 × 10 = 134 368.5 × 10 = 3,685*

```
       27.5
134)3685.0
   − 268
     1005
   −  938
      67 0
    − 67 0
         0
```

Place the decimal point in the quotient.
Divide.

Julie and her family averaged 27.5 miles per gallon.

4 Look Back

Use compatible numbers to estimate the quotient.

368.5 ÷ 13.4 → 360 ÷ 12 = 30

The answer is reasonable, since 27.5 is close to the estimate of 30.

Think and Discuss

1. Tell how the quotient of 48 ÷ 12 is similar to the quotient of 4.8 ÷ 1.2. How is it different?

3 Close

Summarize

Explain how to change the divisor and the dividend to solve 2.4 ÷ 0.9.
(Multiply the divisor, 0.9, by 10 to make it a whole number. Multiply the dividend, 2.4, by 10. Divide: 24 ÷ 9 = 2.6... .)
Remind them that the placement of the decimal point must follow the multiplication by a power of ten.

Answers to Think and Discuss

1. 48 ÷ 12 and 4.8 ÷ 1.2 both have a quotient of 4. The difference is that the first expression contains whole numbers and the second contains decimals.

Reaching All Learners

Through Grouping Strategies

Have students work in groups of three or four. Give each group a set of index cards labeled with digits 0–9 from Chapter 3 Resource Book, p. 111. Have the groups mix and place the cards face down in a pile. Students should then draw three cards to make a dividend and two cards to make a divisor. Have students take turns determining where to place decimal points in each number and then do the division individually. Group members should compare answers and work a problem together if they do not all get the same quotient.

3-8 Exercises

FOR EXTRA PRACTICE
see page 641

internet connect
Homework Help Online
go.hrw.com Keyword: MR4 3-8

GUIDED PRACTICE

See Example 1
Find each quotient.

1. 6.5 ÷ 1.3 **5**
2. 20.7 ÷ 0.6 **34.5**
3. 25.5 ÷ 1.5 **17**
4. 5.4 ÷ 0.9 **6**
5. 13.2 ÷ 2.2 **6**
6. 63.39 ÷ 0.24 **264.125**

See Example 2

7. Marcus drove 354.9 miles in 6.5 hours. On average, how many miles per hour did he drive? **54.6 miles/hour**
8. Anthony spends $87.75 on shrimp. The shrimp cost $9.75 per pound. How many pounds of shrimp does Anthony buy? **9 pounds**

INDEPENDENT PRACTICE

See Example 1
Find each quotient.

9. 3.6 ÷ 0.6 **6**
10. 8.2 ÷ 0.5 **16.4**
11. 18.4 ÷ 2.3 **8**
12. 4.8 ÷ 1.2 **4**
13. 51.2 ÷ 0.24 **213.33...**
14. 32.5 ÷ 2.6 **12.5**
15. 50.9 ÷ 4.5 **$11.3\overline{1}$**
16. 91.6 ÷ 0.45 **203.55...**
17. 6.5 ÷ 1.3 **5**

See Example 2

18. Jen spends $5.98 on ribbon. Ribbon costs $0.92 per meter. How many meters of ribbon does Jen buy? **6.5 meters**
19. Kyle's family drove 329.44 miles. Kyle calculated that the car averaged 28.4 miles per gallon of gas. How many gallons of gas did the car use? **11.6 gallons of gas**
20. Peter is saving $4.95 each week to buy a DVD that costs $24.75, including tax. For how many weeks will he have to save? **5 weeks**

PRACTICE AND PROBLEM SOLVING

Divide.

21. 2.52 ÷ 0.4 **6.3**
22. 12.586 ÷ 0.35 **35.96**
23. 0.5733 ÷ 0.003 **191.1**
24. 10.875 ÷ 1.2 **9.0625**
25. 92.37 ÷ 0.5 **184.74**
26. 8.43 ÷ 0.12 **70.25**

Find the value of each expression.

27. $6.35 \times 10^2 \div 0.5$ **1,270**
28. $8.1 \times 10^2 \div 0.9$ **900**
29. $20.1 \times 10^3 \div 0.1$ **201,000**
30. $2.76 \times 10^2 \div 0.3$ **920**

Evaluate.

31. $0.732 \div n$ for $n = 0.06$ **12.2**
32. $73.814 \div c$ for $c = 1.3$ **56.78**
33. $b \div 0.52$ for $b = 6.344$ **12.2**
34. $r \div 4.17$ for $r = 10.5918$ **2.54**
35. $r \div 3.7$ for $r = 34.928$ **9.44**
36. $45.05 \div a$ for $a = 2.5$ **18.02**

RETEACH 3-8

LESSON 3-8 **Reteach**
Dividing by Decimals

You can use powers of ten to help you divide a decimal by a decimal.

To divide 0.048 by 0.12, first multiply each number by the least power of ten that makes the divisor a whole number.

0.048 ÷ 0.12

$0.12 \cdot 10^2 = 12$ Move the decimal point 2 places to the right.

$0.048 \cdot 10^2 = 4.8$ Move the decimal point 2 places to the right.

Then divide.

4.8 ÷ 12 — **Step 1:** Divide as you would divide a whole number by a whole number.

12)4.8 = 0.4; 4 8; 0 — **Step 2:** Think 48 ÷ 12 = 4.

Step 3: Bring the decimal into the quotient and add a zero placeholder if necessary.

So, 0.048 ÷ 0.12 = 0.4.

Find each quotient.

1. 0.7)0.42 **0.6**
2. 0.08)0.4 **5**
3. 0.5)0.125 **0.25**
4. 0.02)0.3 **15**
5. 0.4)0.08 **0.2**
6. 0.9)0.63 **0.7**
7. 0.008)0.4 **50**
8. 0.04)0.032 **0.8**
9. 0.3)0.06 **0.2**
10. 0.04)0.2 **5**
11. 0.007)4.9 **700**
12. 0.6)0.012 **0.02**

PRACTICE 3-8

LESSON 3-8 **Practice B**
Dividing by Decimals

Find each quotient.

1. 9.0 ÷ 0.9 **10**
2. 29.6 ÷ 3.7 **8**
3. 10.81 ÷ 2.3 **4.7**
4. 10.5 ÷ 1.5 **7**
5. 15.36 ÷ 4.8 **3.2**
6. 9.75 ÷ 1.3 **7.5**
7. 20.4 ÷ 5.1 **4**
8. 37.5 ÷ 2.5 **15**
9. 9.24 ÷ 1.1 **8.4**
10. 16.56 ÷ 6.9 **2.4**
11. 28.9 ÷ 8.5 **3.4**
12. 14.35 ÷ 0.7 **20.5**

Evaluate the expression $x \div 1.2$ for the given value of x.

13. $x = 40.8$ **34**
14. $x = 1.8$ **1.5**
15. $x = 10.8$ **9**
16. $x = 14.4$ **12**
17. $x = 4.32$ **3.6**
18. $x = 0.06$ **0.05**
19. Anna is saving $6.35 a week to buy a computer game that costs $57.15. How many weeks will she have to save to buy the game? **9 weeks**
20. Ben ran a 19.5-mile race last Saturday. His average speed during the race was 7.8 miles per hour. How long did it take Ben to finish the race? **2.5 hours**

3-8 PRACTICE & ASSESS

Students may want to refer back to the lesson examples.

Assignment Guide

If you finished Example 1 assign:
Core 1–6, 9–17, 21–23, 45–54
Enriched 15–17, 21–30, 42–54

If you finished Example 2 assign:
Core 1–20, 31–34, 45–54
Enriched 21–54

Notes

Math Background

To determine whether a quotient will be a terminating or repeating decimal, it is necessary only to examine the ratio of dividend to divisor in lowest whole terms. If the divisor can be expressed as a product of twos and fives, then the quotient will be a terminating decimal; otherwise, it will be a repeating decimal.

For example, the expression 0.04 ÷ 1.6 is translated by the division algorithm to 0.4 ÷ 16. In lowest terms, this is equivalent to the ratio 1:40 (0.4:16 = 4:160 = 1:40). Then, because 40 = 2 × 2 × 2 × 5, the quotient is a terminating decimal.

History LINK

The U.S. Mint was established by the Coinage Act in 1792. The first coins were copper and were made in Philadelphia.

37. ***HISTORY*** The U.S. Treasury first printed paper money in 1862. The paper money we use today is 0.0043 inch thick. Estimate the number of bills you would need to stack to make a pile that is 1 inch thick. If you stacked \$20 bills, what would be the total value of the money in the pile? **about 232 bills; about \$4,640**

38. ***EARTH SCIENCE*** A planet's year is the time it takes that planet to revolve around the Sun. A Mars year is 1.88 Earth years. If you are 13 years old in Earth years, about how old would you be in Mars years? **about 7**

Use the map for Exercises 39–41.

39. Bill drove from Washington, D.C., to Charlotte in 6.5 hours. What was his average speed in miles per hour? **63.5 mi/h**

40. Betty drove a truck from Richmond to Washington, D.C., without stopping. It took her about 2.5 hours. Estimate the average speed she was driving. **Possible range: 40 mi/h to 45 mi/h**

41. How far is it in miles from Washington, D.C., to Baltimore and back? **78.38 miles**

42. ***WHAT'S THE ERROR?*** A student incorrectly answered the division problem below. Explain the error and write the correct quotient.

$$0.004\overline{)53.824} \quad 13.456$$

The decimal point is misplaced. The correct quotient is 13,456.

43. ***WRITE ABOUT IT*** Explain how you know where to place the decimal point in the quotient when you divide by a decimal number. **The decimal point is placed above the decimal point in the dividend.**

44. ***CHALLENGE*** Find the value of a in the division problem.

$$0.4a3\overline{)0.41713} \quad 1.01$$

$a = 1$

Spiral Review

Use mental math to find each sum or product. (Lesson 1-5)

45. $8 \times 5 \times 9$ **360** **46.** $49 + 26 + 11 + 14$ **100** **47.** $4 \times 15 \times 6$ **360**

Solve for y. (Lesson 2-6)

48. $y - 23 = 40$ **$y = 63$** **49.** $14y = 168$ **$y = 12$** **50.** $36 + y = 53$ **$y = 17$**

Order the decimals from greatest to least. (Lesson 3-1)

51. 8.304, 8.009, 8.05 **8.304, 8.05, 8.009**
52. 15.34, 1.589, 5.62 **15.34, 5.62, 1.589**
53. 30.75, 30.211, 30.709 **30.75, 30.709, 30.211**

54. **TEST PREP** Find the product of 1.8×0.541. (Lesson 3-6) **A**

A 0.9738 **B** 9.738 **C** 97.3800 **D** 9.73×10^4

Journal

Have students explain the similarities and differences between dividing whole numbers by whole numbers and decimals by decimals.

Test Prep Doctor

For Exercise 54, encourage students to estimate the answer before looking at the answer choices. 1.8×0.541 can be estimated as $2 \times 0.5 = 1$, eliminating choices **B, C,** and **D.** Make sure students actually do the math before settling on **A** as the answer.

Lesson Quiz

Find each quotient.

1. $49 \div 0.7$ **70**
2. $21.63 \div 2.1$ **10.3**
3. $38.43 \div 6.1$ **6.3**
4. $16.9 \div 5.2$ **3.25**
5. John spent \$13.44 renting 4 videos for the weekend. What was the cost per video? **\$3.36**

Available on Daily Transparency in CRB

CHALLENGE 3-8

LESSON 3-8 **Challenge**
Cutting Decimals

The strips of cloth below need to be cut into equal pieces of given lengths. Draw lines on each strip of cloth to show how many pieces will be cut.

1. Total Length: 9.8 yards Piece Length: 1.4 yards

Students should draw 6 lines across the cloth to cut it into 7 equal pieces.

2. Total Length: 2.5 yards Piece Length: 0.5 yards

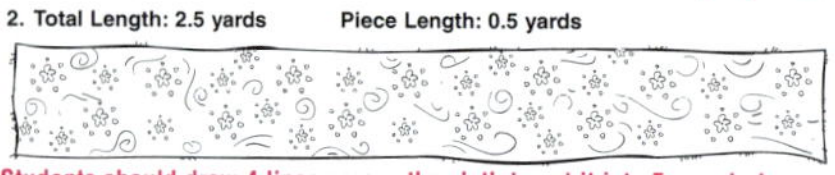

Students should draw 4 lines across the cloth to cut it into 5 equal pieces.

3. Total Length: 10.2 yards Piece Length: 1.7 yards

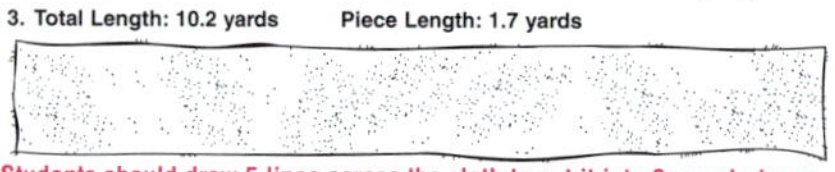

Students should draw 5 lines across the cloth to cut it into 6 equal pieces.

4. Total Length: 6.4 yards Piece Length: 0.8 yards

Students should draw 7 lines across the cloth to cut it into 8 equal pieces.

5. Total Length: 13.6 yards Piece Length: 3.4 yards

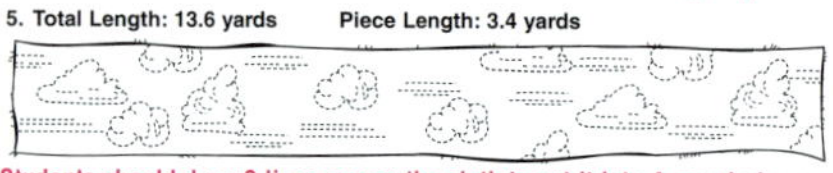

Students should draw 3 lines across the cloth to cut it into 4 equal pieces.

PROBLEM SOLVING 3-8

LESSON 3-8 **Problem Solving**
Dividing by Decimals

Write the correct answer.

1. Jamal spent \$6.75 on wire to build a rabbit hutch. Wire costs \$0.45 per foot. How many feet of wire did Jamal buy?
15 feet

2. Peter drove 195.3 miles in 3.5 hours. On average, how many miles per hour did he drive?
55.8 miles per hour

3. Lisa's family drove 830.76 miles to visit her grandparents. Lisa calculated that they used 30.1 gallons of gas. How many miles per gallon did the car average?
27.6 miles per gallon

4. A chef bought 84.5 pounds of ground beef. He uses 0.5 pound of ground beef for each hamburger. How many hamburgers can he make?
169 hamburgers

5. Mark earned \$276.36 for working 23.5 hours last week. He earned the same amount of money for each hour that he worked. What is Mark's hourly rate of pay?
\$11.76 per hour

6. Alicia wants to cover a section of her wall that is 2 feet wide and 12 feet long with mirrors. Each mirror tile is 2 feet wide and 1.5 feet long. How many mirror tiles does she need to cover that section?
8 tiles

Circle the letter of the correct answer.

7. John ran the city marathon in 196.5 minutes. The marathon is 26.2 miles long. On average, how many miles per hour did John run the race?
A 7 miles per hour
B 6.2 miles per hour
(C) 7.5 miles per hour
D 5.5 miles per hour

8. Shaneeka is saving \$5.75 of her allowance each week to buy a new camera that costs \$51.75. How many weeks will she have to save to have enough money to buy it?
(F) 9 weeks
G 9.5 weeks
H 8.1 weeks
J 8 weeks

3-9 Interpret the Quotient

Problem Solving Skill

Learn to solve problems by interpreting the quotient.

In science lab, Kim learned to make slime from corn starch, water, and food coloring. She has 0.87 kg of corn starch, and the recipe for one bag of slime calls for 0.15 kg. To find the number of bags of slime Kim can make, you need to divide.

EXAMPLE 1 *Measurement Application*

Kim will use 0.87 kg of corn starch to make gift bags of slime for her friends. If each bag requires 0.15 kg of corn starch, how many bags of slime can she make?

The question asks how many whole bags of slime can be made when the corn starch is divided into groups of 0.15 kg.

$0.87 \div 0.15 = ?$

$87 \div 15 = 5.8$

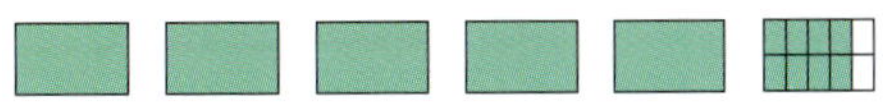

Think: The quotient shows that there is not enough to make 6 bags of slime that are 0.15 kg each. There is only enough for 5 bags. The decimal part of the quotient will not be used in the answer.

Kim can make 5 gift bags of slime.

Remember!

To divide decimals, first write the divisor as a whole number. Multiply the divisor and dividend by the same power of ten.

EXAMPLE 2 *Photography Application*

There are 246 students in the sixth grade. If Ms. Lee buys rolls of film with 24 exposures each, how many rolls will she need to take every student's picture?

The question asks how many whole rolls are needed to take a picture of every one of the students.

$246 \div 24 = 10.25$

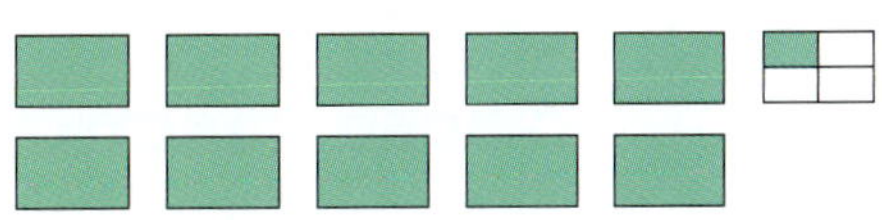

Think: Ten rolls of film will not be enough to take every student's picture. Ms. Lee will need to buy another roll of film. The quotient must be rounded up to the next highest whole number.

Ms. Lee will need 11 rolls of film.

3-9 Organizer

Pacing: Traditional 1 day
Block $\frac{1}{2}$ day

Objective: Students solve problems by interpreting the quotient.

Indiana Academic Standards
6.2.3, 6.7.1, 6.7.2, 6.7.6, 6.7.9, 6.7.10

Warm Up

Divide.

1. 15.264 ÷ 3 **5.088**
2. 3.78 ÷ 3 **1.26**
3. 342 ÷ 7.6 **45**
4. 28.32 ÷ 4.8 **5.9**

Problem of the Day

Divide your age in months by 12. What does the quotient tell you? my age in years

Available on Daily Transparency in CRB

Math Humor

Teacher: Why does your division assignment feel so cold?

Student: A lot of the answers had remainders, and my mom always told me to put leftovers in the refrigerator.

1 Introduce

Alternate Opener

EXPLORATION

3-9 Interpret the Quotient

For each problem, estimate a solution. Then compute with a calculator and interpret the result.

		Estimate	Actual
1.	At Juan's school, each lunch special costs \$3.65. How many lunches can Juan buy with \$20.00?		
2.	Gasoline costs \$1.499 per gallon. How many gallons can Sue buy with \$15.00?		
3.	On Jorge's map, 0.15 cm represents 1 mi. He measures a road which is 7.8 cm. How many mi long is the actual road?		
4.	Ofella makes \$6.79 per hour at her summer job. If she wants to make \$200 per week, how many hours should she work?		

Think and Discuss

5. **Explain** the estimation strategies you used.
6. **Discuss** whether the estimate or the actual answer gives a more reasonable answer to the question each problem asks.

Motivate

Pose some simple problems to get students ready for interpreting quotients. For example: Julia needs enough plates for 25 people. If she buys packs of 10 plates each, how many packs must she buy? (25 ÷ 10 = 2.5; 3 packs) Phil has 19 marbles. He gives an equal number to each of 4 friends. How many marbles does each friend get? (19 ÷ 4 = 4.75; 4 marbles each)

Exploration worksheet and answers on Chapter 3 Resource Book pp. 83 and 129

2 Teach

Lesson Presentation

Guided Instruction

In this lesson, students learn to solve problems by interpreting the quotient. First present the measurement application, in which the decimal part of the quotient is not used to answer the question. Then present Example 2, in which the quotient is rounded up to answer the question. Finally, present the situation in which an exact quotient is needed to answer the question. Explain the reasons for each different interpretation of the quotient.

Additional Examples

Example 1

Suppose Mark wants to make bags of slime. If each bag of slime requires 0.15 kg of corn starch and he has 1.23 kg, how many bags of slime can he make? 8

Example 2

There are 237 students in the seventh grade. If Mr. Jones buys rolls of film with 36 exposures each, how many rolls will he need to take every student's picture? 7

Example 3

Gary has 42.25 meters of rope. If he cuts it into 13 equal pieces, how long is each piece? 3.25 meters

EXAMPLE 3 *Social Studies Application*

Marissa is drawing a time line of the Stone Age. She plans for 6 equal sections, two each for the Paleolithic, Mesolithic, and Neolithic periods. If she has 16.5 meters of paper, how long is each section?

The question asks exactly how long each section will be when the paper is divided into 6 sections.

$16.5 \div 6 = 2.75$ *Think: The question asks for an exact answer, so do not estimate. Use the entire quotient.*

Each section will be **2.75** meters long.

When the question asks	➤ You should
How many whole groups can be made when you divide?	➤ Drop the decimal part of the quotient.
How many whole groups are needed to put all items from the dividend into a group?	➤ Round the quotient up to the next highest whole number.
What is the exact number when you divide?	➤ Use the entire quotient as the answer.

Think and Discuss

1. Tell how you would interpret the quotient: A group of 27 students will ride in vans that carry 12 students each. How many vans are needed?

3-9 PRACTICE & ASSESS

Assignment Guide

If you finished Example 1 assign:
Core 1, 4, 8, 13–18
Enriched 4, 8, 10, 13–18

If you finished Example 2 assign:
Core 1–2, 4–5, 7–8, 13–18
Enriched 1–2, 4–5, 7–8, 10–11, 13–18

If you finished Example 3 assign:
Core 1–9, 13–18
Enriched 4–18

3-9 Exercises

FOR EXTRA PRACTICE see page 641

internet connect
Homework Help Online
go.hrw.com Keyword: MR4 3-9

GUIDED PRACTICE

See Example 1
1. Kay is making beaded belts for her friends from 6.5 meters of cord. One belt uses 0.625 meter of cord. How many belts can she make? 10 belts

See Example 2
2. Julius is supplying cups for a party of 136 people. If cups are sold in packs of 24, how many packs of cups will he need? 6 packs

See Example 3
3. Miranda is decorating for a party. She has 13 balloons and 29.25 meters of ribbon. She wants to tie the same length of ribbon on each balloon. How long will each ribbon be? 2.25 meters

Reaching All Learners

Through Critical Thinking

Have students make up and write a numerical division problem that would have a remainder. Then using the examples as a guide, challenge them to write three different word problems to go along with the division problem, each one requiring a different interpretation of the quotient.

3 Close

Summarize

Direct students' attention to the table following Example 3. Review the different ways to interpret quotients and the questions that signal each action to take.

Answers to Think and Discuss

1. Round up the quotient to the next highest whole number. They will need 3 vans.

INDEPENDENT PRACTICE

See Example 1
4. There are 0.454 kg of corn starch in a container. How many 0.028 kg portions are in one container? **16 portions**

See Example 2
5. Tina needs 36 flowers for her next project. The flowers are sold in bunches of 5. How many bunches will she need? **8 bunches**

See Example 3
6. Bobby's goal is to run 27 miles a week. If he runs the same distance 6 days a week, how many miles would he have to run each day? **4.5 miles**

PRACTICE AND PROBLEM SOLVING

7. Nick wants to write thank-you notes to 15 of his friends. The cards are sold in packs of 6. How many packs does Nick need to buy? **3 packs**
8. The science teacher has 7 packs of seeds and 36 students. If the students should each plant the same number of seeds, how many can each student plant? **4 seeds**

9. A new parking garage at an apartment complex is 16.8 m high. Each floor is 4.2 m high. How many floors are there? **4 floors**
10. ***WRITE A PROBLEM*** Create a problem that is solved by interpreting the quotient.
11. ***WRITE ABOUT IT*** Explain how a calculator shows the remainder when you divide 145 by 8. **Possible answer: A calculator shows the remainder as the decimal part of the quotient 18.125.**
12. ***CHALLENGE*** Leonard wants to place a fence on both sides of a 10-meter walkway. If he puts a post at both ends and at every 2.5 meters in between, how many posts does he use? **10 posts**

10. Possible answer: A cookie recipe calls for 1.75 cups of chocolate pieces. How many batches of cookies can you make with 6 cups of chocolate pieces?

Lesson Quiz

Solve.

1. The cross-country team's goal is to run 26.25 mi next week. If they run only 5 days next week, how many miles would they have to run each day? **5.25 mi**
2. Shannon is having a surprise party for her parents. She wants to invite 22 friends. Invitations come in packages of 8. How many packages does Shannon need to buy? **3**

Available on Daily Transparency in CRB

Spiral Review

Evaluate the expressions to find the missing values in each table. (Lesson 2-1)

13.

x	9x
5	■ **45**
6	■ **54**
7	■ **63**

14.

y	y ÷ 11
121	■ **11**
99	■ **9**
77	■ **7**

Estimate. (Lesson 3-2)

15. 467.32 + 450.64 + 447.9 **1,366**
16. 14.87 × 3.78 **60**
17. 53.67 ÷ 9.18 **6**
18. **TEST PREP** Find the sum of 5.63 + 6.702 + 5.9 + 7.383. (Lesson 3-3) **D**

A 14.707 B 256.15 C 19.774 D 25.615

RETEACH 3-9

LESSON 3-9 **Reteach**
Interpret the Quotient

There are three ways the decimal part of a quotient can be interpreted when you solve a problem.

If the question asks for an exact number, use the entire quotient.
If the question asks how many whole groups are needed to put the dividend into a group, round the quotient up to the next whole number.
If the question asks how many whole groups can be made when you divide, drop the decimal part of the quotient.

To interpret the quotient, decide what the question is asking.

In the school library, there are tables that seat 4 students each. If there are 30 students in a class, how many tables are needed to seat all of the students?

To solve, divide 30 by 4.

30 ÷ 4 = 7.5

The question is asking how many tables (whole groups) are needed to put all of the students in the class (dividend) into a group.

So, round 7.5 up to the next whole number.

8 tables are needed to seat all of the students.

Interpret the quotient to solve each problem.

1. A recipe that serves 6 requires 9 cups of milk. How much milk is needed for each serving?
   1.5 cups are needed for each serving.
2. A storage case holds 24 model cars. Marla has 84 model cars. How many storage cases does she need to store all of her cars?
   Marla needs 4 cases.
3. Kenny has $4.25 to spend at the school carnival. If game tickets are $0.50 each, how many games can Kenny play?
   Kenny can play 8 games.

PRACTICE 3-9

LESSON 3-9 **Practice B**
Interpret the Quotient

Circle the letter of the correct answer.

1. You spent a total of $6.75 for 15 yards of ribbon. How much did the ribbon cost per yard?
   A $0.50
   (B) $0.45
   C $1.35
   D $1.45
2. Buttons come in packs of 12. How many packs should you buy if you need 100 buttons?
   F 10
   G 8
   (H) 9
   J 12
3. Your sewing cabinet has compartments that hold 8 spools of thread each. You have 50 spools of thread. How many compartments can you fill?
   (A) 6
   B 7
   C 5
   D 8
4. You spent a total of $35.75 for velvet cloth. Each yard of the velvet costs $3.25. How many yards did you buy?
   F 10
   G 10.5
   (H) 11
   J 11.5

Write the correct answer.

5. You used a total of 67.5 yards of cotton material to make costumes for the play. Each costume used 11.25 yards of cloth. How many costumes did you make?
   6 costumes
6. You are saving $17.00 each week to buy a new sewing machine that costs $175.50. How many weeks will you have to save to have enough money to buy the sewing machine?
   11 weeks
7. Sequins come in packs of 75. You use 12 sequins on each costume. If you have one pack of sequins, how many costumes can you make?
   6 costumes
8. You pay $26.28 for a subscription to *Sewing Magazine*. You get an issue every month for a year. How much does each issue cost?
   $2.19

CHALLENGE 3-9

LESSON 3-9 **Challenge**
Plan a Party!

You are in charge of buying supplies for the class party. There are 30 students in your class. Use the party supply store advertisement below to plan what to buy. After you pay for all the items, the total cost will be divided evenly among all the students.

Shopping List

Item	Number of Items You Want Per Person	Number of Packs to Buy	Number of Left Over Items	Total Price of Items
Invitations	1	**4**	**2**	**$7.40**
Paper plates	1	**2**	**20**	**$5.60**
Plastic cups	2	**6**	**0**	**$11.70**
Paper napkins	2	**1**	**15**	**$3.65**
Plastic forks	1	**3**	**6**	**$6.15**
			Grand Total Price:	**$34.50**
			Cost Per Student:	**$1.15**

PROBLEM SOLVING 3-9

LESSON 3-9 **Problem Solving**
Interpret the Quotient

Write the correct answer.

1. Five friends split a pizza that costs $16.75. If they shared the bill equally, how much did they each pay?
   $3.35
2. There are 45 choir members going to the recital. Each van can carry 8 people. How many vans are needed?
   6 vans
3. Tara bought 150 beads. She needs 27 beads to make each necklace. How many necklaces can she make?
   5 necklaces
4. Cat food costs $2.85 for five cans. Ben only wants to buy one can. How much will it cost?
   $0.57
5. Tennis balls come in cans of 3. The coach needs 50 tennis balls for practice. How many cans should he order?
   17 cans
6. The rainfall for three months was 4.6 inches, 3.5 inches, and 4.2 inches. What was the average monthly rainfall during that time?
   4.1 inches
7. The students in Daniel's book club have to read a 28-page story. Daniel said he is going to read it in 6 different sittings. About how many pages will Daniel read during each sitting?
   5 pages
8. The gas tank in Brian's car holds 10.5 gallons. His car averages 28.6 miles per gallon of gas. How many times will he have to fill up the gas tank to drive 1,000 miles?
   4 times

Circle the letter of the correct answer.

9. Tom has $15.86 to buy marbles that cost $1.25 each. He wants to know how many marbles he can buy. What should he do after he divides?
   (A) Drop the decimal part of the quotient when he divides.
   B Drop the decimal part of the dividend when he divides.
   C Round the quotient up to the next highest whole number to divide.
   D Use the entire quotient of his division as the answer.
10. Mei needs 135 hot dog rolls for the class picnic. The rolls come in packs of 10. She wants to know how many packs to buy. What should she do after she divides?
   F Drop the decimal part of the quotient when she divides.
   G Drop the decimal part of the dividend when she divides.
   (H) Round the quotient up to the next highest whole number.
   J Use the entire quotient of her division as the answer.

3-10 Organizer

Pacing: Traditional 1 day
Block $\frac{1}{2}$ day

Objective: Students solve equations involving decimals.

Indiana Academic Standards
6.3.1, 6.3.2, 6.5.10, 6.7.1, 6.7.9

Warm Up

Solve.

1. $x - 3 = 11$ $x = 14$
2. $18 = x + 4$ $x = 14$
3. $\frac{x}{7} = 42$ $x = 294$
4. $2x = 52$ $x = 26$
5. $x - 82 = 172$ $x = 254$

Problem of the Day

Find the missing entries in the magic square. 11.25 is the sum of every row, column, and diagonal.

3	6.75	1.5
2.25	3.75	5.25
6	0.75	4.5

Available on Daily Transparency in CRB

Math Humor

An equation with decimals is like a contest between teams of campers. Both sides have *tents*.

3-10 Solving Decimal Equations

Learn to solve equations involving decimals.

Felipe has earned \$45.20 by mowing lawns for his neighbors. He wants to buy inline skates that cost \$69.95. Write and solve an equation to find how much more money Felipe must earn to buy the skates.

Let m be the amount of money Felipe needs. $\$45.20 + m = \$69.95$

You can solve equations with decimals using inverse operations just as you solved equations with whole numbers.

$$\begin{aligned} \$45.20 + m &= \$69.95 \\ -\$45.20 \quad\quad &\quad -\$45.20 \\ m &= \$24.75 \end{aligned}$$

Felipe needs \$24.75 more to buy the inline skates.

EXAMPLE 1 Solving One-Step Equations with Decimals

Remember!
Use inverse operations to get the variable alone on one side of the equation.

Solve each equation. Check your answer.

A $g - 3.1 = 4.5$

$g - 3.1 = 4.5$ *3.1 is subtracted from g.*
$\underline{+3.1} \quad \underline{+3.1}$ *Add 3.1 to both sides to undo the subtraction.*
$g = 7.6$

Check

$g - 3.1 = 4.5$
$7.6 - 3.1 \stackrel{?}{=} 4.5$ *Substitute 7.6 for g in the equation.*
$4.5 \stackrel{?}{=} 4.5$ ✔   *7.6 is the solution.*

B $3k = 8.1$

$3k = 8.1$ *k is multiplied by 3.*
$\frac{3k}{3} = \frac{8.1}{3}$ *Divide both sides by 3 to undo the multiplication.*
$k = 2.7$

Check

$3k = 8.1$
$3(2.7) \stackrel{?}{=} 8.1$ *Substitute 2.7 for k in the equation.*
$8.1 \stackrel{?}{=} 8.1$ ✔   *2.7 is the solution.*

1 Introduce

Alternate Opener

EXPLORATION

3-10 Solving Decimal Equations

For each equation, estimate the solution. Then use a calculator to solve the equation. Compare the calculated solution with your estimated solution.

		Estimate	Actual
1.	$1.25 + x = 10$		
2.	$20 - x = 1.95$		
3.	$6x = 15$		
4.	$\frac{x}{4.5} = 10$		
5.	$\frac{x}{100} = 1.609$		

6. Write a real-world situation for each equation in numbers **1–5**.

Think and Discuss

7. **Explain** which equation it was easiest to estimate a solution for and which equation it was most difficult to estimate a solution for.
8. **Discuss** whether the estimated or the calculated solutions are more reasonable for the real-world situations you wrote in number **6.**

Motivate

Present a simple equation to students, such as $x + 15 = 43$. Have students explain how to solve the equation. Subtract 15 from both sides. $x + 15 - 15 = 43 - 15$, so $x = 28$. Tell students that decimal equations are also solved by using inverse operations.

Exploration worksheet and answers on Chapter 3 Resource Book pp. 92 and 131

2 Teach

Lesson Presentation

Guided Instruction

In this lesson, students learn to solve equations involving decimals. First teach them to solve subtraction, multiplication, and division equations involving decimals using the inverse operations addition, division, and multiplication, respectively. Then teach students to apply the concept to real-world applications.

Teaching Tip If students are intimidated by the use of decimals in equations, have them solve a simpler problem, substituting whole numbers for the decimals. This should give students the reinforcement they need to solve decimal equations.

Solve each equation. Check your answer.

C $\frac{m}{5} = 1.5$

$\frac{m}{5} = 1.5$	*m is divided by 5.*
$\frac{m}{5} \cdot 5 = 1.5 \cdot 5$	*Multiply both sides by 5 to undo the division.*
$m = 7.5$	

Check

$\frac{m}{5} = 1.5$	
$\frac{7.5}{5} \stackrel{?}{=} 1.5$	*Substitute 7.5 for m in the equation.*
$1.5 \stackrel{?}{=} 1.5$ ✔	*7.5 is the solution.*

EXAMPLE 2 *Measurement Application*

Remember!

The area of a rectangle is its length times its width.

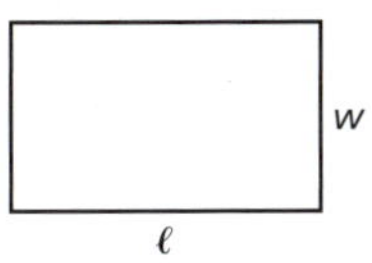

$A = \ell w$

A **The area of the floor in Jonah's bedroom is 28 square meters. If its length is 3.5 meters, what is the width of the bedroom?**

area = length · width

$28 = 3.5 \cdot w$	*Write the equation for the problem.*
$28 = 3.5w$	*Let w be the width of the room.*
$\frac{28}{3.5} = \frac{3.5w}{3.5}$	*w is multiplied by 3.5. Divide both sides by 3.5 to undo the multiplication.*
$8 = w$	

The width of Jonah's bedroom is 8 meters.

B **Jonah puts wall-to-wall carpet in his bedroom. The price of the carpet is $22.50 per square meter. What is the total cost to carpet the bedroom?**

total cost = area · cost of carpet per square meter

$C = 28 \cdot 22.50$	*Let C be the total cost. Write the equation for the problem.*
$C = 630$	*Multiply.*

The cost of carpeting the bedroom is $630.

Think and Discuss

1. **Explain** whether the value of m will be less than or greater than 1 when you solve $5m = 4.5$.
2. **Tell** how you can check the answer in Example 2A.

COMMON ERROR ALERT

Caution students to be sure they are using the inverse operation to get the variable alone on one side of the equation. Remind them that addition and subtraction are inverse operations; so are multiplication and division.

Additional Examples

Example 1

Solve each equation. Check your answers.

A. $k - 6.2 = 9.5$ $k = 15.7$
B. $6k = 7.2$ $k = 1.2$
C. $\frac{m}{7} = 0.6$ $m = 4.2$

Example 2

A. The area of Emily's floor is $33.75\ m^2$. If its length is 4.5 meters, what is its width?
The width of the floor is 7.5 meters.

B. If carpet costs \$23 per m^2, what is the total cost to carpet the floor?
The total cost of carpeting the floor is $776.25.

3 Close

Reaching All Learners
Through Critical Thinking

Ask students to work in pairs, with one demonstrating the solution of a decimal equation to the other. Each step should be shown and explained. Have students reverse roles to solve another decimal equation. Let them know that they can assist their partner if necessary.

Summarize

Have students volunteer to solve decimal equations for the class. As they work, emphasize the inverse operations being used and the placement of the decimal points.

Answers to Think and Discuss

1. Possible answers: The value of m will be less than 1 because $4.5 \div 5$ is less than 1; the value of m will be less than 1 because 4.5 is the product of 5 and m.
2. Substitute 8 for w in the equation $28 = 3.5w$. If both sides of the equation are equal, then 8 is the correct solution.
$28 = 3.5w$
$28 \stackrel{?}{=} 3.5 \cdot 8$
$28 \stackrel{?}{=} 28$ ✔

3-10 PRACTICE & ASSESS

Students may want to refer back to the lesson examples.

Assignment Guide

If you finished Example 1 assign:
Core 1–6, 9–14, 21–25 odd, 42–47
Enriched 15–17, 21–32, 42–47

If you finished Example 2 assign:
Core 1–20, 34, 38, 42–47
Enriched 21–47

Notes

3-10 Exercises

FOR EXTRA PRACTICE see page 641

internet connect: Homework Help Online go.hrw.com Keyword: MR4 3-10

GUIDED PRACTICE

See Example 1 Solve each equation. Check your answer.

1. $a - 2.3 = 4.8$ **$a = 7.1$**
2. $6n = 8.4$ **$n = 1.4$**
3. $\frac{c}{4} = 3.2$ **$c = 12.8$**
4. $8.5 = 2.49 + x$ **$x = 6.01$**
5. $\frac{d}{3.2} = 1.09$ **$d = 3.488$**
6. $1.6 = m \cdot 4$ **$m = 0.4$**

See Example 2

7. The length of a window is 10.5 m, and the width is 5.75 m. Solve the equation $a \div 10.5 = 5.75$ to find the area of the window. **60.375 m^2**
8. The distance around a square picture is 64.8 cm. What is the length of each side? **16.2 cm**

INDEPENDENT PRACTICE

See Example 1 Solve each equation. Check your answer.

9. $b - 5.6 = 3.7$ **$b = 9.3$**
10. $1.6 = \frac{p}{7}$ **$p = 11.2$**
11. $3r = 62.4$ **$r = 20.8$**
12. $9.5 = 5x$ **$x = 1.9$**
13. $a - 4.8 = 5.9$ **$a = 10.7$**
14. $\frac{n}{8} = 0.8$ **$n = 6.4$**
15. $8 + f = 14.56$ **$f = 6.56$**
16. $5.2s = 10.4$ **$s = 2$**
17. $1.95 = z - 2.05$ **$z = 4$**

See Example 2

18. The area of a rectangle is 65.8 square units. The length is 7 units. Solve the equation $7 \cdot w = 65.8$ to find the width of the rectangle. **9.4 units**
19. Irene bought 1.75 kg of grapes for $5.25. What is the price per kilogram of grapes? **$3.00 per kg**
20. Ken placed a fence around his square garden. He used 6.4 meters of fence to enclose all four sides of the garden. How long is each side of his garden? **1.6 meters**

PRACTICE AND PROBLEM SOLVING

Solve each equation and check your answer.

21. $9.8 = t - 42.1$ **$t = 51.9$**
22. $q \div 2.6 = 9.5$ **$q = 24.7$**
23. $45.36 = 5.6 \cdot m$ **$m = 8.1$**
24. $1.3b = 5.46$ **$b = 4.2$**
25. $4.93 = 0.563 + m$ **$m = 4.367$**
26. $\frac{a}{5} = 2.78$ **$a = 13.9$**
27. $w - 64.99 = 13.044$ **$w = 78.034$**
28. $6.205z = 80.665$ **$z = 13$**
29. $74.2 = 38.06 + c$ **$c = 36.14$**
30. $3.7(1.8) = t + 2.9$ **$t = 3.76$**
31. $a - 1.5 = \frac{6.2}{2}$ **$a = 4.6$**
32. $b \cdot 4.4 = 9 + 7.5$ **$b = 3.75$**
33. The shortest side of the triangle is 10 units long.
    a. What are the lengths of the other two sides of the triangle? **19.5 units, 21 units**
    b. What is the perimeter of the triangle? **50.5 units**

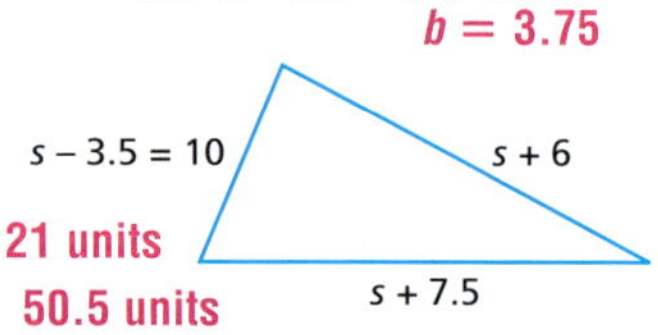

Math Background

As long ago as 2000 B.C., the mathematics of Babylonia included algebra. The methods relied on prose for exposition. The problems included equations of 2nd, 3rd, and 4th degree.

The Rhind papyrus, written by the scribe Ahmes, traces back to Egypt around the year 1650 B.C. It is one of our main sources of knowledge about Egyptian mathematics. It contains 85 problems, some of which relate to the linear form of equation used in this lesson. A drawing of a pair of legs walking from right to left represents + and a drawing of a pair of legs walking from left to right represents –.

RETEACH 3-10

LESSON 3-10 **Reteach**
Decimal Equations

You can write related equations for addition and subtraction equations.
$7.4 + 6.2 = 13.6$ $13.6 - 6.2 = 7.4$

Use related equations to solve each of the following.

A. $x + 4.5 = 7.9$
Think: $7.9 - 4.5 = x$
$x = 3.4$
Check $x + 4.5 = 7.9$
$3.4 + 4.5 \stackrel{?}{=} 7.9$ substitute
$7.9 = 7.9$

B. $x - 0.08 = 6.2$
Think: $6.2 + 0.08 = x$
$x = 6.28$
Check $x - 0.08 = 6.2$
$6.28 - 0.08 \stackrel{?}{=} 6.2$ substitute
$6.2 = 6.2$

Use related facts to solve each equation. Then check each answer.

1. $x + 8.7 = 12.9$ **$x = 4.2$**
2. $x + 8.4 = 16.6$ **$x = 8.2$**
3. $x - 2.65 = 7.8$ **$x = 10.45$**
4. $x - 0.8 = 2.3$ **$x = 3.1$**

You can write related equations for multiplication and division equations.
$3.2 \cdot 2.4 = 7.68$ $7.68 \div 2.4 = 3.2$

Use related equations to solve each of the following.

C. $3x = 1.5$
Think: $1.5 \div 3 = x$
$x = 0.5$
Check: $3x = 1.5$
$3 \cdot 0.5 \stackrel{?}{=} 1.5$ substitute
$1.5 = 1.5$

D. $x \div 6 = 1.2$
Think: $1.2 \cdot 6 = x$
$x = 7.2$
Check: $x \div 6 = 1.2$
$7.2 \div 6 \stackrel{?}{=} 1.2$ substitute
$1.2 = 1.2$

Use related facts to solve each equation. Then check each answer.

5. $x \div 3 = 6.3$ **$x = 18.9$**
6. $x \div 0.2 = 3.4$ **$x = 0.68$**
7. $7x = 4.2$ **$x = 0.6$**
8. $5x = 4.5$ **$x = 0.9$**

PRACTICE 3-10

LESSON 3-10 **Practice B**
Decimal Equations

Solve each equation. Check your answer.

1. $a - 2.7 = 4.8$ **$a = 7.5; 7.5 - 2.7 = 4.8$**
2. $b \div 7 = 1.9$ **$b = 13.3; 13.3 \div 7 = 1.9$**
3. $w - 6.5 = 3.8$ **$w = 10.3; 10.3 - 6.5 = 3.8$**
4. $p \div 0.4 = 1.7$ **$p = 0.68; 0.68 \div 0.4 = 1.7$**
5. $4.5 + x = 8$ **$x = 3.5; 4.5 + 3.5 = 8$**
6. $b \div 3 = 2.5$ **$b = 7.5; 7.5 \div 3 = 2.5$**
7. $7.8 + s = 15.2$ **$s = 7.4; 7.8 + 7.4 = 15.2$**
8. $1.63q = 9.78$ **$q = 6; 1.63 \cdot 6 = 9.78$**
9. $0.05 + x = 2.06$ **$x = 2.01; 0.05 + 2.01 = 2.06$**
10. $1.7n = 2.38$ **$n = 1.4; 1.7 \cdot 1.4 = 2.38$**
11. $t - 6.08 = 12.59$ **$t = 18.67; 18.67 - 6.08 = 12.59$**
12. $9q = 16.2$ **$q = 1.8; 9 \cdot 1.8 = 16.2$**
13. $w - 8.9 = 10.3$ **$w = 19.2; 19.2 - 8.9 = 10.3$**
14. $1.4n = 3.22$ **$n = 2.3; 1.4 \cdot 2.3 = 1.3$**
15. $t - 12.7 = 0.8$ **$t = 13.5; 13.5 - 12.7 = 0.8$**
16. $3.8 + a = 6.5$ **$a = 2.7; 3.8 + 2.7 = 6.5$**
17. The distance around a square photograph is 12.8 centimeters. What is the length of each side of the photograph? **3.2 centimeters**
18. You buy two rolls of film for $3.75 each. You pay with a $10 bill. How much change should you get back? **$2.50**

The London Eye is the world's largest Ferris wheel. Use the table for Exercises 34–38.

Weight of wheel	1,900 metric tons
Time to revolve	30 minutes
Height of wheel	135 meters

34. Write the height of the wheel in kilometers. **0.135 km**

35. There are 1,000 kilograms in a metric ton.

35a. 1,900,000 kg
   a. What is the weight of the wheel in kilograms?
   b. Write the weight in kilograms in scientific notation. 1.9×10^6

36a. 1,800 sec
36. a. How many seconds does it take for the wheel to make one revolution?

36b. 468 m
   b. The wheel moves at a rate of 0.26 meters per second. Use the equation $d \div 0.26 = 1{,}800$ to find the distance of one revolution.

37. 9 capsules
37. Each capsule can hold 25 passengers. How many capsules are needed to hold 210 passengers?

38. Fifteen tickets for the London Eye cost \$112.50. What is the cost for one ticket? **\$7.50**

39. ***WHAT'S THE ERROR?*** When solving the equation $b - 12.98 = 5.03$, a student said that $b = 7.95$. Describe the error. What is the correct value for b? **Possible answer: The student subtracted 5.03 from 12.98 rather than adding; $b = 18.01$**

40. ***WRITE ABOUT IT*** Explain how you solve for the variable in a multiplication equation such as $2.3a = 4.6$. **Possible answer: Divide both sides of the equation by 2.3, so $a = 2$.**

41. ***CHALLENGE*** Solve $1.45n \times 3.2 = 23.942 + 4.13$. **6.05**

Spiral Review

Evaluate each expression. (Lesson 1-4)

42. $6 \times (21 - 15) \div 12$ **3**

43. $72 \div 8 + 2^3 \times 5 - 19$ **30**

Solve each equation. (Lessons 2-5, 2-6, 2-7)

44. $n - 39 = 20$ $n = 59$

45. $3b = 54$ $b = 18$

46. $\frac{c}{3} = 12$ $c = 36$

47. **TEST PREP** Which is the standard form of 1.7565×10^8? (Lesson 3-5) **A**

A 175,650,000 **B** 17,565,000 **C** 1,756,500,000 **D** 17,565,000,000

Journal

Have students write and solve a decimal equation. Then have them place the equation in a real-world context.

Test Prep Doctor

For Exercise 47, students will need to accurately change from scientific to standard notation. Students who answered **B, C,** or **D** multiplied 1.7565 by 10^7, 10^9, or 10^{10} instead of by 10^8.

CHALLENGE 3-10

LESSON 3-10 **Challenge**
Playing Weight

In professional sports, each ball has a maximum, or greatest, weight allowed in play. The lightest official weight for a table tennis ball is only 0.009 ounces. Use the equations below to find the maximum weights, in ounces, of some other sports' balls.

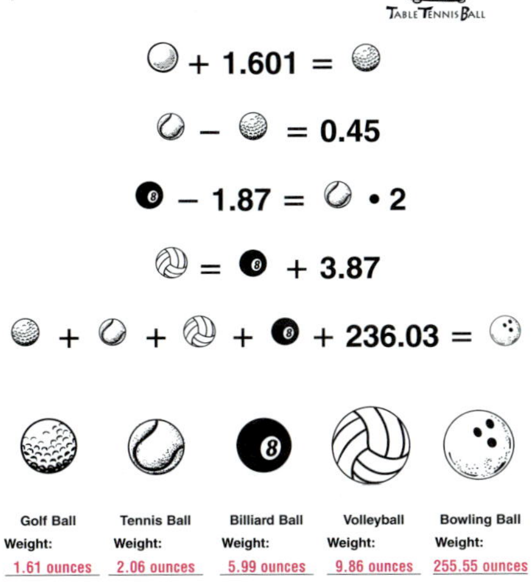

PROBLEM SOLVING 3-10

LESSON 3-10 **Problem Solving**
Decimal Equations

Write the correct answer.

1. Bee hummingbirds weigh only 0.0056 ounces. They have to eat half their body weight every day to survive. How much food does a bee hummingbird have to eat each day?
   0.0028 ounces
2. The desert locust, a type of grasshopper, can jump 10 times the length of its body. The locust is 1.956 inches long. How far can it jump in one leap?
   19.56 inches
3. In 1900, there were about 1.49 million people living in California. In 2000, the population was 33.872 million. How much did the population grow between 1900 and 2000?
   by 32.382 million
4. Juanita has \$567.89 in her checking account. After she deposited her paycheck and paid her rent of \$450.00, she had \$513.82 left in the account. How much was her paycheck?
   \$395.93
5. The average body temperature for people is 98.6°F. The average body temperature for most dogs is 3.4°F higher than for people. The average body temperature for cats is 0.5°F lower than for dogs. What is the normal body temperature for dogs and cats?
   dogs: 102°F; cats: 101.5°F
6. Seattle, Washington, is famous for its rainy climate. Winter is the rainiest season there. From November through December the city gets an average of 5.85 inches of rain each month. Seattle usually gets 6 inches of rain in December. What is the city's average rainfall in November?
   5.7 inches

Circle the letter of the correct answer.

7. The equation to convert from Celsius to Kelvin degrees is K = 273.16 + C. If it is 303.66°K outside, what is the temperature in Celsius degrees?
   A 576.82°C
   (B) 30.5°C
   C 305°C
   D 257.68°C
8. The distance around a square mirror is 6.8 feet. Which of the following equations finds the length of each side of the mirror?
   F $6.8 - x = 4$
   G $x \div 4 = 6.8$
   (H) $4x = 6.8$
   J $6.8 + 4 = x$

Lesson Quiz

Solve each equation. Check your answer.

1. $x - 3.9 = 14.2$ $x = 18.1$

2. $\frac{x}{4} = 8.3$ $x = 33.2$

3. $x - 4.9 = 16.2$ $x = 21.1$

4. $7x = 47.6$ $x = 6.8$

5. The area of the floor in Devon's room is 35.7 m^2. If the width is 4.2 m, what is the length of the bedroom? **8.5 m**

Available on Daily Transparency in CRB

CHAPTER 3
Extension

Pacing: Traditional 1 day
Block $\frac{1}{2}$ day

Objective: Students round measurements to an appropriate number of significant figures.

Indiana Academic Standards
6.5.6

Using the Pages

In Lesson 3-4, students multiplied and divided decimals by powers of ten and converted metric measurements. In this extension, students round measurements to an appropriate number of significant figures. First students determine the number of significant figures in a decimal. Then they use significant figures in addition, subtraction, multiplication, and division.

EXTENSION Significant Figures

Learn to round measurements to an appropriate number of significant figures.

Vocabulary
significant figures

Measurements are approximations due to the limitations of the tools that are used. For example, a meterstick with tick marks in centimeters cannot give a precise measurement in millimeters.

The ruler is marked in centimeters and half centimeters. A measurement in millimeters is only an estimate.

To find out how close your measurement is to an actual measurement, you can use *significant figures.* **Significant figures** are the digits in a measurement that are known with certainty.

Identifying Significant Figures
• Nonzero digits
• Zeros at the end of a number and to the right of the decimal point
• Zeros between significant figures

EXAMPLE 1 Counting Significant Figures

Determine the number of significant figures in each decimal.

A 4.00
4 is a nonzero digit.
The zeros are to the right of the decimal in a number greater than 1.
3 significant figures

B 0.0209
2 and 9 are nonzero digits.
One zero is between nonzero digits.
*Other zeros are **not** significant.*
3 significant figures

When adding or subtracting, find the number in the problem with the least number of *digits* to the right of the decimal point.

1 Introduce

Motivate

Explain to students that science depends on experimentation and experimentation requires measurement. You may use Teaching Transparency T40 in the Chapter 3 Resource Book to illustrate that measurements are never exact, but the number of significant figures in a measurement conveys a sense of the precision of the experiment.

2 Teach

Lesson Presentation

Guided Instruction

In this extension, students round measurements to an appropriate number of significant figures. First teach students to identify significant figures and then determine the number of significant figures in a decimal. Then teach students to find the number of significant figures in the solution of an addition or subtraction problem. Next teach students to find the number of significant figures in the solution of a multiplication or division problem.

EXAMPLE 2 Using Significant Figures in Addition and Subtraction

Write the answer with the appropriate number of significant figures.

35.4 − 7.08

35.4	*1 digit after decimal*
− 7.08	*2 digits after decimal*
28.32	
28.3	*Round the difference to have one digit to the right of the decimal.*

When multiplying and dividing, find the number in the problem with the least number of *significant figures*.

EXAMPLE 3 Using Significant Figures in Multiplication and Division

Write the answer with the appropriate number of significant figures.

0.60 × 1.09

0.60	*2 significant figures*
× 1.09	*3 significant figures*
0.654	
0.65	*Round the product to two significant figures.*

EXTENSION Exercises

Determine the number of significant figures in each decimal.

1. 300.5 **4**
2. 19.050 **5**
3. 0.006 **1**
4. 0.05 **1**
5. 112.15 **5**

Write each answer with the appropriate number of significant figures.

6. 2.5 × 0.6 **2**
7. 11.54 + 2.8 **14.3**
8. 1.20 ÷ 0.8 **2**
9. 22.57 − 4.85 **17.72**

Use the table for Exercises 10 and 11.

10. How many significant figures are in each score? **5**
11. Using significant figures, tell the difference between the gold medal score and the silver medal score. **35.61**

Men's 3 m Synchronized Diving Scores (2000 Summer Olympics)

Medal	Country	Score
Gold	China	365.58
Silver	Russia	329.97
Bronze	Australia	322.86

COMMON ERROR ALERT

In Exercises 6 and 8, students may get the addition and subtraction rule mixed up with the multiplication and division rule. When adding or subtracting, find the number in the problem with the least number of *digits* to the right of the decimal point. When multiplying and dividing, find the number in the problem with the least number of *significant figures*. The answers to Exercise 6 and 8 should have only one significant figure.

Additional Examples

Example 1

Determine the number of significant figures in each decimal.

A. 3.0 2 significant figures

B. 0.003010 4 significant figures

Example 2

Write the answer with the appropriate number of significant figures.

24.7 + 3.561 **28.3**

Example 3

Write the answer with the appropriate number of significant figures.

1.55 ÷ 0.5 **3**

3 Close

Summarize

Explain how to find the answer with the appropriate number of significant figures for 25.62 + 0.1 and 25.62 × 0.1.

Possible answer: For 25.62 + 0.1, add the numbers to get 25.72. Because the least number of digits to the right of the decimal point in each of the addends is one, the sum should be rounded to tenths. The sum is 25.7. For 25.62 × 0.1, multiply the numbers to get 2.562. Because the least number of significant figures in each of the factors is one, the product should have one significant digit. The product is 3.

Teaching Tip Remind students that the rule for finding significant figures for addition and subtraction problems is different from the rule for finding significant figures for multiplication and division problems.

Problem Solving on Location

Indiana

Purpose: *To provide additional practice for problem-solving skills in Chapters 1–3*

Indiana Basketball Hall of Fame

- After problem 4, have students consider the following: Can you conclude that Clyde Lovellette always scored fewer points per game than Oscar Robinson? Explain. Possible answer: No; an average represents all the data. There may have been some games in which Lovellette scored more points than Robinson.
- After problem 5, have students consider the following problem: What is the range of the heights of the four players? 0.1 meter
- After problem 6, ask the following question: A kilogram is about 2.2 pounds. What is the weight in pounds of the heaviest player listed? about 235 pounds

Extension Have students convert the measurements of height and weight in the table to customary measurements. Have them research the players to check their work. Check students' work.

Problem Solving on Location

INDIANA

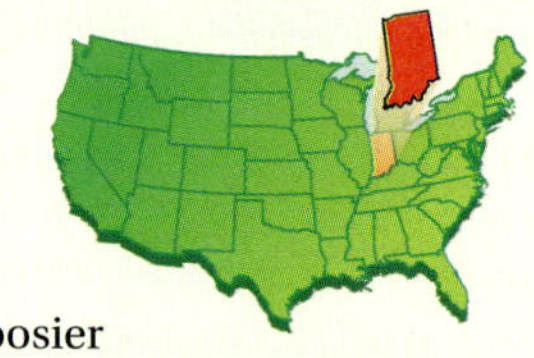

Indiana Basketball Hall of Fame

Visitors to Indiana may not understand "Hoosier Hysteria," the statewide excitement over high-school basketball, but a trip to the Indiana Basketball Hall of Fame may help to explain things. The museum honors Indiana's legendary basketball players and coaches.

After high school, many future Hall of Fame inductees continued to play at the college and professional levels. The table shows professional career averages for some of the players honored in the Hall of Fame.

Indiana Basketball Hall of Fame Inductees

Player (year inducted)	Points per Game	Assists per Game	Rebounds per Game	Height (m)	Mass (kg)
Clyde Lovellette (1982)	17.0	1.6	9.5	2.06	106.69
Oscar Robertson (1982)	25.7	9.5	7.5	1.96	95.34
George McGinnis (1995)	17.2	3.8	9.8	2.03	106.69
Larry Bird (2000)	24.3	6.3	10.0	2.06	99.88

For 1–7, use the table.

1. Which player has the highest points-per-game average? **Oscar Robertson**
2. Order the assists-per-game averages from greatest to least. **9.5, 6.3, 3.8, 1.6**
3. How many more points did Larry Bird average per game than George McGinnis? **7.1**
4. How do you think the points-per-game averages were calculated?
5. Which player is the shortest? **Oscar Robertson**
6. One meter is about 3 feet. About how tall in feet is George McGinnis? **about 6 feet**
7. Order the masses of the players from least to greatest. **95.34, 99.88, 106.69, 106.69**

4. Divide the total number of points for the player by the total number of games.

Lewis and Clark

At Falls of the Ohio State Park in Clarksville, a historical plaque marks the place where Meriwether Lewis and William Clark met before setting out on their famous expedition through the Louisiana Territory and the Pacific Northwest. From the Falls of the Ohio, Lewis and Clark went down the Ohio River to Wood River, Illinois, where they set up camp and trained recruits before crossing the Mississippi River.

The 2003 Lewis and Clark River Festival commemorates the bicentennial anniversary of the start of the expedition. Historical presentations at the Indiana site provide authentic reenactments of the departure and journey west.

Lewis and Clark's travels took them near the present-day cities labeled on the map. The approximate distances, in kilometers, between some of these cities are also shown.

For 1–4, use the map.

1. Lewis and Clark met in Clarksville, Indiana, set up camp in Wood River, Illinois, and then passed through Atchison, Kansas. What distance does the map show for the trail between Clarksville and Atchison? **829.8 km**
2. If the total distance of the trail between Atchison and Bismarck, North Dakota, is 995.9 km, what is the distance of the trail between Yankton and Pierre, South Dakota? **288 km**
3. Between which two sections of the trail is the distance about 200 km? **Between Dillon, MT, and Missoula, MT**
4. The length of the trail between Orofino, Idaho, and Astoria, Oregon, is 582.5 km. Compare the length of the trail between Bismarck and Astoria and the length of the trail between Clarksville and Bismarck. Which is longer? **the distance of the trail between Bismarck and Astoria**

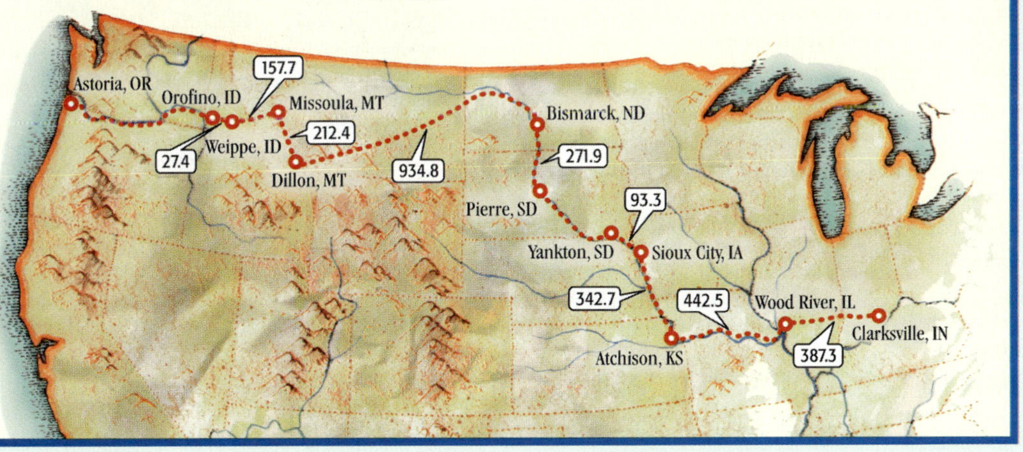

Lewis and Clark

Extension Encourage students to research the journey of Lewis and Clark. Have them write word problems based on information they find in their research. Check students' work.

MATH-ABLES

Game Resources

Puzzles, Twisters & Teasers
Chapter 3 Resource Book

Jumbles

Purpose: *To apply the skill of using operations with decimals to solving a riddle*

Discuss: Discuss with students the best way to order a large group of numbers. Possible answer: Make several different lists (e.g., numbers less than 1, numbers between 1 and 2, 2 and 3, etc.). Order the numbers within each list, and then combine the lists so that the entire set is ordered from least to greatest.

Extend: Challenge students to create a puzzle using decimal operations and their favorite riddle. Have them trade riddles and solve.

Make a Buck

Purpose: *To practice adding decimals by playing a money game*

Discuss: Suppose the sum of the values on the cards in your hand is $0.67. You draw the wild card. What value should you assign the wild card in order to win a point? $0.33

Extend: Have students play the game again, altering the rules so that each player is allowed 8 cards in his or her hand, and the target sum is $2.00.

MATH-ABLES

Jumbles

Do you know what eleven plus two equals?

Use your calculator to evaluate each expression. Keep the letters under the expressions with the answers you get. Then order the answers from least to greatest, and write down the letters in that order. You will spell the answer to the riddle. TWELVE PLUS ONE

$4 - 1.893$	E	(2.107)
$0.21 \div 0.3$	L	(0.7)
$0.443 - 0.0042$	E	(.4388)
$4.509 - 3.526$	V	(.983)
$3.14 \cdot 2.44$	E	(7.6616)
$1.56 \cdot 3.678$	N	(5.73768)
$6.34 \div 2.56$	P	(2.4765625)
$1.19 + 1.293$	L	(2.483)
$8.25 \div 2.5$	U	(3.3)
$7.4 - 2.356$	S	(5.044)
$0.0003 + 0.003$	T	(0.0033)
$0.3 \cdot 0.04$	W	(.012)
$2.17 + 3.42$	O	(5.59)

Make a Buck

The object of the game is to win the most points by adding decimal numbers to make a sum close to but not over $1.00.

Most cards have a decimal number on them representing an amount of money. Others are wild cards: The person who receives a wild card decides its value.

The dealer gives each player four cards. Taking turns, players add the numbers in their hand. If the sum is less than $1.00, a player can either draw a card from the top of the deck or pass.

When each player has taken a turn or passed, the player whose sum is closest to but not over $1.00 scores a point. If players tie for the closest sum, each of those players scores a point. All cards are then discarded and four new cards are dealt to each player.

When all of the cards have been dealt, the player with the most points wins.

internet connect
Go to ***go.hrw.com*** for a complete set of rules and game pieces.
KEYWORD: MR4 Game3

Technology LAB: Scientific Notation

internet connect
Lab Resources Online
go.hrw.com
KEYWORD: MR4 TechLab3

One **septillion** is a very large number. It is 1 followed by 24 zeros, or in exponential notation, 10^{24}.

Enter 1 septillion on your graphing calculator by pressing 1 and then pressing 0 twenty-four times. Press ENTER. What happens?

The calculator displays **1E24**, which means $\mathbf{1 \times 10^{24}}$. A calculator displays very large numbers in **scientific notation.**

Activity

1 Evaluate 248,925 · 259,871 on a calculator. Write the product in standard form.

The product is **6.468838868E10**.
$6.468838868 \times 10^{10} = 64{,}688{,}388{,}680$

The display does not show every digit of the product. The answer is rounded to the nearest ten.

2 Use the **EE** function (the second function of the *comma* key) to enter 2.57×10^7, 5.895×10^{12}, and $3{,}452 \cdot 6.45 \times 10^8$ into a calculator using scientific notation.

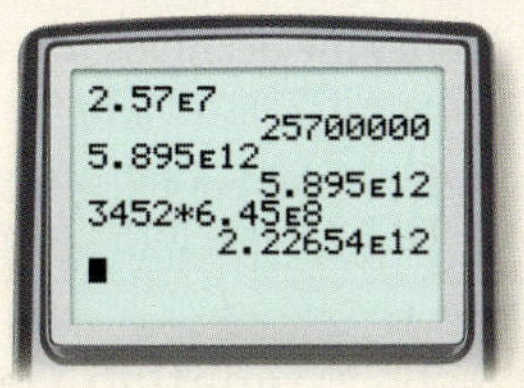

To enter 2.57×10^7, press 2 . 5 7 2nd EE , 7 ENTER.

Notice that, when possible, the calculator displays the results in standard form.

Think and Discuss

1. Tell what happens if you enter a googol into your calculator. A googol is 10^{100}.
On most calculators, you get an "overflow" error or a syntax error—the number is too large for the calculator to interpret.

Try This

Use your calculator to evaluate each expression. Write the results in standard form.

1. 25,468 · 689,234 **17,553,411,510**

2. $5.4 \times 10^4 \cdot 5.6821 \times 10^5$ **30,683,340,000**

3. 1.25 · 2 septillion **2,500,000,000,000,000,000,000,000**

Technology LAB: Scientific Notation

Objective: To use a graphing calculator to study numbers in scientific notation

Materials: Graphing calculator

Lab Resources

Technology Lab Activities . . . . p. 10

Using the Page

This technology activity shows students how to enter and interpret numbers in scientific notation on a graphing calculator. Specific keystrokes may vary, depending on the make and model of the graphing calculator used. The keystrokes given are for a TI-83 model. For keystrokes to other models, visit go.hrw.com.

The Think and Discuss problem is meant to point out the limitations of the calculator. Try This problems 1–3 are meant to help students become familiar with using and interpreting scientific notation on a graphing calculator.

Assessment

Use a calculator to evaluate. Write the results in standard form.

1. $456{,}242 \times 98{,}203$ **44,804,333,130**

2. $(3.25 \times 10^8) \times (2.491 \times 10^7)$ **8,095,750,000,000,000**

Chapter 3 **Study Guide and Review**

Purpose: To help students review and practice concepts and skills presented in Chapter 3

Assessment Resources

Chapter Review
Chapter 3 Resource Book pp. 100–102

***Test and Practice Generator* CD-ROM**

Additional review items in both multiple-choice and free-response format may be generated for any objective in Chapter 3.

Answers

1. front-end estimation
2. scientific notation
3. clustering
4. 5 + 0.6 + 0.08; five and sixty-eight hundredths
5. 1 + 0.007 + 0.0006; one and seventy-six ten-thousandths
6. 1 + 0.2 + 0.003; one and two hundred three thousandths
7. 20 + 3 + 0.005; twenty-three and five thousandths
8. 1.12, 1.2, 1.3
9. 11.07, 11.17, 11.7
10. 0.033, 0.3, 0.303
11. 5.009, 5.5, 5.950
12. 11.32
13. 2.3
14. 80
15. 9

Chapter 3 Study Guide and Review

Vocabulary

Choose the best term from the list above. Words may be used more than once.

1. When you estimate a sum by using only the whole-number part of the decimals, you are using __?__.
2. __?__ is a shorthand method for writing large numbers.
3. __?__ means rounding all the numbers to the same value.

3-1 Representing, Comparing, and Ordering Decimals (pp. 92–95)

EXAMPLES

- **Write 4.025 in expanded form and words.**

  Expanded form: 4 + 0.02 + 0.005

  Word form: four and twenty-five thousandths

- **Order the decimals from least to greatest. 7.8, 7.83, 7.08**

  7.08 < 7.80 < 7.83 *Compare the numbers.*
  7.08, 7.8, 7.83 *Then order the numbers.*

EXERCISES

Write each in expanded form and words.

4. 5.68
5. 1.0076
6. 1.203
7. 23.005

Order the decimals from least to greatest.

8. 1.2, 1.3, 1.12
9. 11.17, 11.7, 11.07
10. 0.3, 0.303, 0.033
11. 5.009, 5.950, 5.5

3-2 Estimating Decimals (pp. 96–99)

EXAMPLES

- **Estimate.**

  5.35 − 0.7904; round to tenths

  5.4 *Align the decimals.*
  − 0.8 *Subtract.*
  4.6

- **Estimate each product or quotient.**

  49.67 × 2.88
  50 × 3 = 150

EXERCISES

Estimate.

12. 8.0954 + 3.218; round to the hundredths
13. 6.8356 − 4.507; round to the tenths

Estimate each product or quotient.

14. 21.19 × 4.23
15. 53.98 ÷ 5.97

3-3 Adding and Subtracting Decimals (pp. 102–105)

EXAMPLE

- **Find the sum or difference.**

7.62 + 0.563

7.620	*Align the decimal points.*
+ 0.563	*Use zeros as placeholders.*
8.183	*Add. Place the decimal point.*

EXERCISES

Find each sum or difference.

16. 7.08 + 4.5 + 13.27 **17.** 6 − 0.7

Evaluate 6.48 − *s* for each value of *s*.

18. $s = 3.9$ **19.** $s = 3.6082$

3-4 Decimals and Metric Measurement (pp. 106–109)

EXAMPLES

- **Multiply or divide.**

326 × 10,000 *Move the decimal point 4 places right.*

= 3,260,000 *Write 4 placeholder zeros.*

- **Convert the measure.**

3,200 g = __?__ kg *(1,000 g = 1 kg)*

3,200 g = 3.2 kg

EXERCISES

Multiply or divide.

20. 12.6×10^4 **21.** $546 \div 10^3$

22. 67×10^5 **23.** $180.6 \div 10^2$

Convert each measure.

24. 8.9 L = __?__ mL

25. 18 cm = __?__ m

3-5 Scientific Notation (pp. 114–117)

EXAMPLES

- **Write the number in scientific notation.**

60,000 *Move the decimal point 4 places to the left.*

$= 6.0 \times 10^4$

- **Write each number in standard form.**

7.18×10^5

= 718,000 *Move the decimal point 5 places right.*

EXERCISES

Write each number in scientific notation.

26. 550,000 **27.** 7,230

28. 1,300,000 **29.** 14.8

Write each number in standard form.

30. 3.02×10^4 **31.** 4.293×10^5

32. 1.7×10^6 **33.** 5.39×10^3

3-6 Multiplying Decimals (pp. 120–123)

EXAMPLE

- **Find the product.**

0.3	*1 decimal place*
× 0.08	*+ 2 decimal places*
0.024	*3 decimal places*

EXERCISES

Find each product.

34. 4 × 2.36 **35.** 0.5 × 1.73

36. 0.6 × 0.012 **37.** 8 × 3.052

Answers

16. 24.85
17. 5.3
18. 2.58
19. 2.8718
20. 126,000
21. 0.546
22. 6,700,000
23. 1.806
24. 8,900
25. 0.18
26. 5.5×10^5
27. 7.23×10^3
28. 1.3×10^6
29. 1.48×10^1
30. 30,200
31. 429,300
32. 1,700,000
33. 5,390
34. 9.44
35. 0.865
36. 0.0072
37. 24.416

Answers

38. 1.03

39. 0.72

40. 3.85

41. 2.59

42. \$3.64

43. 8.1

44. $6.1\overline{6}$

45. $3.87\overline{6}$

46. 52.275

47. 0.75 meter

48. 14 containers

49. 9 cars

50. $a = 13.38$

51. $y = 2.62$

52. $n = 2.29$

53. $p = 60.2$

54. \$5.00

Study Guide and Review

3-7 Dividing Decimals by Whole Numbers (pp. 124–126)

EXAMPLE

■ **Find the quotient.**

Place a decimal point directly above the decimal point in the dividend. Then divide.

$$5\overline{)0.95} = 0.19$$

EXERCISES

Find each quotient.

38. $6.18 \div 6$

39. $2.16 \div 3$

40. $34.65 \div 9$

41. $20.72 \div 8$

42. If four people equally share a bill for \$14.56, how much should each person pay?

3-8 Dividing by Decimals (pp. 127–130)

EXAMPLE

■ **Find the quotient.**

$9.65 \div 0.5$

Make the divisor a whole number.
Place the decimal point in the quotient.

$$5\overline{)96.5} = 19.3$$

EXERCISES

Find each quotient.

43. $4.86 \div 0.6$

44. $1.85 \div 0.3$

45. $34.89 \div 9$

46. $62.73 \div 1.2$

47. Ana cuts some wood that is 3.75 meters long into 5 pieces of equal length. How long is each piece?

3-9 Interpret the Quotient (pp. 131–133)

EXAMPLE

■ **Ms. Ald needs 26 stickers for her preschool class. Stickers are sold in packs of 8. How many packs should she buy?**

$26 \div 8 = 3.25$

3.25 is between 3 and 4.
3 packs will not be enough.

Ms. Ald should buy 4 packs of stickers.

EXERCISES

48. Billy has 3.6 liters of juice. How many 0.25 L containers can he fill?

49. There are 34 people going on a field trip. If each car holds 4 people, how many cars will they need for the field trip?

3-10 Solving Decimal Equations (pp. 134–137)

EXAMPLE

■ **Solve the equation.**

$4x = 20.8$ — *x is multiplied by 4.*

$\frac{4x}{4} = \frac{20.8}{4}$ — *Divide both sides by 4.*

$x = 5.2$

EXERCISES

Solve each equation.

50. $a - 6.2 = 7.18$

51. $3y = 7.86$

52. $n + 4.09 = 6.38$

53. $\frac{p}{7} = 8.6$

54. Jasmine buys 2.25 kg of apples for \$11.25. How much does 1 kg of apples cost?

Chapter Test

Chapter 3

Write each decimal in standard form, expanded form, and words.

1. 3.107 3 + 0.1 + 0.007; three and one hundred seven thousandths

2. Eight and forty-nine thousandths 8.049; 8 + 0.04 + 0.009

Order the decimals from least to greatest.

3. 12.6, 12.07, 12.67 12.07, 12.6, 12.67

4. 3.5, 3.25, 3.08 3.08, 3.25, 3.5

5. 23.84, 23.59, 2.899 2.899, 23.59, 23.84

Estimate by rounding to the indicated place value.

6. 6.178 − 0.2805; hundredths 5.90

7. 7.528 + 6.075; ones 14

Estimate.

8. 21.35 × 3.18 60

9. 98.547 ÷ 4.93 20

10. 11.855 × 8.45 96

Estimate a range for the sum.

11. 3.89 + 42.71 + 12.32 at least 57, but not more than 59.5

12. 20.751 + 2.55 + 17.4 at least 39, but not more than 41

13. 4.987 + 28.27 + 0.098 at least 32, but not more than 33.5

Evaluate.

14. 0.76 + 2.24 3

15. 7 − 0.4 6.6

16. 0.12 × 0.006 0.00072

17. 76 × 10,000 760,000

18. $4.57 \div 10^3$ 0.00457

19. 3.44 ÷ 4 0.86

Convert each measure.

20. 4,700 mL = __?__ L 4.7

21. 3.2 km = __?__ m 3,200

22. 22.8 cm = __?__ m 0.228

Write each number in scientific notation.

23. 16,900 1.69×10^4

24. 180,500 1.805×10^5

25. 3,190,000 3.19×10^6

Write each number in standard form.

26. 3.08×10^5 308,000

27. 1.472×10^6 1,472,000

28. 2.973×10^4 29,730

Solve each equation.

29. $b - 4.7 = 2.1$ $b = 6.8$

30. $5a = 4.75$ $a = 0.95$

31. $\frac{y}{6} = 7.2$ $y = 43.2$

32. The school band is going to a local competition. There are 165 students in the band. If each bus holds 25 students, how many buses will be needed? 7 buses

33. Maria and five friends went shopping. All sweaters at the store were on sale for the same price. Each girl chose a sweater. The total bill was \$126.24. How much did each sweater cost? \$21.04

Chapter Test

Chapter Test

Chapter 3

Purpose: *To assess students' mastery of concepts and skills in Chapter 3*

Assessment Resources

Chapter 3 Tests (Levels A, B, C)
Assessment Resources ..... pp. 43–48

***Test and Practice Generator* CD-ROM**

Additional assessment items in both multiple-choice and free-response format may be generated for any objective in Chapter 3.

Chapter 3 Performance Assessment

Purpose: *To assess students' understanding of concepts in Chapter 3 and combined problem-solving skills*

Assessment Resources

Performance Assessment
Assessment Resources . . . . p. 108

Performance Assessment Teacher Support
Assessment Resources . . . . p. 107

Answers

1–3. See p. A2

4. See Level 3 work sample below.

Scoring Rubric for Problem Solving Item 4

Level 3

Accomplishes the purposes of the task.

Student gives clear explanations, shows understanding of mathematical ideas and processes, and computes accurately.

Level 2

Purposes of the task not fully achieved.

Student demonstrates satisfactory but limited understanding of the mathematical ideas and processes.

Level 1

Purposes of the task not accomplished.

Student shows little evidence of understanding the mathematical ideas and processes and makes computational and/or procedural errors.

Chapter 3 Performance Assessment

Performance Assessment

Show What You Know

Create a portfolio of your work from this chapter. Complete this page and include it with your four best pieces of work from Chapter 3. Choose from your homework or lab assignments, mid-chapter quiz, or any journal entries you have done. Put them together using any design you want. Make your portfolio represent what you consider to be your best work.

Short Response

1. Jocelyn and her 2 sisters went shopping for school supplies. They purchased a box of markers for \$2.99, a pack of pencils for \$2.38, 3 notebooks for \$1.59 each, and 3 videos for \$14.94 each. Find the total cost of their purchases. If the cost is divided equally among the girls, how much should each girl pay? Show your work.
2. Emily wants to rent 6 DVD movies to watch over the weekend. On Friday, she can rent 3 movies from the movie store for \$11.25. On Saturday, she can rent 2 movies for \$6.50. On which day should Emily rent movies? Explain your answer.
3. Michael is making costumes for his school play. Each costume requires 1.25 yards of fabric. If Michael has a total of 12 yards of fabric, how many costumes can he make? Explain how you found your answer.

Extended Problem Solving

4. The large rectangular plot of land in the diagram is being made into a park. The total area of the park is 4.56 square kilometers, and the length of the park is 2.4 kilometers.
   a. Let w represent the width of the park. Write an equation that could be used to find the value of w.
   b. Find the width of the park in kilometers. Show your work.
   c. Find the width of the park in meters, and write it in scientific notation.

Student Work Samples for Item 4

Level 3

a. $4.56 = 2.4 \times w$

b. $\frac{4.56}{2.4} = \frac{2.4}{2.4} \times w$

$1.9 = w$

The width of the park is 1.9 km.

c. 1.9 km = (1.9 × 1,000)m = 1900m

The width of the park in meters is 1.9×10^3.

The student correctly answered parts a, b, and c, and showed understanding of area and scientific notation.

Level 2

a. $4.56 = w \times 2.4$

b. $\frac{4.56}{2.4} = w \times \frac{2.4}{2.4}$

1.9 km = w

c. 1.9 km = (1.9 × 1,000)m = 1900m

The student correctly answered parts a and b, but did not give the answer to part c in scientific notation.

Level 1

a. $w \cdot 2.4 = 4.56$

b. $w = 4.56 \times 2.4$ = 6.144 km

c. 6.144 km = 6,144 m

The student performed an incorrect calculation in part b. The student answered part c incorrectly and did not write the answer in scientific notation.

internet connect
State-Specific Test Practice Online
go.hrw.com Keyword: MR4 TestPrep

Standardized Test Prep

Chapter 3

Cumulative Assessment, Chapters 1–3

1. Which of the following is the standard form for four and seven hundred eighteen ten-thousandths? C
   (A) 4.718 (C) 4.0718
   (B) 4.7018 (D) 4.00718

2. What is the value of $2^2 \times 11 - 7 + 2^3$? F
   (F) 45 (H) 43
   (G) 17 (J) 22

3. What is the product of 0.4×3.25? D
   (A) 1,300 (C) 13.00
   (B) 130 (D) 1.30

TIP! TEST TAKING TIP!
To compare decimal numbers, start from the left and compare digits in the same place-value position. Look for the first place value the digits are different.

4. Which number is the greatest? H
   (F) 18.095 (H) 18.907
   (G) 18.9 (J) 18.75

5. Which equation has 8 as the solution? B
   (A) $y - 3 = 6$
   (B) $5y = 40$
   (C) $\frac{y}{4} = 8$
   (D) $y + 8 = 32$

6. Which is the quotient of $7.89 \div 3$? J
   (F) 263 (H) 0.263
   (G) 26.30 (J) 2.63

7. Which measurement is equivalent to 780 cm? C
   (A) 78 mm (C) 7.8 m
   (B) 0.78 km (D) 7,800 km

8. The area of a rectangle is 83.125 cm^2. Its length is 9.5 cm. Solve $9.5w = 83.125$ to find the width. G
   (F) 7.89 cm
   (G) 8.75 cm
   (H) 9 cm
   (J) 0.875 cm

9. Mr. Myers is placing 6 lights along his driveway. If the length of the driveway is 14.4 meters, how far apart should he place the lights to have them equally spaced along the driveway? D

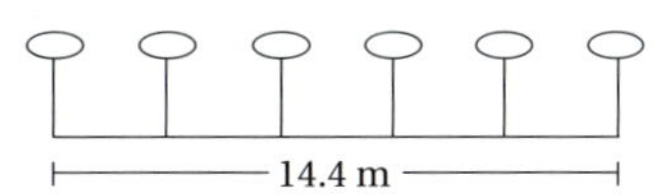

   (A) 86.4 meters
   (B) 2.4 meters
   (C) 20.4 meters
   (D) 2.88 meters

10. ***SHORT RESPONSE*** Explain why 300 cm is the same as 3 m.

11. ***SHORT RESPONSE*** Explain why 54.9×10^6 is not correctly written in scientific notation. What is the correct way to write 54.9×10^6 in scientific notation?

Standardized Test Prep

Chapter 3

Purpose: *To provide review and practice for Chapters 1–3 and standardized tests*

Assessment Resources

Cumulative Tests (Levels A, B, C)
Assessment Resources . . . . pp. 151–162

State-Specific Test Practice Online
KEYWORD: MR4 TestPrep

Test Prep Doctor

In item 3, estimate 0.4×3.25 to be $0.5 \times 3 = 1.5$. Eliminate answer choices **A, B,** and **C** as being too high.

Point out to students that in item 5 they could either solve each equation or substitute 8 into each equation.

Answers

10. There are 100 cm in 1 m. Three groups of 100 cm is 300 cm, so three groups of 1 m is 3 m.

11. The number in front of the multiplication sign must be a number greater than or equal to 1 and less than 10; 5.49×10^7.

Chapter 4

Number Theory and Fractions

Section 4A	Section 4B	Section 4C
Number Theory	**Understanding Fractions**	**Introduction to Fraction Operations**
Lesson 4-1 Divisibility **Lesson 4-2** Factors and Prime Factorization **Lesson 4-3** Greatest Common Factor	**Hands-On Lab 4A** Explore Decimals and Fractions **Lesson 4-4** Decimals and Fractions **Hands-On Lab 4B** Model Equivalent Fractions **Lesson 4-5** Equivalent Fractions **Hands-On Lab 4C** Explore Fraction Measurement **Lesson 4-6** Comparing and Ordering Fractions **Lesson 4-7** Mixed Numbers and Improper Fractions	**Lesson 4-8** Adding and Subtracting with Like Denominators **Lesson 4-9** Multiplying Fractions by Whole Numbers **Extension** Sets of Numbers

Pacing Guide for 45-Minute Classes

Chapter 4

DAY 39	DAY 40	DAY 41	DAY 42	DAY 43
Lesson 4-1	Lesson 4-2	Lesson 4-3	Mid-Chapter Quiz Hands-On Lab 4A	Lesson 4-4
DAY 44	**DAY 45**	**DAY 46**	**DAY 47**	**DAY 48**
Hands-On Lab 4B	Lesson 4-5	Hands-On Lab 4C	Lesson 4-6	Lesson 4-7
DAY 49	**DAY 50**	**DAY 51**	**DAY 52**	**DAY 53**
Mid-Chapter Quiz Lesson 4-8	Lesson 4-9	Extension	Chapter 4 Review	Chapter 4 Assessment

Pacing Guide for 90-Minute Classes

Chapter 4

DAY 19	DAY 20	DAY 21	DAY 22	DAY 23
Chapter 3 Review Lesson 4-1	Chapter 3 Assessment Lesson 4-2	Lesson 4-3 Hands-On Lab 4A	Mid-Chapter Quiz Lesson 4-4 Hands-On Lab 4B	Lesson 4-5 Hands-On Lab 4C
DAY 24	**DAY 25**	**DAY 26**	**DAY 27**	
Lesson 4-6 Lesson 4-7	Mid-Chapter Quiz Lesson 4-8 Lesson 4-9	Extension Chapter 4 Review	Chapter 4 Assessment Hands-On Lab 5A	

Across the Series

HARCOURT GRADE 5

- Use divisibility rules and find the LCM and GCF of a set of numbers.
- Identify prime and composite numbers and write prime factorizations.
- Write equivalent decimals and fractions.
- Compare, order, and simplify fractions.

COURSE 1

- **Use divisibility rules, write prime factorizations of composite numbers, and find the greatest common factor (GCF) of a set of numbers.**
- **Write equivalent decimals, fractions, mixed numbers, and improper fractions.**
- **Compare and order fractions.**
- **Add and subtract fractions with like denominators.**
- **Multiply fractions by whole numbers.**
- **Make Venn diagrams to describe number sets.**

COURSE 2

- Review divisibility rules and prime factorization.
- Find the GCF and LCM of a set of numbers.
- Compare, order, simplify, and write equivalent rational numbers.
- Define and recognize irrational numbers.

Across the Curriculum

LANGUAGE ARTS LINK

Math: Reading and Writing in the Content Area . . . . pp. 25–33

Focus on Problem Solving
Understand the Problem . . . . SE pp. 165, 187

Journal . . . . TE, last page of each lesson

Write About It . . . . SE, last page of each lesson

SOCIAL STUDIES LINK

Social Studies . . . . SE pp. 163, 175, 185, 193

SCIENCE LINK

Astronomy . . . . SE pp. 155, 182

Life Science . . . . SE pp. 159, 163, 170, 185, 188, 191, 195

Physical Science . . . . SE p. 158

TE = *Teacher's Edition* **SE** = *Student Edition*

Interdisciplinary Bulletin Board

Life Science

Although grass uses most of the energy it obtains from the Sun for its own life processes, it stores $\frac{1}{10}$ of the available energy in its tissues. If there are 12,000 units of energy from the Sun available to grass, calculate the units of the energy that the grass will store. Grass stores $\frac{1}{10}$ (12,000) = 1,200 units.

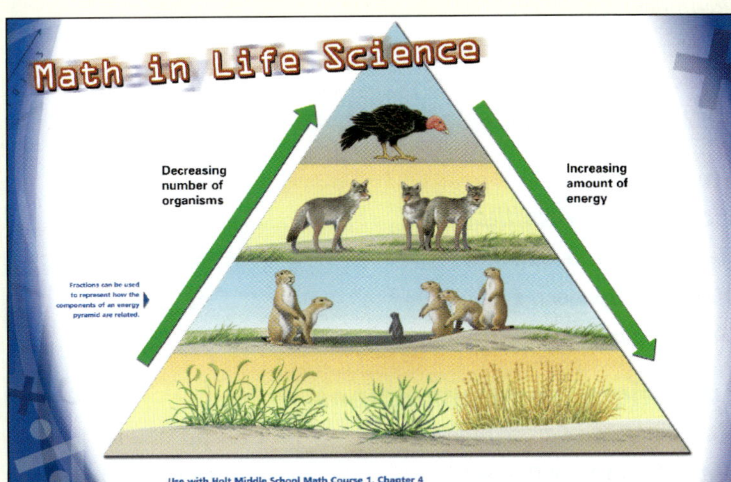

Interdisciplinary posters and worksheets are provided in your resource material.

Chapter 4

Resource Options

Chapter 4 Resource Book

Student Resources

Practice (Levels A, B, C) . . . . pp. 9–11, 19–21, 28–30, 38–40, 48–50, 57–59, 66–68, 76–78, 85–87
Reteach . . . . pp. 12–13, 22, 31, 41–42, 51, 60, 69, 79, 88
Challenge . . . . pp. 14, 23, 32, 43, 52, 61, 70, 80, 89
Problem Solving . . . . pp. 15, 24, 33, 44, 53, 62, 71, 81, 90
Puzzles, Twisters & Teasers . . . . pp. 16, 25, 34, 45, 54, 63, 72, 82, 91
Recording Sheets . . . . pp. 3–4, 8, 18, 27, 37, 47, 56, 65, 75, 84, 95, 99, 102
Chapter Review . . . . pp. 92–94

Teacher and Parent Resources

Chapter Planning and Pacing Guide . . . . p. 5
Section Planning Guides . . . . pp. 6, 35, 73
Parent Letter . . . . pp. 1–2
Teaching Tools . . . . pp. 98–102
Teacher Support for Chapter Project . . . . p. 96
Transparencies . . . . pp. T1–T39
- Daily Transparencies
- Additional Examples Transparencies
- Teaching Transparencies

Reaching All Learners

English Language Learners

Success for English Language Learners . . . . pp. 49–66
Math: Reading and Writing in the Content Area . . . . pp. 25–33
Spanish Homework and Practice . . . . pp. 25–33
Spanish Interactive Study Guide . . . . pp. 25–33
Spanish Family Involvement Activities . . . . pp. 25–36
Multilingual Glossary

Individual Needs

Are You Ready? Intervention and Enrichment . . . . pp. 33–40, 69–72, 225–228, 411–412
Alternate Openers: Explorations . . . . pp. 25–33
Family Involvement Activities . . . . pp. 25–36
Interactive Problem Solving . . . . pp. 25–33
Interactive Study Guide . . . . pp. 25–33
Readiness Activities . . . . pp. 7–8
Math: Reading and Writing in the Content Area . . . . pp. 25–33
Challenge . . . . . CRB pp. 14, 23, 32, 43, 52, 61, 70, 80, 89

Hands-On

Hands-On Lab Activities . . . . pp. 20–26
Technology Lab Activities . . . . pp. 11–15
Alternate Openers: Explorations . . . . pp. 25–33
Family Involvement Activities . . . . pp. 25–36

Applications and Connections

Consumer and Career Math . . . . pp. 13–16
Interdisciplinary Posters . . . . Poster 4, TE p. 150B
Interdisciplinary Poster Worksheets . . . . pp. 10–12

Transparencies

Alternate Openers: Explorations . . . . pp. 25–33
Exercise Answers Transparencies
Chapter 4 Resource Book . . . . pp. T1–T39
- Daily Transparencies
- Additional Examples Transparencies
- Teaching Transparencies

Technology

Teacher Resources

***Lesson Presentations* CD-ROM** . . . . Chapter 4
***Test and Practice Generator* CD-ROM** . . . . Chapter 4
***One-Stop Planner* CD-ROM** . . . . Chapter 4

Student Resources

***Are You Ready? Intervention* CD-ROM**
Skills 6, 7, 15, 54

internet connect go.hrw.com

Homework Help Online	**KEYWORD:** MR4 HWHelp4
Math Tools Online	**KEYWORD:** MR4 Tools
Glossary Online	**KEYWORD:** MR4 Glossary
Chapter Project Online	**KEYWORD:** MR4 PSProject4
Chapter Opener Online	**KEYWORD:** MR4 Ch4
CNN Student News	**KEYWORD:** MR4 CNN4

SE = *Student Edition* **TE** = *Teacher's Edition* **AR** = *Assessment Resources* **CRB** = *Chapter Resource Book* **MK** = *Manipulatives Kit*

Chapter 4

Assessment Options

Assessing Prior Knowledge

Determine whether students have the required prerequisite concepts and skills.

Are You Ready? . . . . SE p. 151
Inventory Test . . . . AR pp. 1–4

Test Preparation

Provide review and practice for chapter and standardized tests.

Standardized Test Prep . . . . SE p. 207
Spiral Review with Test Prep . . . . . SE, last page of each lesson
Study Guide and Review . . . . SE pp. 202–204
Test Prep Tool Kit

Technology

***Test and Practice Generator* CD-ROM**

internet connect

State-Specific Test Practice Online KEYWORD: MR4 TestPrep

Performance Assessment

Assess students' understanding of chapter concepts and combined problem-solving skills.

Performance Assessment . . . . SE p. 206
Includes scoring rubric in TE
Performance Assessment . . . . AR p. 110
Performance Assessment Teacher Support . . . . AR p. 109

Portfolio

Portfolio opportunities appear throughout the Student and Teacher's Editions.

Suggested work samples:

Problem Solving Project . . . . TE p. 150
Performance Assessment . . . . SE p. 206
Portfolio Guide . . . . AR p. xxxv
Journal . . . . TE, last page of each lesson
Write About It . . . . SE, last page of each lesson

Daily Assessment

Obtain daily feedback on students' understanding of concepts.

Spiral Review and Test Prep . . . . . . SE, last page of each lesson

Also Available on Transparency In Chapter 4 Resource Book

Warm Up . . . . TE, first page of each lesson
Problem of the Day . . . . TE, first page of each lesson
Lesson Quiz . . . . TE, last page of each lesson

Student Self-Assessment

Have students evaluate their own work.

Group Project Evaluation . . . . AR p. xxxii
Individual Group Member Evaluation . . . . AR p. xxxiii
Portfolio Guide . . . . AR p. xxxv
Journal . . . . TE, last page of each lesson

Formal Assessment

Assess students' mastery of concepts and skills.

Section Quizzes . . . . AR pp. 11–13
Mid-Chapter Quizzes . . . . SE pp. 164, 186
Chapter Test . . . . SE p. 205
Chapter Tests (Levels A, B, C) . . . . AR pp. 49–54
Cumulative Tests (Levels A, B, C) . . . . AR pp. 163–174
Standardized Test Prep
Cumulative Assessment . . . . SE p. 207
End-of-Year Test . . . . AR pp. 271–274

Technology

***Test and Practice Generator* CD-ROM**

Make tests electronically. This software includes:

- Dynamic practice for Chapter 4
- Customizable tests
- Multiple-choice items for each objective
- Free-response items for each objective
- Teacher management system

SE = *Student Edition* **TE** = *Teacher's Edition* **AR** = *Assessment Resources* **CRB** = *Chapter Resource Book* **MK** = *Manipulatives Kit*

Chapter 4

Chapter 4 Tests

Three levels (A,B,C) of tests are available for each chapter in the *Assessment Resources.*

LEVEL A

Chapter 4 Chapter Test *Form A*

Tell whether each number is divisible by 2, 5, and 10.

1. 40 — 2, 5, 10
2. 75 — 5
3. 130 — 2, 5, 10

Tell whether each number is prime or composite.

4. 13 — prime
5. 84 — composite
6. 21 — composite

List all the factors of each number.

7. 12 — 1, 2, 3, 4, 6, 12
8. 50 — 1, 2, 5, 10, 25, 50
9. 110 — 1, 2, 5, 10, 11, 22, 55, 110

Write the prime factorization of each number.

10. 18 — $2 \cdot 3^2$
11. 24 — $2^3 \cdot 3$
12. 34 — $2 \cdot 17$

Find the GCF of each pair of numbers.

13. 4 and 8 — 4
14. 16 and 40 — 8
15. 8 and 24 — 8

Write each decimal as a fraction or a mixed number in simplest form.

16. 0.5 — $\frac{1}{2}$
17. 1.25 — $1\frac{1}{4}$
18. 0.8 — $\frac{4}{5}$

Write each fraction or mixed number as a decimal.

19. $\frac{1}{10}$ — 0.1
20. $2\frac{1}{4}$ — 2.25
21. $9\frac{3}{8}$ — 9.375

Chapter 4 Chapter Test *Form A, continued*

Write each fraction in simplest form.

22. $\frac{12}{24} =$ $\frac{1}{2}$
23. $\frac{18}{27} =$ $\frac{2}{3}$
24. $\frac{9}{15} =$ $\frac{3}{5}$

Compare. Write <, >, or =.

25. $\frac{5}{12} < \frac{7}{12}$
26. $\frac{1}{3} < \frac{3}{4}$
27. $\frac{2}{3} > \frac{3}{5}$

Order the fractions from least to greatest.

28. $\frac{3}{8}, \frac{1}{8}, \frac{7}{8}$ — $\frac{1}{8}, \frac{3}{8}, \frac{7}{8}$
29. $\frac{1}{2}, \frac{1}{4}, \frac{2}{3}$ — $\frac{1}{4}, \frac{1}{2}, \frac{2}{3}$
30. $\frac{5}{6}, \frac{3}{4}, \frac{4}{5}$ — $\frac{3}{4}, \frac{4}{5}, \frac{5}{6}$

Write each mixed number as an improper fraction.

31. $2\frac{1}{2} =$ $\frac{5}{2}$
32. $3\frac{1}{4} =$ $\frac{13}{4}$
33. $4\frac{3}{8} =$ $\frac{35}{8}$

Multiply. Write your answers in simplest form.

34. $5 \cdot \frac{1}{5}$ — 1
35. $12 \cdot \frac{7}{24}$ — $\frac{7}{2}$
36. $18 \cdot \frac{1}{9}$ — 2

Solve.

37. Beth has a piece of rope $8\frac{7}{8}$ feet long. She is going to cut a piece $1\frac{4}{8}$ feet long. How much rope will she have left? $7\frac{3}{8}$ feet
38. Juan bought $2\frac{1}{2}$ pounds of potato salad and $1\frac{1}{2}$ pounds of coleslaw. How many pounds of food did he buy from the deli? 4 pounds
39. On Wednesday, it rained $1\frac{1}{4}$ inches. On Monday, an additional $\frac{3}{4}$ of an inch fell. How much rain fell altogether? 2 inches
40. A ladies auxiliary served $3\frac{1}{2}$ loaves of banana bread at a luncheon. There were 8 equal slices in every whole loaf. How many slices were served? 28 slices

LEVEL B

Chapter 4 Chapter Test *Form B*

Tell whether each number is divisible by 2, 3, 4, 5, 6, 9, and 10.

1. 840 — 2, 3, 4, 5, 6, 10
2. 875 — 5
3. 1,430 — 2, 5, 10

Tell whether each number is prime or composite.

4. 47 — prime
5. 112 — composite
6. 61 — prime

List all the factors of each number.

7. 49 — 1, 7, 49
8. 100 — 1, 2, 4, 5, 10, 20, 25, 50, 100
9. 144 — 1, 2, 3, 4, 6, 8, 9, 12, 16, 18, 24, 36, 48, 72, 144

Write the prime factorization of each number.

10. 45 — $3^2 \cdot 5$
11. 65 — $5 \cdot 13$
12. 132 — $2^2 \cdot 3 \cdot 11$

Find the GCF of each set of numbers.

13. 54 and 80 — 2
14. 52 and 26 — 26
15. 30, 60, and 90 — 30

Write each decimal as a fraction or a mixed number in simplest form.

16. 0.6 — $\frac{3}{5}$
17. 5.75 — $5\frac{3}{4}$
18. 0.125 — $\frac{1}{8}$

Write each fraction or mixed number as a decimal.

19. $\frac{3}{20}$ — 0.15
20. $6\frac{1}{5}$ — 6.2
21. $9\frac{11}{15}$ — $9.7\overline{3}$

Chapter 4 Chapter Test *Form B, continued*

Write each fraction in simplest form.

22. $\frac{6}{54} =$ $\frac{1}{9}$
23. $\frac{24}{48} =$ $\frac{1}{2}$
24. $\frac{14}{21} =$ $\frac{2}{3}$

Compare. Write <, >, or =.

25. $\frac{5}{8} < \frac{6}{7}$
26. $\frac{3}{4} < \frac{9}{10}$
27. $\frac{8}{15} = \frac{24}{45}$

Order the fractions from least to greatest.

28. $\frac{1}{2}, \frac{4}{5}, \frac{5}{6}$ — $\frac{1}{2}, \frac{4}{5}, \frac{5}{6}$
29. $\frac{5}{9}, \frac{2}{5}, \frac{6}{7}$ — $\frac{2}{5}, \frac{5}{9}, \frac{6}{7}$
30. $\frac{7}{9}, \frac{3}{4}, \frac{11}{15}$ — $\frac{11}{15}, \frac{3}{4}, \frac{7}{9}$

Write each mixed number as an improper fraction.

31. $5\frac{2}{3} =$ $\frac{17}{3}$
32. $3\frac{2}{5} =$ $\frac{17}{5}$
33. $7\frac{5}{9} =$ $\frac{68}{9}$

Multiply. Write your answers in simplest form.

34. $5 \cdot \frac{1}{8}$ — $\frac{5}{8}$
35. $12 \cdot \frac{1}{6}$ — 2
36. $15 \cdot \frac{1}{4}$ — $3\frac{3}{4}$

Solve.

37. On Monday, it snowed $11\frac{1}{2}$ inches. On Tuesday, an additional $1\frac{1}{2}$ inches of snow fell. How much snow fell altogether? 13 inches
38. Linda has a piece of ribbon $15\frac{7}{8}$ inches long. She cuts a piece $3\frac{1}{8}$ inches long. How much ribbon does she have left? $12\frac{3}{4}$ feet
39. Roberto bought of $2\frac{1}{4}$ pounds of ham and $3\frac{5}{8}$ pounds of turkey. How much lunch meat did he buy? $5\frac{7}{8}$ pounds
40. A food service class served $15\frac{1}{2}$ loaves of pumpkin bread at the faculty breakfast. There were 10 equal slices in every whole loaf. How many slices were served? 155 slices

LEVEL C

Chapter 4 Chapter Test *Form C*

Tell whether each number is divisible by 2, 3, 4, 5, 6, 9, and 10.

1. 150 — 2, 3, 5, 6, 10
2. 2,235 — 3, 5
3. 6,720 — 2, 3, 4, 5, 6, 10

Tell whether each number is prime or composite.

4. 213 — composite
5. 4,102 — composite
6. 97 — prime

List all the factors of each number.

7. 94 — 1, 2, 47, 94
8. 86 — 1, 2, 43, 86
9. 1,000 — 1, 2, 4, 5, 8, 10, 20, 25, 40, 50, 100, 125, 200, 250, 500, 1,000

Write the prime factorization of each number.

10. 111 — $3 \cdot 37$
11. 676 — $2^2 \cdot 13^2$
12. 324 — $2^2 \cdot 3^4$

Find the GCF of each set of numbers.

13. 22, 88, and 121 — 11
14. 42, 98, and 112 — 14
15. 32, 54, 72, and 80 — 2

Write each decimal as a fraction or a mixed number in simplest form.

16. 0.7 — $\frac{7}{10}$
17. 12.225 — $12\frac{9}{40}$
18. 4.568 — $4\frac{71}{125}$

Write each fraction or mixed number as a decimal.

19. $\frac{5}{12}$ — $0.41\overline{6}$
20. $8\frac{25}{33}$ — $8.\overline{75}$
21. $2\frac{5}{24}$ — $2.208\overline{3}$

Chapter 4 Chapter Test *Form C, continued*

Write each fraction in simplest form.

22. $\frac{12}{54} =$ $\frac{2}{9}$
23. $\frac{108}{162} =$ $\frac{2}{3}$
24. $\frac{99}{165} =$ $\frac{3}{5}$

Compare. Write <, >, or =.

25. $\frac{1}{8} < \frac{3}{16}$
26. $\frac{38}{57} > \frac{95}{152}$
27. $\frac{84}{105} > \frac{42}{63}$

Order the fractions from least to greatest.

28. $\frac{8}{9}, \frac{7}{12}, \frac{3}{4}$ — $\frac{7}{12}, \frac{3}{4}, \frac{8}{9}$
29. $\frac{19}{21}, \frac{15}{17}, \frac{7}{13}$ — $\frac{7}{13}, \frac{15}{17}, \frac{19}{21}$
30. $\frac{51}{56}, \frac{12}{13}, \frac{19}{20}$ — $\frac{51}{56}, \frac{12}{13}, \frac{19}{20}$

Write each mixed number as an improper fraction.

31. $22\frac{7}{8} =$ $\frac{183}{8}$
32. $34\frac{11}{15} =$ $\frac{521}{15}$
33. $14\frac{18}{19} =$ $\frac{284}{19}$

Multiply. Write your answers in simplest form.

34. $5 \cdot \frac{1}{50}$ — $\frac{1}{10}$
35. $12 \cdot \frac{7}{36}$ — $2\frac{1}{3}$
36. $18 \cdot \frac{1}{81}$ — $\frac{2}{9}$

Solve.

37. Jordan bought $15\frac{5}{8}$ inches of lace from the fabric store on Friday. On Saturday, Jordan bought an additional $14\frac{1}{8}$ inches of lace. How many inches of lace did Jordan buy altogether? $29\frac{3}{4}$ in.
38. Juanita has an oak board $19\frac{11}{16}$ feet long. She is going to cut off a piece of board $6\frac{5}{16}$ feet long. How much of the original board will she have left? $13\frac{3}{8}$ ft
39. Roberto needs $5\frac{1}{2}$ lb of lunch meat for the picnic. He bought $2\frac{1}{4}$ lb of ham and $3\frac{5}{8}$ lb of turkey. How much extra lunch meat did he buy? $\frac{3}{8}$ lb
40. A store has $25\frac{3}{4}$ loaves of bread cut into slices as samples. If there are 12 equal slices in every whole loaf, how many slices are there? 309 slices

***Test and Practice Generator* CD-ROM**

Create and customize multiple versions of the same tests with corresponding answers for any chosen chapter objectives.

Chapter 4 State and Standardized Test Preparation

Test Taking Skill Builder and Standardized Test Practice are provided for each chapter in the *Test Prep Tool Kit.*

Test Taking Skill Builder

Test Taking Strategy **Know How the Test Is Scored**
Chapter 4

Different standardized tests are scored in different ways. Pay attention to the directions the test proctor reads aloud. It helps if you know ahead of time how the test is scored.

- Some multiple choice sections have no penalty for guessing. In this case, be sure to answer every question.
- If the test has a penalty for guessing and you cannot eliminate any answer choices, it is best to leave the question blank.
- Extended response questions are graded using a scoring rubric. Never leave an extended response question blank. Always show each step of your calculations, and explain your thinking process.

Example Extended Response Every 15 days, Jan visits her grandmother. Every 12 days, she visits her uncle. In how many days will Jan visit her grandmother and her uncle on the same day? Show all of your steps and explain your reasoning.

Scoring Rubric
- **3 points:** Student recognizes the need to find the LCM of 12 and 15. Shows calculations to find LCM. Explains reasoning and answers in a complete sentence.
- **2 points:** Student recognizes the need to find LCM but calculates incorrectly, or calculates LCM correctly but does not explain reasoning.
- **1 point:** Student gives a numeric answer but includes no explanation or work.
- **0 points:** Student gives incorrect solution with no explanation or work included.

Solution: A possible 2-point response is shown below. Notice that although the student calculates the LCM as 120 rather than 60, he still receives 2-points for the response.

Find the LCM of 12 and 15.
12: 12, 24, 36, 48, 60, 72, 84, 96, 108, 120
15: 15, 30, 45, 60, 75, 90, 105, 120

The LCM of 12 and 15 is 120. On the 120th day, Sam will visit both his grandmother and his uncle on the same day.

Test Taking Strategy
Chapter 4, continued

Exercises
Use the scoring rubric to answer the following questions.

Scoring Rubric
- **3 points:** Student recognizes the need to find the total number of cars in the lot. Shows calculations, explains reasoning, and answers in a complete sentence.
- **2 points:** Student recognizes the need to find the total number of cars, but calculates incorrectly, or calculates correctly but does not explain reasoning correctly.
- **1 point:** Student gives a numeric answer but includes no explanation or work.
- **0 points:** Student gives incorrect solution with no explanation or work included.

1. A parking lot has space for 1,000 cars but $\frac{2}{5}$ of the spaces are for compact cars. On Tuesday, there were 250 compact cars and some standard size cars in the parking lot. The parking lot was $\frac{3}{4}$ full. How many standard size cars were in the parking lot?

   Solution: 2-point response
   I know that the parking lot is $\frac{3}{4}$ full. $\frac{3}{4} \times 1{,}000 = 750$.
   There are 250 compact cars.
   500 cars

   a. Explain why the student only received 2-points for this response.
   Possible answer: The student failed to show the calculations to arrive at the answer of 500 cars. The student failed to give a complete sentence.

   b. Write a 3-point response.
   I know that the parking lot is $\frac{3}{4}$ full. $\frac{3}{4} \times 1{,}000 = 750$.
   There are 250 compact cars. $750 - 250 = 500$
   There are 500 standard size cars in the parking lot.

   c. What would have happened had the student only given a correct numeric answer?
   The student would have only received 1-point.

Standardized Test Practice

Standardized Test Practice
Chapter 4

Select the best answer for Questions 1–7.

1. For Nina's birthday party, her parents buy a piñata. If there will be a total of nine children at the party, which size package of party favors should her parents buy so that all the children will get an equal amount?
   A 48 favors
   B 97 favors
   C 116 favors
   D 135 favors

2. After two minutes of biking, Brad had biked $\frac{1}{2}$ kilometer, Harold $\frac{3}{4}$ kilometer, Kylie $\frac{4}{5}$ kilometer and Lyle $\frac{2}{3}$ kilometer. Who had biked the farthest?
   F Brad
   G Harold
   H Kylie
   I Lyle

3. Order the fractions and decimals from least to greatest.
   $0.3, \frac{2}{7}, \frac{1}{3}, 0.37$
   A $\frac{2}{7}, 0.3, \frac{1}{3}, 0.37$
   B $0.3, \frac{2}{7}, 0.37, \frac{1}{3}$
   C $\frac{1}{3}, \frac{2}{7}, 0.3, 0.37$
   D $\frac{2}{7}, \frac{1}{3}, 0.3, 0.37$

4. The 6th grade basketball team, have scored 41, 23, 39, and 47 points at their last four games. Which score is a composite number?
   F 23
   G 39
   H 41
   I 47

5. A recipe for your favorite dessert calls for $\frac{2}{3}$ cup of cocoa. How much do you need if you want to make a double batch?
   A $\frac{3}{4}$ cup
   B $1\frac{1}{4}$ cup
   C $1\frac{1}{3}$ cup
   D $1\frac{2}{3}$ cup

6. There are 12 players on the Braves youth baseball team and only $\frac{3}{4}$ of them showed up for practice on Monday night. How many players came to practice?
   F 3 players H 9 players
   G 6 players I 11 players

7. In geometry the fraction $\frac{22}{7}$ is often used as an approximation for pi. What is this improper fraction as a mixed number?
   A $\frac{7}{22}$ C $2\frac{2}{7}$
   B $1\frac{2}{7}$ D $3\frac{1}{7}$

Standardized Test Practice
Chapter 4, continued

Gridded Response
Solve the problems. Use the answer sheet to write and grid-in your answer.

8. What missing number would make the fractions equivalent?
   $\frac{32}{48} = \frac{8}{?}$

9. Nicole is having a wedding shower for her best friend. She is making balloon centerpieces for each table. She has 48 green balloons and 64 white balloons. If she would like the centerpieces to have the greatest equal number of each color balloon, how centerpieces can she make?

10. What number has the prime factorization of $2^3 \times 3^2 \times 7 \times 13$?

Short Response
Solve the problems. Use the answer sheet to write your answers.

11. Eleanor is making candle gift baskets. She has 12 green votives, 16 blue votives, and 24 yellow votives. If she would like the baskets to have the greatest equal number of each color in them, how many baskets will Eleanor be able to make? Explain in words how you determined your answer.

12. In the picture, 6 of the 8 equal pieces are shaded. Determine an equivalent fraction if there are 24 pieces. Draw and shade a picture to represent the equivalent fraction.

Extended Response

13. Lori received the following scores on her math quizzes.
   $\frac{10}{15}, \frac{21}{30}, \frac{5}{5}, \frac{56}{60}, \frac{14}{15}$
   a. Write an equivalent fraction for each score using a denominator of 60.
   b. Order the scores from least to greatest. Show your work.
   c. Lori has another quiz tomorrow worth 30 points. If she wants to score the same on this quiz as she did on the one worth 60 points, how many points does she need? Show your work.

...r Sheet

See Lesson
See Lesson
See Lesson

10. 6552
See Lesson 4-2.

Write your answers in the space provided.

11. 3 baskets; To find the greatest equal number of each color, I found the g... common factor of 12, 16, and 24, which is 4. Because she has 12 green votives, if 4 go into each basket, she can make 3 baskets. (See Les...

12. The equivalent fraction is $\frac{18}{24}$. (See Les...

Extended Response
Write your answers for Problem 13 on the back of this paper.

internet connect
State-Specific Test Practice Online
KEYWORD: MR4 TestPrep

Test Prep Tool Kit

- Standardized Test Prep Workbook
- Countdown to Testing transparencies
- State Test Prep CD-ROM
- Standardized Test Prep Video

Customized answer sheets give students realistic practice for actual standardized tests.

Chapter 4

Number Theory and Fractions

Why Learn This?

Tell students that many formulas used in the real world involve division. Tell them that when they are performing division, it is helpful to understand fractions and the relationship between fractions and decimals. For example, in the plumber's formula, it may be helpful to realize that $\frac{\text{cost of pipe}}{3}$ can be interpreted as (cost of pipe) $\div$ 3 or as $\frac{1}{3} \times$ (cost of pipe).

Using Data

To begin the study of this chapter, have students:

- Estimate the cost of installing a 4 in. × 10 ft pipe with one $\frac{1}{8}$-bend connection. about $203
- Calculate the exact cost of installing a 4 in. × 20 ft pipe with one straight coupling. $369.74
- Round the cost of each connection to the nearest half-dollar and then write each cost as a mixed number. $\frac{1}{4}$ bend: $6\frac{1}{2}$; $\frac{1}{8}$ bend: 6; $\frac{1}{6}$ bend: $7\frac{1}{2}$

Chapter 4

TABLE OF CONTENTS

internet connect
Chapter Opener Online
go.hrw.com
KEYWORD: MR4 Ch4

Number Theory and Fractions

ABS Plastic Drain Pipe

Component	Cost ($)
Pipe 4 in. × 10 ft	11.99
Pipe 4 in. × 20 ft	22.57
Straight coupling	2.19
$\frac{1}{4}$-bend connection	6.49
$\frac{1}{8}$-bend connection	5.99
$\frac{1}{6}$-bend connection	7.49

Career Plumber

Do you like working with your hands to solve problems? If so, you might want to become a skilled trade worker, such as a master plumber.

To calculate the cost of parts and labor, plumbers use basic mathematical formulas. For example, some plumbers might calculate the cost of a new sewer line with a formula like the following:

cost of installed line $= \frac{\text{cost of pipe}}{3} \times 49 + \frac{\text{cost of pipe fittings}}{2}$

PROBLEM SOLVING

Problem Solving Project

Physical Science and Social Studies Connection

Purpose: To solve problems using decimals and fractions

Materials: Plumber's Cost/Profit Analysis worksheet

internet connect
Chapter Project Online: *go.hrw.com*
KEYWORD: MR4 PSProject4

Understand, Plan, Solve, and Look Back

Have students:

- ✓ Complete the Plumber's Cost/Profit Analysis worksheet to learn how to use fractions and decimals to determine prices.
- ✓ Research to find out what the word *plumber* means.
- ✓ Research to find out what an ABS plastic pipe is, why it is used for drainpipe, and what other kinds of plastic pipe are used in building construction.
- ✓ Check students' work.

ARE YOU READY?

Choose the best term from the list to complete each sentence.

denominator
difference
multiples
numerator
product
quotient
add

1. To find the sum of two numbers, you should __?__. add
2. Fractions are written as a __?__ over a __?__. numerator; denominator
3. In the equation $4 \cdot 3 = 12$, 12 is the __?__. product
4. The __?__ of 18 and 10 is 8. difference
5. The numbers 18, 27, and 72 are __?__ of 9. multiples

Complete these exercises to review skills you will need for this chapter.

✓ Write and Read Decimals

Write each decimal in word form.

6. 0.5 five tenths
7. 2.78 two and seventy-eight hundredths
8. 0.125 one hundred twenty-five thousandths
9. 12.8 twelve and eight tenths
10. 125.49 one hundred twenty-five and forty-nine hundredths
11. 8.024 eight and twenty-four thousandths

✓ Identify Sets of Numbers

Determine whether each number is even or odd.

12. 125 odd
13. 29 odd
14. 24 even
15. 127 odd
16. 213 odd
17. 98 even
18. 2 even
19. 17 odd

✓ Multiples

List the first four multiples of each number.

20. 6 — 6, 12, 18, 24
21. 8 — 8, 16, 24, 32
22. 5 — 5, 10, 15, 20
23. 12 — 12, 24, 36, 48
24. 7 — 7, 14, 21, 28
25. 20 — 20, 40, 60, 80
26. 14 — 14, 28, 42, 56
27. 9 — 9, 18, 27, 36

✓ Evaluate Expressions

Evaluate each expression for the given value of the variable.

28. $y + 4.3$ for $y = 3.2$ 7.5
29. $3c$ for $c = 0.75$ 2.25
30. $27.8 - d$ for $d = 9.25$ 18.55
31. $\frac{x}{5}$ for $x = 6.4$ 1.28
32. $a + 4 \div 8$ for $a = 3.75$ 4.25
33. $2.5b$ for $b = 8.4$ 21

Assessing Prior Knowledge

INTERVENTION

Diagnose and Prescribe

Evaluate your students' performance on this page to determine whether intervention is necessary or whether enrichment is appropriate. Options that provide instruction, practice, and a check are listed below.

Resources for Are You Ready?

- ***Are You Ready? Intervention and Enrichment***
- **Recording Sheet for Are You Ready?** *Chapter 4 Resource Book. . . pp. 3–4*

Are You Ready? Intervention **CD-ROM**

ARE YOU READY?

Were students successful with Are You Ready?

NO — INTERVENE

✓ Write and Read Decimals 6–11
Intervention Practice, Skill 15
CD-ROM Intervention Activities, Skill 15

✓ Identify Sets of Numbers. 12–19
Intervention Practice, Skill 6
CD-ROM Intervention Activities, Skill 6

✓ Multiples. 20–27
Intervention Practice, Skill 7
CD-ROM Intervention Activities, Skill 7

✓ Evaluate Expressions. 28–33
Intervention Practice, Skill 54
CD-ROM Intervention Activities, Skill 54

YES — ENRICH

Are You Ready? Enrichment, pp. 413–414

Section 4A

Number Theory

One-Minute Section Planner

Lesson	Materials	Resources
Lesson 4-1 Divisibility **NCTM:** Number and Operations, Communication **NAEP:** Number Properties 5c ☐ SAT-9 ☐ SAT-10 ☑ ITBS ☑ CTBS ☐ MAT ☐ CAT	**Optional** Number cubes with 2, 3, 4, 5, 6, and 9 on the faces *(MK)* Calculators Teaching Transparency T2 *(CRB)*	• *Chapter 4 Resource Book*, pp. 7–16 • Daily Transparency T1, CRB • Additional Examples Transparencies T3–T4, CRB • *Alternate Openers: Explorations*, p. 25
Lesson 4-2 Factors and Prime Factorization **NCTM:** Number and Operations, Communication **NAEP:** Number Properties 5b ☐ SAT-9 ☑ SAT-10 ☐ ITBS ☑ CTBS ☑ MAT ☑ CAT		• *Chapter 4 Resource Book*, pp. 17–25 • Daily Transparency T5, CRB • Additional Examples Transparencies T6–T7, CRB • *Alternate Openers: Explorations*, p. 26
Lesson 4-3 Greatest Common Factor **NCTM:** Number and Operations, Problem Solving, Communication **NAEP:** Number Properties 5d ☑ SAT-9 ☑ SAT-10 ☐ ITBS ☑ CTBS ☑ MAT ☑ CAT	**Optional** Graph Paper *(CRB, p. 98)*	• *Chapter 4 Resource Book*, pp. 26–34 • Daily Transparency T8, CRB • Additional Examples Transparencies T9–T10, CRB • *Alternate Openers: Explorations*, p. 27
Section 4A Assessment		• Mid-Chapter Quiz, SE p. 164 • Section 4A Quiz, AR p. 11 • *Test and Practice Generator* CD-ROM

SAT = *Stanford Achievement Tests* **ITBS** = *Iowa Test of Basic Skills* **CTBS** = *Comprehensive Test of Basic Skills/Terra Nova*
MAT = *Metropolitan Achievement Tests* **CAT** = *California Achievement Test*
NCTM = Complete standards can be found on pages T27–T33. **NAEP** = Complete standards can be found on pages A31–A35.
SE = *Student Edition* **TE** = *Teacher's Edition* **AR** = *Assessment Resources* **CRB** = *Chapter Resource Book* **MK** = *Manipulatives Kit*

Section Overview

Divisibility

Lesson 4-1

Why? You need to find factors of numbers when operating with fractions.

A number is divisible by . . .	Example	Explanation
2 if the last digit is even (0, 2, 4, 6, or 8).	17**6**	**6** is even.
3 if the sum of the digits is divisible by 3.	**525**	**5** + **2** + **5** = 12; 12 is divisible by 3.
4 if the last two digits form a number divisible by 4.	3,5**16**	**16** is divisible by 4.
5 if the last digit is 0 or 5.	11,27**5**	The last digit is **5**.
6 if the number is divisible by both 2 and 3.	**24**	**24** is divisible by both 2 and 3.
9 if the sum of the digits is divisible by 9.	**4,860**	**4** + **8** + **6** + **0** = 18; 18 is divisible by 9.
10 if the last digit is 0.	35,39**0**	The last digit is **0**.

A number is **divisible** by another number if the quotient is a whole number with no remainder (e.g., 15 ÷ 5 = 3).

Factors and Prime Factorization

Lesson 4-2

Why? Prime factorization is used to operate with and simplify fractions.

A **prime number** is greater than 1 and has factors of only 1 and itself: 2, 3, 5, 7, 11, . . .

A **composite number** is greater than 1 and is not prime: 4, 6, 8, 9, 10, . . .

Write the prime factorization of 84.

Use a **factor tree.** Use a **ladder diagram.**

The number 84 is **composite.**

84 = **2** · **2** · **3** · **7** The factors 2, 3, and 7 are **prime.**

Greatest Common Factor

Lesson 4-3

Why? Finding the GCF of a set of numbers is used in operations with fractions.

Find the GCF of 24 and 60.		
Method 1	factors of 24: 1, 2, 3, 4, 6, 8, (12), 24 factors of 60: 1, 2, 3, 4, 5, 6, 10, (12), 15, 20, 30, 60	*List all the factors of each number.* *Circle the greatest common factor.*
Method 2	24 = (2) • (2) • 2 • (3) 60 = (2) • (2) • (3) • 5 2 • 2 • 3 = 12	*Write the prime factorization of each number.* *Circle the common prime factors.* *Find the product of the common prime factors.*

The greatest common factor of 24 and 60 is 12.

4-1 Organizer

Pacing: Traditional 1 day
Block $\frac{1}{2}$ day

Objective: Students use divisibility rules.

Indiana Academic Standards
6.7.1, 6.7.2, 6.7.3, 6.7.4, 6.7.11

Warm Up

Write each number as a product of two whole numbers in as many ways as possible.

1. 20 1×20, 2×10, 4×5

2. 48 1×48, 2×24, 3×16, 4×12, 6×8

Problem of the Day

In this magic square, every row, column, and diagonal has the same sum, 34. Complete the square using the whole numbers from 1 to 16.

4	15	14	1
9	6	7	12
5	10	11	8
16	3	2	13

Available on Daily Transparency in CRB

Math Fact

A horizontal bar is used in mathematics to indicate division. Thus the expression $\frac{63}{7}$ is read "7 divides 63."

4-1 Divisibility

Learn to use divisibility rules.

Vocabulary
divisible
composite number
prime number

This year, 42 girls signed up to play basketball for the Junior Girls League, which has 6 teams. To find whether each team can have the same number of girls, decide if 42 is divisible by 6.

A number is **divisible** by another number if the quotient is a whole number with no remainder.

$$42 \div 6 = 7 \leftarrow \text{Quotient}$$

Since there is no remainder, 42 is divisible by 6. The Junior Girls League can have 6 teams with 7 girls each.

Divisibility Rules

A number is divisible by…	Divisible	Not Divisible
2 if the last digit is even (0, 2, 4, 6, or 8).	3,978	4,975
3 if the sum of the digits is divisible by 3.	315	139
4 if the last two digits form a number divisible by 4.	8,512	7,518
5 if the last digit is 0 or 5.	14,975	10,978
6 if the number is divisible by both 2 and 3.	48	20
9 if the sum of the digits is divisible by 9.	711	93
10 if the last digit is 0.	15,990	10,536

EXAMPLE 1 Checking Divisibility

A Tell whether 610 is divisible by 2, 3, 4, and 5.

2	*The last digit, 0, is even.*	Divisible
3	*The sum of the digits is $6 + 1 + 0 = 7$. 7 is not divisible by 3.*	Not divisible
4	*The last two digits form the number 10. 10 is not divisible by 4.*	Not divisible
5	*The last digit is 0.*	Divisible

So 610 is divisible by 2 and 5.

1 Introduce

Alternate Opener

EXPLORATION

4-1 Divisibility

Some calculators have an INT ÷ key, which returns a quotient and a remainder.

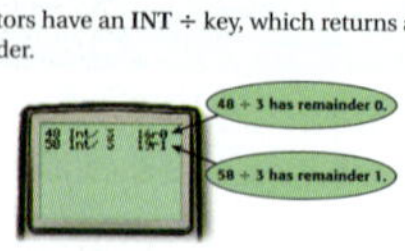

1. Use mental math or a calculator to determine each quotient and remainder. Then add the digits of the dividend.

	Dividend	Divisor	Quotient	Remainder	Sum of Digits
a.	48	3	16	0	$4 + 8 = 12$
b.	58	3	19	1	$5 + 8 = 13$
c.	256	3			
d.	1,011	3			
e.	72	3			
f.	74	3			
g.	129	3			
h.	130	3			

Think and Discuss

2. **Explain** whether 3,129 is divisible by 3.
3. **Describe** the pattern between the remainder and the sum of the digits in the table.

Motivate

Review division facts through $81 \div 9$ with students by using number families or by having students in pairs quiz each other. Point out that the quotients in division facts are all whole numbers without remainders. Ask students which numbers in the division facts you just reviewed are only divided by 1 and themselves (2, 3, 5, 7).

Exploration worksheet and answers on Chapter 4 Resource Book pp. 8 and 103

2 Teach

Lesson Presentation

Guided Instruction

In this lesson, students learn to use divisibility rules. First teach the divisibility rules for the numbers 2–6 and 9–10. Then teach students to use these rules to check for divisibility by these numbers. Finally, explain the terms *prime* and *composite* and how to identify prime and composite numbers.

Teaching Tip Students should know that divisibility rules are most handy when working with very large numbers. Divisibility rules provide information about a number without requiring the use of more complicated division.

B Tell whether 387 is divisible by 6, 9, and 10.

6	*The last digit, 7, is odd, so 387 is not divisible by 2.*	Not divisible
9	*The sum of the digits is 3 + 8 + 7 = 18. 18 is divisible by 9.*	Divisible
10	*The last digit is 7, not 0.*	Not divisible

So 387 is divisible by 9.

Any number greater than 1 is divisible by at least two numbers—1 and the number itself. Numbers that are divisible by more than two numbers are called **composite numbers**.

A **prime number** is divisible by only the numbers 1 and itself. For example, 11 is a prime number because it is divisible by only 1 and 11. The numbers 0 and 1 are neither prime nor composite.

EXAMPLE 2 Identifying Prime and Composite Numbers

Tell whether each number is prime or composite.

A 45
divisible by 1, 3, 5, 9, 15, 45
composite

B 13
divisible by 1, 13
prime

C 19
divisible by 1, 19
prime

D 49
divisible by 1, 7, 49
composite

The prime numbers from 1 through 50 are highlighted below.

1	2	3	4	5	6	7	8	9	10
11	12	13	14	15	16	17	18	19	20
21	22	23	24	25	26	27	28	29	30
31	32	33	34	35	36	37	38	39	40
41	42	43	44	45	46	47	48	49	50

Think and Discuss

1. **Tell** which whole numbers are divisible by 1.
2. **Explain** how you know that 87 is a composite number.
3. **Tell** how the divisibility rules help you identify composite numbers.

Additional Examples

Example 1

A. Tell whether 462 is divisible by 2, 3, 4, and 5. **462 is divisible by 2 and 3.**

B. Tell whether 540 is divisible by 6, 9, and 10. **540 is divisible by 6, 9, and 10.**

Example 2

Tell whether each number is prime or composite.

A. 23 (prime) **B.** 48 (composite)
C. 31 (prime) **D.** 18 (composite)

3 Close

Reaching All Learners
Through Grouping Strategies

Create number cubes with the numbers 2, 3, 4, 5, 6, and 9. Have students work in groups. Generate a list of at least twenty 3-digit numbers. To play the game, students choose a 3-digit number, toss the number cube, state the divisibility rule for the number tossed, and check for divisibility by that number. Toss the number cube again, and check for divisibility by a different number. Have students take turns and repeat the process until all the numbers in the list have been chosen.

Summarize

Review the terms *divisible*, *composite number*, and *prime number* with students. Discuss how the terms relate to each other.

Possible answer: A number is divisible by another number if the quotient is a whole number with no remainder. Composite numbers are divisible by more than two numbers; prime numbers are numbers greater than 1 that are divisible only by 1 and themselves.

Answers to Think and Discuss

1. all whole numbers
2. You know 87 is divisible by 3 because 8 + 7 = 15. Therefore, 87 has factors other than 1 and itself, making it composite.
3. The rules help you quickly check to see if a number is divisible by more than two numbers.

4-1 PRACTICE & ASSESS

Students may want to refer back to the lesson examples.

Assignment Guide

If you finished Example 1 assign:
Core 1–4, 9–16, 29–32, 41, 48–58
Enriched 29–38, 40–58

If you finished Example 2 assign:
Core 1–28, 40, 41, 48–58
Enriched 17–58

Answers

29.		no	no		no	no	
30.		no	no	yes	no	no	yes
31.	yes	no	yes	no	no	no	no
32.	yes	yes	yes	yes	yes	no	yes

4-1 Exercises

FOR EXTRA PRACTICE
see page 642

internet connect
Homework Help Online
go.hrw.com Keyword: MR4 4-1

GUIDED PRACTICE

See Example 1 Tell whether each number is divisible by 2, 3, 4, 5, 6, 9, and 10.

1. 508 **2, 4** 2. 432 **2, 3, 4, 6, 9** 3. 247 **none** 4. 189 **3, 9**

See Example 2 Tell whether each number is prime or composite.

5. 75 **composite** 6. 17 **prime** 7. 27 **composite** 8. 63 **composite**

INDEPENDENT PRACTICE

See Example 1 Tell whether each number is divisible by 2, 3, 4, 5, 6, 9, and 10.

9. 741 **3** 10. 810 **2, 3, 5, 6, 9, 10** 11. 675 **3, 5, 9** 12. 480 **2, 3, 4, 5, 6, 10**
13. 908 **2, 4** 14. 146 **2** 15. 514 **2** 16. 405 **3, 5, 9**

See Example 2 Tell whether each number is prime or composite.

17. 34 **composite** 18. 29 **prime** 19. 61 **prime** 20. 81 **composite**
21. 51 **composite** 22. 23 **prime** 23. 97 **prime** 24. 93 **composite**
25. 77 **composite** 26. 41 **prime** 27. 67 **prime** 28. 39 **composite**

PRACTICE AND PROBLEM SOLVING

Copy and complete the table. Write *yes* if the number is divisible by the given number. Write *no* if it is not.

		2	3	4	5	6	9	10
29.	677	*no*			*no*			*no*
30.	290	*yes*						
31.	1,744							
32.	12,180							

Possible answers:

Replace each box with a digit that will make the number divisible by 3.

33. 74▢ **1, 4, or 7** 34. 8,10▢ **0, 3, 6, or 9** 35. 3,▢41 **1, 4, or 7**
36. ▢,335 **1, 4, or 7** 37. 67,▢11 **0, 3, 6, or 9** 38. 10,0▢1 **1, 4, or 7**

39. Make a table that shows the prime numbers from 50 to 100. **prime numbers from 50 to 100 are 53, 59, 61, 67, 71, 73, 79, 83, 89, and 97.**

40. Tell whether each statement is true or false. Explain your answers.
   a. All even numbers are divisible by 2. **True**
   b. All odd numbers are divisible by 3. **False**
   c. Some even numbers are divisible by 5. **True**

Math Background

The divisibility rule for 9 works because numbers are written in the base-10 system, and every positive power of 10 is 1 greater than a multiple of 9.

Consider the number 243.

$243 = 2 \times 100 + 4 \times 10 + 3$

$243 = 2 \times (99 + 1) + 4 \times (9 + 1) + 3$

$243 = (2 \times 99) + (2 \times 1) + (4 \times 9) + (4 \times 1) + 3$

$243 = (2 \times 99) + (4 \times 9) + (2 \times 1) + (4 \times 1) + 3$

$243 = (2 \times 99) + (4 \times 9) + 2 + 4 + 3$

$(2 \times 99) + (4 \times 9)$ is divisible by 9.

So 243 is divisible by 9 if $2 + 4 + 3$ is divisible by 9.

RETEACH 4-1

LESSON 4-1 **Reteach**
Divisibility

A number is divisible by another number if the quotient is a whole number with no remainder.

15 is divisible by 3 because $15 \div 3 = 5$. There are rules to help you figure out if a number is divisible by another number.

Divisibility Rules
A number is divisible by:
- 2 if the last digit is even.
- 3 if the sum of the digits is divisible by 3.
- 4 if the last two digits form a number that is divisible by 4.
- 5 if the last digit is 0 or 5.
- 6 if the number is divisible by both 2 and 3.
- 9 if the sum of the digits is divisible by 9.
- 10 if the last digit is 0.

To tell whether 315 is divisible by 2, 3, 4, 5, 6, 9, and 10, you can use the divisibility rules listed above.

5 is not an even number, so 315 **is not divisible** by 2.
$3 + 1 + 5 = 9$. 9 is divisible by 3, so 315 **is divisible** by 3.
15 is not divisible by 4, so 315 **is not divisible** by 4.
The last digit is 0 or 5, so 315 **is divisible** by 5.
315 is divisible by 3, but not 2. So, 315 **is not divisible** by 6.
The sum of the digits is 9 and 9 is divisible by 9. So 315 **is divisible** by 9.
The last digit is not 0, so 315 **is not divisible** by 10.

315 is divisible by 3, 5, and 9.

Use divisibility rules to tell whether each number is divisible by 2, 3, 4, 5, 6, 9, and 10.

1. 120 — 2; 3; 4; 5; 6; 10
2. 435 — 3; 5
3. 228 — 2; 3; 4; 6
4. 540 — 2; 3; 4; 5; 6; 9; 10
5. 144 — 2; 3; 4; 6; 9
6. 634 — 2
7. 402 — 2; 3; 6
8. 320 — 2; 4; 5; 10

PRACTICE 4-1

LESSON 4-1 **Practice B**
Divisibility

Tell whether each number is divisible by 2, 3, 4, 5, 6, 9, and 10.

1. 90 — 2; 3; 5; 6; 9; 10
2. 416 — 2; 4
3. 308 — 2; 4
4. 540 — 2; 3; 4; 5; 6; 9; 10
5. 804 — 2; 3; 4; 6
6. 225 — 3; 5; 9
7. 663 — 3
8. 972 — 2; 3; 4; 6; 9
9. 836 — 2; 4

Tell whether each number is prime or composite.

10. 33 — composite
11. 69 — composite
12. 41 — prime
13. 45 — composite
14. 58 — composite
15. 87 — composite
16. 61 — prime
17. 53 — prime
18. 99 — composite

19. Dan counted all the coins in his bank, and he had 72 quarters. Can he exchange the quarters for an even amount of dollar bills? How do you know?
Yes; because there are 4 quarters in 1 dollar, and 72 is divisible by 4.

20. A small town purchased 196 American flags for its Memorial Day parade. Eight locations were selected to display the flags. Can each location have the same number of flags? If no, explain why not. If yes, how many flags will be displayed at each location?
No; because 196 is not divisible by 8.

41. ***ASTRONOMY*** Earth has a diameter of 7,926 miles. Tell whether this number is divisible by 2, 3, 4, 5, 6, 9, and 10. **2, 3, and 6**

42. On which of the bridges in the table could a light fixture be placed every 6 meters so that the first light is at the beginning of the bridge and the last light is at the end of the bridge? **Mackinac Straits**

Golden Gate Bridge

Longest Bridges in the U.S.

Name and State	Length (m)
Verrazano Narrows, NY	1,298
Golden Gate, CA	1,280
Mackinac Straits, MI	1,158
George Washington, NY	1,067

43. A number is between 80 and 100 and is divisible by both 5 and 6. What is the number? **90**

44. ***CHOOSE A STRATEGY*** Find the greatest four-digit number that is divisible by 1, 2, 3, and 4. **9,996**

45. ***WHAT'S THE ERROR?*** To find whether 3,463 is divisible by 4, a student added the digits. The sum, 16, is divisible by 4, so the student stated that 3,463 is divisible by 4. Explain the error.

46. ***WRITE ABOUT IT*** If a number is divisible by both 4 and 9, by what other numbers is it divisible? Explain.

47. ***CHALLENGE*** Find a number that is divisible by 2, 3, 4, 5, 6, and 10, but not 9. **Possible answer: 240**

45. A number is divisible by 4 only if the number formed by the last two digits is divisible by 4, not if the sum of the digits is divisible by 4.

46. 2, 3, and 6; 2 is a factor of 4, 3 is a factor of 9, and if a number is divisible by 2 and 3 it is also divisible by 6.

Spiral Review

Compare. Write < or >. (Lesson 1-1)

48. 10,976 ▒ 100,100 **<**

49. 32,107,120 ▒ 32,170,021 **<**

50. 60,842,250 ▒ 60,847,205 **<**

51. 136,422,190 ▒ 136,242,910 **>**

Write each phrase as a numerical or algebraic expression. (Lesson 2-2)

52. 562 plus t **$562 + t$**

53. the product of n and 16 **$16n$**

54. the quotient of p and 7 **$p \div 7$**

Solve each equation. Check your answer. (Lesson 2-4)

55. $17 + b = 44$ **$b = 27$**

56. $x + 31 = 72$ **$x = 41$**

57. $28 + y = 57$ **$y = 29$**

58. **TEST PREP** Which decimal is greatest? (Lesson 3-1) **B**

A 7.081 B 17.8 C 7.18 D 17.081

Journal

Have students explain the difference between prime and composite numbers using the term *divisible*. Have students give examples of both types of numbers.

Test Prep Doctor

For Exercise 58, students need to line up the decimal points and compare digits in like place-value positions from left to right. **A** and **C** can be eliminated because their greatest digits are in the ones place. Students who answered **D** probably did so because the number has a greater number of decimal places.

CHALLENGE 4-1

LESSON 4-1 **Challenge**
The Chinese Calendar

The Chinese calendar runs in cycles of 12 years. Each year in a cycle is named after an animal. The Chinese believe the animal ruling the year in which you were born has a great effect on your personality. They say, "This is the animal that hides in your heart."

2000 was a year of the dragon.

The Chinese calendar cycle below is for the years 1983–1994. Use the clues to write the last two digits in the year that goes with each animal. Then read the character traits associated with the animals. Which animal hides in your heart?

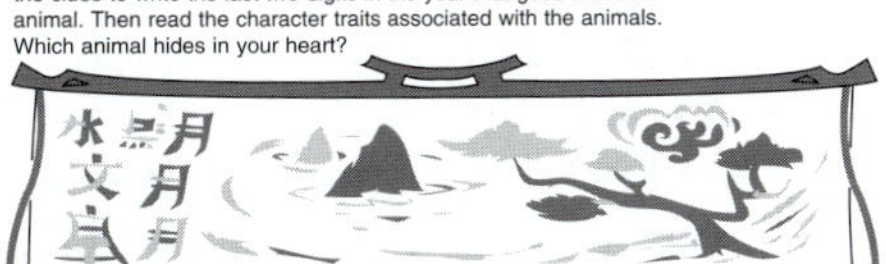

Tiger: 19 **86** **Divisible by:** 2 and 43 **Traits:** bold, adventurous	**Sheep:** 19 **91** **Divisible by:** 7 **Traits:** charming, artistic	**Pig:** 19 **83** **Divisible by:** 1 and 83 **Traits:** loyal, tolerant
Dog: 19 **94** **Divisible by:** 2 and 47 **Traits:** honest, faithful	**Dragon:** 19 **88** **Divisible by:** 2, 4, and 8 **Traits:** fun, energetic	**Rat:** 19 **84** **Divisible by:** 4 and 6 **Traits:** generous, creative
Horse: 19 **90** **Divisible by:** 5 and 9 **Traits:** friendly, hardworking	**Rooster:** 19 **93** **Divisible by:** 3 and 31 **Traits:** decisive, proud	**Ox:** 19 **85** **Divisible by:** 5 **Traits:** confident, leaderlike
Snake: 19 **89** **Divisible by:** 1 and 89 **Traits:** charming, wise	**Monkey:** 19 **92** **Divisible by:** 2 and 4 **Traits:** smart, likable	**Rabbit:** 19 **87** **Divisible by:** 3 and 29 **Traits:** friendly, cooperative

PROBLEM SOLVING 4-1

LESSON 4-1 **Problem Solving**
Divisibility

Use the table to answer the questions.

Subways Around the World

City, Country	Length (mi)
New York, U.S.	247
Mexico City, Mexico	111
Paris, France	125
Moscow, Russia	152
Seoul, South Korea	83
Tokyo, Japan	105

1. Which city's subway has a length that is a prime number of miles?
Seoul, South Korea

2. Which subway could be evenly broken into sections of 2 miles each?
Moscow, Russia

3. Which subways could be evenly broken into sections of 5 miles each?
Paris, France, and Tokyo, Japan

4. Which subways could be evenly divided into sections of track that are each 3 miles long?
Mexico City, Mexico, and Tokyo, Japan

5. Which subway's length is not a prime number, but is also not divisible by 2, 3, 4, 5, 6, or 9?
New York, United States

6. Which subway's length is divisible by 4 miles?
Moscow, Russia

7. Which subway's length is divisible by 7 miles?
Tokyo, Japan

Circle the letter of the correct answer.

8. The subway in Hong Kong, China, has a length that is a prime number of miles. Which of the following is its length?
A 260 miles
B 268 miles
(C) 269 miles
D 265 miles

9. The subway in St. Petersburg, Russia, has a length that is divisible by 3 miles. Which of the following is its length?
(F) 57 miles
G 56 miles
H 55 miles
J 58 miles

Lesson Quiz

Tell whether each number is divisible by 2, 3, 4, 5, 6, 9, and 10.

1. 256 **2, 4**
2. 720 **2, 3, 4, 5, 6, 9, 10**
3. 615 **3, 5**

Tell whether each number is prime or composite.

4. 47 prime
5. 38 composite

Available on Daily Transparency in CRB

4-2 Organizer

Pacing: Traditional 1 day
Block $\frac{1}{2}$ day

Objective: Students write prime factorizations of composite numbers.

Indiana Academic Standards
6.7.1, 6.7.3, 6.7.4, 6.7.11

Warm Up

Identify each number as prime or composite.

1. 19 prime	**2.** 82 composite
3. 57 composite	**4.** 85 composite
5. 101 prime	**6.** 121 composite

Problem of the Day

At the first train stop, 7 people disembarked. At the second stop, 8 people disembarked. At the fourth stop, the last 6 people disembarked. If there were 28 people on the train before the first stop, how many people left at the third stop? 7

Available on Daily Transparency in CRB

Math Humor

Teacher: All prime numbers are odd, with one exception; two is even.

Student: That's odd.

4-2 Factors and Prime Factorization

Learn to write prime factorizations of composite numbers.

Vocabulary
factor
prime factorization

Whole numbers that are multiplied to find a product are called **factors** of that product. A number is divisible by its factors.

$2 \cdot 3 = 6$ (2 and 3: Factors; 6: Product)

$6 \div 3 = 2$
$6 \div 2 = 3$
6 is divisible by 3 and 2.

EXAMPLE 1 Finding Factors

List all of the factors of each number.

A 18

Begin listing factors in pairs.

$18 = 1 \cdot 18$	*1 is a factor.*
$18 = 2 \cdot 9$	*2 is a factor.*
$18 = 3 \cdot 6$	*3 is a factor.*
	4 is not a factor.
	5 is not a factor.
$18 = 6 \cdot 3$	*6 and 3 have already been listed, so stop here.*

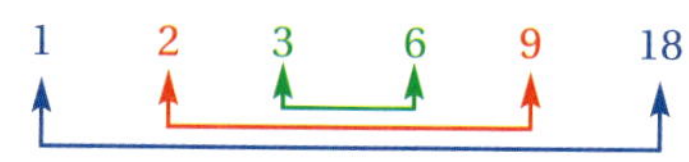

You can draw a diagram to illustrate the factor pairs.

The factors of 18 are 1, 2, 3, 6, 9, and 18.

B 13

$13 = 1 \cdot 13$ *Begin listing factors in pairs. 13 is not divisible by any other whole numbers.*

The factors of 13 are 1 and 13.

Helpful Hint
When the pairs of factors begin to repeat, then you have found all of the factors of the number you are factoring.

You can use factors to write a number in different ways.

Factorization of 12			
$1 \cdot 12$	$2 \cdot 6$	$3 \cdot 4$	$3 \cdot 2 \cdot 2$

Notice that these factors are all prime.

The **prime factorization** of a number is the number written as the product of its prime factors.

1 Introduce
Alternate Opener

EXPLORATION

4-2 Factors and Prime Factorization

1. The yellow rectangle measures 4 units by 6 units and has an area of 24 square units. Use graph paper to draw rectangles that have different whole-number dimensions but still have an area of 24 square units. (*Hint:* 4 × 6 = 24. What factors other than 1 × 24 give you 24?)

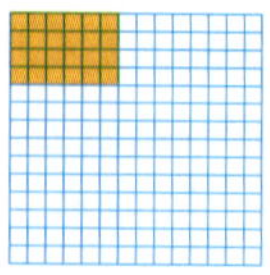

2. The red rectangle measures 3 units by 5 units and has an area of 15 square units. Is it possible to draw rectangles that have whole-number dimensions other than 3 × 5 (and 1 × 15) and still have an area of 15 square units?

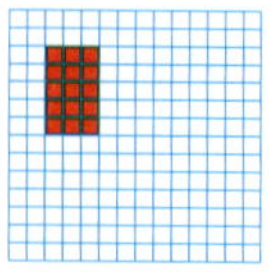

Think and Discuss

3. **Explain** how you can use rectangles to determine factors of numbers.
4. **Explain** why it is possible to draw more than two different rectangles with an area of 24 square units, but it is not possible to draw more than two different rectangles with an area of 15 square units.

Motivate

Ask students to name numbers that can be multiplied to get certain numbers. For example, you might ask them to name numbers that can be multiplied to get 28 (1 and 28; 2 and 14; 4 and 7).

Exploration worksheet and answers on Chapter 4 Resource Book pp. 18 and 105

2 Teach

Lesson Presentation

Guided Instruction

In this lesson, students learn to write prime factorizations of composite numbers. First explain the term *factors* and teach students to list all factors of a number. Then teach them to find the prime factors of a number, using a factor tree and a ladder diagram. Have students practice finding factors and prime factors of other numbers before *Think and Discuss*.

Teaching Tip Some students may prefer using a factor tree, while others may prefer the ladder diagram. Encourage students to use the method that works best for them.

EXAMPLE 2 Writing Prime Factorizations

Write the prime factorization of each number.

A 36

Method 1: Use a factor tree.

Choose any two factors of 36 to begin. Keep finding factors until each branch ends at a prime factor.

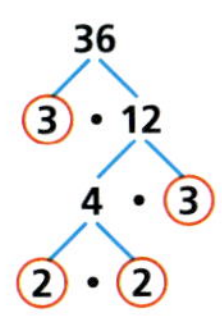

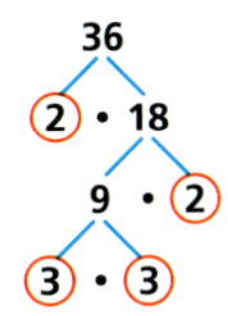

$36 = 3 \cdot 2 \cdot 2 \cdot 3$ $36 = 2 \cdot 3 \cdot 3 \cdot 2$

The prime factorization of 36 is $2 \cdot 2 \cdot 3 \cdot 3$, or $2^2 \cdot 3^2$.

B 54

Method 2: Use a ladder diagram.

Choose a prime factor of 54 to begin. Keep dividing by prime factors until the quotient is 1.

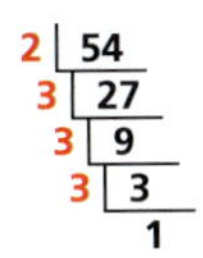

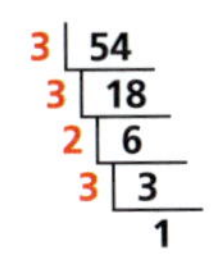

$54 = 2 \cdot 3 \cdot 3 \cdot 3$ $54 = 3 \cdot 3 \cdot 2 \cdot 3$

The prime factorization of 54 is $2 \cdot 3 \cdot 3 \cdot 3$, or $2 \cdot 3^3$.

Helpful Hint

You can use exponents to write prime factorizations. Remember that an exponent tells you how many times the base is a factor.

In Example 2, notice that the prime factors may be written in a different order, but they are still the same factors. Except for changes in the order, there is only one way to write the prime factorization of a number.

Think and Discuss

1. **Tell** how you know when you have found all of the factors of a number.
2. **Tell** how you know when you have found the prime factorization of a number.
3. **Explain** the difference between factors of a number and prime factors of a number.

Additional Examples

Example 1

List all of the factors of each number.

A. 16 The factors of 16 are 1, 2, 4, 8, and 16.

B. 19 The factors of 19 are 1 and 19.

Example 2

Write the prime factorization of each number.

A. 24 $24 = 2 \cdot 2 \cdot 2 \cdot 3$, or $2^3 \cdot 3$

B. 45 $45 = 3 \cdot 3 \cdot 5$, or $3^2 \cdot 5$

3 Close

Reaching All Learners
Through Number Sense

Have students name five numbers with the prime factors 2, 3, and 5. Each factor may be used more than once.

Possible answers:

$30 = 2 \times 3 \times 5$

$60 = 2 \times 2 \times 3 \times 5$, or $2^2 \times 3 \times 5$

$120 = 2 \times 2 \times 2 \times 3 \times 5$, or $2^3 \times 3 \times 5$

$180 = 2 \times 2 \times 3 \times 3 \times 5$, or $2^2 \times 3^2 \times 5$

$900 = 2 \times 2 \times 3 \times 3 \times 5 \times 5$, or $2^2 \times 3^2 \times 5^2$

Summarize

Review how to find factors of a number, and differentiate this process from finding the prime factorization of a number. Have students give examples of each.

Possible responses: factors of 40: 1, 2, 4, 5, 8, 10, 20, 40; prime factorization of 40: $2 \times 2 \times 2 \times 5$, or $2^3 \times 5$

Answers to Think and Discuss

1. Possible answer: when the factor pairs start to repeat
2. Possible answer: when all the factors are prime and their product is the original number
3. *Prime factors* of a number are all prime; *factors* of a number don't have to be prime.

4-2 PRACTICE & ASSESS

Students may want to refer back to the lesson examples.

Assignment Guide

If you finished Example 1 assign:

Core 1–4, 9–16, 25–31 odd, 51–62

Enriched 12–16, 25–32, 42, 44, 49, 51–62

If you finished Example 2 assign:

Core 1–24, 29–35 odd, 42, 43, 45–46, 51–62

Enriched 19–62

Notes

4-2 Exercises

FOR EXTRA PRACTICE see page 642

internet connect
Homework Help Online
go.hrw.com Keyword: MR4 4-2

GUIDED PRACTICE

See Example 1 List all of the factors of each number.

1. 12 1, 2, 3, 4, 6, 12
2. 21 1, 3, 7, 21
3. 52 1, 2, 4, 13, 26, 52
4. 75 1, 3, 5, 15, 25, 75

See Example 2 Write the prime factorization of each number.

5. 48 $2^4 \cdot 3$
6. 20 $2^2 \cdot 5$
7. 66 $2 \cdot 3 \cdot 11$
8. 34 $2 \cdot 17$

INDEPENDENT PRACTICE

See Example 1 List all of the factors of each number.

9. 24 1, 2, 3, 4, 6, 8, 12, 24
10. 37 1, 37
11. 42 1, 2, 3, 6, 7, 14, 21, 42
12. 56 1, 2, 4, 7, 8, 14, 28, 56
13. 67 1, 67
14. 72 1, 2, 3, 4, 6, 8, 9, 12, 18, 24, 36, 72
15. 85 1, 5, 17, 85
16. 92 1, 2, 4, 23, 46, 92

See Example 2 Write the prime factorization of each number.

17. 49 7^2
18. 38 $2 \cdot 19$
19. 76 $2^2 \cdot 19$
20. 60 $2^2 \cdot 3 \cdot 5$
21. 81 3^4
22. 132 $2^2 \cdot 3 \cdot 11$
23. 140 $2^2 \cdot 5 \cdot 7$
24. 87 $3 \cdot 29$

PRACTICE AND PROBLEM SOLVING

Write each number as a product in two different ways. Possible answers:

25. 34 $2 \cdot 17$; $1 \cdot 34$
26. 82 $2 \cdot 41$; $1 \cdot 82$
27. 88 $8 \cdot 11$; $4 \cdot 22$
28. 50 $2 \cdot 25$; $10 \cdot 5$
29. 15 $3 \cdot 5$; $15 \cdot 1$
30. 78 $26 \cdot 3$; $2 \cdot 39$
31. 94 $47 \cdot 2$; $94 \cdot 1$
32. 35 $7 \cdot 5$; $1 \cdot 35$

Find the prime factorization of each number.

33. 99 $3^2 \cdot 11$
34. 249 $3 \cdot 83$
35. 284 $2^2 \cdot 71$
36. 620 $2^2 \cdot 5 \cdot 31$
37. 840 $2^3 \cdot 3 \cdot 5 \cdot 7$
38. 150 $2 \cdot 3 \cdot 5^2$
39. 740 $2^2 \cdot 5 \cdot 37$
40. 402 $2 \cdot 3 \cdot 67$

41. The prime factorization of 50 is $2 \cdot 5^2$. Without dividing or using a diagram, find the prime factorization of 100. $2^2 \cdot 5^2$

42. ***GEOMETRY*** The area of a rectangle is the product of its length and width. Suppose the area of a rectangle is 24 in^2. What are the possible whole number measurements of its length and width?
Possible answers: $1 \cdot 24$; $24 \cdot 1$; $2 \cdot 12$; $12 \cdot 2$; $3 \cdot 8$; $8 \cdot 3$; $4 \cdot 6$; $6 \cdot 4$

43. ***PHYSICAL SCIENCE*** The speed of sound at sea level at 20°C is 343 meters per second. Write the prime factorization of 343. 7^3

44. ***SPORTS*** Little League Baseball began in 1939 in Pennsylvania. When it first started, there were 45 boys on 3 teams.
    a. If the teams were equally sized, how many boys were on each team? 15 boys per team
    b. Name another way the boys could have been divided into equally sized teams. (Remember that a baseball team must have at least 9 players.) 5 teams of 9 players

Math Background

Eratosthenes (275–194 B.C., Greece) developed the following method for finding prime numbers: Make a table from 1 to 100. One is not a prime, so cross it out. Circle 2 because it is a prime. Starting at 2, cross out multiples of 2. Circle 3 because it is a prime. Starting at 3, cross out every third number (even though some may have been crossed out before) in the list. Continue this process until all the numbers in the list have either been circled or crossed out.

Remind students that they are crossing out multiples of each prime. For example, with the prime number 5, every fifth number afterwards (10, 15, 20, 25, …) is a multiple of 5.

RETEACH 4-2

LESSON 4-2 **Reteach**
Factors and Prime Factorization

Factors of a product are the numbers that are multiplied to find that product. A factor is also a whole number that divides the product with no remainder.

To find all of the factors of 24, make a list of multiplication facts.

$1 \cdot 24 = 24$
$2 \cdot 12 = 24$
$3 \cdot 8 = 24$
$4 \cdot 6 = 24$

The factors of 24 are 1, 2, 3, 4, 6, 8, 12, and 24.

Write multiplication facts to find the factors of each number.

1. 20 $1 \cdot 20 = 20$; $2 \cdot 10 = 20$; $4 \cdot 5 = 20$
2. 16 $1 \cdot 16 = 16$; $2 \cdot 8 = 16$; $4 \cdot 4 = 16$
3. 35 $1 \cdot 35 = 35$; $5 \cdot 7 = 35$
4. 31 $1 \cdot 31 = 31$

A number written as the product of prime factors is called the prime factorization of the number.

To write the prime factorization of 24, first write it as product of 2 numbers. Then rewrite each factor as the product of 2 numbers until all of the factors are prime numbers.

$24 = \mathbf{4} \cdot 6$ (Write 24 as the product of 2 numbers.)
$= 2 \cdot 2 \cdot \mathbf{6}$ (Rewrite 4 as the product of 2 prime numbers.)
$= 2 \cdot 2 \cdot 2 \cdot 3$ (Rewrite 6 as the product of 2 prime numbers.)

So, the prime factorization of 24 is $2 \cdot 2 \cdot 2 \cdot 3$ or $2^3 \cdot 3$.

Find the prime factorization of each number.

5. 28 $2^2 \cdot 7$
6. 45 $3^2 \cdot 5$
7. 50 $2 \cdot 5^2$
8. 72 $2^3 \cdot 3^2$

PRACTICE 4-2

LESSON 4-2 **Practice B**
Factors and Prime Factorization

List all the factors for each number.

1. 15 1; 3; 5; 15
2. 24 1; 2; 3; 4; 6; 8; 12; 24
3. 33 1; 3; 11; 33
4. 72 1; 2; 3; 4; 6; 8; 9; 12; 18; 24; 36; 72
5. 48 1; 2; 3; 4; 6; 8; 12; 16; 24; 48
6. 95 1; 5; 19; 95
7. 66 1; 2; 3; 6; 11; 22; 33; 66
8. 87 1; 3; 29; 87
9. 36 1; 2; 3; 4; 6; 9; 12; 18; 36

Write the prime factorization of each number.

10. 44 $2^2 \cdot 11$
11. 56 $2^3 \cdot 7$
12. 42 $2 \cdot 3 \cdot 7$
13. 39 $3 \cdot 13$
14. 36 $2^2 \cdot 3^2$
15. 125 5^3
16. 85 $5 \cdot 17$
17. 100 $2^2 \cdot 5^2$
18. 32 2^5

19. James has an assigned seat for his flight to Denver. The seats on the plane are numbered 1–49. James's seat number is an odd number greater than 10 that is factor of 100. What is his seat number for the flight?
25

20. Linda writes the prime factorization of 40 as $2 \cdot 2 \cdot 2 \cdot 5$ on the board. Phil writes the prime factorization of 40 as $2^3 \cdot 5$. Who is correct?
They both are.

Life Science LINK

Climate changes, habitat destruction, and overhunting can cause animals and plants to die in large numbers. When the entire population of a species begins to die out, the species is considered endangered.

The graph shows the number of endangered species in each category of animal.

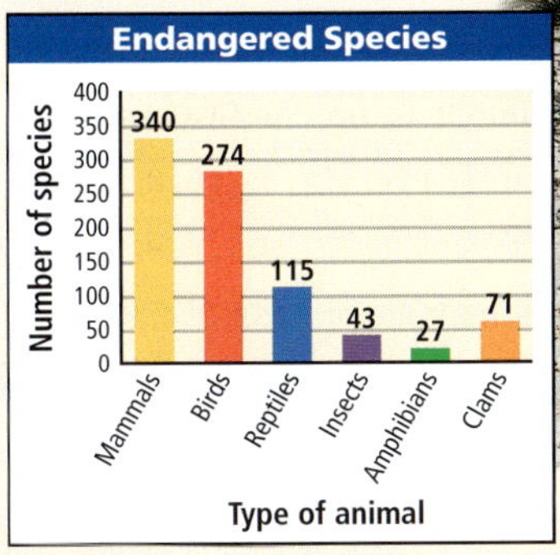

Laysan albatross chicks often die from eating plastic that pollutes the oceans and beaches. Clean-up efforts may prevent the albatross from becoming endangered.

45. How many species of mammals are endangered? Write this number as the product of prime factors. **340; $2^2 \cdot 5 \cdot 17$**

46. Which categories of animals have a prime number of endangered species? **Insects; Clams**

47. How many species of reptiles and amphibians combined are endangered? Write the answer as the product of prime factors. **142; $2 \cdot 71$**

48. ***WHAT'S THE ERROR?*** When asked to write the prime factorization of the number of endangered amphibian species, a student wrote 3×9. Explain the error and write the correct answer. **9 is not a prime number; 3^3**

49. ***WRITE ABOUT IT*** A team of five scientists is going to study endangered insect species. The scientists want to divide the species evenly among them. Will they be able to do this? Why or why not?

49. No, because there are 43 endangered insect species and five scientists and 43 is not divisible by 5.

50. ***CHALLENGE*** Add the number of endangered mammal species to the number of endangered bird species. Find the prime factorization of this number. **$2 \cdot 307$**

go.hrw.com
KEYWORD: MR4 Endangered
CNN Student News

Spiral Review

Estimate each sum or difference by rounding to the place value indicated. (Lesson 1-2)

51. 4,798 + 2,118; thousands **7,000**

52. 6,293 − 3,192; thousands **3,000**

53. 23,978 + 18,164; ten thousands **40,000**

54. 49,169 − 13,919; ten thousands **40,000**

Find each value. (Lesson 1-3)

55. 5^4 **625**

56. 8^3 **512**

57. 9^2 **81**

58. 10^6 **1,000,000**

Solve each equation. (Lessons 2-4 and 2-5)

59. $18 + p = 26$ **$p = 8$**

60. $n - 34 = 177$ **$n = 211$**

61. $7 + b = 84$ **$b = 77$**

62. **TEST PREP** Which number is 7,050,000 expressed in scientific notation? (Lesson 3-5) **D**

A 7.05×10^5 B 70.5×10^5 C 0.705×10^7 D 7.05×10^6

CHALLENGE 4-2

LESSON 4-2 **Challenge**
Prime Shades

For each given number, shade one box in each row of the table to show a prime factor. Then use your shaded boxes to write each number's prime factorization.

1. 12

7	2
3	5
2	11

Prime factorization: $2^2 \cdot 3$

2. 70

5	3
2	11
3	7

Prime factorization: $2 \cdot 5 \cdot 7$

3. 63

3	5
2	3
11	7

Prime factorization: $3^2 \cdot 7$

4. 150

13	5
3	7
2	11
17	5

Prime factorization: $2 \cdot 3 \cdot 5^2$

5. 84

11	7
5	2
13	2
3	17

Prime factorization: $2^2 \cdot 3 \cdot 7$

6. 260

5	11
3	2
17	13
2	7

Prime factorization: $2^2 \cdot 5 \cdot 13$

7. 80

5	7
17	2
2	13
11	2
2	3

Prime factorization: $2^4 \cdot 5$

8. 1,750

17	5
13	7
31	5
5	3
2	11

Prime factorization: $2 \cdot 5^3 \cdot 7$

9. 3,234

5	3
17	7
31	11
2	13
7	23

Prime factorization: $2 \cdot 3 \cdot 7^2 \cdot 11$

PROBLEM SOLVING 4-2

LESSON 4-2 **Problem Solving**
Factors and Prime Factorization

Write the correct answer.

1. The area of a rectangle is the product of its length and width. If a rectangular board has an area of 30 square feet, what are the possible measurements of its length and width?
1, 2, 3, 5, 6, 10, 15, or 30 feet

2. The first-floor apartments in Jenna's building are numbered 100 to 110. How many apartments on that floor are a prime number? What are those apartment numbers?
4 apartments; 101, 103, 107, and 109

3. If a composite number has the first five prime numbers as factors, what is the smallest number it could be? Write that number's prime factorization.
2,310; $2 \cdot 3 \cdot 5 \cdot 7 \cdot 11$

4. Tim's younger brother, Bryant, just had a birthday. Bryant's age only has one factor, and is not a prime number. How old is Bryant?
1 year old

5. A Russian mathematician named Christian Goldbach came up with a theory that every even number greater than 4 can be written as the sum of two odd primes. Test Goldbach's theory with the numbers 6 and 50. **Possible answers:**
6 = 3 + 3; 50 = 19 + 31

6. Mr. Samuels has 24 students in his math class. He wants to divide the students into equal groups, and he wants the number of students in each group to be prime. What are his choices for group sizes? How many groups can he make?
12 groups of 2 students each or 8 groups of 3 students each

Circle the letter of the correct answer.

7. Why is 2 the only even prime number?
A It is the smallest prime number.
(B) All other even numbers are divisible by 2.
C It only has 1 and 2 as factors.
D All odd numbers are prime.

8. What prime numbers are factors of both 60 and 105?
F 2 and 3
G 2 and 5
(H) 3 and 5
J 5 and 7

Interdisciplinary LINK

Life Science

Exercises 45–50 involve using data about endangered species. Protection of endangered species is studied in middle school life science programs such as *HOLT SCIENCE & TECHNOLOGY*.

Journal

Have students explain which is easier to use for finding the prime factorization of a number, a factor tree or a ladder diagram, and why.

Test Prep Doctor

For Exercise 62, encourage students to eliminate answers that are in the wrong format. **B** and **C** can be eliminated because the number multiplied by the power of 10 is not between 1 and 10. Students who answered **A** incorrectly calculated the exponent of 10.

Lesson Quiz

List all the factors of each number.

1. 22 **1, 2, 11, 22**

2. 40 **1, 2, 4, 5, 8, 10, 20, 40**

3. 51 **1, 3, 17, 51**

Write the prime factorization of each number.

4. 32 **2^5**

5. 120 **$2^3 \times 3 \times 5$**

Available on Daily Transparency in CRB

4-3 Organizer

Pacing: Traditional 1 day
Block $\frac{1}{2}$ day

Objective: Students find the greatest common factor (GCF) of a set of numbers.

Indiana Academic Standards
6.1.7, 6.7.1, 6.7.3, 6.7.5, 6.7.9, 6.7.10, 6.7.11

Warm Up

Write the prime factorization of each number.

1. 14 2×7 **3.** 63 $3^2 \times 7$

2. 18 2×3^2 **4.** 54 2×3^3

Problem of the Day

In a parade, there are 15 riders on bicycles and tricycles. In all, there are 34 cycle wheels. How many bicycles and how many tricycles are in the parade? 11 bicycles and 4 tricycles

Available on Daily Transparency in CRB

Math Fact !!

When the greatest common factor of two whole numbers is 1, those numbers are said to be *relatively prime*. The idea is not that they are necessarily prime, but that they are prime *relative to* each other.

4-3 Greatest Common Factor

Learn to find the greatest common factor (GCF) of a set of numbers.

Vocabulary

greatest common factor (GCF)

Factors shared by two or more whole numbers are called common factors. The largest of the common factors is called the **greatest common factor**, or **GCF**.

Factors of 24: 1, 2, 3, 4, 6, 8, 12, 24

Factors of 36: 1, 2, 3, 4, 6, 9, 12, 18, 36

Common factors: 1, 2, 3, 4, 6, (12)

The greatest common factor (GCF) of 24 and 36 is 12.

Example 1 shows three different methods for finding the GCF.

EXAMPLE 1 Finding the GCF

Find the GCF of each set of numbers.

A 16 and 24

Method 1: List the factors.

factors of 16: 1, 2, 4, (8), 16 *List all the factors.*

factors of 24: 1, 2, 3, 4, 6, (8), 12, 24 *Circle the GCF.*

The GCF of 16 and 24 is 8.

B 12, 24, and 32

Method 2: Use prime factorization.

$12 = 2 \cdot 2 \cdot 3$ *Write the prime factorization of each number.*

$24 = 2 \cdot 2 \cdot 2 \cdot 3$

$32 = 2 \cdot 2 \cdot 2 \cdot 2 \cdot 2$ *Find the common prime factors.*

$2 \cdot 2 = 4$ *Find the product of the common prime factors.*

The GCF of 12, 24, and 32 is 4.

C 12, 18, and 60

Method 3: Use a ladder diagram.

2	12	18	60
3	6	9	30
	2	3	10

Begin with a factor that divides into each number. Keep dividing until the three numbers have no common factors.

$2 \cdot 3 = 6$ *Find the product of the numbers you divided by.*

The GCF is 6.

1 Introduce

Alternate Opener

EXPLORATION

4-3 Greatest Common Factor

The sixth-grade band, which has 60 members, and the seventh-grade band, which has 48 members, are getting ready for a parade. How can they march together in blocks with the same number of columns? The model and table below show one possible formation.

1. Use graph paper to draw a model of two other formations that would work.
2. Complete the table to show the number of rows and columns in the other two formations.

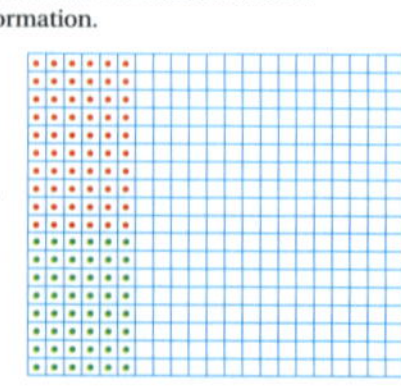

	Formation 1 Rows	Formation 1 Columns	Formation 2 Rows	Formation 2 Columns	Formation 3 Rows	Formation 3 Columns
6th Grade	10	6				
7th Grade	8	6				

Think and Discuss

3. **Discuss** which formation the band director should select if she wants the bands to pass through the parade as quickly as possible.
4. **Explain** why both 48 and 60 must be divisible by the number of columns.

Motivate

Ask students what it means to have something in common with someone (e.g., to have a similar interest, have the same eye color, be the same height, etc.). Give students several numbers and have them list the factors of each number. Have them identify any factors that appear in more than one list.

Exploration worksheet and answers on Chapter 4 Resource Book pp. 27 and 107

2 Teach

Lesson Presentation

Guided Instruction

In this lesson, students learn to find the greatest common factor (GCF) of a set of numbers. Teach the students three methods for finding the GCF: listing factors, using prime factorization, and using a ladder diagram. GCF is used in later lessons involving fractions and algebraic expressions.

Teaching Tip Be sure that students using the prime factorization method realize that the common prime factors will not always line up neatly like they do in Example 1B. Let them know that they can always reorganize the factors.

EXAMPLE 2 **PROBLEM SOLVING APPLICATION**

PROBLEM SOLVING

There are 12 boys and 18 girls in Mr. Ruiz's science class. The students must form lab groups. Each group must have the same number of boys and the same number of girls. What is the greatest number of groups Mr. Ruiz can make if every student must be in a group?

1 Understand the Problem

The **answer** will be the *greatest* number of groups 12 boys and 18 girls can form so that each group has the same number of boys, and each group has the same number of girls.

2 Make a Plan

You can make an organized list of the possible groups.

3 Solve

There are more girls than boys in the class, so there will be more girls than boys in each group.

Boys	Girls	Groups
1	2	(B GG) (B GG) (B GG) (B GG) (B GG) (B GG) (B GG) (B GG) (B GG) 9 boys, 18 girls: There are 3 boys not in groups. ✗
2	3	(BB GGG) (BB GGG) (BB GGG) (BB GGG) (BB GGG) (BB GGG) 12 boys, 18 girls: Every student is in a group. ✓

The greatest number of groups is 6.

4 Look Back

The number of groups will be a common factor of the number of boys and the number of girls. To form the largest number of groups, find the GCF of 12 and 18.

factors of 12: 1, 2, 3, 4, (6), 12 factors of 18: 1, 2, 3, (6), 9, 18

The GCF of 12 and 18 is 6.

Helpful Hint

If more students are put in each group, there will be fewer groups. You need the most groups possible, so put the smallest possible number of students in each team. Start with 1 boy in each group.

Think and Discuss

1. **Explain** what the GCF of two prime numbers is.
2. **Tell** what the least common factor of a group of numbers would be.

COMMON ERROR ALERT

Some students may find *any* common factor, instead of the *greatest* common factor. Remind them to find the *greatest* of the common factors.

Additional Examples

Example 1

Find the GCF of each set of numbers.

A. 28 and 42 14

B. 18, 30, and 24 6

C. 45, 18, and 27 9

Example 2

Jenna has 16 red flowers and 24 yellow flowers. She wants to make bouquets with the same number of each color flower in each bouquet. What is the greatest number of bouquets she can make? 8

3 Close

Reaching All Learners

Through Grouping Strategies

Have students work in groups of three to find the GCF. Each member of the group will use a different method for finding the GCF on the same problem. Each member should explain their method as they compare their answers. For the next problem, the group members will switch methods within the group and proceed as before to find the GCF. Allow group members to assist other group members when necessary.

Summarize

Review the terms *factor* and *greatest common factor*. Have students work in pairs using the prime factorization and ladder diagram methods of finding the GCF of two or more numbers. Then have volunteers illustrate for the class the methods used.

Answers to Think and Discuss

1. The GCF of two prime numbers is 1 because a prime number has only two factors, 1 and the number itself. Two prime numbers will have only the factor 1 in common.
2. The least common factor of a group of numbers is 1.

4-3 PRACTICE & ASSESS

Students may want to refer back to the lesson examples.

Assignment Guide

If you finished Example 1 assign:

Core 1–6, 8–16, 41–50

Enriched 11–16, 19–27, 39–50

If you finished Example 2 assign:

Core 1–18, 34–35, 41–50

Enriched 19–50

Notes

4-3 Exercises

FOR EXTRA PRACTICE see page 642

internet connect — Homework Help Online go.hrw.com Keyword: MR4 4-3

GUIDED PRACTICE

See Example 1 Find the GCF of each set of numbers.

1. 18 and 27 **9**
2. 32 and 72 **8**
3. 21, 42, and 56 **7**
4. 15, 30, and 60 **15**
5. 18, 24, and 36 **6**
6. 9, 36, and 81 **9**

See Example 2

7. Kim is making flower arrangements. She has 16 red roses and 20 pink roses. Each arrangement must have the same number of red roses and the same number of pink roses. What is the greatest number of arrangements Kim can make if every flower is used? **4 arrangements**

INDEPENDENT PRACTICE

See Example 1 Find the GCF of each set of numbers.

8. 10 and 35 **5**
9. 28 and 70 **14**
10. 36 and 72 **36**
11. 26, 48, and 62 **2**
12. 16, 40, and 88 **8**
13. 12, 60, and 68 **4**
14. 30, 45, and 75 **15**
15. 24, 48, and 84 **12**
16. 16, 48, and 72 **8**

See Example 2

17. The local recreation center held a scavenger hunt. There were 15 boys and 9 girls at the event. The group was divided into the greatest number of teams possible with the same number of boys on each team and the same number of girls on each team. How many teams were made if each person was on a team? **3 teams**

18. Ms. Kline makes balloon arrangements. She has 32 blue balloons, 24 yellow balloons, and 16 white balloons. Each arrangement must have the same number of each color. What is the greatest number of arrangements that Ms. Kline can make if every balloon is used? **8 arrangements**

PRACTICE AND PROBLEM SOLVING

Write the GCF of each set of numbers.

19. 60 and 84 **12**
20. 14 and 17 **1**
21. 10, 35, and 110 **5**
22. 21 and 306 **3**
23. 630 and 712 **2**
24. 16, 24, and 40 **8**
25. 75, 225, and 150 **75**
26. 42, 112, and 105 **7**
27. 12, 16, 20, and 24 **4**
28. 16, 24, 30, and 42 **2**
29. 25, 90, 45, and 100 **5**
30. 27, 90, 135, and 72 **9**
31. $2 \times 2 \times 3$ and 2×2 **4**
32. $2 \times 3^2 \times 7$ and $2^2 \times 3$ **6**
33. $3^2 \times 7$ and $2 \times 3 \times 5^2$ **3**

34. Jared has 12 jars of grape jam, 16 jars of strawberry jam, and 24 jars of raspberry jam. He wants to place the jam into the greatest possible number of boxes so that each box has the same number of jars of each kind of jam. How many boxes does he need? **4 boxes**

Math Background

Euclid's work *The Elements* contains a method for determining the greatest common factor of two whole numbers that is known as the Euclidean algorithm.

Here is a simple variation of the Euclidean algorithm: If $a > b$, the GCF of a and b equals the GCF of b and $a - b$. For example, the GCF of 48 and 30 equals the GCF of 30 and 18. This process can be continued until one of the terms divides the other: the GCF of 30 and 18 equals the GCF of 18 and 12; the GCF of 18 and 12 equals the GCF of 12 and 6. Because 6 divides 12, the GCF of 48 and 30 is 6.

RETEACH 4-3

LESSON 4-3 **Reteach**
Greatest Common Factor

The greatest common factor, or GCF, is the largest number that is the factor of any set of at least two numbers.

You can use prime factorization to find the GCF of two or more numbers.

To find the GCF of 18 and 24, first write the prime factorization of each number. Then identify the common prime factors.

$18 = \mathbf{2} \cdot 3 \cdot \mathbf{3}$
$24 = \mathbf{2} \cdot 2 \cdot 2 \cdot \mathbf{3}$

Next, find the product of the common prime factors.
$2 \cdot 3 = 6$

The GCF of 18 and 24 is 6.

Find the GCF of each set of numbers.

1. 32 and 48
$32 = 2^5$
$48 = 2^4 \cdot 3$
16

2. 45 and 81
$45 = 3^2 \cdot 5$
$81 = 3^4$
9

3. 18 and 36
$18 = 2 \cdot 3^2$
$36 = 2^3 \cdot 3^2$
18

4. 14 and 35
$14 = 2 \cdot 7$
$35 = 5 \cdot 7$
7

5. 42 and 72
$42 = 2 \cdot 3 \cdot 7$
$72 = 2^3 \cdot 3^2$
6

6. 56 and 64
$56 = 2^3 \cdot 7$
$64 = 2^6$
8

7. 28, 56, and 84
$28 = 2^2 \cdot 7$
$56 = 2^3 \cdot 7$
$84 = 2^2 \cdot 3 \cdot 7$
28

8. 30, 45, and 75
$30 = 2 \cdot 3 \cdot 5$
$45 = 3^2 \cdot 5$
$75 = 3 \cdot 5^2$
15

9. 36, 45, and 54
$36 = 2^2 \cdot 3^2$
$45 = 3^2 \cdot 5$
$54 = 2 \cdot 3^3$
9

PRACTICE 4-3

LESSON 4-3 **Practice B**
Greatest Common Factor

Find the GCF of each set of numbers.

1. 12 and 15 — 3
2. 18 and 24 — 6
3. 15 and 25 — 5
4. 16 and 24 — 8
5. 36 and 45 — 9
6. 24 and 54 — 6
7. 48 and 64 — 16
8. 27 and 72 — 9
9. 55 and 77 — 11
10. 16, 28, and 48 — 4
11. 15, 35, and 95 — 5
12. 20, 30, and 80 — 10
13. 18, 36, and 54 — 18
14. 27, 36, and 45 — 9
15. 21, 49, and 63 — 7
16. 25, 35, and 45 — 5
17. 28, 42, and 63 — 7
18. 25, 75, and 115 — 5

19. Mr. Thompson's sixth-grade class is competing in the school field day. There are 16 boys and 12 girls in his class. He divided the class into the greatest number of teams possible with the same number of boys on each team and the same number of girls on each team. How many teams were made if each person was on a team? How many girls were on each team? How many boys?

4 teams with 3 girls and 4 boys on each team

20. Barbara is making candy bags for her birthday party. She has 24 lollipops, 12 candy bars, and 42 pieces of gum. She wants each bag to have the same number of each kind of candy. What is the greatest number of bags she can make if all the candy is used? How many pieces of each kind of candy will be in each bag?

6 bags with 4 lollipops, 2 candy bars, and 7 pieces of gum in each bag

Life Science LINK

The biggest flower in the world is the Rafflesia. It grows in the Indonesian rain forest. Blossoms can measure four feet across, weigh about 15 pounds, and contain about 1.5 gallons of water.

35. Mr. Rodriguez is planting 4 types of flowers in his garden. He wants each row to contain the same number of each type of flower. What is the greatest number of rows Mr. Rodriguez can plant if every bulb is used? **6 rows**

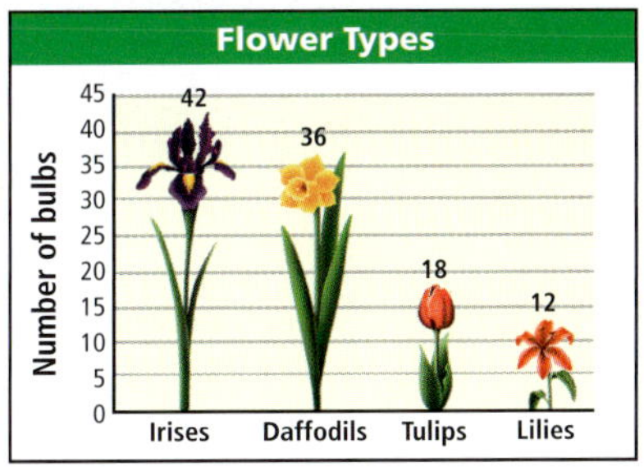

36. Pam is making fruit baskets. She has 30 apples, 24 bananas, and 12 oranges. What is the greatest number of baskets she can make if each type of fruit is distributed equally among the baskets? **6 baskets**

37. In a parade, one school band will march directly behind another school band. All rows must have the same number of students. The first band has 36 students, and the second band has 60 students. What is the greatest number of students who can be in each row? **12**

38. ***SOCIAL STUDIES*** Branches of the U.S. Mint in Denver and Philadelphia make all U.S. coins for circulation. A tiny *D* or *P* on the coin tells you where the coin was minted. Suppose you have 32 *D* quarters and 36 *P* quarters. What is the greatest number of groups you can make with the same number of *D* quarters in each group and the same number of *P* quarters in each group so that every quarter is placed in a group? **4 groups**

39. ***WRITE ABOUT IT*** What method do you like best for finding the GCF? Why? **Check students work.**

40. ***CHALLENGE*** The GCF of three numbers is 9. The sum of the numbers is 90. Find the three numbers. **18, 27, 45 or 9, 27, 54**

Spiral Review

Evaluate each expression. (Lesson 1-4)

41. $5 \times 2^3 + 17 - 3 \times 2$ **51**

42. $85 - (44 + 33) \div 7 + (62 - 12)$ **124**

43. $14 + 4^3 \div 8 \times 2 - 7$ **23**

44. $(76 - 13) \div 3^2 \times 8 + (91 - 47)$ **100**

Solve each equation. (Lesson 2-6)

45. $15n = 45$ $n = 3$

46. $7t = 147$ $t = 21$

47. $6a = 78$ $a = 13$

48. $12b = 216$ $b = 18$

49. **TEST PREP** ___?___ numbers are divisible by more than two numbers. (Lesson 4-1) **C**

A Prime
B Whole
C Composite
D Equivalent

50. **TEST PREP** The numbers 2, 3, 5, and 6 are all factors of which number? (Lesson 4-2) **H**

F 6
G 10
H 30
J 36

COMMON ERROR ALERT

When finding the GCF of a set of three numbers, some students may neglect one of the numbers and find the GCF of only two of the numbers. But what the students can do is find the GCF of the first two numbers and then find the GCF of that GCF and the third number. This will be the GCF of all three numbers.

Journal

Have students write a real-world problem involving GCF similar to the one in Example 2. Have them solve the problem and explain their solution.

Test Prep Doctor

For Exercise 49, students will need to remember definitions of the four math terms given. Prime numbers are divisible by *only* two numbers, so **A** can be eliminated. Whole numbers and equivalent numbers may or may not be divisible by more than two numbers, so **B** and **D** can be eliminated.

For Exercise 50, **F** and **J** can be eliminated because if 5 is a factor, the product ends with 0 or 5. **B** can be eliminated because 6 is not a factor of 10.

CHALLENGE 4-3

LESSON 4-3 **Challenge**
The Greatest Common Flower

A florist made these flower arrangements for a wedding. He used every flower in each crate he had to create the greatest number of arrangements possible. Study the flowers the florist had in each crate on the left. Below each, write the number of arrangements the florist made with those flowers. Then draw a line to the correct arrangement on the right that the florist created with those flowers.

1. 6 18 24
Number of arrangements: 6

2. 12 16 20
Number of arrangements: 4

3. 15 30 35
Number of arrangements: 5

4. 21 42 56
Number of arrangements: 7

5. 9 15 18
Number of arrangements: 3

A.
B.
C.
D.
E.

PROBLEM SOLVING 4-3

LESSON 4-3 **Problem Solving**
Greatest Common Factor

Write the correct answer.

1. Carolyn has 24 bottles of shampoo, 36 tubes of hand lotion, and 60 bars of lavender soap to make gift baskets. She wants to have the same number of each item in every basket. What is the greatest number of baskets she can make without having any of the items left over?
12 baskets

2. There are 40 girls and 32 boys who want to participate in the relay race. If each team must have the same number of girls and boys, what is the greatest number of teams that can race? How many boys and girls will be on each team?
8 teams with 5 girls and 4 boys each

3. Ming has 15 quarters, 30 dimes, and 48 nickels. He wants to group his money so that each group has the same number of each coin. What is the greatest number of groups he can make? How many of each coin will be in each group? How much money will each group be worth?
3 groups with 5 quarters, 10 dimes, and 16 nickels each; $3.05

4. A gardener has 27 tulip bulbs, 45 tomato plants, 108 rose bushes, and 126 herb seedlings to plant in the city garden. He wants each row of the garden to have the same number of each kind of plant. What is the greatest number of rows that the gardener can make if he uses all the plants?
9 rows

Circle the letter of the correct answer.

5. Kim packed 6 boxes with identical supplies. It was the greatest number she could pack and use all the supplies. Which of these is her supply list?
A 24 pencils, 36 pens, 10 rulers
B 12 rulers, 30 pencils, 45 pens
(C) 42 pencils, 18 rulers, 72 pens
D 60 pens, 54 pencils, 32 rulers

6. The sum of three numbers is 60. Their greatest common factor is 4. Which of the following lists shows those three numbers?
F 4, 16, 36
(G) 8, 20, 32
H 14, 16, 30
J 10, 18, 32

Lesson Quiz

Find the greatest common factor of each set of numbers.

1. 18 and 30 **6**

2. 20 and 35 **5**

3. 8, 28, 52 **4**

4. 44, 66, 88 **22**

5. Mrs. Lovejoy makes flower arrangements. She has 36 red carnations, 60 white carnations, and 72 pink carnations. Each arrangement must have the same number of each color. What is the greatest number of arrangements she can make if she uses every carnation? **12 arrangements**

Available on Daily Transparency in CRB

Section 4A

Mid-Chapter Quiz

Purpose: *To assess students' mastery of concepts and skills in Lessons 4-1 through 4-3*

Assessment Resources

Section 4A Quiz

Assessment Resources .... p. 00

***Test and Practice Generator* CD-ROM**

Additional mid-chapter assessment items in both multiple-choice and free-response format may be generated for any objective in Lessons 4-1 through 4-3.

Chapter 4

Mid-Chapter Quiz

LESSON 4-1 (pp. 152–155)

Tell whether each number is divisible by 2, 3, 4, 5, 6, 9, and 10.

1. 708 2, 3, 4, 6 **2.** 514 2 **3.** 470 2, 5, 10 **4.** 338 2

5. 200 2, 4, 5, 10 **6.** 798 2, 3, 6 **7.** 518 2 **8.** 309 3

Tell whether each number is prime or composite.

9. 76 composite **10.** 59 prime **11.** 69 composite **12.** 33 composite

13. 36 composite **14.** 78 composite **15.** 93 composite **16.** 89 prime

LESSON 4-2 (pp. 156–159)

List all of the factors of each number.

17. 26 1, 2, 13, 26 **18.** 32 1, 2, 4, 8 16, 32 **19.** 39 1, 3, 13, 39 **20.** 84 1, 2, 3, 4, 6, 7, 12, 14, 21, 28, 42, 84

21. 54 1, 2, 3, 6, 9, 18, 27, 54 **22.** 85 1, 5, 17, 85 **23.** 27 1, 3, 9, 27 **24.** 29 1, 29

Write the prime factorization of each number.

25. 96 $2 \times 2 \times 2 \times 2 \times 2 \times 3$ **26.** 50 $2 \times 5 \times 5$ **27.** 104 $2 \times 2 \times 2 \times 13$ **28.** 63 $3 \times 3 \times 7$

29. 49 7×7 **30.** 156 $2 \times 2 \times 3 \times 13$ **31.** 62 2×31 **32.** 95 5×19

LESSON 4-3 (pp. 160–163)

Find the GCF of each set of numbers.

33. 16 and 36 4 **34.** 22 and 88 22 **35.** 65 and 91 13

36. 42, 70, and 84 14 **37.** 16, 24, and 48 8 **38.** 20, 55, and 85 5

39. There are 36 sixth-graders and 40 seventh-graders. What is the greatest number of teams that the students can form if each team has the same number of sixth-graders and the same number of seventh-graders and every student must be on a team? 4 teams

40. There are 14 girls and 21 boys in Mrs. Sutter's gym class. To play a certain game, the students must form teams. Each team must have the same number of girls and the same number of boys. What is the greatest number of teams Mrs. Sutter can make if every student is on a team? 7 teams

41. Mrs. Young, an art teacher, is organizing the art supplies. She has 76 red markers, 52 blue markers, and 80 black markers. She wants to divide the markers into boxes with the same number of red, the same number of blue, and the same number of black markers in each box. What is the greatest number of boxes she can have if every marker is placed in a box? 4 boxes

Focus on Problem Solving

Understand the Problem

• Interpret unfamiliar words

You must understand the words in a problem in order to solve it. If there is a word you do not know, try to use context clues to figure out its meaning. Suppose there is a problem about red, green, blue, and chartreuse fabric. You may not know the word *chartreuse*, but you can guess that it is probably a color. To make the problem easier to understand, you could replace *chartreuse* with the name of a familiar color, like *white*.

In some problems, the name of a person, place, or thing might be difficult to pronounce, such as *Mr. Joubert*. When you see a proper noun that you do not know how to pronounce, you can use another proper noun or a pronoun in its place. You could replace *Mr. Joubert* with *he*. You could replace *Koenisburg Street* with *K Street*.

Copy each problem. Underline any words that you do not understand. Then replace each word with a more familiar word.

1. Grace is making flower bouquets. She has 18 chrysanthemums and 42 roses. She wants to arrange them in groups that each have the same number of chrysanthemums and the same number of roses. What is the fewest number of flowers that Grace can have in each group? How many chrysanthemums and how many roses will be in each group?

2. Most marbles are made from glass. The glass is liquefied in a furnace and poured. It is then cut into cylinders that are rounded off and cooled. Suppose 1,200 cooled marbles are put into packages of 8. How many packages could be made? Would there be any marbles left over?

3. In ancient times, many civilizations used calendars that divided the year into months of 30 days. A year has 365 days. How many whole months were in these ancient calendars? Were there any days left over? If so, how many?

4. Mrs. LeFeubre is tiling her garden walkway. It is a rectangle that is 4 feet wide and 20 feet long. Mrs. LeFeubre wants to use square tiles, and she does not want to have to cut any tiles. What is the size of the largest square tile that Mrs. LeFeubre can use?

Answers

1. 10 flowers in each group; 7 roses and 3 chrysanthemums
2. 150 packages; no
3. 12 months; yes; 5
4. Mrs. LeFeubre can use square tiles with side lengths of 4 ft.

Focus on Problem Solving

Purpose: *To focus on understanding the problem by interpreting unfamiliar words*

Problem Solving Resources

Interactive Problem Solving pp. 25–33

Math: Reading and Writing in the Content Area . . . . pp. 25–33

Problem Solving Process

This page focuses on the first step of the problem-solving process: **Understand the Problem**

Discuss

Have students read each problem and list the words they did not understand. Then ask them to give a simpler, or more familiar, replacement word for each of the words they mentioned.

Possible answers:

1. *Bouquets* and *chrysanthemums;* replace *bouquets* with *arrangements* and replace *chrysanthemums* with *carnations.*
2. *Liquefied* and *furnace;* replace *liquefied* with *melted* and replace *furnace* with *fire.*
3. *Ancient* and *civilizations;* replace *ancient* with *old* and replace *civilizations* with *people.*
4. *LeFeubre;* replace *Mrs. LeFeubre* with *Mrs. L.*

Section 4B

Understanding Fractions

One-Minute Section Planner

Lesson	Materials	Resources
Hands-On Lab 4A Explore Decimals and Fractions **NCTM:** Number and Operations, Representation **NAEP:** Number Properties 1b ☐ SAT-9 ☑ SAT-10 ☐ ITBS ☐ CTBS ☑ MAT ☑ CAT	**Required** Decimal squares *Base-ten blocks (MK)*	• *Hands-On Lab Activities,* pp. 20–21
Lesson 4-4 Decimals and Fractions **NCTM:** Number and Operations, Communication, Representation **NAEP:** Number Properties 1j ☑ SAT-9 ☑ SAT-10 ☑ ITBS ☑ CTBS ☑ MAT ☑ CAT	**Optional** Recording Sheet for Reaching All Learners *(CRB, p. 99)* Teaching Transparency T12 *(CRB)* Fraction Bars *(MK, CRB, p. 100)*	• *Chapter 4 Resource Book,* pp. 36–45 • Daily Transparency T11, CRB • Additional Examples Transparencies T13–T15, CRB • *Alternate Openers: Explorations,* p. 28
Hands-On Lab 4B Model Equivalent Fractions **NCTM:** Number and Operations, Reasoning and Proof **NAEP:** Number Properties 1b ☐ SAT-9 ☑ SAT-10 ☐ ITBS ☐ CTBS ☑ MAT ☐ CAT	**Required** Pattern blocks	• *Hands-On Lab Activities,* pp. 22–23, 113
Lesson 4-5 Equivalent Fractions **NCTM:** Number and Operations, Communication **NAEP:** Number Properties 1d ☑ SAT-9 ☑ SAT-10 ☑ ITBS ☑ CTBS ☑ MAT ☑ CAT	**Optional** Fraction bars *(MK, CRB, p. 100)* Number cubes *(MK)*	• *Chapter 4 Resource Book,* pp. 46–54 • Daily Transparency T16, CRB • Additional Examples Transparencies T17–T19, CRB • *Alternate Openers: Explorations,* p. 29
Hands-On Lab 4C Explore Fraction Measurement **NCTM:** Measurement, Connections **NAEP:** Measurement 1e ☑ SAT-9 ☑ SAT-10 ☑ ITBS ☑ CTBS ☑ MAT ☑ CAT	**Required** Inch ruler *(MK, CRB, p. 101)* Paper strips	• *Hands-On Lab Activities,* pp. 24–26, 114
Lesson 4-6 Comparing and Ordering Fractions **NCTM:** Number and Operations, Communication, Connections, Representation **NAEP:** Number Properties 1i ☑ SAT-9 ☑ SAT-10 ☑ ITBS ☑ CTBS ☑ MAT ☑ CAT	**Optional** Fraction bars *(MK, CRB, p. 100)* Teaching Transparency T21 *(CRB)*	• *Chapter 4 Resource Book,* pp. 55–63 • Daily Transparency T20, CRB • Additional Examples Transparencies T22–T23, CRB • *Alternate Openers: Explorations,* p. 30
Lesson 4-7 Mixed Numbers and Improper Fractions **NCTM:** Number and Operations, Communication, Representation **NAEP:** Number Properties 1d ☑ SAT-9 ☑ SAT-10 ☐ ITBS ☐ CTBS ☑ MAT ☑ CAT	**Optional** Inch ruler *(MK, CRB, p. 101)* Teaching Transparency T25 *(CRB)*	• *Chapter 4 Resource Book,* pp. 64–72 • Daily Transparency T24, CRB • Additional Examples Transparencies T26–T27, CRB • *Alternate Openers: Explorations,* p. 31
Section 4B Assessment		• Mid-Chapter Quiz, SE p. 186 • Section 4B Quiz, AR p. 12 • *Test and Practice Generator* CD-ROM

SAT = *Stanford Achievement Tests* **ITBS** = *Iowa Test of Basic Skills* **CTBS** = *Comprehensive Test of Basic Skills/Terra Nova*
MAT = *Metropolitan Achievement Tests* **CAT** = *California Achievement Test*

NCTM = Complete standards can be found on pages T27–T33. **NAEP** = Complete standards can be found on pages A31–A35.

SE = *Student Edition* **TE** = *Teacher's Edition* **AR** = *Assessment Resources* **CRB** = *Chapter Resource Book* **MK** = *Manipulatives Kit*

Section Overview

Decimals and Fractions — Lesson 4-4

Why? To compare and order numbers, it is sometimes necessary to convert between fraction form and decimal form.

Write 0.27 as a fraction.

The place value of the digit farthest to the right is hundredths, so 0.27 is 27 hundredths.

$0.27 = \frac{27}{100}$

Write $\frac{3}{8}$ as a decimal.

Divide 3 by 8 to convert $\frac{3}{8}$ to a decimal.

$8\overline{)3.000}$ = 0.375

$\frac{3}{8} = 0.375$

Order from least to greatest:

$\frac{3}{8}, \frac{1}{3}, 0.35.$

Write all the numbers in decimal form and order the three decimals.

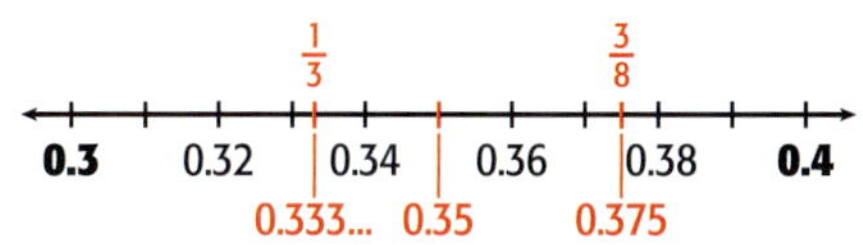

The numbers in order from least to greatest are $\frac{1}{3}$, 0.35, and $\frac{3}{8}$.

Equivalent Fractions, Comparing and Ordering Fractions — Lessons 4-5, 4-6

Why? To compare and order fractions with unlike denominators, write them as equivalent fractions with like denominators.

Equivalent fractions

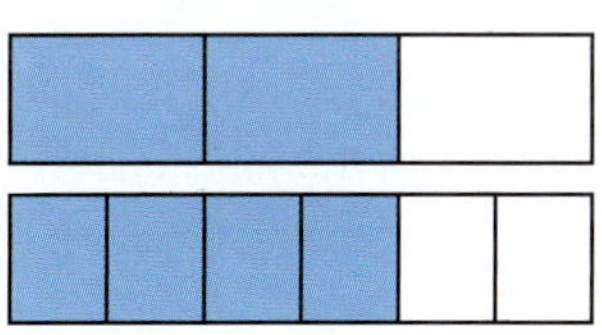

$\frac{2}{3} = \frac{4}{6}$

Find the missing number that makes the fractions equivalent.

$\frac{3}{4} = \frac{■}{20}$

$\frac{4}{5} = \frac{16}{■}$

$\frac{3 \cdot 5}{4 \cdot 5} = \frac{15}{20}$

$\frac{4 \cdot 4}{5 \cdot 4} = \frac{16}{20}$

Order from least to greatest:

$\frac{1}{2}, \frac{5}{12}, \frac{2}{5}.$

Rename with like denominators, and then compare.

$\frac{1 \cdot 30}{2 \cdot 30} = \frac{30}{60}$ $\frac{5 \cdot 5}{12 \cdot 5} = \frac{25}{60}$ $\frac{2 \cdot 12}{5 \cdot 12} = \frac{24}{60}$

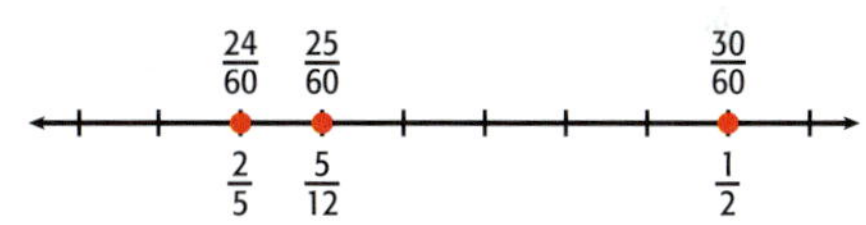

The fractions in order from least to greatest are $\frac{2}{5}, \frac{5}{12}$, and $\frac{1}{2}$.

Mixed Numbers and Improper Fractions — Lesson 4-7

Why? To operate with fractions, you sometimes need to convert between mixed numbers and improper fractions.

Convert the improper fraction $\frac{11}{4}$ to a mixed number:

$$\begin{array}{r} 2\frac{3}{4} \\ 4\overline{)11} \\ -8 \\ \hline 3 \end{array}$$

Divide the numerator by the denominator.

The remainder becomes the numerator, and the divisor remains the denominator.

$\frac{11}{4} = 2\frac{3}{4}$

From a mixed number to an improper fraction:

$2\frac{3}{4} = \frac{4 \times 2 + 3}{4}$

$= \frac{11}{4}$

Hands-On LAB

4A
Explore Decimals & Fractions

Pacing: Traditional 1 day
Block $\frac{1}{2}$ day

Objective: To use decimal grids to show the relationship between decimals and fractions

Materials: Decimal grids

Lab Resources

Hands-On Lab Activities . . . . pp. 20–21

Using the Page

Discuss with students what each small square on the decimal grid represents. Also discuss the decimal represented by each row or column of the grid.

Represent each decimal on a decimal grid.

1. 0.01

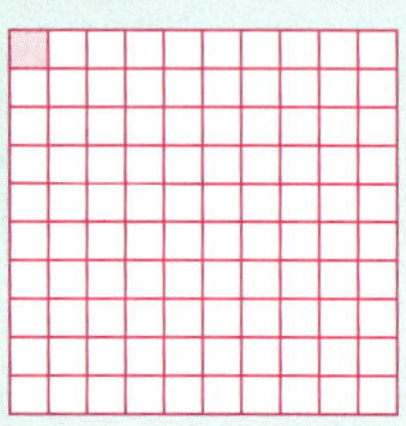

2. 0.4

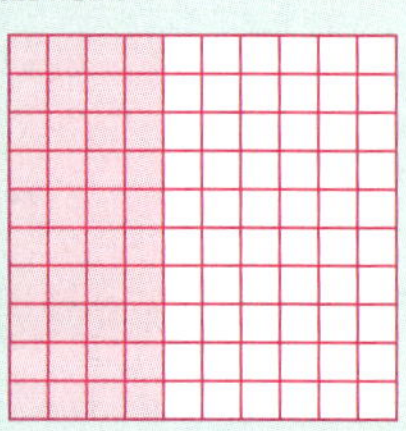

3. 0.64

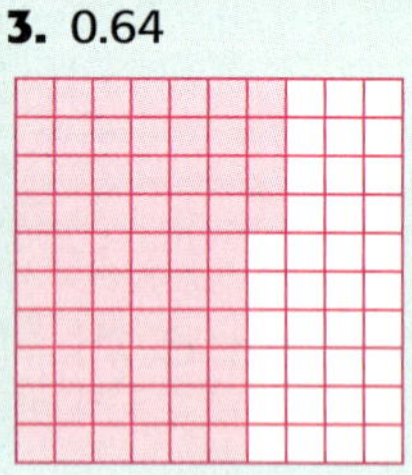

Hands-On LAB 4A

Explore Decimals and Fractions

Use with Lesson 4-4

KEY

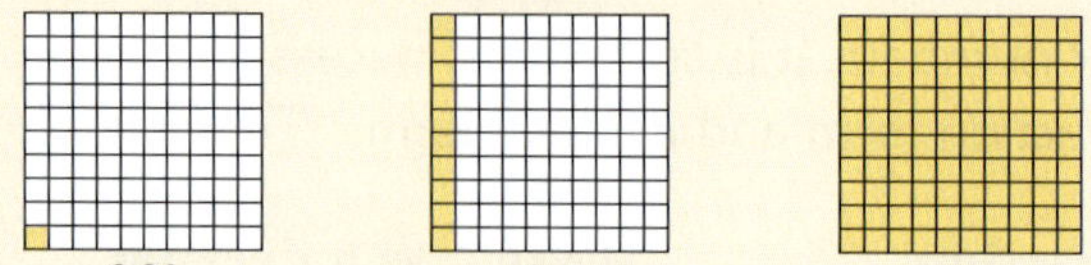

internet connect
Lab Resources Online
go.hrw.com
KEYWORD: MR4 Lab4A

You can use decimal grids to show the relationship between fractions and decimals.

Activity

Write the number represented on each grid as a fraction and as a decimal.

1

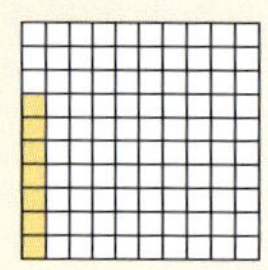

Seven hundredths squares are shaded → 0.07
How many squares are shaded? 7 ← numerator
How many squares are in the whole? 100 ← denominator

$0.07 = \frac{7}{100}$

2

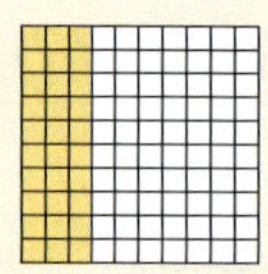

Three tenths columns are shaded → 0.3
How many complete columns are shaded? 3
How many columns are in the whole? 10

$0.3 = \frac{3}{10}$

Think and Discuss

1. Is 0.09 the same as $\frac{9}{10}$? Use decimal grids to support your answer.
Possible answer: No, 0.09 is not the same as $\frac{9}{10}$. $\frac{9}{10}$ is the same as 0.9 (nine tenths), not 0.09 (nine hundredths).

Try This

Use decimal grids to represent each number.

1. 0.8 **2.** $\frac{37}{100}$ **3.** 0.53 **4.** $\frac{1}{10}$ **5.** $\frac{67}{100}$

6. For 1–5, write each decimal as a fraction and each fraction as a decimal.
$\frac{4}{5}$; 0.37; $\frac{53}{100}$; 0.1; 0.67

Answers

Try This

1.

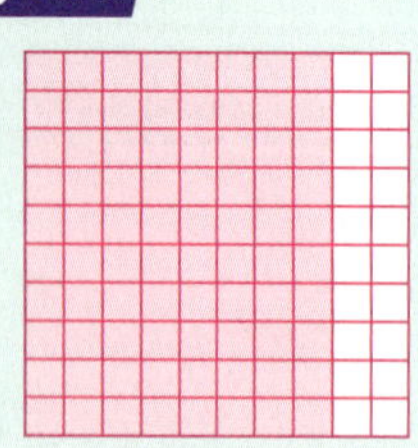

2.

3.

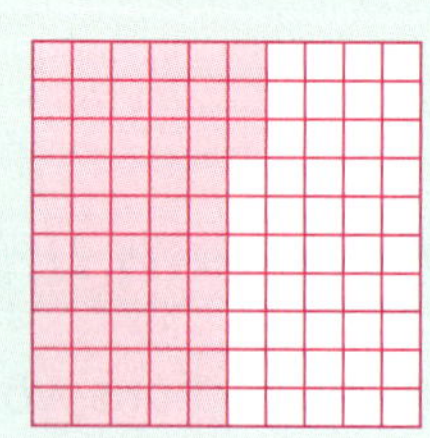

4.

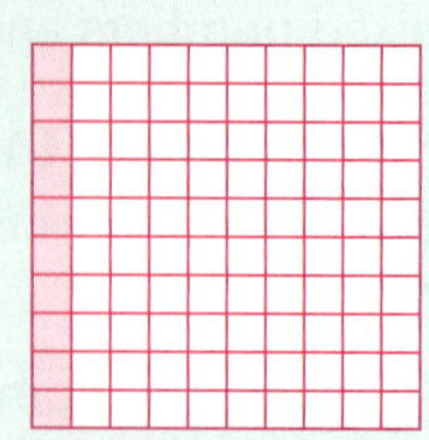

5.

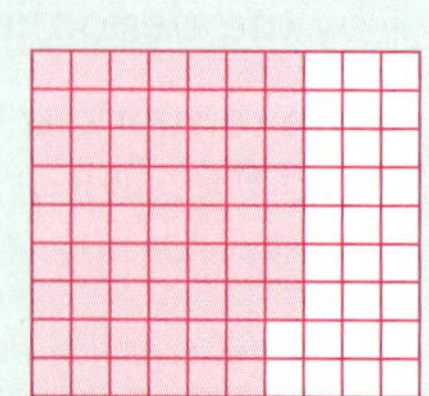

4-4 Decimals and Fractions

Learn to convert between decimals and fractions.

Vocabulary
mixed number
terminating decimal
repeating decimal

Decimals and fractions can often be used to represent the same number.

For example, a baseball player's batting average can be represented as a fraction:

$$\frac{\text{number of hits}}{\text{number of times at bat}}$$

Ty Cobb, at right, holds the record for the highest career batting average in professional baseball. He had 4,189 hits, and he was at bat 11,434 times. His batting average was $\frac{4,189}{11,434}$.

Batting averages are usually written as decimals rounded to three decimal places. To find this decimal, use the fact that the fraction bar means "divided by."

$$4,189 \div 11,434 = 0.3663634773\ldots$$

Ty Cobb's career batting average is reported as .366.

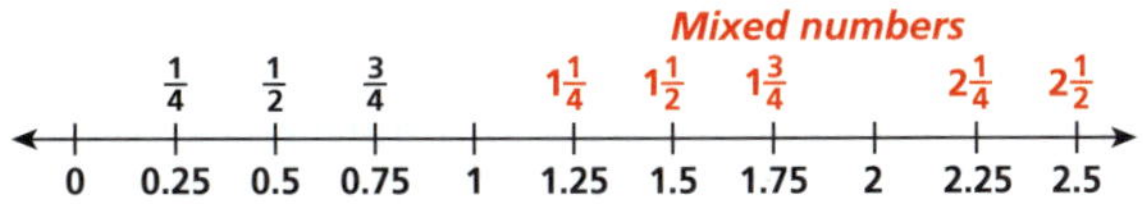

A number that contains both a whole number greater than 0 and a fraction, such as $1\frac{3}{4}$, is called a **mixed number**.

EXAMPLE 1 Writing Decimals as Fractions or Mixed Numbers

Write each decimal as a fraction or mixed number.

A 0.23

0.23 — *Identify the place value of the digit farthest to the right.*

$\frac{23}{100}$ — *The 3 is in the* ***hundred****ths place, so use* ***100*** *as the denominator.*

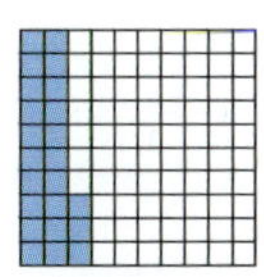

B 1.7

1.7 — *Identify the place value of the digit farthest to the right.*

$1\frac{7}{10}$ — *Write the whole number, 1. The 7 is in the* ***ten****ths place, so use* ***10*** *as the denominator.*

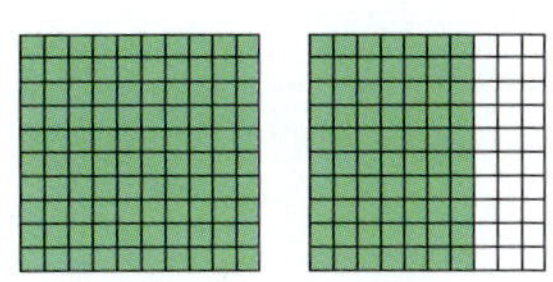

Remember!

Place Value			
Ones	Tenths	Hundredths	Thousandths

4-4 Organizer

Pacing: Traditional 1 day
Block $\frac{1}{2}$ day

Objective: Students convert between decimals and fractions.

Indiana Academic Standards
6.1.3, 6.1.4, 6.1.5

Warm Up

Find the GCF of each set of numbers.

1. 15, 24 **3** **2.** 30, 60 **30**
3. 54, 102 **6** **4.** 12, 16, 24 **4**

Problem of the Day

Write the day of the month as the product of two factors. Possible answer: 12th of the month, 6 × 2

Available on Daily Transparency in CRB

Math Humor

The class was changing decimals to fractions, so the teacher asked, "What's one point four?" The student on the basketball team said, "You get one *point* for making a foul shot."

1 Introduce

Alternate Opener

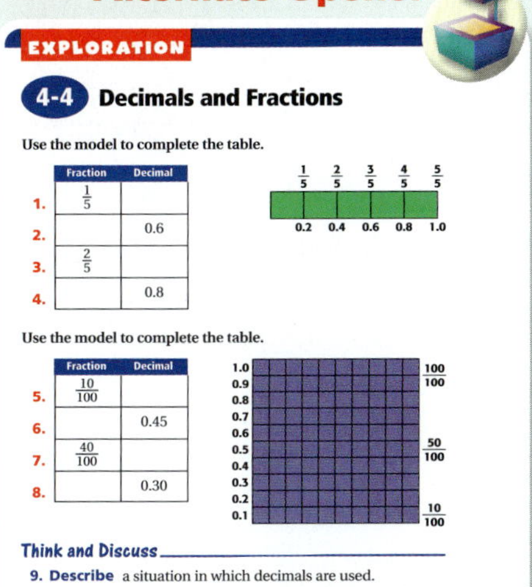

Motivate

Ask students to name different ways to order a set of decimals.

Possible answers: Use a number line; line up decimal points and compare numbers two at a time by comparing digits in the same place-value position, beginning with the greatest position.

Tell students that in this lesson they will order a group of numbers containing both decimals and fractions.

Exploration worksheet and answers on Chapter 4 Resource Book pp. 37 and 109

2 Teach

Lesson Presentation

Guided Instruction

In this lesson, students learn to convert between decimals and fractions. Begin by showing students a number line with mixed numbers and their decimal equivalents. Next teach students to change decimals to fractions and then to change fractions to decimals. Teach them to distinguish between a terminating decimal and a repeating decimal and how to write a repeating decimal. Show students common fractions and their decimal equivalents. Finally, teach how to order a set of numbers containing both fractions and decimals.

Additional Examples

Example 1

Write each decimal as a fraction or mixed number.

A. 0.67 $\frac{67}{100}$ **B.** 5.9 $5\frac{9}{10}$

Example 2

Write each fraction or mixed number as a decimal.

A. $\frac{3}{20}$ 0.15 **B.** $6\frac{1}{3}$ $6.\overline{3}$

Example 3

Order the fractions and decimal from least to greatest.

$\frac{3}{4}$, 0.8, $\frac{7}{10}$ $\frac{7}{10}$, $\frac{3}{4}$, 0.8

Example 3 note: Discuss with students the reason for converting the fractions to decimals instead of converting the decimal to a fraction: Decimals are generally easier to compare than fractions.

EXAMPLE 2 Writing Fractions as Decimals

Write each fraction or mixed number as a decimal.

A $\frac{3}{4}$

$$\begin{array}{r} 0.75 \\ 4\overline{)3.00} \\ -\,28 \\ 20 \\ -\,20 \\ 0 \end{array}$$

Divide 3 by 4.
Add zeros after the decimal point.
The remainder is 0.

$\frac{3}{4} = 0.75$

B $5\frac{2}{3}$

$$\begin{array}{r} 0.666 \\ 3\overline{)2.000} \\ -\,18 \\ 20 \\ -\,18 \\ 20 \\ -\,18 \\ 2 \end{array}$$

Divide 2 by 3.
Add zeros after the decimal point.
The 6 repeats in the quotient.

$5\frac{2}{3} = 5.666... = 5.\overline{6}$

Writing Math

To write a repeating decimal, you can show three dots or draw a bar over the repeating part: 0.666... = $0.\overline{6}$

A **terminating decimal**, such as 0.75, has a finite number of decimal places. A **repeating decimal**, such as 0.666..., has a block of one or more digits that repeat continuously.

Common Fractions and Equivalent Decimals								
$\frac{1}{5}$	$\frac{1}{4}$	$\frac{1}{3}$	$\frac{2}{5}$	$\frac{1}{2}$	$\frac{3}{5}$	$\frac{2}{3}$	$\frac{3}{4}$	$\frac{4}{5}$
0.2	0.25	$0.\overline{3}$	0.4	0.5	0.6	$0.\overline{6}$	0.75	0.8

EXAMPLE 3 Comparing and Ordering Fractions and Decimals

Order the fractions and decimals from least to greatest.

0.5, $\frac{1}{5}$, 0.37

First rewrite the fraction as a decimal. $\frac{1}{5} = 0.2$

Order the three decimals.

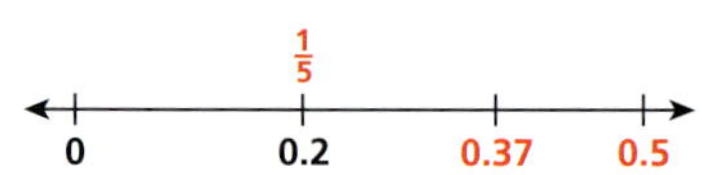

The numbers in order from least to greatest are $\frac{1}{5}$, 0.37, and 0.5.

Think and Discuss

1. **Tell** how reading the decimal 6.9 as "six and nine tenths" helps you to write 6.9 as a mixed number.
2. **Look** at the decimal 0.121122111222.... If the pattern continues, is this a repeating decimal? Why or why not?

Reaching All Learners

Through Number Sense

Have students match common fractions with their decimal equivalents. Knowing these fractions and their decimal equivalents will help students gain confidence in working with fractions.

A recording sheet is provided on Chapter 4 Resource Book p. 99.

3 Close

Summarize

Review the terms *mixed number*, *terminating decimal*, and *repeating decimal* by having students give brief definitions of each.

Possible answer: A mixed number is a number that contains a whole number and a fraction. A terminating decimal is a decimal that has a finite number of decimal places. A repeating decimal repeats the same digit or group of digits without end.

Answers to Think and Discuss

1. Possible answer: By saying it this way you know that you must write 6 first, and then to write the rest as a fraction, you just write 9 over 10.
2. No; Although the digits follow a pattern, the same combination of digits does not repeat.

4-4 Exercises

FOR EXTRA PRACTICE
see page 643

internet connect
Homework Help Online
go.hrw.com Keyword: MR4 4-4

4-4 PRACTICE & ASSESS

Students may want to refer back to the lesson examples.

GUIDED PRACTICE

Accept equivalent fractions.

See Example 1
Write each decimal as a fraction or mixed number.

1. 0.15 $\frac{3}{20}$
2. 1.25 $1\frac{1}{4}$
3. 0.43 $\frac{43}{100}$
4. 2.6 $2\frac{3}{5}$

See Example 2
Write each fraction or mixed number as a decimal.

5. $\frac{2}{5}$ 0.4
6. $2\frac{7}{8}$ 2.875
7. $\frac{1}{8}$ 0.125
8. $4\frac{1}{10}$ 4.1

See Example 3
Order the fractions and decimals from least to greatest.

9. $\frac{2}{3}$, 0.78, 0.21 — 0.21, $\frac{2}{3}$, 0.78
10. $\frac{5}{16}$, 0.67, $\frac{1}{6}$ — $\frac{1}{6}$, $\frac{5}{16}$, 0.67
11. 0.52, $\frac{1}{9}$, 0.3 — $\frac{1}{9}$, 0.3, 0.52

INDEPENDENT PRACTICE

Accept equivalent fractions.

See Example 1
Write each decimal as a fraction or mixed number.

12. 0.31 $\frac{31}{100}$
13. 5.71 $5\frac{71}{100}$
14. 0.13 $\frac{13}{100}$
15. 3.23 $3\frac{23}{100}$
16. 0.5 $\frac{1}{2}$
17. 2.7 $2\frac{7}{10}$
18. 0.19 $\frac{19}{100}$
19. 6.3 $6\frac{3}{10}$

See Example 2
Write each fraction or mixed number as a decimal.

20. $\frac{1}{9}$ $0.\overline{1}$
21. $1\frac{3}{5}$ 1.6
22. $\frac{8}{9}$ $0.\overline{8}$
23. $3\frac{11}{40}$ 3.275
24. $2\frac{5}{6}$ $2.8\overline{3}$
25. $\frac{3}{8}$ 0.375
26. $4\frac{4}{5}$ 4.8
27. $\frac{5}{8}$ 0.625

See Example 3
Order the fractions and decimals from least to greatest.

28. 0.49, 0.82, $\frac{1}{2}$
29. $\frac{3}{8}$, 0.29, $\frac{1}{9}$
30. 0.94, $\frac{4}{5}$, 0.6
31. 0.11, $\frac{1}{10}$, 0.13
32. $\frac{2}{3}$, 0.42, $\frac{2}{5}$
33. $\frac{3}{7}$, 0.76, 0.31

28. 0.49, $\frac{1}{2}$, 0.82 **29.** $\frac{1}{9}$, 0.29, $\frac{3}{8}$ **30.** 0.6, $\frac{4}{5}$, 0.94

31. $\frac{1}{10}$, 0.11, 0.13 **32.** $\frac{2}{5}$, 0.42, $\frac{2}{3}$ **33.** 0.31, $\frac{3}{7}$, 0.76

PRACTICE AND PROBLEM SOLVING

Write each decimal in expanded form and use a whole number or fraction for each place value.

34. 0.81
35. 92.3
36. 13.29
37. 107.17

Write each fraction as a decimal. Tell whether the decimal terminates or repeats.

38. $\frac{7}{9}$ $0.\overline{7}$; repeats
39. $\frac{1}{6}$ $0.1\overline{6}$; repeats
40. $\frac{17}{20}$ 0.85; terminates
41. $\frac{5}{12}$ $0.41\overline{6}$; repeats

Compare. Write <, >, or =.

42. 0.75 ▒ $\frac{3}{4}$ =
43. $\frac{5}{8}$ ▒ 0.5 >
44. 0.78 ▒ $\frac{7}{9}$ >

Assignment Guide

If you finished Example 1 assign:
Core 1–4, 12–19, 54–61
Enriched 12–19, 34–37, 54–61

If you finished Example 2 assign:
Core 1–8, 12–27, 54–61
Enriched 12–27, 34–41, 50–61

If you finished Example 3 assign:
Core 1–33, 48, 49, 54–61
Enriched 20–61

34. $0 + \frac{8}{10} + \frac{1}{100}$
35. $90 + 2 + \frac{3}{10}$
36. $10 + 3 + \frac{2}{10} + \frac{9}{100}$
37. $100 + 7 + \frac{1}{10} + \frac{7}{100}$

RETEACH 4-4

LESSON 4-4 **Reteach**
Decimals and Fractions

You can write decimals as fractions or mixed numbers. A place value chart will help you read the decimal. Remember the decimal point is read as the word "and."

To write 0.47 as a fraction, first think about the decimal in words.

Ones	Tenths	Hundredths	Thousandths	Ten Thousandths
0.	4	7		

0.47 is read "forty-seven hundredths." The place value of the decimal tells you the denominator is 100.

$0.47 = \frac{47}{100}$

To write 8.3 as a mixed number, first think about the decimal in words.

Ones	Tenths	Hundredths	Thousandths	Ten Thousandths
8.	3			

8.3 is read "eight and three tenths." The place value of the decimal tells you the denominator is 10. The decimal point is read as the word "and."

$8.3 = 8\frac{3}{10}$

Write each decimal as a fraction or mixed number.

1. 0.61 $\frac{61}{100}$
2. 3.43 $3\frac{43}{100}$
3. 0.009 $\frac{9}{1,000}$
4. 4.7 $4\frac{7}{10}$
5. 1.5 $1\frac{5}{10}$ or $1\frac{1}{2}$
6. 0.13 $\frac{13}{100}$
7. 5.002 $5\frac{2}{1,000}$ or $5\frac{1}{500}$
8. 0.021 $\frac{21}{1,000}$

PRACTICE 4-4

LESSON 4-4 **Practice B**
Decimals and Fractions

Write each decimal as a fraction or mixed number.

1. 0.23 $\frac{23}{100}$
2. 0.1 $\frac{1}{10}$
3. 3.25 $3\frac{25}{100}$ or $3\frac{1}{4}$
4. $1.\overline{3}$ $1\frac{1}{3}$
5. 5.5 $5\frac{5}{10}$ or $5\frac{1}{2}$
6. 3.7 $3\frac{7}{10}$

Write each fraction or mixed number as a decimal.

7. $\frac{4}{5}$ 0.8
8. $\frac{1}{9}$ $0.\overline{1}$
9. $1\frac{2}{3}$ $1.\overline{6}$
10. $3\frac{3}{5}$ 3.6
11. $2\frac{1}{3}$ $2.\overline{3}$
12. $\frac{8}{9}$ $0.\overline{8}$

Order the fractions and decimals from least to greatest.

13. $\frac{1}{4}$, 0.7, $\frac{3}{5}$ — $\frac{1}{4}$, $\frac{3}{5}$, 0.7
14. 0.25, $\frac{1}{8}$, 0.3 — $\frac{1}{8}$, 0.25, 0.3
15. $\frac{9}{10}$, 0.49, $\frac{1}{2}$ — 0.49, $\frac{1}{2}$, $\frac{9}{10}$

Order the fractions and decimals from greatest to least.

16. 0.13, $\frac{1}{10}$, 0.9 — 0.9, 0.13, $\frac{1}{10}$
17. $\frac{2}{5}$, 0.7, $\frac{2}{3}$ — 0.7, $\frac{2}{3}$, $\frac{2}{5}$
18. 0.65, $\frac{4}{5}$, $\frac{3}{4}$ — $\frac{4}{5}$, $\frac{3}{4}$, 0.65

19. Derrick has a dollar bill and three dimes, Jane has a dollar bill and one quarter, and Kelly has a dollar bill and ten nickels. Who has the most money? the least?
Kelly has the most, and Jane has the least.

20. It rained three and one half inches in April. In May it rained $3\frac{3}{4}$ inches, and in June it rained 3.6 inches. Write the months in order from the greatest to the least amount of rain.
May, June, April

Math Background

As long ago as 1100 B.C., people in China were using the concepts of fractions and mixed numbers. They did not use numerical symbols, but rather words to express these ideas.

Our method of writing fractions probably traces to Hindu mathematicians, although they did not employ the fraction bar. For example, to write the fraction $\frac{1}{2}$, Hindu mathematicians of the first millennium would write $\frac{1}{2}$. The fraction bar was introduced by Arab mathematicians.

Answers

51. Only the 8 repeats, so it should be the only digit with a line over it. $0.3\overline{8}$

52. 0.24 is the decimal for $\frac{6}{25}$. To find this you could multiply 0.04 by 6.

Journal

Tell students to write about one example of a fraction and one example of a decimal number and how they are used in the real world.

Test Prep Doctor

For Exercise 61, students can first eliminate **D** because 9 is only a factor of 18. Students who answered **A** or **B** chose a factor common to all three numbers, but that is not the greatest common factor.

Lesson Quiz

Write each decimal as a fraction or mixed number.

1. 0.24 $\frac{6}{25}$

2. 6.75 $6\frac{3}{4}$

Write each fraction or mixed number as a decimal.

3. $2\frac{3}{5}$ 2.6

4. $\frac{7}{8}$ 0.875

5. Order the fractions and decimals from least to greatest.

0.38, $\frac{3}{4}$, $\frac{1}{5}$ $\frac{1}{5}$, 0.38, $\frac{3}{4}$

6. Jamal ran the following distances on three different days: 0.87 mile, $\frac{2}{3}$ mile, and 0.13 mile. Order the distances from least to greatest.

0.13, $\frac{2}{3}$, 0.87

Available on Daily Transparency in CRB

45. $4\frac{1}{2}$, 4.48, 3.92

46. $10\frac{5}{9}$, 10.5, $10\frac{1}{5}$

47. 125.25, 125.205, $125\frac{1}{5}$

Order the mixed numbers and decimals from greatest to least.

45. 4.48, 3.92, $4\frac{1}{2}$ **46.** $10\frac{5}{9}$, 10.5, $10\frac{1}{5}$ **47.** 125.205, 125.25, $125\frac{1}{5}$

The table shows batting averages for two baseball seasons. Use the table for Exercises 48 and 49.

Player	Season 1	Season 2
Pedro	0.360	$\frac{3}{10}$
Jill	0.380	$\frac{3}{8}$
Lamar	0.290	$\frac{1}{3}$
Britney	0.190	$\frac{3}{20}$

48. Which players had higher batting averages in season 1 than they had in season 2? Pedro, Jill, and Britney

49. Who had the highest batting average in either season? Jill

50. ***LIFE SCIENCE*** Most people with color deficiency (often called color blindness) have trouble distinguishing shades of red and green. About 0.05 of men in the world have color deficiency. What fraction of men have color deficiency? $\frac{1}{20}$

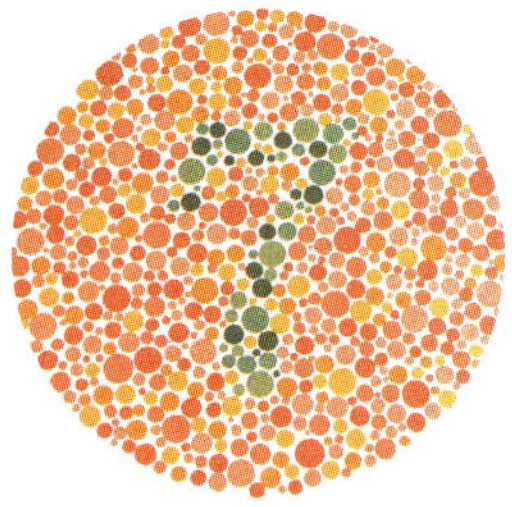

People with normal color vision will see "7" in this color-blindness test.

51. ***WHAT'S THE ERROR?*** A student found the decimal equivalent of $\frac{7}{18}$ to be $0.\overline{38}$. Explain the error. What is the correct answer?

52. ***WRITE ABOUT IT*** The decimal for $\frac{1}{25}$ is 0.04, and the decimal for $\frac{2}{25}$ is 0.08. Without dividing, find the decimal for $\frac{6}{25}$. Explain how you found your answer.

53. ***CHALLENGE*** Write $\frac{1}{999}$ as a decimal. $0.\overline{001}$

Spiral Review

Identify the property illustrated by each equation. (Lesson 1-5)

54. 4 + 5 = 5 + 4 commutative **55.** 3(4 − 1) = 3(4) − 3(1) distributive **56.** (91 + 80) + 72 = 91 + (80 + 72) associative

Find the number of decimal places in each product. Then multiply. (Lesson 3-6)

57. 2.4 · 1.8 2; 4.32 **58.** 19 · 0.5 1; 9.5 **59.** 7.04 · 2.38 4; 16.7552 **60.** 0.4 · 0.1 2; 0.04

61. **TEST PREP** What is the greatest common factor of 12, 18, and 30? (Lesson 4-3) C

A 2 **B** 3 **C** 6 **D** 9

CHALLENGE 4-4

LESSON 4-4 **Challenge**
Fractions of Pizza

Write the fractions as decimals. Then slice and shade pieces of each pizza to represent each fraction. Finally, compare the pizzas in each row by writing <, >, or =.

1. $\frac{3}{4}$ 0.75 $\frac{4}{5}$ 0.8

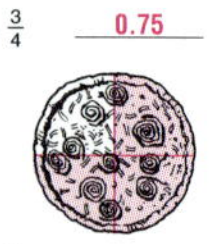 < 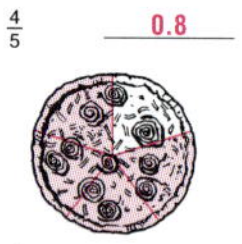

2. $\frac{3}{6}$ 0.5 $\frac{1}{2}$ 0.5

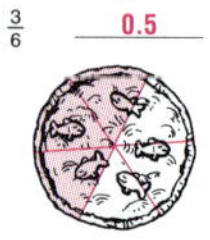 = 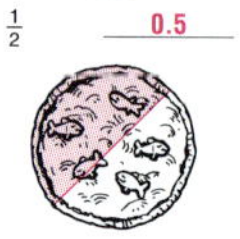

3. $\frac{2}{3}$ $0.\overline{6}$ $\frac{3}{5}$ 0.6

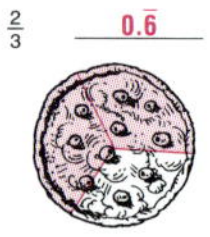 > 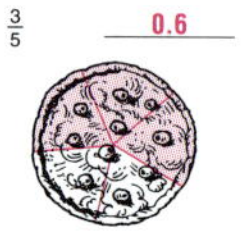

4. $\frac{1}{5}$ 0.2 $\frac{1}{3}$ $0.\overline{3}$

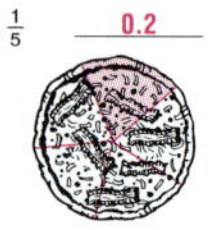 < 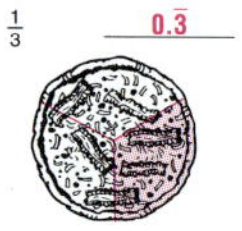

PROBLEM SOLVING 4-4

LESSON 4-4 **Problem Solving**
Decimals and Fractions

Electricity is measured in amperes, or the rate electrical currents flow. A high ampere measurement means that a lot of electricity is being used. The table below shows the average amount of electricity some household appliances use per hour. Use the table to answer the questions.

Electricity Use in the Home

Appliance	Amps per Hour
Blender	$2\frac{1}{2}$
Coffeemaker	$6\frac{2}{3}$
Computer and printer	$1\frac{5}{6}$
Microwave oven	$12\frac{1}{2}$
Popcorn popper	$2\frac{1}{12}$
25-inch television	$1\frac{1}{4}$
VCR	$\frac{1}{3}$

1. How much electricity does an average 25-inch television use each hour? Write your answer as a decimal.
1.25 amperes

2. Which appliance uses an average of 2.5 amps per hour?
blender

3. Which appliance uses the most electricity per hour? Write its ampere measurement as a decimal.
microwave oven;
12.5 amperes

4. How much electricity do most computers and printers use in an hour? Write that measurement as a decimal.
$1.8\overline{3}$ amperes

5. Which appliances have hourly ampere measurements that are repeating decimals?
coffee maker, computer and printer, popcorn popper, and VCR

Circle the letter of the correct answer.

6. In most years, 39.7 percent of the world's energy comes from burning oil. What is this percent written as a fraction?
A $\frac{39}{7}$ percent
B $39\frac{1}{7}$ percent
C $3\frac{9}{7}$ percent
(D) $39\frac{7}{10}$ percent

7. The United States produces about 13.2 percent of the world's hydroelectric power. What fraction of hydroelectric power does the United States produce?
(F) $13\frac{1}{5}$ percent
G $\frac{13}{2}$ percent
H $1\frac{3}{2}$ percent
J $13\frac{1}{2}$ percent

Hands-On LAB 4B Model Equivalent Fractions

Use with Lesson 4-5

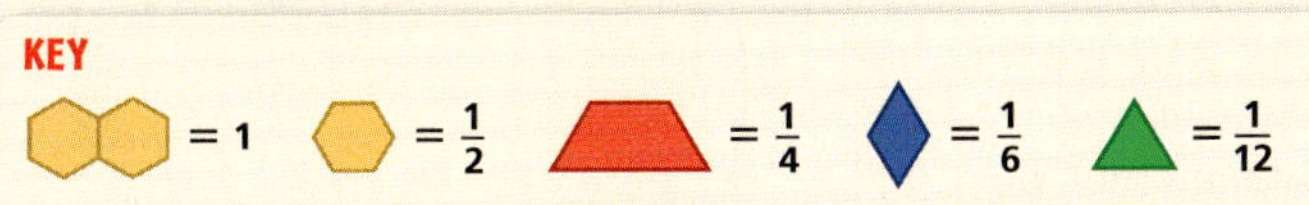

internet connect
Lab Resources Online
go.hrw.com
KEYWORD: MR4 Lab4B

Pattern blocks can be used to model equivalent fractions. To find a fraction that is equivalent to $\frac{1}{2}$, first choose the pattern block that represents $\frac{1}{2}$. Then find all the pieces of one color that will fit evenly on the $\frac{1}{2}$ block. Count these pieces to find the equivalent fraction. You may be able to find more than one equivalent fraction.

$\frac{1}{2} = \frac{2}{4} = \frac{3}{6} = \frac{6}{12}$

Activity

1. Use pattern blocks to find an equivalent fraction for $\frac{8}{12}$.

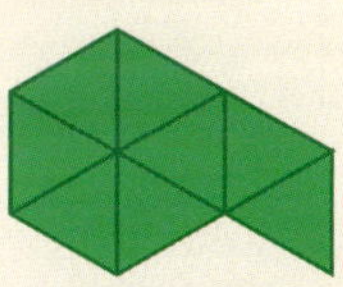

First show $\frac{8}{12}$.

$\frac{8}{12} = \frac{4}{6}$

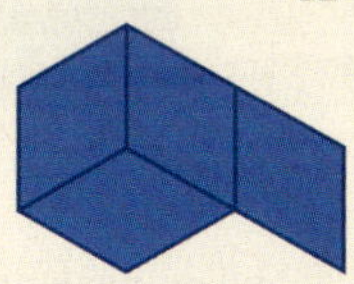

You can cover $\frac{8}{12}$ with four of the $\frac{1}{6}$ pieces.

Think and Discuss

1. Can you find a combination of pattern blocks for $\frac{1}{3}$? Find an equivalent fraction for $\frac{1}{3}$.
2. Are $\frac{9}{12}$ and $\frac{3}{6}$ equivalent? Use pattern blocks to support your answer.

Try This

Write the fraction that is modeled. Then find an equivalent fraction.

1. 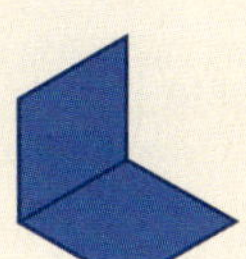

$\frac{2}{6} = \frac{1}{3}$

2. 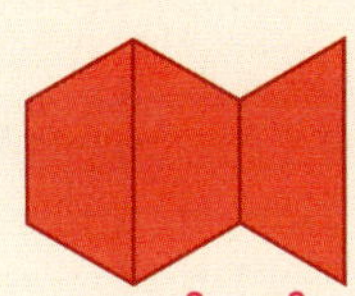

$\frac{3}{4} = \frac{9}{12}$

Hands-On LAB 4B Model Equivalent Fractions

Pacing: Traditional 1 day
Block $\frac{1}{2}$ day

Objective: To use pattern blocks to model equivalent fractions

Materials: Pattern blocks

Lab Resources

Hands-On Lab Activities pp. 22–23, 113

Using the Page

Discuss with students what each pattern block represents.

Model each fraction or mixed number with pattern blocks.

1. 2

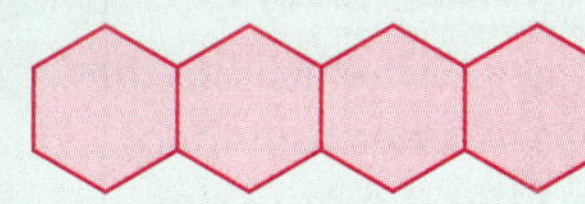

2. $1\frac{1}{4}$

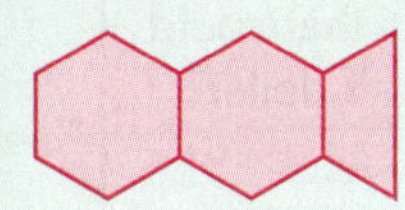

3. $\frac{5}{6}$

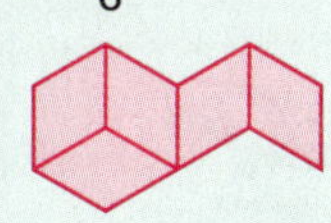

Answers

Think and Discuss

1. Possible answer: Two blue rhombus-shaped pattern blocks are equivalent to $\frac{1}{3}$. $\frac{2}{6}$ is an equivalent fraction for $\frac{1}{3}$.
2. Possible answer: No, $\frac{9}{12}$ and $\frac{3}{6}$ are not equivalent. $\frac{9}{12}$ can be represented by 9 green triangle-shaped pattern blocks, and $\frac{3}{6}$ can be represented by 3 blue rhombus-shaped pattern blocks or 6, not 9, green triangle-shaped pattern blocks.

4-5 Organizer

Pacing: Traditional 1 day
Block $\frac{1}{2}$ day

Objective: Students write equivalent fractions and mixed numbers.

Indiana Academic Standards
6.7.1

Warm Up

List the factors of each number.

1. 8 — 1, 2, 4, 8
2. 10 — 1, 2, 5, 10
3. 16 — 1, 2, 4, 8, 16
4. 20 — 1, 2, 4, 5, 10, 20
5. 30 — 1, 2, 3, 5, 6, 10, 15, 30

Problem of the Day

John has 3 coins, 2 of which are the same. Ellen has 1 fewer coin than John, and Anna has 2 more coins than John. Each girl has only 1 kind of coin. Who has coins that could equal the value of a half dollar?
Ellen and Anna

Available on Daily Transparency in CRB

Math Humor

The fraction one-fifth went to the doctor because it wasn't feeling very well. The doctor said, "You need to calm down. You're too tense (two-tenths)."

4-5 Equivalent Fractions

Learn to write equivalent fractions.

Vocabulary
equivalent fractions
simplest form

Rulers often have marks for inches, $\frac{1}{2}$, $\frac{1}{4}$, and $\frac{1}{8}$ inches.

Notice that $\frac{1}{2}$ in., $\frac{2}{4}$ in., and $\frac{4}{8}$ in. all name the same length. Fractions that represent the same value are **equivalent fractions**. So $\frac{1}{2}$, $\frac{2}{4}$, and $\frac{4}{8}$ are equivalent fractions.

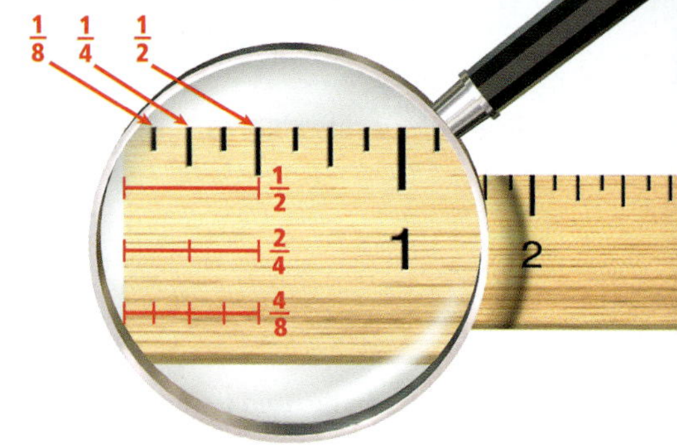

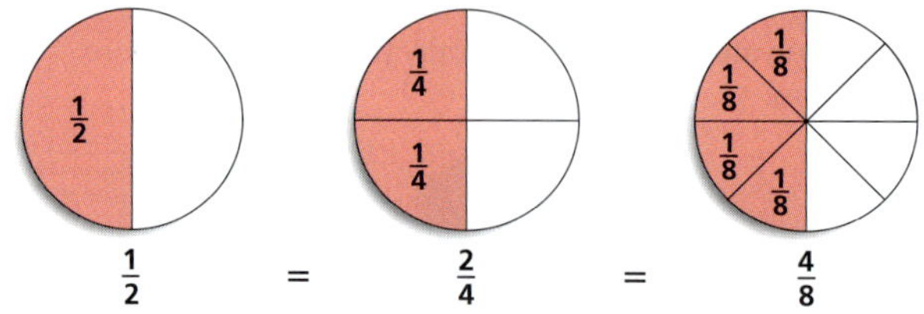

EXAMPLE 1 Finding Equivalent Fractions

Find two equivalent fractions for $\frac{6}{8}$.

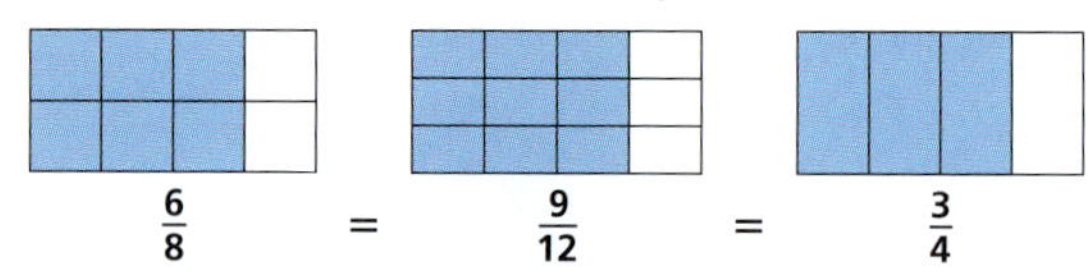

So $\frac{6}{8}$, $\frac{9}{12}$, and $\frac{3}{4}$ are all equivalent fractions.

EXAMPLE 2 Multiplying and Dividing to Find Equivalent Fractions

Find the missing number that makes the fractions equivalent.

A $\frac{2}{3} = \frac{\blacksquare}{18}$

$\frac{2 \cdot 6}{3 \cdot 6} = \frac{12}{18}$ — *In the denominator, 3 is multiplied by 6 to get 18. Multiply the numerator, 2, by the same number, 6.*

So $\frac{2}{3}$ is equivalent to $\frac{12}{18}$.

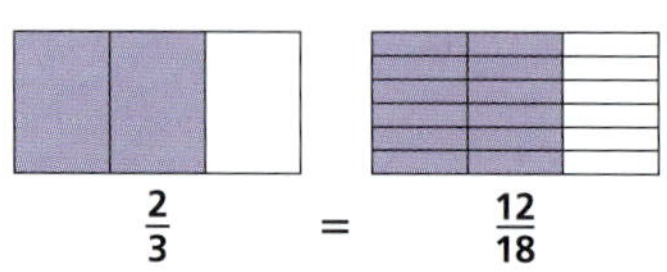

1 Introduce

Alternate Opener

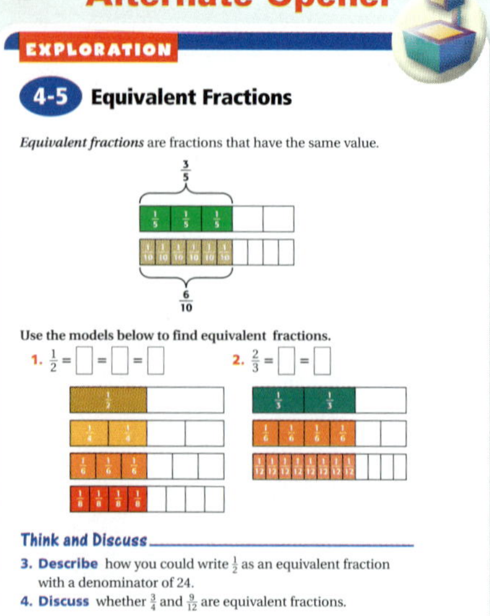
EXPLORATION

4-5 Equivalent Fractions

Equivalent fractions are fractions that have the same value.

Use the models below to find equivalent fractions.

1. $\frac{1}{2} = \square = \square = \square$ 2. $\frac{2}{3} = \square = \square$

Think and Discuss

3. **Describe** how you could write $\frac{1}{3}$ as an equivalent fraction with a denominator of 24.
4. **Discuss** whether $\frac{3}{4}$ and $\frac{9}{12}$ are equivalent fractions.

Motivate

Set up fraction bars in groups of three, with each group representing the same fractional amount (e.g., $\frac{1}{2}$, $\frac{3}{6}$, $\frac{4}{8}$). Ask students to state how the fractions in each group are alike and how they are different.

Possible answer: The same amount is represented in each group. Different numbers of bars were used to represent the same fractional amount.

Exploration worksheet and answers on Chapter 4 Resource Book pp. 47 and 112

2 Teach

Lesson Presentation

Guided Instruction

In this lesson, students learn to write equivalent fractions and mixed numbers. Teach students to use models to find equivalent fractions. Then teach them to multiply and divide to find equivalent fractions and to write a fraction in simplest form using the GCF or a ladder diagram. Students using the GCF method may need to check their final fractions to be sure they are indeed in simplest form.

Teaching Tip Show students how to use a fraction-capable calculator to check whether fractions are in simplest form.

Find the missing number that makes the fractions equivalent.

B $\frac{70}{100} = \frac{7}{■}$

$\frac{70 \div 10}{100 \div 10} = \frac{7}{10}$ *In the numerator, 70 is divided by 10 to get 7. Divide the denominator by the same number, 10.*

So $\frac{70}{100}$ is equivalent to $\frac{7}{10}$.

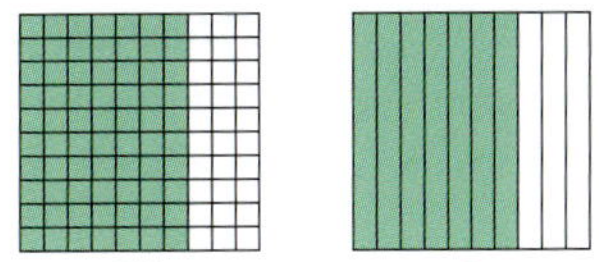

$\frac{70}{100} = \frac{7}{10}$

Every fraction has one equivalent fraction that is called the simplest form of the fraction. A fraction is in **simplest form** when the GCF of the numerator and the denominator is 1.

Example 3 shows two methods for writing a fraction in simplest form.

EXAMPLE 3 Writing Fractions in Simplest Form

Write each fraction in simplest form.

A $\frac{18}{24}$

The GCF of 18 and 24 is 6, so $\frac{18}{24}$ is not in simplest form.

Method 1: Use the GCF.

$\frac{18 \div 6}{24 \div 6} = \frac{3}{4}$ *Divide 18 and 24 by their GCF, 6.*

Method 2: Use a ladder diagram.

2 | 18/24
3 | 9/12
(3/4)

Use a ladder. Divide 18 and 24 by any common factor (except 1) until you cannot divide anymore.

So $\frac{18}{24}$ written in simplest form is $\frac{3}{4}$.

B $\frac{8}{9}$

The GCF of 8 and 9 is 1, so $\frac{8}{9}$ is already in simplest form.

Helpful Hint

Method 2 is useful when you know that the numerator and denominator have common factors, but you are not sure what the GCF is.

Think and Discuss

1. **Explain** whether a fraction is equivalent to itself.
2. **Tell** which of the following fractions are in simplest form: $\frac{9}{21}$, $\frac{20}{25}$, and $\frac{5}{13}$. Explain.
3. **Explain** how you know that $\frac{7}{16}$ is in simplest form.

COMMON ERROR ALERT

Watch for students who add to or subtract from the numerator and denominator instead of multiplying or dividing when making equivalent fractions.

Additional Examples

Example 1

Find two equivalent fractions for $\frac{10}{12}$.

Possible answer: $\frac{10}{12}, \frac{15}{18}, \frac{5}{6}$

Example 2

Find the missing number that makes the fractions equivalent.

A. $\frac{3}{5} = \frac{■}{20}$ 12

B. $\frac{4}{5} = \frac{80}{■}$ 100

Example 3

Write each fraction in simplest form.

A. $\frac{20}{48}$ $\frac{5}{12}$

B. $\frac{7}{10}$ $\frac{7}{10}$ is already in simplest form.

Reaching All Learners

Through Hands-On Experience

Have students work in groups to find equivalent fractions. Groups toss two number cubes (provided in the Manipulatives Kit). The lesser number becomes the numerator of a fraction, and the greater number becomes the denominator. If the numbers are the same, the number is both the numerator and the denominator. Each group member names one equivalent fraction. Then the group decides which one of the two fractions, if either, is in simplest form. If students are having difficulty, encourage them to draw the fraction generated by the number cube and its equivalent.

3 Close

Summarize

Review the terms *equivalent fractions* and *simplest form*. Have the students explain how to find equivalent fractions and how to write fractions in simplest form.

Possible answer: You can find equivalent fractions by multiplying or dividing the numerator and denominator by the same number. You can write a fraction in simplest form by dividing the numerator and denominator by their GCF.

Answers to Think and Discuss

1. Yes, because a fraction would have the same value as itself.
2. $\frac{5}{13}$ is the only fraction whose numerator and denominator have a GCF of 1, so it is the only fraction that is in simplest form.
3. Possible answer: The GCF of 7 and 16 is 1.

4-5 PRACTICE & ASSESS

Students may want to refer back to the lesson examples.

Assignment Guide

If you finished Example 1 assign:
Core 1–4, 12–19, 44–53
Enriched 16–19, 34–38, 41–53

If you finished Example 2 assign:
Core 1–7, 12–25, 35, 37–38, 44–53
Enriched 12–25, 34–38, 41–53

If you finished Example 3 assign:
Core 1–33, 35, 37–38, 40, 44–53
Enriched 8–53

Notes

4-5 Exercises

FOR EXTRA PRACTICE see page 643

internet connect
Homework Help Online
go.hrw.com Keyword: MR4 4-5

GUIDED PRACTICE

See Example 1 Find two equivalent fractions for each fraction. Possible answers:

1. $\frac{4}{6}$ $\frac{2}{3}, \frac{8}{12}$
2. $\frac{3}{12}$ $\frac{1}{4}, \frac{2}{8}$
3. $\frac{3}{6}$ $\frac{1}{2}, \frac{5}{10}$
4. $\frac{6}{16}$ $\frac{3}{8}, \frac{9}{24}$

See Example 2 Find the missing numbers that make the fractions equivalent.

5. $\frac{2}{5} = \frac{10}{\square}$ 25
6. $\frac{7}{21} = \frac{1}{\square}$ 3
7. $\frac{3}{4} = \frac{\square}{28}$ 21

See Example 3 Write each fraction in simplest form.

8. $\frac{2}{10}$ $\frac{1}{5}$
9. $\frac{6}{18}$ $\frac{1}{3}$
10. $\frac{4}{16}$ $\frac{1}{4}$
11. $\frac{9}{15}$ $\frac{3}{5}$

INDEPENDENT PRACTICE

See Example 1 Find two equivalent fractions for each fraction. Possible answers:

12. $\frac{3}{9}$ $\frac{1}{3}, \frac{2}{6}$
13. $\frac{2}{10}$ $\frac{1}{5}, \frac{5}{25}$
14. $\frac{3}{21}$ $\frac{1}{7}, \frac{2}{14}$
15. $\frac{3}{18}$ $\frac{1}{6}, \frac{2}{12}$
16. $\frac{12}{15}$ $\frac{4}{5}, \frac{20}{25}$
17. $\frac{4}{10}$ $\frac{2}{5}, \frac{8}{20}$
18. $\frac{10}{12}$ $\frac{5}{6}, \frac{20}{24}$
19. $\frac{6}{10}$ $\frac{3}{5}, \frac{9}{15}$

See Example 2 Find the missing numbers that make the fractions equivalent.

20. $\frac{3}{7} = \frac{\square}{35}$ 15
21. $\frac{6}{48} = \frac{1}{\square}$ 8
22. $\frac{2}{5} = \frac{28}{\square}$ 70
23. $\frac{2}{7} = \frac{\square}{21}$ 6
24. $\frac{8}{32} = \frac{\square}{4}$ 1
25. $\frac{2}{7} = \frac{40}{\square}$ 140

See Example 3 Write each fraction in simplest form.

26. $\frac{2}{8}$ $\frac{1}{4}$
27. $\frac{10}{15}$ $\frac{2}{3}$
28. $\frac{6}{30}$ $\frac{1}{5}$
29. $\frac{6}{14}$ $\frac{3}{7}$
30. $\frac{12}{16}$ $\frac{3}{4}$
31. $\frac{4}{28}$ $\frac{1}{7}$
32. $\frac{4}{8}$ $\frac{1}{2}$
33. $\frac{10}{35}$ $\frac{2}{7}$

PRACTICE AND PROBLEM SOLVING

34. $\frac{3}{6} = \frac{1}{2}$
35. $\frac{1}{4} = \frac{2}{8}$
36. $\frac{2}{3} = \frac{8}{12}$
37. $\frac{2}{4} = \frac{1}{2}$

Write the equivalent fractions represented by each picture.

34. 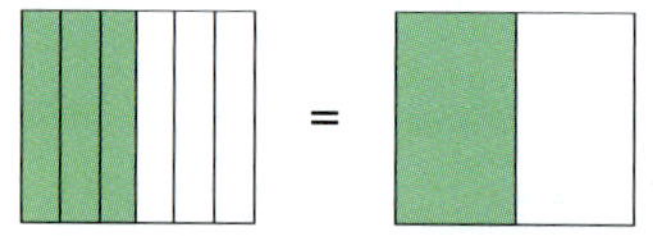

35. 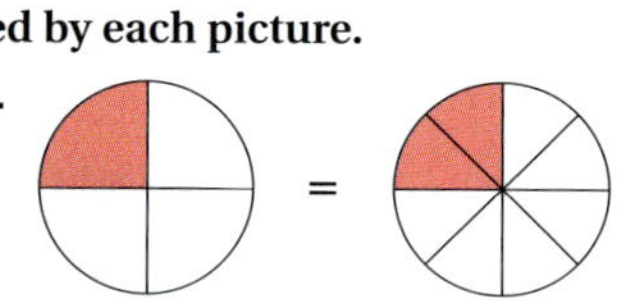

36. 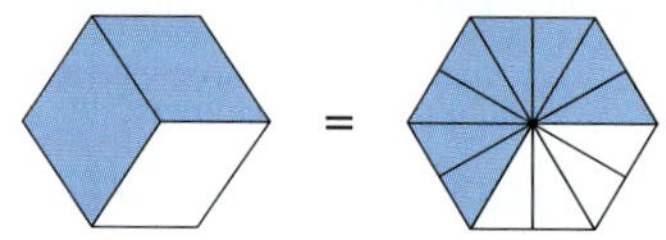

37. 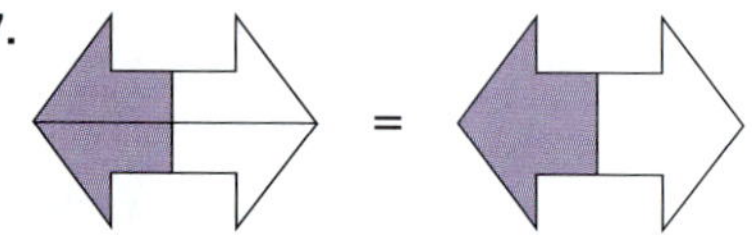

Math Background

In the sixteenth century, before the common acceptance of using places to the right of a decimal point to represent numbers less than 1, fractions were used exclusively. Writing fractions in simplest form made any further computation simpler. Today, with the use of calculators, writing fractions in simplest form is not as necessary.

Writing a fraction in simplest form was once called *abbreviating* the fraction. To abbreviate fractions, people learned the divisibility rules discussed in Lesson 4-1, as well as the Euclidean algorithm discussed in the Math Background of Lesson 4-3.

RETEACH 4-5

LESSON 4-5 **Reteach**
Equivalent Fractions

Fractions that have the same value are equivalent fractions.
$\frac{3}{4} = \frac{6}{8}$

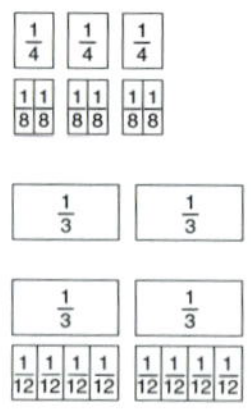

$\frac{3}{4}$ and $\frac{6}{8}$ are equivalent fractions.
You can use fraction strips to help you find equivalent fractions.

To solve $\frac{2}{3} = \frac{?}{12}$, first use fraction strips to model the first fraction.
Then use $\frac{1}{12}$ fraction pieces to make a length as long as the $\frac{2}{3}$ strip.
You need eight $\frac{1}{12}$ pieces, so $\frac{2}{3} = \frac{8}{12}$.

Find the missing number that makes the fractions equivalent.

1. $\frac{3}{4} = \frac{?}{12}$ 9
2. $\frac{8}{10} = \frac{?}{5}$ 4
3. $\frac{6}{9} = \frac{?}{3}$ 2
4. $\frac{5}{6} = \frac{?}{12}$ 10

A fraction is in simplest form when the GCF of the numerator and the denominator is 1.
To write $\frac{4}{6}$ in simplest form, divide the numerator and the denominator by their GCF, 2.
$4 \div 2 = 2$
$6 \div 2 = 3$
So, $\frac{4}{6}$ in simplest form is $\frac{2}{3}$.

Write each fraction in simplest form.

5. $\frac{3}{9}$ $\frac{1}{3}$
6. $\frac{12}{16}$ $\frac{3}{4}$
7. $\frac{14}{18}$ $\frac{7}{9}$
8. $\frac{8}{20}$ $\frac{2}{5}$

PRACTICE 4-5

LESSON 4-5 **Practice B**
Equivalent Fractions

Find two equivalent fractions for the given fraction. Possible answers are given.

1. $\frac{3}{6}$ $\frac{1}{2}, \frac{6}{12}$
2. $\frac{4}{7}$ $\frac{8}{14}, \frac{12}{21}$
3. $\frac{11}{13}$ $\frac{22}{26}, \frac{33}{39}$
4. $\frac{2}{15}$ $\frac{4}{30}, \frac{6}{45}$
5. $\frac{5}{14}$ $\frac{10}{28}, \frac{15}{42}$
6. $\frac{8}{9}$ $\frac{16}{18}, \frac{24}{27}$
7. $\frac{2}{21}$ $\frac{4}{42}, \frac{6}{63}$
8. $\frac{24}{48}$ $\frac{1}{2}, \frac{8}{16}$
9. $\frac{25}{100}$ $\frac{1}{4}, \frac{50}{200}$

Find the missing numbers that makes the fractions equivalent.

10. $\frac{4}{7} = \frac{?}{28}$ 16
11. $\frac{2}{9} = \frac{?}{54}$ 12
12. $\frac{36}{4} = \frac{?}{1}$ 9
13. $\frac{56}{8} = \frac{?}{2}$ 14
14. $1\frac{3}{5} = \frac{?}{25}$ 40
15. $1\frac{4}{7} = \frac{?}{42}$ 66

Write each fraction in simplest form.

16. $\frac{15}{25}$ $\frac{3}{5}$
17. $\frac{8}{36}$ $\frac{2}{9}$
18. $\frac{12}{18}$ $\frac{2}{3}$
19. $\frac{10}{24}$ $\frac{5}{12}$

20. Billy had 24 trading cards. He gave 7 of his cards to Miko and 9 of his cards to Teri. What fraction of his original 24 cards does Billy have left? Write two equivalent fractions for that amount.
$\frac{8}{24}$; possible equivalent fractions: $\frac{4}{12}, \frac{2}{6}$

21. Beth and Kristine ride their bikes to school in the morning. Beth has to ride $1\frac{7}{32}$ miles. Kristine has to ride $\frac{39}{32}$ miles. Who rides the farthest to reach school? Explain.
They ride the same distance, because $1\frac{7}{32} = \frac{39}{32}$ miles.

Social Studies LINK

The Old City Market is a public market in Charleston, South Carolina. Local artists, craftspeople, and vendors display and sell their goods in open-sided booths.

38. You can buy food, such as southern sesame seed cookies, at $\frac{1}{10}$ of the booths. Write two equivalent fractions for $\frac{1}{10}$. Possible answers: $\frac{2}{20}$; $\frac{4}{40}$

39. Handwoven sweetgrass baskets are a regional specialty. About 8 out of every 10 baskets sold are woven at the market. Write a fraction for "8 out of 10." Then write this fraction in simplest form. $\frac{8}{10} = \frac{4}{5}$

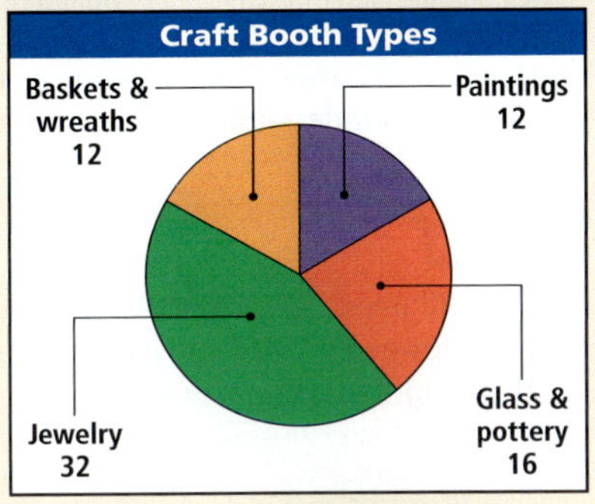

40. Suppose the circle graph shows the number of each kind of craft booth at the Old City Market. For each type of booth, tell what fraction it represents of the total number of craft booths. Write these fractions in simplest form.

41. Customers can buy packages of dried rice and black-eyed peas, which can be made into black-eyed pea soup. One recipe for black-eyed pea soup calls for $\frac{1}{2}$ tsp of basil. How could you measure the basil if you had only a $\frac{1}{4}$ tsp measuring spoon? What if you had only a $\frac{1}{8}$ tsp measuring spoon?

42. *WRITE ABOUT IT* The recipe for soup also calls for $\frac{1}{4}$ tsp of pepper. How many fractions are equivalent to $\frac{1}{4}$? Explain.

43. *CHALLENGE* Silver jewelry is a popular item at the market. Suppose there are 28 bracelets at one jeweler's booth and that $\frac{3}{7}$ of these bracelets have red stones. How many bracelets have red stones? 12

Spiral Review

Order the numbers in each set from greatest to least. (Lesson 1-1)

44. 740, 680, 749, 168
749, 740, 680, 168

45. 204, 271, 640, 644
644, 640, 271, 204

46. 4,192; 4,286; 4,181; 4,287
4,287; 4,286; 4,192; 4,181

Compare. Write <, >, or =. (Lesson 3-1)

47. 4.23 ■ 4.28 <

48. 12.05 ■ 8.79 >

49. 0.45 ■ 0.8 <

50. 50 ■ 0.5 >

51. 14.006 ■ 14.3003 <

52. 23.1945 ■ 23.1928 >

53. **TEST PREP** Which is the prime factorization of 189? (Lesson 4-2) C

A $9 \cdot 3 \cdot 7$ B $7^2 \cdot 3^2$ C $3^3 \cdot 7$ D $3 \cdot 63$

Interdisciplinary LINK

Social Studies

Exercises 38–43 involve using information about a public market in South Carolina. The economy of different regions in the United States is studied in middle school social studies programs such as Holt, Rinehart & Winston's *People, Places, and Change.*

Answers

40. Baskets and wreaths are $\frac{12}{72} = \frac{1}{6}$; jewelry is $\frac{32}{72} = \frac{4}{9}$; glass and pottery are $\frac{16}{72} = \frac{2}{9}$; paintings are $\frac{12}{72} = \frac{1}{6}$.

41. Use two of the $\frac{1}{4}$ tsp measuring spoons; four of the $\frac{1}{8}$ tsp measuring spoons.

42. There are an unlimited number of fractions equivalent to $\frac{1}{4}$. You can continue to multiply the numerator and denominator by the same number to find some equivalent fractions.

Journal

Have students explain which method of writing a fraction in simplest form (GCF or ladder diagram) is the most efficient and why.

Test Prep Doctor

In Exercise 53, students need to remember that the prime factorization of a number includes only prime factors. Students can eliminate answers **A** and **D** because not all the factors are prime. Students who chose **B** selected the prime factorization of 441, not 189.

CHALLENGE 4-5

LESSON 4-5 **Challenge**
Match That Fraction!

Write the fraction that the shaded part of each figure represents. Then match equivalent fractions by drawing lines between numbered and lettered figures.

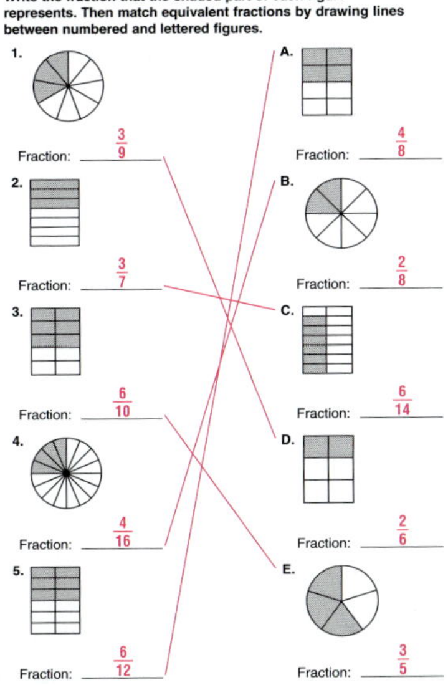

PROBLEM SOLVING 4-5

LESSON 4-5 **Problem Solving**
Equivalent Fractions

About 60 million Americans exercise 100 times or more each year. Their top activities and the fraction of those 60 million people who did them are shown on the circle graph. Use the graph to answer the questions.

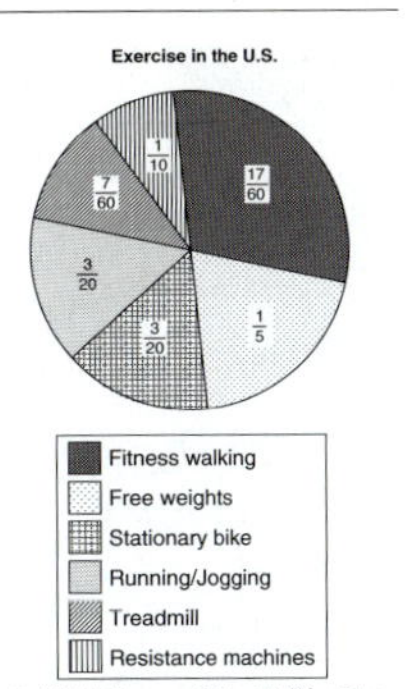

1. Which two activities on the graph did the same number of people use to keep in shape?
stationary bike and running/jogging

2. Which activity did $\frac{3}{15}$ of the people use to exercise?
free weights

3. Which activity had the most participants? Write an equivalent fraction for that activity's participants.
fitness walking; possible answer: $\frac{34}{120}$

4. Which activity had the fewest participants? Write two equivalent fractions for that activity's participants.
resistance machines; possible answers: $\frac{2}{20}$; $\frac{3}{30}$

5. Which fitness activity did [illegible]/300 of the regular exercisers in the United States use to stay healthy?
treadmill

Circle the letter of the correct answer.

6. An average-sized person can burn about $6\frac{1}{2}$ calories a minute while riding a bike. Which of the following is equivalent to that amount?
A $1\frac{2}{2}$ (C) $6\frac{2}{4}$
B $5\frac{6}{2}$ D $6\frac{2}{6}$

7. An average-sized person can burn about 11.25 calories a minute while jogging. Which of the following is not equivalent to that amount?
F $11\frac{1}{4}$ H $11\frac{2}{8}$
(G) $11\frac{1}{2}$ J $11\frac{3}{12}$

Lesson Quiz

Write two equivalent fractions for each given fraction. Possible answers:

1. $\frac{4}{10}$ $\frac{2}{5}$, $\frac{8}{20}$

2. $\frac{7}{14}$ $\frac{1}{2}$, $\frac{14}{28}$

Find the missing number that makes the fractions equivalent.

3. $\frac{2}{7} = \frac{■}{21}$ 6

4. $\frac{4}{15} = \frac{20}{■}$ 75

Write each fraction in simplest form.

5. $\frac{4}{8}$ $\frac{1}{2}$

6. $\frac{7}{49}$ $\frac{1}{7}$

Available on Daily Transparency in CRB

4C Explore Fraction Measurement

Pacing: Traditional 1 day
Block $\frac{1}{2}$ day

Objective: To use handmade rulers to study equivalent fractions

Materials: Strips of paper

Lab Resources

Hands-On Lab Activities pp. 24–26, 114

Using the Pages

Discuss with students how to use each of their rulers.

1. When measuring a line, how should you align the ruler to the line? Make sure the end of the line matches up with 0 on the ruler.
2. If a line measures 7 quarter inches, how long is it? $1\frac{3}{4}$ inches
3. How does a standard ruler take the place of the four rulers you made? A standard ruler has markings to show half-, quarter-, eighth-, and sixteenth-inch measurements.

Explore Fraction Measurement

Use with Lesson 4-5

Look at a standard ruler. It probably has marks for inches, half inches, quarter inches, eighth inches, and sixteenth inches.

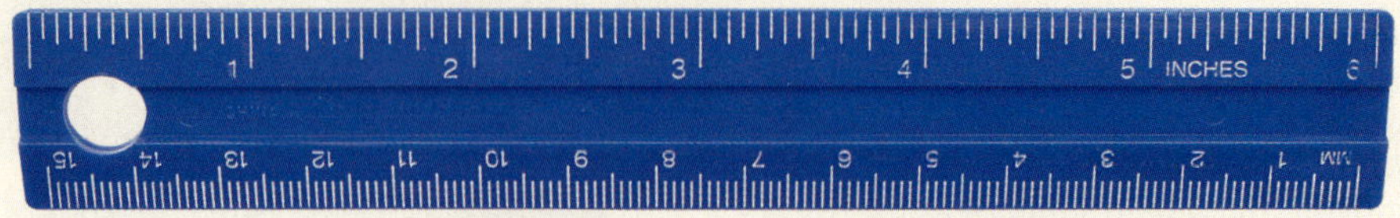

In this activity, you will make some of your own rulers and use them to help you find and understand equivalent fractions.

Activity

1. You will need four strips of paper. On one strip, use your ruler to make a mark for every half inch. Number each mark, beginning with 1. Label this strip "half-inch ruler."

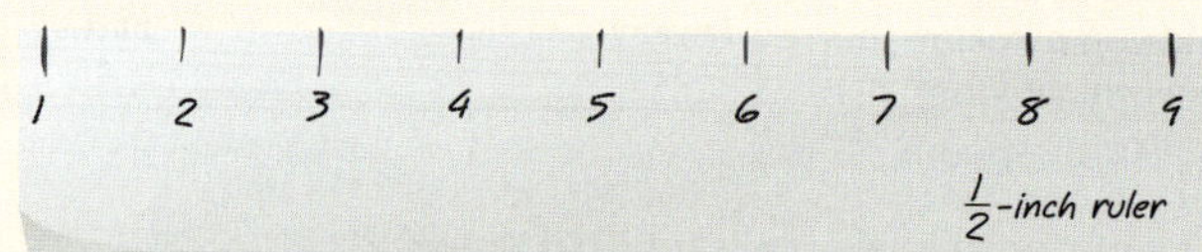

On a second strip, make a mark for every quarter inch. Again, number each mark, beginning with 1. Label this strip "quarter-inch ruler."

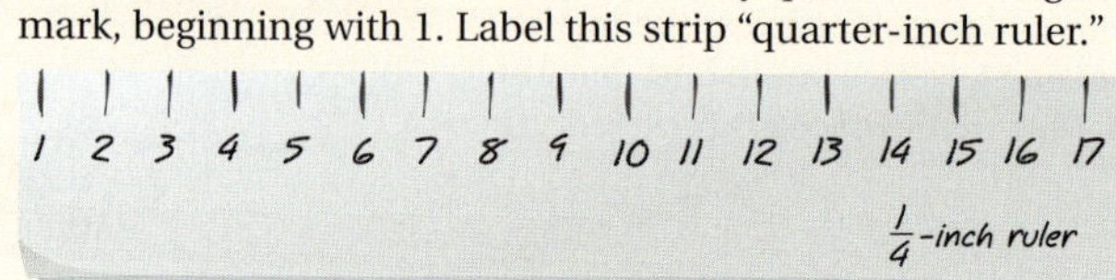

Do the same thing for eighth inches and sixteenth inches.

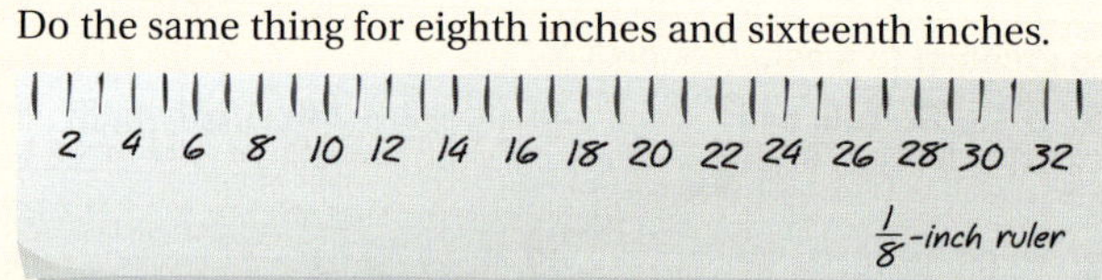

Sandy R. Puckett
Taipei American School
Taipei, Taiwan

Teacher to Teacher

Graphing fractions can be a very powerful way to display equivalent fractions and later to compare fractions.

Draw the positive *x*- and *y*-axes, number equal intervals from 0 to 20, and label the respective axes "numerator" and "denominator." Graph $\frac{1}{2}$ by graphing (1, 2) on the coordinate plane, and label it as $\frac{1}{2}$. Continue graphing and labeling $\frac{2}{4}$, $\frac{3}{6}$, $\frac{4}{8}$, etc. The points for these fractions lie on a straight line that goes through the origin. Every fraction along this line is equivalent to $\frac{1}{2}$. My students graph many common equivalent fractions, using a different color for each line. They then cut out the graphs and tape them in the back of their math books for quick reference.

2 Now use the half-inch ruler you made to measure the line segment at right. How many half inches long is the segment?

Use your quarter-inch ruler to measure the line segment again. How many quarter inches long is the segment?

How many eighth inches long is the segment?

How many sixteenth inches?

Fill in the blanks: $\frac{1}{2} = \frac{\blacksquare}{4} = \frac{\blacksquare}{8} = \frac{\blacksquare}{16}$. **2; 4; 8**

3 Use your quarter-inch ruler to measure the line segment below.

How long is the segment?

Now use your eighth-inch ruler to measure the line segment again. How many eighth inches long is the segment?

How many sixteenth inches?

Fill in the blanks: $\frac{3}{4} = \frac{\blacksquare}{8} = \frac{\blacksquare}{16}$. **6; 12**

Think and Discuss

1. Possible answer: A ruler shows that equivalent fractions have the same value, because the measurements are equivalent. For example, $\frac{1}{2}$ inch is equivalent to $\frac{4}{8}$ inch.

1. How does a ruler show that equivalent fractions have the same value?
2. Look at your lists of equivalent fractions from **2** and **3**. Do you notice any patterns? Describe them. **Possible answer: Both the numerators and the denominators are doubling (being multiplied by 2) from fraction to fraction.**
3. Use your rulers to measure an object longer than 1 inch. Use your measurements to write equivalent fractions. What do you notice about these fractions? **Possible answer: The numerators are greater than the denominators.**

Try This

1. Use your rulers to measure the items below. Use your measurements to write equivalent fractions. **Check students' measurements.**

2. Use your rulers to measure several items in your classroom. Use your measurements to write equivalent fractions. **Check students' work.**

4-6 Organizer

Pacing: Traditional 1 day
Block $\frac{1}{2}$ day

Objective: Students use pictures and number lines to compare and order fractions.

Indiana Academic Standards
6.1.3, 6.7.1

Warm Up

Find the missing number that makes the fractions equivalent.

1. $\frac{3}{5} = \frac{\blacksquare}{15}$ 9
2. $\frac{2}{3} = \frac{\blacksquare}{12}$ 8
3. $\frac{3}{4} = \frac{\blacksquare}{20}$ 15
4. $\frac{5}{8} = \frac{\blacksquare}{24}$ 15

Problem of the Day

From 4:00 to 5:30, Carlos, Lisa, and Toni took turns playing the same computer game. Carlos played for $\frac{1}{2}$ hour, and Lisa played for $\frac{3}{4}$ hour. For how many minutes did Toni play the game? 15 minutes

Available on Daily Transparency in CRB

Math Fact

The word *fraction* comes from the Latin word *frangere*, which means "to break". Thus, a fraction is a broken number, or a piece of a number.

4-6 Comparing and Ordering Fractions

Learn to use pictures and number lines to compare and order fractions.

Vocabulary
like fractions
unlike fractions
common denominator

Rachel and Hannah are making a kind of cookie called *hamantaschen*. They have $\frac{1}{2}$ cup of strawberry jam, but the recipe requires $\frac{1}{3}$ cup.

To determine if they have enough for the recipe, they need to compare the fractions $\frac{1}{2}$ and $\frac{1}{3}$.

When you are comparing fractions, first check their denominators. When fractions have the same denominator, they are called **like fractions**. For example, $\frac{7}{10}$ and $\frac{3}{10}$ are like fractions.

EXAMPLE 1 Comparing Like Fractions

Helpful Hint

When two fractions have the same denominator, the one with the larger numerator is greater.

$\frac{2}{5} < \frac{3}{5}$ $\frac{3}{8} > \frac{1}{8}$

Compare. Write <, >, or =.

A $\frac{7}{10} \blacksquare \frac{3}{10}$

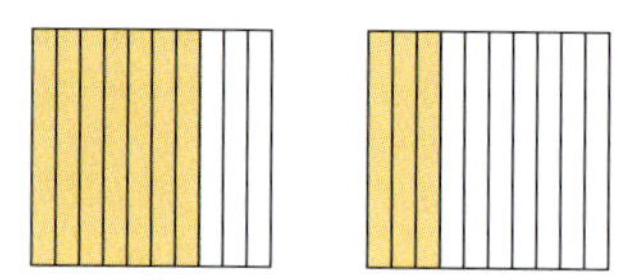

From the model, $\frac{7}{10} > \frac{3}{10}$.

B $\frac{1}{8} \blacksquare \frac{5}{8}$

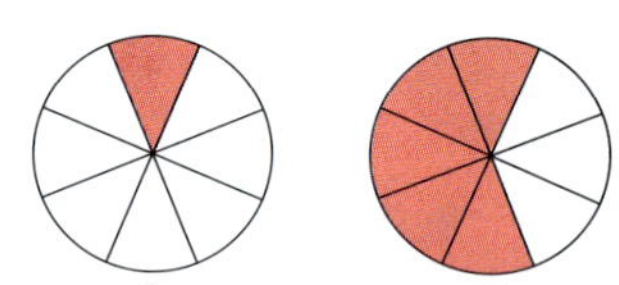

From the model, $\frac{1}{8} < \frac{5}{8}$.

1 Introduce

Alternate Opener

EXPLORATION

4-6 Comparing and Ordering Fractions

Use the model to decide whether the fraction on the left is greater than (>), less than (<), or equal to (=) the fraction on the right.

1. $\frac{1}{2} \square \frac{1}{3}$
2. $\frac{1}{2} \square \frac{1}{4}$
3. $\frac{1}{2} \square \frac{2}{3}$
4. $\frac{1}{3} \square \frac{1}{4}$
5. $\frac{3}{4} \square \frac{2}{3}$
6. $\frac{3}{4} \square \frac{5}{6}$
7. $\frac{1}{6} \square \frac{1}{4}$

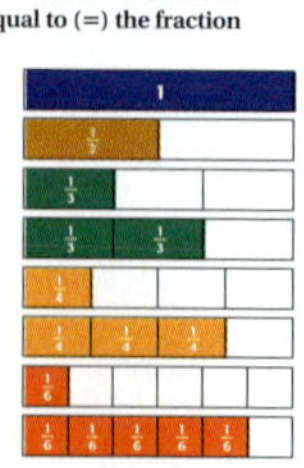

8. Look at the calculator screen to compare $\frac{1}{2}$ and $\frac{2}{3}$. Which fraction is greater? How do you know?

Think and Discuss

9. **Explain** how you could compare a fraction and a decimal.
10. **Explain** why $\frac{1}{17}$ is greater than $\frac{1}{18}$.

Motivate

Ask students to give examples of fractions in different contexts (e.g., $\frac{1}{2}$ gal of ice cream, $\frac{3}{4}$ cup of flour, $\frac{1}{8}$ inch, etc.). Then use the same units of measure, but provide a different fraction (e.g., $\frac{1}{3}$ gal of ice cream, $\frac{2}{3}$ cup of flour, $\frac{1}{16}$ inch, etc.). Explain that, as in the lesson opener, amounts represented as fractions sometimes must be compared in order to make decisions, such as determining whether $\frac{2}{3}$ cup or $\frac{3}{4}$ cup is more.

Exploration worksheet and answers on Chapter 4 Resource Book pp. 56 and 114

2 Teach

Lesson Presentation

Guided Instruction

In this lesson, students learn to use models and number lines to compare and order fractions. Teach them to use models to compare like fractions, then explain how to compare unlike fractions by first changing them to like fractions. Finally, teach students to order three unlike fractions by using a number line.

Teaching Tip

If the denominator of one fraction is a multiple of the denominator of the other fraction, only one of the fractions needs to be changed to compare them.

When two fractions have different denominators, they are called **unlike fractions**. To compare unlike fractions, first rename the fractions so they have the same denominator. This is called finding a **common denominator**.

EXAMPLE 2 *Cooking Application*

Rachel and Hannah have $\frac{1}{2}$ cup of strawberry jam. They need $\frac{1}{3}$ cup to make hamantaschen. Do they have enough strawberry jam for the recipe?

Compare $\frac{1}{2}$ and $\frac{1}{3}$.

Find a common denominator by multiplying the denominators.
$2 \cdot 3 = 6$

Find equivalent fractions with 6 as the denominator.

$\frac{1}{2} = \frac{\square}{6}$ $\quad$ $\frac{1}{3} = \frac{\square}{6}$

$\frac{1 \cdot 3}{2 \cdot 3} = \frac{3}{6}$ $\quad$ $\frac{1 \cdot 2}{3 \cdot 2} = \frac{2}{6}$

$\frac{1}{2} = \frac{3}{6}$ $\quad$ $\frac{1}{3} = \frac{2}{6}$

Compare the like fractions.

$\frac{3}{6} > \frac{2}{6}$, so $\frac{1}{2} > \frac{1}{3}$.

Since $\frac{1}{2}$ cup is more than $\frac{1}{3}$ cup, they have enough strawberry jam.

EXAMPLE 3 Ordering Fractions

Remember!

Numbers increase in value as you move from left to right on a number line.

Order $\frac{3}{7}$, $\frac{3}{4}$, and $\frac{1}{4}$ from least to greatest.

$\frac{3 \cdot 4}{7 \cdot 4} = \frac{12}{28}$ $\quad$ $\frac{3 \cdot 7}{4 \cdot 7} = \frac{21}{28}$ $\quad$ $\frac{1 \cdot 7}{4 \cdot 7} = \frac{7}{28}$ $\quad$ *Rename with like denominators.*

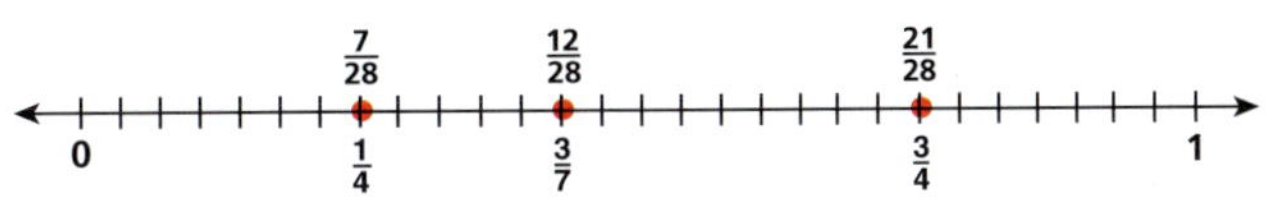

The fractions in order from least to greatest are $\frac{1}{4}$, $\frac{3}{7}$, $\frac{3}{4}$.

Think and Discuss

1. **Tell** whether the values of the fractions change when you rename two fractions so that they have common denominators.
2. **Explain** how to compare $\frac{2}{5}$ and $\frac{4}{5}$.

COMMON ERROR ALERT

When renaming fractions so that they have the same denominator, some students multiply just the denominator and not the numerator. Remind students that they have to multiply both the numerator and the denominator by the same value to create equivalent fractions.

Additional Examples

Example 1

Compare. Write <, >, or =.

A. $\frac{6}{7} \square \frac{4}{7}$ $>$

B. $\frac{1}{9} \square \frac{5}{9}$ $<$

Example 2

Ray has $\frac{2}{3}$ cup of nuts. He needs $\frac{3}{4}$ cup to make cookies. Does he have enough nuts for the recipe?
No

Example 3

Order the fractions from least to greatest. $\frac{4}{5}, \frac{2}{3}, \frac{1}{3}$ $\quad$ $\frac{1}{3}, \frac{2}{3}, \frac{4}{5}$

3 Close

Reaching All Learners

Through Concrete Manipulatives

Have each student or group of students model the fractions being compared in the lesson examples and exercises. Have them use fraction bars provided in the Manipulatives Kit and on Chapter 4 Resource Book p. 100 to verify their results.

Summarize

Have students define *unlike fractions* and explain how to compare them.

Possible answer: Unlike fractions are fractions with different denominators. To compare unlike fractions, change them to equivalent fractions with like denominators and then compare the numerators.

Answers to Think and Discuss

1. No; when you multiply both the numerator and the denominator of a fraction by the same number, the resulting fraction is equivalent.
2. Since the denominators are the same, compare the numerators. $2 < 4$, so $\frac{2}{5} < \frac{4}{5}$.

4-6 PRACTICE & ASSESS

Students may want to refer back to the lesson examples.

Assignment Guide

If you finished Example 1 assign:
Core 1–3, 8–13, 33, 39–53
Enriched 10–13, 21–26, 39–53

If you finished Example 2 assign:
Core 1–4, 8–14, 33–34, 39–53
Enriched 11–14, 33–34, 36–53

If you finished Example 3 assign:
Core 1–20, 33, 35, 39–53
Enriched 21–53

Notes

4-6 Exercises

FOR EXTRA PRACTICE
see page 643

internet connect
Homework Help Online
go.hrw.com Keyword: MR4 4-6

GUIDED PRACTICE

See Example 1

Compare. Write <, >, or =.

1. $\frac{3}{5} \square \frac{2}{5}$ **>**
2. $\frac{1}{9} \square \frac{2}{9}$ **<**
3. $\frac{6}{8} \square \frac{3}{4}$ **=**

See Example 2

4. Arsenio has $\frac{2}{3}$ cup of brown sugar. The recipe he is using requires $\frac{1}{4}$ cup of brown sugar. Does he have enough brown sugar for the recipe? Explain. **yes; possible answer: $\frac{2}{3}$ cup is greater than $\frac{1}{4}$ cup.**

See Example 3

Order the fractions from least to greatest.

5. $\frac{3}{8}, \frac{1}{5}, \frac{2}{3}$ **$\frac{1}{5}, \frac{3}{8}, \frac{2}{3}$**
6. $\frac{1}{4}, \frac{2}{5}, \frac{1}{3}$ **$\frac{1}{4}, \frac{1}{3}, \frac{2}{5}$**
7. $\frac{5}{9}, \frac{1}{8}, \frac{2}{7}$ **$\frac{1}{8}, \frac{2}{7}, \frac{5}{9}$**

INDEPENDENT PRACTICE

See Example 1

Compare. Write <, >, or =.

8. $\frac{2}{5} \square \frac{4}{5}$ **<**
9. $\frac{1}{10} \square \frac{3}{10}$ **<**
10. $\frac{3}{4} \square \frac{15}{20}$ **=**
11. $\frac{4}{5} \square \frac{5}{5}$ **<**
12. $\frac{2}{4} \square \frac{1}{2}$ **=**
13. $\frac{4}{6} \square \frac{16}{24}$ **=**

See Example 2

14. Kelly needs $\frac{2}{3}$ gallon of paint to finish painting her deck. She has $\frac{5}{8}$ gallon of paint. Does she have enough paint to finish her deck? Explain. **no; possible answer: $\frac{5}{8}$ gallon is less than $\frac{2}{3}$ gallon.**

See Example 3

Order the fractions from least to greatest.

15. $\frac{1}{2}, \frac{3}{5}, \frac{3}{7}$ **$\frac{3}{7}, \frac{1}{2}, \frac{3}{5}$**
16. $\frac{1}{6}, \frac{2}{5}, \frac{1}{4}$ **$\frac{1}{6}, \frac{1}{4}, \frac{2}{5}$**
17. $\frac{4}{9}, \frac{3}{8}, \frac{1}{3}$ **$\frac{1}{3}, \frac{3}{8}, \frac{4}{9}$**
18. $\frac{3}{4}, \frac{7}{10}, \frac{2}{3}$ **$\frac{2}{3}, \frac{7}{10}, \frac{3}{4}$**
19. $\frac{13}{18}, \frac{5}{9}, \frac{5}{6}$ **$\frac{5}{9}, \frac{13}{18}, \frac{5}{6}$**
20. $\frac{3}{8}, \frac{1}{4}, \frac{2}{3}$ **$\frac{1}{4}, \frac{3}{8}, \frac{2}{3}$**

PRACTICE AND PROBLEM SOLVING

Compare. Write <, >, or =.

21. $\frac{4}{15} \square \frac{3}{10}$ **<**
22. $\frac{7}{12} \square \frac{13}{30}$ **>**
23. $\frac{5}{9} \square \frac{4}{11}$ **>**
24. $\frac{3}{5} \square \frac{26}{65}$ **>**
25. $\frac{3}{5} \square \frac{2}{21}$ **>**
26. $\frac{24}{41} \square \frac{2}{7}$ **>**

Order the fractions from least to greatest.

27. $\frac{2}{5}, \frac{1}{2}, \frac{3}{10}$ **$\frac{3}{10}, \frac{2}{5}, \frac{1}{2}$**
28. $\frac{3}{4}, \frac{3}{5}, \frac{7}{10}$ **$\frac{3}{5}, \frac{7}{10}, \frac{3}{4}$**
29. $\frac{7}{15}, \frac{2}{3}, \frac{1}{5}$ **$\frac{1}{5}, \frac{7}{15}, \frac{2}{3}$**
30. $\frac{2}{5}, \frac{4}{9}, \frac{11}{15}$ **$\frac{2}{5}, \frac{4}{9}, \frac{11}{15}$**
31. $\frac{7}{12}, \frac{5}{8}, \frac{1}{2}$ **$\frac{1}{2}, \frac{7}{12}, \frac{5}{8}$**
32. $\frac{5}{8}, \frac{3}{4}, \frac{5}{12}$ **$\frac{5}{12}, \frac{5}{8}, \frac{3}{4}$**

Math Background

The relationship between fractions and decimals provides an alternative method for comparing and ordering unlike fractions. By writing each fraction as a decimal, the fractions can be compared without resorting to the introduction of a common denominator.

RETEACH 4-6

LESSON 4-6 **Reteach**
Comparing and Ordering Fractions

$\frac{3}{4}$ and $\frac{2}{3}$ are unlike fractions because they have different denominators. To compare these fractions, graph the fractions on the same number line.

The fraction that is farther to the right on the number line is greater in value.

$\frac{9}{12} > \frac{8}{12}$, so $\frac{3}{4} > \frac{2}{3}$.

Use the number line to compare the fractions. Write < or >.

1. $\frac{1}{3}$ [<] $\frac{5}{6}$
2. $\frac{5}{10}$ [>] $\frac{1}{4}$
3. $\frac{11}{12}$ [>] $\frac{4}{6}$
4. $\frac{6}{9}$ [>] $\frac{1}{6}$

You can use number lines to order fractions from least to greatest.

To order $\frac{1}{2}, \frac{5}{6}$, and $\frac{3}{4}$, graph the values on a number line.

Then list the fractions as they appear from left to right. From least to greatest the fractions are $\frac{1}{2}, \frac{3}{4}$, and $\frac{5}{6}$.

Use the number line to order the fractions from least to greatest.

5. $\frac{7}{12}, \frac{1}{4}, \frac{4}{8}$ **$\frac{1}{4}, \frac{4}{8}, \frac{7}{12}$**
6. $\frac{2}{3}, \frac{7}{12}, \frac{2}{8}$ **$\frac{2}{8}, \frac{7}{12}, \frac{2}{3}$**
7. $\frac{1}{2}, \frac{3}{4}, \frac{3}{9}$ **$\frac{3}{9}, \frac{1}{2}, \frac{3}{4}$**
8. $\frac{5}{12}, \frac{1}{3}, \frac{5}{6}$ **$\frac{1}{3}, \frac{5}{12}, \frac{5}{6}$**

PRACTICE 4-6

LESSON 4-6 **Practice A**
Comparing and Ordering Fractions

Compare. Write <, >, or =.

1. $\frac{1}{6}$ [<] $\frac{2}{6}$
2. $\frac{1}{3}$ [<] $\frac{2}{3}$
3. $\frac{2}{4}$ [=] $\frac{1}{2}$
4. $\frac{1}{2}$ [<] $\frac{3}{4}$
5. $\frac{4}{6}$ [=] $\frac{2}{3}$
6. $\frac{1}{5}$ [<] $\frac{2}{3}$
7. $\frac{1}{2}$ [>] $\frac{2}{5}$
8. $\frac{3}{4}$ [>] $\frac{3}{5}$
9. $\frac{1}{9}$ [<] $\frac{1}{8}$

Order the fractions from least to greatest.

10. $\frac{7}{9}, \frac{1}{9}, \frac{2}{9}$ **$\frac{1}{9}, \frac{2}{9}, \frac{7}{9}$**
11. $\frac{1}{3}, \frac{1}{4}, \frac{2}{3}$ **$\frac{1}{4}, \frac{1}{3}, \frac{2}{3}$**
12. $\frac{2}{5}, \frac{1}{2}, \frac{1}{10}$ **$\frac{1}{10}, \frac{2}{5}, \frac{1}{2}$**

Order the fractions from greatest to least.

13. $\frac{1}{5}, \frac{3}{5}, \frac{2}{5}$ **$\frac{3}{5}, \frac{2}{5}, \frac{1}{5}$**
14. $\frac{2}{7}, \frac{6}{7}, \frac{3}{6}$ **$\frac{6}{7}, \frac{3}{6}, \frac{2}{7}$**
15. $\frac{1}{6}, \frac{2}{3}, \frac{5}{6}$ **$\frac{5}{6}, \frac{2}{3}, \frac{1}{6}$**

16. The Mountain Cats and the Pirates have both played the same number of games. The Mountain Cats have won $\frac{1}{5}$ of its games, and the Pirates have won $\frac{1}{4}$ of its games. Which team has the most wins?
the Pirates

17. Each week, Kelly saves $\frac{1}{2}$ of her allowance. She spends $\frac{1}{3}$ of her allowance on food, and $\frac{1}{6}$ of it on games and movies. What does Kelly do with the largest part of her allowance?
She saves it.

Agriculture LINK

There are over 3,500 different uses for corn products, from ethanol fuel and industrial products to the household items you see above.

33. Kyle operates a hot dog cart in a large city. He spends $\frac{2}{5}$ of his budget on supplies, $\frac{1}{12}$ on advertising, and $\frac{2}{25}$ on taxes and fees. Does Kyle spend more on advertising or more on taxes and fees? **advertising**

34. The Dixon Dragons must win at least $\frac{3}{7}$ of their remaining games to qualify for their district playoffs. If they have 15 games left and they win 7 of them, will the Dragons compete in the playoffs? Explain. **yes; possible answer: $\frac{7}{15}$ is greater than $\frac{3}{7}$**

35. ***AGRICULTURE*** The table shows the fraction of the world's total corn each country produces. List the countries in order from the country that produces the most corn to the country that produces the least corn. **United States, China, Brazil**

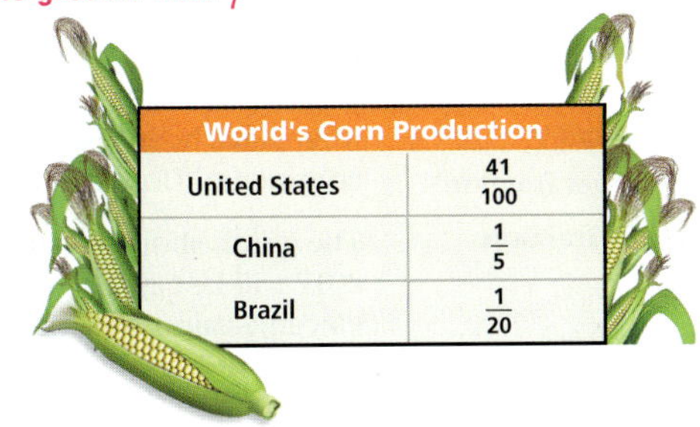

World's Corn Production	
United States	$\frac{41}{100}$
China	$\frac{1}{5}$
Brazil	$\frac{1}{20}$

36. ***WRITE A PROBLEM*** Write a problem that involves comparing two fractions with different denominators.

37. ***WRITE ABOUT IT*** Compare the following fractions.

$\frac{1}{2}$ ■ $\frac{1}{4}$ $\quad$ $\frac{2}{3}$ ■ $\frac{2}{5}$ $\quad$ $\frac{3}{4}$ ■ $\frac{3}{7}$ $\quad$ $\frac{4}{5}$ ■ $\frac{4}{9}$

What do you notice about two fractions that have the same numerator but different denominators? Which one is greater?

38. ***CHALLENGE*** Name a fraction that would make the inequality true.

$\frac{1}{4} >$ ■ $> \frac{1}{5}$ **Possible answer: $\frac{9}{40}$**

Spiral Review

Evaluate each expression when $a = 4$, $b = 2.8$, and $c = 0.9$. (Lesson 3-3)

39. $a + b$ **6.8** **40.** $b - c$ **1.9** **41.** $a + c$ **4.9** **42.** $a - b$ **1.2**

43. $b + c$ **3.7** **44.** $a + b + c$ **7.7** **45.** $a - c$ **3.1** **46.** $a + b - c$ **5.9**

Write each number in scientific notation. (Lesson 3-5)

47. 45 **4.5×10^1** **48.** 820 **8.2×10^2** **49.** 319 **3.19×10^2**

50. 36,000 **3.6×10^4** **51.** 405,000 **4.05×10^5** **52.** 23,000,000 **2.3×10^7**

53. **TEST PREP** Nadine divided a bag of pretzels among 6 friends. Each friend received the same number of pretzels, and 4 extra pretzels remained in the bag. Which number could be the number of pretzels that were in the bag before Nadine shared them? (Lesson 3-9) **D**

A 24 **B** 36 **C** 44 **D** 64

Answers

36. Possible answer: $\frac{2}{7}$ of Hanna's birthday gifts were in bags and $\frac{3}{5}$ were in boxes. Did Hanna receive more gifts in bags or more in boxes?

37. >; >; >; >; Possible answer: When two fractions have the same numerator, the one with the least denominator is greatest.

Journal

Have students write a real-world problem in which unlike fractions are compared. Then have students solve the problem and explain the solution.

Test Prep Doctor

For Exercise 53, students need to find the dividend given the divisor, 6, and the remainder, 4. Students who answered **A** or **B** chose a dividend that would not yield a remainder when divided by 6, while students who answered **C** chose a dividend that would yield a remainder of 2 when divided by 6.

CHALLENGE 4-6

LESSON 4-6 **Challenge**
Light as a Feather

The pygmy shrew is the lightest mammal on Earth. It weighs only 0.05 ounces. That's lighter than most feathers! Use the clues below to complete the table with each tiny animal's weight.

Lightest Mammals

Mammal	Weight (oz)
Pygmy shrew	$\frac{1}{20}$
Pipistrelle bat	$\frac{11}{100}$
Kitti's hog-nosed bat	$\frac{7}{100}$
Harvest mouse	$\frac{9}{50}$
Masked shrew	$\frac{2}{25}$

The smallest pygmy shrews are only $1\frac{1}{2}$ inches long!

1. Thousands of masked shrews live all over North America, but they are rarely seen. Although tiny, they have enormous appetites—often eating more than their body weight each day in bugs, slugs, and other crawly things. A masked shrew weighs more than a kitti's hog-nosed bat, but less than a pipistrelle bat.

2. Several different speices of harvest mice live in the United States. They are excellent climbers and spend much of the night scampering around in search of seeds and bugs to eat. During the day, they stay at home to avoid their worst enemies: hawks, owls, and snakes. A harvest mouse weighs more than any other animal listed in the table.

3. Kitti's hog-nosed bats live only in Asia. They spend most of the day hanging in deep, dark caves. At night, they leave the cave and fly around catching bugs to eat. About the size of a bumblebee, a kitti's hog-nosed bat weighs more than a pygmy shrew, but less than a masked shrew.

4. Pipistrelle bats are the smallest bats in the United States. They are unusual for bats, because they often fly in daylight to catch beetles and other bugs to eat. In winter, they hibernate in caves or other dark places. During that time, the bats are often covered with droplets of water, which sparkle and give them a pearly glow. A pipistrelle bat weighs less than a harvest mouse, but more than a masked shrew.

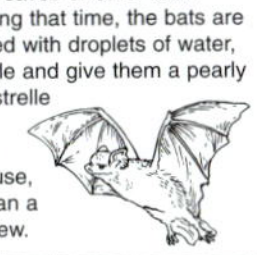

PROBLEM SOLVING 4-6

LESSON 4-6 **Problem Solving**
Comparing and Ordering Fractions

The table shows what fraction of Earth's total land area each of the continents makes up. Use the table to answer the questions.

Earth's Land

Continent	Fraction of Earth's Land
Africa	$\frac{1}{5}$
Antarctica	$\frac{1}{10}$
Asia	$\frac{3}{10}$
Australia	$\frac{1}{20}$
Europe	$\frac{7}{100}$
North America	$\frac{4}{25}$
South America	$\frac{6}{50}$

1. Which continent makes up most of Earth's land?
   Asia
2. Which continent makes up the least part of Earth's land?
   Australia
3. Explain how you would compare the part of Earth's total land area that Australia and Europe make up.
   Change $\frac{1}{20}$ to $\frac{5}{100}$ and then compare it to $\frac{7}{100}$.
4. Which continent covers a larger part of Earth's total land area, North America or South America?
   North America
5. Which continent covers a smaller part of Earth's total land area, Africa or Antarctica?
   Antarctica

Circle the letter of the correct answer.

6. Which of the following lists shows the continents written in order from the greatest part of Earth's total land they cover to the least part?
   (A) Asia, Africa, North America
   B Africa, Asia, North America
   C Asia, South America, North America
   D North America, Asia, South America
7. Which of the following lists shows the continents written in order from the least part of Earth's total land they cover to the greatest part?
   F Antarctica, Europe, South America
   G South America, Antarctica, Europe
   (H) Australia, Europe, Antarctica
   J Antarctica, Europe, Australia

Lesson Quiz

Compare. Write <, >, or =.

1. $\frac{3}{6}$ ■ $\frac{4}{8}$ **=** **2.** $\frac{5}{8}$ ■ $\frac{9}{16}$ **>**

3. You drilled three holes in a piece of wood. The diameters of the holes are $\frac{1}{8}$, $\frac{3}{8}$, and $\frac{3}{16}$ inches. Which hole is the largest? **$\frac{3}{8}$**

Order the fractions from least to greatest.

4. $\frac{7}{8}, \frac{5}{8}, \frac{2}{3}$ **$\frac{5}{8}, \frac{2}{3}, \frac{7}{8}$**

5. $\frac{3}{4}, \frac{5}{8}, \frac{5}{6}$ **$\frac{5}{8}, \frac{3}{4}, \frac{5}{6}$**

Available on Daily Transparency in CRB

4-7 Organizer

Pacing: Traditional 1 day
Block $\frac{1}{2}$ day

Objective: Students convert between mixed numbers and improper fractions.

Indiana Academic Standards
6.1.4, 6.7.1, 6.7.5, 6.7.11

Warm Up

Order the fractions from least to greatest.

1. $\frac{2}{9}, \frac{1}{6}, \frac{2}{3}$ — $\frac{1}{6}, \frac{2}{9}, \frac{2}{3}$
2. $\frac{2}{3}, \frac{7}{12}, \frac{5}{6}$ — $\frac{7}{12}, \frac{2}{3}, \frac{5}{6}$
3. $\frac{5}{8}, \frac{1}{2}, \frac{4}{11}$ — $\frac{4}{11}, \frac{1}{2}, \frac{5}{8}$

Problem of the Day

Two numbers have a product of 48. When the larger number is divided by the smaller, the quotient is 3. What are the numbers? 4 and 12

Available on Daily Transparency in CRB

Math Fact

Many people learn that π equals the mixed number $3\frac{1}{7}$ or the equivalent improper fraction $\frac{22}{7}$. However, these are approximations of the true value of π, because π cannot be expressed as a ratio of whole numbers.

4-7 Mixed Numbers and Improper Fractions

Learn to convert between mixed numbers and improper fractions.

Vocabulary
improper fraction
proper fraction

Reading Math
$\frac{11}{4}$ is read as "eleven-fourths."

Have you ever witnessed a total eclipse of the sun? It occurs when the sun's light is completely blocked out. A total eclipse is rare—only three have been visible in the continental United States since 1963.

The graph shows that the eclipse in 2017 will last $2\frac{3}{4}$ minutes. There are eleven $\frac{1}{4}$-minute sections, so $2\frac{3}{4} = \frac{11}{4}$.

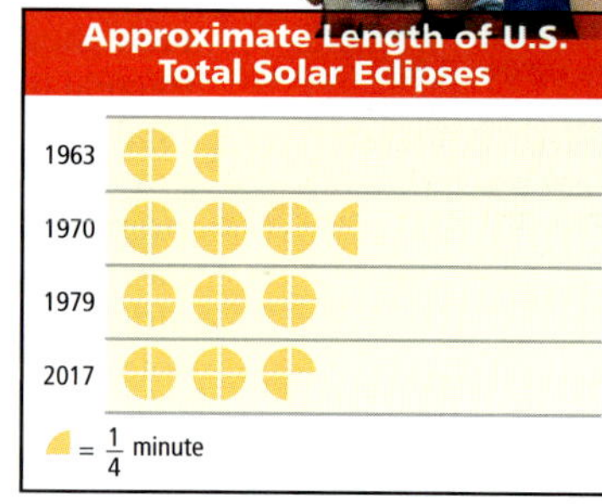

An **improper fraction** is a fraction in which the numerator is greater than or equal to the denominator, such as $\frac{11}{4}$.

Whole numbers can be written as improper fractions. The whole number is the numerator, and the denominator is 1. For example, $7 = \frac{7}{1}$.

When the numerator is less than the denominator, the fraction is called a **proper fraction**.

Improper and Proper Fractions	
Improper Fractions	
• Numerator equals denominator → fraction is equal to 1	$\frac{3}{3} = 1$ $\frac{102}{102} = 1$
• Numerator greater than denominator → fraction is greater than 1	$\frac{9}{5} > 1$ $\frac{13}{1} > 1$
Proper Fractions	
• Numerator less than denominator → fraction is less than 1	$\frac{2}{5} < 1$ $\frac{102}{351} < 1$

You can write an improper fraction as a mixed number.

EXAMPLE 1 *Astronomy Application*

The longest total solar eclipse in the next 200 years will take place in 2186. It will last about $\frac{15}{2}$ minutes. Write $\frac{15}{2}$ as a mixed number.

Method 1: Use a model.

Draw squares divided into half sections. Shade 15 of the half sections.

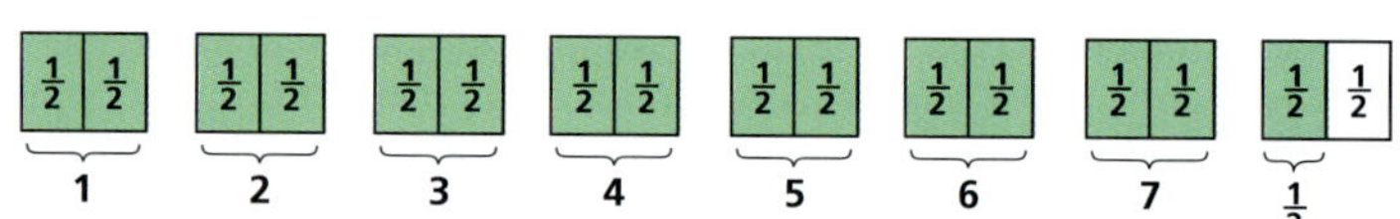

There are 7 whole squares and 1 half square, or $7\frac{1}{2}$ squares, shaded.

1 Introduce

Alternate Opener

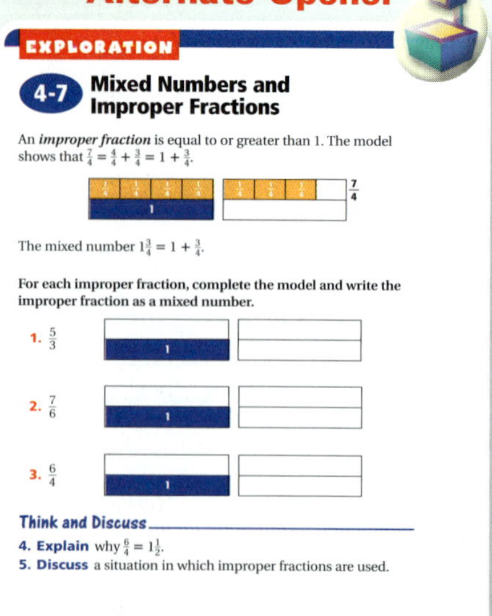
EXPLORATION

4-7 Mixed Numbers and Improper Fractions

An *improper fraction* is equal to or greater than 1. The model shows that $\frac{7}{4} = \frac{4}{4} + \frac{3}{4} = 1 + \frac{3}{4}$.

The mixed number $1\frac{3}{4} = 1 + \frac{3}{4}$.

For each improper fraction, complete the model and write the improper fraction as a mixed number.

1. $\frac{5}{3}$
2. $\frac{7}{6}$
3. $\frac{6}{4}$

Think and Discuss

4. Explain why $\frac{6}{4} = 1\frac{1}{2}$.
5. Discuss a situation in which improper fractions are used.

Motivate

Ask a volunteer to represent, in a drawing, the amount of pizza that 27 students would eat if each student got one slice of pizza and each pizza had been cut into 8 pieces.

Possible answer: 4 circles are divided into eighths, and 3 circles are completely shaded. Only 3 slices in the fourth circle are shaded. Then ask students how much pizza this represents.

Exploration worksheet and answers on Chapter 4 Resource Book pp. 65 and 116

2 Teach

Lesson Presentation

Guided Instruction

In this lesson, students learn to convert between mixed numbers and improper fractions. Define *improper fraction* and *proper fraction.* Teach two ways to convert each type of fraction: using a model and using division (for improper to mixed) and multiplication and addition (for mixed to improper). Students should understand that being able to change the form of a fraction can make working with it easier without changing its value.

Method 2: Use division.

$$2\overline{)15} = 7\tfrac{1}{2}, \quad 15 - 14 = 1$$

Divide the numerator by the denominator.

To form the fraction part of the quotient, use the remainder as the numerator and the divisor as the denominator.

The 2186 eclipse will last about $7\frac{1}{2}$ minutes.

EXAMPLE 2

Writing Mixed Numbers as Improper Fractions

Write $2\frac{1}{5}$ as an improper fraction.

Method 1: Use a model.

You can draw a diagram to illustrate the whole and fractional parts.

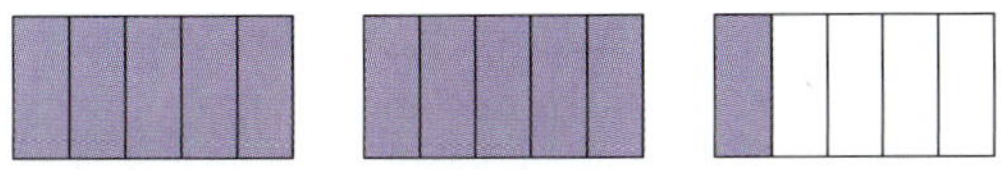

There are 11 fifths, or $\frac{11}{5}$. *Count the fifths in the diagram.*

Method 2: Use multiplication and addition.

When you are changing a mixed number to an improper fraction, spiral clockwise as shown in the picture. The order of operations will help you remember to multiply before you add.

$$2\frac{1}{5} = \frac{(5 \cdot 2) + 1}{5}$$ *Multiply the whole number by the denominator and add the numerator. Keep the same denominator.*

$$= \frac{10 + 1}{5}$$

$$= \frac{11}{5}$$

Think and Discuss

1. **Read** each improper fraction: $\frac{10}{7}, \frac{25}{9}, \frac{31}{16}$.
2. **Tell** whether each fraction is less than 1, equal to 1, or greater than 1: $\frac{21}{21}, \frac{54}{103}, \frac{9}{11}, \frac{7}{3}$.
3. **Explain** why any mixed number written as a fraction will be improper.

COMMON ERROR ALERT

Some students may multiply the numerator by the whole number and add the denominator when changing from a mixed number to an improper fraction. Have students change the resulting improper fraction back to a mixed number and compare with the original mixed number.

Additional Examples

Example 1

Ella hiked for $\frac{9}{4}$ hours yesterday. Write $\frac{9}{4}$ as a mixed number. $2\frac{1}{4}$

Example 2

Write $3\frac{2}{3}$ as an improper fraction. $\frac{11}{3}$

3 Close

Reaching All Learners

Through Hands-On Experience

Have students use inch rulers provided in the Manipulatives Kit and on Chapter 4 Resource Book p. 101 to measure various items in the classroom. For each item, have students record the measurement as both a mixed number and an improper fraction.

Possible answer: pencil: $7\frac{1}{2}$ inches or $\frac{15}{2}$ inches

Summarize

Have students explain how to change from an improper fraction to a mixed number and vice versa and give an example.

Possible answer: To change from an improper fraction to a mixed number, divide the numerator by the denominator. For example, $\frac{8}{5} = 8 \div 5 = 1\frac{3}{5}$. To change from a mixed number to an improper fraction, multiply the denominator by the whole number, add the numerator, and keep the same denominator. For example, $2\frac{1}{3} = \frac{(3 \cdot 2) + 1}{3} = \frac{7}{3}$.

Answers to Think and Discuss

1. ten sevenths; twenty-five ninths; thirty-one sixteenths
2. equal to 1; less than 1; less than 1; greater than 1
3. A mixed number consists of a whole number and a fraction, so it will always be greater than one. Numbers greater than one are improper when written as fractions.

4-7 PRACTICE & ASSESS

Students may want to refer back to the lesson examples.

Assignment Guide

If you finished Example 1 assign:
Core 1, 6, 7, 40, 42, 48–56
Enriched 6, 7, 45–46, 48–56

If you finished Example 2 assign:
Core 1–15, 24–27, 40–44, 48–56
Enriched 24–27, 31–56

Notes

4-7 Exercises

FOR EXTRA PRACTICE
see page 643

internet connect
Homework Help Online
go.hrw.com Keyword: MR4 4-7

GUIDED PRACTICE

See Example 1

1. The fifth largest meteorite found in the United States is named the Navajo. The Navajo weighs $\frac{12}{5}$ tons. Write $\frac{12}{5}$ as a mixed number. $2\frac{2}{5}$

See Example 2

Write each mixed number as an improper fraction.

2. $1\frac{1}{4}$ $\frac{5}{4}$
3. $2\frac{2}{3}$ $\frac{8}{3}$
4. $1\frac{2}{7}$ $\frac{9}{7}$
5. $2\frac{2}{5}$ $\frac{12}{5}$

INDEPENDENT PRACTICE

See Example 1

6. Saturn is the sixth planet from the Sun. It takes Saturn $\frac{59}{2}$ years to revolve around the Sun. Write $\frac{59}{2}$ as a mixed number. $29\frac{1}{2}$
7. Pluto has the lowest surface gravity of all the planets in the solar system. A person who weighs 143 pounds on Earth weighs $\frac{43}{5}$ pounds on Pluto. Write $\frac{43}{5}$ as a mixed number. $8\frac{3}{5}$

See Example 2

Write each mixed number as an improper fraction.

8. $1\frac{3}{5}$ $\frac{8}{5}$
9. $2\frac{2}{9}$ $\frac{20}{9}$
10. $3\frac{1}{7}$ $\frac{22}{7}$
11. $4\frac{1}{3}$ $\frac{13}{3}$
12. $2\frac{3}{8}$ $\frac{19}{8}$
13. $4\frac{1}{6}$ $\frac{25}{6}$
14. $1\frac{4}{9}$ $\frac{13}{9}$
15. $3\frac{4}{5}$ $\frac{19}{5}$

PRACTICE AND PROBLEM SOLVING

Write each improper fraction as a mixed number or whole number. Tell whether your answer is a mixed number or whole number.

16. $\frac{21}{4}$ $5\frac{1}{4}$; mixed number
17. $\frac{32}{8}$ 4; whole number
18. $\frac{20}{3}$
19. $\frac{43}{5}$ $8\frac{3}{5}$; mixed number
20. $\frac{108}{9}$ 12; whole number
21. $\frac{87}{10}$ $8\frac{7}{10}$; mixed number
22. $\frac{98}{11}$
23. $\frac{105}{7}$ 15; whole number

18. $6\frac{2}{3}$; mixed number

22. $8\frac{10}{11}$; mixed number

Write each mixed number as an improper fraction.

24. $9\frac{1}{4}$ $\frac{37}{4}$
25. $4\frac{9}{11}$ $\frac{53}{11}$
26. $11\frac{4}{9}$ $\frac{103}{9}$
27. $18\frac{3}{5}$ $\frac{93}{5}$

Replace each shape with a number that will make the equation correct.

28. $■\frac{2}{5} = \frac{17}{●}$ 3; 5
29. $■\frac{6}{11} = \frac{83}{●}$ 7; 11
30. $■\frac{1}{9} = \frac{118}{●}$ 13;9
31. $■\frac{6}{7} = \frac{55}{●}$ 7; 7
32. $■\frac{9}{10} = \frac{29}{●}$ 2; 10
33. $■\frac{1}{3} = \frac{55}{●}$ 18; 3
34. $21\frac{■}{3} = \frac{65}{●}$ 2; 3
35. $5\frac{■}{15} = \frac{77}{●}$ 2; 15
36. $31\frac{■}{19} = \frac{607}{●}$ 18; 19
37. $14\frac{■}{3} = \frac{45}{●}$ 3; 3
38. $12\frac{■}{10} = \frac{129}{●}$ 9; 10
39. $9\frac{■}{17} = \frac{154}{●}$ 1; 17

Math Background

Improper fractions are really not improper at all. Indeed, they help us to multiply mixed numbers. To multiply $1\frac{1}{4} \times 2\frac{4}{5}$, change the mixed numbers to improper fractions: $\frac{5}{4} \times \frac{14}{5}$.

RETEACH 4-7

LESSON 4-7 **Reteach**
Mixed Numbers and Improper Fractions

A proper fraction is a fraction whose numerator is less than its denominator.
$\frac{2}{3}$, $\frac{1}{4}$, and $\frac{2}{7}$ are examples of proper fractions.

An improper fraction is a fraction whose numerator is greater than or equal to its denominator.
$\frac{3}{2}$, $\frac{8}{3}$, and $\frac{5}{5}$ are examples of improper fractions.

Some improper fractions can be written as mixed numbers.
To write $\frac{7}{4}$ as a mixed number, draw circles divided into $\frac{1}{4}$ sections.
Then shade in 7 of the $\frac{1}{4}$ sections.

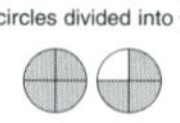

There is one circle and $\frac{3}{4}$ of a circle shaded.
So, $\frac{7}{4} = 1\frac{3}{4}$.

Write each improper fraction as a mixed number.

1. $\frac{14}{3}$ $4\frac{2}{3}$
2. $\frac{11}{2}$ $5\frac{1}{2}$
3. $\frac{15}{4}$ $3\frac{3}{4}$
4. $\frac{19}{6}$ $3\frac{1}{6}$

Mixed numbers can be written as improper fractions.
To write $2\frac{1}{3}$ as an improper fraction, draw 3 circles. Divide each circle into $\frac{1}{3}$ sections. Next, shade in 2 whole circles and one $\frac{1}{3}$ section of the last circle.

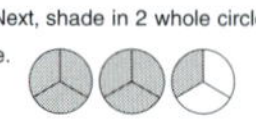

Then find the total number of $\frac{1}{3}$ sections that are shaded.
Seven $\frac{1}{3}$ sections are shaded, so $2\frac{1}{3} = \frac{7}{3}$.

Write each mixed number as an improper fraction.

5. $3\frac{1}{4}$ $\frac{13}{4}$
6. $5\frac{2}{3}$ $\frac{17}{3}$
7. $4\frac{1}{2}$ $\frac{9}{2}$
8. $1\frac{5}{6}$ $\frac{11}{6}$

PRACTICE 4-7

LESSON 4-7 **Practice B**
Mixed Numbers and Improper Fractions

Write each mixed number as an improper fraction.

1. $3\frac{1}{2}$ $\frac{7}{2}$
2. $2\frac{1}{3}$ $\frac{7}{3}$
3. $5\frac{1}{4}$ $\frac{21}{4}$
4. $1\frac{3}{7}$ $\frac{10}{7}$
5. $3\frac{3}{4}$ $\frac{15}{4}$
6. $4\frac{1}{3}$ $\frac{13}{3}$
7. $2\frac{3}{5}$ $\frac{13}{5}$
8. $3\frac{5}{6}$ $\frac{23}{6}$
9. $7\frac{1}{3}$ $\frac{22}{3}$

Write each improper fraction as a mixed number or whole number. Tell whether your answer is a mixed number or whole number.

10. $\frac{17}{3}$ $5\frac{2}{3}$; mixed number
11. $\frac{40}{8}$ 5; whole number
12. $\frac{48}{7}$ $6\frac{6}{7}$; mixed number
13. $\frac{33}{10}$ $3\frac{3}{10}$; mixed number
14. $\frac{50}{8}$ $6\frac{1}{4}$; mixed number
15. $\frac{83}{9}$ $9\frac{2}{9}$; mixed number
16. $\frac{104}{8}$ 13; whole number
17. $\frac{121}{6}$ $20\frac{1}{6}$; mixed number
18. $\frac{78}{11}$ $7\frac{1}{11}$; mixed number

19. The hotel ordered an extra-long rug for a hallway that is $\frac{123}{2}$ feet long. What is the rug's length in feet and inches? Remember, 1 foot = 12 inches.
61 feet and 6 inches

20. During this year's football-throwing contest, John threw the ball $49\frac{2}{3}$ feet. Sharon threw the ball 51 feet. Who threw the ball $\frac{153}{3}$ feet?
Sharon

Life Science LINK

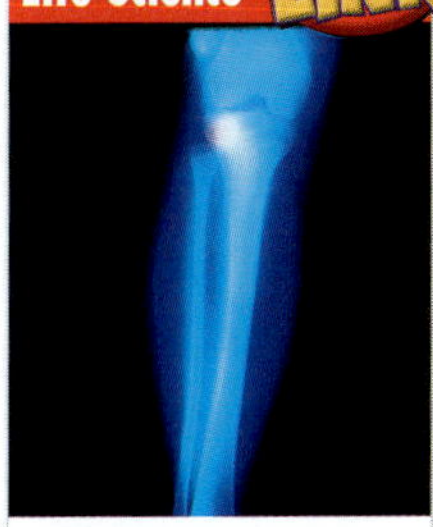

X-rays are used to produce images of bones. Bones absorb different amounts of x-rays than other body tissues. This is what makes them appear on the film.

go.hrw.com
KEYWORD: MR4 X-ray
CNN Student News

40. Daniel is a costume designer for movies and music videos. He recently purchased $\frac{256}{9}$ yards of metallic fabric for space-suit costumes. Write a mixed number to represent the number of yards of fabric Daniel purchased. $28\frac{4}{9}$ **yards**

Use the table for Exercises 41–43.

41. ***LIFE SCIENCE*** Write the length of the ulna as an improper fraction. Then do the same for the length of the humerus. $\frac{141}{5}$; $\frac{73}{2}$

42. ***LIFE SCIENCE*** Write the length of the fibula as a mixed number. Then do the same for the length of the femur. $40\frac{1}{2}$; $50\frac{1}{2}$

43. ***LIFE SCIENCE*** Write the bones in order from longest to shortest.

44. ***SOCIAL STUDIES*** The European country of Monaco, with an area of only $1\frac{4}{5}$ km^2, is one of the smallest countries in the world. Write $1\frac{4}{5}$ as an improper fraction. $\frac{9}{5}$

Longest Human Bones	
Fibula (outer lower leg)	$\frac{81}{2}$ cm
Ulna (inner lower arm)	$28\frac{1}{5}$ cm
Femur (upper leg)	$\frac{101}{2}$ cm
Humerus (upper arm)	$36\frac{1}{2}$ cm
Tibia (inner lower leg)	43 cm

45. ***WHAT'S THE QUESTION?*** The lengths of Victor's three favorite movies are $\frac{11}{4}$ hours, $\frac{9}{4}$ hours, and $\frac{7}{4}$ hours. The answer is $2\frac{1}{4}$ hours. What is the question?

46. ***WRITE ABOUT IT*** Draw models representing $\frac{4}{4}$, $\frac{5}{5}$, and $\frac{9}{9}$. Use your models to explain why a fraction whose numerator is the same as its denominator is equal to 1.

47. ***CHALLENGE*** Write $\frac{65}{12}$ as a decimal. $5.41\overline{6}$

Spiral Review

Find each value. (Lesson 1-3)

48. 3^3 **27** 49. 9^2 **81** 50. 2^6 **64** 51. 4^4 **256**

Solve each equation. (Lesson 2-4)

52. $12 + y = 23$ $y = 11$ 53. $38 + y = 80$ $y = 42$ 54. $y + 76 = 230$ $y = 154$

55. **TEST PREP** Which number is **not** divisible by 3? (Lesson 4-1) **B**

A 240 C 522
B 413 D 735

56. **TEST PREP** Which is the prime factorization of 50? (Lesson 4-2) **F**

F 2×5^2 H 10^5
G 2×5^{10} J 5×10

Answers

43. femur, tibia, fibula, humerus, ulna
45. Possible answer: What mixed number represents the length of the movie that is $\frac{9}{4}$ hours long?
46. because you shade all 4 out of 4, 5 out of 5, and 9 out of 9 pieces in your drawings representing one whole

Journal

Have students write how mixed numbers and improper fractions are used in real-world situations.

Test Prep Doctor

For Exercise 55, students will need to read the question carefully so they know to look for the number that is **not** divisible by 3. Choices **A, C,** and **D** can be eliminated because the sum of the digits in each is divisible by 3.

CHALLENGE 4-7

LESSON 4-7 **Challenge**
Mixed Up Mixed Numbers

Find the whole number and fraction in the box to make a mixed number to match each improper fraction listed below. Cross out each whole number and fraction as you use it. When you have finished, the unused items in the box will form the mixed number that answers the question at the bottom of the page.

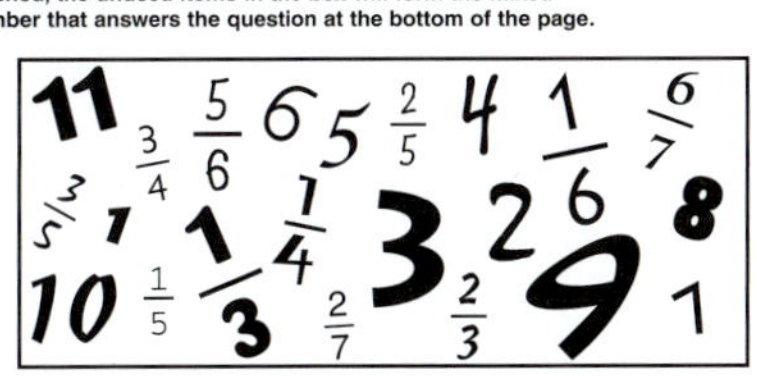

1. $\frac{7}{3}$ = $2\frac{1}{3}$
2. $\frac{20}{3}$ = $6\frac{2}{3}$
3. $\frac{21}{4}$ = $5\frac{1}{4}$
4. $\frac{31}{4}$ = $7\frac{3}{4}$
5. $\frac{41}{5}$ = $8\frac{1}{5}$
6. $\frac{57}{5}$ = $11\frac{2}{5}$
7. $\frac{59}{6}$ = $9\frac{5}{6}$
8. $\frac{19}{6}$ = $3\frac{1}{6}$
9. $\frac{13}{7}$ = $1\frac{6}{7}$
10. $\frac{72}{7}$ = $10\frac{2}{7}$

How old is the moon? $4\frac{3}{5}$ billion years old!

PROBLEM SOLVING 4-7

LESSON 4-7 **Problem Solving**
Mixed Numbers and Improper Fractions

Write the correct answer.

1. If stretched end-to-end, the total length of the blood vessels inside your body could wrap around Earth's equator $\frac{5}{2}$ times! Write this fact as a mixed number.
$2\frac{1}{2}$ **times**

2. In 2000, the average 12-year-old child in the United States earned an allowance of 9 dollars and $\frac{7}{25}$ cents a week. Write this amount as an improper fraction and a decimal.
$\$\frac{232}{25}$; $9.28

3. The normal body temperature for a rattlesnake is between $53\frac{3}{5}$°F and $64\frac{2}{5}$°F. Write this range as improper fractions.
$\frac{268}{5}$°F to $\frac{322}{5}$°F

4. An astronaut who weighs 250 pounds on Earth would weigh $41\frac{1}{2}$ pounds on the moon. Write the astronaut's moon weight as an improper fraction.
$\frac{83}{2}$ **pounds**

5. Adult bees only eat nectar, the substance in flowers used to make honey. A bee could fly 4 million miles on the energy it would get from eating $\frac{9}{2}$ liters of nectar. Write this amount of nectar as a mixed number.
$4\frac{1}{2}$ **liters**

6. A professional baseball can weigh no less than $\frac{45}{9}$ ounces and no more than $\frac{21}{4}$ ounces. Write this range as mixed numbers.
5 ounces to $5\frac{1}{4}$ **ounces**

Circle the letter of the correct answer.

7. Betty needs a piece of lumber that is $\frac{14}{3}$ feet long. Which size should she look for at the hardware store?
A $3\frac{1}{3}$ feet
B $3\frac{1}{4}$ feet
(C) $4\frac{2}{3}$ feet
D $4\frac{1}{4}$ feet

8. What operations are used to change a mixed number to an improper fraction?
(F) multiplication and addition
G division and subtraction
H division and addition
J multiplication and subtraction

Lesson Quiz

Write each mixed number as an improper fraction.

1. $3\frac{3}{5}$ $\frac{18}{5}$
2. $6\frac{1}{9}$ $\frac{55}{9}$
3. $10\frac{1}{2}$ $\frac{21}{2}$
4. $4\frac{1}{8}$ $\frac{33}{8}$
5. Janet watches $\frac{7}{4}$ hours of television per day. Write $\frac{7}{4}$ as a mixed number. $1\frac{3}{4}$

Available on Daily Transparency in CRB

Chapter 4 Mid-Chapter Quiz

Purpose: *To assess students' mastery of concepts and skills in Lessons 4-1 through 4-7*

Assessment Resources

Section 4B Quiz
Assessment Resources .... p. 12

***Test and Practice Generator* CD-ROM**

Additional mid-chapter assessment items in both multiple-choice and free-response format may be generated for any objective in Lessons 4-1 through 4-7.

Chapter 4 Mid-Chapter Quiz

Mid-Chapter Quiz

LESSONS 4-1 AND 4-2 (pp. 152–159)

1. List all of the factors of 38. Is 38 prime or composite? **1, 2, 19, 38—composite**
2. Write the prime factorization of 84. **$2 \times 2 \times 3 \times 7$**

LESSON 4-3 (pp. 160–163)

3. Find the GCF of 25 and 45. **5**
4. Find the GCF of 26, 65, and 91. **13**

LESSON 4-4 (pp. 167–170)

Write each decimal as a fraction.

5. 0.67 $\frac{67}{100}$
6. 0.9 $\frac{9}{10}$
7. 0.43 $\frac{43}{100}$
8. 0.17 $\frac{17}{100}$

Write each fraction as a decimal.

9. $\frac{2}{5}$ **0.4**
10. $\frac{1}{6}$ **$0.1\overline{6}$**
11. $\frac{3}{4}$ **0.75**
12. $\frac{5}{8}$ **0.625**

LESSON 4-5 (pp. 172–175)

Write two equivalent fractions for each fraction.

13. $\frac{9}{12}$ $\frac{3}{4}, \frac{6}{8}$
14. $\frac{18}{42}$ $\frac{9}{21}, \frac{3}{7}$
15. $\frac{25}{30}$ $\frac{5}{6}, \frac{30}{36}$
16. $\frac{8}{20}$ $\frac{4}{10}, \frac{2}{5}$

Write each fraction in simplest form.

17. $\frac{20}{24}$ $\frac{5}{6}$
18. $\frac{14}{49}$ $\frac{2}{7}$
19. $\frac{12}{28}$ $\frac{3}{7}$
20. $\frac{16}{36}$ $\frac{4}{9}$

LESSON 4-6 (pp. 178–181)

Compare. Write $<$, $>$, or $=$.

21. $\frac{3}{4}$ ■ $\frac{2}{3}$ $>$
22. $\frac{7}{9}$ ■ $\frac{5}{6}$ $<$
23. $\frac{4}{9}$ ■ $\frac{4}{7}$ $<$
24. $\frac{5}{11}$ ■ $\frac{3}{5}$ $<$

Order the fractions from least to greatest.

25. $\frac{5}{8}, \frac{1}{2}, \frac{3}{4}$ $\frac{1}{2}, \frac{5}{8}, \frac{3}{4}$
26. $\frac{3}{4}, \frac{3}{5}, \frac{7}{10}$ $\frac{3}{5}, \frac{7}{10}, \frac{3}{4}$
27. $\frac{1}{3}, \frac{3}{8}, \frac{1}{4}$ $\frac{1}{4}, \frac{1}{3}, \frac{3}{8}$
28. $\frac{2}{5}, \frac{4}{9}, \frac{11}{15}$ $\frac{2}{5}, \frac{4}{9}, \frac{11}{15}$

LESSON 4-7 (pp. 182–185)

29. The proboscis bat, with a length of $\frac{19}{5}$ cm, is one of the smallest bats. Write $\frac{19}{5}$ as a mixed number. $3\frac{4}{5}$
30. Write $\frac{15}{2}$ as a mixed number. $7\frac{1}{2}$

Focus on Problem Solving

Understand the Problem

• Write the problem in your own words

One way to understand a problem better is to write it in your own words. Before you do this, you may need to read it over several times, perhaps aloud so that you can hear yourself say the words.

When you write a problem in your own words, try to make the problem simpler. Use smaller words and shorter sentences. Leave out any extra information, but make sure to include all the information you need to answer the question.

Write each problem in your own words. Check that you have included all the information you need to answer the question.

1. Martin is making muffins for his class bake sale. The recipe calls for $2\frac{1}{3}$ cups of flour, but Martin's only measuring cup holds $\frac{1}{3}$ cup. How many of his measuring cups should he use?

2. Mariko sold an old book to a used bookstore. She had hoped to sell it for \$0.80, but the store gave her $\frac{3}{4}$ of a dollar. What is the difference between the two amounts?

3. Koalas of eastern Australia feed mostly on eucalyptus leaves. They select certain trees over others to find the $1\frac{1}{4}$ pounds of food they need each day. Suppose a koala has eaten $1\frac{1}{8}$ pounds of food. Has the koala eaten enough food for the day?

4. The first day of the Tour de France is called the prologue. Each of the days after that is called a stage, and each stage covers a different distance. The total distance covered in the race is about 3,600 km. If a cyclist has completed $\frac{1}{3}$ of the race, how many kilometers has he ridden?

Answers

1. 7
2. \$0.05 (a nickel)
3. No, the koala needs another $\frac{1}{8}$ lb of food.
4. 1,200 km

Focus on Problem Solving

Purpose: *To focus on understanding the problem by writing the problem in your own words*

Problem Solving Resources

Interactive Problem Solving pp. 25–33

Math: Reading and Writing in the Content Area . . . . pp. 25–33

Problem Solving Process

This page focuses on the first step of the problem-solving process: **Understand the Problem**

Discuss

Have students identify the unnecessary information in each problem. Then have them rewrite each problem in their own words and include only the important information.

Possible answers:

1. The muffins are for a bake sale. How many times must Martin fill his $\frac{1}{3}$-cup measuring cup in order to get $2\frac{1}{3}$ cups of flour?
2. Mariko sold an old book to a used bookstore. What is the difference between \$0.80 and $\frac{3}{4}$ of a dollar?
3. Koalas feed mostly on eucalyptus, and they select certain trees over others. If a koala needs $1\frac{1}{4}$ pounds of food and has eaten $1\frac{1}{8}$ pounds of food, has it eaten enough food?
4. The first day of the Tour de France is called the Prologue; each of the days after that is called a stage, and each stage covers a different distance. What is $\frac{1}{3}$ of 3,600 km?

Section 4C

Introduction to Fraction Operations

One-Minute Section Planner

Lesson	Materials	Resources
Lesson 4-8 Adding and Subtracting with Like Denominators **NCTM:** Number and Operations, Reasoning and Proof, Communication, Connections **NAEP:** Number Properties 3g ☑ SAT-9 ☑ SAT-10 ☑ ITBS ☑ CTBS ☑ MAT ☑ CAT	**Optional** Recording Sheet for Reaching All Learners *(CRB, p. 102)* Fraction Bars *(MK, CRB, p. 100)*	• *Chapter 4 Resource Book*, pp. 74–82 • Daily Transparency T28, CRB • Additional Examples Transparencies T29–T31, CRB • *Alternate Openers: Explorations*, p. 32
Lesson 4-9 Multiplying Fractions by Whole Numbers **NCTM:** Number and Operations, Reasoning and Proof, Communication **NAEP:** Number Properties 3g ☑ SAT-9 ☑ SAT-10 ☑ ITBS ☑ CTBS ☐ MAT ☑ CAT	**Optional** Teaching Transparency T33 for Reaching All Learners *(CRB)*	• *Chapter 4 Resource Book*, pp. 83–91 • Daily Transparency T32, CRB • Additional Examples Transparencies T34–T36, CRB • *Alternate Openers: Explorations*, p. 33
Extension Sets of Numbers **NCTM:** Number and Operations, Reasoning and Proof, Connections **NAEP:** Geometry 1b ☐ SAT-9 ☐ SAT-10 ☐ ITBS ☐ CTBS ☐ MAT ☐ CAT	**Optional** Teaching Transparency T37 *(CRB)*	• Additional Examples Transparencies T38–T39, CRB
Section 4C Assessment		• Section 4C Quiz, AR p. 13 • *Test and Practice Generator* CD-ROM

SAT = *Stanford Achievement Tests* **ITBS** = *Iowa Test of Basic Skills* **CTBS** = *Comprehensive Test of Basic Skills/Terra Nova*
MAT = *Metropolitan Achievement Tests* **CAT** = *California Achievement Test*

NCTM = Complete standards can be found on pages T27–T33. **NAEP** = Complete standards can be found on pages A31–A35.

SE = *Student Edition* **TE** = *Teacher's Edition* **AR** = *Assessment Resources* **CRB** = *Chapter Resource Book* **MK** = *Manipulatives Kit*

Section Overview

Adding and Subtracting with Like Denominators — *Lesson 4-8*

Why? Solving real-world problems often involves adding and subtracting fractions and mixed numbers.

When adding or subtracting fractions with like denominators, add or subtract the numerators and keep the same denominator.

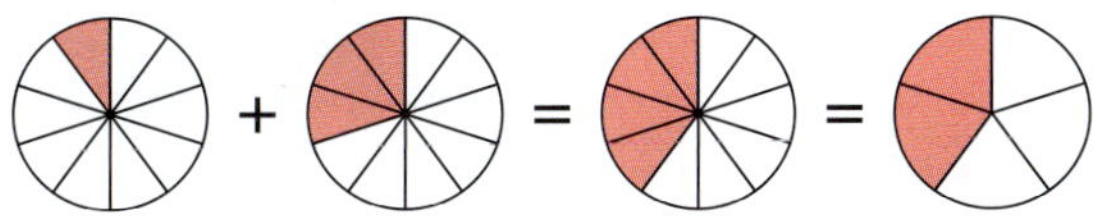

$$\frac{1}{10} + \frac{3}{10} = \frac{4}{10} = \frac{2}{5}$$

Write 1 as a fraction with a denominator of 4, and then subtract.

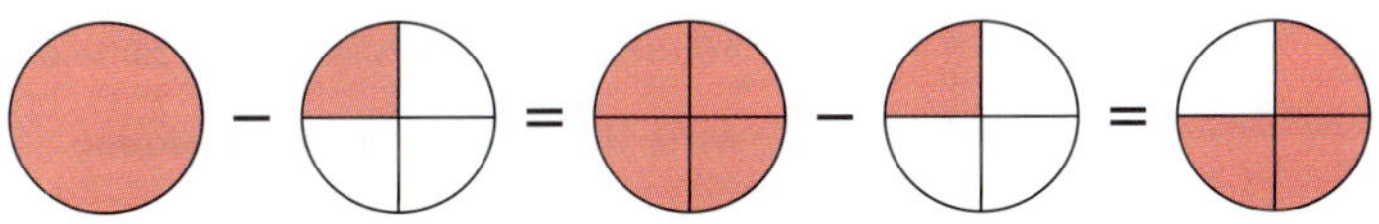

$$1 - \frac{1}{4} = \frac{4}{4} - \frac{1}{4} = \frac{3}{4}$$

Subtract the **fractions** and the **whole numbers** independently.

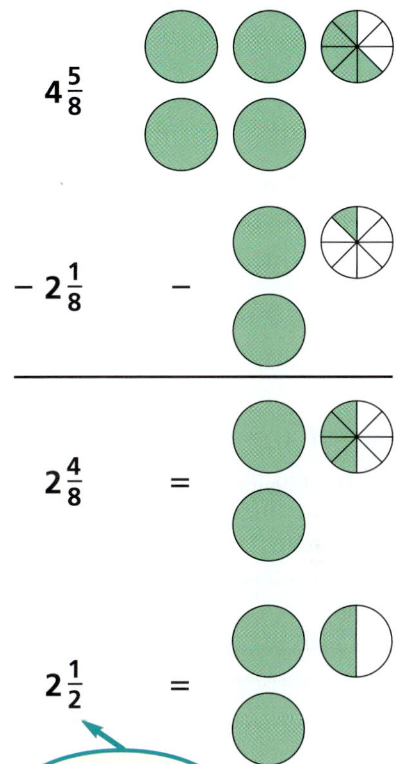

$$4\frac{5}{8}$$

$$-\ 2\frac{1}{8}$$

$$2\frac{4}{8} =$$

$$2\frac{1}{2} =$$

Simplify the answer.

Multiplying Fractions by Whole Numbers — *Lesson 4-9*

Why? Some practical problems require multiplying fractions by whole numbers.

If Jo and Helena each ate $\frac{3}{8}$ of a pizza, how much of the pizza did they eat altogether?

Write **2** as a fraction, and multiply numerators and denominators.

$$2 \cdot \frac{3}{8} = \frac{2}{1} \cdot \frac{3}{8}$$

$$= \frac{2 \cdot 3}{1 \cdot 8}$$

$$= \frac{6}{8}$$

$$= \frac{3}{4}$$

Simplify the answer.

Together, Jo and Helena ate $\frac{3}{4}$ of the pizza.

4-8 Organizer

Pacing: Traditional 1 day
Block $\frac{1}{2}$ day

Objective: Students add and subtract fractions with like denominators.

Indiana Academic Standards
6.2.5, 6.7.1, 6.7.3, 6.7.5, 6.7.11

Warm Up

Write each mixed number as an improper fraction.

1. $6\frac{1}{3}$	$\frac{19}{3}$	**4.** $8\frac{1}{2}$	$\frac{17}{2}$
2. $2\frac{1}{5}$	$\frac{11}{5}$	**5.** $9\frac{4}{5}$	$\frac{49}{5}$
3. $3\frac{2}{7}$	$\frac{23}{7}$	**6.** $4\frac{8}{11}$	$\frac{52}{11}$

Problem of the Day

Marge has six coins that total $1.15. She cannot make change for $1.00, a half dollar, a quarter, a dime, or a nickel. What coins does she have?
half dollar, quarter, 4 dimes

Available on Daily Transparency in CRB

Math Humor

Why do they call that stuff you pour in coffee "half and half"? Why not just call it "one whole"?

4-8 Adding and Subtracting with Like Denominators

Learn to add and subtract fractions with like denominators.

You can estimate the age of an oak tree by measuring around the trunk at four feet above the ground.

The distance around a young oak tree's trunk increases at a rate of approximately $\frac{1}{8}$ inch per month.

EXAMPLE 1 ***Life Science Application***

Sophie plants a young oak tree in her backyard. The distance around the trunk grows at a rate of $\frac{1}{8}$ inch per month. Find how much this distance will increase in two months. Write your answer in simplest form.

$\frac{1}{8} + \frac{1}{8}$

$\frac{1}{8} + \frac{1}{8} = \frac{2}{8}$ *Add the numerators. Keep the same denominator.*

$= \frac{1}{4}$ *Write your answer in simplest form.*

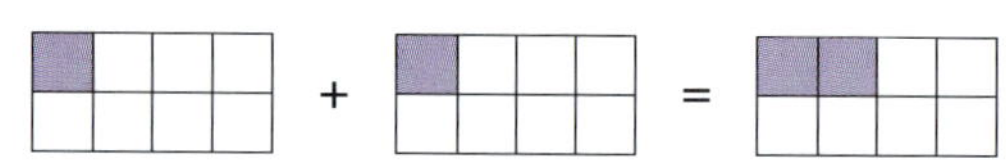

The distance around the trunk will increase by $\frac{1}{4}$ inch.

EXAMPLE 2 **Subtracting Like Fractions and Mixed Numbers**

Remember!
When the numerator equals the denominator, the fraction is equal to 1.
$\frac{3}{3} = 1$ $\frac{173}{173} = 1$

Subtract. Write each answer in simplest form.

A $1 - \frac{2}{3}$

$\frac{3}{3} - \frac{2}{3} = \frac{1}{3}$ *To get a common denominator, rewrite 1 as a fraction with a denominator of 3.* *Subtract the numerators. Keep the same denominator.*

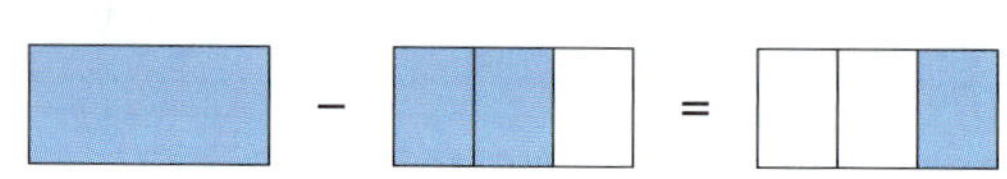

1 Introduce

Alternate Opener

EXPLORATION

4-8 Adding and Subtracting with Like Denominators

Suzanne runs on a $\frac{3}{4}$-mile track. She ran one lap and then decided to run one more lap.

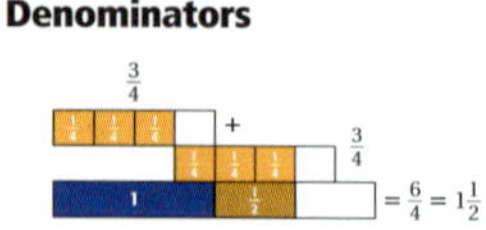

Susan ran $1\frac{1}{2}$ miles.

Draw a model to solve each addition problem.

1. $\frac{3}{5} + \frac{4}{5}$ 2. $\frac{3}{10} + \frac{7}{10}$

Paul is recording a CD and has $\frac{1}{6}$ of the work completed. How much recording is left?

Paul needs to record $\frac{5}{6}$ of the CD. $1 - \frac{1}{6} = \frac{6}{6} - \frac{1}{6} = \frac{5}{6}$

Draw a model to solve each subtraction problem.

3. $\frac{4}{5} - \frac{3}{5}$ 4. $1 - \frac{3}{10}$

Think and Discuss

5. **Explain** how to add and subtract fractions with like denominators.
6. **Explain** how to subtract a fraction from 1.

Motivate

Discuss fractions as they relate to time (e.g., quarter hour and half hour). Ask students how many quarter hours are equal to one half hour. (2 quarter hours) Have students explain how they know this.

Possible answers: Because $\frac{1}{4} + \frac{1}{4} = \frac{1}{2}$, students may refer to a clock face to illustrate their answer.

Exploration worksheet and answers on Chapter 4 Resource Book pp. 75 and 118

2 Teach

Lesson Presentation

Guided Instruction

In this lesson, students learn to add and subtract fractions with like denominators. Teach them to add like fractions by adding the numerators and keeping the same denominator. (Use a pie to model how adding fifths does not change the denominator, only the numerator.) Then teach students to subtract by subtracting the numerators and keeping the same denominator. Finally, have students evaluate expressions involving adding and subtracting fractions with like denominators.

Subtract. Write each answer in simplest form.

B $3\frac{7}{12} - 1\frac{1}{12}$

$3\frac{7}{12} - 1\frac{1}{12}$ *Subtract the fractions.*

$2\frac{6}{12}$ *Then subtract the whole numbers.*

$2\frac{1}{2}$ *Write your answer in simplest form.*

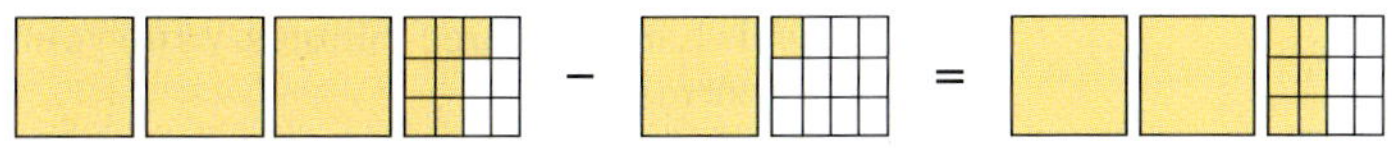

EXAMPLE 3 Evaluating Expressions with Fractions

Evaluate each expression for $x = \frac{3}{8}$. Write each answer in simplest form.

A $\frac{5}{8} - x$

$\frac{5}{8} - x$ *Write the expression.*

$\frac{5}{8} - \frac{3}{8} = \frac{2}{8}$ *Substitute $\frac{3}{8}$ for x and subtract the numerators. Keep the same denominator.*

$= \frac{1}{4}$ *Write your answer in simplest form.*

B $x + 1\frac{1}{8}$

$x + 1\frac{1}{8}$ *Write the expression.*

$\frac{3}{8} + 1\frac{1}{8} = 1\frac{4}{8}$ *Substitute $\frac{3}{8}$ for x. Add the fractions. Then add the whole numbers.*

$= 1\frac{1}{2}$ *Write your answer in simplest form.*

C $x + \frac{7}{8}$

$x + \frac{7}{8}$ *Write the expression.*

$\frac{3}{8} + \frac{7}{8} = \frac{10}{8}$ *Substitute $\frac{3}{8}$ for x and add the numerators. Keep the same denominator.*

$= \frac{5}{4}$ or $1\frac{1}{4}$ *Write your answer in simplest form.*

Helpful Hint

When adding a fraction to a mixed number, you can think of the fraction as having a whole number of 0.

$\frac{3}{8} = 0\frac{3}{8}$

Think and Discuss

1. **Explain** how to add or subtract like fractions.
2. **Tell** why the sum of $\frac{1}{5}$ and $\frac{3}{5}$ is not $\frac{4}{10}$. Give the correct sum.
3. **Describe** how you would add $2\frac{3}{8}$ and $1\frac{1}{8}$. How would you subtract $1\frac{1}{8}$ from $2\frac{3}{8}$?

COMMON ERROR ALERT

Watch for students who add or subtract the numerators *and* the denominators, instead of keeping the same denominator.

Additional Examples

Example 1

Snow was falling at a rate of $\frac{1}{4}$ inch per hour. How much snow fell after two hours? Write your answer in simplest form. $\frac{1}{2}$ inch

Example 2

Subtract. Write your answers in simplest form.

A. $1 - \frac{3}{5}$ $\frac{2}{5}$

B. $5\frac{5}{12} - 2\frac{1}{12}$ $3\frac{1}{3}$

Example 3

Evaluate each expression for $x = \frac{2}{9}$. Write your answers in simplest form.

A. $\frac{5}{9} - x$ $\frac{1}{3}$

B. $x + 2\frac{4}{9}$ $2\frac{2}{3}$, or $\frac{8}{3}$

C. $x + \frac{8}{9}$ $\frac{10}{9}$, or $1\frac{1}{9}$

3 Close

Reaching All Learners

Through Number Sense

Have students complete the magic square, expressing all missing fractions in simplest form. The sum of every row, column, and major diagonal is one.

A recording sheet is provided on Chapter 4 Resource Book p. 102.

$\frac{2}{15}$	$\frac{7}{15}$	$\frac{2}{5}$
$\frac{3}{5}$	$\frac{1}{3}$	$\frac{1}{15}$
$\frac{4}{15}$	$\frac{1}{5}$	$\frac{8}{15}$

Summarize

Have the students explain how to solve $\frac{5}{8} + \frac{7}{8}$ and $\frac{7}{8} - \frac{5}{8}$.

Possible answer: Add 5 and 7 and keep the same denominator. Write the answer as a mixed number in simplest form. $\frac{5}{8} + \frac{7}{8} = \frac{12}{8} = \frac{3}{2} = 1\frac{1}{2}$

Subtract 5 from 7 and keep the same denominator. Write the difference in simplest form. $\frac{7}{8} - \frac{5}{8} = \frac{2}{8} = \frac{1}{4}$

Answers to Think and Discuss

1. Possible answer: First add or subtract the numerators. Then write the answer over the same denominator.
2. When adding like fractions, the denominator stays the same. So the correct sum is $\frac{4}{5}$.
3. Add the fractions first. The like denominators mean the numerators can be added. Then simplify the fraction. Next add the whole numbers. For subtraction, subtract the fractions first. The like denominators mean the numerators can be subtracted. Then subtract the whole numbers.

4-8 PRACTICE & ASSESS

Students may want to refer back to the lesson examples.

Assignment Guide

If you finished Example 1 assign:
Core 1, 10, 19, 21, 23–24, 34, 42–48
Enriched 1, 10, 19, 21, 26–27, 38–39, 42–48

If you finished Example 2 assign:
Core 1–5, 10–14, 19–24, 34, 42–48
Enriched 1–2, 4, 10–14 even, 19–27, 40–48

If you finished Example 3 assign:
Core 1–27, 34, 42–48
Enriched 15–48

Notes

4-8 Exercises

FOR EXTRA PRACTICE
see page 644

internet connect
Homework Help Online
go.hrw.com Keyword: MR4 4-8

GUIDED PRACTICE

See Example 1

1. Marta is filling a bucket with water. The height of the water is increasing $\frac{1}{6}$ foot each minute. Find how much the height of the water will change in three minutes. Write your answer in simplest form. $\frac{1}{2}$ foot

See Example 2

Subtract. Write each answer in simplest form.

2. $2 - \frac{3}{5}$ $1\frac{2}{5}$ **3.** $8 - \frac{6}{7}$ $7\frac{1}{7}$ **4.** $4\frac{2}{3} - 1\frac{1}{3}$ $3\frac{1}{3}$ **5.** $8\frac{7}{12} - 3\frac{5}{12}$ $5\frac{1}{6}$

See Example 3

Evaluate each expression for $x = \frac{3}{10}$. Write each answer in simplest form.

6. $\frac{9}{10} - x$ $\frac{3}{5}$ **7.** $x + \frac{1}{10}$ $\frac{2}{5}$ **8.** $x + \frac{9}{10}$ $\frac{6}{5}$ or $1\frac{1}{5}$ **9.** $x - \frac{1}{10}$ $\frac{1}{5}$

INDEPENDENT PRACTICE

See Example 1

10. Wesley drinks $\frac{2}{13}$ gallon of juice each day. Find the number of gallons of juice Wesley drinks in 5 days. Write your answer in simplest form. $\frac{10}{13}$ gallon

See Example 2

Subtract. Write each answer in simplest form.

11. $1 - \frac{5}{7}$ $\frac{2}{7}$ **12.** $1 - \frac{3}{8}$ $\frac{5}{8}$ **13.** $2\frac{4}{5} - 1\frac{1}{5}$ $1\frac{3}{5}$ **14.** $9\frac{9}{14} - 5\frac{3}{14}$ $4\frac{3}{7}$

See Example 3

Evaluate each expression for $x = \frac{11}{20}$. Write each answer in simplest form.

15. $x + \frac{13}{20}$ $\frac{6}{5}$ or $1\frac{1}{5}$ **16.** $x - \frac{3}{20}$ $\frac{2}{5}$ **17.** $x - \frac{9}{20}$ $\frac{1}{10}$ **18.** $x + \frac{17}{20}$ $\frac{7}{5}$ or $1\frac{2}{5}$

PRACTICE AND PROBLEM SOLVING

Write each sum or difference in simplest form.

19. $\frac{1}{16} + \frac{9}{16}$ $\frac{5}{8}$ **20.** $\frac{15}{26} - \frac{11}{26}$ $\frac{2}{13}$ **21.** $\frac{10}{33} + \frac{4}{33}$ $\frac{14}{33}$

22. $1 - \frac{9}{10}$ $\frac{1}{10}$ **23.** $\frac{26}{75} + \frac{24}{75}$ $\frac{2}{3}$ **24.** $\frac{100}{999} + \frac{899}{999}$ 1

25. $37\frac{13}{18} - 24\frac{7}{18}$ $13\frac{1}{3}$ **26.** $\frac{1}{20} + \frac{7}{20} + \frac{3}{20}$ $\frac{11}{20}$ **27.** $\frac{11}{24} + \frac{1}{24} + \frac{5}{24}$ $\frac{17}{24}$

Evaluate. Write each answer in simplest form.

28. $a + \frac{7}{18}$ for $a = \frac{1}{18}$ $\frac{4}{9}$ **29.** $\frac{6}{13} - j$ for $j = \frac{4}{13}$ $\frac{2}{13}$

30. $c + c$ for $c = \frac{5}{14}$ $\frac{5}{7}$ **31.** $m - \frac{6}{17}$ for $m = 1$ $\frac{11}{17}$

32. $8\frac{14}{15} - z$ for $z = \frac{4}{15}$ $8\frac{2}{3}$ **33.** $13\frac{1}{24} + y$ for $y = 2\frac{5}{24}$ $15\frac{1}{4}$

Math Background

In teaching students to work with fractions, some educators move from addition to subtraction to multiplication and finally to division, the same route used for whole numbers; others begin with multiplication of fractions because it is easier to master than addition and subtraction.

This program begins with addition and subtraction of fractions that have common denominators. This eliminates the need to convert fractions to fractions with a common denominator and allows students to focus on physical representations of addition and subtraction. In the next lesson, students will learn multiplication.

RETEACH 4-8

LESSON 4-8 **Reteach**
Adding and Subtracting with Like Denominators

You can use fraction strips to add like fractions.

To find $\frac{1}{8} + \frac{3}{8}$, first model the expression using $\frac{1}{8}$ fraction pieces.

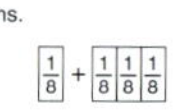

The total number of fraction pieces is 4.
So, $\frac{1}{8} + \frac{3}{8} = \frac{4}{8}$.
Then write the sum in simplest form. $\frac{4}{8} = \frac{1}{2}$.

Write each sum in simplest form.

1. $\frac{1}{4} + \frac{1}{4}$ $\frac{1}{2}$ 2. $\frac{1}{5} + \frac{3}{5}$ $\frac{4}{5}$ 3. $\frac{1}{6} + \frac{5}{6}$ 1 4. $\frac{5}{12} + \frac{1}{12}$ $\frac{1}{2}$

You can use fraction strips to subtract like fractions.

To find $\frac{5}{6} - \frac{1}{6}$, use fraction strips to model the first fraction.

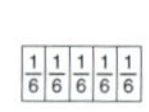

Next, take away fraction strips that represent the second fraction.

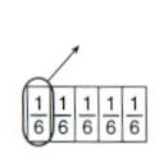

The remaining fraction strips represent the difference.
$\frac{5}{6} - \frac{1}{6} = \frac{4}{6}$

$\frac{4}{6} = \frac{2}{3}$ in simplest form.

Write each difference in simplest form.

5. $\frac{5}{8} - \frac{3}{8}$ $\frac{1}{4}$ 6. $\frac{7}{12} - \frac{5}{12}$ $\frac{1}{6}$ 7. $\frac{3}{4} - \frac{1}{4}$ $\frac{1}{2}$ 8. $\frac{9}{10} - \frac{3}{10}$ $\frac{3}{5}$

PRACTICE 4-8

LESSON 4-8 **Practice B**
Adding and Subtracting with Like Denominators

Subtract. Write your answers in simplest form.

1. $1 - \frac{4}{7}$ $\frac{3}{7}$ 2. $\frac{18}{24} - \frac{10}{24}$ $\frac{1}{3}$ 3. $2\frac{2}{3} - 1\frac{1}{3}$ $1\frac{1}{3}$

4. $8\frac{11}{13} - 5\frac{2}{13}$ $3\frac{9}{13}$ 5. $5 - 3\frac{1}{4}$ $1\frac{3}{4}$ 6. $2 - 1\frac{2}{7}$ $\frac{5}{7}$

7. $6\frac{8}{9} - 4\frac{6}{9}$ $2\frac{2}{9}$ 8. $7\frac{4}{11} - 6\frac{3}{11}$ $1\frac{1}{11}$ 9. $10 - 5\frac{3}{5}$ $4\frac{2}{5}$

Evaluate $\frac{14}{15} - x$ for each value of x. Write your answers in simplest form.

10. $x = \frac{12}{15}$ $\frac{2}{15}$ 11. $x = \frac{2}{15}$ $\frac{4}{5}$ 12. $x = \frac{9}{15}$ $\frac{1}{3}$ 13. $x = \frac{5}{15}$ $\frac{3}{5}$

Write the sum or difference in simplest form.

14. $\frac{17}{21} - \frac{2}{21}$ $\frac{5}{7}$ 15. $\frac{13}{32} + \frac{9}{32}$ $\frac{11}{16}$ 16. $\frac{2}{15} + \frac{8}{15}$ $\frac{2}{3}$

17. $27\frac{76}{100} - 14\frac{26}{100}$ $13\frac{1}{2}$ 18. $\frac{1}{15} + \frac{4}{15} + \frac{5}{15}$ $\frac{2}{3}$ 19. $\frac{9}{26} + \frac{2}{26} + \frac{5}{26}$ $\frac{8}{13}$

20. **Maria has 8 gallons of paint she wants to use in three rooms of her house. If she uses $2\frac{1}{4}$ gallons of the paint in the bedroom and $1\frac{1}{4}$ in the bathroom, how many gallons will she have left to paint the playroom?**
$4\frac{1}{2}$ gallons

21. **Sandy, Ben, and Kwan picked strawberries. Sandy picked $\frac{8}{25}$ of their combined total of strawberries. Ben picked $\frac{7}{25}$ of their strawberries. How much of the strawberries did Kwan pick?**
$\frac{2}{5}$ of the strawberries

34. Carlos had 7 cups of chocolate chips. He used $1\frac{2}{3}$ cups to make a chocolate sauce and $3\frac{1}{3}$ cups to make cookies. How many cups of chocolate chips does Carlos have now? **2 cups**

35. A concert was $2\frac{1}{4}$ hr long. The first musical piece lasted $\frac{1}{4}$ hr. The intermission also lasted $\frac{1}{4}$ hr. How long was the rest of the concert? **$1\frac{3}{4}$ hr**

36. A flight from Washington, D.C., stops in San Francisco and then continues to Seattle. The trip to San Francisco takes $4\frac{5}{8}$ hr. The trip to Seattle takes $1\frac{1}{8}$ hr. What is the total flight time? **$5\frac{6}{8}$ or $5\frac{3}{4}$ hr**

Life Science LINK

The Venus' flytrap, like other carnivorous plants, has adapted to nutrient-poor soil by capturing insects. Their traps spring shut in a fraction of a second, and then enzymes dissolve the insect.

Use the graph for Exercises 37–39.

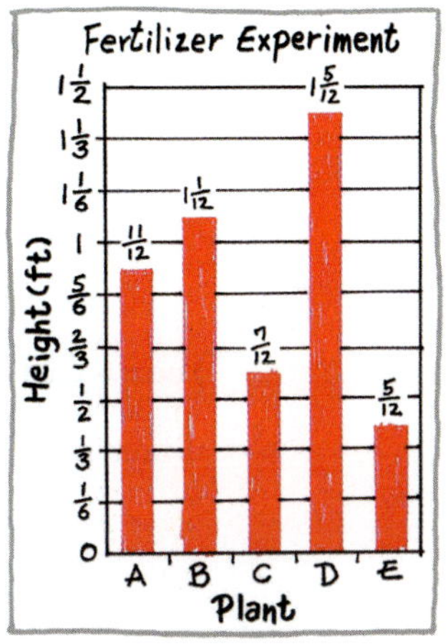

37. ***LIFE SCIENCE*** Sheila performed an experiment to find the most effective plant fertilizer. She used a different fertilizer on each of 5 different plants. The heights of the plants at the end of her experiment are shown in the graph. What is the combined height of plants C and E? **1 foot**

38. ***LIFE SCIENCE*** What is the difference in height between the tallest plant and the shortest plant? **1 foot**

39. ***WHAT'S THE ERROR?*** Sheila found the combined heights of plants B and E to be $1\frac{6}{24}$ feet. Explain the error and give the correct answer in simplest form.

40. ***WRITE ABOUT IT*** When writing 1 as a fraction in a subtraction problem, how do you know what the numerator and denominator should be? Give an example.

41. ***CHALLENGE*** Explain how you might estimate the difference between $\frac{3}{4}$ and $\frac{6}{23}$. **Possible answer: $\frac{6}{23}$ is almost equal to $\frac{1}{4}$. Subtract $\frac{1}{4}$ from $\frac{3}{4}$ to find an estimate of $\frac{2}{4}$, or $\frac{1}{2}$.**

Spiral Review

Evaluate. (Lesson 1-4)

42. $2 + 3 \cdot 4 - 5$ **9**

43. $(9 + 4) \div (3 + 10)$ **1**

44. $23 - 16 + 28$ **35**

Identify a pattern and find the next term. (Lesson 1-7)

45. 3, 10, 17, 24, … **Add 7; 31**

46. 5, 10, 15, 20, … **Add 5; 25**

47. 1, 4, 2, 5, 3, … **Add 3 then subtract 2; 6**

48. **TEST PREP** Paul is seven years younger than his friend Rhonda. If r stands for Rhonda's age, which expression can be used to represent Paul's age? (Lesson 2-2) **B**

A $7 - r$ B $r - 7$ C $7r$ D $r + 7$

Answers

39. Sheila added the numerators and the denominators. She should have added just the numerators, and the answer would have been $1\frac{1}{2}$ feet.

40. The numerator and denominator should be the same as the denominator of the fraction you are subtracting. For example, in $1 - \frac{1}{3}$, you should write 1 as $\frac{3}{3}$ and then subtract the numerators.

Journal

Have students think of and write real-world situations in which like fractions are added and subtracted. Have them describe how to solve each problem.

Test Prep Doctor

For Exercise 48, students need to subtract 7 from r. Students who answered **A** subtracted r from 7, which is incorrect because subtraction is not commutative. Students who answered **C** multiplied r by 7, and students who answered **D** added 7 to r.

CHALLENGE 4-8

LESSON 4-8 **Challenge**
Adding Equals Subtracting?

Use one expression from Box A and one from Box B to write an equation in the pans below. The two expressions must be equal to keep the scale balanced. Use each expression only once. Possible order of students' answers are given.

A		B	
$\frac{1}{4}+\frac{2}{4}$	$\frac{1}{27}+\frac{2}{27}$	$\frac{11}{7}-\frac{6}{7}$	$\frac{11}{12}-\frac{1}{12}$
$\frac{1}{6}+\frac{4}{6}$	$\frac{2}{7}+\frac{3}{7}$	$\frac{4}{8}-\frac{1}{8}$	$1-\frac{8}{9}$
$\frac{3}{16}+\frac{3}{16}$	$\frac{1}{5}+\frac{1}{5}$	$1-\frac{4}{10}$	$2-1\frac{3}{5}$
$\frac{1}{18}+\frac{2}{18}$	$\frac{4}{15}+\frac{5}{15}$	$2-1\frac{1}{4}$	$\frac{5}{6}-\frac{4}{6}$

1. $\frac{1}{4}+\frac{2}{4} = 2-1\frac{1}{4}$

2. $\frac{11}{7}-\frac{6}{7} = \frac{2}{7}+\frac{3}{7}$

3. $\frac{1}{5}+\frac{1}{5} = 2-1\frac{3}{5}$

4. $\frac{1}{6}+\frac{4}{6} = \frac{11}{12}-\frac{1}{12}$

5. $\frac{1}{18}+\frac{2}{18} = \frac{5}{6}-\frac{4}{6}$

6. $\frac{1}{27}+\frac{2}{27} = 1-\frac{8}{9}$

7. $\frac{3}{16}+\frac{3}{16} = \frac{4}{8}-\frac{1}{8}$

8. $1-\frac{4}{10} = \frac{4}{15}+\frac{5}{15}$

PROBLEM SOLVING 4-8

LESSON 4-8 **Problem Solving**
Adding and Subtracting with Like Denominators

Write the answers in simplest form.

1. About $\frac{3}{10}$ of Earth's surface is covered by land, and the rest is water. What fraction of Earth's surface is covered by water? **$\frac{7}{10}$ of Earth's surface**

2. A recipe for cookies calls for $\frac{3}{8}$ cup of chocolate chips. Tameeka wants to double the recipe. How much chocolate chips will she use? **$\frac{3}{4}$ cup**

3. In Mr. Chesterfield's science class, $\frac{2}{9}$ of the boys like his class. Three times as many girls like his science class. How many girls like Mr. Chesterfield's science class? **$\frac{2}{3}$ of the girls**

4. In the United States, $\frac{6}{50}$ of the population is left-handed men and $\frac{5}{50}$ of the population is left-handed women. What part of the population is left-handed? **$\frac{11}{50}$**

5. In the United States, about $\frac{1}{10}$ of the population is born with black hair, and $\frac{7}{10}$ of the population is born with brown hair. What fraction of the total population in the U.S. is born with brown or black hair? **$\frac{4}{5}$ of the population**

6. In the United States, about $\frac{3}{20}$ of the population is born with blond hair, and $\frac{1}{20}$ of the population is born with red hair. What fraction of the total population in the U.S. is born with blond or red hair? **$\frac{1}{5}$ of the population**

Circle the letter of the correct answer.

7. The average height for men in the United States is $5\frac{2}{3}$ feet tall. Bill is $\frac{1}{3}$ foot shorter than average. How tall is Bill?

(A) $5\frac{1}{3}$ feet C 6 feet
B $5\frac{3}{6}$ feet D $5\frac{1}{6}$ feet

8. The average height for women in the United States is $5\frac{1}{3}$ feet tall. Katie is $\frac{2}{3}$ foot taller than average. How tall is Katie?

F $5\frac{1}{3}$ feet (H) 6 feet
G $5\frac{3}{6}$ feet J $5\frac{1}{6}$ feet

Lesson Quiz

Add or subtract. Write each answer in simplest form.

1. $1 - \frac{9}{11}$ **$\frac{2}{11}$**

2. $1 - \frac{8}{9}$ **$\frac{1}{9}$**

3. $2\frac{3}{10} + 4\frac{1}{10}$ **$6\frac{2}{5}$**

4. Evaluate $\frac{16}{21} - x$ for $x = \frac{6}{21}$. **$\frac{10}{21}$**

5. Weston walks $\frac{3}{4}$ mile every day. How far will he have walked in 4 days? **3 miles**

Available on Daily Transparency in CRB

4-9 Organizer

Pacing: Traditional 1 day
Block $\frac{1}{2}$ day

Objective: Students multiply fractions by whole numbers.

Indiana Academic Standards
6.2.4, 6.7.3, 6.7.5, 6.7.11

Warm Up

Multiply.

1. 15 × 2 **30** **2.** 12 × 8 **96**
3. 9 × 16 **144** **4.** 6 × 11 **66**
5. 8 × 15 **120**

Problem of the Day

The lengths of 3 dowels are 13 inches, 27 inches, and 19 inches. How can you use the three dowels to mark off a length of 5 inches?
13 + 19 − 27

Available on Daily Transparency in CRB

Math Fact

European writers of mathematics during medieval times and Renaissance times did not understand how the product of a fraction and a whole number could be less than the whole number.

4-9 Multiplying Fractions by Whole Numbers

Learn to multiply fractions by whole numbers.

Recall that multiplication by a whole number can be represented as repeated addition. For example, $4 \cdot 5 = 5 + 5 + 5 + 5$. You can multiply a whole number by a fraction using the same method.

$$3 \cdot \frac{1}{4} = \frac{1}{4} + \frac{1}{4} + \frac{1}{4} = \frac{3}{4}$$

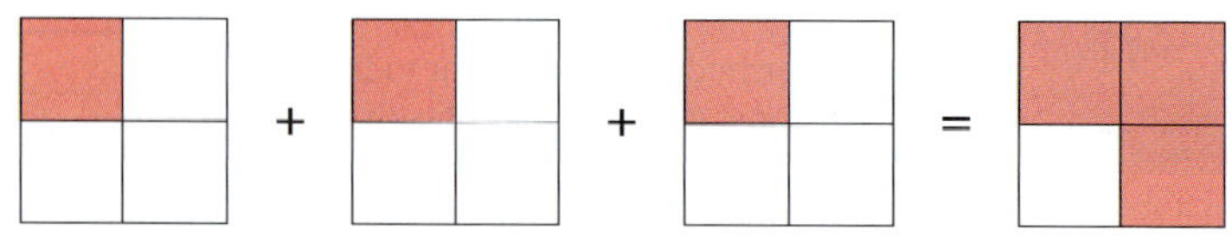

$$3 \cdot \frac{1}{4} = \frac{3}{4}$$

There is another way to multiply with fractions. Remember that a whole number can be written as an improper fraction with 1 in the denominator. So $3 = \frac{3}{1}$.

$$\frac{3}{1} \cdot \frac{1}{4} = \frac{3 \cdot 1}{1 \cdot 4} = \frac{3}{4}$$ ← Multiply numerators. ← Multiply denominators.

EXAMPLE 1 Multiplying Fractions and Whole Numbers

Multiply. Write each answer in simplest form.

A $5 \cdot \frac{1}{8}$

$\frac{5}{1} \cdot \frac{1}{8} = \frac{5 \cdot 1}{1 \cdot 8}$ *Write 5 as a fraction. Multiply numerators and denominators.*

$= \frac{5}{8}$

B $3 \cdot \frac{1}{9}$

$\frac{3}{1} \cdot \frac{1}{9} = \frac{3 \cdot 1}{1 \cdot 9}$ *Write 3 as a fraction. Multiply numerators and denominators.*

$= \frac{3}{9}$

$= \frac{1}{3}$ *Write your answer in simplest form.*

C $4 \cdot \frac{7}{8}$

$\frac{4}{1} \cdot \frac{7}{8} = \frac{4 \cdot 7}{1 \cdot 8}$ *Write 4 as a fraction. Multiply numerators and denominators.*

$= \frac{28}{8}$

$= \frac{7}{2}$ or $3\frac{1}{2}$ *Write your answer in simplest form.*

1 Introduce

Alternate Opener

EXPLORATION

4-9 Multiplying Fractions by Whole Numbers

Rosario requires $\frac{3}{4}$ of a 1-pound bag of clay to make one bowl. How many 1-pound bags of clay will she need to make a set of 6 bowls?

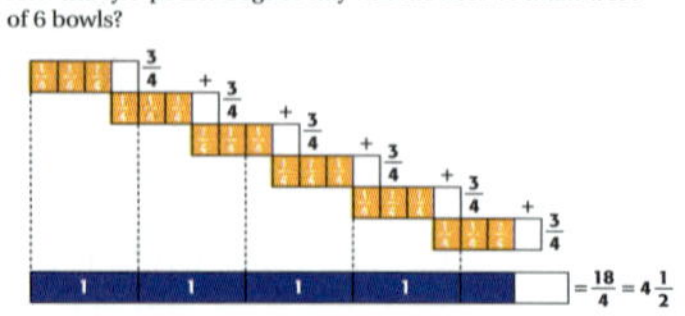

She will need $4\frac{1}{2}$ 1-pound bags of clay.

Draw a model to find each product.

1. $3 \cdot \frac{3}{4}$
2. $4 \cdot \frac{1}{2}$
3. $5 \cdot \frac{2}{3}$
4. $7 \cdot \frac{1}{4}$

Think and Discuss

5. **Explain** why $3 \cdot \frac{3}{4}$ is the same as $\frac{3}{1} \cdot \frac{3}{4} = \frac{3 \cdot 3}{1 \cdot 4}$.

Motivate

Write these two expressions: $\frac{2}{5} + \frac{2}{5} + \frac{2}{5}$ and $\frac{3}{4} + \frac{3}{4} + \frac{3}{4}$. Have a student evaluate each one. Tell the students that these expressions can be represented using multiplication, just as with expressions involving the addition of like whole numbers.

Exploration worksheet and answers on Chapter 4 Resource Book pp. 84 and 120

2 Teach

Lesson Presentation

Guided Instruction

In this lesson, students learn to multiply fractions by whole numbers. Remind students that a whole number can be written as a fraction with a denominator of 1. Teach students to find the product of two fractions by multiplying the numerators to get the new numerator and multiplying the denominators to get the new denominator. Then teach them to evaluate expressions and solve real-world problems in which a fraction and a whole number are multiplied.

EXAMPLE 2 **Evaluating Fraction Expressions**

Evaluate 6x for each value of x. Write each answer in simplest form.

A $x = \frac{1}{8}$

$6x$ — *Write the expression.*

$6 \cdot \frac{1}{8}$ — *Substitute $\frac{1}{8}$ for x.*

$\frac{6}{1} \cdot \frac{1}{8} = \frac{6}{8}$ — *Multiply.*

$= \frac{3}{4}$ — *Write your answer in simplest form.*

B $x = \frac{2}{3}$

$6x$ — *Write the expression.*

$6 \cdot \frac{2}{3}$ — *Substitute $\frac{2}{3}$ for x.*

$\frac{6}{1} \cdot \frac{2}{3} = \frac{12}{3}$ — *Multiply.*

$= \frac{4}{1}$

$= 4$

Sometimes the denominator of an improper fraction will divide evenly into the numerator, as in Example 2B. When this happens, the improper fraction is equivalent to a whole number, not a mixed number.

$$\frac{12}{3} = 4$$

This makes sense if you remember that the fraction bar means "divided by."

EXAMPLE 3 ***Social Studies Application***

Any proposed amendment to the U.S. Constitution must be ratified, or approved, by $\frac{3}{4}$ of the states. When the 13th Amendment abolishing slavery was proposed in 1865, there were 36 states. How many states needed to ratify this amendment in order for it to pass?

To find $\frac{3}{4}$ of 36, multiply.

$\frac{3}{4} \cdot 36 = \frac{3}{4} \cdot \frac{36}{1}$

$= \frac{108}{4}$

$= 27$

For the 13th Amendment to pass, 27 states had to ratify it.

Think and Discuss

1. **Describe** a model you could use to show the product of $4 \cdot \frac{1}{5}$.
2. **Choose** the expression that is correctly multiplied.

$2 \cdot \frac{3}{7} = \frac{6}{7}$ $\qquad$ $2 \cdot \frac{3}{7} = \frac{6}{14}$

COMMON ERROR ALERT

Because students just learned how to add and subtract like fractions, in this lesson some may multiply the numerator, leaving the denominator alone. When practical, have students use repeated addition to check their multiplication.

Additional Examples

Example 1

Multiply. Write your answers in simplest form.

A. $7 \cdot \frac{1}{9}$ $\frac{7}{9}$

B. $6 \cdot \frac{1}{8}$ $\frac{3}{4}$

C. $8 \cdot \frac{2}{3}$ $\frac{16}{3}$, or $5\frac{1}{3}$

Example 2

Evaluate 4x for each value of x. Write your answers in simplest form.

A. $x = \frac{1}{10}$ $\frac{2}{5}$ $\qquad$ **B.** $x = \frac{3}{8}$ $\frac{3}{2}$, or $1\frac{1}{2}$

Example 3

There are 25 students in the music club. Of those students, $\frac{3}{5}$ are also in the band. How many music club students are in the band?
15 students

3 Close

Reaching All Learners

Through Home Connection

Provide students with the following problem, provided on Teaching Transparency T33 in the Chapter 4 Resource Book.

A trail mix recipe called for the following ingredients to make one batch: $\frac{3}{4}$ cup oat cereal, $\frac{2}{3}$ cup wheat cereal, $\frac{1}{3}$ cup raisins, and $\frac{1}{4}$ cup mini pretzels. How much of each ingredient would be needed for 12 batches of the recipe? 9 cups oat cereal, 8 cups wheat cereal, 4 cups raisins, 3 cups mini pretzels

Summarize

Have students explain how multiplying fractions by whole numbers is different from adding and subtracting fractions with like denominators.

Possible answer: When multiplying fractions by whole numbers, you multiply the numerator by the whole number. When adding or subtracting fractions with like denominators, you only add or subtract the numerators, you do not multiply them.

Answers to Think and Discuss

1. Possible answer: 4 circles, each divided into fifths, with one-fifth of each circle shaded
2. The first expression is correct because the whole number was multiplied by the numerator. The second expression is incorrect because the whole number was multiplied by the denominator as well as the numerator.

4-9 PRACTICE & ASSESS

Students may want to refer back to the lesson examples.

Assignment Guide

If you finished Example 1 assign:
Core 1–8, 14–21, 50–58
Enriched 2–8 even, 14–20 even, 36–44, 50–58

If you finished Example 2 assign:
Core 1–12, 14–25, 50–58
Enriched 7–12, 14–25, 27–32, 50–58

If you finished Example 3 assign:
Core 1–26, 45–46, 50–58
Enriched 22–58

Notes

4-9 Exercises

FOR EXTRA PRACTICE see page 644

internet connect — Homework Help Online go.hrw.com Keyword: MR4 4-9

GUIDED PRACTICE

See Example 1 Multiply. Write each answer in simplest form.

1. $8 \cdot \frac{1}{9}$ $\frac{8}{9}$
2. $2 \cdot \frac{1}{5}$ $\frac{2}{5}$
3. $12 \cdot \frac{1}{4}$ 3
4. $7 \cdot \frac{4}{9}$ $3\frac{1}{9}$
5. $3 \cdot \frac{1}{7}$ $\frac{3}{7}$
6. $4 \cdot \frac{2}{11}$ $\frac{8}{11}$
7. $8 \cdot \frac{3}{4}$ 6
8. $18 \cdot \frac{1}{3}$ 6

See Example 2 Evaluate $12x$ for each value of x. Write the answer in simplest form.

9. $x = \frac{2}{3}$ 8
10. $x = \frac{1}{2}$ 6
11. $x = \frac{3}{4}$ 9
12. $x = \frac{5}{6}$ 10

See Example 3

13. The school Community Service Club has 45 members. Of these 45 members, $\frac{3}{5}$ are boys. How many boys are members of the Community Service Club? **27 boys**

INDEPENDENT PRACTICE

See Example 1 Multiply. Write each answer in simplest form.

14. $4 \cdot \frac{1}{10}$ $\frac{2}{5}$
15. $6 \cdot \frac{1}{8}$ $\frac{3}{4}$
16. $3 \cdot \frac{1}{12}$ $\frac{1}{4}$
17. $2 \cdot \frac{2}{5}$ $\frac{4}{5}$
18. $6 \cdot \frac{10}{11}$ $5\frac{5}{11}$
19. $2 \cdot \frac{3}{11}$ $\frac{6}{11}$
20. $15 \cdot \frac{2}{15}$ 2
21. $20 \cdot \frac{1}{2}$ 10

See Example 2 Evaluate $8x$ for each value of x. Write the answer in simplest form.

22. $x = \frac{1}{2}$ 4
23. $x = \frac{3}{4}$ 6
24. $x = \frac{1}{8}$ 1
25. $x = \frac{1}{4}$ 2

See Example 3

26. Kiesha spent 120 minutes completing her homework last night. Of those minutes, $\frac{1}{6}$ were spent on Spanish. How many minutes did Kiesha spend on her Spanish homework? **20 minutes**

PRACTICE AND PROBLEM SOLVING

Evaluate each expression. Write each answer in simplest form.

27. $12b$ for $b = \frac{7}{12}$ 7
28. $20m$ for $m = \frac{1}{20}$ 1
29. $33z$ for $z = \frac{5}{11}$ 15
30. $\frac{2}{3}y$ for $y = 18$ 12
31. $\frac{1}{4}x$ for $x = 20$ 5
32. $\frac{3}{5}a$ for $a = 30$ 18
33. $\frac{4}{5}c$ for $c = 12$ $\frac{48}{5}$ or $9\frac{3}{5}$
34. $14x$ for $x = \frac{3}{8}$ $\frac{21}{4}$ or $5\frac{1}{4}$
35. $\frac{9}{10}n$ for $n = 50$ 45

Compare. Write $<$, $>$, or $=$.

36. $9 \cdot \frac{1}{16}$ ■ $\frac{1}{2}$ $>$
37. $15 \cdot \frac{2}{5}$ ■ 5 $>$
38. $\frac{8}{13}$ ■ $4 \cdot \frac{2}{13}$ $=$
39. $3 \cdot \frac{2}{9}$ ■ $\frac{2}{3}$ $=$
40. $6 \cdot \frac{4}{15}$ ■ $\frac{11}{24}$ $>$
41. 5 ■ $12 \cdot \frac{3}{4}$ $<$
42. $3 \cdot \frac{1}{7}$ ■ $3 \cdot \frac{1}{5}$ $<$
43. $7 \cdot \frac{3}{4}$ ■ $6 \cdot \frac{3}{7}$ $>$
44. $2 \cdot \frac{5}{6}$ ■ $6 \cdot \frac{2}{5}$ $<$

Math Background

Before the seventeenth century, writers of mathematics generally did not try to explain the process of multiplication as it applies to fractions. One exception is Trenchant (1566), who used a square cut into smaller squares to illustrate multiplication.

RETEACH 4-9

LESSON 4-9 **Reteach**
Multiplying Fractions by Whole Numbers

You can use fraction strips to multiply fractions by whole numbers.

To find $3 \cdot \frac{2}{3}$, first think about the expression in words.

$3 \cdot \frac{2}{3}$ means "3 groups of $\frac{2}{3}$."

Then model the expression.

$\frac{1}{3}$	$\frac{1}{3}$		$\frac{1}{3}$	$\frac{1}{3}$		$\frac{1}{3}$	$\frac{1}{3}$

The total number of $\frac{1}{3}$ fraction pieces is 6.

So, $3 \cdot \frac{2}{3} = \frac{6}{3}$. $\frac{6}{3} = 2$ in simplest form.

Use fraction strips to find each product.

1. $4 \cdot \frac{1}{8}$ $\frac{1}{2}$
2. $2 \cdot \frac{2}{5}$ $\frac{4}{5}$
3. $6 \cdot \frac{1}{8}$ $\frac{3}{4}$
4. $8 \cdot \frac{1}{4}$ 2

You can also use counters to multiply fractions by whole numbers.

To find $\frac{1}{2} \cdot 12$, first think about the expression in words.

$\frac{1}{2} \cdot 12 = \frac{12}{2}$, which means "12 divided into 2 equal groups."

Then model the expression.

(○○○○○○)(○○○○○○)

The number of counters in each group is the product.

$\frac{1}{2} \cdot 12 = 6$.

Use counters to find each product.

5. $\frac{1}{3} \cdot 15$ 5
6. $\frac{1}{8} \cdot 24$ 3
7. $\frac{1}{4} \cdot 16$ 4
8. $\frac{1}{12} \cdot 24$ 2

PRACTICE 4-9

LESSON 4-9 **Practice B**
Multiplying Fractions by Whole Numbers

Multiply. Write your answers in simplest form.

1. $5 \cdot \frac{1}{10}$ $\frac{1}{2}$
2. $6 \cdot \frac{1}{18}$ $\frac{1}{3}$
3. $4 \cdot \frac{1}{14}$ $\frac{2}{7}$
4. $3 \cdot \frac{1}{12}$ $\frac{1}{4}$
5. $2 \cdot \frac{1}{8}$ $\frac{1}{4}$
6. $6 \cdot \frac{1}{10}$ $\frac{3}{5}$
7. $3 \cdot \frac{1}{6}$ $\frac{1}{2}$
8. $3 \cdot \frac{5}{12}$ $1\frac{1}{4}$
9. $3 \cdot \frac{2}{7}$ $\frac{6}{7}$
10. $2 \cdot \frac{3}{8}$ $\frac{3}{4}$
11. $10 \cdot \frac{3}{15}$ 2
12. $8 \cdot \frac{2}{14}$ $1\frac{1}{7}$
13. $5 \cdot \frac{2}{10}$ 1
14. $4 \cdot \frac{4}{12}$ $1\frac{1}{3}$
15. $2 \cdot \frac{13}{20}$ $1\frac{3}{10}$

Evaluate $6x$ for each value of x. Write your answers in simplest form.

16. $x = \frac{2}{3}$ 4
17. $x = \frac{2}{8}$ $1\frac{1}{2}$
18. $x = \frac{1}{4}$ $1\frac{1}{2}$
19. $x = \frac{2}{6}$ 2
20. $x = \frac{2}{7}$ $1\frac{5}{7}$
21. $x = \frac{2}{5}$ $2\frac{2}{5}$
22. $x = \frac{3}{11}$ $1\frac{7}{11}$
23. $x = \frac{5}{12}$ $2\frac{1}{2}$

24. Thomas spends 60 minutes exercising. For $\frac{1}{4}$ of that time, he jumps rope. How many minutes does he spend jumping rope? **15 minutes**

25. Kylie made a 4-ounce milk shake. Two-thirds of the milk shake was ice cream. How many ounces of ice cream did Kylie use in the shake? **$2\frac{2}{3}$ ounces**

Life Science LINK

The General Sherman, a giant sequoia tree in California's Sequoia National Park, is one of the largest trees in the world. Its weight is estimated to be equal to the combined weights of 740 elephants.

California also has the nation's tallest grand fir, ponderosa pine, and sugar pine. The table below shows how the heights of these three trees compare with the height of the General Sherman. For example, the height of the grand fir is $\frac{23}{25}$ the height of the General Sherman.

45. The General Sherman tree is 275 ft tall. Find the heights of the three trees in the table. Write your answers as whole numbers or as mixed numbers in simplest form. $253; 225\frac{1}{2}; 231$

Tree Heights Compared with the General Sherman	
Tallest Grand Fir	$\frac{23}{25}$
Tallest Ponderosa Pine	$\frac{41}{50}$
Tallest Sugar Pine	$\frac{21}{25}$

Source: *The Top 10 of Everything 2000*

46. The world's tallest bluegum eucalyptus tree is $\frac{3}{5}$ of the height of the General Sherman tree. How tall is this bluegum eucalyptus? 165 feet tall

47. ***WHAT'S THE QUESTION?*** California is also the location of Joshua Tree National Park. Joshua trees can grow to be 40 ft tall. The answer is $\frac{8}{55}$. What is the question?

48. ***WRITE ABOUT IT*** Find $\frac{1}{5}$ the height of the General Sherman. Then divide the height of the General Sherman by 5. What do you notice? Why does this make sense?

49. ***CHALLENGE*** The world's tallest incense cedar tree is 152 ft tall. What is $\frac{1}{5}$ of $\frac{1}{2}$ of $\frac{1}{4}$ of 152? $3\frac{4}{5}$

Spiral Review

Identify the base and the exponent. (Lesson 1-3)

50. 5^3 base 5, exponent 3
51. 4^8 base 4, exponent 8
52. 9^2 base 9, exponent 2
53. 12 base 12, exponent 1

Solve each equation. (Lesson 2-5)

54. $x - 25 = 40$ 65
55. $c - 18 = 20$ 38
56. $56 = d - 0$ 56
57. $e - 64 = 64$ 128

58. **TEST PREP** Which fraction is **not** equivalent to $\frac{1}{6}$? (Lesson 4-5) A

A $\frac{6}{1}$ **B** $\frac{2}{12}$ **C** $\frac{3}{18}$ **D** $\frac{6}{36}$

Interdisciplinary LINK

Life Science

Exercises 45–49 involve comparing the heights of extremely large trees with the height of the famous General Sherman sequoia tree in California. Plants are studied in middle school life science programs, such as *HOLT SCIENCE & TECHNOLOGY.*

Answers

47. Possible answer: The height of the Joshua tree is what fraction of the height of the General Sherman tree?

48. Possible answer: 55 ft; 55 ft; The answers are the same. One-fifth of a number is the same as dividing a number by 5.

Journal

Have students explain why the product of a fraction and a whole number can be a fraction, a whole number, or a mixed number.

Test Prep Doctor

Exercise 58 involves finding the non-equivalent fraction. Answer **A** is an improper fraction, so it cannot be equivalent to $\frac{1}{6}$. Answers **B, C,** and **D** are equal to $\frac{1}{6}$ when written in simplest form.

CHALLENGE 4-9

LESSON 4-9 **Challenge**
Slowpoke Race

The animals shown below are some of the slowest creatures on Earth. Use their given average speeds to find how far they will travel in the times marked along their racetracks.

Which of these slowpokes traveled the farthest? three-toed sloth

Three-toed sloth
Speed: $\frac{3}{5}$ mi/h

START	2 Hours	3 Hours	5 Hours
	$1\frac{1}{5}$ miles	$1\frac{4}{5}$ miles	3 miles

Earthworm
Speed: $\frac{1}{10}$ mi/h

START	2 Hours	3 Hours	5 Hours
	$\frac{1}{5}$ mile	$\frac{3}{10}$ mile	$\frac{1}{2}$ mile

Tortoise
Speed: $\frac{1}{5}$ mi/h

START	2 Hours	3 Hours	5 Hours
	$\frac{2}{5}$ mile	$\frac{3}{5}$ mile	1 mile

Snail
Speed: $\frac{3}{10}$ mi/h

START	2 Hours	3 Hours	5 Hours
	$\frac{3}{5}$ mile	$\frac{9}{10}$ mile	$1\frac{1}{2}$ miles

PROBLEM SOLVING 4-9

LESSON 4-9 **Problem Solving**
Multiplying Fractions by Whole Numbers

Write the answers in simplest form.

1. Did you know that some people have more bones than the rest of the population? About $\frac{1}{20}$ of all people have an extra rib bone. In a crowd of 60 people, about how many people are likely have an extra rib bone?
   3 people
2. The Appalachian National Scenic Trail is the longest marked walking path in the United States. It extends through 14 states for about 2,000 miles. Last year, Carla hiked $\frac{1}{5}$ of the trail. How many miles of the trail did she hike?
   400 miles
3. Human fingernails can grow up to $\frac{1}{10}$ of a millimeter each day. How much can fingernails grow in one week?
   $\frac{7}{10}$ millimeter
4. Most people dream about $\frac{1}{4}$ of the time they sleep. How long will you probably dream tonight if you sleep for 8 hours?
   2 hours
5. Today, the United States flag has 50 stars—one for each state. The first official U.S. flag was approved in 1795. It had $\frac{3}{10}$ as many stars as today's flag. How many stars were on the first official U.S. flag?
   15 stars
6. The Statue of Liberty is about 305 feet tall from the ground to the tip of her torch. The statue's pedestal makes up about $\frac{1}{2}$ of its height. About how tall is the pedestal of the Statue of Liberty?
   $152\frac{1}{2}$ feet

Circle the letter of the correct answer.

7. The Caldwells own a 60-acre farm. They planted $\frac{3}{5}$ of the land with corn. How many acres of corn did they plant?
   A 12 acres
   (B) 36 acres
   C 20 acres
   D 18 acres
8. Objects on Uranus weigh about $\frac{4}{5}$ of their weight on Earth. If a dog weighs 40 pounds on Earth, how much would it weigh on Uranus?
   (F) 32 pounds
   G 10 pounds
   H 8 pounds
   J 30 pounds

Lesson Quiz

Multiply. Write each answer in simplest form.

1. $10 \cdot \frac{2}{5}$ 4

2. $6 \cdot \frac{1}{10}$ $\frac{3}{5}$

Evaluate 6x for each value of x. Write your answer in simplest form.

3. $x = \frac{1}{4}$ $\frac{3}{2}$, or $1\frac{1}{2}$

4. $x = \frac{5}{6}$ 5

5. Alicia spent 15 minutes making a pizza. Of those minutes, $\frac{1}{3}$ were spent rolling out the crust. How many minutes did Alicia spend rolling out the crust? 5 minutes

Available on Daily Transparency in CRB

CHAPTER 4 Extension

Pacing: Traditional 1 day
Block $\frac{1}{2}$ day

Objective: Students make Venn diagrams to describe number sets.

Using the Pages

In Lesson 4-2, students wrote prime factorizations of composite numbers. In this extension, students make Venn diagrams to describe number sets. Students will identify elements in sets and use Venn diagrams to analyze relationships between those sets.

EXTENSION Sets of Numbers

Learn to make Venn diagrams to describe number sets.

Vocabulary
set
empty set
element
subset
Venn diagram
intersection
union

A group of items is called a **set**. The items in a set are called **elements**. In this chapter, you saw several sets of numbers, such as prime numbers, composite numbers, and factors.

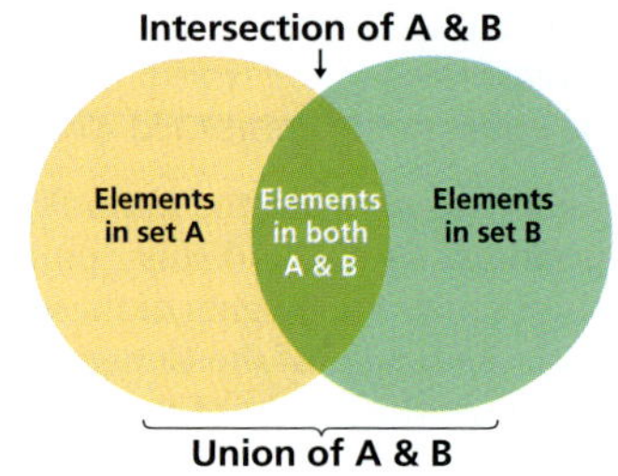

In a **Venn diagram**, circles are used to show relationships between sets. The overlapped region represents elements that are in both set A *and* set B. This set is called the **intersection** of A and B. Elements that are in set A *or* set B make up the **union** of A and B.

EXAMPLE 1 Identifying Elements and Drawing Venn Diagrams

Identify the elements in each set. Then draw a Venn diagram. What is the intersection? What is the union?

A **Set A: prime numbers** **Set B: composite numbers**
Elements of A: 2, 3, 5, 7, …
Elements of B: 4, 6, 8, 9, …

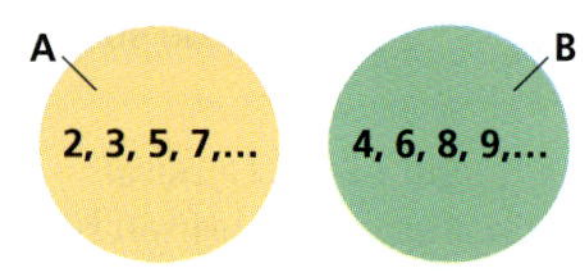

The circles do not overlap because no number is both prime and composite.

Intersection: none. When a set has no elements, it is called an **empty set**. The intersection of A and B is empty.

Union: all numbers that are prime *or* composite—all whole numbers except 0 and 1.

B **Set A: factors of 36** **Set B: factors of 24**
Elements of A: 1, 2, 3, 4, 6, 9, 12, 18, 36
Elements of B: 1, 2, 3, 4, 6, 8, 12, 24

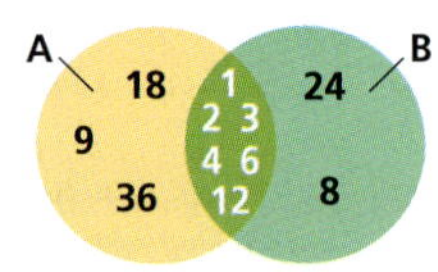

The circles overlap because some factors of 36 are also factors of 24.

Intersection: 1, 2, 3, 4, 6, 12 — *factors of 36* ***and*** *24*
Union: 1, 2, 3, 4, 6, 8, 9, 12, 18, 24, 36 — *factors of 36* ***or*** *24*

1 Introduce

Motivate

Discuss with students how they use graphic organizers in Language Arts, Science, and Social Studies.

Relate how circles, lines, and boxes are used to show the relationships among ideas in the composition of an effective essay.

In science, dot diagrams and three-dimensional drawings are used to represent the way molecules are formed. Lines and arrows are used to illustrate the water cycle and food chain.

In Social studies, graphic organizers such as time lines are used to represent the sequence of events, and organizational flow-charts are used to illustrate the structure of governments and the process a bill goes through to become law.

2 Teach

Lesson Presentation

Guided Instruction

In this extension, students learn to make Venn diagrams to describe number sets. First teach students to identify elements in two sets. Then teach students to draw a Venn diagram of the sets and find the union and intersection of those sets. Then explain that the empty set has no elements, the word *and* indicates intersection, and the word *or* indicates union. Finally, teach students to determine whether one set is a subset of another set.

Identify the elements in each set. Then draw a Venn diagram. What is the intersection? What is the union?

C Set A: factors of 36   Set B: factors of 12

Elements of A: 1, 2, 3, 4, 6, 9, 12, 18, 36

Elements of B: 1, 2, 3, 4, 6, 12

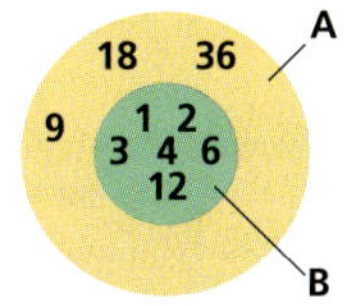

The circle for set B is entirely inside the circle for set A because all factors of 12 are also factors of 36.

Intersection: 1, 2, 3, 4, 6, 12   *factors of 36* ***and*** *12*

Union: 1, 2, 3, 4, 6, 9, 12, 18, 36   *factors of 36* ***or*** *12*

Helpful Hint

To decide whether set B is a subset of set A, ask yourself, "Is every element of B also in A?" If the answer is yes, then B is a subset of A.

Look at Example 1C. When one set is entirely contained in another set, we say the first set is a **subset** of the second set.

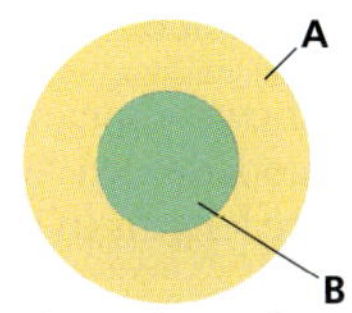

B is a subset of A.

EXTENSION Exercises

Identify the elements in each set. Then draw a Venn diagram. What is the intersection? What is the union?

1. Set A: even numbers
   Set B: odd numbers
2. Set A: factors of 18
   Set B: factors of 40
3. Set A: factors of 72
   Set B: factors of 36
4. Set A: even numbers
   Set B: composite numbers

Tell whether set A is a subset of set B.

5. Set A: whole numbers less than 10
   Set B: whole numbers less than 12
   yes
6. Set A: whole numbers less than 8
   Set B: whole numbers greater than 9
   no
7. Set A: prime numbers
   Set B: odd numbers
   no
8. Set A: numbers divisible by 6
   Set B: numbers divisible by 3
   yes
9. ***WRITE ABOUT IT*** How could you use a Venn diagram to help find the greatest common factor of two numbers? Give an example.
10. ***CHALLENGE*** How could you use a Venn diagram to help find the greatest common factor of three numbers? Give an example.

Additional Examples

Example 1

Identify the elements in each set. Then draw a Venn diagram. What is the intersection? What is the union?

A. Set A: whole numbers
Set B: even numbers

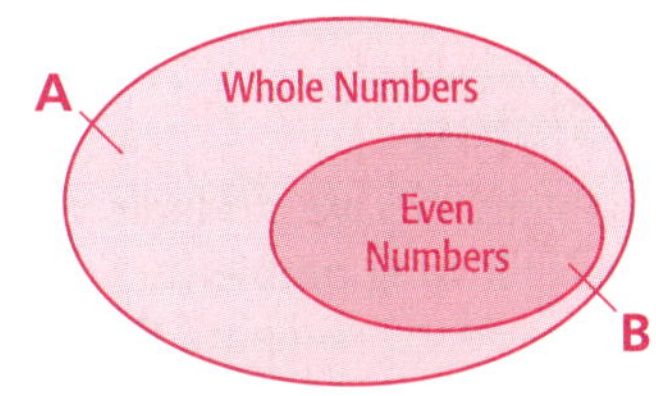

intersection: even number
union: whole numbers

B. Set A: factors of 12
Set B: factors of 18

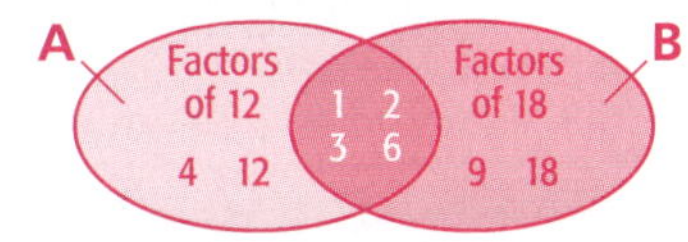

intersection: 1, 2, 3, 6
union: 1, 2, 3, 4, 6, 9, 12, 18

C. Set A: factors of 10
Set B: factors of 7

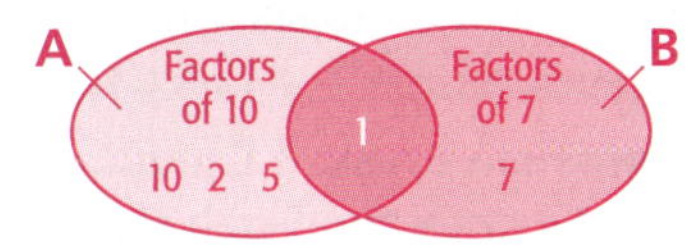

intersection: 1
union: 1, 2, 5, 7, 10

Answers

1–4, 9–10. See p. A2–A3.

3 Close

Summarize

Have the students explain how to find the union and intersection of the set of factors of 6 and the set of factors of 12. Then have students explain whether one of those sets is a subset of the other.

Possible answer: The set of factors of 6 is 1, 2, 3, 6. The set of factors of 12 is 1, 2, 3, 4, 6, 12. The union is all the factors of both 6 and 12. The intersection is only the factors that appear in the diagrams for both 6 and 12. The union is 1, 2, 3, 4, 6, 12. The intersection is 1, 2, 3, 6. Because all the factors of 6 are factors of 12, the factors of six are a subset of the factors of 12.

Teaching Tip

Remind students that in Venn diagrams, *or* means one or the other or both.

Problem Solving on Location

North Carolina

Purpose: *To provide additional practice for problem-solving skills in Chapters 1–4*

Snow Skiing

- After problem 1, have students consider the following problem: Why is it easier to compare the lengths of Snowbird Trail and Rock Island Run than the Lower Omigosh and the Upper Omigosh? The lengths of Snowbird Trail and Rock Island Run have the same denominator, so to compare them you have to compare only the numerators. The lengths of Lower Omigosh and Upper Omigosh have different denominators, so you must find a common denominator in order to compare them.
- After problem 3, ask: How many times do you have to ski Upper Omigosh in order to ski 1 mile? $\frac{1}{5} \cdot 5 = 1$; 5 times
- After problem 6, ask: If Enrique skied Rabbit Hill once and Sarah skied Rock Island Run two times, who skied farther? How much farther? Enrique skied $\frac{1}{9}$ mi farther.

Extension Have students design a ski area. Have them name and label the slopes and write questions for a classmate to solve. Check students' work.

Problem Solving on Location

NORTH CAROLINA

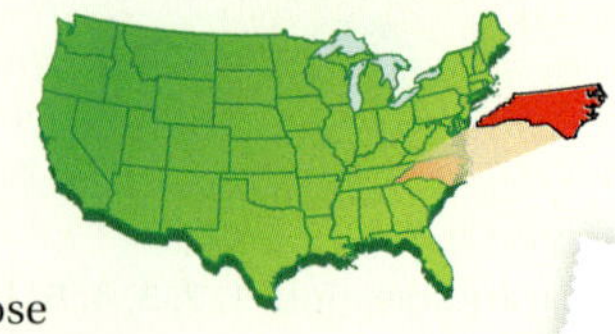

Snow Skiing

Cataloochee Ski Area, in Maggie Valley, North Carolina, has ten ski slopes. Of those ten slopes, $\frac{1}{4}$ of them are for beginners, $\frac{1}{2}$ are for intermediate-level skiers, and $\frac{1}{4}$ are reserved for advanced skiers.

The table shows some of the slopes of the Cataloochee Ski Area and their approximate lengths in miles.

Cataloochee Ski Area Trails

Slope	Length (mi)
Upper Omigosh	$\frac{1}{5}$
Lower Omigosh	$\frac{1}{3}$
Snowbird Trail	$\frac{2}{9}$
Easy Way	$\frac{1}{9}$
Alley Cat	$\frac{1}{6}$
Rabbit Hill	$\frac{1}{3}$
Rock Island Run	$\frac{1}{9}$

For 1–6, use the table.

1. Which is longer, Lower Omigosh or Upper Omigosh? **Lower Omigosh**
2. If you skied Snowbird Trail, Easy Way, and Rock Island Run, how many total miles would you have skied? Write your answer in simplest form. **$\frac{4}{9}$ mi**
3. How many Easy Way slopes would it take to equal the length of one Snowbird Trail slope? **2**
4. If you skied Alley Cat twice, you would have skied a distance equal to the lengths of what other trails? **Lower Omigosh and Rabbit Hill**
5. Write the ski slopes in order from shortest to longest.
6. Raul skied the Snowbird Trail slope, and April skied the Lower Omigosh slope. Who skied farther? **April**

5. Easy Way, Rock Island Run, Alley Cat, Upper Omigosh, Snowbird Trail, Lower Omigosh, Rabbit Hill (Note: Easy Way and Rock Island Run are the same distance, as are Lower Omigosh and Rabbit Hill—they may appear in reverse order.)

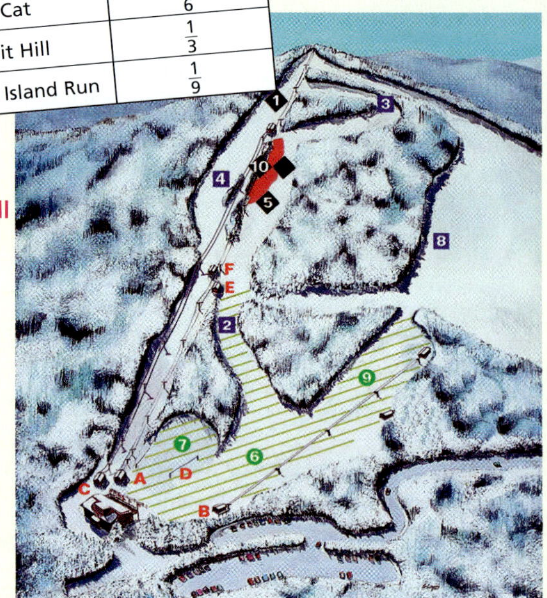

Lighthouses

The $63\frac{2}{5}$-meter-tall Cape Hatteras Lighthouse is the tallest lighthouse in the United States. Located in Buxton, North Carolina, the tower was built of 1.2 million bricks, and its walls are 13 feet thick.

The Cape Hatteras Lighthouse sits over the dangerous shallow sandbars of Diamond Shoals. It projects light more than 50 miles into the Atlantic Ocean, warning travelers away from the area known as the graveyard of the Atlantic.

The following table lists the approximate heights in meters of six operating lighthouses in North Carolina.

For 1–4, use the table.

1. Write the heights of the Bodie Island lighthouse, the Cape Hatteras lighthouse, and the Currituck Beach lighthouse as decimals.
2. Which two lighthouses are the same height?
3. Which lighthouse is the shortest?
4. Write the lighthouses in order from tallest to shortest.
5. North Carolina has 300 miles of coastline. Is it possible for the six lighthouses to be evenly spaced along the coastline so that one lighthouse is at one end and one is at the other end? Explain.

North Carolina Lighthouses

Lighthouse	Height (m)
Bodie Island (Pea Island)	$27\frac{11}{25}$
Cape Hatteras	$63\frac{2}{5}$
Cape Lookout	$51\frac{14}{125}$
Currituck Beach	$49\frac{1}{3}$
Oak Island	$51\frac{14}{125}$
Ocracoke Island	$23\frac{4}{25}$

Lighthouses

- After problem 1, discuss the following: Which of the lighthouses' heights can be converted to decimal form using mental math? Explain. Bodie Island, Cape Hatteras, and Ocracoke Island have heights with denominators that are factors of 100. To convert them to decimals, find the missing factor and multiply the numerator and the denominator by it. For example, for Bodie Island, $\frac{11}{25} \times \frac{4}{4} = \frac{44}{100} = 0.44$. So the Bodie Island lighthouse is 27.44 m tall.
- After problem 3, ask students whether they need to look at the fraction portion of the heights to answer the question. Explain. No; the heights can be compared by looking only at the whole number parts, since the least whole number part is 23, and there is only one height with 23 as its whole number part.

Extension Encourage students to write a problem about North Carolina's lighthouses using the data in the table. Have them challenge a classmate to solve their problem. Check students' work.

Answers

1. Bodie Island: 27.44 m
   Cape Hatteras: 63.4 m
   Currituck Beach: $49.\overline{3}$ m
2. Cape Lookout and Oak Island
3. Ocracoke Island
4. Cape Hatteras, Cape Lookout, Oak Island, Currituck Beach, Bodie Island, Ocracoke Island (Note: Cape Lookout and Oak Island are the same length; they may appear in reverse order.)
5. Yes; if the first lighthouse is at mile 0, then the remaining 5 lighthouses would be spaced evenly over the 300 miles. 300 is divisible by 5, so the lighthouses would be 60 miles apart.

MATH-ABLES

Game Resources

Puzzles, Twisters & Teasers
Chapter 4 Resource Book

Riddle Me This

Purpose: *To apply divisibility rules to solving a riddle*

Discuss: Ask students to explain the divisibility rule for 3. Find the sum of the digits in the number. If the sum is divisible by 3, then the number is divisible by 3.

Extend: Challenge students to create a grid similar to the one in this activity. Have them design the grid so that each square contains a number that a classmate must check for divisibility and must shade according to the answer. Have the results spell out a word or message. Check students' work.

On a Roll

Purpose: *To use knowledge of fractions to play a game*

Discuss: When a student has completed the game board, have the student explain how he or she knows that the squares have been filled in correctly.

Extend: Have students create a new game board using different expressions in each square. For example, they could use the following:

$\frac{\square}{\square} + \frac{\square}{\square} = \frac{\square}{\square}$

Then have them play the game using their new game boards.

MATH-ABLES

Riddle Me This

"When you go from there to here,
you'll find I disappear.
Go from here to there, and then
you'll see me again.
What am I?" **the letter *T***

To solve this riddle, copy the square below. If a number is divisible by 3, color that box red. Remember the divisibility rule for 3. If a number is not divisible by 3, color that box blue.

102	981	210	6,015	72
79	1,204	576	10,019	1,771
548	3,416	12,300	904	1,330
217	2,662	1,746	3,506	15,025
34,351	725	2,352	5,675	6,001

On a Roll

The object is to be the first person to fill in all the squares on your game board.

On your turn, roll a number cube and record the number rolled in any blank square on your game board. Once you have placed a number in a square, you cannot move that number. If you cannot place the number in a square, then your turn is over. The winner is the first player to complete their game board correctly.

internet connect
Go to ***go.hrw.com*** for a complete set of rules and game pieces.
KEYWORD: MR4 Game4

Technology LAB Greatest Common Factor

internet connect
Lab Resources Online
go.hrw.com
KEYWORD: MR4 TechLab4

You can use a graphing calculator to quickly find the greatest common factor (GCF) of two or more numbers. A calculator is particularly useful when you need to find the GCF of large numbers.

Activity

1 Find the GCF of 504 and 3,150.

The GCF is also known as the *greatest common divisor,* or GCD. The GCD function is found on the **MATH** menu.

To find the GCD on a graphing calculator, press MATH. Press ▶ to highlight NUM, and then use ▼ to scroll down and highlight 9:.

Press ENTER 504 , 3150) ENTER.

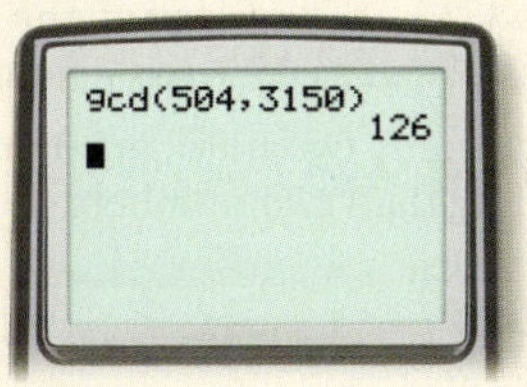

The greatest common factor of 504 and 3,150 is 126.

Think and Discuss

1. Suppose your calculator will not allow you to enter three numbers into the GCD function. How could you still use your calculator to find the GCF of the three following numbers: 4,896; 2,364; and 656? Explain your strategy and why it works. **1. Possible answer: Find the GCF of 4,896 and 2,364. Then find the GCF of 2,364 and 656, and then the GCF of 656 and 4,896. The greatest common factor of these three numbers (12, 4, and 16) would be the greatest common factor of 4,896, 2,364, and 656; in this case, it is 4.**
2. Would you use your calculator to find the GCF of 6 and 18? Why or why not? **Possible answer: No, because it can be done quickly mentally.**

Try This

Find the GCF of each set of numbers.

1. 14, 48 **2**	**2.** 18, 54 **18**	**3.** 99, 121 **11**	**4.** 144, 196 **4**
5. 200, 136 **8**	**6.** 246, 137 **1**	**7.** 72, 860 **4**	**8.** 55, 141, 91 **1**

Technology LAB Greatest Common Factor

Objective: To use a graphing calculator to find the GCF

Materials: Graphing calculator

Lab Resources

Technology Lab Activities . . . . p. 15

Using the Page

This technology activity shows students how to use a graphing calculator to find the greatest common factor of two or more numbers. Specific keystrokes may vary, depending on the make and model of the graphing calculator used. The keystrokes given are for a TI-83 model. For keystrokes to other models, visit go.hrw.com.

The Think and Discuss problems can be used to assess students' understanding of the technology activity. While several of the Try This problems can be done without a graphing calculator, they are meant to help students become familiar with using a graphing calculator to find the GCF.

Assessment

1. What should you enter into the calculator when the display shows **gcd(**? the numbers whose greatest common divisor you are trying to find, separated by a comma
2. Find the greatest common factor of 48 and 88. 8
3. What is the greatest common divisor of 288, 200, and 212? 4

Chapter 4 **Study Guide and Review**

Purpose: *To help students review and practice concepts and skills presented in Chapter 4*

Assessment Resources

Chapter Review

Chapter 4 Resource Book . . . pp. 92–94

***Test and Practice Generator* CD-ROM**

Additional review assessment items in both multiple-choice and free-response format may be generated for any objective in Chapter 4.

Answers

1. improper fraction; mixed number
2. repeating decimal; terminating decimal
3. prime number; composite number
4. 2
5. 2, 3, 5, 6, 9, 10
6. 2, 3, 6, 9
7. 2, 4
8. 2, 5, 10
9. 3
10. composite
11. composite
12. prime
13. composite
14. prime
15. composite
16. composite
17. prime
18. composite
19. prime

Chapter 4 Study Guide and Review

Study Guide and Review

Vocabulary

common denominator	179	mixed number	167
composite number	153	prime factorization	156
divisible	152	prime number	153
equivalent fractions	172	proper fraction	182
factors	156	repeating decimal	168
greatest common factor (GCF)	160	simplest form	173
improper fraction	182	terminating decimal	168
like fractions	178	unlike fractions	179

Complete the sentences below with vocabulary words from the list above. Words may be used more than once.

1. The number $\frac{11}{9}$ is an example of a(n) ___?___, and $3\frac{1}{6}$ is an example of a(n) ___?___.
2. A(n) ___?___, such as 0.3333..., has a block of one or more digits that repeat without end. A(n) ___?___, such as 0.25, has a finite number of decimal places.
3. A(n) ___?___ is divisible by only two numbers, 1 and itself. A(n) ___?___ is divisible by more than two numbers.

4-1 Divisibility (pp. 152–155)

EXAMPLE

■ **Tell whether 210 is divisible by 2, 3, 4, and 6.**

2	*The last digit, 0, is even.*	Divisible
3	*The sum of the digits is divisible by 3.*	Divisible
4	*The number formed by the last two digits is not divisible by 4.*	Not divisible
6	*210 is divisible by 2 and 3.*	Divisible

■ **Tell whether each number is prime or composite.**

17 *only divisible by 1 and 17* prime
25 *divisible by 1, 5, and 25* composite

EXERCISES

Tell whether each number is divisible by 2, 3, 4, 5, 6, 9, and 10.

4. 118
5. 90
6. 342
7. 284
8. 170
9. 393

Tell whether each number is prime or composite.

10. 121
11. 77
12. 13
13. 118
14. 67
15. 93
16. 39
17. 97
18. 85
19. 61

4-2 Factors and Prime Factorization (pp. 156–159)

EXAMPLES

- **List all the factors of 10.**

  $10 = 1 \cdot 10 \qquad 10 = 2 \cdot 5$

  The factors of 10 are 1, 2, 5, and 10.

- **Write the prime factorization of 30.**

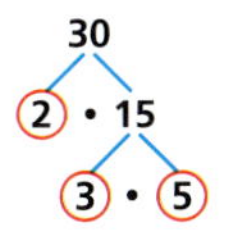

  $30 = 2 \cdot 3 \cdot 5$

EXERCISES

List all the factors of each number.

20. 60 **21.** 72
22. 29 **23.** 56
24. 85 **25.** 71

Write the prime factorization of each number.

26. 65 **27.** 94 **28.** 110
29. 81 **30.** 99 **31.** 76
32. 97 **33.** 55 **34.** 46

4-3 Greatest Common Factor (pp. 160–163)

EXAMPLE

- **Find the GCF of 35 and 50.**

  factors of 35: 1, (5), 7, 35
  factors of 50: 1, 2, (5), 10, 25, 50

  The GCF of 35 and 50 is 5.

EXERCISES

Find the GCF of each set of numbers.

35. 36 and 60
36. 50, 75, and 125
37. 45, 81, and 99

4-4 Decimals and Fractions (pp. 167–170)

EXAMPLES

- **Write 1.29 as a mixed number.**

  $1.29 = 1\frac{29}{100}$

- **Write $\frac{3}{5}$ as a decimal.**

  $5\overline{)3.0}$ = 0.6 $\qquad \frac{3}{5} = 0.6$

EXERCISES

Write as a fraction or mixed number.

38. 0.37 **39.** 1.8 **40.** 0.4

Write as a decimal.

41. $\frac{7}{8}$ **42.** $\frac{2}{5}$ **43.** $\frac{7}{9}$

4-5 Equivalent Fractions (pp. 172–175)

EXAMPLES

- **Find an equivalent fraction for $\frac{4}{5}$.**

  $\frac{4}{5} = \frac{\blacksquare}{15} \qquad \frac{4 \cdot 3}{5 \cdot 3} = \frac{12}{15}$

- **Write $\frac{8}{12}$ in simplest form.**

  $\frac{8 \div 4}{12 \div 4} = \frac{2}{3}$

EXERCISES

Find two equivalent fractions.

44. $\frac{4}{6}$ **45.** $\frac{4}{5}$ **46.** $\frac{3}{12}$

Write each fraction in simplest form.

47. $\frac{14}{16}$ **48.** $\frac{9}{30}$ **49.** $\frac{7}{10}$

Answers

20. 1, 2, 3, 4, 5, 6, 10, 12, 15, 20, 30, 60
21. 1, 2, 3, 4, 6, 8, 9, 12, 18, 24, 36, 72
22. 1, 29
23. 1, 2, 4, 7, 8, 14, 28, 56
24. 1, 5, 17, 85
25. 1, 71
26. $5 \cdot 13$
27. $2 \cdot 47$
28. $2 \cdot 5 \cdot 11$
29. 3^4
30. $3^2 \cdot 11$
31. $2^2 \cdot 19$
32. 97
33. $5 \cdot 11$
34. $2 \cdot 23$
35. 12
36. 25
37. 9
38. $\frac{37}{100}$
38. $1\frac{4}{5}$
40. $\frac{2}{5}$
41. 0.875
42. 0.4
43. $0.\overline{7}$
44. Possible answer: $\frac{2}{3}, \frac{8}{12}$
45. Possible answer: $\frac{8}{10}, \frac{16}{20}$
46. Possible answer: $\frac{1}{4}, \frac{2}{8}$
47. $\frac{7}{8}$
48. $\frac{3}{10}$
49. $\frac{7}{10}$

Answers

50. >

51. >

52. $\frac{3}{8}, \frac{2}{3}, \frac{7}{8}$

53. $\frac{3}{12}, \frac{1}{3}, \frac{4}{6}$

54. $\frac{34}{9}$

55. $\frac{29}{12}$

56. $\frac{37}{7}$

57. $3\frac{5}{6}$

58. $3\frac{2}{5}$

59. $5\frac{1}{8}$

60. 1

61. $\frac{3}{4}$

62. $\frac{3}{5}$

63. $6\frac{5}{7}$

64. $\frac{5}{7}$

65. $\frac{3}{4}$

66. $2\frac{4}{7}$

67. $\frac{8}{9}$

Study Guide and Review

4-6 Comparing and Ordering Fractions (pp. 178–181)

EXAMPLE

- Order from least to greatest.

$\frac{3}{5}, \frac{2}{3}, \frac{1}{3}$ *Rename with like denominators.*

$\frac{3 \cdot 3}{5 \cdot 3} = \frac{9}{15}$ $\frac{2 \cdot 5}{3 \cdot 5} = \frac{10}{15}$ $\frac{1 \cdot 5}{3 \cdot 5} = \frac{5}{15}$

$\frac{1}{3}, \frac{3}{5}, \frac{2}{3}$

EXERCISES

Compare. Write <, >, or =.

50. $\frac{6}{8}$ ■ $\frac{3}{8}$ **51.** $\frac{7}{9}$ ■ $\frac{2}{3}$

Order from least to greatest.

52. $\frac{3}{8}, \frac{2}{3}, \frac{7}{8}$ **53.** $\frac{4}{6}, \frac{3}{12}, \frac{1}{3}$

4-7 Mixed Numbers and Improper Fractions (pp. 182–185)

EXAMPLE

- Write $3\frac{5}{6}$ as an improper fraction.

$3\frac{5}{6} = \frac{(3 \cdot 6) + 5}{6} = \frac{18 + 5}{6} = \frac{23}{6}$

- Write $\frac{19}{4}$ as a mixed number.

$4)\overline{19}$ = 4R3 $\frac{19}{4} = 4\frac{3}{4}$

EXERCISES

Write as an improper fraction.

54. $3\frac{7}{9}$ **55.** $2\frac{5}{12}$ **56.** $5\frac{2}{7}$

Write as a mixed number.

57. $\frac{23}{6}$ **58.** $\frac{17}{5}$ **59.** $\frac{41}{8}$

4-8 Adding and Subtracting with Like Denominators (pp. 188–191)

EXAMPLE

- Subtract $4\frac{5}{6} - 2\frac{1}{6}$. Write your answer in simplest form.

$4\frac{5}{6} - 2\frac{1}{6} = 2\frac{4}{6} = 2\frac{2}{3}$

EXERCISES

Add or subtract. Write each answer in simplest form.

60. $\frac{1}{5} + \frac{4}{5}$ **61.** $1 - \frac{3}{12}$

62. $\frac{9}{10} - \frac{3}{10}$ **63.** $4\frac{2}{7} + 2\frac{3}{7}$

4-9 Multiplying Fractions by Whole Numbers (pp. 192–195)

EXAMPLE

- Multiply $3 \cdot \frac{3}{5}$. Write your answer in simplest form.

$3 \cdot \frac{3}{5} = \frac{3}{1} \cdot \frac{3}{5} = \frac{3 \cdot 3}{1 \cdot 5} = \frac{9}{5}$ or $1\frac{4}{5}$

EXERCISES

Multiply. Write each answer in simplest form.

64. $5 \cdot \frac{1}{7}$ **65.** $2 \cdot \frac{3}{8}$

66. $3 \cdot \frac{6}{7}$ **67.** $4 \cdot \frac{2}{9}$

Chapter Test

Chapter 4

Tell whether each number is divisible by 2, 3, 4, 5, 6, 9, and 10.

1. 384 **2, 3, 4, 6** **2.** 815 **5** **3.** 724 **2, 4** **4.** 624 **2, 3, 4, 6**

List all the factors of each number. Then tell whether each number is prime or composite.

5. 98 **1, 2, 7, 14, 49, 98; composite** **6.** 40 **1, 2, 4, 5, 8, 10, 20, 40; composite** **7.** 45 **1, 3, 5, 9, 15, 45; composite** **8.** 41 **1, 41; prime**

Write the prime factorization of each number.

9. 64 $2 \times 2 \times 2 \times 2 \times 2 \times 2$ **10.** 130 $2 \times 5 \times 13$ **11.** 49 7×7 **12.** 28 $2 \times 2 \times 7$

Find the GCF of each set of numbers.

13. 24 and 108 **12** **14.** 45, 18, and 39 **3** **15.** 49, 77, and 84 **7**

Write each decimal as a fraction or mixed number.

16. 0.37 $\frac{37}{100}$ **17.** 1.9 $1\frac{9}{10}$ **18.** 0.92 $\frac{23}{25}$ **19.** 5.03 $5\frac{3}{100}$

Write each fraction or mixed number as a decimal.

20. $\frac{3}{8}$ **0.375** **21.** $9\frac{3}{5}$ **9.6** **22.** $\frac{2}{3}$ $0.\overline{6}$ **23.** $2\frac{1}{8}$ **2.125**

Write each fraction in simplest form.

24. $\frac{4}{12}$ $\frac{1}{3}$ **25.** $\frac{6}{9}$ $\frac{2}{3}$ **26.** $\frac{3}{15}$ $\frac{1}{5}$ **27.** $\frac{7}{8}$ $\frac{7}{8}$

Write each mixed number as an improper fraction.

28. $4\frac{7}{8}$ $\frac{39}{8}$ **29.** $7\frac{5}{12}$ $\frac{89}{12}$ **30.** $3\frac{5}{7}$ $\frac{26}{7}$ **31.** $1\frac{8}{11}$ $\frac{19}{11}$

Compare. Write <, >, or =.

32. $\frac{5}{6}$ ■ $\frac{3}{6}$ **>** **33.** $\frac{3}{4}$ ■ $\frac{7}{8}$ **<** **34.** $\frac{4}{5}$ ■ $\frac{7}{10}$ **>**

Order the fractions and decimals from least to greatest.

35. 2.17, 2.3, $2\frac{1}{9}$ $2\frac{1}{9}$**, 2.17, 2.3** **36.** 0.1, $\frac{3}{8}$, 0.3 **0.1, 0.3,** $\frac{3}{8}$ **37.** 0.9, $\frac{2}{8}$, 0.35 $\frac{2}{8}$**, 0.35, 0.9**

38. On Monday, it snowed $2\frac{1}{4}$ inches. On Tuesday, an additional $3\frac{3}{4}$ inches of snow fell. How much snow fell altogether on Monday and Tuesday? **6 inches**

Multiply. Write each answer in simplest form.

39. $4 \cdot \frac{1}{3}$ $1\frac{1}{3}$ **40.** $2 \cdot \frac{3}{8}$ $\frac{3}{4}$ **41.** $\frac{1}{4} \cdot 14$ $3\frac{1}{2}$

Chapter Test

Chapter Test

Chapter 4

Purpose: *To assess students' mastery of concepts and skills in Chapter 4*

Assessment Resources

Chapter Tests (Levels A, B, C,)
Assessment Resources ..... pp. 49–54

***Test and Practice Generator* CD-ROM**

Additional assessment items in both multiple-choice and free-response format may be generated for any objective in Chapter 4.

Chapter 4 Performance Assessment

Purpose: *To assess students' understanding of concepts in Chapter 4 and combined problem-solving skills*

Assessment Resources

Performance Assessment
Assessment Resources . . . . p. 110

Performance Assessment Teacher Support
Assessment Resources . . . . p. 109

Answers

1–3. See p. A3.

4. See Level 3 work sample below.

Scoring Rubric for Problem Solving Item 4

Level 3

Accomplishes the purposes of the task.

Student gives clear explanations, shows understanding of mathematical ideas and processes, and computes accurately.

Level 2

Purposes of the task not fully achieved.

Student demonstrates satisfactory but limited understanding of the mathematical ideas and processes.

Level 1

Purposes of the task not accomplished.

Student shows little evidence of understanding the mathematical ideas and processes and makes computational and/or procedural errors.

Chapter 4 Performance Assessment

Performance Assessment

Show What You Know

Create a portfolio of your work from this chapter. Complete this page and include it with your four best pieces of work from Chapter 4. Choose from your homework or lab assignments, mid-chapter quizzes, or any journal entries you have done. Put them together using any design you want. Make your portfolio represent what you consider your best work.

Short Response

1. In Mrs. Matika's class, there are 9 girls and 15 boys. Mrs. Matika wants to divide the class into groups for a project. Each group should have the same number of boys and the same number of girls. What is the greatest number of groups she can make if every student is put in a group? Explain how you determined your answer.

2. Kerry, Janice, Marcos, and Carl ordered a pizza for dinner. Kerry ate $\frac{1}{8}$ of the pizza, Janice ate $\frac{3}{8}$ of the pizza, and Carl ate $\frac{2}{8}$ of the pizza. If there were no leftovers, how much pizza did Marcos eat? Show your work.

3. Find the value of the expression $1\frac{3}{5} + 2\frac{4}{5}$. Write the answer as a mixed number, an improper fraction, and a decimal. Show your work.

Extended Problem Solving

4. Trent plans to purchase the rug in the photograph to place in his den. The rug is 3 yards wide and $5\frac{1}{2}$ yards long.

a. Find the perimeter of the rug by adding the four side measures. Show your work.

b. Find the area of the rug by multiplying the length and width. Write your answer in simplest form.

c. To determine if the rug will fit, Trent measures his den. It is 144 inches wide and 168 inches long. He calculates the area of the den as $18\frac{2}{3}$ square yards. Since the area of the den is greater than the area of the rug, Trent decides to purchase the rug. Do you agree with Trent's decision? Explain your answer.

Student Work Samples for Item 4

Level 3

a. $3 + 5\frac{1}{2} + 3 + 5\frac{1}{2} = 17$ yd

b. $3 \times 5\frac{1}{2} = 16\frac{1}{2}$ yd²

c. The width of the rug is 108 inches which is longer than the length of the den, but the length of the rug is 198 inches which is longer than the length of the den. Trent should not buy the rug.

The student correctly answered parts a and b. The student showed understanding in the explanation for part c.

Level 2

a. $3+3+5\frac{1}{2}+5\frac{1}{2} = 17$ yd

b. $3 \times 5\frac{1}{2} = 15\frac{1}{2}$ yd²

c. Trent should not buy the rug because it will not fit in the room.

The student correctly answered part a, but made an incorrect calculation in part b. The student should explain why the rug will not fit.

Level 1

A. $3 + 5\frac{1}{2} + 5\frac{1}{2} = 14$ yd

B. $3 \times 5\frac{1}{2} = \frac{15}{2} = 7\frac{1}{2}$ yd²

C. TRENT SHOULD BUY THE RUG BECAUSE $7\frac{1}{2}$ IS LESS THAN $18\frac{2}{3}$.

The student added the length of three sides in part a, incorrectly multiplied in part b, and does not understand the concept of area.

internet connect
State-Specific Test Practice Online
go.hrw.com Keyword: MR4 TestPrep

Standardized Test Prep — Chapter 4

Cumulative Assessment, Chapters 1–4

1. Which is a prime number? **C**
   - Ⓐ 9
   - Ⓑ 39
   - Ⓒ 47
   - Ⓓ 51

2. Which number is greatest? **H**
   - Ⓕ 7.056
   - Ⓖ 7.06
   - Ⓗ $7\frac{3}{5}$
   - Ⓙ $7\frac{1}{2}$

3. What is the value of $5^2 \times 3 + 6 - 2^2$? **A**
   - Ⓐ 77
   - Ⓑ 221
   - Ⓒ 125
   - Ⓓ 19

4. Which fraction is **not** equivalent to $\frac{4}{6}$? **J**
   - Ⓕ $\frac{2}{3}$
   - Ⓖ $\frac{10}{15}$
   - Ⓗ $\frac{8}{12}$
   - Ⓙ $\frac{16}{18}$

5. The bar graph shows the four most common kinds of insects and the approximate number of known species of each. Which is the number of true flies written in scientific notation? **C**

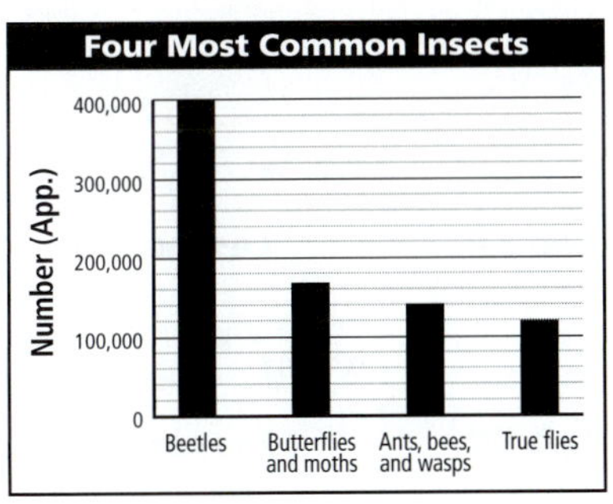

   - Ⓐ 16.5×10^4
   - Ⓑ 1.4×10^5
   - Ⓒ 1.2×10^5
   - Ⓓ 12.0×10^4

6. What is the product of 4 and $\frac{1}{3}$? **G**
   - Ⓕ $\frac{1}{12}$
   - Ⓖ $1\frac{1}{3}$
   - Ⓗ $\frac{3}{4}$
   - Ⓙ $\frac{5}{3}$

7. Which is 2.04 written as a mixed number? **D**
   - Ⓐ $2\frac{1}{4}$
   - Ⓑ $2\frac{2}{5}$
   - Ⓒ $2\frac{2}{10}$
   - Ⓓ $2\frac{1}{25}$

8. Which measure is equivalent to 12 meters? **J**
   - Ⓕ 120 cm
   - Ⓖ 1.2 km
   - Ⓗ 0.12 km
   - Ⓙ 1.2×10^3 cm

TIP! TEST TAKING TIP!
Estimate the answer before solving. Use your estimate to check the reasonableness of your solution.

9. ***SHORT RESPONSE*** At the store, Ryan bought three items that cost \$4.65, \$3.99, and \$2.50. He gave the cashier \$15.00. How much change should Ryan receive? Explain how you found your answer.

10. ***SHORT RESPONSE*** Use prime factorization to find the GCF of 24, 56, and 72. Show your work.

Standardized Test Prep — Chapter 4

Purpose: *To provide review and practice for Chapters 1–4 and standardized tests*

Assessment Resources

Cumulative Tests (Levels A, B, C)
Assessment Resources . . . . . p. 163–174

State-Specific Test Practice Online
KEYWORD: MR4 TestPrep

Test Prep Doctor

For item 1, remind students of the test for divisibility by 3: If the sum of the digits in a number is divisible by 3, then the number is divisible by 3. Each choice in item 1 is divisible by 3 except **C,** which is a prime number.

Point out to students that in item 8, each choice must be converted to meters to see which one is equivalent to 12 meters.

F. The prefix *centi* means 0.01.
$120 \times 0.01 = 1.20$; 120 cm = 1.20 m

G. The prefix *kilo* means 1,000.
$1.2 \times 1{,}000 = 1{,}200$; 1.2 km = 1,200 m

H. $0.12 \times 1{,}000 = 120$; 0.12 km = 120 m

J. $1.2 \times 10^3 = 1{,}200$; $1{,}200 \times 0.01 = 12$;
1.2×10^3 cm = 1,200 cm = 12 m

The correct choice is **J.**

Answers

9. \$3.86; add the cost of the items purchased and subtract from \$15.00; \$4.65 + \$3.99 + \$2.50 = \$11.14, \$15.00 − \$11.14 = \$3.86

10. $24 = 2 \cdot 2 \cdot 2 \cdot 3$
$56 = 2 \cdot 2 \cdot 2 \cdot 7$
$72 = 2 \cdot 2 \cdot 2 \cdot 3 \cdot 3$
$2 \cdot 2 \cdot 2 = 8$
The GCF of 24, 56, and 72 is 8.

Chapter 5

Fraction Operations

Section 5A

Multiplying and Dividing Fractions

Hands-On Lab 5A
Model Fraction Multiplication

Lesson 5-1
Multiplying Fractions

Lesson 5-2
Multiplying Mixed Numbers

Hands-On Lab 5B
Model Fraction Division

Lesson 5-3
Dividing Fractions and Mixed Numbers

Lesson 5-4
Solving Fraction Equations: Multiplication and Division

Section 5B

Adding and Subtracting Fractions

Lesson 5-5
Least Common Multiple

Lesson 5-6
Estimating Fraction Sums and Differences

Hands-On Lab 5C
Model Fraction Addition and Subtraction

Lesson 5-7
Adding and Subtracting with Unlike Denominators

Lesson 5-8
Adding and Subtracting Mixed Numbers

Hands-On Lab 5D
Model Subtraction with Renaming

Lesson 5-9
Renaming to Subtract Mixed Numbers

Lesson 5-10
Solving Fraction Equations: Addition and Subtraction

Pacing Guide for 45-Minute Classes

Chapter 5

DAY 54	DAY 55	DAY 56	DAY 57	DAY 58
Hands-On Lab 5A	Lesson 5-1	Lesson 5-2	Hands-On Lab 5B	Lesson 5-3
DAY 59	**DAY 60**	**DAY 61**	**DAY 62**	**DAY 63**
Lesson 5-4	Mid-Chapter Quiz Lesson 5-5	Lesson 5-6	Hands-On Lab 5C	Lesson 5-7
DAY 64	**DAY 65**	**DAY 66**	**DAY 67**	**DAY 68**
Lesson 5-8	Hands-On Lab 5D	Lesson 5-9	Lesson 5-10	Chapter 5 Review
DAY 69				
Chapter 5 Assessment				

Pacing Guide for 90-Minute Classes

Chapter 5

DAY 27	DAY 28	DAY 29	DAY 30	DAY 31
Chapter 4 Assessment Hands-On Lab 5A	Lesson 5-1 Lesson 5-2	Hands-On Lab 5B Lesson 5-3	Lesson 5-4 Lesson 5-5	Mid-Chapter Quiz Lesson 5-6 Hands-On Lab 5C
DAY 32	**DAY 33**	**DAY 34**	**DAY 35**	
Lesson 5-7 Lesson 5-8	Hands-On Lab 5D Lesson 5-9	Lesson 5-10 Chapter 5 Review	Chapter 5 Assessment Lesson 6-1	

Across the Series

HARCOURT GRADE 5

- Estimate fraction sums and differences.
- Add, subtract, multiply, and divide fractions and mixed numbers.

COURSE 1

- **Estimate fraction operations.**
- **Draw a diagram to solve problems involving common multiples.**
- **Multiply, divide, add, and subtract fractions and mixed numbers.**
- **Solve fraction equations.**

COURSE 2

- Estimate fraction operations.
- Multiply, divide, add, and subtract fractions and mixed numbers.
- Solve fraction equations.

Across the Curriculum

LANGUAGE ARTS LINK

Math: Reading and Writing in the Content Area . . . . pp. 34–43

Focus on Problem Solving

Solve . . . . SE p. 231

Journal . . . . TE, last page of each lesson

Write About It . . . . SE, last page of each lesson

SOCIAL STUDIES LINK

Social Studies . . . . SE pp. 242, 257

Economics . . . . SE p. 254

SCIENCE LINK

Life Science . . . . SE pp. 215, 225, 227, 229, 239, 245, 249

TE = *Teacher's Edition* **SE** = *Student Edition*

Interdisciplinary Bulletin Board

Life Science

For a balanced diet, the FDA recommends the daily requirements shown in the food pyramid. To meet the daily vegetable requirement, you decide to drink vegetable juice. How many cups of juice do you need to drink to meet the minimum requirement?

$2\frac{1}{4}$ cups

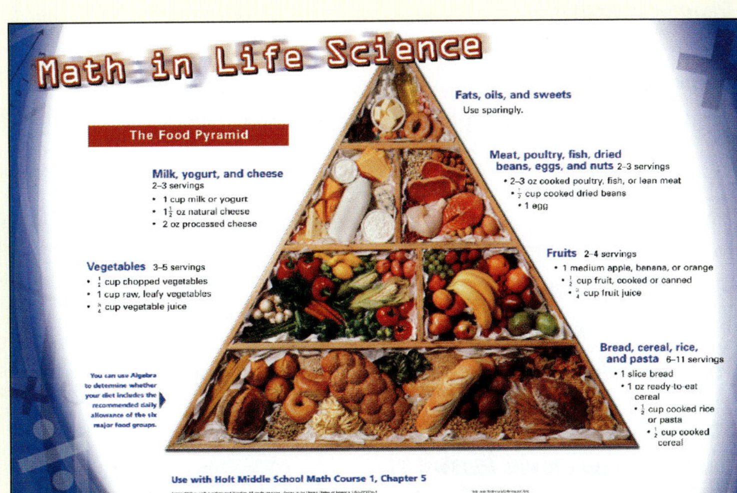

Interdisciplinary posters and worksheets are provided in your resource material.

Chapter 5

Resource Options

Chapter 5 Resource Book

Student Resources

Practice (Levels A, B, C) . . . . pp. 8–10, 17–19, 26–28, 35–37, 45–47, 54–56, 63–65, 72–74, 81–83, 90–92
Reteach . . . . pp. 11, 20, 29, 38, 48, 57, 66, 75, 84, 93
Challenge . . . . pp. 12, 21, 30, 39, 49, 58, 67, 76, 85, 94
Problem Solving . . . . pp. 13, 22, 31, 40, 50, 59, 68, 77, 86, 95
Puzzles, Twisters & Teasers . . . . pp. 14, 23, 32, 41, 51, 60, 69, 78, 87, 96
Recording Sheets . . . . pp. 3, 7, 16, 25, 34, 44, 53, 62, 71, 80, 89, 100, 103, 104, 106, 107
Chapter Review . . . . pp. 97–99

Teacher and Parent Resources

Chapter Planning and Pacing Guide . . . . p. 4
Section Planning Guides . . . . pp. 5, 42
Parent Letter . . . . pp. 1–2
Teaching Tools . . . . pp. 103–109
Teacher Support for Chapter Project . . . . p. 101
Transparencies . . . . pp. T1–T46

- Daily Transparencies
- Additional Examples Transparencies
- Teaching Transparencies

Reaching All Learners

English Language Learners

Success for English Language Learners . . . . pp. 67–86
Math: Reading and Writing in the Content Area . . . . pp. 34–43
Spanish Homework and Practice . . . . pp. 34–43
Spanish Interactive Study Guide . . . . pp. 34–43
Spanish Family Involvement Activities . . . . pp. 37–44
Multilingual Glossary

Individual Needs

Are You Ready? Intervention and Enrichment . . pp. 85–88, 97–100, 153–156, 181–184, 413–414
Alternate Openers: Explorations . . . . pp. 34–43
Family Involvement Activities . . . . pp. 37–44
Interactive Problem Solving . . . . pp. 34–43
Interactive Study Guide . . . . pp. 34–43
Readiness Activities . . . . pp. 9–10
Math: Reading and Writing in the Content Area . . . . pp. 34–43
Challenge . . . . CRB pp. 12, 21, 30, 39, 49, 58, 67, 76, 85, 94

Hands-On

Hands-On Lab Activities . . . . pp. 27–38
Technology Lab Activities . . . . pp. 16–22
Alternate Openers: Explorations . . . . pp. 34–43
Family Involvement Activities . . . . pp. 37–44

Applications and Connections

Consumer and Career Math . . . . pp. 17–20
Interdisciplinary Posters . . . . Poster 5, TE p. 208B
Interdisciplinary Poster Worksheets . . . . pp. 13–15

Transparencies

Alternate Openers: Explorations . . . . pp. 34–43
Exercise Answers Transparencies
Chapter 5 Resource Book . . . . pp. T1–T46

- Daily Transparencies
- Additional Examples Transparencies
- Teaching Transparencies

Technology

Teacher Resources

***Lesson Presentations* CD-ROM** . . . . Chapter 5
***Test and Practice Generator* CD-ROM** . . . . Chapter 5
***One-Stop Planner* CD-ROM** . . . . Chapter 5

Student Resources

***Are You Ready? Intervention* CD-ROM**
Skills 19, 22, 36, 43

internet connect

Homework Help Online	**KEYWORD:** MR4 HWHelp5
Math Tools Online	**KEYWORD:** MR4 Tools
Glossary Online	**KEYWORD:** MR4 Glossary
Chapter Project Online	**KEYWORD:** MR4 PSProject5
Chapter Opener Online	**KEYWORD:** MR4 Ch5
CNN Student News	**KEYWORD:** MR4 CNN5

SE = *Student Edition* **TE** = *Teacher's Edition* **AR** = *Assessment Resources* **CRB** = *Chapter Resource Book* **MK** = *Manipulatives Kit*

Chapter 5

Assessment Options

Assessing Prior Knowledge

Determine whether students have the required prerequisite concepts and skills.

Are You Ready? . . . . SE p. 209
Inventory Test . . . . AR pp. 1–4

Test Preparation

Provide review and practice for chapter and standardized tests.

Standardized Test Prep . . . . SE p. 269
Spiral Review with Test Prep . . . . . SE, last page of each lesson
Study Guide and Review . . . . SE pp. 264–266
Test Prep Tool Kit

Technology

***Test and Practice Generator* CD-ROM**

internet connect

State-Specific Test Practice Online KEYWORD: MR4 TestPrep

go.hrw.com

Performance Assessment

Assess students' understanding of chapter concepts and combined problem-solving skills.

Performance Assessment . . . . SE p. 268
Includes scoring rubric in TE
Performance Assessment . . . . AR p. 112
Performance Assessment Teacher Support . . . . AR p. 111

Portfolio

Portfolio opportunities appear throughout the Student and Teacher's Editions.

Suggested work samples:

Problem Solving Project . . . . TE p. 208
Performance Assessment . . . . SE p. 268
Portfolio Guide . . . . AR p. xxxv
Journal . . . . TE, last page of each lesson
Write About It . . . . SE, last page of each lesson

Daily Assessment

Obtain daily feedback on students' understanding of concepts.

Spiral Review and Test Prep . . . . . . SE, last page of each lesson

Also Available on Transparency In Chapter 5 Resource Book

Warm Up . . . . TE, first page of each lesson
Problem of the Day . . . . TE, first page of each lesson
Lesson Quiz . . . . TE, last page of each lesson

Student Self-Assessment

Have students evaluate their own work.

Group Project Evaluation . . . . AR p. xxxii
Individual Group Member Evaluation . . . . AR p. xxxiii
Portfolio Guide . . . . AR p. xxxv
Journal . . . . TE, last page of each lesson

Formal Assessment

Assess students' mastery of concepts and skills.

Section Quizzes . . . . AR pp. 14, 15
Mid-Chapter Quiz . . . . SE p. 230
Chapter Test . . . . SE p. 267
Chapter Tests (Levels A, B, C) . . . . AR pp. 55–60
Cumulative Tests (Levels A, B, C) . . . . AR pp. 175–186
Standardized Test Prep
Cumulative Assessment . . . . SE p. 269
End-of-Year Test . . . . AR pp. 271–274

Technology

***Test and Practice Generator* CD-ROM**

Make tests electronically. This software includes:

- Dynamic practice for Chapter 5
- Customizable tests
- Multiple-choice items for each objective
- Free-response items for each objective
- Teacher management system

SE = *Student Edition* **TE** = *Teacher's Edition* **AR** = *Assessment Resources* **CRB** = *Chapter Resource Book* **MK** = *Manipulatives Kit*

Chapter 5

Chapter 5 Tests

Three levels (A,B,C) of tests are available for each chapter in the *Assessment Resources.*

Level A

Chapter 5 Chapter Test
Form A

Multiply. Write each answer in simplest form.

1. $\frac{1}{2} \cdot \frac{1}{4}$ — $\frac{1}{8}$
2. $\frac{2}{3} \cdot \frac{3}{7}$ — $\frac{2}{7}$

Evaluate the expression $y \cdot \frac{1}{2}$ for each value of y. Write the answer in simplest form.

3. $y = \frac{2}{3}$ — $\frac{1}{3}$
4. $y = \frac{3}{4}$ — $\frac{3}{8}$

Multiply. Write each answer in simplest form.

5. $\frac{1}{4} \cdot 1\frac{1}{2}$ — $\frac{3}{8}$
6. $2\frac{3}{4} \cdot \frac{1}{8}$ — $\frac{11}{32}$

Find each product. Write the answer in simplest form.

7. $4 \cdot 1\frac{1}{2}$ — 6
8. $1\frac{1}{6} \cdot 2\frac{1}{4}$ — $2\frac{5}{8}$

Find the reciprocal.

9. $\frac{3}{7}$ — $\frac{7}{3}$
10. $\frac{2}{11}$ — $\frac{11}{2}$

Divide. Write each answer in simplest form.

11. $\frac{2}{5} \div 2$ — $\frac{1}{5}$
12. $4\frac{1}{2} \div 1\frac{3}{4}$ — $2\frac{4}{7}$

Solve each equation. Write the answer in simplest form.

13. $\frac{1}{3}a = 2$ — $a = 6$
14. $3t = \frac{1}{4}$ — $t = \frac{1}{12}$
15. $\frac{5y}{6} = 3$ — $y = 3\frac{3}{5}$
16. $3t = \frac{5}{7}$ — $t = \frac{5}{21}$

Find the least common multiple (LCM).

17. 4 and 8 — 8
18. 5 and 7 — 35
19. 4 and 6 — 12
20. 3 and 12 — 12

Estimate each sum or difference by rounding to 0, $\frac{1}{2}$, or 1.

21. $\frac{1}{5} + \frac{5}{8}$ — $\frac{1}{2}$
22. $\frac{7}{8} - \frac{2}{3}$ — 0 or $\frac{1}{2}$
23. $\frac{5}{6} + \frac{7}{8}$ — 2
24. $\frac{7}{8} - \frac{3}{8}$ — $\frac{1}{2}$

Chapter 5 Chapter Test
Form A, continued

Add or subtract. Write each answer in simplest form.

25. $\frac{1}{4} + \frac{1}{3}$ — $\frac{7}{12}$
26. $\frac{1}{5} + \frac{3}{10}$ — $\frac{1}{2}$
27. $\frac{3}{8} + \frac{1}{4}$ — $\frac{5}{8}$
28. $\frac{1}{3} + \frac{2}{5}$ — $\frac{11}{15}$

Find each sum or difference. Write the answer in simplest form.

29. $3\frac{1}{2} - 2\frac{1}{8}$ — $1\frac{3}{8}$
30. $4\frac{1}{2} + 4\frac{2}{3}$ — $9\frac{1}{6}$

Subtract. Write each answer in simplest form.

31. $4\frac{1}{4} - 2\frac{3}{4}$ — $1\frac{1}{2}$
32. $5 - 1\frac{2}{3}$ — $3\frac{1}{3}$
33. $6\frac{7}{10} - 2\frac{4}{5}$ — $3\frac{9}{10}$
34. $10 - 7\frac{2}{5}$ — $2\frac{3}{5}$

Solve each equation. Write the solution in simplest form.

35. $y + 3\frac{1}{2} = 6$ — $y = 2\frac{1}{2}$
36. $2\frac{1}{3} = y - 1\frac{1}{6}$ — $y = 3\frac{1}{2}$
37. $\frac{2}{3}a = 8$ — $a = 12$
38. $n - 2\frac{1}{2} = 5$ — $n = 7\frac{1}{2}$
39. A baker used $2\frac{1}{2}$ pounds of flour for bread, and $1\frac{3}{4}$ pound for cookies. How much flour was used? — $4\frac{1}{4}$ lb
40. As part of a daily fitness program, Lin runs $3\frac{1}{2}$ miles. Today she has run $1\frac{1}{3}$ miles. How much farther does she have to run? — $2\frac{1}{6}$ mi

Level B

Chapter 5 Chapter Test
Form B

Multiply. Write each answer in simplest form.

1. $\frac{5}{7} \cdot \frac{3}{4}$ — $\frac{15}{28}$
2. $\frac{6}{11} \cdot \frac{5}{6}$ — $\frac{5}{11}$

Evaluate the expression $y \cdot \frac{1}{8}$ for each value of y. Write the answer in simplest form.

3. $y = \frac{16}{17}$ — $\frac{2}{17}$
4. $y = \frac{8}{11}$ — $\frac{1}{11}$

Multiply. Write each answer in simplest form.

5. $\frac{2}{3} \cdot 4\frac{1}{2}$ — 3
6. $4\frac{1}{5} \cdot \frac{1}{3}$ — $1\frac{2}{5}$

Find each product. Write the answer in simplest form.

7. $1\frac{1}{2} \cdot 3\frac{1}{6}$ — $4\frac{3}{4}$
8. $3\frac{2}{3} \cdot 5\frac{1}{5}$ — $19\frac{1}{15}$

Find the reciprocal.

9. $\frac{7}{6}$ — $\frac{6}{7}$
10. $\frac{1}{8}$ — 8

Divide. Write each answer in simplest form.

11. $\frac{9}{11} \div 4$ — $\frac{9}{44}$
12. $2\frac{9}{10} \div 3\frac{1}{3}$ — $\frac{87}{100}$

Solve each equation. Write the answer in simplest form.

13. $\frac{2}{3}a = 4$ — $a = 6$
14. $12t = \frac{1}{4}$ — $t = \frac{1}{48}$
15. $\frac{8y}{11} = 6$ — $y = 8\frac{1}{4}$
16. $\frac{1}{2} = \frac{n}{8}$ — $n = 4$

Find the least common multiple (LCM).

17. 6 and 8 — 24
18. 5 and 11 — 55
19. 27, 90, and 84 — 3,780
20. 3, 5, and 8 — 120

Estimate each sum or difference by rounding to 0, $\frac{1}{2}$, or 1.

21. $\frac{1}{12} + \frac{3}{4}$ — 1
22. $\frac{15}{16} - \frac{2}{3}$ — 0 or $\frac{1}{2}$
23. $\frac{17}{20} + \frac{1}{2}$ — $1\frac{1}{2}$
24. $\frac{9}{10} - \frac{7}{8}$ — 0

Chapter 5 Chapter Test
Form B, continued

Add or subtract. Write each answer in simplest form.

25. $\frac{5}{6} - \frac{7}{12}$ — $\frac{1}{4}$
26. $\frac{7}{8} - \frac{5}{12}$ — $\frac{11}{24}$
27. $\frac{15}{24} + \frac{4}{24}$ — $\frac{19}{24}$
28. $\frac{3}{10} + \frac{3}{8}$ — $\frac{27}{40}$

Find each sum or difference. Write the answer in simplest form.

29. $3\frac{3}{4} + 2\frac{1}{8}$ — $5\frac{7}{8}$
30. $9\frac{4}{5} - 2\frac{1}{2}$ — $7\frac{3}{10}$

Subtract. Write each answer in simplest form.

31. $7\frac{1}{8} - 2\frac{5}{8}$ — $4\frac{1}{2}$
32. $9 - 2\frac{2}{5}$ — $6\frac{3}{5}$
33. $15\frac{2}{9} - 7\frac{5}{6}$ — $7\frac{7}{18}$
34. $12 - 7\frac{2}{15}$ — $4\frac{13}{15}$

Solve each equation. Write the solution in simplest form.

35. $y + 4\frac{1}{10} = 7$ — $y = 2\frac{9}{10}$
36. $7\frac{1}{6} = y - 3\frac{2}{3}$ — $y = 10\frac{5}{6}$
37. $\frac{4}{7}a = 6$ — $a = 10\frac{1}{2}$
38. $n - 2\frac{2}{5} = 5\frac{9}{10}$ — $n = 8\frac{3}{10}$
39. Pat has a $5\frac{3}{4}$ pound mixture of pecans and cashews. The mix includes $2\frac{2}{3}$ pounds of cashews. How many pounds are pecans? — $3\frac{1}{12}$ lb
40. At the end of her shift at The Deli Shop, Maria had sold $15\frac{3}{4}$ pounds of sliced turkey and $21\frac{2}{3}$ pounds of ham. What was the total weight of the meat? — $37\frac{5}{12}$ lb

Level C

Chapter 5 Chapter Test
Form C

Multiply. Write each answer in simplest form.

1. $\frac{7}{15} \cdot \frac{5}{6}$ — $\frac{7}{18}$
2. $\frac{11}{18} \cdot \frac{2}{7}$ — $\frac{11}{63}$

Evaluate the expression $z \cdot \frac{2}{9}$ for each value of z. Write the answer in simplest form.

3. $z = \frac{7}{8}$ — $\frac{7}{36}$
4. $z = \frac{18}{25}$ — $\frac{4}{25}$

Multiply. Write each answer in simplest form.

5. $\frac{4}{5} \cdot 3\frac{3}{7}$ — $2\frac{26}{35}$
6. $5\frac{6}{7} \cdot \frac{3}{4}$ — $4\frac{11}{28}$

Find each product. Write the answer in simplest form.

7. $3\frac{7}{8} \cdot 4\frac{3}{8}$ — $16\frac{61}{64}$
8. $4\frac{4}{7} \cdot 2\frac{2}{5}$ — $10\frac{34}{35}$

Find the reciprocal.

9. $\frac{26}{33}$ — $\frac{33}{26}$
10. $\frac{1}{96}$ — 96

Divide. Write each answer in simplest form.

11. $2\frac{5}{9} \div 2$ — $1\frac{5}{18}$
12. $7\frac{1}{8} \div 2\frac{5}{7}$ — $2\frac{5}{8}$

Solve each equation. Write the answer in simplest form.

13. $\frac{7}{4}x = 56$ — $x = 32$
14. $12t = \frac{3}{8}$ — $t = \frac{1}{32}$
15. $\frac{21y}{26} = 7$ — $y = 8\frac{2}{3}$
16. $1\frac{2}{3}n = \frac{9}{10}$ — $n = \frac{27}{50}$

Find the least common multiple (LCM).

17. 18 and 27 — 54
18. 3, 5, and 7 — 105
19. 4, 8, 12, and 18 — 72
20. 3, 12, 18, and 24 — 72

Estimate each sum or difference by rounding to 0, $\frac{1}{2}$, or 1.

21. $\frac{1}{32} + \frac{17}{18}$ — 1
22. $\frac{21}{23} - \frac{44}{47}$ — 0
23. $\frac{19}{39} + \frac{1}{9}$ — $\frac{1}{2}$
24. Compare. Write < or >. $\frac{5}{8} + \frac{3}{5}$ [>] 1

Chapter 5 Chapter Test
Form C, continued

Add or subtract. Write each answer in simplest form.

25. $\frac{7}{8} - \frac{2}{5}$ — $\frac{19}{40}$
26. $\frac{15}{16} - \frac{4}{5}$ — $\frac{11}{80}$
27. $\frac{3}{5} + \frac{4}{9}$ — $1\frac{2}{45}$
28. $\frac{5}{6} - \frac{2}{3} + \frac{5}{12}$ — $\frac{7}{12}$

Find each sum or difference. Write the answer in simplest form.

29. $8\frac{7}{9} - 2\frac{2}{3}$ — $6\frac{1}{9}$
30. $17\frac{5}{9} - 13\frac{3}{7}$ — $4\frac{8}{63}$

Subtract. Write each answer in simplest form.

31. $3\frac{1}{6} - 1\frac{4}{5}$ — $1\frac{11}{30}$
32. $17 - 12\frac{7}{9}$ — $4\frac{2}{9}$
33. $11\frac{7}{8} + 20\frac{5}{6}$ — $32\frac{17}{24}$
34. Evaluate $n - 2\frac{2}{5}$ for $n = 3\frac{7}{10}$. — $1\frac{3}{10}$

Solve each equation. Write the solution in simplest form.

35. $y + 2\frac{3}{5} = 7\frac{5}{12}$ — $y = 4\frac{49}{60}$
36. $5\frac{3}{4} = z - 6\frac{3}{8}$ — $z = 12\frac{1}{8}$
37. $2\frac{4}{7}a = \frac{3}{7}$ — $a = \frac{1}{6}$
38. $3\frac{1}{2} + \frac{1}{4} = n - 2\frac{5}{8}$ — $n = 6\frac{3}{8}$
39. A shelf in a closet measures $3\frac{1}{2}$ feet. If washcloths occupy $1\frac{1}{8}$ feet, how many feet remain for towels? — $2\frac{3}{8}$ ft
40. A radio station plays three songs in a row. If the songs last $3\frac{1}{8}$ minutes, $3\frac{2}{3}$ minutes, and $2\frac{5}{6}$ minutes, how long does the three-song segment last? — $9\frac{5}{8}$ min

Test and Practice Generator CD-ROM

Create and customize multiple versions of the same tests with corresponding answers for any chosen chapter objectives.

Chapter 5 State and Standardized Test Preparation

Test Taking Skill Builder and Standardized Test Practice are provided for each chapter in the *Test Prep Tool Kit.*

TEST TAKING SKILL BUILDER

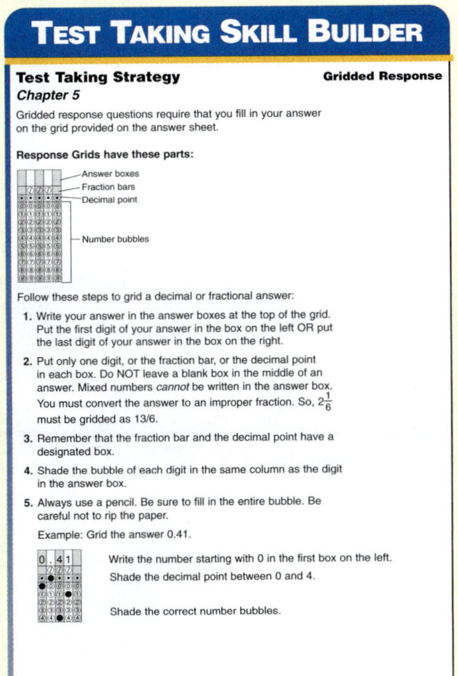
Test Taking Strategy — Gridded Response
Chapter 5

Gridded response questions require that you fill in your answer on the grid provided on the answer sheet.

Response Grids have these parts:

Follow these steps to grid a decimal or fractional answer:

1. Write your answer in the answer boxes at the top of the grid. Put the first digit of your answer in the box on the left OR put the last digit of your answer in the box on the right.
2. Put only one digit, or the fraction bar, or the decimal point in each box. Do NOT leave a blank box in the middle of an answer. Mixed numbers *cannot* be written in the answer box. You must convert the answer to an improper fraction. So, $2\frac{1}{6}$ must be gridded as 13/6.
3. Remember that the fraction bar and the decimal point have a designated box.
4. Shade the bubble of each digit in the same column as the digit in the answer box.
5. Always use a pencil. Be sure to fill in the entire bubble. Be careful not to rip the paper.

Example: Grid the answer 0.41.

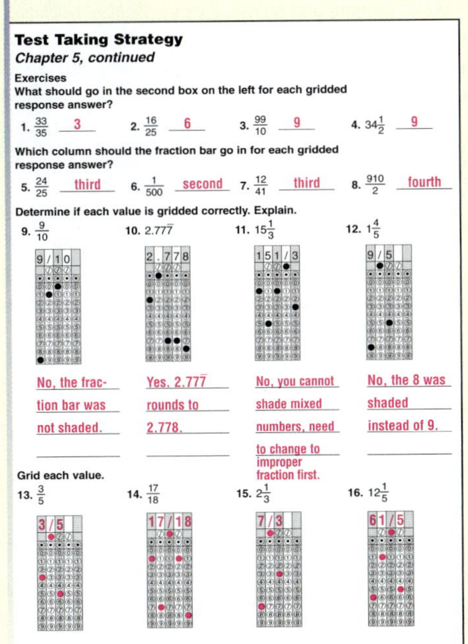
Test Taking Strategy
Chapter 5, continued

Exercises
What should go in the second box on the left for each gridded response answer?

1. $\frac{33}{35}$ 3 2. $\frac{16}{25}$ 6 3. $\frac{99}{10}$ 9 4. $34\frac{1}{2}$ 9

Which column should the fraction bar go in for each gridded response answer?

5. $\frac{24}{25}$ third 6. $\frac{1}{500}$ second 7. $\frac{12}{41}$ third 8. $\frac{910}{2}$ fourth

Determine if each value is gridded correctly. Explain.

9. $\frac{9}{10}$ — No, the fraction bar was not shaded.
10. 2.777 — Yes, 2.777 rounds to 2.778.
11. $15\frac{1}{3}$ — No, you cannot shade mixed numbers, need to change to improper fraction first.
12. $1\frac{4}{5}$ — No, the 8 was shaded instead of 9.

Grid each value.

13. $\frac{3}{5}$ 14. $\frac{17}{18}$ 15. $2\frac{1}{3}$ 16. $12\frac{1}{5}$

STANDARDIZED TEST PRACTICE

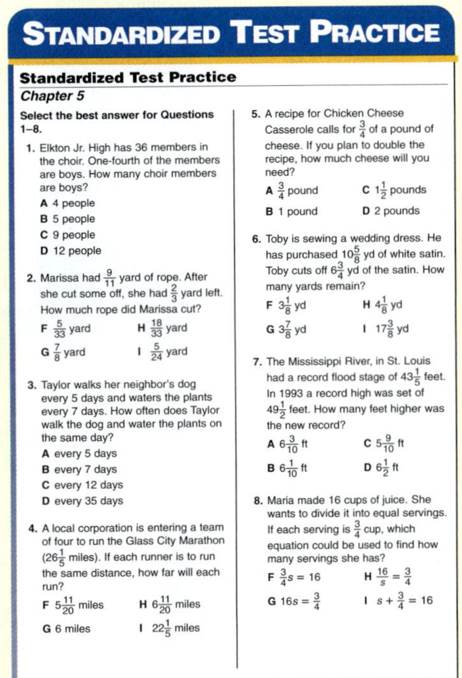
Standardized Test Practice
Chapter 5

Select the best answer for Questions 1–8.

1. Elkton Jr. High has 36 members in the choir. One-fourth of the members are boys. How many choir members are boys?
   A 4 people
   B 5 people
   C 9 people
   D 12 people
2. Marissa had $\frac{9}{11}$ yard of rope. After she cut some off, she had $\frac{2}{3}$ yard left. How much rope did Marissa cut?
   F $\frac{5}{33}$ yard H $\frac{18}{33}$ yard
   G $\frac{7}{8}$ yard I $\frac{5}{24}$ yard
3. Taylor walks her neighbor's dog every 5 days and waters the plants every 7 days. How often does Taylor walk the dog and water the plants on the same day?
   A every 5 days
   B every 7 days
   C every 12 days
   D every 35 days
4. A local corporation is entering a team of four to run the Glass City Marathon ($26\frac{1}{5}$ miles). If each runner is to run the same distance, how far will each run?
   F $5\frac{11}{20}$ miles H $6\frac{11}{20}$ miles
   G 6 miles I $22\frac{1}{5}$ miles
5. A recipe for Chicken Cheese Casserole calls for $\frac{3}{4}$ of a pound of cheese. If you plan to double the recipe, how much cheese will you need?
   A $\frac{3}{4}$ pound C $1\frac{1}{2}$ pounds
   B 1 pound D 2 pounds
6. Toby is sewing a wedding dress. He has purchased $10\frac{5}{8}$ yd of white satin. Toby cuts off $6\frac{3}{4}$ yd of the satin. How many yards remain?
   F $3\frac{1}{8}$ yd H $4\frac{1}{8}$ yd
   G $3\frac{7}{8}$ yd I $17\frac{3}{8}$ yd
7. The Mississippi River, in St. Louis had a record flood stage of $43\frac{1}{5}$ feet. In 1993 a record high was set of $49\frac{1}{2}$ feet. How many feet higher was the new record?
   A $6\frac{3}{10}$ ft C $5\frac{9}{10}$ ft
   B $6\frac{1}{10}$ ft D $6\frac{1}{2}$ ft
8. Maria made 16 cups of juice. She wants to divide it into equal servings. If each serving is $\frac{3}{4}$ cup, which equation could be used to find how many servings she has?
   F $\frac{3}{4}s = 16$ H $\frac{16}{s} = \frac{3}{4}$
   G $16s = \frac{3}{4}$ I $s + \frac{3}{4} = 16$

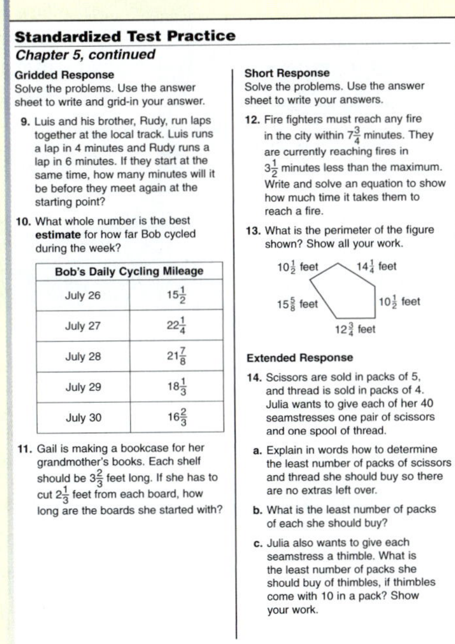
Standardized Test Practice
Chapter 5, continued

Gridded Response
Solve the problems. Use the answer sheet to write and grid-in your answer.

9. Luis and his brother, Rudy, run laps together at the local track. Luis runs a lap in 4 minutes and Rudy runs a lap in 6 minutes. If they start at the same time, how many minutes will it be before they meet again at the starting point?
10. What whole number is the best **estimate** for how far Bob cycled during the week?

Bob's Daily Cycling Mileage	
July 26	$15\frac{1}{2}$
July 27	$22\frac{1}{4}$
July 28	$21\frac{7}{8}$
July 29	$18\frac{1}{3}$
July 30	$16\frac{2}{3}$

11. Gail is making a bookcase for her grandmother's books. Each shelf should be $3\frac{2}{3}$ feet long. If she has to cut $2\frac{1}{3}$ feet from each board, how long are the boards she started with?

Short Response
Solve the problems. Use the answer sheet to write your answers.

12. Fire fighters must reach any fire in the city within $7\frac{3}{4}$ minutes. They are currently reaching fires in $3\frac{1}{2}$ minutes less than the maximum. Write and solve an equation to show how much time it takes them to reach a fire.
13. What is the perimeter of the figure shown? Show all your work.

Extended Response

14. Scissors are sold in packs of 5, and thread is sold in packs of 4. Julia wants to give each of her 40 seamstresses one pair of scissors and one spool of thread.
    a. Explain in words how to determine the least number of packs of scissors and thread she should buy so there are no extras left over.
    b. What is the least number of packs of each she should buy?
    c. Julia also wants to give each seamstress a thimble. What is the least number of packs she should buy of thimbles, if thimbles come with 10 in a pack? Show your work.

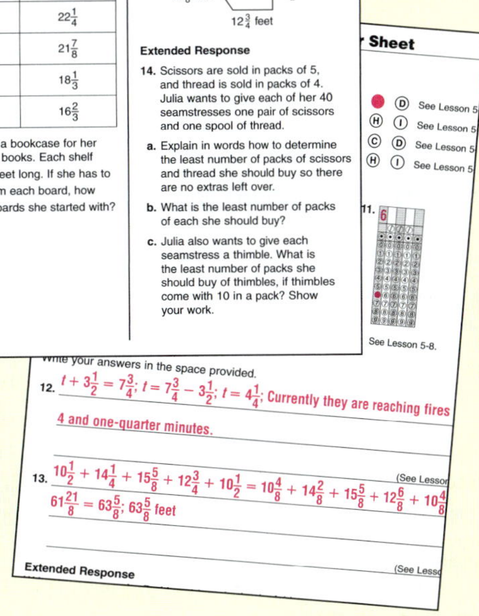

Customized answer sheets give students realistic practice for actual standardized tests.

internet connect

State-Specific Test Practice Online
KEYWORD: MR4 TestPrep

Test Prep Tool Kit

- Standardized Test Prep Workbook
- Countdown to Testing transparencies
- State Test Prep CD-ROM
- Standardized Test Prep Video

Chapter 5

Fraction Operations

Why Learn This?

Tell students that fractions are essential for describing quantities in the real world. Many quantities cannot accurately be described using only whole numbers. Have students look at the chart to see how a painter describes times using fractions. For example, it may be more useful for a painter to describe 30 minutes as $\frac{1}{2}$ hour.

Using Data

To begin the study of this chapter, have students:

- Find the total amount of time it takes to paint a window with oil-based paint and to paint 100 ft of chair rail with latex paint. $1\frac{1}{2}$ hours = 1hr 30 min
- Determine the number of minutes it takes to paint a door with oil-based paint. 30 min
- Compare the numbers of minutes it takes to paint 100 ft of chair rail with latex paint and with stain. It takes 45 minutes with latex paint and 36 minutes with stain.

Chapter 5

Fraction Operations

TABLE OF CONTENTS

internet connect
Chapter Opener Online
go.hrw.com
KEYWORD: MR4 Ch5

Painting Times

Object	Paint	Time (hr)
Wall (100 ft^2)	Oil-based	$\frac{3}{10}$
Wall (100 ft^2)	Latex	$\frac{2}{5}$
Chair rail (100 ft)	Latex	$\frac{3}{4}$
Chair rail (100 ft)	Stain	$\frac{3}{5}$
Door	Oil-based	$\frac{1}{2}$
Window	Oil-based	$\frac{3}{4}$

Career *Painter*

Have you ever wondered how painters estimate how much to charge for a job? Professional painters might paint houses, schools, office buildings, sports stadiums, or even music halls. To estimate how much to charge, many painters use a table that lists the average time it should take to prepare and paint certain objects. The table shows some painting jobs and the amount of time they take to complete.

PROBLEM SOLVING

Problem Solving Project

Social Studies Connection

Purpose: To solve problems using fractions and mixed numbers

Materials: Painter's Estimates worksheet

internet connect
Chapter Project Online: ***go.hrw.com***
KEYWORD: MR4 PSProject5

Understand, Plan, Solve, and Look Back

Have students:

- Complete the Painter's Estimates worksheet to learn to use fractions to calculate the time needed to complete an interior painting project.
- Research to find the difference between oil-based paint and latex paint. Why do we need more than one kind of paint? Why would you choose latex or oil-based paint for a particular job? Why does it take longer to paint with oil-based paint than to paint with latex paint?
- Check students' work.

Are You Ready?

Choose the best term from the list to complete each sentence.

- denominator
- factors
- improper fraction
- like fractions
- multiples
- numerator
- proper fraction
- simplest form
- unlike fractions

1. The first five __?__ of 6 are 6, 12, 18, 24, and 30. The __?__ of 6 are 1, 2, 3, and 6. **multiples; factors**
2. Fractions with the same denominator are called __?__. **like fractions**
3. A fraction is in __?__ when the GCF of the numerator and the denominator is 1. **simplest form**
4. The fraction $\frac{13}{9}$ is a(n) __?__ because the __?__ is greater than the __?__. **improper fraction; numerator; denominator**

Complete these exercises to review skills you will need for this chapter.

✔ Simplify Fractions

Write each fraction in simplest form.

5. $\frac{6}{10}$ **$\frac{3}{5}$**
6. $\frac{5}{15}$ **$\frac{1}{3}$**
7. $\frac{14}{8}$ **$1\frac{3}{4}$**
8. $\frac{8}{12}$ **$\frac{2}{3}$**
9. $\frac{10}{100}$ **$\frac{1}{10}$**
10. $\frac{12}{144}$ **$\frac{1}{12}$**
11. $\frac{33}{121}$ **$\frac{3}{11}$**
12. $\frac{15}{17}$ **$\frac{15}{17}$**

✔ Write Mixed Numbers as Fractions

Write each mixed number as an improper fraction.

13. $1\frac{1}{8}$ **$\frac{9}{8}$**
14. $2\frac{3}{4}$ **$\frac{11}{4}$**
15. $2\frac{4}{5}$ **$\frac{14}{5}$**
16. $1\frac{7}{9}$ **$\frac{16}{9}$**
17. $3\frac{1}{5}$ **$\frac{16}{5}$**
18. $5\frac{2}{3}$ **$\frac{17}{3}$**
19. $4\frac{4}{7}$ **$\frac{32}{7}$**
20. $3\frac{11}{12}$ **$\frac{47}{12}$**

✔ Add and Subtract Like Fractions

Add or subtract. Write each answer in simplest form.

21. $\frac{5}{8} + \frac{1}{8}$ **$\frac{3}{4}$**
22. $\frac{3}{7} + \frac{5}{7}$ **$1\frac{1}{7}$**
23. $\frac{9}{10} - \frac{3}{10}$ **$\frac{3}{5}$**
24. $\frac{5}{9} - \frac{2}{9}$ **$\frac{1}{3}$**

✔ Multiplication Facts

Multiply.

25. 8×11 **88**
26. 7×8 **56**
27. 4×12 **48**
28. 12×7 **84**
29. 10×13 **130**
30. 9×7 **63**
31. 6×8 **48**
32. 11×12 **132**

Assessing Prior Knowledge

INTERVENTION

Diagnose and Prescribe

Evaluate your students' performance on this page to determine whether intervention is necessary or whether enrichment is appropriate. Options that provide instruction, practice, and a check are listed below.

Resources for Are You Ready?

- ***Are You Ready? Intervention and Enrichment***
- **Recording Sheet for Are You Ready?** *Chapter 5 Resource Book . . . . . . p. 3*

Are You Ready? Intervention **CD-ROM**

Are You Ready?

Were students successful with Are You Ready?

NO — INTERVENE

✔ Simplify Fractions 5–12
Intervention Practice, Skill 19
CD-ROM Intervention Activities, Skill 19

✔ Write Mixed Numbers as Fractions 13–20
Intervention Practice, Skill 22
CD-ROM Intervention Activities, Skill 22

✔ Add and Subtract Like Fractions 21–24
Intervention Practice, Skill 43
CD-ROM Intervention Activities, Skill 43

✔ Multiplication Facts 25–32
Intervention Practice, Skill 36
CD-ROM Intervention Activities, Skill 36

YES — ENRICH

Are You Ready? Enrichment, pp. 415–416

Section 5A

Multiplying and Dividing Fractions

One-Minute Section Planner

Lesson	Materials	Resources
Hands-On Lab 5A Model Fraction Multiplication **NCTM:** Number and Operations, Algebra **NAEP:** Number Properties 3d ☐ SAT-9 ☑ SAT-10 ☐ ITBS ☐ CTBS ☑ MAT ☑ CAT	**Required** Grids	• *Hands-On Lab Activities*, pp. 27–30
Lesson 5-1 Multiplying Fractions **NCTM:** Number and Operations, Reasoning and Proof, Communication **NAEP:** Number Properties 3g ☑ SAT-9 ☑ SAT-10 ☑ ITBS ☑ CTBS ☑ MAT ☑ CAT	**Optional** Teaching Transparency T2 *(CRB)*	• *Chapter 5 Resource Book*, pp. 6–14 • Daily Transparency T1, CRB • Additional Examples Transparencies T3–T5, CRB • *Alternate Openers: Explorations*, p. 34
Lesson 5-2 Multiplying Mixed Numbers **NCTM:** Number and Operations, Communication **NAEP:** Number Properties 3g ☐ SAT-9 ☐ SAT-10 ☐ ITBS ☑ CTBS ☑ MAT ☑ CAT	**Optional** Recording Sheet for Reaching All Learners *(CRB, p. 103)*	• *Chapter 5 Resource Book*, pp. 15–23 • Daily Transparency T6, CRB • Additional Examples Transparencies T7–T10, CRB • *Alternate Openers: Explorations*, p. 35
Hands-On Lab 5B Model Fraction Division **NCTM:** Number and Operations, Algebra **NAEP:** Number Properties 3e ☐ SAT-9 ☑ SAT-10 ☐ ITBS ☐ CTBS ☑ MAT ☑ CAT	**Required** Grids	• *Hands-On Lab Activities*, pp. 31–33
Lesson 5-3 Dividing Fractions and Mixed Numbers **NCTM:** Number and Operations, Communication **NAEP:** Number Properties 3g ☑ SAT-9 ☑ SAT-10 ☐ ITBS ☑ CTBS ☑ MAT ☑ CAT	**Optional** Recording Sheet for Reaching All Learners *(CRB, p. 104)*	• *Chapter 5 Resource Book*, pp. 24–32 • Daily Transparency T11, CRB • Additional Examples Transparencies T12–T14, CRB • *Alternate Openers: Explorations*, p. 36
Lesson 5-4 Solving Fraction Equations: Multiplication and Division **NCTM:** Number and Operations, Algebra, Problem Solving, Reasoning and Proof, Communication, Representation **NAEP:** Algebra 4c ☐ SAT-9 ☑ SAT-10 ☐ ITBS ☐ CTBS ☑ MAT ☑ CAT		• *Chapter 5 Resource Book*, pp. 33–41 • Daily Transparency T15, CRB • Additional Examples Transparencies T16–T19, CRB • *Alternate Openers: Explorations*, p. 37
Section 5A Assessment		• Mid-Chapter Quiz, SE p. 230 • Section 5A Quiz, AR p. 14 • *Test and Practice Generator* CD-ROM

SAT = *Stanford Achievement Tests* **ITBS** = *Iowa Test of Basic Skills* **CTBS** = *Comprehensive Test of Basic Skills/Terra Nova*
MAT = *Metropolitan Achievement Tests* **CAT** = *California Achievement Test*

NCTM = Complete standards can be found on pages T27–T33. **NAEP** = Complete standards can be found on pages A31–A35.

SE = *Student Edition* **TE** = *Teacher's Edition* **AR** = *Assessment Resources* **CRB** = *Chapter Resource Book* **MK** = *Manipulatives Kit*

Section Overview

Multiplying Fractions and Mixed Numbers — Lessons 5-1, 5-2

Why? Solving real-world problems often involves multiplying fractions and mixed numbers.

Joe ran two-thirds as far as Adam. If Adam ran $1\frac{4}{5}$ miles, how far did Joe run?

$\frac{2}{3} \cdot 1\frac{4}{5}$

Change the mixed number to an improper fraction.

$\frac{2}{3} \cdot \frac{9}{5} = \frac{18}{15}$

Multiply numerators. Multiply denominators.

$\frac{6}{5}$, or $1\frac{1}{5}$

Joe ran $1\frac{1}{5}$ miles.

Dividing Fractions and Mixed Numbers — Lesson 5-3

Why? Problems that use measurements can be solved by dividing fractions and mixed numbers.

Mary has $2\frac{1}{2}$ yards of ribbon. How many $\frac{1}{4}$ yard lengths of ribbon can she make?

$2\frac{1}{2} \div \frac{1}{4} = \frac{5}{\cancel{2}_1} \cdot \frac{\cancel{4}^2}{1} = \frac{10}{1}$, or 10   Mary can make 10 lengths of ribbon.

Change the mixed number to an improper fraction.

You can simplify before multiplying.

Write division as multiplication by the reciprocal.

Fraction Equations — Lesson 5-4

Why? Application problems can be solved using fraction equations.

Cathy uses 3 cans of paint to paint $\frac{2}{3}$ of her room. How many cans of paint will she use to paint the whole room?

$\frac{2}{3}r = 3$

$\frac{3}{2} \cdot \frac{2}{3}r = \frac{3}{2} \cdot \frac{3}{1}$

Multiply both sides of the equation by $\frac{3}{2}$, the reciprocal of $\frac{2}{3}$.

$r = \frac{9}{2}$, or $4\frac{1}{2}$

Write the solution as a mixed number to represent cans of paint.

Cathy will use $4\frac{1}{2}$ cans of paint.

Hands-On LAB 5A Model Fraction Multiplication

Pacing: Traditional 1 day
Block $\frac{1}{2}$ day

Objective: To use grids to model fraction multiplication

Materials: Pencil and paper

Lab Resources

Hands-On Lab Activities . . . . p. 27

Using the Pages

Discuss with students what each square represents and how to represent fractions with a square.

Represent each fraction with a square.

1. $\frac{2}{3}$ 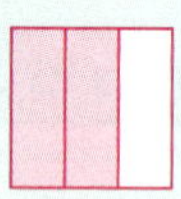

2. $\frac{4}{5}$ 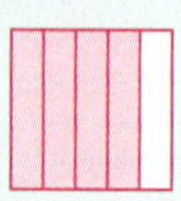

3. $1\frac{3}{4}$ 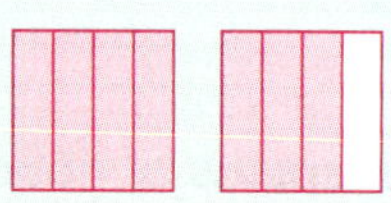

Hands-On LAB 5A Model Fraction Multiplication

Use with Lessons 5-1 and 5-2

internet connect
Lab Resources Online
go.hrw.com
KEYWORD: MR4 Lab5A

You can use grids to help you understand fraction multiplication.

Activity 1

1 Think of $\frac{1}{2} \cdot \frac{1}{3}$ as $\frac{1}{2}$ of $\frac{1}{3}$.

Shade $\frac{1}{3}$ of a square.

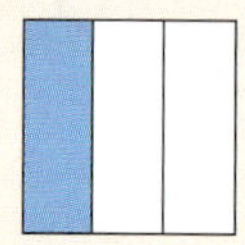

Divide the square into halves.

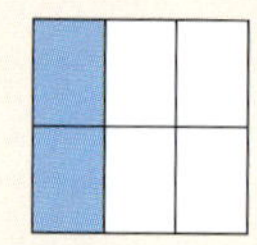

Look at $\frac{1}{2}$ of the part you shaded.

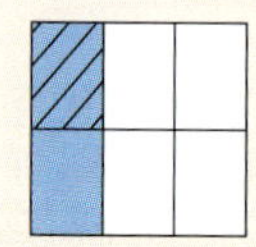

What fraction of the whole is this?

$\frac{1}{2}$ of $\frac{1}{3}$ is $\frac{1}{6}$.

2 Think of $\frac{2}{3} \cdot \frac{1}{2}$ as $\frac{2}{3}$ of $\frac{1}{2}$.

Shade $\frac{1}{2}$ of a square.

Divide the square into thirds.

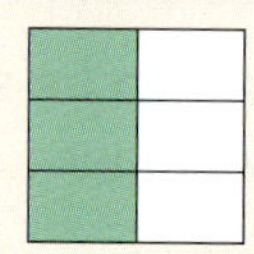

$\frac{2}{3}$ of $\frac{1}{2}$ is $\frac{2}{6}$, or $\frac{1}{3}$.

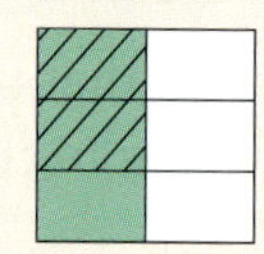

Think and Discuss

1. Tell whether the product is greater than or less than the fractions you started with. **The product will be smaller than both of the fractions.**

Try This

Write the multiplication expression modeled on each grid.

1. 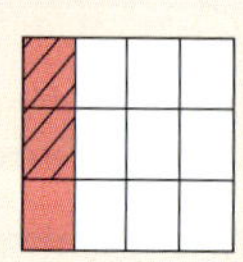 $\frac{2}{3} \cdot \frac{1}{4}$

2. 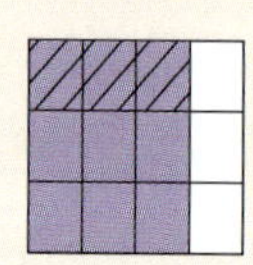 $\frac{1}{3} \cdot \frac{3}{4}$

3. 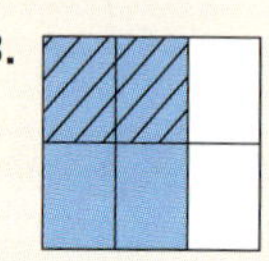 $\frac{1}{2} \cdot \frac{2}{3}$

Use a grid to model each multiplication expression.

4. $\frac{1}{3} \cdot \frac{1}{2}$ $\frac{1}{6}$ 5. $\frac{2}{3} \cdot \frac{1}{3}$ $\frac{2}{9}$ 6. $\frac{1}{4} \cdot \frac{2}{3}$ $\frac{2}{12}$, or $\frac{1}{6}$ 7. $\frac{1}{3} \cdot \frac{3}{4}$ $\frac{3}{12}$, or $\frac{1}{4}$

You can also use grids to model multiplication of mixed numbers.

Activity 2

1. Think of $\frac{1}{2} \cdot 2\frac{1}{2}$ as $\frac{1}{2}$ of $2\frac{1}{2}$.

Shade $2\frac{1}{2}$ squares.

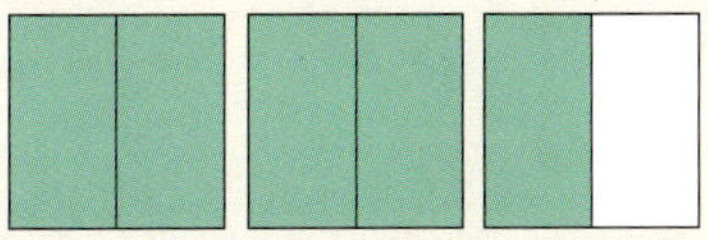

Divide the squares into halves.

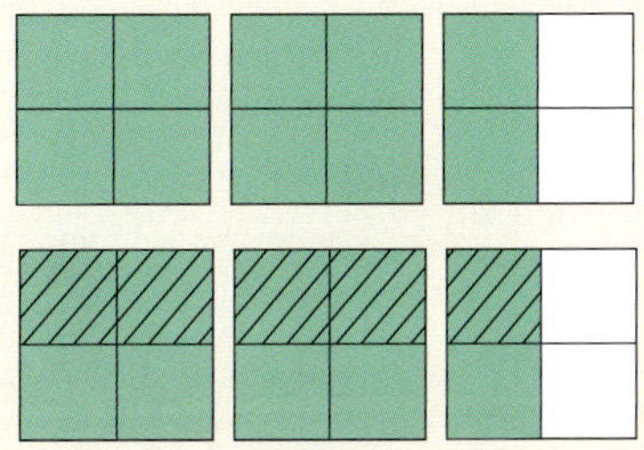

Look at $\frac{1}{2}$ of the part you shaded.

What fraction of the model is this?

$\frac{1}{2}$ of $2\frac{1}{2}$ is $1\frac{1}{4}$.

Think and Discuss

1. Describe how modeling multiplication of mixed numbers is like modeling multiplication of fractions.

Try This

Write the multiplication expression modeled on each grid.

1. 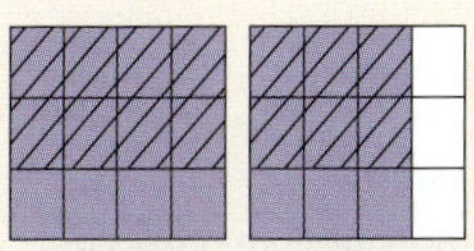 $\frac{2}{3} \cdot 1\frac{3}{4}$

2. 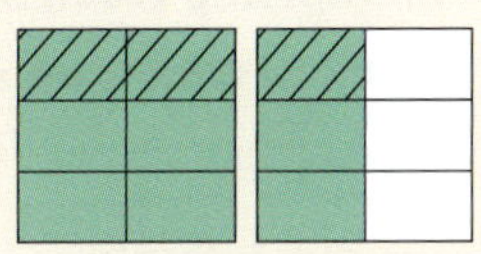 $\frac{1}{3} \cdot 1\frac{1}{2}$

3. 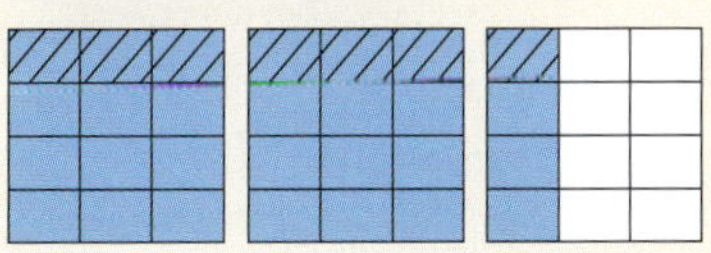 $\frac{1}{4} \cdot 2\frac{1}{3}$

Use a grid to model each multiplication expression.

4. $\frac{1}{3} \cdot 1\frac{1}{2}$ $\frac{3}{6}$, or $\frac{1}{2}$ 5. $\frac{2}{3} \cdot 2\frac{1}{3}$ $\frac{14}{9}$, or $1\frac{5}{9}$ 6. $\frac{1}{4} \cdot 2\frac{2}{3}$ $\frac{8}{12}$, or $\frac{2}{3}$ 7. $\frac{1}{3} \cdot 1\frac{3}{4}$ $\frac{7}{12}$

Answers

Activity 1

Try This

4. 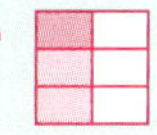

5. 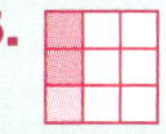

6. 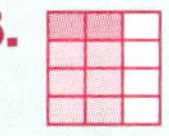

7. 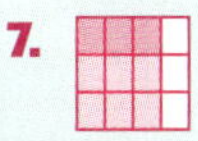

Activity 2

Think and Discuss

1. When you multiply mixed numbers, you should model the second factor first, as you would when you multiply fractions. Divide the model of the second factor into the number of equal groups indicated by the denominator of the first factor. Then shade the portion of the equal groups indicated by the numerator of the first factor.

Try This

4. 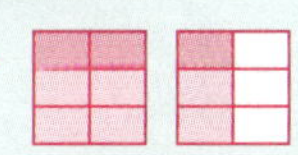

5. 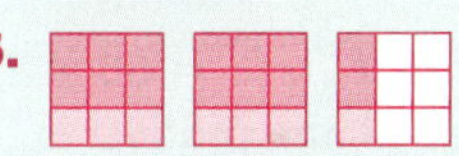

6. 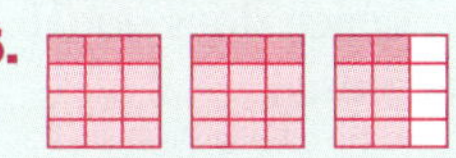

7. 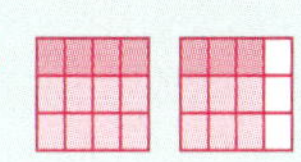

5-1 Organizer

Pacing: Traditional 1 day
Block $\frac{1}{2}$ day

Objective: Students multiply fractions.

Indiana Academic Standards
6.1.7, 6.2.4, 6.7.1, 6.7.5, 6.7.11

Warm Up

1. What is $\frac{1}{2}$ of 12? **6**
2. What is $\frac{1}{2}$ of 100? **50**
3. What is $\frac{1}{2}$ of 120? **60**
4. What is $\frac{1}{4}$ of 100? **25**
5. What is $\frac{1}{4}$ of 480? **120**

Problem of the Day

Your favorite uncle left one-fifth of his estate to each of his three children and the rest to his favorite charity. If his estate was worth \$105,000, how much was given to charity? **\$42,000**

Available on Daily Transparency in CRB

Math Humor

The golfer wondered if $\frac{p}{3} \cdot \frac{ar}{2} = 66$ is a true equation. After all, the product is $\frac{par}{6}$, or *6 under par*, which does equal 66 on most golf courses.

5-1 Multiplying Fractions

Learn to multiply fractions.

On average, people spend $\frac{1}{3}$ of their lives asleep. About $\frac{1}{4}$ of the time they sleep, they dream. What fraction of a lifetime does a person typically spend dreaming?

Your brain keeps working even when you're asleep. It makes sure that you keep breathing and that your heart keeps beating.

One way to find $\frac{1}{4}$ *of* $\frac{1}{3}$ is to make a model.

Find $\frac{1}{4}$ of $\frac{1}{3}$.

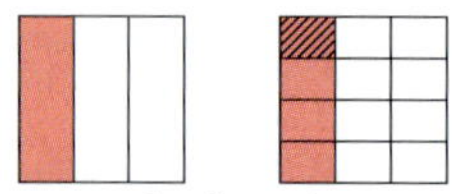

$$\frac{1}{4} \cdot \frac{1}{3} = \frac{1}{12}$$

You can also multiply fractions without making a model.

$\frac{1}{4} \cdot \frac{1}{3} = \frac{1 \cdot 1}{4 \cdot 3}$ ← *Multiply the numerators.* ← *Multiply the denominators.*

$= \frac{1}{12}$ *The answer is in simplest form.*

A person typically spends $\frac{1}{12}$ of his or her lifetime dreaming.

EXAMPLE 1 Multiplying Fractions

Multiply. Write each answer in simplest form.

A $\frac{1}{3} \cdot \frac{3}{5}$

$\frac{1}{3} \cdot \frac{3}{5} = \frac{1 \cdot 3}{3 \cdot 5}$ *Multiply numerators. Multiply denominators.*

$= \frac{3}{15}$ *The GCF of 3 and 15 is 3.*

$= \frac{1}{5}$ *The answer is in simplest form.*

B $\frac{6}{7} \cdot \frac{2}{3}$

$\frac{\overset{2}{\cancel{6}}}{7} \cdot \frac{2}{\underset{1}{\cancel{3}}} = \frac{2}{7} \cdot \frac{2}{1}$ *Use the GCF to simplify the fractions before multiplying. The GCF of 6 and 3 is 3.*

$= \frac{2 \cdot 2}{7 \cdot 1}$ *Multiply numerators. Multiply denominators.*

$= \frac{4}{7}$ *The answer is in simplest form.*

1 Introduce

Alternate Opener

EXPLORATION

5-1 Multiplying Fractions

You can use paper folding to find the product of two fractions. To find $\frac{3}{4}$ of $\frac{1}{2}$, fold the paper in half vertically. Then fold it horizontally into four sections.

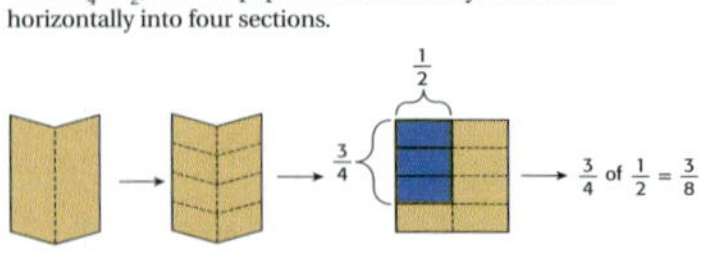

Use paper folding to find each product. Sketch a picture for each product.

1. $\frac{1}{3} \cdot \frac{1}{2}$
2. $\frac{1}{2} \cdot \frac{2}{3}$
3. $\frac{1}{3} \cdot \frac{1}{4}$
4. $\frac{2}{3} \cdot \frac{3}{4}$

Think and Discuss

5. **Explain** how to multiply two fractions.
6. **Discuss** whether it is possible to find the product of three fractions using paper folding.

Motivate

Pose this situation to students:

In this class, $\frac{1}{2}$ of the students are girls. Of those girls, $\frac{1}{3}$ packed a lunch today. What fraction of the class packed a lunch today? $\frac{1}{6}$

Ask students to explain how they would find the answer. Point out that the situation in the lesson opener also involves multiplication of fractions.

Exploration worksheet and answers on Chapter 5 Resource Book pp. 7 and 110

2 Teach

Lesson Presentation

Guided Instruction

In this lesson, students learn to multiply fractions. Show students how to use a model to find the product of two fractions, and then teach them to multiply fractions without using a model. (Teaching Transparency T2 in the Chapter 5 Resource Book) Have students evaluate expressions involving multiplication of fractions.

Teaching Tip Have students use shaded models as they work through the examples.

Multiply. Write each answer in simplest form.

C $\frac{3}{8} \cdot \frac{2}{9}$

$\frac{3}{8} \cdot \frac{2}{9} = \frac{3 \cdot 2}{8 \cdot 9}$ *Multiply numerators. Multiply denominators.*

$= \frac{6}{72}$ *The GCF of 6 and 72 is 6.*

$= \frac{1}{12}$ *The answer is in simplest form.*

EXAMPLE 2 **Evaluating Fraction Expressions**

Evaluate the expression $a \cdot \frac{1}{3}$ for each value of a. Write the answer in simplest form.

A $a = \frac{5}{8}$

$a \cdot \frac{1}{3}$

$\frac{5}{8} \cdot \frac{1}{3}$ *Substitute $\frac{5}{8}$ for a.*

$\frac{5 \cdot 1}{8 \cdot 3}$ *Multiply.*

$\frac{5}{24}$ *The answer is in simplest form.*

B $a = \frac{9}{10}$

$a \cdot \frac{1}{3}$

$\frac{9}{10} \cdot \frac{1}{3}$ *Substitute $\frac{9}{10}$ for a.*

$\frac{\overset{3}{\cancel{9}}}{10} \cdot \frac{1}{\underset{1}{\cancel{3}}}$ *Use the GCF to simplify.*

$\frac{3 \cdot 1}{10 \cdot 1}$ *Multiply.*

$\frac{3}{10}$ *The answer is in simplest form.*

C $a = \frac{3}{4}$

$a \cdot \frac{1}{3}$

$\frac{3}{4} \cdot \frac{1}{3}$ *Substitute $\frac{3}{4}$ for a.*

$\frac{3 \cdot 1}{4 \cdot 3}$ *Multiply numerators. Multiply denominators.*

$\frac{3}{12}$ *The GCF of 3 and 12 is 3.*

$\frac{1}{4}$ *The answer is in simplest form.*

Helpful Hint

You can look for a common factor in a numerator and a denominator to determine whether you can simplify before multiplying.

Think and Discuss

1. **Determine** whether the product of two proper fractions is greater than or less than each factor.
2. **Name** the missing denominator in the equation $\frac{1}{■} \cdot \frac{2}{3} = \frac{2}{21}$.
3. **Tell** how to find the product of $\frac{4}{21} \cdot \frac{6}{10}$ in two different ways.

COMMON ERROR ALERT

When multiplying like fractions, some students may multiply the numerators but not the denominators. Explain that rules for multiplying fractions are different from those for adding and subtracting fractions.

Additional Examples

Example 1

Multiply. Write each answer in simplest form.

A. $\frac{1}{4} \cdot \frac{2}{5}$ $\frac{1}{10}$

B. $\frac{5}{7} \cdot \frac{4}{15}$ $\frac{4}{21}$

C. $\frac{4}{9} \cdot \frac{6}{10}$ $\frac{4}{15}$

Example 2

Evaluate the expression $b \cdot \frac{2}{5}$ for each value of b. Write the answer in simplest form.

A. $b = \frac{1}{3}$ $\frac{2}{15}$

B. $b = \frac{3}{8}$ $\frac{3}{20}$

C. $b = \frac{5}{7}$ $\frac{2}{7}$

3 Close

Reaching All Learners

Through Grouping Strategies

Have students work in groups to practice multiplying fractions. Students toss number cubes to generate fractions to multiply, using the smaller number for the numerator and the larger number for the denominator.

For example, a toss of  would be the fraction $\frac{2}{3}$. Group members work together to multiply the fractions and express the answers in simplest form, taking turns recording the steps they follow.

Summarize

Remind students that the procedure for multiplying fractions differs from that for adding and subtracting fractions with like denominators. Have students explain how they are different.

Possible answers: When adding or subtracting like fractions, you add or subtract the numerators and keep the same denominator. When multiplying fractions, you multiply the numerators and you also multiply the denominators.

Answers to Think and Discuss

1. Possible answer: The product of two proper fractions is less than each factor because you are finding a fraction *of* another fraction.
2. 7
3. First way: Multiply the numerators and the denominators, and then simplify the product. $\frac{4}{21} \cdot \frac{6}{10} = \frac{24}{210} = \frac{4}{35}$ Second way: Use the GCF to simplify the fractions before multiplying. $\frac{\overset{2}{\cancel{4}}}{\underset{7}{\cancel{21}}} \cdot \frac{\overset{2}{\cancel{6}}}{\underset{5}{\cancel{10}}} = \frac{2}{7} \cdot \frac{2}{5} = \frac{4}{35}$

5-1 PRACTICE & ASSESS

Students may want to refer back to the lesson examples.

Assignment Guide

If you finished Example 1 assign:

Core 1–6, 10–18, 25–31 odd, 39–45

Enriched 2–6 even, 10–18 even, 25–32, 35–37, 39–45

If you finished Example 2 assign:

Core 1–24, 25–31 odd, 34, 39–45

Enriched 2–18 even, 19–45

Notes

5-1 Exercises

FOR EXTRA PRACTICE see page 645

internet connect
Homework Help Online
go.hrw.com Keyword: MR4 5-1

GUIDED PRACTICE

See Example 1 Multiply. Write each answer in simplest form.

1. $\frac{1}{2} \cdot \frac{1}{3}$ $\frac{1}{6}$
2. $\frac{2}{5} \cdot \frac{1}{4}$ $\frac{1}{10}$
3. $\frac{4}{7} \cdot \frac{3}{4}$ $\frac{3}{7}$
4. $\frac{5}{6} \cdot \frac{3}{5}$ $\frac{1}{2}$
5. $\frac{4}{9} \cdot \frac{3}{8}$ $\frac{1}{6}$
6. $\frac{2}{11} \cdot \frac{2}{3}$ $\frac{4}{33}$

See Example 2 Evaluate the expression $b \cdot \frac{1}{5}$ for each value of b. Write the answer in simplest form.

7. $b = \frac{2}{3}$ $\frac{2}{15}$
8. $b = \frac{5}{8}$ $\frac{1}{8}$
9. $b = \frac{3}{5}$ $\frac{3}{25}$

INDEPENDENT PRACTICE

See Example 1 Multiply. Write each answer in simplest form.

10. $\frac{1}{3} \cdot \frac{2}{7}$ $\frac{2}{21}$
11. $\frac{1}{3} \cdot \frac{1}{5}$ $\frac{1}{15}$
12. $\frac{5}{6} \cdot \frac{2}{3}$ $\frac{5}{9}$
13. $\frac{1}{3} \cdot \frac{6}{7}$ $\frac{2}{7}$
14. $\frac{3}{10} \cdot \frac{5}{6}$ $\frac{1}{4}$
15. $\frac{7}{9} \cdot \frac{3}{5}$ $\frac{7}{15}$
16. $\frac{1}{2} \cdot \frac{10}{11}$ $\frac{5}{11}$
17. $\frac{3}{5} \cdot \frac{3}{4}$ $\frac{9}{20}$
18. $\frac{8}{9} \cdot \frac{3}{4}$ $\frac{2}{3}$

See Example 2 Evaluate the expression $x \cdot \frac{1}{6}$ for each value of x. Write the answer in simplest form.

19. $x = \frac{4}{5}$ $\frac{2}{15}$
20. $x = \frac{6}{7}$ $\frac{1}{7}$
21. $x = \frac{3}{4}$ $\frac{1}{8}$
22. $x = \frac{8}{9}$ $\frac{4}{27}$
23. $x = \frac{9}{10}$ $\frac{3}{20}$
24. $x = \frac{5}{8}$ $\frac{5}{48}$

PRACTICE AND PROBLEM SOLVING

Find each product. Simplify the answer.

25. $\frac{3}{5} \cdot \frac{4}{9}$ $\frac{4}{15}$
26. $\frac{5}{12} \cdot \frac{9}{10}$ $\frac{3}{8}$
27. $\frac{2}{5} \cdot \frac{2}{7} \cdot \frac{5}{8}$ $\frac{1}{14}$

Compare. Write <, >, or =.

28. $\frac{2}{3} \cdot \frac{1}{4}$ ▒ $\frac{1}{3} \cdot \frac{3}{4}$ $<$
29. $\frac{3}{5} \cdot \frac{3}{4}$ ▒ $\frac{1}{2} \cdot \frac{9}{10}$ $=$
30. $\frac{5}{6} \cdot \frac{2}{3}$ ▒ $\frac{1}{3} \cdot \frac{2}{3}$ $>$

31. A walnut muffin recipe calls for $\frac{3}{4}$ cup walnuts. Mrs. Hooper wants to make $\frac{1}{3}$ of the recipe. What fraction of a cup of walnuts will she need? $\frac{1}{4}$ cup

32. Jim spent $\frac{5}{6}$ of an hour doing chores. He spent $\frac{2}{5}$ of that time washing dishes. What fraction of an hour did he spend washing dishes? $\frac{1}{3}$ of an hour

Math Background

When multiplying fractions, early writers of arithmetic generally did not practice cancellation before multiplication, as we do in this lesson. Thus, they would express a product such as $\frac{6}{11} \cdot \frac{5}{12}$ first as $\frac{30}{132}$ and then reduced as $\frac{5}{22}$, rather than looking for common factors among the numerators and denominators. One can, of course, practice this method today as well, but it is generally easier to simplify before multiplying.

RETEACH 5-1

LESSON 5-1 **Reteach**

Multiplying Fractions

To multiply fractions, multiply the numerators and multiply the denominators.

When multiplying fractions, you can sometimes divide by the GCF to make the problem simpler.

You can divide by the GCF even if the numerator and denominator of the same fraction have a common factor.

$\frac{1}{2} \cdot \frac{2}{3}$

$\frac{1}{\not{2}} \cdot \frac{\not{2}}{3}$

The problem is now $\frac{1}{1} \cdot \frac{1}{3}$.

$\frac{1 \cdot 1}{1 \cdot 3} = \frac{1}{3}$

So, $\frac{1}{2} \cdot \frac{2}{3} = \frac{1}{3}$

Is it possible to simplify before you multiply? If so, what is the GCF?

1. $\frac{1}{4} \cdot \frac{1}{2}$ no
2. $\frac{1}{6} \cdot \frac{3}{4}$ yes; 3
3. $\frac{1}{8} \cdot \frac{2}{3}$ yes; 2
4. $\frac{1}{3} \cdot \frac{2}{5}$ no

Multiply.

5. $\frac{1}{6} \cdot \frac{3}{5}$ $\frac{1}{10}$
6. $\frac{1}{4} \cdot \frac{1}{3}$ $\frac{1}{12}$
7. $\frac{7}{8} \cdot \frac{4}{5}$ $\frac{7}{10}$
8. $\frac{1}{6} \cdot \frac{2}{3}$ $\frac{1}{9}$
9. $\frac{1}{5} \cdot \frac{1}{2}$ $\frac{1}{10}$
10. $\frac{3}{5} \cdot \frac{1}{4}$ $\frac{3}{20}$
11. $\frac{3}{7} \cdot \frac{1}{9}$ $\frac{1}{21}$
12. $\frac{3}{4} \cdot \frac{1}{2}$ $\frac{3}{8}$
13. $\frac{1}{3} \cdot \frac{6}{7}$ $\frac{2}{7}$
14. $\frac{1}{4} \cdot \frac{2}{3}$ $\frac{1}{6}$
15. $\frac{3}{4} \cdot \frac{1}{3}$ $\frac{1}{4}$
16. $\frac{1}{4} \cdot \frac{1}{8}$ $\frac{1}{32}$

PRACTICE 5-1

LESSON 5-1 **Practice B**

Multiplying Fractions

Multiply. Write each answer in simplest form.

1. $\frac{1}{2} \cdot \frac{2}{5}$ $\frac{1}{5}$
2. $\frac{1}{3} \cdot \frac{7}{8}$ $\frac{7}{24}$
3. $\frac{2}{3} \cdot \frac{4}{6}$ $\frac{4}{9}$
4. $\frac{1}{4} \cdot \frac{10}{11}$ $\frac{5}{22}$
5. $\frac{3}{5} \cdot \frac{2}{3}$ $\frac{2}{5}$
6. $\frac{8}{9} \cdot \frac{3}{4}$ $\frac{2}{3}$
7. $\frac{3}{8} \cdot \frac{4}{5}$ $\frac{3}{10}$
8. $\frac{2}{7} \cdot \frac{3}{4}$ $\frac{3}{14}$
9. $\frac{1}{6} \cdot \frac{2}{3}$ $\frac{1}{9}$

Evaluate the expression $x \cdot \frac{1}{5}$ for each value of x. Write each answer in simplest form.

10. $x = \frac{3}{7}$ $\frac{3}{35}$
11. $x = \frac{5}{6}$ $\frac{1}{6}$
12. $x = \frac{2}{3}$ $\frac{2}{15}$
13. $x = \frac{10}{11}$ $\frac{2}{11}$
14. $x = \frac{5}{8}$ $\frac{1}{8}$
15. $x = \frac{4}{5}$ $\frac{4}{25}$

16. A cookie recipe calls for $\frac{2}{3}$ cup of brown sugar. Sarah is making $\frac{1}{4}$ of the recipe. How much brown sugar will she need? $\frac{1}{6}$ cup

17. Nancy spent $\frac{7}{8}$ hour working out at the gym. She spent $\frac{5}{7}$ of that time lifting weights. What fraction of an hour did she spend lifting weights? $\frac{5}{8}$ hour

Life Science LINK

There are about 1,000 species of bats in the world. Bats make up about $\frac{1}{4}$ of the world's mammals.

go.hrw.com KEYWORD: MR4 Bats

CNN Student News

33. A multiplying number machine uses a rule to change one fraction into another fraction. The machine changed $\frac{1}{2}$ into $\frac{1}{8}$, $\frac{1}{5}$ into $\frac{1}{20}$, and $\frac{5}{7}$ into $\frac{5}{28}$.

a. What is the rule? **Multiply by $\frac{1}{4}$**

b. Into what fraction will the machine change $\frac{1}{3}$? $\frac{1}{12}$

34. ***LIFE SCIENCE*** A bat can eat half its weight in insects in one night. If a bat weighing $\frac{3}{4}$ lb eats half its weight in insects, how much do the insects weigh? **$\frac{3}{8}$ lb**

35. The seating plan shows Oak School's theater. The front section has $\frac{3}{4}$ of the seats, and the rear section has $\frac{1}{4}$ of the seats. The school has reserved $\frac{1}{2}$ of the seats in the front section for students. What fraction of the seating is reserved for students? $\frac{3}{8}$

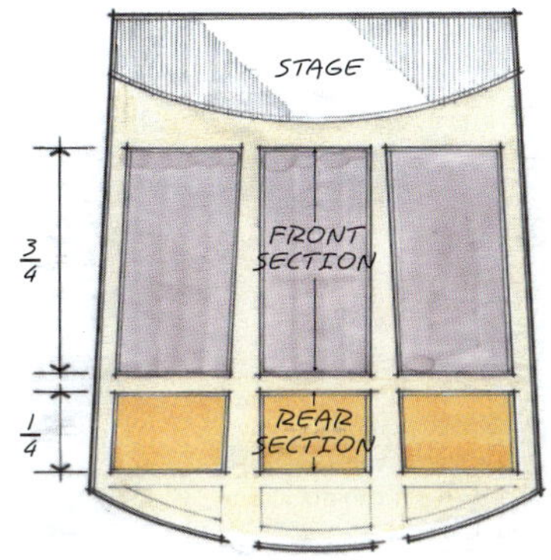

36. ***WRITE A PROBLEM*** Use the seating plan to write a problem in which you need to multiply two fractions. Then solve the problem.

37. ***WRITE ABOUT IT*** Explain how you can use the GCF before multiplying so that the product of two fractions is in simplest form.

38. ***CHALLENGE*** Evaluate the expression. Then simplify your answer.

$$\frac{(2+6)}{5} \cdot \frac{1}{4} \cdot 6 \quad 2\frac{2}{5}$$

Spiral Review

Evaluate each expression. (Lesson 1-4)

39. $3^3 + 12 \div 6 - (5 + 3)$ **21**

40. $30 \div (3 + 2) \times 3 + 12 - 6$ **24**

Compare. Write <, >, or =. (Lesson 3-1)

41. 0.303 ▒ 0.033 **>**

42. 10.17 ▒ 1.701 **>**

43. 3.104 ▒ 3.91 **<**

44. **TEST PREP** A dog's mass is about 9 kg. What is the dog's mass in grams? (Lesson 3-4) **A**

A 9,000 grams **B** 900 grams **C** 90 grams **D** 0.9 grams

45. **TEST PREP** Find the quotient of 3.45 ÷ 0.3. (Lesson 3-8) **J**

F 0.115 **G** 1.15 **H** 10.15 **J** 11.5

COMMON ERROR ALERT

When answering Exercises 28–30, students may assume that they can compare any two fractions. Explain that they must multiply first and then compare the products.

Answers

36. Possible answer: If $\frac{1}{3}$ of the rear seating is reserved for teachers, what fraction of the seating is reserved for teachers? $\frac{1}{12}$

37. Possible answer: You find a common factor in the numerator and denominator. Divide each by the common factor. Repeat the process as many times as possible. Then multiply the fractions. The product will be in simplest form.

Journal

Have students write real-world problems where fractions are multiplied. Have students solve their problems.

Test Prep Doctor

For Exercise 45, students need to remember to move the decimal point in both the divisor and the dividend. Students who answered **F** moved the decimal point in the dividend the wrong direction. Students who answered **G** did not move the decimal point in the dividend. Students who answered **H** may have moved the decimal point correctly, but they divided incorrectly.

CHALLENGE 5-1

LESSON 5-1 Challenge
Fractions of Flowers

For each flower below, shade the two petals whose fractions have a product equal to the fraction written in the center of that flower.

1.

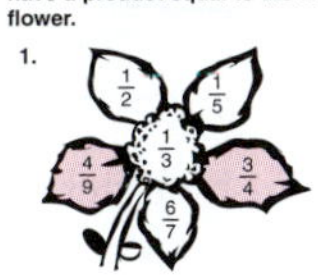

2.

3.

4.

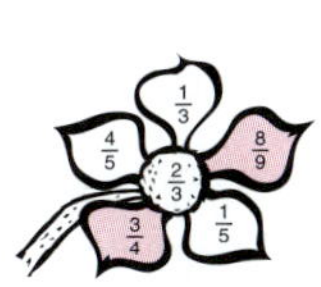

5.

6.

PROBLEM SOLVING 5-1

LESSON 5-1 Problem Solving
Multiplying Fractions

Use the circle graph to answer the questions. Write each answer in simplest form.

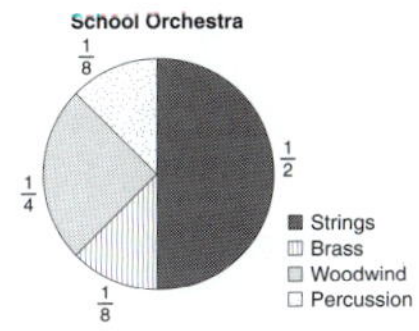

1. Of the students playing stringed instruments, $\frac{3}{4}$ play the violin. What fraction of the whole orchestra is violin players?
   $\frac{3}{8}$ of the orchestra

2. Of the students playing woodwind instruments, $\frac{1}{2}$ play the clarinet. What fraction of the whole orchestra is clarinet players?
   $\frac{1}{8}$ of the orchestra

3. Two-thirds of the students who play a percussion instrument are boys. What fraction of the musicians in the orchestra is boys who play percussion? girls who play percussion?
   boys: $\frac{1}{12}$ of the orchestra;
   girls: $\frac{1}{24}$ of the orchestra

4. The brass section is evenly divided into horns, trumpets, trombones, and tubas. What fraction of the whole orchestra do players of each of those brass instruments make up?
   $\frac{1}{32}$ of the orchestra

Circle the letter of the correct answer.

5. There are 40 students in the orchestra. How many students play either percussion or brass instruments?
   A 5 students
   (B) 10 students
   C 8 students
   D 16 students

6. If 2 more violinists join the orchestra, what fraction of all the musicians would play a stringed instrument?
   (F) $\frac{11}{21}$
   G $\frac{11}{20}$
   H $\frac{1}{20}$
   J $\frac{1}{26}$

Lesson Quiz

Multiply. Write each answer in simplest form.

1. $\frac{1}{4} \cdot \frac{3}{5}$ $\frac{3}{20}$

2. $\frac{3}{10} \cdot \frac{5}{7}$ $\frac{3}{14}$

Evaluate the expression $x \cdot \frac{1}{8}$ for each value of x. Write the answer in simplest form.

3. $x = \frac{4}{5}$ $\frac{1}{10}$

4. $x = \frac{1}{8}$ $\frac{1}{64}$

5. At a particular college $\frac{2}{5}$ of the students take a math class. Of these students $\frac{1}{4}$ take basic algebra. What fraction of all students take basic algebra? $\frac{1}{10}$

Available on Daily Transparency in CRB

5-2 Organizer

Pacing: Traditional 1 day
Block $\frac{1}{2}$ day

Objective: Students multiply mixed numbers.

Indiana Academic Standards
6.1.7, 6.2.4, 6.7.1, 6.7.5, 6.7.11

Warm Up

Multiply.

1. $\frac{1}{3} \cdot \frac{1}{2}$ $\frac{1}{6}$
2. $\frac{4}{5} \cdot \frac{3}{8}$ $\frac{3}{10}$
3. $\frac{3}{11} \cdot 2$ $\frac{6}{11}$
4. $\frac{3}{8} \cdot \frac{5}{8}$ $\frac{15}{64}$

Problem of the Day

Tracy and Zachary are sharing a large pizza. Tracy cuts the pizza in half. Zachary says he cannot eat that much. Zachary cuts his half into fourths and eats one piece. What fraction of the whole pizza does Zachary **not** eat? $\frac{7}{8}$

Available on Daily Transparency in CRB

Math Humor

When you multiply $23\frac{17}{19} \times 14\frac{31}{87}$, what do you get? Tired

5-2 Multiplying Mixed Numbers

Learn to multiply mixed numbers.

Janice and Carlos are making homemade pasta from a recipe that calls for $1\frac{1}{2}$ cups of flour. They want to make $\frac{1}{3}$ of the recipe.

You can find $\frac{1}{3}$ *of* $1\frac{1}{2}$, or multiply $\frac{1}{3}$ by $1\frac{1}{2}$, to find how much flour Janice and Carlos need.

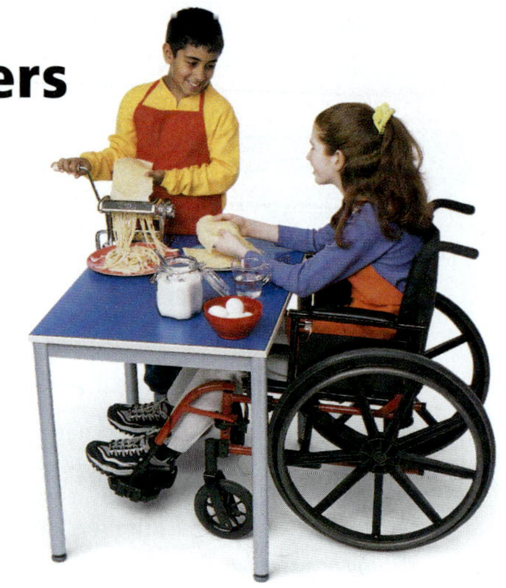

EXAMPLE 1 Multiplying Fractions and Mixed Numbers

Multiply. Write each answer in simplest form.

A $\frac{1}{3} \cdot 1\frac{1}{2}$

$\frac{1}{3} \cdot \frac{3}{2}$ *Write $1\frac{1}{2}$ as an improper fraction. $1\frac{1}{2} = \frac{3}{2}$*

$\frac{1 \cdot 3}{3 \cdot 2}$ *Multiply numerators. Multiply denominators.*

$\frac{3}{6}$

$\frac{1}{2}$ *Write the answer in simplest form.*

B $1\frac{1}{5} \cdot \frac{2}{3}$

$\frac{6}{5} \cdot \frac{2}{3}$ *Write $1\frac{1}{5}$ as an improper fraction. $1\frac{1}{5} = \frac{6}{5}$*

$\frac{6 \cdot 2}{5 \cdot 3}$ *Multiply numerators. Multiply denominators.*

$\frac{12}{15}$

$\frac{4}{5}$ *Write the answer in simplest form.*

C $\frac{3}{4} \cdot 2\frac{1}{3}$

$\frac{3}{4} \cdot \frac{7}{3}$ *Write $2\frac{1}{3}$ as an improper fraction. $2\frac{1}{3} = \frac{7}{3}$*

$\frac{\overset{1}{\cancel{3}}}{4} \cdot \frac{7}{\underset{1}{\cancel{3}}}$ *Use the GCF to simplify before multiplying.*

$\frac{1 \cdot 7}{4 \cdot 1}$

$\frac{7}{4} = 1\frac{3}{4}$ *You can write the answer as a mixed number.*

Remember!

To write a mixed number as an improper fraction, start with the whole number, multiply by the denominator, and add the numerator. Use the same denominator.

$1\frac{1}{5} = \frac{1 \cdot 5 + 1}{5} = \frac{6}{5}$

1 Introduce

Alternate Opener

EXPLORATION

5-2 Multiplying Mixed Numbers

You can use paper folding to find products of mixed numbers. To find $\frac{1}{2} \cdot 1\frac{1}{2}$, first fold two sheets of paper in half vertically to represent $1\frac{1}{2}$. To represent $\frac{1}{2}$ of $1\frac{1}{2}$, fold both sheets in half again horizontally.

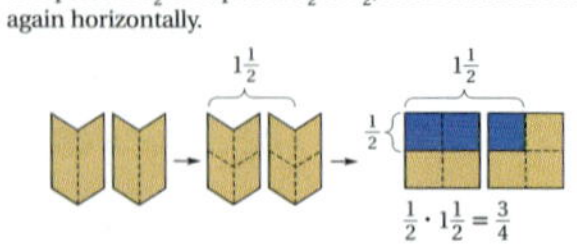

Use paper folding to find each product. Sketch a picture for each product.

1. $\frac{2}{3} \cdot 1\frac{1}{2}$
2. $\frac{1}{3} \cdot 1\frac{3}{4}$

Think and Discuss

3. **Explain** how to multiply a fraction times a mixed number.
4. **Explain** why the product of a fraction and a mixed number is less than the mixed number.

Motivate

To introduce students to multiplying with mixed numbers, show them a model of $2\frac{1}{4}$ like the one below.

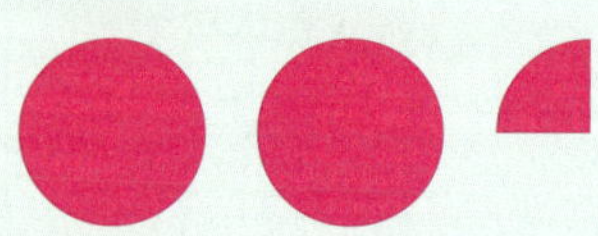

Ask students how they would show $\frac{1}{2}$ of $2\frac{1}{4}$.

Possible answer: Take one-half of the two wholes and one-half of $\frac{1}{4}$. $1 + \frac{1}{8} = 1\frac{1}{8}$

Exploration worksheet and answers on Chapter 5 Resource Book pp. 16 and 112

2 Teach

Lesson Presentation

Guided Instruction

In this lesson, students learn to multiply mixed numbers. Begin by teaching students to multiply a fraction and a mixed number. Then teach students to multiply mixed numbers by mixed numbers and whole numbers by mixed numbers.

Teaching Tip

Have students draw or make models to help conceptualize some of the simpler examples in the lesson.

EXAMPLE 2 Multiplying Mixed Numbers

Find each product. Write the answer in simplest form.

A $2\frac{1}{2} \cdot 1\frac{1}{3}$

$\frac{5}{2} \cdot \frac{4}{3}$ — *Write the mixed numbers as improper fractions. $2\frac{1}{2} = \frac{5}{2}$ $1\frac{1}{3} = \frac{4}{3}$*

$\frac{5 \cdot 4}{2 \cdot 3}$

$\frac{20}{6}$ — *Multiply numerators. Multiply denominators.*

$3\frac{2}{6}$ — *You can write the improper fraction as a mixed number.*

$3\frac{1}{3}$ — *Simplify.*

B $1\frac{1}{4} \cdot 1\frac{1}{3}$

$\frac{5}{4} \cdot \frac{4}{3}$ — *Write the mixed numbers as improper fractions. $1\frac{1}{4} = \frac{5}{4}$ $1\frac{1}{3} = \frac{4}{3}$*

$\frac{5}{\cancel{4}_1} \cdot \frac{\cancel{4}^1}{3}$ — *Use the GCF to simplify before multiplying.*

$\frac{5 \cdot 1}{1 \cdot 3}$ — *Multiply numerators. Multiply denominators.*

$\frac{5}{3}$

$1\frac{2}{3}$ — *You can write the answer as a mixed number.*

C $5 \cdot 3\frac{2}{11}$

$5 \cdot 3\frac{2}{11}$

$5 \cdot \left(3 + \frac{2}{11}\right)$

$(5 \cdot 3) + \left(5 \cdot \frac{2}{11}\right)$ — *Use the Distributive Property.*

$(5 \cdot 3) + \left(\frac{5}{1} \cdot \frac{2}{11}\right)$

$15 + \frac{10}{11}$ — *Multiply.*

$15\frac{10}{11}$ — *Add.*

Think and Discuss

1. **Tell** how you multiply a mixed number by a mixed number.
2. **Explain** two ways you would multiply a mixed number by a whole number.

COMMON ERROR ALERT

Some students may multiply the fraction part by the fraction part and the whole number part by the whole number part separately when multiplying two mixed numbers. Show students that this method results in an incorrect product and emphasize that mixed numbers must first be written as improper fractions before multiplying.

Additional Examples

Example 1

Multiply. Write each answer in simplest form.

A. $\frac{1}{4} \cdot 1\frac{1}{3}$ $\frac{1}{3}$

B. $3\frac{1}{2} \cdot \frac{4}{5}$ $2\frac{4}{5}$

C. $\frac{12}{13} \cdot 2\frac{3}{8}$ $2\frac{5}{26}$

Example 2

Find each product. Write the answer in simplest form.

A. $1\frac{2}{3} \cdot 2\frac{1}{7}$ $3\frac{4}{7}$

B. $1\frac{3}{8} \cdot 2\frac{2}{5}$ $3\frac{3}{10}$

C. $2 \cdot 4\frac{2}{5}$ $8\frac{4}{5}$

3 Close

Reaching All Learners

Through Critical Thinking

Challenge students to use each of the following numbers to form two multiplication number sentences: $\frac{1}{4}, \frac{3}{4}, \frac{7}{8}, 1\frac{7}{8}, 2\frac{1}{2}, 3\frac{1}{2}$.

Possible answer: $\frac{1}{4} \cdot 3\frac{1}{2} = \frac{7}{8}$ and $\frac{3}{4} \cdot 2\frac{1}{2} = 1\frac{7}{8}$

A recording sheet is provided on Chapter 5 Resource Book p. 103.

Summarize

Have the students solve three problems: one in which a fraction and a mixed number are multiplied, one in which two mixed numbers are multiplied, and one in which a whole number and a mixed number are multiplied. After students have solved the problems on their own, work through them together, explaining the steps as you go.

Answers to Think and Discuss

1. Change each mixed number to an improper fraction. Multiply the numerators, and multiply the denominators. Express the product in simplest form, and change any improper fractions to mixed numbers.
2. Write the mixed number as an improper fraction and multiply by the whole number written as a fraction. Or multiply the whole number by the whole number part of the mixed number and by the fraction part of the mixed number. Then add the products.

5-2 PRACTICE & ASSESS

Students may want to refer back to the lesson examples.

Assignment Guide

If you finished Example 1 assign:
Core 1–6, 10–18, 52–55
Enriched 16–18, 25–36, 52–55

If you finished Example 2 assign:
Core 1–24, 37–42, 52–55
Enriched 22–55

Notes

5-2 Exercises

FOR EXTRA PRACTICE see page 645

GUIDED PRACTICE

See Example 1 Multiply. Write each answer in simplest form.

1. $1\frac{1}{4} \cdot \frac{2}{3}$ $\frac{5}{6}$
2. $2\frac{2}{3} \cdot \frac{1}{4}$ $\frac{2}{3}$
3. $\frac{3}{7} \cdot 1\frac{5}{6}$ $\frac{11}{14}$
4. $1\frac{1}{3} \cdot \frac{6}{7}$ $1\frac{1}{7}$
5. $\frac{2}{3} \cdot 1\frac{3}{10}$ $\frac{13}{15}$
6. $2\frac{6}{11} \cdot \frac{2}{7}$ $\frac{8}{11}$

See Example 2 Find each product. Write the answer in simplest form.

7. $1\frac{5}{6} \cdot 1\frac{1}{8}$ $2\frac{1}{16}$
8. $2\frac{2}{5} \cdot 1\frac{1}{12}$ $2\frac{3}{5}$
9. $4 \cdot 5\frac{3}{7}$ $21\frac{5}{7}$

INDEPENDENT PRACTICE

See Example 1 Multiply. Write each answer in simplest form.

10. $1\frac{1}{4} \cdot \frac{3}{4}$ $\frac{15}{16}$
11. $\frac{4}{7} \cdot 1\frac{1}{4}$ $\frac{5}{7}$
12. $1\frac{1}{6} \cdot \frac{2}{5}$ $\frac{7}{15}$
13. $2\frac{1}{6} \cdot \frac{3}{7}$ $\frac{13}{14}$
14. $\frac{5}{9} \cdot 1\frac{9}{10}$ $1\frac{1}{18}$
15. $2\frac{2}{9} \cdot \frac{3}{5}$ $1\frac{1}{3}$
16. $1\frac{3}{10} \cdot \frac{5}{7}$ $\frac{13}{14}$
17. $\frac{3}{5} \cdot 2\frac{2}{9}$ $1\frac{1}{3}$
18. $2\frac{8}{11} \cdot \frac{3}{10}$ $\frac{9}{11}$

See Example 2 Find each product. Write the answer in simplest form.

19. $1\frac{1}{3} \cdot 1\frac{5}{7}$ $2\frac{2}{7}$
20. $1\frac{2}{3} \cdot 2\frac{3}{10}$ $3\frac{5}{6}$
21. $4 \cdot 3\frac{7}{8}$ $15\frac{1}{2}$
22. $6 \cdot 2\frac{1}{3}$ 14
23. $5 \cdot 4\frac{7}{10}$ $23\frac{1}{2}$
24. $2\frac{2}{3} \cdot 3\frac{5}{8}$ $9\frac{2}{3}$
25. $1\frac{1}{2} \cdot 2\frac{2}{5}$ $3\frac{3}{5}$
26. $3\frac{5}{6} \cdot 2\frac{3}{4}$ $10\frac{13}{24}$
27. $2\frac{1}{4} \cdot 1\frac{2}{9}$ $2\frac{3}{4}$

PRACTICE AND PROBLEM SOLVING

Write each product in simplest form.

28. $1\frac{2}{3} \cdot \frac{2}{9}$ $\frac{10}{27}$
29. $3\frac{1}{3} \cdot \frac{7}{10}$ $2\frac{1}{3}$
30. $2 \cdot \frac{5}{8}$ $1\frac{1}{4}$
31. $\frac{3}{8} \cdot \frac{4}{9}$ $\frac{1}{6}$
32. $2\frac{1}{12} \cdot 1\frac{3}{5}$ $3\frac{1}{3}$
33. $3\frac{3}{10} \cdot 4\frac{1}{6}$ $13\frac{3}{4}$
34. $2 \cdot \frac{4}{5} \cdot 1\frac{2}{3}$ $2\frac{2}{3}$
35. $3\frac{5}{6} \cdot \frac{9}{10} \cdot 4\frac{2}{3}$ $16\frac{1}{10}$
36. $1\frac{7}{8} \cdot 2\frac{1}{3} \cdot 4$ $17\frac{1}{2}$

Evaluate each expression.

37. $\frac{1}{2} \cdot c$ for $c = 4\frac{2}{5}$ $2\frac{1}{5}$
38. $1\frac{5}{7} \cdot x$ for $x = \frac{5}{6}$ $1\frac{3}{7}$
39. $1\frac{3}{4} \cdot b$ for $b = 1\frac{1}{7}$ 2
40. $1\frac{5}{9} \cdot n$ for $n = 18$ 28
41. $2\frac{5}{9} \cdot t$ for $t = 4$ $10\frac{2}{9}$
42. $3\frac{3}{4} \cdot p$ for $p = \frac{1}{2}$ $1\frac{7}{8}$
43. $\frac{4}{5} \cdot m$ for $m = 2\frac{2}{3}$ $2\frac{2}{15}$
44. $6y$ for $y = 3\frac{5}{8}$ $21\frac{3}{4}$

Math Background

The expression *improper fractions* carries unfortunate connotations. As this lesson demonstrates, a fraction such as $\frac{25}{9}$ is called improper because its numerator is greater than its denominator. Therefore it is rewritten as $2\frac{7}{9}$. This mixed number has the advantage of making clear the whole numbers between which the fractional expression lies. However, $\frac{25}{9}$ is generally easier to use in arithmetic than $2\frac{7}{9}$. When it comes to deciding how a number should be expressed, students should consider how the number is going to be used.

RETEACH 5-2

LESSON 5-2 **Reteach**
Multiplying Mixed Numbers

To find $\frac{1}{3}$ of $2\frac{1}{2}$, first change $2\frac{1}{2}$ to an improper fraction.

$2\frac{1}{2} = \frac{5}{2}$

Then multiply as you would with two proper fractions.

Check to see if you can divide by the GCF to make the problem simpler. Then multiply the numerators and multiply the denominators.

The problem is now $\frac{1}{3} \cdot \frac{5}{2}$.

$\frac{1 \cdot 5}{3 \cdot 2} = \frac{5}{6}$

So, $\frac{1}{3} \cdot 2\frac{1}{2}$ is $\frac{5}{6}$.

Rewrite each mixed number as an improper fraction. Is it possible to simplify before you multiply? If so, what is the GCF?

1. $\frac{1}{4} \cdot 1\frac{1}{3}$ $= \frac{1}{4} \cdot \frac{4}{3}$ $\frac{1}{3}$
2. $\frac{1}{6} \cdot 2\frac{1}{2}$ $= \frac{1}{6} \cdot \frac{5}{2}$ $\frac{5}{12}$
3. $\frac{1}{8} \cdot 1\frac{1}{2}$ $= \frac{1}{8} \cdot \frac{3}{2}$ $\frac{3}{16}$
4. $\frac{1}{3} \cdot 1\frac{2}{5}$ $= \frac{1}{3} \cdot \frac{7}{5}$ $\frac{7}{15}$
5. $1\frac{1}{3} \cdot 1\frac{2}{3}$ $\frac{4}{3} \cdot \frac{5}{3}$ $2\frac{2}{9}$
6. $1\frac{1}{2} \cdot 1\frac{1}{3}$ $\frac{3}{2} \cdot \frac{4}{3}$ 2
7. $1\frac{3}{4} \cdot 2\frac{1}{2}$ $\frac{7}{4} \cdot \frac{5}{2}$ $4\frac{3}{8}$
8. $1\frac{1}{6} \cdot 2\frac{2}{3}$ $\frac{7}{6} \cdot \frac{8}{3}$ $3\frac{1}{9}$
9. $3\frac{1}{3} \cdot \frac{2}{5}$ $1\frac{1}{3}$
10. $2\frac{1}{2} \cdot \frac{1}{5}$ $\frac{1}{2}$
11. $1\frac{3}{4} \cdot 2\frac{1}{2}$ $4\frac{3}{8}$
12. $3\frac{1}{3} \cdot 1\frac{1}{5}$ 4

PRACTICE 5-2

LESSON 5-2 **Practice B**
Multiplying Mixed Numbers

Multiply. Write each answer in simplest form.

1. $1\frac{2}{3} \cdot \frac{4}{5}$ $1\frac{1}{3}$
2. $1\frac{7}{8} \cdot \frac{4}{5}$ $1\frac{1}{2}$
3. $2\frac{3}{4} \cdot \frac{1}{5}$ $\frac{11}{20}$
4. $2\frac{1}{6} \cdot \frac{2}{3}$ $1\frac{4}{9}$
5. $2\frac{2}{5} \cdot \frac{3}{8}$ $\frac{9}{10}$
6. $1\frac{3}{4} \cdot \frac{5}{6}$ $1\frac{11}{24}$
7. $1\frac{1}{6} \cdot \frac{3}{5}$ $\frac{7}{10}$
8. $\frac{2}{9} \cdot 2\frac{1}{7}$ $\frac{10}{21}$
9. $2\frac{3}{11} \cdot \frac{7}{10}$ $1\frac{13}{22}$

Find each product. Write each answer in simplest form.

10. $\frac{6}{7} \cdot 1\frac{1}{4}$ $1\frac{1}{14}$
11. $\frac{5}{8} \cdot 1\frac{3}{5}$ 1
12. $2\frac{4}{9} \cdot \frac{1}{6}$ $\frac{11}{27}$
13. $1\frac{3}{10} \cdot 1\frac{1}{3}$ $1\frac{11}{15}$
14. $2\frac{1}{2} \cdot 2\frac{1}{2}$ $6\frac{1}{4}$
15. $1\frac{2}{3} \cdot 3\frac{1}{2}$ $5\frac{5}{6}$
16. Dominick lives $1\frac{3}{4}$ miles from his school. If his mother drives him half the way, how far will Dominick have to walk to get to school?
$\frac{7}{8}$ mile
17. Katoni bought $2\frac{1}{2}$ dozen donuts to bring to the office. Since there are 12 donuts in a dozen, how many donuts did Katoni buy?
30 donuts

Computer Science LINK

In a survey, 240 people were asked how many hours per week they spend using the Internet. The circle graph shows which fractions of the people use the Internet for which amounts of time.

Use the graph for Exercises 45–51.

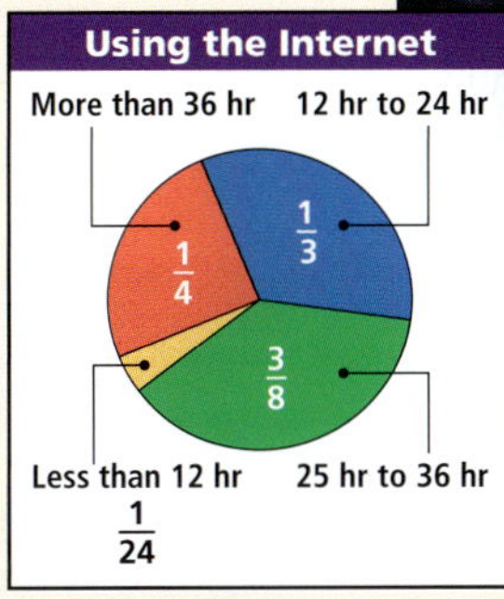

The World Wide Web was developed by Tim Berners-Lee at CERN, in Switzerland. It was designed to help physicists from different parts of the world share information.

go.hrw.com
KEYWORD: MR4 Internet
CNN Student News

45. How many people in all were surveyed? **240 people**

46. Find the number of people who said they use the Internet for 12 hours to 24 hours a week. $\frac{1}{3} \cdot 240 = 80$ **people**

47. Find the number of people who said they use the Internet for 25 hours to 36 hours a week. $\frac{3}{8} \cdot 240 = 90$ **people**

48. Toni's grandfather uses the Internet for $1\frac{1}{2}$ hours each day.

a. How many hours does he use the Internet in one week? (Write the answer as a mixed number.) $1\frac{1}{2} \cdot 7 = 10\frac{1}{2}$ **h**

b. If Toni's grandfather were included in the survey, in which time section of the circle graph would his data be? **Less than 12 h**

49. ***CHOOSE A STRATEGY*** Which set of tallies could represent the number of people who use the Internet for fewer than 12 hours a week? **A**

A 𝍸 𝍸   **C** 𝍸 𝍸 𝍸 𝍸

B 𝍸 𝍸 II   **D** 𝍸 𝍸 𝍸 𝍸 IIII

50. ***WRITE ABOUT IT*** Explain how you can find the number of people surveyed who use the Internet for more than 36 hours a week.

51. ***CHALLENGE*** Five-sixths of the people who use the Internet for 25 hours to 36 hours said they use it for 30 hours each week. Find the number of people who use the Internet for 30 hours each week. **75 people**

Spiral Review

Solve each equation. (Lesson 2-4)

52. $n + 52 = 71$ $n = 19$ **53.** $30 = k - 15$ $k = 45$ **54.** $22 - 18 + c = 30$ $c = 26$

55. TEST PREP Which number is **not** between 7.5 and 8.25? (Lesson 3-1) **D**

A 8.039   **B** 8.219   **C** 7.501   **D** 7.051

Interdisciplinary LINK

Computer Science

Exercises 45–51 involve using data about Internet usage to solve problems involving fraction multiplication.

Answers

50. Possible answer: Find $\frac{1}{4}$ of 240. Use the GCF of 4 and 240 to simplify. $1 \cdot 60 = 60$, so 60 people said they use the Internet for more than 36 hours.

Journal

Have students explain in their own words how to multiply mixed numbers by fractions, by whole numbers, and by mixed numbers.

Test Prep Doctor

Students may wish to review decimal place value before completing Exercise 55. Students who answered **A, B,** or **C** have either forgotten how to compare and order decimals or were careless.

CHALLENGE 5-2

LESSON 5-2 **Challenge**
And They're Off!

Like many sports, horse racing uses a special system of measurement. Horse races are measured in units called *furlongs*. One furlong equals $\frac{1}{8}$ mile. The races described below have different furlong lengths, but they all offer the same prize money to their winners—$1,000,000!

Write the length in miles of each of these horse races in simplest form.

1. Santa Anita Derby, California
Race Length: 9 furlongs   Length in Miles: $1\frac{1}{8}$ miles

2. Kentucky Derby, Kentucky
Race Length: 10 furlongs   Length in Miles: $1\frac{1}{4}$ miles

3. Preakness Stakes, Maryland
Race Length: $9\frac{1}{2}$ furlongs   Length in Miles: $1\frac{3}{16}$ miles

4. Belmont Stakes, New York
Race Length: 12 furlongs   Length in Miles: $1\frac{1}{2}$ miles

5. Breeders' Cup Juvenile, New York
Race Length: $8\frac{1}{2}$ furlongs   Length in Miles: $1\frac{1}{16}$ miles

PROBLEM SOLVING 5-2

LESSON 5-2 **Problem Solving**
Multiplying Mixed Numbers

Use the recipe to answer the questions.

CHOCOLATE CHIP COOKIES Servings: 1 batch
$1\frac{2}{3}$ cups flour
$\frac{3}{4}$ teaspoon baking soda
$\frac{1}{2}$ cup white sugar
$2\frac{1}{3}$ cups semisweet chocolate chips
$\frac{1}{2}$ cup brown sugar
$\frac{3}{4}$ cup butter
1 egg
$1\frac{1}{4}$ teaspoons vanilla

1. If you want to make $2\frac{1}{2}$ batches, how much flour would you need? $4\frac{1}{6}$ cups

2. If you want to make only a $1\frac{1}{2}$ batches, how much chocolate chips would you need? $3\frac{1}{2}$ cups

3. You want to bake $3\frac{1}{4}$ batches. How much vanilla do you need in all? $4\frac{1}{16}$ teaspoons

4. If you make $1\frac{1}{4}$ batches, how much baking soda would you need? $\frac{15}{16}$ teaspoon

5. How many cups of white sugar do you need to make $3\frac{1}{2}$ batches of cookies? $1\frac{3}{4}$ cups

Choose the letter for the best answer.

6. Dan used $2\frac{1}{4}$ cups of butter to make chocolate chip cookies using the above recipe. How many batches of cookies did he make?
(A) 3 batches
B 4 batches
C 5 batches
D 6 batches

7. One bag of chocolate chips holds 2 cups. If you buy five bags, how many cups of chips will you have left over after baking $2\frac{1}{2}$ batches of cookies?
(F) $4\frac{1}{6}$ cups
G $5\frac{5}{6}$ cups
H $2\frac{1}{3}$ cups
J $\frac{1}{3}$ cup

Lesson Quiz

Multiply. Write each answer in simplest form.

1. $3\frac{1}{5} \cdot \frac{5}{8}$ 2   **2.** $1\frac{7}{10} \cdot \frac{5}{7}$ $1\frac{3}{14}$

Find each product. Write the answer in simplest form.

3. $2\frac{1}{5} \cdot 1\frac{1}{4}$ $2\frac{3}{4}$ **4.** $5 \cdot 3\frac{1}{5}$ 16

5. A nurse gave a patient $3\frac{1}{2}$ tablets of a medication. If each tablet contained $\frac{1}{20}$ grain of the medication, how much medication did the patient receive? $\frac{7}{40}$ grain

Available on Daily Transparency in CRB

Hands-On LAB

5B Model Fraction Division

Pacing: Traditional 1 day
Block $\frac{1}{2}$ day

Objective: To use grids to model fraction division

Materials: Pencil and paper

Lab Resources

Hands-On Lab Activities . . . . pp. 31–33

Using the Page

Discuss with students what a grid represents and how to use grids to model fractions.

Model each fraction.

1. $\frac{4}{5}$ 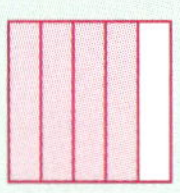

2. $\frac{1}{2}$ 

3. $1\frac{2}{3}$ 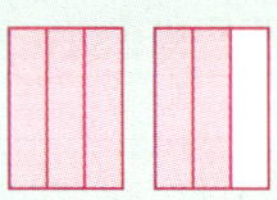

Hands-On LAB 5B

Model Fraction Division

Use with Lesson 5-3

internet connect
Lab Resources Online
go.hrw.com
KEYWORD: MR4 Lab5B

You can use grids to help you understand division of fractions.

Activity 1

1. Think of $4\frac{1}{2} \div 3$ as dividing $4\frac{1}{2}$ into 3 equal groups.

   Shade $4\frac{1}{2}$ squares.

   Divide the squares into 3 equal groups.

   Look at one of the shaded groups.

   What fraction is this?

   $4\frac{1}{2} \div 3$ is $1\frac{1}{2}$.

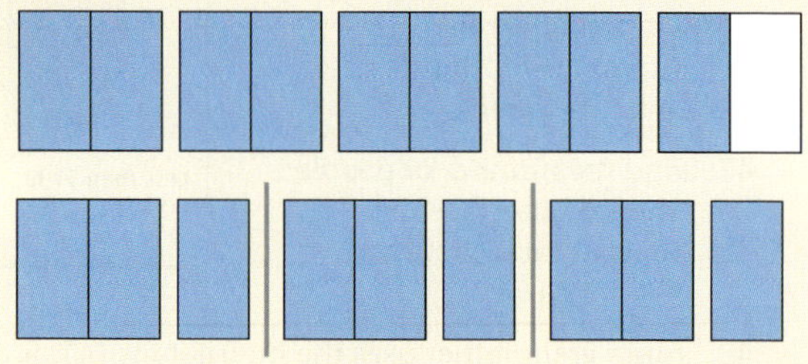

Think and Discuss

1. Explain how you know the number of groups into which you must divide the squares.
   The number of groups is the number you are dividing by.

Try This

Write the division expression modeled on each grid.

1. 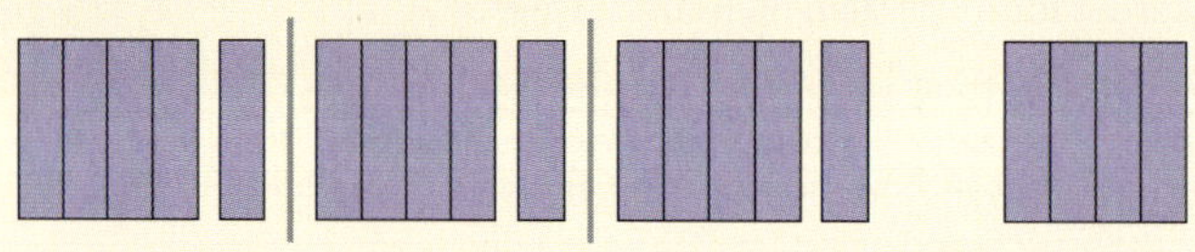 = 1 $3\frac{3}{4} \div 3$

Use grids to model each division expression. Then find the value of the expression.

2. $9\frac{1}{3} \div 4$ $2\frac{1}{3}$
3. $3\frac{3}{4} \div 5$ $\frac{3}{4}$
4. $4\frac{2}{3} \div 2$ $2\frac{1}{3}$
5. $4\frac{1}{5} \div 3$ $1\frac{2}{5}$

Activity 2

1. To find $2\frac{2}{3} \div \frac{2}{3}$, think, "How many groups of $\frac{2}{3}$ are in $2\frac{2}{3}$?"

   Shade $2\frac{2}{3}$ squares.

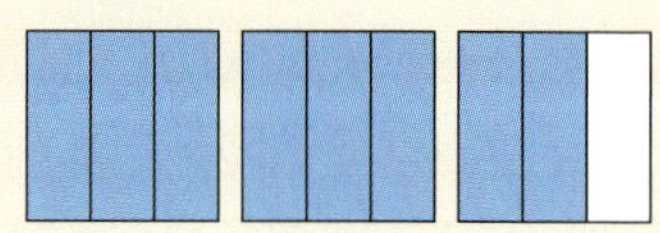

Answers

Activity 1

Try This

2. 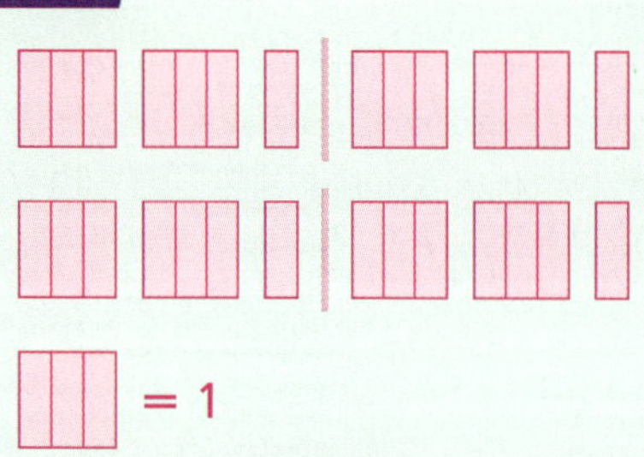

3. 

4. 

5. 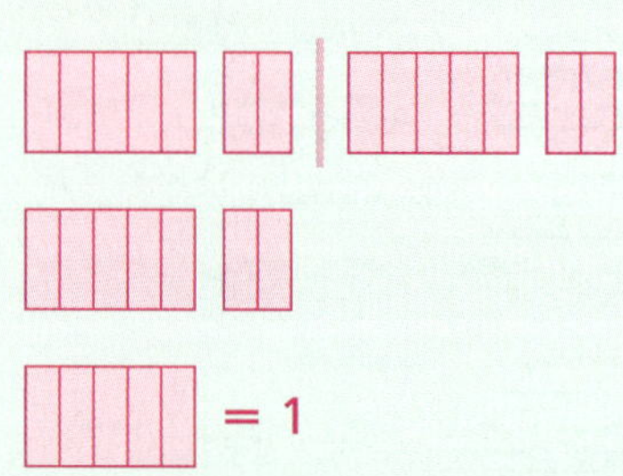

Divide the shaded squares and shaded thirds into equal groups of $\frac{2}{3}$.

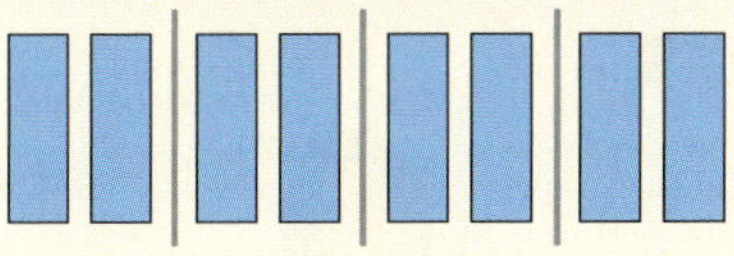

There are 4 groups of $\frac{2}{3}$ in $2\frac{2}{3}$. $2\frac{2}{3} \div \frac{2}{3} = 4$.

2 To find $3 \div \frac{3}{4}$, think, "How many groups of $\frac{3}{4}$ are in 3?"

Shade 3 squares. Then divide the squares into fourths because the denominator of $\frac{3}{4}$ is 4.

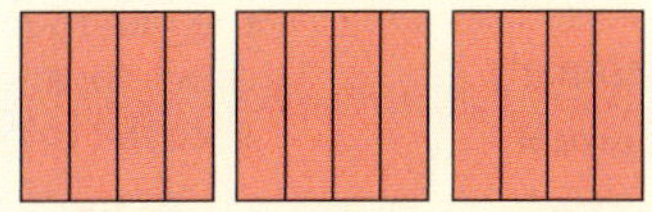

Divide the shaded squares into equal groups of $\frac{3}{4}$.

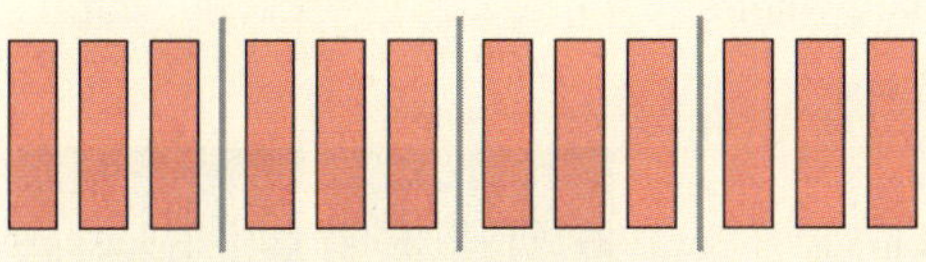

There are 4 groups of $\frac{3}{4}$ in 3. $3 \div \frac{3}{4} = 4$.

Think and Discuss

1. Explain what prediction you can make about the value of $6 \div \frac{3}{4}$ if you know that $3 \div \frac{3}{4}$ is 4. **You can predict that the value would be twice as much, 8, because there will be twice as many groups of $\frac{3}{4}$ in 6 as there were in 3. This is because 3 times 2 is 6.**

Try This

Write the division expression modeled by each grid.

1. 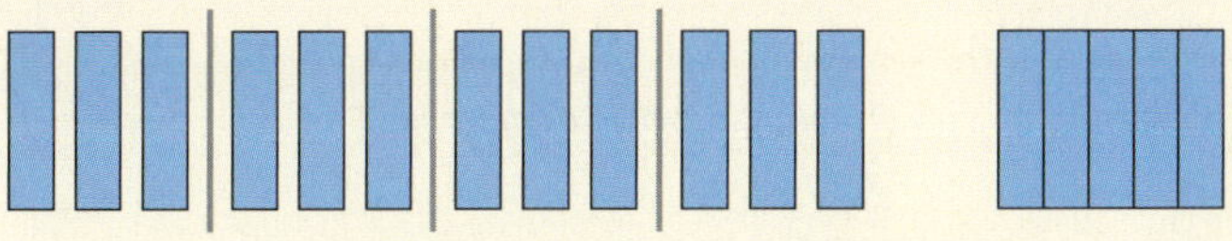

$2\frac{2}{5} \div \frac{3}{5}$

2. 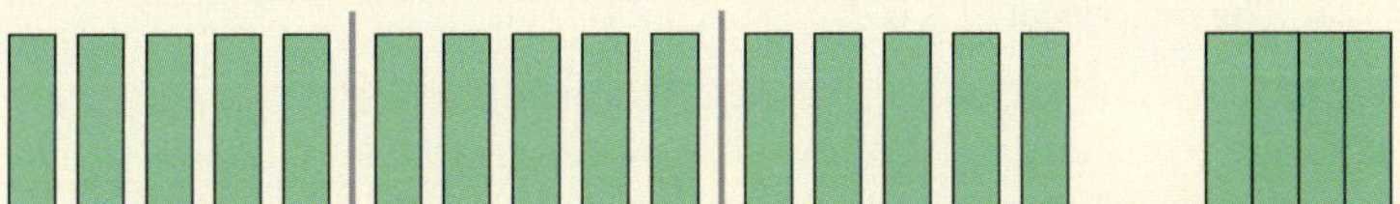

$3\frac{3}{4} \div 1\frac{1}{4}$

Use grids to model each division expression. Then find the value of the expression.

3. $4 \div 1\frac{1}{3}$ **3** **4.** $3\frac{3}{4} \div \frac{3}{4}$ **5** **5.** $5\frac{1}{3} \div \frac{2}{3}$ **8** **6.** $6\frac{2}{3} \div 1\frac{2}{3}$ **4**

Answers

Activity 2

Try This

3. 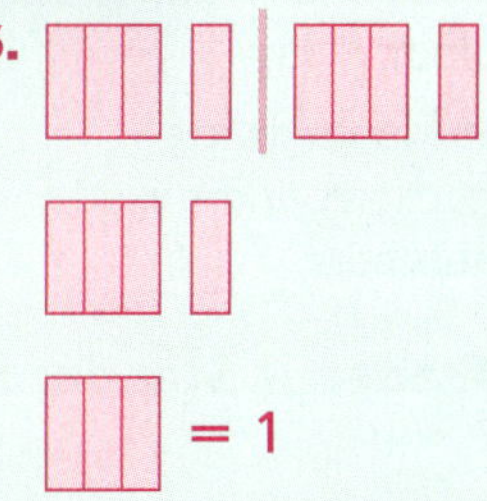

4. 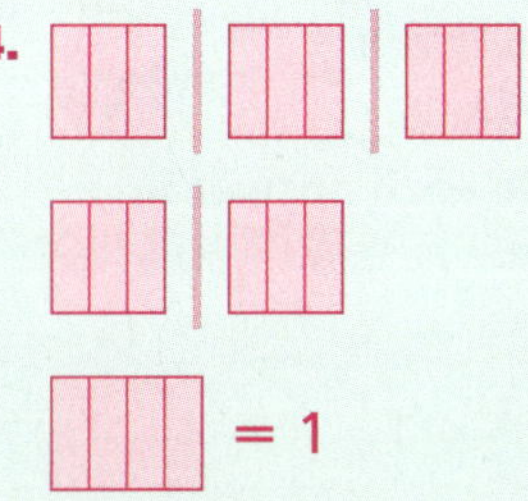

5. 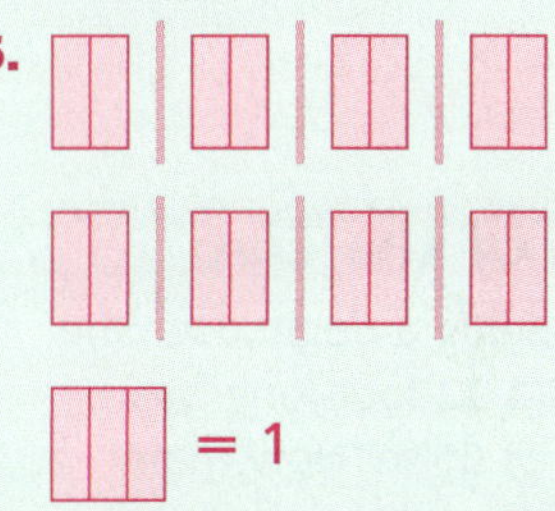

6. 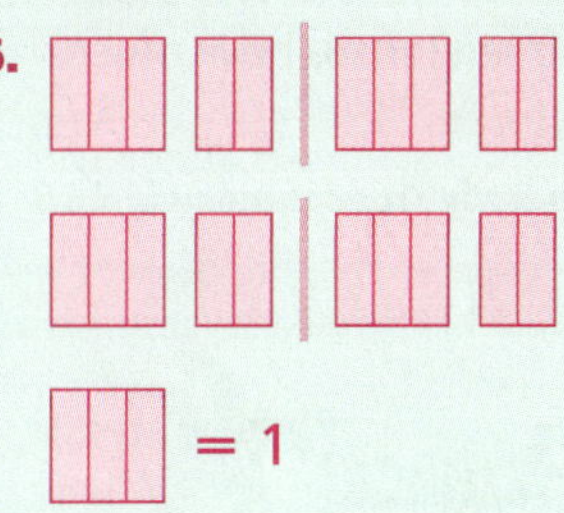

Teacher to Teacher

Dawn Schiller
Temple, Texas

Several of my students have a difficult time dividing fractions. I give each student a set of fraction bars, a Dry-Erase™ board, and a marker. Students model the dividend by drawing a train. Then they draw another train below the first train to represent the divisor. The students then count the number of sections in the second train, which gives them the answer. Fraction bars can be used to model dividing whole numbers by fractions, dividing fractions by fractions, and dividing mixed numbers.

5-3 Organizer

Pacing: Traditional 1 day
Block $\frac{1}{2}$ day

Objective: Students divide fractions and mixed numbers.

Indiana Academic Standards
6.2.4, 6.7.1, 6.7.5, 6.7.11

Warm Up

Multiply. Write each answer in simplest form.

1. $2\frac{1}{2} \cdot 2$ 5
2. $1\frac{1}{2} \cdot 3\frac{1}{2}$ $\frac{21}{4}$ or $5\frac{1}{4}$
3. $3\frac{1}{3} \cdot 3$ 10
4. $2\frac{1}{5} \cdot 1\frac{1}{4}$ $\frac{11}{4}$ or $2\frac{3}{4}$

Problem of the Day

Rachel is building a doghouse. She needs to cut a 12-foot-long board into lengths of 65 in. How many lengths can she cut, and will there be any wood left? If so, how much?
2 cuts; yes, 14 in.

Available on Daily Transparency in CRB

Math Humor

What is the best time to divide a half dollar evenly between two people?
A quarter to two

5-3 Dividing Fractions and Mixed Numbers

Learn to divide fractions and mixed numbers.

Vocabulary
reciprocal

Curtis is making sushi rolls. First, he will place a sheet of seaweed, called *nori,* on the sushi rolling mat. Then, he will use the mat to roll up rice, cucumber, avocado, and crabmeat. Finally, he will slice the roll into smaller pieces.

Curtis has 2 cups of rice and will use $\frac{1}{3}$ cup for each sushi roll. How many sushi rolls can he make?

Think: How many $\frac{1}{3}$ pieces equal 2 wholes?

There are six $\frac{1}{3}$ pieces in 2 wholes.

Curtis can make 6 sushi rolls.

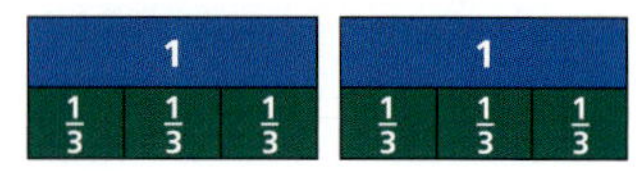

Reciprocals can help you divide by fractions. Two numbers are **reciprocals** if their product is 1.

EXAMPLE 1 Finding Reciprocals

Find the reciprocal.

A $\frac{1}{5}$

$\frac{1}{5} \cdot \blacksquare = 1$ *Think: $\frac{1}{5}$ of what number is 1?*

$\frac{1}{5} \cdot 5 = 1$ *$\frac{1}{5}$ of $\frac{5}{1}$ is 1.*

The reciprocal of $\frac{1}{5}$ is 5.

B $\frac{3}{4}$

$\frac{3}{4} \cdot \blacksquare = 1$ *Think: $\frac{3}{4}$ of what number is 1?*

$\frac{3}{4} \cdot \frac{4}{3} = \frac{12}{12} = 1$ *$\frac{3}{4}$ of $\frac{4}{3}$ is 1.*

The reciprocal of $\frac{3}{4}$ is $\frac{4}{3}$.

C $2\frac{1}{3}$

$\frac{7}{3} \cdot \blacksquare = 1$ *Write $2\frac{1}{3}$ as $\frac{7}{3}$.*

$\frac{7}{3} \cdot \frac{3}{7} = \frac{21}{21} = 1$ *$\frac{7}{3}$ of $\frac{3}{7}$ is 1.*

The reciprocal of $\frac{7}{3}$ is $\frac{3}{7}$.

1 Introduce

Alternate Opener

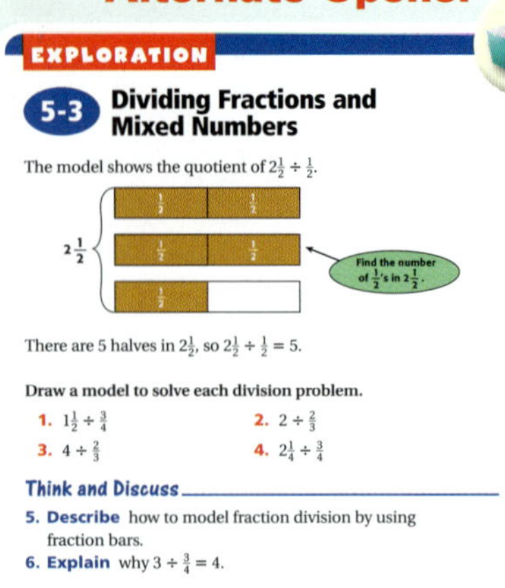
EXPLORATION

5-3 Dividing Fractions and Mixed Numbers

The model shows the quotient of $2\frac{1}{2} \div \frac{1}{2}$.

Find the number of $\frac{1}{2}$'s in $2\frac{1}{2}$.

There are 5 halves in $2\frac{1}{2}$, so $2\frac{1}{2} \div \frac{1}{2} = 5$.

Draw a model to solve each division problem.

1. $1\frac{1}{2} \div \frac{3}{4}$
2. $2 \div \frac{2}{3}$
3. $4 \div \frac{2}{3}$
4. $2\frac{1}{4} \div \frac{3}{4}$

Think and Discuss

5. **Describe** how to model fraction division by using fraction bars.
6. **Explain** why $3 \div \frac{3}{4} = 4$.

Motivate

List the following fractions:

$\frac{2}{3}, \frac{2}{5}, \frac{3}{2}, \frac{3}{4}, \frac{4}{3}, \frac{5}{2}, \frac{5}{6}, \frac{6}{5}$

Have students identify the pairs of fractions with a product of 1.

$\frac{2}{3}$ and $\frac{3}{2}$; $\frac{2}{5}$ and $\frac{5}{2}$; $\frac{3}{4}$ and $\frac{4}{3}$; $\frac{5}{6}$ and $\frac{6}{5}$

When a fraction is inverted, the result is the reciprocal of the original fraction. When reciprocals are multiplied, the product is 1.

Exploration worksheet and answers on Chapter 5 Resource Book pp. 25 and 114

2 Teach

Lesson Presentation

Guided Instruction

In this lesson, students learn to divide fractions and mixed numbers. First explain what a reciprocal is, and have students find reciprocals for fractions and mixed numbers. Then teach students to use a reciprocal and multiplication to divide by a fraction or mixed number.

Look at the relationship between the fractions $\frac{3}{4}$ and $\frac{4}{3}$. If you switch the numerator and denominator of a fraction, you will find its reciprocal. Dividing by a number is the same as multiplying by its reciprocal.

$$24 \div 4 = 6 \qquad 24 \cdot \frac{1}{4} = 6$$

EXAMPLE 2 Using Reciprocals to Divide Fractions and Mixed Numbers

Divide. Write each answer in simplest form.

A $\frac{4}{5} \div 5$

$\frac{4}{5} \div 5 = \frac{4}{5} \cdot \frac{1}{5}$ — *Rewrite as multiplication using the reciprocal of 5, $\frac{1}{5}$.*

$= \frac{4 \cdot 1}{5 \cdot 5}$ — *Multiply by the reciprocal.*

$= \frac{4}{25}$ — *The answer is in simplest form.*

B $\frac{3}{4} \div \frac{1}{2}$

$\frac{3}{4} \div \frac{1}{2} = \frac{3}{4} \cdot \frac{2}{1}$ — *Rewrite as multiplication using the reciprocal of $\frac{1}{2}$, $\frac{2}{1}$.*

$= \frac{3 \cdot \cancel{2}^{1}}{\cancel{4}_{2} \cdot 1}$ — *Simplify before multiplying.*

$= \frac{3}{2}$ — *Multiply.*

$= 1\frac{1}{2}$ — *You can write the answer as a mixed number.*

C $2\frac{2}{3} \div 1\frac{1}{6}$

$2\frac{2}{3} \div 1\frac{1}{6} = \frac{8}{3} \div \frac{7}{6}$ — *Write the mixed numbers as improper fractions. $2\frac{2}{3} = \frac{8}{3}$ and $1\frac{1}{6} = \frac{7}{6}$*

$= \frac{8}{3} \cdot \frac{6}{7}$ — *Rewrite as multiplication.*

$= \frac{8 \cdot \cancel{6}^{2}}{\cancel{3}_{1} \cdot 7}$ — *Simplify before multiplying.*

$= \frac{16}{7}$ — *Multiply.*

$= 2\frac{2}{7}$ — *You can write the answer as a mixed number.*

Think and Discuss

1. **Explain** how you can use mental math to find the value of n in the equation $\frac{5}{8} \cdot n = 1$.
2. **Explain** how to find the reciprocal of $3\frac{6}{11}$.

COMMON ERROR ALERT

Some students may forget that a division problem involving fractions has two parts: changing from division to multiplication and finding the reciprocal of the divisor. Remind them that both steps are necessary to solve the problem correctly.

Additional Examples

Example 1

Find the reciprocal.

A. $\frac{1}{9}$ The reciprocal of $\frac{1}{9}$ is 9.

B. $\frac{2}{3}$ The reciprocal of $\frac{2}{3}$ is $\frac{3}{2}$.

C. $3\frac{1}{5}$ The reciprocal of $\frac{16}{5}$ is $\frac{5}{16}$.

Example 2

Divide. Write each answer in simplest form.

A. $\frac{8}{7} \div 7$ $\frac{8}{49}$

B. $\frac{5}{6} \div \frac{2}{3}$ $1\frac{1}{4}$

C. $2\frac{3}{4} \div 1\frac{1}{12}$ $2\frac{7}{13}$

3 Close

Reaching All Learners

Through Critical Thinking

Challenge students to find the missing numbers in the following division problems. Have them check their answers.

1. $\frac{3}{4} \div$ ■ $= 3\frac{3}{4}$ $\frac{1}{5}$
2. ■ $\div \frac{5}{6} = 1\frac{4}{5}$ $1\frac{1}{2}$
3. $2\frac{1}{2} \div$ ■ $= 1\frac{3}{7}$ $1\frac{3}{4}$
4. ■ $\div 2\frac{1}{10} = 1\frac{2}{3}$ $3\frac{1}{2}$

A recording sheet is provided on Chapter 5 Resource Book p. 104.

Summarize

Review the term *reciprocal*, and have students explain how it is used in division of fractions.

Possible answers: Two numbers are reciprocals if their product is 1. To divide by a fraction or mixed number, multiply the dividend by the reciprocal of the divisor.

Answers to Think and Discuss

1. Possible answer: You know the product of a fraction and its reciprocal is 1. Therefore, n must be the reciprocal of $\frac{5}{8}$, or $\frac{8}{5}$.
2. First change $3\frac{6}{11}$ to an improper fraction, then switch the numerator and denominator. $3\frac{6}{11} = \frac{39}{11}$, so the reciprocal of $3\frac{6}{11}$ is $\frac{11}{39}$.

5-3 PRACTICE & ASSESS

Students may want to refer back to the lesson examples.

Assignment Guide

If you finished Example 1 assign:
Core 1–4, 11–18, 43, 45, 49, 60–64
Enriched 11–18, 43–45, 48–50, 60–64

If you finished Example 2 assign:
Core 1–27, 51–53, 60–64
Enriched 28–64

Notes

Math Background

The use of multiplication by the reciprocal to divide by a fraction can be traced to Hindu and Arab mathematicians in the early Middle Ages.

A second method, which was taught in arithmetic books in the past, is to first rewrite the dividend and divisor using a common denominator. Thus, $\frac{5}{6} \div \frac{7}{8}$ could be rewritten as $\frac{20}{24} \div \frac{21}{24}$. Then the quotient of the numerators is found: $\frac{20}{24} \div \frac{21}{24} = \frac{20}{21}$.

This method has the advantage of fitting in nicely with the algorithms learned for addition and subtraction. Its disadvantage lies in the need for a common denominator.

5-3 Exercises

FOR EXTRA PRACTICE
see page 645

internet connect
Homework Help Online
go.hrw.com Keyword: MR4 5-3

GUIDED PRACTICE

See Example 1 Find the reciprocal.

1. $\frac{2}{7}$ **$\frac{7}{2}$**
2. $\frac{5}{9}$ **$\frac{9}{5}$**
3. $\frac{1}{9}$ **$\frac{9}{1}$, or 9**
4. $\frac{3}{11}$ **$\frac{11}{3}$**

See Example 2 Divide. Write each answer in simplest form.

5. $\frac{5}{6} \div 3$ **$\frac{5}{18}$**
6. $2\frac{1}{7} \div 1\frac{1}{4}$ **$1\frac{5}{7}$**
7. $\frac{5}{12} \div 5$ **$\frac{1}{12}$**
8. $\frac{2}{3} \div \frac{1}{6}$ **4**
9. $\frac{3}{10} \div 1\frac{2}{3}$ **$\frac{9}{50}$**
10. $\frac{4}{7} \div 1\frac{1}{7}$ **$\frac{1}{2}$**

INDEPENDENT PRACTICE

See Example 1 Find the reciprocal.

11. $\frac{7}{8}$ **$\frac{8}{7}$**
12. $\frac{1}{10}$ **$\frac{10}{1}$, or 10**
13. $\frac{3}{8}$ **$\frac{8}{3}$**
14. $\frac{11}{12}$ **$\frac{12}{11}$**
15. $\frac{8}{11}$ **$\frac{11}{8}$**
16. $\frac{5}{6}$ **$\frac{6}{5}$**
17. $\frac{6}{7}$ **$\frac{7}{6}$**
18. $\frac{2}{9}$ **$\frac{9}{2}$**

See Example 2 Divide. Write each answer in simplest form.

19. $\frac{7}{8} \div 4$ **$\frac{7}{32}$**
20. $2\frac{3}{8} \div 1\frac{3}{4}$ **$1\frac{5}{14}$**
21. $\frac{8}{9} \div 12$ **$\frac{2}{27}$**
22. $3\frac{5}{6} \div 1\frac{5}{9}$ **$2\frac{13}{28}$**
23. $\frac{9}{10} \div 3$ **$\frac{3}{10}$**
24. $2\frac{4}{5} \div 1\frac{5}{7}$ **$1\frac{19}{30}$**
25. $\frac{5}{8} \div \frac{1}{2}$ **$1\frac{1}{4}$**
26. $1\frac{1}{2} \div 2\frac{1}{4}$ **$\frac{2}{3}$**
27. $\frac{7}{12} \div 2\frac{5}{8}$ **$\frac{2}{9}$**

PRACTICE AND PROBLEM SOLVING

Multiply or divide. Write each answer in simplest form.

28. $2\frac{3}{4} \div 2\frac{1}{5}$ **$1\frac{1}{4}$**
29. $4\frac{4}{5} \div 2\frac{6}{7}$ **$1\frac{17}{25}$**
30. $\frac{3}{8} \cdot \frac{5}{12}$ **$\frac{5}{32}$**
31. $6 \cdot \frac{7}{9}$ **$4\frac{2}{3}$**
32. $3\frac{1}{7} \div 5$ **$\frac{22}{35}$**
33. $\frac{9}{14} \div \frac{1}{6}$ **$3\frac{6}{7}$**
34. $5\frac{3}{5} \div \frac{4}{7}$ **$9\frac{4}{5}$**
35. $\frac{9}{11} \cdot 2\frac{2}{3}$ **$2\frac{2}{11}$**
36. $2\frac{7}{10} \div 3\frac{3}{5}$ **$\frac{3}{4}$**
37. $\frac{11}{12} \cdot \frac{9}{10} \div 1\frac{1}{4}$ **$\frac{33}{50}$**
38. $2\frac{3}{4} \cdot 1\frac{2}{3} \div 5$ **$\frac{11}{12}$**
39. $1\frac{1}{2} \div \frac{3}{4} \cdot \frac{2}{5}$ **$\frac{4}{5}$**
40. $\frac{3}{4} \cdot \left(\frac{5}{7} \div \frac{1}{2}\right)$ **$1\frac{1}{14}$**
41. $4\frac{2}{3} \div \left(6 \cdot \frac{3}{5}\right)$ **$1\frac{8}{27}$**
42. $5\frac{1}{5} \cdot \left(3\frac{2}{5} \cdot 2\frac{1}{3}\right)$ **$41\frac{19}{75}$**

Decide whether the fractions in each pair are reciprocals. If not, write the reciprocal of each fraction.

43. $\frac{1}{2}, 2$ **yes**
44. $\frac{3}{8}, \frac{16}{6}$ **yes**
45. $\frac{7}{9}, \frac{21}{27}$ **$\frac{9}{7}, \frac{27}{21}$**
46. $\frac{5}{6}, \frac{12}{10}$ **yes**
47. $1\frac{1}{2}, \frac{2}{3}$ **yes**
48. $\frac{2}{5}, \frac{4}{25}$ **$\frac{5}{2}, \frac{25}{4}$**
49. $\frac{3}{7}, 2\frac{1}{3}$ **yes**
50. $5, \frac{5}{1}$ **$\frac{1}{5}, \frac{1}{5}$**

RETEACH 5-3

LESSON 5-3 **Reteach**
Dividing Fractions and Mixed Numbers

Two numbers are reciprocals if their product is 1. $\frac{2}{3}$ and $\frac{3}{2}$ are reciprocals because $\frac{2}{3} \cdot \frac{3}{2} = \frac{6}{6} = 1$.

Dividing by a fraction is the same as multiplying by its reciprocal.

$\frac{1}{4} \div 2 = \frac{1}{8}$ $\qquad$ $\frac{1}{4} \cdot \frac{1}{2} = \frac{1}{8}$

So, you can use reciprocals to divide by fractions.

To find $\frac{2}{3} \div 4$, first rewrite the expression as a multiplication expression using the reciprocal of the divisor, 4.

$\frac{2}{3} \cdot \frac{1}{4}$

Then use canceling to find the product in simplest form.

$\frac{2}{3} \div 4 = \frac{2}{3} \cdot \frac{1}{4} = \frac{1}{3} \cdot \frac{1}{2} = \frac{1}{6}$

To find $3\frac{1}{4} \div 1\frac{1}{2}$, first rewrite the expression using improper fractions.

$\frac{13}{4} \div \frac{3}{2}$

Next, write the expression as a multiplication expression.

$\frac{13}{4} \cdot \frac{2}{3}$

$3\frac{1}{4} \div 1\frac{1}{2} = \frac{13}{4} \div \frac{3}{2} = \frac{13}{4} \cdot \frac{2}{3} = \frac{13}{2} \cdot \frac{1}{3} = \frac{13}{6} = 2\frac{1}{6}$

Divide. Write each answer in simplest form.

1. $\frac{1}{4} \div 3$
   $\frac{1}{4} \div \frac{3}{1}$
   $\frac{1}{4} \cdot \frac{1}{3}$
   $\frac{1}{12}$
2. $1\frac{1}{2} \div 1\frac{1}{4}$
   $\frac{3}{2} \div \frac{5}{4}$
   $\frac{3}{2} \cdot \frac{4}{5}$
   $1\frac{1}{5}$
3. $\frac{3}{8} \div 2$
   $\frac{3}{8} \div \frac{2}{1}$
   $\frac{3}{8} \cdot \frac{1}{2}$
   $\frac{3}{16}$
4. $2\frac{1}{3} \div 1\frac{3}{4}$
   $\frac{7}{3} \div \frac{7}{4}$
   $\frac{7}{3} \cdot \frac{4}{7}$
   $1\frac{1}{3}$
5. $\frac{1}{5} \div 2$ **$\frac{1}{10}$**
6. $1\frac{1}{6} \div 2\frac{2}{3}$ **$\frac{7}{16}$**
7. $\frac{1}{8} \div 4$ **$\frac{1}{32}$**
8. $3\frac{1}{8} \div \frac{1}{2}$ **$6\frac{1}{4}$**

PRACTICE 5-3

LESSON 5-3 **Practice B**
Dividing Fractions and Mixed Numbers

Find the reciprocal.

1. $\frac{5}{7}$ **$\frac{7}{5}$**
2. $\frac{9}{8}$ **$\frac{8}{9}$**
3. $\frac{3}{5}$ **$\frac{5}{3}$**
4. $\frac{1}{10}$ **10**
5. $\frac{4}{9}$ **$\frac{9}{4}$**
6. $\frac{13}{14}$ **$\frac{14}{13}$**
7. $1\frac{1}{3}$ **$\frac{3}{4}$**
8. $2\frac{4}{5}$ **$\frac{5}{14}$**
9. $3\frac{1}{6}$ **$\frac{6}{19}$**

Divide. Write each answer in simplest form.

10. $\frac{5}{6} \div 5$ **$\frac{1}{6}$**
11. $2\frac{3}{4} \div 1\frac{4}{7}$ **$1\frac{3}{4}$**
12. $\frac{7}{8} \div \frac{2}{3}$ **$1\frac{5}{16}$**
13. $3\frac{1}{4} \div 2\frac{3}{4}$ **$1\frac{2}{11}$**
14. $\frac{9}{10} \div 3$ **$\frac{3}{10}$**
15. $\frac{3}{4} \div 9$ **$\frac{1}{12}$**
16. $2\frac{6}{9} \div \frac{6}{7}$ **$3\frac{1}{9}$**
17. $\frac{5}{6} \div 2\frac{3}{10}$ **$\frac{25}{69}$**
18. $2\frac{1}{8} \div 3\frac{1}{4}$ **$\frac{17}{26}$**
19. The rope in the school gymnasium is $10\frac{1}{2}$ feet long. To make it easier to climb, the gym teacher tied a knot in the rope every $\frac{3}{4}$ foot. How many knots are in the rope? **14 knots**
20. Mr. Fulton bought $12\frac{1}{2}$ pounds of ground beef for the cookout. He plans on using $\frac{1}{4}$ pound of beef for each hamburger. How many hamburgers can he make? **50 hamburgers**
21. Mrs. Marks has $9\frac{1}{4}$ ounces of fertilizer for her plants. She plans on using $\frac{3}{4}$ ounce of fertilizer for each plant. How many plants can she fertilizer? **12 plants**

LIFE SCIENCE The bar graph shows the lengths of some species of snakes found in the United States.

Use the bar graph for Exercises 51–53.

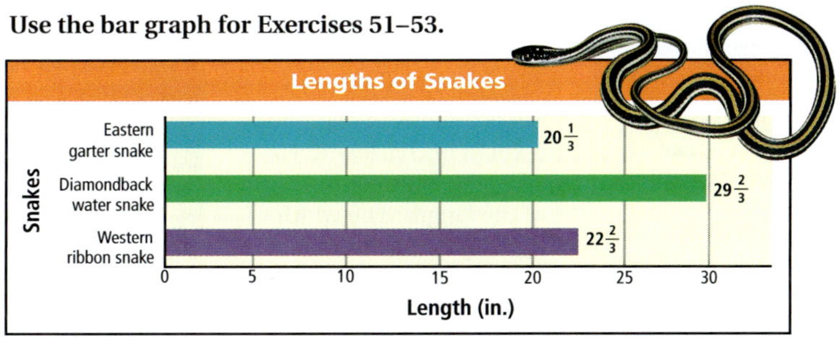

51. Is the length of the eastern garter snake greater than or less than $\frac{1}{2}$ yd? Explain. **Greater than, because $\frac{1}{2}$ yd = 18 in. and $20\frac{1}{3} > 18$.**

52. What is the average length of all the snakes? **$24\frac{2}{9}$ in.**

53. Jim measured the length of a rough green snake. It was $27\frac{1}{3}$ in. long. What would the average length of the snakes be if Jim's measure of a rough green snake were added? **25 in.**

54. At Lina's restaurant, one serving of chili is $1\frac{1}{2}$ cups. The chef makes 48 cups of chili each night. How many servings of chili are in 48 cups? **32 servings**

55. Rhula bought 12 lb of raisins. She packed them into freezer bags so that each bag weighs $\frac{3}{4}$ lb. How many freezer bags did she pack? **16 bags**

56. Lisa had some wood that was $12\frac{1}{2}$ feet long. She cut it into 5 pieces that are equal in length. How long is each piece of wood? **$2\frac{1}{2}$ ft long or 30 in.**

57. ***WHAT'S THE ERROR?*** A student said the reciprocal of $6\frac{2}{3}$ is $6\frac{3}{2}$. Explain the error. Then write the correct reciprocal.

58. ***WRITE ABOUT IT*** Explain how you can divide fractions to find the quotient of $\frac{3}{4} \div 2\frac{1}{3}$.

59. ***CHALLENGE*** Evaluate the expression $\frac{(6-3)}{4} \div \frac{1}{8} \cdot 5$. **30**

Spiral Review

Identify a pattern in each sequence and name the missing term. (Lesson 1-7)

60. 85, 80, 75, 70, 65, ■, … **subtract 5; 60**

61. 1, 4, 7, 10, 13, ■, … **add 3; 16**

62. 2, 6, 5, 9, 8, ■, … **add 4, then subtract 1; 12**

63. **TEST PREP** Which is the most reasonable measure for the length of a pencil? (Lesson 3-4) **B**

A 15 m **B** 15 cm **C** 15 yd **D** 15 oz

64. **TEST PREP** Of which set of numbers is 16 the GCF? (Lesson 4-3) **F**

F 16, 32, 48 **G** 12, 24, 32 **H** 24, 48, 60 **J** 8, 80, 100

Answers

Possible answers:

57. The student did not write $6\frac{2}{3}$ as an improper fraction before switching the denominator and numerator in the fraction. The correct answer is $\frac{3}{20}$.

58. Find the reciprocal of $2\frac{1}{3}$, which is $\frac{3}{7}$. Then multiply $\frac{3}{4} \cdot \frac{3}{7} = \frac{9}{28}$.

Journal

Have students compare and contrast division by fractions and mixed numbers with division by whole numbers and decimals. Have them support their comparisons with examples.

Test Prep Doctor

For Exercise 64, students need to be careful to find the set of numbers that are all divisible by 16. Answers **G** and **J** are incorrect because the first number in each set is smaller than 16 and no number can be a factor of a number that is smaller than itself. Answer **H** is incorrect because the first number, 24, does not have 16 as a factor.

CHALLENGE 5-3

LESSON 5-3 **Challenge**
Inching Across the U.S.A.

You can use a map and a ruler to find the distance between places. On the map below, for example, you measure that 2 inches separate Kansas City, Missouri, and Richmond, Virginia. The map scale shows that $\frac{1}{4}$ inch on the map equals 140 miles. So the distance between Kansas City and Richmond is 1,120 miles.

Calculations:
$2 \div \frac{1}{4} = 8$
$8 \times 140 = 1{,}120$

Use the map and a ruler to find the distance in miles between each pair of cities.

1. Miami, Florida, and New Orleans, Louisiana **770 miles**
2. Denver, Colorado, and Los Angeles, California **1,015 miles**
3. Seattle, Washington, and Minneapolis, Minnesota **1,540 miles**
4. Washington, D.C., and Atlanta, Georgia **665 miles**
5. Oklahoma City, Oklahoma, and Pittsburgh, Pennsylvania **1,120 miles**
6. San Francisco, California, and Boston, Massachusetts **3,080 miles**

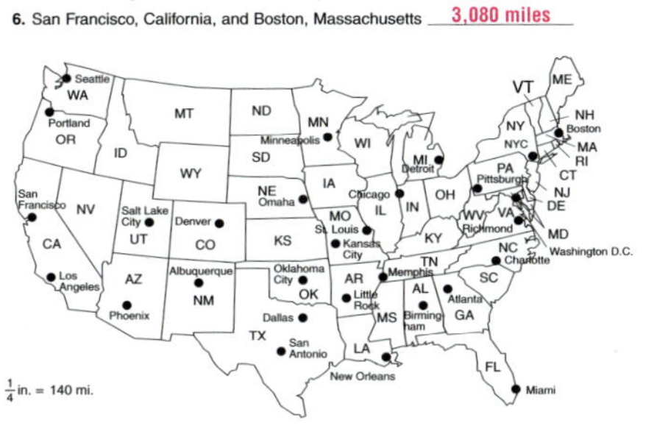

PROBLEM SOLVING 5-3

LESSON 5-3 **Problem Solving**
Dividing Fractions and Mixed Numbers

Write the correct answer in simplest form.

1. Horses are measured in units called *hands*. One inch equals $\frac{1}{4}$ hand. The average Clydesdale horse is $17\frac{1}{5}$ hands high. What is the horse's height in inches? in feet?
$68\frac{4}{5}$ inches; $5\frac{11}{15}$ feet

2. Cloth manufacturers use a unit of measurement called a *finger*. One finger is equal to $4\frac{1}{2}$ inches. If 25 inches are cut off a bolt of cloth, how many fingers of cloth were cut?
$5\frac{5}{9}$ fingers

3. People in England measure weights in units called *stones*. One pound equals $\frac{1}{14}$ of a stone. If a cat weighs $\frac{3}{4}$ stone, how many pounds does it weigh?
$10\frac{1}{2}$ pounds

4. The hiking trail is $\frac{9}{10}$ mile long. There are 6 markers evenly posted along the trail to direct hikers. How far apart are the markers placed?
$\frac{3}{20}$ mile

5. Phyllis bought 14 yards of material to make chair cushions. She cut the material into pieces $1\frac{3}{4}$ yards long to make each cushion. How many cushions did Phyllis make?
8 cushions

6. Dry goods are sold in units called *pecks* and *bushels*. One peck equals $\frac{1}{4}$ bushel. If Peter picks $5\frac{1}{2}$ bushels of peppers, how many pecks of peppers did Peter pick?
22 pecks

Choose the letter for the best answer.

7. A cake recipe calls for $1\frac{1}{2}$ cups of butter. One tablespoon equals $\frac{1}{16}$ cup. How many tablespoons of butter do you need to make the cake?
(A) 24 tablespoons
B 8 tablespoons
C $\frac{3}{32}$ tablespoon
D 9 tablespoons

8. Printed letters are measured in units called *points*. One point equals $\frac{1}{72}$ inch. If you want the title of a paper you are typing on a computer to be $\frac{1}{2}$ inch tall, what type point size should you use?
F 144 point
(G) 36 point
H $\frac{1}{36}$ point
J $\frac{1}{144}$ point

Lesson Quiz

Find the reciprocal.

1. $\frac{1}{11}$ **$\frac{11}{1}$**

2. $\frac{8}{13}$ **$\frac{13}{8}$**

Divide. Write each answer in simplest form.

3. $\frac{4}{7} \div 20$ **$\frac{1}{35}$**

4. $3\frac{1}{2} \div 2\frac{1}{2}$ **$\frac{7}{5}$, or $1\frac{2}{5}$**

5. Rhonda put $2\frac{3}{4}$ pounds of pecans into $\frac{1}{4}$-pound bags. How many bags did Rhonda fill? **11 bags**

Available on Daily Transparency in CRB

5-4 Organizer

Pacing: Traditional 1 day
Block $\frac{1}{2}$ day

Objective: Students solve equations by multiplying and dividing fractions.

Indiana Academic Standards
6.2.4, 6.2.5, 6.3.1, 6.7.1, 6.7.5, 6.7.11

Warm Up

Solve.

1. $x - 5 = 17$ $x = 22$
2. $5x = 125$ $x = 25$
3. $x + 12 = 86$ $x = 74$
4. $9x = 108$ $x = 12$

Problem of the Day

Stephen forgot his locker number, but he remembered that the sum of the digits is 11 and that the digits are all odd numbers. The locker numbers are from 1 to 120. What is Stephen's locker number? 119

Available on Daily Transparency in CRB

Math Humor

Needing 75 cents is like solving the equation $\frac{4}{3}x = 1$. In both cases, the solution is 3 quarters.

5-4 Solving Fraction Equations: Multiplication and Division

Learn to solve equations by multiplying and dividing fractions.

Josef is building a fish pond for koi in his backyard. He makes the width of the pond $\frac{2}{3}$ of the length. The width of the pond is 14 feet. You can use the equation $\frac{2}{3}\ell = 14$ to find the length of the pond.

Small koi in a backyard pond usually grow 2 to 4 inches per year.

EXAMPLE 1 Solving Equations by Multiplying and Dividing

Remember!
Dividing by a number is the same as multiplying by its reciprocal.

Solve each equation. Write the answer in simplest form.

A $\frac{2}{3}\ell = 14$

$$\frac{2}{3}\ell = 14$$

$$\frac{2}{3}\ell \div \frac{2}{3} = 14 \div \frac{2}{3}$$ *Divide both sides of the equation by $\frac{2}{3}$.*

$$\frac{2}{3}\ell \cdot \frac{3}{2} = 14 \cdot \frac{3}{2}$$ *Multiply by $\frac{3}{2}$, the reciprocal of $\frac{2}{3}$.*

$$\ell = 14 \cdot \frac{3}{2}$$

$$\ell = \frac{14 \cdot 3}{1 \cdot 2}$$

$$\ell = \frac{42}{2}, \text{ or } 21$$

B $2x = \frac{1}{3}$

$$2x = \frac{1}{3}$$

$$\frac{2x}{1} \cdot \frac{1}{2} = \frac{1}{3} \cdot \frac{1}{2}$$ *Multiply both sides by the reciprocal of 2.*

$$x = \frac{1 \cdot 1}{3 \cdot 2}$$

$$x = \frac{1}{6}$$ *The answer is in simplest form.*

C $\frac{5x}{6} = 4$

$$\frac{5x}{6} = 4$$

$$\frac{5x}{6} \div \frac{5}{6} = \frac{4}{1} \div \frac{5}{6}$$ *Divide both sides by $\frac{5}{6}$.*

$$\frac{5x}{6} \cdot \frac{6}{5} = \frac{4}{1} \cdot \frac{6}{5}$$ *Multiply by the reciprocal of $\frac{5}{6}$.*

$$x = \frac{24}{5}, \text{ or } 4\frac{4}{5}$$

1 Introduce

Alternate Opener

EXPLORATION

5-4 Solving Fraction Equations: Multiplication and Division

You can use a number line to solve fraction equations. Look at the reasoning used to solve the equation $\frac{1}{2}n = 50$.

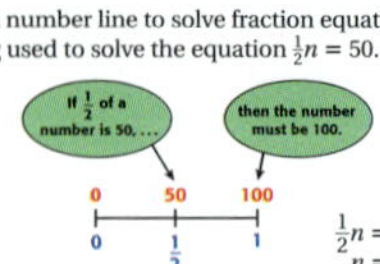

$\frac{1}{2}n = 50$
$n = 100$

Complete the number line to solve each equation.

1. $\frac{1}{2}n = 125$
2. $\frac{1}{3}n = 12$
3. $\frac{3}{4}n = 60$
4. $\frac{2}{3}n = 20$

Think and Discuss

5. **Describe** a real-world situation that could be represented by each equation in numbers 1–4.
6. **Discuss** another way of solving equations that contain fractions.

Motivate

Have students explain how to solve the equation $7b = 35$. Divide both sides of the equation by 7. Then ask students, keeping that in mind, to tell how they would solve the equation $\frac{2}{3}d = 10$. Divide both sides of the equation by $\frac{2}{3}$, or multiply both sides by $\frac{3}{2}$, the reciprocal of $\frac{2}{3}$.

Exploration worksheet and answers on Chapter 5 Resource Book pp. 34 and 116

2 Teach

Lesson Presentation

Guided Instruction

In this lesson, students learn to solve equations by multiplying and dividing fractions. First teach students to solve equations where they multiply by a reciprocal to solve. Then teach students to translate a word problem into an equation that can be solved by multiplying by a reciprocal. Have students check the answers in Example 1 by substituting the values for each variable.

EXAMPLE 2 **PROBLEM SOLVING APPLICATION**

PROBLEM SOLVING

Dexter makes dog biscuits for the animal shelter. He makes $\frac{3}{4}$ of a recipe and uses 15 cups of powdered milk. How many cups of powdered milk are in the recipe?

1 Understand the Problem

The **answer** will be the number of cups of powdered milk in the recipe.

List the **important information:**

- He makes $\frac{3}{4}$ of the recipe.
- He uses 15 cups of powdered milk.

2 Make a Plan

You can write and solve an equation. Let x represent the number of cups in the recipe.

He uses 15 cups, which is three-fourths of the amount in the recipe.
$15 = \frac{3}{4}x$

3 Solve

$15 = \frac{3}{4}x$

$15 \cdot \frac{4}{3} = \frac{3}{4}x \cdot \frac{4}{3}$ *Multiply both sides by $\frac{4}{3}$, the reciprocal of $\frac{3}{4}$.*

$\frac{\overset{5}{\cancel{15}}}{1} \cdot \frac{4}{\underset{1}{\cancel{3}}} = x$ *Simplify. Then multiply.*

$20 = x$

There are 20 cups of powdered milk in the recipe.

4 Look Back

Check $15 = \frac{3}{4}x$

$15 \stackrel{?}{=} \frac{3}{4}(20)$ *Substitute 20 for x.*

$15 \stackrel{?}{=} \frac{\overset{15}{\cancel{60}}}{\underset{1}{\cancel{4}}}$ *Multiply and simplify.*

$15 \stackrel{?}{=} 15$ ✔ *20 is the solution.*

Life Science LINK

No more than $\frac{1}{10}$ of a dog's diet should consist of treats and biscuits.

Think and Discuss

1. **Explain** whether $\frac{2}{3}x = 4$ is the same as $\frac{2}{3} = 4x$.
2. **Tell** how you know which numbers to divide by in the following equations: $\frac{2}{3}x = 4$ and $\frac{4}{5} = 8x$.

COMMON ERROR ALERT

Some students may solve the problem $\frac{2}{5}x = 4$ and get $x = 4 \cdot \frac{2}{5}$. Remind students that they need to multiply by the reciprocal.

Additional Examples

Example 1

Solve each equation. Write the answer in simplest form.

A. $\frac{3}{5}\ell = 25$ $\ell = \frac{125}{3}$, or $41\frac{2}{3}$

B. $7x = \frac{2}{5}$ $x = \frac{2}{35}$

C. $\frac{5y}{8} = 6$ $y = \frac{48}{5}$, or $9\frac{3}{5}$

Example 2

Dexter makes $\frac{2}{3}$ of a recipe, and he uses 12 cups of powdered milk. How many cups of powdered milk are in the recipe? 18 cups

3 Close

Reaching All Learners

Through Grouping Strategies

Have students work in groups to solve equations involving variables multiplied by fractions and mixed numbers. Each student writes an equation on a strip of paper. Then, as a group, students solve and check the solutions to each equation, explaining each step as they work through the problems.

Summarize

Review using reciprocals to solve equations. Explain that solving any equation containing multiplication involves dividing both sides of the equation by the number that is being multiplied by the variable. If that number happens to be a fraction, then the rule for dividing by a fraction applies. In other words, both sides of the equation are multiplied by the reciprocal of that fraction.

Answers to Think and Discuss

1. Possible answer: No, if $\frac{2}{3}x = 4$, $x = 6$. If $\frac{2}{3} = 4x$, $x = \frac{1}{6}$.
2. With $\frac{2}{3}x = 4$, divide by $\frac{2}{3}$ because that's the number that the variable is multiplied by; with $\frac{4}{5} = 8x$, divide by 8 because that's the number that the variable is being multiplied by.

5-4 PRACTICE & ASSESS

Students may want to refer back to the lesson examples.

Assignment Guide

If you finished Example 1 assign:

Core 1–3, 5–10, 13–21 odd, 33–45

Enriched 6–10 even, 13–24, 33–45

If you finished Example 2 assign:

Core 1–16, 22–25, 33–45

Enriched 13–45

Answers

22. $n \cdot \frac{1}{3} = 12$, $n = 36$; Check: $36 \cdot \frac{1}{3} = 12$

23. $n \div 4 = \frac{1}{2}$, $n = 2$; Check: $2 \div 4 = \frac{1}{2}$

24. $n \cdot 1\frac{1}{2} = 9$, $n = 6$; Check: $6 \cdot 1\frac{1}{2} = 9$

Math Background

Consider $\frac{2}{3} \div \frac{3}{5}$. To understand how this can be rewritten as $\frac{2}{3} \cdot \frac{5}{3}$, follow these steps:

Steps	Explanation
$\frac{2}{3} \div \frac{3}{5} = \frac{2}{3} \cdot \frac{5}{3} \cdot \frac{3}{5} \div \frac{3}{5}$	This is valid because $\frac{5}{3} \cdot \frac{3}{5} = \frac{15}{15} = 1$.
$\frac{2}{3} \div \frac{3}{5} = \frac{2}{3} \cdot \frac{5}{3}$	This is valid because $\frac{3}{5} \div \frac{3}{5} = 1$.

5-4 Exercises

FOR EXTRA PRACTICE
see page 645

internet connect
Homework Help Online
go.hrw.com Keyword: MR4 5-4

GUIDED PRACTICE

See Example 1
Solve each equation. Write the answer in simplest form.

1. $\frac{3}{4}z = 12$ $z = 16$ **2.** $4n = \frac{3}{5}$ $n = \frac{3}{20}$ **3.** $\frac{2x}{3} = 5$ $x = 7\frac{1}{2}$

See Example 2
4. In PE class, $\frac{3}{8}$ of the students want to play volleyball. If 9 students want to play volleyball, how many students are in the class? 24

INDEPENDENT PRACTICE

See Example 1
Solve each equation. Write the answer in simplest form.

5. $3t = \frac{2}{7}$ $t = \frac{2}{21}$ **6.** $\frac{1}{3}x = 3$ $x = 9$ **7.** $\frac{3r}{5} = 9$ $r = 15$

8. $\frac{4}{5}a = 1$ $a = 1\frac{1}{4}$ **9.** $\frac{y}{4} = 5$ $y = 20$ **10.** $2b = \frac{6}{7}$ $b = \frac{3}{7}$

See Example 2
11. Jason uses 2 cans of paint to paint $\frac{1}{2}$ of his room. How many cans of paint will he use to paint the whole room? 4 cans

12. Cassandra baby-sits for $\frac{4}{5}$ of an hour and earns \$8. What is her hourly rate? \$10

PRACTICE AND PROBLEM SOLVING

Solve each equation. Write the answer in simplest form.

13. $m = \frac{3}{8} \cdot 4$ $m = 1\frac{1}{2}$ **14.** $\frac{3y}{5} = 6$ $y = 10$ **15.** $4z = \frac{7}{10}$ $z = \frac{7}{40}$

16. $\frac{3}{5}a = \frac{3}{5}$ $a = 1$ **17.** $\frac{1}{6}b = 2\frac{1}{3}$ $b = 14$ **18.** $5c = \frac{2}{3} \div \frac{2}{3}$ $c = \frac{1}{5}$

19. $\frac{1}{2} = \frac{w}{4}$ $w = 2$ **20.** $8 = \frac{2n}{3}$ $n = 12$ **21.** $\frac{1}{4} \cdot \frac{1}{2} = 4d$ $d = \frac{1}{32}$

Write each equation. Then solve, and check the solution.

22. A number n is multiplied by $\frac{1}{3}$ and the product is 12.

23. A number n is divided by 4 and the quotient is $\frac{1}{2}$.

24. A number n is multiplied by $1\frac{1}{2}$ and the product is 9.

25. A recipe for a loaf of bread calls for $\frac{3}{4}$ cup of oatmeal.

a. How much oatmeal do you need if you make half the recipe? $\frac{3}{8}$ cup

b. How much oatmeal do you need if you double the recipe? $1\frac{1}{2}$ cups

26. ***ENTERTAINMENT*** Connie rode the roller coaster at the amusement park. After 3 minutes, the ride was $\frac{3}{4}$ complete. How long did the entire ride take? 4 minutes

RETEACH 5-4

LESSON 5-4 **Reteach**
Solving Fraction Equations: Multiplication and Division

You can write related facts using multiplication and division.

$3 \cdot 4 = 12$ $4 = 12 \div 3$

You can use related facts to solve equations.

A. $\frac{2}{3} \cdot x = 12$

Think: $12 \div \frac{2}{3} = x$

$x = 12 \cdot \frac{3}{2}$

$x = \frac{12}{1} \cdot \frac{3}{2}$

$x = \frac{36}{2}$

$x = 18$

Check: $\frac{2}{3} \cdot x = 12$

$\frac{2}{3} \cdot 18 \stackrel{?}{=} 12$ Substitute

$\frac{2}{3} \cdot \frac{18}{1} \stackrel{?}{=} 12$

$\frac{36}{3} \stackrel{?}{=} 12$

$12 = 12$ ✓

B. $\frac{2x}{5} = 3$

$\frac{2}{5} \cdot x = 3$

Think: $3 \div \frac{2}{5} = x$

$x = 3 \cdot \frac{5}{2}$

$x = \frac{3}{1} \cdot \frac{5}{2}$

$x = \frac{15}{2}$

$x = 7\frac{1}{2}$

Check: $\frac{2x}{5} = 3$

$\frac{2}{5} \cdot x \stackrel{?}{=} 3$

$\frac{2}{5} \cdot \frac{15}{2} \stackrel{?}{=} 3$ Substitute

$\frac{30}{10} \stackrel{?}{=} 3$

$3 = 3$ ✓

Use related facts to solve each equation. Then check each answer.

1. $\frac{1}{4} \cdot x = 3$ $x = 12$ ✓
2. $\frac{3x}{4} = 2$ $x = 2\frac{2}{3}$ ✓
3. $\frac{3}{5} \cdot x = \frac{2}{3}$ $x = 1\frac{1}{9}$ ✓
4. $\frac{1}{3} \cdot x = 6$ $x = 18$ ✓
5. $\frac{2x}{5} = 1$ $x = 2\frac{1}{2}$ ✓
6. $\frac{1}{8} \cdot x = 3$ $x = 24$ ✓

PRACTICE 5-4

LESSON 5-4 **Practice B**
Solving Fraction Equations: Multiplication and Division

Solve each equation. Check your answers.

1. $\frac{1}{4}x = 6$ $x = 24$; $\frac{1}{4} \cdot 24 = 6$ ✓
2. $2t = \frac{4}{7}$ $t = \frac{2}{7}$; $2 \cdot \frac{2}{7} = \frac{4}{7}$ ✓
3. $\frac{3}{5}a = 3$ $a = 5$; $\frac{3}{5} \cdot 5 = 3$ ✓
4. $\frac{r}{6} = 8$ $r = 48$; $\frac{48}{6} = 8$ ✓
5. $\frac{2b}{9} = 4$ $b = 18$; $\frac{(2 \cdot 18)}{9} = 4$ ✓
6. $3y = \frac{4}{5}$ $y = \frac{4}{15}$; $3 \cdot \frac{4}{15} = \frac{4}{5}$ ✓
7. $\frac{2}{3}d = 5$ $d = 7\frac{1}{2}$; $\frac{2}{3} \cdot 7\frac{1}{2} = 5$ ✓
8. $2f = \frac{1}{6}$ $f = \frac{1}{12}$; $2 \cdot \frac{1}{12} = \frac{1}{6}$ ✓
9. $4q = \frac{2}{9}$ $q = \frac{1}{18}$; $4 \cdot \frac{1}{18} = \frac{2}{9}$ ✓
10. $\frac{1}{2}s = 2$ $s = 4$; $\frac{1}{2} \cdot 4 = 2$ ✓
11. $\frac{h}{7} = 5$ $h = 35$; $\frac{35}{7} = 5$ ✓
12. $\frac{1}{4}c = 9$ $c = 36$; $\frac{1}{4} \cdot 36 = 9$ ✓
13. $5g = \frac{5}{6}$ $g = \frac{1}{6}$; $5 \cdot \frac{1}{6} = \frac{5}{6}$ ✓
14. $3k = \frac{1}{9}$ $k = \frac{1}{27}$; $3 \cdot \frac{1}{27} = \frac{1}{9}$ ✓
15. $\frac{3x}{5} = 6$ $x = 10$; $\frac{(3 \cdot 10)}{5} = 6$ ✓
16. It takes 3 buckets of water to fill $\frac{1}{3}$ of a fish tank. How many buckets are needed to fill the whole tank? 9 buckets
17. Jenna got 12, or $\frac{3}{5}$, of her answers on the test right. How many questions were on the test? 20 questions
18. It takes Charles 2 minutes to run $\frac{1}{4}$ of a mile. How long will it take Charles to run a mile? 8 minutes

Life Science LINK

The northwest corner of Madagascar is home to black lemurs. These primates live in groups of 7–10, and they have an average life span of 20–25 years. Much of their habitat is being destroyed by human agricultural activity.

27. **LIFE SCIENCE** Sasha's book report is about animals in Madagascar. She writes 10 pages, which represents $\frac{1}{3}$ of her report, about lemurs. How many more pages does Sasha have to write to complete her book report? **30 pages**

28. Alder cut 3 pieces of fabric from a roll. Each piece of fabric she cut is $1\frac{1}{2}$ yd long. She has 2 yards of fabric left on the roll. How much fabric was on the roll before she cut it? **$6\frac{1}{2}$ yards**

Use the circle graph for Exercises 29 and 30.

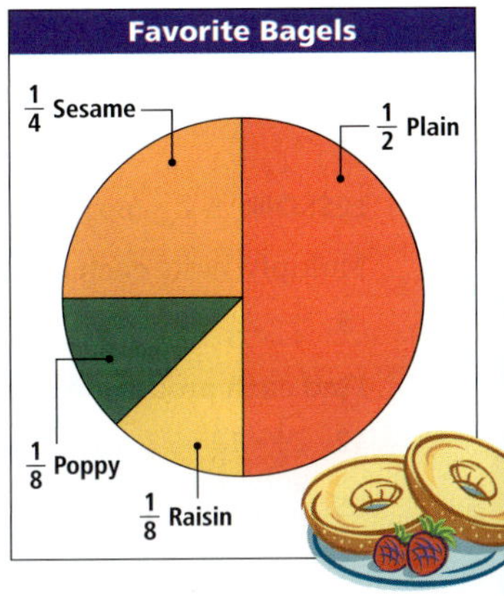

29. The circle graph shows the results of a survey of people who were asked to choose their favorite kind of bagel.
    a. One hundred people chose plain bagels as their favorite kind of bagel. How many people were surveyed in all? **200 people**
    b. One-fifth of the people who chose sesame bagels also chose plain cream cheese as their favorite spread. How many people chose plain cream cheese? (*Hint*: Use the answer to part **a** to help you solve this problem.) **10 people**

30. **WHAT'S THE QUESTION?** If the answer is 25 people, what is the question?

31. **WRITE ABOUT IT** Explain how to solve $\frac{3}{5}x = 4$.

32. **CHALLENGE** Solve.
$2\frac{3}{4}n = \frac{11}{12}$ **$\frac{1}{3}$**

Spiral Review

Estimate each sum or difference by rounding to the thousands place. (Lesson 1-2)

33. 9,074 + 2,123 **11,000**
34. 7,232 − 4,245 **3,000**
35. 9,903 + 5,367 **15,000**
36. 5,234 − 1,901 **3,000**
37. 8,523 + 2,459 **11,000**
38. 15,050 − 10,810 **4,000**

Move the decimal point to multiply or divide by powers of 10. (Lesson 3-4)

39. 718 × 1,000 **718,000**
40. 23.06×10^3 **23,060**
41. 420.34 ÷ 100 **4.2034**
42. 1,000 × 1,500 **1,500,000**
43. 5.03×10^2 **503**
44. 88.5 ÷ 1,000 **0.0885**

45. **TEST PREP** Which fraction is **not** equivalent to $\frac{1}{4}$? (Lesson 4-6) **C**

A $\frac{3}{12}$ B $\frac{4}{16}$ C $\frac{5}{24}$ D $\frac{12}{48}$

Answers

30. Possible answer: How many people chose poppyseed bagels or raisin bagels as their favorite kind of bagel?

31. Possible answer: Divide both sides of the equation by $\frac{3}{5}$. To divide by $\frac{3}{5}$, you can multiply by $\frac{5}{3}$, the reciprocal of $\frac{3}{5}$. Write the answer, $\frac{20}{3}$, in simplest form, $6\frac{2}{3}$.

Journal

Have students explain and give an example of how to solve an equation whose solution involves multiplication of both sides by a reciprocal.

Test Prep Doctor

For Exercise 45, students must remember that for two fractions to be equivalent, both the numerator and denominator of the first fraction are multiplied or divided by the same number to get the second fraction. Students who answered **A, B,** or **D** should recheck their math and they will see that all of those fractions *are* equivalent to $\frac{1}{4}$.

CHALLENGE 5-4

LESSON 5-4 **Challenge**
Crawly Creature Equations

A millipede called the *Illacme plenipes* holds the record for the creature with the most legs—750! However, most millipedes have only 30 legs. Shown below are some other many-legged creatures.

Let [millipede] = the number of legs most millipedes have. Use this information to solve the equations and find how many legs each other crawly creature has.

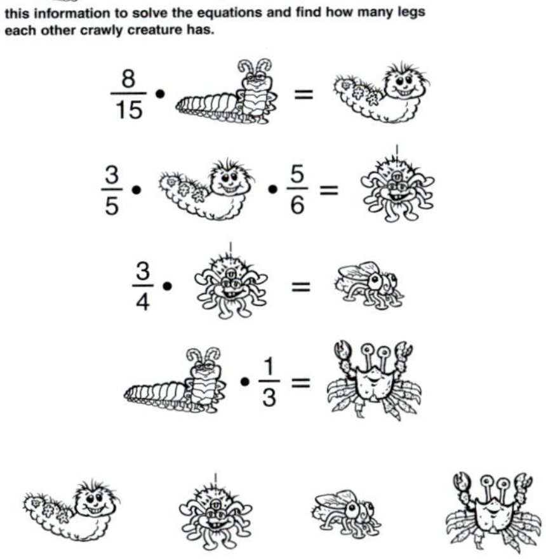

Caterpillars **16 legs**
Spiders **8 legs**
Insects **6 legs**
Crabs **10 legs**

PROBLEM SOLVING 5-4

LESSON 5-4 **Problem Solving**
Solving Fraction Equations: Multiplication and Division

Solve.

1. The number of T-shirts is multiplied by $\frac{1}{2}$ and the product is 18. Write and solve an equation for the number of T-shirts, where *t* represents the number of T-shirts.
   $t \bullet \frac{1}{2} = 18; t = 36$

2. The number of students is divided by 18 and the quotient is $\frac{1}{6}$. Write and solve an equation for the number of students, where *s* represents the number of students.
   $s \div 18 = \frac{1}{6}; s = 3$

3. The number of players is multiplied by $2\frac{1}{2}$ and the product is 25. Write and solve an equation for the number of players, where *p* represents the number of players.
   $p \bullet 2\frac{1}{2} = 25; p = 10$

4. The number of chairs is divided by $\frac{1}{4}$ and the quotient is 12. Write and solve an equation for the number of chairs, where *c* represents the number of chairs.
   $c \div \frac{1}{4} = 12; c = 3$

Circle the letter of the correct answer.

5. Paco bought 10 feet of rope. He cut it into several $\frac{5}{6}$-foot pieces. Which equation can you use to find how many pieces of rope Paco cut?
   A $\frac{5}{6} \div 10 = x$
   B $\frac{5}{6} \div x = 10$
   (C) $10 \div x = \frac{5}{6}$
   D $10x = \frac{5}{6}$

6. Each square on the graph paper has an area of $\frac{4}{9}$ square inch. What is the length and width of each square?
   F $\frac{1}{9}$ inch
   (G) $\frac{2}{3}$ inch
   H $\frac{2}{9}$ inch
   J $\frac{1}{3}$ inch

7. Which operation should you use to solve the equation $6x = \frac{3}{8}$?
   A addition
   B subtraction
   C multiplication
   (D) division

8. A fraction divided by $\frac{2}{3}$ is equal to $1\frac{1}{4}$. What is that fraction?
   F $\frac{1}{3}$
   (G) $\frac{5}{6}$
   H $\frac{1}{4}$
   J $\frac{1}{2}$

Lesson Quiz

Solve each equation. Write the answer in simplest form.

1. $3x = \frac{1}{8}$ **$x = \frac{1}{24}$**
2. $\frac{1}{4}x = 4$ **$x = 16$**
3. $\frac{3}{7}x = 14$ **$x = \frac{98}{3}$ or $32\frac{2}{3}$**
4. $\frac{y}{7} = 9$ **$y = 63$**
5. Rebecca used 3 pt of paint to paint $\frac{1}{4}$ of the trim in her bedroom. How many pints will Rebecca use for the trim in the entire bedroom? **12**

Available on Daily Transparency in CRB

Chapter 5 Mid-Chapter Quiz

Purpose: *To assess students' mastery of concepts and skills in Lessons 5-1 through 5-4*

Assessment Resources

Section 5A Quiz

Assessment Resources . . . . p. 14

***Test and Practice Generator* CD-ROM**

Additional mid-chapter assessment items in both multiple-choice and free-response format may be generated for any objective in Lessons 5-1 through 5-4.

Chapter 5 Mid-Chapter Quiz

Mid-Chapter Quiz

LESSON 5-1 (pp. 212–215)

Multiply. Write each answer in simplest form.

1. $\frac{2}{7} \cdot \frac{3}{4}$ **$\frac{3}{14}$**
2. $\frac{3}{5} \cdot \frac{2}{3}$ **$\frac{2}{5}$**
3. $\frac{7}{12} \cdot \frac{4}{5}$ **$\frac{7}{15}$**
4. $\frac{5}{8} \cdot \frac{9}{10}$ **$\frac{9}{16}$**

Evaluate the expression $t \cdot \frac{1}{8}$ for each value of t. Write the answer in simplest form.

5. $t = \frac{4}{9}$ **$\frac{1}{18}$**
6. $t = \frac{4}{5}$ **$\frac{1}{10}$**
7. $t = \frac{2}{3}$ **$\frac{1}{12}$**
8. $t = \frac{6}{7}$ **$\frac{3}{28}$**

LESSON 5-2 (pp. 216–219)

Multiply. Write each answer in simplest form.

9. $\frac{1}{4} \cdot 2\frac{1}{3}$ **$\frac{7}{12}$**
10. $1\frac{1}{6} \cdot \frac{2}{3}$ **$\frac{7}{9}$**
11. $\frac{7}{8} \cdot 2\frac{2}{3}$ **$2\frac{1}{3}$**

Find each product. Write the answer in simplest form.

12. $2\frac{1}{4} \cdot 1\frac{1}{6}$ **$2\frac{5}{8}$**
13. $1\frac{2}{3} \cdot 2\frac{1}{5}$ **$3\frac{2}{3}$**
14. $3 \cdot 4\frac{2}{7}$ **$12\frac{6}{7}$**
15. $\frac{5}{6}$ of $1\frac{3}{5}$ **$1\frac{1}{3}$**
16. $\frac{1}{5}$ of $2\frac{1}{3}$ **$\frac{7}{15}$**
17. $\frac{3}{4}$ of $1\frac{1}{2}$ **$1\frac{1}{8}$**

LESSON 5-3 (pp. 222–225)

Find the reciprocal.

18. $\frac{2}{7}$ **$\frac{7}{2}$**
19. $\frac{5}{12}$ **$\frac{12}{5}$**
20. $\frac{3}{5}$ **$\frac{5}{3}$**
21. $\frac{1}{10}$ **10**

Divide. Write each answer in simplest form.

22. $\frac{3}{5} \div 4$ **$\frac{3}{20}$**
23. $1\frac{3}{10} \div 3\frac{1}{4}$ **$\frac{2}{5}$**
24. $1\frac{1}{5} \div 2\frac{1}{3}$ **$\frac{18}{35}$**
25. $10 \div 2\frac{1}{2}$ **4**
26. $1\frac{5}{11} \div 1\frac{5}{11}$ **1**
27. $\frac{3}{10} \div \frac{3}{100}$ **10**

LESSON 5-4 (pp. 226–229)

Solve each equation.

28. $\frac{2y}{3} = 10$ **15**
29. $6p = \frac{3}{4}$ **$\frac{1}{8}$**
30. $\frac{2x}{3} = 9$ **$13\frac{1}{2}$**

31. Michael has a black cat and a gray kitten. The black cat weighs 12 pounds. The gray kitten weighs $\frac{3}{5}$ the weight of the black cat. How much does the gray kitten weigh? **$7\frac{1}{5}$ pounds**

32. Ronald rode his bike $7\frac{1}{5}$ miles in an hour. At that speed, how far will he travel in the next $\frac{3}{4}$ hour? **$5\frac{2}{5}$ miles**

33. Amy has some beads that are each $\frac{1}{2}$ inch long. How many beads will she need to make a necklace that is 18 inches long? **36 beads**

Focus on Problem Solving

Solve

Choose the operation: multiplication or division

Read the whole problem before you try to solve it. Determine what action is taking place in the problem. Then decide whether you need to multiply or divide in order to solve the problem.

If you are asked to combine equal groups, you need to multiply. If you are asked to share something equally or to separate something into equal groups, you need to divide.

Action	Operation	
Combining equal groups	Multiplication	
Sharing things equally or separating into equal groups	Division	

Read each problem, and determine the action taking place. Choose an operation to solve the problem. Then solve, and write the answer in simplest form.

1. Jason picked 30 cups of raspberries. He put them in freezer bags with $\frac{3}{4}$ cup in each bag. How many bags does he have?

2. When the cranberry flowers start to open in June, cranberry growers usually bring in about $1\frac{1}{2}$ beehives per acre of cranberries to pollinate the flowers. A grower has 36 acres of cranberries. About how many beehives does she need?

3. A recipe that makes 3 cranberry banana loaves calls for 4 cups of cranberries. Linh wants to make only 1 loaf. How many cups of cranberries does she need?

4. Clay wants to double a recipe for blueberry muffins that calls for $\frac{3}{4}$ cup of blueberries. How many blueberries will he need?

Focus on Problem Solving

Purpose: *To focus on choosing an operation to solve the problem*

Problem Solving Resources

Interactive Problem Solving pp. 34–43

Math: Reading and Writing in the Content Area . . . . pp. 34–43

Problem Solving Process

This page focuses on the third step of the problem-solving process: **Solve**

Discuss

Have students identify the key words or phrases that indicate which operation to use. Determine the action taking place and name the operation.

Possible answers:

1. $\frac{3}{4}$ cup *in each* bag; separating into equal groups; division
2. $1\frac{1}{2}$ beehives *per* acre; combining equal groups; multiplication
3. recipe that makes 3 loaves. . . *only* wants 1; separating into equal groups; division
4. *double* a recipe; combining equal groups; multiplication

Answers

1. 40 bags
2. about 54 beehives
3. $1\frac{1}{3}$ cups of cranberries
4. $1\frac{1}{2}$ cups blueberries

Section 5B

Adding and Subtracting Fractions

One-Minute Section Planner

Lesson	Materials	Resources
Lesson 5-5 Least Common Multiple **NCTM:** Number and Operations, Communication **NAEP:** Number Properties 5b ☑ SAT-9 ☑ SAT-10 ☐ ITBS ☑ CTBS ☑ MAT ☑ CAT	**Optional** Spinner *(MK, CRB, p. 105)*	• *Chapter 5 Resource Book*, pp. 43–51 • Daily Transparency T20, CRB • Additional Examples Transparencies T21–T23, CRB • *Alternate Openers: Explorations*, p. 38
Lesson 5-6 Estimating Fraction Sums and Differences **NCTM:** Number and Operations, Communication **NAEP:** Number Properties 2a ☑ SAT-9 ☑ SAT-10 ☐ ITBS ☐ CTBS ☑ MAT ☐ CAT	**Optional** Teaching Transparency T25 *(CRB)* Fraction calculators	• *Chapter 5 Resource Book*, pp. 52–60 • Daily Transparency T24, CRB • Additional Examples Transparencies T26–T28, CRB • *Alternate Openers: Explorations*, p. 39
Hands-On Lab 5C Model Fraction Addition and Subtraction **NCTM:** Number and Operations, Algebra **NAEP:** Number Properties 3e ☐ SAT-9 ☑ SAT-10 ☐ ITBS ☐ CTBS ☑ MAT ☑ CAT	**Required** Fraction bars *(MK, CRB, p. 109)*	• *Hands-On Lab Activities*, pp. 34–36
Lesson 5-7 Adding and Subtracting with Unlike Denominators **NCTM:** Number and Operations, Reasoning and Proof, Communication **NAEP:** Number Properties 3a ☑ SAT-9 ☑ SAT-10 ☑ ITBS ☑ CTBS ☑ MAT ☑ CAT	**Optional** Recording Sheet for Motivate *(CRB, p. 106)* Teaching Transparency T30 *(CRB)* Recording Sheet for Reaching All Learners *(CRB, p. 107)* Fraction bars *(MK, CRB, p. 109)*	• *Chapter 5 Resource Book*, pp. 61–69 • Daily Transparency T29, CRB • Additional Examples Transparencies T31–T33, CRB • *Alternate Openers: Explorations*, p. 40
Lesson 5-8 Adding and Subtracting Mixed Numbers **NCTM:** Number and Operations, Communication **NAEP:** Number Properties 3g ☐ SAT-9 ☐ SAT-10 ☐ ITBS ☑ CTBS ☑ MAT ☑ CAT	**Optional** Rulers *(MK, CRB, p. 108)*	• *Chapter 5 Resource Book*, pp. 70–78 • Daily Transparency T34, CRB • Additional Examples Transparencies T35–T37, CRB • *Alternate Openers: Explorations*, p. 41
Hands-On Lab 5D Model Subtraction with Renaming **NCTM:** Number and Operations, Algebra **NAEP:** Number Properties 3e ☐ SAT-9 ☑ SAT-10 ☐ ITBS ☐ CTBS ☑ MAT ☑ CAT	**Required** Fraction bars *(MK, CRB, p. 109)*	• *Hands-On Lab Activities*, pp. 37–38
Lesson 5-9 Renaming to Subtract Mixed Numbers **NCTM:** Number and Operations, Communication, Connections **NAEP:** Number Properties 3a ☐ SAT-9 ☑ SAT-10 ☐ ITBS ☑ CTBS ☑ MAT ☑ CAT	**Optional** Fraction bars *(MK, CRB, p. 109)* Teaching Transparency T39 *(CRB)*	• *Chapter 5 Resource Book*, pp. 79–87 • Daily Transparency T38, CRB • Additional Examples Transparencies T40–T42, CRB • *Alternate Openers: Explorations*, p. 42
Lesson 5-10 Solving Fraction Equations: Addition and Subtraction **NCTM:** Number and Operations, Algebra, Reasoning and Proof, Communication, Representation **NAEP:** Algebra 4c ☐ SAT-9 ☑ SAT-10 ☐ ITBS ☐ CTBS ☑ MAT ☑ CAT		• *Chapter 5 Resource Book*, pp. 88–96 • Daily Transparency T43, CRB • Additional Examples Transparencies T44–T46, CRB • *Alternate Openers: Explorations*, p. 43
Section 5B Assessment		• Section 5B Quiz, AR p. 15 • *Test and Practice Generator* CD-ROM

SAT = *Stanford Achievement Tests* **ITBS** = *Iowa Test of Basic Skills* **CTBS** = *Comprehensive Test of Basic Skills/Terra Nova*
MAT = *Metropolitan Achievement Tests* **CAT** = *California Achievement Test*

NCTM = Complete standards can be found on pages T27–T33. **NAEP** = Complete standards can be found on pages A31–A35.

SE = *Student Edition* **TE** = *Teacher's Edition* **AR** = *Assessment Resources* **CRB** = *Chapter Resource Book* **MK** = *Manipulatives Kit*

Section Overview

Estimating Fraction Sums and Differences — Lesson 5-6

Why? Estimation skills are often used in daily life.

Kristi is looking for a two-bedroom apartment. In one apartment, the master bedroom is $14\frac{3}{4}$ feet wide and the other bedroom is $12\frac{1}{3}$ feet wide. About how much wider is the master bedroom than the other bedroom?

$14\frac{3}{4}$
The numerator is about the same as the denominator, so the fraction is closest to 1. You can round this number to 15.

$12\frac{1}{3}$
The numerator is about half the denominator, so the fraction is closest to $\frac{1}{2}$. You can round this number to $12\frac{1}{2}$.

$$14\frac{3}{4} - 12\frac{1}{3} \approx 15 - 12\frac{1}{2} \approx 2\frac{1}{2}$$

The master bedroom is about $2\frac{1}{2}$ feet wider than the other bedroom.

Adding and Subtracting Fractions — Lessons 5-5, 5-7, 5-8

Why? The least common denominator (LCD) is the least common multiple (LCM) of the denominators. You can use the LCD to add and subtract fractions and mixed numbers.

To add or subtract unlike fractions, first rewrite them as equivalent fractions with a common denominator.

$$\frac{1}{6} + \frac{7}{10}$$

$$\frac{1 \cdot 5}{6 \cdot 5} + \frac{7 \cdot 3}{10 \cdot 3}$$

$$\frac{5}{30} + \frac{21}{30}$$

$$\frac{26}{30}$$

$$\frac{13}{15}$$

To find a common denominator, find the LCM of 6 and 10.

List the multiples.
6: 6, 12, 18, 24, 30, . . .
10: 10, 20, 30, . . .

Use prime factorization.
$6 = 2 \cdot 3$
$10 = 2 \cdot 5$
$\text{LCM} = 2 \cdot 3 \cdot 5 = 30$

Renaming to Subtract and Fraction Equations — Lessons 5-9, 5-10

Why? In order to solve fraction equations, you may need to rename to subtract.

Subtract $1\frac{3}{4}$ from both sides to undo the addition.

$$x + 1\frac{3}{4} = 3\frac{1}{4}$$

$$x + 1\frac{3}{4} - 1\frac{3}{4} = 3\frac{1}{4} - 1\frac{3}{4}$$

$$x = 1\frac{2}{4}, \text{ or } 1\frac{1}{2}$$

$$\begin{array}{r} 3\frac{1}{4} \\ -\,1\frac{3}{4} \\ \hline \end{array} \longrightarrow \begin{array}{r} 2\frac{5}{4} \\ -\,1\frac{3}{4} \\ \hline 1\frac{2}{4} \end{array}$$

Rename $3\frac{1}{4}$ as $2\frac{5}{4}$ for subtracting numerators.

5-5 Organizer

Pacing: Traditional 1 day
Block $\frac{1}{2}$ day

Objective: Students find the least common multiple (LCM) of a group of numbers.

Indiana Academic Standards
6.1.7, 6.7.1, 6.7.3, 6.7.11

Warm Up

Write the first five multiples of each number.

1. 5 — 5, 10, 15, 20, 25
2. 6 — 6, 12, 18, 24, 30
3. 10 — 10, 20, 30, 40, 50
4. 12 — 12, 24, 36, 48, 60

Problem of the Day

Greg, Sam, and Mary all work at the same high school. One of them is a principal, one of them is a teacher, and one of them is a janitor. Sam is older than Mary. Mary does not live in the same town as the principal. The teacher, the oldest of the three, often plays golf with Greg. What is each person's job? Greg, principal; Sam, teacher; Mary, janitor

Available on Daily Transparency in CRB

5-5 Least Common Multiple

Learn to find the least common multiple (LCM) of a group of numbers.

Vocabulary
least common multiple (LCM)

After games in Lydia's soccer league, one player's family brings snacks for both teams to share. This week Lydia's family will provide juice boxes and granola bars for 24 players.

You can make a model to help you find the least number of juice and granola packs Lydia's family should buy. Use colored counters, drawings, or pictures to illustrate the problem.

EXAMPLE 1 ***Consumer Application***

Juice comes in packs of 6, and granola bars in packs of 8. If there are 24 players, what is the least number of packs needed so that each player has a drink and granola bar and there are none left over?

Draw juice boxes in groups of 6. Draw granola bars in groups of 8. Stop when you have drawn the same number of each.

There are 24 juice boxes and 24 granola bars.

Lydia's family should buy 4 packs of juice and 3 packs of granola bars.

The smallest number that is a multiple of two or more numbers is the **least common multiple (LCM)**. In Example 1, the LCM of 6 and 8 is 24.

1 Introduce

Alternate Opener

EXPLORATION

5-5 Least Common Multiple

1. Sarah and Jane enter a walkathon for charity. They start together, but Sarah completes one lap every 6 minutes while Jane completes one lap every 8 minutes.

Number of Laps Completed	Sarah's Time (min)	Jane's Time (min)
1	$6 \cdot 1 = 6$	$8 \cdot 1 = 8$
2	$6 \cdot 2 = 12$	$8 \cdot 2 = 16$
3	$6 \cdot 3 = 18$	$8 \cdot 3 = 24$
4	$6 \cdot 4 = 24$	$8 \cdot 4 = 32$
5	$6 \cdot 5 = 30$	$8 \cdot 5 = 40$
6	$6 \cdot 6 = 36$	$8 \cdot 6 = 48$
7	$6 \cdot 7 = 42$	$8 \cdot 7 = 56$
8	$6 \cdot 8 = 48$	$8 \cdot 8 = 64$

a. After how many minutes will Sarah and Jane meet at the start again?

b. When will they meet the next time?

Think and Discuss

2. Discuss the solution to number **1a** using the term *common multiple.*

3. Compare the solution to number **1a** with the solution to number **1b**, and describe these solutions using the terms *common multiple* and *least common multiple.*

Motivate

Have students work on the following problem. Two faucets are dripping. One faucet will drip every 4 seconds and the other faucet drips every 9 seconds. If a drop of water falls from both faucets at the same time, how many seconds will it be before you see the faucets drip at the same time again? 36 seconds

Exploration worksheet and answers on Chapter 5 Resource Book pp. 44 and 118

2 Teach

Lesson Presentation

Guided Instruction

In this lesson, students learn to find the least common multiple (LCM) for a group of numbers. First show them a real-world application of the LCM. Then teach them three methods for finding the LCM of a group of numbers: use a number line, use a list, and use prime factorization.

Teaching Tip For Example 2, have students use each of the three methods for finding the LCM for each given set of numbers.

EXAMPLE 2 Using Multiples to Find the LCM

Find the least common multiple (LCM).

Method 1: Use a number line.

A 6 and 9

Use a number line to skip count by 6 and 9.

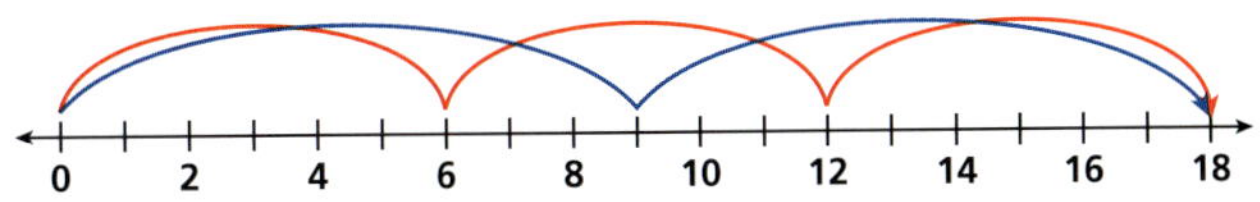

The least common multiple (LCM) of 6 and 9 is 18.

Method 2: Use a list.

B 3, 5, and 6

3: 3, 6, 9, 12, 15, 18, 21, 24, 27, **30**, 33, . . .
5: 5, 10, 15, 20, 25, **30**, 35, . . .
6: 6, 12, 18, 24, **30**, 36, . . .

List multiples of 3, 5, and 6. Find the smallest number that is in all the lists.

LCM: 30

Method 3: Use prime factorization.

C 8 and 12

$8 = 2 \cdot 2 \cdot 2$
$12 = 2 \cdot 2 \cdot \quad 3$
↓ ↓ ↓ ↓
$2 \cdot 2 \cdot 2 \cdot 3$
$2 \cdot 2 \cdot 2 \cdot 3 = 24$

Write the prime factorization of each number. Line up the common factors. To find the LCM, multiply one number from each column.

LCM: 24

D 12, 10, and 15

$12 = 2^2 \cdot 3$
$10 = 2 \quad \cdot \quad 5$
$15 = \quad 3 \cdot 5$

Write the prime factorization of each number in exponential form.

↓ ↓ ↓
$2^2 \cdot 3 \cdot 5$
$2^2 \cdot 3 \cdot 5 = 60$

To find the LCM, multiply each prime factor once with the greatest exponent used in any of the prime factorizations.

LCM: 60

Remember!
The prime factorization of a number is the number written as a product of its prime factors.

Think and Discuss

1. **Explain** why you cannot find a greatest common multiple for a group of numbers.
2. **Tell** whether the LCM of a set of numbers can ever be smaller than any of the numbers in the set.

COMMON ERROR ALERT

Some students may confuse LCM with GCF. Tell them to concentrate on the words *multiple* and *factor* to keep their meanings straight. Finding the GCF and LCM for the same set of numbers helps students see the difference.

Additional Examples

Example 1

English muffins come in packs of 8, and eggs come in cartons of 12. If there are 24 students, what is the least number of packs and cartons needed so that each student has a muffin sandwich with one egg?
Three packs of English muffins and 2 cartons of eggs are needed.

Example 2

Find the least common multiple (LCM).

A. 3 and 4 **12**
B. 4, 5, and 8 **40**
C. 6 and 20 **60**
D. 15, 6, and 4 **60**

3 Close

Reaching All Learners
Through Concrete Manipulatives

Give students spinners labeled with the digits 1–9 (provided in the Manipulatives Kit and on Chapter 5 Resource Book p. 105). Have students spin their spinners to generate sets of three or four 1- and 2-digit numbers. Have students use each of the three given methods to find the LCM for each set of numbers.

Summarize

Review the term *least common multiple* (LCM) and have students explain how the LCM differs from the GCF.

Possible answer: The LCM is the *least* number that is a *multiple* of all numbers in a given set. The GCF is the *greatest* number that is a *factor* of all numbers in a given set (or by which all numbers in a set are divisible). The GCF is smaller than the LCM unless all of the numbers are equal.

Answers to Think and Discuss

1. Possible answer: For any common multiple found, you can always create a greater common multiple by continuing to multiply the common multiple by any whole number.
2. No, the LCM of a set of numbers can be equal to one of the numbers in the set, but never smaller than any number in the set.

5-5 PRACTICE & ASSESS

Students may want to refer back to the lesson examples.

Assignment Guide

If you finished Example 1 assign:
Core 1, 14, 34, 38–48
Enriched 1, 14, 34–35, 38–48

If you finished Example 2 assign:
Core 1–26, 34, 38–48
Enriched 11–48

Notes

Math Background

An alternative method for finding the least common multiple of two numbers is to divide their product by their greatest common factor. For example, because the greatest common factor of 10 and 12 is 2, the least common multiple of 10 and 12 is $(10 \cdot 12) \div 2 = 60$. This method has the advantage of not requiring the enumeration of multiples.

5-5 Exercises

FOR EXTRA PRACTICE see page 646

internet connect
Homework Help Online
go.hrw.com Keyword: MR4 5-5

GUIDED PRACTICE

See Example 1

1. Pencils are sold in packs of 12, and erasers in packs of 9. Mr. Joplin wants to give each of 36 students a pencil and an eraser. What is the least number of packs he should buy so there are none left over? **3 packs of pencils and 4 packs of erasers**

See Example 2

Find the least common multiple (LCM).

2. 3 and 5 **15**
3. 4 and 9 **36**
4. 2, 3, and 6 **6**
5. 2, 4, and 5 **20**
6. 4 and 12 **12**
7. 6 and 16 **48**
8. 4, 6, and 8 **24**
9. 2, 5, and 8 **40**
10. 6 and 10 **30**
11. 21 and 63 **63**
12. 3, 5, and 9 **45**
13. 5, 6, and 25 **150**

INDEPENDENT PRACTICE

See Example 1

14. String-cheese sticks are sold in packs of 10, and celery sticks in packs of 15. Ms. Sobrino wants to give each of 30 students one string-cheese stick and one celery stick. What is the least number of packs she should buy so there are none left over? **3 packs of string cheese sticks and 2 packs of celery sticks**

See Example 2

Find the least common multiple (LCM).

15. 2 and 8 **8**
16. 3 and 7 **21**
17. 4 and 10 **20**
18. 3 and 9 **9**
19. 3, 6, and 9 **18**
20. 4, 8, and 10 **40**
21. 4, 6, and 12 **12**
22. 4, 6, and 7 **84**
23. 3, 8, and 12 **24**
24. 3, 7, and 10 **210**
25. 2, 6, and 11 **66**
26. 2, 3, 6, and 9 **18**
27. 2, 4, 5, and 6 **60**
28. 10 and 11 **110**
29. 4, 5, and 7 **140**
30. 2, 3, 6, and 8 **24**

PRACTICE AND PROBLEM SOLVING

31. What is the LCM of 6 and 12? **12**
32. What is the LCM of 5 and 11? **55**
33. **Find the missing numbers in the diagram.**
    a. a two-digit multiple of 4 that is not a multiple of 6 **Possible answers: 16, 20, 28**
    b. a two-digit multiple of 6 that is not a multiple of 4 **Possible answers: 18, 30, 42**
    c. the LCM of 4 and 6 **12**
    d. a three-digit common multiple of 4 and 6 **120, 144, 168, and 192**

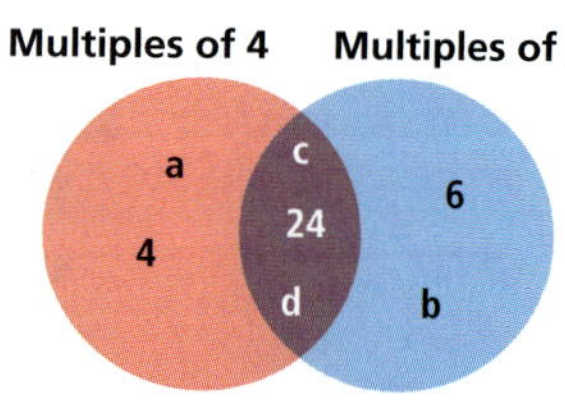

RETEACH 5-5

LESSON 5-5 **Reteach**
Least Common Multiple

The smallest number that is a multiple of two or more numbers is called the least common multiple (LCM).

To find the least common multiple of 3, 6, and 8, list the multiples for each number and put a circle around the LCM in the three list.

Multiples of 3: 3, 6, 9, 12, 15, 18, 21, 24
Multiples of 6: 6, 12, 18, 24, 30, 36, 42
Multiples of 8: 8, 16, 24, 32, 40, 48, 56

So 24 is the LCM of 3, 6, and 8.

List the multiples of each number to help you find the least common multiple of each group.

1. 3 and 4
   Multiples of 3: ______
   Multiples of 4: ______
   LCM: **12**
2. 5 and 7
   Multiples of 5: ______
   Multiples of 7: ______
   LCM: **35**
3. 8 and 12
   Multiples of 8: ______
   Multiples of 12: ______
   LCM: **24**
4. 2 and 9
   Multiples of 2: ______
   Multiples of 9: ______
   LCM: **18**
5. 4 and 6
   Multiples of 4: ______
   Multiples of 6: ______
   LCM: **12**
6. 4 and 10
   Multiples of 4: ______
   Multiples of 10: ______
   LCM: **20**
7. 2, 5, and 6
   Multiples of 2: ______
   Multiples of 5: ______
   Multiples of 6: ______
   LCM: **30**
8. 3, 4, and 9
   Multiples of 3: ______
   Multiples of 4: ______
   Multiples of 9: ______
   LCM: **36**
9. 8, 10, and 12
   Multiples of 8: ______
   Multiples of 10: ______
   Multiples of 12: ______
   LCM: **120**

PRACTICE 5-5

LESSON 5-5 **Practice B**
Least Common Multiple

Find the least common multiple (LCM).

1. 2 and 5 **10**
2. 4 and 3 **12**
3. 6 and 4 **12**
4. 6 and 8 **24**
5. 5 and 9 **45**
6. 4 and 5 **20**
7. 10 and 15 **30**
8. 8 and 12 **24**
9. 6 and 10 **30**
10. 3, 6, and 9 **18**
11. 2, 5, and 10 **10**
12. 4, 7, and 14 **28**
13. 3, 5, and 9 **45**
14. 2, 5, and 8 **40**
15. 3, 9, and 12 **36**
16. Mr. Stevenson is ordering shirts and hats for his Boy Scout troop. There are 45 scouts in the troop. Hats come in packs of 3, and shirts come in packs of 5. What is the least number of packs of each he should order to so that each scout will have 1 hat and 1 shirt, and none will be left over?
    15 packs of hats and 9 packs of shirts
17. Tony wants to make 36 party bags. Glitter pens come in packs of 6. Stickers come in sheets of 4, and balls come in packs of 3. What is the least number of each package he should buy to have 1 of each item in every party bag, and no supplies left over?
    6 packs of pens, 9 sheets of stickers, and 12 packs of balls
18. Glenda is making 30 school supply baskets. Notepads come in packs of 5. Erasers come in packs of 15, and markers come in packs of 3. What is the least number of each package she should buy to have 1 of each item in every basket, and no supplies left over?
    6 packs of notepads, 2 packs of erasers, and 10 packs of markers

34. During its grand opening weekend, a restaurant gave every eighth customer a complimentary appetizer, every twelfth customer a complimentary beverage, and every fifteenth customer a complimentary dish of frozen yogurt.
   a. Which customer was the first to receive all three complimentary items? **120**
   b. Which customer was the first to receive a complimentary appetizer and frozen yogurt? **120**
   c. If the restaurant served 500 customers that weekend, how many of those customers received all three complimentary items? **4**

35. ***CHOOSE A STRATEGY*** Sophia gave $\frac{1}{2}$ of her semi-precious-rock collection to her son. She gave $\frac{1}{2}$ of what she had left to her grandson. Then she gave $\frac{1}{2}$ of what she had left to her great-grandson. She kept 10 rocks for herself. How many rocks did she have in the beginning? **B**

   A 40 **B** 80 **C** 100 **D** 160

36. ***WRITE ABOUT IT*** Explain the steps you can use to find the LCM of two numbers. Choose two numbers to show an example of your method.

37. ***CHALLENGE*** Find the LCM of each pair of numbers.

   a. 4 and 6 **b.** 8 and 9 **c.** 5 and 7 **d.** 8 and 10

   When is the LCM of two numbers equal to the product of the two numbers? **a. 12 b. 72 c. 35 d. 40; when they have no common prime factors**

Spiral Review

Order the numbers from least to greatest. (Lesson 1-1)

38. 1,235; 354; 846 **354; 846; 1,235**
39. 978; 679; 879 **679; 879; 978**
40. 1,264; 1,098; 1,104 **1,098; 1,104; 1,264**

Determine whether the given value of each variable is a solution. (Lesson 2-3)

41. $y + 37 = 64$ for $y = 27$ **27 is a solution**
42. $43 - c = 19$ for $c = 24$ **24 is a solution**
43. $72 \div z = 9$ for $z = 7$ **7 is not a solution**

Write each number in standard form. (Lesson 3-5)

44. 6.479×10^3 **6,479** **45.** 0.208×10^2 **20.8** **46.** 13.507×10^4 **135,070** **47.** 7.1×10^5 **710,000**

48. TEST PREP The prime factorization of which number is $2 \times 2 \times 3 \times 5$? (Lesson 4-2) **B**

A 100 **B** 60 **C** 30 **D** 20

Answers

36. Possible answer: Choose 3 and 4. List multiples of the first number: 3, 6, 9, 12, 15, ... List multiples of the second number: 4, 8, 12, 16, ... Find the smallest number they have in common. The LCM is 12.

Journal

Have students list three numbers and find their LCM. Have students explain two methods they could use for finding the LCM.

Test Prep Doctor

For Exercise 48, students may not realize that finding the answer is as simple as multiplying together the given prime factors. Students who answered **C** or **D** may have multiplied three of the four factors. Students who answered **A** probably just guessed, incorrectly, at the answer.

CHALLENGE 5-5

LESSON 5-5 **Challenge**
Moons Over Neptune

We measure one month by our moon's orbital period, or the time it takes the Moon to travel once around Earth, which is about 30 days. But what if you lived on Neptune? It has 8 moons! How could you pick just one moon to measure your months? One possible solution is to calculate one month based on when two of Neptune's moons are in conjunction at some arbitrary starting point in the sky, or appear to be in the same place in the sky. The diagram below shows some of the moons you could use to measure your months on Neptune.

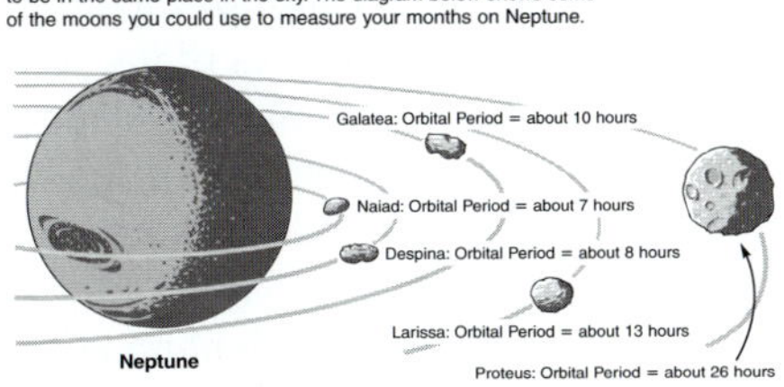

Use the diagram and least common multiples to complete the chart below. For each row, write how long your month on Neptune would be if you used those moons in conjunction as the length of one month.

Neptune Moons to Use	Lenth of One Neptune Month
Naiad and Despina	about 56 hours
Larissa and Proteus	about 26 hours
Galatea and Despina	about 40 hours
Despina and Proteus	about 104 hours

PROBLEM SOLVING 5-5

LESSON 5-5 **Problem Solving**
Least Common Multiple

Use the table to answer the questions.

Party Supplies

Item	Number per Pack
Invitations	12
Balloons	30
Paper plates	10
Paper napkins	24
Plastic cups	15
Noise makers	5

1. You want to have an equal number of plastic cups and paper plates. What is the least number of packs of each you can buy?
   3 packs of plates and 2 packs of cups
2. You want to invite 48 people to a party. What is the least number of packs of invitations and napkins you should buy to have one for each person and none left over?
   4 packs of invitations and 2 packs of napkins
3. You want to have an equal number of noisemakers and balloons at your party. What is the least number of packs of each you can buy?
   1 pack of balloons and 6 packs of noise makers
4. You bought an equal number of packs of plates and cups so that each of your 20 guests would have 3 cups and 2 plates. How many packs of each item did you buy?
   4 packs of cups and 4 packs of plates

Circle the letter of the correct answer.

5. The LCM for three items listed in the table is 60 packs. Which of the following are those three items?
   A balloons, plates, noise makers
   (B) noise makers, invitations, balloons
   C napkins, cups, plates
   D balloons, napkins, plates
6. To have one of each item for 120 party guests, you buy 10 packs of one item and 24 packs of the other. What are those two items?
   F plates and invitations
   G balloons and cups
   H napkins and plates
   (J) invitations and noise makers

Lesson Quiz

Find the least common multiple (LCM).

1. 6, 14 **42**
2. 9, 12 **36**
3. 5, 6, 10 **30**
4. 12, 16, 24, 36 **144**
5. Two students in Mrs. Albring's preschool class are stacking blocks, one on top of the other. Reece's blocks are 4 cm high, and Maddy's blocks are 9 cm high. How tall will their stacks be when they are the same height for the first time? **36 cm**

Available on Daily Transparency in CRB

5-6 Organizer

Pacing: Traditional 1 day
Block $\frac{1}{2}$ day

Objective: Students estimate sums and differences of fractions and mixed numbers.

Indiana Academic Standards
6.2.10, 6.7.1, 6.7.7, 6.7.11

Warm Up

Write each sum or difference in simplest form.

1. $\frac{3}{8} + \frac{5}{8}$ 1 **2.** $\frac{9}{10} - \frac{3}{10}$ $\frac{3}{5}$

3. $2\frac{1}{4} + 1\frac{1}{4}$ $3\frac{1}{2}$ **4.** $5\frac{8}{9} - 1\frac{2}{9}$ $4\frac{2}{3}$

Problem of the Day

Students at a school dance formed equal teams to play a game. When they formed teams of 3, 4, 5, or 6, there was always one person left. What is the smallest number of people that there could have been at the dance? 61, the LCM of all the numbers, plus 1

Available on Daily Transparency in CRB

Math Humor

What was the movie about?
About $2\frac{1}{2}$ hours

5-6 Estimating Fraction Sums and Differences

Learn to estimate sums and differences of fractions and mixed numbers.

Members of the Nature Club went mountain biking in Canyonlands National Park, Utah. They biked $10\frac{3}{10}$ miles on Monday and $9\frac{3}{4}$ miles on Tuesday.

Canyonlands National Park, Utah, is a 337,570-acre park that has many canyons, mesas, arches, and spires.

You can estimate fractions by rounding to 0, $\frac{1}{2}$, or 1.

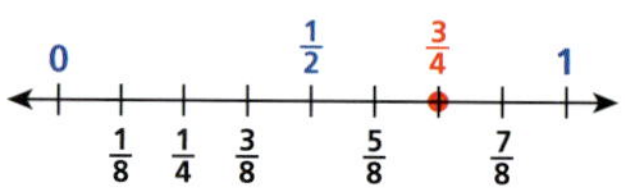

The fraction $\frac{3}{4}$ rounds to 1.

You can round fractions by comparing the numerator and denominator.

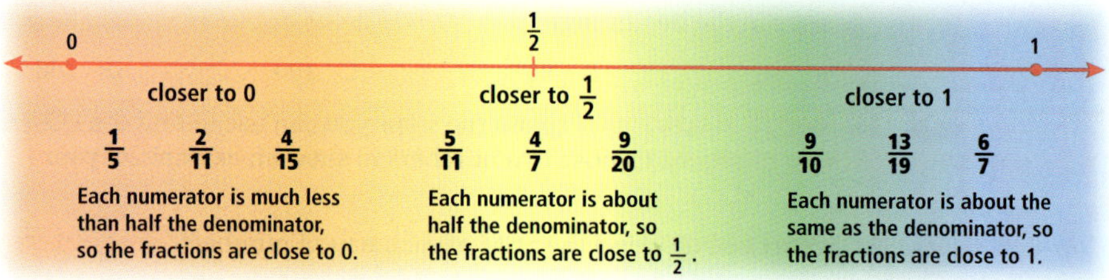

EXAMPLE 1 Estimating Fractions

Estimate each sum or difference by rounding to 0, $\frac{1}{2}$, or 1.

A $\frac{8}{9} + \frac{2}{11}$

$\frac{8}{9} + \frac{2}{11}$ *Think: $\frac{8}{9}$ rounds to 1 and $\frac{2}{11}$ rounds to 0.*

$1 + 0 = 1$

$\frac{8}{9} + \frac{2}{11}$ is about 1.

B $\frac{7}{12} - \frac{8}{15}$

$\frac{7}{12} - \frac{8}{15}$ *Think: $\frac{7}{12}$ rounds to $\frac{1}{2}$ and $\frac{8}{15}$ rounds to $\frac{1}{2}$.*

$\frac{1}{2} - \frac{1}{2} = 0$

$\frac{7}{12} - \frac{8}{15}$ is about 0.

1 Introduce

Exploration

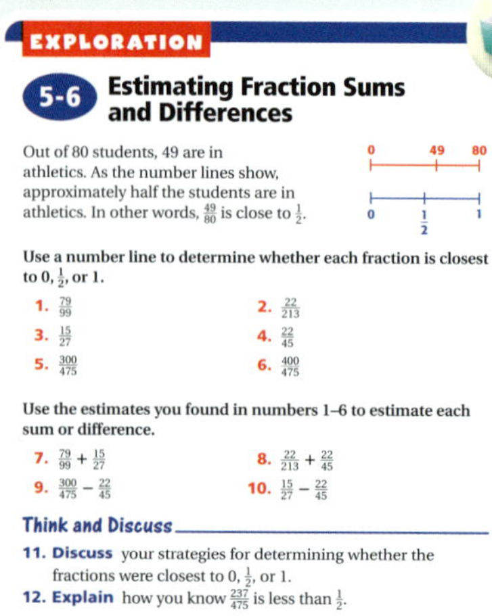
EXPLORATION

5-6 Estimating Fraction Sums and Differences

Out of 80 students, 49 are in athletics. As the number lines show, approximately half the students are in athletics. In other words, $\frac{49}{80}$ is close to $\frac{1}{2}$.

0 49 80
0 $\frac{1}{2}$ 1

Use a number line to determine whether each fraction is closest to 0, $\frac{1}{2}$, or 1.

1. $\frac{79}{99}$ 2. $\frac{22}{213}$
3. $\frac{15}{27}$ 4. $\frac{22}{45}$
5. $\frac{300}{475}$ 6. $\frac{400}{475}$

Use the estimates you found in numbers 1–6 to estimate each sum or difference.

7. $\frac{79}{99} + \frac{15}{27}$ 8. $\frac{22}{213} + \frac{22}{45}$
9. $\frac{300}{475} - \frac{22}{45}$ 10. $\frac{15}{27} - \frac{22}{45}$

Think and Discuss

11. **Discuss** your strategies for determining whether the fractions were closest to 0, $\frac{1}{2}$, or 1.
12. **Explain** how you know $\frac{237}{475}$ is less than $\frac{1}{2}$.

Motivate

Ask students what it means to *estimate*. Possible answer: to find an answer that is close to the exact answer. Have students estimate whole number sums and differences and explain how they estimated each.

Exploration worksheet and answers on Chapter 5 Resource Book pp. 53 and 120

2 Teach

Lesson Presentation

Guided Instruction

In this lesson, students learn to estimate sums and differences of fractions and mixed numbers. First teach students the rule for rounding fractional parts. If the numerator is much less than half the denominator, the fraction is close to 0. If the numerator is about half the denominator, the fraction is close to $\frac{1}{2}$. If the numerator is about the same as the denominator, the fraction is close to 1. You may use Teaching Transparency T25 in the Chapter 5 Resource Book. Then teach them to estimate fraction sums and differences and then mixed number sums and differences.

You can also estimate by rounding mixed numbers. You compare each mixed number to the two nearest whole numbers and the nearest $\frac{1}{2}$.

Does $10\frac{3}{10}$ round to 10, $10\frac{1}{2}$, or 11?

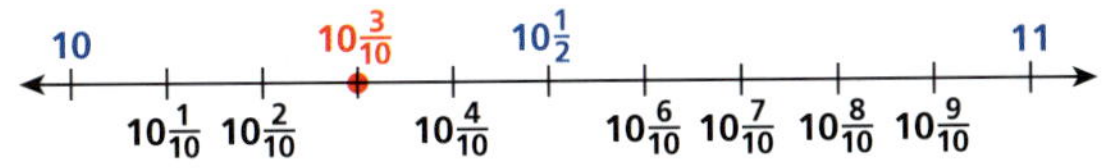

The mixed number $10\frac{3}{10}$ rounds to $10\frac{1}{2}$.

EXAMPLE 2 *Sports Application*

The table shows the distances the Nature Club biked in Utah.

Nature Club's Biking Distances	
Day	**Distances (mi)**
Monday	$10\frac{3}{10}$
Tuesday	$9\frac{3}{4}$
Wednesday	$12\frac{1}{4}$
Thursday	$4\frac{7}{10}$

A **About how far did the Nature Club ride on Monday and Tuesday?**

$10\frac{3}{10} + 9\frac{3}{4}$

$10\frac{1}{2} + 10 = 20\frac{1}{2}$

They rode **about** $20\frac{1}{2}$ miles.

B **About how much farther did the Nature Club ride on Wednesday than on Thursday?**

$12\frac{1}{4} - 4\frac{7}{10}$

$12\frac{1}{2} - 5 = 7\frac{1}{2}$

They rode **about** $7\frac{1}{2}$ miles farther on Wednesday than on Thursday.

C **Estimate the total distance that the Nature Club rode on Monday, Tuesday, and Wednesday.**

$10\frac{3}{10} + 9\frac{3}{4} + 12\frac{1}{4}$

$10\frac{1}{2} + 10 + 12\frac{1}{2} = 33$

They rode **about** 33 miles.

Think and Discuss

1. **Tell** whether each fraction rounds to 0, $\frac{1}{2}$, or 1: $\frac{5}{6}, \frac{2}{15}, \frac{7}{13}$.
2. **Explain** how to round mixed numbers to the nearest whole number.
3. **Determine** whether the Nature Club met their goal to ride at least 35 total miles.

COMMON ERROR ALERT

Students may be confused by the concept of rounding to a fraction, because they have more practice with rounding to whole numbers. Spend extra time explaining rounding to a fraction as needed.

Additional Examples

Example 1

Estimate each sum or difference by rounding to 0, $\frac{1}{2}$, or 1.

A. $\frac{6}{7} + \frac{3}{8}$ about $1\frac{1}{2}$

B. $\frac{9}{10} - \frac{7}{8}$ about 0

Example 2

Tosha's Walking Distances	
Day	**Distance (miles)**
Tuesday	$5\frac{1}{10}$
Thursday	$4\frac{7}{8}$
Saturday	$6\frac{3}{7}$
Sunday	$8\frac{9}{10}$

A. About how far did Tosha walk on Tuesday and Thursday? about 10 miles

B. About how much farther did Tosha walk on Sunday than on Thursday? about 4 miles farther

C. Estimate the total distance Tosha walked on Thursday, Saturday, and Sunday. about $20\frac{1}{2}$ miles

3 Close

Summarize

Close the lesson by reviewing with students how to estimate fraction sums and differences: Round each fraction to 0, $\frac{1}{2}$, or 1, and then add or subtract the rounded fractions. Have students estimate a sum and a difference to make sure they understand the process.

Reaching All Learners

Through Curriculum Integration

Language Arts Discuss with students that many words in the English language have Latin origins. The word *fraction* comes from the Latin word *frangere*, which means "to break." The word *estimate* comes from the Latin word *aestimatus*, which means "to value." Encourage students to research the origins of other math terms and to report to the class what they discover.

Answers to Think and Discuss

1. 1; 0; $\frac{1}{2}$
2. If the fraction part of the mixed number is less than $\frac{1}{2}$, drop the fraction to round to the whole number. If the fraction part of the mixed number is $\frac{1}{2}$ or greater, round up to the next whole number.
3. Yes, they rode a total of about 10 + 10 + 12 + 5 = 37 miles.

5-6 PRACTICE & ASSESS

5-6 Exercises

FOR EXTRA PRACTICE
see page 646

internet connect
Homework Help Online
go.hrw.com Keyword: MR4 5-6

Students may want to refer back to the lesson examples.

GUIDED PRACTICE

See Example 1 Estimate each sum or difference by rounding to 0, $\frac{1}{2}$, or 1.

1. $\frac{8}{9} + \frac{1}{6}$ about 1
2. $\frac{11}{12} - \frac{4}{9}$ about $\frac{1}{2}$
3. $\frac{3}{7} + \frac{1}{12}$ about $\frac{1}{2}$
4. $\frac{6}{13} - \frac{2}{5}$ about 0

See Example 2 Use the table for Exercises 5 and 6.

5. About how far did Mark run during week 1 and week 2? 16 miles
6. About how much farther did Mark run during week 2 than during week 3? 1 mile

Mark's Running Distances

Week	Distance (mi)
1	$8\frac{3}{4}$
2	$7\frac{1}{5}$
3	$5\frac{5}{6}$

Assignment Guide

If you finished Example 1 assign:
Core 1–4, 7–10, 20–22, 33–42
Enriched 14–25, 33–42

If you finished Example 2 assign:
Core 1–13, 20–22, 26–29, 33–42
Enriched 13–42

INDEPENDENT PRACTICE

See Example 1 Estimate each sum or difference by rounding to 0, $\frac{1}{2}$, or 1.

7. $\frac{7}{8} - \frac{3}{8}$ about $\frac{1}{2}$
8. $\frac{3}{10} + \frac{3}{4}$ about $1\frac{1}{2}$
9. $\frac{5}{6} - \frac{7}{8}$ about 0
10. $\frac{7}{10} + \frac{1}{6}$ about 1

See Example 2 Use the table for Exercises 11–13.

11. About how much do the meteorites in Brenham and Goose Lake weigh together? 4 tons
12. About how much more does the meteorite in Willamette weigh than the meteorite in Norton County? $15\frac{1}{2}$ tons
13. About how much do the two meteorites in Kansas weigh together? $3\frac{1}{2}$ tons

Meteorites in the United States

Location	Weight (tons)
Willamette, AZ	$16\frac{1}{2}$
Brenham, KS	$2\frac{3}{5}$
Goose Lake, CA	$1\frac{3}{10}$
Norton County, KS	$1\frac{1}{10}$

Notes

PRACTICE AND PROBLEM SOLVING

Estimate each sum or difference to compare. Write $<$ or $>$.

14. $\frac{5}{6} + \frac{7}{9}$ ■ 3 $<$
15. $2\frac{8}{15} - 1\frac{1}{11}$ ■ 1 $>$
16. $1\frac{2}{21} + \frac{3}{7}$ ■ 2 $<$
17. $1\frac{7}{13} - \frac{8}{9}$ ■ 1 $<$
18. $3\frac{2}{10} + 2\frac{2}{5}$ ■ 6 $<$
19. $4\frac{6}{9} - 2\frac{3}{19}$ ■ 2 $>$

Estimate.

20. $\frac{7}{8} + \frac{4}{7} + \frac{7}{13}$ 2
21. $\frac{6}{11} + \frac{9}{17} + \frac{3}{5}$ $1\frac{1}{2}$
22. $\frac{8}{9} + \frac{3}{4} + \frac{9}{10}$ 3
23. $1\frac{5}{8} + 2\frac{1}{15} + 2\frac{12}{13}$ $6\frac{1}{2}$
24. $4\frac{11}{12} + 3\frac{1}{19} + 5\frac{4}{7}$ $13\frac{1}{2}$
25. $10\frac{1}{9} + 8\frac{5}{9} + 11\frac{13}{14}$ $30\frac{1}{2}$

Math Background

To estimate the sum of many mixed numbers or decimal numbers, the following strategy is both efficient and effective: If there are n addends in all, add the whole parts only and then add $\frac{n}{2}$ to the sum of the whole parts.

The rationale behind the method is as follows: If there are many addends, the fractional or decimal parts should be distributed in a fairly uniform manner between 0 and 1. Therefore the average value of a fractional or decimal part of a number should be near $\frac{1}{2}$. Because there are n addends in all, the sum of these parts should be near $n \cdot \frac{1}{2} = \frac{n}{2}$.

RETEACH 5-6

LESSON 5-6 **Reteach**
Estimating Fraction Sums and Differences

You can use number lines to help you estimate fraction sums and differences.

To estimate the sum of $\frac{5}{6}$ and $\frac{1}{3}$, locate each fraction on a number line. Then round each fraction to 0, $\frac{1}{2}$, or 1.

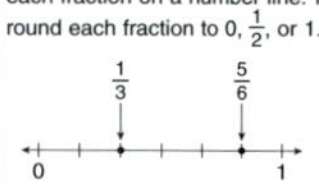

$\frac{5}{6} + \frac{1}{3} \approx$

$1 + \frac{1}{2} = 1\frac{1}{2}$

So, $\frac{5}{6} + \frac{1}{3}$ is about $1\frac{1}{2}$.

To estimate the difference between $\frac{7}{8}$ and $\frac{2}{4}$, locate each fraction on a number line. Then round each fraction to 0, $\frac{1}{2}$, or 1.

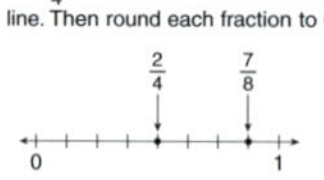

$\frac{7}{8} - \frac{2}{4} \approx$

$1 - \frac{1}{2} = \frac{1}{2}$

So, $\frac{7}{8} - \frac{2}{4}$ is about $\frac{1}{2}$.

Use the number line to round each fraction to 0, $\frac{1}{2}$, or 1 to estimate each sum or difference.

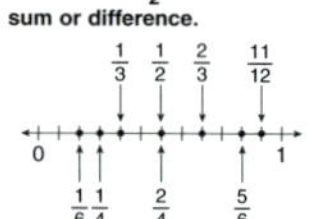

1. $\frac{5}{6} + \frac{1}{6}$ 1
2. $\frac{11}{12} - \frac{1}{2}$ $\frac{1}{2}$
3. $\frac{2}{3} + \frac{2}{4}$ 1
4. $\frac{1}{4} - \frac{1}{3}$ 0

Use the number line to round each fraction to 0, $\frac{1}{2}$, or 1 to estimate each sum or difference.

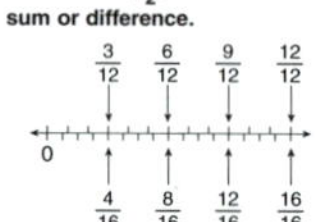

5. $\frac{7}{12} + \frac{2}{6}$ 1
6. $\frac{5}{6} - \frac{3}{8}$ $\frac{1}{2}$
7. $\frac{1}{4} + \frac{2}{6}$ $\frac{1}{2}$
8. $\frac{7}{8} + \frac{14}{16}$ 2

PRACTICE 5-6

LESSON 5-6 **Practice B**
Estimating Fraction Sums and Differences

Estimate each sum or difference by rounding to 0, $\frac{1}{2}$, or 1. Possible answers:

1. $\frac{5}{6} + \frac{3}{10}$ $1\frac{1}{2}$
2. $\frac{7}{9} - \frac{4}{5}$ 0
3. $\frac{9}{10} - \frac{3}{7}$ $\frac{1}{2}$
4. $\frac{4}{9} + \frac{1}{4}$ 1
5. $\frac{1}{8} + \frac{1}{6}$ 0
6. $\frac{7}{8} - \frac{4}{5}$ 0
7. $\frac{5}{8} + \frac{2}{7}$ 1
8. $\frac{7}{10} + \frac{11}{12}$ $1\frac{1}{2}$
9. $\frac{8}{9} + \frac{4}{7}$ $1\frac{1}{2}$
10. $\frac{5}{11} + 1\frac{2}{7}$ 2
11. $1\frac{6}{11} - \frac{2}{5}$ 1
12. $4\frac{2}{7} - 1\frac{7}{9}$ $2\frac{1}{2}$

Use the table for Exercises 13–15.

Fruit Punch

Ingredient	Amount (cups)
Orange juice	$4\frac{3}{5}$
Cranberry juice	$2\frac{1}{4}$
Ginger ale	$\frac{7}{8}$

13. About how much more orange juice than ginger ale is used in the punch? about $3\frac{1}{2}$ cups
14. About how much juice is used in the punch? about 7 cups
15. About how many cups of fruit punch does this recipe make? about 8 cups
16. Damonte rolled the medicine ball $9\frac{3}{4}$ feet. Zachary rolled it $9\frac{7}{12}$ feet. Who rolled the medicine ball the farthest? About how much farther? Damonte; about $\frac{1}{2}$ foot farther
17. Sara ran $5\frac{6}{7}$ miles on Monday and $4\frac{1}{4}$ miles on Tuesday. About how many miles did she run in all during those two days? about $10\frac{1}{2}$ miles

Use an inch ruler for Exercises 26–29. Measure to the nearest $\frac{1}{4}$ inch.

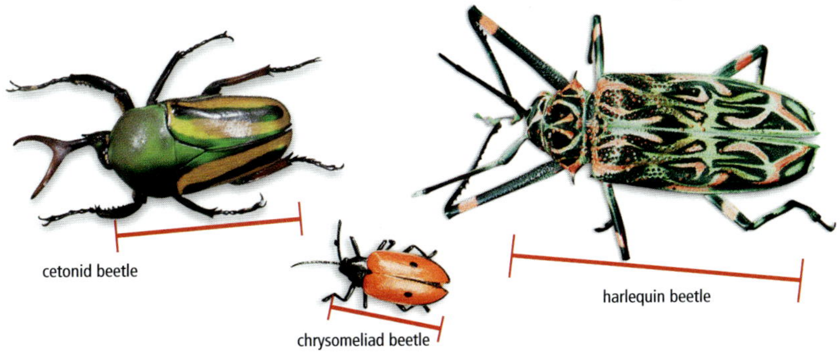

26. About how long is the chrysomeliad beetle? $\frac{3}{4}$ in.

27. About how long is the cetonid beetle? $1\frac{1}{4}$ in.

28. About how much longer is the harlequin beetle than the cetonid beetle? $\frac{1}{2}$ in.

29. About how much longer is the harlequin beetle than the chrysomeliad beetle? 1 in.

30. ***WRITE A PROBLEM*** Write a problem about a trip that can be solved by estimating fractions. Exchange with a classmate and solve.

31. ***WRITE ABOUT IT*** Explain how to estimate the sum of two mixed numbers. Give an example to explain your answer.

32. ***CHALLENGE*** Estimate. $5\frac{1}{2}$

$$\left[5\frac{7}{8} - 2\frac{3}{20}\right] + 1\frac{4}{7}$$

Spiral Review

Write each expression in exponential form. (Lesson 1-3)

33. $8 \times 8 \times 8 \times 8 \times 8$ 8^5 34. $4 \times 4 \times 4 \times 4$ 4^4 35. $7 \times 7 \times 7 \times 7 \times 7 \times 7$ 7^6

Find the missing values in each table. (Lesson 2-1)

36.

x	6	7	9
$x^2 - 5$	■	■	■

31 44 76

37.

a	12	10	8
■	28	24	20

$a \times 2 + 4$

Evaluate $4y$ for each value of y. (Lesson 3-6)

38. $y = 2.13$ 8.52 39. $y = 4.015$ 16.06 40. $y = 3.6$ 14.4 41. $y = 0.78$ 3.12

42. **TEST PREP** Which number is **not** a factor of 42? (Lesson 4-2) C

A 2 B 6 C 4 D 21

Answers

30. Possible answer: Bob will drive 426 miles to visit his grandmother. He drives $\frac{3}{8}$ of the trip and stops to buy gasoline. Then Bob drives $\frac{1}{5}$ of the trip before he makes a rest stop. About how much of the trip has Bob driven?

31. Possible answer: Estimate the fraction part of each mixed number. Add the fractions, and then add the whole numbers. $5\frac{1}{3} + 8\frac{7}{8}$ Round $5\frac{1}{3}$ to $5\frac{1}{2}$ because $\frac{1}{3}$ is closer to $\frac{1}{2}$ than to 0. Round $8\frac{7}{8}$ to 9 because $\frac{7}{8}$ is closer to 1 than to $\frac{1}{2}$. The estimated sum is $14\frac{1}{2}$.

Journal

Have students describe situations when fraction sums and differences may be estimated as well as situations when an exact sum or difference would be needed.

Test Prep Doctor

For Exercise 42, students need to identify the number by which 42 is **not** divisible. If students answered **A, B,** or **D,** have them check their math using the division algorithm. They will find that each of those numbers **is** a factor of 42.

CHALLENGE 5-6

LESSON 5-6 **Challenge**
What's on the Menu?

The menu below shows the average amounts of food most Americans eat each day. Use the menu to complete the diet estimation chart that follows.

Average American Daily Menu

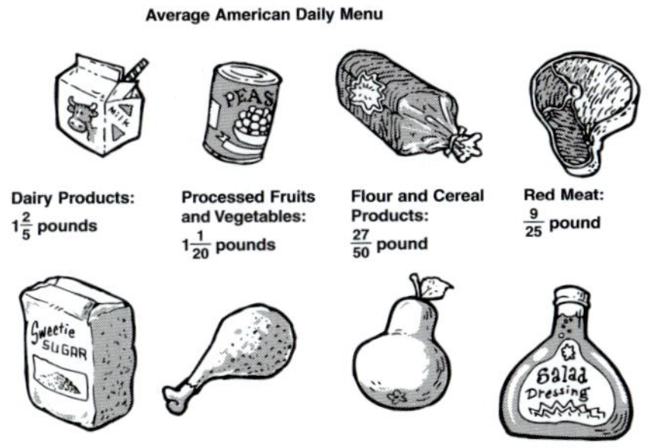

Dairy Products: $1\frac{2}{5}$ pounds
Processed Fruits and Vegetables: $1\frac{1}{20}$ pounds
Flour and Cereal Products: $\frac{27}{50}$ pound
Red Meat: $\frac{9}{25}$ pound
Sweetener Products: $\frac{21}{50}$ pound
Poultry: $\frac{9}{50}$ pounds
Fresh Fruits and Vegetables: $\frac{9}{10}$ pounds
Fats and Oils: $\frac{9}{50}$ pounds

Estimated American Diet

Foods	Estimated Daily Consumption
Red meat and poultry	about $\frac{1}{2}$ pound
Fruits and vegetables	about 2 pounds
Sweetener products and fats and oils	about $\frac{1}{2}$ pound
Dairy and flour and cereal products	about 2 pounds
Total Daily Food Consumption:	about 5 pounds

PROBLEM SOLVING 5-6

LESSON 5-6 **Problem Solving**
Estimating Fraction Sums and Differences

Use the table to answer the questions. Possible answers:

Portland, Oregon, Average Monthly Rainfall

Month	Jan	Feb	Mar	Apr	May	Jun	Jul	Aug	Sep	Oct	Nov	Dec
Rain (in.)	$5\frac{2}{5}$	$3\frac{9}{10}$	$3\frac{3}{5}$	$2\frac{2}{5}$	$2\frac{1}{10}$	$1\frac{1}{2}$	$\frac{3}{5}$	$1\frac{1}{10}$	$1\frac{4}{5}$	$2\frac{7}{10}$	$5\frac{3}{10}$	$6\frac{1}{10}$

1. About how much does it rain in Portland in January and February combined? about $9\frac{1}{2}$ inches

2. About how much more does it rain in Portland in October than in September? about 1 inch more

3. In most years, about how much rain does Portland receive from May through July? about 4 inches

4. What is the difference between Portland's average rainfall in March and May? about $1\frac{1}{2}$ inches

5. What is the difference in rainfall between Portland's rainiest and driest months? about $5\frac{1}{2}$ inches

6. About how much rain does Portland receive in most years all together? about $36\frac{1}{2}$ inches

Circle the letter of the correct answer.

7. About how much rain does Portland receive during its three rainiest months all together?
   (A) about 17 inches
   B about 16 inches
   C about 18 inches
   D about 15 inches

8. In which month in Portland can you expect about $\frac{1}{2}$ inch less rainfall than in June?
   F May
   G July
   H September
   (J) August

Lesson Quiz

Estimate each sum or difference by rounding to 0, $\frac{1}{2}$, or 1.

1. $\frac{9}{10} - \frac{2}{5}$ $\frac{1}{2}$ 2. $\frac{3}{8} + \frac{8}{9}$ $1\frac{1}{2}$

3. $\frac{10}{11} - \frac{8}{9}$ 0 4. $\frac{1}{4} + \frac{8}{15}$ 1

5. The conservation club picked up trash along the road for three weeks. The table shows the number of pounds of trash that was collected. About how many pounds did they collect in weeks 2 and 3?

Week	Pounds Picked Up
1	$18\frac{1}{2}$
2	$16\frac{1}{3}$
3	$20\frac{9}{10}$

37 pounds

Available on Daily Transparency in CRB

Hands-On LAB 5C

Model Fraction Addition and Subtraction

Pacing: Traditional 1 day
Block $\frac{1}{2}$ day

Objective: To use fraction bars to model addition and subtraction of fractions

Materials: Fraction bars

Lab Resources

Hands-On Lab Activities . . . . pp. 34–36

Using the Pages

Discuss with students what each fraction bar represents.

Represent each fraction using fraction bars.

1. $\frac{3}{8}$

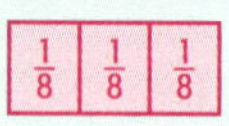

2. $\frac{5}{12}$

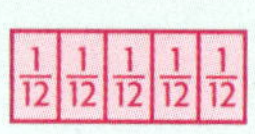

3. How can you use $\frac{1}{6}$ bars to represent $\frac{1}{2}$? Use three $\frac{1}{6}$ bars.

Hands-On LAB 5C Model Fraction Addition and Subtraction

Use with Lessons 5-7 and 5-8

When fractions have different denominators, you need to find a common denominator before you can add or subtract them. Write equivalent fractions with the same denominator, and then perform the operation.

internet connect
Lab Resources Online
go.hrw.com
KEYWORD: MR4 Lab5C

Activity 1

1. Find $\frac{1}{8} + \frac{1}{4}$.

Use fraction bars to represent both fractions.

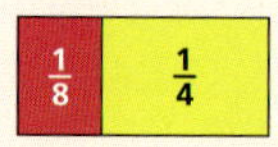

Which fractions fit exactly across $\frac{1}{8}$ and $\frac{1}{4}$?

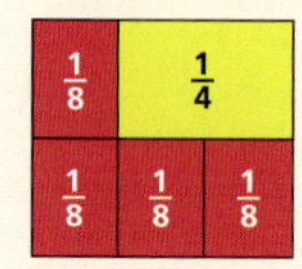

$\frac{1}{4} = \frac{2}{8}$ $\quad$ $\frac{1}{8} + \frac{2}{8} = \frac{3}{8}$

2. Find $\frac{2}{3} + \frac{1}{2}$.

Use fraction bars to represent both fractions.

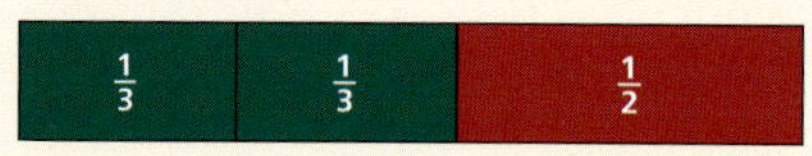

$\frac{2}{3} + \frac{1}{2}$

Which fractions fit exactly across $\frac{2}{3}$ and $\frac{1}{2}$?

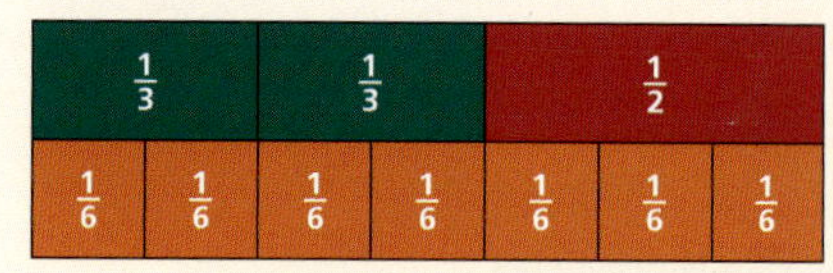

$\frac{2}{3} = \frac{4}{6}$ $\quad$ $\frac{1}{2} = \frac{3}{6}$

$\frac{4}{6} + \frac{3}{6} = \frac{7}{6} = 1\frac{1}{6}$

Think and Discuss

1. Explain what the denominators of $\frac{1}{6}$, $\frac{1}{4}$, $\frac{2}{3}$, and $\frac{1}{2}$ have in common. (*Hint:* Think of common multiples.) The denominators 6, 4, 3, and 2 have an LCM of 12.

Try This

Model each addition expression with fraction bars, and find the sum.

1. $\frac{1}{4} + \frac{1}{2}$ $\frac{3}{4}$
2. $\frac{3}{8} + \frac{1}{4}$ $\frac{5}{8}$
3. $\frac{1}{2} + \frac{2}{5}$ $\frac{9}{10}$
4. $\frac{3}{4} + \frac{1}{6}$ $\frac{11}{12}$
5. $\frac{1}{3} + \frac{1}{6}$ $\frac{3}{6}$, or $\frac{1}{2}$
6. $\frac{7}{8} + \frac{3}{4}$ $\frac{13}{8}$, or $1\frac{5}{8}$
7. $\frac{2}{3} + \frac{1}{4}$ $\frac{11}{12}$
8. $\frac{5}{8} + \frac{1}{4}$ $\frac{7}{8}$

Answers

Activity 1

Try This

1.

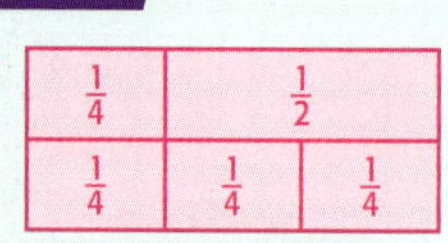

2.

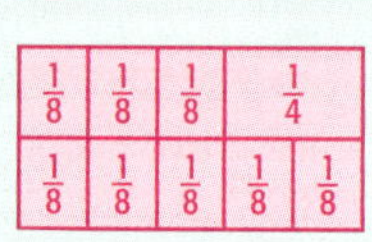

3.

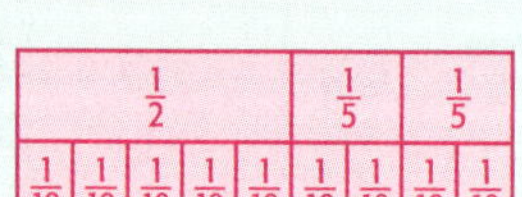

4.

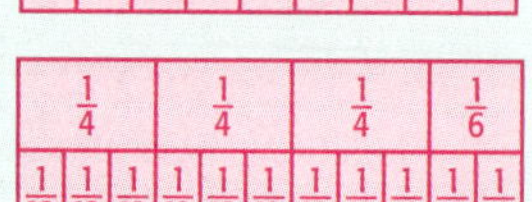

5.

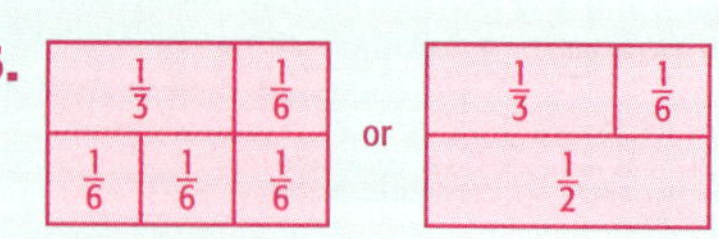

6.

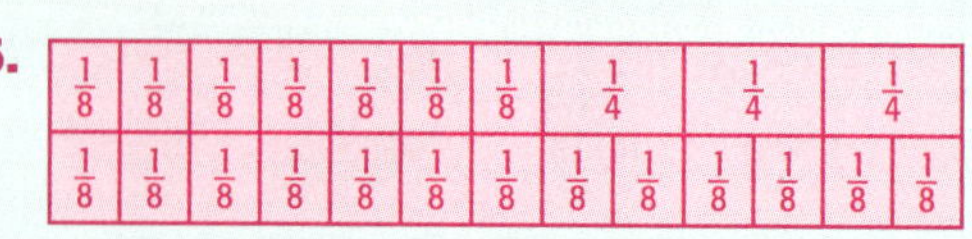

7.

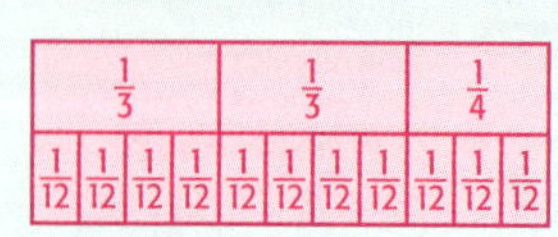

8.

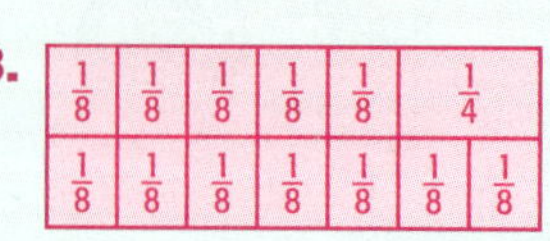

Activity 2

1 Find $\frac{1}{3} - \frac{1}{6}$.

Use fraction bars to represent both fractions.

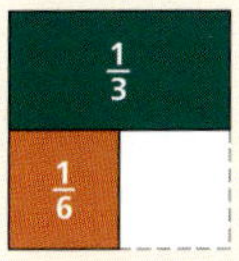

Which fractions fit exactly across $\frac{1}{3}$ and $\frac{1}{6}$?

$\frac{1}{3} = \frac{2}{6}$

Subtract $\frac{1}{6}$ from $\frac{2}{6}$.

$\frac{2}{6} - \frac{1}{6} = \frac{1}{6}$

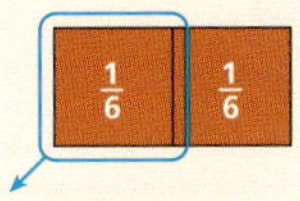

2 Find $\frac{1}{2} - \frac{1}{3}$.

Use fraction bars to represent both fractions.

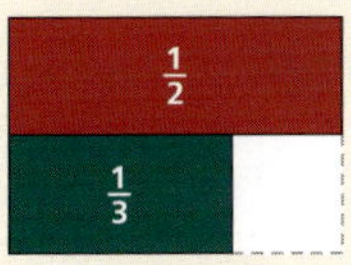

Which fractions fit exactly across $\frac{1}{2}$ and $\frac{1}{3}$?

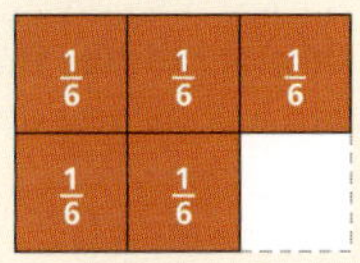

$\frac{1}{2} = \frac{3}{6}$

$\frac{1}{3} = \frac{2}{6}$

Subtract $\frac{2}{6}$ from $\frac{3}{6}$.

$\frac{3}{6} - \frac{2}{6} = \frac{1}{6}$

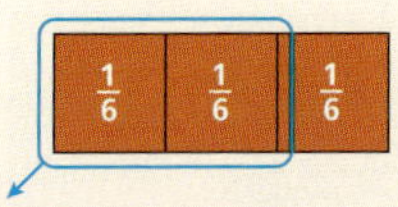

Think and Discuss

1. Explain what the area surrounded by a dashed line represents.
the difference between the two fractions, or the value of the expression

Try This

Model each subtraction expression with fraction bars, and find the difference.

1. $\frac{3}{4} - \frac{1}{3}$ **$\frac{5}{12}$**
2. $\frac{1}{3} - \frac{1}{4}$ **$\frac{1}{12}$**
3. $\frac{1}{2} - \frac{2}{5}$ **$\frac{1}{10}$**
4. $\frac{5}{6} - \frac{1}{3}$ **$\frac{3}{6} = \frac{1}{2}$**
5. $\frac{1}{2} - \frac{5}{12}$ **$\frac{1}{12}$**
6. $\frac{7}{8} - \frac{3}{4}$ **$\frac{1}{8}$**
7. $\frac{1}{4} - \frac{1}{8}$ **$\frac{1}{8}$**
8. $\frac{1}{4} - \frac{1}{6}$ **$\frac{1}{12}$**

Answers

Activity 2

Try This

1.

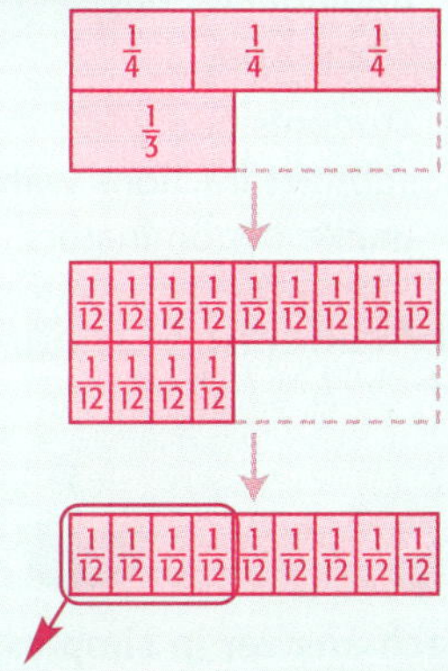

2.

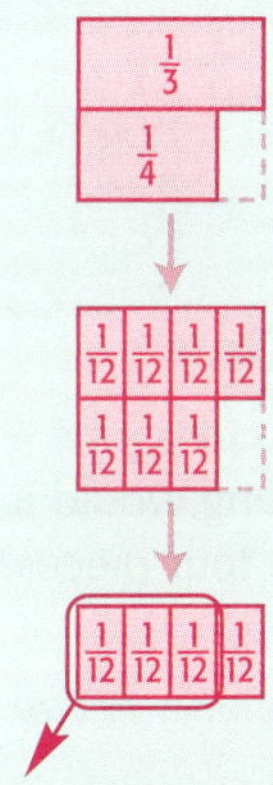

3.

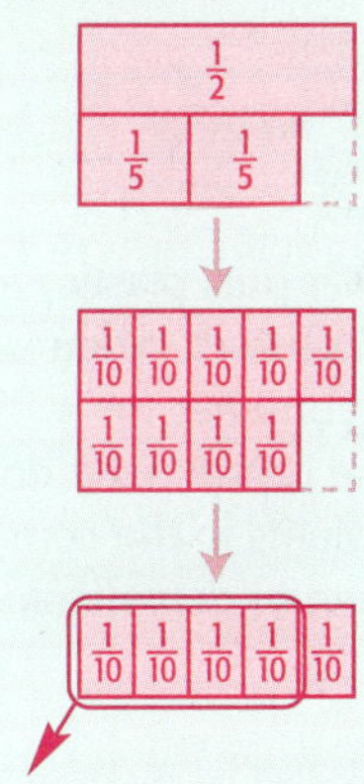

4.

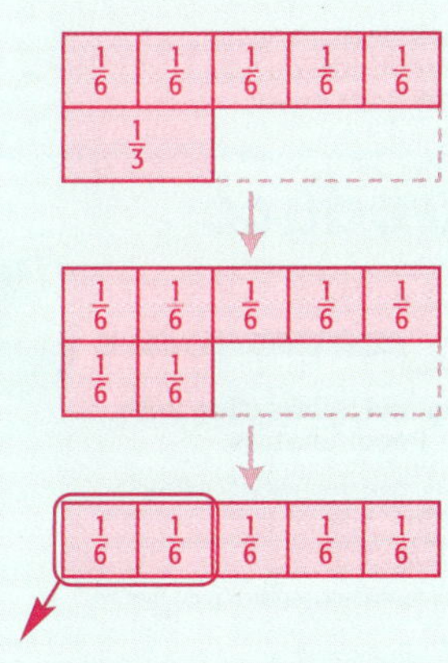

5.

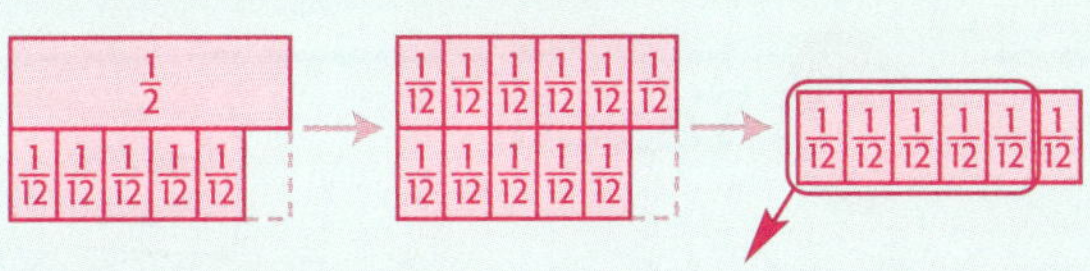

6.

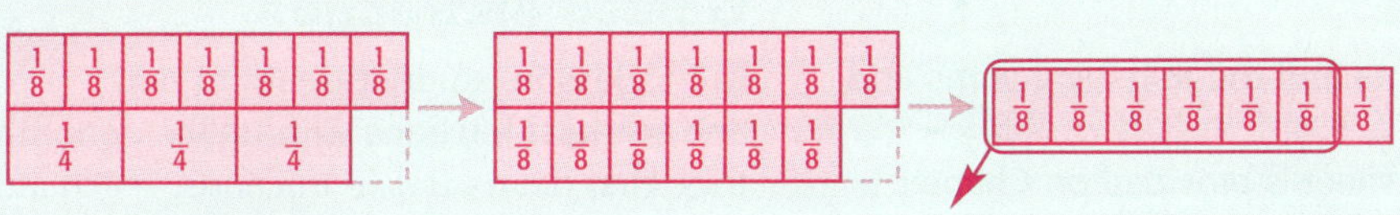

7.

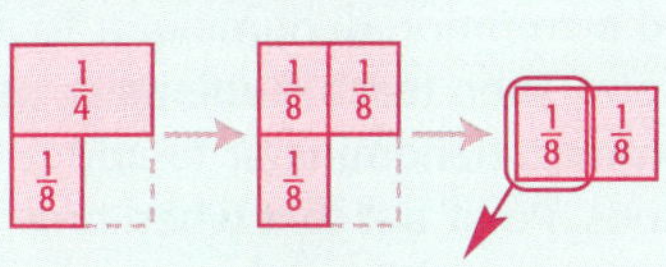

8.

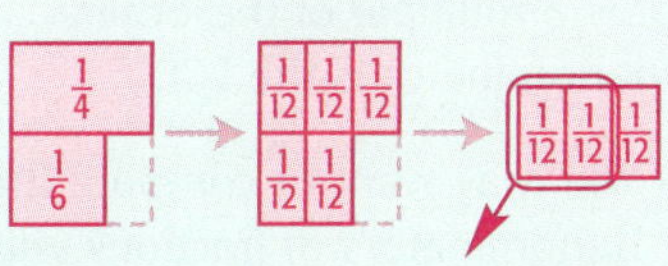

5-7 Organizer

Pacing: Traditional 1 day
Block $\frac{1}{2}$ day

Objective: Students add and subtract fractions with unlike denominators.

Indiana Academic Standards
6.1.7, 6.2.10, 6.7.1, 6.7.3, 6.7.5, 6.7.11

Warm Up

Add. Write each answer in simplest form.

1. $\frac{1}{7} + \frac{1}{7}$ $\frac{2}{7}$
2. $\frac{8}{15} + \frac{3}{15}$ $\frac{11}{15}$
3. $\frac{7}{9} + \frac{2}{9}$ 1
4. $\frac{11}{20} + \frac{4}{20}$ $\frac{3}{4}$

Problem of the Day

If it takes 12 minutes to cut a pipe into three pieces, how long would it take to cut the pipe into four pieces?
18 minutes

Available on Daily Transparency in CRB

Math Humor

Imagine a company that puts plastic coatings on cards and posters. When things go wrong, someone has to remove the plastic. The person who does this for all the departments in the company is called the *common delaminator.*

5-7 Adding and Subtracting with Unlike Denominators

Learn to add and subtract fractions with unlike denominators.

Vocabulary
least common denominator (LCD)

Remember!
Fractions that represent the same value are equivalent.

The Pacific Ocean covers $\frac{1}{3}$ of Earth's surface. The Atlantic Ocean covers $\frac{1}{5}$ of Earth's surface. To find the fraction of Earth's surface that is covered by both oceans, you can add $\frac{1}{3}$ and $\frac{1}{5}$, which are unlike fractions.

To add or subtract unlike fractions, first rewrite them as equivalent fractions with a common denominator.

EXAMPLE 1 Social Studies Application

What fraction of Earth's surface is covered by the Atlantic and Pacific Oceans?

Add $\frac{1}{3}$ and $\frac{1}{5}$.

$$\frac{1}{3} + \frac{1}{5}$$ *Find a common denominator for 3 and 5.*

Write equivalent fractions with 15 as the common denominator.

$$\frac{1}{3} \rightarrow \frac{5}{15}$$
$$+\frac{1}{5} \rightarrow \frac{3}{15}$$
$$\frac{8}{15}$$ *Add the numerators. Keep the common denominator.*

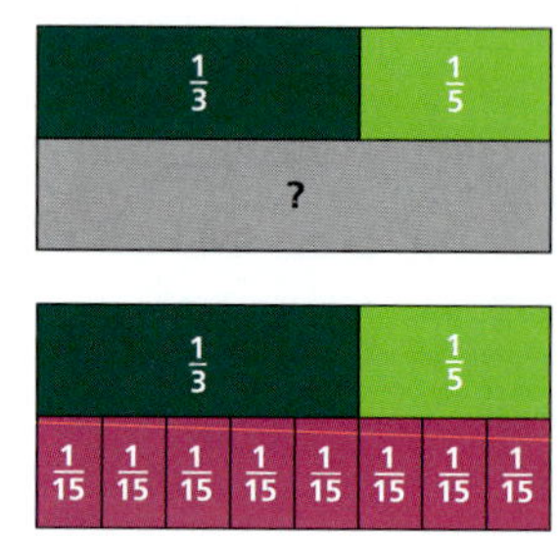

The Pacific and Atlantic Oceans cover $\frac{8}{15}$ of Earth's surface.

You can use *any* common denominator or the *least common denominator* to add and subtract unlike fractions. The **least common denominator (LCD)** is the least common multiple of the denominators.

1 Introduce

Alternate Opener

EXPLORATION

5-7 Adding and Subtracting with Unlike Denominators

Fractions are pieces of a whole. When you add or subtract fractions with unlike denominators, you are usually adding or subtracting pieces of different sizes. Look at the model used to solve the problem below.

Phil combines $\frac{1}{4}$ gallon of paint with $\frac{1}{2}$ gallon of paint. How much paint does he have now?

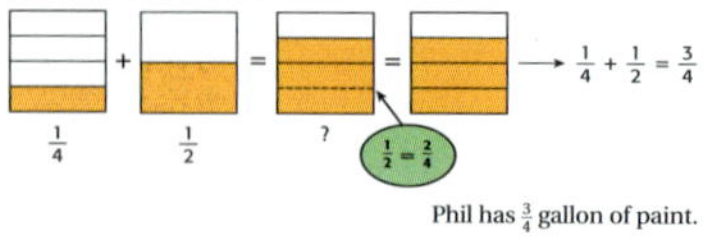

Phil has $\frac{3}{4}$ gallon of paint.

1. Draw a model to show that $1 - \frac{1}{4} = \frac{3}{4}$.

Draw a model to solve each problem. Simplify your answers.

2. $\frac{1}{2} + \frac{1}{3}$
3. $\frac{4}{5} - \frac{1}{2}$

Think and Discuss

4. **Explain** how to add and subtract fractions with unlike denominators.
5. **Draw** a model to show $\frac{1}{2} + \frac{2}{3} = 1\frac{1}{6}$.

Motivate

Have students find the following sums and differences: $\frac{2}{7} + \frac{3}{7}$; $\frac{1}{8} + \frac{3}{8}$; $\frac{10}{11} - \frac{4}{11}$; $\frac{11}{12} - \frac{7}{12}$. A recording sheet is provided on Chapter 5 Resource Book p. 106. $\frac{5}{7}$, $\frac{4}{8}$, or $\frac{1}{2}$, $\frac{6}{11}$, $\frac{4}{12}$, or $\frac{1}{3}$

Have students compare the denominators in each expression and tell what they notice. The denominators are the same in each expression.

Explain that when you add or subtract fractions with different denominators, you have to rewrite them as equivalent fractions with like denominators.

Exploration worksheet and answers on Chapter 5 Resource Book pp. 62 and 122

2 Teach

Lesson Presentation

Guided Instruction

In this lesson, students learn to add and subtract fractions with unlike denominators. First define *unlike fractions*. Teach students to find a common denominator to add unlike fractions. Then teach students to use the least common denominator to add unlike fractions. Point out that when one denominator is a multiple of the other, the greater denominator is the LCD.

Teaching Tip

You may want to compare the denominators of fractions with units of measure. For example, just as you cannot add 2 in. and 3 cm, you cannot add $\frac{2}{5}$ and $\frac{3}{10}$ without first writing them with a common denominator.

EXAMPLE 2 Adding and Subtracting Unlike Fractions

Add or subtract. Write each answer in simplest form.

Method 1: Multiply denominators.

A $\frac{9}{10} - \frac{7}{8}$

$\frac{9}{10} - \frac{7}{8}$ *Multiply the denominators. 10 · 8 = 80*

$\frac{72}{80} - \frac{70}{80}$ *Write equivalent fractions.*

$\frac{2}{80}$ *Subtract.*

$\frac{1}{40}$ *Write the answer in simplest form.*

Method 2: Use the LCD.

B $\frac{9}{10} - \frac{7}{8}$

$\frac{9}{10} - \frac{7}{8}$ *The LCD is 40.*

$\frac{36}{40} - \frac{35}{40}$ *Write equivalent fractions.*

$\frac{1}{40}$ *Subtract.*

Method 3: Use mental math.

C $\frac{5}{12} + \frac{1}{6}$

$\frac{5}{12} + \frac{1}{6}$ *Think: 12 is a multiple of 6, so the LCD is 12.*

$\frac{5}{12} + \frac{2}{12}$ *Rewrite $\frac{1}{6}$ with a denominator of 12.*

$\frac{7}{12}$ *Add.*

D $\frac{5}{12} - \frac{1}{6}$

$\frac{5}{12} - \frac{1}{6}$ *Think: 12 is a multiple of 6, so the LCD is 12.*

$\frac{5}{12} - \frac{2}{12}$ *Rewrite $\frac{1}{6}$ with a denominator of 12.*

$\frac{3}{12}$ *Subtract.*

$\frac{1}{4}$ *Write the answer in simplest form.*

Think and Discuss

1. **Explain** an advantage of using the least common denominator (LCD) when adding unlike fractions.
2. **Tell** when the least common denominator (LCD) of two fractions is the product of their denominators.
3. **Explain** how you can use mental math to subtract $\frac{1}{12}$ from $\frac{3}{4}$.

COMMON ERROR ALERT

Some students may forget to find a common denominator for the fractions, and add or subtract both numerators and denominators. Watch for this especially when they are adding denominators.

Additional Examples

Example 1

Mark made a pizza with pepperoni covering $\frac{1}{4}$ of the pizza and onions covering another $\frac{1}{3}$. What fraction of the pizza is covered by pepperoni and onions? $\frac{7}{12}$

Example 2

Add or subtract. Write each answer in simplest form.

A. $\frac{7}{10} - \frac{3}{8}$ $\frac{13}{40}$

B. $\frac{11}{12} - \frac{3}{8}$ $\frac{13}{24}$

C. $\frac{5}{16} + \frac{1}{8}$ $\frac{7}{16}$

D. $\frac{5}{16} - \frac{1}{8}$ $\frac{3}{16}$

3 Close

Reaching All Learners

Through Concrete Manipulatives

Have students work in groups using fraction bars to complete addition and subtraction squares (provided on a recording sheet on Chapter 5 Resource Book p. 107). Work across and down; the sum of the first and second will equal the third.

The addition square is shown below.

$\frac{1}{4}$	$\frac{1}{12}$	$\frac{1}{3}$
$\frac{1}{3}$	$\frac{1}{6}$	$\frac{1}{2}$
$\frac{7}{12}$	$\frac{1}{4}$	$\frac{5}{6}$

Summarize

Review the terms *unlike fractions* and *least common denominator* (LCD). Discuss how the terms relate to each other.

Possible answers: *Unlike fractions* are fractions with different denominators. The *least common denominator* is the least common multiple of the denominators of unlike fractions. Use the LCD or any common multiple of the denominators when adding and subtracting unlike fractions.

Answers to Think and Discuss

1. Possible answer: Using the LCD will give the smallest possible fractions to work with, which may reduce the number of steps needed to express the answer in simplest form.
2. When the denominators have no factors other than 1 in common, the LCD is a product of the two denominators.
3. Because 12 is a multiple of 4, the LCD is 12. $\frac{9}{12} - \frac{1}{12} = \frac{8}{12}$, or $\frac{2}{3}$

5-7 PRACTICE & ASSESS

Students may want to refer back to the lesson examples.

Assignment Guide

If you finished Example 1 assign:
Core 1, 6–7, 17–29 odd, 44–49
Enriched 31–49

If you finished Example 2 assign:
Core 1–15, 17–29 odd, 36, 38–39, 44–49
Enriched 19–49

Notes

5-7 Exercises

FOR EXTRA PRACTICE
see page 646

internet connect
Homework Help Online
go.hrw.com Keyword: MR4 5-7

GUIDED PRACTICE

See Example 1

1. A trailer hauling wood weighs $\frac{2}{3}$ ton. The trailer weighs $\frac{1}{4}$ ton without the wood. What is the weight of the wood? $\frac{5}{12}$ ton

See Example 2

Add or subtract. Write each answer in simplest form.

2. $\frac{1}{3}+\frac{1}{9}$ $\frac{4}{9}$
3. $\frac{7}{10}-\frac{2}{5}$ $\frac{3}{10}$
4. $\frac{2}{3}-\frac{2}{5}$ $\frac{4}{15}$
5. $\frac{1}{2}+\frac{3}{7}$ $\frac{13}{14}$

INDEPENDENT PRACTICE

See Example 1

6. Approximately $\frac{1}{5}$ of the world's population lives in China. The people of India make up about $\frac{1}{6}$ of the world's population. What fraction of the world's people live in either China or India? $\frac{11}{30}$
7. Cedric is making an Italian dish using a recipe that calls for $\frac{2}{3}$ cup of grated mozarella cheese. If Cedric has grated $\frac{1}{2}$ cup of mozarella cheese, how much more does he need to grate? $\frac{1}{6}$ cup

See Example 2

Add or subtract. Write each answer in simplest form.

8. $\frac{3}{4}-\frac{1}{2}$ $\frac{1}{4}$
9. $\frac{1}{6}+\frac{5}{12}$ $\frac{7}{12}$
10. $\frac{5}{6}-\frac{3}{4}$ $\frac{1}{12}$
11. $\frac{1}{5}+\frac{1}{4}$ $\frac{9}{20}$
12. $\frac{7}{10}+\frac{1}{8}$ $\frac{33}{40}$
13. $\frac{1}{3}+\frac{4}{5}$ $1\frac{2}{15}$
14. $\frac{8}{9}-\frac{2}{3}$ $\frac{2}{9}$
15. $\frac{5}{8}+\frac{1}{2}$ $1\frac{1}{8}$

PRACTICE AND PROBLEM SOLVING

Find each sum or difference. Write your answer in simplest form.

16. $\frac{3}{10}+\frac{1}{2}$ $\frac{4}{5}$
17. $\frac{4}{5}-\frac{1}{3}$ $\frac{7}{15}$
18. $\frac{5}{8}-\frac{1}{6}$ $\frac{11}{24}$
19. $\frac{1}{6}+\frac{2}{9}$ $\frac{7}{18}$
20. $\frac{2}{7}+\frac{2}{5}$ $\frac{24}{35}$
21. $\frac{7}{12}-\frac{1}{4}$ $\frac{1}{3}$
22. $\frac{7}{8}-\frac{2}{3}$ $\frac{5}{24}$
23. $\frac{2}{11}+\frac{2}{3}$ $\frac{28}{33}$

Evaluate each expression for $b=\frac{1}{2}$. Write your answer in simplest form.

24. $b+\frac{1}{3}$ $\frac{5}{6}$
25. $\frac{8}{9}-b$ $\frac{7}{18}$
26. $b-\frac{2}{11}$ $\frac{7}{22}$
27. $\frac{2}{7}+b$ $\frac{11}{14}$
28. $b+b$ $\frac{2}{2}$ *or* 1
29. $b-b$ 0

Evaluate. Write each answer in simplest form.

30. $\frac{1}{3}+\frac{1}{9}+\frac{1}{3}$ $\frac{7}{9}$
31. $\frac{9}{10}-\frac{2}{10}-\frac{1}{5}$ $\frac{1}{2}$
32. $\frac{1}{2}+\frac{1}{4}-\frac{1}{8}$ $\frac{5}{8}$
33. $\frac{5}{6}-\frac{2}{3}+\frac{7}{12}$ $\frac{3}{4}$
34. $\frac{2}{3}+\frac{1}{4}-\frac{1}{6}$ $\frac{3}{4}$
35. $\frac{2}{9}+\frac{1}{6}+\frac{1}{3}$ $\frac{13}{18}$

36. Bailey spent $\frac{2}{3}$ of his monthly allowance at the movies and $\frac{1}{5}$ of it on baseball cards. What fraction of Bailey's allowance is left? $\frac{2}{15}$

Math Background

Example 2 makes clear that both the method of using the LCD and the method of multiplying denominators are suitable for adding or subtracting fractions. The LCD method generally has the advantage of working with simpler numbers, but the method of multiplying denominators is actually the method found most often historically. This method can be found in use as early as the twelfth century.

The use of the LCD is found occasionally in the work of some fifteenth and sixteenth century arithmeticians. This method did not gain general acceptance until the seventeenth century.

RETEACH 5-7

LESSON 5-7 **Reteach**
Adding and Subtracting with Unlike Denominators

Unlike fractions have different denominators. To add and subtract fractions, you must have a common denominator. The least common denominator (LCD) is the least common multiple of the denominators.

To add or subtract unlike fractions, first find the LCD of the fractions.

A. $\frac{2}{3}+\frac{1}{4}$
Multiples of 4: 4, 8, **12**,...
Multiples of 3: 3, 6, 9, **12**,...
The LCD is 12.

Next, use fraction strips to find equivalent fractions.

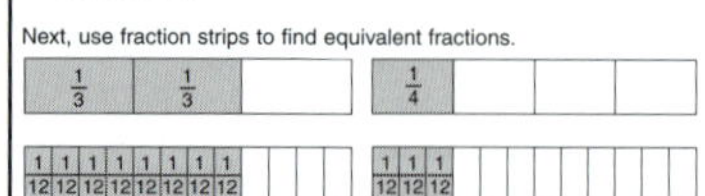

Then use fraction strips to find the sum or difference.

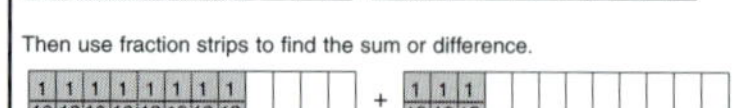

$\frac{8}{12}+\frac{3}{12}=\frac{11}{12}$

So, $\frac{2}{3}+\frac{1}{4}=\frac{11}{12}$.

Use fraction strips to find each sum or difference. Write your answer in simplest form.

1. $\frac{1}{4}+\frac{1}{8}$ $\frac{3}{8}$
2. $\frac{5}{6}-\frac{2}{3}$ $\frac{1}{6}$
3. $\frac{3}{4}-\frac{1}{3}$ $\frac{5}{12}$
4. $\frac{3}{5}+\frac{3}{10}$ $\frac{9}{10}$
5. $\frac{3}{4}+\frac{1}{6}$ $\frac{11}{12}$
6. $\frac{1}{2}+\frac{3}{8}$ $\frac{7}{8}$
7. $\frac{2}{3}-\frac{1}{6}$ $\frac{1}{2}$
8. $\frac{1}{3}-\frac{1}{4}$ $\frac{1}{12}$

PRACTICE 5-7

LESSON 5-7 **Practice B**
Adding and Subtracting with Unlike Denominators

Add or subtract. Write your answers in simplest form.

1. $\frac{6}{7}+\frac{1}{3}$ $1\frac{4}{21}$
2. $\frac{3}{7}-\frac{2}{5}$ $\frac{1}{35}$
3. $\frac{1}{4}+\frac{3}{8}$ $\frac{5}{8}$
4. $\frac{7}{8}-\frac{2}{3}$ $\frac{5}{24}$
5. $\frac{1}{6}+\frac{3}{5}$ $\frac{23}{30}$
6. $\frac{5}{6}-\frac{2}{3}$ $\frac{1}{6}$
7. $\frac{5}{9}-\frac{1}{3}$ $\frac{2}{9}$
8. $\frac{7}{8}+\frac{3}{4}$ $1\frac{5}{8}$
9. $\frac{5}{12}-\frac{1}{6}$ $\frac{1}{4}$
10. $\frac{4}{5}-\frac{7}{11}$ $\frac{9}{55}$
11. $\frac{4}{9}+\frac{5}{6}$ $1\frac{5}{18}$
12. $\frac{5}{8}+\frac{2}{3}$ $1\frac{7}{24}$

Evaluate each expression for $b=\frac{1}{3}$. Write your answers in simplest form.

13. $b+\frac{5}{8}$ $\frac{23}{24}$
14. $\frac{7}{9}-b$ $\frac{4}{9}$
15. $\frac{2}{7}+b$ $\frac{13}{21}$
16. $b+b$ $\frac{2}{3}$
17. $\frac{11}{12}-b$ $\frac{7}{12}$
18. $\frac{3}{4}-b$ $\frac{5}{12}$

19. There are three grades in Kyle's middle school—sixth, seventh, and eighth. One-third of the students are in sixth grade and $\frac{1}{4}$ are in seventh grade. What fraction of the schools' students are in eighth grade?
$\frac{5}{12}$ of the students

20. Sarah is making a dessert that calls for $\frac{4}{5}$ cup of crushed cookies. If she has already crushed $\frac{7}{10}$ cup, how much more does she need?
$\frac{1}{10}$ cup

Life Science LINK

The red lorikeet, galah cockatoo, and green-cheeked Amazon are three very colorful birds. The African grey parrot is known for its ability to mimic sounds it hears. In fact, one African grey named Prudle had a vocabulary of almost 1,000 words.

37. What is the average weight of the red lorikeet, African grey parrot, and the galah cockatoo? $\frac{2}{3}$ lb

38. Which bird weighs more, the green-cheeked Amazon or the red lorikeet? green-cheeked Amazon

39. What is the difference in weights between the green-cheeked Amazon and the red lorikeet? $\frac{9}{40}$ lb

40. Does the red lorikeet weigh more or less than $\frac{1}{2}$ lb? Explain.

41. ***WHAT'S THE ERROR?*** A student found the difference in weight between the African grey parrot and the galah cockatoo to be 1 lb. Explain the error. Then find the correct difference between the weights of these birds.

42. ***WRITE ABOUT IT*** Explain how you find the difference in weight between the galah cockatoo and green-cheeked Amazon.

43. ***CHALLENGE*** Find the average weight of the birds. $\frac{13}{20}$ lb

Red lorikeet, $\frac{3}{8}$ lb, $11\frac{2}{5}$ in.

Galah cockatoo, $\frac{3}{4}$ lb, $13\frac{9}{10}$ in.

Green-cheeked Amazon, $\frac{3}{5}$ lb, $13\frac{1}{5}$ in.

African grey parrot, $\frac{7}{8}$ lb, 13 in.

Spiral Review

Find each value. (Lesson 1-3)

44. 4^2 16 **45.** 2^4 16 **46.** 10^5 100,000 **47.** 7^3 343

48. **TEST PREP** Which list is ordered from least to greatest? (Lesson 3-1) D

A 75.4, 75.09, 75.28
B 75.28, 75.09, 75.4
C 75.09, 75.4, 75.28
D 75.09, 75.28, 75.4

49. **TEST PREP** Of which pair of numbers is 1 the only common factor? (Lesson 4-3) G

F 12 and 9 G 7 and 16 H 21 and 15 J 3 and 6

Interdisciplinary LINK

Life Science

Exercises 37–43 involve information about the weights of some birds. Students study mammals and birds in middle school life science programs such as *Holt Science & Technology*.

Answers

40. Possible answer: Less than $\frac{1}{2}$ lb, because $\frac{4}{8}$ lb $= \frac{1}{2}$ lb and $\frac{3}{8} < \frac{4}{8}$.

41. Possible answer: The student subtracted the numerators and denominators in $\frac{7}{8} - \frac{3}{4}$ and got a difference of $\frac{4}{4} = 1$. The correct difference is $\frac{1}{8}$ lb.

42. Possible answer: Subtract $\frac{3}{4} - \frac{3}{5}$. Use the LCD of 20, so $\frac{15}{20} - \frac{12}{20} = \frac{3}{20}$ lb.

Journal

Have students explain how adding or subtracting unlike fractions and adding or subtracting like fractions are different and how they are alike.

Test Prep Doctor

For Exercise 49, students can eliminate answer choices in which the pair of numbers has a common factor other than 1. **F, H,** and **J** can all be eliminated because each pair has 3 as a common factor.

CHALLENGE 5-7

LESSON 5-7 **Challenge**
Egyptian Fractions

Did you know that ancient Egyptians used fractions 5,000 years ago? Some of their fractions were like the ones we use today. However, the Egyptians only used **unit fractions,** or fractions with a numerator of 1. All other fractions had to be written as a sum of unit fractions. And no sum could repeat the same unit fraction! For example, the Egyptians would write $\frac{3}{4}$ as $\frac{1}{2} + \frac{1}{4}$. They would not write $\frac{1}{4} + \frac{1}{4} + \frac{1}{4}$.

Ancient Egyptians did not have paper. They recorded their math work on papyrus, or thin strips of dried plants. Study the Egyptian fractions recorded on the papyrus scrolls below. Then write each fraction the way we do today.

1. $\frac{1}{2} + \frac{1}{6}$ $\frac{2}{3}$
2. $\frac{1}{2} + \frac{1}{4} + \frac{1}{20}$ $\frac{4}{5}$
3. $\frac{1}{2} + \frac{1}{3}$ $\frac{5}{6}$
4. $\frac{1}{4} + \frac{1}{8}$ $\frac{3}{8}$
5. $\frac{1}{2} + \frac{1}{3} + \frac{1}{15}$ $\frac{9}{10}$

PROBLEM SOLVING 5-7

LESSON 5-7 **Problem Solving**
Adding and Subtracting with Unlike Denominators

Use the circle graph to answer the questions. Write each answer in simplest form.

World Population, 2001

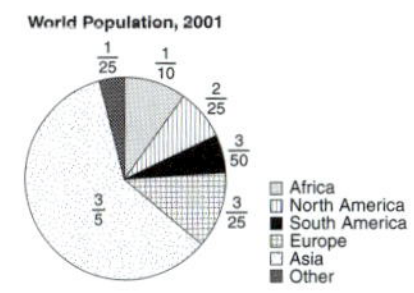

1. On which two continents do most people in the world live? How much of the total population do they make up together?
Asia and Europe; $\frac{18}{25}$ of the population

2. How much of the world's population live in either North America or South America?
$\frac{7}{50}$ of the population

3. How much more of the world's total population lives in Asia than in Africa?
$\frac{1}{2}$ of the population

4. How much of Earth's total population do people in Asia and Africa make up all together?
$\frac{7}{10}$ of the population

5. What is the difference between North America's part of the total population and Africa's part?
Africa has $\frac{1}{50}$ more.

Circle the letter of the correct answer.

6. How much more of the population lives in Europe than in North America?
(A) $\frac{1}{25}$ of the population
B $\frac{1}{5}$ of the population
C $\frac{1}{15}$ of the population
D $\frac{1}{10}$ of the population

7. How much of the world's population lives in North America and Europe?
F $\frac{1}{25}$ of the population
G $\frac{1}{15}$ of the population
(H) $\frac{1}{5}$ of the population
J $\frac{1}{20}$ of the population

Lesson Quiz

Add or subtract. Write each answer in simplest form.

1. $\frac{3}{8} + \frac{1}{6}$ $\frac{13}{24}$ **2.** $\frac{1}{2} - \frac{1}{8}$ $\frac{3}{8}$

3. $\frac{5}{12} - \frac{1}{6}$ $\frac{1}{4}$ **4.** $\frac{1}{6} + \frac{1}{8}$ $\frac{7}{24}$

5. Bonnie is making oatmeal bars, and the recipe calls for $\frac{3}{4}$ cup of brown sugar. If she has $\frac{1}{3}$ cup of brown sugar, how much more does she need? $\frac{5}{12}$ cup

Available on Daily Transparency in CRB

5-8 Organizer

Pacing: Traditional 1 day
Block $\frac{1}{2}$ day

Objective: Students add and subtract mixed numbers with unlike denominators.

Indiana Academic Standards
6.1.7, 6.2.5

Warm Up

Add or subtract. Write each answer in simplest form.

1. $\frac{7}{12} - \frac{1}{2}$ $\frac{1}{12}$
2. $\frac{3}{11} + \frac{2}{3}$ $\frac{31}{33}$
3. $\frac{1}{6} + \frac{3}{4}$ $\frac{11}{12}$
4. $\frac{11}{12} - \frac{1}{3}$ $\frac{7}{12}$

Problem of the Day

The sum of every row, column, and diagonal is the same. Complete the magic square.

$\frac{3}{4}$	$1\frac{1}{3}$	$\frac{11}{12}$
$1\frac{1}{6}$	1	$\frac{5}{6}$
$1\frac{1}{12}$	$\frac{2}{3}$	$1\frac{1}{4}$

Available on Daily Transparency in CRB

Math Fact

Two-by-four boards measure $1\frac{1}{2}$ inches by $3\frac{1}{2}$ inches. Because of this, carpenters must add and subtract using mixed numbers.

5-8 Adding and Subtracting Mixed Numbers

Learn to add and subtract mixed numbers with unlike denominators.

Chameleons can change color at any time to camouflage themselves. They live high in trees and are seldom seen on the ground.

A Parsons chameleon, which is the largest kind of chameleon, can extend its tongue $1\frac{1}{2}$ times the length of its body. This allows the chameleon to capture food it otherwise would not be able to reach.

The chameleon is the only animal capable of moving each eye independently of the other. A chameleon can turn its eyes about 360°.

EXAMPLE 1 Adding and Subtracting Mixed Numbers

Find each sum or difference. Write the answer in simplest form.

A $2\frac{3}{4} + 1\frac{1}{6}$

$$2\frac{3}{4} \rightarrow 2\frac{18}{24}$$
$$+1\frac{1}{6} \rightarrow +1\frac{4}{24}$$
$$3\frac{22}{24} = 3\frac{11}{12}$$

Multiply the denominators. 4 • 6 = 24
Write equivalent fractions with a denominator of 24.
Add the fractions and then the whole numbers, and simplify.

B $4\frac{5}{6} - 2\frac{2}{9}$

$$4\frac{5}{6} \rightarrow 4\frac{15}{18}$$
$$-2\frac{2}{9} \rightarrow -2\frac{4}{18}$$
$$2\frac{11}{18}$$

The LCD is 18.
Write equivalent fractions with a denominator of 18.
Subtract the fractions and then the whole numbers.

C $2\frac{2}{3} + 1\frac{3}{4}$

$$2\frac{2}{3} \rightarrow 2\frac{8}{12}$$
$$+1\frac{3}{4} \rightarrow +1\frac{9}{12}$$
$$3\frac{17}{12} = 4\frac{5}{12}$$

The LCD is 12.
Write equivalent fractions with a denominator of 12.
Add the fractions and then the whole numbers. $3\frac{17}{12} = 3 + 1\frac{5}{12}$

1 Introduce

Alternate Opener

EXPLORATION

5-8 Adding and Subtracting Mixed Numbers

The graph shows typical rainfall levels for a city in the Southwest for the first 5 months of the year. What is the approximate total rainfall from January through May?

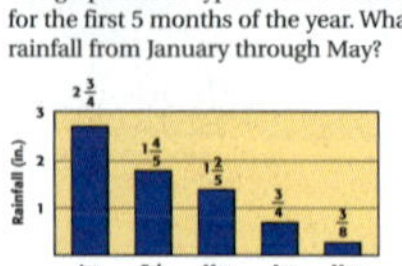

A mixed number contains a whole number and a fraction. To estimate with mixed numbers, round each mixed number to the nearest whole number.

Actual $2\frac{3}{4} + 1\frac{4}{5} + 1\frac{2}{5} + \frac{3}{4} + \frac{3}{8}$

Estimated $3 + 2 + 1 + 1 + 0 = 7$ in.

The total rainfall is about 7 inches.

Estimate each sum or difference.

1. $13\frac{1}{2} - 9\frac{27}{32}$
2. $1\frac{1}{2} + 9\frac{3}{8} - 2\frac{1}{4}$
3. $17\frac{7}{8} + 19\frac{1}{10}$
4. $4\frac{1}{8} - 1\frac{3}{10} + 3\frac{1}{4}$

Think and Discuss

5. **Discuss** the estimation strategies you used.
6. **Describe** a real-world situation in which mixed numbers are added or subtracted.

Motivate

Have students give examples of situations when they would need to add or subtract mixed numbers. Ask students what they needed to do in the previous lesson to add or subtract fractions with unlike denominators. Write equivalent fractions with common denominators. Explain that the same rule applies to mixed number addition and subtraction.

Exploration worksheet and answers on Chapter 5 Resource Book pp. 71 and 124

2 Teach

Lesson Presentation

Guided Instruction

In this lesson, students learn to add and subtract mixed numbers with unlike denominators. Teach them to find sums and differences by writing equivalent fractions with common denominators. Then apply the concept to a measurement situation.

Find each sum or difference. Write the answer in simplest form.

D $8\frac{2}{5} - 6\frac{3}{10}$

$$\begin{array}{rcr} 8\frac{2}{5} & \longrightarrow & 8\frac{4}{10} \\ -\,6\frac{3}{10} & \longrightarrow & -\,6\frac{3}{10} \\ \hline & & 2\frac{1}{10} \end{array}$$

Think: 10 is a multiple of 5, so 10 is the LCD.

Write equivalent fractions with a denominator of 10.

Subtract the fractions and then the whole numbers.

EXAMPLE 2 *Measurement Application*

The length of a Parsons chameleon's body is $23\frac{1}{2}$ inches. The chameleon can extend its tongue $35\frac{1}{4}$ inches. What is the total length of its body and its tongue?

Add $23\frac{1}{2}$ and $35\frac{1}{4}$.

$$\begin{array}{rcr} 23\frac{1}{2} & \longrightarrow & 23\frac{2}{4} \\ +\,35\frac{1}{4} & \longrightarrow & +\,35\frac{1}{4} \\ \hline & & 58\frac{3}{4} \end{array}$$

Find a common denominator. Write equivalent fractions with the LCD, 4, as the denominator.

Add the fractions and then the whole numbers.

The total length of the chameleon's body and tongue is $58\frac{3}{4}$ inches.

Helpful Hint

You can use mental math to find an LCD. *Think:* 4 is a multiple of 2 and 4.

Think and Discuss

1. **Tell** what mistake was made when subtracting $2\frac{1}{2}$ from $5\frac{3}{5}$ gave the following result: $5\frac{3}{5} - 2\frac{1}{2} = 3\frac{2}{3}$.
2. **Explain** why you first find equivalent fractions when adding $1\frac{1}{5}$ and $1\frac{1}{2}$.
3. **Tell** how you know that $5\frac{1}{2} - 3\frac{1}{4}$ is more than 2.

COMMON ERROR ALERT

When renaming mixed numbers using a common denominator, some students, concentrating on the fractions, may overlook keeping the whole number as part of the newly renamed mixed number. Remind students that the whole number part of this mixed number should not be separated from the part that is a fraction.

Additional Examples

Example 1

Find each sum or difference. Write the answer in simplest form.

A. $3\frac{1}{8} + 1\frac{5}{6}$ $4\frac{23}{24}$

B. $5\frac{2}{3} - 1\frac{1}{4}$ $4\frac{5}{12}$

C. $2\frac{1}{2} + 4\frac{4}{5}$ $7\frac{3}{10}$

D. $6\frac{2}{3} - 3\frac{2}{9}$ $3\frac{4}{9}$

Example 2

The length of Jen's kitten's body is $10\frac{1}{4}$ inches long. Its tail is $5\frac{1}{8}$ inches long. What is the total length of its body and tail? $15\frac{3}{8}$ inches

3 Close

Reaching All Learners

Through Hands-On Experience

Have students measure, in inches, several objects around the classroom and then record their findings. (Rulers are provided in the Manipulatives Kit and on Chapter 5 Resource Book p. 108.) Then move some of the objects next to one another. Have students use their findings to calculate the combined width of adjacent objects and to check the results by making an actual measurement. Ask the students to explain any discrepancies.

Summarize

Review adding and subtracting mixed numbers with unlike denominators. Have students explain how they would add $12\frac{3}{4} + 3\frac{5}{6}$.

Possible answer: Write equivalent fractions with a denominator of 12. Add the fractions, and then add the whole numbers.

$12\frac{9}{12} + 3\frac{10}{12} = 15\frac{19}{12} = 16\frac{7}{12}$

Answers to Think and Discuss

1. Instead of using equivalent fractions with a common denominator, the numerators were subtracted and the denominators were subtracted.
2. Possible answer: You need to have common denominators before you can add the numerators.
3. 5 minus 3 is 2 and $\frac{1}{2}$ is greater than $\frac{1}{4}$, so the answer would be greater than 2.

5-8 PRACTICE & ASSESS

Students may want to refer back to the lesson examples.

Assignment Guide

If you finished Example 1 assign:

Core 1–4, 6–9, 11–21 odd, 45–53

Enriched 2–8 even, 11–25 odd, 43–53

If you finished Example 2 assign:

Core 1–10, 23–28, 37–41 odd, 45–53

Enriched 29–53

Notes

5-8 Exercises

FOR EXTRA PRACTICE see page 646

internet connect Homework Help Online go.hrw.com Keyword: MR4 5-8

GUIDED PRACTICE

See Example 1 Find each sum or difference. Write the answer in simplest form.

1. $7\frac{1}{12} + 3\frac{1}{3}$ $10\frac{5}{12}$
2. $2\frac{1}{6} + 2\frac{3}{8}$ $4\frac{13}{24}$
3. $8\frac{5}{6} - 2\frac{3}{4}$ $6\frac{1}{12}$
4. $6\frac{6}{7} - 1\frac{1}{2}$ $5\frac{5}{14}$

See Example 2

5. A sea turtle traveled $7\frac{3}{4}$ hours in two days. It traveled $3\frac{1}{2}$ hours on the first day. How many hours did it travel on the second day? $4\frac{1}{4}$

INDEPENDENT PRACTICE

See Example 1 Find each sum or difference. Write the answer in simplest form.

6. $3\frac{9}{10} - 1\frac{2}{5}$ $2\frac{1}{2}$
7. $2\frac{1}{6} + 4\frac{5}{12}$ $6\frac{7}{12}$
8. $5\frac{9}{11} + 5\frac{1}{3}$ $11\frac{5}{33}$
9. $9\frac{3}{4} - 3\frac{1}{2}$ $6\frac{1}{4}$

See Example 2

10. The drama club rehearsed $1\frac{3}{4}$ hours Friday and $3\frac{1}{6}$ hours Saturday. How many total hours did the students rehearse? $4\frac{11}{12}$

PRACTICE AND PROBLEM SOLVING

Add or subtract. Write each answer in simplest form.

11. $6\frac{3}{10} + 3\frac{2}{5}$ $9\frac{7}{10}$
12. $10\frac{2}{3} - 2\frac{1}{12}$ $8\frac{7}{12}$
13. $14\frac{3}{4} - 6\frac{5}{12}$ $8\frac{1}{3}$
14. $19\frac{1}{10} + 10\frac{1}{2}$ $29\frac{3}{5}$
15. $15\frac{5}{6} + 18\frac{2}{3}$ $34\frac{1}{2}$
16. $17\frac{1}{6} + 12\frac{1}{4}$ $29\frac{5}{12}$
17. $23\frac{9}{10} - 20\frac{3}{9}$ $3\frac{51}{90}$
18. $32\frac{5}{7} - 13\frac{2}{5}$ $19\frac{11}{35}$
19. $28\frac{11}{12} - 8\frac{5}{9}$ $20\frac{13}{36}$
20. $12\frac{2}{11} + 20\frac{2}{3}$ $32\frac{28}{33}$
21. $36\frac{5}{8} - 24\frac{5}{12}$ $12\frac{5}{24}$
22. $48\frac{9}{11} + 2\frac{1}{4}$ $51\frac{3}{44}$

Evaluate. Write each answer as a fraction in simplest form.

23. $0.3 + \frac{2}{5}$ $\frac{7}{10}$
24. $\frac{4}{5} + 0.9$ $\frac{17}{10}$ *or* $1\frac{7}{10}$
25. $5\frac{4}{5} - 3.2$ $2\frac{3}{5}$
26. $6.3 + \frac{4}{5}$ $7\frac{1}{10}$
27. $23\frac{3}{4} - 10.5$ $13\frac{1}{4}$
28. $18.9 - 6\frac{1}{2}$ $12\frac{2}{5}$

Evaluate each expression for $n = 2\frac{1}{3}$. Write your answer in simplest form.

29. $2\frac{2}{3} + n$ 5
30. $5 - \left(1\frac{2}{3} + n\right)$ 1
31. $n - 1\frac{1}{4}$ $1\frac{1}{12}$
32. $n + 5\frac{7}{9}$ $8\frac{1}{9}$
33. $6\left(3\frac{4}{9} + n\right)$ $34\frac{2}{3}$
34. $2n - n$ $2\frac{1}{3}$

35. ***MEASUREMENT*** Kyle's backpack weighs $14\frac{7}{20}$ lb. Kirsten's backpack weighs $12\frac{1}{4}$ lb.
    a. How much do the backpacks weigh together? $26\frac{3}{5}$ pounds
    b. How much more does Kyle's backpack weigh than Kirsten's backpack? $2\frac{1}{10}$ lb
    c. Kyle takes his $3\frac{1}{4}$ lb math book out of his backpack. How much does his backpack weigh now? $11\frac{1}{10}$ lb

Math Background

In Lesson 5-2, students learned to multiply mixed numbers by first rewriting each as a fraction. The same method could be used here to add mixed numbers, but this is not the most efficient way for this to be done.

The Associative and Commutative Properties of Addition allow one to conceive of the sum of two mixed numbers as two separate sums: the sum of their whole parts and the sum of their fractional parts. By treating the sums separately, addition of mixed numbers is reduced to two algorithms already mastered: addition of whole numbers and addition of fractions.

RETEACH 5-8

LESSON 5-8 **Reteach**

Adding and Subtracting Mixed Numbers

You can use what you know about improper fractions to add and subtract mixed numbers.

To find the sum or difference of mixed numbers, first write the mixed numbers as improper fractions.

A. $3\frac{1}{4} + 2\frac{1}{3}$
$= \frac{13}{4} + \frac{7}{3}$

B. $4\frac{1}{2} - 2\frac{2}{3}$
$= \frac{9}{2} - \frac{8}{3}$

Next, find equivalent fractions with a least common denominator.

$\frac{13}{4} + \frac{7}{3}$
$= \frac{39}{12} + \frac{28}{12}$

$\frac{9}{2} - \frac{8}{3}$
$= \frac{27}{6} - \frac{16}{6}$

Then add or subtract the like fractions.

$\frac{39}{12} + \frac{28}{12}$
$= \frac{67}{12}$

$\frac{27}{6} - \frac{16}{6}$
$= \frac{11}{6}$

Write the answer as a mixed number in simplest form.

$\frac{67}{12}$
$= 5\frac{7}{12}$

$\frac{11}{6}$
$= 1\frac{5}{6}$

So, $3\frac{1}{4} + 2\frac{1}{3} = 5\frac{7}{12}$.

So, $4\frac{1}{2} - 2\frac{2}{3} = 1\frac{5}{6}$.

Find each sum or difference. Write your answer in simplest form.

1. $1\frac{1}{4} + 1\frac{1}{2}$ $= \frac{5}{4} + \frac{3}{2}$ $= \frac{5}{4} + \frac{6}{4}$ $2\frac{3}{4}$
2. $3\frac{1}{6} + 1\frac{2}{3}$ $= \frac{19}{6} + \frac{5}{3}$ $= \frac{19}{6} + \frac{10}{6}$ $4\frac{5}{6}$
3. $2\frac{1}{8} + 4\frac{1}{2}$ $= \frac{17}{8} + \frac{9}{2}$ $= \frac{17}{8} + \frac{36}{8}$ $6\frac{5}{8}$
4. $4\frac{1}{3} + 1\frac{1}{2}$ $= \frac{13}{3} + \frac{3}{2}$ $= \frac{26}{6} + \frac{9}{6}$ $5\frac{5}{6}$
5. $2\frac{3}{5} + 1\frac{1}{10}$ $3\frac{7}{10}$
6. $3\frac{1}{6} + 1\frac{1}{12}$ $4\frac{1}{4}$
7. $2\frac{5}{8} - 1\frac{1}{4}$ $1\frac{3}{8}$
8. $5\frac{2}{3} - 2\frac{1}{4}$ $3\frac{5}{12}$

PRACTICE 5-8

LESSON 5-8 **Practice B**

Adding and Subtracting Mixed Numbers

Find each sum or difference. Write each answer in simplest form.

1. $4\frac{3}{8} + 5\frac{1}{4}$ $9\frac{5}{8}$
2. $11\frac{2}{5} - 8\frac{1}{3}$ $3\frac{1}{15}$
3. $7\frac{1}{3} + 3\frac{2}{9}$ $10\frac{5}{9}$
4. $22\frac{5}{6} - 17\frac{1}{4}$ $5\frac{7}{12}$
5. $32\frac{4}{7} - 14\frac{1}{3}$ $18\frac{5}{21}$
6. $12\frac{1}{4} + 5\frac{1}{12}$ $17\frac{1}{3}$
7. $29\frac{1}{3} - 14\frac{1}{6}$ $15\frac{1}{6}$
8. $5\frac{3}{4} - 1\frac{7}{11}$ $4\frac{5}{44}$
9. $21\frac{1}{6} + 1\frac{3}{8}$ $22\frac{13}{24}$
10. $15\frac{7}{12} - 14\frac{3}{8}$ $1\frac{5}{24}$
11. $5\frac{6}{15} + 4\frac{3}{10}$ $9\frac{7}{10}$
12. $25\frac{1}{7} + 25\frac{2}{5}$ $50\frac{19}{35}$
13. $3\frac{2}{5} + 1\frac{1}{3}$ $4\frac{11}{15}$
14. $1\frac{2}{5} - 1\frac{2}{10}$ $\frac{1}{5}$
15. $3\frac{3}{5} - 2\frac{1}{2}$ $1\frac{1}{10}$
16. $6\frac{3}{4} - 3\frac{3}{10}$ $3\frac{9}{20}$
17. $4\frac{4}{5} + 2\frac{1}{10}$ $6\frac{9}{10}$
18. $32\frac{1}{2} + 5\frac{1}{3}$ $37\frac{5}{6}$
19. Donald is making a party mix. He bought $2\frac{1}{4}$ pounds of pecans and $3\frac{1}{5}$ pounds of walnuts. How many pounds of nuts did Donald buy in all? $5\frac{9}{20}$ pounds
20. Mrs. Watson's cookie recipe calls for $3\frac{4}{7}$ cups of sugar. Mr. Clark's cookie recipe calls for $4\frac{2}{3}$ cups of sugar. How much more sugar does Mr. Clark's recipe use? $1\frac{2}{21}$ cups more
21. Tasha's cat weighs $15\frac{5}{12}$ lb. Naomi's cat weighs $11\frac{1}{3}$ lb. Can they bring both of their cats to the vet in a carrier that can hold up to 27 pounds? Explain.
    Yes; because the cats' combined weight is $26\frac{3}{4}$ pounds, which is less the 27 pounds

Life Science LINK

Research shows that elephants hear through their feet. The feet act as drums, sending vibrations through the bones to the elephant's ears.

36. ***LIFE SCIENCE*** Elephants can communicate through low-frequency infrasonic rumbles. Such sounds can travel from $\frac{1}{8}$ km to $9\frac{1}{2}$ km. Find the difference between these two distances. $9\frac{3}{8}$ km

37. The route Jo usually takes to work is $4\frac{2}{5}$ mi. After heavy rains, when that road is flooded, she must take a different route that is $4\frac{9}{10}$ mi. How much longer is Jo's alternate route? $\frac{1}{2}$ mi

38. Mr. Hansley used $1\frac{2}{3}$ c of flour to make muffins and $4\frac{1}{2}$ c to make bread. If he has $3\frac{5}{6}$ c left, how much flour did Mr. Hansley have before making the muffins and bread? 10 c

Use the drawing for Exercises 39–42.

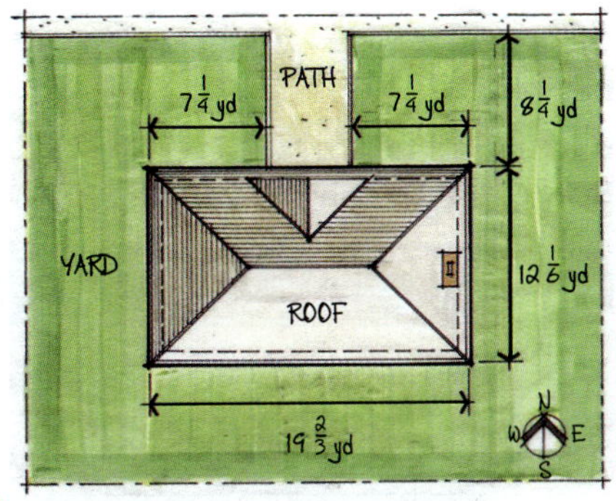

39. Sarah is a landscape architect designing a garden. Based on her drawing, how much longer is the south side of the building than the west side?

40. Sarah needs to determine how many azalea bushes she can plant along both sides of the path. What is the sum of the lengths of the two sides of the path? $16\frac{1}{2}$ yards

41. How wide is the path? $5\frac{1}{6}$ yards

42. ***WHAT'S THE QUESTION?*** The answer is $63\frac{2}{3}$ yd. What is the question?

43. ***WRITE ABOUT IT*** Explain how you would use the sum of $\frac{2}{5}$ and $\frac{1}{3}$ to find the sum of $10\frac{2}{5}$ and $6\frac{1}{3}$. Possible answer: Find the sum of the whole numbers and add it to the sum of $\frac{2}{5}$ and $\frac{1}{3}$.

44. ***CHALLENGE*** Find each missing numerator.

a. $3\frac{x}{9} + 4\frac{2}{9} = 7\frac{7}{9}$ $x = 5$

b. $1\frac{3}{10} + 9\frac{x}{2} = 10\frac{4}{5}$ $x = 1$

Spiral Review

Move the decimal point to multiply or divide by powers of 10. (Lesson 3-4)

45. $1{,}465 \times 100$ 146,500
46. $32.09 \div 10^2$ 0.3209
47. $209{,}467 \times 1{,}000$ 209,467,000
48. $1.7 \div 10^4$ 0.00017

Find each product. (Lesson 5-1)

49. $\frac{2}{3} \times \frac{1}{5}$ $\frac{2}{15}$
50. $\frac{3}{7} \times \frac{1}{4}$ $\frac{3}{28}$
51. $\frac{2}{9} \times \frac{3}{8}$ $\frac{1}{12}$
52. $\frac{1}{4} \times \frac{6}{7}$ $\frac{3}{14}$

53. **TEST PREP** What is the least common multiple of 5 and 8? (Lesson 5-5) A

A 40   B 20   C 80   D 60

Answers

39. $7\frac{1}{2}$ yd

42. What is the distance around, or the perimeter of, the house?

Journal

Have students outline the steps involved in adding and subtracting mixed numbers with unlike denominators. Have them give an example for each operation.

Test Prep Doctor

For Exercise 53, students must identify the number that is the **least** common multiple of 5 and 8, not just any common multiple. Students who answered **B** or **D** identified numbers that are only multiples of 5. Students who answered **C** identified a common multiple, but not the least common multiple.

CHALLENGE 5-8

LESSON 5-8 **Challenge**
Maximum Snakes

The bar graph below shows the maximum lengths for the longest snakes in the world. Use the graph to find how much each of the snakes in the City Zoo is below its maximum length.

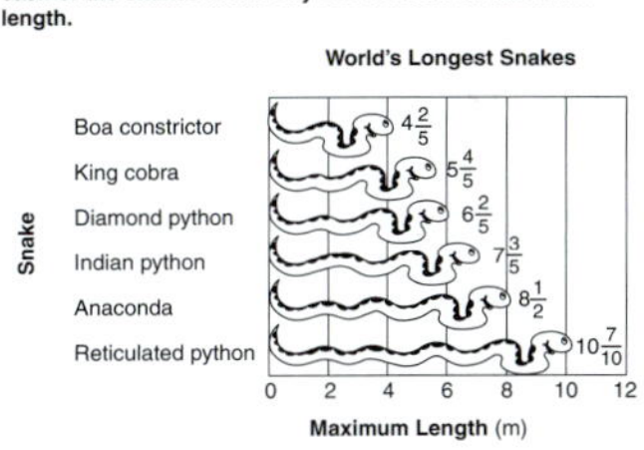

Snakes in the City Zoo

Snake	Length (in meters)	Difference from Maximum Length
Kevin (king cobra)	$3\frac{1}{2}$	$2\frac{3}{10}$ meters
Annie (anaconda)	$5\frac{1}{3}$	$3\frac{1}{6}$ meters
Bob (boa constrictor)	$3\frac{1}{4}$	$1\frac{3}{20}$ meters
Ivy (Indian python)	$4\frac{2}{7}$	$3\frac{11}{35}$ meters
Reggie (reticulated python)	$8\frac{3}{5}$	$2\frac{1}{10}$ meters
Diana (diamond python)	$4\frac{3}{8}$	$2\frac{1}{40}$ meters

PROBLEM SOLVING 5-8

LESSON 5-8 **Problem Solving**
Adding and Subtracting Mixed Numbers

Write the correct answer in simplest form.

1. Of the planets in our solar system, Jupiter and Neptune have the greatest surface gravity. Jupiter's gravitational pull is $2\frac{16}{25}$ stronger than Earth's, and Neptune's is $1\frac{1}{5}$ stronger. What is the difference between Jupiter's and Neptune's surface gravity levels?
Jupiter's is $1\frac{11}{25}$ higher.

2. Escape velocity is the speed a rocket must attain to overcome a planet's gravitational pull. Earth's escape velocity is $6\frac{9}{10}$ miles per second! The Moon's escape velocity is $5\frac{2}{5}$ miles per second slower. How fast does a rocket have to launch to escape the moon's gravity?
$1\frac{1}{2}$ miles per second

3. The two longest total solar eclipses occurred in 1991 and 1992. The first one lasted $6\frac{5}{6}$ minutes. The eclipse of 1992 lasted $5\frac{1}{3}$ minutes. How much longer was 1991's eclipse?
$1\frac{1}{2}$ minutes

4. The two largest meteorites found in the U.S. landed in Canyon Diablo, Arizona, and Willamette, Oregon. The Arizona meteorite weighs $33\frac{1}{10}$ tons! Oregon's weighs $16\frac{1}{2}$ tons. How much do the two meteorites weigh in all?
$49\frac{3}{5}$ tons

Circle the letter of the correct answer.

5. Not including the Sun, Proxima Centauri is the closest star to Earth. It is $4\frac{11}{50}$ light years away! The next closest star is Alpha Centauri. It is $\frac{13}{100}$ light years farther than Proxima. How far is Alpha Centauri from Earth?
(A) $4\frac{7}{20}$ light years
B $4\frac{13}{100}$ light years
C $4\frac{6}{25}$ light years
D $4\frac{1}{50}$ light years

6. It takes about $5\frac{1}{3}$ minutes for light from the Sun to reach Earth. The Moon is closer to Earth, so its light reaches Earth faster—about $5\frac{19}{60}$ minutes faster than from the Sun. How long does light from the Moon take to reach Earth?
F $\frac{3}{10}$ of a minute
(G) $\frac{1}{60}$ of a minute
H $\frac{1}{3}$ of a minute
J $\frac{4}{15}$ of a minute

Lesson Quiz

Find each sum or difference. Write the answer in simplest form.

1. $7\frac{11}{12} - 4\frac{2}{3}$ $3\frac{1}{4}$
2. $9\frac{1}{6} + 7\frac{3}{8}$ $16\frac{13}{24}$
3. $6\frac{7}{8} + 1\frac{3}{10}$ $8\frac{7}{40}$
4. $10\frac{2}{5} - 4\frac{1}{6}$ $6\frac{7}{30}$
5. Miles worked $5\frac{3}{4}$ hours on Friday and $8\frac{2}{3}$ hours on Saturday. How many total hours did he work? $14\frac{5}{12}$ hours

Available on Daily Transparency in CRB

Hands-On LAB 5D Model Subtraction with Renaming

Pacing: Traditional 1 day
Block $\frac{1}{2}$ day

Objective: To use fraction bars to model subtracting fractions by renaming them

Materials: Fraction bars

Lab Resources

Hands-On Lab Activities . . . . pp. 37–38

Using the Pages

Discuss with students what each fraction bar represents and how to model subtraction using fraction bars.

Model each subtraction problem using fraction bars.

1. $\frac{1}{2} - \frac{1}{3}$

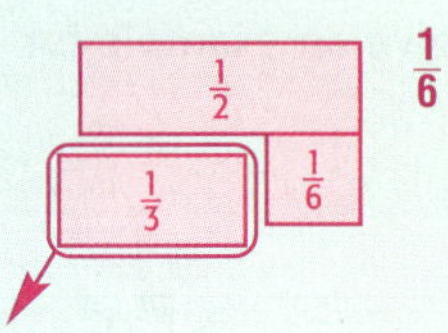

2. $\frac{5}{12} - \frac{1}{3}$

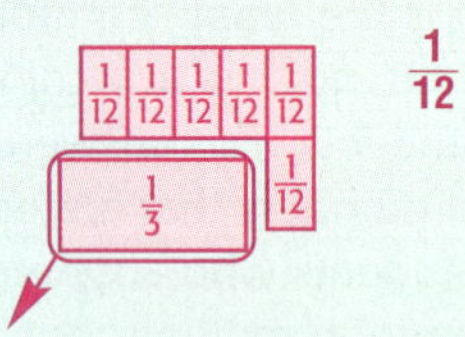

3. $\frac{7}{8} - \frac{1}{4}$

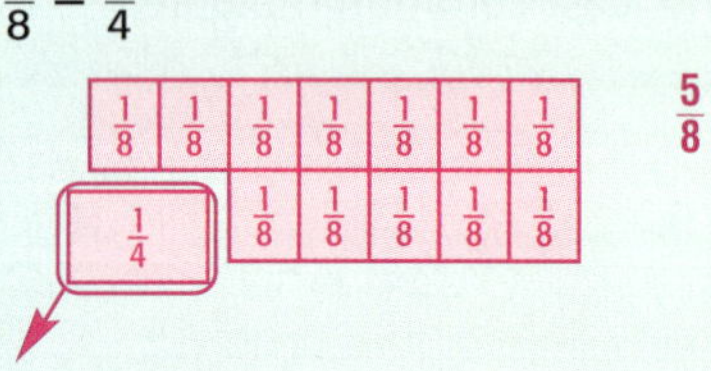

Hands-On LAB 5D Model Subtraction with Renaming

Use with Lesson 5-9

Sometimes you need to rename a mixed number before you can subtract. To rename a mixed number, divide one or more of the whole numbers into fractional parts.

internet connect
Lab Resources Online
go.hrw.com
KEYWORD: MR4 Lab5D

Activity

1 Find $2\frac{1}{3} - 1\frac{2}{3}$.

Use fraction bars to model $2\frac{1}{3}$.

You cannot subtract $\frac{2}{3}$ from $\frac{1}{3}$. You need to rename $2\frac{1}{3}$.

You can subtract $1\frac{2}{3}$ from $1\frac{4}{3}$.

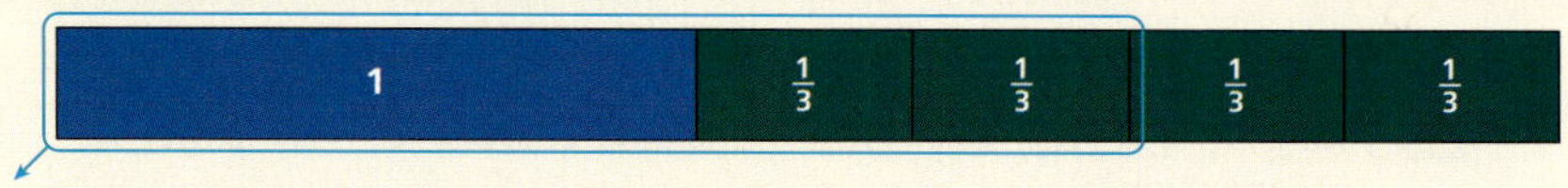

$2\frac{1}{3} - 1\frac{2}{3} = 1\frac{4}{3} - 1\frac{2}{3} = \frac{2}{3}$

2 Find $2\frac{1}{6} - 1\frac{5}{12}$.

Use fraction bars to model $2\frac{1}{6}$.

Write an equivalent fraction with a denominator of 12.

Still, you cannot subtract $\frac{5}{12}$ from $\frac{2}{12}$. You need to rename $2\frac{2}{12}$.

You can subtract $1\frac{5}{12}$ from $1\frac{14}{12}$.

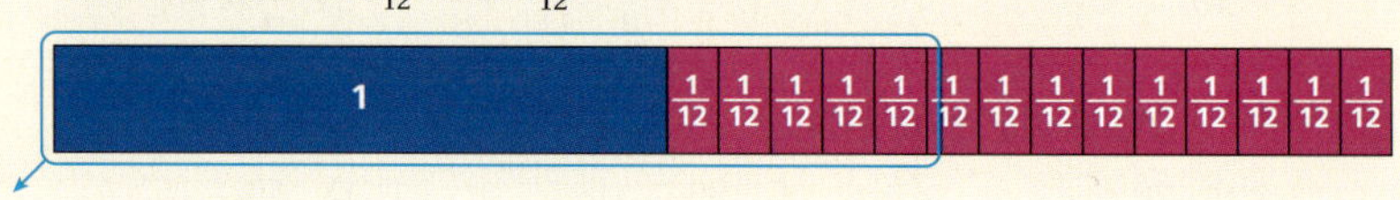

$2\frac{1}{6} - 1\frac{5}{12} = 1\frac{14}{12} - 1\frac{5}{12} = \frac{9}{12}$, or $\frac{3}{4}$

3 Find $2\frac{1}{4} - 1\frac{3}{8}$.

Use fraction bars to model $2\frac{1}{4}$.

Write an equivalent fraction.

Still, you cannot subtract $\frac{3}{8}$ from $\frac{2}{8}$. You need to rename $2\frac{2}{8}$.

You can subtract $1\frac{3}{8}$ from $1\frac{10}{8}$.

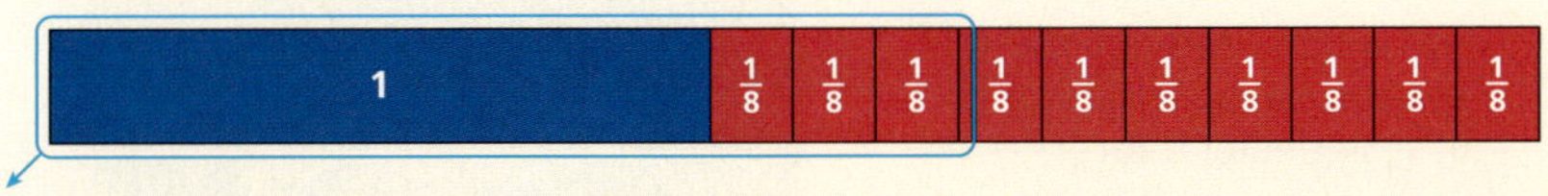

$2\frac{1}{4} - 1\frac{3}{8} = 1\frac{10}{8} - 1\frac{3}{8} = \frac{7}{8}$

Think and Discuss

1. Tell whether you need to rename before you subtract $3\frac{3}{8} - 1\frac{1}{8}$. **No; you can subtract $\frac{1}{8}$ from $\frac{3}{8}$.**
2. Tell whether you need to rename before you subtract $4\frac{3}{5} - 1\frac{7}{10}$.
Yes; you cannot subtract $\frac{7}{10}$ from $\frac{3}{5}$ or the equivalent fraction $\frac{6}{10}$.

Try This

Give the subtraction expression that is modeled.

1. 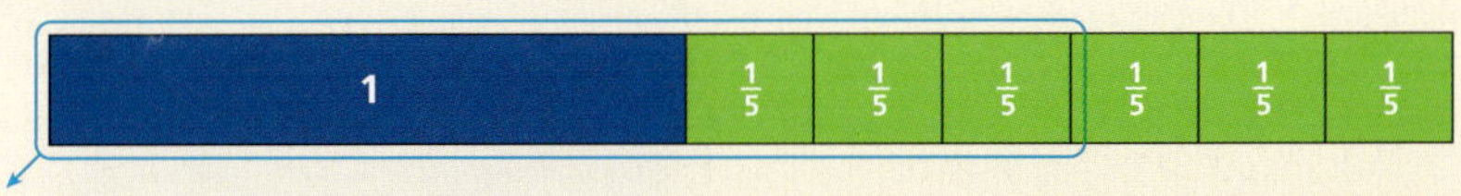 $2\frac{1}{5} - 1\frac{3}{5}$

2. 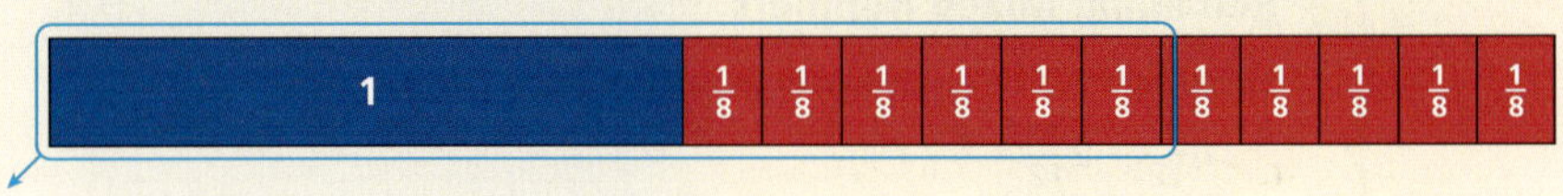 $2\frac{3}{8} - 1\frac{3}{4}$

3. 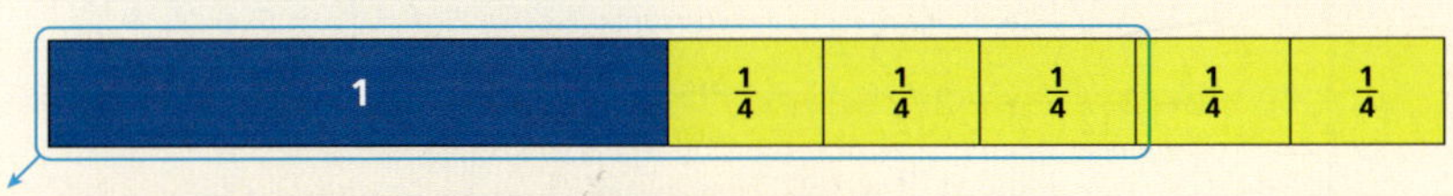 $2\frac{1}{4} - 1\frac{3}{4}$

4. 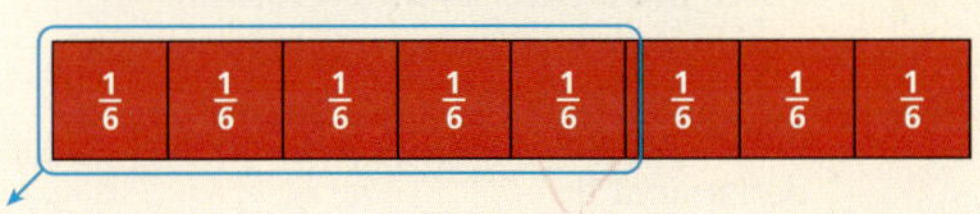 $1\frac{1}{3} - \frac{5}{6}$

Use fraction bars to subtract. Draw each model.

5. $2\frac{3}{8} - 1\frac{1}{2}$ **$\frac{7}{8}$**
6. $3\frac{1}{3} - 1\frac{2}{3}$ **$1\frac{2}{3}$**
7. $4\frac{1}{4} - 2\frac{5}{6}$ **$1\frac{5}{12}$**
8. $3\frac{1}{6} - 1\frac{1}{4}$ **$1\frac{11}{12}$**

Answers

Try This

5. 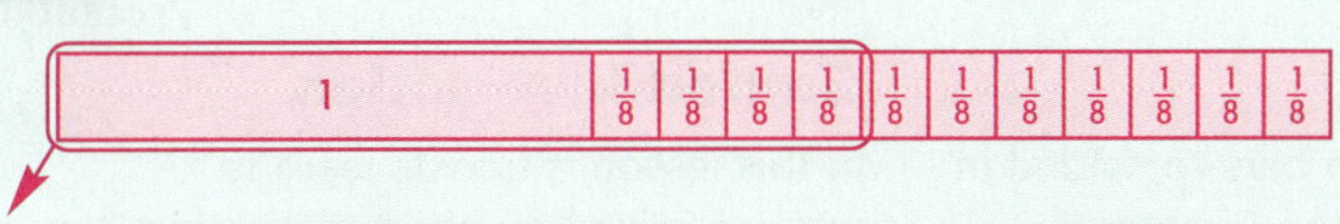 $\frac{7}{8}$

6. 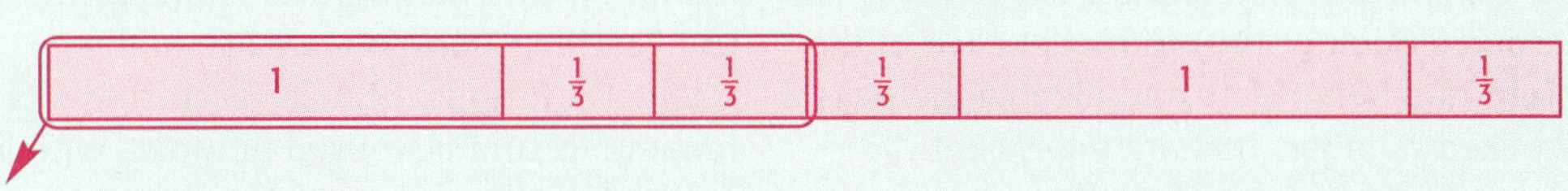 $1\frac{2}{3}$

7. 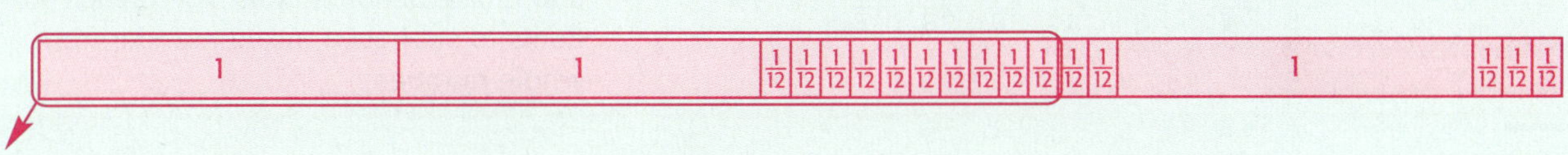 $1\frac{5}{12}$

8. 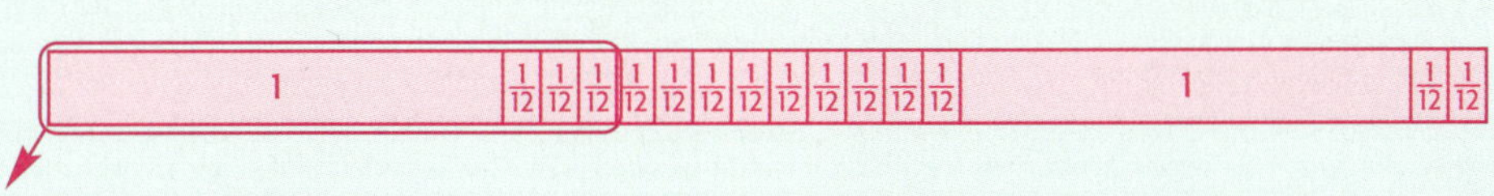 $1\frac{11}{12}$

5-9 Organizer

Pacing: Traditional 1 day
Block $\frac{1}{2}$ day

Objective: Students rename mixed numbers to subtract.

Indiana Academic Standards
6.1.4, 6.1.7, 6.2.5

Warm Up

Add or subtract.

1. $3\frac{2}{5} + 1\frac{3}{5}$ 5
2. $7\frac{9}{11} + \frac{11}{11}$ $8\frac{9}{11}$
3. $8\frac{4}{5} - 2\frac{1}{4}$ $6\frac{11}{20}$
4. $6\frac{1}{3} - 1\frac{1}{9}$ $5\frac{2}{9}$

Problem of the Day

Complete the magic square so that every row, column, and diagonal has the same sum. Answer:

$1\frac{7}{8}$	$\frac{5}{8}$	$\frac{1}{2}$	$2\frac{1}{4}$
$1\frac{1}{4}$	$1\frac{1}{2}$	$1\frac{5}{8}$	$\frac{7}{8}$
$1\frac{3}{4}$	1	$1\frac{1}{8}$	$1\frac{3}{8}$
$\frac{3}{8}$	$2\frac{1}{8}$	2	$\frac{3}{4}$

Available on Daily Transparency in CRB

Math Humor

Teacher: Why did you change every mixed number on the subtraction quiz to zero?

Student: You told us to rename mixed numbers to make subtraction easier.

5-9 Renaming to Subtract Mixed Numbers

Learn to rename mixed numbers to subtract.

Jimmy and his family planted a tree when it was $1\frac{3}{4}$ ft tall. Now the tree is $2\frac{1}{4}$ ft tall. How much has the tree grown since it was planted?

The difference in the heights is $2\frac{1}{4} - 1\frac{3}{4}$.

You will need to rename $2\frac{1}{4}$ because the fraction in $1\frac{3}{4}$ is greater than $\frac{1}{4}$.

Divide *one whole* of $2\frac{1}{4}$ into fourths.

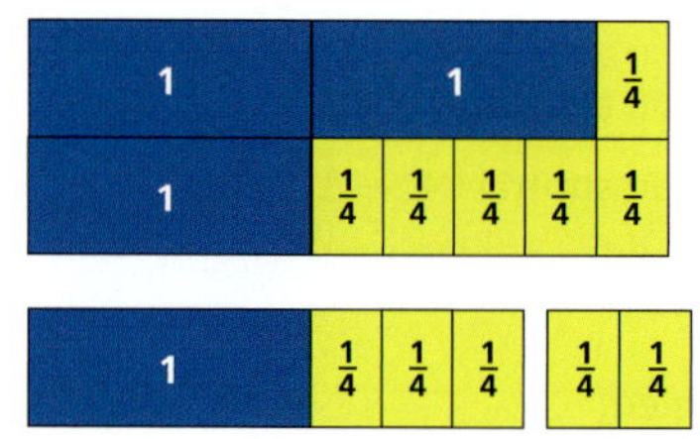

Rename $2\frac{1}{4}$ as $1\frac{5}{4}$.

$$\begin{array}{r} 2\frac{1}{4} \rightarrow \ \ 1\frac{5}{4} \\ -1\frac{3}{4} \rightarrow -1\frac{3}{4} \\ \hline \frac{2}{4} = \frac{1}{2} \end{array}$$

The tree has grown $\frac{1}{2}$ ft since it was planted.

EXAMPLE 1 Renaming Mixed Numbers

Subtract. Write each answer in simplest form.

A $6\frac{5}{12} - 2\frac{7}{12}$

$$\begin{array}{r} 6\frac{5}{12} \rightarrow \ \ 5\frac{17}{12} \\ -2\frac{7}{12} \rightarrow -2\frac{7}{12} \\ \hline 3\frac{10}{12} = 3\frac{5}{6} \end{array}$$

Rename $6\frac{5}{12}$ as $5 + 1\frac{5}{12} = 5 + \frac{12}{12} + \frac{5}{12}$.

Subtract the fractions and then the whole numbers.

Write the answer in simplest form.

B $7\frac{2}{3} - 2\frac{5}{6}$

$$\begin{array}{r} 7\frac{4}{6} \rightarrow \ \ 6\frac{10}{6} \\ -2\frac{5}{6} \rightarrow -2\frac{5}{6} \\ \hline 4\frac{5}{6} \end{array}$$

6 is a multiple of 3, so 6 is a common denominator.

Rename $7\frac{4}{6}$ as $6 + 1\frac{4}{6} = 6 + \frac{6}{6} + \frac{4}{6}$.

Subtract the fractions and then the whole numbers.

1 Introduce

Exploration

EXPLORATION

5-9 Renaming to Subtract Mixed Numbers

A baker starts the day with $2\frac{1}{4}$ lemon cakes and sells $1\frac{1}{2}$ lemon cakes during the day. How much cake is left over?

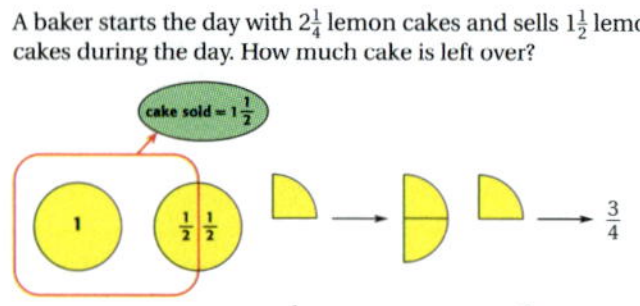

initial amount of cake = $2\frac{1}{4}$ cake left = $\frac{3}{4}$

There is $\frac{3}{4}$ of a cake left over.

Use a model to solve each subtraction problem.

1. $2 - 1\frac{1}{4}$
2. $4 - 1\frac{1}{2}$
3. $5\frac{1}{4} - 3\frac{1}{2}$
4. $1\frac{1}{8} - \frac{3}{4}$

Think and Discuss

5. **Discuss** your method for subtracting mixed numbers.
6. **Explain** why the method used to solve the problem about the lemon cakes is called renaming.

Motivate

Have students use fraction bars (provided in the Manipulatives Kit and on Chapter 5 Resource Book p. 109) to model each of the mixed numbers below and identify the pairs of mixed numbers that are equivalent.

$1\frac{11}{8}$ $1\frac{13}{8}$ $2\frac{3}{8}$ $2\frac{1}{8}$ $2\frac{5}{8}$ $1\frac{9}{8}$

$1\frac{9}{8}$ and $2\frac{1}{8}$; $1\frac{11}{8}$ and $2\frac{3}{8}$; $1\frac{13}{8}$ and $2\frac{5}{8}$

Exploration worksheet and answers on Chapter 5 Resource Book pp. 80 and 126

2 Teach

Lesson Presentation

Guided Instruction

In this lesson, students learn to rename mixed numbers to subtract. You may use Teaching Transparency T39 in the Chapter 5 Resource Book. Teach students to rename to subtract mixed numbers with like and unlike denominators. Then teach students to subtract a mixed number from a whole number.

Subtract. Write each answer in simplest form.

C $8\frac{1}{4} - 5\frac{2}{3}$

$8\frac{3}{12} \longrightarrow 7\frac{15}{12}$
$-5\frac{8}{12} \longrightarrow -5\frac{8}{12}$
$2\frac{7}{12}$

The LCM of 4 and 3 is 12.
Rename $8\frac{3}{12}$ as $7 + 1\frac{3}{12} = 7 + \frac{12}{12} + \frac{3}{12}$.
Subtract the fractions and then the whole numbers.

D $8 - 5\frac{3}{4}$

$8 \longrightarrow 7\frac{4}{4}$
$-5\frac{3}{4} \longrightarrow -5\frac{3}{4}$
$2\frac{1}{4}$

Write 8 as a mixed number with a denominator of 4. Rename 8 as $7 + \frac{4}{4}$.
Subtract the fractions and then the whole numbers.

EXAMPLE 2 *Measurement Application*

Dave is re-covering an old couch and cushions. He determines that he needs 17 yards of fabric for the job.

A Dave has $1\frac{2}{3}$ yards of fabric. How many more yards does he need?

$17 \longrightarrow 16\frac{3}{3}$
$-1\frac{2}{3} \longrightarrow -1\frac{2}{3}$
$15\frac{1}{3}$

Write 17 as a mixed number with a denominator of 3. Rename 17 as $16 + \frac{3}{3}$.
Subtract the fractions and then the whole numbers.

Dave needs another $15\frac{1}{3}$ yards of material.

B If Dave uses $9\frac{5}{6}$ yards of fabric to cover the couch frame, how much of the 17 yards will he have left?

$17 \longrightarrow 16\frac{6}{6}$
$-9\frac{5}{6} \longrightarrow -9\frac{5}{6}$
$7\frac{1}{6}$

Write 17 as a mixed number with a denominator of 6. Rename 17 as $16 + \frac{6}{6}$.
Subtract the fractions and then the whole numbers.

Dave will have $7\frac{1}{6}$ yards of material left.

Think and Discuss

1. **Explain** why you rename 2 as $1\frac{8}{8}$ instead of $1\frac{3}{3}$ when you find $2 - 1\frac{3}{8}$.
2. **Give an example** of a subtraction expression in which you would need to rename the first mixed number to subtract.

COMMON ERROR ALERT

Some students may have difficulty renaming the mixed number(s) being subtracted. Have these students use fraction models to help them with this step.

Additional Examples

Example 1

Subtract. Write each answer in simplest form.

A. $7\frac{1}{6} - 2\frac{5}{6}$ $4\frac{1}{3}$
B. $8\frac{2}{5} - 6\frac{7}{10}$ $1\frac{7}{10}$
C. $6 - 3\frac{2}{3}$ $2\frac{1}{3}$
D. $5\frac{1}{3} - 2\frac{3}{4}$ $2\frac{7}{12}$

Example 2

Li is making a quilt. She needs 15 yards of fabric.

A. Li has $2\frac{3}{4}$ yards of fabric. How many more yards does she need? another $12\frac{1}{4}$ yards

B. If Li uses $11\frac{1}{6}$ yards of fabric, how much of the 15 yards will she have left? $3\frac{5}{6}$ yards

3 Close

Reaching All Learners
Through Number Sense

Challenge students to identify a rule and complete the following subtraction pattern.

$10\frac{3}{5}, 8\frac{4}{5}, 7, 5\frac{1}{5}$, ▬, ▬, . . .

Subtract $1\frac{4}{5}$ from each term to get the next one; $3\frac{2}{5}$, $1\frac{3}{5}$.

Have students make up their own mixed number subtraction patterns for a partner to solve.

Summarize

Review the steps for subtracting mixed numbers:

- Rename the mixed numbers so that they have like denominators.
- Rename the mixed number being subtracted from, if its fraction is less than the fraction in the mixed number being subtracted.
- Subtract the fractions and then the whole numbers.
- Write the answer in simplest form.

Answers to Think and Discuss

1. Possible answer: You need to use a denominator of 8 so that you can subtract $\frac{3}{8}$ from it.
2. Answers should show a mixed number subtracted from a second mixed number, where the fraction in the number being subtracted is greater than the fraction in the number from which it is subtracted.

5-9 PRACTICE & ASSESS

Students may want to refer back to the lesson examples.

Assignment Guide

If you finished Example 1 assign:
Core 1–4, 6–13, 15–21 odd, 27, 29, 43–49
Enriched 2–12 even, 15–35 odd, 43–49

If you finished Example 2 assign:
Core 1–14, 27–30, 35–39 odd, 43–49
Enriched 23–49

Notes

5-9 Exercises

FOR EXTRA PRACTICE
see page 646

internet connect
Homework Help Online
go.hrw.com Keyword: MR4 5-9

GUIDED PRACTICE

See Example 1
Subtract. Write each answer in simplest form.

1. $2\frac{1}{2} - 1\frac{3}{4}$ $\frac{3}{4}$
2. $8\frac{2}{9} - 2\frac{7}{9}$ $5\frac{4}{9}$
3. $3\frac{2}{6} - 1\frac{2}{3}$ $1\frac{2}{3}$
4. $7\frac{1}{4} - 4\frac{11}{12}$ $2\frac{1}{3}$

See Example 2
5. Mr. Jones purchased a 4-pound bag of flour. He used $1\frac{2}{5}$ pounds of flour to make bread. How many pounds of flour are left? $2\frac{3}{5}$

INDEPENDENT PRACTICE

See Example 1
Subtract. Write each answer in simplest form.

6. $6\frac{3}{11} - 3\frac{10}{11}$ $2\frac{4}{11}$
7. $9\frac{2}{5} - 5\frac{3}{5}$ $3\frac{4}{5}$
8. $4\frac{3}{10} - 3\frac{3}{5}$ $\frac{7}{10}$
9. $10\frac{1}{2} - 2\frac{5}{8}$ $7\frac{7}{8}$
10. $11\frac{3}{4} - 9\frac{1}{8}$ $2\frac{5}{8}$
11. $7\frac{5}{9} - 2\frac{5}{6}$ $4\frac{13}{18}$
12. $6 - 2\frac{2}{3}$ $3\frac{1}{3}$
13. $5\frac{7}{10} - 3\frac{1}{2}$ $2\frac{1}{5}$

See Example 2
14. A standard piece of notebook paper has a length of 11 inches and a width of $8\frac{1}{2}$ inches. What is the difference between these two measures? $2\frac{1}{2}$ inches

PRACTICE AND PROBLEM SOLVING

Find each difference. Write the answer in simplest form.

15. $8 - 6\frac{4}{7}$ $1\frac{3}{7}$
16. $13\frac{1}{9} - 11\frac{2}{3}$ $1\frac{4}{9}$
17. $10\frac{3}{4} - 6\frac{1}{2}$ $4\frac{1}{4}$
18. $13 - 4\frac{2}{11}$ $8\frac{9}{11}$
19. $15\frac{2}{5} - 12\frac{3}{4}$ $2\frac{13}{20}$
20. $17\frac{5}{9} - 6\frac{1}{3}$ $11\frac{2}{9}$
21. $18\frac{1}{4} - 14\frac{3}{8}$ $3\frac{7}{8}$
22. $20\frac{1}{6} - 7\frac{4}{9}$ $12\frac{13}{18}$

Evaluate each expression. Write the answer in simplest form.

23. $4\frac{2}{3} + 5\frac{1}{3} - 7\frac{1}{8}$ $2\frac{7}{8}$
24. $12\frac{5}{9} - 6\frac{2}{3} + 1\frac{4}{9}$ $7\frac{1}{3}$
25. $7\frac{4}{11} - 2\frac{8}{11} - \frac{10}{11}$ $3\frac{8}{11}$
26. $8\frac{1}{3} - 5\frac{8}{9} + 8\frac{1}{2}$ $10\frac{17}{18}$

Evaluate each expression for $a = 6\frac{2}{3}$, $b = 8\frac{1}{2}$, and $c = 1\frac{3}{4}$. Write the answer in simplest form.

27. $a - c$ $4\frac{11}{12}$
28. $b - c$ $6\frac{3}{4}$
29. $b - a$ $1\frac{5}{6}$
30. $10 - b$ $1\frac{1}{2}$
31. $b - (a + c)$ $\frac{1}{12}$
32. $c + (b - a)$ $3\frac{7}{12}$
33. $(a + b) - c$ $13\frac{5}{12}$
34. $(10 - c) - a$ $1\frac{7}{12}$

35. ***ECONOMICS*** A single share of stock in a company cost $\$23\frac{2}{5}$ on Monday. By Tuesday, the cost of a share in the company had fallen to $\$19\frac{1}{5}$. By how much did the price of a share fall? $\$4\frac{1}{5}$

Math Background

The method taught here for subtracting mixed numbers is directly analogous to the method for subtracting whole numbers and decimal numbers. The only new element is that the manipulation involves increasing a digit not by 10 but by a fraction that equals 1.

Similar renaming occurs when finding elapsed time. To find the elapsed time from 10:40 to 11:10, you would subtract 11:10 from 10:40. Because 1 hour = 60 minutes, when you rename, you get 10:70 − 10:40 = 0:30. The elapsed time is 30 minutes.

RETEACH 5-9

LESSON 5-9 **Reteach**
Renaming to Subtract Mixed Numbers

You can use fraction strips to rename to subtract mixed numbers.
To find $3\frac{1}{4} - 1\frac{3}{4}$, first model the first mixed number in the expression.

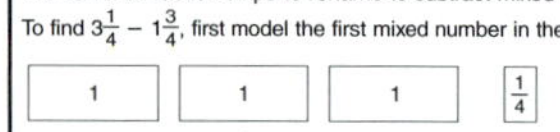

There are not enough $\frac{1}{4}$ pieces to subtract, so you have to rename.
Trade one one-whole strip for four $\frac{1}{4}$ pieces, because $\frac{4}{4} = 1$.

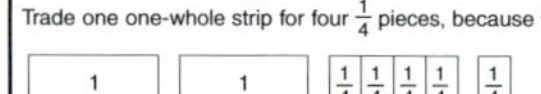

Now there are enough $\frac{1}{4}$ pieces to subtract. Take away $1\frac{3}{4}$.

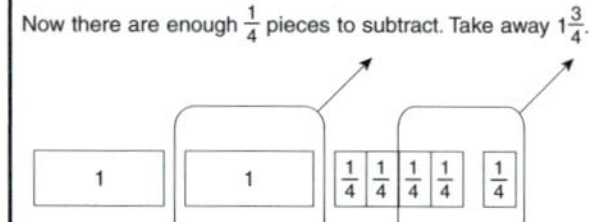

The remaining pieces represent the difference. Write the difference in simplest form.

$3\frac{1}{4} - 1\frac{3}{4} = 1\frac{2}{4} = 1\frac{1}{2}$

Use fraction strips to find each difference. Write your answer in simplest form.

1. $3\frac{1}{4} - 2\frac{3}{4}$ $\frac{1}{2}$
2. $3\frac{1}{6} - 1\frac{5}{6}$ $1\frac{1}{3}$
3. $4\frac{3}{8} - 1\frac{7}{8}$ $2\frac{1}{2}$
4. $3\frac{1}{3} - 2\frac{2}{3}$ $\frac{2}{3}$
5. $5\frac{5}{12} - 2\frac{7}{12}$ $2\frac{5}{6}$
6. $3\frac{3}{10} - 1\frac{9}{10}$ $1\frac{2}{5}$
7. $5\frac{1}{8} - 1\frac{5}{8}$ $3\frac{1}{2}$
8. $4 - 1\frac{1}{3}$ $2\frac{2}{3}$
9. $3\frac{1}{8} - 1\frac{3}{8}$ $1\frac{3}{4}$
10. $2\frac{1}{8} - 1\frac{7}{8}$ $\frac{1}{4}$
11. $3 - 1\frac{1}{4}$ $1\frac{3}{4}$
12. $6\frac{3}{8} - 2\frac{5}{8}$ $3\frac{3}{4}$

PRACTICE 5-9

LESSON 5-9 **Practice B**
Renaming to Subtract Mixed Numbers

Subtract. Write the answer in simplest form.

1. $4 - 2\frac{3}{8}$ $1\frac{5}{8}$
2. $5\frac{1}{6} - 2\frac{2}{3}$ $2\frac{1}{2}$
3. $14 - 8\frac{2}{9}$ $5\frac{7}{9}$
4. $19\frac{1}{7} - 5\frac{1}{3}$ $13\frac{17}{21}$
5. $7\frac{1}{4} - 3\frac{5}{8}$ $3\frac{5}{8}$
6. $10\frac{1}{5} - 5\frac{7}{10}$ $4\frac{1}{2}$
7. $1\frac{1}{6} - \frac{7}{9}$ $\frac{7}{18}$
8. $9\frac{1}{4} - 1\frac{7}{16}$ $7\frac{13}{16}$
9. $6\frac{1}{5} - 3\frac{1}{4}$ $2\frac{19}{20}$

Evaluate each expression for $a = 1\frac{1}{2}$, $b = 2\frac{1}{3}$, $c = \frac{1}{4}$, and $d = 3$. Write each answer in simplest form.

10. $b - a$ $\frac{5}{6}$
11. $a - c$ $1\frac{1}{4}$
12. $b - c$ $2\frac{1}{12}$
13. $d - a$ $1\frac{1}{2}$
14. $d - b$ $\frac{2}{3}$
15. $d - c$ $2\frac{3}{4}$

16. Tim had 6 feet of wrapping paper for Kylie's birthday present. He used $3\frac{3}{8}$ feet of the paper to wrap her gift. How much paper did Tim have left? $2\frac{5}{8}$ feet of paper
17. At his last doctor's visit, Pablo was $60\frac{1}{2}$ inches tall. At today's visit, he measured $61\frac{1}{6}$ inches. How much did Pablo grow between visits? $\frac{2}{3}$ inch
18. Yesterday, Danielle rode her bike for $5\frac{1}{2}$ miles. Today, she rode her bike for $6\frac{1}{4}$ miles. How much farther did Danielle ride her bike today? $\frac{3}{4}$ mile

36. Octavio used a brand new 6-hour tape to record some television shows. He recorded a movie that is $1\frac{1}{2}$ hours long and a cooking show that is $1\frac{1}{4}$ hours long. How much time is left on the tape? $3\frac{1}{4}$

Use the table for Exercises 37–40.

37. Gustavo is working at a gift wrap center. He has 2 yd^2 of wrapping paper to wrap a small box. How much wrapping paper will be left after he wraps the gift? $1\frac{1}{12}$ yards^2

38. Gustavo must now wrap two extra-large boxes. If he has 6 yd^2 of wrapping paper, how much more wrapping paper will he need to wrap the two gifts? $\frac{2}{9}$ yards^2

39. To wrap a large box, Gustavo used $\frac{3}{4}$ yd^2 less wrapping paper than the amount listed in the table. How many square yards did he use to wrap the gift? $1\frac{11}{12}$ yards^2

Gustavo's Gift Wrap Table

Gift Size	Paper Needed (yd^2)
Small	$\frac{11}{12}$
Medium	$1\frac{5}{9}$
Large	$2\frac{2}{3}$
X-large	$3\frac{1}{9}$

40. ***WHAT'S THE ERROR?*** Gustavo calculated the difference between the amount needed to wrap an extra-large box and the amount needed to wrap a medium box to be $2\frac{4}{9}$ yd^2. Explain his error and find the correct answer.

41. ***WRITE ABOUT IT*** Explain why you write equivalent fractions before you rename them. Explain why you do not rename them first.

42. ***CHALLENGE*** Fill in the box with a mixed number that makes the inequality true.

$$12\frac{1}{2} - 8\frac{3}{4} > 10 - \blacksquare$$ Any number $> 6\frac{1}{4}$

Spiral Review

Solve. (Lesson 3-9)

43. Gina made 4 dozen enchiladas for a dinner party. If she made 3 enchiladas for each guest, how many guests were at Gina's party? 16

Find each product. Write your answers in simplest form. (Lesson 5-1)

44. $\frac{2}{3} \times \frac{1}{5}$ $\frac{2}{15}$
45. $\frac{3}{7} \times \frac{1}{4}$ $\frac{3}{28}$
46. $\frac{2}{9} \times \frac{3}{8}$ $\frac{1}{12}$
47. $\frac{1}{4} \times \frac{6}{7}$ $\frac{3}{14}$

48. **TEST PREP** Which is the correct value of $\frac{3}{4}$ of $\frac{3}{4}$? (Lesson 5-1) **B**

A 1 **B** $\frac{9}{16}$ **C** $\frac{1}{3}$ **D** 3

49. **TEST PREP** Solve $\frac{3}{4}x = 9$. (Lesson 5-4) **J**

F 6 **G** 8 **H** 10 **J** 12

Answers

40. Gustavo did not rename $3\frac{1}{9}$ and he subtracted 1 from 5 instead of 5 from 1. The correct difference is $1\frac{5}{9}$ yd^2.

41. Possible answer: It is easier to determine if you need to rename after you find equivalent fractions; unnecessary steps may be involved if you rename before finding equivalent fractions.

Journal

Have students explain the steps they follow when subtracting mixed numbers. Have them address situations when the fraction part of the first mixed number is less than the fraction part of the mixed number being subtracted.

Test Prep Doctor

For Exercise 48, students will need to know that multiplication is used to find a fraction *of* another fraction. Students who answered **A** divided $\frac{3}{4}$ by $\frac{3}{4}$. Students who answered **C** or **D** probably attempted to name a reciprocal, though incorrectly.

CHALLENGE 5-9

LESSON 5-9 **Challenge**
Renaming to Subtract Mixed Numbers

What are the most popular first names in the United States?

Rename fractions or mixed numbers to solve each problem below. Write your answers in simplest form. Then, in the box at the bottom of the page, write each problem's letter in the blanks above its solution. When you have solved all the problems, you will have found the answer to the question.

$8\frac{7}{12} - 7\frac{3}{4}$ $\frac{5}{6}$ A

$9\frac{1}{8} - 8\frac{3}{4}$ $\frac{3}{8}$ E

$10\frac{1}{3} - 9\frac{2}{3}$ $\frac{2}{3}$ J

$6\frac{1}{2} - 5\frac{4}{5}$ $\frac{7}{10}$ M

$5\frac{1}{5} - 4\frac{4}{5}$ $\frac{2}{5}$ R

$7\frac{2}{9} - 6\frac{2}{3}$ $\frac{5}{9}$ S

$12\frac{2}{5} - 11\frac{1}{2}$ $\frac{9}{10}$ Y

#1 Name For American Men: J ($\frac{2}{3}$) A ($\frac{5}{6}$) M ($\frac{7}{10}$) E ($\frac{3}{8}$) S ($\frac{5}{9}$)

#1 Name For American Women: M ($\frac{7}{10}$) A ($\frac{5}{6}$) R ($\frac{2}{5}$) Y ($\frac{9}{10}$)

PROBLEM SOLVING 5-9

LESSON 5-9 **Problem Solving**
Renaming to Subtract Mixed Numbers

Write the correct answer in simplest form.

1. The average person in the United States eats $6\frac{13}{16}$ pounds of potato chips each year. The average person in Ireland eats $5\frac{15}{16}$ pounds. How much more potato chips do Americans eat a year than people in Ireland?
$\frac{7}{8}$ pound more

2. The average person in the United States eats $270\frac{1}{16}$ pounds of meat each year. The average person in Australia eats $238\frac{1}{2}$ pounds. How much more meat do Americans eat a year than people in Australia?
$31\frac{9}{16}$ pounds more

3. The average Americans eats $24\frac{1}{2}$ pounds of ice cream every year. The average person in Israel eats $15\frac{4}{5}$ pounds. How much more ice cream do Americans eat each year?
$8\frac{7}{10}$ pounds more

4. People in Switzerland eat the most chocolate—26 pounds a year per person. Most Americans eat $12\frac{9}{16}$ pounds each year. How much more chocolate do the Swiss eat?
$13\frac{7}{16}$ pounds more

5. The average person in the United States chews $1\frac{9}{16}$ pounds of gum each year. The average person in Japan chews $\frac{7}{8}$ pound. How much more gum do Americans chew?
$\frac{11}{16}$ pound more

6. Norwegians eat the most frozen foods—$78\frac{1}{2}$ pounds per person each year. Most Americans eat $35\frac{15}{16}$ pounds. How much more frozen foods do people in Norway eat?
$42\frac{9}{16}$ pounds more

Circle the letter of the correct answer.

7. Most people around the world eat $41\frac{7}{8}$ pounds of sugar each year. Most Americans eat $66\frac{3}{4}$ pounds. How much more sugar do Americans eat than the world's average?
A $25\frac{7}{8}$ pounds more
B $25\frac{1}{8}$ pounds more
(C) $24\frac{7}{8}$ pounds more
D $24\frac{1}{8}$ pounds more

8. The average person eats 208 pounds of vegetables and $125\frac{5}{8}$ pounds of fruit each year. How much more vegetables do most people eat than fruit?
F $83\frac{5}{8}$ pounds more
(G) $82\frac{3}{8}$ pounds more
H $123\frac{5}{8}$ pounds more
J $83\frac{3}{8}$ pounds more

Lesson Quiz

Subtract. Write each answer in simplest form.

1. $7\frac{4}{7} - 3\frac{6}{7}$ $3\frac{5}{7}$
2. $8\frac{7}{10} - 3\frac{4}{5}$ $4\frac{9}{10}$
3. $7 - 5\frac{2}{9}$ $1\frac{7}{9}$
4. $10\frac{3}{8} - 4\frac{9}{16}$ $5\frac{13}{16}$
5. Irena put 10 cups of food in the bird feeder. The birds ate $3\frac{3}{8}$ cups. How many cups are left? $6\frac{5}{8}$ cups

Available on Daily Transparency in CRB

5-10 Organizer

Pacing: Traditional 1 day
Block $\frac{1}{2}$ day

Objective: Students solve equations by adding and subtracting fractions.

Indiana Academic Standards
6.1.7, 6.2.5, 6.3.1, 6.7.5, 6.7.11

Warm Up

Solve.

1. $x - 15 = 9$ $\quad x = 24$
2. $x + 21 = 34$ $\quad x = 13$
3. $17 = x - 11$ $\quad x = 28$
4. $22 = x - 34$ $\quad x = 56$

Problem of the Day

If a newborn baby weighs two pounds plus three-fourths its own weight, how much does it weigh?
8 pounds

Available on Daily Transparency in CRB

Math Humor

Two crooked tailors made pleats in a skirt to sew in some stolen money. One made four pleats, and the other made only one. Both were arrested, but the second tailor went free because in court he *pleated the fifth*.

5-10 Solving Fraction Equations: Addition and Subtraction

Learn to solve equations by adding and subtracting fractions.

Sugarcane is the main source of the sugar we use to sweeten our foods. It grows in tropical areas, such as Costa Rica and Haiti.

In one year, the average person in Costa Rica consumes $24\frac{1}{4}$ lb less sugar than the average person in the United States consumes.

This painting depicts the landscape of Haiti, a tropical area where sugarcane grows.

EXAMPLE 1 Solving Equations by Adding and Subtracting

Solve each equation. Write the solution in simplest form.

A $x + 6\frac{2}{3} = 11$

$$x + 6\frac{2}{3} = 11$$

$$\underline{-6\frac{2}{3}} \quad \underline{-6\frac{2}{3}}$$ *Subtract $6\frac{2}{3}$ from both sides to undo the addition.*

$$x = 10\frac{3}{3} - 6\frac{2}{3}$$ *Rename 11 as $10\frac{3}{3}$.*

$$x = 4\frac{1}{3}$$ *Subtract.*

B $2\frac{1}{4} = x - 3\frac{1}{2}$

$$2\frac{1}{4} = x - 3\frac{1}{2}$$

$$\underline{+3\frac{1}{2}} \quad \underline{+3\frac{1}{2}}$$ *Add $3\frac{1}{2}$ to both sides to undo the subtraction.*

$$2\frac{1}{4} + 3\frac{2}{4} = x$$ *Find a common denominator.*

$$5\frac{3}{4} = x$$ *Add.*

C $5\frac{3}{5} = m + \frac{7}{10}$

$$5\frac{3}{5} = m + \frac{7}{10}$$

$$\underline{-\frac{7}{10}} \quad \underline{-\frac{7}{10}}$$ *Subtract $\frac{7}{10}$ from both sides to undo the addition.*

$$5\frac{6}{10} - \frac{7}{10} = m$$ *Find a common denominator.*

$$4\frac{16}{10} - \frac{7}{10} = m$$ *Rename $5\frac{6}{10}$ as $4\frac{10}{10} + \frac{6}{10}$.*

$$4\frac{9}{10} = m$$ *Subtract.*

1 Introduce

Exploration

EXPLORATION

5-10 Solving Fraction Equations: Addition and Subtraction

You can use mental math to solve addition and subtraction equations that contain fractions. Look at the reasoning used to solve $\frac{7}{9} - x = \frac{2}{9}$.

Since the denominators of the fractions are the same, you can rewrite $\frac{7}{9} - x = \frac{2}{9}$ as a simpler equation: $7 - ■ = 2$.

$7 - ■ = 2$	*Think:* 7 minus what number equals 2?
$7 - 5 = 2$	Use mental math.
$\frac{7}{9} - \frac{5}{9} = \frac{2}{9}$	Write the equation, using the denominator.
$x = \frac{5}{9}$	Write the value of x.

Using the example above as a guide, complete the table below.

	Equation	Simpler Equation	Value of x
1.	$\frac{1}{2} + x = \frac{7}{2}$	$1 + ■ = 7$	
2.	$x - \frac{2}{5} = \frac{1}{5}$	$■ - 2 = 1$	
3.	$x + \frac{1}{3} = \frac{2}{3}$	$■ + 1 = 2$	
4.	$x - \frac{1}{3} = \frac{2}{3}$	$■ - 1 = 2$	

Think and Discuss

5. **Explain** how you could use mental math to solve the equation $\frac{1}{3} + x = \frac{5}{6}$.

Motivate

Review how to solve equations. Ask students what operation undoes addition. subtraction Ask what operation undoes subtraction. addition Have students explain how to solve an equation where a whole number is added to a variable. Subtract the whole number added to the variable from both sides of the equation.

Exploration worksheet and answers on Chapter 5 Resource Book pp. 89 and 128

2 Teach

Lesson Presentation

Guided Instruction

In this lesson, students learn how to solve equations by adding and subtracting fractions. After students solve addition and subtraction equations that require renaming and finding common denominators, have them apply the concept to a real-world situation. Remind students that addition and subtraction are inverse operations.

Solve each equation. Write the solution in simplest form.

D $w - \frac{1}{2} = 2\frac{3}{4}$

$$w - \frac{1}{2} = 2\frac{3}{4}$$
$$\underline{+\frac{1}{2} \quad +\frac{1}{2}}$$ *Add $\frac{1}{2}$ to both sides to undo the subtraction.*
$$w = 2\frac{3}{4} + \frac{1}{2}$$
$$w = 2\frac{3}{4} + \frac{2}{4}$$ *Find a common denominator.*
$$w = 2\frac{5}{4}$$ *Add.*
$$w = 3\frac{1}{4}$$ *$2\frac{5}{4} = 2 + 1\frac{1}{4}$*

EXAMPLE 2 *Social Studies Application*

On average, a person in Costa Rica consumes $132\frac{1}{4}$ lb of sugar per year. If the average person in Costa Rica consumes $24\frac{1}{4}$ lb less than the average person in the U.S., what is the average sugar consumption per year by a person in the U.S.?

$$u - 24\frac{1}{4} = 132\frac{1}{4}$$ *Let u represent the average amount of sugar consumed in the U.S.*
$$\underline{+24\frac{1}{4} \quad +24\frac{1}{4}}$$ *Add $24\frac{1}{4}$ to both sides to undo the subtraction.*
$$u = 156\frac{2}{4} = 156\frac{1}{2}$$ *Simplify.*

Check

$$u - 24\frac{1}{4} = 132\frac{1}{4}$$
$$156\frac{1}{2} - 24\frac{1}{4} \stackrel{?}{=} 132\frac{1}{4}$$ *Substitute $156\frac{1}{2}$ for u.*
$$156\frac{2}{4} - 24\frac{1}{4} \stackrel{?}{=} 132\frac{1}{4}$$ *Find a common denominator.*
$$132\frac{1}{4} \stackrel{?}{=} 132\frac{1}{4} \checkmark$$ *$156\frac{1}{2}$ is the solution.*

On average, a person in the U.S. consumes $156\frac{1}{2}$ lb of sugar per year.

Think and Discuss

1. **Explain** how renaming a mixed number when subtracting is similar to regrouping when subtracting whole numbers.
2. **Give an example** of an addition equation with a solution that is a fraction between 3 and 4.

COMMON ERROR ALERT

Students may use the wrong operation when solving equations, for example, using addition to solve an addition equation. Remind them that the *inverse* operation isolates the variable.

Additional Examples

Example 1

Solve each equation. Write the solution in simplest form.

A. $x + 5\frac{3}{5} = 14$ $\quad x = 8\frac{2}{5}$

B. $3\frac{2}{9} = x - 4\frac{1}{3}$ $\quad 7\frac{5}{9} = x$

C. $6\frac{1}{6} = m + \frac{7}{12}$ $\quad 5\frac{7}{12} = m$

D. $w - \frac{4}{5} = 3\frac{3}{10}$ $\quad w = 4\frac{1}{10}$

Example 2

Linda's dog weighs $85\frac{1}{4}$ pounds. If Linda's dog weighs $17\frac{1}{2}$ pounds less than Ian's dog, how much does Ian's dog weigh? $102\frac{3}{4}$ pounds

3 Close

Reaching All Learners

Through Home Connection

Have students look through cookbooks at home for recipes involving mixed numbers that can be placed into addition and subtraction equations. Have students bring in their problems to share with the class.

Possible answer: Steve's chicken recipe calls for $1\frac{1}{2}$ teaspoons of garlic. The amount of garlic is $1\frac{1}{4}$ teaspoons more than the amount of salt called for. How much salt does the recipe call for?

$s + 1\frac{1}{4} = 1\frac{1}{2}$; $s = \frac{1}{4}$

The recipe requires $\frac{1}{4}$ teaspoon of salt.

Summarize

Work through the following problem with the class to review the lesson.

$$y + \frac{3}{5} = 7\frac{1}{2}$$
$$y + \frac{3}{5} - \frac{3}{5} = 7\frac{1}{2} - \frac{3}{5}$$
$$y = 7\frac{5}{10} - \frac{6}{10}$$
$$y = 6\frac{15}{10} - \frac{6}{10}$$
$$y = 6\frac{9}{10}$$

Answers to Think and Discuss

1. Possible answer: You must regroup one whole from the whole numbers place to add to the fractions place, just like borrowing 1 group of 10 from the tens place to add to the ones place.
2. Possible answer: $a + \frac{1}{2} = 4$, $a = 3\frac{1}{2}$

5-10 PRACTICE & ASSESS

Students may want to refer back to the lesson examples.

Assignment Guide

If you finished Example 1 assign:

Core 1–4, 6–9, 11–15 odd, 38–42

Enriched 2–8 even, 11–18, 38–42

If you finished Example 2 assign:

Core 1–10, 11–15 odd, 30, 33, 38–42

Enriched 23–42

Notes

5-10 Exercises

FOR EXTRA PRACTICE see page 646

internet connect — Homework Help Online go.hrw.com Keyword: MR4 5-10

GUIDED PRACTICE

See Example 1 Solve each equation. Write the solution in simplest form.

1. $x + 2\frac{1}{2} = 7$ $4\frac{1}{2}$
2. $3\frac{1}{3} = x - 5\frac{1}{9}$ $8\frac{4}{9}$
3. $9\frac{3}{4} = x + 4\frac{1}{8}$ $5\frac{5}{8}$
4. $x + 1\frac{1}{5} = 5\frac{3}{10}$ $4\frac{1}{10}$

See Example 2

5. A tailor increased the length of a robe by $2\frac{1}{4}$ inches. The new length of the robe is 60 inches. What was the original length? $57\frac{3}{4}$ in.

INDEPENDENT PRACTICE

See Example 1 Solve each equation. Write the solution in simplest form.

6. $x - 4\frac{3}{4} = 1\frac{1}{12}$ $5\frac{5}{6}$
7. $x + 5\frac{3}{8} = 9$ $3\frac{5}{8}$
8. $3\frac{1}{2} = 1\frac{3}{10} + x$ $2\frac{1}{5}$
9. $4\frac{2}{3} = x - \frac{1}{6}$ $4\frac{5}{6}$

See Example 2

10. Robert is taking a movie-making class in school. He edited his short video and cut $3\frac{2}{5}$ minutes. The new length of the video is $12\frac{1}{10}$ minutes. How long was his video before he cut it? $15\frac{1}{2}$ minutes

PRACTICE AND PROBLEM SOLVING

Find the solution to each equation. Check your answers.

11. $y + 8\frac{2}{4} = 10$ $1\frac{1}{2}$
12. $p - 1\frac{2}{5} = 3\frac{7}{10}$ $5\frac{1}{10}$
13. $6\frac{2}{3} + n = 7\frac{5}{6}$ $1\frac{1}{6}$
14. $5\frac{3}{5} = s - 2\frac{3}{10}$ $7\frac{9}{10}$
15. $k - 8\frac{1}{4} = 2\frac{2}{3} - 1\frac{1}{3}$ $9\frac{7}{12}$
16. $\frac{5}{6} + \frac{1}{8} = c + \frac{5}{8}$ $\frac{1}{3}$
17. $m + 4 = 6\frac{3}{8} - 1\frac{1}{4}$ $1\frac{1}{8}$
18. $12\frac{1}{6} - 10\frac{1}{9} + 2\frac{2}{3} = y$ $4\frac{13}{18}$
19. $3\frac{2}{9} - 1\frac{1}{3} = p - 5\frac{1}{2}$ $7\frac{7}{18}$
20. $q - 4\frac{1}{4} = 1\frac{1}{6} + 1\frac{1}{2}$ $6\frac{11}{12}$
21. $h = 9\frac{3}{11} - 6\frac{2}{3} + 2\frac{1}{11}$ $4\frac{23}{33}$
22. $a + 5\frac{1}{4} + 2\frac{1}{2} = 13\frac{1}{6}$ $5\frac{5}{12}$
23. $6\frac{2}{9} = n - 2\frac{3}{8} - 1\frac{1}{9}$ $9\frac{17}{24}$
24. $d - 20\frac{1}{4} + 2\frac{1}{10} = 12\frac{3}{10}$ $30\frac{9}{20}$
25. $4\frac{1}{2} + \frac{1}{5} = z - 5\frac{1}{5}$ $9\frac{9}{10}$
26. $11\frac{2}{7} = w + 3\frac{1}{2} - 1\frac{1}{7}$ $8\frac{13}{14}$
27. $4\frac{1}{8} + 2\frac{3}{4} + 5\frac{1}{2} = r$ $12\frac{3}{8}$
28. $9 - 5\frac{7}{8} = x - 1\frac{1}{8}$ $4\frac{1}{4}$
29. The difference between Cristina's and Erin's heights is $\frac{1}{2}$ foot. Erin's height is $4\frac{1}{4}$ feet, and she is shorter than Cristina. How tall is Cristina? $4\frac{3}{4}$ feet
30. ***MEASUREMENT*** Lori used $2\frac{5}{8}$ ounces of shampoo to wash her dog. When she was finished, the bottle contained $13\frac{3}{8}$ ounces of shampoo. How many ounces of shampoo were in the bottle before Lori washed her dog? 16 ounces

Math Background

Although the equations in this lesson are no different in structure from those encountered earlier involving whole numbers, the introduction of fractions and mixed numbers could be enough to confuse some students.

For example, some students may find Example 2 difficult as written, but they would have no problem with it if the numbers were whole. Therefore, a good strategy for students to use is *Solve a Simpler Problem*. Here, students could set up the equation, imagining whole numbers instead of mixed numbers. Then, once solved, the equation can be set up again and solved using the original mixed numbers.

RETEACH 5-10

LESSON 5-10 **Reteach**

Solving Fraction Equations: Addition and Subtraction

You can write related facts using addition and subtraction.

$3 + 4 = 7$ $\quad$ $7 - 4 = 3$

You can use related facts to solve equations.

A. $x + 2\frac{1}{2} = 4$

Think: $4 - 2\frac{1}{2} = x$

$x = 4 - 2\frac{1}{2}$

$x = 3\frac{2}{2} - 2\frac{1}{2}$ Rename 4 as $3\frac{2}{2}$.

$x = 1\frac{1}{2}$

B. $x - 4\frac{1}{3} = 3\frac{1}{2}$

Think: $3\frac{1}{2} + 4\frac{1}{3} = x$

$x = 3\frac{1}{2} + 4\frac{1}{3}$

$x = \frac{7}{2} + \frac{13}{3}$ Write the mixed numbers as improper fractions.

$x = \frac{21}{6} + \frac{26}{6}$ Write the fractions using a common denominator.

$x = \frac{47}{6}$

$x = 7\frac{5}{6}$ Write the sum as a mixed number.

Use related facts to solve each equation.

1. $x + 3\frac{1}{3} = 7$
   $x = 7 - 3\frac{1}{3}$
   $x = 6\frac{3}{3} - 3\frac{1}{3}$
   $x = 3\frac{2}{3}$
2. $x - 2\frac{1}{4} = 4\frac{1}{2}$
   $x = 4\frac{1}{2} + 2\frac{1}{4}$
   $x = \frac{9}{2} + \frac{9}{4}$
   $x = \frac{18}{4} + \frac{9}{4}$
   $x = 6\frac{3}{4}$
3. $x + \frac{3}{8} = 5\frac{1}{4}$
   $x = 5\frac{1}{4} - \frac{3}{8}$
   $x = \frac{21}{4} - \frac{3}{8}$
   $x = \frac{42}{8} - \frac{3}{8}$
   $x = 4\frac{7}{8}$
4. $x - \frac{5}{12} = 2\frac{1}{2}$
   $x = 2\frac{1}{2} + \frac{5}{12}$
   $x = \frac{5}{2} + \frac{5}{12}$
   $x = \frac{30}{12} + \frac{5}{12}$
   $x = 2\frac{11}{12}$
5. $x - 1\frac{3}{4} = 7\frac{1}{2}$ $\quad x = 9\frac{1}{4}$
6. $x - 3\frac{2}{3} = 1\frac{1}{3}$ $\quad x = 5$
7. $x + 3\frac{1}{2} = 6\frac{1}{4}$ $\quad x = 2\frac{3}{4}$
8. $x - 2\frac{2}{5} = 1\frac{3}{10}$ $\quad x = 3\frac{7}{10}$

PRACTICE 5-10

LESSON 5-10 **Practice B**

Solving Fraction Equations: Addition and Subtraction

Solve each equation. Write the answer in simplest form. Check your answers.

1. $k + 3\frac{3}{4} = 5\frac{2}{3} - 1\frac{1}{3}$ $\quad k = \frac{7}{12}$
2. $a - 2\frac{2}{11} = 2\frac{5}{22} - 1\frac{2}{11}$ $\quad a = 3\frac{5}{22}$
3. $2\frac{2}{7} = n - 4\frac{2}{3} - 1\frac{1}{3}$ $\quad n = 8\frac{2}{7}$
4. $6\frac{1}{4} = z + 1\frac{5}{8}$ $\quad z = 4\frac{5}{8}$
5. $5\frac{1}{4} = x + \frac{7}{16}$ $\quad x = 4\frac{13}{16}$
6. $r + 6 = 9\frac{2}{5} - 2\frac{1}{2}$ $\quad r = \frac{9}{10}$
7. $11\frac{2}{5} = q - 4\frac{2}{7} + 2\frac{1}{7}$ $\quad q = 13\frac{19}{35}$
8. $4\frac{2}{5} - 2\frac{1}{2} = p + \frac{3}{10}$ $\quad p = 1\frac{3}{5}$
9. $\frac{3}{8} + \frac{1}{6} = c - 4\frac{5}{6}$ $\quad c = 5\frac{3}{8}$
10. $2\frac{1}{4} + c = 2\frac{1}{3} + 1\frac{1}{6}$ $\quad c = 1\frac{1}{4}$
11. A seamstress raised the hem on Helen's skirt by $1\frac{1}{3}$ inches. The skirt's original length was 16 inches. What is the new length?
    $14\frac{2}{3}$ inches
12. The bike trail is $5\frac{1}{4}$ miles long. Jessie has already cycled $2\frac{5}{8}$ miles of the trail. How much farther does she need to go to finish the trail?
    $2\frac{5}{8}$ miles

31. ***SPORTS*** Jack decreased his best time in the 400-meter race by $1\frac{3}{10}$ seconds. His new best time is $52\frac{3}{5}$ seconds. What was Jack's old time in the 400-meter race? **$53\frac{9}{10}$**

32. ***CRAFTS*** Juan makes bracelets to sell at his mother's gift shop. He alternates between green and blue beads.

What is the length of the green bead? **$\frac{3}{8}$ in.**

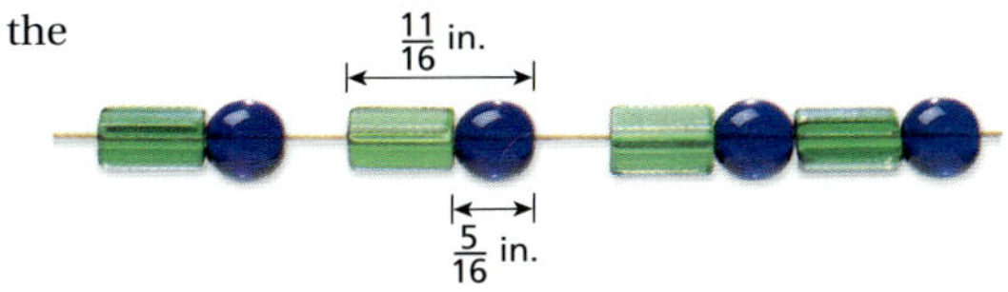

33. ***MUSIC*** A string quartet is performing Antonio Vivaldi's *The Four Seasons.* The concert is scheduled to last 45 minutes.

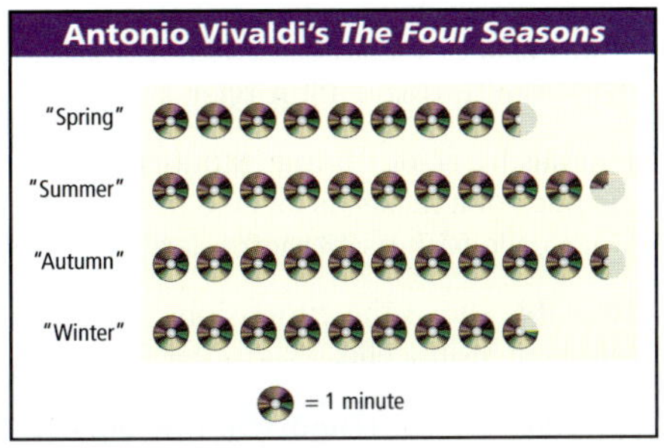

a. After playing "Spring," "Summer," and "Autumn," how much time will be left in the concert?

b. Is the concert long enough to play the four movements and another piece that is $6\frac{1}{2}$ minutes long? Explain.

34. ***WRITE A PROBLEM*** Use the pictograph to write a subtraction problem with two mixed numbers. **Possible answer: What is the difference in time between Summer and Spring?**

35. ***CHOOSE A STRATEGY*** How can you draw a line that is 5 inches long using only one sheet of $8\frac{1}{2}$ in. × 11 in. notebook paper?

36. ***WRITE ABOUT IT*** Explain how you know whether to add a number to or subtract a number from both sides of an equation in order to solve the equation.

37. ***CHALLENGE*** Use the numbers 1, 2, 3, 4, 5, and 6 to write a subtraction problem with two mixed numbers that have a difference of $4\frac{13}{20}$. **$6\frac{2}{5} - 1\frac{3}{4}$**

Spiral Review

Order the numbers from least to greatest. (Lesson 1-1)

38. 1,497; 2,560; 1,038 **1,038; 1,497; 2,560**

39. 10,462; 9,198; 11,320 **9,198; 10,462; 11,320**

40. 4,706; 11,765; 1,765 **1,765; 4,706; 11,765**

41. TEST PREP Which expression contains a variable? (Lesson 2-1) **A**

A $19p$ **B** $\frac{1}{4}$ **C** $0.25\overline{5}$ **D** 8^2

42. TEST PREP What is the standard form of 10 + 4 + 0.2 + 0.06 + 0.003? (Lesson 3-1) **F**

F 14.263 **G** 14,263 **H** 142.63 **J** 1.4263

Answers

33. a. $15\frac{3}{4}$ minutes left in the concert

b. Yes. The total time for the four movements and another piece is $44\frac{1}{2}$ minutes.

35. Fold the top right corner to the left side of the page. The distance from the top right corner to the bottom left corner is 2.5 inches (11−8.5). Draw a line segment that is twice this length.

36. You would do the opposite of the operation that was in front of the number you wanted to move.

Journal

Have students write algebraic equations involving adding and subtracting fractions. Have them solve the equations and outline the steps they followed to solve one of the addition equations and one of the subtraction equations.

Test Prep Doctor

For Exercise 42, have students think back to decimal place value. While each answer choice contains all of the correct digits in the correct order, only **F** has the decimal point positioned correctly.

CHALLENGE 5-10

LESSON 5-10 **Challenge**

You Read My Mind!

Here's a trick you can use to amaze your friends and family. Start by asking your friends to think of a number—any number. Pretend you are reading their minds while you write the number 6 on a piece of paper. (Don't show it to them.) Then use the steps below to tell them what to do. The fraction $\frac{2}{5}$ is used as an example choice, but the trick works for any chosen fraction, mixed number, decimal, or whole number.

Step	Example
1. Double your number.	$\frac{2}{5} + \frac{2}{5} = \frac{4}{5}$
2. Add 12 to your sum.	$\frac{4}{5} + 12 = 12\frac{4}{5}$
3. Divide your new sum by 2.	$12\frac{4}{5} \div 2 = 6\frac{2}{5}$
4. Subtract your chosen number from that quotient.	$6\frac{2}{5} - \frac{2}{5} = 6$

Now amaze your friends by showing that you wrote the same number they ended with. No matter what number is chosen, this trick always ends in 6. Equations explain why it works—but don't tell your friends this part.

Let x = the chosen number.

STEP 1	STEP 2	STEP 3	STEP 4
$2x$	$2x + 12$	$(2x + 12) \div 2 = x + 6$	$x + 6 - x = 6$

Before you try the trick, practice it on the fractions below. Use the equations for each step and show all your work.

1. Chosen Number: $\frac{7}{9}$

STEP 1: $2 \bullet \frac{7}{9} = \frac{14}{9} = 1\frac{5}{9}$

STEP 2: $1\frac{5}{9} + 12 = 13\frac{5}{9}$

STEP 3: $13\frac{5}{9} \div 2 = \frac{122}{18} = 6\frac{14}{18} = 6\frac{7}{9}$

STEP 4: $6\frac{7}{9} - \frac{7}{9} = 6$

2. Chosen Number: $3\frac{1}{4}$

STEP 1: $2 \bullet 3\frac{1}{4} = \frac{26}{4} = 6\frac{2}{4} = 6\frac{1}{2}$

STEP 2: $6\frac{1}{2} + 12 = 18\frac{1}{2}$

STEP 3: $18\frac{1}{2} \div 2 = \frac{37}{4} = 9\frac{1}{4}$

STEP 4: $9\frac{1}{4} - 3\frac{1}{4} = 6$

PROBLEM SOLVING 5-10

LESSON 5-10 **Problem Solving**

Solving Fraction Equations: Addition and Subtraction

Write the correct answer in simplest form.

1. It usually takes Brian $1\frac{1}{2}$ hours to get to work from the time he gets out of bed. His drive to the office takes $\frac{3}{4}$ hour. How much time does he spend getting ready for work?

   $\frac{3}{4}$ of an hour

2. Before she went to the hairdresser, Sheila's hair was $7\frac{1}{4}$ inches long. When she left the salon, it was $5\frac{1}{2}$ inches long. How much of her hair did Sheila get cut off?

   $1\frac{3}{4}$ inches

3. One lap around the gym is $\frac{1}{3}$ mile long. Kim has already run 5 times around. If she wants to run 2 miles total, how much farther does she have to go?

   $\frac{1}{3}$ mile more

4. Darius timed his speech at $5\frac{1}{6}$ minutes. His time limit for the speech is $4\frac{1}{2}$ minutes. How much does he need to cut from his speech?

   $\frac{2}{3}$ minute

5. Mei and Alex bought the same amount of food at the deli. Mei bought $1\frac{1}{4}$ pounds of turkey and $1\frac{1}{3}$ pound of cheese. Alex bought $1\frac{1}{2}$ pounds of turkey. How much cheese did Alex buy?

   $1\frac{1}{12}$ pound

6. When Lynn got her dog, Max, he weighed $10\frac{1}{2}$ pounds. During the next 6 months, he gained $8\frac{4}{5}$ pounds. At his one-year check-up he had gained another $4\frac{1}{3}$ pounds. How much did Max weigh when he was 1 year old?

   $23\frac{19}{30}$ pounds

Circle the letter of the correct answer.

7. Charlie picked up 2 planks of wood at the hardware store. One is $6\frac{1}{4}$ feet long and the other is $5\frac{5}{8}$ feet long. How much should he cut from the first plank to make them the same length?

   (A) $\frac{5}{8}$ foot C $1\frac{3}{8}$ feet

   B $\frac{1}{2}$ foot D $1\frac{5}{8}$ feet

8. Carmen used $3\frac{3}{4}$ cups of flour to make a cake. She had $\frac{1}{2}$ cup of flour left over. Which equation can you use to find how much flour she had before baking the cake?

   F $x + \frac{1}{2} = 3\frac{3}{4}$ H $3\frac{3}{4} - \frac{1}{2} = x$

   (G) $x - 3\frac{3}{4} = \frac{1}{2}$ J $3\frac{3}{4} - x = \frac{1}{2}$

Lesson Quiz

Solve each equation. Write the solution in simplest form.

1. $x - 3\frac{3}{4} = 2\frac{5}{12}$ **$x = 6\frac{1}{6}$**

2. $x + 6\frac{5}{8} = 11$ **$x = 4\frac{3}{8}$**

3. $1\frac{7}{10} = x - 2\frac{4}{5}$ **$x = 4\frac{1}{2}$**

4. $5\frac{3}{4} = x - \frac{1}{2}$ **$x = 6\frac{1}{4}$**

5. Marco needs to work 40 hours per week. He has worked $31\frac{5}{8}$ hours so far this week. How many hours does he need to work on Friday to meet his 40-hour requirement? **$8\frac{3}{8}$ hours**

Available on Daily Transparency in CRB

Problem Solving on Location

Virginia

***Purpose:** To provide additional practice for problem-solving skills in Chapters 1–5*

Presidents

- After problem 1, have students consider the following problem: What is the difference in the rooms' heights in inches? 6 in.
- After problem 4, have students guess which room has the greatest perimeter. Have them find the perimeter. entrance hall; $103\frac{1}{3}$ ft. How did you determine whether the entrance hall or the parlor had the greater perimeter? Possible answer: Their perimeters are close, but if you use rounding to estimate, $\frac{11}{12}$ is closer to 1 than $\frac{1}{4}$. So the length of the entrance is closer to 28 ft, and the length of the parlor is closer to 27 ft. And, since the widths both round to 24 ft, the entrance hall must have the greater perimeter.

Extension Have students convert the measurements in the table to inches and feet. Remind them that there are 12 inches in 1 foot.

Room	Length	Width	Height
Entrance hall	27 ft 11 in.	23 ft 9 in.	18 ft 2 in.
Library	15 ft 3 in.	14 ft 10 in.	10 ft
Cabinet	18 ft 6 in.	11 ft 10 in.	10 ft
Bedroom	18 ft 7 in.	13 ft 5 in.	18 ft 8 in.
Parlor	27 ft 3 in.	23 ft 8 in.	18 ft 2 in.
Tea room	15 ft 1 in.	11 ft 2 in.	17 ft 11 in.

Problem Solving on Location

VIRGINIA

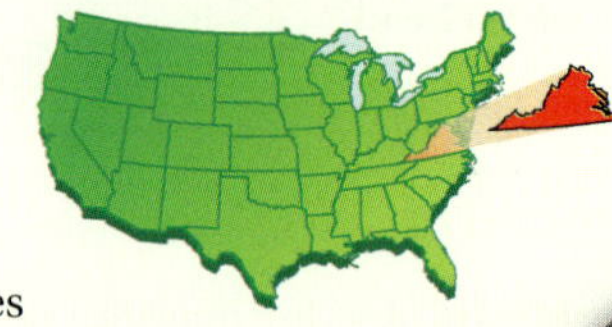

Presidents

Eight presidents were born in Virginia: George Washington, Thomas Jefferson, James Monroe, James Madison, William Henry Harrison, Benjamin Harrison, John Tyler, and Woodrow Wilson.

Thomas Jefferson's home, Monticello, which is pictured on the back of a nickel, is located in Charlottesville. The area of Monticello with its 43 rooms, is about 11,000 ft^2.

The table shows the dimensions of some of the first-floor rooms of Monticello.

Room	Length (ft)	Width (ft)	Height (ft)
Entrance hall	$27\frac{11}{12}$	$23\frac{3}{4}$	$18\frac{1}{6}$
Library	$15\frac{1}{4}$	$14\frac{5}{6}$	10
Cabinet	$18\frac{1}{2}$	$11\frac{5}{6}$	10
Bedroom	$18\frac{7}{12}$	$13\frac{5}{12}$	$18\frac{2}{3}$
Parlor	$27\frac{1}{4}$	$23\frac{2}{3}$	$18\frac{1}{6}$
Tea room	$15\frac{1}{12}$	$11\frac{1}{6}$	$17\frac{11}{12}$

For 1–5, use the table above.

1. Find the difference between the height of Jefferson's bedroom and the height of the entrance hall. $\frac{1}{2}$ ft
2. What is the difference between the length of the tea room and its width? $3\frac{11}{12}$ ft
3. Jefferson used the cabinet room, which was next to his bedroom, as a place to read, write, and draft plans. What was the total length of his bedroom and the cabinet? $37\frac{1}{12}$ ft
4. Without actually multiplying the length times the width to find the area of the rooms, decide which two rooms you think have the greatest area. Explain. **The entrance hall and the parlor are the greatest because they have the greatest dimensions.**
5. Estimate the area of the library by rounding its length and width and then multiplying them. $15\frac{1}{4}$ rounds to 15 and $14\frac{5}{6}$ rounds to 15; $15 \cdot 15 = 225$ ft^2

Apple Crops

Virginia is ranked sixth in the United States for apple production. It produces about 8 to 10 million bushels of apples per year. Many popular varieties of apples, such as Red Delicious, Golden Delicious, Rome, Gala, Granny Smith, Jonathan, Fuji, and Ginger Gold, are grown in Virginia.

Here is the recipe for a Virginia sliced apple pie.

For 1–3, use the recipe.

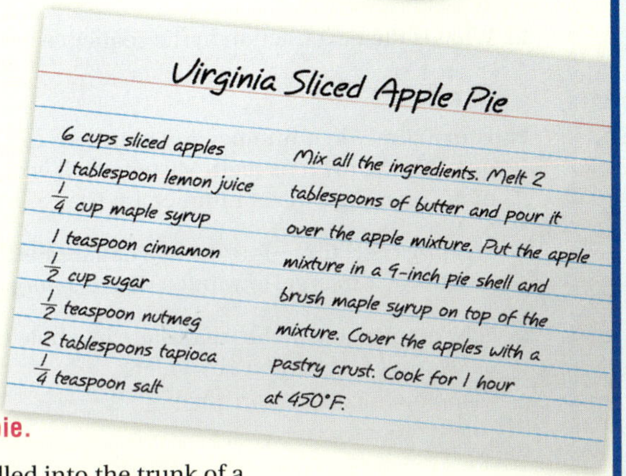

Virginia Sliced Apple Pie

6 cups sliced apples
1 tablespoon lemon juice
$\frac{1}{4}$ cup maple syrup
1 teaspoon cinnamon
$\frac{1}{2}$ cup sugar
$\frac{1}{2}$ teaspoon nutmeg
2 tablespoons tapioca
$\frac{1}{4}$ teaspoon salt

Mix all the ingredients. Melt 2 tablespoons of butter and pour it over the apple mixture. Put the apple mixture in a 9-inch pie shell and brush maple syrup on top of the mixture. Cover the apples with a pastry crust. Cook for 1 hour at 450°F.

1. Jeannie wants to make three Virginia sliced apple pies. How much of each ingredient will she need?
2. Rod wants to make 9 pies for his family reunion. If a bag of sugar holds 4 cups, how many bags of sugar will he need to buy?
3. How much more sugar than maple syrup is in the apple pie recipe? $\frac{1}{4}$ **cup**
4. When Jason visited his grandmother, she had three-fourths of an apple pie left. If Jason ate half of what she had left, how much of the pie did he eat? **He ate $\frac{3}{8}$ of the pie.**
5. To make maple syrup, first a hole must be drilled into the trunk of a tree and the sap collected. It takes about 50 gallons of sap to make 1 gallon of maple syrup. How many gallons of sap are needed to make $2\frac{1}{2}$ gallons of maple syrup? **125 gallons**

1. **18 cups sliced apples; 3 tbsp lemon juice; $\frac{3}{4}$ cup syrup; 3 tsp cinnamon; $\frac{3}{2}$, or $1\frac{1}{2}$, cups sugar; $\frac{3}{2}$, or $1\frac{1}{2}$ tsp nutmeg; 6 tbsp tapioca; $\frac{3}{4}$ tsp salt**

2. **$\frac{1}{2} \cdot 9 = \frac{9}{2}$, or $4\frac{1}{2}$, cups of sugar; So he will need 2 bags.**

Apple Crops

- After problem 2, ask students to explain the steps necessary to solve the problem. Determine the amount of sugar needed to make 9 pies: $9 \times \frac{1}{2}$ cup $= 4\frac{1}{2}$ cups; $4\frac{1}{2} > 4$, so Rod will need to buy 2 bags of sugar.
- After problem 4, have students draw a diagram or use fraction bars to model the problem. Check students' models.

Extension Have students find a different apple pie recipe and compare it to the Virginia sliced apple pie recipe. Have them write sentences such as "My recipe calls for $\frac{1}{2}$ teaspoon more cinnamon than the Virginia sliced apple pie recipe" to compare the quantities of the ingredients. Check students' work.

MATH-ABLES

Game Resources

Puzzles, Twisters & Teasers
Chapter 5 Resource Book

Fraction Riddles

Purpose: *To apply using operations with fractions to solving riddles*

Discuss: Ask students to explain how the riddle in problem 1 works. Is the problem as difficult as it sounds? Explain.
Possible answer:
When you multiply the first nine factors, $\frac{1}{2} \cdot \frac{2}{3} \cdot \frac{3}{4} \cdot \frac{4}{5} \cdot \frac{5}{6} \cdot \frac{7}{8} \cdot \frac{8}{9} \cdot \frac{9}{10}$, you find that the product simplifies to $\frac{1}{10}$ because each factor in the denominator, except for 10, is also in the numerator. $\frac{1}{10}$ of 1000 is 100.

Extend: Challenge students to create their own fraction riddles. Have them use problems 1–5 as models. Check students' work.

Fraction Bingo

Purpose: *To practice using fraction operations and equivalent fractions in a game format*

Discuss: When students get a "bingo," have them demonstrate for the class how each winning solution was obtained.

Extend: Have students create new expression cards for each solution on their bingo card. Use the new cards to play again.

MATH-ABLES

Fraction Riddles

1. What is the value of one-half of two-thirds of three-fourths of four-fifths of five-sixths of six-sevenths of seven-eighths of eight-ninths of nine-tenths of one thousand? **100**
2. What is the next fraction in the sequence below?
   $\frac{1}{12}, \frac{1}{6}, \frac{1}{4}, \frac{1}{3}, \ldots$ **$\frac{5}{12}$; The sequence is $\frac{1}{12}, \frac{2}{12} = \frac{1}{6}, \frac{3}{12} = \frac{1}{4}, \frac{4}{12} = \frac{1}{3}, \ldots$**
3. I am a three-digit number. My hundreds digit is one-third of my tens digit. My tens digit is one-third of my ones digit. What number am I? **139**
4. A *splorg* costs three-fourths of a dollar plus three-fourths of a *splorg*. How much does a *splorg* cost? **\$3**
5. How many cubic inches of dirt are in a hole that measures $\frac{1}{3}$ feet by $\frac{1}{4}$ feet by $\frac{1}{2}$ feet?
   None; there is no dirt in a hole.

Fraction Bingo

The object is to be the first player to cover five squares in a row horizontally, vertically, or diagonally.

One person is the caller. On each of the caller's cards, there is an expression containing fractions. When the caller draws a card, he or she reads the expression aloud for the players.

The players must find the value of the expression. If a square on the player's card has that value or a fraction equivalent to that value, they cover the square.

The first player to cover five squares in a row is the winner. Take turns being the caller. A variation can be played in which the winner is the first person to cover all their squares.

internet connect
Go to ***go.hrw.com*** to print out cards for Fraction Bingo.
KEYWORD: MR4 Game5

Technology LAB

Fraction Operations

internet connect
Lab Resources Online
go.hrw.com
KEYWORD: MR4 TechLab5

To enter a fraction, like $\frac{1}{8}$, into a graphing calculator, use [÷]. When you press [ENTER], the fraction is converted to a decimal.
To convert a decimal number, such as 0.625, to a fraction, enter the number on your calculator. Press [MATH], and then press [ENTER] to select Frac.

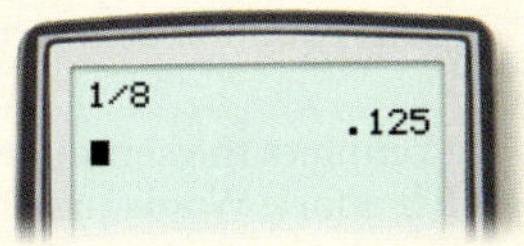

When you press [ENTER] again, the calculator displays the fraction equivalent.

The decimal 0.625 is equivalent to $\frac{5}{8}$.

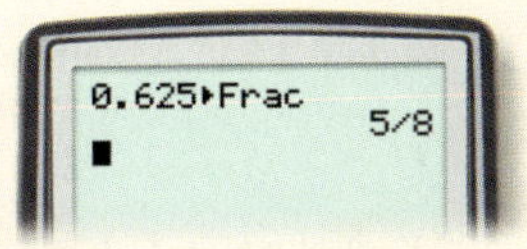

Activity

1 Use your calculator to find the value of $\frac{2}{5} + \frac{3}{7}$. Write the sum as a fraction.

Step 1: To find the sum, enter $\frac{2}{5} + \frac{3}{7}$, and then press [ENTER]. The calculator displays the sum as a decimal.

Step 2: To convert the decimal to a fraction, press [MATH], press [ENTER] to select Frac, and then press [ENTER] again.

$$\frac{2}{5} + \frac{3}{7} = \frac{29}{35}$$

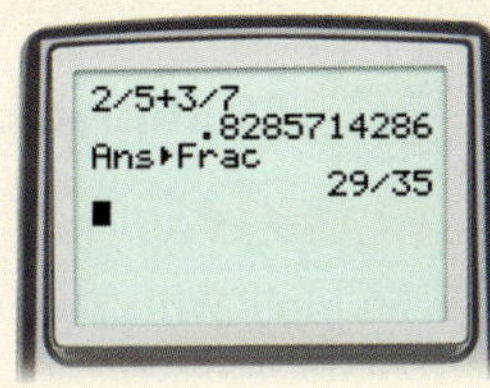

Think and Discuss

1. Without using a calculator, write 0.10001 as a fraction. Then use a graphing calculator to convert 0.10001 to a fraction. What happens? Why do you think this happened? $\frac{10,001}{100,000}$; The calculator will not convert 0.10001 to a fraction because its denominator is too large.

Try This

Add or subtract. Write the answer as a fraction or a mixed number and as a decimal rounded to the nearest hundredth.

1. $\frac{1}{4} + \frac{2}{3}$ $\frac{11}{12}$; 0.92 **2.** $\frac{2}{3} - \frac{1}{4}$ $\frac{5}{12}$; 0.42 **3.** $\frac{3}{5} + \frac{5}{3}$ $\frac{34}{15}$; 2.27 **4.** $\frac{4}{3} - \frac{3}{4}$ $\frac{7}{12}$; 0.58

Technology LAB

Fraction Operations

Objective: To use a graphing calculator to perform fraction operations

Materials: Graphing calculator

Lab Resources

Using the Page

This technology activity shows students how to perform fraction operations and convert between fraction and decimal form on a graphing calculator. Specific keystrokes may vary, depending on the make and model of the graphing calculator used. The keystrokes given are for a TI-83 model. For keystrokes to other models, visit go.hrw.com.

The Think and Discuss problem is meant to point out the limitations of the graphing calculator. While Try This problems 1–4 can be done without a graphing calculator, they are meant to help students become familiar with performing fraction operations on a graphing calculator.

Assessment

1. Use a graphing calculator to convert the decimal 0.532 to a fraction in simplest form. $\frac{133}{250}$
2. How would you find the simplest form of the sum $1\frac{1}{2} + 2\frac{2}{3}$ on a graphing calculator? Press 1 [+] 1 [÷] 2 [+] 2 [+] 2 [÷] 3 [MATH] [ENTER] [ENTER] to convert the sum to a fraction. Then convert it to a mixed number $\left(\frac{25}{6} = 4\frac{1}{6}\right)$.

Chapter 5 Study Guide and Review

Purpose: *To help students review and practice concepts and skills presented in Chapter 5*

Assessment Resources

Chapter Review

Chapter 5 Resource Book . . . pp. 97–99

***Test and Practice Generator* CD-ROM**

Additional review items in both multiple-choice and free-response format may be generated for any objective in Chapter 5.

Answers

1. reciprocals
2. least common denominator
3. $\frac{1}{3}$
4. $\frac{15}{28}$
5. $\frac{1}{10}$
6. $\frac{7}{25}$
7. $\frac{5}{81}$
8. $\frac{3}{14}$
9. $\frac{9}{10}$
10. $1\frac{1}{4}$
11. 2
12. $\frac{4}{21}$
13. $\frac{3}{20}$
14. $\frac{5}{9}$

Chapter 5 Study Guide and Review

Vocabulary

Complete the sentences below with vocabulary words from the list above. Words may be used more than once.

1. Two numbers are __?__ if their product is 1.
2. The __?__ is the smallest number that is a common multiple of two or more denominators.

5-1 Multiplying Fractions (pp. 212–215)

EXAMPLE

■ **Multiply. Write the answer in simplest form.**

$\frac{3}{4} \cdot \frac{1}{3}$ *Multiply. Then simplify.*

$\frac{3 \cdot 1}{4 \cdot 3} = \frac{3}{12} = \frac{1}{4}$

EXERCISES

Multiply. Write each answer in simplest form.

3. $\frac{5}{6} \cdot \frac{2}{5}$
4. $\frac{5}{7} \cdot \frac{3}{4}$
5. $\frac{4}{5} \cdot \frac{1}{8}$
6. $\frac{7}{10} \cdot \frac{2}{5}$
7. $\frac{1}{9} \cdot \frac{5}{9}$
8. $\frac{1}{4} \cdot \frac{6}{7}$

5-2 Multiplying Mixed Numbers (pp. 216–219)

EXAMPLE

■ **Multiply. Write the answer in simplest form.**

$\frac{2}{5} \cdot 1\frac{2}{3} = \frac{2}{5} \cdot \frac{5}{3} = \frac{10}{15} = \frac{2}{3}$

EXERCISES

Multiply. Write each answer in simplest form.

9. $\frac{2}{5} \cdot 2\frac{1}{4}$
10. $\frac{3}{4} \cdot 1\frac{2}{3}$
11. $3\frac{1}{3} \cdot \frac{3}{5}$

5-3 Dividing Fractions and Mixed Numbers (pp. 222–225)

EXAMPLE

■ **Divide. Write the answer in simplest form.**

$\frac{3}{4} \div 6 = \frac{3 \cdot 1}{4 \cdot 6} = \frac{3}{24} = \frac{1}{8}$

EXERCISES

Divide. Write each answer in simplest form.

12. $\frac{4}{7} \div 3$
13. $\frac{3}{10} \div 2$
14. $1\frac{1}{3} \div 2\frac{2}{5}$

5-4 Solving Fraction Equations: Multiplication and Division

(pp. 226–229)

EXAMPLE

- Solve the equation.

$\frac{4}{5}n = 12$

$\frac{4}{5}n \div \frac{4}{5} = 12 \div \frac{4}{5}$ *Divide both sides by $\frac{4}{5}$.*

$\frac{4}{5}n \cdot \frac{5}{4} = 12 \cdot \frac{5}{4}$ *Multiply by the reciprocal.*

$n = \frac{60}{4} = 15$

EXERCISES

Solve each equation.

15. $4a = \frac{1}{2}$ **16.** $\frac{3b}{4} = 1\frac{1}{2}$

17. $\frac{2m}{7} = 5$ **18.** $6g = \frac{4}{5}$

19. $\frac{5}{6}r = 9$ **20.** $\frac{s}{8} = 6\frac{1}{4}$

5-5 Least Common Multiple

(pp. 232–235)

EXAMPLE

- Find the least common multiple (LCM) of 4, 6, and 8.

4: 4, 8, 12, 16, 20, 24, 28, …
6: 6, 12, 18, 24, 30, …
8: 8, 16, 24, 32, …
LCM: 24

EXERCISES

Find the least common multiple (LCM).

21. 3, 5, and 10 **22.** 6, 8, and 16

23. 3, 9, and 27 **24.** 4, 12, and 30

25. 25 and 45 **26.** 12, 22, and 30

5-6 Estimating Fraction Sums and Differences

(pp. 236–239)

EXAMPLE

- Estimate the sum or difference by rounding fractions to 0, $\frac{1}{2}$, or 1.

$\frac{7}{8} + \frac{1}{7}$ *Think: 1 + 0.*

$\frac{7}{8} + \frac{1}{7}$ is about 1.

EXERCISES

Estimate each sum or difference by rounding fractions to 0, $\frac{1}{2}$, or 1.

27. $\frac{3}{5} + \frac{3}{7}$ **28.** $\frac{6}{7} - \frac{5}{9}$

29. $4\frac{9}{10} + 6\frac{1}{5}$ **30.** $7\frac{5}{11} - 4\frac{3}{4}$

5-7 Adding and Subtracting with Unlike Denominators

(pp. 242–245)

EXAMPLE

- $\frac{7}{9} + \frac{2}{3}$

$\frac{7}{9} + \frac{2}{3}$ *Write equivalent fractions. Add.*

$\frac{7}{9} + \frac{6}{9} = \frac{13}{9} = 1\frac{4}{9}$

EXERCISES

Add or subtract. Write each answer in simplest form.

31. $\frac{1}{5} + \frac{5}{8}$ **32.** $\frac{1}{6} + \frac{7}{12}$

33. $\frac{13}{15} - \frac{4}{5}$ **34.** $\frac{7}{8} - \frac{2}{3}$

Study Guide and Review

Answers

15. $a = \frac{1}{8}$

16. $b = 2$

17. $m = 17\frac{1}{2}$

18. $g = \frac{2}{15}$

19. $r = 10\frac{4}{5}$

20. $s = 50$

21. 30

22. 48

23. 27

24. 60

25. 225

26. 660

27. about 1

28. about $\frac{1}{2}$

29. about 11

30. about $2\frac{1}{2}$

31. $\frac{33}{40}$

32. $\frac{3}{4}$

33. $\frac{1}{15}$

34. $\frac{5}{24}$

Answers

35. $4\frac{7}{10}$

36. $3\frac{1}{18}$

37. $11\frac{43}{60}$

38. $4\frac{1}{12}$

39. $1\frac{13}{20}$

40. $2\frac{3}{8}$

41. $\frac{11}{30}$ gal

42. $3\frac{2}{3}$

43. $1\frac{1}{2}$

44. $5\frac{2}{3}$

45. $2\frac{5}{8}$

46. $6\frac{13}{14}$

47. $1\frac{1}{8}$

48. $4\frac{3}{4}$ feet

49. $30\frac{3}{20}$

50. $14\frac{11}{12}$

51. $5\frac{5}{12}$

52. $3\frac{4}{9}$

53. $5\frac{7}{15}$

54. 7 oz

Study Guide and Review

5-8 Adding and Subtracting Mixed Numbers (pp. 246–249)

EXAMPLE

- **Find the difference. Write the answer in simplest form.**

$5\frac{5}{8} - 3\frac{1}{6}$

$5\frac{15}{24} - 3\frac{4}{24}$ *Write equivalent fractions.*

$2\frac{11}{24}$ *Subtract.*

EXERCISES

Find each sum or difference. Write the answer in simplest form.

35. $1\frac{3}{10} + 3\frac{2}{5}$ **36.** $4\frac{5}{9} - 1\frac{1}{2}$

37. $5\frac{5}{12} + 6\frac{3}{10}$ **38.** $2\frac{1}{4} + 1\frac{5}{6}$

39. $2\frac{9}{10} - 1\frac{1}{4}$ **40.** $6\frac{3}{4} - 4\frac{3}{8}$

41. Angela had $\frac{7}{10}$ gallon of paint. She used $\frac{1}{3}$ gallon for a project. How much paint did she have left?

5-9 Renaming to Subtract Mixed Numbers (pp. 252–255)

EXAMPLE

- **Subtract.**

$4\frac{7}{10} - 2\frac{9}{10}$

$3\frac{17}{10} - 2\frac{9}{10}$ *Rename $4\frac{7}{10}$. Subtract.*

$1\frac{8}{10}$

$1\frac{4}{5}$

EXERCISES

Subtract. Write each answer in simplest form.

42. $7\frac{2}{9} - 3\frac{5}{9}$ **43.** $3\frac{1}{5} - 1\frac{7}{10}$

44. $8\frac{7}{12} - 2\frac{11}{12}$ **45.** $5\frac{3}{8} - 2\frac{3}{4}$

46. $11\frac{6}{7} - 4\frac{13}{14}$ **47.** $10 - 8\frac{7}{8}$

48. Georgette needs 8 feet of ribbon to decorate gifts. She has $3\frac{1}{4}$ feet of ribbon. How many more feet of ribbon does Georgette need?

5-10 Solving Fraction Equations: Addition and Subtraction (pp. 256–259)

EXAMPLE

- **Solve $n + 2\frac{5}{7} = 8$.**

$n + 2\frac{5}{7} - 2\frac{5}{7} = 8 - 2\frac{5}{7}$

$n = 8 - 2\frac{5}{7}$

$n = 7\frac{7}{7} - 2\frac{5}{7}$

$n = 5\frac{2}{7}$

EXERCISES

Solve each equation. Write the solution in simplest form.

49. $x - 12\frac{3}{4} = 17\frac{2}{5}$ **50.** $t + 6\frac{11}{12} = 21\frac{5}{6}$

51. $3\frac{2}{3} = m - 1\frac{3}{4}$ **52.** $5\frac{2}{3} = p + 2\frac{2}{9}$

53. $y - 1\frac{2}{3} = 3\frac{4}{5}$

54. Jon poured $1\frac{1}{2}$ oz of lemon juice onto a salad. He has $5\frac{1}{2}$ oz lemon juice left in the bottle. How many ounces of lemon juice were in the bottle before Jon poured some on the salad?

Chapter Test

Chapter 5

Find the reciprocal.

1. $\frac{3}{5}$ $\frac{5}{3}$
2. $\frac{7}{11}$ $\frac{11}{7}$
3. $\frac{5}{9}$ $\frac{9}{5}$
4. $\frac{1}{8}$ 8

Find the least common multiple (LCM).

5. 10 and 15 30
6. 4, 6, and 18 36
7. 9, 10, and 12 180
8. 6, 15, and 20 60

Estimate each sum or difference by rounding to 0, $\frac{1}{2}$, or 1.

9. $\frac{1}{8} + \frac{4}{7}$ about $\frac{1}{2}$
10. $\frac{11}{12} - \frac{4}{9}$ about $\frac{1}{2}$
11. $\frac{4}{5} + \frac{1}{9}$ about 1

Evaluate each expression. Write the answer in simplest form.

12. $4\frac{1}{9} - 2\frac{4}{9}$ $1\frac{2}{3}$
13. $\frac{2}{5} \cdot \frac{5}{6}$ $\frac{1}{3}$
14. $2\frac{1}{3} \div \frac{5}{6}$ $2\frac{4}{5}$
15. $1\frac{7}{10} + 3\frac{3}{4}$ $5\frac{9}{20}$
16. $\frac{3}{7} \cdot \frac{4}{9}$ $\frac{4}{21}$
17. $\frac{2}{3} - \frac{3}{8}$ $\frac{7}{24}$
18. $1\frac{3}{8} \cdot \frac{6}{11}$ $\frac{3}{4}$
19. $3\frac{1}{8} \div 1\frac{1}{4}$ $2\frac{1}{2}$
20. $\frac{7}{8} \div 2$ $\frac{7}{16}$
21. $\frac{3}{8} \cdot \frac{3}{4}$ $\frac{9}{32}$
22. $3\frac{1}{3} \div 1\frac{5}{12}$ $2\frac{6}{17}$
23. $4 \cdot 2\frac{2}{7}$ $9\frac{1}{7}$
24. $2\frac{1}{4} \cdot 2\frac{2}{3}$ 6
25. $\frac{1}{12} + \frac{5}{6}$ $\frac{11}{12}$
26. $\frac{4}{5} \cdot 1\frac{1}{3}$ $1\frac{1}{15}$
27. $\frac{3}{8} \cdot \frac{2}{3}$ $\frac{1}{4}$

Evaluate the expression $n \cdot \frac{1}{4}$ for each value of n. Write the answer in simplest form.

28. $n = \frac{7}{8}$ $\frac{7}{32}$
29. $n = \frac{2}{5}$ $\frac{1}{10}$
30. $n = \frac{8}{9}$ $\frac{2}{9}$
31. $n = \frac{4}{11}$ $\frac{1}{11}$

Solve each equation. Write the solution in simplest form.

32. $3r = \frac{9}{10}$ $\frac{3}{10}$
33. $n + 3\frac{1}{6} = 12$ $8\frac{5}{6}$
34. $5\frac{5}{6} = x - 3\frac{1}{4}$ $9\frac{1}{12}$
35. $\frac{2}{5}t = 9$ $22\frac{1}{2}$
36. $\frac{4}{5}m = 7$ $8\frac{3}{4}$
37. $y - 15\frac{3}{5} = 2\frac{1}{3}$ $17\frac{14}{15}$
38. Jessica purchased a bag of cat food. She feeds her cat 1 cup of cat food each day. After 7 days, she has fed her cat $\frac{2}{3}$ of the food in the bag. How many cups of food were in the bag of cat food when Jessica bought it? $10\frac{1}{2}$ cups
39. On Saturday, Cecelia ran $3\frac{3}{7}$ miles. On Sunday, she ran $4\frac{5}{6}$ miles. About how much farther did Cecelia run on Sunday than on Saturday? about $1\frac{1}{2}$ miles farther
40. Michael studied social studies for $\frac{3}{4}$ of an hour, Spanish for $1\frac{1}{2}$ hours, and math for $1\frac{1}{4}$ hours. How many hours did Michael spend studying all three subjects? $3\frac{1}{2}$ hours

Chapter Test

Chapter Test

Chapter 5

Purpose: *To assess students' mastery of concepts and skills in Chapter 5*

Assessment Resources

Chapter 5 Tests (Levels A, B, C)
Assessment Resources ..... pp. 55–60

***Test and Practice Generator* CD-ROM**

Additional assessment items in both multiple-choice and free-response format may be generated for any objective in Chapter 5.

Chapter 5 Performance Assessment

Purpose: To assess students' understanding of concepts in Chapter 5 and combined problem solving skills

Assessment Resources

Performance Assessment

Performance Assessment Teacher Support

Answers

1–3. See p. A3.

4. See Level 3 work sample below.

Scoring Rubric for Problem Solving Item 4

Level 3

Accomplishes the purposes of the task.

Student gives clear explanations, shows understanding of mathematical ideas and processes, and computes accurately.

Level 2

Purposes of the task not fully achieved.

Student demonstrates satisfactory but limited understanding of the mathematical ideas and processes.

Level 1

Purposes of the task not accomplished.

Student shows little evidence of understanding the mathematical ideas and processes and makes computational and/or procedural errors.

Chapter 5 Performance Assessment

Show What You Know

Create a portfolio of your best work from this chapter. Complete this page and include it with your four best pieces of work from Chapter 5. Choose from your homework or lab assignments, mid-chapter quiz, or any journal entries you have done. Put them together using any design you want. Make your portfolio represent what you consider your best work.

Short Response

1. Use prime factorization to find the least common multiple of 7, 12, and 15. Show your work.
2. Daphne will distribute cereal samples with pamphlets about good nutrition. The samples come in packages of 15. The pamphlets come in packages of 20. What is the least number of cereal samples and pamphlets that Daphne can get to have the same number of each? How many packages of each will she have? Show your work.
3. Estimate the sum of $\frac{4}{5}$ and $\frac{9}{10}$. Is your answer an overestimate or an underestimate? Explain.

CLOSE TO HOME JOHN McPHERSON

Bobby's excitement about going to summer camp faded as soon as he read the sign.

Extended Problem Solving

4. During the summer, Garrett attends a day camp for 6 hours each day. The circle graph shows what fraction of each day he spends doing different activities.
   a. How long does Garrett spend doing each activity? Write the activities in order from longest to shortest.
   b. Sports activities and playground games are all held on the camp fields. What fraction of the day does Garrett spend on the fields? Write your answer in simplest form.
   c. Lunch and crafts are held in the cafeteria. How many hours does Garrett spend in the cafeteria during a 5-day week at day camp? Write your answer in simplest form, and show the work necessary to determine the correct answer.

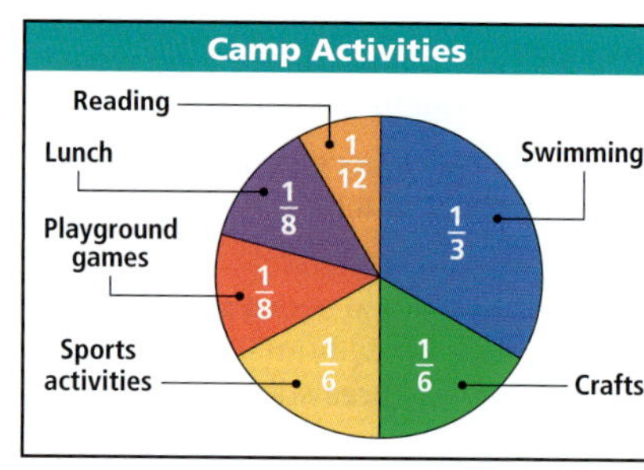

Student Work Samples for Item 4

Level 3

a. Swimming: $\frac{1}{3}(6) = 2$ hr
Crafts / Sports: $\frac{1}{6}(6) = 1$ hr
Playground / Lunch: $\frac{1}{8}(6) = \frac{3}{4}$ hr = 45 min.
Reading: $\frac{1}{12}(6) = \frac{1}{2}$ hr = 30 min.

b. $\frac{1}{6} + \frac{1}{8} = \frac{4}{24} + \frac{3}{24} = \frac{7}{24}$ day

c. 1 hr + $\frac{3}{4}$ hr = $1\frac{3}{4}$ hr $1\frac{3}{4}(5) = 8\frac{3}{4}$
$8\frac{3}{4}$ hr or 8 hr 45 min.

The student answered parts a, b, and c correctly. The student showed all the steps in part c.

Level 2

a. Swimming − $\frac{1}{3}(6) = 2$ hours
Crafts − $\frac{1}{6}(6) = 1$ hour
Sports − $\frac{1}{6}(6) = 1$ hour
Playground − $\frac{1}{8}(6) = \frac{3}{4}$ hour
Lunch − $\frac{1}{8}(6) = \frac{3}{4}$ hour
Reading − $\frac{1}{12}(6) = \frac{1}{2}$ hour

b. 1 hour + $\frac{3}{4}$ hour = $1\frac{3}{4}$ hour
$1\frac{3}{4} \div 6 = \frac{6}{4} \cdot \frac{1}{6} = \frac{1}{4}$ day

c. 1 hour + $\frac{3}{4}$ hour = $1\frac{3}{4}$ hour
$1\frac{3}{4} \cdot 5 = \frac{6}{4} \times 5 = \frac{30}{4} = 7\frac{1}{2}$ hours

The student correctly answered part a. The student incorrectly wrote $1\frac{3}{4}$ as $\frac{6}{4}$, resulting in incorrect calculations for parts b and c.

Level 1

a. Swimming, Crafts, Sports, Playground, Lunch, Reading

b. $\frac{1}{6} + \frac{1}{8} = \frac{2}{14} = \frac{1}{7}$ day

c. $\frac{1}{7} \times 5 = \frac{5}{7}$ hr

The student did not answer the question posed in part a. The student incorrectly added fractions in part b and did not understand part c.

internet connect
State-Specific Test Practice Online
go.hrw.com Keyword: MR4 TestPrep

Standardized Test Prep

Chapter 5

Cumulative Assessment, Chapters 1–5

1. Which number is greater than $\frac{4}{5}$? **D**

Ⓐ $\frac{1}{2}$ Ⓒ $\frac{3}{4}$

Ⓑ $\frac{5}{6}$ Ⓓ $\frac{2}{3}$

2. Which number is divisible by 2, 3, 6, and 9 but **not** by 4, 5, or 10? **F**

Ⓕ 882 Ⓗ 684

Ⓖ 768 Ⓙ 180

3. Which is the standard form of thirty-one and twenty-two thousandths? **C**

Ⓐ 31.22 Ⓒ 31.022

Ⓑ 31,022 Ⓓ 31,022,000

TIP! TEST TAKING TIP!

To multiply mixed numbers, write the mixed number as an improper fraction. Multiply the numerators, and then multiply the denominators.

4. Find the product. $2\frac{2}{3} \cdot 3\frac{1}{2}$ **F**

Ⓕ $6\frac{1}{3}$ Ⓗ $\frac{1}{3}$

Ⓖ 2 Ⓙ $9\frac{1}{3}$

5. For which equation is $n = 5$ a solution? **D**

Ⓐ $7n = 25 + 3$

Ⓑ $\frac{n}{8} = 5$

Ⓒ $47 - n = (8 \times 5)$

Ⓓ $\frac{n}{6} \cdot \frac{1}{3} = \frac{5}{18}$

6. What is the LCM of 4, 7, and 14? **H**

Ⓕ 2 Ⓗ 28

Ⓖ 14 Ⓙ 56

7. Which is a common denominator of $\frac{1}{4}$, $\frac{5}{6}$, and $\frac{3}{8}$? **D**

Ⓐ 2 Ⓒ 15

Ⓑ 12 Ⓓ 24

8. What is the distance around the rectangular picture frame shown? **F**

Ⓕ $5\frac{1}{4}$ in. Ⓗ $2\frac{5}{8}$ in.

Ⓖ 4 in. Ⓙ $3\frac{7}{8}$ in.

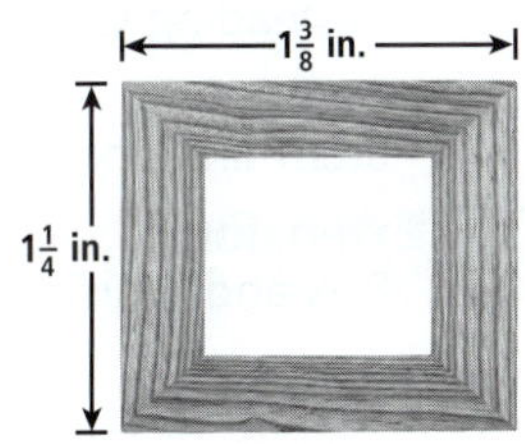

9. ***SHORT RESPONSE*** Lucy has a total of $\frac{5}{6}$ yard of ribbon to wrap gifts for her friends. The bow on each gift requires $\frac{1}{6}$ yard of ribbon. How can you determine how many bows b Lucy can make? Find the correct value for b.

10. ***SHORT RESPONSE*** Mr. Frost is painting his garage, which has a surface area of 500 square feet. He has $1\frac{3}{4}$ gallons of paint. If one gallon of paint covers 150 square feet, how much more paint must Mr. Frost buy? Show your work.

Standardized Test Prep

Chapter 5

Purpose: *To provide review and practice for Chapters 1–5 and standardized tests*

Assessment Resources

Cumulative Tests (Levels A, B, C)
Assessment Resources. . . . pp. 175–186

State-Specific Test Practice Online
KEYWORD: MR4 TestPrep

Test Prep Doctor

For item 5, remind students to substitute 5 for n to check if each equation balances.

In item 7, point out that students may find the least common denominator or any common denominator.

Answers

9. $\frac{1}{6}b = \frac{5}{6}$

$b = \frac{5}{6} \div \frac{1}{6}$

$= \frac{5}{6} \times \frac{6}{1}$

$= 5$

Lucy has enough ribbon for five bows.

10. $\frac{500 \text{ ft}^2}{150 \text{ ft}^2 \text{ per gal}} = 3\frac{1}{3}$ gal

Mr. Frost has $1\frac{3}{4}$ gal.

$3\frac{1}{3} - 1\frac{3}{4} = 1\frac{7}{12}$ gal

Mr. Frost needs $1\frac{7}{12}$ gallons of paint, so he must buy 2 gallons.

Chapter 6

Collect and Display Data

Section 6A

Organizing Data

Lesson 6-1
Make a Table

Lesson 6-2
Range, Mean, Median, and Mode

Lesson 6-3
Additional Data and Outliers

Section 6B

Displaying and Interpreting Data

Lesson 6-4
Bar Graphs

Technology Lab 6A
Create Bar Graphs

Lesson 6-5
Frequency Tables and Histograms

Lesson 6-6
Ordered Pairs

Lesson 6-7
Line Graphs

Lesson 6-8
Misleading Graphs

Lesson 6-9
Stem-and-Leaf Plots

Extension
Box-and-Whisker Plots

Pacing Guide for 45-Minute Classes

Chapter 6

DAY 70	DAY 71	DAY 72	DAY 73	DAY 74
Lesson 6-1	Lesson 6-2	Lesson 6-3	Mid-Chapter Quiz Lesson 6-4	Technology Lab 6A
DAY 75	**DAY 76**	**DAY 77**	**DAY 78**	**DAY 79**
Lesson 6-5	Lesson 6-6	Lesson 6-7	Lesson 6-8	Lesson 6-9
DAY 80	**DAY 81**	**DAY 82**		
Extension	Chapter 6 Review	Chapter 6 Assessment		

Pacing Guide for 90-Minute Classes

Chapter 6

DAY 35	DAY 36	DAY 37	DAY 38	DAY 39
Chapter 5 Assessment Lesson 6-1	Lesson 6-2 Lesson 6-3	Mid-Chapter Quiz Lesson 6-4 Technology Lab 6A	Lesson 6-5 Lesson 6-6	Lesson 6-7 Lesson 6-8
DAY 40	**DAY 41**	**DAY 42**		
Lesson 6-9 Extension	Chapter 6 Review Lesson 7-1	Chapter 6 Assessment Lesson 7-2		

Across the Series

HARCOURT GRADE 5

- Organize, display, and analyze data in tables, line plots, bar graphs, line graphs, histograms, and circle graphs.
- Find measures of central tendency.
- Graph ordered pairs.

COURSE 1

- **Find measures of central tendency and range.**
- **Organize, display, and analyze data in tables, histograms, bar graphs, line graphs, stem-and-leaf plots, and box-and-whisker plots.**
- **Graph ordered pairs.**
- **Recognize misleading graphs and statistics.**
- **Use a spreadsheet or graphing calculator to create graphs.**

COURSE 2

- Identify populations and choose samples.
- Find measures of central tendency and range.
- Organize, display, and analyze data in tables, bar graphs, histograms, line graphs, stem-and-leaf plots, box-and-whisker plots, and scatter plots.
- Recognize misleading graphs and statistics.
- Use a spreadsheet or graphing calculator to create graphs.

Across the Curriculum

LANGUAGE ARTS LINK

Math: Reading and Writing in the Content Area . . . . pp. 44–52

Focus on Problem Solving
Make a Plan . . . . SE p. 283

Journal . . . . TE pp. 281, 287, 293, 300, 304

Write About It . . . . SE pp. 274, 281, 287, 293, 296, 300, 304

SOCIAL STUDIES LINK

Social Studies . . . . SE pp. 278, 293

SCIENCE LINK

Life Science . . . . SE p. 300

Earth Science . . . . SE p. 281

Health . . . . SE p. 304

TE = *Teacher's Edition* **SE** = *Student Edition*

Interdisciplinary Bulletin Board

Social Studies

Most of the countries in Southwest Asia are between 500 and 4,000 meters above sea level. Name the type of graph you would use to show the elevation and area of each country in Southwest Asia. double-bar graph

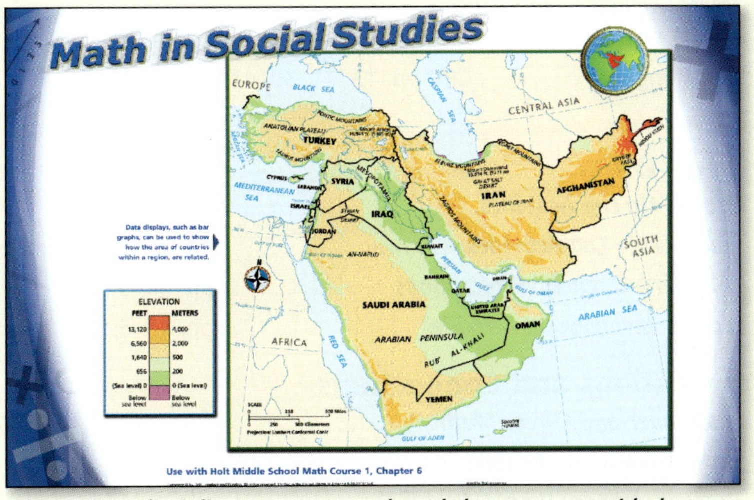

Interdisciplinary posters and worksheets are provided in your resource material.

Chapter 6

Resource Options

Chapter 6 Resource Book

Student Resources

Practice (Levels A, B, C) . . . . pp. 8–10, 17–19, 26–28, 36–38, 46–48, 56–58, 65–67, 75–77, 84–86

Reteach . . . . pp. 11, 20, 29, 39–40, 49–50, 59, 68–69, 78, 87

Challenge . . . . pp. 12, 21, 30, 41, 51, 60, 70, 79, 88

Problem Solving . . . . pp. 13, 22, 31, 42, 52, 61, 71, 80, 89

Puzzles, Twisters & Teasers . . . . pp. 14, 23, 32, 43, 53, 62, 72, 81, 90

Recording Sheets . . . . pp. 3, 7, 16, 25, 35, 45, 55, 64, 74, 83, 94

Chapter Review . . . . pp. 91–93

Teacher and Parent Resources

Chapter Planning and Pacing Guide . . . . p. 4

Section Planning Guides . . . . pp. 5, 33

Parent Letter . . . . pp. 1–2

Teaching Tools . . . . p. 97

Teacher Support for Chapter Project . . . . p. 95

Transparencies . . . . pp. T1–T36

- Daily Transparencies
- Additional Examples Transparencies
- Teaching Transparencies

Reaching All Learners

English Language Learners

Success for English Language Learners . . . . pp. 87–104

Math: Reading and Writing in the Content Area . . . . pp. 44–52

Spanish Homework and Practice . . . . pp. 44–52

Spanish Interactive Study Guide . . . . pp. 44–52

Spanish Family Involvement Activities . . . . pp. 45–52

Multilingual Glossary

Individual Needs

Are You Ready? Intervention and Enrichment . . . pp. 13–16, 25–32, 113–116, 253–256, 415–416

Alternate Openers: Explorations . . . . pp. 44–52

Family Involvement Activities . . . . pp. 45–52

Interactive Problem Solving . . . . pp. 44–52

Interactive Study Guide . . . . pp. 44–52

Readiness Activities . . . . pp. 11–12

Math: Reading and Writing in the Content Area . . . . pp. 44–52

Challenge . . . . CRB pp. 12, 21, 30, 41, 51, 60, 70, 79, 88

Hands-On

Hands-On Lab Activities . . . . pp. 36–42

Technology Lab Activities . . . . pp. 22–24

Alternate Openers: Explorations . . . . pp. 44–52

Family Involvement Activities . . . . pp. 45–52

Applications and Connections

Consumer and Career Math . . . . pp. 21–24

Interdisciplinary Posters . . . . Poster 6, TE p. 270B

Interdisciplinary Poster Worksheets . . . . pp. 16–18

Transparencies

Alternate Openers: Explorations . . . . pp. 44–52

Exercise Answers Transparencies

Chapter 6 Resource Book . . . . pp. T1–T36

- Daily Transparencies
- Additional Examples Transparencies
- Teaching Transparencies

Technology

Teacher Resources

***Lesson Presentations* CD-ROM** . . . . Chapter 6

***Test and Practice Generator* CD-ROM** . . . . Chapter 6

***One-Stop Planner* CD-ROM** . . . . Chapter 6

Student Resources

***Are You Ready? Intervention* CD-ROM**
Skills 1, 4, 5, 26, 61

internet connect go.hrw.com

Homework Help Online	**KEYWORD:** MR4 HWHelp6
Math Tools Online	**KEYWORD:** MR4 Tools
Glossary Online	**KEYWORD:** MR4 Glossary
Chapter Project Online	**KEYWORD:** MR4 PSProject6
Chapter Opener Online	**KEYWORD:** MR4 Ch6
CNN Student News™	**KEYWORD:** MR4 CNN6

SE = *Student Edition* **TE** = *Teacher's Edition* **AR** = *Assessment Resources* **CRB** = *Chapter Resource Book* **MK** = *Manipulatives Kit*

Assessment Options

Assessing Prior Knowledge

Determine whether students have the required prerequisite concepts and skills.

Are You Ready? . . . . SE p. 271
Inventory Test . . . . AR pp. 1–4

Test Preparation

Provide review and practice for chapter and standardized tests.

Standardized Test Prep . . . . SE p. 319
Spiral Review with Test Prep . . . . . SE, last page of each lesson
Study Guide and Review . . . . SE pp. 314–316
Test Prep Tool Kit

Technology

***Test and Practice Generator* CD-ROM**

internet connect

State-Specific Test Practice Online KEYWORD: MR4 TestPrep

go.hrw.com

Performance Assessment

Assess students' understanding of chapter concepts and combined problem-solving skills.

Performance Assessment . . . . SE p. 318
Includes scoring rubric in TE
Performance Assessment . . . . AR p. 114
Performance Assessment Teacher Support . . . . AR p. 113

Portfolio

Portfolio opportunities appear throughout the Student and Teacher's Editions.

Suggested work samples:

Problem Solving Project . . . . TE p. 270
Performance Assessment . . . . SE p. 318
Portfolio Guide . . . . AR p. xxxv
Journal . . . . TE pp. 281, 287, 293, 300, 304
Write About It . . . . . . SE pp. 274, 281, 287 293, 296, 300, 304

Daily Assessment

Obtain daily feedback on students' understanding of concepts.

Spiral Review and Test Prep . . . . . . SE, last page of each lesson

Also Available on Transparency In Chapter 6 Resource Book

Warm Up . . . . TE, first page of each lesson
Problem of the Day . . . . TE, first page of each lesson
Lesson Quiz . . . . TE, last page of each lesson

Student Self-Assessment

Have students evaluate their own work.

Group Project Evaluation . . . . AR p. xxxii
Individual Group Member Evaluation . . . . AR p. xxxiii
Portfolio Guide . . . . AR p. xxxv
Journal . . . . TE pp. 281, 287, 293, 300, 304

Formal Assessment

Assess students' mastery of concepts and skills.

Section Quizzes . . . . AR pp. 16–17
Mid-Chapter Quiz . . . . SE p. 282
Chapter Test . . . . SE p. 317
Chapter Tests (Levels A, B, C) . . . . AR pp. 61–66
Cumulative Tests (Levels A, B, C) . . . . AR pp. 187–198
Standardized Test Prep
Cumulative Assessment . . . . SE p. 319
End-of-Year Test . . . . AR pp. 271–274

Technology

***Test and Practice Generator* CD-ROM**

Make tests electronically. This software includes:

- Dynamic practice for Chapter 6
- Customizable tests
- Multiple-choice items for each objective
- Free-response items for each objective
- Teacher management system

SE = *Student Edition* **TE** = *Teacher's Edition* **AR** = *Assessment Resources* **CRB** = *Chapter Resource Book* **MK** = *Manipulatives Kit*

Chapter 6

Chapter 6 Tests

Three levels (A,B,C) of tests are available for each chapter in the *Assessment Resources.*

LEVEL A

CHAPTER 6 Chapter Test
Form A

Monday the closing price was $12.00. Tuesday it was $14.00. Wednesday it was $16.00. Thursday it was $18.00. Friday the closing price was $20.00.

1. Make a table.

Day	Price
Monday	$12.00
Tuesday	$14.00
Wednesday	$16.00
Thursday	$18.00
Friday	$20.00

2. Find the range, mean, median, and mode of the data set.
Quiz Scores: 10 8 9 8 7

range: 3; mean: 8.4;

median: 8; mode: 8

The table shows the number of days employees took for vacation.

10	5	9	14	14	12	10	14	12	14

3. Find the mean, median and mode of the data.

mean: 11.4; median: 12;

mode: 14

4. Which of the mean, median, and mode best describes the data?

Possible answer: mean or median

5. Make a bar graph.

Number of TV Advertisements

Food	15	Health	12
Movies	6	TV previews	9
Games/toys	3		

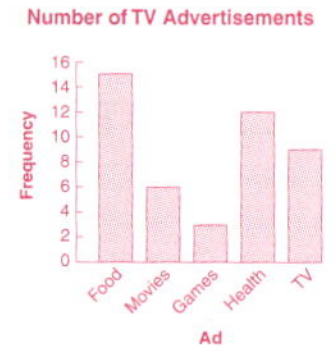

1	2	3	4	2
3	5	1	1	2
2	3	3	3	4
2	1	1	0	3
3	4	5	1	2

6. The data in the table shows the number of videos rented by 25 customers on a given day. Complete the frequency table with the given intervals.

Videos	Frequency
0–1	7
2–3	13
4–5	5

CHAPTER 6 Chapter Test
Form A, continued

7. Name the ordered pairs for each location on the grid.

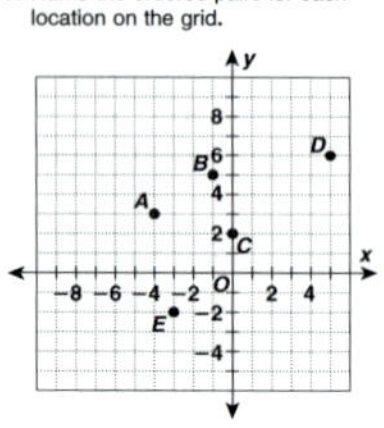

A(−4, 3), *B*(−1, 5), *C*(0, 2)

D(5, 6), *E*(−3, −2)

8. Make a line graph for the depths of the river over 6 weeks.

Week	1	2	3	4	5	6
Depth (in in.)	15	12	13	14	16	18

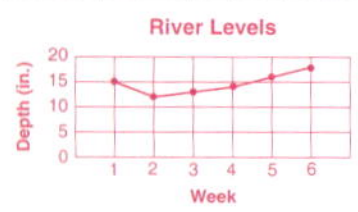

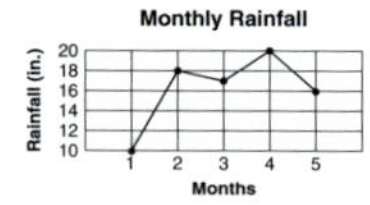

9. Why is this graph misleading?

The vertical scale does not start at 0.

10. What might people believe from the misleading graph?

Possible answer: People might believe that there was no rainfall the first month.

11. Make a stem-and-leaf plot.
10 31 26 30 22 17 23 12 31 28

Stem	Leaves
1	0 2 7
2	2 3 6 8
3	0 1 1

LEVEL B

CHAPTER 6 Chapter Test
Form B

In 1965, 520,000 children were enrolled in preschool. In 1975, 1,748,000 were enrolled. In 1985, 2,491,000 were enrolled. In 1995, 4,399,000 were enrolled, and in 2000, 4,481,000 children were enrolled.

1. Make a table of the data.

Year	Number of students
1965	520,000
1975	1,748,000
1985	2,491,000
1995	4,399,000
2000	4,481,000

2. Find the range, mean, median, and mode. 89 99 77 94 86 89

range: 22; mean: 89; median: 89; mode: 89

Attendance at Weekly Sales Meetings

Date	Number in Attendance
July 3	48
July 10	50
July 17	36
July 24	47
July 31	53
August 7	50
August 14	38

3. Find the mean, median, and mode.

mean: 46; median: 48; mode: 50

4. Which of the mean, median, and mode best describes the data?

median

5. Make a bar graph.

Number of students in each class

Math	23	Gym	26
Spanish	12	Science	24
English	25	History	19

12	3	9	7	10
10	3	8	10	5
5	8	6	0	11
11	9	7	4	5
3	3	6	13	2
14	2	3	12	5

6. The data in the table are the numbers of responses by the Swanton Rescue Squad for each day in April. Make a frequency table with intervals. Possible answer:

Responses	Frequency
0–2	3
3–5	10
6–8	6
9–11	7
12–14	4

CHAPTER 6 Chapter Test
Form B, continued

7. Name the ordered pairs for each location on the grid.

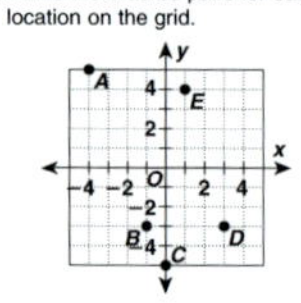

A (−4, 5), *B* (−1, −3),
C (0, −5), *D* (3, −3),
E (1, 4)

8. Make a line graph.

Year	Sales (in thousands)
1996	$114
1997	$130
1998	$144
1999	$140
2000	$152
2001	$154

Possible answer:

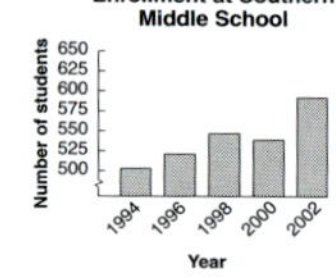

9. Why is this graph misleading?

The vertical scale is broken.

10. What might people believe from the misleading graph?

People might believe that enrollment has more than doubled since 1994.

11. Make a stem-and-leaf plot. 46 61 76 67 53 46 46 53 48 64 74

Stem	Leaves
4	6 6 6 8
5	3 3
6	1 4 7
7	4 6

LEVEL C

CHAPTER 6 Chapter Test
Form C

1. In 1995, 31.7% of U.S. households had a computer. In 1996, 35.5% had a computer. In 1997, 39.2% had a computer. In 1998, 42.6% had a computer. In 1999, 48.2% had a computer, and in 2000, 53.0% of households had a computer.

Year	Percentage
1995	31.7
1996	35.5
1997	39.2
1998	42.6
1999	48.2
2000	53.0

2. The daily high temperatures for one week in August are listed. Find the range, mean, median, and mode to the nearest tenth.
86 92 88 91 95 84 86

range: 11; mean: 88.9;

median: 88; mode: 86

Books Read by 6th Graders

Sept.	Oct.	Nov.	Dec.	Jan.
29	30	30	15	28

3. Find the mean, median, and mode.

mean: 26.4; median: 29;

mode: 30

4. Which of the mean, median, and mode best describes the data?

median

5. Make a double bar graph.

Number of Bushels of Apples Picked

	Week 1	Week 2	Week 3	Week 4
Red	21	15	32	38
Yellow	25	18	27	36

100	108	111	117
111	113	103	103
115	116	117	100
102	109	107	114
102	103	106	102
116	115	111	115

6. Make a frequency table of the data above.
Possible answer:

Interval	Frequency
100–104	8
105–109	4
110–114	5
115–119	7

CHAPTER 6 Chapter Test
Form C, continued

7. Name the ordered pairs for each location.

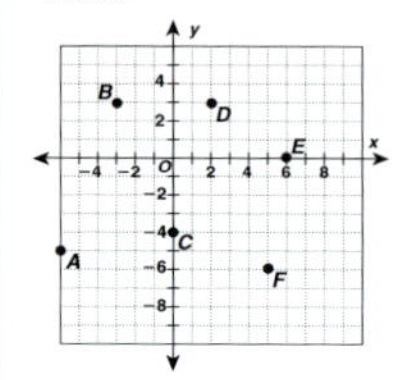

A(−6, −5), *B*(−3, 3), *C*(0, −4),
D(2, 3), *E*(6, 0), *F*(5, −6)

8. Make a double line graph.

Population of Orlando and Ft. Myers

	Orlando	Ft. Myers
1970	523,000	105,000
1980	805,000	205,000
1990	1,239,000	339,000
2000	1,535,000	401,000

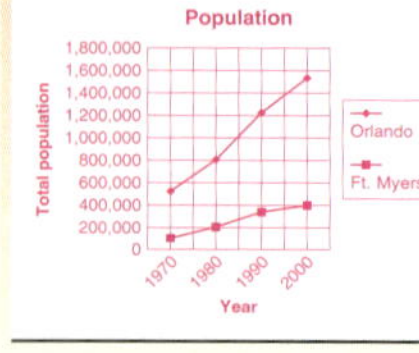

9. Use the data given to construct a misleading bar graph showing that the company's business is falling quickly.

Monthly Sales

June	$25,000
July	$24,000
August	$23,000

Possible answer:

10. Explain why your bar graph is misleading.

Possible answer: The lower part of the vertical scale is missing.

11. Make a stem-and-leaf plot.
100, 103, 112, 120, 116, 127, 104, 109, 124, 118, 116 129, 119

Stem	Leaves
10	0 3 4 9
11	2 6 6 8 9
12	0 4 7 9

Test and Practice Generator
CD-ROM

Create and customize multiple versions of the same tests with corresponding answers for any chosen chapter objectives.

Chapter 6 State and Standardized Test Preparation

Test Taking Skill Builder and Standardized Test Practice are provided for each chapter in the *Test Prep Tool Kit.*

Test Taking Skill Builder

Test Taking Strategy — **Context-Based Response Questions**

Chapter 6

Use the information provided in a table, a diagram, or a graph to answer a context-based response question. To receive full credit for this type of question, you need to explain your reasoning and show all of your work in detail. Be sure to check your answer for sense once you complete the question.

Example Wesley displays the following graph in his science class. He then makes the statement that there are more known beetle species than in all of the other types of species combined. Do you support Wesley's statement? Why or why not?

Subgroups of the Animal Kingdom

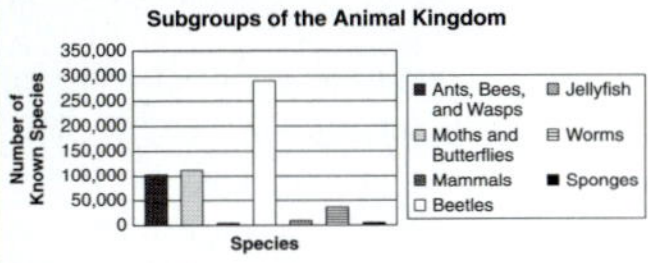

To receive full credit for this question, work through this type of process.

First read over the question again: What information does the question statement provide? Plan to use this information to solve the problem.

Given: Data from the graph can be used to find the number of known species of beetles as well as the total number of known species of all of the other types of species. Then you can compare the two values to determine if you support Wesley's statement.

Find the number of species: Beetles: There are about 290,000 types of beetle species. All other species: 103,000 + 112,000 + 4,000 + 9,000 + 36,000 + 5,000 = 269,000
There are about 269,000 types of other species shown.

Answer in a complete sentence: I support Wesley's statement because 290,000 is greater than 269,000.

Check that your answer is reasonable: By looking at the graph, you can see that the number of known beetle species is far greater than any of the other types of species. It is a reasonable conclusion that Wesley's statement is correct.

Standardized Test Practice

Standardized Test Practice

Chapter 6

Select the best answer for Questions 1–5.

1. If you add 5 to every item in a data set what will happen to the mean?
   A It is multiplied by 5.
   B It is increased by 5.
   C It is divided by 5.
   D The mean does not change.

Use the bar graph for Questions 2 and 3.

American Kennel Club Registration

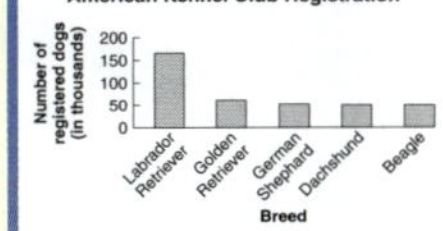

2. A flea collar company is planning a new advertisement campaign. They plan to use the most popular breed of puppy in their ad. Which breed of puppy should they use?
   F Beagle
   G German Shephard
   H Golden Retriever
   I Labrador Retriever

3. What was the number of registered golden retrievers in 2001?
   A 40,000
   B 60,000
   C 50,000
   D 160,000

4. What are the coordinates of point *A*?

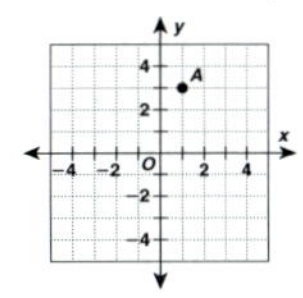

   F (−1, 3)
   G (3, 1)
   H (3, −1)
   I (1, 3)

5. What makes this bar graph misleading?

Sales Profits

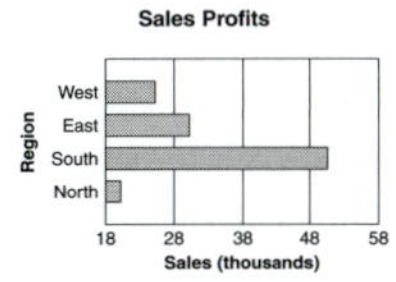

   A The title is incorrect.
   B The horizontal scale does not start at zero.
   C The vertical scale does not start at zero.
   D The space between the data values is not the same.

internet connect

State-Specific Test Practice Online

KEYWORD: MR4 TestPrep

Test Prep Tool Kit

- Standardized Test Prep Workbook
- Countdown to Testing transparencies
- State Test Prep CD-ROM
- Standardized Test Prep Video

Test Taking Strategy

Chapter 6, continued

Exercises

Answer each question.

1. Data for 6 out of the 8 planets is shown below. Maria says the mean length of day in Earth hours is greater than that of the median. Is she correct? Explain.

Planetary Days

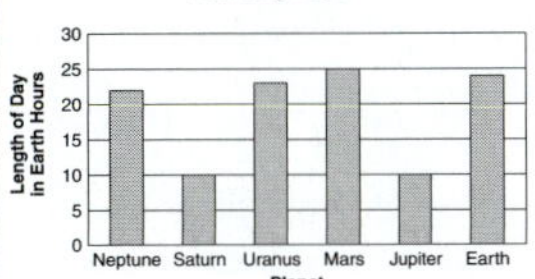

   a. What information does the problem statement provide? Write a plan for how you will use this information to answer the question.
   a bar graph; I will use the values on the graph to calculate the mean and median of the data. I will then compare the mean and median to determine if Maria is correct.
   b. Use your plan of solve the problem. Show your work.
   mean = 22 + 10 + 23 + 25 + 10 + 24 = 114 ÷ 6 = 19 Earth hours, median = 22.5 Earth hours; No Maria is not correct. The mean is not greater than the mode.
   c. Is your answer reasonable? Explain.
   Possible answer: Yes, by looking at the graph there are more larger values than smaller values, so the median would be a larger number. The mean would not be as large as the median because of the two smaller values.

Standardized Test Practice

Chapter 6, continued

Gridded Response

Solve the problems. Use the answer sheet to write and grid-in your answer.

6. Identify the outlier of the following set of test scores.
   60, 64, 59, 63, 88, 67, 70

7. How many servings will you get from 6 cartons of rainbow sherbet?

Sherbet Punch	
Cartons of Sherbet	Number of Servings
2	8
3	12
4	16
5	?
6	?

8. How many points total did Kayla have for all three volleyball games?

	Martin	Keegan	Kayla	Madison
Game 1	II	III	I	IIII
Game 2	~~IIII~~	II	I	II
Game 3	II	III	III	II

Short Response

Solve the problems. Use the answer sheet to write your answers.

9. Use the given data to construct a line graph. Be sure to include all parts of the graph.

Ticket Prices at the Zoo	
1992	$3.75
1994	$5.00
1996	$5.00
1998	$6.00
2000	$6.50
2002	$7.00

10. Explain how to use the table in Question 9 to determine the ticket price at the zoo in 2004.

11. Explain the different ways a graph can be misleading.

Extended Response

12.

Stem	Leaf
2	1 1 2 3 4 4
3	1 2
4	0 0 5 5 6
5	7 8 9
6	1 3 3 4 4 5

Key: 2 | 3 means 23

   a. What is the least number in the data set? What is the greatest number? Show how to determine the range of the data.
   b. Find the mean, median and mode of the data set. Show your work.
   c. Is it easier to find the mean, median, or mode from a stem-and-leaf plot? Explain why.

Sheet

Ⓗ ● See Lesson 6

Ⓒ Ⓓ See Lesson 6

8.

See Lesson 6-5.

9.

Zoo Ticket Prices

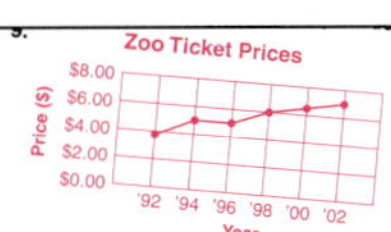

See Lesson 6-7.

10. The table shows that each year the price of a ticket at the zoo increas by 50 cents. So the price of a ticket in 2003 will be 7.00 + 0.50 = $7

(See Less

11. A graph can be misleading if a broken axis or distorted pictures are u

(See Less

Customized answer sheets give students realistic practice for actual standardized tests.

Chapter 6

Collect and Display Data

Why Learn This?

Tell students that weather statistics are collected every day. For example, the high and low temperatures, the amount of precipitation, and the time of sunrise and sunset are recorded daily. Meteorologists compare this data to past weather and also use it to predict future weather.

Using Data

To begin the study of this chapter, have students:

- Describe what is meant by *average high temperature.* The average high temperature for each month is the average of the highest recorded temperatures for each day of the month.
- Identify which national park has the highest average high temperature during July. Badlands, SD Which has the lowest? Crater Lake, OR
- Find the difference in the average high temperatures of the national park with the highest and the national park with the lowest average high temperatures during June. 32°C–19°C = 13°C

Chapter 6

TABLE OF CONTENTS

internet connect
Chapter Opener Online
go.hrw.com
KEYWORD: MR4 Ch6

Collect and Display Data

	Average High Temperatures (°C)		
National Park	Jun	Jul	Aug
Badlands, SD	27	33	32
Big Bend, TX	32	31	31
Crater Lake, OR	19	25	24
Everglades, FL	31	32	32

Career *Meteorologist*

Weather affects our daily activities, and weather information is useful and often necessary. Businesses such as farms, ski resorts, and airlines need to know weather conditions.

This information comes from meteorologists, people who study and forecast the weather. They gather data such as temperature, wind speed, and rainfall. They then study this data and make predictions.

The table lists the average daily high temperatures during the summer in some popular national parks.

Problem Solving Project

Earth Science and Social Studies Connection

Purpose: To solve problems by collecting and displaying data

Materials: National Park Climate worksheet, thermometers

internet connect
Chapter Project Online: *go.hrw.com*
KEYWORD: MR4 PSProject6

Understand, Plan, Solve, and Look Back

Have students:

- ✓ Complete the National Park Climate worksheet to learn about the temperature conditions in our national parks.
- ✓ Make a table comparing the high temperatures of the five parks they most want to visit. Then use the table to create a graph.
- ✓ Take a poll of their class or of the whole school to discover how many different national parks the students have visited. Then have students create a graph of the data they collected.
- ✓ Check students' work.

ARE YOU READY?

Choose the best term from the list to complete each sentence.

place value
horizontally
vertically
sum
quotients

1. The answer to an addition problem is called the __?__. sum
2. The __?__ of the 6 in 5,672 is hundreds. place value
3. When you move __?__, you move left or right. When you move __?__, you move up or down. horizontally; vertically

Complete these exercises to review skills you will need for this chapter.

✔ Place Value

Write the digit in the tens place of each number.

4. 718 1
5. 989 8
6. 55 5
7. 7,709 0

✔ Compare and Order Whole Numbers

Order the numbers from least to greatest.

8. 40, 32, 51, 78, 26, 43, 27
26, 27, 32, 40, 43, 51, 78
9. 132, 150, 218, 176, 166
132, 150, 166, 176, 218
10. 92, 91, 84, 92, 87, 90
84, 87, 90, 91, 92, 92
11. 23, 19, 33, 27, 31, 31, 28, 18
18, 19, 23, 27, 28, 31, 31, 33

Find the greatest number in each set.

12. 452, 426, 502, 467, 530, 512 530
13. 711, 765, 723, 778, 704, 781 781
14. 143, 122, 125, 137, 140, 118, 139 143
15. 1,053; 1,106; 1,043; 1,210; 1,039; 1,122 1,210

✔ Write Fractions as Decimals

Write each fraction as a decimal.

16. $\frac{1}{4}$ 0.25
17. $\frac{5}{8}$ 0.625
18. $\frac{1}{6}$ $0.1\overline{6}$
19. $\frac{2}{5}$ 0.4

✔ Locate Points on a Number Line

Name the point on the number line that corresponds to each given value.

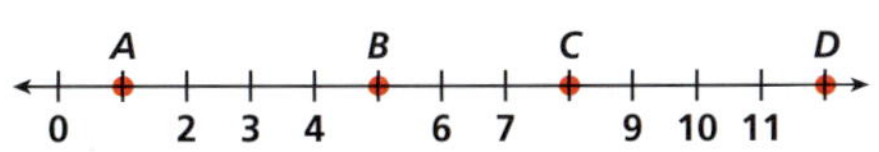

20. 5 B
21. 12 D
22. 8 C
23. 1 A

Assessing Prior Knowledge

INTERVENTION

Diagnose and Prescribe

Evaluate your students' performance on this page to determine whether intervention is necessary or whether enrichment is appropriate. Options that provide instruction, practice, and a check are listed below.

Resources for Are You Ready?

- ***Are You Ready? Intervention and Enrichment***
- **Recording Sheet for Are You Ready?** *Chapter 6 Resource Book . . . . . . p. 3*

Are You Ready? Intervention **CD-ROM**

ARE YOU READY?

Were students successful with Are You Ready?

NO INTERVENE

✔ Place Value 4–7
Intervention Practice, Skill 1
CD-ROM Intervention Activities, Skill 1

✔ Compare and Order Whole Numbers 8–15
Intervention Practice, Skills 4, 5
CD-ROM Intervention Activities, Skills 4, 5

✔ Write Fractions as Decimals . . . 16–19
Intervention Practice, Skill 26
CD-ROM Intervention Activities, Skill 26

✔ Locate Points on a Number Line 20–23
Intervention Practice, Skill 61
CD-ROM Intervention Activities, Skill 61

YES ENRICH

Are You Ready? Enrichment, pp. 417–418

Section 6A

Organizing Data

One-Minute Section Planner

Lesson	Materials	Resources
Lesson 6-1 Make a Table **NCTM:** Data Analysis and Probability, Problem Solving, Communication, Connections, Representation **NAEP:** Data Analysis and Probability 1a ☑ SAT-9 ☑ SAT-10 ☐ ITBS ☑ CTBS ☑ MAT ☐ CAT	**Optional** Almanacs	• *Chapter 6 Resource Book*, pp. 6–14 • Daily Transparency T1, CRB • Additional Examples Transparencies T2–T3, CRB • *Alternate Openers: Explorations*, p. 44
Lesson 6-2 Range, Mean, Median, and Mode **NCTM:** Number and Operations, Data Analysis and Probability, Communication **NAEP:** Data Analysis and Probability 2a ☑ SAT-9 ☑ SAT-10 ☐ ITBS ☐ CTBS ☑ MAT ☑ CAT		• *Chapter 6 Resource Book*, pp. 15–23 • Daily Transparency T4, CRB • Additional Examples Transparencies T5–T6, CRB • *Alternate Openers: Explorations*, p. 45
Lesson 6-3 Additional Data and Outliers **NCTM:** Data Analysis and Probability, Communication **NAEP:** Data Analysis and Probability 2c ☑ SAT-9 ☑ SAT-10 ☐ ITBS ☐ CTBS ☐ MAT ☐ CAT	**Optional** Magazines or newspapers containing data sets	• *Chapter 6 Resource Book*, pp. 24–32 • Daily Transparency T7, CRB • Additional Examples Transparencies T8–T10, CRB • *Alternate Openers: Explorations*, p. 46
Section 6A Assessment		• Mid-Chapter Quiz, SE p. 282 • Section 6A Quiz, AR p. 16 • *Test and Practice Generator* CD-ROM

SAT = *Stanford Achievement Tests* **ITBS** = *Iowa Test of Basic Skills* **CTBS** = *Comprehensive Test of Basic Skills/Terra Nova*
MAT = *Metropolitan Achievement Tests* **CAT** = *California Achievement Test*
NCTM = Complete standards can be found on pages T27–T33. **NAEP** = Complete standards can be found on pages A31–A35.
SE = *Student Edition* **TE** = *Teacher's Edition* **AR** = *Assessment Resources* **CRB** = *Chapter Resource Book* **MK** = *Manipulatives Kit*

Section Overview

Make a Table — *Lesson 6-1*

Why? Making a table helps you to organize and interpret data.

At 1 P.M., the temperature was 72°F. At 2 P.M. it was 74°F. At 3 P.M., it was 76°F. At 4 P.M., it was 73°F.

Time (P.M.)	Temperature (°F)
1	72
2	74
3	76
4	73

Range, Mean, Median, and Mode — *Lesson 6-2*

Why? The mean, median, and mode are measures used to represent values of a data set.

Consider the following data set: **9, 2, 2, 4, 8, 2, 8.**

Mean

The **mean (average)** is the sum of all the items, divided by the number of items in the set.

$$9 + 2 + 2 + 4 + 8 + 2 + 8 = 35$$

$$35 \div 7 = 5$$

The mean is 5.

Median

The **median** is the middle value when the data are in numerical order or the mean of the two middle values if there is an even number of items.

2, 2, 2, 4, 8, 8, 9

The median is 4.

Mode

The **mode** is the value or values that occur most often. There may be more than one mode for a data set. If all values occur an equal number of times, the data set has no mode.

The number 2 occurs most often. The mode is 2.

Additional Data and Outliers — *Lesson 6-3*

Why? Data values that are not close to most other values in the data set can greatly change the mean.

Add the value 47 to the data set above to form the new data set: **9, 2, 2, 4, 8, 2, 8, 47.** Find the mean, median, and mode of the new data set.

$$9 + 2 + 2 + 4 + 8 + 2 + 8 + 47 = 82$$

$$82 \div 8 = 10.25$$

The mean is 10.25

2, 2, 2, 4, 8, 8, 9, 47

The mean of 4 and 8 is 6. The median is 6.

9, 2, 2, 4, 8, 2, 8, 47

The number 2 still occurs most often. The mode is 2.

In this case, the mean is not very representative of the data set because it is greater than all the data values except 47. The mean best represents a data set when the values are close and there are not outliers.

6-1 Organizer

Pacing: Traditional 1 day
Block $\frac{1}{2}$ day

Objective: Students use tables to record and organize data.

Indiana Academic Standards
6.6.1, 6.7.1, 6.7.11

Warm Up

Write the values in simplest form.

1. $\frac{1}{3} + \frac{5}{8}$ $\frac{23}{24}$
2. $\frac{5}{8} - \frac{5}{12}$ $\frac{5}{24}$
3. $\frac{7}{8} \div \frac{1}{3}$ $2\frac{5}{8}$
4. $5\frac{5}{6} \cdot 2\frac{1}{4}$ $13\frac{1}{8}$

Problem of the Day

If February 1 falls on a Tuesday, then March 1 falls on what day of the week? Tuesday or Wednesday, depending on whether or not it is a leap year

Available on Daily Transparency in CRB

Math Humor

I'm learning how to solve math problems by making a table. Can I borrow your tools?

6-1 Make a Table

Problem Solving Strategy

Learn to use tables to record and organize data.

Weather forecasters collect data about weather. By organizing and interpreting this data, they can often warn people of severe weather before it happens. This advance warning can save lives.

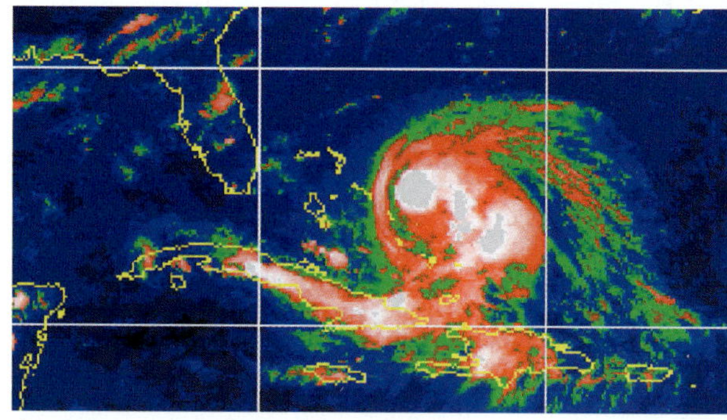

This satellite image shows a hurricane approaching Florida's coastline.

One way to organize data is to make a table. By looking at a table, you may see patterns and relationships.

EXAMPLE 1 *Weather Application*

Use the data about Hurricane Mitch to make a table. Then use your table to describe how the hurricane's strength changed over time.

On October 24, 1998, Hurricane Mitch's wind speed was 90 mi/h. On October 26, its wind speed was 130 mi/h. On October 27, its wind speed was 150 mi/h. On October 31, its wind speed was 40 mi/h. On November 1, its wind speed was 30 mi/h.

Date (1998)	Wind Speed
October 24	90 mi/h
October 26	130 mi/h
October 27	150 mi/h
October 31	40 mi/h
November 1	30 mi/h

Make a table. Write the dates in order so that you can see how the hurricane's strength changed over time.

From the table, you can see that Hurricane Mitch became stronger from October 24 to October 27 and then weakened from October 27 to November 1.

Weather LINK

The National Weather Service estimated that Mitch's wind speed reached 180 mi/h. This made Mitch a Category 5 hurricane, which is the strongest type.

go.hrw.com KEYWORD: MR4 Hurricane

CNN Student News

1 Introduce

Alternate Opener

EXPLORATION

6-1 Make a Table

The table shows the number of medals awarded to the top 13 medal-winning countries during the 2002 Winter Olympics.

1. Compute the total number of medals won by each country.

Country	Gold	Silver	Bronze	Total
Germany	12	16	7	
USA	10	13	11	
Norway	11	7	6	
Canada	6	3	8	
Austria	2	4	10	
Russia	6	6	4	
Italy	4	4	4	
France	4	5	2	
Switzerland	3	2	6	
China	2	2	4	
Netherlands	3	5	0	
Finland	4	2	1	
Sweden	0	2	4	

Think and Discuss

2. **Explain** why the table is set up the way it is.
3. **Describe** a different way to organize this data.

Motivate

Have students look through various almanacs and describe how different data is organized. Possible answers: in graphs, in tables Explain that representing data in an organized way makes it easier to compare and draw conclusions from the data.

Exploration worksheet and answers on Chapter 6 Resource Book pp. 7 and 98

2 Teach

Lesson Presentation

Guided Instruction

In this lesson, students learn to use tables to record and organize data. Explain to students how the data in the Examples are easier to describe when organized in the tables than when in paragraph form.

EXAMPLE 2 **Organizing Data in a Table**

Use the temperature data to make a table. Then use your table to find a pattern in the data and draw a conclusion.

At 10 A.M., the temperature was 62°F. At noon, it was 65°F. At 2 P.M., it was 68°F. At 4 P.M., it was 70°F. At 6 P.M., it was 66°F.

Time	Temperature (°F)
10 A.M.	62
Noon	65
2 P.M.	68
4 P.M.	70
6 P.M.	66

The temperature rose until 4 P.M., and then it dropped. One conclusion is that the high temperature on this day was at least 70°F.

Think and Discuss

1. **Tell** how a table helps you organize data.
2. **Explain** why the data in Example 2 was arranged from earliest to latest time instead of from lowest to highest temperature.

6-1 Exercises

FOR EXTRA PRACTICE
see page 647

internet connect
Homework Help Online
go.hrw.com Keyword: MR4 6-1

GUIDED PRACTICE

See Example 1
1. On Monday, the high temperature was 72°F. On Tuesday, the high was 75°F. On Wednesday, the high was 68°F. On Thursday, the high was 62°F. On Friday, the high was 55°F. Use this data to make a table.

See Example 2
2. Use your table from Exercise 1 to find a pattern in the data and draw a conclusion. Possible answer: The daily high peaked on Tuesday and then dropped for the remainder of the week. The temperature will continue to drop over the weekend.

INDEPENDENT PRACTICE

See Example 1
3. On his first math test, Joe made a grade of 70. On the second test, Joe made a grade of 75. On the third test, Joe made a grade of 80. On the fourth test, Joe made a grade of 85. On the fifth test, Joe made a grade of 90. Use this data to make a table.

See Example 2
4. Use your table from Exercise 3 to find a pattern in the data and draw a conclusion.

Additional Examples

Example 1

Use the audience data to make a table. Then use your table to describe how attendance changed over time.

On May 1, there were 275 people. On May 2, there were 302 people. On May 3, there were 322 people.

Example 2

Use the temperature data to make a table. Then use your table to find a pattern in the data and draw a conclusion.

At 3 A.M., the temperature was 53°F. At 5 A.M., it was 52°F. At 7 A.M., it was 50°F. At 9 A.M., it was 53°F. At 11 A.M., it was 57°F.

Answers on Additional Examples Transparencies T2–T3 in CRB

6-1 PRACTICE & ASSESS

Assignment Guide

If you finished Example 1 assign:
Core 1, 3, 9–15
Enriched 3, 7, 9–15

If you finished Example 2 assign:
Core 1–5, 9–15
Enriched 3–4, 6–15

Answers

1, 3–4. See p. A3.

3 Close

Reaching All Learners
Through Home Connection

Have students gather data at home and organize the data in tables. Ideas for the data include the amount of soft drinks consumed weekly, weekly gasoline consumption, hours of TV watched per week, etc.

Summarize

Give students an unorganized set of data and have them organize it in a table. For example, you could list the number of sixth-grade students enrolled in your school in the past 5 years.

If you use the sixth-grade enrollment example, students may organize the data by year and describe the trends, if any, in enrollment.

Answers to Think and Discuss

1. Possible answer: By using a table, you can easily line up data and reduce the number of words so you can notice the patterns in data.
2. Possible answer: so you can see how the temperature changed throughout the day

Lesson Quiz

1. Humans have the following approximate heart rates at the ages given: newborn, 135 beats per minute (bpm); 2 years old, 110 bpm; 6 years old, 95 bpm; 10 years old, 87 bpm; 20 years old, 71 bpm; 40 years old, 72 bpm; and 60 years old, 74 bpm. Use this data to make a table.
   See p. A8.
2. Use the data from problem 1 to estimate how many times per minute an 8-year-old's heart beats. 91

Available on Daily Transparency in CRB

Answers

5. See p. A3.

6. The cities in the table are ordered alphabetically, not by population. Jakarta has the smallest population.

7. The table on the left would be useful when you are interested in how temperature changes over time. The table on the right would be useful if you wanted to know when it would be a certain temperature.

8. Jeffery is in sixth grade. Victoria is in seventh, and Arthur is in eighth.

PRACTICE AND PROBLEM SOLVING

5. For ice-skating on a frozen pond to be safe, the ice should be at least 7 inches thick. Use the data below to make a table, and estimate the date on which it first became safe to ice-skate.

On December 3, the ice was 1 in. thick. On December 18, the ice was 2 in. thick. On January 3, the ice was 5 in. thick. On January 18, the ice was 11 in. thick. On February 3, the ice was 17 in. thick.

6. ***WHAT'S THE ERROR?*** A student read this table about the populations of large cities and decided that Buenos Aires had the smallest population. Why might the student have made this mistake? Which city does have the smallest population?

City	Population
Buenos Aires	13,430,000
Calcutta	13,400,000
Jakarta	13,300,000
Mexico City	19,430,000
Mumbai	16,630,000

7. ***WRITE ABOUT IT*** The tables below were made using identical data that have been organized differently. When might each table be useful?

Time	Temperature (°F)
6 A.M.	55
10 A.M.	68
2 P.M.	75
6 P.M.	62
10 P.M.	58

Time	Temperature (°F)
2 P.M.	75
10 A.M.	68
6 P.M.	62
10 P.M.	58
6 A.M.	55

8. ***CHALLENGE*** Arthur, Victoria, and Jeffrey are in the sixth, seventh, and eighth grades, although not necessarily in that order. Victoria is not in eighth grade. The sixth-grader is in choir with Arthur and in band with Victoria. Which student is in which grade? Use a yes/no table like the one at right to help you answer this question.

	Arthur	Victoria	Jeffrey
6th			
7th			
8th		No	

Spiral Review

Write each number in scientific notation. (Lesson 3-5)

9. 5,234,000 5.234×10^6 **10.** 23 2.3×10^1 **11.** 12.078 1.2078×10^1

Multiply. (Lesson 3-6)

12. $0.3 \cdot 0.1$ 0.03 **13.** $0.16 \cdot 0.5$ 0.08 **14.** $1.2 \cdot 0.2$ 0.24

15. **TEST PREP** Find the quotient: $0.64 \div 8$. (Lesson 3-7) D

A 80 **B** 8 **C** 0.8 **D** 0.08

RETEACH 6-1

LESSON 6-1 **Reteach**
Make a Table

You can make a table to organize data. Then you can use the table to see patterns and draw conclusions.

During the week-long book fair, 324 books were sold. On Monday, 45 books were sold. On Tuesday, students bought 58 books. On Wednesday, 79 books were sold. Sixty-two books were sold on Thursday, and students bought 51 books on Friday.

Day	Books Sold
Monday	45
Tuesday	58
Wednesday	79
Thursday	62
Friday	51

To make a table, arrange the information in order by days so you can see patterns over time. Remember to make headings for each column of the table.

From the table, you can see that the number of books sold increased from Monday to Wednesday, and decreased from Wednesday to Friday.

Use the data to make a table. Then use the table to find a pattern in the data and draw a conclusion.

1. During the championship series, the school basketball team earned 24 points in the first game, 28 points in the second game, 33 points in the third game, 42 points in the fourth game, and 49 points in the last game.

Game	Points Earned
1	24
2	28
3	33
4	42
5	49

The team earned more points each time that it played a game. The team earned at least 24 points each game.

2. In the sixth grade, 18 students study Spanish, 35 students study French, 11 students study Latin, and 5 students study no foreign language at all.

Foreign Language	Students
French	35
Spanish	18
Latin	11
None	5

Most of the sixth grade students study French as a foreign language.

PRACTICE 6-1

LESSON 6-1 **Practice B**
Make a Table

Complete each activity and answer each question.

1. Pizza Express sells different-sized pizzas. The jumbo pizza has 20 slices. The extra large has 16 slices. The large has 12 slices. There are 8 slices in a medium, and 6 slices in a small. A personal-sized pizza has 4 slices. Use this data to complete the table at right, from largest to smallest pizza.

Pizza Express Pizza Sizes

Size	Slices
Jumbo	20
Extra large	16
Large	12
Medium	8
Small	6
Personal	4

2. What pattern do you see in the table's data?
   The first 4 sizes decrease by 4 slices each; the last 3 sizes decrease by 2 slices each.
3. A plain large pizza at Pizza Express costs \$13.75. A large pizza with one topping costs \$14.20. A 2-topping large pizza costs \$14.65, and a 3-topping large pizza costs \$15.10. If you want 4 toppings on your large pizza, it will cost you \$15.55. Use this data to complete the table at right.

Pizza Express Pizza Prices

Large Pizza	Price
Plain	\$13.75
1 Topping	\$14.20
2 Toppings	\$14.65
3 Toppings	\$15.10
4 Toppings	\$15.55

4. What pattern do you see in the table's data?
   The price increases by \$0.45 for each additional topping.
5. How much does each slice of a 1-topping large pizza from Pizza Express cost? Round your answer to the nearest hundredth of a dollar.
   \$1.18 per slice
6. You and three friends buy two large pizzas from Pizza Express. One pizza has pepperoni and onions, and one pizza is plain. If you equally share the total price, how much will you each pay? How many slices will you each get?
   \$7.10 each; 6 slices each

CHALLENGE 6-1

LESSON 6-1 **Challenge**
Liberty Logic

You can use tables and logic to organize information and solve problems. For example, you have some measurements for different parts of the Statue of Liberty, but you do not know which measurement goes with which part. To solve the problem, first organize all the possibilities in a logic table. Then use the clues to fill out the table.

Because each part has only one measurement, there can be only one **YES** in each row and column of your logic table.

Three measurements for parts of the statue's face are 30 inches, 36 inches, and 54 inches. Which of those measurements are for the width of her mouth, the length of her nose, and the width of each of her eyes?

Clue 1: Her mouth is wider than each of her eyes.
Clue 2: The length of her nose is greater than the width of her mouth.

	30 inches	36 inches	54 inches
Width of Mouth	NO	YES	NO
Length of Nose	NO	NO	YES
Width of each Eye	YES	NO	NO

Three measurements for the tablet she holds are 24 inches, 163 inches, and 283 inches. Use the clues and logic table below to find the length, width, and thickness of the Statue of Liberty tablet.

Clue 1: The tablet is longer than it is wide.
Clue 2: The tablet is less than 100 inches thick.

	24 inches	163 inches	283 inches
Length	NO	NO	YES
Width	NO	YES	NO
Thickness	YES	NO	NO

PROBLEM SOLVING 6-1

LESSON 6-1 **Problem Solving**
Make a Table

Complete each activity and answer each question.

1. In January, the normal temperature in Atlanta, Georgia, is 41°F. In February, the normal temperature in Atlanta is 45°F. In March, the normal temperature in Atlanta is 54°F, and in April, it is 62°F. Atlanta's normal temperature in May is 69°F. Use this data to complete the table at right.

Atlanta Normal Temperatures

Month	Temperature (°F)
January	41
February	45
March	54
April	62
May	69

2. In which month given does Atlanta have the highest temperature? the lowest?
   in May; in January
3. Use your table from Exercise 1 to find a pattern in the data and draw a conclusion about the temperature in June.
   Pattern: The normal temperature in Atlanta increases each month from January to May. Possible conclusion: Atlanta's normal temperature in June is higher than 69°F.
4. In what other ways could you organize Atlanta's temperature data in a table?
   Possible answers: by temperature from lowest to highest, or from highest to lowest

Circle the letter of the correct answer.

5. Which of these statements about Atlanta's temperature data from January to May is true?
   (A) It is always higher than 40°F.
   B It is always lower than 60°F.
   C It is hotter in March than in April.
   D It is cooler in February than in January.
6. Between which two months in Atlanta does the normal temperature change the most?
   F January and February
   (G) February and March
   H March and April
   J April and May

6-2 Range, Mean, Median, and Mode

Learn to find the range, mean, median, and mode of a data set.

Vocabulary
range
mean
median
mode

Players on a volleyball team measured how high they could jump. The results in inches are recorded in the table.

13	23	21	20	21	24	18

Some descriptions of a set of data are called the *range, mean, median,* and *mode.*

- The **range** is the difference between the least and greatest values in the set.
- The **mean** is the sum of all the items, divided by the number of items in the set. (The mean is sometimes called the *average.*)
- The **median** is the middle value when the data are in numerical order, or the mean of the two middle values if there are an even number of items.
- The **mode** is the value or values that occur most often. There may be more than one mode for a data set. When all values occur an equal number of times, the data set has no mode.

EXAMPLE 1 Finding the Range, Mean, Median, and Mode of a Data Set

Find the range, mean, median, and mode of each data set.

A

Heights of Vertical Jumps (in.)						
13	23	21	20	21	24	18

Start by writing the data in numerical order.

13, 18, 20, 21, 21, 23, 24

range: $24 - 13 = 11$ *Subtract least value from greatest value.*

mean: $13 + 18 + 20 + 21 + 21 + 23 + 24 = 140$ *Add all values.*

$140 \div 7 = 20$ *Divide the sum by the number of items.*

median: 21 *There are an odd number of items, so find the middle value.*

mode: 21 *21 occurs most often.*

The range is 11 in.; the mean is 20 in.; the median is 21 in.; and the mode is 21 in.

6-2 Organizer

Pacing: Traditional 1 day
Block $\frac{1}{2}$ day

Objective: Students find the range, mean, median, and mode of a data set.

Indiana Academic Standards
6.6.3, 6.7.1, 6.7.5, 6.7.11

Warm Up

Order the numbers from least to greatest.

1. 8, 6, 25, 7, 4, 12 **4, 6, 7, 8, 12, 25**

2. 60, 11, 27, 45, 32 **11, 27, 32, 45, 60**

Divide.

3. $720 \div 4$ **180** **4.** $760 \div 10$ **76**

Problem of the Day

Ms. Red, Ms. Blue, and Ms. Green attended a holiday party. They each wore a different-colored dress, red, blue, or green. Ms. Red said to the lady wearing the blue dress, "Did you notice that none of us is wearing a dress with the color that corresponds to our name?" Who wore which color dress?

Ms. Red: green; Ms. Blue: red; Ms. Green: blue

Available on Daily Transparency in CRB

1 Introduce

Alternate Opener

EXPLORATION

6-2 Range, Mean, Median, and Mode

The table shows an ordered list of all the times in the men's 1,500-meter speed-skating competition in the 2002 Winter Olympics. The fastest time was 1:43.95, or 1 minute 43.95 seconds. Notice that all of the times are only seconds apart.

1:43.95	1:44.57	1:45.26	1:45.34	1:45.41	1:45.51	1:45.63	1:45.82
1:45.86	1:45.97	1:45.98	1:46.00	1:46.04	1:46.29	1:46.38	1:46.40
1:46.75	1:46.99	1:47.04	1:47.21	1:47.26	1:47.63	1:47.64	1:47.72
1:47.78	1:47.83	1:48.02	1:48.13	1:48.20	1:48.27	1:48.40	1:48.57
1:48.58	1:48.76	1:49.24	1:49.42	1:49.45	1:49.50	1:49.57	1:50.15
1:50.26	1:50.70	1:51.02	1:51.02	1:51.81	1:52.01	1:52.87	

1. How many seconds behind the winner was the second-place skater? the third-place skater?
2. Find the range. (Subtract the fastest time from the slowest time.)
3. Find the median. (The median is the number in the middle of the data set.)
4. Find the mean. (The mean is the average of all the times.)

Think and Discuss

5. **Discuss** whether you think the range, median, or mean best represents the data.
6. **Explain** how the mean compares with the median.

Motivate

To give students a meaningful context for studying data, discuss how report card grades are figured. Explain that when teachers add a student's scores and divide by the number of scores, they are finding the *average*, or *mean*, score for that student. Tell students that *range*, *median*, and *mode* are other statistics that can be used to help describe data such as test scores.

Exploration worksheet and answers on Chapter 6 Resource Book pp. 16 and 100

2 Teach

Lesson Presentation

Guided Instruction

In this lesson, students learn to find the range, mean, median, and mode of a set of data. First teach them to find these values for a set containing an odd number of items, and then teach them to find the median for a set containing an even number of items. Be sure to discuss the three possibilities for mode: one, more than one, and none. Help students to understand that range, mean, median, and mode are ways to summarize or describe a data set.

Additional Examples

Example 1

Find the range, mean, median, and mode of each data set.

A. depths of puddles (in.)

5 8 3 5 4 2 1

range: 7 in.
mean: 4 in.
median: 4 in.
mode: 5 in.

B. number of points scored

96 75 84 7

range: 89 points
mean: 65.5 points
median: 79.5 points
mode: none

Find the range, mean, median, and mode of each data set.

B

NFL Career Touchdowns			
Marcus Allen	145	Franco Harris	100
Jim Brown	126	Walter Payton	125

Write the data in numerical order: 100, 125, 126, 145

range: $145 - 100 = 45$

mean: $\frac{145 + 126 + 100 + 125}{4}$

$= 124$

median: 100, (125, 126), 145

There are an even number of items, so find the mean of the two middle values.

$\frac{125 + 126}{2} = 125.5$

mode: none

The range is 45 touchdowns; the mean is 124 touchdowns; the median is 125.5 touchdowns; and there is no mode.

Think and Discuss

1. Describe what you can say about the values in a data set if the set has a small range.

2. Tell how many modes are in the following data set. Explain your answer. 15, 12, 13, 15, 12, 11

6-2 PRACTICE & ASSESS

6-2 Exercises

FOR EXTRA PRACTICE
see page 647

internet connect
Homework Help Online
go.hrw.com Keyword: MR4 6-2

Students may want to refer back to the lesson examples.

GUIDED PRACTICE

See Example 1 Find the range, mean, median, and mode of the data set.

1.

Heights of Students (in.)	51	67	63	52	49	48	48

range = 19, mean = 54, median = 51, mode = 48

INDEPENDENT PRACTICE

See Example 1 Find the range, mean, median, and mode of each data set.

2.

Ages of Students (yr)	14	16	15	17	16	12

2. range = 5, mean = 15, median = 15.5, mode = 16

Assignment Guide

If you finished Example 1 assign:
Core 1–5, 8–15
Enriched 1, 4–15

Teach

Reaching All Learners

Through Curriculum Integration

Have students gather data related to other subject areas and find the range, mean, median, and mode for the data. For example, science: heights and weights of a certain species of animal; social studies: sizes of states or countries; language arts: ages of famous authors.

3 Close

Summarize

Have students write brief definitions of new vocabulary in the lesson: *range*, *mean*, *median*, and *mode*.

Possible answer: For a given set of data, the range is the difference between the least and greatest values. The mean is the sum of all the items, divided by the number of items. The median is the middle value when the data are in numerical order. The mode is the value or values that occur most often.

Answers to Think and Discuss

1. Possible answer: If the set of data has a small range, then there is not very much difference between the values of the individual items.
2. Two modes; the values 12 and 15 both occur twice, which is more than any other values occur.

Find the range, mean, median, and mode of each data set.

3.

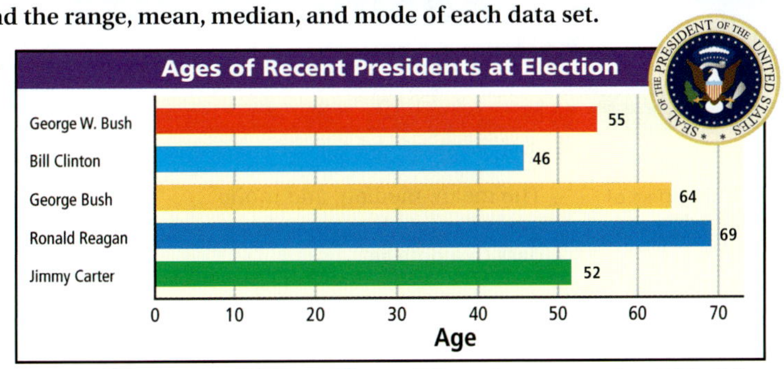

range = 23, mean = 57.2, median = 55, mode = no mode of this data set

PRACTICE AND PROBLEM SOLVING

4. Frank has 3 nickels, 5 dimes, and 2 quarters. Find the range, mean, median, and mode of the values of Frank's coins. range = 20 cents, mean = 11.5 cents, median = 10 cents, mode = 10 cents

5. ***EDUCATION*** For the six New England states, the mean scores on the math section of the SAT one year were as follows: Connecticut, 509; Maine, 500; Massachusetts, 513; New Hampshire, 519; Rhode Island, 500; and Vermont, 508. Create a table using this data. Then find the range, mean, median, and mode.

6. ***WHAT'S THE QUESTION?*** On an exam, three students scored 75, four students scored 82, three students scored 88, four students scored 93, and one student scored 99. If the answer is 88, what is the question?

7. ***CHALLENGE*** In the Super Bowls from 1997 to 2002, the winning team won by a mean of $12\frac{1}{6}$ points. By how many points did the Green Bay Packers win in 1997? 14

Year	Super Bowl Champion	Points Won By
2002	New England Patriots	3
2001	Baltimore Ravens	27
2000	St. Louis Rams	7
1999	Denver Broncos	15
1998	Denver Broncos	7
1997	Green Bay Packers	■

5. range = 19, mean = 508.2, median = 508.5, mode = 500

6. What was the median score on the exam?

Spiral Review

Order the fractions from least to greatest. (Lesson 4-6)

8. $\frac{3}{7}, \frac{5}{4}, \frac{2}{6}$ $\frac{2}{6}, \frac{3}{7}, \frac{5}{4}$

9. $\frac{2}{3}, \frac{4}{11}, \frac{5}{8}$ $\frac{4}{11}, \frac{5}{8}, \frac{2}{3}$

10. $\frac{3}{10}, \frac{3}{8}, \frac{1}{3}$ $\frac{3}{10}, \frac{1}{3}, \frac{3}{8}$

Multiply. Write your answers in simplest form. (Lesson 5-1)

11. $\frac{3}{5} \cdot \frac{6}{7}$ $\frac{18}{35}$

12. $\frac{2}{3} \cdot \frac{4}{5}$ $\frac{8}{15}$

13. $\frac{7}{9} \cdot \frac{3}{4}$ $\frac{7}{12}$

14. $\frac{1}{7} \cdot \frac{1}{2}$ $\frac{1}{14}$

15. **TEST PREP** Write the product in simplest form: $3\frac{6}{7} \cdot \frac{1}{3}$. (Lesson 5-2) B

A $1\frac{27}{21}$ B $1\frac{2}{7}$ C $1\frac{5}{7}$ D $1\frac{3}{4}$

Lesson Quiz

Use the following data set: 18, 20, 56, 47, 30, 18, 21.

1. Find the range. 38
2. Find the mean. 30
3. Find the median. 21
4. Find the mode. 18
5. Bonnie ran a mile in 8 minutes, 8 minutes, 7 minutes, 9 minutes, and 8 minutes. What was her mean time? 8 minutes

Available on Daily Transparency in CRB

Answers

5. Complete answer on p. A3

RETEACH 6-2

LESSON 6-2 **Reteach**

Range, Mean, Median, and Mode

You can find the range, mean, median, and mode to describe a set of data.

Terry's Test Scores	76	81	94	81	78

The **range** is the difference between the greatest and least values in the set of data.

94 − 76 = 18 Use subtraction to find the range.

The range is 18 points.

The **mean** or average is the sum of the items divided by the number of items.

76 + 81 + 94 + 81 + 78 = 410 First, find the sum of the values.

410 ÷ 5 = 82 Then divide the sum by the number of values in the set of data.

The mean is 82 points.

The **median** is the middle value of an ordered set of data. If there are two middle values, the median is the mean of those two values.

76, 78, **81**, 81, 94 Put the values in order first.

The median is 81 points.

The **mode** is the value that occurs most often in a set of data.

The mode is 81 points.

Find the range, mean, median, and mode of each set of values.

1. 23, 78, 45, 22
   range: 56
   mean: 42
   median: 34
   mode: no mode

2. 102, 79, 82, 103, 79
   range: 24
   mean: 89
   median: 82
   mode: 79

3. 56, 99, 112, 112, 56
   range: 56
   mean: 87
   median: 99
   mode: 56, 112

PRACTICE 6-2

LESSON 6-2 **Practice B**

Range, Mean, Median, and Mode

Find the range, mean, median, and mode of each data set.

1.

School Sit-Up Records (sit-ups per minute)	31	28	30	31	30

range: 3 sit-ups; mean: 30 sit-ups; median: 30 sit-ups; mode: 30 and 31 sit-ups

2.

Brian's Math Test Scores	86	90	93	85	79	92

range: 14; mean: 87.5; median: 88; mode: none

3.

Heights of Basketball Players (in.)	72	75	78	72	73

range: 6 in.; mean: 74 in.; median: 73 in.; mode: 72 in.

4.

Team Heart Rates (beats per min)	70	68	70	72	68	66

range: 6 bpm; mean: 69 bpm; median: 69 bpm; mode: 68 and 70 bpm

5.

Daily Winter Temperatures (°F)	45	50	47	52	53	45	51

range: 8°F; mean: 49°F; median: 50°F; mode: 45°F

6.

Daily Theater Ticket Sales	68	74	71	69	74	78	70

range: 10; mean: 72; median: 71; mode: 74

7. Anita has two sisters and three brothers. The mean of all their ages is 6 years. What will their mean age be 10 years from now? Twenty years from now?

16 years; 26 years

8. In a class of 28 sixth graders, all but one of the students are 12 years old. Which two data measurements are the same for the student's ages? What are those measurements?

the median and mode; 12 years

CHALLENGE 6-2

LESSON 6-2 **Challenge**

Speedy Data

Match each set of data with its description.

Mammal Speeds	
Antelope	54 mi/h
Cheetah	65 mi/h
Greyhound	42 mi/h
Horse	43 mi/h

C, H, M, N

Range Descriptions	
A Range	4 mi/h
B Range	29 mi/h
C Range	23 mi/h
D Range	6 mi/h

Insect Speeds	
Bumble bee	11 mi/h
Honey bee	7 mi/h
Hornet	13 mi/h
Horsefly	9 mi/h

D, G, J, N

Mean Descriptions	
E Mean	49 mi/h
F Mean	42 mi/h
G Mean	10 mi/h
H Mean	51 mi/h

Fish Speeds	
Bluefin tuna	47 mi/h
Bonefish	40 mi/h
Sailfish	69 mi/h
Swordfish	40 mi/h

B, E, L, O

Median Descriptions	
J Median	10 mi/h
K Median	42 mi/h
L Median	43.5 mi/h
M Median	48.5 mi/h

Bird Speeds	
Crane	42 mi/h
Goose	42 mi/h
Mallard	40 mi/h
Swan	44 mi/h

A, F, K, P

Mode Descriptions	
N Mode	none
O Mode	40 mi/h
P Mode	42 mi/h
Q Mode	44 mi/h

PROBLEM SOLVING 6-2

LESSON 6-2 **Problem Solving**

Range, Mean, Median, and Mode

Write the correct answer.

1. Use the table at right to find the range, mean, median, and mode of the data set.

range: 2 wins; mean: 4 wins; median: 4 wins; mode: 3 and 5 wins

World Series Winners

Team	Number of Wins
Baltimore Orioles	3
Boston Red Sox	5
Detroit Tigers	4
Minnesota Twins	3
Pittsburgh Pirates	5

2. When you use the data for only 2 of the teams in the table, the mean, median, and mode for the data are the same. Which teams are they?

Orioles and Twins or Pirates and Red Sox

3. The states that border the Gulf of Mexico are Alabama, Florida, Louisiana, Mississippi, and Texas. What are the range, mean, median, and mode for the number of letters in those states' names?

range: 6 letters; mean: 7.8 letters; median: 7 letters; mode: 7 letters

4. There are 5 whole numbers in a data set. The mean of the data is 10. The median and mode are both 9. The least number in the data set is 7, and the greatest is 14. What are the numbers in the data set?

7, 9, 9, 11, and 14

Circle the letter of the correct answer.

5. If the mean of two numbers is 2.5, what is true about the data?
   A Both numbers are greater than 5.
   B One of the numbers is less than 2.
   C One of the numbers is 2.5.
   (D) The sum of the data is not divisible by 2.

6. Tom wants to find the average height of the students in his class. Which measurement should he find?
   F the range
   (G) the mean
   H the median
   J the mode

6-3 Organizer

Pacing: Traditional 1 day
Block $\frac{1}{2}$ day

Objective: Students learn the effect of additional data and outliers.

Indiana Academic Standards
6.5.10, 6.6.3, 6.7.1, 6.7.5, 6.7.11

Warm Up

Use the numbers to answer the questions.

146, 161, 114, 178, 150, 134, 172, 131, 128

1. What is the greatest number? 178
2. What is the least number? 114
3. How can you find the median? Order the numbers and find the middle value.

Problem of the Day

Ms. Green has 6 red gloves and 10 blue gloves in a box. She closes her eyes and picks some gloves. What is the least number of gloves Ms. Green will have to pick to ensure 2 gloves of the same color? 3

Available on Daily Transparency in CRB

Math Humor

No one believed the baseball player who popped up for the fourteenth time in a row. He was an *out and outlier.*

6-3 Additional Data and Outliers

Learn the effect of additional data and outliers.

Vocabulary
outlier

The mean, median, and mode may change when you add data to a data set.

USA's Jim Shea in Men's Skeleton at the 2002 Winter Olympics

EXAMPLE 1 ***Sports Application***

A **Find the mean, median, and mode of the data in the table.**

U.S. Winter Olympic Medals Won								
Year	2002	1998	1994	1992	1988	1984	1980	1976
Medals	34	13	13	11	6	8	12	10

mean = 13.375 mode = 13 median = 11.5

B **The United States also won 8 medals in 1972 and 5 medals in 1968. Add this data to the data in the table and find the mean, median, and mode.**

mean = 12 modes = 8, 13 median = 10.5

The mean decreased by 1.375, there is an additional mode, and the median decreased by 1.

An **outlier** is a value in a set that is very different from the other values.

EXAMPLE 2 ***Social Studies Application***

In 2001, 64-year-old Sherman Bull became the oldest person to reach the top of Mount Everest. Other climbers to reach the summit that day were 33, 31, 31, 32, 33, and 28 years old. Find the mean, median, and mode without and with Bull's age.

Data without Bull's age: mean ≈ 31.3 modes = 31, 33 median = 31.5

Data with Bull's age: mean = 36 modes = 31, 33 median = 32

When you add Bull's age, the mean increases by 4.7, the modes stay the same, and the median increases by 0.5. The mean is the most affected by the outlier—notice that it is greater than every age in the set except Bull's. The median is closer to most of the climbers' ages.

Helpful Hint
Sherman Bull's age is an outlier because he is much older than the others in the group.

1 Introduce

Alternate Opener

EXPLORATION

6-3 Additional Data and Outliers

Casey Fitzrandolph won the men's 500-meter speed-skating competition in the 2002 Olympics with a time of 69.23 seconds. The table lists the top 32 times in the 500-meter race.

69.23	69.26	69.47	69.49	69.59	69.60	69.60	69.81
69.86	69.89	70.10	70.11	70.28	70.32	70.33	70.44
70.57	70.75	70.84	70.88	70.97	71.27	71.39	71.54
71.96	72.07	72.49	72.58	72.64	72.69	72.93	74.81

1. Find the range, which is the difference between the fastest and the slowest time.
2. Find the median, which is the number in the middle of the data set.
3. Find the mean with a calculator.
4. The table excluded three more times. These times are 108.46, 117.41, and 133.57. Calculate the range, median, and mean including these three additional times.

Think and Discuss

5. **Discuss** why the last three times in number 4 were excluded from the table above.
6. **Describe** how the additional three times affect the range, median, and mean.

Motivate

Give students the following list of CD prices. Have them describe the data set and tell you anything that they notice about the data set.

\$13, \$11, \$15, \$15, \$29, \$15, \$12, \$16

Possible answer: The mode is \$15. Most of the prices are from \$11 to \$16, but one is much higher, \$29.

Exploration worksheet and answers on Chapter 6 Resource Book pp. 25 and 102

2 Teach

Lesson Presentation

Guided Instruction

In this lesson, students learn the effect of additional data and outliers. First teach them how adding data to an existing set of data can change the mean, median, and mode. Then teach students how an outlier can affect the mean, median, and mode of a data set. Have them determine which measure of central tendency best describes a given data set.

Sometimes one or two data values can greatly affect the mean, median, or mode. When one of these values is affected like this, you should choose a different value to best describe the data set.

EXAMPLE 3 **Describing a Data Set**

The Seawells are shopping for a DVD player. They found ten DVD players with the following prices:

\$175, \$180, \$130, \$150, \$180, \$500, \$160, \$180, \$150, \$160

What are the mean, median, and mode of this data set? Which one best describes the data set?

mean = \$196.50 mode = \$180 median = \$167.50

The median price is the best description of the prices. Most of the DVD players cost *about* \$167.50.

The mean is higher than most of the prices because of the \$500 player, and the mode is higher because of the three players that cost \$180 each.

Some data sets do not contain numbers. For example, the circle graph shows the results of a survey to find people's favorite color.

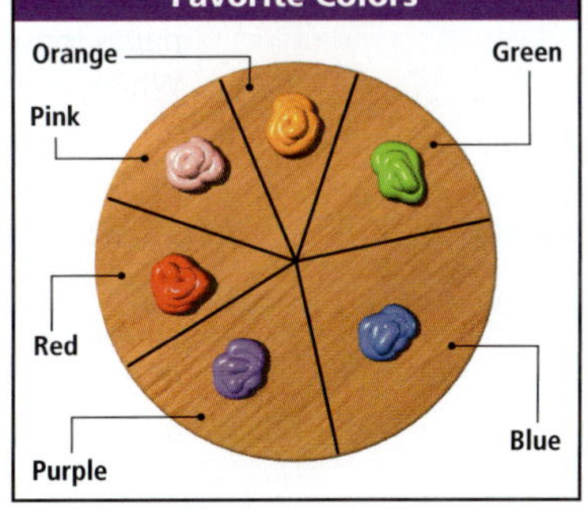

When it does not contain numbers, the only way to describe the data set is with the mode. You cannot find a mean or a median for a set of colors.

The mode for this data set is blue. Most people in this survey chose blue as their favorite color.

Think and Discuss

1. **Explain** how an outlier with a large value will affect the mean of a data set. What is the effect of a small outlier value?
2. **Explain** why the mean would not be a good description of the following high temperatures that occurred over 7 days: 72°F, 73°F, 70°F, 68°F, 70°F, 71°F, and 39°F.
3. **Give an example** of a data set that could be described only by its mode.

Additional Examples

Example 1

A. Find the mean, median, and mode of the data in the table.

EMS Football Games Won					
Year	'98	'99	'00	'01	'02
Games	11	5	7	5	7

mean = 7; modes = 5, 7; median = 7

B. EMS also won 13 games in 1997 and 8 games in 1996. Add this data to the data in the table and find the mean, median, and mode.

mean = 8; modes = 5, 7; median = 7

Example 2

Ms. Gray is 25 years old. She took a class with students who were 55, 52, 59, 61, 63, and 58 years old. Find the mean, median, and mode with and without Ms. Gray's age.

with:

mean ≈ 53.3; no mode; median = 58

without:

mean = 58; no mode; median = 58.5

Example 3

The Yorks found 8 pairs of skates with the following prices: \$35, \$42, \$75, \$40, \$47, \$34, \$45, and \$40. What are the mean, median, and mode of this data set? Which statistic best describes the data set?

mean = \$44.75 mode = \$40

median = \$41

The median price is the best description of the prices.

3 Close

Summarize

Have the students explain what an *outlier* is and how it can affect a set of data.

Possible answer: An outlier is a value in a set that is very different from the other values. It can affect the mean of a set of data by making it much higher or lower than it would have been without the outlier.

Reaching All Learners

Through Grouping Strategies

Have groups do research (for example, in newspapers and magazines) to find data sets with outliers. You may want to choose a particular data set to save time. Have students find the mean, median, and mode of the data and determine which is the best description of the data set. Then ask students to find the mean, median, and mode without including the outliers.

Answers to Think and Discuss

1. Possible answers: The mean will be greater than it would have been without the outlier with a large value. An outlier with a small value will make the mean less than it would have been without the outlier.
2. Possible answer: because with the outlier of 39° the mean will be too low
3. Possible answer: results of a survey to find people's favorite singer or any data set with nonnumeric data

6-3 PRACTICE & ASSESS

Students may want to refer back to the lesson examples.

Assignment Guide

If you finished Example 1 assign:
Core 1, 4, 7, 12–20
Enriched 4, 7, 9, 11–20

If you finished Example 2 assign:
Core 1–2, 4–5, 7, 12–20
Enriched 4–5, 7–9, 11–20

If you finished Example 3 assign:
Core 1–7, 12–20
Enriched 5–20

Answers

1. a. mean = 4.75, median = 5, no mode
   b. mean = 10, median = 7, no mode
2. with: mean = 45.4, median = 42, no mode; without: mean = 40.2, median = 40, no mode
5. with: mean = 710.4, median = 788, no mode; without: mean = 877.75, median = 868, no mode
6. mean = 22.7, median = 18, mode = 32; mean

6-3 Exercises

FOR EXTRA PRACTICE see page 647

internet connect
Homework Help Online
go.hrw.com Keyword: MR4 6-3

GUIDED PRACTICE

See Example 1

1. The graph shows how many times some countries have won the Davis Cup in tennis from 1900 to 2000.
   a. Find the mean, median, and mode of the data.
   b. The United States won 31 Davis Cups between 1900 and 2000. Add this number to the data in the graph and find the mean, median, and mode.

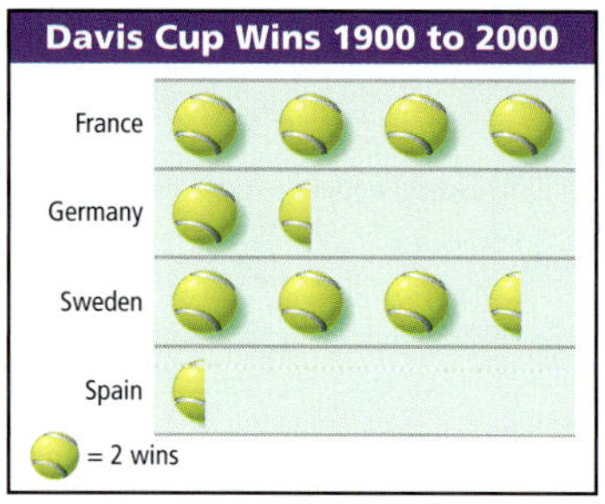

See Example 2

2. In 1998, 77-year-old John Glenn became the oldest person to travel into space. Other astronauts traveling on that same mission were 43, 37, 38, 46, 35, and 42 years old. Find the mean, median, and mode of all their ages with and without Glenn's age.

See Example 3

3. Kate read books that were 240, 450, 180, 160, 195, 170, 240, and 165 pages long. What are the mean, median, and mode of this data set? Which one best describes the data set? mean = 225, median = 187.5, mode = 240; median

INDEPENDENT PRACTICE

See Example 1

4. The graph shows the ages of the 10 youngest signers of the Declaration of Independence.
   a. Find the mean, median, and mode of the data. mean ≈ 30.9, median = 32, mode = 33
   b. Benjamin Franklin was 70 years old when he signed the Declaration of Independence. Add his age to the data in the graph and find the mean, median, and mode. mean ≈ 34.5, median = 33, mode = 33

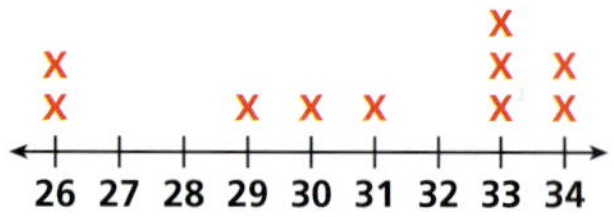

Ages of 10 Youngest Signers of Declaration of Independence

See Example 2

5. The map shows the population densities of several states along the Atlantic coast. Find the mean, median, and mode of the data with and without Maine's population density.

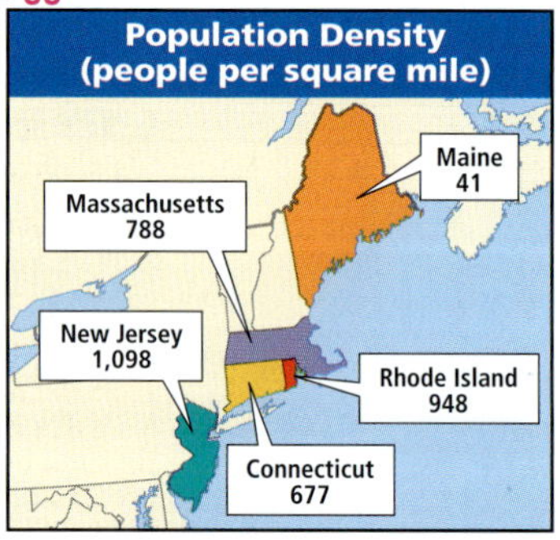

See Example 3

6. The passengers in a van are 16, 19, 17, 18, 15, 14, 32, 32, and 41 years old. What are the mean, median, and mode of this data set? Which one best describes the data set?

Math Background

A group of five workers with annual incomes of $22,000, $35,000, $40,000, $48,000, and $60,000 have a mean income of $41,000 and a median income of $40,000. The inclusion of an individual with annual income of $1,500,000 increases the median only to $44,000 but inflates the mean to over $280,000.

Income data presented by the United States Census Bureau is based upon a much larger data set than six individuals. Nevertheless, incomes that are outliers significantly skew the mean. Therefore the median income is the standard *average* used to report income in the United States.

RETEACH 6-3

LESSON 6-3 **Reteach**
Additional Data and Outliers

An **outlier** is a value in a set of data that is much greater or much less than the other values.

Number of Minutes Spent on Homework

Mon	Tue	Wed	Thurs	Fri
47	42	45	46	10

The outlier is 10 minutes, because it is much less than the other values in the set.

An outlier may affect the mean, median, or mode.

Data without Friday's value: mean = 45 median = 45.5 no mode
Data with Friday's value: mean = 38 median = 45 no mode

When Friday's value is included, the mean decreases by 7 minutes, the median decreases by 0.5 minutes, and the mode stays the same. The mean is most affected by the outlier because it is less than every value except for the outlier itself.

Find the mean, median, and mode for the set of data with and without the outlier.

1. 22, 25, 48, 26, 21, 27, 26, 29

   With outlier: mean: 28; median: 26; mode: 26

   Without outlier: mean: 25.1; median: 26; mode: 26

When an outlier affects the mean, median, or mode, choose a value that best describes the data.

In the example above, the median best describes the data because 45 minutes is closer to most of the data values in the set.

Find the mean, median, and mode. Then decide which best describes the set of data.

2. 16, 12, 14, 17, 81, 18, 13, 19, 14, 19

   mean: 22.3; median: 16.5; mode: 14 and 19. The median best describes the data.

PRACTICE 6-3

LESSON 6-3 **Practice B**
Additional Data and Outliers

Use the table to answer Exercises 1–2.

Population Densities

State	People (per mi²)
Idaho	16
Nevada	18
New Mexico	15
North Dakota	9
South Dakota	10

1. The table shows population data for some of the least-crowded states. Find the mean, median, and mode of the data.

   mean: 13.6; median: 15; mode: none

2. Alaska has the lowest population density of any state. Only about 1 person per square mile lives there. Add this number to the data in the table and find the mean, median, and mode.

   mean: 11.5; median: 12.5; mode: none

Use the table to answer Exercises 3–4.

State Counties

State	Number of Counties
Illinois	102
Iowa	99
North Carolina	100
Tennessee	95
Virginia	95

3. The table shows some of the states with the most counties. Find the mean, median, and mode of the data.

   mean: 98.2; median: 99; mode: 95

4. With 254 counties, Texas has more counties than any other state. Add this number to the data in the table and find the mean, median, and mode.

   mean: $124.1\overline{6}$; median: 99.5; mode: 95

5. In Exercise 1, which measurement best describes the data? Why is Alaska's population density an outlier for that data set?

   the mean; because it is much lower than the other data

6. In Exercise 4, why is the number of counties in Texas an outlier for the data set? Which measurement best describes the data set with Texas included?

   because Texas has many more counties; the median

Earth Science LINK

On September 13, 1922, the temperature in El Azizia, Libya, reached 136°F, the record high for the planet. (*Source: The World Almanac and Book of Facts*)

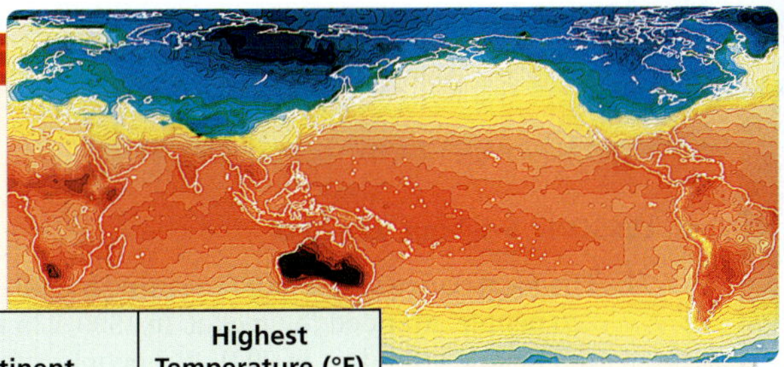

This satellite map shows the world's surface temperature. The dark blue areas are coldest, and the deep red areas are hottest.

Continent	Highest Temperature (°F)
Africa	136
Antarctica	59
Asia	129
Australia	128
Europe	122
North America	134
South America	120

go.hrw.com
KEYWORD: MR4 Heat
CNN Student News

7. What are the mean, median, and mode of the highest recorded temperatures on each continent? mean ≈ 118.29, median = 128, no mode

8. a. Which temperature is an outlier? 59°

   b. What are the mean, median, and mode of the temperatures if the outlier is not included? mean ≈ 128.17, median = 128.5, no mode

9. **WHAT'S THE ERROR?** A student stated that the median temperature would rise to 120.6°F if a new record high of 75°F were recorded in Antarctica. Explain the error. How would the median temperature actually be affected if a high of 75°F were recorded in Antarctica? The mean of the data would rise to 120.6°F. The median would remain unchanged.

10. **WRITE ABOUT IT** Is the data in the table best described by the mean, median, or mode? Explain. Possible answer: median; Antarctica's temperature of 59°F lowers the mean.

11. **CHALLENGE** Suppose a new high temperature were recorded in Europe, and the new mean temperature became 120°F. What is Europe's new high temperature? 134°F

Spiral Review

List all the factors of each number. (Lesson 4-2)

12. 57 1, 3, 19, 57
13. 36 1, 2, 3, 4, 6, 9, 12, 18, 36
14. 54 1, 2, 3, 6, 9, 18, 27, 54

Find the GCF of each set of numbers. (Lesson 4-3)

15. 6 and 15 3
16. 18 and 56 2
17. 12, 16, and 32 4
18. 24, 63, and 81 3

19. **TEST PREP** What is the least common multiple of 4, 12, and 15? (Lesson 5-5) B

    A 30　B 60　C 45　D 90

20. **TEST PREP** Over 5 days, Pedro jogged 6 mi, 5 mi, 2 mi, 2 mi, and 4 mi. Find the mean distance that Pedro jogged. (Lesson 6-2) H

    F 4 mi　G 2 mi　H 3.8 mi　J 4.75 mi

Interdisciplinary LINK

Earth Science

Exercises 7–11 require students to use the table of record high temperatures of continents in the world. Students study weather in middle school Earth science programs such as *Holt Science and Technology*.

Journal

Have students give an example of a data set with an outlier. Have them find the mean, median, and mode of the data without and with the outlier and explain how the outlier changes these measures of central tendency.

Test Prep Doctor

For Exercise 19, students must find the least multiple that is common to all three numbers (4, 12, and 15). Choices **A, C,** and **D** can be eliminated because the numbers given are multiples of 15 only.

CHALLENGE 6-3

LESSON 6-3 **Challenge**
Outer Space Outlier

You have been chosen to train as an astronaut! The statisticians at NASA are not happy, because you are a major outlier for their data. Use the information below to find how you will affect their astronaut data. Some answers depend on student ages. Sample answers are given for age 12.

Youngest Astronauts

In 1970, Russian astronaut Gherman S. Titov became the youngest person to travel into space. He was 25 years old at liftoff. The ages of some of the other youngest astronauts of all time were 26, 29, 28, 27, 26, and 28.

Data Without Your Age:	Data With Your Age:
Mean age: 27	Mean age: 25.125
Median age: 27	Median age: 26.5
Mode age: 26 and 28	Mode age: 26 and 28

Oldest Astronauts

In 1998, American astronaut John H. Glenn became the oldest person to travel into space. He was 77 years old at liftoff. The ages of some of the other oldest astronauts of all time were 54, 59, 61, 56, 58, and 55.

Data Without Your Age:	Data With Your Age:
Mean age: 60	Mean age: 54
Median age: 58	Median age: 57
Mode age: no mode	Mode age: no mode

PROBLEM SOLVING 6-3

LESSON 6-3 **Problem Solving**
Additional Data and Outliers

Use the table to answer the questions.

Successful Films in the U.S.

Film	U.S. Earnings for first release (million $)
E.T. the Extra-Terrestrial	400
Forrest Gump	330
Independence Day	305
Jurassic Park	357
The Lion King	313

1. Find the mean, median, and mode of the earnings data.
   mean: $341 million; median: $330 million; mode: none

2. *Titanic* earned more money in the United States than any other film—a total of $600 million! Add this figure to the data and find the mean, median, and mode. Round your answer for the mean to the nearest whole million.
   mean: $384 million; median: $343.5 million; mode: none

3. A person in Canada watches an average of 74 minutes of television each day. In Germany, a person watches an average of 68 minutes of TV a day. In France, people watch TV an average of 67 minutes a day. In Spain, the average time watching TV is 91 minutes a day, and in Ireland it is 74 minutes a day. Find the mean, median, and mode of the data.
   mean: 74.8 minutes; median: 74 minutes; mode: 74 minutes

4. People in the United States watch more television than in any other country. Americans watch an average of 118 minutes a day! Add this number to the data and find the mean, median, and mode.
   mean: 82 minutes; median: 74 minutes; mode: 74 minutes

Circle the letter of the correct answer.

5. In Exercise 2, which data measurement changed the least with the addition of *Titanic*'s earnings?
   A the range
   B the mean
   (C) the median
   D the upper extreme

6. In Exercise 4, which measurements best describe the data?
   F mean and median
   G range and mean
   (H) median and mode
   J range and mode

Lesson Quiz

At the college bookstore, your brother buys 6 textbooks at the following prices: $21, $58, $68, $125, $36, and $140.

1. Find the mean. $74.67
2. Find the median. $63
3. Find the mode. none
4. Your brother signs up for an additional class, and the textbook costs $225. Recalculate the mean, including the extra book. $96.14

Available on Daily Transparency in CRB

Chapter 6 Mid-Chapter Quiz

Purpose: *To assess students' mastery of concepts and skills in Lessons 6-1 through 6-3*

Assessment Resources

Section 6A Quiz

Assessment Resources .... p. 16

***Test and Practice Generator* CD-ROM**

Additional mid-chapter assessment items in both multiple-choice and free-response format may be generated for any objective in Lessons 6-1 through 6-3.

Answers

1.

Year	Attendance
1998	220
1999	235
2000	250
2001	242
2002	258

Chapter 6 Mid-Chapter Quiz

Mid-Chapter Quiz

LESSON 6-1 (pp. 272–274)

1. The local dance studio holds a spring recital each year. In 1998, 220 people attended the recital. In 1999, 235 people attended. In 2000, 250 people attended. In 2001, 242 people attended. In 2002, 258 people attended. Use the attendance data to make a table. Then use your table to describe how attendance changed over time.
   The attendance increased from 1998 to 2000, decreased in 2001, and increased again in 2002.

LESSON 6-2 (pp. 275–277)

Find the range, mean, median, and mode of each data set.

2.

Distance (mi)					
5	6	4	7	3	5

range: 4, mean: 5, median: 5, mode: 5

3.

Test Scores				
78	80	86	92	90

range: 14, mean: 85.2, median: 86, no mode

4.

Ages of Students (yr)							
11	13	12	12	12	13	9	14

range: 5, mean: 12, median: 12, mode: 12

5.

Number of Pages in Each Book						
145	119	156	158	125	128	135

range: 39, mean: 138, median: 135, no mode

LESSON 6-3 (pp. 278–281)

6. The table shows the number of people who attended each monthly meeting from January to May.

Number of People Attending				
Jan	Feb	Mar	Apr	May
27	26	32	30	30

   a. Find the mean, median, and mode of the attendances. mean: 29; median: 30; mode: 30
   b. In June, 39 people attended the meeting, and in July, 26 people attended the meeting. Add this data to the table and find the mean, median, and mode with the new data. mean: 30, median: 30, modes: 26 and 30

7. The four states with the longest coastlines are Alaska, Florida, California, and Hawaii. Alaska's coastline is 6,640 miles. Florida's coastline is 1,350 miles. California's coastline is 840 miles, and Hawaii's coastline is 750 miles. Find the mean, median, and mode of the lengths with and without Alaska's. with Alaska—mean: 2,395, median: 1,095, no mode; without Alaska—mean: 980, median: 840, no mode

8. The daily snowfall amounts for the first ten days of December are listed below.

   2 in., 5 in., 0 in., 0 in., 15 in., 1 in., 0 in., 3 in., 1 in., 4 in.

   What are the mean, median, and mode of this data set? Which one best describes the data set? mean: 3.1 in., median: 1.5 in., mode: 0 in; median

Focus on Problem Solving

Make a Plan

• Prioritize and sequence information

Some problems give you a lot of information. Read the entire problem carefully to be sure you understand all of the facts. You may need to read it over several times, perhaps aloud so that you can hear yourself say the words.

Then decide which information is most important (prioritize). Is there any information that is absolutely necessary to solve the problem? This information is important.

Finally, put the information in order (sequence). Use comparison words like *before, after, longer, shorter,* and so on to help you. Write the sequence down before you try to solve the problem.

Read the problems below and answer the questions that follow.

1. The compact disc (CD) was invented 273 years after the piano. The tape recorder was invented in 1898. Thomas Edison invented the phonograph 21 years before the tape recorder and 95 years before the compact disc. What is the date of each invention?

   a. Which invention's date can you use to find the dates of all the others?

   b. Can you solve the problem without this date? Explain.

   c. List the inventions in order from earliest invention to latest invention.

2. Jon recorded the heights of his family members. There are 4 people in Jon's family, including Jon. Jon's mother is 2 inches taller than Jon's father. Jon is 56 inches tall. Jon's sister is 4 inches taller than Jon and 5 inches shorter than Jon's father. What are the heights of Jon and his family members?

   a. Whose height can you use to find the heights of all the others?

   b. Can you solve the problem without this height? Explain.

   c. List Jon's family members in order from shortest to tallest.

?

1898

?

?

Answers

1. piano: 1699
phonograph: 1877
tape recorder: 1898
CD: 1972

2. Jon: 56 in.
Jon's sister: 60 in.
Jon's father: 65 in.
Jon's mother: 67 in.

Focus on Problem Solving

Purpose: *To focus on making a plan to solve the problem by prioritizing and sequencing the information in the problem*

Problem Solving Resources

Interactive Problem Solving pp. 44–52

Math: Reading and Writing in the Content Area . . . . pp. 44–52

Problem Solving Process

This page focuses on the second step of the problem-solving process:
Make a Plan

Discuss

Have students discuss their answers for parts **a–c** for each problem and then discuss the strategies they used to order the information.

Possible answers:

1. a. the tape recorder's invention date

b. If the invention date of any of the other inventions were given instead, you could still find the other dates, but without at least one date, the problem could not be solved.

c. piano, phonograph, tape recorder, CD

Possible strategy:

Use a timeline to position the inventions relative to each other. Label the time elapsed between inventions. Use the fact that the tape recorder was invented in 1898 to find the other information.

2. a. Jon's height

b. If the height of any one of the family members were given instead, you could find the heights of all of the family members. But without at least one height, the problem could not be solved.

c. Jon, Jon's sister, Jon's father, Jon's mother

Possible strategy:

Position the family members' heights relative to each other on a vertical number line, using tick marks to show inches. Use Jon's height to find the heights of the other family members.

Section 6B

Displaying and Interpreting Data

One-Minute Section Planner

Lesson	Materials	Resources
Lesson 6-4 Bar Graphs **NCTM:** Data Analysis and Probability, Problem Solving, Communication, Connections, Representation **NAEP:** Data Analysis and Probability 1b ☑ SAT-9 ☑ SAT-10 ☑ ITBS ☑ CTBS ☑ MAT ☑ CAT	**Optional** Graph paper *(CRB, p. 97)* Connecting cubes *(MK)*	• *Chapter 6 Resource Book*, pp. 34–43 • Daily Transparency T11, CRB • Additional Examples Transparencies T12–T15, CRB • *Alternate Openers: Explorations*, p. 47
Technology Lab 6A Create Bar Graphs **NCTM:** Data Analysis and Probability, Representation **NAEP:** Data Analysis and Probability 1b ☐ SAT-9 ☐ SAT-10 ☐ ITBS ☐ CTBS ☐ MAT ☐ CAT	**Required** Spreadsheet software	• *Technology Lab Activities,* pp. 23–24
Lesson 6-5 Frequency Tables and Histograms **NCTM:** Data Analysis and Probability, Communication, Representation **NAEP:** Data Analysis and Probability 1b ☐ SAT-9 ☑ SAT-10 ☐ ITBS ☑ CTBS ☑ MAT ☐ CAT	**Optional** Graph paper *(CRB, p. 97)*	• *Chapter 6 Resource Book*, pp. 44–53 • Daily Transparency T16, CRB • Additional Examples Transparencies T17–T20, CRB • *Alternate Openers: Explorations*, p. 48
Lesson 6-6 Ordered Pairs **NCTM:** Number and Operations, Geometry, Communication, Connections **NAEP:** Algebra 2c ☑ SAT-9 ☑ SAT-10 ☐ ITBS ☑ CTBS ☑ MAT ☐ CAT	**Required** Graph paper *(CRB, p. 97)* **Optional** Road maps	• *Chapter 6 Resource Book*, pp. 54–62 • Daily Transparency T21, CRB • Additional Examples Transparencies T22–T23, CRB • *Alternate Openers: Explorations*, p. 49
Lesson 6-7 Line Graphs **NCTM:** Data Analysis and Probability, Communication, Connections, Representation **NAEP:** Data Analysis and Probability 1b ☑ SAT-9 ☑ SAT-10 ☐ ITBS ☐ CTBS ☑ MAT ☑ CAT	**Required** Graph paper *(CRB, p. 97)* **Optional** Line graphs from newspapers or magazines	• *Chapter 6 Resource Book*, pp. 63–72 • Daily Transparency T24, CRB • Additional Examples Transparencies T25–T27, CRB • *Alternate Openers: Explorations*, p. 50
Lesson 6-8 Misleading Graphs **NCTM:** Data Analysis and Probability, Reasoning and Proof, Communication, Representation **NAEP:** Data Analysis and Probability 1d ☐ SAT-9 ☑ SAT-10 ☐ ITBS ☐ CTBS ☐ MAT ☐ CAT	**Optional** Teaching Transparency T29 *(CRB)* Graph paper *(CRB, p. 97)*	• *Chapter 6 Resource Book*, pp. 73–81 • Daily Transparency T28, CRB • Additional Examples Transparencies T30–T32, CRB • *Alternate Openers: Explorations*, p. 51
Lesson 6-9 Stem-and-Leaf Plots **NCTM:** Data Analysis and Probability, Communication, Representation **NAEP:** Data Analysis and Probability 1b ☐ SAT-9 ☐ SAT-10 ☐ ITBS ☐ CTBS ☑ MAT ☐ CAT		• *Chapter 6 Resource Book*, pp. 82–90 • Daily Transparency T33, CRB • Additional Examples Transparencies T34–T35, CRB • *Alternate Openers: Explorations*, p. 52
Extension Box-and-Whisker Plots **NCTM:** Data Analysis and Probability, Representation **NAEP:** Data Analysis and Probability 1b ☐ SAT-9 ☐ SAT-10 ☐ ITBS ☐ CTBS ☐ MAT ☐ CAT		• Additional Examples Transparency T33, CRB
Section 6B Assessment		• Section 6B Quiz, AR p. 17 • *Test and Practice Generator* CD-ROM

SAT = *Stanford Achievement Tests* **ITBS** = *Iowa Test of Basic Skills* **CTBS** = *Comprehensive Test of Basic Skills/Terra Nova*
MAT = *Metropolitan Achievement Tests* **CAT** = *California Achievement Test*

NCTM = Complete standards can be found on pages T27–T33. **NAEP** = Complete standards can be found on pages A31–A35.

SE = *Student Edition* **TE** = *Teacher's Edition* **AR** = *Assessment Resources* **CRB** = *Chapter Resource Book* **MK** = *Manipulatives Kit*

Section Overview

Bar Graphs, Frequency Tables, and Histograms — *Lessons 6-4 and 6-5*

Why? Data is easier to interpret when it is organized into a table or a graph.

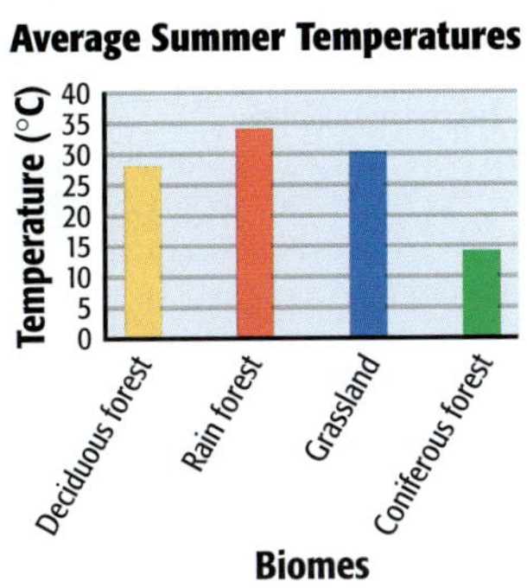

Use a **bar graph** to display countable data that are grouped in categories.

Data Item	Tally	Frequency	Cumulative Frequency
Item 1	|||	3	3
Item 2	||||	4	7
Item 3	||	2	9
Item 4	卌	5	14

A **frequency table** tells the number of times an event, category, or group occurs. The **cumulative frequency** column shows a running total of all frequencies.

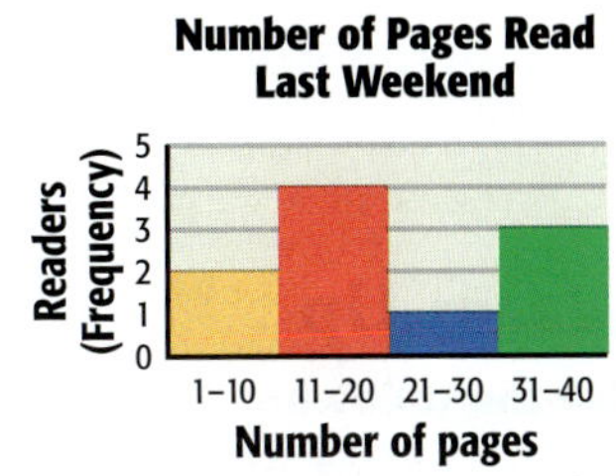

A **histogram** is a bar graph that shows the number of data items that occur within each interval. The bars of a histogram should touch but not overlap.

Line Graphs — *Technology Lab 6A, Lessons 6-6 through 6-8*

Why? On Wall Street, line graphs are used to keep track of stock prices.

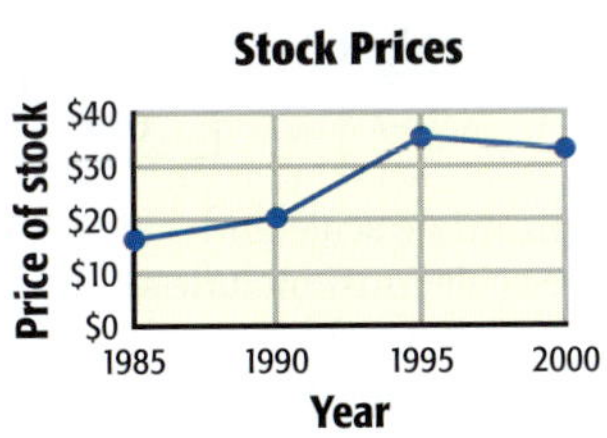

A **line graph** displays a set of data, using line segments. Data that shows change over time is best displayed in a line graph.

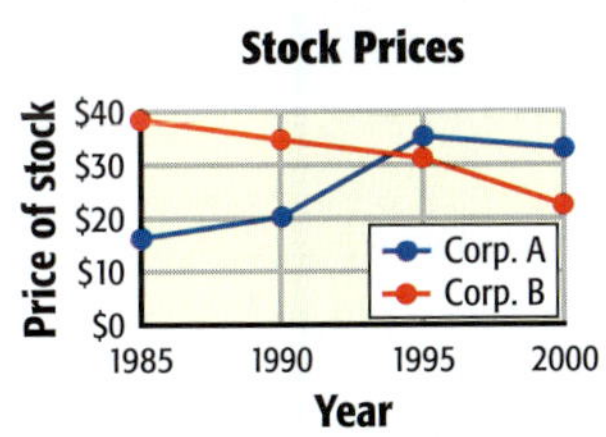

Line graphs that display two sets of data are called **double-line graphs**.

BEWARE! Misleading bar graphs or line graphs can be created by adjusting the scale of a graph to hide or emphasize a feature of the numerical data.

Stem-and-Leaf Plots — *Lesson 6-9*

Why? You can use a stem-and-leaf plot when you want to display data in an organized way that allows you to see each value.

Stems	Leaves
1	1 2 3
2	2 2 3 6 7
3	1 1 3 3 8 9
4	7 8
5	4 6 9

Key: 3|8 means 38

A **stem-and-leaf plot** arranges data by place value.

6-4 Organizer

Pacing: Traditional 1 day
Block $\frac{1}{2}$ day

Objective: Students display and analyze data in bar graphs.

Indiana Academic Standards
6.6.1, 6.7.1, 6.7.11

Warm Up

Use the following data set.

45 55 58 63 63 37 76 46 34

1. What is the mean of the data? 53
2. What is the median of the data? 55
3. What is the mode of the data? 63

Problem of the Day

The distance around the bases is 4 × 90 feet. How many runs does a baseball team need to score before the scoring base runners have covered a mile? (1 mile = 5,280 feet)
15 runs

Available on Daily Transparency in CRB

Math Fact

Even though it is sometimes used as a singular noun, the word *data* is actually the plural of the Latin word *datum*, which means "a gift".

6-4 Bar Graphs

Learn to display and analyze data in bar graphs.

Vocabulary
bar graph
double-bar graph

A biome is a large region characterized by a specific climate. There are ten land biomes on Earth. Some are pictured at right. Each gets a different amount of rainfall.

A *bar graph* can be used to display and compare data about rainfall. A **bar graph** displays data with vertical or horizontal bars.

EXAMPLE 1 Reading a Bar Graph

Use the bar graph to answer each question.

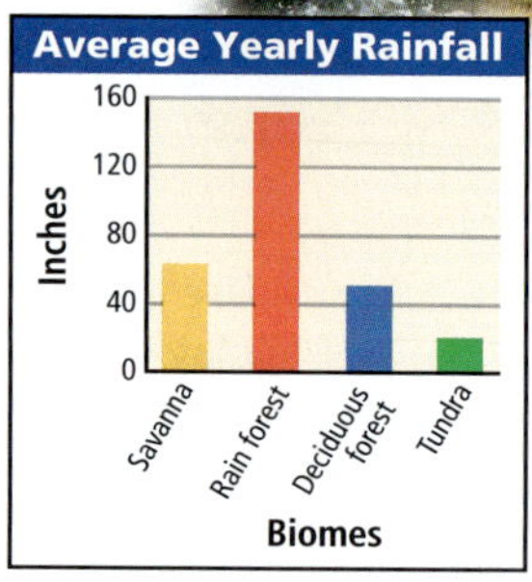

A **Which biome in the graph has the most rainfall?**

Find the highest bar.

The rain forest has the most rainfall.

B **Which biomes in the graph have an average yearly rainfall less than 40 inches?**

Find the bar or bars whose heights measure less than 40.

The tundra has an average yearly rainfall less than 40 inches.

EXAMPLE 2 Making a Bar Graph

Use the given data to make a bar graph.

Coal Reserves (billion metric tons)		
Asia	Europe	Africa
695	404	66

Step 1: Find an appropriate scale and interval. The scale must include all of the data values. The interval separates the scale into equal parts.

Step 2: Use the data to determine the lengths of the bars. Draw bars of equal width. The bars cannot touch.

Step 3: Title the graph and label the axes.

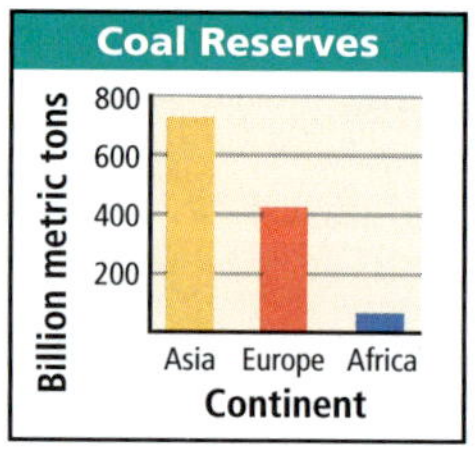

1 Introduce

Alternate Opener

EXPLORATION

6-4 Bar Graphs

The bar graph shows the medal totals of the four countries that won the most medals at the 2002 Winter Olympic Games.

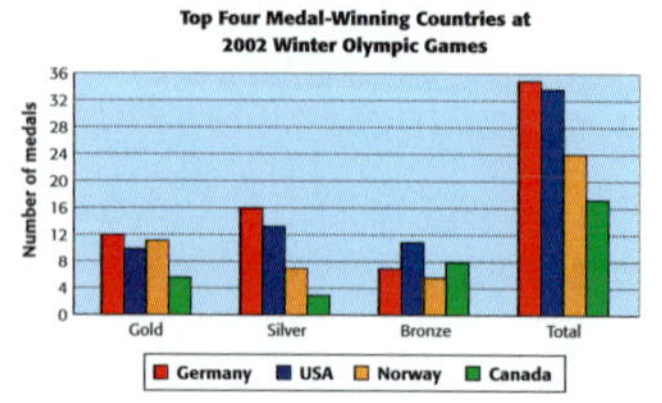

1. Name the country that won
   a. the most gold medals.
   b. the most silver medals.
   c. the most bronze medals.
   d. the most medals overall.
2. What is the total number of medals won by each country?

Think and Discuss

3. **Explain** a different way to display the data.
4. **Discuss** information that is not shown by the bar graph.

Motivate

Have students discuss places where they have seen graphs and what the graphs represent. Possible answers: newspapers, magazines, television; stock performance, survey results Explain that this lesson focuses on *bar graphs,* which display data with vertical or horizontal bars.

Exploration worksheet and answers on Chapter 6 Resource Book pp. 35 and 104

2 Teach

Lesson Presentation

Guided Instruction

In this lesson, students learn to display and analyze data in bar graphs. First teach students to read a bar graph. Then teach them to make bar graphs and double-bar graphs, given sets of data. Have students think of other examples of when a double-bar graph would be appropriate.

A **double-bar graph** shows two sets of related data.

EXAMPLE 3 PROBLEM SOLVING APPLICATION

PROBLEM SOLVING

Make a double-bar graph to compare the data in the table.

Life Expectancies in Atlantic South America				
	Brazil	Argentina	Uruguay	Paraguay
Male (yr)	59	71	73	70
Female (yr)	69	79	79	74

1 Understand the Problem

You are asked to use a graph to compare the data given in the table. You will need to use all of the information given.

2 Make a Plan

You can make a double-bar graph to display the two sets of data.

3 Solve

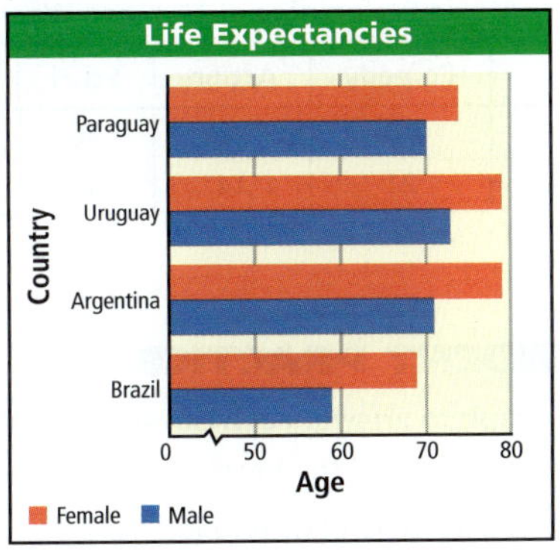

Determine appropriate scales for both sets of data.

Use the data to determine the lengths of the bars. Draw bars of equal width. Bars should be in pairs. Use a different color for male ages and female ages.

Title the graph and label both axes.

Include a key to show what each bar represents.

4 Look Back

You could make two separate graphs, one of male ages and one of female ages. However, it is easier to compare the two data sets when they are on the same graph.

Reading Math

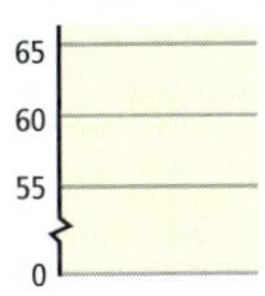

This symbol means there is a break in the scale. Some numbers were left out because they were not needed for the graph.

Think and Discuss

1. **Give** comparisons you can make by looking at a bar graph.
2. **Describe** the kind of data you would display in a bar graph.
3. **Tell** why the graph in Example 3 needs a key.

Additional Examples

Example 1

Use the bar graph to answer each question.

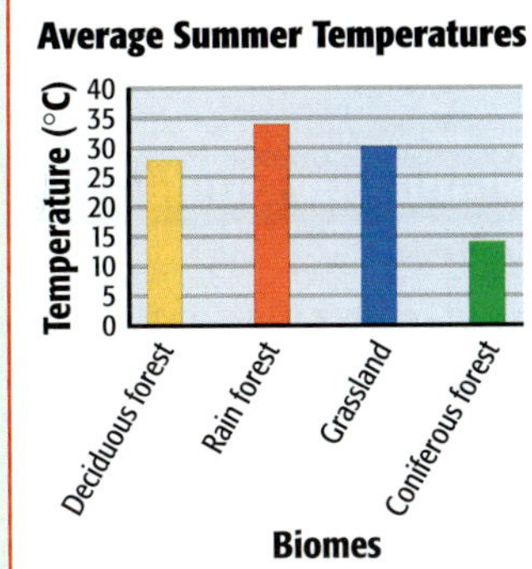

A. Which biome in the graph has the least average summer temperature?

B. Which biomes in the graph have an average summer temperature of 30°C or greater?

Example 2

Use the data to make a bar graph.

Magazine Subscriptions Sold		
Grade 6	Grade 7	Grade 8
258	597	374

Example 3

Make a bar graph to compare the data in the table.

Club Memberships			
Club	Art	Music	Science
Boys	12	6	16
Girls	8	14	4

Answers on Additional Examples Transparencies T12–T15 in CRB

3 Close

Reaching All Learners

Through Concrete Manipulatives

Some students may benefit from building double-bar graph bars with connecting cubes before drawing the graphs on paper. Give these students cubes in two colors, one for each set of data. Allow students to construct the bars for each data set separately and to then combine them to make the double-bar graph. Students can refer to the connecting cube model as they draw their graphs on paper. To represent large numbers, cubes could be 10's or 100's.

Summarize

Review the difference between a bar graph and a double-bar graph, and review the parts of each.

Answers to Think and Discuss

1. Possible answers: least and greatest value; values that are less than or greater than a given value
2. countable data that are grouped in categories
3. The key lets you know which color bar represents males and which color represents females.

6-4 PRACTICE & ASSESS

Students may want to refer back to the lesson examples.

Assignment Guide

If you finished Example 1 assign:
Core 1–2, 5, 9, 11, 16–23
Enriched 2, 6, 9–12, 16–23

If you finished Example 2 assign:
Core 1–3, 5–11 odd, 16–23
Enriched 2, 6–7, 9–23

If you finished Example 3 assign:
Core 1–11, 16–23
Enriched 2, 6–23

Answers

3.

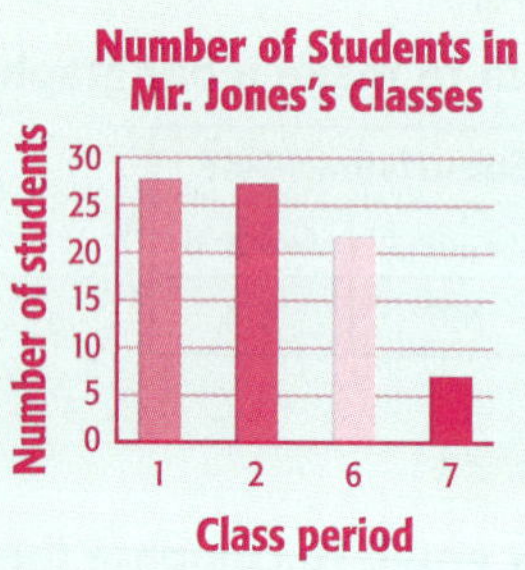

4, 7–8. See p. A3.

6-4 Exercises

FOR EXTRA PRACTICE
see page 648

internet connect
Homework Help Online
go.hrw.com Keyword: MR4 6-4

GUIDED PRACTICE

See Example 1

Use the bar graph to answer each question.

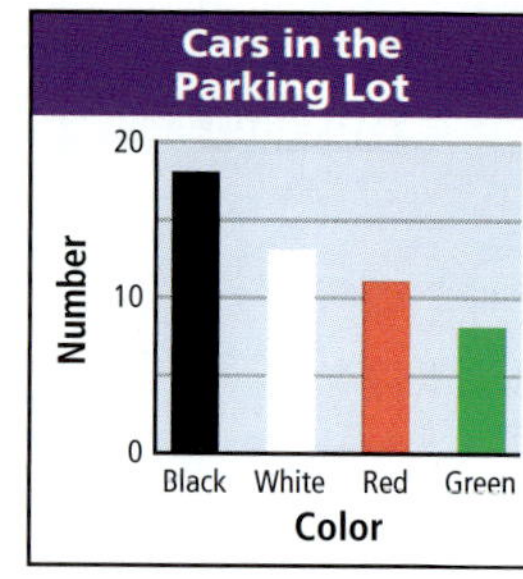

1. Which color was the least common among the cars in the parking lot? **green**

2. Which colors appeared more than ten times in the parking lot? **black, white, red**

See Example 2

3. Use the given data to make a bar graph.

Students in Mr. Jones's History Classes			
Period 1	28	Period 6	22
Period 2	27	Period 7	7

See Example 3

4. Make a bar graph to compare the data in the table.

Movie Preferences of Men and Women Polled at the Mall						
	Comedy	Action	Sci-Fi	Horror	Drama	Other
Men	16	27	16	23	12	6
Women	21	14	8	18	30	9

INDEPENDENT PRACTICE

See Example 1

Use the bar graph to answer each question.

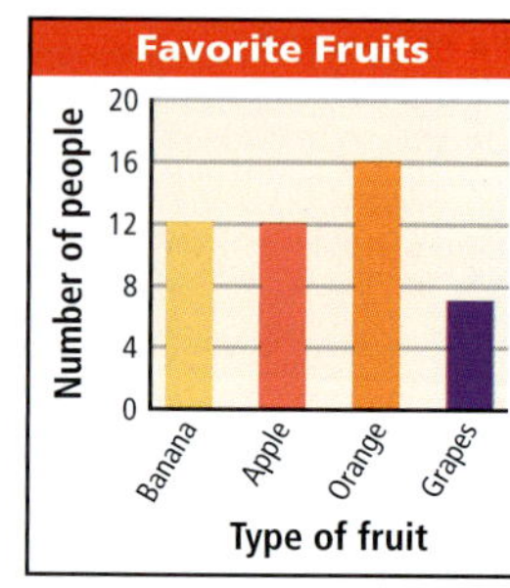

5. Which fruit was liked the best? **orange**

6. Which fruits were liked by equal numbers of people? **banana, apple**

See Example 2

7. Use the given data to make a bar graph.

Days with Rainfall			
January	14	March	16
February	12	April	23

See Example 3

8. Make a bar graph to compare the data in the table.

Heart Rates Before and After Exercise (beats per minute)						
	Jason	Jamal	Ray	Tonya	Peter	Brenda
Before	60	62	61	65	64	65
After	131	140	128	140	135	120

Math Background

Studying and constructing bar graphs is a way for students to begin their approach to ordered pairs and coordinate graphing. In a vertical bar graph, the location of a bar corresponds to the *x*-coordinate of a point, and the height of a bar corresponds to the *y*-coordinate of a point. This connection between data presentation and coordinate graphing will be carried one step further in Lesson 6-8, in which line graphs are studied.

RETEACH 6-4

LESSON 6-4 **Reteach**
Bar Graphs

You can make a bar graph to compare amounts.

To make a bar graph using the data in the table, first choose a scale that includes all of the data values. Next, separate the scale into equal parts, called intervals.

Then draw bars to match the data. The bars should be of equal width and should not touch. Give your graph a title and label its axes.

Annual Read-a-thon Totals			
Grade	6	7	8
Books Read	86	42	98

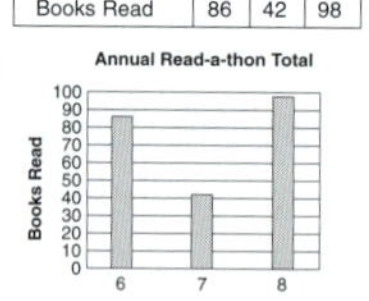

Use the data to make a bar graph.

1.

Canned Food Drive Totals			
Grade	6	7	8
Cans Collected	96	74	62

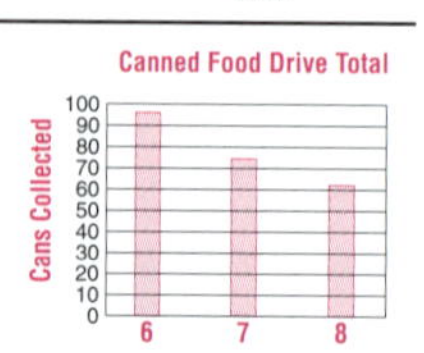

Look at the bar graph for the Read-a-thon above. Which grade read almost twice as many books as the seventh grade?

The bar for the sixth grade is about twice a long as the bar for the seventh grade. So the sixth grade read almost twice as many books as the seventh grade.

Use the bar graph you made in Exercise 1.

2. How many more items did the sixth grade collect than the eighth grade?

The sixth grade collected 34 more items.

PRACTICE 6-4

LESSON 6-4 **Practice B**
Bar Graphs

Use the bar graph to answer each question.

1. In which country did people spend the most money on toys in 2000?

the United States

2. In which two countries did people spend the same amount of money on toys in 2000? How much did they each spend?

France and Germany; $3 million each

3. In which country did people spend $9 million on toys in 2000?

Japan

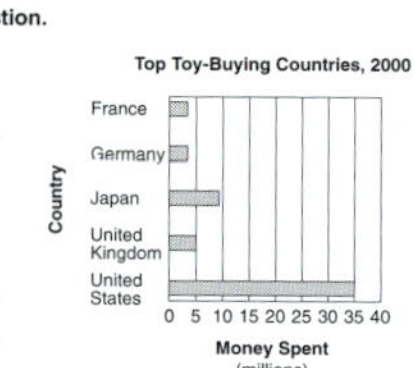

Make a double-bar graph to compare the data in the table.

Female Groups with the Most Top 10 and Top 20 Hits

Top 10		Top 20	
The Supremes	20	The Supremes	24
The Pointer Sisters	7	The Pointer Sisters	13
TLC	9	TLC	11
En Vogue	5	En Vogue	7
Spice Girls	4	Spice Girls	7

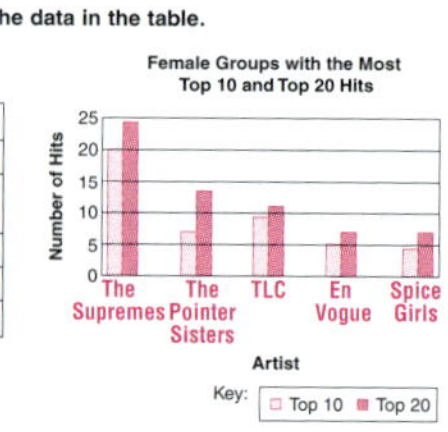

PRACTICE AND PROBLEM SOLVING

Use the bar graph for Exercises 9–12.

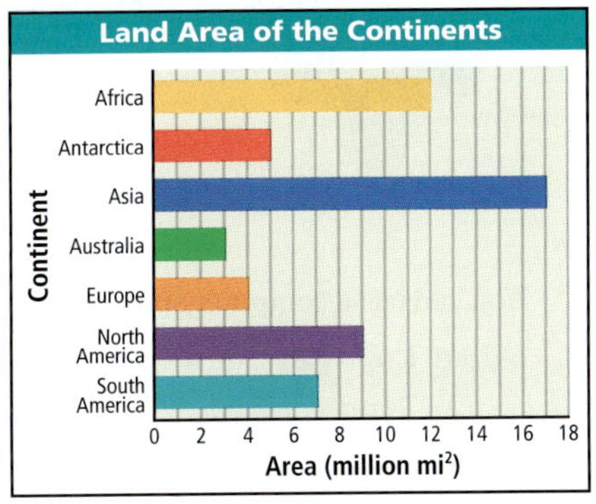

9. What is the range of the land area of the continents? **14 million mi²**
10. What is the mode of the land area of the continents? **no mode**
11. What is the mean of the land area of the continents? **≈ 8.14 million mi²**
12. What is the median of the land area of the continents? **7 million mi²**

13. ***CHOOSE A STRATEGY*** The heights of Maria, Glenn, Carol, and Luis are shown in the graph, but the labels are missing.
    - Maria is neither the tallest nor the shortest.
    - Glenn is taller than Carol.
    - There is only one person taller than Luis.

    Which student's name should go with each bar in the graph?
    From left to right: Maria, Carol, Glenn, Luis

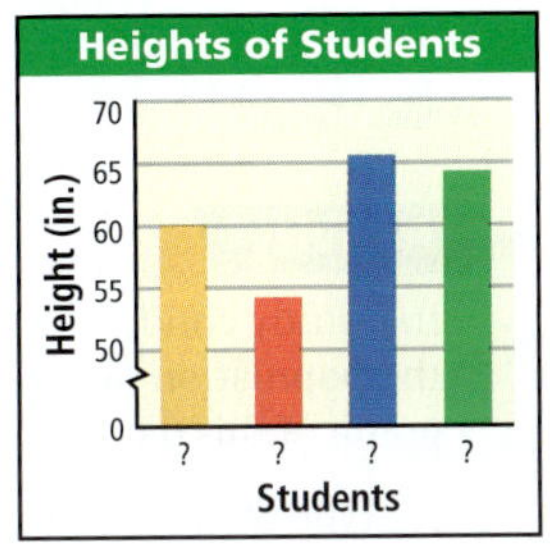

14. ***WRITE ABOUT IT*** Explain how you would make a bar graph of the five most populated cities in the United States.

15. ***CHALLENGE*** Create a bar graph displaying the number of A's, B's, C's, D's, and F's in Ms. Walker's class if the grades were the following: 81, 87, 80, 75, 77, 98, 52, 78, 75, 82, 74, 95, 76, 52, 76, 53, 86, 77, 90, 83, 96, 83, 74, 67, 90, 65, 69, 93, 68, and 76.

Grading System	
A	90–100
B	80–89
C	70–79
D	60–69
F	0–59

Spiral Review

Write each phrase as a numerical or algebraic expression. (Lesson 2-2)

16. 739 minus 103 **739 − 103**
17. the product of 7 and z **$7z$**
18. the difference of 12 and n **$12 - n$**

Write each mixed number as an improper fraction. (Lesson 4-7)

19. $2\frac{2}{5}$ **$\frac{12}{5}$**
20. $1\frac{3}{4}$ **$\frac{7}{4}$**
21. $4\frac{1}{7}$ **$\frac{29}{7}$**
22. $3\frac{1}{3}$ **$\frac{10}{3}$**

23. **TEST PREP** Which of the following is the solution to $4x = \frac{3}{4}$? (Lesson 5-4) **A**

    A $x = \frac{3}{16}$ **B** $x = \frac{3}{4}$ **C** $x = 3$ **D** $x = 5\frac{1}{3}$

Answers

14. Possible answer: Find the five most populated cities in an up-to-date reference book. Draw the axes. Label the vertical axis "Population." Choose an appropriate scale. Mark off equal intervals from 0 to your maximum. Label the horizontal axis "Cities," and list the cities. For each city, draw a bar. Give the graph a title.

15.

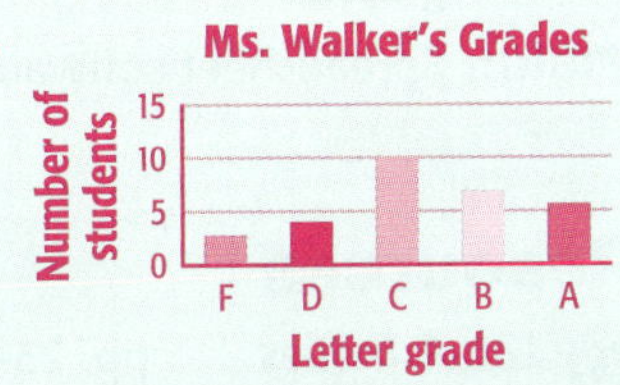

Journal

Have students think of and describe a situation in which a double-bar graph would be used to represent data. Have them come up with data and make the graph.

Test Prep Doctor

For Exercise 23, you may wish to review solving a multiplication equation and dividing a fraction by a whole number. Students who answered **B** might have just guessed. Students who answered **C** multiplied $\frac{3}{4} \times 4$. Students who answered **D** multiplied $4 \times \frac{4}{3}$.

Lesson Quiz

Use the bar graph to answer each question.

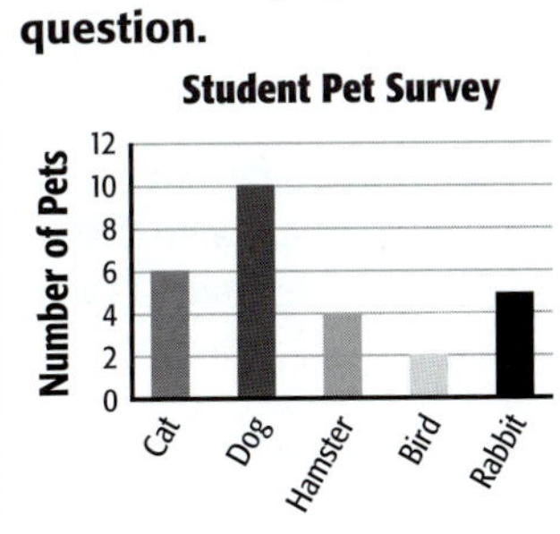

1. Which animal was least popular among students? bird
2. Which pet was more popular to twice as many students as rabbits were? dog
3. Make a bar graph of this data.
   Number of Daily Servings
   Grains = 6 Fruit = 2
   Meat = 2 Milk = 3
   Vegetables = 3

See p. A8.

Available on Daily Transparency in CRB

CHALLENGE 6-4

LESSON 6-4 **Challenge**
Picture a Bar Graph

Sometimes people use illustrations instead of bars to display data on bar graphs. For example, a bar graph showing the sizes of some forests might use trees for the graph's bars.

Use the data given in each table below to make an illustrated bar graph. Make sure the pictures you choose for your bars relate to the subject of each graph.

Fastest U.S. Roller Coasters

Roller Coaster, Location	Speed (mi/h)
Desperado, NV	80
Goliath, CA	85
Millennium Force, OH	92
Superman: The Escape, CA	100
Titan, TX	85

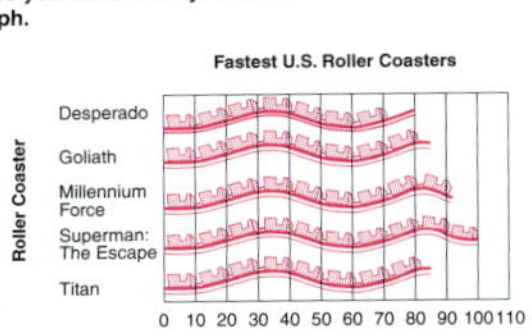

Check students' graphs. Ask students to explain why they chose their illustrations for each graph's bars.

Surfers with the Most World Championship Wins

Surfer, Country	Wins
Tom Carroll, Australia	2
Tom Curren, United States	3
Damien Hardman, Australia	2
Mark Richards, Australia	4
Kelly Slater, United States	6

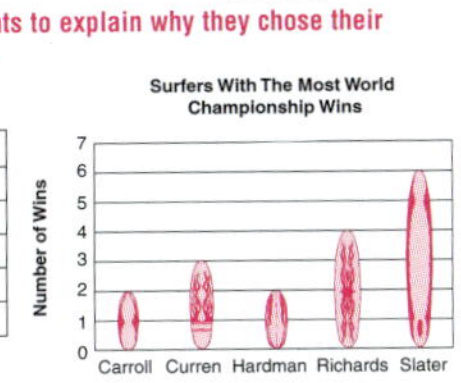

PROBLEM SOLVING 6-4

LESSON 6-4 **Problem Solving**
Bar Graphs

Use the bar graph for Exercises 1–4.

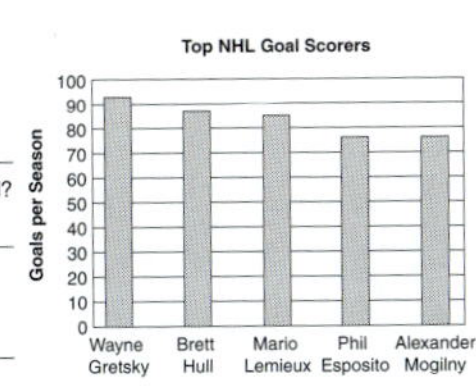

1. What is the range of the goals the hockey players scored per season? **16 goals**
2. What is the mode of the goals scored? **76 goals**
3. What is the mean number of goals the players scored? **83 goals**

Use the bar graph for Exercises 5–8.

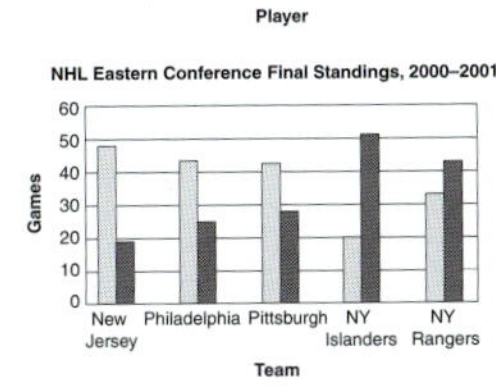

4. Which team won the most games that season? **New Jersey**
5. Which team lost the most games that season? **NY Islanders**
6. What was the mean number of games won? **37.4 games**
7. What was the mean number of games lost? **33.2 games**

Circle the letter of the correct answer.

8. Which hockey team had the greatest difference between the number of games won and lost?
   A New Jersey
   (B) New York Islanders
   C Philadelphia
   D Pittsburgh

9. How do you know the mode of a data set by looking at a bar graph?
   (F) The mode has two or more bars on the graph with the same height.
   G The mode has the tallest bar.
   H The mode has the lowest bar.
   J The bar for the mode is in the middle of the graph.

Technology LAB

6A Create Bar Graphs

Pacing: Traditional 1 day
Block $\frac{1}{2}$ day

Objective: To use a spreadsheet program to create bar graphs

Materials: Spreadsheet software

Lab Resources

Technology Lab Activities . . . . pp. 23–24

Using the Pages

This technology activity shows students how to use spreadsheet software to create bar graphs. Specific instructions may vary, depending on the spreadsheet software used. The instructions given are for Microsoft Excel. For instructions for other software, visit go.hrw.com.

The Think and Discuss problem is meant to be a critical-thinking exercise. While Try This problems 1 and 2 can be done without a computer, they are meant to help students use spreadsheet software to create bar graphs.

Assessment

1. What pattern do you see in the projected population from 2000 to 2035? The populations will increase steadily every 5 years.
2. How can you change the title of the graph? Go to **Chart Options**. Select the **Titles** tab, and change the **Chart Title**.
3. In Excel, what is the difference between the **Bar** option and the **Column** option in the **Chart Type** screen? Selecting **Column** yields a vertical bar graph; selecting **Bar** yields a horizontal bar graph.

Technology LAB 6A

Create Bar Graphs

Use with Lesson 6-4

internet connect
Lab Resources Online
go.hrw.com
KEYWORD: MR4 TechLab6A

You can use a computer spreadsheet to draw bar graphs. The Chart Wizard icon, , on a spreadsheet menu looks like a bar graph. The Chart Wizard allows you to create different types of graphs.

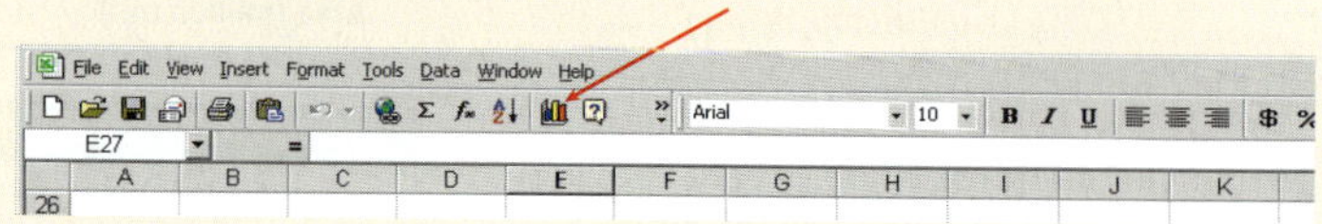

Activity

In a study conducted in December 2001 at Texas A&M University, the population of Texas through 2035 was projected. Make a bar graph of this data.

Texas Population	
Year	**Population**
2000	20,851,820
2005	23,207,929
2010	25,897,018
2015	28,971,283
2020	32,427,282
2025	36,273,829
2030	40,538,290
2035	45,283,746

1. Type the titles *Year* and *Population* into cells A1 and B1. Then type the data into columns A (year) and B (population).
2. Select the cells containing the titles and the data. Do this by placing your pointer in A1, clicking and holding the mouse button, and dragging the pointer down to B9.
3. Click the Chart Wizard icon. Highlight **Column** to make a vertical bar graph. Click **Next**.

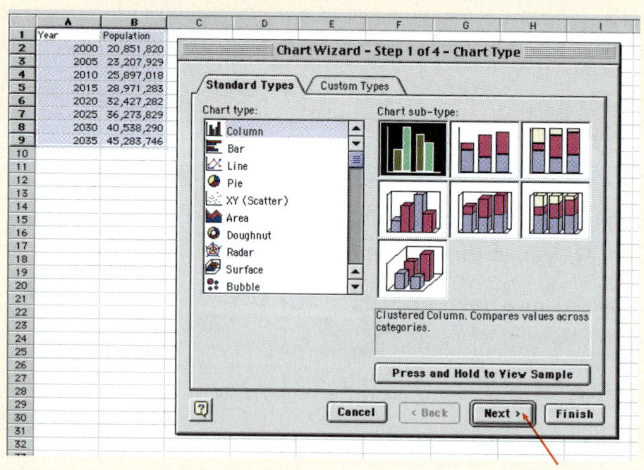

4 The next screen shows where the data from the graph comes from. Click **Next**.

5 Title your graph and both axes. Click the **Legend** tab. Click the box next to **Show Legend** to turn off the key. (You would need a key if you were making a double-bar graph.) Click **Next** when you are finished.

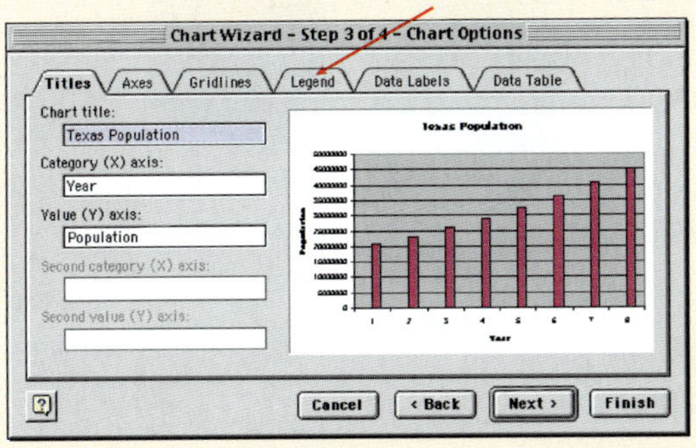

6 The next screen asks you where you want to place your chart. Click **Finish** to place it in your spreadsheet.

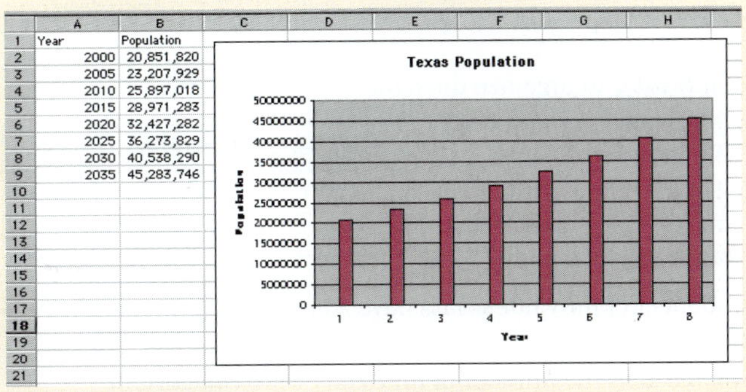

Think and Discuss

1. Do you think the population of Texas will be 32,427,282 in the year 2020 as shown in the graph? Explain. **Probably not; the number is only a projection and cannot be assumed to be completely accurate.**

Try This

1. Redraw the bar graph in the activity to show the population of Texas as 39,000,000 in 2035 and 33,000,000 in 2040.

2. The table shows the number of countries in which some common languages are spoken. Make a bar graph of the data.

Language	English	Arabic	Spanish	French
Number of Countries	54	24	21	33

Answers

Try This

1.

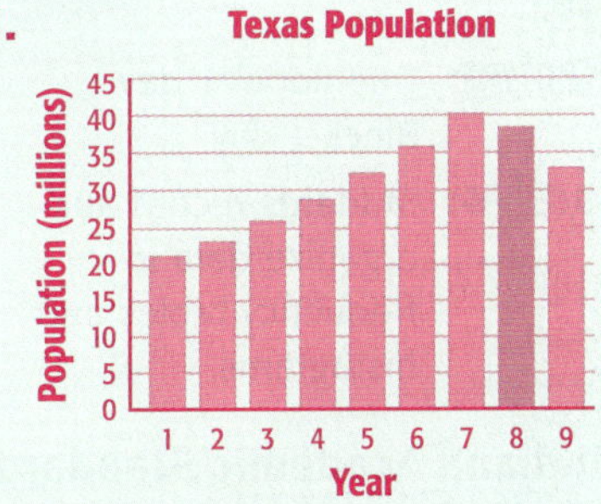

2.

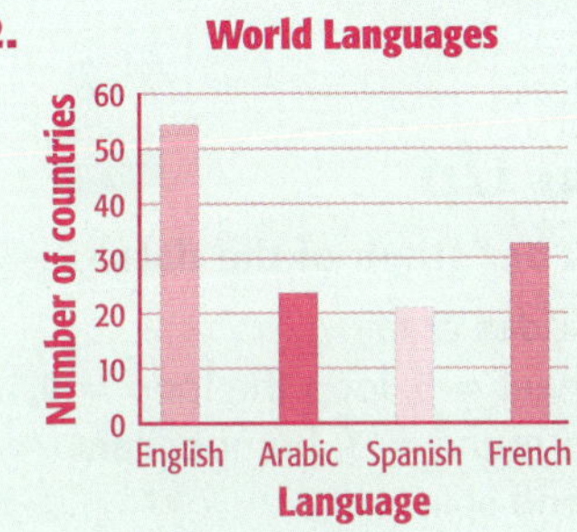

6-5 Organizer

Pacing: Traditional 1 day
Block $\frac{1}{2}$ day

Objective: Students record and organize data in frequency tables and histograms.

Indiana Academic Standards
6.6.1, 6.6.2, 6.7.1, 6.7.5, 6.7.11

Warm Up

Create a bar graph of the data.

Favorite rides at fair:
Ferris wheel = 5, loop the loop = 4, merry-go-round = 3, bumper cars = 7, sit and spin = 9

Problem of the Day

A set of 7 numbers has a mean of 36, a median of 37, a mode of 37, and a range of 6. What could the 7 numbers be? Possible answer: 33, 33, 36, 37, 37, 37, 39

Answers on Chapter 6 Resource Book p. T16

Available on Daily Transparency in CRB

Math Humor

What kind of a message do you send to a snake? A histogram

6-5 Frequency Tables and Histograms

Learn to organize data in frequency tables and histograms.

Vocabulary
frequency table
cumulative frequency
histogram

Your fingerprints are unlike anyone else's. Even identical twins have slightly different fingerprint patterns.

All fingerprints have one of three patterns: whorl, arch, or loop.

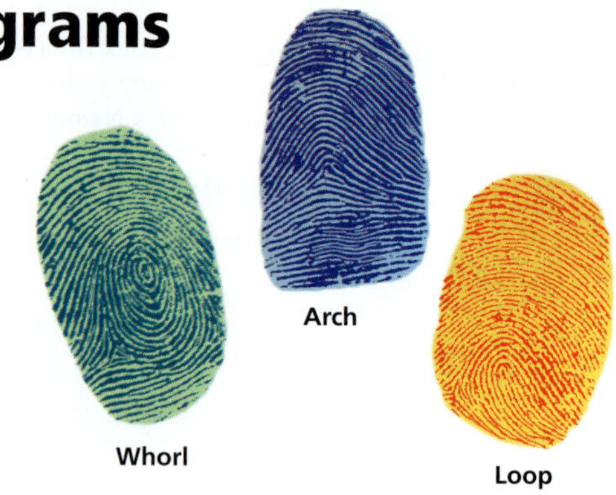

EXAMPLE 1 Making a Tally Table

Each student in Mrs. Choe's class recorded their fingerprint pattern. Which type do most students in Mrs. Choe's class have?

whorl	loop	loop	loop	loop	arch	loop
whorl	arch	loop	arch	loop	arch	whorl

Make a *tally table* to organize the data.

Step 1: Make a column for each fingerprint pattern.

Step 2: For each fingerprint, make a tally mark in the appropriate column.

Number of Fingerprint Patterns		
Whorl	Arch	Loop
///	////	~~////~~ //

Most students in Mrs. Choe's class have a loop fingerprint pattern.

Reading Math

A group of four tally marks with a line through it means *five*.

~~////~~ = 5
~~////~~ ~~////~~ = 10

A **frequency table** tells the number of times an event, category, or group occurs. The **cumulative frequency** column shows a running total of all frequencies.

EXAMPLE 2 Making a Cumulative Frequency Table

Use the tally table above to make a cumulative frequency table.

Step 1: Make a row for each pattern.

Step 2: The frequency is how many times each pattern occurred.

Step 3: Find the cumulative frequency for each row by adding all frequency values above or in that row.

Number of Fingerprint Patterns		
Fingerprint Pattern	Frequency	Cumulative Frequency
Whorl	3	3
Arch	4	7
Loop	7	14

1 Introduce

Alternate Opener

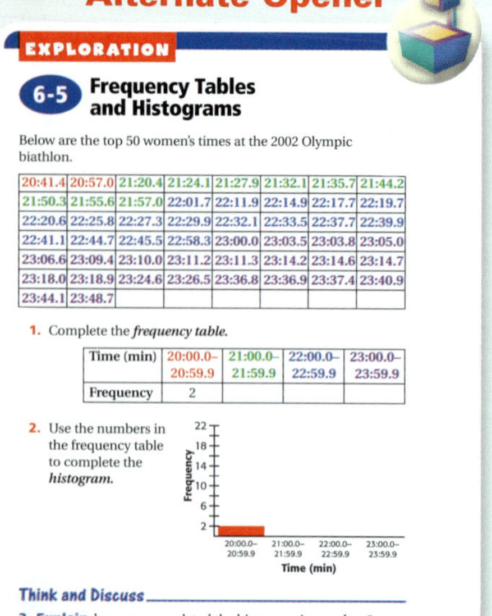

EXPLORATION

6-5 Frequency Tables and Histograms

Below are the top 50 women's times at the 2002 Olympic biathlon.

20:41.4	20:57.0	21:20.4	21:24.1	21:27.9	21:32.1	21:35.7	21:44.2
21:50.3	21:55.6	21:57.0	22:01.7	22:11.9	22:14.9	22:17.7	22:19.7
22:20.6	22:25.8	22:27.3	22:29.9	22:32.1	22:33.5	22:37.7	22:39.9
22:41.1	22:44.7	22:45.5	22:58.3	23:00.0	23:03.5	23:03.8	23:05.0
23:06.6	23:09.4	23:10.0	23:11.2	23:11.3	23:14.2	23:14.6	23:14.7
23:18.0	23:18.9	23:24.6	23:26.5	23:36.8	23:36.9	23:37.4	23:40.9
23:44.1	23:48.7						

1. Complete the *frequency table.*

Time (min)	20:00.0–20:59.9	21:00.0–21:59.9	22:00.0–22:59.9	23:00.0–23:59.9
Frequency	2			

2. Use the numbers in the frequency table to complete the *histogram.*

Think and Discuss

3. **Explain** how you completed the histogram in number 2.

Motivate

Using the photos in the lesson opener, have students determine which type of fingerprint pattern they have. Determine which type most students in the class have.

Exploration worksheet and answers on Chapter 6 Resource Book pp. 45 and 106

2 Teach

Lesson Presentation

Guided Instruction

In this lesson, students learn to record and organize data in frequency tables and histograms. Teach students to make a tally table and a cumulative frequency table. Then teach them to make a frequency table with intervals, and a histogram. You can add your class's data to the data for Mrs. Choe's class, or you can have your students create new tally and cumulative frequency tables based on their data. Discuss how histograms are different from bar graphs. Histograms compare data in intervals; bar graphs compare data in categories.

EXAMPLE 3 Making a Frequency Table with Intervals

Social Studies LINK

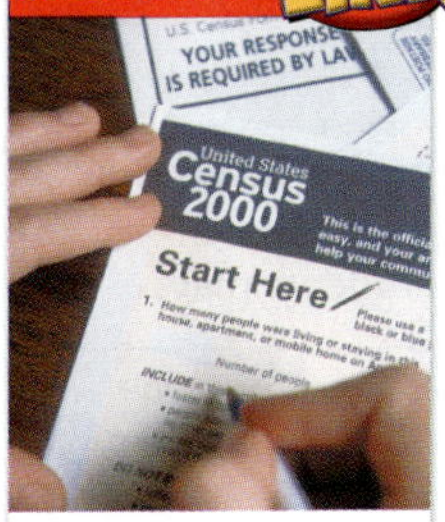

The number of representatives a state has depends on its population. A state can gain or lose representatives after a population count called a census.

Use the data in the table to make a frequency table with intervals.

Number of Representatives per State in the U.S. House of Representatives												
7	1	6	4	52	6	6	1	1	23	11	2	2
20	10	5	4	6	7	2	8	10	16	8	5	9
1	3	2	2	13	3	31	12	1	19	6	5	21
2	6	1	9	30	3	1	11	9	3	9		

Step 1: Choose equal intervals.

Step 2: Find the number of data values in each interval. Write these numbers in the "Frequency" row.

Number of Representatives per State in the U.S. House of Representatives									
Number	0–5	6–11	12–17	18–23	24–29	30–35	36–41	42–47	48–53
Frequency	22	18	3	4	0	2	0	0	1

This table shows that 22 states have between 0 and 5 representatives, 18 states have between 6 and 11 representatives, and so on.

A **histogram** is a bar graph that shows the number of data items that occur within each interval.

EXAMPLE 4 Making a Histogram

Use the frequency table in Example 3 to make a histogram.

Step 1: Choose an appropriate scale and interval.

Step 2: Draw a bar for the number of states in each interval. The bars should touch but not overlap.

Step 3: Title the graph and label the axes.

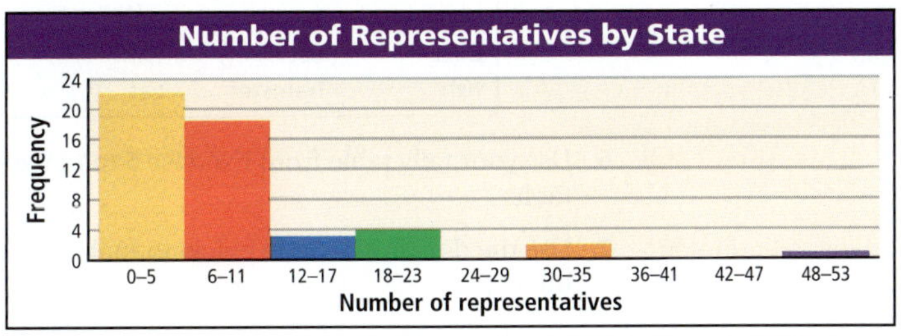

Think and Discuss

1. Tell how to find cumulative frequency.

Additional Examples

Example 1

Students in Mr. Ray's class recorded their fingerprint patterns. Which type of pattern do most students in Mr. Ray's class have?

whorl loop whorl loop
arch arch loop whorl
loop arch whorl arch
arch whorl arch loop

arch

Example 2

Use the tally table from Additional Example 1 to make a cumulative frequency table.

Example 3

Use the data in the table below to make a frequency table with intervals.

Pages Read Last Weekend				
12	15	40	19	7
5	22	34	37	18

Example 4

Use the frequency table from Additional Example 3 to make a histogram.

Answers on Additional Examples Transparencies T17–T20 in CRB

Reaching All Learners

Through Curriculum Integration

Sports: Have the students do research and make a frequency table of the number of Super Bowls won by the teams in the National Football League or the number of gold medals won by the top five countries in the Olympics.

3 Close

Summarize

Review the procedures for completing a cumulative frequency table and a frequency table based on intervals. Then review how an interval-based frequency table is related to a histogram.

Answers to Think and Discuss

1. Possible answer: For each row, add all frequency values above and in that row.

6-5 PRACTICE & ASSESS

Students may want to refer back to the lesson examples.

Assignment Guide

If you finished Example 1 assign:
Core 1, 5, 9, 15–23
Enriched 1, 5, 9, 15–23

If you finished Example 2 assign:
Core 1–2, 5–6, 9–10, 15–23
Enriched 5–6, 9–10, 13, 15–23

If you finished Example 3 assign:
Core 1–3, 5–7, 9–10, 15–23
Enriched 5–7, 9–11, 13–23

If you finished Example 4 assign:
Core 1–12, 15–23
Enriched 3–23

Answers

1–8. See p. A3.

6-5 Exercises

FOR EXTRA PRACTICE see page 648

internet connect
Homework Help Online
go.hrw.com Keyword: MR4 6-5

GUIDED PRACTICE

See Example 1

1. Each student in the band recorded the type of instrument he or she plays. The results are shown in the box. Make a tally table to organize the data. Which instrument do the fewest students play?

trumpet	tuba	French horn	drums	trombone
drums	trombone	trombone	trumpet	trumpet
trumpet	French horn	trumpet	French horn	French horn

See Example 2

2. Use your tally table from Exercise 1 to make a cumulative frequency table.

See Example 3

3. Use the data in the table below to make a frequency table with intervals.

Length of Each U.S. Presidency (yr)																				
8	4	8	8	8	4	8	4	0	4	4	1	3	4	4	4	4	8	4	0	4
4	4	4	4	8	4	8	2	6	4	12	8	8	2	6	5	3	4	8	4	8

See Example 4

4. Use your frequency table from Exercise 3 to make a histogram.

INDEPENDENT PRACTICE

See Example 1

5. Students recorded the type of pet they own. The results are shown in the box. Make a tally table. Which type of pet do most students own?

cat	cat	bird	dog	dog
dog	bird	dog	bird	fish
bird	cat	fish	dog	cat
fish	hamster	cat	hamster	dog

See Example 2

6. Use your tally table from Exercise 5 to make a cumulative frequency table.

See Example 3

7. Use the data in the table below to make a frequency table with intervals.

Number of Olympic Medals Won by 25 Countries													
8	88	59	12	11	57	38	17	14	28	28	26	25	23
18	8	29	34	14	17	13	13	58	12	97	10	9	

See Example 4

8. Use your frequency table from Exercise 7 to make a histogram.

Math Background

A tally was originally a stick on which notches were made with a sharp object to indicate numbers. The tallies we use in a frequency table today correspond to the notches on a tally stick. The word *tally* has the same root as the word *tailor*. The connection between the two is the idea of cutting: a tally originally cut notches onto a stick, and a tailor cuts cloth.

Evidence of civilizations keeping numerical records on a stick can be traced as far back as 1350 B.C.

RETEACH 6-5

LESSON 6-5 **Reteach**
Frequency Tables and Histograms

Julie picked the following cards from a deck.

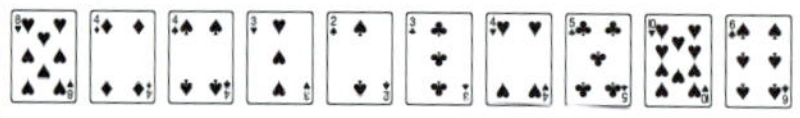

You can make a tally table to organize the data. Make a row for each suit of cards. Then for each card, make a tally mark in the appropriate row.

Julie's Cards			
Clubs	Spades	Hearts	Diamonds
|	|||	||||	|

1. Make a tally table to organize the data.

Rolls of a Number Cube						
2	3	6	5	1	4	1
3	3	5	1	6	1	4

Rolls of a Number Cube					
1	2	3	4	5	6
||||	|	|||	||	||	||

A frequency table tells the number of times an event, category, or group occurs. The cumulative frequency is the running total of all of the frequencies.

To make a frequency table of Julie's data, make a row for each type of card. Then use the tally table to find the frequency of picking each type of card.

	Julie's Cards			
Card	Clubs	Spades	Hearts	Diamonds
Frequency	2	3	4	1
Cumulative Frequency	2	5	9	10

2. Make a frequency table for the tally table you made Exercise 1.

	Roll of a Number Cube					
Roll	1	2	3	4	5	6
Frequency	4	1	3	2	2	2
Cumulative Frequency	4	5	8	10	12	14

PRACTICE 6-5

LESSON 6-5 **Practice B**
Frequency Tables and Histograms

Mr. Allen said that his students could choose one day a week to not have homework assigned. He recorded their votes in the box below.

Monday	Friday	Thursday	Friday	Wednesday
Friday	Tuesday	Wednesday	Friday	Monday
Tuesday	Monday	Friday	Monday	Friday

1. Make a tally table to organize the data. Which day got the most votes?

Friday

Tally Table for Homework Votes

Mon	||||
Tues	||
Wed	||
Thurs	|
Fri	~~||||~~ |

2. Use your tally table from Exercise 1 to make a cumulative frequency table. How many students voted in all?

15 students

Number of Votes

Day	Frequency	Cumulative Frequency
Mon	4	4
Tues	2	6
Wed	2	8
Thurs	1	9
Fri	6	15

3. Use the data in the box below to make a frequency table with intervals.

Class Social Studies Test Scores									
78	95	81	83	75	68	100	73	92	85
59	70	88	92	99	87	75	67	89	84

Possible answer:

Class Social Studies Test Scores					
Scores	51–60	61–70	71–80	81–90	91–100
Frequency	1	3	4	7	5

4. In which range of scores did most of the students' tests fall?

81–90

PRACTICE AND PROBLEM SOLVING

9. a. Students in a gym class recorded their favorite sports. Use their data to make a tally table.

basketball football soccer	hockey track and field track and field	hockey football football	soccer football baseball	tennis basketball track and field

b. Use your tally table to make a cumulative frequency table.

10. ***SOCIAL STUDIES*** The map shows the populations of Australia's states and territories. Use the data to make a frequency table with intervals.

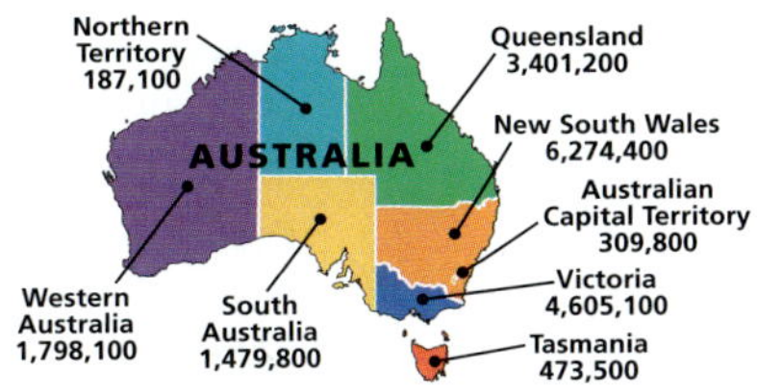

11. ***SOCIAL STUDIES*** Use your frequency table from Exercise 10 to make a histogram.

12. ***WRITE ABOUT IT*** Choose one of the histograms you made for this lesson and redraw it using different intervals. How did the histogram change? Explain.

13. ***WHAT'S THE ERROR?*** Describe the error in the frequency table.

Age of Consumers

Age	Frequency	Cumulative Frequency
Child	3	
Teenager	2	15
Adult	10	

13. The cumulative frequency column is incorrect; it should have three rows and the numbers should be 3, 5, and 15.

14. ***CHALLENGE*** Can you find the mean, median, and mode price using this frequency table? If so, find them. If not, explain why not.

Cost of Video Game Rentals at Different Stores

Price	\$2.00–\$2.99	\$3.00–\$3.99	\$4.00–\$4.99	\$5.00–\$5.99
Frequency	5	12	8	5

14. You cannot find them. You do not know the actual data points, only the ranges they fall in.

Spiral Review

Write each decimal in expanded form and word form. (Lesson 3-1)

15. 1.23   16. 0.45   17. 26.07   18. 80.002

Find the reciprocal of each number. (Lesson 5-3)

19. 6 $\frac{1}{6}$ 20. $\frac{4}{7}$ $\frac{7}{4}$ 21. $\frac{2}{9}$ $\frac{9}{2}$ 22. $\frac{11}{5}$ $\frac{5}{11}$

23. **TEST PREP** The __?__ of a data set is always a value in the set. (Lesson 6-2) C

A range   B median   C mode   D mean

Answers

9–11. See p. A3.

12. Possible answer:

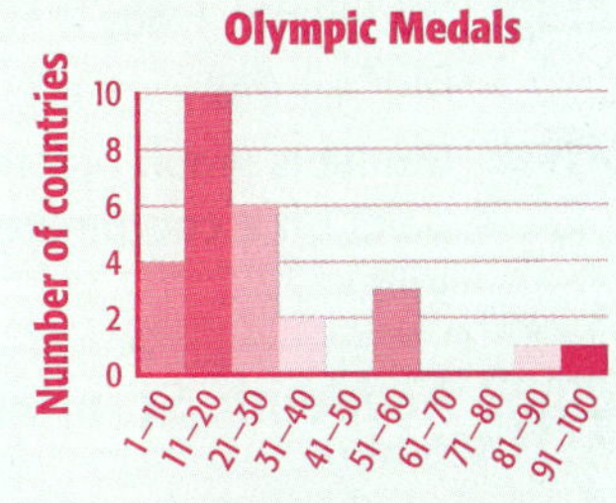

The number of intervals has doubled, and the histogram displays more detailed information.

15. 1 + 0.2 + 0.03; one and twenty-three hundredths

16. 0.4 + 0.05; forty-five hundredths

17. 20 + 6 + 0.07; twenty-six and seven-hundredths

18. 80 + 0.002; eighty and two-thousandths

Journal

Have students explain why a histogram is more appropriate than a bar graph for displaying the data in Example 4.

Test Prep Doctor

For Exercise 23, encourage students to test each answer choice and to eliminate those that don't make the sentence true. Choices **A** and **D** can be eliminated because they are always found by using computation. Choice **B** can be eliminated because when there is an even number of items in a data set, the median is found by using computation.

CHALLENGE 6-5

LESSON 6-5 **Challenge**
Write Often

What letter is used more than any other letter in the English language? e

The box below contains the six English letters that are used most often. Use the box to complete the tally table and frequency table at the bottom of the page. Your completed tables will show the answer to the question.

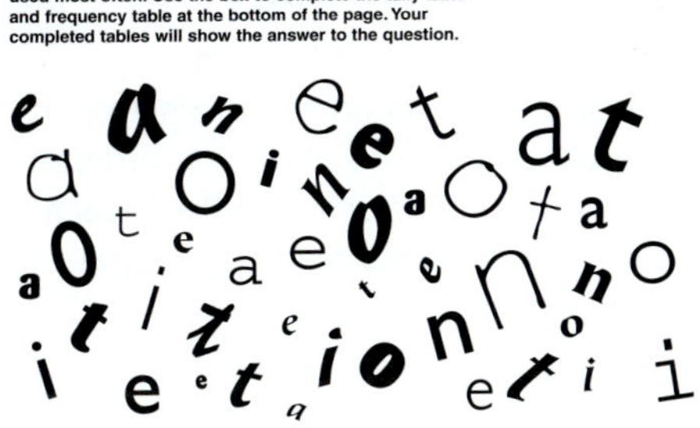

Tally Table for the Number of Letters

A	𝍸 III
E	𝍸 𝍸
I	𝍸 I
N	𝍸
O	𝍸 II
T	𝍸 IIII

Number of Letters

Letter	Frequency	Cumulative Frequency
A	8	8
E	10	18
I	6	24
N	5	29
O	7	36
T	9	45

PROBLEM SOLVING 6-5

LESSON 6-5 **Problem Solving**
Frequency Tables and Histograms

The sixth grade class voted on their favorite ice cream flavors. The results of the vote are shown below.

chocolate vanilla chocolate	vanilla chocolate strawberry	strawberry vanilla vanilla	vanilla chocolate vanilla	vanilla strawberry chocolate

1. Use the data to make a cumulative frequency table. How many students voted in all?
15 students

Ice Cream Flavor Votes

Flavor	Frequency	Cumulative Frequency
Chocolate	5	5
Vanilla	7	12
Strawberry	3	15

2. Which flavor got the most votes?
vanilla

Use the histogram for Exercises 3–5.

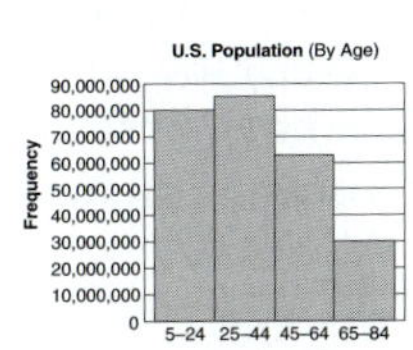

3. How many years make up each age interval on the histogram?
20 years

4. Which range of ages on the histogram has the highest population?
25–44

5. Which range of ages has the lowest population?
65–84

Circle the letter of the correct answer.

6. Which of the following cannot be used to make a cumulative frequency table?
A histogram
B tally table
C frequency table with intervals
(D) double-bar graph

7. Which question can be answered by using the histogram above?
A How many people in the United States are younger than 5 years?
B What is the mean age of all people in the United States?
C How many people in the United States are older than 84 years old?
(D) How many people in the United States are age 25 to 64?

Lesson Quiz

1. Students listed the number of days they spent on vacation in one year. Make a tally table with intervals of 5.

2, 18, 5, 15, 7, 10, 1, 10, 4, 16, 7, 11, 17, 3, 8, 14, 13, 10

2. Use your tally table from problem 1 to make a frequency and cumulative frequency table.

Answers on p. A8.

Available on Daily Transparency in CRB

6-6 Organizer

Pacing: Traditional 1 day
Block $\frac{1}{2}$ day

Objective: Students graph ordered pairs on a coordinate grid.

Indiana Academic Standards
6.3.7, 6.7.1, 6.7.5, 6.7.11

Warm Up

24 35 47 67 36 22 80 41

Use the set of data above to find the following:

1. the mean 44 **2.** the median 38.5
3. the mode none **4.** the range 58

Problem of the Day

What is the largest 6-digit number with each digit different and no digit a prime number? 986,410

Available on Daily Transparency in CRB

Additional Examples

Example 1

Name the ordered pair for each location.

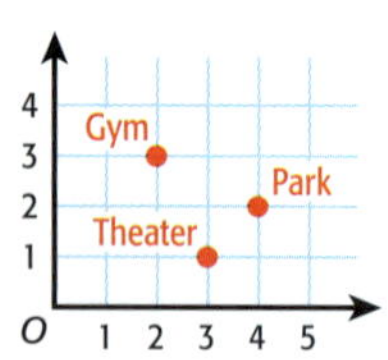

A. Gym (2, 3)
B. Theater (3, 1)
C. Park (4, 2)

6-6 Ordered Pairs

Learn to graph ordered pairs on a coordinate grid.

Vocabulary
coordinate grid
ordered pair

Cities, towns, and neighborhoods are often laid out on a grid. This makes it easier to map and find locations.

A **coordinate grid** is formed by horizontal and vertical lines and is used to locate points.

Each point on a coordinate grid can be located by using an **ordered pair** of numbers, such as (4, 6). The starting point is (0, 0).

San Diego, CA. Image courtesy of spaceimaging.com.

- The first number tells how far to move horizontally from (0, 0).
- The second number tells how far to move vertically.

EXAMPLE 1 **Identifying Ordered Pairs**

Name the ordered pair for each location.

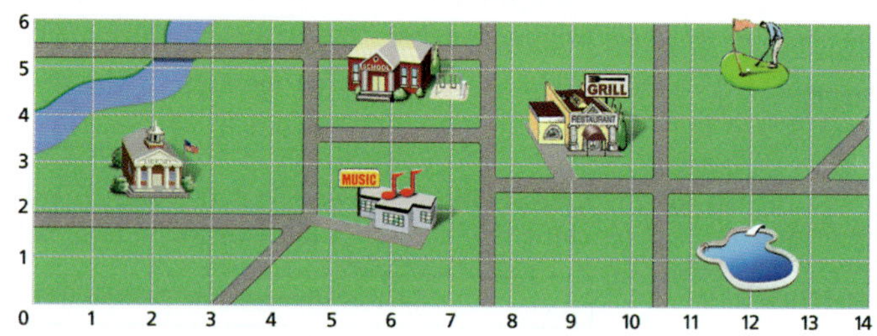

A Library

Start at (0, 0). Move right 2 units and then up 3 units.

The library is located at (2, 3).

B School

Start at (0, 0). Move right 6 units and then up 5 units.

The school is located at (6, 5).

C Pool

Start at (0, 0). Move right 12 units and up 1 unit.

The pool is located at (12, 1).

1 Introduce

Alternate Opener

EXPLORATION

6-6 Ordered Pairs

The table shows the number of faces and vertices of the five regular polyhedrons.

Faces	4	8	20	6	12
Vertices	4	6	12	8	20

This data can be represented on a graph using *ordered pairs*. Each ordered pair is composed of the number of faces and the number of vertices of the polyhedron.

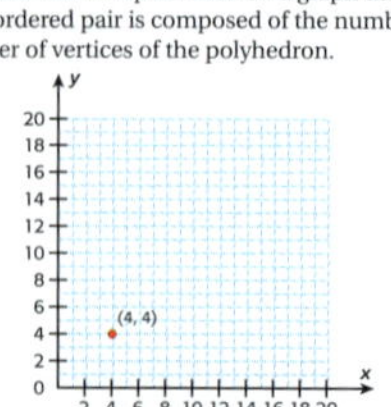

1. Plot and label the remaining ordered pairs. First find the number of faces on the horizontal line. From the number of faces, move up to find the number of vertices on the vertical line.

Think and Discuss

2. **Explain** how to plot ordered pairs.
3. **Discuss** what the point (6, 8) means.

Motivate

Have students look at maps that contain horizontal and vertical columns and rows (e.g., state road maps). Discuss the reasons why the lines are there. After students have shared their ideas, point out the letters and numbers at the ends of the grid lines and explain how those are there to help people find locations on the map.

Exploration worksheet and answers on Chapter 6 Resource Book pp. 55 and 108

2 Teach

Lesson Presentation

Guided Instruction

In this lesson, students learn to graph ordered pairs on a coordinate grid. First teach students to name the ordered pairs for given locations, and then teach them to graph points on the grid, given the ordered pairs.

EXAMPLE 2 **Graphing Ordered Pairs**

Graph and label each point on a coordinate grid.

A $Q(4, 6)$ *Start at (0, 0).* *Move right 4 units.* *Move up 6 units.*

B $S(0, 4)$ *Start at (0, 0).* *Move right 0 units.* *Move up 4 units.*

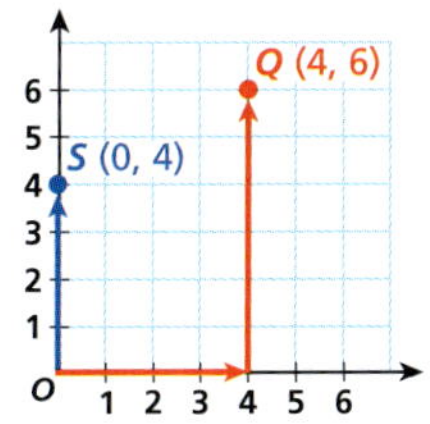

Think and Discuss

1. **Tell** what point is the starting location when you are graphing on a coordinate grid.
2. **Describe** how to graph (2, 8) on a coordinate grid.

Additional Examples

Example 2

Graph and label each point on a coordinate grid.

A. $L(3, 5)$ **B.** $M(4, 0)$

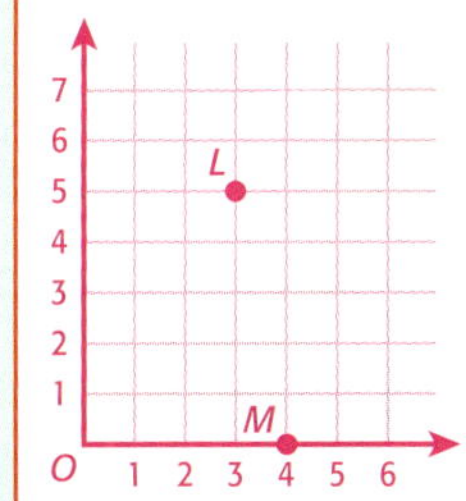

6-6 Exercises

FOR EXTRA PRACTICE
see page 648

internet connect
Homework Help Online
go.hrw.com Keyword: MR4 6-6

GUIDED PRACTICE

See Example 1 Name the ordered pair for each location.

1. school (2, 3)
2. store (0, 7)
3. hospital (7, 6)
4. mall (9, 1)
5. office (4, 5)
6. hotel (11, 4)

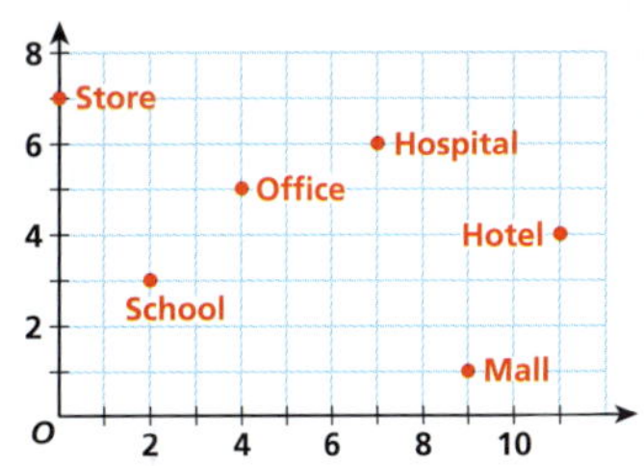

See Example 2 Graph and label each point on a coordinate grid.

7. $T(3, 4)$
8. $S(2, 8)$
9. $U(5, 5)$
10. $V(4, 1)$

INDEPENDENT PRACTICE

See Example 1 Name the ordered pair for each location.

11. diner (3, 0)
12. library (6, 6)
13. store (1, 4)
14. bank (10, 4)
15. theater (11, 7)
16. town hall (1, 7)

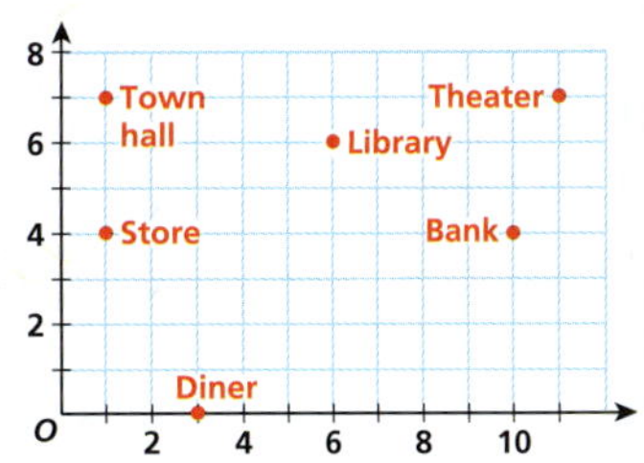

6-6 PRACTICE & ASSESS

Assignment Guide

If you finished Example 1 assign:
Core 1–6, 11–16, 29–33 odd, 39–44
Enriched 2–6 even, 12–16 even, 24–28 even, 36, 38–44

If you finished Example 2 assign:
Core 1–22, 23–35 odd, 39–44
Enriched 11–44

Graph paper for Exercises 7–10, 17–22, and 38 is provided on Chapter 6 Resource Book p. 97.

Answers

7–10. See p. A4.

3 Close

Summarize

Review the terms *coordinate grid* and *ordered pair*. Have students explain how the terms are related.

Possible answer: Ordered pairs are used to name locations on a coordinate grid. The first number in an ordered pair tells how far to move horizontally from (0, 0). The second number tells how far to move vertically from (0, 0).

Reaching All Learners

Through Grouping Strategies

Have students work with partners. Each partner draws axes on graph paper and plots 5–10 locations on the grid. Partners switch coordinate grids, and each student writes the ordered pair for each location shown on his or her partner's grid. Partners switch papers again and check each other's work, working together to fix any errors that were found.

Answers to Think and Discuss

1. at (0, 0), in the bottom left corner
2. Start at (0, 0). Move 2 units right and then 8 units up.

Lesson Quiz

Give the ordered pair for each point.

1. A (4, 6)
2. B (6, 1)
3. C (1, 4)
4. D (2, 1)

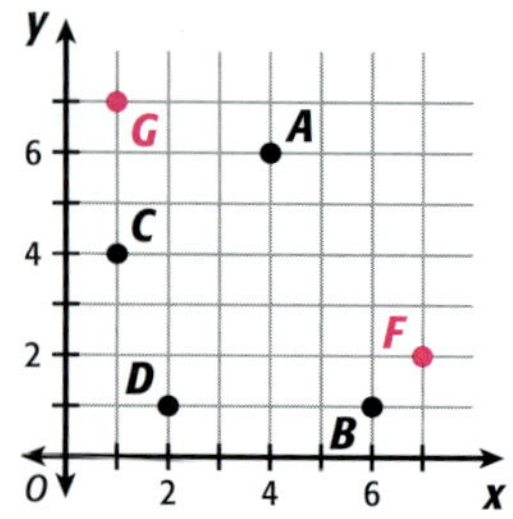

Graph and label each point on a coordinate grid.

5. F(7, 2)
6. G(1, 7)

Available on Daily Transparency in CRB

Answers

17–22, 36. See p. A4.

See Example 2 — Graph and label each point on a coordinate grid.

17. P(5, 1) **18.** R(2, 4) **19.** Q(3, 2)

20. V(6, 5) **21.** X(1, 3) **22.** Y(7, 4)

PRACTICE AND PROBLEM SOLVING

Use the coordinate grid for Exercises 23–34.

Name the point found at each location.

23. (1, 7) A **24.** (5, 9) G **25.** (3, 3) C

26. (4, 7) N **27.** (7, 4) P **28.** (7, 7) F

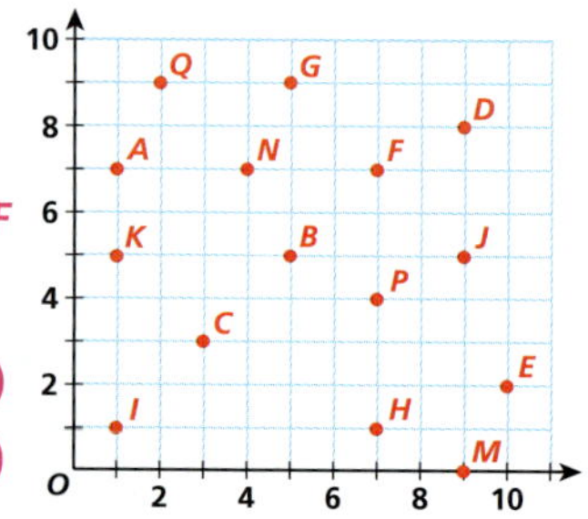

Give the ordered pair for each point.

29. D (9, 8) **30.** H (7, 1) **31.** K (1, 5)

32. Q (2, 9) **33.** M (9, 0) **34.** B (5, 5)

35. The Spirit Club marches in a large grid formation at sporting events. To spell words, some students hold red cards while others hold gold cards. Give five ordered pairs where students are holding red cards in the formation at right.

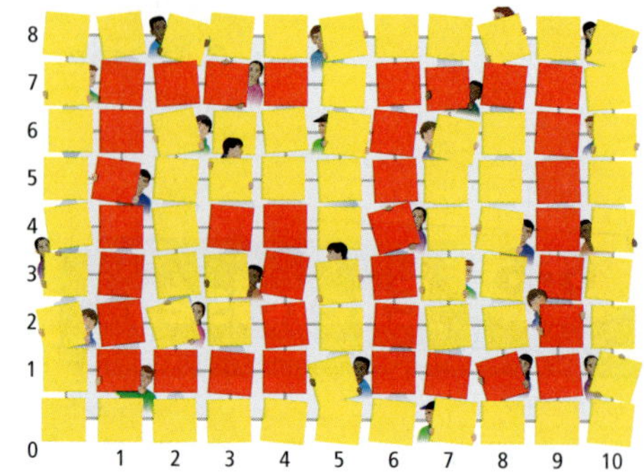

35. Possible answers: (1, 1), (3, 4), (4, 3), (6, 5), (7, 7)

36. ***WRITE ABOUT IT*** Explain the difference between the points (3, 2) and (2, 3).

37. ***WHAT'S THE QUESTION?*** If the answer is "Start at (0, 0) and move 3 units to the right," what is the question?
Possible answer: How do you graph the point (3, 0)?

38. ***CHALLENGE*** Locate and graph points that can be connected to form your initials. What are the ordered pairs for these points?

38. Possible answer: LYN; L: (1, 4), (1, 1) and (3, 1); Y(4, 4), (5, 3), (6, 4), and (5, 1); N:(7, 1), (7, 4), (9,1) and (9,4).

Spiral Review

Find the prime factorization of each number. (Lesson 4-2)

39. 18 $2 \cdot 3^2$ **40.** 20 $2^2 \cdot 5$ **41.** 33 $3 \cdot 11$ **42.** 50 $2 \cdot 5^2$

43. The marching band's halftime show was $10\frac{5}{6}$ minutes long. Then the director added a song. The new length of the show is $12\frac{1}{3}$ minutes. How long is the song that was added? (Lesson 5-10) $1\frac{1}{2}$ minutes

44. **TEST PREP** Which number, if any, is an outlier in the set 0, 1, 4, 0, 3, 4, 2, and 1? (Lesson 6-3) D

A 0 **B** 1 **C** 4 **D** No outlier

RETEACH 6-6

LESSON 6-6 Reteach
Ordered Pairs

A coordinate plane is formed by horizontal and vertical lines and is used to locate points.

An ordered pair names the location of a point by using two numbers.

The ordered pair (2, 5) gives the location of point A on the coordinate plane.

The first number, 2, tells the horizontal distance from the starting point (0, 0).

The second number, 5, tells the vertical distance.

To find the ordered pair for point B, start at (0, 0). Then move 6 units right and 3 units up. The coordinates of point B are (6, 3).

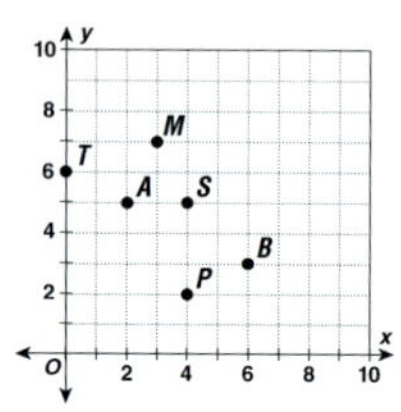

Give the ordered pair for each point shown on the coordinate plane above.

1. P (4, 2)
2. T (0, 6)
3. M (3, 7)
4. S (4, 5)

You can plot points in a coordinate plane.

To plot F (6, 4), start at (0, 0). Then move 6 units right and 4 units up.

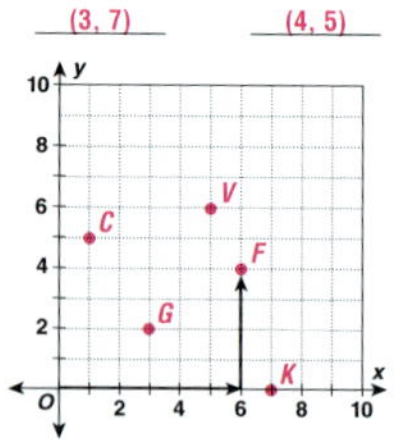

Plot each point in the coordinate plane above.

5. V (5, 6)
6. G (3, 2)
7. K (7, 0)
8. C (1, 5)

PRACTICE 6-6

LESSON 6-6 Practice B
Ordered Pairs

Name the ordered pair for each location on the plane.

1. gym (1, 2)
2. dining hall (0, 4)
3. offices (3, 3)
4. library (4, 1)
5. classrooms (4, 5)
6. dormitories (2, 1)

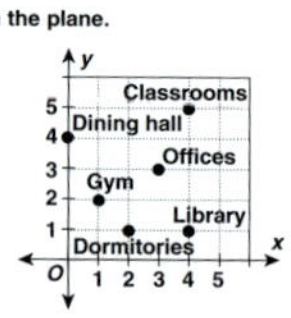

Graph and label each point on the coordinate plane.

7. A (5, 1)
8. B (2, 2)
9. C (1, 3)
10. D (4, 3)
11. E (5, 5)
12. F (2, 4)

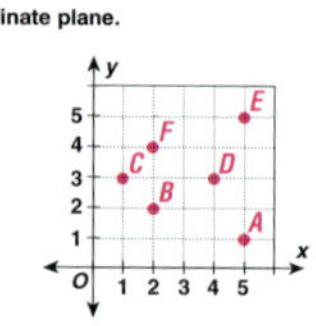

13. On a map of his neighborhood, Mark's house is located at point (7, 3). His best friend, Cheryl, lives 2 units west and 1 unit south of him. What ordered pair describes the location of Cheryl's house on their neighborhood map?
(5, 2)

14. Quan used a coordinate plane map of the zoo during his visit. Starting at (0, 0), he walked 3 units up and 4 units to the right to reach the tiger cages. Then he walked 1 unit down and 1 unit left to see the pandas. Describe the directions Quan should walk to get back to his starting point.
walk 2 units down and 3 units to the left

CHALLENGE 6-6

LESSON 6-6 Challenge
Treasure Island

According to legend, a pirate named Blackbeard buried his stolen treasures somewhere on the Outer Banks, off the coast of North Carolina. While on vacation there, you found an old sea chest buried on the beach. Inside it was a map that just may lead you to Blackbeard's hidden treasures!

Follow the map's clues to each location on the island. For each clue, name the location and the ordered pair that describes its point on the map. Graph each of those points on the map. Then draw an X where you think the treasure is buried.

1. Land your ship 7 units east and 6 units north of the (0, 0) starting point. Pirates Cove; (7, 6)
2. Walk 5 units west. Fool's Forest; (2, 6)
3. Walk 5 units south and 1 unit east. Skipper's Village; (3, 1)
4. Walk 2 units east and 2 units north. Terror Hills; (5, 3)
5. Walk 2 units north and 1 unit east. Silver Lake; (6, 5)
6. Walk 1 unit east and 3 units north. Deadman's Swamp; (7, 8)
7. Walk 3 units west and 1 unit north to find my treasure. Shipwreck Rocks; (4, 9)

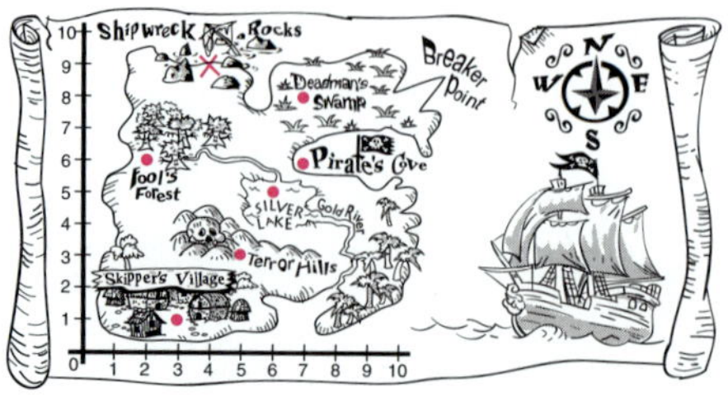

PROBLEM SOLVING 6-6

LESSON 6-6 Problem Solving
Ordered Pairs

Use the coordinate plane to answer each question.

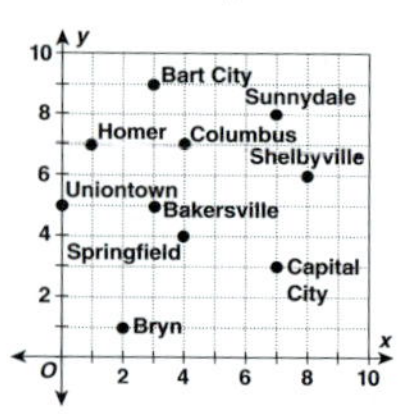

1. What city is located at point (4, 4) on the map?
Springfield

2. Which city is located at point (8, 6) on the map?
Shelbyville

3. Which city's location is given by an ordered pair that includes a 0?
Uniontown

4. What ordered pair describes the location of Capital City?
(7, 3)

5. If you started at (0, 0) and moved 1 unit north and 2 units east, which city would you reach?
Bryn

6. Which two cities on the map are both located 4 units to the right of (0, 0)?
Springfield and Columbus

Circle the letter of the correct answer.

7. If you started in Bart City and moved 2 units south and 2 units west, which city would you reach?
A Columbus
B Sunnydale
Ⓒ Homer
D Bakersville

8. Starting at (0, 0), which of the following directions would lead you to Capital City?
Ⓕ Go 7 units east and 3 units north.
G Go 5 units north and 3 units east.
H Go 3 unit east and 7 units north.
J Go 8 units east and 6 units north.

6-7 Line Graphs

Learn to display and analyze data in line graphs.

Vocabulary

line graph

double-line graph

A New England Dame School, 1713

The first permanent English settlement in the New World was founded in 1607. It contained 104 colonists. Population increased quickly as more and more immigrants left Europe for North America.

The table shows the estimated population of English American colonies from 1650 to 1700.

Population of American Colonies				
Year	1650	1670	1690	1700
Population	50,400	111,900	210,400	250,900

Data that shows change over time is best displayed in a *line graph.* A **line graph** displays a set of data using line segments.

EXAMPLE 1 **Making a Line Graph**

Use the data in the table above to make a line graph.

Step 1: Place *years* on the horizontal axis and *population* on the vertical axis. Label the axes.

Step 2: Determine an appropriate scale and interval for each axis.

Step 3: Mark a point for each data value. Connect the points with straight lines.

Step 4: Title the graph.

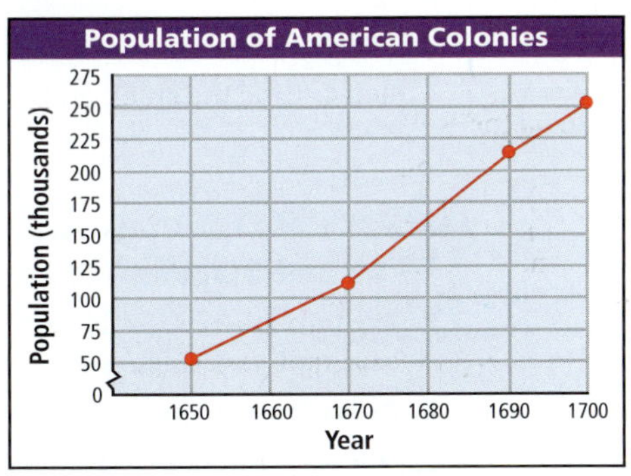

Helpful Hint

Because time passes whether or not the population changes, time is *independent* of population. Always put the independent quantity on the horizontal axis.

6-7 Organizer

Pacing: Traditional 1 day
Block $\frac{1}{2}$ day

Objective: Students display and analyze data in line graphs.

Indiana Academic Standards
6.6.1, 6.7.1, 6.7.5, 6.7.11

Warm Up

Describe how to graph each point on a coordinate grid.

1. (4, 5) right 4, up 5
2. (0, 2) up 2
3. (3, 0) right 3

Problem of the Day

Study the first two columns to determine a pattern to help fill in the blank square at the bottom.

4	49	6
11	134	12
7	85	6

Available on Daily Transparency in CRB

Math Humor

When the student dropped her graph on the floor, the teacher sent her to the principal's office for stepping over the line.

1 Introduce

Alternate Opener

EXPLORATION

6-7 Line Graphs

Luis works at a record store. His manager asked him to graph the number of CDs returned each day during one week.

Day	Sun	Mon	Tue	Wed	Thu	Fri	Sat
CDs Returned	10	3	2	4	7	11	14

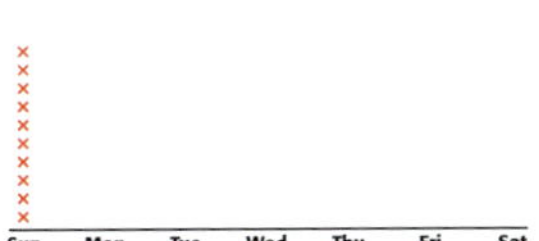

1. Complete the graph.
2. Draw a vertical line to the left of Sun (Sunday) to form the y-axis. Number this line from 1 to 14.
3. Delete all the ×'s in each stack except the one at the top.
4. Connect the ×'s with line segments. You have constructed a ***line graph.***

Think and Discuss

5. **Explain** how to construct a line graph.
6. **Discuss** some advantages of displaying data on a line graph rather than in a table.

Motivate

Display a line graph from a magazine or newspaper. Discuss what the graph shows, what is shown along the horizontal axis, and what is shown along the vertical axis. Explain that line graphs are the best graphs to show change over time.

Exploration worksheet and answers on Chapter 6 Resource Book pp. 64 and 110

2 Teach

Lesson Presentation

Guided Instruction

In this lesson, students display and analyze data in line graphs. First teach students to make a line graph, given data in a table. Then have students answer questions about a given line graph. Next, teach students to make a double-line graph, given data in a table.

Teaching Tip

Discuss why line graphs are better than bar graphs when representing data that show change over time. On a line graph, you can estimate values for times that fall between the given times.

Additional Examples

Example 1

Use the data in the table to make a line graph.

Population of New Hampshire	
Year	**Population**
1650	1,300
1670	1,800
1690	4,200
1700	5,000

Example 2

Use the line graph to answer each question.

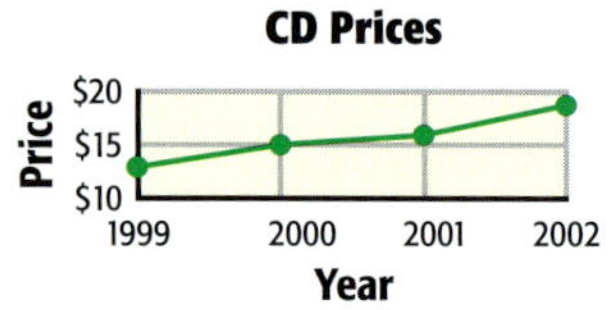

A. In which year did CDs cost the most? 2002

B. About how much did CDs cost in 2000? about $15

C. Did CD prices increase or decrease from 1999 through 2002? They increased.

Example 3

Use the data in the table to make a double-line graph.

Stock Prices				
	1985	**1990**	**1995**	**2000**
Corp. A	$16	$20	$34	$33
Corp. B	$38	$35	$31	$21

Answers on Additional Examples Transparencies T25–T27 in CRB

EXAMPLE 2 Reading a Line Graph

Use the line graph to answer each question.

A In which year did mountain bikes cost the least? 1997

B About how much did mountain bikes cost in 1999? about $300

C Did mountain bike prices increase or decrease from 1997 through 2001? They increased.

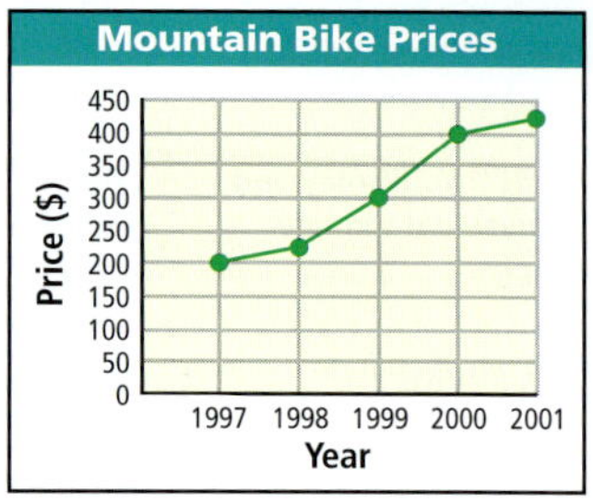

Line graphs that display two sets of data are called **double-line graphs**.

EXAMPLE 3 Making a Double-Line Graph

Use the data in the table to make a double-line graph.

Life Expectancy in the United States							
	1970	**1975**	**1980**	**1985**	**1990**	**1995**	**2000**
Male (yr)	67	69	70	71	72	73	74
Female (yr)	71	77	77	78	79	79	80

Helpful Hint

Use different colors of lines to connect the male and female values so you will easily be able to tell the data apart.

Step 1: Determine an appropriate scale and interval.

Step 2: Mark a point for each male value and connect the points.

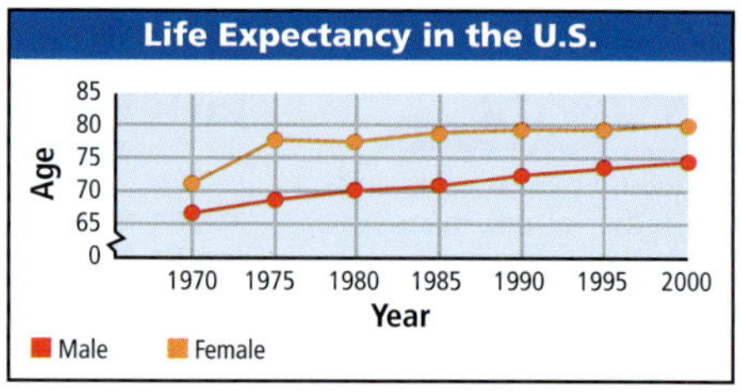

Step 3: Mark a point for each female value and connect the points.

Step 4: Title the graph and label both axes. Include a key.

Think and Discuss

1. **Explain** when it would be helpful to use a line graph instead of a bar graph to display data.
2. **Describe** how you might use a line graph to make predictions.
3. **Tell** why the graph in Example 3 needs a key.

Teach

Reaching All Learners

Through Curriculum Integration

Economics: Have students look in newspapers, magazines, and/or online for line graphs that depict stock market performance for various companies. Have students explain what each graph shows and then write and answer several questions about the graph.

3 Close

Summarize

Have students write brief descriptions for *line graphs* and *double-line graphs*. Have students give an example of data that would best be represented by a line graph.

Possible answer: Line graphs represent data values as points connected by lines. A double-line graph shows two sets of data. You could show in a line graph how the number of houses in a given town changes over the course of several years.

Answers to Think and Discuss

1. when the data show change over a period of time
2. Possible answer: You could look for a trend in the data, for example, increasing or decreasing, and then use that information to predict what will happen in the future.
3. so you can tell which line represents males and which represents females

6-7 Exercises

FOR EXTRA PRACTICE
see page 648

internet connect
Homework Help Online
go.hrw.com Keyword: MR4 6-7

6-7 PRACTICE & ASSESS

GUIDED PRACTICE

See Example 1

1. Use the data in the table to make a line graph.

School Enrollment				
Year	2000	2001	2002	2003
Students	2,000	2,500	2,750	3,500

Students may want to refer back to the lesson examples.

See Example 2

Use the line graph to answer each question.

2. In which year did the most students participate in the science fair? **2000**

3. Did the number of students increase or decrease from 2000 to 2001? **decrease**

Participants in Science Fair

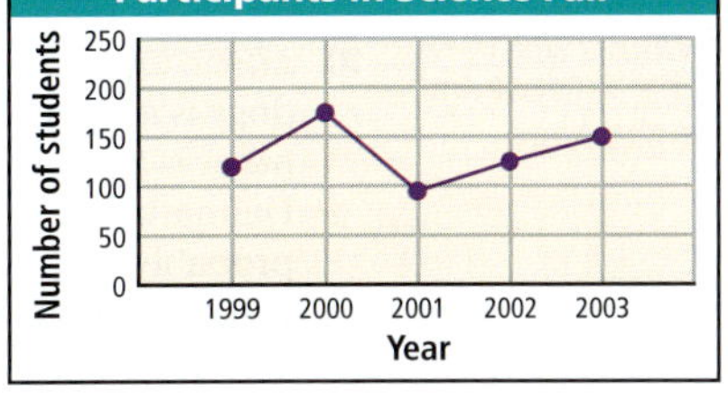

See Example 3

4. Use the data in the table to make a double-line graph.

	January	February	March	April	May
Stock A	$10	$12	$20	$25	$22
Stock B	$8	$8	$12	$20	$30

INDEPENDENT PRACTICE

See Example 1

5. Use the data in the table to make a line graph.

Winning Times in the Iditarod Dog Sled Race							
Year	1995	1996	1997	1998	1999	2000	2001
Time (hr)	219	222	225	222	231	217	236

See Example 2

Use the line graph to answer each question.

6. About how many personal computers were in use in the United States in 1996? **70 million**

7. When was the number of personal computers in use about 105 million? **1999**

Personal Computer Use in the U.S.

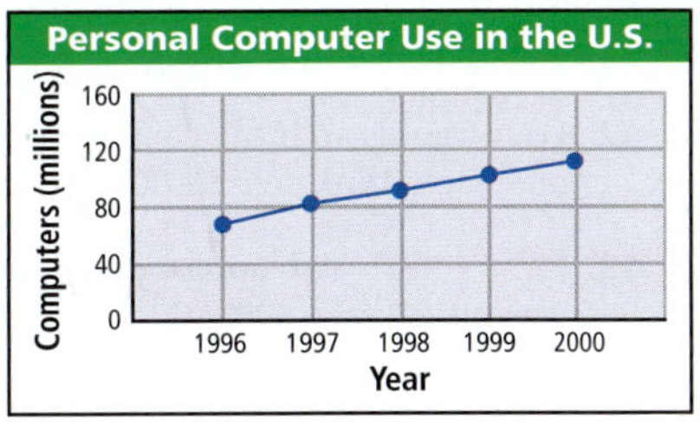

Assignment Guide

If you finished Example 1 assign:
Core 1, 5, 14–18
Enriched 1, 5, 12, 14–18

If you finished Example 2 assign:
Core 1–3, 5–7, 9, 11, 14–18
Enriched 1, 5–6, 9–10, 12–18

If you finished Example 3 assign:
Core 1–11, 14–18
Enriched 3–18

Answers

1. School Enrollment

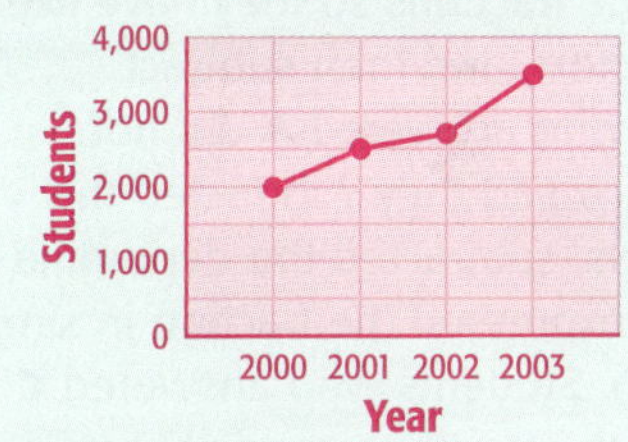

4. Comparison of Stock Prices

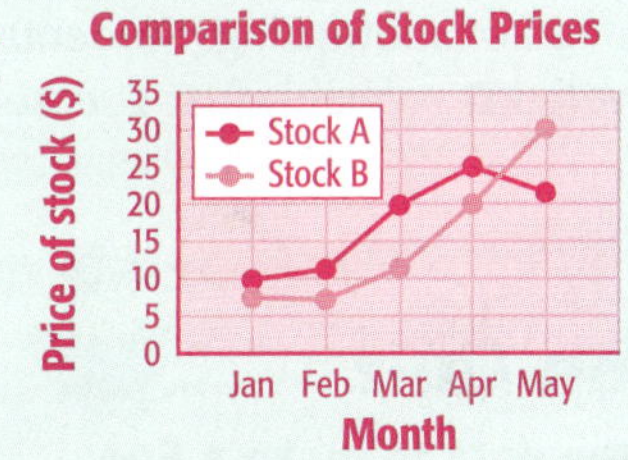

5. See p. A4.

RETEACH 6-7

LESSON 6-7 **Reteach**
Line Graphs

A line graph shows change over time.

You can represent data by making a line graph.

Stock Sales (millions)

Mon	Tue	Wed	Thurs	Fri
1.5	2.0	2.25	1.75	0.5

To make a line graph, make "days" the horizontal axis and "sales" the vertical axis. Label the axes.

Then determine an appropriate scale and interval for each axis.

Think of the data in the table as ordered pairs. Mark a point for each ordered pair. Then connect the points with straight segments.

Make sure your line graph has a title.

Stock Sales

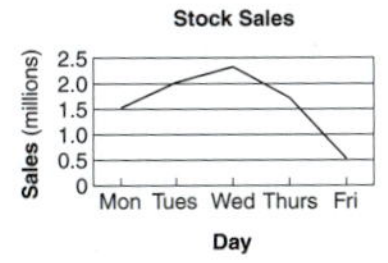

Use the table to make a line graph.

1. **Millie's Savings Account**

Jan	Feb	March	April	May
30	40	35	45	25

Millie's Savings Account

PRACTICE 6-7

LESSON 6-7 **Practice B**
Line Graphs

Use the line graph to answer each question.

1. In which year were the average weekly earnings in the United States the highest?
2000

2. In general, how did average weekly earnings in the United States change between 1970 and 2000?
The earnings increased.

3. In which year did the average United States worker earn about $350 a week?
1990

Average U.S. Weekly Earnings

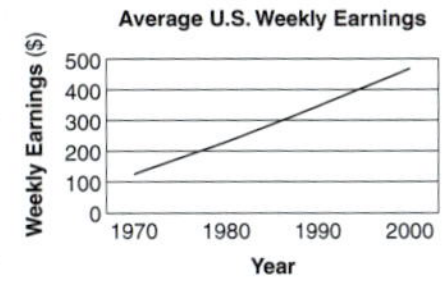

4. Use the given data to make a line graph.

U.S. Minimum Wage

Year	Hourly Rate
1970	$1.60
1980	$3.10
1990	$3.80
2000	$5.15

U.S. Minimum Wage

5. Between which two years shown on the graph did the U.S. minimum wage change the least?
1980 and 1990

6. How has the hourly minimum wage changed in the U.S. since 1970?
It has increased.

Math Background

When you are deciding whether to display data using a line graph or a bar graph, it is sometimes unclear which to choose. The general choice of a line graph to show change over time and a bar graph to show noncontinuous data is a good rule of thumb, but cannot be followed in every instance. For example, if a graph is to show tourist population at a beach resort over the four seasons of a year, a bar graph would be a better choice, even though the graph shows change over time.

Answers

8. Soccer Team Fund-Raising Efforts

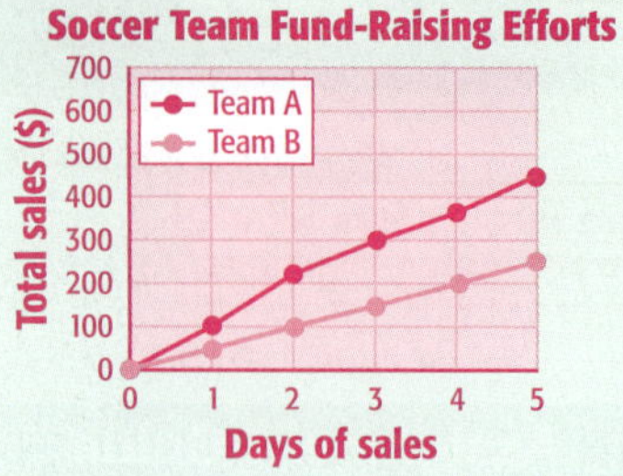

11–12. See p. A4.

Journal

Have students make up and solve a word problem for the graphs they made for Exercise 8.

Test Prep Doctor

For Exercise 18, students need to rename the fractions so they have like denominators, and then subtract. Students who answered **A** did not rename; instead they subtracted the numerators, subtracted the denominators, and expressed the fraction in simplest form. Students who answered **C** renamed the fractions correctly but added instead of subtracting. Students who answered **D** probably just guessed.

Lesson Quiz

1. Use the data to make a line graph. See p. A8.

Number of Aluminum Cans Collected				
Mon	Tue	Wed	Thu	Fri
100	150	200	125	175

Use the line graph to answer each question.

Plant Growth

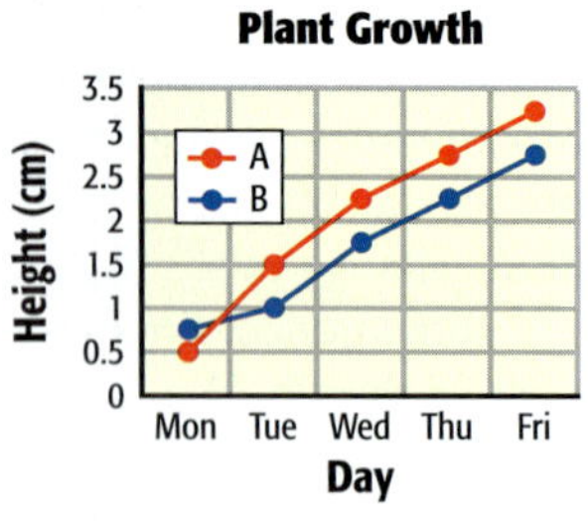

2. Which plant was taller on Tuesday? A

3. Which plant grew more between Thursday and Friday? Each grew the same amount.

4. Which plant grew the most in one week? A

Available on Daily Transparency in CRB

See Example 3

8. Use the data in the table to make a double-line graph.

Soccer Team's Total Fund-Raising Sales						
Day	0	1	2	3	4	5
Team A	\$0	\$100	\$225	\$300	\$370	\$450
Team B	\$0	\$50	\$100	\$150	\$200	\$250

PRACTICE AND PROBLEM SOLVING

Use the line graph for Exercises 9 and 10.

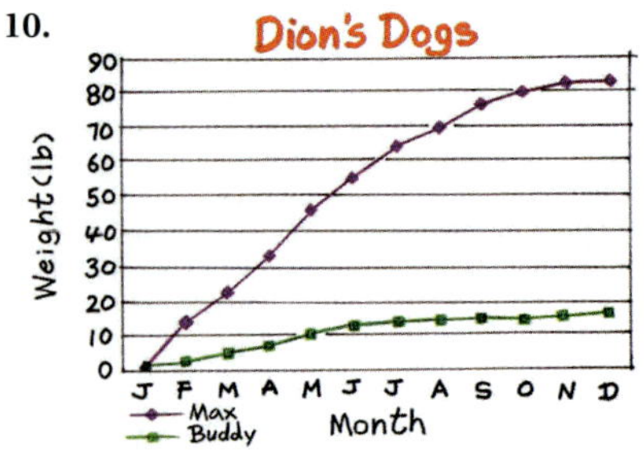

9. ***LIFE SCIENCE*** Estimate the difference in the dogs' weights in March.

9. about 17 pounds

10. ***LIFE SCIENCE*** One of Dion's dogs is a Great Dane, and the other is a miniature Dalmatian. Which dog is probably the Great Dane? Justify your answer.

10. Max is probably the Great Dane because he weighs more.

11. ***LIFE SCIENCE*** The table shows the weights in pounds for Sara Beth's two pets. Use the data to make a line graph that is similar to Dion's.

	Jan	Feb	Mar	Apr	May	Jun	Jul	Aug	Sep	Oct	Nov	Dec
Ginger	3	9	15	21	24	25	26	25	26	27	26	28
Toto	4	8	13	17	24	26	27	29	25	26	28	28

12. ***WRITE ABOUT IT*** Suppose you have a bowl of soup with lunch. Draw a line graph that could represent the changes in the soup's temperature during lunch. Explain.

13. ***CHALLENGE*** Describe a situation that this graph could represent.
Possible answer: the speed of a runner in a 1-mile race

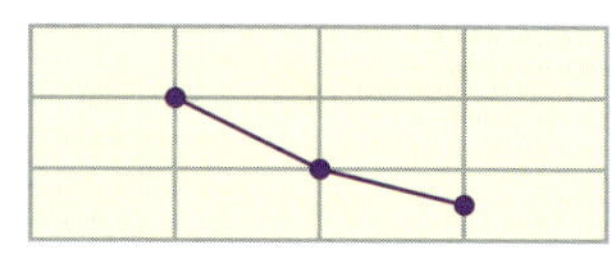

Spiral Review

Identify the property that is illustrated by each equation. (Lesson 1-5)

14. $3 + (4 + 5) = (3 + 4) + 5$ associative

15. $19(24) = 19(20) + 19(4)$ distributive

16. $(2)(13) = (13)(2)$ commutative

17. Four friends split the cost of a pizza and four drinks. The pizza cost \$12, and each drink cost \$2.00. How much did each person pay? (Lesson 3-8) \$5.00

18. **TEST PREP** Two apples weigh $\frac{1}{4}$ lb and $\frac{3}{16}$ lb. Find the difference in their weights. (Lesson 5-7) B

A $\frac{1}{6}$ lb **B** $\frac{1}{16}$ lb **C** $\frac{7}{16}$ lb **D** $\frac{1}{4}$ lb

CHALLENGE 6-7

LESSON 6-7 **Challenge**
A Trendy Park

Because line graphs show changes over time, you can use them to make predictions based on trends, or patterns. United States park rangers make line graphs to look for trends. They count the number of people who visit their parks each month. Then the park rangers analyze the data on line graphs to look for trends and predict how many visitors to expect each month in the coming years. This data helps the rangers schedule workers and provide services for their visitors.

Great Smoky Mountains National Park in Tennessee and North Carolina receives more visitors each year than any other national park. Imagine you are a park ranger there. Use the line graph below to identify trends and make predictions about the number of visitors the park will receive in the future.

Visitors At Great Smoky Mountains National Park, 2000

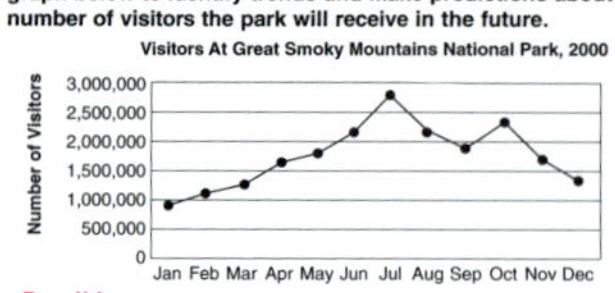

Possible answers:

1. In which month next year should you plan for the most visitors at your park? July

2. What can you expect next year at the park between October and January?
Each month the number of visitors will decrease.

3. You are in charge of deciding how many rangers should be scheduled to work at Great Smoky Mountains National Park each month next year. How will the number of park rangers you schedule change each month from January to June?
Each month, I will schedule more park rangers to work than the month before.

4. Which month next year would be best for you to take time off from your park ranger job and go on your own vacation? Explain.
January; because that is the month when the park has the least amount of visitors.

PROBLEM SOLVING 6-7

LESSON 6-7 **Problem Solving**
Line Graphs

Use the line graphs to answer each question.

U.S. Farm Population

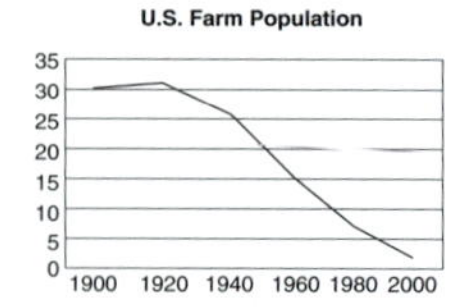

Size of U.S. Farms

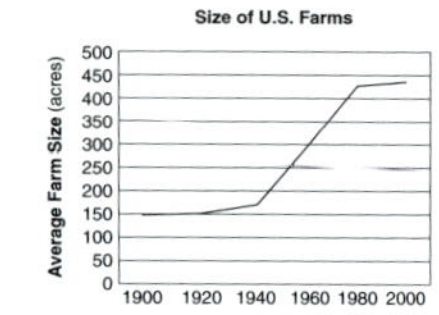

1. In which year was the U.S. farm population the highest? the lowest?
1920; 2000

2. In which year was the size of the average U.S. farm the largest? the smallest?
2000; 1900

3. How many people lived on farms in the United States in 1940?
26 million

4. How many acres did the average farm in the United States cover in 1980?
426 acres

5. In general, how has the U.S. farm population changed in the last 100 years?
The population has decreased.

6. In general, how has the size of the average U.S. farm changed in the last 100 years?
The average size has increased.

Circle the letter of the correct answer.

7. Between which two years did the U.S. farm population increase?
(A) 1900 and 1920
B 1920 and 1940
C 1940 and 1960
D 1960 and 1980

8. Between which two years did the average size of farms in the United States change the least?
(F) 1900 and 1920
G 1920 and 1940
H 1960 and 1980
J 1980 and 2000

6-8 Misleading Graphs

Learn to recognize misleading graphs.

Data can be displayed in many different ways. Sometimes people who make graphs choose to display data in a misleading way.

This bar graph was created by a group of students who believe their school should increase support of the football team. How could this bar graph be misleading?

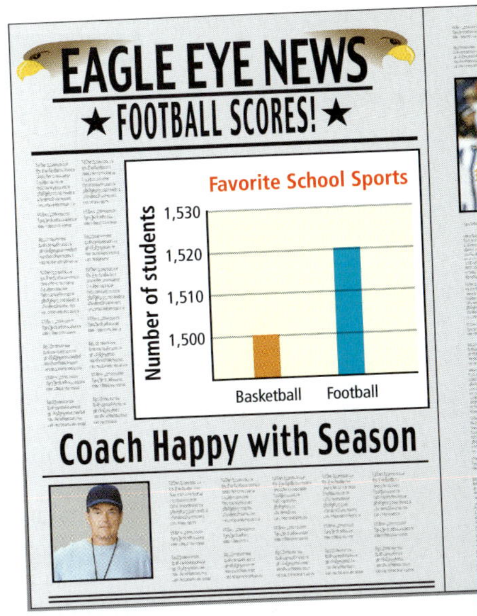

At a glance, you might conclude that about three times as many students prefer football to basketball. But if you look at the values of the bars, you can see that only 20 more students chose football over basketball.

EXAMPLE 1 Misleading Bar Graphs

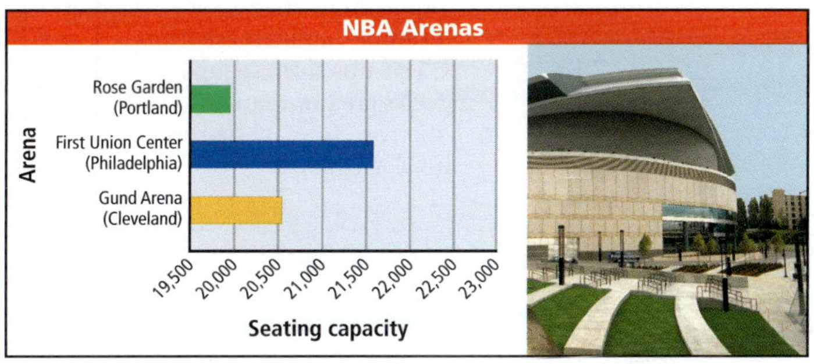

A **Why is this bar graph misleading?**

Because the lower part of the horizontal scale is missing, the differences in seating capacities are exaggerated.

B **What might people believe from the misleading graph?**

People might believe that the First Union Center holds 2–4 times as many people as Gund Arena and the Rose Garden. In reality, the First Union Center holds only one to two thousand more people than the other two arenas.

6-8 Organizer

Pacing: Traditional 1 day
Block $\frac{1}{2}$ day

Objective: Students recognize misleading graphs.

Indiana Academic Standards
6.6.1, 6.7.1, 6.7.5, 6.7.10, 6.7.11

Warm Up

Use the data below to answer each question.

20 21 23 24 27 33 34 35 36
38 40 41 42 43 46 52 53

1. What is the median? 36
2. What is the mode? none
3. What is the range? 33

Problem of the Day

Nine students in a group found that their mean score was 86 for the first math test. On the next test, each student in the group scored 7 points higher than on the first test. What was their mean score for the two tests? 89.5

Available on Daily Transparency in CRB

Math Humor

The bankrupt restaurant was like a statistician waiting for data. There were lots of empty tables.

1 Introduce

Alternate Opener

EXPLORATION

6-8 Misleading Graphs

The graph shows the total number of medals won by four countries at the 2002 Winter Olympics.

1. According to the height of each bar, which country appears to have won approximately half the number of medals won by the United States?
2. Look at the numbers on the left to estimate the number of medals won by each country.
3. Use the estimates in number 2 to determine whether the answer to number 1 is accurate.

Think and Discuss

4. **Discuss** how the graph is misleading.
5. **Explain** how you could modify the graph to represent the data more accurately.

Motivate

Show students the graph at the beginning of the lesson (Teaching Transparency T29 in the Chapter 6 Resource Book). Discuss how the graph is misleading and what information people might believe from the graph. The scale does not begin at zero, and people might believe that 3 times as many students prefer football to basketball.

Exploration worksheet and answers on Chapter 6 Resource Book pp. 74 and 113

2 Teach

Lesson Presentation

Guided Instruction

In this lesson, students learn to recognize misleading graphs. First teach students to analyze a misleading bar graph. Then teach them to analyze line graphs that are misleading. Emphasize the importance of analyzing scales to decide whether graphs are misleading.

Teaching Tip Have students describe how they could change the graphs in Examples 1 and 2 to make them less misleading.

Additional Examples

Example 1

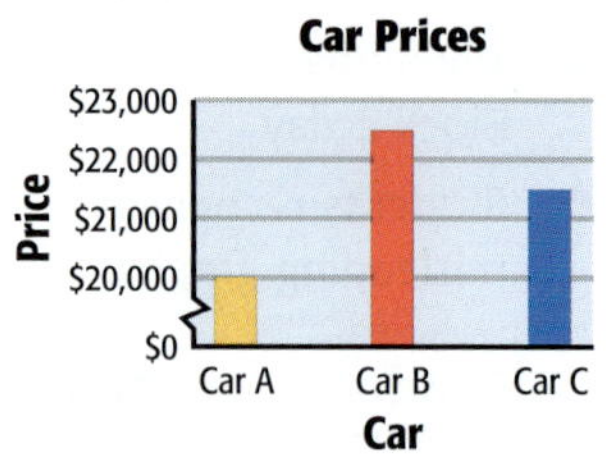

A. Explain why this bar graph is misleading.

B. What might people believe from the misleading graph?

Example 2

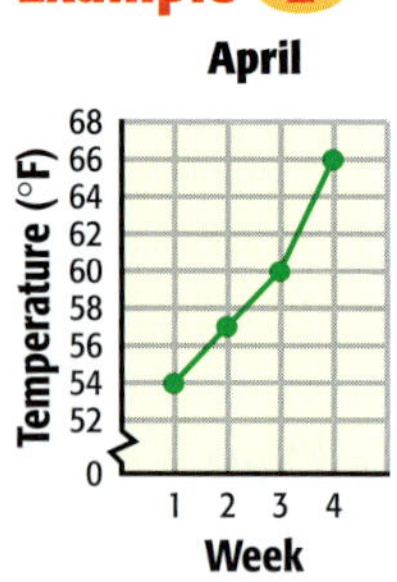

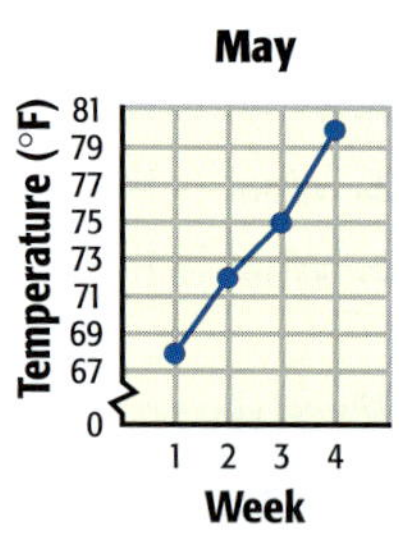

A. Explain why these line graphs are misleading.

B. What might people believe from these misleading graphs?

C. Explain why this graph is misleading.

Answers on Additional Examples Transparencies T30–T32 in CRB

EXAMPLE 2 **Misleading Line Graphs**

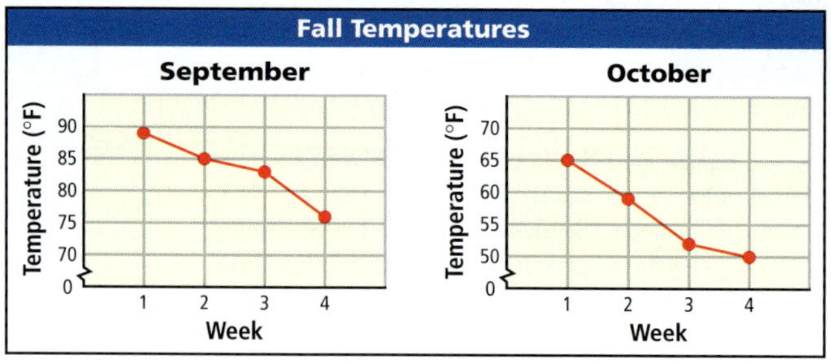

A **Why are these line graphs misleading?**

If you look at the scale for each graph, you will notice that the September graph goes from 75°F to 90°F and the October graph goes from 50°F to 65°F.

B **What might people believe from these misleading graphs?**

People might believe that the temperatures in October were about the same as the temperatures in September. In reality, the temperatures in September were 20–30 degrees higher.

C **Why is this line graph misleading?**

The scale does not have equal intervals. So, for example, an increase from 35 sit-ups to 40 sit-ups looks greater than an increase from 30 sit-ups to 35 sit-ups.

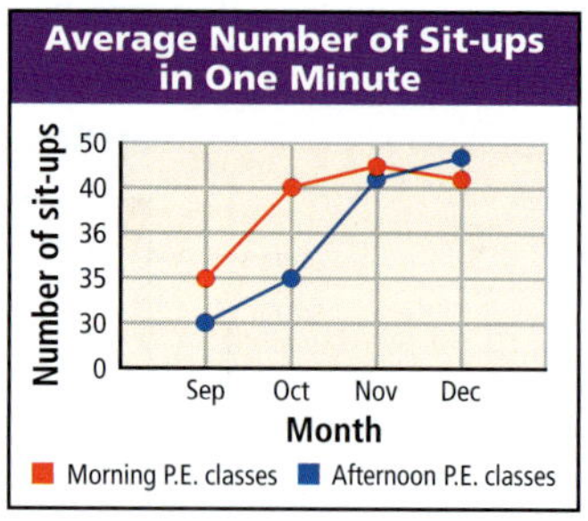

Think and Discuss

1. **Give an example** of a situation in which you think someone would intentionally try to make a graph misleading.
2. **Tell** who might have made the misleading graph in Example 2C.
3. **Tell** how you could change the graph in Example 2C so that it is not misleading.

Teach

Reaching All Learners

Through Critical Thinking

Have students work in groups of four to make misleading graphs and graphs that are not misleading. Have each group decide on a set of data to graph. Half of the group then graphs the data in a way that is misleading, while the other half graphs it in a way that is not misleading.

3 Close

Summarize

Review with students that graphs can be misleading, whether intentional or not. Remind students to carefully analyze all graphs before making conclusions about the data presented.

Answers to Think and Discuss

1. Possible answer: in advertising, to try to convince consumers to buy one product instead of another
2. Possible answer: The PE teacher may have made the graph as a way to convince the school principal to do away with afternoon PE classes.
3. Possible answer: Change the scale on the number of sit-ups to increments of 5.

6-8 Exercises

FOR EXTRA PRACTICE
see page 648

internet connect
Homework Help Online
go.hrw.com Keyword: MR4 6-8

GUIDED PRACTICE

See Example 1

1. Why is this bar graph misleading?
2. What might people believe from the misleading graph?
that the community center had more volunteers in the past than now

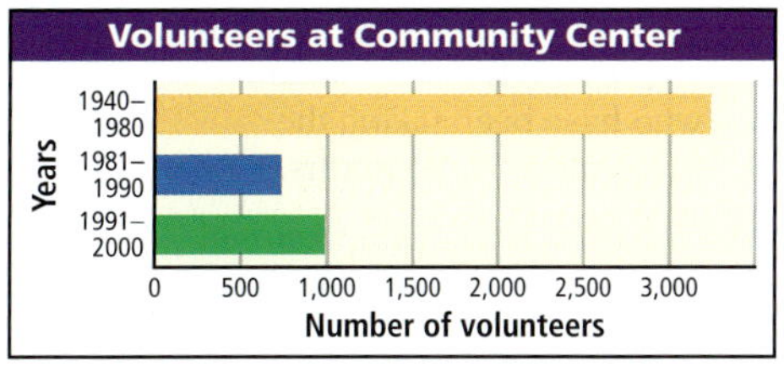

See Example 2

3. Why is this line graph misleading?
4. What might people believe from the misleading graph?
It appears that Kerry biked farther in the 30-minute period when actually she did not.

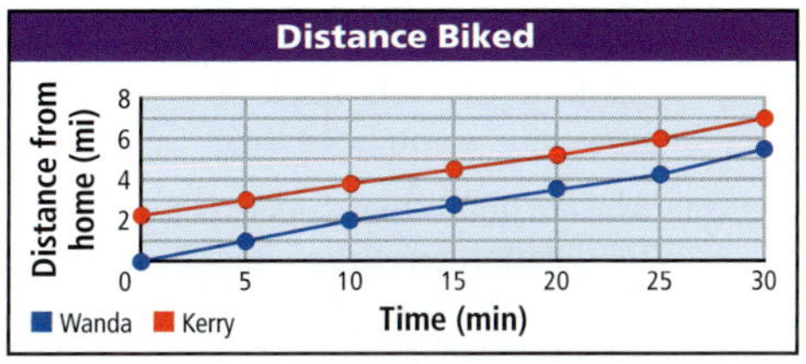

INDEPENDENT PRACTICE

See Example 1

5. Why is this bar graph misleading?
6. What might people believe from the misleading graph?
that the sixth graders have read 7 times more books than the seventh graders

See Example 2

7. Why is this line graph misleading?
8. What might people believe from the misleading graph?
Possible answer: The minimum wage is not increasing at as fast of a rate as it was several years ago

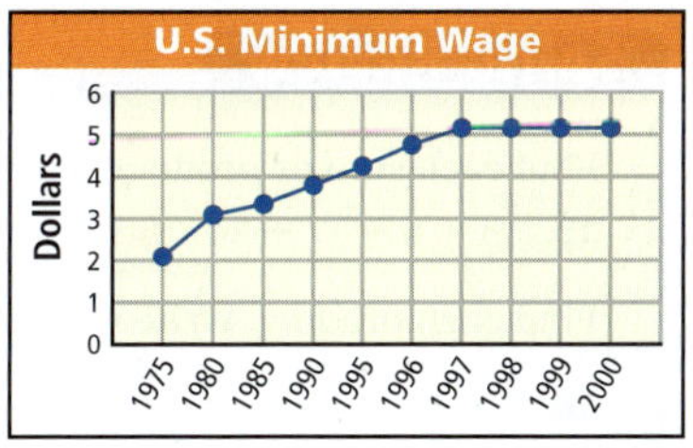

6-8 PRACTICE & ASSESS

Assignment Guide

If you finished Example 1 assign:
Core 1–2, 5–6, 15–22
Enriched 5–6, 12, 14–22

If you finished Example 2 assign:
Core 1–11, 15–22
Enriched 3–4, 7–22

Answers

1. Possible answer: The years for the top bar represent 40 years, but the other columns only represent 10 years.
3. Possible answer: Kerry does not begin biking from home.
5. The vertical axis begins at 430 rather than zero.
7. the yearly increments changed

RETEACH 6-8

LESSON 6-8 **Reteach**
Misleading Graphs

Graphs are often made to influence you. When you look at a graph, you need to figure out if the graph is accurate or if it is misleading.

Look at the graph below.

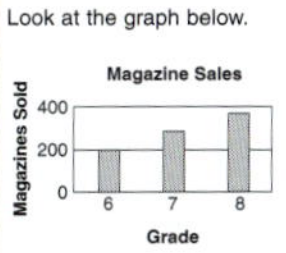

The graph is misleading because the intervals for the scale are so great. When you first look at the graph it appears that each grade sold about the same number of magazines.

Look at each graph. Then explain why each graph is misleading.

1.

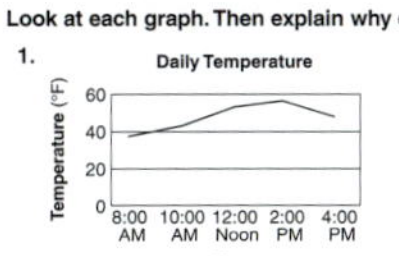

Possible answers are given.
The line graph is misleading because the intervals on the scale are so great. The graph leads you to think that the temperature changed very little throughout the day.

2.

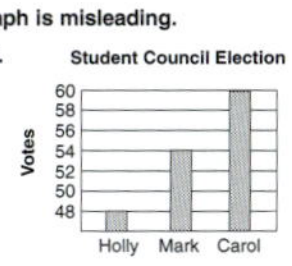

The bar graph is misleading because of the scale. The graph leads you to believe that the winner won the election by a large margin when it was actually a close race.

PRACTICE 6-8

LESSON 6-8 **Practice B**
Misleading Graphs

Use the graph to answer each question. Possible answers are given.

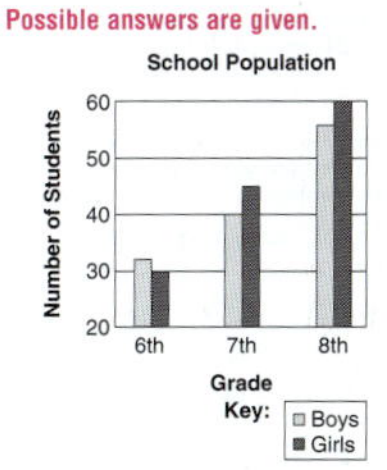

1. Explain why this bar graph is misleading.
Because the lower part of the vertical scale is missing, the differences in grades are exaggerated.
2. What might people believe from the misleading graph?
There are 4 times as many students in the 8th grade than the 6th grade.

Use the graph to answer each question.

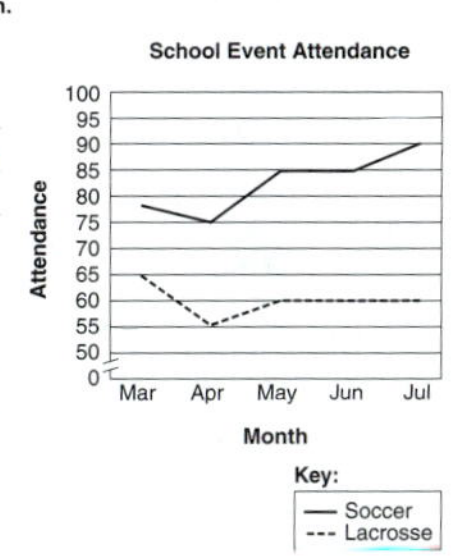

3. Explain why this line graph is misleading.
Because there is a break in the vertical scale, the differences in attendance seem greater than they really are.
4. What might people believe from the misleading graph?
In some months, 3 times more people attended soccer games than lacrosse games.

Math Background

Thanks to advances in computers and printing technology, graphs that include artwork, shading, color, perspective changes, and three-dimensional effects are now easy to produce. While they often have a pleasing appearance, they are also potentially misleading. The important thing to remember is that every graph must be examined with a critical eye.

More information on this issue can be found in Stephen M. Kosslyn's *Elements of Graph Design*.

Answers

9. Possible answer: line graph; because you easily see the changes from month to month

10. Possible answer:

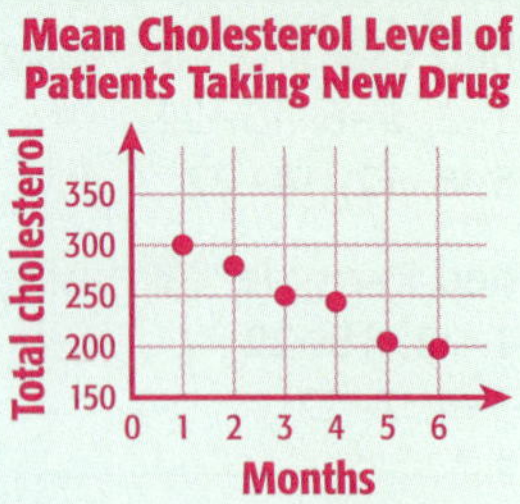

The vertical scale begins at 150.

11. Possible answer:

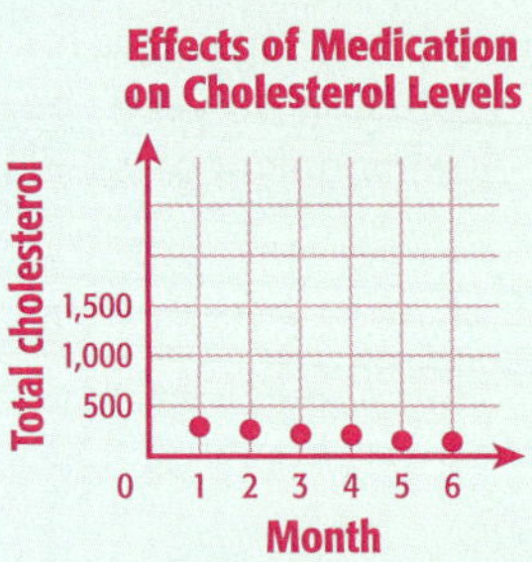

The vertical scale is far too large.

14. See p. A4.

Journal

Have students write about a situation in which they might want to use a misleading graph.

Test Prep Doctor

For Exercise 22, encourage students to eliminate wrong answer choices. Choice **C** can be eliminated right away because a tally table is not a graph. Choices **A** and **B** can be eliminated because the graph needs to show two sets of data.

Lesson Quiz

1. Why might this line graph be misleading?

The scale does not start at zero.

2. What might people believe from the graph? Possible answer: that there were hardly any visitors on Monday

Available on Daily Transparency in CRB

Health LINK

A research company has developed a new medication for lowering cholesterol levels. The table shows the mean monthly cholesterol levels for patients who have been taking the medication for 6 months.

Mean Cholesterol Level of Patients Taking New Medicine	
Month	Total Cholesterol
1	300
2	275
3	240
4	230
5	210
6	190

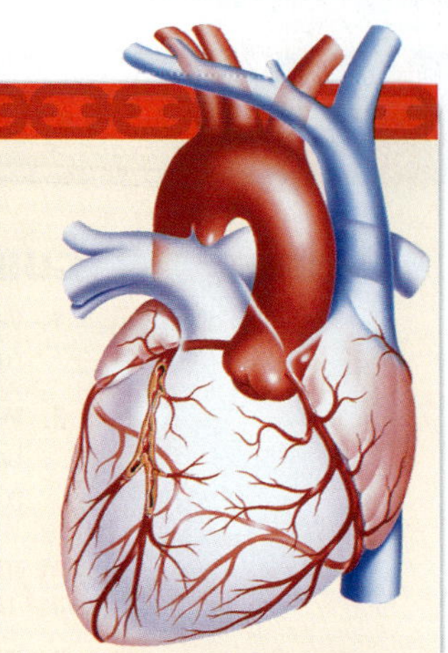

A heart with coronary artery disease, caused by buildup of fatty deposits

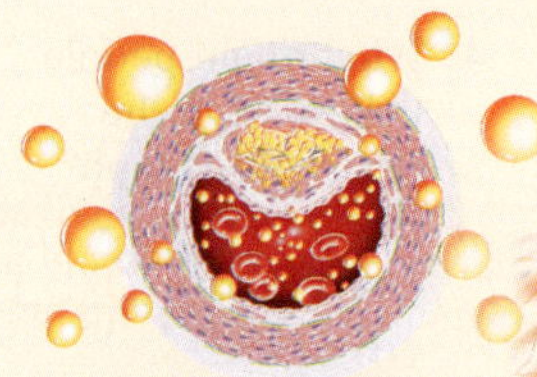

An artery that has been narrowed by high levels of blood cholesterol

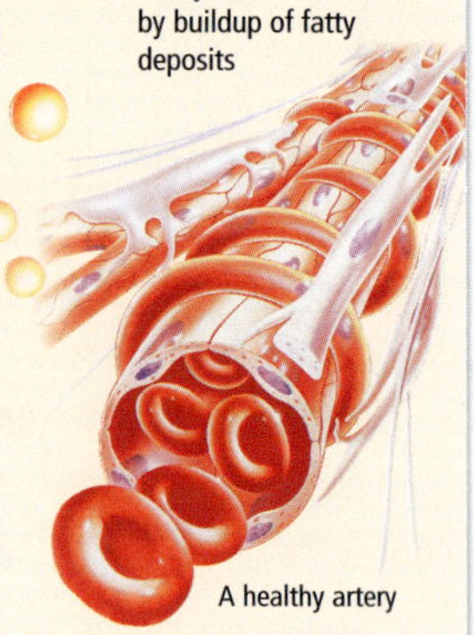

A healthy artery

9. What kind of graph would you make to display this data? Why?
10. Make a graph that suggests the medication greatly reduces cholesterol levels. Explain how your graph does this.
11. Make a graph that suggests the medication has little effect on cholesterol levels. Explain how your graph does this.
12. ***WHAT'S THE QUESTION?*** Look at the entries in the table. If the answer is 110, what is the question? What is the range of the data in the table?
13. ***WRITE ABOUT IT*** Suppose you saw your graph from Exercise 11 in an advertisement. What do you think it might be an advertisement for? Explain. for a competitor's drug
14. ***CHALLENGE*** What additional information could the research company gather and use to make a double-line graph that shows how its medication affects cholesterol levels?

Spiral Review

Find each sum or product. (Lesson 1-5)

15. 13 + 6 + 17 + 24 **60**
16. 4 · 11 · 3 **132**
17. 45 + 12 + 35 + 28 **120**

Find each product. Write your answers in simplest form. (Lesson 5-1)

18. $\frac{2}{3} \cdot \frac{1}{5}$ $\frac{2}{15}$
19. $\frac{3}{7} \cdot \frac{1}{4}$ $\frac{3}{28}$
20. $\frac{2}{9} \cdot \frac{3}{8}$ $\frac{1}{12}$
21. $\frac{1}{4} \cdot \frac{6}{7}$ $\frac{3}{14}$

22. **TEST PREP** Which type of graph would you use to display two sets of data that change over time? (Lesson 6-7) **D**

A Bar graph **B** Line graph **C** Tally table **D** Double-line graph

CHALLENGE 6-8

LESSON 6-8 Challenge
Graph Detective

You are a police detective in Capital City. A gang of criminals there is distributing misleading graphs to convince people that your city does not need to hire more police officers. It's your job to catch these graph crooks.

Search the graphs below for evidence of misleading displays of data. Then use your detective skills to explain why each graph is misleading.

Possible answers:

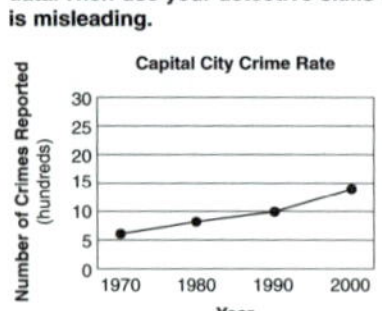

Why is this line graph misleading?
The intervals on the y-axis are so large, it looks like the crime rate has changed very little.

What might people believe from this misleading graph?
The city's crime rate is low and has not changed very much since 1970, so the city does not need to hire more police officers.

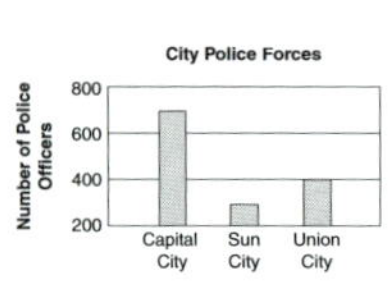

Why is this bar graph misleading?
Because the lower part of the vertical scale is missing, the differences in the cities' police forces are exaggerated.

What might people believe from this misleading graph?
Capital City has 5 times as many police officers as Sun City and $2\frac{1}{2}$ times as many as Union City, so Capital City does not need to hire more police officers.

PROBLEM SOLVING 6-8

LESSON 6-8 Problem Solving
Misleading Graphs

Use the graphs to answer each question. Possible answers:

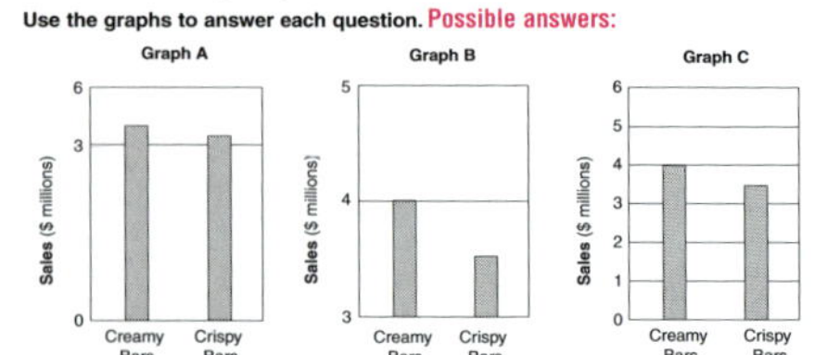

1. Why is Graph A misleading?
The vertical scale intervals are not equal, which makes the data look closer than it actually is.
2. Why is Graph B misleading?
The lower part of the vertical scale is missing; differences in sales are exaggerated.
3. What might people believe from reading Graph A?
About the same number of Crispy Bars and Creamy Bars were sold.
4. What might people believe from reading Graph B?
Creamy Bar sales were twice the sales of Crispy Bar sales.
5. Which graph do you think was made by the company that sells Crispy Bars? the company that sells Creamy Bars?
Graph A; Graph B
6. If you were writing a newspaper article about candy bar sales, which graph would be best to use? Why?
Graph C; it accurately displays the data.

Circle the letter of the correct answer.

7. Which of the following information is different on all three graphs above?
(A) the vertical scale
B the Crispy Bars sales data
C the Creamy Bars sales data
D the horizontal scale

8. Which of the following is a way that graphs can be misleading?
F breaks in scales
G uneven scales
H missing parts of scales
(J) all of the above

6-9 Stem-and-Leaf Plots

Learn to make and analyze stem-and-leaf plots.

Vocabulary

stem-and-leaf plot

A **stem-and-leaf plot** shows data arranged by place value. You can use a stem-and-leaf plot when you want to display data in an organized way that allows you to see each value.

The Explorer Scouts had a competition to see who could build the highest card tower. The table shows the number of levels reached by each scout.

Bryan Berg and his card model of the Iowa State Capitol

Number of Card-Tower Levels									
12	23	31	50	14	17	25	44	51	20
23	18	35	15	19	15	23	42	21	13

EXAMPLE 1 Creating Stem-and-Leaf Plots

Use the data in the table above to make a stem-and-leaf plot.

Step 1: Group the data by tens digits.

Step 2: Order the data from least to greatest.

Step 3: List the tens digits of the data in order from least to greatest. Write these in the "stems" column.

Step 4: For each tens digit, record the ones digits of each data value in order from least to greatest. Write these in the "leaves" column.

Step 5: Title the graph and add a key.

12 13 14 15 15 17 18 19
20 21 23 23 23 25
31 35
42 44
50 51

Number of Card Tower Levels

Stems	Leaves
1	2 3 4 5 5 7 8 9
2	0 1 3 3 3 5
3	1 5
4	2 4
5	0 1

Key: 1|5 means 15

Helpful Hint

To write 42 in a stem-and-leaf plot, write each digit in a separate column.

4|2
Stem Leaf

6-9 Organizer

Pacing: Traditional 1 day
Block $\frac{1}{2}$ day

Objective: Students make and analyze stem-and-leaf plots.

Indiana Academic Standards
6.6.1, 6.6.3, 6.7.1, 6.7.3, 6.7.5, 6.7.11

Warm Up

A set of data ranges from 12 to 86. What intervals would you use to display this data in a histogram with four intervals? Possible answer: 10–29, 30–49, 50–69, 70–89

Problem of the Day

What is the least number that can be divided evenly by each of the numbers 1 through 12? 27,720

Available on Daily Transparency in CRB

Math Fact

The stem-and-leaf plot was invented by Professor Jon Tukey, of Princeton University, in the 1960s.

1 Introduce

Alternate Opener

EXPLORATION

6-9 Stem-and-Leaf Plots

The table shows the times in seconds and hundredths of seconds for the women's 500-meter speed-skating competition in the 2002 Winter Olympics.

74.75	74.94	75.19	75.37	75.39	75.64	75.64	76.17	76.20	76.31
76.37	76.42	76.62	76.73	76.73	76.86	76.92	77.10	77.37	77.60
77.60	77.71	78.26	78.63	78.79	78.89	79.28	79.45	79.45	

You can organize the data by seconds and hundredths of seconds. Notice how the times are grouped using different colors.

1. Complete the ***stem-and-leaf plot.*** The stems represent seconds and the leaves represent hundredths of seconds. Notice how the colors correspond to the colors used in the table.

Stems	Leaves
74	75 94
75	
76	
77	
78	
79	

Key: 74 | 75 means 74.75

Think and Discuss

2. **Explain** what it means in this case for a stem to have the most number of leaves.
3. **Explain** what it means in this case for a stem to have the least number of leaves.

Motivate

Give students a series of 2-digit numbers in random order. Have students group the numbers by tens digits. Then have them order the numbers in each group from least to greatest. Explain that they just completed the first two steps of making a *stem-and-leaf plot.*

Exploration worksheet and answers on Chapter 6 Resource Book pp. 83 and 115

2 Teach

Lesson Presentation

Guided Instruction

In this lesson, students make and analyze stem-and-leaf plots. Teach students to make a stem-and-leaf plot, given a set of unordered data. Then teach students to analyze a stem-and-leaf plot to identify least and greatest values, mean, median, mode, and range.

Additional Examples

Example 1

Use the data to make a stem-and-leaf plot.

75 86 83 91 94 88 84 99 79 86

Example 2

Find the least value, greatest value, mean, median, mode, and range of the data.

Stems	Leaves
4	0 0 1 5 7
5	1 1 2 4
6	3 3 3 5 9 9
7	0 4 4
8	3 6 7
9	1 4

Key: 4|0 means 40

Answers on Additional Examples Transparencies T34–T35 in CRB

EXAMPLE 2 Reading Stem-and-Leaf Plots

Find the least value, greatest value, mean, median, mode, and range of the data.

Stems	Leaves
5	8
6	8 9
7	2 4 8
8	0 4 5 6 8
9	0 0 2 3 6 7 8
10	
11	7

Key: 5|8 means 58

The least stem and least leaf give the least value, 58.

The greatest stem and greatest leaf give the greatest value, 117.

Use the data values to find the mean.
(58 + . . . + 117) ÷ 19 = 85

The median is the middle value in the table, 86.

To find the mode, look for the number that occurs most often in a row of leaves. Then identify its stem. The mode is 90.

The range is the difference between the greatest and least value.
117 − 58 = 59

Helpful Hint

If a stem has no leaves, there are no data points with that stem. In the stem-and-leaf plot in Example 2, there are no data values between 100 and 109.

Think and Discuss

1. Describe how to show 25 on a stem-and-leaf plot.

6-9 PRACTICE & ASSESS

Students may want to refer back to the lesson examples.

Assignment Guide

If you finished Example 1 assign:
Core 1, 8, 17, 20–25
Enriched 8, 17, 19–25

If you finished Example 2 assign:
Core 1–8, 9–17 odd, 20–25
Enriched 6–25

Answers

1, 8. See p. A4.

6-9 Exercises

FOR EXTRA PRACTICE
see page 648

internet connect
Homework Help Online
go.hrw.com Keyword: MR4 6-9

GUIDED PRACTICE

See Example 1

1. Use the data in the table to make a stem-and-leaf plot.

Daily High Temperatures (°F)	45	56	40	39	37	48	51

See Example 2

Find each value of the data.

2. smallest value 10
3. largest value 44
4. mean 27.8
5. median 32
6. mode no mode
7. range 34

Stems	Leaves
1	0 2
2	
3	2
4	1 4

Key: 1|0 means 10

INDEPENDENT PRACTICE

See Example 1

8. Use the data in the table to make a stem-and-leaf plot.

Heights of Plants (cm)	30	12	27	28	15	47	37	28	40	20

Teach

Reaching All Learners

Through Grouping Strategies

Have students work in groups to make a stem-and-leaf plot for which all of the following are true:

- range = 35
- mode = 25
- median = 30

Possible answer:

Stems	Leaves
2	5 5 8
3	0 2 5
4	
5	
6	0

Key: 2|5 means 25

3 Close

Summarize

Have students find the least and greatest values, median, mode, and range for the stem-and-leaf plot in Example 1. Then have students tell how they would find the mean of the data.

Smallest = 12; largest = 51; median = 22; mode = 23; range = 39; Possible answer: Add all the numbers and divide by the number of numbers.

Answers to Think and Discuss

1. 2 | 5, where 2 is the stem and 5 is a leaf

See Example 2

Find each value of the data.

Stems	Leaves
4	1 2 2
5	1 3
6	7 8

Key: 4|1 means 41

9. least value 41
10. greatest value 68
11. mean 52
12. median 51
13. mode 42
14. range 27

PRACTICE AND PROBLEM SOLVING

18. It must be incorrect because it is significantly smaller than all of the numbers in the data set. The classmate found the median of the leaves only. The mean is ≈ 95.9.

For Exercises 15 and 16, write the letter of the stem-and-leaf plot described.

A.

Stems	Leaves
1	0 3 4
2	0 0 1 1 1 3
3	4 5 9
4	8
5	

Key: 1|0 means 10

B.

Stems	Leaves
1	6
2	2 3
3	0 1 4
4	1 4 8
5	8 8 8

Key: 1|6 means 16

C.

Stems	Leaves
1	4
2	
3	
4	3 6 8
5	2 2 4

Key: 1|4 means 14

15. The data set has a mode of 58. B
16. The data set has a median of 48. C

Use the table for Exercises 17 and 18.

Cars with Only One Passenger					
82	103	95	125	88	94
89	92	94	99	87	80
109	101	100	83	124	81

17. Karla recorded the number of cars with only one passenger that came through a toll booth each day. Use Karla's data to make a stem-and-leaf plot.

18. ***WHAT'S THE ERROR?*** Karla's classmate looked at the stem-and-leaf plot and said that the mean number of cars with only one passenger is 4. Explain Karla's classmate's error. What is the correct mean?

19. ***CHALLENGE*** Josh is the second youngest of 4 teenage boys, all 2 years apart in age. Josh's mother is 3 times as old as Josh is, and she is 24 years younger than her father. Make a stem-and-leaf plot to show the ages of Josh, his brothers, his mother, and his grandfather.

Spiral Review

20. Holly read 128 pages on Monday, 239 pages on Tuesday, and 152 pages each day on Wednesday through Friday. Estimate the number of pages Holly read to the nearest ten. (Lesson 1-2) 820 pages

Find each sum or difference. (Lesson 3-3)

21. 12.56 + 8.91 21.47
22. 19.05 − 2.27 16.78
23. 5 + 8.25 + 10.2 23.45
24. 40 − 20.66 19.34

25. **TEST PREP** Which value is **not** always a number in the data set it represents? (Lesson 6-2) D

A Mode B Lowest value C Highest value D Mean

Lesson Quiz

1. **Make a stem-and-leaf plot of the data.** See p. A8.

42 36 40 31 29 49 21 28 52 27 22 35 30 46 34 34

Find each value using the stem-and-leaf plot.

2. What is the least value? 21
3. What is the mean? 34.75
4. What is the median? 34
5. What is the mode? 34

Available on Daily Transparency in CRB

Answers

17, 19. See p. A4.

RETEACH 6-9

LESSON 6-9 **Reteach**
Stem-and-Leaf Plots

You can use place value to make a stem-and-leaf plot.

Points Earned in Games During Basketball Season						
27	16	34	29	48	12	33
20	18	42	51	27	32	41

Write the numbers in order from least to greatest.

12 16 18 20 27 27 29 32 33 34 41 42 48 51

List the tens digits in order from least to greatest in the first, or stem, column. Then, for each tens digit, record the ones digit for each data value in order from least to greatest in the second, or leaves, column.

Points Earned

Stem	Leaves
1	2 6 8
2	0 7 7 9
3	2 3 4
4	1 2 8
5	1

Key: 1 | 2 = 12

Make sure your graph has a title and a key.

Use the data to make a stem-and-leaf plot.

1.

Valerie's Test Scores				
62	84	93	88	89
76	68	81	91	88

Valerie's Test Scores

Stem	Leaves
6	2 8
7	6
8	1 4 8 8 9
9	1 3

Key: 6 | 2 = 62

2. What is the range? The range is 31 points.
3. What is the median? The median is 86 points.
4. What is the mode? The mode is 88 points.

PRACTICE 6-9

LESSON 6-9 **Practice B**
Stem-and-Leaf Plots

Complete each activity and answer the questions.

1. Use the data in the table to complete the stem-and-leaf plot below.

Richmond, Virginia, Monthly Normal Temperatures (°F)											
Jan	Feb	Mar	April	May	June	July	Aug	Sep	Oct	Nov	Dec
37	39	48	57	74	78	77	76	70	59	50	40

Stem	Leaves
3	7 9
4	0 8
5	0 7 9
6	
7	0 4 6 7 8

Key: 1 | 2 = 12°F

Find each value of the data set.

Stem	Leaves
6	1 4
7	1 6
8	2 2
9	0 1 8

Key: 6 | 5 = 65

2. smallest value 61
3. largest value 98
4. mean $79.\overline{4}$
5. median 82
6. mode 82
7. range 37
8. Look at the stem-and-leaf plot you made for Exercise 1. How many months in Richmond have a normal temperature above 70°F? 4 months
9. How would you display a data value of 100 on the stem-and-leaf plot above? Use 10 for the stem and 0 for the leaf.

CHALLENGE 6-9

LESSON 6-9 **Challenge**
A Plot of Trees

As part of your job with the National Forest System, you must compile a report on the tallest trees in the United States. You have already collected the data and displayed it on the bar graph below. To complete the report, you need to make a stem-and-leaf plot of the data. Use the hundreds digits of the data for your stems. Then analyze the data.

National Forest System Report, 2000

Tallest Trees in the U.S.

Stem	Leaves
1	78 94
2	06 19 27 32 72 75 81
3	21

Key: 1 | 78 = 178

Data Analysis

Range of heights: 143 feet
Mean height: 240.5 feet
Median height: 229.5 feet
Mode height: none

Tallest Trees in the United States

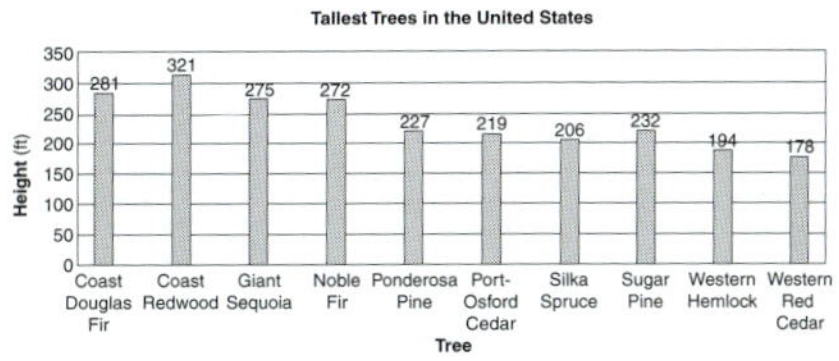

PROBLEM SOLVING 6-9

LESSON 6-9 **Problem Solving**
Stem-and-Leaf Plots

Use the Texas stem-and-leaf plots to answer each question.

Dallas Normal Monthy Temperatures

Stem	Leaves
4	3 7 8
5	6 7
6	6 7
7	3 7
8	1 5 5

Key: 4 | 3 = 43°F

Houston Normal Monthy Temperatures

Stem	Leaves
5	0 4 4
6	1 1 8
7	0 5 8
8	0 2 3

Key: 5 | 0 = 50°F

1. Which city's temperature data has a mode of 85°F? Dallas
2. Which city's temperature data has a range of 33°F? Houston
3. Which city has the lowest data value? What is that value? Dallas; 43°F
4. Which city has the highest data value? What is that value? Dallas; 85°F
5. Which city's temperature data has a mean of 68°F? Houston
6. Which city's temperature data has a median of 69°F? Houston

Circle the letter of the correct answer.

7. What do the data values 54°F and 61°F represent for the plots above?
   - A the ranges of normal temperatures in Dallas and Houston
   - (B) the mode normal temperatures for Houston
   - C the mean and median normal temperatures for Dallas
   - D the lowest normal temperatures for Dallas and Houston
8. Which of the following would be the best way to display the Dallas and Houston temperature data?
   - (F) on a line graph
   - G in a tally table
   - H on a bar graph
   - J on a coordinate plane

CHAPTER 6 Extension

Pacing: Traditional 1 day
Block $\frac{1}{2}$ day

Objective: Students make and read box-and-whisker plots.

Using the Pages

In Lesson 6-9, students made and analyzed stem-and-leaf plots. In this extension, students will learn to represent data using another descriptive organizer, the box-and-whisker plot.

EXTENSION Box-and-Whisker Plots

Learn to make and read box-and-whisker plots.

Vocabulary
box-and-whisker plot
lower extreme
lower quartile
upper quartile
upper extreme

A **box-and-whisker plot** shows how data is distributed. To make a box-and-whisker plot of a data set, you need to know the following five values:

- the **lower extreme**, the least value
- the **lower quartile**, the median of the lower half of the data
- the median of the data
- the **upper quartile**, the median of the upper half of the data
- the **upper extreme**, the greatest value

EXAMPLE 1 Making a Box-and-Whisker Plot

Use the data in the table to make a box-and-whisker plot.

Cans Collected (lb)	10	23	15	17	26	27	21	22	19	11	16

10, 11, 15, 16, 17, 19, 21, 22, 23, 26, 27 *Order from least to greatest.*
10, 11, 15, 16, 17, 19, 21, 22, 23, 26, (27) *Find the upper extreme.*
(10), 11, 15, 16, 17, 19, 21, 22, 23, 26, 27 *Find the lower extreme.*
10, 11, 15, 16, 17, (19), 21, 22, 23, 26, 27 *Find the median.*
10, 11, (15), 16, 17, 19, 21, 22, 23, 26, 27 *Find the lower quartile.*
10, 11, 15, 16, 17, 19, 21, 22, (23), 26, 27 *Find the upper quartile.*

Step 1: Make a box using the median and the upper and lower quartiles.

Step 2: Place a dot at the upper and lower extremes.

Step 3: Connect the dots to the box with segments called whiskers.

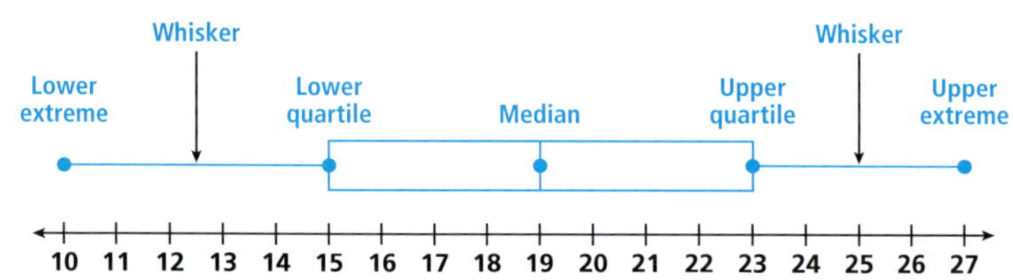

1 Introduce

Motivate

It is easier to compare data sets if the data sets are represented graphically. The box-and-whisker plot is a good method of showing how data is distributed.

2 Teach

Lesson Presentation

Guided Instruction

In this extension, students make and read box-and-whisker plots. Teach students the five-value summary of a data set: the lower extreme, the lower quartile, the median, the upper quartile, and the upper extreme. Teach students to find the five values and to use those values to make a box-and-whisker plot. Lastly, discuss with students the range and the interquartile range.

The range is the difference between the upper and lower extremes.
$27 - 10 = 17$

The interquartile range is the difference between the upper and lower quartiles. $23 - 15 = 8$

EXTENSION Exercises

Use the box-and-whisker plot to find each value.

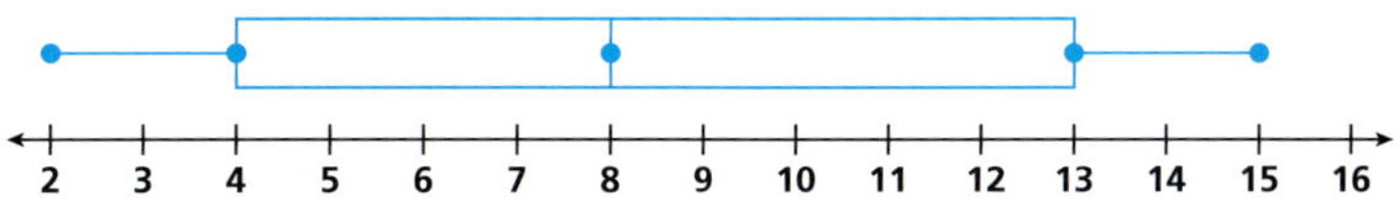

1. lower extreme **2**
2. median **8**
3. upper quartile **13**
4. lower quartile **4**
5. upper extreme **15**
6. interquartile range **9**
7. Use the data in the table to create a box-and-whisker plot.

Waiting Times for Movie Tickets (min)					
0	0	5	5	2	9
4	4	1	8	20	3

The two box-and-whisker plots represent the test scores for two different math classes. Use the box-and-whisker plots for Exercises 8–12.

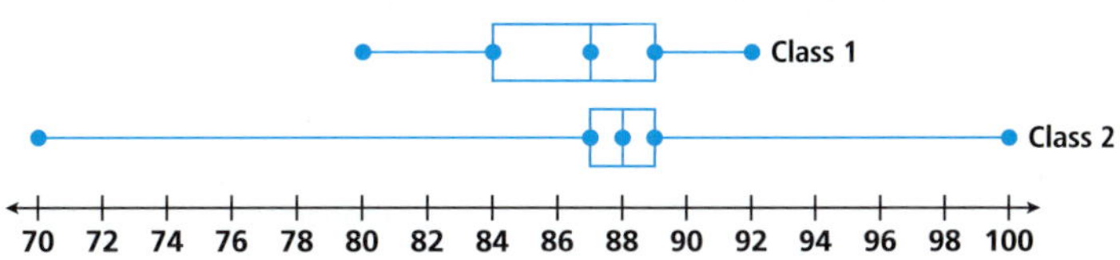

8. What is each class's median score? **Class 1: 87, Class 2: 88**
9. Which class's interquartile range is the greatest? **Class 1**
10. What is the highest score from either class? **100**
11. What is the difference between the ranges of the two sets of scores? **18**
12. Can you use these plots to determine which class has more students? Explain why or why not. **No; The plots do not indicate how many pieces of data were used to create them.**

COMMON ERROR ALERT

In the data set 2, 4, 4, 9, 10, students may include the underlined 4 when trying to find the lower quartile. Remind students that the lower half of the data set is 2, 4 and doesn't include the underlined 4. The lower quartile of this data set is 3.

Additional Example

Example 1

Use the data in the table to make a box-and-whisker plot.

High Temperatures (°F)							
72	75	78	65	68	70	73	68
67	71	74					

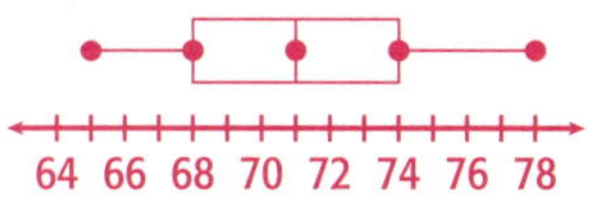

Answer

7.
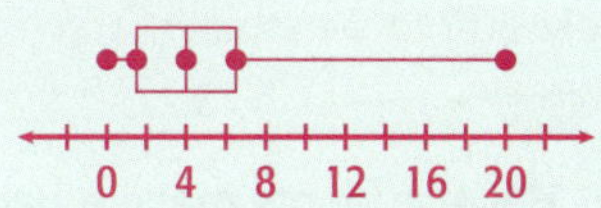

3 Close

Summarize

Have a student volunteer explain how to make a box-and-whisker plot of the following data set: 7, 8, 3, 9, 6, 5, 1, 2, and 4.

Possible answer: Order the data set: 1, 2, 3, 4, 5, 6, 7, 8, 9; upper extreme = 9; lower extreme = 1; median = 5; lower quartile = 2.5; upper quartile = 7.5; make a box using the median and the lower and upper quartiles; place a dot at the upper and lower extremes; connect the dots to the box with segments called whiskers.

Teaching Tip Remind students that the data should be ordered from least to greatest or from greatest to least when finding the five values.

Problem Solving on Location

Michigan

Purpose: *To provide additional practice for problem-solving skills in Chapters 1–6*

Shipwrecks

- After problem 4, have students discuss how finding the median was different in problems 3 and 4. In problem 3, the median lies halfway between 65 and 105. You must find the average of 65 and 105 to find the median. In problem 4, the two middle data values are both 65, so 65 is the median.
- After problem 5, have students speculate as to why there are so many modes. Why are there multiple shipwrecks with the same depths? Possible answer: There may be certain areas where shipwrecks are more likely due to weather, water conditions, etc.

Extension Have students make a bar graph to compare the resting depths of the steamers.

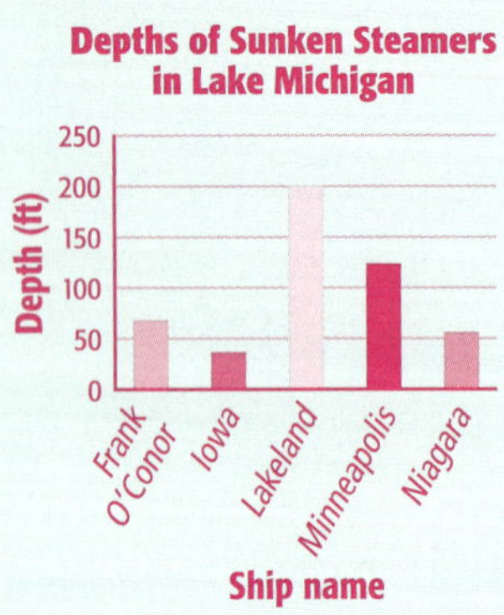

Problem Solving on Location

MICHIGAN

Shipwrecks

Over the past two hundred years, a large number of shipwrecks have occurred on Lake Michigan. One of the most famous shipwrecks was that of the *S.S. Milwaukee* on October 22, 1929. The *Milwaukee* was a steel steam-car ferry that was thought to be unsinkable because of its reinforced hulls and its massive size.

Today, the *S.S. Milwaukee* sits upright at a depth of 125 feet underwater. It is still intact, and all of its cargo is accounted for and in place.

The table lists other shipwrecks that have occurred on Lake Michigan and the depths to which the ships sank.

Lake Michigan Shipwrecks		
Ship's Name	**Type of Ship**	**Depth (ft)**
A. P. Nichols	Schooner	40
Forest	Schooner	40
Frank O'Connor	Steamer	65
Grace A. Channon	Schooner	200
Hetty Taylor	Schooner	105
Illinois	Barge	35
Iowa	Steamer	35
J. S. Crouse	Freighter	15
Jenny Bell	Schooner	125
Lakeland	Steamer	200
Linda E.	Barge	250
Lumberman	Schooner	65
Minneapolis	Steamer	125
Niagara	Steamer	52

For 1–5, use the table.

1. Find the range of the depths of the sunken ships. 235 ft
2. Find the mean of the depths of the sunken ships. Round your answer to the nearest hundredth. 96.57 ft
3. Find the median of the depths of the sunken schooners. 85 ft
4. Find the median of the depths of all the sunken ships. 65 ft
5. Find the mode(s) of the depths of the sunken ships. 35 ft, 40 ft, 65 ft, 125 ft, 200 ft

Butterflies

Every spring, the Frederik Meijer Gardens, in Grand Rapids, Michigan, hosts the largest temporary butterfly exhibit in the nation. The exhibit contains over 5,000 butterflies from various parts of the world. The butterflies are housed in a 15,000-square-foot tropical environment in the Gardens' conservatory, where visitors can walk among them.

The graph shows some of the most common types of butterflies at the exhibit.

For 1–5, use the graph.

1. Which type of butterfly is most common at the exhibit? **blue morpho**
2. The exhibit contains more than 400 of which type(s) of butterfly? **blue morpho, cydno longwing, monarch**
3. Estimate how many more Cydno longwings are at the exhibit than blue wings. **about 230**
4. About how many of the exhibit's butterflies are from the types shown in the graph? About what fraction is this of the entire butterfly population at the exhibit? (*Hint*: Remember there are about 5,000 butterflies total.) **about 2,500; $\frac{1}{2}$**
5. Order the types of butterflies in the graph from greatest number at the exhibit to least number at the exhibit.

blue morpho, cydno longwing, monarch, julia longwing, blue wing, zebra longwing, tropical swallowtail

Butterflies

- After problem 2, ask students to look at the data from the graph in a different way by making a frequency table and histogram that show the number of different types of butterflies that have 0–199, 200–399, 400–599 or 600–799, butterflies at the exhibit.

Number of Butterflies per Type				
Number	0–199	200–399	400–599	600–799
Frequency by type	1	3	2	1

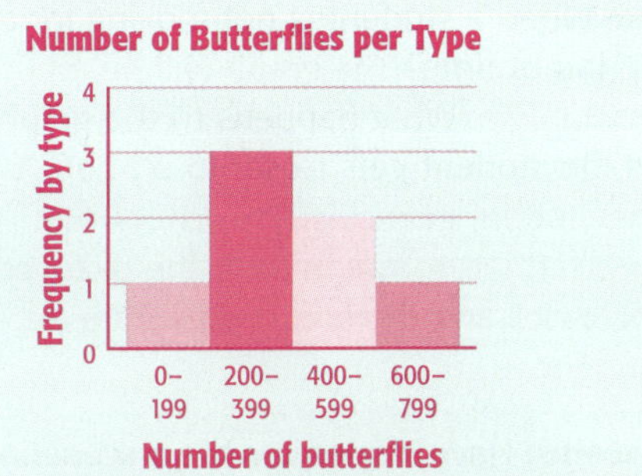

- After problem 5, have the students estimate the number of blue morpho butterflies that the graph indicates are at the exhibit. about 610 Then ask them to determine how many more blue morpho butterflies are at the exhibit than tropical swallowtails. about 435 What data vocabulary term best describes this difference? This difference is the *range* of the data displayed by the graph.

Extension Have students research different types of butterflies and collect pictures of, and interesting facts about, several species. Then have them create and conduct a survey to find out which butterfly is most interesting to their classmates. They should display the results of their surveys as both a data table and a graph. Check students' work.

MATH-ABLES

Game Resources

Puzzles, Twisters & Teasers
Chapter 6 Resource Book

A Thousand Words

Purpose: *To apply the problem-solving skill of interpreting graphs to a story problem*

Discuss: Ask: What will the graph look like when a student is not moving for a period of time? The graph will be a horizontal line. What happens to the graph as the student gets closer to school? As the student gets closer to school, he gets farther from home, so the distance increases and the graph rises from left to right.

Extend: Have students write a scenario similar to those presented in the problem. Have them make several different graphs and challenge a classmate to determine which graph matches their scenario. Check students' work.

Spinnermeania

Purpose: *To practice calculating mean, median, and mode in a game format*

Discuss: Discuss with students the range of answers they can expect for mean, median, and mode. What are the highest mean, median, and mode values possible? Are those values likely? The highest value possible for each measure is 10. To earn a mean, median, or mode of 10, the spinner would have to land on four 10's, which is not likely.

Extend: Have students create a new spinner using ten numbers chosen randomly from 0 to 100. Have them play the game again using the new board. Check students' work.

MATH-ABLES

A Thousand Words

Did you ever hear the saying "A picture is worth a thousand words"? A graph can be worth a thousand words too!

Each of the graphs below tells a story about a student's trip to school. Read each story and think about what each graph is showing. Can you match each graph with its story?

Kyla:
I rode my bike to school at a steady pace. I had to stop and wait for the light to change at two intersections. **B**

Tom:
I walked to my bus stop and waited there for the bus. After I boarded the bus, it was driven straight to school. **C**

Megan:
On my way to school, I stopped at my friend's house. She wasn't ready yet, so I waited for her. Then we walked to school. **A**

Graph A

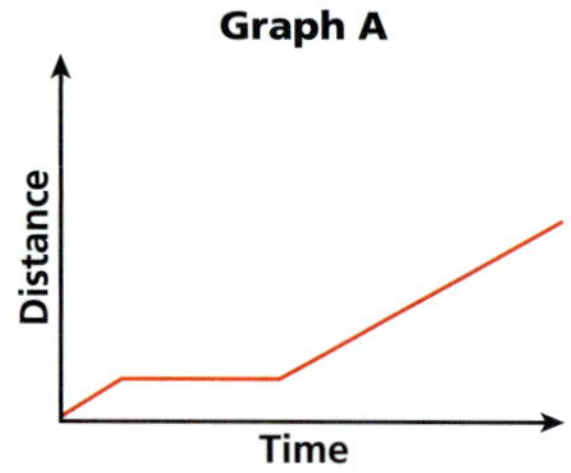

Graph B

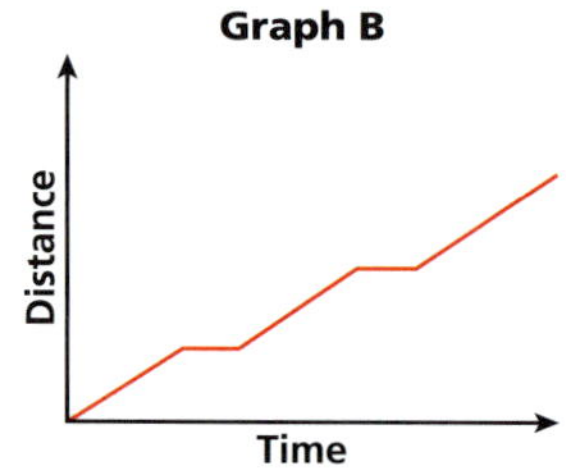

Graph C

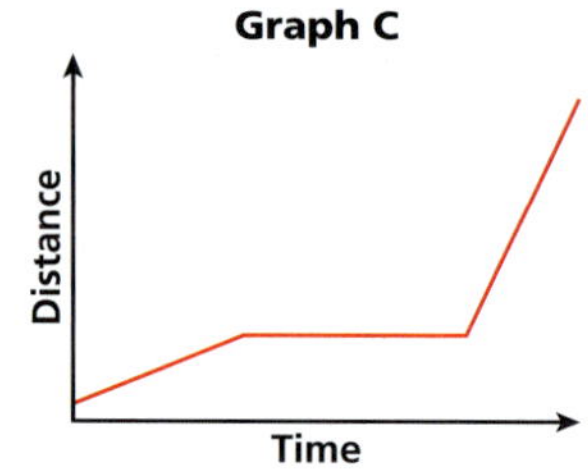

Spinnermeania

Round 1: On your turn, spin the spinner four times and record the results. After everyone has had a turn, find the mean, median, and mode of your results. For every category in which you have the highest number, you get one point. If there is a tie in a category, each player with that number gets a point. If your data set has more than one mode, use the greatest one.

Spin five times in round 2, eight times in round 3, ten times in round 4, and twelve times in round 5. The player with the highest score at the end of five rounds is the winner.

internet connect
Go to ***go.hrw.com*** for a complete set of game pieces.
KEYWORD: MR4 Game6

Technology LAB: Create Line Graphs

internet connect
Lab Resources Online
go.hrw.com
KEYWORD: MR4 TechLab6

Activity

The table shows the tide heights at various times. Use the data to make a line graph.

Time (A.M.)	9:00	9:06	9:12	9:18	9:24
Height (ft)	2.41	2.31	2.21	2.10	2.00

1. Enter the data. Put times in the **L1** list and heights in the **L2** list. You cannot enter times, so use 0 for 9:00, 1 for 9:06, and so on.

Times: STAT ENTER 0 ENTER 1 ENTER 2 ENTER 3 ENTER 4 ENTER

Heights: ▶ 2.41 ENTER 2.31 ENTER 2.21 ENTER 2.10 ENTER 2.00 ENTER

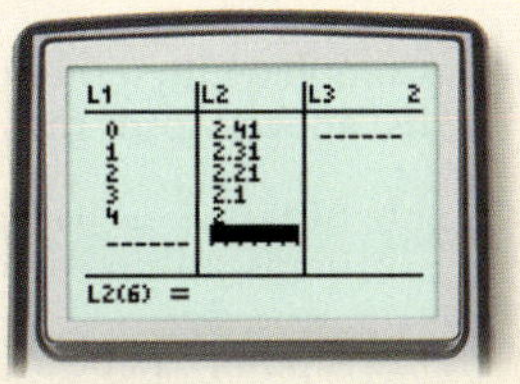

2. Choose a scale. Press WINDOW. The calculator uses **X** for the horizontal axis (time) and **Y** for the vertical axis (height). Set **Xmin** to 0, **Xmax** to 5, **Ymin** to 1, and **Ymax** to 3.

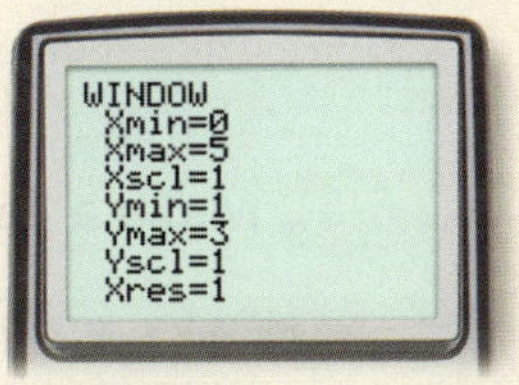

3. Make the graph.

Select Plot 1: 2nd PLOT Y= ENTER ENTER.
Use the arrow keys to highlight the line graph icon and press ENTER. Then press GRAPH.

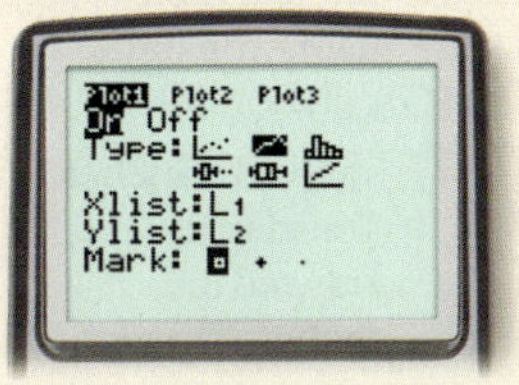

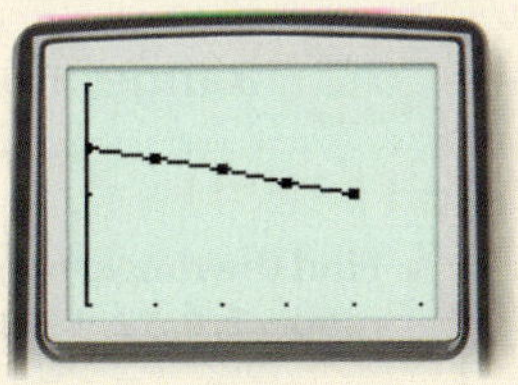

Think and Discuss

1. What does the line graph tell you about the data? What do you think was happening when the measurements were taken? **The tide was decreasing steadily. The tide was probably going out when the measurements were taken.**

Try This

1. The table shows amounts of snowfall during one winter. Use the data to make a line graph.

Month	Nov	Dec	Jan	Feb	Mar
Snowfall (in.)	2	6	9	6	2

Answers

Try This

1.

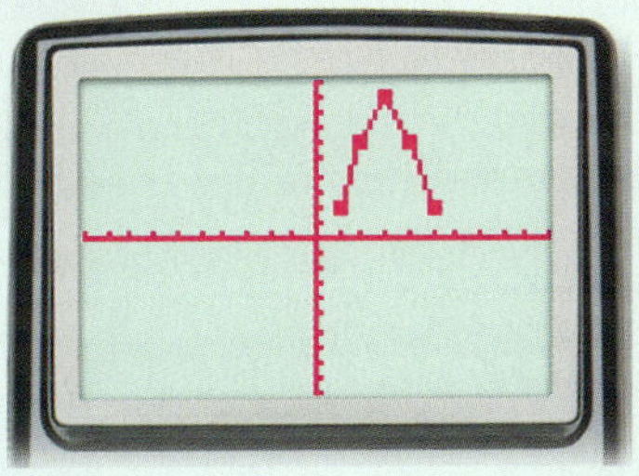

Technology LAB: Create Line Graphs

Objective: To use a graphing calculator to create line graphs

Materials: Graphing calculator

Lab Resources

Technology Lab Activities . . . . p. 25

Using the Page

This technology activity shows students how to create line graphs on a graphing calculator. Specific keystrokes may vary, depending on the make and model of the graphing calculator used. The keystrokes given are for a TI-83 model. For keystrokes to other models, visit go.hrw.com.

The Think and Discuss problem is meant to encourage students to interpret the line graph. While the Try This problem can be done without a graphing calculator, it is meant to help students use a graphing calculator to create line graphs.

Assessment

1. Suppose the tide's height was 1.94 ft at 9:30 A.M. How could you add this data to the graph? Press STAT ENTER. Use the arrow keys to move down and add 5 to L1. Use the arrow keys to move right and add 1.94 to L2. Press GRAPH to see the new graph.
2. Set the window to standard zoom by pressing ZOOM 6. How does changing the window change the impression that the graph gives? The standard zoom shows a larger area, so the line looks flatter. The decrease in the tide's height is not as obvious as it is in the first graph.

Chapter 6 Study Guide and Review

Purpose: *To help students review and practice concepts and skills presented in Chapter 6*

Assessment Resources

Chapter Review
Chapter 6 Resource Book . . . pp. 91–93

***Test and Practice Generator* CD-ROM**

Additional review items in both multiple-choice and free-response format may be generated for any objective in Chapter 6.

Answers

1. histogram
2. ordered pair
3. mode
4.

Snake Lengths (ft)	
Anaconda	35 ft
Diamond python	21 ft
King cobra	19 ft
Boa constrictor	16 ft

5. Range: 7; mean: 37; median: 38 mode: 39

Chapter 6 Study Guide and Review

Study Guide and Review

Vocabulary

bar graph 284
coordinate grid 294
cumulative frequency . . . 290
double-bar graph 285
double-line graph 298
frequency table 290
histogram 291
line graph 297
mean 275
median 275
mode 275
ordered pair 294
outlier 278
range 275
stem-and-leaf plot 305

Complete the sentences below with vocabulary words from the list above. Words may be used more than once.

1. A(n) __?__ uses vertical or horizontal bars to show the number of items within each interval.
2. A point can be located by using a(n) __?__ of numbers such as (3, 5).
3. In a data set, the __?__ is the value or values that occur most often.

6-1 Make a Table (pp. 272–274)

EXAMPLES

■ **Make a table using the data.**

Monday it snowed 2 inches. Tuesday it snowed 3.5 inches. Thursday it snowed 4.25 inches.

Day	Snowfall
Mon	2 in.
Tue	3.5 in.
Thu	4.25 in.

EXERCISES

4. Make a table using the data on snake lengths.

An anaconda can be up to 35 ft long. A diamond python can be up to 21 ft long. A king cobra can be up to 19 ft long. A boa constrictor can be up to 16 ft long.

6-2 Range, Mean, Median, and Mode (pp. 275–277)

EXAMPLE

■ **Find the range, mean, median, and mode. 7, 8, 12, 10, 8**

range: $12 - 7 = 5$
mean: $7 + 8 + 8 + 10 + 12 = 45$
$45 \div 5 = 9$
median: 8
mode: 8

EXERCISES

Find the range, mean, median, and mode.

5.

Hours Worked Each Week						
32	39	39	38	36	39	36

6-3 Additional Data and Outliers (pp. 278–281)

EXAMPLE

- **Find the mean, median, and mode with and without the outlier.**

10, 4, 7, 8, 34, 7, 7, 12, 5, 8 *The outlier is 34.*
Without: **mean** = 10.2, **mode** = 7,
median = 7.5
With: **mean** ≈ 7.555, **mode** = 7, **median** = 7

EXERCISES

Find the mean, median, and mode of each data set with and without the outlier.

6. 12, 11, 9, 38, 10, 8, 12
7. 34, 12, 32, 45, 32
8. 16, 12, 15, 52, 10, 13

6-4 Bar Graphs (pp. 284–287)

EXAMPLE

- **Which grades have more than 200 students?** 6th grade and 8th grade

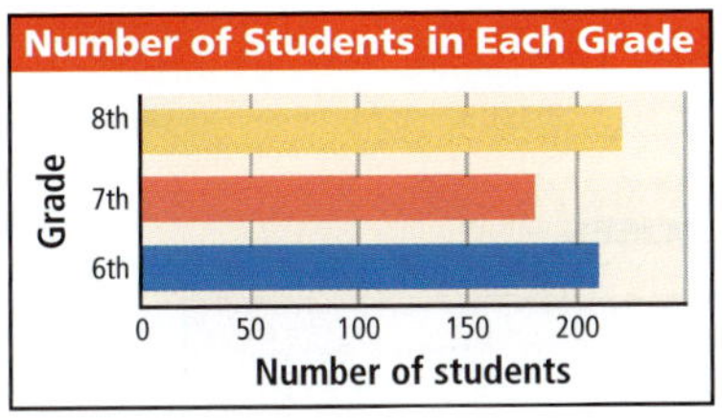

EXERCISES

Use the bar graph at left for Exercise 9.

9. Which grade has the most students?
10. Use the data to make a bar graph.

Test	Math	English	History	Science
Grade	95	85	90	80

6-5 Frequency Tables and Histograms (pp. 290–293)

EXAMPLE

- **Make a frequency table with intervals.**

Ages of people at Irene's birthday party: 37, 39, 18, 15, 13

Ages of People at Irene's Birthday Party				
Ages	13–19	20–26	27–33	34–40
Frequency	3	0	0	2

EXERCISES

11. Make a frequency table with intervals.

Points Scored					
6	4	5	4	7	10

12. Use the frequency table from Exercise 11 to make a histogram.

6-6 Ordered Pairs (pp. 294–296)

EXAMPLE

- **Name the ordered pair for *A*.**

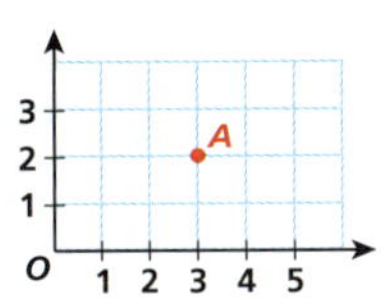

A is at (3, 2).

EXERCISES

Name the ordered pair for each location.

13. Bob's house
14. toy store

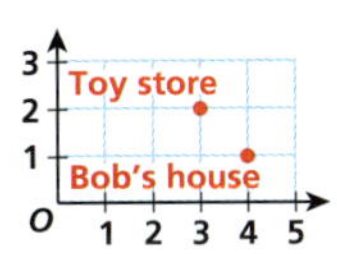

Answers

6. With outlier: mean ≈ 14.29; median = 11; mode = 12; without outlier: mean ≈ 10.33; median = 10.5; mode = 12

7. With outlier: mean = 31; median = 32; mode = 32; without outlier: mean = 35.75; median = 33; mode = 32

8. With outlier: mean ≈ 19.67; median = 14; mode = none; without outlier: mean = 13.2; median = 13; mode = none

9. 8th grade

10.

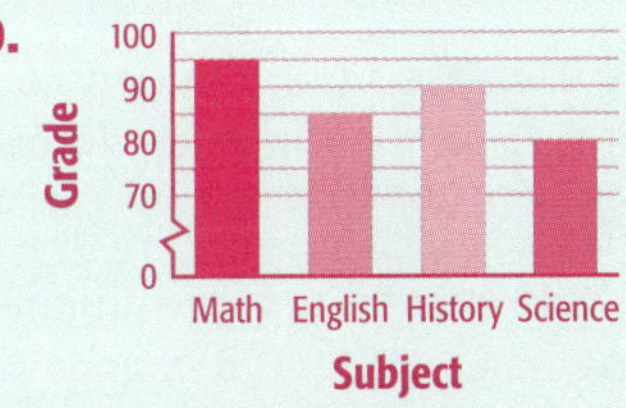

11.

Points Scored			
Points (Intervals)	1–4	5–8	9–12
Frequency	2	3	1

12.

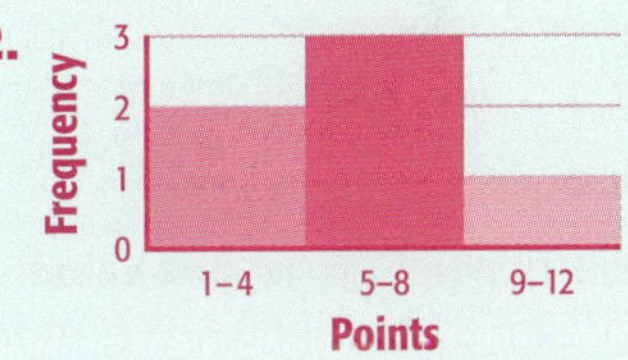

13. (4, 1)

14. (3, 2)

6-7 Line Graphs (pp. 297–300)

EXAMPLE

- **Use the data to make a line graph.**

Temperature Recording (°F)				
Day 1	Day 2	Day 3	Day 4	Day 5
32	36	38	40	36

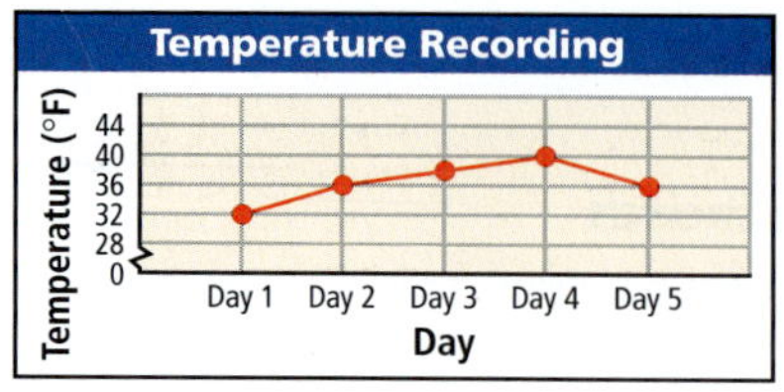

EXERCISES

15. Use the data to make a line graph.

Bookstore Sales			
Jan	Feb	Mar	Apr
$425	$320	$450	$530

Use your line graph from Exercise 15.

16. When were bookstore sales the greatest?
17. Describe the trend in bookstore sales over the four months.

6-8 Misleading Graphs (pp. 301–304)

EXAMPLES

- **Why is this graph misleading?**

The lower part of the scale is missing.

EXERCISES

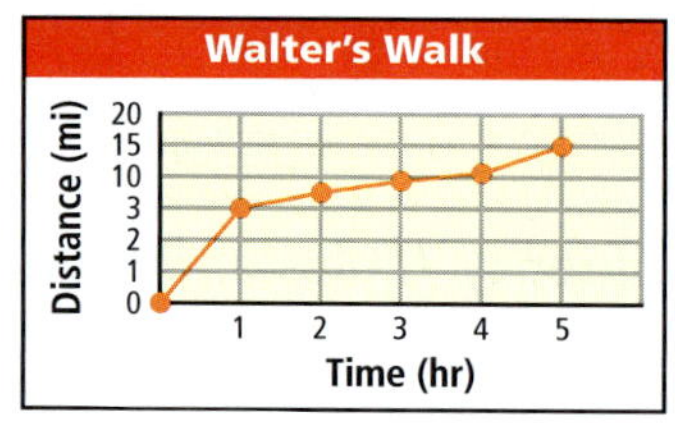

18. Explain why this graph is misleading.

6-9 Stem-and-Leaf Plots (pp. 305–307)

EXAMPLE

- **Make a stem-and-leaf plot.**

Test Scores							
66	72	80	92	88	86	85	94

Stems	Leaves
6	6
7	2
8	0 5 6 8
9	2 4

Key: 6|6 means 66

EXERCISES

19. Make a stem-and-leaf plot.

Basketball Scores							
22	26	34	46	20	44	40	28

20. List the least value, greatest value, mean, median, mode, and range of the data in the stem-and-leaf plot from Exercise 19.

Study Guide and Review

Answers

15. Bookstore Sales

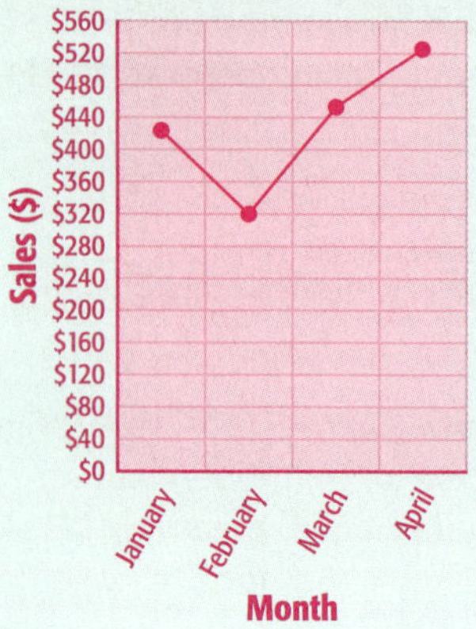

16. April

17. Sales decreased from January to February and then increased from February to April.

18. The scale starts out in increments of one mile and then it changes to 5 miles.

19. **Basketball Scores**

Stems	Leaves
2	0 2 6 8
3	4
4	0 4 6

Key: 2|0 means 20

20. smallest value: 20, largest value: 46, mean: 32.5, median: 31, mode: none, range: 26

Chapter Test

Chapter 6

1. Use the data about sound to make a table.

   The loudness of a sound is measured by the size of its vibrations. The unit of measurement is the decibel (dB). A soft whisper is 30 dB. Conversation is 60 dB. A loud shout is 100 dB. The pain threshold for humans is 130 dB. An airplane takeoff at 100 ft is 140 dB.

Use the bar graph for Exercises 2–5.

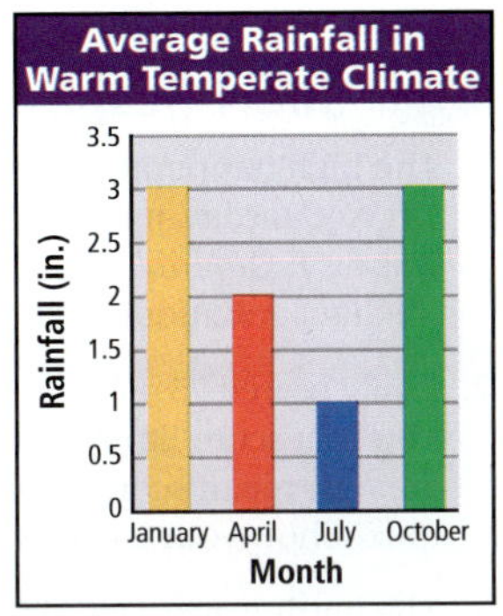

2. Find the mean, median, and mode of the rainfall amounts. **mean: 2.25, median: 2.5, mode: 3**
3. Which month had the lowest average rainfall? **July**
4. Which months had rainfall amounts greater than 2 inches? **January and October**
5. In a tropical climate, the rainfall in January was 0 in. In April, it was 2 in. In July, it was 22 in. In October, it was 7 in. Make a double-bar graph to compare the rainfall in the tropical climate and the warm temperate climate.
6. Use the book report data to make a frequency table. Then use the frequency table to make a histogram.

Scores on Book Reports							
82	88	60	75	95	92	71	82
78	93	87	76	90	80	70	85

Name the ordered pair for each point on the grid.

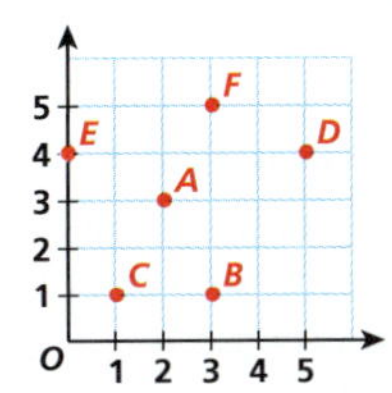

7. *A* **(2, 3)** 8. *B* **(3, 1)** 9. *C* **(1, 1)**

10. *D* **(5, 4)** 11. *E* **(0, 4)** 12. *F* **(3, 5)**

Graph and label the following points on a coordinate grid.

13. *T*(3, 4) 14. *M*(0, 6) 15. *P*(5, 1) 16. *S*(3, 2)

17. Make a double-line graph of the height data at right. How could you alter your graph to magnify the difference between males and females?

Average Heights				
	Birth	4 yr	12 yr	18 yr
Male	50 cm	100 cm	150 cm	180 cm
Female	50 cm	100 cm	152 cm	164 cm

18. Make a stem-and-leaf plot of the push-up data. Use your stem-and-leaf plot to find the mean, median, and mode of the data.

Number of Push-ups Performed						
35	33	25	45	52	21	18
41	27	35	40	53	24	38

Chapter Test

Chapter Test

Chapter 6

Purpose: *To assess students' mastery of concepts and skills in Chapter 6*

Assessment Resources

Chapter Tests (Levels A, B, C,)
Assessment Resources . . . . . pp. 61–66

***Test and Practice Generator* CD-ROM**

Additional assessment items in both multiple-choice and free-response format may be generated for any objective in Chapter 6.

Answers

1.

Sound Data	
Sound	**Loudness**
Soft whisper	30 dB
Conversation	60 dB
Loud shout	100 dB
Pain threshold for humans	130 dB
Airplane takeoff at 100 ft	140 dB

5.

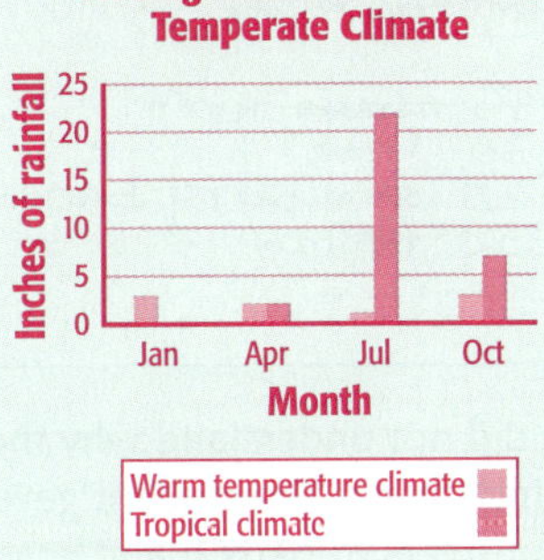

6.

Scores on Book Report	
Score	**Frequency**
60–69	1
70–79	5
80–89	6
90–99	4

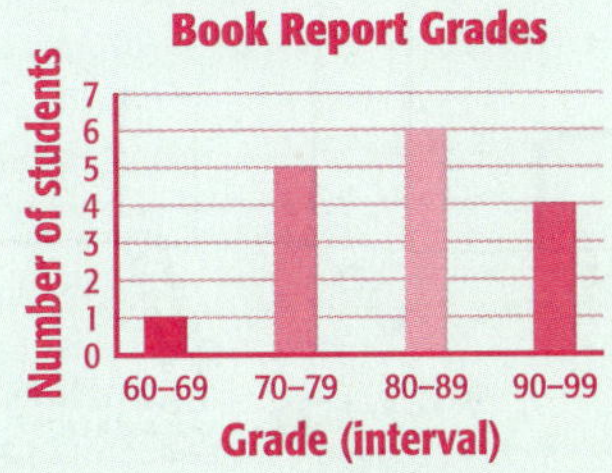

13–16.

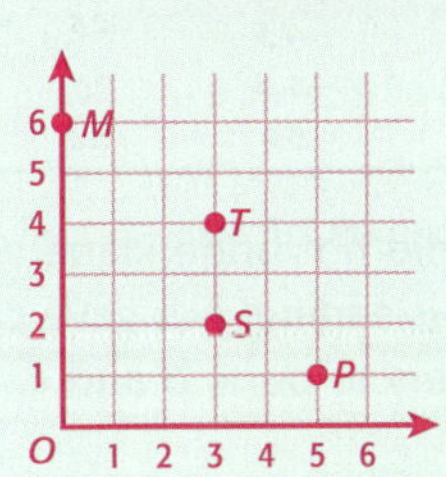

17.

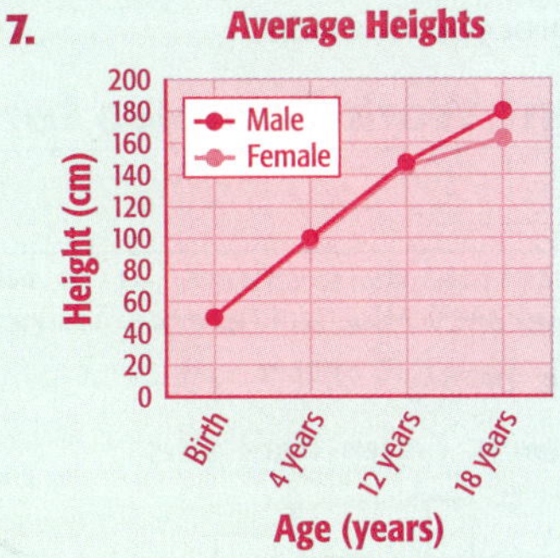

To magnify the difference, you could start the scale at 50.

18. **Number of Push-ups Performed**

Stems	Leaves
1	8
2	1 4 5 7
3	3 5 5 8
4	0 1 5
5	2 3

Key: 1|8 means 18

mean ≈ 34.79, median: 35, mode: 35

Chapter 6 Performance Assessment

Purpose: *To assess students' understanding of concepts in Chapter 6 and combined problem-solving skills*

Assessment Resources

Performance Assessment
Assessment Resources . . . . p. 114

Performance Assessment Teacher Support
Assessment Resources . . . . p. 113

Answers

1–3. See p. A4.

4. See Level 3 work sample below.

Scoring Rubric for Problem Solving Item 4

Level 3
Accomplishes the purposes of the task.
Student gives clear explanations, shows understanding of mathematical ideas and processes, and computes accurately.

Level 2
Purposes of the task not fully achieved.
Student demonstrates satisfactory but limited understanding of the mathematical ideas and processes.

Level 1
Purposes of the task not accomplished.
Student shows little evidence of understanding the mathematical ideas and processes and makes computational and/or procedural errors.

Chapter 6 Performance Assessment

Show What You Know

Create a portfolio of your work from this chapter. Complete this page and include it with your four best pieces of work from Chapter 6. Choose from your homework or lab assignments, mid-chapter quiz, or any journal entries you have done. Put them together using any design you want. Make your portfolio represent what you consider your best work.

Short Response

1. The high temperature on Monday was 54°F. On Tuesday, it was 62°F. On Wednesday, it was 65°F. On Thursday, it was 60°F. On Friday, it was 62°F. Organize this data in a table. Find the range, mean, median, and mode of the data. Show your work.
2. Emily scored 75, 85, 35, 85, 70, and 10 on her first six math quizzes. The best score Emily can make on a math quiz is 100. What is the greatest mean score Emily could have after taking the seventh math quiz? Show your work.
3. The batting averages for the players on two softball teams are given below. Assuming each player had the same number of at bats, tell which team has the higher batting average. Show your work.

Team A	.213	.138	.115	.152	.297	.101	.198	.176	.189
Team B	.145	.313	.103	.228	.184	.183	.261	.149	.163

Extended Problem Solving

4. A group of people at a shopping mall were surveyed to determine their favorite frozen fruit bars. The graph at right shows the results of the survey.
   a. Explain why the graph is misleading.
   b. Use the same data to construct a graph that is not misleading.
   c. Create a cumulative frequency table of the data. How many people at the mall participated in the survey?

Student Work Samples for Item 4

Level 3

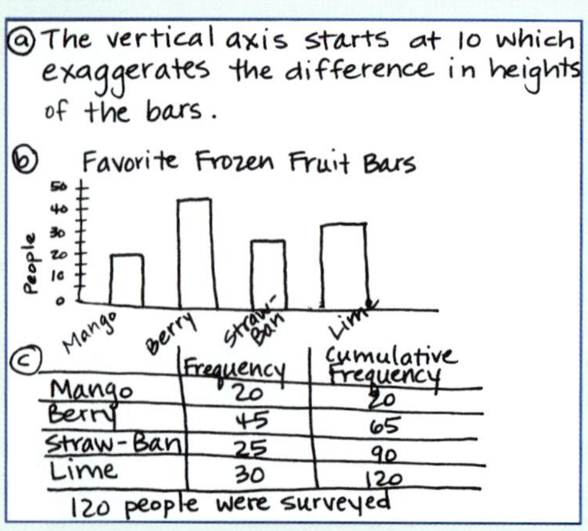

The student gave complete, correct answers.

Level 2

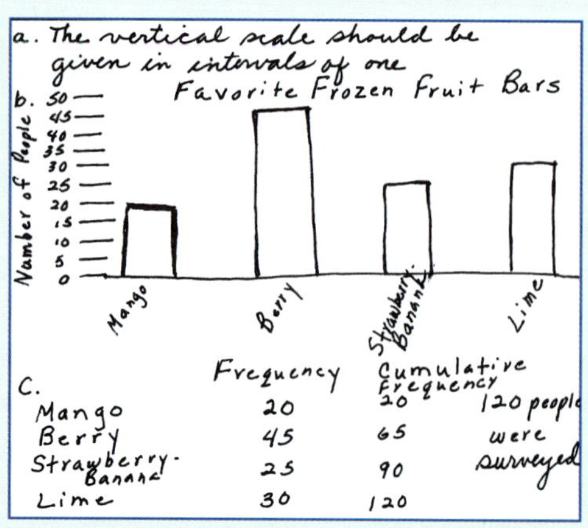

The student did not understand why the graph was misleading, but gave complete, correct answers to parts b and c.

Level 1

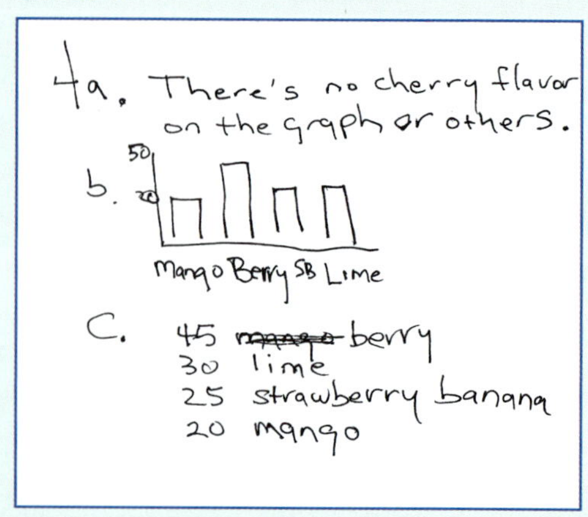

The student did not understand why the graph was misleading. The student gave incomplete answers to parts b and c.

internet connect
State-Specific Test Practice Online
go.hrw.com Keyword: MR4 TestPrep

Standardized Test Prep

Chapter 6

Cumulative Assessment, Chapters 1–6

1. What is the mode of the following data set?
17, 13, 14, 13, 21, 18, 16, 19 **A**

(A) 13 (C) 16.5
(B) 16 (D) 16.375

2. What is the LCM of 6, 8, and 12? **H**

(F) 2 (H) 24
(G) 12 (J) 48

3. Which expression has the greatest value? **A**

(A) $5.35 \cdot 1.6$ (C) $35.7 \div 6.8$
(B) $12\frac{2}{3} \div 2$ (D) $2\frac{1}{3} \cdot 3\frac{3}{5}$

4. Which is a type of graph that uses bars and intervals to display data? **G**

(F) Stem-and-leaf plot
(G) Histogram
(H) Double-line graph
(J) Cumulative frequency

5. $7\frac{5}{10} + 4\frac{3}{4}$ **D**

(A) $11\frac{4}{7}$ (C) $12\frac{3}{4}$
(B) $11\frac{1}{4}$ (D) $12\frac{1}{4}$

6. Which decimal is equivalent to $4\frac{3}{8}$? **G**

(F) 4.125 (H) 4.38
(G) 4.375 (J) 4.8

TIP! **TEST TAKING TIP!**

Eliminate possibilities. If there are an odd number of items in the data set, the median will be a value from the set.

7. Find the mean and median of the following data set: 25, 30, 27, 26, 32, 32, and 24. **B**

(A) Mean: 28; median: 32
(B) Mean: 28; median: 27
(C) Mean: 27; median: 28
(D) Mean: 28; median: 28

8. What is the outlier of the following data set: 55, 62, 71, 64, 28, 64? **J**

(F) 64 (H) 36
(G) 63 (J) 28

9. ***SHORT RESPONSE*** Terry and Carl collect baseball cards. Terry has 151 fewer cards than Carl. Let *c* represent the number of cards Carl has and write an expression for the number of cards Terry has. Explain how to find the number of cards Carl has if Terry has 728.

10. ***SHORT RESPONSE*** Name the ordered pair for each point on the coordinate grid. Suppose a new point *F* is two units to the left and three units above point *B*. Explain how to find the ordered pair for point *F*.

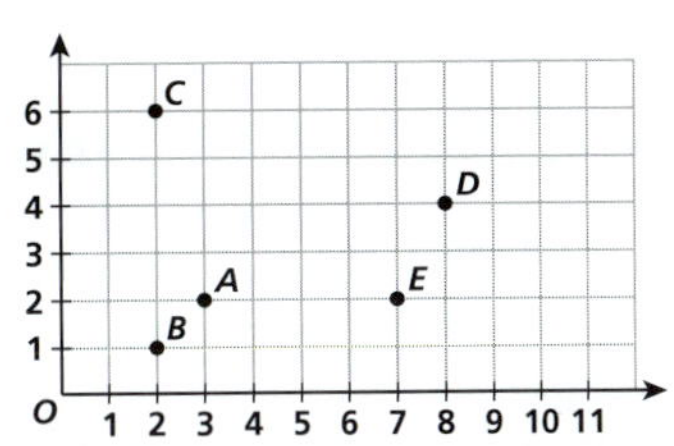

Standardized Test Prep

Standardized Test Prep

Chapter 6

Purpose: *To provide review and practice for Chapters 1–6 and standardized tests*

Assessment Resources

Cumulative Tests (Levels A, B, C)
Assessment Resources . . . . pp. 187–198

State-Specific Test Practice Online
KEYWORD: MR4 TestPrep

Test Prep Doctor

In item 8, eliminate answer choice **A** because there are 3 values in the sixties in the data set. Eliminate answer choices **B** and **C** because they are not in the data set.

Point out to students that in Item 5, answer choices **F** and **G** are too small because the sum has to be at least 12. Answer choice **H** is too large because $8 + 4\frac{3}{4} = 12\frac{3}{4}$.

Answers

9. $c - 151$;

$$\begin{aligned} c - 151 &= 728 \\ +151 &\quad +151 \\ c &= 879 \end{aligned}$$

Carl has 879 cards.

10. *A*(3, 2); *B*(2, 1); *C*(2, 6); *D*(8, 4) *E*(7, 2); move 2 units left of *B* and 3 units above *B*, *F*(0, 4)

Chapter 7

Plane Geometry

Section 7A

Lines and Angles

Lesson 7-1
Points, Lines, and Planes

Lesson 7-2
Angles

Hands-On Lab 7A
Construct Congruent Segments and Angles

Lesson 7-3
Angle Relationships

Lesson 7-4
Classifying Lines

Hands-On Lab 7B
Parallel Line Relationships

Section 7B

Polygons

Lesson 7-5
Triangles

Lesson 7-6
Quadrilaterals

Lesson 7-7
Polygons

Lesson 7-8
Geometric Patterns

Section 7C

Polygon Relationships

Lesson 7-9
Congruence

Lesson 7-10
Transformations

Lesson 7-11
Symmetry

Lesson 7-12
Tessellations

Hands-On Lab 7C
Create Tessellations

Extension
Compass and Straightedge Constructions

Pacing Guide for 45-Minute Classes

Chapter 7

DAY 83	DAY 84	DAY 85	DAY 86	DAY 87
Lesson 7-1	Lesson 7-2	Hands-On Lab 7A	Lesson 7-3	Lesson 7-4
DAY 88	**DAY 89**	**DAY 90**	**DAY 91**	**DAY 92**
Hands-On Lab 7B	Mid-Chapter Quiz Lesson 7-5	Lesson 7-6	Lesson 7-7	Lesson 7-8
DAY 93	**DAY 94**	**DAY 95**	**DAY 96**	**DAY 97**
Mid-Chapter Quiz Lesson 7-9	Lesson 7-10	Lesson 7-11	Lesson 7-12	Hands-On Lab 7C
DAY 98	**DAY 99**	**DAY 100**		
Extension	Chapter 7 Review	Chapter 7 Assessment		

Pacing Guide for 90-Minute Classes

Chapter 7

DAY 41	DAY 42	DAY 43	DAY 44	DAY 45
Chapter 6 Review Lesson 7-1	Chapter 6 Assessment Lesson 7-2	Hands-On Lab 7A Lesson 7-3	Lesson 7-4 Hands-On Lab 7B	Mid-Chapter Quiz Lesson 7-5 Lesson 7-6

DAY 46	DAY 47	DAY 48	DAY 49	DAY 50	DAY 51
Lesson 7-7 Lesson 7-8	Mid-Chapter Quiz Lesson 7-9 Lesson 7-10	Lesson 7-11 Lesson 7-12	Hands-On Lab 7C Extension	Chapter 7 Review Lesson 8-1	Chapter 7 Assessment Technology Lab 8A

Across the Series

HARCOURT GRADE 5

- Classify and construct lines, angles, and polygons.
- Draw and identify circles and their parts.
- Identify congruent and similar figures and lines of symmetry.
- Identify solid figures and draw them from different views.
- Transform a figure on the coordinate plane.

COURSE 1

- **Classify and construct plane figures, congruent figures, and bisectors.**
- **Understand and use angle relationships and line relationships.**
- **Use congruence to solve problems.**
- **Recognize, describe, and extend patterns of geometric figures.**
- **Describe and show geometric transformations, symmetry, and tessellations.**

COURSE 2

- Classify and construct plane figures.
- Understand and use angle relationships and line relationships.
- Classify and find unknown angle measures in polygons.
- Use congruence to solve problems.
- Describe and show geometric transformations, symmetry, and tessellations.

Across the Curriculum

LANGUAGE ARTS LINK

Math: Reading and Writing in the Content Area . . . . pp. 53–64

Focus on Problem Solving

Solve . . . . SE p. 343

Make a Plan . . . . SE p. 361

Journal . . . . TE pp. 325, 329, 335, 339, 347, 351, 355, 359, 368, 372

Write About It . . . . SE, last page of each lesson

Language Arts . . . . SE p. 368

SOCIAL STUDIES LINK

Social Studies . . . . SE pp. 347, 359, 370

Geography . . . . SE p. 325

SCIENCE LINK

Science . . . . SE p. 337

SE = *Student Edition* **TE** = *Teacher's Edition*

Interdisciplinary Bulletin Board

Earth Science

Latitude is the distance north or south, measured in degrees, from the equator. *Longitude* is the distance east or west, measured in degrees, from the prime meridian. Which contain parallel lines, the lines of latitude or the lines of longitude?

lines of latitude

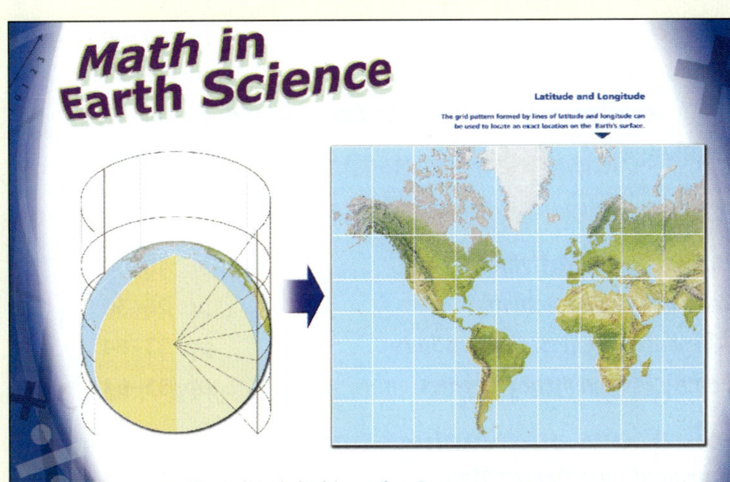

Interdisciplinary posters and worksheets are provided in your resource material.

Chapter 7

Resource Options

Chapter 7 Resource Book

Student Resources

Practice (Levels A, B, C) .... pp. 8–10, 17–19, 27–29, 37–39, 47–49, 57–59, 66–68, 75–77, 85–87, 94–96 104–106, 113–115

Reteach .... pp. 11, 20–21, 30–31, 40, 50–51, 60, 69, 78, 88, 97–98, 107, 116

Challenge .... pp. 12, 22, 32, 41, 52, 61, 70, 79, 89, 99, 108, 117

Problem Solving .... pp. 13, 23, 33, 42, 53, 62, 71, 80, 90, 100, 109, 118

Puzzles, Twisters & Teasers .... pp. 14, 24, 34, 43, 54, 63, 72, 81, 91, 101, 110, 119

Recording Sheets .... pp. 3, 7, 16, 26, 36, 46, 56, 65, 74, 84, 93, 103, 112, 123, 127–128, 130, 137–139

Chapter Review .... pp. 120–122

Teacher and Parent Resources

Chapter Planning and Pacing Guide .... p. 4

Section Planning Guides .... pp. 5, 44, 82

Parent Letter .... pp. 1–2

Teaching Tools .... pp. 126–139

Teacher Support for Chapter Project .... p. 124

Transparencies .... pp. T1–T58

- Daily Transparencies
- Additional Examples Transparencies
- Teaching Transparencies

Reaching All Learners

English Language Learners

Success for English Language Learners ...... pp. 105–128

Math: Reading and Writing in the Content Area. pp. 53–64

Spanish Homework and Practice .... pp. 53–64

Spanish Interactive Study Guide .... pp. 53–64

Spanish Family Involvement Activities .... pp. 53–64

Multilingual Glossary

Individual Needs

Are You Ready? Intervention and Enrichment. pp. 301–304, 313–316, 333–336, 417–418

Alternate Openers: Explorations .... pp. 53–64

Family Involvement Activities .... pp. 53–64

Interactive Problem Solving .... pp. 53–64

Interactive Study Guide .... pp. 53–64

Readiness Activities .... pp. 13–14

Math: Reading and Writing in the Content Area. pp. 53–64

Challenge .... CRB pp. 12, 22, 32, 41, 52, 61, 70, 79, 89, 99, 108, 117

Hands-On

Hands-On Lab Activities .... pp. 48–60

Technology Lab Activities .... pp. 26–30

Alternate Openers: Explorations .... pp. 53–64

Family Involvement Activities .... pp. 53–64

Applications and Connections

Consumer and Career Math .... pp. 25–28

Interdisciplinary Posters .... Poster 7, TE p. 320B

Interdisciplinary Poster Worksheets .... pp. 19–21

Transparencies

Alternate Openers: Explorations .... pp. 53–64

Exercise Answers Transparencies

Chapter 7 Resource Book .... pp. T1–T58

- Daily Transparencies
- Additional Examples Transparencies
- Teaching Transparencies

Technology

Teacher Resources

Lesson Presentations **CD-ROM** .... Chapter 7

Test and Practice Generator **CD-ROM** .... Chapter 7

One-Stop Planner **CD-ROM** .... Chapter 7

Student Resources

Are You Ready? Intervention **CD-ROM**
Skills 73, 76, 81

internet connect

go.hrw.com

Homework Help Online	**KEYWORD:** MR4 HWHelp7
Math Tools Online	**KEYWORD:** MR4 Tools
Glossary Online	**KEYWORD:** MR4 Glossary
Chapter Project Online	**KEYWORD:** MR4 PSProject7
Chapter Opener Online	**KEYWORD:** MR4 Ch7
CNN Student News	**KEYWORD:** MR4 CNN7

SE = *Student Edition* **TE** = *Teacher's Edition* **AR** = *Assessment Resources* **CRB** = *Chapter Resource Book* **MK** = *Manipulatives Kit*

Chapter 7

Assessment Options

Assessing Prior Knowledge

Determine whether students have the required prerequisite concepts and skills.

Are You Ready? . . . . SE p. 321
Inventory Test . . . . AR pp. 1–4

Test Preparation

Provide review and practice for chapter and standardized tests.

Standardized Test Prep . . . . SE p. 389
Spiral Review with Test Prep . . . . . SE, last page of each lesson
Study Guide and Review . . . . SE pp. 384–386
Test Prep Tool Kit

Technology

***Test and Practice Generator* CD-ROM**

internet connect

State-Specific Test Practice Online KEYWORD: MR4 TestPrep

Performance Assessment

Assess students' understanding of chapter concepts and combined problem-solving skills.

Performance Assessment . . . . SE p. 388
Includes scoring rubric in TE
Performance Assessment . . . . AR p. 116
Performance Assessment Teacher Support . . . . AR p. 115

Portfolio

Portfolio opportunities appear throughout the Student and Teacher's Editions.

Suggested work samples:

Problem Solving Project . . . . TE p. 320
Performance Assessment . . . . SE p. 388
Portfolio Guide . . . . AR p. xxxv
Journal . . . . TE pp. 325, 329, 335, 339, 347, 351, 355, 359, 368, 372
Write About It . . . . SE, last page of each lesson

Daily Assessment

Obtain daily feedback on students' understanding of concepts.

Spiral Review and Test Prep . . . . . . SE, last page of each lesson

Also Available on Transparency In Chapter 7 Resource Book

Warm Up . . . . TE, first page of each lesson
Problem of the Day . . . . TE, first page of each lesson
Lesson Quiz . . . . TE, last page of each lesson

Student Self-Assessment

Have students evaluate their own work.

Group Project Evaluation . . . . AR p. xxxii
Individual Group Member Evaluation . . . . AR p. xxxiii
Portfolio Guide . . . . AR p. xxxv
Journal . . . . TE pp. 325, 329, 335, 339, 347, 351, 355, 359, 368, 372

Formal Assessment

Assess students' mastery of concepts and skills.

Section Quizzes . . . . AR pp. 18–20
Mid-Chapter Quiz . . . . SE pp. 342, 360
Chapter Test . . . . SE p. 387
Chapter Tests (Levels A, B, C) . . . . AR pp. 67–72
Cumulative Tests (Levels A, B, C) . . . . AR pp. 199–210
Standardized Test Prep
Cumulative Assessment . . . . SE p. 389
End-of-Year Test . . . . AR pp. 271–274

Technology

***Test and Practice Generator* CD-ROM**

Make tests electronically. This software includes:

- Dynamic practice for Chapter 7
- Customizable tests
- Multiple-choice items for each objective
- Free-response items for each objective
- Teacher management system

SE = *Student Edition* **TE** = *Teacher's Edition* **AR** = *Assessment Resources* **CRB** = *Chapter Resource Book* **MK** = *Manipulatives Kit*

Chapter 7 Tests

Three levels (A,B,C) of tests are available for each chapter in the *Assessment Resources.*

Level A

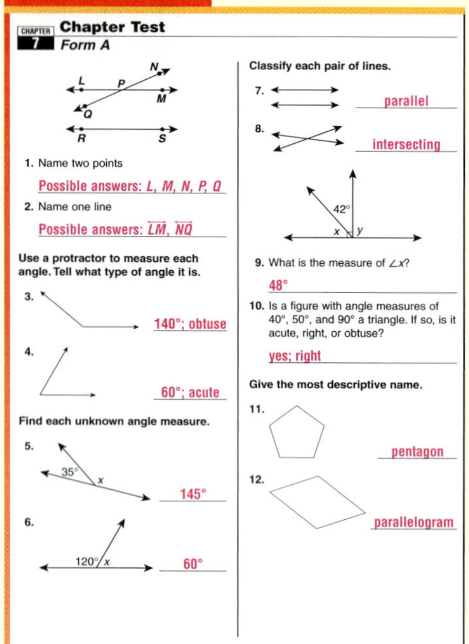
Chapter 7 Chapter Test
Form A

1. Name two points
Possible answers: L, M, N, P, Q
2. Name one line
Possible answers: $\overleftrightarrow{LM}$, $\overleftrightarrow{NQ}$

Use a protractor to measure each angle. Tell what type of angle it is.
3. 140°; obtuse
4. 60°; acute

Find each unknown angle measure.
5. 145°
6. 60°

Classify each pair of lines.
7. parallel
8. intersecting
9. What is the measure of ∠x?
48°
10. Is a figure with angle measures of 40°, 50°, and 90° a triangle. If so, is it acute, right, or obtuse?
yes; right

Give the most descriptive name.
11. pentagon
12. parallelogram

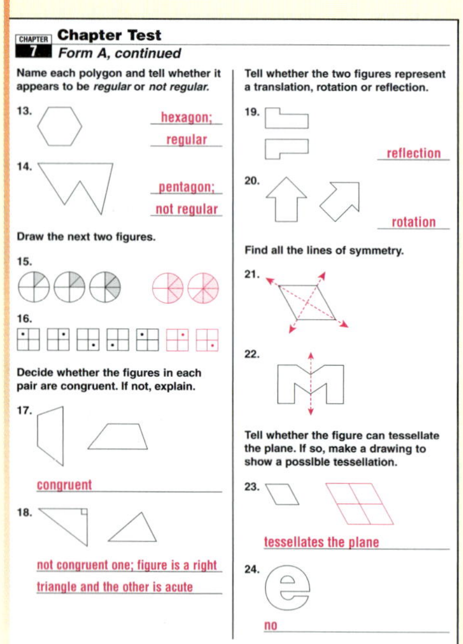
Chapter 7 Chapter Test
Form A, continued

Name each polygon and tell whether it appears to be *regular* or *not regular.*
13. hexagon; regular
14. pentagon; not regular

Draw the next two figures.
15.
16.

Decide whether the figures in each pair are congruent. If not, explain.
17. congruent
18. not congruent one; figure is a right triangle and the other is acute

Tell whether the two figures represent a translation, rotation or reflection.
19. reflection
20. rotation

Find all the lines of symmetry.
21.
22.

Tell whether the figure can tessellate the plane. If so, make a drawing to show a possible tessellation.
23. tessellates the plane
24. no

Level B

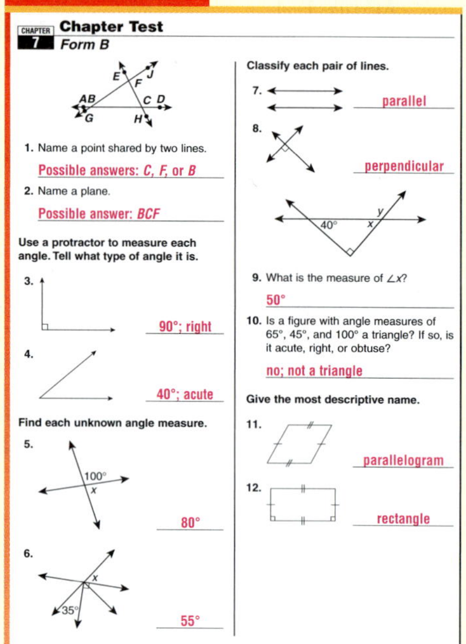
Chapter 7 Chapter Test
Form B

1. Name a point shared by two lines.
Possible answers: C, F, or B
2. Name a plane.
Possible answer: BCF

Use a protractor to measure each angle. Tell what type of angle it is.
3. 90°; right
4. 40°; acute

Find each unknown angle measure.
5. 80°
6. 55°

Classify each pair of lines.
7. parallel
8. perpendicular
9. What is the measure of ∠x?
50°
10. Is a figure with angle measures of 65°, 45°, and 100° a triangle? If so, is it acute, right, or obtuse?
no; not a triangle

Give the most descriptive name.
11. parallelogram
12. rectangle

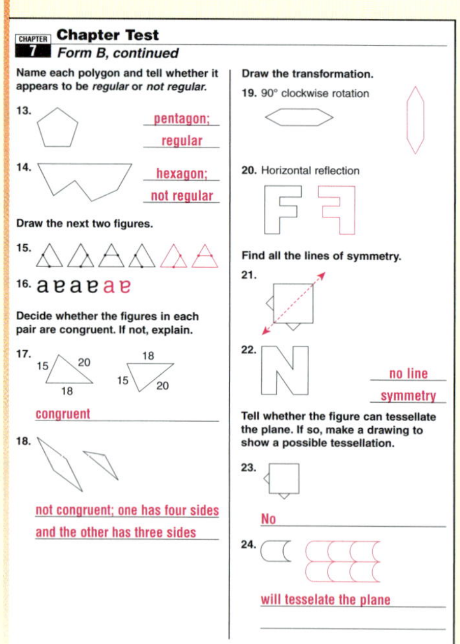
Chapter 7 Chapter Test
Form B, continued

Name each polygon and tell whether it appears to be *regular* or *not regular.*
13. pentagon; regular
14. hexagon; not regular

Draw the next two figures.
15.
16.

Decide whether the figures in each pair are congruent. If not, explain.
17. congruent
18. not congruent; one has four sides and the other has three sides

Draw the transformation.
19. 90° clockwise rotation
20. Horizontal reflection

Find all the lines of symmetry.
21.
22. no line symmetry

Tell whether the figure can tessellate the plane. If so, make a drawing to show a possible tessellation.
23. No
24. will tesselate the plane

Level C

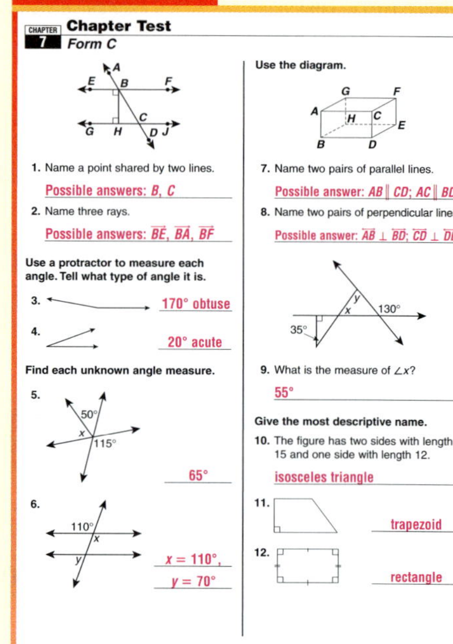
Chapter 7 Chapter Test
Form C

1. Name a point shared by two lines.
Possible answers: B, C
2. Name three rays.
Possible answers: $\overrightarrow{BE}$, $\overrightarrow{BA}$, $\overrightarrow{BF}$

Use a protractor to measure each angle. Tell what type of angle it is.
3. 170° obtuse
4. 20° acute

Find each unknown angle measure.
5. 65°
6. x = 110°, y = 70°

Use the diagram.
7. Name two pairs of parallel lines.
Possible answer: $AB \parallel CD$; $AC \parallel BD$
8. Name two pairs of perpendicular lines.
Possible answer: $AB \perp BD$; $CD \perp DE$
9. What is the measure of ∠x?
55°

Give the most descriptive name.
10. The figure has two sides with length 15 and one side with length 12.
isosceles triangle
11. trapezoid
12. rectangle

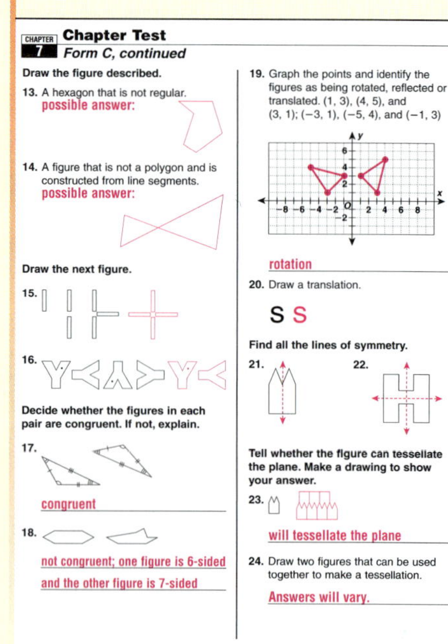
Chapter 7 Chapter Test
Form C, continued

Draw the figure described.
13. A hexagon that is not regular.
possible answer:
14. A figure that is not a polygon and is constructed from line segments.
possible answer:

Draw the next figure.
15.
16.

Decide whether the figures in each pair are congruent. If not, explain.
17. congruent
18. not congruent; one figure is 6-sided and the other figure is 7-sided

19. Graph the points and identify the figures as being rotated, reflected or translated. (1, 3), (4, 5), and (3, 1); (−3, 1), (−5, 4), and (−1, 3)
rotation
20. Draw a translation.

Find all the lines of symmetry.
21.
22.

Tell whether the figure can tessellate the plane. Make a drawing to show your answer.
23. will tessellate the plane
24. Draw two figures that can be used together to make a tessellation.
Answers will vary.

Test and Practice Generator CD-ROM

Create and customize multiple versions of the same tests with corresponding answers for any chosen chapter objectives.

Chapter 7 State and Standardized Test Preparation

Test Taking Skill Builder and Standardized Test Practice are provided for each chapter in the *Test Prep Tool Kit.*

Test Taking Skill Builder

Test Taking Strategy **Short Response—Drawing Diagrams**
Chapter 7

Some short response questions require you to draw and label a diagram. You need to make sure you draw the figure as described in the problem statement and provide all markings and labeling as appropriate. Short response questions often have multiple correct solutions. So, your score is based on a scoring rubric.

Example 1
Short Response Sketch and label ∠*ABC* and ∠*BCD*, a pair of adjacent, supplementary angles.

Scoring Rubric
- **2 points:** Student correctly sketches a pair of adjacent, supplementary angles, and correctly labels them *ABC* and *BCD*.
- **1 points:** Student correctly sketches the angles, but labels them incorrectly, or sketches adjacent angles that are not supplementary.
- **1 point:** Student sketches two angles that are neither adjacent nor supplementary.
- **0 points:** Student sketches one angle or has no response.

Solution
2-point response:
See the sketch at the right for one possible response. Notice that the angles are adjacent to one another. The angles are supplementary, meaning they form a 180° angle, and points *A*, *B*, *C*, and *D* are labeled.

1-point response:
See the sketch at the right for one possible response. Notice that the angles are adjacent to one another, however, they are NOT supplementary, meaning they do NOT form a 180° angle. Although the angles are labeled, they are not labeled correctly. The angles in the diagram are ∠*ACB* and ∠*BCD*, not ∠*ABC* and ∠*BCD*.

0-point response:
See the sketch at the right for one possible response. Notice that only one angle is drawn and labeled. The student did not complete the required diagram.

Test Taking Strategy
Chapter 7, continued

Exercises Possible answers are given.
Use the scoring rubric for each question.

Scoring Rubric
- **2 points:** Student sketches diagram as required, shows all work and makes no errors in computations. The question is answered in a complete sentence.
- **1 point:** Student shows most work and makes no errors. The student answers the question but fails to draw a diagram.
- **0 points:** There is no response or it is completely incorrect.

1. Draw a regular 8-sided polygon Find the sum of the angles measures in the figure and find the measure of each angle.

1-point response	0-point response
180(8 − 2) = 1,080 There are 1,080 degrees in a regular 8-sided polygon. $\frac{1,080}{8}$ = 135 degrees The measure of each angle is 135°.	90(8) = 720 degrees $\frac{720}{10}$ = 72 degrees

a. Name three things you would have to add to the 1-point response to make it a 2-point response.
Add a diagram of an 8-sided regular polygon; write the formula for the sum of the measures of the angles in a polygon, and show all the work.

b. Why did the 0-point response only receive 0-points?
Possible answer: The response is completely incorrect.

2. Sketch and label two right triangles that are congruent. Write two congruence statements. Write a 2-point response.
△*ABC* ≅ *DEF*; ∠*A* ≅ ∠*D*; ∠*C* ≅ ∠*E*

Standardized Test Practice

Standardized Test Practice
Chapter 7

Select the best answer for Questions 1–8.

1. Which figure has four congruent sides?
   A hexagon
   B rhombus
   C triangle
   D trapezoid
2. What is the missing length of the congruent triangles?
   F 2 cm
   G 6 cm
   H 8 cm
   I 10 cm
3. The figure shown appears to be an example of what kind of triangle?
   A acute
   B right
   C obtuse
   D equilateral
4. Which letter does NOT have a line of symmetry?
   F H
   G S
   H O
   I W
5. The figure shown is what type of transformation?
   A translation
   B glide reflection
   C reflection
   D rotation
6. What is the name of the shape?
   F quadrilateral
   G pentagon
   H hexagon
   I octagon
7. Which of the following cannot tessellate the plane?
   A hexagon
   B square
   C octagon
   D equilateral triangle
8. Which is a good estimation of the measure of the angle?
   F 16°
   G 25°
   H 58°
   I 131°

Standardized Test Practice
Chapter 7, continued

Gridded Response
Solve the problems. Use the answer sheet to write and grid-in your answer.

9. What is the sum of the angles of a hexagon?

Use the figure to answer Questions 10 and 11.

10. What is the measure of ∠*BOD*?
11. What is the measure of ∠*COB*?

Short Response
Solve the problems. Use the answer sheet to write your answers.

12. Explain in words the pattern. Then draw the next figure.
13. Compare and contrast perpendicular lines and intersecting lines. Draw an example of each. What symbol do you use to show that two lines are perpendicular?

Extended Response
14. Use the diagram.
   a. Explain in words the relationship between ∠*MOQ* and ∠*PON*. Find the measure of ∠*PON*.
   b. Classify ∠*ROQ*. What is its measure?
   c. Explain in words how to find the measure of ∠*MOR* and ∠*NOS*.

Answer Sheet (partial):
Write your answers in the space provided.
12. The pattern is three shapes: a circle, a square, and a triangle. The first series has the three shapes with no dots. The second series has three shapes with one dot, and the third series has three shapes with two dots, etc. The next shape is a square with two dots.
13. Perpendicular lines are lines that intersect at a 90° angle. Intersecting lines can intersect at any angle. You use a small square to represent the two lines are perpendicular.
Intersecting lines Perpendicular lines

internet connect
State-Specific Test Practice Online
KEYWORD: MR4 TestPrep

Test Prep Tool Kit

- Standardized Test Prep Workbook
- Countdown to Testing transparencies
- State Test Prep CD-ROM
- Standardized Test Prep Video

Customized answer sheets give students realistic practice for actual standardized tests.

Chapter 7

Plane Geometry

Why Learn This?

Tell students that the language of geometry is essential in describing real-world objects. For example, the shapes of classroom objects, such as desktops, floor tiles, blackboards, and book covers can be described as types of plane figures. Students are likely to be familiar with basic figures, such as triangles, circles, and quadrilaterals. Other names are needed to describe more complex figures.

Using Data

To begin the study of this chapter, have students:

- Sketch an example of a nonagon. Check students' sketches. Possible answers:

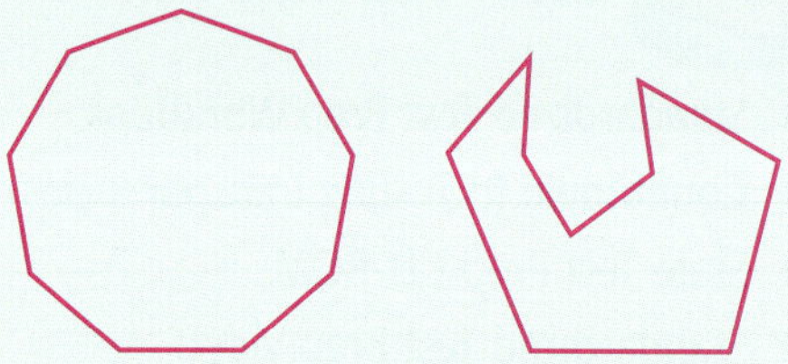

- Find the total number of sides in a heptagon and hexagon. 13
- Name a common object shaped like a regular octagon. stop sign

Chapter 7

Plane Geometry

TABLE OF CONTENTS

internet connect
Chapter Opener Online
go.hrw.com
KEYWORD: MR4 Ch7

Name of Figure	Number of Sides
Pentagon	5
Hexagon	6
Heptagon	7
Octagon	8
Nonagon	9
Decagon	10
Undecagon	11
Dodecagon	12

Career *Artist*

Artists help us to see our world in new ways. They use their creativity in many different kinds of careers. Artists might design graphics for Web sites, draw cartoons, design textiles and furniture, paint murals, or even illustrate courtroom scenes. Artists work with many materials, such as different kinds of paints, paper, stone, metal, stained glass, and tile. The table shows some geometric figures that an artist might use in a design.

PROBLEM SOLVING

Problem Solving Project

Art and Social Studies Connection

Purpose: To solve problems using angles, lines, and their relationships

Materials: Tile Artistry worksheet, ruler, compass, tiling materials, overhead projector, polygon pieces

internet connect
Chapter Project Online: *go.hrw.com*
KEYWORD: MR4 PSProject7

Understand, Plan, Solve, and Look Back

Have students:

- ✓ Complete the Tile Artistry worksheet to solve problems using polygons.
- ✓ Discover the similarities between regular and irregular polygons that have the same number of sides.
- ✓ Use or create a set of polygon pieces. Have students copy and/or create figures.
- ✓ Check students' work.

ARE YOU READY?

Choose the best term from the list to complete each sentence.

- protractor
- ruler
- triangle
- quadrilateral
- horizontal
- vertical
- clockwise
- counterclockwise

1. A closed figure with three sides is a ___?___, and a closed figure with four sides is a ___?___. triangle; quadrilateral
2. A ___?___ is used to measure and draw angles. protractor

3. 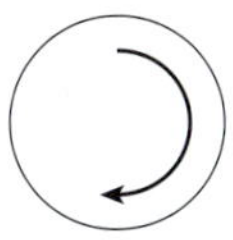

The arrow inside the circle is moving ___?___. clockwise

4.

A line that extends left to right is ___?___. horizontal

Complete these exercises to review skills you will need for this chapter.

✔ Measure with Customary and Metric Units

Use an inch ruler to measure each line to the nearest $\frac{1}{2}$ in.

5. 2 in.
6. $1\frac{1}{2}$ in.

Use a centimeter ruler to measure each line to the nearest centimeter.

7. 3 cm
8. 1.5 cm

✔ Identify Polygons

Tell how many sides and angles each figure has.

9. 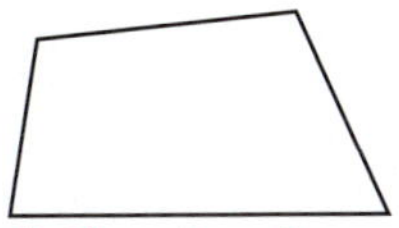
4 sides, 4 angles

10. 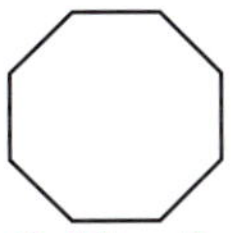
8 sides, 8 angles

11. 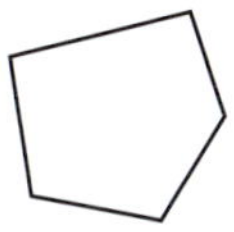
5 sides, 5 angles

✔ Identify Congruent Figures

Which two figures are exactly the same size and shape but in different positions? B

12. 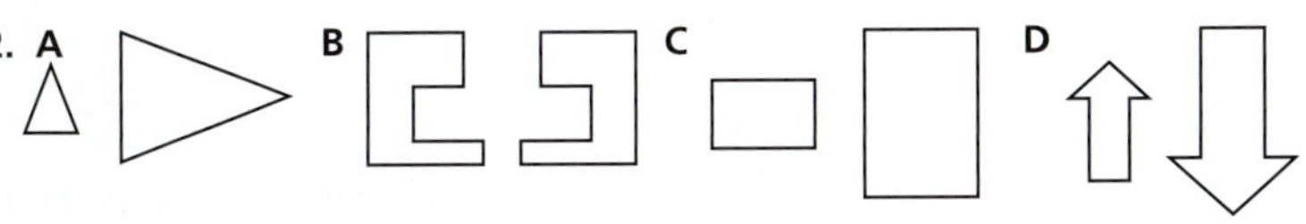

Assessing Prior Knowledge

INTERVENTION

Diagnose and Prescribe

Evaluate your students' performance on this page to determine whether intervention is necessary or whether enrichment is appropriate. Options that provide instruction, practice, and a check are listed below.

Resources for Are You Ready?

- ***Are You Ready? Intervention and Enrichment***
- **Recording Sheet for Are You Ready?** *Chapter 7 Resource Book. . . . . . p. 3*

Are You Ready? Intervention **CD-ROM**

ARE YOU READY?

Were students successful with Are You Ready?

NO INTERVENE

✔ Measure with Customary and Metric Units. 5–8
Intervention Practice, Skill 73
CD-ROM Intervention Activities, Skill 73

✔ Identify Polygons 9–11
Intervention Practice, Skill 76
CD-ROM Intervention Activities, Skill 76

✔ Identify Congruent Figures 12
Intervention Practice, Skill 81
CD-ROM Intervention Activities, Skill 81

YES ENRICH

Are You Ready? Enrichment, pp. 419–420

Section 7A

Lines and Angles

One-Minute Section Planner

Lesson	Materials	Resources
Lesson 7-1 Points, Lines, and Planes **NCTM:** Geometry, Communication **NAEP:** Geometry 1a ☑ SAT-9 ☑ SAT-10 ☐ ITBS ☐ CTBS ☐ MAT ☐ CAT	**Optional** Teaching Transparency T2 *(CRB)* Index cards *(CRB, p. 126)*	• *Chapter 7 Resource Book,* pp. 6–14 • Daily Transparency T1, CRB • Additional Examples Transparencies T3–T4, CRB • *Alternate Openers: Explorations,* p. 53
Lesson 7-2 Angles **NCTM:** Geometry, Measurement, Communication **NAEP:** Measurement 1b ☑ SAT-9 ☑ SAT-10 ☐ ITBS ☑ CTBS ☑ MAT ☑ CAT	**Required** Protractor *(MK, CRB, p. 129)* Straightedge *(MK)* Recording Sheet for Exercises 1–3, 8–16, 25–27, and 37 *(CRB, pp. 127–128)* **Optional** Teaching Transparency T6 *(CRB)*	• *Chapter 7 Resource Book,* pp. 15–24 • Daily Transparency T5, CRB • Additional Examples Transparencies T7–T9, CRB • *Alternate Openers: Explorations,* p. 54
Hands-On Lab 7A Construct Congruent Segments and Angles **NCTM:** Geometry **NAEP:** Geometry 3b ☐ SAT-9 ☐ SAT-10 ☐ ITBS ☐ CTBS ☐ MAT ☐ CAT	**Required** Compass *(MK)* Straightedge *(MK)*	• *Hands-On Lab Activities,* pp. 48–50
Lesson 7-3 Angle Relationships **NCTM:** Geometry, Communication **NAEP:** Geometry 3b ☑ SAT-9 ☑ SAT-10 ☑ ITBS ☑ CTBS ☑ MAT ☑ CAT	**Optional** Teaching Transparency T11 *(CRB)* Protractor *(MK, CRB, p. 129)* Cardboard angles *(MK)*	• *Chapter 7 Resource Book,* pp. 25–34 • Daily Transparency T10, CRB • Additional Examples Transparencies T12–T13, CRB • *Alternate Openers: Explorations,* p. 55
Lesson 7-4 Classifying Lines **NCTM:** Geometry, Communication **NAEP:** Geometry 4b ☑ SAT-9 ☑ SAT-10 ☑ ITBS ☑ CTBS ☑ MAT ☑ CAT	**Optional** Teaching Transparency T15 *(CRB)* Recording Sheet for Reaching All Learners *(CRB, p. 130)*	• *Chapter 7 Resource Book,* pp. 35–43 • Daily Transparency T14, CRB • Additional Examples Transparencies T16–T18, CRB • *Alternate Openers: Explorations,* p. 56
Hands-On Lab 7B Parallel Line Relationships **NCTM:** Geometry, Reasoning and Proof **NAEP:** Geometry 3g ☑ SAT-9 ☑ SAT-10 ☑ ITBS ☑ CTBS ☑ MAT ☑ CAT	**Required** Protractor *(MK, CRB, p. 129)* Straightedge *(MK)*	• *Hands-On Lab Activities,* p. 51
Section 7A Assessment		• Mid-Chapter Quiz, SE p. 342 • Section 7A Quiz, AR p. 18 • *Test and Practice Generator* CD-ROM

SAT = *Stanford Achievement Tests* **ITBS** = *Iowa Test of Basic Skills* **CTBS** = *Comprehensive Test of Basic Skills/Terra Nova*
MAT = *Metropolitan Achievement Tests* **CAT** = *California Achievement Test*
NCTM = Complete standards can be found on pages T27–T33. **NAEP** = Complete standards can be found on pages A31–A35.
SE = *Student Edition* **TE** = *Teacher's Edition* **AR** = *Assessment Resources* **CRB** = *Chapter Resource Book* **MK** = *Manipulatives Kit*

Section Overview

Points, Lines, and Planes — *Lesson 7-1*

Why? Points, lines, and planes are the foundation of geometry.

A **line** is a straight path that extends without end in opposite directions. $\overleftrightarrow{PQ}$ and $\overleftrightarrow{NR}$ are lines.

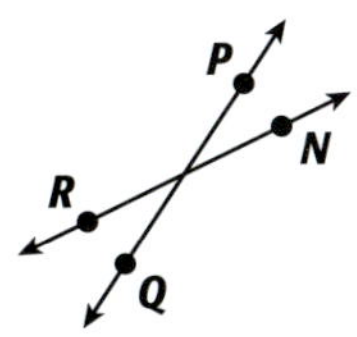

A **point** is an exact location. *P, Q, R, M,* and *N* are points.

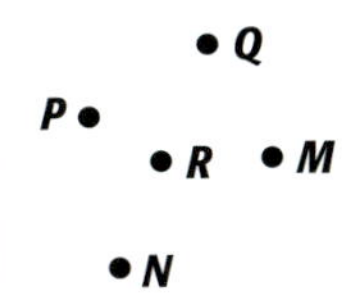

Line **segments** and **rays** are parts of lines. $\overline{PQ}$ is a line segment. $\overrightarrow{RS}$ is a ray.

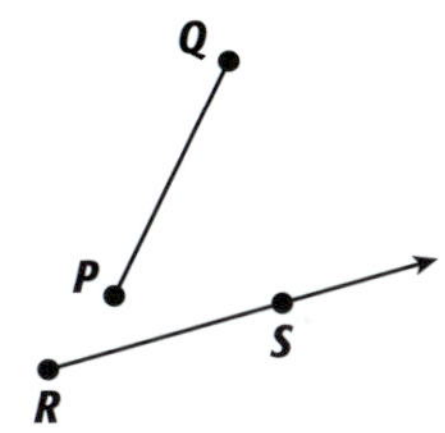

A **plane** is a flat surface that extends without end in all directions. Plane *MNP* is a plane.

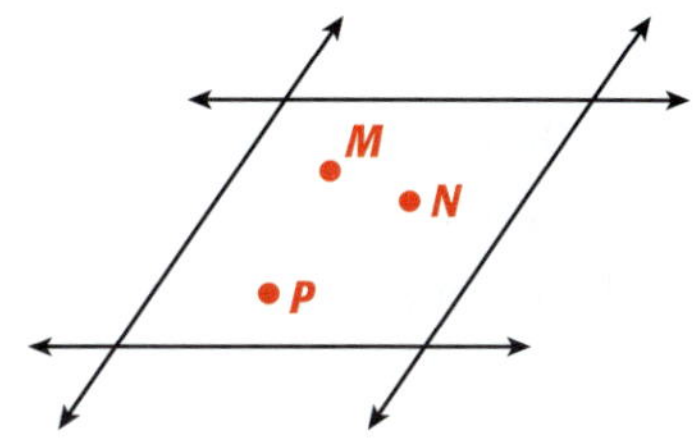

Angles and Angle Relationships — *Lessons 7-2, 7-3*

Why? Many geometric figures contain angles.

An **angle** is formed by two rays with a common end-point, called the **vertex**.

acute angle $<$ 90° $<$ obtuse angle $<$ 180°

straight angle = 180° right angle = 90°

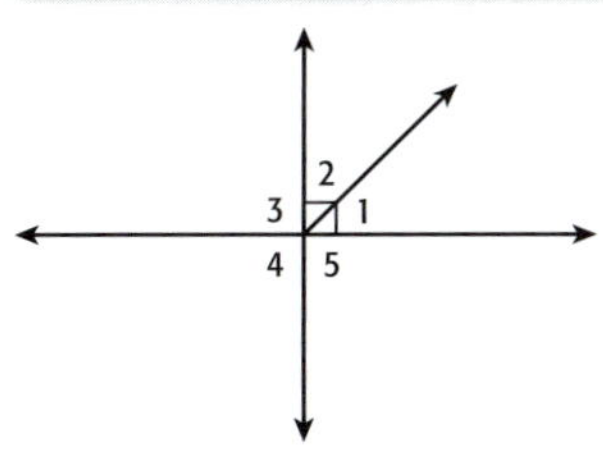

Complementary Angles	Supplementary Angles	Adjacent Angles	Vertical Angles
$\angle 1$ and $\angle 2$	$\angle 3$ and $\angle 4$ $\angle 4$ and $\angle 5$	$\angle 1$ and $\angle 2$ $\angle 2$ and $\angle 3$ $\angle 3$ and $\angle 4$ $\angle 4$ and $\angle 5$ $\angle 5$ and $\angle 1$	$\angle 3$ and $\angle 5$

Classifying Lines — *Lesson 7-4*

Why? Parallel and perpendicular lines are used in construction, among other applications.

Intersecting lines are lines that cross at one common point.

Line *YZ* intersects line *WX*.

$\overleftrightarrow{YZ}$ intersects $\overleftrightarrow{WX}$.

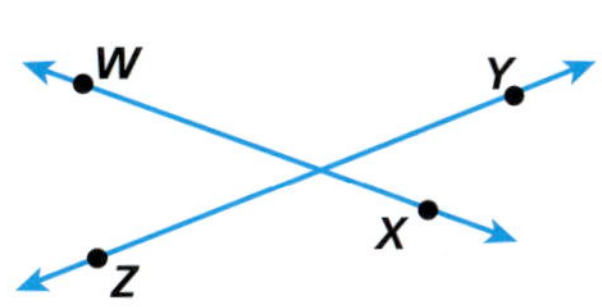

Parallel lines are lines in the same plane that never intersect.

Line *AB* is parallel to line *ML*.

$\overleftrightarrow{AB} \parallel \overleftrightarrow{ML}$

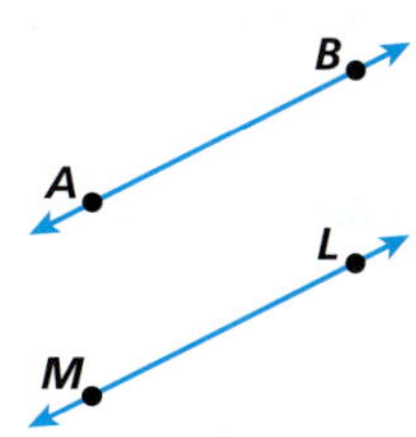

Perpendicular lines intersect to form 90° angles, or right angles.

Line *RS* is perpendicular to line *TU*. $\overleftrightarrow{RS} \perp \overleftrightarrow{TU}$

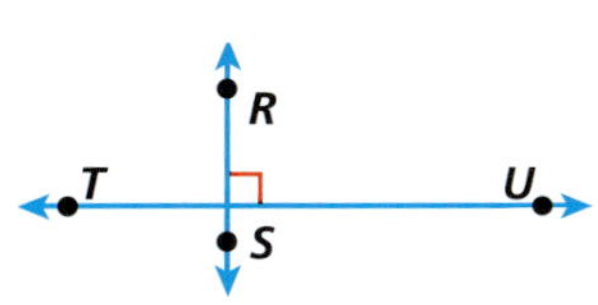

Skew lines are lines that lie in different planes. They are neither parallel nor intersecting.

Line *AB* and line *ML* are skew.

$\overleftrightarrow{AB}$ and $\overleftrightarrow{ML}$ are skew.

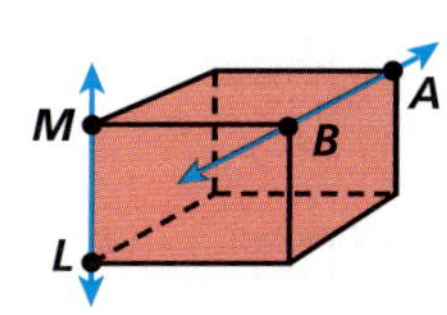

7-1 Organizer

Pacing: Traditional 1 day
Block $\frac{1}{2}$ day

Objective: Students describe figures by using the terms of geometry.

Indiana Academic Standards
6.7.2, 6.7.5, 6.7.11

Warm Up

What geometry term might you associate with each object?

1. a string on a guitar line segment
2. a window plane or rectangle
3. the tip of a pencil point
4. a sheet of paper plane or rectangle

Problem of the Day

Draw a clock face that includes the numerals 1–12. Draw two lines that do not intersect and that separate the clock face into three parts so that the sums of the numbers on each part are the same.

Available on Daily Transparency in CRB

Math Humor

What did the circle say to the line?
I'll be around if you need me.

7-1 Points, Lines, and Planes

Learn to describe figures by using the terms of geometry.

Vocabulary
point
line
plane
line segment
ray

The building blocks of geometry are *points, lines,* and *planes.*

A **point** is an exact location.

point *P*, *P*

A point is named by a capital letter.

A **line** is a straight path that extends without end in opposite directions.

line *AB*, $\overleftrightarrow{AB}$, line *BA*, $\overleftrightarrow{BA}$

A line is named by two points on the line.

A **plane** is a flat surface that extends without end in all directions.

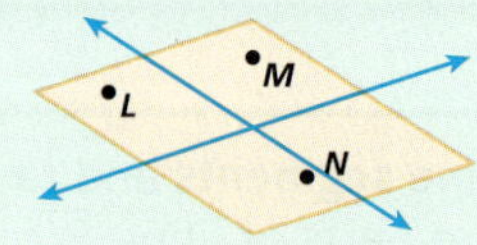

plane *LMN*, plane *MLN*, plane *NLM*

A plane is named by three points on the plane that are not on the same line.

EXAMPLE 1 Identifying Points, Lines, and Planes

Use the diagram to name each geometric figure.

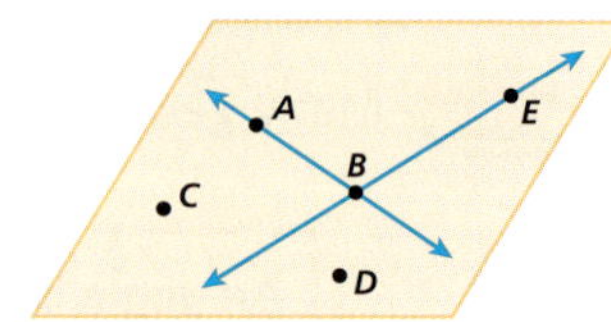

A three points
A, *C*, and *D*
Five points are labeled: points A, B, C, D, and E.

B two lines
$\overleftrightarrow{AB}$ and $\overleftrightarrow{BE}$
You can also write $\overleftrightarrow{BA}$ *and* $\overleftrightarrow{EB}$.

C a point shared by two lines
point *B*
Point B is a point on $\overleftrightarrow{AB}$ *and* $\overleftrightarrow{BE}$.

D a plane
plane *ADC*
Use any three points in the plane that are not on the same line. Write the three points in any order.

1 Introduce

Alternate Opener

EXPLORATION

7-1 Points, Lines, and Planes

Geometry can be used to describe the physical world around us. Check the box of the geometry term that each real-world item represents.

		Point	Line Segment	Plane
1.	A freckle			
2.	A strand of hair			
3.	A poster			
4.	A pixel on your calculator screen			
5.	A period at the end of a sentence			
6.	A guitar string			
7.	The minute hand of a clock			
8.	A computer screen			

Think and Discuss

9. **Describe** the characteristics of the items that you classified as *points* in the table above.
10. **Describe** the characteristics of the items that you classified as *line segments* in the table above.

Motivate

Ask students to give examples of symbols that are used instead of words (e.g., pictures on men's and women's restrooms). Have them explain why symbols are sometimes used instead of words. Explain that in this lesson, symbols are used to name lines, line segments, and rays.

Exploration worksheet and answers on Chapter 7 Resource Book pp. 7 and 141

2 Teach

Lesson Presentation

Guided Instruction

In this lesson, students learn to describe figures by using the terms of geometry. Use the diagrams to teach students the meanings of *point, line, plane, line segment,* and *ray* (Teaching Transparency T2 in the Chapter 7 Resource Book). Point out that when naming a ray the order of the points *does* matter—the endpoint is always listed first.

Teaching Tip Remind students that line segments have two endpoints, rays have one endpoint, and lines have no endpoints.

Line segments and rays are parts of lines. Use points on a line to name line segments and rays.

A **line segment** is made of two endpoints and all the points between the endpoints.

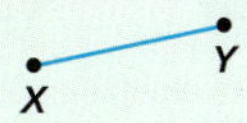

line segment XY, $\overline{XY}$,
line segment YX, $\overline{YX}$

A line segment is named by its endpoints.

A **ray** has one endpoint. From the endpoint, the ray extends without end in one direction only.

ray JK, $\overrightarrow{JK}$

A ray is named by its endpoint first followed by another point on the ray.

EXAMPLE 2 Identifying Line Segments and Rays

Use the diagram to give a possible name to each figure.

A three different line segments

$\overline{TU}$, $\overline{UV}$, and $\overline{TV}$

You can also write $\overline{UT}$, $\overline{VU}$, and $\overline{VT}$.

B three ways to name the line

$\overleftrightarrow{UT}$, $\overleftrightarrow{VU}$, and $\overleftrightarrow{VT}$

You can also write $\overleftrightarrow{TU}$, $\overleftrightarrow{UV}$, and $\overleftrightarrow{TV}$.

C six different rays

$\overrightarrow{TU}$, $\overrightarrow{TV}$, $\overrightarrow{VT}$, $\overrightarrow{VU}$, $\overrightarrow{UV}$, and $\overrightarrow{UT}$

D another name for ray TU

$\overrightarrow{TV}$

T is still the endpoint. V is another point on the ray.

Think and Discuss

1. **Name** the geometric figure suggested by each of the following: a page of a book; a dot (also called a *pixel*) on a computer screen; the path of a jet across the sky.
2. **Explain** how $\overrightarrow{XY}$ is different from $\overleftrightarrow{XY}$.
3. **Explain** how $\overline{AB}$ is different from $\overleftrightarrow{AB}$.

Additional Examples

Example 1

Use the diagram to name each geometric figure.

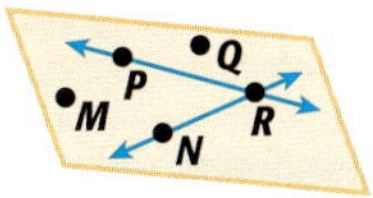

A. three points
M, N, and P (also Q and R)

B. two lines
$\overleftrightarrow{PR}$ and $\overleftrightarrow{NR}$ or $\overleftrightarrow{RP}$ and $\overleftrightarrow{RN}$

C. a location shared by two lines
point R

D. a plane
plane QRM

Example 2

Use the diagram to give a possible name to each figure.

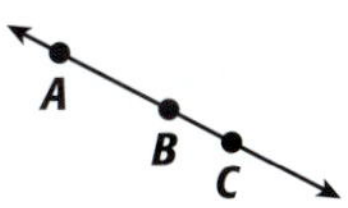

A. three different line segments
$\overline{AB}$, $\overline{BC}$, $\overline{AC}$, or $\overline{BA}$, $\overline{CB}$, $\overline{CA}$

B. three ways to name the line
$\overleftrightarrow{AB}$, $\overleftrightarrow{BC}$, $\overleftrightarrow{AC}$ or $\overleftrightarrow{BA}$, $\overleftrightarrow{CB}$, $\overleftrightarrow{CA}$

C. six different rays
$\overrightarrow{AB}$, $\overrightarrow{AC}$, $\overrightarrow{BC}$, $\overrightarrow{CB}$, $\overrightarrow{CA}$, $\overrightarrow{BA}$

D. another name for ray AB
$\overrightarrow{AC}$

3 Close

Reaching All Learners

Through Concrete Manipulatives

Have students work with a partner to play a geometry memory game. Give each pair a set of ten index cards (Chapter 7 Resource Book, p. 126). On five of the cards have students write the terms *point*, *line*, *plane*, *line segment*, and *ray* (one per card). On the other five cards, have them draw pictures representing each term (again, one per card). Have students mix the cards, place them face down, and take turns turning over pairs of cards to make a match. This continues until all matches have been made. (These cards can be kept and added to as students progress through the chapter.)

Summarize

Draw a diagram like the one shown in Example 1, and have volunteers identify a *point*, a *line*, a *plane*, a *line segment*, and a *ray* on the diagram.

Answers to Think and Discuss

1. plane; point; line segment
2. Possible answer: $\overrightarrow{XY}$ is a ray with endpoint X that passes through point Y; $\overleftrightarrow{XY}$ is a line that passes through points X and Y.
3. Possible answer: $\overline{AB}$ is a line segment with endpoints A and B; $\overleftrightarrow{AB}$ is a line that passes through points A and B.

7-1 PRACTICE & ASSESS

Students may want to refer back to the lesson examples.

Assignment Guide

If you finished Example 1 assign:
Core 1–4, 8–11, 20, 25–30
Enriched 15–19, 21–30

If you finished Example 2 assign:
Core 1–21, 25–30
Enriched 4–30

Answers

Possible answers:

5. $\overleftrightarrow{AC}$ and $\overleftrightarrow{AB}$
6. $\overrightarrow{AC}$, $\overrightarrow{BC}$, $\overrightarrow{BA}$, and $\overrightarrow{CA}$
13. $\overrightarrow{WX}$, $\overrightarrow{XY}$, $\overrightarrow{YZ}$, $\overrightarrow{ZY}$, $\overrightarrow{YW}$, and $\overrightarrow{ZX}$
17. $\overrightarrow{CA}$ and $\overrightarrow{CB}$
18. $\overrightarrow{AC}$
19. $\overleftrightarrow{BC}$ and $\overleftrightarrow{CB}$

7-1 Exercises

FOR EXTRA PRACTICE
see page 649

internet connect
Homework Help Online
go.hrw.com Keyword: MR4 7-1

GUIDED PRACTICE

See Example 1 **Use the diagram to name each geometric figure.**

1. two points Possible answer: *M* and *N*
2. a line Possible answer: $\overleftrightarrow{KN}$
3. a point shared by two lines *K*
4. a plane Possible answer: *JKL*

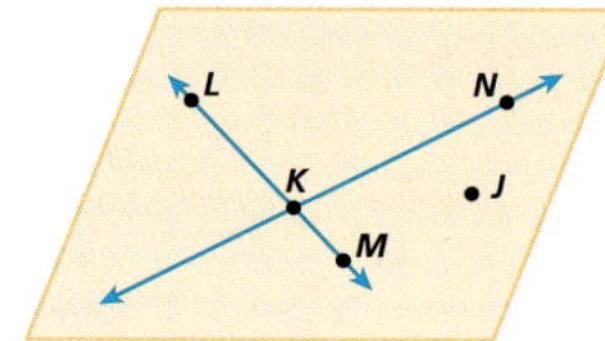

See Example 2 **Use the diagram to give a possible name to each figure.**

5. two different ways to name the line
6. four different names for rays
7. another name for $\overrightarrow{AC}$ $\overrightarrow{AB}$

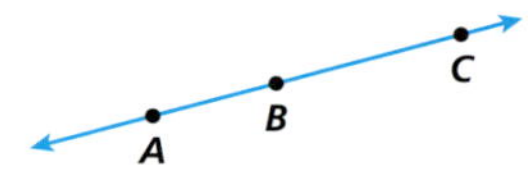

INDEPENDENT PRACTICE

See Example 1 **Use the diagram to name each geometric figure.**

8. three points Possible answer: *D*, *E*, and *F*
9. two lines Possible answer: $\overleftrightarrow{DF}$; $\overleftrightarrow{ED}$
10. a point shared by a line and a ray *E*
11. a plane Possible answer: *FGH*

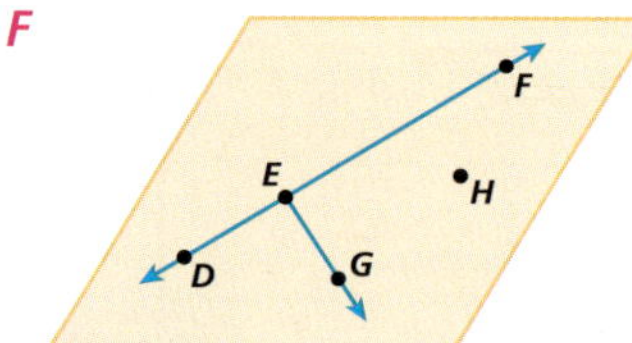

See Example 2 **Use the diagram to give a possible name to each figure.**

12. two different line segments Possible answer: $\overline{WX}$ and $\overline{YZ}$
13. six different names for rays
14. another name for $\overrightarrow{YX}$ $\overrightarrow{YW}$

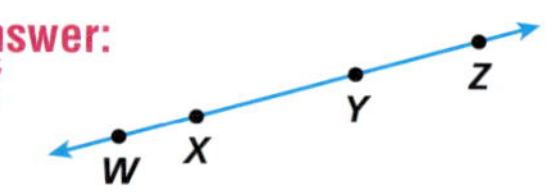

PRACTICE AND PROBLEM SOLVING

Use the diagram to find a name for each geometric figure described.

15. a point shared by three lines *C*
16. two points on the same line Possible answer: *C* and *F*
17. two different rays
18. another name for $\overrightarrow{AD}$
19. two different names for the same line

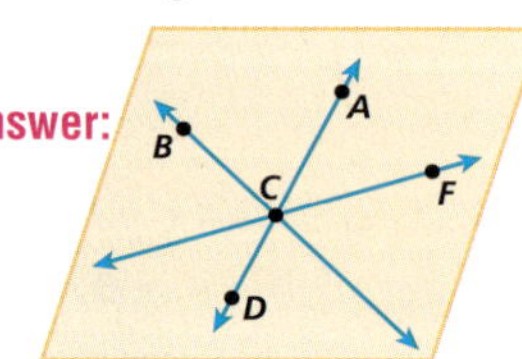

Math Background

When Euclid wrote *The Elements*, the book to which the modern study of geometry can trace its roots, he began with definitions of the terms *point* and *line*. To define a point, Euclid wrote "that which has no part."

In modern developments of geometry as a logical system, one must begin with undefined terms. When your students encounter geometry in high school, the terms *point*, *line*, and *plane* will be treated as undefined terms. These terms will then form the basis for defining more-complicated geometric figures.

RETEACH 7-1

LESSON 7-1 **Reteach**
Points, Lines, and Planes

Here are terms that can help you understand geometry.

A point is an exact location. A point has no size.

•*A* point *A* or *A* Use a capital letter to name a point.

A line is a straight path that extends without end in opposite directions. A line has infinite length, but no width.

A *B* $\overleftrightarrow{AB}$ or $\overleftrightarrow{BA}$ Two points name a line.

A plane is a flat surface that extends without end in all directions. A plane has infinite length and width, but no depth.

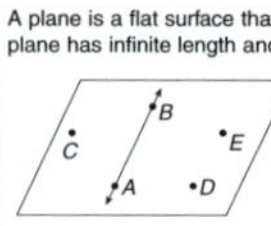

CDE *EDC* A plane is named by 3
DEC *CED* points on the plane that
ECD *DCE* are not on a line.

A line segment has two endpoints. The length of a line segment can be measured.

J *K* $\overline{JK}$ or $\overline{KJ}$ A line segment is named by its endpoints.

X *Y* $\overrightarrow{XY}$

A ray has one endpoint and extends without end in one direction. A ray is named by its endpoint first and another point on the ray.

Use the diagram to name each geometric figure. Possible answers are given.

1. two points *R*, *S*
2. two lines $\overleftrightarrow{PQ}$ and $\overleftrightarrow{RQ}$
3. a plane *STU*
4. two line segments $\overline{PQ}$ and $\overline{QR}$
5. two rays $\overrightarrow{QP}$ and $\overrightarrow{QR}$
6. a point shared by two lines *Q*

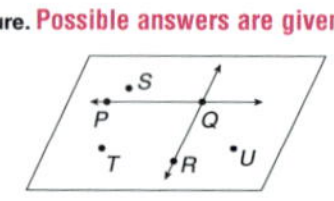

PRACTICE 7-1

LESSON 7-1 **Practice B**
Points, Lines, and Planes

Use the diagram to name each geometric figure. Possible answers are given.

1. two points *A* and *B*
2. a plane plane *ABD*
3. a line segment $\overline{BD}$
4. a point shared by two lines *A*
5. a line $\overleftrightarrow{CD}$

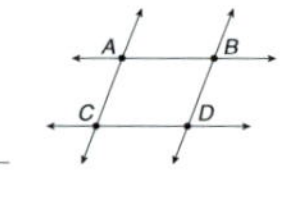

Use the diagram to give a possible name for each figure. Possible answers are given.

6. two different ways to name the line
line *XY* and $\overleftrightarrow{XY}$
7. four different names for rays
ray *PY*, ray *PX*, $\overrightarrow{PY}$, and $\overrightarrow{PX}$
8. another name for $\overline{QP}$
$\overline{PQ}$

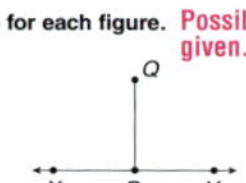

9. Is the following statement always true, sometimes true, or never true? Explain your reasoning. A line is longer than a line segment.
It is always true, because a line segment only extends between two endpoints, but a line extends without end in opposite directions.
10. Using endpoints as your basis, explain how a line, a line segment, and a ray are different.
A line has no endpoints, a ray has one endpoint, and a line segment has two endpoints.

Geography LINK

Mapmakers often must use a kind of code to give information on a map. To aid with interpreting this code, they also usually include a *compass rose,* a *scale,* and a *legend.*

The compass rose has arrows that point north, south, east, and west. The scale tells you the real-world distance that is represented by a distance on the map. The legend explains what each symbol on the map represents.

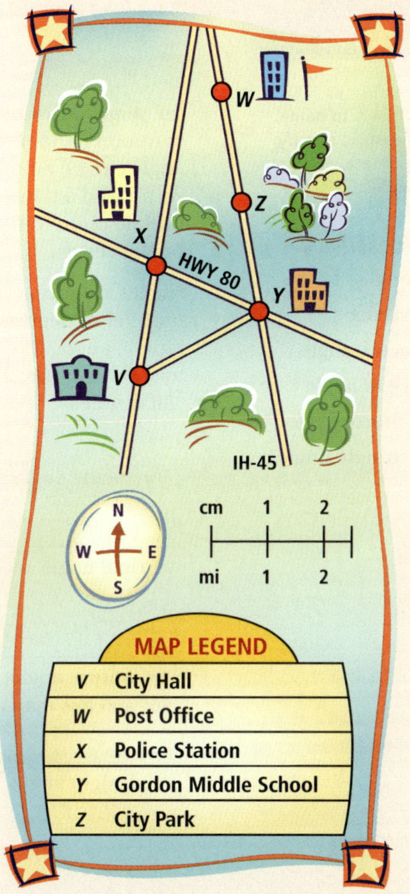

20. Name the geometric figure suggested by each part of the map.

- **a.** City Hall and Gordon Middle School **points**
- **b.** Highway 80 **line**
- **c.** the road from the park to the post office **line segment**

21. Use a centimeter ruler to measure the route from the police station to Gordon Middle School. Use the scale to find the actual distance your measurement represents. **2 miles**

22. ***WHAT'S THE QUESTION?*** The answer is the route from City Hall to Gordon Middle School. What is the question?

23. ***WRITE ABOUT IT*** Explain why the road from City Hall that goes past the police station suggests a ray named $\overrightarrow{VX}$ rather than a ray named $\overrightarrow{XV}$.

24. ***CHALLENGE*** What are all the possible names for the line suggested by IH-45?
$\overleftrightarrow{YZ}$, $\overleftrightarrow{YW}$, $\overleftrightarrow{ZY}$, $\overleftrightarrow{ZW}$, $\overleftrightarrow{WY}$, and $\overleftrightarrow{WZ}$

Spiral Review

25. A cocker spaniel weighs seven times more than her puppy. Write an algebraic expression for the weight of the mother. Use *p* to represent the weight of the puppy. (Lesson 2-2) **7*p***

Find the quotient. (Lessons 3-7, 3-8)

26. $45.5 \div 5$ **9.1** **27.** $103.7 \div 2$ **51.85** **28.** $35 \div 2.5$ **14** **29.** $4.25 \div 0.25$ **17**

30. **TEST PREP** Which sum is the greatest? (Lesson 5-7) **C**

A $\frac{1}{3} + \frac{1}{4}$ **B** $\frac{1}{6} + \frac{5}{12}$ **C** $\frac{1}{4} + \frac{7}{12}$ **D** $\frac{1}{12} + \frac{1}{2}$

Interdisciplinary LINK

Geography

Exercises 20–24 involve reading and interpreting a map. Students learn to work with maps in middle school social studies programs, such as Holt, Rinehart, and Winston's *People, Places, and Change.*

Answers

22. Possible answer: Which route is longer than 2 miles?

23. The first letter in the name of a ray should be the name of the end-point. *V* is the endpoint of the ray.

Journal

Have students draw a diagram. Then have them name four points, two lines, two line segments, three rays, and a plane shown in the diagram.

Test Prep Doctor

For Exercise 30, students need to use common denominators to find each sum. Because each pair of fractions has an LCD of 12, the sums will be easy to compare. **A, B,** and **D** all have the same sum, $\frac{7}{12}$, which is less than the sum for **C,** $\frac{10}{12}$. So the correct answer is **C.**

CHALLENGE 7-1

LESSON 7-1 **Challenge**
Points of Light

Astronomers have divided the sky into 88 constellations, or groups of stars. To map these constellations, astronomers use a point for each star and draw imaginary lines between the points. One of the easiest groups of stars to see is called the Big Dipper. The Big Dipper is not officially a constellation. It is only part of a large constellation called Ursa Major, or Great Bear.

Imagine the stars of Ursa Major are labeled as shown below. Follow these instructions to make a star map of Ursa Major.

1. Use line segments to connect points *A–G* in alphabetical order, and then draw $\overline{DG}$.
2. Use line segments to connect points *G–L* in alphabetical order, and then draw $\overline{FJ}$.
3. Use line segments to connect points *M–O* in alphabetical order. Then draw $\overline{EM}$ and $\overline{MP}$.
4. Label the star grouping formed by the 7 points *A–G* as the "Big Dipper."
5. Then label each point of the Big Dipper with its proper star name: *A*: Alkaid; *B*: Mizar; *C*: Alioth; *D*: Megrez; *E*: Phecda; *F*: Merak; *G*: Dubhe

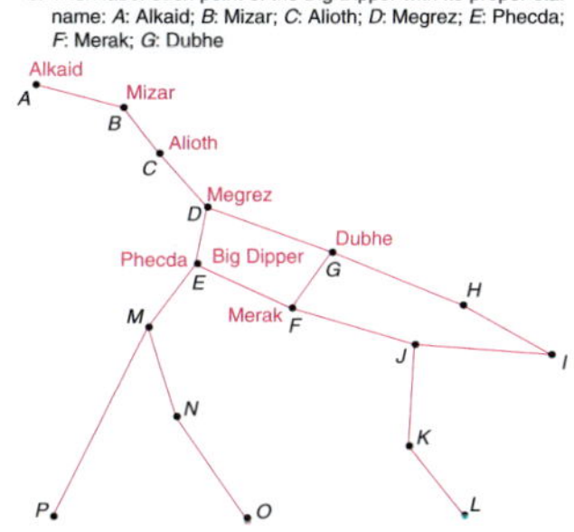

PROBLEM SOLVING 7-1

LESSON 7-1 **Problem Solving**
Points, Lines, and Planes

Place your hand down flat on a sheet of paper. Draw a point at the tip of your thumb, the tip of your middle finger, and the tip of your pinky.

1. Label the thumb point *A*, the middle finger point *B*, and the pinky point *C*.
   Check students' drawings.
2. Name all the planes you possibly can with points *A*, *B*, and *C*.
   plane *ABC*
3. Draw and name all the lines you can make with points *A*, *B*, and *C*.
   $\overleftrightarrow{AB}$, $\overleftrightarrow{AC}$, and $\overleftrightarrow{BC}$
4. Name all the line segments possible using points *A*, *B*, and *C*.
   $\overline{AB}$, $\overline{AC}$, and $\overline{BC}$
5. Name all the rays possible using points *A*, *B*, and *C*.
   $\overrightarrow{AB}$, $\overrightarrow{AC}$, $\overrightarrow{BA}$, $\overrightarrow{BC}$, $\overrightarrow{CA}$, and $\overrightarrow{CB}$
6. Choose one line that you drew. Give all the different possible names for that line.
   Possible answer: $\overleftrightarrow{AB}$, $\overleftrightarrow{BA}$, **line *AB*, and line *BA***

Classify each statement as true or false. If it is false, explain why.

7. An infinite number of lines can be drawn through one point.
   true
8. Exactly one line can be drawn between two points.
   true
9. A line contains exactly one ray.
   False; any point on a line defines another ray on the line.
10. If points *A* and *B* are on a line, then line segment *AB* and line segment *BA* are the same.
   true

Circle the letter of the correct answer.

11. Which of the following has exactly one endpoint?
   - **A** $\overrightarrow{OP}$
   - **B** $\overline{AB}$
   - **C** $\overleftrightarrow{TR}$
   - **(D)** $\overrightarrow{SM}$
12. Which of the following is a straight path that extends without end in opposite directions?
   - **F** a point
   - **(G)** a line
   - **H** a ray
   - **J** a line segment

Lesson Quiz

Use the diagram to name each geometric figure.

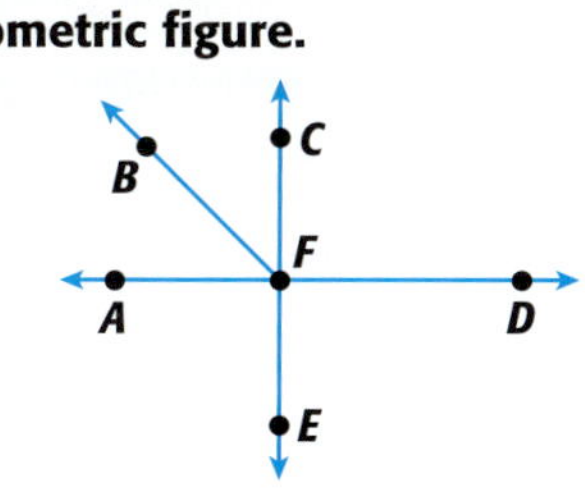

1. three points Possible answer: *A, B, C*

2. two lines Possible answer: $\overleftrightarrow{AD}$, $\overleftrightarrow{CE}$

3. a point shared by a line and a ray Possible answer: *F*

4. a plane Possible answer: plane *AEC*

Available on Daily Transparency in CRB

7-2 Organizer

Pacing: Traditional 1 day
Block $\frac{1}{2}$ day

Objective: Students measure, classify, estimate, and draw angles.

Indiana Academic Standards
6.5.1, 6.7.2, 6.7.5, 6.7.11

Warm Up

1. Draw two points. Label one point *A* and the other point *B*.
2. Draw a line through points *A* and *B*.
3. Draw a ray with *A* as an endpoint and *C* as a point on the ray.
4. Name all the rays in your drawing.
$\overrightarrow{AB}$, $\overrightarrow{BA}$, and $\overrightarrow{AC}$

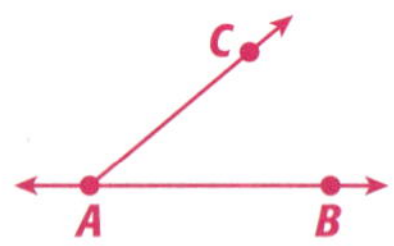

Problem of the Day

The measure of Jack's angle is twice that of Amy's and half that of Nate's. The sum of the measures of Amy's and Trisha's angles is equal to the sum of the measures of Jack's and Nate's angles. The sum of the measures of all the angles is equal to 180°. What is the measure of each student's angle?
Jack's angle: 30°; Nate's angle: 60°; Amy's angle: 15°; Trisha's angle: 75°

Available on Daily Transparency in CRB

7-2 Angles

Learn to name, measure, classify, estimate, and draw angles.

Vocabulary
angle
vertex
acute angle
right angle
obtuse angle
straight angle

An **angle** is formed by two rays with a common endpoint, called the **vertex**. An angle can be named by its vertex or by its vertex and a point from each ray. The middle point in the name should always be the vertex.

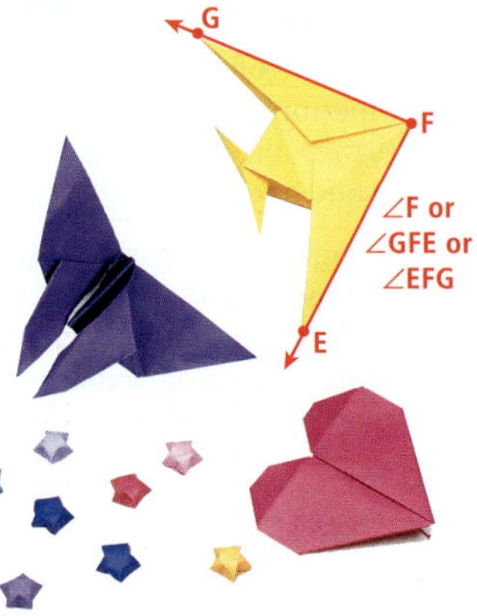

Angles are measured in degrees. The number of degrees determines the type of angle. Use the symbol ° to show degrees: 90° means "90 degrees."

An **acute angle** measures less than 90°.

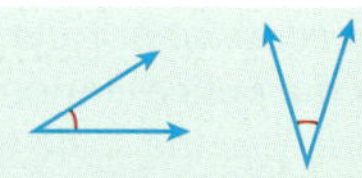

A **right angle** measures exactly 90°.

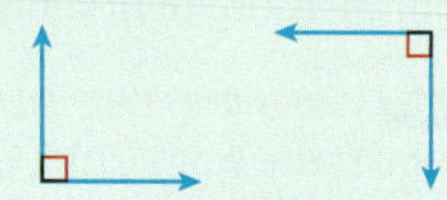

An **obtuse angle** measures more than 90° and less that 180°.

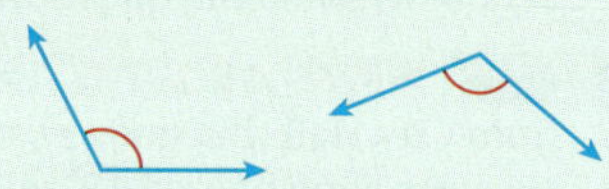

A **straight angle** measures exactly 180°.

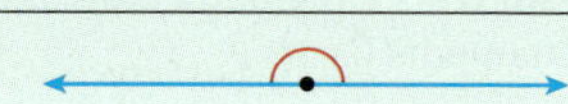

EXAMPLE 1 **Measuring an Angle with a Protractor**

Use a protractor to measure the angle. Tell what type of angle it is.

- Place the center point of the protractor on the vertex of the angle.
- Place the protractor so that ray *YZ* passes through the 0° mark.
- Using the scale that starts with 0° along ray *YZ*, read the measure where ray *YX* crosses.

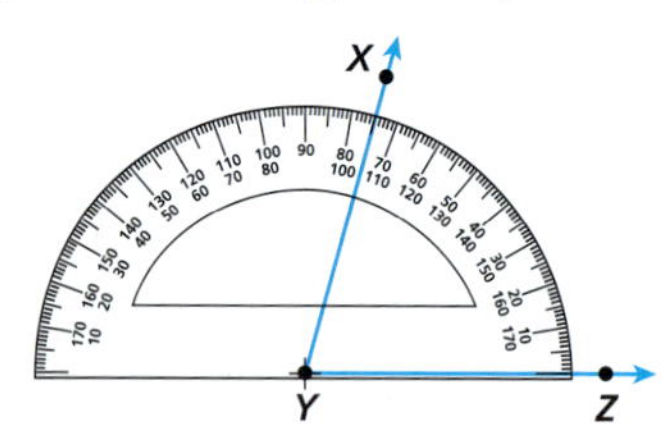

- The measure of $\angle XYZ$ is 75°. Write this as $m\angle XYZ = 75°$.
- Since $75° < 90°$, the angle is acute.

1 Introduce

Alternate Opener

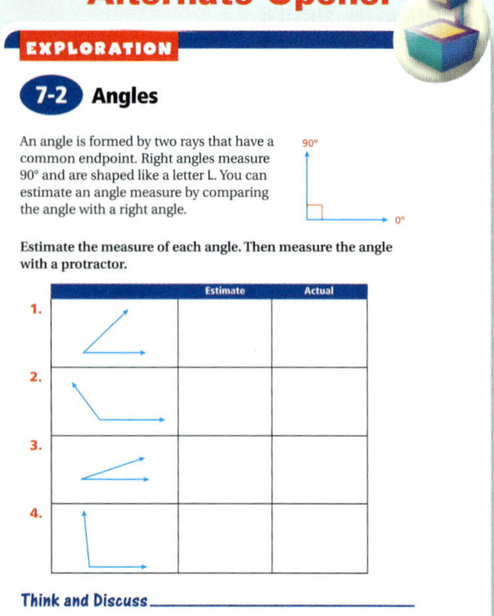
EXPLORATION

7-2 Angles

An angle is formed by two rays that have a common endpoint. Right angles measure 90° and are shaped like a letter L. You can estimate an angle measure by comparing the angle with a right angle.

Estimate the measure of each angle. Then measure the angle with a protractor.

		Estimate	Actual
1.			
2.			
3.			
4.			

Think and Discuss

5. Discuss how you estimated the angle measures.

Motivate

To get students familiar with protractors, distribute a protractor to each student. Have students tell you things they notice about the protractor, such as the following:

- There is a flat edge and a rounded edge.
- There are two sets of numbers along the rounded edge. Each set goes from 0 to 180.

Exploration worksheet and answers on Chapter 7 Resource Book pp. 16 and 143

2 Teach

Lesson Presentation

Guided Instruction

In this lesson, students learn to name, measure, classify, estimate, and draw angles. First teach students to classify angles as acute, right, obtuse, or straight (Teaching Transparency T6 in the Chapter 7 Resource Book). Then teach students to measure angles and to draw angles, given their measure, using a protractor (Manipulatives Kit). Finally teach them to estimate angle measures using the benchmarks of 45°, 90°, and 180°.

Teaching Tip Tell students that all measurements have errors. The angles they construct with a protractor may not be exact.

EXAMPLE 2 Drawing an Angle with a Protractor

Use a protractor to draw an angle that measures 150°.

- Draw a ray on a sheet of paper.
- Place the center point of the protractor on the endpoint of the ray.
- Place the protractor so that the ray passes through the 0° mark.
- Make a mark at 150° above the scale on the protractor.
- Use a straightedge to draw a ray from the endpoint of the first ray through the mark you made at 150°.

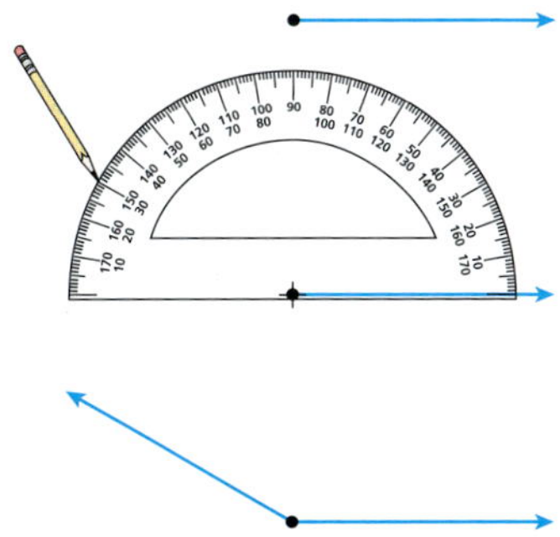

To estimate the measure of an angle, compare it with an angle whose measure you already know. A right angle has half the measure of a straight angle. A 45° angle has half the measure of a right angle.

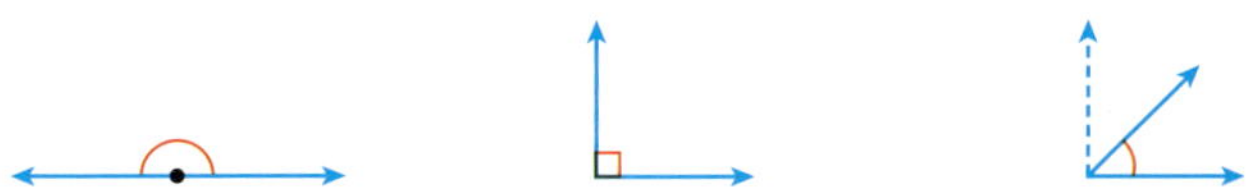

180° 90° 45°

EXAMPLE 3 Estimating Angle Measures

Estimate the measure of the angle, and then use a protractor to check the reasonableness of your estimate.

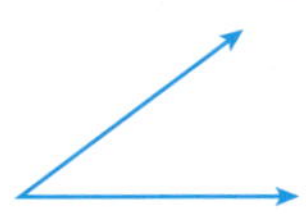

Think: The measure of the angle is close to 45°, but it is a little less. A good estimate would be about 35°.

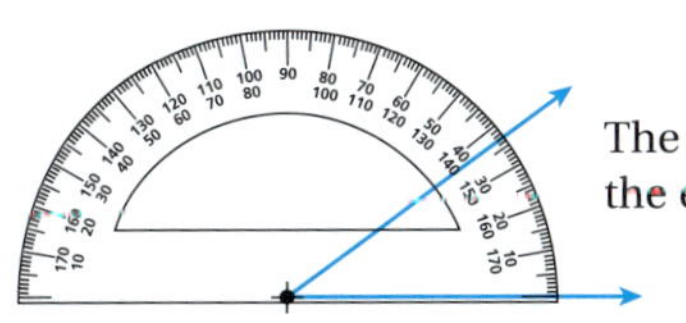

The angle measures 37°, so the estimate is reasonable.

Think and Discuss

1. **Tell** what types of angles are in Examples 2 and 3.
2. **Explain** how you know which scale to read on a protractor.

Additional Examples

Example 1

Use a protractor to measure the angle. Tell what type of angle it is.

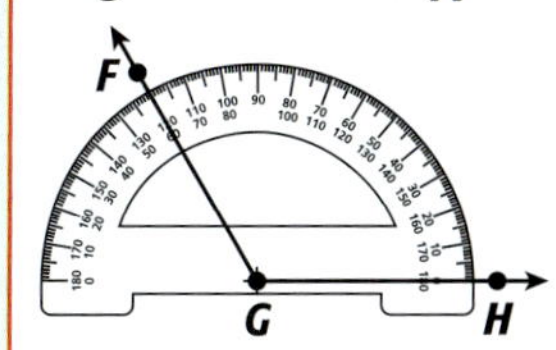

$m\angle FGH = 120°$; obtuse

Example 2

Use a protractor to draw an angle that measures 80°.

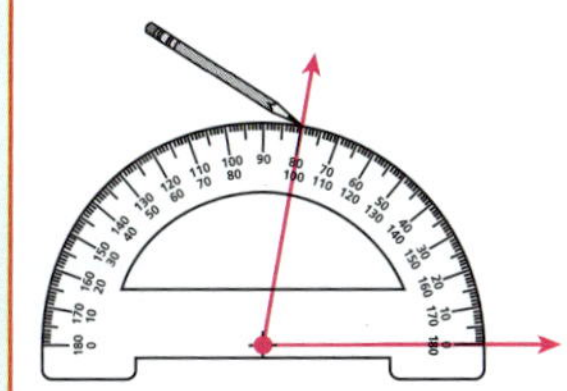

Example 3

Estimate the measure of the angle, and then use a protractor to check the reasonableness of your estimate.

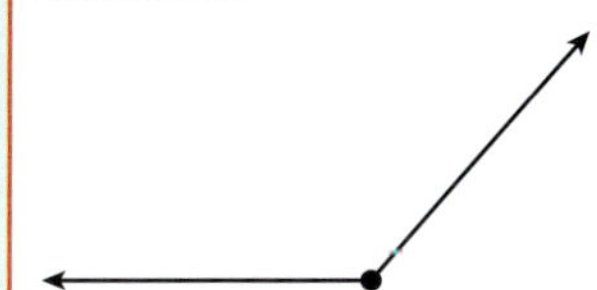

between 90° and 180°; estimate: 135°

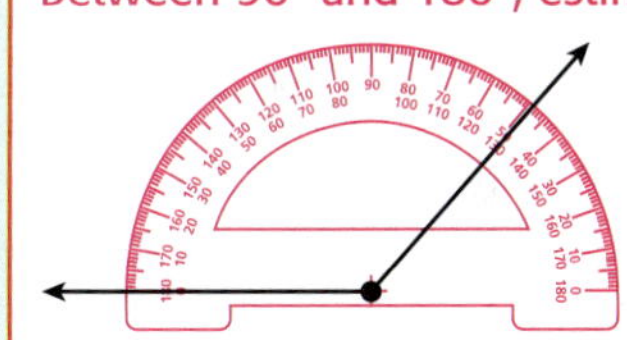

The angle measures 131°.

3 Close

Reaching All Learners

Through Curriculum Integration

Art Have students do research to find examples of geometric art. Have them measure and classify the angles they see in these examples. Encourage students to create their own geometric artwork, and discuss the angles that they used.

Summarize

Display examples of acute, right, obtuse, and straight angles. Have students classify the angles. Then review the steps to measure an angle: place the center of the protractor over the vertex of the angle; line up one ray and a 0° mark; read the measure from that 0° mark to where the other ray crosses the scale.

Answers to Think and Discuss

1. Example 2: obtuse angle; Example 3: acute angle
2. Begin reading the scale where one of the rays passes through the 0° mark.

7-2 PRACTICE & ASSESS

Students may want to refer back to the lesson examples.

Assignment Guide

If you finished Example 1 assign:
Core 1–3, 11–16, 32–34, 40–48
Enriched 1–3, 11–16, 33–35, 40–48

If you finished Example 2 assign:
Core 1–7, 11–24, 40–48
Enriched 11–24, 32–38, 40–48

If you finished Example 3 assign:
Core 1–27, 35, 36, 40–48
Enriched 11–48

You may use the Recording Sheet on Chapter 7 Resource Book pp. 127–128 for Exercises 1–3, 8–16, 25–27, and 37.

Answers

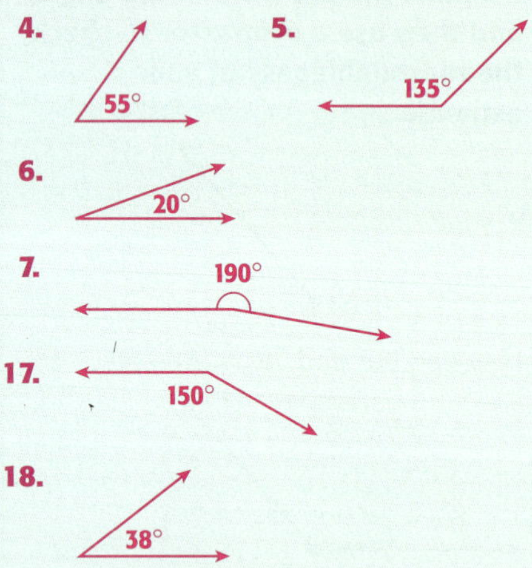

19–24. See p. A4.

7-2 Exercises

FOR EXTRA PRACTICE see page 649

internet connect
Homework Help Online
go.hrw.com Keyword: MR4 7-2

GUIDED PRACTICE

See Example 1 Use a protractor to measure each angle. Tell what type of angle it is.

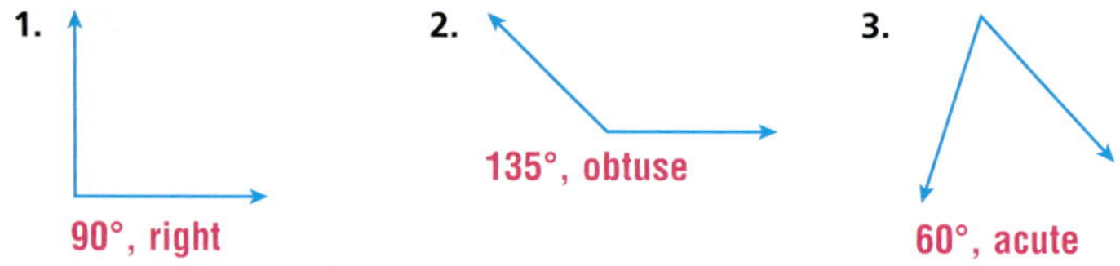

See Example 2 Use a protractor to draw an angle with each given measure.

4. 55° 5. 135° 6. 20° 7. 190°

See Example 3 Estimate the measure of each angle, and then use a protractor to check the reasonableness of your estimate.

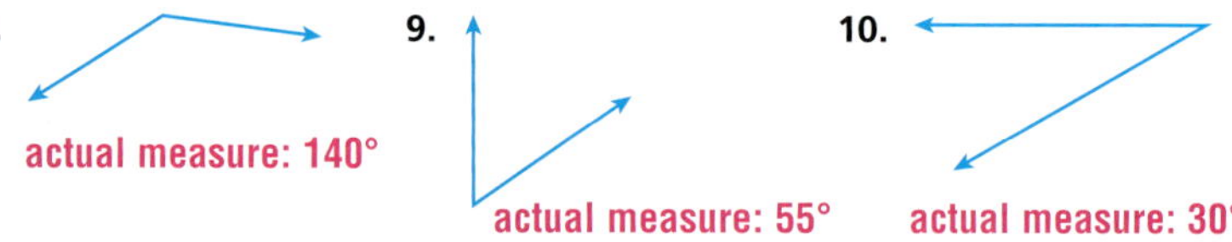

INDEPENDENT PRACTICE

See Example 1 Use a protractor to measure each angle. Tell what type of angle it is.

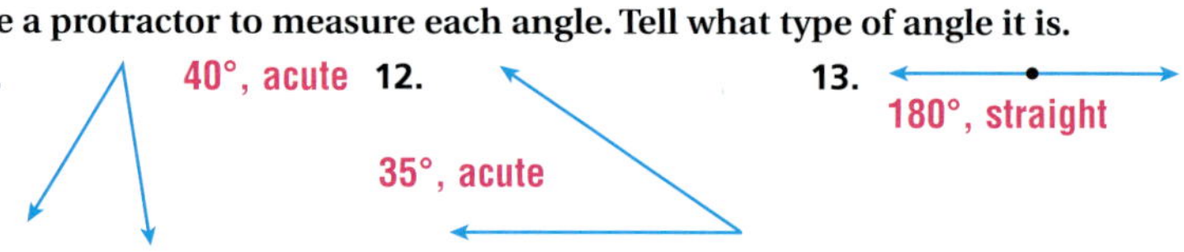

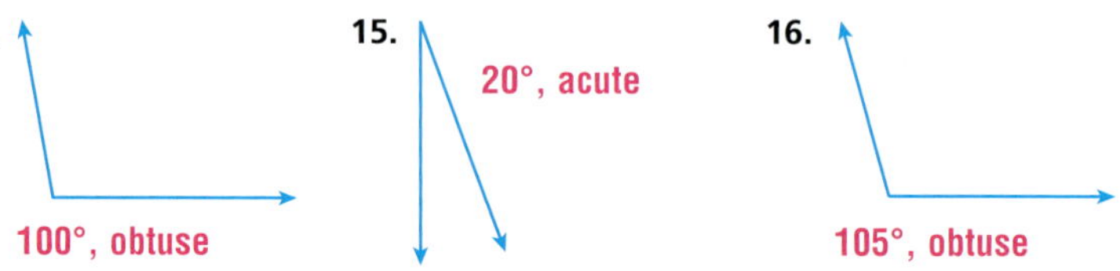

See Example 2 Use a protractor to draw an angle with each given measure.

17. 150° 18. 38° 19. 90° 20. 72°
21. 112° 22. 180° 23. 19° 24. 45°

See Example 3 Estimate the measure of each angle, and then use a protractor to check the reasonableness of your estimate.

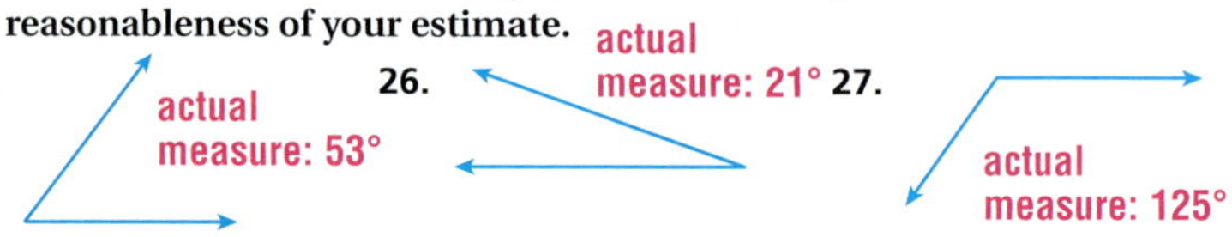

Math Background

The measurement system for angles shown in this lesson comes from the division of a circle into 360 equal parts, or degrees. That idea can be traced to the ancient Babylonians.

In some areas of mathematics, instead of degrees, angles are measured in radians. To find the radian measure of an angle, place the angle's vertex at the center of a circle of radius 1. The radian measure of the angle equals the length of the arc of the circle intercepted by the angle. For example, because a circle of radius 1 has circumference 2π, a 90° angle has a radian measure of $\frac{2\pi}{4} = \frac{\pi}{2}$. Radians are used in trigonometry and analysis.

RETEACH 7-2

LESSON 7-2 **Reteach**
Angles

An angle is formed by two rays with a common endpoint, called the vertex.

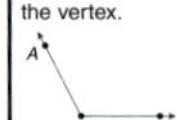

You can name an angle in different ways.

By its vertex $\angle B$ OR By its vertex and one point from each ray. $\angle ABC$, $\angle CBA$ Vertex letter in the middle.

Give three ways to name each angle.

1. $\angle Q$, $\angle PQR$, $\angle RQP$
2. $\angle M$, $\angle GMZ$, $\angle ZMG$

You can use a protractor to measure an angle. Then you can classify the angle by its measure.

To measure the angle, place the center point of the protractor on the vertex, K. Make sure that the protractor is placed so that $\overrightarrow{KL}$ passes through 0°. The measure of the angle is where $\overrightarrow{KJ}$ crosses the protractor.

The measure of the angle is 60 degrees. Write $m\angle K = 60°$. $\angle JKL$ is an acute angle because its measure is less than 90°.

If the measure is:	> 90°	= 90°	= 180°
Then the angle is:	obtuse	right	straight

Use a protractor to measure the angle and classify it.

3. $m\angle Y = 120°$, obtuse
4. $m\angle L = 60°$, acute

PRACTICE 7-2

LESSON 7-2 **Practice B**
Angles

Use a protractor to measure each angle. Tell what type of angle it is.

1. 35°; acute
2. 90°; right
3. 120°; obtuse

Use a protractor to draw an angle with each given measure.

4. 70° 5. 115° 6. 45°

Estimate the measure of each angle, and then use a protractor to check the reasonableness of your estimate. Possible estimates are given.

7. 80°; 82°
8. 130°; 128°
9. 45°; 47°

10. The frame for a kite has two angles that together form a right angle. What type of angle is each of those angles? Explain.
They are both acute angles; The sum of the two angles is a right angle, so they each must measure less than 90°.

11. What kinds of angles are in each of the letters in this word? **TAXI**
The *T* has all right angles; the *A* has acute and obtuse angles; the *X* has acute and obtuse angles; and the *I* has one straight angle.

PRACTICE AND PROBLEM SOLVING

Use a protractor to draw each angle.

28. an acute angle whose measure is less than 45°

29. an obtuse angle whose measure is between 100° and 160°

30. a right angle

31. two acute angles that together form a right angle

Name the smallest angle formed by the hands on each clock.

32.

straight

33.

obtuse

34.

acute

35. ***AVIATION*** Classify the angle the body of an airplane makes with the runway as the airplane takes off. acute angle

36. ***SPORTS*** The quarterback of a football team throws a long pass, and the angle the path of the ball makes with the ground is 30°. Draw an angle with this measurement.

37. ***WHAT'S THE ERROR?*** A student wrote that the measure of this angle is 156°. Explain the error the student may have made, and give the correct measure of the angle. How can the student avoid making the same mistake again?

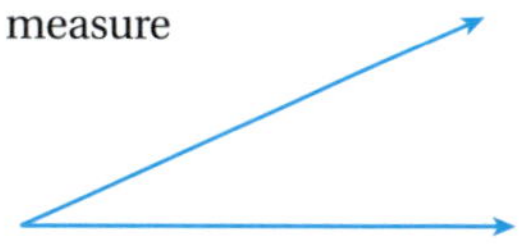

38. ***WRITE ABOUT IT*** Describe how an acute angle and an obtuse angle are different.

39. ***CHALLENGE*** How many times during the day do the hands of a clock form a straight angle? 24

37. The student read the wrong scale on the protractor. The student should estimate the measure of the angle before measuring it.

38. The measure of an acute angle is less than 90°, while the measure of an obtuse angle is greater than 90° and less than 180°.

Spiral Review

Find each value. (Lesson 1-3)

40. 9^2 81 **41.** 2^6 64 **42.** 3^3 27 **43.** 1^{12} 1

Find each sum or difference. (Lesson 3-3)

44. 2.1 + 0.9 3.0 **45.** 4 − 1.2 2.8 **46.** 1.5 + 3.2 4.7 **47.** 5 − 2.6 2.4

48. **TEST PREP** Which fraction is not equivalent to two-thirds? (Lesson 4-6) B

A $\frac{8}{12}$ **B** $\frac{4}{9}$ **C** $\frac{2}{3}$ **D** $\frac{10}{15}$

Answers

28. Possible answer:

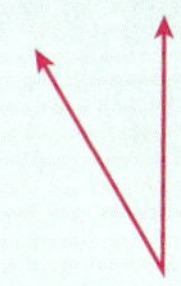

29. Possible answer:

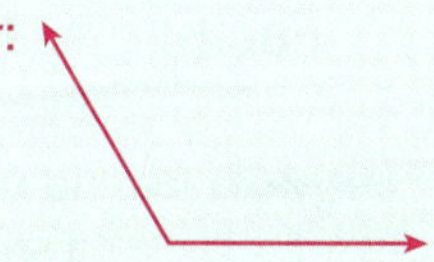

30.

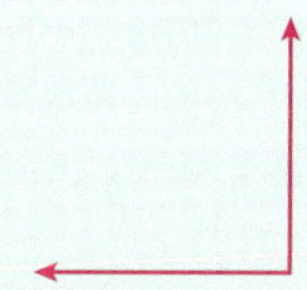

31. Possible answer:

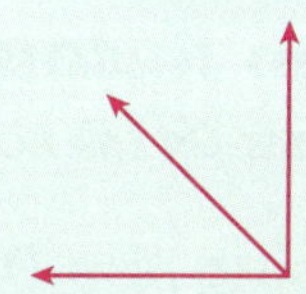

36.

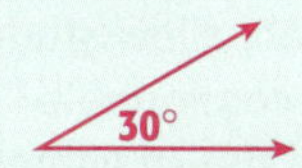

Journal

Have students draw examples of an acute, a right, and an obtuse angle. Have students use a protractor to measure each angle.

Test Prep Doctor

For Exercise 48, students need to remember that a fraction is equivalent to another if the numerator and denominator of the first fraction are both multiplied or divided by the same amount to get the second fraction. Because **A, C,** and **D** all name fractions equivalent to $\frac{2}{3}$, those choices can be eliminated.

CHALLENGE 7-2

LESSON 7-2 **Challenge**
I'll Meet You in 30°

No one knows for sure when people invented clocks to measure time. The first clocks were probably sundials, which measure time by the sun's shadow moving around a circle. Although today's clocks no longer use the sun to measure time, many people still use a circle to display time. A circle has 360°, and a clock face has 12 hours marked. So, each hour on the clock represents 30° of the circle.

Write the time shown on each clock below. Then use a protractor to measure each angle formed by the clock's hands and tell what kind of angle it is.

1. 
3:00; 90°; right angle

2. 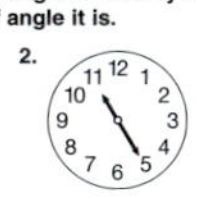
11:25; 180°; straight angle

3. 
9:05; 120°; obtuse angle

4. 
5:10; 90°; right angle

5. 
8:00; 120°; obtuse angle

6. 
4:30; 60°; acute angle

7. 
2:35; 150°; obtuse angle

8. 
6:00; 180°; straight angle

9. 
10:50; 30°; acute angle

PROBLEM SOLVING 7-2

LESSON 7-2 **Problem Solving**
Angles

Write the correct answer.

1. When a patient is lying flat in a hospital bed, what type of angle does the patient's body form? What is the measurement of that angle?
straight angle; 180°

2. When a patient is sitting straight up in a hospital bed, the upper body has been raised to what angle? What type of angle is that?
90° angle; right angle

3. Most hospital beds have a setting for the Fowler position. In this position, the patient's upper body is raised to form a 60° to 70° angle from a flat position. What types of angles are these?
They are both acute.

4. What are the greatest and least differences between the straight-up position and the Fowler position in a hospital bed?
least: 20°; greatest: 30°

5. Medical technicians often set the handles of crutches so that the patient's elbow is at a 30° angle. What type of angle is this?
acute angle

6. By law, wheelchair ramps in public places cannot be greater than 45 degrees. Which type of angle does a wheelchair ramp in public form with the ground?
acute angle

Circle the letter of the correct answer.

7. Physical therapists use a goniometer to measure the extension of a sitting patient's knee. Resting is 90°, and full extension is 180°. What angle does the goniometer measure if the patient's knee is at $\frac{1}{2}$ extension?
A 45°
B 90°
(C) 135°
D 0°

8. The Q-angle is measured between two points on a patient's hip joint and one point on the knee joint. A normal Q-measure from men is 14° plus or minus 3 degrees. What type of angle is any normal Q-angle for men?
F straight
G obtuse
H right
(J) acute

Lesson Quiz

Use a protractor to draw an angle with the given measure. Tell what type of angle it is.

1. 140°
Check student drawings; obtuse.

2. 20° Check student drawings; acute.

3. Draw a right angle.
Check student drawings.

4. Is the angle shown closer to 30° or 120°? 30°

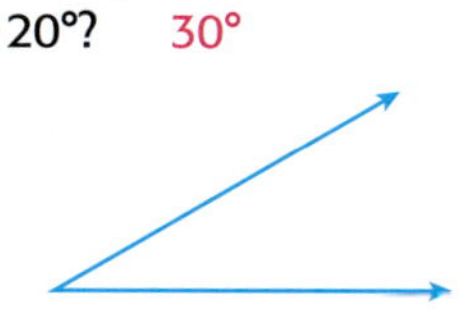

Available on Daily Transparency in CRB

Hands-On LAB 7A Construct Congruent Segments and Angles

Pacing: Traditional 1 day
Block $\frac{1}{2}$ day

Objective: To use a compass and a straightedge to construct congruent segments and angles

Materials: Compass, straightedge

Lab Resources

Hands-On Lab Activities . . . . pp. 48–50

Using the Pages

Before beginning the lab, allow students several minutes to get comfortable using the compass and the straightedge.

1. Construct a circle and label its center *C*.

2. Use the straightedge to draw a line segment about 4 inches long. Label the segment $\overline{AB}$. Open the compass to the length of segment $\overline{AB}$ and construct a circle that has center *A* and that passes through point *B*.

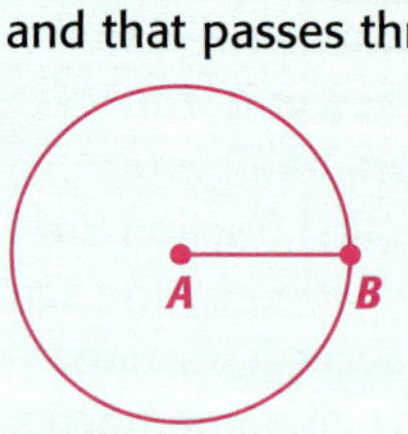

Hands-On LAB 7A Construct Congruent Segments and Angles

Use with Lesson 7-3

Line segments are congruent if they are the same length. You can use a compass and straightedge to construct congruent line segments.

internet connect
Lab Resources Online
go.hrw.com
KEYWORD: MR4 Lab7A

Activity 1

1. Draw a line segment congruent to line segment *AM*.

Draw a ray, and label the endpoint *B*.

Place your compass point on point *A*. Open the compass to the length of $\overline{AM}$. Use the same opening to draw an arc that intersects the ray. Label the intersection point *Y*.

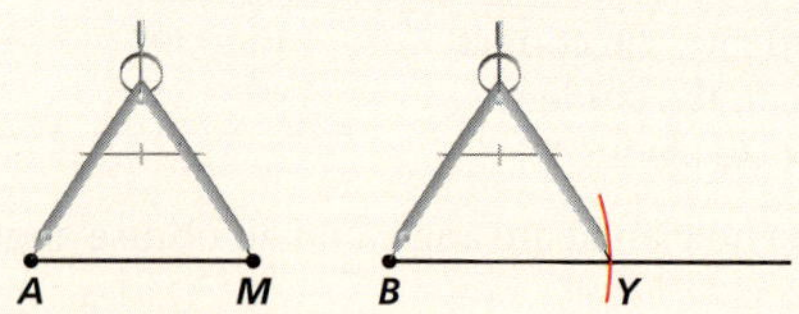

$\overline{AM} \cong \overline{BY}$

Think and Discuss

1. Tell whether it is more accurate to use a ruler or a compass to construct congruent line segments. Explain your answer.

Try This

Construct a line segment that is congruent to the given line segment. **Check students' work.**

1. 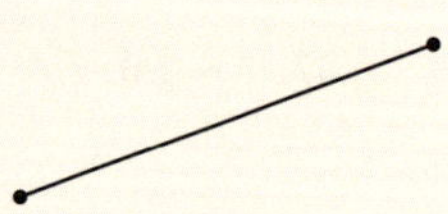

2. 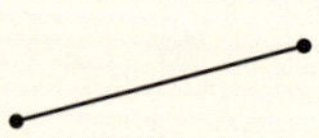

3. 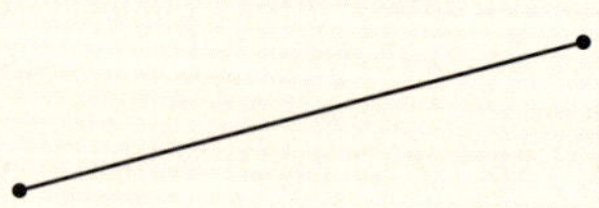

Answers

Activity 1

Think and Discuss

Possible answers:

1. The intersection of the arc and the segment gives the second endpoint of the congruent segment. If the segment were too short, the arc would not intersect it.
2. Compass; as long as the compass is spread accurately and the setting does not change, a compass is the better tool. There is room for error when reading measurements from a ruler, and the ruler may not have markings precise enough to be accurate.

Angles are congruent if they have the same measure in degrees. You can use a compass and straightedge to construct congruent angles.

Activity 2

1 Draw an angle congruent to ∠C.
Draw a ray, and label the endpoint *P*.

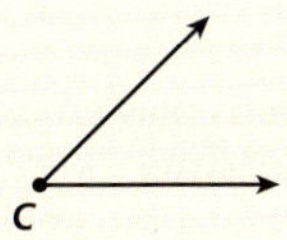

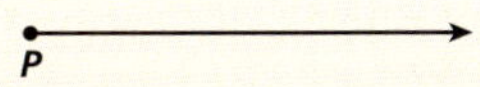

Place your compass point on point *C* and draw an arc through ∠*C*. Label the points of intersection *D* and *E*. Place the compass point on *P*, and draw a similar arc through the ray. Label the point of intersection *R*.

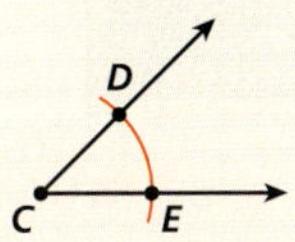

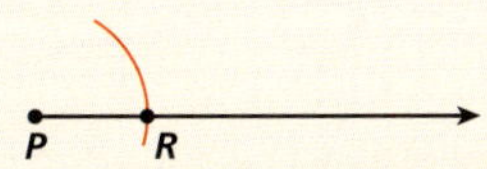

Use the compass to measure the arc in ∠*DCE*. Then place the compass point on *R* and draw an arc that intersects the first one. Label the point of intersection *Q*. Draw $\overrightarrow{PQ}$.

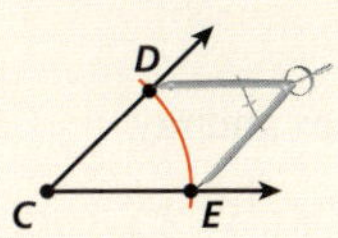

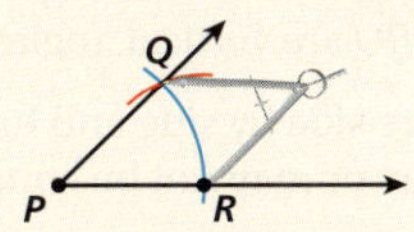

∠*DCE* ≅ ∠*QPR*

Think and Discuss

1. Explain how drawing and constructing are different.

Try This

Construct a congruent angle for each given angle. **Check students' work.**

1.

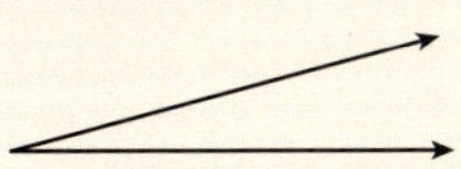

2.

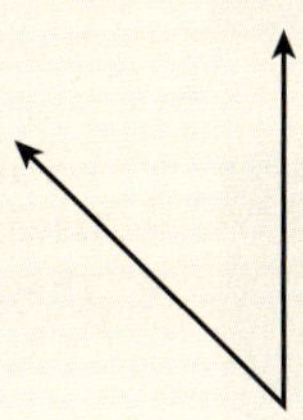

3.

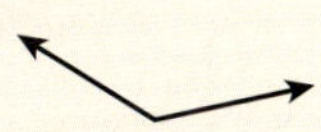

Answers

Activity 2

Think and Discuss

1. Possible answer: Any tools may be used in drawing. Constructions use a compass and straightedge.

7-3 Organizer

Pacing: Traditional 1 day
Block $\frac{1}{2}$ day

Objective: Students understand relationships of angles.

Indiana Academic Standards
6.3.5, 6.4.1, 6.4.2, 6.7.1, 6.7.2, 6.7.5

Warm Up

Identify the type of angle.

1. 70° acute
2. 90° right
3. 140° obtuse
4. 180° straight

Problem of the Day

A line forms an angle of 57° with the vertical axis. What angle does the line form with the horizontal axis?
33° or 147°

Available on Daily Transparency in CRB

Math Humor

Two wrongs don't make a right, but

- two rights make a straight angle,
- two writes make a second draft,
- two rites make a double ceremony, and
- two Wrights make an airplane.

7-3 Angle Relationships

Learn to understand relationships of angles.

Vocabulary
congruent
vertical angles
adjacent angles
complementary angles
supplementary angles

Angle relationships play an important role in many sports and games. Miniature-golf players must understand angles to know where to aim the ball. In the miniature-golf hole shown, m∠1 = m∠2, m∠3 = m∠4, and m∠5 = m∠6.

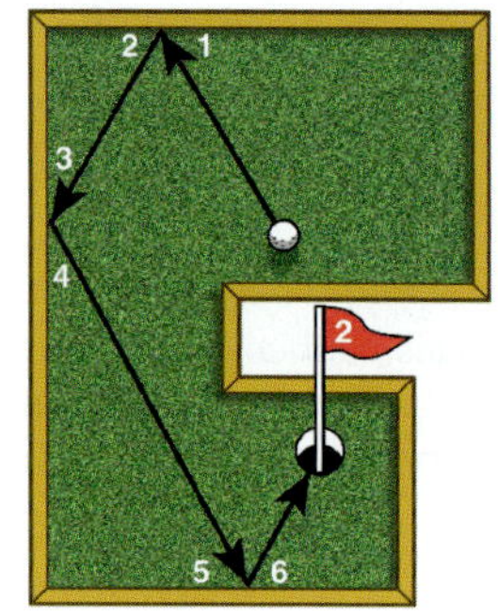

When angles have the same measure, they are said to be **congruent**.

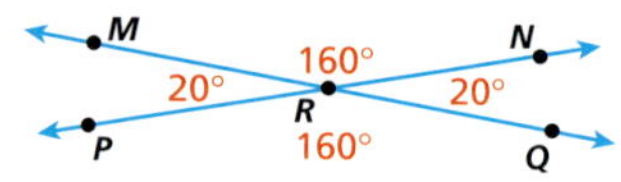

Reading Math
m∠1 is read "the measure of angle 1."

Vertical angles are formed opposite each other when two lines intersect. Vertical angles have the same measure, so they are always congruent.

∠MRP and *∠NRQ* are vertical angles.
∠MRN and *∠PRQ* are vertical angles.

Adjacent angles are side by side and have a common vertex and ray. Adjacent angles may or may not be congruent.

∠MRN and *∠NRQ* are adjacent angles. They share vertex *R* and $\overrightarrow{RN}$.
∠NRQ and *∠QRP* are adjacent angles. They share vertex *R* and $\overrightarrow{RQ}$.

EXAMPLE 1 Identifying Types of Angle Pairs

Identify the type of each angle pair shown.

A

∠1 and ∠2 are opposite each other and are formed by two intersecting lines.
They are vertical angles.

B

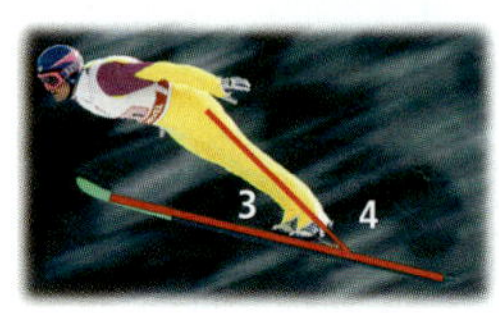

∠3 and ∠4 are side by side and have a common vertex and ray.
They are adjacent angles.

1 Introduce

Alternate Opener

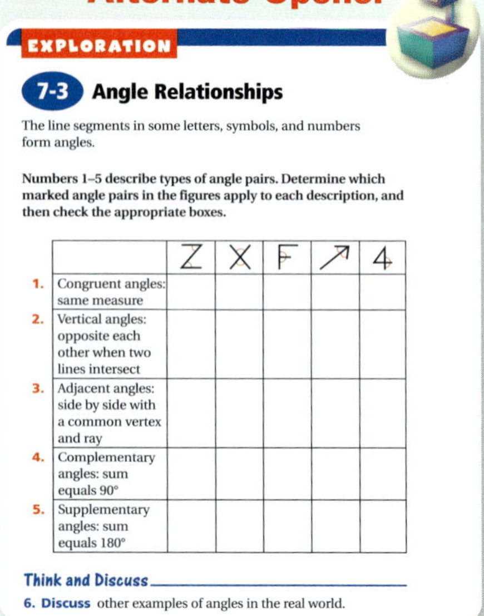

EXPLORATION

7-3 Angle Relationships

The line segments in some letters, symbols, and numbers form angles.

Numbers 1–5 describe types of angle pairs. Determine which marked angle pairs in the figures apply to each description, and then check the appropriate boxes.

		Z	X	F	↗	4
1.	Congruent angles: same measure					
2.	Vertical angles: opposite each other when two lines intersect					
3.	Adjacent angles: side by side with a common vertex and ray					
4.	Complementary angles: sum equals 90°					
5.	Supplementary angles: sum equals 180°					

Think and Discuss

6. Discuss other examples of angles in the real world.

Motivate

To introduce angle relationships, draw two intersecting lines as below.

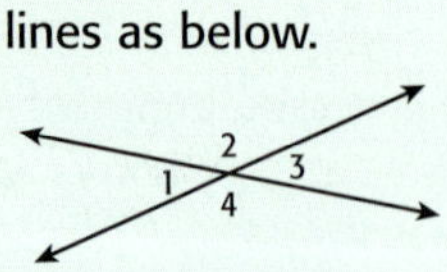

Ask students to tell you what they notice about the angles. For example, ∠1 and ∠3 are opposite each other, and so are ∠4 and ∠2; ∠1 and ∠2 are side by side and have a common vertex and ray, and so do ∠2 and ∠3, ∠3 and ∠4, and ∠4 and ∠1.

Exploration worksheet and answers on Chapter 7 Resource Book pp. 26 and 145

2 Teach

Lesson Presentation

Guided Instruction

In this lesson, students learn to describe relationships of angles. Explain the term *congruent*, and show students examples of *vertical* and *adjacent* angles. Then show students examples of *complementary* and *supplementary* angles, and teach how to find the measure of an unknown angle, given the measure of its complementary or supplementary angle.

Teaching Tip Remind students that complementary angles and supplementary angles do not have to be adjacent.

Complementary angles are two angles whose measures have a sum of 90°.

65° + 25° = 90°
∠*LMN* and ∠*NMP* are complementary.

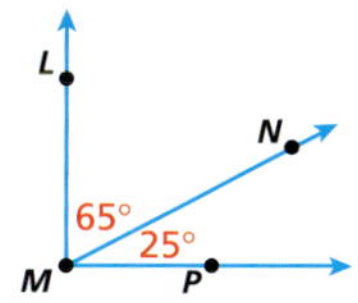

Supplementary angles are two angles whose measures have a sum of 180°.

65° + 115° = 180°
∠*GHK* and ∠*KHJ* are supplementary.

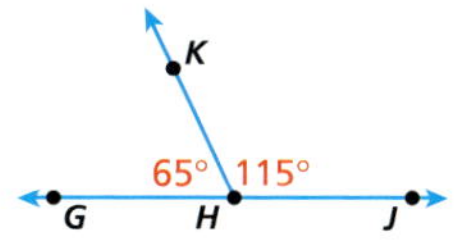

EXAMPLE 2 Identifying an Unknown Angle Measure

Find each unknown angle measure.

A The angles are complementary.

$$\begin{aligned} 55° + a &= 90° \\ -55° \quad & \quad -55° \\ a &= 35° \end{aligned}$$

The sum of the measures is 90°.

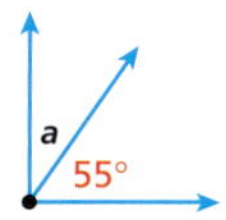

B The angles are supplementary.

$$\begin{aligned} 75° + b &= 180° \\ -75° \quad & \quad -75° \\ b &= 105° \end{aligned}$$

The sum of the measures is 180°.

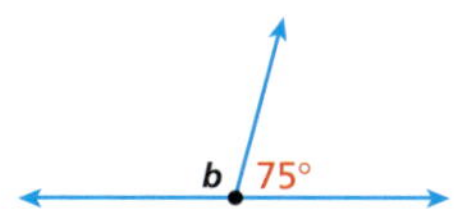

C The angles are vertical angles.

$c = 51°$ *Vertical angles are congruent.*

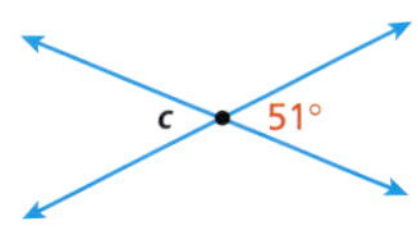

D Angles *JGF* and *KGH* are congruent.

$$\begin{aligned} d + e + 136° &= 180° \\ -136° \quad & \quad -136° \\ d + e &= 44° \end{aligned}$$

$d = 22°$ and $e = 22°$

The sum of the measures is 180°. Each angle measures half of 44°.

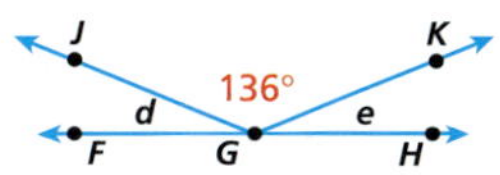

Think and Discuss

1. **Give** the measure of ∠2 if ∠1 and ∠2 are vertical angles and m∠1 = 40°.
2. **Give** the measure of ∠3 if ∠3 and ∠4 are supplementary and m∠4 = 150°.
3. **Tell** whether the angles in Example 1B are supplementary or complementary.

Additional Examples

Example 1

Identify the type of each angle pair shown.

A.

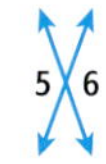

vertical

B.

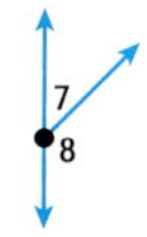

supplementary

Example 2

Find each unknown angle measure.

A. The angles are complementary.

m∠1 = 19°

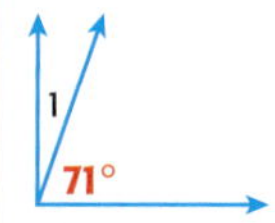

B. The angles are supplementary.

m∠2 = 55°

C. The angles are vertical angles.

m∠3 = 82°

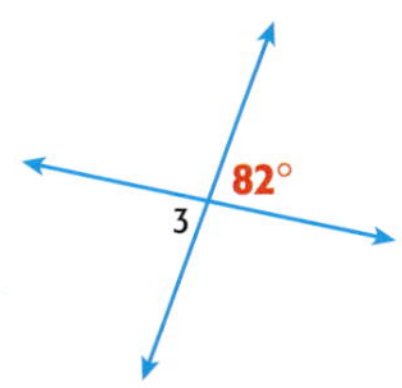

D. ∠*JKL* and ∠*MKN* are congruent.

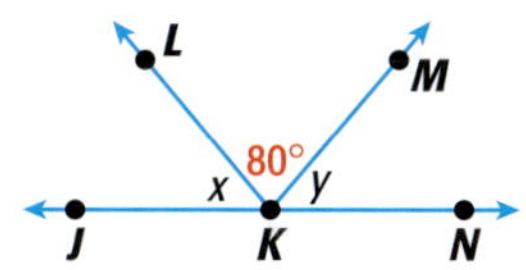

m∠*JKL* = 50° and m∠*MKN* = 50°

3 Close

Reaching All Learners

Through Hands-On Experience

Give each pair of students a protractor and two strips of cardboard fastened at the center so they intersect (provided in the Manipulatives Kit). Show students that the fastener is the vertex of the angles formed by the lines on the strips of cardboard. Have students move the strips to make vertical angles of varying sizes. Have pairs measure one angle and use what they know to calculate the measures of the remaining angles. Repeat with other sets of angles.

Summarize

Review the new vocabulary in the lesson: *congruent*, *vertical angles*, *adjacent angles*, *complementary angles*, and *supplementary angles*. Have students volunteer to draw examples of each.

Check students' drawings.

Answers to Think and Discuss

1. m∠2 = 40°
2. m∠3 = 30°
3. supplementary

7-3 PRACTICE & ASSESS

Students may want to refer back to the lesson examples.

Assignment Guide

If you finished Example 1 assign:
Core 1–2, 5–6, 9–10, 31–32
Enriched 1–2, 5–6, 9–10, 31–32

If you finished Example 2 assign:
Core 1–24, 31–32
Enriched 7–32

Notes

7-3 Exercises

FOR EXTRA PRACTICE see page 649

internet connect
Homework Help Online
go.hrw.com Keyword: MR4 7-3

GUIDED PRACTICE

See Example 1 Identify the type of each angle pair shown.

1. 

adjacent

2. 

adjacent

See Example 2 Find each unknown angle measure.

3. The angles are complementary.
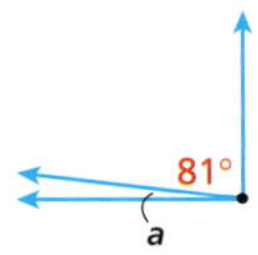

$m\angle a = 9°$

4. The angles are supplementary.
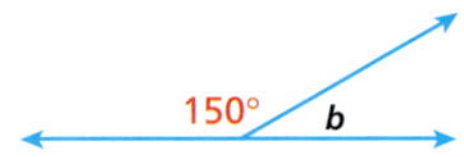

$m\angle b = 30°$

INDEPENDENT PRACTICE

See Example 1 Identify the type of each angle pair shown.

5. 

adjacent

6. 

vertical

See Example 2 Find each unknown angle measure.

7. The angles are vertical angles.
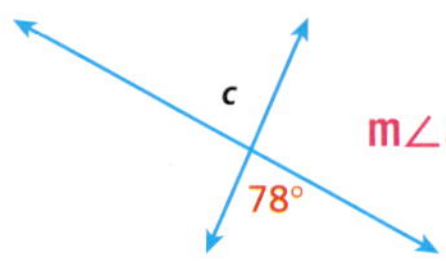

$m\angle c = 78°$

8. The angles are supplementary.
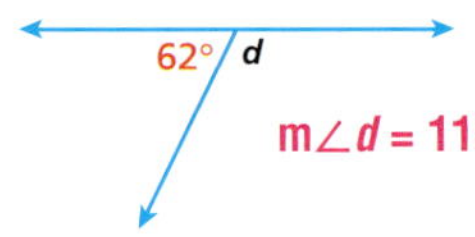

$m\angle d = 118°$

PRACTICE AND PROBLEM SOLVING

Use the figure for Exercises 9–12.

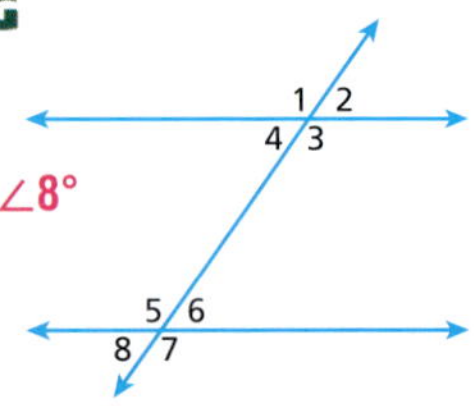

9. Which angles are not adjacent to ∠3? angles 1, 5, 6, 7, and 8
10. Name all the pairs of vertical angles that include ∠8. ∠6 and ∠8°
11. If the m∠6 is 72°, what are the measures of ∠5, ∠7, and ∠8? 108°, 108°, 72°
12. What is the sum of the measures of ∠1, ∠2, ∠3, and ∠4? 360°

Math Background

If ∠1 and ∠2 are supplementary, then m∠1 + m∠2 = 180°. If ∠2 and ∠3 are supplementary, then m∠2 + m∠3 = 180°. Therefore, m∠1 + m∠2 = m∠2 + m∠3. Subtracting m∠2 from each side, m∠1 = m∠3. So, if two angles are supplementary to the same angle, their measures are equal. In a similar fashion, it can be shown that complements of equal angles are equal.

RETEACH 7-3

LESSON 7-3 **Reteach**
Angle Relationships

When two lines intersect, two pairs of opposite angles, called vertical angles, are formed. Vertical angles are congruent.

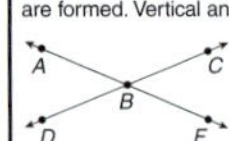

∠ABC and ∠DBF are vertical angles.
∠ABD and ∠CBF are vertical angles.

Adjacent angles have a common side and vertex.

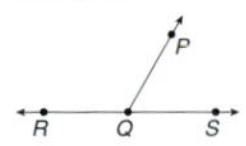

∠PQR and ∠PQS are adjacent angles.

Identify the type of each angle pair shown.

1. 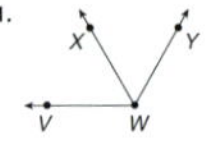

adjacent angles

2. 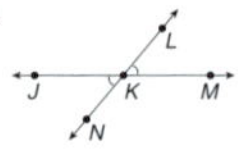

vertical angles

If the sum of the measures of two angles is 90°, then the angles are complementary angles.

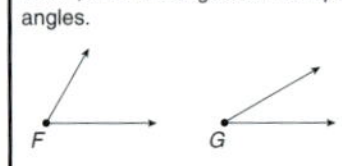

∠F and ∠G are complementary angles.

If the sum of the measures of two angles is 180°, then the angles are supplementary angles.

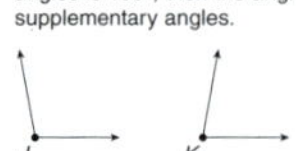

∠J and ∠K are supplementary angles.

Identify the type of each angle pair shown.

3. 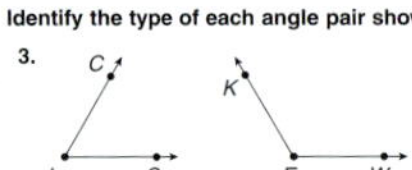

supplementary angles

4. 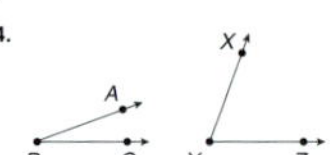

complementary angles

PRACTICE 7-3

LESSON 7-3 **Practice B**
Angle Relationships

Identify the type of each angle pair shown.

1. 

vertical angles

2. 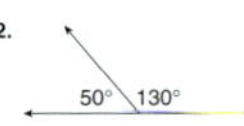

supplementary angles

3. 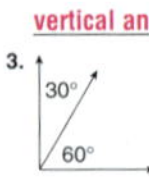

complementary angles

4. 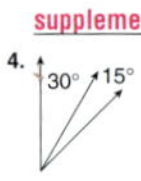

adjacent angles

Find each unknown angle measure.

5. The angles are supplementary.
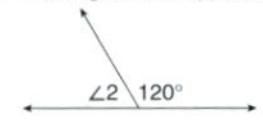

$m\angle 2 = 60°$

6. The angles are complementary.
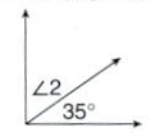

$m\angle 2 = 55°$

7. Anita says the plus sign + forms 2 pairs of vertical angles. Charles says it forms 2 pairs of congruent angles. Who is correct? Explain.
They are both correct. The intersecting lines of the plus sign form pairs of vertical angles. Vertical angles always have the same measure, so they are always congruent.

8. Is the following statement always true, sometimes true, or never true? Explain your reasoning. Two congruent angles that are complementary both measure 45°.
It is always true. Congruent angles have the same measure, and the sum of two complementary angles is 90°; 90° ÷ 2 = 45°.

Use the figure for Exercises 13–16.

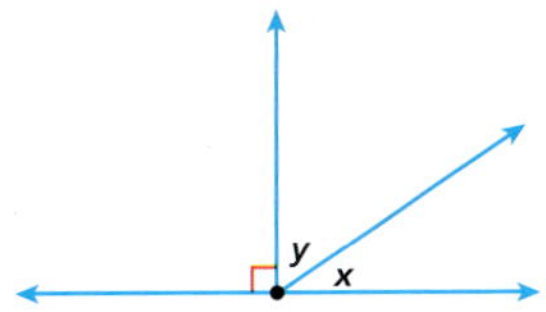

13. If $y = 51°$, what does x equal? **39°**

14. If $x = 64°$, what does y equal? **26°**

15. If $x = 40.09°$, what does y equal? **49.91°**

16. If $y = 27\frac{1}{3}°$, what does x equal? **$62\frac{2}{3}°$**

Find the measure of the angle that is complementary to each given angle. Use a protractor to draw both angles.

17. 47° **43°** **18.** 62° **28°** **19.** 55° **35°** **20.** 31° **59°**

Find the measure of the angle that is supplementary to each given angle. Use a protractor to draw both angles.

21. 75° **105°** **22.** 102° **78°** **23.** 136° **44°** **24.** 81° **99°**

25. Angles A and B are complementary. If the measure of angle A equals the measure of angle B, what is the measure of each angle? **45°**

26. Angles C and D are each complementary to angle F. How are angle C and angle D related? **angles C and D are congruent angles**

27. The measure of angle 1 is 43°. Angle 2 is complementary to angle 1, and angle 3 is supplementary to angle 1.

a. Give m∠2 and m∠3. **47°, 137°**

b. Use a protractor to draw ∠1, ∠2, and ∠3.

28. ***WRITE A PROBLEM*** Draw a pair of adjacent supplementary angles. Write a problem in which the measure of one of the angles must be found.

29. ***WRITE ABOUT IT*** Two angles are supplementary to the same angle. Explain the relationship between the measures of these angles.

29. The angles will have the same measure, because they are supplementary to the same angle.

30. ***CHALLENGE*** The measure of angle A is 38°. Angle B is complementary to angle A. Angle C is supplementary to angle B. What is the measure of angle C? **128°**

Spiral Review

31. Tami worked 4 hours on Saturday at the city pool. She spent $1\frac{3}{4}$ hours cleaning the pool and the remaining time working as a lifeguard. How many hours did Tami spend working as a lifeguard? (Lesson 5-9) **$2\frac{1}{4}$ hours**

32. **TEST PREP** Which type of data would not be appropriate for a line graph? (Lesson 6-7) **D**

A Average rainfall over a ten-year period

B A tennis player's earnings during his career

C The value of a share of stock during the last 14 months

D The results of a survey on favorite television programs

Answers

17.

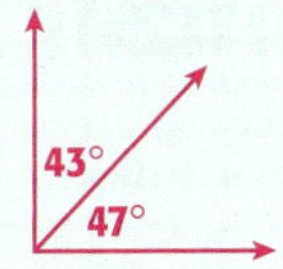

18.

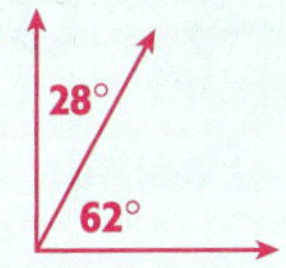

19.

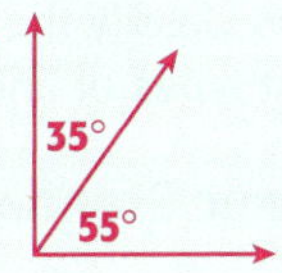

20.

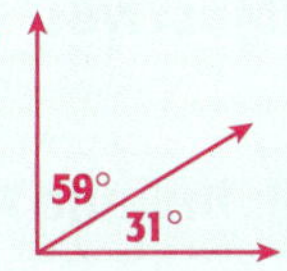

21.

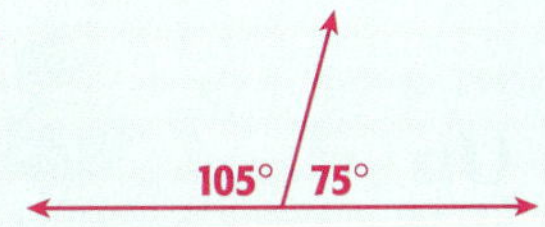

22.

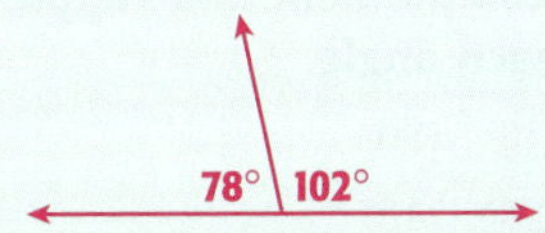

23.

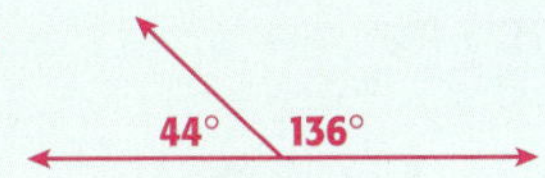

24, 27–28. See p. A4.

Journal

Have students explain the methods used to compute the measures of complementary and supplementary angles.

Test Prep Doctor

For Exercise 32, encourage students to first think about the kinds of data that are best displayed in line graphs and in bar graphs. As students look over the answer choices, they should notice that **A, B,** and **C** all describe data that change over time and *are* appropriate for a line graph.

CHALLENGE 7-3

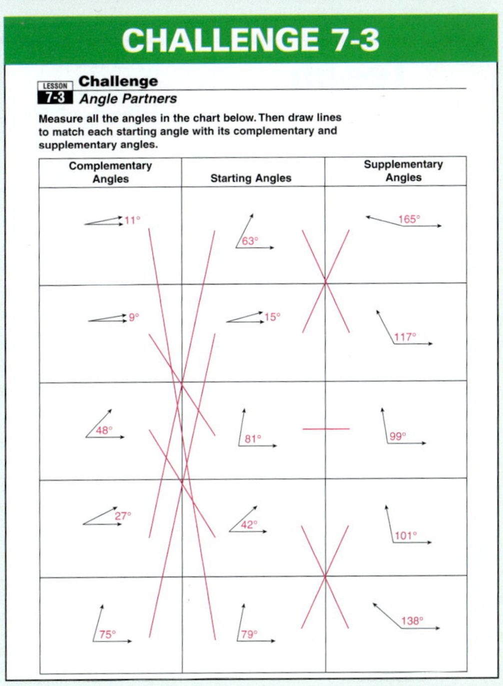

LESSON 7-3 **Challenge**

Angle Partners

Measure all the angles in the chart below. Then draw lines to match each starting angle with its complementary and supplementary angles.

Complementary Angles	Starting Angles	Supplementary Angles
11°	63°	165°
9°	15°	117°
48°	81°	99°
27°	42°	101°
75°	79°	138°

PROBLEM SOLVING 7-3

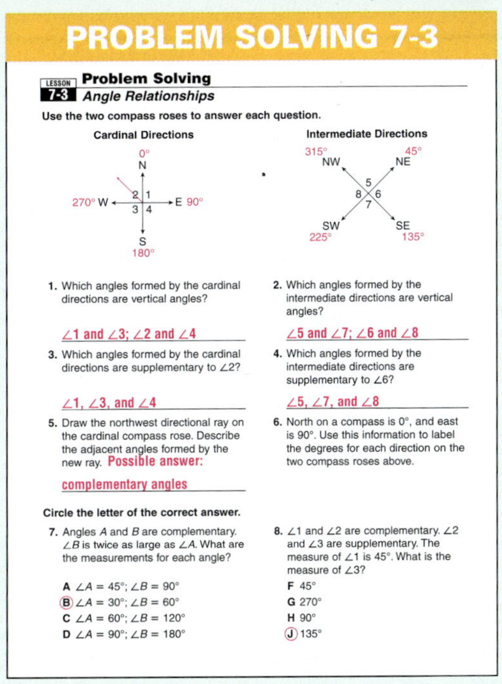

LESSON 7-3 **Problem Solving**

Angle Relationships

Use the two compass roses to answer each question.

1. Which angles formed by the cardinal directions are vertical angles? **∠1 and ∠3; ∠2 and ∠4**

2. Which angles formed by the intermediate directions are vertical angles? **∠5 and ∠7; ∠6 and ∠8**

3. Which angles formed by the cardinal directions are supplementary to ∠2? **∠1, ∠3, and ∠4**

4. Which angles formed by the intermediate directions are supplementary to ∠6? **∠5, ∠7, and ∠8**

5. Draw the northwest directional ray on the cardinal compass rose. Describe the adjacent angles formed by the new ray. **Possible answer: complementary angles**

6. North on a compass is 0°, and east is 90°. Use this information to label the degrees for each direction on the two compass roses above.

Circle the letter of the correct answer.

7. Angles A and B are complementary. ∠B is twice as large as ∠A. What are the measurements for each angle?

A ∠A = 45°; ∠B = 90°
(B) ∠A = 30°; ∠B = 60°
C ∠A = 60°; ∠B = 120°
D ∠A = 90°; ∠B = 180°

8. ∠1 and ∠2 are complementary. ∠2 and ∠3 are supplementary. The measure of ∠1 is 45°. What is the measure of ∠3?

F 45°
G 270°
H 90°
(J) 135°

Lesson Quiz

Give the complement of each angle.

1. 70° **20°**

2. 42° **48°**

Give the supplement of each angle.

3. 120° **60°**

4. 17° **163°**

5. Identify the type of angle pair shown. **adjacent**

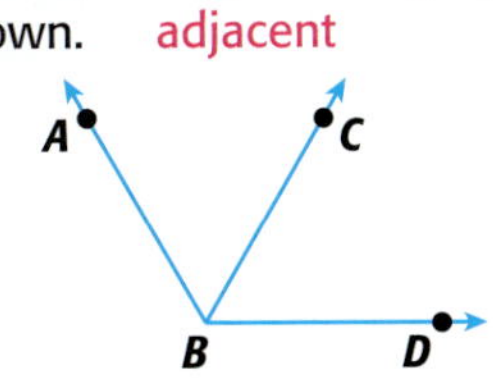

Available on Daily Transparency in CRB

7-4 Organizer

Pacing: Traditional 1 day
Block $\frac{1}{2}$ day

Objective: Students classify the different types of lines.

Indiana Academic Standards
6.7.1, 6.7.2, 6.7.5, 6.7.11

Warm Up

Give the complement and supplement of each angle.

1. 80° 10°, 100°
2. 64° 26°, 116°
3. 15° 75°, 165°

Problem of the Day

Draw three points that are not in a straight line. Label them *A, B,* and *C.* How many different lines can you draw that contains two of the points? Name the lines.
3; $\overleftrightarrow{AB}$, $\overleftrightarrow{AC}$, $\overleftrightarrow{BC}$

Available on Daily Transparency in CRB

Math Fact

Another word that is sometimes used to describe skew lines is *noncoplanar*, which means that they cannot lie in the same plane.

7-4 Classifying Lines

Learn to classify the different types of lines.

Vocabulary
parallel lines
perpendicular lines
skew lines

The photograph of the houses and the table below show some of the ways that lines can relate to each other. The yellow lines are intersecting. The purple lines are parallel. The green lines are perpendicular. The white lines are skew.

Writing Math
The square inside a right angle shows that the rays of the angle are perpendicular.

Intersecting lines are lines that cross at one common point.	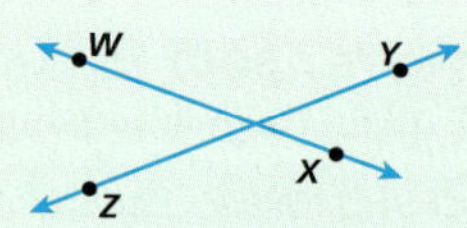	Line *YZ* intersects line *WX*. $\overleftrightarrow{YZ}$ intersects $\overleftrightarrow{WX}$.
Parallel lines are lines in the same plane that never intersect.	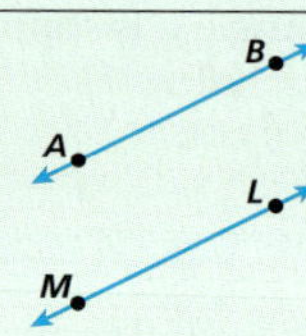	Line *AB* is parallel to line *ML*. $\overleftrightarrow{AB} \parallel \overleftrightarrow{ML}$
Perpendicular lines intersect to form 90° angles, or right angles.	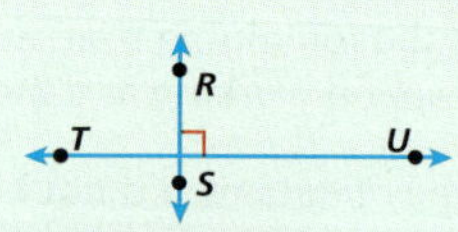	Line *RS* is perpendicular to line *TU*. $\overleftrightarrow{RS} \perp \overleftrightarrow{TU}$
Skew lines are lines that lie in different planes. They are neither parallel nor intersecting.	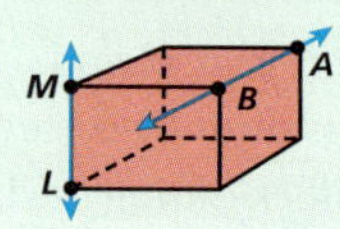	Line *AB* and line *ML* are skew. $\overleftrightarrow{AB}$ and $\overleftrightarrow{ML}$ are skew.

1 Introduce

Alternate Opener

EXPLORATION

7-4 Classifying Lines

In the diagram below, two *parallel lines* are intersected by a third line.

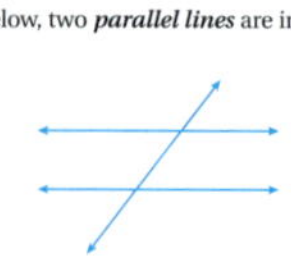

1. How many angles are formed when the third line intersects the two parallel lines? Label these angles.
2. Congruent angles have the same measure. Which angles in the diagram are congruent?

Think and Discuss

3. **Discuss** any relationships you see among the locations of congruent angles.
4. **Explain** which pairs of angles in the diagram have measures that add to 180°.

Motivate

Fold a sheet of paper in half lengthwise, and then fold it in half the other direction. Unfold the paper and ask students what they notice about the lines formed. They intersect. Ask the students what kinds of angles are formed where the lines intersect. right angles Explain that lines that intersect to form right angles are called *perpendicular* lines.

Exploration worksheet and answers on Chapter 7 Resource Book pp. 36 and 147

2 Teach

Lesson Presentation

Guided Instruction

In this lesson students learn to classify different types of lines. Show students the photo in the lesson opener, which provides a real context for illustrating intersecting, parallel, perpendicular, and skew lines. Have students work through the examples to practice classifying pairs of lines. Have students think of other examples for each type of line relationship.

Teaching Tip Understanding the concept of skew lines calls for thinking in three dimensions. You may wish to use yardsticks to show students a three-dimensional model of skew lines.

EXAMPLE 1 Classifying Pairs of Lines

Classify each pair of lines.

A

The lines are in the same plane. They do not appear to intersect.
They are parallel.

B

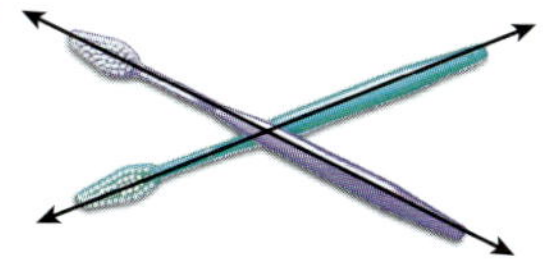

The lines cross at one common point.
They are intersecting.

C

The lines intersect to form right angles.
They are perpendicular.

D

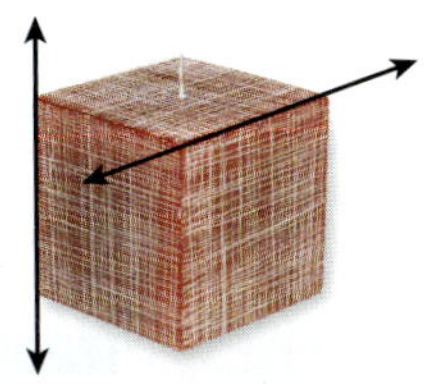

The lines are in different planes and are not parallel or intersecting.
They are skew.

EXAMPLE 2 *Science Application*

The particles in a transverse wave move up and down as the wave travels to the right. What type of line relationship does this represent?

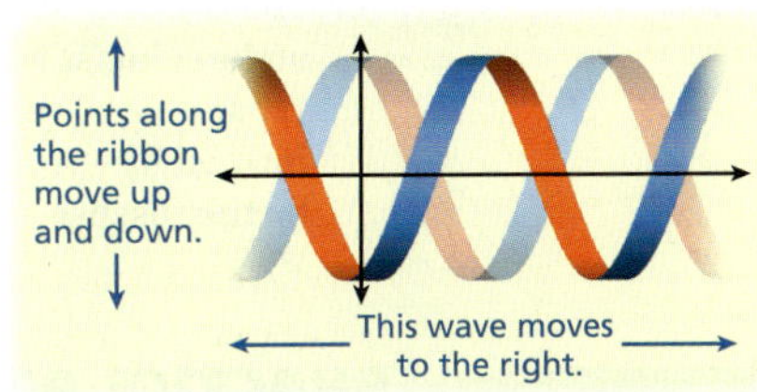

The direction that the particles move forms a right angle with the direction that the wave is traveling. The lines are perpendicular.

Think and Discuss

1. **Give an example** of intersecting, parallel, perpendicular, and skew lines or line segments in your classroom.
2. **Determine** whether two lines must be parallel if they do not intersect. Explain.

COMMON ERROR ALERT

Students may confuse intersecting and perpendicular lines. Remind them that perpendicular lines are a special kind of intersection.

Additional Examples

Example 1

Classify each pair of lines.

A.

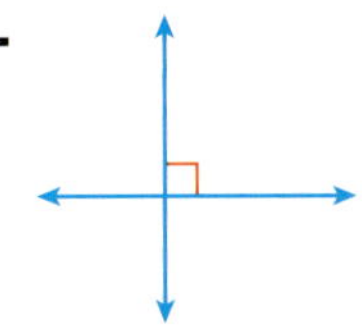

perpendicular

B.

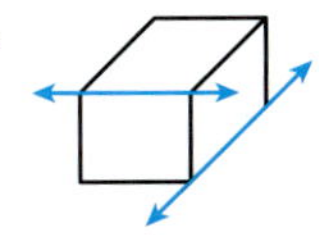

skew

C.

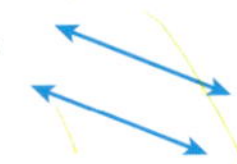

parallel

D.

intersecting

Example 2

The handrails on an escalator are in the same plane. What type of line relationship do they represent?

The lines are parallel.

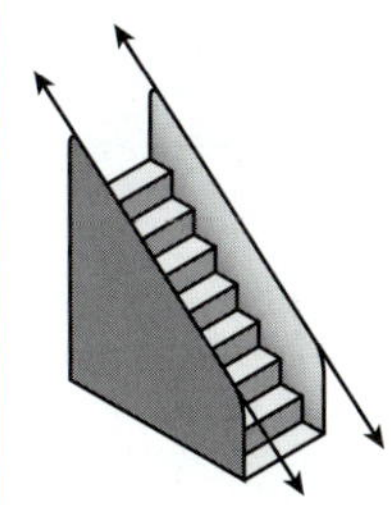

3 Close

Reaching All Learners

Through Cognitive Strategies

Give students old road maps on which they can identify several examples of parallel, intersecting, and perpendicular lines. You may use the Recording Sheet on Chapter 7 Resource Book p. 130. Have students write questions about the line relationships they find on the map and exchange maps and questions with other students.

Possible answers: What street is perpendicular to Main Street? What street intersects Stagecoach Boulevard? What street runs parallel to Alcove Drive?

Summarize

Ask students these questions to review the new vocabulary in the lesson:

- What do you call lines that intersect to form 90° angles? perpendicular
- What do you call lines in the same plane that never intersect? parallel
- What do you call lines that do not lie in the same plane and never intersect? skew

Answers to Think and Discuss

1. Possible answers: intersecting, lines between the ceiling tiles; parallel, the top and bottom of the door; perpendicular, the top corner of the window; skew, the line where the wall and the ceiling meet in front of the room and the line where the wall and the floor meet on the right side of the room
2. Possible answer: No; two lines that don't intersect are skew if they do not and cannot lie in the same plane.

7-4 PRACTICE & ASSESS

Students may want to refer back to the lesson examples.

Assignment Guide

If you finished Example 1 assign:
Core 1–2, 4–5, 7–11, 13, 24–32
Enriched 11–19, 21, 24–32

If you finished Example 2 assign:
Core 1–19, 24–32
Enriched 5–32

Notes

7-4 Exercises

FOR EXTRA PRACTICE see page 649

internet connect
Homework Help Online
go.hrw.com Keyword: MR4 7-4

GUIDED PRACTICE

See Example 1 Classify each pair of lines.

1. 
intersecting

2. 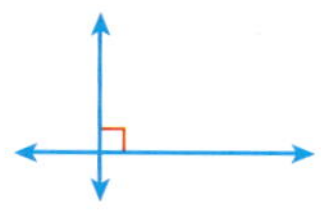
perpendicular

See Example 2

3. Jamal dropped a fishing line from a pier, as shown in the drawing. What type of relationship is formed by the lines? perpendicular

INDEPENDENT PRACTICE

See Example 1 Classify each pair of lines.

4. 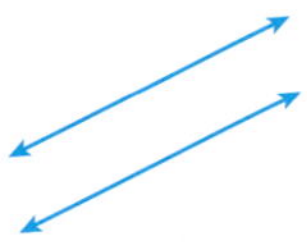
parallel

5. 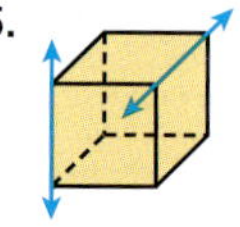
skew

See Example 2

6. The drawing shows where an archaeologist found two fossils. What type of relationship is formed by the lines suggested by the fossils? skew

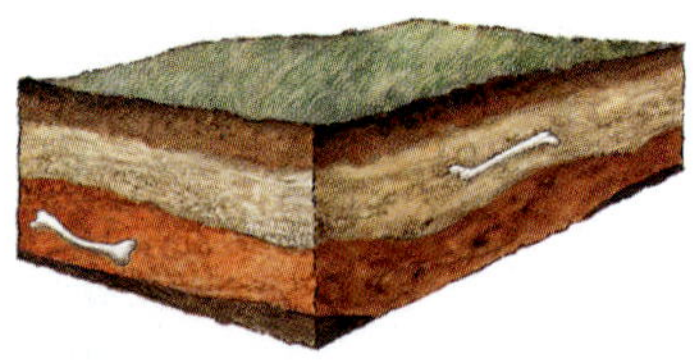

PRACTICE AND PROBLEM SOLVING

Describe each pair of lines as parallel, skew, intersecting, or perpendicular.

7. 
intersecting

8. 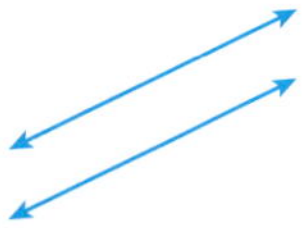
parallel

9. 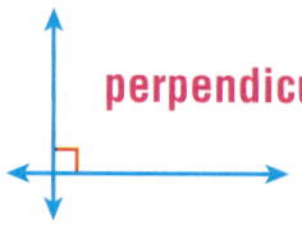
perpendicular

10. 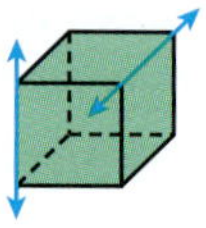
skew

Math Background

Based on our experiences in the real world, it seems reasonable to assume that parallel lines exist. This assumption is made explicit in Euclid's fifth postulate, now known as the Parallel Postulate. This postulate, set forth as a statement to be accepted without proof, offers the idea that through any point not on a given line, there exists exactly one line that can be drawn parallel to the given line.

In the nineteenth century, mathematicians came to realize that new geometries could be created that use alternative versions of the Parallel Postulate. These are known as non-Euclidean geometries.

RETEACH 7-4

LESSON 7-4 **Reteach**
Classifying Lines

Some lines have relationships.

Intersecting lines cross each other at one point.

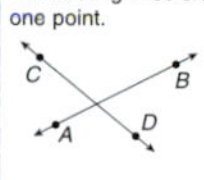

$\overleftrightarrow{AB}$ intersects $\overleftrightarrow{CD}$.

Perpendicular lines intersect to form right angles.

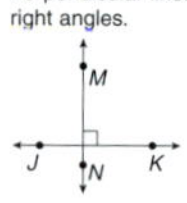

$\overleftrightarrow{JK} \perp \overleftrightarrow{MN}$

Parallel lines lie in the same plane but never intersect.

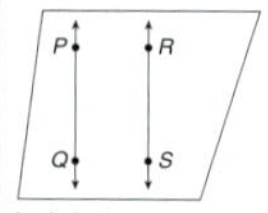

$\overleftrightarrow{PQ} \parallel \overleftrightarrow{RS}$

Skew lines lie in different planes and do not intersect.

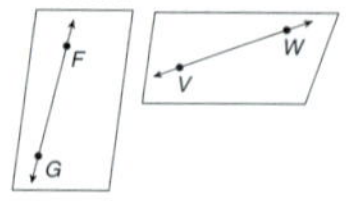

$\overleftrightarrow{FG}$ and $\overleftrightarrow{VW}$ are skew.

Classify each pair of lines.

1. 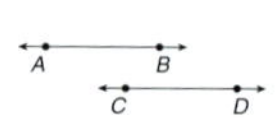
parallel

2. 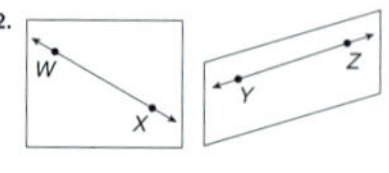
skew

3. 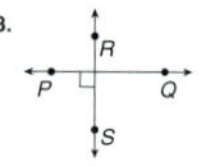
perpendicular

4. 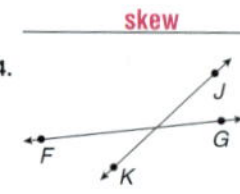
intersecting

PRACTICE 7-4

LESSON 7-4 **Practice B**
Classifying Lines

Classify each pair of lines.

1. 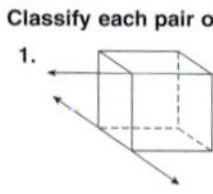
skew lines

2. 
intersecting lines

3. 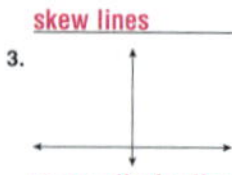
perpendicular lines

4. 
parallel lines

Match each description with its correct classification.

5. $\overleftrightarrow{AB}$ and $\overleftrightarrow{EF}$ lie on the same plane and never intersect. B
6. $\overleftrightarrow{AB}$ and $\overleftrightarrow{EF}$ cross each other at one common point. A
7. $\overleftrightarrow{AB}$ and $\overleftrightarrow{EF}$ lie on different planes and are neither parallel nor intersecting. C
8. $\overleftrightarrow{AB}$ and $\overleftrightarrow{EF}$ intersect to form right angles. D

A. $\overleftrightarrow{AB}$ intersects $\overleftrightarrow{EF}$.
B. $\overleftrightarrow{AB} \parallel \overleftrightarrow{EF}$
C. $\overleftrightarrow{AB}$ and $\overleftrightarrow{EF}$ are skew.
D. $\overleftrightarrow{AB} \perp \overleftrightarrow{EF}$

9. Oak Street runs parallel to Elm Street in a flat section of town. Tom tells you to meet him at the intersection of Oak and Elm. Explain why these instructions are impossible to follow.

Because Oak and Elm are parallel streets on the same plane, they will never intersect.

10. Look around your classroom. Name a pair of parallel lines and a pair of perpendicular lines that you see.

Answers will vary. Possible parallel lines: the 2 sides of my desk; Possible perpendicular lines: the top and side of the chalkboard

The lines in the figure intersect to form a rectangular box.

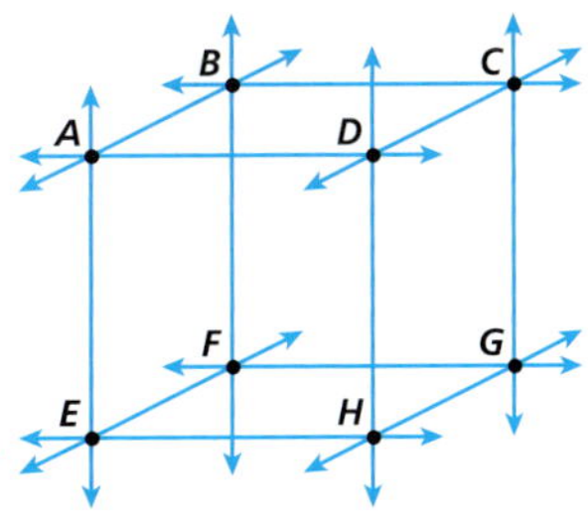

11. Name all the lines that are parallel to $\overleftrightarrow{AD}$. $\overleftrightarrow{BC}$, $\overleftrightarrow{FG}$, and $\overleftrightarrow{EH}$

12. Name all the lines that are perpendicular to $\overleftrightarrow{FG}$. $\overleftrightarrow{BF}$, $\overleftrightarrow{GH}$, $\overleftrightarrow{EF}$, $\overleftrightarrow{CG}$

13. Name a pair of lines that are skew. Possible answer: $\overleftrightarrow{AD}$ and $\overleftrightarrow{GH}$

14. Name all the lines that are not parallel to and do not intersect $\overleftrightarrow{DH}$. $\overleftrightarrow{AB}$, $\overleftrightarrow{EF}$, $\overleftrightarrow{FG}$, $\overleftrightarrow{BC}$

Tell whether each statement is *always*, *sometimes*, or *never* true.

15. Intersecting lines are parallel. never

16. Intersecting lines are perpendicular. sometimes

17. Perpendicular lines are intersecting. always

18. Parallel lines are in the same plane. always

19. Parallel lines are skew. never

20. Capitol Street intersects 1st, 2nd, and 3rd Avenues, which are parallel to each other. West Street and East Street are perpendicular to 2nd Avenue.
    a. Draw a map showing the six streets.
    b. Suppose East and West Streets were perpendicular to Capitol Street rather than 2nd Avenue. Draw a map showing the streets.

21. ***WHAT'S THE ERROR?*** A student drew two lines and claimed that the lines were both parallel and intersecting. Explain the error.

22. ***WRITE ABOUT IT*** Explain the similarities and differences between perpendicular and intersecting lines.

23. ***CHALLENGE*** Lines x, y, and z are in a plane. If lines x and y are parallel and line z intersects line x, does line z intersect line y? Explain. Yes; line z will intersect both lines x and y if it intersects line x.

Spiral Review

Multiply. Write each answer in simplest form. (Lesson 4-9)

24. $5 \cdot \frac{1}{10}$ $\frac{1}{2}$
25. $21 \cdot \frac{1}{3}$ 7
26. $\frac{2}{7} \cdot 14$ 4
27. $\frac{5}{12} \cdot 2$ $\frac{5}{6}$

Subtract. Write each answer in simplest form. (Lesson 5-9)

28. $5\frac{2}{3} - 4\frac{5}{6}$ $\frac{5}{6}$
29. $12\frac{4}{7} - 3\frac{6}{7}$ $8\frac{5}{7}$
30. $9\frac{7}{12} - 2\frac{1}{3}$ $7\frac{1}{4}$
31. $11\frac{5}{8} - 5\frac{1}{4}$ $6\frac{3}{8}$

32. **TEST PREP** Which is the outlier of the data set 24, 76, 28, 24, 35, 31, 28, 24, 27? (Lesson 6-3) D

A 24 B 33 C 28 D 76

Answers

20. a, b. See pp. A4 and A5.

21. By definition, parallel lines do not intersect. It is impossible to draw lines that are both parallel and intersecting.

22. Both perpendicular and intersecting lines cross at exactly one point. Perpendicular lines must form right angles, but intersecting lines may not.

Journal

Have students draw and label intersecting, parallel, and perpendicular lines and write a brief explanation of each. Also have students explain how they drew skew lines on their paper.

Test Prep Doctor

For Exercise 32, students will have to think back to Chapter 6, where they studied measures of central tendency. Students who answered **A** confused *outlier* with *mode*. Students who answered **B** confused *outlier* with *mean*, and students who answered **C** confused *outlier* with *median*.

CHALLENGE 7-4

LESSON 7-4 **Challenge**
Capital Streets

Washington, D.C., became the capital of the United States in 1800. The city's planners designed wide streets reaching out from two main centers—the White House and the Capitol Building. The President works and lives in the White House, and Congress works in the Capitol.

The map below shows the section of Washington, D.C, where the White House and Capitol are located. Use the following clues to label the streets on the map. Then name all the pairs of perpendicular streets on the map.

1. Madison Drive, Jefferson Drive, and Independence Avenue are parallel streets. Madison Drive is north of Jefferson Drive and Jefferson Drive is north of Independence Avenue.
2. 14th Street and 3rd Street are parallel streets. 14th Street is west of 3rd Street.
3. Pennsylvania Avenue intersects 14th Street near the White House.
4. The National Mall lies between Madison Drive to the north and Jefferson Drive to the south.

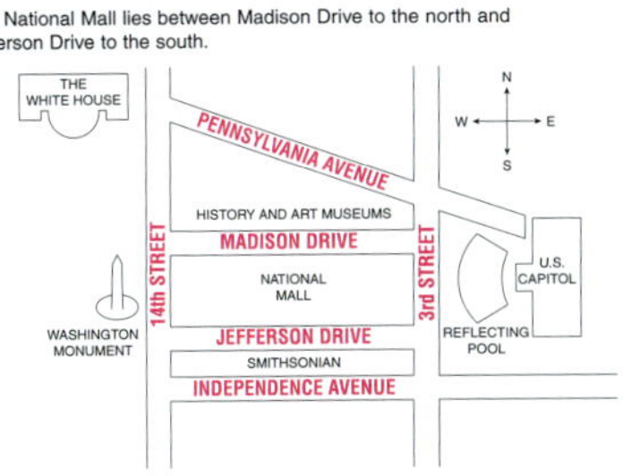

14th ⊥ Madison; 14th ⊥ Jefferson; 14th ⊥ Independence;
3rd ⊥ Madison; 3rd ⊥ Jefferson; 3rd ⊥ Independence

PROBLEM SOLVING 7-4

LESSON 7-4 **Problem Solving**
Classifying Lines

Use the map to answer each question.

1. The area where the borders of Utah, Colorado, Arizona, and New Mexico meet is sometimes called the Four Corners. What kind of lines are formed where the borders meet?
   perpendicular lines
2. Which borderlines on the map are skew lines?
   none
3. What kinds of lines are suggested by the eastern and western borders of New Mexico?
   parallel lines
4. Which three western borderlines intersect near the Grand Canyon?
   Nevada, Utah, and Arizona

Western U.S. States

5. Which two western states seem to have congruent borderlines?
   Colorado and Wyoming

Circle the letter of the correct answer.

6. Which of the following do not appear to be parallel to the western borderline of Nevada?
   (A) the western borderline of California
   B the western borderline of Wyoming
   C the eastern borderline of Montana
   D the eastern borderline of Arizona
7. Which of these western states do not have borderlines that intersect near Great Salt Lake?
   F Utah and Nevada
   G Utah and Idaho
   H Utah and Wyoming
   (J) Utah and Colorado

Lesson Quiz

1. Sketch a pair of perpendicular lines.

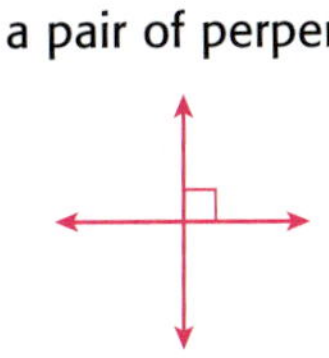

2. Sketch a pair of parallel lines.

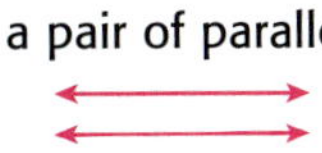

Use the figure to classify the lines.

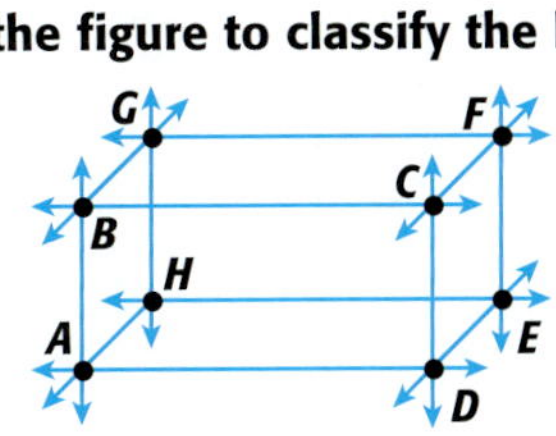

3. $\overleftrightarrow{AD}$ and $\overleftrightarrow{BC}$ parallel
4. $\overleftrightarrow{AD}$ and $\overleftrightarrow{CF}$ skew
5. $\overleftrightarrow{AB}$ and $\overleftrightarrow{BG}$ perpendicular

Available on Daily Transparency in CRB

Hands-On LAB 7B Parallel Line Relationships

Pacing: Traditional 1 day
Block $\frac{1}{2}$ day

Objective: To use a compass, a straightedge, and a protractor to explore parallel line relationships

Materials: Compass, straightedge, protractor

Lab Resources

Hands-On Lab Activities . . . . p. 51

Using the Pages

Remind students how to use the materials to draw geometric figures.

Draw each figure using the tool given.

1. 52° angle; protractor

Check students' work.

2. Line segment $\overline{CD}$; straightedge

Check students' work.

3. Point P and two lines that intersect at P; straightedge

Check students' work.

Hands-On LAB 7B Parallel Line Relationships

Use with Lesson 7-4

Parallel lines are in the same plane and never intersect. You can use a straightedge and protractor to draw parallel lines.

internet connect
Lab Resources Online
go.hrw.com
KEYWORD: MR4 Lab7B

Activity

1 Draw a line on your paper. Label two points A and B.

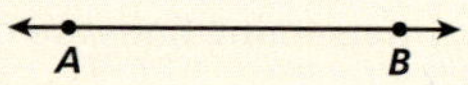

Use your protractor to measure and mark a 90° angle at each point.

Draw rays with endpoints A and B through the marks you made with the protractor.

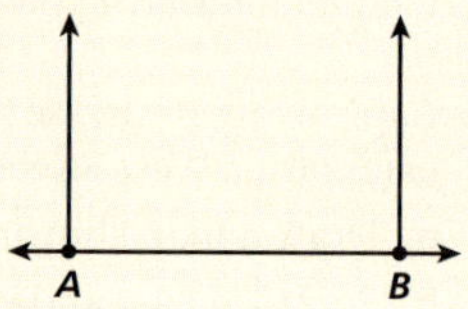

Place the point of your compass on point A, and draw an arc through the ray.

Use the same compass opening to draw an arc through the ray at point B.

Label the points of intersection X and Y.

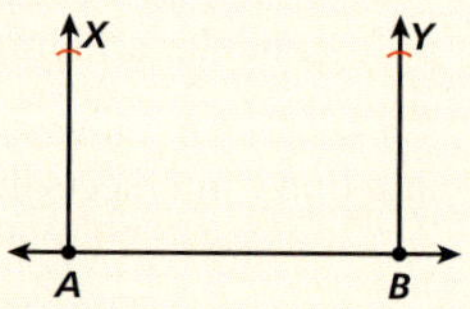

Now use your straightedge to draw a line through X and Y.

$\overline{AB} \parallel \overline{XY}$

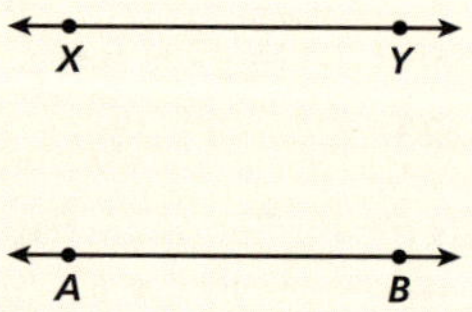

When a pair of parallel lines is intersected by a third line, the angles formed have special relationships.

2 Draw a pair of parallel lines and a third line that intersects them. Label the angles 1 through 8, as shown.

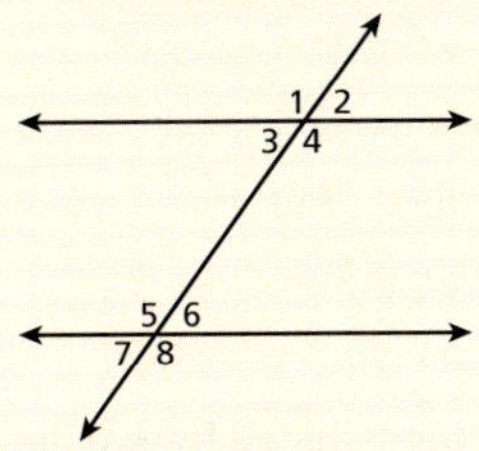

Angles inside the parallel lines are called *interior angles*. The interior angles here are angles 3, 4, 5, and 6.

Angles outside the parallel lines are called *exterior angles*. The exterior angles here are angles 1, 2, 7, and 8.

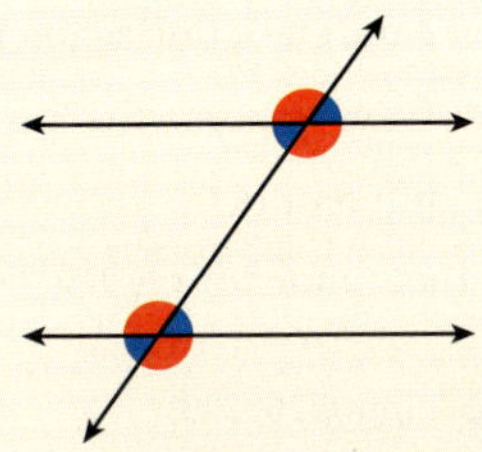

Measure each angle, and write its measurement inside the angle.

Shade angles with the same measure with the same color.

The interior angles with the same measure are called *alternate interior angles*. They are angles 3 and 6 and angles 4 and 5.

The exterior angles with the same measure are called *alternate exterior angles*. They are angles 1 and 8 and angles 2 and 7.

Angles in the same position when the third line intersects the parallel lines are called *corresponding angles*.

Think and Discuss

1. Name three pairs of corresponding angles. **1 and 5, 2 and 6, 3 and 7, or 4 and 8**
2. Tell the relationship between the measure of interior angles and the measure of exterior angles. **The sum of the measures of an interior angle and an adjacent exterior angle at the same intersection is 180°.**

Try This

Follow the steps to construct and label the diagram.

1. Draw a pair of parallel lines, and draw a third line intersecting them where one angle measures 75°.
2. Label each angle on the diagram using the measure you know.

Answers

Try This

1.

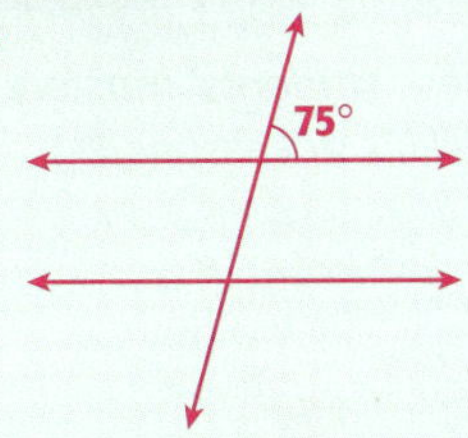

2.

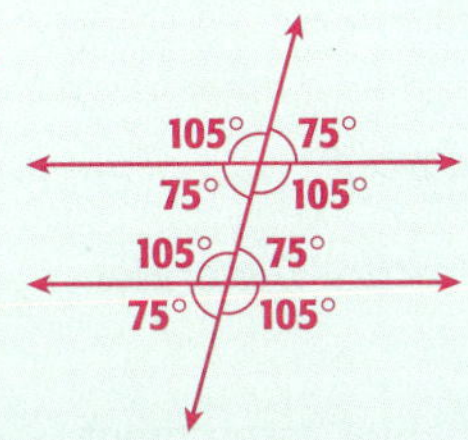

Chapter 7 Mid-Chapter Quiz

Purpose: *To assess students' mastery of concepts and skills in Lessons 7-1 through 7-4*

Assessment Resources

Section 7A Quiz
Assessment Resources . . . . p. 18

***Test and Practice Generator* CD-ROM**

Additional mid-chapter assessment items in both multiple-choice and free-response format may be generated for any objective in Lessons 7-1 through 7-4.

Chapter 7 Mid-Chapter Quiz

Mid-Chapter Quiz

LESSON 7-1 (pp. 322–325)

Use the diagram. Possible answers:

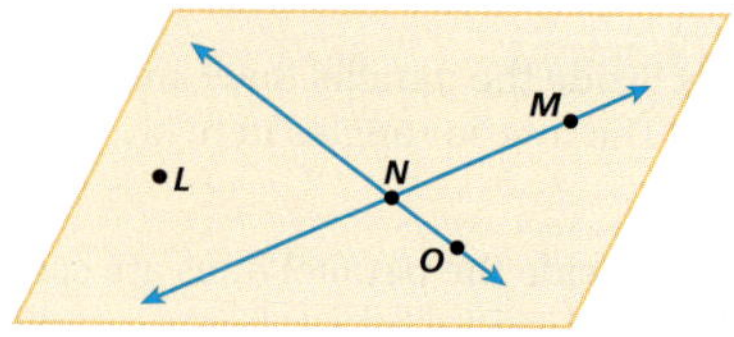

1. Name three points. L, M, N
2. Name two lines. $\overleftrightarrow{NO}$; $\overleftrightarrow{NM}$
3. Name a point shared by two lines. N
4. Name a plane. MOL

Use the diagram. Possible answers:

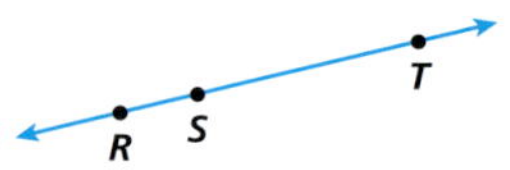

5. Name three different line segments. $\overline{RS}$; $\overline{ST}$; $\overline{RT}$
6. Give three ways to name the line. $\overleftrightarrow{TR}$; $\overleftrightarrow{ST}$; $\overleftrightarrow{RS}$
7. Name six different rays. $\overrightarrow{RS}$; $\overrightarrow{SR}$; $\overrightarrow{ST}$; $\overrightarrow{TS}$; $\overrightarrow{RT}$; $\overrightarrow{TR}$
8. Give another name for ray *RS*. $\overrightarrow{RT}$

LESSON 7-2 (pp. 326–329)

Estimate the measure of each angle, and use a protractor to check the reasonableness of your estimate. Tell what type each angle is.

9. 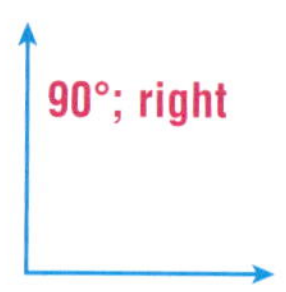
90°; right

10. 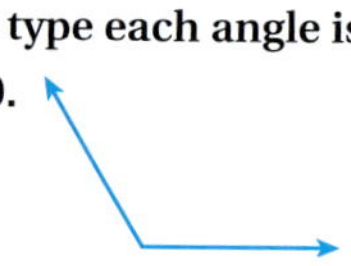
120°; obtuse

11. 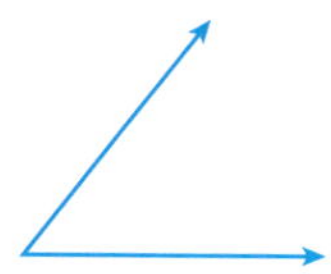
52°; acute

12. 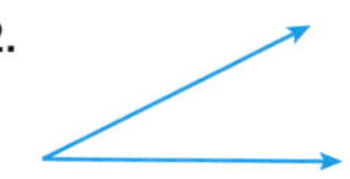
27°; acute

LESSON 7-3 (pp. 332–335)

Find each unknown angle measure.

13. $a = 45°$
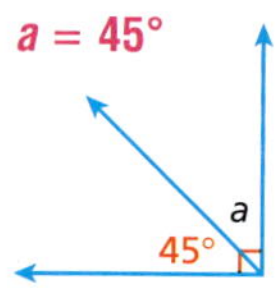

14. 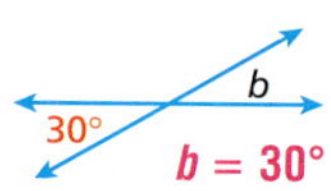

$b = 30°$

15. 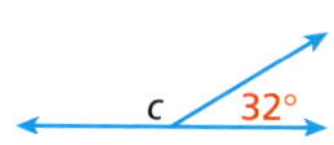

$c = 148°$

16. 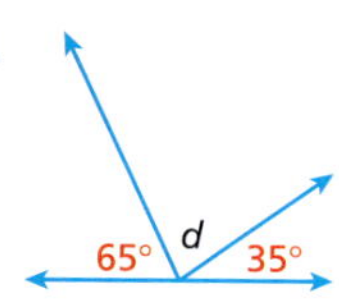

$d = 80°$

LESSON 7-4 (pp. 336–339)

Classify each pair of lines.

17. 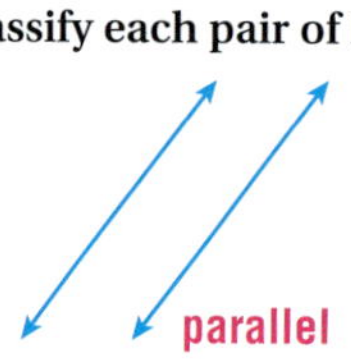
parallel

18. 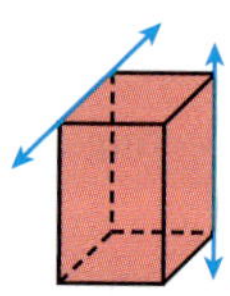
skew

19. 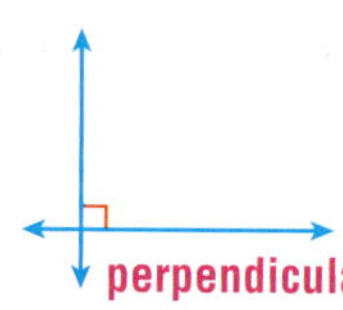
perpendicular

20. 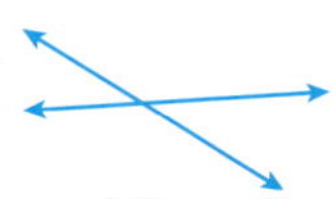
intersecting

Focus on Problem Solving

Solve

• Eliminate answer choices

Sometimes, when a problem has multiple answer choices, you can eliminate some of the choices to help you solve the problem.

For example, a problem reads, "The missing shape is not a red triangle." If one of the answer choices is a red triangle, you can eliminate that answer choice.

Read each problem, and look at the answer choices. Determine whether you can eliminate any of the answer choices before solving the problem. Then solve.

Smileys are letters and symbols that look like faces if you turn them around. When you write an e-mail to someone, you can use smileys to show how you are feeling.

Smileys	
Symbol	**Meaning**
:-(	Frown
:-D	Laugh
:-)	Smile
:-o	Shout
;-)	Wink

For 1–3, use the table.

1. Dora made a pattern with smileys. Which smiley will she probably use next?

   :-D :-) :-D :-) :-D :-) :-D :-) :-D ▢

   A :-D
   B :-)
   C :-)
   D :-D

2. Troy made a pattern with smileys. Identify a pattern. Which smiley is missing?

   :-(;-) :-o :-(;-) :-o :-(▢ :-o

   F :-(
   G :-o
   H ;-)
   J ;-)

3. To end an e-mail, Mya typed four smileys in a row. The shout is first. The wink is between the frown and the smile. The smile is not last. In which order did Mya type the smileys?

   A :-o :-(;-) :-)
   B :-o :-) ;-) :-(
   C :-) ;-) :-o :-(
   D :-o ;-) :-(:-)

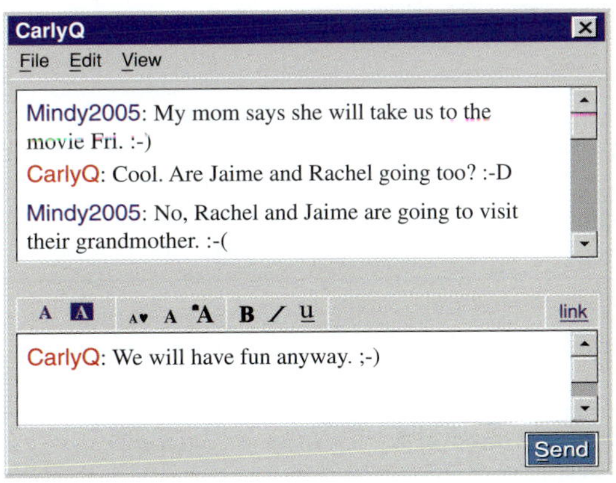

Answers

1. C
2. J
3. B

Focus on Problem Solving

Purpose: *To focus on solving a problem by eliminating answer choices*

Problem Solving Resources

Interactive Problem Solving pp. 53–64

Math: Reading and Writing in the Content Area . . . . pp. 53–64

Problem Solving Process

This page focuses on the third step of the problem-solving process:
Solve

Discuss

Have students discuss which answers can be eliminated immediately and why. Then identify the pattern when possible to help solve the problem.

1. Eliminate **B** and **D** because green always precedes red, and **B** and **D** are blue.
   smileys: laugh, smile, laugh, smile, laugh . . .
   colors: red, blue, green, red, blue, green, red . . .
2. Eliminate **F** and **H** because there are never two of the same color in a row.
   smileys: frown, wink, shout, frown, wink, shout, frown . . .
   colors: red, blue, red, blue, red . . .
3. Eliminate **A** and **D** because the smile is not last, and eliminate **C** because the shout is first.

Section 7B

Polygons

One-Minute Section Planner

Lesson	Materials	Resources
Lesson 7-5 Triangles **NCTM:** Geometry, Reasoning and Proof, Communication **NAEP:** Geometry 1b ☐ SAT-9 ☐ SAT-10 ☑ ITBS ☑ CTBS ☐ MAT ☑ CAT	**Optional** Teaching Transparency T20 *(CRB)* Protractor *(MK, CRB, p. 129)* Straightedge *(MK)* Geoboards *(MK)* Centimeter ruler *(MK, CRB, p. 131)*	• *Chapter 7 Resource Book,* pp. 45–54 • Daily Transparency T19, CRB • Additional Examples Transparencies T21–T23, CRB • *Alternate Openers: Explorations,* p. 57
Lesson 7-6 Quadrilaterals **NCTM:** Geometry, Reasoning and Proof, Communication **NAEP:** Geometry 3b ☑ SAT-9 ☑ SAT-10 ☑ ITBS ☑ CTBS ☐ MAT ☑ CAT	**Optional** Teaching Transparencies T25–T26 *(CRB)* Geoboards *(MK)* Cut-out quadrilaterals *(CRB, p. 133)*	• *Chapter 7 Resource Book,* pp. 55–63 • Daily Transparency T24, CRB • Additional Examples Transparencies T27–T28, CRB • *Alternate Openers: Explorations,* p. 58
Lesson 7-7 Polygons **NCTM:** Geometry, Problem Solving, Reasoning and Proof, Communication **NAEP:** Geometry 5a ☑ SAT-9 ☑ SAT-10 ☑ ITBS ☑ CTBS ☐ MAT ☑ CAT	**Optional** Teaching Transparency T30 *(CRB)*	• *Chapter 7 Resource Book,* pp. 64–72 • Daily Transparency T29, CRB • Additional Examples Transparencies T31–T33, CRB • *Alternate Openers: Explorations,* p. 59
Lesson 7-8 Geometric Patterns **NCTM:** Geometry, Reasoning and Proof, Communication, Connections **NAEP:** Algebra 1a ☑ SAT-9 ☑ SAT-10 ☑ ITBS ☑ CTBS ☑ MAT ☑ CAT	**Optional** Teaching Transparency T35 *(CRB)* Pattern blocks *(MK, CRB, p. 134)*	• *Chapter 7 Resource Book,* pp. 73–81 • Daily Transparency T34, CRB • Additional Examples Transparencies T36–T38, CRB • *Alternate Openers: Explorations,* p. 60
Section 7B Assessment		• Mid-Chapter Quiz, SE p. 360 • Section 7B Quiz, AR p. 19 • *Test and Practice Generator* CD-ROM

SAT = *Stanford Achievement Tests* **ITBS** = *Iowa Test of Basic Skills* **CTBS** = *Comprehensive Test of Basic Skills/Terra Nova*
MAT = *Metropolitan Achievement Tests* **CAT** = *California Achievement Test*

NCTM = Complete standards can be found on pages T27–T33. **NAEP** = Complete standards can be found on pages A31–A35.

SE = *Student Edition* **TE** = *Teacher's Edition* **AR** = *Assessment Resources* **CRB** = *Chapter Resource Book* **MK** = *Manipulatives Kit*

Section Overview

Triangles — Lesson 7-5

Why? Triangles are often used in construction to provide structural support.

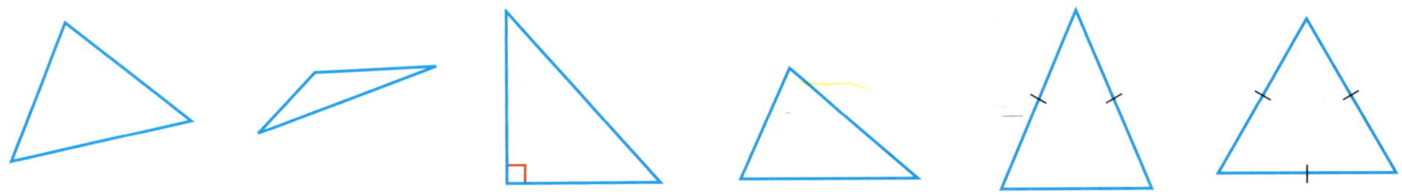

Acute triangle | Obtuse triangle | Right triangle | Scalene triangle | Isosceles triangle | Equilateral triangle

The sum of the measures of the angles in any triangle is 180°.

Quadrilaterals and Polygons — Lessons 7-6, 7-7

Why? Polygons are used in many architectural designs.

Quadrilaterals

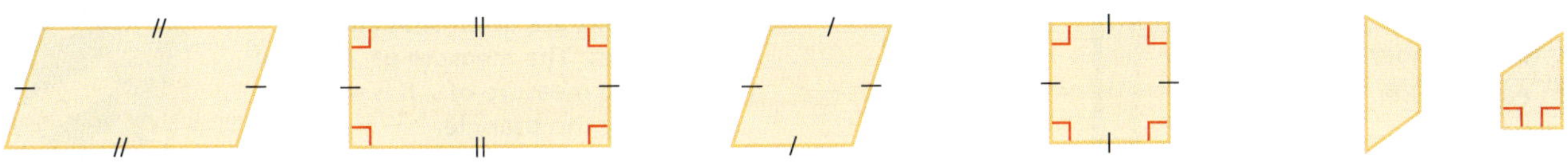

Parallelogram
Opposite sides are parallel and congruent. Opposite angles are congruent.

Rectangle
Parallelogram with four right angles

Rhombus
Parallelogram with four congruent sides

Square
Rectangle with four congruent sides

Trapezoid
Quadrilateral with exactly two parallel sides; may have two right angles.

Polygons

Name	Triangle	Quadrilateral	Pentagon	Hexagon	Octagon
Sides and Angles	3	4	5	6	8
Regular	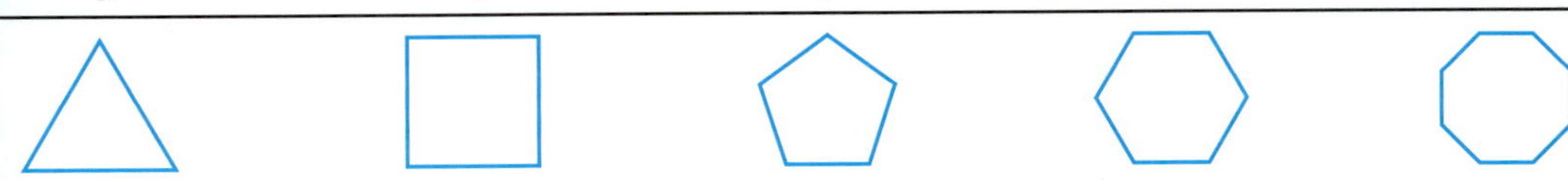				
Not Regular	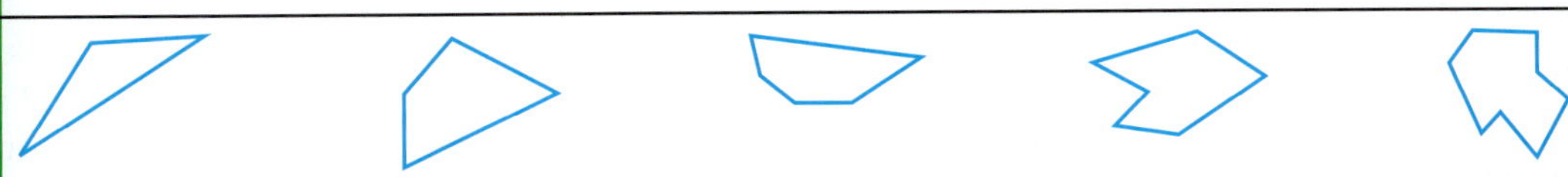				

Geometric Patterns — Lesson 7-8

Why? Many designs and works of art involve geometric patterns.

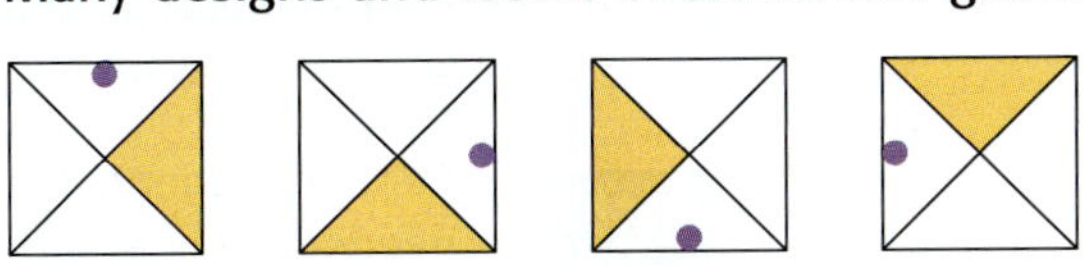

Perfect squares, such as 2^2, 3^2, and 4^2 are also called square numbers because they can be modeled as a square array.

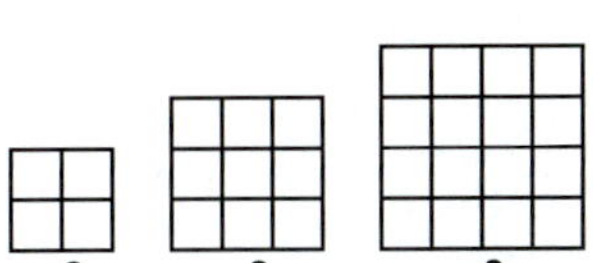

2^2 3^2 4^2

7-5 Organizer

Pacing: Traditional 1 day
Block $\frac{1}{2}$ day

Objective: Students classify triangles and solve problems involving angle and side measures of triangles.

Indiana Academic Standards
6.3.1, 6.3.5, 6.4.3, 6.4.4, 6.7.5, 6.7.11

Warm Up

1. What are two angles whose sum is 90°? complementary angles
2. What are two angles whose sum is 180°? supplementary angles
3. A part of a line between two points is called a _______. segment
4. Two lines that intersect at 90° are _________. perpendicular

Problem of the Day

Find the total number of shaded triangles in each figure. 3, 6, 10

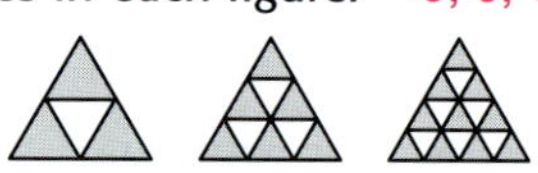

Available on Daily Transparency in CRB

Math Humor

The scalene triangle confused everyone on the baseball field. When it stood on a different base, it had a different height.

7-5 Triangles

Learn to classify triangles and solve problems involving angle and side measures of triangles.

Vocabulary
acute triangle
obtuse triangle
right triangle
scalene triangle
isosceles triangle
equilateral triangle

A triangle is a closed figure with three line segments and three angles. Triangles can be classified by the measures of their angles. An **acute triangle** has only acute angles. An **obtuse triangle** has one obtuse angle. A **right triangle** has one right angle.

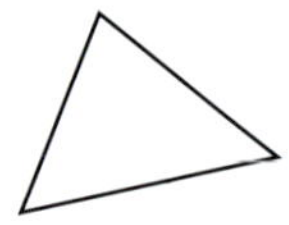

Acute triangle

Obtuse triangle

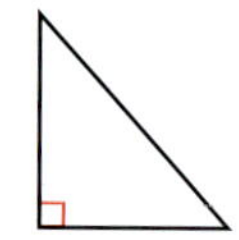

Right triangle

To decide whether a triangle is acute, obtuse, or right, you need to know the measures of its angles.

The sum of the measures of the angles in any triangle is 180°. You can see this if you tear the corners from a triangle and arrange them around a point on a line.

By knowing the sum of the measures of the angles in a triangle, you can find unknown angle measures.

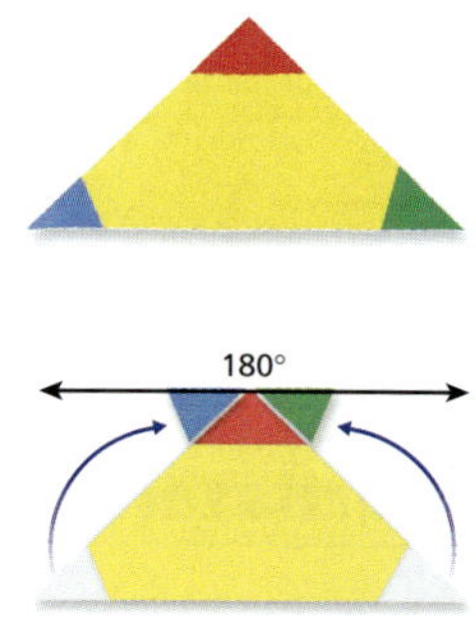

EXAMPLE 1 *Sports Application*

Boat sails are often shaped like triangles. The measure of $\angle A$ is 70°, and the measure of $\angle B$ is 45°. Classify the triangle.

To classify the triangle, find the measure of $\angle C$ on the sail.

$m\angle C = 180° - (70° + 45°)$
$m\angle C = 180° - 115°$ *Subtract the sum of the known angle measures from 180°.*
$m\angle C = 65°$

So the measure of $\angle C$ is 65°. Because $\triangle ABC$ has only acute angles, the boat sail is an acute triangle.

You can use what you know about vertical, adjacent, complementary, and supplementary angles to find the missing measures of angles.

1 Introduce

Alternate Opener

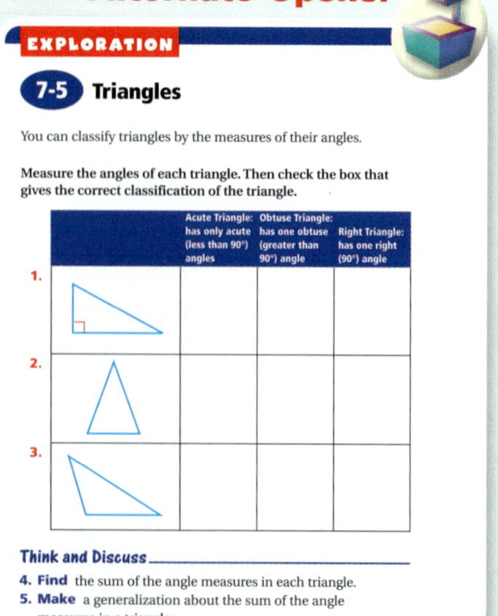

EXPLORATION

7-5 Triangles

You can classify triangles by the measures of their angles.

Measure the angles of each triangle. Then check the box that gives the correct classification of the triangle.

	Acute Triangle: has only acute (less than 90°) angles	Obtuse Triangle: has one obtuse (greater than 90°) angle	Right Triangle: has one right (90°) angle
1.			
2.			
3.			

Think and Discuss

4. Find the sum of the angle measures in each triangle.
5. Make a generalization about the sum of the angle measures in a triangle.

Motivate

Have each student use a straightedge to draw a triangle. Have students measure the angles in their triangles and count the number of right, acute, and obtuse angles. Have them measure the sides of their triangles, and note how many sides (if any) are congruent. Students can refer back to these drawings as you teach the different types of triangles. Straightedges, protractors, and rulers are provided in the Manipulatives Kit.

Exploration worksheet and answers on Chapter 7 Resource Book pp. 46 and 149

2 Teach

Lesson Presentation

Guided Instruction

In this lesson, students learn to classify triangles and solve problems involving angle and side measures of triangles. First teach students to use angles to classify triangles and to use the sum of the measures of the angles in a triangle to find the measure of an unknown angle. Then teach them to classify triangles by the lengths of the sides. Have students classify the acute, obtuse, and right triangles as scalene, isosceles, or equilateral, and vice versa.

Teaching Tip Remind students that a triangle needs only one obtuse angle to be an obtuse triangle, but three acute angles to be an acute triangle.

EXAMPLE 2 Using Properties of Angles to Label Triangles

Remember!

Vertical angles are congruent. The sum of the measures of complementary angles is 90°. The sum of the measures of supplementary angles is 180°.

Use the diagram to find the measure of each indicated angle.

A $\angle BDE$

$\angle BDE$ and $\angle ADC$ are vertical angles, so $m\angle BDE = m\angle ADC$.

$$m\angle ADC = 180° - (30° + 35°)$$
$$= 180° - 65°$$
$$= 115°$$

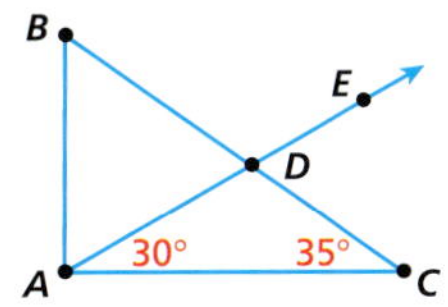

$$m\angle BDE = 115°$$

B $\angle ADB$

The sum of $m\angle BDE$ and $m\angle ADB$ is 180°.

$$m\angle ADB = 180° - 115°$$
$$= 65°$$
$$m\angle ADB = 65°$$

Triangles can be classified by the lengths of their sides. A **scalene triangle** has no congruent sides. An **isosceles triangle** has at least two congruent sides. An **equilateral triangle** has three congruent sides.

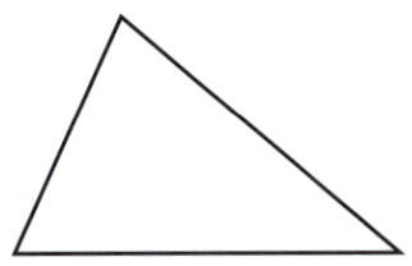

Scalene triangle

Isosceles triangle

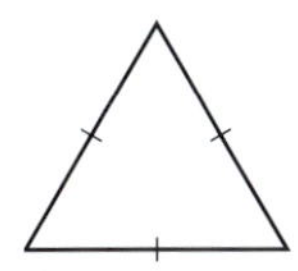

Equilateral triangle

EXAMPLE 3 Classifying Triangles by Lengths of Sides

Classify the triangle. The sum of the lengths of the sides is 7.8 cm.

$$a + (3.8 + 2) = 7.8$$
$$a + 5.8 = 7.8$$
$$a + 5.8 - 5.8 = 7.8 - 5.8$$
$$a = 2$$

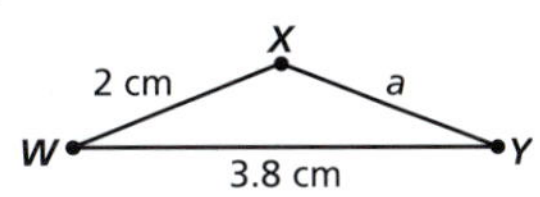

Side a is 2 centimeters long. Because $\triangle WXY$ has at least two sides, but not three, that are the same length, it is an isosceles triangle.

Think and Discuss

1. Explain why a triangle cannot have two obtuse angles.

2. Tell whether a right triangle can also be an acute triangle. Explain.

Additional Examples

Example 1

Sara designed this triangular trophy. The measure of $\angle E$ is 38°, and the measure of $\angle F$ is 52°. Classify the triangle.

right triangle

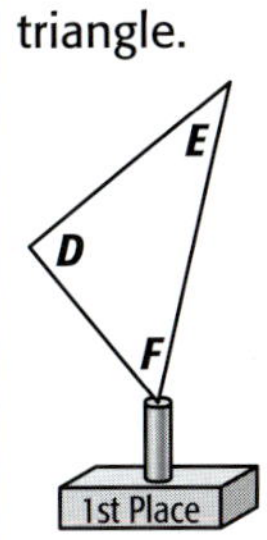

Example 2

Use the diagram to find the measure of each indicated angle.

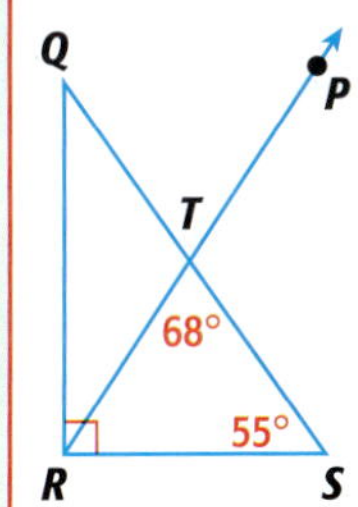

A. $\angle QTR$ 112°

B. $\angle QRT$ 33°

Example 3

Classify the triangle. The sum of the lengths of the sides is 19.5 in.

equilateral

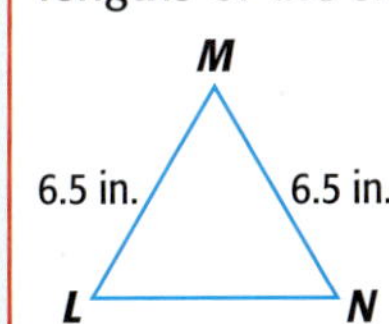

3 Close

Reaching All Learners

Through Hands-On Experience

Have students use geoboards (Manipulatives Kit) to model acute, right, obtuse, scalene, isosceles, and equilateral triangles. Have them draw their triangles on dot paper (Chapter 7 Resource Book p. 132) and classify each according to measures of angles and lengths of sides.

Summarize

Discuss the following questions to review the vocabulary in the lesson:

- Can an obtuse triangle also be an acute triangle? Why or why not? No; obtuse triangles have two acute angles, but acute triangles must have three acute angles.
- Can a right triangle also be an isosceles triangle? Why or why not? Yes; if the right triangle has two sides that are equal lengths, then the triangle is isosceles.
- Can an acute triangle also be a scalene triangle? Why or why not? Yes; if the sides are all different lengths, then the triangle is scalene.

Answers to Think and Discuss

1. Possible answer: The sum of the measures in a triangle is 180°. Because an obtuse angle has a measure greater than 90°, two obtuse angles would have a measure greater than 180°.
2. No; a right triangle has one 90° angle. In an acute triangle, all three angles measure less than 90°.

7-5 PRACTICE & ASSESS

Students may want to refer back to the lesson examples.

Assignment Guide

If you finished Example 1 assign:
Core 1, 6, 11, 13, 28–36
Enriched 1, 6, 25, 26, 28–36

If you finished Example 2 assign:
Core 1–3, 6–8, 11, 13, 28–36
Enriched 1–3, 6–8, 25, 26, 28–36

If you finished Example 3 assign:
Core 1–16, 23, 28–36
Enriched 11–36

Notes

7-5 Exercises

FOR EXTRA PRACTICE
see page 650

internet connect
Homework Help Online
go.hrw.com Keyword: MR4 7-5

GUIDED PRACTICE

See Example 1
1. Three stars form a triangular constellation. Two of the angles measure 20° and 50°. Classify the triangle. **obtuse triangle**

See Example 2
Use the diagram to find the measure of each indicated angle.

2. $\angle XZV$ **98°**

3. $\angle VZW$ **82°**

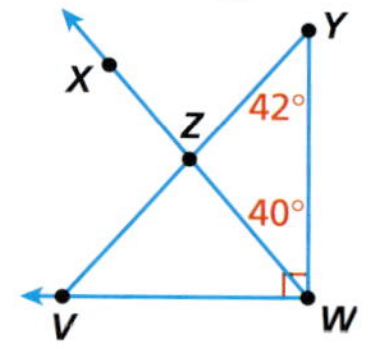

See Example 3
Classify each triangle using the given information.

4. The sum of the lengths of the three sides is 24 cm. **equilateral**

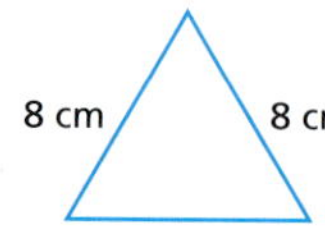

5. The sum of the lengths of the three sides is 30 ft. **isosceles**

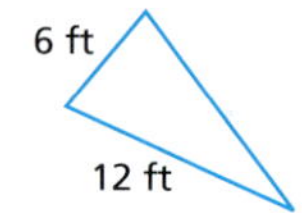

INDEPENDENT PRACTICE

See Example 1
6. Interstate highways connecting towns *R*, *S*, and *T* form a triangle. Two of the angles measure 40° and 42°. Classify the triangle. **obtuse triangle**

See Example 2
Use the diagram to find the measure of each indicated angle.

7. $\angle KNJ$ **60°**

8. $\angle LKM$ **70°**

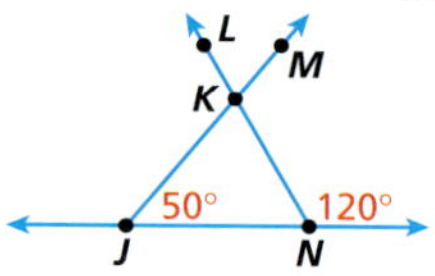

See Example 3
Classify each triangle using the given information.

9. The sum of the lengths of the three sides is 10.5 in. **scalene**

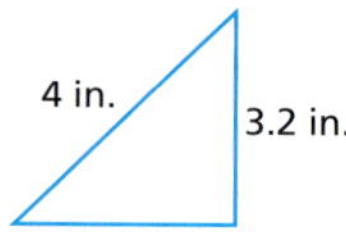

10. The sum of the lengths of the three sides is 231 km. **scalene**

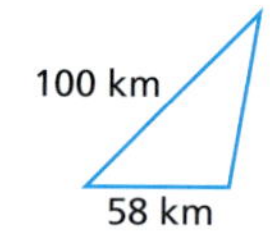

PRACTICE AND PROBLEM SOLVING

If the angles can form a triangle, classify it as acute, obtuse, or right.

11. 45°, 90°, 45° **yes, right**

12. 51°, 88°, 41° **yes, acute**

13. 55°, 102°, 33° **no**

14. 37°, 40°, 103° **yes, obtuse**

Math Background

Although isosceles and equilateral triangles are defined in terms of side lengths, it turns out that their angles have special properties as well. In an isosceles triangle, the angles opposite the congruent sides are congruent. In an equilateral triangle, all three angles are congruent and measure 60°.

Students can explore each of these propositions by folding paper or by using protractors. In a high school geometry course, where the first result is known as the Isosceles Triangle Theorem, both propositions are easily proven.

RETEACH 7-5

LESSON 7-5 **Reteach**
Triangles

You can classify triangles by the measures of their angles.

An acute triangle has only acute angles.
A right triangle has one right angle.
An obtuse triangle has one obtuse angle.

To classify the triangle below by its angles, you first need to find the measure of all the angles.

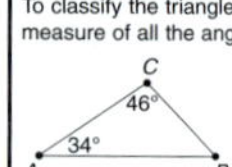

The sum of the angles of a triangle is 180°.

To find the unknown angle, first find the sum of the two known angles.

34 + 46 = 80

Then subtract the sum from 180.

180 − 80 = 100

The difference is the measure of the third angle.

The measures of the angles are 34°, 46°, and 100°. So the triangle is obtuse.

Find each unknown angle. Then classify each triangle.

1. 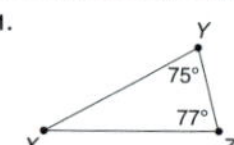

28°, acute

2. 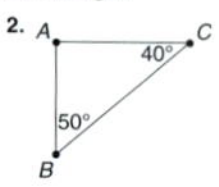

90°, right

3. 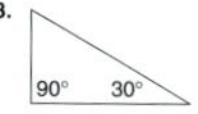

60°, right

4. 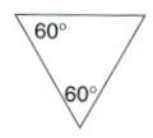

60°, acute

PRACTICE 7-5

LESSON 7-5 **Practice B**
Triangles

Use the diagram to find the measure of each indicated angle.

1. $\angle CBD$ **90°**
2. $\angle DAC$ **45°**
3. $\angle ABC$ **90°**
4. $\angle EBA$ **90°**
5. $\angle ACB$ **45°**

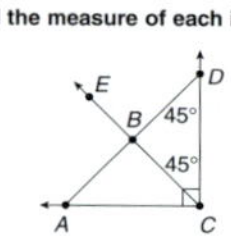

Classify each triangle using the given information.

6. The sum of the lengths of the three sides is 30 in.

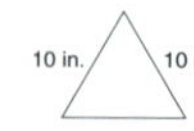

equilateral

7. The sum of the lengths of the three sides is 15 cm.

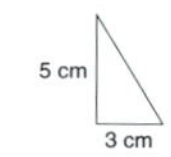

scalene

8. The sum of the lengths of the three sides is 22 ft.

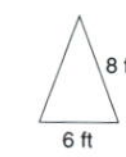

isosceles

9. The angles of a triangular sail measure 90°, 30°, and 60°. Its sides measure approximately 2 feet, 3.5 feet, and 4 feet. Classify the triangular shape of the sail in two different ways.

It is a right triangle; It is a scalene triangle.

10. Two angles in one triangle are congruent to two angles in another triangle. What can you conclude about the third angle in both triangles?

They are congruent.

Social Studies LINK

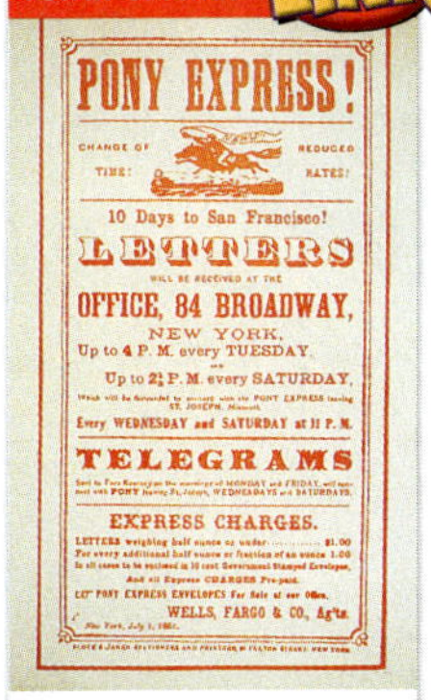

The Pony Express was established in 1860. This system of horseback messengers carried information across the United States before our modern postal service.

The lengths of two sides are given for $\triangle ABC$. Use the sum of the lengths of the three sides to calculate the length of the third side and classify each triangle.

15. $AB = 7$ cm; $BC = 7$ cm; sum = 15.9 cm **1.9 cm, isosceles**

16. $AB = 24$ in.; $BC = 30$ in.; sum = 92 in. **38 in., scalene**

17. $AB = 1\frac{1}{6}$ ft; $BC = 1\frac{1}{6}$ ft; sum = $3\frac{1}{2}$ ft **$1\frac{1}{6}$ ft, equilateral**

18. $AB = 9.5$ m; $BC = 9.5$ m; sum = 38 m **19 m, isosceles**

Draw an example of each triangle described.

19. a scalene acute triangle

20. an isosceles right triangle

21. an isosceles obtuse triangle

22. a scalene right triangle

23. ***SOCIAL STUDIES*** Some triangular stamps are made by dividing a rectangle into two parts. Classify the triangle that is made by cutting on a line that connects one corner of a rectangle to the opposite corner. **right triangle**

24. ***MEASUREMENT*** Use a centimeter ruler to measure each side of triangle A. Add the lengths of any two sides and compare the sum to the length of the third side. Add a different pair of lengths and compare the sum to the third side. Do the same for triangles B and C. What do you notice? **Possible answer: The sum of two sides must be greater than the third side to form a triangle.**

25. ***CHOOSE A STRATEGY*** How many triangles are in the figure at right? **27**

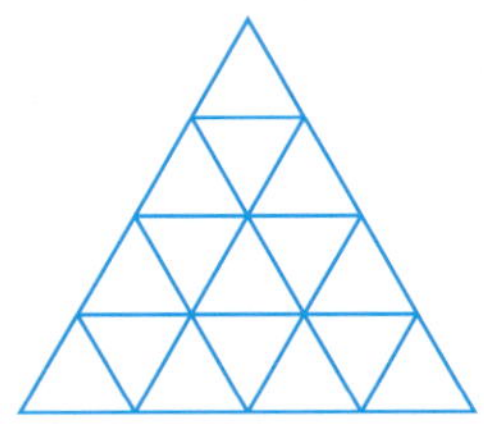

26. ***WRITE ABOUT IT*** Explain why a triangle cannot have two right angles.

26. The sum of the angles of a triangle is 180°. The sum of the measures of two right angles is 180°. The measure of the third angle would be 0°, which is not possible.

27. ***CHALLENGE*** Find the sum of the angles of a square. (*Hint:* Divide the square into two triangles.) **360°**

Spiral Review

Find the greatest common factor of each set of numbers. (Lesson 4-3)

28. 12, 36 **12** **29.** 15, 24 **3** **30.** 18, 24, 42 **6** **31.** 5, 14, 17 **1**

Plot each point on a coordinate plane. (Lesson 6-6)

32. A (3, 5) **33.** B (6, 2) **34.** C (0, 4) **35.** D (1, 0)

36. **TEST PREP** Which number could represent the measure of an acute angle? (Lesson 7-2) **A**

A 71° **B** 90° **C** 112° **D** 180°

Answers

Possible answers 19–22:

19.

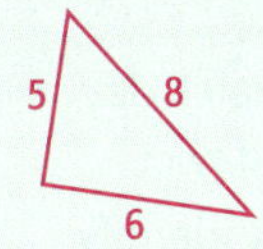

20.

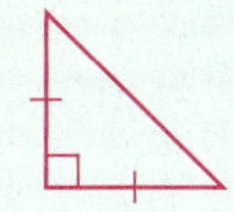

21.

22.

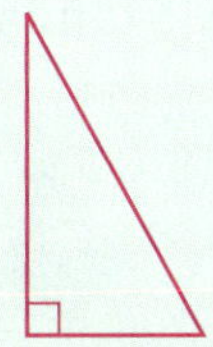

32–35.

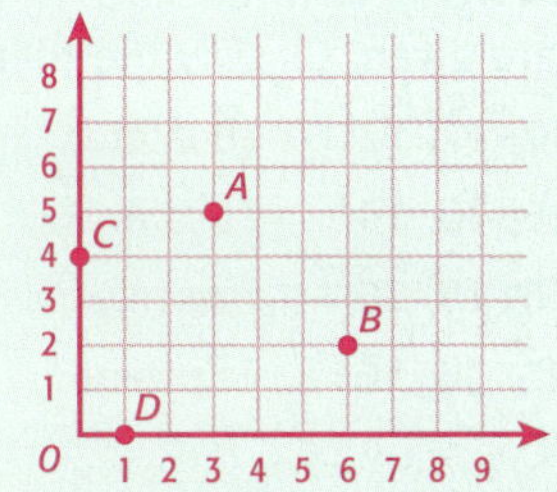

Journal

Have students draw examples of the six types of triangles presented in the lesson. For each triangle, have students explain why it is that type of triangle.

Test Prep Doctor

For Exercise 36, encourage students to eliminate incorrect answer choices before settling on a final answer. Students who answered **B** may have confused acute angles with right angles. Students who answered **C** confused acute angles with obtuse angles, and students who answered **D** confused acute angles with straight angles.

CHALLENGE 7-5

LESSON 7-5 **Challenge** *Square Legs*

About 2,400 years ago a Greek philosopher and mathematician named Pythagoras proved a very important rule for triangles. Today people all over the world study and use this rule named in his honor—the Pythagorean Theorem. It states the relationship between the side lengths of a right triangle.

Pythagoras

PYTHAGOREAN THEOREM

In a right triangle, the square of the length of the hypotenuse is equal to the sum of the squares of the lengths of the legs.

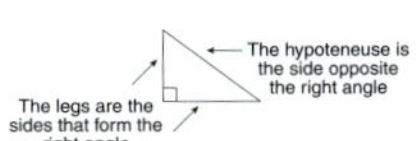

(leg)^2 + (leg)^2 = (hypotenuse)^2

$a^2 + b^2 = c^2$

$3^2 + 4^2 = 5^2$

$9 + 16 = 25$

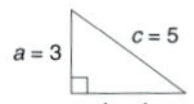

Use the Pythagorean Theorem to find the length of the missing side for each right triangle below.

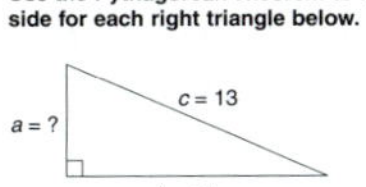

a = 5

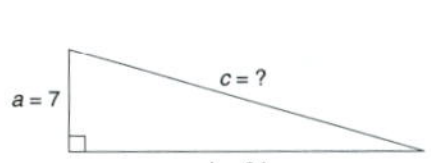

c = 25

PROBLEM SOLVING 7-5

LESSON 7-5 **Problem Solving** *Triangles*

Use the triangle diagram to answer each question.

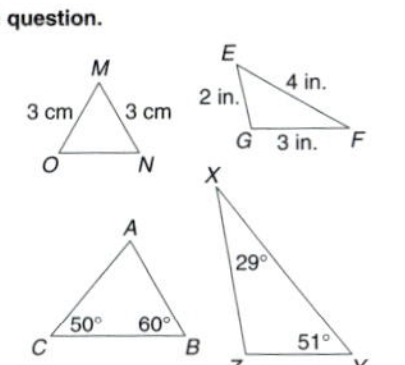

1. Classify triangle ABC. What is the measure of the missing angle?
   $\triangle ABC$ is an acute triangle, 70°
2. Classify triangle XYZ. What is the measure of the missing angle?
   $\triangle XYZ$ is an obtuse triangle, 100°
3. If triangle MNO is an equilateral triangle, what is the measure of the missing side?
   $\overline{ON}$ = 3 cm
4. What is the complement of $\angle XYZ$?
   39°
5. Classify triangle EFG.
   $\triangle EFG$ is a scalene triangle
6. Explain how you can find the measure of $\angle CAB$.
   Because the sum of the measures of the angles in any triangle is 180°, subtract the measure of $\angle ACB$ and $\angle ABC$ from 180°.
7. What is the supplement of $\angle ABC$?
   120°

Circle the letter of the correct answer.

8. Which of the following statements is always true?
   A A right triangle is a scalene triangle.
   (B) An equilateral triangle is an isosceles triangle.
   C An isosceles triangle is an obtuse triangle.
   D A right triangle is an acute triangle.
9. Which of the following is not true of all right triangles?
   F The sum of the measures of the angles is 180°.
   (G) Two of its angles are complementary angles.
   H At least two of its angles are acute.
   J The side with the greatest length is opposite the right angle.

Lesson Quiz

If the angles can form a triangle, classify the triangle as acute, obtuse, or right.

1. 37°, 53°, 90° right

2. 65°, 110°, 25° not a triangle

3. 61°, 78°, 41° acute

4. 115°, 25°, 40° obtuse

The lengths of three sides of a triangle are given. Classify the triangle.

5. 12, 16, 25 scalene

6. 10, 10, 15 isosceles

Available on Daily Transparency in CRB

7-6 Organizer

Pacing: Traditional 1 day
Block $\frac{1}{2}$ day

Objective: Students identify, classify, and compare quadrilaterals.

Indiana Academic Standards
6.4.3, 6.7.1, 6.7.2, 6.7.5, 6.7.11

Warm Up

The lengths of three sides of a triangle are given. Classify the triangle.

1. 12, 12, 12 equilateral
2. 18, 10, 14 scalene
3. 15, 15, 26 isosceles
4. 23, 36, 16 scalene

Problem of the Day

How many different rectangles are in the figure?

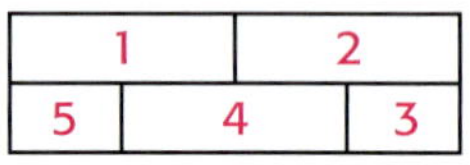

10; if the rectangles are marked 1–5 clockwise from upper left, they are 1; 2; 3; 4; 5; 1-2; 3-4; 4-5; 3-5; 1-2-3-4-5.

Available on Daily Transparency in CRB

Math Fact !

The figure formed by joining the consecutive midpoints of the sides of a quadrilateral is a parallelogram.

7-6 Quadrilaterals

Learn to identify, classify, and compare quadrilaterals.

Vocabulary
quadrilateral
parallelogram
rectangle
rhombus
square
trapezoid

A **quadrilateral** is a plane figure with four sides and four angles.

Five special types of quadrilaterals and their properties are shown in the table below. The same mark on two or more sides of a figure indicates that the sides are congruent.

Parallelogram		Opposite sides are parallel and congruent. Opposite angles are congruent.
Rectangle	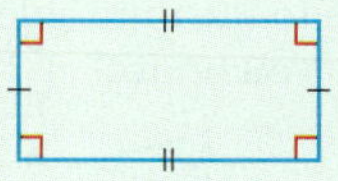	Parallelogram with four right angles
Rhombus	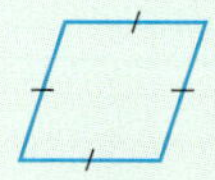	Parallelogram with four congruent sides
Square	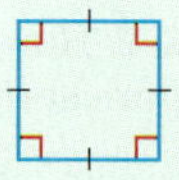	Rectangle with four congruent sides
Trapezoid	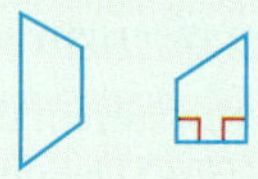	Quadrilateral with exactly two parallel sides May have two right angles

EXAMPLE 1 Naming Quadrilaterals

Give the most descriptive name for each figure.

A 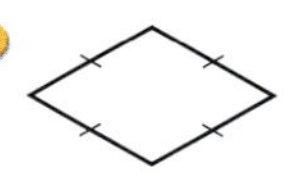

The figure is a quadrilateral, a parallelogram, and a rhombus.

Rhombus is the most descriptive name.

B 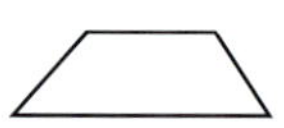

The figure is a quadrilateral and a trapezoid.

Trapezoid is the most descriptive name.

1 Introduce

Alternate Opener

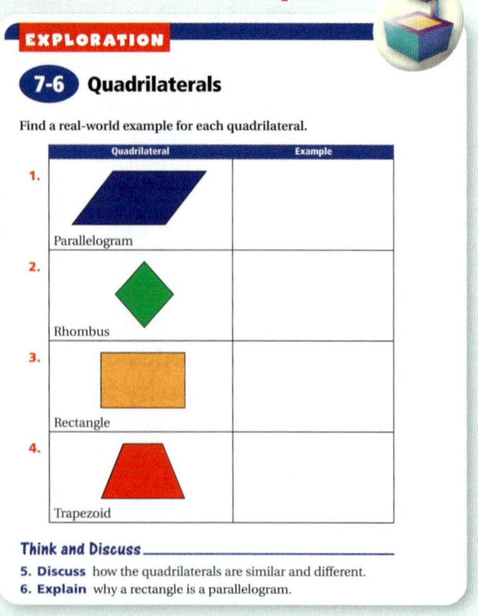
EXPLORATION

7-6 Quadrilaterals

Find a real-world example for each quadrilateral.

Quadrilateral	Example
1. Parallelogram	
2. Rhombus	
3. Rectangle	
4. Trapezoid	

Think and Discuss

5. **Discuss** how the quadrilaterals are similar and different.
6. **Explain** why a rectangle is a parallelogram.

Motivate

Have students examine a set of cut-out quadrilaterals: parallelogram, rectangle, rhombus, square, and trapezoid (provided in Chapter 7 Resource Book p. 133). Have students name similarities and differences between the figures.

Possible answers: They all have four sides. Some have all right angles. Some have parallel sides. Some have all equal sides.

Exploration worksheet and answers on Chapter 7 Resource Book pp. 56 and 152

2 Teach

Lesson Presentation

Guided Instruction

In this lesson, students learn to identify, classify, and compare quadrilaterals. Show students the chart illustrating the types of quadrilaterals and describing the properties of each (Teaching Transparency T25 in CRB). Teach students to find the most descriptive name for a given quadrilateral, and have students complete statements comparing quadrilaterals.

Teaching Tip Have students use geoboards (Manipulatives Kit) to model each quadrilateral as you define each type.

Give the most descriptive name for each figure.

C 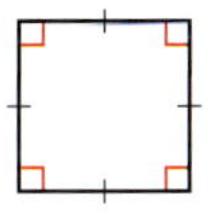

The figure is a quadrilateral, parallelogram, rectangle, rhombus, and square.

Square is the most descriptive name.

D 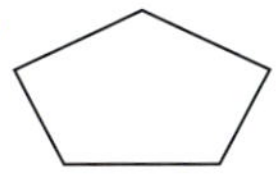

This figure is a plane figure, but it has more than 4 sides.

The figure is not a quadrilateral.

You can draw a diagram to classify quadrilaterals based on their properties.

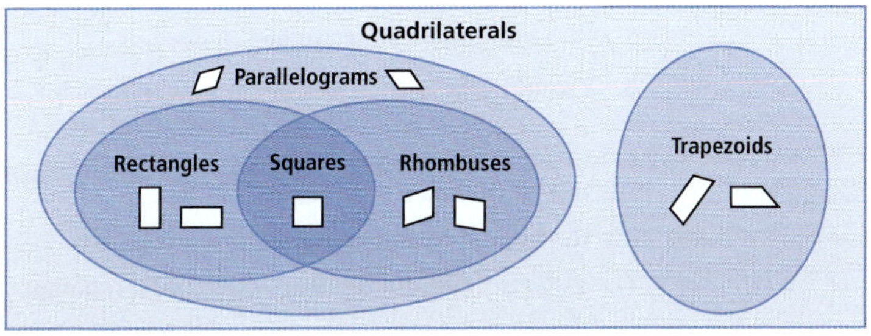

EXAMPLE 2 Classifying Quadrilaterals

Complete each statement.

A **A rhombus that is a rectangle is also a ___?___.**

A rhombus has four congruent sides, and the opposite sides are parallel. If it is a rectangle, it has four right angles, which makes it a **square**.

B **A square can also be called a ___?___, ___?___, and ___?___.**

A square has opposite sides that are parallel; it can be called a **parallelogram**.

A square has four congruent sides; it can be called a **rhombus**.

A square has four right angles; it can be called a **rectangle**.

Think and Discuss

1. **Tell** whether all squares are rhombuses and whether all rhombuses are squares.
2. **Compare** a trapezoid with a rectangle.

COMMON ERROR ALERT

Students may be confused by the fact that some quadrilaterals can be described by more than one name. Give students several examples of these quadrilaterals to illustrate the point.

Additional Examples

Example 1

Give the most descriptive name for each figure.

A. 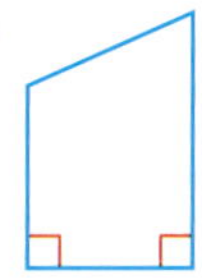 trapezoid

B.  rectangle

C. 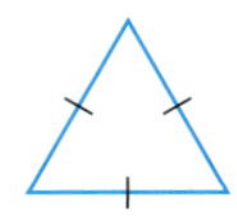 equilateral triangle

D. 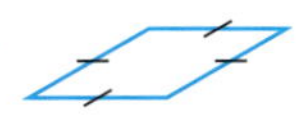 rhombus

Example 2

Complete each statement.

A. A rectangle can also be called a ___?___. parallelogram

B. A parallelogram cannot be a ___?___. trapezoid

3 Close

Reaching All Learners

Through Home Connection

Have students provide examples or drawings of the different types of quadrilaterals they may find in their homes. These could include clocks, mirrors, pictures cut out from magazines, patterns on bedspreads, etc. Ask them to give all the possible names for each quadrilateral.

Summarize

Display a parallelogram, rectangle, rhombus, square, and trapezoid. Have students use the most descriptive name to classify each figure. Ask students what all of these figures are called. quadrilaterals

Answers to Think and Discuss

1. All squares are rhombuses because they have four congruent sides. Not every rhombus is a square, because the angles in a rhombus do not have to be right angles.
2. Possible answer: Both have four sides. A rectangle has two sets of parallel sides and a trapezoid has only one set of parallel sides. A trapezoid may have two right angles. A rectangle has four right angles.

7-6 PRACTICE & ASSESS

Students may want to refer back to the lesson examples.

Assignment Guide

If you finished Example 1 assign:
Core 1–3, 7–9, 13–15, 34–36
Enriched 1–3, 7–9, 13–15, 34–36

If you finished Example 2 assign:
Core 1–23, 34–36
Enriched 10–36

Answers

13. quadrilateral, parallelogram, rectangle, rhombus, (square)

14. quadrilateral, parallelogram, (rectangle)

15. quadrilateral, parallelogram, (rhombus)

Math Background

A theorem in geometry states that a line segment joining the midpoints of two sides of a triangle is parallel to the third side and equal to one-half the length of the third side.

This theorem can be used to prove the math fact of this lesson.

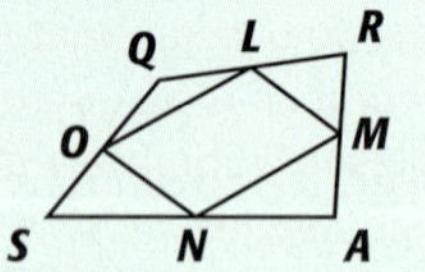

Draw diagonal $\overline{QA}$. Because $\overline{LM}$ joins the midpoints of two sides of a triangle, $\overline{LM} \parallel \overline{QA}$ and $LM = \frac{1}{2}QA$. Because $\overline{ON}$ joins the midpoints of two sides of a triangle, $\overline{ON} \parallel \overline{QA}$ and $ON = \frac{1}{2}QA$. So, $\overline{LM} \parallel \overline{ON}$ and $LM = ON$. In a similar manner, it can be shown that $\overline{LO} \parallel \overline{MN}$ and $LO = MN$. Therefore $LMNO$ is a parallelogram.

7-6 Exercises

FOR EXTRA PRACTICE see page 650

internet connect
Homework Help Online
go.hrw.com Keyword: MR4 7-6

GUIDED PRACTICE

See Example 1 Give the most descriptive name for each figure.

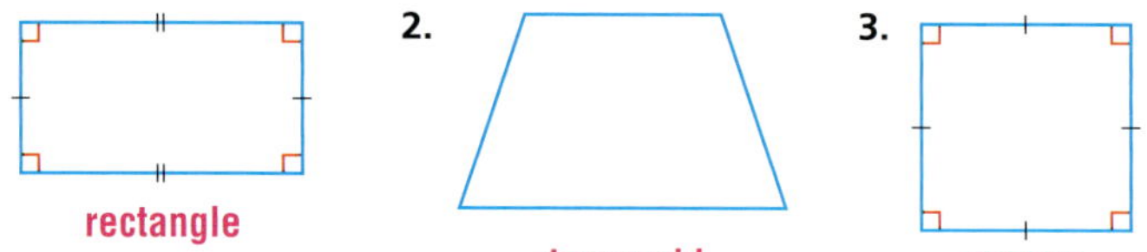

1. rectangle
2. trapezoid
3. square

See Example 2 Complete each statement.

4. A trapezoid is also a __?__. quadrilateral
5. All __?__ are also rectangles. squares
6. A square has four __?__ angles. right

INDEPENDENT PRACTICE

See Example 1 Give the most descriptive name for each figure.

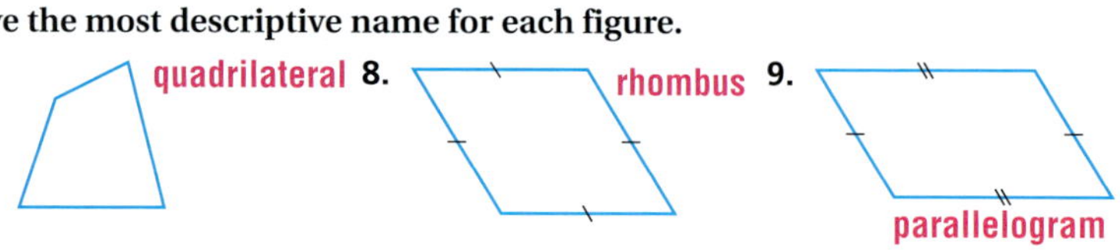

7. quadrilateral
8. rhombus
9. parallelogram

See Example 2 Complete each statement.

10. A rhombus with four right angles is a __?__. square
11. A parallelogram cannot be a __?__. trapezoid
12. A quadrilateral with four congruent sides and no right angles can be called a __?__ and a __?__. rhombus, parallelogram

PRACTICE AND PROBLEM SOLVING

Give all of the possible names for each figure. Circle the most exact name.

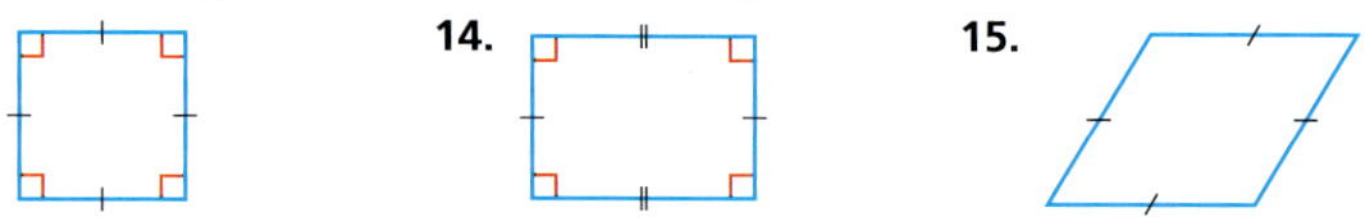

Determine if the given statements are *sometimes, always,* or *never* true.

16. A square is a rectangle. always
17. A trapezoid is a parallelogram. never
18. A rhombus is a square. sometimes
19. A parallelogram is a quadrilateral. always
20. A rectangle is a rhombus. sometimes
21. Four-sided figures are parallelograms. sometimes
22. All rectangles are squares. sometimes
23. All trapezoids have one right angle. sometimes

RETEACH 7-6

LESSON 7-6 **Reteach**
Quadrilaterals

A quadrilateral is a plane figure with four sides and four angles.

There are special types of quadrilaterals.

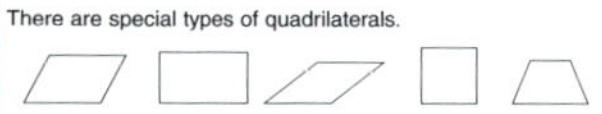

parallelogram rectangle rhombus square trapezoid

A parallelogram has opposite sides that are parallel and congruent. Opposite angles are also congruent.

A rectangle is a parallelogram that has four right angles.

A rhombus is a parallelogram with four congruent sides.

A square is a rectangle with four congruent sides.

A trapezoid is a quadrilateral with exactly two parallel sides.

The best description for the figure below is rectangle because it is a parallelogram with four right angles.

Give the best name for each figure.

1. trapezoid
2. square
3. parallelogram
4. rhombus
5. rectangle
6. trapezoid

PRACTICE 7-6

LESSON 7-6 **Practice B**
Quadrilaterals

Give the most descriptive name for each figure.

1. rhombus
2. trapezoid
3. parallelogram
4. quadrilateral
5. square
6. rectangle

Complete each statement.

7. All rectangles are also parallelograms.
8. A rhombus is sometimes a square.
9. All trapezoids are also quadrilaterals.
10. A quadrilateral is any plane figure with four straight sides and four angles.
11. A quadrilateral with two sets of parallel lines, but does not have 90° angles is called a rhombus.
12. Devon made a table top in the shape of a quadrilateral. All of its angles measure 90°. What could the shape of Devon's table top be?
a rectangle or a square
13. The perimeter of a rhombus is 64 inches. What is the length of each side of the rhombus? Explain.
16 inches, All four sides of a rhombus are congruent, and $64 \div 4 = 16$.
14. Explain why a trapezoid is a quadrilateral, but a quadrilateral is not always a trapezoid.
A trapezoid has four sides, always making it a quadrilateral. A quadrilateral can,but not always, have two parallel sides, thus a quatrilateral is not always a trapezoid.

The first baseball game played at Shea Stadium was on April 17, 1964. In September 1974, the Mets played a 25-inning game against the Cardinals, almost three times the length of a regulation game.

go.hrw.com KEYWORD: MR4 Baseball

CNN Student News

Draw each quadrilateral as described. If it is not possible to draw, explain why.

24. a rectangle that is also a square

25. a rhombus that is also a trapezoid **not possible: A rhombus has two pairs of parallel sides, and a trapezoid only has one pair.**

26. a parallelogram that is not a rectangle

27. a square that is not a rhombus **not possible: A square has four congruent sides. It must always be a rhombus.**

28. ***SPORTS*** A baseball diamond is in the shape of a square. The distance from home plate to first base is 90 ft. What is the distance around the baseball diamond? **360 ft**

29. A rectangular picture frame is 3 in. wider than it is tall. The total length of the four sides of the frame is 38 in.

a. The dimensions of the frame could be 10 in. by 13 in. because one dimension is 3 in. longer than the other. Explain how you know the frame is not 10 in. by 13 in.

b. How can you use your answer from part **a** to find the dimensions of the frame?

c. Using parts **a** and **b,** what are the dimensions of the frame? **8 in. by 11 in.**

30. Anika drew a quadrilateral. Then she drew a line segment connecting one pair of opposite corners. She saw that she had divided the original quadrilateral into two right isosceles triangles. Classify the quadrilateral she began with. **square**

31. ***WHAT'S THE ERROR?*** A student said that any quadrilateral with two right angles and a pair of parallel sides is a rectangle. What is the error in the statement?

32. ***WRITE ABOUT IT*** Explain why a square is also a rectangle and a rhombus.

33. ***CHALLENGE*** Part of a quadrilateral is hidden. What are the possible types of quadrilaterals that the figure could be? **trapezoid, parallelogram, rectangle, rhombus, square**

Spiral Review

34. A reporter interviewed 100 drivers and asked them how many times each had received a speeding ticket. Create a bar graph that displays the data. (Lesson 6-4)

Tickets	Number of Drivers
0	48
1	34
2	10
3	5
4+	3

35. Angles M and N are supplementary. If the measure of angle M is 33°, what is the measure of angle N? (Lesson 7-3) **147°**

36. **TEST PREP** ___?___ are lines in the same plane that never intersect. (Lesson 7-4) **C**

A Skew lines
B Intersecting lines
C Parallel lines
D Perpendicular lines

Answers

24. 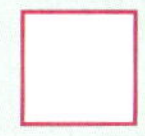 **26.** 

29. a. If the frame is 10 in. by 13 in., the total length of the sides is 46 in., not 38 in.

b. Possible answer: The dimensions, 10 in. by 13 in., were too long, so try shorter lengths. Try 9 in. and 12 in., since 12 in. is 3 in. more than 9 in. This is also too large, so you can try shorter lengths.

31–32. See p. A5.

34.

Journal

Have students write brief descriptions and draw pictures of the five types of quadrilaterals discussed in this lesson.

Test Prep Doctor

For Exercise 36, students can eliminate **B** and **D** because they *do* intersect. Students who answered **A** may not have read the sentence carefully enough to understand that the lines in question must be *in the same plane*.

CHALLENGE 7-6

LESSON 7-6 **Challenge**
What's the Point?

For each grid below, name the ordered pair of the missing fourth point that could be used to complete the given quadrilateral.

1. 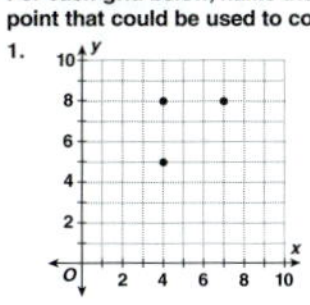
square
(7, 5)

2. 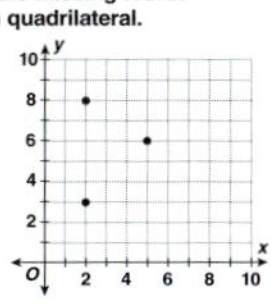
trapezoid
Possible answer: (5, 5)

3. 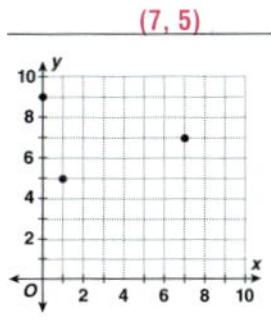
quadrilateral
Possible answer: (3, 9)

4. 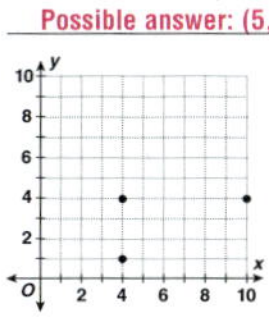
rectangle
(10, 1)

5. 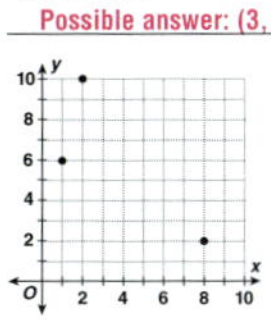
parallelogram
(9, 6)

6. 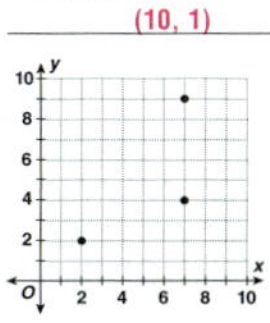
rhombus
(2, 7)

PROBLEM SOLVING 7-6

LESSON 7-6 **Problem Solving**
Quadrilaterals

Write the correct answer.

1. Fill in this Venn diagram using the terms quadrilaterals, squares, rectangles, rhombuses, parallelograms, and trapezoids.

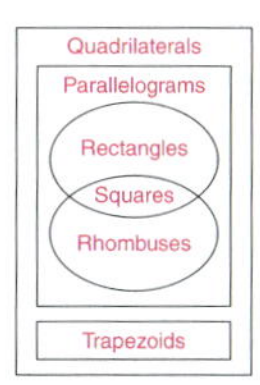

2. Part of this quadrilateral is hidden. What could it possibly be?

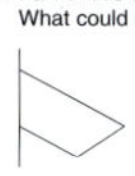

a trapezoid, a parallelogram

3. How could you make a trapezoid from a rectangle using only one cut?

Cut a right triangle from one side

4. An engineer wants to build a building with a parallelogram base. He wants the four corners to be right angles and the four sides congruent. What type of base does the engineer want?

square

Circle the letter of the correct answer.

5. Each side of a quadrilateral-shaped picture frame has the same length. Which of the following is not a possible shape for the frame?
A a rhombus
B a square
(C) a trapezoid
D a parallelogram

6. The total length of the four sides of the picture frame from Exercise 5 is 4 feet, 8 inches. What is the length of each of its sides?
(F) 14 inches
G 1 foot, 3 inches
H 12 inches
J 2 inches

Lesson Quiz

Complete each statement.

1. A quadrilateral with four right angles is a ___?___.
square or rectangle

2. A parallelogram with four right angles and four congruent sides is a ___?___. square

3. A figure with 4 sides and 4 angles is a ___?___. quadrilateral

4. Give the most descriptive name for this quadrilateral. trapezoid

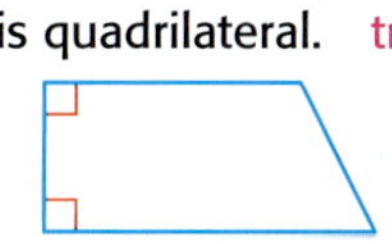

Available on Daily Transparency in CRB

7-7 Organizer

Pacing: Traditional 1 day
Block $\frac{1}{2}$ day

Objective: Students identify regular and not regular polygons and find the angle measures of regular polygons.

Indiana Academic Standards
6.4.4, 6.7.1, 6.7.2, 6.7.5, 6.7.11

Warm Up

True or false?

1. Some trapezoids are parallelograms. false
2. Some figures with 4 right angles are squares. true

Problem of the Day

Four square tables pushed together can seat either 8 or 10 people. How many people could 12 square tables pushed together seat?

14, 16, or 26 people

Available on Daily Transparency in CRB

Math Humor

What did the parrot owner say when he saw his birdcage door open and the cage empty? Polygon

7-7 Polygons

Learn to identify regular and not regular polygons and to find the angle measures of regular polygons.

Vocabulary
polygon
regular polygon

Triangles and quadrilaterals are examples of polygons. A **polygon** is a closed plane figure formed by three or more line segments. A **regular polygon** is a polygon in which all sides are congruent and all angles are congruent.

Polygons are named by the number of their sides and angles.

	Triangle	Quadrilateral	Pentagon	Hexagon	Octagon
Sides and Angles	3	4	5	6	8
Regular					
Not Regular					

EXAMPLE 1 Identifying Polygons

Tell whether each shape is a polygon. If so, give its name and tell whether it appears to be regular or not regular.

A

There are 4 sides and 4 angles.
quadrilateral
The sides and angles appear to be congruent.
regular

B

There are 4 sides and 4 angles.
quadrilateral
All 4 sides do not appear to be congruent.
not regular

The sum of the interior angle measures in a triangle is 180°, so the sum of the interior angle measures in a quadrilateral is 360°.

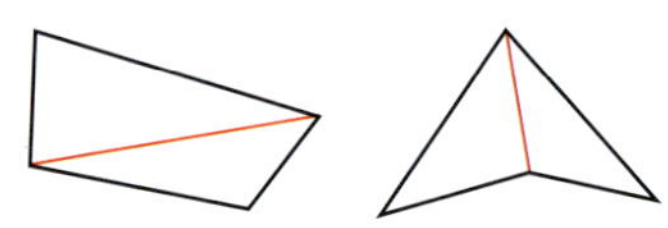

1 Introduce

Alternate Opener

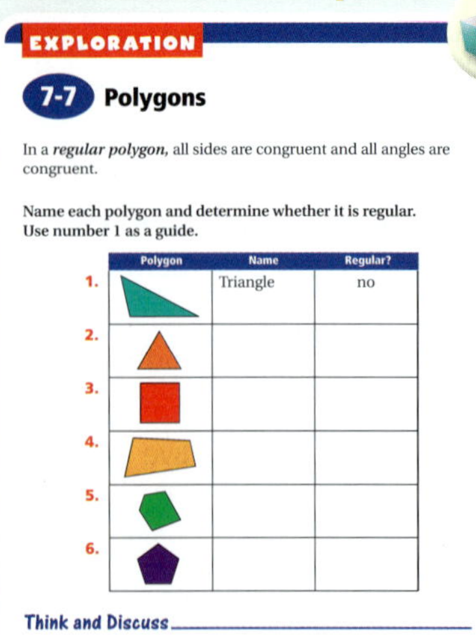
EXPLORATION

7-7 Polygons

In a *regular polygon,* all sides are congruent and all angles are congruent.

Name each polygon and determine whether it is regular. Use number 1 as a guide.

	Polygon	Name	Regular?
1.		Triangle	no
2.			
3.			
4.			
5.			
6.			

Think and Discuss

7. Explain how you classified each polygon in numbers 2–6.

Motivate

Have students draw closed plane figures with 3, 4, 5, 6, and 8 sides. Have students identify the figures they can name. They should at least be able to name *triangle* and *quadrilateral.* Tell them that in this lesson they will learn the names of 5-, 6-, and 8-sided polygons.

Exploration worksheet and answers on Chapter 7 Resource Book pp. 65 and 154

2 Teach

Lesson Presentation

Guided Instruction

In this lesson, students learn to identify regular and not regular polygons and to find the angle measures of regular polygons. First teach the difference between regular and not regular polygons and the names for 5-, 6-, and 8-sided polygons. You may use Teaching Transparency T30 in the Chapter 7 Resource Book. Then teach students to find the sum of the measures of the interior angles of an octagon.

Teaching Tip Direct students' attention to the Reading Math box explaining the polygon prefixes.

EXAMPLE 2 PROBLEM SOLVING APPLICATION

PROBLEM SOLVING

A stop sign is in the shape of a regular octagon. What is the measure of each angle of the stop sign?

1 Understand the Problem

The **answer** will be the measure of each angle in a regular octagon. List the **important information:**

- A regular octagon has 8 congruent sides and 8 congruent angles.

2 Make a Plan

Make a table to look for a pattern using regular polygons.

3 Solve

Draw some regular polygons and divide each into triangles.

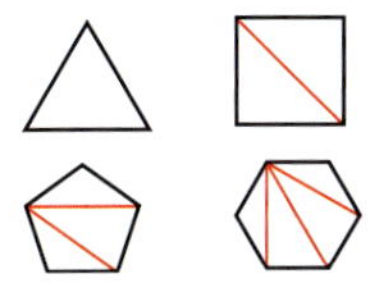

Polygon	Sides	Triangles	Sum of Angle Measures
Triangle	3	1	180°
Quadrilateral	4	2	2 × 180° = 360°
Pentagon	5	3	3 × 180° = 540°
Hexagon	6	4	4 × 180° = 720°

The number of triangles is always 2 fewer than the number of sides. An octagon can be divided into $8 - 2 = 6$ triangles.

The sum of the interior angle measures in an octagon is $6 \times 180° = 1{,}080°$.

So the measure of each angle is $1{,}080° \div 8 = 135°$.

4 Look Back

Each angle in a regular octagon is obtuse. 135° is a reasonable answer, because an obtuse angle is between 90° and 180°.

Reading Math

The prefixes in the names of the polygons tell you how many sides and angles there are.

tri- = three
quad- = four
penta- = five
hexa- = six
octa- = eight

Think and Discuss

1. **Classify** the angles in each figure: a regular triangle, a regular hexagon, and a rectangle.
2. **Name** an object that is in the shape of a pentagon and an object that is in the shape of an octagon.

COMMON ERROR ALERT

Some students may not understand the reasoning for finding the sums of angle measures in various polygons. Help them to see that the sum of the interior angles of the polygon is the same as the sum of all the angles in all the triangles in the polygon's interior.

Additional Examples

Example 1

Tell whether each shape is a polygon. If so, give its name and tell whether it appears to be regular or not regular.

A.

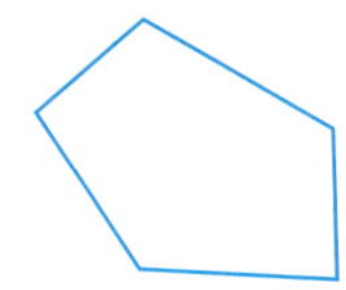

polygon
pentagon
not regular

B.

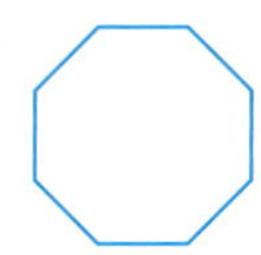

polygon
octagon
regular

Example 2

Malcolm designed a wall hanging that was a regular 9-sided polygon (called a *nonagon*). What is the measure of each angle of the nonagon? 140°

3 Close

Reaching All Learners
Through Number Sense

Have students research the names of other polygons. Have them find the sum of the angle measures for each polygon and the measure of each angle if the polygon is regular (to the nearest tenth of a degree).

Possible answers:
7-sided: heptagon; 900°; 128.6°
10-sided: decagon; 1,440°; 144°
11-sided: undecagon; 1,620°; 147.3°
12-sided: dodecagon; 1,800°; 150°
15-sided: pentadecagon; 2,340°; 156°

Summarize

Review the polygons in the lesson. Ask students to tell you the difference between a polygon that is regular and one that is not regular.

A regular polygon has all congruent sides and all congruent angles. A polygon that is not regular does not have all congruent sides and all congruent angles.

Answers to Think and Discuss

1. acute; obtuse; right
2. Possible answers: the front of a small birdhouse; a stop sign

7-7 PRACTICE & ASSESS

Students may want to refer back to the lesson examples.

Assignment Guide

If you finished Example 1 assign:
Core 1–3, 5–7, 9–18, 23–29
Enriched 5–19, 23–29

If you finished Example 2 assign:
Core 1–19, 23–29
Enriched 4–29

Notes

7-7 Exercises

FOR EXTRA PRACTICE
see page 650

internet connect
Homework Help Online
go.hrw.com Keyword: MR4 7-7

GUIDED PRACTICE

See Example 1

Tell whether each shape is a polygon. If so, give its name and tell whether it appears to be regular or not regular.

1.  polygon, hexagon, regular

2.  polygon, quadrilateral, not regular

3.  polygon, triangle, regular

See Example 2

4. A flower bed in the park is in the shape of a rhombus. The distance around the flower bed is 160 meters. What is the length of one side of the flower bed? **40 meters**

INDEPENDENT PRACTICE

See Example 1

Tell whether each shape is a polygon. If so, give its name and tell whether it appears to be regular or not regular.

5. 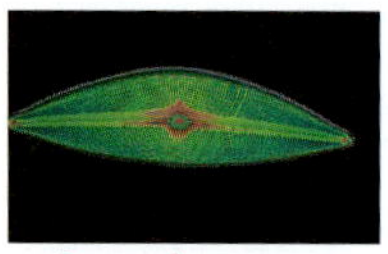 not a polygon

6.  polygon, triangle, regular

7. 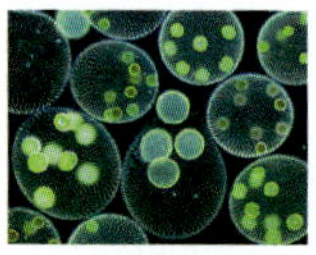 not a polygon

See Example 2

8. Janet made a sign for her room in the shape of a regular pentagon. What is the measure of each angle of the pentagon? **108°**

PRACTICE AND PROBLEM SOLVING

Explain why each shape is *not* a polygon.

9. 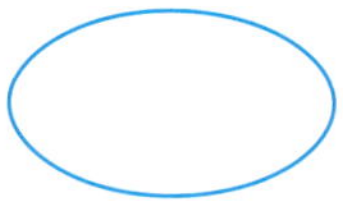 not formed by line segments

10.  not a closed figure

11.  not formed by line segments

Name each polygon.

12.  octagon

13. 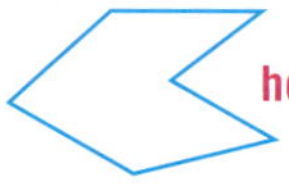 hexagon

14.  pentagon

Math Background

If you extend a side of a polygon past a vertex, you will form an exterior angle. The sum of the measures of the interior and exterior angle at that vertex is 180°. For a polygon with n sides, the sum of all the interior and exterior angle pairs is $180n$. The sum of the measures of the interior angles of a polygon with n sides is $180(n - 2)$. Thus, the sum of the exterior angles, one at each vertex, of a polygon of n sides is $180n - 180(n - 2) = 180n - 180n + 360 = 360$.

RETEACH 7-7

LESSON 7-7 **Reteach**
Polygons

A polygon is a closed plane figure formed by three or more line segments.

Polygons are named by the number of their sides and angles.

Sides and Angles	3	4	5	6	8
Polygon	triangle	quadrilateral	pentagon	hexagon	octagon

A regular polygon has all congruent sides and angles.

Each angle is 90 degrees. Each side is 3 centimeters long.

If the angles and the sides are not all congruent, then the polygon is not regular.

Name each polygon and tell whether it is regular or not regular.

1.  regular

2.  not regular

To find the angle measures of a regular hexagon, first divide the polygon into triangles by drawing connecting segments from one vertex to all of its nonadjacent vertices.

There are 4 triangles.

Next, multiply the number of triangles by 180.

$4 \cdot 180 = 720$ The sum of the angle measures of a triangle is 180°. Divide the product by 6, the number of sides, to find the measure of each angle.

$\frac{720}{6} = 120$ All of the angles of a regular polygon are congruent.

Each angle is 120°.

Find the angle measure of each regular polygon.

3. regular quadrilateral 90°
4. regular pentagon 108°

PRACTICE 7-7

LESSON 7-7 **Practice B**
Polygons

Name each polygon and tell whether it appears to be *regular* or *not regular*.

1.  triangle; not regular

2.  hexagon; regular

3.  octagon; regular

4.  quadrilateral; regular

5.  pentagon; not regular

6. 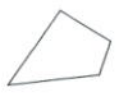 quadrilateral; not regular

7. The public swimming pool is in the shape of a regular hexagon. Each side of the pool measures 5 feet. What is the distance around the entire pool?

30 feet

8. In the space below, draw a regular quadrilateral. Now draw one diagonal of that quadrilateral. Describe the two polygons that are formed.

They are both triangles.

Classify each of the following polygons as either *always* regular, *sometimes* regular, or *never* regular.

		Always	Sometimes	Never
15.	Equilateral triangle	?	?	?
16.	Trapezoid	?	?	?
17.	Right triangle	?	?	?
18.	Parallelogram	?	?	?

MEASUREMENT A *diagonal* is a line segment that connects two nonadjacent vertices of a polygon. One diagonal is shown in each figure.

19. a. How many diagonals does a rectangle have? 2

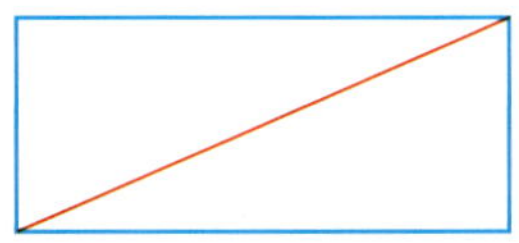

b. How many diagonals does a pentagon have? 5

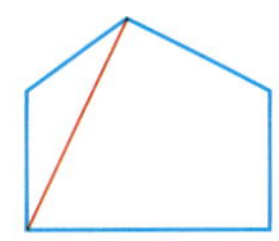

20. ***WHAT'S THE ERROR?*** A student said a rectangle is never a regular polygon because the lengths of all the sides are not congruent. What error did the student make? Explain why a rectangle is sometimes a regular polygon.

21. ***WRITE ABOUT IT*** What polygon is formed when two equilateral triangles are placed side by side, with one upside down? Draw examples, and explain whether the polygon formed by the two triangles is regular.

22. ***CHALLENGE*** A figure is formed by placing 6 equilateral triangle tiles around a regular hexagon tile. The distance around the regular hexagon is 60 cm. A snail moves along the sides of the figure. How far will the snail travel until it gets back to its starting point? 120 cm

Spiral Review

Use the pattern to write the first five terms of the sequence. (Lesson 1-7)

23. Start with 10; add 3.
10, 13, 16, 19, 22

24. Start with 4; multiply by 4.
4, 16, 64, 256, 1,024

Write each decimal as a fraction in simplest form. (Lesson 4-4)

25. 0.9 $\frac{9}{10}$ **26.** 0.71 $\frac{71}{100}$ **27.** 0.20 $\frac{1}{5}$ **28.** 0.88 $\frac{22}{25}$

29. **TEST PREP** Which is the least common multiple (LCM) of 4, 6, and 12? (Lesson 5-5) C

A 2 **B** 3 **C** 12 **D** 24

Answers

15. always

16. never

17. never

18. sometimes

20. A square is always a rectangle, and a square has congruent sides and angles, so it is a regular polygon.

21. Possible answer: Two equilateral triangles form a parallelogram. It is not regular because not all angles are congruent.

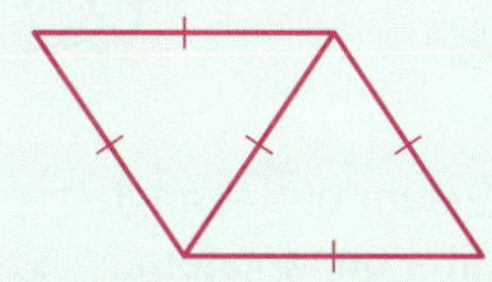

Journal

Have students explain the difference between polygons that are regular and polygons that are not regular.

Test Prep Doctor

For Exercise 29, encourage students to eliminate incorrect answers. **A** is the GCF, not the LCM. **B** is a common factor of two of the numbers, not the LCM. **D** is a common multiple, but not the *least* common multiple. **C** is correct because 12 is the smallest number that is a multiple of 4, 6, and 12.

CHALLENGE 7-7

LESSON 7-7 **Challenge**
Sign Language

Traffic signs are symbols used to convey information. In the United States, certain shapes are used only for certain traffic signs. As a result, you can often understand the message of a traffic sign simply by classifying its geometric shape.

Tell whether each traffic sign below is a polygon. If it is, name the polygon that best describes its shape, and tell whether it appears to be *regular* or *not regular*.

1. 

yes, octagon, regular

2. 

not a polygon

3. 

yes, triangle, regular

4. 
yes, pentagon, not regular

5. 

yes, quadrilateral, not regular

6. 

not a polygon

7. 

yes, trapezoid, not regular

8. 
yes, quadrilateral, regular

9. 

yes, triangle, not regular

PROBLEM SOLVING 7-7

LESSON 7-7 **Problem Solving**
Polygons

Write the correct answer.

1. Name each polygon in this figure.
1, 2, 4, 5, and 6 are triangles; 3 is a parallelogram; 7 is a square (some students may name the entire figure a square)

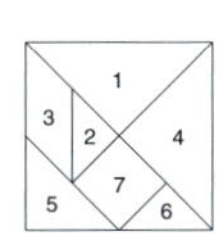

2. How could you use the sum of the angles inside a triangle to find the sum of the angles inside a heptagon?
Five triangles are needed to fill a heptagon; $5 \times 180° = 900°$. So, the sum of the angles inside a heptagon is 900°.

3. How could you use the sum of the angles inside a triangle to find the sum of the angles inside a decagon?
Eight triangles are needed to fill a decagon; $8 \times 180° = 1{,}440°$. So, the sum of the angles inside a decagon is 1,440°.

4. In the space below, draw a rectangle and a parallelogram with side lengths congruent to the rectangle's. Now draw the diagonals for each of those polygons. What new polygons are formed by the diagonals in each quadrilateral?

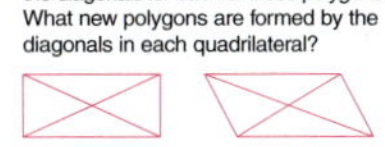

triangles

5. In Exercise 4, what is true of the diagonals in the rectangle that isn't true of the diagonals of the parallelogram?
The diagonals in the rectangle are congruent.

Circle the letter of the correct answer.

6. The total length of the sides of a regular hexagon is $13\frac{1}{2}$ inches. What is the length of each side?
A $2\frac{7}{10}$ inches
(B) $2\frac{1}{4}$ inches
C $3\frac{3}{4}$ inches
D $1\frac{11}{16}$ inches

7. Which of the following statements is sometimes false?
(F) A plane figure is a polygon.
G Each side of a polygon intersects exactly two other sides.
H A polygon is a closed figure.
J A polygon has straight sides.

Lesson Quiz

1. Name each polygon and tell whether it appears to be regular or not regular.

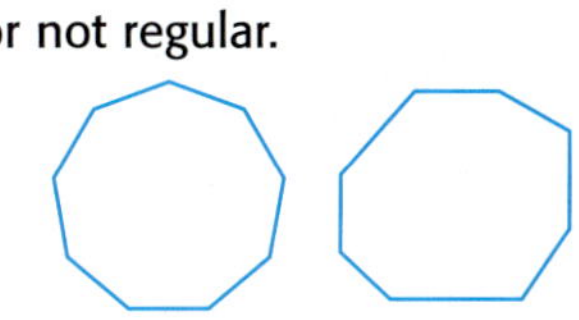

nonagon, regular; octagon, not regular

2. What is the measure of each angle in a regular dodecagon (12-sided figure)? 150°

Available on Daily Transparency in CRB

7-8 Organizer

Pacing: Traditional 1 day
Block $\frac{1}{2}$ day

Objective: Students recognize, describe, and extend geometric patterns.

Indiana Academic Standards
6.7.1, 6.7.2, 6.7.5, 6.7.11

Warm Up

Divide.

1. What is the sum of the angle measures in a quadrilateral? 360°
2. What is the sum of the angle measures in a hexagon? 720°
3. What is the measure of each angle in a regular octagon? 135°

Problem of the Day

Which three letters come next in the following series: *W, T, L, C, N, I, . . .*
T, F, S; the letters are the initial letters of the words in the question.

Available on Daily Transparency in CRB

Math Fact

The use of geometric patterns can be found in works of art and craft throughout history. For example, Egyptian bowls and Etruscan vases dating back more than 2,000 years incorporate geometric patterns.

7-8 Geometric Patterns

Learn to recognize, describe, and extend geometric patterns.

Native American art often involves geometric patterns. The patterns are based on the shape, color, size, position, or number of geometric figures.

This blanket has a geometric pattern. The first row with a complete figure has a parallelogram with a horse in its center. The next row has two parallelograms with cows in the centers. This pattern continues. If the weaver wanted to make a longer blanket, the next row would be two parallelograms with pictures of cows.

This Navajo blanket was made in the late seventeenth century.

EXAMPLE 1 **Extending Geometric Patterns**

Identify a possible pattern. Use the pattern to draw the next figure.

A

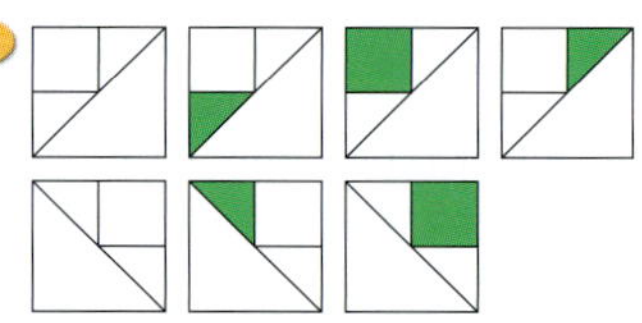

The small shapes within the figure are shaded one at a time from bottom to top. Then the figure is rotated and the top triangle is shaded.

So the next figure might look like this:

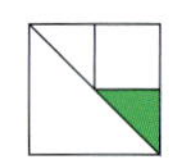

B

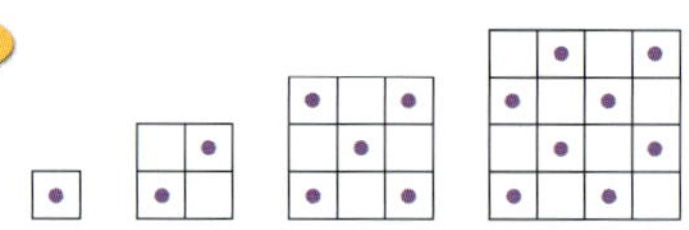

The figures from left to right are a 1 × 1 square, a 2 × 2 square, a 3 × 3 square, and a 4 × 4 square.

So the next figure might look like this:

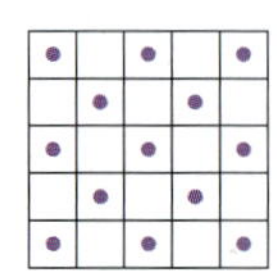

Remember!

Perfect squares, such as 2^2, 3^2, and 4^2, are also called "square numbers" because they can be modeled as a square array.

1 Introduce

Alternate Opener

EXPLORATION

7-8 Geometric Patterns

Identify the pattern, and draw the next three figures in the sequence.

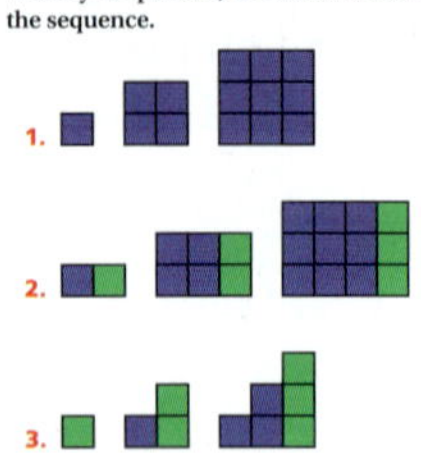

Think and Discuss

4. **Explain** how the sequence in number 2 is built on the sequence in number 1.
5. **Describe** in words the sequence in number 3.

Motivate

Discuss possible rules for each number pattern below, and have students find the next two numbers in each pattern.

Possible answers:

1, 3, 5, 7, 9, . . . Add 2; 11, 13.

5, 10, 15, 20, 25, . . . Add 5; 30, 35.

4, 14, 9, 19, 14, 24, . . . Add 10, and then subtract 5; 19, 29.

Have students make up other patterns.

Exploration worksheet and answers on Chapter 7 Resource Book pp. 74 and 156

2 Teach

Lesson Presentation

Guided Instruction

In this lesson, students learn to recognize, describe, and extend geometric patterns. Teach students to describe patterns, to use the descriptions to extend the patterns, and to identify missing elements in the patterns. Encourage students to give other possible ways to identify each pattern, because there is not just one way to describe a pattern.

Teaching Tip Point out patterns that are different from those that simply repeat (e.g., A, B, A, B , . . .). Example 2 illustrates the concept of a growing pattern.

EXAMPLE 2 Completing Geometric Patterns

Identify a possible pattern. Use the pattern to draw the missing figure.

A

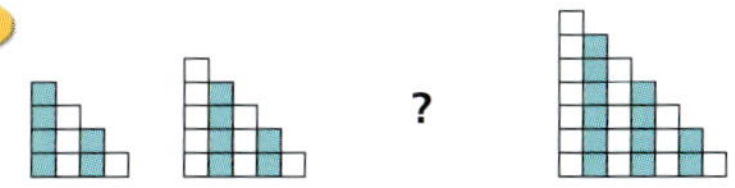

The first figure from the bottom row to the top has 4 squares and then 3, 2, and 1 square. The next figure has 5 squares in the bottom row and then 4, 3, 2, and 1.

So the missing figure might look like this:

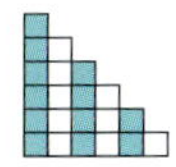

B

Each figure is an equilateral triangle. The first figure has 3 red triangles along the base. The third figure has 5 red triangles, and the last figure has 6.

So the missing figure might look like this:

EXAMPLE 3 *Art Application*

Dan is painting a clay pot. Identify a pattern in Dan's design and tell what the finished pot might look like.

The pattern from bottom to top is narrow stripe, wide stripe, narrow stripe, wide stripe. The color pattern from bottom to top is blue, green, yellow, blue, green.

If this pattern is followed, the finished pot might look like the pot at left.

Think and Discuss

1. **Explain** how you can use a pattern to find the number of squares in the next, or fifth, figure in Example 2A.
2. **Tell** how you can use a pattern to find the number of small red triangles in the sixth figure in Example 2B.

Additional Examples

Example 1

Identify a possible pattern. Use the pattern to draw the next figure.

A.

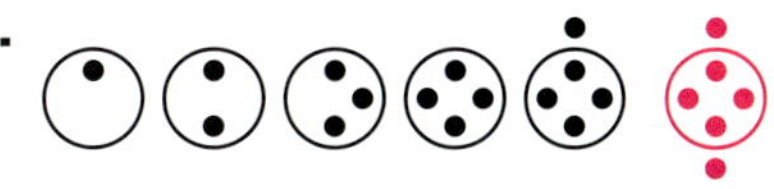

B.

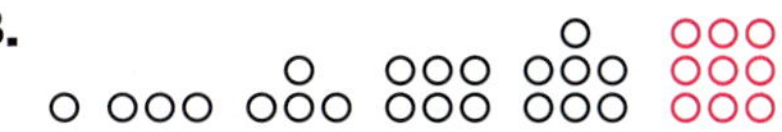

Example 2

Identify a possible pattern. Use the pattern to draw the missing figure.

A.

B.

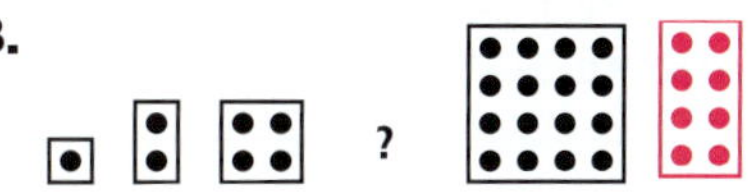

Example 3

Travis is painting a platter. Identify a pattern that Travis is using and draw what the finished platter might look like.

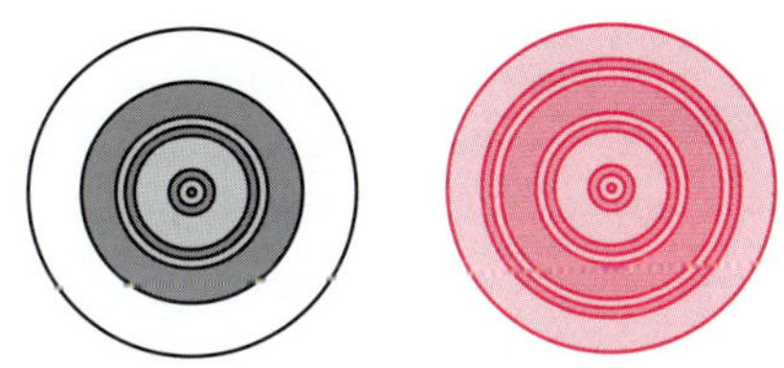

3 Close

Reaching All Learners
Through Modeling

Have students build geometric patterns with pattern blocks (Manipulatives Kit). Ask them to draw and describe their patterns. If time allows, share patterns with the class, and have students tell what figure might come next in the pattern.

Summarize

Have students make patterns that follow a given rule. For example, each figure is twice as wide as the figure to its left, or each figure has one more triangle than the one before it.

Check students' patterns.

Answers to Think and Discuss

1. Possible answer: Each figure has one more square in the bottom row than the one to its left. So the fifth figure will have from the bottom row to the top the following number of squares 8, 7, 6, 5, 4, 3, 2, and finally 1.
2. Possible answer: The second figure has 4 more red triangles than the first, the third has 5 more than the second, and the fourth has 6 more than the third. So the fifth figure would have 7 more red triangles than the fourth ($21 + 7 = 28$), and the sixth figure would have 8 more than the fifth ($28 + 8 = 36$ red triangles).

7-8 PRACTICE & ASSESS

Students may want to refer back to the lesson examples.

Assignment Guide

If you finished Example 1 assign:
Core 1, 4, 7, 12–15
Enriched 1, 4, 7–8, 11–15

If you finished Example 2 assign:
Core 1–2, 4–5, 7, 9, 12–15
Enriched 1–2, 4–5, 7–9, 11–15

If you finished Example 3 assign:
Core 1–9, 12–15
Enriched 3–15

Answers

1. Possible answer: Rotate figure 90° clockwise.

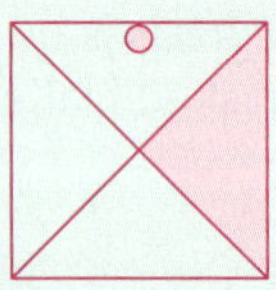

2. Possible answer: Move dot and triangle one position counterclockwise.

3. Possible answer: purple, purple, red, yellow, green, yellow; red, yellow, green, yellow, purple

4–5, 7. See next page.

7-8 Exercises

FOR EXTRA PRACTICE see page 650

internet connect
Homework Help Online
go.hrw.com Keyword: MR4 7-8

GUIDED PRACTICE

See Example 1 Identify a possible pattern. Use the pattern to draw the next figure.

1.

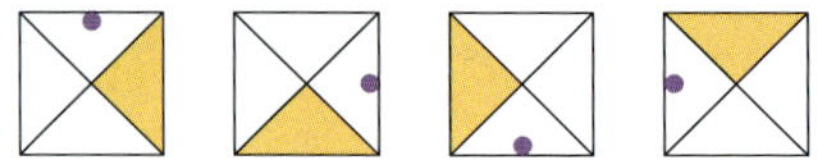

See Example 2 Identify a possible pattern. Use the pattern to draw the missing figure.

2.

See Example 3 **3.** Oscar is making a beaded necklace. Identify a pattern in Oscar's design. Then tell which five beads Oscar will probably use next.

INDEPENDENT PRACTICE

See Example 1 Identify a possible pattern. Use the pattern to draw the next figure.

4.

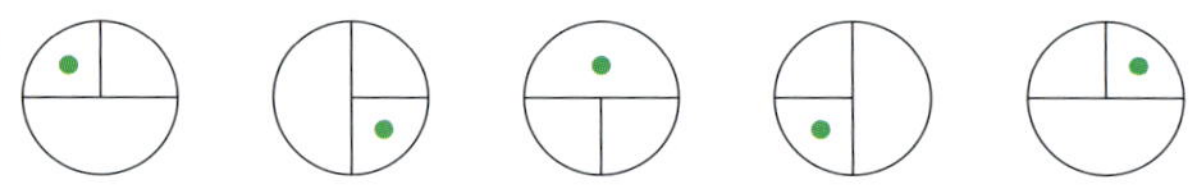

See Example 2 Identify a possible pattern. Use the pattern to draw the missing figure.

5.

See Example 3 **6.** Tamara is planting flowers in her garden. She makes groups of purple flowers and groups of pink flowers.

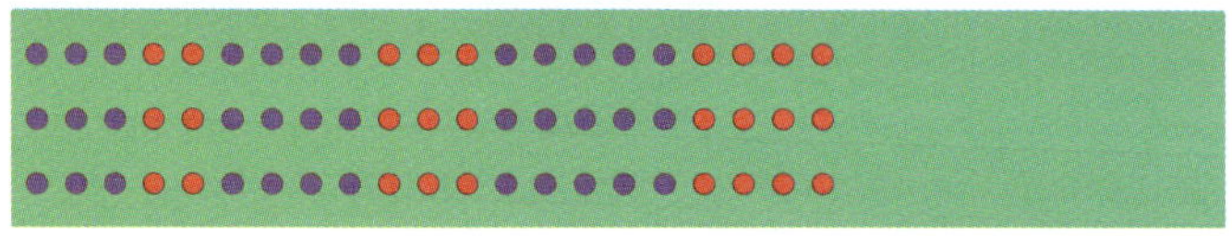

If she continues this pattern, how many flowers might Tamara plant in the next group of purple flowers? How many flowers might she plant in the next group of pink flowers? **18 purple flowers; 15 pink flowers**

PRACTICE AND PROBLEM SOLVING

Draw the next figure in the pattern.

7.

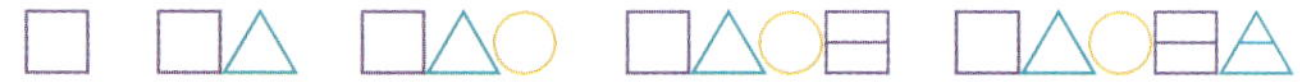

Math Background

A formula for the sum of the first n counting numbers can be found with the help of geometric patterns. For example, the sum of $1 + 2 + 3 + 4$ can be represented as a pattern of squares (below left). The diagram on the right shows the sum twice. The two copies of the sum fit together to form a rectangle containing $4 \cdot 5 = 20$ squares. So the sum equals $\frac{4 \cdot 5}{2} = 10$.

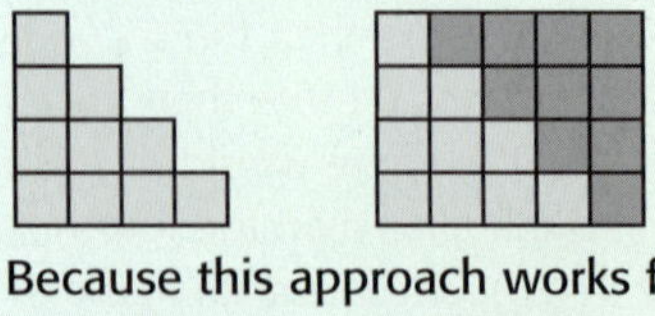

Because this approach works for any value of n, the sum of the first n counting numbers equals $\frac{n(n+1)}{2}$.

RETEACH 7-8

LESSON 7-8 **Reteach**
Geometric Patterns

Sometimes, patterns involve geometric shapes and figures. The pattern can be based on shape, color, size, position, or quantity.

Before you draw a missing figure in a pattern, you first have to look at the other figures to recognize a relationship among the figures.

In the pattern above, each figure has one more side than the figure that proceeds it.

To continue the pattern, the next figure would be a hexagon.

Identify a pattern for each. Then draw the missing figure.

1.

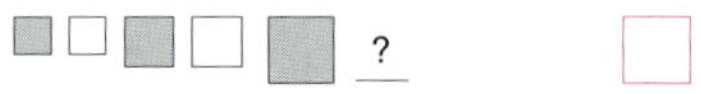

2.

3.

PRACTICE 7-8

LESSON 7-8 **Practice B**
Geometric Patterns

Identify a possible pattern. Use the pattern to draw the next figure.

1.

Pattern: A bottom row is added to each figure with 1 more dot than the row above it.

2.

Pattern: One more square is shaded than the figure before it.

3.

Pattern: Repeating groups of 1 right triangle, 1 equilateral triangle, and 1 scalene triangle

4.

Pattern: One more diagonal is drawn from the same vertex in each hexagon in the pattern.

5. Use triangles to create a geometric pattern. Describe your pattern.

Accept all geometric patterns using triangles. Students' descriptions should match their patterns.

Social Studies LINK

The art of decorating houses is practiced throughout Africa. In South Africa, Ndebele and Basotho people paint their houses with brightly colored patterns made up of geometric shapes.

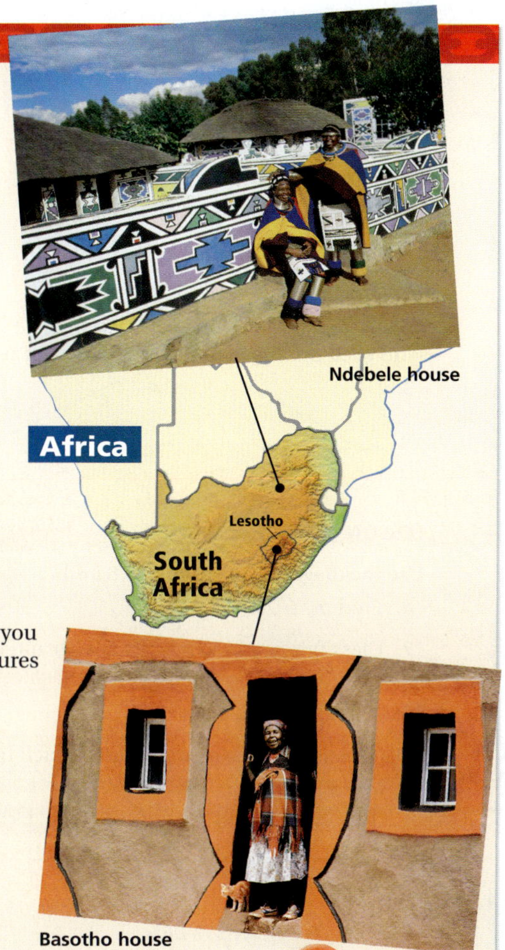

Basotho house

8. Look at the shapes found on the wall surrounding the Ndebele house. Identify a possible pattern that was used to paint the top band of the wall. Use the pattern to draw the shapes hidden by the Ndebele people. (You do not need to make color part of the pattern.)

9. Look at the design on the Basotho house. Identify a possible pattern. Use the pattern to draw a picture of what the house might look like.

10. ***WRITE ABOUT IT*** Look closely at the Ndebele house. Draw four geometric figures you see painted on the house. Then use those figures to make a pattern. Describe your pattern.

11. ***CHALLENGE*** Look at the designs below, which were made using an African motif. Identify a possible pattern. If the pattern continues, how many motifs will be in the sixth design? If there are 45 motifs, what will the design number be? **28 motifs, Design 8**

go.hrw.com
KEYWORD: MR4 Patterns
CNN Student News

Spiral Review

Write each expression in exponential form. (Lesson 4-2)

12. $3 \times 3 \times 3 \times 5 \times 5$ $3^3 \times 5^2$

13. $7 \times 7 \times 4 \times 4$ $7^2 \times 4^2$

14. $2 \times 2 \times 2 \times 2 \times 3 \times 3 \times 5$ $2^4 \times 3^2 \times 5$

15. **TEST PREP** Multiply $\frac{4}{5} \cdot \frac{3}{5}$. Write the answer in simplest form. (Lesson 5-1) **C**

A $1\frac{2}{5}$ **B** $\frac{3}{5}$ **C** $\frac{12}{25}$ **D** $1\frac{1}{3}$

Answers

4. Possible answer: Rotate figure 90° clockwise, and move dot to next clockwise section.

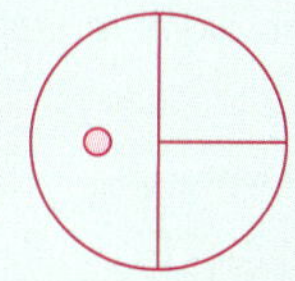

5. Possible answer: The number of objects doubles each time.

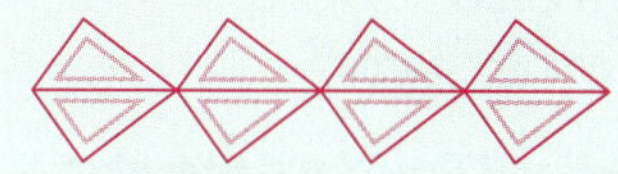

7. 

8. Possible answer: A pattern is triangle, white stripe, upside down triangle, white stripe, triangle, etc.

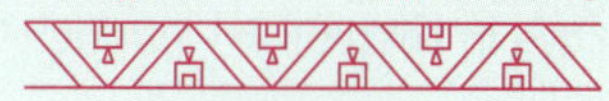

9. Possible answer:

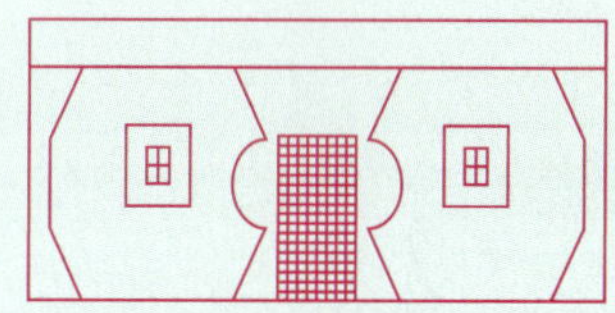

10. Check that students' work includes figures from the house.

Journal

Have students make up their own geometric patterns and describe each pattern that they make. Encourage them to make both repeating and growing patterns.

Test Prep Doctor

For Exercise 15, encourage students to estimate the answer before looking at the answer choices. Because $1 \cdot \frac{1}{2}$ can be used to estimate $\frac{4}{5} \cdot \frac{3}{5}$, the answer will be close to $\frac{1}{2}$. This eliminates choices **A** and **D** right away. Because **B** is one of the factors, that choice can also be eliminated.

CHALLENGE 7-8

LESSON 7-8 **Challenge**
Polygon Patterns

Look for patterns to complete this chart and discover two rules for polygons.

Regular Polygon	Number of Sides, n	Number of Triangles to Fill	Sum of the Interior Angles
Triangle	$n = 3$	1	1(180°) = 180°
Quadrilateral	$n = 4$	2	2(180°) = 360°
Pentagon	$n = 5$	3	3(180°) = 540°
Hexagon	$n = 6$	4	4(180°) = 720°

1. Use the patterns to write an expression for the number of triangles needed to fill any regular polygon.
   $n - 2$
2. Use the patterns to write an expression for the sum of all the angles inside any regular polygon.
   $(n - 2)180°$
3. Using these expressions, how many triangles are needed to fill any regular octagon? What is the sum of a regular octagon's interior angles?
   6; 1,080°

PROBLEM SOLVING 7-8

LESSON 7-8 **Problem Solving**
Geometric Patterns

Complete this chart and look for patterns. Then answer the questions. Possible answers are given for 1–4.

	Number of Points on the Line	Draw and Label the Line and Points	Number of Different Line Segments in the Line
	2	A B	1
1.	3	A B C	3
2.	4	A B C D	6
3.	5	A B C D E	10
4.	6	A B C D E F	15

Circle the letter of the correct answer.

5. If n = the number of points on a line, which of the following expressions shows the number of different line segments on that line?
   A $2n - 3$
   B (circled) $(n^2 - n) \div 2$
   C $(n \div 2) \cdot 5$
   D $10n \div 2$

6. Using the pattern in the table and your answer to Exercise 5, how many different line segments will be on a line if there are 10 points on the line?
   F 17 line segments
   G 25 line segments
   H (circled) 45 line segments
   J 50 line segments

Lesson Quiz

Identify a possible pattern. Use the pattern to draw the next figure.

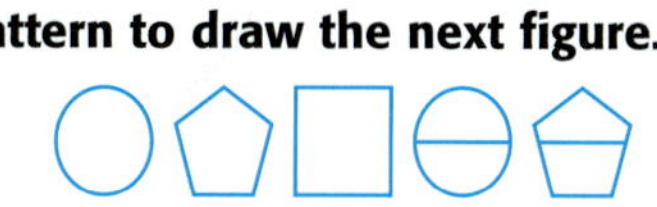

Possible answer: 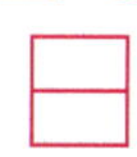

Available on Daily Transparency in CRB

Chapter 7 Mid-Chapter Quiz

Purpose: *To assess students' mastery of concepts and skills in Lessons 7-1 through 7-8*

Assessment Resources

Section 7B Quiz

Assessment Resources . . . . p. 19

***Test and Practice Generator* CD-ROM**

Additional mid-chapter assessment items in both multiple-choice and free-response format may be generated for any objective in Lessons 7-1 through 7-8.

Answers

16. The shaded part of the triangle is moving clockwise around each section.

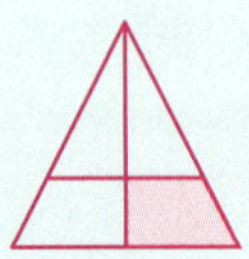

17. Divide each square into four smaller squares.

Chapter 7 Mid-Chapter Quiz

LESSON 7-1 (pp. 322–325)

Use the diagram for problems 1 and 2.   Possible answers given.

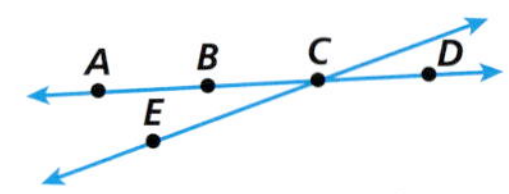

1. Name two lines. $\overleftrightarrow{AD}$, $\overleftrightarrow{CE}$
2. Name three different rays. $\overrightarrow{BA}$, $\overrightarrow{BD}$, and $\overrightarrow{CE}$

LESSON 7-2 and 7-3 and 7-4 (pp. 326–339)

Find each unknown angle measure, and tell what type of angle it is. Classify lines ℓ and m.

3. 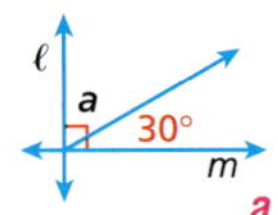

$a = 60°$; acute; perpendicular

4. 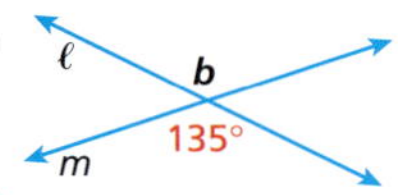

$b = 135°$; obtuse; intersecting

5. 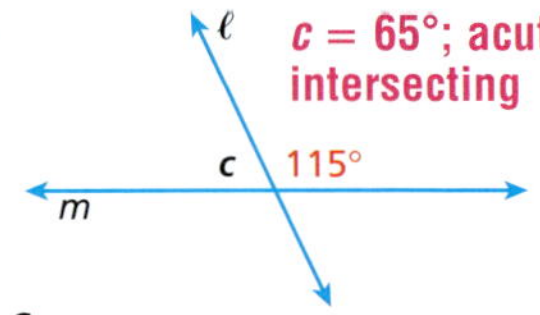

$c = 65°$; acute; intersecting

LESSON 7-5 (pp. 344–347)

Use the diagram for problems 6 and 7.

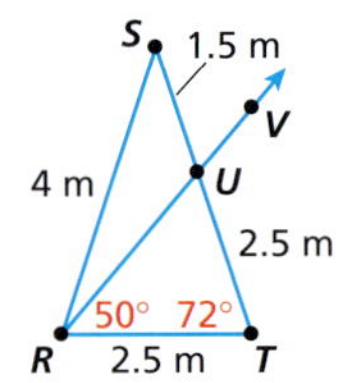

6. Find m∠*SUV*. 58°
7. Classify triangle *STR* by its angles and by its sides. acute isosceles triangle

LESSON 7-6 (pp. 348–351)

Give the most descriptive name for each figure.

8. 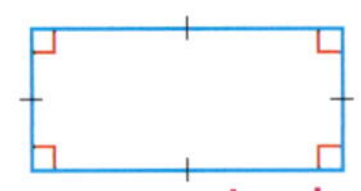 rectangle
9. 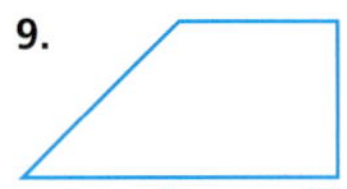 trapezoid
10. 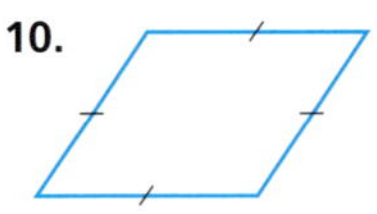 rhombus
11. 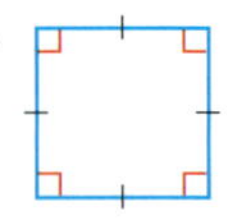 square

LESSON 7-7 (pp. 352–355)

Name each polygon, and tell whether it appears to be regular or not regular.

12. 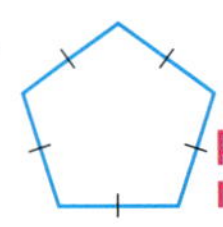 pentagon; regular
13.  hexagon; not regular
14. 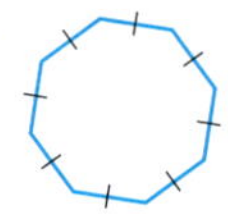 octagon; regular
15. 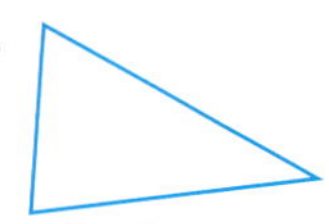 triangle; not regular

LESSON 7-8 (pp. 356–359)

Identify a possible pattern. Use the pattern to draw the missing figure.

16. 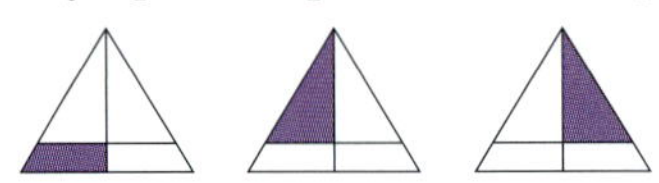

17. 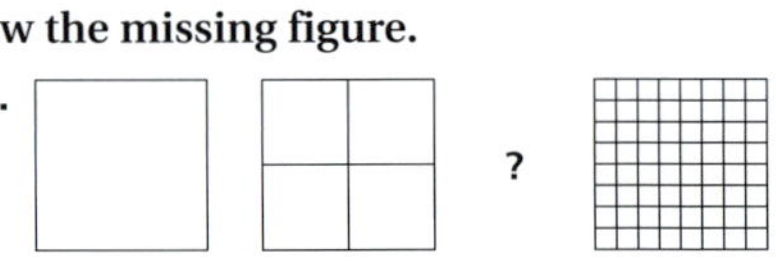

Focus on Problem Solving

Make a Plan

- **Draw a diagram**

Sometimes a problem seems difficult because it is described in words only. You can draw a diagram to help you picture the problem. Try to label all the information you are given on your diagram. Then use the diagram to solve the problem.

Read each problem. Draw a diagram to help you solve the problem. Then solve.

1. Bob used a ruler to draw a quadrilateral. First he drew a line 3 in. long and labeled it $\overline{AB}$. From B, he drew a line 2 in. long and labeled the endpoint C. From A, he drew a line $2\frac{1}{2}$ in. long and labeled the endpoint D. What is the length of $\overline{CD}$ if the perimeter of Bob's quadrilateral is $12\frac{1}{2}$ in?

2. Karen has a vegetable garden that is 12 feet long and 10 feet wide. She plans to plant tomatoes in one-half of the garden. She will divide the other half of the garden equally into three beds, where she'll grow cabbage, pumpkin, and radishes.
   - **a.** What are the possible whole number dimensions of the tomato bed?
   - **b.** What fraction of the garden will Karen use to grow cabbage?

3. Pam draws three parallel lines that are an equal distance apart. The two outside lines are 8 cm apart. How far apart is the middle line from the outside lines?

4. Jan connected the following points on a coordinate grid: (2, 4), (4, 6), (6, 6), (6, 2), (3, 2), and (2, 4).
   - **a.** What figure did Jan draw?
   - **b.** How many right angles does the figure have?

5. Triangle ABC is isoceles. The measure of angle B is equal to the measure of angle C. The measure of angle B equals 50°. What is the measure of angle A?

Answers

1. 5 inches

2. a. 6 ft × 10 ft or 5 ft × 12 ft
b. $\frac{1}{6}$

3. 4 cm

4. a. pentagon
b. 2

5. 80°

Focus on Problem Solving

Purpose: *To focus on making a plan to help solve the problem by drawing a diagram*

Problem Solving Resources

Interactive Problem Solving pp. 53–64

Math: Reading and Writing in the Content Area . . . . pp. 53–64

Problem Solving Process

This page focuses on the second step of the problem-solving process: **Make a Plan**

Discuss

Possible answer: For problems 1–5, have students identify what must be labeled in each diagram, and then draw a diagram to help them solve the problem.

1. names of vertices; side lengths

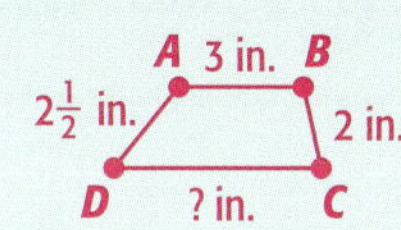

2. length and width of garden; length and width of each bed

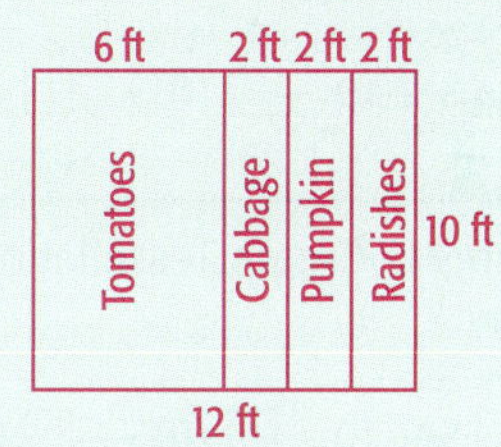

3. distance between outside lines

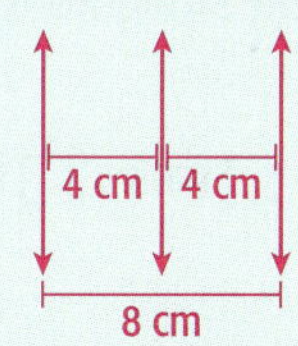

4. names of points

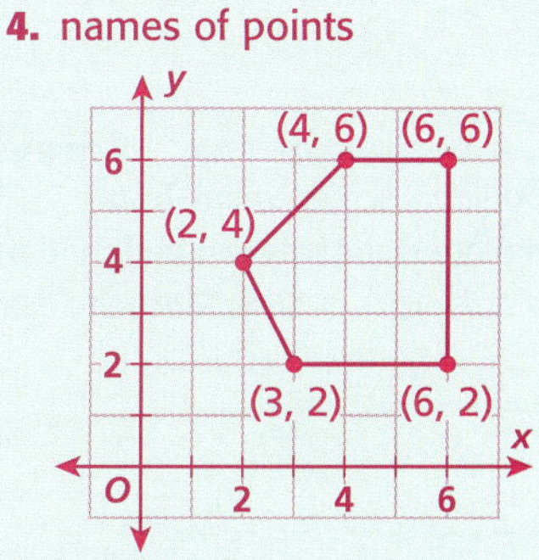

5. measures of ∠B and ∠C

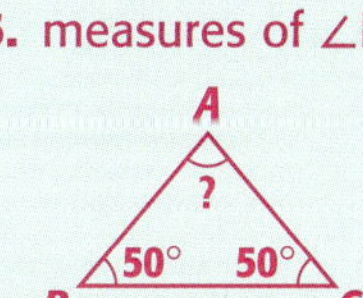

Section 7C

Polygon Relationships

One-Minute Section Planner

Lesson	Materials	Resources
Lesson 7-9 Congruence **NCTM:** Geometry, Communication **NAEP:** Geometry 2e ☐ SAT-9 ☐ SAT-10 ☐ ITBS ☑ CTBS ☐ MAT ☐ CAT	**Optional** Inch ruler *(MK, CRB, p. 131)* Cutout polygons *(CRB, p. 135)* Graph paper *(CRB, p. 136)*	• *Chapter 7 Resource Book,* pp. 83–91 • Daily Transparency T39, CRB • Additional Examples Transparencies T40–T42, CRB • *Alternate Openers: Explorations,* p. 61
Lesson 7-10 Transformations **NCTM:** Geometry, Communication **NAEP:** Geometry 2c ☑ SAT-9 ☑ SAT-10 ☐ ITBS ☑ CTBS ☑ MAT ☑ CAT	**Required** Recording Sheet for Exercises 4, 5, 9–14, 17 *(CRB, p. 137)* **Optional** Teaching Transparency T44 *(CRB)* Graph paper *(CRB, p. 136)*	• *Chapter 7 Resource Book,* pp. 92–101 • Daily Transparency T43, CRB • Additional Examples Transparencies T45–T46, CRB • *Alternate Openers: Explorations,* p. 62
Lesson 7-11 Symmetry **NCTM:** Geometry, Communication **NAEP:** Geometry 2a ☑ SAT-9 ☑ SAT-10 ☐ ITBS ☑ CTBS ☐ MAT ☐ CAT	**Required** Recording Sheet for Exercises 4–8 and 12–16 *(CRB, p. 138)* **Optional** Teaching Transparency T48 *(CRB)*	• *Chapter 7 Resource Book,* pp. 102–110 • Daily Transparency T47, CRB • Additional Examples Transparencies T49–T51, CRB • *Alternate Openers: Explorations,* p. 63
Lesson 7-12 Tessellations **NCTM:** Geometry, Communication, Connections **NAEP:** Geometry 2d ☑ SAT-9 ☑ SAT-10 ☐ ITBS ☐ CTBS ☑ MAT ☑ CAT	**Optional** Teaching Transparency T53 *(CRB)*	• *Chapter 7 Resource Book,* pp. 111–119 • Daily Transparency T52, CRB • Additional Examples Transparencies T54–T55, CRB • *Alternate Openers: Explorations,* p. 64
Hands-On Lab 7C Create Tessellations **NCTM:** Geometry **NAEP:** Geometry 2d ☐ SAT-9 ☐ SAT-10 ☐ ITBS ☐ CTBS ☐ MAT ☐ CAT	**Required** Scissors *(MK)* Tape	• *Hands-On Lab Activities,* pp. 59–60
Extension Compass and Straightedge Construction **NCTM:** Geometry **NAEP:** Geometry 3b ☐ SAT-9 ☐ SAT-10 ☐ ITBS ☐ CTBS ☐ MAT ☐ CAT	**Required** Recording Sheet for Exercises 4, 5 *(CRB, p. 139)* Compass *(MK)* Straightedge *(MK)*	• Additional Examples Transparencies T56–T58, CRB
Section 7C Assessment		• Section 7C Quiz, AR p. 20 • *Test and Practice Generator* CD-ROM

SAT = *Stanford Achievement Tests* **ITBS** = *Iowa Test of Basic Skills* **CTBS** = *Comprehensive Test of Basic Skills/Terra Nova*
MAT = *Metropolitan Achievement Tests* **CAT** = *California Achievement Test*

NCTM = Complete standards can be found on pages T27–T33. **NAEP** = Complete standards can be found on pages A31–A35.

SE = *Student Edition* **TE** = *Teacher's Edition* **AR** = *Assessment Resources* **CRB** = *Chapter Resource Book* **MK** = *Manipulatives Kit*

Section Overview

Congruence — Lesson 7-9

Why? Manufacturers use congruent figures in mass production.

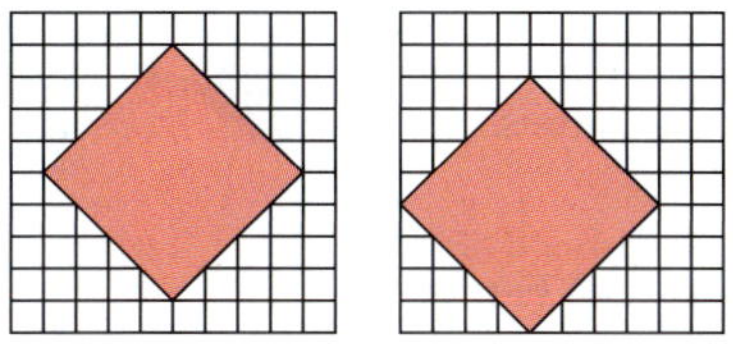

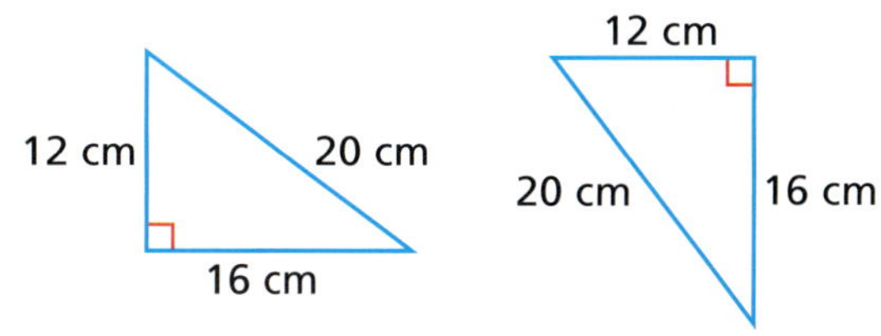

Congruent figures have the same shape and size.

Transformations — Lesson 7-10

Why? Transformations can be used to alter the size, shape, or position of geometric figures.

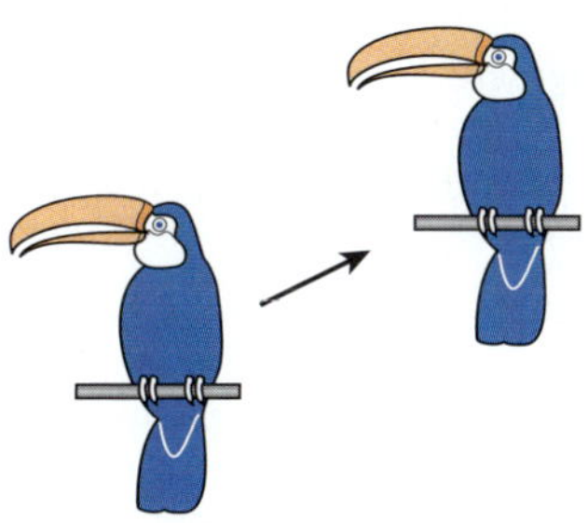

Translation

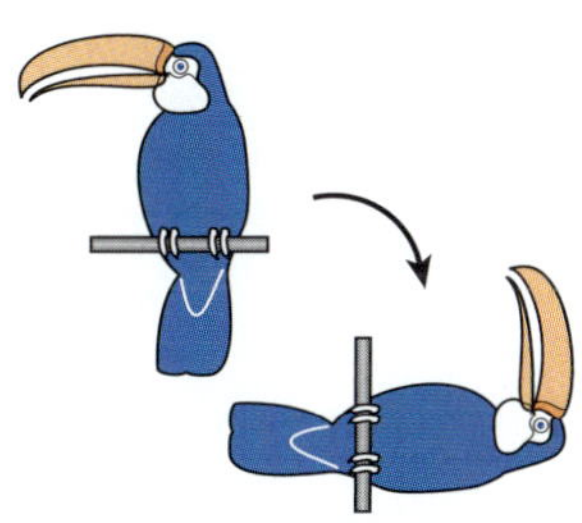

Rotation

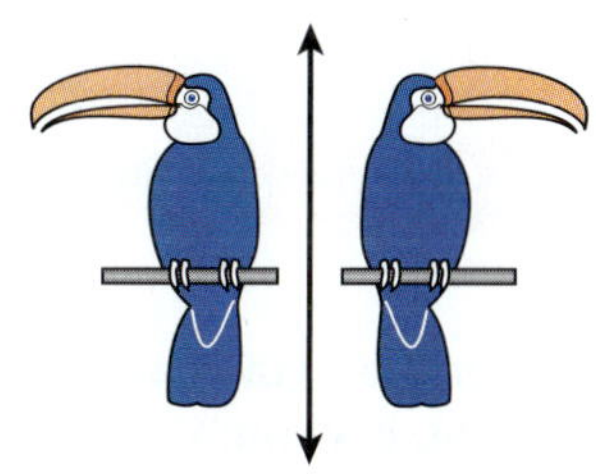

Reflection

Symmetry, Tessellations — Lessons 7-11, 7-12

Why? Tessellations are used in works of art.

A figure has **line symmetry** if it can be folded or reflected so that the two parts of the figures match, or are congruent. The line of reflection is called the **line of symmetry.**

A tessellation

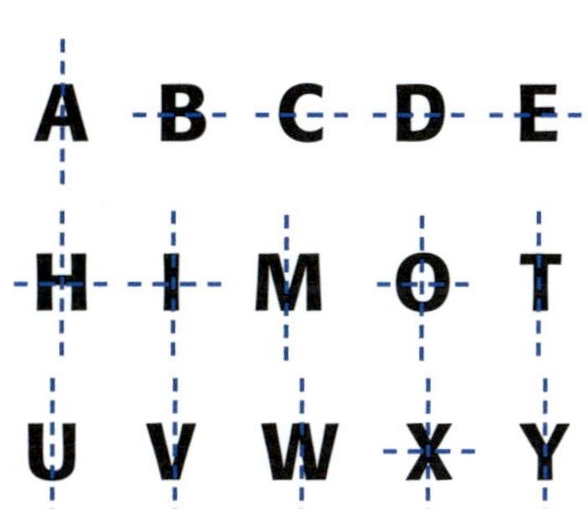

Lines of Symmetry

7-9 Organizer

Pacing: Traditional 1 day
Block $\frac{1}{2}$ day

Objective: Students identify congruent figures and use congruence to solve problems.

Indiana Academic Standards
6.7.1, 6.7.2, 6.7.5, 6.7.10, 6.7.11

Warm Up

1. A straight angle has what degree measure? 180°
2. How do you name a line? Name any two points on a line.

Problem of the Day

The sum of two decimals is 9.3; their difference is 4.3, and their product is 17.00. What are they? 2.5, 6.8

Available on Daily Transparency in CRB

Additional Example

Example 1

Decide whether the figures are congruent. If not, explain.

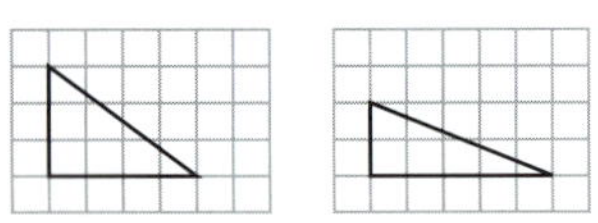

no; triangles are not the same size

7-9 Congruence

Learn to identify congruent figures and to use congruence to solve problems.

You know that angles that have the same measure are congruent. Figures that have the same shape and same size are also congruent.

You can use stencils to decorate pages of a scrapbook. The stencil helps you draw congruent figures.

EXAMPLE 1 **Identifying Congruent Figures**

Decide whether the figures in each pair are congruent. If not, explain.

A

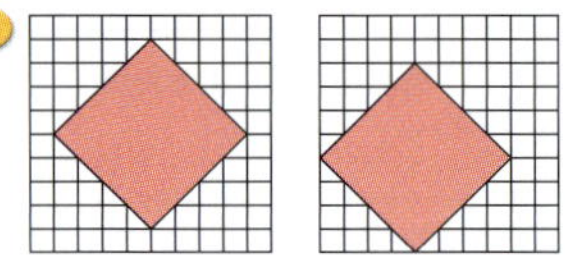

The figures are congruent.

These figures have the same shape and size.

B

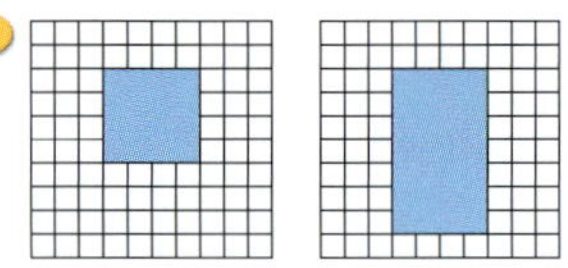

The figures are not congruent.

These figures are both quadrilaterals. But they are neither the same size nor the same shape.

C

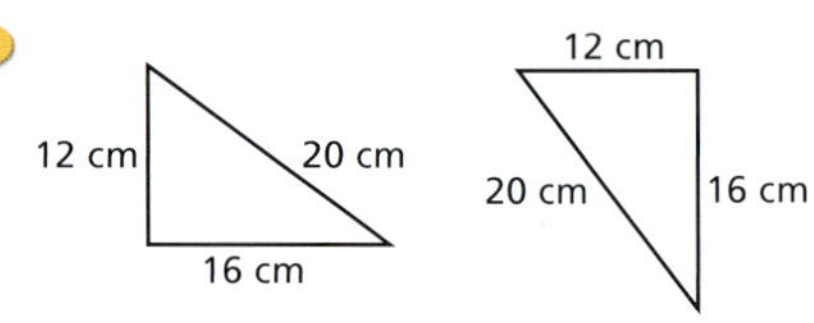

The triangles are congruent.

Each triangle has a 12 cm side, a 16 cm side, and a 20 cm side.

D

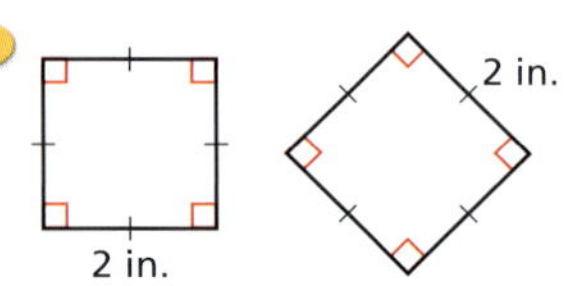

The figures are congruent.

Each figure is a square. Each side of each square measures 2 inches.

1 Introduce

Alternate Opener

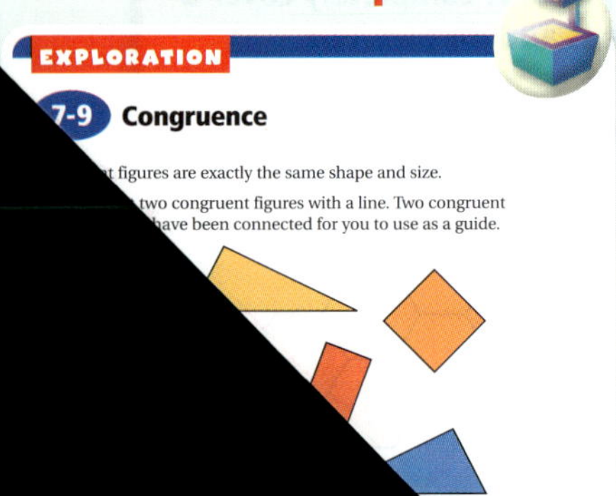

Motivate

Provide students with pairs of cutout polygons (provided on Chapter 7 Resource Book p. 135). Some of the polygon pairs should be congruent, and some should be the same shape but not the same size. Have students match the polygons that are the same size and shape. Explain that polygons, like line segments and angles, can be congruent.

Exploration worksheet and answers on Chapter 7 Resource Book pp. 84 and 158

2 Teach

Lesson Presentation

Guided Instruction

In this lesson, students learn to identify congruent figures and to use congruence to solve problems. Explain that congruent figures need not be oriented in the same way. Have students draw the figures in Examples 1A and 1B on graph paper (provided on Chapter 7 Resource Book p. 136) and cut them out to test congruence. Discuss reasons for needing to know whether figures are congruent, as illustrated in Example 2.

EXAMPLE 2 ***Consumer Application***

Landra needs a ground cloth that is congruent to the tent floor. Which ground cloth should she buy?

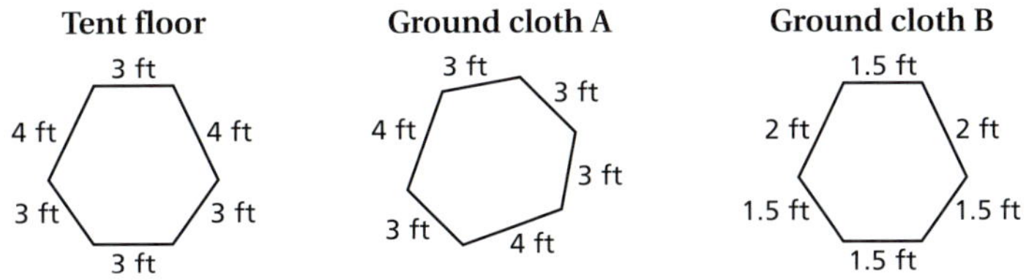

Which ground cloth is the same size and shape as the tent floor?
Both cloths are hexagons. Only Cloth A is the same size as the floor.

Cloth A is congruent to the tent floor.

Think and Discuss

1. **Explain** whether you can determine that figures are congruent just by looking at them.
2. **Tell** what information you would need to know about two rectangles to determine whether they are congruent.

Additional Examples

Example 2

Jodi needs a sleeping pad that is congruent to her sleeping bag. Which pad should she buy?

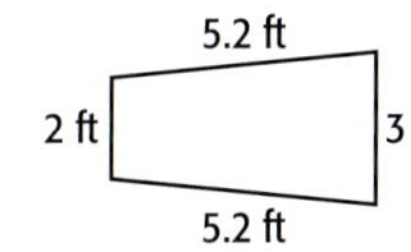

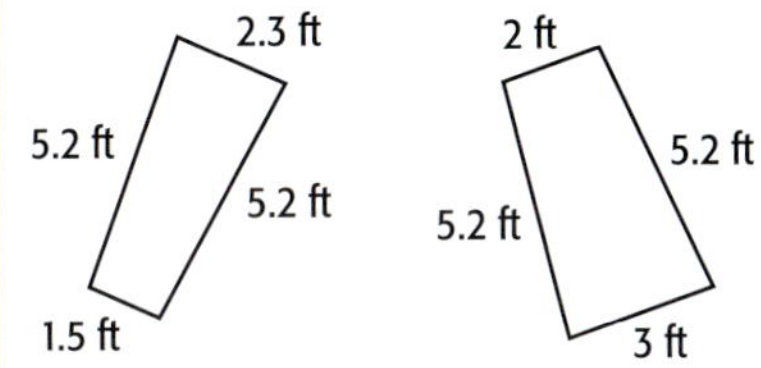

sleeping pad B

7-9 Exercises

FOR EXTRA PRACTICE
see page 651

internet connect
Homework Help Online
go.hrw.com Keyword: MR4 7-9

7-9 PRACTICE & ASSESS

Students may want to refer back to the lesson examples.

GUIDED PRACTICE

See Example 1 Decide whether the figures in each pair are congruent. If not, explain.

1. 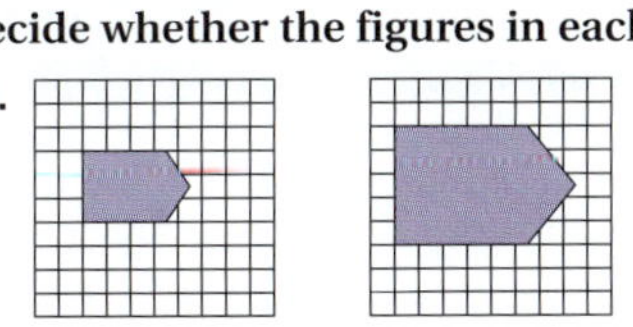

not congruent; different sizes

2. 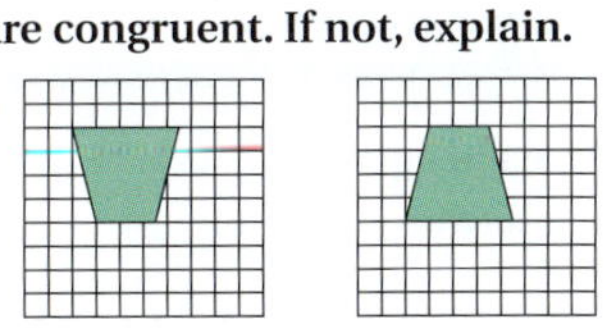

congruent

See Example 2 3. Which quadrilateral is congruent to the bottom of the box? **Figure A**

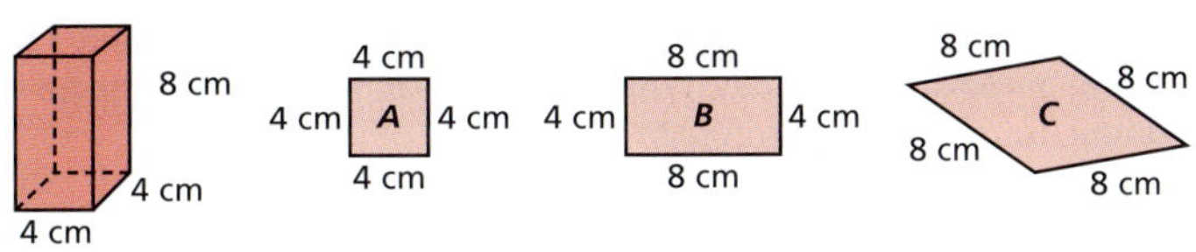

Assignment Guide

If you finished Example 1 assign:
Core 1, 7, 11–17
Enriched 1–2, 4–5, 8, 11–17

If you finished Example 2 assign:
Core 1–8, 11–17
Enriched 3–17

3 Close

Summarize

Ask students how they can tell whether two figures are congruent.

Possible answer: Two figures are congruent if they are the same shape and the same size.

Reaching All Learners
Through Grouping Strategies

Give groups of students sheets of graph paper (provided on Chapter 7 Resource Book p. 136). The student leader of each group draws a triangle, a quadrilateral, and a pentagon on his or her paper. Then the other group members draw figures that are congruent. Have the group discuss why some, if any, of the figures are not congruent. This continues until each group member has had a chance to be the leader.

Answers to Think and Discuss

1. Possible answer: No; two figures may look like they are the same size, but you cannot be certain unless you measure the corresponding parts of the figures.
2. Possible answer: If the lengths and widths of two rectangles are equal, then the rectangles are congruent.

Lesson Quiz

True or false?

1. If two figures have sides of the same length, they are congruent. false
2. Congruent figures have angle measures that are equal. true
3. Decide whether the figures are congruent. If not, explain.

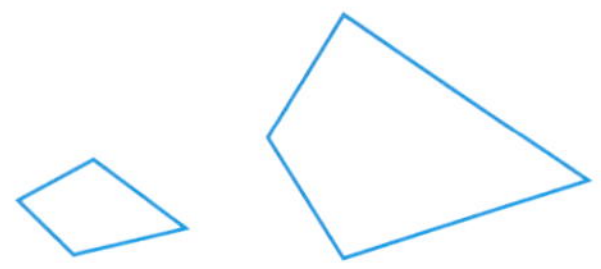

No, both are quadrilaterals but they are not the same size or shape.

Available on Daily Transparency in CRB

Answers

7–8. See p. A5.

INDEPENDENT PRACTICE

See Example 1 Decide whether the figures in each pair are congruent. If not, explain.

4. 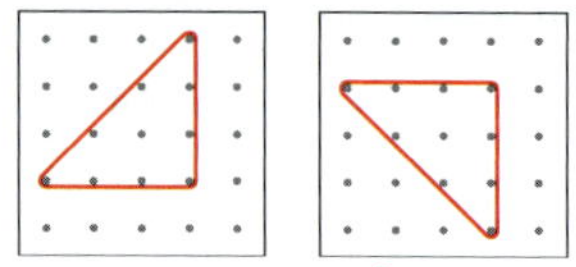

congruent

5. 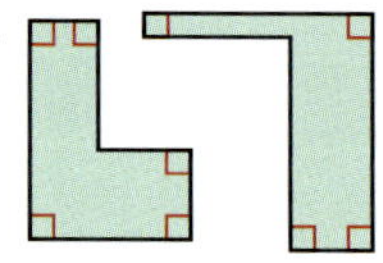

The figures are irregular hexagons that are not congruent.

See Example 2 6. Which puzzle piece will fit into the empty space? Piece a

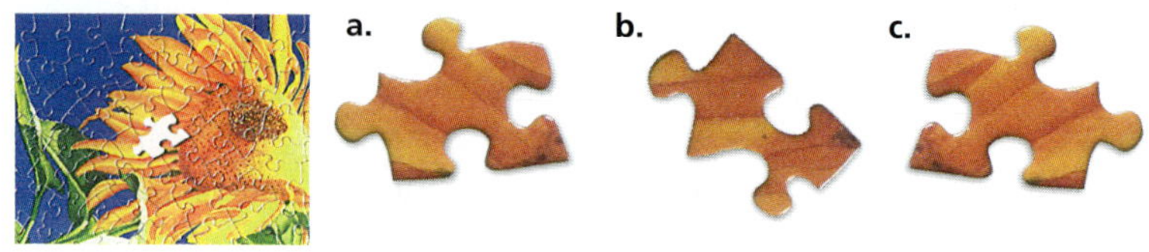

PRACTICE AND PROBLEM SOLVING

7. Copy the dot grid. Then draw three figures congruent to the given figure. The figures can have common sides but should not overlap.

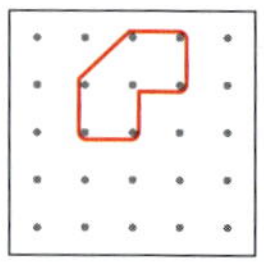

8. ***MEASUREMENT*** Use an inch ruler to draw two congruent rectangles with side lengths that are longer than 2 in. and shorter than 6 in. Label each side length.
9. ***WRITE ABOUT IT*** Explain how to tell whether two polygons are congruent. Possible answer: You can measure the sides and angles to make sure they are the same size and shape.
10. ***CHALLENGE*** Two quadrilaterals have side lengths 2 cm, 2 cm, 5 cm, and 5 cm. Are the two quadrilaterals congruent? Explain. Not necessarily; One may be a rectangle with 90° angles and one may be a parallelogram with angles less than and greater than 90°.

Spiral Review

Solve each equation. (Lessons 2-6 and 2-7)

11. $5t = 45$ $t = 9$ **12.** $72 = 3n$ $n = 24$ **13.** $\frac{s}{6} = 8$ $s = 48$

Find the GCF of each set of numbers. (Lesson 4-3)

14. 12, 18, 24 6 **15.** 15, 18, 30 3 **16.** 16, 24, 42 2

17. **TEST PREP** Last week Leo spent 2 hours, 3 hours, 2 hours, and 1 hour doing homework each day. Which is the mean amount of time he spent doing homework? (Lesson 6-2) C

A 8 hours **B** $2\frac{1}{2}$ hours **C** 2 hours **D** $1\frac{1}{2}$ hours

RETEACH 7-9

LESSON 7-9 Reteach
Congruence

Two figures are congruent if they have the same size and shape.

Look at the two figures.

They are congruent because they are both triangles and they are the same size.

Now look at these two figures.

These figures are not congruent. They are both rectangles but they are not the same size.

Decide whether the figures in each pair are congruent. If not, explain.

1. 

The figures are congruent.

2. 

The figures are not congruent; they are not the same shape.

3. 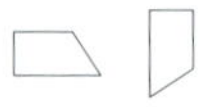

The figures are congruent.

PRACTICE 7-9

LESSON 7-9 Practice B
Congruence

Decide whether the figures in each pair are congruent. If not, explain.

1. 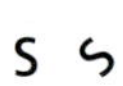

congruent

2. 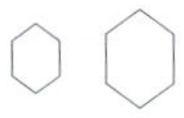

not congruent; they have different sizes

3. 

not congruent; they have different lengths

4. 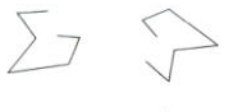

congruent

Use the diagram for Exercises 5–7.

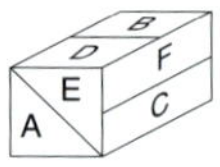

5. Which part of the figure is congruent to A? E
6. Which part of the figure is congruent to D? B
7. Which part of the figure is congruent to F? C
8. Name two parts of your body that appear to be congruent.

Possible answers: eyes, ears, arms, legs, feet, hands

9. Square *ABCD* is congruent to square *FGHJ*. The total length of the sides of square *ABCD* is 12 meters. What is the length of each side of square *FGHJ*?

3 meters

CHALLENGE 7-9

LESSON 7-9 Challenge
Tangram

Tangram is an ancient Chinese puzzle made by cutting a square into 7 pieces—2 congruent small triangles, 2 congruent large triangles, 1 medium triangle, a square, and a parallelogram. To solve a Tangram, you need to rearrange those seven pieces to exactly reproduce a given image.

The seven Tangram puzzle pieces are shown below. Trace the pieces and cut them out. Then use them to solve each Tangram that follows.

1. 

2. 

PROBLEM SOLVING 7-9

LESSON 7-9 Problem Solving
Congruence

Write the correct answer.

1. Similar figures have the same shape but may have different sizes. How are similar figures different from congruent figures?

Congruent figures must have the same shape and size.

2. Pentagon A and Pentagon B are congruent regular polygons. If the total length of the sides of Pentagon B is 68.5 feet, what is the length of each side of Pentagon A?

13.7 feet

3. Is the following statement always true, sometimes true, or never true? Two congruent figures are similar figures. Explain.

Always true; Possible answer: Congruent figures have the same shape, which is the definition of similar figures.

4. Draw a figure congruent to this line segment. Explain how you drew your congruent figure.

I measured the line segment and drew my segment the same length.

5. Which word makes this statement true? Corresponding parts of congruent figures are ________.

congruent

6. If two angles of a right triangle are congruent, what are the measures of each angle in the triangle?

45°, 45°, and 90°

Circle the letter of the correct answer.

7. Which of the following polygons do not always have all congruent sides?
   - **A** a square
   - **B** an equilateral triangle
   - **C** a rhombus
   - **(D)** a pentagon
8. If ∠*A* of rectangle *ABCD* is congruent to ∠*X* of triangle *XYZ*, which of these statements must be true?
   - **F** Rectangle *ABCD* is also a square.
   - **(G)** Triangle *XYZ* is a right triangle.
   - **H** Rectangle *ABCD* is a regular polygon.
   - **J** Triangle *XYZ* is an acute triangle.

7-10 Transformations

Learn to use translations, reflections, and rotations to transform geometric shapes.

Vocabulary
- transformation
- translation
- rotation
- reflection
- line of reflection

A rigid **transformation** moves a figure without changing its size or shape. So the original figure and the transformed figure are always congruent.

The illustrations of the alien show three transformations: a translation, a rotation, and a reflection. Notice the transformed alien does not change in size or shape.

A **translation** is the movement of a figure along a straight line.

Only the location of the figure changes with a translation.

A **rotation** is the movement of a figure around a point. A point of rotation can be on or outside a figure.

The location and position of a figure can change with a rotation.

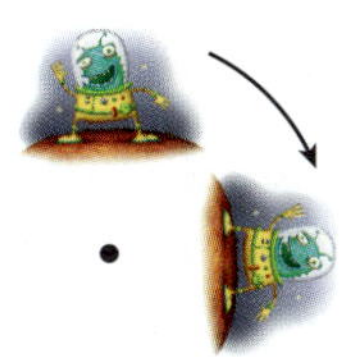

When a figure flips over a line, creating a mirror image, it is called a **reflection**. The line the figure is flipped over is called the **line of reflection**.

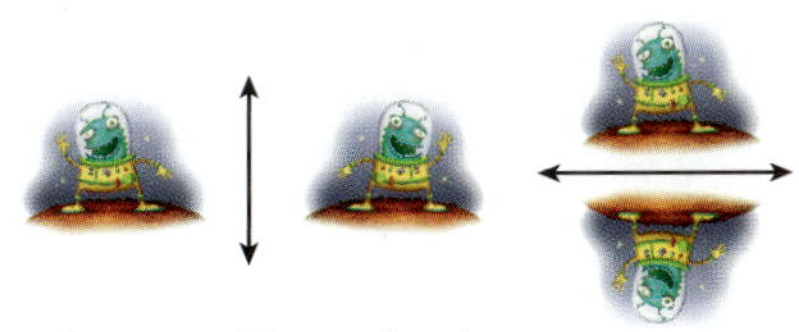

The location and position of a figure change with a reflection.

EXAMPLE 1 Identifying Transformations

Tell whether each is a translation, rotation, or reflection.

A

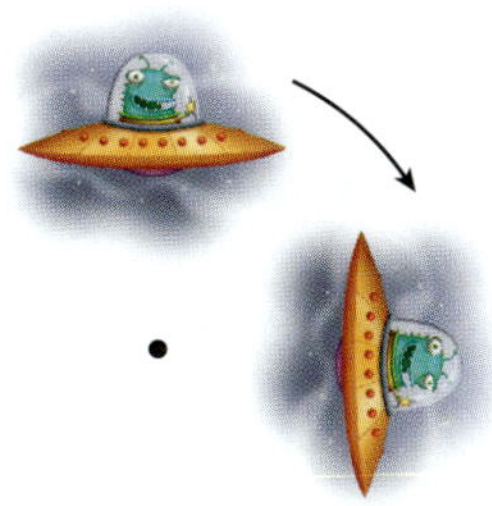

The figure moves around a point.

It is a rotation.

7-10 Organizer

Pacing: Traditional 1 day
Block $\frac{1}{2}$ day

Objective: Students use translations, reflections, and rotations to transform geometric shapes.

Indiana Academic Standards
6.4.6, 6.7.5, 6.7.11

Warm Up

Tell whether the figures described are congruent.

1. two triangles, each with sides measuring 24 cm, 32 cm, and 40 cm congruent
2. a square with sides of length 22 cm and a rectangle with side lengths 12 cm and 11 cm not congruent

Problem of the Day

Imagine that each figure is folded so that point A lies on point B. What figures would be formed?

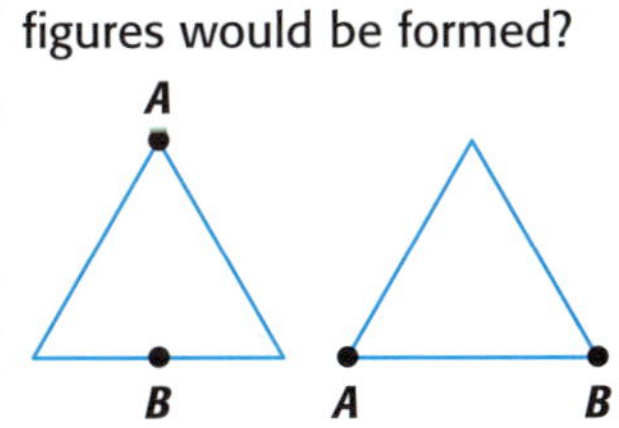

trapezoid, right triangle

Available on Daily Transparency in CRB

1 Introduce

Alternate Opener

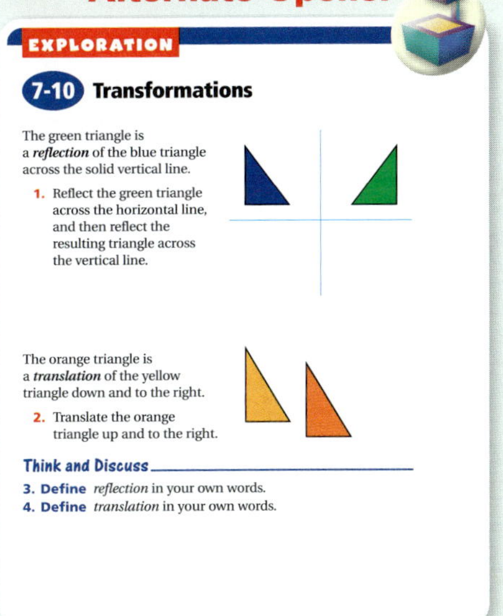
EXPLORATION

7-10 Transformations

The green triangle is a *reflection* of the blue triangle across the solid vertical line.

1. Reflect the green triangle across the horizontal line, and then reflect the resulting triangle across the vertical line.

The orange triangle is a *translation* of the yellow triangle down and to the right.

2. Translate the orange triangle up and to the right.

Think and Discuss

3. **Define** *reflection* in your own words.
4. **Define** *translation* in your own words.

Motivate

To introduce students to transformations, show them the pictures on Teaching Transparency T44 in the Chapter 7 Resource Book, and have them explain the relationship between the figures in each.

Possible answers: The triangles are on the same line as each other. The hearts are mirror images of each other. The cat heads are turned in different directions.

Exploration worksheet and answers on Chapter 7 Resource Book pp. 93 and 160

2 Teach

Lesson Presentation

Guided Instruction

In this lesson, students learn to use translations, reflections, and rotations to transform geometric shapes. Teach students first to identify transformations and then to draw transformations.

Teaching Tip Have students trace and cut out the figures shown in the examples so they can physically perform the transformations shown.

Review the meaning of *clockwise* and *counterclockwise* for Example 2.

Additional Examples

Example 1

Tell whether each is a translation, rotation, or reflection.

A.

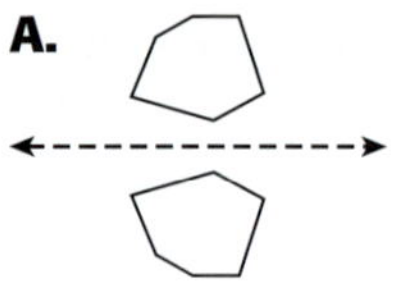

B.

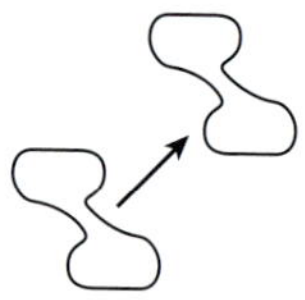

reflection translation

C.

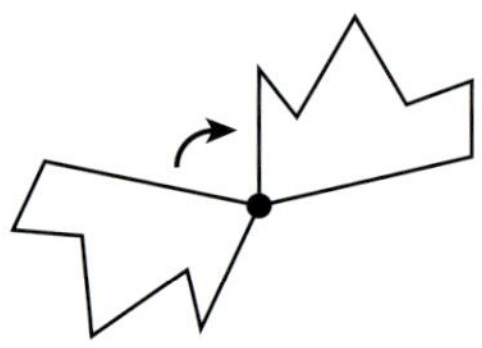

rotation

Example 2

Draw each transformation.

A. Draw a 180° rotation about the point shown.

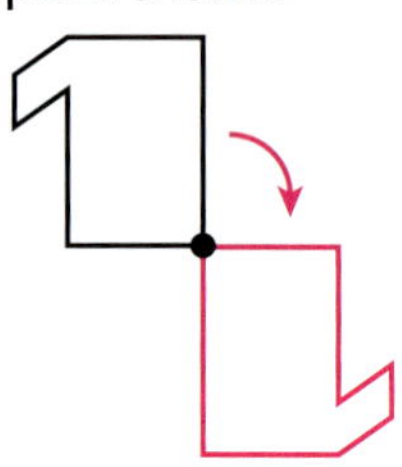

B. Draw a horizontal reflection.

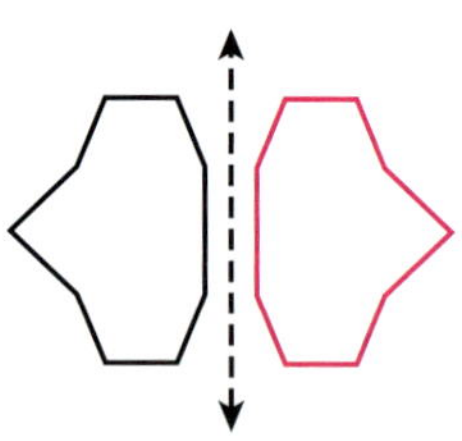

Tell whether each is a translation, rotation, or reflection.

B

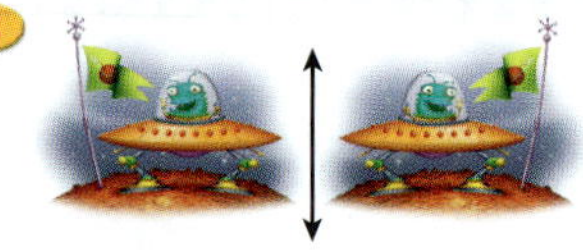

The figure is flipped over a line.

It is a reflection.

C

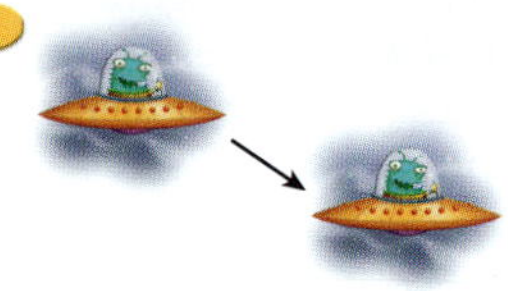

The figure is moved along a line.

It is a translation.

A full turn is a 360° rotation. So a $\frac{1}{4}$ turn is 90°, and a $\frac{1}{2}$ turn is 180°.

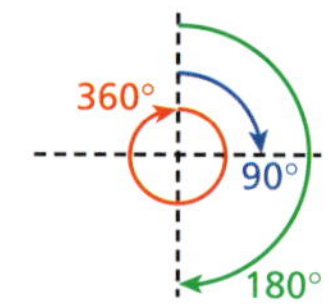

EXAMPLE 2 **Drawing Transformations**

Draw each transformation.

A

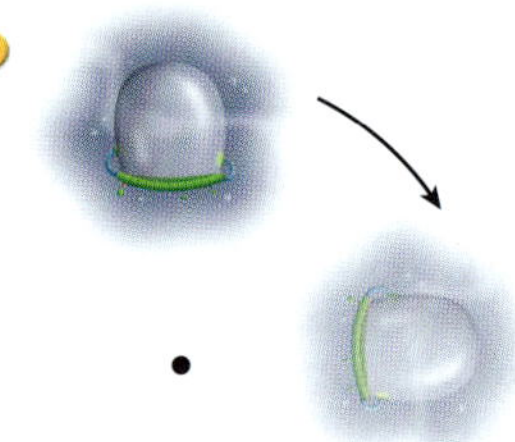

Draw a 90° clockwise rotation about the point shown.

Trace the figure and the point of rotation.

Place your pencil on the point of rotation.

Rotate the figure clockwise 90°.

Trace the figure in its new location.

B

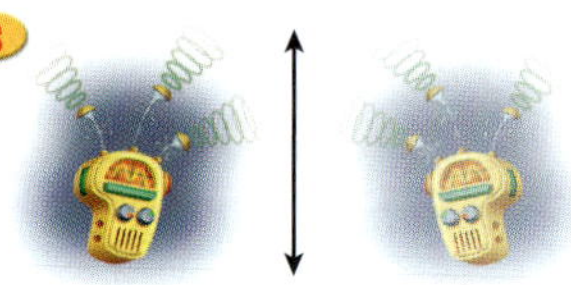

Draw a horizontal reflection.

Trace the figure and the line of reflection.

Fold along the line of reflection.

Trace the figure in its new location.

Think and Discuss

1. **Give examples** of reflections that occur in the real world.
2. **Name** a figure that can be rotated so that it will land on top of itself.

3 Close

Reaching All Learners

Through Hands-On Experience

Have each student draw some unusual plane figures on graph paper (provided on Chapter 7 Resource Book p. 136). Then have them translate, rotate, and reflect each figure.

Have students mount their drawings on poster board and label each transformation.

Summarize

Review the vocabulary from the lesson. In a rigid *transformation* a figure is moved without changing its size or shape. In a *rotation* a figure is turned around a point. In a *reflection* a mirror image is created across a line. In a *translation* a figure is moved along a line. Provide examples of each.

Answers to Think and Discuss

1. Possible answers: wallpaper designs, reflections in a mirror, reflections of buildings or trees in a lake
2. Possible answer: any figure rotated 360° about any point

FOR EXTRA PRACTICE
see page 651

internet connect
Homework Help Online
go.hrw.com Keyword: MR4 7-10

GUIDED PRACTICE

See Example 1 Tell whether each is a translation, rotation, or reflection.

1. 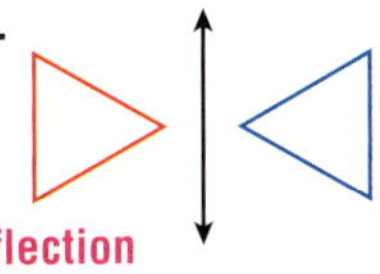
reflection

2. 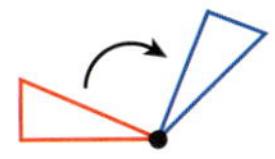
rotation

3. 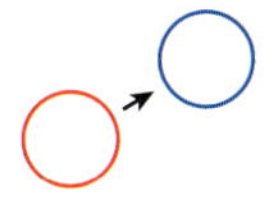
translation

See Example 2 Draw each transformation.

4. Draw a 180° clockwise rotation about the point shown.

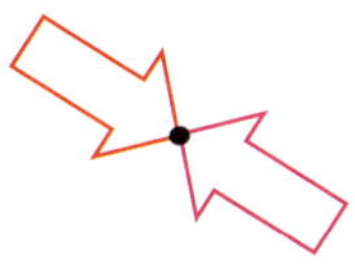

5. Draw a horizontal reflection across the line.

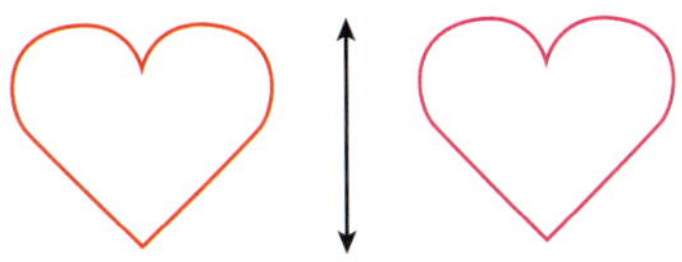

INDEPENDENT PRACTICE

See Example 1 Tell whether each is a translation, rotation, or reflection.

6. 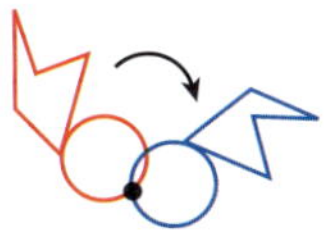
rotation

7. 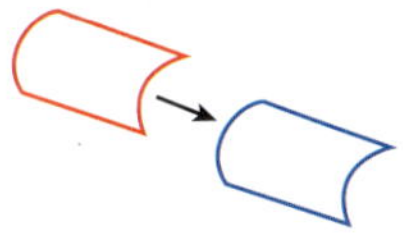
translation

8. 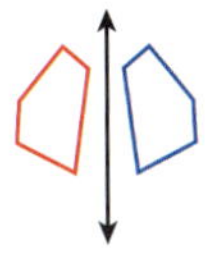
reflection

See Example 2 Draw each transformation.

9. Draw a vertical reflection across the line.

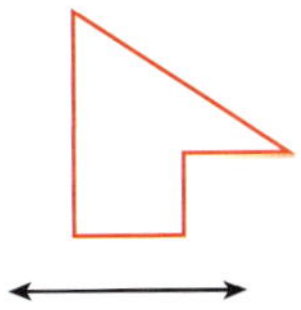

10. Draw a 90° counterclockwise rotation about the point.

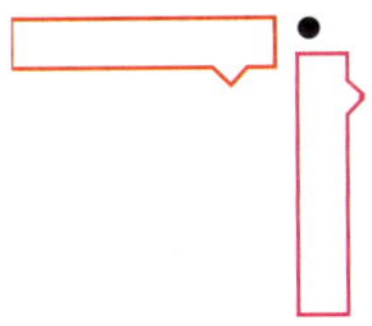

11. Draw a translation.

12. Draw a translation.

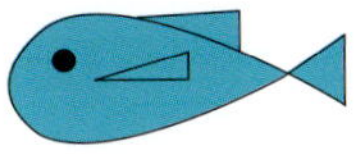

13. Draw a 90° clockwise rotation about the point.

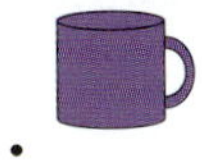

14. Draw a horizontal reflection across the line.

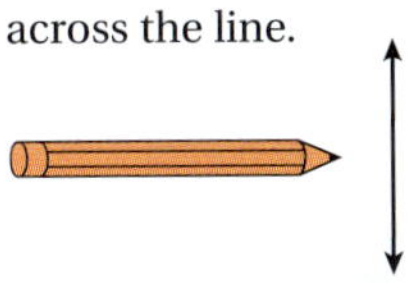

7-10 PRACTICE & ASSESS

Students may want to refer back to the lesson examples.

Assignment Guide

If you finished Example 1 assign:
Core 1–3, 6–8, 15, 21–27
Enriched 2–3, 6–8, 15–16, 21–27

If you finished Example 2 assign:
Core 1–16, 21–27
Enriched 5–27

You may use the recording sheet on Chapter 7 Resource Book p. 137 for Exercises 4, 5, 9–14, and 17.

Answers

9. 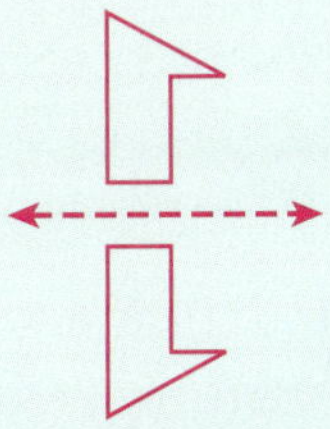

12. 

13. 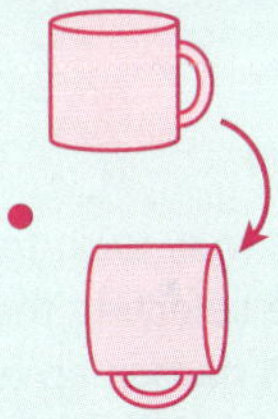

14. 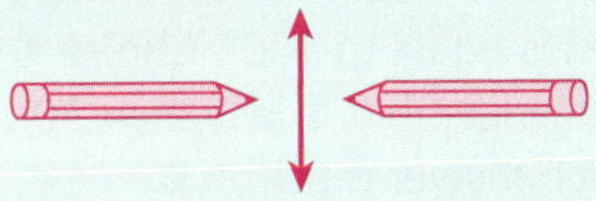

RETEACH 7-10

LESSON 7-10 **Reteach**
Transformations

A transformation moves a figure without changing its size or shape. So, a figure and its transformation are always congruent.

A translation is the movement of a figure along a straight line. The location is the only thing that changes with a translation.

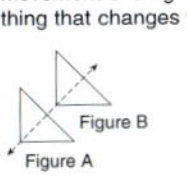

Figure B is a translation of Figure A.

A rotation is the movement of a figure around a point. The location and position of a figure can change with a rotation.

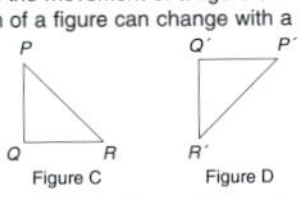

Figure D is a rotation of Figure C around point *R*.

When a figure flips over a line creating a mirror image, it is called a reflection. The line the figure flips over is called the line of reflection. The location and position of a figure change with a reflection.

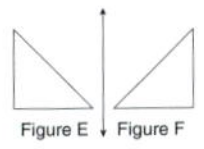

Figure F is a reflection of Figure E.

Identify each transformation.

1. 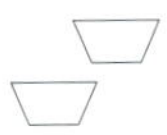
translation

2. 
rotation

PRACTICE 7-10

LESSON 7-10 **Practice B**
Transformations

Tell whether each is a translation, rotation, or reflection.

1. 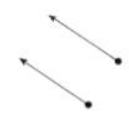
translation

2. 

reflection

3. 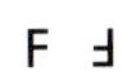

rotation

Draw each transformation.

4. Draw a 180° clockwise rotation about the point.

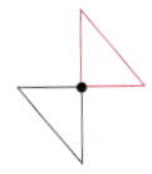

5. Draw a vertical reflection across the dotted line.

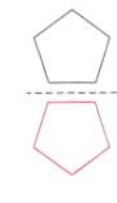

6. Without using reflections, how can you get this △ to look like this ▽?
Use the top vertex of △ as the point of rotation and rotate the triangle 180° clockwise or counterclockwise.

7. Describe a horizontal reflection of the word **MOM**. Can you think of any other words that would have a similar horizontal reflection? Possible answer:
The word MOM will look exactly the same if it is a horizontal reflection left or right. Other possible words: TOOT, HAH

Math Background

You can combine two or more transformations to form a *composition*. An example of a composition is the glide reflection. A glide reflection is created by reflecting a figure across a given line and then translating the resulting figure along that same line.

An example of a transformation that changes size (but not shape) is a dilation. When a geometric figure undergoes a dilation, the resulting figure is similar to the original figure.

Answers

16. horizontally: A, H, I, M, O, T, U, V, W, X, Y; vertically: B, C, D, E, H, I, O, X

17.

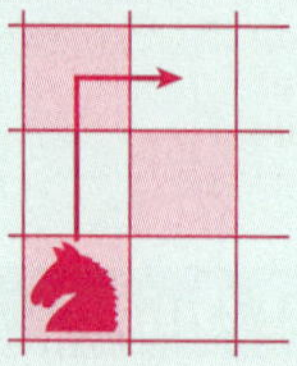

19. Possible answer: To translate the piece, move it along a straight line; to rotate the piece, move it in a circular path; to reflect the piece, flip it over a line to create a mirror image.

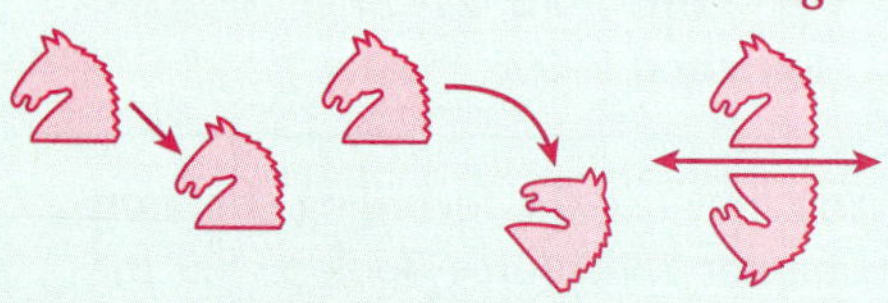

20.

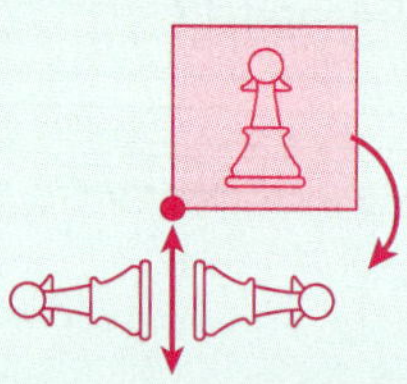

Journal

Have students create a design using transformations. Have them explain what kinds of transformations they used.

Test Prep Doctor

For Exercise 27, students may wish to eliminate choices before selecting the correct answer. Choice **A** names an angle that is equal to 90°. Choice **B** names an angle that is equal to 180°. Choice **C** names an angle that is less than 90°. So the correct answer must be **D**.

Lesson Quiz

1. Tell whether the figure is translated, rotated, or reflected. rotated

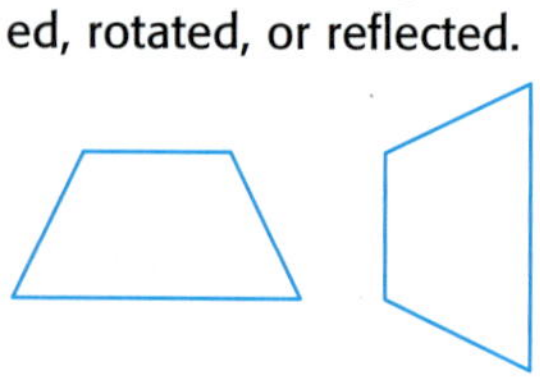

2. Draw a vertical reflection of the first figure in problem 1.

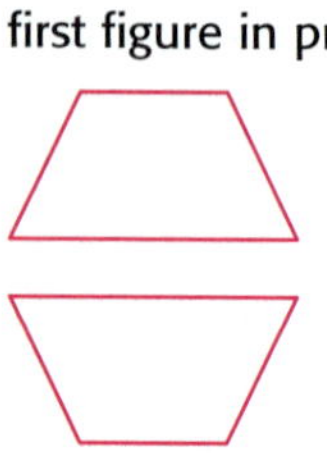

Available on Daily Transparency in CRB

PRACTICE AND PROBLEM SOLVING

Hobbies LINK

In a game of chess, each player has 318,979,564,000 possible ways to make the first four moves.

15. Which is a horizontal reflection of this red arrow? **A**

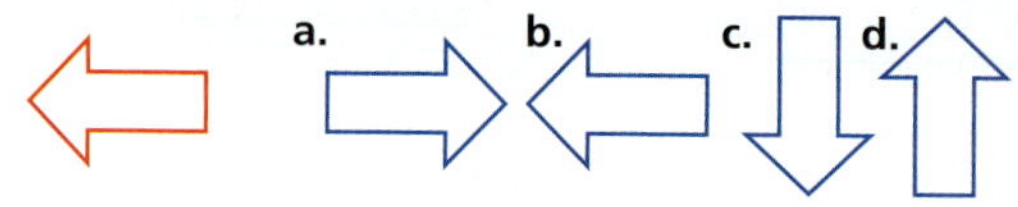

16. ***LANGUAGE ARTS*** Which letters in the alphabet can be horizontally reflected and still look the same? Which letters can be vertically reflected and still look the same?

Use the chessboard for Exercises 17–20.

HOBBIES Chess is a game of skill that is played on a board divided into 64 squares. Each chess piece is moved differently.

17. Copy the lower left corner of the chessboard. Then show the indicated knight moving in a translation of two forward and one right.

Knight King Pawn

18. ***CHOOSE A STRATEGY*** If the knight, king, and pawn are placed in a straight line, how many ways can they be arranged? **C**

A 3 **B** 4 **C** 6 **D** 12

19. ***WRITE ABOUT IT*** Draw one of the chess pieces. Then draw a translation, rotation, and reflection of that piece. Describe each transformation.

20. ***CHALLENGE*** Draw one of the chess pieces rotated 90° clockwise around the vertex of a square and then horizontally reflected.

Spiral Review

Find the least common multiple (LCM). (Lesson 5-5)

21. 4 and 12 **12** **22.** 7, 14, and 21 **42** **23.** 6, 9, and 24 **72**

Use the stem-and-leaf plot. (Lessons 6-2, 6-9)

24. Find the median. **36**

25. Find the mode. **36**

26. Find the range. **33**

Stems	Leaves
2	0 1 3
3	2 5 6 6 6 7
4	5 8 9
5	2 3 3

Key: 2|3 means 23

27. **TEST PREP** Which type of angle has a measure between 90° and 180°? (Lesson 7-2) **D**

A Right **B** Straight **C** Acute **D** Obtuse

CHALLENGE 7-10

LESSON 7-10 **Challenge**
Rotation Ride

The Ferris wheel shown below is turning clockwise around its rotation point at the center of the wheel. For each person on the Ferris wheel, describe how many degrees he or she must rotate to be able to get off the ride.

1. Darius 90°
2. Jenna 225°
3. Paul 0°
4. Sarah 315°
5. Mei 270°
6. Ben 135°
7. Mindy 45°
8. Carlos 180°

PROBLEM SOLVING 7-10

LESSON 7-10 **Problem Solving**
Transformations

Write the correct answer.

1. If the rotation point of a circle is its center, how will all rotations affect the circle?
The circle will never change.

2. What transformation could make an arrow pointing east become an arrow pointing north?
a 90° counterclockwise rotation

3. What transformation could make the number 9 become the number 6?
a vertical and a horizontal reflection or a 180° rotation

4. What transformation could make the letter P look like the letter b?
a vertical reflection

5. On the coordinate plane at right, graph Triangle A with vertices (3, 1), (6, 1), and (3, 5). Then graph Triangle B with vertices (3, 6), (6, 6), and (3, 10). What transformation best describes the change from Triangle A to Triangle B?
vertical translation

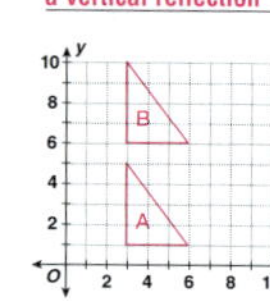

Circle the letter of the correct answer.

6. Which transformation best describes the figure on the right?

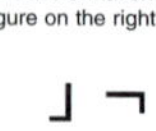

A 90° clockwise rotation
B horizontal reflection
(C) 90° counterclockwise rotation
D horizontal translation

7. Which transformation best describes the figure on the left?

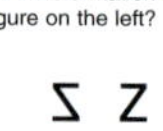

(F) horizontal reflection
G 180° counterclockwise rotation
H 90° counterclockwise rotation
J horizontal translation

7-11 Symmetry

Learn to identify line symmetry.

Vocabulary
line symmetry
line of symmetry

A figure has **line symmetry** if it can be folded or reflected so that the two parts of the figure match, or are congruent. The line of reflection is called the **line of symmetry**.

You could draw a line of symmetry on this windmill. The shape of the building and the position of the blades are symmetrical.

EXAMPLE 1 Identifying Lines of Symmetry

Determine whether each dashed line appears to be a line of symmetry.

A

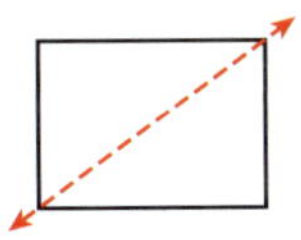

The two parts of the figure are congruent, but they do not match exactly when folded or reflected across the line.

The line does not appear to be a line of symmetry.

B

The two parts of the figure appear to match exactly when folded or reflected across the line.

The line appears to be a line of symmetry.

Some figures have more than one line of symmetry.

EXAMPLE 2 Finding Multiple Lines of Symmetry

Find all of the lines of symmetry in each regular polygon.

A

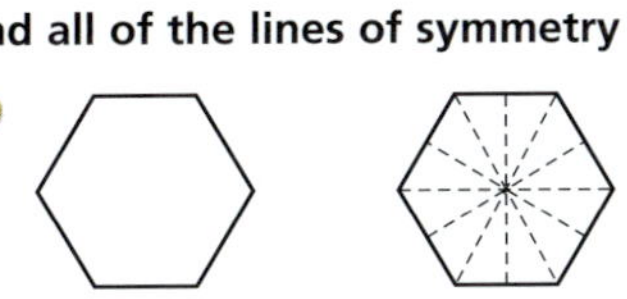

Trace each figure and cut it out.
Fold the figure in half in different ways.
Count the lines of symmetry.

6 lines of symmetry

7-11 Organizer

Pacing: Traditional 1 day
Block $\frac{1}{2}$ day

Objective: Students identify line symmetry.

Indiana Academic Standards
6.7.1, 6.7.5, 6.7.11

Warm Up

True or false?

1. In a vertical reflection the original figure is flipped over a line of symmetry. true
2. In a translation the original figure slides along a straight line and is flipped. false
3. In rotation a figure is moved around a point. true

Problem of the Day

Name four plane figures that can be rotated 180° around a center point and look the same after the rotation as they did before. Possible answer: rectangle, square, parallelogram, regular hexagon

Available on Daily Transparency in CRB

Math Fact

When written in block letters, *H, I, O,* and *X* have both vertical and horizontal lines of symmetry.

1 Introduce

Alternate Opener

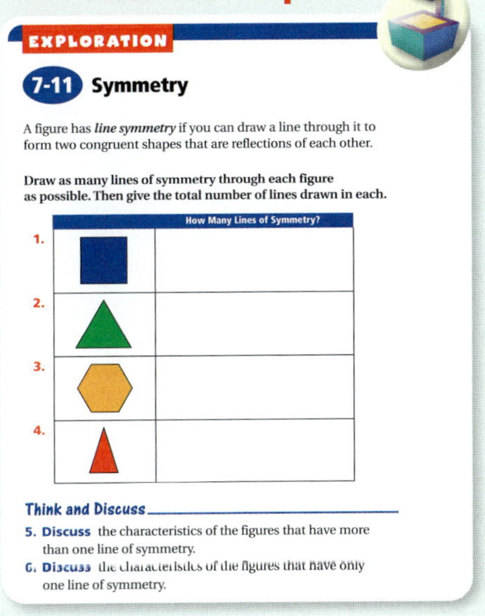

EXPLORATION

7-11 Symmetry

A figure has *line symmetry* if you can draw a line through it to form two congruent shapes that are reflections of each other.

Draw as many lines of symmetry through each figure as possible. Then give the total number of lines drawn in each.

		How Many Lines of Symmetry?
1.		
2.		
3.		
4.		

Think and Discuss

5. **Discuss** the characteristics of the figures that have more than one line of symmetry.
6. **Discuss** the characteristics of the figures that have only one line of symmetry.

Motivate

Have each student fold a piece of paper in half and cut an unusual shape from it. Then have them unfold the shapes they cut out. Ask students to tell you what they notice about the parts of the shape on either side of the fold line. They are identical mirror images of each other. Explain that the fold line is a *line of symmetry*.

Exploration worksheet and answers on Chapter 7 Resource Book pp. 103 and 162

2 Teach

Lesson Presentation

Guided Instruction

In this lesson, students learn to identify lines of symmetry. Teach students how to determine whether a given line dividing a figure is a line of symmetry. Explain that figures can have zero, one, or more than one line of symmetry. Have students find multiple lines of symmetry in polygons and find symmetry in flags.

Additional Examples

Example 1

Determine whether the dashed line appears to be a line of symmetry.

A. 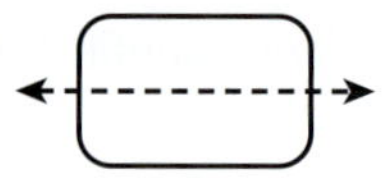

yes

B. 

no

Example 2

Find all the lines of symmetry in each regular polygon.

A. 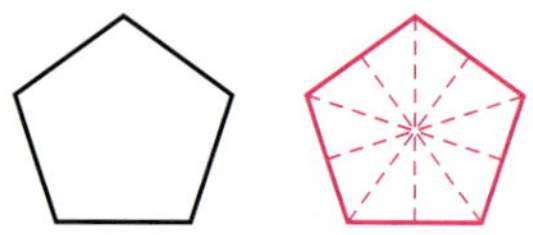

B. 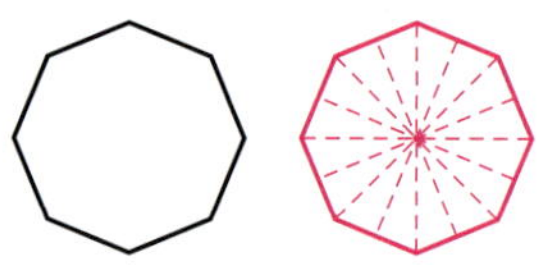

Example 3

Find all the lines of symmetry in the design on each flag.

A. Alaska

B. Arizona

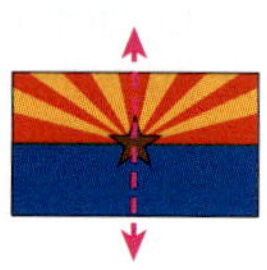

no lines of symmetry

C. Colorado

D. New Mexico

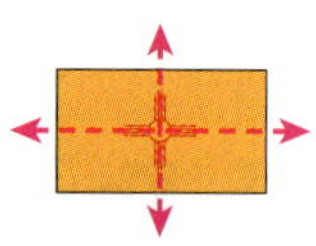

Find all of the lines of symmetry in each regular polygon.

B 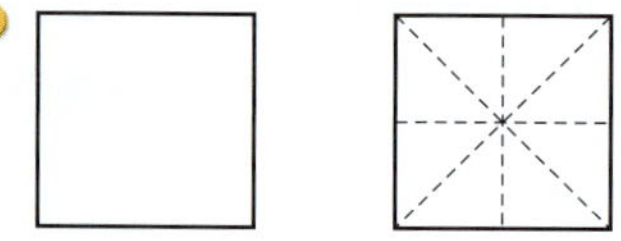

4 lines of symmetry

Count the lines of symmetry.

C 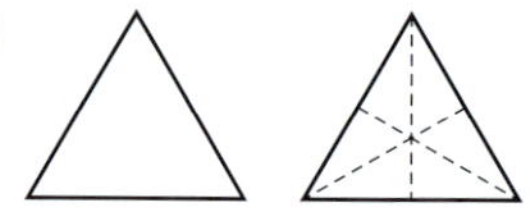

3 lines of symmetry

Count the lines of symmetry.

EXAMPLE 3 *Social Studies Application*

Find all of the lines of symmetry in each flag design.

A Antigua and Barbuda

1 line of symmetry

B Macedonia

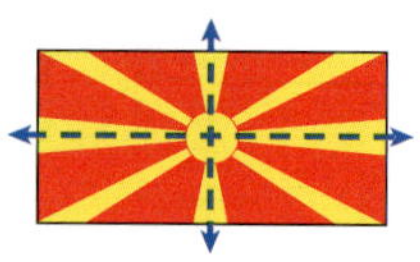

2 lines of symmetry

C Norway

1 line of symmetry

D Lesotho

There are no lines of symmetry.

Think and Discuss

1. **Explain** how you can use your knowledge of reflection to create a figure that has a line of symmetry.
2. **Determine** whether all hexagons have six lines of symmetry.
3. **Name** objects with line symmetry in your classroom. Tell how many lines of symmetry each of these objects has.

Reaching All Learners

Through Curriculum Integration

Social Studies Have students research flags from other countries. Have them categorize the flags as having zero, one, or two lines of symmetry.

3 Close

Summarize

Have the students write a brief definition of *line symmetry* and draw an example of a figure with its line or lines of symmetry drawn.

Possible answer: A figure has line symmetry if a line can divide it into two parts that match when the figure is folded along the line. Check students' drawings.

Answers to Think and Discuss

1. Possible answer: If a figure is reflected over a line that touches it, the line of reflection becomes the line of symmetry for the resulting figure.
2. Possible answer: No, only regular hexagons have six lines of symmetry.
3. Possible answers: a rectangular chalkboard; two lines of symmetry

7-11 Exercises

FOR EXTRA PRACTICE
see page 651

internet connect
Homework Help Online
go.hrw.com Keyword: MR4 7-11

GUIDED PRACTICE

See Example 1 Determine whether each dashed line appears to be a line of symmetry.

1. The line is a line of symmetry.
2. The line is not a line of symmetry.
3. The line is a line of symmetry.

See Example 2 Find all of the lines of symmetry in each regular polygon.

4. 4 lines of symmetry
5. 3 lines of symmetry
6. 5 lines of symmetry

See Example 3 Find all of the lines of symmetry in each design.

7. 

4 lines of symmetry

8. 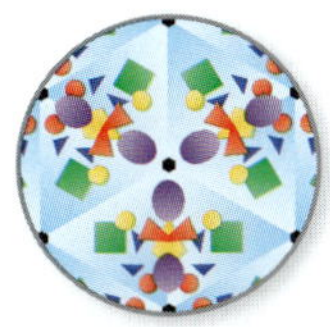

3 lines of symmetry

INDEPENDENT PRACTICE

See Example 1 Determine whether each dashed line appears to be a line of symmetry.

9. The line is not a line of symmetry.
10. The line is a line of symmetry.
11. The line is not a line of symmetry.

See Example 2 Find all of the lines of symmetry in each regular polygon.

12. 4 lines of symmetry
13. 8 lines of symmetry
14. 7 lines of symmetry

See Example 3 Find all of the lines of symmetry in each object.

15. 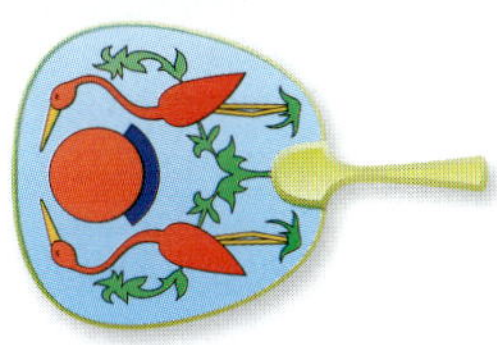

1 line of symmetry

16. 

no lines of symmetry

7-11 PRACTICE & ASSESS

Students may want to refer back to the lesson examples.

Assignment Guide

If you finished Example 1 assign:
Core 1–3, 9–11, 21–26
Enriched 1–3, 9–11, 21–26

If you finished Example 2 assign:
Core 1–6, 9–14, 17, 21–26
Enriched 2–6, 9–14, 17, 18, 21–26

If you finished Example 3 assign:
Core 1–17, 21–26
Enriched 4–26

A recording sheet for Exercises 4–8 and 12–16 is provided on Chapter 7 Resource Book p. 138.

RETEACH 7-11

LESSON 7-11 Reteach
Symmetry

A figure has line symmetry if it can be folded or reflected so that the two overlapping parts of the figure are congruent. The line of reflection is called the line of symmetry.

To figure out if a dashed line is a line of symmetry, test to see if the two parts match exactly when folded or reflected across the line.

The two parts match, so the line is a line of symmetry.

Decide if each line is a line of symmetry.

1. no
2. yes

To find all of the lines of symmetry of the regular octagon, first trace the polygon and cut it out.

Then fold the polygon in half in different ways to find the lines of symmetry.

A regular octagon has 8 lines of symmetry.

Find all of the lines of symmetry for each figure. Write how many lines of symmetry each figure has.

3. 1 line
4. 4 lines

PRACTICE 7-11

LESSON 7-11 Practice B
Symmetry

Determine whether each dashed line appears to be a line of symmetry.

1. no
2. yes
3. yes

Find all the lines of symmetry in each regular polygon.

4.
5.
6.

Draw each cut-out figure as it would look unfolded.

7.
8.

9. Which has more lines of symmetry, a square or a rectangle?
a square

10. Of the numbers 1 through 9, which numbers can have lines of symmetry?
3, and 8

Math Background

The study of mathematics includes the study of rotational symmetry in addition to the study of line symmetry. When a figure can be rotated about a point between 0° and 360°, and the resulting figure matches the original figure point for point, the figure has rotational symmetry. For example, a square has 90° rotational symmetry about the point where its diagonals meet.

Rotational symmetry abounds in nature, for example, in snowflakes and flower petals.

Interdisciplinary LINK

Music

Exercises 17–20 focus on some unique instruments from various cultures.

Answers

19. Possible answer: Trace the figure and cut it out. Fold the figure in half in as many ways as possible. If the halves match, then each fold line is a line of symmetry.

21. One hundred one and twenty-five hundredths

22. Three million, four thousand, five hundred six

23. Twelve billion, thirty million, nine hundred twenty-one thousand

24. Forty-seven and three hundred five ten-thousandths

Journal

Have students describe how they determine whether a figure has zero, one, or two lines of symmetry.

Test Prep Doctor

For Exercise 25, students can use mathematical reasoning to eliminate answers. The fraction equivalent to $10\frac{3}{4}$ will have a numerator that is about 10 times as great as its denominator (because the numerator divided by the denominator will equal $10\frac{3}{4}$), so **A** and **D** cannot be correct. Using the conventional method for writing a mixed number as a fraction, $10\frac{3}{4} = \frac{43}{4}$. Because $\frac{43}{4} \cdot \frac{2}{2} = \frac{86}{8}$, the correct answer is **C**.

Lesson Quiz

Does the described figure have line symmetry?

1. right isosceles triangle   no

2. rectangle   yes

Do the following capital letters of the alphabet have symmetry? If so, is the line of symmetry horizontal or vertical?

3. H   yes, vertical and horizontal

4. R   no

Available on Daily Transparency in CRB

Music LINK

Music is an art form enjoyed by many cultures. Some cultures play music on unique instruments. You might hear the sun drum or turtle drum in Native American music. In music made by people from the Appalachian Mountains, you might hear the strains of a dulcimer. The photo shows young musicians playing sitars, instruments heard in north Indian classical music.

17. Determine whether the dashed line in each drawing is a line of symmetry.

a.

The line is a line of symmetry.

b.

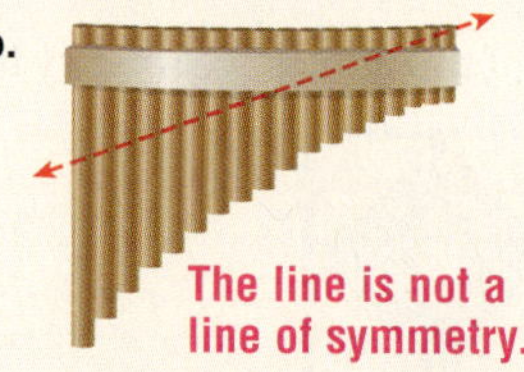

The line is not a line of symmetry.

18. The triangle is a percussion instrument formed by a rod of steel or chrome that is bent into the shape of an equilateral triangle. It is open at one corner to allow the notes to resonate. How many lines of symmetry can you find in an equilateral triangle?   3

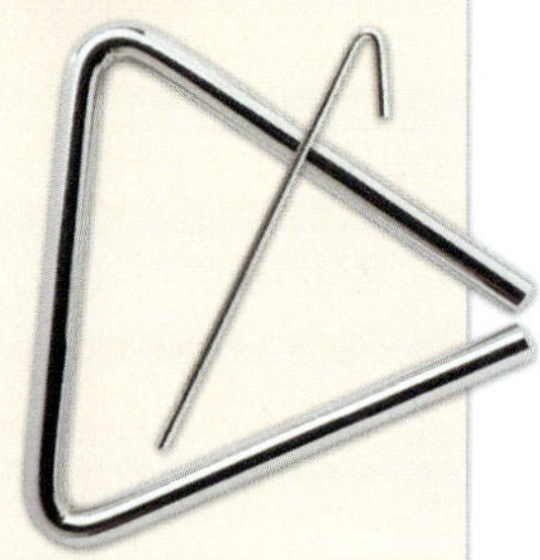

19. *WRITE ABOUT IT* The turtle drum is a regular octagon. How can you find all of the lines of symmetry in a regular polygon?

20. *CHALLENGE* A student drew a drum in the shape of an octagon on a grid. What are the coordinates of the vertices of the unfolded half of the drum drawing if the fold shown is a line of symmetry?   (5, 4), (6, 3), (6, 2), (5, 1)

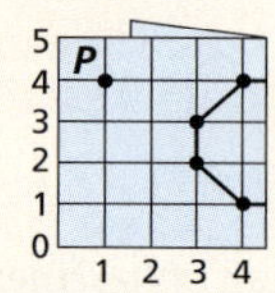

Spiral Review

Write each number in word form. (Lessons 1-1 and 3-1)

21. 101.25   **22.** 3,004,506   **23.** 12,030,921,000   **24.** 47.0305

25. **TEST PREP** Which fraction is equivalent to $10\frac{3}{4}$? (Lesson 4-7)   C

A $\frac{166}{8}$ **B** $\frac{83}{8}$ **C** $\frac{86}{8}$ **D** $\frac{33}{4}$

26. **TEST PREP** Look at the coordinate grid in Exercise 20. Which ordered pair best represents point *P*? (Lesson 6-6)   G

F (4, 1)   **G** (1, 4)   **H** (0, 3)   **J** (3, 0)

CHALLENGE 7-11

LESSON 7-11 **Challenge**
Alphabetical Symmetry

Tell whether each capital letter has a horizontal line of symmetry, a vertical line of symmetry, both, or neither. Draw the lines of symmetry to illustrate.

A vertical	B horizontal	C horizontal	D horizontal
E horizontal	F neither	G neither	H both
I both	J neither	K horizontal	L neither
M vertical	N neither	O both	P neither
Q neither	R neither	S neither	T vertical
U vertical	V vertical	W vertical	X both
Y vertical	Z neither		

PROBLEM SOLVING 7-11

LESSON 7-11 **Problem Solving**
Symmetry

Write the correct answer.

1. Do your body and face appear to have a vertical line of symmetry or a horizontal line of symmetry?
   vertical

2. Which letter of the alphabet has an infinite, or endless, number of lines of symmetry?
   the letter o

3. Ted says the diagonals of a rectangle are also its lines of symmetry. Do you agree? Explain.
   No, because when folded or reflected along a diagonal, the two parts of the rectangle do not match.

4. Using the digits 0 through 9 and not repeating any digits, write a 3-digit number that has a horizontal line of symmetry.
   3-digit numbers only choices are: 083, 038, 803, 830, 308, 380

5. Draw a line of symmetry for this word.

   DOCK

6. Draw the lines of symmetry for this star.

Circle the letter of the correct answer.

7. How many lines of symmetry does this hexagon have?
   A 4
   B 8
   (C) 6
   D 2

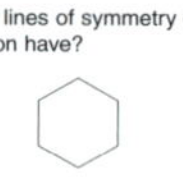

8. How many lines of symmetry does this flower have?
   F 3
   (G) 4
   H 5
   J 6

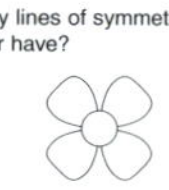

7-12 Tessellations

Learn to identify tessellations and shapes that can tessellate.

Vocabulary
tessellation

A **tessellation** is a repeating arrangement of one or more shapes that completely covers a plane with no gaps and no overlaps.

In the design shown, the shape of a fish is used to make a tessellation.

The shape of the fish was copied over and over until the design completely covered an area with no gaps or overlaps.

Although most tessellations are made by humans, a few occur in nature. Honeycombs are naturally occuring tessellations of regular hexagons.

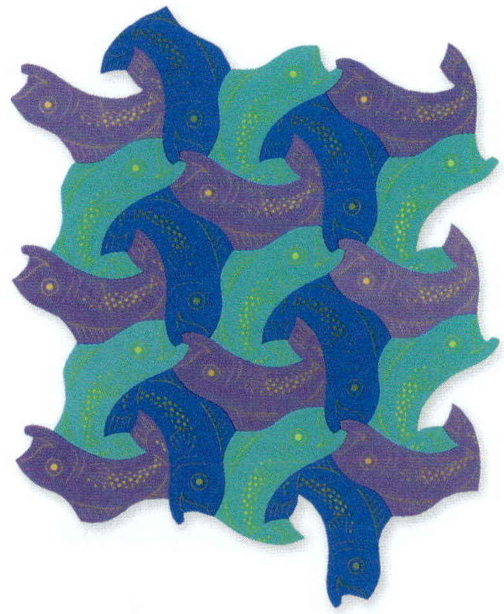

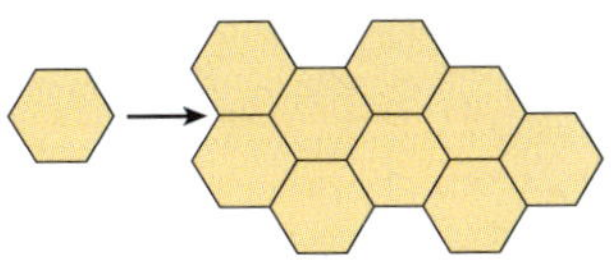

EXAMPLE 1 Identifying Polygons That Tessellate the Plane

Identify whether each polygon can tessellate the plane. Make a drawing to show your answer.

A

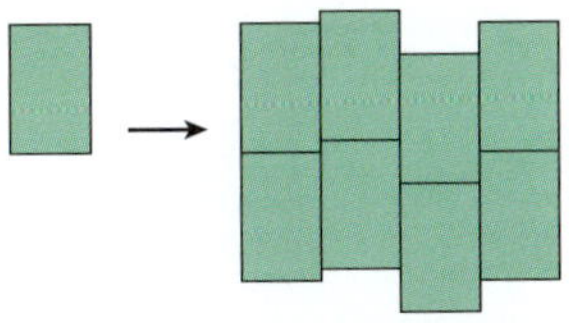

The rectangles cover the plane without any gaps or overlaps.

The rectangle can tessellate the plane.

B

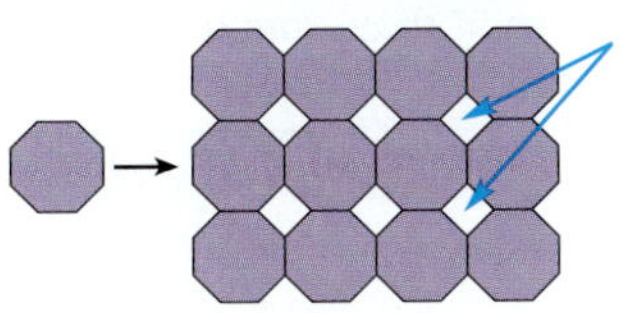

There are gaps between the octagons.

The octagon cannot tessellate the plane.

7-12 Organizer

Pacing: Traditional 1 day
Block $\frac{1}{2}$ day

Objective: Students identify tessellations and shapes that can tessellate.

Indiana Academic Standards
6.7.1, 6.7.5, 6.7.10, 6.7.11

Warm Up

Do the following letters of the alphabet have symmetry, and if so, is the line of symmetry horizontal or vertical?

1. X yes; both
2. P no
3. C yes; horizontal
4. T yes; vertical

Problem of the Day

If either diagonal is drawn in a figure, the figure is divided into two congruent triangles. What is the most general classification of this figure? parallelogram

Available on Daily Transparency in CRB

Math Humor

A flight attendant decorated a jet with colored cords from graduation caps. He wanted to tassel-late a plane.

1 Introduce

Alternate Opener

EXPLORATION

7-12 Tessellations

Copy each figure repeatedly to see if you can cover an area with no gaps or overlaps. Use the first arrangement as a guide.

	Figure	Arrangement	Covers Area as Specified?
1.			Yes
2.			
3.			
4.			
5.			
6.			

Think and Discuss

7. **Discuss** the characteristics of the figures with which you could cover an area with no gaps or overlaps.
8. **Discuss** the characteristics of the figures with which you could not cover an area with no gaps or overlaps.

Motivate

Have students look at the honeycomb from the lesson (provided on Teaching Transparency T53 in the Chapter 7 Resource Book). Ask them to identify the shape and the transformation used in the honeycomb to cover the plane. regular hexagon; translated Have students name other designs in which shapes completely cover a plane.

Possible answer: floor tiles

Exploration worksheet and answers on Chapter 7 Resource Book pp. 112 and 164

2 Teach

Lesson Presentation

Guided Instruction

In this lesson, students learn to identify tessellations and shapes that can tessellate. Show students human-made and natural tessellations. Help students identify regular polygons and other shapes that can and cannot tessellate a plane.

Additional Examples

Example 1

Identify whether the polygon can tessellate the plane. Make a drawing to show your answer.

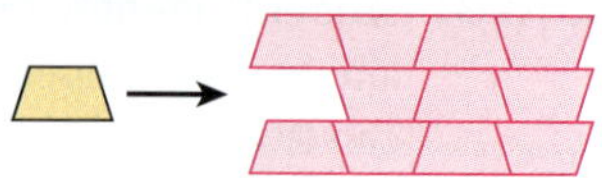

yes

Example 2

Identify whether the shape can tessellate the plane. Make a drawing to show your answer.

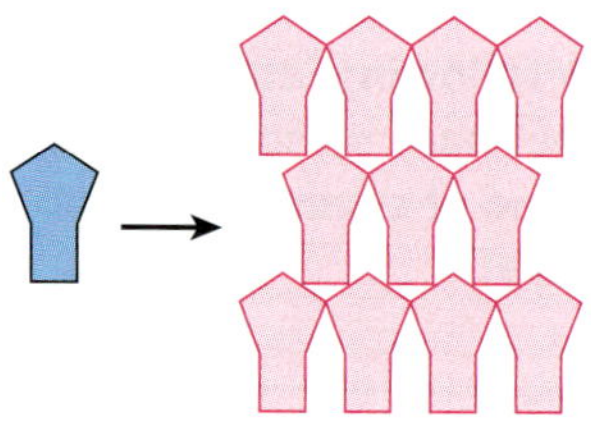

no

7-12 PRACTICE & ASSESS

Assignment Guide

If you finished Example 1 assign:
Core 1–3, 7–9, 13, 17–21
Enriched 3, 7–9, 13–15, 17–21

If you finished Example 2 assign:
Core 1–14, 17–21
Enriched 3–21

Answers

1–6. Complete answers on p. A5.

EXAMPLE 2 Identifying Nonpolygons That Tessellate the Plane

Identify whether each shape can tessellate the plane. Make a drawing to show your answer.

A

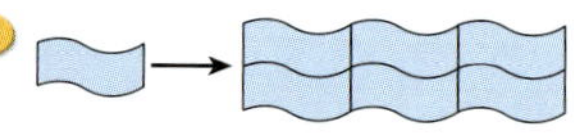

The shapes cover the plane without any gaps or overlaps.

This shape can tessellate the plane.

B

There are gaps between the shapes.

This shape cannot tessellate the plane.

Think and Discuss

1. Explain how you know when a pattern of shapes forms a tessellation.

2. Give an example of a shape that cannot form a tessellation.

7-12 Exercises

FOR EXTRA PRACTICE see page 651

internet connect Homework Help Online go.hrw.com Keyword: MR4 7-12

GUIDED PRACTICE

See Example 1 Identify whether each polygon can tessellate the plane. Make a drawing to show your answer.

1. 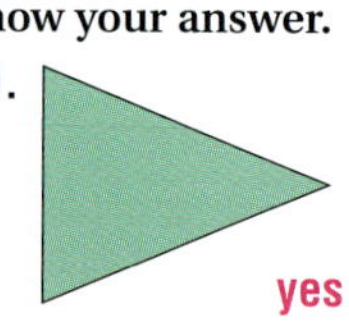 yes

2. 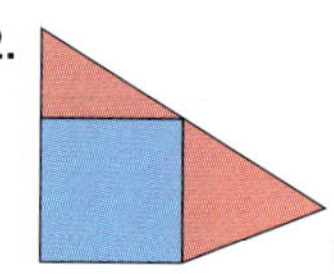 yes

3. 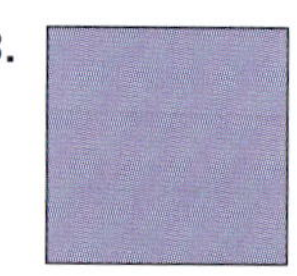 yes

See Example 2 Identify whether each shape can tessellate the plane. Make a drawing to show your answer.

4.  no

5. 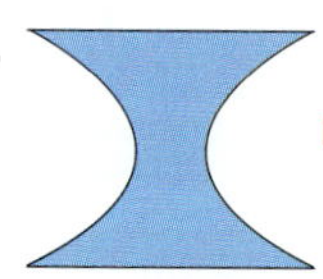 no

6. 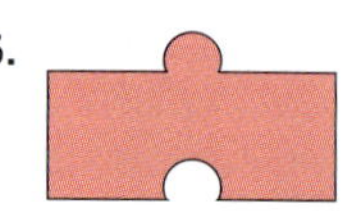 yes

Reaching All Learners

Through Curriculum Integration

Art Have students do research about the artwork of M. C. Escher. Have them choose a favorite work of his and describe how he used tessellations in the piece.

3 Close

Summarize

Review the term *tessellation*. Remind students that there must not be any gaps or overlaps when a figure tessellates a plane.

Answers to Think and Discuss

1. If you can cover a space with the shape without any gaps or overlaps, then it forms a tessellation.

2. Answers will vary. Possible answer: a regular pentagon

INDEPENDENT PRACTICE

See Example 1

Identify whether each polygon can tessellate the plane. Make a drawing to show your answer.

7. 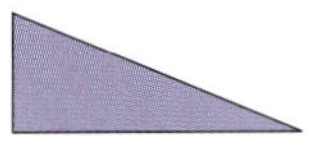 yes

8. 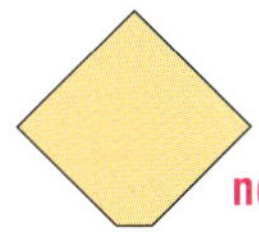 no

9.  yes

See Example 2

Identify whether each shape can tessellate the plane. Make a drawing to show your answer.

10. 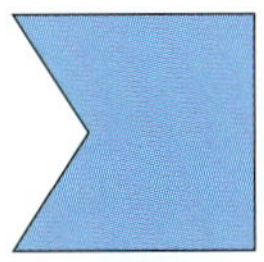 yes

11. 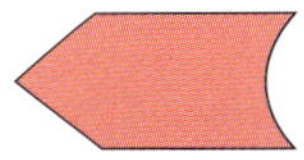 no

12.  yes

PRACTICE AND PROBLEM SOLVING

13. ***CRAFTS*** This quilt design is a tessellation. Triangular pieces of fabric have been sewn together. Which piece of fabric is missing from the design? B

A 

B 

C 

14. ***CRAFTS*** Determine whether the design on the Japanese wallpaper at right is a tessellation. If not, explain. yes

15. ***WRITE ABOUT IT*** What does it mean to tessellate a plane?

16. ***CHALLENGE*** Draw three figures that can be used together to make a tessellation.

Spiral Review

Write two word phrases for each expression. (Lesson 2-2)

17. $b + 13$ *b* plus 13, 13 more than *b*

18. $(2)(12)$ 2 times 12, 12 multiplied by 2

19. $26 - c$ 26 minus *c*, *c* less than 26

20. $m \div 3$ *m* divided by 3, the quotient of *m* and 3

21. **TEST PREP** Which is the reciprocal of $2\frac{2}{3}$? (Lesson 5-3) D

A $\frac{3}{2}$ B $2\frac{3}{2}$ C $\frac{8}{3}$ D $\frac{3}{8}$

Lesson Quiz

True or false?

1. A regular hexagon can tessellate the plane. true
2. A regular octagon can tessellate the plane. false
3. A square can tessellate the plane. true
4. Tell whether the shape can tessellate the plane. Make a drawing to show your answer. yes

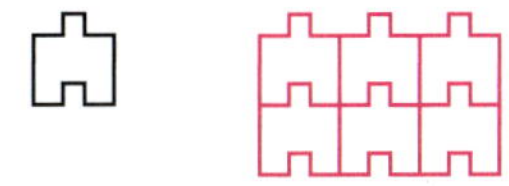

Available on Daily Transparency in CRB

Answers

7–12, 15, 16. Complete answers on p. A5.

RETEACH 7-12

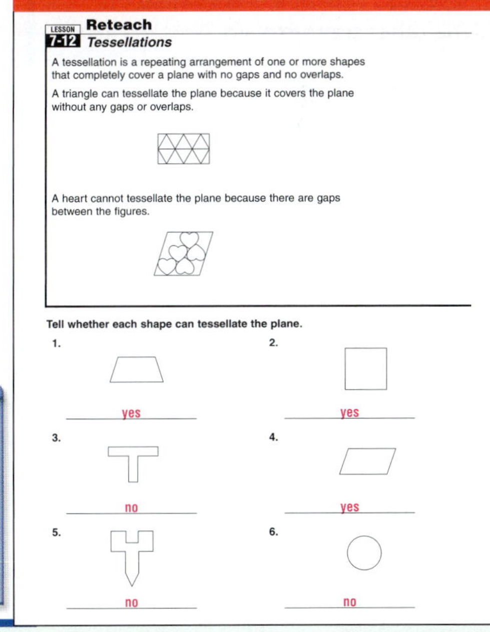

LESSON 7-12 Reteach
Tessellations

A tessellation is a repeating arrangement of one or more shapes that completely cover a plane with no gaps and no overlaps.

A triangle can tessellate the plane because it covers the plane without any gaps or overlaps.

A heart cannot tessellate the plane because there are gaps between the figures.

Tell whether each shape can tessellate the plane.

1. yes
2. yes
3. no
4. yes
5. no
6. no

PRACTICE 7-12

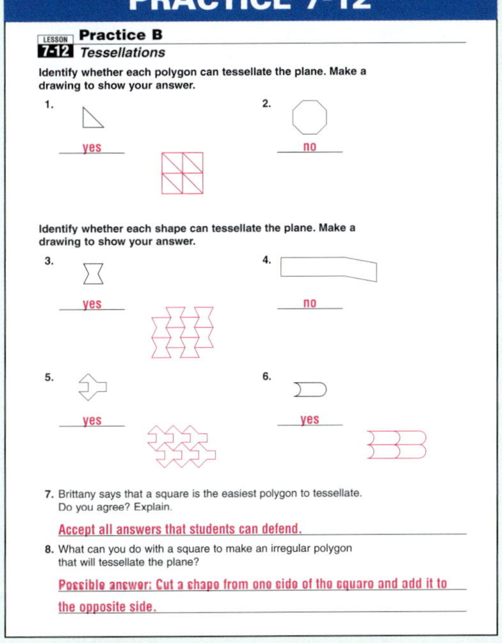

LESSON 7-12 Practice B
Tessellations

Identify whether each polygon can tessellate the plane. Make a drawing to show your answer.

1. yes
2. no

Identify whether each shape can tessellate the plane. Make a drawing to show your answer.

3. yes
4. no
5. yes
6. yes

7. Brittany says that a square is the easiest polygon to tessellate. Do you agree? Explain.
Accept all answers that students can defend.

8. What can you do with a square to make an irregular polygon that will tessellate the plane?
Possible answer: Cut a shape from one side of the square and add it to the opposite side.

CHALLENGE 7-12

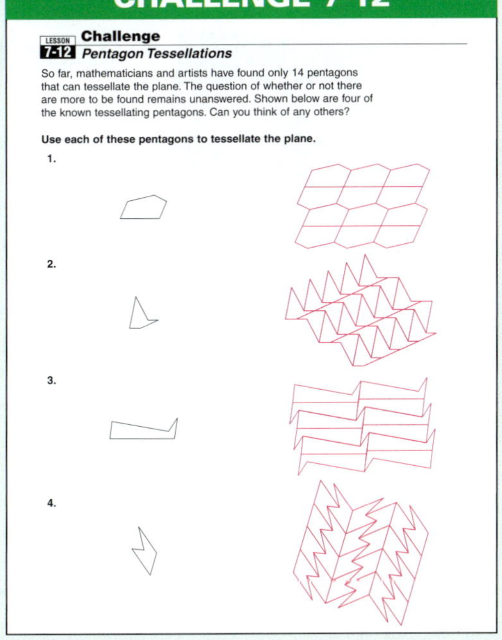

LESSON 7-12 Challenge
Pentagon Tessellations

So far, mathematicians and artists have found only 14 pentagons that can tessellate the plane. The question of whether or not there are more to be found remains unanswered. Shown below are four of the known tessellating pentagons. Can you think of any others?

Use each of these pentagons to tessellate the plane.

1.
2.
3.
4.

PROBLEM SOLVING 7-12

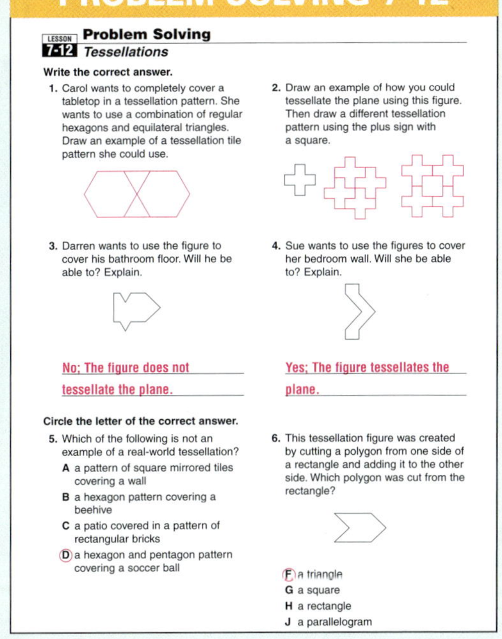

LESSON 7-12 Problem Solving
Tessellations

Write the correct answer.

1. Carol wants to completely cover a tabletop in a tessellation pattern. She wants to use a combination of regular hexagons and equilateral triangles. Draw an example of a tessellation tile pattern she could use.

2. Draw an example of how you could tessellate the plane using this figure. Then draw a different tessellation pattern using the plus sign with a square.

3. Darren wants to use the figure to cover his bathroom floor. Will he be able to? Explain.
No; The figure does not tessellate the plane.

4. Sue wants to use the figures to cover her bedroom wall. Will she be able to? Explain.
Yes; The figure tessellates the plane.

Circle the letter of the correct answer.

5. Which of the following is not an example of a real-world tessellation?
A a pattern of square mirrored tiles covering a wall
B a hexagon pattern covering a beehive
C a patio covered in a pattern of rectangular bricks
(D) a hexagon and pentagon pattern covering a soccer ball

6. This tessellation figure was created by cutting a polygon from one side of a rectangle and adding it to the other side. Which polygon was cut from the rectangle?
(F) a triangle
G a square
H a rectangle
J a parallelogram

Hands On LAB 7C Create Tessellations

Pacing: Traditional 1 day
Block $\frac{1}{2}$ day

Objective: To use paper, scissors, and tape to create tessellations

Materials: Paper, scissors, tape

Lab Resources

Hands-On Lab Activities . . . . . pp. 59–60

Using the Pages

Discuss with students what must be done in order for their shapes to tessellate the plane.

1. After cutting out a shape from one side, can you tape it anywhere on the other side? Explain. No; you have to tape it exactly opposite from where it was cut out so that it will fit into the space left by cutting out the same shape from the next tile.
2. How many cutouts can you make when forming the basic shape for the tessellation? Explain. You can cut out and tape as many shapes as possible while still being able to position them accurately. Shapes can be cut out from any side of the square.

Hands On LAB 7C Create Tessellations

Use with Lesson 7-12

A repeating arrangement of one or more shapes that completely covers a plane, with no gaps or overlaps, is called a *tessellation*. You can make your own tessellations using paper, scissors, and tape.

internet connect
Lab Resources Online
go.hrw.com
KEYWORD: MR4 Lab7C

Activity

1 Start with a square.

Use scissors to cut out a shape from one side of the square.

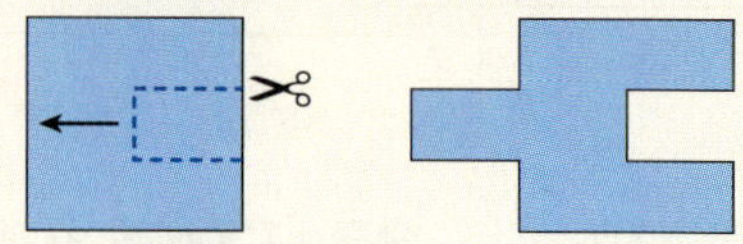

Translate the shape you cut out to the opposite side of the square and tape the two pieces together.

Trace this new shape to form at least two rows of a tessellation. You will need to translate, rotate, or reflect the shape.

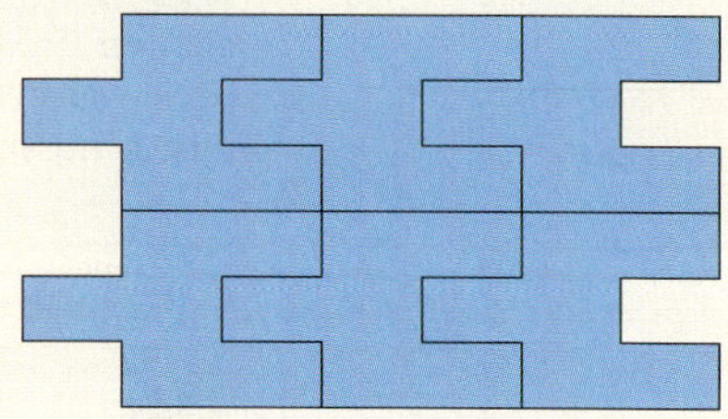

2 Start again with a square.

Use scissors to cut out shapes and move them to the opposite sides of the square.

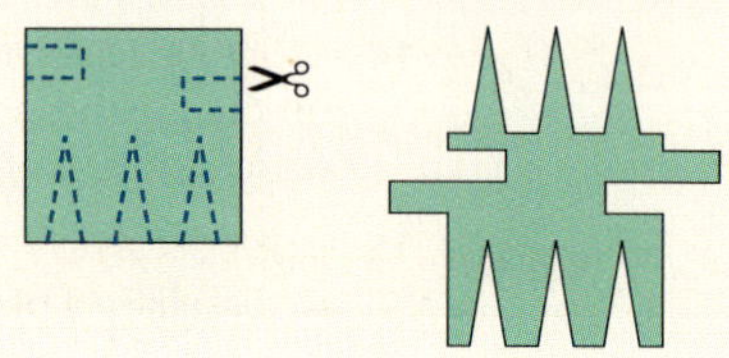

Trace this new shape to form at least two rows of a tessellation. You will need to translate, rotate, or reflect the shape.

Teacher to Teacher

Kendra Vaught
San Gabriel, California

When we work through the activity, I like to highlight the vocabulary and reinforce it for my ELD students. Once they have successfully made their original template and traced it a few times, I challenge them to color it creatively and see whether they can make a picture.

When they have done the translation, we go through the same process with the rotations. When the projects are finished, they look great on the bulletin board. I like to challenge them to tell me the rule that polygons must satisfy to make a successful tessellation. (The sum of the angles must equal 360°.)

3. You can base a tessellating shape on other polygons.

Try starting with a hexagon.

Use scissors to cut out a shape from one side of the hexagon. Translate the shape to the opposite side of the hexagon.

Try repeating these steps on other sides of the hexagon.

Trace the new shape to form a tessellation. You will need to translate, rotate, or reflect the shape.

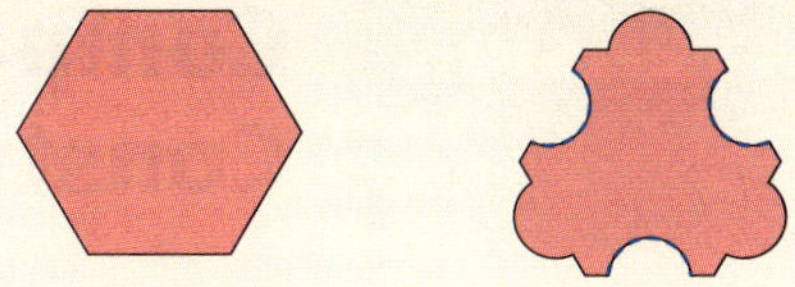

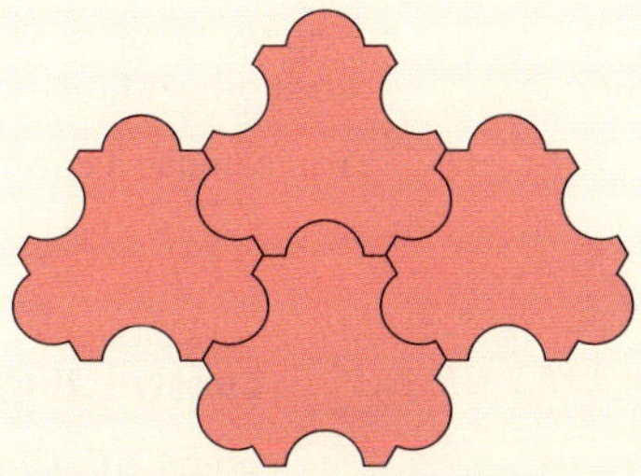

Think and Discuss

1. Tell whether you can make a tessellation out of circles. **No. Circles have no edges that allow them to fit together.**
2. Tell whether any polygon can make a tessellation. **No. Some polygons do not fit together without leaving gaps.**

Try This

Make each tessellation shape described. Then form two rows of a tessellation. **Check students' work.**

1. 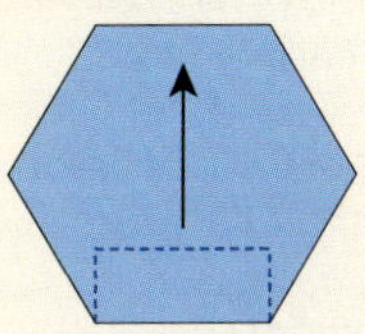

2. 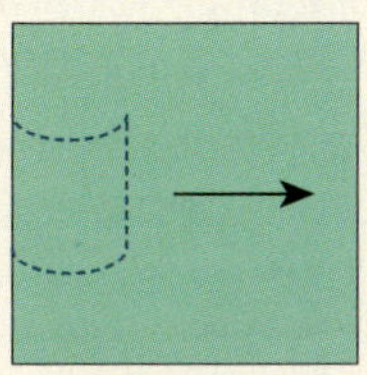

3. 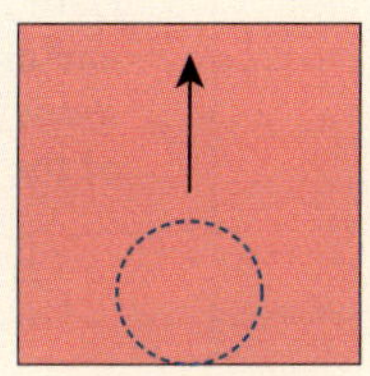

Tell whether each shape can be used to form a tessellation.

4.  **yes**

5. 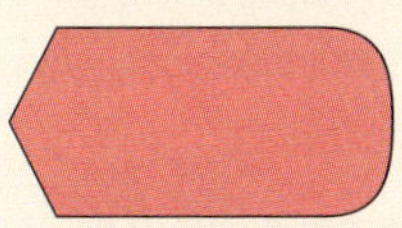 **no**

6. 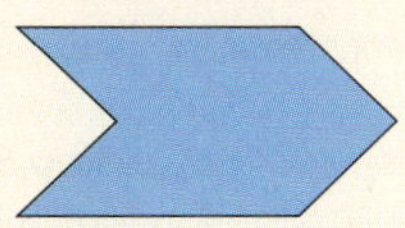 **yes**

7. Cut out a polygon, and then change it by cutting out a part of one side. Translate the cut-out part to the opposite side. Can your shape form a tessellation? Make a drawing to show your answer. **Check students' work.**

Answers

Try This

1. 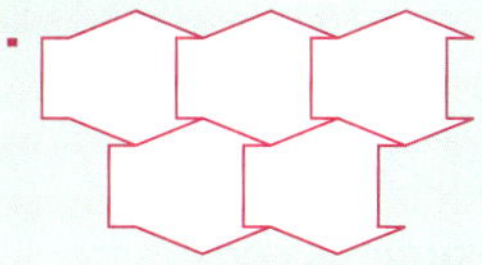

2. 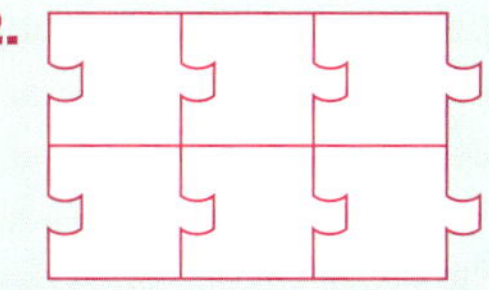

3. 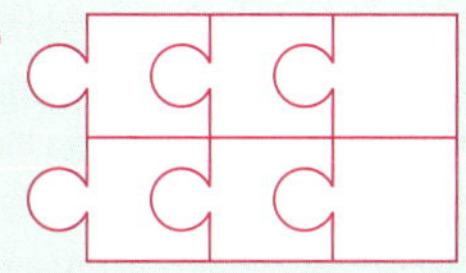

CHAPTER 7 Extension

Pacing: Traditional 1 day
Block $\frac{1}{2}$ day

Objective: Students construct perpendicular bisectors and angle bisectors.

Using the Pages

In Lab 7A, students constructed congruent segments and congruent angles. In this extension, students construct perpendicular bisectors and angle bisectors.

EXTENSION Compass and Straightedge Constructions

Learn to construct perpendicular bisectors and angle bisectors.

Vocabulary
bisect

To **bisect** something means to divide it into two congruent parts. The perpendicular bisector of a line segment divides it into two line segments of equal length and is perpendicular to the original segment. The midpoint of a segment is the point halfway between the endpoints of the segment. The midpoint bisects the line segment.

You can use a compass and straightedge to bisect a line segment.

EXAMPLE 1 Constructing a Perpendicular Bisector

Given $\overline{AC}$, construct its perpendicular bisector.

Step 1
Open your compass to a distance greater than half of $\overline{AC}$ and do not adjust it throughout the construction.

Step 2
Place your compass point on A and draw an arc as shown.

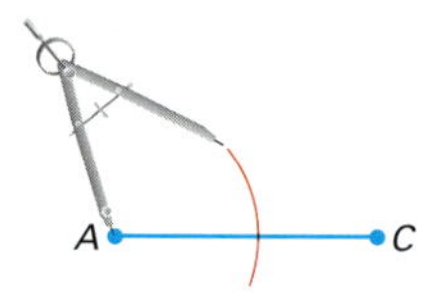

Step 3
Place your compass point on C and draw a second arc as shown. Label the intersections of the two arcs B and D.

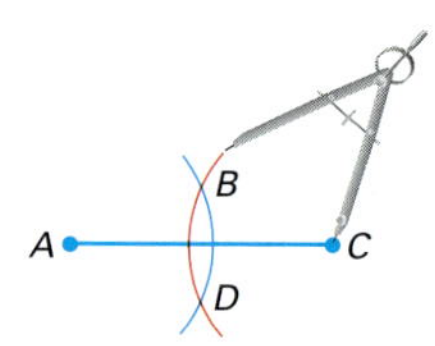

Step 4
Use a straightedge to draw $\overleftrightarrow{BD}$, the perpendicular bisector of $\overline{AC}$.

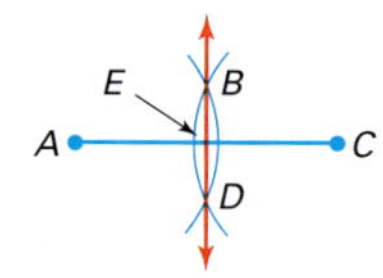

Step 5 *Check*
Use a ruler to measure the length of $\overline{AE}$ and $\overline{EC}$. $\overline{AE} \cong \overline{EC}$.
Use a protractor to measure $\angle AEB$ and $\angle BEC$.
$m\angle AEB = 90°$ and $m\angle BEC = 90°$.
$\overline{AC} \perp \overleftrightarrow{BD}$

1 Introduce

Motivate

Engineers, architects, fashion designers, landscape designers, builders, and artists use accurate constructions to perform their crafts. A geometric construction is a mathematical drawing done with the use of only a straightedge and a compass. Two basic geometric constructions are constructing a perpendicular bisector and bisecting an angle.

2 Teach

Lesson Presentation

Guided Instruction

In this extension, students construct perpendicular bisectors and angle bisectors. Define *bisect* and review *perpendicular.* Teach students to construct the perpendicular bisector of a segment. Then teach students the steps in constructing the bisector of an angle.

Teaching Tip If some students are using a traditional compass, with a metal point, suggest they put several layers of paper under their worksheets to help keep the compass from slipping.

The bisector of an angle divides the angle into two angles of equal measure.

EXAMPLE 2 Constructing an Angle Bisector

Given $\angle A$, construct its bisector.

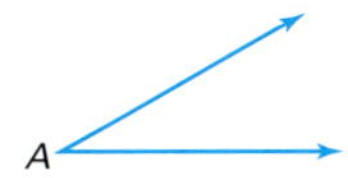

Step 1

Place your compass point on A and draw an arc through the rays of the angle as shown. Label the intersection points B and C.

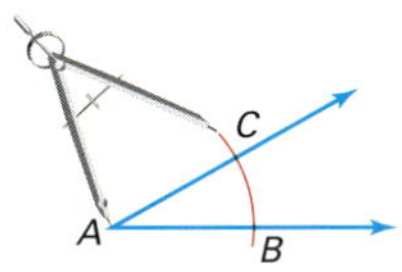

Step 2

Place your compass point on B and draw an arc inside of $\angle A$. Then, without adjusting your compass, place the compass point on C and draw an arc that intersects the first arc. Label the intersection of the arcs F.

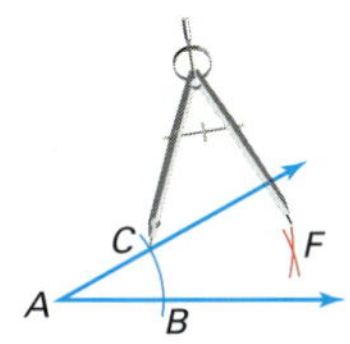

Step 3

Draw a ray from A through F. $\overrightarrow{AF}$ is the bisector of $\angle BAC$.

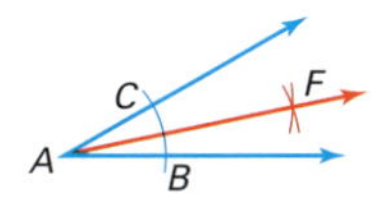

EXTENSION Exercises

1. Draw line segment DH 4 centimeters long. Then construct $\overleftrightarrow{XY}$, a perpendicular bisector of $\overline{DH}$.
2. Draw $\angle L$, and then construct a bisector $\overrightarrow{LG}$. $m\angle L = 120°$.
3. Draw $\angle S$, and then construct a bisector $\overrightarrow{SF}$. $m\angle S = 60°$.

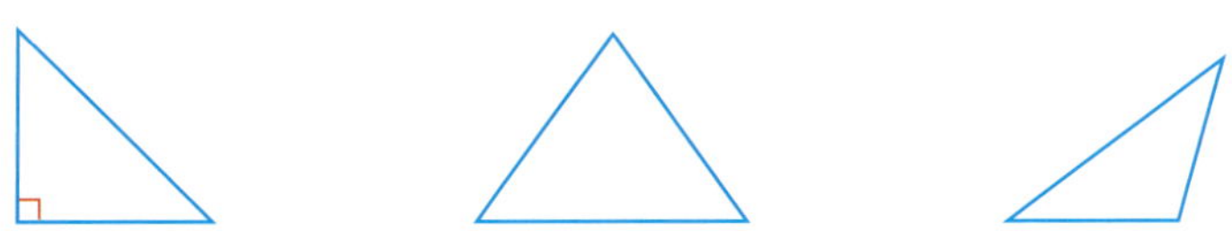

4. Trace each of the triangles above. For each triangle, construct the perpendicular bisector of each side. What do you notice?
5. Trace each of the triangles above. For each triangle, construct the bisector of each angle. What do you notice?

COMMON ERROR ALERT

When constructing a perpendicular bisector, some students may not open their compasses wide enough on the first step. Remind them to open their compasses to a distance greater than half of the line segment.

Additional Examples

Example 1

Given $\overline{XY}$, construct its perpendicular bisector.

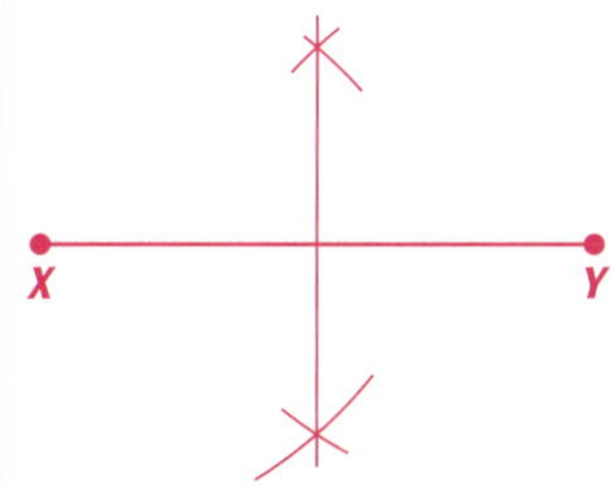

Example 2

Given $\angle B$, construct its bisector.

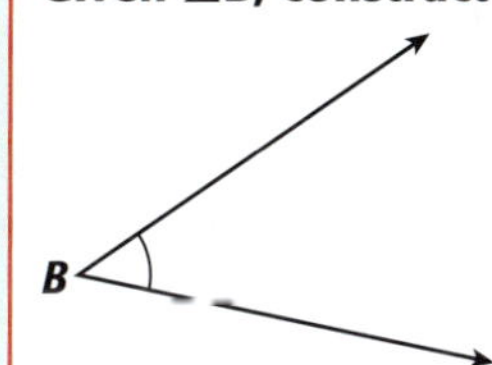

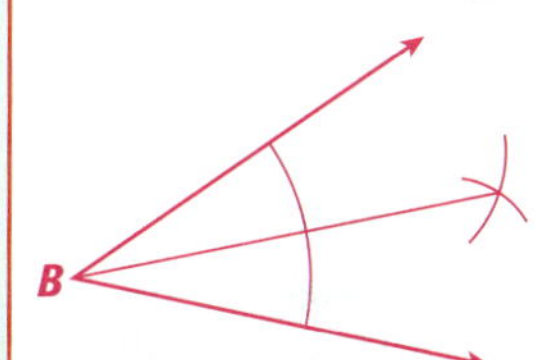

You may use the recording sheet on Chapter 7 Resource Book p. 139 for Exercises 4 and 5.

3 Close

Summarize

Have the students demonstrate how to construct the perpendicular bisector of a segment. Then have students demonstrate how to construct an angle bisector.

Answers

1.

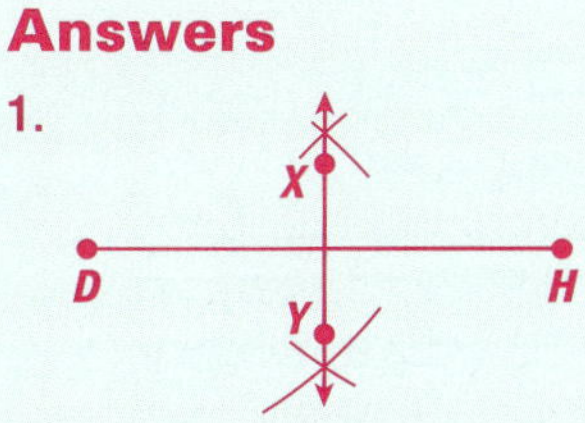

2.

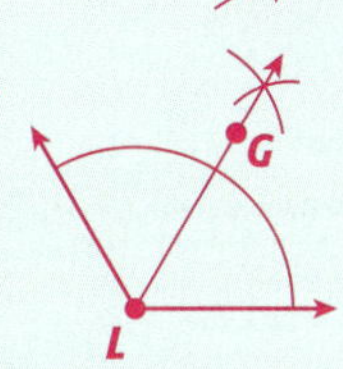

3.

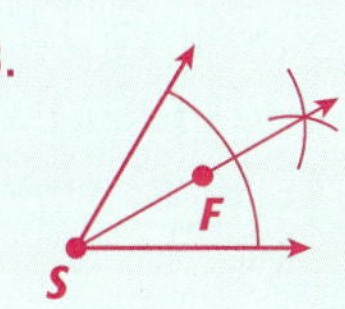

4–5. See p. A5.

Problem Solving on Location

Missouri

Purpose: *To provide additional practice for problem-solving skills in Chapters 1–7*

The Gateway Arch

- After problem 3, have students use any geometric vocabulary they can to describe the triangles that make up the legs of the arch.

  Possible answer: Each triangle is an acute equilateral triangle. The triangles are all similar. Each triangle is a transformation (dilation and translation) of the others.

- After problem 4, have students extend their thinking about height by considering the following: Does the elevator that carries tourists to the top of the Gateway Arch travel exactly 630 feet? Explain.

  Possible answer: The elevator will travel more than 630 feet. The distance from the bottom of one leg to the top of the arch is greater than 630 feet because of the structure's curve.

Extension Have students make a model of the Gateway Arch using clay or modeling dough. Make sure the legs of the arch have the triangular shape. Check students' work.

Problem Solving on Location

MISSOURI

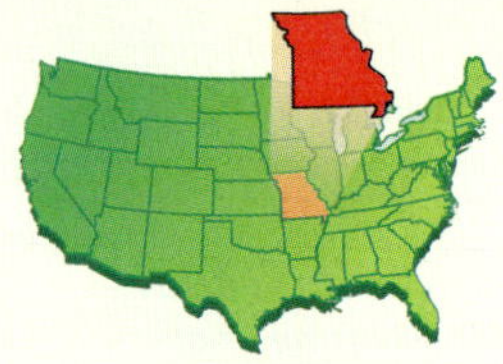

The Gateway Arch

The St. Louis waterfront on the Mississippi River is the site of the Gateway Arch. Under the Arch is a museum that tells the story of the opening of the West in the 1800s.

The Arch is 630 ft tall. The ends of the Arch are 630 ft apart, and they get narrower as the arch goes up. The arch is made up of triangular sections. Each section is an equilateral triangle. At ground level, the triangular sections have sides 54 ft long, and at the top, the sections have sides 17 ft long.

1. What is the distance around a triangular section at the bottom of one of the ends of the Arch? **162 ft**
2. What is the distance around a triangular section at the top of the Arch? **51 ft**
3. Give a possible side length for a triangular section that is above ground level but not at the top of the Arch.
4. One story is about 10 feet tall. About how many stories tall is the Arch? **63 stories**

Forty city blocks of St. Louis had to be cleared to build the Arch. Below is a partial map of St. Louis in 1804.

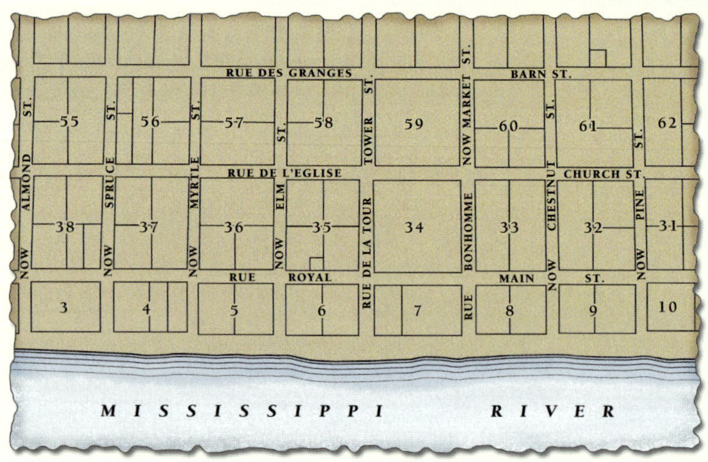

For 5–7, use the map.

5. Name two streets that appear parallel to each other.
6. Name two streets that appear perpendicular to each other.
7. Name one street that appears to be parallel to the Mississippi River.

Answers

Possible answers:

3. Answer should be a length between 17 ft and 54 ft.
5. Barn St. and Church St.
6. Barn St. and Tower St.
7. Main St.

The Climatron

The Climatron, a geodesic dome, is a major attraction at the Missouri Botanical Garden in St. Louis. A geodesic dome is a strong structure made up of short, straight, lightweight bars that form a grid of polygons. The Climatron is formed out of aluminum and acrylic hexagons.

The Climatron houses a rainforest, streams, waterfalls, and 1,200 species of plants, including banana trees, cacao trees, coffee trees, passion flowers, and orchids. The dome was named *Climatron* to emphasize the controlled climate in the greenhouse.

1. What is the measure of one angle in a regular hexagon? **120°**

2. Name some polygons other than hexagons that you see in the Climatron. **triangles; parallelograms**

3. There are 72 different triangular glass panes in the Climatron. Draw five different triangles, and classify each triangle. **Check students' work.**

Look at the pictures to see how the Climatron compares in size with some other famous domes.

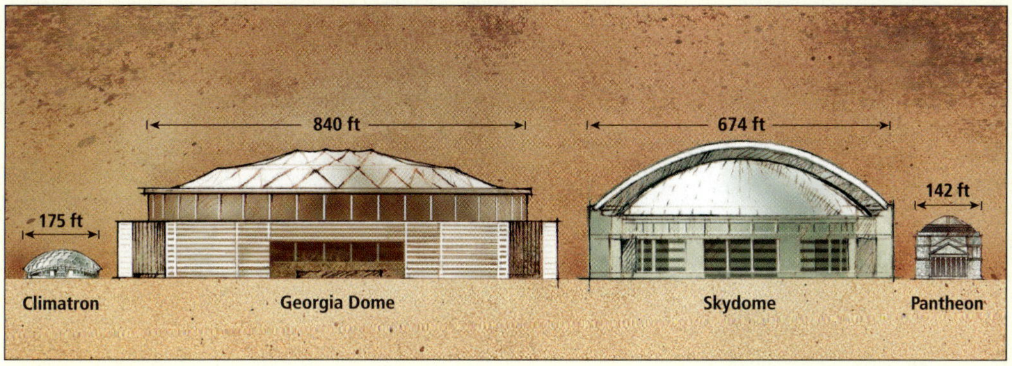

For 4–6, use the pictures of the domes.

4. How much longer is the distance across the base of the Climatron than the distance across the base of the Pantheon? **33 ft**

5. How much shorter is the distance across the base of the Climatron than the distance across the base of the Georgia Dome? **665 ft**

6. Order the domes from longest distance across the base to shortest distance across the base. **Georgia Dome, Skydome, Climatron, Pantheon**

The Climatron

- After problem 1, have students show how they can divide a regular hexagon into six equilateral triangles in order to find the measure of each interior angle.

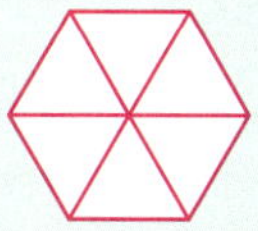

Each angle of each triangle measures 60°. Each angle of the hexagon is made up of two adjacent 60° angles, so each interior angle of the hexagon measures 120°.

- After problem 3, have students consider how many possible triangle classifications there are. Have them list all the possibilities.
  There are 7 possibilities:
  equilateral
  scalene acute
  isosceles acute
  isosceles right
  scalene right
  isosceles obtuse
  scalene obtuse

Extension Encourage students to construct a geodesic dome using toothpicks joined by small balls of clay. Check students' work.

MATH-ABLES

Game Resources

Puzzles, Twisters & Teasers
Chapter 7 Resource Book

Tangrams

Purpose: *To apply knowledge of geometry to visualizing and creating shapes*

Discuss: Ask students to describe the shapes that make up the tangram puzzle. two large isosceles triangles; one medium-sized isosceles triangle; two small isosceles triangles; one square; one parallelogram. What fraction of the area of the square made from all of the tangram pieces is one large isosceles triangle? $\frac{1}{4}$ What fraction of the large isosceles triangle's area is the medium-sized isosceles triangle? $\frac{1}{2}$

Extend: Have students make a figure using all seven tangram pieces. Have them trace around their shape to form an outline. Then have them challenge a classmate to re-create the shape using only the outline. Check students' work.

MATH-ABLES

Tangrams

A tangram is an ancient Chinese puzzle. The seven shapes that make this square can be arranged to make many other figures. Copy the shapes that make this square, and then cut them apart. See if you can arrange the pieces to make the figures below.

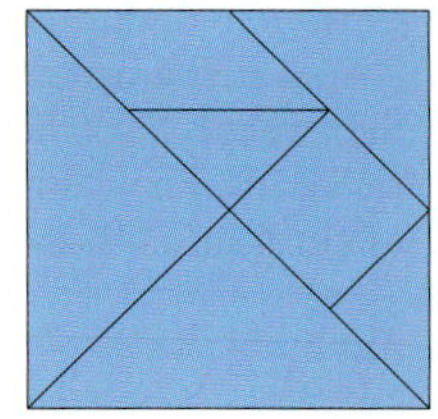

internet connect
Go to ***go.hrw.com*** to print out Tangram puzzle pieces.
KEYWORD: MR4 Game7

Answers

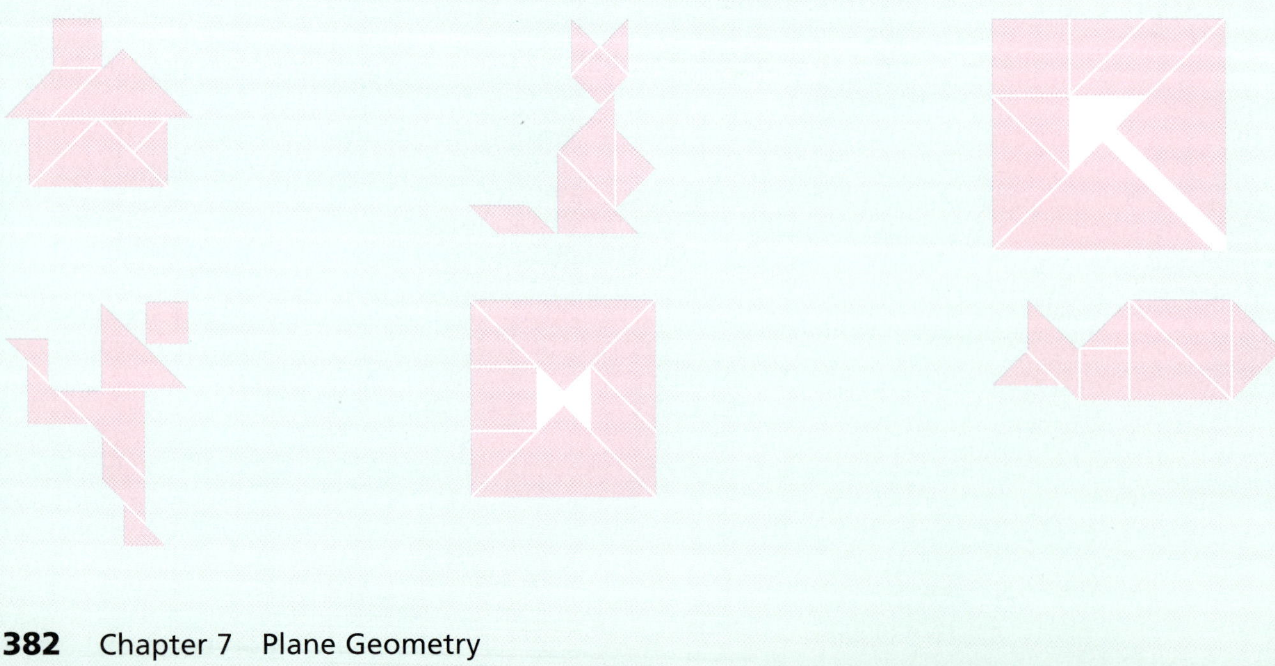

Technology LAB: Angles in Triangles

internet connect
Lab Resources Online
go.hrw.com
KEYWORD: MR4 TechLab7

The sum of the angle measures is the same for any triangle. You can use geometry software to find this sum and to check that the sum is the same for many different triangles.

Activity

1. Use the geometry software to make triangle *ABC*. Then use the angle measure tool to measure ∠*B*.

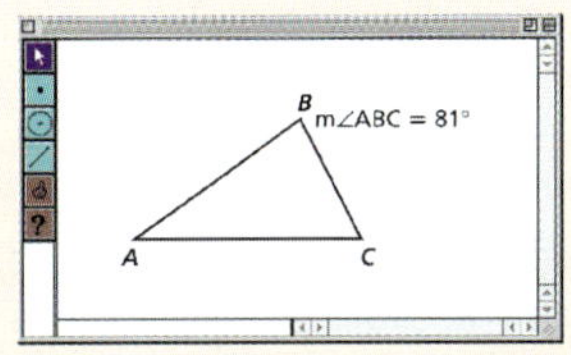

2. Use the angle measure tool to measure ∠*C* and ∠*A*. Then use the calculator tool to add the measures of the three angles. Notice that the sum is 180°.

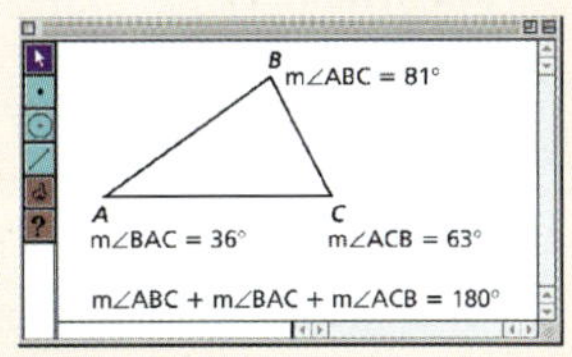

3. Select vertex *A* and drag it around to change the shape of triangle *ABC*. Watch the angle sum. Change the shape of the triangle again and then again. Be sure to make acute and obtuse triangles.

   Notice that the sum of the angle measures is always 180°, regardless of the triangle's shape.

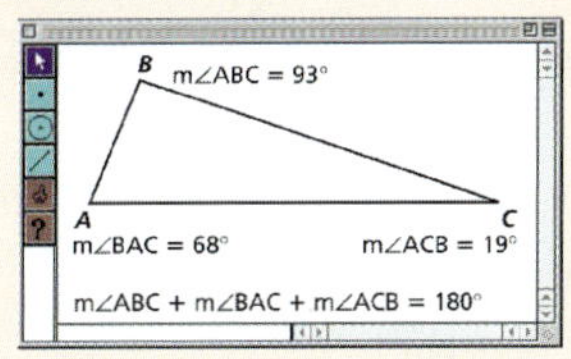

Think and Discuss

1. Can you use geometry software to draw a triangle with two obtuse angles? Explain. **No, the sum of the angles cannot be greater than 180°.**

Try This

Solve. Then use geometry software to check each answer.

1. In triangle *ABC*, m∠*B* = 49.15° and m∠*A* = 113.75°. Find m∠*C*. **m∠C = 17.1°**
2. Use geometry software to construct an acute triangle *XYZ*. Give the measures of its angles, and check that their sum is 180°. **Check students' work.**

Technology LAB: Angles in Triangles

Objective: To use geometry software to explore the angles in a triangle

Materials: Geometry software

Lab Resources

Technology Lab Activities . . . . p. 30

Using the Page

This technology activity shows students how to use geometry software to explore the sum of the measures of the angles of a triangle. Specific instructions may vary, depending on the geometry software used. The instructions given are for Geometer's Sketch Pad. For instructions to other software, visit go.hrw.com.

The Think and Discuss problem can be used to assess students' understanding of the technology activity. While Try This problems 1 and 2 can be done without geometry software, they are meant to help students become familiar with using geometry software to measure the angles of a triangle.

Assessment

1. Use the software to construct a triangle with one obtuse angle. Label the triangle *MNP*. What is the measure of each angle of the triangle? Check students' work.
2. In triangle *XYZ*, m∠*X* = 150° and m∠*Y* = 25°. What is m∠*Z*? 5°

Chapter 7 Study Guide and Review

Purpose: *To help students review and practice concepts and skills presented in Chapter 7*

Assessment Resources

Chapter Review

Chapter 7 Resource Book . pp. 120–122

***Test and Practice Generator* CD-ROM**

Additional review items in both multiple-choice and free-response format may be generated for any objective in Chapter 7.

Answers

1. trapezoid
2. polygon
3. Possible answer: $\overleftrightarrow{ED}$, $\overleftrightarrow{AD}$
4. acute
5. obtuse
6. acute

Chapter 7 Study Guide and Review

Study Guide and Review

Vocabulary

acute angle 326	obtuse triangle 344	right triangle 344
acute triangle 344	parallel lines 336	rotation 365
adjacent angles 332	parallelogram 348	scalene triangle 345
angle 326	perpendicular lines 336	skew lines 336
complementary angles 332	plane 322	square 348
congruent 332	point 322	straight angle 326
equilateral triangle 345	polygon 352	supplementary angles 332
isosceles triangle 345	quadrilateral 348	tessellation 373
line 322	ray 323	transformation 365
line of reflection 365	rectangle 348	translation 365
line of symmetry 369	reflection 365	trapezoid 348
line segment 323	regular polygon 352	vertex 326
line symmetry 369	rhombus 348	vertical angles 332
obtuse angle 326	right angle 326	

Choose the best term from the list above. Words may be used more than once.

1. A quadrilateral with exactly two parallel sides is called a(n) __?__.
2. A(n) __?__ is a closed plane figure formed by three or more line segments.

7-1 Points, Lines, and Planes (pp. 322–325)

EXAMPLE

■ **Use the diagram.**

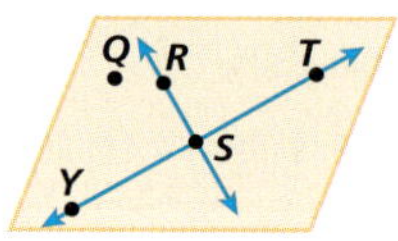

Name a line. $\overleftrightarrow{RS}$
Name a line segment. $\overline{ST}$

EXERCISE

Use the diagram.

3. Name two lines.

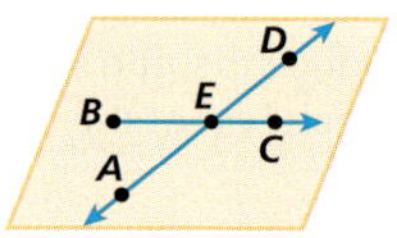

7-2 Angles (pp. 326–329)

EXAMPLE

■ **Classify each angle as acute, right, obtuse, or straight.**

$m\angle A = 80°$
$80° < 90°$, so $\angle A$ is acute.

EXERCISES

Classify each angle as acute, right, obtuse, or straight.

4. $m\angle x = 60°$
5. $m\angle x = 100°$
6. $m\angle x = 45°$

7-3 Angle Relationships (pp. 332–335)

EXAMPLE

■ **Find the unknown angle measure.**

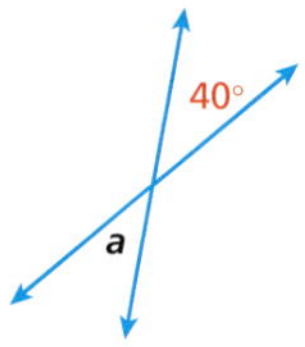

m∠a = 40° *Vertical angles are congruent.*

EXERCISES

Find each unknown angle measure.

7. 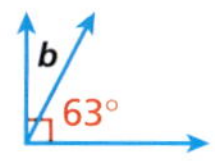

8. 

7-4 Classifying Lines (pp. 336–339)

EXAMPLE

■ **Classify the lines.**

The red lines are parallel.
The blue lines are perpendicular.

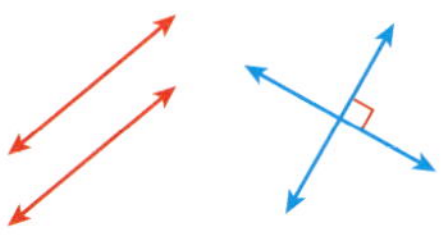

EXERCISES

Classify each pair of lines.

9. 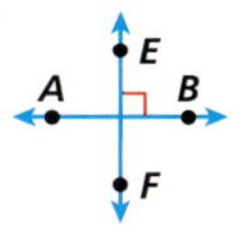

10. 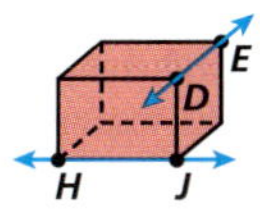

7-5 Triangles (pp. 344–347)

EXAMPLE

■ **Classify the triangle using the given information.**

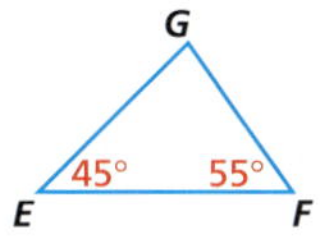

m∠G + 45° + 55° = 180°
m∠G = 80°, so △EFG is an acute triangle.

EXERCISES

Classify the triangle using the given information.

11. 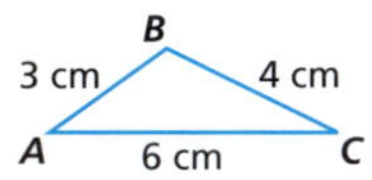

7-6 Quadrilaterals (pp. 348–351)

EXAMPLE

■ **Give the most exact name for the figure.**

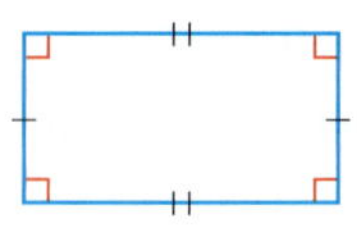

The most exact name is rectangle.

EXERCISES

Give the most exact name for the figure.

12. 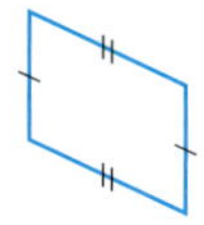

7-7 Polygons (pp. 352–355)

EXAMPLE

■ **Name the polygon and tell whether it appears to be regular or not regular.**

It is a regular octagon.

EXERCISES

Name each polygon and tell whether it appears to be regular or not regular.

13. 

14. 

Study Guide and Review

Answers

7. $b = 27°$
8. $d = 98°$
9. perpendicular
10. skew
11. obtuse scalene
12. parallelogram
13. triangle; not regular
14. rectangle; not regular

Answers

15. Add 1 shaded and 1 white triangle each time.

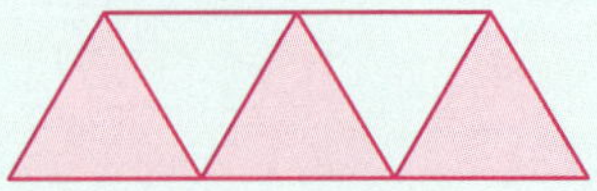

16. not congruent; different sizes

17. congruent

18. translation

19. The line is a line of symmetry.

20. The figure can tessellate the plane.

Study Guide and Review

7-8 Geometric Patterns (pp. 356–359)

EXAMPLE

■ **Identify a possible pattern. Use the pattern to draw the missing figure.**

The missing figure might be 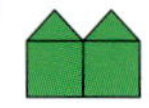.

EXERCISE

Identify a possible pattern. Use the pattern to draw the missing figure.

15. 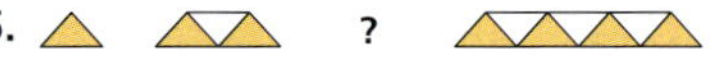

7-9 Congruence (pp. 362–364)

EXAMPLE

■ **Decide whether the figures are congruent. If not, explain.**

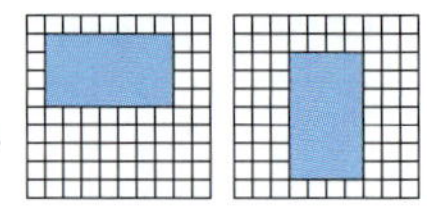

These figures are congruent.

EXERCISES

Decide whether the figures in each pair are congruent. If not, explain.

16. 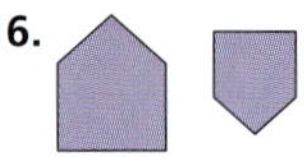

17. 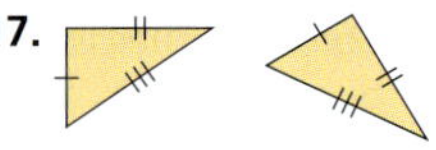

7-10 Transformations (pp. 365–368)

EXAMPLE

■ **Tell whether the transformation is a translation, rotation, or reflection.**

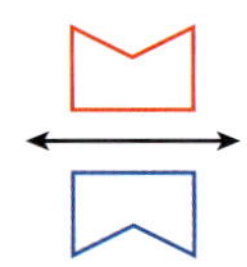

The transformation is a reflection.

EXERCISE

Tell whether the transformation is a translation, rotation, or reflection.

18. 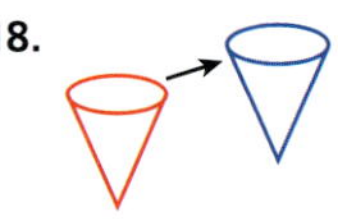

7-11 Symmetry (pp. 369–372)

EXAMPLE

■ **Determine whether the dashed line appears to be a line of symmetry.**

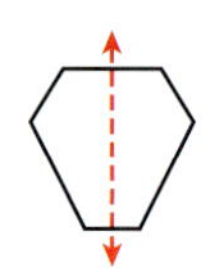

The line appears to be a line of symmetry.

EXERCISE

Determine whether the dashed line appears to be a line of symmetry.

19. 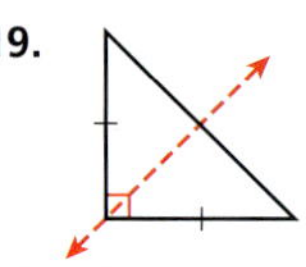

7-12 Tessellations (pp. 373–375)

EXAMPLE

■ **Identify whether the polygon can tessellate the plane.**

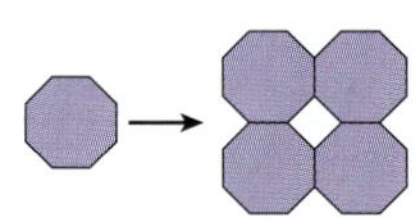

The octagon cannot tessellate the plane.

EXERCISES

Identify whether the polygon can tessellate the plane.

20. 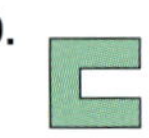

Chapter Test

Chapter 7

Classify each pair of angles or lines.

1. 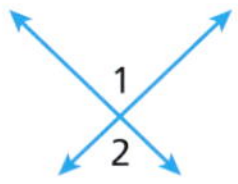

vertical angles

2. 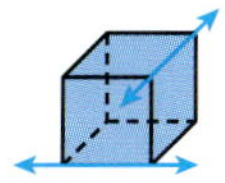
skew

3. 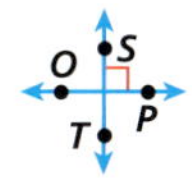

perpendicular

Classify the triangles by angle and side measures. Possible answers given.

4. 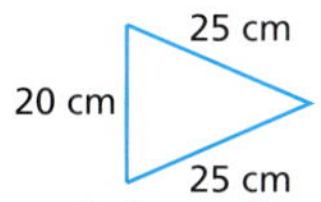

acute isosceles

5. 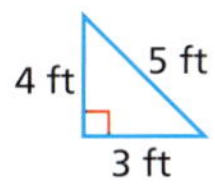

right scalene

6. 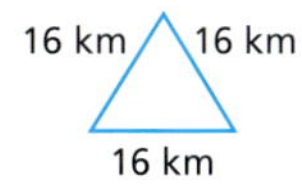

acute equilateral

Find the unknown angle measure.

7. 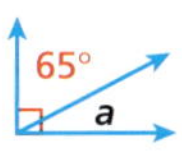

$a = 25°$

8. 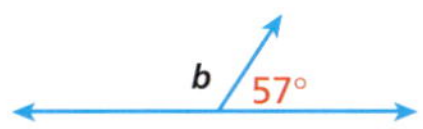

$b = 123°$

9. 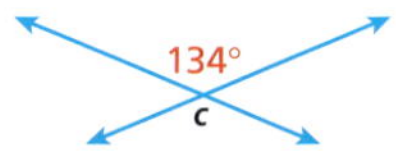

$c = 134°$

10. Triangle ABC has sides of equal length. The measure of $\angle A$ is 60°, and the measure of $\angle B$ is 60°. What is the measure of $\angle C$? Classify the triangle based on the measures of the angles and lengths of the sides.
$\angle C$ is 60°; The triangle is an acute, equilateral triangle.

Draw the indicated transformation.

11. Reflect across the line.

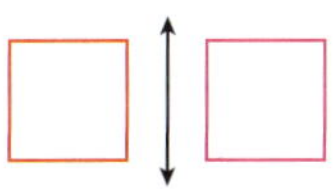

12. Rotate 270° clockwise about the point.

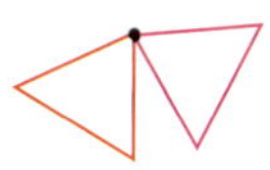

13. Translate $\frac{3}{4}$ in. right.

Make a drawing to show whether the figure can tessellate the plane.

14. 
does tessellate

15. 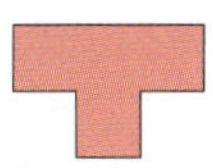
does tessellate

16. 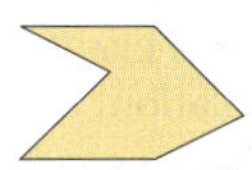
does tessellate

Decide whether the figures in each pair are congruent. If not, explain.

17. 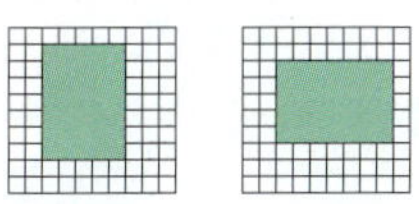
congruent

18. 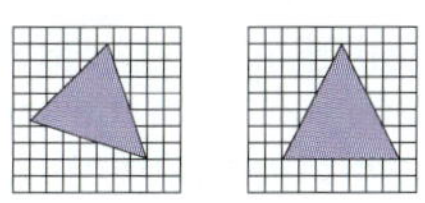
congruent

Identify a possible pattern. Use your pattern to draw the next figure.

19. 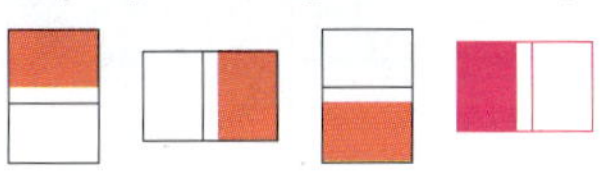
rotate 90° clockwise

20. 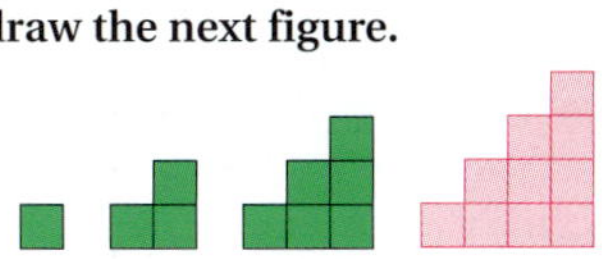
increase the number of squares to the right by 1

Chapter Test

Chapter 7

Purpose: *To assess students' mastery of concepts and skills in Chapter 7*

Assessment Resources

Chapter 7 Tests (Levels A, B, C)
Assessment Resources ..... pp. 67–72

***Test and Practice Generator* CD-ROM**

Additional assessment items in both multiple-choice and free-response format may be generated for any objective in Chapter 7.

Answers

14. does tessellate

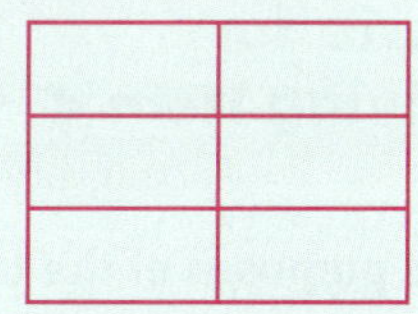

15. does tessellate

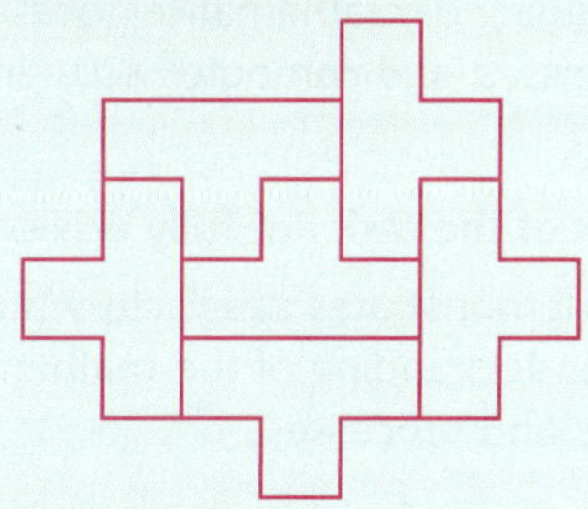

16. does tessellate

Chapter 7 **Performance Assessment**

Purpose: To assess students' understanding of concepts in Chapter 7 and combined problem-solving skills

Assessment Resources

Performance Assessment
Assessment Resources . . . . p. 116

Performance Assessment Teacher Support
Assessment Resources . . . . p. 115

Use the recording sheet on Chapter 7 Resource Book p. 140 for item **3b**.

Answers

1–3. See p. A6.

4. See Level 3 work sample below.

Scoring Rubric for Problem Solving Item 4

Level 3

Accomplishes the purposes of the task.

Student gives clear explanations, shows understanding of mathematical ideas and processes, and computes accurately.

Level 2

Purposes of the task not fully achieved.

Student demonstrates satisfactory but limited understanding of the mathematical ideas and processes.

Level 1

Purposes of the task not accomplished.

Student shows little evidence of understanding the mathematical ideas and processes and makes computational and/or procedural errors.

Chapter 7 **Performance Assessment**

Performance Assessment

Show What You Know

Create a portfolio of your work from this chapter. Complete this page and include it with your four best pieces of work from Chapter 7. Choose from your homework or lab assignments, mid-chapter quizzes, or any journal entries you have done. Put them together using any design you want. Make your portfolio represent what you consider your best work.

Short Response

1. Monica drew a quadrilateral. Then she drew a line segment connecting one pair of opposite corners. She saw that she had divided the quadrilateral into two congruent right isosceles triangles. Draw and classify all of the possible quadrilaterals she could have drawn.

2. José drew two intersecting lines. Of the four angles he formed, two adjacent angles are congruent. What can you conclude about the relationship between the two lines and the measures of each angle? Explain.

3. Nancy drew a quadrilateral with a right angle. The other three angles in her figure are congruent. Let x represent the measure of the other three angles. Write and solve an equation to find the value of x. Name the kind of quadrilateral Nancy drew.

Extended Problem Solving

4. Roger Penrose, a famous mathematician, used plane figures to create interesting patterns like the one below through tessellation.

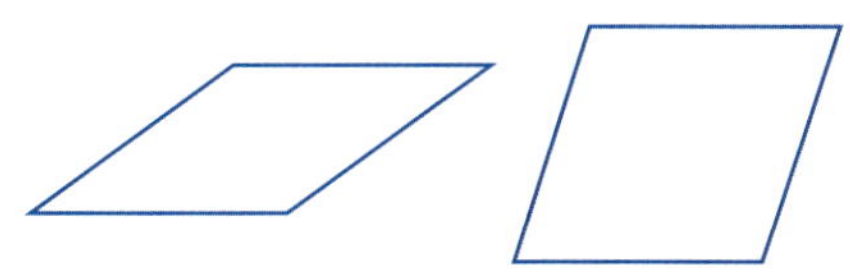

   a. Give the name that best describes each of the two plane figures at right.
   b. The obtuse angle in the first figure has a measure of 144°. Write an equation that could be used to determine the measure m of the acute angle in this figure. Solve for m. Show your work.
   c. Trace the two figures to create your own Penrose tiling.

Student Work Samples for Item 4

Level 3

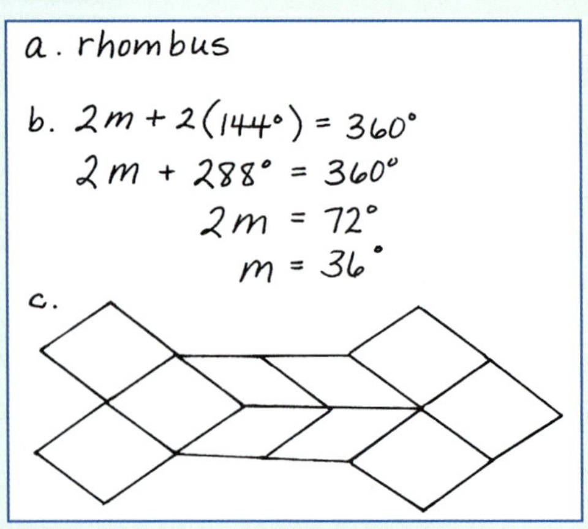

The student gave complete, correct answers and showed all the steps in part **b.**

Level 2

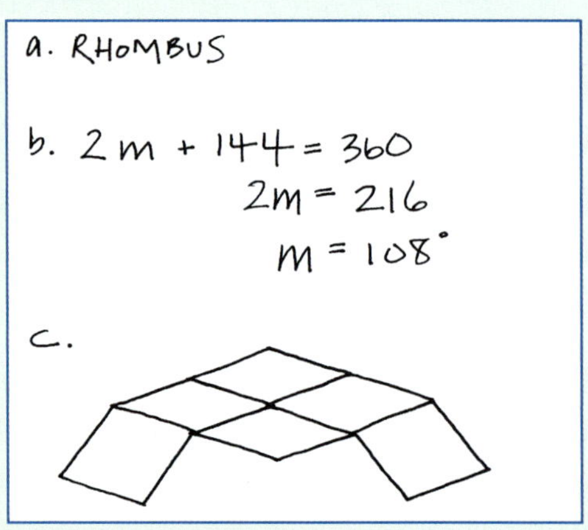

The student made an error when writing an equation for part **b,** which resulted in an incorrect answer.

Level 1

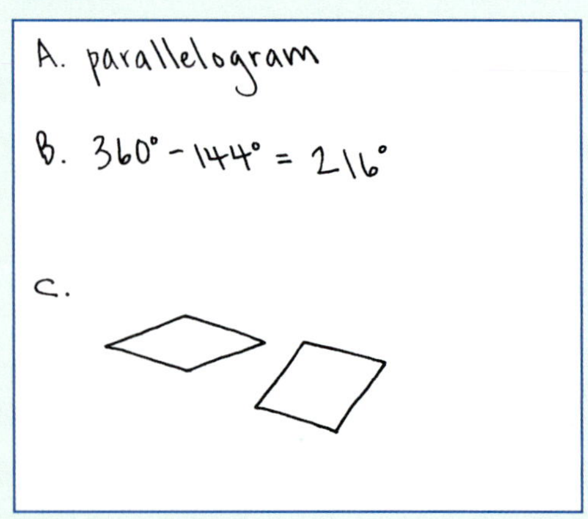

The student gave incorrect or incomplete answers and did not show an understanding of quadrilaterals.

internet connect
State-Specific Test Practice Online
go.hrw.com Keyword: MR4 TestPrep

Standardized Test Prep

Chapter 7

Cumulative Assessment, Chapters 1–7

1. Which of the following is an acute angle? C

Ⓐ

Ⓒ 

Ⓑ 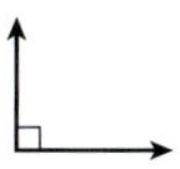

Ⓓ 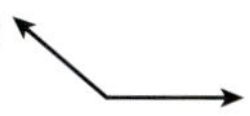

2. Two angles whose measures have a sum of 180° are what type of angles? J

Ⓕ Congruent

Ⓖ Vertical

Ⓗ Complementary

Ⓙ Supplementary

3. In which equation does $n = 3$? D

Ⓐ $n + 13 = 17$ Ⓒ $4.8 - n = 1.2$

Ⓑ $4 = 12n$ Ⓓ $\frac{5}{6} \cdot n = 2\frac{1}{2}$

4. Which of the following numbers is divisible by 2, 3, and 6, but **not** 9? F

Ⓕ 312 Ⓗ 288

Ⓖ 306 Ⓙ 256

5. Subtract. $6\frac{3}{4} - 3\frac{5}{8}$ B

Ⓐ $3\frac{1}{2}$ Ⓒ $2\frac{3}{4}$

Ⓑ $3\frac{1}{8}$ Ⓓ $2\frac{1}{4}$

6. Which expression has the greatest value? J

Ⓕ $2^3 + 20 \div 4$

Ⓖ $3.85 \div 0.25$

Ⓗ $15\frac{1}{3} \div 2$

Ⓙ $4\frac{1}{4} \cdot 3\frac{2}{3}$

7. What term is used to describe quadrilaterals that have opposite sides that are parallel and congruent and opposite angles that are congruent? C

Ⓐ Rectangle

Ⓑ Tessellation

Ⓒ Parallelogram

Ⓓ Trapezoid

8. What type of transformation is shown in the picture? F

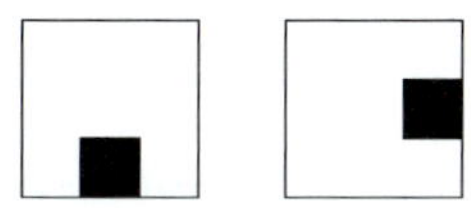

Ⓕ 90° counterclockwise rotation

Ⓖ Translation to the right

Ⓗ Horizontal reflection

Ⓙ 180° counterclockwise rotation

TIP!

TEST TAKING TIP!

The sum of the measures of the angles in a triangle is 180°.

9. ***SHORT RESPONSE*** For triangle ABC, $\angle A$ measures 35° and $\angle B$ measures 30°. Show the steps necessary to find the measure of $\angle C$, and classify the triangle by its angles.

10. ***SHORT RESPONSE*** Find the median and mode of the following data set. Explain how you found your answers.

13, 8, 9, 10, 12, 11, 13, 13, 10, 18

Standardized Test Prep

Chapter 7

Purpose: *To provide review and practice for Chapter 1–7 and standardized tests*

Assessment Resources

Cumulative Tests (Levels A, B, C)
Assessment Resources. . . . pp. 199–210

State-Specific Test Practice Online
KEYWORD: MR4 Testprep

Test Prep Doctor

In item 8, eliminate answer choices **H** and **J** because the angle measures do not total 180°. Eliminate answer choice **F** because a 115° angle is not acute.

Answers

9. $m\angle A + m\angle B + m\angle C = 180°$
$35° + 30° + m\angle C = 180°$
$65° + m\angle C = 180°$
$m\angle C = 180° - 65°$
$m\angle C = 115°$;
obtuse

10. 8, 9, 10, 10, 11, 12, 13, 13, 13, 18

median $= \frac{11 + 12}{2} = 11.5$

mode = 13 (appears 3 times)

mean $= \frac{8 + 9 + 10 + 10 + 11 + 12 + 13 + 13 + 13 + 18}{10}$
$= 11.7$

Chapter 8

Ratio, Proportion, and Percent

Section 8A

Ratios and Proportions

Lesson 8-1
Ratios and Rates

Hands-On Lab 8A
Explore Proportions

Lesson 8-2
Proportions

Lesson 8-3
Proportions and Customary Measurement

Lesson 8-4
Similar Figures

Lesson 8-5
Indirect Measurement

Lesson 8-6
Scale Drawings and Maps

Section 8B

Percents

Lesson 8-7
Percents

Lesson 8-8
Percents, Decimals, and Fractions

Lesson 8-9
Percent Problems

Hands-On Lab 8B
Construct Circle Graphs

Lesson 8-10
Using Percents

Extension
Simple Interest

Pacing Guide for 45-Minute Classes

Chapter 8

DAY 101	DAY 102	DAY 103	DAY 104	DAY 105
Lesson 8-1	Hands-On Lab 8A	Lesson 8-2	Lesson 8-3	Lesson 8-4
DAY 106	**DAY 107**	**DAY 108**	**DAY 109**	**DAY 110**
Lesson 8-5	Lesson 8-6	Mid-Chapter Quiz Lesson 8-7	Lesson 8-8	Lesson 8-9
DAY 111	**DAY 112**	**DAY 113**	**DAY 114**	**DAY 115**
Hands-On Lab 8B	Lesson 8-10	Extension	Chapter 8 Review	Chapter 8 Assessment

Pacing Guide for 90-Minute Classes

Chapter 8

DAY 50	DAY 51	DAY 52	DAY 53	DAY 54
Chapter 7 Review Lesson 8-1	Chapter 7 Assessment Hands-On Lab 8A	Lesson 8-2 Lesson 8-3	Lesson 8-4 Lesson 8-5	Lesson 8-6 Lesson 8-7
DAY 55	**DAY 56**	**DAY 57**	**DAY 58**	
Mid-Chapter Quiz Lesson 8-8 Lesson 8-9	Hands-On Lab 8B Lesson 8-10	Extension Chapter 8 Review	Chapter 8 Assessment Lesson 9-1	

Across the Series

HARCOURT GRADE 5

- Write ratios and proportions.
- Interpret scale drawings.
- Relate fractions, decimals, and percents.
- Find the percent of a number.
- Interpret circle graphs.

COURSE 1

- **Write ratios, find unit rates, and solve proportions.**
- **Identify similar figures, find unknown measures, and convert measurements.**
- **Make scale drawings and maps.**
- **Relate fractions, decimals, and percents.**
- **Solve problems including those involving discounts, tips, sales tax, and simple interest.**
- **Construct circle graphs.**

COURSE 2

- Write ratios, find unit rates, and solve proportions.
- Identify similar figures, find unknown measures, and convert measurements.
- Make scale drawings, models, and maps.
- Relate fractions, decimals, and percents.
- Solve problems involving percents of change and simple interest.
- Construct circle graphs.

Across the Curriculum

LANGUAGE ARTS LINK

Math: Reading and Writing in the Content Area . . . . pp. 65–74

Focus on Problem Solving

Make a Plan . . . . SE p. 417

Journal . . . . TE pp. 395, 401, 408, 415, 421, 425, 429, 435

Write About It . . . . SE, last page of each lesson

SOCIAL STUDIES LINK

Social Studies . . . . SE pp. 401, 415, 435

SCIENCE LINK

Science . . . . SE p. 398

Life Science . . . . SE p. 419

Earth Science . . . . SE pp. 395, 423

Physical Science . . . . SE p. 395

Astronomy . . . . SE p. 413

Chemistry . . . . SE p. 429

TE = *Teacher's Edition* **SE** = *Student Edition*

Interdisciplinary Bulletin Board

Earth Science

A grade of coal with a higher carbon content, such as anthracite, contains more usable energy and releases less pollution than a grade with a lower carbon content, such as lignite. How much more carbon is in a 20-gram sample of anthracite than in a 20-gram sample of lignite?

4 grams of carbon

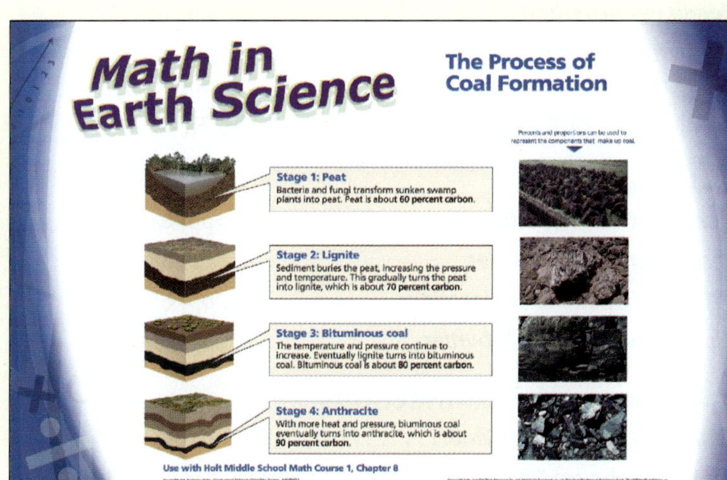

Interdisciplinary posters and worksheets are provided in your resource material.

Chapter 8

Resource Options

Chapter 8 Resource Book

Student Resources

Practice (Levels A, B, C) . . . . pp. 8–10, 18–20, 27–29, 36–38, 45–47, 54–56, 64–66, 73–75, 82–84, 91–93

Reteach . . . . pp. 11–12, 21, 30, 39, 48, 57, 67, 76, 85, 94

Challenge . . . . pp. 13, 22, 31, 40, 49, 58, 68, 77, 86, 95

Problem Solving . . . . pp. 14, 23, 32, 41, 50, 59, 69, 78, 87, 96

Puzzles, Twisters & Teasers . . . . pp. 15, 24, 33, 42, 51, 60, 70, 79, 88, 97

Recording Sheets . . . . pp. 3, 7, 17, 26, 35, 44, 53, 63, 72, 81, 90, 101, 104, 109

Chapter Review . . . . pp. 98–100

Teacher and Parent Resources

Chapter Planning and Pacing Guide . . . . p. 4

Section Planning Guides . . . . pp. 5, 61

Parent Letter . . . . pp. 1–2

Teaching Tools . . . . pp. 104–109

Teacher Support for Chapter Project . . . . p. 102

Transparencies . . . . pp. T1–T48

- Daily Transparencies
- Additional Examples Transparencies
- Teaching Transparencies

Reaching All Learners

English Language Learners

Success for English Language Learners . . . . . . pp. 129–148

Math: Reading and Writing in the Content Area . . . . pp. 65–74

Spanish Homework and Practice . . . . pp. 65–74

Spanish Interactive Study Guide . . . . pp. 65–74

Spanish Family Involvement Activities . . . . pp. 65–72

Multilingual Glossary

Individual Needs

Are You Ready? Intervention and Enrichment . . pp. 65–68, 85–88, 105–108, 113–116, 169–172, 419–420

Alternate Openers: Explorations . . . . pp. 65–74

Family Involvement Activities . . . . pp. 65–72

Interactive Problem Solving . . . . pp. 65–74

Interactive Study Guide . . . . pp. 65–74

Readiness Activities . . . . pp. 15–16

Math: Reading and Writing in the Content Area . . . . pp. 65–74

Challenge . . CRB pp. 13, 22, 31, 40, 49, 58, 68, 77, 86, 95

Hands-On

Hands-On Lab Activities . . . . pp. 61–80

Technology Lab Activities . . . . pp. 31–36

Alternate Openers: Explorations . . . . pp. 65–74

Family Involvement Activities . . . . pp. 65–72

Applications and Connections

Consumer and Career Math . . . . pp. 29–32

Interdisciplinary Posters . . . . Poster 8, TE p. 390B

Interdisciplinary Poster Worksheets . . . . pp. 22–24

Transparencies

Alternate Openers: Explorations . . . . pp. 65–74

Exercise Answers Transparencies

Chapter 8 Resource Book . . . . pp. T1–T48

- Daily Transparencies
- Additional Examples Transparencies
- Teaching Transparencies

Technology

Teacher Resources

***Lesson Presentations* CD-ROM** . . . . Chapter 8

***Test and Practice Generator* CD-ROM** . . . . Chapter 8

***One-Stop Planner* CD-ROM** . . . . Chapter 8

Student Resources

***Are You Ready? Intervention* CD-ROM**
Skills 14, 19, 24, 26, 40

internet connect

Homework Help Online	**KEYWORD:** MR4 HWHelp8
Math Tools Online	**KEYWORD:** MR4 Tools
Glossary Online	**KEYWORD:** MR4 Glossary
Chapter Project Online	**KEYWORD:** MR4 PSProject8
Chapter Opener Online	**KEYWORD:** MR4 Ch8
CNN Student News	**KEYWORD:** MR4 CNN8

SE = *Student Edition* **TE** = *Teacher's Edition* **AR** = *Assessment Resources* **CRB** = *Chapter Resource Book* **MK** = *Manipulatives Kit*

Chapter 8 Assessment Options

Assessing Prior Knowledge

Determine whether students have the required prerequisite concepts and skills.

Are You Ready? . . . . SE p. 391
Inventory Test . . . . AR pp. 1–4

Test Preparation

Provide review and practice for chapter and standardized tests.

Standardized Test Prep . . . . SE p. 447
Spiral Review with Test Prep . . . . . SE, last page of each lesson
Study Guide and Review . . . . SE pp. 442–444
Test Prep Tool Kit

Technology

***Test and Practice Generator* CD-ROM**

internet connect

State-Specific Test Practice Online KEYWORD: MR4 TestPrep

Performance Assessment

Assess students' understanding of chapter concepts and combined problem-solving skills.

Performance Assessment . . . . SE p. 446
Includes scoring rubric in TE
Performance Assessment . . . . AR p. 118
Performance Assessment Teacher Support . . . . AR p. 117

Portfolio

Portfolio opportunities appear throughout the Student and Teacher's Editions.

Suggested work samples:

Problem Solving Project . . . . TE p. 390
Performance Assessment . . . . SE p. 446
Portfolio Guide . . . . AR p. xxxv
Journal . . . . TE pp. 395, 401, 408, 415, 421, 425, 429, 435
Write About It . . . . SE, last page of each lesson

Daily Assessment

Obtain daily feedback on students' understanding of concepts.

Spiral Review and Test Prep . . . . . . SE, last page of each lesson

Also Available on Transparency In Chapter 8 Resource Book

Warm Up . . . . TE, first page of each lesson
Problem of the Day . . . . TE, first page of each lesson
Lesson Quiz . . . . TE, last page of each lesson

Student Self-Assessment

Have students evaluate their own work.

Group Project Evaluation . . . . AR p. xxxii
Individual Group Member Evaluation . . . . AR p. xxxiii
Portfolio Guide . . . . AR p. xxxv
Journal . . . . TE pp. 395, 401, 408, 415, 421, 425, 429, 435

Formal Assessment

Assess students' mastery of concepts and skills.

Section Quizzes . . . . AR pp. 21–22
Mid-Chapter Quiz . . . . SE p. 416
Chapter Test . . . . SE p. 445
Chapter Tests (Levels A, B, C) . . . . AR pp. 73–78
Cumulative Tests (Levels A, B, C) . . . . AR pp. 211–222
Standardized Test Prep
Cumulative Assessment . . . . SE p. 447
End-of-Year Test . . . . AR pp. 271–274

Technology

***Test and Practice Generator* CD-ROM**

Make tests electronically. This software includes:

- Dynamic practice for Chapter 8
- Customizable tests
- Multiple-choice items for each objective
- Free-response items for each objective
- Teacher management system

SE = *Student Edition* **TE** = *Teacher's Edition* **AR** = *Assessment Resources* **CRB** = *Chapter Resource Book* **MK** = *Manipulatives Kit*

Chapter 8

Chapter 8 Tests

Three levels (A,B,C) of tests are available for each chapter in the *Assessment Resources.*

LEVEL A

CHAPTER 8 Chapter Test *Form A*

Use the shapes pictured to write the following ratios.

1. stars to circles 3:2
2. squares to total shapes 4:9
3. Write 3 equivalent ratios to compare the number of circles with stars. Possible answers: 2:3, 4:6, 6:9
4. On a rose bush there are 8 roses in bloom and 6 buds yet to open. What is the ratio of blooms to buds? 4:3

Find the missing value in each proportion.

5. $\frac{8}{y} = \frac{1}{5}$ $y = 40$
6. $\frac{2}{8} = \frac{x}{20}$ $x = 5$
7. $\frac{16}{3} = \frac{48}{y}$ $y = 9$
8. The Johnson family will spend three weeks on vacation this summer. How many days is this? 21 days
9. A newborn baby at Bayside Hospital weighs 128 ounces. How many pounds is this? 8 lb
10. The two triangles are similar. Find the missing measurements.

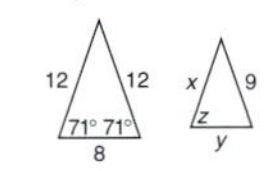

$x = 9$; $y = 6$; $z = 71°$

11. The two rectangles are similar. Find the missing measurements.

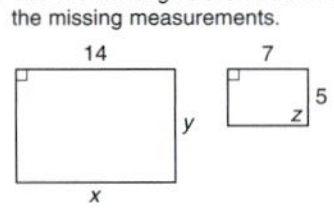

$x = 14$; $y = 10$; $z = 90°$

12. A flagpole casts a shadow that is 14 feet long. At the same time a 4-foot-tall bench casts a shadow that is 5.6 feet long. How tall is the flagpole? 10 ft

CHAPTER 8 Chapter Test *Form A, continued*

13. A 20-foot monument casts a shadow 32 feet long. How long of a shadow would a 6-foot man cast at the same time? 9.6 feet
14. A scale on a map is 1 inch = 15 miles. If it is 60 miles from the Indiana state line to Toledo, Ohio, how many inches is this on the map? 4 in.
15. A model boat is constructed using the scale 1 foot:16 feet. The actual length of the boat is 48 feet. Find the length of the scale model boat. 3 ft

Write each percent as a fraction in simplest form and as a decimal.

16. 25% $\frac{1}{4}$; 0.25
17. 20% $\frac{1}{5}$; 0.20
18. 75% $\frac{3}{4}$; 0.75
19. 72% $\frac{18}{25}$; 0.72
20. Order from least to greatest: 25%, $\frac{33}{100}$, 0.14, $\frac{21}{50}$. 0.14, 25%, $\frac{33}{100}$, $\frac{21}{50}$

Write each decimal or fraction as a percent.

21. 0.63 63%
22. 0.08 8%
23. $\frac{17}{50}$ 34%
24. $\frac{4}{25}$ 16%

Find each percent.

25. 60% of 95 57
26. 12% of 50 6
27. 20% of 500 100
28. When Sarah and Jane go out to lunch, Sarah offers to buy if Jane will leave the tip. If the bill was for $14.00 and Jane wanted to leave a 15% tip, how much tip should she leave? $2.10
29. Karen wants to buy a sweater that costs $34.00. If the sales tax is 6%, what is the total cost of the purchase? $36.04

LEVEL B

CHAPTER 8 Chapter Test *Form B*

Use the shapes pictured to write the following ratios.

1. squares to octagons 3:2
2. octagons to diamonds 4:3
3. Write 3 equivalent ratios to compare the number of circles with the number of squares. Possible answers: 5:6, 10:12, 15:18
4. If there are 11 boys on a soccer team and 4 of them scored goals at the last game, what is the ratio of boys that scored to boys that did not score? 4:7

Find the missing value in each proportion.

5. $\frac{5}{y} = \frac{9}{27}$ $y = 15$
6. $\frac{6}{24} = \frac{x}{9}$ $x = 2.25$
7. $\frac{14}{5} = \frac{35}{y}$ $y = 12.5$
8. A bag of apples weighs 3 pounds. How many ounces is this? 48 oz
9. A punch bowl holds 2 gallons of punch. How many 8-ounce cups is this? 32 cups
10. The two triangles are similar. Find the missing measurements.

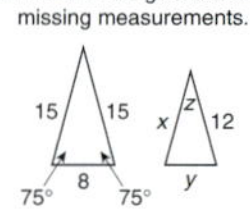

$x = 12$; $y = 6.4$; $z = 30°$

11. The two triangles are similar. Find the missing measurements.

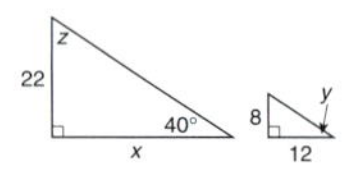

$x = 33$; $y = 40°$; $z = 50°$

12. A house casts a shadow that is 18 feet long. At the same time, a 4-foot-tall child casts a shadow that is 3.2 feet long. How tall is the house? 22.5 ft

CHAPTER 8 Chapter Test *Form B, continued*

13. A 7-foot ladder casts a shadow 12 feet long. How long of a shadow would a 15-foot cherry tree cast at the same time? Round your answer to the nearest tenth. 25.7 ft
14. A scale model of a space shuttle is constructed using the scale 1 inch:146 inches. The actual length of the space shuttle is $194\frac{2}{3}$ feet. Find the length of the scale model in inches. 16 in.
15. A model car is constructed using the scale 1 foot:12 feet. The actual length of the car is 18 feet. Find the length of the scale model car. 1.5 ft

Write each percent as a fraction in simplest form and as a decimal.

16. 54% $\frac{27}{50}$; 0.54
17. 60% $\frac{3}{5}$; 0.60
18. 35% $\frac{7}{20}$; 0.35
19. 44% $\frac{11}{25}$; 0.44
20. Order from least to greatest: 21.7%, $\frac{1}{5}$, 0.21, $\frac{2}{9}$. $\frac{1}{5}$, 0.21, 21.7%, $\frac{2}{9}$

Write each decimal or fraction as a percent. Round to the nearest tenth of a percent.

21. 0.817 81.7%
22. 0.021 2.1%
23. $\frac{23}{25}$ 92%
24. $\frac{7}{9}$ 77.8%

Find each percent.

25. 57% of 45 25.65
26. 13% of 97 12.61
27. 9% of 450 40.5
28. A compact disc costs $17.97. The sales tax rate is 5%. How much will the total cost be for this compact disc? $18.87
29. Carly has a coupon for 15% off the price of a camera. The camera has an original cost of $234.00. How much will the camera cost after the discount? $198.90

LEVEL C

CHAPTER 8 Chapter Test *Form C*

Use the shapes pictured to write the following ratios.

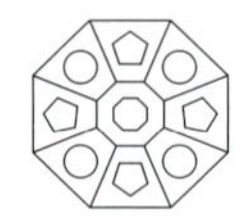

1. circles to octagons 4:3
2. trapezoids to octagons 8:3
3. Write 3 equivalent ratios to compare the number of pentagons with the number of octagons. Possible answers: 4:3, 8:6, 12:9
4. On a bus, 6 passengers are wearing headphones and 7 are not. What is the ratio of passengers wearing headphones to total passengers? 6:13

Find the missing value in each proportion.

5. $\frac{0.6}{y} = \frac{3}{14}$ $y = 2.8$
6. $\frac{8.7}{9.6} = \frac{x}{4}$ $x = 3.625$
7. $\frac{1.6}{3} = \frac{7.2}{y}$ $y = 13.5$
8. If you drank 10 pints of water a day, how many gallons would you drink in one week? 8.75 gal
9. Jake grew an average of 0.5 cm per month last year. How many inches did he grow in the year? (hint: 2.5 cm = 1 inch) 2.4 in.
10. The two regular pentagons are similar. Find the missing measurements.

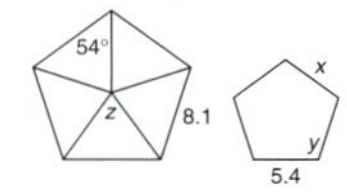

$x = 5.4$; $y = 108°$; $z = 72°$

11. The two trapezoids are similar. Find the missing measurements.

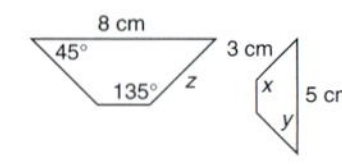

$x = 135°$; $y = 45°$; $z = 4.8$ cm

CHAPTER 8 Chapter Test *Form C, continued*

12. A 6-foot-tall woman casts a shadow that is 18 feet long. At the same time, a flower has a shadow of 2 feet. How many inches tall is the flower? 8 in.
13. A 3-meter statue casts a shadow that is 5 meters long. At the same time the rock next to the statue casts a shadow that is 15 cm long. How tall is the rock? 9 cm
14. If a scale of a map is 1 inch:75 miles and it is $3\frac{1}{2}$ inches from St. Louis, Missouri to Indianapolis, Indiana, how long would it take to travel from one city to the other if you were traveling at 55 miles per hour? about 5 hours
15. In a biology poster, a scale drawing of a beetle is 9 cm:2 cm. If the drawing on the poster is 28 cm, how long is the actual beetle? 6.2 cm

Write each percent as a fraction in simplest form and as a decimal.

16. 94.2% $\frac{471}{500}$; 0.942
17. 9% $\frac{9}{100}$; 0.09
18. 0.5% $\frac{1}{200}$; 0.005
19. 265% $\frac{53}{20}$; 2.65
20. Order from least to greatest: 88.2%, $\frac{22}{25}$, 0.875, $\frac{6}{7}$. $\frac{6}{7}$, 0.875, $\frac{22}{25}$, 88.2%

Write each decimal or fraction as a percent.

21. 1.025 102.5%
22. 0.007 0.7%
23. $\frac{5}{8}$ 62.5%
24. $\frac{7}{30}$ 23.3%

Find each percent.

25. 32% of 19.5 6.24
26. 1,753% of 125 2,191.25
27. 1.5% of 750 11.25
28. A coat costs $159.99 and is on clearance for 50% off. When you get to the register you get an additional 20% off all clearance merchandise. To the nearest cent, how much will the coat cost? $64.00
29. A set of luggage costs $578.00, but is on sale this week for 15% off. If sales tax is $6\frac{1}{4}$%, what is the total price for the luggage to the nearest cent? $522.01

Test and Practice Generator CD-ROM

Create and customize multiple versions of the same tests with corresponding answers for any chosen chapter objectives.

Chapter 8 State and Standardized Test Preparation

Test Taking Skill Builder and Standardized Test Practice are provided for each chapter in the *Test Prep Tool Kit.*

Test Taking Skill Builder

Test Taking Strategy **Response Questions**
Chapter 8

Short response questions require you to find the solution to a problem but do not provide answer choices or a grid. To get full credit, you need to show each step of your calculations, provide your reasoning, and when appropriate, answer in a complete sentence.

Example 1 Short Response George painted walls at his aunt's house over the weekend. She paid him $39. He worked for 6 hours. How much did George earn per hour?

Solution
Reasoning: Find the unit rate.

Show all of your work: $\frac{39}{6} = 6.5$

Answer the question in a complete sentence: George earned $6.50 per hour.

Example 2 Short Response For a birthday party, Ellen and Abby want to enlarge a 4 in. by 6 in. picture of the birthday girl so that the length of the photo is 15 in. What is the width of the enlargement?

Solution
Reasoning: Set up and solve a proportion.
Show all of your work:

$\frac{4}{6} = \frac{x}{15}$

$4 \cdot 15 = 6 \cdot x$

$60 = 6x$

$\frac{60}{6} = \frac{6x}{6}$

$10 = x$

Answer the question in a complete sentence:
The width of the enlargement is 10 in.

Test Taking Strategy
Chapter 8, continued

Exercises Possible answers are given.

Answer each question.

1. Terrance is standing on a sidewalk next to a lamppost. The lamppost is 9 ft tall and casts a 3-ft shadow. Terrance is 6 ft tall. How long is his shadow?

Response:

$\frac{9 \text{ ft}}{3 \text{ ft}} = \frac{6 \text{ ft}}{x \text{ ft}}$

$9 \cdot x = 3 \cdot 6$

$9x = 18$

$\frac{9x}{9} = \frac{18x}{9}$

$2 = x$

Terrance's shadow is 2 ft long.

a. The response to the question did not receive full credit. Why?

The response did not include the student's reasoning.

b. Complete the response so that it receives full credit.

Write a proportion and solve for x, the length of Terrance's shadow.

2. Marcus takes his sister out to dinner for landing the lead role in the school play. The cost of the meal is $28. Since the service was good, he wants to leave a 20% tip. What is his total bill after he adds the tip?

Response: Calculate the amount of the tip and then add the amount of the tip to the cost of the meal.
The total bill after the tip is added is $33.60.

The response to the question provides a correct answer. Do you think the student received full credit for the response? Explain.

No, the student did not show any of the work.

Standardized Test Practice

Standardized Test Practice
Chapter 8

Select the best answer for Questions 1–7.

1. At Video Mania, 35% of all the rental movies are comedies. What is this percentage as a fraction in simplest form?

A $\frac{7}{20}$ C $\frac{13}{20}$
B $\frac{17}{50}$ D $\frac{35}{100}$

2. A car salesman has sold 15 vehicles over the past 4 months. If 20% of the vehicles sold have been trucks, how many trucks has the salesman sold during this time?

F 3 trucks
G 6 trucks
H 9 trucks
I 187 trucks

3. Which is NOT an equivalent ratio to compare the number of suns to moons?

A $\frac{18}{24}$ C $\frac{9}{12}$
B $\frac{3}{6}$ D $\frac{24}{32}$

4. Your three dogs eat 25 pounds of dog food in 2 weeks. How many pounds will they eat in one year?

F 300 lb
G 600 lb
H 650 lb
I 1,300 lb

5. What is the missing length and the measure of $\angle J$ in the similar triangles below?

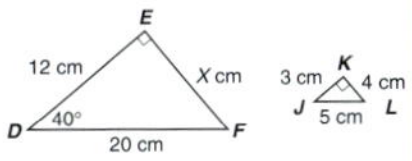

A 12 cm, 90°
B 16 cm, 40°
C 16 cm, 50°
D 20 cm, 40°

6. A beverage stand at a movie theater sells 4 different sizes of drinks. Which is a better deal?

F 12 oz for $1.25
G 16 oz for $1.65
H 20 oz for $2.15
I 32 oz for $3.00

7. To make 36 oz of homemade lemonade, you combine 3 cups of cold water with 1 cup of lemon juice and $\frac{1}{2}$ cup of sugar. If you only have enough lemons for a $\frac{1}{2}$ cup of juice, how much sugar will you need?

A $\frac{1}{8}$ cup
B $\frac{1}{4}$ cup
C $\frac{3}{4}$ cup
D 1 cup

Standardized Test Practice
Chapter 8, continued

Gridded Response
Solve the problems. Use the answer sheet to write and grid-in your answer.

8. A father builds his daughter a dollhouse that is modeled after his own home. The dollhouse measures 24 inches long and 16 inches wide. If the actual house is 40 feet wide, how long is the actual house?

9. A 30 ft building casts a 45 ft shadow and a tree next to the building casts a 24 ft shadow. How tall is the tree?

10. On a map, the distance from your house to the capital building is 4.2 inches. If the scale of the map is 1 inch = 5 miles, how far is the actual distance from your house to the capital building?

Short Response
Solve the problems. Use the answer sheet to write your answers.

11. Explain in words how to write 0.38 as a fraction and a percent.

12. In the 2001, Virginia elections, approximately 46% of the 4,109,127 registered voters actually cast their ballot. How many voters voted? Round to the nearest person. How many registered voters did not cast their ballot?

13. You bought a pair of shoes for $36 and four pairs of socks for $16. You pay a 6% sales tax. How much is the sales tax? How much is the total?

14. Out of 350 6th graders, 73% of the students raised enough money to go to Space Camp. Explain in words how to write this percent as a decimal.

Extended Response

15. During a fundraiser, potential customers had a choice of buying a book cover, a coffee mug with the school logo, or a school pennant. Of the 400 items that were sold, 12% were coffee mugs.

a. Write a proportion to find how many coffee mugs were sold.

b. If 240 book covers were sold, what percent of the items sold were book covers?

c. For next year's sale, the committee has decided to sell only two items. Which two items would you suggest to the committee and why?

Sheet

Ⓒ Ⓓ See Lesson 8
Ⓒ Ⓓ See Lesson 8
Ⓗ ● See Lesson 8

10. 21

See Lesson 8-6.

Write your answers in the space provided.

11. To write 0.38 as a percent, multiply the number by 100. 0.38 = 38% To write 0 as a fraction, write it over 100 and simplify: $\frac{38}{100} = \frac{19}{50}$.

(See Less

12. $0.46 \times 4,109,127 = 1,890,198$ people; $4,109,127 - 1890,198 = 2,218,929$ pe

(See Lesso

13. $36 + 16 = 52$; $52 \times 0.06 = 3.12$; You pay $3.12 in sales tax. $52 + 3.12 = \$55$ You pay $55.12 total.

(See Less

14. Write the percent as a fraction with a denominator of 100. Then divide.

Customized answer sheets give students realistic practice for actual standardized tests.

internet connect
State-Specific Test Practice Online
KEYWORD: MR4 TestPrep

Test Prep Tool Kit

- Standardized Test Prep Workbook
- Countdown to Testing transparencies
- State Test Prep CD-ROM
- Standardized Test Prep Video

Chapter 8

Ratio, Proportion, and Percent

Why Learn This?

Tell students that a small sample is often used to gain information about a larger group. For example, a fisheries biologist may find the number of tagged fish in a sample of 100 fish. Then, since the total number of tagged fish is known, the fisheries biologist can use the fraction of tagged fish out of recaptured fish to estimate the total number of fish in the lake.

Using Data

To begin the study of this chapter, have students:

- Write a fraction to show the portion of the fish recaptured in Duck Lake that were tagged. $\frac{23}{96}$
- Write in simplest form the fraction of the fish recaptured in Los Dos Perros Lake that were tagged. $\frac{32}{40} = \frac{4}{5}$
- Use the information about Robyn Lake to fill in the tag, release, and recapture formula. $\frac{18}{26} = \frac{75}{x}$
- Use the information in the table and the tag, release, and recapture formula to estimate the total number of fish in each lake. Duck: 451; Los Dos Perros: 70; Robyn: 109

Chapter 8

TABLE OF CONTENTS

internet connect
Chapter Opener Online
go.hrw.com
KEYWORD: MR4 Ch8

Ratio, Proportion, and Percent

Career — Fisheries Biologist

A fisheries biologist interacts with nature and with people. Fisheries biologists complete surveys, improve habitats, monitor water conditions, and work with land developers.

Fisheries biologists often must determine the number of fish in a lake or pond. They use the tag, release, and recapture method to estimate this number.

$$\frac{\text{tagged number in recapture}}{\text{total number recaptured}} = \frac{\text{number originally tagged}}{\text{total number in lake}}$$

Lake	Tagged Number in Recapture	Total Number Recaptured	Number Originally Tagged
Duck	23	96	108
Los Dos Perros	32	40	56
Robyn	18	26	75

PROBLEM SOLVING

Problem Solving Project

Life Science Connection

Purpose: To solve problems using ratio, proportion, and percent

Materials: Counting the Unknown worksheet, paper bags or cans, Unifix© cubes, centimeter cubes, multiple colors of cardboard squares or other similar objects, marking pens

internet connect
Chapter Project Online: *go.hrw.com*
KEYWORD: MR4 PSProject8

Understand, Plan, Solve, and Look Back

Have students:

- Complete the Counting the Unknown worksheet.
- Examine the chart and explain what they observe. What does the data suggest about the number of fish in each lake? Is it easy to tell which lake has the most fish? Why or why not?
- Make up data for a new lake, exchange data with their classmates, and calculate the estimated populations.
- Think of other situations in nature and in other areas where tagging and collecting specimens might be useful.
- Check students' work.

ARE YOU READY?

Choose the best term from the list to complete each sentence.

1. A(n) __?__ is a three-sided polygon, and a(n) __?__ is a four-sided polygon. **triangle; quadrilateral**
2. A(n) __?__ is used to name a part of a whole. **fraction**
3. When two numbers have the same value, they are said to be __?__. **equivalent**
4. When writing 0.25 as a fraction, 25 is the __?__ and 100 is the __?__. **numerator; denominator**

fraction
numerator
denominator
equivalent
angle
triangle
quadrilateral
pentagon

Complete these exercises to review skills you will need for this chapter.

✔ Simplify Fractions

Write each fraction in simplest form.

5. $\frac{6}{10}$ $\frac{3}{5}$
6. $\frac{9}{12}$ $\frac{3}{4}$
7. $\frac{8}{6}$ $1\frac{1}{3}$ or $\frac{4}{3}$

✔ Write Equivalent Fractions

Write three equivalent fractions for each given fraction. **Possible answers:**

8. $\frac{4}{16}$ $\frac{1}{4}, \frac{2}{8}, \frac{3}{12}$
9. $\frac{5}{10}$ $\frac{1}{2}, \frac{10}{20}, \frac{25}{50}$
10. $\frac{5}{6}$ $\frac{10}{12}, \frac{15}{18}, \frac{20}{24}$

✔ Write Fractions as Decimals

Write each fraction as a decimal.

11. $\frac{3}{10}$ 0.3
12. $\frac{3}{4}$ 0.75
13. $\frac{5}{8}$ 0.625
14. $\frac{11}{12}$ $0.91\overline{6}$

✔ Write Decimals as Fractions

Write each decimal as a fraction in simplest form.

15. 0.5 $\frac{1}{2}$
16. 0.35 $\frac{7}{20}$
17. 0.08 $\frac{2}{25}$
18. 0.12 $\frac{3}{25}$

✔ Multiply Decimals

Multiply.

19. $0.42 \cdot 10$ 4.2
20. $0.3 \cdot 52$ 15.6
21. $20.5 \cdot 0.25$ 5.125
22. $6.75 \cdot 0.40$ 2.7
23. $9.8 \cdot 0.2$ 1.96
24. $0.8 \cdot 7.4$ 5.92
25. $0.52 \cdot 0.64$ 0.3328
26. $0.75 \cdot 8.9$ 6.675

Assessing Prior Knowledge

INTERVENTION

Diagnose and Prescribe

Evaluate your students' performance on this page to determine whether intervention is necessary or whether enrichment is appropriate. Options that provide instruction, practice, and a check are listed below.

Resources for Are You Ready?

- ***Are You Ready? Intervention and Enrichment***
- **Recording Sheet for Are You Ready?** *Chapter 8 Resource Book . . . . . . p. 3*

Are You Ready? Intervention **CD-ROM**

ARE YOU READY?

Were students successful with Are You Ready?

NO INTERVENE

✔ **Simply Fractions 5–7**
Intervention Practice, Skill 19
CD-ROM Intervention Activities, Skill 19

✔ **Write Equivalent Fractions 8–10**
Intervention Practice, Skill 24
CD-ROM Intervention Activities, Skill 24

✔ **Write Fractions as Decimals 11–14**
Intervention Practice, Skill 26
CD-ROM Intervention Activities, Skill 26

✔ **Write Decimals as Fractions 15–18**
Intervention Practice, Skill 14
CD-ROM Intervention Activities, Skill 14

✔ **Multiply Decimals 19–26**
Intervention Practice, Skill 40
CD-ROM Intervention Activities, Skill 40

YES ENRICH

Are You Ready? Enrichment, pp. 421–422

Section **8A**

Ratios and Proportions

One-Minute Section Planner

Lesson	Materials	Resources
Lesson 8-1 Ratios and Rates **NCTM:** Number and Operations, Communication **NAEP:** Number Properties 4a ☑ SAT-9 ☑ SAT-10 ☑ ITBS ☑ CTBS ☑ MAT ☑ CAT	**Optional** Teaching Transparency T2 *(CRB)*	• *Chapter 8 Resource Book,* pp. 6–15 • Daily Transparency T1, CRB • Additional Examples Transparencies T3–T4, CRB • *Alternate Openers: Explorations,* p. 65
Hands-On Lab 8A Explore Proportions **NCTM:** Number and Operations, Reasoning and Proof **NAEP:** Number Properties 4b ☐ SAT-9 ☑ SAT-10 ☐ ITBS ☑ CTBS ☑ MAT ☐ CAT	**Required** Two-color counters *(MK)*	• *Hands-On Lab Activities,* pp. 67–70
Lesson 8-2 Proportions **NCTM:** Number and Operations, Reasoning and Proof, Communication **NAEP:** Number Properties 4b ☑ SAT-9 ☑ SAT-10 ☑ ITBS ☑ CTBS ☑ MAT ☑ CAT	**Optional** Recording Sheet for Reaching All Learners *(CRB, p. 104)* Teaching Transparency T6 *(CRB)*	• *Chapter 8 Resource Book,* pp. 16–24 • Daily Transparency T5, CRB • Additional Examples Transparencies T7–T9, CRB • *Alternate Openers: Explorations,* p. 66
Lesson 8-3 Proportions and Customary Measurement **NCTM:** Measurement, Reasoning and Proof, Communication, Connections **NAEP:** Measurement 2b ☑ SAT-9 ☑ SAT-10 ☐ ITBS ☐ CTBS ☑ MAT ☐ CAT	**Optional** Rulers *(MK and CRB p. 105)* Teaching Transparency T11 *(CRB)*	• *Chapter 8 Resource Book,* pp. 25–33 • Daily Transparency T10, CRB • Additional Examples Transparencies T12–T13, CRB • *Alternate Openers: Explorations,* p. 67
Lesson 8-4 Similar Figures **NCTM:** Number and Operations, Geometry, Problem Solving, Communication, Connections **NAEP:** Geometry 2f ☑ SAT-9 ☑ SAT-10 ☑ ITBS ☑ CTBS ☑ MAT ☑ CAT	**Optional** Rulers *(MK)* Protractors *(MK)* Similar Polygons for Reaching All Learners (*CRB* p. 107) Teaching Transparency T15 *(CRB)*	• *Chapter 8 Resource Book,* pp. 34–42 • Daily Transparency T14, CRB • Additional Examples Transparencies T16–T18, CRB • *Alternate Openers: Explorations,* p. 68
Lesson 8-5 Indirect Measurement **NCTM:** Number and Operations, Geometry, Communication, Connections, Representation **NAEP:** Measurement 1k ☑ SAT-9 ☑ SAT-10 ☐ ITBS ☐ CTBS ☑ MAT ☑ CAT	**Optional** Teaching Transparency T20 *(CRB)*	• *Chapter 8 Resource Book,* pp. 43–51 • Daily Transparency T19, CRB • Additional Examples Transparencies T21–T22, CRB • *Alternate Openers: Explorations,* p. 69
Lesson 8-6 Scale Drawings and Maps **NCTM:** Number and Operations, Geometry, Measurement, Communication, Connections, Representation **NAEP:** Measurement 2f ☑ SAT-9 ☑ SAT-10 ☑ ITBS ☑ CTBS ☑ MAT ☑ CAT	**Optional** Teaching Transparencies T24–T25 *(CRB)* Maps and atlas	• *Chapter 8 Resource Book,* pp. 52–60 • Daily Transparency T23, CRB • Additional Examples Transparencies T26–T28, CRB • *Alternate Openers: Explorations,* p. 70
Section 8A Assessment		• Mid-Chapter Quiz, SE p. 416 • Section 8A Quiz, AR p. 21 • *Test and Practice Generator* CD-ROM

SAT = *Stanford Achievement Tests* **ITBS** = *Iowa Test of Basic Skills* **CTBS** = *Comprehensive Test of Basic Skills/Terra Nova*
MAT = *Metropolitan Achievement Tests* **CAT** = *California Achievement Test*

NCTM = Complete standards can be found on pages T27–T33. **NAEP** = Complete standards can be found on pages A31–A35.

SE = *Student Edition* **TE** = *Teacher's Edition* **AR** = *Assessment Resources* **CRB** = *Chapter Resource Book* **MK** = *Manipulatives Kit*

Section Overview

Ratios, Rates, and Proportions

Lessons 8-1, 8-2, 8-3

Why? You can use ratios to compare quantities or describe rates.

A **ratio** is a comparison of two quantities that uses division. A ratio can be **written 3 ways.**

One molecule of water contains 2 hydrogen atoms and 1 oxygen atom. So the ratio of hydrogen to oxygen in water can be written the following ways:

$\frac{2}{1}$ 2 to 1 2:1

A **proportion** shows that two ratios are equivalent.

Cross products in a proportion are equal.

$$\frac{2}{1} = \frac{6}{3} \quad \frac{2}{1} \times \frac{6}{3} \Rightarrow 2 \times 3 = 1 \times 6$$

Proportions can be used to convert units within the customary system.

Convert 4 yards to feet.

$$\frac{1 \text{ yd}}{3 \text{ft}} = \frac{4 \text{ yd}}{x \text{ ft}}$$

$$1 \cdot x = 3 \cdot 4$$

$$x = 12$$

So 4 yards is 12 feet.

Similar Figures, Scale Drawings

Lessons 8-4, 8-5, 8-6

Why? Blueprints are scale drawings used in construction.

Similar figures have the same shape, but not necessarily the same size.

For similar figures:

- Corresponding sides have lengths that are proportional.
- Corresponding angles are congruent.

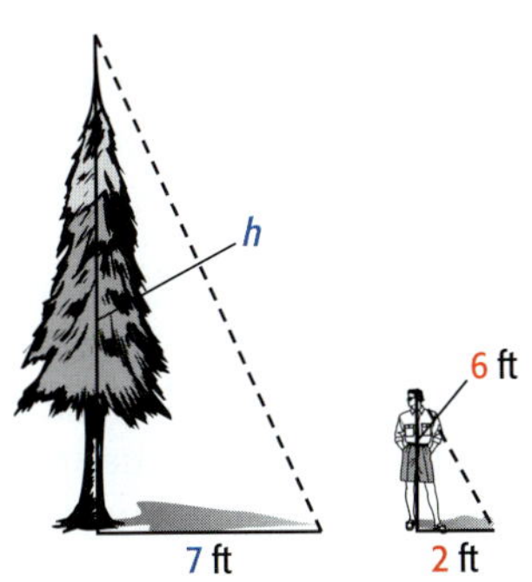

$$\frac{6}{h} = \frac{2}{7}$$

$$2 \cdot h = 6 \cdot 7$$

$$h = 21$$

A **scale** is a ratio between two sets of measurements.

A **scale drawing**, such as a map, is a drawing of a real object that is proportionally smaller or larger than the real object.

8-1 Organizer

Pacing: Traditional 1 day
Block $\frac{1}{2}$ day

Objective: Students write ratios and rates and find unit rates.

Indiana Academic Standards
6.1.6, 6.2.6, 6.7.1, 6.7.5, 6.7.11

Warm Up

Write each fraction in simplest form.

1. $\frac{2}{6}$ $\frac{1}{3}$
2. $\frac{4}{16}$ $\frac{1}{4}$
3. $\frac{25}{70}$ $\frac{5}{14}$
4. $\frac{6}{52}$ $\frac{3}{26}$

Problem of the Day

What three consecutive odd numbers are factors of 105? 3, 5, 7

Available on Daily Transparency in CRB

Math Fact!

A colon was first used to indicate a ratio by Gottfried Leibniz, the coinventor of calculus, in 1639.

8-1 Ratios and Rates

Learn to write ratios and rates and to find unit rates.

Vocabulary
ratio
equivalent ratios
rate
unit rate

For a time, the Boston Symphony Orchestra was made up of 95 musicians.

Violins	29	Violas	12
Cellos	10	Basses	9
Flutes	5	Trumpets	3
Double reeds	8	Percussion	5
Clarinets	4	Harp	1
Horns	6	Trombones	3

You can compare the different groups by using ratios. A **ratio** is a comparison of two quantities using division.

Reading Math
Read the ratio $\frac{29}{12}$ as "twenty-nine to twelve."

For example, you can use a ratio to compare the number of violins with the number of violas. This ratio can be written in three ways.

Terms ⟶ $\frac{29}{12}$ 29 to 12 29:12

Notice that the ratio of violins to violas, $\frac{29}{12}$, is different from the ratio of violas to violins, $\frac{12}{29}$. The order of the terms is important.

Ratios can be written to compare a part to a part, a part to the whole, or the whole to a part.

EXAMPLE 1 Writing Ratios

Use the table above to write each ratio.

A flutes to clarinets

$\frac{5}{4}$ *or* 5 to 4 *or* 5:4 *Part to part*

B trumpets to total instruments

$\frac{3}{95}$ *or* 3 to 95 *or* 3:95 *Part to whole*

C total instruments to basses

$\frac{95}{9}$ *or* 95 to 9 *or* 95:9 *Whole to part*

Equivalent ratios are ratios that name the same comparison. You can find an equivalent ratio by multiplying or dividing both terms of a ratio by the same number.

1 Introduce

Alternate Opener

EXPLORATION

8-1 Ratios and Rates

A TV network offers the numbers of shows each week shown in the table.

You can compare the numbers of TV shows by using ratios. A *ratio* is a comparison of two quantities that uses division. For example, the ratio of science fiction shows to drama shows is $\frac{3}{14}$, which can also be written 3:14 or 3 to 14.

Type of TV Show	Number of Shows
Comedy	14
Drama	14
Science fiction	3
Game show	7
Talk show	15
News	14
Morning show	10
Late-night show	5
Sports	6

Find each ratio.

1. comedy shows to game shows
2. game shows to news shows
3. morning shows to late-night shows
4. talk shows to sports shows

Think and Discuss

5. **Discuss** whether the ratios in numbers 1–4 compare part to part, part to whole, or whole to part.
6. **Discuss** whether order is important when calculating ratios. (*Hint:* Is $\frac{\text{news}}{\text{sports}}$ equivalent to $\frac{\text{sports}}{\text{news}}$?)

Motivate

Ask students if they have ever heard the terms *rate* and *ratios* and, if so, in what context.

Possible answers: win/loss ratios, heart rate, rate of pay, rate of speed, etc.

Exploration worksheet and answers on Chapter 8 Resource Book pp. 7 and 110

2 Teach

Lesson Presentation

Guided Instruction

In this lesson, students learn to write ratios and rates and to find unit rates. First teach what a ratio is and the three ways to write ratios. Then teach students that a ratio can compare a part to a part, a part to the whole, or the whole to a part. Finally teach students to find equivalent ratios. Have students apply a unit rate in a consumer application.

EXAMPLE 2 Writing Equivalent Ratios

Write three equivalent ratios to compare the number of stars with the number of moons in the pattern.

$\frac{\text{number of stars}}{\text{number of moons}} = \frac{4}{6}$ *There are 4 stars and 6 moons.*

$\frac{4}{6} = \frac{4 \div 2}{6 \div 2} = \frac{2}{3}$ *There are 2 stars for every 3 moons.*

$\frac{4}{6} = \frac{4 \cdot 2}{6 \cdot 2} = \frac{8}{12}$ *If you double the pattern, there will be 8 stars and 12 moons.*

So $\frac{4}{6}$, $\frac{2}{3}$, and $\frac{8}{12}$ are equivalent ratios.

A **rate** compares two quantities that have different units of measure.

Suppose a 2-liter bottle of soda costs $1.98.

$$\text{rate} = \frac{\text{price}}{\text{number of liters}} = \frac{\$1.98}{2 \text{ liters}} \qquad \$1.98 \text{ for 2 liters}$$

When the comparison is to one unit, the rate is called a **unit rate**.

Divide both terms by the second term to find the unit rate.

$$\text{unit rate} = \frac{\$1.98}{2} = \frac{\$1.98 \div 2}{2 \div 2} = \frac{\$0.99}{1} \qquad \$0.99 \text{ for 1 liter}$$

When the prices of two or more items are compared, the item with the lowest unit rate is the best deal.

EXAMPLE 3 *Consumer Application*

A 2-liter bottle of soda costs $2.02. A 3-liter bottle of the same soda costs $2.79. Which is the better deal?

2-liter bottle		3-liter bottle	
$\frac{\$2.02}{2 \text{ liters}}$	*Write the rate.*	$\frac{\$2.79}{3 \text{ liters}}$	*Write the rate.*
$\frac{\$2.02 \div 2}{2 \text{ liters} \div 2}$	*Divide both terms by 2.*	$\frac{\$2.79 \div 3}{3 \text{ liters} \div 3}$	*Divide both terms by 3.*
$\frac{\$1.01}{1 \text{ liter}}$	*$1.01 for 1 liter*	$\frac{\$0.93}{1 \text{ liter}}$	*$0.93 for 1 liter*

The 3-liter bottle is the better deal.

Think and Discuss

1. **Explain** why the ratio 2 boys:5 girls is different from the ratio 5 girls:2 boys.
2. **Describe** how to determine what number to divide by when finding a unit rate.

COMMON ERROR ALERT

Some students may think that the order of the terms in a ratio is unimportant. Explain that using the wrong order completely changes the meaning of the ratio.

Additional Examples

Example 1

Use the table to write each ratio.

Animals at the Vet	
Cats	5
Dogs	7
Rabbits	2

A. cats to rabbits $\frac{5}{2}$

B. dogs to total number of pets $\frac{7}{14}$

C. total number of pets to cats $\frac{14}{5}$

Example 2

Write three equivalent ratios to compare the number of diamonds to the number of spades in the pattern.

♦♠♠♦♠♠♦♠♠

$\frac{3}{6}$, $\frac{1}{2}$, and $\frac{9}{18}$ are equivalent ratios.

Example 3

A 3-pack of paper towels costs $2.79. A 6-pack costs $5.46. Which is the better deal?

The 6-pack is the better deal.

3 Close

Reaching All Learners

Through Home Connection

Provide quantity pricing (e.g., 3 for $1.46) for several different items from two different grocery stores. Have students work to find the unit rates for the items and compare their findings with those of their classmates. Which store had the best unit rates for each item?

Summarize

You may wish to review the terms *ratio*, *equivalent ratios*, *rate*, and *unit rate*. Have students explain how to find an equivalent ratio.

Possible answer: Multiply or divide both terms of a ratio by the same number to find an equivalent ratio.

Answers to Think and Discuss

1. Possible answer: because even though the terms are the same, they are in a different order
2. Possible answer: After the rate is written as a ratio, divide both terms by the lower term to find the unit rate.

8-1 PRACTICE & ASSESS

Students may want to refer back to the lesson examples.

Assignment Guide

If you finished Example 1 assign:
Core 1–3, 6–8, 14–15, 25–32
Enriched 3, 7, 14–18, 23, 25–32

If you finished Example 2 assign:
Core 1–4, 6–9, 14–15, 25–32
Enriched 14–19, 21–32

If you finished Example 3 assign:
Core 1–20, 25–32
Enriched 6–32

Answers

11. 10 to 7, 10:7, $\frac{10}{7}$

12. 24 to 11, 24:11, twenty-four to eleven

13. four to thirty, 4:30, $\frac{4}{30}$

8-1 Exercises

FOR EXTRA PRACTICE
see page 652

internet connect
Homework Help Online
go.hrw.com Keyword: MR4 8-1

GUIDED PRACTICE

See Example 1
Use the table to write each ratio.

Jacqueline's Software Collection	
Educational games	16
Word processing	2
Art programs	10
Arcade games	10
Music programs	3

1. music programs to art programs **3:10**
2. arcade games to entire collection **10:41**
3. entire collection to educational games **41:16**

See Example 2
4. Write three equivalent ratios to compare the number of red hearts in the picture with the total number of hearts. **Possible answer: $\frac{2}{9}$, $\frac{4}{18}$, $\frac{6}{27}$**

See Example 3
5. An 8-ounce bag of sunflower seeds costs $1.68. A 4-ounce bag of sunflower seeds costs $0.88. Which is the better deal? **the 8-ounce bag**

INDEPENDENT PRACTICE

See Example 1
Use the table to write each ratio.

	Redbirds	Blue Socks
Left-Handed Batters	8	3
Right-Handed Batters	11	19

6. Redbirds to Blue Socks **19:22**
7. right-handed Blue Socks to left-handed Blue Socks **19:3**
8. left-handed Redbirds to total Redbirds **$\frac{8}{19}$**

See Example 2
9. Write three equivalent ratios to compare the number of stars in the picture with the number of stripes. **Possible answer: 6:9, 2:3, 12:18**

See Example 3
10. Gina charges $28 for 3 hours of swimming lessons. Hector charges $18 for 2 hours of swimming lessons. Which instructor offers a better deal? **Hector**

PRACTICE AND PROBLEM SOLVING

Write each ratio three different ways.

11. ten to seven
12. $\frac{24}{11}$
13. 4 to 30

Math Background

Ratio is an ancient topic in mathematics. For example, ratios were used in early Greek works in arithmetic, geometry, and music.

The word *ratio* comes from the Latin word *ratus,* which means "calculation." During the Middle Ages, the word *ratio* was used to mean "computation." During that time, the word *proportio* was used to indicate what we mean by a ratio today.

RETEACH 8-1

LESSON 8-1 **Reteach**
Ratios and Rates

A ratio is a comparison of two quantities by division.

To compare the number of times vowels are used to the number of time consonants are used in the word "mathematics," first find each quantity.

Number of times vowels are used: 4
Number of times consonants are used: 7

Then write the comparison as a ratio, using the quantities in the same order as they appear in the word expression. There are three ways to write a ratio.

$\frac{4}{7}$ 4 to 7 4:7

Write each ratio.

1. days in May to days in a year **31 to 365**
2. sides of triangle to sides of a square **3 to 4**

Equivalent ratios are ratios that name the same comparison.
The ratio of inches in a foot to inches in a yard is $\frac{12}{36}$. To find equivalent ratios, divide or multiply the numerator and denominator by the same number.

$\frac{12}{36} = \frac{12 \div 3}{36 \div 3} = \frac{4}{12}$ $\frac{12}{36} = \frac{12 \cdot 2}{36 \cdot 2} = \frac{24}{72}$

So, $\frac{12}{36}$, $\frac{4}{12}$, and $\frac{24}{72}$ are equivalent ratios.

Write three equivalent ratios to compare each of the following.
Possible answers are given.

3. 8 triangles to 12 circles **2:3, 4:6, 6:9**
4. 20 pencils to 25 erasers **4:5, 8:10, 12:15**
5. 5 girls to 6 boys **10:12, 15:18, 20:24**
6. 10 pants to 14 shirts **5:7, 15:21, 20:28**

PRACTICE 8-1

LESSON 8-1 **Practice B**
Ratios and Rates

Use the table to write each ratio.

Animals in the Zoo	
Elephants	12
Giraffes	8
Lions	9
Seals	10
Otters	16

1. lions to elephants **9:12 or 3:4**
2. giraffes to otters **8:16 or 1:2**
3. lions to seals **9:10**
4. seals to elephants **10:12 or 5:6**
5. elephants to lions **12:9 or 4:3**
6. Write three equivalent ratios to compare the number of diamonds with the number of spades in the box. **Possible answer: 6:9, 2:3, 12:18**

Use the table to write each ratio as a fraction.

Baseball Team Stats		
	Titans	Orioles
Wins	12	9
Losses	14	15

7. Titans wins to Titans losses **$\frac{12}{14}$ or $\frac{6}{7}$**
8. Orioles losses to Orioles wins **$\frac{15}{9}$ or $\frac{5}{3}$**
9. Titans losses to Orioles losses **$\frac{14}{15}$**
10. Orioles wins to Titans wins **$\frac{9}{12}$ or $\frac{3}{4}$**
11. A 6-ounce bag of raisins costs $2.46. An 8-ounce bag of raisins costs $3.20. Which is the better deal? **the 8-ounce bag**
12. Barry earns $36.00 for 6 hours of yard work. Henry earns $24.00 for 3 hours of yard work. Who has the better hourly rate of pay? **Henry**

Use the diagram of an oxygen atom and a boron atom for Exercises 14–17. Find each ratio. Then give two equivalent ratios.

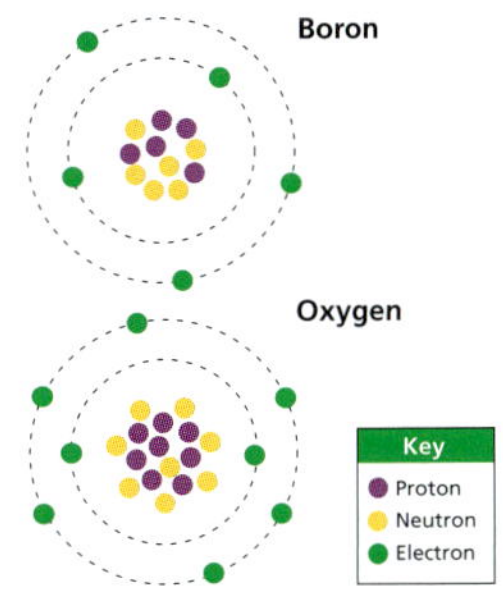

14. oxygen protons to boron protons **8:5, Possible answer: 16:10, 24:15**
15. boron neutrons to boron protons **6:5, Possible answer: 12:10, 18:15**
16. boron electrons to oxygen electrons **5:8, Possible answer: 10:16, 15:24**
17. oxygen electrons to oxygen protons **8:8, Possible answer: 4:4, 1:1**
18. A lifeguard-training program includes 16 hours of instruction in basic first aid and 8 hours of instruction in cardiopulmonary resuscitation (CPR). Write the ratio of hours of CPR instruction to hours of first aid instruction. **8:16**

19. Cassandra has three pictures on her desk. The pictures measure 4 in. long by 6 in. wide, 24 mm long by 36 mm wide, and 6 cm long by 7 cm wide. Which photos have a length-to-width ratio equivalent to 2:3? **4 × 6 and 24 × 36**
20. On which day did Alfonso run faster? **Wednesday**

Alfonso's Runs		
Day	Distance (m)	Time (min)
Monday	1,020	6
Wednesday	1,554	9

21. ***EARTH SCIENCE*** Water rushes over Niagara Falls at the rate of 180 million cubic feet every 30 minutes. How much water goes over the falls in 1 minute? **6 million ft^3**

22. ***WRITE ABOUT IT*** How are equivalent ratios like equivalent fractions?

23. ***WHAT'S THE QUESTION?*** The ratio of total students in Mr. Avalon's class to students in the class who have a blue backpack is 3 to 1. The answer is 1:2. What is the question?

24. ***CHALLENGE*** There are 36 performers in a dance recital. The ratio of men to women is 2:7. How many men are in the dance recital? **8 men**

Spiral Review

Solve each equation. (Lesson 3-10)

25. $5.5 = 5c$ **1.1**
26. $d + 4.96 = 9$ **4.04**
27. $t - 12.5 = 39.04$ **51.54**

Write each improper fraction as a mixed number. (Lesson 4-7)

28. $\frac{13}{4}$ **$3\frac{1}{4}$**
29. $\frac{70}{9}$ **$7\frac{7}{9}$**
30. $\frac{41}{3}$ **$13\frac{2}{3}$**
31. $\frac{75}{6}$ **$12\frac{1}{2}$**

32. **TEST PREP** How many lines of symmetry are in a rectangle that is not a square? (Lesson 7-11) **B**

A 1 **B** 2 **C** 4 **D** 6

Answers

22. You can find equivalent ratios just like you find equivalent fractions: Divide or multiply both terms by the same number.

23. Possible answer: What is the ratio of students with blue backpacks to students without blue backpacks?

Journal

Have students define *ratio,* and have them give examples of and explain how to find equivalent ratios.

Test Prep Doctor

For Exercise 32, encourage students to draw a rectangle and all of its lines of symmetry. You may wish to remind them that when a figure is folded along a line of symmetry, both sides match exactly. Students who answered **A** neglected to find all of the lines of symmetry for the rectangle. Students who answered **C** included the diagonals. Students who answered **D** probably counted the sides of the rectangle as well as the lines of symmetry.

CHALLENGE 8-1

LESSON 8-1 **Challenge**
The Golden Ratio

For centuries, people all over the world have considered a certain rectangle to be one of the most beautiful shapes. Which of these rectangles do you find the most attractive?

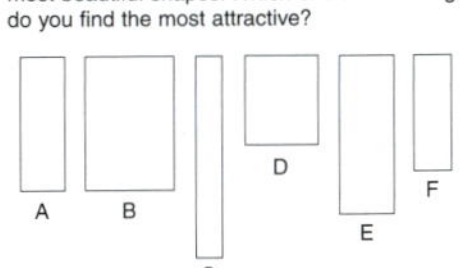

Golden Ratio

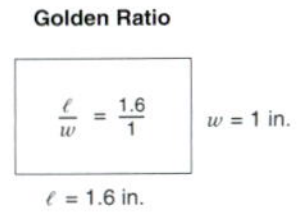

If you are like most people, you chose rectangle B. Why? It's a golden rectangle, of course! In a golden rectangle, the ratio of the length to the width is called the **golden ratio**—about 1.6 to 1.

The golden ratio pops up all over the place—in music, sculptures, the Egyptian pyramids, seashells, paintings, pinecones, and of course in rectangles.

To create your own golden rectangle, just write a ratio equivalent to the golden ratio. This will give you the length and width of another golden rectangle.

Use a ruler to draw a new golden rectangle in the space below. Then draw several non-golden rectangles around it. Now conduct a survey of your family and friends to see if they choose the golden rectangle as their favorite.

PROBLEM SOLVING 8-1

LESSON 8-1 **Problem Solving**
Ratios and Rates

Use the table to answer each question.

Atomic Particles of Elements

Element	Protons	Neutrons	Electrons
Gold	79	118	79
Iron	26	30	26
Neon	10	10	10
Platinum	78	117	78
Silver	47	61	47
Tin	50	69	50

1. What is the ratio of gold protons to silver protons? **79:47**
2. What is the ratio of gold neutrons to platinum protons? **118:78 or 59:39**
3. What is the ratio of platinum neutrons to neon neutrons? **117:10**
4. What is the ratio of iron electrons to tin electrons? **26:50 or 13:25**
5. What are two equivalent ratios of the ratio of neon protons to tin protons? **Possible answer: 10:50 and 1:5**
6. What are two equivalent ratios of the ratio of iron protons to iron neutrons? **Possible answer: 26:30 and 13:15**
7. A ratio of one element's neutrons to another element's electrons is equivalent to 3 to 5. What are those two elements? **iron neutrons to tin electrons**
8. The ratio of two elements' protons is equivalent to 3 to 1. What are those two elements? **platinum to iron**

Circle the letter of the correct answer.

9. Which element in the table has a ratio of 1 to 1, no matter what parts you are comparing in the ratio?
   A iron
   (B) neon
   C tin
   D silver
10. If the ratio for any element is 1:1, which two parts is the ratio comparing?
    F protons to neutrons
    G electrons to neutrons
    (H) protons to electrons
    J neutrons to electrons

Lesson Quiz

Flower Bulbs Planted	
Crocuses	20
Tulips	9
Daffodils	17
Lilies	5

Use the table to write each ratio.

1. tulips to daffodils **9:17**
2. crocuses to total number of bulbs **20:51**
3. total bulbs to lilies **51:5**
4. A dozen eggs cost $1.25 at one market. At a competing market, $1\frac{1}{2}$ dozen eggs costs $2.00. Which is the better buy? **eggs from the first market**

Available on Daily Transparency in CRB

Hands-On LAB 8A Explore Proportions

Pacing: Traditional 1 day
Block $\frac{1}{2}$ day

Objective: To use two-color counters to explore proportions

Materials: Two-color counters

Lab Resources

Hands-On Lab Activities . . . . pp. 67–70

Using the Pages

Discuss with students how to use two-color counters to model ratios.

Represent each ratio using two-color counters.

1. 3 red counters to 4 yellow counters.

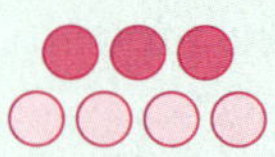

2. 5 red counters to 3 yellow counters.

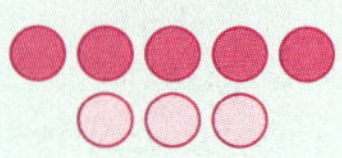

Write the ratio represented by the counters.

3.

4 red counters to 2 yellow counters

Hands-On LAB 8A Explore Proportions

Use with Lesson 8-2

internet connect
Lab Resources Online
go.hrw.com
KEYWORD: MR4 Lab8A

You can use counters to model equivalent ratios.

Activity 1

Find three ratios that are equivalent to $\frac{6}{12}$.

1. Show 6 red counters and 12 yellow counters.

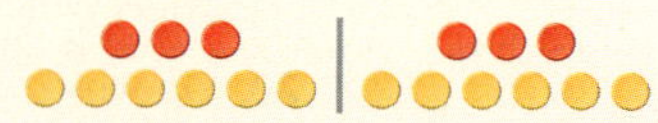

2. Separate the red counters into two equal groups. Then separate the yellow counters into two equal groups.

3. Write the ratio of red counters in each group to yellow counters in each group.

$$\frac{3 \text{ red counters}}{6 \text{ yellow counters}} = \frac{3}{6}$$

4. Now separate the red counters into three equal groups. Then separate the yellow counters into three equal groups.

5. Write the ratio of red counters in each group to yellow counters in each group.

$$\frac{2 \text{ red counters}}{4 \text{ yellow counters}} = \frac{2}{4}$$

6. Now separate the red counters into six equal groups. Then separate the yellow counters into six equal groups.

7. Write the ratio of red counters in each group to yellow counters in each group.

$$\frac{1 \text{ red counter}}{2 \text{ yellow counters}} = \frac{1}{2}$$

The three ratios you wrote are equivalent to $\frac{6}{12}$.

$$\frac{6}{12} = \frac{3}{6} = \frac{2}{4} = \frac{1}{2}$$

When you write an equation showing equivalent ratios, that equation is called a **proportion.**

Think and Discuss

1. How do the models show that the ratios are equivalent?
The number of counters is the same in each model.

Try This

Use models to determine whether the ratios form a proportion.

1. $\frac{1}{3}$ and $\frac{4}{12}$ yes   2. $\frac{3}{4}$ and $\frac{6}{9}$ no   3. $\frac{4}{10}$ and $\frac{2}{5}$ yes

Activity 2

Write a proportion in which one of the ratios is $\frac{1}{3}$.

1. You must find a ratio that is equivalent to $\frac{1}{3}$. First show one red counter and three yellow counters.

2. Show one more group of one red counter and three yellow counters.

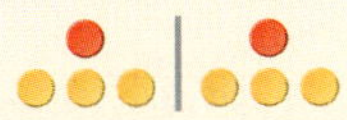

3. Write the ratio of red counters to yellow counters for the two groups.

$$\frac{2 \text{ red counters}}{6 \text{ yellow counters}} = \frac{2}{6}$$

4. The two ratios are equivalent. Write the proportion $\frac{1}{3} = \frac{2}{6}$.

You can find more equivalent ratios by adding more groups of one red counter and three yellow counters. Use your models to write proportions.

$$\frac{3 \text{ red counters}}{9 \text{ yellow counters}} = \frac{3}{9}$$

$$\frac{3}{9} = \frac{1}{3}$$

$$\frac{4 \text{ red counters}}{12 \text{ yellow counters}} = \frac{4}{12}$$

$$\frac{4}{12} = \frac{1}{3}$$

Think and Discuss

1. The models above show that $\frac{1}{3}$, $\frac{2}{6}$, $\frac{3}{9}$, and $\frac{4}{12}$ are equivalent ratios. Do you see a pattern in this list of ratios? Every time the first term increases by 1, the second term increases by 3.
2. Use counters to find another ratio that is equivalent to $\frac{1}{3}$.
Possible answer: $\frac{5}{15}$ is equivalent to $\frac{1}{3}$.

Try This

Use counters to write a proportion containing each given ratio. Possible answers:

1. $\frac{1}{4}$ $\frac{1}{4} = \frac{3}{12}$ 2. $\frac{1}{5}$ $\frac{1}{5} = \frac{2}{10}$ 3. $\frac{3}{7}$ $\frac{3}{7} = \frac{6}{14}$ 4. $\frac{1}{6}$ $\frac{1}{6} = \frac{3}{18}$ 5. $\frac{4}{9}$ $\frac{4}{9} = \frac{8}{18}$

8-2 Organizer

Pacing: Traditional 1 day
Block $\frac{1}{2}$ day

Objective: Students write and solve proportions.

Indiana Academic Standards
6.1.6, 6.2.7, 6.7.1, 6.7.5, 6.7.11

Warm Up

Use the table to write each ratio.

Brown bears	3
Giraffes	2
Monkeys	17
Polar bears	4

1. giraffes to monkeys **2:17**
2. polar bears to all bears **4:7**
3. monkeys to all animals **17:26**
4. all animals to all bears **26:7**

Problem of the Day

A carpenter can build one doghouse in one day. How many doghouses can 12 carpenters build in 20 days? **240**

Available on Daily Transparency in CRB

Math Fact

In a proportion of the form $\frac{a}{b} = \frac{b}{c}$, b is called the *geometric mean* of a and c.

8-2 Proportions

Learn to write and solve proportions.

Vocabulary
proportion

Have you ever heard water called H_2O? H_2O is the scientific formula for water. One molecule of water contains two hydrogen atoms (H_2) and one oxygen atom (O). No matter how many molecules of water you have, hydrogen and oxygen will always be in the ratio 2 to 1.

Water Molecules	1	2	3	4
$\frac{\text{Hydrogen}}{\text{Oxygen}}$	$\frac{2}{1}$	$\frac{4}{2}$	$\frac{6}{3}$	$\frac{8}{4}$

Notice that $\frac{2}{1}$, $\frac{4}{2}$, $\frac{6}{3}$, and $\frac{8}{4}$ are equivalent ratios.

A **proportion** is an equation that shows two equivalent ratios.

$\frac{2}{1} = \frac{4}{2}$ $\quad\frac{4}{2} = \frac{8}{4}$ $\quad\frac{2}{1} = \frac{6}{3}$

Read the proportion $\frac{2}{1} = \frac{4}{2}$ as "two is to one as four is to two."

EXAMPLE 1 Modeling Proportions

Write a proportion for the model.

First write the ratio of triangles to circles.

$$\frac{\text{number of triangles}}{\text{number of circles}} = \frac{4}{2}$$

Next separate the triangles and the circles into two equal groups.

Now write the ratio of triangles to circles in each group.

$$\frac{\text{number of triangles in each group}}{\text{number of circles in each group}} = \frac{2}{1}$$

A proportion shown by the model is $\frac{4}{2} = \frac{2}{1}$.

1 Introduce

Alternate Opener

EXPLORATION

8-2 Proportions

An automobile assembly line finishes 3 cars every 2 hours.

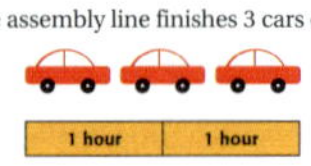

1. Use the diagram to determine how many cars are finished each hour.
2. Use the diagram to determine approximately how long it takes to finish 1 car.
3. If it takes 2 hours to finish 3 cars, how many hours does it take to finish
   a. 6 cars? b. 9 cars? c. 12 cars?
   d. 4 cars? e. 8 cars? f. 16 cars?

Think and Discuss

4. **Discuss** how you used the diagram to solve numbers 1 and 2.
5. **Explain** how you solved numbers 3a–3f.

Motivate

To review how to express ratios as fractions and how to find equivalent ratios, pose this problem:

Ethan is mixing orange juice. For every 1 can of frozen concentrate, he adds 3 cans of water. What is the ratio of concentrate to water? Give the ratio as a fraction. Write an equivalent ratio to show how much water Ethan will need if he uses 3 cans of concentrate. $\frac{1}{3}$; $\frac{1}{3} = \frac{3}{9}$; He will need 9 cans of water. This will get students ready to learn about *proportions*.

Exploration worksheet and answers on Chapter 8 Resource Book pp. 17 and 112

2 Teach

Lesson Presentation

Guided Instruction

In this lesson, students write and solve proportions. Teach students the format for writing proportions as equivalent ratios, and have them practice doing this. Then teach them to use cross products to find missing values in proportions. Have students use cross products to find missing values in a measurement context.

Teaching Tip You may wish to allow students to use two-color counters to model some of the proportions in the lesson.

CROSS PRODUCTS

Cross products in proportions are equal.

$\frac{4}{8} = \frac{2}{4}$	$\frac{3}{5} = \frac{9}{15}$	$\frac{9}{6} = \frac{3}{2}$	$\frac{14}{7} = \frac{2}{1}$
$8 \cdot 2 = 4 \cdot 4$	$5 \cdot 9 = 3 \cdot 15$	$6 \cdot 3 = 9 \cdot 2$	$7 \cdot 2 = 14 \cdot 1$
$16 = 16$	$45 = 45$	$18 = 18$	$14 = 14$

EXAMPLE 2 Using Cross Products to Complete Proportions

Find the missing value in the proportion $\frac{3}{4} = \frac{n}{16}$.

$\frac{3}{4} = \frac{n}{16}$ *Find the cross products.*

$4 \cdot n = 3 \cdot 16$ *The cross products are equal.*

$4n = 48$ *n is multiplied by 4.*

$\frac{4n}{4} = \frac{48}{4}$ *Divide both sides by 4 to undo the multiplication.*

$n = 12$

EXAMPLE 3 *Measurement Application*

Helpful Hint

In a proportion, the units must be in the same order in both ratios.

$\frac{tsp}{lb} = \frac{tsp}{lb}$ or $\frac{lb}{tsp} = \frac{lb}{tsp}$

The label from a bottle of pet vitamins shows recommended dosages. What dosage would you give an adult dog that weighs 15 lb?

Pet Vitamins

- **Adult dogs:** 1 tsp per 20 lb body weight
- **Puppies, pregnant dogs, or nursing dogs:** 1 tsp per 10 lb body weight
- **Cats:** 1 tsp per 12 lb body weight

$\frac{1 \text{ tsp}}{20 \text{ lb}} = \frac{v}{15 \text{ lb}}$ *Let v be the amount of vitamins for a 15 lb dog.*

$\frac{1 \text{ tsp}}{20 \text{ lb}} = \frac{v}{15 \text{ lb}}$ *Write a proportion.*

$20 \cdot v = 1 \cdot 15$ *The cross products are equal.*

$20v = 15$ *v is multiplied by 20.*

$\frac{20v}{20} = \frac{15}{20}$ *Divide both sides by 20 to undo the multiplication.*

$v = \frac{3}{4}$ tsp *Write your answer in simplest form.*

You should give $\frac{3}{4}$ tsp of vitamins to a 15 lb dog.

Think and Discuss

1. **Tell** whether $\frac{7}{8} = \frac{4}{14}$ is a proportion. How do you know?
2. **Give an example** of a proportion. Tell how you know that it is a proportion.

COMMON ERROR ALERT

Some students may try to cancel the numerator on one side of the equation with the denominator on the other side of the equation, as when multiplying fractions, instead of *cross multiplying*. Stress that they can't cancel across the equal sign.

Additional Examples

Example 1

Write a proportion for the model.

♥♥♥♥☆☆☆☆☆☆☆☆

♥♥☆☆☆☆

♥♥☆☆☆☆

A proportion shown by the model is $\frac{4}{8} = \frac{2}{4}$.

Example 2

Find the missing value in the proportion.

$\frac{5}{6} = \frac{n}{18}$ $n = 15$

Example 3

According to the label, 1 tablespoon of plant fertilizer should be used per 6 gallons of water. How many tablespoons of fertilizer would you use for 4 gallons of water?

You would use $\frac{2}{3}$ tbsp of fertilizer.

3 Close

Summarize

Review the fact that a *proportion* is an equation that shows two equivalent ratios. Demonstrate how to find a missing value in a proportion by using cross products.

Reaching All Learners

Through Curriculum Integration

Science The lesson opener deals with H_2O. You may use the recording sheet on CRB p. 104 to investigate other chemical formulations. Use the table to show equivalent ratios based on a number of molecules for each compound, and write two or three proportions based on the equivalent ratios in the table.

Answers to Think and Discuss

1. Possible answer: No; the cross products (32 and 98) are not equal.
2. Possible answer: $\frac{3}{5} = \frac{6}{10}$; I know it is a proportion because $\frac{3}{5}$ and $\frac{6}{10}$ are equivalent ratios, and also because $3 \times 10 = 5 \times 6$.

8-2 PRACTICE & ASSESS

Students may want to refer back to the lesson examples.

Assignment Guide

If you finished Example 1 assign:
Core 1, 6, 26–29
Enriched 1, 6, 26–29

If you finished Example 2 assign:
Core 1–4, 6–9, 11–17, 26–29
Enriched 3, 4, 7–9, 11–19, 26–29

If you finished Example 3 assign:
Core 1–15, 20, 26–29
Enriched 10–29

Notes

8-2 Exercises

FOR EXTRA PRACTICE
see page 652

internet connect
Homework Help Online
go.hrw.com Keyword: MR4 8-2

GUIDED PRACTICE

See Example 1
1. Write a proportion for the model. Possible answer: $\frac{6}{3} = \frac{2}{1}$

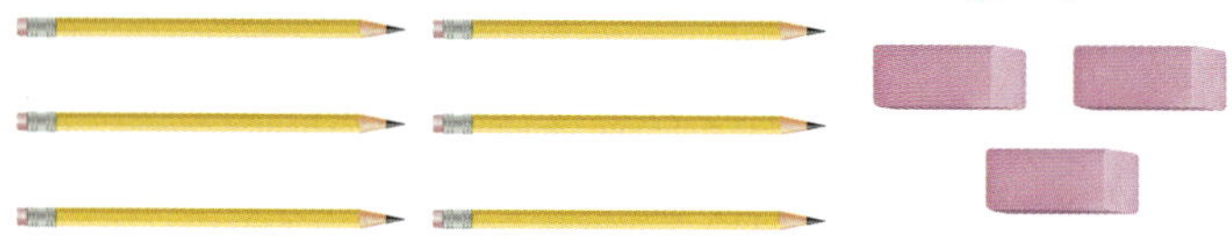

See Example 2
Find the missing value in each proportion.

2. $\frac{12}{9} = \frac{n}{3}$ 4
3. $\frac{t}{5} = \frac{28}{20}$ 7
4. $\frac{1}{c} = \frac{6}{12}$ 2

See Example 3
5. Ursula is entering a bicycle race for charity. Her mother pledges \$0.75 for every 0.5 mile she bikes. If Ursula bikes 17.5 miles, how much will her mother donate? \$26.25

INDEPENDENT PRACTICE

See Example 1
6. Write a proportion for the model. Possible answer: $\frac{12}{9} = \frac{4}{3}$

See Example 2
Find the missing value in each proportion.

7. $\frac{3}{2} = \frac{24}{d}$ 16
8. $\frac{p}{40} = \frac{3}{8}$ 15
9. $\frac{6}{14} = \frac{x}{7}$ 3

See Example 3
10. According to Ty's study guidelines, how many minutes of science reading should he do if his science class is 90 minutes long? 30 minutes

Ty's Study Guidelines

Class	Reading Time
Literature	35 minutes for every 50 minutes of class time
Science	20 minutes for every 60 minutes of class time
History	30 minutes for every 55 minutes of class time

PRACTICE AND PROBLEM SOLVING

Find the value of p in each proportion.

11. $\frac{18}{6} = \frac{6}{p}$ 2
12. $\frac{4}{p} = \frac{48}{60}$ 5
13. $\frac{p}{10} = \frac{15}{50}$ 3
14. $\frac{21}{15} = \frac{p}{5}$ 7
15. $\frac{3}{6} = \frac{p}{8}$ 4
16. $\frac{15}{5} = \frac{9}{p}$ 3
17. $\frac{150}{2} = \frac{p}{1}$ 75
18. $\frac{1}{12} = \frac{0.8}{p}$ 9.6
19. $\frac{8.1}{27} = \frac{p}{5}$ 1.5

Math Background

The statement that cross products in a proportion are equal is true because you are multiplying both sides of the proportion by the denominators of the fractions. Consider the following demonstration:

For real numbers a, b, c, and d,

$\frac{a}{b} = \frac{c}{d}$,

$\frac{a}{b} = \frac{ad}{bd}$, and $\frac{c}{d} = \frac{cb}{db}$.

Since the fractions to the right of the equal signs have the same denominator, they are equal if and only if their numerators are equal (i.e., if and only if $ad = cb$).

RETEACH 8-2

LESSON 8-2 **Reteach**
Proportions

A proportion is an equation that shows two equivalent ratios.

$\frac{3}{4} = \frac{9}{12}$ is an example of a proportion.

$3 \cdot 12 = 36$ and $4 \cdot 9 = 36$. The cross products of proportions are equal.

You can use cross products to find the missing value in a proportion.

$\frac{3}{x} = \frac{12}{48}$

$12 \cdot x = 3 \cdot 48$ To find x, first find the cross products.

$12x = 144$

Think: $144 \div 12 = x$ Then use a related math sentence to solve the equation.

$x = 12$

So, $\frac{3}{12} = \frac{12}{48}$.

Find the cross products to solve each proportion.

1. $\frac{x}{8} = \frac{3}{4}$ $x \cdot 4 =$ $8 \cdot 3$ $x = 6$
2. $\frac{2}{3} = \frac{x}{6}$ $2 \cdot 6 =$ $3 \cdot x$ $x = 4$
3. $\frac{2}{5} = \frac{4}{x}$ $2 \cdot x =$ $5 \cdot 4$ $x = 10$
4. $\frac{6}{x} = \frac{1}{3}$ $6 \cdot 3 =$ $x \cdot 1$ $x = 18$
5. $\frac{3}{8} = \frac{12}{x}$ $x = 32$
6. $\frac{3}{5} = \frac{6}{x}$ $x = 10$
7. $\frac{x}{8} = \frac{2}{16}$ $x = 1$
8. $\frac{2}{9} = \frac{4}{x}$ $x = 18$
9. $\frac{3}{4} = \frac{15}{x}$ $x = 20$
10. $\frac{1}{2} = \frac{x}{30}$ $x = 15$
11. $\frac{x}{5} = \frac{24}{30}$ $x = 4$
12. $\frac{25}{35} = \frac{5}{x}$ $x = 7$

PRACTICE 8-2

LESSON 8-2 **Practice B**
Proportions

Find the missing value in each proportion.

1. $\frac{24}{8} = \frac{n}{2}$ $n = 6$
2. $\frac{4}{9} = \frac{20}{n}$ $n = 45$
3. $\frac{n}{36} = \frac{5}{6}$ $n = 30$
4. $\frac{n}{5} = \frac{4}{10}$ $n = 2$
5. $\frac{3}{9} = \frac{2}{n}$ $n = 6$
6. $\frac{6}{n} = \frac{3}{7}$ $n = 14$
7. $\frac{5}{3} = \frac{n}{6}$ $n = 10$
8. $\frac{9}{6} = \frac{6}{n}$ $n = 4$
9. $\frac{2}{130} = \frac{1}{n}$ $n = 65$

Write a proportion for each model.

10. 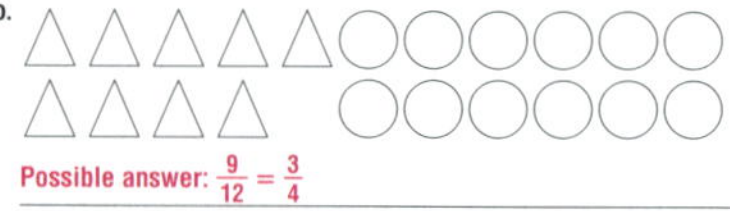

Possible answer: $\frac{9}{12} = \frac{3}{4}$

11. 

Possible answer: $\frac{16}{4} = \frac{4}{1}$

12. Shane's neighbor pledged \$1.25 for every 0.5 miles that Shane swims in the charity swim-a-thon. If Shane swims 3 miles, how much money will his neighbor donate?

\$7.50

13. Barbara's goal is to practice piano 20 minutes for every 5 minutes of lessons she takes. If she takes a 20 minute piano lesson this week, how many minutes should she practice this week?

80 minutes

Social Studies LINK

The value of the U.S. dollar as compared to the values of currencies from other countries changes every day. The graph shows the recent value of various currencies compared to the U.S. dollar. Use the graph for Exercises 20–25.

20. What is the value of 9.72 European euros in U.S. dollars? **$8.55**

21. You have $100 in U.S. dollars. Determine how much money this is in euros, Canadian dollars, renminbi, shekels, and Mexican pesos.

22. A watch in Israel costs 82 shekels. In the U.S., the watch costs $25. In which country does the watch cost less? **The watch cost less in Israel.**

23. ***WHAT'S THE ERROR?*** A student set up the proportion $\frac{1}{8.28} = \frac{x}{30}$ to determine the value of 30 U.S. dollars in China. What is wrong with this proportion? Write the correct proportion, and find the missing value.

24. ***WRITE ABOUT IT*** Would you prefer to have five U.S. dollars or five Canadian dollars? Why?

25. ***CHALLENGE*** A dime is worth about how many Mexican pesos? **about 0.91**

go.hrw.com
KEYWORD: MR4 Value
CNN Student News

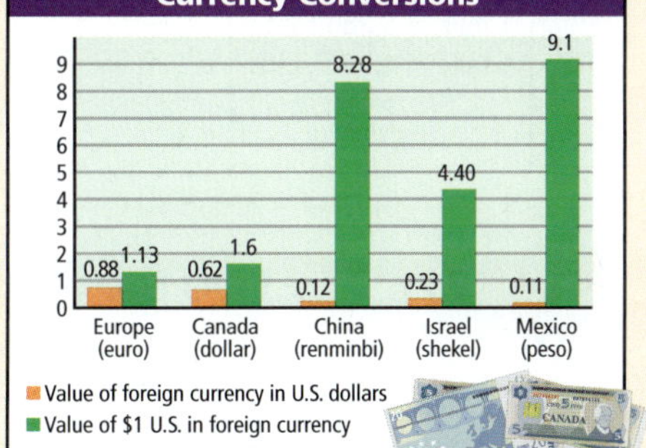

Spiral Review

26. The lengths of Ali's three jumps in the long jump were $9\frac{2}{3}$ feet, $9\frac{1}{2}$ feet, and $9\frac{5}{6}$ feet. Order these lengths from greatest to least. (Lesson 4-7) $9\frac{5}{6}, 9\frac{2}{3}, 9\frac{1}{2}$

Name the measure of central tendency described. (Lesson 6-2)

27. the number that appears the most often in a data set **mode**

28. the number in the middle of a data set that is in order from least to greatest **median**

29. **TEST PREP** Which polygon has the greatest number of sides? (Lesson 7-7) **C**

A Trapezoid **B** Quadrilateral **C** Hexagon **D** Triangle

Interdisciplinary LINK

Social Studies

Exercises 20–25 involve comparing the value of the U.S. dollar with values of currencies from other countries. Students learn more about these other countries in middle school social studies programs, such as Holt, Rinehart & Winston's *People, Places, and Change*.

Answers

21. 113 euros, 160 Canadian dollars, 828 renminbi, 440 shekels, and 910 Mexican pesos.

23. Possible answer: They inverted the x and the 30. The proportion should be $\frac{1}{8.28} = \frac{30}{x}$, and $30 U.S. is equivalent to 248.4 of China's renminbi.

24. Possible answer: U.S. dollars; they are worth more than Canadian dollars.

Journal

Have students write an explanation of how proportions and ratios are related. Have them write a proportion with a missing value and use cross products to solve the proportion.

Test Prep Doctor

For Exercise 30, students will need to remember the names of polygons learned in the previous chapter. Students who answered **A** or **B** incorrectly chose figures with 4 sides. Students who answered **D** may have been looking for the polygon with the *least* number of sides.

CHALLENGE 8-2

LESSON 8-2 **Challenge**
Patriotic Proportions

On August 21, 1959, President Eisenhower signed an order that established the official proportions of the United States flag. No matter what size the flag is, it must match those proportions to be used officially.

Official Proportions for the United States Flag	
Width of flag	1
Length of flag	$1\frac{9}{10}$
Width of union	$\frac{7}{13}$
Length of union	$\frac{19}{25}$
Width of each stripe	$\frac{1}{13}$

The union is the blue area.
The 50 stars represent the 50 states.

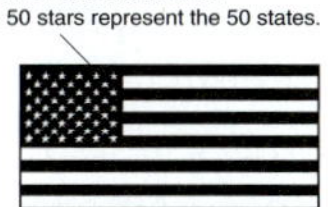

The 13 stripes represent the first 13 states.

Use the official proportions to find the missing dimension of each flag.

1. Length of flag = 10 feet; Width of flag = $5\frac{5}{19}$ feet
2. Width of flag = 57 yards; Length of flag = $108\frac{3}{10}$ yards
3. Width of flag = 13 centimeters; Width of Union = 7 centimeters
4. Width of flag = 260 inches; Width of each stripe = 20 inches
5. Length of flag = 25 meters; Length of Union = 10 meters

Choose a width in inches for a United States flag. Then use a ruler to draw your flag with the official proportional length in the space below.

Check students' flag widths and lengths for the correct width-to-length ratio of 1 inch to 1.9 inches.

PROBLEM SOLVING 8-2

LESSON 8-2 **Problem Solving**
Proportions

Write the correct answer.

1. For most people, the ratio of the length of their head to their total height is 1:7. Use proportions to test your measurements and see if they match this ratio.
   Answers will vary but should test the 1:7 head to height ratio measurements.
2. The ratio of an object's weight on Earth to its weight on the Moon is 6:1. The first person to walk on the Moon was Neil Armstrong. He weighed 165 pound on Earth. How much did he weigh on the Moon?
   27.5 pounds
3. It has been found that the distance from a person's eye to the end of the fingers of his outstretched hand is proportional to the distance between his eyes at a 10:1 ratio. If the distance between your eyes is 2.3 inches, what should the distance from your eye to your outstretched fingers be?
   23 inches
4. Chemists write the formula of ordinary sugar as $C_{12}H_{22}O_{11}$, which means that the ratios of one molecule of sugar are always 12 carbon atoms to 22 hydrogen atoms to 11 oxygen atoms. If there are four sugar molecules, how many atoms of each element will there be in 4 molecules of sugar?
   48 carbon, 88 hydrogen, 44 oxygen
5. According to doctors, a healthy diet should follow the ratio for meat to vegetables of 2.5 servings to 4 servings. If you eat 7 servings of meat a week, how many servings of vegetables should you eat?
   11.2 servings
6. A 150-pound person will burn 100 calories while sitting still for one hour. Following this ratio, how many calories will a 100-pound person burn while sitting still for one hour?
   $66\frac{2}{3}$ calories

Circle the letter of the correct answer.

7. Recently, 1 U.S. dollar was worth 1.58 in euros. If you exchanged $25 at that rate, how many euros would you get?
   (A) 39.50 euros
   B 15.82 euros
   C 26.58 euros
   D 23.42 euros
8. Recently, 1 United States dollar was worth 0.69 English pounds. If you exchanged 500 English pounds, how many dollars would you get?
   F 345 U.S. dollars
   (G) 725 U.S. dollars
   H 500.69 U.S dollars
   J 499.31 U.S. dollars

Lesson Quiz

1. Write a proportion for the model.
   □□□□□□◊◊ $\frac{2}{6} = \frac{1}{3}$

Find the missing value in each proportion.

2. $\frac{9}{5} = \frac{45}{x}$ $x = 25$
3. $\frac{p}{36} = \frac{5}{6}$ $p = 30$
4. The label on a bottle of salad dressing states that there are 3 grams of fat per tablespoon. If you use 3 tablespoons, how many grams of fat would you be getting? **9 g**

Available on Daily Transparency in CRB

8-3 Organizer

Pacing: Traditional 1 day
Block $\frac{1}{2}$ day

Objective: Students use proportions to make conversions within the customary system.

Indiana Academic Standards
6.2.7, 6.7.1, 6.7.2, 6.7.11

Warm Up

Find the missing value in each proportion.

1. $\frac{4}{1} = \frac{24}{x}$ $x = 6$ **2.** $\frac{x}{30} = \frac{5}{6}$ $x = 25$

3. $\frac{14}{16} = \frac{x}{40}$ $x = 35$ **4.** $\frac{8}{x} = \frac{12}{18}$ $x = 12$

Problem of the Day

There are 4 ounces in a gill. There are 4 gills in a pint. There are 8 pints in a gallon. How many ounces are the same as the total of 3 gallons, 3 pints, 3 gills, and 3 ounces? 447

Available on Daily Transparency in CRB

Math Fact

Julius Caesar used a calendar system that made March (*Martius* in Latin) the first month of the year. Thus September, October, November, and December were the seventh, eighth, ninth, and tenth months of his year. This is why those months' names contain prefixes meaning *seven, eight, nine,* and *ten.*

8-3 Proportions and Customary Measurement

Learn to use proportions to make conversions within the customary system.

The Washington Monument is about 185 yards tall. This is almost equal to the length of two football fields. How many feet is this length?

You can use the information in the table below to write a proportion that will help you answer this question.

Common Customary Measurements			
Length	**Weight**	**Time**	**Capacity**
1 foot = 12 inches 1 yard = 36 inches 1 yard = 3 feet 1 mile = 5,280 feet 1 mile = 1,760 yards	1 pound = 16 ounces 1 ton = 2,000 pounds	1 minute = 60 seconds 1 hour = 60 minutes 1 day = 24 hours 1 week = 7 days 1 year = 12 months 1 year = 365 days 1 leap year = 366 days	1 cup = 8 fluid ounces 1 pint = 2 cups 1 quart = 2 pints 1 quart = 4 cups 1 gallon = 4 quarts 1 gallon = 16 cups

EXAMPLE 1 Using Proportions to Convert Measurements

A **Find the height in feet of the Washington Monument.**

$\frac{1 \text{ yd}}{3 \text{ ft}} = \frac{185 \text{ yd}}{x \text{ ft}}$ *1 yard is 3 feet. Write a proportion. Use a variable for the value you are trying to find.*

$3 \cdot 185 = 1 \cdot x$ *The cross products are equal.*

$555 = x$

The Washington Monument is 555 feet tall.

B **In March 1994, a rainbow was visible for 360 minutes over parts of the United Kingdom. How many hours was it visible?**

$\frac{1 \text{ hr}}{60 \text{ min}} = \frac{x \text{ hr}}{360 \text{ min}}$ *1 hour is 60 minutes. Write a proportion. Use a variable for the value you are trying to find.*

$60 \cdot x = 1 \cdot 360$ *The cross products are equal.*

$\frac{60x}{60} = \frac{360}{60}$ *x is multiplied by 60. Divide both sides by 60 to undo the multiplication.*

$x = 6$

The rainbow was visible for 6 hours.

1 Introduce

Alternate Opener

EXPLORATION

8-3 Proportions and Customary Measurement

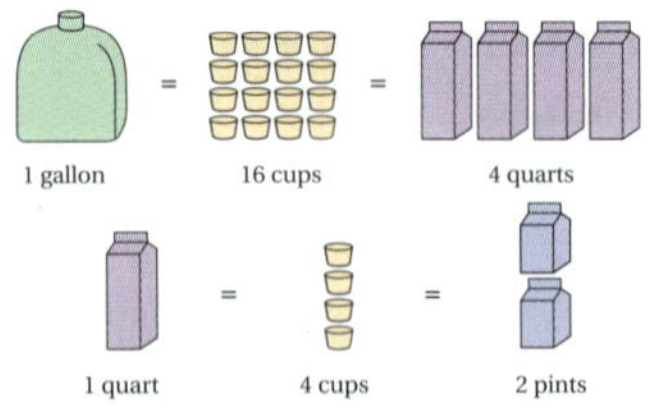

Use the equivalent measurements above to convert each measurement.

1. 2 gallons = ____ cups
2. 12 quarts = ____ cups
3. 3 quarts = ____ cups
4. 1 gallon = ____ pints
5. 6 pints = ____ quarts
6. 20 cups = ____ gallons

Think and Discuss

7. **Explain** how you converted each measurement.
8. **Discuss** situations in which measurement conversions are used.

Motivate

Ask students how many centimeters are equal to 1 meter. 100 centimeters Then ask how many centimeters are equal to 4 meters. 400 centimeters Demonstrate that a proportion can be set up using these equivalent measures:

$\frac{100 \text{ cm}}{1 \text{ m}} = \frac{400 \text{ cm}}{4 \text{ m}}$

Explain that proportions can be used to convert customary measurements.

Exploration worksheet and answers on Chapter 8 Resource Book pp. 26 and 114

2 Teach

Lesson Presentation

Guided Instruction

In this lesson, students use proportions to make conversions within the customary system. Review common customary measurements with students (Teaching Transparency T11 in Chapter 8 Resource Book), and then teach them to use proportions to find equivalent measures. You may wish to have students check their answers by substituting the solution for the variable in the proportion and determining whether the cross products are equal.

C The world's largest ice cream sundae weighed about 55,000 pounds. How many tons did it weigh?

$\frac{1 \text{ ton}}{2{,}000 \text{ lb}} = \frac{x \text{ tons}}{55{,}000 \text{ lb}}$ — *1 ton is 2,000 pounds. Write a proportion. Use a variable for the value you are trying to find.*

$2{,}000 \cdot x = 1 \cdot 55{,}000$ — *The cross products are equal.*

$2{,}000x = 55{,}000$ — *x is multiplied by 2,000.*

$\frac{2{,}000x}{2{,}000} = \frac{55{,}000}{2{,}000}$ — *Divide both sides by 2,000 to undo the multiplication.*

$x = 27.5$

The ice cream sundae weighed about 27.5 tons.

Think and Discuss

1. **Describe** a situation in which you would need to convert measurements.
2. **Explain** how to set up a proportion to convert miles to yards.
3. **Tell** what is wrong with this proportion: $\frac{24 \text{ hr}}{1 \text{ day}} = \frac{x \text{ days}}{168 \text{ hr}}$. Then tell the correct way to write it.

Additional Example

Example 1

A. Sonja went hiking for 4 hours. For how many minutes did she go hiking? 240 minutes

B. Mr. Lee is 72 inches tall. Find his height in feet. 6 feet

C. Hunter used 56 cups of water to wash his car. How many gallons of water did he use? 3.5 gallons

8-3 Exercises

FOR EXTRA PRACTICE see page 652

internet connect — Homework Help Online go.hrw.com Keyword: MR4 8-3

8-3 PRACTICE & ASSESS

Students may want to refer back to the lesson examples.

GUIDED PRACTICE

See Example 1

1. Linda cut off 1.5 feet of her hair to donate to an organization that makes wigs for children with cancer. How many inches of hair did she cut off? **18 inches**
2. An adult male of average size normally has about 6 quarts of blood in his body. Approximately how many cups of blood does the average adult male have in his body? **24 cups**

INDEPENDENT PRACTICE

See Example 1

3. The steel used to make the Statue of Liberty weighs about 125 tons. How many pounds of steel were used to make the Statue of Liberty? **250,000 pounds**
4. Ky will spend 105 days in Nepal as an exchange student. How many weeks will Ky spend in Nepal? **15 weeks**
5. Lake Superior is about 1,302 feet deep at its deepest point. Find this depth in yards. **434 yards**

Assignment Guide

If you finished Example 1 assign:
Core 1–11, 13–17 odd, 23–27
Enriched 2, 4, 6–27

3 Close

Reaching All Learners
Through Hands-On Experience

Have students measure several items in the classroom (Rulers are available in Manipulatives Kit and Chapter 8 Resource Book p. 105). Have them set up a proportion for each item to determine an equivalent customary measure (e.g., door height is 96 in. and $\frac{12 \text{ in.}}{1 \text{ ft}} = \frac{96 \text{ in.}}{x \text{ ft}}$). Students can then measure again, using the new unit of measurement, to check their answers.

Summarize

You may wish to review some of the more common customary measurement equivalents with students and post them in your room for easy access. Ask them how to write and solve a proportion to find how many feet a football player traveled if he ran 7 yards.

$\frac{1 \text{ yd}}{3 \text{ ft}} = \frac{7 \text{ yd}}{x \text{ ft}}$

$3 \cdot 7 = 1 \cdot x$

$21 = x$

The football player ran 21 feet.

Answers to Think and Discuss

1. Possible answer: when you are asked to provide a measurement in feet but your tape measures only inches
2. Use the relationship 1 mile equals 1,760 yards to set up the proportion.
3. Possible answer: The upper terms must have the same units, and the lower terms must have the same units. So in the second ratio, *168 hr* should be the upper term and *x days* should be the lower term.

Lesson Quiz

Find the missing value.

1. ▢ in. = 6 yd 216
2. 24 pt = ▢ gal 3

Compare. Write <, >, or =.

3. 42 oz ▢ $2\frac{1}{2}$ lb >
4. 7,920 ft ▢ 1.5 mi =
5. If your family is going to take a 3-week vacation, how many days will you be gone? 21 days

Available on Daily Transparency in CRB

PRACTICE AND PROBLEM SOLVING

Fill in each missing value.

6. ▢ c = 72 fl oz 9
7. ▢ in. = 4 yd 144
8. 14,000 lb = ▢ tons 7
9. ▢ yd = 93 ft 31
10. 20 min = ▢ s 1,200
11. 98 days = ▢ weeks 14

Compare. Write <, >, or =.

12. 18 ft ▢ 220 in. <
13. 24 lb ▢ 388 oz <
14. 21 hr ▢ $\frac{5}{6}$ day >
15. $\frac{1}{2}$ pt ▢ 1 c =

16. If an object travels 66 inches per minute, how many feet does it travel per minute? $5\frac{1}{2}$
17. If you drink 14 quarts of water per week, on average, how many pints do you drink per day? 4 pints
18. ***ART*** In Paris, the sculpture *Long-Term Parking,* created by Armand Fernandez, contains 60 cars embedded in 3.5 million pounds of concrete. How many tons of concrete is this? 1750 tons
19. ***SPORTS*** The width of a tennis court is 27 feet.
    a. How many yards wide is a tennis court? 9 yd
    b. How many inches wide is a tennis court? 324 in.
    c. An inch is equal to about 2.5 centimeters. Estimate the width in centimeters of a tennis court. about 810 cm
20. ***CHOOSE A STRATEGY*** A customer wanted a 25-foot piece of wire. The clerk incorrectly measured the wire with a yardstick that was 2 inches too short. How many inches were missing from the customer's piece of wire?
21. ***WRITE ABOUT IT*** Explain how to compare a weight given in ounces with a weight given in pounds.
22. ***CHALLENGE*** Human hair grows at an average rate of 12 cm per year. How many inches per month does the average human hair grow? (*Hint:* 2.5 cm is about 1 in.) 0.4 inches per month

Art LINK

Long-Term Parking is 65 feet tall and stands in front of a parking lot in Paris.

20. Possible answer: Draw a diagram to figure out that 16 inches were missing from the piece of wire.

21. First convert either the pounds to ounces or the ounces to pounds so that they have the same units. Then compare.

Spiral Review

Solve each equation. (Lessons 2-4, 2-5)

23. $d + 24 = 40$ 16
24. $x - 15 = 5$ 20
25. $9 + c = 44$ 35
26. A new box of dominoes is opened, and 7 dominoes are added to it. The box now has 33 dominoes in it. Write an equation that can be used to find the number of dominoes in a new box. (Lesson 2-4) Possible answer: $b + 7 = 33$
27. **TEST PREP** Which length is the longest? (Lesson 3-4) B

    A 30 cm **B** 3,000 mm **C** 0.03 m **D** 3 cm

RETEACH 8-3

LESSON 8-3 **Reteach**
Proportions and Customary Measurement

You can use the table below to solve proportions that involve customary measures.

Length	Weight	Time	Capacity
1 foot = 12 inches 1 yard = 36 inches 1 yard = 3 feet 1 mile = 5,280 feet 1 mile = 1,760 yards	1 pound = 16 ounces 1 ton = 2,000 pounds	1 minute = 60 seconds 1 hour = 60 minutes 1 day = 24 hours 1 week = 7 days 1 year = 365 days 1 leap year = 366 days	1 cup = 8 fluid ounces 1 pint = 2 cups 1 quart = 2 pints 1 quart = 4 cups 1 gallon = 4 quarts 1 gallon = 16 cups

To figure out how many hours are in three days, set up a proportion, where the first ratio uses 24 hours is 1 day, and the second ratio uses a variable for the value you are trying to find.

$$\frac{24 \text{ hours}}{1 \text{ day}} = \frac{72 \text{ hours}}{x \text{ days}}$$

Then solve the proportion.

$$\frac{24 \text{ hours}}{1 \text{ day}} = \frac{72 \text{ hours}}{x \text{ days}}$$ First, find the cross products.

$$24x = 72$$

Think: $72 \div 24 = x$ Then, use a related math sentence to solve the equation.

$$x = 3$$

So, there are 72 hours in 3 days.

Use the table above to set up a proportion. Then find each of the values.

1. the number of pounds in 80 ounces
   $\frac{1}{16} = \frac{x}{80}$
   5 pounds
2. the number of quarts in 6 gallons
   $\frac{4}{1} = \frac{x}{6}$
   24 quarts
3. the number of yards in 5 miles
   8,800 yards
4. the number of minutes in 20 hours
   1,200 minutes

PRACTICE 8-3

LESSON 8-3 **Practice B**
Proportions and Customary Measurement

Find each missing value.

1. 3 yards = 108 inches
2. 29 yards = 87 feet
3. 13 cups = 104 fluid ounces
4. 2 years = 104 weeks
5. 4 pounds = 64 ounces
6. 48 hours = 2 days
7. 540 minutes = 9 hours
8. 3 gallons = 48 cups
9. 8 cups = 4 pints
10. 36 inches = 1 yards

Compare. Write <, >, or =.

11. 4 quarts < 24 cups
12. 2.5 feet < 32 inches
13. 250 seconds > 4 minutes
14. 5 cups = 40 fluid ounces
15. 56 ounces = 3.5 pounds
16. 38 hours > $1\frac{1}{2}$ days
17. 1.5 miles > 2,500 yards
18. $3\frac{1}{2}$ tons > 6,000 pounds
19. Cassandra drank $8\frac{1}{2}$ cups of water during the mountain hike. How many fluid ounces of water did she drink?
    68 fluid ounces
20. Stan cut a wooden plank into 4 pieces. Each piece was 18 inches long. How long was the plank before Stan cut it?
    72 inches or 6 feet long

CHALLENGE 8-3

LESSON 8-3 **Challenge**
Pro-portional Basketball

Use proportions to convert each professional basketball measurement.

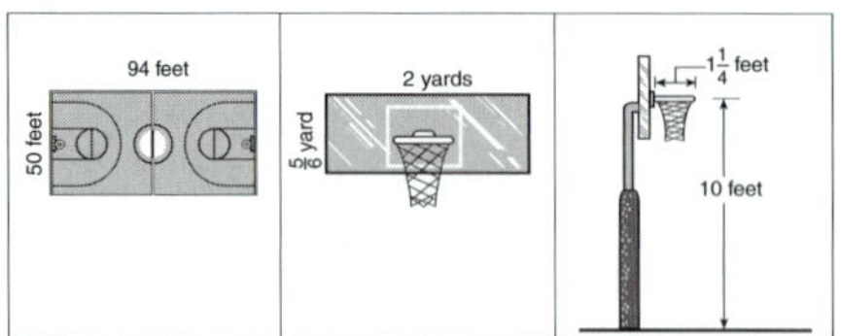

Court	Backboard	Basket
length: $31\frac{1}{3}$ yards	length: 6 feet	Height: 120 inches
width: $16\frac{2}{3}$ yards	width: $2\frac{1}{2}$ feet	Diameter: 18 inches

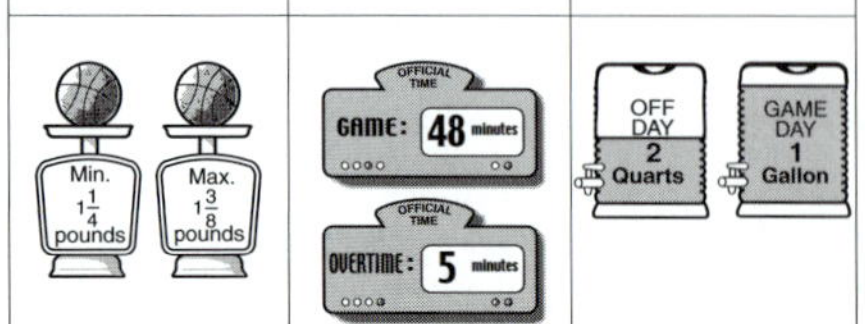

Ball Weights	Official Times	Water Intake
Minimum: 20 ounces	Game: $\frac{4}{5}$ hour	Off Day 8 cups
Maximum: 22 ounces	Overtime: $\frac{1}{12}$ hour	Game Day 16 cups

PROBLEM SOLVING 8-3

LESSON 8-3 **Problem Solving**
Proportions and Customary Measurement

Write the correct answer.

1. Each side of a professional baseball base must measure 15 inches. What is the base's side length in feet?
   $1\frac{1}{4}$ feet
2. In the NBA, any shot made from 22 feet or more from the basket is worth 3 points. How many yards from the basket is that?
   $7\frac{1}{3}$ yards
3. The maximum weight for a professional bowling ball is 16 pounds. What is the maximum weight in ounces?
   256 ounces
4. A professional hockey goal is 6 feet wide and 4 feet high. What is the area of the goal in square yards?
   $2\frac{2}{3}$ square yards
5. An NFL football field is 120 yards long. How many times would you have to run across the field to run 1 mile?
   $14\frac{2}{3}$ times
6. The halftime break in an NBA game is 15 minutes. How many seconds long is the break? How many hours?
   900 seconds; $\frac{1}{4}$ hour
7. The official length for a marathon race is 26.2 miles. How many yards long is a marathon? How many feet?
   46,112 yards;
   138,336 ft
8. The fastest time for a marathon race was set in 1999. The runner completed the race in 7,542 seconds. What is the record in hours?
   2.095 hours

Circle the letter of the correct answer.

9. An NFL football can be no less than $\frac{87}{96}$ feet long. What is the minimum length for an official football in inches?
   (**A**) $10\frac{7}{8}$ inches
   B $1\frac{3}{32}$ inches
   C $\frac{87}{1152}$ inches
   D $2\frac{69}{96}$ inches
10. An official Olympic-sized swimming pool holds 880,000 gallons of water! How many fluid ounces of water is that?
    F 1,4080,000 fluid ounces
    G 7,040,000 fluid ounces
    (**H**) 112,640,000 fluid ounces
    J 1,760,000 fluid ounces

8-4 Similar Figures

Learn to use ratios to identify similar figures.

Vocabulary

similar

corresponding sides

corresponding angles

Two or more figures are **similar** if they have exactly the same shape. Similar figures may be different sizes.

Similar figures have corresponding sides and corresponding angles.

- **Corresponding sides** have lengths that are proportional.
- **Corresponding angles** are congruent.

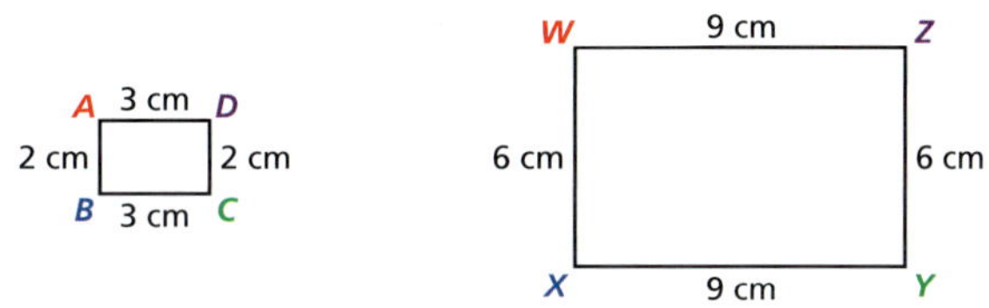

Corresponding sides:

$\overline{AB}$ corresponds to $\overline{WX}$.
$\overline{BC}$ corresponds to $\overline{XY}$.
$\overline{CD}$ corresponds to $\overline{YZ}$.
$\overline{AD}$ corresponds to $\overline{WZ}$.

Corresponding angles:

$\angle A$ corresponds to $\angle W$.
$\angle B$ corresponds to $\angle X$.
$\angle C$ corresponds to $\angle Y$.
$\angle D$ corresponds to $\angle Z$.

In the rectangles above, one proportion is $\frac{AB}{WX} = \frac{AD}{WZ}$, or $\frac{2}{6} = \frac{3}{9}$.

If you cannot use corresponding side lengths to write a proportion, or if corresponding angles are not congruent, then the figures are not similar.

EXAMPLE 1 Finding Missing Measures in Similar Figures

The two triangles are similar. Find the missing length x and the measure of $\angle A$.

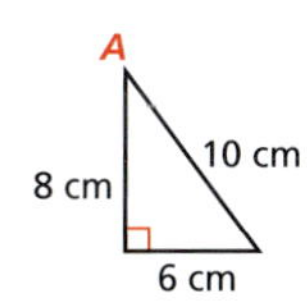

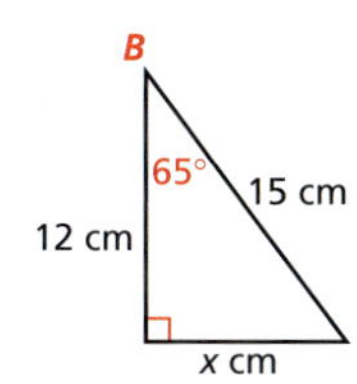

$\frac{8}{12} = \frac{6}{x}$ *Write a proportion using corresponding side lengths.*

$12 \cdot 6 = 8 \cdot x$ *The cross products are equal.*

$72 = 8x$ *x is multiplied by 8.*

$\frac{72}{8} = \frac{8x}{8}$ *Divide both sides by 8 to undo the multiplication.*

$9 \text{ cm} = x$

Angle A is congruent to angle B, and $m\angle B = 65°$.
$m\angle A = 65°$

8-4 Organizer

Pacing: Traditional 1 day
Block $\frac{1}{2}$ day

Objective: Students use ratios to identify similar figures.

Indiana Academic Standards
6.2.7, 6.4.5, 6.7.1, 6.7.11

Warm Up

Fill in the missing value.

1. ▢ c = 2 qt **8**
2. 180 in. = ▢ yd **5**
3. 3 tons = ▢ lb **6,000**
4. ▢ min = 2,760 s **46**

Problem of the Day

How many 8 in. by 10 in. rectangular tiles would be needed to cover a 16 ft by 20 ft floor? **576**

Available on Daily Transparency in CRB

Math Humor

A class in Texas and a class in California did the same science fair experiment. Their results were so similar that the judges knew the sides must have been corresponding.

1 Introduce

Alternate Opener

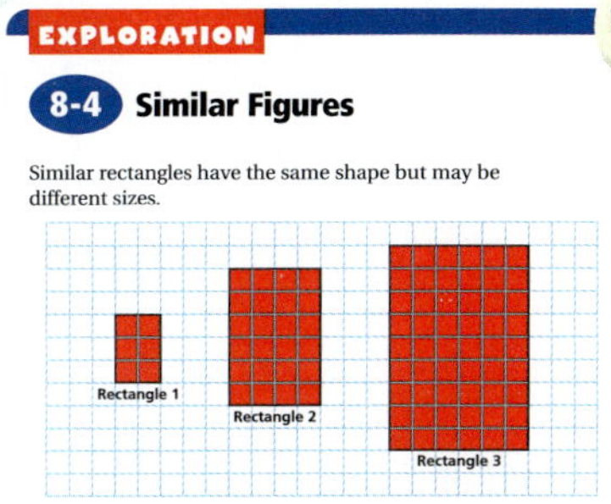

EXPLORATION

8-4 Similar Figures

Similar rectangles have the same shape but may be different sizes.

Measure the length and width of each rectangle and find each ratio.

	Ratio 1	Ratio 2
1.	length of rectangle 1 / length of rectangle 2 =	width of rectangle 1 / width of rectangle 2 =
2.	length of rectangle 2 / length of rectangle 3 =	width of rectangle 2 / width of rectangle 3 =
3.	length of rectangle 1 / length of rectangle 3 =	width of rectangle 1 / width of rectangle 3 =

Think and Discuss

4. **Describe** the pattern between ratio 1 and ratio 2.
5. **Explain** why the three rectangles are similar.

Motivate

Have students use a protractor to draw two angles that are congruent to each other. Ask how they know the angles are congruent. Their measures are the same. Then draw and label two pairs of line segments, a 1-inch and a 2-inch, and a 3-inch and a 6-inch. Ask students whether the lengths of these segments can form a proportion. yes; $\frac{1}{2} = \frac{3}{6}$ Remind students to measure carefully because measurements will never be exact. Protractors and rulers are provided in the Manipulatives Kit.

Exploration worksheet and answers on Chapter 8 Resource Book pp. 35 and 116

2 Teach

Lesson Presentation

Guided Instruction

In this lesson, students learn to use ratios to identify similar figures. Define the terms *similar*, *corresponding sides*, and *corresponding angles*. Teach students to use a proportion to find the length of a missing side, given two similar polygons. Then have students apply the concept in a problem-solving context.

Additional Examples

Example 1

The two triangles are similar. Find the missing length y and the measure of $\angle D$.

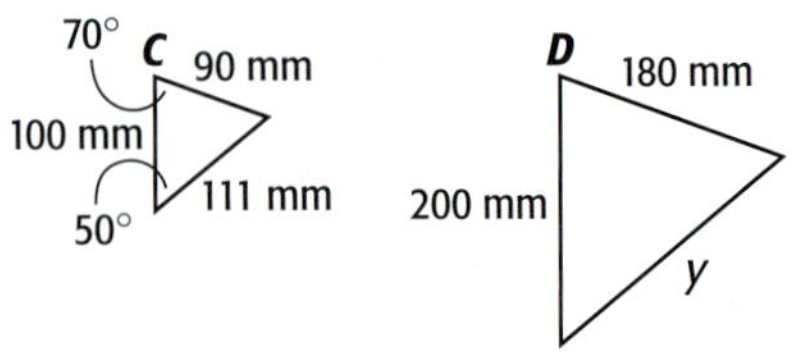

$y = 222$ mm $\quad m\angle D = 70°$

Example 2

This reduction is similar to a picture that Katie painted. The height of the actual painting is 54 centimeters. What is the width of the actual painting?

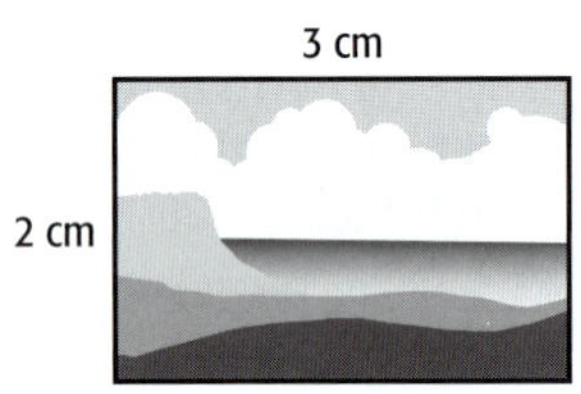

81 cm

EXAMPLE 2 **PROBLEM SOLVING APPLICATION**

The Boating Party **was painted by American artist Mary Cassatt. This reduction is similar to the actual painting. The height of the actual painting is 90.2 cm. To the nearest centimeter, what is the width of the actual painting?**

1 Understand the Problem

The **answer** will be the width of the actual painting.

List the **important information:**

- The actual painting and the reduction above are similar.
- The reduced painting is 4.6 cm tall and 6 cm wide.
- The actual painting is 90.2 cm tall.

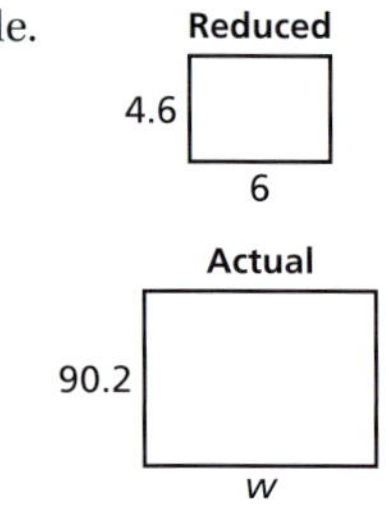

2 Make a Plan

Draw a diagram to represent the situation.
Use the corresponding sides to write a proportion.

3 Solve

$\frac{4.6 \text{ cm}}{90.2 \text{ cm}} = \frac{6 \text{ cm}}{w \text{ cm}}$ *Write a proportion.*

$90.2 \cdot 6 = 4.6 \cdot w$ *The cross products are equal.*

$541.2 = 4.6w$ *w is multiplied by 4.6.*

$\frac{541.2}{4.6} = \frac{4.6w}{4.6}$ *Divide both sides by 4.6 to undo the multiplication.*

$118 \approx w$ *Round to the nearest centimeter.*

Remember!

The symbol ≈ means "is approximately equal to."

The width of the actual painting is about 118 cm.

4 Look Back

Estimate to check your answer. The ratio of the heights is about 5:90, or 1:18. The ratio of the widths is about 6:120, or 1:20. Since these ratios are close to each other, 118 cm is a reasonable answer.

Think and Discuss

1. **Name** two items in your classroom that appear to be similar figures.
2. **Describe** how similar figures are different from congruent figures.

Reaching All Learners

Through Concrete Manipulatives

You may have students cut out the similar polygons provided on CRB p. 107 and have them explore the proportionality of corresponding sides. They can place the cutout figures on top of each other to prove that the corresponding angles are congruent. If time is limited, use this as a reinforcing activity to be completed at home or when time allows.

3 Close

Summarize

Have the students discuss how the new vocabulary terms in the lesson (*similar*, *corresponding sides*, and *corresponding angles*) relate to each other.

Possible answer: Two figures are similar if they have the same shape. Corresponding sides of similar figures are proportional. Corresponding angles of similar figures are congruent.

Answers to Think and Discuss

1. Possible answer: the window pane and the window
2. Congruent figures are figures with the same size and shape; similar figures have the same shape, but they can be different sizes as long as they are in proportion to each other.

8-4 Exercises

FOR EXTRA PRACTICE see page 652

internet connect
Homework Help Online
go.hrw.com Keyword: MR4 8-4

8-4 PRACTICE & ASSESS

Students may want to refer back to the lesson examples.

GUIDED PRACTICE

See Example 1

1. The two triangles are similar. Find the missing length x and the measure of $\angle G$. **The length of the missing side is 4 cm. m$\angle G = 37°$**

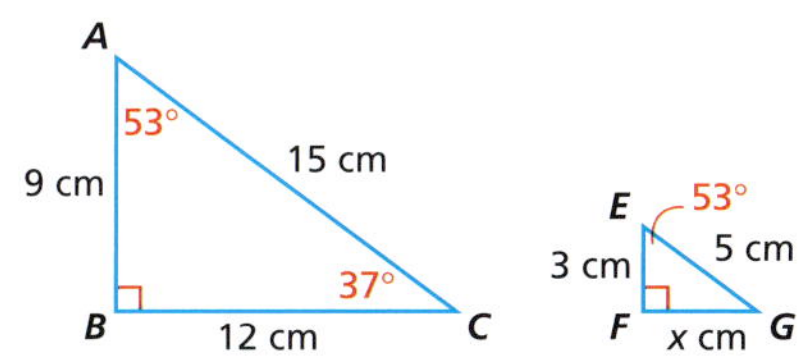

See Example 2

2. Pat's school photo package includes one large photo and several smaller photos. The large photo is similar to the photo at right. If the height of the large photo is 10 in., what is its width? **7.5 in.**

INDEPENDENT PRACTICE

See Example 1

3. The two triangles are similar. Find the missing length n and the measure of $\angle M$. **The length of the missing side, n, is 3 inches. m$\angle M = 110°$**

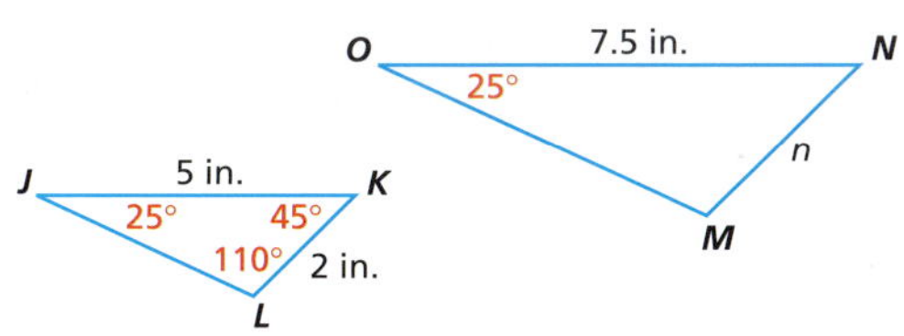

See Example 2

4. LeJuan swims in a pool that is similar to an Olympic-sized pool. LeJuan's pool is 30 m long by 8 m wide. The length of an Olympic-sized pool is 50 m. To the nearest meter, what is the width of an Olympic-sized pool? **13 m**

PRACTICE AND PROBLEM SOLVING

Name the corresponding sides and angles for each pair of similar figures.

5.

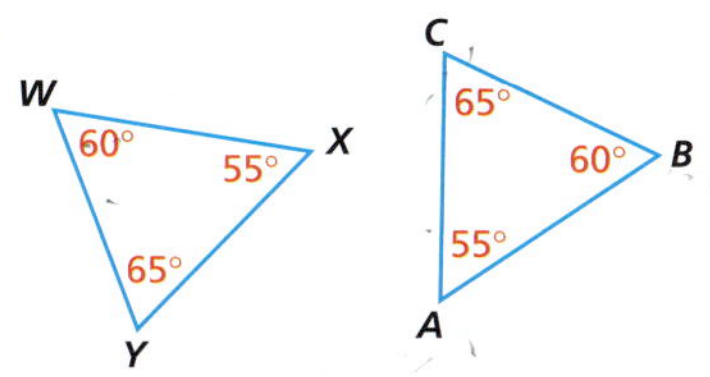

6.

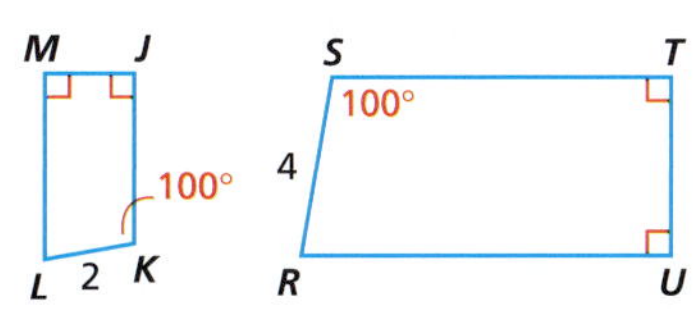

Assignment Guide

If you finished Example 1 assign:
Core 1, 3, 7, 8, 15–19
Enriched 1, 3, 7, 8, 15–19

If you finished Example 2 assign:
Core 1–8, 15–19
Enriched 7–19

Answers

5. sides: $\overline{AC}$ and $\overline{XY}$; $\overline{XW}$ and $\overline{AB}$; $\overline{BC}$ and $\overline{WY}$; angles: X and A; W and B; Y and C
6. sides: $\overline{KJ}$ and $\overline{ST}$; $\overline{KL}$ and $\overline{SR}$; $\overline{LM}$ and $\overline{RU}$; $\overline{JM}$ and $\overline{TU}$; angles: S and K; L and R; J and T; M and U

RETEACH 8-4

LESSON 8-4 **Reteach**
Similar Figures

Two figures are similar if they have the same shape but are different sizes.

Similar figures have corresponding sides and corresponding angles. Corresponding sides are proportional. Corresponding angles are congruent.

Look at the similar triangles below.

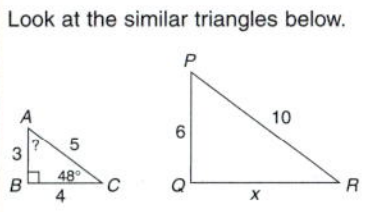

$\overline{AB}$ corresponds to $\overline{PQ}$. $\angle A$ corresponds to $\angle P$.
$\overline{BC}$ corresponds to $\overline{QR}$. $\angle B$ corresponds to $\angle Q$.
$\overline{AC}$ corresponds to $\overline{PR}$. $\angle C$ corresponds to $\angle R$.

What is the length of $\overline{QR}$?

$\frac{AB}{BC} = \frac{PQ}{QR}$ Set up a proportion.
$\frac{3}{4} = \frac{6}{x}$ Substitute the values.
$3 \cdot x = 4 \cdot 6$ The cross products are equal.
$3x = 24$ x is multiplied by 3.
$\frac{3x}{3} = \frac{24}{3}$ Divide both sides by 3.
$x = 8$

So, the length of $\overline{QR}$ is 8 units.

Find each missing length.

1.

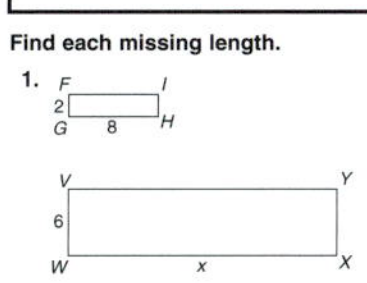

x = 24 units

2.

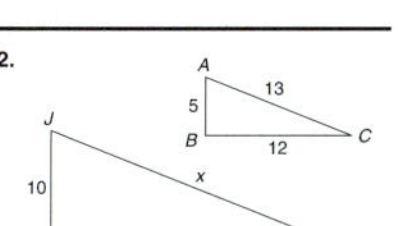

x = 26 units

PRACTICE 8-4

LESSON 8-4 **Practice B**
Similar Figures

Write the correct answers.

1. The two triangles are similar. Find the missing length x and the measure of $\angle A$. **x = 18 ft; m$\angle A = 80°$**

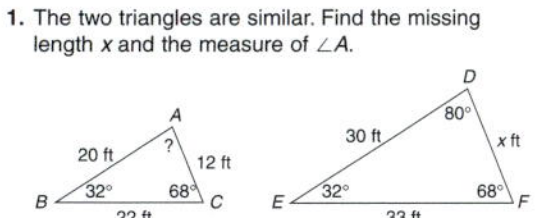

2. The two triangles are similar. Find the missing length x and the measure of $\angle J$. **x = 8m; m$\angle J = 23°$**

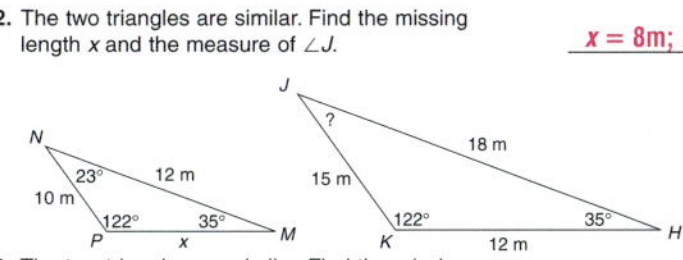

3. The two triangles are similar. Find the missing length x and the measure of $\angle N$. **x = 24 cm; m$\angle N = 53°$**

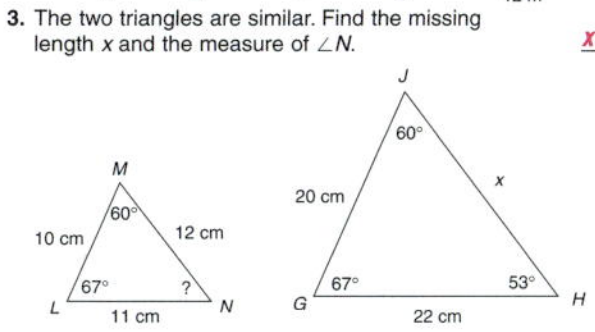

4. Juanita planted two flower gardens in similar square shapes. What are the measures of all the angles in each garden? Explain how you know.

They are all 90°; All squares have all right angles.

Math Background

As with congruence, triangles can be proven similar without having to prove that all pairs of corresponding angles are congruent and all pairs of corresponding sides are in proportion.

If either of the following situations occurs, you can conclude that two triangles are similar: All corresponding sides have lengths that are proportional, or two pairs of corresponding angles are congruent.

Answers

7. m∠E = 78°, m∠L = 78°, m∠M = 51°; the length of $\overline{ML}$ is 21 in.
8. m∠H = 80°, m∠J = 80°, m∠I = 100°, m∠Z = 100°; the length of $\overline{WX}$ is 5.5 yd, the length of $\overline{ZY}$ is 5.5 yd, and the length of $\overline{WZ}$ is 4 yd.
9. Yes; the corresponding angles are equal and the corresponding sides are in proportion.
10. No; the corresponding sides are not in proportion.

Journal

Have students write an explanation for the following statement: If two figures are congruent, they must also be similar.

Test Prep Doctor

For Exercise 19, students need to remember that parallelograms have two sets of parallel sides. Students should know that **A, B,** and **C** fit this description. Only **D** is not a parallelogram. Students who seem to think there is no correct answer have probably confused *parallelogram* with *quadrilateral*.

Lesson Quiz

These two triangles are similar.

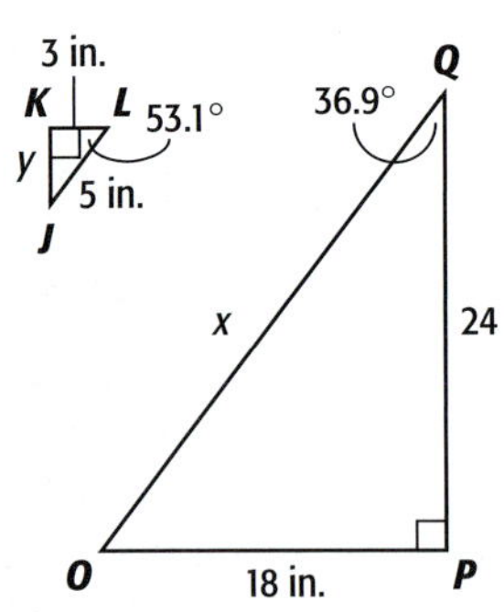

1. Find the missing length *x*. **30 in.**
2. Find the measure of ∠J. **36.9°**
3. Find the missing length *y*. **4 in.**
4. Find the measure of ∠P. **90°**
5. Susan is making a wood deck from plans for an 8 ft by 10 ft deck. However, she is going to increase its size proportionally. If the length is to be 15 ft, what will the width be? **12 ft**

Available on Daily Transparency in CRB

The figures in each pair are similar. Find the unknown measures.

7.

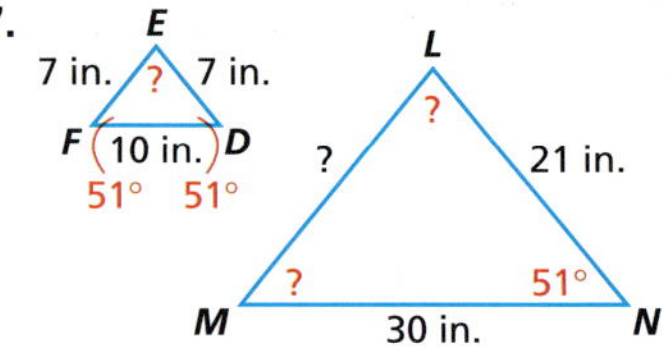

8.

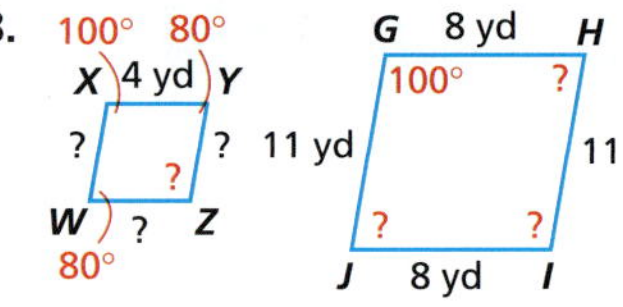

Tell whether the figures in each pair are similar. Explain your answers.

9.

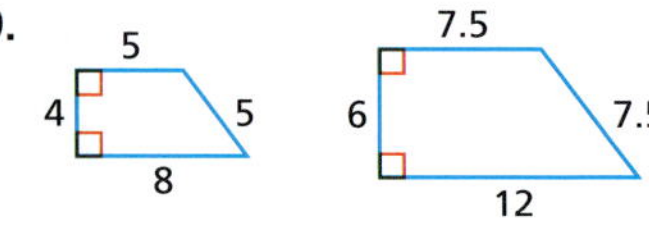

10.

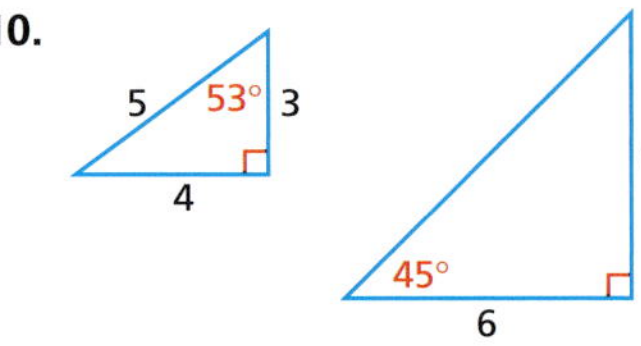

11. ***GRAPHIC ART*** Lenny designs billboards. He sketches his billboards before he paints them. The sketch and the billboard are similar. If the height of the billboard is 30 ft, what is the width to the nearest foot of the billboard? **50 feet**

12. ***WHAT'S THE ERROR?*** A student drew two rectangles. The dimensions of the rectangles are 10 in. by 9 in. and 5 in. by 3 in. The student said that the rectangles are similar. What's the error?

12. The lengths of the sides are not proportional, because $\frac{10}{5}$ does not equal $\frac{9}{3}$.

13. No; All right triangles have one angle that measures 90 degrees, but not all right triangles have the same angle measures for the other angles.

13. ***WRITE ABOUT IT*** Are all right triangles similar? Explain your answer.
14. ***CHALLENGE*** Draw two similar right triangles whose sides are in a ratio of 5:2. **Possible answer:**

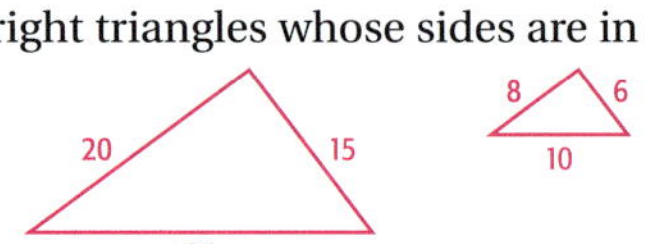

Spiral Review

Write each decimal in standard form. (Lesson 3-1)

15. four tenths **0.4**
16. thirty-one hundredths **0.31**
17. ten and seven thousandths **10.007**
18. A chef has $10\frac{1}{3}$ cups of flour. He uses $2\frac{1}{6}$ cups to make banana bread. How much flour does the chef have after making the bread? (Lesson 5-8) **$8\frac{1}{6}$ cups**
19. **TEST PREP** Which quadrilateral is **not** a parallelogram? (Lesson 7-6) **D**

A Rhombus **B** Rectangle **C** Square **D** Trapezoid

CHALLENGE 8-4

LESSON 8-4 **Challenge**
You Won't Believe Your Eyes!

Answer each question by looking at the drawings below. Then use what you know about similar and congruent figures to verify your answers.

1. 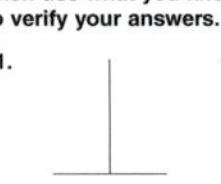

Are the two line segments congruent?

yes

2. 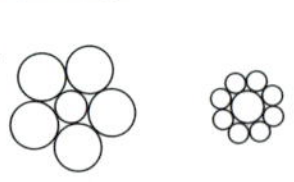

Are the two center circles similar or congruent?

congruent

3. 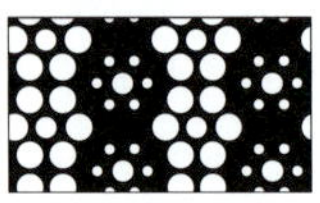

Are any of these circles similar?

yes; all the circles

4. 

Are any of these line segments congruent?

yes; horizontal lines

5. 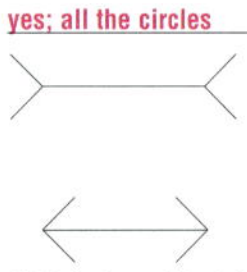

Which horizontal line is longer?

Neither, they are congruent.

6. 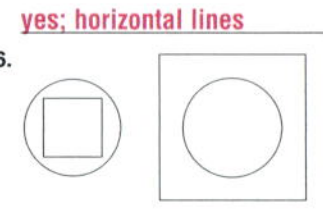

Which two figures are congruent? Which two figures are similar?

congruent circles; similar squares

PROBLEM SOLVING 8-4

LESSON 8-4 **Problem Solving**
Similar Figures

Write the correct answer.

1. The map at right shows the dimensions of the Bermuda Triangle, a region of the Atlantic Ocean where many ships and airplanes have disappeared. If a theme park makes a swimming pool in a similar figure, and the longest side of the pool is 0.5 mile long, about how long would the other sides of the pool have to be?

0.403 mile

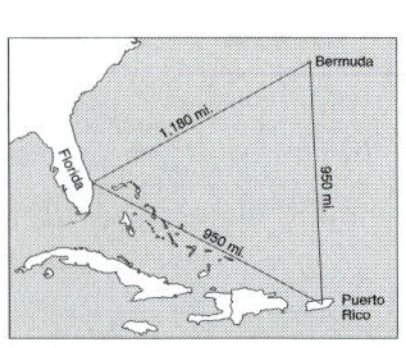

2. Completed in 1883, *The Battle of Gettysburg* is one of the largest paintings in the world. It is 410 feet long and 70 feet tall. A museum shop sells a print of the painting that is similar to the original. The print is 2.05 feet long. How tall is the print?

0.35 ft

3. *Panorama of the Mississippi* was the largest painting ever created in the United States. It was 12 feet tall and 5,000 feet long! If you wanted to make a copy similar to the original that was 2 feet tall, how many feet long would the copy have to be?

$833\frac{1}{3}$ feet

4. Two tables shaped like triangles are similar. The measure of one of the larger table's angles is 38°, and another angle is half that size. What are the measures of all the angles in the smaller table?

38°, 19°, and 123°

5. Two rectangular gardens are similar. The area of the larger garden is 8.28 m^2, and its length is 6.9 m. The smaller garden is 0.6 m wide. What is the smaller garden's length and area?

length = 3.45 m; area = 2.07 m^2

Circle the letter of the correct answer.

6. Which of the following is not always true if two figures are similar?
   - A They have the same shape.
   - **(B)** They have the same size.
   - C Their corresponding sides have proportional lengths.
   - D Their corresponding angles are congruent.
7. Which of the following figures are always similar?
   - F two rectangles
   - G two triangles
   - **(H)** two squares
   - J two pentagons

8-5 Indirect Measurement

Learn to use proportions and similar figures to find unknown measures.

Vocabulary
indirect measurement

Residents of Maine spent 14 days in 1999 building this enormous snowman. How could you measure the height of this snowman?

One way to find a height that you cannot measure directly is to use similar figures and proportions. This method is called **indirect measurement**.

Suppose that on a sunny day, the snowman cast a shadow that was 228 feet long. A 6-foot-tall person standing by the snowman cast a 12-foot-long shadow.

Both the person and the snowman form right angles with the ground, and their shadows are cast at the same angle. This means we can form two similar right triangles and use proportions to find the missing height.

EXAMPLE 1 Using Indirect Measurement

Use the similar triangles above to find the height of the snowman.

$\frac{6}{h} = \frac{12}{228}$	*Write a proportion using corresponding sides.*
$12 \cdot h = 6 \cdot 228$	*The cross products are equal.*
$12h = 1{,}368$	*h is multiplied by 12.*
$\frac{12h}{12} = \frac{1{,}368}{12}$	*Divide both sides by 12 to undo the multiplication.*
$h = 114$	

The snowman was 114 feet tall.

8-5 Organizer

Pacing: Traditional 1 day
Block $\frac{1}{2}$ day

Objective: Students use proportions and similar figures to find unknown measures.

Indiana Academic Standards
6.2.7, 6.3.1, 6.7.1, 6.7.11

Warm Up

Find the missing value in each proportion.

1. $\frac{6}{t} = \frac{18}{45}$ $t = 15$
2. $\frac{k}{19} = \frac{20}{76}$ $k = 5$
3. $\frac{6}{8} = \frac{42}{n}$ $n = 56$
4. $\frac{21}{11} = \frac{x}{44}$ $x = 84$

Problem of the Day

Bryce, Kate, and Annie have drawn rectangles. Each side of Bryce's rectangle is twice the size of one side of Kate's. The same side of Kate's rectangle is congruent to one side of Annie's. One side of Annie's rectangle is congruent to one side of Bryce's. Which two rectangles could be congruent? Annie's and Kate's

Available on Daily Transparency in CRB

1 Introduce

Alternate Opener

EXPLORATION

8-5 Indirect Measurement

The heights of very tall structures can be measured indirectly using similar figures and proportions. This method is called ***indirect measurement.***

Augustine and Carmen want to measure the height of the school's flagpole. To do this, they go outside and hold a meterstick upright. The meterstick casts a shadow that measures 50 cm.

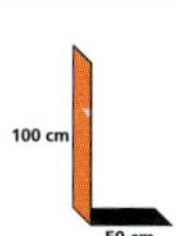

1. If the shadow of the flagpole measures 6 meters, how tall is the flagpole? To answer this question, follow these steps:
   a. Draw a sketch of the flagpole and its shadow next to the sketch of the meterstick and its shadow.
   b. Label the height of the flagpole x.
   c. Write the proportion $\frac{\text{height of flagpole}}{\text{shadow of flagpole}} = \frac{\text{height of meterstick}}{\text{shadow of meterstick}}$, and substitute the values for the given measurements.
   d. Solve the proportion for x.

Think and Discuss

2. **Explain** how you solved the proportion for x in number **1d.**
3. **Explain** whether you could have solved the problem by writing the proportion as $\frac{\text{shadow of flagpole}}{\text{height of flagpole}} = \frac{\text{shadow of meterstick}}{\text{height of meterstick}}$.

Motivate

Have students brainstorm a list of items that are too tall for them to measure, such as buildings or tall masts on large sailboats. Challenge students to hypothesize how they could use proportions to find the heights of very tall items.

Exploration worksheet and answers on Chapter 8 Resource Book pp. 44 and 118

2 Teach

Lesson Presentation

Guided Instruction

In this lesson, students learn to use proportions and similar figures to find unknown measures. Teach the indirect measurement method. Have students solve the question presented in the lesson opener and then a similar question. You may use Teaching Transparency T20 in Chapter 8 Resource Book. Teach students to use cross products to check each solution.

Additional Examples

Example 1

Use the similar triangles to find the height of the tree.

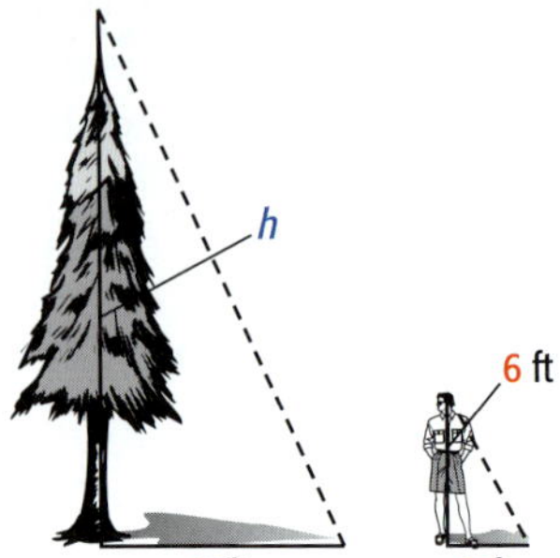

21 feet

Example 2

A rocket casts a shadow that is 91.5 feet long. A 4-foot model rocket casts a shadow that is 3 feet long. How tall is the rocket? 122 feet

EXAMPLE 2 *Measurement Application*

A lighthouse casts a shadow that is 36 m long when a meterstick casts a shadow that is 3 m long. How tall is the lighthouse?

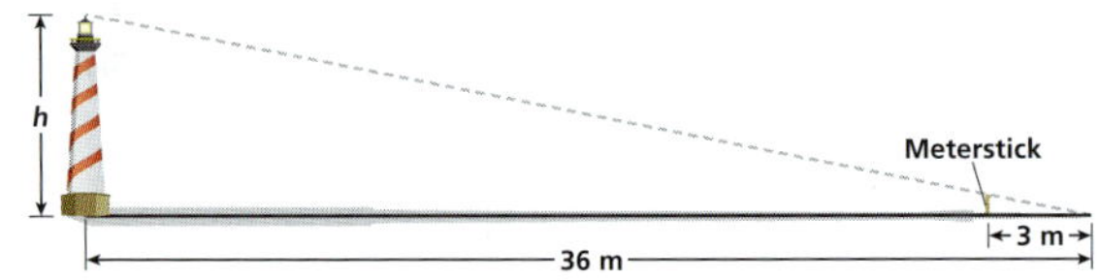

$\frac{h}{1} = \frac{36}{3}$ — *Write a proportion using corresponding sides.*

$1 \cdot 36 = 3 \cdot h$ — *The cross products are equal.*

$36 = 3h$ — *h is multiplied by 3.*

$\frac{36}{3} = \frac{3h}{3}$ — *Divide both sides by 3 to undo the multiplication.*

$12 = h$

The lighthouse is 12 m tall.

Think and Discuss

1. **Name** two items for which it would make sense to use indirect measurement to find their heights.
2. **Name** two items for which it would **not** make sense to use indirect measurement to find their heights.

8-5 PRACTICE & ASSESS

Assignment Guide

If you finished Example 1 assign:
Core 1, 3, 7, 10–15
Enriched 1, 3, 8, 10–15

If you finished Example 2 assign:
Core 1–7, 10–15
Enriched 3–15

8-5 Exercises

FOR EXTRA PRACTICE see page 652

internet connect
Homework Help Online
go.hrw.com Keyword: MR4 8-5

GUIDED PRACTICE

See Example 1

1. Use the similar triangles to find the height of the flagpole. 15 ft

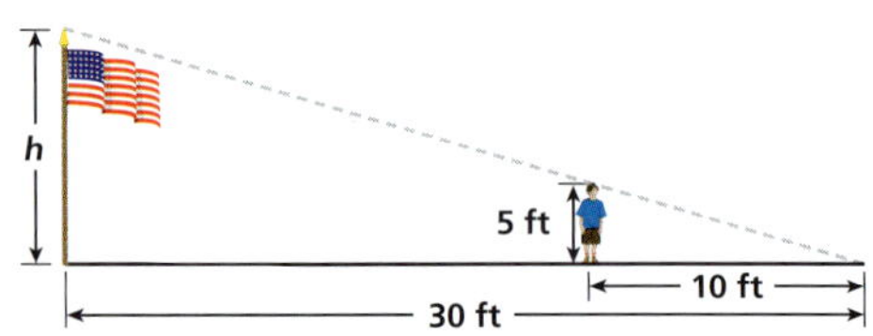

See Example 2

2. A tree casts a shadow that is 26 ft long. At the same time, a yardstick casts a shadow that is 4 ft long. How tall is the tree? 19.5 ft

Teach

Reaching All Learners
Through World Math

Have students do research to find heights of various landmarks around the world. (You may wish to assign a different landmark to each student.) Then have them write an indirect measurement problem about their landmarks. (Model how to write such a problem.) Display these problems on a bulletin board and have students solve some of the problems written by the other students.

3 Close

Summarize

You may wish to review the teaching examples, explaining step by step what each number represents in each proportion.

Answers to Think and Discuss

1. Possible answers: a flagpole, a tall tree
2. Possible answers: a pencil, a can of soup

INDEPENDENT PRACTICE

See Example 1

3. Use the similar triangles to find the height of the lamppost. **18 ft**

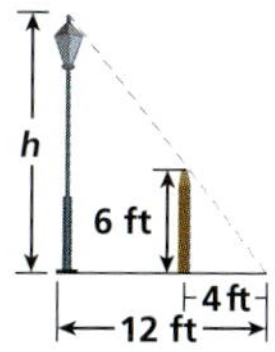

See Example 2

4. On a sunny day, the Eiffel Tower cast a shadow that was 328 feet long. A 6-foot-tall person standing by the tower cast a 2-foot-long shadow. How tall is the Eiffel Tower? **984 ft**

PRACTICE AND PROBLEM SOLVING

8. Possible answer: Using indirect measurements allows you to find the heights of tall objects that would otherwise be difficult to measure.

Find the unknown heights.

5. 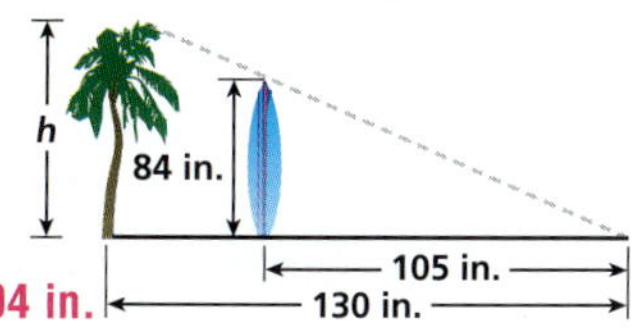

104 in.

6. 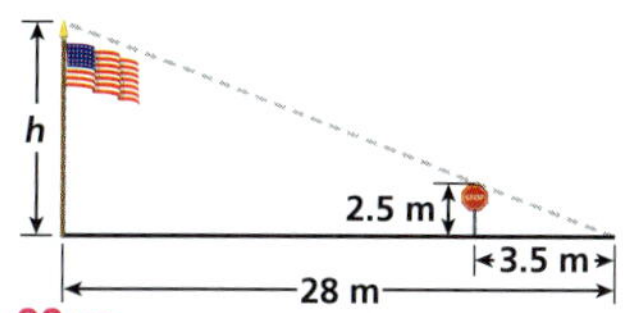

20 m

7. A statue casts a shadow that is 360 m long. At the same time, a person who is 2 m tall casts a shadow that is 6 m long. How tall is the statue? **120 m**

8. ***WRITE ABOUT IT*** How are indirect measurements useful?

9. ***CHALLENGE*** A 5.5-foot-tall girl stands so that her shadow lines up with the shadow of a telephone pole. The tip of her shadow is even with the tip of the pole's shadow. If the length of the pole's shadow is 40 feet and the girl is standing 27.5 feet away from the pole, how tall is the telephone pole? **17.6 ft**

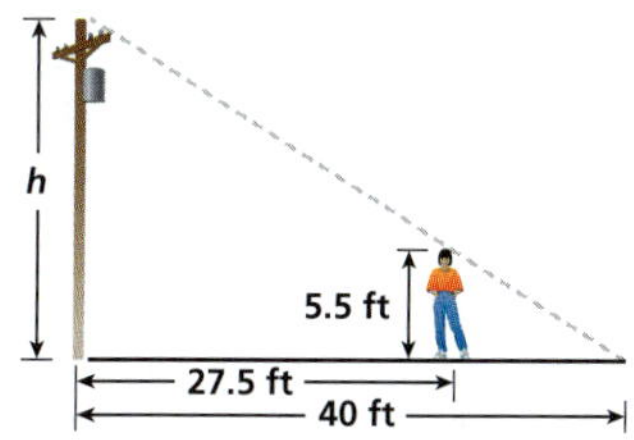

Spiral Review

Evaluate. (Lesson 1-4)

10. $(2 + 7 - 5) \div 2$ **2**
11. $10(6 - 3)$ **30**
12. $5 + 8 \cdot 7 - 1$ **60**
13. $5 + (8 + 2) - 3$ **12**

14. Make a stem-and-leaf plot of the following data: 85, 102, 89, 86, 104, 92, 103, 97, 91, 100. (Lesson 6-9)

15. **TEST PREP** A triangle has one right angle. What could be the measures of the other two angles? (Lesson 7-5) **B**

A 30° and 15° B 70° and 20° C 60° and 120° D 100° and 90°

Lesson Quiz

1. On a sunny day, a telephone pole casts a shadow 21 ft long. A 5-foot-tall mailbox next to the pole casts a shadow 3 ft long. How tall is the pole? **35 ft**

2. On a sunny afternoon, a goalpost casts a 75 ft shadow. A 6.5 ft football player next to the goal post has a shadow 19.5 ft long. How tall is the goalpost? **25 feet**

Available on Daily Transparency in CRB

Answers

14. See p. A6.

RETEACH 8-5

LESSON 8-5 **Reteach**
Indirect Measurement

If you cannot measure a length directly, you can use indirect measurement. Indirect measurement uses similar figures and proportions to find lengths.

The small tree is 8 feet high and it casts a 12-foot shadow. The large tree casts a 36-foot shadow.

The triangles formed by the trees and the shadows are similar. So, their heights are proportional.

To find the height of the large tree, first set up a proportion. Use a variable to stand for the height of the large tree.

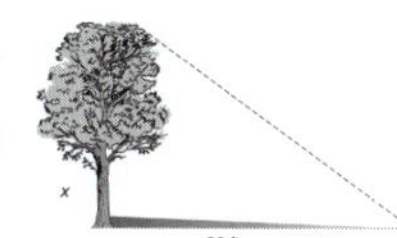

$\frac{8}{12} = \frac{x}{36}$ Write a proportion using corresponding sides.

$8 \cdot 36 = 12 \cdot x$ The cross products are equal.

$12x = 288$ x is multiplied by 12.

$\frac{12x}{12} = \frac{288}{12}$ Divide both sides by 12.

$x = 24$

So, the height of the tall tree is 24 feet.

Use indirect measurement to find the height of the statue.

1. 

150 feet

2. 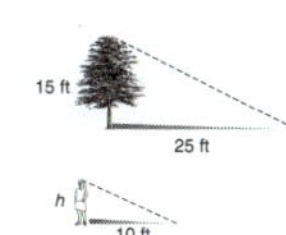

6 feet

PRACTICE 8-5

LESSON 8-5 **Practice B**
Indirect Measurement

Write the correct answer.

1. Use similar triangles to find the height of the building. **$h = 24$ m**

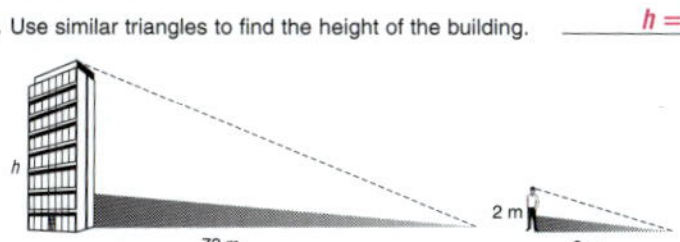

2. Use similar triangles to find the height of the taller tree. **5 meters**

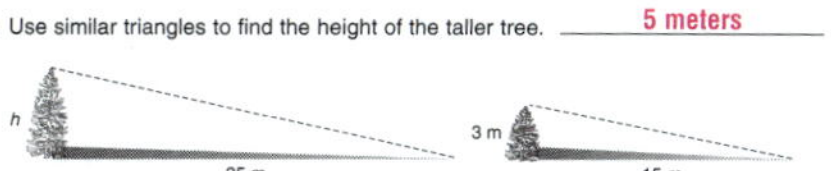

3. A lamppost casts a shadow that is 35 yards long. A 3-foot-tall mailbox casts a shadow that is 5 yards long. How tall is the lamppost? **21 feet**

4. A 6-foot-tall scarecrow in a farmer's field casts a shadow that is 21 feet long. A dog standing next to the scarecrow is 2 feet tall. How long is the dog's shadow? **7 feet**

5. A building casts a shadow that is 348 meters long. At the same time, a person who is 2 meters tall casts a shadow that is 6 meters long. How tall is the building? **116 meters**

6. On a sunny day, a tree casts a shadow that is 146 feet long. At the same time, a person who is 5.6 feet tall standing beside the tree casts a shadow that is 11.2 feet long. How tall is the tree? **73 feet**

7. In the early afternoon, a tree casts a shadow that is 2 feet long. A 4.2-foot-tall boy standing next to the tree casts a shadow that is 0.7 feet long. How tall is the tree? **12 feet**

8. Steve's pet parakeet is 100 mm tall. It casts a shadow that is 250 mm long. A cockatiel sitting next to the parakeet casts a shadow that is 450 mm long. How tall is the cockatiel? **180 millimeters**

CHALLENGE 8-5

LESSON 8-5 **Challenge**
Mirror Measurements

When it is noon, nighttime, a cloudy day, or when you are inside, there are hardly any shadows to use for indirect measurement. Instead, you can use mirrors to measure in the following way.

Place a mirror on the floor. Move back until you see the reflection of the top of the object you want to measure in the mirror. This creates two similar triangles. You can then use proportions to find the unknown height:

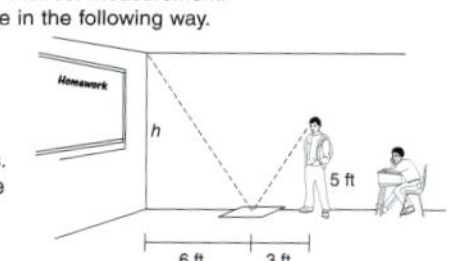

$$\frac{h}{5} = \frac{6}{3}$$
$$h \cdot 3 = 5 \cdot 6$$
$$3h = 30$$
$$\frac{3h}{3} = \frac{30}{3}$$
$$h = 10$$

So, the height of the classroom is 10 feet.

Find the missing height in each drawing to the nearest whole foot.

1. 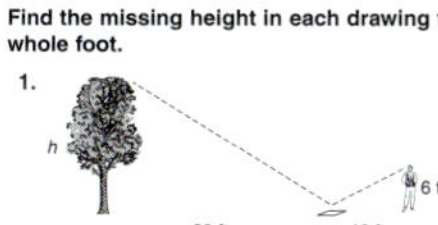

$h = 18$ feet

2. 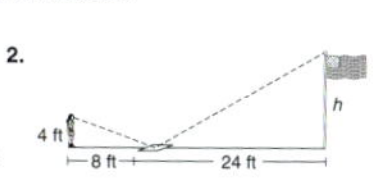

$h = 12$ feet

3. 

$h = 19$ feet

4. 

$h = 25$ feet

PROBLEM SOLVING 8-5

LESSON 8-5 **Problem Solving**
Indirect Measurement

Use the table to answer each question.

1. The Petronas Towers in Malaysia are the tallest buildings in the world. On a sunny day, the Petronas Towers cast shadows that are 4,428 feet long. A 6-foot-tall person standing by one building casts an 18-foot-long shadow. How tall are the Petronas Towers? **1,476 feet**

2. The Sears Tower in Chicago is the tallest building in the United States. On a sunny day, the Sears Tower casts a shadow that is 2,908 feet long. A 5-foot-tall person standing by the building casts a 10-foot-long shadow. How tall is the Sears Tower? **1,454 feet**

3. The world's tallest man cast a shadow that was 535 inches long. At the same time, a woman who was 5 feet 4 inches tall cast a shadow that was 320 inches long. How tall was the world's tallest man in feet and inches? **8 feet 11 inches**

4. Hoover Dam on the Colorado River casts a shadow that is 2,904 feet long. At the same time, an 18-foot-tall flagpole next to the dam casts a shadow that is 72 feet long. How tall is Hoover Dam? **726 feet**

5. An NFL goalpost casts a shadow that is 170 feet long. At the same time, a yardstick casts a shadow that is 51 feet long. How tall is an NFL goalpost? **10 feet**

6. A gorilla casts a shadow that is 600 centimeters long. A 92-centimeter-tall chimpanzee casts a shadow that is 276 centimeters long. What is the height of the gorilla in meters? **2 meters**

Circle the letter of the correct answer.

7. A 6-foot-tall man casts a shadow that is 30 feet long. If a boy standing next to the man casts a shadow that is 12 feet long, how tall is the boy?

A 2.2 feet **(C) 2.4 feet**
B 5 feet D 2 feet

8. An ostrich is 108 inches tall. If its shadow is 162 inches, and an emu standing next to it casts a 90-inch shadow, how tall is the emu?

F 162 inches **(H) 60 inches**
G 90 inches J 194.4 inches

8-6 Organizer

Pacing: Traditional 1 day
Block $\frac{1}{2}$ day

Objective: Students read and use map scales and scale drawings.

Indiana Academic Standards
6.2.7, 6.3.1, 6.7.1, 6.7.2, 6.7.5, 6.7.11

Warm Up

Find the unknown heights.

1. A tower casts a 56 ft shadow. A 5 ft girl next to it casts a 3.5 ft shadow. How tall is the tower? 80 ft
2. On a sunny day, a 50 ft silo casts a 10 ft shadow. The barn next to the silo casts a shadow that is 4 ft long. How tall is the barn? 20 ft

Problem of the Day

Hal runs 4 miles in 32 minutes. Julie runs 5 miles more than Hal runs. If Julie runs at the same rate as Hal, for how many minutes will Julie run? 72 minutes

Available on Daily Transparency in CRB

Math Humor

What do you call people who are in favor of tractors? Protractors

8-6 Scale Drawings and Maps

Learn to read and use map scales and scale drawings.

Vocabulary
scale drawing
scale

The map of Yosemite National Park shown above is a *scale drawing*. A **scale drawing** is a drawing of a real object that is proportionally smaller or larger than the real object. In other words, measurements on a scale drawing are in proportion to the measurements of the real object.

A **scale** is a ratio between two sets of measurements. In the map above, the scale is 1 in:2 mi. This ratio means that 1 inch on the map represents 2 miles in Yosemite National Park.

EXAMPLE 1 Finding Actual Distances

On the map, the distance between El Capitan and Panorama Cliff is 2 inches. What is the actual distance?

$\frac{1 \text{ in.}}{2 \text{ mi}} = \frac{2 \text{ in.}}{x \text{ mi}}$ *Write a proportion using the scale. Let x be the actual number of miles from El Capitan to Panorama Cliff.*

$2 \cdot 2 = 1 \cdot x$ *The cross products are equal.*

$4 = x$

The actual distance from El Capitan to Panorama Cliff is 4 miles.

Helpful Hint

In Example 1, think "1 inch is 2 miles, so 2 inches is how many miles?" This approach will help you set up proportions in similar problems.

1 Introduce

Alternate Opener

EXPLORATION

8-6 Scale Drawings and Maps

A *scale* is a ratio between two sets of measurements. For example, the scale 2 in:1 mi means that 2 inches on a scale drawing represents 1 mile.

1. Each letter of the Hollywood sign measures 50 ft tall and 30 ft wide. Use the rectangles below to sketch a scale drawing of the Hollywood sign. The side lengths of each square inside each rectangle represent 10 ft.

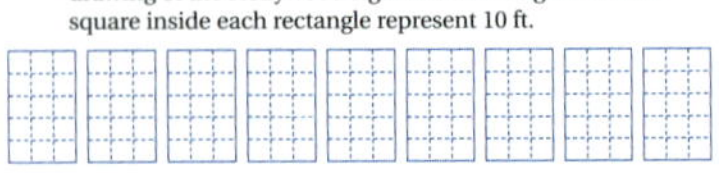

2. If the total width of the Hollywood sign is approximately 450 ft, what is the approximate distance between each pair of neighboring letters?

Think and Discuss

3. **Explain** how you found the answer in number 2.
4. **Discuss** other examples of scale drawings.

Motivate

Have a class discussion about maps. Ask students to name features of maps, such as roads, mountains, bodies of water, towns/cities, and so on. Point out the scale on a map and ask students if they know what it is and what it is there for. You may want to use Teaching Transparency T25.

Exploration worksheet and answers on Chapter 8 Resource Book pp. 53 and 120

2 Teach

Lesson Presentation

Guided Instruction

In this lesson, students learn to read and use map scales and scale drawings. Teach students to set up proportions using a map scale and a given map distance to find the actual distance. Then teach students to determine the map distance, given an actual distance.

Teaching Tip Point out that map scales vary depending on the size of the area being mapped and the size of a map. An inch on an 11 × 17 inch map of a small town would represent a smaller distance than an inch on an 11 × 17 inch map of a state or country.

EXAMPLE 2 *Astronomy Application*

A **What is the actual distance from Mercury to Earth?**

Use your inch ruler to measure the distance from the center of Mercury to the center of Earth on the drawing. Mercury and Earth are about 3 inches apart.

$\frac{1 \text{ in.}}{30 \text{ million km}} = \frac{3 \text{ in.}}{x \text{ million km}}$ *Write a proportion. Let x be the actual distance from Mercury to Earth.*

$30 \cdot 3 = 1 \cdot x$ *The cross products are equal.*

$90 = x$

The actual distance from Mercury to Earth is about 90 million km.

B **The actual distance from Mercury to Venus is 50 million kilometers. How far apart should Mercury and Venus be drawn?**

$\frac{1 \text{ in.}}{30 \text{ million km}} = \frac{x \text{ in.}}{50 \text{ million km}}$ *Write a proportion. Let x be the distance from Mercury to Venus on the drawing.*

$30 \cdot x = 1 \cdot 50$ *The cross products are equal.*

$30x = 50$ *x is multiplied by 30.*

$\frac{30x}{30} = \frac{50}{30}$ *Divide both sides by 30 to undo the multiplication.*

$x = 1\frac{2}{3}$

Mercury and Venus should be drawn $1\frac{2}{3}$ inches apart.

Think and Discuss

1. **Give an example** of when you would use a scale drawing.
2. **Suppose** that you are going to make a scale drawing of your classroom with a scale of 1 inch:3 feet. Select a distance in your classroom and measure it. What will this distance be on your drawing?

COMMON ERROR ALERT

Some students may confuse the order of the measurements in the scale. The distance on the drawing always precedes the actual distance.

Additional Examples

Example 1

The scale on a map is 4 in:1 mi. On the map, the distance between two towns is 20 in. What is the actual distance?

The actual distance is 5 miles.

Example 2

A. If a drawing of the planets were made using the scale 1 in:30 million km, the distance from Mars to Jupiter on the drawing would be about 18.3 in. What is the actual distance from Mars to Jupiter?

The actual distance from Mars to Jupiter is about 549 million km.

B. The actual distance from Earth to Mars is about 78 million kilometers. How far apart should they be drawn?

Earth and Mars should be $2\frac{3}{5}$ inches apart on the drawing.

3 Close

Reaching All Learners
Through Curriculum Integration

Social Studies Give each student a map or atlas. Have them use the map's scale to calculate actual distances between cities or other landmarks. Have students show how they wrote and solved each proportion to find each distance.

Summarize

You may wish to have students explain the terms *scale* and *scale drawing*. Ask how a scale on a map is used to find actual distances.

Possible answers: A scale is a ratio between two sets of measurements, and a scale drawing is a drawing of a real object that is similar to the real object. You can set up a proportion using a map scale and a measured distance on the map to find the actual distance.

Answers to Think and Discuss

1. Possible answer: to draw the floor plan of a house
2. Possible answer: The distance from the door to the teacher's desk is 9 feet. So on my scale drawing that distance should be 3 inches.

8-6 PRACTICE & ASSESS

Students may want to refer back to the lesson examples.

Assignment Guide

If you finished Example 1 assign:
Core 1, 4, 13–16, 21–27
Enriched 13–17, 19, 21–27

If you finished Example 2 assign:
Core 1–6, 13–16, 21–27
Enriched 7–12, 18, 19, 21–27

Notes

Math Background

The mathematical problem of mapping Earth lies in the fact that its surface is curved and the surface of a map is flat. There is no way to maintain both the correct shapes of the various objects on Earth and the relative sizes of those objects when making a world map. For this reason, maps historically have made some parts of the world appear larger than they actually are (e.g., Europe) while making other parts appear smaller than they actually are (e.g., Africa). More recent mapmakers have sought to correct this distortion of relative size, but as a result they have created maps in which the shapes of continents do not resemble their true shapes.

8-6 Exercises

FOR EXTRA PRACTICE
see page 652

internet connect
Homework Help Online
go.hrw.com Keyword: MR4 8-6

GUIDED PRACTICE

See Example 1

1. On the map, the distance between the post office and the fountain is 6 cm. What is the actual distance? **300 ft**

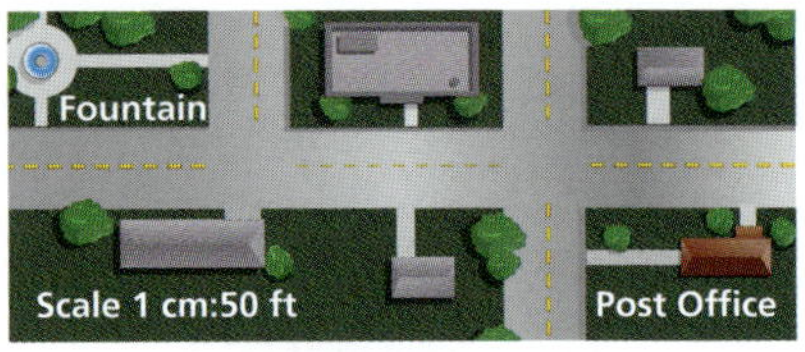

See Example 2

2. What is the actual length of the car? **3 m**
3. The actual height of the car is 1.6 meters. Is the car's height in the drawing correct? **No; the height should be about 2 cm**

Scale: 1 cm:0.8 m

INDEPENDENT PRACTICE

See Example 1

4. On the map of California, Los Angeles is 1.25 inches from Malibu. Find the actual distance from Los Angeles to Malibu. **25 miles**

See Example 2

5. Riverside, California, is 50 miles from Los Angeles. On the map, how far should Riverside be from Los Angeles? **2.5 inches**
6. A paramecium is a one-celled organism. The scale drawing at right is larger than an actual paramecium. Find the actual length of the paramecium. **0.009375 inches**

Scale: 1 in:20 mi

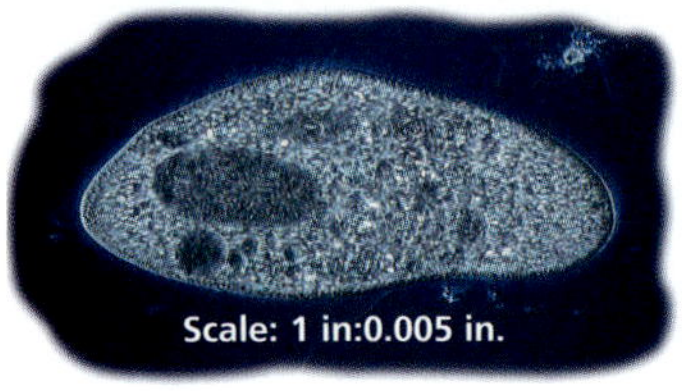
Scale: 1 in:0.005 in.

PRACTICE AND PROBLEM SOLVING

Suppose you are asked to make a scale drawing of a room. The scale is 1 in:4 ft. Use the actual lengths below to find the lengths in the drawing.

7. north wall: 12 ft **3 in.**
8. south wall: 8 ft **2 in.**
9. east wall: 5 ft **1.25 in.**
10. west wall: 10 ft **2.5 in.**
11. door width: 3.5 ft **.875 in.**
12. window width: 2.5 ft **.625 in.**

RETEACH 8-6

LESSON 8-6 **Reteach**
Scale Drawings and Maps

A scale drawing is a drawing of a real object that is proportionally smaller or larger than the real object.

A scale is a ratio between two sets of measurements. In the map below, the scale is 2 cm: 0.5 km. This means that each centimeter on the map represents 0.25 kilometers.

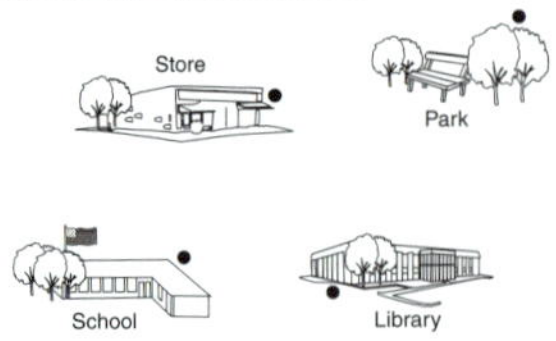

To find the actual distance from school to the library, first measure the distance on the map using a ruler.

The distance on the map is 3 centimeters.

$\frac{2\text{ cm}}{0.5\text{ km}} = \frac{3\text{ cm}}{x\text{ km}}$	Write a proportion using the scale.
$2 \cdot x = 0.5 \cdot 3$	The cross products are equal.
$2x = 1.5$	x is multiplied by 2.
$\frac{2x}{2} = \frac{1.5}{2}$	Divide both sides by 2.
$x = 0.75$	

The distance from school to the library is 0.75 kilometers.

Use the map to find each actual distance.

1. from the store to the library **1 kilometer**
2. from the park to the store **1.25 kilometers**
3. from the park to the library **1.625 kilometers**
4. from the park to the school **2 kilometers**

PRACTICE 8-6

LESSON 8-6 **Practice B**
Scale Drawings and Maps

Use the map to answer the questions.

1. On the map, the distance between Big Cypress Swamp and Lake Okeechobee is $\frac{1}{4}$ inch. What is the actual distance? **25 miles**
2. On the map, the distance between Key West and Cuba is $\frac{9}{10}$ inch. What is the actual distance? **90 miles**
3. Use a ruler to measure the distance between Key West and Key Largo on the map. What is the actual distance? **100 miles**
4. The Overseas Highway connects Key West to mainland Florida. It is 110 miles long. If it were shown on this map, how many inches long would it be? **$1\frac{1}{10}$ inches**

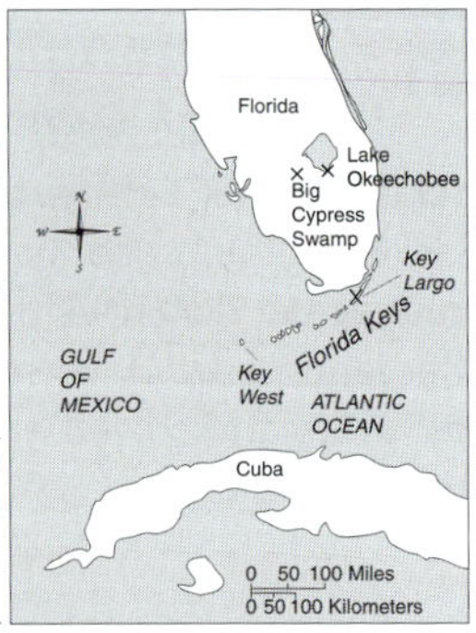

Use the scale drawing to answer each question.

5. This scale drawing is of the lighthouse on Key West, originally built in 1825. What is the actual height of the lighthouse? **85 feet**
6. The original lighthouse was 66 feet tall. It was rebuilt at its present height after a hurricane destroyed it in 1846. How tall would the original lighthouse be in this scale drawing? **$1\frac{13}{20}$ inches**

1 inch = 40 feet

Social Studies LINK

Texas is the second largest state in the country and is the largest state in the lower 48 states. It is more than 1,120 kilometers across. There is even a ranch in Texas that is larger than Rhode Island!

Scale: 1 cm:119 km

13. What is the distance in kilometers from Houston to Dallas? **357 km**

14. What is the distance in kilometers from Corpus Christi to San Antonio? **226.1 km**

15. What is the distance in kilometers from Austin to Dallas? **297.5 km**

16. Name two cities on the map that are more than 200 kilometers apart. **Possible answer: Corpus Christi and Amarillo**

17. Wichita Falls is about 480 kilometers from San Antonio.

- **a.** About how far apart should these two cities be on the map? **about 4 cm**
- **b.** What else would you need to know to be able to place Wichita Falls on the map? **Possible answer: You would have to know which direction Wichita Falls is from San Antonio.**

18. ***WRITE A PROBLEM*** Write a problem using the map and its scale. **Possible answer: About how far is Fort Worth from Dallas?**

19. ***WRITE ABOUT IT*** Explain how to find the actual distance between two cities if you know the distance on a map and the scale of the map. **Write a proportion using the map distance and the scale. Solve the proportion.**

20. ***CHALLENGE*** If you drive at a constant speed of 100 kilometers per hour, about how long will it take you to drive from Amarillo to San Antonio? **about 7 hours**

Spiral Review

21. Patrice had \$5.89 in her pocket and \$9.81 in her purse. She spent \$12.99 on a new CD. How much money does Patrice have after her purchase? (Lesson 3-3) **\$2.71**

Divide. Write each answer in simplest form. (Lesson 5-3)

22. $\frac{2}{3} \div \frac{1}{3}$ **2**   **23.** $\frac{9}{10} \div \frac{3}{4}$ $1\frac{1}{5}$ **24.** $2\frac{3}{8} \div \frac{1}{4}$ $9\frac{1}{2}$ **25.** $1\frac{1}{4} \div 2\frac{1}{3}$ $\frac{15}{28}$

26. **TEST PREP** A ___?___ has two endpoints. (Lesson 7-1) **B**

A plane   **B** line segment   **C** line   **D** ray

27. **TEST PREP** Which ratio is equivalent to 3:2? (Lesson 8-1) **H**

F 5:1   **G** 2:3   **H** 6:4   **J** 4:3

Interdisciplinary LINK

Social Studies

Exercises 13–20 involve using a scale to determine actual distances represented on a map. Students learn to read maps and use scales in middle school social studies programs, such as Holt, Rinehart & Winston's *People, Places, and Change.*

Journal

Have students write about how the scale on a map is useful. If the scale is 1 in. = 5 mi, what does this mean, and how can this information be used?

Test Prep Doctor

For Exercise 26, students need to remember some basic geometry terms. A *plane* is a flat surface; it doesn't have endpoints. A *line* extends in two opposite directions and doesn't have endpoints. A *ray* has one endpoint and extends in one direction. Only **B,** *line segment,* has two endpoints.

CHALLENGE 8-6

LESSON 8-6 **Challenge**
Solar System String

Distances in outer space are usually measured in millions of miles. Understanding or comparing such huge measurements can be difficult, and it is impossible to map or draw them in their actual scale. Here's an activity that can help you understand the vast scale of our solar system. Identify the 1-millimeter mark on your ruler. This tiny distance represents 1,000,000 miles in space! You will use it as the scale for your model: 1 millimeter = 1 million miles.

Make a scale model of our solar system.

1. Cut a piece of string 4 meters long. Tape a small piece of paper at one end of the string and label it "Sun."
2. From the sun, measure 3.6 cm. Tape a "Mercury" label there.
3. From Mercury, measure another 3.1 cm. Tape a "Venus" label there.
4. From Venus, measure another 2.6 cm. Tape an "Earth" label there.
5. From Earth, measure another 4.9 mm. Tape a "Mars" label there.
6. From Mars, measure another 34.2 cm. Tape a "Jupiter" label there.
7. From Jupiter, measure another 40.2 cm. Tape a "Saturn" label there.
8. From Saturn, measure another 89.8 cm. Tape a "Uranus" label there.
9. From Uranus, measure another 1.01 m. Tape a "Neptune" label there.
10. From Neptune, measure another 88.1 cm. Tape a "Pluto" label there.

Now use the scale and your model to find the actual distance of each planet from Earth. For example, the distance from Earth to the sun on the string measures 93 mm, so the actual distance is 93 million miles.

Earth to Mercury: **57 million miles**
Earth to Venus: **26 million miles**
Earth to Mars: **49 million miles**
Earth to Jupiter: **391 million miles**
Earth to Saturn: **793 million miles**
Earth to Uranus: **1,691 million miles**
Earth to Neptune: **2,701 million miles**
Earth to Pluto: **3,582 million miles**

PROBLEM SOLVING 8-6

LESSON 8-6 **Problem Solving**
Scale Drawings and Maps

Write the correct answer.

1. About how many kilometers long is the northern border of California along Oregon?
   about 300 kilometers
2. What is the distance in kilometers from Los Angeles to San Francisco?
   about 600 kilometers
3. How many kilometers would you have to drive to get from San Diego to Sacramento?
   about 800 kilometers
4. At its longest point, about how many kilometers long is Death Valley National Park?
   about 250 kilometers
5. Approximately what is the distance, in kilometers, between Redwood National Park and Yosemite National Park?
   about 500 kilometers

Circle the letter of the correct answer.

6. Which of the following two cities in California are about 200 kilometers apart?
   - **(A)** San Diego and Los Angeles
   - **B** Monterey and Los Angeles
   - **C** San Francisco and Fresno
   - **D** Palm Springs and Bakersfield
7. Joshua Tree National Park is about 200 kilometers from Sequoia National Park. How many centimeters should separate those parks on this map?
   - **F** 220 cm
   - **G** 22 cm
   - **(H)** 2 cm
   - **J** 0.22 cm

Lesson Quiz

On a map of the Great Lakes, 2 cm = 45 km. Find the actual distance of the following, given their distances on the map.

1. Detroit to Cleveland = 12 cm **270 km**
2. Duluth to Nipigon = 20 cm **450 km**
3. Buffalo to Syracuse = 10 cm **225 km**
4. Sault Ste. Marie to Toronto = 33 cm **742.5 km**

Available on Daily Transparency in CRB

Chapter 8 **Mid-Chapter Quiz**

Purpose: *To assess students' mastery of concepts and skills in Lessons 8-1 through 8-6*

Assessment Resources

Section 8A Quiz
Assessment Resources .... p. 21

Test and Practice Generator **CD-ROM**

Additional mid-chapter assessment items in both multiple-choice and free-response format may be generated for any objective in Lessons 8-1 through 8-6.

Chapter 8 Mid-Chapter Quiz

Mid-Chapter Quiz

LESSON 8-1 (pp. 392–395)

Use the table to write the following ratios.

Types of CDs in Mark's Music Collection			
Classical	4	Jazz	3
Country	9	Pop	14
Dance	8	Rock	10

1. classical CDs to rock CDs **4:10**
2. dance CDs to jazz CDs **8:3**
3. country to total CDs **9:48**
4. Write three equivalent ratios to compare the number of jazz CDs with the number of country CDs. $\frac{3}{9}, \frac{1}{3}, \frac{6}{18}$
5. A package containing 6 pairs of socks costs \$6.89. A package containing 4 pairs of socks costs \$4.64. Which is the better deal? **the 6 pair-pack of socks**

LESSON 8-2 (pp. 398–401)

Find the missing value in each proportion.

6. $\frac{1}{4} = \frac{n}{12}$ $n = 3$ 7. $\frac{3}{n} = \frac{15}{25}$ $n = 5$ 8. $\frac{n}{4} = \frac{18}{6}$ $n = 12$ 9. $\frac{10}{4} = \frac{5}{n}$ $n = 2$

LESSON 8-3 (pp. 402–404)

10. You have 26 feet of intestine in your body. How many inches of intestine do you have? **312 inches**
11. Death Valley is 282 feet below sea level. How many yards below sea level is it? **94 yards**

LESSON 8-4 (pp. 405–408)

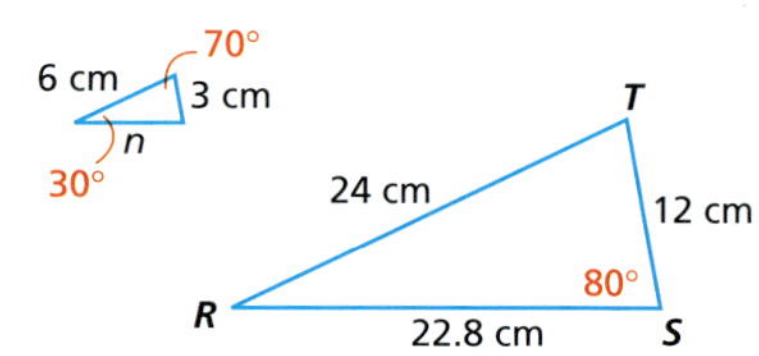

12. The two triangles are similar. Find the missing length n and the measure of $\angle R$.
$n = 5.7$ cm; $m\angle R = 30°$

LESSON 8-5 (pp. 409–411)

13. A tree casts a shadow that is 18 feet long. At the same time, a 5-foot-tall person casts a shadow that is 3.6 feet long. How tall is the tree?
about 25 feet tall

LESSON 8-6 (pp. 412–415)

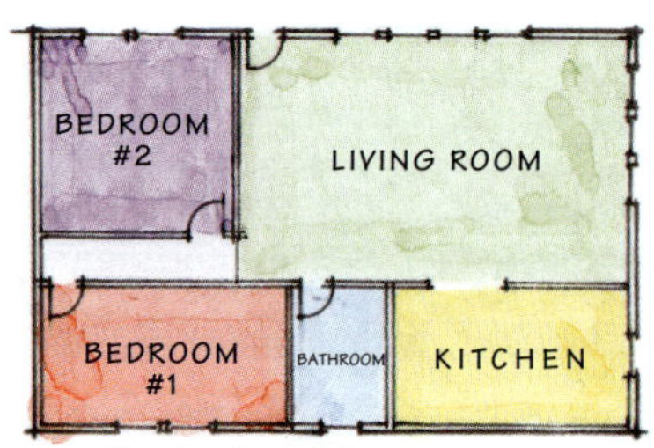

Scale: 1cm:8ft

Use the scale drawing to answer each question.

14. What is the actual length of the kitchen?
20 ft
15. What are the actual length and width of bedroom 1? **20 ft by 12 ft**

Focus on Problem Solving

Make a Plan

• Estimate or find an exact answer

Sometimes an estimate is all you need to solve a problem, and sometimes you need to find an exact answer.

One way to decide whether you can estimate is to see if you can rewrite the problem using the words *at most, at least,* or *about.* For example, suppose Laura has $30. Then she could spend *at most* $30. She would not have to spend *exactly* $30. Or, if you know it takes 15 minutes to get to school, you must leave your house *at least* (not exactly) 15 minutes before school starts.

Read the problems below. Decide whether you can estimate or whether you must find the exact answer. How do you know?

1. Alex is a radio station disc jockey. He is making a list of songs that should last no longer than 30 minutes total when played in a row. His list of songs and their playing times are given in the table. Does Alex have the right amount of music?

Song Title	Length (min)
Color Me Blue	4.5
Hittin' the Road	7.2
Stand Up, Shout	2.6
Top Dog	3.6
Kelso Blues	4.3
Smile on Me	5.7
A Long Time Ago	6.4

2. For every 10 minutes of music, Alex has to play 1.5 minutes of commercials. If Alex plays the songs on the list, how much time does he need to allow for commercials?

3. If Alex must play the songs on the list and the commercials in 30 minutes, how much music time does he need to cut to allow for commercials?

Focus on Problem Solving

Purpose: *To focus on making a plan to solve a problem*

Problem Solving Resources

Interactive Problem Solving . . pp. 65–74

Math: Reading and Writing in the Content Area . . . . pp. 65–74

Problem Solving Process

This page focuses on the second step of the problem-solving process: **Make a Plan**

Discuss

Have students rewrite each problem and use the words *at most, at least,* or *exactly* to determine whether they can estimate or whether they must find an exact answer.

Possible answers:

1. Alex's songs should last *at most* 30 minutes, but they can last less than 30 minutes, so you can estimate.
2. Alex has to play *exactly* 1.5 minutes of commercials for every 10 minutes of music, so you need an exact answer.
3. Alex must know *exactly* how many minutes of music must be cut in order for the songs and commercials to last *exactly* 30 minutes, so you need an exact answer.

Answers

1. No; by rounding the length of each song to the nearest minute you find that Alex has about 35 minutes of music, which is over 30 minutes by about 5 minutes.
2. $\frac{1.5}{10} = \frac{x}{34.3}$; $x = 5.145$ minutes
3. Alex must play songs and commercials for exactly 30 minutes. For every 11.5 minutes of air time there should be 1.5 minutes of commercials. Solving the proportion $\frac{11.5}{1.5} = \frac{30}{x}$ indicates that there should be 3.9 minutes of commercials for 30 minutes of airtime. Since Alex's song list already exceeds 30 minutes by 4.3 minutes, Alex needs to cut 3.9 + 4.3 = 8.2 minutes of music.

Section **8B**

Percents

One-Minute Section Planner

Lesson	Materials	Resources
Lesson 8-7 Percents **NCTM:** Number and Operations, Communication, Representation **NAEP:** Number Properties 1b ☑ SAT-9 ☑ SAT-10 ☑ ITBS ☑ CTBS ☑ MAT ☑ CAT	**Optional** Teaching Transparency T30 *(CRB)* 10 × 10 grid *(CRB, p. 108)* Number cubes *(MK)*	• *Chapter 8 Resource Book,* pp. 62–70 • Daily Transparency T29, CRB • Additional Examples Transparencies T31–T32, CRB • *Alternate Openers: Explorations,* p. 71
Lesson 8-8 Percents, Decimals, and Fractions **NCTM:** Number and Operations, Communication, Representation **NAEP:** Number Properties 1e ☑ SAT-9 ☑ SAT-10 ☑ ITBS ☑ CTBS ☑ MAT ☑ CAT	**Optional** Teaching Transparency T34 *(CRB)* Recording Sheet for Reaching All Learners *(CRB, p. 109)*	• *Chapter 8 Resource Book,* pp. 71–79 • Daily Transparency T33, CRB • Additional Examples Transparencies T35–T37, CRB • *Alternate Openers: Explorations,* p. 72
Lesson 8-9 Percent Problems **NCTM:** Number and Operations, Communication, Connections **NAEP:** Number Properties 3g ☑ SAT-9 ☑ SAT-10 ☑ ITBS ☑ CTBS ☑ MAT ☑ CAT		• *Chapter 8 Resource Book,* pp. 80–88 • Daily Transparency T38, CRB • Additional Examples Transparencies T39–T41, CRB • *Alternate Openers: Explorations,* p. 73
Hands-On Lab 8B Construct Circle Graphs **NCTM:** Data Analysis and Probability, Representation **NAEP:** Data Analysis and Probability 1b ☐ SAT-9 ☐ SAT-10 ☐ ITBS ☐ CTBS ☐ MAT ☐ CAT	**Required** Compasses *(MK)* Straightedges *(MK)* Protractors *(MK)*	• *Hands-On Lab Activities,* pp. 78–80
Lesson 8-10 Using Percents **NCTM:** Number and Operations, Communication **NAEP:** Number Properties 4d ☑ SAT-9 ☑ SAT-10 ☑ ITBS ☑ CTBS ☑ MAT ☑ CAT	**Optional** Newspaper advertisement Teaching Transparency T43 *(CRB)*	• *Chapter 8 Resource Book,* pp. 89–97 • Daily Transparency T42, CRB • Additional Examples Transparencies T44–T46, CRB • *Alternate Openers: Explorations,* p. 74
Extension Simple Interest **NCTM:** Number and Operations, Connections **NAEP:** Number Properties 3g ☐ SAT-9 ☑ SAT-10 ☐ ITBS ☐ CTBS ☑ MAT ☑ CAT	**Optional** Teaching Transparency T47 *(CRB)*	• Additional Examples Transparency T48, CRB
Section 8B Assessment		• Section 8B Quiz, AR p. 22 • *Test and Practice Generator* CD-ROM

SAT = *Stanford Achievement Tests* **ITBS** = *Iowa Test of Basic Skills* **CTBS** = *Comprehensive Test of Basic Skills/Terra Nova*
MAT = *Metropolitan Achievement Tests* **CAT** = *California Achievement Test*
NCTM = Complete standards can be found on pages T27–T33. **NAEP** = Complete standards can be found on pages A31–A35.
SE = *Student Edition* **TE** = *Teacher's Edition* **AR** = *Assessment Resources* **CRB** = *Chapter Resource Book* **MK** = *Manipulatives Kit*

Section Overview

Percents, Decimals, and Fractions — *Lessons 8-7, 8-8*

Why? In calculations, percents are changed to decimals or fractions.

A percent is a ratio of a number to 100.

Percents to Decimals

$45\% = \frac{45}{100} = 0.45$

Percents to Fractions

$65\% = \frac{65 \div 5}{100 \div 5} = \frac{13}{20}$

Decimals to Percents

$0.27 = \frac{27}{100} = 27\%$

Fractions to Percents

$\frac{3}{5} = \frac{3 \times 20}{5 \times 20} = \frac{60}{100} = 60\%$

Percent Problems — *Lesson 8-9*

Why? Statistics, such as those in sports, are sometimes reported as percents.

Three Types of Percent Problems

20% of 80 is ▢.	▢ % of 80 is 16.	20% of ▢ is 16.
$0.20 \cdot 80 = x$	$x \cdot 80 = 16$	$0.20 \cdot x = 16$
$x = 0.20 \cdot 80$	$80x = 16$	$0.20x = 16$
$x = 16$	$x = \frac{16}{80} = \frac{1}{5} = 20\%$	$x = \frac{16}{0.20} = 80$
20% of 80 is 16.	20% of 80 is 16.	20% of 80 is 16.

Using Percents — *Lesson 8-10*

Why? Percents are used in calculating discounts, tips, and sales tax.

Common Uses of Percents	
Discounts	A **discount** is an amount that is subtracted from the regular price of an item. discount = regular price • discount rate
Tips	A **tip** is an amount added to a bill. tip = total bill • tip rate
Sales Tax	**Sales tax** is an amount added to the price of an item. sales tax = purchase price • sales tax rate

8-7 Organizer

Pacing: Traditional 1 day
Block $\frac{1}{2}$ day

Objective: Students write percents as decimals and as fractions.

Indiana Academic Standards
6.1.4, 6.7.1, 6.7.5, 6.7.11

Warm Up

Write each fraction as a decimal.

1. $\frac{3}{4}$ 0.75 **2.** $\frac{9}{10}$ 0.9

Write each decimal as a fraction.

3. 0.375 $\frac{3}{8}$ **4.** 0.05 $\frac{1}{20}$

Problem of the Day

Wally wanted to change the scale of a drawing from 1 in. = 2 ft to 1 in. = 10 ft. The scale height of a building in the first drawing is 25 in. How high is the building in the new drawing? 5 in.

Available on Daily Transparency in CRB

Math Humor

Coach: You're not giving 100%.

Player: I am! I give 50% when I'm on offense and 50% when I'm on defense.

8-7 Percents

Learn to write percents as decimals and as fractions.

Vocabulary
percent

Most states charge sales tax on items you purchase. Sales tax is a percent of the item's price. A **percent** is a ratio of a number to 100.

You can remember that *percent* means "per hundred." For example, 8% means "8 per hundred," or "8 out of 100."

If a sales tax rate is 8%, the following statements are true:

- For every \$1.00 you spend, you pay \$0.08 in sales tax.
- For every \$10.00 you spend, you pay \$0.80 in sales tax.
- For every \$100 you spend, you pay \$8 in sales tax.

At a sales tax rate of 8%, the tax on this guitar and amplifier would be \$36.56.

Because *percent* means "per hundred," 100% means "100 out of 100." This is why 100% is often used to mean "all" or "the whole thing."

EXAMPLE 1 Modeling Percents

Use a 10-by-10-square grid to model 8%.

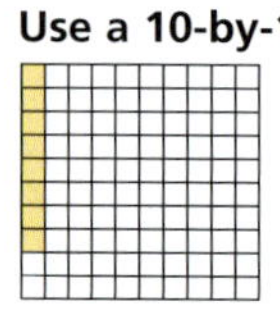

A 10-by-10-square grid has 100 squares.
8% means "8 out of 100," or $\frac{8}{100}$.
Shade 8 squares out of 100 squares.

EXAMPLE 2 Writing Percents as Fractions

Write 40% as a fraction in simplest form.

$40\% = \frac{40}{100}$ *Write the percent as a fraction with a denominator of 100.*

$\frac{40 \div 20}{100 \div 20} = \frac{2}{5}$ *Write the fraction in simplest form.*

Written as a fraction, 40% is $\frac{2}{5}$.

1 Introduce

Alternate Opener

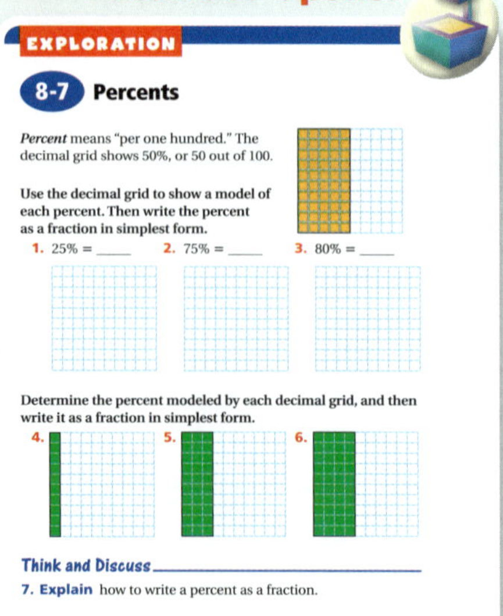
EXPLORATION

8-7 Percents

Percent means "per one hundred." The decimal grid shows 50%, or 50 out of 100.

Use the decimal grid to show a model of each percent. Then write the percent as a fraction in simplest form.

1. 25% = ____ 2. 75% = ____ 3. 80% = ____

Determine the percent modeled by each decimal grid, and then write it as a fraction in simplest form.

4. 5. 6.

Think and Discuss

7. Explain how to write a percent as a fraction.

Motivate

Ask students to give you examples of ways they have seen percents used. If necessary, show them what the percent symbol (%) looks like. Possible answers: chance of rain; real juice content of drinks; sales signs in stores; grades on tests

Exploration worksheet and answers on Chapter 8 Resource Book pp. 63 and 122

2 Teach

Lesson Presentation

Guided Instruction

In this lesson, students learn to write percents as decimals and as fractions. Teach students to use a 10-by-10 grid (available on Teaching Transparency T30 and on Chapter 8 Resource Book p. 108) to model a percent as a part of 100. Then teach students to write percents as fractions and as decimals. Examples 3 and 5 provide real-world settings for the use of percents.

EXAMPLE 3 *Life Science Application*

Up to 55% of the heat lost by your body can be lost through your head. Write 55% as a fraction in simplest form.

$55\% = \frac{55}{100}$ — *Write the percent as a fraction with a denominator of 100.*

$\frac{55 \div 5}{100 \div 5} = \frac{11}{20}$ — *Write the fraction in simplest form.*

Written as a fraction, 55% is $\frac{11}{20}$.

EXAMPLE 4 **Writing Percents as Decimals**

Remember!

To divide by 100, move the decimal point two places to the left.

$24 \div 100 = 0.24$

Write 24% as a decimal.

$24\% = \frac{24}{100}$ — *Write the percent as a fraction with a denominator of 100.*

Write the fraction as a decimal.

$$\begin{array}{r} 0.24 \\ 100\overline{)24.00} \\ -\underline{200} \\ 400 \\ -\underline{400} \\ 0 \end{array}$$

Written as a decimal, 24% is 0.24.

EXAMPLE 5 *Earth Science Application*

The water frozen in glaciers makes up almost 75% of the world's fresh water supply. Write 75% as a decimal.

$75\% = \frac{75}{100}$ — *Write the percent as a fraction with a denominator of 100.*

$75 \div 100 = 0.75$ — *Write the fraction as a decimal.*

Written as a decimal, 75% is 0.75.

Think and Discuss

1. **Give an example** of a situation in which you have seen percents.
2. **Tell** how much sales tax you would have to pay on \$1, \$10, and \$100 if your state had a 5% sales tax rate.
3. **Explain** how to write a percent as a fraction.
4. **Write** 100% as a decimal and as a fraction.

COMMON ERROR ALERT

When changing from a percent to a decimal, some students may divide 100 by the percent instead of the other way around. Remind them to divide by the denominator.

Additional Examples

Example 1

Use a 10-by-10-square grid to model 17%.

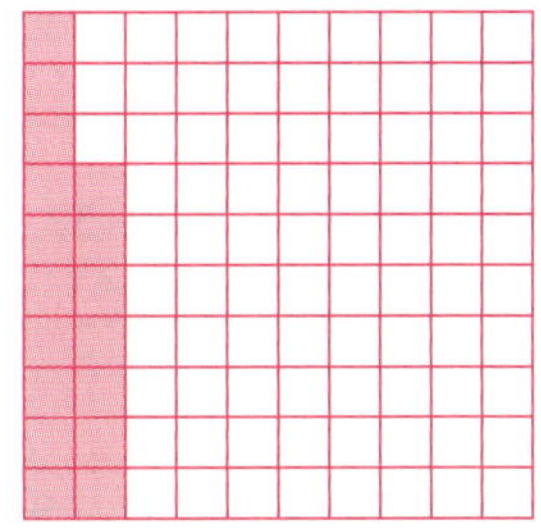

Example 2

Write 35% as a fraction in simplest form. $\frac{7}{20}$

Example 3

Janell has 20% body fat. Write 20% as a fraction in simplest form. $\frac{1}{5}$

Example 4

Write 56% as a decimal. **0.56**

Example 5

Water made up 85% of the fluids that Kirk drank yesterday. Write 85% as a decimal. **0.85**

3 Close

Reaching All Learners

Through Grouping Strategies

Have students work in pairs to model percents and write them as decimals and fractions. Give each pair a set of 10-by-10 grids (Chapter 8 Resource Book p. 108) and two number cubes (Manipulatives Kit) that go up to at least 9. One student rolls one or two number cubes and writes the total shown as a percent. The other student shades a grid to model the percent and writes the percent as a decimal and as a fraction. The first student checks the other's work. Students continue, taking turns.

Summarize

Model a percent using a 10-by-10 grid, and show the percent as an equivalent decimal and fraction.

Answers to Think and Discuss

1. Possible answer: A store has a 25% off sale.
2. 5 cents; 50 cents; 500 cents, or \$5.00
3. Convert the percent to a fraction with a denominator of 100, and then simplify the fraction.
4. 100% written as a decimal and as a fraction is 1 because 1.00 = 1 and $\frac{100}{100} = 1$.

8-7 PRACTICE & ASSESS

Students may want to refer back to the lesson examples.

Assignment Guide

If you finished Example 1 assign:
Core 1–3, 12–14, 44, 49–54
Enriched 1–3, 12–14, 44, 49–54

If you finished Example 2 assign:
Core 1–6, 12–20, 44, 49–54
Enriched 29–44, 49–54

If you finished Example 3 assign:
Core 1–7, 12–21, 29–38, 49–54
Enriched 16–21, 29–40, 41–45, 49–54

If you finished Example 4 assign:
Core 1–10 12–27, 41, 49–54
Enriched 1–10, 29–54

If you finished Example 5 assign:
Core 1–28, 41–42, 49–54
Enriched 17–54

Answers

1–3, 12–14. See p. A6.

37. $\frac{47}{50}$, 0.94

40. $\frac{13}{25}$, 0.52

8-7 Exercises

FOR EXTRA PRACTICE see page 653

internet connect
Homework Help Online
go.hrw.com Keyword: MR4 8-7

GUIDED PRACTICE

See Example 1 Use a 10-by-10-square grid to model each percent.

1. 45% **2.** 3% **3.** 61%

See Example 2 Write each percent as a fraction in simplest form.

4. 25% $\frac{1}{4}$ **5.** 80% $\frac{4}{5}$ **6.** 54% $\frac{27}{50}$

See Example 3 **7.** Belize is a country in Central America. Of the land in Belize, 92% is made up of forests and woodlands. Write 92% as a fraction in simplest form. $\frac{23}{25}$

See Example 4 Write each percent as a decimal.

8. 72% 0.72 **9.** 4% 0.04 **10.** 90% 0.9

See Example 5 **11.** About 64% of the runways at airports in the United States are not paved. Write 64% as a decimal. 0.64

INDEPENDENT PRACTICE

See Example 1 Use a 10-by-10-square grid to model each percent.

12. 14% **13.** 98% **14.** 36%

See Example 2 Write each percent as a fraction in simplest form.

15. 20% $\frac{1}{5}$ **16.** 75% $\frac{3}{4}$ **17.** 11% $\frac{11}{100}$

18. 5% $\frac{1}{20}$ **19.** 64% $\frac{16}{25}$ **20.** 31% $\frac{31}{100}$

See Example 3 **21.** Nikki must answer 80% of the questions on her final exam correctly to pass her class. Write 80% as a fraction in simplest form. $\frac{4}{5}$

See Example 4 Write each percent as a decimal.

22. 44% 0.44 **23.** 13% 0.13 **24.** 29% 0.29

25. 60% 0.6 **26.** 92% 0.92 **27.** 7% 0.07

See Example 5 **28.** Brett was absent 2% of the school year. Write 2% as a decimal. 0.02

PRACTICE AND PROBLEM SOLVING

Write each percent as a fraction in simplest form and as a decimal.

29. 23% $\frac{23}{100}$, 0.23 **30.** 1% $\frac{1}{100}$, 0.01 **31.** 49% $\frac{49}{100}$, 0.49 **32.** 70% $\frac{7}{10}$, 0.7

33. 37% $\frac{37}{100}$, 0.37 **34.** 85% $\frac{17}{20}$, 0.85 **35.** 8% $\frac{2}{25}$, 0.08 **36.** 63% $\frac{63}{100}$, 0.63

37. 94% **38.** 100% 1, 1 **39.** 0% 0, 0 **40.** 52%

Math Background

The percent symbol, %, has evolved over time. Manuscripts are available from the fifteenth century showing percent symbolized as "per c̊," which is an abbreviation of the Latin phrase *per cento.* By the middle of the seventeenth century, percent had begun to be symbolized by "per $\frac{o}{o}$." Following this, the word *per* was dropped, leaving only the symbol $\frac{o}{o}$. The modern form of the symbol is a direct descendant of $\frac{o}{o}$.

Another symbol, ‰, is used to indicate *per thousand.* It is not commonly used in the United States.

RETEACH 8-7

LESSON 8-7 **Reteach**
Percents

A percent is a ratio of a number to 100. Percent means "per hundred."

To write 38% as a fraction, write a fraction with a denominator of 100.

$\frac{38}{100}$

Then write the fraction in simplest form.

$\frac{38}{100} = \frac{38 \div 2}{100 \div 2} = \frac{19}{50}$

So, $38\% = \frac{19}{50}$.

Write each percent as a fraction in simplest form.

1. 43% $\frac{43}{100}$ 2. 72% $\frac{18}{25}$ 3. 88% $\frac{22}{25}$ 4. 35% $\frac{7}{20}$

To write 38% as a decimal, first write it as fraction.

$38\% = \frac{38}{100}$

$\frac{38}{100}$ means "38 divided by 100."

$$\begin{array}{r} 0.38 \\ 100\overline{)38.00} \\ -300 \\ 800 \\ -800 \\ 0 \end{array}$$

So, 38% = 0.38.

Write each percent as a decimal.

5. 64% 0.64 6. 92% 0.92 7. 73% 0.73 8. 33% 0.33

PRACTICE 8-7

LESSON 8-7 **Practice B**
Percents

Write each percent as a fraction in simplest form.

1. 30% $\frac{3}{10}$ 2. 42% $\frac{21}{50}$ 3. 18% $\frac{9}{50}$

4. 35% $\frac{7}{20}$ 5. 100% $\frac{1}{1}$ or 1 6. 29% $\frac{29}{100}$

7. 56% $\frac{14}{25}$ 8. 70% $\frac{7}{10}$ 9. 25% $\frac{1}{4}$

Write each percent as a decimal.

10. 19% 0.19 11. 45% 0.45 12. 3% 0.03

13. 80% 0.8 14. 24% 0.24 15. 6% 0.06

Order the percents from least to greatest.

16. 89%, 42%, 91%, 27% 27%, 42%, 89%, 91%

17. 2%, 55%, 63%, 31% 2%, 31%, 55%, 63%

18. Sarah correctly answered 84% of the questions on her math test. What fraction of the test questions did she answer correctly? Write your answer in simplest form.
$\frac{21}{25}$

19. Chloe swam 40 laps in the pool, but this was only 50% of her total swimming workout. How many more laps does she still need to swim?
40 more laps

Music LINK

The circle graph shows the percent of radio stations around the world that play each type of music listed. Use the graph for Exercises 41–48.

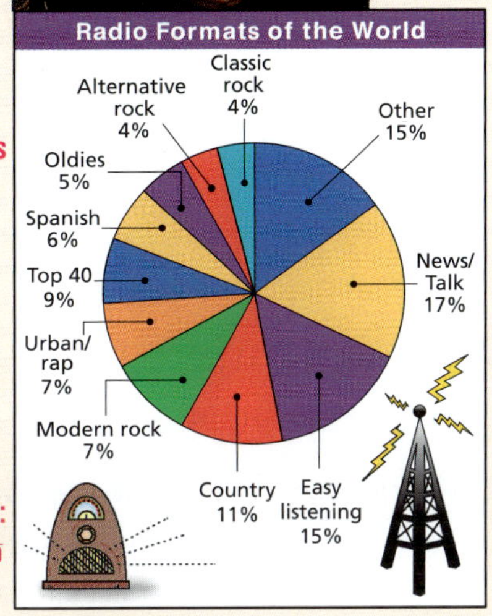

Source: Scholastic Kid's Almanac for the 21st Century

41. What fraction of the radio stations play easy listening music? Write this fraction in simplest form.

42. Use a 10-by-10-square grid to model the percent of radio stations that play country music. Then write this percent as a decimal. **11% = 0.11**

43. Which type of music makes up $\frac{1}{20}$ of the graph? **Oldies**

44. Someone reading the graph said, "More than $\frac{1}{10}$ of the radio stations play top 40 music." Do you agree with this statement? Why or why not?

45. Suppose you converted all of the percents in the graph to decimals and added them. Without actually doing this, tell what the sum would be. Explain.

46. ***WRITE A PROBLEM*** Write a question about the circle graph that involves changing a percent to a fraction. Then answer your question. **Possible Answer: What fraction of the radio stations play country music? $\frac{11}{100}$**

47. ***WRITE ABOUT IT*** How does the percent of radio stations that play Spanish music compare with the fraction $\frac{1}{6}$? Explain.

48. ***CHALLENGE*** Name a fraction that is greater than the percent of radio stations that play Spanish music but less than the percent of radio stations that play urban/rap music. **Possible answer: $\frac{61}{1000}$**

Spiral Review

49. The heights of four plants are 139 cm, 208 cm, 144 cm, and 165 cm. Estimate the sum of the heights by rounding to the nearest ten. (Lesson 1-2) **660 cm**

Find the range, mean, median, and mode of each data set. (Lesson 6-2)

50. 22, 24, 22, 29, 33, 14 **51.** 87, 16, 19, 21, 23 **52.** 365, 180, 360, 720, 59

53. **TEST PREP** Which is the measure of an obtuse angle? (Lesson 7-2) **D**

A 90° B 0° C 45° D 125°

54. **TEST PREP** Squares *ABCD* and *WXYZ* are congruent. The length of $\overline{AB}$ is 5 in. What is the perimeter of square *WXYZ*? (Lesson 7-9) **H**

F 5 in. G 9 in. H 20 in. J 25 in.

CHALLENGE 8-7

LESSON 8-7 **Challenge**
Per State

To show a percent, you can shade a 10-by-10 grid in any design that you want. For each percent below, try to shade the grid to look like the state it describes.

1. California has the largest population of any state. About 12% of all Americans live in California.
   Possible 12% shading for California
2. Florida is the top tourist state. About 26% of all visitors to the United States choose Florida for their vacations.
   Possible 26% shading for Florida
3. Nevada is the fastest-growing state. Its population has grown about 66% in the last ten years.
   Possible 66% shading for Nevada
4. Alaska is the largest state. It makes up about 15% of the total area of the United States.
   Possible 15% shading for Alaska
5. Washington produces the most apples. About 50% of all the apples grown in the U.S. come from Washington.
   Possible 50% shading for Washington
6. Texas is the top oil-producing state. About 21% of all the oil produced in the United States comes from Texas.
   Possible 21% shading for Texas

PROBLEM SOLVING 8-7

LESSON 8-7 **Problem Solving**
Percents

Use the circle graph to answer each question in simplest form.

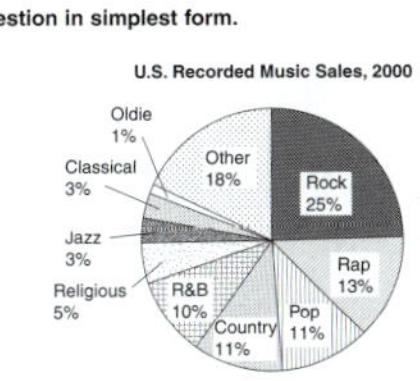

1. What fraction of the total 2000 music sales in the United States were rock recordings? $\frac{1}{4}$
2. On this grid, model the percent of total United States music sales that were rap recordings. Then write that percent as a decimal. 0.13
3. What kind of music made up $\frac{1}{20}$ of the total U.S. music recording sales? religious
4. What fraction of the United States music sales were country recordings? $\frac{11}{100}$

Circle the letter of the correct answer.

5. What fraction of all United States recording sales did jazz and classical music make up together?
   A $\frac{6}{10}$ **(B) $\frac{3}{50}$** C $\frac{1}{5}$ D $\frac{11}{100}$
6. What kind of music made up $\frac{1}{10}$ of the total music recording sales in the United States in 2000?
   F Pop G Jazz **(H) R&B** J Oldies

Answers

41. $\frac{15}{100} = \frac{3}{20}$

42. See p. A6.

44. No; $\frac{1}{10}$ is equivalent to 10%, and only 9% play Top 40.

45. 1; the sum of the percents would be 100%, which is equivalent to 1.

47. 6% is equal to $\frac{6}{100}$, or $\frac{3}{50}$, which is smaller than $\frac{1}{6}$.

50. range: 19, mean: 24, median: 23, mode: 22

51. range: 71, mean: 33.2, median: 21, mode: none

52. range: 661, mean: 336.8, median: 360, mode: none

Journal

Have students write three different percents, including percents with both 1 and 2 digits. For each percent, have them draw and shade a 10-by-10-square grid to model it, then have them write it as a decimal and as a fraction in simplest form. For one of the percents, have them explain why all three forms are equivalent.

Test Prep Doctor

Exercise 54 requires knowledge of three facts: congruent figures have congruent corresponding sides, a square has 4 congruent sides, and the perimeter of a square equals 4*s*, where *s* is the length of one side. Students who answered **F** found the length of a corresponding side of square *WXYZ*. Students who answered **J** found the area of the square. Students who answered **G** may have added the number of sides, 4, to the length of one side, 5.

Lesson Quiz

Write each percent as a fraction in simplest form.

1. 52% $\frac{13}{25}$

2. 29% $\frac{29}{100}$

Write each percent as a decimal.

3. 17% 0.17

4. 86% 0.86

5. A store clerk has an 8% sales increase. Write the increase as a fraction in simplest form and as a decimal. $\frac{2}{25}$, 0.08

Available on Daily Transparency in CRB

8-8 Organizer

Pacing: Traditional 1 day
Block $\frac{1}{2}$ day

Objective: Students write decimals and fractions as percents.

Indiana Academic Standards
6.1.4, 6.1.5, 6.7.1, 6.7.5, 6.7.11

Warm Up

Write each percent as a fraction in simplest form.

1. 37% $\frac{37}{100}$ **2.** 78% $\frac{39}{50}$

Write each percent as a decimal.

3. 59% 0.59 **4.** 7% 0.07

Problem of the Day

Jennifer gave Karen 50% of her comic books. Karen gave Jack 50% of the books she got from Jennifer. Jack gave Lucas 50% of the books he got from Karen. Lucas got 4 comic books. What percent of Jennifer's comic books does Lucas have? 12.5%

Available on Daily Transparency in CRB

Math Humor

The student was so unfamiliar with football that he thought a quarterback was a 25-cent refund.

8-8 Percents, Decimals, and Fractions

Learn to write decimals and fractions as percents.

Percents, decimals, and fractions appear in newspapers, on television, and on the Internet. To fully understand the data you see in your everyday life, you should be able to change from one number form to another.

Andrew Toos/Cartoon Resource.com

"Oh yes, a one-half of one percent allowance increase is quite a bit."

EXAMPLE 1 Writing Decimals as Percents

Write each decimal as a percent.

Method 1: Use place value.

A 0.3

$0.3 = \frac{3}{10}$ *Write the decimal as a fraction.*

$\frac{3 \cdot 10}{10 \cdot 10} = \frac{30}{100}$ *Write an equivalent fraction with 100 as the denominator.*

$\frac{30}{100} = 30\%$ *Write the numerator with a percent symbol.*

B 0.43

$0.43 = \frac{43}{100}$ *Write the decimal as a fraction.*

$\frac{43}{100} = 43\%$ *Write the numerator with a percent symbol.*

Method 2: Multiply by 100.

C 0.7431

$0.7431 \cdot 100$ *Multiply by 100.*

74.31% *Add the percent symbol.*

D 0.023

$0.023 \cdot 100$ *Multiply by 100.*

2.3% *Add the percent symbol.*

1 Introduce

Alternate Opener

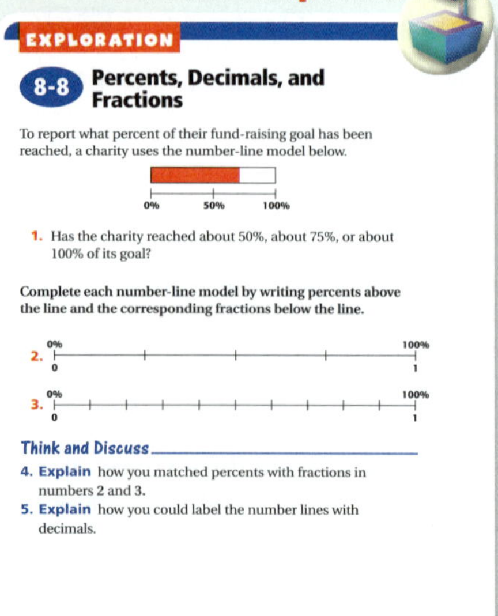

EXPLORATION

8-8 Percents, Decimals, and Fractions

To report what percent of their fund-raising goal has been reached, a charity uses the number-line model below.

1. Has the charity reached about 50%, about 75%, or about 100% of its goal?

Complete each number-line model by writing percents above the line and the corresponding fractions below the line.

2.

3.

Think and Discuss

4. **Explain** how you matched percents with fractions in numbers 2 and 3.
5. **Explain** how you could label the number lines with decimals.

Motivate

Have students explain how to convert from a percent to a fraction and from a percent to a decimal. Have them hypothesize how to do the opposite, changing a fraction or decimal to a percent.

Exploration worksheet and answers on Chapter 8 Resource Book pp. 72 and 124

2 Teach

Lesson Presentation

Guided Instruction

In this lesson, students learn to write decimals and fractions as percents. Teach them two methods for writing a decimal as a percent—one using place value and one involving multiplying by 100. Then teach students to use equivalent fractions or division to change a fraction to a percent. Discuss the table of common equivalent fractions, decimals, and percents.

EXAMPLE 2 Writing Fractions As Percents

Write each fraction as a percent.

Method 1: Write an equivalent fraction with a denominator of 100.

> **Helpful Hint**
>
> When the denominator is a factor of 100, it is often easier to use method 1. When the denominator is not a factor of 100, it is usually easier to use method 2.

A $\frac{4}{5}$

$\frac{4 \cdot 20}{5 \cdot 20} = \frac{80}{100}$ *Write an equivalent fraction with a denominator of 100.*

$\frac{80}{100} = 80\%$ *Write the numerator with a percent symbol.*

Method 2: Use division to write the fraction as a decimal.

B $\frac{3}{8}$

$8\overline{)3.000}$ = 0.375 *Divide the numerator by the denominator.*

$0.375 = 37.5\%$ *Multiply by 100 by moving the decimal point right two places. Add the percent symbol.*

EXAMPLE 3 Earth Science Application

About $\frac{39}{50}$ of Earth's atmosphere is made up of nitrogen. About what percent of the atmosphere is nitrogen?

$\frac{39}{50}$

$\frac{39 \cdot 2}{50 \cdot 2} = \frac{78}{100}$ *Write an equivalent fraction with a denominator of 100.*

$\frac{78}{100} = 78\%$ *Write the numerator with a percent symbol.*

About 78% of Earth's atmosphere is made up of nitrogen.

Common Equivalent Fractions, Decimals, and Percents

Fraction	$\frac{1}{5}$	$\frac{1}{4}$	$\frac{1}{3}$	$\frac{2}{5}$	$\frac{1}{2}$	$\frac{3}{5}$	$\frac{2}{3}$	$\frac{3}{4}$	$\frac{4}{5}$
Decimal	0.2	0.25	$0.\overline{3}$	0.4	0.5	0.6	$0.\overline{6}$	0.75	0.8
Percent	20%	25%	$33.\overline{3}\%$	40%	50%	60%	$66.\overline{6}\%$	75%	80%

Think and Discuss

1. **Tell** which method you prefer for converting decimals to percents—using equivalent fractions or multiplying by 100. Why?
2. **Give** two different ways to write three-tenths.
3. **Explain** how to write fractions as percents using two different methods.

COMMON ERROR ALERT

When writing decimals as percents, students may simply drop the decimal point and write a percent symbol. For example, they may think that 0.3 = 3% and 0.435 = 435%. Reinforce the idea of multiplying by 100 and *moving* the decimal point two spaces right rather than simply ignoring the decimal point.

Additional Examples

Example 1

Write each decimal as a percent.

A. 0.7 70%
B. 0.16 16%
C. 0.4118 41.18%
D. 0.067 6.7%

Example 2

Write each fraction as a percent.

A. $\frac{9}{25}$ 36%
B. $\frac{3}{20}$ 15%

Example 3

One year, $\frac{7}{25}$ of people with home offices were self-employed. What percent of people with home offices were self-employed?

28% of people with home offices were self-employed.

3 Close

Reaching All Learners

Through Number Sense

Have students divide the following list into sets of three equivalent numbers:

$\frac{1}{4}$, 75%, 0.40, $\frac{1}{8}$, 40%, 0.50, $\frac{2}{5}$, 25%, 0.75, $\frac{1}{2}$, 12.5%, 0.25, $\frac{3}{4}$, 50%, and 0.125.

Have students work in pairs to check results. A recording sheet is available on Chapter 8 Resource Book p. 109.

$\frac{1}{4}$, 25%, 0.25

75%, 0.75, $\frac{3}{4}$

0.40, 40%, $\frac{2}{5}$

$\frac{1}{8}$, 12.5%, 0.125

0.50, $\frac{1}{2}$, 50%

Summarize

Review each of the two methods presented for changing decimals to percents and fractions to percents. Have students discuss the advantages of each method.

Answers to Think and Discuss

1. Possible answer: multiplying by 100, because it is easier to move the decimal point two places right
2. decimal: 0.3; fraction: $\frac{3}{10}$
3. Possible answer: Write an equivalent fraction with a denominator of 100 and write the numerator with a percent symbol. Divide the numerator by the denominator, multiply by 100 by moving the decimal point two places right, and add a percent symbol.

8-8 PRACTICE & ASSESS

Students may want to refer back to the lesson examples.

Assignment Guide

If you finished Example 1 assign:
Core 1–4, 10–17, 60–68
Enriched 1–4, 10–13, 27–30, 60–68

If you finished Example 2 assign:
Core 1–8, 10–25, 31–38, 60–68
Enriched 10–25, 31–38, 46–51, 55, 56, 60–68

If you finished Example 3 assign:
Core 1–26, 31–38, 46–51, 60–68
Enriched 20–68

Notes

8-8 Exercises

FOR EXTRA PRACTICE
see page 653

internet connect
Homework Help Online
go.hrw.com Keyword: MR4 8-8

GUIDED PRACTICE

See Example 1 Write each decimal as a percent.

1. 0.39 **39%** 2. 0.125 **12.5%** 3. 0.8 **80%** 4. 0.112 **11.2%**

See Example 2 Write each fraction as a percent.

5. $\frac{11}{25}$ **44%** 6. $\frac{7}{8}$ **87.5%** 7. $\frac{7}{10}$ **70%** 8. $\frac{1}{2}$ **50%**

See Example 3 9. Patti spent $\frac{3}{4}$ of her allowance on a new backpack. What percent of her allowance did she spend? **75%**

INDEPENDENT PRACTICE

See Example 1 Write each decimal as a percent.

10. 0.6 **60%** 11. 0.55 **55%** 12. 0.34 **34%** 13. 0.308 **30.8%**
14. 0.941 **94.1%** 15. 0.01 **1%** 16. 0.62 **62%** 17. 0.02 **2%**

See Example 2 Write each fraction as a percent.

18. $\frac{3}{5}$ **60%** 19. $\frac{3}{10}$ **30%** 20. $\frac{24}{25}$ **96%** 21. $\frac{9}{20}$ **45%**
22. $\frac{1}{8}$ **12.5%** 23. $\frac{11}{16}$ **68.75%** 24. $\frac{37}{50}$ **74%** 25. $\frac{2}{5}$ **40%**

See Example 3 26. About $\frac{1}{125}$ of the people in the United States have the last name *Johnson*. What percent of people in the United States have this last name? **0.8%**

PRACTICE AND PROBLEM SOLVING

Write each decimal as a percent and a fraction.

27. 0.04 **4%, $\frac{1}{25}$** 28. 0.32 **32%, $\frac{8}{25}$** 29. 0.45 **45%, $\frac{9}{20}$** 30. 0.59 **59%, $\frac{59}{100}$**
31. 0.81 **81%, $\frac{81}{100}$** 32. 0.6 **60%, $\frac{3}{5}$** 33. 0.39 **39%, $\frac{39}{100}$** 34. 0.14 **14%, $\frac{7}{50}$**

Write each fraction as a percent and as a decimal. Round to the nearest hundredth, if necessary.

35. $\frac{4}{5}$ **80%, 0.8** 36. $\frac{1}{3}$ **33.33%, 0.33** 37. $\frac{5}{6}$ **83.33%, 0.83** 38. $\frac{7}{12}$ **58.33%, 0.58**
39. $\frac{2}{30}$ **6.67%, 0.07** 40. $\frac{1}{25}$ **4%, 0.04** 41. $\frac{8}{11}$ **72.73%, 0.73** 42. $\frac{4}{15}$ **26.67%, 0.27**

Compare. Write $<$, $>$, or $=$.

43. 70% ▒ $\frac{3}{4}$ **$<$** 44. $\frac{5}{8}$ ▒ 6.25% **$>$** 45. 0.2 ▒ $\frac{1}{5}$ **$=$**
46. 0.7 ▒ 7% **$>$** 47. $\frac{9}{10}$ ▒ 0.3 **$>$** 48. 37% ▒ $\frac{3}{7}$ **$<$**

Math Background

Another method of writing a fraction as a percent is to use a proportion. For example, to write $\frac{4}{5}$ as a percent, use the proportion $\frac{4}{5} = \frac{x}{100}$, since *percent* means "per hundred." Solving for x yields the percent.

$$\frac{4}{5} = \frac{x}{100}$$
$$5x = 400$$
$$x = 80$$

So $\frac{4}{5}$ is 80%.

RETEACH 8-8

LESSON 8-8 **Reteach**
Percents, Decimals, and Fractions

You can write decimals as percents.

To write 0.5 as a percent, multiply the decimal by 100%.

$0.5 \cdot 100\% = 50\%$

To multiply a number by 100, move the decimal point two places to the right.

0.50

So, 0.5 = 50%.

Write each decimal as a percent.

1. 0.8 **80%** 2. 0.64 **64%** 3. 0.075 **7.5%** 4. 0.29 **29%**

You can solve a proportion to write a fraction as percent.

To write $\frac{3}{4}$ as a percent, first set up a proportion.

$\frac{3}{4} = \frac{x}{100}$

$3 \cdot 100 = 4 \cdot x$ The cross products are equal.

$300 = 4x$ x is multiplied by 4.

$\frac{4x}{4} = \frac{300}{4}$ Divide both sides by 4.

$x = 75$

So, $\frac{3}{4} = \frac{75}{100}$

$\frac{75}{100} = 75\%$, So, $\frac{3}{4} = 75\%$.

Write each fraction as a percent.

5. $\frac{4}{5}$ **80%** 6. $\frac{9}{10}$ **90%** 7. $\frac{1}{8}$ **12.5%** 8. $\frac{7}{25}$ **28%**
9. $\frac{1}{4}$ **25%** 10. $\frac{5}{6}$ **83.3%** 11. $\frac{3}{4}$ **75%** 12. $\frac{1}{5}$ **20%**

PRACTICE 8-8

LESSON 8-8 **Practice B**
Percents, Decimals, and Fractions

Write each decimal as a percent.

1. 0.03 **3%** 2. 0.92 **92%** 3. 0.18 **18%**
4. 0.49 **49%** 5. 0.7 **70%** 6. 0.09 **9%**
7. 0.26 **26%** 8. 0.11 **11%** 9. 1.0 **100%**

Write each fraction as a percent.

10. $\frac{2}{5}$ **40%** 11. $\frac{1}{5}$ **20%** 12. $\frac{7}{10}$ **70%**
13. $\frac{1}{20}$ **5%** 14. $\frac{1}{50}$ **2%** 15. $\frac{4}{50}$ **8%**

Compare. Write $<$, $>$, or $=$.

16. 60% **$<$** $\frac{2}{3}$ 17. 0.4 **$=$** $\frac{2}{5}$
18. 0.5 **$>$** 5% 19. $\frac{1}{100}$ **$<$** 0.03
20. $\frac{7}{9}$ **$>$** 72% 21. $\frac{3}{10}$ **$<$** 35%

22. Bradley completed $\frac{3}{5}$ of his homework. What percent of his homework does he still need to complete?
40%

23. After reading a book for English class, 100 students were asked whether or not they enjoyed it. Nine twenty-fifths of the students did not like the book. How many students liked the book?
64 students

Order the numbers from least to greatest.

$\frac{21}{50}$, 0.43, 45% **49.** 45%, $\frac{21}{50}$, 0.43

50. $\frac{7}{8}$, 90%, 0.098 0.098, $\frac{7}{8}$, 90%

51. 0.7, 26%, $\frac{1}{4}$

52. 38%, $\frac{7}{25}$, 0.21

53. $\frac{9}{20}$, 14%, 0.125

54. 0.605, 17%, $\frac{5}{9}$ 17%, $\frac{5}{9}$, 0.605

55. ***ENTERTAINMENT*** About 97 million households in the United States have at least one television. Use the table below to answer the questions that follow.

Entertainment LINK

This photo from 1953 shows one of the first color television cameras.

go.hrw.com KEYWORD: MR4 TV

CNN Student News

Television in the United States	
Fraction of households with at least one television	$\frac{49}{50}$
Percent of televisions that are color	99%
Fraction of households with three televisions	$\frac{19}{50}$
Percent of television owners with a VCR	82%
Fraction of television owners with basic cable	$\frac{2}{3}$

a. About what percent of television owners have basic cable? about 67%

b. Write a decimal to express the percent of television owners who have color televisions. 0.99

c. What percent of television owners have three televisions? 38%

56. A record-company official estimates that 3 out of every 100 albums released become hits. Model this number on a 10-by-10-square grid. What percent of albums do not become hits?

57. ***WHAT'S THE QUESTION?*** Out of 25 students, 12 prefer to take their test on Monday, and 5 prefer to take their test on Tuesday. The answer is 32%. What is the question?

58. ***WRITE ABOUT IT*** Explain why 0.8 is equal to 80% and not 8%.

59. ***CHALLENGE*** The dimensions of a rectangle are 0.5 yard and 24% of a yard. What is the area of the rectangle? Write your answer as a fraction in simplest form. $\frac{3}{25}$ square yard

Spiral Review

Solve each equation. (Lesson 2-6)

60. $7c = 77$ 11 **61.** $12j = 228$ 19 **62.** $22m = 176$ 8 **63.** $41z = 205$ 5

Draw each geometric figure. (Lesson 7-1)

64. $\overleftrightarrow{CD}$ **65.** $\overrightarrow{GM}$ **66.** $\overline{XY}$ **67.** point A

68. **TEST PREP** The sides of a number cube are numbered from 1 through 6. What is the ratio of multiples of 3 to multiples of 4? (Lesson 8-1) B

A 1:1 **B** 2:1 **C** 1:2 **D** 2:2

COMMON ERROR ALERT

Students may be confused about Exercises 49–55. You may wish to help students by telling them to convert each number to the same form before comparing and ordering.

Answers

51. $\frac{1}{4}$, 26%, 0.7 **52.** 0.21, $\frac{7}{25}$, 38%

53. 0.125, 14%, $\frac{9}{20}$

56.  97%

57. What percent of the students prefer neither Monday nor Tuesday for their test day?

58. To express 0.8 as a percent, you multiply by 100 by moving the decimal point two places to the right and adding the percent symbol, making it 80%, not 8%.

64–67. See p. A6.

Journal

Have students explain which method of changing decimals to percents and fractions to percents they prefer and why.

Test Prep Doctor

For Exercise 68, students need to remember what a multiple is and they need to be sure to write the ratio in the correct order. Students who answered **A** found the ratio of 3's to 4's. Students who answered **C** found the ratio of multiples of 4 to multiples of 3. Students who answered **D** probably need to review the concept of ratio.

Lesson Quiz

Write each decimal as a percent.

1. 0.26 26%

2. 0.419 41.9%

Write each fraction as a percent.

3. $\frac{1}{5}$ 20%

4. $\frac{9}{16}$ 56.25%

5. About $\frac{1}{16}$ of all the students at a local high school own their own car. What percent is this? 6.25%

Available on Daily Transparency in CRB

CHALLENGE 8-8

LESSON 8-8 **Challenge**

Trash or Treasure?

People in the United States produce about 208 million tons of garbage every year! We recycle about 56 million tons of that garbage, or about 27% of the total.

Complete the chart at right. Then display the percents on the circle graph below. Remember to give your graph a title. Label each section of the graph with the material and the percent of the total garbage recycled that each section represents. You may wish to color each section differently or add illustrations.

United States Recycling

Material	Total Garbage Recycled	
	Fraction	Percent
Metals	$\frac{1}{10}$	10%
Yard Waste	$\frac{17}{100}$	17%
Glass	$\frac{3}{50}$	6%
Paper	$\frac{29}{50}$	58%
Plastics	$\frac{1}{50}$	2%
All Other Materials	$\frac{7}{100}$	7%

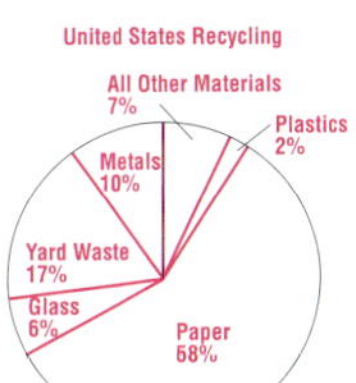

PROBLEM SOLVING 8-8

LESSON 8-8 **Problem Solving**

Percents, Decimals, and Fractions

Write the correct answer.

1. Deserts cover about $\frac{1}{7}$ of all the land on Earth. About what percent of Earth's land is made up of deserts?

   about 14%

2. The Sahara is the largest desert in the world. It covers about 3% of the total area of Africa. What decimal expresses this percent?

   0.03

3. Cactus plants survive in deserts by storing water in their thick stems. In fact, water makes up $\frac{3}{4}$ of the saguaro cactus's total weight. What percent of its weight is water?

   75%

4. Daytime temperatures in the Sahara can reach 130°F! At night, however, the temperature can drop by 62%. What decimal expresses this percent?

   0.62

5. About $\frac{4}{5}$ of all the water in the southwestern United States is used for irrigation to grow crops in the desert. What percent of that region's water is used for desert irrigation?

   80%

6. About 2,000 years ago, Native Americans built irrigation canals across deserts that carried water to nearly $\frac{1}{3}$ of Arizona's total area. About what percent of Arizona's area is that?

   about 33%

7. The desert nation of Saudi Arabia is the world's largest oil producer. About $\frac{1}{4}$ of all the oil imported to the United States is shipped from Saudi Arabia. What percent of our nation's oil is that?

   25%

8. About $\frac{2}{5}$ of all the food produced on Earth is grown on irrigated cropland. What percent of the world's food production relies on irrigation? What is the percent written as a decimal?

   40%; 0.4

Circle the letter of the correct answer.

9. About $\frac{3}{25}$ of all the freshwater in the United States is used for drinking, washing, and other domestic purposes. What percent of our fresh water resources is that?

   A 3% (C) 12%

   B 25% D $\frac{1}{5}$

10. Factories and other industrial users account for about $\frac{23}{50}$ of the total water usage in the United States. Which of the following show that amount as a percent and decimal?

    (F) 46% and 0.46 H 50% and 0.5

    G 23% and 0.23 J 46% and 4.6

8-9 Organizer

Pacing: Traditional 1 day
Block $\frac{1}{2}$ day

Objective: Students find the missing value in a percent problem.

Indiana Academic Standards
6.2.3, 6.2.7, 6.2.8, 6.7.1, 6.7.5, 6.7.11

Warm Up

Write each decimal as a percent and fraction.

1. 0.38   38%, $\frac{19}{50}$
2. 0.06   6%, $\frac{3}{50}$
3. 0.2   20%, $\frac{1}{5}$

Problem of the Day

Lucky Jim won $16,000,000 in a lottery. Every year for 10 years he spent 50% of what was left. How much did Lucky Jim have after 10 years? $15,625

Available on Daily Transparency in CRB

Math Humor

What does a house full of happy cats smell like? Purr scent

8-9 Percent Problems

Learn to find the missing value in a percent problem.

The frozen-yogurt stand in the mall sells 420 frozen-yogurt cups per day, on average. Forty-five percent of the frozen-yogurt cups are sold to teenagers. On average, how many frozen-yogurt cups are sold to teenagers each day?

To answer this question, you will need to find 45% of 420.

To find the percent one number is of another, use this proportion:

$$\frac{\%}{100} = \frac{\text{is}}{\text{of}}$$

Because you are looking for 45% of 420, 45 replaces the percent sign and 420 replaces "of." The first denominator, 100, always stays the same. The "is" part is what you have been asked to find.

EXAMPLE 1 *Consumer Application*

How many frozen-yogurt cups are sold to teenagers each day?

First estimate your answer. Think: $45\% = \frac{45}{100}$, which is close to $\frac{1}{2}$.

So about $\frac{1}{2}$ of the 420 yogurt cups are sold to teenagers.

$\frac{1}{2} \cdot 420 = 210$ ← *This is the estimate.*

Now solve:

$\frac{45}{100} = \frac{y}{420}$ *Let y represent the number of yogurt cups sold to teenagers.*

$100 \cdot y = 45 \cdot 420$ *The cross products are equal.*

$100y = 18{,}900$ *y is multiplied by 100.*

$\frac{100y}{100} = \frac{18{,}900}{100}$ *Divide both sides of the equation by 100 to undo the multiplication.*

$y = 189$

Since 189 is close to your estimate of 210, 189 is a reasonable answer. About 189 yogurt cups per day are sold to teenagers.

Helpful Hint

Think: "45 out of 100 is how many out of 420?"

1 Introduce

Alternate Opener

EXPLORATION

8-9 Percent Problems

You can use a number line to find the percent of a number.

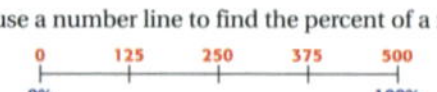

Use the number-line model above to complete each problem.

1. $\frac{125}{500} =$ ____% 2. $\frac{250}{500} =$ ____% 3. $\frac{375}{500} =$ ____%

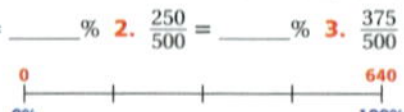

Label the number-line model above to find each percent of 640.

4. 50% of 640 is ____.
5. 25% of 640 is ____.
6. 75% of 640 is ____.

Think and Discuss

7. **Discuss** what it means to find 100% of a number. (*Hint:* What is 100% of 640?)
8. **Explain** how you can use number-line models to solve percent problems.

Motivate

Review estimation with students. Ask them why it can be a good idea to estimate an answer before solving a problem. Possible answer: so you will know if the answer you get is reasonable Also review some common decimal to fraction equivalencies.

Exploration worksheet and answers on Chapter 8 Resource Book pp. 81 and 126

2 Teach

Lesson Presentation

Guided Instruction

In this lesson, students learn to find the missing value in a percent problem. Teach students to set up a proportion to answer questions involving percent. Then teach them to multiply by a decimal to find a percent of a given number.

EXAMPLE 2 ***Technology Application***

Heather is downloading a file from the Internet. So far, she has downloaded 75% of the file. If 30 minutes have passed since she started, how long will it take her to download the rest of the file?

$\frac{\%}{100} = \frac{\text{is}}{\text{of}}$

$\frac{75}{100} = \frac{30}{m}$ *75% of the file has downloaded, so 30 minutes **is** 75% **of** the total time needed.*

$100 \cdot 30 = 75 \cdot m$ *The cross products are equal.*

$3{,}000 = 75m$ *m is multiplied by 75.*

$\frac{3{,}000}{75} = \frac{75m}{75}$ *Divide both sides by 75 to undo the multiplication.*

$40 = m$

The time needed to download the entire file is 40 min. So far, the file has been downloading for 30 min. Because $40 - 30 = 10$, the remainder of the file will be downloaded in 10 min.

Instead of using proportions, you can also multiply to find a percent of a number.

EXAMPLE 3 **Multiplying to Find a Percent of a Number**

A **Find 20% of 150.**

20% = 0.20 *Write the percent as a decimal.*

$0.20 \cdot 150$ *Multiply using the decimal.*

30

So 30 is 20% of 150.

B **Find 5% of 90.**

5% = 0.05 *Write the percent as a decimal.*

$0.05 \cdot 90$ *Multiply using the decimal.*

4.5

So 4.5 is 5% of 90.

Think and Discuss

1. **Explain** why you must subtract 30 from 40 in Example 2.
2. **Give an example** of a time when you would need to find a percent of a number.

COMMON ERROR ALERT

Some students may multiply by the percent instead of by the equivalent decimal when finding percent of a number (e.g., multiply 5 × 25 to find 5% of 25). Explain that the percent must be written as a decimal or fraction before computing with it.

Additional Examples

Example 1

There are 560 students in Ella's school. If 35% of the students participate in after-school sports, how many students participate in after-school sports?

196 students participate in after-school sports.

Example 2

Johan is 25% of the way through his exercises. If he has exercised for 20 minutes so far, how much longer does he have to work out?

He still has another 60 min to go.

Example 3

A. Find 36% of 50. 18

B. Find 17% of 95. 16.15

Example 3 note: The *of* in each problem can be replaced with ×. So 20% of 150 means 20% × 150, and 5% of 90 means 5% × 90.

3 Close

Reaching All Learners
Through World Math

Have students do research to find real data about population percents and write problems involving the percents. Have students exchange problems to solve.

Possible answer:

47% of the people in Smithville are males. The population of Smithville is 100,000. How many males live in Smithville?

Answer: Find 47% of 100,000.

$0.47 \times 100{,}000 = 47{,}000$

47,000 males live in Smithville.

Summarize

Review the proportion presented in the lesson opener:

$$\frac{\%}{100} = \frac{\text{is}}{\text{of}}$$

Have students tell you what replaces %, *of,* and *is* in the proportion.

The number before the percent sign replaces %, the total number replaces *of,* and the number that is the percentage of the total replaces *is.*

Answers to Think and Discuss

1. Possible answer: You must find the remaining minutes. So, you subtract the time it took for 75% of the files to download (30 min) from the amount of time it should take 100% of the files to download (40 min).
2. Possible answer: when you want to find the amount of discount when you know the regular price and the percent of discount, for instance, finding the amount of money saved when saving 5% off the regular price of $50

8-9 PRACTICE & ASSESS

Students may want to refer back to the lesson examples.

Assignment Guide

If you finished Example 1 assign:
Core 1, 7, 8, 25, 29–38
Enriched 7, 8, 24, 25, 29–38

If you finished Example 2 assign:
Core 1–2, 7–10, 23, 25, 29–38
Enriched 7–10, 24–27, 29–38

If you finished Example 3 assign:
Core 1–18, 29–38
Enriched 11–38

Notes

Math Background

The proportion shown on the first page of the lesson can be rewritten using variables and cross products to assume one of the forms shown below:

Given $\frac{A}{100} = \frac{C}{B}$,

$\frac{A}{100} \cdot B = C$, $\frac{100}{A} \cdot C = B$, and $\frac{100}{B} \cdot C = A$.

Students can then substitute for two of the three unknown quantities and solve for the third quantity. This method allows students to find a missing percent (A), the percent of a number (C), or a number when a percent of it is known (B).

8-9 Exercises

FOR EXTRA PRACTICE see page 653

internet connect
Homework Help Online
go.hrw.com Keyword: MR4 8-9

GUIDED PRACTICE

See Example 1

1. Members of the drama club sold T-shirts for their upcoming musical. Of the 80 T-shirts sold, 55% were size medium. How many of the T-shirts sold were size medium? **44 T-shirts**

See Example 2

2. Loni has read 25% of a book. If she has been reading for 5 hours, how many more hours will it take her to complete the book? **15 hours**

See Example 3

3. Find 12% of 56. **6.72**
4. Find 65% of 240. **156**
5. Find 85% of 115. **97.75**
6. Find 70% of 54. **37.8**

INDEPENDENT PRACTICE

See Example 1

7. Tamara collects porcelain dolls. Of the 24 dolls that she has, 25% have blond hair. How many of her dolls have blond hair? **6 dolls**
8. Mr. Green has a garden. Of the 40 seeds he planted, 35% were vegetable seeds. How many vegetable seeds did he plant? **14 seeds**

See Example 2

9. Kevin has mowed 40% of the lawn. If he has been mowing for 20 minutes, how long will it take him to mow the rest of the lawn? **30 minutes**
10. Maggie ordered a painting. She paid 30% of the total cost when she ordered it, and she will pay the remaining amount when it is delivered. If she has paid $15, how much more does she owe? **$35**

See Example 3

11. Find 22% of 130. **28.6**
12. Find 78% of 350. **273**
13. Find 9% of 50. **4.5**
14. Find 45% of 210. **94.5**

PRACTICE AND PROBLEM SOLVING

Find the percent of each number.

15. 6% of 38 **2.28**
16. 20% of 182 **36.4**
17. 32% of 205 **65.6**
18. 14% of 88 **12.32**
19. 78% of 52 **40.56**
20. 31% of 345 **106.95**
21. 10% of 50 **5**
22. 1.5% of 800 **12**

23. ***GEOMETRY*** The width of a rectangular room is 75% of the length of the room. The room is 12 feet long.
   a. How wide is the room? **9 feet**
   b. The area of a rectangle is the product of the length and the width. What is the area of the room? **108 square feet**

RETEACH 8-9

LESSON 8-9 **Reteach**
Percent Problems

You can use proportions to solve percent problems.

To find 25% of 72, first set up a proportion.

$\frac{25}{100} = \frac{x}{72}$

$25 \cdot 72 = 100 \cdot x$ Next, find cross products.

$1{,}800 = 100x$

$\frac{100x}{100} = \frac{1{,}800}{100}$ Then solve the equation.

$x = 18$

So, 18 is 25% of 72.

Use a proportion to find each number.

1. Find 3% of 75. **2.25**
2. Find 15% of 85. **12.75**
3. Find 20% of 50. **10**
4. Find 6% of 90. **5.4**

You can use multiplication to solve percent problems.

To find 9% of 70, first write the percent as a decimal.
9% = 0.09

Then multiply using the decimal.

$0.09 \cdot 70 = 6.3$

So, 9% of 70 = 6.3.

Use multiplication to find each number.

5. Find 80% of 48. **38.4**
6. Find 6% of 30. **1.8**
7. Find 40% of 120. **48**
8. Find 20% of 98. **19.6**
9. Find 70% of 70. **49**
10. Find 35% of 120. **42**
11. Find 9% of 50. **4.5**
12. Find 40% of 150. **60**

PRACTICE 8-9

LESSON 8-9 **Practice B**
Percent Problems

Find the percent of each number.

1. 8% of 40 **3.2**
2. 105% of 80 **84**
3. 35% of 300 **105**
4. 13% of 66 **8.58**
5. 64% of 50 **32**
6. 51% of 445 **226.95**
7. 14% of 56 **7.84**
8. 98% of 72 **70.56**
9. 24% of 230 **55.2**
10. 35% of 225 **78.75**
11. 44% of 89 **39.16**
12. 3% of 114 **3.42**
13. 70% of 68 **47.6**
14. 1.5% of 300 **4.5**
15. 85% of 240 **204**
16. 47% of 13 **6.11**
17. 20% of 522 **104.4**
18. 2.5% of 400 **10**

19. Jenna ordered 28 shirts for her soccer team. Seventy-five percent of those shirts were size large. How many large shirts did Jenna order?
21 large shirts

20. Douglas sold 125 sandwiches to raise money for his boy scout troop. Eighty percent of those sandwiches were sold in his neighborhood. How many sandwiches did Douglas sell in his neighborhood?
100 sandwiches

21. Samuel has run for 45 minutes. If he has completed 60% of his run, how many minutes will Samuel run in all?
75 minutes

Technology LINK

Someday you may do all of your schoolwork on a computer. You may even carry a computer to class with you, if your school becomes a digital school, like the one pictured above.

24. ***TECHNOLOGY*** Students were asked in a school survey about how they use their computers. The circle graph shows the results.

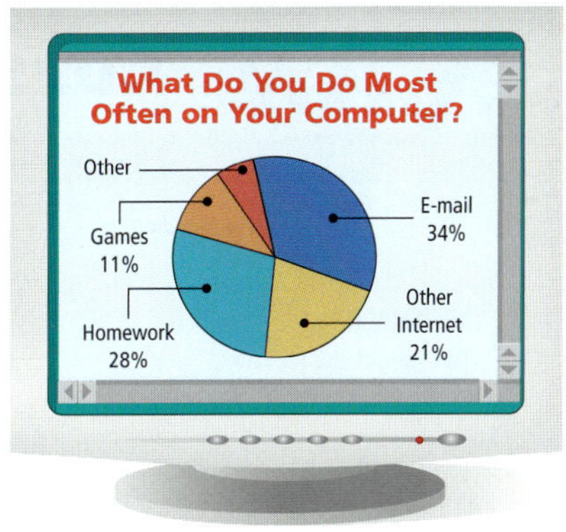

   a. If there are 850 students in the school, how many spend most of their computer time using e-mail? **289**
   b. Fifty-one students selected "other." What percent of the school population does this represent? **6%**
   c. Which choices were selected by more than 200 students? **Email, homework**
   d. How many more students chose Internet than chose playing games? **85**

25. ***CHEMISTRY*** Glucose is a type of sugar. A glucose molecule is composed of 24 atoms. Hydrogen atoms make up 50% of the molecule, carbon atoms make up 25% of the molecule, and oxygen atoms make up the other 25%. How many of each atom are in a molecule of glucose? **12 atoms of hydrogen, 6 atoms of carbon, and 6 atoms of oxygen**

26. ***WHAT'S THE ERROR?*** To find 80% of 130, a student set up the proportion $\frac{80}{100} = \frac{130}{x}$. Explain the error. Write the correct proportion, and find the missing value.

27. ***WRITE ABOUT IT*** Suppose you were asked to find 48% of 300 and your answer was 6.25. Would your answer be reasonable? How do you know? What is the correct answer?

28. ***CHALLENGE*** Mrs. Peterson makes ceramic figurines. She recently made 25 figurines. Of those figurines, 16 are animals. What percent of the figurines are **not** animals? **36%**

Spiral Review

Find each quotient. (Lesson 3-8)

29. 5.6 ÷ 0.8 **7** **30.** 30.8 ÷ 1.4 **22** **31.** 254.1 ÷ 0.35 **726** **32.** 11.5 ÷ 0.05 **230**

Write the prime factorization of each number. (Lesson 4-2)

33. 38 $2 \cdot 19$ **34.** 50 $2 \cdot 5^2$ **35.** 120 $2^3 \cdot 3 \cdot 5$ **36.** 214 $2 \cdot 107$

37. TEST PREP What is the least common denominator for $\frac{1}{2}$ and $\frac{5}{6}$? (Lesson 5-7) **B**

A 12 **B** 6 **C** 2 **D** 4

38. TEST PREP Which is the movement of a figure around a point? (Lesson 7-10) **H**

F Translation **G** Reflection **H** Rotation **J** Tessellation

Answers

26. The student set up the proportion for "80% of what number is 130?" instead of "what is 80% of 130?" The proportion should be $\frac{80}{100} = \frac{x}{130}$. The correct answer is 104.

27. Possible answer: No, the answer is not reasonable because 48% is close to half, and 6.25 is not anywhere close to half of 300; 144.

Journal

Have students think of and write real-world situations in which percents are used.

Test Prep Doctor

For Exercise 38, students can eliminate the following incorrect answers: **A,** because it describes a slide along a line; **B,** because it doesn't describe a flip across a line; and **D,** because it doesn't name a transformation. Remind students that *rotate* means "move about a point."

CHALLENGE 8-9

LESSON 8-9 **Challenge**
Pet Percentages

The United States Census Bureau counts all the people in the United States—but they do not count our pets! So, veterinarians use the percents shown in the table below to estimate pet populations. Their estimated U.S. pet population data is based on the 2000 census, which counted about 106 million households in the United States.

U.S. Pet Census, 2000

Pet	Percent of all Households	Estimated U.S. Pet Population
Dogs	53%	56,180,000
Cats	60%	63,600,000
Birds	13%	13,780,000
Horses	4%	4,240,000

Use the percents to estimate the number of pets that your class owns altogether, and the number of pets that your school owns altogether. Let each student in your class and each student in your school represent 1 household.

My Class and School Pet Population

Pet	Estimated Class Pet Population	Estimated School Pet Population
Dogs	15.9	106
Cats	18	120
Birds	3.9	26
Horses	1.2	8

Answers will vary depending on the number of students in the class and the number of students in the school. Example answers are given for a 30-student class and a 200-student school.

PROBLEM SOLVING 8-9

LESSON 8-9 **Problem Solving**
Percent Problems

In 2000, the population of the United States was about 280 million people.

Use this information to answer each question.

1. About 20% of the total United States population is 14 years old or younger. How many people is that? **56 million people**
2. About 6% of the total United States population is 75 years old or older. How many people is that? **16.8 million people**
3. About 50% of Americans live in states that border the Atlantic or Pacific Ocean. How many people is that? **140 million people**
4. About 12% of all Americans live in California. What is the population of California? **33.6 million people**
5. About 7.5% of all Americans live in the New York City metropolitan area. What is the population of that region? **21 million people**
6. About 12.3% of all Americans have Hispanic ancestors. What is the Hispanic American population here? **34.44 million people**
7. Males make up about 49% of the total population of the United States. How many males live here? **137.2 million men**
8. About 75% of all Americans live in urban areas. How many Americans live in or near large cities? **210 million Americans**

Circle the letter of the correct answer.

9. About 7.4% of all Americans live in Texas. What is the population of Texas?
   A 74 million C 7.4 million
   (B) 20.72 million D 2.072 million
10. Between 1990 and 2000, the population of the United States grew by about 12%. What was the U.S. population in 1990?
   (F) 250 million H 313.6 million
   G 33.6 million J 268 million

Lesson Quiz

1. Find 28% of 310. **86.8**
2. Find 70% of 542. **379.4**
3. Martha is taking a 100-question test. She has completed 60% of the test in 45 minutes. How much longer will it take her to finish the test? **30 min**
4. Crystal has a collection of 72 pennies. If 25% of them are Canadian, how many Canadian pennies does she have? **18**

Available on Daily Transparency in CRB

8B Construct Circle Graphs

Pacing: Traditional 1 day
Block $\frac{1}{2}$ day

Objective: To use a compass, straightedge, and protractor to construct circle graphs

Materials: Compass, straightedge, protractor

Lab Resources

Hands-On Lab Activities . . . . pp. 78–80

Using the Pages

Have students practice using a compass and protractor to construct circles and angles.

Use the compass to draw a circle with the given radius. Mark the center and label it *C*. Use a straightedge to draw a radius.

1. $r = \frac{1}{2}$

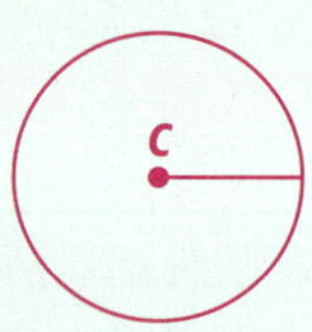

Use a protractor to draw angles with the given measurements.

2. 32°

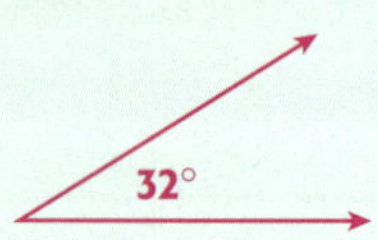

3. 115°

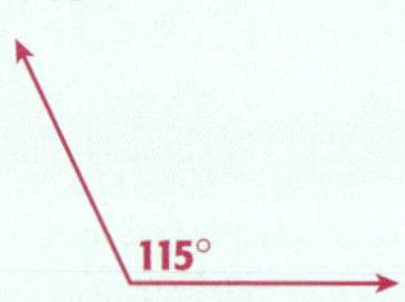

Construct Circle Graphs

Use with Lesson 8-9

REMEMBER
The sum of the measures of the angles in any circle is 360°.

internet connect
Lab Resources Online
go.hrw.com
KEYWORD: MR4 Lab8B

A circle graph shows parts of a whole. If you think of a complete circle as 100%, you can express sections of a circle graph as percents.

- Ms. Shipley's class earned \$400 at the school fair. What fraction of the \$400 did the class earn at the bake sale?
- What percent of the \$400 did the class earn at the bake sale?

Activity

At Mazel Middle School, students were surveyed about their favorite types of TV programs. Make a circle graph to represent the results.

Students' Favorite Programs

Type of Program	Number of Students
Science	25
Cooking	15
Sports	50
Sitcoms	150
Movies	60
Cartoons	200

1. Find the total number of students surveyed.
   $25 + 15 + 50 + 150 + 60 + 200 = 500$
2. Find the percent of the total represented by students who like science programs.
   $$\frac{25}{500} = 5\%$$
3. Since there are 360° in a circle, multiply 5% by 360°. This will give you an angle measure in degrees.
   $$0.05 \cdot 360° = 18°$$
4. Use a compass to draw a circle. Mark the center and use a straightedge to draw a line from the center to the edge of the circle.

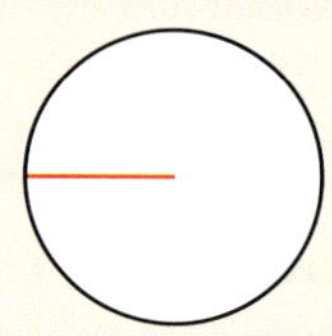

5 Use your protractor to draw an angle measuring 18°. The vertex of the angle will be the center of the circle, and one side will be the line that you drew. The section formed represents the percent of students who prefer science programs.

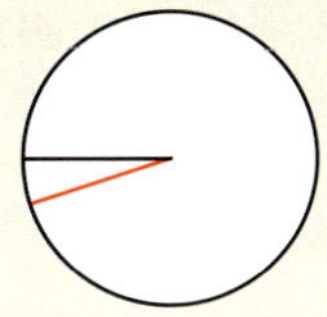

6 Repeat 2 through 5 for each type of program. Label each section, and give the graph a title.

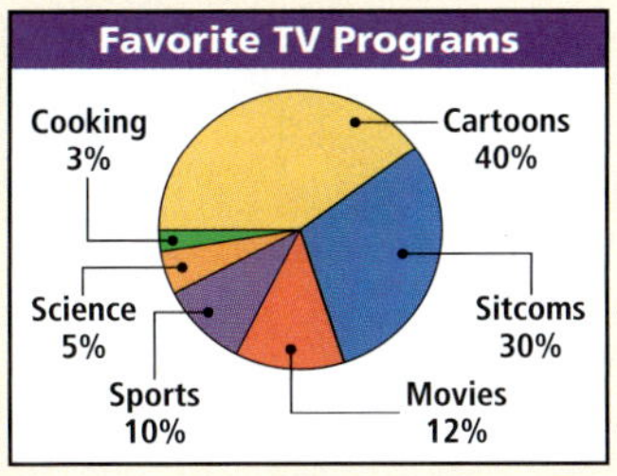

Think and Discuss

1. Looking at your circle graph, discuss five pieces of information you have learned about the TV habits of students at Mazel Middle School.
2. What does the whole circle represent? **all of the students surveyed**
3. Why do you need to know that there are 360° in a circle?
4. How does the size of each section of your circle graph relate to the percent that it represents?
Larger percentages have larger sections; smaller percentages have smaller sections.

Try This

1. People at a mall were surveyed about their favorite pets. Make a circle graph to display the results of the survey.

Favorite Pets	
Type of Pet	Number of People
Dog	225
Fish	150
Bird	112
Cat	198
Other	65

2. Collect data from your classmates about their favorite colors. Use the data to make a circle graph with no more than five sections.
Check students' work.
3. The circle graph shows the results of a survey about what people in the United States like to eat for breakfast. If this survey included 1,500 people, how many people said they like to eat cereal for breakfast? **390**

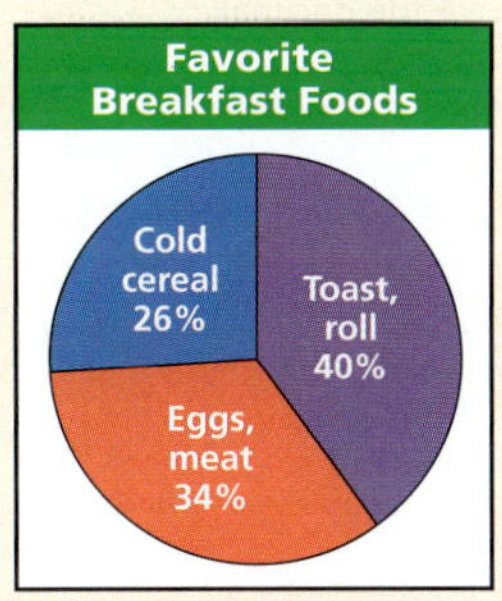

Answers

Think and Discuss

1. Possible answer: The most popular kind of TV show is cartoons and the least popular is cooking shows. 4 out of 10 students like cartoons the best, 3 out of 10 students like sitcoms the best, and 1 out of 10 like sports the best.
3. To find the number of degrees each section should have, you must multiply the percent the section represents by the total number of degrees in a circle.

Try This

1. Favorite Pets

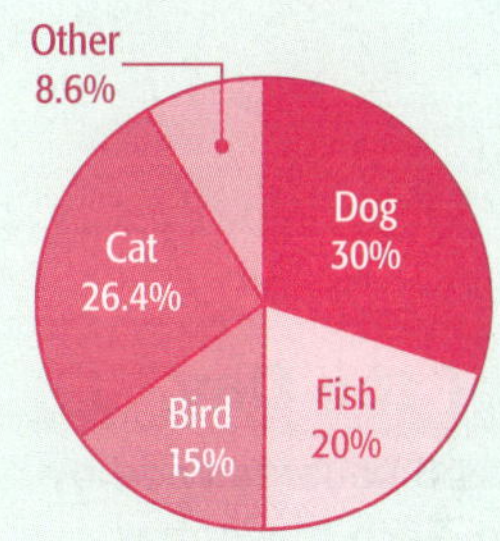

8-10 Organizer

Pacing: Traditional 1 day
Block $\frac{1}{2}$ day

Objective: Students solve percent problems that involve discounts, tips, and sales tax.

Indiana Academic Standards
6.2.8, 6.5.10, 6.7.1, 6.7.2, 6.7.3, 6.7.5, 6.7.11

Warm Up

Find the percent of each number.

1. 75% of 300 — 225
2. 93% of 56 — 52.08
3. 32% of 128 — 40.96
4. 9% of 60 — 5.4

Problem of the Day

A chessboard is 8 squares wide by 8 squares long. Each player has 8 pawns, 1 king, and 7 other pieces. At the start of a game, all the pieces are on the board, 1 piece per square. What percent of the total number of squares have a chess piece? 50%

Available on Daily Transparency in CRB

Math Humor

The Internet shopper thought she found a bargain when the ad promised, "Designer jeans 50% off." A week later, she opened the package and found one trouser leg.

8-10 Using Percents

Learn to solve percent problems that involve discounts, tips, and sales tax.

Vocabulary
discount
tip
sales tax

Percents show up often in daily life. Think of examples that you have seen of percents—sales at stores, tips in restaurants, and sales tax on purchases. You can estimate percents such as these to find amounts of money.

Common Uses of Percents	
Discounts	A **discount** is an amount that is subtracted from the regular price of an item. discount = price · discount rate total cost = price − discount
Tips	A **tip** is an amount added to a bill for service. tip = bill · tip rate total cost = bill + tip
Sales tax	**Sales tax** is an amount added to the price of an item. sales tax = price · sales tax rate total cost = price + sales tax

EXAMPLE 1 Finding Discounts

A music store sign reads "10% off the regular price." If Nichole wants to buy a CD whose regular price is $14.99, about how much will she pay for her CD after the discount?

Remember!
To multiply by 0.10, move the decimal point one place left.

Step 1: First round $14.99 to $15.

Step 2: Find 10% of $15 by multiplying 0.10 · $15. (**Hint:** Moving the decimal point one place left is a shortcut.)

10% of 15 = 0.10 · $15 = $1.50

The approximate discount is $1.50. Subtract this amount from $15.00 to estimate the cost of the CD.

$15.00 − $1.50 = $13.50

Nichole will pay about $13.50 for the CD.

1 Introduce

Alternate Opener

EXPLORATION

8-10 Using Percents

Stores that go out of business often offer big discounts on purchases. In such situations, 50% off sales are common.

Estimate the discount for each item at 50% off. Then calculate the actual discount.

	Item	Price	Estimated Discount	Actual Discount
1.	Shirt	$39.95		
2.	DVD player	$288.99		
3.	Speakers	$239.95		
4.	TV	$1,035.29		
5.	MP3 player	$247.99		

Think and Discuss

6. **Discuss** the estimation strategies you used.
7. **Explain** whether a one-time 50% discount is equivalent to two consecutive 25% discounts. (*Hint:* Use $100.00 as the base amount.)

Motivate

Ask students to explain what sales tax is and what, if any, strategies they use to figure out how much tax they will have to pay when they buy something at the store. Then discuss tips and how to figure out how much of a tip to leave at a restaurant.

Exploration worksheet and answers on Chapter 8 Resource Book pp. 90 and 128

2 Teach

Lesson Presentation

Guided Instruction

In this lesson, students learn to solve percent problems that involve discounts, tips, and sales tax. Define the new vocabulary for the lesson, and then teach students to compute each. Explain that it's a good idea when making purchases to estimate the total price after sales tax or discount to be sure you are charged the correct amount. (Teaching Transparency T43 is available in CRB)

Teaching Tip You may wish to have students make a personal "tip chart" showing 15% and 20% tips for various amounts.

When estimating percents, use percents that you can calculate mentally.

- You can find 10% of a number by moving the decimal point one place to the left.
- You can find 1% of a number by moving the decimal point two places to the left.
- You can find 5% of a number by finding one-half of 10% of the number.

EXAMPLE 2 Finding Tips

Leslie's lunch bill is $13.95. She wants to leave a tip that is 15% of the bill. About how much should her tip be?

Step 1: First round $13.95 to $14.

Step 2: Think: 15% = 10% + 5%
10% of $14 = 0.10 · $14 = $1.40

Step 3: 5% = 10% ÷ 2
= $1.40 ÷ 2 = $0.70

Step 4: 15% = 10% + 5%
= $1.40 + $0.70 = $2.10

Leslie should leave about $2.10 as a tip.

EXAMPLE 3 Finding Sales Tax

Marc is buying a scooter for $79.65. The sales tax rate is 6%. About how much will the total cost of the scooter be?

Step 1: First round $79.65 to $80.

Step 2: Think: 6% = 6 · 1%
1% of $80 = 0.01 · $80 = $0.80

Step 3: 6% = 6 · 1%
= 6 · $0.80 = $4.80

The approximate sales tax is $4.80. Add this amount to $80 to estimate the total cost of the scooter.

$80 + $4.80 = $84.80

Marc will pay about $84.80 for the scooter.

Think and Discuss

1. **Tell** when it would be useful to estimate the percent of a number.
2. **Explain** how to estimate to find the sales tax of an item.

COMMON ERROR ALERT

When finding the price after a discount, students may multiply by the discount rate and neglect to subtract that amount from the original price. Explain that a *discount* is the amount saved, and the *sale price* is the final cost after subtracting the discount.

Additional Examples

Example 1

A clothing store is having a 10% off sale. If Angela wants to buy a sweater whose regular price is $19.95, about how much will she pay for the sweater after the discount?

Angela will pay about $18 for the sweater.

Example 2

Ben's dinner bill is $7.85. He wants to leave a tip that is 15% of the bill. About how much should his tip be?

Ben should leave about $1.20 as a tip.

Example 3

Ann is buying a $29.75 dog bed. The sales tax rate is 7%. About how much will the total cost be?

Ann will pay about $32.10 for the dog bed.

3 Close

Summarize

Have students explain how *discounts* are different from *sales tax* and *tips*.

Possible answer: Discounts are subtracted from the price of an item. Sales tax and tips are added to the total cost.

Reaching All Learners
Through World Math

Have students use examples of percents from the advertising flyers in newspapers. Have them use the original prices and the percent of discount to find the sale prices.

Answers to Think and Discuss

1. Possible answer: When you are buying something, you can use the estimate to check that you are being charged the correct sales tax or given the correct discount.
2. Possible answer: Round the cost of the item to the nearest dollar, and then multiply by the sales tax rate.

8-10 PRACTICE & ASSESS

Students may want to refer back to the lesson examples.

Assignment Guide

If you finished Example 1 assign:
Core 1, 4–5, 10, 17–26
Enriched 1, 4, 10, 14, 17–26

If you finished Example 2 assign:
Core 1–2, 4–7, 10, 17–26
Enriched 1–2, 4, 6, 10, 14, 16–26

If you finished Example 3 assign:
Core 1–10, 13, 17–26
Enriched 6–26

Answers

11. Yes; the total cost with tax is $31.92. $32.50 is greater than $31.92.

8-10 Exercises

FOR EXTRA PRACTICE see page 653

internet connect
Homework Help Online
go.hrw.com Keyword: MR4 8-10

GUIDED PRACTICE

See Example 1

1. Norine wants to buy a beaded necklace that is on sale for 10% off the marked price. If the marked price is $8.49, about how much will the necklace cost after the discount? about $7.65

See Example 2

2. Alice and Wagner ordered a pizza to be delivered. The total bill was $12.15. They want to give the delivery person a tip that is 20% of the bill. About how much should the tip be? about $2.40

See Example 3

3. A bicycle sells for $139.75. The sales tax rate is 8%. About how much will the total cost of the bicycle be? about $151.20

INDEPENDENT PRACTICE

See Example 1

4. Peter has a coupon for 15% off the price of any item in a sporting goods store. He wants to buy a pair of sneakers that are priced at $36.99. About how much will the sneakers cost after the discount? about $31.45
5. All DVDs are discounted 25% off the original price. The DVD that Marissa wants to buy was originally priced at $24.98. About how much will the DVD cost after the discount? about $18.75

See Example 2

6. Michael's breakfast bill came to $7.65. He wants to leave a tip that is 15% of the bill. About how much should he leave for the tip? about $1.20
7. Betty and her family went out for dinner. Their bill was $73.82. Betty's parents left a tip that was 15% of the bill. About how much was the tip that they left? about $11.00

See Example 3

8. A computer game costs $36.85. The sales tax rate is 6%. About how much will the total cost be for this computer game? about $39.00
9. Irene is buying party supplies. The cost of her supplies is $52.75. The sales tax rate is 5%. About how much will the total cost of her party supplies be? about $55.65

PRACTICE AND PROBLEM SOLVING

10. An electronics store is going out of business. The sign on the door reads "All items on sale for 60% off the ticketed price." A computer has a ticketed price of $649, and a printer has a ticketed price of $199. What is the total cost of both items after the discount? $339.20
11. Jackie has $32.50 to buy a new pair of jeans. The pair she likes costs $38 but are marked "20% off ticketed price." The sales tax rate is 5%. Does Jackie have enough money to buy the jeans? Explain.

Math Background

Taxation is one of the oldest applications of arithmetic. Records of it are found in the histories of people from around the globe. The word *tax* is derived from the Latin word *taxare,* which means "to estimate or evaluate."

Tipping is a widely varying phenomenon. In some areas, it is standard practice to tip a certain percentage of a bill. On the other hand, there are establishments and cultures in which tipping is either expressly forbidden or considered unnecessary. A person traveling in a place with a different culture is advised to become acquainted with the conventions of tipping there.

RETEACH 8-10

LESSON 8-10 **Reteach**
Using Percents

There are many uses for percents.

Common Uses of Percents

Discounts	A **discount** is an amount that is subtracted from the regular price of an item. discount = regular price • discount rate
Tips	A **tip** is an amount added to a bill. tip = total bill • tip rate
Sales Tax	**Sales tax** is an amount added to the price of an item. sales tax = purchase price • sales tax rate

Rachel is buying a sweater that costs $42. The sales tax rate is 5%. About how much will the total cost of the sweater be?

You can use fractions to find the amount of sales tax.

First round $42 to $40.

Think: 5% is equal to $\frac{1}{20}$.

So, the amount of tax is about $\frac{1}{20} \cdot \$40$.

The tax is about $2.00.

Then find the sum of the price of the sweater and the tax.

$42 + $2.00 = $44.00

Rachel will pay about $44.00 for the sweater.

Solve each problem. Estimates may vary.

1. About how much would you pay for a meal that costs $29.75 if you left a 15% tip?
$34.50
2. About how much do you save if a book whose regular price is $25.00 is on sale for 10% off?
$2.50
3. About how much would you pay for a box of markers whose price is $5.99 with a sales tax rate of 9.5%?
$6.60

PRACTICE 8-10

LESSON 8-10 **Practice B**
Using Percents

Write the correct answer.

1. Carl and Rita ate breakfast at the local diner. Their bill came to $11.48. They gave their waitress a tip that was 25% of the bill. How much money did they give the waitress for her tip?
$2.87
2. The school's goal for the charity fundraiser was $3,000. They exceeded the goal by 22%. How much money for charity did the school raise at the event?
$3,660
3. Rob had a 15% off coupon for the sporting goods store. He bought a tennis racket that had a regular ticket price of $94.00. How much did Rob spend on the racket after using his coupon?
$79.90
4. Lisa's family ordered sandwiches to be delivered. The total bill was $21.85. They gave the delivery person a tip that was 20% of the bill. How much did they tip the delivery person?
$4.37
5. A portable CD player costs $118.26. The sales tax rate is 7%. About how much will it cost to buy the CD player?
$126.54
6. Kathy bought two CDs that each cost $14.95. The sales tax rate was 5%. About how much did Kathy pay in all?
$31.40
7. Tom bought $65.86 worth of books at the book fair. He got a 12% discount since he volunteered at the fair. About how much did Tom's books cost after the discount?
$57.96
8. Sawyer bought a T-shirt for $12.78 and shorts for $17.97. The sales tax rate was 6%. About how much money did Sawyer spend altogether?
$32.60
9. Melody buys a skateboard that costs $79.81 and a helmet that costs $26.41. She uses a 45% off coupon on the purchase. If Melody pays with a $100 bill, about how much change should she get back?
$41.58
10. Bruce saved $35.00 to buy a new video game. The game's original price was $42.00, but it was on sale for 30% off. The sales tax rate was 5%. Did Bruce have enough money to buy the game? Explain.
Yes; with the discount and sales tax, the total cost was $30.87.

12. Lenny, Robert, and Katrina went out for lunch. The items they ordered are listed on the receipt. The sales tax rate was 7%, and they left a tip that was 15% of the total bill. How much did the three friends spend in all? **$27.49**

**** Thank you ****

Chicken Sandwich - 1	$5.95
Hamburger - 1	$4.75
Roast Beef Sandwich - 1	$7.35
Milk - 2	$2.40
Iced Tea - 1	$1.89

13. ***SOCIAL STUDIES*** The table shows the sales tax rate in some states.

State	Sales Tax Rate
Georgia	4%
Kentucky	6%
New York	4%
North Carolina	4.5%

a. A shirt costs $18.95. Will the shirt cost more after sales tax in Georgia or in Kentucky? About how much more?

b. A video game in North Carolina costs $59.75. The same video game in New York costs $60. After sales tax, in which state will the video game cost less? How much less?

14. ***WHAT'S THE ERROR?*** The original price of an item was $48.65. The item was discounted 40%. A customer calculated the price after the discount to be $19.46. What's the error? Give the correct price after the discount.

15. ***WRITE ABOUT IT*** Discuss the difference between a discount, sales tax, and a tip, in relation to the total cost. How does each affect the total cost? Give examples of situations in which each one is used.

16. ***CHALLENGE*** Suppose a jacket is discounted 50% off the original price and then discounted an additional 20%. Is this the same as discounting the jacket 70% off the original price? Explain, and give an example to support your answer.

Spiral Review

Give all the factors of each number. (Lesson 4-2)

17. 20 **1, 2, 4, 5, 10, 20**
18. 40 **1, 2, 4, 5, 8, 10, 20, 40**
19. 59 **1, 59**
20. 85 **1, 5, 17, 85**

Find the mean, median, and mode for each set of data. (Lesson 6-2)

21. 23, 24, 25, 22, 23, 28, 30, 21, 20 **mean: 24; median: 23; mode: 23**

22. 70, 80, 78, 82, 90, 96, 74, 80 **mean: 81.25; median: 80; mode: 80**

Use a protractor to draw an angle with each measure. (Lesson 7-2)

23. 120°
24. 35°
25. 90°

26. **TEST PREP** Which fraction is equivalent to 4%? (Lesson 8-7) **B**

A $\frac{2}{5}$ B $\frac{1}{25}$ C $\frac{1}{4}$ D $\frac{4}{1}$

CHALLENGE 8-10

LESSON 8-10 **Challenge**
Shop Smart

The Sport Zone and Sport City are both competing for customers by offering big discounts. To be a smart customer, you need to decide which store is offering the better price on each item.

For each item, write the store offering the best deal and the price you will pay there to the nearest whole cent.

The Sport Zone | Sport City

1.

The Sport Zone; $118.99

2.

Sport City; $30.30

3.

The Sport Zone; $15.50

4.

Sport City; $42.20

PROBLEM SOLVING 8-10

LESSON 8-10 **Problem Solving**
Using Percents

Use the table to answer each question.

Federal Income Tax Rates, 2001

Single Income	Tax Rate	Married Joint Income	Tax Rate
$0 to $27,050	15%	$0 to $45,200	15%
$27,051 to $65,550	27.5%	$45,201 to $109,250	27.5%
$65,551 to $136,740	30.5%	$109,251 to $166,500	30.5%
$136,741 to $297,350	35.5%	$166,501 to $297,350	35.5%
More than $297,350	39.1%	More than $297,350	31.5%

1. If a single person makes $25,000 a year, how much federal income tax will he or she have to pay? **$3,750**

2. If a married couple makes $148,000 together, how much federal income tax will they have to pay? **$45,140**

3. The average American with a college degree earns $33,365 a year. About how much federal income tax does he or she have to pay at a single rate? **$9,175.38**

4. The governor of New York makes a higher salary than in any other state—$179,000 a year. How much federal income tax does that governor have to pay at a single rate? **$63,545**

5. The average salary for a public school teacher in the United States is $42,898. If two teachers are married, what is the average amount of federal income taxes they have to pay together? **$23,593.90**

6. In 2002 President George W. Bush received an annual salary of $400,000. Vice President Dick Cheney got $186,300. How much federal income tax do they each have to pay on their salary? **Bush: $126,000; Cheney: $66,136.50**

Circle the letter of the correct answer.

7. Members of the U.S. Congress each earn $145,100 a year. How much federal income tax does each pay on their salary?
   (A) $51,510.50 C $21,765
   B $44,255.50 D $39,902.50

8. A married couple each working a minimum-wage job will earn an average of $21,424 together a year. How much income tax will they pay?
   F $5,891.60 H $321.36
   (G) $3,213.60 J $6,534.32

Answers

13. a. Kentucky; about $0.38
    b. New York; $0.04 less than in North Carolina

14–16. See p. A6.

23.
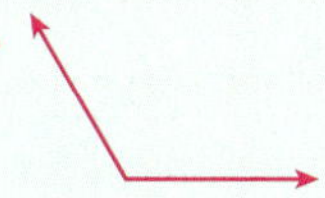

24.
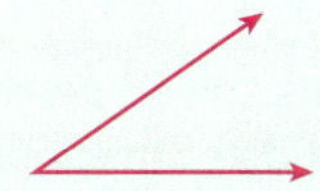

25.
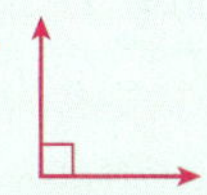

Journal

Have students estimate and solve the problem below. Have them explain each step in their solution.

Tom's Toys is having a 25% off sale. The regular price of a badminton set is $39.80. How much will the set cost after the discount, including 6% sales tax? **$31.64**

Test Prep Doctor

For Exercise 26, students need to remember that percents, when expressed as fractions, have a denominator of 100. This means students need to find the fraction equivalent of $\frac{4}{100}$. $\frac{4 \div 4}{100 \div 4} = \frac{1}{25}$, so the correct answer is **B.**

Lesson Quiz

1. Sean's new jeans are priced at $29.97, but the sale sign reads, "Take 15% off." About how much will the jeans cost after the discount? about $25.50

2. The bill for a family dinner comes to $56.78. About how much would a 20% tip be? about $11.50

3. The price on a book is $12.99. If sales tax is 4%, about how much will its total cost be? about $13.50

4. Megan wants a new bike. She is happy to see a sign that reads, "All bikes 10% off." If the original price of the bike was $159.90 and sales tax is 6%, about how much will the total cost of the bike be? about $150

Available on Daily Transparency in CRB

CHAPTER 8

Extension

Pacing: Traditional 1 day
Block $\frac{1}{2}$ day

Objective: Students find simple interest.

Using the Pages

In Lesson 8-10, students solved percent problems that involved discounts, tips, and sales tax. In this extension, students find simple interest.

EXTENSION Simple Interest

Learn to find simple interest.

Vocabulary
interest
principal
simple interest

When you save money in a savings account, you earn money that the bank adds to your account. The added money is called **interest**. The original amount you put into the account is the **principal**. Interest is a percentage of the principal.

One type of interest is called *simple interest*. **Simple interest** is a fixed percentage of the original principal and is often paid over a certain time period. For example, simple interest may be paid once per year or several times per year. In this section, we will assume that simple interest is paid once per year.

Simple Interest

$$I = P \cdot r \cdot t$$

Interest (I)
Principal (P)
Annual interest rate (as a decimal) (r)
Time that the money earns interest (in years) (t)

Note that interest rates are usually given as percents, but you must convert the rates to decimals when you use the simple interest formula above.

EXAMPLE 1 Finding Simple Interest

Alyssa put $250 in a savings account at a simple interest rate of 6% per year.

A If she does not add money to or take money from her account, how much interest will she have earned at the end of 3 years?

$I = P \cdot r \cdot t$ — *P = $250, r = 0.06, t = 3 years*
$I = 250 \cdot 0.06 \cdot 3$ — *Multiply.*
$I = \$45$

Alyssa will earn $45 in interest in 3 years.

B How much money will be in her account after 3 years?

To find the total amount in Alyssa's account after three years, add the interest to the principal.

$$\$250 + \$45 = \$295$$

Alyssa will have $295 in her account after 3 years.

1 Introduce

Motivate

Money doesn't grow on trees, but it can grow if you save and invest wisely. Earning simple interest on your money is one way to make your money grow.

2 Teach

Lesson Presentation

Guided Instruction

In this extension, students find simple interest. Define *interest, principal,* and *simple interest.* Show students the formula for simple interest, $I = Prt$. Have students find simple interest using the formula. Then have students add the interest to the principal to get the new balance.

EXTENSION Exercises

1. Tamara put $425 in a savings account at a simple interest rate of 7% per year. How much interest will she have earned after 5 years? **$148.75**

2. Jerome put $75 in a savings account at a simple interest rate of 3% per year. How much interest will Jerome have earned after 1 year? How much money will he have in his account after 1 year? **$2.25; $77.25**

Use the equation $I = P \cdot r \cdot t$ to find the missing amount.

3. principal = $320
   interest rate = 5% per year
   time = 2 years
   interest = ▒ **$32**

4. principal = $150
   interest rate = 2% per year
   time = 7 years
   interest = ▒ **$21**

5. principal = ▒ **$250**
   interest rate = 4% per year
   time = 3 years
   interest = $30

6. principal = $456
   interest rate = 6% per year
   time = ▒ **4 years**
   interest = $109.44

7. Mr. Bruckner is saving to go on a vacation. He put $340 in a savings account at a simple interest rate of 4% per year. How much money will he have in the savings account after 2 years? **$367.20**

8. When you borrow money, the amount borrowed is the principal. Instead of receiving interest, you pay interest on the principal. Kendra borrowed $1,500 from the bank to buy a home computer. The bank is charging her a simple interest rate of 7% per year. How much interest will Kendra owe the bank after 1 year? **$105.00**

9. Mr. Pei paid $7,500 in interest over 20 years at 1% per year on a loan. How much money did he borrow? **$37,500**

10. Hunter put $165 in a savings account at a simple interest rate of 6% per year. Nicholas put $145 in a savings account at a simple interest rate of 7% per year. Who will have earned more interest after 3 years? How much more? **Nicholas; $0.75 more**

11. ***WRITE ABOUT IT*** Explain the difference between principal and interest.

12. ***WRITE ABOUT IT*** Would you prefer a high or low interest rate when you are borrowing money? When you are saving money? Explain.

13. ***CHALLENGE*** Madison put $200 in a savings account at an interest rate of 5%. Each year the interest is added to the principal, and then the new amount of interest is calculated. If Madison does not add money to or take money out of the account, how much will she have after 3 years? **$231.53**

COMMON ERROR ALERT

In Exercise 5, some students may multiply the interest by the number of years by the interest rate to get the principal. Remind students that they need to substitute the given values into the formula $I = Prt$ and then solve for P.

Additional Example

Example 1

Ricardo put $300 into a savings account at a simple interest rate of 7% per year.

A. If he does not add money to, or take money from, his account, how much interest will he have earned at the end of 2 years? **$42**

B. How much money will be in his account after 2 years? **$342**

Answers

11. The principal is the amount of money you originally have or borrow, and interest is the amount of money you earn or have to pay on the principal after a specified number of years at a certain rate.

12. When you borrow money, you want a low interest rate because you will have to pay back less. When you are saving money, you want a high interest rate because that way you will earn more money.

3 Close

Summarize

Review the terms *interest*, *principal*, and *simple interest*. Have the students tell what each of the variables represents in the formula for simple interest, $I = Prt$.

I = interest earned, P = principal, r = interest rate per year written as a decimal, and t = time in years.

Teaching Tip Make sure that the students convert the interest rate to a decimal when they use the simple interest formula.

Problem Solving on Location

New York

Purpose: *To provide additional practice for problem-solving skills in Chapters 1–8*

Ellis Island

- After problem 3, have students estimate the ratio of the number of Russian immigrants to the number of Italian immigrants. 19:25 Have students estimate the ratio of the combined number of Italian and Russian immigrants to the total number of immigrants. 44:107
- After problem 4, have students consider the following problem: About how many people per day passed through Ellis Island in its first 40 years of operation? about 730 people per day

Extension Have students use the information in the exercise to make a circle graph to show the immigration to Ellis Island during this period. Have them use the categories Italy, Russia, and other.

Country of Origin for Ellis Island Immigrants (1892–1932)

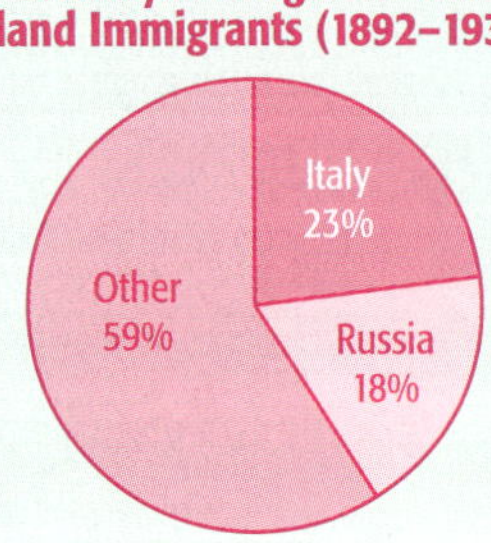

Problem Solving on Location

NEW YORK

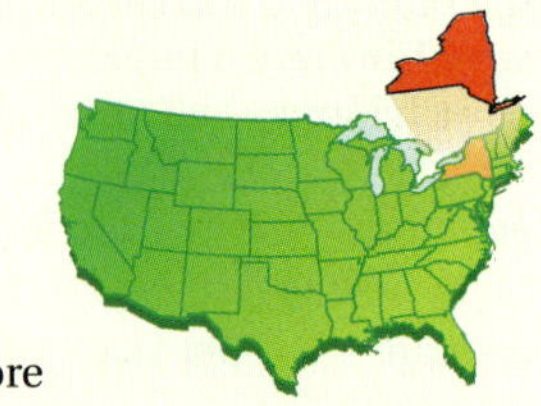

Ellis Island

From 1892 to 1954, Ellis Island, located in New York Harbor, was the first stop for many immigrants to the United States. More than 40 percent of all U.S. citizens alive today have an ancestor who came through Ellis Island as an immigrant.

More than 10,700,000 immigrants passed through Ellis Island in its first 40 years of operation. The table shows the number of immigrants from the two countries that had the greatest number of people come through Ellis Island.

Number of Italian and Russian Immigrants Through Ellis Island in its First 40 Years	
Country	**Number of Immigrants**
Italy	2,500,000
Russia	1,900,000

For 1–4, use the table.

1. About what percent of the total number of immigrants came from Italy? **about 23%**
2. About what percent of the total number of immigrants came from Russia? **about 18%**
3. During this period, about 500,000 immigrants came to Ellis Island from Ireland. Estimate the ratio of the immigrants that came from Ireland to the total number of immigrants that came during this period. **about 1:21**
4. Estimate the ratio of the immigrants that came from a country other than Italy or Russia during the 40-year period to the total number of immigrants that came during the same time. **about 6:11**

The Statue of Liberty

As immigrants sailed toward New York Harbor, their first sight in the distance was the Statue of Liberty. A gift from France in 1886, the statue stands directly across from Ellis Island in New York Harbor.

1. If you visit the Statue of Liberty, you can climb 22 stories to the crown. In all, you would walk up 354 steps. Find the number of steps per story. Round your answer to the nearest whole number. about 16 steps per story
2. To get to the Statue of Liberty, you take a 2-mile ferry ride from Battery Park in New York City. If the ferry ride takes 15 minutes, what is your speed in miles per hour? 7.5 mi/h

Models of the Statue of Liberty are popular souvenirs in New York. Suppose the scale of a particular model is 1 in:8 ft. The actual dimensions of the statue are given in the table.

For 3–6, use the table.

3. Explain what the scale 1 in:8 ft means. 1 in. on the model represents 8 ft on the actual statue.
4. What is the length of the model statue's index finger? 1 in.
5. What is the length of the model's hand? $2\frac{1}{16}$ in.
6. What is the distance from the bottom of the model's base to the tip of its torch? $38\frac{1}{8}$ in.

Statue of Liberty	
Height of Base	154 ft
Height of Statue	151 ft
Length of Hand	16.5 ft
Length of Index Finger	8 ft

Statue of Liberty

- After problem 1, discuss the following: In 1999, more than 5.3 million people visited the Statue of Liberty. About how many people per day visited the Statue of Liberty in 1999? about 14,500 people per day
- After problem 5, have students consider the following questions: Compare the Statue of Liberty's index finger length and its hand length. What is the ratio of the lengths? about 1:2 Does the ratio of index finger length to hand length seem reasonable to you? Explain. Possible answer: Yes; the index finger is about half as long as the hand. Looking at my own hand, that seems reasonable.

Extension Encourage students to write a problem about a visit to the Statue of Liberty, using the data in the table. Then have them exchange problems and solve. Check students' work.

MATH-ABLES

Game Resources

Puzzles, Twisters & Teasers
Chapter 8 Resource Book

The Golden Rectangle

Purpose: *To apply the skill of writing proportions to identifying golden rectangles*

Discuss: Show students how to test a 3-by-5-inch index card to see whether it approximates a golden rectangle. Let $\ell = 3$ in. and $w = 5$ in. Ask: What proportion can you write to test the rectangle? $\frac{3}{5} = \frac{5}{8}$ Is the proportion true? **no** Have students write each fraction as a decimal to show that the ratios are nearly equal. **0.60 ≈ 0.625**

Extend: Have students measure visually appealing rectangles in their environment (e.g., flags, windows, notebook paper) to determine whether they approximate golden rectangles. Check students' work.

Triple Play

Purpose: *To practice finding equivalent fractions, decimals, and percents in a game format.*

Discuss: After a student wins the game, have the winner show each set of cards to the other players so they can verify that each set shows an equivalent fraction, decimal, and percent.

Extend: For each set of three cards, have students create an additional card showing a fraction strip representing the fraction. Have them repeat the game, with four cards now needed to create a set.

Example: fourth card for the set 0.1, 10%, $\frac{1}{10}$:

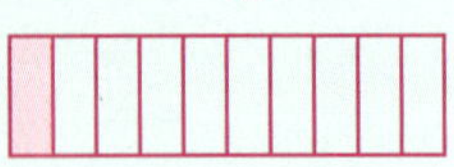

MATH-ABLES

The Golden Rectangle

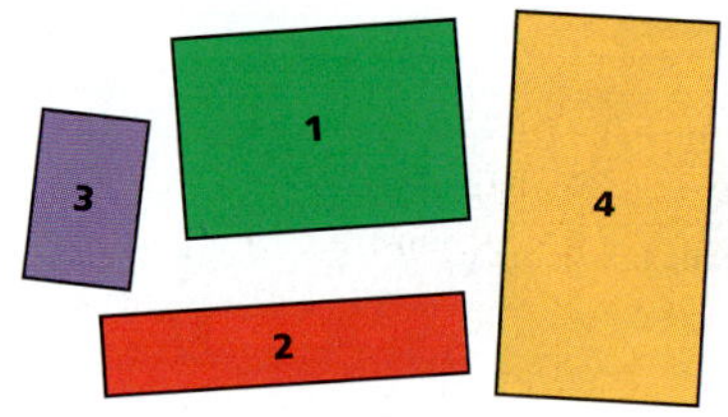

Which rectangle do you find most visually pleasing?

Did you choose rectangle 3? If so, you agree with artists and architects throughout history. Rectangle 3 is a golden rectangle. Golden rectangles are said to be the most pleasing to the human eye.

In a golden rectangle, the ratio of the length of the longer side to the length of the shorter side is approximately equal to 1.6. In other words,

$$\frac{\text{length of longer side}}{\text{length of shorter side}} \approx \frac{1.6}{1}$$

Measure the length and width of each rectangle below. Which could be golden rectangles? Are they the most pleasing to your eye?

golden rectangle; not a golden rectangle; golden rectangle

Triple Play

Number of players: 3–5

Deal five cards to each player. Place the remaining cards in a pile facedown. At any time, you may remove *triples* from your hand. A *triple* is a fraction card, a decimal card, and a percent card that are all equivalent.

On your turn, ask any other player for a specific card. For example, if you have the $\frac{3}{5}$ card, you might ask another player if he or she has the 60% card. If so, he or she must give it to you, and you repeat your turn. If not, take the top card from the deck, and your turn is over.

The first player to get rid of his or her cards is the winner.

internet connect
Go to ***go.hrw.com*** for a complete set of rules and game pieces.
KEYWORD: MR4 Game8

Technology LAB

Fractions, Decimals, and Percents

internet connect
Lab Resources Online
go.hrw.com
KEYWORD: MR4 TechLab8

You can use your calculator to quickly change between fractions, decimals, and percents.

Activity

1. To write a decimal as a fraction on a graphing calculator, use the **FRAC** command from the **MATH** menu.

   Find the fraction equivalent of 0.225 by pressing 0.225 MATH 1 ENTER.

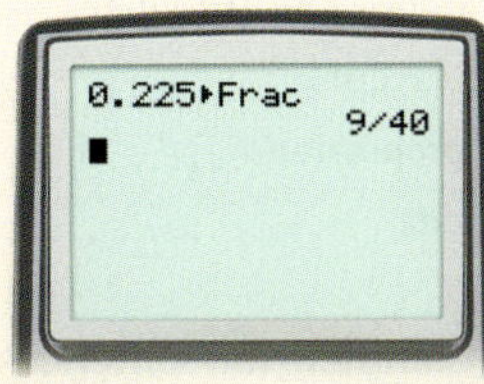

2. To write a percent as a fraction, first write the percent as a fraction whose denominator is 100. Then use the **FRAC** command to find the simplest form of the fraction.

   Find the fraction equivalent of 65% by pressing 65 ÷ 100 MATH 1 ENTER.

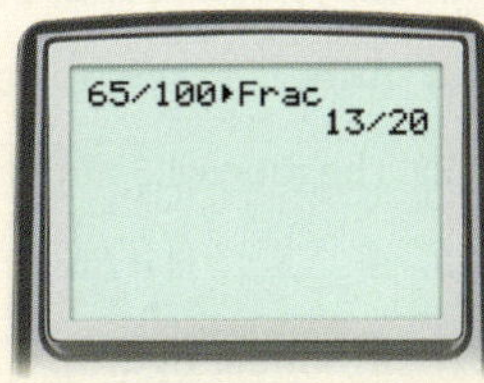

3. To write a fraction as a percent, multiply the fraction by 100.

   Find the percent equivalent of $\frac{11}{25}$ by pressing 11 ÷ 25 × 100 ENTER.

   $\frac{11}{25} = 44\%$

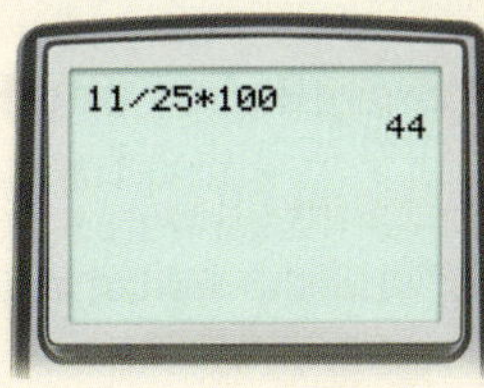

Think and Discuss

1. Use the **FRAC** command on a graphing calculator to find the fraction equivalent of 0.1428571429 by pressing 0.1428571429 MATH 1 ENTER. Describe what happens. **Possible answer: Some graphing calculators cannot convert decimals with more than 4 digits to fractions. Others will accept decimal approximations for common fractions and will return $\frac{1}{7}$ in this case.**

Try This

1. Write each percent as a fraction.
   a. 57.5% $\frac{23}{40}$ **b.** 32.5% $\frac{13}{40}$ **c.** 3.25% $\frac{13}{400}$ **d.** 1.65% $\frac{33}{2,000}$ **e.** 81.25% $\frac{13}{16}$
2. Write each fraction as a percent.
   a. $\frac{7}{40}$ **17.5%** **b.** $\frac{3}{8}$ **37.5%** **c.** $\frac{19}{25}$ **76%** **d.** $\frac{3}{16}$ **18.75%** **e.** $\frac{17}{20}$ **85%**

Technology LAB

Fractions, Decimals, and Percents

Objective: To use a graphing calculator to change between fractions, decimals, and percents

Materials: Graphing calculator

Lab Resources

Technology Lab Activities . . . pp. 35–36

Using the Page

This technology activity shows students how to change between the fraction, decimal, and percent forms of a number on a graphing calculator. Specific keystrokes may vary, depending on the make and model of the graphing calculator used. The keystrokes given are for a TI-83 model. For keystrokes to other models, visit go.hrw.com.

The Think and Discuss problem is meant to point out the limitations of the graphing calculator. While Try This problems 1 and 2 can be done without a graphing calculator, they are meant to help students become familiar with changing between fractions, decimals, and percents on a graphing calculator.

Assessment

1. What keystrokes are needed to convert the fraction $\frac{3}{26}$ to a decimal?
   3 ÷ 26 ENTER
2. What keystrokes are needed to convert the decimal 0.082 to a fraction?
   0.082 MATH 1 ENTER
3. What keystrokes are needed to convert the percent 32.4% to a fraction?
   32.4 ÷ 100 MATH 1 ENTER

Chapter 8

Study Guide and Review

Purpose: *To help students review and practice concepts and skills presented in Chapter 8*

Assessment Resources

Chapter Review

Chapter 8 Resource Book . . pp. 98–100

***Test and Practice Generator* CD-ROM**

Additional review items in both multiple-choice and free-response format may be generated for any objective in Chapter 8.

Answers

1. discount
2. equivalent ratios
3. percent
4. corresponding angles
5. Possible answers: 2:4; 3:6; 6:12
6. 12 oz for \$2.64
7. $n = 9$
8. $n = 3$
9. $n = 14$
10. $n = 2$

Chapter 8

Study Guide and Review

Study Guide and Review

Vocabulary

corresponding angles	405	ratio	392
corresponding sides	405	sales tax	432
discount	432	scale	412
equivalent ratios	392	scale drawing	412
indirect measurement	409	similar	405
percent	418	tip	432
proportion	398	unit rate	393
rate	393		

Complete the sentences below with vocabulary words from the list above. Words may be used more than once.

1. A(n) __?__ is an amount subtracted from the regular price of an item.
2. The ratios 4:3 and 8:6 are __?__ because they name the same comparison.
3. A __?__ is a ratio of a number to 100.
4. In similar figures, __?__ are congruent.

8-1 Ratios and Rates (pp. 392–395)

EXAMPLE

■ **Write the ratio of hearts to diamonds.**

♥ ♥ ♥ ♥

♦ ♦ ♦ ♦ ♦ ♦ ♦ ♦

$\frac{\text{hearts}}{\text{diamonds}} = \frac{4}{8}$

EXERCISES

5. Write three equivalent ratios for 4:8.
6. Which is the better deal—an 8 oz package of pretzels for \$1.92 or a 12 oz package of pretzels for \$2.64?

8-2 Proportions (pp. 398–401)

EXAMPLE

■ **Find the value of n in $\frac{5}{6} = \frac{n}{12}$.**

$6 \cdot n = 5 \cdot 12$ *Cross products are equal.*

$\frac{6n}{6} = \frac{60}{6}$ *Divide both sides by 6.*

$n = 10$

EXERCISES

Find the value of n in each proportion.

7. $\frac{3}{5} = \frac{n}{15}$
8. $\frac{1}{n} = \frac{3}{9}$
9. $\frac{7}{8} = \frac{n}{16}$
10. $\frac{n}{4} = \frac{8}{16}$

8-3 Proportions and Customary Measurement (pp. 402–404)

EXAMPLE

- **How many inches are in 5 feet?**

$\frac{12 \text{ in.}}{1 \text{ ft}} = \frac{x \text{ in.}}{5 \text{ ft}}$ *1 foot is 12 inches.*

$1 \cdot x = 12 \cdot 5$ *Cross products are equal.*

$x = 60$ There are 60 in. in 5 ft.

EXERCISES

11. Marc made 3 gallons of punch. How many cups of punch did he make?

12. Pam spent 150 minutes researching a project. How many hours is this?

8-4 Similar Figures (pp. 405–408)

EXAMPLE

- **The triangles are similar. Find *b*.**

$\frac{1}{32} = \frac{2}{b}$ *Write a proportion.*

$32 \cdot 2 = 1 \cdot b$ *Cross products are equal.*

$64 \text{ cm} = b$

EXERCISES

13. The shapes are similar. Find n and $m\angle A$.

8-5 Indirect Measurement (pp. 409–411)

EXAMPLE

- **A tree casts a 12 ft shadow when a 6 ft man casts a 4 ft shadow. How tall is the tree?**

$\frac{h}{6} = \frac{12}{4}$ *Write a proportion.*

$6 \cdot 12 = 4 \cdot h$ *The cross products are equal.*

$\frac{72}{4} = \frac{4h}{4}$ *Divide both sides by 4.*

$18 = h$ The tree is 18 ft tall.

EXERCISES

14. Find the height of the building.

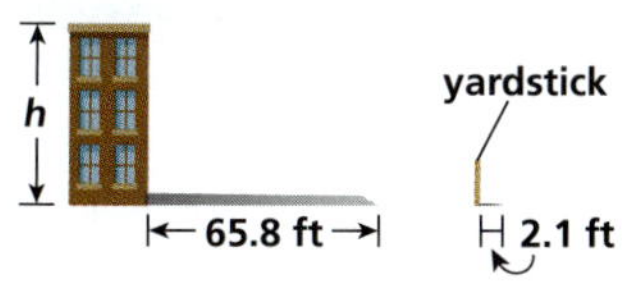

8-6 Scale Drawings and Maps (pp. 412–415)

EXAMPLE

- **Find the actual distance from *A* to *B*.**

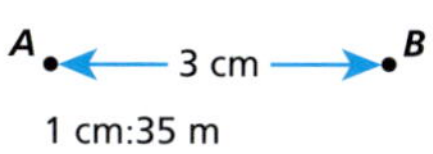

$\frac{1 \text{ cm}}{35 \text{ m}} = \frac{3 \text{ cm}}{x \text{ m}}$ *Write a proportion.*

$35 \cdot 3 = 1 \cdot x$ *Cross products are equal.*

$105 = x$

The actual distance is 105 m.

EXERCISES

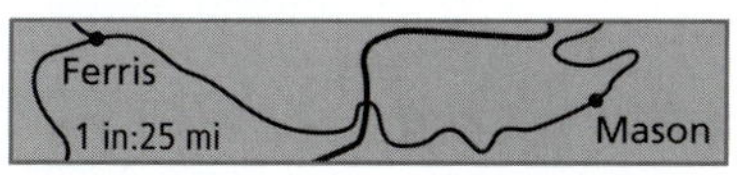

15. Find the actual distance from Ferris to Mason.

16. Renfield is 75 mi from Mason. About how far apart should Renfield and Mason be on the map?

Study Guide and Review

Answers

11. 48 cups

12. $2\frac{1}{2}$, or 2.5

13. n = 11 inches; $m\angle A = 90°$

14. 94 ft

15. 43.75 miles

16. 3 inches

Answers

17. $\frac{3}{4}$

18. $\frac{3}{50}$

19. $\frac{3}{10}$

20. 0.08

21. 0.65

22. 0.2

23. 89.6%

24. 70%

25. 5.7%

26. 12%

27. 70%

28. 25%

29. 87.5%

30. 80%

31. 6.25%

32. 12

33. 5.94

34. 117 tickets

35. about $19.00

36. about $4.35

37. about $1.08

Study Guide and Review

8-7 Percents (pp. 418–421)

EXAMPLE

- **Write 48% as a fraction in simplest form.**

  $48\% = \frac{48}{100}$ $\quad \frac{48 \div 4}{100 \div 4} = \frac{12}{25}$

EXERCISES

Write each as a fraction in simplest form.

17. 75% **18.** 6% **19.** 30%

Write each percent as a decimal.

20. 8% **21.** 65% **22.** 20%

8-8 Percents, Decimals, and Fractions (pp. 422–425)

EXAMPLES

- **Write 0.365 as a percent.**

  0.365 = 36.5% *Multiply by 100.*

- **Write $\frac{3}{5}$ as a percent.**

  $\frac{3 \cdot 20}{5 \cdot 20} = \frac{60}{100} = 60\%$

EXERCISES

Write each decimal or fraction as a percent.

23. 0.896 **24.** 0.70 **25.** 0.057

26. 0.12 **27.** $\frac{7}{10}$ **28.** $\frac{3}{12}$

29. $\frac{7}{8}$ **30.** $\frac{4}{5}$ **31.** $\frac{1}{16}$

8-9 Percent Problems (pp. 426–429)

EXAMPLE

- **Find 30% of 85.**

  30% = 0.30 *Write 30% as a decimal.*

  0.30 · 85 = 25.5 *Multiply.*

EXERCISES

32. Find 25% of 48. **33.** Find 33% of 18.

34. A total of 325 tickets were sold for the school concert, and 36% of these were sold to students. How many tickets were sold to students?

8-10 Using Percents (pp. 432–435)

EXAMPLE

- **A DVD costs $24.98. The sales tax is 5%. About how much is the tax?**

  Step 1: Round $24.98 to $25.

  Step 2: 5% = 5 · 1%
  1% of $25 = 0.01 · $25 = $0.25

  Step 3: 5% = 5 · 1%
  = 5 · $0.25 = $1.25

  The tax is about $1.25.

EXERCISES

35. A sweater is marked 40% off the original price. The original price was $31.75. About how much is the sweater after the discount?

36. Barry and his friends went out for lunch. The bill was $28.68. About how much should they leave for a 15% tip?

37. Ana is purchasing a book for $17.89. The sales tax rate is 6%. About how much will she pay in sales tax?

Chapter Test — Chapter 8

Use the table to write the ratios.

Types of Videos in Richard's Collection			
Comedy	5	Cartoon	7
Drama	6	Exercise	3
Music	3	Documentary	2

1. three equivalent ratios to compare dramas to documentaries **6:2, 12:4, 3:1**
2. documentaries to total videos **2:26**
3. music videos to exercise videos **3:3**
4. Which is a better deal—5 videos for \$29.50 or 3 videos for \$17.25? **3 videos for \$17.25**

Find the value of n in each proportion.

5. $\frac{5}{6} = \frac{n}{24}$ **$n = 20$**
6. $\frac{8}{n} = \frac{12}{3}$ **$n = 2$**
7. $\frac{n}{10} = \frac{3}{6}$ **$n = 5$**
8. $\frac{3}{9} = \frac{4}{n}$ **$n = 12$**
9. A cocoa recipe calls for 4 tbsp cocoa mix to make an 8 oz serving. How many tbsp of cocoa mix are needed to make a 15 oz serving? **$7\frac{1}{2}$ tbsp**
10. Among states that have a shoreline, Pennsylvania has the shortest one. It is 89 miles long. How many yards long is it? **156,640 yards**
11. A 3-foot-tall mailbox casts a shadow that is 1.8 feet long. At the same time, a nearby street lamp casts a shadow that is 12 feet long. How tall is the street lamp? **20 ft**

Use the scale drawing for Exercises 12 and 13.

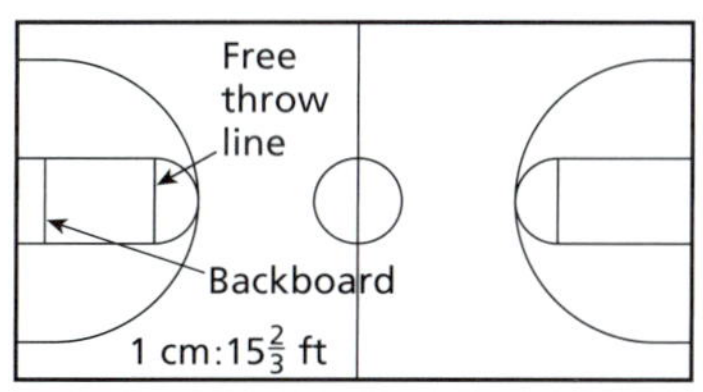

12. The length of the court in the drawing is 6 cm. How long is the actual court? **94 ft**
13. The free-throw line is always 15 feet from the backboard. Is the distance between the backboard and the free-throw line correct in the drawing? Explain.

Write each percent as a fraction in simplest form and as a decimal.

14. 66% **$\frac{33}{50}$, 0.66**
15. 90% **$\frac{9}{10}$, 0.9**
16. 5% **$\frac{1}{20}$, 0.05**
17. 18% **$\frac{9}{50}$, 0.18**

Write each decimal or fraction as a percent.

18. 0.546 **54.6%**
19. 0.092 **9.2%**
20. $\frac{14}{25}$ **56%**
21. $\frac{1}{8}$ **12.5%**

Find each percent.

22. 55% of 218 **119.9**
23. 30% of 310 **93**
24. 25% of 78 **19.5**
25. A bookstore sells paperback books at 20% off the listed price. If Brandy wants to buy a paperback book whose listed price is \$12.95, about how much will she pay for the book after the discount? **about \$10.40**

Chapter Test

Chapter Test — Chapter 8

Purpose: *To assess students' mastery of concepts and skills in Chapter 8*

Assessment Resources

Chapter 8 Tests (Levels A, B, C)
Assessment Resources . . . . . pp. 73–78

***Test and Practice Generator* CD-ROM**

Additional assessment items in both multiple-choice and free-response format may be generated for any objective in Chapter 8.

Answers

13. Yes; the distance from the backboard to the free-throw line is correct. It should be a little less than 1 cm, and it is.

Chapter 8 Performance Assessment

Purpose: *To assess students' understanding of concepts in Chapter 8 and combined problem-solving skills*

Assessment Resources

Performance Assessment
Assessment Resources . . . . p. 118

Performance Assessment Teacher Support
Assessment Resources . . . . p. 117

Answers

1–2. See p. A6.

3. See Level 3 work sample below.

Scoring Rubric for Problem Solving Item 3

Level 3
Accomplishes the purposes of the task.
Student gives clear explanations, shows understanding of mathematical ideas and processes, and computes accurately.

Level 2
Purposes of the task not fully achieved.
Student demonstrates satisfactory but limited understanding of the mathematical ideas and processes.

Level 1
Purposes of the task not accomplished.
Student shows little evidence of understanding the mathematical ideas and processes and makes computational and/or procedural errors.

Chapter 8 Performance Assessment

Show What You Know

Create a portfolio of your work from this chapter. Complete this page and include it with your four best pieces of work from Chapter 8. Choose from your homework or lab assignments, mid-chapter quiz, or any journal entries you have done. Put them together using any design you want. Make your portfolio represent what you consider your best work.

Short Response

1. Find the unit rate for each of the following. Show your work. Which is the best deal?

3 for \$7.80   5 for \$13.25   10 for \$25.00   8 for \$19.84

2. Georgette purchased a sweater on sale for \$19.60. The original price was \$28.00. What percent of the original price was discounted? Show your work.

Extended Problem Solving

3. The small purple rectangle in the drawing is 8 millimeters wide and 18 millimeters tall. The larger purple rectangle is 18 millimeters wide and 25 millimeters tall.

a. Are the two purple rectangles similar? Explain your answer.

b. A third rectangle is similar to the smaller purple rectangle. The width of the third rectangle is 14 millimeters. Let x represent the length of the third rectangle. Write an equation that could be used to find x.

c. Find the length of the third rectangle. Show your work.

Student Work Samples for Item 3

Level 3

a. No. Corresponding angles are congruent, but corresponding sides aren't proportional. $\frac{8}{18} \neq \frac{18}{25}$

b. $\frac{8}{18} = \frac{14}{x}$

c. $8x = 18 \cdot 14$
$8x = 252$
$x = 31.5$ mm

The student demonstrated an understanding of similar figures and proportions.

Level 2

a. No. Side lengths are not proportional.

b. $\frac{8}{18} = \frac{x}{14}$

c. $8(14) = 18x$
$112 = 18x$
$x = 6.22$ mm

The student gave a correct answer for part **a**, but did not correctly set up a proportion to answer parts **b** and **c.**

Level 1

a. no, they don't look the same

b. $18 - 8 = 10$
$14 + 10 = 24$

c. 24 mm

The student gave an incorrect explanation for part **a** and does not understand proportions.

internet connect
State-Specific Test Practice Online
go.hrw.com Keyword: MR4 TestPrep

Standardized Test Prep

Chapter 8

Cumulative Assessment, Chapters 1–8

1. A quadrilateral whose opposite sides are parallel and that has four right angles is a __?__. A
   - (A) Rectangle
   - (B) Rhombus
   - (C) Parallelogram
   - (D) Trapezoid

TIP! TEST TAKING TIP!
Estimate the correct answer before solving. You can often use your estimate to eliminate answer choices.

2. Find 8% of 215. H
   - (F) 1,720
   - (G) 172
   - (H) 17.2
   - (J) 1.72

3. Which is an obtuse angle? B
   - (A) 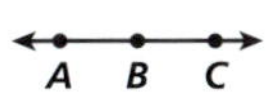

   - (B) 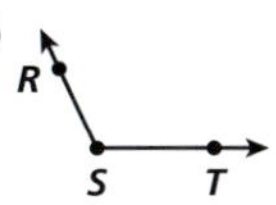

   - (C) 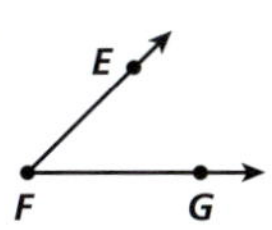

   - (D) 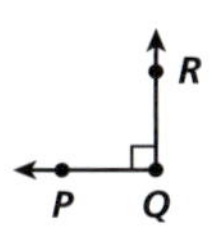

4. $3\frac{2}{7} + 2\frac{2}{3}$ G
   - (F) $5\frac{2}{5}$
   - (G) $5\frac{20}{21}$
   - (H) $5\frac{4}{7}$
   - (J) $5\frac{4}{21}$

5. Which is greatest? D
   - (A) $\frac{3}{8}$
   - (B) 5%
   - (C) $\frac{2}{5}$
   - (D) 0.5

6. Find the GCF of 6, 18, and 30. H
   - (F) 30
   - (G) 18
   - (H) 6
   - (J) 3

7. Which ratio could be used to compare the number of triangles to the number of circles in the pattern? C

   ▲▲▲●▲▲●●▲▲▲●

   - (A) 3:9
   - (B) 3:6
   - (C) 2:1
   - (D) 3:1

8. Use your ruler to help you find the actual distance between *R* and *T*. J
   - (F) $\frac{3}{5}$ mile
   - (G) 3 miles
   - (H) 5 miles
   - (J) 15 miles

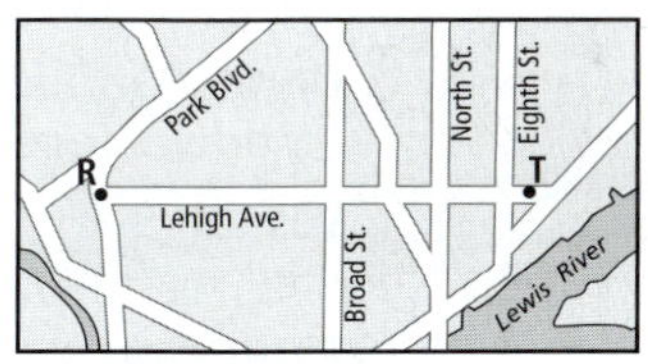

Scale 1 in:10 mi

9. ***SHORT RESPONSE*** Show how to write $\frac{3}{5}$ as a decimal and as a percent.

10. ***SHORT RESPONSE*** Marta purchased 12 gallons of gas for $13.08. Explain the term *unit rate* and give the unit rate for Marta's purchase.

Standardized Test Prep

Standardized Test Prep

Chapter 8

Purpose: *To provide review and practice for Chapter 1–8 and standardized tests*

Assessment Resources

Cumulative Tests (Levels A, B, C)
Assessment Resources . . . . pp. 211–222

State-Specific Test Practice Online
KEYWORD: MR4 TestPrep

Test Prep Doctor

In item 9, the ratio is triangles to circles; because there are more triangles than circles, **A** and **B** are incorrect. There are twice as many triangles as circles, so **C** is incorrect.

Answers

9. $\frac{3}{5} = \frac{60}{100}$

   $\frac{60}{100} = 60\%$

   $\frac{60}{100} = \frac{60 \div 10}{100 \div 10} = \frac{6}{10} = 0.6$

10. A unit rate compares a measured quantity with one unit of another measure.

    $\frac{\$13.08}{12 \text{ gal}} = \frac{\$13.08 \div 12}{12 \text{ gal} \div 12} = \frac{\$1.09}{\text{gal}}$

Chapter 9

Integers

Section 9A

Integers

Lesson 9-1
Understanding Integers

Lesson 9-2
Comparing and Ordering Integers

Lesson 9-3
The Coordinate Plane

Section 9B

Integer Operations

Hands-On Lab 9A
Model Integer Addition

Lesson 9-4
Adding Integers

Hands-On Lab 9B
Model Integer Subtraction

Lesson 9-5
Subtracting Integers

Lesson 9-6
Multiplying Integers

Lesson 9-7
Dividing Integers

Hands-On Lab 9C
Model Integer Equations

Lesson 9-8
Solving Integer Equations

Extension
Integer Exponents

Pacing Guide for 45-Minute Classes

Chapter 9

DAY 116	DAY 117	DAY 118	DAY 119	DAY 120
Lesson 9-1	Lesson 9-2	Lesson 9-3	Mid-Chapter Quiz Hands-On Lab 9A	Lesson 9-4
DAY 121	**DAY 122**	**DAY 123**	**DAY 124**	**DAY 125**
Hands-On Lab 9B	Lesson 9-5	Lesson 9-6	Lesson 9-7	Hands-On Lab 9C
DAY 126	**DAY 127**	**DAY 128**	**DAY 129**	
Lesson 9-8	Extension	Chapter 9 Review	Chapter 9 Assessment	

Pacing Guide for 90-Minute Classes

Chapter 9

DAY 58	DAY 59	DAY 60	DAY 61	DAY 62
Chapter 8 Assessment Lesson 9-1	Lesson 9-2 Lesson 9-3	Mid-Chapter Quiz Hands-On Lab 9A Lesson 9-4	Hands-On Lab 9B Lesson 9-5	Lesson 9-6 Lesson 9-7
DAY 63	**DAY 64**	**DAY 65**		
Hands-On Lab 9C Lesson 9-8	Extension Chapter 9 Review	Chapter 9 Assessment Lesson 10-1		

Across the Series

HARCOURT GRADE 5

- Use a number line to find opposites and absolute values of integers.
- Use a number line to compare and order integers.
- Use counters and a number line to add and subtract integers.

COURSE 1

- **Identify and graph integers, find opposites, and find the absolute value of an integer.**
- **Compare and order integers.**
- **Use ordered pairs to locate and graph points on a coordinate plane.**
- **Use counters and a number line to add and subtract integers.**
- **Multiply and divide integers.**
- **Solve equations containing integers.**
- **Recognize negative exponents by examining patterns.**

COURSE 2

- Compare, order, add, subtract, multiply, and divide integers, and determine absolute value.
- Graph and identify ordered pairs on a coordinate plane.
- Solve one-step equations with integers.

Across the Curriculum

LANGUAGE ARTS LINK

Math: Reading and Writing in the Content Area . . . . pp. 75–82

Focus on Problem Solving
Understand the Problem . . . . SE p. 463

Journal . . . . TE pp. 453, 457, 461, 468, 479, 485

Write About It . . . . SE, last page of each lesson

SOCIAL STUDIES LINK

Social Studies . . . . SE pp. 461, 485

Geography . . . . SE p. 457

History . . . . SE p. 468

SCIENCE LINK

Life Science . . . . SE pp. 479, 485

Earth Science . . . . SE pp. 453, 457, 466, 468, 472, 475

TE = *Teacher's Edition* **SE** = *Student Edition*

Interdisciplinary Bulletin Board

Social Studies

A record high temperature of 58°C (136°F) was recorded in El Azizia, Libya, in 1922. In 1983, in Vostok, Antarctica, a record low temperature of −89°C (−192°F) was recorded. What is the difference in degrees Celsius and in degrees Fahrenheit?

$58 - (-89) = 147°C$; $136 - (-192) = 328°F$

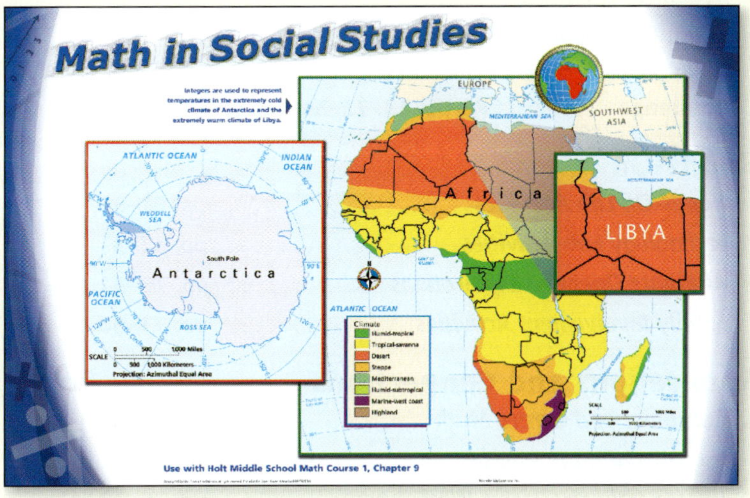

Interdisciplinary posters and worksheets are provided in your resource material.

Chapter 9

Resource Options

Chapter 9 Resource Book

Student Resources

Practice (Levels A, B, C) . . . . pp. 9–11, 18–20, 27–29, 37–39, 47–49, 56–58, 65–67, 75–77

Reteach . . . . pp. 12, 21, 30, 40, 50, 59, 68, 78

Challenge . . . . pp. 13, 22, 31, 41, 51, 60, 69, 79

Problem Solving . . . . pp. 14, 23, 32, 42, 52, 61, 70, 80

Puzzles, Twisters & Teasers . . . . pp. 15, 24, 33, 43, 53, 62, 71, 81

Recording Sheets . . . . pp. 3, 7–8, 17, 26, 36, 45–46, 55, 64, 73–74, 84, 89

Chapter Review . . . . pp. 82–83

Teacher and Parent Resources

Chapter Planning and Pacing Guide . . . . p. 4

Section Planning Guides . . . . pp. 5, 34

Parent Letter . . . . pp. 1–2

Teaching Tools . . . . pp. 87–90

Teacher Support for Chapter Project . . . . p. 85

Transparencies . . . . pp. T1–T31

- Daily Transparencies
- Additional Examples Transparencies
- Teaching Transparencies

Reaching All Learners

English Language Learners

Success for English Language Learners . . . . . . pp. 149–164

Math: Reading and Writing in the Content Area . . . . pp. 75–82

Spanish Homework and Practice . . . . pp. 75–82

Spanish Interactive Study Guide . . . . pp. 75–82

Spanish Family Involvement Activities . . . . pp. 73–80

Multilingual Glossary

Individual Needs

Are You Ready? Intervention and Enrichment . . pp. 25–28, 145–148, 225–228, 285–288, 421–422

Alternate Openers: Explorations . . . . pp. 75–82

Family Involvement Activities . . . . pp. 73–80

Interactive Problem Solving . . . . pp. 75–82

Interactive Study Guide . . . . pp. 75–82

Readiness Activities . . . . pp. 17–18

Math: Reading and Writing in the Content Area . . . . pp. 75–82

Challenge . . . . CRB pp. 13, 22, 31, 41, 51, 60, 69, 79

Hands-On

Hands-On Lab Activities . . . . pp. 81–84

Technology Lab Activities . . . . pp. 37–41

Alternate Openers: Explorations . . . . pp. 75–82

Family Involvement Activities . . . . pp. 73–80

Applications and Connections

Consumer and Career Math . . . . pp. 33–36

Interdisciplinary Posters . . . . Poster 9, TE p. 448B

Interdisciplinary Poster Worksheets . . . . pp. 25–27

Transparencies

Alternate Openers: Explorations . . . . pp. 75–82

Exercise Answers Transparencies

Chapter 9 Resource Book . . . . pp. T1–T31

- Daily Transparencies
- Additional Examples Transparencies
- Teaching Transparencies

Technology

Teacher Resources

***Lesson Presentations* CD-ROM** . . . . Chapter 9

***Test and Practice Generator* CD-ROM** . . . . Chapter 9

***One-Stop Planner* CD-ROM** . . . . Chapter 9

Student Resources

***Are You Ready? Intervention* CD-ROM**
Skills 4, 34, 54, 69

internet connect go.hrw.com

Homework Help Online	**KEYWORD:** MR4 HWHelp9
Math Tools Online	**KEYWORD:** MR4 Tools
Glossary Online	**KEYWORD:** MR4 Glossary
Chapter Project Online	**KEYWORD:** MR4 PSProject9
Chapter Opener Online	**KEYWORD:** MR4 Ch9
CNN Student News™	**KEYWORD:** MR4 CNN9

SE = *Student Edition* **TE** = *Teacher's Edition* **AR** = *Assessment Resources* **CRB** = *Chapter Resource Book* **MK** = *Manipulatives Kit*

Chapter 9

Assessment Options

Assessing Prior Knowledge

Determine whether students have the required prerequisite concepts and skills.

Are You Ready? . . . . SE p. 449
Inventory Test . . . . AR pp. 1–4

Test Preparation

Provide review and practice for chapter and standardized tests.

Standardized Test Prep . . . . SE p. 497
Spiral Review with Test Prep . . . . . SE, last page of each lesson
Study Guide and Review . . . . SE pp. 492–494
Test Prep Tool Kit

Technology

***Test and Practice Generator* CD-ROM**

internet connect

State-Specific Test Practice Online **KEYWORD:** MR4 TestPrep

Performance Assessment

Assess students' understanding of chapter concepts and combined problem-solving skills.

Performance Assessment . . . . SE p. 496
Includes scoring rubric in TE
Performance Assessment . . . . AR p. 120
Performance Assessment Teacher Support . . . . AR p. 119

Portfolio

Portfolio opportunities appear throughout the Student and Teacher's Editions.

Suggested work samples:

Problem Solving Project . . . . TE p. 448
Performance Assessment . . . . SE p. 496
Portfolio Guide . . . . AR p. xxxv
Journal . . . . TE pp. 453, 457, 461, 468, 479, 485
Write About It . . . . SE, last page of each lesson

Daily Assessment

Obtain daily feedback on students' understanding of concepts.

Spiral Review and Test Prep . . . . . . SE, last page of each lesson

Also Available on Transparency in Chapter 9 Resource Book

Warm Up . . . . TE, first page of each lesson
Problem of the Day . . . . TE, first page of each lesson
Lesson Quiz . . . . TE, last page of each lesson

Student Self-Assessment

Have students evaluate their own work.

Group Project Evaluation . . . . AR p. xxxii
Individual Group Member Evaluation . . . . AR p. xxxiii
Portfolio Guide . . . . AR p. xxxv
Journal . . . . TE pp. 453, 457, 461, 468, 479, 485

Formal Assessment

Assess students' mastery of concepts and skills.

Section Quizzes . . . . AR pp. 23, 24
Mid-Chapter Quiz . . . . SE p. 462
Chapter Test . . . . SE p. 495
Chapter Tests (Levels A, B, C) . . . . AR pp. 79–84
Cumulative Tests (Levels A, B, C) . . . . AR pp. 223–234
Standardized Test Prep
Cumulative Assessment . . . . SE p. 497
End-of-Year Test . . . . AR pp. 271–274

Technology

***Test and Practice Generator* CD-ROM**

Make tests electronically. This software includes:

- Dynamic practice for Chapter 9
- Customizable tests
- Multiple-choice items for each objective
- Free-response items for each objective
- Teacher management system

Chapter 9

Chapter 9 Tests

Three levels (A,B,C) of tests are available for each chapter in the *Assessment Resources.*

Level A

Chapter 9 Chapter Test
Form A

Name a positive or negative number to represent each situation.

1. John lost 5 yards on a play. −5
2. Tara has 3 apples. 3

Write the opposite of each integer.

3. 7 −7
4. −10 10
5. −1 1
6. 14 −14

Find the absolute value.

7. |1| = 1
8. |−7| = 7
9. |−11| = 11
10. |6| = 6

Compare. Write < or >.

11. −10 < 13
12. 7 > −1
13. −10 > −11
14. 0 > −2

Order each set of integers from least to greatest.

15. 10, −7, 2 — −7, 2, 10
16. −5, −20, 4, 21 — −20, −5, 4, 21
17. 1, −6, 4, −12 — −12, −6, 1, 4

Name the quadrant and give the coordinates where each point is located.

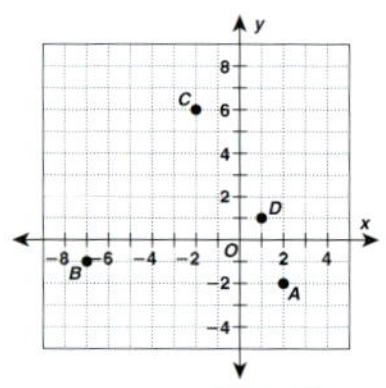

18. *A* — IV (2, −2)
19. *B* — III (−7, −1)
20. *C* — II (−2, 6)
21. *D* — I (1, 1)

Chapter 9 Chapter Test
Form A, continued

Graph each point in the coordinate plane.

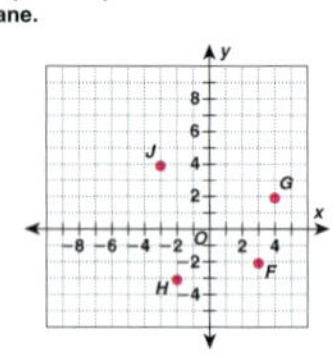

22. *F* (3, −2)
23. *G* (4, 2)
24. *H* (−2, −3)
25. *J* (−3, 4)

Find each sum.

26. 3 + 4 = 7
27. 0 + (−5) = −5
28. −2 + (−1) = −3
29. −7 + 3 = −4

Find each difference.

30. 6 − 2 = 4
31. 5 − (−2) = 7
32. −4 − 5 = −9
33. −3 − (−2) = −1

Find each product.

34. 6 • 3 = 18
35. 4 • (−1) = −4
36. −2 • (−3) = 6
37. −5 • 3 = −15

Find each quotient.

38. 12 ÷ 2 = 6
39. 21 ÷ (−3) = −7
40. −28 ÷ 7 = −4

Evaluate each expression for the given value of the variable.

41. (−10) ÷ *w*; *w* = 2 — −5
42. *m* • 4; *m* = −3 — −12
43. *x* + 6; *x* = −2 — 4
44. *h* − 3; *h* = 1 — −2

Solve each equation.

45. *p* − 7 = 3 — *p* = 10
46. 3*x* = −9 — *x* = −3
47. *q* + 10 = 6 — *q* = −4
48. $\frac{x}{4} = -2$ — *x* = −8
49. *a* + 3 = 3 — *a* = 0
50. −5*y* = 20 — *y* = −4

Level B

Chapter 9 Chapter Test
Form B

Name a positive or negative number to represent each situation.

1. Tony has $104 in his bank account. 104
2. Abby noticed that it is 4° below zero outside. −4

Write the opposite of each integer.

3. +107 −107
4. −12 12
5. +99 −99
6. −1 1

Find the absolute value.

7. |34| = 34
8. |−17| = 17
9. |120| = 120
10. |−653| = 653

Compare. Write < or >.

11. 9 > −7
12. −5 < 4
13. −12 > −16
14. −100 < 1

Order each set of integers from least to greatest.

15. −10, 2, −3 — −10, −3, 2
16. 16, 17, −18, −16 — −18, −16, 16, 17
17. −1, 1, 0, −3, −2 — −3, −2, −1, 0, 1

Name the quadrant and give the coordinates where each point is located.

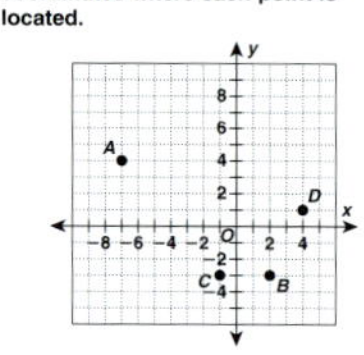

18. *A* — II (−7, 4)
19. *B* — IV (2, −3)
20. *C* — III (−1, −3)
21. *D* — I (4, 1)

Chapter 9 Chapter Test
Form B, continued

Graph each point in the coordinate plane.

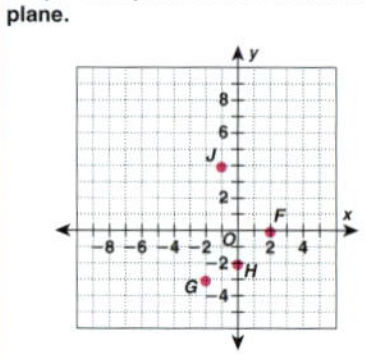

22. *F* (2, 0)
23. *G* (−2, −3)
24. *H* (0, −2)
25. *J* (−1, 4)

Find each sum.

26. 9 + 2 = 11
27. −5 + (−3) = −8
28. 3 + (−12) = −9
29. −17 + 3 = −14

Find each difference.

30. 9 − 2 = 7
31. −3 − 8 = −11
32. 4 − (−11) = 15
33. −3 − (−12) = 9

Find each product.

34. −2 • 6 = −12
35. −9 • (−1) = 9
36. 6 • (−6) = −36
37. −8 • 3 = −24

Find each quotient.

38. 21 ÷ 7 = 3
39. 50 ÷ (−2) = −25
40. −30 ÷ (−5) = 6

Evaluate each expression for the given value of the variable.

41. (−20) ÷ *w*; *w* = 5 — −4
42. *m* • 9; *m* = −3 — −27
43. *x* + 32; *x* = −12 — 20
44. *h* − (−4); *h* = −8 — −4

Solve each equation.

45. *p* + 17 = −2 — *p* = −19
46. −4*x* = −16 — *x* = 4
47. *q* − 15 = −7 — *q* = 8
48. $\frac{x}{3} = -9$ — *x* = −27
49. $-\frac{z}{2} = -1$ — *z* = 2
50. 8*y* = −64 — *y* = −8

Level C

Chapter 9 Chapter Test
Form C

Name a positive or negative number to represent each situation.

1. Haley's checking account is overdrawn by $52. −52
2. Josh lost $3 on a bet and won $8 on his next bet. 5

Write the opposite of the value of each expression.

3. 0 — 0
4. −7 + 2 — 5
5. 5 − 9 — 4
6. −3 + 5 — −2

Find the absolute value.

7. |−7 + 5| = 2
8. |4 + (−2)| = 2
9. |−8 + 0| = 8
10. |−6 − 3| = 9

Compare. Write < or >.

11. 11 − 3 > −1 + 7
12. −5 + 16 > −12 + 1
13. −1 − 5 < −7 + 2
14. −17 − 6 < 14 − 7

Order each set of integers from least to greatest.

15. −363, −365, −362, −370 — −370, −365, −363, −362
16. −23, |−23|, −14, 0, |14| — −23, −14, 0, |14|, |−23|
17. 54,012, 59,361, 54,021, −57,011, −59,631, −57,110 — −59,631, −57,110, −57,011, 54,012, 54,021, 59,361

Name the quadrant and give the coordinates where each point is located.

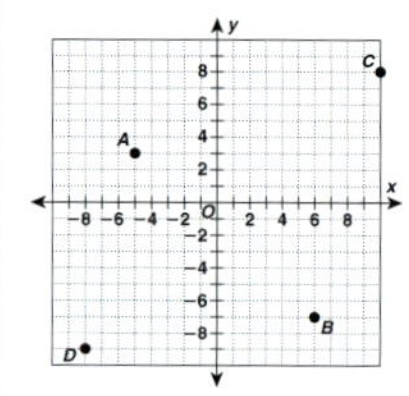

18. *A* — II (−5, 3)
19. *B* — IV (6, −7)
20. *C* — I (10, 8)
21. *D* — III (−8, −9)

Chapter 9 Chapter Test
Form C, continued

Graph each point in the coordinate plane.

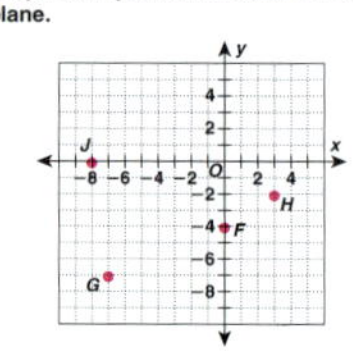

22. *F* (0, −4)
23. *G* (−7, −7)
24. *H* (3, −2)
25. *J* (−8, 0)

Find each sum.

26. |3| + |−11| = 14
27. −16 + (−15) = −31
28. 47 + (−39) = 8
29. |−7| + |3| = 10

Find each difference.

30. −5 − (−15) = 10
31. −32 − 36 = −68
32. −107 − (−204) = 97
33. |−3| − |−2| = 1

Find each product.

34. −11 • 12 = −132
35. |−30| • (−7) = −210
36. −|−8| • (−13) = 104
37. |−5| • |3| = 15

Find each quotient.

38. $\frac{144}{-12}$ = −12
39. −75 ÷ (−25) = 3
40. $\frac{|460|}{|-2|}$ = 230

Evaluate each expression for the given value of the variable.

41. $\frac{-w}{6}$; *w* = −18 — 3
42. −67 + *m*; *m* = −24 — −91
43. *x* − 38; *x* = |−47| — 9
44. 15 • (−*h*); *h* = 5 — −75

Solve each equation.

45. 5*p* = −105 — *p* = −21
46. −20 + *x* = −40 — *x* = −20
47. $\frac{-4}{-q} = 4$ — *q* = 1
48. 164 − *x* = 178 — *x* = −14
49. $\frac{z}{16} = -4$ — *z* = −64
50. |−6|*y* = |102| — *y* = 17

***Test and Practice Generator* CD-ROM**

Create and customize multiple versions of the same tests with corresponding answers for any chosen chapter objectives.

Chapter 9 State and Standardized Test Preparation

Test Taking Skill Builder and Standardized Test Practice are provided for each chapter in the *Test Prep Tool Kit.*

TEST TAKING SKILL BUILDER

Test Taking Strategy — **Multiple Choice Questions—Working Backwards**

Chapter 9

There will be times that you may not know how to solve a multiple choice test question. If the test does not penalize you for guessing, then you need to provide an answer to every question. One method to help you to make an educated guess is to use the answer choices provided and work backwards to solve the question.

Example 1 Find the value of x that makes the equation true.

$x + 4 = -9$

A $x = 4$ **B** $x = -4$ **C** $x = -13$ **D** $x = 13$

Work backwards to find the correct solution.

Try Choice A: $x = 4$; $x + 4 = -9$
$4 + 4 \stackrel{?}{=} -9$ Substitute 4 into the equation.
$8 \neq -9$ $x = 4$ is not the correct solution.

Try Choice B: $x = -4$; $x + 4 = -9$
$-4 + 4 \stackrel{?}{=} -9$ Substitute −4 into the equation.
$0 \neq -9$ $x = -4$ is not the correct solution.

Try Choice C: $x = -13$; $x + 4 = -9$
$-13 + 4 \stackrel{?}{=} -9$ Substitute −13 into the equation.
$-9 = -9$ $x = -13$ is the correct solution.

The correct answer is Choice C.

Example 2 Solve for x. $\frac{x}{3} = -12$

F 36 **G** 4 **H** −4 **I** −36

Try Choice F: $x = 36$
$\frac{36}{3} \stackrel{?}{=} -12$ Substitute 36 into the equation.
$12 \neq -12$ $x = 36$ is not a correct solution.

Try Choice I next since the only thing wrong with Choice F was the negative sign.
$\frac{-36}{3} \stackrel{?}{=} -12$ Substitute −36 into the equation.
$-12 = -12$ $x = -36$ is the correct solution.

The correct answer is Choice I.

Test Taking Strategy

Chapter 9, continued

Exercises Possible answers are given.

1. Find the value of x that makes the equation true. $x + 5 = -4$

   A $x = 9$ **B** $x = 1$ **C** $x = -1$ **D** $x = -9$

   A student worked backwards to answer the question. The following is the student's work.

   Try Choice A: Substitute 9 for x.
   $(9) + 5 \stackrel{?}{=} -4$;
   $14 \neq -4$

   Try Choice C: Substitute −1 for x.
   $(-1) + 5 \stackrel{?}{=} -4$;
   $4 \neq -4$

   Try Choice B: Substitute 1 for x.
   $(1) + 5 \stackrel{?}{=} -4$;
   $6 \neq -4$

   Try Choice D: Substitute −9 for x.
   $(-9) + 5 \stackrel{?}{=} -4$;
   $-4 = -4$

   a. After the student tried Choice A, why did she try another answer choice?
   The value in Choice A did not make the equation a true statement.

   b. The student selected Choice C as the correct answer. Do you agree with her selection? If not, what selection would you have made? Explain.
   No, the student should have chosen Choice D. The only value that makes the equation a true statement is given in Choice D.

2. Find the value of x that makes the equation true.

   $6x + 3 = -21$

   F $x = 6$ **G** $x = 3$ **H** $x = -4$ **I** $x = -6$

   a. Explain how you can work backwards to answer this question.
   Substitute the given value in each answer choice into the equation and see if it results in a true statement.

   b. Once you find one answer choice that is correct, do you need to check the other answer choices? Why?
   No, because there is only one correct answer to this question.

STANDARDIZED TEST PRACTICE

Standardized Test Practice

Chapter 9

Select the best answer for Questions 1–8.

1. Evaluate $w \div (-4)$ for $w = -44$.
   A −11 **C** 10
   B −10 **D** 11

2. The temperature outside is −4°C. It drops 6° at night. What is the new temperature?
   F −10° **G** −2°
   H 2° **I** 10°

3. Mrs. Lee kept a record of the changes in the water level as the tide changed. At what time did the water level increase the most?

Time	Change in Water Level (inches)
6:00 A.M.	−6
8:00 A.M.	−4
10:00 A.M.	0
noon	+3
2:00 P.M.	+6
4:00 P.M.	+11

   A between 6:00 A.M. and 8:00 A.M.
   B between 8:00 A.M. and 10:00 A.M.
   C between noon and 2:00 P.M.
   D between 2:00 P.M. and 4:00 P.M.

4. At the Fall Festival, Mr. Moulter's class had expenses of \$5, \$2, and \$3. The sales at the end of the festival were \$6, \$8, and \$3. What was their total profit?
   F −\$7 **H** \$10
   G \$7 **I** \$17

5. Chelsea and 3 other friends went golfing. Use the table to determine who won. (Hint: Low score wins.)

Player	Score
Chelsea	−2
Bryan	−5
Laurie	2
Todd	5

   A Chelsea
   B Bryan
   C Laurie
   D Todd

6. On a coordinate plane, the point where the axes intersect is called the ______.
   F quadrant
   G origin
   H ordered pair
   I coordinate point

7. In a board game, Trina made the following moves: 10 spaces forward, 3 spaces back, 2 spaces forward, and 5 spaces back. Where is Trina's position from where she started?
   A back 4 spaces
   B back 8 spaces
   C forward 4 spaces
   D forward 6 spaces

8. What number must 12 be multiplied by to get a product of −12?
   F 0 **H** −1
   G 1 **I** −12

Standardized Test Practice

Chapter 9, continued

Gridded Response
Solve the problems. Use the answer sheet to write and grid-in your answer.

9. In one week the temperature rose 16° to reach 10°F. The temperature was originally negative ____°.

10. You spent \$25 on groceries and \$31 at the bookstore. You had \$150 before you began shopping. How much do you have left?

Short Response
Solve the problems. Use the answer sheet to write your answers.

11. Graph and label each point on the coordinate plane.
    A (3, 4)
    B (−2, −4)
    C (0, −3)
    D (2, 0)

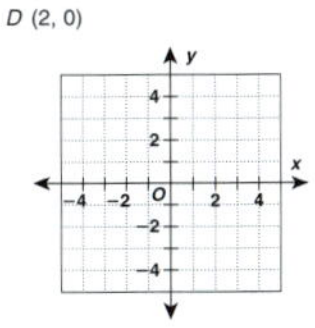

12. A submarine starts at sea level and dives 500 feet, rises 50 feet, but then has to rise 175 feet more to reach periscope depth. After receiving a report from the base, the submarine dives by an amount equal to 3 times its first rise. Write an expression to represent the submarine's change in sea level and then find the final depth of the submarine.

13. Explain how positive and negative integers represent transactions with money.

Extended Response

14. Larry has been investing his money in the stock market. He bought 25 shares of a mutual fund for \$530. He sold the fund for a profit of \$120.

    a. Write and solve an equation to find the price per share at which Larry sold his fund.

    b. The price of the fund dropped \$80 the day after Larry sold it. Write and solve an equation to find the price Larry would have sold it if he had waited until the next day.

    c. Would Larry have earned a profit if he had waited until the next day to sell the stock? Explain.

Sheet

Ⓒ Ⓓ See Lesson 9
Ⓗ Ⓘ See Lesson 9
● Ⓓ See Lesson 9
● Ⓘ See Lesson 9

nse
wers in the space pro

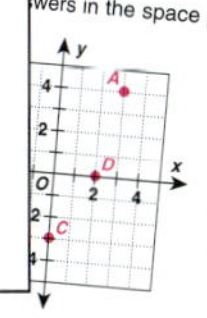

(See Lesson 9-3.)

12. 0 − 500 + 50 + 175 − 3(50); The final depth of the submarine is −4

13. Positive numbers represent money saved or earned and negative numbers represent money spent. (See Less

Extended Response (See Less
Write your answers for Problem 14 on the back of this paper.
See Lesson 9-8.

internet connect go.hrw.com

State-Specific Test Practice Online

KEYWORD: MR4 TestPrep

Test Prep Tool Kit

- Standardized Test Prep Workbook
- Countdown to Testing transparencies
- State Test Prep CD-ROM
- Standardized Test Prep Video

Customized answer sheets give students realistic practice for actual standardized tests.

Chapter 9

Integers

Why Learn This?

Tell students that many everyday situations make positive and negative numbers necessary. For example, elevation above and below sea level can be described by positive and negative integers, as shown in the table. Integers are used in many other fields besides geography. For instance, in finance, positive and negative integers are used to describe monetary changes.

Using Data

To begin the study of this chapter, have students:

- Interpret the meaning of –411, the lowest point in Asia. 411 meters below sea level
- Identify the continents with the lowest and the highest points. Antarctica; Asia
- Identify the elevation of the lowest point in North America. 86 meters below sea level

Chapter 9

TABLE OF CONTENTS

internet connect

Chapter Opener Online
go.hrw.com
KEYWORD: MR4 Ch9

Integers

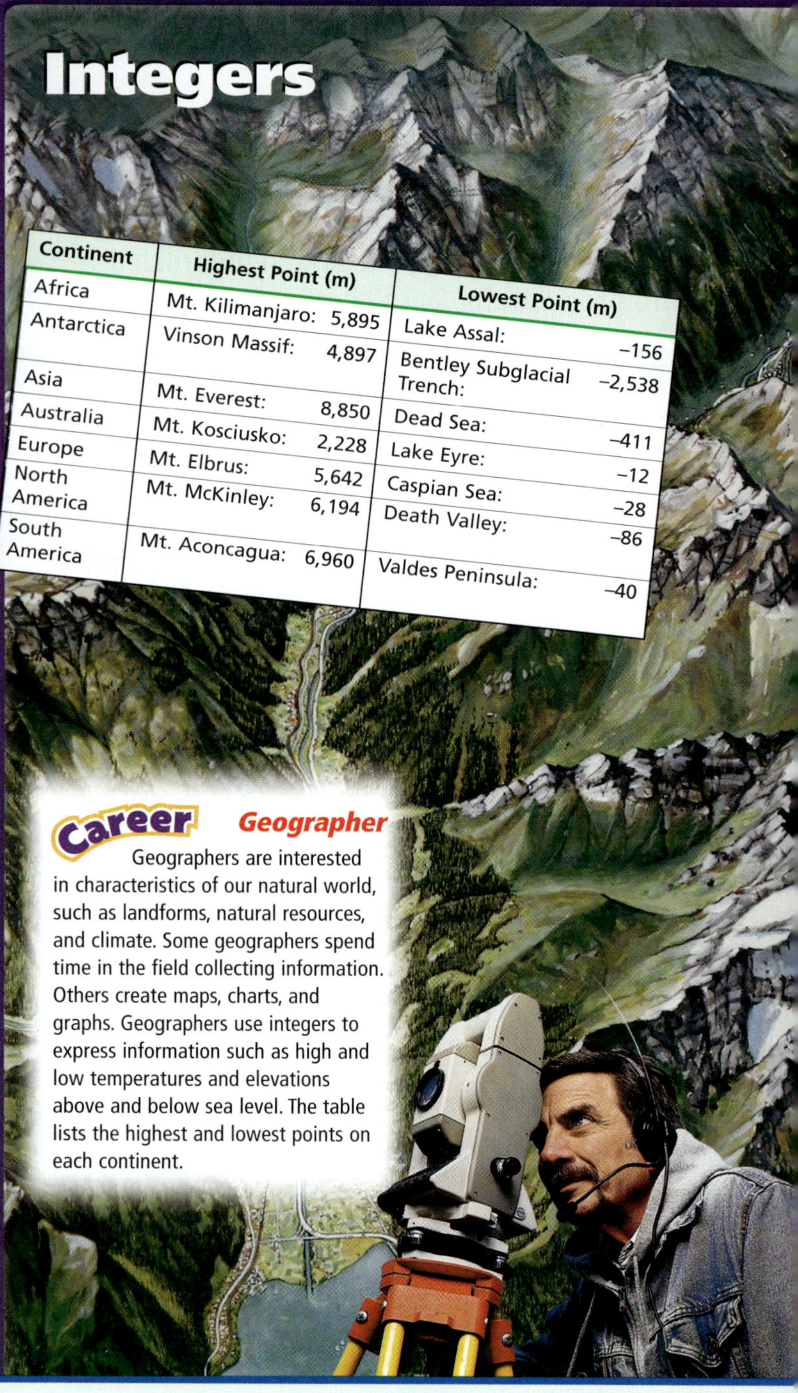

Continent	Highest Point (m)		Lowest Point (m)	
Africa	Mt. Kilimanjaro:	5,895	Lake Assal:	–156
Antarctica	Vinson Massif:	4,897	Bentley Subglacial Trench:	–2,538
Asia	Mt. Everest:	8,850	Dead Sea:	–411
Australia	Mt. Kosciusko:	2,228	Lake Eyre:	–12
Europe	Mt. Elbrus:	5,642	Caspian Sea:	–28
North America	Mt. McKinley:	6,194	Death Valley:	–86
South America	Mt. Aconcagua:	6,960	Valdes Peninsula:	–40

Career ***Geographer***

Geographers are interested in characteristics of our natural world, such as landforms, natural resources, and climate. Some geographers spend time in the field collecting information. Others create maps, charts, and graphs. Geographers use integers to express information such as high and low temperatures and elevations above and below sea level. The table lists the highest and lowest points on each continent.

PROBLEM SOLVING

Problem Solving Project

Social Studies and Earth Science Connection

Purpose: To solve problems using integers

Materials: Continental Ups and Downs worksheet, modeling materials

internet connect

Chapter Project Online: ***go.hrw.com***
KEYWORD: MR4 PSProject9

Understand, Plan, Solve, and Look Back

Have students:

- ✓ Complete the Continental Ups and Downs worksheet to discover how integers can help them understand the topography of the earth.
- ✓ Create a scale model of a continent showing the high and low points. Compare their continent with the models of others, including those who chose different continents.
- ✓ Research the locations of the high and low points in each continent. Ask them if they can identify any patterns.
- ✓ Check students' work.

ARE YOU READY?

Choose the best term from the list to complete each sentence.

1. When you __?__ a numerical expression, you find its value. evaluate
2. __?__ are the set of numbers 0, 1, 2, 3, 4, ... whole numbers
3. A(n) __?__ is an exact location in space. point
4. A(n) __?__ is a mathematical statement that two quantities are equal. equation

equation
evaluate
exponents
less than
point
whole numbers

Complete these exercises to review skills you will need for this chapter.

✔ Compare Whole Numbers

Write <, >, or = to compare the numbers.

5. 9 ■ 2 >	**6.** 4 ■ 5 <	**7.** 8 ■ 1 >	**8.** 3 ■ 3 =
9. 412 ■ 214 >	**10.** 1,076 ■ 1,074 >	**11.** 502 ■ 520 <	**12.** 9,123 ■ 9,001 >

✔ Whole Number Operations

Add or subtract.

13. 8 + 3 11	**14.** 10 − 2 8	**15.** 7 + 6 13	**16.** 15 − 8 7
17. 129 + 30 159	**18.** 32 − 25 7	**19.** 72 + 93 165	**20.** 120 − 87 33

Multiply or divide.

21. 3 · 9 27	**22.** 16 ÷ 4 4	**23.** 6 · 7 42	**24.** 25 ÷ 5 5
25. 119 · 5 595	**26.** 156 ÷ 6 26	**27.** 249 · 44 10,956	**28.** 275 ÷ 25 11

✔ Graph Ordered Pairs

Graph each ordered pair.

29. (1, 3)	**30.** (0, 5)	**31.** (3, 2)	**32.** (4, 0)
33. (6, 4)	**34.** (2, 5)	**35.** (0, 1)	**36.** (1, 0)

✔ Evaluate Expressions

Evaluate $n + 4$ for each value of n.

37. $n = 10$ 14	**38.** $n = 5$ 9	**39.** $n = 16$ 20	**40.** $n = 27$ 31
41. $n = 0$ 4	**42.** $n = 4$ 8	**43.** $n = 19$ 23	**44.** $n = 33$ 37

Assessing Prior Knowledge

INTERVENTION

Diagnose and Prescribe

Evaluate your students' performance on this page to determine whether intervention is necessary or whether enrichment is appropriate. Options that provide instruction, practice, and a check are listed below.

Resources for Are You Ready?

- ***Are You Ready? Intervention and Enrichment***
- **Recording Sheet for Are You Ready?** *Chapter 11 Resource Book . . . . . p. 3*
- ***Are You Ready? Intervention* CD-ROM**

Answers 29–36.

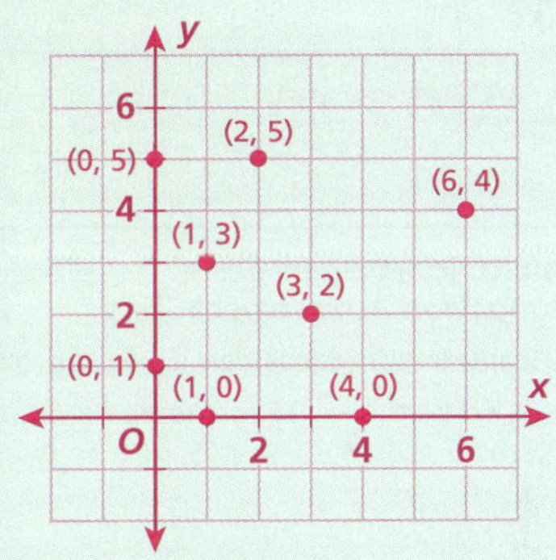

ARE YOU READY?

Were students successful with Are You Ready?

NO INTERVENE

YES ENRICH

✔ Compare Whole Numbers 5–12
Intervention Practice, Skill 4
CD-ROM
Intervention Activities, Skill 4

✔ Whole Number Operations 13–28
Intervention Practice, Skill 34
CD-ROM
Intervention Activities, Skill 34

✔ Graph Ordered Pairs 29–36
Intervention Practice, Skill 69
CD-ROM
Intervention Activities, Skill 69

✔ Evaluate Expressions 37–44
Intervention Practice, Skill 54
CD-ROM
Intervention Activities, Skill 54

Are You Ready? Enrichment, pp. 423–424

Section **9A**

Integers

One-Minute Section Planner

Lesson	Materials	Resources
Lesson 9-1 Understanding Integers **NCTM:** Number and Operations, Communication, Connections **NAEP:** Number Properties 1g ☑ SAT-9 ☑ SAT-10 ☑ ITBS ☑ CTBS ☑ MAT ☑ CAT	**Optional** Two-color Counters *(MK)*	• *Chapter 9 Resource Book,* pp. 6–15 • Daily Transparency T1, CRB • Additional Examples Transparencies T2–T3, CRB • *Alternate Openers: Explorations,* p. 75
Lesson 9-2 Comparing and Ordering Integers **NCTM:** Number and Operations, Problem Solving, Communication **NAEP:** Number Properties 1j ☐ SAT-9 ☑ SAT-10 ☐ ITBS ☐ CTBS ☑ MAT ☑ CAT	**Optional** Large sticky notes	• *Chapter 9 Resource Book,* pp. 16–24 • Daily Transparency T4, CRB • Additional Examples Transparencies T5–T7, CRB • *Alternate Openers: Explorations,* p. 76
Lesson 9-3 The Coordinate Plane **NCTM:** Algebra, Geometry, Communication **NAEP:** Algebra 2c ☑ SAT-9 ☑ SAT-10 ☑ ITBS ☑ CTBS ☑ MAT ☑ CAT	**Optional** Teaching Transparency T9 *(CRB)* State maps Coordinate plane *(CRB p. 87)*	• *Chapter 9 Resource Book,* pp. 25–33 • Daily Transparency T8, CRB • Additional Examples Transparencies T10–T11, CRB • *Alternate Openers: Explorations,* p. 77
Section 9A Assessment		• Mid-Chapter Quiz, SE p. 462 • Section 9A Quiz, AR p. 23 • *Test and Practice Generator* CD-ROM

SAT = *Stanford Achievement Tests* **ITBS** = *Iowa Test of Basic Skills* **CTBS** = *Comprehensive Test of Basic Skills/Terra Nova*
MAT = *Metropolitan Achievement Tests* **CAT** = *California Achievement Test*

NCTM = Complete standards can be found on pages T27–T33. **NAEP** = Complete standards can be found on pages A31–A35.

SE = *Student Edition* **TE** = *Teacher's Edition* **AR** = *Assessment Resources* **CRB** = *Chapter Resource Book* **MK** = *Manipulatives Kit*

Section Overview

Understanding, Comparing, and Ordering Integers — *Lessons 9-1, 9-2*

Why? By including integers, we can solve equations such as $x + 4 = 2$ and evaluate subtraction expressions such as $4 - 9$.

The **integers** are the set of whole numbers and their **opposites**.

Opposites are the same distance from 0, but on opposite sides of 0. The opposite of 0 is itself, 0.

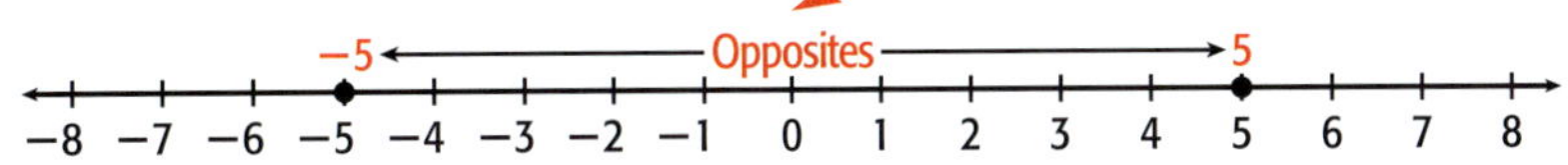

The **absolute value** of an integer is the integer's distance from zero on a number line.

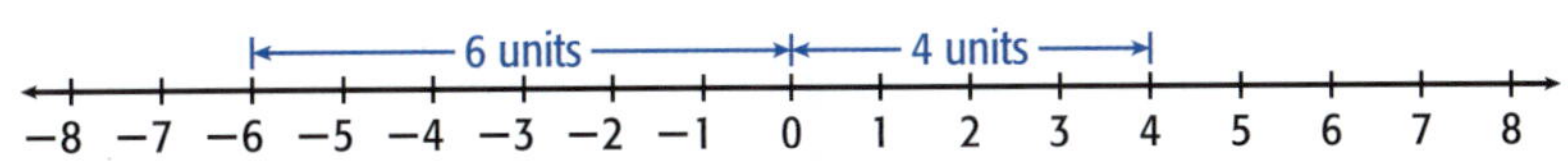

Because distance is always nonnegative, the absolute value of an integer is always nonnegative.

Integers are **ordered from least to greatest** as you move left to right along the number line.

To order the integers −12, 14, −28, 77, 0, 51, and −79 from least to greatest, consider their relative positions on a number line.

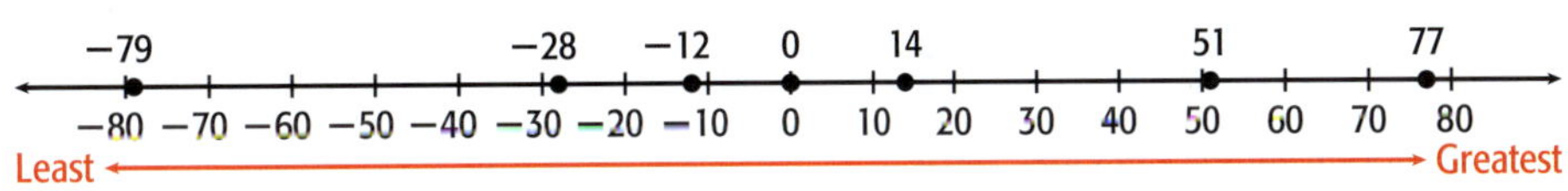

The numbers written from least to greatest are −79, −28, −12, 0, 14, 51, and 77.

Graphing on a Coordinate Plane — *Lesson 9-3*

Why? When a coordinate plane includes integers, the horizontal and vertical axes divide the plane into four quadrants.

Each point on the coordinate plane is identified by an **ordered pair** of numbers: an *x*-coordinate and a *y*-coordinate.

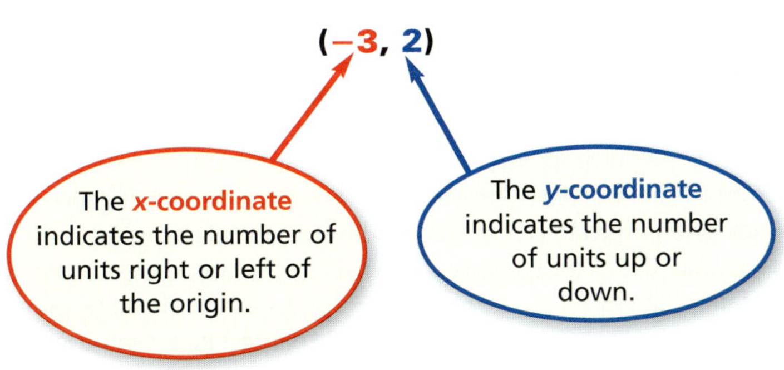

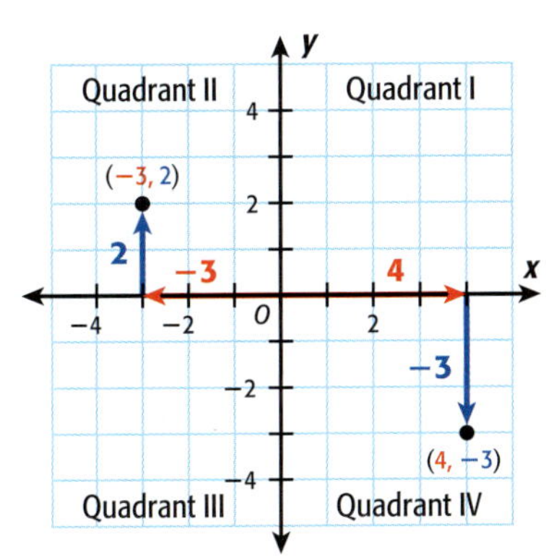

9-1 Organizer

Pacing: Traditional 1 day
Block $\frac{1}{2}$ day

Objective: Students identify and graph integers, find opposites, and find the absolute value of an integer.

Indiana Academic Standards
6.1.1, 6.1.2, 6.7.1, 6.7.5, 6.7.11

Warm Up

Add or subtract.

1. 16 + 25 41 **2.** 84 − 12 72

3. Graph the even numbers from 1 to 10 on a number line.

Problem of the Day

Carlo uses a double-pan balance and three different weights to weigh bird seed. If his weights are 1 lb, 2 lb, and 5 lb, what whole pound amounts is he able to weigh?
1, 2, 3, 5, 6, 7, and 8 lb

Available on Daily Transparency in CRB

Math Fact

Business people traditionally use black ink to represent profits and red ink to represent losses. That is why you are *in the red* when you are losing money.

9-1 Understanding Integers

Learn to identify and graph integers, find opposites, and find the absolute value of an integer.

Vocabulary
positive number
negative number
opposites
integer
absolute value

The highest temperature recorded in the United States is 134°F, in Death Valley, California. The lowest recorded temperature is 80° below 0°F, in Prospect Creek, Alaska.

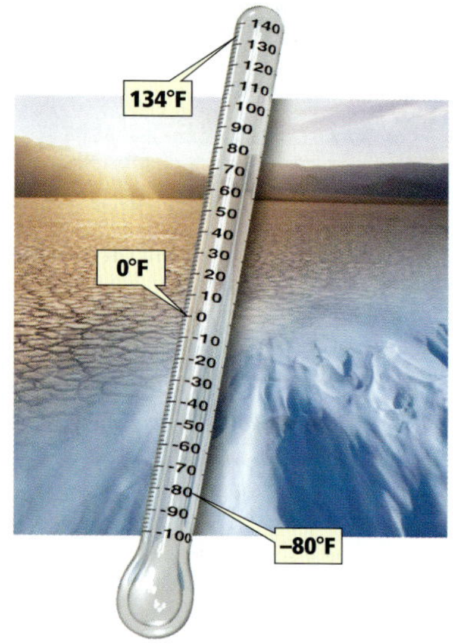

Positive numbers are greater than 0. They may be written with a positive sign (+), but they are usually written without it. So, the highest temperature can be written as +134°F or 134°F.

Negative numbers are less than 0. They are always written with a negative sign (−). So, the lowest temperature is written as −80°F.

EXAMPLE 1 Identifying Positive and Negative Numbers in the Real World

Name a positive or negative number to represent each situation.

A **a gain of 20 yards in football**
Positive numbers can represent *gains* or *increases.*
+20

B **spending $75**
Negative numbers can represent *losses* or *decreases.*
−75

C **10 feet below sea level**
Negative numbers can represent values *below* or *less than* a certain value.
−10

Remember!
The set of whole numbers includes zero and the counting numbers.
{0, 1, 2, 3, 4, ...}

You can graph positive and negative numbers on a number line.

On a number line, **opposites** are the same distance from 0 but on different sides of 0.

Integers are the set of all whole numbers and their opposites.

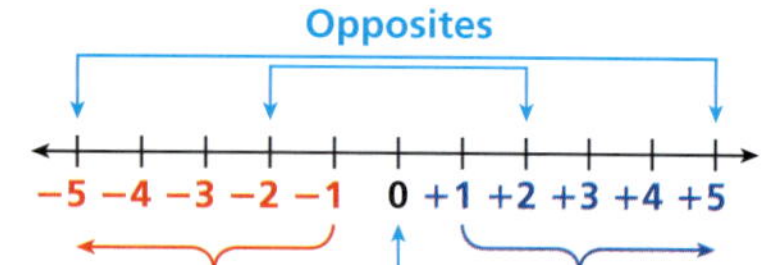

1 Introduce

Alternate Opener

EXPLORATION

9-1 Understanding Integers

Integers can be represented with two-color counters. A red counter represents −1, and a yellow counter represents 1. Each red and yellow pair is called a zero pair.

A zero pair has a value of zero. To find the value on an integer mat, remove zero pairs.

The value on this mat is −1.

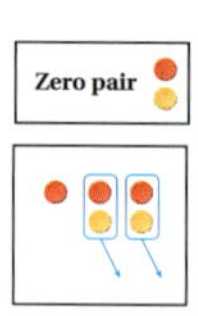

Find the value on each mat.

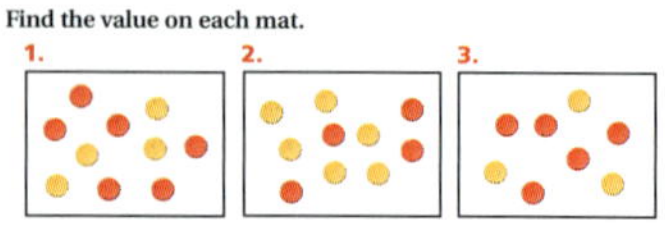

4. Create four different mats that show a value of −1.

Think and Discuss

5. **Discuss** how you can tell whether the value on a mat will be positive or negative.
6. **Describe** the strategies you used to create different mats that show a value of −1.

Motivate

Show students examples of negative numbers. (Collect and display temperatures, bank and credit card statements, stock reports, etc.) Explain that the negative sign (−) indicates that a number is negative, or less than zero. You may want to mention that negative amounts of money often indicate debt, or money owed. (You may use the two-color counters in the Manipulatives Kit.)

Exploration worksheet and answers on Chapter 9 Resource Book pp. 7–8 and 91

2 Teach

Lesson Presentation

Guided Instruction

In this lesson, students learn to identify and graph positive and negative integers, find opposites, and find the absolute value of an integer. First, introduce integer number lines and teach students how to graph positive and negative integers. Then discuss and demonstrate opposites. Next, show absolute values on a number line. Emphasize that absolute value is a distance and that distance can never be negative.

Teaching Tip

Explain to students that the number −2 can be read "negative two" or "the opposite of 2." So −(−2) is "the opposite of −2," or 2.

EXAMPLE 2 Graphing Integers

Graph each integer and its opposite on a number line.

A -4

−5 −4 −3 −2 −1 0 +1 +2 +3 +4 +5

+4 is the same distance from 0 as −4.

B $+3$

−4 −3 −2 −1 0 +1 +2 +3 +4

−3 is the same distance from 0 as +3.

C 0

−4 −3 −2 −1 0 +1 +2 +3 +4

Zero is its own opposite.

The **absolute value** of an integer is its distance from 0 on a number line. The symbol for absolute value is $| \; |$.

Reading Math

Read $|3|$ as "the absolute value of 3."

Read $|-3|$ as "the absolute value of negative 3."

$|-3| = 3 \quad |3| = 3$

←3 units→←3 units→

−5 −4 −3 −2 −1 0 +1 +2 +3 +4 +5

- Absolute values are never negative.
- Opposite integers have the same absolute value.
- $|0| = 0$

EXAMPLE 3 Finding Absolute Value

Use the number line to find the absolute value of each integer.

−10 −9 −8 −7 −6 −5 −4 −3 −2 −1 0 +1 +2 +3 +4 +5 +6 +7 +8 +9 +10

A $|5|$

5 — *5 is 5 units from 0, so $|5| = 5$.*

B $|-7|$

7 — *−7 is 7 units from 0, so $|-7| = 7$.*

Think and Discuss

1. Is –3.2 an integer? Why or why not?
2. **Give** the opposite of 14. What is the opposite of –11?
3. **Name** all the integers with an absolute value of 12.

COMMON ERROR ALERT

Caution students not to confuse *absolute value* with *opposite*. Absolute values are never negative. An opposite can be positive or negative.

Additional Examples

Example 1

Name a positive or negative number to represent each situation.

A. a jet climbing to an altitude of 20,000 feet **+20,000**

B. taking $15 out of the bank **−15**

C. 7 degrees below zero **−7**

Example 2

Graph each integer and its opposite on a number line.

A. +2

−3 −2 −1 0 1 2 3

B. −5

−6 −4 −2 0 2 4 6

C. +1

−3 −2 −1 0 1 2 3

Example 3

Use the number line to find the absolute value of each integer.

−4 −2 0 2 4

A. $|-4|$ 4

B. $|2|$ 2

3 Close

Reaching All Learners
Through Curriculum Integration

Language Arts To help students understand the concept of opposite integers, discuss what *opposite* means in contexts other than math. Have students generate a list of opposite terms (e.g., black and white, up and down, tall and short). Then have students name some pairs of opposite integers (e.g., 1 and −1, −13 and 13).

Summarize

You may wish to have the students state for the class brief definitions of the new vocabulary in the lesson: *positive, negative, opposites, integer*, and *absolute value*. Discuss how the terms relate to each other.

Possible answers: A positive number is greater than 0. A negative number is less than 0. Opposites are positive and negative numbers that are the same distance from 0 on a number line (except for 0, which is its own opposite). Integers are the set of all whole numbers and their opposites. Absolute value is an integer's distance from 0 on a number line.

Answers to Think and Discuss

1. No; only whole numbers and their opposites are integers.
2. a. −14
   b. 11
3. −12 and +12

9-1 PRACTICE & ASSESS

Students may want to refer back to the lesson examples.

Assignment Guide

If you finished Example 1 assign:
Core 1–2, 11–14, 39–43, 46–50
Enriched 11–14, 23–24, 39–43, 46–50

If you finished Example 2 assign:
Core 1–6, 11–18, 24, 26, 39–43, 46–50
Enriched 2, 4, 12–18 even, 23–30, 39–43, 46–50

If you finished Example 3 assign:
Core 1–30, 39–42, 46–50
Enriched 2–10 even, 19–50

Answers

3. −4 −2 0 2 4
4. −3 −2 −1 0 1 2 3
5. −7 −5 −3 −1 1 3 5 7
6. −9 −6 −3 0 3 6 9
15. −5 −3 −1 1 3 5
16. −7 −5 −3 −1 1 3 5 7
17. −4 −2 0 2 4
18. −4 −2 0 2 4

9-1 Exercises

FOR EXTRA PRACTICE see page 654

internet connect
Homework Help Online
go.hrw.com Keyword: MR4 9-1

GUIDED PRACTICE

See Example 1 Name a positive or negative number to represent each situation.

1. an increase of 5 points +5 **2.** a loss of 15 yards −15

See Example 2 Graph each integer and its opposite on a number line.

3. −2 **4.** 1 **5.** −6 **6.** 9

See Example 3 Use the number line to find the absolute value of each integer.

−4 −3 −2 −1 0 +1 +2 +3 +4

7. $|-3|$ 3 **8.** $|4|$ 4 **9.** $|1|$ 1 **10.** $|2|$ 2

INDEPENDENT PRACTICE

See Example 1 Name a positive or negative number to represent each situation.

11. earning $50 +50 **12.** 20° below zero −20
13. 7 feet above sea level +7 **14.** a decrease of 39 points −39

See Example 2 Graph each integer and its opposite on a number line.

15. −5 **16.** 6 **17.** 2 **18.** −3

See Example 3 Use the number line to find the absolute value of each integer.

−5 −4 −3 −2 −1 0 1 2 3 4 5

19. $|-4|$ 4 **20.** $|-1|$ 1 **21.** $|3|$ 3 **22.** $|-2|$ 2

PRACTICE AND PROBLEM SOLVING

Write a situation that each integer could represent. Possible answers:

23. +49 a gain of 49 yards **24.** −83 spending $83 **25.** −7 7 degrees below zero **26.** +15 earning $15

Write the opposite of each integer.

27. −92 +92 **28.** +75 −75 **29.** −25 +25 **30.** 0 0

Find the absolute value.

31. $|419|$ 419 **32.** $|-189|$ 189 **33.** $|723|$ 723 **34.** $|-806|$ 806
35. $|35|$ 35 **36.** $|150|$ 150 **37.** $|-295|$ 295 **38.** $|-80|$ 80

Math Background

The concept of negative numbers can be traced to Hindu mathematicians. They used negative numbers to represent debts, as we do today, and formulated rules for the arithmetic of integers. Their ideas were acquired by Arab mathematicians, who passed the ideas on to European scientists over time.

RETEACH 9-1

LESSON 9-1 **Reteach**
Understanding Integers

−5 −4 −3 −2 −1 0 1 2 3 4 5
negative numbers positive numbers

Opposites are the same distance from zero on a number line, but in different directions. −3 and 3 are opposites because each number is 3 units from zero on a number line.

Integers are the set of all whole numbers and their opposites.

Graph each integer and its opposite on a number line.

−10 −9 −8 −7 −6 −5 −4 −3 −2 −1 0 1 2 3 4 5 6 7 8 9 10

1. −1 −1, 1 2. 9 −9, 9 3. 6 −6, 6 4. −5 −5, 5

The distance an integer is from zero on a number line is the absolute value of the integer.

Find the absolute value of −4.

4 units
−5 −4 −3 −2 −1 0 1 2 3 4 5

−4 is 4 units from zero on a number line, so $|-4| = 4$.

Find the absolute value of each integer.

−10 −9 −8 −7 −6 −5 −4 −3 −2 −1 0 1 2 3 4 5 6 7 8 9 10

5. $|-7|$ 7 6. $|8|$ 8 7. $|-6|$ 6 8. $|10|$ 10
9. $|-5|$ 5 10. $|-3|$ 3 11. $|9|$ 9 12. $|2|$ 2

PRACTICE 9-1

LESSON 9-1 **Practice B**
Understanding Integers

Name a positive or negative number to represent each situation.

1. depositing $85 in a bank account +85 or 85
2. riding an elevator down 3 floors −3
3. the foundation of a house sinking 5 inches −5
4. a temperature of 98° above zero +98 or 98

Graph each integer and its opposite on the number line. Check student's graphs.

−6 −5 −4 −3 −2 −1 0 1 2 3 4 5 6

5. −2 −2 and +2 6. +3 −3 and +3 7. −5 −5 and +5 8. +1 −1 and +1

Use the number line to find the absolute value of each integer.

−6 −5 −4 −3 −2 −1 0 1 2 3 4 5 6

9. $|-1|$ 1 10. $|+4|$ 4 11. $|-3|$ 3 12. $|-5|$ 5
13. $|2|$ 2 14. $|0|$ 0 15. $|6|$ 6 16. $|-6|$ 6

17. The highest point in the state of Louisiana is Driskall Mountain. It rises 535 feet above sea level. Write the elevation of Driskall Mountain as an integer. +535

18. The lowest point in the state of Louisiana is New Orleans. This city's elevation is 8 feet below sea level. Write the elevation of New Orleans as an integer. −8

Earth Science LINK

Underwater vehicles called submersibles are used to explore the deepest parts of the ocean. The *Alvin*, pictured above, was used to photograph and explore the *Titanic* in 1987.

39. ***EARTH SCIENCE*** The Mariana Trench is the deepest part of the Pacific Ocean, reaching a depth of 10,924 meters. Write the depth in meters of the Mariana Trench as an integer. **−10,924**

40. ***SPORTS*** When the Mountain Lions football team returned the kickoff, they gained 45 yards. Write an integer to represent this situation. **+45**

41. ***EARTH SCIENCE*** From June 21 to December 21, most of the United States loses 1 to 2 minutes of daylight each day. But on December 21, most of the country begins to gain 1 to 2 minutes of daylight each day. What integer could you write for a gain of 2 minutes? a loss of 2 minutes? **+2; −2**

42. Match each temperature with the correct point on the thermometer.

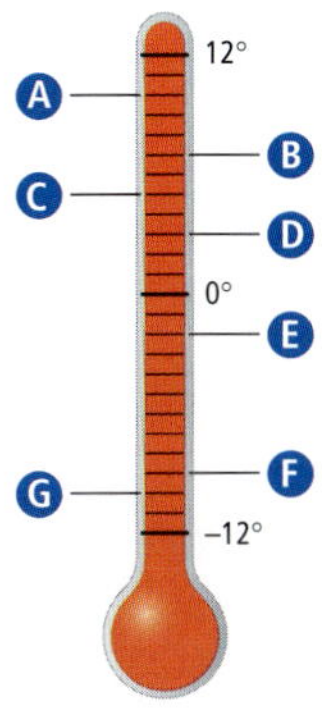

a. −10°F **G**
b. 5°F **C**
c. 10°F **A**
d. −2°F **E**
e. −9°F **F**
f. 7°F **B**
g. 3°F **D**

43. Which cannot be represented by −8? **C**

A a temperature drop of 8°F
B a depth of 8 meters
C a growth of 8 centimeters
D a time 8 years ago

44. ***WRITE ABOUT IT*** Why do opposites have the same absolute value? **Because opposites are the same distance from 0 and the absolute value of an integer is its distance from zero.**

45. ***CHALLENGE*** What is the value of $|19 - 2|$? of $|2 - 11|$? **17; 9**

Spiral Review

Order the numbers in each set from least to greatest. (Lesson 1-1)

46. 1,945; 2,649; 1,495; 2,609
1,495 < 1,945 < 2,609 < 2,649

47. 17,465; 17,509; 17,395; 17,498
17,395 < 17,465 < 17,498 < 17,509

Solve each equation. (Lesson 2-4)

48. $n + 10 = 25$ **15**

49. $28 = 4 + x$ **24**

50. **TEST PREP** Which fraction is **not** in simplest form? (Lesson 4-5) **C**

A $\frac{2}{3}$ **B** $\frac{17}{31}$ **C** $\frac{3}{9}$ **D** $\frac{1}{8}$

COMMON ERROR ALERT

Listen for students using the term *plus* when they mean "positive" and *minus* when they mean "negative." Have the class use the terms *positive* and *negative* correctly.

Journal

Have students write about a situation in which they might use negative numbers.

Test Prep Doctor

For Exercise 50, remind students to read carefully. Point out that they need to identify the fraction that is **not** in simplest form. Students can then eliminate answer choices that are in simplest form (fractions in which the numerator and denominator do not share a common factor other than 1). Because 3 and 9 have a common factor, 3, the correct answer must be **C.**

CHALLENGE 9-1

LESSON 9-1 **Challenge**
Boiling Up and Freezing Down

Draw a line on the thermometer to show each given temperature record. Then label each state's name at its correct line on the thermometer.

1. The lowest temperature in California was recorded on January 20, 1937. It was −45°F.
2. The highest temperature in North Carolina was recorded on August 21, 1983. It was 110°F.
3. The lowest temperature in Georgia was recorded on January 27, 1940. It was −17°F.
4. The highest temperature in Maine was recorded on July 10, 1911. It was 105°F.
5. The lowest temperature in Missouri was recorded on February 13, 1905. It was −40°F.
6. The lowest temperature in Texas was recorded on February 8, 1933. It was −23°F.
7. The highest temperature in Maryland was recorded on July 10, 1936. It was 109°F.
8. The lowest temperature in Massachusetts was recorded on January 12, 1981. It was −35°F.
9. The highest temperature in Oklahoma was recorded on June 27, 1994. It was 120°F.
10. The highest temperature in Mississippi was recorded on July 29, 1930. It was 115°F.

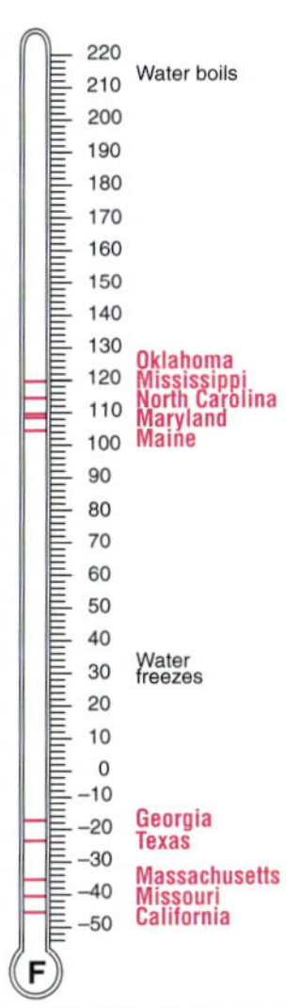

PROBLEM SOLVING 9-1

LESSON 9-1 **Problem Solving**
Understanding Integers

Write the correct answer.

1. The element mercury is used in thermometers because it expands as it is heated. Mercury melts at 38°F below zero. Write this temperature as an integer.
−38°F

2. Denver, Colorado, earned the nickname "Mile High City" because of its elevation of 5,280 feet above sea level. Write Denver's elevation as an integer in feet and miles.
+5,280 feet or +1 mile

3. The lowest temperature recorded in San Francisco was 20°F. Buffalo's lowest recorded temperature was the opposite of San Francisco's. What was Buffalo's record temperature?
−20°F

4. Greenland holds the record for the lowest temperature recorded on Earth. The absolute value of that temperature in degrees Fahrenheit is 65. What is Earth's lowest recorded temperature?
−65°F

5. In 1960, explorers on the submarine *Trieste 2* set the world record for the deepest dive. The ship reached 35,814 feet below sea level. Write this depth as an integer.
−35,814 feet

6. In 1960, Joseph W. Kittinger, Jr., set the record for the highest parachute jump. He jumped from an air balloon at 102,800 feet above sea level. Write this altitude as an integer.
+102,800 feet

Circle the letter of the correct answer.

7. Which situation cannot be represented by the integer −10?
A an elevation of 10 feet below sea level
(B) a temperature increase of 10°F
C a golf score of 10 under par
D a bank withdrawal of $10

8. Paper was invented in China one thousand, nine hundred years ago. Which integer represents this date?
F 1,900
G 900
(H) −1,900
J −1,000

9. The elevation of the Dead Sea is about 1,310 feet below sea level. Which integer represents this elevation?
(A) −1,310
B −131
C 131
D 1,310

10. The quarterback had a 10-yard loss and then a 25-yard gain. Which integer represents a 25-yard gain?
F −25
G −10
(H) 25
J 10

Lesson Quiz

Name a positive or negative number to represent each situation.

1. saving $15 **+15**
2. 12 feet below sea level **−12**
3. What is the opposite of −6? **6**
4. What is the absolute value of −12? **12**
5. When the Swanton Bulldogs football team passed the football, they gained 25 yards. Write an integer to represent this situation. **+25**

Available on Daily Transparency in CRB

9-2 Organizer

Pacing: Traditional 1 day
Block $\frac{1}{2}$ day

Objective: Students compare and order integers.

Indiana Academic Standards
6.1.3, 6.7.1, 6.7.3, 6.7.5, 6.7.11

Warm Up

Compare. Write <, >, or =.

1. 8,426 ▇ 8,246 >
2. 9,625 ▇ 6,852 >
3. 2,071 ▇ 2,171 <
4. 2,250 ▇ 2,250 =

Problem of the Day

Four friends are waiting in line at the amusement park. Jenna is in front of Kyle. Kyle is behind Gary and in front of Maggie. Gary is first. In what order are they waiting? Gary, Jenna, Kyle, Maggie

Available on Daily Transparency in CRB

Math Humor

Why was 6 afraid of 7?
Because seven eight nine!

9-2 Comparing and Ordering Integers

Learn to compare and order integers.

The table shows three golfers' scores from a 2001 tournament.

Player	Score
David Berganio	+6
Sergio Garcia	−16
Tiger Woods	−4

In golf, the player with the lowest score wins the game. You can compare integers to find the winner of the tournament.

Sergio Garcia

EXAMPLE 1 Comparing Integers

Use the number line to compare each pair of integers. Write < or >.

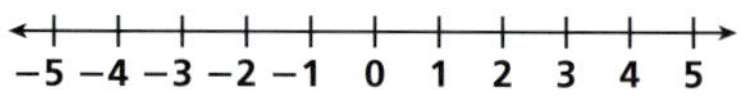

Remember!
Numbers on a number line increase in value as you move from left to right.

A −4 ▇ 2
$-4 < 2$ *−4 is to the left of 2 on the number line.*

B −3 ▇ −5
$-3 > -5$ *−3 is to the right of −5 on the number line.*

C 0 ▇ −4
$0 > -4$ *0 is to the right of −4 on the number line.*

EXAMPLE 2 Ordering Integers

Order the integers in each set from least to greatest.

A 4, −2, 1

Graph the integers on the same number line.

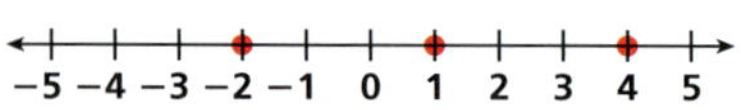

Then read the numbers from left to right: −2, 1, 4.

1 Introduce

Alternate Opener

EXPLORATION

9-2 Comparing and Ordering Integers

1. The completed table shows the average January temperatures in degrees Fahrenheit and degrees Celsius for some U.S. cities. Complete the other table by ordering the cities from warmest to coolest.

	°F	°C
Juneau, AK	24	−4
Phoenix, AZ	54	12
Atlanta, GA	41	5
Des Moines, IA	19	−7
Bismarck, ND	9	−13
Houston, TX	50	10
Boston, MA	29	−2
Kansas City, MO	26	−3

Source: *Statistical Abstract of the United States*

Warmest

	°F	°C

Coolest

Boston is colder than Houston, since $29 < 50$ in degrees Fahrenheit and $-2 < 10$ in degrees Celsius.

Houston is warmer than Boston, since $50 > 29$ in degrees Fahrenheit and $10 > -2$ in degrees Celsius.

2. Use inequality symbols to compare the Kansas City temperature in degrees Celsius with each of the other temperatures in degrees Celsius.

Think and Discuss

3. **Describe** your method for ordering the cities from warmest to coolest.

Motivate

Draw and label a number line from 10 to 50, with intervals of 5. Have students graph 20, 25, 45, 35, 30, 10, and 40 on the number line. Teach students to compare pairs of numbers using the terms *greater than* and *less than* and their symbols (e.g., 10 is less than 40; 25 > 20). Discuss how a number line helps to identify which number is greater.

Exploration worksheet and answers on Chapter 9 Resource Book pp. 17 and 93

2 Teach

Lesson Presentation

Guided Instruction

In this lesson, students learn to compare and order integers. First, teach students to compare integers by graphing the integers on a number line. Then teach how sets of 3, 4, or more integers can be ordered by graphing the set on a number line. Emphasize that the greater integer is always to the right of the lesser integer.

Teaching Tip Remind students that in a comparison the inequality sign always "points to" the lesser of two numbers.

Order the integers in each set from least to greatest.

B −2, 0, 2, −5

Graph the integers on the same number line.

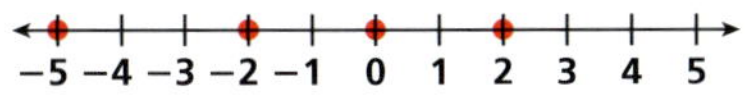

Then read the numbers from left to right: −5, −2, 0, 2.

EXAMPLE 3 PROBLEM SOLVING APPLICATION

PROBLEM SOLVING

At a 2001 golf tournament, David Berganio scored +6, Sergio Garcia scored −16, and Tiger Woods scored −4. One of these three players was the winner of the tournament. Who won the tournament?

1 Understand the Problem

The **answer** will be the player with the *lowest* score.
List the **important information:**

- David Berganio scored +6.
- Sergio Garcia scored −16.
- Tiger Woods scored −4.

2 Make a Plan

You can draw a diagram to order the scores from least to greatest.

3 Solve

Draw a number line and graph each player's score on it.

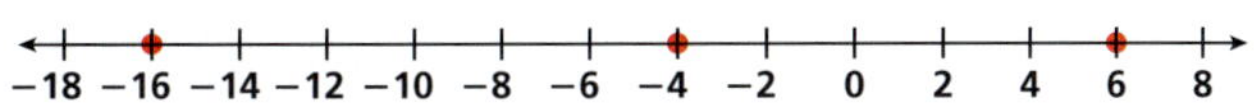

Sergio Garcia's score, –16, is farthest to the left, so it is the lowest score. Sergio Garcia won this tournament.

4 Look Back

Negative integers are always less than positive integers, so David Berganio cannot be the winner. Since Sergio Garcia's score of −16 is less than Tiger Woods's score of −4, Sergio Garcia won.

Sports LINK

In golf, *par* is the standard number of strokes needed to hit the ball into the hole. A score of −12 means "12 strokes under par." A score of +2 means "2 strokes over par."

go.hrw.com
KEYWORD: MR4 Golf
CNN Student News.

Think and Discuss

1. **Tell** which is greater, a negative or a positive integer. Explain.
2. **Tell** which is greater, 0 or a negative integer. Explain.
3. **Explain** how to tell which of two negative integers is greater.

COMMON ERROR ALERT

Some students may think that the integer with the greater absolute value is greater. Remind students to use a number line for comparing integers.

Additional Examples

Example 1

Use the number line to compare each pair of integers. Write $<$ or $>$.

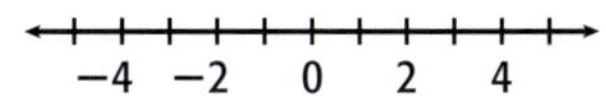

A. −2 ▢ 2 $<$
B. 3 ▢ −5 $>$
C. −1 ▢ −4 $>$

Example 2

Order the integers in each set from least to greatest.

A. −2, 3, −1 **−2, −1, 3**
B. 4, −3, −5, 2 **−5, −3, 2, 4**

Example 3

In a golf match, Craig scored +2, Cameron scored +3, and Rob scored −1. Who won the golf match?
Rob won the golf match.

Example 2B note: Point out to students that no matter how many integers are in the set, the method for ordering them is the same: graph the integers on the same number line, then read them from left to right.

3 Close

Reaching All Learners
Through Hands-On Learning

For a hands-on approach, have students write the integers being compared on sticky notes and arrange them on a large number line on the board. Have them explain why they placed the numbers in the position they did, and encourage them to rearrange the notes if placed incorrectly. Chips may also be used.

Summarize

Have students list two integers that are greater than −10 and two integers that are less than −10. Then have students list the integers, including −10, in order from least to greatest, and discuss how a number line shows the integers' relationships to −10.

Possible answers: Greater than −10: −5, 2; less than −10: −11, −20;
least to greatest: −20, −11, −10, −5, 2;
A number line shows that −20 and −11 are less than −10 because they are to the left of −10 on the number line and that −5 and 2 are greater than −10 because they are to the right of −10 on the number line.

Answers to Think and Discuss

1. A positive integer; Possible answer: Positive integers are located to the right of negative integers on a number line. Because numbers on a number line increase in value from left to right, positive integers are always greater than negative integers.
2. 0; It is greater than all negative integers because it is to the right of all negative integers on a number line.
3. Possible answer: Graph the integers on a number line. The integer that is farther to the right is the greater of the two integers.

9-2 PRACTICE & ASSESS

Students may want to refer back to the lesson examples.

Assignment Guide

If you finished Example 1 assign:
Core 1–3, 8–10, 18–20, 36–47
Enriched 18–23, 31–33, 36–47

If you finished Example 2 assign:
Core 1–6, 8–16, 30, 36–47
Enriched 20–47

If you finished Example 3 assign:
Core 1–17, 30–32, 36–47
Enriched 16–47

Answers

27. −25, −4, 15, 31
28. −73, −67, −10, 20, 82
29. −50, −27, 38, 42, 69

9-2 Exercises

FOR EXTRA PRACTICE see page 654

internet connect
Homework Help Online
go.hrw.com Keyword: MR4 9-2

GUIDED PRACTICE

See Example 1

Use the number line to compare each pair of integers. Write < or >.

−5 −4 −3 −2 −1 0 1 2 3 4 5

1. −4 ■ −5 >
2. −2 ■ 0 <
3. −1 ■ 3 <

See Example 2

Order the integers in each set from least to greatest.

4. 9, 0, −2 −2, 0, 9
5. 7, −4, 3, −5 −5, −4, 3, 7
6. 8, −6, −1, 10 −6, −1, 8, 10

See Example 3

7. At what time was the temperature the lowest? 3:30 A.M.

Time	Temperature (°F)
10:00 P.M.	1
Midnight	−4
3:30 A.M.	−6
6:00 A.M.	1

INDEPENDENT PRACTICE

See Example 1

Use the number line to compare each pair of integers. Write < or >.

−5 −4 −3 −2 −1 0 1 2 3 4 5

8. 0 ■ 2 <
9. 4 ■ −4 >
10. −3 ■ −1 <

See Example 2

Order the integers in each set from least to greatest.

11. 11, −6, −3 −6, −3, 11
12. 15, −8, 7 −8, 7, 15
13. 5, −12, 0, 1 −12, 0, 1, 5
14. −9, 13, −1, −16 −16, −9, −1, 13
15. 24, −6, 7, −10, 4 −10, −6, 4, 7, 24
16. 22, 0, −19, 8, −3 −19, −3, 0, 8, 22

See Example 3

17. The table shows the depths of the world's three largest oceans. Which ocean is the deepest? Pacific

Ocean	Depth (ft)
Pacific	−36,200
Atlantic	−30,246
Indian	−24,442

PRACTICE AND PROBLEM SOLVING

Compare. Write < or >.

18. −30 ■ 25 <
19. 0 ■ −49 >
20. −16 ■ −51 >
21. −64 ■ −15 <
22. 77 ■ 300 <
23. −28 ■ 1 <

Order the integers in each set from least to greatest.

24. −39, 14, 21 −39, 14, 21
25. −18, −9, −31 −31, −18, −9
26. 0, −26, 43, −12 −26, −12, 0, 43
27. 15, −25, −4, 31
28. −67, 82, −73, −10, 20
29. 42, −27, 69, −50, 38

Math Background

There is an interesting connection between integers and their absolute values. While $-16 < -2$ is true for the integers, $|-16| > |-2|$ is true for their absolute values. Therefore, if a and b are negative and $a < b$, then $|a| > |b|$. A similar connection with opposites exists. The negative integer with the lesser opposite is greater. This is also true for all negative real numbers.

RETEACH 9-2

LESSON 9-2 **Reteach**
Comparing and Ordering Integers

You can use a number line to compare and order integers.

As you move right on a number line, the values of the integers increase. As you move left on a number line, the values of the integers decrease.

Compare −4 and 2.

−5 −4 −3 −2 −1 0 1 2 3 4 5

−4 is to the left of 2, so −4 < 2.

Compare the integers. Write < or >.

1. −5 −4 −3 −2 −1 0 1 2 3 4 5

1 > −4

2. −5 −4 −3 −2 −1 0 1 2 3 4 5

−5 < −2

3. −3 < 2
4. −1 > −4
5. 5 > 0
6. −2 < 3

Order −3, 4 and −1 from least to greatest.

−5 −4 −3 −2 −1 0 1 2 3 4 5

List the numbers as they appear from left to right.

The integers in order from least to greatest are −3, −1, 4.

Order the integers from least to greatest.

7. −2, −5, −1 −5, −2, −1
8. 0, −5, 5 −5, 0, 5
9. −4, 2, −3 −4, −3, 2
10. 3, −1, −4 −4, −1, 3

PRACTICE 9-2

LESSON 9-2 **Practice B**
Comparing and Ordering Integers

Use the number line to compare each pair of integers. Write < or >.

−10 −9 −8 −7 −6 −5 −4 −3 −2 −1 0 1 2 3 4 5 6 7 8 9 10

1. 10 > −2
2. 0 < 3
3. −5 < 0
4. −7 < 6
5. −6 > −9
6. −8 > −10

Order the integers in each set from least to greatest.

7. 5, −2, 6 −2, 5, 6
8. 0, 9, −3 −3, 0, 9
9. −1, 6, 1 −1, 1, 6
10. −8, −9, 9 −9, −8, 9
11. 15, 1, −5 −5, 1, 15
12. −4, −7, −2 −7, −4, −2

Order the integers in each set from greatest to least.

13. 8, −6, 4 8, 4, −6
14. −2, 1, 2 2, 1, −2
15. 0, 7, −8 7, 0, −8
16. −1, 1, 0 1, 0, −1
17. −12, 2, 1 2, 1, −12
18. −10, −12, −11 −10, −11, −12

19. The lowest point in the Potomac River is 1 foot above sea level. The lowest point in the Colorado River is 70 feet above sea level. The lowest point in the Delaware River is sea level. Write the names of these three rivers in order from the lowest to the highest elevation.

Delaware River, Potomac River, Colorado River

20. The lowest recorded temperature in Alabama was 27°F below zero. In Florida, the lowest recorded temperature was 2°F below zero. The lowest temperature ever recorded in Hawaii was 12°F above zero. Write the names of these three states in order from the highest to the lowest recorded temperatures.

Hawaii, Florida, Alabama

30. Which set of integers is written in order from greatest to least? **B**

A 0, −4, −3, −1 **B** 9, −9, −10, −15

C 2, −4, 8, −16 **D** −8, −7, −6, −5

31. ***EARTH SCIENCE*** The normal high temperature in January for Barrow, Alaska, is −7°F. The normal high temperature in January for Los Angeles is 68°F. Compare the two temperatures using < or >. **68 > −7 or −7 < 68**

32. ***GEOGRAPHY*** The table shows elevations for several natural features. Write the features in order from the least elevation to the greatest elevation. **San Augustin Cave, Dead Sea, Mt. Rainier, Kilimanjaro, Mt. Everest**

Elevations of Natural Features

Feature	Elevation
Mt. Everest	29,022 ft
Mt. Rainier	14,410 ft
Kilimanjaro	19,000 ft
San Augustin Cave	−2,189 ft
Dead Sea	−1,296 ft

33. ***WHAT'S THE ERROR?*** Your classmate says that 0 < −91. Explain why this is incorrect. **all neg. numbers are less than 0**

34. ***WRITE ABOUT IT*** Explain how you would order from least to greatest three numbers that include a positive number, a negative number, and zero. **Negative numbers are less than 0. Positive numbers are greater than 0. So, the order would be negative, 0, positive.**

35. ***CHALLENGE*** Write < or >. |−4| ▒ |−3| **>**

Spiral Review

Evaluate each expression. (Lesson 1-4)

36. 2 • (17 − 5) ÷ 4 **6** **37.** (11 + 7) ÷ 3 − 5 **1** **38.** 3^2 ÷ 3 + (5 • 3) **18**

39. 8 • (14 ÷ 2) − 2^3 **48** **40.** 8^2 + (36 ÷ 4) − 5 **68** **41.** 4^2 ÷ (5 − 3) − 8 **0**

Divide. (Lesson 3-7)

42. 1.40 ÷ 2 **0.70** **43.** 3.3 ÷ 3 **1.1** **44.** 0.85 ÷ 5 **0.17** **45.** 0.375 ÷ 3 **0.125**

46. **TEST PREP** An isosceles triangle has at least __?__ congruent sides. (Lesson 7-5) **C**

A zero **B** one **C** two **D** three

47. **TEST PREP** What is 35% written as a decimal? (Lesson 8-8) **J**

F 35.0 **G** 3.5 **H** 0.035 **J** 0.35

COMMON ERROR ALERT

Point out to students that in Exercise 30 they are asked to find the set of integers ordered from greatest to least.

Journal

Have students think of and write about real-world situations in which sets of positive and negative integers are used.

Test Prep Doctor

For Exercise 47, students should remember that percent means "per hundred" and that 35% means "35 parts out of 100." If students answered **F,** then they assumed that a percent has the same value as the whole number written before the percent symbol. If they answered **G** or **H,** then they either incorrectly divided 35 by 100 or divided 35 by 10 or 1,000.

CHALLENGE 9-2

LESSON 9-2 **Challenge**
Integer Maze

During winter, bears and many other animals hibernate, or go into a sleeplike state. When they do so, their body temperatures greatly decrease. How much can their temperatures change during hibernation?

Write the following integers along the maze so they are increasing from start to finish. The shaded box in the maze will have the answer to the question.

5, −14, −39, −3, 61, −60, −23, −72, −48, −11, 100, −45, 10, −57, −1, −29, −64, −37, 0, −65, 74, 98, −28, −7, −63, −49, −21, −54, 27, 53, −9, −32, −16, 35, −30, −18, −52, 86, −46, 42, −56, −41, −22, −43, 19

PROBLEM SOLVING 9-2

LESSON 9-2 **Problem Solving**
Comparing and Ordering Integers

Use the table below to answer each question.

Continental Elevation Facts

Continent	Highest Point	Elevation (ft) above sea level	Lowest Point	Elevation (ft) below sea level
Africa	Mount Kilimanjaro	19,340	Lake Assal	−512
Antarctica	Vinson Massif	16,066	Bentley Subglacial Trench	−8,327
Asia	Mount Everest	29,035	Dead Sea	−1,349
Australia	Mount Kosciusko	7,310	Lake Eyre	−52
Europe	Mount Elbrus	18,510	Caspian Sea	−92
North America	Mount McKinley	20,320	Death Valley	−282
South America	Mount Aconcagua	22,834	Valdes Peninsula	−131

1. What is the highest point on Earth? What is its elevation?
Mount Everest; 29,035 feet above sea level

2. What is the lowest point on Earth? What is its elevation?
Bentley Subglacial Trench; 8,327 feet below sea level

3. Which point on Earth is higher, Mount Elbrus or Mount Kilimanjaro?
Mount Kilimanjaro

4. Which point on Earth is lower, the Caspian Sea or Lake Eyre?
Caspian Sea

Circle the letter of the correct answer.

5. Which continent has a higher elevation than North America?
A Antarctica
(B) South America
C Europe
D Australia

6. Which continent has a lower elevation than Africa?
F Australia
G Europe
(H) Asia
J South America

7. Write the continents in order by their highest points, from highest elevation to lowest elevation.
Asia, South America, North America, Africa, Europe, Antarctica, Australia

Lesson Quiz

Order the integers in each set from least to greatest.

1. −3, 7, 4 **−3, 4, 7**

2. −11, 2, 5, −15 **−15, −11, 2, 5**

Compare. Write <, >, or =.

3. −3 ▒ −4 **>**

4. −12 ▒ −10 **<**

5. A location in Carlsbad Caverns is 752 ft below sea level, and another location is 910 ft below sea level. Which location is closer to sea level? **the location at −752 feet**

Available on Daily Transparency in CRB

9-3 Organizer

Pacing: Traditional 1 day
Block $\frac{1}{2}$ day

Objective: Students locate and graph points on a coordinate plane.

Indiana Academic Standards
6.3.7, 6.7.1, 6.7.5, 6.7.11

Warm Up

Use the number line to compare each pair of integers. Write < or >.

−10 −5 0 5 10

1. 7 ▢ −7 > **2.** −8 ▢ −3 <
3. 0 ▢ −4 > **4.** −2 ▢ −5 >

Problem of the Day

While delivering pizza, Christian drove 4 miles south, 6 miles west, 2 miles north, 8 miles east, and then 2 miles north. How far is Christian from where he started? 2 miles

Available on Daily Transparency in CRB

Math Fact

The numbers assigned to each point on a coordinate plane are called *Cartesian coordinates*. This name is in honor of French mathematician René Descartes.

9-3 The Coordinate Plane

Learn to locate and graph points on the coordinate plane.

Vocabulary
coordinate plane
axes
x-axis
y-axis
quadrants
origin
coordinates
x-coordinate
y-coordinate

A **coordinate plane** is formed by two number lines in a plane that intersect at right angles. The point of intersection is the zero on each number line.

- The two number lines are called the **axes**.
- The horizontal axis is called the **x-axis**.
- The vertical axis is called the **y-axis**.
- The two axes divide the coordinate plane into four **quadrants**.
- The point where the axes intersect is called the **origin**.

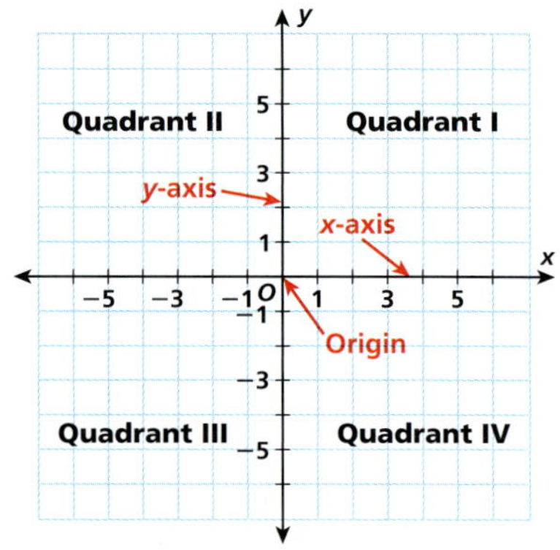

EXAMPLE 1 Identifying Quadrants

Name the quadrant where each point is located.

A *M*
Quadrant I

B *J*
Quadrant IV

C *R*
x-axis
no quadrant

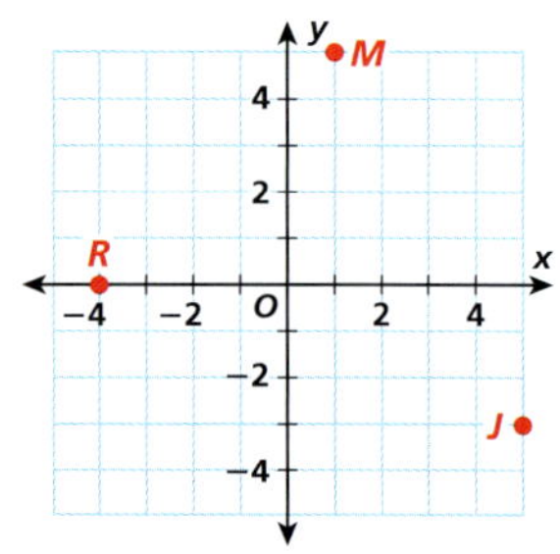

Helpful Hint
Points on the axes are not in any quadrant.

An ordered pair gives the location of a point on a coordinate plane. The first number tells how far to move right (positive) or left (negative) from the origin. The second number tells how far to move up (positive) or down (negative).

The numbers in an ordered pair are called **coordinates**. The first number is called the ***x*-coordinate**. The second number is called the ***y*-coordinate**.

The ordered pair for the origin is (0, 0).

1 Introduce

Alternate Opener

EXPLORATION

9-3 The Coordinate Plane

On the *coordinate plane* below, the color of the first number in each ordered pair matches the color of the *x-axis*, and the color of the second number in each ordered pair matches the color of the *y-axis*.

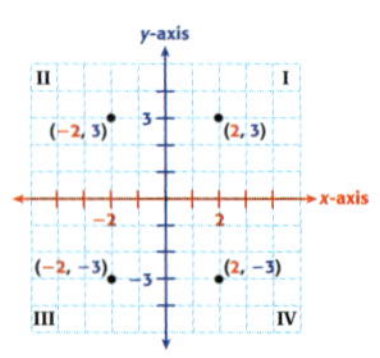

1. The four points graphed are labeled with their ordered pairs. How are these four ordered pairs alike? How are they different?
2. Plot the points (3, 4), (−3, 4), (−3, −4), and (3, −4) on the same coordinate plane.

Think and Discuss

3. **Describe** what each number in an ordered pair tells you.
4. **Explain** how a negative sign indicates which direction to move when plotting points on a coordinate plane.

Motivate

To introduce students to the four-quadrant coordinate plane, give the following two examples. (1) In order to reach a second-floor window, you move the ladder directly below the window and then climb up. (2) To dig for treasure, you walk to the spot and then dig down to the treasure. Both examples involve a horizontal move followed by a vertical move.

Exploration worksheet and answers on Chapter 9 Resource Book pp. 26 and 95

2 Teach

Lesson Presentation

Guided Instruction

In this lesson, students learn to locate and graph points on a coordinate plane. First, identify the four quadrants of the plane (Teaching Transparency T9, CRB) and teach students that the first number in an ordered pair tells where to move horizontally from the origin, and the second number tells where to move vertically. Then teach students to give the coordinates of points in different quadrants. Finally, teach them to graph points, given the coordinates.

Teaching Tip
Remind students to always begin at the origin (0, 0) when finding coordinates or graphing points on a coordinate plane.

EXAMPLE 2 Locating Points on a Coordinate Plane

Give the coordinates of each point.

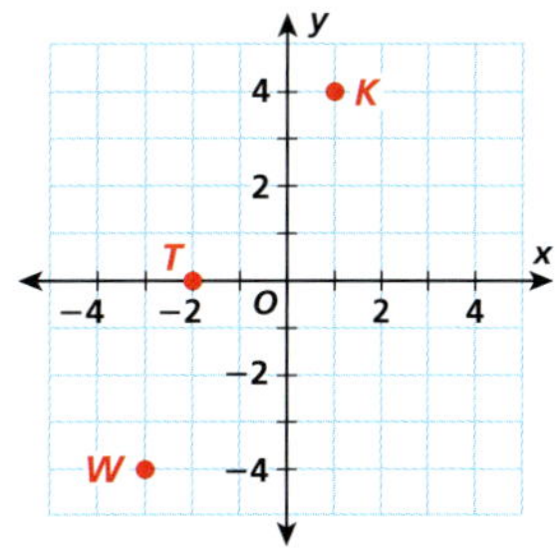

A K

From the origin, K is 1 unit right and 4 units up.

(1, 4)

B T

From the origin, T is 2 units left on the x-axis.

(−2, 0)

C W

From the origin, W is 3 units left and 4 units down.

(−3, −4)

EXAMPLE 3 Graphing Points on a Coordinate Plane

Graph each point on a coordinate plane.

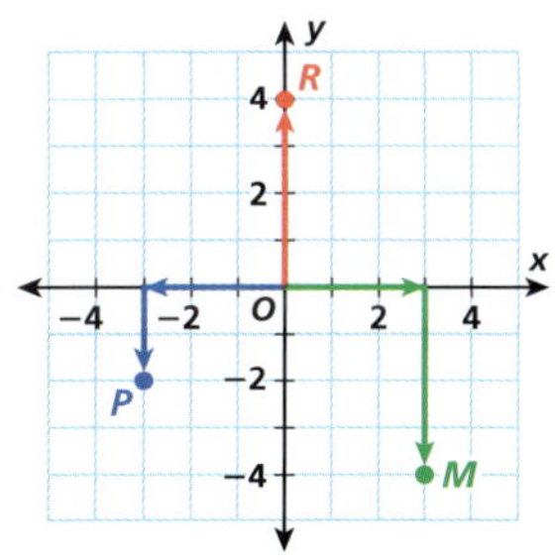

A $P(-3, -2)$

From the origin, move 3 units left and 2 units down.

B $R(0, 4)$

From the origin, move 4 units up.

C $M(3, -4)$

From the origin, move 3 units right and 4 units down.

Think and Discuss

1. Which number in an ordered pair tells you how far to move left or right from the origin? up or down?
2. **Describe** how graphing the point (5, 4) is similar to graphing the point (5, −4). How is it different?
3. **Tell** why it is important to start at the origin when you are graphing points.

COMMON ERROR ALERT

Some students may forget what the first and second numbers in an ordered pair mean. To help them, remind students of the two examples used to motivate them in the introduction.

Additional Examples

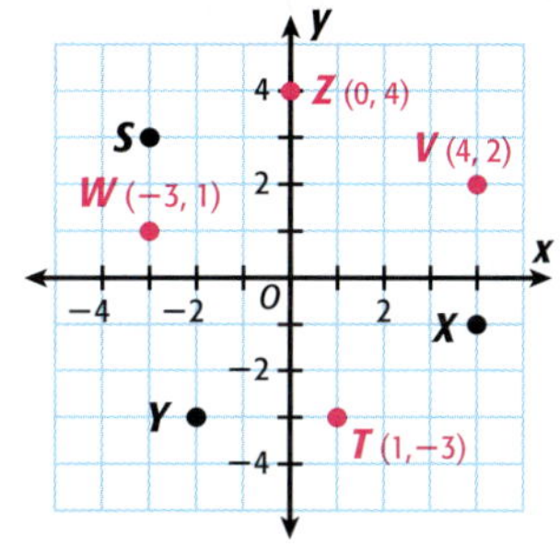

Example 1

Name the quadrant where each point is located.

A. X Quadrant IV

B. Y Quadrant III

C. S Quadrant II

Example 2

Give the coordinates of each point.

A. X (4, −1) **B.** Y (−2, −3) **C.** S (−3, 3)

Example 3

Graph each point on a coordinate plane. (See answers above.)

A. $V(4, 2)$ **B.** $W(-3, 1)$

C. $Z(0, 4)$ **D.** $T(1, -3)$

3 Close

Reaching All Learners

Through Curriculum Integration

Geography Have students work in groups to draw coordinate grid lines on maps of your state. Instruct students to draw the x- and y-axes through the state capital and the other lines at $\frac{1}{2}$-inch increments above, below, to the left, and to the right of the axes. Have the students label the grid lines, beginning with the axes, with the appropriate numbers, and give coordinates for various cities and towns on the map.

Summarize

To review new vocabulary, have the students draw coordinate planes and label the *x*-axis, *y*-axis, quadrants I–IV, and the origin. Then have students graph a point in each quadrant and identify each point's *x*-coordinate and *y*-coordinate.

Answers to Think and Discuss

1. **a.** the first number
   b. the second number
2. Possible answer: Move 5 units to the right on the *x*-axis for both points. But you move up 4 units for (5, 4) and down 4 units for (5, −4).
3. Possible answer: because coordinates describe a location in relation to the orgin

9-3 PRACTICE & ASSESS

9-3 Exercises

FOR EXTRA PRACTICE
see page 654

Internet connect
Homework Help Online
go.hrw.com Keyword: MR4 9-3

Students may want to refer back to the lesson examples.

GUIDED PRACTICE

Use the coordinate plane for Exercises 1–4.

See Example 1 Name the quadrant where each point is located.

1. T **III**
2. U **no quadrant**

See Example 2 Give the coordinates of each point.

3. A **(−2, 0)**
4. B **(1, 2)**

See Example 3 Graph each point on a coordinate plane.

5. $E(4, 2)$
6. $F(-1, -4)$

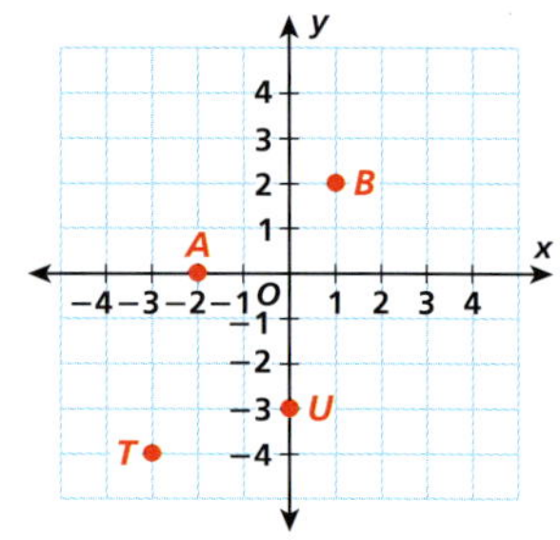

INDEPENDENT PRACTICE

Use the coordinate plane for Exercises 7–14.

See Example 1 Name the quadrant where each point is located.

7. Q **no quadrant**
8. X **II**
9. Y **I**
10. Z **II**

See Example 2 Give the coordinates of each point.

11. P **(3, −4)**
12. R **(−2, 4)**
13. T **(4, 4)**
14. H **(−2, −3)**

See Example 3 Graph each point on a coordinate plane.

15. $L(0, 3)$
16. $M(3, -3)$
17. $V(-4, 3)$
18. $N(-2, -1)$

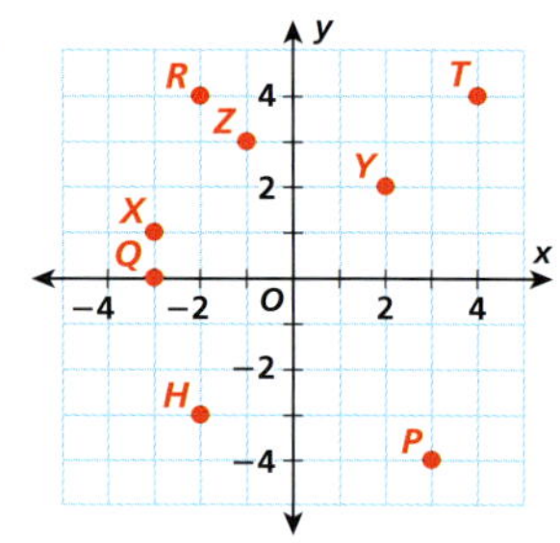

PRACTICE AND PROBLEM SOLVING

Name the quadrant where each ordered pair is located.

19. (3, −1) **IV**
20. (2, 1) **I**
21. (−2, 3) **II**

Graph each ordered pair.

22. (0, −5)
23. (−4, −4)
24. (5, 0)
25. (−2, 2)
26. (0, −3)
27. (1, −4)

Assignment Guide

If you finished Example 1 assign:
Core 1–2, 7–10, 34–40
Enriched 7–10, 19–21, 34–40

If you finished Example 2 assign:
Core 1–4, 7–14, 19–21, 34–40
Enriched 7–14, 19–21, 29, 30, 33, 34–40

If you finished Example 3 assign:
Core 1–18, 28–31, 34–40
Enriched 7–40

Answers

5 and 6.

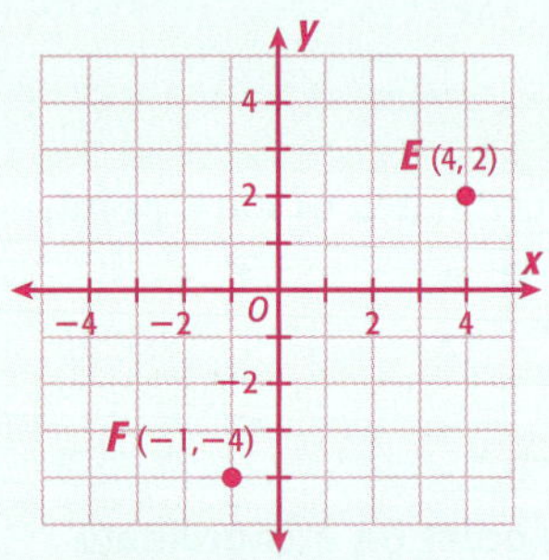

15–18.

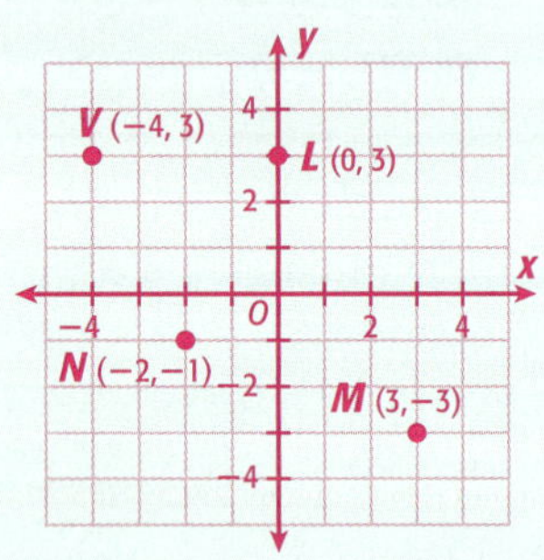

22–27.

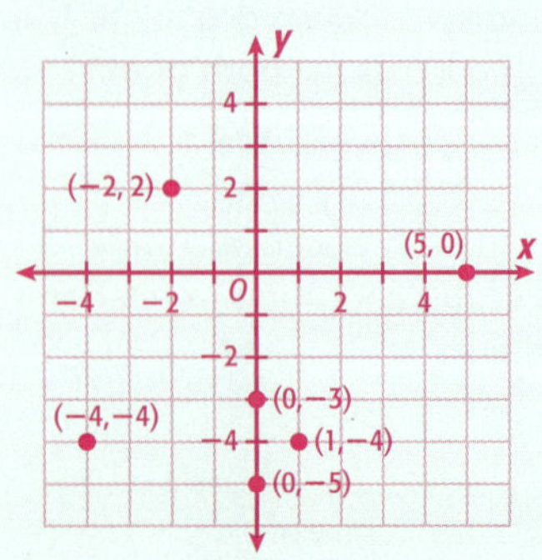

RETEACH 9-3

LESSON 9-3 **Reteach**
The Coordinate Plane

The coordinate plane is divided into four quadrants. They are numbered I, II, III, and IV.

An ordered pair tells the location of a point. The x-coordinate tells you how far to move right or left. The y-coordinate tells you how far to move up or down.

The coordinates of point A are (5, −4) because it is 5 units to the right of the origin and 4 units down. It is located in quadrant IV.

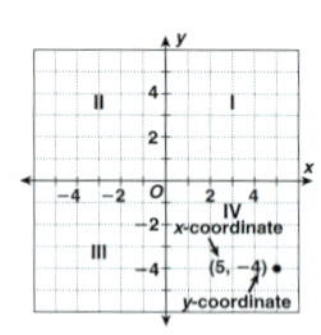

Name the quadrant or axis where each point is located. Then give the coordinates of each point.

1. D II; (−4, 1)
2. Q III; (−3, −6)
3. F x-axis; (6, 0)
4. T I; (5, 5)
5. P IV; (4, −5)
6. W y-axis; (0, −4)

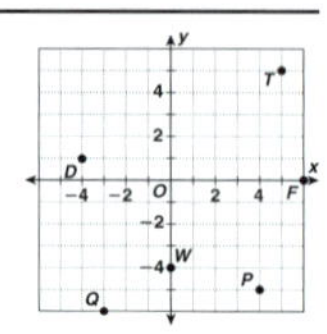

Graph each point in the coordinate plane.

7. B (−1, 6)
8. R (8, −5)
9. V (−3, −4)
10. Z (0, −6)

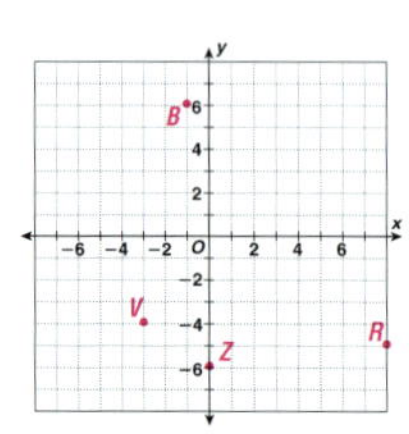

PRACTICE 9-3

LESSON 9-3 **Practice B**
The Coordinate Plane

Use the coordinate plane to answer questions 1–12.

Name the quadrant where each point is located.

1. D II
2. P II
3. Y III
4. B IV
5. C I
6. X I

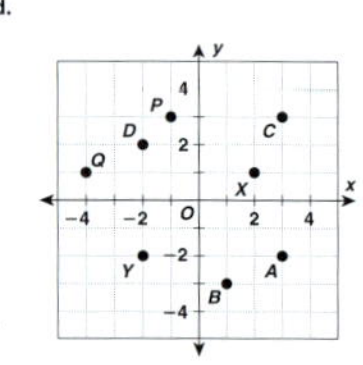

Give the coordinates of each point.

7. X (2, 1)
8. A (3, −2)
9. P (−1, 3)
10. Q (−4, 1)
11. Y (−2, −2)
12. D (−2, 2)

Graph each point on the coordinate plane at right. Check students' graphs.

13. X (3, 1)
14. T (−2, −2)
15. C (1, −2)
16. U (0, −3)
17. P (2, 0)
18. A (−4, −1)

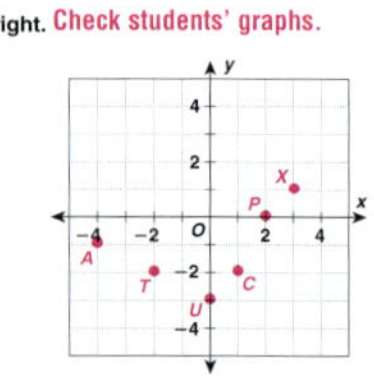

19. Does every point lie in a quadrant? Explain.

No, if a point is on either axis it does not lie in a quadrant.

20. When a point lies on the x-axis, what do you know about its y-coordinate? When a point lies on the y-axis, what do you know about its x-coordinate?

Its y-coordinate is 0; its x-coordinate is 0.

Social Studies LINK

We use a coordinate system on Earth to find exact locations. The *equator* is like the *x*-axis, and the *prime meridian* is like the *y*-axis.

The lines that run east-west are *lines of latitude.* They are measured in degrees north and south of the equator.

The lines that run north-south are *lines of longitude.* They are measured in degrees east and west of the prime meridian.

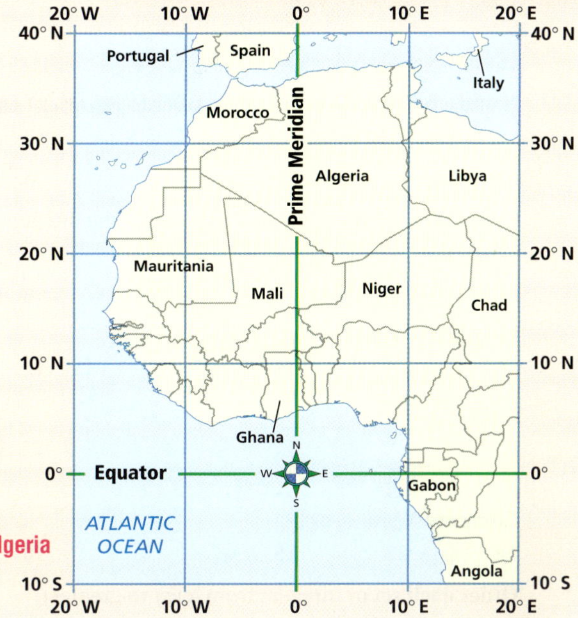

28. In what country is the location 0° latitude, 10° E longitude? **Gabon**

29. Give the coordinates of a location in Algeria. **Possible answer: 30° N latitude, 0° longitude**

30. Name two countries that lie along the 30° N line of latitude. **Possible answers: Morocco, Libya, Algeria**

31. Where would you be if you were located at 10° S latitude, 10° W longitude? **Atlantic Ocean**

32. ***WRITE ABOUT IT*** How is the coordinate system we use to locate places on Earth different from the coordinate plane? How is it similar?

go.hrw.com
KEYWORD: MR4 Africa
CNN Student News

33. ***CHALLENGE*** Begin at 10° S latitude, 20° E longitude. Travel 40° north and 20° west. What country would you be in now? **Algeria**

Spiral Review

Multiply. Write each answer in simplest form. (Lesson 5-1)

34. $\frac{2}{3} \cdot \frac{4}{7}$ $\frac{8}{21}$

35. $\frac{1}{5} \cdot \frac{3}{8}$ $\frac{3}{40}$

36. $\frac{3}{4} \cdot \frac{1}{2}$ $\frac{3}{8}$

Give the most exact name for each figure. (Lesson 7-6)

37. 
parallelogram

38. 
rhombus

39. 
trapezoid

40. **TEST PREP** A __?__ compares a number to 100. (Lesson 8-7) **C**

A rate **B** ratio **C** percent **D** proportion

Interdisciplinary LINK

Social Studies

Exercises 28–33 involve using the coordinate system of Earth's latitude and longitude for naming locations on a map. These mapping skills are used in middle school social studies programs, such as Holt, Rinehart & Winston's *People, Places, and Change.*

Answers

32. The coordinate plane has *x*- and *y*-coordinates and the Earth's coordinate system has longitude and latitude. They are similar in that both have a grid system.

Journal

Have students write about how giving someone directions is like writing an ordered pair.

Test Prep Doctor

For Exercise 40, encourage students to eliminate answer choices before choosing an answer. You may want to advise them to pick the most exact answer. Students who answered **A** or **B** are on the right track but need some additional reinforcement. A *rate* and a *ratio* compare two numbers, but the comparison is not restricted to 100. A *proportion* is an equation showing two equivalent ratios. The correct answer is **C.**

CHALLENGE 9-3

LESSON 9-3 **Challenge**
Plot and See

Graph each point below in the order given. Connect the points as you graph them to see a creature that lives most of its life 50 feet below sea level, or −50 feet.

START: (0, 20), (1, 19), (3, 18), (5, 15), (6, 15), (6, 12), (10, 8), (12, 8), (11, 6), (9, 6), (9, 7), (3, 11), (0, 11), (−2, 12), (3, 4), (2, 1), (3, −4), (−1, −13), (0, −17), (4, −18), (7, −17), (7, −16), (5, −16), (6, −14), (8, −15),(9, −18), (6, −20), (3, −20), (−2, −19), (−4, −13), (−2, −4), (−3, −1), (−4, −2), (−6, 2), (−5, 5), (−10, 12), (−6, 18), (0, 20) STOP!

Check students' graphs. The picture should be a seahorse.

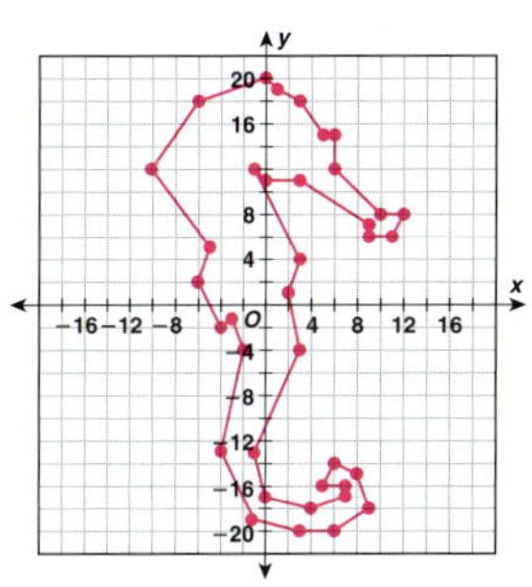

PROBLEM SOLVING 9-3

LESSON 9-3 **Problem Solving**
The Coordinate Plane

Use the coordinate plane on the map of Texas below to answer each question.

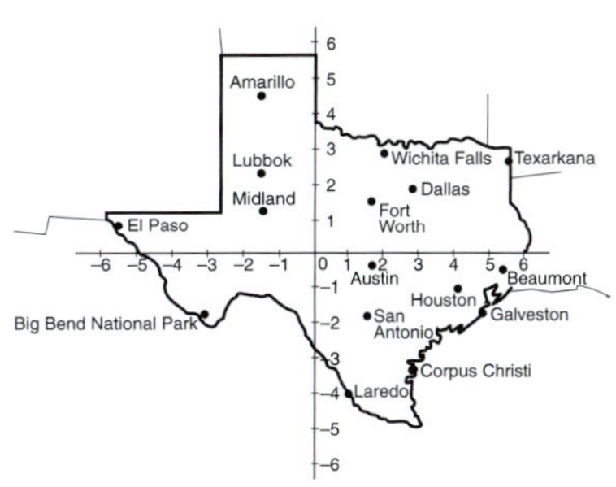

1. Which location in Texas is closest to the ordered pair (5, −2)? Galveston
2. What ordered pair best describes the location of Dallas, Texas? (3, 2)
3. Which location in Texas is closest to the ordered pair (−6, 1)? El Paso
4. Which location in Texas is located in Quadrant III of this coordinate plane? Big Bend National Park
5. Which three locations in Texas all have positive *y*-coordinates and nearly the same *x*-coordinate? Midland, Lubbock, and Amarillo
6. Which cities on this map of Texas have locations with *y*-coordinates less than −3? Laredo and Corpus Christi

Lesson Quiz

Name the quadrant where each ordered pair is located.

1. (3, −5) IV **2.** (−4, −2) III

3. (6, 2) I **4.** (−7, 9) II

Give the coordinates of each point.

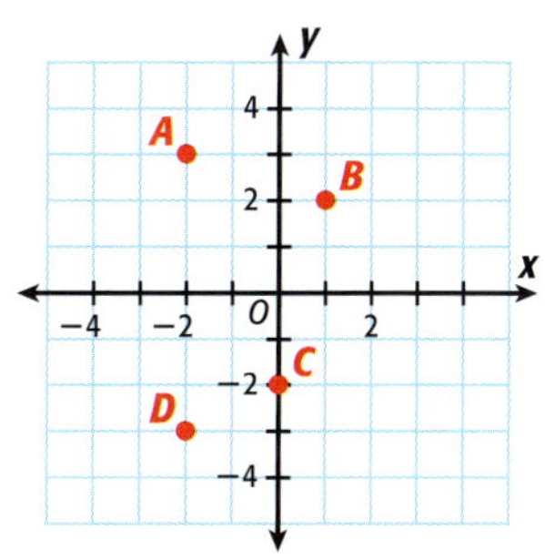

5. *A* (−2, 3) **6.** *B* (1, 2)

7. *C* (0, −2) **8.** *D* (−2, −3)

Available on Daily Transparency in CRB

Chapter 9 Mid-Chapter Quiz

Purpose: *To assess students' mastery of concepts and skills in Lessons 9-1 through 9-3*

Assessment Resources

Section 9A Quiz

Assessment Resources . . . . p. 23

***Test and Practice Generator* CD-ROM**

Additional mid-chapter assessment items in both multiple-choice and free-response format may be generated for any objective in Lessons 9-1 through 9-3.

Answers

5.

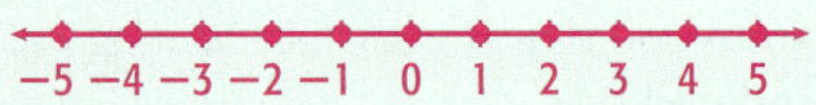

30–35.

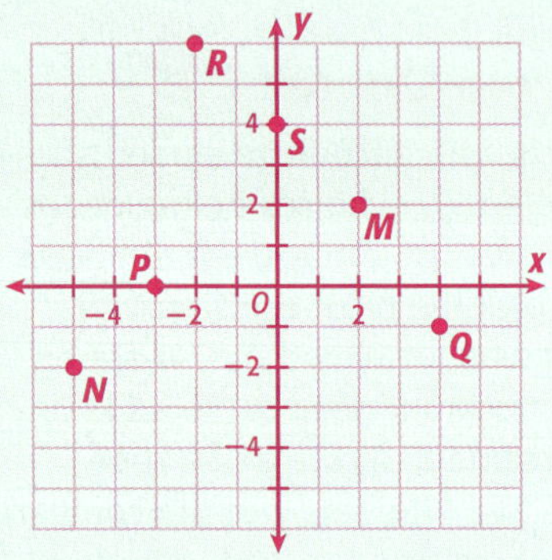

Chapter 9 Mid-Chapter Quiz

LESSON 9-1 (pp. 450–453)

Name a positive or negative number to represent each situation.

1. a gain of 10 yards **+10**

2. 45 feet below sea level **−45**

3. 5 degrees below zero **−5**

4. earning \$50 **+50 or 50**

5. Draw a number line and show all of the integers from −5 to 5.

Write the opposite of each integer.

6. 9 **−9** **7.** −17 **+17** **8.** 1 **−1** **9.** −20 **+20**

Find the absolute value.

10. $|5|$ **5** **11.** $|-16|$ **16** **12.** $|2|$ **2** **13.** $|-13|$ **13**

LESSON 9-2 (pp. 454–457)

Compare. Write < or >.

14. 9 ■ −22 **>** **15.** 4 ■ −7 **>** **16.** −8 ■ 2 **<** **17.** −10 ■ −19 **>**

Order each set of integers from least to greatest.

18. 2, −7, 14 **−7, 2, 14**

19. 25, −9, 4, −21 **−21, −9, 4, 25**

20. 10, 0, −23, −17, 8 **−23, −17, 0, 8, 10**

21. During an archaeological dig, the farther down an object is found, the older it is. If pieces of jewelry are found at −7 ft, −17 ft, −4 ft, and −9 ft, which piece is oldest? **the one found at −17 ft.**

Mid-Chapter Quiz

LESSON 9-3 (pp. 458–461)

Use the coordinate plane for problems 22–29.

Name the quadrant where each point is located.

22. *A* **no quadrant** **23.** *Y* **III**

24. *J* **I** **25.** *C* **II**

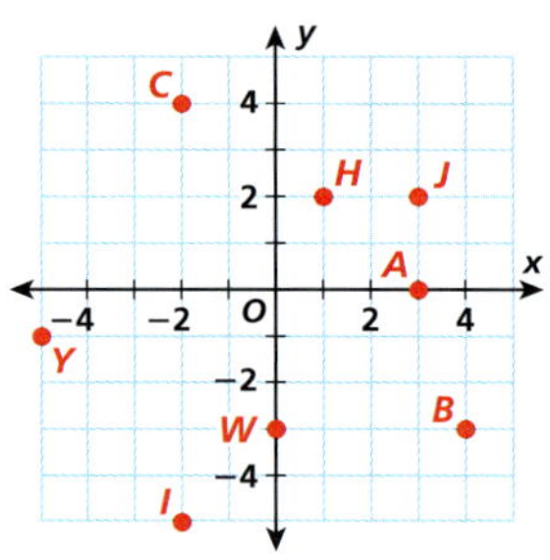

Give the coordinates of each point.

26. *H* **(1, 2)** **27.** *I* **(−2, −5)**

28. *W* **(0, −3)** **29.** *B* **(4, −3)**

Graph each point on a coordinate plane.

30. $N(-5, -2)$ **31.** $M(2, 2)$

32. $P(-3, 0)$ **33.** $Q(4, -1)$

34. $R(-2, 6)$ **35.** $S(0, 4)$

Focus on Problem Solving

Understand the Problem

• Restate the question

After reading a real-world problem (perhaps several times), look at the question in the problem. Rewrite the question as a statement in your own words. For example, if the question is "How much money did the museum earn?" you could write, "Find the amount of money the museum earned."

Now you have a simple sentence telling you what you must do. This can help you understand and remember what the problem is about. This can also help you find the necessary information in the problem.

Read the problems below. Rewrite each question as a statement in your own words.

1. Israel is one of the hottest countries in Asia. A temperature of 129°F was once recorded there. This is the opposite of the coldest recorded temperature in Antarctica. How cold has it been in Antarctica?

2. The average recorded temperature in Fairbanks, Alaska, in January is about −10°F. In February, the average temperature is about −4°F. Is the average temperature lower in January or in February?

3. The south pole on Mars is made of frozen carbon dioxide, which has a temperature of −193°F. The coldest day recorded on Earth was −129°F, in Antarctica. Which temperature is lower?

4. The pirate Blackbeard's ship, the *Queen Anne's Revenge,* sank at Beauford Inlet, North Carolina, in 1718. In 1996, divers discovered a shipwreck believed to be the *Queen Anne's Revenge*. The ship's cannons were found 21 feet below the water's surface, and the ship's bell was found 20 feet below the surface. Were the cannons or the bell closer to the surface?

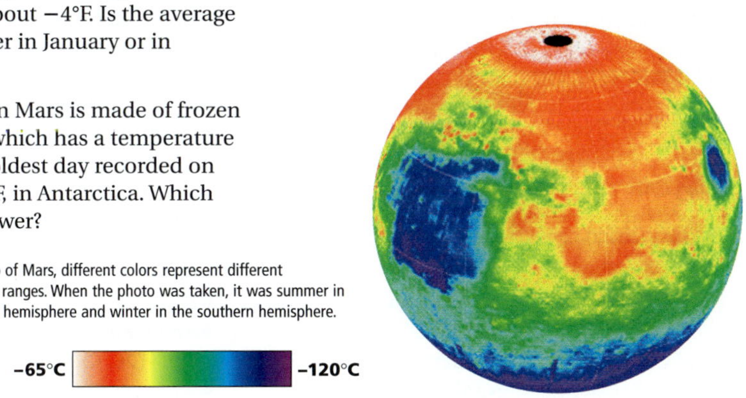

In this photo of Mars, different colors represent different temperature ranges. When the photo was taken, it was summer in the northern hemisphere and winter in the southern hemisphere.

Answers

1. −129°F
2. January
3. −193°F
4. The bell was closer to the surface.

Focus on Problem Solving

Purpose: *To focus on understanding the problem*

Problem Solving Resources

Interactive Problem Solving pp. 75–82

Math: Reading and Writing in the Content Area . . . . pp. 75–82

Problem Solving Process

This page focuses on the first step of the problem-solving process: **Understand the Problem**

Discuss

Have students identify the quantities being compared in the problems, and then simplify the problems using their own words.

1. the hottest temperature in Israel and the coldest temperature in Antarctica; Find the opposite of 129°F.
2. −10°F and −4°F; tell which is smaller, −10 or −4.
3. −193°F and −129°F; tell which is lower, −193 or −129.
4. 20 feet below the water's surface and 21 feet below the water's surface; tell which is closer to 0, −21 or −20.

Section 9B

Integer Operations

One-Minute Section Planner

Lesson	Materials	Resources
Hands-On Lab 9A Model Integer Addition **NCTM:** Number and Operations, Algebra **NAEP:** Number Properties 3f ☐ SAT-9 ☑ SAT-10 ☐ ITBS ☑ CTBS ☑ MAT ☐ CAT	**Required** Two-color counters *(MK)*	• *Hands-On Lab Activities,* p. 81
Lesson 9-4 Adding Integers **NCTM:** Number and Operations, Communication **NAEP:** Number Properties 3a ☑ SAT-9 ☑ SAT-10 ☑ ITBS ☑ CTBS ☑ MAT ☑ CAT	**Optional** Two-color counters *(MK)* Number lines *(CRB p. 88)*	• *Chapter 9 Resource Book,* pp. 35–43 • Daily Transparency T12, CRB • Additional Examples Transparencies T13–T15, CRB • *Alternate Openers: Explorations,* p. 78
Hands-On Lab 9B Model Integer Subtraction **NCTM:** Number and Operations, Algebra **NAEP:** Number Properties 3f ☐ SAT-9 ☑ SAT-10 ☐ ITBS ☑ CTBS ☑ MAT ☐ CAT	**Required** Two-color counters *(MK)*	• *Hands-On Lab Activities,* p. 82
Lesson 9-5 Subtracting Integers **NCTM:** Number and Operations, Communication **NAEP:** Number Properties 3a ☑ SAT-9 ☑ SAT-10 ☑ ITBS ☑ CTBS ☑ MAT ☑ CAT	**Optional** Number lines *(CRB p. 88)*	• *Chapter 9 Resource Book,* pp. 44–53 • Daily Transparency T16, CRB • Additional Examples Transparencies T17–T18, CRB • *Alternate Openers: Explorations,* p. 79
Lesson 9-6 Multiplying Integers **NCTM:** Number and Operations, Communication **NAEP:** Number Properties 3a ☑ SAT-9 ☑ SAT-10 ☑ ITBS ☑ CTBS ☑ MAT ☑ CAT	**Optional** Teaching Transparency T19 *(CRB)* Number cubes *(MK)* Two-color counters *(MK)*	• *Chapter 9 Resource Book,* pp. 54–62 • Daily Transparency T19, CRB • Additional Examples Transparencies T21–T22, CRB • *Alternate Openers: Explorations,* p. 80
Lesson 9-7 Dividing Integers **NCTM:** Number and Operations, Communication **NAEP:** Number Properties 3a ☑ SAT-9 ☑ SAT-10 ☑ ITBS ☑ CTBS ☑ MAT ☑ CAT	**Optional** Teaching Transparency T22 *(CRB)* Recording sheet for RAL *(CRB, p. 89)*	• *Chapter 9 Resource Book,* pp. 63–71 • Daily Transparency T23, CRB • Additional Examples Transparencies T25–T26, CRB • *Alternate Openers: Explorations,* p. 81
Hands-On Lab 9C Model Integer Equations **NCTM:** Number and Operations, Algebra **NAEP:** Algebra 4b ☐ SAT-9 ☑ SAT-10 ☐ ITBS ☑ CTBS ☑ MAT ☐ CAT	**Required** Algebra tiles *(MK, CRB p. 90)*	• *Hands-On Lab Activities,* pp. 83–84
Lesson 9-8 Solving Integer Equations **NCTM:** Number and Operations, Algebra, Reasoning and Proof, Communication, Representation **NAEP:** Algebra 4a ☐ SAT-9 ☐ SAT-10 ☐ ITBS ☐ CTBS ☐ MAT ☐ CAT	**Optional** Algebra tiles *(MK, CRB p. 90)*	• *Chapter 9 Resource Book,* pp. 72–81 • Daily Transparency T27, CRB • Additional Examples Transparencies T28–T30, CRB • *Alternate Openers: Explorations,* p. 82
Extension Integer Exponents **NCTM:** Number and Operations, Reasoning and Proof **NAEP:** Number Properties 9j ☐ SAT-9 ☐ SAT-10 ☐ ITBS ☐ CTBS ☐ MAT ☐ CAT		• Additional Examples Transparency T31, CRB
Section 9B Assessment		• Section 9B Quiz, AR p. 24 • *Test and Practice Generator* CD-ROM

SAT = *Stanford Achievement Tests* **ITBS** = *Iowa Test of Basic Skills* **CTBS** = *Comprehensive Test of Basic Skills/Terra Nova*
MAT = *Metropolitan Achievement Tests* **CAT** = *California Achievement Test*

NCTM = Complete standards can be found on pages T27–T33. **NAEP** = Complete standards can be found on pages A31–A35.

SE = *Student Edition* **TE** = *Teacher's Edition* **AR** = *Assessment Resources* **CRB** = *Chapter Resource Book* **MK** = *Manipulatives Kit*

Section Overview

Integer Operations *Hands-On Labs 9A, 9B, Lessons 9-4 through 9-7*

Why? When you know how to operate with integers, you can solve equations and problems involving integers.

Addition	Rule	Examples
Add integers with like signs.	Find the sum of their absolute values. Then use the sign of the integers.	$5 + 4 = 9$ $(-6) + (-2) = -8$
Add integers with unlike signs.	Find the difference of their absolute values. Then use the sign of the integer with the greater absolute value.	$9 + (-3) = 6$ $-8 + 7 = -1$

Subtraction	Rule	Examples
Subtract integers.	To subtract an integer, add its opposite.	$2 - (-3) = 2 + 3$ $= 5$ $-7 - 1 = -7 + (-1)$ $= -8$

Multiplication and Division	Rule	Examples
Multiply or divide integers with like signs.	Find the product or quotient of their absolute values. The answer will be positive.	$4 \cdot 5 = 20$ $-24 \div -6 = 4$
Multiply or divide integers with unlike signs.	Find the product or quotient of their absolute values. The answer will be negative.	$-8 \cdot 3 = -24$ $36 \div -4 = -9$

Integer Equations *Lesson 9-8*

Why? Solving equations is necessary in many problem-solving situations.

> When solving equations with integers, the goal is the same as with whole numbers—use the inverse of the operation on the variable to isolate the variable on one side of the equation.

A scuba diver begins to ascend. He rises 10 feet to a depth of −54 feet. At what depth did the scuba diver begin his ascent?

$$\begin{aligned} x + 10 &= -54 \\ \underline{-10} &\quad \underline{-10} \\ x &= -64 \end{aligned}$$

The scuba diver began the ascent at −64 feet, or 64 feet below sea level.

A submarine descends 175 feet at a time. How many times will it need to repeat a 175 ft descent to reach a depth of 14,000 feet below sea level?

$$-175x = -14{,}000$$

$$\frac{-175x}{-175} = \frac{-14{,}000}{-175}$$

$$x = 80$$

The submarine will repeat the descent 80 times.

Hands-On LAB 9A Model Integer Addition

Pacing: Traditional 1 day
Block $\frac{1}{2}$ day

Objective: To use two-color counters to model addition of integers

Materials: Two-color counters

Lab Resources

Hands-On Lab Activities .... p. 71

Using the Page

Discuss with students what each red or yellow counter represents.

Represent each expression with two-color counters.

1. −4
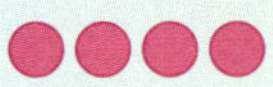

2. 3

3. 3 + (−1)
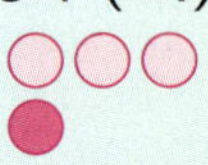

4. −2 + (−2)
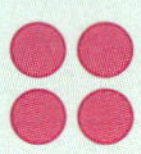

Answers

Try This

1.

2.
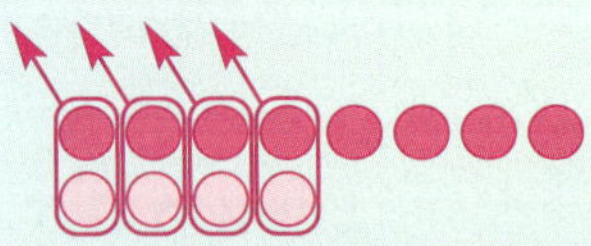

3.
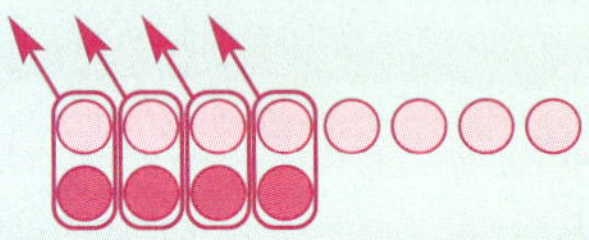

4.
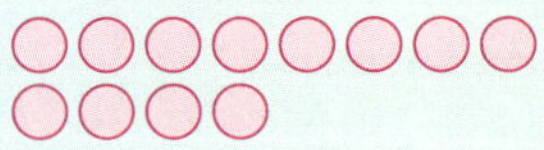

Hands-On LAB 9A Model Integer Addition

Use with Lesson 9-4

KEY (yellow) = 1 (red) = −1

REMEMBER Subtracting zero from a number does not change the number's value.

internet connect Lab Resources Online *go.hrw.com* **KEYWORD:** MR4 Lab9A

Two-color counters can be used to represent integers. Yellow counters represent positive numbers and red counters represent negative numbers.

Activity

Model with two-color counters.

1 3 + 4
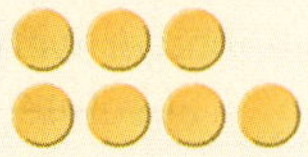
3 + 4 = 7

2 −5 + (−3)
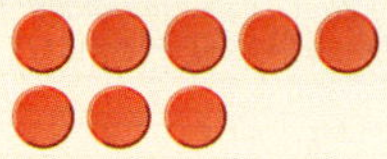
−5 + (−3) = −8

One red and one yellow counter together equal zero. Whenever you have a pair of red and yellow counters, you can remove them without changing the value of the model.

3 3 + (−4)
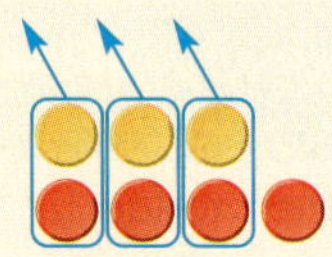
3 + (−4) = −1

Think and Discuss

1. When adding integers, would changing the order in which you add them affect the answer? Explain. **Changing the order of the integers in an addition problem does not affect the answer. For example, −6 + 4 = 2 and 4 + (−6) = 2.**
2. When can you remove counters from an addition model? **You may remove counters from an addition problem when you have a pair (one red and one yellow) of counters representing 0.**

Try This

Model with two-color counters.

1. −8 + (−4) **−12** **2.** −8 + 4 **−4** **3.** 8 + (−4) **+4** **4.** 8 + 4 **+12**

Teacher to Teacher

Karen Smith
Duxbury, Massachusetts

Game for Combining Integers

Divide the class into pairs. I use two-sided discs; the red side is negative, and the yellow side is positive. Each pair of students should have ten discs and a pencil and paper to mark off each round. Tell the students that one red and one yellow disc will cancel each other out because of the Inverse Property of Addition. Students take turns tossing all ten discs onto a desk and then use the cancellation rule to determine their score (e.g., 6 red and 4 yellow = −2 points). The students add scores from one round to the next. At the end of ten rounds, the student whose score is the closest to zero wins.

9-4 Adding Integers

Learn to add integers.

One of the world's most active volcanoes is Kilauea, in Hawaii. Kilauea's base is 9 km below sea level. The top of Kilauea is 10 km above the base of the mountain.

You can add the integers −9 and 10 to find the height of Kilauea above sea level.

Adding Integers on a Number Line

Move **right** on a number line to add a **positive** integer.
Move **left** on a number line to add a **negative** integer.

EXAMPLE 1 **Writing Integer Addition**

Write the addition modeled on each number line.

A

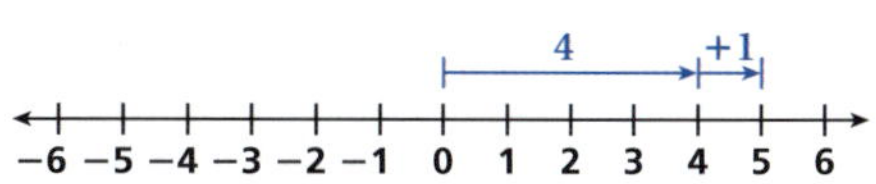

The addition modeled is $4 + 1 = 5$.

B

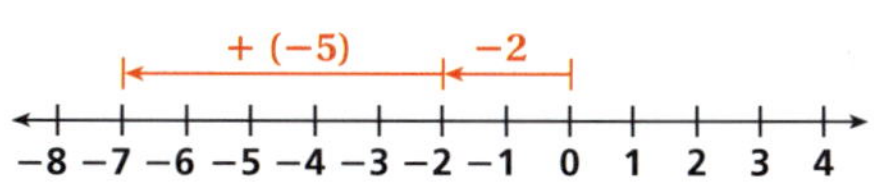

The addition modeled is $-2 + (-5) = -7$.

C

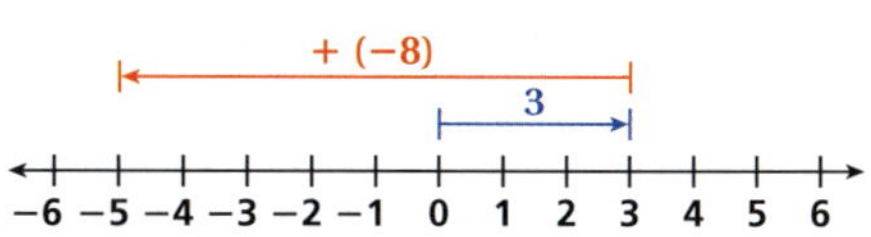

The addition modeled is $3 + (-8) = -5$.

Writing Math

Parentheses are used to separate addition, subtraction, multiplication, and division signs from negative integers.
$-2 + (-5) = -7$

9-4 Organizer

Pacing: Traditional 1 day
Block $\frac{1}{2}$ day

Objective: Students add integers.

Indiana Academic Standards
6.2.1, 6.7.5, 6.7.11

Warm Up

Add.

1. $2 + 3$ 5 **2.** $7 + 4$ 11

3. $8 + 5$ 13 **4.** $17 + 12$ 29

Name the integer that corresponds to each point.

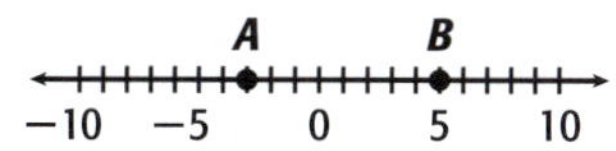

5. *A* −3 **6.** *B* 5

Problem of the Day

Karen earned 40 points on part I of a test, 26 points on part II, and 25 points on part III. What number of points did Karen earn on parts I and III combined? 65

Available on Daily Transparency in CRB

Math Fact

It is not possible to reach a temperature lower than absolute zero. To convert absolute zero to the Fahrenheit scale, you must add a negative 459° (−459°).

1 Introduce

Alternate Opener

EXPLORATION

9-4 Adding Integers

You can use a thermometer to model addition of integers.

1. Suppose the temperature starts at −10°F and increases 30° during the day. Complete the addition statement to show the new temperature.
   −10° + 30° = ____
2. Suppose the temperature starts at 20°F and drops 40° overnight. Complete the addition statement to show the new temperature.
   20° + (−40°) = ____
3. Draw a thermometer and show 20° + (−10°). Find the sum.

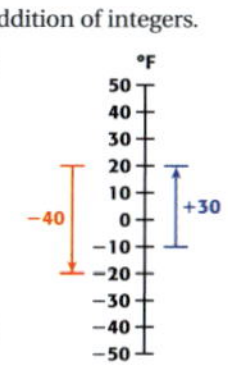

Think and Discuss

4. **Describe** how to add a positive integer using a thermometer.
5. **Describe** how to add a negative integer using a thermometer.

Motivate

Ask students to help write rules for a game called *Pegs & Holes*, in which *P* stands for pegs and *H* stands for holes. (Adding pegs and holes follows the same rules as adding integers.) Give examples, such as $4P + 5P = 9P$ and $2H + 3H = 5H$. Adding *P*'s to *H*'s is different. For example, $5P + 3H = 2P$, $4P + 4H = 0$, and $1P + 6H = 5H$. Create more examples. See if the students can state a rule for each case.

Exploration worksheet and answers on Chapter 9 Resource Book pp. 36 and 97

2 Teach

Lesson Presentation

Guided Instruction

In this lesson, students learn to add integers. First, teach the rules for adding integers on a number line (Chapter 9 Resource Book p. 88) and the rules for adding integers using two-color counters (Manipulatives Kit). Then teach addition using a number line and counters. Next, have students model addition of integers on a number line as well as model addition using counters.

Additional Examples

Example 1

Write the addition modeled on each number line.

A.

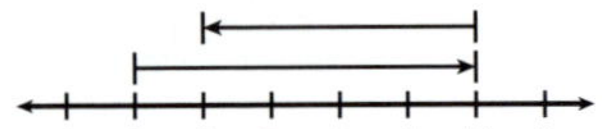

5 + (−4) = 1

B.

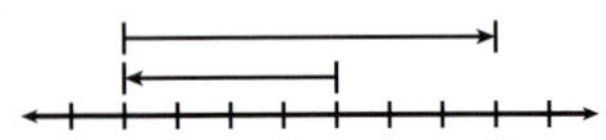

−4 + 7 = 3

C.

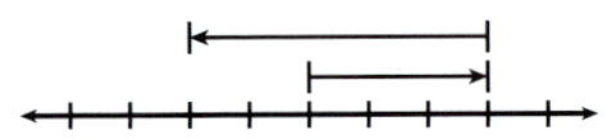

3 + (−5) = −2

Example 2

Find each sum.

A. −3 + (−2) −5

B. 6 + (−8) −2

Example 3

Evaluate *y* + (−2) for each value of *y*.

A. *y* = 7 5 **B.** *y* = 1 −1

Example 4

A sunken ship is 12 m below sea level. A search plane flies 35 m above the sunken ship. How far above the sea is the plane?

23 m above the sea

EXAMPLE 2 Adding Integers

Find each sum.

A $6 + (-5)$

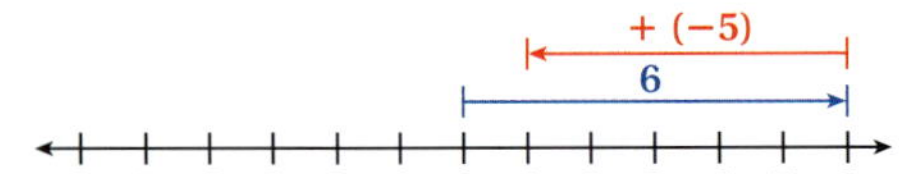

Think:

$6 + (-5) = 1$

B $-7 + 4$

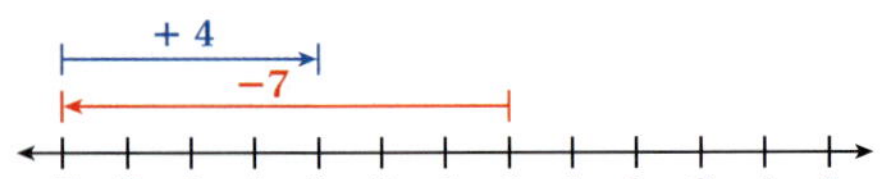

Think:

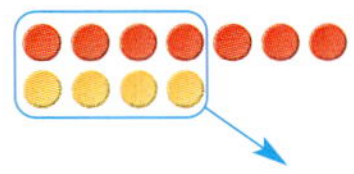

$-7 + 4 = -3$

EXAMPLE 3 Evaluating Integer Expressions

Evaluate *x* + 3 for each value of *x*.

A $x = 1$

$x + 3$	*Write the expression.*
$1 + 3$	*Substitute 1 for x.*
4	*Add.*

B $x = -9$

$x + 3$	*Write the expression.*
$-9 + 3$	*Substitute −9 for x.*
-6	*Add.*

EXAMPLE 4 *Earth Science Application*

The base of Kilauea is 9 km below sea level. The top is 10 km above the base. How high above sea level is Kilauea?

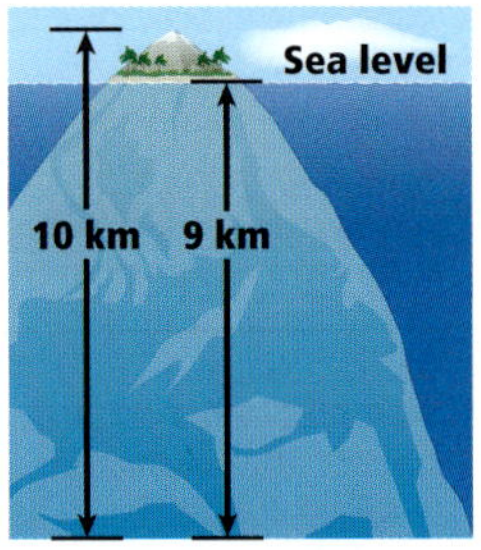

The base is 9 km below sea level and the top is 10 km above the base.

$-9 + 10$

1

Kilauea is 1 km above sea level.

Think and Discuss

1. Tell if the sum of a positive integer and −8 is greater than −8 or less than −8. Explain.

2. Give the sum of a number and its opposite.

3 Close

Reaching All Learners

Through Concrete Manipulatives

Some students may not be ready to make the transition from concrete models (counters) to pictorial models (number lines) when adding positive and negative integers. Allow these students to use color counters provided in the Manipulatives Kit when working through the examples.

Summarize

Review the procedures for adding integers. Discuss how to use counters and how to use a number line when adding integers with opposite signs. Ask students to tell how to use each method to find the sum of −7 + 3.

Possible answers:

a. Counters: Use 7 red counters and 3 yellow counters. Make pairs of red and yellow counters (there will be 3 pairs), and then remove them. There are 4 red counters left.
So −7 + 3 = −4.

b. Number line: Count 7 units to the left of 0, to −7. Then count 3 units right from −7, to −4. So −7 + 3 = −4.

Answers to Think and Discuss

1. Possible answer: Greater than; When you add a positive integer to any integer, you move to the right on the number line. Because the sum of −8 and a positive integer will be to the right of −8, even if the sum is negative, it will still be greater than −8.

2. zero

9-4 Exercises

FOR EXTRA PRACTICE
see page 655

internet connect
Homework Help Online
go.hrw.com Keyword: MR4 9-4

GUIDED PRACTICE

See Example 1 Write the addition modeled on the number line.

1. 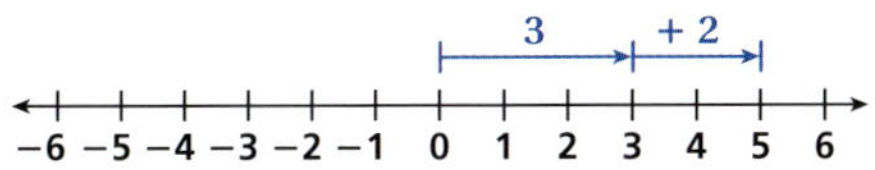

3 + 2 = 5

See Example 2 Find each sum.

2. $-5 + 9$ **4** **3.** $-3 + (-2)$ **−5** **4.** $8 + (-7)$ **1**

See Example 3 Evaluate $n + (-2)$ for each value of n.

5. $n = -10$ **−12** **6.** $n = 2$ **0** **7.** $n = -2$ **−4**

See Example 4 **8.** A submarine at the water's surface dropped down 100 ft. After thirty minutes at that depth, it dove an additional 500 ft. What was its depth after the second dive? **−600 ft**

INDEPENDENT PRACTICE

See Example 1 Write the addition modeled on each number line.

9. 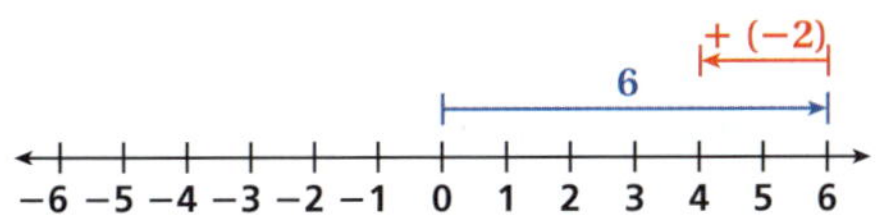

6 + (−2) = 4

10. 

−4 + (−2) = −6

See Example 2 Find each sum.

11. $4 + 7$ **11** **12.** $2 + (-12)$ **−10** **13.** $9 + (-9)$ **0**

14. $-8 + 2$ **−6** **15.** $-2 + 8$ **6** **16.** $-1 + (-6)$ **−7**

See Example 3 Evaluate $-6 + a$ for each value of a.

17. $a = -10$ **−16** **18.** $a = 7$ **1** **19.** $a = -2$ **−8**

20. $a = 4$ **−2** **21.** $a = -9$ **−15** **22.** $a = 8$ **2**

See Example 4 **23.** Jon works on a cruise ship and sleeps in a cabin that is 6 feet below sea level. The main deck is 35 feet above Jon's cabin. How far above sea level is the main deck? **29 feet**

9-4 PRACTICE & ASSESS

Students may want to refer back to the lesson examples.

Assignment Guide

If you finished Example 1 assign:
Core 1, 9–10, 24–26, 44–52
Enriched 24–26, 41–52

If you finished Example 2 assign:
Core 1–4, 9–16, 44–52
Enriched 14–16, 27–32, 41–52

If you finished Example 3 assign:
Core 1–7, 9–22, 44–52
Enriched 24–52

If you finished Example 4 assign:
Core 1–32, 44–52
Enriched 12–52

Notes

RETEACH 9-4

LESSON 9-4 **Reteach**
Adding Integers

Two-color counters can help you understand integers.

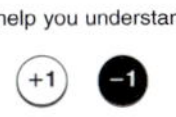

One positive counter and one negative counter is called a zero pair because together they have a value of zero.

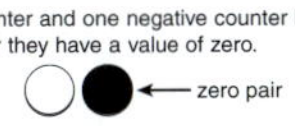

1 + (−1) = 0

You can use two-color counters to find the sum of integers.

To use counters to find the sum of −3 and 5, model the addition expression first.

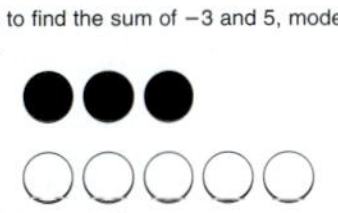

Then remove all the zero pairs.

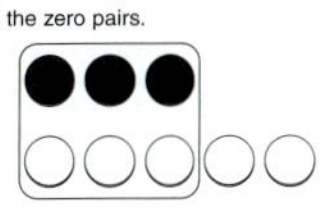

The remaining counters represent the sum.
−3 + 5 = 2

Use counters to find each sum.

1. −4 + 7 **3**   2. 4 + (−8) **−4**   3. −3 + −6 **−9**   4. −3 + 1 **−2**

5. −2 + 7 **5**   6. 5 + (−6) **−1**   7. −8 + (−3) **−11**   8. 10 + 2 **12**

9. 9 + −12 **−3**   10. 4 + (−10) **−6**   11. −8 + −5 **−13**   12. 6 + −5 **1**

PRACTICE 9-4

LESSON 9-4 **Practice B**
Adding Integers

Write the addition modeled on each number line.

1. 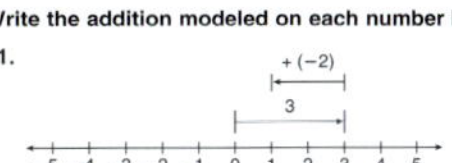

3 + (−2)

2. 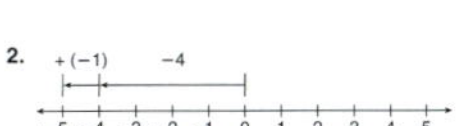

−4 + (−1)

3. 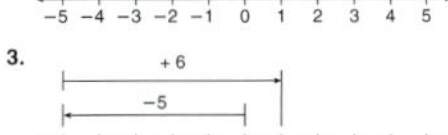

−5 + 6

Find each sum.

4. 5 + (−1) **4**   5. −3 + 2 **−1**   6. −8 + (−4) **−12**

7. −2 + (−1) **−3**   8. 9 + (−6) **3**   9. −10 + 5 **−5**

10. 12 + (−3) **9**   11. 0 + (−7) **−7**   12. 17 + (−9) **8**

Evaluate $n + (-1)$ for each value of n.

13. $n = 2$ **1**   14. $n = -4$ **−5**   15. $n = 5$ **4**

19. When Calvin played golf today, he scored a +1 on the first hole, a −2 on the second hole, a −1 on the third, and a +4 on the fourth. What was Calvin's total score for the first four holes?
+2 or 2

20. The average temperature for February was 4°F below zero. By March, the average temperature had increased 11 degrees. What was the average temperature in March?
7°F

Math Background

Color counters and number lines are useful tools for understanding addition of integers with opposite signs. Mastering these tools prepares students to better understand the more formal method:

- Find the absolute value of each addend.
- Subtract the smaller absolute value from the larger absolute value.
- Attach the sign of the addend that has the greater absolute value to the difference from the last step.

Answers

24.

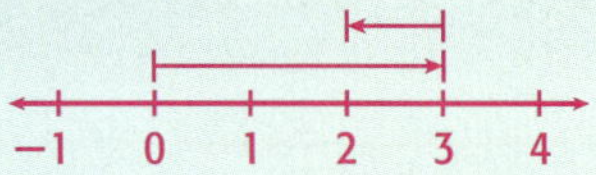

25.

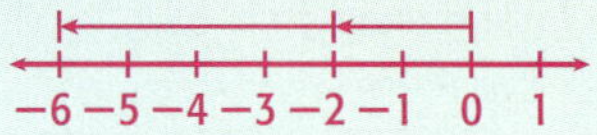

26.

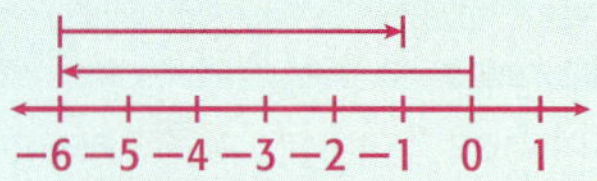

42. a. Positive; when you add two integers with the same signs, the sum will share the sign.

b. Negative; when you add two integers with the same sign, the sum will share the sign.

c. Sometimes positive, sometimes negative; the sign of the sum of two integers with different signs is the sign of the integer with the greater absolute value.

Journal

Have students write about why the rules are different for the addition of positive integers and the addition of negative integers.

Test Prep Doctor

For Exercise 52, encourage students to eliminate incorrect answer choices. **A** is sometimes true, but not always. **B** and **C** are never true. Therefore, the correct answer is **D.**

Lesson Quiz

Find each sum.

1. $7 + (-3)$ 4

2. $-5 + 2$ −3

3. $-8 + (-4)$ −12

4. Evaluate $x + 5$ for $x = -4$. 1

5. At midnight on a winter night, the temperature was −12°F. By 10 A.M. the temperature had risen 42°F. What was the new temperature? 30°F

Available on Daily Transparency in CRB

PRACTICE AND PROBLEM SOLVING

Model each addition problem on a number line.

24. $3 + (-1)$ **25.** $-2 + (-4)$ **26.** $-6 + 5$

Find each sum.

27. $-18 + 25$ 7 **28.** $8 + (-2)$ 6 **29.** $-5 + -6$ −11

30. $-6 + (-3)$ −9 **31.** $4 + (-1)$ 3 **32.** $20 + (-3)$ 17

Evaluate each expression for the given value of the variable.

33. $x + (-3); x = 7$ 4 **34.** $-9 + n; n = 7$ −2 **35.** $a + 5; a = -6$ −1

36. $m + (-2); m = -4$ −6 **37.** $-10 + x; x = -7$ −17 **38.** $n + 19; n = -5$ 14

History LINK

Augustus was originally named Octavian, but the Roman senate gave him the title Augustus, meaning "revered one." He ruled the Roman Empire for more than 40 years.

39. ***EARTH SCIENCE*** The temperature at midnight was −2°F. During the next 4 hours, a decrease of 4°F was recorded. What was the temperature at 4 A.M.? −6°F

40. ***SPORTS*** In the 2001 U.S. Women's Open, Cristie Kerr had the following scores for the four rounds of golf: −1, +3, +1, and 0. What was her total score? +3

41. ***CHOOSE A STRATEGY*** The first Roman emperor, Augustus, was born in 63 B.C. and died in A.D. 14. How many years did he live? (*Hint*: Years B.C. are like negative numbers. Years A.D. are like positive numbers. There was no year 0.) 76 yrs

42. ***WRITE ABOUT IT*** When adding two integers, what will the sign of the answer be when

a. both integers are positive? positive

b. both integers are negative? negative

c. one integer is positive and the other is negative? sometimes positive, sometimes negative

Explain your answers.

43. ***CHALLENGE*** Evaluate $-3 + (-2) + (-1) + 0 + 1 + 2 + 3 + 4$. Then use this pattern to find the sum of the integers from −10 to 11 and from −100 to 101. 4; 11; 101

Spiral Review

Find the GCF of each pair of numbers. (Lesson 4-3)

44. 12, 18 6 **45.** 12, 36 12 **46.** 25, 33 1 **47.** 45, 27 9

Find the LCM of each pair of numbers. (Lesson 5-5)

48. 9, 15 45 **49.** 4, 30 60 **50.** 17, 3 51 **51.** 10, 25 50

52. **TEST PREP** __?__ are less than zero. (Lesson 9-1) D

A Integers B Positive numbers C Absolute values D Negative numbers

CHALLENGE 9-4

LESSON 9-4 **Challenge**
Time Adds Up

Solve the addition problems below to find the date each toy or game was invented. Then use those dates to label the time line at the bottom of the page.

	Invention	Addition	Date
1.	Frisbee	50 + (−2)	1948
2.	Pogo Stick	−11 + 32	1921
3.	Crossword Puzzle	−1 + 17 + (−3)	1913
4.	Skateboard	−43 + 101	1958
5.	Yo-Yo	38 + (−9)	1929
6.	Rollerblades	−25 + 105	1980
7.	Teddy Bear	−2 + 8 + (−3)	1903
8.	Slinky	50 + (−4)	1946

Check students' time lines.

Time Line of Toys and Games

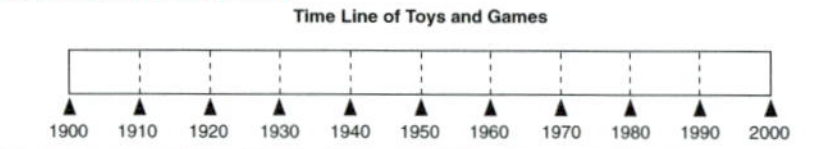

PROBLEM SOLVING 9-4

LESSON 9-4 **Problem Solving**
Adding Integers

In 1997, Tiger Woods became the youngest golfer ever to win the Masters Tournament. There are four rounds of 18 holes in the Masters Tournament. Use Woods's scorecard to answer questions 1–6.

Tiger Woods																		
Hole	1	2	3	4	5	6	7	8	9	10	11	12	13	14	15	16	17	18
Rd. 1	1	0	0	1	0	0	0	1	1	−1	0	−1	−1	0	−2	0	−1	0
Rd. 2	0	−1	1	0	−1	0	0	−1	0	0	0	0	−2	−1	−1	0	0	0
Rd. 3	0	−1	0	0	−1	0	−1	−1	0	0	−1	0	0	0	−1	0	0	−1
Rd. 4	0	−1	0	0	1	0	1	−1	0	0	−1	0	−1	−1	0	0	0	0

1. What was Woods's total score for round 1 of the tournament? −2

2. What was his total score for the second round of the tournament? −6

3. What was his total score for the third round of the tournament? −7

4. What was his total score for the fourth round of the tournament? −3

5. Woods's final score in 1997 was the lowest in the history of the Masters Tournament. What was Woods's record-breaking final score? −18

6. Tom Kite placed second in the 1997 Masters Tournament. His final score was 12 strokes higher than Tiger Woods's final score. What was Kite's final score? −6

Circle the letter of the correct answer.

7. Which of the following is the sum of Woods's scores on the 8th hole?
A 2
B 1
C −1
(D) −2

8. Which of the following is the sum of Woods's scores on the 15th hole?
F 4
(G) −4
H 0
J 1

Hands-On LAB 9B Model Integer Subtraction

Use with Lesson 9-5

KEY

(yellow) = 1 (red) = −1

REMEMBER

Adding zero to a number does not change the number's value.

(yellow) + (red) = 0

internet connect
Lab Resources Online
go.hrw.com
KEYWORD: MR4 Lab9B

Activity

Model with two-color counters.

1. $3 - 2$

$3 - 2 = 1$

2. $-3 - (-2)$

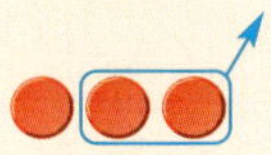

$-3 - (-2) = -1$

3. $3 - (-2)$

You do not have any red counters, so you cannot subtract −2. Add pairs of red and yellow counters until you have enough red counters to subtract.

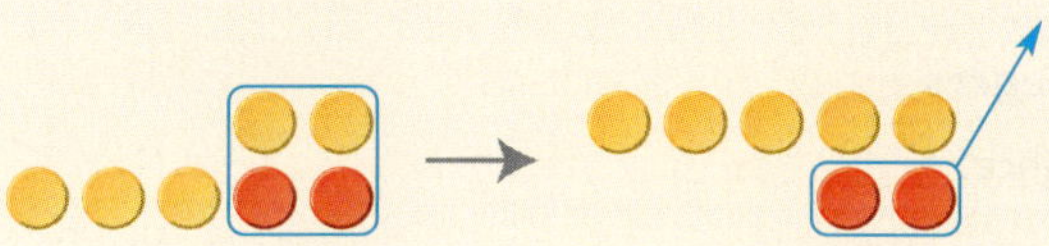

Add these. *Now you can subtract −2.*

$3 - (-2) = 5$

Think and Discuss

Possible answers:

1. How do you show subtraction with counters? **by removing counters**
2. Why can you add pairs of red and yellow counters to a subtraction model? **Adding a pair of red and yellow counters is like adding 0, which doesn't change the value of the number.**

Try This

Model with two-color counters.

1. $5 - 4$ **1**
2. $4 - (-5)$ **9**
3. $-4 - 5$ **−9**
4. $-4 - (-5)$ **1**

Hands-On LAB 9B Model Integer Subtraction

Pacing: Traditional 1 day
Block $\frac{1}{2}$ day

Objective: To use two-color counters to model subtraction of integers

Materials: Two-color counters

Lab Resources

Using the Page

Discuss with students what each side of the two-color counter represents. Discuss what is represented by a pair of counters in which one is red and one is yellow.

Represent each expression with two-color counters.

1. -2

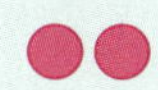

2. 4

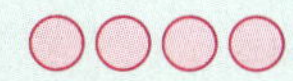

3. $5 - 3$

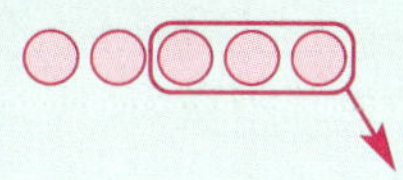

4. $-2 - (4)$

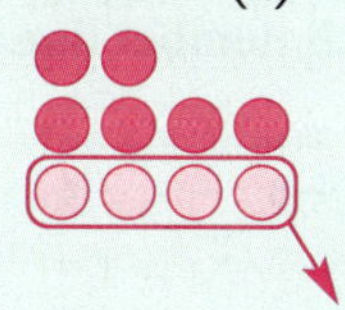

Answers

Try This

1.

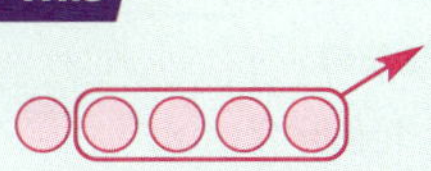

2.

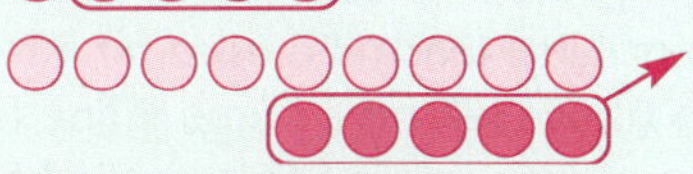

3.

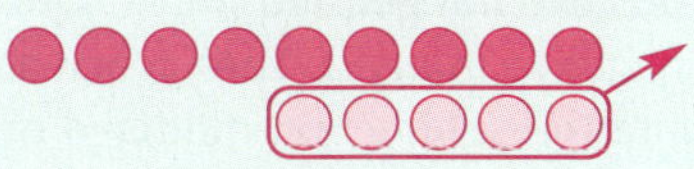

4.

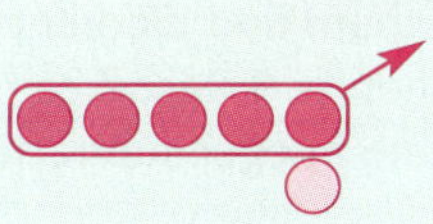

9-5 Organizer

Pacing: Traditional 1 day
Block $\frac{1}{2}$ day

Objective: Students subtract integers.

Indiana Academic Standards
6.2.1, 6.7.5, 6.7.11

Warm Up

Find each sum.

1. $-4 + 6$ 2   **2.** $5 + 12$ 17

3. $-3 + 3$ 0   **4.** $-8 + 9$ 1

Give the opposite of each number.

5. 7 -7 **6.** -8 8

7. -3 3   **8.** 1 -1

Problem of the Day available on Daily Transparency in CRB

Additional Example

Example 1

Write the subtraction modeled on each number line.

A.

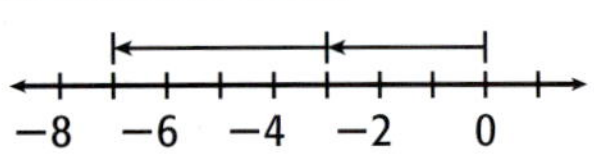

$-3 - 4 = -7$

B.

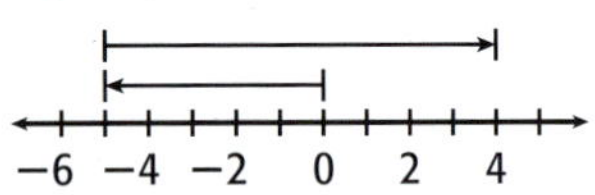

$-5 - (-9) = 4$

9-5 Subtracting Integers

Learn to subtract integers.

On a number line, integer subtraction is the opposite of integer addition. Integer subtraction "undoes" integer addition.

Subtracting Integers on a Number Line
Move **left** on a number line to subtract a **positive** integer.
Move **right** on a number line to subtract a **negative** integer.

EXAMPLE 1 Writing Integer Subtraction

Write the subtraction modeled on each number line.

A

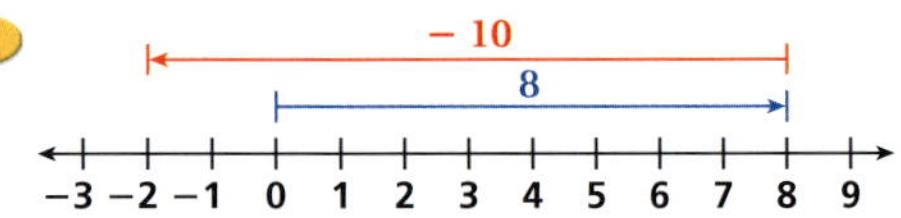

The subtraction modeled is $8 - 10 = -2$.

B

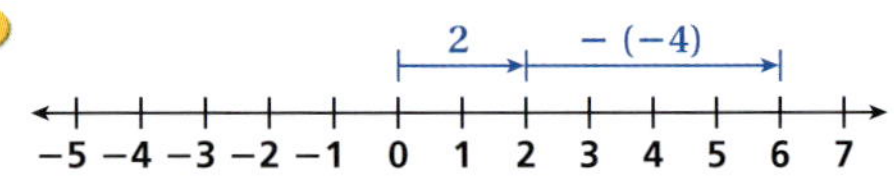

The subtraction modeled is $2 - (-4) = 6$.

EXAMPLE 2 Subtracting Integers

Find each difference.

A $7 - 4$

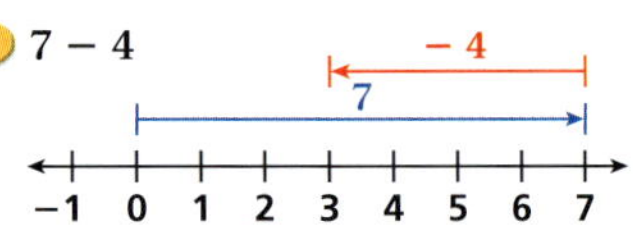

Think:

$7 - 4 = 3$

B $-8 - (-2)$

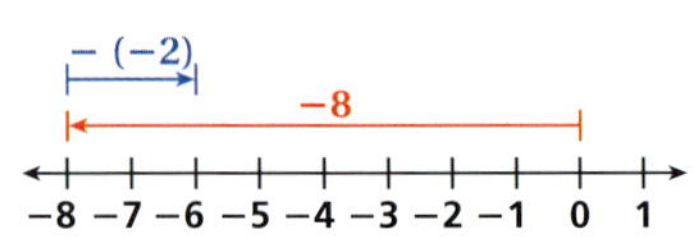

Think:

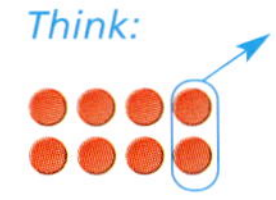

$-8 - (-2) = -6$

1 Introduce

Alternate Opener

EXPLORATION

9-5 Subtracting Integers

You can use a number line to model subtracting integers.

To subtract 20 from 50, begin at the number being subtracted, 20, and count the number of units to the number 50.

The direction is **right,** so the difference is **positive.**

$50 - 20 = 30$

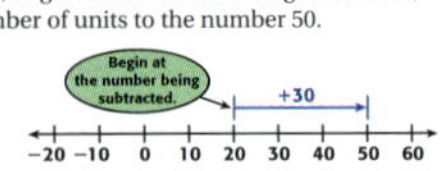

To subtract -20 from -60, begin at the number being subtracted, -20, and count the number of units to the number -60.

The direction is **left,** so the difference is **negative.**

$-60 - (-20) = -40$

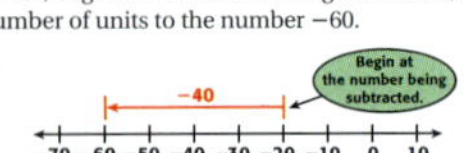

Use a number line to find each difference.

1. $12 - 10 =$ ____   **2.** $15 - 20 =$ ____

3. $-8 - 5 =$ ____   **4.** $-6 - (-4) =$ ____

5. $2 - (-4) =$ ____   **6.** $7 - (-2) =$ ____

Think and Discuss

7. Describe how to use a number line to model subtraction.

Motivate

Have students look at addition models on a number line, such as $6 + 2$ and $6 + (-2)$. Show students that adding -2 and subtracting 2 give the same result.

Exploration worksheet and answers on Chapter 9 Resource Book pp. 45–46 and 99

2 Teach

Lesson Presentation

Guided Instruction

In this lesson, students learn to subtract integers. First, teach the rules for integer subtraction (move left on a number line to subtract a positive integer and move right to subtract a negative integer). Then teach students to write the subtraction modeled on number lines. Next let them find differences using number lines (Chapter 9 Resource Book p. 88) and evaluate integer subtraction expressions.

EXAMPLE 3 **Evaluating Integer Expressions**

Evaluate $x - (-4)$ for each value of x.

A $x = -4$

$x - (-4)$ — *Write the expression.*

$-4 - (-4)$ — *Substitute −4 for x.*

0 — *Subtract.*

B $x = -5$

$x - (-4)$ — *Write the expression.*

$-5 - (-4)$ — *Substitute −5 for x.*

-1 — *Subtract.*

Think and Discuss

1. In which direction do you move to add a positive integer? In which direction do you move to subtract a positive integer?
2. How do your answers to Example 1 help show that addition and subtraction are inverses?

9-5 Exercises

FOR EXTRA PRACTICE see page 655

internet connect
Homework Help Online
go.hrw.com Keyword: MR4 9-5

GUIDED PRACTICE

See Example 1

1. Write the subtraction modeled on the number line. $6 - 5 = 1$

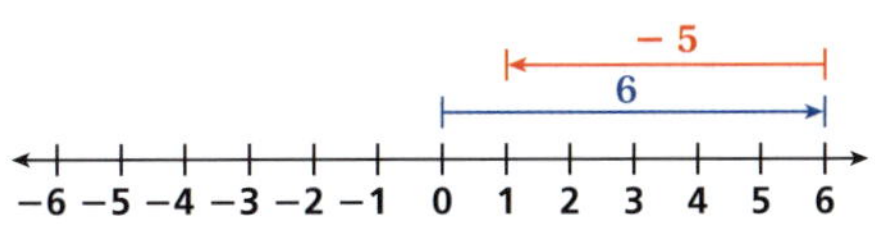

See Example 2

Find each difference.

2. $6 - 3$ **3**
3. $3 - 6$ **−3**
4. $10 - (-4)$ **14**

See Example 3

Evaluate $n - (-6)$ for each value of n.

5. $n = -4$ **2**
6. $n = 2$ **8**
7. $n = -15$ **−9**

INDEPENDENT PRACTICE

See Example 1

8. Write the subtraction modeled on the number line. $-2 - (-3) = 1$

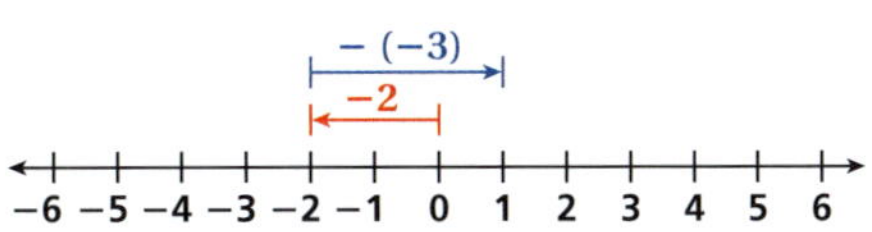

See Example 2

Find each difference.

9. $3 - 7$ **−4**
10. $-4 - 9$ **−13**
11. $2 - (-9)$ **11**

Additional Examples

Example 2

Find each difference.

A. $4 - 6$ **−2** **B.** $3 - (-3)$ **6**

Example 3

Evaluate $a - 4$ for each value of a.

A. $a = 2$ **−2** **B.** $a = 8$ **4**

9-5 PRACTICE & ASSESS

Students may want to refer back to the lesson examples.

Assignment Guide

If you finished Example 1 assign:
Core 1, 8, 18–20, 33–37
Enriched 21–23, 31–37

If you finished Example 2 assign:
Core 1–4, 8–11, 27, 33–37
Enriched 18–23, 30–37

If you finished Example 3 assign:
Core 1–28, 33–37
Enriched 5–37

3 Close

Reaching All Learners
Through Curriculum Integration

Social Studies Have students find and list the record high and low temperatures for various countries. Then have students subtract to find the difference between the high and low temperatures for a given country.

Summarize

Emphasize to students that subtraction of an integer on a number line looks the same as addition of that integer's opposite.

Answers to Think and Discuss

1. right; left
2. Possible answer: To add a positive integer, you move in the opposite direction than you would when you subtract a positive integer.

Lesson Quiz

Find each difference.

1. 8 − 5 3 **2.** −9 − (−3) −6

Evaluate *x* − (−7) for each value of *x*.

3. *x* = −2 5 **4.** *x* = 5 12

5. The peak of Mt. Everest is approximately 26,544 ft above sea level. The shore of the Dead Sea is about 1,200 ft below sea level. How much higher is the peak of Mt. Everest than the shore of the Dead Sea? 27,744 ft

Available on Daily Transparency in CRB

See Example 3 Evaluate $m - (-3)$ for each value of m.

12. $m = -1$ 2 **13.** $m = 7$ 10 **14.** $m = -8$ −5

15. $m = 4$ 7 **16.** $m = -9$ −6 **17.** $m = -15$ −12

PRACTICE AND PROBLEM SOLVING

Find each difference.

18. −12 − (−6) −6 **19.** 7 − (−3) 10 **20.** −4 − (−3) −1

21. 13 − (−8) 21 **22.** 4 − 12 −8 **23.** 2 − (−3) 5

Evaluate each expression for the given value of the variable.

24. $n - (-10)$, $n = 2$ 12 **25.** $-6 - m$, $m = -9$ 3 **26.** $x - 2$, $x = 6$ 4

27. ***EARTH SCIENCE*** The surface of an underground water supply was 10 m below sea level. After one year, the depth of the water supply has decreased by 9 m. How far below sea level is the water's surface now? −19 m

28. ***CONSTRUCTION*** A 200-foot column holds an oil rig platform above the ocean's surface. The column rests on the ocean floor 175 feet below sea level. How high is the platform above sea level? 25 ft

29. ***EARTH SCIENCE*** During summer 1997, NASA landed the *Pathfinder* on Mars. On July 9, *Pathfinder* reported a temperature of −1°F on the planet's surface. On July 10, it reported a temperature of 8°F. Find the difference between the temperature on July 10 and the temperature on July 9. 9

30. ***WHAT'S THE ERROR?*** Your friend says that 0 − (−4) = −4. Explain why this is incorrect.

31. ***WRITE ABOUT IT*** Will the difference between two negative numbers ever be positive? Use examples to support your answer.

32. ***CHALLENGE*** This pyramid was built by subtracting integers. Two integers are subtracted from left to right, and their difference is centered above them. Find the missing numbers.

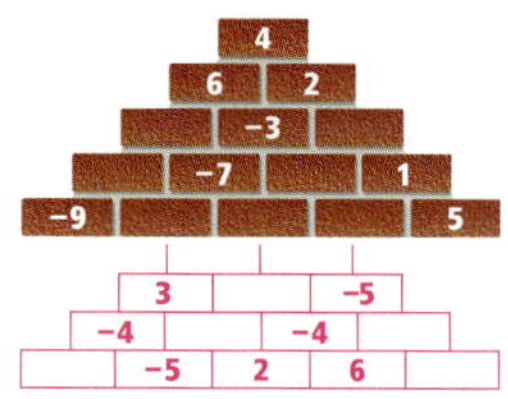

30. Zero minus a negative number is the same as zero plus a positive number and that sum will always be positive.

31. Yes, the difference may be negative, positive, or zero. Examples: −2 − (−4) = 2, −6 − (−6) = 0.

Spiral Review

Evaluate. (Lesson 1-3)

33. 3^3 27 **34.** 4^2 16 **35.** 5^3 125 **36.** 2^5 32

37. **TEST PREP** Two numbers are __?__ if their product is 1. (Lesson 5-3) A

A reciprocals **C** greatest common factors

B opposites **D** least common multiples

RETEACH 9-5

LESSON 9-5 **Reteach**
Subtracting Integers

Two-color counters can help you understand integers.

(+1) (−1)

One positive counter and one negative counter is called a zero pair because together they have a value of zero.

← zero pair

1 + (−1) = 0

You can use two-color counters to find the difference between integers. 5 − 7
The first integer tells you how many counters to start off with.

If there are not enough counters to subtract, add zero pairs.

Then take away the number of counters given by the second integer.

The remaining counters represent the difference. 5 − 7 = −2

Use counters to find each difference.

1. −3 − (5) −8 **2.** 5 − 8 −3 **3.** −9 − (−7) −2 **4.** 1 − 6 −5

5. −4 − (−3) −1 **6.** 2 − (−5) 7 **7.** −6 − 4 −10 **8.** 5 − (−4) 9

9. 2 − (−3) 5 **10.** −7 − (−3) −4 **11.** 10 − 8 2 **12.** −4 − (−5) 1

PRACTICE 9-5

LESSON 9-5 **Practice B**
Subtracting Integers

Write the subtraction modeled on each number line.

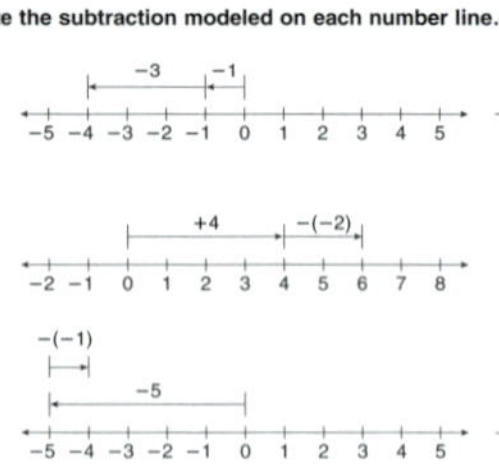

1. −1 − 3

2. 4 − (−2)

3. −5 − (−1)

Find each difference.

4. 8 − (−1) 9 **5.** −5 − 2 −7 **6.** −10 − (−3) −7

7. −2 − (−1) −1 **8.** 4 − (−6) 10 **9.** −9 − (−5) −4

10. 15 − (−8) 23 **11.** 0 − (−6) 6 **12.** −20 − (−9) −11

Evaluate *n* − (−2) for each value of *n*.

13. *n* = 2 4 **14.** *n* = −4 −2 **15.** *n* = 5 7

16. *n* = −3 −1 **17.** *n* = 1 3 **18.** *n* = 0 2

19. In a golf tournament, Sarah scored a +2 on the first round and a −4 on the second round. What was the difference between her scores on the first two rounds? 6

20. Washington, D.C., has an elevation of 1 foot above sea level. The elevation of New Orleans is 8 feet below sea level. What is the difference in the two cities' elevations? 9 feet

CHALLENGE 9-5

LESSON 9-5 **Challenge**
Going Up or Down?

Five friends parked in an underground garage, and they cannot remember where they left their car! So they split up and searched the garage. Read the descriptions of their trips between levels. Use integers to show on which level each person started and the change in their locations. Then, name the level where they exited the elevator and label their names at that level on the diagram. Check students' labels on the diagram.

1. Tammy got on the elevator on the ground level. Then she rode down seven levels and got off the elevator. 0; −7, −7 Level

2. Jamal got on the elevator on the ninth underground level. He rode up seven levels and got off there. −9; +7, −2 Level

3. Maria entered the elevator on the sixth underground level. She rode up six levels and then down one level before getting off the elevator. −6; +5, −1 Level

4. Shaneeka started on the fifth level underground. She rode the elevator down three levels, up two more levels, and then got off. −5; −3, +2, −6 Level

5. Bill got on the elevator on the fourth underground level. He rode down four levels. Then he rode up three more levels, got off the elevator, and found the car! Now he just has to find his friends. −4; −4, +3, −5 Level

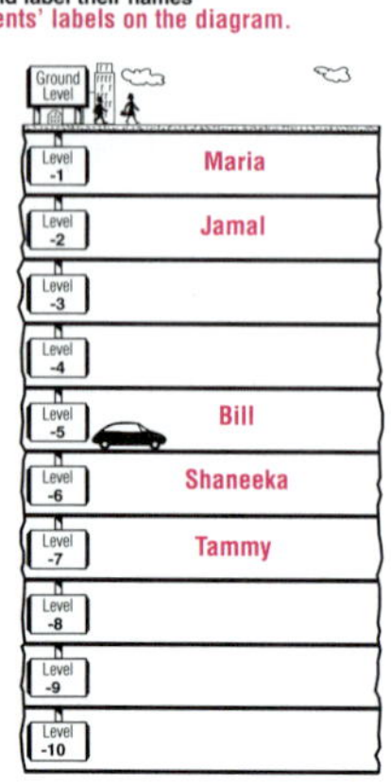

PROBLEM SOLVING 9-5

LESSON 9-5 **Problem Solving**
Subtracting Integers

Write the correct answer.

1. The average surface temperature on Earth is 59°F. The average surface temperature on Mars is 126°F lower than on Earth. What is the average surface temperature on Mars? −67°F

2. The average surface temperature on Saturn is 46°F colder than on Jupiter. Jupiter's average surface temperature is 162°F below zero. What is the average surface temperature on Saturn? −208°F

3. Venus has the hottest average surface temperature at 854°F. Mercury, the planet closest to the Sun, has an average surface temperature that is 522°F colder than Venus's. What is Mercury's average surface temperature? 332°F

4. Pluto, the planet farthest from the Sun, has an average surface temperature of 355°F below zero. Neptune, its closest neighbor, has the coldest average surface temperature. It is 10°F colder on Neptune than on Pluto. What is the average surface temperature on Neptune? −365°F

Circle the letter of the correct answer.

5. Which of the following is the difference between 247°F below zero and 221°F above zero?
A −26°F
B 129°F
(C) −468°F
D 468°F

6. Which of the following is the difference between 806°C above zero and 328°C below zero?
(F) 1,134°C
G 478°C
H −478°C
J −1,134°C

7. Which of the following is the difference between −40°C and −30°C?
(A) −10°C
B 708C
C −120°C
D −1°C

8. Which of the following is the difference between 8,700°F and −344°F?
F 8,356°F
G 900°F
H −9°F
(J) 9,044°F

9-6 Multiplying Integers

Learn to multiply integers.

You have seen that you can multiply whole numbers to count items in equally sized groups.

There are three sets of twins in the sixth grade. How many sixth graders are twins?

A set of twins is 2 people.

$3 \cdot 2 = 6$ *3 sets of 2 is 6.*

So 6 students in the sixth grade are twins.

Multiplying with integers is similar.

Numbers	$3 \cdot 2$	$-3 \cdot 2$	$3 \cdot (-2)$	$-3 \cdot (-2)$
Words	3 groups of 2	the opposite of 3 groups of 2	3 groups of −2	the opposite of 3 groups of −2
Addition	$2 + 2 + 2$	$-(2 + 2 + 2)$	$(-2) + (-2) + (-2)$	$-[(-2) + (-2) + (-2)]$
Product	6	−6	−6	6

EXAMPLE 1 Multiplying Integers

Find each product.

A $4 \cdot 3$

$4 \cdot 3 = 12$ *Think: 4 groups of 3*

B $2 \cdot (-4)$

$2 \cdot (-4) = -8$ *Think: 2 groups of −4*

C $-5 \cdot 2$

$-5 \cdot 2 = -10$ *Think: **the opposite of** 5 groups of 2*

D $-3 \cdot (-4)$

$-3 \cdot (-4) = 12$ *Think: **the opposite of** 3 groups of −4*

Remember!

To find the opposite of a number, change the sign. The opposite of 6 is −6. The opposite of −4 is 4.

9-6 Organizer

Pacing: Traditional 1 day
Block $\frac{1}{2}$ day

Objective: Students multiply integers.

Indiana Academic Standards
6.2.2, 6.7.5, 6.7.11

Warm Up

Find each product.

1. $8 \cdot 4$ 32
2. $7 \cdot 12$ 84
3. $3 \cdot 9$ 27
4. $6 \cdot 5$ 30
5. $80 \cdot 6$ 480
6. $50 \cdot 6$ 300
7. $40 \cdot 90$ 3,600
8. $20 \cdot 700$ 14,000

Problem of the Day

Catherine has $14.00 and earns $12.00 for each lawn she mows. If she mows 4 lawns and buys 5 DVDs that cost $11.95 each, including tax, how much money does she have left? $2.25

Available on Daily Transparency in CRB

Math Fact

Because the product of two numbers with the same sign can never be negative, no negative number has a real square root.

1 Introduce

Alternate Opener

EXPLORATION

9-6 Multiplying Integers

Complete each table.

1.

$2 \cdot 3 = 6$	$-2 \cdot (3) = -6$	$2 \cdot (-3) = -6$	$-2 \cdot (-3) = 6$
$2 \cdot 2 =$	$-2 \cdot (2) =$	$2 \cdot (-2) =$	$-2 \cdot (-2) =$
$2 \cdot 1 =$	$-2 \cdot (1) =$	$2 \cdot (-1) =$	$-2 \cdot (-1) =$

2.

$3 \cdot 3 =$	$-3 \cdot (3) =$	$3 \cdot (-3) =$	$-3 \cdot (-3) =$
$3 \cdot 2 =$	$-3 \cdot (2) =$	$3 \cdot (-2) =$	$-3 \cdot (-2) =$
$3 \cdot 1 =$	$-3 \cdot (1) =$	$3 \cdot (-1) =$	$-3 \cdot (-1) =$

3.

$4 \cdot 3 =$	$-4 \cdot (3) =$	$4 \cdot (-3) =$	$-4 \cdot (-3) =$
$4 \cdot 2 =$	$-4 \cdot (2) =$	$4 \cdot (-2) =$	$-4 \cdot (-2) =$
$4 \cdot 1 =$	$-4 \cdot (1) =$	$4 \cdot (-1) =$	$-4 \cdot (-1) =$

Think and Discuss

4. Describe the patterns you notice in each of the tables.

Motivate

Discuss with students the meaning of multiplication. Have students write examples of situations that involve multiplication. (For example, 3 classes of 25 students each equal how many students in all?) Challenge students to think of situations in which the result could be represented by a negative integer. (Possible situations: a scuba diver descending 25 feet at a time 3 times; an automatic withdrawal of $15 per month from a bank account for 6 months; etc.)

Exploration worksheet and answers on Chapter 9 Resource Book pp. 55 and 101

2 Teach

Lesson Presentation

Guided Instruction

In this lesson, students learn to multiply integers (Teaching Transparency T20, CRB). First, review the four possible situations for integer multiplication, $(+) \cdot (+)$, $(+) \cdot (-)$, $(-) \cdot (-)$, and $(-) \cdot (+)$. Then teach the rules for multiplying integers. Give special attention to the rule $(-) \cdot (-) = (+)$. Next, teach students to evaluate expressions that involve the multiplication of integers.

Additional Examples

Example 1

Find each product.

A. 5 · 2 **10**

B. 4 · (−5) **−20**

C. −3 · 2 **−6**

D. −2 · (−4) **8**

Example 2

Evaluate −7*x* for each value of *x*.

A. *x* = −3 **21**

B. *x* = 5 **−35**

MULTIPLYING INTEGERS

If the signs are the same, the product is positive.

$4 \cdot 3 = 12$ $-6 \cdot (-3) = 18$

If the signs are different, the product is negative.

$-2 \cdot 5 = -10$ $7 \cdot (-8) = -56$

The product of any number and 0 is 0.

$0 \cdot 9 = 0$ $(-12) \cdot 0 = 0$

EXAMPLE 2 Evaluating Integer Expressions

Remember!
5x means 5 · x.

Evaluate 5*x* for each value of *x*.

A $x = -4$

$5x$	*Write the expression.*
$5 \cdot (-4)$	*Substitute −4 for x.*
-20	*The signs are different, so the answer is negative.*

B $x = 0$

$5x$	*Write the expression.*
$5 \cdot 0$	*Substitute 0 for x.*
0	*Any number times 0 is 0.*

Think and Discuss

1. Explain how multiplying integers is similar to multiplying whole numbers. How is it different?

9-6 PRACTICE & ASSESS

Assignment Guide

If you finished Example 1 assign:
Core 1–6, 13–18, 25–32, 45–53
Enriched 1–6, 13–18, 25–29, 42–53

If you finished Example 2 assign:
Core 1–32, 42, 45–53
Enriched 13–53

9-6 Exercises

FOR EXTRA PRACTICE
see page 655

internet connect
Homework Help Online
go.hrw.com Keyword: MR4 9-6

GUIDED PRACTICE

See Example 1 Find each product.

1. $6 \cdot 4$ **24** **2.** $5 \cdot (-2)$ **−10** **3.** $-3 \cdot 7$ **−21**

4. $-9 \cdot (-1)$ **9** **5.** $13 \cdot 0$ **0** **6.** $-8 \cdot (-2)$ **16**

See Example 2 Evaluate $3n$ for each value of n.

7. $n = 3$ **9** **8.** $n = -2$ **−6** **9.** $n = 11$ **33**

10. $n = -8$ **−24** **11.** $n = -12$ **−36** **12.** $n = 6$ **18**

Reaching All Learners
Through Grouping Strategies

Have students work in groups to generate positive and negative numbers by simultaneously rolling a number cube and flipping a two-color counter (provided in the Manipulatives Kit). Let red indicate negative and yellow indicate positive. The students record pairs of numbers and find their products. After five different products have been found, have them order the products from least to greatest. It would be helpful to model how to play this game before assigning it to students.

3 Close

Summarize

Give the students four multiplication situations (positive times positive, positive times negative, negative times positive, and negative times negative), and have a volunteer give an expression as an example of each situation. Have another student evaluate the expression and state the rule for multiplying that pair of integers.

Possible answer:
Rules: Positive times negative or negative times positive gives a negative product; negative times negative or positive times positive gives a positive product.

Answer to Think and Discuss

1. Possible answer:

Similar: Both can be shown as repeated addition; for both, you are finding the total, given the number of groups and the number in each group.

Different: You have to decide whether the product is positive or negative when multiplying integers.

INDEPENDENT PRACTICE

See Example 1 — Find each product.

13. $5 \cdot 9$ **45**
14. $-7 \cdot 6$ **−42**
15. $8 \cdot (-4)$ **−32**
16. $-13 \cdot (-3)$ **39**
17. $4 \cdot 12$ **48**
18. $6 \cdot (-12)$ **−72**

See Example 2 — Evaluate $-4a$ for each value of a.

19. $a = 6$ **−24**
20. $a = 12$ **−48**
21. $a = 3$ **−12**
22. $a = -10$ **40**
23. $a = 7$ **−28**
24. $a = -15$ **60**

PRACTICE AND PROBLEM SOLVING

Multiply.

25. $-2 \cdot 3$ **−6**
26. $-4(9)$ **−36**
27. $-6 \cdot (-6)$ **36**
28. $-5(-8)$ **40**
29. $-12 \cdot 2$ **−24**
30. $-9(9)$ **−81**
31. $-6 \cdot 7$ **−42**
32. $-6(25)$ **−150**

Evaluate each expression for the given value of the variable.

33. $n \cdot (-7)$; $n = -2$ **14**
34. $-6 \cdot m$; $m = 4$ **−24**
35. $9x$; $x = 6$ **54**
36. $-5m$; $m = 5$ **−25**
37. $x \cdot 10$; $x = -9$ **−90**
38. $-8 \cdot n$; $n = -1$ **8**
39. $-15 \cdot x$; $x = 6$ **−90**
40. $-13n$; $n = -4$ **52**
41. $m \cdot 14$; $m = -3$ **−42**

42. ***EARTH SCIENCE*** When the moon, the sun, and Earth are in a straight line, spring tides occur on Earth. Spring tides may cause high and low tides to be two times as great as normal. If high tides at a certain location are usually 2 ft and low tides are usually −2 ft, what might the spring tides be? **4 ft to −4 ft**

43. ***WRITE ABOUT IT*** What is the sign of the product when you multiply three negative integers? four negative integers? Use examples to explain your answers. **negative; positive. possible examples: $-2 \cdot -2 \cdot -2 = -8$; $-2 \cdot -2 \cdot -2 \cdot -2 = 16$**

44. ***CHALLENGE*** Name 2 integers whose product is −36 and whose sum is 0. **6 and −6**

Spiral Review

Solve each equation. Check your answers. (Lesson 2-6)

45. $9y = 81$ **$y = 9$**
46. $70 = 10x$ **$x = 7$**
47. $6 \cdot 8 = n$ **$n = 48$**
48. $60 = 12m$ **$m = 5$**

Write two equivalent ratios. (Lesson 8-1) **Possible answers:**

49. $\frac{1}{2}$ **$\frac{2}{4}$, $\frac{4}{8}$**
50. $\frac{3}{12}$ **$\frac{1}{4}$, $\frac{6}{24}$**
51. $\frac{2}{3}$ **$\frac{4}{6}$, $\frac{6}{9}$**
52. $\frac{5}{15}$ **$\frac{1}{3}$, $\frac{10}{30}$**

53. **TEST PREP** On a coordinate plane, the point where the axes intersect is called the __?__. (Lesson 9-3) **C**

A quadrant B 0-axis C origin D coordinate point

Lesson Quiz

Find each product.

1. $6 \cdot (4)$ **24**
2. $3 \cdot (-2)$ **−6**
3. $-9 \cdot (-2)$ **18**
4. $-6 \cdot 5$ **−30**
5. Evaluate $3y$ for $y = -7$. **−21**
6. During a football game, Raymond's team lost 6 yards on each of 3 plays and gained 8 yards on each of two plays. What integer represents the total change in the team's position? **−2**

Available on Daily Transparency in CRB

RETEACH 9-6

LESSON 9-6 **Reteach**
Multiplying Integers

You can use two-color counters to multiply integers.

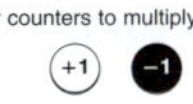

Multiply −4 by 2.

First, think about the numerical expression in words.

−4 • 2 means "the opposite of 4 groups of 2."

Then use counters to represent the expression.

4 groups of 2

4 groups of 2 are 8.

The opposite of 8 is −8.

−4 • 2 = −8.

Use counters to find each product.

1. 3 • (−3) **−9**
2. −5 • (−1) **5**
3. 2 • 3 **6**
4. −3 • 4 **−12**
5. −4 • (4) **−16**
6. 0 • (−3) **0**
7. −11 • (1) **−11**
8. −6 • −2 **12**
9. −2 • (−4) **8**
10. 7 • (−2) **−14**
11. 5 • 3 **15**
12. −8 • (−2) **16**

PRACTICE 9-6

LESSON 9-6 **Practice B**
Multiplying Integers

Write the sign of each product.

1. 7 • 8 **positive**
2. 5 • (−9) **negative**
3. −4 • 12 **negative**
4. −6 • (−11) **positive**
5. −3 • 8 **negative**
6. −12 • (−18) **positive**

Find each product.

7. 5 • (−7) **−35**
8. −4 • 3 **−12**
9. −8 • (−2) **16**
10. −9 • (−1) **9**
11. 5 • (−6) **−30**
12. −10 • (−4) **40**
13. 6 • (−8) **−48**
14. 0 • (−3) **0**
15. 7 • (−9) **−63**

Evaluate 4*n* for each value of *n*.

16. *n* = 2 **8**
17. *n* = −4 **−16**
18. *n* = −7 **−28**
19. *n* = −3 **−12**
20. *n* = 11 **44**
21. *n* = 0 **0**

Evaluate −3*n* for each value of *n*.

22. *n* = −5 **15**
23. *n* = 0 **0**
24. *n* = 6 **−18**
25. *n* = −8 **24**
26. *n* = 7 **−21**
27. *n* = −1 **3**

28. Last month, Tyler made five withdrawals of $25 each from his bank account and no deposits. What multiplication expression models Tyler's bank transactions last month? **5 • (−25)**

29. The Atlantic Ocean is sinking 4 inches every 100 years. Write a multiplication expression that models how much the Atlantic Ocean will sink in 300 years. How many inches will it sink in that time? **3 • (−4); 12 inches**

CHALLENGE 9-6

LESSON 9-6 **Challenge**
Factor Machines

The integer at the top of each factor machine is the product of the first integer in each row and some unknown factor. Help the machines work by loading the correct missing factors.

1.

2.
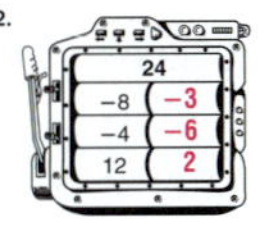

3.
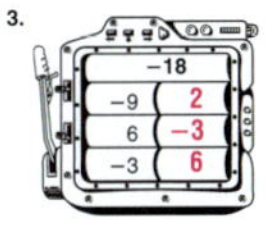

4.
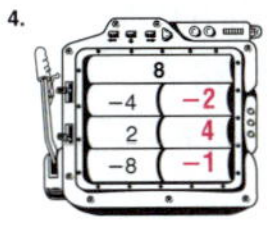

5.
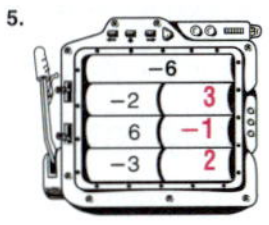

6.
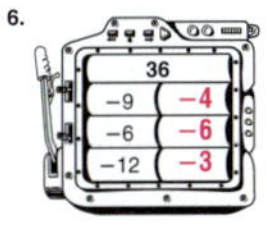

7.
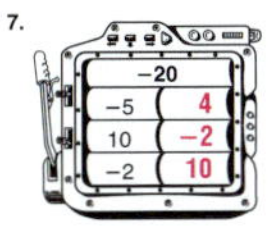

8.
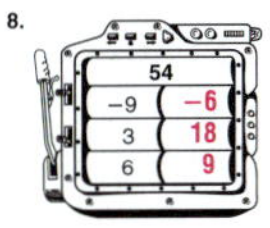

9.
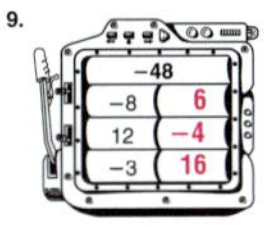

PROBLEM SOLVING 9-6

LESSON 9-6 **Problem Solving**
Multiplying Integers

Write the correct answer.

1. The coldest temperature ever recorded in Rhode Island was 25°F below zero. Though Nevada lies much farther south, its coldest temperature was twice as cold as Rhode Island's. What was Nevada's record cold temperature? **−50°F**

2. Tom and Kim made up a game in which black tiles equal +5 points each, and red tiles equal −3 points each. The person with the most points wins. At the end of the game Tom had 6 red tiles and 4 black tiles, and Kim had 4 red tiles and 3 black tiles. Who won? **Kim**

3. During a month-long drought, the amount of water in the family's well changed −4 gallons a day. How much did the amount of water in the well change after one week? **−28 gallons**

4. Sperm whales dive deeper than any other mammals. They regularly dive to 3,937 feet below sea level. But they sometimes dive to twice this depth! To what elevation can sperm whales dive? **−7,874 feet**

5. On Monday morning, the value of LCM stock was $15 a share. Then the value of the stock changed by −3 dollars a day for 4 days in a row. What was the value of one share of LCM stock after the fourth day? **$3**

6. Lake Manitoba and Lake Winnipeg are two of the largest lakes in Canada. The greatest depth of Lake Manitoba is 12 feet. Lake Winnipeg is 5 times deeper than Lake Manitoba. What is the greatest depth of Lake Winnipeg? **60 feet**

Circle the letter of the best answer.

7. Which addition expression could be used to check the product of 5 • (−3)?
   - A 5 + 5 + 5
   - B −3 + (−3) + (−3)
   - C 5 + 5 + 5 + 5 + 5
   - **(D)** −3 + (−3) + (−3) + (−3) + (−3)

8. Which property allows you to rewrite −2 • (−4) as −4 • (−2)?
   - **(F)** Commutative Property
   - G Distributive Property
   - H Integer Property
   - J Associative Property

9-7 Dividing Integers

Learn to divide integers.

Mona is a biologist studying an endangered species of wombat. Each year she records the change in the wombat population.

Year	Change in Population
1	−10
2	−5
3	−1
4	+4

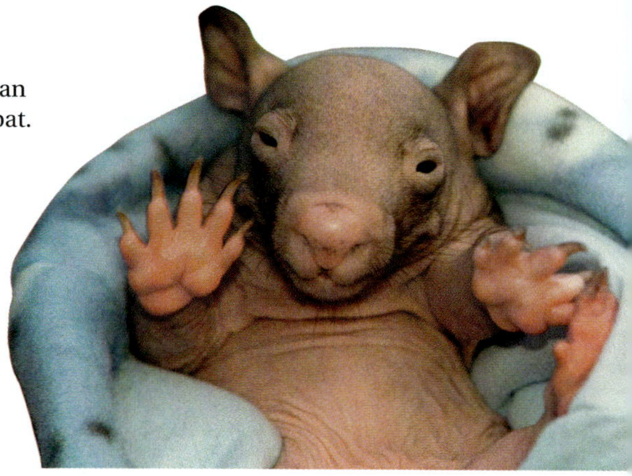

Baby Australian wombat

One way to describe the change in the wombat population over time is to find the mean of the data in the table.

Remember!

To find the mean of a list of numbers:
1. Add all the numbers together.
2. Divide by how many numbers are in the list.

$$\frac{-10 + (-5) + (-1) + 4}{4} = \frac{-12}{4} = -12 \div 4 = \blacksquare$$

Multiplication and division are inverse operations. To solve a division problem, think of the related multiplication.

To solve −12 ÷ 4, think: What number times 4 equals −12?

$$-3 \cdot 4 = -12, \text{ so } -12 \div 4 = -3$$

The mean change in the wombat population is **−3**. So on average, the population **decreased by 3 wombats** per year.

EXAMPLE 1 Dividing Integers

Find each quotient.

A $12 \div (-3)$

Think: What number times −3 equals 12?

$-4 \cdot (-3) = 12$, so $12 \div (-3) = -4$.

B $-15 \div (-3)$

Think: What number times −3 equals −15?

$5 \cdot (-3) = -15$, so $-15 \div (-3) = 5$.

9-7 Organizer

Pacing: Traditional 1 day
Block $\frac{1}{2}$ day

Objective: Students divide integers.

Indiana Academic Standards
6.2.2, 6.7.5, 6.7.11

Warm Up

Find each quotient.

1. 18 ÷ 2 **9** **2.** 42 ÷ 7 **6**
3. 56 ÷ 8 **7** **4.** 24 ÷ 6 **4**
5. 3,600 ÷ 4 **900** **6.** 540 ÷ 60 **9**

Problem of the Day

Hank wanted to record the number of marbles he lost to his friend Marcus each day for 5 days. He forgot to record one day, but for the other days he wrote 8, 2, 3, and 4. The average number he lost was 4. What number did he forget to write? **3**

Available on Daily Transparency in CRB

Math Humor

Student: I love doing gazintas. They're my favorite!

Math teacher: I don't understand. What do you mean by *gazintas*?

Student: You know. Like, three gazinta 12 four times; two gazinta 18 nine times…

1 Introduce

Alternate Opener

EXPLORATION

9-7 Dividing Integers

For each multiplication statement, you can write two related division statements.

Multiplication statement	Division statements
2 · 3 = 6	6 ÷ 3 = 2 and 6 ÷ 2 = 3

Complete each table.

1.

Multiply	4 · (−3) =	−4 · (−3) =	−4 · 3 =
Divide	−12 ÷ 4 =	12 ÷ (−4) =	−12 ÷ (−4) =
	−12 ÷ (−3) =	12 ÷ (−3) =	−12 ÷ 3 =

2.

Multiply	2 · (−5) =	−2 · (−5) =	−2 · 5 =
Divide	−10 ÷ 2 =	10 ÷ (−2) =	−10 ÷ (−2) =
	−10 ÷ (−5) =	10 ÷ (−5) =	−10 ÷ 5 =

3.

Multiply	8 · (−3) =	−8 · (−3) =	−8 · 3 =
Divide	−24 ÷ 8 =	24 ÷ (−8) =	−24 ÷ (−8) =
	−24 ÷ (−3) =	24 ÷ (−3) =	−24 ÷ 3 =

Think and Discuss

4. Describe what you think the sign rules are for dividing a positive integer by a negative integer, a negative integer by a positive integer, and a negative integer by a negative integer.

Motivate

Remind students that division is the inverse of multiplication by writing 5 • 2 = 10; therefore, 10 ÷ 5 = 2 and 10 ÷ 2 = 5. Tell them that with this information they can discover the rules for dividing integers. Give examples, such as −3 • −4 = 12 and 2 • −3 = −6. See if anyone discovers the rules for dividing integers.

Exploration worksheet and answers on Chapter 9 Resource Book pp. 64 and 103

2 Teach

Lesson Presentation

Guided Instruction

In this lesson, students learn to divide integers. (Teaching Transparency T24, CRB) First, review multiplication of integers and state the rules. Then write related division statements. For example, −2 • 3 = −6, so −6 ÷ −2 = 3 and −6 ÷ 3 = −2. Point out that the sign rules for dividing integers are the same as the sign rules for multiplying integers.

Because division is the inverse of multiplication, the rules for dividing integers are the same as the rules for multiplying integers.

DIVIDING INTEGERS

If the signs are the same, the quotient is positive.

$24 \div 3 = 8$ $\quad -6 \div (-3) = 2$

If the signs are different, the quotient is negative.

$-20 \div 5 = -4$ $\quad 72 \div (-8) = -9$

Zero divided by any integer equals 0.

$\frac{0}{14} = 0$ $\quad \frac{0}{-11} = 0$

You cannot divide any integer by 0.

EXAMPLE 2 Evaluating Integer Expressions

Remember! $\frac{x}{3}$ means $x \div 3$.

Evaluate $\frac{x}{3}$ for each value of x.

A $x = 6$

$\frac{x}{3}$	*Write the expression.*
$\frac{6}{3} = 6 \div 3$	*Substitute 6 for x.*
$= 2$	*The signs are the same, so the answer is positive.*

B $x = -18$

$\frac{x}{3}$	*Write the expression.*
$\frac{-18}{3} = -18 \div 3$	*Substitute −18 for x.*
$= -6$	*The signs are different, so the answer is negative.*

C $x = -12$

$\frac{x}{3}$	*Write the expression.*
$\frac{-12}{3} = -12 \div 3$	*Substitute −12 for x.*
$= -4$	*The signs are different, so the answer is negative.*

Think and Discuss

Complete each sentence.

1. The quotient of two integers with like signs is __?__.

2. The quotient of two integers with unlike signs is __?__.

COMMON ERROR ALERT

Some students may think the quotient of two negative numbers is negative. Remind students that the rules for adding integers are different from the rules for multiplying and dividing integers.

Additional Examples

Example 1

Find each quotient.

A. $-30 \div 6$ $\quad -5$

B. $-42 \div (-7)$ $\quad 6$

Example 2

Evaluate $\frac{d}{4}$ for each value of d.

A. $d = 16$ $\quad 4$

B. $d = -24$ $\quad -6$

C. $d = -12$ $\quad -3$

Example 2 note: You may wish to have students evaluate some expressions with the variable in the denominator. For example:

Evaluate $\frac{27}{x}$ for $x = -3$.

$\frac{27}{x} = \frac{27}{-3} = -9$

Reaching All Learners
Through Critical Thinking

Have students place the correct signs in front of the dividend and the divisor to make the sentence true. You may use the recording sheet provided in the Chapter 9 Resource Book on p. 89. Have them list all possible answers.

1. ▢ 8 ÷ ▢ 8 = −1 $\quad$ +,− or −,+

2. ▢ 36 ÷ ▢ 4 = 9 $\quad$ +,+ or −,−

3. ▢ 10 ÷ ▢ 2 = −5 $\quad$ +,− or −,+

4. ▢ 63 ÷ ▢ 7 = 9 $\quad$ +,+ or −,−

5. ▢ 12 ÷ ▢ 4 = −3 $\quad$ +,− or −,+

3 Close

Summarize

Summarize the rules for adding, subtracting, multiplying, and dividing integers. Also give expressions as examples to illustrate each rule.

Answers to Think and Discuss

1. positive

2. negative

9-7 PRACTICE & ASSESS

Students may want to refer back to the lesson examples.

Assignment Guide

If you finished Example 1 assign:
Core 1–3, 7–12, 19–24, 31, 37–44
Enriched 9–12, 19–24, 31–44

If you finished Example 2 assign:
Core 1–18, 31–34, 37–44
Enriched 15–44

Notes

9-7 Exercises

FOR EXTRA PRACTICE see page 655

internet connect — Homework Help Online go.hrw.com Keyword: MR4 9-7

GUIDED PRACTICE

See Example 1 Find each quotient.

1. $64 \div 8$ **8**
2. $10 \div (-2)$ **−5**
3. $-21 \div (-7)$ **3**

See Example 2 Evaluate $\frac{m}{2}$ for each value of m.

4. $m = -4$ **−2**
5. $m = 20$ **10**
6. $m = -30$ **−15**

INDEPENDENT PRACTICE

See Example 1 Find each quotient.

7. $45 \div 9$ **5**
8. $-42 \div 6$ **−7**
9. $32 \div (-4)$ **−8**
10. $-60 \div (-10)$ **6**
11. $-75 \div 15$ **−5**
12. $22 \div 11$ **2**

See Example 2 Evaluate $\frac{n}{4}$ for each value of n.

13. $n = 4$ **1**
14. $n = -32$ **−8**
15. $n = 12$ **3**
16. $n = 64$ **16**
17. $n = -92$ **−23**
18. $n = 56$ **14**

PRACTICE AND PROBLEM SOLVING

Divide.

19. $-12 \div 2$ **−6**
20. $\frac{16}{-4}$ **−4**
21. $-6 \div (-6)$ **1**
22. $\frac{-30}{-3}$ **10**
23. $-45 \div 9$ **−5**
24. $\frac{-35}{5}$ **−7**

Evaluate each expression for the given value of the variable.

25. $n \div (-7)$; $n = -21$ **3**
26. $\frac{m}{3}$; $m = -15$ **−5**
27. $\frac{x}{4}$; $x = 32$ **8**
28. $\frac{a}{3}$; $a = -9$ **−3**
29. $w \div (-2)$; $w = -18$ **9**
30. $-48 \div n$; $n = -8$ **6**

31. The graph shows the low temperatures for 5 days in Fairbanks, Alaska.
    a. Find the mean low temperature for Monday, Tuesday, and Wednesday. **−3**
    b. Find the mean low temperature for all 5 days. **−2**

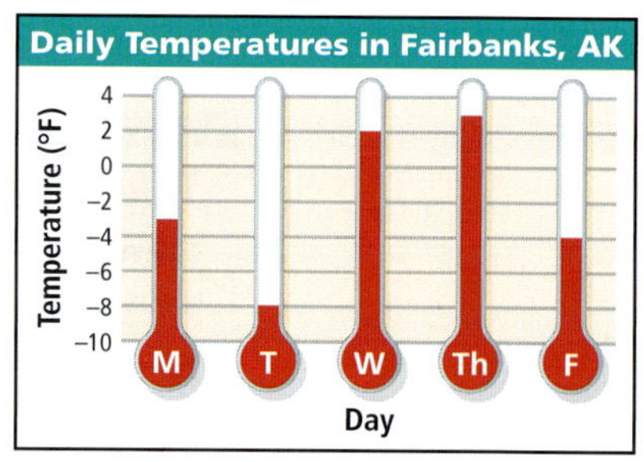

Math Background

Another direct connection can be made between the rules of multiplication and division. Because division can be written as multiplication by the reciprocal, the rules for multiplication can be applied directly to any division problem. Examples:

$$-6 \div 2 = -6 \times \frac{1}{2}$$
$$6 \div -2 = 6 \times -\frac{1}{2}$$
$$-6 \div -2 = -6 \times -\frac{1}{2}$$

RETEACH 9-7

LESSON 9-7 **Reteach**
Dividing Integers

You can use two-color counters to divide integers.

(+1) (−1)

Divide −8 by 2.

First, think about the numerical expression in words.

$-8 \div 2$ means "−8 divided into 2 equal groups."

Then use counters to represent the expression.

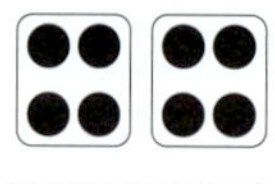

There are 4 negative counters in each group.

$-8 \div 2 = -4$.

Use counters to find each quotient.

1. $-15 \div 3$ **−5**
2. $-12 \div 2$ **−6**
3. $9 \div 3$ **3**
4. $16 \div 4$ **4**
5. $-11 \div 1$ **−11**
6. $-6 \div 3$ **−2**
7. $-20 \div 4$ **−5**
8. $21 \div 3$ **7**
9. $-14 \div 7$ **−2**
10. $-7 \div 7$ **−1**
11. $-18 \div 3$ **−6**
12. $12 \div 4$ **3**
13. $4 \div 2$ **2**
14. $-18 \div 9$ **−2**
15. $-5 \div 5$ **−1**
16. $-20 \div 5$ **−4**

PRACTICE 9-7

LESSON 9-7 **Practice B**
Dividing Integers

Write the sign of each quotient.

1. $56 \div 8$ **positive**
2. $-45 \div (-9)$ **positive**
3. $36 \div (-12)$ **negative**
4. $54 \div (-6)$ **negative**
5. $-84 \div 7$ **negative**
6. $-225 \div (-15)$ **positive**

Find each quotient.

7. $-45 \div 9$ **−5**
8. $15 \div (-3)$ **−5**
9. $-56 \div 8$ **−7**
10. $-10 \div (-5)$ **2**
11. $28 \div (-7)$ **−4**
12. $-36 \div (-6)$ **6**
13. $81 \div 9$ **9**
14. $-72 \div 9$ **−8**
15. $-121 \div (-11)$ **11**

Evaluate $\frac{n}{-3}$ for each value of n.

16. $n = 6$ **−2**
17. $n = -18$ **6**
18. $n = -24$ **8**
19. $n = -36$ **12**
20. $n = 30$ **−10**
21. $n = -21$ **7**

Evaluate $n \div 2$ for each value of n.

22. $n = -14$ **−7**
23. $n = 20$ **10**
24. $n = -24$ **−12**
25. $n = 8$ **4**
26. $n = -18$ **−9**
27. $n = -22$ **−11**

28. What two division equations can you use to check the answer to the problem $6 \cdot (-4) = -24$?
    $-24 \div 6 = -4$ or $-24 \div (-4) = 6$
29. Why are the rules for dividing integers similar to the rules for multiplying integers?
    because division is the inverse of multiplication
30. What two multiplication equations can you use to check the answer to the problem $-32 \div 8 = -4$?
    $8 \cdot (-4) = -32$ or $(-4) \cdot 8 = -32$
31. Name two integers whose product is −18 and whose quotient is −2.
    6 and −3 or −6 and 3

Life Science LINK

The Mediterranean monk seal is one of the world's rarest mammals. Monk seals have become endangered largely because divers hunt them for their skin and disturb their habitat.

Annette found this table in a science article about monk seals.

Changes in Population of Monk Seals							
Years	1966–1970	1971–1975	1976–1980	1981–1985	1986–1990	1991–1995	1996–2000
Change	−250	550	−300	−150	−50	100	200

32a. According to the table, what was the change in the monk seal population from 1966 to 1970? **−250**

b. What does this number mean? **a decrease of 250 seals from 1966 to 1970**

33. Find the mean change per year from 1971 to 1975. (*Hint:* This is a range of 5 years, so divide by 5.) What does your answer mean? **110; an increase of 110 seals each year from 1971 to 1975**

34. Find the mean change per year from 1981 to 1990. What does your answer mean? **−20; a decrease of 20 seals per year from 1981 to 1990**

35. ***WRITE ABOUT IT*** Why is it important to use both positive and negative numbers when tracking the changes in a population?

36. ***CHALLENGE*** Suppose that there were 500 monk seals in 1966. How many were there in 2000? **600**

Spiral Review

Solve each equation. Check your answers. (Lesson 3-10)

37. $4.2 + n = 6.7$ **$n = 2.5$** **38.** $x - 2.3 = 1.6$ **$x = 3.9$** **39.** $1.5w = 3.6$ **$w = 2.4$**

Solve each equation. Check your answers. (Lesson 5-10)

40. $\frac{1}{2} + m = 2$ **$1\frac{1}{2}$ or $\frac{3}{2}$** **41.** $3 - \frac{2}{3} = n$ **$2\frac{1}{3}$ or $\frac{7}{3}$** **42.** $\frac{1}{5} + 6 = x$ **$6\frac{1}{5}$ or $\frac{31}{5}$**

43. TEST PREP ___?___ angles are always congruent. (Lesson 7-3) **C**

- **A** Complementary
- **B** Adjacent
- **C** Vertical
- **D** Supplementary

44. TEST PREP Which of the following are equivalent ratios? (Lesson 8-1) **G**

F $\frac{2}{3}, \frac{3}{2}$ **G** $\frac{1}{2}, \frac{4}{8}$ **H** $\frac{1}{5}, \frac{4}{5}$ **J** $\frac{3}{4}, \frac{5}{6}$

CHALLENGE 9-7

LESSON 9-7 **Challenge**
Divide and Answer

Long ago, people in India began using negative numbers. They were the first known people to do so. When did the people of India start using negative numbers?

Solve each division problem below. Then in the box at the bottom of the page, write each problem's letter in the blank above its quotient. When you have solved all the problems, you will have found the answer to the question.

- A $18 \div (-3)$ **−6**
- D $-27 \div (-9)$ **3**
- E $-81 \div 9$ **−9**
- F $24 \div (-3)$ **−8**
- G $-35 \div (-5)$ **7**
- H $24 \div (-2)$ **−12**
- N $-40 \div (-8)$ **5**
- O $32 \div (-8)$ **−4**
- R $-17 \div 17$ **−1**
- S $-48 \div (-8)$ **6**
- T $22 \div (-2)$ **−11**
- U $-100 \div (-10)$ **10**
- Y $16 \div (-8)$ **−2**

O	N	E		T	H	O	U	S	A	N	D,
−4	5	−9		−11	−12	−4	10	6	−6	5	3

F	O	U	R		H	U	N	D	R	E	D
−8	−4	10	−1		−12	10	5	3	−1	−9	3

Y	E	A	R	S		A	G	O
−2	−9	−6	−1	6		−6	7	−4

PROBLEM SOLVING 9-7

LESSON 9-7 **Problem Solving**
Dividing Integers

Use the table below to answer questions 1–6.

Temperatures for Barrow, Alaska

	JAN	FEB	MAR	APRIL	MAY	JUNE	JULY	AUG	SEPT	OCT	NOV	DEC
Temp (°F)	−13	−18	−15	−2	19	34	39	38	31	14	−2	−11

1. What is the average temperature in Barrow for December and January? **−12°F**
2. What is the average temperature in Barrow for March and July? **12°F**
3. Which month's average temperature is half as warm as August's? **May**
4. What is the average temperature in Barrow for October and November? **6°F**
5. What is the average temperature in Barrow for January through April? **−12°F**
6. What is the city's average temperature for September through December? **8°F**

Circle the letter of the best answer.

7. A submarine dove to a depth of 168 feet in 7 minutes. What was the average rate of change in its location?
   - A 24 feet
   - B 168 feet
   - **(C)** −24 feet
   - D −168 feet
8. In its first 4 months of business, Skyscraper Records reported its losses as −$1,520. What was the company's average monthly loss?
   - F −$1,520
   - **(G)** −$380
   - H −$38
   - J $380
9. Which of these expressions checks the solution to the division problem $-8 \div (-2) = 4$?
   - A $-8 \cdot (-2)$
   - B $4 \cdot 4$
   - C $-2 \cdot (2)$
   - **(D)** $4 \cdot (-2)$
10. A glacier is melting 3 in^3 a year. At that rate, how long will it take for the glacier to change by −24 in^3?
    - F 72 years
    - G 6 years
    - **(H)** 8 years
    - J 24 years

Interdisciplinary LINK

Life Science

Exercises 32–36 involve changes in the population of an endangered species. Protection of endangered species is studied in middle school life science programs such as *Holt Science & Technology*.

Answers

35. so you know whether the population is growing or shrinking

Journal

Have students describe in writing how integer division can be used in the real world.

Test Prep Doctor

For Exercise 43, all of the answer choices describe angles. Review these terms without giving away the answer to the question. Stress to students that the important word in the sentence is *always*. Answers **A, B,** and **D** would be acceptable answers only if *always* were replaced with *sometimes*.

Lesson Quiz

Find each quotient.

1. $18 \div (-3)$ **−6** **2.** $-36 \div (-6)$ **6**

3. $-64 \div (-8)$ **8** **4.** $-45 \div 3$ **−15**

Evaluate $\frac{x}{4}$ for each value of x.

5. $x = -12$ **−3** **6.** $x = 20$ **5**

7. Carmen's mom has $3,200 in her bank account. If she uses the money to make monthly rent payments of $375, how many whole payments can she make? **8**

Available on Daily Transparency in CRB

Hands-On LAB 9C Model Integer Equations

Pacing: Traditional 1 day
Block $\frac{1}{2}$ day

Objective: To use algebra tiles to model and solve integer equations

Materials: Algebra tiles, equation mat

Lab Resources

Hands-On Lab Activities . . . . pp. 83–84

Using the Pages

Discuss with students what each algebra tile represents. Also discuss what the two sides of the equation mat represent.

1. How can you represent the equation $x - 5 = -2$ with algebra tiles?

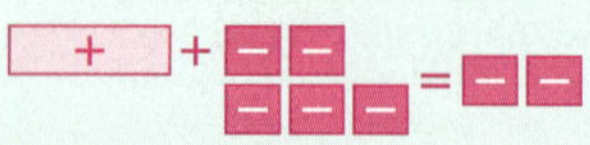

2. How do you solve the equation using the algebra tiles? What must you do first? Add 3 positive-negative pairs to the right side of the equation so that you can remove 5 negative tiles from each side.

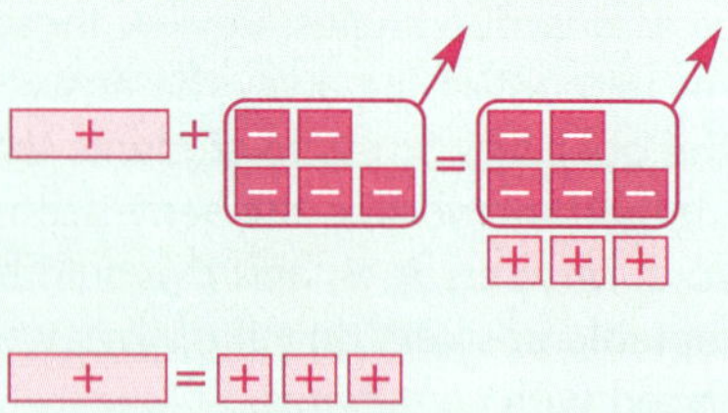

Hands-On LAB 9C Model Integer Equations

Use with Lesson 9-8

KEY

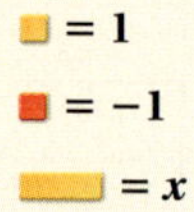

REMEMBER

You can add or subtract the same number on both sides of an equation.

Adding or subtracting zero does not change a number's value.

internet connect
Lab Resources Online
go.hrw.com
KEYWORD: MR4 Lab9C

You can use algebra tiles to model equations. An equation mat represents the two sides of an equation. To find the value of the variable, get the x-tile by itself on one side of the mat. You may remove the same number of yellow tiles or the same number of red tiles from both sides.

Activity

Use algebra tiles to model and solve each equation.

1 $x + 2 = 6$

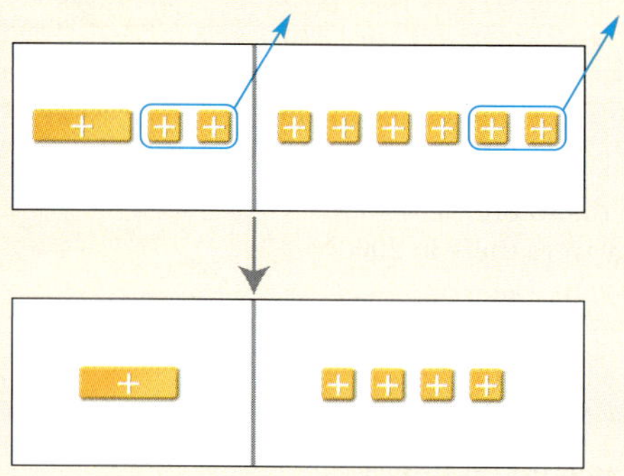

Remove 2 yellow tiles from both sides of the mat.

$x = 4$

2 $x - 3 = -5$

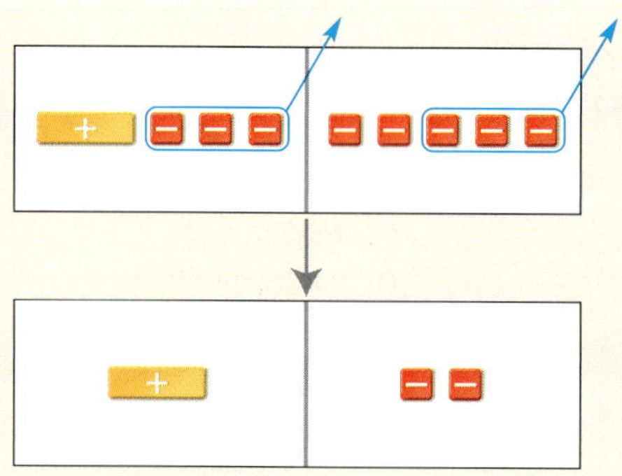

Use red tiles to model subtraction. Remove 3 red tiles from both sides of the mat.

$x = -2$

3 $x + 6 = 2$

You do not have enough yellow tiles on the right side to remove 6 from both sides. Add pairs of red and yellow tiles to the right side until you have enough yellow tiles to subtract.

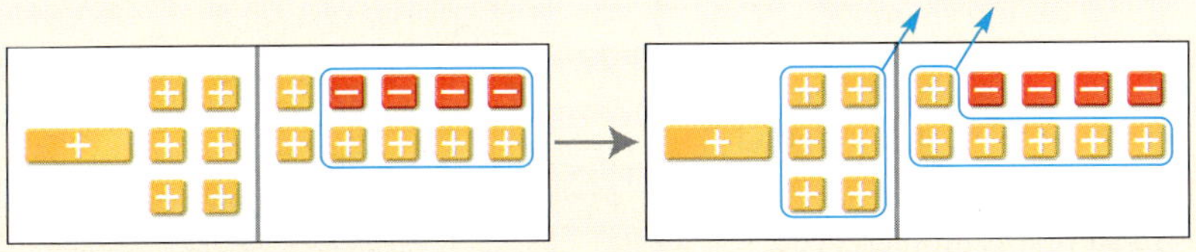

$x = -4$

Add these tiles. *Now you can remove 6 yellow tiles from both sides of the mat.*

4 $3x = -9$

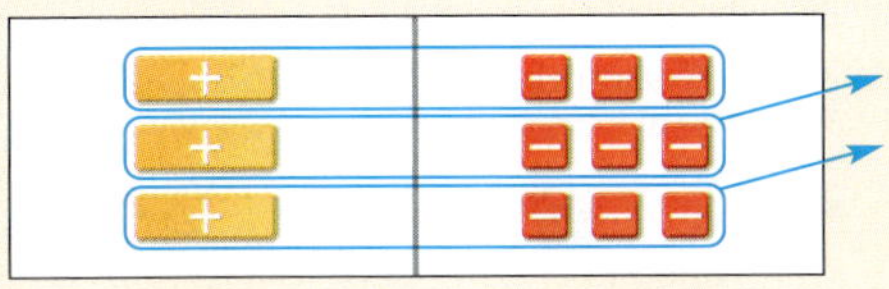

Divide each side into 3 equal groups. Remove all but one of the groups.

$x = -3$

Think and Discuss

1. In **4**, why did you divide both sides into 3 groups? **You divided both sides into groups in order to isolate the bar representing *x*.**
2. Why can you add pairs of red and yellow tiles to an equation mat? Why is it not necessary to add them to both sides? **Because a pair of red and yellow tiles represents 0, it doesn't change the value of the equation.**
3. When you add zero to an equation, how do you know the number of red and yellow tiles to add? **You add as many as you need in order to have enough tiles to subtract an equal number from both sides.**
4. How can you use algebra tiles to check your answers? **You can work the problem backward using algebra tiles to check your work.**

Try This

Use algebra tiles to model and solve each equation.

1. $x + 6 = 3$ **−3** **2.** $x - 1 = -8$ **−7** **3.** $2x = 14$ **7** **4.** $4x = -8$ **−2**

Answers

Try This

1. $x + 6 = 3$

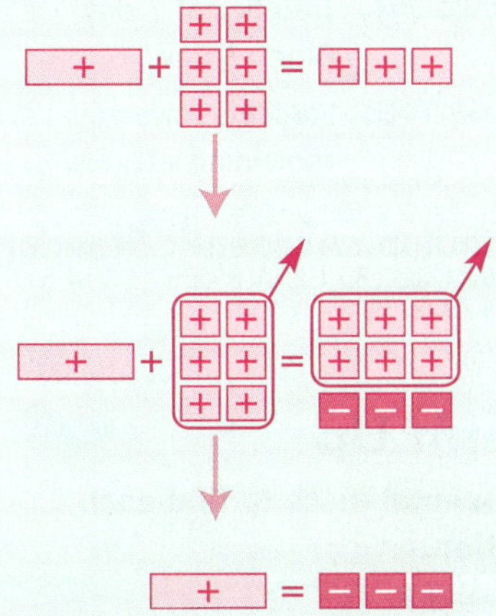

$x = -3$

2. $x - 1 = -8$

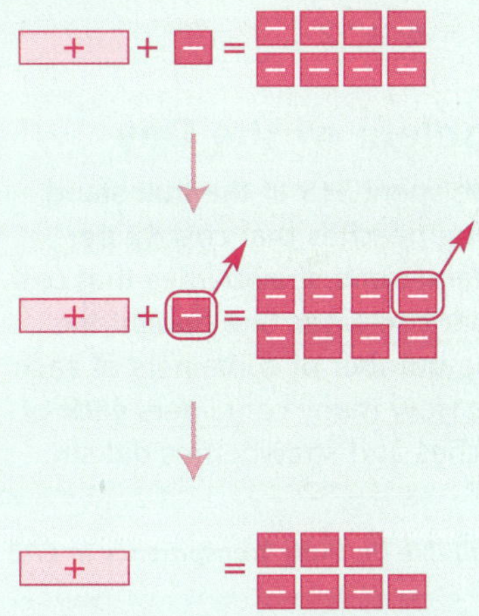

$x = -7$

3. $2x = 14$

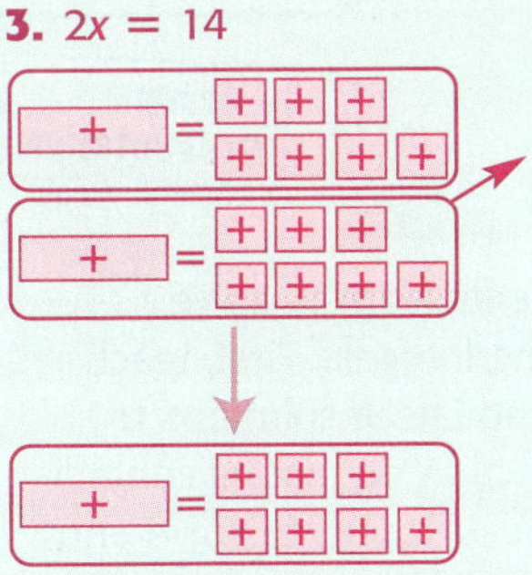

$x = 7$

4. $4x = -8$

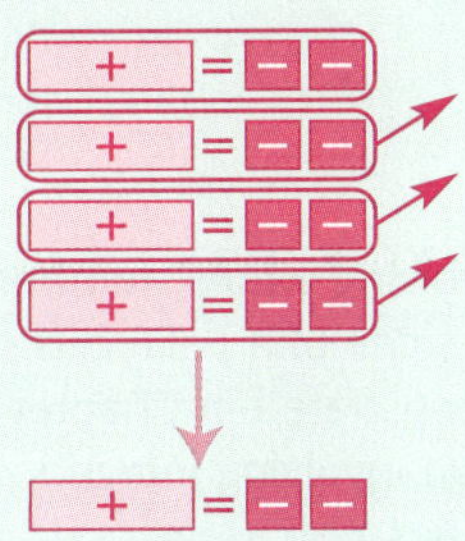

$x = -2$

9-8 Organizer

Pacing: Traditional 1 day
Block $\frac{1}{2}$ day

Objective: Students solve equations containing integers.

Indiana Academic Standards
6.2.1, 6.2.2, 6.3.1, 6.7.1, 6.7.2, 6.7.3, 6.7.4, 6.7.5, 6.7.9, 6.7.11

Warm Up

Use mental math to find each solution.

1. $6 + x = 12$ $x = 6$
2. $3x = 15$ $x = 5$
3. $\frac{x}{5} = 4$ $x = 20$
4. $x - 8 = 12$ $x = 20$

Problem of the Day

Marie spent $15 at the fruit stand, buying peaches that cost $2 per container and strawberries that cost $3 per container. She bought the same number of containers of each fruit. How many containers each of peaches and strawberries did she buy? 3

Available on Daily Transparency in CRB

Math Fact

French philosopher and mathematician René Descartes was the first to use *a*, *b*, and *c* to represent known quantities and *x*, *y*, and *z* to represent unknown quantities.

9-8 Solving Integer Equations

Learn to solve equations containing integers.

The entrance to the Great Pyramid of Khufu is 55 ft above ground. The underground chamber is 102 ft below ground. From the entrance, what is the distance to the underground chamber?

To solve this problem, you can use an equation containing integers.

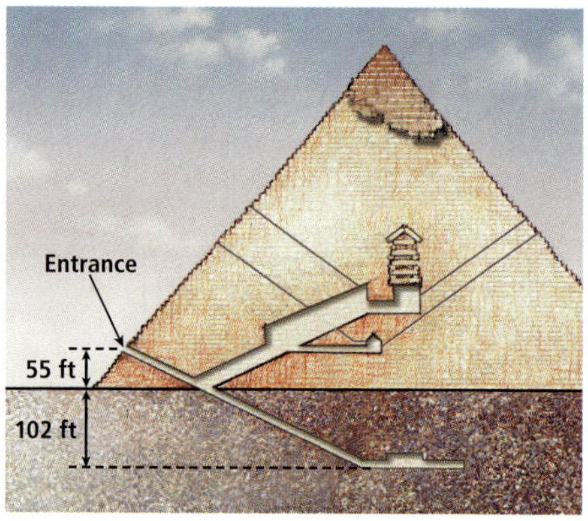

height of entrance	+	distance to underground chamber	=	height of underground chamber
55	**+**	***d***	**=**	**-102**

$$55 + d = -102 \qquad \textit{Write the equation.}$$
$$\underline{-55} \quad \underline{-55} \qquad \textit{Subtract 55 from both sides.}$$
$$d = -157$$

It is -157 ft from the entrance to the underground chamber. The sign is negative, which means you go down 157 ft.

EXAMPLE 1 Adding and Subtracting to Solve Equations

Helpful Hint

To solve this equation using algebra tiles, you can add four red tiles to both sides and then remove pairs of red and yellow tiles. This is because subtracting a number is the same as adding its opposite.

A Solve $4 + x = -2$. Check your answer.

$$4 + x = -2 \qquad \textit{4 is added to x.}$$
$$\underline{-4} \quad \underline{-4} \qquad \textit{Subtract 4 from both sides to undo the addition.}$$
$$x = -6$$

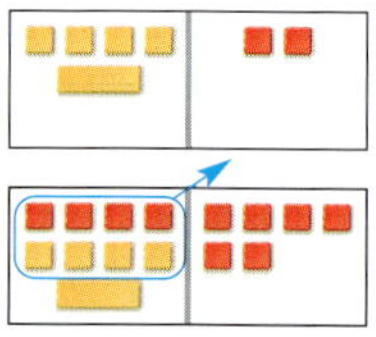

Check

$$4 + x = -2 \qquad \textit{Write the equation.}$$
$$4 + (-6) \stackrel{?}{=} -2 \qquad \textit{Substitute } -6 \textit{ for x.}$$
$$-2 \stackrel{?}{=} -2 \checkmark \qquad -6 \textit{ is a solution.}$$

1 Introduce

Alternate Opener

EXPLORATION

9-8 Solving Integer Equations

You can use algebra tiles to model solving integer equations.

The equation $x - 3 = 5$ is modeled.

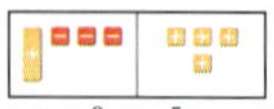

To get x alone on one side, add three positive tiles to each side of the mat. This allows you to remove three zero pairs from the left side.

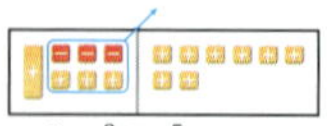

The solution is 8.

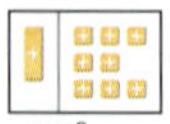

Use algebra tiles to solve each equation.

1. $x + 5 = 9$
2. $x - 6 = 2$
3. $x + 4 = -1$
4. $6 = x - 7$
5. $8 = x + 2$
6. $3 = x - 9$

Think and Discuss

7. **Explain** how you know when to add zero pairs.

Motivate

Solve an equation involving whole numbers by operating on both sides of a balanced set of scales, as was demonstrated in Chapter 2. Then solve an equation involving integers and show the similarities.

Exploration worksheet and answers on Chapter 9 Resource Book pp. 73–74 and 105–106

2 Teach

Lesson Presentation

Guided Instruction

In this lesson, students learn to solve equations containing integers. First, teach students to solve and check solutions to addition and subtraction equations containing integers. Then teach them to solve and check solutions to multiplication and division equations. Remind students that to solve an equation, they must "undo" the operation being performed on the variable.

Teaching Tip Remind students to think of the rules for adding, subtracting, multiplying, and dividing integers when they are solving for a variable. Have them check the solution each time to catch any mistakes.

B Solve $y - 6 = -5$. Check your answer.

$y - 6 = -5$	*6 is subtracted from y.*
$\underline{+6 \quad +6}$	*Add 6 to both sides to*
$y \quad = \quad 1$	*undo the subtraction.*

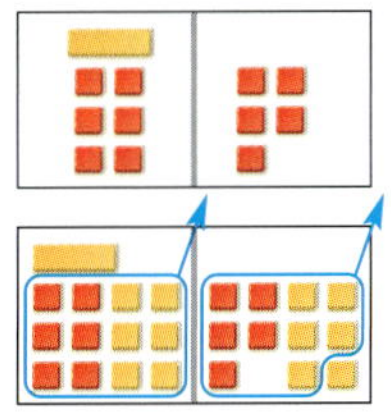

Check

$y - 6 = -5$	*Write the equation.*
$1 - 6 \stackrel{?}{=} -5$	*Substitute 1 for y.*
$-5 \stackrel{?}{=} -5$ ✓	*1 is a solution.*

EXAMPLE 2 Multiplying and Dividing to Solve Equations

Solve each equation. Check your answers.

A $-3a = 15$

$\frac{-3a}{-3} = \frac{15}{-3}$	*a is multiplied by −3. Divide both sides by −3 to undo the multiplication.*
$a = -5$	

Check

$-3a = 15$	*Write the equation.*
$-3(-5) \stackrel{?}{=} 15$	*Substitute −5 for a.*
$15 \stackrel{?}{=} 15$ ✓	*−5 is a solution.*

B $\frac{b}{-4} = -2$

$-4 \cdot \frac{b}{-4} = -4 \cdot (-2)$	*b is divided by −4. Multiply both sides by −4 to undo the division.*
$b = 8$	

Check

$\frac{b}{-4} = -2$	*Write the equation.*
$8 \div (-4) \stackrel{?}{=} -2$	*Substitute 8 for b.*
$-2 \stackrel{?}{=} -2$ ✓	*8 is a solution.*

Think and Discuss

1. **Tell** what operation you would use to solve $x + 12 = -32$.
2. **Tell** whether the solution to $-9t = -27$ will be positive or negative without actually solving the equation.
3. **Explain** how to check your answer to an integer equation.

COMMON ERROR ALERT

Some students may have difficulty affixing signs to integers when writing equations to solve word problems. Have students search for key words that would indicate the use of negative integers. For example, in the lesson opener, "The underground chamber is 102 feet **below** ground."

Additional Examples

Example 1

Solve each equation. Check your answers.

A. $-8 + y = -13$ $\quad y = -5$

B. $n - 2 = -8$ $\quad n = -6$

Example 2

Solve each equation.

A. $4m = -20$ $\quad m = -5$

B. $\frac{x}{3} = -7$ $\quad x = -21$

3 Close

Reaching All Learners

Through Cognitive Strategies

While students are solving examples or exercises, have them circle the operation and the integer that is operating on the variable. Then directly below the circled operation and integer, have students write the inverse operation and the same integer that was circled. Remind students to keep the equation balanced by writing the inverse operation and the same integer that was circled on the other side of the equation.

Summarize

Discuss the steps for finding and checking the solution to addition, subtraction, multiplication, and division equations with integers. Be sure to include and spend time on equations involving subtracting negatives and equations involving multiplying and dividing by negatives.

Answers to Think and Discuss

1. subtraction
2. positive, because a negative integer multiplied by a positive integer has a negative product
3. In the original equation, substitute the solution for the variable and follow the order of operations. The solution is correct if the left side of the equation equals the right side.

9-8 PRACTICE & ASSESS

Students may want to refer back to the lesson examples.

Assignment Guide

If you finished Example 1 assign:
Core 1–3, 7–12, 27, 28, 35, 36, 49–63
Enriched 7–12, 27, 28, 35, 36, 42, 43, 49–63

If you finished Example 2 assign:
Core 1–18, 40, 41, 49–63
Enriched 28–63

Notes

9-8 Exercises

FOR EXTRA PRACTICE see page 655

internet connect
Homework Help Online
go.hrw.com Keyword: MR4 9-8

GUIDED PRACTICE

See Example 1
Solve each equation. Check your answers.

1. $m - 3 = 9$ $m = 12$
2. $a - 8 = -13$ $a = -5$
3. $z - 12 = -3$ $z = 9$

See Example 2
4. $-4b = 32$ $b = -8$
5. $\frac{w}{3} = 18$ $w = 54$
6. $5c = -35$ $c = -7$

INDEPENDENT PRACTICE

See Example 1
Solve each equation. Check your answers.

7. $g - 9 = -5$ $g = 4$
8. $v - 7 = 19$ $v = 26$
9. $t - 13 = -27$ $t = -14$
10. $x + 2 = -12$ $x = -14$
11. $y + 9 = -10$ $y = -19$
12. $20 + w = 10$ $w = -10$

See Example 2
13. $6j = 48$ $j = 8$
14. $7s = -49$ $s = -7$
15. $\frac{a}{-2} = 26$ $a = -52$
16. $\frac{m}{-12} = 4$ $m = -48$
17. $\frac{k}{5} = -4$ $k = -20$
18. $u \div 6 = -10$ $u = -60$

PRACTICE AND PROBLEM SOLVING

Solve each equation. Check your answers.

19. $x - 12 = 5$ $x = 17$
20. $w - 3 = -2$ $w = 1$
21. $-7k = 28$ $k = -4$
22. $\frac{m}{-3} = 5$ $m = -15$
23. $a - 10 = 9$ $a = 19$
24. $n - 19 = -22$ $n = -3$
25. $13g = -39$ $g = -3$
26. $s \div 6 = -3$ $s = -18$
27. $24 + f = 16$ $f = -8$
28. $d - 26 = 7$ $d = 33$
29. $-6c = 54$ $c = -9$
30. $h \div (-4) = 21$ $h = -84$
31. $b - 17 = 15$ $b = 32$
32. $u - 82 = -7$ $u = 75$
33. $-8a = -64$ $a = 8$
34. $\frac{t}{11} = -5$ $t = -55$
35. $31 + j = -14$ $j = -45$
36. $c + 23 = 10$ $c = -13$
37. $15n = -60$ $n = -4$
38. $z \div (-5) = -9$ $z = 45$
39. $j - 20 = -23$ $j = -3$

40. A submarine captain sets the following diving course: dive 200 ft, stop, and then dive another 200 ft. If this pattern is continued, how many dives will be necessary to reach a location 14,000 ft below sea level? **70 dives**

41. While exploring a cave, Lin noticed that the temperature dropped 4°F for every 30 ft that she descended. What is Lin's depth if the temperature is 8° lower than the temperature at the surface? **−60 ft**

42. ***SPORTS*** After two rounds in the 2001 LPGA Champions Classic, Wendy Doolan had a score of –12. Her score in the second round was –8. What was her score in the first round? **−4**

Math Background

Before students studied this chapter, they may not have been able to solve or understand the solution to an equation such as $x + 1 = 0$. Through the study of integers, students can now solve equations of the form $x + a = b$, $x - a = b$, $ax = b$, and $\frac{x}{a} = b$, where a and b are integers. The one exception is that equations in this lesson of the form $ax = b$ are restricted to cases in which $\frac{b}{a}$ is an integer.

RETEACH 9-8

LESSON 9-8 **Reteach**
Solving Integer Equations

You can write related equations between addition and subtraction equations.
$7 + 6 = 13$ $13 - 6 = 7$

You can write related equations between multiplication and division equations.
$3 \cdot -4 = -12$ $-12 \div -4 = 3$

Use related equations to solve each of the following.

A. $x + 4 = -2$
Think: $-2 - 4 = x$
$x = -6$
Check: $x + 4 = -2$
$-6 + 4 \stackrel{?}{=} -2$ Substitute.
$-2 = -2$

B. $-3a = 15$
Think: $15 \div -3 = a$
$a = -5$
Check: $-3a = 15$
$-3 \cdot (-5) \stackrel{?}{=} 15$ Substitute.
$15 = 15$

C. $m - 6 = 1$
Think: $1 + 6 = m$
$m = 7$
Check: $m - 6 = 1$
$7 - 6 \stackrel{?}{=} 1$ Substitute.
$1 = 1$

D. $b \div 4 = 6$
Think: $6 \cdot 4 = b$
$b = 24$
Check: $b \div 4 = 6$
$24 \div 4 \stackrel{?}{=} 6$ Substitute.
$6 = 6$

Use related facts to solve each equation. Then check each answer.

1. $x + 8 = 12$ $x = 4$
2. $x \div 9 = 4$ $x = 36$
3. $x - 12 = 7$ $x = 19$
4. $5x = 40$ $x = 8$
5. $x \div 3 = 6$ $x = 18$
6. $x + (-12) = 16$ $x = 28$
7. $7x = -42$ $x = -6$
8. $x - (-8) = 23$ $x = 15$
9. $x - 3 = -9$ $x = -6$
10. $-4x = 28$ $x = -7$
11. $x + (-9) = -13$ $x = -4$
12. $x \div (-5) = -4$ $x = 20$

PRACTICE 9-8

LESSON 9-8 **Practice B**
Solving Integer Equations

Write what you should do to solve each equation.

1. $x - 4 = -10$ Add 4 to both sides.
2. $-2x = 8$ Divide both sides by −2.
3. $-3 + x = 12$ Subtract −3 from both sides.
4. $x \div 6 = -9$ Multiply both sides by 6.
5. $-7 + x = -15$ Subtract −7 from both sides.
6. $35 = -5x$ Divide both sides by −5.

Solve each equation.

7. $-45 \div x = -5$ $x = 9$
8. $x - 9 = -1$ $x = 8$
9. $36 \div x = -6$ $x = -6$
10. $x - 10 = -12$ $x = -2$
11. $-8x = 56$ $x = -7$
12. $x \div 7 = -9$ $x = -63$
13. $3x = -36$ $x = -12$
14. $15 + x = 21$ $x = 6$
15. $-4x = 64$ $x = -16$
16. $x \div (-3) = -5$ $x = 15$
17. $x \div -12 = -5$ $x = 60$
18. $x - 13 = -9$ $x = 4$
19. $-7 + x = 4$ $x = 11$
20. $-9x = 54$ $x = -6$
21. $49 \div x = -7$ $x = -7$

22. If you multiply a value x by −2 and the product is −14, what sign is the value of x? Explain.
positive; because the product of a negative and a positive value is negative

23. You separate an amount into 3 equal groups of −6. Write and solve an division equation to model this situation.
$x \div 3 = -6$; $x = -18$

Social Studies LINK

Bolivia actually has two capitals. La Paz is Bolivia's chief industrial city, and the congress meets there. However, the supreme court meets in Sucre, farther south.

Use the graph for Exercises 43 and 44.

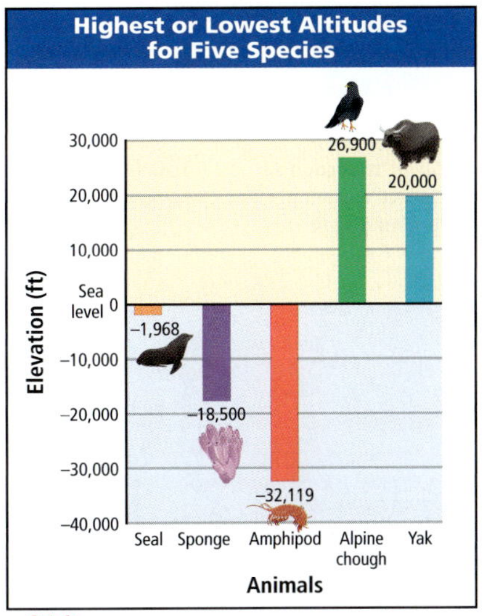

43. ***LIFE SCIENCE*** Scientists have found live bacteria at elevations of 135,000 ft. This is 153,500 ft above one of the animals in the graph. Which one? (*Hint:* Solve $x + 153{,}500 = 135{,}000$.) $x = -18{,}500$; sponge

44. The world's highest capital city is La Paz, Bolivia, with an elevation of 11,808 ft. The highest altitude that a yak has been found at is how much higher than La Paz? (*Hint:* Solve $11{,}808 + x = 20{,}000$.) 8,192 ft.

45. Carla is a diver. On Friday, she dove 5 times as deep as she dove on Monday. If she dove to −120 ft on Friday, how deep did she dive on Monday? −24 ft.

46. ***WRITE A PROBLEM*** Write a word problem that could be solved using the equation $x - 3 = -15$.

47. ***WRITE ABOUT IT*** Is the solution to $3n = -12$ positive or negative? How could you tell without solving the equation?

48. ***CHALLENGE*** Find each answer.

 a. $12 \div (-3 \cdot 2) \div 2$ −1 b. $12 \div (-3 \cdot 2 \div 2)$ −4

 Why are the answers different even though the numbers are the same? order of operations

Spiral Review

Write the prime factorization using exponents. (Lesson 4-2)

49. 76 $2^2 \times 19$ **50.** 12 $2^2 \times 3$ **51.** 16 2^4 **52.** 18 2×3^2

53. 21 3×7 **54.** 128 2^7 **55.** 156 $2^2 \times 3 \times 13$ **56.** 49 7^2

Add or subtract. Write each answer in simplest form. (Lesson 5-7)

57. $\frac{1}{2} + \frac{3}{4}$ $1\frac{1}{4}$ or $\frac{5}{4}$ **58.** $\frac{2}{3} - \frac{1}{5}$ $\frac{7}{15}$ **59.** $\frac{2}{5} + \frac{1}{2}$ $\frac{9}{10}$

60. $\frac{5}{6} - \frac{1}{3}$ $\frac{1}{2}$ **61.** $\frac{3}{7} + \frac{1}{4}$ $\frac{19}{28}$ **62.** $\frac{8}{9} - \frac{1}{6}$ $\frac{13}{18}$

63. TEST PREP A __?__ is a rectangle with four congruent sides. (Lesson 7-6) B

A triangle **B** square **C** rhombus **D** trapezoid

Answers

46. Possible answer: Three less than a number is −15. What is that number?

47. Negative; a negative number divided by a positive number gives a negative quotient.

Journal

Have students write about why there is a need for integers in our number system.

Test Prep Doctor

For Exercise 63, all of the answer choices are names of polygons. **A** is not a quadrilateral. **C** is a quadrilateral with four congruent sides, but it does not have four right angles. **D** is a quadrilateral, but it has neither four congruent sides nor four right angles.

CHALLENGE 9-8

LESSON 9-8 **Challenge**
Balancing Act

Use one expression from Box A and one from Box B to write an equation in the pans below. The two expressions must be equal to keep the scale balanced. Use each expression only once. Possible answers are given.

A		B	
−7 + 1	−6 • (−3)	−2 • (−5)	−6 − (−10)
−16 ÷ (−4)	−5 + (−2)	−4 • 3	25 + (−7)
−15 • 0	−8 − (−5)	2 • (−3)	−27 ÷ 9
−6 + (−6)	17 + (−7)	−12 + 12	14 ÷ (−2)

1. −7 + 1 = 2 • (−3)
2. −16 ÷ (−4) = −6 − (−10)
3. −6 • (−3) = 25 + −7
4. −6 + (−6) = −4 • 3
5. −5 + (−2) = 14 ÷ (−2)
6. 17 + (−7) = −2 • (−5)
7. −8 − (−5) = −27 ÷ 9
8. −15 • 0 = −12 + 12

PROBLEM SOLVING 9-8

LESSON 9-8 **Problem Solving**
Solving Integer Equations

For questions 1–8, the temperatures found are in °F.

1. The highest recorded temperature in Africa is the solution to $x \div (-4) = -34$. What is Africa's highest recorded temperature? 136°F
2. The lowest recorded temperature in Australia is the solution to $7x = -56$. What is Australia's lowest recorded temperature? −8°F
3. To find Africa's lowest recorded temperature, solve the following equation: $80 - x = 91$. −11°F
4. To find Europe's highest recorded temperature, solve the following equation: $x \div -2 = -61$. 122°F
5. The solution to $-2x = -116$ is the highest recorded temperature in Antartica. What is Antartica's highest recorded temperature? 58°F
6. The solution to $x + (-23) = -90$ is the lowest recorded temperature in Europe. What is Europe's lowest recorded temperature? −67°F

Circle the letter of the correct answer.

7. Which of the following is a solution to $x + (-11) = -140$?
 A 12
 (B) −129
 C −151
 D −1,540
8. Which of the following is a solution to $-110 + x = 19$?
 F 91
 (G) 129
 H −5
 J −2,090
9. Which of the following is a solution to $5x = -75$?
 A −375
 B −80
 C −70
 (D) −15
10. Which of the following is a solution to $-270 \div x = -30$?
 F 8,100
 G −300
 (H) 9
 J −240

Lesson Quiz

Solve each equation.

1. $5 + x = -6$ $x = -11$

2. $y - 9 = -7$ $y = 2$

3. $-6a = 24$ $a = -4$

4. $\frac{x}{5} = -9$ $x = -45$

5. A submarine captain sets the following diving course: dive 300 ft, stop, and then dive another 300 ft. If this pattern is continued, how many dives will be necessary to reach a location 3,000 ft below sea level? 10

Available on Daily Transparency in CRB

CHAPTER 9

Extension

Pacing: Traditional 1 day
Block $\frac{1}{2}$ day

Objective: Students recognize negative exponents by examining patterns.

Using this Page

In Lesson 1-3, students learned to represent numbers by using exponents. In this extension, students will develop a rule to calculate numbers raised to negative powers by examining patterns in negative exponents.

EXTENSION

Integer Exponents

Learn to recognize negative exponents by examining patterns.

You have already learned about positive exponents. Exponents can be negative, too. To determine the values of negative powers, write some positive powers and look for a pattern.

Remember!

Exponent

$10^3 = 10 \cdot 10 \cdot 10$

Base

EXAMPLE 1 Finding Patterns in Exponents

Find a pattern in the table.

Power	10^3	10^2	10^1	10^0	10^{-1}	10^{-2}
Value	1,000	100	10	1	$\frac{1}{10}$	$\frac{1}{100}$

÷ 10 ÷ 10 ÷ 10 ÷ 10 ÷ 10

One possible pattern is "divide by 10."

EXAMPLE 2 Using Patterns in Exponents

Find each value: 2^0, 2^{-1}, 2^{-2}, 2^{-3}.

Make a table like the one in Example 1. Write some powers of 2 that you know, and look for a pattern.

Power	2^3	2^2	2^1	2^0	2^{-1}	2^{-2}	2^{-3}
Value	8	4	2				

÷ 2 ÷ 2 ÷ 2 ÷ 2 ÷ 2 ÷ 2

One possible pattern is "divide by 2."

$2^0 = 2 \div 2 = 1 \quad 2^{-1} = 1 \div 2 = \frac{1}{2} \quad 2^{-2} = \frac{1}{2} \div 2 = \frac{1}{4} \quad 2^{-3} = \frac{1}{4} \div 2 = \frac{1}{8}$

Look at the table in Example 2. There is another pattern.

$$2^{-1} = \frac{1}{2^1} = \frac{1}{2} \qquad 2^{-2} = \frac{1}{2^2} = \frac{1}{4} \qquad 2^{-3} = \frac{1}{2^3} = \frac{1}{8}$$

This pattern works for all negative exponents. A number raised to a negative exponent equals 1 divided by that number raised to the opposite (positive) exponent.

1 Introduce

Motivate

Have students complete this geometric sequence.

8, 4, 2, 1, ___, ___, ___, ... $\frac{1}{2}, \frac{1}{4}, \frac{1}{8}$

Write the first four terms in exponential form. $2^3, 2^2, 2^1, 2^0$, ___, ___, ___, ...

Then ask students to complete the sequence. $2^{-1}, 2^{-2}, 2^{-3}$

2 Teach

Lesson Presentation

Guided Instruction

In this extension, students learn to recognize negative exponents by examining patterns. Show students tables of decreasing powers of numbers. Then point out the pattern. Next, help them see the following rule: A number raised to a negative exponent equals 1 divided by that number raised to the opposite (positive) exponent.

EXTENSION Exercises

Complete each table by extending the pattern.

1.

Power	3^3	3^2	3^1	3^0	3^{-1}	3^{-2}
Value	27	9	3	■ 1	■ $\frac{1}{3}$	■ $\frac{1}{9}$

2.

Power	5^{-2}	5^{-1}	5^0	5^1	5^2	5^3
Value	■ $\frac{1}{25}$	■ $\frac{1}{5}$	■ 1	5	25	125

3.

Power	6^3	6^2	6^1	6^0	6^{-1}	6^{-2}
Value	216	36	6	■ 1	■ $\frac{1}{6}$	■ $\frac{1}{36}$

Find the missing exponent.

4. $81 = 9^{■}$ 9^2 **5.** $\frac{1}{7} = 7^{■}$ 7^{-1} **6.** $64 = 4^{■}$ 4^3 **7.** $\frac{1}{64} = 8^{■}$ 8^{-2}

8. $49 = 7^{■}$ 7^2 **9.** $\frac{1}{3} = 3^{■}$ 3^{-1} **10.** $25 = 5^{■}$ 5^2 **11.** $\frac{1}{49} = 7^{■}$ 7^{-2}

12. $64 = 2^{■}$ 2^6 **13.** $\frac{1}{16} = 4^{■}$ 4^{-2} **14.** $\frac{1}{64} = 4^{■}$ 4^{-3} **15.** $\frac{1}{81} = 3^{■}$ 3^{-4}

Find each value.

16. 8^3 512 **17.** 3^{-3} $\frac{1}{27}$ **18.** 6^3 216 **19.** 9^{-3} $\frac{1}{729}$

20. 7^{-3} $\frac{1}{343}$ **21.** 4^4 256 **22.** 1^{-8} 1 **23.** 8^{-2} $\frac{1}{64}$

24. 1^2 1 **25.** 5^{-3} $\frac{1}{125}$ **26.** 4^2 16 **27.** 1^{-3} 1

28. For each row of the table, find the number that is not equal to the other three.

a. 10	10	10^{-1}	$\frac{1}{10}$	0.1
b. $\frac{1}{3}$	27	3^3	$\frac{1}{3}$	$3 \cdot 3 \cdot 3$
c. −25	$\frac{1}{25}$	5^{-2}	0.04	−25

29. What do you think is the value of any number raised to the 0 power?
Any number raised to the zero power is 1.

30. ***WRITE ABOUT IT*** What is the value of 1 raised to a negative exponent? Use examples to support your answer.

31. ***WRITE ABOUT IT*** You cannot raise 0 to a negative exponent. Why?

COMMON ERROR ALERT

Some students may think that any number with a negative exponent is negative. Remind them that a number raised to a negative power equals 1 divided by that number raised to the opposite (positive) power. Thus $3^{-2} = \frac{1}{3^2}$ is a positive number.

Additional Examples

Example 1

Find a pattern in the table.

Power	4^3	4^2	4^1	4^0	4^{-1}	4^{-2}
Value	64	16	4	1	$\frac{1}{4}$	$\frac{1}{16}$

One possible pattern is "divide by 4."

Example 2

Find each value: $3^0, 3^{-1}, 3^{-2}, 3^{-3}$

Power	3^2	3^1	3^0	3^{-1}	3^{-2}	3^{-3}
Value	9	3	1	$\frac{1}{3}$	$\frac{1}{9}$	$\frac{1}{27}$

Answers

30. The value of 1 raised to a negative power is 1; Possible answer: $1^{-5} = \frac{1}{1^5} = 1, 1^{-2} = \frac{1}{1^2} = 1.$

31. Raising 0 to a negative power results in 0 being in the denominator of a fraction, which is not defined.

3 Close

Summarize

Have the students explain zero as an exponent and positive and negative exponents.

Possible answer: Positive exponents tell how many times the base is used as a factor. Any number raised to the 0 power, except 0, equals 1. Zero raised to the 0 power is undefined. A number raised to a negative power equals 1 divided by that number raised to the opposite (positive) power.

Teaching Tip Point out to students that a negative number raised to a negative power may be negative.
Example: $(-2)^{-3} = \frac{1}{(-2)^3} = \frac{1}{-8} = -\frac{1}{8}$

Problem Solving on Location

Ohio

Purpose: *To provide additional practice for problem-solving skills in Chapters 1–9*

Temperatures

- After problem 1, have students identify the key word that signaled which operation to perform. difference
- After problem 4, have students consider the following: According to the rule, what change in elevation will cause the temperature to drop 1°F? 250 feet
- After problem 5, ask students the following question: If the wind speed is 15 mi/h and the air temperature feels like 5°F, what is a good estimation of the actual air temperature? about 27°F

Extension Have students research the record high and low temperatures for a nearby city for November and December. Have them determine the difference between each month's record high and low temperatures. Check students' work.

Problem Solving on Location

OHIO

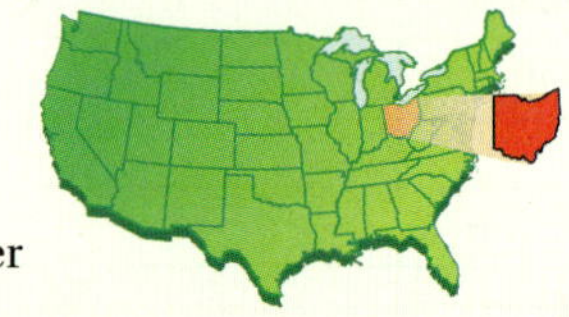

Temperatures

Ohio has experienced some extreme weather in the past 150 years. On July 21, 1934, the temperature hit a record 113°F in Gallipolis. The residents of Milligan braved temperatures of −39°F on February 10, 1899.

1. Find the difference between Ohio's record high temperature and its record low temperature. **152°F**

For 2–3, use the table.

Monthly Record Temperatures in Columbus		
	November	December
Low	−4°F	−17°F
High	80°F	76°F

2. By how many degrees did Columbus's record high November temperature exceed its record low November temperature? **84°F**
3. Find the difference between Columbus's record high November temperature and its record high December temperature. **4°F**
4. Here's a useful rule for estimating temperature change: Temperature decreases by about 4°F for every 1,000 feet that you climb. Suppose that you hike 500 feet up to a campsite. By about how many degrees will the temperature have decreased by the time you reach your campsite? **about 2°F**

Because of the *wind-chill factor,* air temperatures feel colder when the wind is blowing. The stronger the wind, the colder the air feels. The table below shows how much colder it feels.

For 5–7, use the table.

Wind-Chill Factor				
Wind Speed	Air Temperature (°F)			
	35	30	25	20
5 mi/h	32	27	22	16
10 mi/h	22	16	10	3
15 mi/h	16	9	2	−5
20 mi/h	12	4	−3	−10

5. Suppose the wind is blowing at 10 mi/h. How many degrees colder will the air feel if the air temperature is 30°F? **30 − 16 = 14°F colder**
6. Suppose the wind is blowing at 20 mi/h. How many degrees colder will the air feel if the air temperature is 25°F? **25 − (−3) = 28°F colder**
7. Describe any pattern you see in the row for 15 mi/h. **Possible answer: When there is a 15 mi/h wind, for each 5°F drop in temperature, it feels like the temperature has dropped by 7°F.**

Elevations

Ohio consists mostly of rolling plains that slope downward, toward the southwestern corner of the state. Ohio's highest point is Campbell Hill, at 1,549 feet above sea level. Its lowest point is the Ohio River, at 455 feet above sea level. The lowest point in the United States is Death Valley, California, where the elevation is 282 feet below sea level (−282 ft). New Orleans, Louisiana, has an elevation of 8 feet below sea level (−8 ft).

Campbell Hill

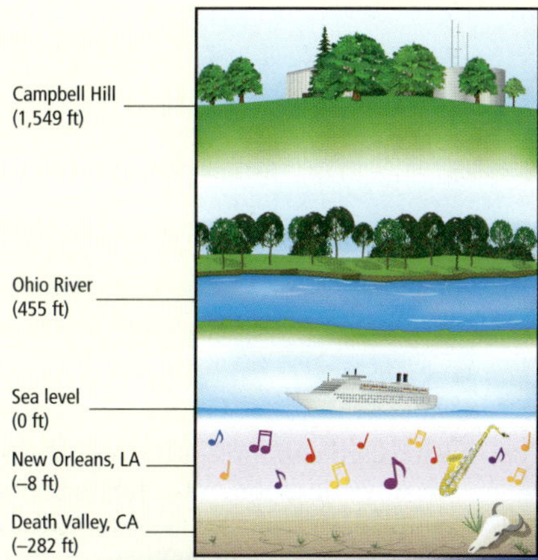

For 1–4, use the diagram.

1. What is the difference in elevation between the lowest point in Ohio and the highest point in Ohio? **1,094 ft**
2. What is the difference in elevation between Death Valley and the top of Campbell Hill? **1,831 ft**
3. What is the difference in elevation between Death Valley and the lowest point in Ohio? **737 ft**
4. The elevation of Death Valley is about how many times the elevation of New Orleans? **about 35 times**

Elevations

- After problem 1, ask the following question: What integer can be used to represent the elevation of Campbell Hill? +1,549
- After problem 3, have students draw a number line to show the difference between −8 and −282. Check students' work
- After problem 4, have students explain how they estimated. 8 × 30 = 240, and 8 × 40 = 320. 282 is about halfway between 240 and 320, so the estimate is 35.

Extension Have students create a properly scaled number line to show the elevations given in this activity. Have them research another state's highest and lowest elevations and plot them on the number line. Check students' work.

MATH-ABLES

Game Resources

Puzzles, Twisters & Teasers
Chapter 9 Resource Book

A Math Riddle

Purpose: *To apply the skill of graphing points to solve a riddle*

Discuss: Ask students to identify common mistakes they may encounter when doing this activity. Ask students what they can do to avoid those mistakes.

Possible answer: Confusing *x*- and *y*-coordinates: be sure to plot points in the correct order; leaving out a pair of points: cross off pairs of points as you plot them; not connecting in the right order: connect points as you plot each pair.

Extend: Have students write their names on graph paper, making the letters using straight lines. Have students identify the ordered pairs and create a puzzle like the one in this activity. Then have them randomly trade papers and complete the puzzle to determine whose paper they have.

Zero Sum

Purpose: *To practice adding integers in a game format*

Discuss: When a student wins a round, have him or her write a number sentence to show that his or her sum is closest to zero.

Extend: Have students play the game with a different objective—the winning player is the one with the sum farthest from 0.

MATH-ABLES

A Math Riddle

What coin doubles in value when half is subtracted?
a half dollar

To find the answer, graph each set of points. Connect each pair of points with a straight line.

1. (−8, 3) (−6, 3) **2.** (−9, 1) (−7, 5) **3.** (−7, 5) (−5, 1) **4.** (−3, 1) (−3, 5)
5. (−1, 1) (−1, 5) **6.** (−3, 3) (−1, 3) **7.** (1, 1) (3, 5) **8.** (3, 5) (5, 1)
9. (2, 3) (4, 3) **10.** (6, 1) (6, 5) **11.** (6, 1) (8, 1) **12.** (9, 1) (9, 5)
13. (9, 5) (11, 5) **14.** (9, 3) (11, 3) **15.** (−9, −5) (−9, −1) **16.** (−9, −1) (−7, −3)
17. (−7, −3) (−9, −5) **18.** (−6, −1) (−6, −5) **19.** (−6, −5) (−4, −5) **20.** (−4, −5) (−4, −1)
21. (−4, −1) (−6, −1) **22.** (−3, −1) (−3, −5) **23.** (−3, −5) (−1, 5) **24.** (1, −1) (1, −5)
25. (1, −5) (3, −5) **26.** (4, −5) (6, −1) **27.** (6, −1) (8, −5) **28.** (5, −3) (7, −3)
29. (9, −5) (9, −1) **30.** (9, −1) (11, −3) **31.** (11, −3) (9, −3) **32.** (9, −3) (11, −5)

Zero Sum

Each card contains either a positive number, a negative number, or 0. The dealer deals three cards to each player. On your turn, you may exchange one or two of your cards for new ones, or you may keep your three original cards. After everyone has had a turn, the player whose sum is closest to 0 wins the round and receives everyone's cards. The dealer deals a new round and the game continues until the dealer runs out of cards. The winner is the player with the most cards at the end of the game.

internet connect
Go to ***go.hrw.com*** for a complete set of rules and game pieces.
KEYWORD: MR4 Game9

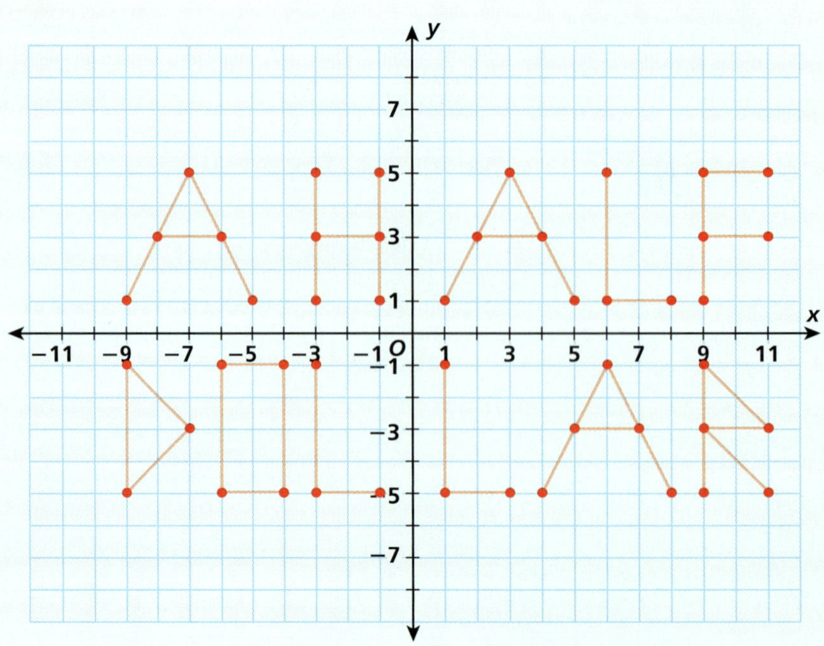

Technology LAB

Graph Points

internet connect
Lab Resources Online
go.hrw.com
KEYWORD: MR4 TechLab9

To graph on your calculator, you must first set the viewing window. To do this, press [WINDOW].

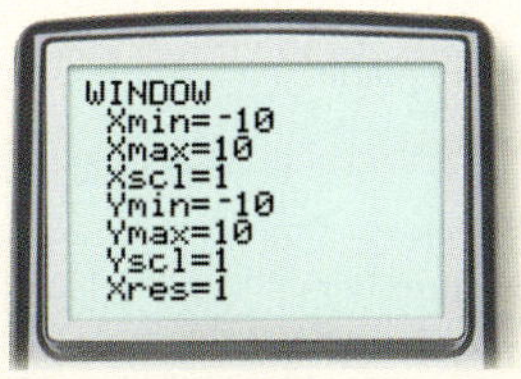

The calculator automatically uses the **standard window** shown at right unless you change the settings.

The x-values and y-values go from −10 to 10. These are set by **Xmin**, **Xmax**, **Ymin**, and **Ymax**.

Xscl and **Yscl** give the distance between tick marks. In the standard window, tick marks are 1 unit apart.

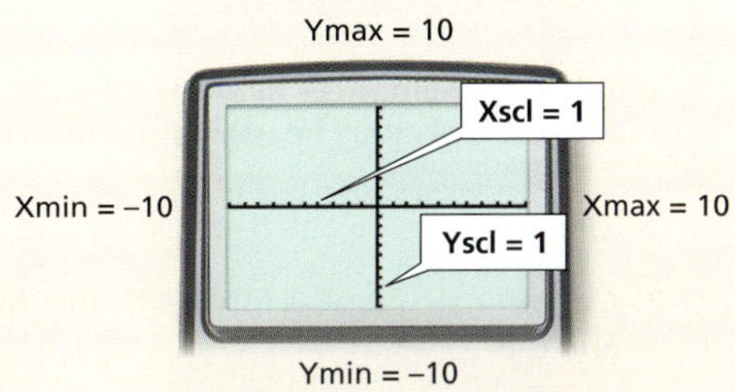

Activity

Graph the points (2, 5), (4, 3), (−4, 6), and (5, −7) on your calculator.

1. Access the **DRAW** menu by pressing [2nd] [PRGM] (DRAW).
2. Press [▶] to highlight [POINTS]. Make sure [1:] is highlighted to select **Pt-On**. Press [ENTER].
3. The calculator gives you the opening parenthesis for the ordered pair. To plot (2, 5), press 2 [,] 5 [)] [ENTER].
4. To graph the next point, press [2nd] [MODE] (QUIT) to quit the coordinate plane. Then repeat steps 1–3.

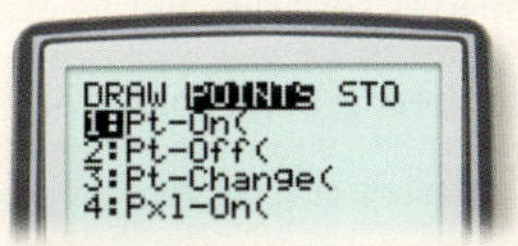

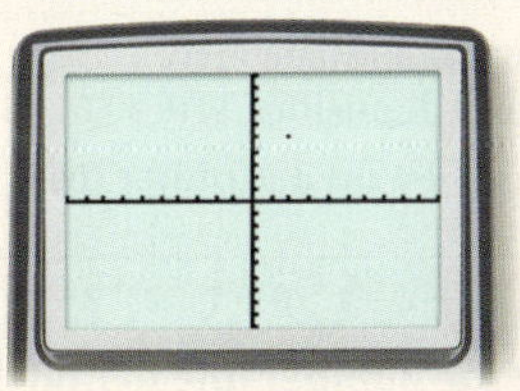

Think and Discuss

1. Suppose you want to graph (9, 16) on a graphing calculator. Should you use the standard window? Why or why not? **No, because the y-value is 16 and the Ymax on the standard window is 10.**

Try This

1. Graph (1, 3), (5, 3), (−5, −2), (3, −6), and (−8, 1) on your calculator.

Answers

Activity

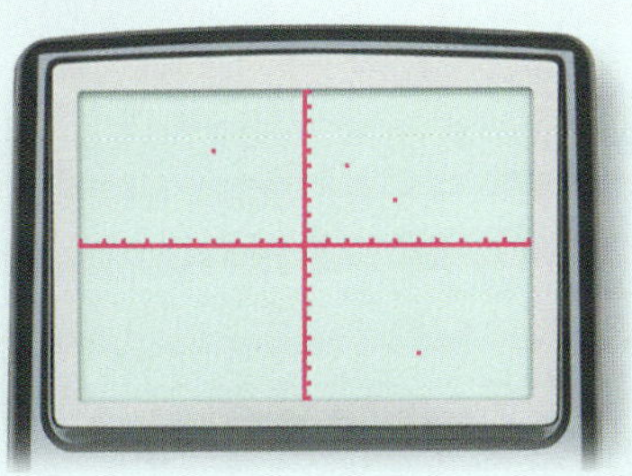

Try This

1.

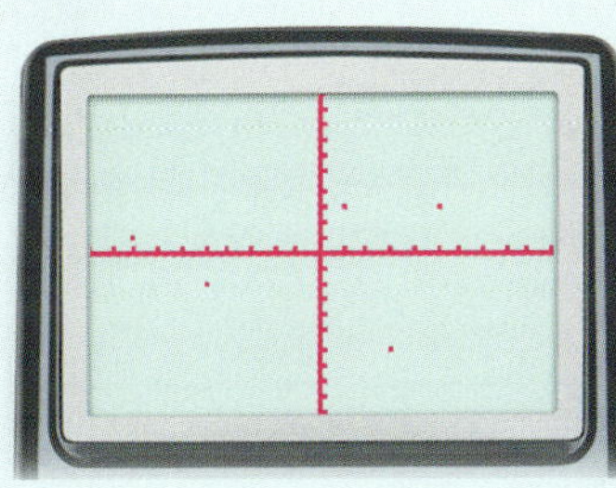

Technology LAB

Graph Points on a Calculator

Objective: To use a graphing calculator to graph points

Materials: Graphing calculator

Lab Resources

Using the Page

This technology activity shows students how to graph points. Specific keystrokes may vary, depending on the make and model of the graphing calculator used. The keystrokes given are for a TI-83 model. For keystrokes to other models, visit go.hrw.com.

The Think and Discuss problems can be used to assess students' understanding of the technology activity. While Try This problem 1 can be done without a graphing calculator, it is meant to help students become familiar with using a graphing calculator to graph points.

Assessment

1. How could you change the WINDOW screen to plot the point (−4, 20)?

Possible answer:

Xmin = −10	Ymin = 0
Xmax = 0	Ymax = 25
Xscl = 1	Yscl = 5

2. How do you enter a comma in the ordered pair? press [,]
3. Plot the points (−3, −4) and (4, 6) on a graphing calculator.

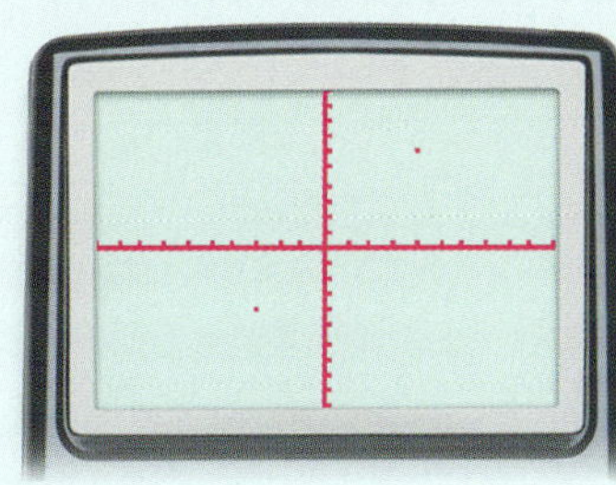

Chapter 9 Study Guide and Review

Purpose: *To help students review and practice concepts and skills presented in Chapter 9*

Assessment Resources

Chapter Review

Chapter 9 Resource Book . . pp. 82–83

***Test and Practice Generator* CD-ROM**

Additional review items in both multiple-choice and free-response format may be generated for any objective in Chapter 9.

Answers

1. opposites

2. *x*-axis, *y*-axis

3. coordinate plane, quadrants

4. positive, negative

5. +10

6. −50

7.

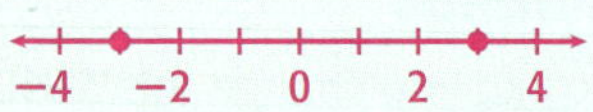

8.

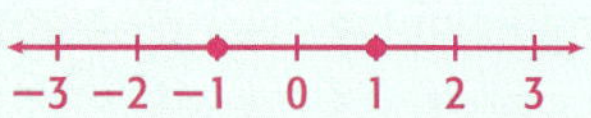

9.

10. 2

11. 1

12. 0

Chapter 9 Study Guide and Review

Vocabulary

Complete the sentences below with vocabulary words from the list above. Words may be used more than once.

1. The numbers −6 and 6 are called __?__.

2. A coordinate plane is formed by the intersection of the __?__ and the __?__.

3. The axes separate the __?__ into four __?__.

4. Numbers greater than 0 are __?__ and numbers less than 0 are __?__.

9-1 Understanding Integers (pp. 450–453)

EXAMPLE

Name a positive or negative number to represent each situation.

- 15 feet below sea level **−15**
- a bank deposit of \$10 **+10**
- **Graph +4 on a number line.**

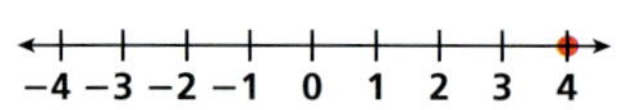

- **Use the number line to find |−3|.**

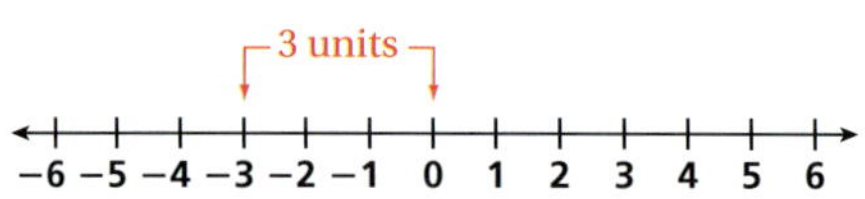

−3 is 3 units from 0, so |−3| = 3.

EXERCISES

Name a positive or negative number to represent each situation.

5. a raise of \$10 **6.** a loss of \$50

Graph each integer and its opposite on a number line.

7. −3 **8.** 1 **9.** 0

Use the number line to find the absolute value of each integer.

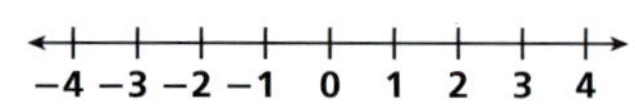

10. 2 **11.** −1 **12.** 0

9-2 Comparing and Ordering Integers (pp. 454–457)

EXAMPLE

- **Compare −2 and 3. Write < or >.**

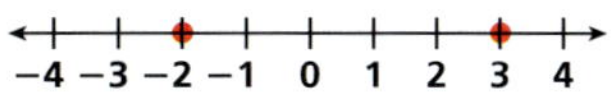

$-2 < 3$ *−2 is left of 3 on the number line.*

- **Order 3, −2, and 0 from least to greatest.**

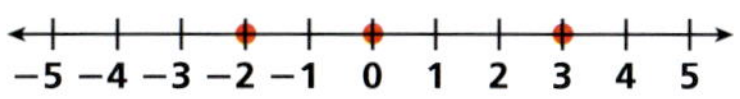

Read from left to right: −2, 0, 3.

EXERCISES

Compare. Write < or >.

13. 3 ■ 4 **14.** −2 ■ 5 **15.** 0 ■ 6

16. −5 ■ −7 **17.** 8 ■ −11 **18.** −4 ■ 0

Order each set of integers from least to greatest.

19. 2, −1, 4 **20.** −3, 0, 4

21. −3, 1, −2, 0 **22.** −6, −8, 0

23. 7, −4, −7 **24.** −1, 7, 3, −5

9-3 The Coordinate Plane (pp. 458–461)

EXAMPLE

- **Give the coordinates of *A* and identify the quadrant in which it lies.**

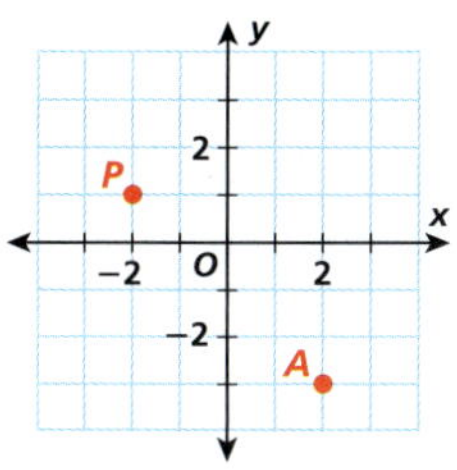

A is in the fourth quadrant. Its coordinates are (2, −3).

- **Graph *P*(−2, 1) on a coordinate plane.**
From (0, 0), move 2 units left and 1 unit up.

EXERCISES

Give the coordinates of each point.

25. *A* **26.** *C*

Give the quadrant in which each point lies.

27. *A* **28.** *B*

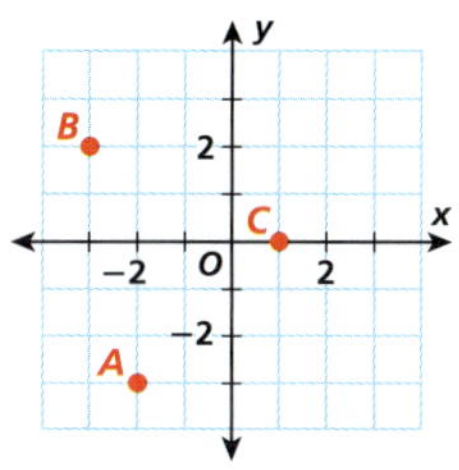

Graph each point on a coordinate plane.

29. *D*(−2, −1) **30.** *E*(0, 0) **31.** *F*(3, 4)

9-4 Adding Integers (pp. 465–468)

EXAMPLE

- **Find the sum: 3 + (−2).**

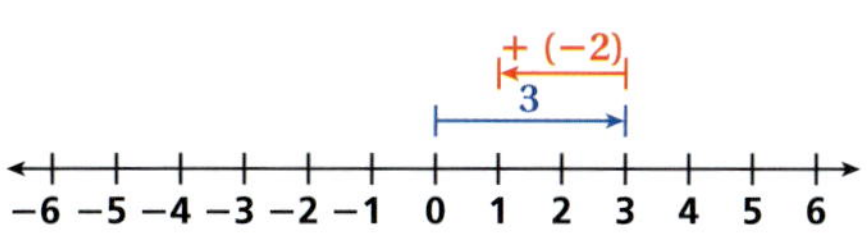

3 + (−2) = 1

- **Evaluate $x + (-2)$ when $x = 9$.**
9 + (−2) = 7

EXERCISES

Find each sum.

32. −4 + 2 **33.** 4 + (−4)

34. 3 + (−2) **35.** −3 + (−2)

Evaluate $x + 3$ for the following values.

36. $x = -20$ **37.** $x = 5$

Study Guide and Review

Answers

13. <
14. <
15. <
16. >
17. >
18. <
19. −1, 2, 4
20. −3, 0, 4
21. −3, −2, 0, 1
22. −8, −6, 0
23. −7, −4, 7
24. −5, −1, 3, 7
25. (−2, −3)
26. (1, 0)
27. III
28. II
29–31.

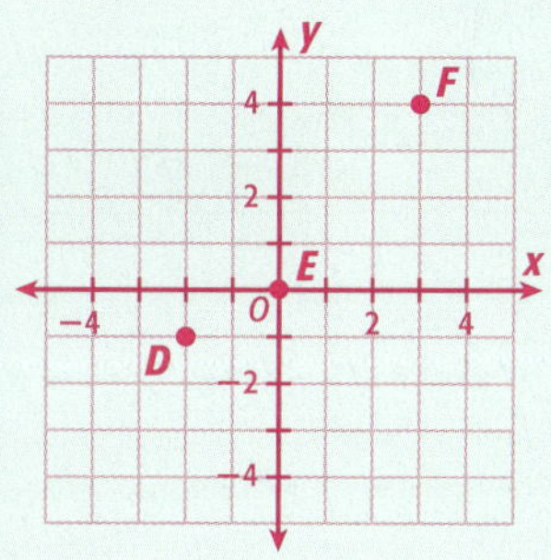

32. −2
33. 0
34. 1
35. −5
36. −17
37. 8

Answers

38. −8
39. 9
40. 10
41. 0
42. 13
43. −6
44. −10
45. 6
46. 6
47. −8
48. −16
49. 45
50. −3
51. 3
52. 2
53. −2
54. −12
55. −9
56. $w = 4$
57. $a = -12$
58. $q = -7$
59. $x = -5$

Study Guide and Review

9-5 Subtracting Integers (pp. 470–472)

EXAMPLE

- **Subtract the integers: $3 - (-2)$.**

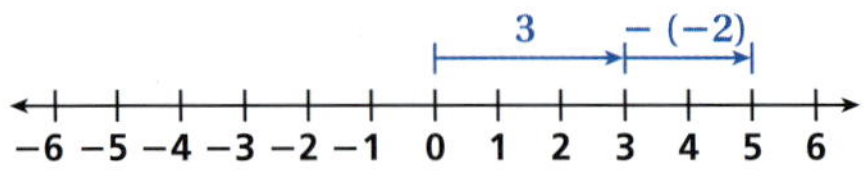

$3 - (-2) = 5$

- **Evaluate $n - 4$ for $n = -1$.**

$(-1) - 4 = -5$

EXERCISES

Find each difference.

38. $-6 - 2$
39. $5 - (-4)$
40. $5 - (-5)$
41. $-3 - (-3)$

Evaluate $x - (-1)$ for each value of x.

42. $x = 12$
43. $x = -7$

9-6 Multiplying Integers (pp. 473–475)

EXAMPLE

- **Find the product: $3 \cdot (-2)$.**

Think: 3 groups of −2

$3 \cdot (-2) = -6$

- **Evaluate $-2x$ for $x = -4$.**

$-2(-4) = 8$

EXERCISES

Find each product.

44. $5 \cdot (-2)$
45. $3 \cdot 2$
46. $-3 \cdot (-2)$
47. $-4 \cdot 2$

Evaluate each expression for the given value of the variable.

48. $n \cdot (-8);\ n = 2$
49. $-9y;\ y = -5$

9-7 Dividing Integers (pp. 476–479)

EXAMPLE

- **$-24 \div 4$**

Think: $-6 \cdot 4 = -24$

$-24 \div 4 = -6$

- **Evaluate $\frac{x}{-2}$ for $x = 14$.**

$\frac{14}{-2} = -7$

EXERCISES

Find each quotient.

50. $6 \div (-2)$
51. $9 \div 3$
52. $-14 \div (-7)$
53. $-4 \div 2$

Evaluate each expression for the given value of the variable.

54. $n \div 2;\ n = -24$
55. $\frac{x}{-3};\ x = 27$

9-8 Solving Integer Equations (pp. 482–485)

EXAMPLE

- **Solve the equation.**

$x + 4 = 18$

$$\begin{aligned} x + 4 &= 18 \\ -4 \quad &\ \ -4 \quad \textit{Subtract 4 from both sides.} \\ x \quad &= 14 \end{aligned}$$

EXERCISES

Solve each equation. Check your answers.

56. $w - 5 = -1$
57. $\frac{a}{-4} = 3$
58. $2q = -14$
59. $x + 3 = -2$

Chapter Test

Chapter 9

Name a positive or negative number to represent each situation.

1. 30° below zero **−30**
2. a bank deposit of $75 **+75**
3. an increase of 10 points **+10**
4. a loss of 5 yards **−5**

Write the opposite of each integer.

5. -3 **3**
6. 2 **−2**
7. -19 **19**
8. 0 **0**

Find the absolute value.

9. $|-6|$ **6**
10. $|7|$ **7**
11. $|-20|$ **20**
12. $|11|$ **11**

Compare. Write < or >.

13. -4 ■ 4 **<**
14. 2 ■ -9 **>**
15. -10 ■ 8 **<**
16. -2 ■ -12 **>**

Order each set of integers from least to greatest.

17. 21, −19, 34 **−19, 21, 34**
18. −16, −2, 13, 46 **−16, −2, 13, 46**
19. −10, 0, 25, −7, 18 **−10, −7, 0, 18, 25**

Graph each point on a coordinate plane.

20. $A(2, 3)$
21. $C(-1, 3)$
22. $E(0, 1)$
23. $B(3, -2)$
24. $D(2, 0)$
25. $F(-1, -2)$

Add, subtract, multiply, or divide.

26. $-4 + 4$ **0**
27. $-2 - 9$ **−11**
28. $-3 \cdot 8$ **−24**
29. $12 \div (-3)$ **−4**
30. $-48 \div (-4)$ **12**
31. $13 + (-9)$ **4**
32. $8 - (-11)$ **19**
33. $-7 \cdot (-6)$ **42**
34. $7 \cdot (-9)$ **−63**
35. $-42 \div 2$ **−21**
36. $-15 + (-10)$ **−25**
37. $-31 - (-16)$ **−15**

Evaluate each expression for the given value of the variable.

38. $n + 3,\ n = -10$ **−7**
39. $9 - x,\ x = -9$ **18**
40. $m \cdot 4,\ m = -6$ **−24**
41. $\frac{15}{a},\ a = -3$ **−5**
42. $(-11) + z,\ z = 28$ **17**
43. $w - (-8),\ w = 13$ **21**
44. $-7c,\ c = 13$ **−91**
45. $n \div 4,\ n = -32$ **−8**
46. $p + (-14),\ p = -22$ **−36**

Solve each equation.

47. $\frac{b}{7} = -3$ **$b = -21$**
48. $-9 \cdot f = -81$ **$f = 9$**
49. $r - 14 = -32$ **$r = -18$**
50. $y + 17 = -2$ **$y = -19$**

Chapter Test

Chapter Test

Chapter 9

Purpose: *To assess students' mastery of concepts and skills in Chapter 9*

Assessment Resources

Chapter 9 Tests (Levels A, B, C)
Assessment Resources ..... pp. 79–84

***Test and Practice Generator* CD-ROM**

Additional assessment items in both multiple-choice and free-response format may be generated for any objective in Chapter 9.

Answers

20–25.

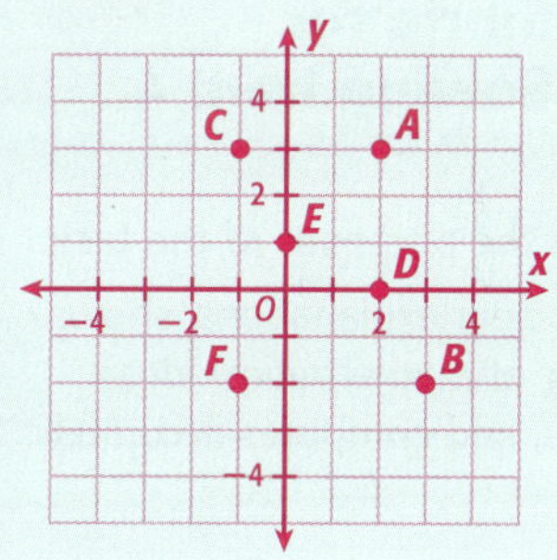

Chapter 9 Performance Assessment

Purpose: *To assess students' understanding of concepts in Chapter 9 and combined problem solving skills*

Assessment Resources

Performance Assessment
Assessment Resources . . . . p. 120

Performance Assessment Teacher Support
Assessment Resources . . . . p. 119

Answers

1–3. See p. A6

4. See Level 3 work sample below.

Scoring Rubric for Problem Solving Item 4

Level 3

Accomplishes the purposes of the task.

Student gives clear explanations, shows understanding of mathematical ideas and processes, and computes accurately.

Level 2

Purposes of the task not fully achieved.

Student demonstrates satisfactory but limited understanding of the mathematical ideas and processes.

Level 1

Purposes of the task not accomplished.

Student shows little evidence of understanding the mathematical ideas and processes and makes computational and/or procedural errors.

Chapter 9 Performance Assessment

Show What You Know

Create a portfolio of your work from this chapter. Complete this page and include it with your four best pieces of work from Chapter 9. Choose from your homework or lab assignments, mid-chapter quiz, or any journal entries you have done. Put them together using any design you want. Make your portfolio represent what you consider your best work.

Short Response

1. The high temperatures in Nome, Alaska, for one week were 5°F, 4°F, −2°F, −3°F, −1°F, and 2°F. What was the average high temperature in Nome for that week? Show all your steps.
2. Lionel finished four rounds of a golf tournament with a score of −10. His scores on the first three rounds were −4, −2, and −6. What was his score on the last round? Show your work.
3. Mount McKinley is 20,320 feet above sea level. Death Valley is 20,602 feet lower than Mount McKinley. Write and solve an equation to find the elevation of Death Valley.

Extended Problem Solving

4. Asheka has a checking account and a savings account. There is no monthly service fee for her checking account, but she must pay $4.00 a month to maintain her savings account.
   a. Asheka's checking account had a balance of $50. She then deposited $50 and wrote checks for $40, $24, and $18. What is Asheka's new checking account balance?
   b. During June, Asheka deposited $1,250 in her savings account, withdrew $575, and paid the monthly fee. Write an equation that could be used to find the balance b at the beginning of June if the balance at the end of June was $968.
   c. Find Asheka's savings account balance at the beginning of June.

Student Work Samples for Item 4

Level 3

a. $50 + $50 = $100
   $40 + $24 + $18 = $82
   $100 − $82 = $18

b. b + $1250 − $575 − $4 = $968

c. b + $671 = $968
   b = $297

The student gave correct answers and correctly set up and solved an equation.

Level 2

A. $50 + $50 − $40 − $24 − $18 = $18

B. b + $968 = $1250 − $575 − $4

C. b + $968 = $671
   b = $−297

The student correctly answered part **a**, but incorrectly set up the equation.

Level 1

a. 50 + 50 + 40 + 24 + 18 = $182

b. 1250 + 575 + 968 = $2793

c. $50

The student did not understand the concepts and incorrectly answered all parts.

internet connect
State-Specific Test Practice Online
go.hrw.com Keyword: MR4 TestPrep

Standardized Test Prep

Chapter 9

Cumulative Assessment, Chapters 1–9

1. Which is the correct value of 3^3? D

Ⓐ 6 Ⓑ 9 Ⓒ 12 Ⓓ 27

TEST TAKING TIP!

You can solve the equation, or you can substitute each answer choice in the equation to check whether it is a solution.

2. What is the solution to $18 + g = 72$? G

Ⓕ $g = 4$ Ⓖ $g = 54$ Ⓗ $g = 90$ Ⓙ $g = 1{,}296$

3. Which is the correct value of $18 - (2 + 4) \cdot 4$? A

Ⓐ −6 Ⓑ 10 Ⓒ 42 Ⓓ 48

4. Which is the greatest common factor of 24, 32, and 40? H

Ⓕ 4 Ⓖ 6 Ⓗ 8 Ⓙ 10

5. Which decimal is equivalent to 0.035? B

Ⓐ 0.35 Ⓑ 0.0350 Ⓒ 0.305 Ⓓ 0.0305

6. Which is the value of $|-62|$? J

Ⓕ −62 Ⓖ −4 Ⓗ 0 Ⓙ 62

7. Which set of numbers has a mean of 7? A

Ⓐ 4, 8, 9 Ⓑ 3, 7, 8 Ⓒ 7, 7, 10 Ⓓ 7, 9, 11

8. Which decimal and fraction are equivalent to 35%? G

Ⓕ 35.0, $\frac{7}{20}$ Ⓖ 0.35, $\frac{7}{20}$ Ⓗ 0.35, $\frac{35}{10}$ Ⓙ 3.5, $\frac{35}{10}$

9. ***SHORT RESPONSE*** Darius drinks $1\frac{1}{2}$ cups of water three times a day. How many cups of water does Darius drink per week? Show your work.

10. ***SHORT RESPONSE*** The line graph below shows the temperature from 8 A.M. to noon. What was the mean change per hour in temperature from 8 A.M. to noon? Explain how you found your answer.

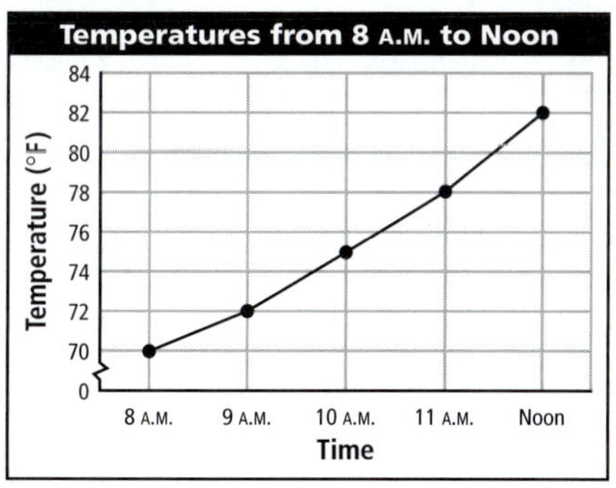

Standardized Test Prep

Standardized Test Prep

Chapter 9

Purpose: *To provide review and practice for Chapters 1–9 and standardized tests*

Assessment Resources

Cumulative Tests (Levels A, B, C)
Assessment Resources. . . . pp. 223–234

State-Specific Test Practice Online
KEYWORD: MR4 TestPrep

Test Prep Doctor

Expand on the test-taking tip given for item 2 by reminding students to first eliminate unreasonable answers. Answer **F** is too low. Answers **H** and **J** are too high. **G** has to be the correct choice.

Point out to students that all incorrect choices in item 7 may be eliminated by estimating. Because $1\frac{1}{2} \cdot 3 \approx 5$ and because $5 \cdot 7 = 35$, the correct answer must be close to 35.

Answers

9. $1\frac{1}{2}$ cups $\times 3 = 4\frac{1}{2}$ cups
$4\frac{1}{2}$ cups $\times 7 = 31\frac{1}{2}$ cups

10. 8–9: 2°
9–10: 3°
10–11: 3°
11–12: 4°
$\frac{2° + 3° + 3° + 4°}{4} = 3°F$

Chapter 10

Perimeter, Area, and Volume

Section 10A

Perimeter, Area, and Circumference

Lesson 10-1
Finding Perimeter

Lesson 10-2
Estimating and Finding Area

Lesson 10-3
Break into Simpler Parts

Lesson 10-4
Comparing Perimeter and Area

Hands-On Lab 10A
Discover Properties of Circles

Lesson 10-5
Circles

Section 10B

Volume and Surface Area

Hands-On Lab 10B
Draw Views of Solid Figures

Lesson 10-6
Solid Figures

Hands-On Lab 10C
Model Solid Figures

Lesson 10-7
Surface Area

Lesson 10-8
Finding Volume

Lesson 10-9
Volume of Cylinders

Pacing Guide for 45-Minute Classes

Chapter 10

DAY 130	DAY 131	DAY 132	DAY 133	DAY 134
Lesson 10-1	Lesson 10-2	Lesson 10-3	Lesson 10-4	Hands-On Lab 10A
DAY 135	**DAY 136**	**DAY 137**	**DAY 138**	**DAY 139**
Lesson 10-5	Mid-Chapter Quiz Hands-On Lab 10B	Lesson 10-6	Hands-On Lab 10C	Lesson 10-7
DAY 140	**DAY 141**	**DAY 142**	**DAY 143**	
Lesson 10-8	Lesson 10-9	Chapter 10 Review	Chapter 10 Assessment	

Pacing Guide for 90-Minute Classes

Chapter 10

DAY 65	DAY 66	DAY 67	DAY 68	DAY 69
Chapter 9 Assessment Lesson 10-1	Lesson 10-2 Lesson 10-3	Lesson 10-4 Hands-On Lab 10A	Lesson 10-5 Hands-On Lab 10B	Mid-Chapter Quiz Lesson 10-6 Hands-On Lab 10C
DAY 70	**DAY 71**	**DAY 72**		
Lesson 10-7 Lesson 10-8	Lesson 10-9 Chapter 10 Review	Chapter 10 Assessment Lesson 11-1		

Across the Series

HARCOURT GRADE 5

- Find the perimeter of polygons and the circumference of circles.
- Find the area of polygons.
- Identify and draw solids.
- Find the surface area and volume of prisms.

COURSE 1

- **Find the perimeter and area of polygons.**
- **Explore how area and perimeter are affected by changing the dimensions.**
- **Identify parts of circles; find the area and circumference of circles.**
- **Identify, draw, and build solids.**
- **Find the surface area of prisms, pyramids, and cylinders.**
- **Find the volume of prisms and cylinders.**

COURSE 2

- Find the perimeter, circumference, and area of geometric figures and irregular figures.
- Draw, identify, and build solids.
- Find the volume of prisms, cylinders, pyramids, cones, and spheres.
- Find the surface area of prisms, cylinders, and spheres.
- Find the surface area, volume, and weight of proportional solids.

Across the Curriculum

LANGUAGE ARTS LINK

Math: Reading and Writing in the Content Area . . . . pp. 83–91

Focus on Problem Solving

Solve . . . . SE p. 521

Journal . . . . TE pp. 503, 507, 519, 527, 533, 537, 541

Write About It . . . . SE, last page of each lesson

SOCIAL STUDIES LINK

Social Studies . . . . SE pp. 507, 510

History . . . . SE pp. 519, 538

SCIENCE LINK

Physical Science . . . . SE p. 537

Science . . . . SE p. 541

TE = *Teacher's Edition* **SE** = *Student Edition*

Interdisciplinary Bulletin Board

Architecture

The Pentagon, headquarters of the United States Department of Defense, is one of the world's largest office buildings. The total length of the corridors within the Pentagon is 17.5 miles. How many feet is this? 92,050 ft

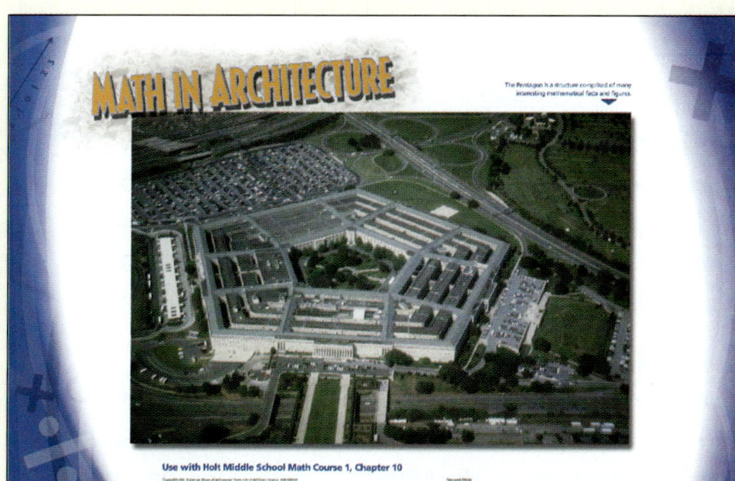

Interdisciplinary posters and worksheets are provided in your resource material.

Chapter 10

Resource Options

Chapter 10 Resource Book

Student Resources

Practice (Levels A, B, C) .... pp. 8–10, 17–19, 27–29, 36–38, 45–47, 56–58, 65–67, 74–76, 83–85
Reteach........ pp. 11, 20–21, 30, 39, 48–49, 59, 68, 77, 86
Challenge .... pp. 12, 22, 31, 40, 50, 60, 69, 78, 87
Problem Solving .... pp. 13, 23, 32, 41, 51, 61, 70, 79, 88
Puzzles, Twisters & Teasers .... pp. 14, 24, 33, 42, 52, 62, 71, 80, 89
Recording Sheets .... pp. 3, 7, 16, 26, 35, 44, 55, 64, 73, 82, 93, 99, 102
Chapter Review........................... pp. 90–92

Teacher and Parent Resources

Chapter Planning and Pacing Guide.................. p. 4
Section Planning Guides .... pp. 5, 53
Parent Letter .... pp. 1–2
Teaching Tools.......................... pp. 96–104
Teacher Support for Chapter Project .... p. 94
Transparencies .... pp. T1–T38
- Daily Transparencies
- Additional Examples Transparencies
- Teaching Transparencies

Reaching All Learners

English Language Learners

Success for English Language Learners ...... pp. 165–182
Math: Reading and Writing in the Content Area .... pp. 83–91
Spanish Homework and Practice .... pp. 83–91
Spanish Interactive Study Guide .... pp. 83–91
Spanish Family Involvement Activities......... pp. 81–88
Multilingual Glossary

Individual Needs

Are You Ready? Intervention and Enrichment. pp. 145–148, 169–172, 177–180, 285–288, 313–316, 423–424
Alternate Openers: Explorations .... pp. 83–91
Family Involvement Activities .... pp. 81–88
Interactive Problem Solving................. pp. 83–91
Interactive Study Guide .... pp. 83–91
Readiness Activities .... pp. 19–20
Math: Reading and Writing in the Content Area .... pp. 83–91
Challenge ..... CRB pp. 12, 22, 31, 40, 50, 60, 69, 78, 87

Hands-On

Hands-On Lab Activities.................. pp. 85–100
Technology Lab Activities.................. pp. 42–44
Alternate Openers: Explorations .... pp. 83–91
Family Involvement Activities .... pp. 81–88

Applications and Connections

Consumer and Career Math................. pp. 37–40
Interdisciplinary Posters .... Poster 10, TE p. 498B
Interdisciplinary Poster Worksheets .... pp. 28–30

Transparencies

Alternate Openers: Explorations................ pp. 83–91
Exercise Answers Transparencies
Chapter 10 Resource Book.................... pp. T1–T38
- Daily Transparencies
- Additional Examples Transparencies
- Teaching Transparencies

Technology

Teacher Resources

***Lesson Presentations* CD-ROM**........... Chapter 10
***Test and Practice Generator* CD-ROM** ...... Chapter 10
***One-Stop Planner* CD-ROM** .... Chapter 10

Student Resources

***Are You Ready? Intervention* CD-ROM**
Skills 34, 40, 42, 72, 76

internet connect go.hrw.com

Homework Help Online	**KEYWORD:** MR4 HWHelp10
Math Tools Online	**KEYWORD:** MR4 Tools
Glossary Online	**KEYWORD:** MR4 Glossary
Chapter Project Online	**KEYWORD:** MR4 PSProject10
Chapter Opener Online	**KEYWORD:** MR4 Ch10
CNN Student News™	**KEYWORD:** MR4 CNN10

SE = *Student Edition* **TE** = *Teacher's Edition* **AR** = *Assessment Resources* **CRB** = *Chapter Resource Book* **MK** = *Manipulatives Kit*

Chapter 10

Assessment Options

Assessing Prior Knowledge

Determine whether students have the required prerequisite concepts and skills.

Are You Ready? . . . . SE p. 499
Inventory Test . . . . AR pp. 1–4

Test Preparation

Provide review and practice for chapter and standardized tests.

Standardized Test Prep . . . . SE p. 551
Spiral Review with Test Prep . . . . . SE, last page of each lesson
Study Guide and Review . . . . SE pp. 546–548
Test Prep Tool Kit

Technology

***Test and Practice Generator* CD-ROM**

internet connect

State-Specific Test Practice Online KEYWORD: MR4 TestPrep

Performance Assessment

Assess students' understanding of chapter concepts and combined problem-solving skills.

Performance Assessment . . . . SE p. 550
Includes scoring rubric in TE
Performance Assessment . . . . AR p. 122
Performance Assessment Teacher Support . . . . AR p. 121

Portfolio

Portfolio opportunities appear throughout the Student and Teacher's Editions.

Suggested work samples:

Problem Solving Project . . . . TE p. 498
Performance Assessment . . . . SE p. 550
Portfolio Guide . . . . AR p. xxxv
Journal . . . . TE pp. 503, 507, 519, 527, 533, 537, 541
Write About It . . . . SE, last page of each lesson

Daily Assessment

Obtain daily feedback on students' understanding of concepts.

Spiral Review and Test Prep . . . . . . SE, last page of each lesson

Also Available on Transparency In Chapter 10 Resource Book

Warm Up . . . . TE, first page of each lesson
Problem of the Day . . . . TE, first page of each lesson
Lesson Quiz . . . . TE, last page of each lesson

Student Self-Assessment

Have students evaluate their own work.

Group Project Evaluation . . . . AR p. xxxii
Individual Group Member Evaluation . . . . AR p. xxxiii
Portfolio Guide . . . . AR p. xxxv
Journal . . . . TE pp. 503, 507, 519, 527, 533, 537, 541

Formal Assessment

Assess students' mastery of concepts and skills.

Section Quizzes . . . . AR pp. 25–26
Mid-Chapter Quiz . . . . SE p. 520
Chapter Test . . . . SE p. 549
Chapter Tests (Levels A, B, C) . . . . AR pp. 85–90
Cumulative Tests (Levels A, B, C) . . . . AR pp. 235–246
Standardized Test Prep
Cumulative Assessment . . . . SE p. 551
End-of-Year Test . . . . AR pp. 271–274

Technology

***Test and Practice Generator* CD-ROM**

Make tests electronically. This software includes:

- Dynamic practice for Chapter 10
- Customizable tests
- Multiple-choice items for each objective
- Free-response items for each objective
- Teacher management system

SE = *Student Edition* **TE** = *Teacher's Edition* **AR** = *Assessment Resources* **CRB** = *Chapter Resource Book* **MK** = *Manipulatives Kit*

Chapter 10

Chapter 10 Tests

Three levels (A,B,C) of tests are available for each chapter in the *Assessment Resources.*

Level A

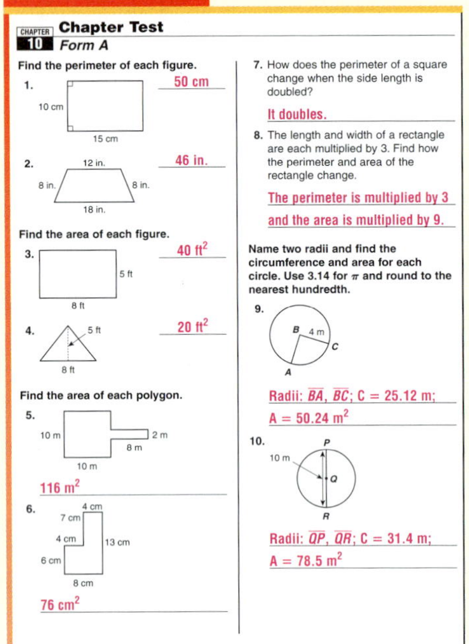
CHAPTER 10 Chapter Test Form A

Find the perimeter of each figure.

1. 50 cm
2. 46 in.

Find the area of each figure.

3. 40 ft^2
4. 20 ft^2

Find the area of each polygon.

5. 116 m^2
6. 76 cm^2

7. How does the perimeter of a square change when the side length is doubled? It doubles.

8. The length and width of a rectangle are each multiplied by 3. Find how the perimeter and area of the rectangle change. The perimeter is multiplied by 3 and the area is multiplied by 9.

Name two radii and find the circumference and area for each circle. Use 3.14 for π and round to the nearest hundredth.

9. Radii: $\overline{BA}$, $\overline{BC}$; C = 25.12 m; A = 50.24 m^2
10. Radii: $\overline{QP}$, $\overline{QR}$; C = 31.4 m; A = 78.5 m^2

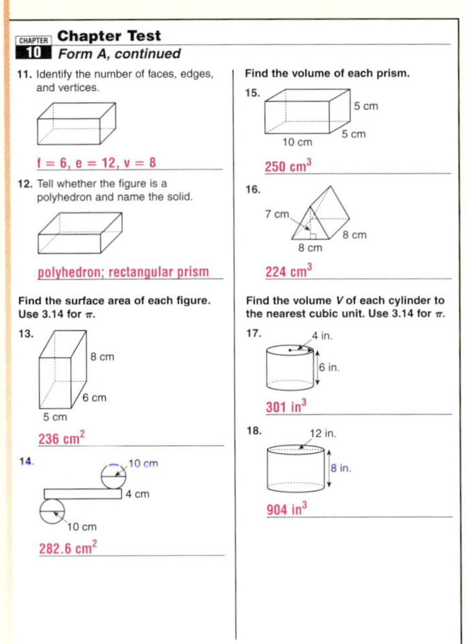
CHAPTER 10 Chapter Test Form A, continued

11. Identify the number of faces, edges, and vertices. f = 6, e = 12, v = 8
12. Tell whether the figure is a polyhedron and name the solid. polyhedron; rectangular prism

Find the surface area of each figure. Use 3.14 for π.

13. 236 cm^2
14. 282.6 cm^2

Find the volume of each prism.

15. 250 cm^3
16. 224 cm^3

Find the volume *V* of each cylinder to the nearest cubic unit. Use 3.14 for π.

17. 301 in^3
18. 904 in^3

Level B

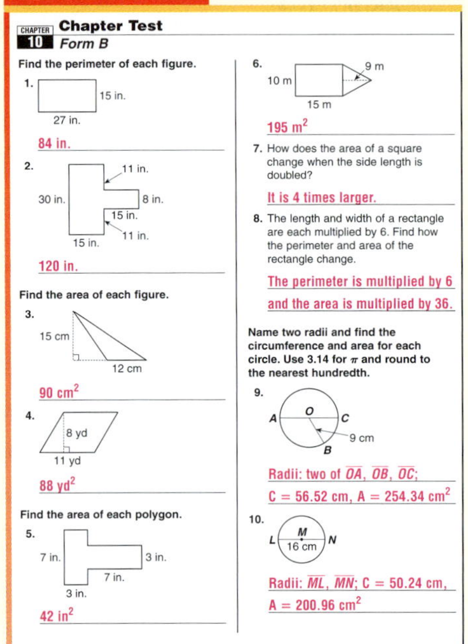
CHAPTER 10 Chapter Test Form B

Find the perimeter of each figure.

1. 84 in.
2. 120 in.

Find the area of each figure.

3. 90 cm^2
4. 88 yd^2

Find the area of each polygon.

5. 42 in^2
6. 195 m^2

7. How does the area of a square change when the side length is doubled? It is 4 times larger.

8. The length and width of a rectangle are each multiplied by 6. Find how the perimeter and area of the rectangle change. The perimeter is multiplied by 6 and the area is multiplied by 36.

Name two radii and find the circumference and area for each circle. Use 3.14 for π and round to the nearest hundredth.

9. Radii: two of $\overline{OA}$, $\overline{OB}$, $\overline{OC}$; C = 56.52 cm, A = 254.34 cm^2
10. Radii: $\overline{ML}$, $\overline{MN}$; C = 50.24 cm, A = 200.96 cm^2

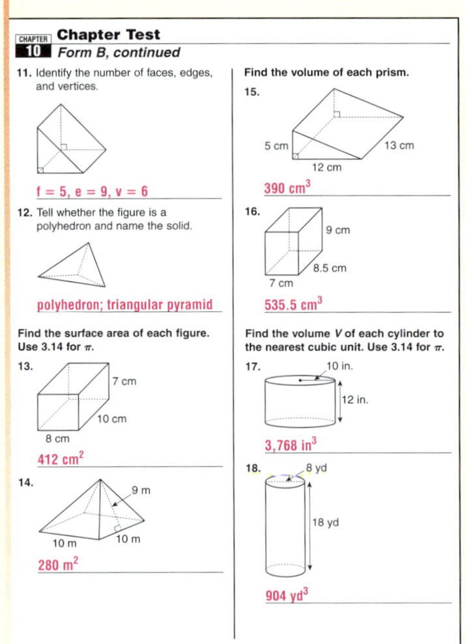
CHAPTER 10 Chapter Test Form B, continued

11. Identify the number of faces, edges, and vertices. f = 5, e = 9, v = 6
12. Tell whether the figure is a polyhedron and name the solid. polyhedron; triangular pyramid

Find the surface area of each figure. Use 3.14 for π.

13. 412 cm^2
14. 280 m^2

Find the volume of each prism.

15. 390 cm^3
16. 535.5 cm^3

Find the volume *V* of each cylinder to the nearest cubic unit. Use 3.14 for π.

17. 3,768 in^3
18. 904 yd^3

Level C

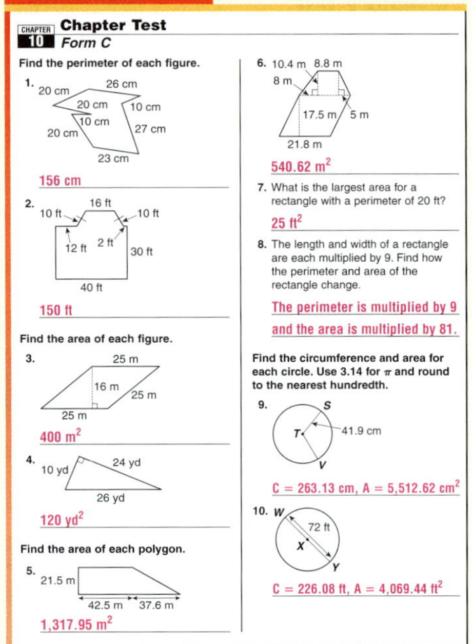
CHAPTER 10 Chapter Test Form C

Find the perimeter of each figure.

1. 156 cm
2. 150 ft

Find the area of each figure.

3. 400 m^2
4. 120 yd^2

Find the area of each polygon.

5. 1,317.95 m^2
6. 540.62 m^2

7. What is the largest area for a rectangle with a perimeter of 20 ft? 25 ft^2

8. The length and width of a rectangle are each multiplied by 9. Find how the perimeter and area of the rectangle change. The perimeter is multiplied by 9 and the area is multiplied by 81.

Find the circumference and area for each circle. Use 3.14 for π and round to the nearest hundredth.

9. C = 263.13 cm, A = 5,512.62 cm^2
10. C = 226.08 ft, A = 4,069.44 ft^2

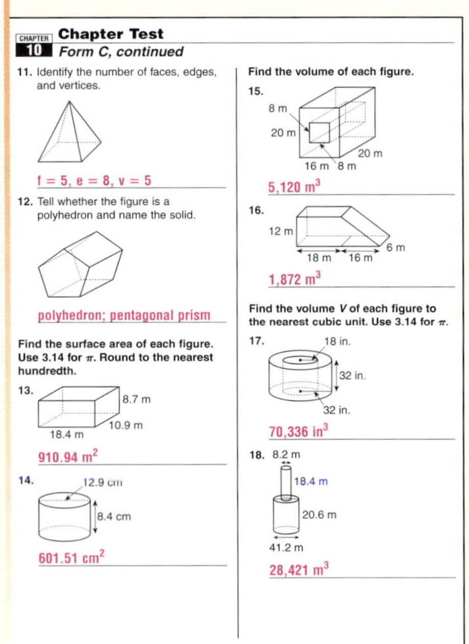
CHAPTER 10 Chapter Test Form C, continued

11. Identify the number of faces, edges, and vertices. f = 5, e = 8, v = 5
12. Tell whether the figure is a polyhedron and name the solid. polyhedron; pentagonal prism

Find the surface area of each figure. Use 3.14 for π. Round to the nearest hundredth.

13. 910.94 m^2
14. 601.51 cm^2

Find the volume of each figure.

15. 5,120 m^3
16. 1,872 m^3

Find the volume *V* of each figure to the nearest cubic unit. Use 3.14 for π.

17. 70,336 in^3
18. 28,421 m^3

***Test and Practice Generator* CD-ROM**

Create and customize multiple versions of the same tests with corresponding answers for any chosen chapter objectives.

Chapter 10 State and Standardized Test Preparation

Test Taking Skill Builder and Standardized Test Practice are provided for each chapter in the *Test Prep Tool Kit.*

Test Taking Skill Builder

Test Taking Strategy — **Sketch a Picture or Diagram**
Chapter 10

Sketching a picture or diagram can help you find the solution to some problems.

Example 1 Short Response Daniel is planning to fence in a circular garden. The distance from the center of the garden to the edge of the garden is 10 ft. How many feet of fencing should Daniel purchase?

Solution: It helps to draw a diagram with the given information. This way you can visualize the problem.

Sketch a circle with radius labeled "10 ft".

Now calculate the circumference of the circle. Use $\pi \approx 3.14$.

$$C = 2\pi r$$
$$= 2 \cdot \pi \cdot 10$$
$$\approx 62.8$$

Daniel needs to buy 62.8 ft of fencing.

Example 2 Multiple Choice Joseph wants to make a poster shaped like a house. He has a piece of cardboard shaped like a rectangle. The length of the cardboard is 24 in. and its width is 15 in. He also has a piece of cardboard shaped like an isosceles triangle. The base of this cardboard measures 24 in. and the lengths of the other two sides measure 17 in. each. What is the perimeter of the figure after he connects the cardboard together?

A 71 in. B 88 in. C 112 in. D 136 in.

Solution: Draw a diagram to help you understand the situation. Sketch a rectangle with side lengths 15 and 24. Attach to it an isosceles triangle with side lengths 17, 17, and 24.

The perimeter of Joseph's poster is calculated as follows:
17 + 17 + 15 + 24 + 15 = 88.

The perimeter of Joseph's poster is 88 in. The correct answer is Choice B.

Test Taking Strategy
Chapter 10, continued

Exercises

1. A circle with a radius of 3 inches is inscribed in a square. What is the area between the circle and the square?
   A 36 in² B 19.26 in² C 7.74 in² D 6 in²
   a. How would you draw a circle *inscribed* in a square?
   Draw a circle inside of a square.
   b. Sketch a diagram of the problem.
   c. How does the diagram help you to solve the problem?
   Once I drew the diagram I could see the area that needed to be calculated.
2. Maria is wrapping a birthday present. The shipping box measures 10 cm by 5 cm by 5 cm. Inside the shipping box is a small gift box that measures 2 cm by 2 cm by 2 cm. How much space is left inside the shipping box for packaging material?
   F 250 cm² G 242 cm² H 8 cm² I 6 cm²
   a. What geometric shape is the shipping box?
   a rectangular prism
   b. Sketch and label a diagram of the problem.
   c. Explain how to determine the correct answer. What is the correct answer?
   I subtracted the volume of the cube from the volume of the box. The correct answer is Choice G.

Standardized Test Practice

Standardized Test Practice
Chapter 10

Select the best answer for Questions 1–8.

1. Identify the figure shown.
   A square pyramid
   B cube
   C cylinder
   D cone
2. What is the volume of a 5 feet long, 3 feet wide, and 2 feet deep rectangular toy chest?
   F 10 ft³ H 30 ft³
   G 15 ft³ I 60 ft³
3. Find the volume of the cylinder. Use 3.14 for π.
   A 508.7 yd³ C 169.6 yd³
   B 339.1 yd³ D 162 yd³
4. What is the volume of the figure shown?
   F 288 cm² H 48 cm²
   G 144 cm² I 12 cm²
5. What is the perimeter of the polygon?
   A 50 cm C 42 cm
   B 48.9 cm D 37.1 cm
6. Mrs. Albright has a rug for her classroom floor. The rug is 20 feet long. What else must she know to calculate how many square feet the rug will occupy?
   F The number of rugs in the room.
   G The width of the classroom.
   H The width of the rug.
   I The height of the classroom.
7. If square floor tile costs $2 per square foot, what is the cost of tiling a rectangular floor 12 feet by 15 feet?
   A $360 C $108
   B $180 D $54
8. Estimate the area of the figure shown.
   F 150 cm² H 189.25 cm²
   G 175 cm² I 228.5 cm²

Standardized Test Practice
Chapter 10, continued

Gridded Response
Solve the problems. Use the answer sheet to write and grid-in your answer.

9. Kara wants to cover the surface of the pyramid shown with brown paper for a social studies project. How many square centimeters of paper will she need?
10. What is the area in square inches that a ceiling fan covers if the length of one of the fan blades extends 15 in. from the center? Use 3.14 for π and round to the nearest inch.
11. The circumference at the base of the Tower of Pisa is 48.6 m. What is the radius in meters of the Tower of Pisa? Use 3.14 for π.

Short Response
Solve the problems. Use the answer sheet to write your answers.

12. The top of a shed has the shape of a square pyramid. The base edges are 12 ft and the slant height is 8 feet. How much plastic is needed to cover the top of the shed? (Hint: the base of the pyramid is not included.)
13. An architect wants to change the dimensions of a rectangular room he had drafted. In the new draft, each dimension is twice as large as in the first drawing. What happens to the area of the second room in relation to the first room? Find the area of each room.

Extended Response

14. The state of Kansas is rectangular in shape. It is approximately 401 mi long and 204 mi wide.
    a. Find the area of Kansas.
    b. In 2000, the census calculated that there were 2,688,418 people living in Kansas. Estimate the population per square mile.
    c. If you can cycle 121 miles per day, how many days would it take you to cycle the borders of Kansas? Explain your answer.

Answer Sheet

12. $4\left(\frac{1}{2} \cdot 12 \cdot 8\right) = 192$; 192 square feet of plastic is needed.

13. The area of the second room is four times as large. Area of the first room is 80 sq ft. Area of the second room is 320 sq ft.

Extended Response
Write your answers for Problem 14 on the back of this paper.

internet connect
State-Specific Test Practice Online
KEYWORD: MR4 TestPrep

Test Prep Tool Kit

- Standardized Test Prep Workbook
- Countdown to Testing transparencies
- State Test Prep CD-ROM
- Standardized Test Prep Video

Customized answer sheets give students realistic practice for actual standardized tests.

Chapter 10

Perimeter, Area, and Volume

Why Learn This?

Tell students that the study of mathematics is essential to many career fields. *Pure* mathematicians study mathematical ideas and try to create and prove new rules about the way values and figures behave in certain situations. *Applied* mathematicians use the mathematical concepts discovered by pure mathematicians to evaluate and solve real-world situations. The study of π is of interest to both pure and applied mathematicians.

Using Data

To begin the study of this chapter, have students:

- Identify the value of π to the nearest hundredth. 3.14
- Use words to explain what π represents. In any circle, π is the ratio of the circumference to the diameter.

Chapter 10

Perimeter, Area, and Volume

TABLE OF CONTENTS

internet connect
Chapter Opener Online
go.hrw.com
KEYWORD: MR4 Ch10

Career ***Mathematician***

Some mathematicians apply their knowledge in areas such as airplane scheduling, medical safety, and automobile and industrial research. Other mathematicians prefer to study the concepts behind mathematics.

For hundreds of years, mathematicians have studied the relationship between the circumference and the diameter of a circle. This ratio is called *pi* and is represented by the Greek letter π.

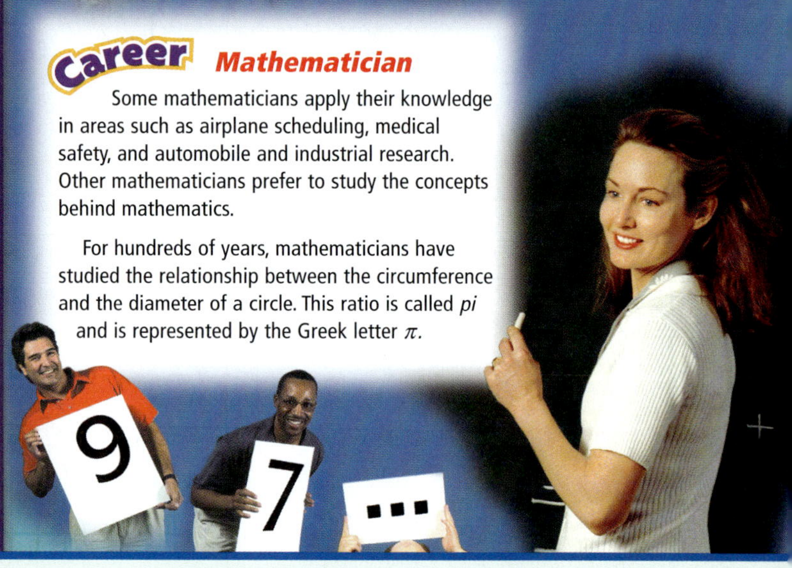

Problem Solving Project

Social Studies and Geometry Connection

Purpose: To solve problems relating to circles, area, and *pi*

Materials: Discovering *Pi* worksheet, cans and other objects, strings, and rulers or other measuring devices

internet connect
Chapter Project Online: ***go.hrw.com***
KEYWORD: MR4 PSProject10

Understand, Plan, Solve, and Look Back

Have students:

- ✔ Complete the Discovering *Pi* worksheet.
- ✔ Explain whether $A = \pi r^2$ can be used to find the exact area of a circle.
- ✔ Research ways to find the area of a circle without using the equation $A = \pi r^2$ and create drawings to demonstrate them.
- ✔ Research the history of *pi* and then make a wall chart to display the known decimal places of *pi*.
- ✔ Check students' work.

ARE YOU READY?

Choose the best term from the list to complete each sentence.

1. A(n) __?__ is a quadrilateral with opposite sides that are parallel and congruent. parallelogram
2. Some customary units of length are __?__ and __?__. Some metric units of length are __?__ and __?__. inches; feet; centimeters; meters
3. A(n) __?__ is a quadrilateral with side lengths that are all congruent and four right angles. square
4. A(n) __?__ is a polygon with six sides. hexagon

square
feet
cube
meters
liters
hexagon
inches
parallelogram
trapezoid
centimeters
sphere

Complete these exercises to review skills you will need for this chapter.

✔ Add and Multiply Whole Numbers, Fractions, and Decimals

Find each sum or product.

5. $1.5 + 2.4 + 3.6 + 2.5$ 10
6. $2 \cdot 3.5 \cdot 4$ 28
7. $\frac{22}{7} \cdot 21$ 66
8. $\frac{1}{2} \cdot 5 \cdot 4$ 10
9. $3.2 \cdot 5.6$ 17.92
10. $\frac{1}{2} \cdot 10 \cdot 3$ 15
11. $(2 \cdot 5) + (6 \cdot 8)$ 58
12. $2(3.5) + 2(1.5)$ 10
13. $9(20 + 7)$ 243

✔ Estimate Metric Lengths

Use a centimeter ruler to measure each line to the nearest centimeter.

14. ______________ 6 cm
15. __________ 4 cm

✔ Identify Polygons

Name each polygon. Determine whether it appears to be regular or not regular.

16. 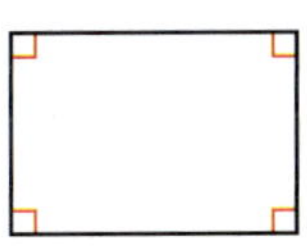
rectangle; not regular

17. 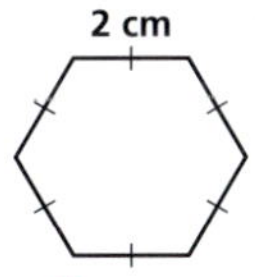

octagon; regular

18. 

pentagon; not regular

Assessing Prior Knowledge

INTERVENTION

Diagnose and Prescribe

Evaluate your students' performance on this page to determine whether intervention is necessary or whether enrichment is appropriate. Options that provide instruction, practice, and a check are listed below.

Resources for Are You Ready?

- ***Are You Ready? Intervention and Enrichment***
- **Recording Sheet for Are You Ready?** *Chapter 10 Resource Book. . . . . p. 3*
- ***Are You Ready? Intervention* CD-ROM**

ARE YOU READY?

Were students successful with Are You Ready?

NO — INTERVENE

✔ Add and Multiply Whole Numbers, Fractions, and Decimals. 5–13
Intervention Practice, Skill 34, 40, 42
CD-ROM
Intervention Activities, Skill 34, 40, 42

✔ Estimate Metric Lengths 14–15
Intervention Practice, Skill 72
CD-ROM
Intervention Activities, Skill 72

✔ Identify Polygons 16–18
Intervention Practice, Skill 76
CD-ROM
Intervention Activities, Skill 76

YES — ENRICH

Are You Ready? Enrichment, pp. 425–426

Section 10A

Perimeter, Area, and Circumference

One-Minute Section Planner

Lesson	Materials	Resources
Lesson 10-1 Finding Perimeter **NCTM:** Geometry, Measurement, Communication, Connections **NAEP:** Measurement 1h ☑ SAT-9 ☑ SAT-10 ☑ ITBS ☑ CTBS ☑ MAT ☑ CAT	**Optional** Teaching Transparency T2 *(CRB)* Cut out polygons *(CRB p. 96)* Geoboards *(MK)* Dot paper *(CRB p. 97)*	• *Chapter 10 Resource Book*, pp. 6–14 • Daily Transparency T1, CRB • Additional Examples Transparencies T3–T5, CRB • *Alternate Openers: Explorations*, p. 83
Lesson 10-2 Estimating and Finding Area **NCTM:** Geometry, Measurement, Communication, Connections **NAEP:** Measurement 1c ☑ SAT-9 ☑ SAT-10 ☑ ITBS ☑ CTBS ☑ MAT ☑ CAT	**Optional** Teaching Transparency T7 *(CRB)* Recording Sheet for Motivate *(CRB p. 99)* Australian map *(CRB p. 100)* Graph paper *(CRB p. 101)*	• *Chapter 10 Resource Book*, pp. 15–24 • Daily Transparency T6, CRB • Additional Examples Transparencies T8–T9, CRB • *Alternate Openers: Explorations*, p. 84
Lesson 10-3 Break into Simpler Parts **NCTM:** Geometry, Measurement, Communication **NAEP:** Measurement 1g ☐ SAT-9 ☑ SAT-10 ☐ ITBS ☑ CTBS ☑ MAT ☑ CAT		• *Chapter 10 Resource Book*, pp. 25–33 • Daily Transparency T10, CRB • Additional Examples Transparencies T11–T13, CRB • *Alternate Openers: Explorations*, p. 85
Lesson 10-4 Comparing Perimeter and Area **NCTM:** Geometry, Measurement, Reasoning and Proof, Communication **NAEP:** Measurement 1b ☐ SAT-9 ☐ SAT-10 ☐ ITBS ☐ CTBS ☐ MAT ☐ CAT		• *Chapter 10 Resource Book*, pp. 34–42 • Daily Transparency T14, CRB • Additional Examples Transparencies T15–T16, CRB • *Alternate Openers: Explorations*, p. 86
Hands-On Lab 10A Discover Properties of Circles **NCTM:** Geometry, Measurement, Reasoning and Proof **NAEP:** Geometry 1d ☐ SAT-9 ☐ SAT-10 ☐ ITBS ☐ CTBS ☐ MAT ☐ CAT	**Required** Compasses *(MK)* Rulers *(MK)*	• *Hands-On Lab Activities*, pp. 90–91, 114
Lesson 10-5 Circles **NCTM:** Geometry, Measurement, Communication **NAEP:** Measurement 1g ☑ SAT-9 ☑ SAT-10 ☑ ITBS ☑ CTBS ☑ MAT ☑ CAT	**Optional** Recording Sheet for Reaching All Learners *(CRB p. 102)*	• *Chapter 10 Resource Book*, pp. 43–52 • Daily Transparency T17, CRB • Additional Examples Transparencies T18–T20, CRB • *Alternate Openers: Explorations*, p. 87
Section 10A Assessment		• Mid-Chapter Quiz, SE p. 520 • Section 10A Quiz, AR p. 25 • *Test and Practice Generator* CD-ROM

SAT = *Stanford Achievement Tests* **ITBS** = *Iowa Test of Basic Skills* **CTBS** = *Comprehensive Test of Basic Skills/Terra Nova*
MAT = *Metropolitan Achievement Tests* **CAT** = *California Achievement Test*

NCTM = Complete standards can be found on pages T27–T33. **NAEP** = Complete standards can be found on pages A31–A35.

SE = *Student Edition* **TE** = *Teacher's Edition* **AR** = *Assessment Resources* **CRB** = *Chapter Resource Book* **MK** = *Manipulatives Kit*

Section Overview

Finding Perimeter — Lesson 10-1

Why? You would need to find the perimeter of your backyard to know how much fencing is needed to enclose it.

The **perimeter** of a figure is the distance around it.

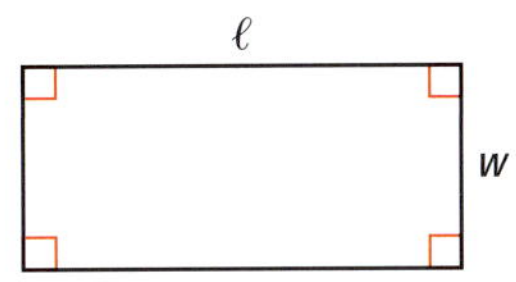

The formula for the **perimeter of a rectangle** is
$P = 2\ell + 2w$.

Area, Comparing Area and Perimeter — Lessons 10-2, 10-3, 10-4

Why? You would need to find the area of your backyard to know how much sod is needed to cover it.

The **area** of a figure is the amount of surface it covers.

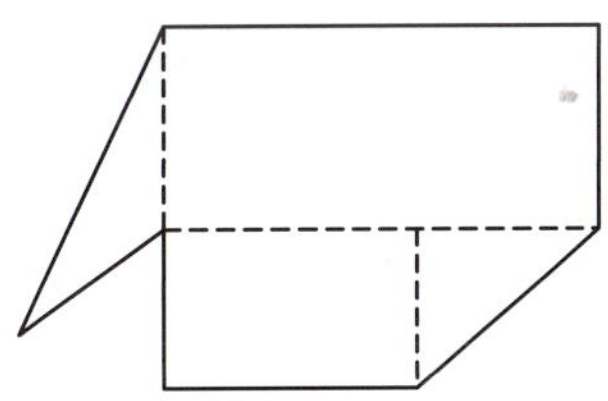

To find the area of a composite figure, break the figure apart into rectangles, parallelograms, and triangles.

Area Formulas	
Rectangle	$A = \ell w$
Parallelogram	$A = bh$
Triangle	$A = \frac{1}{2}bh$
Square	$A = s^2$

Circles — Hands-On Lab 10A, Lesson 10-5

Why? The shape of a bicycle wheel is a circle. The size of tires and inner tubes are given by their diameters.

Circumference of a Circle
the distance around a circle
$C = \pi d$, or $C = 2\pi r$
Find the circumference of a circle with radius 5 cm.
$C = 2\pi r$
$C = 2\pi 5$
$C = 10\pi$ cm

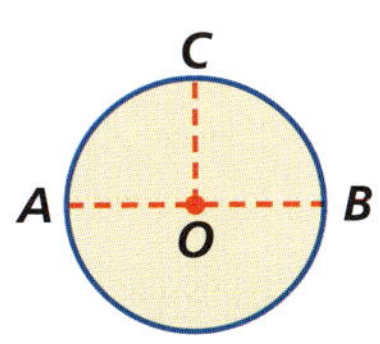

The circle is circle O.
$\overline{AB}$ is a diameter.
$\overline{OA}$, $\overline{OB}$, and $\overline{OC}$ are radii.

Area of a Circle
the amount of surface it covers
$A = \pi r^2$
Find the area of a circle with radius 5 cm.
$A = \pi r^2$
$A = \pi 5^2$
$A = 25\pi$ cm^2

Pi is the ratio of the circumference to the diameter, $\frac{C}{d}$, for any circle. This ratio is represented by the Greek letter π, which is read as "**pi**." $\frac{C}{d} = \pi$

The decimal representation of **pi** starts with 3.14159265 . . . and goes on forever without a repeating pattern. We approximate **pi** using either 3.14 or $\frac{22}{7}$.

10-1 Organizer

Pacing: Traditional 1 day
Block $\frac{1}{2}$ day

Objective: Students find the perimeter and missing side lengths of a polygon.

Indiana Academic Standards
6.3.2, 6.3.5, 6.5.1, 6.7.2, 6.7.5, 6.7.11

Warm Up

1. What figure has four equal sides and four right angles? square
2. What figure has eight sides? octagon
3. What figure has five sides pentagon

Problem of the Day

Find the greatest perimeter possible when 6 identical unit squares are arranged to form a closed figure. Adjacent squares must share an entire side. 14 units

Available on Daily Transparency in CRB

Math Fact !!

The word *perimeter* comes from the Greek words *peri*, which means "around," and *metron*, which means "to measure." So the perimeter is the distance around a figure.

10-1 Finding Perimeter

Learn to find the perimeter and missing side lengths of a polygon.

Vocabulary
perimeter

One of the biggest finger paintings ever painted is *Ten Fingers, Ten Toes*. It is 8.53 meters wide and 10.66 meters long.

The **perimeter** of a figure is the distance around it. To find the perimeter of the painting you can add the lengths of the sides.

8.53 + 10.66 + 8.53 + 10.66 = 38.38

The perimeter of the painting is 38.38 meters.

EXAMPLE 1 Finding the Perimeter of a Polygon

Find the perimeter of the figure.

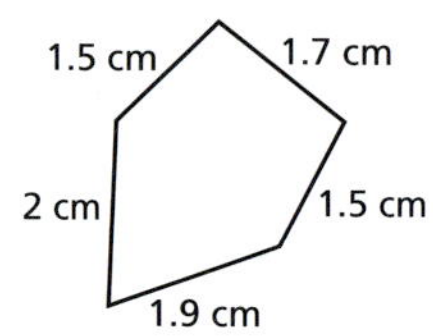

1.5 + 1.7 + 1.5 + 1.9 + 2 = 8.6

Add all the side lengths.

The perimeter is 8.6 cm.

PERIMETER OF A RECTANGLE

The opposite sides of a rectangle are equal in length. Find the perimeter of a rectangle by using the formula, in which ℓ is the length and w is the width.

$P = 2\ell + 2w$

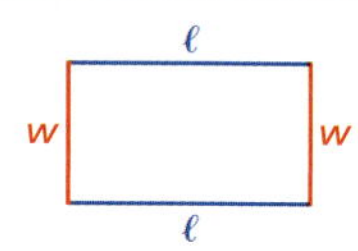

$P = \ell + \ell + w + w$

EXAMPLE 2 Using a Formula to Find Perimeter

Find the perimeter *P* of the rectangle.

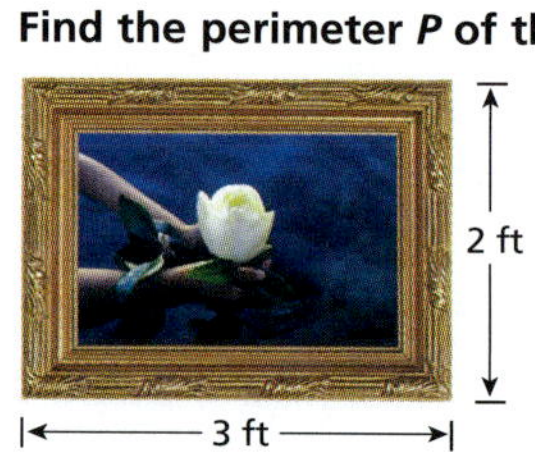

$P = 2\ell + 2w$

$P = (2 \cdot 3) + (2 \cdot 2)$ *Substitute 3 for ℓ and 2 for w.*

$P = 6 + 4$ *Multiply.*

$P = 10$ *Add.*

The perimeter is 10 feet.

1 Introduce

Alternate Opener

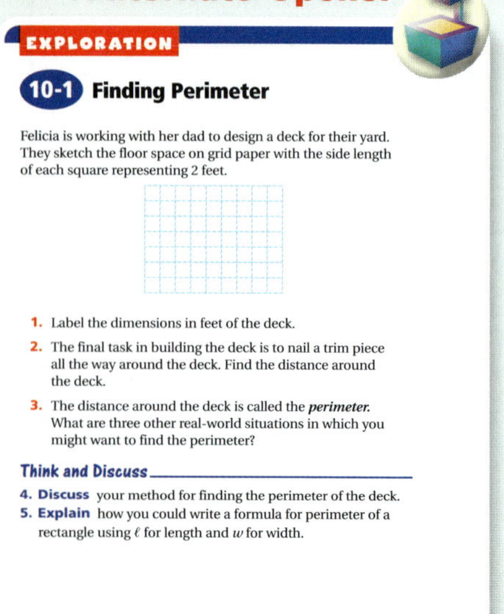

EXPLORATION

10-1 Finding Perimeter

Felicia is working with her dad to design a deck for their yard. They sketch the floor space on grid paper with the side length of each square representing 2 feet.

1. Label the dimensions in feet of the deck.
2. The final task in building the deck is to nail a trim piece all the way around the deck. Find the distance around the deck.
3. The distance around the deck is called the *perimeter*. What are three other real-world situations in which you might want to find the perimeter?

Think and Discuss

4. **Discuss** your method for finding the perimeter of the deck.
5. **Explain** how you could write a formula for perimeter of a rectangle using ℓ for length and w for width.

Motivate

Give students various simple cutout polygons (provided on Chapter 10 Resource Book p. 96). Have students use rulers to measure the distance around each shape. Discuss strategies used, such as recording the length of each side and then adding to find the total distance around. Explain that the distance around a figure is its *perimeter*.

Exploration worksheet and answers on Chapter 10 Resource Book pp. 7 and 105

2 Teach

Lesson Presentation

Guided Instruction

In this lesson, students learn to find the perimeter and missing side lengths of a polygon. Teach students to find the perimeter of a figure by adding the lengths of all the sides. Then teach them to use a formula to find perimeter of a rectangle and to use an equation to find the length of a missing side.

EXAMPLE 3 Finding Unknown Side Lengths and the Perimeter of a Polygon

Find each unknown measure.

A **What is the length of side a if the perimeter equals 105 m?**

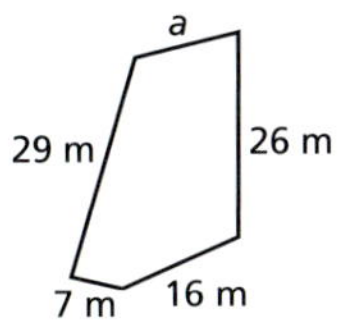

P = sum of side lengths

$105 = a + 26 + 16 + 7 + 29$ *Use the values you know.*

$105 = a + 78$ *Add the known lengths.*

$105 - 78 = a + 78 - 78$ *Subtract 78 from both sides.*

$27 = a$

Side a is 27 m long.

B **What is the perimeter of the polygon?**

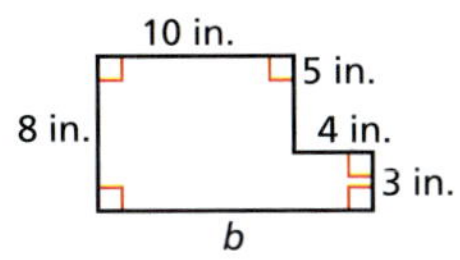

First find the unknown side length.

Find the sides opposite side b.

The length of side b = 10 + 4.

Side b is 14 in. long.

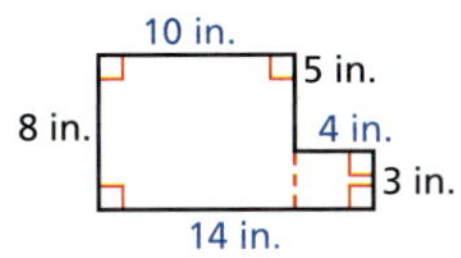

Find the perimeter.

$P = 14 + 8 + 10 + 5 + 4 + 3$

$P = 44$

The perimeter of the polygon is 44 in.

C **The width of a rectangle is 12 cm. What is the perimeter of the rectangle if the length is 3 times the width?**

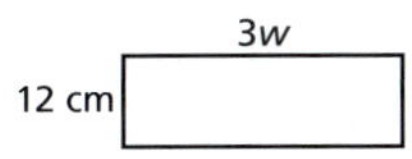

$\ell = 3w$ *Find the length.*

$\ell = (3 \cdot 12)$ *Substitute 12 for w.*

$\ell = 36$ *Multiply.*

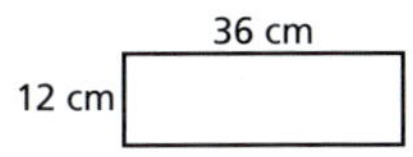

$P = 2\ell + 2w$ *Use the formula for the perimeter of a rectangle.*

$P = 2(36) + 2(12)$ *Substitute 36 and 12.*

$P = 72 + 24$ *Multiply.*

$P = 96$ *Add.*

The perimeter of the rectangle is 96 cm.

Think and Discuss

1. **Explain** how to find the perimeter of a regular pentagon if you know the length of one side.
2. **Tell** what formula you can use to find the perimeter of a square.

Additional Examples

Example 1

Find the perimeter of the figure.

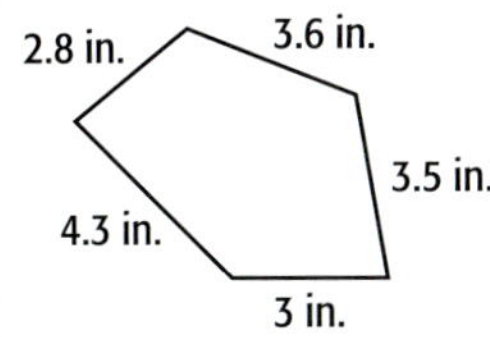

17.2 in.

Example 2

Find the perimeter *P* of the rectangle.

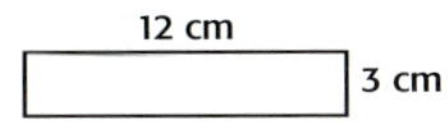

30 cm

Example 3

Find each unknown measure.

A. What is the length of side a if the perimeter equals 1,471 mm?

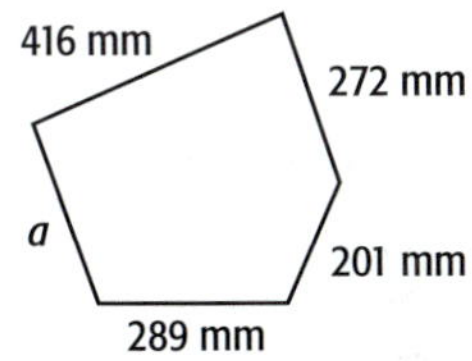

293 mm

B. What is the perimeter of the polygon?

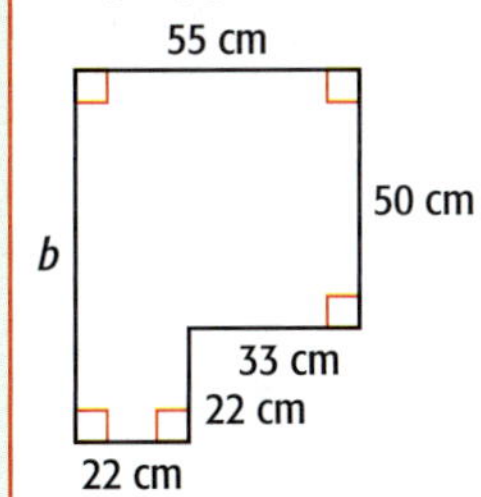

254 cm

C. The width of a rectangle is 19 cm. What is the perimeter of the rectangle if the length is 4 times the width? 190 cm

3 Close

Reaching All Learners
Through Concrete Manipulatives

Have each student make several polygons on a geoboard (provided in the Manipulatives Kit) and draw each polygon on dot paper (provided on Chapter 10 Resource Book p. 97). Have students measure the perimeter of each polygon. This activity reinforces the meaning of perimeter, maintains measuring skills, and requires the students to apply the correct fraction operation.

Summarize

Review the term *perimeter* and have students explain how to find the perimeter of a polygon.

Possible answer: Measure each side of the polygon. Add the lengths of the sides to find the perimeter.

Answers to Think and Discuss

1. Possible answer: Multiply the length of the known side by 5.
2. $P = 4s$, where s is the length of each side

10-1 PRACTICE & ASSESS

Students may want to refer back to the lesson examples.

Assignment Guide

If you finished Example 1 assign:
Core 1–2, 6–7, 13, 16–17, 23–30
Enriched 6–7, 13, 16–19, 23–30

If you finished Example 2 assign:
Core 1–4, 6–10, 16, 17, 23–30
Enriched 3, 4, 6–10, 16–19, 23–30

If you finished Example 3 assign:
Core 1–13, 16, 17, 23–30
Enriched 8–30

Notes

Math Background

The study of perimeter allows students to develop some simple and easily understandable formulas. Besides the formula in the text for the perimeter of a rectangle, a formula for the perimeter of a regular polygon can be stated as follows: $P = ns$, where n is the number of sides in the polygon and s is the length of each side.

10-1 Exercises

FOR EXTRA PRACTICE see page 656

internet connect
Homework Help Online
go.hrw.com Keyword: MR4 10-1

GUIDED PRACTICE

See Example 1 Find the perimeter of each figure.

1. 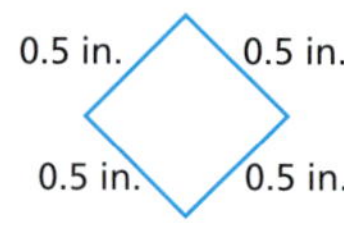

2 in.

2. 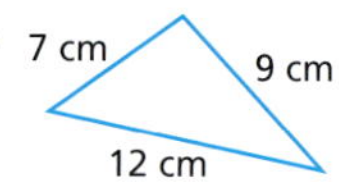

28 cm

See Example 2 Find the perimeter P of each rectangle.

3. 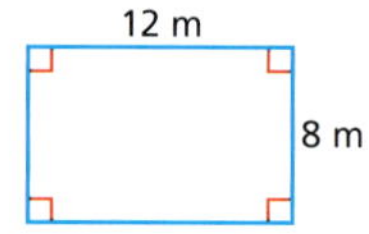

40 m

4. 

22.6 in.

See Example 3 Find the unknown measure.

5. What is the length of side b if the perimeter equals 21 yd?
7 yd

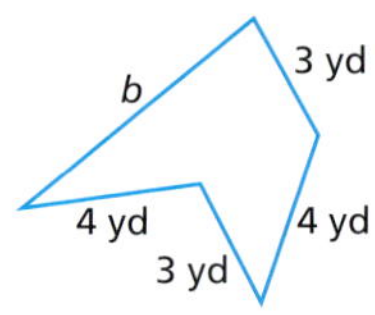

INDEPENDENT PRACTICE

See Example 1 Find the perimeter of each figure.

6. 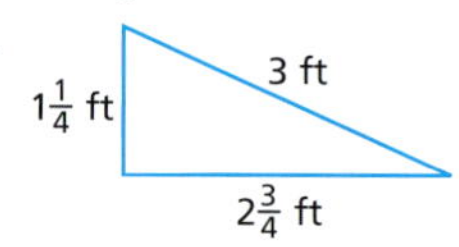

7 ft

7. regular octagon
96 in.

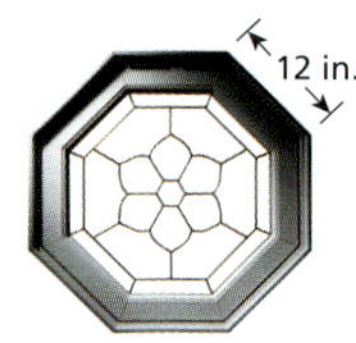

See Example 2 Find the perimeter P of each rectangle.

8. 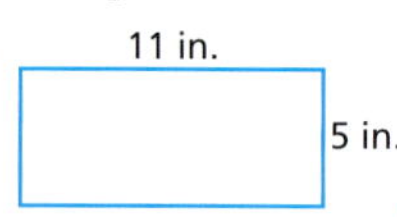

32 in.

9. 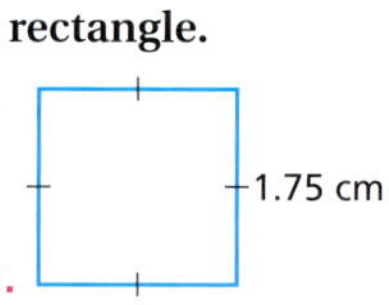

7 cm

10. 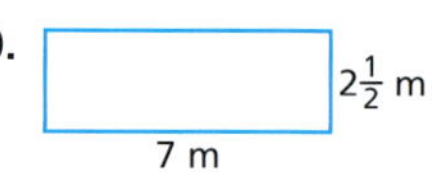

19 m

See Example 3 Find each unknown measure.

11. What is the perimeter of the polygon?
42 m

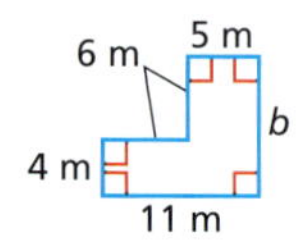

12. The width of a rectangle is 15 ft. What is the perimeter of the rectangle if the length is 5 ft longer than the width? **70 ft**

RETEACH 10-1

LESSON 10-1 **Reteach**
Finding Perimeter

The perimeter of a figure is the distance around it. To find the perimeter of a figure, find the sum of the lengths of its sides.

Look at the figure below.

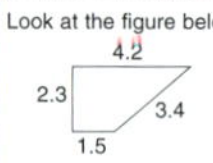

The perimeter is the sum of the lengths of its sides.

$P = 2.3 + 1.5 + 3.4 + 4.2$
$= 11.4$

So, the perimeter of the figure is 11.4 units.

Find the perimeter of each figure.

1. 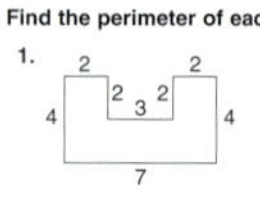

26 units

2. 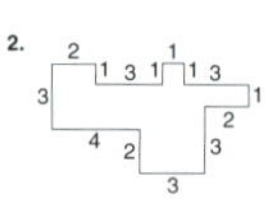

30 units

If you know the perimeter of a figure, you can find the unknown length of a side.

The perimeter of the figure below is 94 units.

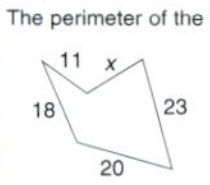

To find the missing length x first find the sum of the known lengths.

$23 + 18 + 20 + 11 = 72$

Then subtract the sum from the perimeter.

$94 - 72 = 22$ So, side x is 22 units long.

Find the missing length for each figure.

3. 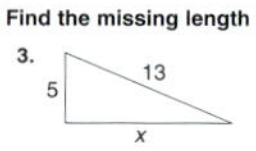

Perimeter = 30 units
$x = 12$ units

4. 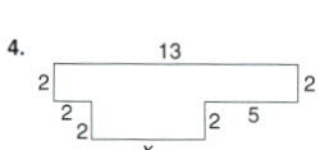

Perimeter = 34 units
$x = 6$ units

PRACTICE 10-1

LESSON 10-1 **Practice B**
Finding Perimeter

Find the perimeter of each figure.

1. 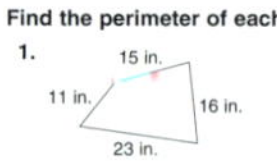

65 in.

2. 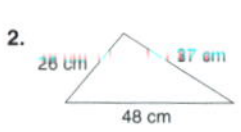

111 cm

3. 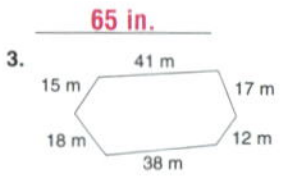

141 m

4. 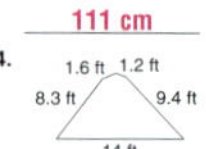

34.5 ft

Find the perimeter P of each rectangle.

5. 

80 yd

6. 

102 mi

7. 

9 ft

Find the unknown measure.

8. What is the length of side b if the perimeter equals 47 in.?

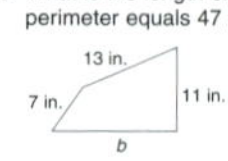

$b = 16$ in.

9. What is the length of side s if the perimeter equals 119 yd?

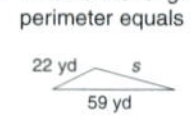

$s = 38$ yd

10. Benjamin is putting a fence around his rectangular-shaped yard. The yard is 38 feet long and 27 feet wide. How many feet of fencing does Benjamin need to surround his entire yard? 130 ft of fencing

11. If you drove from Bakersville to Salem and then to San Mateo, your entire 81-mile journey would form a triangle. The distance from Salem to San Mateo is 24 miles. The distance from Bakersville to San Mateo is 40 miles. How many miles is it from Salem to Bakersville? 17 miles

PRACTICE AND PROBLEM SOLVING

Use the figure *ACDEFG* for Exercises 13–15.

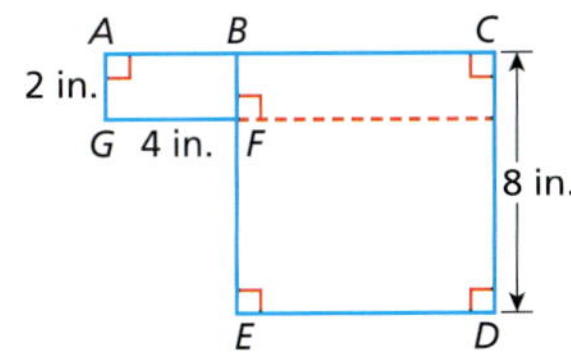

13. What is the length of side *FE*? **6 in.**
14. If the perimeter of rectangle *BCDE* is 34 in., what is the length of side *BC*? **9 in.**
15. Use your answer from Exercise 14 to find the perimeter of figure *ACDEFG*. **42 in.**

Find the perimeter of each figure.

16. a triangle with side lengths 6 in., 8 in., and 10 in. **24 in.**
17. a regular pentagon with side length $\frac{2}{5}$ km **2 km**
18. ***SPORTS*** The diagram shows one-half of a badminton court.
    a. What are the dimensions of the whole court? **44 ft × 20 ft**
    b. What is the perimeter of the whole court? **128 ft**

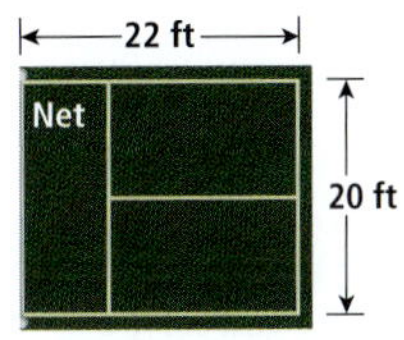

19. ***MEASUREMENT*** Use the map and a centimeter ruler to find the perimeter of the triangle formed by the three cities. **8.5 cm**

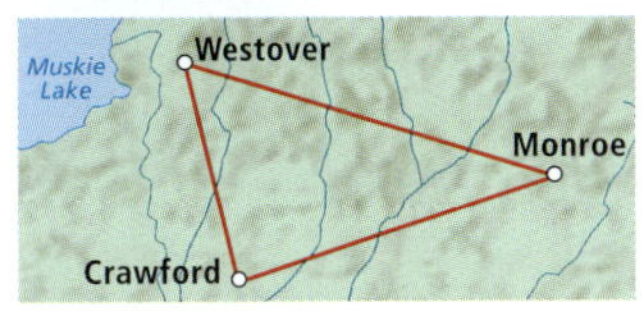

20. ***WHAT'S THE ERROR?*** A student found the perimeter of a 10-inch-by-13-inch rectangle to be 23 inches. Explain the student's error. Then find the correct perimeter.
21. ***WRITE ABOUT IT*** Explain how to find the unknown length of a side of a triangle that has a perimeter of 24 yd and two sides that measure 6 yd and 8 yd.
22. ***CHALLENGE*** The perimeter of a regular octagon is 20 m. What is the length of one side of the octagon? **2.5 m**

Spiral Review

Find each sum or difference. (Lesson 3-3)

23. 30 − 5.32 **24.68**
24. 80.37 + 15.125 **95.495**
25. 100 − 25.65 **74.35**
26. 200.6 + 62.78 **263.38**

Solve each proportion. (Lesson 8-2)

27. $\frac{9}{15} = \frac{x}{5}$ **3**
28. $\frac{a}{20} = \frac{3}{15}$ **4**
29. $\frac{1}{7} = \frac{6}{k}$ **42**

30. **TEST PREP** Which decimal is equivalent to 85%? (Lesson 8-7) **C**

A 85.0 **B** 8.5 **C** 0.85 **D** 0.085

Answers

20. The student did not multiply the length and width by 2 before adding. The correct perimeter is 46 inches.
21. Subtract the lengths of the known sides from the perimeter. The answer is the length of the third side.

Journal

Have students draw polygons and find their perimeters. Students should explain the steps they followed to find each perimeter.

Test Prep Doctor

For Exercise 30, students need to divide the percent by 100 to find the decimal equivalent. Students who answered **A** did not divide; they just removed the % symbol. Students who answered **B** divided by 10, and students who answered **D** divided by 1,000.

CHALLENGE 10-1

LESSON 10-1 **Challenge**
Perimeters Around the World

Find the perimeter of the base of each famous site.

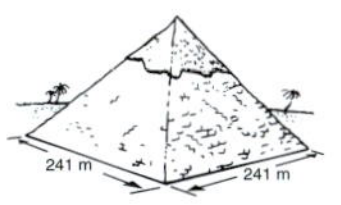

Great Pyramid of Khufu, Giza, Egypt

Perimeter: **964 m**

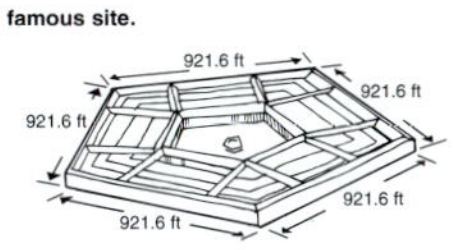

Pentagon, Arlington, Virginia

Perimeter: **4,608 ft**

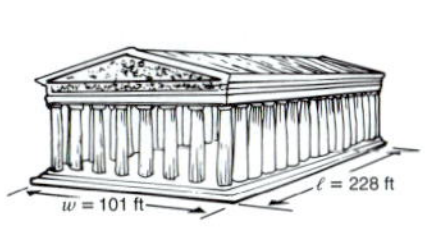

Parthenon, Athens, Greece

Perimeter: **658 ft**

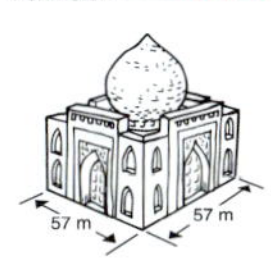

Taj Mahal, Agra, India

Perimeter: **228 m**

Red Square, Moscow, Russia

Perimeter: **1,100 m**

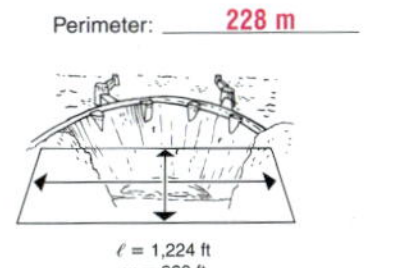

Hoover Dam, Arizona and Nevada

Perimeter: **3,768 ft**

PROBLEM SOLVING 10-1

LESSON 10-1 **Problem Solving**
Finding Perimeter

Write the correct answer.

1. Use a ruler to find the perimeter of your math textbook in inches. **39.5 inches**
2. Use a ruler to find the perimeter of your desk in feet and inches. **Answer depends on size of desk.**
3. The world's largest flag weighs 3,000 pounds and requires at least 500 people to set up! This United States flag is 505 feet long and 255 feet wide. What is the perimeter of this United States flag? **1,520 feet**
4. Students in Lisbon, Ohio, built the world's largest mousetrap in 1998. The mousetrap is 9 feet 10 inches long and 4 feet 5 inches wide—and it actually works! What is the perimeter of the mousetrap in feet and inches? **28 feet 6 inches**
5. The giant ball dropped every New Year's Eve in New York City is covered with 504 crystal equilateral triangles. The average perimeter of each triangle is $15\frac{3}{4}$ inches. What is the average side length of each crystal triangle on the ball? **$5\frac{1}{4}$ inches**
6. United States dollar bills are 2.61 inches wide and 6.14 inches long. Larger notes in circulation before 1919 measured 3.125 inches wide by 7.4218 inches long. What is the difference between the old and new dollar bill perimeters? **3.5936 inches**

Circle the letter of the correct answer.

7. The perimeter of regular octagon-shaped swimming pool is 42 feet. What is the length of each side of the pool?
   A 5 feet C 5 feet 2 inches
   (B) 5 feet 3 inches D 5.2 feet
8. Each Scrabble® tile is 1.8 centimeters wide and 2.1 centimeters tall. If the tiles spell the word LOVE, what is the perimeter of the entire word?
   F 7.8 cm H 12 cm
   (G) 18.6 cm J 31.2 cm

Lesson Quiz

Find each perimeter.

1. 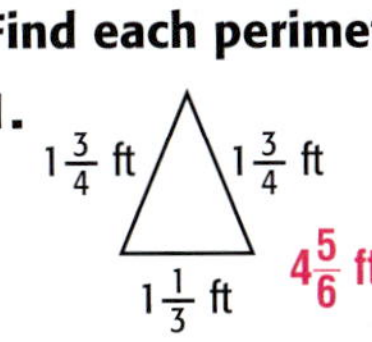

$4\frac{5}{6}$ ft

2. 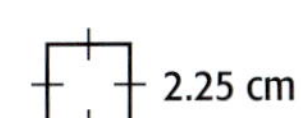

9 cm

3. What is the perimeter of a polygon with side lengths of 15 cm, 18 cm, 21 cm, 32 cm, and 26 cm? **112 cm**
4. What is the perimeter of a rectangle with length 22 cm and width 8 cm? **60 cm**

Available on Daily Transparency in CRB

10-2 Organizer

Pacing: Traditional 1 day
Block $\frac{1}{2}$ day

Objective: Students estimate the area of irregular figures and find the area of rectangles, triangles, and parallelograms.

Indiana Academic Standards
6.3.2, 6.3.5, 6.7.1, 6.7.2, 6.7.5, 6.7.6, 6.7.7, 6.7.11

Warm Up

1. What is the perimeter of a square with side lengths of 15 in.? **60 in.**
2. What is the perimeter of a rectangle with length 16 cm and width 11 cm? **54 cm**

Problem of the Day

Two wibbles equal four wabbles, and eight wabbles equal two bibbles. If a square has a perimeter of 16 wibbles, what is its area in square bibbles?
4 square bibbles

Available on Daily Transparency in CRB

Math Fact

Ancient Egyptian surveyors found the area of a four-sided field by finding one-half the sum of each pair of opposite sides and then finding their product. If the field is rectangular, the method is exact, but if not, it leads to an overestimate.

10-2 Estimating and Finding Area

Learn to estimate the area of irregular figures and to find the area of rectangles, triangles, and parallelograms.

Vocabulary
area

When colonists settled the land that would become the United States, ownership boundaries were sometimes natural landmarks such as rivers, trees, and hills. Landowners who wanted to know the size of their property needed to find the areas of irregular shapes.

The **area** of a figure is the amount of surface it covers. We measure area in square units.

EXAMPLE 1 Estimating the Area of an Irregular Figure

Estimate the area of the figure.

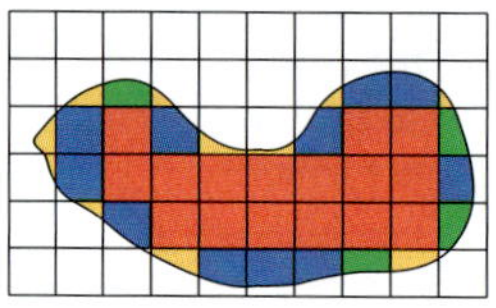

□ = 1 mi^2

Count full squares: 16 red squares.
Count almost-full squares: 11 blue squares.
Count squares that are about half-full: 4 green squares ≈ 2 full squares.
Do not count almost empty yellow squares.
Add. 16 + 11 + 2 = 29

The area of the figure is about 29 mi^2.

AREA OF A RECTANGLE

To find the area of a rectangle, multiply the length by the width.

$$A = \ell w$$
$$A = 4 \cdot 3 = 12$$

The area of the rectangle is 12 square units.

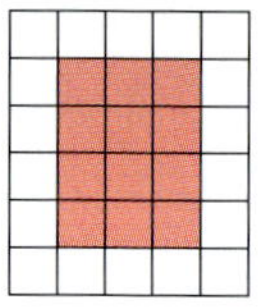

EXAMPLE 2 Finding the Area of a Rectangle

Find the area of the rectangle.

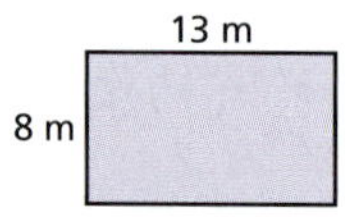

$A = \ell w$ *Write the formula.*

$A = 13 \cdot 8$ *Substitute 13 for ℓ.*

$A = 104$ *Substitute 8 for w.*

The area is 104 m^2.

1 Introduce

Alternate Opener

EXPLORATION

10-2 Estimating and Finding Area

Mr. and Mrs. Domínguez want to have the bottom of their pool refinished. A sketch of the pool is shown below with the side length of each square representing 1 yard. Before they begin the refinishing project, they have to estimate the area in square yards of the bottom of the pool.

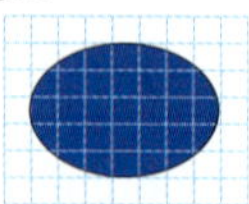

1. Estimate the area, in square yards, of the bottom of the pool.
2. Compare your estimate with the estimates of others in your class, and then average your estimates.
3. What are three other real-world situations in which you might want to estimate area?

Think and Discuss

4. **Discuss** the strategies you used for estimating the area of the bottom of the pool.
5. **Explain** how you could use squares to help you estimate the areas of irregular shapes.

Motivate

Provide students with figures on graph paper (provided on a recording sheet on Chapter 10 Resource Book p. 99) and have them find the perimeter of each figure. Then have them count the number of squares inside each figure. Explain that this is the *area* of the figure and that area is measured in square units.

Exploration worksheet and answers on Chapter 10 Resource Book pp. 16 and 107

2 Teach

Lesson Presentation

Guided Instruction

In this lesson, students learn to estimate the area of irregular figures and find the area of rectangles, triangles, and to parallelograms. Review the vocabulary associated with these figures. Teach students to estimate the area of an irregular figure on a grid by counting full, almost-full, and half-full squares. Then teach students the formulas for finding the area of rectangles, parallelograms, and triangles.

You can use the formula for the area of a rectangle to write a formula for the area of a parallelogram. Imagine cutting off the triangle drawn in the parallelogram and sliding it to the right to form a rectangle.

The area of a parallelogram = bh. The area of a rectangle = ℓw.

The **base** of the parallelogram is the **length** of the rectangle.
The **height** of the parallelogram is the **width** of the rectangle.

EXAMPLE 3 Finding the Area of a Parallelogram

Find the area of the parallelogram.

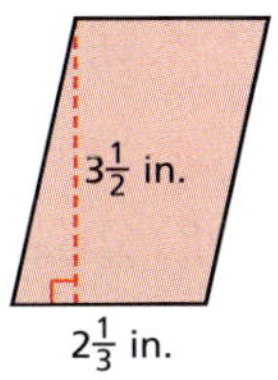

$A = bh$	*Write the formula.*
$A = 2\frac{1}{3} \cdot 3\frac{1}{2}$	*Substitute $2\frac{1}{3}$ for b and $3\frac{1}{2}$ for h.*
$A = \frac{7}{3} \cdot \frac{7}{2}$	*Multiply.*
$A = \frac{49}{6}$, or $8\frac{1}{6}$	The area is $8\frac{1}{6}$ in^2.

You can make a parallelogram out of two congruent triangles.

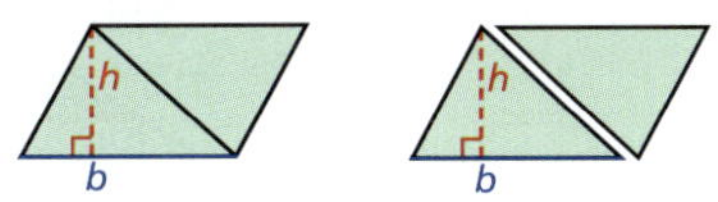

The area of each triangle is half the area of the parallelogram, so the formula for the area of a triangle is $A = \frac{1}{2}bh$.

EXAMPLE 4 Finding the Area of a Triangle

Find the area of the triangle.

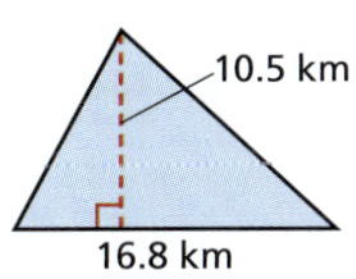

$A = \frac{1}{2}bh$	*Write the formula.*
$A = \frac{1}{2}(16.8 \cdot 10.5)$	*Substitute 16.8 for b.* *Substitute 10.5 for h.*
$A = \frac{1}{2}(176.4)$	*Multiply.*
$A = 88.2$	The area is 88.2 km^2.

Think and Discuss

1. **Explain** how the area of a triangle and the area of a rectangle that have the same base and the same height are related.
2. **Give** a formula for the area of a square.

Additional Examples

Example 1

Estimate the area of the figure.

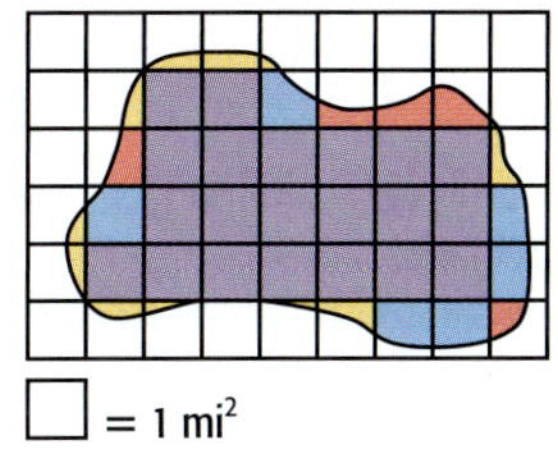

30 mi^2

Example 2

Find the area of the rectangle.

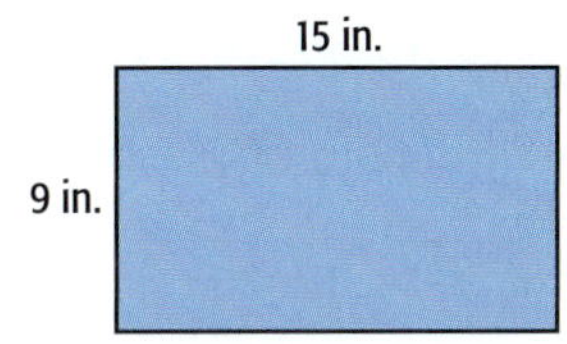

135 in^2

Example 3

Find the area of the parallelogram.

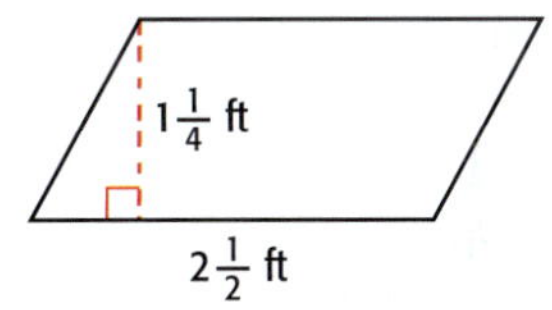

$3\frac{1}{8}$ ft^2

Example 4

Find the area of the triangle.

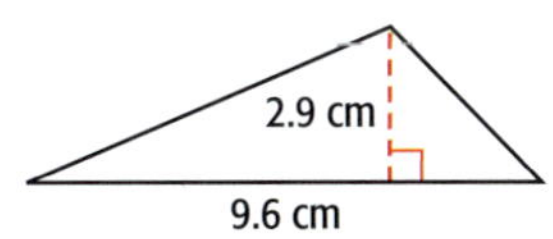

13.92 cm^2

3 Close

Reaching All Learners

Through Curriculum Integration

Social Studies Have students use graph paper (provided on Chapter 10 Resource Book p. 101) to trace outlines of states on a map of Australia (provided on Chapter 10 Resource Book p. 100). Then have them use the method shown in Example 1 to estimate the area of various states.

Summarize

Review the formulas for area of a rectangle, parallelogram, and triangle. Help students to recite them from memory.

Answers to Think and Discuss

1. Possible answer: The area of the rectangle is twice the area of the triangle.
2. Possible answer: $A = s^2$, where s is the length of each side

10-2 PRACTICE & ASSESS

Students may want to refer back to the lesson examples.

Assignment Guide

If you finished Example 1 assign:
Core 1–2, 9–10, 17, 21–26
Enriched 2, 9–10, 17–18, 21–26

If you finished Example 2 assign:
Core 1–4, 9–12, 17, 21–26
Enriched 3–4, 9–12, 17–19, 21–26

If you finished Example 3 assign:
Core 1–6, 9–14, 17, 21–26
Enriched 4–6, 9–14, 17–26

If you finished Example 4 assign:
Core 1–17, 21–26
Enriched 4–26

Notes

10-2 Exercises

FOR EXTRA PRACTICE
see page 656

internet connect
Homework Help Online
go.hrw.com Keyword: MR4 10-2

GUIDED PRACTICE

See Example 1 Estimate the area of each figure.

1. 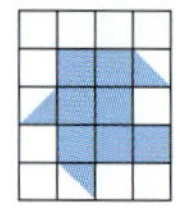 about 8.5 square units

2. 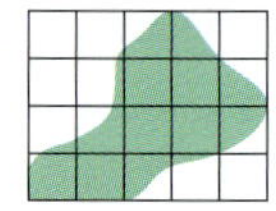 about 9 square units

See Example 2 Find the area of each rectangle.

3. 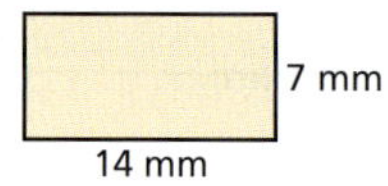

98 mm^2

4. 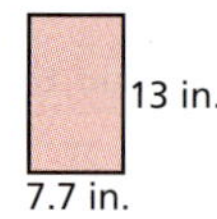

100.1 in^2

See Example 3 Find the area of each parallelogram.

5. 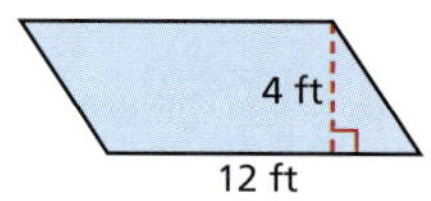

48 ft^2

6. 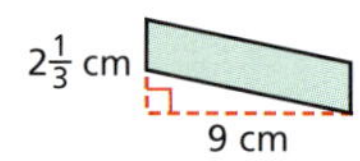

21 cm^2

See Example 4 Find the area of each triangle.

7. 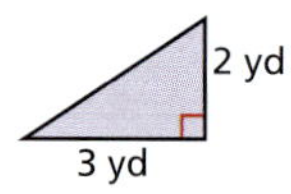

3 yd^2

8. 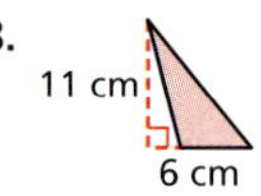

33 mm^2

INDEPENDENT PRACTICE

See Example 1 Estimate the area of each figure.

9. 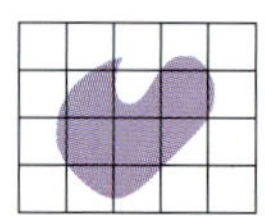 about 6 square units

10. 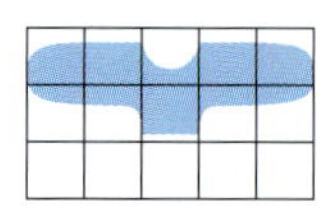 about 5 square units

See Example 2 Find the area of each rectangle.

11. 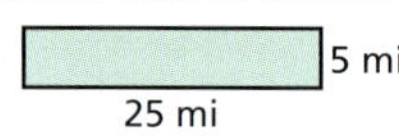

125 mi^2

12. 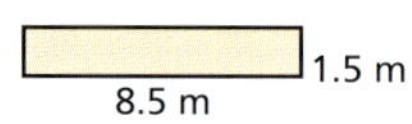

12.75 m^2

See Example 3 Find the area of each parallelogram.

13. 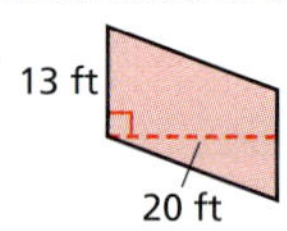

260 ft^2

14. 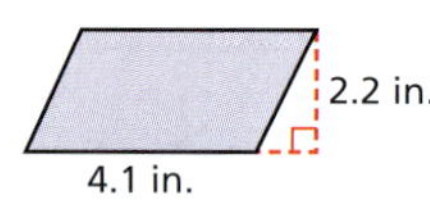

9.02 in^2

See Example 4 Find the area of each triangle.

15. 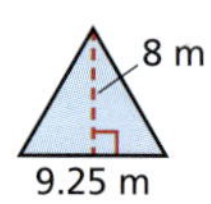

37 m^2

16. 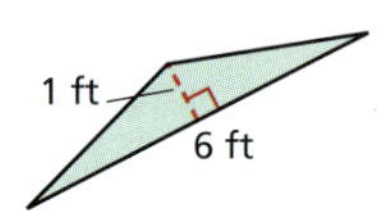

3 ft^2

Math Background

A formula that does not require the use of height exists for triangles and for a special class of quadrilaterals.

Heron's formula is so named because the Greek mathematician Heron proved it around the year 75 B.C. According to Heron's formula, the area of any triangle with sides of length a, b, and c equals $\sqrt{s(s-a)(s-b)(s-c)}$. The semiperimeter s equals one-half the perimeter of a figure.

RETEACH 10-2

LESSON 10-2 Reteach
Estimating and Finding Area

You can use grid paper to estimate and find area of figures. The area of a figure is the amount of surface it covers. Area is measured in square units.

To estimate the area of the figure below, first find the number of fully shaded squares. Next, find the number of nearly full squares. Then, find the number of half or nearly half-shaded squares.

There are 6 fully shaded squares.
There are 2 nearly full squares.
There are 8 half-shaded squares.
$(8 \cdot \frac{1}{2} = 4)$

Find the estimated total number of squares shaded.
6 + 2 + 4 = 12. The area of the figure is about 12 square units.

Estimate the area of each figure. Possible answers:

1. 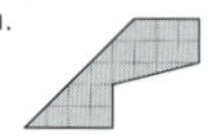 18 units

2. 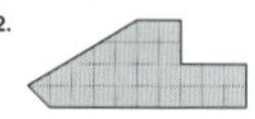 27 units

To find the area of a rectangle, find the total number of square units.

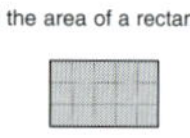

There are 3 rows of 5 squares.
3 • 5 = 15
So, the area of the rectangle is 15 square units.

Find the area of each rectangle.

3.  12 square units

4. 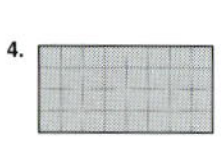 32 square units

PRACTICE 10-2

LESSON 10-2 Practice B
Estimating and Finding Area

Estimate the area of each figure.

1. 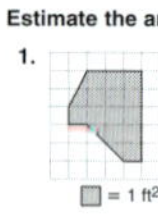

about 15 square feet

2. 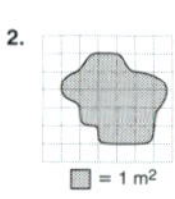

about 19 square meters

Find the area of each rectangle.

3. 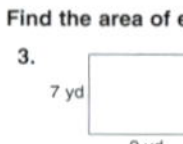

63 yd^2

4. 

96 mi^2

Find the area of each parallelogram.

5. 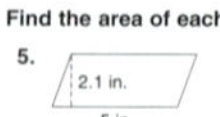

10.5 in^2

6. 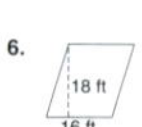

288 ft^2

Find the area of each triangle.

7. 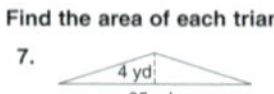

50 yd^2

8. 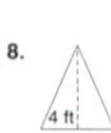

7 ft^2

9. A section of a stained-glass window is shaped like a parallelogram. Its base is 6.5 inches, and its height is 4 inches. How much glass is needed to cover the section completely?
26 in^2

10. Your rectangular yard is 10 feet wide and 26 feet long. How many square feet of grass do you need to plant if you want to cover the entire yard?
260 ft^2

Social Studies LINK

Iceland has many active volcanoes and frequent earthquakes. There are more hot springs in Iceland than in any other country in the world. One spring, the Great Geysir, is capable of releasing about 250 liters of boiling water per second. The word *geyser* comes from the Great Geysir, which, though it rarely erupts anymore, can spray hot water up to 60 meters high.

Sightseers watch the eruption of Geyser Namafjall in the Myvatn Region of North Iceland.

Use the map for Exercises 17–18.

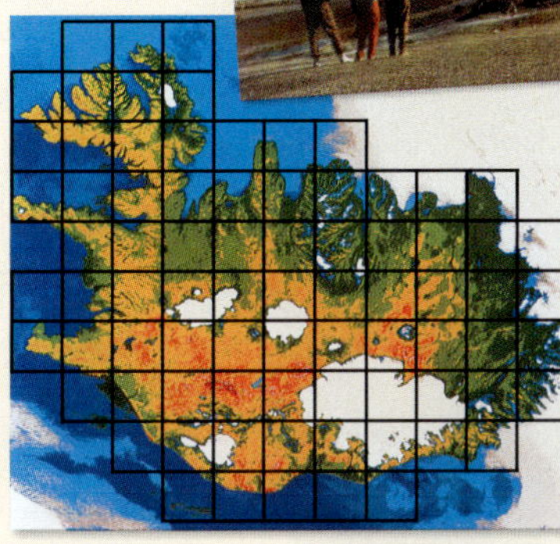

17. One square on the map represents 1,700 km^2. Which is a reasonable estimate for the area of Iceland? **B**

A Less than 65,000 km^2

B Between 90,000 and 105,000 km^2

C Between 120,000 and 135,000 km^2

D Greater than 150,000 km^2

18. About 10% of the area of Iceland is covered with glaciers. Estimate the area covered by glaciers.
Possible answer: Between 9,000 and 10,500 km^2

19. ***WRITE ABOUT IT*** The House is Iceland's oldest building. When it was built in 1765, the builders measured length in *ells*. The House is 14 ells wide and 20 ells long. Explain how to find the area in ells of the House.
Multiply 14 by 20, which is 280 square ells

20. ***CHALLENGE*** The length of one ell varied from country to country. In England, one ell was equal to $1\frac{1}{4}$ yd. Suppose the House were measured in English ells. Find the area in yards of the House. **$437\frac{1}{2}$ yd^2**

go.hrw.com
KEYWORD: MR4 Iceland
CNN Student News

Spiral Review

21. Damien's favorite song is 4.2 minutes long. Jan's favorite song is 2.89 minutes long. Estimate the difference in the lengths of the songs. (Lesson 3-2) **1 minute**

Find each product. (Lesson 5-2)

22. $2\frac{2}{3} \times \frac{1}{8}$ **$\frac{1}{3}$**   **23.** $\frac{1}{4} \times 3\frac{1}{2}$ **$\frac{7}{8}$**   **24.** $1\frac{1}{4} \times 1\frac{2}{5}$ **$1\frac{3}{4}$**   **25.** $2\frac{1}{5} \times 2\frac{2}{3}$ **$5\frac{13}{15}$**

26. TEST PREP Which ratio is equivalent to $\frac{1}{20}$? (Lesson 8-1) **A**

A 9:180   B 180 to 9   C 4 to 100   D 100:4

Interdisciplinary LINK

Social Studies

Exercises 17–20 involve the estimation and calculation of area as it relates to Iceland. Students study about Iceland in middle school social studies programs, such as Holt, Rinehart & Winston's *People, Places, and Change.*

Journal

Have students think of and write real-world situations in which areas of rectangles, triangles, and parallelograms must be found. Have them find the area in each situation, using the appropriate area formula.

Test Prep Doctor

The key ideas students need to know in order to answer Exercise 26 are that in a ratio the order matters and that equivalent ratios are found by multiplying both terms in a ratio by the same factor. Students who answered **B** found a ratio equivalent to $\frac{20}{1}$, not $\frac{1}{20}$. Students who answered **C** or **D** multiplied the terms by different factors; students who answered **D** also switched the order of the terms.

CHALLENGE 10-2

LESSON 10-2 **Challenge**
Chewing Gum Archaeology

Archaeologists found the oldest piece of chewing gum in Sweden—it was 9,000 years old! The scientists studied the gum to learn about the people who may have made and chewed it. You can be a chewing gum archaeologist, too. How? When a standard-sized wad of chewed gum is stepped on, you can study the area of the flattened gum to find the weight of the person who squished it!

Estimate the area covered by each flattened piece of gum below. Then use the table to find the gum squisher's weight.

Bubble gum was invented in Pennsylvania in 1928.

Estimated Area of Squished Gum	Estimated Weight of Gum Squisher
about 25 cm^2	about 50 lb
about 50 cm^2	about 100 lb
about 75 cm^2	about 150 lb
about 100 cm^2	about 200 lb

1. 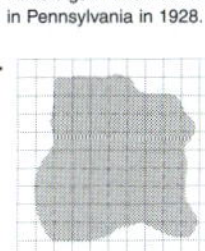

Estimated Area: **about 65 cm^2**
Estimated Weight: **about 150 lb**

2. 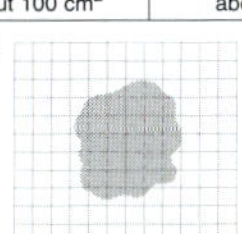

Estimated Area: **about 27 cm^2**
Estimated Weight: **about 50 lb**

3. 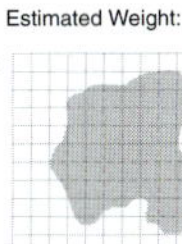

Estimated Area: **about 59 cm^2**
Estimated Weight: **about 100 lb**

4. 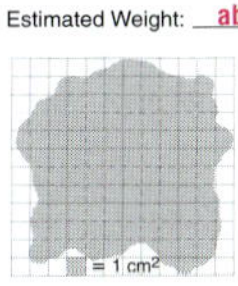

Estimated Area: **about 105 cm^2**
Estimated Weight: **about 200 lb**

PROBLEM SOLVING 10-2

LESSON 10-2 **Problem Solving**
Estimating and Finding Area

Use the table to answer each question.

State Information

State	Approx. Width (mi)	Approx. length (mi)	Water Area (mi^2)
Colorado	280	380	376
Kansas	210	400	462
New Mexico	343	370	234
North Dakota	211	340	1,724
Pennsylvania	160	283	1,239

1. New Mexico is the 5th largest state in the United States. What is its approximate total area?
126,910 mi^2

2. Kansas is the 15th largest state in the United States. What is its approximate total area?
84,000 mi^2

3. What is the difference between North Dakota's land area and water area?
70,016 mi^2

4. What is Pennsylvania's approximate land area?
45,280 mi^2

5. What is the difference between Colorado's land area and Pennsylvania's land area?
61,120 mi^2

6. About what percent of the total area of Pennsylvania is covered by land?
about 97%

Circle the letter of the correct answer.

7. Rhode Island is the smallest state. Its total land area is approximately 1,200 mi^2. Rhode Island is approximately 40 miles long. About how wide is Rhode Island?
A about 20 mi
B about 40 mi
C about 50 mi
(D) about 30 mi

8. The entire United States covers 3,794,085 square miles of North America. About how much of that area is not made up of the 5 states in the chart?
F 2,537,470 mi^2
(G) 3,359,755 mi^2
H 3,686,525 mi^2
J 3,1310,818 mi^2

Lesson Quiz

Find the area of each figure.

1. 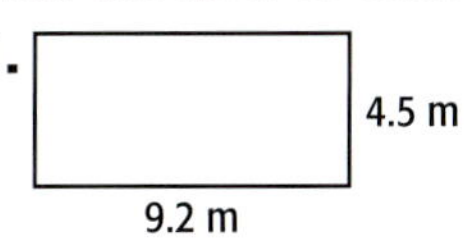

41.4 m^2

2. 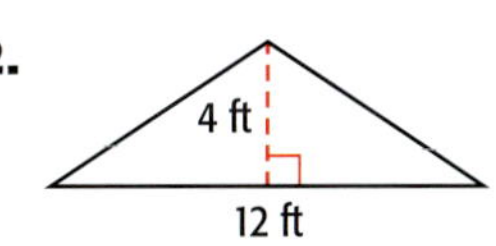

24 ft^2

3. What is the area of a parallelogram with base 16 in. and height 10 in.? **160 in^2**

4. What is the area of a triangle with base 10 in. and height 7 in.? **35 in^2**

Available on Daily Transparency in CRB

10-3 Organizer

Pacing: Traditional 1 day
Block $\frac{1}{2}$ day

Objective: Students break a polygon into simpler parts to find its area.

Indiana Academic Standards
6.3.2, 6.3.5, 6.7.3, 6.7.4, 6.7.11

Warm Up

1. What is the area of a rectangle with length 10 cm and width 4 cm? **40 cm²**
2. What is the area of a parallelogram with base 18 ft and height 12 ft? **216 ft²**
3. What is the area of a triangle with base 16 cm and height 8 cm? **64 cm²**

Problem of the Day

Four squares are stacked in a tower. The bottom square is 12 inches on a side. The perimeter of each of the other squares is half of the one below it. What is the perimeter of the combined figure? **69 in.**

Available on Daily Transparency in CRB

10-3 Break into Simpler Parts

Problem Solving Skill

Learn to break a polygon into simpler parts to find its area.

You know how to find the area of rectangles, parallelograms, and triangles. To help you find the areas of other polygons, first solve a simpler problem by breaking the polygons apart into rectangles, parallelograms, and triangles.

EXAMPLE 1 Finding Areas of Composite Figures

Find the area of each polygon.

A

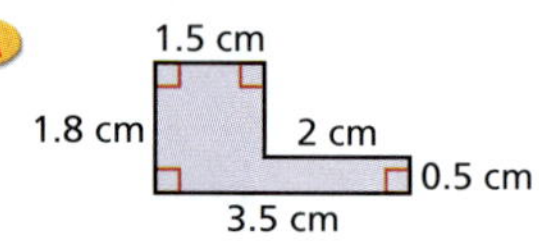

Think: Break the polygon apart into rectangles.

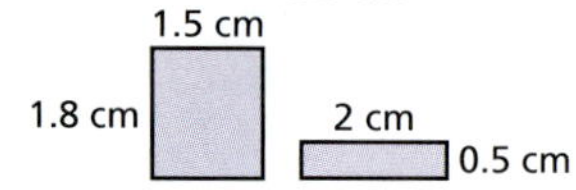

Find the area of each rectangle.

$A = \ell w$ | $A = \ell w$ — *Write the formula for the area of a rectangle.*

$A = 1.8 \cdot 1.5$ | $A = 2 \cdot 0.5$

$A = 2.7$ | $A = 1$

$2.7 + 1 = 3.7$ — *Add to find the total area.*

The area of the polygon is 3.7 cm².

B

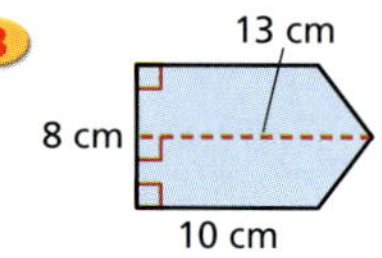

Think: Break the figure apart into a triangle and a rectangle.

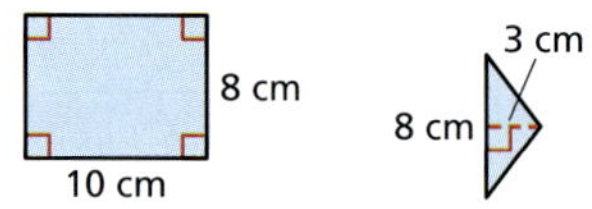

$A = \ell w$ | $A = \frac{1}{2}bh$ — *Find the area of each polygon.*

$A = 8 \cdot 10$ | $A = \frac{1}{2} \cdot 8 \cdot 3$

$A = 80$ | $A = 12$

$80 + 12 = 92$ — *Add to find the total area of the figure.*

The area of the figure is 92 cm².

1 Introduce

Alternate Opener

EXPLORATION

10-3 Break into Simpler Parts

Phil and Louise are planning to sod their backyard. Sod is sold in square yards. They sketched their yard on a piece of graph paper, where the side length of each square represents 1 yard.

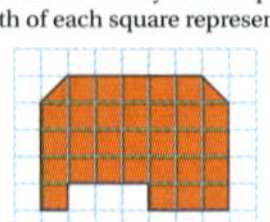

1. a. Use graph paper to divide the figure into several figures.
   b. Name each figure from number 1.
   c. Find the area in square yards of each figure.
   d. Add the areas in part c.

Think and Discuss

2. **Discuss** the strategies you used for finding area.
3. **Explain** how you could use squares to help you find the areas of different shapes.

Motivate

Review using formulas to find the area of rectangles and triangles. Ask students what formula they would use to find the area of a 4 in. × 7 in. rectangle $A = \ell w$; $A = 4 \cdot 7 = 28$; $A = 28$ in² and a triangle with a base of 3 cm and a height of 4 cm. $A = \frac{1}{2}bh$; $A = \frac{1}{2}(3 \cdot 4) = \frac{1}{2}(12) = 6$; $A = 6$ cm²

Exploration worksheet and answers on Chapter 10 Resource Book pp. 26 and 109

2 Teach

Lesson Presentation

Guided Instruction

In this lesson, students learn to break a polygon into simpler parts to find the area of the polygon. Teach students how to break a composite figure into rectangles, triangles, and parallelograms. Explain that the areas of the smaller figures can be added to find the area of the composite figure.

Teaching Tip Counting squares can be easier if students draw lines that begin and end at the intersection of the grid lines on the graph paper.

EXAMPLE 2 *ART APPLICATION*

Stan made a wall hanging. All the sides are 6 inches long, except for two longer sides that are each 12 inches. All the angles are right angles. What is the area of the wall hanging?

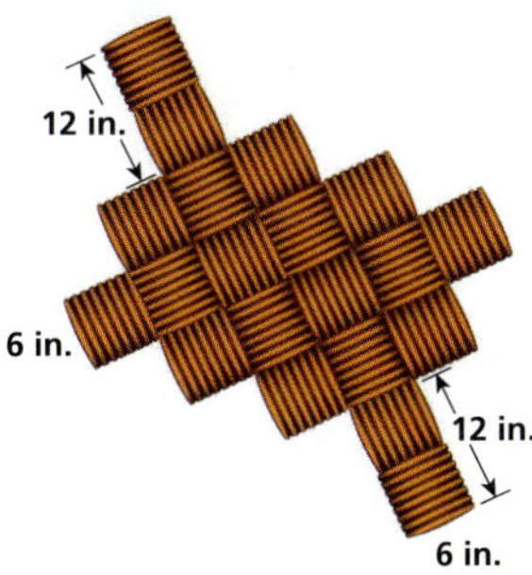

Think: Divide the wall hanging into 20 squares.

Find the area of one square that has a side length of 6 in.

$A = \ell w$ *Write the formula.*

$A = 6 \cdot 6 = 36$

$20 \cdot 36 = 720$ *Multiply to find the area of the 20 squares.*

The area of the wall hanging is 720 in^2.

Helpful Hint

You can also use the formula $A = s^2$, where s is the length of a side, to find the area of a square.

Think and Discuss

1. **Explain** how you can find the area of a regular octagon by breaking it apart into congruent triangles, if you know the area of one triangle.
2. **Explain** one way you can find the area of a trapezoid.

Additional Examples

Example 1

Find the area of each polygon.

A.

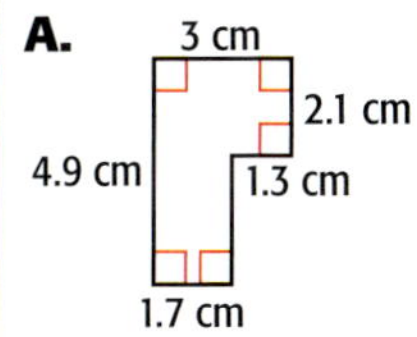

B.

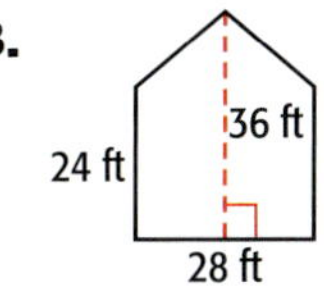

11.06 cm^2 840 ft^2

Example 2

Patrick made a design. All the sides are 5 inches long, except for two longer sides that are each 20 inches. All the angles are right angles. What is the area of the design? 250 in^2

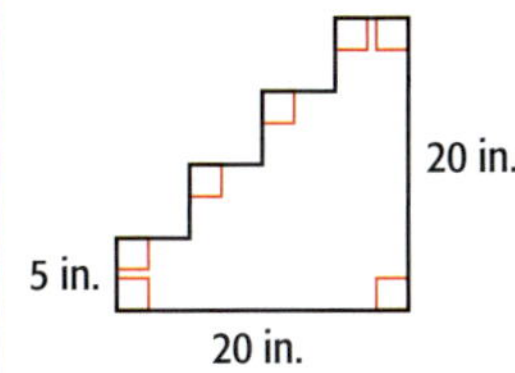

10-3 Exercises

FOR EXTRA PRACTICE
see page 656

internet connect
Homework Help Online
go.hrw.com Keyword: MR4 10-3

GUIDED PRACTICE

See Example 1

Find the area of each polygon.

1. 32 in^2

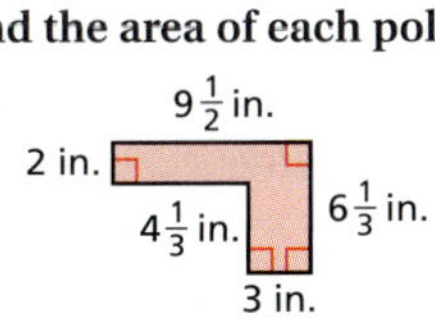

2. 18.84 cm^2

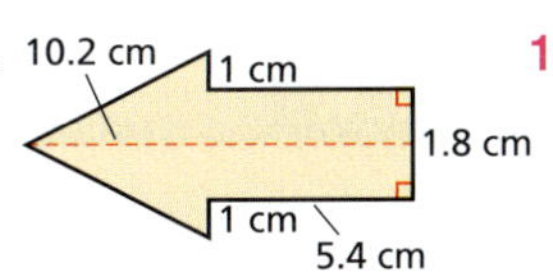

See Example 2

3. Gina used tiles to create a design. The lengths of all the sides are 3 cm, except for two longer sides that are 9 cm. What is the area of Gina's design? 54 cm^2

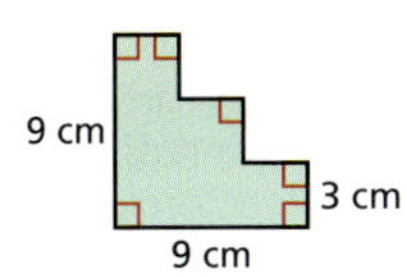

10-3 PRACTICE & ASSESS

Assignment Guide

If you finished Example 1 assign:
Core 1–2, 4–5, 7a–7b, 10–14
Enriched 1–2, 4–5, 7, 10–14

If you finished Example 2 assign:
Core 1–7b, 10–14
Enriched 3–14

3 Close

Reaching All Learners
Through Grouping Strategies

Have students work in pairs to find areas of composite figures. Each student draws a figure on a sheet of graph paper (provided on Chapter 10 Resource Book p. 101). Students trade figures with their partners; divide the figures into rectangles, triangles, and parallelograms; find the areas of the individual figures; and find the total area of each composite figure.

Summarize

Have students explain how they would divide the figure below in order to find its area.

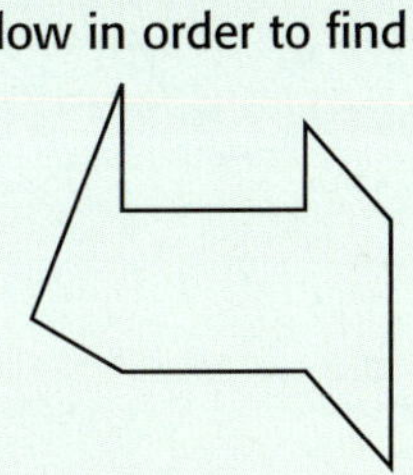

Possible answer: Divide the figure from left to right into a triangle, a rectangle, and a parallelogram. Then find the sum of the three areas.

Answers to Think and Discuss

1. Possible answer: A regular octagon can be divided into 8 congruent triangles. If you already know the area of one, you can multiply that by 8 to find the area of the octagon.
2. Possible answers: Break it apart into a parallelogram and a triangle, find the area of each figure, and add the areas. Or break it apart into a rectangle and two triangles, find the area of each figure, and add the areas. The formula for the area of a trapezoid is $\frac{1}{2}h(b_1 + b_2)$.

Lesson Quiz

Find the area of the figure shown.

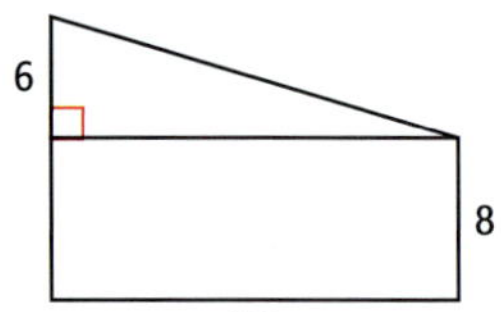

220 units^2

Available on Daily Transparency in CRB

Answers

7–8. See p. A6.

INDEPENDENT PRACTICE

See Example 1

Find the area of each polygon.

4. 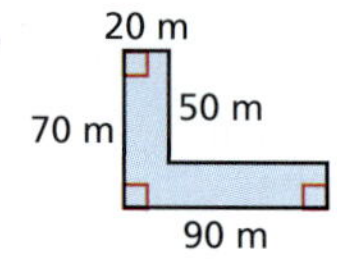

2,800 m^2

5. 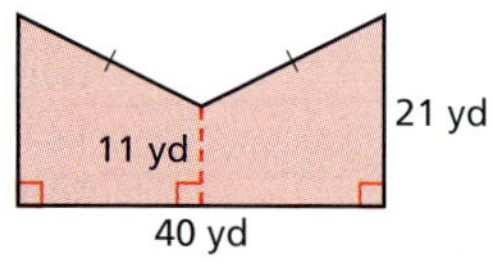

640 yd^2

See Example 2

6. Edgar plants daffodils around a rectangular pond. The yellow part of the diagram shows where the daffodils are planted. What is the area of the yellow part of the diagram? 70.75 m^2

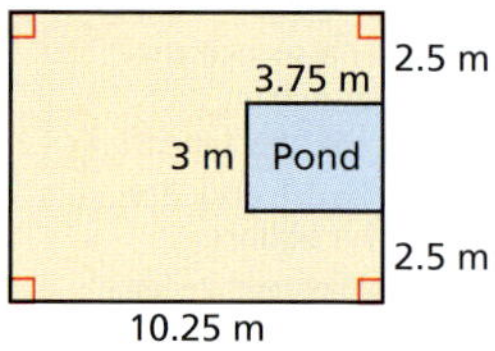

PRACTICE AND PROBLEM SOLVING

7. ***SOCIAL STUDIES*** The map shows the approximate dimensions of the state of South Australia.

a. Look at the red figure outlining South Australia. Divide the figure into a right triangle and a rectangle.

b. Find the total area of the rectangle and triangle.

c. The total area of Australia is about 7.7 million km^2. About what fraction of the total area of Australia is the area of the state of South Australia?

8. ***WRITE ABOUT IT*** Draw a figure that can be broken up into two rectangles. Label the lengths of each side. Explain how you can find the area of the figure. Then find the area.

9. ***CHALLENGE*** The perimeter of this figure is 42.5 cm. Find the area of this figure. 72.25 cm^2

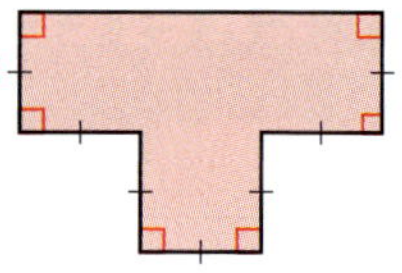

Spiral Review

Solve each equation. (Lesson 5-4)

10. $\frac{1}{4}x = 12$ 48 **11.** $5\frac{1}{3}y = 9$ $1\frac{11}{16}$ **12.** $\frac{2n}{5} = 3$ 7.5 **13.** $3a = \frac{1}{7}$ $\frac{1}{21}$

14. **TEST PREP** In which quadrant on a coordinate plane is the point (–1, 2) located? (Lesson 9-3) B

A Quadrant I **B** Quadrant II **C** Quadrant III **D** Quadrant IV

RETEACH 10-3

LESSON 10-3 **Reteach**
Break into Simpler Parts

Sometimes you can use area formulas you know to help you find the area of other figures.

To find the area of the figure below, first divide the figure into figures you know.

The figure is made up of a triangle, a parallelogram, and a rectangle.

Next, find the area of each figure.

Triangle	Parallelogram	Rectangle
$A = \frac{1}{2}bh$	$A = bh$	$A = \ell w$
$= \frac{1}{2}(3 \cdot 4)$	$= 3 \cdot 4$	$= 4 \cdot 5$
$= 6$	$= 12$	$= 20$

Then, find the sum of all of the areas.

6 + 12 + 20 = 38 The area of the figure is 38 square units.

Find the area of each figure.

1. 15 square units
2. 25 square units
3. 21 square units
4. 34 square units

PRACTICE 10-3

LESSON 10-3 **Practice B**
Break into Simpler Parts

Find the area of each polygon.

1. 36 in^2
2. 80 cm^2
3. 22.5 ft^2
4. 12 yd^2
5. 8.75 mi^2
6. 108 m^2

7. Three paintings are shaped like an 8-foot square, a 7-foot by 4-foot rectangle, and a triangle with a 6-foot base and a height of 7 feet. If those paintings are hung together on the outside of a building, how much of the building's wall will they cover altogether?
113 ft^2

8. Two diagonals divide a square carpet into 4 congruent triangles. The base of each triangle is 5 feet and the height is 2.5 feet. What is the area of the entire carpet?
25 ft^2

CHALLENGE 10-3

LESSON 10-3 **Challenge**
Area Combinations

Combine these polygons to draw figures with the given areas. You must use all the polygons, and each can be used only once. Sample answers given

1. Total Area = 26 cm^2
2. Total Area = 30 cm^2
3. Total Area = 22 cm^2
4. Total Area = 16 cm^2

PROBLEM SOLVING 10-3

LESSON 10-3 **Problem Solving Strategy**
Break into Simpler Parts

Write the correct answer.

1. The shape of Nevada can almost be divided into a perfect rectangle and a perfect triangle. About how many square miles does Nevada cover?
about 109,600 mi^2

2. The shape of Oklahoma can almost be divided into 2 perfect rectangles and 1 triangle. About how many square miles does Oklahoma cover?
about 63,800 mi^2

3. The front side of an apartment building is a rectangle 60 feet tall and 25 feet wide. Bricks cover its surface, except for a door and 10 windows. The door is 7 feet tall and 3 feet wide. Each window is 4 feet tall and 2 feet wide. How many square feet of bricks cover the front side of the building?
1,399 ft^2

4. Each side of a square garden is 12 meters long. A hedge wall 1 meter wide surrounds the garden. What is the area of the entire garden including the hedge wall? How many square meters of land does the hedge wall cover alone?
entire garden = 196 m^2; hedge wall = 52 m^2

Circle the letter of the correct answer.

5. A figure is formed by a square and a triangle. Its total area is 32.5 m^2. The area of the triangle is 7.5 m^2. What is the length of each side of the square?
A 5 meters (circled) **C** 15 meters
B 25 meters **D** 16.25 meters

6. A rectangle is formed by two congruent right triangles. The area of each triangle is 6 in^2. If each side of the rectangle is a whole number of inches, which of these could not be its perimeter?
F 26 inches **H** 24 inches (circled)
G 16 inches **J** 14 inches

10-4 Comparing Perimeter and Area

Learn to make a model to explore how area and perimeter are affected by changes in the dimensions of a figure.

Ms. Cohn wants to enlarge this photo by doubling its length and width.

You can make a model on grid paper to see how the area and the perimeter of a figure change when its dimensions change.

The original photo is a 3 in. × 2 in. rectangle.	The enlarged photo will be a 6 in. × 4 in. rectangle.
perimeter = 10 in.	perimeter = 20 in.
area = 6 in^2	area = 24 in^2

If Ms. Cohn doubles the dimensions of the photo, the **perimeter** will also be doubled, and the **area** will be four times greater than the area of the original photo.

EXAMPLE 1 Changing Dimensions

Find how the perimeter and the area of the figure change when its dimensions change.

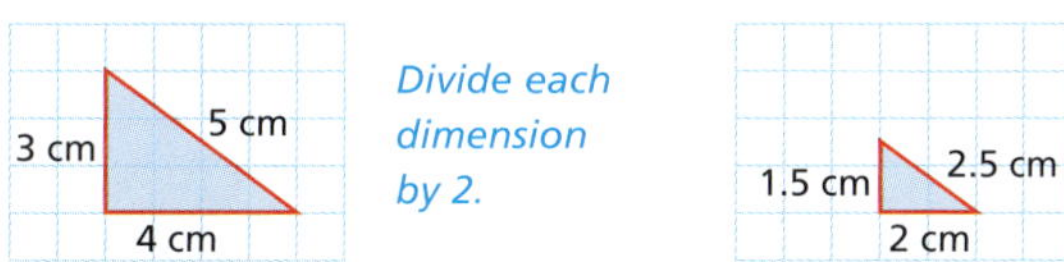

$P = 12$ cm
$A = 6$ cm^2

$P = 6$ cm
$A = 1.5$ cm^2

When the dimensions of the triangle are divided by 2, the **perimeter** is divided by **2**, and the **area** is divided by **4**, or 2^2.

10-4 Organizer

Pacing: Traditional 1 day
Block $\frac{1}{2}$ day

Objective: Students make a model to explore how area and perimeter are affected by changes in the dimensions of a figure.

Indiana Academic Standards
6.3.5, 6.3.9, 6.5.1, 6.7.1, 6.7.2, 6.7.5, 6.7.11

Warm Up

1. What is the area of a figure made up of a rectangle with length 12 cm and height 4 cm and a parallelogram with length 12 cm and height 6 cm? **120 cm^2**
2. What is the area of a figure consisting of a triangle sitting on top of a rectangle? The triangle has a base of 12 in. and height of 9 in., and the rectangle has a base of 12 in. and a height of 5 in. **114 in^2**

Problem of the Day

If sixteen people sit, evenly spaced, in a circle for story time, who sits directly across from person 5? **13**

Available on Daily Transparency in CRB

Math Humor

As a student was blowing up a balloon, he noticed that as it got more perimeter, it also got *airier*.

1 Introduce

Alternate Opener

EXPLORATION

10-4 Comparing Perimeter and Area

Suzanne is enlarging a color copy of a 3 in. by 5 in. photograph to 6 in. by 10 in. A model is shown below.

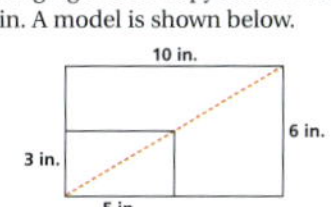

Find the perimeter of each color copy.

	Color Copy	Perimeter = 2 · Length + 2 · Width
1.	3 in. by 5 in.	
2.	6 in. by 10 in.	

3. How do the perimeters compare?

Find the area of each color copy.

	Color Copy	Area = Length · Width
4.	3 in. by 5 in.	
5.	6 in. by 10 in.	

6. How do the areas compare?

Think and Discuss

7. **Explain** how you compared the perimeters.
8. **Explain** how you compared the areas.

Motivate

Review the terms *perimeter* and *area* and the formulas for finding perimeter and area of a rectangle.

Perimeter is the distance around a figure, and area is the amount of surface a figure covers. The formula for perimeter of a rectangle is $P = 2\ell + 2w$. The formula for area of a rectangle is $A = \ell w$.

Exploration worksheet and answers on Chapter 10 Resource Book pp. 35 and 111

2 Teach

Lesson Presentation

Guided Instruction

In this lesson, students learn to make a model to explore how area and perimeter are affected by changes in the dimensions of a figure. Teach students what happens to the perimeter and area of a figure when its dimensions are doubled, halved, and tripled.

Additional Example

Example 1

Find how the perimeter and the area of the figure change when its dimensions change.

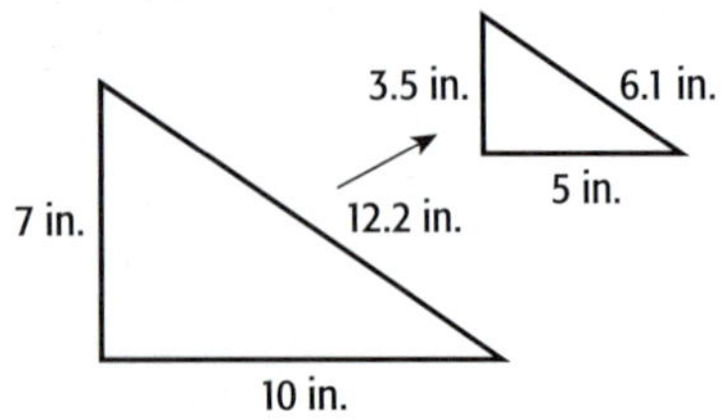

The perimeter is divided by 2, and the area is divided by 4.

Example 2

Draw a rectangle whose dimensions are 4 times as large as the given rectangle. How do the perimeter and area change?

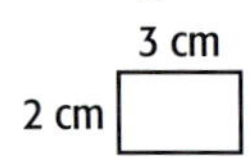

The perimeter is multiplied by 4, and the area is multiplied by 16.

EXAMPLE 2 ***Measurement Application***

Use a centimeter ruler to measure the photo. Draw a rectangle whose sides are 3 times as long to enlarge the photo. How do the perimeter and the area change?

$P = 6$ cm
$A = 2$ cm^2

Multiply each dimension by 3.

$P = 18$ cm
$A = 18$ cm^2

When the dimensions of the rectangle are multiplied by 3, the perimeter is multiplied by 3, and the area is multiplied by 9, or 3^2.

Think and Discuss

1. **Explain** how the perimeter of a triangle changes when all the side lengths are doubled.
2. **Tell** how the area of a rectangle changes when all the side lengths are divided in half.

10-4 PRACTICE & ASSESS

Assignment Guide

If you finished Example 1 assign:
Core 1, 3, 6, 11–14
Enriched 1, 3, 9, 11–14

If you finished Example 2 assign:
Core 1–5, 7, 11–14
Enriched 5–14

10-4 Exercises

FOR EXTRA PRACTICE
see page 656

internet connect
Homework Help Online
go.hrw.com Keyword: MR4 10-4

GUIDED PRACTICE

See Example 1

1. Find how the perimeter and the area of the figure change when its dimensions change.

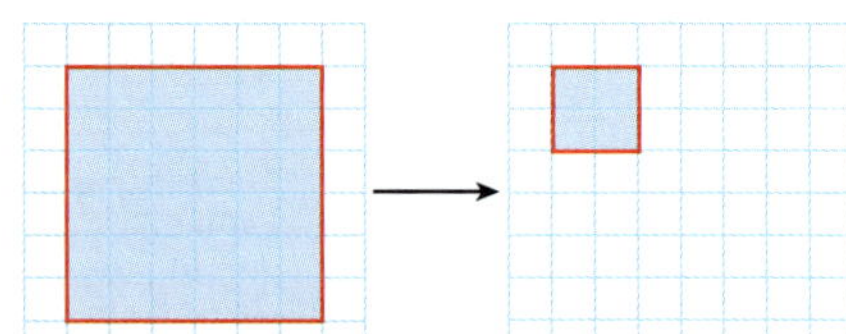

When the dimensions of the square are divided by 3, the perimeter is divided by 3, and the area is divided by 9 or 3^2.

See Example 2

2. Use a centimeter ruler to measure the rectangle. Then draw another rectangle with dimensions that are 2 times greater than the given rectangle. How do the perimeter and the area change when the dimensions change?

The perimeter is multiplied by 2 and the area is multiplied by 4, or 2^2.

3 Close

Summarize

Review how perimeter and area of a figure change when the scale changes. (Perimeter changes linearly with scale, and area changes with the square of the scale factor.)

Answers to Think and Discuss

1. The perimeter of the triangle doubles.
2. The area becomes one-fourth the area of the original figure.

Reaching All Learners

Through Hands-On Experience

Have students find the perimeters and areas of figures in and around their classroom. Then have them figure out what the perimeters and areas of those items would be if the dimensions were halved and tripled. Students can record their findings and share them with the class.

INDEPENDENT PRACTICE

See Example 1

3. Find how the perimeter and the area of the figure change when its dimensions change.

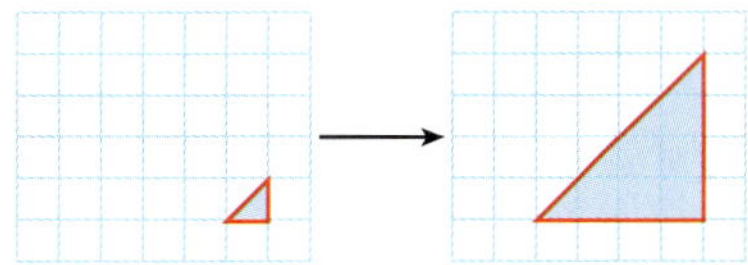

See Example 2

4. Use a centimeter ruler to measure the triangle. Then draw another triangle with dimensions that are half as great as the given triangle. How do the perimeter and the area change when the dimensions change?

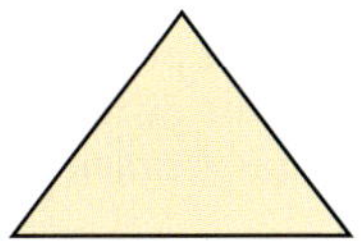

PRACTICE AND PROBLEM SOLVING

5. The schoolyard is a rectangle with a length of 120 ft and a width of 80 ft. The PE teacher plans to make a field in the schoolyard by dividing either the length or the width of the schoolyard in half.
   a. What will the area of the field be if she divides only one of the dimensions in half? **4,800 ft^2**
   b. What will the perimeter of the field be if she divides only the length in half? the width in half? **280 ft, 320 ft**

Use the table for Exercises 6–8.

Photo Frames (in.)	
6 × 8	12 × 16
8 × 10	16 × 20
9 × 12	18 × 24

6. Give an example of two frames, one with dimensions that are double the dimensions of the other. **6 × 8 and 12 × 16, 9 × 12 and 18 × 24, 8 × 10 and 16 × 20**
7. George has a 3 in. × 4 in. photo. Which frame has dimensions that are 6 times greater than the dimensions of George's photo? **18 × 24**
8. If George enlarges a 3 in. × 4 in. photo so that it is 12 in. × 16 in., how will its area change? **The area will be 4^2 times greater.**
9. ***WRITE ABOUT IT*** What happens to the area and the perimeter of a rectangle when the length and width are multiplied by 4?
10. ***CHALLENGE*** A rectangle has a perimeter of 24 meters. If its length and width are whole numbers, what is its greatest possible area? **36 m^2**

9. When the dimensions are multiplied by 4 the area will be 4^2 greater and the perimeter will be 4 times as great.

Spiral Review

Write each phrase as a numerical or algebraic expression. (Lesson 2-2)

11. 19 times 3 **19 × 3**
12. the quotient of *g* and 6 **$g \div 6$**
13. the sum of 5 and 9 **5 + 9**
14. **TEST PREP** Angle *A* and Angle *B* are supplementary. What is the measure of $\angle B$ if the measure of $\angle A$ is 75°? (Lesson 7-3) **C**

A 15° B 25° C 105° D 150°

Lesson Quiz

Find how the perimeter and area of the triangle change when its dimensions change.

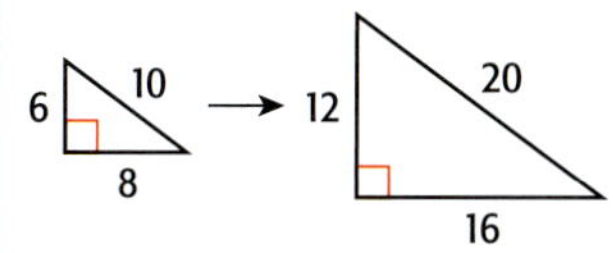

The perimeter is multiplied by 2, and the area is multiplied by 4; perimeter = 24, area = 24; perimeter = 48, area = 96.

Available on Daily Transparency in CRB

Answers

2–4. Complete answers on p. A6

RETEACH 10-4

LESSON 10-4 **Reteach**

Comparing Perimeter and Area

When the dimensions of a figure change, the perimeter and area of the figure change.

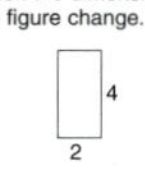

$P = 2\ell + 2w$ $= (2 \cdot 4) + (2 \cdot 2)$ $= 8 + 4$ $= 12$ units

$A = \ell w$ $= 4 \cdot 2$ $= 8$ square units

The dimensions are $\ell = 4$ and $w = 2$.

If each dimension of the rectangle is divided by 2, the perimeter and area change.

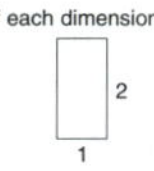

$P = 2\ell + 2w$ $= (2 \cdot 2) + (2 \cdot 1)$ $= 4 + 2$ $= 6$ units

$A = \ell w$ $= 2 \cdot 1$ $= 2$ square units

The new dimensions are $\ell = 2$ and $w = 1$.

When the dimensions of a rectangle are divided by 2, the perimeter is divided by 2 and the area is divided by 4.

Write how the perimeter and area change when the dimensions change.

1. 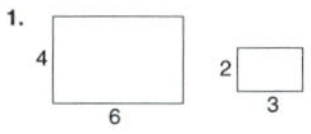

When the dimensions of a rectangle are divided by 2 the perimeter is divided by 2 and the area is divided by 4.

2. 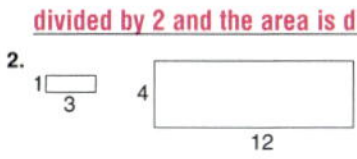

When the dimensions of a rectangle are multiplied by 4 the perimeter is multiplied by 4 and the area is multiplied by 16.

PRACTICE 10-4

LESSON 10-4 **Practice B**

Comparing Perimeter and Area

Write how the perimeter and the area of the figure change when its dimensions change.

1. 

$P =$ 26 in. $P =$ 52 in.

$A =$ 40 in^2 $A =$ 160 in^2

When the dimensions of the rectangle are doubled, the perimeter is doubled, and the area is 4 times greater.

2. Use a centimeter ruler to measure the triangle. Then draw another triangle with dimensions that are half as great as the given triangle. How do the perimeter and the area change when the dimensions change?

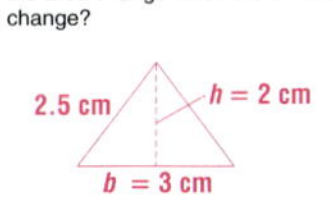

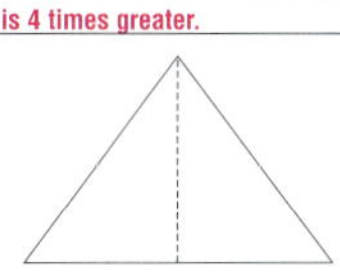

When the dimensions of the triangle are half as great, the perimeter is divided by 2, and the area is divided by 4.

3. Nina wants to make a smaller version of a painting she saw in a museum. The museum painting was a square with each side measuring 6.4 feet. If Nina makes her copy half the size of the original painting, how much space will it cover on her wall?

10.24 square feet

4. How many feet of wood will Nina need to make a frame for her painting from Exercise 3?

12.8 feet of wood

CHALLENGE 10-4

LESSON 10-4 **Challenge**

Hooray for Hollywood!

The "Hollywood" sign in Los Angeles, California, is one of the most recognized landmarks of the United States. Each of the sign's letters covers a rectangular space 9 meters wide and 12 meters tall.

Built in 1923, the original sign said "Hollywoodland."

Find these measurements of the "Hollywood" sign.

1. What is the perimeter of each letter's rectangle? 42 m
2. What is the area of each letter's rectangle? 108 m^2
3. What is the perimeter of the rectangle formed by the entire sign, including only the letters? 186 meters
4. What is the area the rectangle formed by the entire sign, including only the letters? 972 m^2

Reduce the dimensions of each letter in the "Hollywood" sign by one-half.

5. What is the perimeter of each reduced letter's rectangle? 21 m
6. What is the area of each reduced letter's rectangle? 27 m^2
7. What is the perimeter of the rectangle formed by the entire reduced sign, including only the letters? 93 m
8. What is the area of the rectangle formed by the entire reduced sign, including only the letters? 243 m^2
9. How does the perimeter of each letter change when its dimensions change?

   When each letter is reduced by one-half, its perimeter is divided by 2.
10. How does the area of the entire sign change when its dimensions change?

    When the entire sign is reduced by one-half its area is divided by 4.

PROBLEM SOLVING 10-4

LESSON 10-4 **Problem Solving**

Comparing Perimeter and Area

Write the correct answer.

1. Fiona's school photograph is 6 inches long and 5 inches wide. If she orders a triple enlargement how would this affect the area of the photo? How would the enlargement affect the frame she would need for the photo?

   The area of the enlarged photo will be 9 times larger than the original. She will need a frame 3 times larger than the frame for the original photo.
2. The Whitman's kitchen is 8 feet long and 6 feet wide. They are planning on renovating the kitchen to have more space. If they double just the width, how will it affect the area of the room? if they double just the length? if they double both measurements?

   If they double just 1 measurement, the area would double. If they double both measurements, the room's area would be 4 times larger.
3. Kent saw a table in a magazine that was 3 feet wide and 4 feet long. If he wants to make a similar version of the table with an area 4 times larger, what dimensions should he use? How will the perimeter of Kent's table differ from the table in the magazine?

   6 ft wide and 8 ft long; The perimeter of Kent's table will be 2 times larger than the magazine's table.
4. The triangular sail on Shakeera's boat is 8 meters wide and 10 meters tall. She wants to make a model of the boat that is $\frac{1}{20}$ of its actual size. How much canvas will Shakeera use for the model boat's sail? How does that amount compare to the canvas used for the real boat's sail?

   0.1 m^2 of canvas; The real sail uses 400 times more canvas than the model.

Circle the letter of the correct answer.

5. A triangle is 6.4 cm long and 8.2 cm tall. If you triple its dimensions, what would be the area of the enlarged triangle?

   A 78.72 cm^2 (C) 236.16 cm^2
   B 157.44 cm^2 D 472.32 cm^2
6. The dimensions of a regular pentagon are doubled. The perimeter of the enlarged pentagon is 25 yards. What was the length of each side of the original pentagon?

   (F) 2.5 yards H 5 yards
   G 12. yards J 16.25 yards

10A Discover Properties of Circles

Pacing: Traditional 1 day
Block $\frac{1}{2}$ day

Objective: To use a compass, string, ruler, scissors, and a calculator to explore circles

Materials: Compass, string, ruler, scissors, calculator

Lab Resources

Hands-On Lab Activities pp. 90–91, 114

Using the Pages

Discuss with students items in the classroom that could be used to create circles of different diameters. Encourage students to use as many different diameters as they possibly can.

Guide students as they measure each circle's diameter and circumference.

1. How can you use string to measure a circle's circumference? Wrap the string around the circle, then unwrap the string and measure to the correct length.
2. How can you measure a circle's diameter? Place the ruler so that it passes through the center, and measure from one side of the circle to the other side.

Guide students as they transform a circle into a parallelogram to find its area.

3. How do you draw a circle with a 2-inch radius? Open the compass so that the width between the points measures 2 inches on a ruler. Tighten the hinge and draw a circle.
4. Why is the base of the parallelogram equal to πr? The base of the parallelogram is made up of half the circle's circumference. Because $C = \pi d$, half the circumference is $\frac{1}{2} \times \pi \times d = \frac{1}{2} \times \pi \times 2r = \pi r$.

Hands-On LAB 10A Discover Properties of Circles

Use with Lesson 10-5

Circles are not polygons because they are not made of line segments. The distance around a circle is not called the perimeter; it is called the *circumference.*

Activity 1

1. Use a compass to draw several different-sized circles, or trace around circular objects such as lids, cups, and plates.
2. For each circle, use a ruler to measure the distance across it through its center. Record this as the *diameter.*
3. Lay a piece of string around the circle and mark the string where it meets itself. Measure the length, and record it as the circumference.
4. Use a calculator to find the relationship between the circumference C, and the diameter d. Round this value to the nearest hundredth, and record it in the table.

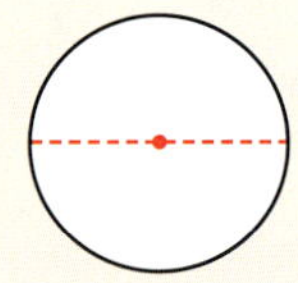

Diameter

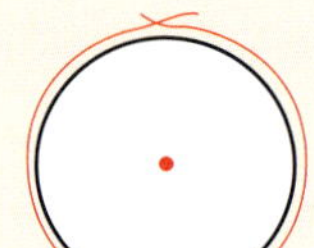

Circumference

Possible answers:

	Circle 1	Circle 2	Circle 3	Circle 4
Circumference C	20 cm	15.2 cm	12.3 cm	5.7 cm
Diameter d	6.4 cm	4.5 cm	3.8 cm	2 cm
$\frac{C}{d}$	3.13	3.38	3.24	2.85

The ratio of C to d is called *pi*, which is represented by the Greek letter π. You can write the equation for circumference as $C = \pi d$.

Think and Discuss

1. Give an approximation of π using the data in your table. **Possible answer: The average of the 4 approximations of π is about 3.15.**

Try This

Use your approximation of π to find the circumference of each circle. Compare your answer to the given value of C. **Possible answers:**

1. 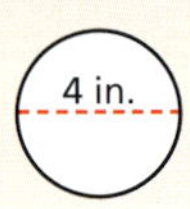

$C = 12.57$ in.
$C = 12.6$ in.; 12.57 in. rounded to the nearest tenth is 12.6 in., so the answers are very close to each other.

2. 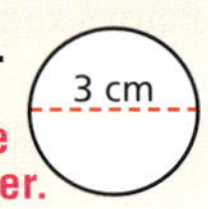

$C = 9.42$ cm
$C = 9.45$ cm; 9.45 is only 0.03 more than 9.42 so the two answers are very close to each other.

You can use your approximation of π to learn about the area of a circle.

Activity 2

1. The *radius* of a circle is half of its diameter. Use a compass to draw a circle with a 2-inch radius. Cut your circle out and fold it three times as shown.

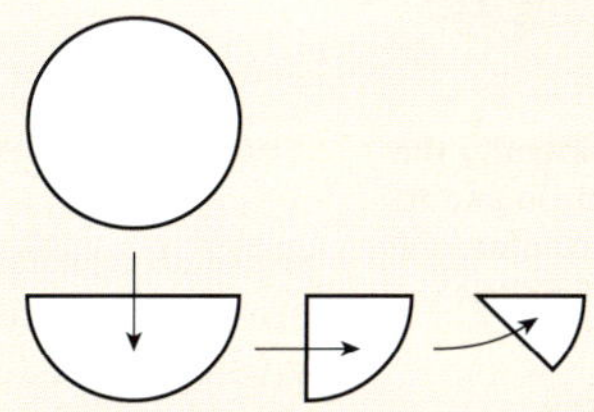

2. Unfold the circle, trace the folds, and shade one-half of the circle.

3. Cut along the folds, and fit the pieces together to make a figure that looks approximately like a parallelogram.

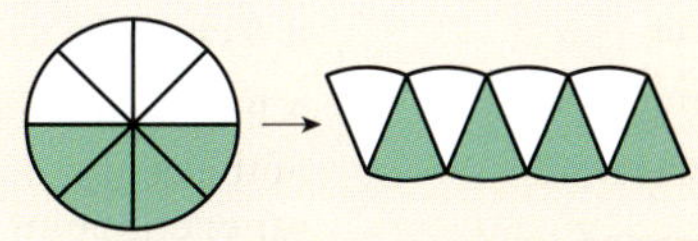

Think of this figure as a parallelogram. The base and height of the parallelogram relate to the parts of the circle.

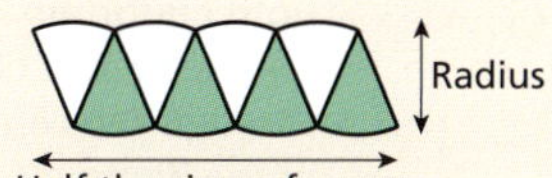

base (b) = $\frac{1}{2}$ the circumference of the circle, or πr

height (h) = the radius of the circle, or r

To find the area of a parallelogram, use the equation $A = bh$.

To find the area of a circle, use the equation $A = \pi r(r) = \pi r^2$.

Think and Discuss

1. Compare the lengths of all the diameters of a circle. **They have the same measure.**
2. Compare the lengths of all the radii of a circle. **They have the same measure.**

Try This

Find the area of each circle with the given measure.

1. $r = 4$ yd **$A = 16\pi$ yd$^2 \approx 50.24$ yd^2**
2. $r = 3$ in. **$A = 9\pi$ in$^2 \approx 28.26$ in^2**
3. $d = 10$ m **$A = 25\pi$ m$^2 \approx 78.5$ m^2**

Find the area of each circle.

4. 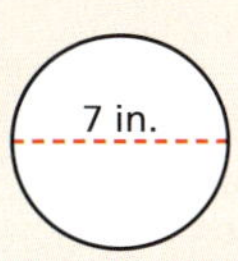

$A = 12.25\pi$ in$^2 \approx 38.465$ in^2

5. 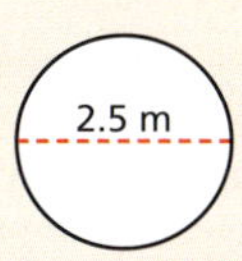

$A = 1.5625\pi$ m$^2 \approx 4.906$ m^2

6. 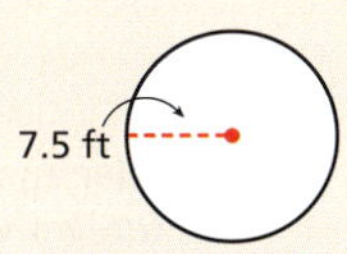

$A = 56.25\pi$ ft$^2 \approx 176.625$ ft^2

10-5 Organizer

Pacing: Traditional 1 day
Block $\frac{1}{2}$ day

Objective: Students identify the parts of a circle and find the circumference and area of a circle.

Indiana Academic Standards
6.3.2, 6.3.5, 6.5.4, 6.5.5

Warm Up

The length and width of a rectangle are each multiplied by 5. Find how the perimeter and area of the rectangle change. The perimeter is multiplied by 5, and the area is multiplied by 25.

Problem of the Day

When using a calculator to find the height of a rectangle whose length one knew, a student accidentally multiplied by 20 when she should have divided by 20. The answer displayed was 520. What is the correct height? 1.3

Available on Daily Transparency in CRB

Math Humor

Teacher: "Pi *r* squared. . ."
Student: "No, pie are round."

10-5 Circles

Learn to identify the parts of a circle and to find the circumference and area of a circle.

Vocabulary
circle
center
radius (radii)
diameter
circumference
pi

The shape of an inline skate wheel is a *circle*. A **circle** is the set of all points in a plane that are the same distance from a given point, called the **center**. At the edge of the wheel, every point is *5 cm* from the center.

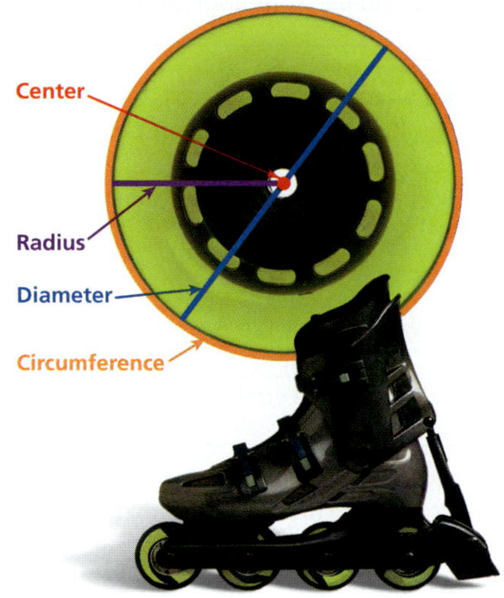

A line segment with one endpoint at the center of the circle and the other endpoint on the circle is a **radius** (plural: *radii*).

A *chord* is a line segment with both endpoints on a circle. A **diameter** is a chord that passes through the center of the circle. The length of the diameter is twice the length of the radius.

The radius of the wheel = *5 cm.*

The diameter of the wheel = 2 · *5 cm* = *10 cm.*

Like a polygon, a circle is a plane figure. But a circle is not a polygon because it is not made of line segments.

EXAMPLE 1 Naming Parts of a Circle

Name the circle, a diameter, and three radii.

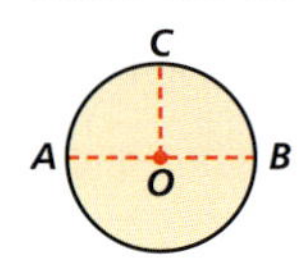

The circle is circle *O*.
$\overline{AB}$ is a diameter.
$\overline{OA}$, $\overline{OB}$, and $\overline{OC}$ are radii.

The distance around a circle is called the **circumference**.

The ratio of the circumference to the diameter, $\frac{C}{d}$, is the same for any circle. This ratio is represented by the Greek letter π, which is read "***pi***."

$$\frac{C}{d} = \pi$$

The decimal representation of *pi* starts with 3.14159265 . . . and goes on forever without repeating. We estimate *pi* using either 3.14 or $\frac{22}{7}$.

1 Introduce

Alternate Opener

EXPLORATION

10-5 Circles

People in ancient civilizations learned to estimate the distance around a circle (***circumference***) by multiplying the distance across (***diameter***) by three.

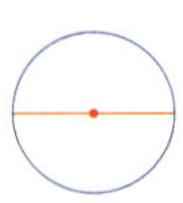

1. Does the distance around the circle above look like three times the diameter?
2. The table below contains actual measurements, to the nearest tenth of a centimeter, of two cans. Use a calculator to find the ratio $\frac{\text{circumference}}{\text{diameter}}$.

Object	Diameter (cm)	Circumference (cm)	$\frac{\text{Circumference}}{\text{Diameter}}$
Juice can	5.2	16.3	
Coffee can	15.7	49.3	

3. Are the $\frac{\text{circumference}}{\text{diameter}}$ ratios you found in number 2 reasonably close to 3?

Think and Discuss

4. **Discuss** whether the ratio $\frac{\text{circumference}}{\text{diameter}}$ changes according to the size of the circle.

Motivate

Have students name real-world items that are shaped like circles.
Possible answers: wheels, CDs, plates, clocks

Explain that, as for other plane figures, there are formulas for finding the distance around a circle and the area of a circle.

Exploration worksheet and answers on Chapter 10 Resource Book pp. 44 and 113

2 Teach

Lesson Presentation

Guided Instruction

In this lesson, students learn to identify parts of a circle and find the circumference and area of a circle. Teach students parts of a circle (center, radius, diameter). Then teach the formulas for finding a circle's circumference and area.

Teaching Tip

You may wish to use a CD to further illustrate the parts of a circle presented in the lesson.

The formula for the circumference of a circle is $C = \pi d$, or $C = 2\pi r$.

EXAMPLE 2 Using the Formula for the Circumference of a Circle

Find each missing value to the nearest hundredth. Use 3.14 for *pi*.

A

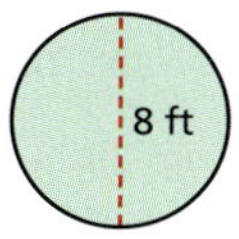

$d = 8$ ft; $C = ?$

$C = \pi d$ — *Write the formula.*

$C \approx 3.14 \cdot 8$ — *Replace π with 3.14 and d with 8.*

$C \approx 25.12$ ft

B

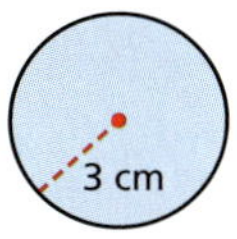

$r = 3$ cm; $C = ?$

$C = 2\pi r$ — *Write the formula.*

$C \approx 2 \cdot 3.14 \cdot 3$ — *Replace π with 3.14 and r with 3.*

$C \approx 18.84$ cm

C $C = 37.68$ in.; $d = ?$

$C = \pi d$ — *Write the formula.*

$37.68 \approx 3.14d$ — *Replace C with 37.68, and π with 3.14.*

$\frac{37.68}{3.14} \approx \frac{3.14d}{3.14}$ — *Divide both sides by 3.14.*

$12.00 \text{ in.} \approx d$

The formula for the area of a circle is $A = \pi r^2$.

EXAMPLE 3 Using the Formula for the Area of a Circle

Find the area of the circle. Use $\frac{22}{7}$ for *pi*.

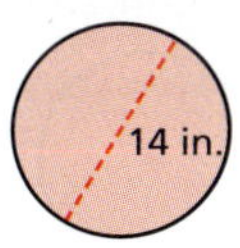

$d = 14$ in.; $A = ?$

$A = \pi r^2$ — *Write the formula to find the area.*

$r = d \div 2$ — *The length of the diameter is twice the length of the radius.*

$r = 14 \div 2 = 7$

$A \approx \frac{22}{7} \cdot 7^2$ — *Replace π with $\frac{22}{7}$ and r with 7.*

$A \approx \frac{22}{\cancel{7}_1} \cdot \cancel{49}^7$ — *Use the GCF to simplify.*

$A \approx 154 \text{ in}^2$ — *Multiply.*

Think and Discuss

1. **Explain** how to find the radius in Example 2C.
2. **Tell** whether the approximation 3.14 is less than or greater than π.
3. **Express** $\frac{22}{7}$ as a decimal rounded to the nearest hundredth. Compare this number with 3.14.

COMMON ERROR ALERT

When using the formula for area, some students may multiply the radius by 2 instead of squaring it. To prevent this, have them write the formula as $A = \pi \times r \times r$.

Additional Examples

Example 1

Name the circle, a diameter, and three radii.

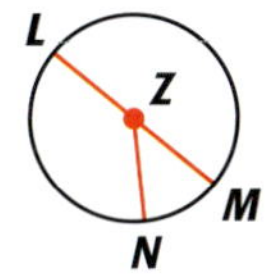

The circle is circle Z. $\overline{LM}$ is a diameter. $\overline{ZL}$, $\overline{ZM}$, and $\overline{ZN}$ are radii.

Example 2

Find the missing value to the nearest hundredth. Use 3.14 for *pi*.

A.

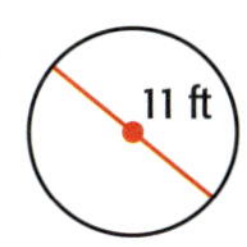

$C \approx 34.54$ ft

$d = 11$ ft; $C =$ ▢

B.

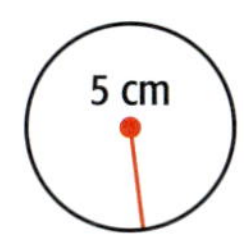

$C \approx 31.4$ cm

$r = 5$ cm; $C =$ ▢

C. $C = 21.98$ in.; $d =$ ▢ $d \approx 7$ in.

Example 3

Find the area of the circle. Use $\frac{22}{7}$ for *pi*.

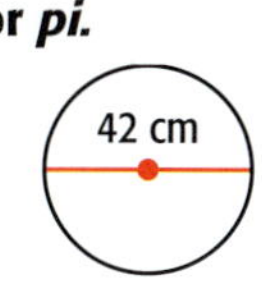

$A = 1{,}386 \text{ cm}^2$

3 Close

Reaching All Learners

Through Number Sense

Use the recording sheet provided on Chapter 10 Resource Book p. 102 to observe the influence of the different representations of pi. For each circle have students find the area and circumference using three different values for pi: 3.14, $\frac{22}{7}$, and the π button on a calculator.

Summarize

Review the new vocabulary in the lesson: *circle*, *center*, *radius*, *diameter*, *circumference*, and *pi*. Discuss how the terms relate to each other.

Answers to Think and Discuss

1. $12 \div 2 = 6$; the diameter is two times the length of the radius.
2. less than π, because $\pi = 3.14159265...$
3. 3.14; When rounded to the nearest hundredth, it is 3.14.

10-5 PRACTICE & ASSESS

Students may want to refer back to the lesson examples.

Assignment Guide

If you finished Example 1 assign:
Core 1, 6, 16, 23–30
Enriched 1, 6, 16, 23–30

If you finished Example 2 assign:
Core 1–3, 6–9, 13–15, 17, 23–30
Enriched 1–3, 6–9, 16–17, 19–20, 23–30

If you finished Example 3 assign:
Core 1–17, 23–30
Enriched 8–30

Answers

13. 14 m, 43.96 m, 153.86 m^2

14. 5.75 ft, 36.11 ft, 103.81 ft^2

15. 2.25 cm, 1.13 cm, 3.97 cm^2

10-5 Exercises

FOR EXTRA PRACTICE
see page 656

GUIDED PRACTICE

See Example 1

1. Point G is the center of the circle. Name the circle, a diameter, and three radii.
circle G, diameter $\overline{EF}$, and radii $\overline{GF}$, $\overline{GE}$, and $\overline{GD}$

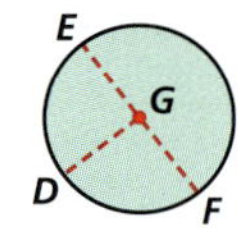

See Example 2

Find each missing value to the nearest hundredth. Use 3.14 for *pi*.

2. $C = \underline{?}$
31.4 mm

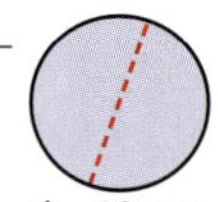
$d = 10$ mm

3. $C = \underline{?}$
12.56 in.

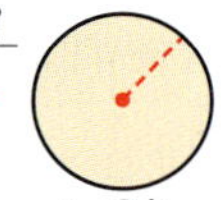
$r = 2$ in.

See Example 3

Find the area of each circle to the nearest hundredth. Use $\frac{22}{7}$ for *pi*.

4. $A = \underline{?}$
154 ft^2

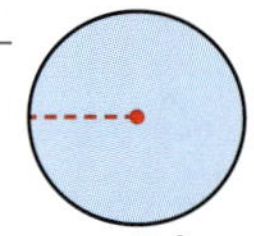
$r = 7$ ft

5. $A = \underline{?}$
616 cm^2

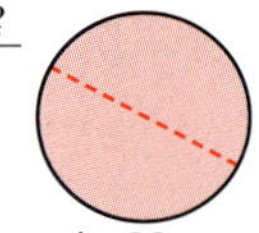
$d = 28$ cm

INDEPENDENT PRACTICE

See Example 1

6. Point P is the center of the circle. Name the circle, a diameter, and three radii.
circle P, diameter $\overline{RS}$, and radii $\overline{PR}$, $\overline{PS}$, and $\overline{PQ}$

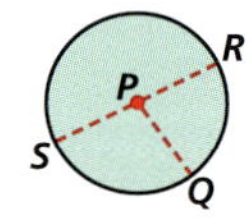

See Example 2

Find each missing value to the nearest hundredth. Use 3.14 for *pi*.

7. $C = \underline{?}$
4.71 m

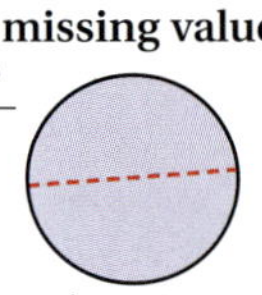
$d = 1.5$ m

8. $C = \underline{?}$
5.02 cm

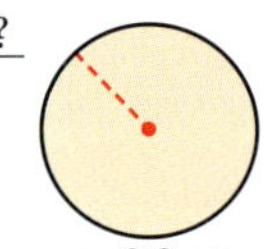
$r = 0.8$ cm

9. $d = \underline{?}$
0.5 in.

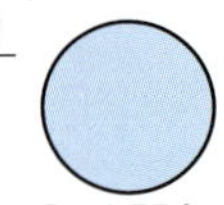
$C = 1.57$ in.

See Example 3

Find the area of each circle to the nearest hundredth. Use $\frac{22}{7}$ for *pi*.

10. $A = \underline{?}$
38.5 yd^2

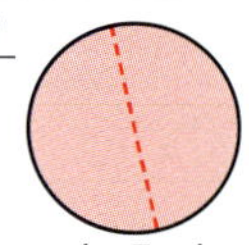
$d = 7$ yd

11. $A = \underline{?}$
9.63 cm^2

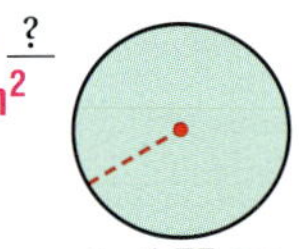
$r = 1.75$ cm

12. $A = \underline{?}$
2,464 ft^2

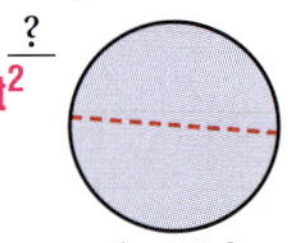
$d = 56$ ft

PRACTICE AND PROBLEM SOLVING

Fill in the blanks. Use 3.14 for *pi* and round to the nearest hundredth.

13. If $r = 7$ m, then $d = \underline{?}$, $C = \underline{?}$, and $A = \underline{?}$.

14. If $d = 11.5$ ft, then $r = \underline{?}$, $C = \underline{?}$, and $A = \underline{?}$.

15. If $C = 7.065$ cm, then $d = \underline{?}$, $r = \underline{?}$, and $A = \underline{?}$.

Math Background

Around 225 B.C., the mathematician Archimedes wrote a short work called *Measurement of a Circle*. In it, he presented the result we now write as $A = \pi r^2$.

Archimedes worked by inscribing regular polygons within a circle and circumscribing regular polygons about the circle. But as the number of sides in the regular polygons increase, the areas of the inscribed and circumscribed polygons move toward each other. In modern terminology, the two areas approach a limit. Archimedes was able to show that their common limit, the area of the circle, is πr^2.

RETEACH 10-5

LESSON 10-5 **Reteach**
Circles

A circle is the set of all points in a plane that are the same distance from a given point, called the center.

A line segment with an endpoint at the center and an endpoint at the circle is called the radius r. A line segment that passes through the center and has both endpoints on the circle is called the diameter d. The length of the diameter is twice the radius.

This is circle P.
$\overline{QR}$ is a diameter.
$\overline{PQ}$ and $\overline{PR}$ are radii.

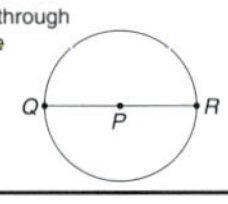

Identify each part of the circle below.

1. circle $\underline{A}$
2. diameter $\underline{\overline{BC}}$
3. two radii $\underline{\overline{AB} \text{ and } \overline{AC}}$

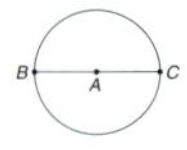

The distance around a circle is called the circumference. The ratio of the circumference of a circle to its diameter is always the same. This relationship is called pi and is represented by the Greek letter π.

Pi is a decimal number that has no end and never repeats. 3.14 or $\frac{22}{7}$ is used to estimate pi.

To find the circumference C of a circle, use a formula.

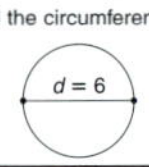

$C = \pi d$
$\approx 3.14 \cdot 6$
≈ 18.84
The circumference is about 18.84 units.

Find the circumference. Use 3.14 for π.

4.
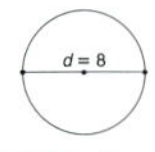

$C \approx 25.12$ units

5.

$C \approx 12.56$ units

PRACTICE 10-5

LESSON 10-5 **Practice B**
Circles

Use the circle to answer each question.

1. Name the circle. circle X
2. Name two diameters.
$\overline{YW}$ and $\overline{VZ}$
3. Name four radii.
$\overline{XV}$, $\overline{XW}$, $\overline{XZ}$, and $\overline{XY}$

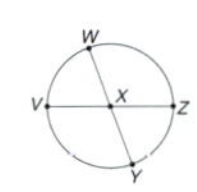

Find each missing value to the nearest hundredth. Use 3.14 for π.

4.
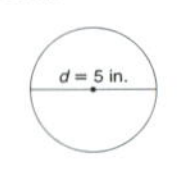

$C =$ 15.7 in.

5.
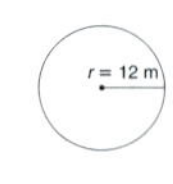

$C =$ 75.36 m

Find the area of each circle. Use $\frac{22}{7}$ for π.

6.
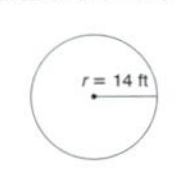

$A =$ 616 ft^2

7.
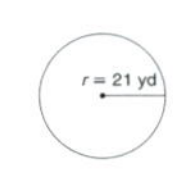

$A =$ 1,386 yd^2

8. The first Ferris wheel was built in 1893 in Chicago. Its diameter was 250 feet. How many feet did the Ferris wheel rotate with each complete turn? Use 3.14 for π.
785 feet

9. Stonehenge, a circle of large carved stones in England, was built more than 1,000 years ago. The circle of stones has a diameter of 108 feet. About how many square feet of land does Stonehenge cover? Use $\frac{22}{7}$ for π.
9,164.57 square feet

History LINK

Performers like Patsy Rosales have learned to twirl multiple Hula Hoops around their bodies. The world record for most Hula Hoops kept in revolution is held by Katja Lange, who kept 90 of them spinning.

16. ***HISTORY*** The first Hula Hoop® was introduced in 1958. It is one of the most popular toys in U.S. history. What is the circumference of a Hula Hoop with a 3 ft diameter? (Use 3.14 for π. Round the answer to the nearest hundredth.) **9.42 ft**

17. ***MEASUREMENT*** Draw a circle with center O and radius $\overline{AO}$ that is 2 inches long.
    a. Draw the diameter $\overline{AB}$ and give its length.
    b. What is the circumference of circle O? (Use 3.14 for π. Round to the nearest hundredth.) **12.56 in.**
    c. What is the area of circle O? **12.56 in²**

18. ***SPORTS*** The diameter of the circle that a shot-putter stands in is 7 ft. What is the area of the circle? (Use $\frac{22}{7}$ for π.) **$38\frac{1}{2}$ ft²**

19. a. Estimate the difference between the area of a pizza with a 6 in. diameter and a pizza with a 12 in. diameter. (Use 3.14 for π. Round to the nearest whole number.)
    b. Is the area of the 12 in. pizza about twice as large as the area of the 6 in. pizza? Explain.

19a. Possible answer: The difference is about 85 in².
19b. No, the area of the 12-in. pizza is about 4 times as large as the 6-in. pizza.

20. ***WRITE A PROBLEM*** Write a problem that involves finding the circumference and the area of a circle when given the diameter. Give the answer to your problem.

21. ***WRITE ABOUT IT*** The circumference of a circle is 3.14 m. Explain how you can find the diameter and radius of the circle.

22. ***CHALLENGE*** What is the area of the shaded part of the figure? (Use 3.14 for π. Round the answer to the nearest hundredth.) **0.43 m²**

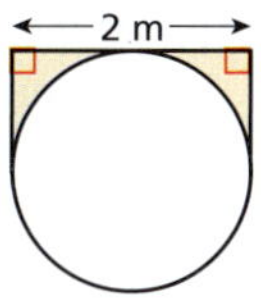

Spiral Review

Find each quotient. (Lesson 3-7)

23. 25.5 ÷ 5 **5.1**
24. 44.7 ÷ 3 **14.9**
25. 96.48 ÷ 6 **16.08**
26. 0.0378 ÷ 9 **0.0042**

Order the fractions from greatest to least. (Lesson 4-6)

27. $\frac{1}{2}, \frac{3}{8}, \frac{5}{8}$ **$\frac{5}{8}, \frac{1}{2}, \frac{3}{8}$**
28. $\frac{3}{4}, \frac{10}{12}, \frac{1}{12}$ **$\frac{10}{12}, \frac{3}{4}, \frac{1}{12}$**
29. $\frac{3}{10}, \frac{3}{5}, \frac{7}{10}$ **$\frac{7}{10}, \frac{3}{5}, \frac{3}{10}$**

30. **TEST PREP** Which number is 20% of 360? (Lesson 8-9) **B**

A 18 B 72 C 380 D 7,200

Answers

17. a.

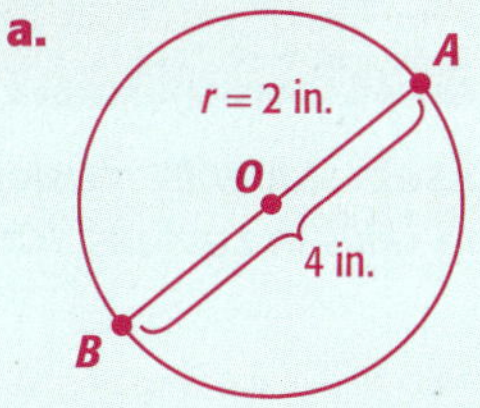

20. Possible answer: The surface of a disk has a diameter of 1 ft. What are the circumference and area of the disk? 3.14 ft and 0.785 ft²

21. Use the formula for the circumference of a circle, $C = \pi d$. So $3.14 = 3.14d$, $d = 1$ in., and $r = 0.5$ in.

Journal

Have students use a compass to draw a circle. Have students label the center, a radius, and a diameter on the circle. Have students also find the circumference and area of the circle.

Test Prep Doctor

For Exercise 30, students need to multiply 360 by 0.20. Students who answered **A** divided 360 by 20. Students who answered **C** added 20 to 360. Students who answered **D** multiplied 360 by 20 instead of by 0.20.

CHALLENGE 10-5

LESSON 10-5 **Challenge**
Pies Are Squared

Have you ever heard the saying, "You can't fit a square peg in a round hole"? Well, sometimes you can fit a round pie in a square dish—if you squeeze really hard and don't mind messy areas.

Find the area of each pie and pie dish. Use 3.14 for π, and round your measurements to the nearest hundredth. Then imagine you could squish the pies. Which square pan would each round pie best fit in? Draw a line from the pie to the appropriate pie dish.

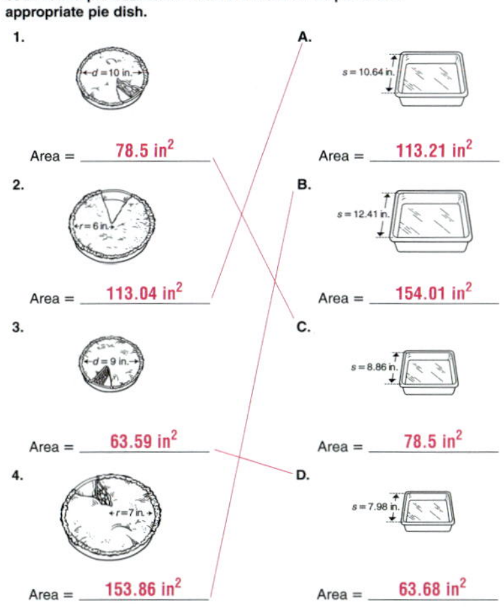

PROBLEM SOLVING 10-5

LESSON 10-5 **Problem Solving**
Circles

Use the table to answer each question. Use 3.14 for π.

Official U.S. Coin Sizes

Coin	Diameter (rounded to nearest mm)
Penny	19
Nickel	21
Dime	18
Quarter	24
Half Dollar	31
Dollar	27

1. Which coin has the smallest radius? How long is that coin's radius?
   dime; 9 mm
2. What is the circumference of a nickel?
   65.94 mm
3. What is the area of a quarter?
   452.16 mm²
4. Which coin has a greater area, a dollar or half dollar? What is the difference in their areas?
   half dollar; 182.12 mm²
5. If you rolled a dollar coin on its edge, how far would it go with each complete turn?
   84.78 mm
6. Which U.S. coins will fit in a vending machine coin slot that is 2 centimeters wide?
   penny and dime
7. An engraving of Thomas Jefferson's home, Monticello, covers about $\frac{1}{3}$ of a nickel's tails side. What is the area of the Monticello etching?
   about 115.395 mm²
8. The engraved words "United States of America" run about one-half the circumference of all U.S. coins. On which coin will the words run about 38 mm?
   quarter

Circle the letter of the correct answer.

9. A dime has 118 ridges evenly spaced along its circumference. About how wide is each ridge?
   A about 0.24 mm
   (B) about 0.48 mm
   C about 0.15 mm
   D about 0.08 mm
10. Your two coins together cover an area of about 540 mm². How much money do you have?
   (F) $0.11 H $1.10
   G $1.50 J $0.35

Lesson Quiz

Find the circumference and area of each circle. Use 3.14 for π.

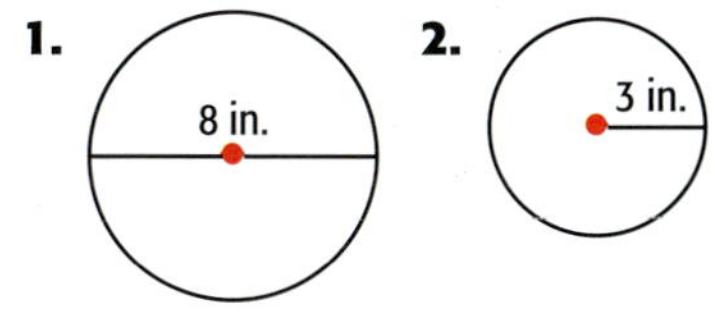

1. **$C = 25.12$ in.**
   $A = 50.24$ in²
2. **$C = 18.84$ in.**
   $A = 28.26$ in²

3. Find the area of a circle with a diameter of 20 feet. Use 3.14 for π.
   314 ft²

Available on Daily Transparency in CRB

Chapter 10 Mid-Chapter Quiz

Purpose: *To assess students' mastery of concepts and skills in Lessons 10-1 through 10-5*

Assessment Resources

Section 10A Quiz

Assessment Resources .... p. 25

***Test and Practice Generator* CD-ROM**

Additional mid-chapter assessment items in both multiple-choice and free-response format may be generated for any objective in Lessons 10-1 through 10-5.

Answers

11. $\overline{AC}$ and $\overline{AE}$ are radii; $C = 43.96$ cm; $A = 153.86$ cm^2.

12. $\overline{DH}$ and $\overline{DG}$ are radii; $C = 9.42$ in.; $A = 7.07$ in^2.

13. $\overline{IG}$ and $\overline{IF}$ are radii; $C = 51.81$ km; $A = 213.72$ km^2.

14. $\overline{KJ}$, $\overline{KM}$, and $\overline{KL}$ are radii; $C = 131.88$ cm; $A = 1{,}384.74$ cm^2.

18. $A \approx 28\frac{2}{7}$ cm^2

19. $A \approx 616$ ft^2

20. $A \approx 9\frac{5}{8}$ in^2

Chapter 10 Mid-Chapter Quiz

Mid-Chapter Quiz

LESSON 10-1 (pp. 500–503)

Find the perimeter of each figure.

1. 44 cm

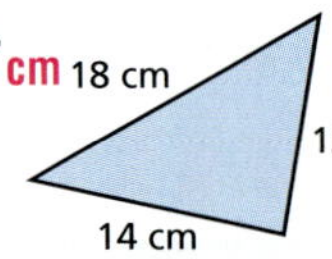

2.

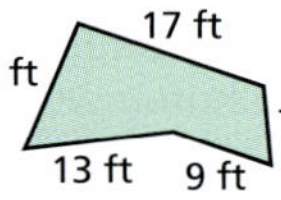

58 ft

3.

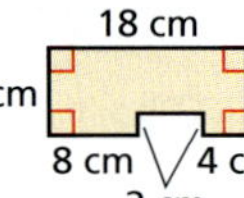

56 cm

LESSON 10-2 (pp. 504–507)

Find the area of each figure.

4.

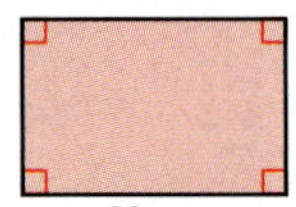

2,542 cm^2

5.

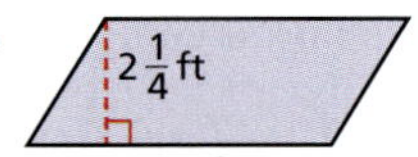

12 ft^2

6.

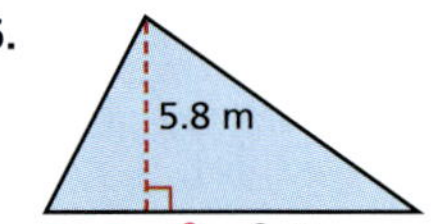

23.2 m^2

LESSON 10-3 (pp. 508–510)

Find the area of each polygon.

7.

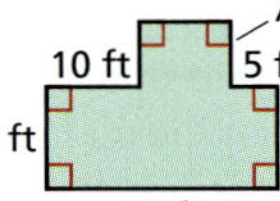

345 ft^2

8.

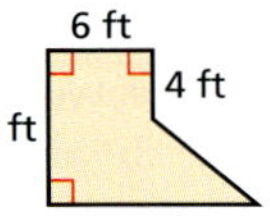

69 ft^2

9.

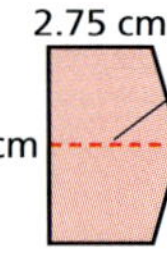

16.41 cm^2

LESSON 10-4 (pp. 511–513)

10. The length and width of a rectangle are each multiplied by 4. Find how the perimeter and the area of the rectangle change.
The perimeter is multiplied by 4, and the area is multiplied by 4^2, or 16.

LESSON 10-5 (pp. 516–519)

Name two radii and find the circumference and area for each circle. Use 3.14 for *pi* and round to the nearest hundredth.

11.

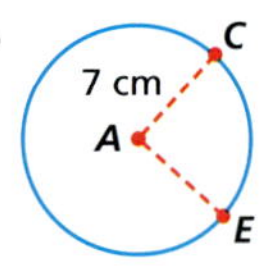

12.

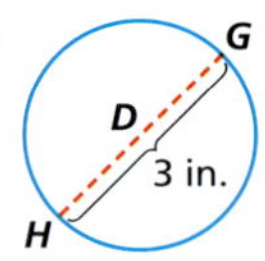

13.

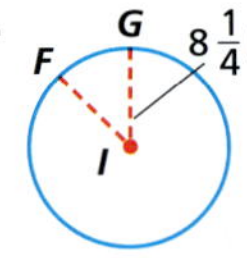

14.

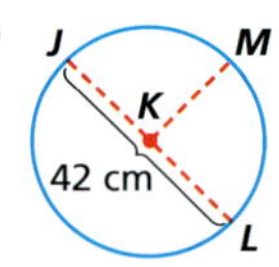

Find each missing value to the nearest hundredth. Use 3.14 for *pi*.

15. $r = 9$ in.; $C = \underline{?}$ 56.52 in.

16. $d = 20$ m; $C = \underline{?}$ 62.8 m

17. $C = 37.68$ ft; $d = \underline{?}$ 12 ft

Find the area of each circle. Use $\frac{22}{7}$ for pi.

18. $d = 6$ cm; $A = \underline{?}$

19. $r = 14$ ft; $A = \underline{?}$

20. $r = 1\frac{3}{4}$ in.; $A = \underline{?}$

Focus on Problem Solving

Solve
• Choose the operation

Read the whole problem before you try to solve it. Determine what action is taking place in the problem. Then decide whether you need to add, subtract, multiply, or divide in order to solve the problem.

Action	Operation
Combining or putting together	Add
Removing or taking away Comparing or finding the difference	Subtract
Combining equal groups	Multiply
Sharing equally or separating into equal groups	Divide

Read each problem and determine the action taking place. Choose an operation, and then solve the problem.

1. There are 3 lily ponds in the botanical gardens. They are identical in size and shape. The total area of the ponds is 165 ft^2. What is the area of each lily pond?

2. The greenhouse is made up of 6 rectangular rooms with an area of 4,800 ft^2 each. What is the total area of the greenhouse?

3. A shady area with 17 different varieties of magnolia trees, which bloom from March to June, surrounds the plaza in Magnolia Park. In the center of the plaza, there is a circular bed of shrubs as shown in the chart. If the total area of the park is 625 ft^2, what is the area of the plaza?

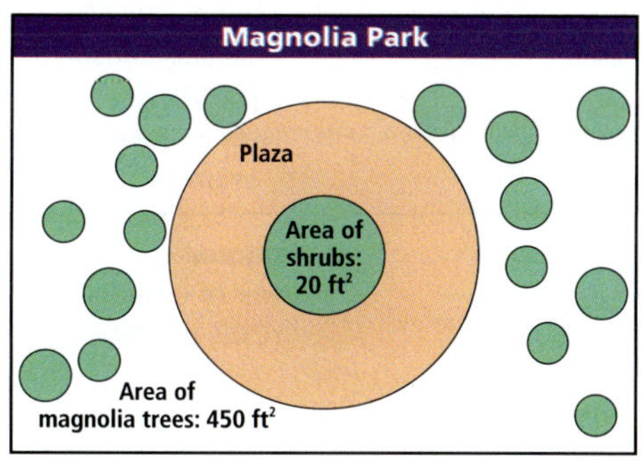

Answers

1. 55 ft^2
2. 28,800 ft^2
3. 155 ft^2

Focus on Problem Solving

Purpose: *To focus on choosing an operation to solve a problem*

Problem Solving Resources

Interactive Problem Solving . . pp. 83–91

Math: Reading and Writing in the Content Area . . . . pp. 83–91

Problem Solving Process

This page focuses on the third step of the problem-solving process: **Solve**

Discuss

Have students discuss the words or phrases in each exercise that signaled which operation to choose.

Possible answers:

1. The *three ponds* are *identical,* so *each* pond has the same area. Therefore, you want to separate the *total area* into 3 equal groups; divide.
2. *Each* of the *6* rooms are equal in area. To find the *total area,* combine 6 equal groups; multiply.
3. To find the *area of the plaza,* you must find the difference between the *total area* of the park and the area that *surrounds the plaza;* subtract.

Section 10B

Volume and Surface Area

One-Minute Section Planner

Lesson	Materials	Resources
Hands-On Lab 10B Draw Views of Solid Figures **NCTM:** Geometry, Reasoning and Proof **NAEP:** Geometry 1e ☐ SAT-9 ☐ SAT-10 ☐ ITBS ☐ CTBS ☐ MAT ☐ CAT	**Required** Centimeter cubes *(MK)*	• *Hands-On Lab Activities,* pp. 92–94
Lesson 10-6 Solid Figures **NCTM:** Geometry, Communication **NAEP:** Geometry 1c ☐ SAT-9 ☐ SAT-10 ☑ ITBS ☐ CTBS ☑ MAT ☑ CAT	**Optional** Teaching Transparency T22 *(CRB)* Nets for Motivate *(CRB, pp. 103–104)*	• *Chapter 10 Resource Book*, pp. 54–62 • Daily Transparency T21, CRB • Additional Examples Transparencies T23–T24, CRB • *Alternate Openers: Explorations*, p. 88
Hands-On Lab 10C Model Solid Figures **NCTM:** Geometry, Reasoning and Proof **NAEP:** Geometry 1f ☐ SAT-9 ☐ SAT-10 ☐ ITBS ☐ CTBS ☐ MAT ☐ CAT		• *Hands-On Lab Activities,* pp. 97–98
Lesson 10-7 Surface Area **NCTM:** Geometry, Measurement, Communication **NAEP:** Measurement 1j ☑ SAT-9 ☑ SAT-10 ☐ ITBS ☑ CTBS ☑ MAT ☑ CAT	**Optional** Graph paper *(CRB p. 101)* Nets for Reaching All Learners *(CRB pp. 103–104)*	• *Chapter 10 Resource Book*, pp. 63–71 • Daily Transparency T25, CRB • Additional Examples Transparencies T26–T29, CRB • *Alternate Openers: Explorations*, p. 89
Lesson 10-8 Finding Volume **NCTM:** Geometry, Measurement, Problem Solving, Communication **NAEP:** Measurement 1j ☐ SAT-9 ☑ SAT-10 ☑ ITBS ☑ CTBS ☑ MAT ☑ CAT	**Optional** Teaching Transparency T31 *(CRB)* Nets for Reaching All Learners *(CRB, p. 103)*	• *Chapter 10 Resource Book*, pp. 72–80 • Daily Transparency T30, CRB • Additional Examples Transparencies T32–T34, CRB • *Alternate Openers: Explorations*, p. 90
Lesson 10-9 Volume of Cylinders **NCTM:** Geometry, Measurement, Communication, Connections **NAEP:** Measurement 1j ☐ SAT-9 ☐ SAT-10 ☐ ITBS ☐ CTBS ☐ MAT ☐ CAT	**Optional** Net of Cylinder *(CRB, p. 104)*	• *Chapter 10 Resource Book*, pp. 81–89 • Daily Transparency T35, CRB • Additional Examples Transparencies T36–T38, CRB • *Alternate Openers: Explorations*, p. 91
Section 7B Assessment		• Section 10B Quiz, AR p. 26 • *Test and Practice Generator* CD-ROM

SAT = *Stanford Achievement Tests* **ITBS** = *Iowa Test of Basic Skills* **CTBS** = *Comprehensive Test of Basic Skills/Terra Nova*
MAT = *Metropolitan Achievement Tests* **CAT** = *California Achievement Test*
NCTM = Complete standards can be found on pages T27–T33. **NAEP** = Complete standards can be found on pages A31–A35.
SE = *Student Edition* **TE** = *Teacher's Edition* **AR** = *Assessment Resources* **CRB** = *Chapter Resource Book* **MK** = *Manipulatives Kit*

Section Overview

Solid Figures — *Hands-On Labs 10B, 10C, Lesson 10-6*

Why? Buildings are solid figures.

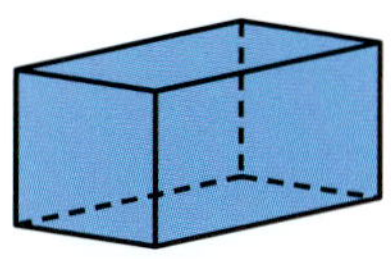

Rectangular prism

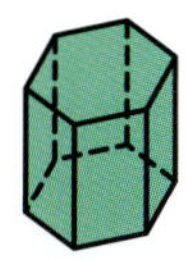

Hexagonal prism

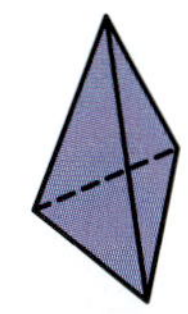

Triangular pyramid

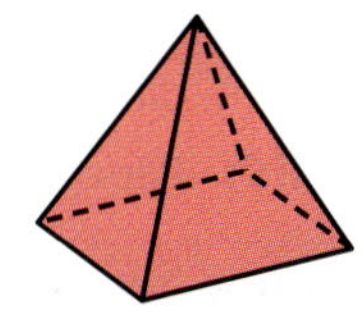

Square pyramid

A **prism** is a polyhedron with two congruent, parallel bases and other faces that are all parallelograms.

A **pyramid** has one polygon-shaped base, and the other faces are triangles that come to a point.

Cylinder

A **cylinder** has two congruent, parallel circular bases. A cylinder is not a polyhedron because not all of its surfaces are polygons.

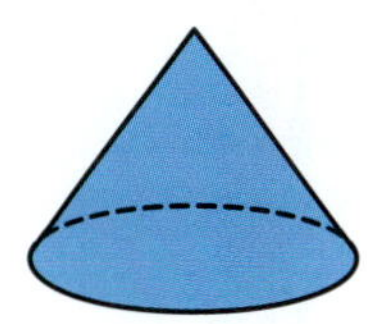

Cone

A **cone** has a circular base and a curved surface that comes to a point.

Surface Area — *Lesson 10-7*

Why? Find the surface area of your room to determine how much paint you will need to paint your room.

The **surface area** of a solid figure is the sum of the areas of its surfaces.

A **net** is a flat pattern that shows all the surfaces of a solid figure.

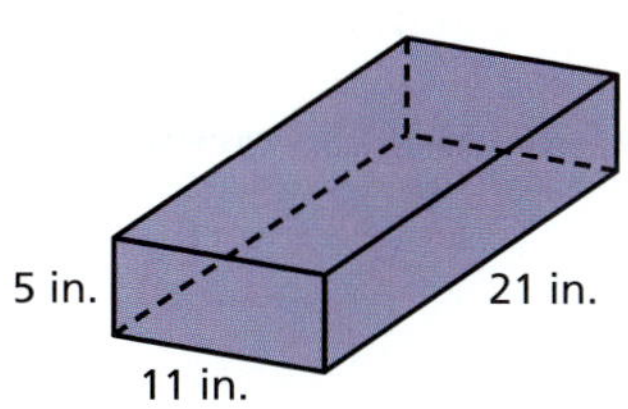

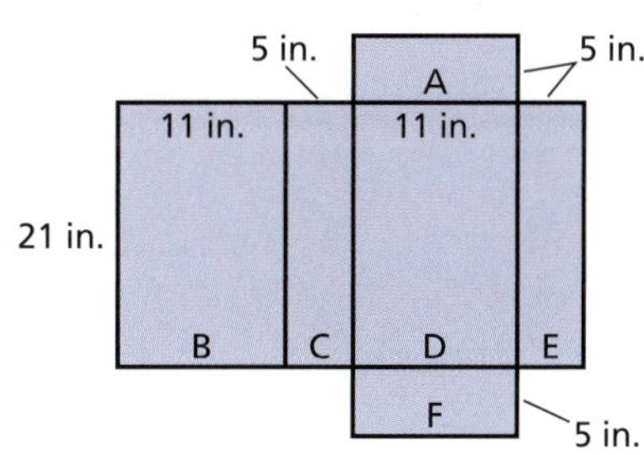

Find the surface area S of the prism.

Use the formula $A = \ell w$ to find the area of each face.

A: $A = 11 \times 5 = 55$
B: $A = 21 \times 11 = 231$
C: $A = 21 \times 5 = 105$
D: $A = 21 \times 11 = 231$
E: $A = 21 \times 5 = 105$
F: $A = 11 \times 5 = 55$

Add the areas of the faces.

$S = 55 + 231 + 105 + 231 + 105 + 55 = 782$

The surface area is 782 in^2.

Volume — *Lessons 10-8, 10-9*

Why? Sometimes we need to answer the question, how much will it hold?

Volume is the number of cubic units needed to fill a space.

To find the **volume of a prism**, use the formula $V = Bh$, where B is the area of the base, and h is the height.

To find the **volume of a cylinder**, use the formula $V = Bh = \pi r^2 h$, where B is the area of the base, and h is the height.

Hands-On LAB 10B Draw Views of Solid Figures

Pacing: Traditional 1 day
Block $\frac{1}{2}$ day

Objective: To use drawings of solid figures to study different views

Materials: Paper, pencil

Indiana Academic Standards
6.4.7, 6.7.1, 6.7.2, 6.7.3, 6.7.4, 6.7.5, 6.7.11

Lab Resources

Hands-On Lab Activities . . . . p. 85

Using the Pages

Discuss with students how to find different views of a solid figure. Take a textbook and lay it on the table. Stand so that your eyes are directly over the book and draw a sketch of the view on the board. Next, bend down so that your eyes are directly to the textbook's right, and repeat. Finally, bend down so that your eyes are directly in front of the textbook and draw a sketch of the view.

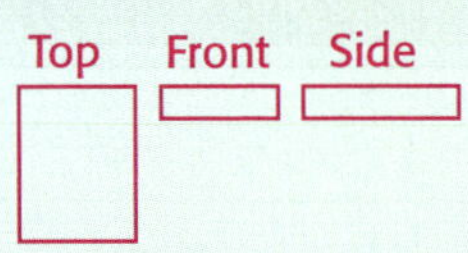

Answers

Activity 1

Try This

1.
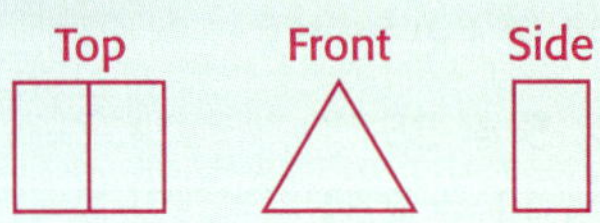

2.
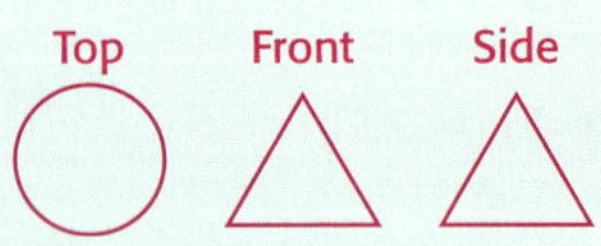

3.
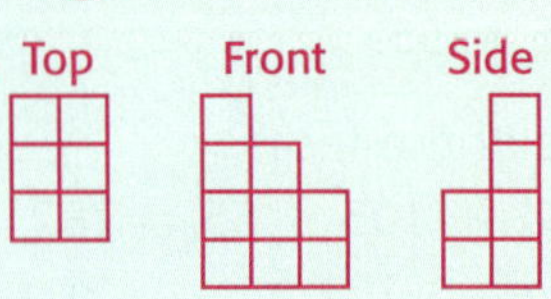

Hands-On LAB 10B Draw Views of Solid Figures

Use with Lesson 10-6

internet connect
Lab Resources Online
go.hrw.com
KEYWORD: MR4 Lab10B

Activity 1

1. Draw a rectangular prism. Imagine that you are looking at the top of the prism, and draw what you would see. Draw the front and side views of the prism.

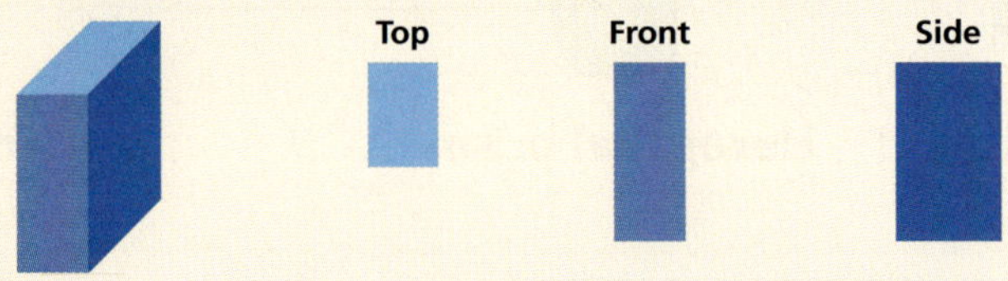

All the faces of a rectangular prism are rectangles.

2. Stack centimeter cubes to make the solid figure shown. Draw the top, front, and side views.

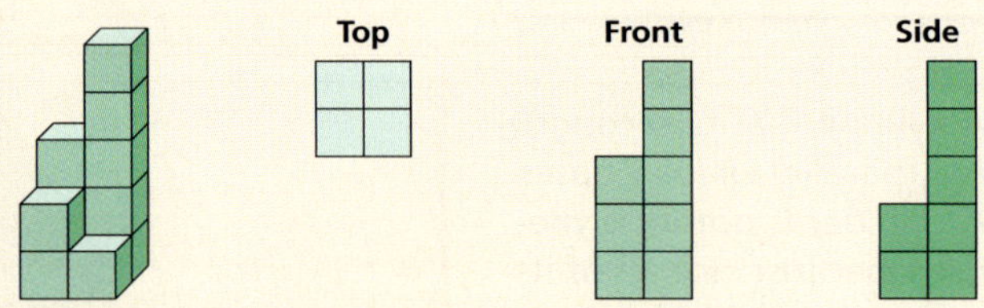

Each view shows a different configuration of squares representing the number of cubes you see.

Think and Discuss

1. Explain why a side view of a figure might change if you look at a different side. **Possible answer: Some blocks may be hidden from certain side views and visible from others.**

Try This

Draw the top, front, and side views of each solid figure.

1. 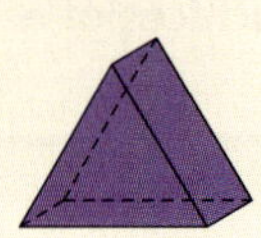

2. 

3. 

You can use different views of a solid to identify the figure.

Activity 2

1 Name the solid figure that has the given views.

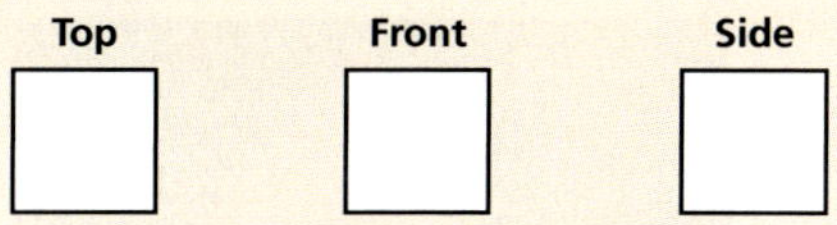

Each view of the solid shows a square.

The solid is a cube.

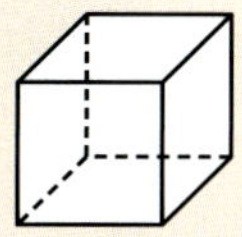

2 Name the solid figure that has the given views.

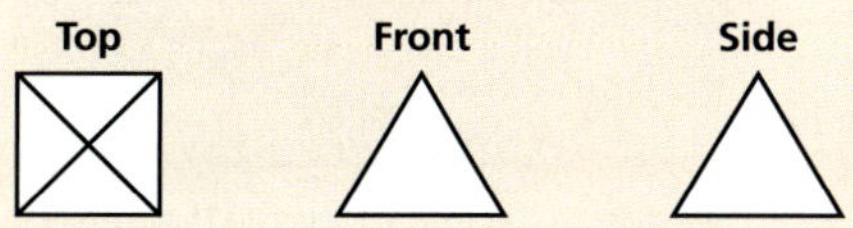

The top view shows that the base is a square. It also shows that the other faces come together at a point.

The front and side views show that the other faces are triangles.

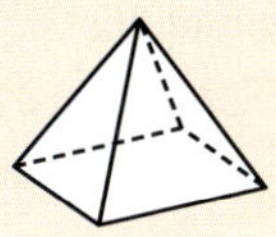

The solid is a square pyramid.

Think and Discuss

1. Explain which views show how tall a solid figure is.
The front and side views best show how tall a solid figure is.

Try This

Name each solid figure that has the given views.

1.

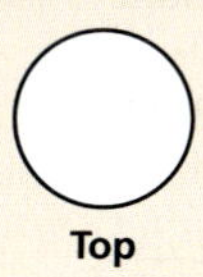

cylinder

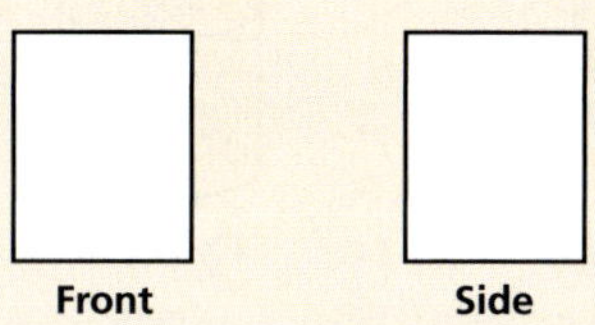

2.

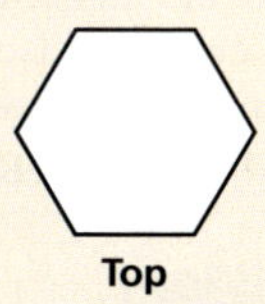

hexagonal prism

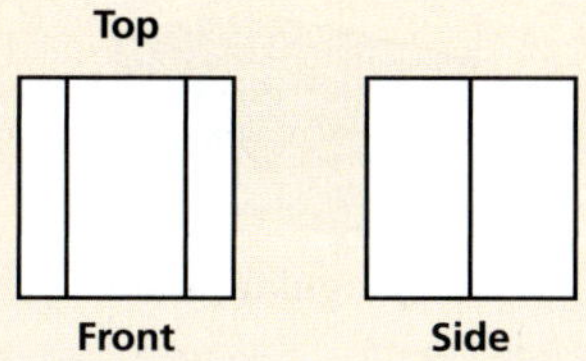

10-6 Organizer

Pacing: Traditional 1 day
Block $\frac{1}{2}$ day

Objective: Students name solid figures.

Indiana Academic Standards
6.7.1, 6.7.2, 6.7.3, 6.7.5, 6.7.11

Warm Up

Solve. Use 3.14 for π.

1. The diameter of a circle is 12 in. What is its circumference? **37.68 in.**
2. The radius of a circle is 9 cm. What is its circumference? **56.52 cm**
3. Find the area of a circle with a 12 ft radius. **452.16 ft^2**

Problem of the Day

To measure the perimeter of her square patio, Becky used an old bicycle wheel with a 22 in. diameter. She rolled the wheel from one corner of the patio along the edge to the next. The wheel made 6.75 revolutions. What is the perimeter in feet of the patio? Use 3.14 for π. **155.43 ft**

Available on Daily Transparency in CRB

Math Fact

The base of a pyramid may contain any number of sides greater than two, but the lateral faces will all be triangles.

10-6 Solid Figures

Learn to name solid figures.

Vocabulary
polyhedron
face
edge
vertex
prism
base
pyramid
cylinder
cone

A **polyhedron** is a three-dimensional object, or solid figure, with flat surfaces, called **faces**, that are polygons.

When two faces of a solid figure share a side, they form an **edge**. On a solid figure, a point at which three or more edges meet is a **vertex** (plural: *vertices*).

A cube is formed by 6 square faces. It has 8 vertices and 12 edges. The sculpture in front of this building is based on a cube. The artist's work is not a polyhedron because of the hole cut through the middle.

This sculpture, *Red Cube*, in front of the Marine Midland Bank in New York City was done by Isamu Noguchi.

EXAMPLE 1 Identifying Faces, Edges, and Vertices

Identify the number of faces, edges, and vertices on each solid figure.

A

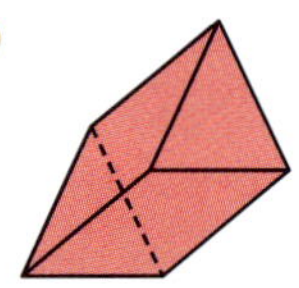

5 faces
9 edges
6 vertices

B

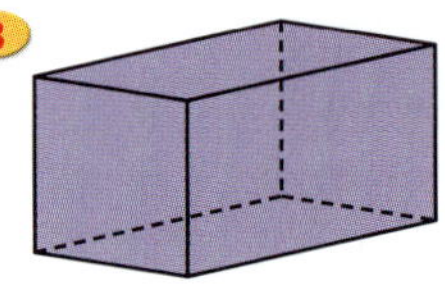

6 faces
12 edges
8 vertices

A **prism** is a polyhedron with two congruent, parallel **bases**, and other faces that are all parallelograms. A prism is named for the shape of its bases. A **cylinder** also has two congruent, parallel bases, but bases of a cylinder are circular. Cylinders are not polyhedra because not every surface is a polygon.

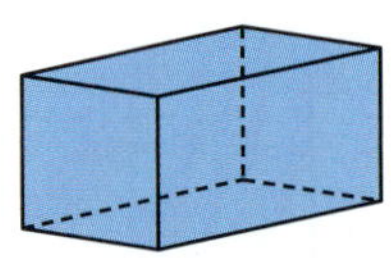

Rectangular prism

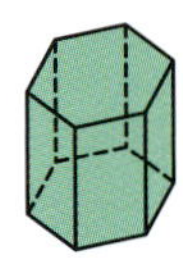

Hexagonal prism

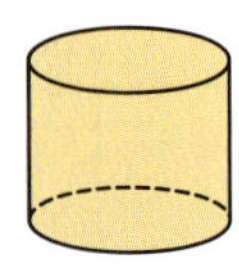

Cylinder

1 Introduce

Alternate Opener

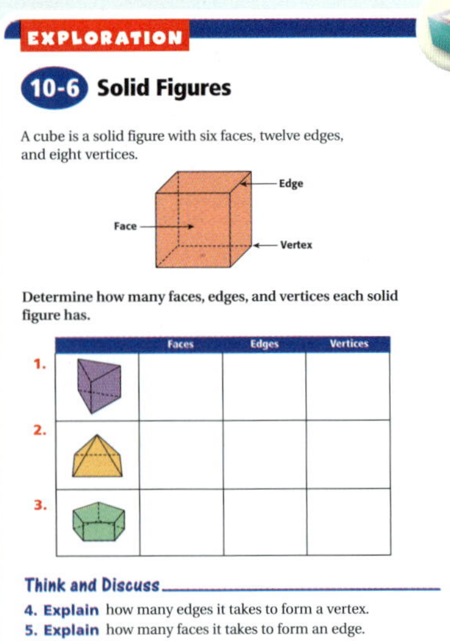

EXPLORATION

10-6 Solid Figures

A cube is a solid figure with six faces, twelve edges, and eight vertices.

Edge
Face
Vertex

Determine how many faces, edges, and vertices each solid figure has.

	Faces	Edges	Vertices
1.			
2.			
3.			

Think and Discuss
4. **Explain** how many edges it takes to form a vertex.
5. **Explain** how many faces it takes to form an edge.

Motivate

Display a collection of solid figures from everyday life or utilize the models that can be assembled on Chapter 10 Resource Book pp. 103–104. With the class, separate the figures into solids with no curved surfaces and solids with one or more curved surface. Have students name any of the solid figures that they are familiar with.

Exploration worksheet and answers on Chapter 10 Resource Book pp. 55 and 115

2 Teach

Lesson Presentation

Guided Instruction

In this lesson, students learn to name solid figures. Teach the new vocabulary words in the lesson. Then teach students how to count faces, edges, and vertices on pictures of solid figures. Then explain how to decide whether or not a solid figure is a polyhedron.

Helpful Hint

The point of a cone is called its *vertex*.

A **pyramid** has one polygon shaped base, and the other faces are triangles that come to a point. A pyramid is named for the shape of its base. A **cone** has a circular base and a curved surface that comes to a point. Cones are not polyhedra because not every face is a polygon.

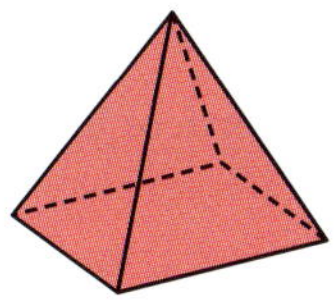

Square pyramid

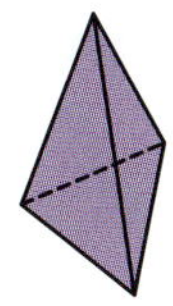

Triangular pyramid

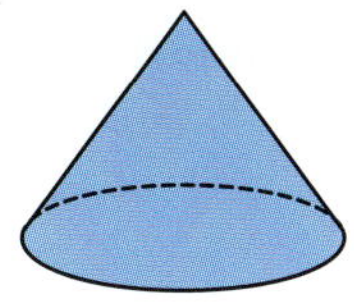

Cone

EXAMPLE 2 Naming Solid Figures

Name the solid figure represented by each object.

A

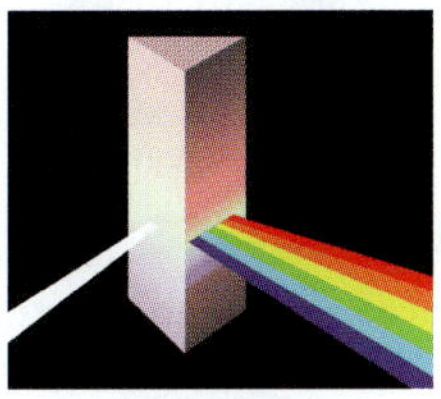

All the faces are flat and are polygons.

The figure is a polyhedron.

There are two congruent, parallel bases, so the figure is a prism.

The bases are triangles.

The figure is a triangular prism.

B

There is a curved surface.

The figure is not a polyhedron.

There is a flat, circular base.

The lateral surface comes to a point.

The figure represents a cone.

C

All the faces are flat and are polygons.

The figure is a polyhedron.

It has one base and the other faces are triangles that meet at a point, so the figure is a pyramid.

The base is a square.

The figure is a square pyramid.

Think and Discuss

1. **Tell** how a cylinder and a prism are alike and how they are different.
2. **Explain** how a cone and a pyramid are alike and how they are different.
3. **Name** a prism that has squares as its bases and as its faces.

Additional Examples

Example 1

Identify the number of faces, edges, and vertices on each solid figure.

A.

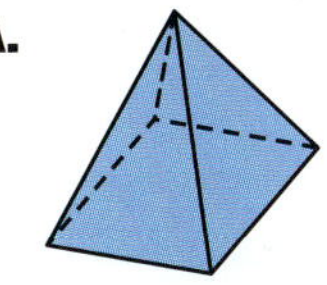

5 faces
8 edges
5 vertices

B.

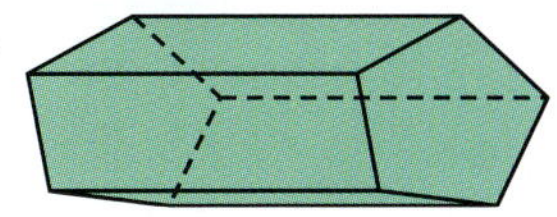

7 faces 15 edges 10 vertices

Example 2

Name the solid represented by each object.

A.

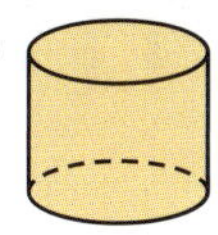

cylinder

B.

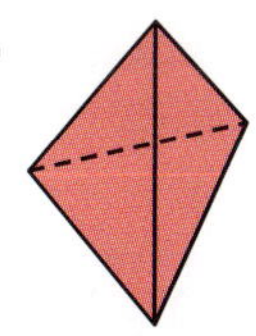

triangular pyramid

C.

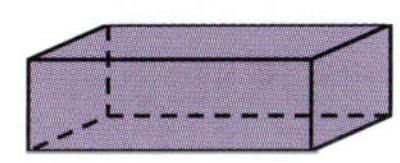

rectangular prism

3 Close

Reaching All Learners

Through Curriculum Integration

Social Studies Ask students in groups to research one famous architectural site, for example, the Great Wall of China, the Roman Colosseum, or the Great Pyramid. Tell each group to find where it is located, when and why it was constructed, and what it was used for. Also have the students name the geometric solids used in the construction and why they think those shapes were used. Have each group share a picture of the structure and their research with the class.

Summarize

You may display the collection of solid figures on Teaching Transparency T22 in the CRB. Have students identify each figure and tell whether it is a *polyhedron*. For each polyhedron have students tell the number of *faces*, *edges*, and *vertices*.

Answers to Think and Discuss

1. Possible answer: Alike: They both have two parallel, flat bases. Different: A cylinder has circular bases, and a prism has polygon-shaped bases.
2. Possible answer: Alike: They both have one flat base and both come to a point. Different: A cone has a circular base, and a pyramid has a polygon-shaped base.
3. cube

10-6 PRACTICE & ASSESS

Students may want to refer back to the lesson examples.

Assignment Guide

If you finished Example 1 assign:
Core 1–3, 7–9, 20–23, 29–36
Enriched 1–3, 7–9, 20–23, 29–36

If you finished Example 2 assign:
Core 1–19, 29–36
Enriched 10–36

Notes

Math Background

The vertices, edges, and faces of the polyhedrons discussed in this lesson satisfy a relationship known as Euler's formula. If V stands for the number of vertices, E for the number of edges, and F for the number of faces, then Euler's formula states that $V - E + F = 2$.

A regular polyhedron is a solid figure in which all the faces are identical regular polygons. The ancient Greeks suspected, but were unable to prove, that there are only five regular polyhedra. What the Greeks suspected was proven using Euler's formula.

Polyhedrons that satisfy Euler's formula are called *simply connected*.

10-6 Exercises

FOR EXTRA PRACTICE see page 657

internet connect
Homework Help Online
go.hrw.com Keyword: MR4 10-6

GUIDED PRACTICE

See Example 1 Identify the number of faces, edges, and vertices in each solid figure.

1. 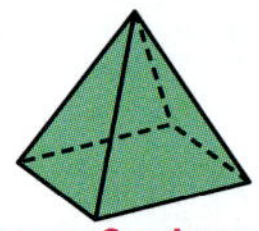
5 faces, 8 edges, 5 vertices

2. 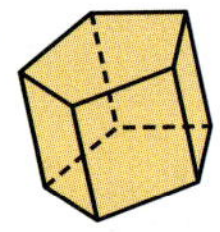
7 faces, 15 edges, 10 vertices

3. 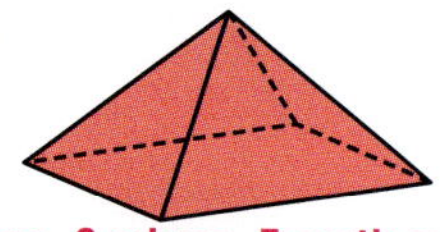
5 faces, 8 edges, 5 vertices

See Example 2 Name the solid figure represented by each object.

4. 
hexagonal prism

5. 
square pyramid

6. 
cube

INDEPENDENT PRACTICE

See Example 1 Identify the number of faces, edges, and vertices in each solid figure.

7. 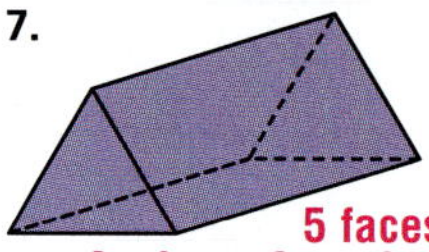
5 faces, 9 edges, 6 vertices

8. 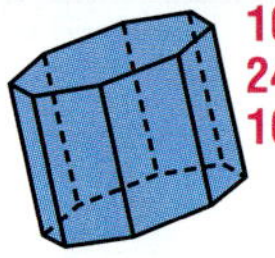
10 faces, 24 edges, 16 vertices

9. 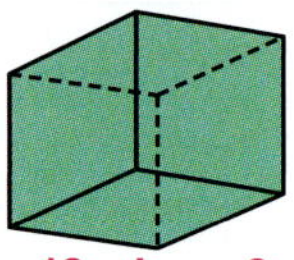
6 faces, 12 edges, 8 vertices

See Example 2 Name the solid figure represented by each object.

10. 
cylinder

11. 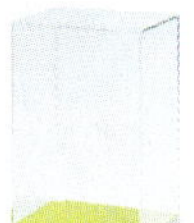
rectangular prism

12. 
cone

PRACTICE AND PROBLEM SOLVING

Name each figure and tell whether it is a polyhedron.

13. 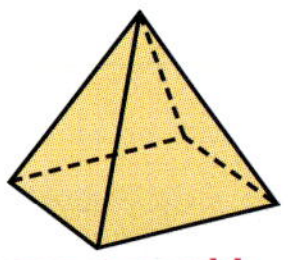
square pyramid, yes

14. 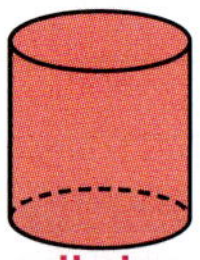
cylinder, no

15. 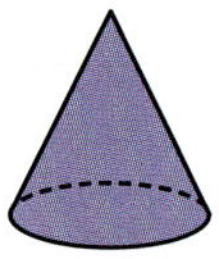
cone, no

Write the letter of each figure described.

16. prism A and D
17. has triangular faces B, C and D
18. has 6 faces A
19. has 5 vertices B

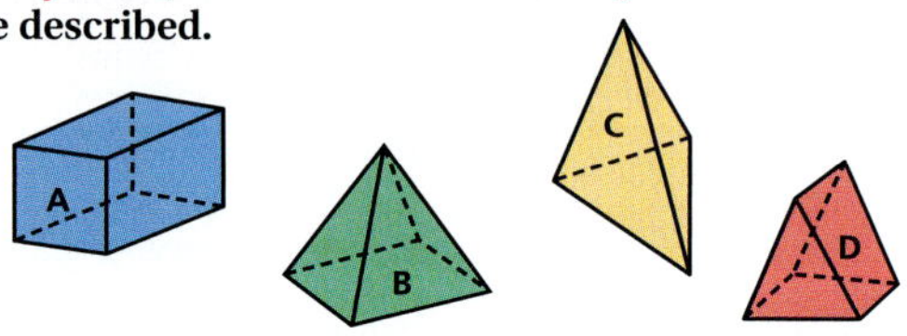

RETEACH 10-6

LESSON 10-6 **Reteach**
Solid Figures

A polyhedron is a solid figure. A polyhedron has flat surfaces, called faces. Faces that share a side form an edge. The point at which three or more edges meet is called a vertex.

This polyhedron has 4 faces, 4 vertices, and 6 edges.

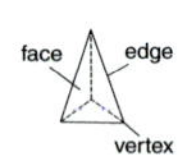

Identify the number of faces, edges, and vertices on each solid figure.

1. 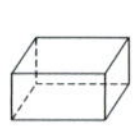
6, 12, 8

2. 
5, 8, 5

A prism is a solid figure with two congruent parallel bases. The other faces are all parallelograms.

A pyramid has one base. The other faces are triangles.

A cylinder has two circular parallel bases joined by a curved surface.

A cone has one circular base and a curved surface that comes to a point.

Pyramids and prisms are polyhedrons because each face is a polygon. Cylinders and cones are not polyhedrons because they have curved surfaces.

Tell whether each figure is a polyhedron and name each solid.

3. 
no, cone

4. 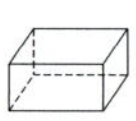
yes, rectangular prism

PRACTICE 10-6

LESSON 10-6 **Practice B**
Solid Figures

Identify the number of faces, edges, and vertices in each solid figure.

1. 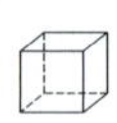
6 faces; 12 edges; 8 vertices

2. 
4 faces 6 edges; 4 vertices

3. 
5 faces; 8 edges; 5 vertices

Tell whether each figure is a polyhedron and name the solid.

4. 
no; cone

5. 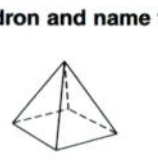
yes; rectangular pyramid

6. 
yes; triangular prism

7. Kelly wants to make a box in the shape of a cube. How many pieces of wood does she need? In what shape should she cut them? Explain.

She needs 6 square pieces of wood because a cube has 6 square faces.

8. Kwan made a sculpture in the shape of a polyhedron. It only has one base that is a triangle. What solid figure is her sculpture? Explain your reasoning.

It is a triangular pyramid, because a pyramid has only 1 base, and the shape of that base defines what kind of pyramid it is.

Write *true* or *false* for each statement.

20. A cone does not have a flat surface. **false**

21. The bases of a cylinder are congruent. **true**

22. All pyramids have five or more vertices. **false**

23. All of the edges of a cube are congruent. **true**

24. ***ARCHITECTURE*** Name the solid figure represented by each building.

a. 
pyramid

b. 
rectangular prism

c. 
cylinder

25. ***HOBBIES*** Li makes candles with her mother. She made a candle in the shape of a pyramid that had 9 faces. How many sides did the base of the candle have? Name the polyhedron formed by the candle. **8; octagonal pyramid**

26. ***WHAT'S THE ERROR?*** A student says that any polyhedron can be named if the number of faces it has is known. What is the student's error? **Possible answer: The number of faces does not tell what shape they have or what solid is formed.**

27. ***WRITE ABOUT IT*** How are a cone and cylinder alike? How are they different? **Both a cone and a cylinder have a circular base. A cone has a vertex and a cylinder does not.**

28. ***CHALLENGE*** The top of a square pyramid is cut off, and the cut is made parallel to the base of the pyramid. What are the shapes of the faces of the new figure? **squares and trapezoids**

Spiral Review

Order the numbers from greatest to least. (Lesson 1-1)

29. 108, 24, 89, 75, 5, 91 **108, 91, 89, 75, 24, 5**

30. 246, 235, 241, 36, 240 **246, 241, 240, 235, 36**

31. 19, 18, 15, 17, 13 **19, 18, 17, 15, 13**

Find each product. (Lesson 3-6)

32. 1.2×8 **9.6**

33. 0.05×0.6 **0.03**

34. 14×0.02 **0.28**

35. 22.1×22.1 **488.41**

36. **TEST PREP** Which of the following types of data should not be displayed using a line graph? (Lesson 6-7) **D**

A The temperature each hour in one day

B A child's height on each of her birthdays

C The price of a computer from 1990 to 2000

D Students' favorite foods

Journal

Have students write brief definitions for the following new vocabulary: *polyhedron*, *face*, *edge*, and *vertex*. Have them draw a picture for each term.

Test Prep Doctor

Encourage students to read Exercise 36 carefully before choosing the answer; they need to identify the type of data that should *not* be displayed in a line graph. Choices **A, B,** and **C** describe data that change over time and therefore *should* be displayed in line graphs.

CHALLENGE 10-6

LESSON 10-6 **Challenge**
Polyhedron Patterns

Complete these charts to discover the polyhedron patterns.

	Triangular Prism	Rectangular Prism	Pentagonal Prism	Hexagonal Prism
Base's Number of Sides	3	4	5	6
Faces	5	6	7	8
Vertices	6	8	10	12
Edges	9	12	15	18

PRISM PATTERNS: If n = the number of sides on the base of a prism, what three expressions show that prism's number of faces, vertices, and edges?

faces $= n + 2$; vertices $= 2n$; edges $= 3n$

	Triangular Pyramid	Rectangular Pyramid	Pentagonal Pyramid	Hexagonal Pyramid
Base's Number of Sides	3	4	5	6
Faces	4	5	6	7
Vertices	4	5	6	7
Edges	6	8	10	12

PYRAMID PATTERNS: If n = the number of sides on the base of a pyramid, what three expressions show that pyramid's number of faces, vertices, and edges?

faces $= n + 1$; vertices $= n + 1$; edges $= 2n$

PROBLEM SOLVING 10-6

LESSON 10-6 **Problem Solving**
Solid Figures

Write the correct answer.

1. Pamela folded an origami figure that has 5 faces, 8 edges, and 5 vertices. What kind of solid figure could Pamela have created?
a rectangular or square pyramid

2. Look at your classroom chalkboard. What kind of solid figure is the board eraser? What kind of solid figure is the chalk?
eraser: rectangular prism; chalk: cylinder

3. If you cut a cylinder in half between its two bases, what two solid figures are formed?
2 cylinders

4. You have two hexagons. How many rectangles do you need to create a hexagonal prism?
6 rectangles

5. A museum needs to ship a sculpture that has a curved lateral surface and one flat circular base. In what shape box should they mail the sculpture?
a cone

6. A glass prism reflects white light as a multicolored band of light called a spectrum. The prism has 5 glass faces with 9 edges and 6 vertices. What kind of prism it it?
a triangular prism

7. All of the faces of a paperweight are triangles. Is this enough information to classify this solid figure? Explain.
Yes, It is a triangular pyramid.

8. Paulo says that if you know the number of faces a pyramid has, you also know how many vertices it has. Do you agree? Explain.
Yes; A pyramid always has the same number of faces and vertices.

Circle the letter of the correct answer.

9. How is a triangular prism different from a triangular pyramid?
(A) The prism has 2 bases.
B The pyramid has 2 bases.
C All of the prism's faces are triangles.
D The pyramid has 5 faces.

10. Which of these statements is not true about a cylinder?
F It has 2 circular bases.
G It has a curved lateral surface.
H It is a solid figure.
(J) It is a polyhedron.

Lesson Quiz

1. Identify the number of faces, edges, and vertices in the figure shown. 8 faces, 18 edges, and 12 vertices

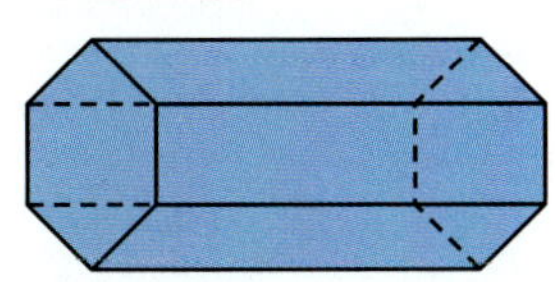

Identify the figure described.

2. two congruent circular faces connected by a curved surface cylinder

3. one flat circular face and a curved lateral surface that comes to a point cone

Available on Daily Transparency in CRB

10C Model Solid Figures

Pacing: Traditional 1 day
Block $\frac{1}{2}$ day

Objective: To use paper, scissors, and tape to make nets and form them into solid figures

Materials: Heavy paper, scissors, tape

Lab Resources

Hands-On Lab Activities . . . . p. 86

Using the Pages

Discuss with students how to create a rectangular prism out of paper. Tell students that their cutouts must be precise for the prism to fit together correctly.

1. Why are the rectangles always cut out in matching pairs? You need two of each rectangle because opposite sides of a rectangular prism are congruent rectangles.
2. What shapes would you need to cut out to build a triangular prism out of paper? 2 triangles, 3 rectangles
3. What shapes would you need to cut out to build a cube? 6 identical squares

Model Solid Figures

Use with Lesson 10-6

You can build a solid figure by cutting its faces from paper, taping them together, and then folding them to form the solid. A pattern of shapes that can be folded to form a solid figure is called a *net*.

Activity 1

1. To make a pattern for a rectangular prism follow the steps below.

   a. Draw the following rectangles and cut them out:

   Two 2 in. × 3 in. rectangles

   Two 1 in. × 3 in. rectangles

   Two 1 in. × 2 in. rectangles

   b. Tape the pieces together to form the prism.

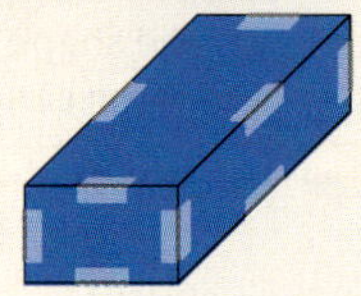

   c. Remove the tape from some of the edges so that the pattern lies flat.

2. Name the solid figure that can be formed by the net shown.

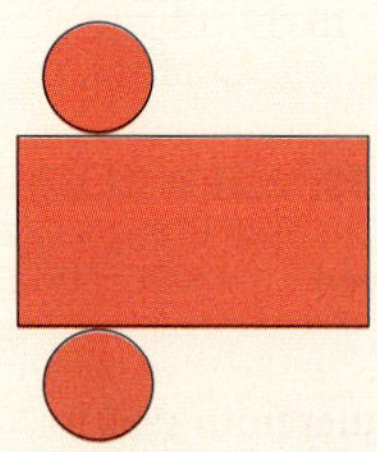

There are circular faces, so the solid is not a prism or a pyramid. Because there are two circular faces, the solid is not a cone. It could be a cylinder.

The net shows one rectangular face that can be curved to form a cylinder.

The net can form a cylinder.

Teacher to Teacher

Sara Höfler
Orlando, Florida

Give each student a black-line net on colored paper. Then have each student use the net to work through each step of the problems listed under Try This. Have them use the scientific method to analyze the steps they take to solve the problems. As students go through this process, they can write a detailed account of their actions and thought processes. This activity is a great way to integrate science and language arts into the math curriculum.

3 Tell whether the net can form a prism.

The net shows two pentagonal-shaped bases that appear to be congruent.

All other faces are rectangles. The net might form a pentagonal prism.

The base has five sides, but there are only four rectangular faces.

The net cannot form a prism.

Think and Discuss

1. Compare the nets for a rectangular prism and a cube.
2. Tell what shapes will always appear in a net for a pyramid. **triangles**
3. Tell what shapes will always appear in a net for a prism. **rectangles**

Try This

Tell whether each net can be folded to form a cube. If not, explain.

1. 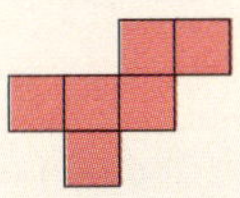 **yes**

2. 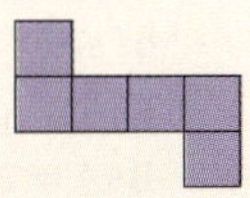 **yes**

3. 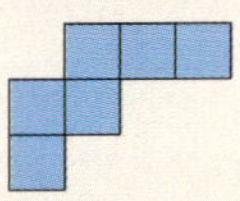 **no**

4. 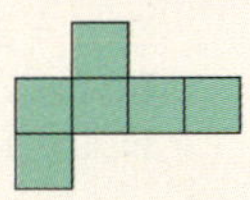 **yes**

Name the solid that can be formed from each net.

5. 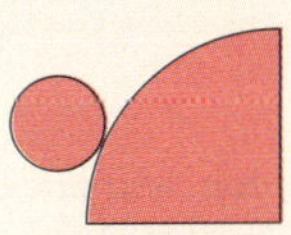 **cone**

6. 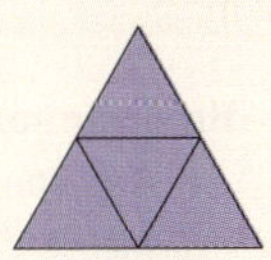 **triangular pyramid**

7. 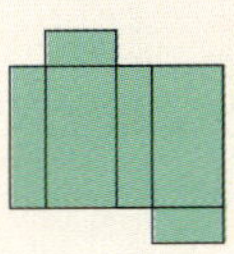 **rectangular prism**

8. 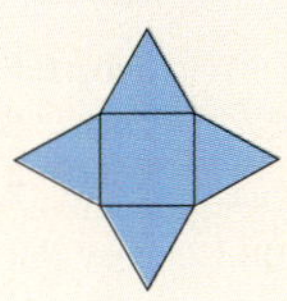 **square pyramid**

Answers

Think and Discuss

1. Both cubes and rectangular prisms have six faces, so the same number of polygons must be used to construct each. But a cube is made from 6 identical squares, while a rectangular prism can have up to 3 pairs of different-sized rectangles for its sides.

10-7 Organizer

Pacing: Traditional 1 day
Block $\frac{1}{2}$ day

Objective: Students find the surface areas of prisms, pyramids, and cylinders.

Indiana Academic Standards
6.3.2, 6.3.4, 6.3.5, 6.5.4, 6.5.5, 6.5.8

Warm Up

Identify the figure described.

1. two parallel congruent faces, with the other faces being parallelograms prism
2. a polyhedron that has a vertex and a face at opposite ends, with the other faces being triangles pyramid

Problem of the Day

Which figure has the longer side and by how much, a square with an area of 81 ft^2 or a square with perimeter of 84 ft? A square with a perimeter of 84 ft; by 12 ft

Available on Daily Transparency in CRB

Math Humor

Teacher: A net can help you calculate the surface area of a polyhedron.

Student: Annette who?

10-7 Surface Area

Learn to find the surface areas of prisms, pyramids, and cylinders.

Vocabulary
surface area
net

Katie made a toy for her cat to scratch by attaching carpet to the faces of a wooden box. The amount of carpet needed to cover the box is equal to the surface area of the box.

The **surface area** of a solid figure is the sum of the areas of its surfaces. To help you see all the surfaces of a solid figure, you can use a *net*. A **net** is the pattern made when the surface of a solid is layed out flat showing each face of the figure.

EXAMPLE 1 Finding the Surface Area of a Prism

Find the surface area *S* of each prism.

A **Method 1: Use a net.**

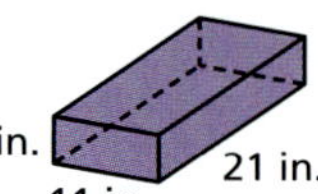

Draw a net to help you see each face of the prism.

Use the formula $A = \ell w$ to find the area of each face.

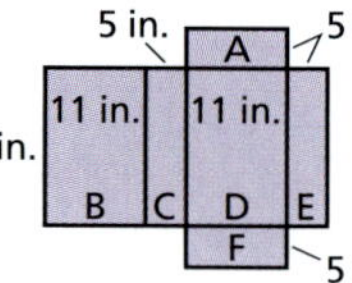

A: $A = 11 \times 5 = 55$
B: $A = 21 \times 11 = 231$
C: $A = 21 \times 5 = 105$
D: $A = 21 \times 11 = 231$
E: $A = 21 \times 5 = 105$
F: $A = 11 \times 5 = 55$

$S = 55 + 231 + 105 + 231 + 105 + 55 = 782$ *Add the areas of each face.*

The surface area is 782 in^2.

B **Method 2: Use a three-dimensional drawing.**

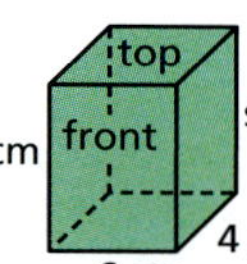

Find the area of the front, top, and side, and multiply each by 2 to include the opposite faces.

Front: $6 \times 8 = 48 \longrightarrow 48 \times 2 = 96$
Top: $6 \times 4 = 24 \longrightarrow 24 \times 2 = 48$
Side: $4 \times 8 = 32 \longrightarrow 32 \times 2 = 64$

$S = 96 + 48 + 64 = 208$ *Add the areas of the faces.*

The surface area is 208 cm^2.

1 Introduce

Alternate Opener

EXPLORATION

10-7 Surface Area

You can use grid paper to make nets that cover boxes, or rectangular solids. The area of the net is the ***surface area of the solid.***

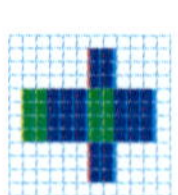

1. Find the combined area of the blue rectangles (the sides of the box).
2. Find the combined area of the green rectangles (the top and bottom of the box).
3. Add the areas you found in numbers 1 and 2. This is the surface area of the box.
4. On the grid below, draw a different net that can cover a box, and find its surface area.

Think and Discuss

5. **Explain** how you can use a net to find surface area.

Motivate

Have students examine polyhedra and name the polygon faces of each. For example, a square pyramid has one square face and four triangular faces. You may use Teaching Transparency T22 in the Chapter 10 Resource Book. Ask students to explain how to find the area of each individual face. Review formulas for area presented earlier in this chapter. Point out to students that finding the surface area of hexagonal prisms, cones, and other solid figures will be studied in a later geometry course.

Exploration worksheet and answers on Chapter 10 Resource Book pp. 64 and 117

2 Teach

Lesson Presentation

Guided Instruction

In this lesson, students learn to find the surface area of prisms, pyramids, and cylinders. Teach students to use nets and three-dimensional drawings to identify all of the faces of the figures.

Teaching Tip Discuss why the curved surface of a cylinder is a rectangle. Point out that the dimensions of the rectangle are the circumference of the cylinder and the height of the cylinder.

The surface area of a pyramid equals the sum of the area of the base and the areas of the triangular faces. To find the surface area of a pyramid, think of its net.

EXAMPLE 2 Finding the Surface Area of a Pyramid

Find the surface area S of the pyramid.

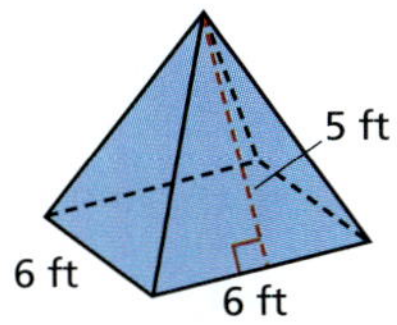

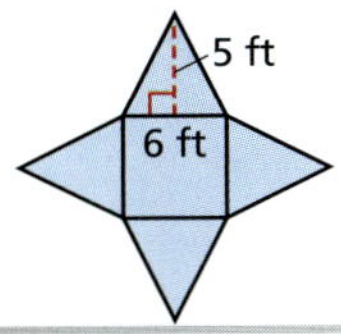

S = area of square + 4 × (area of triangular face)

$S = s^2 + 4 \times \left(\frac{1}{2}bh\right)$

$S = 6^2 + 4 \times \left(\frac{1}{2} \times 6 \times 5\right)$ *Substitute.*

$S = 36 + 4 \times 15$

$S = 36 + 60$

$S = 96$

The surface area is 96 ft^2.

The surface area of a cylinder equals the sum of the area of its bases and the area of its curved surface.

EXAMPLE 3 Finding the Surface Area of a Cylinder

Find the surface area S of the cylinder. Use 3.14 for π, and round to the nearest hundredth.

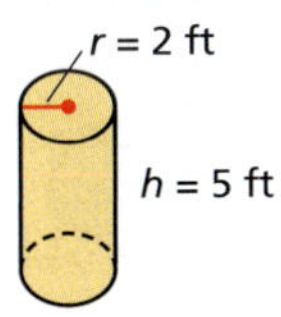

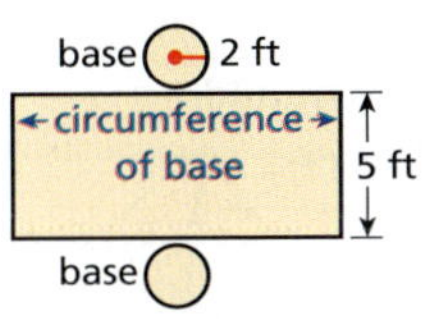

S = area of lateral surface + 2 × (area of each base)

$S = h \times (2\pi r) + 2 \times (\pi r^2)$

$S = 5 \times (2 \times \pi \times 2) + 2 \times (\pi \times 2^2)$ *Substitute.*

$S = 5 \times 4\pi + 2 \times 4\pi$

$S \approx 5 \times 4(3.14) + 2 \times 4(3.14)$ *Use 3.14 for π.*

$S \approx 5 \times 12.56 + 2 \times 12.56$

$S \approx 62.8 + 25.12$

$S \approx 87.92$

The surface area is about 87.92 ft^2.

Helpful Hint

To find the area of the curved surface of a cylinder, multiply its height by the circumference of the base.

Think and Discuss

1. **Describe** how to find the surface area of a pentagonal prism.
2. **Tell** how to find the surface area of a cube if you know the area of one face.

COMMON ERROR ALERT

Students may omit faces when finding the surface areas of solid figures. To remedy this, have them begin by listing the faces and checking them off as they find the area of each.

Additional Examples

Example 1

Find the surface area S of each prism.

A.

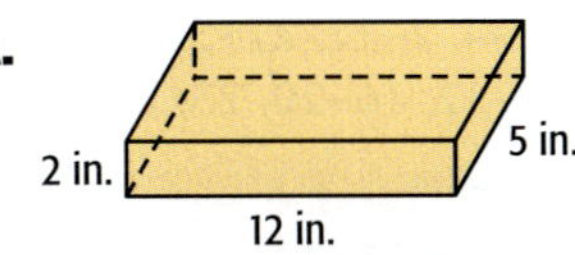

188 in^2

B.

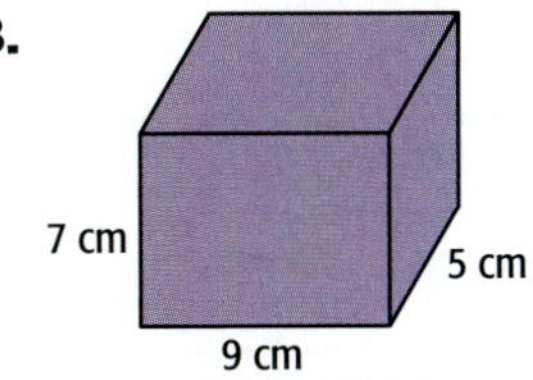

286 cm^2

Example 2

Find the surface area S of the pyramid.

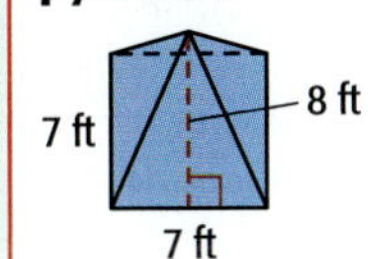

161 ft^2

Example 3

Find the surface area S of the cylinder.

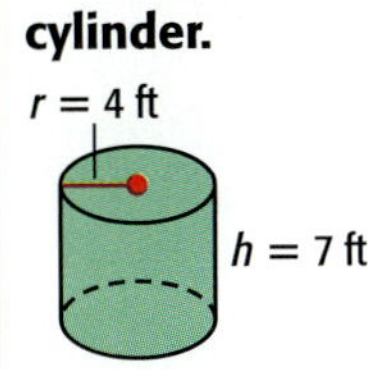

276.32 ft^2

3 Close

Reaching All Learners

Through Concrete Manipulatives

Give students the nets provided on pp. 103–104 in Chapter 10 Resource Book. Have them cut out the nets and assemble them to make models of the solid figures. Students can then use the graph paper squares to estimate surface area before computing the actual surface area. Comparing their actual answers with the estimates will help them determine if their answers are reasonable.

Summarize

Review the new vocabulary terms *surface area* and *net.* Ask students how a net can help when finding surface area of a solid figure.

Possible answer: A net can help you see each face of the solid figure, and you need to add the areas of the faces to find the surface area.

Answers to Think and Discuss

1. Find the sum of the areas of its two pentagonal bases and its five rectangular faces.
2. Multiply the area of the face by 6.

10-7 PRACTICE & ASSESS

Students may want to refer back to the lesson examples.

Assignment Guide

If you finished Example 1 assign:
Core 1–2, 7–9, 18–19, 27–32
Enriched 8, 18–19, 23–24, 26–32

If you finished Example 2 assign:
Core 1–4, 7–12, 27–32
Enriched 9–12, 16–20, 23, 27–32

If you finished Example 3 assign:
Core 1–15, 27–32
Enriched 13–32

Notes

10-7 Exercises

FOR EXTRA PRACTICE
see page 657

internet connect
Homework Help Online
go.hrw.com Keyword: MR4 10-7

GUIDED PRACTICE

See Example 1 Find the surface area S of each prism.

1. 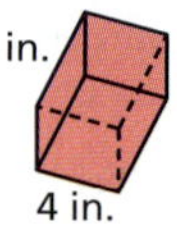

94 in^2

2. 

112 m^2

See Example 2 Find the surface area S of each pyramid.

3. 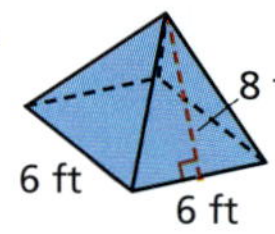

132 ft^2

4. 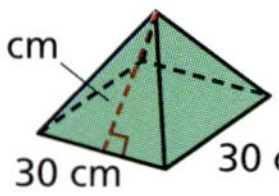

$2{,}640 \text{ cm}^2$

See Example 3 Find the surface area S of each cylinder. Use 3.14 for π, and round to the nearest hundredth.

5. 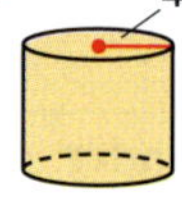

326.56 ft^2

6. 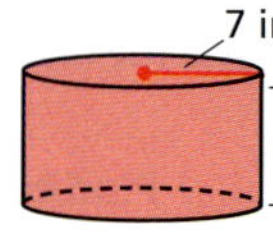

747.32 in^2

INDEPENDENT PRACTICE

See Example 1 Find the surface area S of each prism.

7. 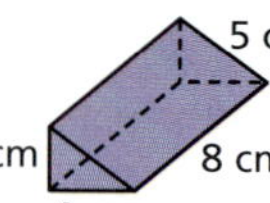

108 cm^2

8. 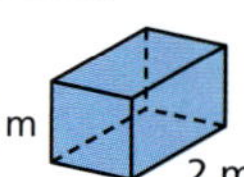

$16\frac{1}{2} \text{ m}^2$

9. 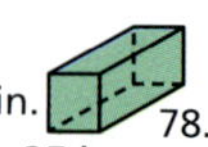

$14{,}650.75 \text{ in}^2$

See Example 2 Find the surface area S of each pyramid.

10. 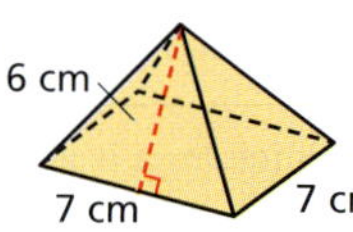

133 cm^2

11. 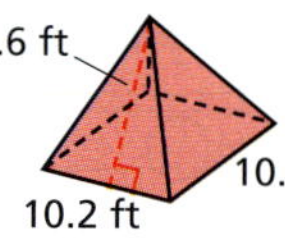

381.48 ft^2

12. 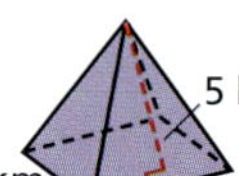

11 km^2

See Example 3 Find the surface area S of each cylinder. Use 3.14 for π, and round to the nearest hundredth.

13. 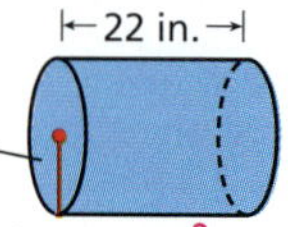

$1{,}274.84 \text{ in}^2$

14. 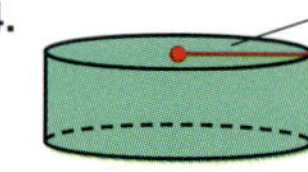

712.717 m^2

15. 

126.39 in^2

Math Background

Just as formulas for the perimeter and area of polygons were developed in Lessons 10-1 and 10-2, nets and hands-on materials can be used here to develop formulas for the surface area of various solid figures.

Among the formulas you and your class can develop are the following:

- a cube with sides s: $S = 6s^2$
- a rectangular prism: $2\ell h + 2wh$, or $2(\ell w + \ell h + wh)$
- a square pyramid with base of length s and lateral faces of height ℓ: $S = s^2 + 2s\ell$, or $s(s + 2\ell)$
- a cylinder with radius r and height h: $S = 2\pi r^2 + 2\pi rh$, or $2\pi r(r + h)$

RETEACH 10-7

LESSON 10-7 **Reteach**
Surface Area

You can use what you know about finding the area of polygon to find the surface area of a solid figure.

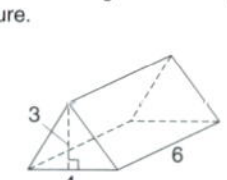

To find the surface area of the prism above, first find the area of each face.

2 congruent triangular bases — 3 rectangular faces

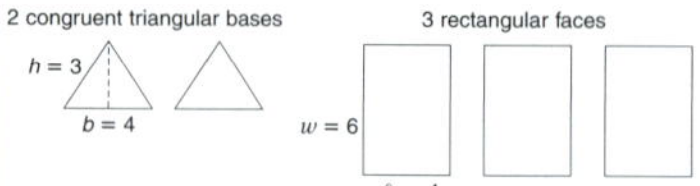

$A = \frac{1}{2}bh$
$= \frac{1}{2}(4 \cdot 3)$
$= 6$ square units

$A = \ell w$
$= 4 \cdot 6$
$= 24$ square units

Then, find the sum of all of the faces of the prism.
$SA = 6 + 6 + 24 + 24 + 24$
$= 84$ square units

Find the surface area of each figure.

1. 

24 square units

2. 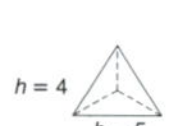

40 square units

PRACTICE 10-7

LESSON 10-7 **Practice B**
Surface Area

Find the surface area S of each prism.

1. 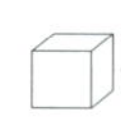

$S = 600 \text{ in}^2$

2. 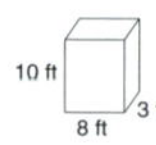

$S = 268 \text{ ft}^2$

Find the surface area S of each pyramid.

3. 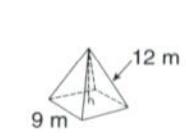

$S = 297 \text{ ft}^2$

4. 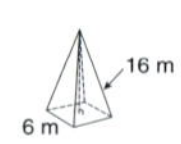

$S = 228 \text{ m}^2$

Find the surface area S of each cylinder. Use 3.14 for π.

5. 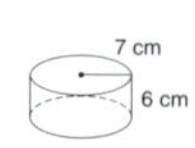

$S \approx 571.48 \text{ cm}^2$

6. 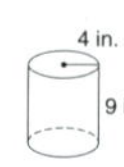

$S \approx 326.56 \text{ in}^2$

7. Why can you find an exact surface area measurement for a prism and pyramid but not for a cylinder?
because to find the area of a circle, you use pi, which is an estimated, not an exact value

8. The surface area of a rectangular prism is 48 square feet. The area of its front is 4 square feet, and the area of one side is 10 square feet. What is the area of the top of the prism?
10 ft^2

PRACTICE AND PROBLEM SOLVING

Find the surface area of each figure.

16. 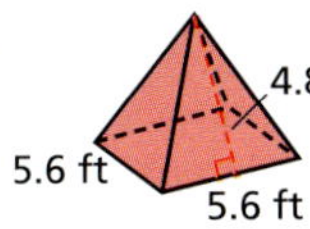

85.12 ft^2

17. 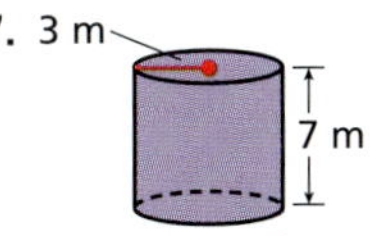

188.4 m^2

18. 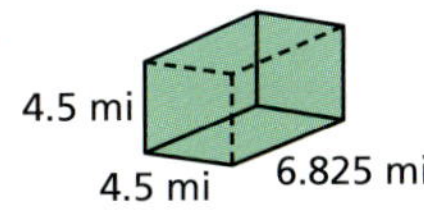

163.35 mi^2

Find the surface area of each solid figure with the given measurements.

19. cube; $s = 1\frac{1}{2}$ km $13\frac{1}{2}\text{ km}^2$

20. square pyramid; base side = 12 m; triangular face height = 8 m 336 m^2

21. cylinder; $d = 10$ in.; $h = 6$ in. 345.4 in^2

22. ***ARCHITECTURE*** The entrance to the Louvre Museum is a glass-paned square pyramid. The width of the base is 34.2 m, and the height of the triangular sides is 27 m. What is the surface area of the glass? $1{,}846.8\text{ m}^2$

23. Find the length, height, and surface area of each rectangular prism.
    a. The length is half the width. The height is half the length. The width is 20 m. $l = 10$ m; $h = 5$ m; 700 m^2
    b. The length is three times the height. The height is one-fourth the width. The width is 12 in. $l = 9$ in.; $h = 3$ in.; 342 in^2

24. ***WHAT'S THE QUESTION?*** The surface area of a cube is 150 cm^2. The answer is 5 cm. What is the question? Possible answer: What is the length of each side of the cube?

25. ***WRITE ABOUT IT*** How is finding the surface area of a rectangular pyramid different from finding the surface area of a triangular prism?

26. ***CHALLENGE*** This cube is made of 27 smaller cubes whose sides measure 1 in.
    a. What is the surface area of the large cube? 54 in^2
    b. Remove one small cube from each of the eight corners of the larger cube. What is the surface area of the solid formed? 54 in^2

Architecture LINK

I. M. Pei is the architect of the pyramid-shaped addition to the Louvre in Paris, France.

go.hrw.com KEYWORD: MR4 Pei

CNN Student News

Spiral Review

27. On a sunny day, a 4-foot-tall girl casts a shadow that is 7.2 feet long. She is standing near a tree that casts a shadow 25.56 feet long. How tall is the tree? (Lesson 8-5) 14.2 ft

Compare. Write <, >, or =. (Lesson 9-2)

28. 0 ▆ −4 >
29. −345 ▆ 7 <
30. −12 ▆ −6 <
31. 14 ▆ 18 <

32. **TEST PREP** Which of the following is a solution to the equation $-7a = 42$? (Lesson 9-8) D

A $a = 6$ B $a = 294$ C $a = 49$ D $a = -6$

Answers

25. The difference is the number of rectangular and triangular faces. A rectangular pyramid has 5 faces, 1 rectangle, and 4 triangles, and a triangular prism has 5 face, 2 triangles, and 3 rectangles.

Journal

Assign each student a prism, pyramid, or cylinder. Have the student draw a picture of the figure, draw its net, label its dimensions, and find its surface area.

Test Prep Doctor

For Exercise 32, students need to remember that division undoes multiplication and that a positive integer divided by a negative integer has a negative quotient. Students who answered **A** divided both sides by 7, not −7. Students who answered **B** multiplied 42 by 7. Students who answered **C** multiplied 7 by 7.

CHALLENGE 10-7

LESSON 10-7 **Challenge**
A Monumental Paint Job

Completed in 1927, the Lincoln Memorial in Washington, D.C., honors President Abraham Lincoln. The monument was built in the style of Greek temples, with 36 columns surrounding the outside of the monument and 8 columns surrounding the statue of Lincoln inside.

It's your job to paint the columns of the Lincoln Memorial. You will paint its exterior columns blue, and its interior columns red. Use the diagram to figure out how much paint you should buy and how much the paint will cost.

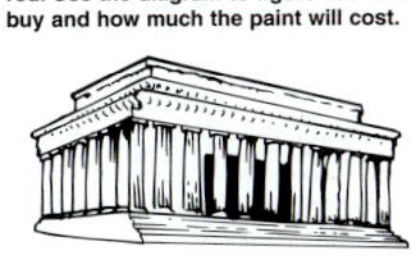

each outside column:
$h = 44$ ft
$r = 3.7$ ft

each inside column:
$h = 50$ ft
$r = 3.75$ ft

Surface Areas to Paint	Gallons of Paint Needed	Cost of Paint
Remember to consider the parts of each column that will **not** be painted.	One gallon of paint will cover 350 square feet. Round **up** to the nearest whole gallon.	Each gallon of paint costs $16.99.
Exterior Columns: $36{,}805.824\text{ ft}^2$	Exterior Columns: 106 gallons	Blue Paint: $1,800.94
Interior Columns: $9{,}420\text{ ft}^2$	Interior Columns: 27 gallons	Red Paint: $458.73

PROBLEM SOLVING 10-7

LESSON 10-7 **Problem Solving**
Surface Area

Write the correct answer.

1. The world's largest cookie was baked in Wisconsin in 1992. Its diameter was 34 feet and contained about 4 million chocolate chips! If the cookie was a cylinder 1 foot tall, and you wanted to cover it with icing, how many square inches would you have to ice? Use 3.14 for π.
   $276{,}721.92\text{ in}^2$

2. The top of the Washington Monument is a square pyramid covered with white marble. Each triangular face is 58 feet tall and 34 feet wide. About how many square feet of marble covers the top of the monument? (The base is hollow.)
   about $3{,}944\text{ ft}^2$ of marble

3. The Parthenon, a famous temple in Greece, is surrounded by large stone columns. Each column is 10.4 meters tall and has a diameter of 1.9 meters. To the nearest whole square meter, what is the surface area of each column (not including the top and bottom)?
   62 m^2

4. The tablet that the Statue of Liberty holds is 7.2 meters long, 4.1 meters wide, and 0.6 meters thick. The tablet is covered with thin copper sheeting. If the tablet was freestanding, how many square meters of copper covers the statue's tablet?
   72.6 m^2 of copper

5. The largest Egyptian pyramid is called the Great Pyramid of Khufu. It has a 756-foot square base and a slant height of 481 feet. What is the total surface area of the faces of the Pyramid of Khufu?
   $727{,}272\text{ ft}^2$

6. A glass triangular prism for a telescope is 5.5 inches tall. Each side of the triangular base is 4 inches long, with a 3-inch height. How much glass covers the surface of the prism?
   78 in^2 of glass

Circle the letter of the correct answer.

7. A can of frozen orange juice is 7.5 inches tall, and its base diameter is 3.5 inches. What size strip of paper is used for its label?
   (A) 84.43 in^2 C 576.98 in^2
   B 26.25 in^2 D 101.66 in^2

8. Tara made fuzzy cubes to hang in her car. Each side of the 2 cubes is 4 inches long. How much fuzzy material did Tara use to make both cubes?
   F 96 in^2 H 16 in^2
   (G) 192 in^2 J 128 in^2

Lesson Quiz

Find the surface area of each figure. Use 3.14 for π.

1. rectangular prism with base length 6 ft, width 5 ft, and height 7 ft 214 ft^2
2. cylinder with radius 3 ft and height 7 ft 188.4 ft^2
3. Find the surface area of the figure shown. 208 ft^2

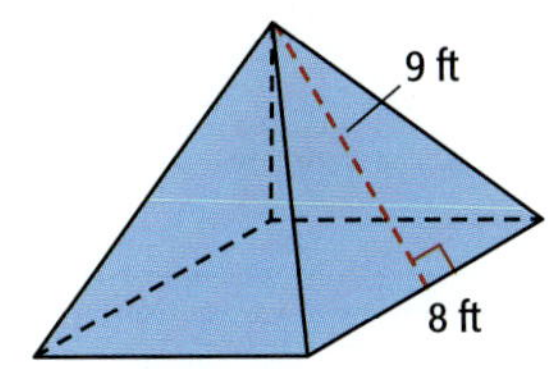

Available on Daily Transparency in CRB

10-8 Organizer

Pacing: Traditional 1 day
Block $\frac{1}{2}$ day

Objective: Students estimate and find the volumes of rectangular prisms and triangular prisms.

Indiana Academic Standards
6.3.2, 6.3.5, 6.5.8

Warm Up

Find the surface area of each figure. Use 3.14 for π.

1. rectangular prism with base length 8 in., width 6 in., and height 12 in. 432 in^2
2. cylinder with diameter 8 ft and height 5 ft 226.08 ft^2

Problem of the Day

A rectangular park is bordered by a 3-foot-wide sidewalk. The park, including the sidewalk, measures 125 ft by 180 ft. What is the area of the park, not including the sidewalk? 20,706 ft^2

Available on Daily Transparency in CRB

Math Humor

Teacher: Use *volume* in a sentence.

Student: My parents are always telling me to turn down the volume on my CD player.

10-8 Finding Volume

Learn to estimate and find the volumes of rectangular prisms and triangular prisms.

Vocabulary
volume

Volume is the number of cubic units needed to fill a space.

It takes 10, or 5 · 2, centimeter cubes to cover the bottom layer of this rectangular prism.

There are 3 layers of 10 cubes each. It takes 30, or 5 · 2 · 3, cubes to fill the prism.

The volume of the prism is 5 cm · 2 cm · 3 cm = 30 cm^3.

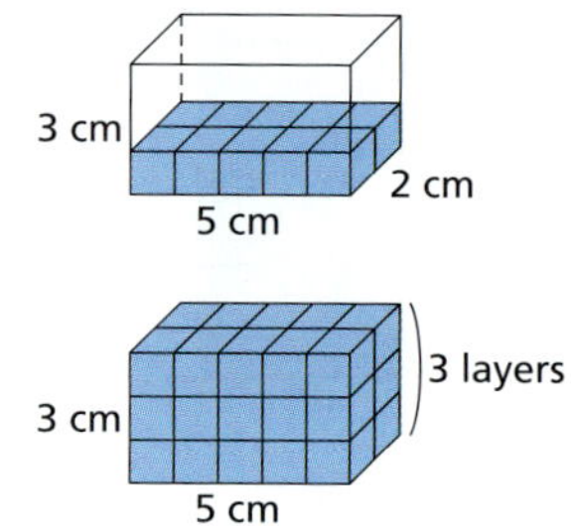

EXAMPLE 1 Finding the Volume of a Rectangular Prism

Find the volume of the rectangular prism.

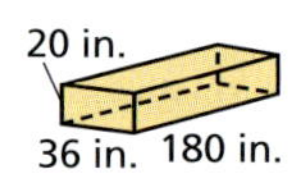

$V = \ell wh$ *Write the formula.*

$V = 180 \cdot 36 \cdot 20$ *$\ell = 180$; $w = 36$; $h = 20$*

$V = 129{,}600 \text{ in}^3$ *Multiply.*

To find the volume of any prism, you can use the formula $V = Bh$, where B is the area of the base, and h is the prism's height. So, to find the volume of a triangular prism, B is the area of the triangular base and h is the height of the prism.

EXAMPLE 2 Finding the Volume of a Triangular Prism

Find the volume of each triangular prism.

A

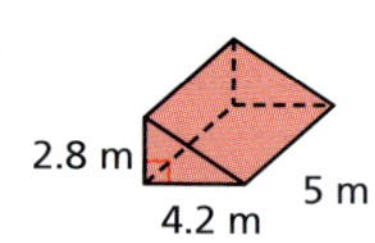

$V = Bh$ *Write the formula.*

$V = \left(\frac{1}{2} \cdot 2.8 \cdot 4.2\right) \cdot 5$ *$B = \frac{1}{2} \cdot 2.8 \cdot 4.2$; $h = 5$*

$V = 29.4 \text{ m}^3$ *Multiply.*

B

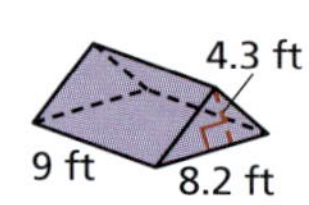

$V = Bh$ *Write the formula.*

$V = \left(\frac{1}{2} \cdot 8.2 \cdot 4.3\right) \cdot 9$ *$B = \frac{1}{2} \cdot 8.2 \cdot 4.3$; $h = 9$*

$V = 158.67 \text{ ft}^3$ *Multiply.*

1 Introduce

Alternate Opener

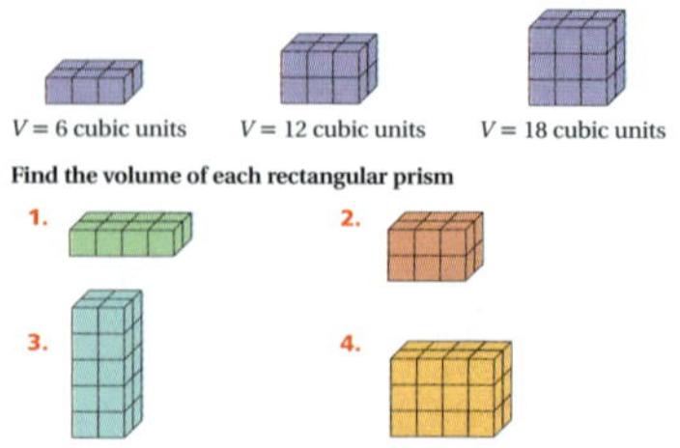

EXPLORATION

10-8 Finding Volume

Volume is the number of cubic units that fill a space. Notice how the volume of a rectangular prism increases as the height increases.

V = 6 cubic units V = 12 cubic units V = 18 cubic units

Find the volume of each rectangular prism

1. 2. 3. 4.

Think and Discuss

5. **Explain** how you found the volume of each rectangular prism.
6. **Discuss** why the formulas V = base · height and V = length · width · height are equivalent.

Motivate

Review that prisms have two congruent bases connected by rectangular faces. Show students a rectangular prism. Ask students to name the prism and explain their answer. It is a rectangular prism because the bases are shaped like rectangles. Show them a triangular prism. Ask them to name the prism and explain their answer. It is a triangular prism because the bases are shaped like triangles.

Exploration worksheet and answers on Chapter 10 Resource Book pp. 73 and 119

2 Teach

Lesson Presentation

Guided Instruction

In this lesson, students learn to estimate and find the volumes of rectangular prisms and triangular prisms. Explain and teach the formula for finding volume of a rectangular prism and the formula for finding volume of a triangular prism. Then teach students to make a model to find dimensions that would yield a given volume.

EXAMPLE 3 PROBLEM SOLVING APPLICATION

PROBLEM SOLVING

A craft supplier ships 12 cubic trinket boxes in a case. What are the possible dimensions for a case of the trinket boxes?

1 Understand the Problem

The **answer** will be all possible dimensions for a case of 12 cubic boxes.

List the **important information:**

- There are 12 trinket boxes in a case.
- The boxes are cubic, or square prisms.

2 Make a Plan

You can make models using cubes to find the possible dimensions for a case of 12 trinket boxes.

3 Solve

Make different arrangements of 12 cubes.

The possible dimensions for a case of 12 cubic trinket boxes are the following: $1 \cdot 1 \cdot 12$, $1 \cdot 2 \cdot 6$, $1 \cdot 3 \cdot 4$, and $2 \cdot 2 \cdot 3$.

4 Look Back

Notice that each dimension is a factor of 12. Also, the product of the dimensions (length · width · height) is 12, showing that the volume of each case is 12 cubes.

Think and Discuss

1. **Explain** how to find the height of a rectangular prism if you know its length, width, and volume.
2. **Tell** how to use mental math strategies to find the volume of a rectangular prism with dimensions 2 cm, 15 cm, and 5 cm.

Additional Examples

Example 1

Find the volume of the rectangular prism.

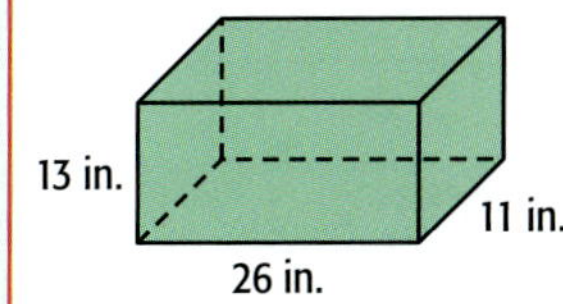

3,718 in^3

Example 2

Find the volume of each triangular prism.

A.

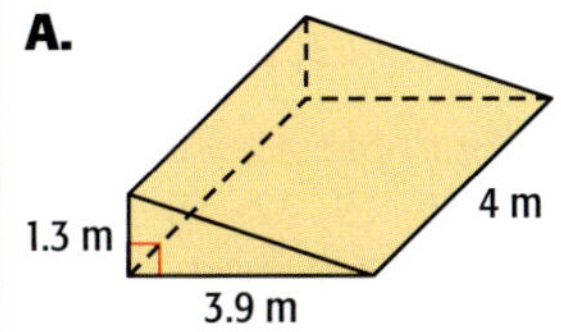

10.14 m^3

B.

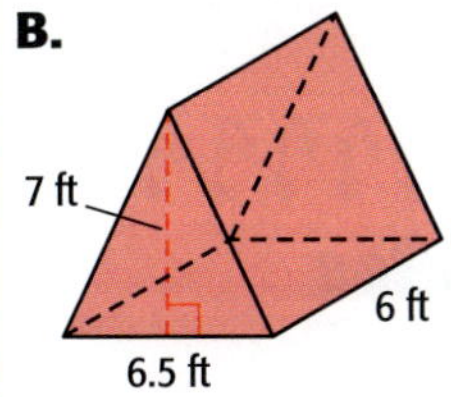

136.5 ft^3

Example 3

Suppose a facial tissue company ships 16 cubic boxes of tissue in each case. What are the possible dimensions for a case of tissue?

$16 \times 1 \times 1$, $8 \times 2 \times 1$, $4 \times 4 \times 1$, and $4 \times 2 \times 2$

3 Close

Reaching All Learners

Through Grouping Strategies

Have students construct rectangular and triangular prisms from nets provided on Chapter 10 Resource Book p. 103. With a partner, students should measure and record the dimensions of their prisms. Then have students compute the volumes of both prisms independently and compare answers with each other. If the answers do not agree, partners should work together to find the correct answers.

Summarize

Have students explain the new vocabulary term, *volume*. Review the formulas for finding the volume of a rectangular prism ($V = \ell wh$) and the volume of a triangular prism ($V = Bh$, where B is the area of the base).

Possible answer: Volume is the number of cubic units needed to fill a space.

Answers to Think and Discuss

1. Possible answer: Divide its volume by the product of its length and width.
2. Possible answer: First multiply $2 \times 5 = 10$, then multiply $10 \times 15 = 150$. So the volume is 150 cm^3.

10-8 PRACTICE & ASSESS

Students may want to refer back to the lesson examples.

Assignment Guide

If you finished Example 1 assign:
Core 1–3, 8–10, 27–33
Enriched 8–10, 18–20, 27–33

If you finished Example 2 assign:
Core 1–6, 8–13, 27–33
Enriched 11–13, 15–23, 27–33

If you finished Example 3 assign:
Core 1–14, 18–19, 27–33
Enriched 11–33

Notes

10-8 Exercises

FOR EXTRA PRACTICE see page 657

internet connect
Homework Help Online
go.hrw.com Keyword: MR4 10-8

GUIDED PRACTICE

See Example 1 Find the volume of each rectangular prism.

1. 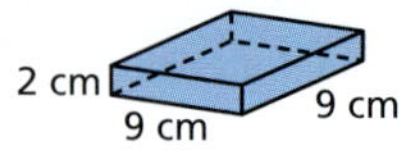

162 cm³

2. 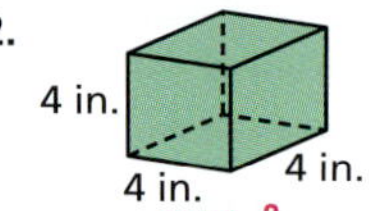

64 in³

3. 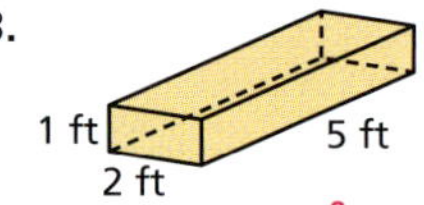

10 ft³

See Example 2 Find the volume of each triangular prism.

4. 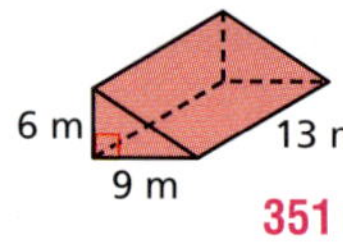

351 m³

5. 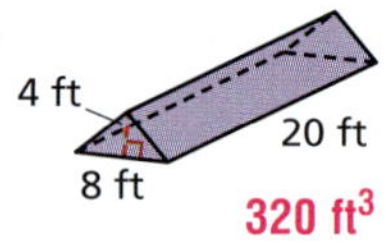

320 ft³

6. 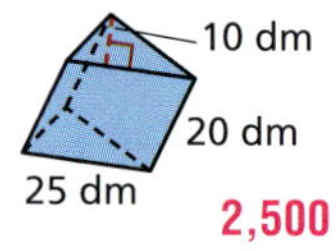

2,500 dm³

See Example 3

7. A toy company packs 10 cubic boxes of toys in a case. What are the possible dimensions for a case of toys? 1 × 1 × 10 and 2 × 5 × 1

INDEPENDENT PRACTICE

See Example 1 Find the volume of each rectangular prism.

8. 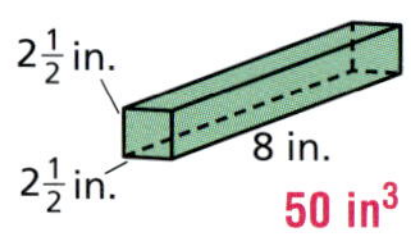

50 in³

9. 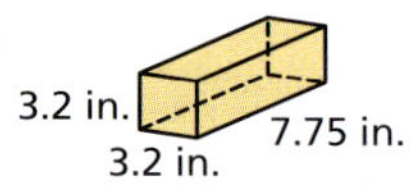

79.36 in³

10. 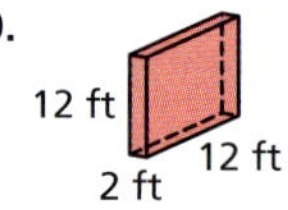

288 ft³

See Example 2 Find the volume of each triangular prism.

11. 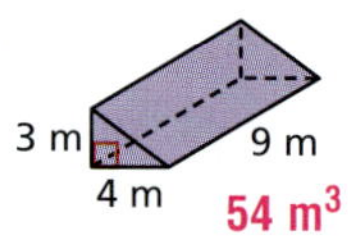

54 m³

12. 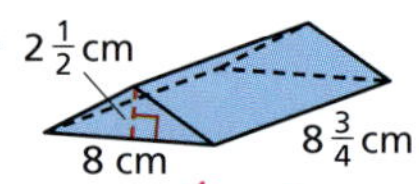

$87\frac{1}{2}$ cm³

13. 71.72 ft³ 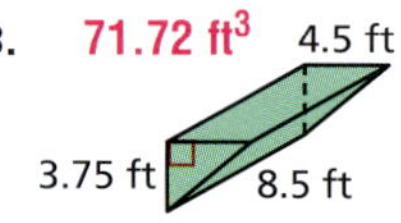

See Example 3

14. A printing company packs 15 cubic boxes of business cards in a larger shipping box. What are the possible dimensions for the shipping box? 1 × 1 × 15, 1 × 3 × 5

PRACTICE AND PROBLEM SOLVING

Find the volume of each figure.

15. 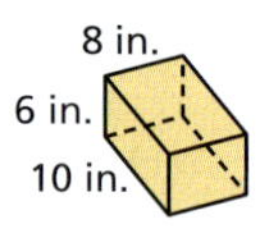

480 in³

16. 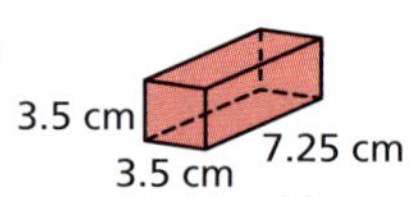

88.81 cm³

17. 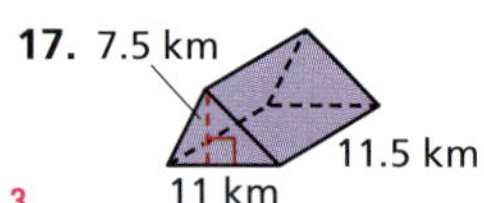

474.375 km³

Find the missing measurement for each rectangular prism.

18. $\ell = \underline{\quad ? \quad}$; $w = 25$ m; $h = 4$ m; $V = 300\text{ m}^3$ 3 m

19. $\ell = 9$ ft; $w = \underline{\quad ? \quad}$; $h = 5$ ft; $V = 900\text{ ft}^3$ 20 ft

Math Background

When students first learned about the exponent 2, they were told that an expression like 5^2 is read "five squared." That is related to the fact that a square with sides of length 5 units has an area of 5^2, or 25 square units.

A similar idea can be explained for this lesson. An expression like 6^3 can be related to a cube whose sides have a length of 6 units. By the formula for the volume of a rectangular prism, this cube has a volume of 6 × 6 × 6, or 6^3 cubic units. Just as the model of a square illustrates why we refer to 5^2 as "five squared," the model of a cube illustrates why we refer to 6^3 as "six cubed."

RETEACH 10-8

LESSON 10-8 **Reteach**
Finding Volume

Volume is the number of cubic units needed to fill a space. To find the volume of a rectangular prism, first find the area of the base.

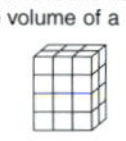

length = 3 units
width = 2 units
$A = \ell w = 3 \cdot 2 = 6$ square units.

The area of the base tells you how many cubic units are in the first layer of the prism.

Next, multiply the result by the number of layers in the prism. The prism has 4 layers, so multiply 6 by 4.

$6 \cdot 4 = 24$

So, the volume of the rectangular prism is 24 cubic units.

Find the volume of each rectangular prism.

1. 
16 square units

2. 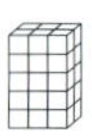
30 square units

To find the area of a triangular prism, first find the area of the base.

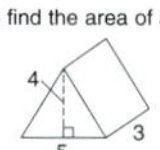

$A = \frac{1}{2}bh$
$= \frac{1}{2}(5 \cdot 4)$
$= 10$ square units

Then multiply the result by the height of the prism.

$10 \cdot 3 = 30$

The volume of the triangular prism is 30 cubic units.

Find the volume of each triangular prism.

3. 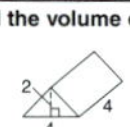

16 square units

4. 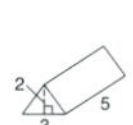

15 square units

PRACTICE 10-8

LESSON 10-8 **Practice B**
Finding Volume

Find the volume of each rectangular prism.

1. 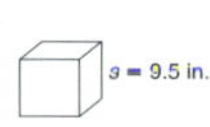

857.375 in³

2. 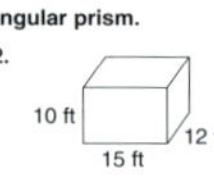

1,800 ft³

3. 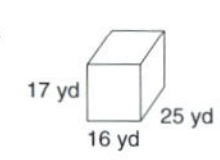

6,800 yd³

4. 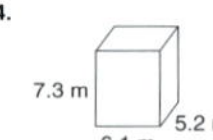

231.556 m³

5. 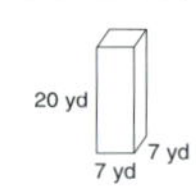

980 yd³

6. 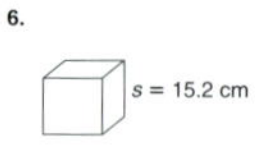

3,511.808 cm³

Find the volume of each triangular prism.

7. 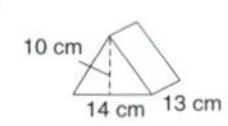

910 cm³

8. 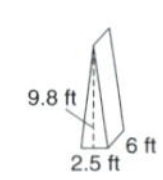

73.5 ft³

9. 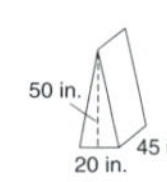

22,500 in³

10. Fawn built a sandbox that is 6 feet long, 5 feet wide, and $\frac{1}{2}$ foot tall. How many cubic feet of sand does she need to fill the box?
15 ft³

11. Unfinished lumber is sold in units called board feet. A board foot is the volume of lumber contained in a board 1 inch thick, 1 foot wide, and 1 foot long. How many cubic inches of wood are in 1 board foot?
144 cubic inches of wood

Physical Science LINK

The density of a substance is a measure of its mass per unit of volume. The density of a particular substance is always the same. The formula for density is the mass of a substance divided by its volume, or $D = \frac{m}{V}$.

Iron filings are attracted by a magnet.

Copper is used in color-coded telephone wires.

20. Find the volume of each substance in the table. 10 cm³, 1 cm³, 3.5 cm³, 300 cm³, 20 cm³

21. Calculate the density of each substance in the table. 8.96 g/cm³, 19.32 g/cm³, 5.02 g/cm³, 0.4 g/cm³, 10.5 g/cm³

22. Water has a density of 1 g/cm³. A substance whose density is less than that of water will float. Which of the substances in the table will float in water? pine

23. A fresh egg has a density of approximately 1.2 g/cm³. A spoiled egg has a density of about 0.9 g/cm³. How can you tell whether an egg is fresh without cracking it open?

24. Alicia has a solid rectangular prism of a substance she believes is gold. The dimensions of the prism are 2 cm by 1 cm by 2 cm, and the mass is 20.08 g.

a. Find the volume of Alicia's substance.

b. Is the substance that Alicia has gold? Explain.

Rectangular Prisms

Substance	Length (cm)	Width (cm)	Height (cm)	Mass (g)
Copper	2	1	5	89.6
Gold	$\frac{2}{3}$	$\frac{3}{4}$	2	19.32
Iron pyrite	0.25	2	7	17.57
Pine	10	10	3	120
Silver	2.5	4	2	210

25. *WRITE ABOUT IT* In a science lab, you are given a prism of copper. You determine that its dimensions are 4 cm, 2 cm, and 6 cm. Without weighing the prism, how can you determine its mass? Explain your answer.

Gold is used to make many pieces of jewelry.

26. *CHALLENGE* A solid rectangular prism of silver has a mass of 84 g. What are some possible dimensions of the prism?
Possible answers: 2 cm × 4 cm × 1 cm, 1 cm × 1 cm × 8 cm, 2 cm × 2 cm × 2 cm

Spiral Review

Find the mean of each set. (Lesson 6-2)

27. 0, 5, 2, −3, 7, 1 2

28. 6, 6, 6, 6, 6, 6, 6, 6, 6, 6, 6, 6 6

Find 20% of each number. (Lesson 8-9)

29. 200 40 **30.** 50 10 **31.** 15 3 **32.** 3,000 600

33. TEST PREP Which sum is negative? (Lesson 9-4) D

A −9 + 10 B 17 + (−4) C 0 + 5 D −3 + (−5)

Answers

23. Check to see whether the egg floats in water; if it does then the egg is spoiled.

24 **a.** 4 cm³
b. Alicia does not have gold because the density of Alicia's substance is 5.02 g/cm³, and the density of gold is 19.32 g/cm³.

25. Find the density of the prism of copper in the table. Because the density of a substance remains the same, multiply the volume of your prism by the density of copper to find your prism's mass.

Journal

Have students write the formulas for volumes of rectangular and triangular prisms and explain what the variables in each formula stand for. Have them draw pictures to demonstrate their understanding.

Test Prep Doctor

For Exercise 33, students need to remember the rules for adding positive and negative integers: the sum takes the sign of the integer with the greater absolute value. Students who answered **A** or **B** probably confused the rules for multiplication with those for addition. Answer **C** can be eliminated because neither addend is negative.

CHALLENGE 10-8

LESSON 10-8 **Challenge**
Fish Tank Math

Did you know that most fish grow according to the size of their tank? In fact, a freshwater fish needs about 1 gallon of water for every inch of its body. The volume of 1 gallon of water is about 231 in³. So, a fish tank with a volume of a 700 in³ will hold 700 ÷ 231 ≈ 3 gallons of water. This size tank would be the healthiest home for a 3-inch-long fish.

Find the volume of each fish tank. Then match each fish to the best tank for it.

Angelfish $\ell = 6$ in | Koi Goldfish $\ell = 24$ in | Redtail Shark $\ell = 5$ in | Clown Loach $\ell = 12$ in

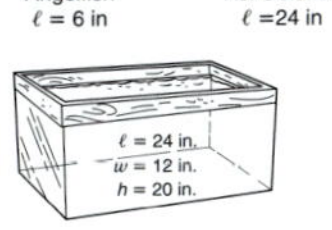

Volume of Tank: 5,760 in³
Gallons of Water: about 25
Fish: koi goldfish

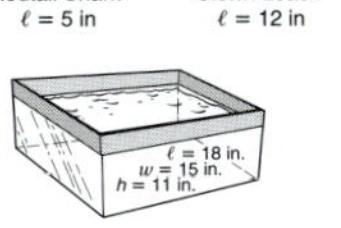

Volume of Tank: 2,970 in³
Gallons of Water: about 13
Fish: clown loach

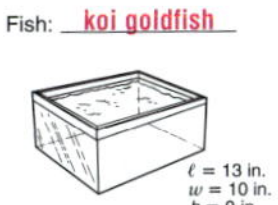

$\ell = 13$ in.
$w = 10$ in.
$h = 9$ in.

Volume of Tank: 1,170 in³
Gallons of Water: about 5
Fish: redtail shark

$\ell = 25$ in.
$w = 8$ in.
$h = 8$ in.

Volume of Tank: 1,600 in³
Gallons of Water: about 7
Fish: angelfish

PROBLEM SOLVING 10-8

LESSON 10-8 **Problem Solving**
Finding Volume

Write the correct answer.

1. At 726 feet tall, Hoover Dam is one of the world's largest concrete dams. In fact, it holds enough concrete to pave a two-lane highway from New York City to San Francisco! The dam is shaped like a rectangular prism with a base 1,224 feet long and 660 feet wide. About how much concrete forms Hoover Dam?
about 586,491,840 ft³ of concrete

2. The Vietnam Veterans Memorial in Washington, D.C., is a 493.5-foot-long wall made of polished black granite engraved with the names of soldiers who died in the war. The wall is 0.25 feet thick and has an average height of 9 feet. About how many cubic feet of black granite was used in the Vietnam Veterans Memorial?
about 1,110.375 ft³ of black granite

3. Benitoite, a triangular prism crystal, is the official state gem of California. One benitoite crystal found in California is 1.2 cm tall, with a base width of 2 cm and a base height of 1.3 cm. How many cubic centimeters of benitoite are in that crystal?
1.56 cm³ of benitoite

4. The Flatiron Building in New York City is a triangular prism. A solid bronze souvenir model of the building is 5 inches tall, with a base height of 1.5 inches and a base width of 2.5 inches. How much bronze was used to make the model?
9.375 in³ of bronze

5. Individual slices of pizza are sold in 2-inch-tall triangular prism boxes. The box base is 8 inches wide, with a 7-inch height. How many cubic inches of pizza will fit in each box?
56 in³ of pizza

6. The world's largest chocolate bar is a huge rectangular prism weighing more than a ton! The bar is 9 feet long, 4 feet tall, and 1 foot wide. How many cubic feet of chocolate does it have?
36 ft³ of chocolate

Circle the letter of the correct answer.

7. A box can hold 175 cubic inches of cereal. If the box is 7 inches long and 2.5 inches wide, how tall is it?
A 25 in. C 17.5 in.
(B) 10 in. D 9.5 in.

8. A triangular prism used to reflect light is made of 120 cm³ of glass. If the prism is 5 centimeters tall, what is the area of each of its triangular bases?
F 24 cm
G 12 cm
H 12 cm²
(J) 24 cm²

Lesson Quiz

Find the volume of each figure.

1. rectangular prism with length 20 cm, width 15 cm, and height 12 cm 3,600 cm³
2. triangular prism with a height of 12 cm and a triangular base with base length 7.3 cm and height 3.5 cm 153.3 cm³
3. Find the volume of the figure shown. 38.13 cm³

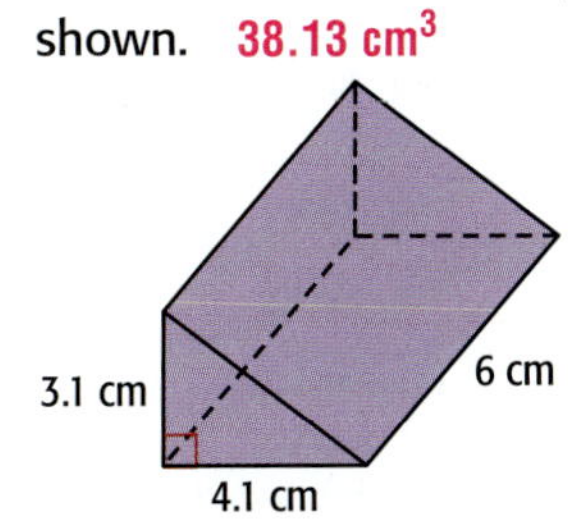

Available on Daily Transparency in CRB

10-9 Organizer

Pacing: Traditional 1 day
Block $\frac{1}{2}$ day

Objective: Students find volumes of cylinders.

Indiana Academic Standards
6.3.2, 6.3.5, 6.5.4, 6.5.5, 6.5.8

Warm Up

Find the volume of each figure described.

1. rectangular prism with length 12 cm, width 11 cm, and height 10 cm **1,320 cm^3**
2. triangular prism with height 11 cm and triangular base with base length 10.2 cm and height 6.4 cm **359.04 cm^3**

Problem of the Day

The height of a box is half its width. The length is 12 in. longer than its width. If the volume of the box is 28 in^3, what are the dimensions of the box? **1 in. × 2 in. × 14 in.**

Available on Daily Transparency in CRB

Math Humor

To get students ready for the lesson on curved solid figures, the math teacher read them the fairy tale *Cylinder-ella.*

10-9 Volume of Cylinders

Learn to find volumes of cylinders.

Thomas Edison invented the first phonograph in 1877. The main part of this phonograph was a cylinder with a 4-inch diameter and a height of $3\frac{3}{8}$ inches.

To find the volume of a cylinder, you can use the same method as you did for prisms: Multiply the area of the base by the height.

volume of a cylinder = area of base × height

The area of the circular base is πr^2, so the formula is $V = Bh = \pi r^2 h$.

EXAMPLE 1 Finding the Volume of a Cylinder

Find the volume V of each cylinder to the nearest cubic unit.

A

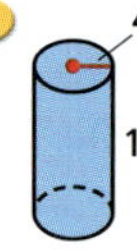

$V = \pi r^2 h$ *Write the formula.*

$V \approx 3.14 \times 4^2 \times 15$ *Replace π with 3.14, r with 4, and h with 15.*

$V \approx 753.6$ *Multiply.*

The volume is about 754 in^3.

B

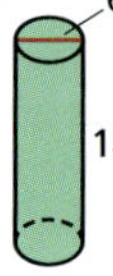

6 ft ÷ 2 = 3 ft *Find the radius.*

$V = \pi r^2 h$ *Write the formula.*

$V \approx 3.14 \times 3^2 \times 18$ *Replace π with 3.14, r with 3, and h with 18.*

$V \approx 508.68$ *Multiply.*

The volume is about 509 ft^3.

C

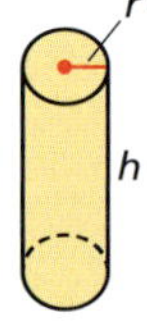

$r = \frac{h}{6} + 1$ *Find the radius.*

$r = \frac{24}{6} + 1 = 5$ *Substitute 24 for h.*

$V = \pi r^2 h$ *Write the formula.*

$V \approx 3.14 \times 5^2 \times 24$ *Replace π with 3.14, r with 5, and h with 24.*

$V \approx 1{,}884$ *Multiply.*

The volume is about 1,884 cm^3.

1 Introduce

Alternate Opener

EXPLORATION

10-9 Volume of Cylinders

The area of the base of a soup can is 4.9 in^2, and the height is 4 in. To find the volume of this can, multiply the area of the base times the height.

volume = area of base · height ($V = Bh$)

$V = 4.9 \cdot 4 = 19.6$ in^3

The soup can has a volume of 19.6 in^3.

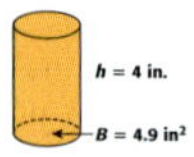

Find the volume of each cylinder.

	Area of Base	Height	Volume = Area of Base · Height
1.	12.6 in^2	8 in.	
2.	28.3 cm^2	10 cm	
3.	3.14 ft^2	2 ft	
4.	113.1 in^2	12 in.	
5.	176.7 cm^2	25 cm	

Think and Discuss

6. **Explain** how to find the volume of a cylinder.
7. **Discuss** why the formulas $V = B \cdot h$ and $V = \text{p} \cdot r^2 \cdot h$ are equivalent (r = radius).

Motivate

Ask students to tell you the formula for surface area of a cylinder. $S = 2\pi rh + 2\pi r^2$ Tell students that the formula for volume of a cylinder involves multiplying the area of the base by the height. To exercise their mathematical reasoning, see whether students can come up with the formula on their own before looking in the text $V = \pi r^2 h$

Exploration worksheet and answers on Chapter 10 Resource Book pp. 82 and 121

2 Teach

Lesson Presentation

Guided Instruction

In this lesson, students learn to find volumes of cylinders. Teach students to use the formula for volume of a cylinder, and have students compare volumes of cylinders. Remind students that they can substitute 3.14 or $\frac{22}{7}$ as an estimated value for *pi.*

EXAMPLE 2 ***Music Application***

The cylinder in Edison's first phonograph had a 4 in. diameter and a height of about 3 in. The standard phonograph manufactured 21 years later had a 2 in. diameter and a height of 4 in. Estimate the volume of each cylinder to the nearest cubic inch.

A **Edison's first phonograph**

4 in. ÷ 2 = 2 in.	*Find the radius.*
$V = \pi r^2 h$	*Write the formula.*
$V \approx 3.14 \times 2^2 \times 3$	*Replace π with 3.14, r with 2, and h with 3.*
$V \approx 37.68$	*Multiply.*

The volume of Edison's first phonograph was about 38 in^3.

B **Edison's standard phonograph**

2 in. ÷ 2 = 1 in.	*Find the radius.*
$V = \pi r^2 h$	*Write the formula.*
$V \approx \frac{22}{7} \times 1^2 \times 4$	*Replace π with $\frac{22}{7}$, r with 1, and h with 4.*
$V \approx \frac{88}{7} = 12\frac{4}{7}$	*Multiply.*

The volume of the standard phonograph was about 13 in^3.

EXAMPLE 3 **Comparing Volumes of Cylinders**

Find which cylinder has the greater volume.

Cylinder 1: $V = \pi r^2 h$

$V \approx 3.14 \times 6^2 \times 12$

$V \approx 1{,}356.48 \text{ cm}^3$

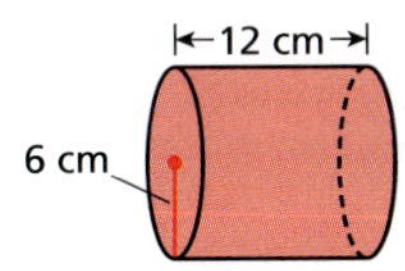

Cylinder 2: $V = \pi r^2 h$

$V \approx 3.14 \times 4^2 \times 16$

$V \approx 803.84 \text{ cm}^3$

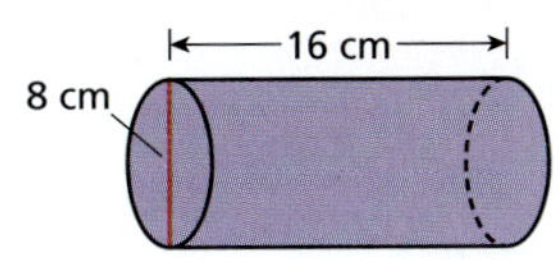

Cylinder 1 has the greater volume because $1{,}356.48 \text{ cm}^3 > 803.84 \text{ cm}^3$.

Think and Discuss

1. **Explain** how the formula for the volume of a cylinder is similar to the formula for the volume of a rectangular prism.
2. **Explain** which parts of a cylinder are represented by πr^2 and h in the formula $V = \pi r^2 h$.

Additional Examples

Example 1

Find the volume V of each cylinder to the nearest cubic unit.

A. $r = 4$ ft, $h = 7$ ft about 352 ft^3

B. $d = 10$ cm, $h = 11$ cm about 864 cm^3

C. $r = \frac{h}{3} + 4$, $h = 9$ in. about 1,385 in^3

Example 2

Ali has a cylinder-shaped pencil holder with a 3 in. diameter and a height of 5 in. Scott has a cylinder-shaped pencil holder with a 4 in. diameter and a height of 6 in. Estimate the volume of each cylinder to the nearest cubic inch.

A. Ali's pencil holder about 35 in^3

B. Scott's pencil holder about 75 in^3

Example 3

Find which cylinder has the greater volume.

Cylinder 1:

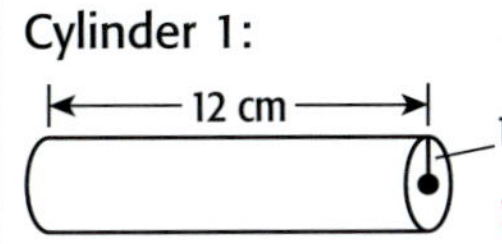

$V \approx 84.78 \text{ cm}^3$

Cylinder 2:

$V \approx 169.56 \text{ cm}^3$

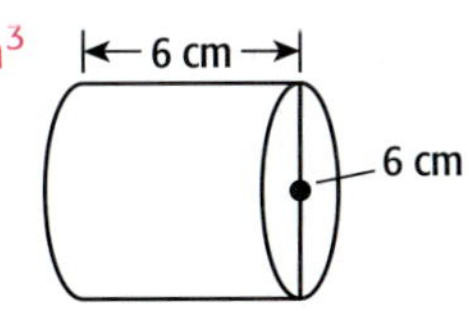

Cylinder 2 has the greater volume because $169.56 \text{ cm}^3 > 84.78 \text{ cm}^3$.

3 Close

Reaching All Learners

Through Home Connection

Discuss common household cylinders, for example, paper towel tubes, toilet paper tubes, oatmeal containers, shampoo bottles, etc. Have students measure and record the radius and height and compute the volume for some common household cylinders. Have students share their findings with the class.

Summarize

Review the formula for volume of a cylinder and work through finding the volume of a cylinder that is 4 inches tall and has a radius of 2 inches.

$V = \pi r^2 h$

$V \approx 3.14 \cdot 2^2 \cdot 4$

$V \approx 50.24 \text{ in}^3$

Answers to Think and Discuss

1. Possible answer: For both, you multiply the height by the area of the base.
2. πr^2 represents the area of the base of the cylinder (a circle); h represents the height of the cylinder.

10-9 PRACTICE & ASSESS

Students may want to refer back to the lesson examples.

Assignment Guide

If you finished Example 1 assign:
Core 1–3, 6–8, 27–32
Enriched 6–8, 21–23, 27–32

If you finished Example 2 assign:
Core 1–4, 6–9, 27–32
Enriched 16–23, 27–32

If you finished Example 3 assign:
Core 1–10, 14–17, 27–32
Enriched 14–32

Notes

10-9 Exercises

FOR EXTRA PRACTICE
see page 657

internet connect
Homework Help Online
go.hrw.com Keyword: MR4 10-9

GUIDED PRACTICE

See Example 1 Find the volume V of each cylinder to the nearest cubic unit.

1. 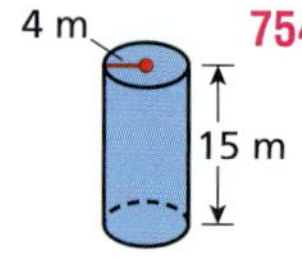

754 m^3

2. 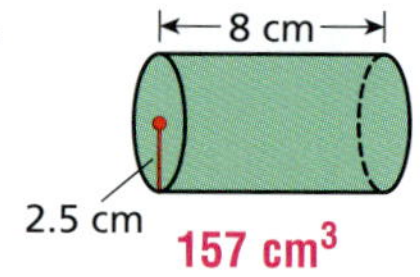

157 cm^3

3. 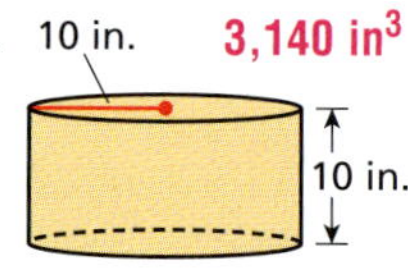

3,140 in^3

See Example 2 **4.** A cylindrical bucket with a diameter of 4 inches is filled with rainwater to a height of 2.5 inches. Estimate the volume of the rainwater to the nearest cubic inch. 31 in^3

See Example 3 **5.** Find which cylinder, A or B, in the diagram has the greater volume. Cylinder B

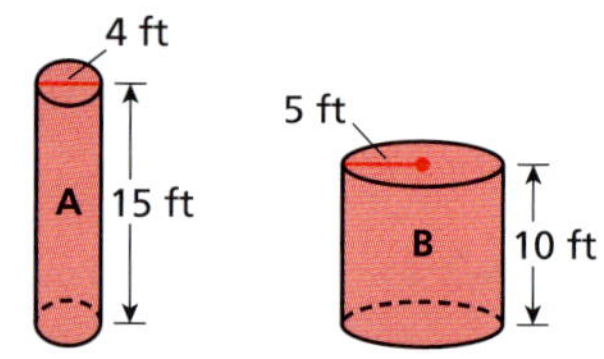

INDEPENDENT PRACTICE

See Example 1 Find the volume V of each cylinder to the nearest cubic unit.

6. 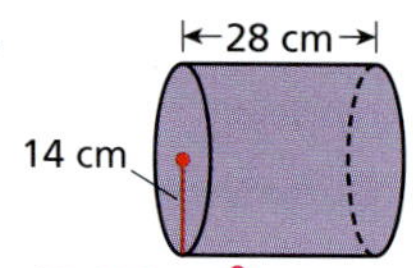

17,232 cm^3

7. 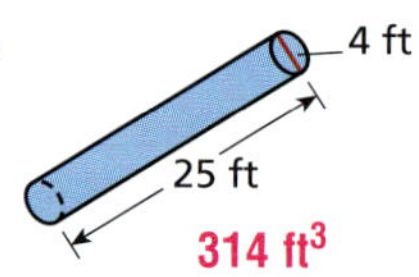

314 ft^3

8. 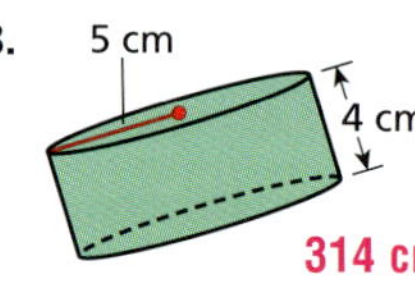

314 cm^3

See Example 2 **9.** Wooden dowels are solid cylinders of wood. One dowel has a radius of 1 cm, and another dowel has a radius of 3 cm. Both dowels have a height of 10 cm. Estimate the volume of each dowel to the nearest cubic cm.
31 cm^3 and 283 cm^3

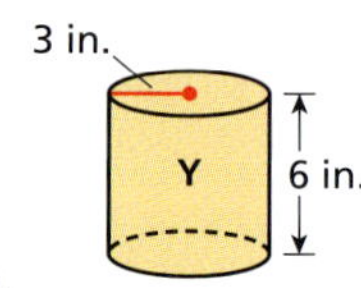

See Example 3 **10.** Find which cylinder, X or Y, in the diagram has the greater volume.
Cylinder X

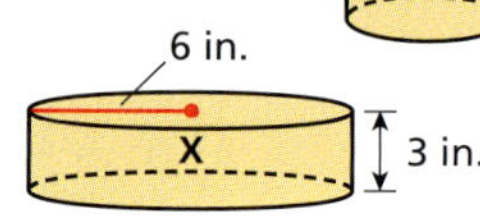

PRACTICE AND PROBLEM SOLVING

Find the volume of each cylinder to the nearest cubic unit.

11. 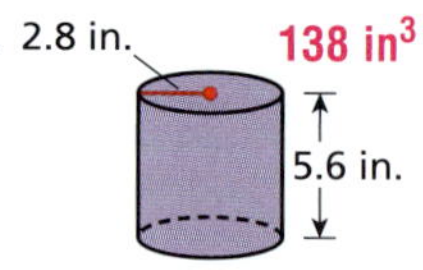

138 in^3

12. 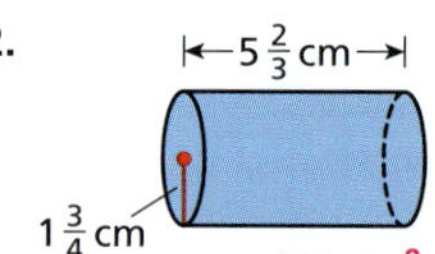

54 cm^3

13. 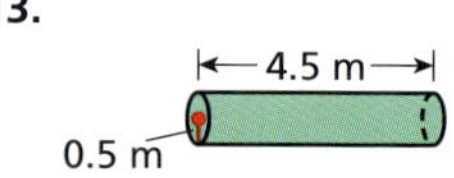

4 m^3

Find the volume of each cylinder using the information given.

14. $r = 6$ cm; $h = 6$ cm 678.24 cm^3 **15.** $d = 4$ in.; $h = 8$ in. 100.48 in^3

16. $r = 7.5$ ft; $h = 11.25$ ft 1,987.03 ft^3 **17.** $d = 12\frac{1}{4}$ yd; $h = 5\frac{3}{5}$ yd 659.67 yd^3

Math Background

The ancient Greek mathematician Archimedes wrote a two-volume work entitled *On the Sphere and the Cylinder*. In it, he proved that a cylinder and sphere with equal heights and diameters have volumes, as well as surface areas, in the ratio of 3:2.

According to the historian Plutarch, Archimedes was so pleased with his proof that he asked his friends to inscribe it on his tomb. And Cicero claims to have found the inscription when he discovered Archimedes's grave more than 100 years later.

RETEACH 10-9

LESSON 10-9 **Reteach**
Volume of Cylinders

You can use what you know about area to help you find the volume of a cylinder.

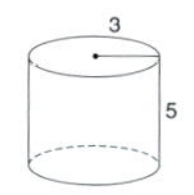

To find the volume of a cylinder, first find the area of the circular base.

$A = \pi r^2$ Use 3.14 for π.
$\approx 3.14 \cdot 3^2$
$\approx 3.14 \cdot 9$
≈ 28.26 square units

The area of the circular base is about 28.26 square units.

Next, multiply your answer by the height of the cylinder. The height of the cylinder is 5 units, so multiply 28.26 by 5.

$28.26 \cdot 5 = 141.3$

So, the volume of the cylinder is about 141.3 cubic units.

Find the volume of each cylinder.

1. (1, 3) 9.42 cubic units
2. (4, 2) 100.48 cubic units
3. (3 cm, 6 cm) 169.56 cubic units
4. (6 ft, 10 ft) 1,130.4 cubic units

PRACTICE 10-9

LESSON 10-9 **Practice B**
Volume of Cylinders

Find the volume V of each cylinder to the nearest whole cubic unit.

1. (6 in., 12 in.) $V \approx 1,356$ in^3
2. (4 ft, 11 ft) $V \approx 553$ ft^3
3. (3 yd, 20 yd) $V \approx 565$ yd^3
4. (2 m, 7.5 m) $V \approx 94$ m^3
5. (1.3 cm, 10 cm) $V \approx 53$ cm^3
6. (2.7 yd, 5.9 yd) $V \approx 135$ yd^3
7. (10 cm, 13 cm) $V \approx 4,082$ cm^3
8. (16 yd, 27 yd) $V \approx 21,704$ yd^3
9. (5 ft, 8 ft) $V \approx 628$ ft^3

10. A cylindrical package of oatmeal is 20 centimeters tall. The diameter of its base is 10 centimeters. About how much oatmeal does the package hold?
about 1,570 cubic centimeters of oatmeal

11. The volume of a can is about 50.24 in^3. The radius of its base is 2 inches. How tall is the can?
4 inches

Find the volume of each interior cylinder to the nearest cubic unit.

18. 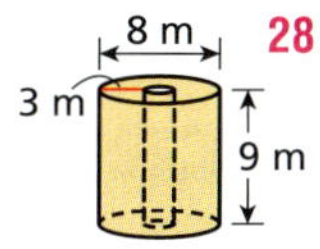

28 m^3

19. 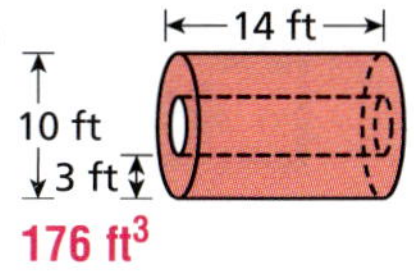

176 ft^3

20. 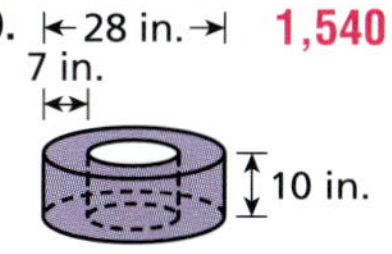

$1{,}540 \text{ in}^3$

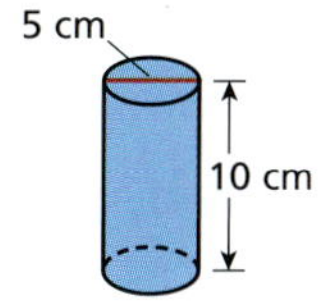

21. ***MEASUREMENT*** Could this blue can hold 200 cm^3 of juice? How do you know? It cannot hold 200 cm^3 of juice because it only has a volume of 196.25 cm^3.

22. ***GARDENING*** Kyle wants to fill a cylindrical planter with soil that costs \$0.01 per cubic inch. The diameter of the planter is 8 inches and the height is 6 inches. About how much will Kyle spend on soil? \$3.00

23. ***SCIENCE*** A scientist filled a cylindrical beaker with 942 mm^3 of a chemical solution. The area of the base of the cylinder is 78.5 mm^2. What is the height of the solution? 12 mm

24. ***CHOOSE A STRATEGY*** Fran, Gene, Helen, and Ira have cylinders with different volumes. Gene's cylinder holds more than Fran's. Ira's cylinder holds more than Helen's, but less than Fran's. Whose cylinder has the largest volume? What color cylinder does each person have?

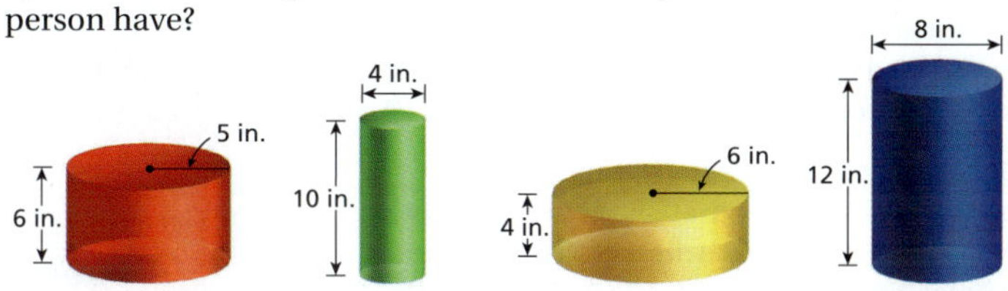

25. ***WRITE ABOUT IT*** Explain why volume is expressed in cubic units of measurement.

26. ***CHALLENGE*** Find the volume of the shaded area. 35.74 cm^3

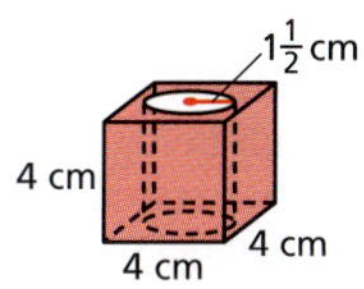

Spiral Review

Determine whether the given value is a solution to each equation. (Lesson 2-3)

27. $2x + 3 = 10$; $x = 4$ no

28. $5(b - 3) = 25$; $b = 8$ yes

Write a fraction whose value is between the given fractions. (Lesson 4-6)

29. $\frac{1}{2}$ and $\frac{5}{6}$ Possible answer: $\frac{4}{6}$

30. $\frac{2}{3}$ and $\frac{5}{12}$ Possible answer: $\frac{7}{12}$

31. $\frac{3}{5}$ and $\frac{9}{10}$ Possible answer: $\frac{3}{4}$

32. **TEST PREP** Which of the following is the best deal? (Lesson 8-1) B

A 2 lb for \$8.40 B 3 lb for \$12.50 C 4 lb for \$17.00 D 5 lb for \$21.00

Answer:

24. Gene has the largest cylinder; Gene–blue Fran–red, Ira–yellow, Helen–green

25. Volume records how much 3-dimensional space an object occupies.

Journal

Have students write an explanation of how to find the volume of a cylinder.

Test Prep Doctor

Students need to compare unit costs to answer Exercise 32. Remind them to divide the total cost by the number of units to find the unit cost. Choice **C** is the worst deal, with a unit cost of \$4.25/lb. Choices **A** and **D** both have unit costs of \$4.20/lb. At \$4.17/lb, choice **B** is the best deal.

CHALLENGE 10-9

LESSON 10-9 **Challenge**
Champion Trees

Trees are among Earth's largest and oldest living things. The National Register of Big Trees measures trees throughout the United States to find the largest of each species—the champions. The four huge trees described below are the only champions that have been listed in the register since the program began in 1940.

Use the register's data to find the circumference and volume of each champion American tree trunk to the nearest foot and cubic foot.

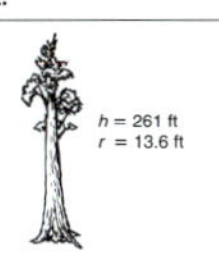

$h = 261$ ft, $r = 13.6$ ft

"General Sherman"
Giant Sequoia, California
Circumference: 85 ft
Volume: $151{,}582 \text{ ft}^3$

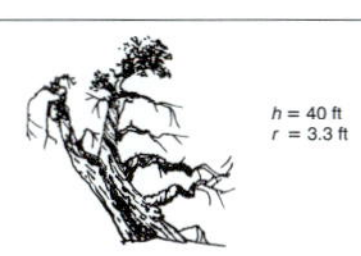

$h = 40$ ft, $r = 3.3$ ft

"Jardine Juniper"
Rocky Mountain Juniper, Utah
Circumference: 21 ft
Volume: $1{,}368 \text{ ft}^3$

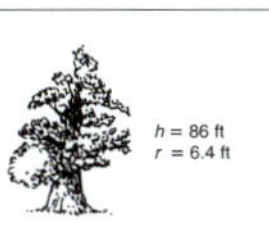

$h = 86$ ft, $r = 6.4$ ft

"Bennett Juniper"
Western Juniper, California
Circumference: 40 ft
Volume: $11{,}061 \text{ ft}^3$

$h = 96$ ft, $r = 5.1$ ft

"Wye Oak"
White Oak, Maryland
Circumference: 32 ft
Volume: $7{,}840 \text{ ft}^3$

PROBLEM SOLVING 10-9

LESSON 10-9 **Problem Solving**
Volume of Cylinders

Write the correct answer.

1. The Hubble Space Telescope was launched into space in 1990. Shaped like a cylinder, the telescope is 15.9 meters long, with a diameter of 4.2 meters. To the nearest whole cubic foot, what is the volume of the Hubble Space Telescope?
about 220 ft^3

2. The Living Color aquarium in Bermuda is the largest freestanding cylindrical aquarium in the Western Hemisphere. With a 10-foot diameter and an 18-foot height, the aquarium holds 10,400 gallons of water! What is the aquarium's volume in cubic feet?
$1{,}413 \text{ ft}^3$

3. In 1902 an American music company built the world's largest music recording cylinder. Nicknamed "Brutus," the cylinder is 5 feet tall, with a 2-foot diameter. What is the volume of the "Brutus" cylinder?
15.7 ft^3

4. The world's largest glass of orange juice was filled in Florida in 1998. At 8 feet tall and with a 2-foot radius, the glass held about 700 gallons of orange juice. What was the volume of that huge glass of orange juice?
100.48 ft^3

5. A large can of soda is 7.5 inches tall and has a 3-inch diameter. A small can of soda is 5 inches tall with a 2.5-inch diameter. To the nearest whole cubic inch, how much more soda does the large can hold?
28 in^3 more soda

6. The maximum length for an official professional baseball bat is 36 inches. Its maximum diameter is 2.6 inches. To the nearest whole cubic inch, what is the maximum volume of a professional baseball bat?
191 in^3

Circle the letter of the correct answer.

7. A cylindrical candle is tightly packed in a rectangular box with a volume of 144 in^3. Which of these could be the dimensions of the candle?
A $h = 6$ in.; $r = 3$ in.
B $h = 2$ in.; $r = 5$ in.
(C) $h = 4$ in.; $r = 3$ in.
D $h = 3$ in.; $r = 4$ in.

8. A can of tennis balls is 21 centimeters tall and has a diameter of 8 centimeters. What is the volume of the tennis ball can?
F 17,408.16 cm^3 H 527.52 cm^3
(G) 1,055.04 cm^3 J 263.76 cm^3

Lesson Quiz

Find the volume of each cylinder to the nearest cubic unit. Use 3.14 for π.

1. radius = 9 ft, height = 4 ft
1017 ft^3

2. radius = 3.2 ft, height = 6 ft
193 ft^3

3. Which cylinder has a greater volume? cylinder B

a. radius 5.6 ft and height 12 ft
$1{,}181.64 \text{ ft}^3$

b. radius 9.1 ft and height 6 ft
$1{,}560.14 \text{ ft}^3$

Available on Daily Transparency in CRB

Problem Solving on Location

Illinois

***Purpose:** To provide additional practice for problem-solving skills in Chapters 1–10*

Fish, Whales, and More

- After problem 1, have students consider the following:
  If the turtle's nest is 1.5 feet deep, what volume of dirt did the turtle have to dig up to make the nest?
  about 10.6 ft^3
- After problem 3, have students determine what portion of a 2,500 lb Beluga whale's weight is *not* blubber.
  1,500 pounds
- After problems 5–7, have students consider the ratio of the diameter of a giant squid's eye to its overall length of 55 ft. Then consider an adult human that is about 68 inches tall. Find the ratio of the diameter of the human's eye to its overall height. How do the two ratios compare? For the giant squid, the ratio is about 16:55, or 1:3.4. For the human, the ratio is about 1:68. The diameter of the giant squid's eye compared with its total length is proportionally much larger than the diameter of the human's eye compared with its height.

Extension Have students make scale models to compare the sizes of the eyes of humans, giant squids, and blue whales. Have them present their work on posters. Check students' work. A student might use a large marble to illustrate the size of their own eye, a child-size basketball for the size of a blue whale's eye, and a large balloon for the size of a giant squid's eye.

Problem Solving on Location

ILLINOIS

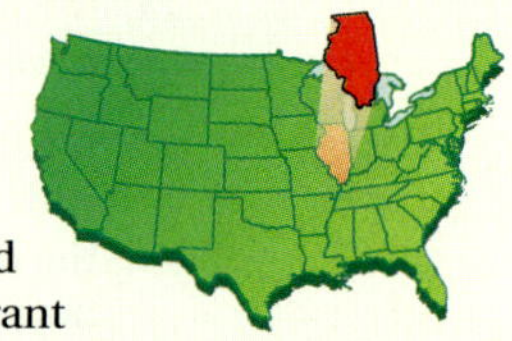

Fish, Whales, and More

The Chicago Bears football team plays at Chicago's Soldier Field, and turtles, sharks, and whales play at the nearby Shedd Aquarium in Grant Park. Since the aquarium opened in 1930, more than 100 million visitors have walked through its doors to view its spectacular exhibits of sea life.

1. If a giant river turtle digs a circular nest 3 feet in diameter, what is the circumference of the nest? (Use 3.14 for π and round to the nearest hundredth.) $C = 9.42$ ft
2. If a turtle swims 24 feet in a circle, what is the radius and diameter of the circle? (Use 3.14 for π and round to the nearest tenth.) $r = 3.8$ ft; $d = 7.6$ ft
3. Beluga whales live in icy habitats. Their bodies have two separate layers of blubber to help them keep warm. If 40% of a whale's body weight is made up of blubber, how heavy is the blubber in a whale that weighs 2,000 pounds? $2{,}000 \times 0.4 = 800$ lb
4. The animals at the aquarium eat about 650 pounds of fish a day. An adult beluga whale eats about 50 pounds of fish a day. In one day, about what fraction of the 650 pounds of fish is eaten by an adult beluga whale? about $\frac{1}{13}$

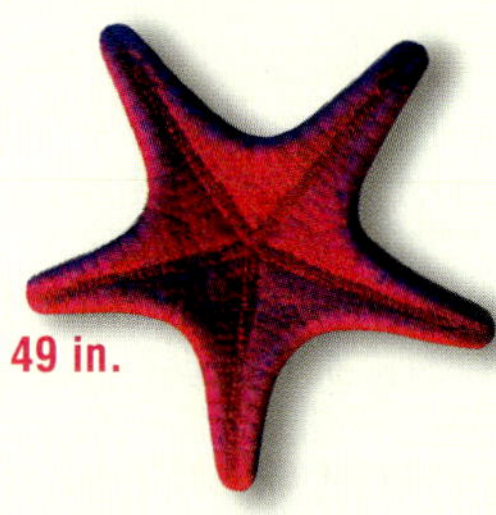

For 5–7, use 3.14 for π, and round to the nearest whole number.

5. The giant squid has the largest eye of any living animal. One giant squid that measures 55 feet in length has an eye with a diameter of 15.74 inches. What is the circumference of the giant squid's eye? about 49 in.
6. One blue whale's eye has a diameter of 4.7 inches. How much greater is the circumference of the giant squid's eye than the circumference of the blue whale's eye? about 34 in.
7. A human's eye has a diameter of 0.94 inches. How much greater is the circumference of the giant squid's eye than the circumference of a human's eye? about 46 in.

Caribbean Reef

The Shedd Aquarium's popular Caribbean Reef exhibit holds 90,000 gallons of water. Inside the tank there are about 250 sea creatures, representing 70 different species.

For 1–4, use the drawing, which shows a typical aquarium window.

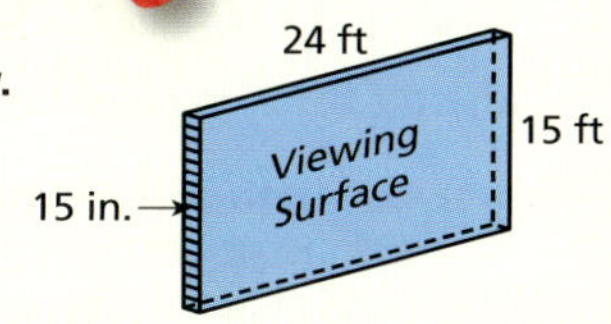

1. What is the perimeter of the viewing surface? **78 ft**
2. What is the area of the viewing surface? **360 ft^2**
3. Classify the solid figure formed by the window. **rectangular prism**
4. What is the volume of the window? **450 ft^3**
5. Suppose the tank in the Caribbean Reef exhibit is shaped like a square prism with a base whose sides are each 30 feet and a height of 16 feet.
   - **a.** What would the area of the viewing surface be? **480 ft^2**
   - **b.** What would the volume of the tank be? **14,400 ft^3**
   - **c.** How many cubic inches of water would the tank hold? **24,883,200 in^3**
6. Suppose the tank in the Caribbean Reef exhibit is shaped like a cylinder with a diameter of 30 feet and a height of 16 feet.
   - **a.** What would the volume of the tank be? (Use 3.14 for π.) **11,304 ft^3**
   - **b.** How many cubic inches of water would the tank hold? **19,533,312 in^3**
7. A gallon of water has a volume of 231 cubic inches. Would either the square-prism tank in problem 5 or the cylinder tank in problem 6 be large enough to hold the water in the tank of the actual Caribbean Reef exhibit? Explain.

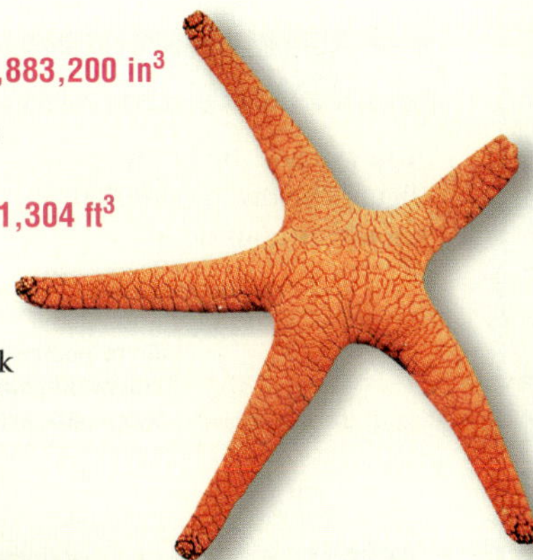

Caribbean Reef

- After problem 5c, have students explain how they arrived at their answers. The volume of the tank is 14,400 ft^3. There are 12^3 = 1,728 cubic inches in a cubic foot, so there are 14,400 × 1,728 cubic inches in the tank.
- After problem 6a, have students compare the volume of the square prism tank in problem 5 with the volume of the cylindrical tank. Point out that both tanks are 30 feet wide at the base and 16 feet tall. Why is the volume of the cylindrical tank less? A circle with diameter 30 feet has less area than a square with side length 30 feet. (This can be illustrated by showing a circle of diameter x inscribed in a square with side length x, and shading the area outside the circle.) And the volume formula, Bh, gives a smaller value when either value, B or h, is reduced.

Extension Have students design an aquarium exhibit, giving dimensions for a tank and viewing windows. Have students write questions about their designs and trade with classmates to solve. Check students' work.

Answers

7. The prism tank in problem 5 will hold 24,883,200 ÷ 231 = 107,719 gal; the cylinder tank in problem 6 will hold 19,533,312 ÷ 231 = 84,560 gal; the prism tank holds more than 90,000 gal, so it would be large enough to hold the water in the Caribbean Reef exhibit.

MATH-ABLES

Game Resources
Puzzles, Twisters & Teasers
Chapter 10 Resource Book

Polygon Hide-and-Seek

Purpose: *To apply the skill of identifying polygons to a puzzle*

Discuss: As students identify each figure, have them justify each choice. Possible answer: For an obtuse scalene triangle, I choose $\triangle ABF$. It is an obtuse triangle because one of its angles, $\angle A$, measures greater than 90°. It is a scalene triangle because none of its three sides are congruent.

Extend: Challenge students to create their own figures like the one in this lesson. Have them make a list of polygons for a classmate to find. Remind students to be sure that each polygon exists in their drawings. Check students' work.

Poly-Cross Puzzle

Purpose: *To practice identifying solid figures in a crossword puzzle*

Discuss: Because spelling is important in this exercise, have students refer to the lessons in Chapter 10 to check spelling, if necessary.

Extend: Have students create a matching card game, making one card for a figure's name and another for a sketch of the figure. They place all cards face down and take turns pulling pairs of cards. They return the cards if they are not a match. If they are, they keep the cards and take another turn.

MATH-ABLES

Polygon Hide-and-Seek

Use the figure to name each polygon described.

1. an obtuse scalene triangle $\triangle ABF$
2. a right isosceles triangle $\triangle BJC$
3. a parallelogram with no right angles *FBJE*
4. a trapezoid with two congruent sides *ABCF*
5. a pentagon with three congruent sides *EFABJ*

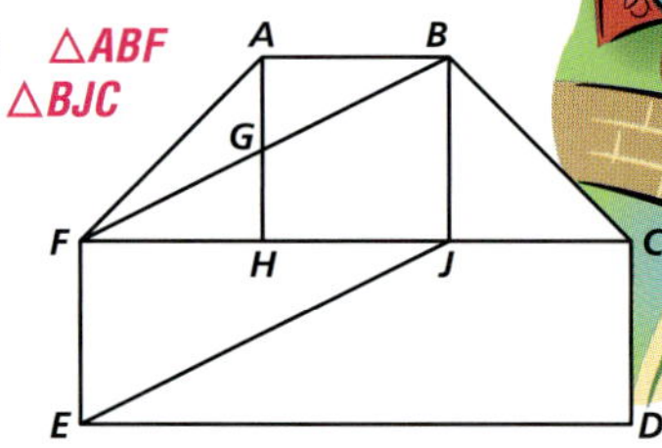

Poly-Cross Puzzle

You will use the names of the figures below to complete a crossword puzzle.

internet connect
Go to ***go.hrw.com*** for a blank crossword puzzle.
KEYWORD: MR4 Game10

ACROSS

1. 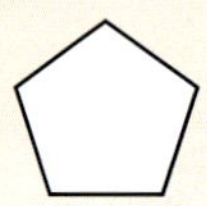 pentagon

2.  square

3. 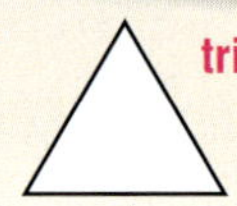 triangle

4. 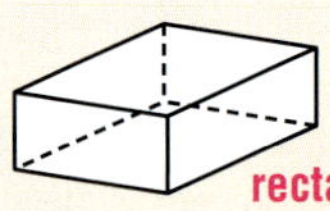 rectangular prism

5. 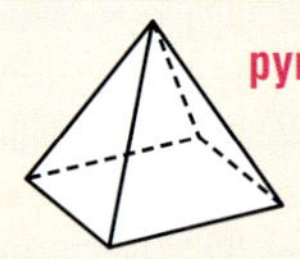 pyramid

6. 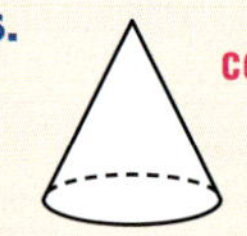 cone

DOWN

1. 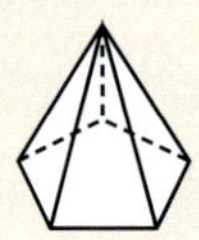 pentagonal pyramid

7.  cylinder

8. 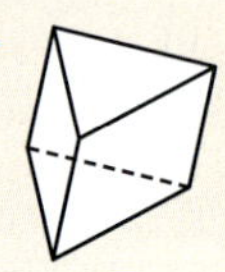 triangular prism

Technology LAB

Area and Perimeter

internet connect
Lab Resources Online
go.hrw.com
KEYWORD: MR4 TechLab10

Geometry software can be used to explore geometric formulas.

Activity

1 Use your geometry software to explore the formula for the area of a rectangle, $A = \ell \cdot w$.

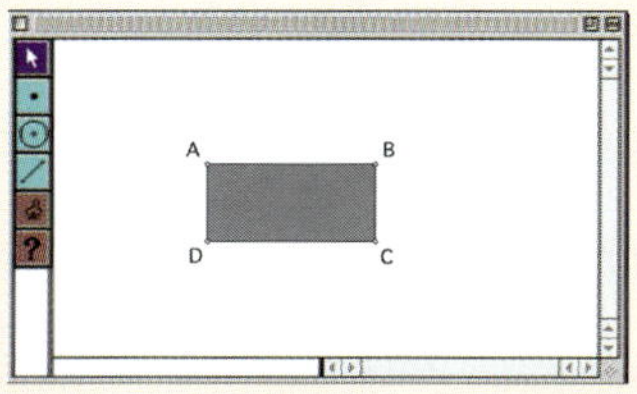

a. Construct a rectangle *ABCD*. Choose four points and connect them with line segments, making sure the opposite sides are parallel.

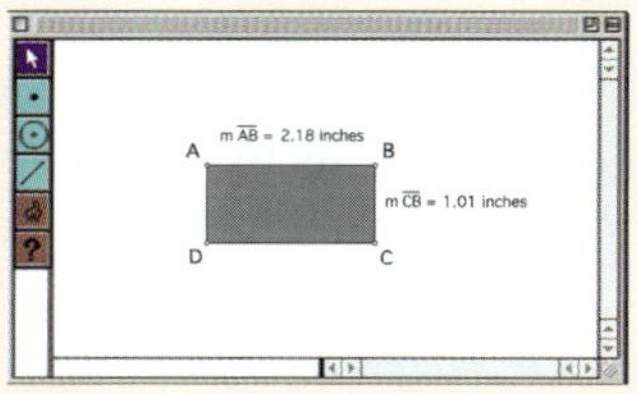

b. Use the distance tool to measure the length of sides $\overline{AB}$ and $\overline{CB}$.

Select the interior of the rectangle, and then use the area tool to measure the area.

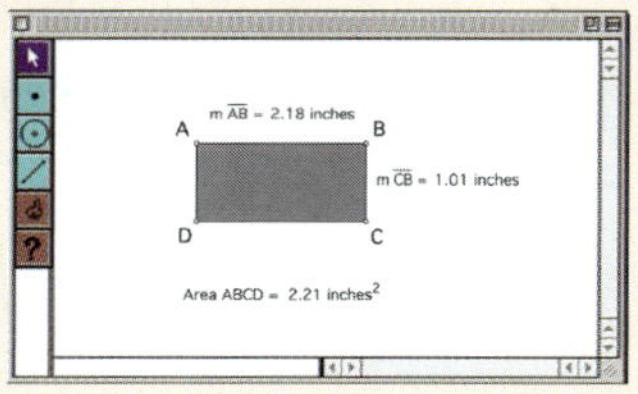

c. Use a calculator or paper and pencil to find the product of the side lengths. Round to the hundredths place.
$2.18 \cdot 1.01 \approx 2.20$

Notice that the geometry software rounds the product to 2.21, which is close to 2.20.
So $Area = AB \cdot CB = \ell \cdot w$.

Think and Discuss

1. Tell whether the perimeter P of rectangle *ABCD* is equal to $2 \cdot (AB + CB)$. **yes**
2. Determine whether the area of rectangle *ABCD* divided by 2 is equal to the perimeter. **No; half of the area of rectangle *ABCD* is not equal to its perimeter.**

Try This

1. Use geometry software to construct a triangle *ABC* where $m\angle B = 90°$.
   a. Measure the area of the triangle and the lengths of sides $\overline{AB}$ and $\overline{CB}$. Find $\frac{1}{2} \cdot AB \cdot CB$. **Check students' work.**
   b. Drag angle *A*, making sure $m\angle B = 90°$. Do this three more times to construct triangles with different areas and side lengths. For each triangle, find $\frac{1}{2} \cdot AB \cdot CB$. What do you conclude? **The area of each triangle is equal to $\frac{1}{2} \cdot AB \cdot CB$.**

Technology LAB Area and Perimeter

Objective: To use geometry software to explore area and perimeter

Materials: Computer, geometry software

Lab Resources

Technology Lab Activities .... p. 44

Using the Page

This technology activity shows students how to use geometry software to measure side lengths and compute area.

The Think and Discuss problems can be used to assess students' ability to interpret the results given by the software. The Try This problem provides students with an opportunity to use geometry software to find information and draw conclusions.

Assessment

Discuss how to use the software to create and explore figures.

1. Create triangle *ABC*, and use the software to find its area. Students will choose three points and connect them with line segments and then click inside the triangle and use the area tool. Areas will vary. Check students' work.
2. Create square *WXYZ* and use a calculator to find its area. Students will create four congruent segments that meet at right angles. They will use the distance tool to measure each side and then use the formula $A = s^2$ to find the square's area.

Chapter 10 Study Guide and Review

Purpose: *To help students review and practice concepts and skills presented in Chapter 10*

Assessment Resources

Chapter Review

Chapter 10 Resource Book . . pp. 90–92

***Test and Practice Generator* CD-ROM**

Additional review items in both multiple-choice and free-response format may be generated for any objective in Chapter 10.

Answers

1. polyhedron
2. volume
3. perimeter; circumference
4. diameter
5. 33.9 in.
6. 6 ft

Chapter 10 Study Guide and Review

Study Guide and Review

Vocabulary

area	504	**net**	530
base	524	**perimeter**	500
center	516	***pi***	516
circle	516	**polyhedron**	524
circumference	516	**prism**	524
cone	525	**pyramid**	525
cylinder	524	**radius (radii)**	516
diameter	516	**surface area**	530
edge	524	**vertex**	524
face	524	**volume**	534

Complete the sentences below with vocabulary words from the list above. Words may be used more than once.

1. A __?__ is a three-dimensional object with flat faces that are polygons.
2. The number of cubic units needed to fill a space is called __?__.
3. The distance around a figure is called the __?__, and the distance around a circle is called the __?__.
4. A line segment that passes through the center of a circle and has both endpoints on the circle is a __?__.

10-1 Finding Perimeter (pp. 500–503)

EXAMPLE

- **Find the perimeter of the figure.**

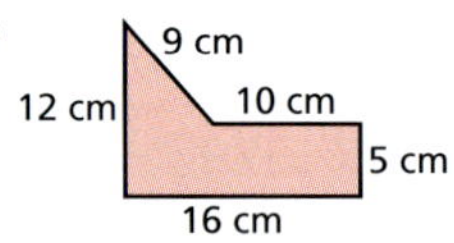

$P = 9 + 10 + 5 + 16 + 12$ *Add all the side lengths.*
$P = 52$
The perimeter is 52 cm.

EXERCISES

5. Find the perimeter.

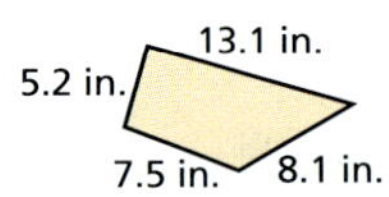

6. What is the length of n if the perimeter is 20 ft?

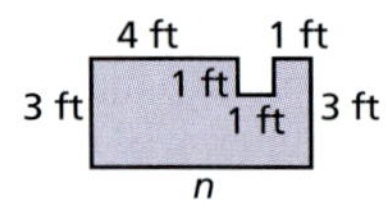

10-2 Estimating and Finding Area (pp. 504–507)

EXAMPLE

- **Find the area of the rectangle.**

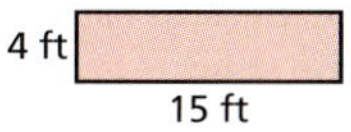

$A = \ell w$
$A = 15 \cdot 4 = 60$
The area is 60 ft^2.

EXERCISES

Find the area of each figure.

7.
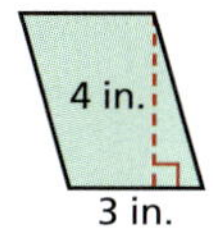

8.
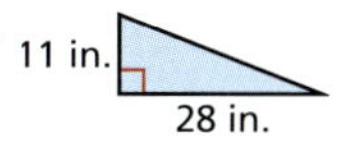

10-3 Break into Simpler Parts (pp. 508–510)

EXAMPLE

- **Find the area of the polygon.**

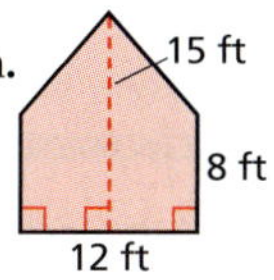

$A = 8 \cdot 12 = 96$
$A = \frac{1}{2} \cdot 12 \cdot 7 = 42$
The area of the figure is
42 ft^2 + 96 ft^2 = 138 ft^2.

EXERCISES

Find the area of each polygon.

9.
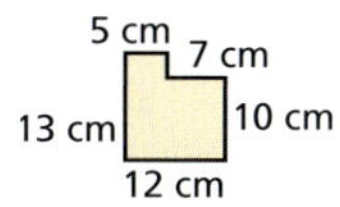

10.
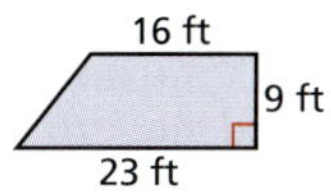

10-4 Comparing Perimeter and Area (pp. 511–513)

EXAMPLE

- **Find how the perimeter and area of a rectangle change when the dimensions change.**
When the dimensions of the rectangle are multiplied by x, the perimeter is multiplied by x, and the area is multiplied by x^2.

EXERCISES

Find how the perimeter and area change when the dimensions change.

11.
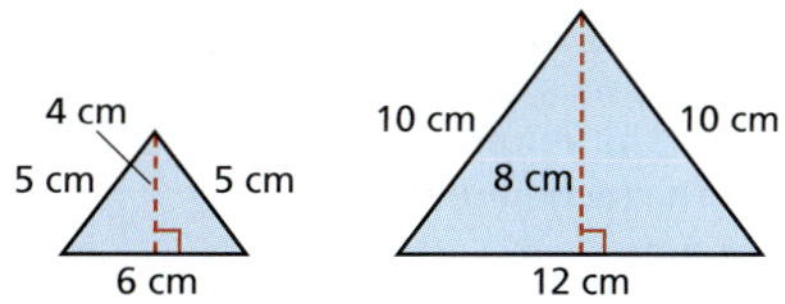

10-5 Circles (pp. 516–519)

EXAMPLE

- **Find the circumference and the area of the circle. Use 3.14 for *pi*. Round to the nearest hundredth.**

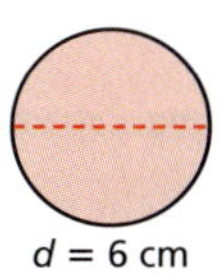

$C = \pi d$
$C \approx 3.14 \cdot 6$
$C \approx 18.84$ cm

$A = \pi r^2$
$r = d \div 2 = 6 \div 2 = 3$
$A \approx 3.14 \cdot 3^2$
$A \approx 3.14 \cdot 9 \approx 28.26$ cm^2

EXERCISES

Find each missing value to the nearest hundredth. Use 3.14 for *pi*.

12. $d = 10$ ft; $C = \underline{?}$
13. $r = 8$ cm; $C = \underline{?}$
14. $C = 28.26$ m; $d = \underline{?}$
15. $C = 69.08$ ft; $r = \underline{?}$
16. Find the area of a circle with radius 7 cm. Use $\frac{22}{7}$ for *pi.*
17. Find the area of a circle with diameter 28 cm. Use $\frac{22}{7}$ for *pi.*

Study Guide and Review

Answers

7. 12 ft^2
8. 154 in^2
9. 135 cm^2
10. 175.5 ft^2
11. The perimeter is multiplied by 2, and the area is multiplied by 4, or 2^2.
12. 31.4 ft
13. 50.24 cm
14. 9 m
15. 11 ft
16. 154 cm^2
17. 616 cm^2

Answers

18. 5 faces, 8 edges, 5 vertices; square pyramid

19. 6 faces, 12 edges, 8 vertices; rectangular prism

20. 125 m^2

21. 102 cm^2

22. 384 cm^3

23. 6,300 in^3

24. 353 m^3

25. 2,308 ft^3

10-6 Solid Figures (pp. 524–527)

EXAMPLE

- **Identify the number of faces, edges, and vertices in the solid figure. Then name the solid.**

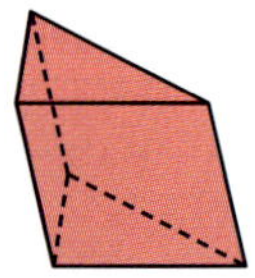

5 faces; 9 edges; 6 vertices
There are two congruent parallel bases, so the figure is a prism. The bases are triangles.
The solid is a triangular prism.

EXERCISES

Identify the number of faces, edges, and vertices in each solid figure. Then name the solid.

18.

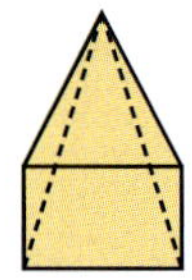

19.

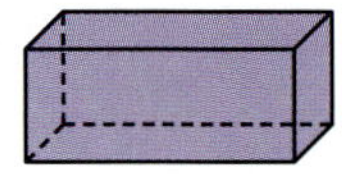

10-7 Surface Area (pp. 530–533)

EXAMPLE

- **Find the surface area S of the cylinder.**

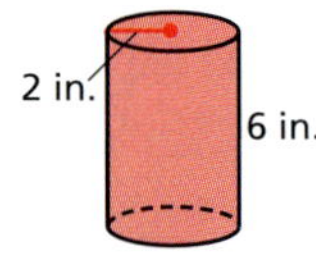

$S = h \cdot (2\pi r) + 2 \cdot (\pi r^2)$
$S \approx 6 \cdot (2 \cdot 3.14 \cdot 2) + 2 \cdot (3.14 \cdot 2^2)$
$S \approx 100.48 \text{ in}^2$

EXERCISES

Find the surface area S of each solid.

20.

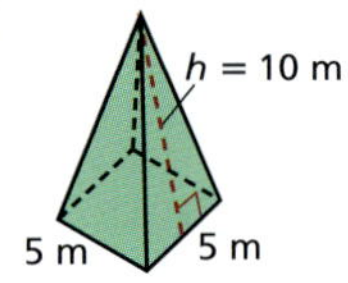

21.

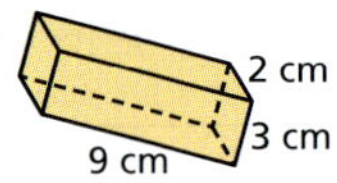

10-8 Finding Volume (pp. 534–537)

EXAMPLE

- **Find the volume of the rectangular prism.**

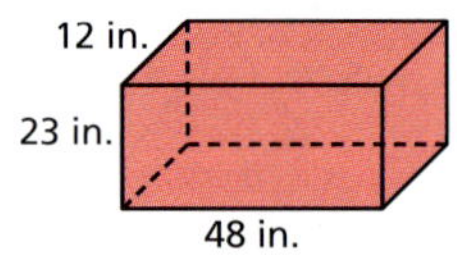

$V = \ell wh$
$V = 48 \cdot 12 \cdot 23$
$V = 13{,}248 \text{ in}^3$

EXERCISES

Find the volume of each prism.

22.

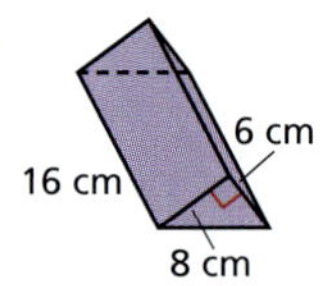

23.

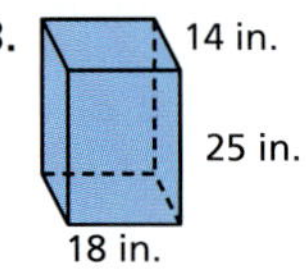

10-9 Volume of Cylinders (pp. 538–541)

EXAMPLE

- **Find the volume of the cylinder to the nearest cubic unit.**

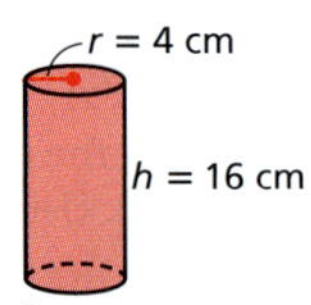

$V \approx 3.14 \cdot 4^2 \cdot 16$
$V \approx 803.84 \text{ cm}^3$
The volume is about 804 cm^3.

EXERCISES

Find the volume of each cylinder to the nearest cubic unit.

24.

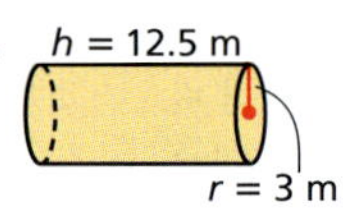

25.

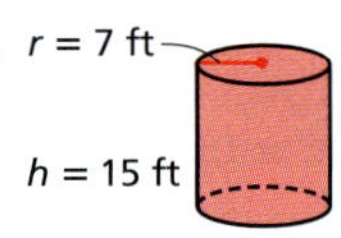

Chapter Test

Chapter 10

Find the perimeter and area of each figure.

1. 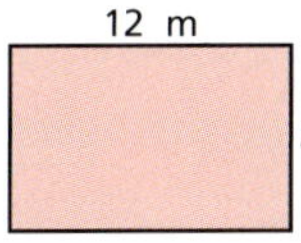

40 m, 96 m^2

2. 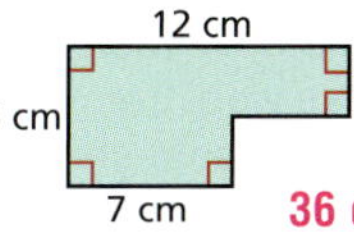

36 cm, 57 cm^2

3. 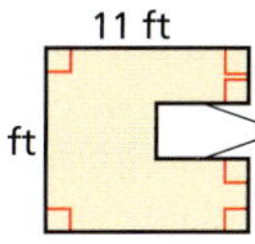

52 ft, 95 ft^2

4. Find how the perimeter and the area of a rectangle change when the length and width are doubled.
The perimeter is multiplied by 2, and the area is multiplied by 4, or 2^2

Name the circle and two radii for each circle. Then find the area and the circumference of each circle.

5. 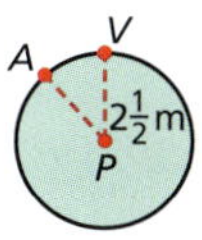

6. 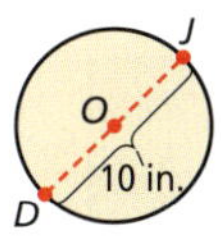

7. 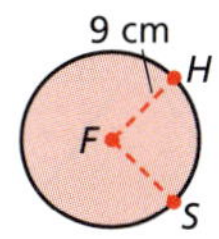

Identify the number of faces, edges, and vertices in each solid figure. Then tell whether each figure is a polyhedron and name the solid.

8. 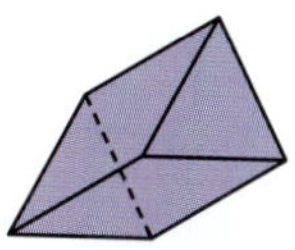

9. 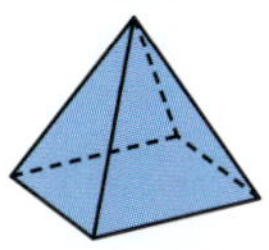

10. 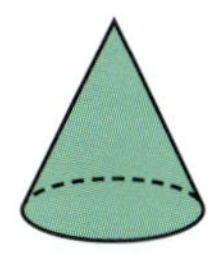

Find the surface area S of each solid.

11. 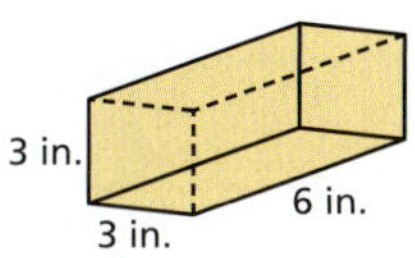

90 in^2

12. 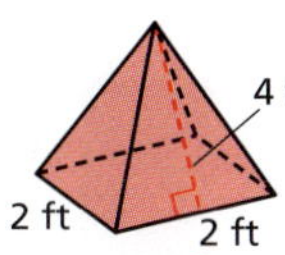

20 ft^2

13. 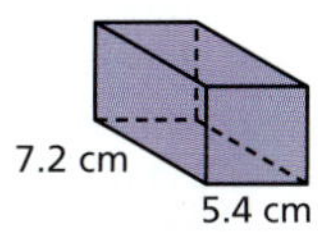

208.8 cm^2

Find the volume V of each solid.

14. 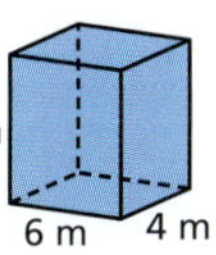

192 m^3

15. 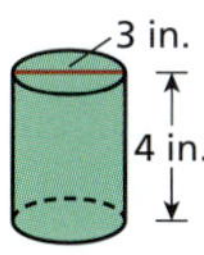

28.26 in^3

16. 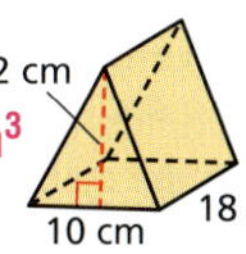

1,080 cm^3

17. Patricia has two cylinder-shaped jars. Jar A has a radius of 6 cm and a height of 9 cm. Jar B has a diameter of 8 cm and a height of 17 cm. Which jar has the greater volume? How much greater? Jar A; 163.28 cm^3

Chapter Test

Chapter 10

Purpose: *To assess students' mastery of concepts and skills in Chapter 10*

Assessment Resources

Chapter 10 Tests (Levels A, B, C)
Assessment Resources ..... pp. 85–90

***Test and Practice Generator* CD-ROM**

Additional assessment items in both multiple-choice and free-response format may be generated for any objective in Chapter 10.

Answers

5. Circle P; $\overline{PA}$ and $\overline{PV}$ are radii; C = 15.7 m; A = 19.625 m^2.
6. Circle O; $\overline{DJ}$ is a diameter; $\overline{OJ}$ and $\overline{OD}$ are radii; C = 31.4 in.; A = 78.5 in^2.
7. Circle F; $\overline{FS}$ and $\overline{FH}$ are radii; C = 56.52 m; A = 254.34 m^2.
8. 5 faces; 9 edges; 6 vertices; yes; triangular prism
9. 5 faces; 8 edge; 5 vertices; yes; square pyramid
10. 2 faces; 1 edges, 1 vertex; no; cone

Chapter 10 Performance Assessment

Purpose: *To assess students' understanding of concepts in Chapter 10 and combined problem-solving skills*

Assessment Resources

Performance Assessment
Assessment Resources . . . . p. 122

Performance Assessment Teacher Support
Assessment Resources . . . . p. 121

Answers

1–2. See p. A6.
3. See Level 3 work sample below.

Scoring Rubric for Problem Solving Item 3

Level 3
Accomplishes the purposes of the task.
Student gives clear explanations, shows understanding of mathematical ideas and processes, and computes accurately.

Level 2
Purposes of the task not fully achieved.
Student demonstrates satisfactory but limited understanding of the mathematical ideas and processes.

Level 1
Purposes of the task not accomplished.
Student shows little evidence of understanding the mathematical ideas and processes and makes computational and/or procedural errors.

Chapter 10 Performance Assessment

Show What You Know

Create a portfolio of your work from this chapter. Complete this page and include it with your four best pieces of work from Chapter 10. Choose from your homework or lab assignments, mid-chapter quiz, or any journal entries you have done. Put them together using any design you want. Make your portfolio represent what you consider your best work.

Short Response

1. Find the perimeter and area of each figure. Show your work. Which figure has the greatest perimeter? Which has the greatest area?

a.
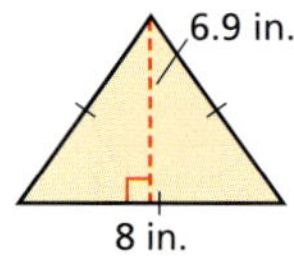

b.
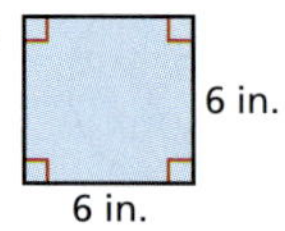

c.
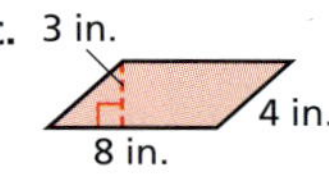

2. George will use 32 feet of fencing to enclose a rectangular garden. George wants the greatest possible area for his garden. What dimensions, in whole feet, will his garden be? Explain how you found your answer.

Extended Problem Solving

3. The ancient Sumerians kept records by etching onto clay tablets. After etching the tablet, they rolled cylindrical stones over the clay. The stones were carved with unique images whose impressions were then transferred onto the tablets. These impressions proved that the writings were originals.

a. Explain the relationship between the circumference of the base of the cylindrical stone and the indicated length of the clay impression.

b. Find the indicated length of the clay impression. Use 3.14 for π, and round your answer to the nearest hundredth. Show your work.

c. Find the volume of the cylindrical stone. Use 3.14 for π, round your answer to the nearest hundredth, and show your work.

Student Work Samples for Item 3

Level 3

a. The circumference of the base is the length indicated on the impression.

b. $C = \pi d$ = length of impression
$= 3.14(1.44)$
$= 4.52$ cm

c. $V = \pi r^2 h$ $r = \frac{1.44}{2} = 0.72$
$= 3.14(0.72)^2(3)$
$= 4.88 \text{ cm}^3$

The student correctly and completely answered all parts of the question.

Level 2

a. The length of the impression is equal to the circumference.

b. $C = \pi r^2$
$= (3.14)\left(\frac{1.44}{2}\right)^2$
≈ 1.63cm

c. $V = \pi dh$
$= (3.14)(1.44)(3)$
$= 13.56 \text{ cm}^3$

The student correctly answered part **a,** but confused the formulas for area and volume.

Level 1

a. circumference = half length

b. $3 \times 1.44 = 4.32$ cm

c. $3(3.14) = 9.42 \text{ cm}^3$

The student did not understand the concepts of circumference, area, and volume.

internet connect
State-Specific Test Practice Online
go.hrw.com Keyword: MR4 TestPrep

Standardized Test Prep

Chapter 10

Cumulative Assessment, Chapters 1–10

TIP!
TEST TAKING TIP!
Perimeter is the distance around a figure, and area is the amount of surface a figure covers.

1. What is the perimeter and the area of a rectangle with length 8 cm and width 6 cm? **B**
   (A) $P = 14$ cm, $A = 48$ cm^2
   (B) $P = 28$ cm, $A = 48$ cm^2
   (C) $P = 48$ cm, $A = 14$ cm^2
   (D) $P = 48$ cm, $A = 28$ cm^2

2. Which expression has the greatest value? **J**
   (F) 20% · 150 (H) $4\frac{1}{2} \cdot 5\frac{2}{3}$
   (G) $2^2 \cdot 5 + 8$ (J) $4 \div 0.12$

3. Which value is equivalent to 521 cm? **C**
   (A) 5,210 m (C) 5.21 m
   (B) 0.521 m (D) 0.0521 km

4. Which is a polyhedron? **H**
   (F) 
   (H) 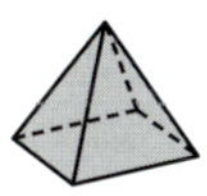
   (G) 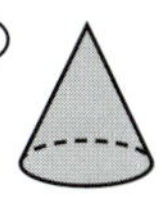
   (J) 

5. Evaluate $3\frac{1}{3} \cdot 2\frac{4}{5} + 1\frac{2}{3}$. **C**
   (A) $7\frac{14}{15}$ (C) 11
   (B) $4\frac{1}{3}$ (D) $5\frac{1}{2}$

6. What is the measure of angle *a*? **F**

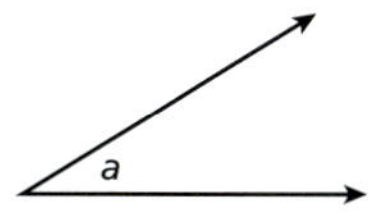

   (F) 32° (H) 68°
   (G) 58° (J) 148°

7. What is the prime factorization of 126? **D**
   (A) $2 \cdot 63$ (C) $1 \cdot 2 \cdot 3 \cdot 21$
   (B) $2 \cdot 7 \cdot 9$ (D) $2 \cdot 3^2 \cdot 7$

8. Which figure below has the greatest volume? **J**
   (F) F (H) H
   (G) G (J) J

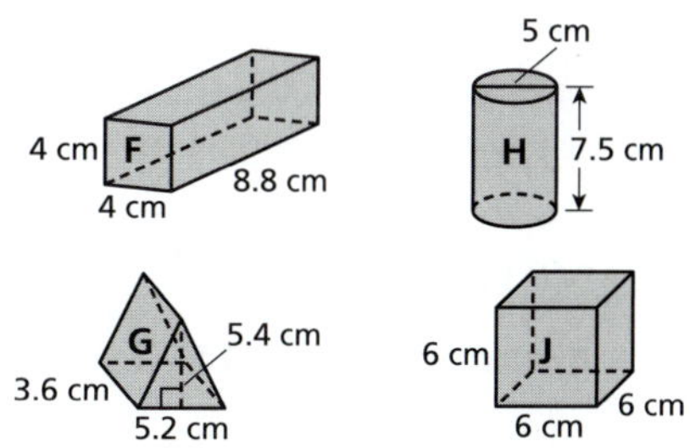

9. ***SHORT RESPONSE*** Aaron plans to buy a shirt that is on sale for 40% off the ticketed price. The ticketed price is \$24.00, and sales tax is 7%. How much will Aaron pay for the shirt? Show your work.

10. ***SHORT RESPONSE*** Use prime factorization to find the LCM of 10, 12, and 15. Show your work.

Standardized Test Prep

Standardized Test Prep

Chapter 10

Purpose: *To provide review and practice for Chapters 1–10 and standardized tests*

Assessment Resources

Cumulative Tests (Levels A, B, C)
Assessment Resources . . . . pp. 235–246

State-Specific Test Practice Online
KEYWORD: MR4 TestPrep

Test Prep Doctor

For item 8, remind students of the meaning of *LCM* (Least Common Multiple). The key is the word *multiple*; the answer must be a multiple of all three numbers. **F** can be eliminated because 30 is not a multiple of 12. **H** and **J** can be eliminated because they are not the least of the common multiples of the numbers.

Answers

9. 40% of \$24.00 = \$9.60
\$24.00 − \$9.60 = \$14.40
7% of \$14.40 ≈ \$1.01
\$14.40 + \$1.01 ≈ \$15.41

10. 10: 2 × 5
12: 2 × 2 × 3
15: 3 × 5
↓ ↓ ↓ ↓
2 × 2 × 3 × 5 = 60

Chapter 11

Probability

Section 11A

Understanding Probability

Lesson 11-1
Introduction to Probability

Lesson 11-2
Experimental Probability

Hands-On Lab 11A
Simulations

Lesson 11-3
Theoretical Probability

Section 11B

Using Probability

Lesson 11-4
Make an Organized List

Lesson 11-5
Compound Events

Hands-On Lab 11B
Explore Permutations and Combinations

Lesson 11-6
Making Predictions

Extension
Odds

Pacing Guide for 45-Minute Classes

Chapter 11

DAY 144	DAY 145	DAY 146	DAY 147	DAY 148
Lesson 11-1	Lesson 11-2	Hands-On Lab 11A	Lesson 11-3	Mid-Chapter Quiz Lesson 11-4
DAY 149	**DAY 150**	**DAY 151**	**DAY 152**	**DAY 153**
Lesson 11-5	Hands-On Lab 11B	Lesson 11-6	Extension	Chapter 11 Review
DAY 154				
Chapter 11 Assessment				

Pacing Guide for 90-Minute Classes

Chapter 11

DAY 72	DAY 73	DAY 74	DAY 75	DAY 76
Chapter 10 Assessment Lesson 11-1	Lesson 11-2 Hands-On Lab 11A	Lesson 11-3 Lesson 11-4	Mid-Chapter Quiz Lesson 11-5 Hands-On Lab 11B	Lesson 11-6 Extension
DAY 77	**DAY 78**			
Chapter 11 Review Lesson 12-1	Chapter 11 Assessment Hands-On Lab 12A			

Across the Series

HARCOURT GRADE 5

- Predict and write outcomes of probability experiments.
- Express probabilities as fractions.
- Compare probabilities.

COURSE 1

- **Estimate likelihood, and find experimental and theoretical probabilities.**
- **Use simulation to model experiments.**
- **Use an organized list to find all possible outcomes, including compound events, and find theoretical probabilities.**
- **Explore permutations and combinations.**
- **Use data to estimate probabilities and make predictions.**
- **Find odds for and against specified outcomes.**

COURSE 2

- Estimate likelihood and probability.
- Use an organized list to find all possible outcomes—including compound, independent, and dependent events—and find theoretical probabilities.
- Use Pascal's Triangle to find probabilities.
- Find probabilities involving permutations and combinations.
- Find odds for and against specified outcomes.

Across the Curriculum

LANGUAGE ARTS LINK

Math: Reading and Writing in the Content Area . . . . pp. 92–97

Focus on Problem Solving

Look Back. . . . . SE p. 569

Journal . . . . TE, last page of each lesson

Write About It . . . . SE, last page of each lesson

SOCIAL STUDIES LINK

Social Studies . . . . SE pp. 567, 583

SCIENCE LINK

Life Science . . . . SE pp. 557, 570, 577

TE = *Teacher's Edition* **SE** = *Student Edition*

Interdisciplinary Bulletin Board

Life Science

A Punnett square is a diagram used to show all possible combinations of alleles, or gene forms, in offspring. Which Punnett square shows a 100% probability that the offspring will always have the same alleles? the cross of a true-breeding purple flower with a true-breeding white flower

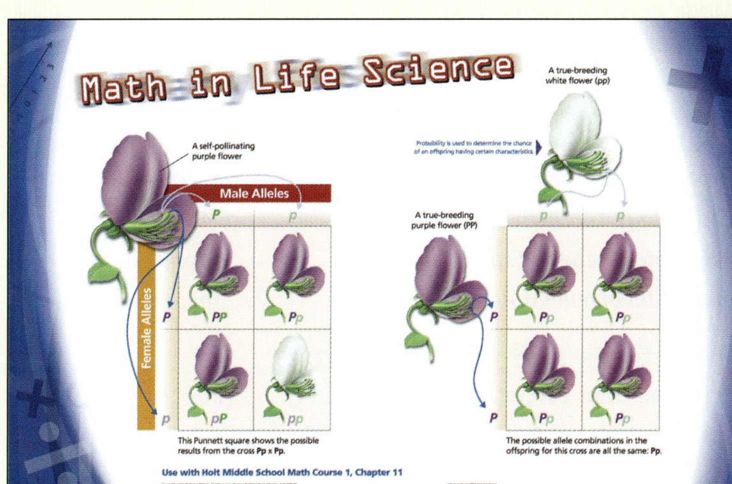

Interdisciplinary posters and worksheets are provided in your resource material.

Chapter 11

Resource Options

Chapter 11 Resource Book

Student Resources

Practice (Levels A, B, C) .... pp. 8–10, 18–20, 27–29, 37–39, 46–48, 55–57
Reteach .... pp. 11–12, 21, 30, 40, 49, 58
Challenge .... pp. 13, 22, 31, 41, 50, 59
Problem Solving .... pp. 14, 23, 32, 42, 51, 60
Puzzles, Twisters & Teasers .... pp. 15, 24, 33, 43, 52, 61
Recording Sheets .... pp. 3, 7, 17, 26, 36, 45, 54, 64
Chapter Review .... pp. 62–63

Teacher and Parent Resources

Chapter Planning and Pacing Guide .... p. 4
Section Planning Guides .... pp. 5, 34
Parent Letter .... pp. 1–2
Teaching Tools .... p. 67
Teacher Support for Chapter Project .... p. 65
Transparencies .... pp. T1–T29
- Daily Transparencies
- Additional Examples Transparencies
- Teaching Transparencies

Reaching All Learners

English Language Learners

Success for English Language Learners ...... pp. 183–194
Math: Reading and Writing in the Content Area .... pp. 92–97
Spanish Homework and Practice .... pp. 92–97
Spanish Interactive Study Guide .... pp. 92–97
Spanish Family Involvement Activities .... pp. 89–96
Multilingual Glossary

Individual Needs

Are You Ready? Intervention and Enrichment . . . pp. 81–84, 113–120, 137–140, 425–426
Alternate Openers: Explorations .... pp. 92–97
Family Involvement Activities .... pp. 89–96
Interactive Problem Solving .... pp. 92–97
Interactive Study Guide .... pp. 92–97
Readiness Activities .... pp. 21–22
Math: Reading and Writing in the Content Area .... pp. 92–97
Challenge .... CRB pp. 13, 22, 31, 41, 50, 59

Hands-On

Hands-On Lab Activities .... pp. 102–106
Technology Lab Activities .... pp. 45–48, 53
Alternate Openers: Explorations .... pp. 92–97
Family Involvement Activities .... pp. 89–96

Applications and Connections

Consumer and Career Math .... pp. 41–44
Interdisciplinary Posters .... Poster 11, TE p. 552B
Interdisciplinary Poster Worksheets .... pp. 31–33

Transparencies

Alternate Openers: Explorations .... pp. 92–97
Exercise Answers Transparencies
Chapter 11 Resource Book .... pp. T1–T29
- Daily Transparencies
- Additional Examples Transparencies
- Teaching Transparencies

Technology

Teacher Resources

***Lesson Presentations* CD-ROM** .... Chapter 11
***Test and Practice Generator* CD-ROM** ...... Chapter 11
***One-Stop Planner* CD-ROM** .... Chapter 11

Student Resources

***Are You Ready? Intervention* CD-ROM**
Skills 18, 26, 27, 32

internet connect go.hrw.com

Homework Help Online	**KEYWORD:** MR4 HWHelp11
Math Tools Online	**KEYWORD:** MR4 Tools
Glossary Online	**KEYWORD:** MR4 Glossary
Chapter Project Online	**KEYWORD:** MR4 PSProject11
Chapter Opener Online	**KEYWORD:** MR4 Ch11
CNN Student News™	**KEYWORD:** MR4 CNN11

SE = *Student Edition* **TE** = *Teacher's Edition* **AR** = *Assessment Resources* **CRB** = *Chapter Resource Book* **MK** = *Manipulatives Kit*

Chapter 11

Assessment Options

Assessing Prior Knowledge

Determine whether students have the required prerequisite concepts and skills.

Are You Ready? . . . . SE p. 553
Inventory Test . . . . AR pp. 1–4

Test Preparation

Provide review and practice for chapter and standardized tests.

Standardized Test Prep . . . . SE p. 595
Spiral Review with Test Prep . . . . . SE, last page of each lesson
Study Guide and Review . . . . SE pp. 590–592
Test Prep Tool Kit

Technology

***Test and Practice Generator* CD-ROM**

internet connect

State-Specific Test Practice Online KEYWORD: MR4 TestPrep

go.hrw.com

Performance Assessment

Assess students' understanding of chapter concepts and combined problem-solving skills.

Performance Assessment . . . . SE p. 594
Includes scoring rubric in TE
Performance Assessment . . . . AR p. 124
Performance Assessment Teacher Support . . . . AR p. 123

Portfolio

Portfolio opportunities appear throughout the Student and Teacher's Editions.

Suggested work samples:

Problem Solving Project . . . . TE p. 552
Performance Assessment . . . . SE p. 594
Portfolio Guide . . . . AR p. xxxv
Journal . . . . TE, last page of each lesson
Write About It . . . . SE, last page of each lesson

Daily Assessment

Obtain daily feedback on students' understanding of concepts.

Spiral Review and Test Prep . . . . . . SE, last page of each lesson

Also Available on Transparency In Chapter 11 Resource Book

Warm Up . . . . TE, first page of each lesson
Problem of the Day . . . . TE, first page of each lesson
Lesson Quiz . . . . TE, last page of each lesson

Student Self-Assessment

Have students evaluate their own work.

Group Project Evaluation . . . . AR p. xxxii
Individual Group Member Evaluation . . . . AR p. xxxiii
Portfolio Guide . . . . AR p. xxxv
Journal . . . . TE, last page of each lesson

Formal Assessment

Assess students' mastery of concepts and skills.

Section Quizzes . . . . AR pp. 27–28
Mid-Chapter Quiz . . . . SE p. 568
Chapter Test . . . . SE p. 593
Chapter Tests (Levels A, B, C) . . . . AR pp. 91–96
Cumulative Tests (Levels A, B, C) . . . . AR pp. 247–258
Standardized Test Prep
Cumulative Assessment . . . . SE p. 595
End-of-Year Test . . . . AR pp. 271–274

Technology

***Test and Practice Generator* CD-ROM**

Make tests electronically. This software includes:

- Dynamic practice for Chapter 11
- Customizable tests
- Multiple-choice items for each objective
- Free-response items for each objective
- Teacher management system

SE = *Student Edition* **TE** = *Teacher's Edition* **AR** = *Assessment Resources* **CRB** = *Chapter Resource Book* **MK** = *Manipulatives Kit*

Chapter 11

Chapter 11 Tests

Three levels (A,B,C) of tests are available for each chapter in the *Assessment Resources.*

LEVEL A

CHAPTER 11 Chapter Test
Form A

Write impossible, unlikely, likely, as likely as not, or certain to describe each event.

1. You toss a coin and it comes up tails.
as likely as not

2. You spin a 6 on the spinner shown.

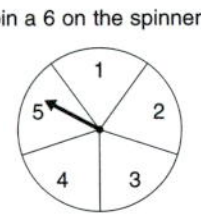

impossible

3. The weather report indicates that there is a 60% chance of rain for Saturday. Write this probability as a decimal and fraction.
0.60 and $\frac{3}{5}$

4. For the experiment, identify the outcome shown and the sample space.

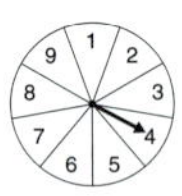

Outcome: 4; Sample space: {1, 2, 3, 4, 5, 6, 7, 8, 9}.

Outcome	Frequency
up	13
down	7

5. A bottle cap is tossed 20 times and the results are recorded in the table above. What is the probability that the bottle cap will land up?
$\frac{13}{20}$

6. Find the experimental probability of the bottle cap landing down.
$\frac{7}{20}$

7. Find the experimental probability of the bottle cap landing up or down.
1

Outcome	A	B	C	D	E
Frequency	𝍸	I	II	III	II

8. Cass recorded the number of times a spinner landed on each letter. Based on Cass' experiment, on which letter is the spinner most likely to land?
A

9. On which letter is the spinner least likely to land?
B

LEVEL B

CHAPTER 11 Chapter Test
Form B

Write impossible, unlikely, likely, as likely as not, or certain to describe each event.

1. You win the lottery.
unlikely

2. The month of December has 31 days.
certain

3. The chance of Colleen winning a new DVD player is 0.05. Write this probability as a fraction and as a percent.
5% and $\frac{1}{20}$

4. For the experiment, identify the outcome shown and the sample space.

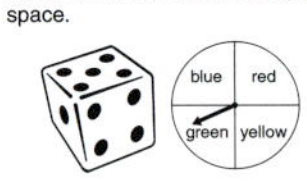

Outcome: 5, green; Sample Space: All possible combinations of {1, 2, 3, 4, 5, 6} and {red, blue, yellow, green}.

Outcome	Frequency
Lands on Red	5
Lands on Black	7
Lands on both colors	13

5. A dime is tossed onto a checkerboard 25 times and the results are shown in the table above. Find the experimental probability of tossing a dime and having it land on red.
$\frac{5}{25} = \frac{1}{5}$

6. Find the experimental probability of tossing a dime and having it land on red or on black.
$\frac{12}{25}$

7. Find the experimental probability of tossing a dime and having it land on both colors.
$\frac{13}{25}$

Outcome	A	B	C	D	E
Frequency	I	IIII	𝍸	III	IIII

8. Samantha recorded the number of times a spinner landed on each letter. Based on Samantha's experiment, on which letter is the spinner most likely to land?
C

9. On which letter is the spinner least likely to land?
A

LEVEL C

CHAPTER 11 Chapter Test
Form C

Write impossible, unlikely, likely, as likely as not, or certain to describe each event.

1. You toss a quarter and it lands on heads.
as likely as not

2. You roll a 7 on a standard number cube.
impossible

3. There is a $\frac{15}{40}$ chance of selecting a green M & M out of a package. Write this probability as a decimal and percent.
0.375 and 37.5%

4. For the experiment, identify the outcome shown and the sample space.

Outcome: HHT; Sample Space: {HHH, HHT, HTH, HTT, THH, THT, TTH, TTT}

	Up	Down	Side
Sue	6	7	7
Jeremy	4	6	10
Louis	5	8	7

5. Sue, Jeremy, and Louis conducted an experiment by each tossing a paper cup 20 times to see if it would land up, down, or on its side. The results are shown in the table. What is the probability that it will land up for the entire group?
$\frac{15}{60} = \frac{1}{4}$

6. Find the experimental probability of the cup landing down for Jeremy.
$\frac{6}{20} = \frac{3}{10}$

7. Find the experimental probability of the cup landing on its side for Sue.
$\frac{7}{20}$

	A	B	C
Frequency	III	IIII	𝍸
	D	E	F
Frequency	III	𝍸	III

8. Samantha recorded the number of times a spinner landed on each letter. Based on her experiment, on which letter is the spinner most likely to land?
C or E

9. On which letter(s) is the spinner least likely to land?
A, D, and F

CHAPTER 11 Chapter Test
Form A, continued

10. What is the probability of this spinner landing on A?

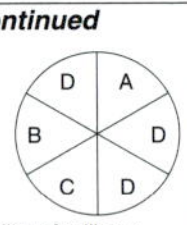

$\frac{1}{6}$

11. What is the probability of rolling a number greater than 6 on a fair number cube?
0

12. Raymond has a 95% chance of winning a prize in the spelling contest. What is the probability that he will not win a prize in the spelling contest?
5%

13. Jessica is preparing lunch for school. She has a choice of chicken or roast beef, one fruit choice (banana or orange), and one cookie (chocolate chip or oatmeal). How many different lunches could she make?
8 choices

14. If Jessica chooses at random, find the probability of selecting chicken, a banana, and an oatmeal cookie.
$\frac{1}{8}$

15. You go to the local restaurant for lunch. You have 3 choices of pasta, 3 choices for sauce, and either soup or salad. How many different choices are there for lunch?
18 choices

Solve, using the standard number cube and the spinner shown.

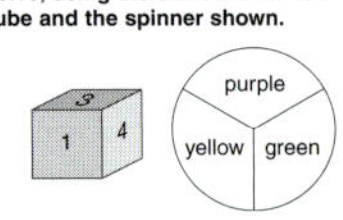

16. The number cube is rolled and the spinner is spun. What is the probability that you will roll a 7 and spin white?
0

17. What is the probability that you will roll 2 and spin yellow?
$\frac{1}{18}$

18. What is the probability that you will roll an even number, and spin purple?
$\frac{1}{6}$

19. About 5% of the items produced by a company are defective. Out of 1,000 items, how many would you predict will be defective?
50 items

20. If you roll a fair 6-sided number cube 100 times, how many times would you expect to roll an odd number?
50 times

CHAPTER 11 Chapter Test
Form B, continued

10. What is the probability of this spinner landing on D?

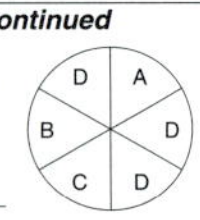

$\frac{1}{2}$

11. What is the probability of rolling a number less than 5 on a fair number cube?
$\frac{2}{3}$

12. Rebecca has a 5% chance of not completing the marathon. What is the probability that she will complete the marathon?
95%

13. Danica can choose an outfit from the following clothes: two pairs of slacks (navy or black), four blouses (red, blue, white, and striped), and three pairs of shoes (blue, black, and sandals). How many different choices of outfits does she have?
24 choices

14. If Danica chooses an outfit at random, find the probability of her selecting the black slacks, red blouse, and black shoes.
$\frac{1}{24}$

15. You go to the local restaurant for lunch. You have 3 choices of meat (chicken, fish, or beef), 2 choices for a vegetable (peas or beans), and 2 choices for soup (tomato or vegetable). How many different choices are there for lunch?
12 choices

Solve using the standard number cube and the spinner shown.

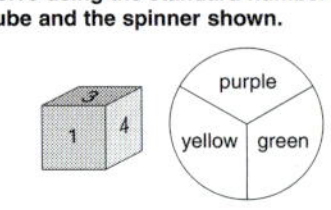

16. The number cube is rolled and the spinner is spun. What is the probability that you will roll a 2 and spin purple?
$\frac{1}{18}$

17. What is the probability that you will roll 1 or 3 and spin yellow?
$\frac{1}{9}$

18. What is the probability that you will roll an even number, and spin purple or green?
$\frac{1}{3}$

19. About 4% of the items produced by a company are defective. Out of 8000 items, how many would you predict will be defective?
320 items

20. If you roll a fair 12-sided die 300 times, how many times can you expect to roll an odd number?
150 times

CHAPTER 11 Chapter Test
Form C, continued

10. What is the probability of the spinner landing on A?

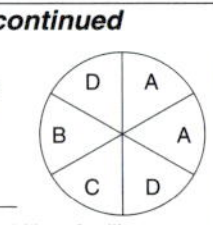

$\frac{1}{3}$

11. What is the probability of rolling a number less than 6 on a fair number cube?
$\frac{5}{6}$

12. A mixture of trail mix is 40% granola, 10% cashews, 20% M & M's, 15% pecans, and 15% raisins. What is the probability of not selecting a pecan?
85%

13. A local print shop offers 4 color choices, 6 different font styles, 3 different RSVP cards, and 2 different types of wording choices for wedding invitations. How many different wedding invitations do you have to choose from?
144 choices

14. If the print shop discontinues one particular font style, how many different choices are still available?
120 choices

15. A store stocks 8 different styles of shorts. Each style comes in 9 sizes and 5 colors. How many pairs of shorts must the store carry to have at least one pair of each size, style, and color?
360 pairs

Solve using the standard number cube and the spinner shown.

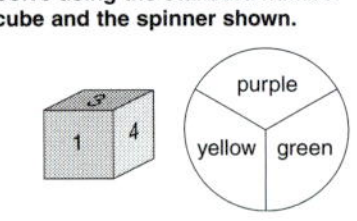

16. The number cube is rolled and the spinner is spun. What is the probability that you will roll a 4 and spin yellow?
$\frac{1}{18}$

17. What is the probability that you will roll an odd number and spin yellow?
$\frac{1}{6}$

18. What is the probability that you will not roll a 4 and will spin green?
$\frac{5}{18}$

19. About 0.5% of the items produced by a company are defective. Out of 1200 items inspected, how many can you predict will be defective?
6 items

20. If you spin a fair spinner with 10 equal outcomes numbered 1 through 10 300 times, how many times can you expect to spin a 5?
about 30 times

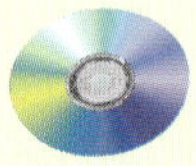

Test and Practice Generator
CD-ROM

Create and customize multiple versions of the same tests with corresponding answers for any chosen chapter objectives.

Chapter 11 State and Standardized Test Preparation

Test Taking Skill Builder and Standardized Test Practice are provided for each chapter in the *Test Prep Tool Kit.*

Test Taking Skill Builder

Test Taking Strategy **Gridded Response**

Chapter 11

Some questions on standardized tests require you to grid in your answer on a grid. Be sure you know how to correctly grid your answer.

Example Gridded Response There are 33 students in art class. The janitor walks into the class and chooses one student to help him clean the kiln. Eleven of the students are involved in creating a self-portrait. The rest of the class is working on a sculpture. What is the probability of the janitor selecting a student working on the sculpture?

Solution:
There are 33 students in the class and 11 of them are working on a self-portrait. Find the number of students working on the sculpture.
33 − 11 = 22 students

Next, find the probability of selecting a student working on the sculpture.

P(selecting a student working on the sculpture) = $\frac{22}{33} = \frac{2}{3}$

- You now need to grid in the answer $\frac{2}{3}$.
- When your answer is in the form of a fraction, such as $\frac{2}{3}$, in order to grid your answer correctly, you need to either grid the fraction $\frac{2}{3}$, or convert the fraction to its decimal equivalent, $\frac{2}{3} = 0.\overline{6} \approx 0.666 \approx 0.667$. If a fraction converts to a repeating decimal, every space in the grid must be filled in. You can correctly bubble either 0.666 or 0.667
- Four ways to correctly grid in the answer $\frac{2}{3}$ are shown below.

2/3 | 2/3 | 0.666 | 0.667

Test Taking Strategy

Chapter 11, continued

Exercises
What should go in the first box on the left for each gridded response answer?

1. 2.67 __2__ 2. $\frac{7}{10}$ __7__ 3. 519 __5__ 4. $4\frac{1}{2}$ __9__

If you grid your answer starting in the far left column, which column should the decimal or fraction bar go in for each answer below?

5. $\frac{2}{7}$ __second__ 6. 15.24 __third__ 7. $3\frac{2}{8}$ __third__ 8. 895 __none__

Tell what error was made in each gridded response below.

9. 126 — The fourth and fifth columns have zero shaded, they should be blank.
10. 21/4 — Cannot shade mixed numbers, need to change to improper fraction first.
11. 6/25 — 6 and the fraction bar were shaded in the same column.
12. 0.55 — The fourth column should have a 5 shaded in.

Grid each answer.

13. Elizabeth saved money to purchase a new television. She can also choose two of the following options: VCR, DVD, or CD player. How many options does Elizabeth have?
Answer: 3

14. Billy rolls a number cube and flips a coin. Find the probability Billy will roll a 5 and the coin will land on "tails".
Answer: $\frac{1}{12}$

Standardized Test Practice

Standardized Test Practice

Chapter 11

Select the best answer for Questions 1–7.

1. Find the probability of getting at least 2 tails when a coin is tossed three times.
A $\frac{1}{4}$ C $\frac{1}{2}$
B $\frac{3}{8}$ D $\frac{5}{8}$

2. A antique dealer predicts that 40% of its sales are from antique linens. Predict how much money the dealer has made if she sold $500 worth of linens.
F $40
G $100
H $200
I $300

3. Veronica spun the arrow on a spinner 60 times. The results are shown in the table. Which of these spinners did Veronica most likely spin?

Shape	Circle	Triangle	Square	Total Spins
Number of Times	22	20	18	60

A B C D

4. There are 30 jellybeans in a bag. Five are red, 6 yellow, 5 black, 3 orange, 4 green, and 7 purple. If Leona selects one jellybean from the bag, what is the probability that it will be black?
F 5
G $\frac{1}{5}$
H $\frac{5}{30}$
I 30%

5. You roll a fair number cube 36 times. How many times would you expect to roll a number that is five or greater?
A 3 C 12
B 6 D 18

6. Which describes the chances that Mr. Fredrickson will be late for his 4:00 meeting if he stuck in traffic 30 miles away and it is 3:50?
F impossible
G unlikely
H likely
I certain

7. A student randomly guessed the answers to three questions on a true-false test. What is the probability that the student correctly answered all three questions?
A $\frac{1}{8}$ C $\frac{1}{3}$
B $\frac{1}{6}$ D $\frac{1}{2}$

Standardized Test Practice

Chapter 11, continued

Gridded Response
Solve the problems. Use the answer sheet to write and grid-in your answer.

8. If there is a 28% chance of snow today, what is the percent chance that it will NOT snow?

9. Gail flips a fair coin and spins a fair spinner numbered from 1 to 6. What is the probability of landing on a number divisible by 3 and the coin showing heads?

10. An artist chooses 3 colors from 6 standard colors for a mural. How many different combinations of colors can the artist chose from?

Short Response
Solve the problems. Use the answer sheet to write your answers.

11. Firefighters can enter a burning building through one of four entrances. They can exit the building through any one of the six exits. How many ways is it possible for them to enter and exit the building? Draw a tree diagram to illustrate how you determined your answer.

12. Katie is running late for work and she still has to put on her shoes and socks. She has 10 pairs of shoes in her closet, of which 1 pair are loafers, 4 pairs are tennis shoes, 2 pairs are sandals, and 3 pairs are high-heels. If Katie randomly picks 1 tennis shoe from the closet, what is the probability that Katie will randomly pick out another tennis shoe? Show your work.

13. What is the probability that a week of the year picked at random will have 7 days? Explain in words how you determined your answer.

Extended Response

14. When a customer enters the Vet Pet Store, an employee randomly hands the customer a pet to hold. Today, the store has 8 puppies, 8 kittens, 12 birds, 5 lizards, and 2 bunnies for sale.
a. What is the probability that an employee will hand a puppy to a customer?
b. It has been the manager's experience that an employee gives a customer a lizard to hold 1 out of 10 times. How does this compare with the theoretical probability of the employee choosing a lizard?
c. Compare the likelihood of a customer receiving a kitten to hold with the likelihood of a customer receiving each of the other animals.

Sheet

See Lesson 11
See Lesson 11
See Lesson 11

10. 20
See Lesson 11-4.

Write your answers in the space provided.
11. 24 ways (See Lesson 11-4.)
12. P(choosing a second tennis shoe) = $\frac{7}{19}$ (See Lesson 11-3.)
13. The probability is 1 because every week has 7 days. (See Lesson 11-1.)

Extended Response
Write your answers for Problem 14 on the back of this paper.
See Lessons 11-2 and 11-3.

Entrances Exits A B C

internet connect
State-Specific Test Practice Online
KEYWORD: MR4 TestPrep

go.hrw.com

Test Prep Tool Kit

- Standardized Test Prep Workbook
- Countdown to Testing transparencies
- State Test Prep CD-ROM
- Standardized Test Prep Video

Customized answer sheets give students realistic practice for actual standardized tests.

Chapter 11

Probability

Why Learn This?

Tell students that probability is used to determine the likelihood that an event will happen. One way to find the probability that an event will happen is to study past trials and use information about the past to predict the future. For example, if a coin is tossed 10 times and lands heads up 4 times, you could predict that if you toss the same coin 100 times, it will land heads up 40 times. A financial advisor uses an investment fund's past performance to predict its future performance.

Using Data

To begin the study of this chapter, have students:

- Find the interest earned on a $100 investment over 5 years at 8%. $47
- Estimate the interest rate on a $100 investment that returned $18 after 2 years. about 8.5%
- Estimate the amount of time it will take a $100 investment to earn $100 in interest at 9%. about 8 years

Chapter 11

Probability

TABLE OF CONTENTS

internet connect
Chapter Opener Online
go.hrw.com
KEYWORD: MR4 Ch11

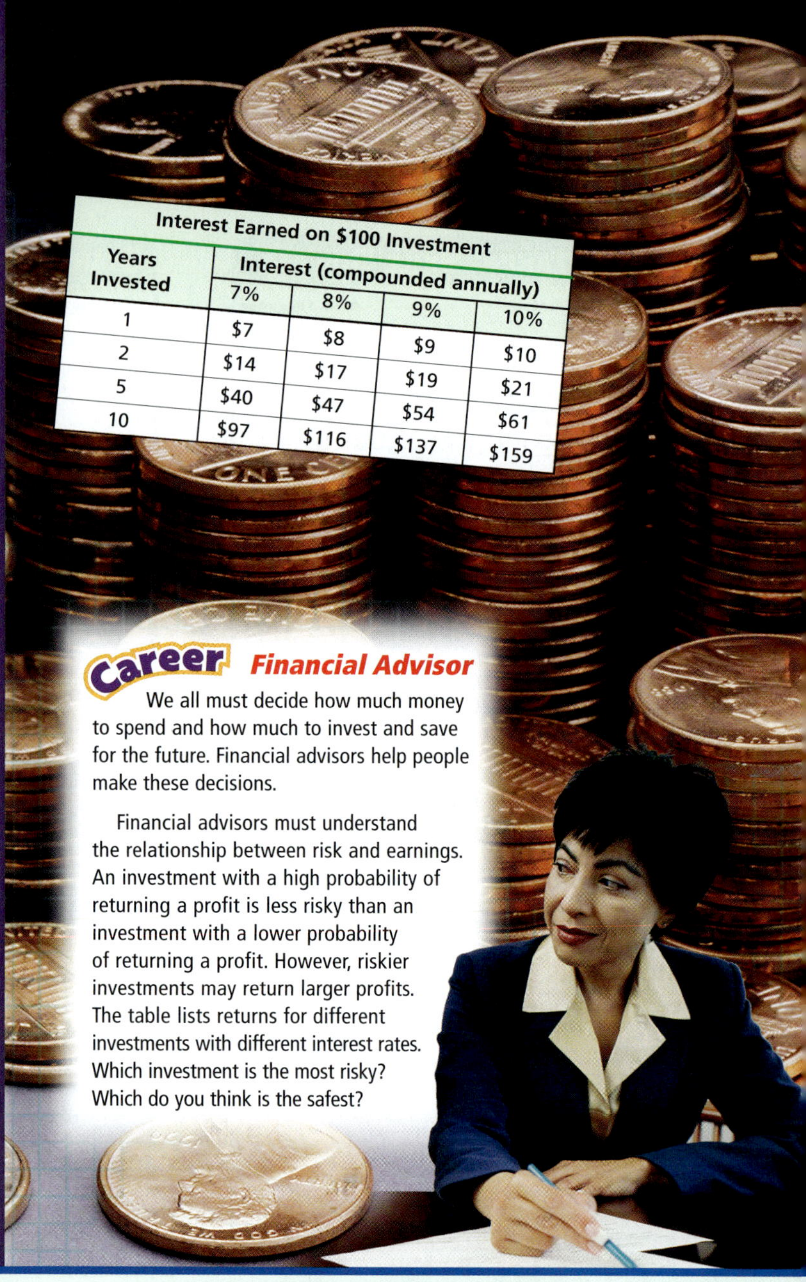

Interest Earned on $100 Investment

Years Invested	Interest (compounded annually)			
	7%	8%	9%	10%
1	$7	$8	$9	$10
2	$14	$17	$19	$21
5	$40	$47	$54	$61
10	$97	$116	$137	$159

Career ***Financial Advisor***

We all must decide how much money to spend and how much to invest and save for the future. Financial advisors help people make these decisions.

Financial advisors must understand the relationship between risk and earnings. An investment with a high probability of returning a profit is less risky than an investment with a lower probability of returning a profit. However, riskier investments may return larger profits. The table lists returns for different investments with different interest rates. Which investment is the most risky? Which do you think is the safest?

Problem Solving Project

Social Studies and Economics Connection

Purpose: To solve problems using probability

Materials: Financial Planning worksheet

internet connect
Chapter Project Online: ***go.hrw.com***
KEYWORD: MR4 PSProject11

Understand, Plan, Solve, and Look Back

Have students:

- ✓Examine the chart and explain what interest rate they would prefer. Which investment do they consider the safest?
- ✓Complete the Financial Planning worksheet to learn more about probability.
- ✓Calculate the amount earned at each interest rate over a 20-year period.
- ✓Research to find out what a typical investment fund earned 10 years ago and 5 years ago, and what it is earning today. Based on this research, predict the fund's performance 5 years from now.
- ✓Check students' work.

ARE YOU READY?

Choose the best term from the list to complete each sentence.

ratio
fraction
percent
frequency
tree diagram
table
part
whole

1. The denominator of a fraction represents the __?__, and the numerator represents the __?__. **whole; part**
2. A __?__ can be used to show all the possible combinations. **tree diagram**
3. A __?__ is a comparison of two quantities by division. **ratio**
4. Tally marks in a table show the __?__, or total, for each result. **frequency**
5. A ratio of a number to 100 is called a __?__. **percent**

Complete these exercises to review skills you will need for this chapter.

✔ Model Fractions

Write the fraction in simplest form that represents the shaded portion.

6. 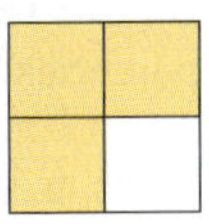 $\frac{3}{4}$

7. 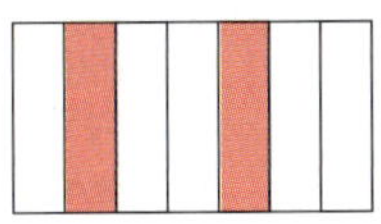 $\frac{2}{7}$

8. 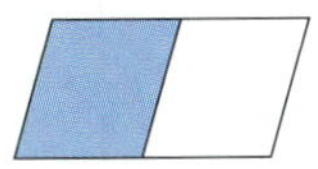 $\frac{1}{2}$

✔ Write Fractions as Decimals

Write each fraction as a decimal.

9. $\frac{9}{10}$ **0.9**
10. $\frac{1}{2}$ **0.5**
11. $\frac{12}{25}$ **0.48**
12. $\frac{11}{20}$ **0.55**

✔ Compare Fractions, Decimals, and Percents

Compare. Write <, >, or =.

13. 0.35 ■ 0.4 **<**
14. 0.25 ■ 25% **=**
15. $\frac{3}{5}$ ■ 0.7 **<**
16. 0.5 ■ $\frac{23}{50}$ **>**

✔ Write Ratios

Write each ratio.

17. blue circles to total circles **3:7**

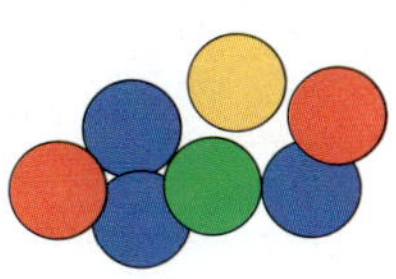

18. squares to triangles **5:4**

Assessing Prior Knowledge

INTERVENTION

Diagnose and Prescribe

Evaluate your students' performance on this page to determine whether intervention is necessary or whether enrichment is appropriate. Options that provide instruction, practice, and a check are listed below.

Resources for Are You Ready?

- ***Are You Ready? Intervention and Enrichment***
- **Recording Sheet for Are You Ready?** *Chapter 11 Resource Book . . . . . p. 3*

***Are You Ready? Intervention* CD-ROM**

ARE YOU READY?

Were students successful with Are You Ready?

NO — INTERVENE

✔ **Model Fractions 6–8**
Intervention Practice, Skill 18
CD-ROM Intervention Activities, Skill 18

✔ **Write Fractions as Decimals 9–12**
Intervention Practice, Skill 26
CD-ROM Intervention Activities, Skill 26

✔ **Compare Fractions, Decimals, and Percents 13–16**
Intervention Practice, Skill 32
CD-ROM Intervention Activities, Skill 32

✔ **Write Ratios 17–18**
Intervention Practice, Skill 27
CD-ROM Intervention Activities, Skill 27

YES — ENRICH

Are You Ready? Enrichment, pp. 427–428

Section 11A

Understanding Probability

One-Minute Section Planner

Lesson	Materials	Resources
Lesson 11-1 Introduction to Probability **NCTM:** Data Analysis and Probability, Reasoning and Proof, Communication, Connections **NAEP:** Data Analysis and Probability 4a ☑ SAT-9 ☑ SAT-10 ☐ ITBS ☑ CTBS ☑ MAT ☑ CAT	**Optional** Number cubes *(MK)* Teaching Transparency T2 *(CRB)*	• *Chapter 11 Resource Book*, pp. 6–15 • Daily Transparency T1, CRB • Additional Examples Transparencies T3–T5, CRB • *Alternate Openers: Explorations*, p. 92
Lesson 11-2 Experimental Probability **NCTM:** Data Analysis and Probability, Reasoning and Proof, Communication **NAEP:** Data Analysis and Probability 4c ☐ SAT-9 ☐ SAT-10 ☐ ITBS ☐ CTBS ☐ MAT ☐ CAT	**Optional** Spinners *(MK, CRB, p. 67)* Teaching Transparency T7 *(CRB)*	• *Chapter 11 Resource Book*, pp. 16–24 • Daily Transparency T6, CRB • Additional Examples Transparencies T8–T10, CRB • *Alternate Openers: Explorations*, p. 93
Hands-On Lab 11A Simulations **NCTM:** Data Analysis and Probability, Representation **NAEP:** Data Analysis and Probability 4c ☐ SAT-9 ☐ SAT-10 ☐ ITBS ☐ CTBS ☐ MAT ☐ CAT	**Required** Number cubes *(MK)* Coins	• *Hands-On Lab Activities*, pp. 101–104
Lesson 11-3 Theoretical Probability **NCTM:** Data Analysis and Probability, Reasoning and Proof, Communication **NAEP:** Data Analysis and Probability 4d ☐ SAT-9 ☑ SAT-10 ☐ ITBS ☑ CTBS ☑ MAT ☑ CAT	**Optional** Teaching Transparency T12 *(CRB)*	• *Chapter 11 Resource Book*, pp. 25–33 • Daily Transparency T11, CRB • Additional Examples Transparencies T13–T14, CRB • *Alternate Openers: Explorations*, p. 94
Section 11A Assessment		• Mid-Chapter Quiz, SE p. 568 • Section 11A Quiz, AG p. 27 • *Test and Practice Generator* CD-ROM

SAT = *Stanford Achievement Tests* **ITBS** = *Iowa Test of Basic Skills* **CTBS** = *Comprehensive Test of Basic Skills/Terra Nova*
MAT = *Metropolitan Achievement Tests* **CAT** = *California Achievement Test*

NCTM = Complete standards can be found on pages T27–T33. **NAEP** = Complete standards can be found on pages A31–A35.

SE = *Student Edition* **TE** = *Teacher's Edition* **AR** = *Assessment Resources* **CRB** = *Chapter Resource Book* **MK** = *Manipulatives Kit*

Section Overview

Introduction to Probability — *Lesson 11-1*

Why? Probability is used in making predictions, such as predictions about the weather.

Probabilities are written as fractions or decimals from 0 to 1 or percents from 0% to 100%. The higher an event's probability, the more likely that event is to happen.

Probability is the measure of how likely an event is to occur.

Probability	Likelihood
0%	Never happen
50%	Same chance of happening as of not happening
100%	Always happen

Experimental Probability, Simulations — *Lesson 11-2, Hands-On Lab 11A*

Why? Performing an experiment is one way to estimate the probability of an event.

Vocabulary	Definition	Example
Experiment	An activity involving chance that can have different results	Two coins are tossed.
Outcome	A result of an experiment	HT (H = heads, T = tails)
Sample space	The set of all possible outcomes	HH, HT, TH, TT
Experimental probability	$\frac{\text{Number of times event occurs}}{\text{Total number of trials}}$	If you toss two coins 100 times, and they both come up heads on 24 of those trials, then the experimental probability of both coming up heads, based on this experiment, is as follows: probability $= \frac{24}{100} = \frac{6}{25} = 0.24 = 24\%$.

A **simulation** is a model of a real situation. Random numbers can be used to simulate random events in real situations.

Theoretical Probability — *Lesson 11-3*

Why? Theoretical methods can be used to find probability without performing experiments or simulations.

One situation in which you can use theoretical probability is when all outcomes have the same chance of occurring, such as tossing fair coins or number cubes. In other words, the outcomes are equally likely.

Theoretical Probability

$$\frac{\text{number of ways event can occur}}{\text{total number of possible outcomes}}$$

11-1 Organizer

Pacing: Traditional 1 day
Block $\frac{1}{2}$ day

Objective: Students estimate the likelihood of an event and write and compare probabilities.

Indiana Academic Standards
6.6.5, 6.6.6, 6.7.1, 6.7.2, 6.7.5, 6.7.11

Warm Up

Write each fraction as a decimal and as a percent.

1. $\frac{3}{4}$ 0.75; 75%
2. $\frac{1}{2}$ 0.50, 50%
3. $\frac{2}{9}$ $0.\overline{2}$; 22.2%
4. $\frac{3}{8}$ 0.375; 37.5%

Problem of the Day

What fraction of the numbers from 0 to 99 are divisible by 3? $\frac{34}{100}$, or $\frac{17}{50}$

Available on Daily Transparency in CRB

Math Humor

Did you hear the joke about the statistician? probably

11-1 Introduction to Probability

Learn to estimate the likelihood of an event and to write and compare probabilities.

Vocabulary
probability

The weather report gives a 5% chance of rain today. Will you wear your raincoat? What if the report gives a 95% chance of rain?

In this situation, you are using probability to help make a decision. **Probability** is the measure of how likely an event is to occur. In this case, both 5% and 95% are probabilities of rain.

Probabilities are written as fractions or decimals from 0 to 1 or as percents from 0% to 100%. The higher an event's probability, the more likely that event is to happen.

- Events with a probability of 0, or 0%, never happen.
- Events with a probability of 1, or 100%, always happen.
- Events with a probability of 0.5, or 50%, have the same chance of happening as of not happening.

Impossible	Unlikely	As likely as not	Likely	Certain
0 0%		0.5 $\frac{1}{2}$ 50%		1 100%

A 95% chance of rain means rain is highly likely. A 5% chance of rain means rain is highly unlikely.

EXAMPLE 1 Estimating the Likelihood of an Event

Write *impossible*, *unlikely*, *as likely as not*, *likely*, or *certain* to describe each event.

A **The month of June has 30 days.**
certain

B **A coin toss comes up heads.**
as likely as not

C **You roll a 9 on a standard number cube.**
impossible

D **This spinner lands on red.**
likely

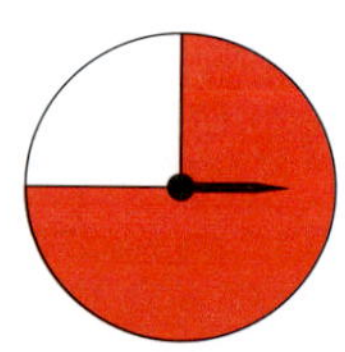

Helpful Hint
A standard number cube is numbered from 1 to 6.

1 Introduce

Alternate Opener

EXPLORATION

11-1 Introduction to Probability

Probability describes how likely it is that an event will occur. For example, it is likely that a dog will eat a treat, and it is unlikely that a human being will live 200 years.

Tell whether each event is likely or unlikely to happen.

	Event	Probability
1.	Having a blackout during a thunderstorm	
2.	Losing your money in an old vending machine	
3.	Winning the lottery	
4.	Finding a \$100 bill on the street	
5.	Passing a test for which you studied very hard	
6.	Being in a traffic jam at 5:00 P.M. on Monday	
7.	Hearing your favorite song when you first turn on the radio	

Think and Discuss

8. **Name** an event that is impossible.
9. **Name** an event that is certain to happen.

Motivate

To introduce students to probability, ask students to roll a number cube 6 times and record each of the 6 outcomes. Discuss and compare the outcomes to introduce the concept of probability. If the activity is done in small groups, collect outcome data from each group for comparison.

Exploration worksheet and answers on Chapter 11 Resource Book pp. 7 and 68

2 Teach

Lesson Presentation

Guided Instruction

In this lesson, students learn to estimate the likelihood of an event and to write and compare probabilities. Define the term *probability,* and explain the terms *impossible*, *unlikely*, *as likely as not*, *likely*, and *certain*. Teach students to express given probabilities as decimals, fractions, and percents. Then teach students how to compare given probabilities to determine which is more likely.

EXAMPLE 2 Writing Probabilities

A **The weather report gives a 35% chance of rain for tomorrow. Write this probability as a decimal and as a fraction.**

$35\% = 0.35$ *Write as a decimal.*

$35\% = \frac{35}{100} = \frac{7}{20}$ *Write as a fraction in simplest form.*

B **The chance that Ethan is chosen to represent his class in the student council is 0.6. Write this probability as a fraction and as a percent.**

$0.6 = \frac{6}{10} = \frac{3}{5}$ *Write as a fraction in simplest form.*

$0.6 = 60\%$ *Write as a percent.*

C **There is a $\frac{9}{25}$ chance of getting a green gumball out of a certain machine. Write this probability as a decimal and as a percent.**

$\frac{9}{25} = 9 \div 25 = 0.36$ *Write as a decimal.*

$\frac{9}{25} = \frac{9 \cdot 4}{25 \cdot 4} = \frac{36}{100} = 36\%$ *Write as a percent.*

Helpful Hint

In Example 2C, after you find the decimal form of $\frac{9}{25}$, you can use it to find the percent.

$0.36 = 36\%$

EXAMPLE 3 Comparing Probabilities

A **On a flowering plant called the four o'clock, there is a 50% chance the flowers will be pink, a 25% chance the flowers will be white, and a 25% chance the flowers will be red. Is it more likely that the flowers will be pink or white?**

Compare: $50\% > 25\%$

The flowers are more likely to be pink than white.

B **When you spin this spinner, there is a 25% chance that it will land on red, a 50% chance that it will land on yellow, and a 25% chance that it will land on blue. Is it more likely to land on red or on blue?**

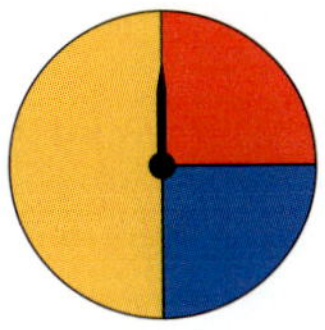

Compare: $25\% = 25\%$

It is as likely to land on red as on blue.

Think and Discuss

1. Give an example of a situation that involves probability.

2. Name events that can be described by each of the following terms: *impossible, likely, as likely as not, unlikely,* and *certain.*

Additional Examples

Example 1

Write *impossible, unlikely, as likely as not, likely,* or *certain* to describe each event.

A. You roll an even number on a standard number cube. as likely as not

B. February has 28 days. certain

C. A spinner is equally divided into green, blue, yellow, and red. The spinner lands on blue. unlikely

D. A spinner with sectors labeled 2, 4, 6, 8, and 10 lands on an odd number. impossible

Example 2

A. Write the probability 75% as a decimal and as a fraction. **0.75, $\frac{3}{4}$**

B. Write the probability 0.8 as a fraction and as a percent. **$\frac{4}{5}$, 80%**

C. Write the probability $\frac{7}{50}$ as a decimal and as a percent. **0.14, 14%**

Example 3

A. On a standard number cube, there is a 50% chance of rolling a multiple of 2 and a $33\frac{1}{3}\%$ chance of rolling a multiple of 3. Is it more likely to roll a multiple of 2 or a multiple of 3? a multiple of 2

B. When you spin a certain spinner, it has a 15% chance of landing on yellow, a 15% chance of landing on green, and a 70% chance of landing on purple. Is it more likely to land on purple or green? land on purple

3 Close

Reaching All Learners
Through Grouping Strategies

Have students in groups discuss probabilities of events in other school subjects. Examples: "It is certain that we will have a spelling test on Friday." "It is likely that we will run laps in physical education." "It is impossible for humans to live without H_2O."

Summarize

Review probability concepts presented in the lesson and the likelihood of different events.

Answers to Think and Discuss

1. Possible answer: playing a board game
2. Possible answers: impossible: a cat will fly; likely: I will get an A on my report card; as likely as not: I flip a coin and it lands heads up; unlikely: it will snow in Florida; certain: the sun will rise tomorrow.

11-1 PRACTICE & ASSESS

Students may want to refer back to the lesson examples.

Assignment Guide

If you finished Example 1 assign:
Core 1–2, 5–8, 11–14, 21–26
Enriched 1–2, 5–8, 11–14, 21–26

If you finished Example 2 assign:
Core 1–3, 5–9, 11–14, 21–26
Enriched 5–9, 11–17, 21–26

If you finished Example 3 assign:
Core 1–15, 18, 21–26
Enriched 5–26

Notes

Math Background

The study of probability has its origins in problems that arise in games of chance. Most famous among these is known as the "problem of the points."

The problem of the points involves the question of how to divide the stakes in an interrupted game of chance between two equally skilled players, knowing the scores of the players and the number of points needed to win. In 1654, Blaise Pascal and Pierre de Fermat established a correspondence in which they discussed the problem and solved it correctly, but differently. This correspondence is considered the foundation of the mathematical study of probability.

11-1 Exercises

FOR EXTRA PRACTICE see page 658

internet connect
Homework Help Online
go.hrw.com Keyword: MR4 11-1

GUIDED PRACTICE

See Example 1

Write *impossible, unlikely, as likely as not, likely,* or *certain* to describe each event.

1. This year has 12 months. **certain**
2. You win the lottery. **unlikely**

See Example 2

3. Suppose that the chance of reaching into a bag of coins and selecting a quarter is 40%. Write this probability as a decimal and as a fraction. **0.4, $\frac{2}{5}$**

See Example 3

4. If there are two children in a family, there is a 25% chance that both children are boys, a 25% chance that both children are girls, and a 50% chance that one child is a boy and the other is a girl. Which is more likely, that both children are boys or that one child is a boy and the other is a girl? **boy and girl**

INDEPENDENT PRACTICE

See Example 1

Write *impossible, unlikely, equally likely, likely,* or *certain* to describe each event.

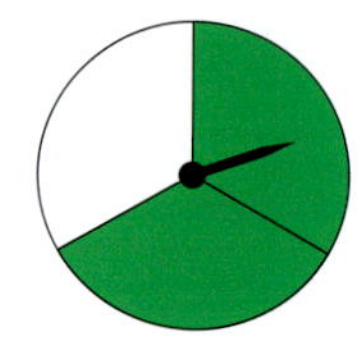

5. The spinner at right lands on green. **likely**
6. The spinner at right lands on blue. **impossible**
7. You guess one winning number between 1 and 500. **unlikely**
8. You correctly guess one of eight winning numbers between 1 and 10. **likely**

See Example 2

9. The chance of Jill's missing a free throw is $\frac{3}{10}$. Write this probability as a decimal and as a percent. **0.3, 30%**

See Example 3

10. If you choose from a bag of mixed nuts, there is a 45% chance of choosing a peanut, a 20% chance of choosing a pecan, a 15% chance of choosing a cashew, and a 20% chance of choosing a walnut. Is it less likely that you will choose a pecan or a cashew from the bag? **cashew**

PRACTICE AND PROBLEM SOLVING

Describe the events as *impossible, unlikely, as likely as not, likely,* or *certain.*

11. The probability of winning a game is $\frac{2}{3}$. **likely**
12. The probability of being chosen for a team is 0.09. **unlikely**
13. There is a 50% chance of snow today. **as likely as not**
14. Your chances of being struck by lightning are $\frac{1}{2{,}000{,}000}$. **unlikely**

RETEACH 11-1

LESSON 11-1 **Reteach**
Introduction to Probability

Probability is the measure of how likely it is that an event will occur.

You can write the probability of an event as a fraction or decimal from 0 to 1, or as a percent from 0 to 100%, inclusive.

The probability of tossing a penny and it landing on heads is $\frac{1}{2}$.

To write the probability as a decimal, divide the numerator by the denominator.

$1 \div 2 = 0.5$ The probability of landing on heads is 0.5.

To write the probability as a percent, first write the fraction as a decimal. Then move the decimal point two places to the right.

$1 \div 2 = 0.5 = 50\%$ The probability of landing on heads is 50%.

Write the probability.

1. The probability of Joy winning the race is 0.4. Write this probability as a fraction and a percent.
$\frac{2}{5}$, 40%

The higher the probability, the more likely the event is to occur.

Events with a probability of 0 or 0% never happen.
Events with a probability of 1 or 100% always happen.

Events with a probability of $\frac{1}{2}$, 0.5, or 50%, have the same chance of happening as of not happening.

If you have a bag of red marbles and blue marbles, the probability of pulling out a green marble is 0. The event is impossible because there are no green marbles in the bag.

Write *impossible, unlikely, as likely as not, likely,* or *certain* to describe each event.

2. A week has seven days. **certain**
3. Rolling a 7 on a standard number cube. **impossible**

PRACTICE 11-1

LESSON 11-1 **Practice B**
Introduction to Probability

Write *impossible, unlikely, as likely as not, likely,* or *certain* to describe each event.

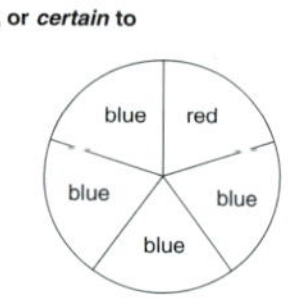

1. landing on blue **likely**
2. landing on green **impossible**
3. landing on red **unlikely**
4. landing on blue or red **certain**
5. You will spin the spinner clockwise. **as likely as not**

Write each probability as a decimal and as a fraction.

6. There is a 10% chance of rain tomorrow. **0.1, $\frac{1}{10}$**
7. There is a 75% chance of snow tomorrow. **0.75, $\frac{3}{4}$**
8. There is a 25% chance of hail tomorrow. **0.25, $\frac{1}{4}$**

Compare probabilities.

Prize Winning Probabilities	
Color TV	17%
DVD player	22%
Watch	13%
Stereo	21%
Diamond ring	27%

9. Are you more likely to win a color TV or a watch? **a color TV**
10. Are you more likely to win a DVD player or a stereo? **a DVD player**
11. Are you more likely to win a diamond ring, a DVD player, or a stereo? **a diamond ring**
12. A bag has 4 red marbles, 3 blue marbles, 4 green marbles, and 1 black marble. Which term best describes the probability of picking a black marble from the bag: impossible, likely, as likely as not, unlikely, or impossible? **unlikely**

Life Science LINK

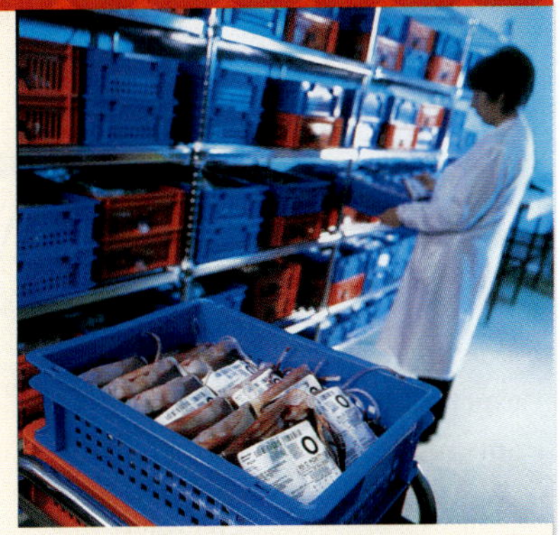
The donated blood in these bags is Type O.

Each year, millions of people donate blood. This blood is given to patients with certain bleeding disorders, people who have been in serious accidents, or patients who lose a lot of their own blood during surgery.

There are eight different human blood types, which are shown in the chart, along with the percent of people who have each type. It is very important that people receive the right type of blood. If they do not, their bodies will not recognize the foreign blood cells and will attack the cells.

15. How would you describe the probability of a person having AB positive blood: impossible, unlikely, as likely as not, likely, or certain? Explain. unlikely; 3 people out of 100 could have AB positive blood.

16. If a person is randomly chosen, which blood type is he or she most likely to have? O positive

17. If a person is randomly chosen, which blood type is he or she least likely to have? AB negative

18. Write the probability that a randomly chosen person will have A negative blood as a decimal and as a fraction in simplest form. 0.06; $\frac{3}{50}$

19. ***WRITE ABOUT IT*** Blood banks especially encourage people with certain types of blood to donate. Which blood types do you think these are? Explain.

20. ***CHALLENGE*** A person with AB positive blood can safely receive O, A, B, or AB blood. What is the probability that a randomly chosen person could donate blood to a person with AB positive blood? probability of 1

Blood Types in the United States

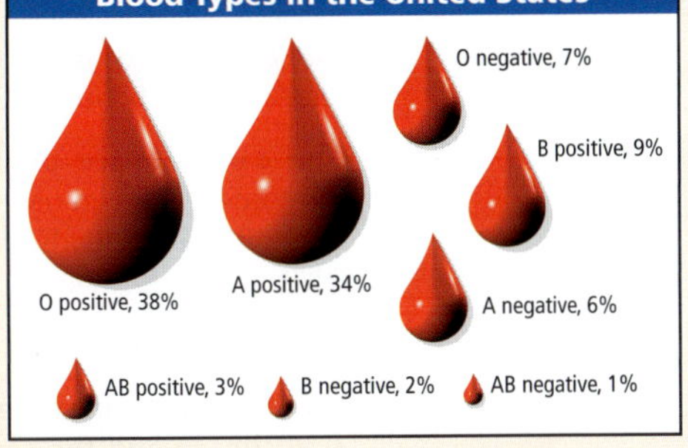

go.hrw.com
KEYWORD: MR4 Blood
CNN Student News

Spiral Review

21. Juanita needs $\frac{3}{4}$ cup of pineapple juice to make punch. She has only half that amount. How much pineapple juice does Juanita have? (Lesson 5-1) $\frac{3}{8}$

Name the quadrant on a coordinate plane where each point is located. (Lesson 9-3)

22. (4, −6) IV **23.** (−1, 5) II **24.** (2, 3) I **25.** (−2, −4) III

26. **TEST PREP** The perimeter of a square is 24 m. What is the length of one side of the square? (Lesson 10-1) A

A 6 m **B** 8 m **C** 12 m **D** 16 m

CHALLENGE 11-1

LESSON 11-1 **Challenge**
Shark!

If you are like most people, you think sharks are scary. People have feared these ancient creatures for thousands of years.

But we actually have little to fear—statistically speaking, that is. Sharks rarely attack people. In fact, you have a greater chance of being killed by a bee than by a shark! Still, knowing the probability of where most sharks attack might make your next trip to the beach a little more relaxing. Or will it?

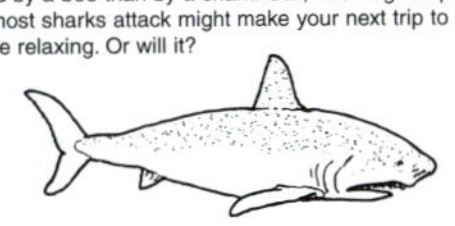
The Great White shark is one of the 25 species of sharks known to attack people. It has about 3,000 razor-sharp teeth!

Complete the tables below to describe the probabilities of where shark attacks are most likely to occur.

Where Sharks Attack People

Location of Attack (feet from shore)	Probability Fraction	Probability Percent
0–50	$\frac{31}{100}$	31%
51–100	$\frac{11}{100}$	11%
101–150	$\frac{3}{50}$	6%
151–200	$\frac{3}{100}$	3%
201–5,200	$\frac{27}{100}$	27%
Farther than 5,201	$\frac{11}{50}$	22%

1. How many feet from shore do most sharks attack people?
   0–50 feet
2. Why do you think this is the most likely location for attacks?
   Possible answer: because most people swim close to shore

PROBLEM SOLVING 11-1

LESSON 11-1 **Problem Solving**
Introduction to Probability

Floods are categorized by their probability of occurrence. For example, a flood categorized as a 20-year flood means it has a 1 in 20 chance of occurring in any given year. Complete the flood probability chart below. Then use it to answer the questions. Write answers in simplest form.

Flood Probabilities of Occurrence

	Category	Probability Fraction	Probability Decimal	Probability Percent
1.	2-year flood	$\frac{1}{2}$	0.5	50%
2.	5-year flood	$\frac{1}{5}$	0.2	20%
3.	10-year flood	$\frac{1}{10}$	0.1	10%
4.	50-year flood	$\frac{1}{50}$	0.02	2%
5.	100-year flood	$\frac{1}{100}$	0.01	1%

6. Which flood category in the table is the most likely to occur in a given year? The least likely?
   most: 2-year flood; least: 100-year flood
7. Following the naming system in the table, what category name would you use for a flood that is certain to occur in any given year?
   a 1-year flood
8. The Yukon River in Alaska had a 100-year flood in 1992. Does this mean that another 100-year flood could not occur on the Yukon River until 2092? Explain.
   No; The probability of occurrence is the same every year—1%.
9. During the Mississippi River system flood of 1993, about 8 million acres of land were flooded! It was a rare 500-year flood. What is the percent of probability that another 500-year flood will occur on the Mississippi River system next year?
   0.2%

Interdisciplinary LINK

Life Science

Exercises 15–20 involve questions about the distribution of human blood types. Students learn about blood types in middle school science programs such as *Holt Science and Technology.*

Answers

19. Possible answer: all types that are not O positive or A positive, because when needed, these would be the hardest to locate

Journal

Have students write a description of a spinner that, when spun, has an outcome that is always certain.

Test Prep Doctor

For Exercise 26, students need to know that the perimeter of a square is equal to 4 times the length of a side. So students must divide the given perimeter by 4 to find the length of one side. Students who answered **B** divided by 3, and students who answered **C** divided by 2. Students who answered **D** multiplied by $\frac{2}{3}$.

Lesson Quiz

Write *impossible, unlikely, equally likely, likely,* or *certain* to describe each event.

1. The sun will rise tomorrow. certain
2. You will roll 13 when rolling two dice. impossible
3. There is a 0.125 chance of picking the winning ticket. Write this probability as a fraction and as a percent. $\frac{1}{8}$, 12.5%
4. At Hamburger Hut, there is a 20% chance of getting a plastic dinosaur cup and a 35% chance of getting a plastic rabbit cup. Is it less likely that you will receive a rabbit cup or dinosaur cup? dinosaur cup

Available on Daily Transparency in CRB

11-2 Organizer

Pacing: Traditional 1 day
Block $\frac{1}{2}$ day

Objective: Students find the experimental probability of an event.

Indiana Academic Standards
6.6.5, 6.6.6, 6.7.1, 6.7.2, 6.7.3, 6.7.4, 6.7.5, 6.7.10, 6.7.11

Warm Up

Write *impossible, unlikely, equally likely, likely,* or *certain* to describe each event.

1. A particular person's birthday falls on the first of a month. unlikely
2. You roll an odd number on a fair number cube. equally likely
3. There is a 0.14 probability of picking the winning ticket. Write this as a fraction and as a percent. $\frac{7}{50}$, 14%

Problem of the Day

Max picks a letter out of this problem at random. What is the probability that the letter is in the first half of the alphabet? $\frac{57}{101}$

Available on Daily Transparency in CRB

Math Fact !!

Ars Conjectandi was the first book devoted entirely to probability. It was written by Jacques Bernoulli in 1713.

11-2 Experimental Probability

Learn to find the experimental probability of an event.

Vocabulary
experiment
outcome
sample space
experimental probability

Four Possibilities

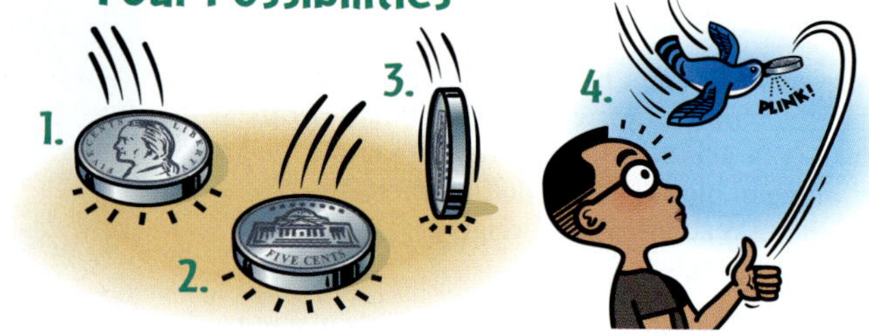

An **experiment** is an activity involving chance that can have different results. Flipping a coin and rolling a number cube are examples of experiments.

The different results that can occur are called **outcomes** of the experiment. If you are flipping a coin, heads is one possible outcome.

The **sample space** of an experiment is the set of all possible outcomes. You can use { } to show sample spaces. When a coin is being flipped, {heads, tails} is the sample space.

EXAMPLE 1 Identifying Outcomes and Sample Spaces

For each experiment, identify the outcome shown and the sample space.

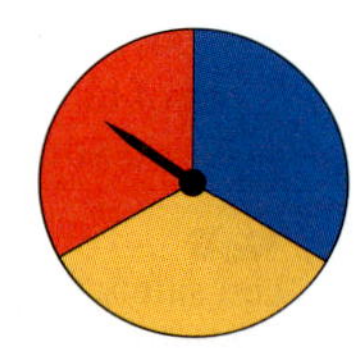

A spinning a spinner
outcome shown: red
sample space: {red, blue, yellow}

B tossing two coins
outcomes shown: heads, tails (H, T)
sample space: {HH, HT, TH, TT}

Performing an experiment is one way to estimate the probability of an event. If an experiment is repeated many times, the **experimental probability** of an event is the ratio of the number of times the event occurs to the total number of times the experiment is performed.

EXPERIMENTAL PROBABILITY

$$\text{probability} \approx \frac{\text{number of times the event occurs}}{\text{total number of trials}}$$

1 Introduce

Alternate Opener

EXPLORATION

11-2 Experimental Probability

You can find the ***experimental probability*** of an event by dividing the number of times an event occurs by the total number of times the experiment is performed.

$$\text{probability} = \frac{\text{number of times an event occurs}}{\text{total number of trials}}$$

1. The data in the table show the number of free throws five players made in a season. Find the $\frac{\text{made}}{\text{attempts}}$ ratio for each player.

	Bo	Jack	Ali	Kim	José
Free Throws Made	30	32	15	36	24
Attempts	48	64	25	48	49
$\frac{\text{Made}}{\text{Attempts}}$					

2. Which player has the best chance of making a free throw?
3. Which player has the worst chance of making a free throw?

Think and Discuss

4. **Discuss** how you determined the answers for numbers 2 and 3.
5. **Explain** how to write each $\frac{\text{made}}{\text{attempts}}$ ratio as a percent.

Motivate

Review the term *probability* and the different likelihoods of events (e.g., unlikely, certain, etc.). Remind students that probability can be expressed as a fraction, decimal, or percent. Have them express $\frac{1}{4}$ as a percent and as a decimal. 25%; 0.25

Exploration worksheet and answers on Chapter 11 Resource Book pp. 17 and 70

2 Teach

Lesson Presentation

Guided Instruction

In this lesson, students learn to find the experimental probability of an event. Introduce the new vocabulary for the lesson: *experiment*, *outcome*, *sample space*, and *experimental probability*. Emphasize that finding experimental probability requires you to do an experiment and collect data or use existing data. Teach students to express experimental probability as a fraction and to compare experimental probabilities.

Teaching Tip When students are identifying sample spaces, remind them to use { } to enclose their answer.

EXAMPLE 2 Finding Experimental Probability

Writing Math

The probability of an event can be written as *P*(event). *P*(blue) means "the probability that blue will be the outcome."

For one month, Tosha recorded the time at which her school bus arrived. She organized her results in a frequency table.

Time	7:00–7:04	7:05–7:09	7:10–7:15
Frequency	8	9	3

A **Find the experimental probability that the bus will arrive between 7:00 and 7:04.**

$$P(\text{between 7:00 and 7:04}) \approx \frac{\text{number of times the event occurs}}{\text{total number of trials}}$$

$$= \frac{8}{20} = \frac{2}{5}$$

B **Find the experimental probability that the bus will arrive before 7:10.**

$$P(\text{before 7:10}) \approx \frac{\text{number of times the event occurs}}{\text{total number of trials}}$$

$$= \frac{8 + 9}{20}$$ *Before 7:10 includes 7:00–7:04 and 7:05–7:09.*

$$= \frac{17}{20}$$

EXAMPLE 3 Comparing Experimental Probabilities

Ian tossed a cone 30 times and recorded whether it landed on its base or on its side. Based on Ian's experiment, which way is the cone more likely to land?

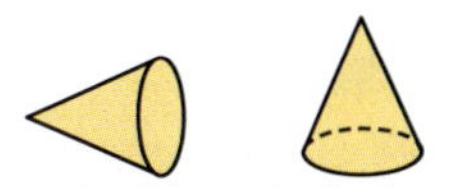

On its side On its base

Outcome	On its base	On its side
Frequency	𝍸 II	𝍸 𝍸 𝍸 𝍸 III

$$P(\text{base}) \approx \frac{\text{number of times the event occurs}}{\text{total number of trials}} = \frac{7}{30}$$

$$P(\text{side}) \approx \frac{\text{number of times the event occurs}}{\text{total number of trials}} = \frac{23}{30}$$

Find the experimental probability of each outcome.

$$\frac{7}{30} < \frac{23}{30}$$ *Compare the probabilities.*

It is more likely that the cone will land on its side.

Think and Discuss

1. **Explain** whether you and a friend will get the same experimental probability for an event if you perform the same experiment.
2. **Tell** why it is important to repeat an experiment many times.

Additional Examples

Example 1

Identify the outcome shown and the sample space.

spinning two spinners

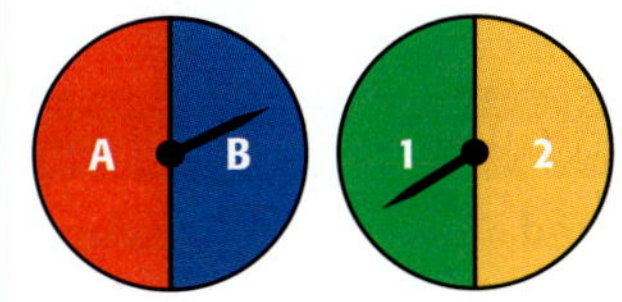

B1; {A1, A2, B1, B2}

Example 2

For one month, Mr. Crowe recorded the time at which his train arrived. He organized his results in a frequency table.

Time	6:50–6:52	6:53–6:56	6:57–7:00
Freq.	7	8	5

Find the experimental probability of the train arriving before 6:57. $\frac{3}{4}$

Example 3

Erika tossed a cylinder 30 times and recorded whether it landed on one of its bases or on its side. According to Erika's experiment, which way is the cylinder more likely to land?

Outcome	On a base	On its side
Freq.	𝍸 IIII	𝍸 𝍸 𝍸 𝍸 I

more likely to land on its side

3 Close

Reaching All Learners
Through Hands-On Experience

Give students spinners divided into 5 equal sections labeled 1–5 (provided in the Manipulatives Kit and on Chapter 11 Resource Book p. 67). Have each student conduct 20 trials with the spinner, recording the outcomes in a tally table. Then have them write the experimental probability for each outcome in the sample space {1, 2, 3, 4, 5} and then compare the probabilities to determine which, if any, outcome is more likely.

Summarize

Give the students examples of the *sample space* for a given experiment, and review how to find *experimental probability*.

Answers to Think and Discuss

1. Possible answer: Not necessarily; even though the outcomes have the same chances of happening, they may not occur the same number of times.
2. Possible answer: Each time you repeat the experiment you will probably get slightly different results. The more times you repeat the experiment, the closer your experiments probability will get to the actual probability.

11-2 PRACTICE & ASSESS

Students may want to refer back to the lesson examples.

Assignment Guide

If you finished Example 1 assign:
Core 1, 4–5, 9, 11, 17–21
Enriched 1, 4, 9, 10, 12, 17–21

If you finished Example 2 assign:
Core 1–2, 4–7, 9, 11, 17–21
Enriched 5–7, 9–14, 17–21

If you finished Example 3 assign:
Core 1–9, 14, 17–21
Enriched 6–21

Notes

11-2 Exercises

FOR EXTRA PRACTICE see page 658

internet connect
Homework Help Online
go.hrw.com Keyword: MR4 11-2

GUIDED PRACTICE

See Example 1

1. Identify the outcome and the sample space shown on the spinner. 6; {2, 4, 6, 8}

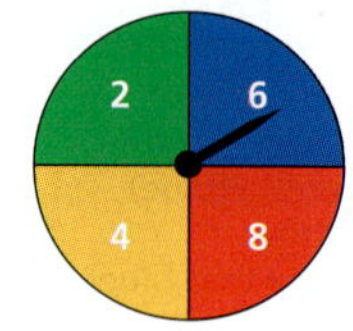

Josh recorded the number of hits his favorite baseball player made in each of 15 games. He organized his results in a frequency table.

Number of Hits	0	1	2	3
Frequency	4	8	2	1

See Example 2

2. Find the experimental probability that this player will get one hit in a game. $\frac{8}{15}$

See Example 3

3. Based on Josh's results, is this player more likely to get two hits in a game or no hits in a game? How many hits will this player most likely get in a game? no hits; 1 hit

INDEPENDENT PRACTICE

See Example 1

For each experiment, identify the outcome shown and the sample space.

4. 

4; {1, 2, 3, 4, 5, 6}

5. 

HTH; {HHH, HHT, HTH, THH, HTT, THT, TTH, TTT}

Jennifer has a bag of marbles. She removed one marble, recorded the color, and placed it back in the bag. She repeated this process several times and recorded her results in the table.

Color	Frequency
White	𝍸
Red	///
Yellow	𝍸
Black	𝍸 𝍸 //

See Example 2

6. Find the experimental probability that a marble selected from the bag will be red. $\frac{3}{25}$

7. Find the experimental probability that a marble selected from the bag will not be black. $\frac{13}{25}$

See Example 3

8. Based on Jennifer's experiment, which color marble is she most likely to select from the bag? black

Math Background

When you are calculating experimental probability, it is crucial that you repeat the experiment many times. When an experiment is repeated under essentially unchanged conditions, *statistical regularity* occurs. Under statistical regularity, the experimental probability of an event varies less and less as the number of trials in an experiment increases.

An example can help make this clear. If a coin were tossed 10 times, it would not be unusual for it to land heads up 7 times. However, if the same coin were tossed 1,000 times and it landed heads up 700 times, you would probably conclude that the coin was unfair.

RETEACH 11-2

LESSON 11-2 **Reteach**
Experimental Probability

An experiment is an activity involving chance that can have different results. Rolling a standard number cube is an experiment.

The different results that are possible are called the outcomes of an experiment. The sample space of an experiment is the set of all possible outcomes. When rolling a standard number cube, the sample space is {1, 2, 3, 4, 5, 6}.

Identify the outcome shown and the sample space.

1. spinning a spinner

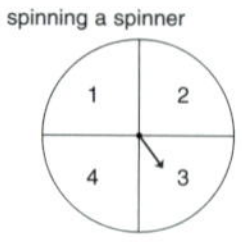

3; {1, 2, 3, 4}

2. tossing a coin

heads; {heads, tails}

If you perform an experiment many times, you can estimate the probability of an event.

To find the experimental probability of an event, use a formula.

$$\text{probability} \approx \frac{\text{number of times the event occurs}}{\text{total number of trials}}$$

Suppose you rolled a standard number cube 20 times and it landed on 5 six times. The probability of rolling a 5 is

$P(\text{rolling } 5) \approx \frac{6}{20} = \frac{3}{10}$

Find the experimental probability of each event.

Pulling Marbles Out of a Bag

Red	Blue	Green
𝍸 ||||	𝍸 |	𝍸

3. The experimental probability of pulling out a red marble. $\frac{9}{20}$

4. The experimental probability of pulling out a green marble. $\frac{1}{4}$

PRACTICE 11-2

LESSON 11-2 **Practice B**
Experimental Probability

For each experiment, identify the outcome shown and the sample space.

1. 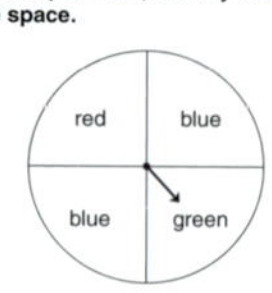

outcome: green
sample space: {red, blue, green}

2. 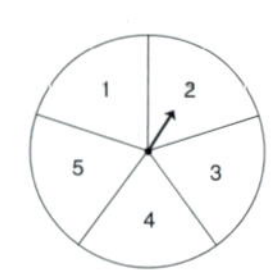

outcome: 2
sample space: {1, 2, 3, 4, 5}

Amanda has a standard deck of playing cards. She picked one card, recorded the suit, and placed it back in the deck. She repeated this process several times and recorded her results in the table.

Heart	𝍸 ||
Diamond	||||
Spade	𝍸
Club	𝍸 |||

3. Find the experimental probability that a card selected from the deck will be a spade. $\frac{5}{24}$

4. Find the experimental probability that a card selected from the deck will be a diamond. $\frac{1}{6}$

5. Based on Amanda's experiment, which card suit is she most likely to select from the deck? clubs

6. Based on Amanda's experiment, which card suit is she least likely to select from the deck? diamonds

7. In 28 coin tosses, John got tails up 14 times. What is the experimental probability that John will make get tails up on his next toss? $\frac{1}{2}$

PRACTICE AND PROBLEM SOLVING

Identify the sample space for each situation.

9. Fe has three clean T-shirts—a yellow one, a red one, and a green one. Without looking, she pulls one from her laundry basket. {yellow T-shirt, red T-shirt, green T-shirt}

10. You roll a number cube whose sides are all numbered 7. {7, 7, 7, 7, 7, 7}

11. The principal will choose two of the three finalists in the science fair to attend a banquet. The three finalists are Anna, Joel, and Roseann. {Anna & Joel, Roseann & Anna, Joel & Roseann}

12. 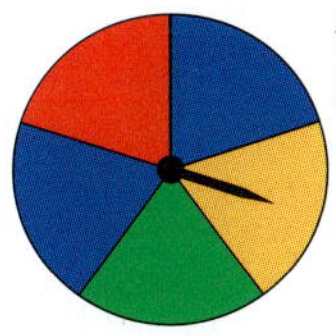 {blue, blue, red, green, yellow}

13. 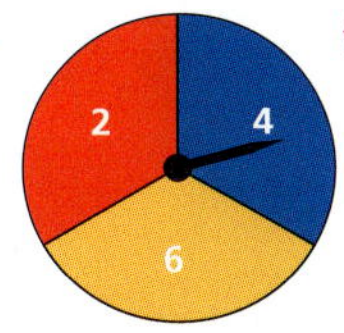

{2, 4, 6}

14. ***WEATHER*** Janet recorded the high temperature every day in January. She recorded her results in a frequency table.

Temperature (°F)	26–35	36–45	46–55	56–65
Number of Days	10	9	11	1

a. According to Janet's results, what is the probability that a day in January will be warmer than 55°F? $\frac{1}{31}$

b. Describe this probability as certain, likely, as likely as not, unlikely, or impossible. unlikely

15. ***WRITE ABOUT IT*** Conduct an experiment in which you toss a coin 100 times. Keep a tally of the number of times the coin shows heads. According to your results, what is the experimental probability that it will show heads? Compare your results with a classmate. Did you both get the same experimental probability? Why or why not? Check students' work.

16. ***CHALLENGE*** Suppose you roll two number cubes and add the two numbers that come up. What do you think the most likely sum would be? (*Hint:* Perform an experiment.) Check students' work.

Spiral Review

Find the next three numbers in each sequence. (Lesson 1-7)

17. 1, 3, 5, 7, … 9, 11, 13

18. 2, 4, 6, 8, 10, … 12, 14, 16

19. 1, 4, 9, 16, … 25, 36, 49

20. **TEST PREP** Which expression is **not** equal to half of n? (Lesson 8-8) B

A $0.5n$ **B** 5% of n **C** $\frac{n}{2}$ **D** $n \div 2$

21. **TEST PREP** Choose the greatest amount. (Lesson 8-9) G

F 45% of 200 **G** 60% of 190 **H** 50% of 150 **J** 100% of 110

Journal

Have students write a real-world problem similar to the one in Example 3.

Test Prep Doctor

Encourage students to read Exercise 20 carefully; they need to find the expression that is *not* equal to half of *n*. Because choices **A, C,** and **D** all name values that *are* equal to half of *n*, the correct answer is **B.**

Lesson Quiz

1. The spinner below was spun. Identify the outcome shown and the sample space. outcome: green; sample space: {red, blue, green, purple, yellow}

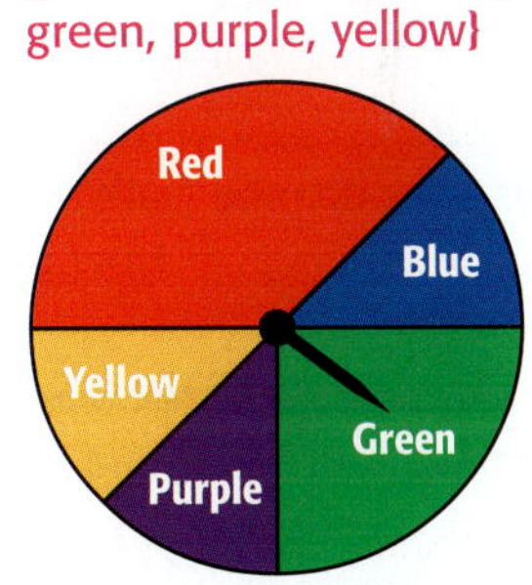

Sandra spun the spinner above several times and recorded the results in the table.

Color	Red	Blue	Green	Purple	Yellow
Freq.	4	2	1	2	0

2. Find the experimental probability that the spinner will land on blue. $\frac{2}{9}$

3. Find the experimental probability that the spinner will land on red. $\frac{4}{9}$

4. Based on the experiment, on which color will the spinner most likely land? red

Available on Daily Transparency in CRB

CHALLENGE 11-2

LESSON 11-2 **Challenge**
Batter Up!

A batting average is the probability of a player getting a hit. The sample space is the total number of at bats a player has had, not including walks. A hit is a favorable outcome.

How do you find this probability? Say Janet gets 30 hits at 100 at bats. Then her hitting probability is $\frac{30}{100}$, or 30%. This means that each time Janet steps up to the plate, she has a 30% chance of hitting the ball. For baseball, you multiply the percentages by 10 to get a batting average. So, Janet's hitting probability of 30% means her batting average is 300. Because very few professional baseball players hit above 300, you can see how hard it is to get a hit in the major leagues!

Complete the table below to find which major league player had the all-time highest career batting average. Round the decimals to the nearest thousandth.

Players with the Highest Career Batting Averages

Player	At Bats	Hits	Average
Babe Ruth	8,399	2,873	342
Ty Cobb	11,434	4,189	366
Harry Heilmann	7,787	2,660	341
Ed Delahanty	7,505	2,597	346
Ted Williams	7,706	2,654	344
Rogers Hornsby	8,173	2,930	358
Billy Hamilton	6,268	2,158	344
Joe Jackson	4,981	1,772	355
Dan Brouthers	6,711	2,296	342
Tris Speaker	10,195	3,514	344

PROBLEM SOLVING 11-2

LESSON 11-2 **Problem Solving**
Experimental Probability

Write the correct answer. Write answers in simplest form.

1. Brandy tossed a fair coin several times. She recorded in this table the result of each toss. What is the experimental probability that Brandy's next toss will land heads up?

Heads Up	卌 卌 卌 I
Tails Up	卌 卌 IIII

$\frac{8}{15}$

2. In this table, Charles recorded the gender of each person who shopped at his store this morning. What is the experimental probability that his next customer will be a woman?

Male	卌 卌 卌 卌 II
Female	卌 卌 卌 III

$\frac{9}{20}$

3. Nita packed 4 pairs of shorts for her beach vacation—a blue pair, a white pair, a denim pair, and a black pair. Without looking, she pulls out the blue pair from her suitcase. What are the outcome and the sample space for this experiment?

outcome: blue shorts; sample space: {blue shorts, white shorts, denim shorts, black shorts}

4. Mick rolled two number cubes at the same time. Each cube is numbered 1 through 6. The cubes showed a sum of 7. What are the outcome and sample space of this experiment?

outcome: 7; sample space: {2, 3, 4, 5, 6, 7, 8, 9, 10, 11, 12}

Abdul recorded the number of free throws his favorite basketball player made in each of 24 games. He organized his results in this frequency table. Circle the letter of the correct answer.

Free Throws Made	0	1	2	3	4
Frequency	1	4	7	9	3

5. What is the experimental probability that this player will make 1 free throw in the next game?

A $\frac{1}{24}$ C $\frac{7}{24}$
(B) $\frac{1}{6}$ D $\frac{1}{8}$

6. Based on Abdul's experiment, how many free throws will this player most likely make in any given game?

(F) 3 H 0
G 4 J 2

Hands-On LAB

11A Simulations

Pacing: Traditional 1 day
Block $\frac{1}{2}$ day

Objective: To use a number cube and coin to simulate probability experiments

Materials: Number cube, coin

Lab Resources

Hands-On Lab Activities . . . . p. 90

Using the Pages

Discuss what each roll of the number cube represents and how students should record their results in the table. Then discuss what each coin toss represents and how students should record their results.

1. What outcome does rolling a 3 represent? buying a box of cereal containing the letter *U*
2. A student conducted the simulation. The first 10 rolls were 4, 2, 5, 3, 4, 2, 4, 5, 2, and 1. What does her simulation tell about the actual contest? Does the simulation model a winning combination? It is going to take more than 10 boxes to win a prize. This player has the letters *Y, O, U, W,* and *I,* but does not have an *N,* because no 6 was rolled.
3. What does the outcome "tails" represent? Amy missing a shot How is it recorded in the table? as a *T*
4. How will you be able to tell whether Amy has made four shots in a row? Look for four consecutive *H*'s in a single row of the table.

Hands-On LAB 11A

Simulations

Use with Lesson 11-2

internet connect
Lab Resources Online
go.hrw.com
KEYWORD: MR4 Lab11A

A **simulation** is a model of an experiment that would be difficult or inconvenient to actually perform. In this lab, you will conduct simulations.

Activity 1

A cereal company is having a contest. To win a prize, you must collect six different cards that spell *YOU WIN.* One of the six letters is put into each cereal box. The letters are divided equally among the boxes. How many boxes do you think you will have to buy to collect all six cards?

1. Since there are six different cards that are evenly distributed, you can use a number cube to simulate collecting the letters. Each of the numbers from 1 to 6 will represent a letter. A roll of the number cube will simulate buying one box of cereal, and the number rolled will represent the letter inside that box.

1	2	3	4	5	6
Y	O	U	W	I	N
/	~~////~~ /	////	//	//	/

2. Roll the number cube, and keep track of the numbers you roll. Continue to roll the number cube until you have rolled every number at least once.

Think and Discuss

1. Look at the results in the table above. What was the last number rolled? How do you know?
2. How many rolls did it take to get all six numbers in your simulation? **Possible answer: 15**
3. How many boxes of cereal do you think you would have to buy to get all six letters? If you bought this many boxes, would you be sure to win? Explain.

Try This

1. Repeat the simulation three more times. Record your results. **Check students' work.**
2. Combine your data with the data of 5 of your classmates. Find the mean number of rolls from all 6 sets of data. **Possible answer:** $\frac{9 + 12 + 15 + 6 + 13 + 17}{6} = 12$
3. How many boxes of cereal do you think you would have to buy to get all six letters? Is this number different from what you thought after the first simulation? Explain.

Answers

Activity 1

Think and Discuss

1. 1 or 6; since 1 and 6 were each rolled only once, and the simulation is complete once all numbers have been rolled, then either 1 or 6 was the last number rolled.
3. Possible answer: 15; every simulation will have a different result, so even if you use results from your experiment to estimate the number of boxes you should buy, you can never be sure that you will win.

Try This

3. Possible answer: 12; yes, the results of my first two simulations were 15 and 9.

Activity 2

Amy is a basketball player who usually makes $\frac{1}{2}$ of the baskets that she attempts. Suppose she makes 20 shots in each game. If she plays ten games, in how many games do you think she will make at least four baskets in a row?

Trial	Results
1	THTHHTTHTTHTHTTHHHTT
2	HHTTTHTHTHHHHHTTHTHT
3	HTTTHTTTHTHTTTHTTHTT
4	HTHTHTHTHTTHTHTTTTTT
5	THTTTTHHTHTHHTHTTHTT
6	HTTHTHHHHHTHHHHHHHHH
7	TTHHTTHHHTHTHHTTHTTT
8	HTTHTTHTTTHHTTHTTHTT
9	HHHTTTTTHHHHHTHHTHHT
10	HTTHHTTHHHTHHTHTHHHH

1. There are two possible outcomes every time Amy shoots the ball—either she will make the basket or she will miss. Since Amy makes $\frac{1}{2}$ of her shots, you can toss a coin to simulate one shot. Let heads represent making the basket, and let tails represent missing.
2. Toss the coin 20 times to simulate one game. Keep track of your results.
3. Repeat 2 nine more times to simulate ten games.

Think and Discuss

1. Why does tossing a coin 20 times represent only one trial? **Each trial represents one game and Amy makes 20 shots in a game.**
2. Do any of your sequences contain four or more heads in a row? How many? **Possible answer: yes; 4**
3. In how many games do you think Amy will make at least four baskets in a row? Out of *every* ten games, will Amy always make at least four baskets in a row this number of times?
4. You can use your simulation to find the experimental probability that Amy will make at least four or more baskets in a row. Divide the number of trials in which the coin came up heads at least four times in a row by the total number of trials. What is the experimental probability that Amy will make at least four baskets in a row? **Possible answer: $\frac{4}{10} = 0.4$**
5. Suppose Amy made only $\frac{1}{3}$ of her shots. Would you still be able to use a coin as a simulation? Why or why not? **No, because the probability of a coin landing on a particular side is $\frac{1}{2}$; there is no way to simulate $\frac{1}{3}$ with a coin.**

Try This

1. In a group of ten families that each have four children, how many families do you think will have two girls and two boys? Make a prediction, and then design and carry out a simulation to answer this question. (Assume that having a boy and having a girl are equally likely events.) Was your prediction close? **Check students' work.**
2. Use your results from the previous problem to give the experimental probability that a family with four children will have two girls and two boys. **Check students' work.**
3. Think of an experiment, and design your own simulation to model it. **Check students' work.**

Answers

Activity 2

Think and Discuss

3. Possible answer: 4; every simulation will have different results, and the actual game results are dependent on outside factors that the experiment does not account for. So even if you use results from your experiment to estimate the number of baskets in a row that she will make, there is no guarantee that this will occur in every game.

11-3 Organizer

Pacing: Traditional 1 day
Block $\frac{1}{2}$ day

Objective: Students find the theoretical probability of an event.

Indiana Academic Standards
6.6.4, 6.6.5, 6.6.6, 6.7.1, 6.7.2, 6.7.5, 6.7.10, 6.7.11

Warm Up

Tim took one marble from a bag, recorded the color, and returned it to the bag. He repeated this several times and recorded the results.

Green	Red	Yellow	Purple
卌 卌 II	III	IIII	I

1. Find the experimental probability that a marble selected from the bag will be green. $\frac{3}{5}$
2. Find the experimental probability that a marble selected from the bag will not be yellow. $\frac{4}{5}$

Problem of the Day

What is the probability that the sum of four consecutive whole numbers is divisible by 4? 0

Available on Daily Transparency in CRB

Math Fact !

Dice have been found in Egyptian tombs dating from 2000 B.C.

11-3 Theoretical Probability

Learn to find the theoretical probability of an event.

Vocabulary
theoretical probability
equally likely
fair

Another way to estimate probability of an event is to use **theoretical probability**. One situation in which you can use theoretical probability is when all outcomes have the same chance of occurring. In other words, the outcomes are **equally likely**.

Equally likely outcomes

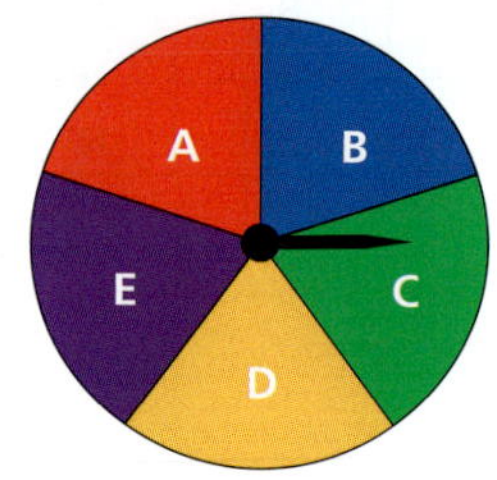

There is the same chance that the spinner will land on any of these letters.

Not equally likely outcomes

There is a greater chance that the spinner will land on 1 than on any other number.

An experiment with equally likely outcomes is said to be **fair**. You can usually assume that experiments involving items such as coins and number cubes are fair.

THEORETICAL PROBABILITY

$$\text{probability} \approx \frac{\text{number of ways event can occur}}{\text{total number of possible outcomes}}$$

EXAMPLE 1 Finding Theoretical Probability

A What is the probability that a fair coin will land heads up?

There are two possible outcomes when flipping a coin, heads or tails. Both are equally likely because the coin is fair.

$$P(\text{heads}) = \frac{\square}{\text{2 possible outcomes}}$$

There is only one way for the coin to land heads up.

$$P(\text{heads}) = \frac{\text{1 way event can occur}}{\text{2 possible outcomes}}$$

$$P(\text{heads}) = \frac{\text{1 way event can occur}}{\text{2 possible outcomes}} = \frac{1}{2}$$

1 Introduce

Alternate Opener

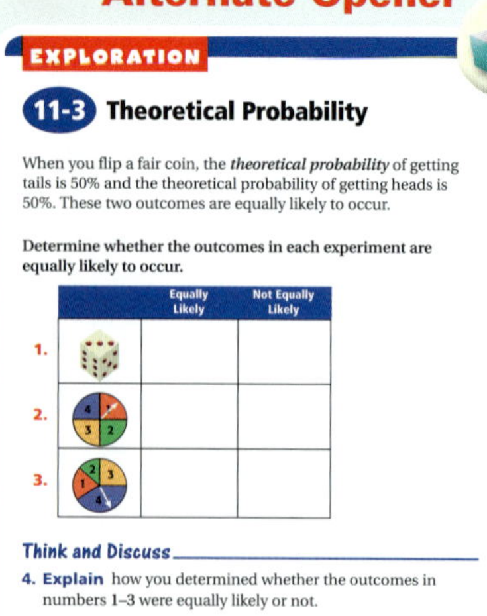

EXPLORATION

11-3 Theoretical Probability

When you flip a fair coin, the *theoretical probability* of getting tails is 50% and the theoretical probability of getting heads is 50%. These two outcomes are equally likely to occur.

Determine whether the outcomes in each experiment are equally likely to occur.

		Equally Likely	Not Equally Likely
1.			
2.			
3.			

Think and Discuss

4. **Explain** how you determined whether the outcomes in numbers 1–3 were equally likely or not.

Motivate

Ask students to explain experimental probability and how to write experimental probability as a fraction. Tell them that although experimental probability can only be found if an experiment has been conducted, theoretical probability can be found without conducting an experiment.

Exploration worksheet and answers on Chapter 11 Resource Book pp. 26 and 72

2 Teach

Lesson Presentation

Guided Instruction

In this lesson, students learn to find the theoretical probability of an event. Define the new vocabulary for the lesson: *theoretical probability*, *equally likely*, and *fair*. Teach students to write the theoretical probabilities of events happening and not happening. Point out that the sum of the probability of an event happening and the probability of that event not happening is 1, or 100%.

B **What is the probability of rolling a number less than 5 on a fair number cube?**

There are six possible outcomes when a fair number cube is rolled: 1, 2, 3, 4, 5, or 6. All are equally likely.

$$P(\text{less than } 5) = \frac{\blacksquare}{\text{6 possible outcomes}}$$

There are 4 ways to roll a number less than 5: 1, 2, 3, or 4.

$$P(\text{less than } 5) = \frac{\text{4 ways event can occur}}{\text{6 possible outcomes}}$$

$$P(\text{less than } 5) = \frac{\text{4 ways event can occur}}{\text{6 possible outcomes}} = \frac{4}{6} = \frac{2}{3}$$

Think about a single experiment, such as tossing a coin. There are two possible outcomes, heads or tails. What is $P(\text{heads}) + P(\text{tails})$?

Experimental Probability (coin tossed 10 times)			Theoretical Probability
H	T		
卌 I	IIII	$P(\text{heads}) = \frac{6}{10}$ $P(\text{tails}) = \frac{4}{10}$ $\frac{6}{10} + \frac{4}{10} = \frac{10}{10} = 1$	$P(\text{heads}) = \frac{1}{2}$ $P(\text{tails}) = \frac{1}{2}$ $\frac{1}{2} + \frac{1}{2} = \frac{2}{2} = 1$

No matter how you determine the probabilities, their sum is 1.

This is true for any experiment—the probabilities of the individual outcomes add to 1 (or 100%, if the probabilities are given as percents).

EXAMPLE 2 Finding Probabilities of Events Not Happening

Suppose there is a 10% chance of rain today. What is the probability that it will not rain?

In this situation, there are two possible outcomes, either it will rain or it will not rain.

$P(\text{rain}) + P(\text{not rain}) = 100\%$

$10\% + P(\text{not rain}) = 100\%$

$\underline{-10\%} \qquad \underline{-10\%}$ *Subtract 10% from each side.*

$P(\text{not rain}) = 90\%$

Think and Discuss

1. **Give an example** of a fair experiment. Give an example of an unfair experiment.
2. Suppose there is a 50% chance of snow and a 10% chance of sleet. What is the probability that neither one will occur?

COMMON ERROR ALERT

When calculating probability and there is more than one favorable outcome, some students may incorrectly find the probability of only one of two or more favorable outcomes occurring. Have these students find each probability separately and then add them together.

Additional Examples

Example 1

A. What is the probability of this fair spinner landing on 3? $\frac{1}{3}$

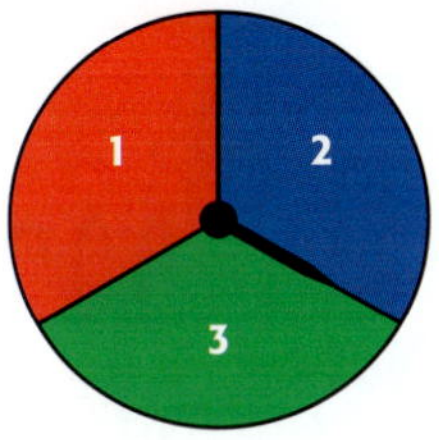

B. What is the probability of rolling a number greater than 4 on a fair number cube? $\frac{1}{3}$

Example 2

Suppose there is a 45% chance of snow tomorrow. What is the probability that it will not snow? 55%

3 Close

Reaching All Learners
Through Curriculum Integration

Life Science Gregor Mendel crossbred plants that had round-pea genes with plants that had wrinkled-pea genes. The first-generation offspring all had round peas. When these plants were bred with each other, 25% of the second generation had wrinkled peas, and the rest had round peas. Have students use probability to write a hypothesis about getting round peas in the second generation.

Possible answer: The probability of getting round peas in the second generation is $\frac{3}{4}$, so round peas are likely in the second generation.

Summarize

Review the new vocabulary in the lesson: *theoretical probability*, *equally likely*, and *fair*. Discuss how *equally likely* and *fair* relate to each other.

Possible response: Outcomes with the same chance of happening are said to be *equally likely*. An experiment or game with equally likely outcomes is *fair*.

Answers to Think and Discuss

1. Possible answers: You pull a tile from a bag with 5 red, 5 blue, 5 green, and 5 white tiles. You spin a spinner that is $\frac{3}{5}$ yellow and $\frac{2}{5}$ pink.
2. $50\% + 10\% + P(\text{no snow/sleet}) = 100\%$
   $P(\text{no snow/sleet}) = 100\% - 60\% = 40\%$

11-3 PRACTICE & ASSESS

Students may want to refer back to the lesson examples.

Assignment Guide

If you finished Example 1 assign:
Core 1–2, 5–7, 10–13, 37–42
Enriched 5–7, 18–21, 30–31, 37–42

If you finished Example 2 assign:
Core 1–13, 18–19, 24–25, 37–42
Enriched 20–42

Notes

11-3 Exercises

FOR EXTRA PRACTICE

see page 658

internet connect
Homework Help Online
go.hrw.com Keyword: MR4 11-3

GUIDED PRACTICE

See Example 1

1. What is the probability that a fair coin will land tails up? $\frac{1}{2}$

2. What is the probability of randomly choosing a vowel from the letters *A*, *B*, *C*, *D*, and *E*? $\frac{2}{5}$

See Example 2

3. The probability that a spinner will land on blue is 26%. What is the probability that it will not land on blue? 74%

4. Suppose you have an unfair number cube and the probability of rolling a 2 is 0.7. What is the probability that you will not roll a 2? 0.3

INDEPENDENT PRACTICE

See Example 1

5. What is the probability of rolling the number 3 on a fair number cube? $\frac{1}{6}$

6. What is the probability of rolling a number that is a multiple of 3 on a fair number cube? $\frac{1}{3}$

7. Find the probability that a yellow marble will be chosen from a bag that contains 3 green marbles, 2 red marbles, and 4 yellow marbles. $\frac{4}{9}$

See Example 2

8. Suppose there is an 81% chance of snow today. What is the probability that it will not snow? 19%

9. On a game show, the chance that the spinner will land on the winning color is 0.04. Find the probability that it will not land on the winning color. 0.96

PRACTICE AND PROBLEM SOLVING

A standard number cube is rolled. Find each probability.

10. $P(4)$ $\frac{1}{6}$

11. P(not 3) $\frac{5}{6}$

12. P(1, 2, or 3) $\frac{1}{2}$

13. P(number greater than 0) 1

14. P(odd number) $\frac{1}{2}$

15. P(number divisible by 5) $\frac{1}{6}$

16. P(prime number) $\frac{1}{2}$

17. P(negative number) 0

Nine pieces of paper with the numbers 1, 2, 2, 3, 4, 4, 5, 6, and 6 printed on them are placed in a bag. A student chooses one without looking. Compare the probabilities. Write <, >, or =.

18. $P(1)$ ■ $P(5)$ =

19. $P(3)$ ■ $P(2)$ <

20. $P(4)$ ■ P(5 or 6) <

21. P(3 or 5) ■ $P(6)$ =

22. P(even number) ■ P(odd number) >

23. P(multiple of 3) ■ P(number less than 4) <

Math Background

The symbol ≈ is used with both experimental and theoretical probability. Any calculation of probability is always an estimate. The symbol = is used in the theoretical probability after the assumption of a fair experiment.

RETEACH 11-3

LESSON 11-3 **Reteach**
Theoretical Probability

You can use theoretical probability to estimate the probability of an event.

To find the theoretical probability of an event, first find the number of ways the event can occur. Then divide that number by the total number of possible outcomes.

Think about a standard number cube. To find the theoretical probability of rolling a number greater than 2, find the number of possible outcomes that are greater than 2.

3, 4, 5, and 6 are greater than 2. So there are 4 outcomes that are greater than 2.

There are 6 possible outcomes, so divide 4 by 6.

$$\text{Probability} = \frac{\text{number of ways event can occur}}{\text{total number of possible outcomes}}$$

$P(\text{rolling a number greater than 2}) = \frac{4}{6} = \frac{2}{3}$

Find the theoretical probability of each event.

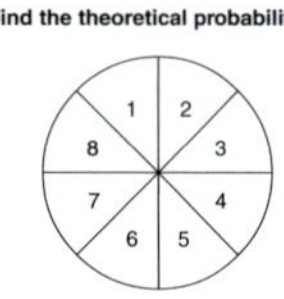

1. P(landing on an even number) $\frac{1}{2}$
2. P(landing on a prime number) $\frac{1}{2}$
3. P(landing on a number divisible by 3) $\frac{1}{4}$
4. P(landing on a number with 3 factors) $\frac{1}{8}$
5. P(landing on a number greater than 8) 0
6. P(landing on a number less than 9) 1

PRACTICE 11-3

LESSON 11-3 **Practice B**
Theoretical Probability

Find the probability of each event using the spinner.

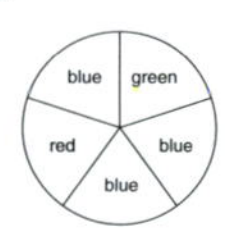

1. landing on blue $\frac{3}{5}$
2. landing on red $\frac{1}{5}$
3. landing on green $\frac{1}{5}$
4. not landing on blue $\frac{2}{5}$

Find the probability of each event using the bag of marbles.

5. picking a black marble $\frac{4}{9}$
6. picking a striped marble $\frac{1}{3}$
7. picking a white marble $\frac{2}{9}$
8. not picking a white marble $\frac{7}{9}$

A standard number cube is rolled. Find each probability.

9. $P(2)$ $\frac{1}{6}$
10. P(even number) $\frac{1}{2}$
11. P(4 or 5) $\frac{1}{3}$
12. P(odd number) $\frac{1}{2}$
13. Out of 10 fair coin tosses, a coin landed tails up 4 times. How does this experimental probability of a fair coin landing tails up compare to the theoretical probability of the same event?
The experimental probability, 40%, is less than the theoretical probability, 50%.
14. The probability of a spinner landing on blue is $\frac{3}{4}$. What is the probability of it not landing on blue written as a percent?
25%

Games LINK

Mah Jong was developed in China. Although the game's exact history is unknown, Mah Jong may be over 2,000 years old. Today the tiles are often made of plastic, but traditional tiles are still made of bamboo, bone, and ivory.

go.hrw.com
KEYWORD: MR4 Game

CNN Student News

For Exercises 24–29, *A* represents an event. The probability that *A* will happen is given. Find the probability that *A* will not happen.

24. $P(A) = 47\%$ 53% **25.** $P(A) = 0.9$ 0.1 **26.** $P(A) = \frac{7}{12}$ $\frac{5}{12}$

27. $P(A) = \frac{5}{8}$ $\frac{3}{8}$ **28.** $P(A) = 0.23$ 0.77 **29.** $P(A) = 100\%$ 0%

30. ***GAMES*** Mah Jong is a traditional Chinese game played with 144 decorated tiles—36 Bamboo tiles, 36 Circle tiles, 36 Character tiles, 16 Wind tiles, 12 Dragon tiles, and 8 bonus tiles. The tiles are the same shape and size, and are all blank on the back. Suppose the tiles are all placed face down and you choose one. What is the probability that you will choose a Wind tile? Write your answer as a fraction in simplest form. $\frac{1}{9}$

31. ***GEOMETRY*** This net can be folded to make a solid figure. The solid figure can then be rolled like a number cube. Give the probability of rolling each number with the solid figure. $\frac{1}{8}$

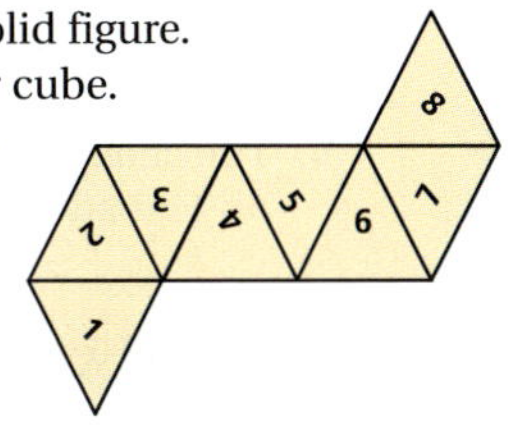

32. ***SOCIAL STUDIES*** In a recent presidential election, the probability that an eligible person voted was about 45%. Is it more likely that an eligible person voted or did not vote?
It is more likely that an eligible person did not vote.

33. Molly has a bag with eight marbles in it: four red, two green, and two yellow. Without looking, she draws a green marble and does not put it back into the bag. What is the probability of her drawing a yellow marble after the green marble has been removed? $\frac{2}{7}$

34. ***WHAT'S THE ERROR?*** If you toss a cylinder, it can land on its top, on its bottom, or on its side. Your friend says that $P(\text{top}) = \frac{1}{3}$. What mistake did your friend make? The error is that landing on its top, bottom, or side are not equally likely events.

35. ***WRITE ABOUT IT*** Toss a coin 20 times and record your results. According to your experiment, what is the probability that the coin shows tails? What is the theoretical probability that it shows tails? How do the two probabilities compare? Repeat the experiment, but this time toss the coin 50 times. Now how do the probabilities compare?
Check students' work.

36. ***CHALLENGE*** Suppose you perform an experiment in which you toss a fair coin and roll a fair number cube. Find the theoretical probability that heads *and* 3 will be the outcomes. $\frac{1}{12}$

Spiral Review

Find each length in meters. (Lesson 3-4)

37. 20 cm 0.2 m **38.** 4 mm 0.004 m **39.** 9,000 km 9,000,000 m **40.** 100 km 100,000 m

41. Make a table that shows the number of days in each month of the year. (Lesson 6-1)

42. **TEST PREP** Which quotient is greatest? (Lesson 9-7) B

A $-10 \div 5$ **B** $-8 \div (-2)$ **C** $15 \div (-5)$ **D** $-10 \div (-5)$

CHALLENGE 11-3

LESSON 11-3 **Challenge**
Plant Probabilities

Botanists often develop new plants by crossing two "parent" plants. Each parent plant gives one gene to each seedling, or "child" plant. Imagine the first parent has two red genes (RR), which produce red flowers. The second parent has two white genes (WW), which produce white flowers. Complete this chart to organize the probabilities of producing a new plant with pink flowers (RW).

		Red-flowering Plant	
		R	R
White-flowering Plant	W	RW	RW
	W	RW	RW

Now imagine crossing two of the new pink-flowering plants. Complete this chart to see the results. Then use both charts to answer the questions that follow.

		Pink-flowering Plant	
		R	W
Pink-flowering Plant	R	RR	RW
	W	RW	WW

1. What is the probability that a plant in the first crossing will be pink? 1
2. What is the probability that a plant in the second crossing will be pink? $\frac{1}{2}$
3. What is the probability that a plant in the second crossing will be red? $\frac{1}{4}$
4. What is the probability that a plant in the second crossing will not be white? $\frac{3}{4}$

PROBLEM SOLVING 11-3

LESSON 11-3 **Problem Solving**
Theoretical Probability

Each time a letter is drawn, it is returned to the bag. Write the correct answer. Write answers in simplest form.

1. At the beginning of a game, each player picks letter tiles from a bag without looking. What is the probability that a player will pick a blank tile?
$\frac{1}{50}$
2. Which letter are you most likely to pick from the bag? Write this probability as a fraction, decimal, and percent.
E; $\frac{3}{25}$, 0.12, 12%
3. Which letters are you least likely to pick from the bag? What is the probability that you will pick any one of those letters? Write this probability as a fraction, decimal, and percent.
J, K, Q, X, and Z; $\frac{1}{100}$; 0.01; 1%
4. The probability of randomly picking a letter is $\frac{3}{50}$. What could that letter possibly be?
N, T, or R

Numbers of Tiles for Each Letter

Letter	Tiles	Letter	Tiles
A	9	O	8
B	2	P	2
C	2	Q	1
D	4	R	6
E	12	S	4
F	2	T	6
G	3	U	4
H	2	V	2
I	9	W	2
J	1	X	1
K	1	Y	2
L	4	Z	1
M	2	BLANK	2
N	6		

5. The probability of randomly picking a letter is $\frac{1}{25}$. What could that letter possibly be?
D, L, S, or U

Circle the letter of the correct answer.

6. What is the probability that you will select a vowel tile from the bag?
A $\frac{9}{100}$ (C) $\frac{11}{25}$
B $\frac{26}{49}$ D $\frac{21}{50}$
7. Most words with a *Q* must also have a *U*. What is the probability that you will select a *U*?
F $\frac{1}{100}$ H $\frac{1}{20}$
(G) $\frac{1}{25}$ D $\frac{1}{300}$

Answers

41.

Month	Number of Days
January	31
February	28
March	31
April	30
May	31
June	30
July	31
August	31
September	30
October	31
November	30
December	31

Journal

Have students write an explanation for how theoretical probability is different from experimental probability. Have them give examples of each type of probability.

Test Prep Doctor

Students need to remember two key ideas to solve Exercise 42. First, they must know that division of integers with like signs gives a positive quotient and division of integers with different signs gives a negative quotient. Also, they must know that positive integers are always greater than negative integers. The answers to **B** and **D** are positive. Since **A** and **C** have negative answers, they can be eliminated. The correct answer is **B.**

Lesson Quiz

Use the spinner shown for problems 1–3.

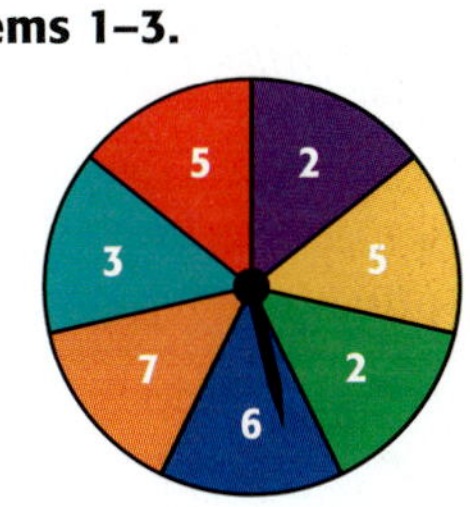

1. $P(2)$ $\frac{2}{7}$
2. $P(\text{odd number})$ $\frac{4}{7}$
3. $P(\text{factor of }6)$ $\frac{4}{7}$
4. Suppose there is a 2% chance of spinning the winning number at a carnival game. What is the probability of not winning? 98%

Available on Daily Transparency in CRB

Chapter 11 Mid-Chapter Quiz

Purpose: *To assess students' mastery of concepts and skills in Lessons 11-1 through 11-3*

Assessment Resources

Section 11A Quiz
Assessment Resources .... p. 27

***Test and Practice Generator* CD-ROM**

Additional mid-chapter assessment items in both multiple-choice and free-response format may be generated for any objective in Lessons 11-1 through 11-3.

Chapter 11 Mid-Chapter Quiz

Mid-Chapter Quiz

LESSON 11-1 (pp. 554–557)

For problems 1 and 2, write *impossible, unlikely, as likely as not, likely,* or *certain* to describe the event.

1. This spinner lands on blue. impossible
2. You roll an even number on a fair number cube. as likely as not
3. Mitch entered a contest to win concert tickets. The chance that Mitch will win is 0.15. Write this probability as a fraction and as a percent. $\frac{3}{20}$; 15%
4. The chance of rain is 33% on Tuesday, 45% on Wednesday, and 35% on Thursday. On which day is it most likely to rain? Wednesday

LESSON 11-2 (pp. 558–561)

For each experiment, identify the outcome shown and the sample space.

5.  1; {1, 2, 3, 4, 5, 6}

6.  yellow; {green, yellow, red, blue}

7. Gregory surveyed students to determine their favorite colors. His results are recorded in the table. Find the experimental probability that a student's favorite color is blue. $\frac{8}{31}$

Color	Red	Yellow	Blue	Purple
Frequency	13	11	16	22

8. Find the experimental probability that a student's favorite color is red or yellow. $\frac{12}{31}$
9. Jeremy recorded the number of times a spinner landed on each number. Based on Jeremy's experiment, on which number is the spinner most likely to land? 2

Outcome	1	2	3
Frequency	𝍸 II	𝍸 𝍸 II	𝍸 I

LESSON 11-3 (pp. 564–567)

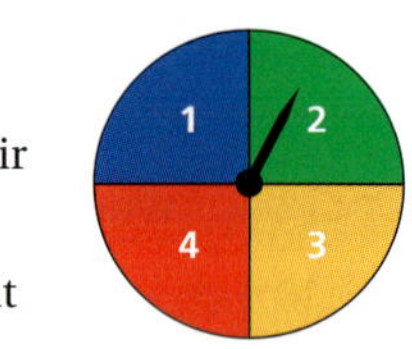

10. What is the probability that this spinner will land on 2? $\frac{1}{4}$
11. What is the probability of rolling a number less than 3 on a fair number cube? $\frac{1}{3}$
12. Kirk has a 33% chance of scoring in the basketball game. What is the probability that Kirk will **not** score in the game? 67%

Focus on Problem Solving

Look Back

• Estimate to check that your answer is reasonable

When you have finished solving a problem, take a minute to reread the problem. See if your answer makes sense. Make sure that your answer is reasonable given the situation in the problem.

One way to do this is to estimate the answer before you begin solving the problem. Then when you get your final answer, compare it with your original estimate. If your answer is not close to your estimate, check your work again.

Each problem below has an answer given, but it is not right. How do you know that the answer is not reasonable? Give your own estimate of the correct answer.

1. A rental car agency has 55 blue cars, 32 red cars, and 70 white cars. A customer is given a car at random. How many color outcomes are possible?

   Answer: 2,100

2. A box has 120 marbles. If the probability of drawing a blue marble is $\frac{3}{8}$ and the probability of drawing a red marble is $\frac{5}{8}$, how many of each color are in the box?

   Answer: 100 blue marbles and 20 red marbles

3. A store manager decides to survey one out of every ten shoppers. How many would be surveyed out of 350 shoppers?

   Answer: 3 shoppers

4. Sue has just started to collect old dimes. She has six dimes from 1941, five dimes from 1932, and one dime from 1930. If she chooses one dime at random, what is the probability that it is from before 1932?

   Answer: 50%

Answers

1. 3

2. 45 blue marbles and 75 red marbles

3. 35 shoppers

4. $\frac{1}{12}$

Focus on Problem Solving

Purpose: *To focus on looking back to check that your answer is reasonable*

Problem Solving Resources

Interactive Problem Solving . pp. 92–97

Math: Reading and Writing in the Content Area . . . . pp. 92–97

Problem Solving Process

This page focuses on the last step of the problem-solving process: **Look Back**

Discuss

Have students explain how they knew each answer was unreasonable.

1. For each customer, there are only three possible colors they could get. So 2,100 is too large to be the correct answer.

2. The numbers 100 and 20 are not divisible by 3, and the probability of drawing a blue marble is 3 out of 8, which means that the number of blue marbles must be divisible by 3.

3. 350 divided by 10 is much larger than 3, since 3 × 10 = 30.

4. There is only 1 dime that is older than 1932, and a 50% probability would suggest that half of the time, or 6 out of 12 times, you would draw the same dime.

Section **11B**

Using Probability

One-Minute Section Planner

Lesson	Materials	Resources
Lesson 11-4 Make an Organized List **NCTM:** Number and Operations, Data Analysis and Probability, Problem Solving, Reasoning and Proof, Communication **NAEP:** Data Analysis and Probability 4f ☐ SAT-9 ☑ SAT-10 ☐ ITBS ☐ CTBS ☐ MAT ☑ CAT	**Optional** 4 boxes 2 items that fit inside the boxes	• *Chapter 11 Resource Book*, pp. 35–43 • Daily Transparency T15, CRB • Additional Examples Transparencies T16–T18, CRB • *Alternate Openers: Explorations*, p. 95
Lesson 11-5 Compound Events **NCTM:** Data Analysis and Probability, Communication **NAEP:** Data Analysis and Probability 4b ☐ SAT-9 ☐ SAT-10 ☐ ITBS ☐ CTBS ☐ MAT ☑ CAT	**Optional** Coins Number cubes *(MK)* Spinners *(MK, CRB, p. 67)*	• *Chapter 11 Resource Book*, pp. 44–52 • Daily Transparency T19, CRB • Additional Examples Transparencies T20–T21, CRB • *Alternate Openers: Explorations*, p. 96
Hands-On Lab 11B Explore Permutations and Combinations **NCTM:** Number and Operations, Data Analysis and Probability, Reasoning and Proof **NAEP:** Data Analysis and Probability 4e ☑ SAT-9 ☑ SAT-10 ☐ ITBS ☐ CTBS ☐ MAT ☐ CAT	**Required** Index cards	• *Hands-On Lab Activities,* pp. 105–106
Lesson 11-6 Making Predictions **NCTM:** Data Analysis and Probability, Problem Solving, Communication, Connections **NAEP:** Data Analysis and Probability 4g ☑ SAT-9 ☑ SAT-10 ☐ ITBS ☐ CTBS ☑ MAT ☑ CAT		• *Chapter 11 Resource Book*, pp. 53–61 • Daily Transparency T22, CRB • Additional Examples Transparencies T23–T26, CRB • *Alternate Openers: Explorations*, p. 97
Extension Odds **NCTM:** Data Analysis and Probability **NAEP:** Data Analysis and Probability 4j ☐ SAT-9 ☐ SAT-10 ☐ ITBS ☐ CTBS ☐ MAT ☐ CAT	**Optional** Teaching Transparency T27 *(CRB)*	• Additional Examples Transparencies T28–T29, CRB
Section 11B Assessment		• Section 11B Quiz, AG p. 28 • *Test and Practice Generator* CD-ROM

SAT = *Stanford Achievement Tests* **ITBS** = *Iowa Test of Basic Skills* **CTBS** = *Comprehensive Test of Basic Skills/Terra Nova*
MAT = *Metropolitan Achievement Tests* **CAT** = *California Achievement Test*

NCTM = Complete standards can be found on pages T27–T33. **NAEP** = Complete standards can be found on pages A31–A35.

SE = *Student Edition* **TE** = *Teacher's Edition* **AR** = *Assessment Resources* **CRB** = *Chapter Resource Book* **MK** = *Manipulatives Kit*

Section Overview

Making an Organized List — Lesson 11-4

Why? When you have to find many possibilities, one way to find them all is to make an organized list.

A tree diagram is one way to organize information.

Charles is having soup and salad for lunch. He can choose from onion, tomato, and potato soup. He can have a chef, garden, or spinach salad. How many different lunches could Charles have?

There are 9 different lunches that Charles could have.

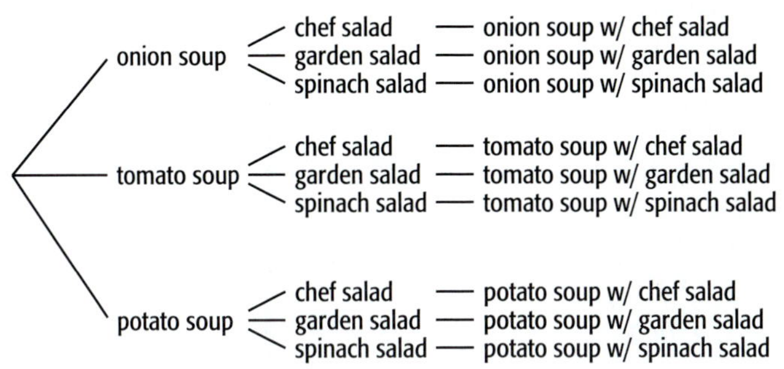

Compound Events — Lesson 11-5

Why? There are counting methods to help determine the probabilities of multiple events.

Assume that the births of boys and girls are equally likely. If a family is going to have 4 children, what is the probability that exactly 2 will be girls? (B = boy, G = girl)

> A **compound event** consists of two or more single events.

There are **16** possible ways for the gender birth order to occur.

BBBB BBBG BBGB BBGG
BGBB BGBG BGGB BGGG
GBBB GBBG GBGB GBGG
GGBB GGBG GGGB GGGG

There are **6** possible ways that there could be exactly 2 girls.

BBGG BGBG BGGB
GGBB GBGB GBBG

The probability of exactly 2 girls out of 4 births is as follows:

$$P(\text{exactly two girls}) = \frac{\text{6 ways event can occur}}{\text{16 possible outcomes}} = \frac{3}{8} = 37.5\%$$

Making Predictions — Lesson 11-6

Why? Insurance companies use probabilities to make predictions about life expectancy.

If you roll a number cube 24 times, how many times can you expect to roll a 5?

$P(\text{rolling a 5}) = \frac{1}{6}$ Use the probability to set up a proportion.

$$\frac{1}{6} = \frac{x}{24}$$
$$6 \cdot x = 1 \cdot 24$$
$$6x = 24$$
$$x = 4$$

You can expect to roll a 5 about 4 times.

11-4 Organizer

Pacing: Traditional 1 day
Block $\frac{1}{2}$ day

Objective: Students make an organized list to find all possible outcomes.

Indiana Academic Standards
6.6.4, 6.7.1, 6.7.2, 6.7.4, 6.7.5, 6.7.11

Warm Up

A game die with eight sides numbered 1 through 8 is rolled. Find each probability.

1. $P(1, 2, \text{or } 3)$ $\frac{3}{8}$
2. $P(\text{even number})$ $\frac{1}{2}$
3. $P(\text{number greater than } 9)$ 0

Problem of the Day

Sam and Pam can have an apple, an orange, or a pear. What is the probability that they will pick the same snack? $\frac{1}{3}$

Available on Daily Transparency in CRB

Math Fact

A tree diagram is so named because it starts at one point and repeatedly branches outward, similar to the way a tree develops branches.

11-4 Make an Organized List

Problem Solving Strategy

Learn to make an organized list to find all possible outcomes.

DNA is a substance found in cells. It is part of what makes up chromosomes, where genetic information is stored.

DNA contains four chemical bases, which are abbreviated A, T, G, and C. Sequences of three bases code for specific amino acids. How could you figure out how many different sequences are possible?

When you have to find many possibilities, one way to find them all is to make an organized list. A tree diagram is one way to organize information.

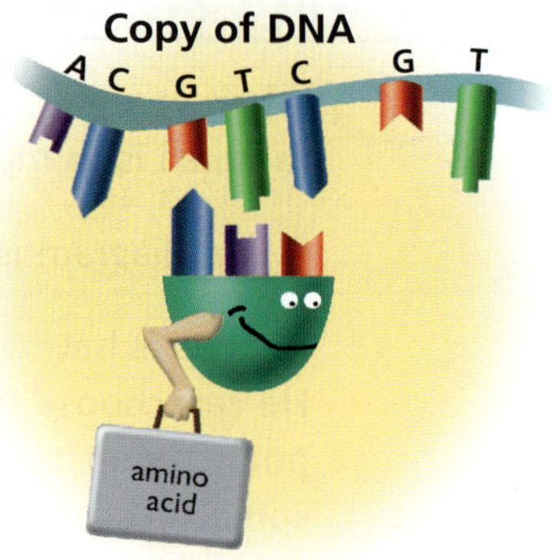

EXAMPLE 1 **Using a Tree Diagram**

At a circus, the clowns have two choices of clown suits—polka dots or stripes. They have three choices of wigs—pigtails, rainbow curly hair, or blue hair. How many different costumes can the clowns wear?

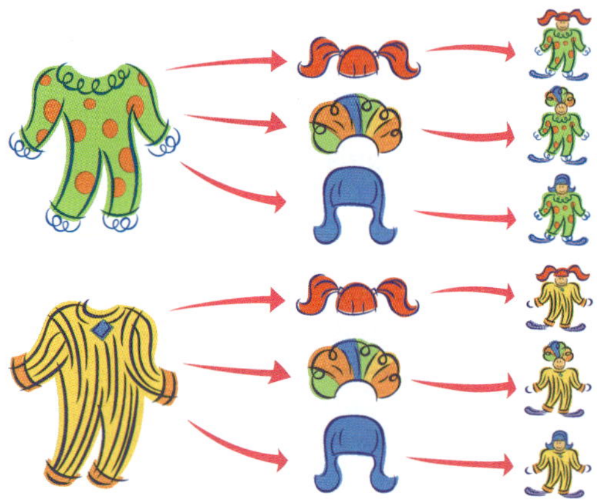

Follow each branch on the tree diagram to find all of the possible outcomes. There are 6 different costume combinations.

1 Introduce

Alternate Opener

EXPLORATION

11-4 Make an Organized List

A teacher made a quiz with three true or false questions. You can make an organized list to determine all of the answer possibilities for the three questions.

1. Complete the tree diagram to list all possible answer outcomes.

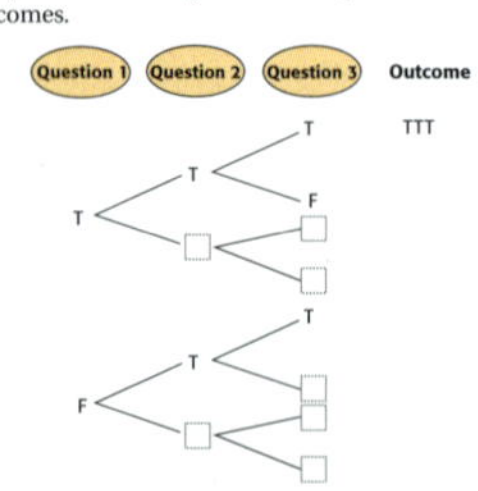

2. How many outcomes are possible?

Think and Discuss

3. **Discuss** your method for organizing the list of outcomes.
4. **Explain** how you could determine the number of outcomes possible with four questions.

Motivate

Display 4 different boxes and 2 different items (such as a pencil and a stapler) that could fit in any of the boxes. Ask students to name all of the possible ways one item could be put inside of one box. stapler in box 1, pencil in box 1, stapler in box 2, pencil in box 2, stapler in box 3, pencil in box 3, stapler in box 4, pencil in box 4

Exploration worksheet and answers on Chapter 11 Resource Book pp. 36 and 74

2 Teach

Lesson Presentation

Guided Instruction

In this lesson, students make an organized list to find all possible outcomes. First teach students to make a tree diagram to show all possible outcomes and then to make an organized list without the help of a tree diagram. Have students choose their preferred methods and explain their choices.

EXAMPLE 2 PROBLEM SOLVING APPLICATION

PROBLEM SOLVING

DNA contains four bases, A, T, G, and C. A sequence of three bases codes for an amino acid. A base may appear more than once in a sequence. For example, CCC codes for the amino acid proline. The order of the bases is also important. CAT codes for the amino acid histidine, but TAC codes for tyrosine, and CTA codes for leucine.

How many different sequences of three can be formed from the four bases?

1 Understand the Problem

List the **important information:**

- There are four bases, A, T, G, and C.
- A base may repeat in a sequence.
- The order of the bases is important.

2 Make a Plan

You can make an organized list to keep track of the sequences.

3 Solve

First, find all the sequences that begin with A.

- List all that begin with AA. *AAA, AAT, AAG, AAC*
- List all that begin with AT. *ATA, ATT, ATG, ATC*
- List all that begin with AG. *AGA, AGT, AGG, AGC*
- Finally, list all that begin with AC. *ACA, ACT, ACG, ACC*

There are 16 sequences that begin with A. To find the sequences that begin with T, replace the beginning A in each sequence above with T. There are 16 sequences that begin with T. The same is true for sequences that begin with C and G.

$$16 + 16 + 16 + 16 = 64$$

There are 64 sequences of three bases that can be made.

4 Look Back

You could have continued to list all 64 possibilities. But looking for a pattern in the list shortened the amount of work needed.

Science LINK

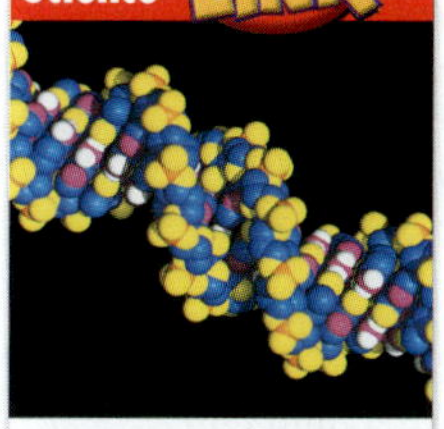

There are 64 sequences of three bases, called *codons,* but cells use only 20 amino acids. Most amino acids have more than one codon. For example, TAT and TAC both code for the amino acid tyrosine.

Think and Discuss

1. **Explain** the advantages of an organized list over a random list.
2. **Describe** how you can check whether your list is accurate.

COMMON ERROR ALERT

Some students may leave out entire groups of entries when completing their tree diagrams. Explain that each of the second choices must be listed beside each of the first choices.

Additional Examples

Example 1

Matt wants to take a 3-day weekend trip to visit his grandparents. He can take either Friday or Monday off from work, and he can either fly, drive, take a train, or take a bus. How many options are available to Matt? 8

Example 2

One girl and one boy will be chosen to go to the state science fair. The girl finalists are Alia, Brenda, Cathy, Deb, and Erika. The boy finalists are Frank, Greg, and Hal. How many different pairs of one girl and one boy can be formed? 15

3 Close

Reaching All Learners
Through Home Connection

Have students work in pairs to make tree diagrams to organize the possible outcomes of choices they make at home. Possible ideas include sandwiches (choices of breads and meats), outfits for school (choices of pants/skirts and tops), and ways to make the bed (choices of sheets and blankets).

Summarize

Have students name a situation in which a tree diagram could be used to find possible combinations. Work through the process of making a tree diagram with students.

Answers to Think and Discuss

1. Possible answer: When you make an organized list, you are less likely to forget information or leave anything out.
2. Possible answer: Organize the list in a different way, listing the information in a different order. Compare the lists to see whether they have the same items.

11-4 PRACTICE & ASSESS

Students may want to refer back to the lesson examples.

Assignment Guide

If you finished Example 1 assign:
Core 1, 3–4, 6, 8, 15–22
Enriched 6–10, 15–22

If you finished Example 2 assign:
Core 1–8, 15–22
Enriched 7–22

Notes

11-4 Exercises

FOR EXTRA PRACTICE see page 659

internet connect
Homework Help Online
go.hrw.com Keyword: MR4 11-4

GUIDED PRACTICE

See Example 1

1. Carl can choose turkey, tacos, or pasta for his main dish at lunch. His choices for a side dish are fruit and salad. How many different lunch combinations are available to him? **6**

See Example 2

2. Patrice, Jason, Kenya, Leon, and Brice are auditioning for the school play. The director has two roles available, a doctor and a teacher. Each can be played by either a boy or a girl. How many different ways can the two roles be assigned? **20**

INDEPENDENT PRACTICE

See Example 1

3. Mr. Li is offering a make-up science test. He can give the test on Monday, Tuesday, or Thursday, before school, during lunch, or after school. How many different times can Mr. Li give his make-up test? **9**

4. The Outdoor Club is planning its annual Spring Festival. The members must vote to choose the day of the event and the main activity. The event can take place on Saturday or Sunday. The main activity can be a foot race, a bicycle race, a hike, a swim, or a scavenger hunt. How many different combinations for the day and activity are there? **10**

See Example 2

5. Greta's apartment building is protected by a security system that requires a pass code to let in residents. The code is made up of numbers from 1 to 3. The code is three digits long, and a digit can repeat. How many different pass codes are possible? **27**

PRACTICE AND PROBLEM SOLVING

6. Keisha will choose a shirt and a skirt or a pair of pants from her closet to wear to school. Find the number of different outfits she can make if she has
   a. 3 shirts, 3 pants, and 3 skirts. **18**
   b. 7 shirts, 5 pants, and 3 skirts. **56**
   c. 4 shirts, 2 pants, and 2 skirts. **16**
   d. 5 shirts, 2 pants, and 6 skirts. **40**
   e. 2 shirts, 4 pants, and 3 skirts. **14**
   f. 6 shirts, 7 pants, and 4 skirts. **66**

7. A middle school is purchasing new basketball jerseys. Each jersey will have a two-digit number on it. The possible digits are 0, 1, 2, 3, 4, and 5, and a digit may appear twice on a jersey. How many different two-digit numbers are possible? **36**

8. One burger restaurant offers a single, double, or triple burger on either a plain or sesame seed bun. Another restaurant offers single and double burgers and four different choices of buns. Does the first or second restaurant offer more possible burger combinations? **the second restaurant**

Math Background

When three number cubes are rolled, the sums of 9 and 10 can each be produced using six different combinations of addends. But, in practice, 10 appears as a sum more often than 9 when three number cubes are rolled. By analyzing the problem using the methods of this lesson, it can be shown that there are 216 different ways in which three number cubes can be rolled, with 27 ways leading to a sum of 10, and 25 ways leading to a sum of 9.

RETEACH 11-4

LESSON 11-4 **Reteach**
Make an Organized List

You can make an organized list to help you solve problems.

Look at the sandwich choices posted at the deli.

Bread	Meat
Wheat	Ham
Rye	Turkey
White	

To find the possible sandwiches that can be made with one type of bread and one type of meat, use a tree diagram to make an organized list of the sandwiches.

Bread	Meat	Combination
wheat	ham	ham on wheat
	turkey	turkey on wheat
rye	ham	ham on rye
	turkey	turkey on rye
white	ham	ham on white
	turkey	turkey on white

Then follow the branches to find the total number of outcomes.

There are 6 possible sandwiches.

Solve.

1. Erin can choose a waffle cone, sugar cone, or chocolate-dipped cone at the carnival. Her choices for ice cream are chocolate, vanilla, strawberry, or cookies and cream. How many different ice cream cones combinations are available to her?
   12 different cones

2. Mrs. Baylus is a parent involved in a neighborhood car pool. She can drive on Monday, Wednesday, or Friday, before school or after school. How many different times can Mrs. Baylus drive for the car pool?
   6 different times

PRACTICE 11-4

LESSON 11-4 **Practice B**
Make an Organized List

Make an organized list to answer each question.

1. Brian wants to buy a new bicycle. He can choose a 10-speed or 3-speed bike. The bikes come in red, blue, black, and purple. How many different bikes can Brian choose from?

Speed	Color	Combination
10-speed	red	red 10-speed bike
	blue	blue 10-speed bike
	black	black 10-speed bike
	purple	purple 10-speed bike
3-speed	red	red 3-speed bike
	blue	blue 3-speed bike
	black	black 3-speed bike
	purple	purple 3-speed bike

8 different bikes

2. Mr. Simon can leave for Miami on Monday, Tuesday, or Wednesday. He can fly, drive, or take a train. How many different travelling options does Mr. Simon have?

Day	Method	Combination
Monday	fly	fly on Monday
	drive	drive on Monday
	train	take train on Monday
Tuesday	fly	fly on Tuesday
	drive	drive on Tuesday
	train	take train on Tuesday
Wednesday	fly	fly on Wednesday
	drive	drive on Wednesday
	train	take train on Wednesday

9 different options

3. The marching band is choosing new uniforms. They can select black or white pants. They can choose a blue, red, green, or black shirt. From how many different uniforms can the band choose?

Pants	Shirt	Combination
black	blue	black pants, blue shirt
	red	black pants, red shirt
	green	black pants, green shirt
	black	black pants, black shirt
white	blue	white pants, blue shirt
	red	white pants, red shirt
	green	white pants, green shirt
	black	white pants, black shirt

8 different uniforms

4. Sara, Jimmy, and Chantall are sitting beside one another on a bench. In how many different orders could they possibly be sitting from left to right?

1st	2nd	3rd
Sara	Jimmy	Chantall
	Chantall	Jimmy
Jimmy	Sara	Chantall
	Chantall	Sara
Chantall	Sara	Jimmy
	Jimmy	Sara

6 different orders

9. Omar is redecorating his bedroom. He can choose one paint color, one border, and one type of brush.

   a. How many different combinations of paint, border, and brush are possible? **12**

   b. If Omar found another brush that he could use, how many different combinations would be possible? **18**

10. Japanese children play a game called *Jan-Ken-Pon.* You may know it as Rock, Paper, Scissors. Two players shout at the same time, "*jan-ken-pon!*" On "*pon!*" each player shows one of three hand positions—closed fist (*gu*), open hand palm down (*pa*), or index and middle finger extended to form a V (*choki*). How many different outcomes are possible in this game? **9**

11. **CHOOSE A STRATEGY** At a meeting, each person shook hands with every other person exactly one time. There were a total of 28 handshakes. How many people were at the meeting? **8**

12. **WRITE A PROBLEM** Write a question that involves making an organized list. Then answer your question.

13. **WRITE ABOUT IT** Suppose you are going to choose one boy and one girl from your class for a group project. How can you find the number of possible combinations? Explain.

14. **CHALLENGE** A sailor has five flags: blue, green, red, orange, and yellow. Suppose she wants to fly three flags, but their order is not important; red, orange, yellow is the same as yellow, orange, red. How many different combinations of flags are possible? **10**

Spiral Review

Evaluate. (Lesson 1-3)

15. 13^2 **169** **16.** 4^3 **64** **17.** 2^5 **32** **18.** 3^4 **81**

Find each sum or difference. (Lesson 5-7)

19. $\frac{1}{9} + \frac{1}{3}$ $\frac{4}{9}$ **20.** $\frac{11}{12} - \frac{5}{6}$ $\frac{1}{12}$ **21.** $\frac{1}{5} + \frac{3}{10} - \frac{1}{15}$ $\frac{13}{30}$

22. TEST PREP Choose the circle with the greatest circumference. (Lesson 10-5) **C**

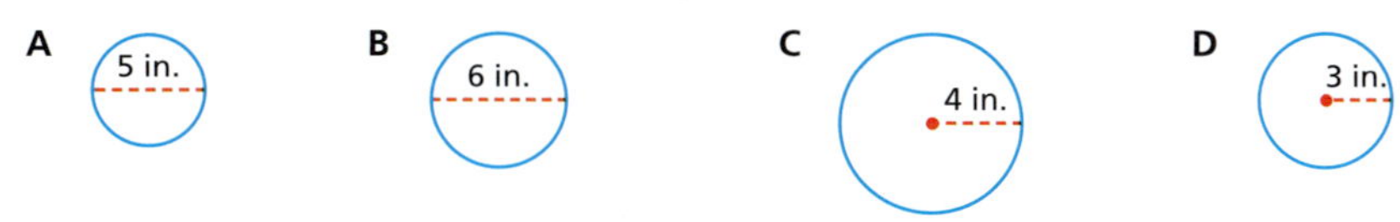

Answers

12. Possible answer: You can choose either a thin or a thick crust pizza. You can have 1 of the following toppings: peppers or onions. How many different pizzas can you make? 4

13. You can make a tree diagram. Start with all the boys' names and match them with each of the girls' names.

Journal

Have students explain a situation in which a tree diagram or organized list can be used to find all the possible outcomes.

Test Prep Doctor

For Exercise 22, students can use the formulas $C = \pi d$ or $C = 2\pi r$ to compute the circumferences of the circles. Some students may realize that they can simply identify the circle with the greatest diameter to find the circle with the greatest circumference. Students who answered **A** or **B** probably assumed that the radius measurements in **C** and **D** were actually diameter measurements. The correct answer is **C.**

CHALLENGE 11-4

LESSON 11-4 **Challenge**
Pascal's Triangle

A mathematician named Blaise Pascal is often called the father of probability theory. In 1654, Pascal investigated the chances of getting different values when rolling dice. Pascal's triangle, shown below, was one result of Pacal's probability work.

Each number in Pascal's triangle is the sum of the two numbers just above it. The triangle has an infinite number or rows—it can go on forever. The first row, Row 0, is always 1. The first entry in each row (Entry 0) is always 1, too.

You can use Pascal's triangle to find combinations. For example, say you want to choose 2 different pizza topping from the 5 toppings available. How many different choices do you have? To answer this question, all you have to do is look at Row 5 (number in group), Entry 2 (number in combination) in Pascal's triangle—10.

Row 0	1
Row 1	1 1
Row 2	1 2 1
Row 3	1 3 3 1
Row 4	1 4 6 4 1
Row 5	1 5 **10** 10 5 1
Row 6	1 6 15 20 15 6 1
Row 7	1 7 21 35 35 21 7 1
Row 8	1 8 28 56 70 56 28 8 1

Entry 2, Row 5

So, you could choose 10 different 2-topping combinations for your pizza.

Use Pascal's triangle to answer each question. To find some combinations, you may have to add rows to the triangle.

1. The ice cream parlor offers 8 different sundae toppings. You want to choose 6 different toppings. How many different 6-topping combinations can you choose?
   28 combinations

2. There are 7 musical notes in the major scale—A, B, C, D, E, F, and G. Combinations of 3 notes played together are called chords. How many different chords can you play without repeating notes?
   35 chords

3. There are 6 people going scuba diving. For safety, they are grouped in pairs. How many different scuba-diving pairs can be formed?
   15 pairs

4. You must choose 3 different digits for your computer password. How many different passwords can you choose from the digits 0 through 9?
   120 passwords

PROBLEM SOLVING 11-4

LESSON 11-4 **Problem Solving**
Make an Organized List

Write the correct answer.

1. Computer spreadsheet programs use letter-number combinations to name cells. How many different cells can a spreadsheet have where its name has 1 English letter followed by 1 digit?
   260 different cells

2. An airline has five different flights to San Francisco today. Each flight offers first-class or coach seats. From how many different tickets to San Francisco can you choose today?
   10 different tickets

3. On Friday, the school cafeteria is serving pizza, hamburgers, chicken, milk, chocolate milk, and juice. From how many different meal-drink combinations can you choose?
   9 combinations

4. Tanya packed 4 T-shirts, 6 pairs of shorts, and 2 pairs of shoes for her vacation. How many different short-shirt-shoes outfit combinations can she wear?
   48 different outfits

5. There are 4 people at a meeting. Every person shakes hands with each other person once. How many handshakes are done in all?
   6 handshakes

6. There are 3,628,800 different ways to arrange the digits 0 through 9! How many different ways can you arrange the digits 1, 2, and 3?
   6 different ways

Circle the letter of the correct answer.

7. A spinner has 6 equal sections labeled *A, B, C, D, E,* and *F*. A second spinner has 5 equal sections colored red, blue, green, yellow, and black. If you spin both spinners at the same time, how many different possible outcomes are there?

   A 5 C 11
   B 6 (D) 30

8. How many different ways can you get from point *A* to point *G*?

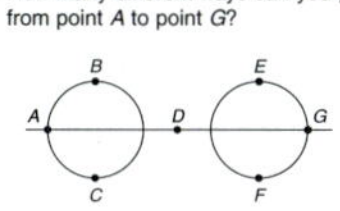

   F 4 H 5
   (G) 9 J 12

Lesson Quiz

1. A baseball coach has 4 pitchers, 3 catchers, and 2 shortstops on his team. How many different combinations of players can he use for the positions? **24**

2. You are taking a 5-question true/false test. How many possible combinations of answers are there? **10**

3. You are planning a small game booth at the local street fair. You have a choice of 3 games and 4 different prizes. How many combinations of games and prizes are there? **12**

Available on Daily Transparency in CRB

11-5 Organizer

Pacing: Traditional 1 day
Block $\frac{1}{2}$ day

Objective: Students list all the outcomes and find the theoretical probability of a compound event.

Indiana Academic Standards
6.6.4, 6.6.6, 6.7.1, 6.7.5, 6.7.11

Warm Up

A cafe offers a soup-and-sandwich combination lunch. You can choose tomato soup, chicken noodle soup, or clam chowder. You can choose a turkey, ham, veggie, or tuna sandwich. How many lunch combinations are there? 12

Problem of the Day

Rory dropped a quarter, a nickel, a dime, and a penny. What is the probability that all four landed tails up? $\frac{1}{16}$

Available on Daily Transparency in CRB

Math Fact

When two events cannot occur at the same time, they are called *disjoint* or *mutually exclusive*. So the probability of two mutually exclusive events occurring is 0.

11-5 Compound Events

Learn to list all the outcomes and find the theoretical probability of a compound event.

Vocabulary
compound event

If a family is going to have four children, there are 16 possibilities for the birth order of the children based on gender (boy, B, or girl, G).

BBBB, BBBG, BBGB, BBGG,
BGBB, BGBG, BGGB, BGGG,
GBBB, GBBG, GBGB, GBGG,
GGBB, GGBG, GGGB, GGGG

A **compound event** consists of two or more single events. For example, the birth of one child is a single event. The births of four children make up a compound event.

EXAMPLE 1 Finding Probabilities of Compound Events

Theresa rolls a fair number cube and then flips a fair coin.

A **Find the probability that the number cube will show an odd number and that the coin will show tails.**

First find all of the possible outcomes.

		Number Cube					
		1	2	3	4	5	6
Coin	H	1, H	2, H	3, H	4, H	5, H	6, H
	T	1, T	2, T	3, T	4, T	5, T	6, T

There are 12 possible outcomes, and all are equally likely.
Three of the outcomes have an odd number and tails:
1, T; 3, T; and 5, T.

$$P(\text{odd, tails}) = \frac{3 \text{ ways event can occur}}{12 \text{ possible outcomes}}$$
$$= \frac{3}{12}$$
$$= \frac{1}{4}$$

Write your answer in simplest form.

B **Find the probability that the number cube will show a 2 and that the coin will show heads.**

Only one outcome is 2, H.

$$P(2, \text{H}) = \frac{1 \text{ way event can occur}}{12 \text{ possible outcomes}}$$
$$= \frac{1}{12}$$

1 Introduce

Alternate Opener

EXPLORATION

11-5 Compound Events

A *compound event* consists of two or more single events.

A bat had two babies. The birth of each baby bat is a single event. The birth of both bats is a compound event.

		First bat born	
		Female	Male
Second bat born	Female	FF	MF
	Male	FM	MM

1. List all possible outcomes for two offspring.
2. Divide 1 (FF) by the total number of outcomes to find the probability that both bats are female.
3. Divide 1 (MM) by the total number of outcomes to find the probability that both bats are male.
4. Divide 2 (MF and FM) by the total number of outcomes to find the probability that one bat is female and the other is male.

Think and Discuss

5. **Explain** how to find the probability that three males would be born among four births.

Motivate

Have students give you examples of the use of the term *compound*, outside of math.

Possible answer: A compound word is a single word, such as *airplane*, made up of two or more single words. Explain that compound events consist of two or more single events.

Exploration worksheet and answers on Chapter 11 Resource Book pp. 45 and 76

2 Teach

Lesson Presentation

Guided Instruction

In this lesson, students learn to list all the outcomes and find the theoretical probability of a compound event. Define the term *compound event*. Teach students to use tables and tree diagrams to list the possible outcomes of compound events. Have students make a tree diagram to display the possible outcomes in Example 1A and a table to display the possible outcomes in Example 1C.

C **The following experiment is going to be performed.**

Step 1: Toss a fair coin. **Step 2:** Spin the spinner. **Step 3:** Choose a marble.

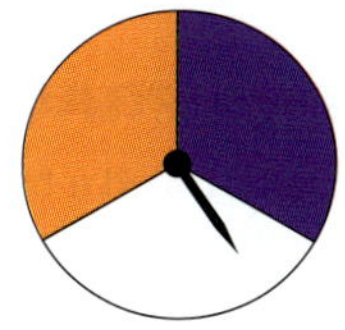

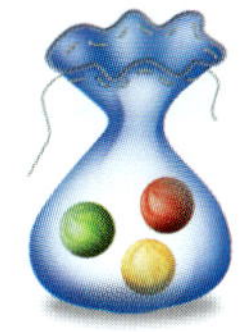

What is the probability that the coin will show heads, the spinner will land on orange, and a red marble will be chosen?

Coin	Spinner	Marble	Outcome
Heads	purple	red	→ heads, purple, red
		yellow	→ heads, purple, yellow
		green	→ heads, purple, green
	orange	red	→ heads, orange, red
		yellow	→ heads, orange, yellow
		green	→ heads, orange, green
	white	red	→ heads, white, red
		yellow	→ heads, white, yellow
		green	→ heads, white, green
Tails	purple	red	→ tails, purple, red
		yellow	→ tails, purple, yellow
		green	→ tails, purple, green
	orange	red	→ tails, orange, red
		yellow	→ tails, orange, yellow
		green	→ tails, orange, green
	white	red	→ tails, white, red
		yellow	→ tails, white, yellow
		green	→ tails, white, green

There are 18 equally likely outcomes.

$$P(\text{heads, orange, red}) = \frac{\text{1 way event can occur}}{\text{18 possible outcomes}} = \frac{1}{18}$$

Think and Discuss

1. **Give an example** of a compound event.
2. **Explain** any pattern you noticed while finding the number of possible outcomes in a compound event.

COMMON ERROR ALERT

When finding the number of possible outcomes for a compound event, students may add the number of outcomes for each independent event. However, the correct operation to use is multiplication.

Additional Example

Example 1

Jerome spins the spinner and rolls a fair number cube.

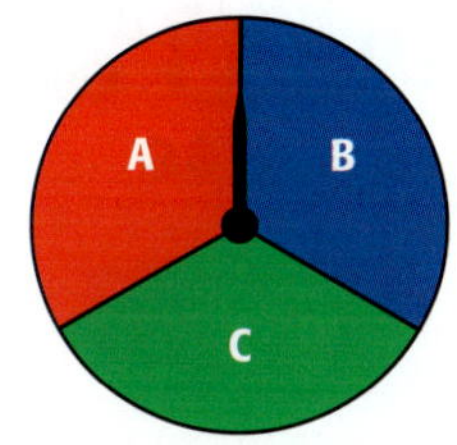

A. Find the probability of the number cube showing an even number and the spinner showing B. $\frac{1}{6}$

B. Find the probability of the number cube showing 4 and the spinner showing A. $\frac{1}{18}$

C. In the experiment proposed in the student book, what is the probability of the coin showing tails, the spinner showing purple, and a green marble being chosen? $\frac{1}{18}$

3 Close

Reaching All Learners

Through Grouping Strategies

Give each group a coin, a number cube, and a spinner. (Number cubes and spinners are provided in the Manipulatives Kit) Have the groups design and conduct experiments involving compound events. Each group should first compute the theoretical probability of a given outcome and then conduct the experiment and find the experimental probability of that outcome. Have group members discuss the relationship between their theoretical and experimental probabilities.

Summarize

Review the term *compound event* and discuss ways to find the possible outcomes of a compound event. Also review how to write a theoretical probability as a fraction.

Answers to Think and Discuss

1. Possible answer: rolling a number cube and pulling a marble out of a bag
2. Possible answer: Multiply the number of possible outcomes for each independent event to find the number of possible outcomes for the compound event.

11-5 PRACTICE & ASSESS

Students may want to refer back to the lesson examples.

Assignment Guide

If you finished Example 1 assign:
Core 1–15, 24–28
Enriched 9–28

Notes

11-5 Exercises

FOR EXTRA PRACTICE
see page 659

internet connect
Homework Help Online
go.hrw.com Keyword: MR4 11-5

GUIDED PRACTICE

See Example 1

1. Patrick rolled a fair number cube twice. Find the probability that the number cube will show an even number both times. $\frac{1}{4}$
2. A boy and a girl each flip a coin. What is the probability that the boy's coin will show heads and the girl's coin will show tails? $\frac{1}{4}$

INDEPENDENT PRACTICE

See Example 1

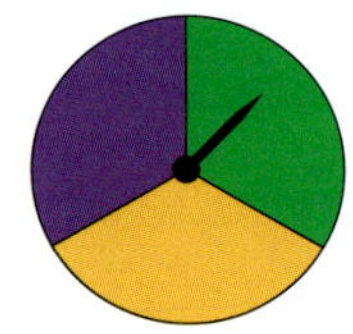

3. If you spin the spinner twice, what is the probability that it will land on green on the first spin and on purple on the second spin? $\frac{1}{9}$
4. What is the probability that the spinner will land on either green or purple on the first spin and yellow on the second spin? $\frac{2}{9}$
5. What is the probability that the spinner will land on the same color twice in a row? $\frac{1}{3}$

PRACTICE AND PROBLEM SOLVING

An experiment involves spinning each spinner once. Find each probability.

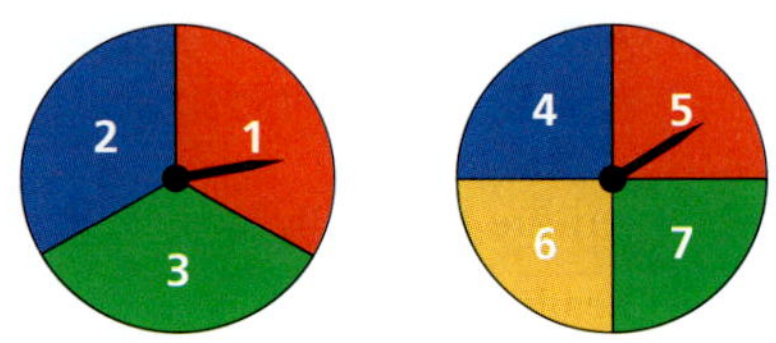

6. P(2 on spinner 1 and 5 on spinner 2) $\frac{1}{12}$
7. P(not 1 on spinner 1 and not 7 on spinner 2) $\frac{1}{2}$
8. P(even number on both spinners) $\frac{1}{6}$
9. P(odd number on spinner 1 and even number on spinner 2) $\frac{1}{3}$
10. P(number on spinner 2 is greater than number on spinner 1) 1
11. P(same number on both spinners) 0
12. P(a multiple of 3 on both spinners) $\frac{1}{12}$
13. P(different number on each spinner) 1

Math Background

If events A and B are independent, then the probability of both A and B occurring equals $P(A) \cdot P(B)$. Here, independence of two events means that the occurrence of one had no relation to the occurrence of the other. With this formula, Example 1A yields the result $P(\text{odd, tails}) = P(\text{odd}) \cdot P(\text{tails}) = \frac{3}{6} \cdot \frac{1}{2} = \frac{1}{4}$. Example 1B yields the result $P(2, \text{H}) = P(2) \cdot P(\text{H}) = \frac{1}{6} \cdot \frac{1}{2} = \frac{1}{12}$. Example 1C yields the result $P(\text{heads, orange, red}) = P(\text{heads}) \cdot P(\text{orange}) \cdot P(\text{red}) = \frac{1}{2} \cdot \frac{1}{3} \cdot \frac{1}{3} = \frac{1}{18}$.

RETEACH 11-5

LESSON 11-5 **Reteach**
Compound Events

A compound event consists of two or more single events.

To find the probability of landing on a prime number and black, first make a list of all the possibilities.

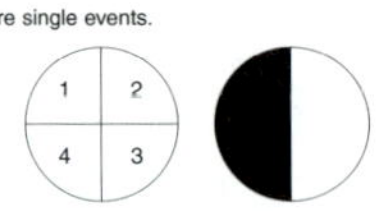

There are 8 possible outcomes.

Number Spinner

Color Spinner	1	2	3	4
Black	1, B	2, B	3, B	4, B
White	1, W	2, W	3, W	4, W

Next, find the number of outcomes that have a prime number and black.

There are two outcomes that have a prime number and black.

Number Spinner

Color Spinner	1	2	3	4
Black	1, B	(2, B)	(3, B)	4, B
White	1, W	2, W	3, W	4, W

Then find the probability.

$P(\text{prime number, black}) = \frac{2 \text{ ways the event can occur}}{8 \text{ possible outcomes}}$

$= \frac{2}{8} = \frac{1}{4}$

Use the spinners to find the probability of each compound event.

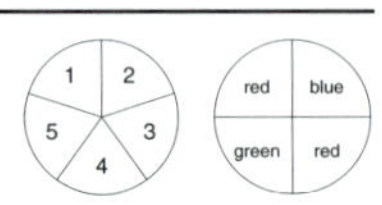

1. P(2, red) $\frac{1}{10}$
2. P(odd number, blue) $\frac{3}{20}$
3. P(even number, red) $\frac{1}{5}$
4. P(number greater than 4, green) $\frac{1}{20}$
5. P(number less than 4, red) $\frac{3}{10}$

PRACTICE 11-5

LESSON 11-5 **Practice B**
Compound Events

This spinner is spun twice.

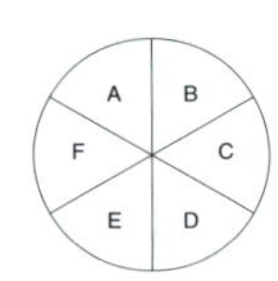

1. What is the probability of the spinner landing on B both times? $\frac{1}{36}$
2. What is the probability of the spinner landing on B, then C? $\frac{1}{36}$
3. What is the probability of the spinner landing on a vowel and then a consonant? $\frac{2}{9}$
4. What is the probability of not spinning D either time? $\frac{25}{36}$

A coin is tossed three times.

5. How many possible outcomes are there? 8 outcomes
6. What is the probability of the coin landing heads up three times? $\frac{1}{8}$
7. What is the probability of the coin landing heads up twice and tails up once? $\frac{3}{8}$

There are five cards numbered 1, 2, 3, 4, and 5 in a bag. Each time a card is drawn, it is replaced. Find the probability of each event.

8. P(1, then 2) $\frac{1}{25}$
9. P(4, then 4) $\frac{1}{25}$
10. P(even, then odd) $\frac{6}{25}$
11. P(odd, then odd) $\frac{9}{25}$
12. What is the probability of a coin landing on heads and a number cube landing on an even number? $\frac{1}{4}$

A fair number cube is rolled, and a fair coin is tossed. Compare the probabilities. Write <, >, or =.

14. P(3 and tails) ▒ P(5 and heads) =

15. P(even number and tails) ▒ P(odd number and heads) =

16. P(number less than 3 and tails) ▒ P(odd number and tails) <

17. P(number greater than 5 and heads) ▒ P(prime number and tails) <

18. ***LIFE SCIENCE*** If a cat has 5 kittens, what is the probability that they are all female? What is the probability that they are all male? (Assume that having a male and having a female are equally likely events.) $\frac{1}{32}$; $\frac{1}{32}$

19. A jar contains tiles that are numbered 1, 2, 3, 4, and 5. Danny removes a tile from the jar, replaces the tile, and draws a second tile. What is the probability that Danny will draw the same number both times? $\frac{1}{25}$

20. The students in Jared's class have ID numbers made up of two digits from 1 through 6. The same digit can be used twice. In fact, the same digit is used twice in Jared's ID number. If Jared rolls 2 number cubes, what is the probability that he does **not** roll his ID number? $\frac{35}{36}$

21. ***WHAT'S THE ERROR?*** One of your classmates said, "If you flip a coin and roll a number cube, the probability of getting heads and a 3 is $\frac{1}{2} + \frac{1}{6} = \frac{2}{3}$." What mistake did your classmate make? Explain how to find the correct answer. the classmate added; should have multiplied

22. ***WRITE ABOUT IT*** Describe a situation that involves a compound event. Possible answer: rolling a fair number cube and flipping a coin

23. ***CHALLENGE*** You roll a number cube six times. What is the probability of rolling the numbers 1 through 6 in order? $\frac{1}{46{,}656}$

Spiral Review

Solve each equation. (Lesson 5-10)

24. $g + \frac{3}{10} = \frac{2}{5}$ $g = \frac{1}{10}$

25. $7m = \frac{1}{2}$ $m = \frac{1}{14}$

26. $\frac{2}{3}p = \frac{1}{6}$ $p = \frac{1}{4}$

27. Angles A and B are supplementary. The measure of angle A is 38 degrees. What is the measure of angle B? (Lesson 7-3) 142 degrees

28. **TEST PREP** Which of the following is always negative? (Lesson 9-1) D

A The opposite of a number

B An integer

C The absolute value of a number

D A number less than zero

Journal

Have students describe an experiment involving compound events. Have them find the theoretical probabilities of a few different outcomes of the experiment.

Test Prep Doctor

For Exercise 28, encourage students to eliminate answer choices before choosing the correct one. Choice **A** can be eliminated because the opposite of a negative number is positive, choice **B** can be eliminated because integers can be positive, and choice **C** can be eliminated because the absolute value of a number is always positive. The correct answer is **D.**

CHALLENGE 11-5

LESSON 11-5 **Challenge**
Give Probability a Hand

Have you ever made a decision by playing Rock, Paper, Scissors? Did you know that people all over the world have played this game for hundreds of years? In it, a player makes his or her hand into the shape of three possible throws at the same time—Rock, Paper, or Scissors. Rock beats Scissors, Scissors beats Paper, and Paper beats Rock.

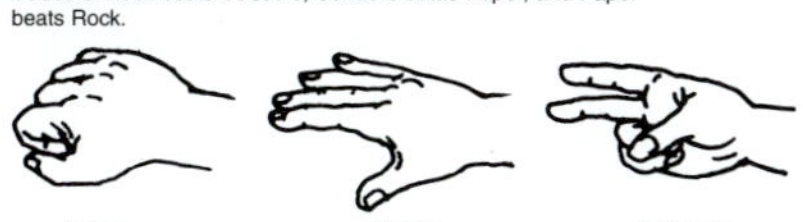

Imagine you are playing Rock, Paper, Scissors with a friend. You are Player 1, and your friend is Player 2. Find each probability.

1. P(1 rock, 2 paper) = $\frac{1}{9}$
2. P(both scissors) = $\frac{1}{9}$
3. P(1 not rock, 2 scissors) = $\frac{2}{9}$
4. P(1 not scissors, 2 not paper) = $\frac{4}{9}$
5. Which throw has the highest chance of winning each round of the game? Explain.
   Possible answer: None; all of the throws have an equal probability of winning because each throw beats one other throw.

Imagine you are playing Rock, Paper, Scissors with two friends. You are Player 1, and your friends are Player 2 and Player 3. Find each probability.

6. P(1 rock, 2 paper, 3 scissors) = $\frac{1}{27}$
7. P(all paper) = $\frac{1}{27}$
8. P(1 not scissors, 2 scissors, 3 not paper) = $\frac{4}{27}$

PROBLEM SOLVING 11-5

LESSON 11-5 **Problem Solving**
Compound Events

You have two decks of playing cards. You draw one card from each deck at the same time. Write the correct answer.

Standard Deck of Playing Cards

Suit	Color	Number
Spades	Black	13
Hearts	Red	13
Clubs	Black	13
Diamonds	Red	13

1. What is the probability that you will draw a black card from Deck 1 and a red card from Deck 2? $\frac{1}{4}$
2. What is the probability that you will draw a club card from both decks? $\frac{1}{16}$
3. What is the probability that you will draw a heart from Deck 1 and a black card from Deck 2? $\frac{1}{8}$

You roll two standard number cubes at the same time. Write the correct answer.

4. What is the probability that you roll doubles, or the same two numbers? $\frac{1}{6}$
5. What is the probability of rolling a sum less than 6? $\frac{5}{18}$
6. Which sums are you least likely to get? What is the probability of rolling either of those sums? 2 or 12; $\frac{1}{18}$
7. Which sum are you most likely to get? What is the probability of rolling that sum? 7; $\frac{1}{6}$

Circle the correct answer.

8. Spinner 1 has 5 equal sections labeled 1 through 5. Spinner 2 has 4 equal sections labeled 1 through 4. What is the probability that both spinners will land on an odd number at the same time?
   A $\frac{1}{5}$ C $\frac{1}{4}$
   B $\frac{4}{9}$ (D) $\frac{3}{10}$
9. Charlie hit the bull's-eye of a dartboard 80 times in 100 shots. Jim hit the bull's-eye 90 times in 100 shots. What are the chances that they will both hit the bull's-eye in the next shot?
   F 89% H 17%
   (G) 72% J 98%

Lesson Quiz

An experiment involves one spin of the spinner and one flip of the coin. Find each probability.

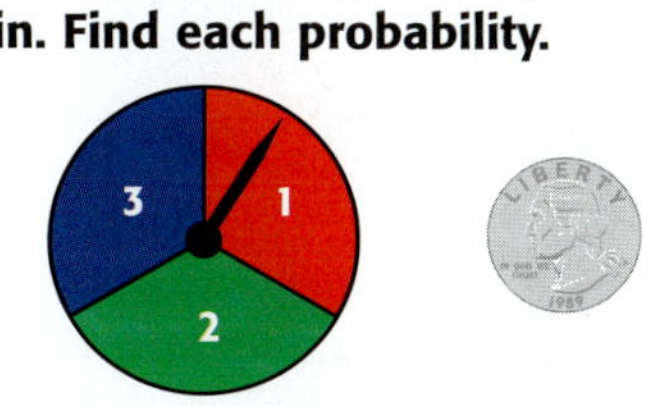

1. What is the probability of the spinner landing on red and the coin landing tails up? $\frac{1}{6}$

2. What is the probability of the spinner landing on an odd number and the coin landing heads up? $\frac{1}{3}$

3. What is the probability of the spinner landing on an even number and the coin landing tails up? $\frac{1}{6}$

Available on Daily Transparency in CRB

Hands-On LAB

11B Explore Permutations and Combinations

Pacing: Traditional 1 day
Block $\frac{1}{2}$ day

Objective: To use index cards to model permutations and combinations

Materials: Index cards

Lab Resources

Hands-On Lab Activities. . . pp. 105–106

Using the Pages

Discuss with students what each arrangement of index cards represents.

Represent each arrangement with index cards. Then use the situations described in the activities to answer the questions about each arrangement.

1. Susan, Ellen, Jeffrey Check students' work.
2. Ellen, Susan, Jeffrey Check students' work.
3. Do the two models represent different outcomes? Explain. Yes. The order of the students is important in Activity 1.
4. Cora, Babe Check students' work.
5. Babe, Cora Check students' work.
6. Do the two models represent different outcomes? Explain. No. The order of the students is not important in Activity 2.

Hands-On LAB 11B

Explore Permutations and Combinations

Use with Lesson 11-5

For a compound event, you often must count the arrangements of individual outcomes. To do this, you must know whether the order of the outcomes in these arrangements matters. With three outcomes *A*, *B*, and *C*, when is *A-B-C* different from *C-B-A*, and when is it considered to be the same?

Activity 1

In how many different arrangements can Ellen, Susan, and Jeffrey sit in a row?

1. Write each name on 6 index cards. You will have a total of 18 cards. Show all the different ways the cards can be arranged in a row.

Arrangement	1	2	3	4	5	6
First Seat	Ellen	Ellen	Susan	Susan	Jeffrey	Jeffrey
Second Seat	Susan	Jeffrey	Jeffrey	Ellen	Susan	Ellen
Third Seat	Jeffrey	Susan	Ellen	Jeffrey	Ellen	Susan

There are 6 different ways that these three people can sit in a row.

Notice that the order of the students in the different arrangements is important. "Ellen, Susan, Jeffrey" is different from "Ellen, Jeffrey, Susan." An arrangement in which order is important is called a **permutation.**

Think and Discuss

1. Think of another situation in which the order in an arrangement is important. Can you think of a situation in which the order would **not** be important? Explain.

Try This

1. Cindy, Laurie, Marty, and Joel are running for president of their class. The person who gets the second greatest amount of votes will be the vice president. How many different ways can the election turn out? 24

Activity 2

1. Abe, Babe, Cora, and Dora are going to work on a project in groups of 2. How many different ways can they pair off?

Write each name on 3 index cards. You will have a total of 12 cards. Show all pairings.

Abe	Babe	Babe	Cora	Cora	Dora
Abe	Cora	Babe	Dora		
Abe	Dora				

There are 6 different possible pairs.

Notice that in this situation, the order in the pairs is not important. "Abe, Cora" is the same as "Cora, Abe." When order is not important, the arrangements are called **combinations.**

Think and Discuss

Tell whether each of the following is a permutation or a combination. Explain.

1. There are 20 horses in a race. Ribbons are given for first, second, and third place. How many possible ways can the ribbons be awarded? Permutation; order is important.
2. There are 20 violin players trying out for the school band and 6 players will be chosen. How many different ways could students be selected for the band? Combination; order is not important.
3. Connie has 10 different barrettes. She wears 2 each day. How many ways can she choose 2 barrettes each morning? Combination; order is not important.
4. Yoko belongs to a book club, and she has just received 25 new books. How many possible ways are there for them to be placed on the shelf? Permutation; order is important.

Try This

1. The video club is sponsoring a double feature. How many ways can club members choose 2 movies from a list of 6 possibilities? 15
2. Ms. Baker must pick a team of 3 students to send to the state mathematics competition. She has decided to choose 3 students from the 5 with the highest grades in her class. Ms. Baker can either send 3 equal representatives, or she can send a captain, an assistant captain, and a secretary. Which choice results in more possible teams? Explain. Find the number of teams possible for each choice.

Answers

Activity 1

Think and Discuss

1. Possible answer: The order is important when conducting a systematic sample in which every third person will be surveyed. Even if the first three people through the door are the same in each case, the order in which they enter affects the survey results. The order is not important in a situation in which the first three people into a store win a prize. All three people will win, regardless of who was first, second, or third to enter.

Activity 2

Try This

2. Choosing three equal team members is a combination problem and results in 10 possible teams; choosing a captain, an assistant captain, and a secretary is a permutation problem and results in 60 possible teams. The permutation problem has more possible teams because the order in which team members are chosen is important.

11-6 Organizer

Pacing: Traditional 1 day
Block $\frac{1}{2}$ day

Objective: Students use probability to predict future events.

Indiana Academic Standards
6.6.5, 6.6.6, 6.7.1, 6.7.2, 6.7.11

Warm Up

1. Zachary rolled a fair number cube twice. Find the probability of the number cube showing an odd number both times. $\frac{1}{4}$
2. Larissa rolled a fair number cube twice. Find the probability of the number cube showing the same number both times. $\frac{1}{36}$

Problem of the Day

The average of three numbers is 45. If the average of the first two numbers is 47, what is the third number? 41

Available on Daily Transparency in CRB

Math Humor

Teacher: Does anyone know the weather prediction for tomorrow?
Student: It will rain nickels and dimes.
Teacher: Now why would you say that?
Student: I heard on the radio that there would be change in the weather.

11-6 Making Predictions

Learn to use probability to predict events.

Vocabulary
prediction

The Old Farmer's Almanac, first published in 1792, predicts weather, sunrise and sunset times, and tides.

A **prediction** is a guess about something in the future. The predictions in *The Old Farmer's Almanac* are based on several factors, such as the cycles of the Sun and Moon. Another way to make a prediction is to use probability.

EXAMPLE 1 Using Probability to Make Predictions

A **An airline claims that its flights have a 92% probability of being on time. Out of 1,000 flights, how many would you predict will be on time?**

You can write a proportion. Remember that *percent* means "per hundred."

$\frac{92}{100} = \frac{x}{1{,}000}$	*Think: 92 out of 100 is how many out of 1,000?*
$100 \cdot x = 92 \cdot 1{,}000$	*The cross products are equal.*
$100x = 92{,}000$	*x is multiplied by 100.*
$\frac{100x}{100} = \frac{92{,}000}{100}$	*Divide both sides by 100 to undo the multiplication.*
$x = 920$	

You can predict that about 920 of 1,000 flights will be on time.

B **If you roll a number cube 24 times, how many times do you expect to roll a 5?**

$P(\text{rolling a 5}) = \frac{1}{6}$

$\frac{1}{6} = \frac{x}{24}$	*Think: 1 out of 6 is how many out of 24?*
$6 \cdot x = 1 \cdot 24$	*The cross products are equal.*
$6x = 24$	*x is multiplied by 6.*
$\frac{6x}{6} = \frac{24}{6}$	*Divide both sides by 6 to undo the multiplication.*
$x = 4$	

You can expect to roll a 5 about 4 times.

1 Introduce

Alternate Opener

EXPLORATION

11-6 Making Predictions

You can use probabilities to make ***predictions***. For example, the probability of rolling a 1 on a number cube is $\frac{1}{6}$. If you roll the cube 12 times, how many times do you predict you will roll a 1?

Each time you roll the cube, the probability of rolling a 1 is $\frac{1}{6}$.

Number of Trials → Probability → Predicted Value

$12 \cdot \frac{1}{6} = 2$

So you expect to roll a 1 twice when you roll a cube 12 times.

Use the model as a guide to make each prediction.

	Event	Probability	Number of Trials	Predicted Value
1.	Getting heads when flipping a coin	$\frac{1}{2}$	Flip the coin 20 times.	
2.	Spinning a 2 on a spinner divided into 4 equal sections	$\frac{1}{4}$	Spin the spinner 24 times.	

Think and Discuss

3. **Explain** how to find a predicted value.
4. **Discuss** whether it is certain that you will roll a 1 twice when you roll a number cube 12 times.

Motivate

To get students ready for the lesson, review how to solve proportions. Work together to solve the proportion $\frac{2}{3} = \frac{x}{6}$.

1. The cross products are equal.
   $3x = 12$
2. x is multiplied by 3, so divide both sides by 3 to undo the multiplication.
   $\frac{3x}{3} = \frac{12}{3}$
3. Solve.
   $x = 4$

Exploration worksheet and answers on Chapter 11 Resource Book pp. 54 and 78

2 Teach

Lesson Presentation

Guided Instruction

In this lesson, students learn to use probability to predict future events. Define the term *prediction*. Teach students to set up proportions to make predictions involving probability. Have students verify each answer by substituting the value for x in each proportion.

EXAMPLE 2 **PROBLEM SOLVING APPLICATION**

PROBLEM SOLVING

A stadium sells yearly parking passes. If you have a parking pass, you can park at the stadium for any event during that year.

The managers of the stadium estimate that the probability that a person with a pass will attend any one event is 80%. The parking lot has 300 spaces. If the managers want the lot to be full at every event, how many passes should they sell?

1 Understand the Problem

The **answer** will be the number of parking passes they should sell.

List the **important information:**

- *P*(person with pass attends event) = 80%
- There are 300 parking spaces.

2 Make a Plan

The managers want to fill all 300 spaces. But, on average, only 80% of parking pass holders will attend. So 80% of pass holders must equal 300. You can write an equation to find this number.

3 Solve

$\frac{80}{100} = \frac{300}{x}$ — *Think: 80 out of 100 is 300 out of how many?*

$100 \cdot 300 = 80 \cdot x$ — *The cross products are equal.*

$30{,}000 = 80x$ — *x is multiplied by 80.*

$\frac{30{,}000}{80} = \frac{80x}{80}$ — *Divide both sides by 80 to undo the multiplication.*

$375 = x$

The managers should sell 375 parking passes.

4 Look Back

If the managers sold only 300 passes, the parking lot would not usually be full because only about 80% of the people with passes will attend any one event. The managers should sell more than 300 passes, so 375 is a reasonable answer.

Think and Discuss

1. Tell whether you expect to be exactly right if you make a prediction based on probability. Explain your answer.

COMMON ERROR ALERT

Students may set up their proportions incorrectly. If *x* represents the number of times a favorable outcome occurs, it should be in the numerator. If it represents the number of possible outcomes, then it belongs in the denominator.

Additional Examples

Example 1

A. A store claims that 78% of shoppers end up buying something. Out of 1,000 shoppers, how many would you predict will buy something? about 780

B. If you roll a number cube 30 times, how many times do you expect to roll a number greater than 2? about 20 times

Example 2

Suppose the managers of a second stadium, like the one in the student book, also sell yearly parking passes. The managers of the second stadium estimate that the probability of a person with a pass attending any one event is 50%. The parking lot has 400 spaces. If the managers want the lot to be full at every event, how many passes should they sell? **800**

3 Close

Summarize

Review that a *prediction* is a guess about something in the future. Remind students that probability and proportions can be used to make predictions.

Reaching All Learners

Through Grouping Strategies

Have students working in groups research airlines' Web sites and find the percent of flights that are on time. (This information is usually updated monthly.) You may wish to assign a different airline to each group. Have them predict the number of on-time flights out of 1,000 flights (or out of 750 flights to make it a little more challenging).

Answers to Think and Discuss

1. No; a prediction, by definition, is just a guess. Probability allows you to guess more likely outcomes, but they are still just guesses.

11-6 PRACTICE & ASSESS

Students may want to refer back to the lesson examples.

Assignment Guide

If you finished Example 1 assign:
Core 1–2, 4–6, 8–9, 15–22
Enriched 4–6, 8–11, 15–22

If you finished Example 2 assign:
Core 1–10, 15–22
Enriched 4–22

Notes

11-6 Exercises

FOR EXTRA PRACTICE see page 659

internet connect
Homework Help Online
go.hrw.com Keyword: MR4 11-6

GUIDED PRACTICE

See Example 1

1. A local newspaper states that 12% of the city's residents have volunteered at an animal shelter. Out of 5,000 residents, how many would you predict have volunteered at the animal shelter? **600**

2. If you roll a fair number cube 30 times, how many times would you expect to roll a number that is a multiple of 3? **10**

See Example 2

3. Airlines routinely overbook flights, which means that they sell more tickets than there are seats on the planes. They do this because ticketed customers sometimes do not show up for flights, and the airlines want to fill the planes. Suppose an airline estimates that 93% of customers will show up for a particular flight. If the plane seats 186 people, how many tickets should the airline sell? **200 tickets**

INDEPENDENT PRACTICE

See Example 1

4. The U.S. Bureau of Engraving and Printing prints paper money. The bureau estimates that 45% of the bills printed on any given day are $1 bills. Out of 500 bills printed in one day, how many would you predict are $1 bills? **225**

5. If you flip a coin 64 times, how many times do you expect the coin to show tails? **32**

6. A bag contains 2 black chips, 5 red chips, and 4 white chips. You pick a chip from the bag, record its color, and put the chip back in the bag. If you repeat this process 99 times, how many times do you expect to remove a red chip from the bag? **45**

See Example 2

7. The director of a blood bank is eager to increase his supply of O negative blood, because O negative blood can be given to people with any blood type. The probability that a person has O negative blood is 7%. The director would like to have 9 O negative donors each day. How many total donors does the director need to find each day to reach his goal of O negative donors? **129 donors**

PRACTICE AND PROBLEM SOLVING

8. A random survey of 50 people in Harrisburg indicates that 10 of them know the name of the mayor of their neighboring city.

a. Out of 5,500 Harrisburg residents, how many would you expect to know the name of the mayor of the neighboring city? **1,100**

b. Out of 600 Harrisburg residents, how many would you predict do not know the name of the mayor of the neighboring city? **480**

Math Background

The idea of using data and probability to analyze and make predictions about real-world events can be traced to seventeenth century England.

John Graunt and his friend Sir William Petty collected and analyzed data concerning birth rates and death rates in the mid-seventeenth century.

Today data collection and analysis are central to the operation of governments and insurance companies, and statistical methods play an important role in both sciences and social sciences.

RETEACH 11-6

LESSON 11-6 **Reteach**
Making Predictions

A prediction is a guess about something in the future. You can use theoretical probability to make predictions.

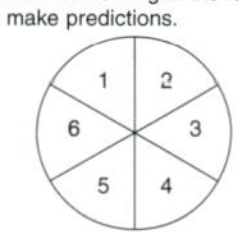

Look at the spinner above. To find the number of times you expect to land on a composite number in 27 spins, first find the theoretical probability of the event.

$P(\text{composite number}) \approx \frac{2}{6} = \frac{1}{3}$

This means that the spinner will land on a composite number about $\frac{1}{3}$ of the time.

Then multiply the theoretical probability by the number of times that you carry out the experiment.

Find $\frac{1}{3}$ of 27 times.

$\frac{1}{3} \cdot 27$
$= \frac{27}{3}$
$= 9$

You can predict that the spinner will land on a composite number about 9 times.

Look at the spinner above. Use theoretical probability to make a prediction.

1. Landing on an odd number in 42 spins.
   21 times
2. Landing on a number that is neither prime nor composite in 30 spins.
   5 times
3. Landing on a number that is less than 5 in 18 spins.
   12 times

PRACTICE 11-6

LESSON 11-6 **Practice B**
Making Predictions

Use the sample survey to make predictions.

Favorite Sports

Sport	Number of Students
Football	28
Basketball	35
Soccer	20
Baseball	45
Hockey	15
Other	7

1. If you randomly selected a person, what is the probability that his or her favorite sport is basketball?
   $\frac{7}{30}$
2. In a group of 200 people, how many do you predict would choose baseball as their favorite sport?
   60 people
3. In a class of 45 students, how many students do you predict would choose soccer as their favorite sport?
   6 students
4. In a group of 100 people, how many do you predict would choose hockey as their favorite sport?
   10 people
5. A local newspaper states that 75% of all the city's voters turned out for the city council elections. If you randomly selected 200 people in that city, how many do you predict would have voted in the election?
   150 people
6. If you roll a fair number cube 30 times, how many times would you expect to roll an odd number?
   15 times
7. A company claims that 8% of its customers were unhappy with the DVD players they bought. If the company sold DVD players to 2,000 people last year, how many of those customers do you predict were unhappy with their purchases?
   160 customers
8. If you toss a fair coin 48 times, how many times do you predict it will land tails up?
   24 times

Social Studies LINK

Canada's population consists of several diverse groups. The Native Canadians lived in Canada before the Europeans arrived. The French were the first Europeans to settle successfully in Canada. There were also British settlers, and after the American Revolution, many Americans who had remained loyal to the king of England fled to Canada.

The graph shows the results of a survey of 400 Canadian citizens.

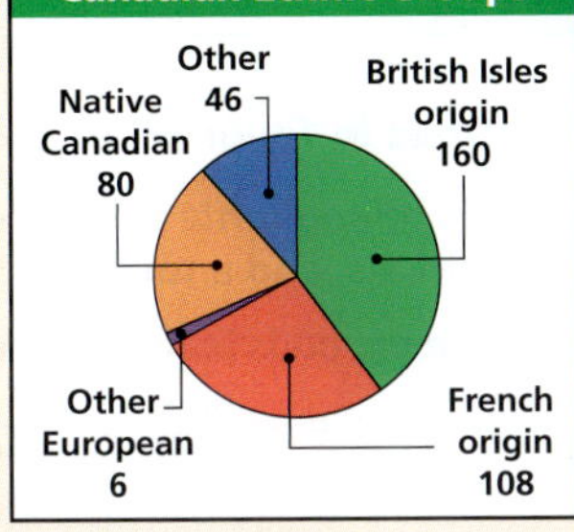

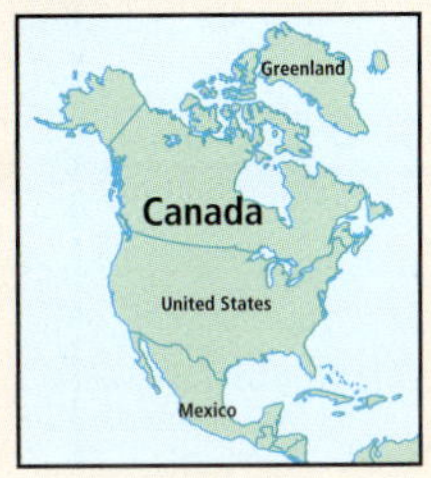

9. Out of 75 Canadians, how many would you predict are of French origin? about 20

10. A random group of Canadians includes 18 Native Canadians. How many total Canadians would you predict are in the group? about 90

11. Predict the number of people of British origin in a group of 50 Canadians. about 20

12. ***WHAT'S THE ERROR?*** A student said that in any group of Canadians, 20 of them will be Native Canadians. What mistake did this student make?

13. ***WRITE ABOUT IT*** How could you predict the number of people of French *or* Native Canadian origin in a group of 150 Canadians?

14. ***CHALLENGE*** In a group of Canadians, 15 are in the Other European origin category. Predict how many Canadians in the same group are **not** in that category. 985

Spiral Review

Solve each equation. (Lesson 9-8)

15. $\frac{y}{-10} = 12$ $y = -120$ **16.** $\frac{p}{25} = -4$ $p = -100$ **17.** $\frac{j}{-3} = -15$ $j = 45$

Find the GCF of each set of numbers. (Lesson 4-3)

18. 8 and 12 4 **19.** 18 and 42 6 **20.** 5 and 80 5 **21.** 2, 9, and 13 1

22. **TEST PREP** Which solid figure has the greatest number of faces? (Lesson 10-6) D

A Cube **B** Triangular prism **C** Cone **D** Octagonal prism

CHALLENGE 11-6

LESSON 11-6 **Challenge**
Predictions in a Flash

Lightning strikes somewhere on Earth almost 9 million times a day! However, very few people are struck by lightning. In fact, your chance of being struck by lightning each year in the United States is only 1 in 600,000. But wait! This is not a simple probability. Where you live, the time of day, and even your hobbies affect your chances of being hit. To be on the safe side, stay indoors during a storm, and avoid open spaces, trees, poles and other tall objects that can attract lightning.

The people who live in the states listed below have the highest probabilities of being struck by lightning. Use their given probabilities and populations to predict how many people in each state will likely be struck by lightning next year.

1.

Florida; 16 million; 0.0003%
48 people

2.

North Carolina; 8 million; 0.0002%
16 people

3.
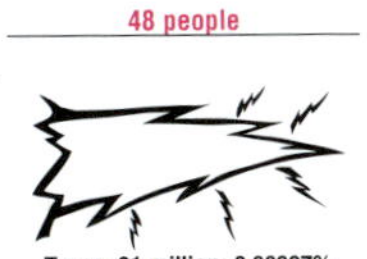
Texas; 21 million; 0.00007%
about 15 people

4.

New York; 19 million; 0.00009%
about 17 people

PROBLEM SOLVING 11-6

LESSON 11-6 **Problem Solving**
Making Predictions

Write the correct answer.

U.S. Public High School Graduation Rates, Top 5 States

State	Number of Students	Percent that Graduate
Iowa	497,301	83.2%
Minnesota	854,034	84.7%
Nebraska	288,261	87.9%
North Dakota	112,751	84.5%
Utah	480,255	83.7%

1. In which state are students most likely to graduate from public high school? About how many of the students who are enrolled in that state now do you predict will graduate?
Nebraska; about 253,381 students

2. About how many students enrolled in North Dakota public high schools now do you predict will graduate?
about 95,275 students

3. About how many students enrolled in Minnesota public high schools now do you predict will graduate?
about 723,367 students

4. In which state do you predict more students in public high schools will graduate—Iowa or Utah? How many more?
Iowa; about 11,781 more students

Circle the letter of the correct answer.

5. The total U.S. high school graduation rate is 68.1%. There are 48,857,321 students enrolled in public schools. About how many of those students do you predict will graduate?
A about 332 million students
B about 20 million students
(C) about 33 million students
D about 16 million students

6. About 11% of all students in the U.S. are enrolled in private schools. There are more than 48 million students in the U.S. About how many do you predict will go to private schools?
(F) about 5,280,000 students
G about 6 million students
H about 52,800 students
J about 528,000 students

Interdisciplinary LINK

Social Studies

Exercises 9–14 focus on the history and culture of Canada.

Answers

12. The student should have said 20%, not 20, of the group will be native Canadians.

13. First find what percent of the Canadians in the sample group are of French origin. Then find what percent of the Canadians in the sample group are of native Canadian origin. Multiply both percents by 150 and add the products.

Journal

Have students explain in their own words how proportions and probability can be used to make predictions. Have students give examples.

Test Prep Doctor

For Exercise 22, you may wish to have students sketch each figure and count its faces before determining the answer. A cube has 6 faces, a triangular prism has 5 faces, a cone has 1 flat base and 1 curved surface, and an octagonal prism has 10 faces. So the correct answer is **D.**

Lesson Quiz

1. The owner of a local pizzeria estimates that 72% of his customers order pepperoni on their pizza. Out of 250 orders taken in one day, how many would you predict to have pepperoni? 180

2. A bag contains 9 red chips, 4 blue chips, and 7 yellow chips. You pick a chip from the bag, record its color, and put the chip back in the bag. If you do this 100 times, how many times do you expect to remove a yellow chip from the bag? 35

3. A quality-control inspector has determined that 3% of the items he checks are defective. If the company he works for produces 3,000 items per day, how many does the inspector predict will be defective? 90

Available on Daily Transparency in CRB

CHAPTER 11 Extension

Pacing: Traditional 1 day
Block $\frac{1}{2}$ day

Objective: Students find the odds for and against a specified outcome.

Using the Pages

In Lesson 11-3, students found the theoretical probability of an event. In this extension, students find the odds for and against a specified outcome.

EXTENSION Odds

Learn to find the odds for and against a specified outcome.

Vocabulary
odds

Odds in favor of an event are written as a ratio of the number of ways the event can happen to the number of ways the event can fail to happen.

ODDS IN FAVOR OF AN EVENT

$$\text{odds in favor} = \frac{\text{number of ways event can happen}}{\text{number of ways event can fail to happen}}$$

EXAMPLE 1 Finding Odds in Favor of an Event

The band is selling raffle tickets for \$2.00 each. Suppose you bought 3 tickets, and a total of 500 raffle tickets were sold.

A What are the odds in favor of your winning?

$$\text{odds in favor} = \frac{\text{number of ways event can happen}}{\text{number of ways event can fail to happen}}$$

$$= \frac{3}{497}$$ *You have 3 tickets that could be drawn. There are 497 other tickets.*

The odds in favor of your winning the raffle are 3 to 497.

B What is the probability that you will win the raffle?

$$P(\text{win}) = \frac{\text{number of ways event can happen}}{\text{number of possible outcomes}} = \frac{3}{500}$$

You can also calculate the odds against an event happening.

ODDS AGAINST AN EVENT

$$\text{odds against} = \frac{\text{number of ways event can fail to happen}}{\text{number of ways event can happen}}$$

EXAMPLE 2 Finding Odds Against an Event

There are 615 students in Carl's class, and Mr. Rosenweig will draw one name at random to attend the governor's lunch.

A What are the odds against Carl's being chosen?

$$\text{odds against} = \frac{\text{number of ways event can fail to happen}}{\text{number of ways event can happen}}$$

$$= \frac{614}{1}$$ *There are 614 other students who can be chosen. There is 1 way Carl can be chosen.*

The odds against Carl's being chosen are 614 to 1.

Helpful Hint

The two numbers given as the odds will add up to the total number of possible outcomes.

1 Introduce

Motivate

Every year before the Major League Baseball season starts, the odds of each team winning the World Series are calculated.

2 Teach

Lesson Presentation

Guided Instruction

In this extension, students find the odds for and against a specified outcome. Define *odds in favor of an event* and give an example that shows both the odds and the probability for the event. Then define *odds against an event* and give an example that shows both the odds and the probability against the event.

B **What is the probability that Carl will not be chosen?**

$$P(\text{not chosen}) = \frac{\text{number of ways event can happen}}{\text{number of possible outcomes}} = \frac{614}{615}$$

EXTENSION Exercises

There are 30 students in Darian's class. His teacher put each student's name in a hat and chose one name.

1. What are the odds in favor of Darian's name being chosen? 1 to 29
2. What is the probability that Darian's name will be chosen? $\frac{1}{30}$
3. What are the odds against Darian's name being chosen? 29 to 1
4. What is the probability that Darian's name will not be chosen? $\frac{29}{30}$

A fair number cube is rolled. Find the odds in favor of and against each of the following outcomes.

5. rolling a 4 1 to 5; 5 to 1
6. rolling a number greater than 2 4 to 2; 2 to 4
7. rolling a number divisible by 3 2 to 4; 4 to 2
8. rolling a factor of 12 5 to 1; 1 to 5
9. rolling an odd number 1 to 1; 1 to 1
10. rolling a 1, a 2, a 3, a 4, or a 6 5 to 1; 1 to 5
11. The surface of Earth is about 70% water and 30% land. If you stopped a spinning globe with your finger, what are the odds in favor of your finger landing on a body of water? 7 to 3

The circle graph shows the origins of bananas in the United States. Use the graph for Exercises 12 and 13.

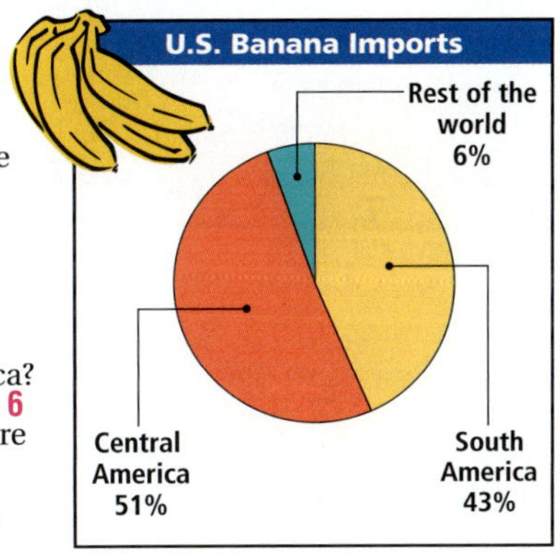

12. If you are eating a banana, what are the odds in favor of its being from Central America? 51 to 49
13. What are the odds against the banana's being from somewhere other than Central or South America? 94 to 6
14. The odds against winning a prize are 2 to 9. Are you more likely to win a prize or not to win a prize? Explain your answer.
15. ***CHALLENGE*** The probability of winning a game is 0.25. What are the odds against winning the game? 3 to 1

COMMON ERROR ALERT

In Exercise 3, some students may think that the odds of 29 to 1 are incorrect since probability cannot be greater than 1. Remind them that odds and probability are different.

Additional Examples

Example 1

In the sixth-grade class, 211 students are right-handed and 137 students are left-handed. What are the odds that a student chosen at random is left-handed? $\frac{137}{211}$

Example 2

Mr. Rosenweig must choose a student from Carl's school to attend the governer's lunch. There are 301 boys and 314 girls in Carl's school, and Mr. Rosenweig will draw one name at random. What are the odds against a boy being chosen? $\frac{314}{301}$

Answers

14. You are more likely to win. Possible answer: The odds against winning are 2 to 9. That means the odds in favor of winning are 9 to 2.

3 Close

Summarize

Have the students find the odds for and against rolling an even number on a fair number cube. 1 to 1; 1 to 1

Teaching Tip Stress the fact that odds should be read as a ratio and not as a fraction. For example, odds may be written as $\frac{3}{4}$ but should be read as 3 to 4, because odds compare two quantities. Odds should not be written as a decimal or percent.

Problem Solving on Location

Kansas

Purpose: *To provide additional practice for problem-solving skills in Chapters 1–11*

State Parks

- After problem 2, have students find the probability that the Porters will visit a park that does not begin with the letter *E*. Have them discuss two ways to find the probability and to verify that both methods give the same probability. $\frac{2}{3}$; use the ratio: $\frac{\text{number of parks } not \text{ beginning with } E}{\text{total number of parks}}$, or subtract the probability of visiting a park that does begin with *E* from 1.
- After problem 3, have students consider the following problem:
  If the Porter family decided to visit all 6 parks, in how many different ways could they visit the parks? 720 How many times as many options does the Porter family have for choosing an order to visit 6 parks than the Hernandez family has for choosing an order to visit 4 parks? The Porter family has 30 times as many options as the Hernandez family.

Extension Have students use the information in the map to write a problem in which the solver must make a prediction. Have students trade papers to solve. Check students' work.

Problem Solving on Location

KANSAS

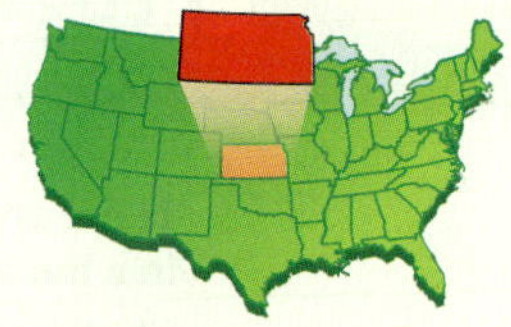

State Parks

Kansas is home to over twenty state parks. Many of these parks exist for the preservation of wildlife in Kansas, while others allow hunting and fishing with the proper permits. The map shows the locations of some of the state parks in Kansas.

1. Each member of the Porter family wants to visit a different park. They decide that each of the 6 family members will write the name of the park they want to visit on a piece of paper. Then they will draw from a hat to decide which park to visit. What is the probability that the Porters will visit Mushroom Rock? $\frac{1}{6}$
2. What is the probability that the Porters will visit a park that begins with the letter *E*? $\frac{1}{3}$
3. The Hernandez family wants to visit the following state parks on their summer vacation to Kansas: Prairie Dog, Webster, Cedar Bluff, and Meade. In how many different orders could they visit all of these parks? 24

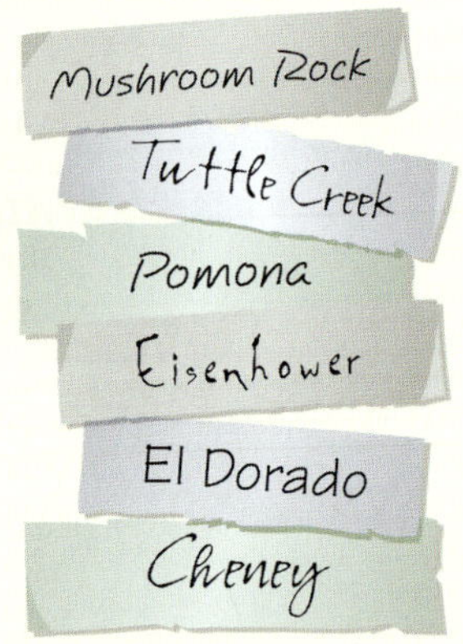

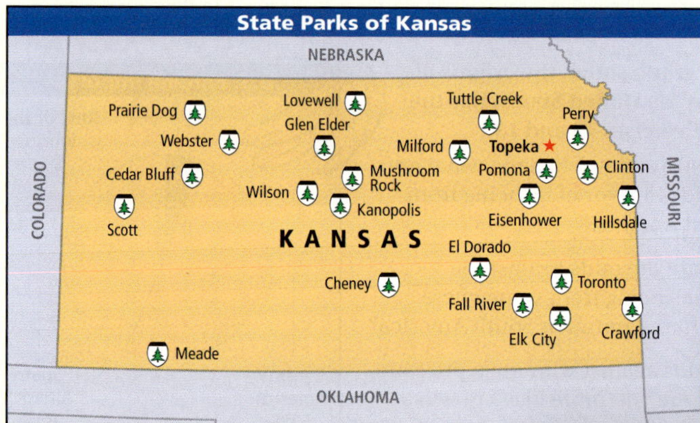

Mushroom Rock State Park

Tornadoes

Compared with the other states, Kansas ranks third in the frequency of tornado occurrences. The table shows the tornado statistics for Wichita County.

The F-scale is the Fujita-Pearson Tornado Scale, which measures the severity of tornadoes. F0 is the weakest tornado rating, and F5 is the strongest. F0 tornadoes have winds ranging from 40–72 mi/h and do minor damage, often to trees and chimneys. F5 tornadoes have winds ranging from 261–318 mi/h and can lift homes off foundations and throw automobiles as far as 100 meters.

Tornadoes in Wichita County (1950–1995)		
Date	**Time**	**F-Scale Rating**
June 16, 1950	3:30 P.M.	F1
May 8, 1951	6:00 P.M.	F1
June 20, 1951	10:00 P.M.	F2
June 12, 1955	7:00 P.M.	F1
July 12, 1956	7:20 P.M.	F2
July 16, 1959	1:30 P.M.	F0
May 13, 1968	1:40 P.M.	F3
May 28, 1969	5:34 P.M.	F0
May 27, 1988	5:30 P.M.	F0
June 10, 1989	7:15 P.M.	F0
June 10, 1989	7:15 P.M.	F0
May 15, 1991	5:27 P.M.	F0
May 15, 1991	5:27 P.M.	F0
May 15, 1991	5:27 P.M.	F0
May 15, 1991	5:27 P.M.	F0
August 12, 1993	8:00 P.M.	F0
March 25, 1995	3:03 P.M.	F0

For 1–3, use the table.

1. Based on the data in the table, what is the experimental probability that a tornado will be F0? F3? F5? $\frac{11}{17}$; $\frac{1}{17}$; 0
2. According to the data, what is the experimental probability of a tornado occurring in May? in December? $\frac{8}{17}$; 0
3. Would you say that the probability of a tornado occurring in the morning is impossible, unlikely, as likely as not, likely, or certain? Explain.

Tornadoes

- After problem 1, discuss whether the data ensures that Wichita County will never have an F5 tornado. No. The fact that Wichita County has never had an F5 tornado does not mean that one could not happen in the future.
- After problem 3, ask students to use the data in the table to describe what they think is the time of day during which a tornado is most likely. Have them justify their opinions. Possible answer: Tornadoes are most likely to occur during the late afternoon and evening hours. Of the 17 tornadoes in the chart, 12 occurred between 3:30 P.M. and 7 P.M.

Extension Have students research the system used to classify tornadoes. Have them present the information to their classmates. Check students' work.

Answers

3. Unlikely; none of the tornadoes listed in the table occurred before noon. This does not mean it is impossible for a tornado to occur in the morning; it just means that it is very rare, since, over a period of 45 years, there has never been a tornado in the morning.

MATH-ABLES

Game Resources

Puzzles, Twisters & Teasers
Chapter 11 Resource Book

Probability Brain Teasers

Purpose: *To apply probability ideas to solving riddles*

Discuss: Have students use models to act out problems 2 and 3. For problem 2, have them use 2-color counters to represent black and white socks. For problem 3, have them use index cards labeled with the names Dale, Melvin, Carter, and Ken and the phrases "Dale's food," "Melvin's food," etc.

Extend: Give students the following brain teaser: Four students are playing a game in which they draw a card that shows a number and then roll a pair of number cubes and find the sum. If the sum matches the number on a player's card, the player scores. The four cards are shown:

7 4 11 10

Amy has the greatest probability of scoring. Ben and Kurt have half the probability of scoring that Amy does. The number on Kurt's card is 3 more than the number on Amy's card. Match the player with his or her card. Amy: 7; Ben: 4; Sam: 11; Kurt: 10

Round and Round and Round

Purpose: *To practice probability in a game format*

Discuss: Review the rules of play with students. Have students discuss reasons why one spinner could win more often than the others.

Possible answers: if one spinner has higher numbers overall than the others, or if one spinner has more of the higher numbers and just a few low numbers, while the other spinners have more low numbers and just a few high numbers

Extend: Have students create new spinners for the game. Challenge them to create a set of spinners in which no spinner is certain to beat another spinner. Check students' work.

MATH-ABLES

Probability Brain Teasers

Can you solve these riddles that involve probability? Watch out—some of them are tricky!

1. In Wade City, 5% of the residents have unlisted phone numbers. If you selected 100 people at random from the town's phone directory, how many of them would you predict have unlisted numbers?
   None; the people were chosen from the phone book.
2. Amanda has a drawer that contains 24 black socks and 18 white socks. If she reaches into the drawer without looking, how many socks does she have to pull out in order to be *certain* that she will have two socks of the same color? 3
3. Dale, Melvin, Carter, and Ken went out to eat. Each person ordered something different. When the food came, the waiter could not remember who had ordered what, so he set the plates down at random in front of the four friends. What is the probability that exactly three of the boys got what they ordered?
   0; if 3 boys got the correct food, then so did the 4th boy.

Round and Round and Round

This is a game for two players.

The object of this game is to determine which of the three spinners is the winning spinner (lands on the greater number most often).

Both players choose a spinner and spin at the same time. Record which spinner lands on the greater number. Repeat this 19 times, keeping track of which spinner wins each time. Repeat this process until you have played spinner A against spinner B, spinner B against spinner C, and spinner A against spinner C. Spin each pair of spinners 20 times and record the results.

Which spinner wins more often, A or B?
Which spinner wins more often, B or C?
Which spinner wins more often, A or C?
Is there anything surprising about your results?

internet connect
Go to ***go.hrw.com*** for a complete set of rules and game pieces.
KEYWORD: MR4 Game11

Technology LAB: Random Numbers

internet connect
Lab Resources Online
go.hrw.com
KEYWORD: MR4 TechLab11

Your calculator can randomly generate numbers using the **randInt** function. This is helpful when you cannot actually perform an experiment. Instead, you can use random numbers to simulate it.

Activity

A dodecahedron is a 12-sided solid figure in which all faces are congruent. Imagine a dodecahedron whose faces are numbered from 1 to 12. Roll this dodecahedron 25 times and give your experimental probability of rolling a 6.

You may not have a dodecahedron handy to perform this experiment, but your calculator can simulate it.

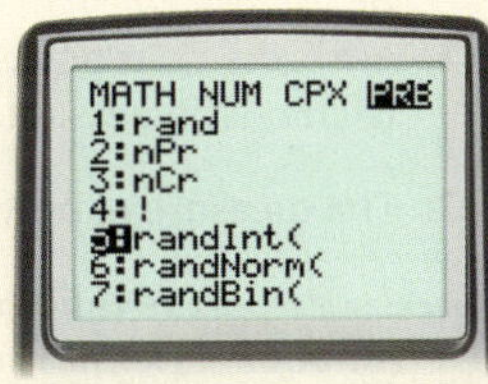

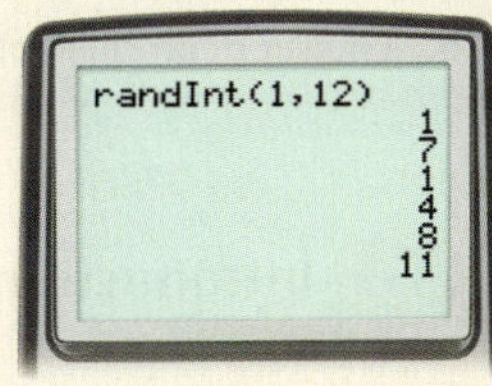

1. Press MATH and use the right arrow key to highlight PRB. Use the down arrow key to highlight 5:. Press ENTER.
2. To have the calculator give you a number from 1 to 12, press 1 , 12) ENTER. The number that appears represents the number you have "rolled" on the dodecahedron.
3. To generate another random number, press ENTER. Continue to press ENTER as many times as you wish to represent rolling the dodecahedron.

Think and Discuss

1. How could you use your calculator to simulate rolling a standard number cube? flipping a coin? Generate random integers from 1 to 6; generate random integers from 1 to 2.
2. What kinds of experiments do you think random numbers would be most useful to simulate? experiments that are expensive or very time-consuming

Try This

1. Simulate rolling a dodecahedron 50 times, and give the experimental probability of rolling a 9. Check students' work.
2. Design your own experiment, and describe how you would use your calculator to simulate it. Then perform the simulation several times, and give your experimental probabilities of each outcome. Check students' work.

Technology LAB: Random Numbers

Objective: To use a graphing calculator to generate random numbers

Materials: Graphing calculator

Lab Resources

Technology Lab Activities . . . . p. 53

Using the Page

This technology activity shows students how to generate random numbers on a graphing calculator. Specific keystrokes may vary, depending on the make and model of the graphing calculator used. The keystrokes given are for a TI-83 model. For keystrokes to other models, visit go.hrw.com.

The Think and Discuss problems can be used to assess students' understanding of the technology activity. Although the experiments in Try This problems 1–2 can actually be performed without a calculator, they are meant to help students become familiar with using a graphing calculator to generate random numbers.

Assessment

1. What sequence of keystrokes is needed to simulate tossing a coin 5 times?

   MATH ▶ ▶ ▶ 5

   1 , 2) ENTER

   ENTER ENTER ENTER ENTER

2. Simulate rolling a number cube 5 times on the graphing calculator. Possible answer: 6, 2, 5, 6, 2

Chapter 11 Study Guide and Review

Purpose: *To help students review and practice concepts and skills presented in Chapter 11*

Assessment Resources

Chapter Review

Chapter 11 Resource Book . . pp. 62–63

***Test and Practice Generator* CD-ROM**

Additional review items in both multiple-choice and free-response format may be generated for any objective in Chapter 11.

Answers

1. equally likely
2. experiment; outcome
3. probability
4. theoretical probability
5. sample space
6. certain
7. 0.75, $\frac{3}{4}$
8. white

Chapter 11 Study Guide and Review

Vocabulary

compound event	574	outcome	558
equally likely	564	prediction	580
experiment	558	probability	554
experimental probability	558	sample space	558
fair	564	theoretical probability	564

Complete the sentences below with vocabulary words from the list above. Words may be used more than once.

1. When all outcomes have the same probability of occurring, the outcomes are __?__.
2. A(n) __?__ is an activity involving chance that can have different results. Each possible result is called a(n) __?__.
3. The measure of how likely an event is to occur is the event's __?__.
4. __?__ is the ratio of the number of ways an event can occur to the total number of possible outcomes.
5. The set of all possible outcomes for an experiment is the __?__.

Study Guide and Review

11-1 Introduction to Probability (pp. 554–557)

EXAMPLE

- **Is it impossible, unlikely, as likely as not, likely, or certain that the spinner will land on yellow?**

Half of the spinner is yellow, so it is as likely to land on yellow as not.

EXERCISES

6. Is it impossible, unlikely, as likely as not, likely, or certain that next week will have 7 days?
7. There is a 75% chance that George will win a race. Write this probability as a decimal and as a fraction.
8. Barry has a 30% chance of picking a black sock and a 50% chance of picking a white sock from his drawer. Which color sock is he more likely to pick?

11-2 Experimental Probability (pp. 558–561)

EXAMPLE

- **Margie recorded the number of times a spinner landed on each color. Based on Margie's experiment, on which color is the spinner most likely to land?**

Outcomes	Red	Blue	Green
Frequency	卌 卌 ||||	||||	卌 ||

$P(\text{red}) \approx \frac{14}{25}$ $P(\text{blue}) \approx \frac{4}{25}$ $P(\text{green}) \approx \frac{7}{25}$

The spinner will most likely land on red.

EXERCISES

9. One day, the cafeteria supervisor recorded the number of students who chose each type of beverage. She organized her results in a table. Find the experimental probability that a student will choose juice.

Beverage	Juice	Milk	Water
Frequency	20	37	18

11-3 Theoretical Probability (pp. 564–567)

EXAMPLE

- **What is the probability of rolling a 4 on a fair number cube?**

There are six possible outcomes when a number cube is rolled: 1, 2, 3, 4, 5, or 6. All are equally likely because the number cube is fair.

$P = \frac{\text{number of ways event can occur}}{\text{total number of possible outcomes}}$

$P(4) = \frac{\text{1 way event can occur}}{\text{6 possible outcomes}} = \frac{1}{6}$

EXERCISES

10. What is the probability that the spinner will land on yellow?

11. What is the probability of rolling a number greater than 3 on a fair number cube?

12. There is a 25% chance of choosing a purple marble from a bag. Find the probability of choosing a marble that is **not** purple.

11-4 Make an Organized List (pp. 570–573)

EXAMPLE

- **Liz is wrapping a gift. She can use gold or silver paper and either a red or white ribbon. From how many different combinations can Liz choose?**

Gold < red / white

Silver < red / white

Follow each branch to find all outcomes.

There are 4 different combinations.

EXERCISES

13. The local restaurant has a lunch special in which you can pick an appetizer, a sandwich, and a drink. How many different lunch-special combinations are there if you have the following choices?

appetizers: soup or salad

sandwiches: turkey, roast beef, or ham

drinks: juice, milk, or iced tea

Answers

9. $\frac{20}{75} = \frac{4}{15}$

10. $\frac{1}{4}$

11. $\frac{1}{2}$

12. 75%

13. 18 combinations

Answers

14. $\frac{1}{6}$

15. $\frac{1}{8}$

16. 100 items

17. 25 times

18. 1,575 teenagers

19. 100 students

11-5 Compound Events (pp. 574–577)

EXAMPLE

■ **What is the probability of spinning red or blue and having the coin land heads up?**

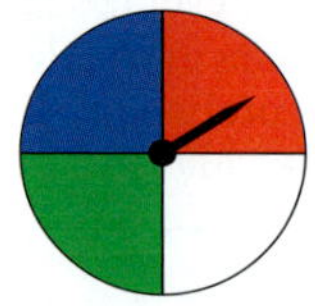

	Red	Blue	Green	White
Heads	red, H	blue, H	green, H	white, H
Tails	red, T	blue, T	green, T	white, T

There are 8 possible outcomes, and all are equally likely.

$$P(\text{red or blue, H}) = \frac{\text{2 ways event can occur}}{\text{8 possible outcomes}}$$
$$= \frac{2}{8} = \frac{1}{4}$$

EXERCISES

14. Find the probability that a blue marble will be chosen, the first coin will show heads, and the second coin will show tails.

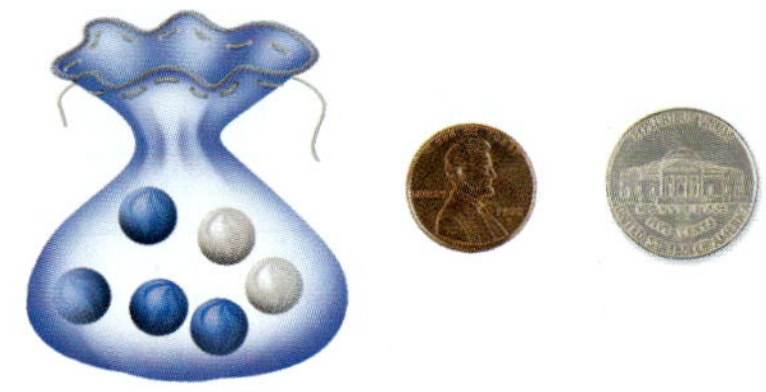

15. Jacob rolled a fair number cube, flipped a fair penny, and then flipped a fair quarter. Find the probability that the number cube will show an even number and both coins will show heads.

11-6 Making Predictions (pp. 580–583)

EXAMPLE

■ **If you spin the spinner 30 times, how many times do you expect it to land on red?**

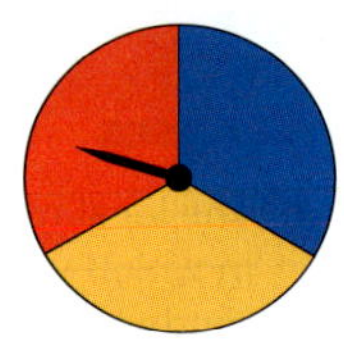

$P(\text{red}) = \frac{1}{3}$

$\frac{1}{3} = \frac{x}{30}$

$3 \cdot x = 1 \cdot 30$ — *The cross products are equal.*

$3x = 30$ — *x is multiplied by 3.*

$\frac{3x}{3} = \frac{30}{3}$ — *Divide both sides by 3 to undo the multiplication.*

$x = 10$

You can expect it to land on red about 10 times.

EXERCISES

16. About 2% of the items produced by a company are defective. Out of 5,000 items, how many can you predict will be defective?

17. If you roll a fair number cube 50 times, how many times can you expect to roll an even number?

18. A survey of 500 teenagers indicated that 175 of them use their computers regularly. Out of 4,500 teenagers, predict how many use their computers regularly.

19. A survey of 100 sixth-grade students indicated that 20 of them take music lessons. Out of 500 sixth-grade students, predict how many take music lessons.

Chapter Test

For Exercises 1 and 2, write *impossible, unlikely, as likely as not, likely,* or *certain* to describe each event.

1. You roll a 3 on a fair number cube. **unlikely**
2. You pick a blue marble from a bag of 5 white marbles and 20 blue marbles. **likely**
3. There is a 12% chance of rain tomorrow. Write this probability as a decimal and as a fraction. **0.12; $\frac{3}{25}$**
4. The probability that Mark will be selected for a scholarship is 0.8. Write this probability as a percent and as a fraction. **80%; $\frac{4}{5}$**
5. Iris asked 60 students what time they go to bed. Her results are in the table. Estimate the probability that a student chosen at random goes to bed at 8:30 P.M. **$\frac{2}{5}$**

Time (P.M.)	8:00	8:30	9:00	9:30
Frequency	12	24	18	6

6. Estimate the probability that a student chosen at random goes to bed before 8:30 P.M. **$\frac{1}{5}$**
7. Josh threw darts at a dartboard 10 times. Assume that he threw the darts randomly and did not aim. Based on his results, what is the probability that a dart will land in the center circle? **$\frac{1}{10}$**

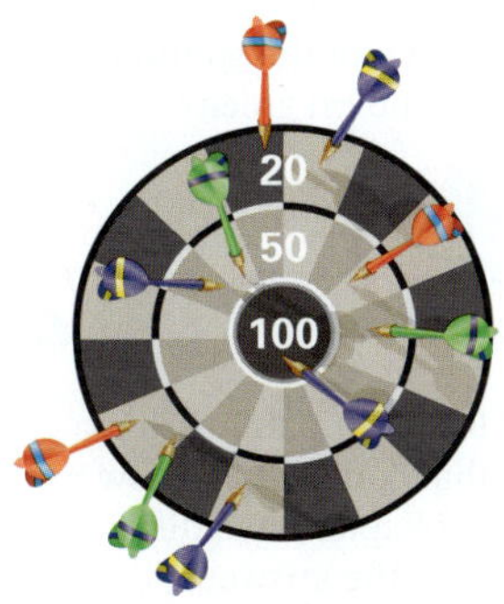

8. What is the probability of rolling an even number greater than 2 on a fair number cube? **$\frac{1}{3}$**
9. The baseball game has a 64% chance of being rained out. What is the probability that it will **not** be rained out? **36%**
10. Peter has four photos to arrange in a frame. How many different ways can he arrange the photos? **24 ways**

11. Marsha can wear jeans or black pants with a red, blue, or white shirt. How many different outfits can she choose from? **6 different outfits**
12. If you roll a number cube 36 times, how many times do you expect to roll an even number? **18**
13. Find the probability that you will pick a blue marble from both bags and that the spinner will land on blue. **$\frac{1}{24}$**

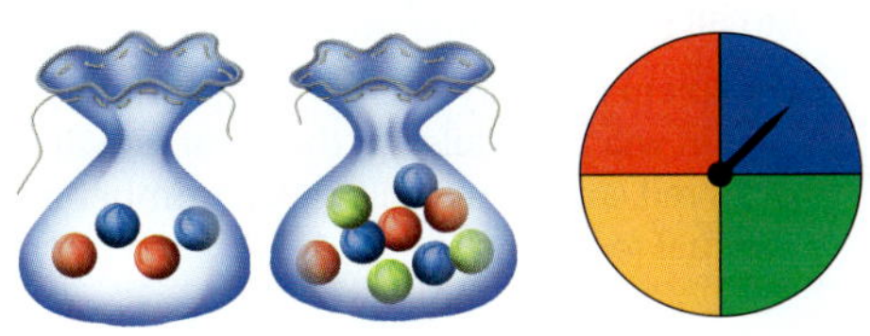

Chapter Test — Chapter 11

Purpose: *To assess students' mastery of concepts and skills in Chapter 11*

Assessment Resources

Chapter 11 Tests (Levels A, B, C)
Assessment Resources ..... pp. 91–96

***Test and Practice Generator* CD-ROM**

Additional assessment items in both multiple-choice and free-response format may be generated for any objective in Chapter 11.

Chapter 11 Performance Assessment

Purpose: *To assess students' understanding of concepts in Chapter 11 and combined problem-solving skills*

Assessment Resources

Performance Assessment
Assessment Resources . . . . p. 124

Performance Assessment Teacher Support
Assessment Resources . . . . p. 123

Answers

1–3. See p. A6.

4. See Level 3 work sample below.

Scoring Rubric for Problem Solving Item 4

Level 3

Accomplishes the purposes of the task.

Student gives clear explanations, shows understanding of mathematical ideas and processes, and computes accurately.

Level 2

Purposes of the task not fully achieved.

Student demonstrates satisfactory but limited understanding of the mathematical ideas and processes.

Level 1

Purposes of the task not accomplished.

Student shows little evidence of understanding the mathematical ideas and processes and makes computational and/or procedural errors.

Chapter 11 Performance Assessment

Performance Assessment

Show What You Know

Create a portfolio of your work from this chapter. Complete this page and include it with your four best pieces of work from Chapter 11. Choose from your homework or lab assignments, mid-chapter quiz, or any journal entries you have done. Put them together using any design you want. Make your portfolio represent what you consider your best work.

Short Response

1. A restaurant offers a choice of roast beef, chicken, or fish, with broccoli, carrots, or corn, and a soup or salad. Make an organized list to determine how many different meal combinations the restaurant offers.

2. Find the theoretical probability that the spinner will land on each color. Which of the spinners is fair? Explain. Which spinner is most likely to land on green?

3. What is the probability that a green marble will not be chosen from a bag that contains 2 yellow marbles, 2 green marbles, and 3 red marbles? Explain two different methods that you can use to determine the answer.

Extended Problem Solving

4. Jamie is performing an experiment in which she tosses the coin, spins the spinner, and then chooses a card.
   a. How many possible outcomes are there?
   b. What is the probability that the coin will land heads up, the spinner will land on red, and Jamie will pick a number greater than 5?
   c. Explain two ways to find the probability that the coin will land tails up, the spinner will not land on red, and Jamie will pick a number less than or equal to 5.

Student Work Samples for Item 4

Level 3

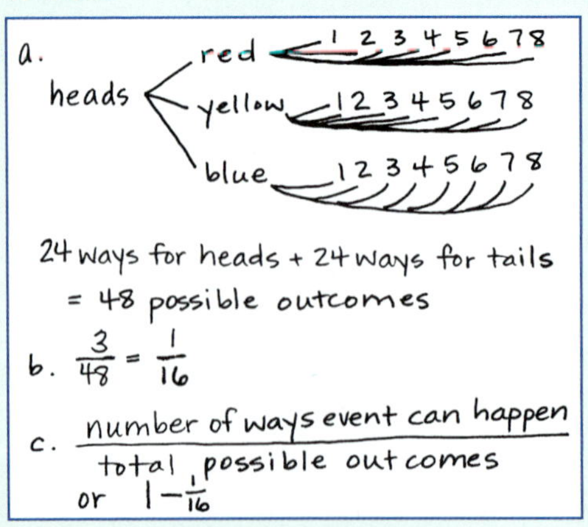

The student correctly answered all parts of the question and used a tree diagram.

Level 2

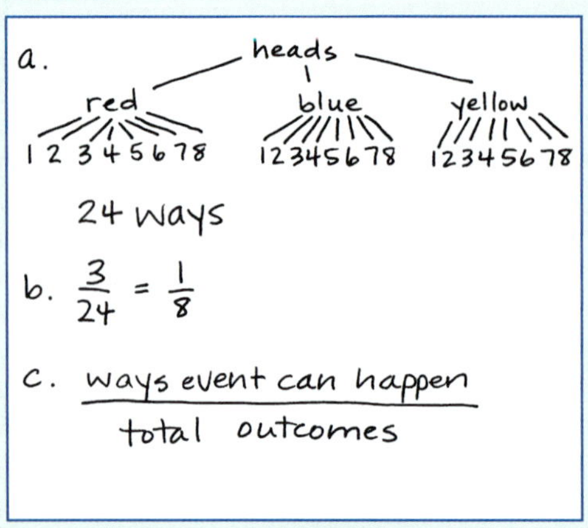

The student gave a partial answer to parts **a** and **c** and an incorrect answer to part **b.**

Level 1

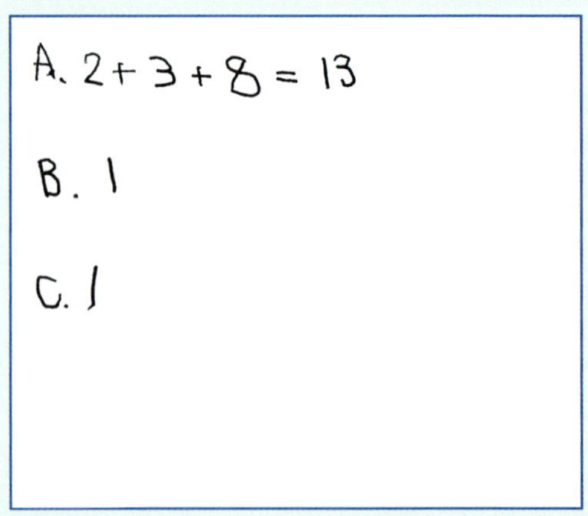

The student did not show understanding of probability or combinations.

internet connect
State-Specific Test Practice Online
go.hrw.com Keyword: MR4 TestPrep

Standardized Test Prep

Chapter 11

Cumulative Assessment, Chapters 1–11

TIP! TEST TAKING TIP!
Do not waste a lot of time on a question that is giving you trouble. Skip it and come back to it later if you have time.

1. What is the probability that the spinner lands on 1 or 2? **A**

Ⓐ $\frac{1}{4}$ Ⓒ $\frac{1}{8}$

Ⓑ $\frac{1}{2}$ Ⓓ $\frac{3}{8}$

2. Which set is ordered from least to greatest? **G**

Ⓕ −4, −8, 0, 2 Ⓗ 3, 1, −4, −5

Ⓖ −6, −2, 1, 5 Ⓙ −2, −5, 6, 8

3. Which data set has a mean of 15? **C**

Ⓐ 5, 17, 16, 16, 15, 7, 8

Ⓑ 11, 15, 15, 17, 14, 12

Ⓒ 21, 10, 17, 16, 14, 16, 11

Ⓓ 25, 10, 10, 24, 18, 12

4. What is the GCF of 8, 16, and 20? **G**

Ⓕ 2 Ⓗ 160

Ⓖ 4 Ⓙ 20

5. Which expression has a value of 12? **C**

Ⓐ 6 · (−2) Ⓒ −3 · (−4)

Ⓑ −15 + (−3) Ⓓ 22 − (−10)

6. Mr. Rodriguez is remodeling his den. He can choose hardwood, carpet, or tiles for the floor and wallpaper, paint, or paneling for the walls. From how many different combinations can he choose? **J**

Ⓕ 5 Ⓗ 8

Ⓖ 6 Ⓙ 9

7. A computer is on sale for 25% off the ticketed price of \$986.00. How much will the computer cost after the discount? **B**

Ⓐ \$246.50 Ⓒ \$961.00

Ⓑ \$739.50 Ⓓ \$961.35

8. What is the name for the transformation shown? **H**

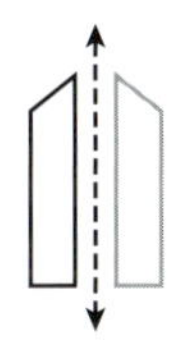

Ⓕ translation Ⓗ reflection

Ⓖ rotation Ⓙ tessellation

9. ***SHORT RESPONSE*** Lily is purchasing shelves for her new bookcase. She has 45 books to place in the bookcase. If 12 books fit on a shelf, how many shelves should she purchase? Explain your answer.

10. ***SHORT RESPONSE*** Sam flipped a fair coin and rolled a fair number cube. What is the probability that the coin will show heads and the cube will show a number greater than 2? Show how you found your answer.

Standardized Test Prep
Chapter 11

Purpose: *To provide review and practice for Chapters 1–11 and standardized tests*

Assessment Resources

Cumulative Tests (Levels A, B, C)
Assessment Resources . . . . pp. 247–258

State-Specific Test Practice Online
KEYWORD: MR4 TestPrep

Test Prep Doctor

For item 8, remind students that a percent must be changed to a decimal or a fraction before being used in a calculation. Choice **F** can be eliminated because \$246.50 represents the savings but not the sale price. Choice **H** can be eliminated because \$25 was subtracted. Choice **J** can be eliminated because \$24.65 was subtracted rather than \$246.50.

Answers

9. 45 ÷ 12 = 3.75
Lily cannot purchase part of a shelf, so she needs to buy 4 shelves.

10. P(heads) $= \frac{1}{2}$
P(3, 4, 5, or 6) $= \frac{2}{3}$
$\frac{1}{2} \cdot \frac{2}{3} = \frac{2}{6} = \frac{1}{3}$

Chapter 12

Functions and Coordinate Geometry

Section 12A

Introduction to Functions

Lesson 12-1
Tables and Functions

Hands-On Lab 12A
Explore Linear and Nonlinear Relationships

Lesson 12-2
Graphing Functions

Section 12B

Coordinate Geometry

Lesson 12-3
Graphing Translations

Lesson 12-4
Graphing Reflections

Lesson 12-5
Graphing Rotations

Lesson 12-6
Stretching and Shrinking

Pacing Guide for 45-Minute Classes

Chapter 12

DAY 155	DAY 156	DAY 157	DAY 158	DAY 159
Lesson 12-1	Hands-On Lab 12A	Lesson 12-2	Mid-Chapter Quiz Lesson 12-3	Lesson 12-4
DAY 160	**DAY 161**	**DAY 162**	**DAY 163**	
Lesson 12-5	Lesson 12-6	Chapter 12 Review	Chapter 12 Assessment	

Pacing Guide for 90-Minute Classes

Chapter 12

DAY 77	DAY 78	DAY 79	DAY 80	DAY 81
Chapter 11 Review Lesson 12-1	Chapter 11 Assessment Hands-On Lab 12A	Lesson 12-2 Lesson 12-3	Mid-Chapter Quiz Lesson 12-4 Lesson 12-5	Chapter 12 Review
DAY 82				
Chapter 12 Assessment				

Across the Series

HARCOURT GRADE 5

- Identify, write, and graph ordered pairs on a coordinate plane.
- Transform a figure on a coordinate plane, and use the plane to graph geometric relationships.
- Use a rule to complete a function table, and plot the ordered pairs.

COURSE 1

- **Use a function table to write an equation for a function.**
- **Use tiles to explore linear and nonlinear relationships.**
- **Represent linear functions with ordered pairs and graphs.**
- **Use translations, reflections, and rotations individually to change positions of figures on a coordinate plane.**
- **Visualize and show the results of stretching or shrinking a figure.**

COURSE 2

- Write an equation for a function table.
- Interpret graphs of functions that represent situations.
- Identify and graph linear functions, and explore nonlinear relations.
- Use translations, reflections, and rotations individually to change positions of figures on a coordinate plane.

Across the Curriculum

LANGUAGE ARTS LINK

Math: Reading and Writing in the Content Area . . . . pp. 98–103

Focus on Problem Solving

Solve . . . . SE p. 609

Journal . . . . TE pp. 601, 607, 619, 623

Write About It . . . . SE, last page of each lesson

SOCIAL STUDIES LINK

Social Studies . . . . SE p. 623

SCIENCE LINK

Physical Science. . . . . SE p. 607

TE = *Teacher's Edition* **SE** = *Student Edition*

Interdisciplinary Bulletin Board

Earth Science

There are three temperature regions in oceans. Notice how the water temperature of an ocean decreases as the depth of the water increases. What is the difference in water temperature between a depth of 300 meters and a depth of 700 meters? about 20°C

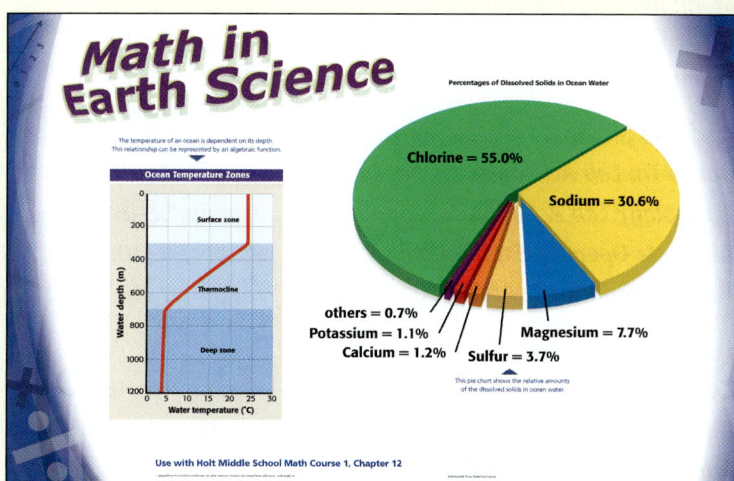

Interdisciplinary posters and worksheets are provided in your resource material.

Chapter 12

Resource Options

Chapter 12 Resource Book

Student Resources

Teacher and Parent Resources

Reaching All Learners

English Language Learners

Individual Needs

Hands-On

Applications and Connections

Transparencies

Technology

Teacher Resources

***Lesson Presentations* CD-ROM** .... Chapter 12
***Test and Practice Generator* CD-ROM** ...... Chapter 12
***One-Stop Planner* CD-ROM** .... Chapter 12

Student Resources

***Are You Ready? Intervention* CD-ROM**
Skills 58, 59, 60, 68, 87

internet connect

Homework Help Online	**KEYWORD:** MR4 HWHelp12
Math Tools Online	**KEYWORD:** MR4 Tools
Glossary Online	**KEYWORD:** MR4 Glossary
Chapter Project Online	**KEYWORD:** MR4 PSProject12
Chapter Opener Online	**KEYWORD:** MR4 Ch12
CNN Student News	**KEYWORD:** MR4 CNN12

SE = *Student Edition* **TE** = *Teacher's Edition* **AR** = *Assessment Resources* **CRB** = *Chapter Resource Book* **MK** = *Manipulatives Kit*

Chapter 12

Assessment Options

Assessing Prior Knowledge

Determine whether students have the required prerequisite concepts and skills.

Are You Ready? . . . . SE p. 597
Inventory Test . . . . AR pp. 1–4

Test Preparation

Provide review and practice for chapter and standardized tests.

Standardized Test Prep . . . . SE p. 633
Spiral Review with Test Prep . . . . . SE, last page of each lesson
Study Guide and Review . . . . SE pp. 628–630
Test Prep Tool Kit

Technology

***Test and Practice Generator* CD-ROM**

internet connect

State-Specific Test Practice Online KEYWORD: MR4 TestPrep

go.hrw.com

Performance Assessment

Assess students' understanding of chapter concepts and combined problem-solving skills.

Performance Assessment . . . . SE p. 632
Includes scoring rubric in TE
Performance Assessment . . . . AR p. 126
Performance Assessment Teacher Support . . . . AR p. 125

Portfolio

Portfolio opportunities appear throughout the Student and Teacher's Editions.

Suggested work samples:

Problem Solving Project . . . . TE p. 596
Performance Assessment . . . . SE p. 632
Portfolio Guide . . . . AR p. xxxv
Journal . . . . TE pp. 601, 607, 619, 623
Write About It . . . . SE, last page of each lesson

Daily Assessment

Obtain daily feedback on students' understanding of concepts.

Spiral Review and Test Prep . . . . . . SE, last page of each lesson

Also Available on Transparency In Chapter 12 Resource Book

Warm Up . . . . TE, first page of each lesson
Problem of the Day . . . . TE, first page of each lesson
Lesson Quiz . . . . TE, last page of each lesson

Student Self-Assessment

Have students evaluate their own work.

Group Project Evaluation . . . . AR p. xxxii
Individual Group Member Evaluation . . . . AR p. xxxiii
Portfolio Guide . . . . AR p. xxxv
Journal . . . . TE pp. 601, 607, 619, 623

Formal Assessment

Assess students' mastery of concepts and skills.

Section Quizzes . . . . AR pp. 29–30
Mid-Chapter Quiz . . . . SE p. 608
Chapter Test . . . . SE p. 631
Chapter Tests (Levels A, B, C) . . . . AR pp. 97–102
Cumulative Tests (Levels A, B, C) . . . . AR pp. 259–270
Standardized Test Prep
Cumulative Assessment . . . . SE p. 633
End-of-Year Test . . . . AR pp. 271–274

Technology

***Test and Practice Generator* CD-ROM**

Make tests electronically. This software includes:

- Dynamic practice for Chapter 12
- Customizable tests
- Multiple-choice items for each objective
- Free-response items for each objective
- Teacher management system

SE = *Student Edition* **TE** = *Teacher's Edition* **AR** = *Assessment Resources* **CRB** = *Chapter Resource Book* **MK** = *Manipulatives Kit*

Chapter 12 Tests

Three levels (A,B,C) of tests are available for each chapter in the *Assessment Resources.*

LEVEL A

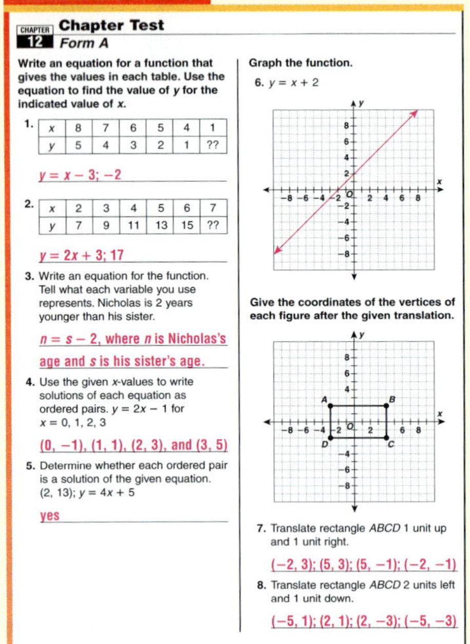

Chapter 12 Chapter Test
Form A

Write an equation for a function that gives the values in each table. Use the equation to find the value of *y* for the indicated value of *x*.

1.

x	8	7	6	5	4	1
y	5	4	3	2	1	??

$y = x - 3$; -2

2.

x	2	3	4	5	6	7
y	7	9	11	13	15	??

$y = 2x + 3$; 17

3. Write an equation for the function. Tell what each variable you use represents. Nicholas is 2 years younger than his sister.

$n = s - 2$, where n is Nicholas's age and s is his sister's age.

4. Use the given x-values to write solutions of each equation as ordered pairs. $y = 2x - 1$ for $x = 0, 1, 2, 3$

(0, −1), (1, 1), (2, 3), and (3, 5)

5. Determine whether each ordered pair is a solution of the given equation. (2, 13); $y = 4x + 5$

yes

Graph the function.

6. $y = x + 2$

Give the coordinates of the vertices of each figure after the given translation.

7. Translate rectangle *ABCD* 1 unit up and 1 unit right.

(−2, 3); (5, 3); (5, −1); (−2, −1)

8. Translate rectangle *ABCD* 2 units left and 1 unit down.

(−5, 1); (2, 1); (2, −3); (−5, −3)

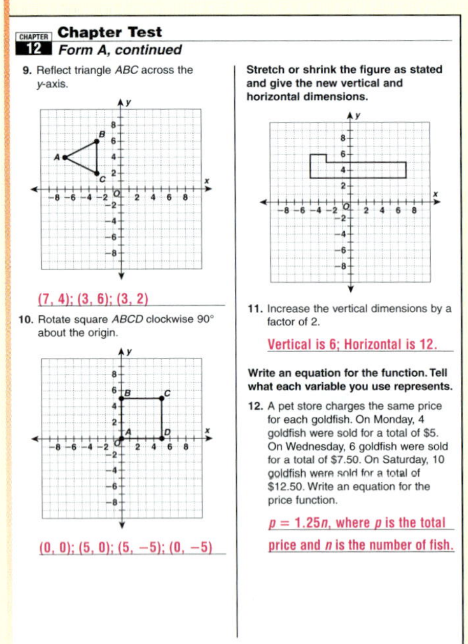

Chapter 12 Chapter Test
Form A, continued

9. Reflect triangle *ABC* across the *y*-axis.

(7, 4); (3, 6); (3, 2)

10. Rotate square *ABCD* clockwise 90° about the origin.

(0, 0); (5, 0); (5, −5); (0, −5)

Stretch or shrink the figure as stated and give the new vertical and horizontal dimensions.

11. Increase the vertical dimensions by a factor of 2.

Vertical is 6; Horizontal is 12.

Write an equation for the function. Tell what each variable you use represents.

12. A pet store charges the same price for each goldfish. On Monday, 4 goldfish were sold for a total of \$5. On Wednesday, 6 goldfish were sold for a total of \$7.50. On Saturday, 10 goldfish were sold for a total of \$12.50. Write an equation for the price function.

$p = 1.25n$, where p is the total price and n is the number of fish.

LEVEL B

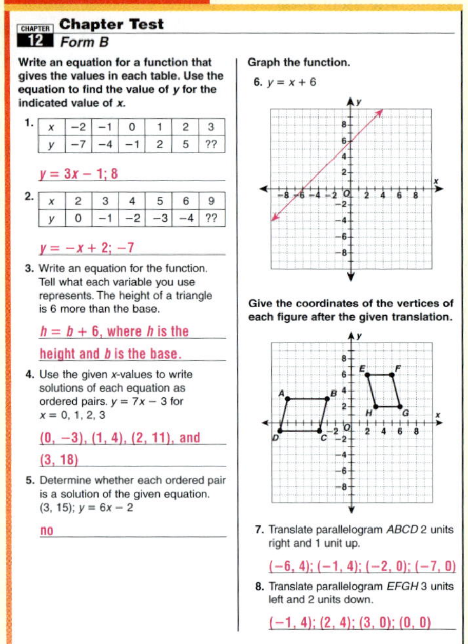

Chapter 12 Chapter Test
Form B

Write an equation for a function that gives the values in each table. Use the equation to find the value of *y* for the indicated value of *x*.

1.

x	−2	−1	0	1	2	3
y	−7	−4	−1	2	5	??

$y = 3x - 1$; 8

2.

x	2	3	4	5	6	9
y	0	−1	−2	−3	−4	??

$y = -x + 2$; -7

3. Write an equation for the function. Tell what each variable you use represents. The height of a triangle is 6 more than the base.

$h = b + 6$, where h is the height and b is the base.

4. Use the given x-values to write solutions of each equation as ordered pairs. $y = 7x - 3$ for $x = 0, 1, 2, 3$

(0, −3), (1, 4), (2, 11), and (3, 18)

5. Determine whether each ordered pair is a solution of the given equation. (3, 15); $y = 6x - 2$

no

Graph the function.

6. $y = x + 6$

Give the coordinates of the vertices of each figure after the given translation.

7. Translate parallelogram *ABCD* 2 units right and 1 unit up.

(−6, 4); (−1, 4); (−2, 0); (−7, 0)

8. Translate parallelogram *EFGH* 3 units left and 2 units down.

(−1, 4); (2, 4); (3, 0); (0, 0)

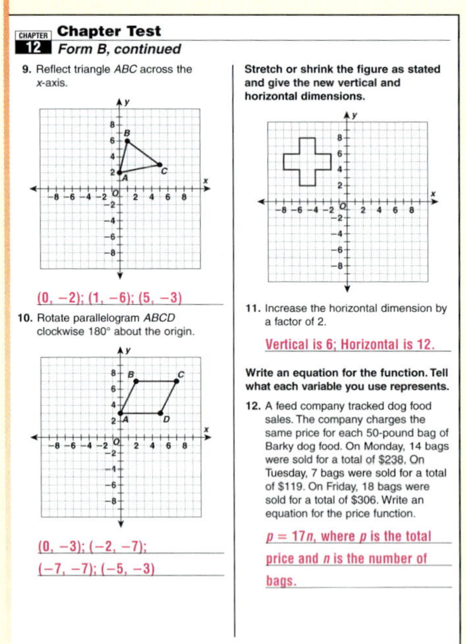

Chapter 12 Chapter Test
Form B, continued

9. Reflect triangle *ABC* across the *x*-axis.

(0, −2); (1, −6); (5, −3)

10. Rotate parallelogram *ABCD* clockwise 180° about the origin.

(0, −3); (−2, −7); (−7, −7); (−5, −3)

Stretch or shrink the figure as stated and give the new vertical and horizontal dimensions.

11. Increase the horizontal dimension by a factor of 2.

Vertical is 6; Horizontal is 12.

Write an equation for the function. Tell what each variable you use represents.

12. A feed company tracked dog food sales. The company charges the same price for each 50-pound bag of Barky dog food. On Monday, 14 bags were sold for a total of \$238. On Tuesday, 7 bags were sold for a total of \$119. On Friday, 18 bags were sold for a total of \$306. Write an equation for the price function.

$p = 17n$, where p is the total price and n is the number of bags.

LEVEL C

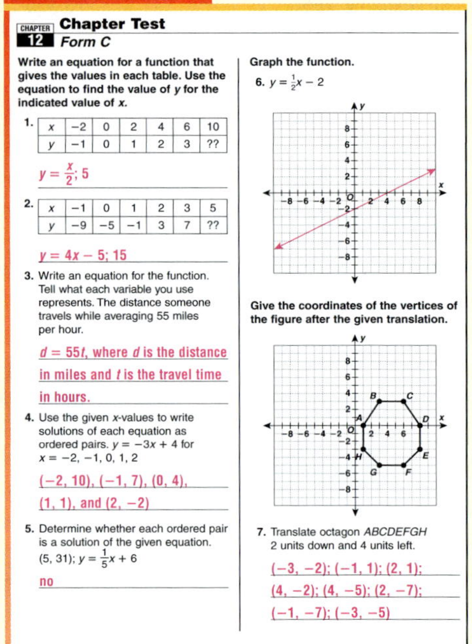

Chapter 12 Chapter Test
Form C

Write an equation for a function that gives the values in each table. Use the equation to find the value of *y* for the indicated value of *x*.

1.

x	−2	0	2	4	6	10
y	−1	0	1	2	3	??

$y = \frac{x}{2}$; 5

2.

x	−1	0	1	2	3	5
y	−9	−5	−1	3	7	??

$y = 4x - 5$; 15

3. Write an equation for the function. Tell what each variable you use represents. The distance someone travels while averaging 55 miles per hour.

$d = 55t$, where d is the distance in miles and t is the travel time in hours.

4. Use the given x-values to write solutions of each equation as ordered pairs. $y = -3x + 4$ for $x = -2, -1, 0, 1, 2$

(−2, 10), (−1, 7), (0, 4), (1, 1), and (2, −2)

5. Determine whether each ordered pair is a solution of the given equation. (5, 31); $y = \frac{1}{5}x + 6$

no

Graph the function.

6. $y = \frac{1}{2}x - 2$

Give the coordinates of the vertices of the figure after the given translation.

7. Translate octagon *ABCDEFGH* 2 units down and 4 units left.

(−3, −2); (−1, 1); (2, 1); (4, −2); (4, −5); (2, −7); (−1, −7); (−3, −5)

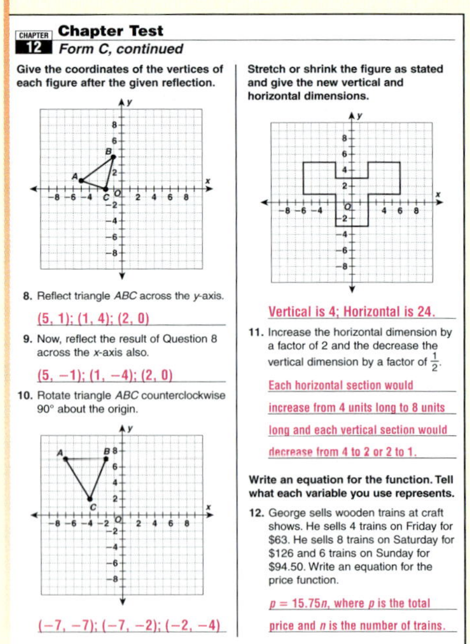

Chapter 12 Chapter Test
Form C, continued

Give the coordinates of the vertices of each figure after the given reflection.

8. Reflect triangle *ABC* across the *y*-axis.

(5, 1); (1, 4); (2, 0)

9. Now, reflect the result of Question 8 across the *x*-axis also.

(5, −1); (1, −4); (2, 0)

10. Rotate triangle *ABC* counterclockwise 90° about the origin.

(−7, −7); (−7, −2); (−2, −4)

Stretch or shrink the figure as stated and give the new vertical and horizontal dimensions.

Vertical is 4; Horizontal is 24.

11. Increase the horizontal dimension by a factor of 2 and the decrease the vertical dimension by a factor of $\frac{1}{2}$.

Each horizontal section would increase from 4 units long to 8 units long and each vertical section would decrease from 4 to 2 or 2 to 1.

Write an equation for the function. Tell what each variable you use represents.

12. George sells wooden trains at craft shows. He sells 4 trains on Friday for \$63. He sells 8 trains on Saturday for \$126 and 6 trains on Sunday for \$94.50. Write an equation for the price function.

$p = 15.75n$, where p is the total price and n is the number of trains.

Test and Practice Generator CD-ROM

Create and customize multiple versions of the same tests with corresponding answers for any chosen chapter objectives.

Chapter 12 State and Standardized Test Preparation

Test Taking Skill Builder and Standardized Test Practice are provided for each chapter in the *Test Prep Tool Kit.*

TEST TAKING SKILL BUILDER

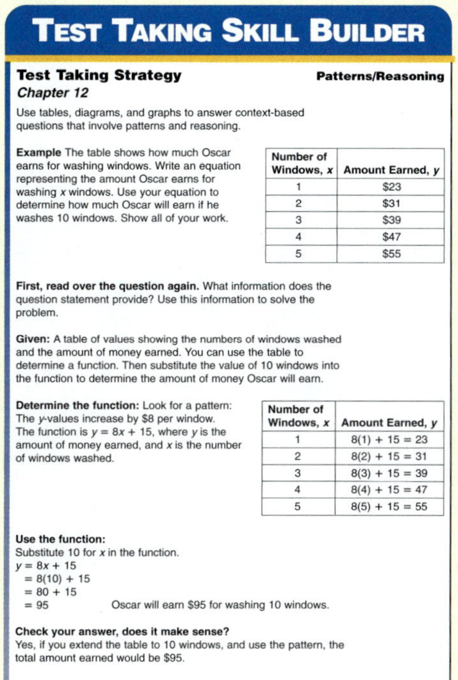
Test Taking Strategy — Patterns/Reasoning
Chapter 12

Use tables, diagrams, and graphs to answer context-based questions that involve patterns and reasoning.

Example The table shows how much Oscar earns for washing windows. Write an equation representing the amount Oscar earns for washing x windows. Use your equation to determine how much Oscar will earn if he washes 10 windows. Show all of your work.

Number of Windows, x	Amount Earned, y
1	$23
2	$31
3	$39
4	$47
5	$55

First, read over the question again. What information does the question statement provide? Use this information to solve the problem.

Given: A table of values showing the numbers of windows washed and the amount of money earned. You can use the table to determine a function. Then substitute the value of 10 windows into the function to determine the amount of money Oscar will earn.

Determine the function: Look for a pattern: The y-values increase by $8 per window. The function is $y = 8x + 15$, where y is the amount of money earned, and x is the number of windows washed.

Number of Windows, x	Amount Earned, y
1	$8(1) + 15 = 23$
2	$8(2) + 15 = 31$
3	$8(3) + 15 = 39$
4	$8(4) + 15 = 47$
5	$8(5) + 15 = 55$

Use the function:
Substitute 10 for x in the function.
$y = 8x + 15$
$= 8(10) + 15$
$= 80 + 15$
$= 95$ Oscar will earn $95 for washing 10 windows.

Check your answer, does it make sense?
Yes, if you extend the table to 10 windows, and use the pattern, the total amount earned would be $95.

STANDARDIZED TEST PRACTICE

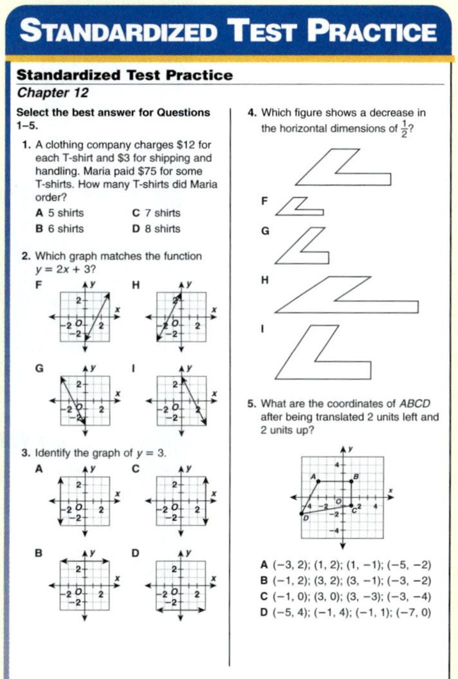
Standardized Test Practice
Chapter 12

Select the best answer for Questions 1–5.

1. A clothing company charges $12 for each T-shirt and $3 for shipping and handling. Maria paid $75 for some T-shirts. How many T-shirts did Maria order?
   A 5 shirts C 7 shirts
   B 6 shirts D 8 shirts
2. Which graph matches the function $y = 2x + 3$?
   F G H I
3. Identify the graph of $y = 3$.
   A B C D
4. Which figure shows a decrease in the horizontal dimensions of $\frac{1}{2}$?
   F G H I
5. What are the coordinates of $ABCD$ after being translated 2 units left and 2 units up?
   A (−3, 2); (1, 2); (1, −1); (−5, −2)
   B (−1, 2); (3, 2); (3, −1); (−3, −2)
   C (−1, 0); (3, 0); (3, −3); (−3, −4)
   D (−5, 4); (−1, 4); (−1, 1); (−7, 0)

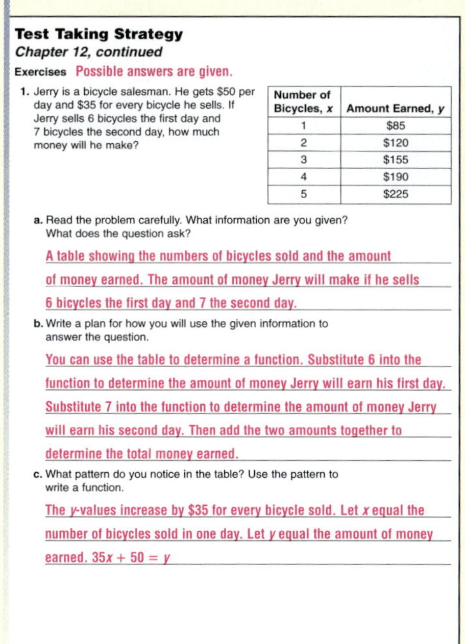
Test Taking Strategy
Chapter 12, continued

Exercises Possible answers are given.

1. Jerry is a bicycle salesman. He gets $50 per day and $35 for every bicycle he sells. If Jerry sells 6 bicycles the first day and 7 bicycles the second day, how much money will he make?

Number of Bicycles, x	Amount Earned, y
1	$85
2	$120
3	$155
4	$190
5	$225

a. Read the problem carefully. What information are you given? What does the question ask?
A table showing the numbers of bicycles sold and the amount of money earned. The amount of money Jerry will make if he sells 6 bicycles the first day and 7 the second day.

b. Write a plan for how you will use the given information to answer the question.
You can use the table to determine a function. Substitute 6 into the function to determine the amount of money Jerry will earn his first day. Substitute 7 into the function to determine the amount of money Jerry will earn his second day. Then add the two amounts together to determine the total money earned.

c. What pattern do you notice in the table? Use the pattern to write a function.
The y-values increase by $35 for every bicycle sold. Let x equal the number of bicycles sold in one day. Let y equal the amount of money earned. $35x + 50 = y$

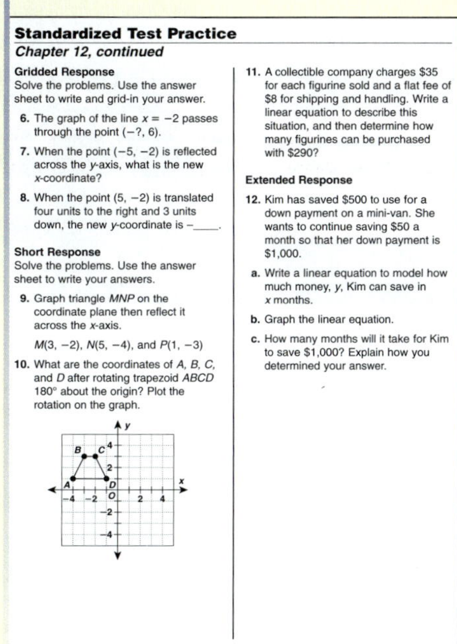
Standardized Test Practice
Chapter 12, continued

Gridded Response
Solve the problems. Use the answer sheet to write and grid-in your answer.

6. The graph of the line $x = -2$ passes through the point (−?, 6).
7. When the point (−5, −2) is reflected across the y-axis, what is the new x-coordinate?
8. When the point (5, −2) is translated four units to the right and 3 units down, the new y-coordinate is −____.

Short Response
Solve the problems. Use the answer sheet to write your answers.

9. Graph triangle MNP on the coordinate plane then reflect it across the x-axis.
   $M(3, -2)$, $N(5, -4)$, and $P(1, -3)$
10. What are the coordinates of A, B, C, and D after rotating trapezoid $ABCD$ 180° about the origin? Plot the rotation on the graph.
11. A collectible company charges $35 for each figurine sold and a flat fee of $8 for shipping and handling. Write a linear equation to describe this situation, and then determine how many figurines can be purchased with $290?

Extended Response

12. Kim has saved $500 to use for a down payment on a mini-van. She wants to continue saving $50 a month so that her down payment is $1,000.
    a. Write a linear equation to model how much money, y, Kim can save in x months.
    b. Graph the linear equation.
    c. How many months will it take for Kim to save $1,000? Explain how you determined your answer.

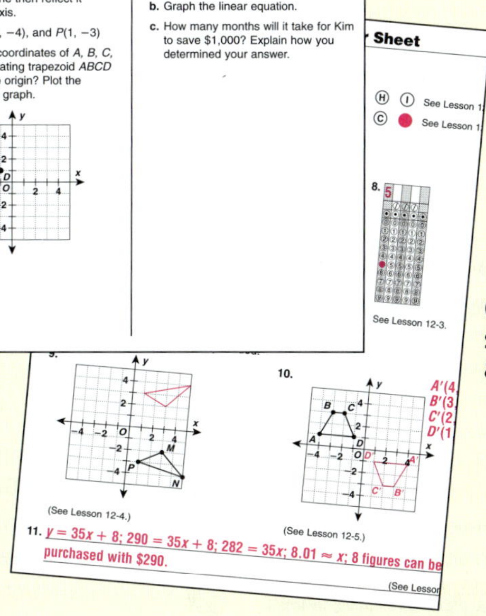

internet connect
State-Specific Test Practice Online
KEYWORD: MR4 TestPrep

Test Prep Tool Kit

- Standardized Test Prep Workbook
- Countdown to Testing transparencies
- State Test Prep CD-ROM
- Standardized Test Prep Video

Customized answer sheets give students realistic practice for actual standardized tests.

Chapter 12

Functions and Coordinate Geometry

Why Learn This?

Tell students that functions describe relationships in which one quantity depends on another quantity. In the equation $y = 2x$, the value of y depends on the value of x. A sports physiologist knows that the number of calories burned is a function of the speed that the exerciser is biking, walking, or running. The sports physiologist can use a function to produce the number of calories burned (the output) when a certain speed (the input) is used.

Using Data

To begin the study of this chapter, have students:

- Estimate the rate of energy use for walking at 2.5 m/s. about 315 calories/hr
- Determine the number of calories used during a 4-hour cycling tour at an average speed of 3 m/s. 540
- Determine the speed of a runner who used 810 calories on a 1.5-hour training run. 4 m/s

Chapter 12

Functions and Coordinate Geometry

TABLE OF CONTENTS

internet connect

Chapter Opener Online
go.hrw.com
KEYWORD: MR4 Ch12

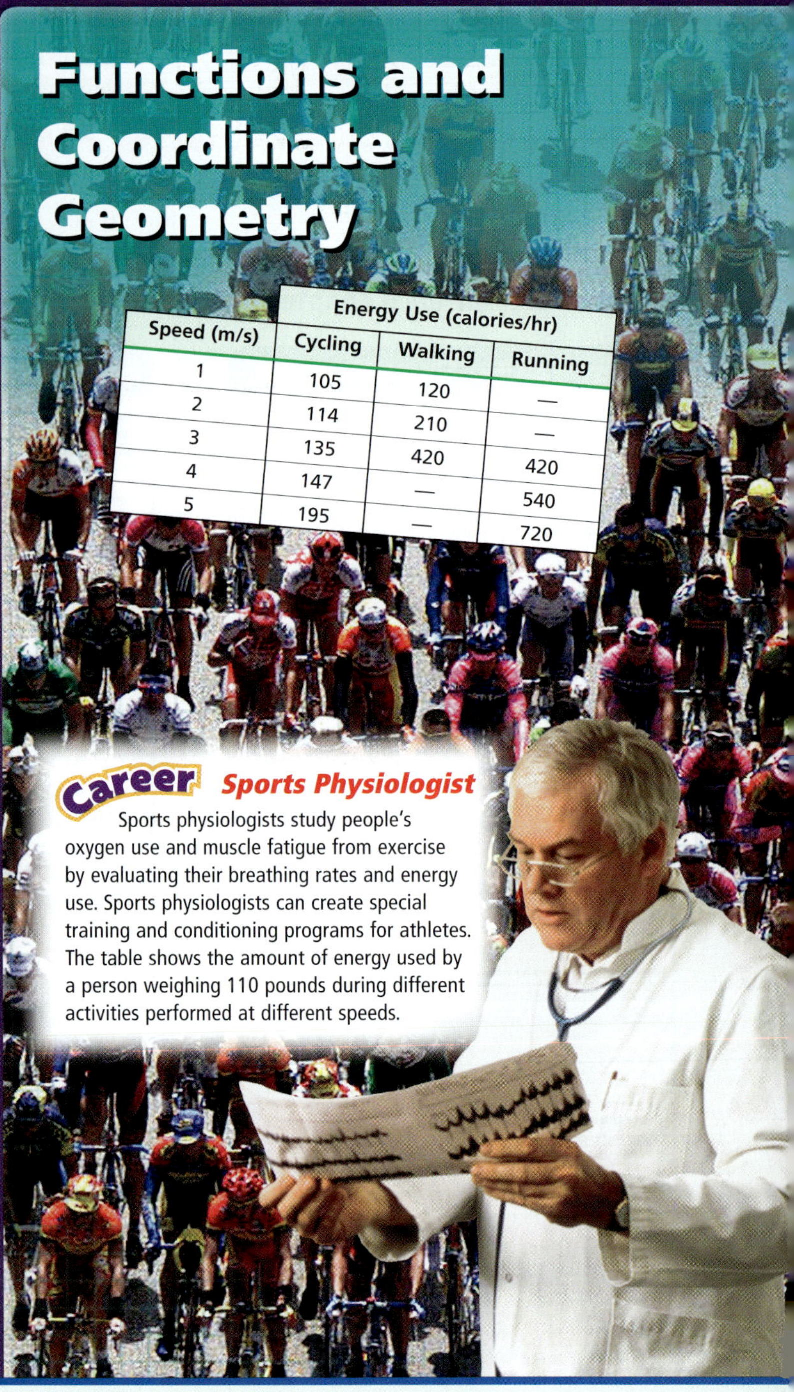

Speed (m/s)	Energy Use (calories/hr)		
	Cycling	Walking	Running
1	105	120	—
2	114	210	—
3	135	420	420
4	147	—	540
5	195	—	720

Career ***Sports Physiologist***

Sports physiologists study people's oxygen use and muscle fatigue from exercise by evaluating their breathing rates and energy use. Sports physiologists can create special training and conditioning programs for athletes. The table shows the amount of energy used by a person weighing 110 pounds during different activities performed at different speeds.

PROBLEM SOLVING

Problem Solving Project

Health and Life Science Connection

Purpose: To solve problems using data, linear functions, and graphs

Materials: Counting on Calories worksheet

internet connect

Chapter Project Online: ***go.hrw.com***
KEYWORD: MR4 PSProject12

Understand, Plan, Solve, and Look Back

Have students:

- Complete the Counting on Calories worksheet.
- Create a measured course. Have students time each other walking and running. Have them create a graph of the class results. Have them estimate their calorie use for an hour's activity at that speed. Do they think they could continue that pace for an hour? Why or why not?
- Estimate how many calories a world-class marathoner would use in setting the world record.
- Check students' work.

ARE YOU READY?

Choose the best term from the list to complete each sentence.

1. The x-axis on a coordinate grid is the __?__ axis, and the y-axis is the __?__ axis. horizontal; vertical
2. The point (1, 2) is written as a(n) __?__. ordered pair
3. A mathematical statement that says two quantities are equal is a(n) __?__. equation
4. A(n) __?__ is a mathematical phrase that includes only numbers and operation symbols. numerical expression
5. A(n) __?__ is formed by two number lines in a plane that intersect at right angles. coordinate plane

coordinate plane
equation
horizontal
numerical expression
ordered pair
vertical
origin

Complete these exercises to review skills you will need for this chapter.

✔ Equations

Solve each equation.

6. $4t = 32$ $t = 8$
7. $2b + 4 = 12$ $b = 4$
8. $15 = 6r - 3$ $r = 3$
9. $3x = 72$ $x = 24$
10. $23 = 4a - 5$ $a = 7$
11. $12m + 3 = 63$ $m = 5$

✔ Ordered Pairs

Give the coordinates of each point.

12. A (1, 2)
13. Z (−3, 0)
14. C (−2, −3)
15. G (2, −1)
16. M (0, 0)
17. H (−1, 3)

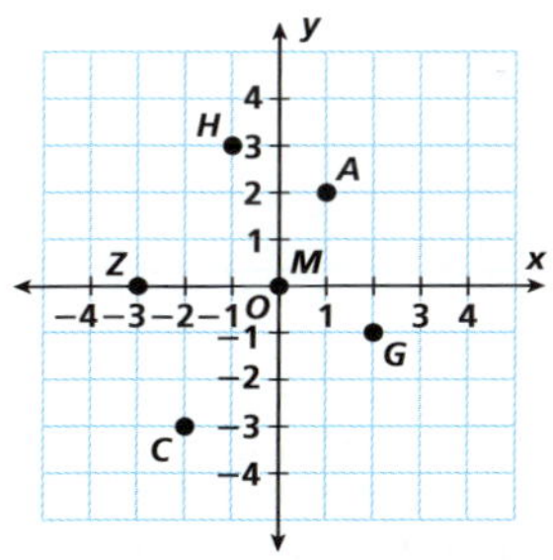

✔ Transformations

Tell whether each transformation is a translation, a rotation, or a reflection.

18.
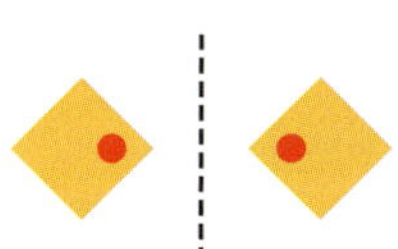
reflection

19.
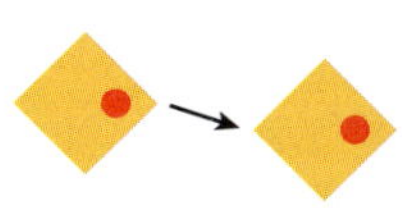
translation

20.
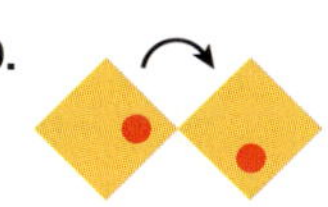
rotation

Assessing Prior Knowledge

INTERVENTION

Diagnose and Prescribe

Evaluate your students' performance on this page to determine whether intervention is necessary or whether enrichment is appropriate. Options that provide instruction, practice, and a check are listed below.

Resources for Are You Ready?

- ***Are You Ready? Intervention and Enrichment***
- **Recording Sheet for Are You Ready?** *Chapter 12 Resource Book. . . . . p. 3*

Are You Ready? Intervention **CD-ROM**

ARE YOU READY?

Were students successful with Are You Ready?

NO INTERVENE | **YES ENRICH**

✔Equations 6–11
Intervention Practice, Skill 58, 59, 60
CD-ROM
Intervention Activities, Skill 58, 59, 60

✔Ordered Pairs 12–17
Intervention Practice, Skill 68
CD-ROM
Intervention Activities, Skill 68

✔Transformations 18–20
Intervention Practice, Skill 87
CD-ROM
Intervention Activities, Skill 87

Are You Ready? Enrichment, pp. 429–430

Section

12A Introduction to Functions

One-Minute Section Planner

Lesson	Materials	Resources
Lesson 12-1 Tables and Functions **NCTM:** Algebra, Problem Solving, Communication, Representation **NAEP:** Algebra 1b ☑ SAT-9 ☑ SAT-10 ☐ ITBS ☑ CTBS ☐ MAT ☑ CAT		• *Chapter 12 Resource Book*, pp. 6–14 • Daily Transparency T1, CRB • Additional Examples Transparencies T2–T3, CRB • *Alternate Openers: Explorations*, p. 98
Hands-On Lab 12A Explore Linear and Nonlinear Relationships **NCTM:** Algebra, Representation **NAEP:** Algebra 1e ☑ SAT-9 ☐ SAT-10 ☐ ITBS ☑ CTBS ☐ MAT ☐ CAT	**Required** Square tiles or graph paper *(MK, CRB, p. 68)*	• *Hands-On Lab Activities*, pp. 107–108
Lesson 12-2 Graphing Functions **NCTM:** Algebra, Communication **NAEP:** Algebra 2b ☐ SAT-9 ☐ SAT-10 ☐ ITBS ☐ CTBS ☐ MAT ☐ CAT	**Required** Graph paper *(CRB, p. 68)*	• *Chapter 12 Resource Book*, pp. 15–24 • Daily Transparency T4, CRB • Additional Examples Transparencies T5–T7, CRB • *Alternate Openers: Explorations*, p. 99
Section 12A Assessment		• Mid-Chapter Quiz, SE p. 608 • Section 12A Quiz, AR p. 29 • *Test and Practice Generator* CD-ROM

SAT = *Stanford Achievement Tests* **ITBS** = *Iowa Test of Basic Skills* **CTBS** = *Comprehensive Test of Basic Skills/Terra Nova*
MAT = *Metropolitan Achievement Tests* **CAT** = *California Achievement Test*

NCTM = Complete standards can be found on pages T27–T33. **NAEP** = Complete standards can be found on pages A31–A35.

SE = *Student Edition* **TE** = *Teacher's Edition* **AR** = *Assessment Resources* **CRB** = *Chapter Resource Book* **MK** = *Manipulatives Kit*

Section Overview

Tables and Functions
Lesson 12-1

Why? You can use a function to convert between customary units and metric units.

A **function** is a rule that relates two variables such that each **input** value of one variable corresponds to exactly one **output** value of the other variable.

A function table shows some of the values for a function.

Input x	3	4	5	6	10
Output y	7	9	11	13	■

You can write an equation for the function that gives these values and use the equation to find the missing output value.

Each **output** is 1 more than 2 times the **input.**

y = 2 times x plus 1

$y = 2x + 1$

$y = 2(10) + 1$

$y = 20 + 1$

$y = 21$

When x is 10, y is 21.

Graphing Functions
Lesson 12-2

Why? You can use the graph of the linear relationship between degrees Celsius and kelvins to convert temperatures.

Graph the function described by the equation $y = 2x + 1$.

Make a function table.

x	$2x + 1$	y
−3	2(−3) + 1	−5
0	2(0) + 1	1
1	2(1) + 1	3

Write the ordered pairs.

(x, y)
(−3, −5)
(0, 1)
(1, 3)

Graph the ordered pairs on a coordinate plane.
Draw a line through the points.

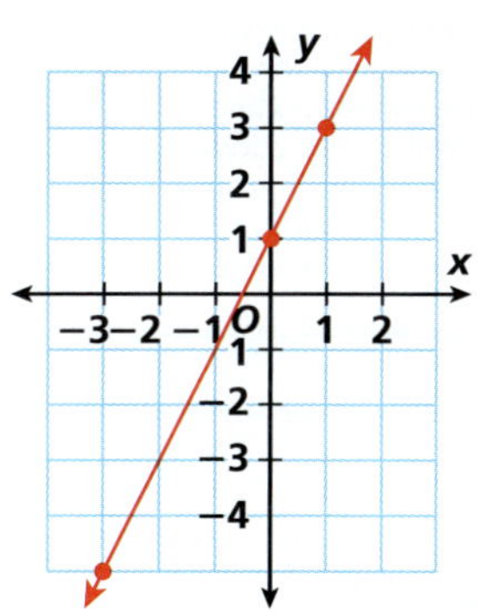

Every point on the line represents a solution to the equation.

When you graph the ordered pairs of a function, the points may form a straight line. An equation for such a function is called a **linear equation.**

12-1 Organizer

Pacing: Traditional 1 day
Block $\frac{1}{2}$ day

Objective: Students use data in a table to write an equation for a function and use the equation to find a missing value.

Indiana Academic Standards
6.3.5, 6.3.9, 6.7.1, 6.7.2, 6.7.3, 6.7.4, 6.7.5, 6.7.9, 6.7.10, 6.7.11

Warm Up

Evaluate each expression for the given value of the variable.

1. $4x - 1$ for $x = 2$ 7
2. $7y + 3$ for $y = 5$ 38
3. $\frac{1}{2}x + 2$ for $x = -6$ -1
4. $8y - 3$ for $y = -2$ -19

Problem of the Day

Maria rented an electric car at a rate of $40 per day and $0.15 per mile. She returned the car the same day, gave the rental clerk a $100 bill, and got $21.75 back in change. How far did Maria drive the car? 255 miles

Available on Daily Transparency in CRB

Math Humor

When x was equal to zero, no one listened to its opinion. They all agreed its input had no value.

12-1 Tables and Functions

Learn to use data in a table to write an equation for a function and to use the equation to find a missing value.

Vocabulary
function
input
output

A baseball pitch thrown too high, low, or wide is considered outside the strike zone. A pitcher threw a ball 4 inches too low. How far in centimeters was the ball outside the strike zone? Make a table to show how the number of centimeters increases as the number of inches increases.

"Come on, ump, that pitch was at least four centimeters outside!"

Andrew Toos/Cartoon Resource.com

	Inches	Centimeters	
	1	2.54	
+1	2	5.08	+2.54
+1	3	7.62	+2.54
+1	4	10.16	+2.54

The number of centimeters is 2.54 times the number of inches. Let x represent the number of inches and y represent the number of centimeters. Then the equation $y = 2.54x$ relates centimeters to inches.

A **function** is a rule that relates two quantities so that each **input** value corresponds exactly to one **output** value.

When the input is 4 in., the output is 10.16 cm. So the ball was 10.16 centimeters outside the strike zone.

You can use a function table to show some of the values for a function.

EXAMPLE 1 Writing Equations from Function Tables

Write an equation for a function that gives the values in the table. Use the equation to find the value of y for the indicated value of x.

x	3	4	5	6	7	10
y	7	9	11	13	15	■

y is 2 times x + 1. *Compare x and y to find a pattern.*
$y = 2x + 1$ *Use the pattern to write an equation.*
$y = 2(10) + 1$ *Substitute 10 for x.*
$y = 20 + 1 = 21$ *Use your function rule to find y when x = 10.*

Helpful Hint
When all the y-values are greater than the corresponding x-values, use addition and/or multiplication in your equation.

1 Introduce

Alternate Opener

EXPLORATION

12-1 Tables and Functions

A school has scheduled a trip for 210 students to a theme park. The school can rent up to six buses for $200 each. Each bus seats a maximum of 60 students. Tickets to the theme park cost $25 per student.

1. Use the first example as a guide to complete the table.

Group Number	Number of Students	Cost of Bus Rental and Tickets	Total Cost
1	10	$200 + 25 \cdot 10 = 200 + 250$	$450.00
2	20		
3	30		
4	40		
5	50		
6	60		

2. What numbers in the table remain constant?
3. What numbers in the table vary?
4. Use the cost of the bus ($200), the cost of each ticket ($25), and the number of students (x) to write an equation for the total cost (c). (*Hint:* Look at the middle column of the table to write the equation.)

Think and Discuss

5. **Explain** what makes the total cost of a group vary.
6. **Discuss** possible ways of reducing the total cost of taking 210 students to the theme park.

Motivate

To introduce students to functions, have them complete a table with familiar data, such as the following:

Tables	1	2	3	4
Legs	4	8	12	16

Ask students to discuss any patterns they notice in the table.

Possible answer: The number of tables keeps increasing by 1. The number of legs keeps increasing by 4. The number of legs is 4 times the number of tables.

Exploration worksheet and answers on Chapter 12 Resource Book pp. 7 and 69

2 Teach

Lesson Presentation

Guided Instruction

In this lesson, students learn to use data in a table to write an equation for a function and to use the equation to find a missing value. Define *function*, *input*, and *output*. First teach students to write equations for functions shown in tables and described with words. Then work through the steps for solving a real-world problem involving functions.

You can write equations for functions that are described in words.

EXAMPLE 2 **Translating Words into Math**

Write an equation for the function. Tell what each variable you use represents.

The length of a rectangle is 5 times its width.

ℓ = length of rectangle *Choose variables for the equation.*

w = width of rectangle

$\ell = 5w$ *Write an equation.*

EXAMPLE 3 **PROBLEM SOLVING APPLICATION**

PROBLEM SOLVING

Car washers tracked the number of cars they washed and the total amount of money they earned. They charged the same price for each car they washed. They earned \$60 for 20 cars, \$66 for 22 cars, and \$81 for 27 cars. Write an equation for the function.

1 Understand the Problem

The **answer** will be an equation that describes the relationship between the number of cars washed and the money earned.

2 Make a Plan

You can make a table to display the data.

3 Solve

Let c be the number of cars. Let m be the amount of money earned.

c	20	22	27
m	60	66	81

m is equal to 3 times c. *Compare c and m.*

$m = 3c$ *Write an equation.*

4 Look Back

Substitute the c and m values in the table to check that they are solutions of the equation $m = 3c$.

$m = 3c$ (20, 60) | $m = 3c$ (22, 66) | $m = 3c$ (27, 81)

$60 \stackrel{?}{=} 3 \cdot 20$ | $66 \stackrel{?}{=} 3 \cdot 22$ | $81 \stackrel{?}{=} 3 \cdot 27$

$60 \stackrel{?}{=} 60$ ✓ | $66 \stackrel{?}{=} 66$ ✓ | $81 \stackrel{?}{=} 81$ ✓

Think and Discuss

1. Explain how you find the y-value when the x-value is 20 for the function $y = 5x$.

COMMON ERROR ALERT

Some students may write an equation for a relationship using only x-values or only y-values instead of the relationship between x- *and* y-values. To prevent this error, you may wish to have students color code their tables, using one color for all x-values and one color for all y-values.

Additional Examples

Example 1

Write an equation for a function that gives the values in the table. Use the equation to find the value of y for the indicated value of x.

x	3	4	5	6	7	10
y	13	16	19	22	25	■

$y = 3x + 4$; when $x = 10$, $y = 34$.

Example 2

Write an equation for the function. Tell what each variable you use represents.

The height of a painting is 7 times its width.

$h = 7w$; h = height, w = width

Example 3

The school choir tracked the number of tickets sold and the total amount of money received. The choir members received \$80 for 20 tickets, \$88 for 22 tickets, and \$108 for 27 tickets. If each ticket costs the same, write an equation for the function.

$m = 4t$

Reaching All Learners

Through Home Connection

Have students make tables of values that can be expressed as functions. For example, students might make a table showing the relationship of plates to silverware at the dinner table:

Plates	1	2	3	4
Silverware	3	6	9	12

Have students write equations for the functions that represent the values in their tables.

For the example above, the equation might be $y = 3x$.

3 Close

Summarize

Review the new vocabulary in the lesson: *function*, *input*, and *output*. Ask students to explain how the terms relate to each other.

Possible answer: A *function* is a rule that relates two quantities such that each *input* value corresponds exactly to one *output* value.

Answers to Think and Discuss

1. Substitute 20 for x, and then multiply. When x is 20, y is 100.

12-1 PRACTICE & ASSESS

Students may want to refer back to the lesson examples.

Assignment Guide

If you finished Example 1 assign:
Core 1–2, 5–6, 10–11, 20–24
Enriched 1–2, 5–6, 10–11, 20–24

If you finished Example 2 assign:
Core 1–3, 5–8, 10–13, 20–24
Enriched 1–3, 5–8, 10–13, 20–24

If you finished Example 3 assign:
Core 1–13, 15, 20–24
Enriched 5–24

Answers

10. $y = -4x + 2$; -54

11. $y = x + 3.4$; 2.4, 8.4

12-1 Exercises

FOR EXTRA PRACTICE see page 660

internet connect Homework Help Online go.hrw.com Keyword: MR4 12-1

GUIDED PRACTICE

See Example 1

Write an equation for a function that gives the values in each table. Use the equation to find the value of y for the indicated value of x.

1.

x	1	2	3	6	9
y	−5	−4	−3	0	

$y = x - 6$; 3

2.

x	3	4	5	6	10
y	16	21	26	31	

$y = 5x + 1$; 51

See Example 2

Write an equation for the function. Tell what each variable you use represents. Possible answers:

3. Jen is 6 years younger than her brother.
j = Jen's age, b = brother's age, $j = b - 6$

See Example 3

4. Brenda sells balloon bouquets. She charges the same price for each balloon in a bouquet. The cost of a bouquet with 6 balloons is \$3, with 9 balloons is \$4.50, and with 12 balloons is \$6. Write an equation for the function.
c = cost of bouquet, b = number of balloons, $c = \$0.50b$

INDEPENDENT PRACTICE

See Example 1

Write an equation for a function that gives the values in each table. Use the equation to find the value of y for the indicated value of x.

5.

x	0	1	2	5	7
y	0	4	8	20	

$y = 4x$; 28

6.

x	4	5	6	7	12
y	−2	0	2	4	

$y = 2x - 10$; 14

See Example 2

Write an equation for the function. Tell what each variable you use represents. Possible answers:

7. The cost of a case of bottled juices is \$2 less than the cost of twelve individual bottles.
c = cost of a case, s = cost of 1 bottle, $c = 12s - 2$

8. The population of New York is twice as large as the population of Michigan.
n = population of New York, m = population of Michigan, $n = 2m$

See Example 3

9. Oliver is playing a video game. He earns the same number of points for each prize he captures. He earned 1,050 points for 7 prizes, 1,500 points for 10 prizes, and 2,850 points for 19 prizes. Write an equation for the function.
p = number of points, m = number of prizes, $p = 150m$

PRACTICE AND PROBLEM SOLVING

Write an equation for a function that gives the values in each table, and then find the missing terms.

10.

x	2	3	5	9	11	14
y	−6	−10	−18	−34	−42	

11.

x	−1	0	1	2	5	7
y		3.4	4.4	5.4		10.4

Math Background

The function concept is used today throughout mathematics. The definition given in the present lesson corresponds to the definition devised by the mathematician Lejeune Dirichlet in the nineteenth century.

When y is a function of x, x is called the *independent variable,* and y is called the *dependent variable*. Whenever a value is assigned to x, a value is automatically assigned to y by some rule or correspondence. The set of all possible values of x is called the *domain* of a function. The set of all possible values of y is called the *range* of a function.

RETEACH 12-1

LESSON 12-1 **Reteach**
Tables and Functions

A function is a rule that relates two quantities so that each input value corresponds to exactly one output value. In the table below, the x-values are the input and the y-values are the output.

x	0	1	2	3	4	5	6	7
y	4	5	6	7	8	9	10	?

To write an equation for a table of values, first compare the x- and y-values to find a pattern.

Each y-value is 4 more than its corresponding x-value.

Then use the pattern to write a rule for the table.

$y = x + 4$

You can use the rule to find a missing value in a table. To find the value of y in table above when $x = 7$, substitute 7 for x in the equation.

$y = x + 4$
$y = 7 + 4$
$y = 11$
So y is 11 when x is 7.

Write an equation for a function that gives the values in each table. Use the equation to find the value of y for the indicated value of x.

1.

x	1	2	3	4	5	6
y	3	6	9	12	15	?

$y = 3x$, $y = 18$

2.

x	18	17	16	15	14	13
y	15	14	13	?	11	10

$y = x - 3$, $y = 12$

You can also write equations for functions that are described in words.

The length of the pool is 6 times the width of the pool.

ℓ = length of pool, w = width of pool — Choose variables for the equation.

$\ell = 6w$ — Write an equation.

Write an equation for the function. Tell what each variable you use represents.

3. Todd is 6 inches taller than Scott.
t = Todd's height
s = Scott's height
$t = s + 6$

4. Alana is 4 times as old as Tracey.
a = Alana's age
t = Tracey's age
$a = 4t$

PRACTICE 12-1

LESSON 12-1 **Practice B**
Tables and Functions

Write an equation for a function that gives the values in each table. Use the equation to find the value of y for the indicated value of x.

1.

x	1	2	3	4	5
y	7	14	21	28	◆

$y = 7x$
$y = 35$

2.

x	2	3	4	5	6
y	−3	−2	−1	0	◆

$y = x - 5$
$y = 1$

3.

x	20	16	12	8	4
y	10	8	6	4	◆

$y = x \div 2$
$y = 2$

4.

x	7	8	9	10	11
y	11	12	13	14	◆

$y = x + 4$
$y = 15$

Write an equation for the function. Tell what each variable you use represents.

5. Amanda is 7 years younger than her cousin.
Possible answer: $y = x - 7$; y = Amanda's age; x = her cousin's age

6. The population of North Carolina is twice as large as the population of South Carolina.
Possible answer: $n = 2s$; n = population of North Carolina; s = population of South Carolina

7. An Internet book company charges \$7 for each paperback book, plus \$2.75 for shipping and handling per order.
Possible answer: $y = 7x + 2.75$; y = total price of order; x = number of books purchased

8. Henry records how many days he rides his bike and how far he rides each week. He rides the same distance each time. He rode 18 miles in 3 days, 24 miles in 4 days, and 42 miles in 7 days. Write an equation for the function.
$m = 6d$; m = miles, and d = days

Write an equation for each function. Define the variables that you use.

12. 

13. 

Career LINK

Graphic designers work in media such as magazines, brochures, product packaging, and the Internet to visually communicate the appeal of a company or its products.

go.hrw.com
KEYWORD: MR4 Designer
CNN Student News

14. The height of a triangle is 5 centimeters more than twice the length of its base. Write an equation relating the height of the triangle to the length of its base. Find the height when the base is 20 centimeters long. b = base of triangle, h = height of triangle, $h = 2b + 5$; $h = 45$

15. Georgia earns $6.50 per hour at a part-time job. She wants to buy a sweater that costs $58.50. Write an equation relating the number of hours she works to the amount of money she earns. Find how many hours Georgia needs to work to buy the sweater. h = number of hours worked, $58.50 = 6.5h$; 9 hours

Use the table for Exercises 16–18.

Width (pixels)	Length (pixels)
30	95
40	125
50	155
60	185

16. ***GRAPHIC DESIGN*** Margo is designing a Web page displaying similar rectangles. She uses a linear pattern to determine the dimensions of each rectangle. Use the table to write an equation relating the width of a rectangle to the length of a rectangle. Find the length of a rectangle that has a width of 250 pixels. $l = 3w + 5$; 755 pixels

17. ***WHAT'S THE ERROR?*** Margo predicted that the length of a rectangle with a width of 100 pixels would be 310 pixels. Explain the error she made. Then find the correct length.

18. ***WRITE ABOUT IT*** Explain how to write an equation for the data in the table.

19. ***CHALLENGE*** Write an equation that would give the same y-values as $y = 2x + 1$ for $x = 0, 1, 2, 3$. Possible answer: $2y = 4x + 2$

Spiral Review

Evaluate each expression for $x = 5$. (Lesson 2-1)

20. $x + 7$ 12 **21.** $-4x$ −20 **22.** $3x + 6$ 21 **23.** $x^2 + 5$ 30

24. TEST PREP Jazmine played a game of chance 100 times and won 24 times. Which of the following is the best estimate of the experimental probability of winning the game? (Lesson 11-6) **B**

A 5% **B** 25% **C** 50% **D** 75%

Answers

12. Let c be the total cost and h be the number of hours.
$c = \$125 + \$55h$

13. Let f be the total cab fare and m be the number of miles.
$f = \$2.50 + \$0.90m$

17. Possible answer: The student may have tripled the width and added 10 instead of 5. The correct answer is 305 pixels.

18. Possible answer: each ℓ value is greater than its w value, so try multiplying or adding. When w is multiplied by 3, and 5 is added, the result is ℓ. $\ell = 3w + 5$. This works for all the pairs.

Journal

Have students think of and write real-world situations in which a function is used to find the relationship between two quantities. Have them write the equation for the function and find the y-values for any three x-values.

Test Prep Doctor

For Exercise 24, stress to students that they are to find the best estimate of the experimental probability. Otherwise, they may be confused because 24% does not appear as an answer choice. Students who answered **A** or **C** may need to review the term *experimental probability*. Students who answered **D** found the approximate experimental probability of losing the game.

CHALLENGE 12-1

LESSON 12-1 **Challenge**
Multifunctional Measurements

Any function can be written as two different equations—one for each variable. Write two different equations for the function that gives the values in each table. Then choose either equation to answer the question that follows.

A **stone** is a British unit of weight.

stones (s)	1	2	3	4	5
pounds (p)	14	28	42	56	70

1. $p =$ $14 \times s$ 2. $s =$ $p \div 14$

3. In a competition called Stones of Strength, athletes must lift five huge rocks onto platforms. The first rock weighs 245 pounds. How many stone does it weigh?

$17\frac{1}{2}$ stone

A **fathom** is a unit sailors use to measure the depth of water.

fathoms (m)	4	6	8	10	12
feet (f)	24	36	48	60	72

4. $f =$ $6 \times m$ 5. $m =$ $f \div 6$

6. Years ago, sailors on riverboats shouted "mark twain" when the river's depth was 2 fathoms—barely deep enough for a boat to navigate safely. How many feet deep was the water at mark twain?

12 feet

An **acre** is a unit for measuring areas of land.

acre (a)	1,280	1,920	2,560	3,200	3,840
square miles (s)	2	3	4	5	6

7. $a =$ $640 \times s$ 8. $s =$ $a \div 640$

9. King Ranch in Texas covers about 1,300 square miles—larger than the entire state of Rhode Island! How many acres of land does King Ranch cover?

832,000 acres

PROBLEM SOLVING 12-1

LESSON 12-1 **Problem Solving**
Tables and Functions

Use the tables to answer each question.

Table 1

miles	kilometers
2	3.22
3	4.83
4	6.44
5	8.05

Table 2

ounces	grams
1	28.35
2	56.7
3	85.05
4	113.4

Table 3

gallons	liters
5	18.95
10	37.9
15	56.85
20	75.8

1. Write an equation for a function that gives the values in table 1. Define the variables you use. Use your equation to find the missing term in the table.

Possible answer: $k = 1.61m$; k = kilometers; m = miles

2. Write an equation for a function that gives the values in table 2. Define the variables you use. Use your equation to find the missing term in the table.

Possible answer: $g = 28.35z$; g = grams; z = ounces

3. Write an equation for a function that gives the values in table 3. Define the variables you use. Use your equation to find the missing term in the table.

Possible answer: $l = 3.79g$; l = liters; g = gallons

4. There are 4 quarts in a gallon. Write an equation for a function relating quarts to liters. Then use your equation to find how many liters of oil a 50-quart barrel can hold.

Possible answer: $l = 0.9475q$; l = liters; q = quarts; 47.375 liters of oil

Circle the letter of the correct answer.

5. The Rocky Mountains stretch 3,750 miles across North America. What is this length in kilometers?
A 2,329.2 kilometers
B 1,164.6 kilometers
(C) 6,037.5 kilometers
D 12,075 kilometers

6. A hummingbird egg only weighs 0.25 grams! How many ounces does the egg weigh?
F about 7.0875 ounces
(G) about 0.009 ounces
H about 28 ounces
J about 9 ounces

Lesson Quiz

1. Write an equation for a function that gives the values in the table below. Use the equation to find the value for y for the indicated value of x. $y = 3x$; 21

x	0	1	3	5	7
y	0	3	9	15	■

2. Write an equation for the function. Tell what each variable you use represents. The height of a round can is 2 times its radius. $h = 2r$, where h is the height and r is the radius.

Available on Daily Transparency in CRB

Hands-On LAB 12A Explore Linear and Nonlinear Relationships

Pacing: Traditional 1 day
Block $\frac{1}{2}$ day

Objective: To use graph-paper drawings to explore linear and nonlinear relationships

Materials: Graph paper

Lab Resources

Hands-On Lab Activities .... p. 96

Using the Pages

Discuss with students how to use a graph-paper model to find ordered pairs. Have students do the following:

1. Explain how to form the next figure in each pattern. First example: At each stage, make a new column of squares on the right side, 2 units higher than the previous column. Second example: At each stage, make a square using the specified number of blocks.
2. Explain how you can use the graph to show that the function $y = \sqrt{x}$ is nonlinear. Possible answer: The points on the graph do not lie on a line.
3. Explain how you can use the table to show that the function $y = \sqrt{x}$ is nonlinear. The ratio of the change in y to the change in x is not constant.

Answers

Activity

1. a.

Stage 4 Stage 5 Stage 6

Hands-On LAB 12A Explore Linear and Nonlinear Relationships

Use with Lesson 12-2

internet connect
Lab Resources Online
go.hrw.com
KEYWORD: MR4 Lab12A

You can learn about linear and nonlinear relationships by looking at patterns.

Activity

1. This model shows stage 1 to stage 3 of a pattern.

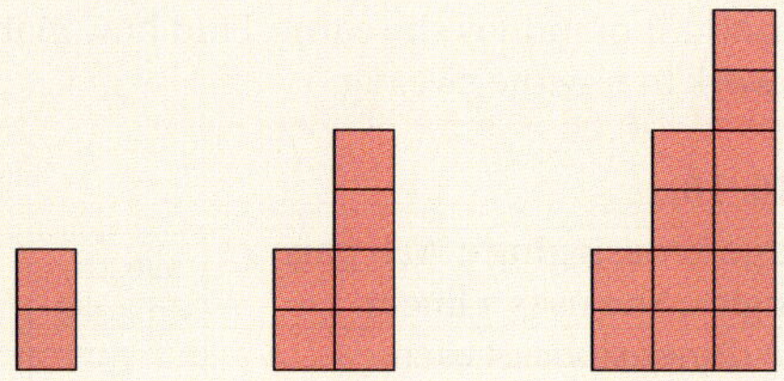

Stage 1 Stage 2 Stage 3

a. Use square tiles or graph paper to model stages 4, 5, and 6.

b. Record each stage and the perimeter of each figure in a table.

c. Graph the ordered pairs (x, y) from the table on a coordinate plane.

Stage (x)	Perimeter (y)
1	6
2	12
3	18
4	24
5	30
6	36

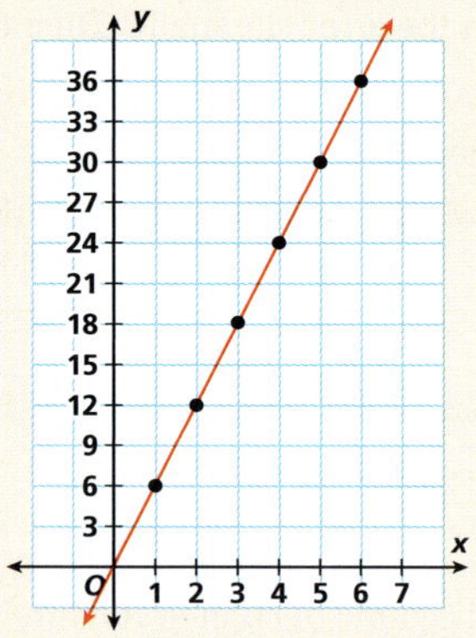

If you connected the points you graphed, you would draw a straight line. This shows that the relationship between the stage and the perimeter of the figure is linear. The equation for this line is $y = 3x$.

2 This table shows the ordered pairs for stages 1, 4, and 9 of a pattern.

Stage (x)	Square Root (y)
1	1
4	2
9	3
16	4
25	5
36	6

$1 = 1 \cdot 1$ or $1 = 1^2$

$4 = 2 \cdot 2$ or $4 = 2^2$

$9 = 3 \cdot 3$ or $9 = 3^2$

The square root of 4 is 2, which is written like this: $\sqrt{4} = 2$.

a. You can model this by arranging squares in 1-by-1, 2-by-2, and 3-by-3 blocks.

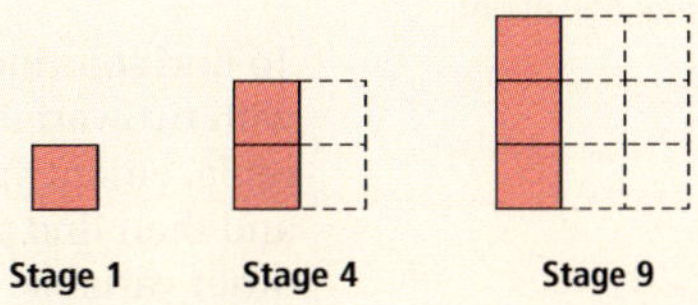

b. Record each stage number and the number's square root in a table. Graph the ordered pairs (x, y) from the table on a coordinate plane.

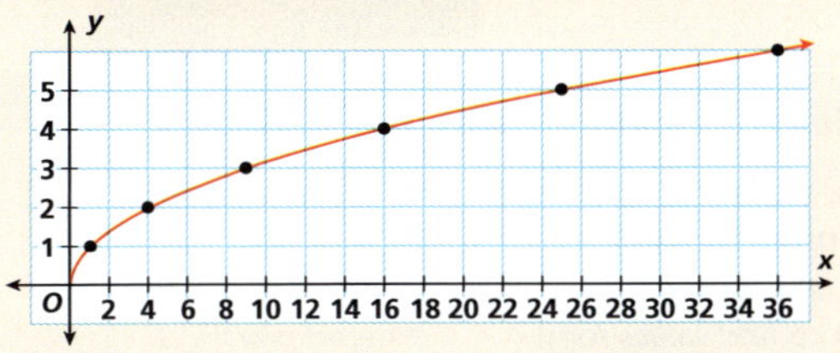

If you connect the graphed points, you draw a curved line. This shows that the relationship between the stage number and that number's square root is nonlinear. The equation for this curve is $y = \sqrt{x}$.

Think and Discuss

1. Explain what pattern you see in the y-values of the ordered pairs from the graph above.
The y-values are changing at a constant rate of 1.

Try This

Use the x-values 1, 2, 3, and 4 to find ordered pairs for each equation. Then graph the equation. Tell whether the relationship between x and y is linear or nonlinear.

1. $y = 2 + x$ linear

2. $y = 4x$ linear

3. $y = x^3$ nonlinear

4. $y = x + 4$ linear

5. $y = x(2 + x)$ nonlinear

6. $y = x + x$ linear

Answers

Try This

1. (1, 3), (2, 4), (3, 5), (4, 6)

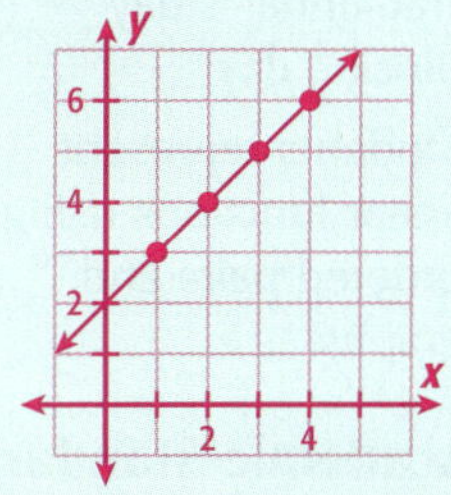

2. (1, 4), (2, 8), (3, 12), (4, 16)

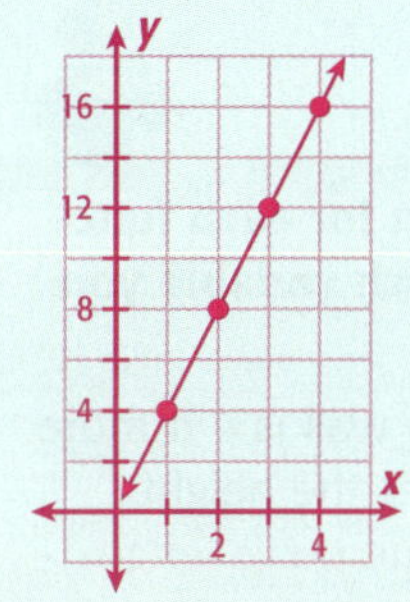

3. (1, 1), (2, 8), (3, 27), (4, 64)

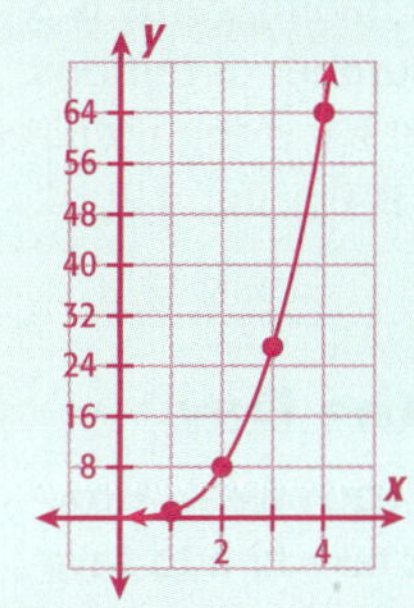

4. (1, 5), (2, 6), (3, 7), (4, 8)

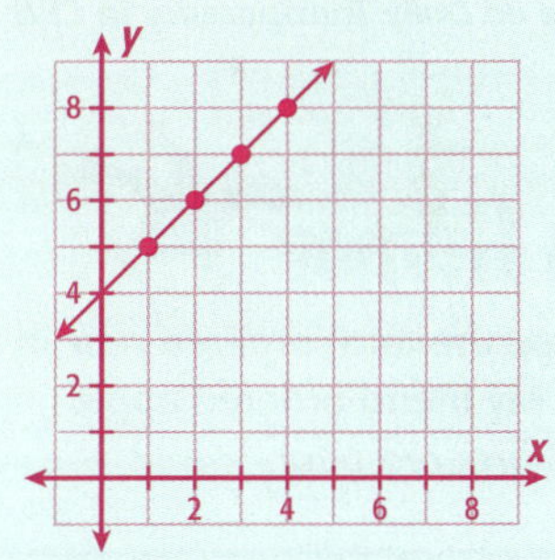

5. (1, 3), (2, 8), (3, 15), (4, 24)

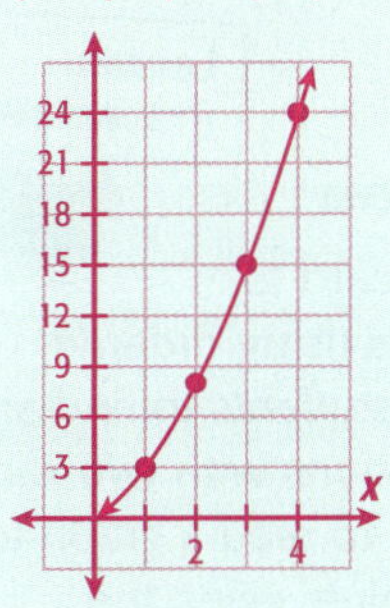

6. (1, 2), (2, 4), (3, 6), (4, 8)

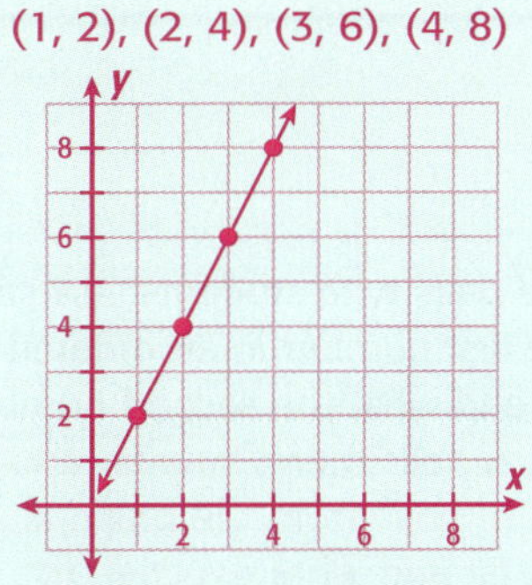

12-2 Organizer

Pacing: Traditional 1 day
Block $\frac{1}{2}$ day

Objective: Students represent linear functions using ordered pairs and graphs.

Indiana Academic Standards
6.3.7, 6.3.8, 6.3.9, 6.7.8

Warm Up

Write an equation for each function. Tell what each variable you use represents.

1. The length of a wall is 4 ft more than three times the height. $\ell = 3h + 4$, where ℓ is length and h is height.
2. The number of trading cards is 3 less than the number of buttons. $c = b - 3$, where c is the number of cards and b is the number of buttons.

Problem of the Day

Steve saved $1.50 each week. How many weeks did it take him to save enough to buy a $45 skateboard? 30

Available on Daily Transparency in CRB

Math Humor

Can you use the term *ordered pair* in a sentence? My friend ordered apple juice, but I *ordered pear*.

12-2 Graphing Functions

Learn to represent linear functions using ordered pairs and graphs.

Vocabulary
linear equation

Christa is ordering CDs online. Each CD costs $16, and the shipping and handling charge is $6 for the whole order.

The total cost, y, depends on the number of CDs, x. This function is described by the equation $y = 16x + 6$.

To find solutions of an equation with two variables, first choose a replacement value for one variable and then find the value of the other variable.

EXAMPLE 1 Finding Solutions of Equations with Two Variables

Use the given x-values to write solutions of the equation $y = 16x + 6$ as ordered pairs.

Make a function table by using the given values for x to find values for y. *Write these solutions as ordered pairs.*

x	$16x + 6$	y	(x, y)
1	$16(1) + 6$	22	(1, 22)
2	$16(2) + 6$	38	(2, 38)
3	$16(3) + 6$	54	(3, 54)
4	$16(4) + 6$	70	(4, 70)

Check if an ordered pair is a solution of an equation by putting the x and y values into the equation to see if they make it a true statement.

EXAMPLE 2 Checking Solutions of Equations with Two Variables

Determine whether the ordered pair is a solution to the given equation.

$(8, 16); y = 2x$

$y = 2x$ *Write the equation.*

$16 \stackrel{?}{=} 2(8)$ *Substitute 8 for x and 16 for y.*

$16 \stackrel{?}{=} 16$ ✔

So (8, 16) is a solution of $y = 2x$.

1 Introduce

Alternate Opener

EXPLORATION

12-2 Graphing Functions

To graph a ***linear equation,*** you can plot ordered pairs from a table of x- and y-values as points on a coordinate grid.

The table and graph at right model the function $y = x + 2$.

x	y
0	2
1	3
2	4
3	5

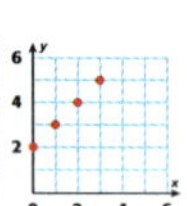

Graph each set of ordered pairs on the coordinate grid.

1. $y = x - 2$

x	y
2	0
3	1
4	2
5	3

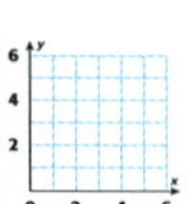

2. $y = x + 1$

x	y
1	2
2	3
3	4
4	5

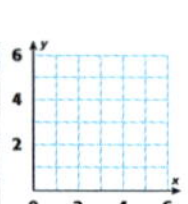

Use the points on each graph to complete each table.

3.

x	y

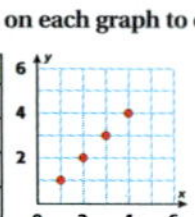

4.

x	y

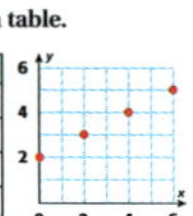

Think and Discuss

5. **Explain** how to use the points on the graph of a linear equation to write ordered pairs.

Motivate

Review ordered pairs with students. Ask students what the first number in an ordered pair tells them and what the second number in an ordered pair tells them. how far and what direction to move from the origin along the x-axis; how far and what direction to move parallel to the y-axis Have students explain how to graph the ordered pairs (2, 3) and (−4, −2) From the origin, move 2 spaces to the right and 3 spaces up. From the origin, move 4 spaces to the left and 2 spaces down.

Exploration worksheet and answers on Chapter 12 Resource Book pp. 16 and 71

2 Teach

Lesson Presentation

Guided Instruction

In this lesson, students learn to represent linear functions using ordered pairs and graphs. Teach students to find and check solutions for equations with two variables. Demonstrate how to read a graph to find an approximate y-value, given the x-value. Finally, teach students to graph a function described by a particular equation.

Teaching Tip After working through all of the examples, have students graph the functions described by the equations in Examples 1 and 2.

You can also graph the solutions of an equation on a coordinate plane. When you graph the ordered pairs of some functions, they form a straight line. The equations that express these functions are called **linear equations**.

EXAMPLE 3 **Reading Solutions on Graphs**

Use the graph of the linear function to find the value of *y* for the given value of *x*.

$x = 1$

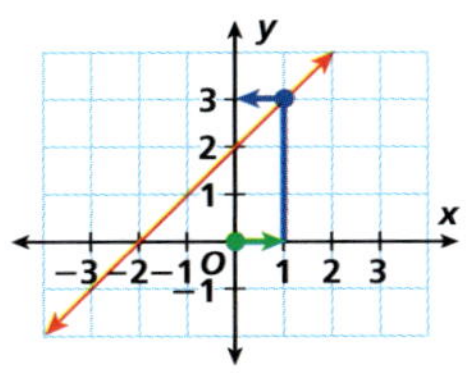

Start at the origin and move 1 unit right. Move up until you reach the graph. Move left to find the y-value on the y-axis.

When $x = 1$, $y = 3$. The ordered pair is $(1, 3)$.

EXAMPLE 4 **Graphing Linear Functions**

Graph the function described by the equation.

$y = 2x + 1$

Make a function table. *Write the solutions as ordered pairs.*

x	$2x + 1$	y	(x, y)
−1	2(−1) + 1	−1	(−1, −1)
0	2(0) + 1	1	(0, 1)
1	2(1) + 1	3	(1, 3)

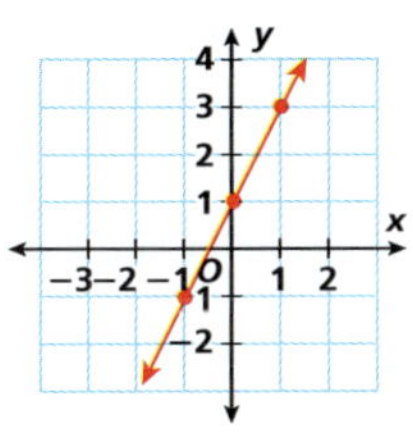

Graph the ordered pairs on a coordinate plane. Draw a line through the points to represent all the values of x you could have chosen and the corresponding values of y.

Think and Discuss

1. **Explain** why the points in Example 4 are not the only points on the graph. Name two points that you did not plot.
2. **Tell** whether the equation $y = 10x - 5$ describes a linear function.

Additional Examples

Example 1

Use the given *x*-values to write solutions of the equation as ordered pairs.

$y = 4x + 2;\ x = 1, 2, 3, 4$

(1, 6), (2, 10), (3, 14), (4, 18)

Example 2

Determine whether the ordered pair is a solution to the given equation.

$(3, 21);\ y = 7x$ (3, 21) is a solution.

Example 3

Use the graph of the linear function to find the value of *y* for the given value of *x*.

$x = 4$

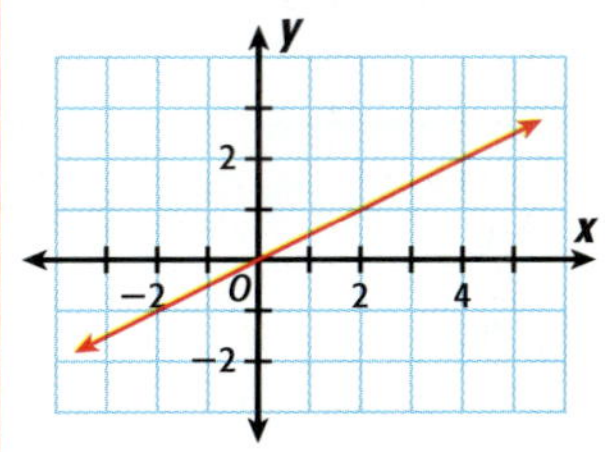

When $x = 4$, $y = 2$.

Example 4

Graph the function described by the equation.

$y = -x - 2$

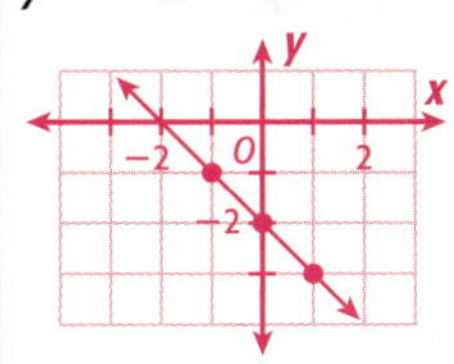

3 Close

Reaching All Learners
Through Grouping Strategies

Have students work in pairs to write an equation with two variables. Equations should be similar in form to the ones in Examples 1, 2, and 4. Collect students' equations, and randomly redistribute them so that each pair of students has a different pair's equation. Have the students make a function table for the equations and write solutions of the equations as ordered pairs. Then have the students graph the functions described by the equations.

Summarize

Review how to represent an equation using ordered pairs and how to graph a function described by an equation. Remind students that when the ordered pairs of a function form a straight line, the equation representing the function is called a *linear equation*.

Answers to Think and Discuss

1. Possible answer: By extending the table, an infinite number of values of x and y can be found; (2, 5) and (−2, −3).
2. Possible answer: Yes; when you graph the solutions to this equation the ordered pairs form a straight line.

12-2 PRACTICE & ASSESS

Students may want to refer back to the lesson examples.

Assignment Guide

If you finished Example 1 assign:
Core 1–2, 9–10, 22, 29–37
Enriched 1–2, 9–10, 22, 29–37

If you finished Example 2 assign:
Core 1–4, 9–12, 21–22, 29–37
Enriched 1–4, 9–12, 21–22, 29–37

If you finished Example 3 assign:
Core 1–6, 9–16, 21–22, 29–37
Enriched 1–6, 9–16, 21–22, 29–37

If you finished Example 4 assign:
Core 1–18, 22–24, 29–37
Enriched 7–37

Answers

7.

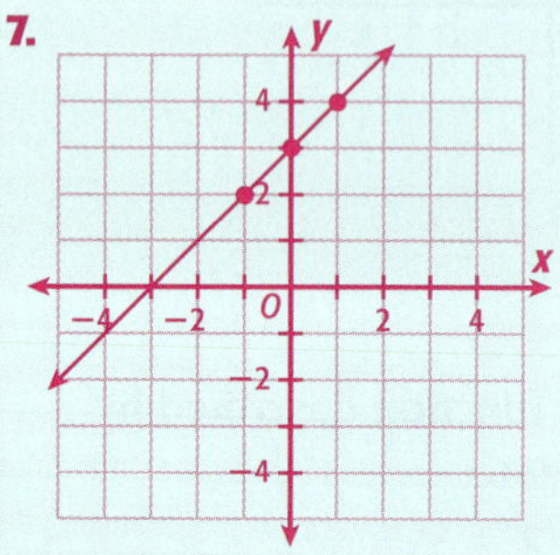

8, 17–20 See p. A7.

12-2 Exercises

FOR EXTRA PRACTICE see page 660

internet connect
Homework Help Online
go.hrw.com Keyword: MR4 12-2

GUIDED PRACTICE

See Example 1 Use the given x-values to write solutions of each equation as ordered pairs.

1. $y = 6x + 2$ for $x = 1, 2, 3, 4$ (1, 8); (2, 14); (3, 20); (4, 26)

2. $y = -2x$ for $x = 1, 2, 3, 4$ (1, −2); (2, −4); (3, −6); (4, −8)

See Example 2 Determine whether each ordered pair is a solution of the given equation.

3. (2, 12); $y = 4x$ no

4. (5, 9); $y = 2x - 1$ yes

See Example 3 Use the graph of the linear function to find the value of y for each given value of x.

5. $x = 1$ 2

6. $x = 0$ 1

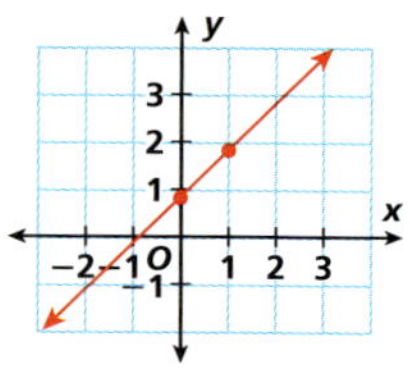

See Example 4 Graph the function described by each equation.

7. $y = x + 3$

8. $y = 3x - 1$

INDEPENDENT PRACTICE

See Example 1 Use the given x-values to write solutions of each equation as ordered pairs.

9. $y = -4x + 1$ for $x = 1, 2, 3, 4$ (1, −3); (2, −7); (3, −11); (4, −15)

10. $y = 5x - 5$ for $x = 1, 2, 3, 4$ (1, 0); (2, 5); (3, 10); (4, 15)

See Example 2 Determine whether each ordered pair is a solution of the given equation.

11. (3, −10); $y = -6x + 8$ yes

12. (−8, 1); $y = 7x - 15$ no

See Example 3 Use the graph of the linear function to find the value of y for each given value of x.

13. $x = -2$ 1

14. $x = 1$ 4

15. $x = 0$ 3

16. $x = -1$ 2

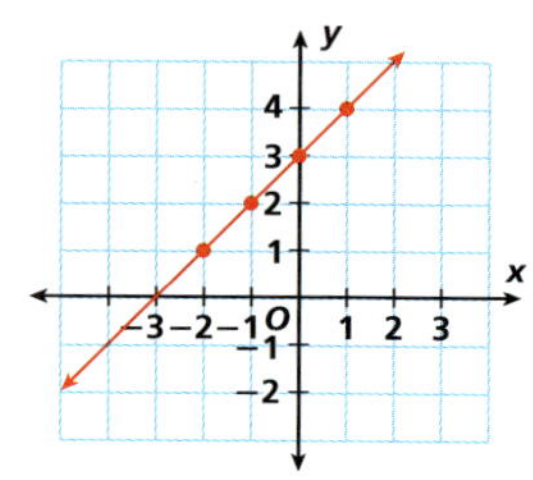

See Example 4 Graph the function described by each equation.

17. $y = 4x + 1$

18. $y = -x - 2$

PRACTICE AND PROBLEM SOLVING

Complete each table, and then use the table to graph the function.

19. $y = x - 2$ −3; −2; −1; 0

x	−1	0	1	2
y	■	■	■	■

20. $y = 2x - 4$ −6; −4; −2; 0

x	−1	0	1	2
y	■	■	■	■

21. Which of the ordered pairs below is not a solution of $y = 4x + 9$?
(1, 14), (0, 9), (−1, 5), (−2, 1), (2, 17) (1, 14)

Math Background

Exact values cannot be read from an actual graph since the line has thickness and there are measurement errors. A graph can only abstractly determine x- and y-values.

RETEACH 12-2

LESSON 12-2 **Reteach**
Graphing Functions

You can express solutions to equations as ordered pairs.

To express solutions to $y = x - 3$ as ordered pairs, substitute the given values for x in the equation. Then write the solutions as ordered pairs.

x	$x - 3$	y	ordered pair (x, y)
9	9 − 3	6	(9, 6)
7	7 − 3	4	(7, 4)
5	5 − 3	2	(5, 2)
3	3 − 3	0	(3, 0)

Use the given x-values to write solutions of the equation as ordered pairs.

1.

x	$2x$	y	ordered pair (x, y)
1	2 • 1	2	(1, 2)
2	2 • 2	4	(2, 4)
3	2 • 3	6	(3, 6)
4	2 • 4	8	(4, 8)

You can check whether an ordered pair is a solution to an equation by substituting the x- and y-values into the equation.

Is (2, 3) a solution to $y = x + 1$?

$y = x + 1$ Write the equation.
$3 = 2 + 1$ Substitute 3 for y and 2 for x.
$3 = 3$ ✓

So (2, 3) is a solution to $y = x + 1$.

Determine whether each ordered pair is a solution to the given equation.

2. (1, 4); $y = 4x$ yes

3. (3, 1); $y = 2x - 3$ no

4. (2, 8); $y = 2x + 4$ yes

PRACTICE 12-2

LESSON 12-2 **Practice B**
Graphing Functions

Use the given x-values to write solutions of each equation as ordered pairs.

1. $y = 5x + 3$ for $x = 1, 2, 3$ (1, 8); (2, 13); (3, 18)

2. $y = -4x$ for $x = 3, 5, 7$ (3, −12); (5, −20), (7, −28)

Determine whether each ordered pair is a solution of the given equation.

3. (6, 4); $y = 2x - 8$ yes

4. (8, 72); $y = x \div 9$ no

5. (−3, −18); $y = -6x$ no

6. (5, 64); $y = 12x + 4$ yes

Use the graph of the linear function to find the value of y for each given value of x.

7. $x = 2$ $y = 4$

8. $x = 1$ $y = 2$

9. $x = 0$ $y = 0$

10. $x = -1$ $y = -2$

11. $x = -2$ $y = -4$

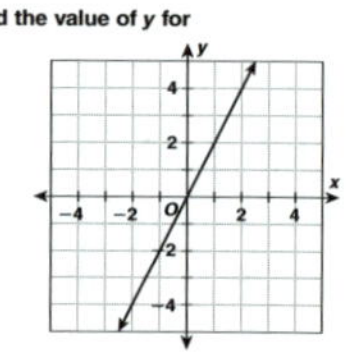

Graph the function described by each equation.

12. $y = x + 1$

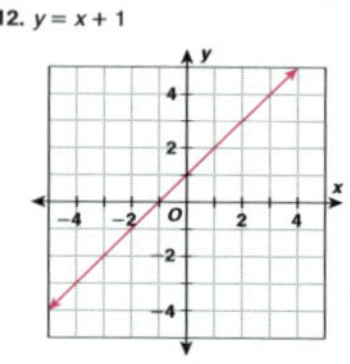

13. $y = 3 - x$

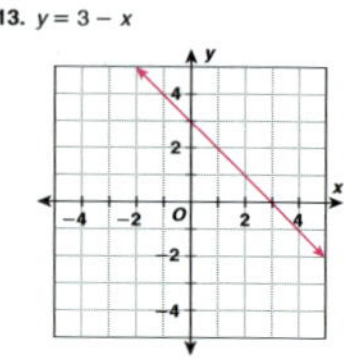

Physical Science LINK

Temperature can be expressed according to different scales. The Kelvin scale is divided into units called kelvins, and the Celsius scale is divided into degrees Celsius.

The table shows several temperatures recorded in degrees Celsius and their equivalent measures in kelvins.

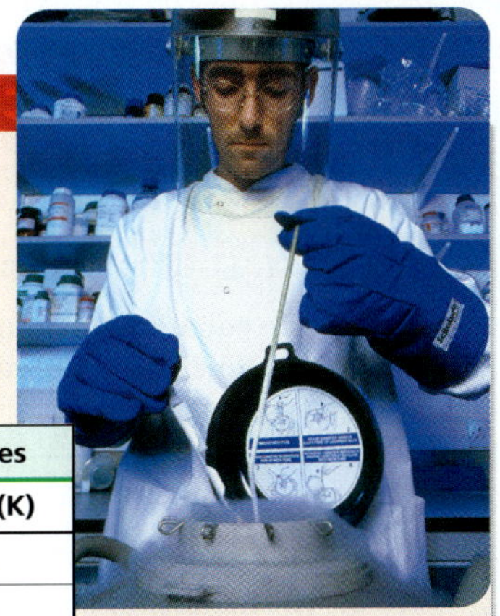
A technician preserves brain cells in this tank of liquid nitrogen, which is at −196°C, for later research. Scientists hope to understand more about how the brain works and how some diseases start.

Equivalent Temperatures

Celsius (°C)	Kelvin (K)
−100	173
−50	223
0	273
50	323
100	373

22. Write an equation for a function that gives the values in the table. Define the variables that you use. $K = C + 273$, K = kelvins, C = degrees Celsius

23. Graph the function described by your equation.

24. Use your equation to find the equivalent Kelvin temperature for −54°C. 219 kelvins

25. Use your equation to find the equivalent Celsius temperature for 77 kelvins. −196°C

26. *WHAT'S THE QUESTION?* The answer is −273°C. What is the question? **26.** What temperature on the Celsius scale is equivalent to 0 kelvins?

27. *WRITE ABOUT IT* Explain how to use your equation to determine whether 75°C is equivalent to 345 kelvins. Then determine whether the temperatures are equivalent.

28. *CHALLENGE* How many ordered-pair solutions exist for the equation you wrote in Exercise 22? an infinite number of ordered pairs

go.hrw.com
KEYWORD: MR4 Temp
CNN Student News

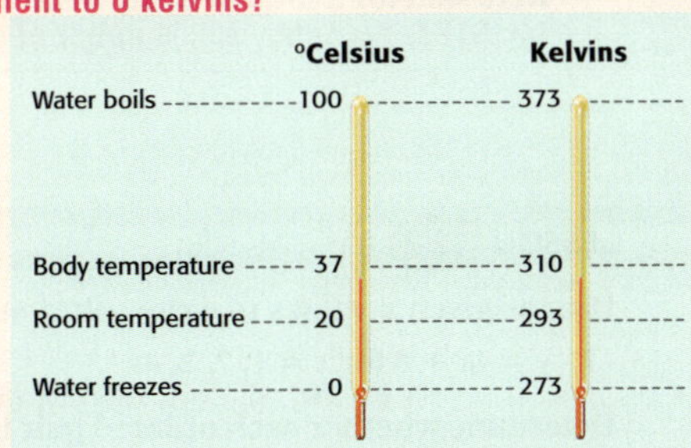

Spiral Review

Write each percent as a fraction in simplest form. (Lesson 8-7)

29. 25% $\frac{1}{4}$ **30.** 78% $\frac{39}{50}$ **31.** 40% $\frac{2}{5}$ **32.** 99% $\frac{99}{100}$

In which quadrant would you find each point? (Lesson 9-3)

33. (5, −1) IV **34.** (2, 14) I **35.** (−6, 3) II **36.** (−9, −15) III

37. **TEST PREP** The dimensions of four rectangular prisms are given below. Which prism has the greatest volume? (Lesson 10-8) A

A 9 cm × 1.5 cm × 1.5 cm

B 3 cm × 3 cm × 2 cm

C 2 cm × 9 cm × 1.1 cm

D 7 cm × 2 cm × 1 cm

Answers

23.

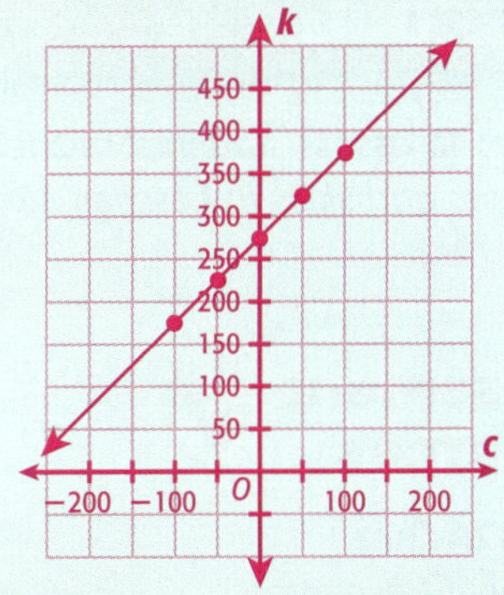

27. When $c = 75$ and $k = 345$, check to see whether the equation is balanced. It is not, so the temperatures are not the same. 75 degrees Celsius is equivalent to 348 kelvins.

Journal

Have students graph the functions represented by the equations they wrote about for the Lesson 12-1 Journal activity.

Test Prep Doctor

For Exercise 37, students need to remember the formula for finding the volume of a rectangular prism: $V = \ell wh$. Prism **A** has a volume of 20.25 cm^3, Prism **B** has a volume of 18 cm^3, Prism **C** has a volume of 19.8 cm^3, and Prism **D** has a volume of 14 cm^3. Since $20.25 > 19.8 > 18 > 14$, the correct answer is **A**.

CHALLENGE 12-2

LESSON 12-2 Challenge
Cricket Thermometers

How can you tell the temperature outside when you don't have a thermometer? Just ask a cricket!

As temperatures increase, crickets chirp faster. When the first frost occurs, the chirping stops because the crickets are too cold to move.

To use a cricket thermometer, count how many times you hear a cricket chirp in 15 seconds (c).

Then use this equation to estimate the temperature outside in degrees Fahrenheit (t).

$t = \frac{1}{4}c + 40$

Only male crickets chirp. Scientists once thought the chirp was a mating call, but it's not. Female crickets are deaf!

Use the cricket temperature equation to complete the function table. Then use the table to graph the cricket temperature function.

Function table		Ordered Pairs
c	t	(c, t)
0	40	(0, 40)
20	45	(20, 45)
40	50	(40, 50)
60	55	(60, 55)
80	60	(80, 60)
100	65	(100, 65)

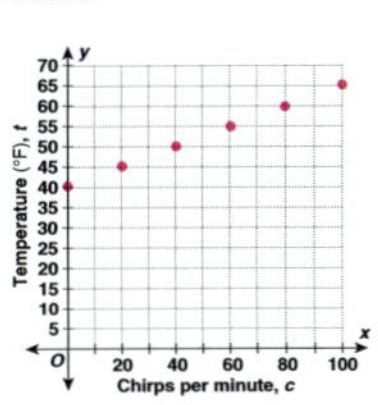

PROBLEM SOLVING 12-2

LESSON 12-2 Problem Solving
Graphing Functions

Use the table to answer each question.

Equivalent Temperatures

Celsius (°C)	Fahrenheit (°F)
−20	−4
−10	14
0	32
10	50
20	68

1. $F = \frac{9}{5}C + 32$ is an equation for the function that gives the values in the table. What does each variable represent in the equation? Use the equation to complete the table.
   F = degrees Fahrenheit;
   C = degrees Celsius
2. Write a different equation for a function that gives the values in the table.
   $C = \frac{5}{9}(F - 32)$
3. Is the ordered pair (30, 86) a solution for either equation? Why or why not? What does each value in the ordered pair represent?
   Yes, the ordered pair is (C, F); $86 = \frac{9}{5}(30) + 32$
4. Graph the function described by either equation on the graph at right.
   Check students' graphs.

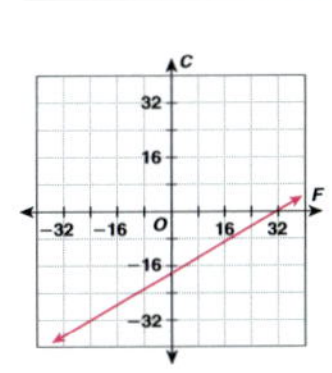

5. Use your graph to find the equivalent Fahrenheit temperature for −8°C.
   46°F

Circle the letter of the correct answer.

6. What Celsius temperature is equivalent to −58°F?
   (A) −50°C C 50°C
   B 14.4°C D −40°C
7. Which of these ordered pairs is not a solution for the equation in Exercise 1?
   F (100, 212) (H) (−40, 104)
   G (0, 32) J (60, 140)

Lesson Quiz

1. Use the given x-values to write solutions as ordered pairs to the equation $y = -3x + 1$ for $x = 0, 1, 2,$ and 3. (0, 1), (1, −2), (2, −5), (3, −8)
2. Determine whether (4, −2) is a solution to the equation $y = -5x + 3$. no, $-2 \neq -5(4) + 3$
3. Graph the function described by the equation $y = -x + 3$.

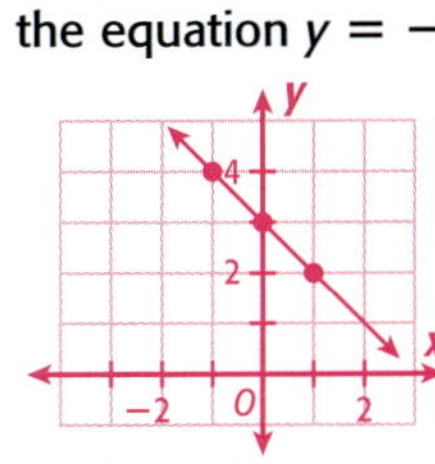

Available on Daily Transparency in CRB

Chapter 12 Mid-Chapter Quiz

Purpose: *To assess students' mastery of concepts and skills in Lessons 12-1 through 12-2*

Assessment Resources

Section 12A Quiz

Assessment Resources .... p. 29

***Test and Practice Generator* CD-ROM**

Additional mid-chapter assessment items in both multiple-choice and free-response format may be generated for any objective in Lessons 12-1 through 12-2.

You may use the graph paper on Chapter 12 Resource Book p. 68 for items 17–20.

Chapter 12 Mid-Chapter Quiz

Mid-Chapter Quiz

LESSON 12-1 (pp. 598–601)

Write an equation for a function that gives the values in each table. Use the equation to find the value of *y* for each indicated value of *x*.

1.

x	2	3	4	5	8
y	7	9	11	13	■

$y = 2x + 3$; 19

2.

x	1	4	5	6	8
y	■	18	23	28	38

$y = 5x - 2$; 3

Write an equation for the function. Tell what each variable you use represents.

3. The number of plates is 5 less than 3 times the number of cups. $p = 3c - 5$; p = plates, c = cups
4. The time Rodney spends running is 10 minutes more than twice the time he spends stretching. $r = 2s + 10$; r = time spent running, s = time spent stretching
5. The height of a triangle is twice the length of its base. $h = 2b$; h = height, b = base
6. A store manager tracked T-shirt sales. The store charges the same price for each T-shirt. On Monday, 5 shirts were sold for a total of \$60. On Tuesday, 8 shirts were sold for a total of \$96. On Wednesday, 11 shirts were sold for a total of \$132. Write an equation for the function. $t = \$12s$; t = total, s = number of shirts

LESSON 12-2 (pp. 604–607)

Use the given *x*-values to write solutions of each equation as ordered pairs.

7. $y = 4x + 6$ for $x = 1, 2, 3, 4$ (1, 10), (2, 14), (3, 18), (4, 22)
8. $y = 10x - 7$ for $x = 2, 3, 4, 5$ (2, 13), (3, 23), (4, 33), (5, 43)

Determine whether each ordered pair is a solution of the given equation.

9. (3, 7); $y = 2x + 1$ yes
10. (5, 1); $y = x - 5$ no
11. (4, 8); $y = 5x - 12$ yes
12. (9, 6); $y = \frac{1}{3}x + 4$ no

Use the graph of the linear function at right to find the value of *y* for each given value of *x*.

13. $x = 3$ −5
14. $x = 0$ 1
15. $x = -1$ 3
16. $x = -2$ 5

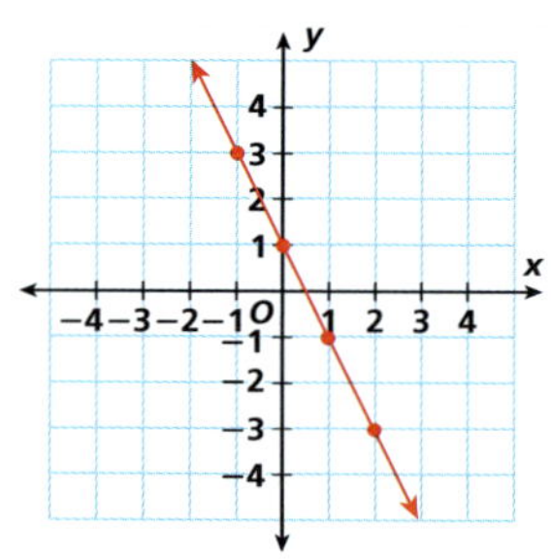

Graph the function described by each equation.

17. $y = x + 5$
18. $y = 3x + 2$
19. $y = x - 3$
20. $y = -2x$

Answers

17.

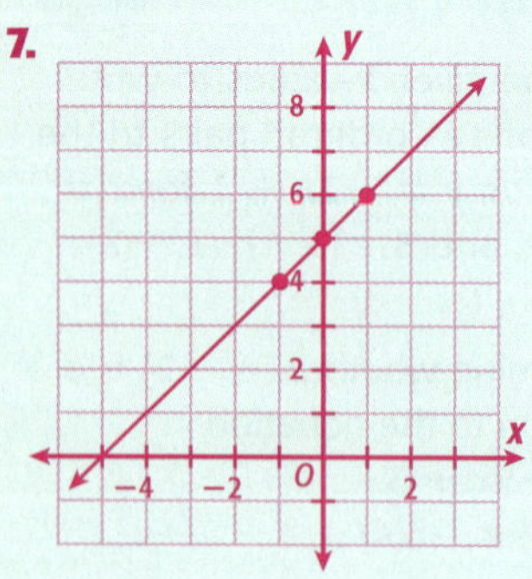

18.

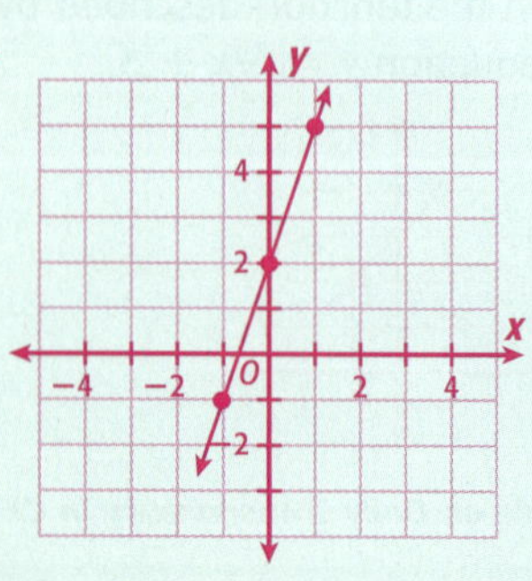

19.

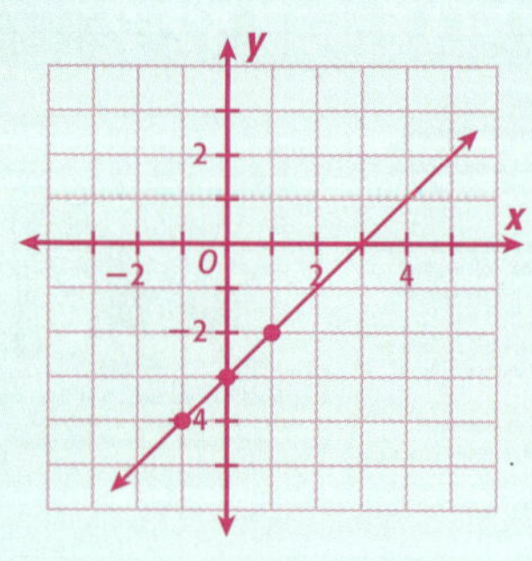

20.

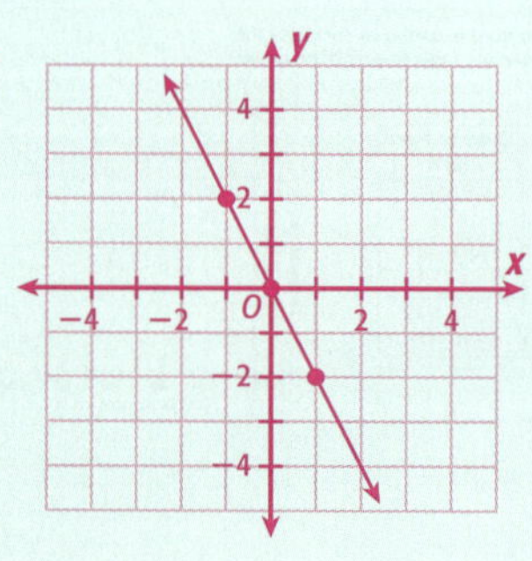

Focus on Problem Solving

Look Back

• Check that the question is answered

Sometimes a problem asks you to go through a series of steps and then give the answer. When you read a question, ask yourself what you need to find to answer it. After you have solved the problem, read the question again and make sure you have completely answered it.

Read each problem. Follow all the directions in the problem and perform each required step. Then check that you have completely answered the question.

1 If all the x- and y-coordinates of A, B, C, and D were multiplied by 3, graphed, and connected, what would the area of the new figure be?

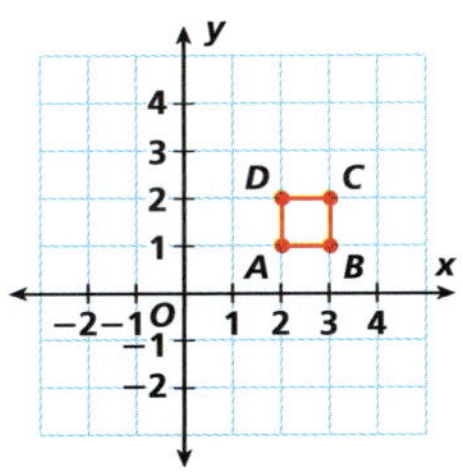

2 If all the x- and y-coodinates of P, Q, and R were multiplied by 4, graphed, and connected, how many units long would side $\overline{PQ}$ be in the new figure?

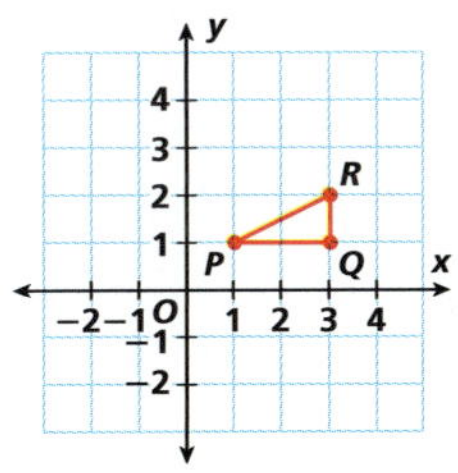

3 Vicky drew a cat's face on the coordinate grid at right. Then she stretched the cat's face vertically by a factor of 2 to create a new face. Give the new coordinates of A, B, C, D, and E in the new cat's face.

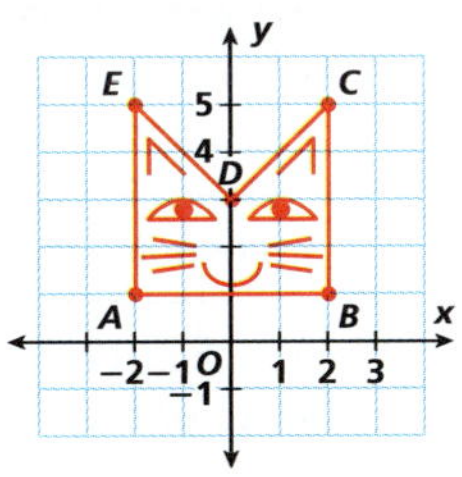

4 Ralph transformed the cat face below. The new coordinates are $A'(-2, -1)$, $B'(2, -1)$, $C'(2, -5)$, $D'(0, -3)$, and $E'(-2, -5)$. Graph $A'B'C'D'E'$ to create the new cat's face. What kind of transformations did Ralph use to create the new cat's face?

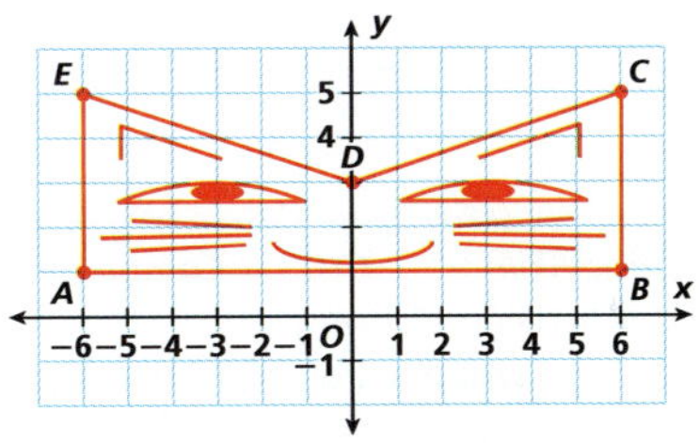

Answers

1. 9 units^2

2. 8 units

3. $A(-2, 2)$, $B(2, 2)$, $C(2, 10)$, $D(0, 6)$, and $E(-2, 10)$

4. He reflected the cat's face across the x-axis and scaled it horizontally by a factor of $\frac{1}{3}$.

Focus on Problem Solving

Purpose: *To focus on looking back to check that the question has been answered*

Problem Solving Resources

Interactive Problem Solving . . pp. 98–103

Math: Reading and Writing in the Content Area . . . . pp. 98–103

Problem Solving Process

This page focuses on the last step of the problem-solving process: **Look Back**

Discuss

Have students identify all the steps that must be taken to completely answer the question.

1. Multiply each coordinate by 3, graph the new coordinates, and connect. Finally, find the area of the new figure.

2. Multiply each coordinate by 4, graph the new coordinates, and connect. Finally, find the length of side PQ in the new figure.

3. Find the coordinates of points A, B, C, D, and E on the graph of the cat's face. Then multiply the y-coordinate of each point by 2, since the y-coordinate is the vertical coordinate.

4. Use the new coordinates to graph the cat's face and then identify the type of transformation that occurred.

Section 12B

Coordinate Geometry

One-Minute Section Planner

Lesson	Materials	Resources
Lesson 12-3 Graphing Translations **NCTM:** Algebra, Geometry, Communication, Connections **NAEP:** Geometry 2c ☑ SAT-9 ☑ SAT-10 ☐ ITBS ☑ CTBS ☑ MAT ☑ CAT	**Required** Graph paper *(CRB, p. 68)*	• *Chapter 12 Resource Book*, pp. 26–34 • Daily Transparency T8, CRB • Additional Examples Transparencies T9–T10, CRB • *Alternate Openers: Explorations*, p. 100
Lesson 12-4 Graphing Reflections **NCTM:** Algebra, Geometry, Communication, Connections **NAEP:** Geometry 2c ☑ SAT-9 ☑ SAT-10 ☐ ITBS ☑ CTBS ☑ MAT ☑ CAT	**Required** Graph paper *(CRB, p. 68)*	• *Chapter 12 Resource Book*, pp. 35–43 • Daily Transparency T11, CRB • Additional Examples Transparencies T12–T13, CRB • *Alternate Openers: Explorations*, p. 101
Lesson 12-5 Graphing Rotations **NCTM:** Algebra, Geometry, Communication, Connections **NAEP:** Geometry 2c ☐ SAT-9 ☐ SAT-10 ☐ ITBS ☐ CTBS ☑ MAT ☑ CAT	**Required** Graph paper *(CRB, p. 68)* **Optional** Teaching Transparency T15 *(CRB)* Scissors Graph paper *(CRB, p. 68)*	• *Chapter 12 Resource Book*, pp. 44–52 • Daily Transparency T14, CRB • Additional Examples Transparencies T16–T18, CRB • *Alternate Openers: Explorations*, p. 102
Lesson 12-6 Stretching and Shrinking **NCTM:** Algebra, Geometry, Communication, Connections **NAEP:** Geometry 4d ☐ SAT-9 ☐ SAT-10 ☐ ITBS ☐ CTBS ☑ MAT ☑ CAT	**Required** Graph paper *(CRB, p. 68)*	• *Chapter 12 Resource Book*, pp. 53–61 • Daily Transparency T19, CRB • Additional Examples Transparencies T20–T23, CRB • *Alternate Openers: Explorations*, p. 103
Section 12B Assessment		• Section 12B Quiz, AR p. 30 • *Test and Practice Generator* CD-ROM

SAT = *Stanford Achievement Tests* **ITBS** = *Iowa Test of Basic Skills* **CTBS** = *Comprehensive Test of Basic Skills/Terra Nova*
MAT = *Metropolitan Achievement Tests* **CAT** = *California Achievement Test*

NCTM = Complete standards can be found on pages T27–T33. **NAEP** = Complete standards can be found on pages A31–A35.

SE = *Student Edition* **TE** = *Teacher's Edition* **AR** = *Assessment Resources*
CRB = *Chapter Resource Book* **MK** = *Manipulatives Kit*

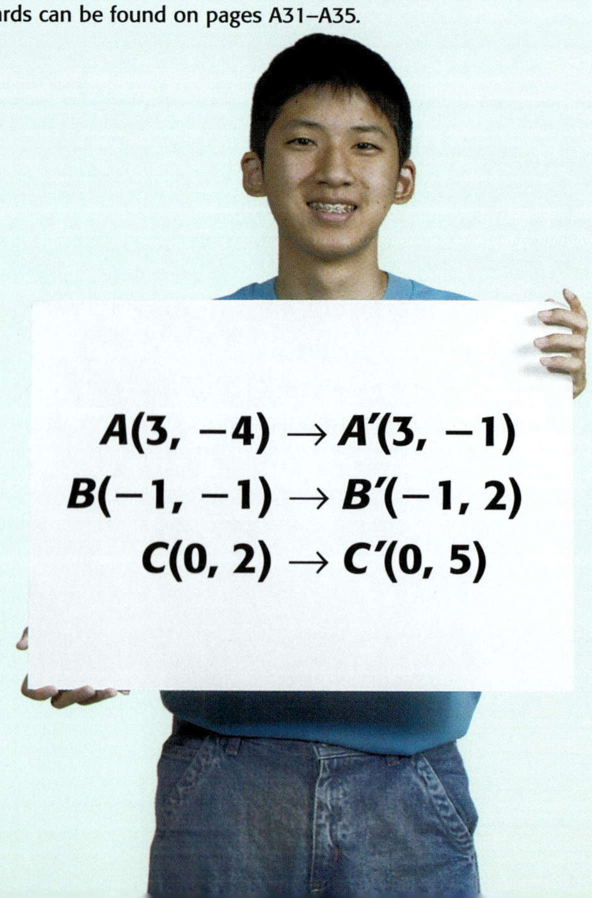

Section Overview

Translations, Reflections, and Rotations *Lessons 12-3, 12-4, 12-5*

Why? Elevators and escalators translate people between floors.

A **translation** is a movement or slide of a figure along a straight line.

When a figure is flipped over a line, the new image is a **reflection** of the figure.

A **rotation** is the movement of a figure about a point, called the center of rotation.

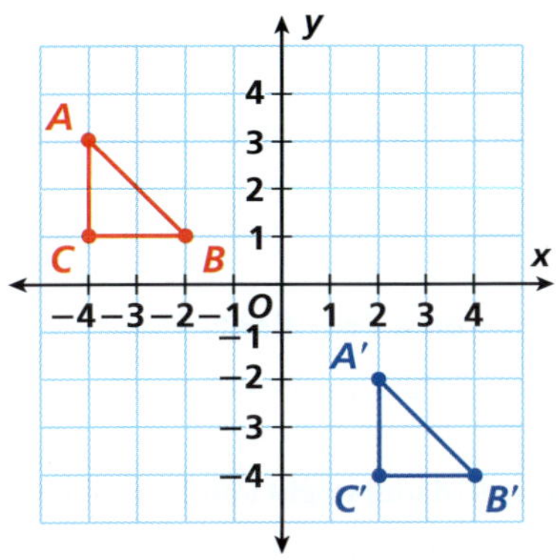

Triangle *ABC* is translated 6 units right and 5 units down.

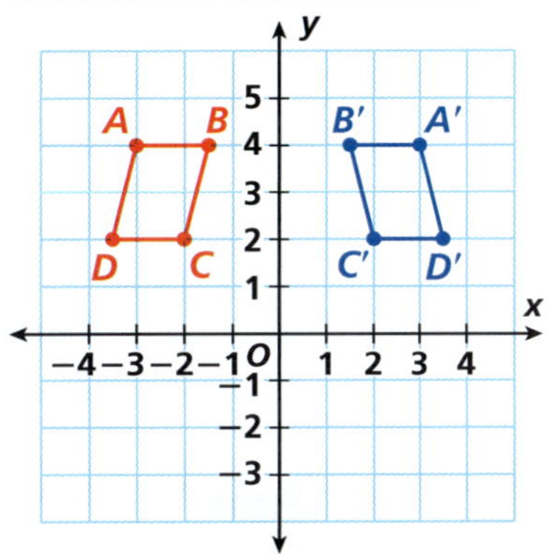

Parallelogram *ABCD* is reflected across the *y*-axis.

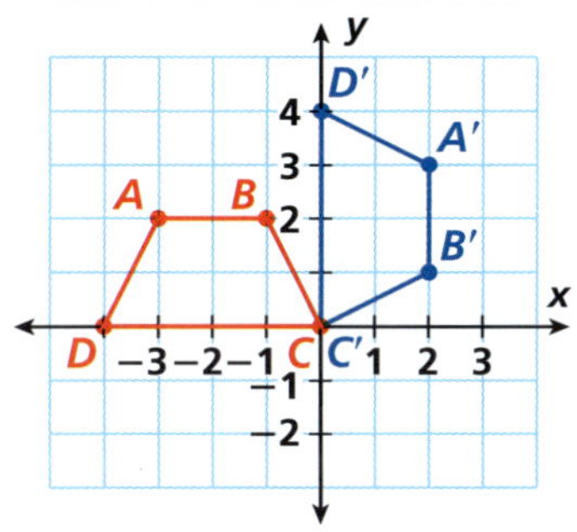

Trapezoid *ABCD* is rotated clockwise 90° about the origin.

Stretching and Shrinking *Lesson 12-6*

Why? Photographic images can be stretched or shrunk.

By increasing or decreasing one dimension of a figure, an artist can change a design.

Stretching

Original face

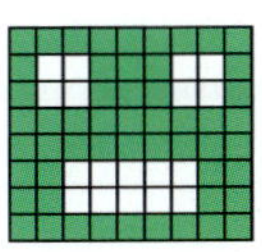

Stretched horizontally by a factor of 3

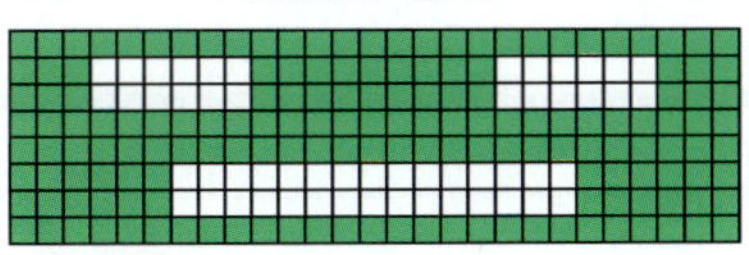

Stretched vertically by a factor of 2

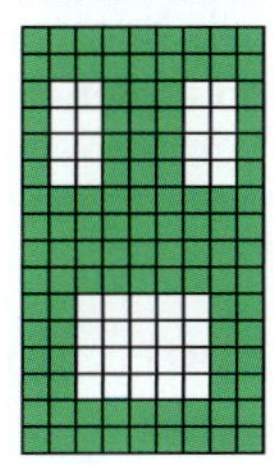

Shrinking

Original N

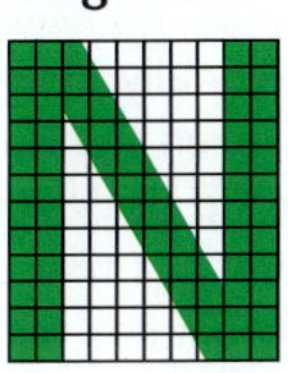

Shrunk vertically by a factor of $\frac{1}{3}$

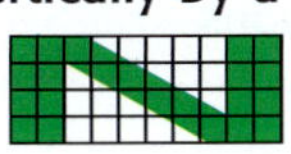

Shrunk horizontally by a factor of $\frac{1}{2}$

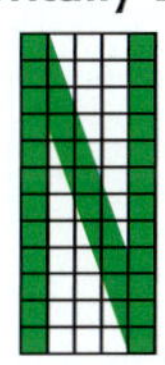

12-3 Organizer

Pacing: Traditional 1 day
Block $\frac{1}{2}$ day

Objective: Students use translations to change the positions of figures on a coordinate plane.

Indiana Academic Standards
6.3.7, 6.4.6, 6.7.1, 6.7.5, 6.7.11

Warm Up

1. Use the given x-values to write solutions of the following equation as ordered pairs.
$y = 6x - 2$ for $x = 0, 1, 2, 3$
(0, −2), (1, 4), (2, 10), (3, 16)

2. Determine whether (3, −13) is a solution to the equation
$y = -4x - 1$. yes

Problem of the Day

Samantha's house is 3 blocks east and 5 blocks south of Tyra. If Tyra walks straight south and then straight east to Samantha's house, does she walk more blocks east or more blocks south? How many more?
south; 2 blocks

Available on Daily Transparency in CRB

Math Humor

The crew asked the bilingual pilot to slide a square along the x-axis because he knew how to translate on a plane.

12-3 Graphing Translations

Learn to use translations to change the positions of figures on a coordinate plane.

Have you ever seen a marching band that stays in formation while moving forward, backward, or sideways? The moves may have been planned using a coordinate plane.

A translation is a movement of a figure along a straight line. You can translate a figure on a coordinate plane by sliding it horizontally, vertically, or diagonally.

Reading Math

After a transformation is performed on a figure, *A′* labels the image, or new location, of *A*. Read *A′* as "*A* prime."

EXAMPLE 1 Translating Figures on a Coordinate Plane

Give the coordinates of the vertices of the figure after the given translation.

Translate triangle ABC 6 units right and 5 units down.

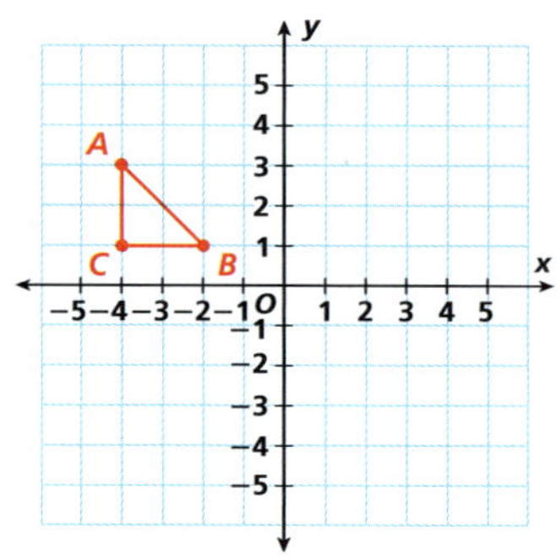

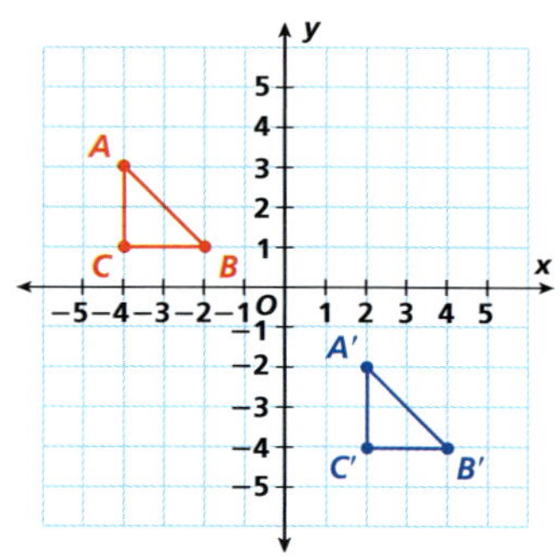

To move the triangle 6 units right, add 6 to each of the x-coordinates.
To move the triangle 5 units down, subtract 5 from each of the y-coordinates.

ABC				$A'B'C'$
$A(-4, 3)$	→	$A'(-4 + 6, 3 - 5)$	→	$A'(2, -2)$
$B(-2, 1)$	→	$B'(-2 + 6, 1 - 5)$	→	$B'(4, -4)$
$C(-4, 1)$	→	$C'(-4 + 6, 1 - 5)$	→	$C'(2, -4)$

1 Introduce

Alternate Opener

EXPLORATION

12-3 Graphing Translations

The graph shows the *translation* (slide) of the red triangle to the blue triangle.

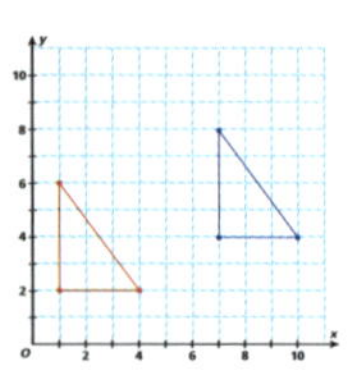

1. Label the coordinates of the vertices of each triangle.
2. Are the two triangles congruent?
3. Examine the blue translated triangle.
   a. How far is it shifted to the right?
   b. How far is it shifted up?

Think and Discuss

4. **Explain** how to find the coordinates of the vertices of the translated triangle by using the coordinates of the vertices of the original triangle.
5. **Explain** how you determined whether the triangles are congruent.

Motivate

To introduce students to the concept of graphing translations, review the term *translation*. Ask students to define the term and give several examples of translations.

Possible response: A translation is the movement of a figure along a straight line. An example of a translation is sliding a figure along a line, such as someone sliding down a straight slide at a park. They are translated from the top of the slide to the bottom of the slide.

Exploration worksheet and answers on Chapter 12 Resource Book pp. 27 and 73

2 Teach

Lesson Presentation

Guided Instruction

In this lesson, students learn to use translations to change the positions of figures on a coordinate plane. Remind students that figures can be translated horizontally, vertically, or diagonally. Teach students to translate a figure, given the number of units right or left and up or down that the figure is to move. Point out that addition is used to find new coordinates when the figure is moved up or right, and subtraction is used when the figure is moved down or left.

EXAMPLE 2 *Music Application*

Members of a marching band begin in a square formation, represented by the square *FGHJ*. Then they move 4 steps left and 3 steps forward. Give the coordinates of the vertices of the square after such a translation.

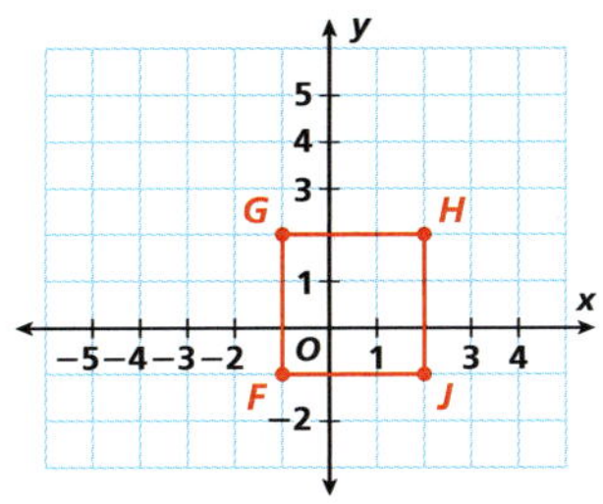

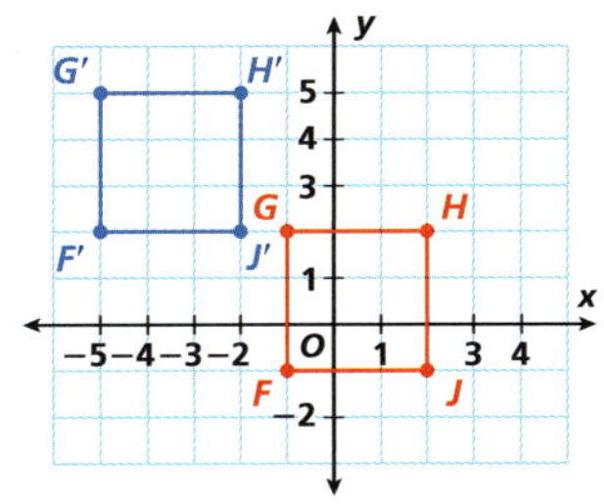

To move 4 steps left, subtract 4 from the x-coordinates.
To move 3 steps forward, add 3 to the y-coordinates.

FGHJ				*F′G′H′J′*
$F(-1, -1)$	→	$F'(-1 - 4, -1 + 3)$	→	$F'(-5, 2)$
$G(-1, 2)$	→	$G'(-1 - 4, 2 + 3)$	→	$G'(-5, 5)$
$H(2, 2)$	→	$H'(2 - 4, 2 + 3)$	→	$H'(-2, 5)$
$J(2, -1)$	→	$J'(2 - 4, -1 + 3)$	→	$J'(-2, 2)$

Think and Discuss

1. **Tell** how the location of a point changes when the x-coordinate increases.
2. **Tell** how the location of a point changes when the y-coordinate decreases.

Additional Examples

Example 1

Give the coordinates of the vertices of the figure after the given translation.

Translate triangle *DEF* with vertices $D(1, 4)$, $E(4, 2)$, and $F(-3, -3)$ 4 units left and 3 units up.
$D'(-3, 7)$, $E'(0, 5)$, $F'(-7, 0)$

Example 2

Members of a marching band begin in a trapezoid formation that can be represented by trapezoid *KLMN* with vertices $K(-2, 1)$, $L(1, 1)$, $M(3, -3)$, and $N(-4, -3)$. Then they move 5 steps right and 4 steps down. Give the coordinates of the vertices of trapezoid *KLMN* after such a translation.
$K'(3, -3)$, $L'(6, -3)$, $M'(8, -7)$, $N'(1, -7)$

You may use graph paper on Chapter 12 Resource Book p. 68 for Exercises 1–8.

12-3 Exercises

FOR EXTRA PRACTICE
see page 661

internet connect
Homework Help Online
go.hrw.com Keyword: MR4 12-3

GUIDED PRACTICE

See Example 1

Give the coordinates of the vertices of each figure after the given translation.

1. Translate triangle *RST* 3 units up.
$R'(2, 5)$ $S'(3, 2)$ $T'(-1, 2)$

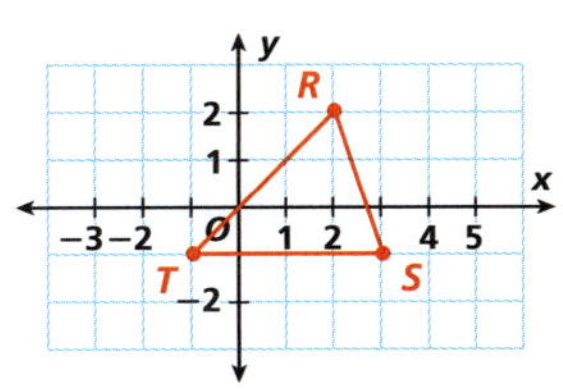

12-3 PRACTICE & ASSESS

Assignment Guide

If you finished Example 1 assign:
Core 1, 3–4, 7, 8, 11–14
Enriched 4, 6–9, 11–14

If you finished Example 2 assign:
Core 1–5, 7, 9, 11–14
Enriched 4–14

3 Close

Reaching All Learners
Through Concrete Manipulatives

Have each student draw and cut out a figure on a coordinate grid. Then have each student place his or her figure on a new coordinate plane and list the coordinates for each vertex of the figure. Have each student translate the figure (e.g., move the figure 3 units right and 7 units down). Students then find the coordinates of the vertices of the translated figure. Encourage them to see the relationship between how the figure was translated and how the vertices were changed.

Summarize

Review the procedures for finding the new x- and y-coordinates of a translated figure (i.e., using addition for a figure that moves right or up and using subtraction for a figure that moves left or down.)

Answers to Think and Discuss

1. It moves to the right.
2. It moves down.

Lesson Quiz

Give the coordinates of the vertices of triangle *ABC*, with vertices *A*(−5, −4), *B*(−3, 2), and *C*(1, −3), after the given translations.

1. Translate triangle *ABC* 3 units up and 2 units right.
   A′(−3, −1), *B*′(−1, 5), and *C*′(3, 0)
2. Translate triangle *ABC* 5 units down and 3 units left.
   A′(−8, −9), *B*′(−6, −3), and *C*′(−2, −8)
3. Translate triangle *ABC* 2 units down and 4 units right.
   A′(−1, −6), *B*′(1, 0), and *C*′(5, −5)

Available on Daily Transparency in CRB

See Example 2

2. The graph shows the position of a fountain on the plans for a new park. The architect wants to move it 2 units right and 3 units down. Give the coordinates of the vertices of the fountain after this translation.
   M′(0, −1) *N*′(1, 0) *O*′(2, −1) *P*′(2, −3) *Q*′(0, −3)

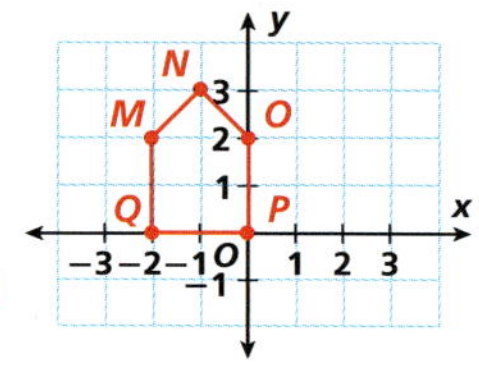

INDEPENDENT PRACTICE

See Example 1

Give the coordinates of the vertices of each figure after the given translation.

3. Translate triangle *RST* 5 units up and 2 units left. *R*′(0, 7) *S*′(1, 4) *T*′(−3, 4)
4. Translate triangle *RST* 6 units right and 1 unit down.
   R′(8, 1) *S*′(9, −2) *T*′(5, −2)

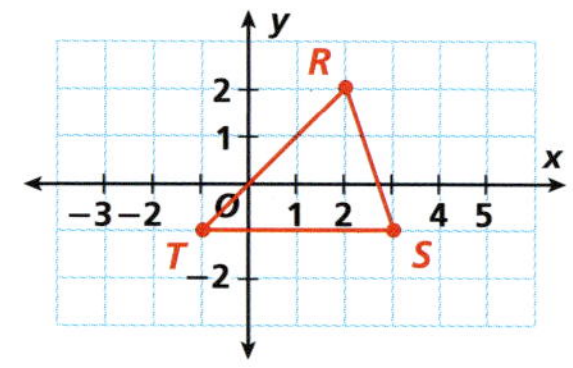

See Example 2

5. The placement of a rug in a room is represented on the graph. The rug is moved 7 units left and 4 units down. Give the coordinates of the vertices of the rug after this translation.
   A′(−9, − 3) *B*′(−4, −3) *C*′(−4, −5) *D*′(−9, −5)

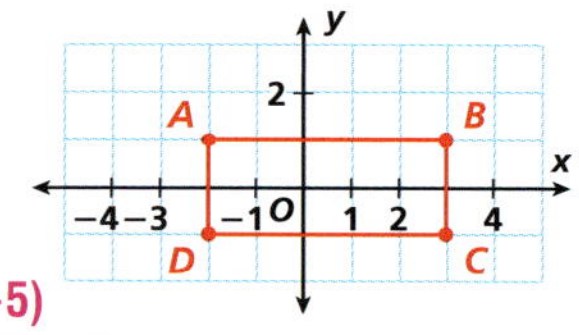

PRACTICE AND PROBLEM SOLVING

Give the coordinates of the trapezoid after each given translation.

6. Translate the trapezoid 1 unit up and 8 units left. *P*′(−9, 1) *Q*′(−5, 2) *R*′(−5, −2) *S*′(−9, −1)
7. Translate the trapezoid 2 units down and 2 units right. *P*′(1, −2) *Q*′(5, −1) *R*′(5, −5), *S*′(1, −4)

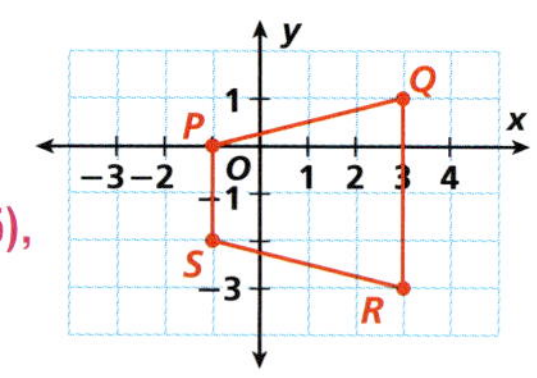

8. Rectangle *ABCD* was translated 2 units down and 5 units left. The new coordinates for the rectangle are *A*′(1, 2), *B*′(4, 2), *C*′(4, 1), and *D*′(1, 1). What were the coordinates of *A*, *B*, *C*, and *D*? *A*(6, 4), *B*(9, 4), *C*(9, 3), *D*(6, 3)
9. ***WRITE ABOUT IT*** Explain how to find the new coordinates of a point that is translated 3 units right and 2 units down.
   Add 3 to the *x*-coordinate and subtract 2 from the *y*-coordinate.
10. ***CHALLENGE*** Give examples of how you could translate rectangle *A*′*B*′*C*′*D*′ from Exercise 8 so that the entire figure is in quadrant IV.
    Possible answer: Translate the figure down 4 units.

Spiral Review

Solve for *x*. (Lesson 9-8)

11. $x + 10 = -2$ −12
12. $x - 20 = -5$ 15
13. $-9x = 45$ −5
14. **TEST PREP** What is the area of a triangle with base 8 in. and height 2.5 in? (Lesson 10-2) B

   A 10.5 in^2 **B** 10 in^2 **C** 13 in^2 **D** 20 in^2

RETEACH 12-3

LESSON 12-3 **Reteach**
Graphing Translations

A translation is a movement of a figure along a straight line.

Translate the triangle 2 units left and 3 units up.

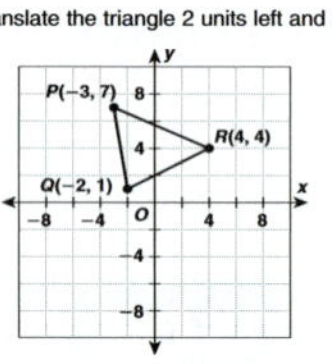

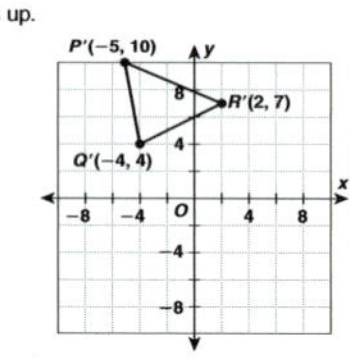

To find the coordinate of the vertices of the translated figure, subtract 2 from each *x*-coordinate and add 3 to each *y*-coordinate.

P(−3, 7) ⟶ *P*′(−3 − 2, 7 + 3) ⟶ *P*′(−5, 10)
Q(−2, 1) ⟶ *Q*′(−2 − 2, 1 + 3) ⟶ *Q*′(−4, 4)
R(4, 4) ⟶ *R*′(4 − 2, 4 + 3) ⟶ *R*′(2, 7)

Give the coordinates of the vertices of the figure after the given translation.

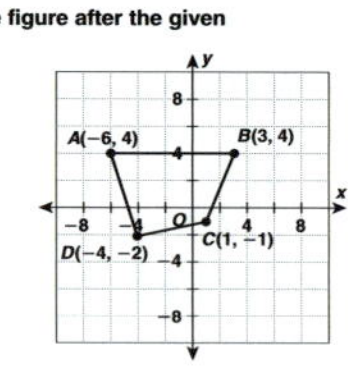

1. Translate quadrilateral *ABCD* 4 units right and 3 units down.
   A′(−2, 1); *B*′(7, 1); *C*′(5, −4); *D*′(0, −5)
2. Translate quadrilateral *ABCD* 2 units left and 1 unit up.
   A′(−8, 5); *B*′(1, 5); *C*′(−1, 0); *D*′(−6, −1)
3. Translate quadrilateral *ABCD* 3 units right and 2 units up.
   A′(−3, 6); *B*′(6, 6); *C*′(4, 1); *D*′(−1, 0)
4. Translate quadrilateral *ABCD* 1 unit left and 4 units down.
   A′(−7, 0); *B*′(2, 0); *C*′(0, −5); *D*′(−5, −6)

PRACTICE 12-3

LESSON 12-3 **Practice B**
Graphing Translations

Give the coordinates of the vertices of each figure after the given translation.

1. Translate $\overline{FG}$ 4 units down and 2 units right
   F′(−2, 0); *G*′(0, −6)
2. Translate $\overline{FG}$ 3 units up and 1 unit left.
   F′(−5, 7); *G*′(−3, 1)

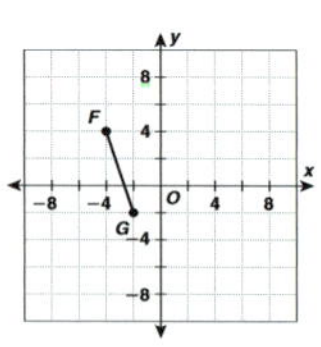

3. Translate △*XYZ* 1 unit up and 2 units right.
   X′(4, −1); *Y*′(8, −2); *Z*′(6, −5)
4. Translate △*XYZ* 5 units up and 3 units left.
   X′(−1, 3); *Y*′(3, 2); *Z*′(1, −1)

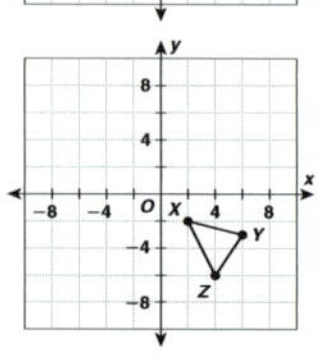

5. Translate △*ABC* 7 units up and 4 units right.
   A′(−1, 6); *B*′(3, 6); *C*′(3, 2)
6. The coordinates of the vertices of triangle *ABC* after translation are *A*′(−4, 0), *B*′(0, 0), and *C*′(0, −4). Describe this translation.
   It was translated 1 unit up and 1 unit right.

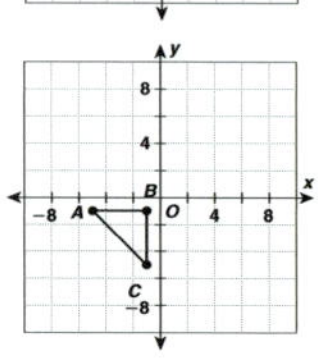

CHALLENGE 12-3

LESSON 12-3 **Challenge**
Chess Translations

The game of chess is played on a board with 64 squares. Each square and each movement of a chess piece during a game is described by coordinates on the board grid (letter, number). There are six different kinds of chess pieces—king, rook, bishop, queen, knight, and pawn. Each piece has its own way that it can move. For example:

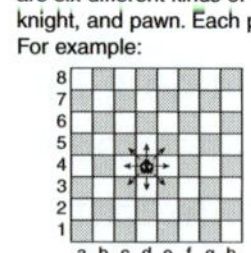

A **king** can move one square in any direction.

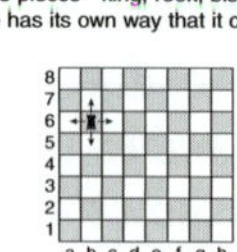

A **rook** can move any number of squares up, down, left, or right.

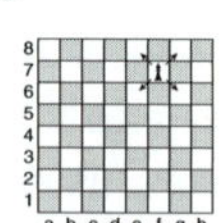

A **bishop** can move any number of squares diagonally.

Give the coordinates of each chess piece after the given translation.

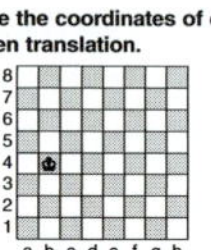

Translation: 1 unit up, 1 unit right
New coordinates: c5

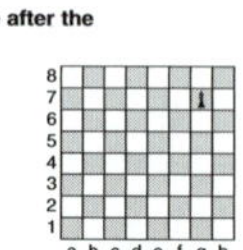

Translation: 3 units down, 3 units left
New coordinates: d4

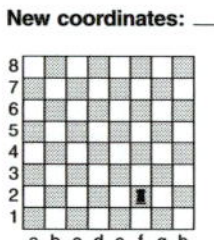

Translation: 5 units up
New coordinates: f7

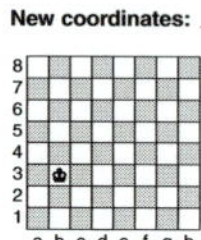

Translation: 1 unit down, 1 unit left
New coordinates: a2

PROBLEM SOLVING 12-3

LESSON 12-3 **Problem Solving**
Graphing Translations

Use the room floor plan to answer each question. Each square on the floor plan equals two feet.

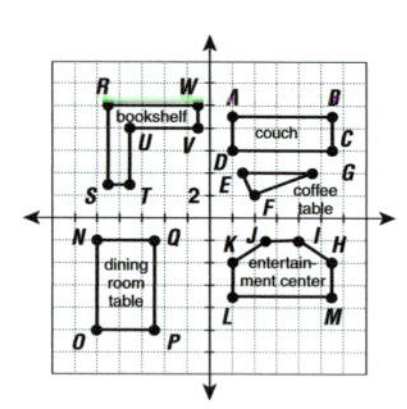

1. If you move the couch 2 feet up and 1 foot left, what will be the coordinates of each corner of the couch after the translation?
   A′(1, 11); *B*′(10, 11); *C*′(10, 8); *D*′(1, 8);
2. If you move the dining room table 1 foot down and 4 feet right, what will be the coordinates of each corner of the table after the translation?
   N′(−6, −3); *O*′(−6, −11); *P*′(−1, −11); *Q*′(−1, −3);
3. You move the coffee table so that its translated vertices are *E*′(5, 2), *F*′(6, 0), and *G*′(11, 2). Describe the translation of the coffee table.
   It was moved 2 feet right and 2 feet down.
4. You move the book shelf so that its translated vertices are *R*′(−6, 10), *S*′(−6, 3), *T*′(−4, 3), *U*′(−4, 8), *V*′(2, 8), and *W*′(2, 10). Describe the translation of the coffee table.
   It was moved 3 feet right.

Circle the letter of the correct answer.

5. You move the entertainment center 4 feet down and 1 foot left. Which of the following are **not** coordinates for one of the vertices of the translated entertainment center?
   A (1, −8) **C** (4, −6)
   B (10, −11) **(D)** (12, −8)
6. The coordinates of each corner of the couch after a translation are *A*′(3, 4), *B*′(12, 4), *C*′(12, 1), and *D*′(3, 1). Which of the following describes this translation of the couch?
   F down 5 feet, left 1 foot
   G up 1 foot, left 5 feet
   (H) down 5 feet, right 1 foot
   J up 1 foot, right 5 feet

12-4 Graphing Reflections

Learn to use reflections to change the positions of figures on a coordinate plane.

Textile designers create patterns for printed, woven, or knitted fabrics. The design might include a shape and the image of that shape's reflection.

To design a fabric pattern, a designer might use a coordinate plane. A figure can be reflected on a coordinate plane across the *x*-axis or the *y*-axis.

Remember!

When a figure is flipped over a line, the new image is a reflection of the original.

EXAMPLE 1 Reflecting Figures on a Coordinate Plane

Give the coordinates of the figure after the given reflection.

Reflect parallelogram *ABCD* across the *y*-axis.

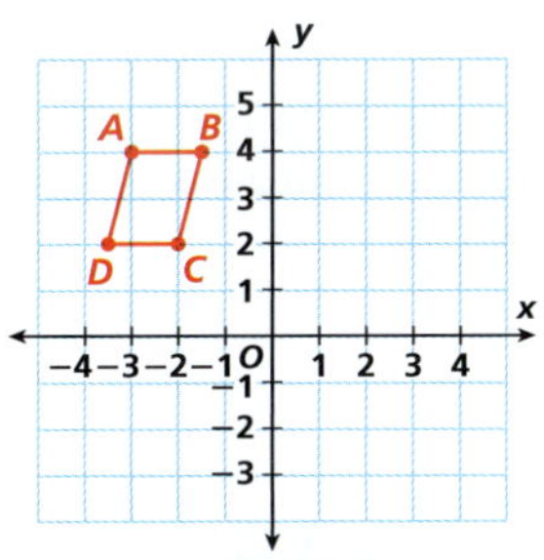

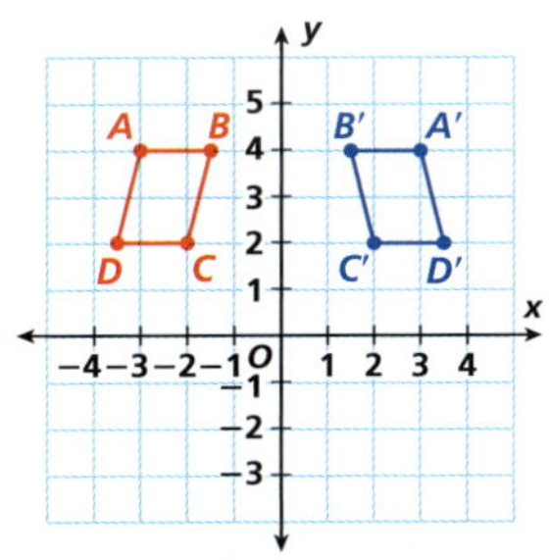

To reflect the parallelogram across the y-axis, write the opposites of the x-coordinates.

The y-coordinates do not change.

ABCD		*A′B′C′D′*
$A(-3, 4)$	→	$A'(3, 4)$
$B\left(-1\frac{1}{2}, 4\right)$	→	$B'\left(1\frac{1}{2}, 4\right)$
$C(-2, 2)$	→	$C'(2, 2)$
$D\left(-3\frac{1}{2}, 2\right)$	→	$D'\left(3\frac{1}{2}, 2\right)$

12-4 Organizer

Pacing: Traditional 1 day
Block $\frac{1}{2}$ day

Objective: Students use reflections to change the positions of figures on a coordinate plane.

Indiana Academic Standards
6.3.7, 6.4.6, 6.7.1, 6.7.3, 6.7.4, 6.7.5, 6.7.8, 6.7.11

Warm Up

A parallelogram has vertices (−4, 1), (0, 1), (−3, 4), and (1, 4). What are its vertices after it has been translated 4 units right and 3 units up?
(0, 4), (4, 4), (1, 7), (5, 7)

Problem of the Day

These are *rits*: 24042, 383, and 4994. These are not *rits*: 39239, 28, and 5505. Which of these are *rits:* 39883, 4040, and 101? Why?
101 is a *rit* because it is the same forward and backward.

Available on Daily Transparency in CRB

Math Humor

The actor wondered and wondered why he had messed up the last sentence in the play. Finally he went into a rage about it. It's only natural that someone who reflects that much will eventually flip over a line.

1 Introduce

Alternate Opener

EXPLORATION

12-4 Graphing Reflections

The graph shows a *reflection* of the red triangle across the *y*-axis.

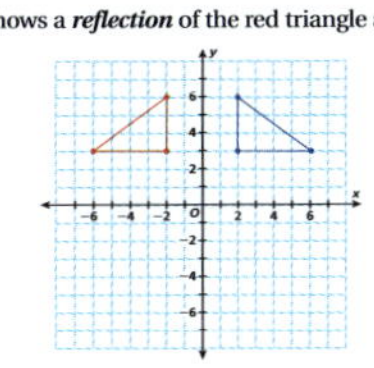

1. Label the coordinates of the vertices of each triangle.
2. Are the two triangles congruent?
3. Examine the reflection and the coordinates of its vertices.
   a. How are the coordinates different from the original coordinates?
   b. How are the coordinates similar to the original coordinates?

Think and Discuss

4. **Explain** how to find the coordinates of the vertices of the reflected triangle by using the coordinates of the vertices of the original triangle.
5. **Explain** how to reflect the red triangle across the *x*-axis.

Motivate

To introduce students to the concept of graphing reflections, review the term *reflection*. Ask students to define the term and site examples of reflections.

Possible answer: A reflection is the movement of a figure by flipping it over a line, such as a person's reflection in a mirror.

Exploration worksheet and answers on Chapter 12 Resource Book pp. 36 and 75

2 Teach

Lesson Presentation

Guided Instruction

In this lesson, students learn to use reflections to change the positions of figures on a coordinate plane. First, teach students that when reflecting a figure across one axis of a coordinate plane, the coordinates for that axis stay the same. Finally, show students examples of a reflection across the *y*-axis and a reflection across the *x*-axis.

Additional Examples

Example 1

Give the coordinates of the figure after the given reflection.

Reflect parallelogram *EFGH* with vertices *E*(−6, 1), *F*(−2, 1), *G*(−4, −2), and *H*(−8, −2) across the *y*-axis.
E′(6, 1), *F*′(2, 1), *G*′(4, −2), and *H*′(8, −2)

Example 2

A designer is using a stencil that is shaped like a quadrilateral with vertices *P*(1, 7), *Q*(2, 1), *R*(−1, 1), and *S*(−3, 3). A pattern is made by reflecting the figure across the *x*-axis. Give the coordinates of the vertices of the figure after the reflection.
P′(1, −7), *Q*′(2, −1), *R*′(−1, −1), and *S*′(−3, −3)

You may use graph paper on Chapter 12 Resource Book p. 68 for the Exercises.

EXAMPLE 2 Design Application

A designer is using a stencil that is shaped like the figure below. A pattern is made by reflecting the figure across the *x*-axis. Give the coordinates of the vertices of the figure after the given reflection.

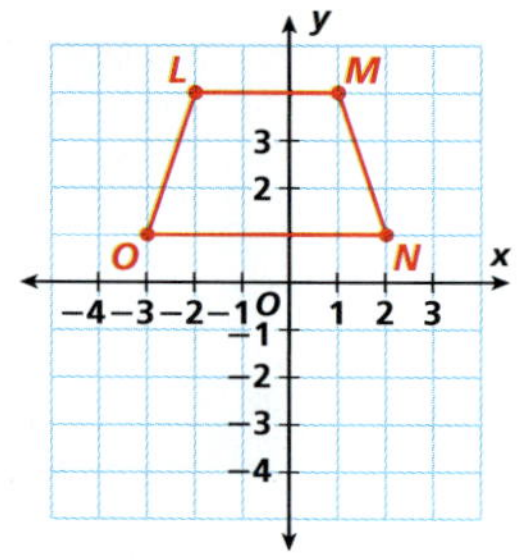

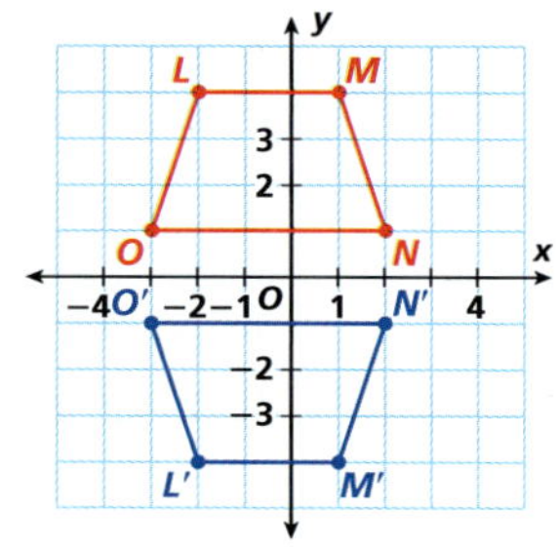

To reflect the trapezoid across the x-axis, write the opposites of the y-coordinates. The x-coordinates do not change.

LMNO		*L′M′N′O′*
L(−2, 4)	→	*L*′(−2, −4)
M(1, 4)	→	*M*′(1, −4)
N(2, 1)	→	*N*′(2, −1)
O(−3, 1)	→	*O*′(−3, −1)

Think and Discuss

1. **Tell** why the *x*-coordinates of a figure do not change when the figure is reflected across the *x*-axis.
2. **Decribe** how the *x*-coordinates of a figure change when the figure is reflected across the *y*-axis.

12-4 PRACTICE & ASSESS

Assignment Guide

If you finished Example 1 assign:
Core 1, 3, 5, 10–13
Enriched 3, 5–6, 10–13

If you finished Example 2 assign:
Core 1–5, 7, 10–13
Enriched 4–13

12-4 Exercises

FOR EXTRA PRACTICE see page 661

internet connect
Homework Help Online
go.hrw.com Keyword: MR4 12-4

GUIDED PRACTICE

See Example 1

Give the coordinates of the vertices of each figure after the given reflection.

1. Reflect triangle *PQR* across the *y*-axis.
P′(3, −1) *Q*′(1, −1) *R*′(1, −4)

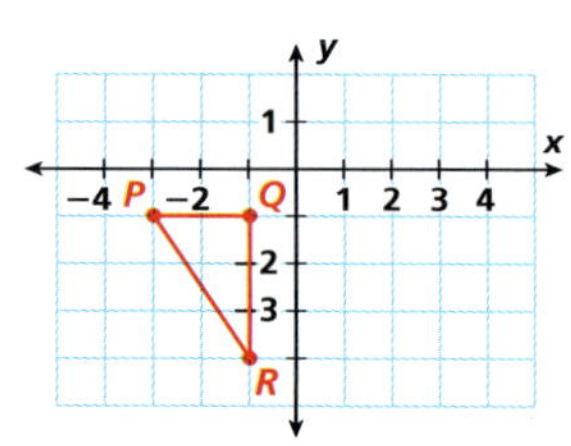

3 Close

Reaching All Learners
Through Curriculum Integration

Art Have students find examples of reflections in artwork. Then have students discuss where the line of symmetry, or axis of reflection, exists in their examples.

Summarize

Review the procedures for finding the new *x*- and *y*-coordinates of a reflected figure (i.e., keeping the coordinates the same for the axis over which the figure was reflected and writing the opposites of the other coordinates).

Answer to Think and Discuss

1. Possible answer: because the figure remains the same number of units away from the *y*-axis and stays on the same side of the *y*-axis
2. The new *x*-coordinates are the opposites of the original *x*-coordinates.

See Example 2

2. Some cheerleaders stood in a formation similar to trapezoid *ABCD*. Then, they moved to form a second figure represented by reflecting *ABCD* across the *x*-axis. Give the coordinates of the vertices of the new trapezoid.
A′(−2, −2) *B*′(0, −2) *C*′(1, 0) *D*′(−3, 0)

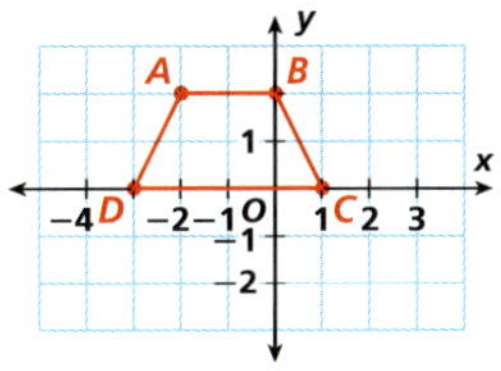

INDEPENDENT PRACTICE

See Example 1

Give the coordinates of the vertices of each figure after the given reflection.

3. Reflect triangle *NQP* across the *x*-axis.
N′(0, −4) *Q*′(0, −1) *P*′(2, 0)

See Example 2

4. Patricia created a plan for her new garden by graphing the parallelogram *JKLM*. She changed her mind and decided to reflect the figure across the *y*-axis. Give the new coordinates of the vertices of the figure.
J′(1, 2) *K*′(−4, 2) *L*′(−3, −1) *M*′(2, −1)

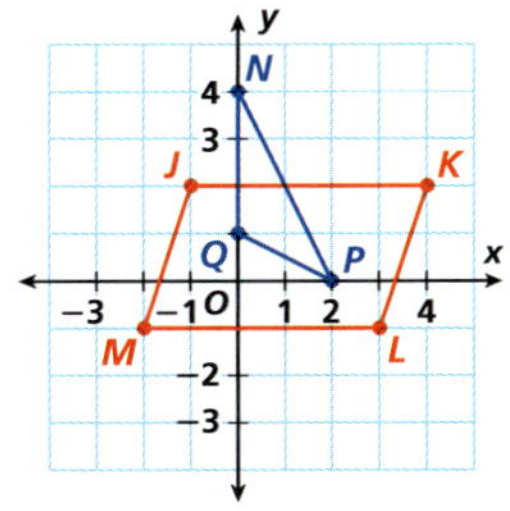

PRACTICE AND PROBLEM SOLVING

Give the coordinates of the square after each given reflection.

5. Reflect square *ABCD* across the *y*-axis.
A′(−1, 3) *B*′(−3, 3) *C*′(−3, 1) *D*′(−1, 1)
6. Reflect square *ABCD* across the *x*-axis.
A′(1, −3) *B*′(3, −3) *C*′(3, −1) *D*′(1, −1)

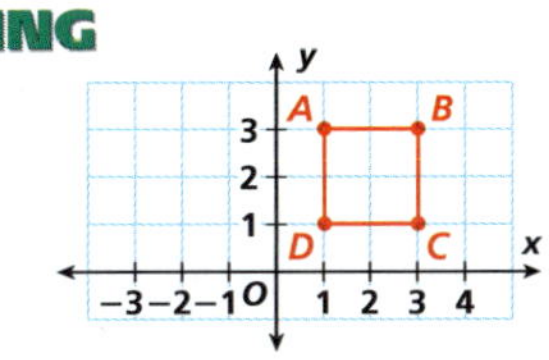

7. Tell what word is spelled when the cross-stitch piece at right is reflected across the *x*-axis. WOW

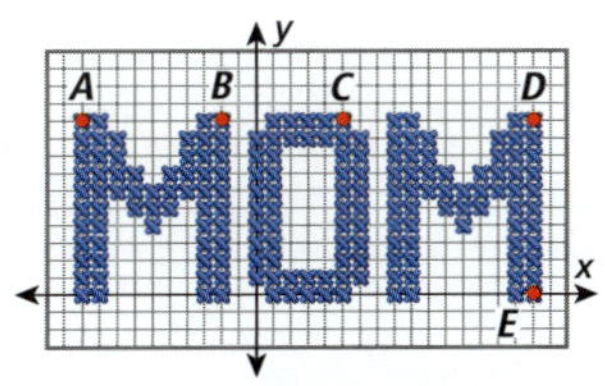

8. The *x*-coordinates will be the opposite of what they were in the original image.

8. ***WRITE ABOUT IT*** Explain how the coordinates of an image change when the image is reflected across the *y*-axis.

9. ***CHALLENGE*** Reflect square *ABCD* from Exercises 5 and 6 across the line $\overline{BD}$. Give the new coordinates of the vertices.
A′(3, 1), *B*′(3, 3), *C*′(1, 3), *D*′(1, 1)

Spiral Review

Write each number in scientific notation. (Lesson 3-5)

10. 2,345 2.345×10^3 11. 100 1×10^2 12. 56,700 5.67×10^4

13. **TEST PREP** The probability of a spinner with three sections landing on green is 56%, and the probability of it landing on yellow is 0.24. What is the probability of it landing on blue if blue is the other color? (Lesson 11-3) A

A 20% B 24% C 30% D 44%

Lesson Quiz

Give the coordinates of the vertices of each figure after the given reflection.

1. Reflect triangle *ABC* with vertices *A*(5, 1), *B*(2, 3), and *C*(4, 6) across the *x*-axis. *A*′(5, −1), *B*′(2, −3), *C*′(4, −6)
2. Reflect parallelogram *PQRS* with vertices *P*(−4, −4), *Q*(1, −4), *R*(−2, 2), and *S*(3, 2) across the *y*-axis. *P*′(4, −4), *Q*′(−1, −4), *R*′(2, 2) and *S*′(−3, 2)

Available on Daily Transparency in CRB

RETEACH 12-4

LESSON 12-4 **Reteach**
Graphing Reflections

A reflection is an imaged flipped over a line. A figure can be reflected across the *x*-axis or the *y*-axis on a coordinate plane.

Reflect the triangle across the *x*-axis.

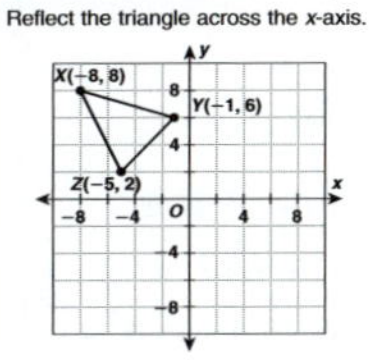

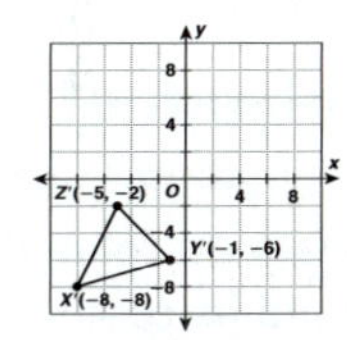

X (−8, 8)	→ X′(−8, −8)
Y (−1, 6)	→ Y′(−1, −6)
Z (−5, 2)	→ Z′(−5, −2)

To find the coordinate of the vertices of the reflected figure, follow these rules. The *x*-coordinates is the same and the *y*-coordinates changes to its opposite.

Give the coordinates of the vertices of each figure after the given reflection.

1. Reflect figure *ABCD* across the *x*-axis.
A′(6, −9); *B*′(7, −1);
C′(4, 4); *D*′(2, −4)
2. Reflect figure *ABCD* across the *y*-axis.
A′(−6, 9); *B*′(−7, 1);
C′(−4, −4); *D*′(−2, 4)

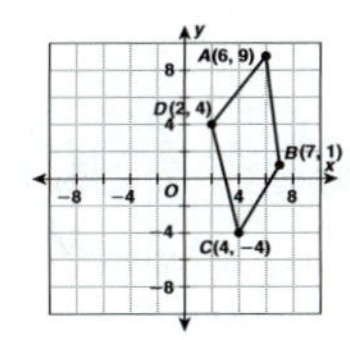

PRACTICE 12-4

LESSON 12-4 **Practice B**
Graphing Reflections

Give the coordinates of the vertices of each figure after the given reflection.

1. Reflect △*XYZ* across the *x*-axis.
X′(2, 1); *Y*′(2, 7);
Z′(6, 2)
2. Reflect △*XYZ* across the *y*-axis.
X′(−2, −1); *Y*′(−2, −7);
Z′(−6, −2)
3. Reflect △*ABC* across the *x*-axis.
A′(−6, 4); *B*′(−2, 1);
C′(−4, 7)
4. Reflect △*ABC* across the *y*-axis.
A′(6, −4); *B*′(2, −1);
C′(4, −7)
5. Reflect square *FGHJ* across the *y*-axis.
F′(4, 4); *G*′(4, 7);
H′(1, 7); *J*′(1, 4)
6. Reflect square *FGHJ* across the *x*-axis.
F′(−4, −4); *G*′(−4, −7);
H′(−1, −7); *J*′(−1, −4)

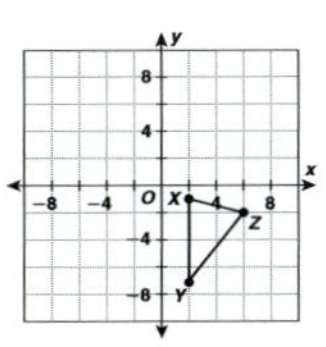

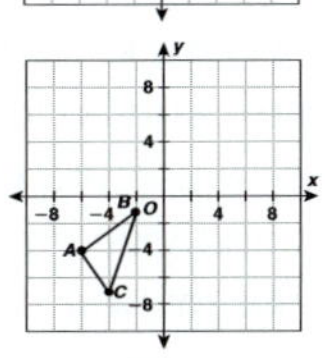

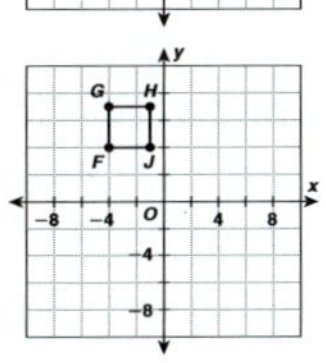

CHALLENGE 12-4

LESSON 12-4 **Challenge**
Math Reflections

STEP 1: Use the given endpoints to graph these 12 line segments on the coordinate plane below.

STEP 2: Reflect the line segments across the *x*-axis.

STEP 3: Reflect the new line segments across the *y*-axis.

STEP 4: Reflect the new line segments across the *x*-axis.

(1, 4), (1, 8)	(4, 5), (6, 5)	(8, 8), (8, 4)	(3, 8), (3, 4)
(10, 4), (10, 8)	(7, 8), (9, 8)	(1, 8), (2, 6)	(10, 6), (12, 6)
(5, 8), (6, 4)	(2, 6), (3, 8)	(12, 4), (12, 8)	(4, 4), (5, 8)

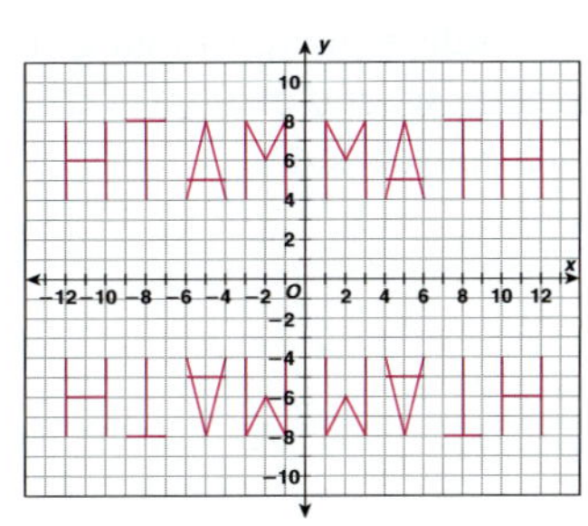

PROBLEM SOLVING 12-4

LESSON 12-4 **Problem Solving**
Graphing Reflections

Write the correct answer.

1. Does the word MOM have line symmetry? If yes, is the line horizontal or vertical?
yes; vertical line symmetry
2. Name the coordinates for the endpoints of each line segment that make up the graphed word MOM.
A(2, 6); *B*(2, 2); *C*(3, 5); *D*(4, 6);
E(4, 2); *F*(5, 2); *G*(5, 6); *H*(7, 6);
I(7, 2); *J*(8, 2); *K*(8, 6); *L*(9, 5);
M(10, 6); *N*(10, 2)

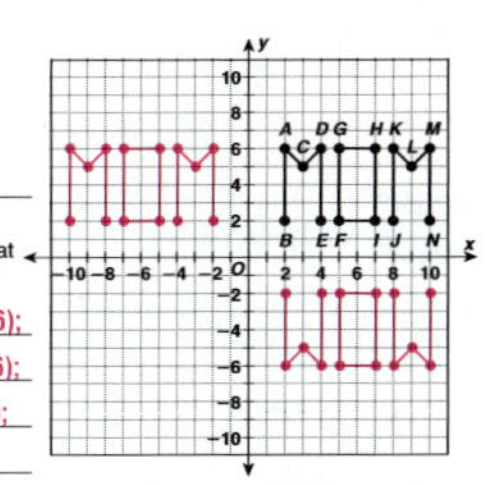

3. Reflect the word MOM across the *x*-axis. Give the coordinates of each endpoint after this reflection.
A′(2, −6); *B*′(2, −2); *C*′(3, −5);
D′(4, −6); *E*′(4, −2); *F*′(5, −2);
G′(5, −6); *H*′(7, −6); *I*′(7, −2);
J′(8, −2); *K*′(8, −6); *L*′(9, −5);
M′(10, −6); *N*′(10, −2)
4. Reflect the word MOM across the y-axis. Give the coordinates of each endpoint after this reflection.
A′(−2, 6); *B*′(−2, 2); *C*′(−3, 5);
D′(−4, 6); *E*′(−4, 2); *F*′(−5, 2);
G′(−5, 6); *H*′(−7, 6); *I*′(−7, 2);
J′(−8, 2); *K*′(−8, 6); *L*′(−9, 5);
M′(−10, 6); *N*′(−10, 2)

Circle the letter of the correct answer.

5. The vertices of triangle *ABC* are *A*(−8, −2), *B*(−2, −2), and *C*(−2, −8). What are the coordinates of the vertices if the triangle is reflected across the *x*-axis?
A *A*′(8, 2), *B*′(2, 2), *C*′(2, 8)
B *A*′(8, −2), *B*′(2, −2), *C*′(2, −8)
(C) *A*′(−8, 2), *B*′(−2, 2), *C*′(−2, 8)
D *A*′(−2, −8), *B*′(−2, −2), *C*′(−8, −2)
6. Which of the following describes a reflection of any point (*a*, *b*) across the *y*-axis?
(F) (*a*, *b*) → (−*a*, *b*)
G (*a*, *b*) → (*a*, −*b*)
H (*a*, *b*) → (−*b*, *a*)
J (*a*, *b*) → (−*a*, −*b*)

12-5 Organizer

Pacing: Traditional 1 day
Block $\frac{1}{2}$ day

Objective: Students use rotations to change positions of figures on a coordinate plane.

Indiana Academic Standards
6.3.7, 6.7.1, 6.7.3, 6.7.4, 6.7.5, 6.7.8, 6.7.9, 6.7.11

Warm Up

Parallelogram *ABCD* has vertices (−4, 1), (0, 1), (−3, 4), and (1, 4).

1. What are the vertices of *ABCD* after it has been reflected across the *x*-axis? (−4, −1), (0, −1), (−3, −4), (1, −4)
2. What are the vertices after it has been reflected across the *y*-axis? (4, 1), (0, 1), (3, 4), (−1, 4)

Problem of the Day

If each of the capital letters of the alphabet is rotated a half turn around its center, which will look the same? H, I, N, O, S, X, Z

Available on Daily Transparency in CRB

Math Fact

Rotating a figure 180° about the origin is equivalent to reflecting the figure twice: once across the *x*-axis and once across the *y*-axis, in either order.

12-5 Graphing Rotations

Learn to use rotations to change positions of figures on a coordinate plane.

Swimmers on synchronized swimming teams perform routines that involve making designs in the water with their bodies. They can rotate a design by changing their positions in the water.

You can rotate a figure about the origin or another point on a coordinate plane.

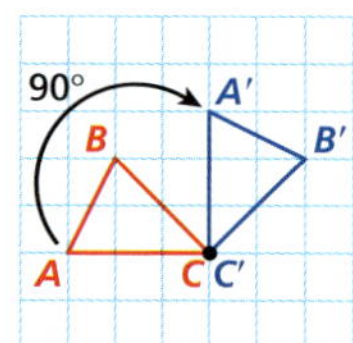

90° rotation

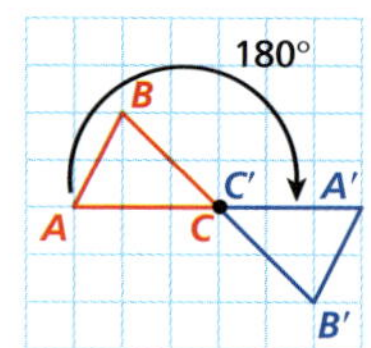

180° rotation

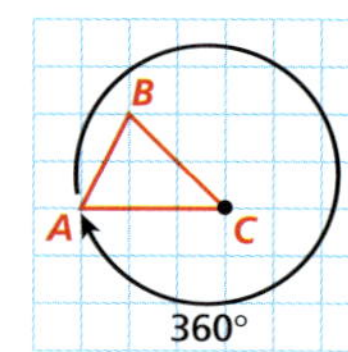

360° rotation

EXAMPLE 1 Rotating Figures on a Coordinate Plane

Give the coordinates of the vertices of the figure after the given rotation.

> **Remember!**
> A rotation is the movement of a figure about a point. Rotating a figure "about the origin" means that the origin is the center of rotation.

Rotate trapezoid *ABCD* clockwise 90° about the origin.

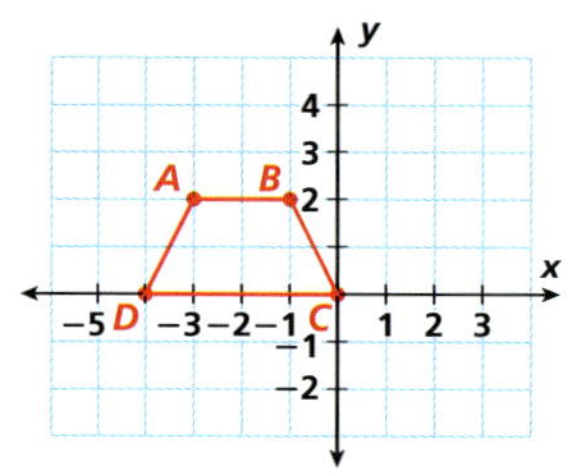

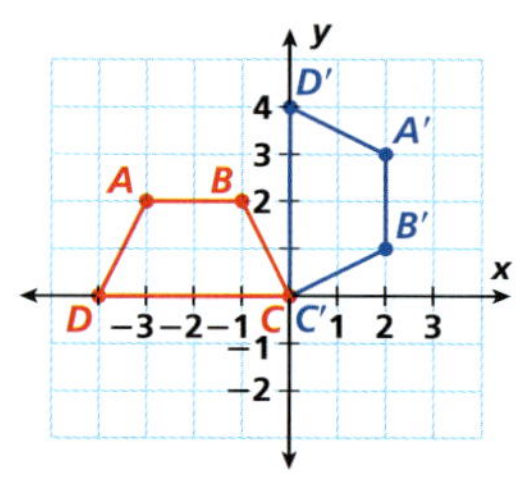

The new x-coordinates are the old y-coordinates.

The new y-coordinates are the opposites of the old x-coordinates.

ABCD		*A′B′C′D′*
A(−3, 2)	→	*A′*(2, 3)
B(−1, 2)	→	*B′*(2, 1)
C(0, 0)	→	*C′*(0, 0)
D(−4, 0)	→	*D′*(0, 4)

1 Introduce

Alternate Opener

EXPLORATION

12-5 Graphing Rotations

The graph shows the blue triangle as a 180° rotation of the red triangle about the point (5, 6).

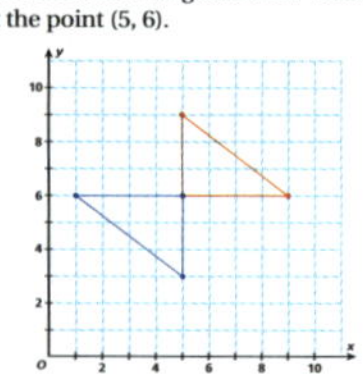

1. Label the coordinates of the vertices of each triangle.
2. Are the two triangles congruent?
3. Examine the rotation and the coordinates of its vertices.
   a. How are the coordinates different from the original coordinates?
   b. How are the coordinates similar to the original coordinates?

Think and Discuss

4. **Describe** some real-world examples of rotations.
5. **Explain** how you determined whether the two triangles are congruent.

Motivate

To introduce students to the concept of graphing rotations, review the term *rotation*. Ask students to define the term and give examples of rotations and examples of figures that are not rotated.

Exploration worksheet and answers on Chapter 12 Resource Book pp. 45 and 77

2 Teach

Lesson Presentation

Guided Instruction

In this lesson, students learn to use rotations to change positions of figures on a coordinate plane.

You may want to review terms like *origin, clockwise, counterclockwise,* etc. Teach students to use the origin as the point of rotation and give examples of figures that have been rotated 90°, 180°, and 360°. You may want to use Teaching Transparency T15 in the Chapter 12 Resource Book. Explain that figures can be rotated clockwise (to the right) or counterclockwise (to the left) and explain how the coordinates change for different degrees of rotation.

EXAMPLE 2 *Sports Application*

A synchronized swimming team forms this figure. The swimmers rotate the figure without changing its size or shape.

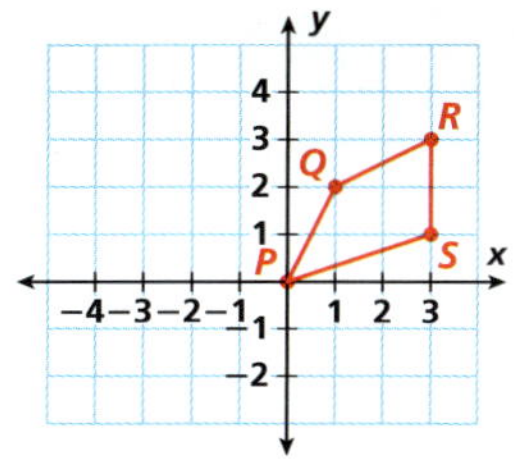

A **Give the coordinates of the vertices of the figure after a clockwise rotation of 90° about the origin.**

The new x-coordinates are the old y-coordinates.

The new y-coordinates are the opposites of the old x-coordinates.

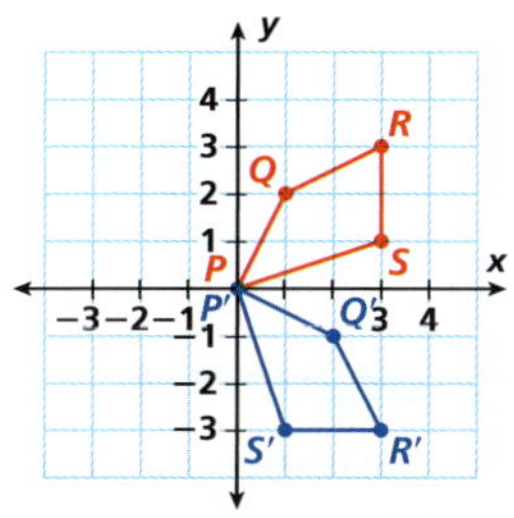

PQRS		*P'Q'R'S'*
P(0, 0)	→	*P'*(0, 0)
Q(1, 2)	→	*Q'*(2, −1)
R(3, 3)	→	*R'*(3, −3)
S(3, 1)	→	*S'*(1, −3)

B **Give the coordinates of the vertices of the figure after a counterclockwise rotation of 180° about the origin.**

The new x-coordinates are the opposites of the old x-coordinates.

The new y-coordinates are the opposites of the old y-coordinates.

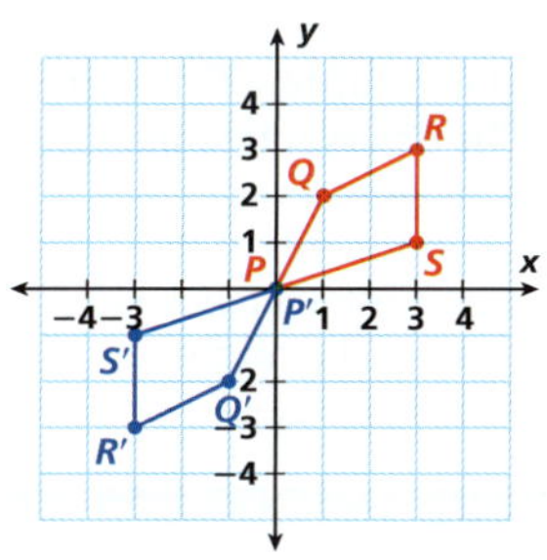

PQRS		*P'Q'R'S'*
P(0, 0)	→	*P'*(0, 0)
Q(1, 2)	→	*Q'*(−1, −2)
R(3, 3)	→	*R'*(−3, −3)
S(3, 1)	→	*S'*(−3, −1)

Think and Discuss

1. **Tell** how the *x*- and *y*-coordinates would change if you rotated figure *ABCD* from Example 1 counterclockwise 90° about the origin.
2. **Tell** what the coordinates would be if you rotated figure *P'Q'R'S'* from Example 2A clockwise 90° about the origin.

COMMON ERROR ALERT

Some students may try to change the coordinates of the vertex located at the origin (0, 0). To prevent this error, tell students to imagine that a tack is holding that vertex on the origin, and the figure is rotated without moving the tack.

Additional Examples

Example 1

Give the coordinates of the vertices of the figure after the given rotation.

Rotate parallelogram *EFGH* with vertices *E*(−4, 0), *F*(0, 0), *G*(−2, −3), and *H*(−6, −3) clockwise 90° about the origin.

E'(0, 4), *F'*(0, 0), *G'*(−3, 2), *H'*(−3, 6)

Example 2

A figure with vertices *A*(0, 0), *B*(1, 4), *C*(3, 5), *D*(7, 3), and *E*(6, 0) is rotated without changing its size or shape.

A. Give the coordinates of the vertices of the figure after a clockwise rotation of 90° about the origin.
A'(0, 0), *B'*(4, −1), *C'*(5, −3), *D'*(3, −7), *E'*(0, −6)

B. Give the coordinates of the vertices of the figure after a counterclockwise rotation of 180° about the origin.
A'(0, 0), *B'*(−1, −4), *C'*(−3, −5), *D'*(−7, −3), *E'*(−6, 0)

3 Close

Reaching All Learners

Through Concrete Manipulatives

Tell each student to draw a triangle with coordinates *A*(−7, 0), *B*(−2, 3), and *C*(0, 0) and label the vertices. Next have students trace the triangle and cut out the tracing. You may use the graph paper provided on Chapter 12 Resource Book p. 68. Ask students to place the cut-out triangle on top of triangle *ABC* and then rotate the cut-out triangle 90° clockwise, keeping vertex *C* at the origin. Have students identify the coordinates of the new vertices. Repeat the steps above for a 180° rotation and a 270° rotation.
A'(0, 7), *B'*(3, 2), *C'*(0, 0); *A''*(7, 0), *B''*(2, −3), *C''*(0, 0); *A'''*(0, −7), *B'''*(−3, −2), *C'''*(0, 0)

Summarize

Review different ways figures can be rotated about the origin (e.g., number of degrees and direction). Have students give examples of 90°, 180°, and 270° clockwise and counterclockwise rotations.

Answers to Think and Discuss

1. The new *y*-coordinates would be the old *x*-coordinates, and the new *x*-coordinates would be the opposites of the old *y*-coordinates.
2. *P''*(0, 0), *Q''*(−1, −2), *R''*(−3, −3), *S''*(−3, −1)

12-5 PRACTICE & ASSESS

Students may want to refer back to the lesson examples.

Assignment Guide

If you finished Example 1 assign:
Core 1–2, 4–5, 7–8, 17–21
Enriched 7–12, 17–21

If you finished Example 2 assign:
Core 1–8, 11, 17–21
Enriched 7–21

You may use the graph paper on Chapter 12 Resource Book p. 68 for the exercises.

Answers

7. $A'(3, 0)$, $B'(3, -3)$, $C'(0, -3)$, $D'(0, 0)$

8. $A'(3, 0)$, $B'(3, -3)$, $C'(0, -3)$, $D'(0, 0)$

9. $B'(3, -3)$, $C'(3, 0)$, $E'(4, 0)$

10. $A'(-5, 5)$, $B'(-2, 5)$, $E'(-1, 2)$, $D'(-5, 2)$

12-5 Exercises

FOR EXTRA PRACTICE see page 661

internet connect
Homework Help Online
go.hrw.com Keyword: MR4 12-5

GUIDED PRACTICE

See Example 1 **Give the coordinates of the vertices of each figure after the given rotation.**

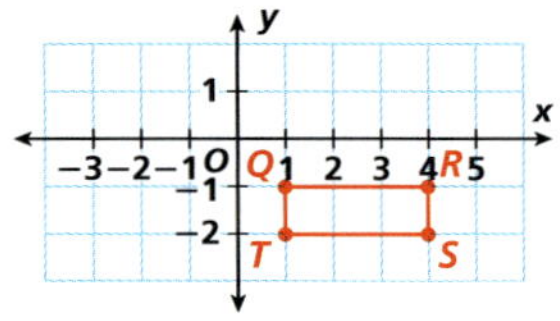

1. Rotate rectangle $QRST$ clockwise 180° about the origin. $Q'(-1, 1)$ $R'(-4, 1)$ $S'(-4, 2)$ $T'(-1, 2)$
2. Rotate rectangle $QRST$ counterclockwise 90° about the origin. $Q'(1, 1)$ $R'(1, 4)$ $S'(2, 4)$ $T'(2, 1)$

See Example 2

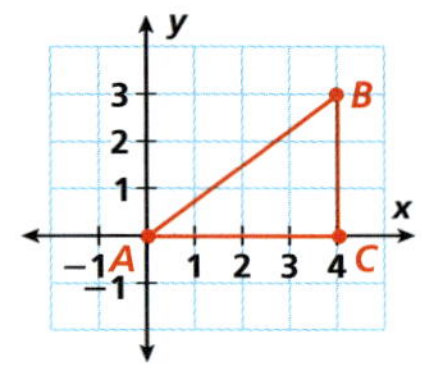

3. Sean is using triangular tiles that look like the triangle on the graph. If he rotates the tiles, he can create a pattern. Give the coordinates of the vertices of triangle ABC after a clockwise rotation of 90° about point A. $A'(0, 0)$ $B'(3, -4)$ $C'(0, -4)$

INDEPENDENT PRACTICE

See Example 1 **Give the coordinates of the vertices of each figure after the given rotation.**

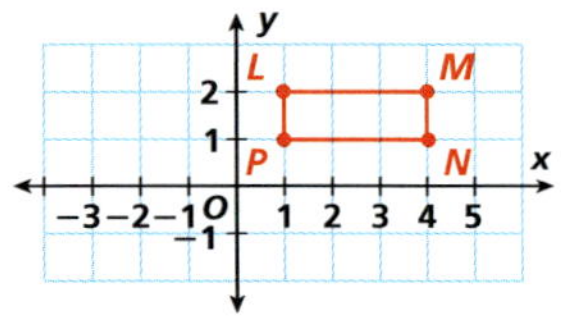

4. Rotate rectangle $LMNP$ counterclockwise 180° about the origin. $P'(-1, -1)$ $N'(-4, -1)$ $M'(-4, -2)$ $L'(-1, -2)$
5. Rotate rectangle $LMNP$ clockwise 90° about the origin. $P'(1, -1)$ $N'(1, -4)$ $M'(2, -4)$ $L'(2, -1)$

See Example 2

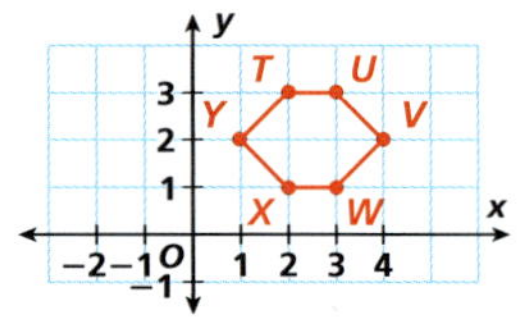

6. A group of sky divers forms a figure like this in the air. The divers rotate the figure without changing its size or shape. Give the coordinates of the vertices of the figure after a clockwise rotation of 180° about the origin. $T'(-2, -3)$ $U'(-3, -3)$ $V'(-4, -2)$ $W'(-3, -1)$ $X'(-2, -1)$ $Y'(-1, -2)$

PRACTICE AND PROBLEM SOLVING

Use the graph for Exercises 7–10. Give the coordinates of the vertices of each figure after the given transformation.

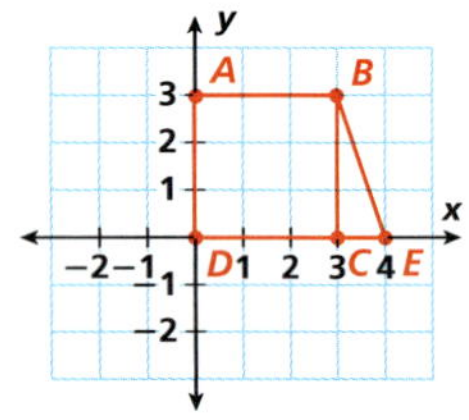

7. Rotate square $ABCD$ clockwise 90° about the origin.
8. Rotate square $ABCD$ counterclockwise 270° about the origin.
9. Reflect triangle BCE across the x-axis.
10. Translate trapezoid $ABED$ up 2 units and left 5 units.

Math Background

A figure that has rotational symmetry coincides exactly with an image of itself after a rotation about its center of a certain measurement less than 360°. An equilateral triangle has rotational symmetry of 120°, and a square has rotational symmetry of 90°. In general, any regular polygon (i.e., a polygon with all sides equal in length and all angles equal in measure) with n sides has rotational symmetry of $\left(\frac{360}{n}\right)°$.

A regular polygon with n sides also has n lines of symmetry. Rectangles exhibit rotational symmetry and have line symmetry. They are symmetrical about the two lines through the center parallel to the sides.

RETEACH 12-5

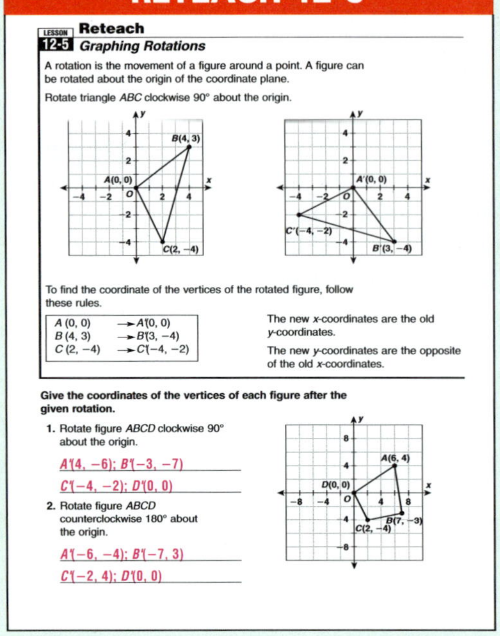

LESSON 12-5 Reteach
Graphing Rotations

A rotation is the movement of a figure around a point. A figure can be rotated about the origin of the coordinate plane.

Rotate triangle ABC clockwise 90° about the origin.

To find the coordinate of the vertices of the rotated figure, follow these rules.

$A(0, 0) \rightarrow A'(0, 0)$
$B(4, 3) \rightarrow B'(3, -4)$
$C(2, -4) \rightarrow C'(-4, -2)$

The new x-coordinates are the old y-coordinates.
The new y-coordinates are the opposite of the old x-coordinates.

Give the coordinates of the vertices of each figure after the given rotation.

1. Rotate figure $ABCD$ clockwise 90° about the origin.
$A'(4, -6)$; $B'(-3, -7)$
$C'(-4, -2)$; $D'(0, 0)$
2. Rotate figure $ABCD$ counterclockwise 180° about the origin.
$A'(-6, -4)$; $B'(-7, 3)$
$C'(-2, 4)$; $D'(0, 0)$

PRACTICE 12-5

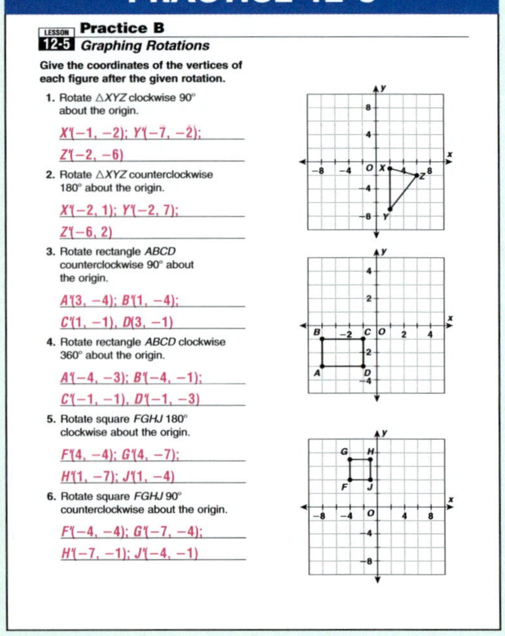

LESSON 12-5 Practice B
Graphing Rotations

Give the coordinates of the vertices of each figure after the given rotation.

1. Rotate $\triangle XYZ$ clockwise 90° about the origin.
$X'(-1, -2)$; $Y'(-7, -2)$;
$Z'(-2, -6)$
2. Rotate $\triangle XYZ$ counterclockwise 180° about the origin.
$X'(-2, 1)$; $Y'(-2, 7)$;
$Z'(-6, 2)$
3. Rotate rectangle $ABCD$ counterclockwise 90° about the origin.
$A'(3, -4)$; $B'(1, -4)$;
$C'(1, -1)$, $D'(3, -1)$
4. Rotate rectangle $ABCD$ clockwise 360° about the origin.
$A'(-4, -3)$; $B'(-4, -1)$;
$C'(-1, -1)$, $D'(-1, -3)$
5. Rotate square $FGHJ$ 180° clockwise about the origin.
$F'(4, -4)$; $G'(4, -7)$;
$H'(1, -7)$; $J'(1, -4)$
6. Rotate square $FGHJ$ 90° counterclockwise about the origin.
$F'(-4, -4)$; $G'(-7, -4)$;
$H'(-7, -1)$; $J'(-4, -1)$

Art LINK

Some works of art, like *A Fish Story*, by Gustave Verbeek, are optical illusions. When this picture is rotated 180°, a different image appears.

11. **ART** Use the points labeled on the graph at right to rotate the design 180° clockwise about the origin. Tell what word is spelled after this rotation. math

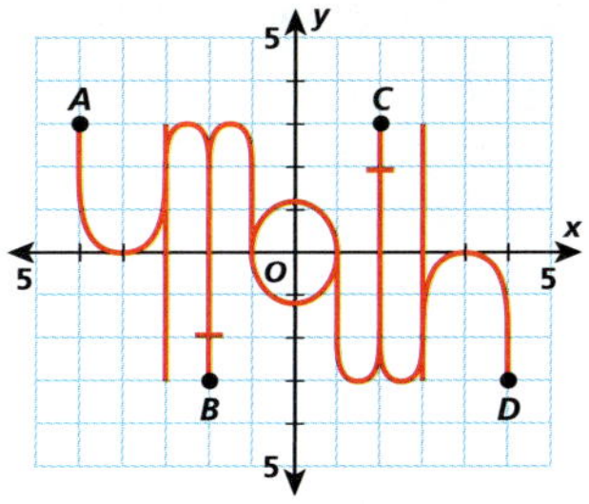

12. Graph the points $A(2, 6)$, $B(-3, 6)$, $C(-3, 1)$, and $D(2, 1)$. Then join the points to form rectangle $ABCD$.
    a. Translate the rectangle 3 units up. What are the coordinates of the vertices of $A'B'C'D'$?
    b. Rotate $A'B'C'D'$ 360° about the origin. Now what are the coordinates of its vertices? $A'(2, 9)$, $B'(-3, 9)$, $C'(-3, 4)$, $D'(2, 4)$

12a. $A'(2, 9)$, $B'(-3, 9)$, $C'(-3, 4)$, $D'(2, 4)$

Use the graph of the rhombus for Exercises 13–16.

13. Lara translated the rhombus 2 units right and then rotated it clockwise 180° about the origin. Mike rotated the original rhombus clockwise 180° about the origin and then translated it 2 units right. Are the new coordinates of Lara's rhombus the same as the new coordinates of Mike's rhombus? Explain.

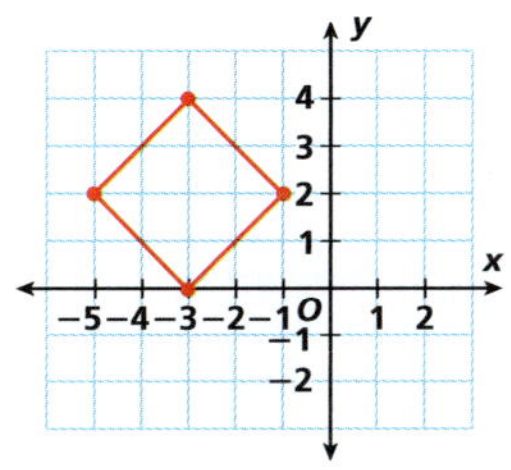

14. **WRITE A PROBLEM** Write a problem about rotation using the rhombus. Solve your problem.
15. **WRITE ABOUT IT** Describe a translation, a reflection, and a rotation that can be performed on the rhombus.
16. **CHALLENGE** Describe two different transformations that would not change the coordinates of the rhombus.

Spiral Review

Order each set of numbers from least to greatest. (Lesson 3-1)

17. 1.2, 0.445, 1.06, 0.9
    0.445, 0.9, 1.06, 1.2
18. 2.45, 2.678, 2.007, 2.02
    2.007, 2.02, 2.45, 2.678
19. 7.99, 7.999, 7.9, 7.09
    7.09, 7.9, 7.99, 7.999
20. **TEST PREP** Choose the coordinates that are the farthest to the right of the origin on a coordinate plane. (Lesson 9-3) **C**

    A (0, 12) **B** (−19, 7) **C** (7, 0) **D** (4, 15)

21. **TEST PREP** How many different 4-digit numbers can be made using the digits 5, 3, 2, and 7? (Lesson 11-4) **J**

    F 4 **G** 12 **H** 16 **J** 24

Answers

13. No; Mike's figure will be four units to the right of Lara's.
14. Possible answer: What are the coordinates of the vertices of the rhombus after it has been rotated 90° clockwise about the origin? (0, 3), (2, 1), (2, 5), (4, 3)
15. Check students' work.
16. Possible answer: Rotate the image 360° about the origin, or translate the image up two units and then down two units.

Journal

Have students draw a figure on a coordinate grid, labeling and giving the coordinates of its vertices. On a second grid, have students draw the original figure rotated 90° counterclockwise and rotated 180° clockwise.

Test Prep Doctor

For Exercise 21, it is important that students realize that order does matter. Most students will probably tackle this problem by making an organized list. Students who answered **F, G,** or **H** should go back and check their work. Here are the 24 possible 4-digit numbers:

5327	3527	2537	7532
5372	3572	2573	7523
5237	3257	2357	7352
5273	3275	2375	7325
5732	3752	2753	7253
5723	3725	2735	7235

CHALLENGE 12-5

LESSON 12-5 Challenge
Coded Rotations

Before modern technology, the United States Navy sent messages by Semaphore Code. In this code, a messenger rotates two flags into different positions to indicate letters of the alphabet.

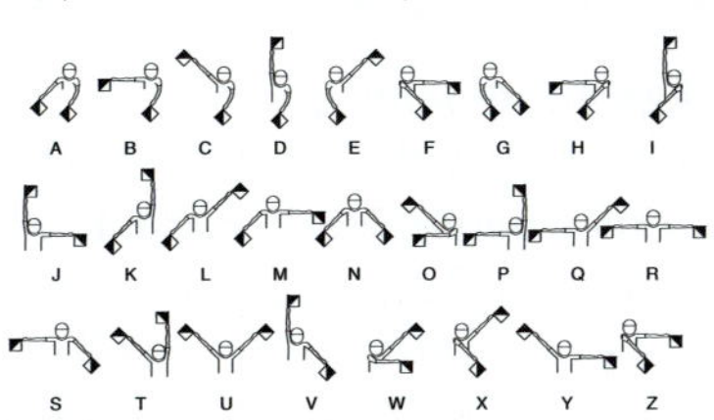

Complete the chart by describing how to rotate one flag about the origin to change each starting Semaphore Code letter to show each signaled letter. Answers are given from perspective of person facing us.

Starting Letter	Rotation	Signaled Letter
M	Rotate the left flag 90° clockwise.	A
D	Rotate the right flag 90° counterclockwise.	B
C	Rotate the left flag 180° counterclockwise.	T
V	Rotate the right flag 180° counterclockwise.	G
R	Rotate the right flag 90° clockwise	J

PROBLEM SOLVING 12-5

LESSON 12-5 Problem Solving
Graphing Rotations

Write the correct answer.

1. Graph points $A(-8, 2)$, $B(-8, 7)$, $C(-5, 7)$, and $D(-5, 2)$. Then join the points to form rectangle $ABCD$.

   Check students' graphs.

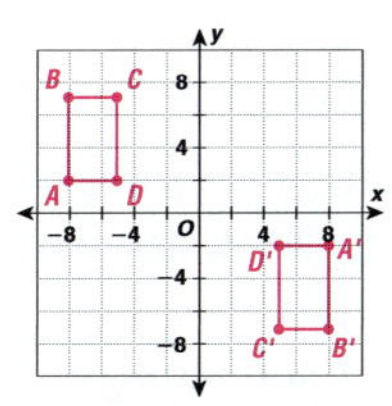

2. Rotate $ABCD$ 180° counterclockwise about the origin. What are the coordinates of the rotated rectangle?

   Check students' graphs;
   $A'(8, -2)$; $B'(8, -7)$;
   $C'(5, -7)$; $D'(5, -2)$

3. Reflect $ABCD$ across the y-axis. Then reflect the reflected $ABCD$ across the x-axis. What are the coordinates of the double-reflected rectangle?

   Check students' graphs;
   $A''(8, -2)$; $B''(8, -7)$;
   $C''(5, -7)$; $D''(5, -2)$

4. What conclusion can you make about the relationship between a double reflection of a figure over both axes and a 180° rotation of that figure about the origin?

   They are the same transformation of the figure.

Circle the letter of the correct answer.

5. The vertices of triangle ABC are $A(5, -3)$, $B(10, -7)$, and $C(1, -7)$. What are the coordinates of the vertices if it is rotated 90° counterclockwise about the origin?
   A $A'(-5, -3)$, $B'(-10, -7)$, $C'(-1, -7)$
   B $A'(-5, 3)$, $B'(-10, 3)$, $C'(-1, 7)$
   (C) $A'(3, 5)$, $B'(7, 10)$, $C'(7, 1)$
   D $A'(3, -5)$, $B'(7, -10)$, $C'(7, -1)$

6. How do you find the coordinates of the vertices of a figure when it is rotated 180° clockwise about the origin?
   (F) Change the sign of both coordinates.
   G Change the sign of the x-coordinate and reverse the coordinates.
   H Change the sign of the y-coordinate and reverse the coordinates.
   J Change the sign of the x-coordinate.

Lesson Quiz

Give the coordinates of the vertices of the trapezoid after the given rotation.

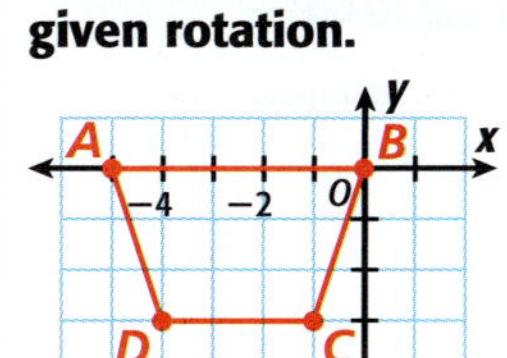

1. Rotate trapezoid $ABCD$ counterclockwise 180° about the origin. $A'(5, 0)$, $B'(0, 0)$, $C'(1, 3)$, $D'(4, 3)$
2. Rotate trapezoid $ABCD$ clockwise 90° about the origin. $A'(0, 5)$, $B'(0, 0)$, $C'(-3, 1)$, $D'(-3, 4)$

Available on Daily Transparency in CRB

12-6 Organizer

Pacing: Traditional 1 day
Block $\frac{1}{2}$ day

Objective: Students visualize and show the results of stretching or shrinking a figure.

Indiana Academic Standards
6.7.1, 6.7.2, 6.7.3, 6.7.4, 6.7.5, 6.7.8, 6.7.10, 6.7.11

Warm Up

ABCD has vertices (−4, 1), (0, 1), (−3, 4), and (1, 4). What are the vertices of *ABCD* after it has been rotated clockwise 180° about the origin?
(4, −1), (0, −1), (3, −4), (−1, −4)

Problem of the Day

Alice was 4 feet tall. She took a bite of one side of the caterpillar's mushroom and became 5 times as tall! Then she took a bite of the other side of the mushroom and became $\frac{1}{4}$ times as tall. She took a bite from each side two more times. How tall was Alice then? $7\frac{13}{16}$ feet

Available on Daily Transparency in CRB

Math Fact

Examples of reflections, rotations, and translations are found in M. C. Escher's famous works of art.

12-6 Stretching and Shrinking

Learn to visualize and show the results of stretching or shrinking a figure.

By increasing or decreasing the size of one dimension of a figure, the look of a design can change.

A funhouse mirror distorts your reflection because of its curved surfaces. The parts that curve inward stretch your image, and the parts that curve outward shrink your image.

Remember!
Vertical means "straight up or down." *Horizontal* means "straight across," such as from left to right.

EXAMPLE 1 Stretching Figures

Write the dimensions of each part of the figure. Stretch the figure as stated, and give the new dimensions of each part.

A Increase the horizontal dimensions of the face, eyes, and mouth by a factor of 3.

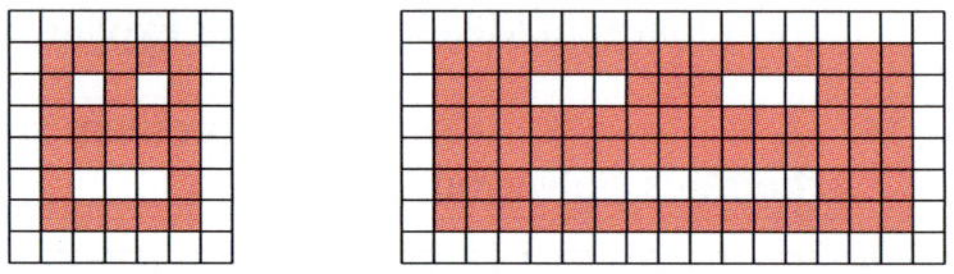

	Original Dimensions	New Dimensions
Face	Vertical 6 Horizontal 5	Vertical 6 Horizontal 15
Eyes	Vertical 1 Horizontal 1	Vertical 1 Horizontal 3
Mouth	Vertical 1 Horizontal 3	Vertical 1 Horizontal 9

B Increase the vertical dimensions of the face, eyes, and mouth by a factor of 2.

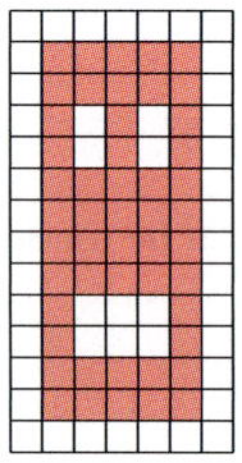

	Original Dimensions	New Dimensions
Face	Vertical 6 Horizontal 5	Vertical 12 Horizontal 5
Eyes	Vertical 1 Horizontal 1	Vertical 2 Horizontal 1
Mouth	Vertical 1 Horizontal 3	Vertical 2 Horizontal 3

1 Introduce

Alternate Opener

EXPLORATION

12-6 Stretching and Shrinking

You can stretch the triangle in the coordinate plane.

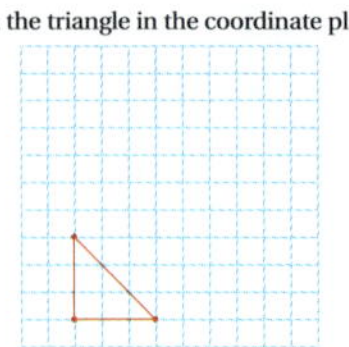

1. Label the height and width of the triangle.
2. Create a new triangle by multiplying the dimensions by 2.
3. Explain how the height and width of the triangle changed after the dimensions were multiplied by 2.

Think and Discuss

4. **Discuss** real-world applications of stretches.
5. **Explain** how many times the original triangle fits into the triangle you drew in number 2.

Motivate

Discuss the words *stretch* and *shrink*. Ask students: What happens when you stretch something? It gets bigger. What happens when you shrink something? It gets smaller. Explain that this lesson deals with stretching and shrinking figures horizontally and vertically.

Exploration worksheet and answers on Chapter 12 Resource Book pp. 54 and 79

2 Teach

Lesson Presentation

Guided Instruction

In this lesson, students learn to visualize and show the results of stretching or shrinking a figure. Teach students what happens when a figure is stretched or shrunk horizontally or vertically. Point out that only the horizontal dimensions change when the figure stretches or shrinks horizontally and only the vertical dimensions change when the figure stretches or shrinks vertically.

EXAMPLE 2 Shrinking Figures

Write the dimensions of each figure. Shrink the figure as stated and give the new dimensions.

A Decrease the vertical dimensions by multiplying by $\frac{1}{3}$.

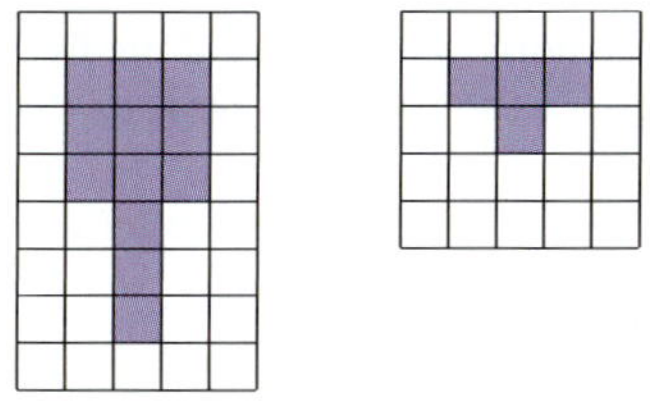

Original Dimensions		New Dimensions	
Vertical	6	Vertical	2
Horizontal	3	Horizontal	3

B Decrease the horizontal dimensions by multiplying by $\frac{1}{2}$.

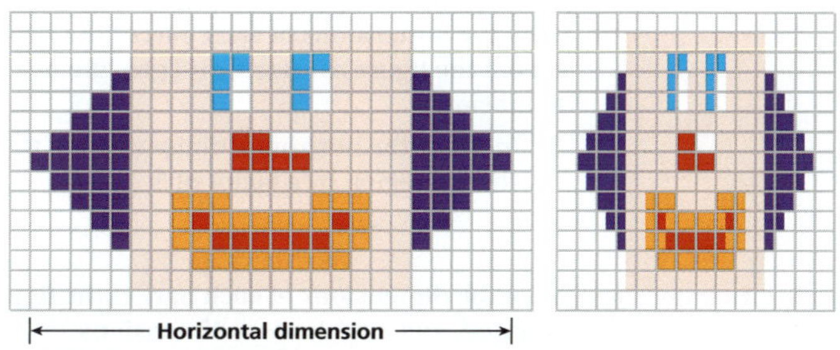

Original Dimensions		New Dimensions	
Vertical	13	Vertical	13
Horizontal	24	Horizontal	12

Think and Discuss

1. **Describe** what happens to a figure when the horizontal dimension is stretched.
2. **Tell** whether a figure whose vertical dimension has been shrunk by being multiplied by $\frac{1}{3}$ is similar to the original figure. Explain.

Additional Examples

Example 1

Write the dimensions of the figure. Stretch the figure as stated and give the new dimensions.

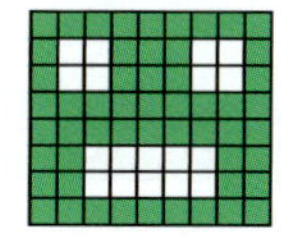

A. Increase the horizontal dimensions by a factor of 3.

vertical 8 → no change
horizontal 9 → 27

B. Increase the vertical dimensions by a factor of 2.

vertical 8 → 16
horizontal 9 → no change

Example 2

Write the dimensions of the figure. Shrink the figure as stated and give the new dimensions.

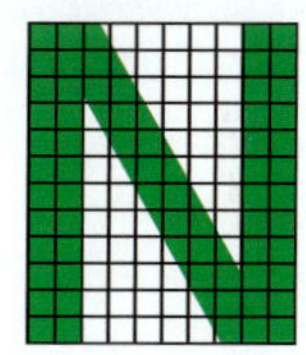

A. Decrease the vertical dimensions by multiplying by $\frac{1}{3}$.

vertical 12 → 4
horizontal 10 → no change

B. Decrease the horizontal dimensions by multiplying by $\frac{1}{2}$.

vertical 12 → no change
horizontal 10 → 5

Complete answers on Additional Examples Transparencies in CRB

Reaching All Learners
Through Critical Thinking

Explain to students that when a figure is enlarged or reduced by the same factor in all directions, it is called a *dilation*. Have students find real-world examples of dilations, e.g., different-sized maps of the same location. Challenge students to determine the factor by which the dimensions were increased or decreased for the dilations.

3 Close

Summarize

Have the students explain what happens when a figure is stretched and what happens when a figure is shrunk.

Possible answer: A figure gets wider when stretched horizontally and taller when stretched vertically. A figure gets narrower when shrunk horizontally and shorter when shrunk vertically.

Answers to Think and Discuss

1. It stretches from side to side and becomes wider.
2. Possible answer: No; for the figures to be similar, the vertical and the horizontal dimensions would have to be multiplied by the same amount, because similar figures are the same shape but different sizes. Changing only one dimension changes the shape.

12-6 PRACTICE & ASSESS

Students may want to refer back to the lesson examples.

Assignment Guide

If you finished Example 1 assign:
Core 1–2, 5, 9, 14–18
Enriched 2, 5, 9–10, 14–18

If you finished Example 2 assign:
Core 1–9, 14–18
Enriched 5–18

You may use the graph paper on Chapter 12 Resource Book p. 68 for the exercises.

Answers

5–7. See p. A7.

12-6 Exercises

FOR EXTRA PRACTICE
see page 661

internet connect
Homework Help Online
go.hrw.com Keyword: MR4 12-6

GUIDED PRACTICE

See Example 1

Write the dimensions of each part of the figure. Stretch the figure as stated, and give the new dimensions.

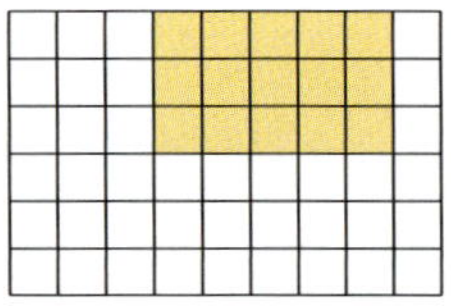

1. Increase the horizontal dimension by a factor of 3.
   vertical: 3, horizontal: 5; vertical: 3, horizontal: 15
2. Increase the vertical dimension by a factor of 10.
   vertical: 3, horizontal: 5; vertical: 30, horizontal: 5

See Example 2

Write the dimensions of each part of the figure above. Shrink the figure as stated, and give the new dimensions.

3. Decrease the horizontal dimensions by multiplying by $\frac{1}{5}$.
   vertical: 3, horizontal: 5; vertical 3, horizontal: 1
4. Decrease the vertical dimensions by multiplying by $\frac{1}{3}$.
   vertical: 3, horizontal: 5; vertical: 1, horizontal: 5

INDEPENDENT PRACTICE

See Example 1

Write the dimensions of each part of the figure. Stretch the figure as stated, and give the new dimensions.

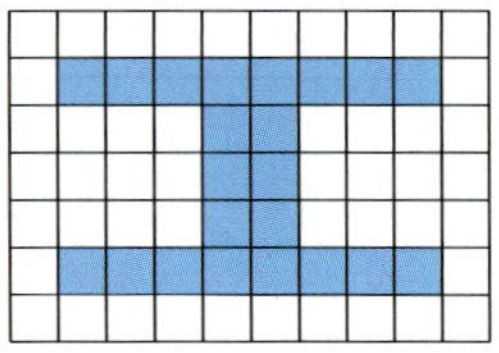

5. Increase the horizontal dimensions of the shaded region by a factor of 5.

See Example 2

Write the dimensions of each part of the figure above. Shrink the figure as stated, and give the new dimensions.

6. Decrease the vertical dimensions of the shaded region by multiplying by $\frac{1}{3}$.
7. Decrease the horizontal dimensions of the shaded region by multiplying by $\frac{1}{2}$.

PRACTICE AND PROBLEM SOLVING

8. When Craig put his basketball jersey in the dryer, it looked like the picture.
   a. What was the length of the jersey before he put it in the dryer? 36 inches
   b. When Craig took the jersey out of the dryer, he noticed that it had shrunk in length only and was now only $\frac{5}{6}$ as long as it was. Find how many inches the jersey shrank. 6 inches

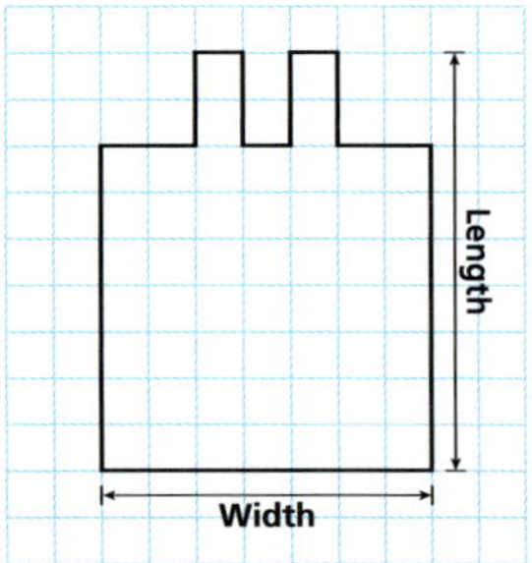

1 square = 4 in.

Math Background

Unlike the transformations discussed in the three previous lessons, stretching or shrinking does not create an image that is congruent to the original figure. When both the vertical and horizontal dimensions of a figure are stretched or shrunk by the same factor, however, the image will be similar to the original figure.

RETEACH 12-6

LESSON 12-6 **Reteach**
Stretching and Shrinking

You can stretch or shrink a figure by increasing or decreasing the size of one dimension.

Look at the figure below.

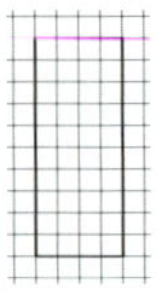

The horizontal dimension is 4 units.
The vertical dimension is 10 units

To stretch its horizontal dimension by a factor of 2, multiply that dimension by 2.

$4 \cdot 2 = 8$

So the new dimensions are, horizontal: 8 units and vertical: 10 units.

To shrink its vertical dimension by a factor of 2, multiply the dimension by $\frac{1}{2}$.

$10 \cdot \frac{1}{2} = 5$

So the new dimensions are, horizontal: 4 units and vertical: 5 units.

Stretch or shrink the figure as stated and give the new dimensions of each part.

1. Decrease the horizontal dimension by a factor of 3.

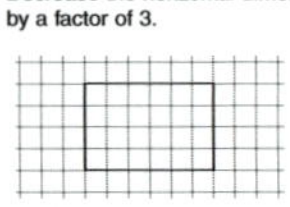

vertical: 4 units;
horizontal: 2 units

2. Increase the vertical dimension by a factor of 4.

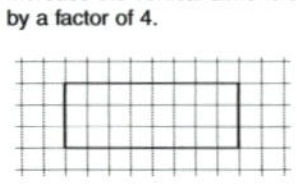

vertical: 12 units;
horizontal: 8 units

PRACTICE 12-6

LESSON 12-6 **Practice B**
Stretching and Shrinking

Write the dimensions of each part of the figure. Stretch the figure as stated and give the new dimensions of each part.

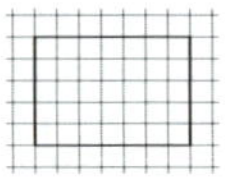

1. Original dimensions:
   horizontal: 7 squares
   and vertical: 5 squares
2. Increase the vertical dimension of the shaded region by a factor of 4.
   horizontal: 7 squares
   vertical: 20 squares
3. Increase the horizontal dimension of the shaded region by a factor of 2.
   horizontal: 14 squares
   and vertical: 5 squares

Write the dimensions of each part of the figure. Shrink the figure as stated and give the new dimensions of each part.

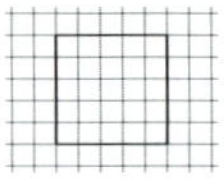

4. Original dimensions:
   horizontal: 5 squares
   and vertical: 5 squares
5. Decrease the vertical dimension by multiplying by $\frac{1}{3}$.
   horizontal: 5 squares
   and vertical: $\frac{5}{3}$ squares
6. Decrease the horizontal dimension by multiplying by $\frac{1}{5}$.
   horizontal: 1 square
   and vertical: 5 squares
7. A painting is 36 inches wide and 60 inches tall. If you decrease its vertical dimension by a factor of 12, what will the new area be?
   The area will be smaller: 180 in^2
8. If you increase its horizontal dimension by a factor of 3, what will the new perimeter be?
   The perimeter will be larger: 336 in.

Social Studies LINK

The United Nations, or U.N., is an international organization established in 1945. The U.N. now includes more than 180 nations. Samoa and Tonga are Pacific island nations that are members of the U.N. Samoa joined on December 15, 1976, and Tonga became a member on September 14, 1999.

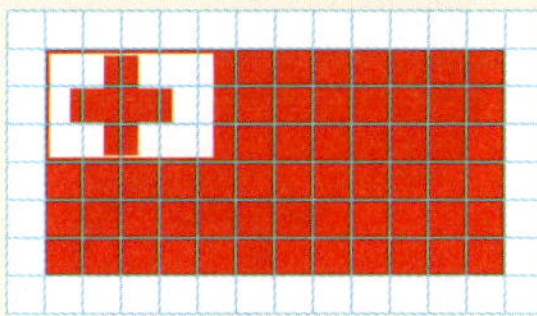

Tonga

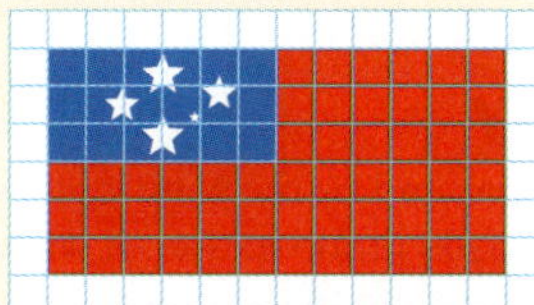

Samoa

9. a. Draw the flag of Tonga, and increase both the horizontal and vertical dimensions by a factor of 4.
   b. What is the perimeter of the cross on the Tongan flag before and after the change? **12 units; 48 units**

10. The horizontal dimension of the Tongan flag was increased by a factor of 4, and the vertical dimension was increased by a factor of 2. Is the cross on the new flag similar to the cross in the original flag? Explain.

11. ***CHOOSE A STRATEGY*** Which of the following would give the Samoan flag an area of 36 square units? **C**
    - **A** vertical increase by a factor of 2
    - **B** horizontal decrease by a factor of 3
    - **C** vertical decrease by a factor of 2

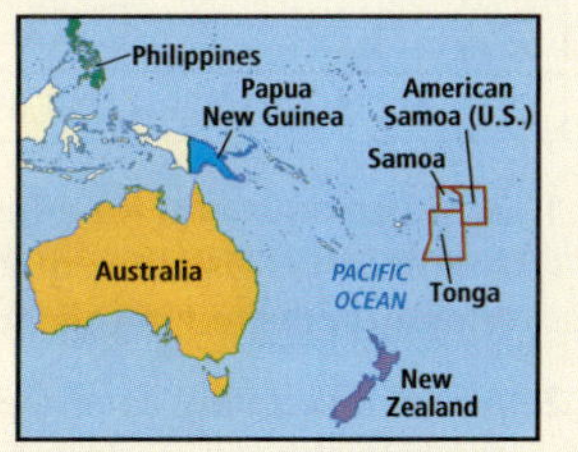

12. ***WRITE ABOUT IT*** Explain which dimensions of the Tongan flag you could change so that the new flag from Exercise 10 is similar to the original flag.

13. ***CHALLENGE*** A Samoan flag was made with a perimeter of 180 units. By what whole-number factor were the horizontal and vertical dimensions increased? **The dimensions were increased by a factor of 5.**

Spiral Review

Determine whether each number is divisible by 2, 3, or 5. (Lesson 4-1)

14. 155 **5** **15.** 14 **2** **16.** 99 **3** **17.** 2,345 **5**

18. **TEST PREP** A circle has a diameter of 12 in. What is the circumference of the circle rounded to the nearest hundredth? (Use 3.14 for π.) (Lesson 10-5) **B**

A 15.14 in^2 **B** 37.68 in^2 **C** 113.04 in^2 **D** 150.72 in^2

Answers

9.a. See pp. A7.

10. No. The horizontal dimensions and the vertical dimensions were not increased by the same factor.

12. Possible answer: Multiply the horizontal dimension by $\frac{1}{2}$ or multiply the vertical dimension by 2.

Journal

Have students draw a figure on a grid. Tell students to stretch the figure vertically by a factor of 4, drawing the new figure. Have students shrink the original figure horizontally by a factor of $\frac{1}{3}$, drawing the new figure. Ask students to list the vertical and horizontal dimensions of all three figures.

Test Prep Doctor

For Exercise 18, students need to know the formula for circumference of a circle: $C = \pi d$. An estimate for the answer is 36. Eliminate answer **A** because it is too small. Eliminate answers **C** and **D** because they are too large. The correct answer is **B.**

Lesson Quiz

Determine the vertical and horizontal dimensions of the figure. Stretch the figure as stated and give the new dimensions.

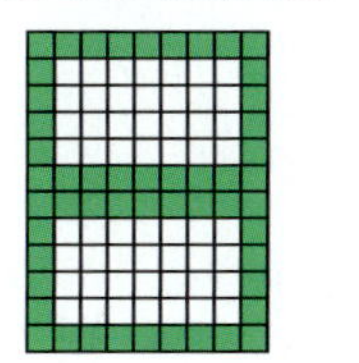

1. Increase the vertical dimensions of the shaded region by a factor of 4. vertical 12→48; horizontal 9→no change

Write the dimensions of the figure above. Shrink the figure as stated and give the new dimensions.

2. Decrease the horizontal dimensions of the shaded region by multiplying by $\frac{1}{3}$. vertical 12→no change; horizontal 9→3

Available on Daily Transparency in CRB

CHALLENGE 12-6

LESSON 12-6 Challenge
Dilations

When you enlarge or reduce a figure to produce a similar figure, the transformation is called a dilation. Just like other transformations, you can graph dilations. When a figure is dilated with its center at the origin and a scale factor k, the result for each point of the figure is $(a, b) \longrightarrow (ka, kb)$.

Graph and label each figure after the given dilation with its center at the origin. Next to each dilation, write whether it is an enlargement or a reduction.

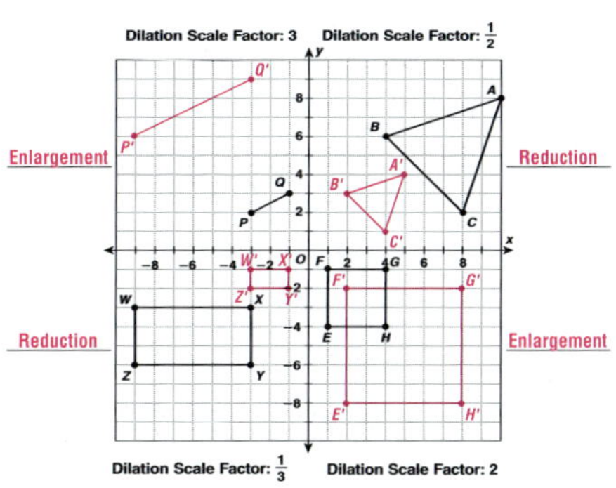

PROBLEM SOLVING 12-6

LESSON 12-6 Problem Solving
Stretching and Shrinking

Write the correct answer.

1. An NBA basketball court is 94 feet long and 50 feet wide. When there are not enough players, people sometimes play half-court basketball, using only one basket. How do they change the dimensions of the playing court? What are its new dimensions?

   Its length is decreased by multiplying by $\frac{1}{2}$; It is 47 feet long and 50 feet wide.

2. Gold is one of the most malleable, or changeable, elements on Earth. A one-ounce square of gold can be pounded into a thin wire that stretches 60 miles! To accomplish this stretch, how do the dimensions of the gold square have to change? Explain.

   One dimension has to decrease for the thinness of the wire, and the other has to increase for the length of the wire.

3. Photo researchers review thumbnail, or reduced, photos to choose which ones to buy. One thumbnail photo is 2 inches tall and $1\frac{1}{3}$ inches wide. The actual photo is similar to the thumbnail, but its dimensions are increased by a factor of 3. What are the dimensions of the actual photo?

   6 inches tall and 4 inches wide

4. If you change all the dimensions of a figure by a factor greater than 1, how is the figure changed? By a factor less than 1? By a factor of 1?

   If the factor is greater than 1, the figure is enlarged; less than 1: the figure is reduced; factor of 1: the figure does not change.

Circle the letter of the correct answer.

5. A billboard is 20 feet wide and 5 feet tall. Which of the following would give the billboard an area of 300 square feet?
   - **A** Horizontal increase by a factor of 10.
   - **(B)** Vertical increase by a factor of 3.
   - **C** Horizontal increase by a factor of 2.
   - **D** Vertical increase by a factor of 6.

6. The perimeter of a rectangle is 14 feet. After one of its dimensions is decreased by multiplying it by $\frac{1}{2}$, its perimeter is 10 feet. Which of the following could be the original dimensions of the rectangle?
   - **F** $\ell = 6$ feet, $w = 1$ foot
   - **G** $\ell = 3.5$ feet, $w = 3.5$ feet
   - **H** $\ell = 5$ feet, $w = 2$ feet
   - **(J)** $\ell = 4$ feet, $w = 3$ feet

Problem Solving on Location

New Jersey

Purpose: *To provide additional practice for problem-solving skills in Chapters 1–12*

Garden State Parkway

- After problem 1, have students identify what each variable in their equation stands for. Possible answer: t is the total amount of money collected in tolls; c is the number of cars. Which variable is a function of the other? The total amount of money collected is a function of the number of cars.
- After problem 2, have students discuss how the answer can be found either by using the equation from problem 1 or by using a ratio. Use the equation $t = 0.35c$ with $t = \$1{,}000$. Solve for c to get 2,857. Round to 3,000. About \$1 in tolls is collected for every 3 cars, so the proportion $\frac{\$1}{3\text{ cars}} = \frac{\$1{,}000}{c\text{ cars}}$ can be used.

Extension Have students write equations to find the amount of money m going to maintenance and improvement for a number of cars c and the amount of money m going to maintenance and improvement for a given amount of toll money t collected. Then have them write questions based on the equations and trade papers to solve. $m = \frac{7}{35}t$; $m = 0.07c$; Check students' work.

Problem Solving on Location

NEW JERSEY

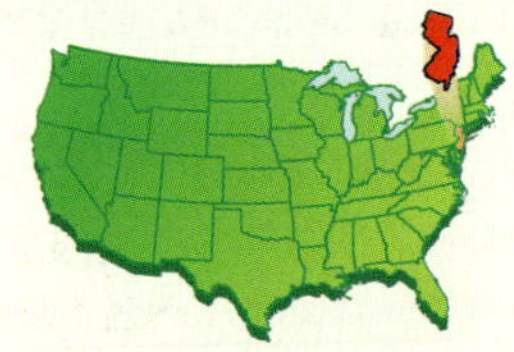

Garden State Parkway

The Garden State Parkway, a toll road operated by the New Jersey Highway Authority, is one of the safest roads in the nation. It was built in 1952 and runs from one end of the state to the other—from the New York state line to Cape May. Drivers must pay a 35¢ toll to drive on the Garden State Parkway. The New Jersey Highway Authority designates 7¢ from each toll to go to maintenance and improvement of the road.

The table shows the relationship between the number of cars c and the amount of money collected from tolls t.

For 1–3, use the table.

c	1	2	3	4	5
t	\$0.35	\$0.70	\$1.05	\$1.40	\$1.75

1. Write an equation for a function that gives the values in this table. Use the equation to find the amount of money collected in tolls from 7 cars. $t = 0.35c$; \$2.45
2. About how many cars need to pass through the toll stations for the New Jersey Highway Authority to collect \$1,000? C

   A 30 **C** 3,000
   B 300 **D** 30,000
3. The New Jersey Highway Authority estimates that it costs about 2.2 cents per mile to drive on the Garden State Parkway. Write an equation for a function that gives the values in the table below. Use the equation to find how much it costs to travel 15 miles on the Parkway. $t = 2.2m$; \$0.33

m	1	2	3	4	5	6
t	2.2¢	4.4¢	6.6¢	8.8¢	11¢	13.2¢

Birds and Butterflies in Cape May

When summer is ending, the migration of birds and butterflies to the New Jersey shore is just beginning. Hundreds of thousands of birds and butterflies pass through Cape May as they journey to warmer climates. Early October is a great time to spot butterflies, when there are about 100 different butterfly species around the wetlands of Cape May.

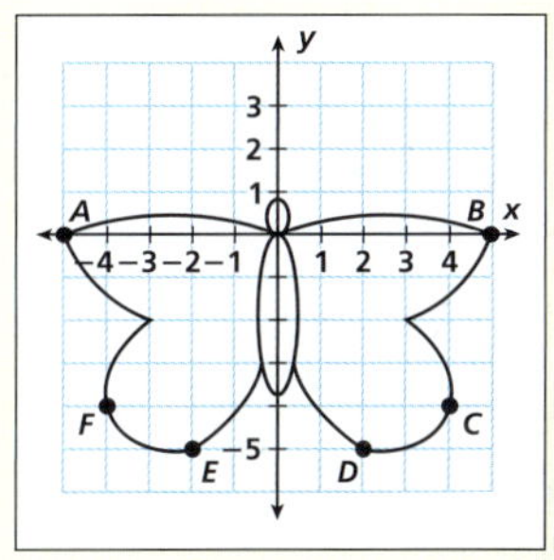

For 1–3, use the graph of the butterfly.

1. Which axis appears to be a line of symmetry? **the *y*-axis**
2. Reflect points *A, B, C, D, E,* and *F* across the *x*-axis. Give the new coordinates of the points. **$A(-5, 0)$, $B(5, 0)$, $C(4, 4)$, $D(2, 5)$, $E(-2, 5)$, $F(-4, 4)$**
3. Draw a similar, larger butterfly by increasing the vertical and horizontal dimensions by a factor of 2.
4. The osprey population is declining in New Jersey. Volunteers have built nesting platforms to help increase the survival rate of osprey chicks. The platforms are 2 feet by 3 feet. What is the area of the platforms? **6 ft^2**
5. New Jersey is also home to thousands of wild turkeys that flourish in the woodlands. A bird-watcher recorded the number of eggs she saw in different nests. Is the relationship of the number of eggs to nests a function? Explain.

Number of Nests	1	2	3	4	5
Number of Eggs	9	12	15	16	1

6. Let *x* be the number of nests and *y* be the number of eggs. Graph the ordered pairs given in the table. Do you think it makes sense to connect the points on the graph? Explain.

Birds and Butterflies in Cape May

- After problem 1, have students explain how to determine which axis is a line of symmetry. A line of symmetry is one that, if each point in the figure is reflected over it, the transformed figure would lie exactly where the original figure lay before the transformation.
- After problem 5, have students consider what happens to the image of the bird if each dimension is transformed by a factor of −12. The image will stretch by a factor of 12 and will be rotated 180° about the origin.

Extension Have students draw a simple image and identify the ordered pairs used. Next, have them transform the image by multiplying each coordinate by a factor of −3. Check students' work.

Answers

3.

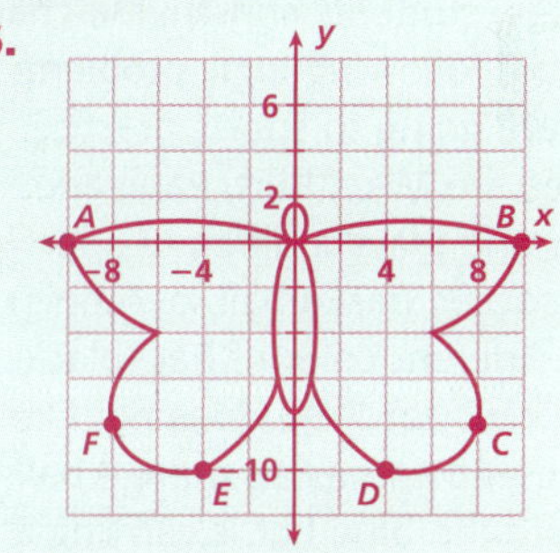

5. The relationship is a function. Each input value has a unique output value.

6. It does not make sense to connect the points on the graph because there can be only a whole number of eggs and nests.

MATH-ABLES

Game Resources

Puzzles, Twisters & Teasers
Chapter 12 Resource Book

Logic Puzzle

Purpose: *To apply problem-solving skills to a logic puzzle*

Discuss: Ask students to explain how the chart works. How can you use the chart to solve the puzzle? Use each clue to place *O*'s and *X*'s in the appropriate squares. When a row or column of squares has 4 *X*'s, the fifth square must contain an *O*. Clues sometimes give information in a way that is not straightforward. What can you conclude about Angela's choice of restaurant, using clues 1 and 3? Angela did not choose tacos, because she did not participate on Friday.

Extend: Have students construct a chart and solve the following logic problem: Four boys are going to sports practice. Their names are Mike, Kurt, Paul, and Andy. Each one plays a different sport (baseball, soccer, basketball, or tennis) and has a different color of hat (black, red, blue, or white). Paul does not like to wear red and carries an orange ball. The boy with the blue hat is carrying a racket. Kurt does not play soccer. Mike uses a bat and wears a black hat. The soccer player wears a red hat.

Mike: baseball, black
Kurt: tennis, blue
Paul: basketball, white
Andy: soccer, red
Check students' charts.

MATH-ABLES

Logic Puzzle

Each day from Monday through Friday, Mayuri, Naomi, Brett, Thomas, and Angela took turns picking a restaurant for lunch. They ate at restaurants that serve either Chinese food, hamburgers, pizza, seafood, or tacos. Use the clues below to determine which student picked the restaurant on each day and which restaurant the student picked.

1. Angela skipped Friday's lunch to play in a basketball game.
2. Brett picked the restaurant on Wednesday.
3. The students ate tacos on Friday.
4. Naomi is allergic to seafood and volunteered to pick the first restaurant.
5. Thomas picked a hamburger restaurant on the day before another student chose a pizza restaurant.

You can use a chart like the one below to help you solve this puzzle. Place an *O* in a square for something that is true and an *X* in a square for something that cannot be true. Remember that when you place an *O* in a square, you can put *X*'s in the rest of the squares in that row and column. The information from the first two clues has been entered for you.

		Student					Restaurant				
		Mayuri	Naomi	Brett	Thomas	Angela	Seafood	Pizza	Hamburger	Chinese	Tacos
Day	Monday	X	O	X	X	X	X	X	X	O	X
	Tuesday	X	X	X	O	X	X	X	O	X	X
	Wednesday	X	X	O	X	X	X	O	X	X	X
	Thursday	X	X	X	X	O	O	X	X	X	X
	Friday	O	X	X	X	X	X	X	X	X	O
Restaurant	Seafood	X	X	X	X	O					
	Pizza	X	X	O	X	X					
	Hamburgers	X	X	X	O	X					
	Chinese	X	O	X	X	X					
	Tacos	O	X	X	X	X					

internet connect
Go to ***go.hrw.com*** for a blank logic puzzle chart.
KEYWORD: MR4 Game12

Answers

Mayuri: tacos, Friday
Naomi: Chinese, Monday
Brett: pizza, Wednesday
Thomas: hamburgers, Tuesday
Angela: seafood, Thursday

Technology LAB: Use Graphs to Estimate Solutions

internet connect
Lab Resources Online
go.hrw.com
KEYWORD: MR4 TechLab12

You can use graphs to estimate solutions to equations.

Activity

1 Use a graphing calculator to estimate the solution to the equation $x - 3 = 4$.

a. Press [Y=] and enter $x - 3$ for **Y1** and 4 for **Y2.** These are the left and right sides of the equation $x - 3 = 4$.
Press [ZOOM] 6 to select **ZStandard.** This sets the view of the x-axis and y-axis from –10 to 10.

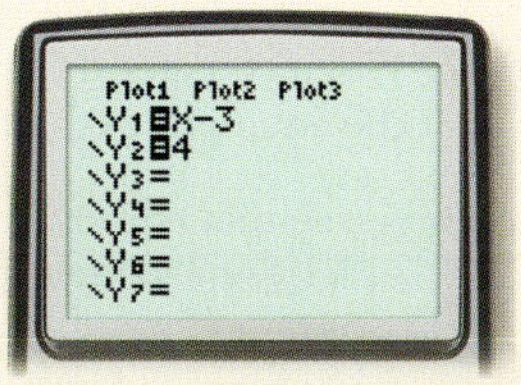

b. There is one graph of a line representing $y = x - 3$ and a second graph of a line representing $y = 4$.

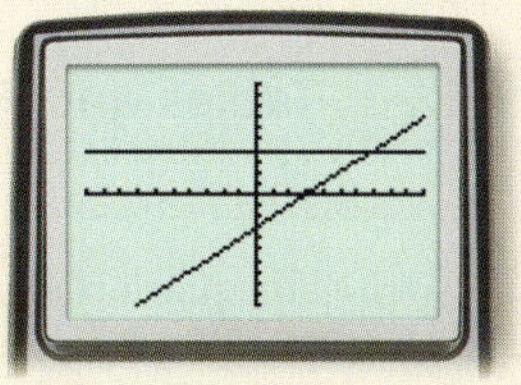

c. The expression $x - 3$ and the number 4 have equal values at the point where their graphs intersect.

To find the coordinates of the point of intersection, press [2nd] [TRACE] (CALC) 5. A flashing cursor appears. Use the arrow keys to move the cursor near the intersection and press [ENTER]. Do this for both graphs.

At the bottom of the window, a guess is shown for the value. Press [ENTER] again to see the coordinates of the point.

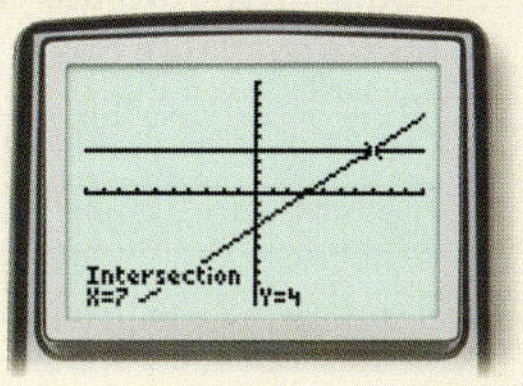

The point of intersection is (7, 4).

Think and Discuss

1. Tell how you would solve $2x + 5 = 6x - 3$ by using a graphing calculator.
Enter [Y=] 2 [X,T,θ,n] [+] 5 [ENTER] 6 [X,T,θ,n] [–] 3 [ZOOM] 6. Find the point of intersection (2, 9). Check this solution by substituting 2 for x into the equation to see if both sides are equal.

Try This

Use a graphing calculator to estimate the solution to each equation.

1. $x - 5 = 2$ $x = 7$
2. $x + 3 = -3$ $x = -6$
3. $3\frac{1}{2}x = 7\frac{1}{4}$ $x = 2\frac{1}{14}$
4. $x - 1.75 = 6.35$ $x = 8.1$

Technology LAB: Use Graphs to Estimate Solutions

Objective: To use a graphing calculator to estimate solutions

Materials: Graphing calculator

Lab Resources

Technology Lab Activities .... p. 54

Using the Page

This technology activity shows students how to estimate solutions using graphs, which can be done on any graphing calculator. Specific keystrokes may vary, depending on the make and model of the graphing calculator used. The keystrokes given are for a TI-83 model. For keystrokes to other models, visit go.hrw.com.

The Think and Discuss problem can be used to assess students' understanding of the technology activity. Although Try This problems 1–4 can be done without a graphing calculator, they are meant to provide students with experiences that help them become familiar with how to estimate solutions using a graphing calculator.

Assessment

1. What should be entered in Y1 and Y2 to estimate the solution to the equation $x + 4 = -4$?
Y1 = $x + 4$
Y2 = -4
2. Use the graphing calculator to estimate the solution to the equation $x - 3 = -2x + 3$. $x = 2$
3. How can you check the solution you found in **2**? Substitute 2 for x and check that the equation is true.
$2 - 3 = -2(2) + 3$; $-1 = -1$

Chapter 12 Study Guide and Review

Purpose: To help students review and practice concepts and skills presented in Chapter 12

Assessment Resources

Chapter Review

Chapter 12 Resource Book . . . pp. 62–64

***Test and Practice Generator* CD-ROM**

Additional review items in both multiple-choice and free-response format may be generated for any objective in Chapter 12.

You may use graph paper on Chapter 12 Resource Book p. 68 for items 13–19.

Answers

1. output, input
2. linear equation
3. function
4. $y = 2x + 2$; $y = 18$ when $x = 8$
5. $y = x \div 2 + 1$; $y = 4$ when $x = 6$
6. $y = 5x - 2$; $y = 53$ when $x = 11$
7. $\ell = 4w$; w = width, ℓ = length

Chapter 12 Study Guide and Review

Study Guide and Review

Vocabulary

function 598
input 598
linear equation 605
output 598

Complete the sentences below with vocabulary words from the list above. Words may be used more than once.

1. For the equation $y = 3x$, the __?__ is 12 when the __?__ is 4.
2. When you graph the ordered pairs of a function and a straight line is formed, the equation of the function is called a __?__.
3. A rule that relates two quantities so that each value of x corresponds with exactly one value of y is called a __?__.

12-1 Tables and Functions (pp. 598–601)

EXAMPLE

- **Write an equation for a function that gives the values in the table. Use the equation to find the value of y for the indicated value of x.**

x	2	3	4	5	6	12
y	5	8	11	14	17	■

y is 3 times x minus 1. *Compare x and y to find a pattern.*

$y = 3x - 1$ *Use the pattern to write an equation.*

$y = 3(12) - 1$ *Substitute 12 for x.*

$y = 36 - 1$

$y = 35$ *Use your function rule to find y when x = 12.*

When x is 12, y is 35.

EXERCISES

Write an equation for a function that gives the values in each table. Use the equation to find the value of y for each indicated value of x.

4.

x	2	3	4	5	6	8
y	6	8	10	12	14	■

5.

x	20	18	16	14	12	6
y	11	10	9	8	7	■

6.

x	1	3	5	7	9	11
y	3	13	23	33	43	■

Write an equation to describe the function. Tell what each variable you use represents.

7. The length of a rectangle is 4 times its width.

12-2 Graphing Functions (pp. 604–607)

EXAMPLE

- **Graph the function described by the equation $y = 3x + 4$.**

Make a function table.

Write the solutions as ordered pairs.

x	$3x + 4$	y	(x, y)
−2	3(−2) + 4	−2	(−2, −2)
−1	3(−1) + 4	1	(−1, 1)
0	3(0) + 4	4	(0, 4)

Graph the ordered pairs on a coordinate plane.

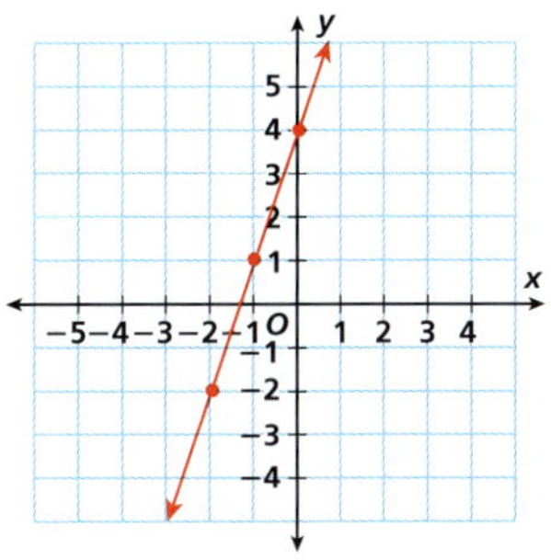

EXERCISES

Use the given x-values to write solutions of each equation as ordered pairs.

8. $y = 2x - 5$ for $x = 1, 2, 3, 4$
9. $y = x + 7$ for $x = 1, 2, 3, 4$

Determine whether each ordered pair is a solution to the given equation.

10. (3, 12); $y = 5x - 3$
11. (6, 14); $y = 2x + 3$
12. Use the graph of the linear function to find the value of y when x is 2.

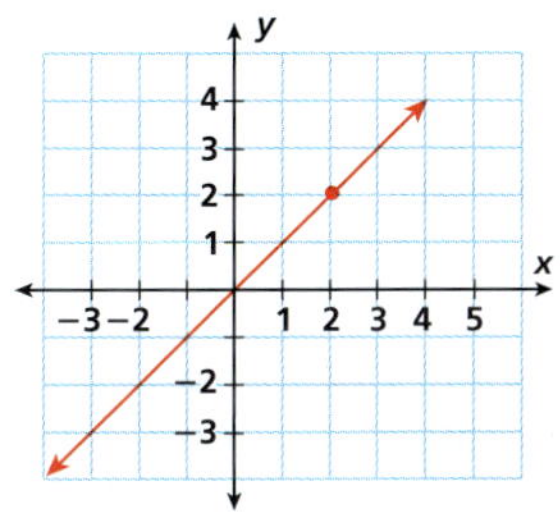

13. Graph the function described by the equation $y = 2x + 1$.

12-3 Graphing Translations (pp. 610–612)

EXAMPLE

- **Give the coordinates of the figure after the given translation.**

Translate triangle RST 3 units left and 1 unit up.

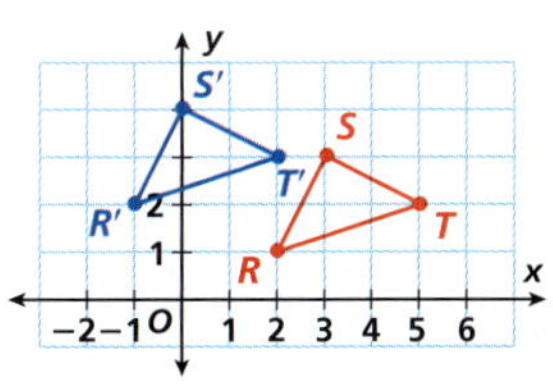

$R(2, 1) \rightarrow R'(2 - 3, 1 + 1) \rightarrow R'(-1, 2)$
$S(3, 3) \rightarrow S'(3 - 3, 3 + 1) \rightarrow S'(0, 4)$
$T(5, 2) \rightarrow T'(5 - 3, 2 + 1) \rightarrow T'(2, 3)$

EXERCISES

Give the coordinates of the figure after the given translation.

14. Translate square $ABCD$ 3 units right and 2 units down.
15. Translate triangle FGH 1 unit left and 4 units up.
16. Translate triangle FGH 2 units right and 1 unit down.

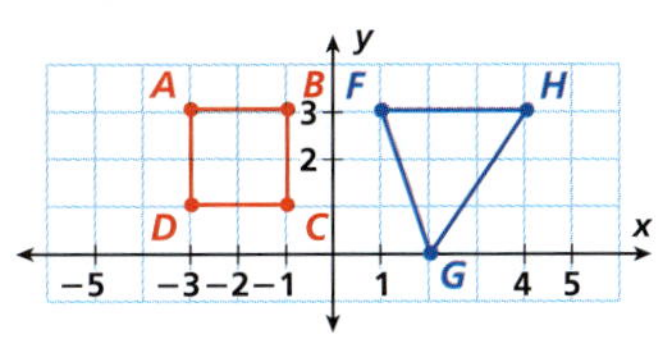

Answers

8. (1, −3); (2, −1); (3, 1); (4, 3)
9. (1, 8); (2, 9); (3, 10); (4, 11)
10. yes
11. no
12. $y = 2$
13.

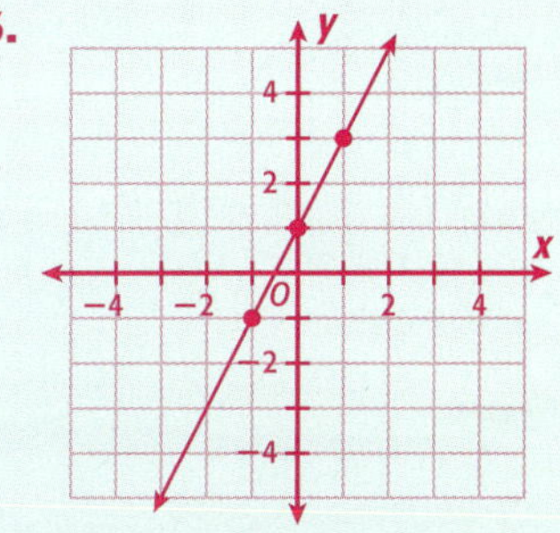

14. $A'(0, 1)$, $B'(2, 1)$, $C'(2, -1)$ $D'(0, -1)$
15. $F'(0, 7)$, $G'(1, 4)$, $H'(3, 7)$
16. $F'(3, 2)$, $G'(4, -1)$, $H'(6, 2)$

Answers

17. $A'(-1,-2)$, $B'(-2,-1)$, $C'(-5,-3)$, $D'(-4,-4)$

18. $A'(1,-1)$, $B'(3,-2)$, $C'(3,-5)$, $D'(1,-4)$

19. original dimensions: vertical: 6, horizontal: 9; new dimensions: vertical: 12, horizontal: 9

12-4 Graphing Reflections (pp. 613–615)

EXAMPLE

- **Give the coordinates of the vertices of the figure after the given reflection.**

  Reflect figure *LMNO* across the *y*-axis.

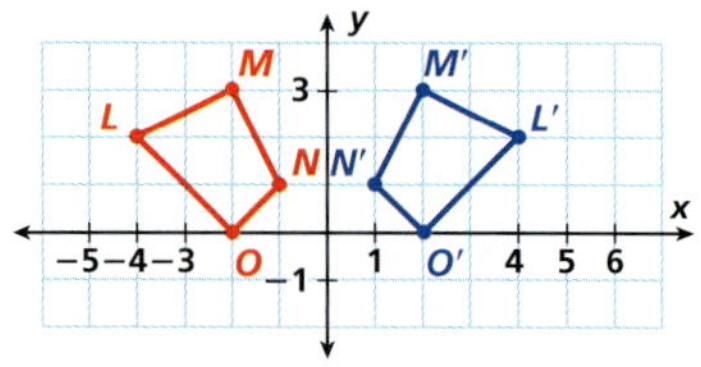

$L(-4, 2) \rightarrow L'(4, 2)$ $N(-1, 1) \rightarrow N'(1, 1)$
$M(-2, 3) \rightarrow M'(2, 3)$ $O(-2, 0) \rightarrow O'(2, 0)$

EXERCISES

Give the coordinates of the vertices of the figure after the given reflection.

17. Reflect parallelogram *ABCD* across the *y*-axis.

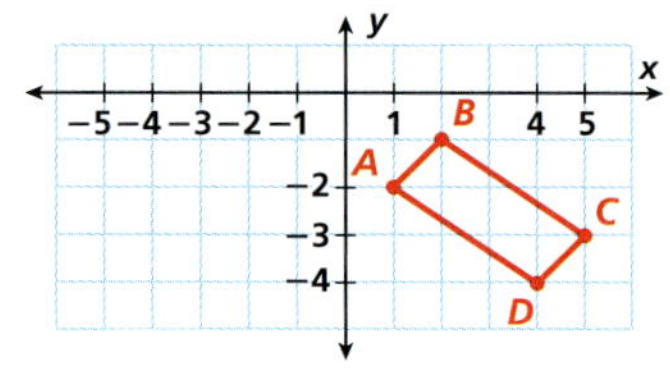

12-5 Graphing Rotations (pp. 616–619)

EXAMPLE

- **Give the coordinates of the figure after the given rotation.**

  Rotate triangle *ABC* clockwise 90° about the origin.

  $A(2, 1) \rightarrow A'(1, -2)$
  $B(2, 3) \rightarrow B'(3, -2)$
  $C(4, 1) \rightarrow C'(1, -4)$

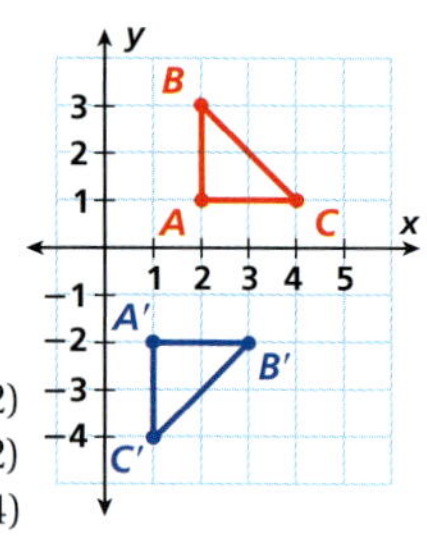

EXERCISES

Give the coordinates of the figure after the given rotation.

18. Rotate parallelogram *ABCD* clockwise 90° about the origin.

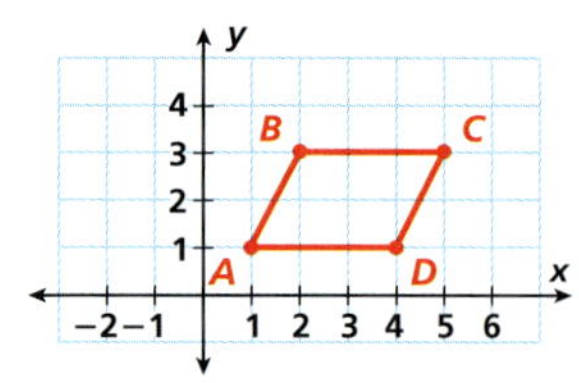

12-6 Stretching and Shrinking (pp. 620–623)

EXAMPLE

- **Write the dimensions of the figure. Stretch the figure horizontally by a factor of 3.**

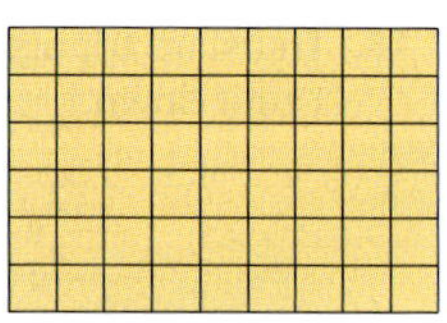

horizontally: 9 squares → 27 squares
vertically: 6 squares → 6 squares

EXERCISES

Write the dimensions of each part of the figure. Stretch or shrink the figure as stated, and give the new dimensions.

19. Increase the vertical dimensions of the figure in the example by a factor of 2.

Chapter 12 Chapter Test

Write an equation for a function that gives the values in each table. Use the equation to find the value of y for each indicated value of x.

1.

x	2	3	4	5	6	7
y		8	11	14	17	20

$y = 3x - 1$; 5

2.

x	1	2	3	4	5	9
y	8	10	12	14	16	

$y = 2x + 6$; 24

Write an equation to describe the function. Tell what each variable you use represents.

3. The number of buttons on the jacket is 4 more than the number of zippers. $b = z + 4$; b = buttons, z = zippers

4. The length of a parallelogram is 2 in. more than twice the height. $l = 2h + 2$; l = length, h = height

5. The number of cards is 6 less than the number of envelopes. $c = e - 6$; c = cards, e = envelopes

6. The width of the rectangle is 4 cm less than the length. $w = l - 4$; l = length, w = width

Use the given x-values to write solutions of each equation as ordered pairs. Then graph the equation.

7. $y = 5x - 3$ for $x = 1, 2, 3, 4$ (1, 2), (2, 7), (3, 12), (4, 17)

8. $y = 2x - 3$ for $x = 0, 1, 2, 3$ (0, −3), (1, −1), (2, 1), (3, 3)

Determine whether each ordered pair is a solution of the given equation.

9. (2, 5); $y = 3x - 1$ yes

10. (0, 6); $y = 6x$ no

11. (−3, −5); $y = 2x + 1$ yes

Use the graph of the linear function to find the value of y for each indicated value of x.

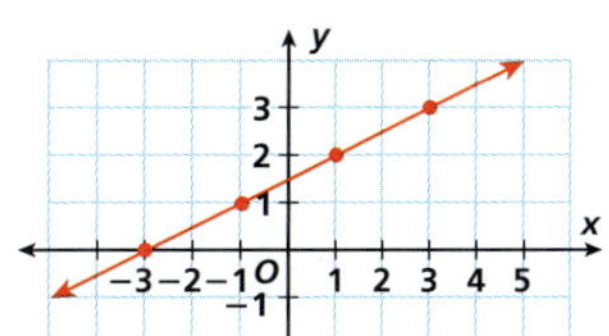

12. $x = 1$ $y = 2$

13. $x = -1$ $y = 1$

14. $x = 3$ $y = 3$

15. $x = -3$ $y = 0$

Give the coordinates of the vertices of each figure after the given transformation.

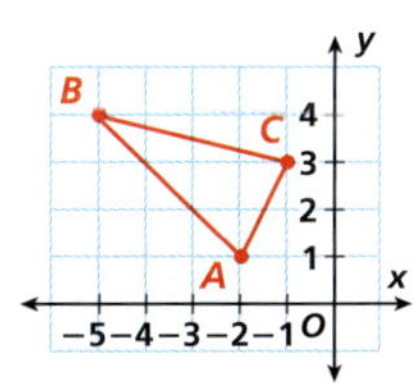

16. Translate triangle ABC 3 units right and 2 units down. $A'(1, -1)$, $B'(-2, 2)$, $C'(2, 1)$

17. Reflect triangle ABC across the y-axis. $A'(2, 1)$, $B'(5, 4)$, $C'(1, 3)$

18. Rotate triangle ABC clockwise 90° about the origin. $A'(1, 2)$, $B'(4, 5)$, $C'(3, 1)$

Write the dimensions of each part of the figure. Stretch or shrink the figure as stated, and give the new dimensions.

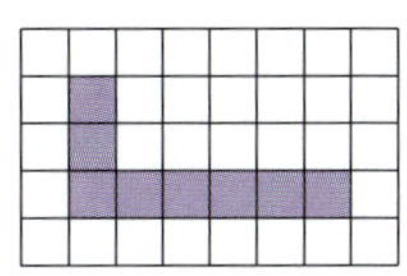

19. Increase the horizontal dimensions by a factor of 2.

20. Decrease the vertical dimensions by multiplying by $\frac{1}{3}$.

Chapter 12 Chapter Test

Purpose: *To assess students' mastery of concepts and skills in Chapter 12*

Assessment Resources

Chapter 12 Tests (Levels A, B, C)
Assessment Resources .... pp. 97–102

***Test and Practice Generator* CD-ROM**

Additional assessment items in both multiple-choice and free-response format may be generated for any objective in Chapter 12.

You may use graph paper on Chapter 12 Resource Book p. 68 for items 7–8.

Answers

7. (1, 2), (2, 7), (3, 12), (4, 17);

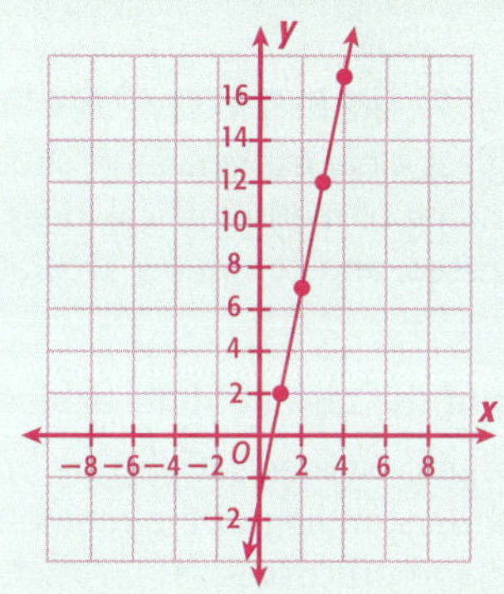

8. (0, −3), (1, −1), (2, 1), (3, 3);

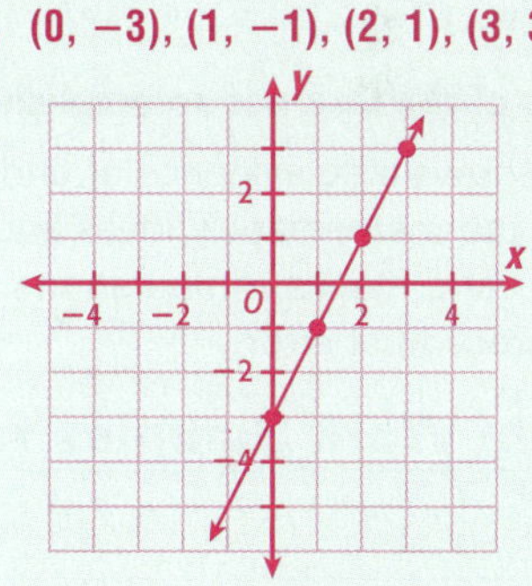

19. vertical: 3, horizontal: 6; vertical: 3, horizontal: 12

20. vertical: 3, horizontal: 6; vertical: 1, horizontal: 6

Chapter 12 Performance Assessment

Purpose: To assess students' understanding of concepts in Chapter 12 and combined problem-solving skills

Assessment Resources

Performance Assessment
Assessment Resources . . . . p. 126

Performance Assessment Teacher Support
Assessment Resources . . . . p. 125

Answers

1–2. See p. A7.

3. See Level 3 work sample below.

Scoring Rubric for Problem Solving Item 3

Level 3
Accomplishes the purposes of the task.
Student gives clear explanations, shows understanding of mathematical ideas and processes, and computes accurately.

Level 2
Purposes of the task not fully achieved.
Student demonstrates satisfactory but limited understanding of the mathematical ideas and processes.

Level 1
Purposes of the task not accomplished.
Student shows little evidence of understanding the mathematical ideas and processes and makes computational and/or procedural errors.

Chapter 12 Performance Assessment

Show What You Know

Create a portfolio of your work from this chapter. Complete this page and include it with your four best pieces of work from Chapter 12. Choose from your homework or lab assignments, mid-chapter quiz, or any journal entries you have done. Put them together using any design you want. Make your portfolio represent what you consider your best work.

Short Response

1. Draw quadrilateral *ABCD* on a coordinate plane with *A*(1,2), *B*(1,5), *C*(5,5), and *D*(5,2). What is the area of the quadrilateral? Describe two different methods you could use to find the answer.

2. On a coordinate plane, draw quadrilateral *EFGH* with *E*(1,2), *F*(2,5), *G*(4,4), and *H*(3,1). Reflect the figure across the *x*-axis. Explain how to find the coordinates of the vertices of the figure after the reflection.

Extended Problem Solving

3. A store sold 52 folk-art masks in September for a total price of $624. In October, sales totaled $492 for 41 masks. In November, sales totaled $456 for 38 masks. All of the masks cost the same.

a. Make a table to display the data, and then graph the data. Is the function linear?

b. Write an equation to represent the function. Indicate what each variable represents.

c. In December, the store sold 67 masks. What was the total sales amount in December?

Student Work Samples for Item 3

Level 3

A.

masks	$
38	$456
41	$492
52	$624

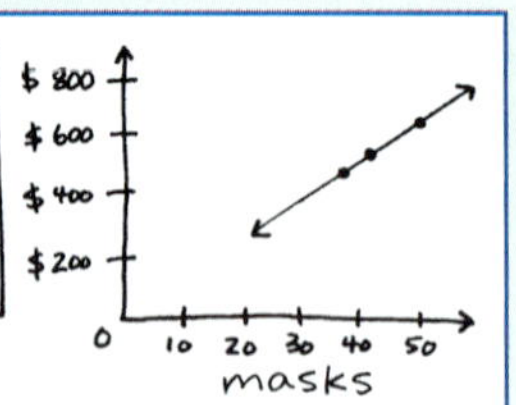

yes

B. m = masks d = dollars
d = 12m

C. d = 12(67)
d = $804

The student made a correct table, graph, and equation, and answered all parts completely and correctly.

Level 2

a.

masks	38	41	52
$	$456	$492	$624

b. x = masks y = dollars
x = 12y

c. 67 = 12y
y = $5.58

The student gave an incomplete answer to part **a** and wrote an incorrect equation in part **b.**

Level 1

a. yes

b. y = x y = masks

c. $670

The student did not show understanding of linear equations or graphs.

internet connect
State-Specific Test Practice Online
go.hrw.com Keyword: MR4 TestPrep

Standardized Test Prep

Chapter 12

Cumulative Assessment, Chapters 1–12

1. Which ordered pair is a solution of the equation $y = 3x + 2$? C

   Ⓐ (1, 6) Ⓒ (2, 8)
   Ⓑ (5, 1) Ⓓ (−2, 4)

2. What is 15% of 130? F

   Ⓕ 19.5 Ⓗ 1.35
   Ⓖ 13 Ⓙ 1.95

TIP! TEST TAKING TIP!
To reflect a figure across the *y*-axis, write the opposites of the *x*-coordinates. The *y*-coordinates do not change.

3. A figure has the following vertices: $A(1, -2)$, $B(2, -5)$, $C(6, -3)$. What are the coordinates after the figure is reflected across the *y*-axis? D

   Ⓐ $A'(1, 2)$, $B'(2, 5)$, $C'(6, 3)$
   Ⓑ $A'(-2, -1)$, $B'(-5, -2)$, $C'(-3, -6)$
   Ⓒ $A'(-1, 2)$, $B'(-2, 5)$, $C'(-6, 3)$
   Ⓓ $A'(-1, -2)$, $B'(-2, -5)$, $C'(-6, -3)$

4. What kind of transformation is shown? F

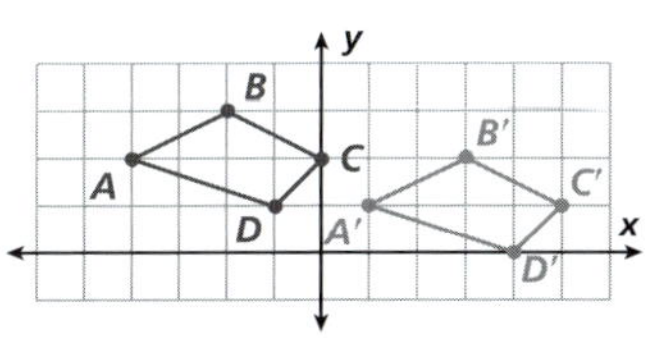

   Ⓕ Translation Ⓗ Reflection
   Ⓖ Rotation Ⓙ Tessellation

5. What is the probability of rolling a number less than 3 on a fair number cube? B

   Ⓐ $\frac{1}{2}$ Ⓒ $\frac{2}{3}$
   Ⓑ $\frac{1}{3}$ Ⓓ $\frac{1}{6}$

6. Multiply $3\frac{5}{6} \cdot 1\frac{2}{3}$. H

   Ⓕ $3\frac{5}{9}$ Ⓗ $6\frac{7}{18}$
   Ⓖ $5\frac{1}{2}$ Ⓙ $4\frac{7}{9}$

7. Which set is in order from least to greatest? C

   Ⓐ 4, 2, −1, 0 Ⓒ −3, −1, 2, 4
   Ⓑ 0, −1, 3, 5 Ⓓ −3, −5, 1, 4

8. Which of the following numbers is divisible by 2, 3, 4, and 6, but **not** by 9? J

   Ⓕ 216 Ⓗ 822
   Ⓖ 414 Ⓙ 912

9. ***SHORT RESPONSE*** A cylinder has a diameter of 6 in. and a height of 8 in. What is the cylinder's volume? Use 3.14 for π, and round to the nearest hundredth. Show your work.

10. ***SHORT RESPONSE*** Explain how to find the coordinates of triangle *ABC* after it is rotated clockwise 90° about the origin.

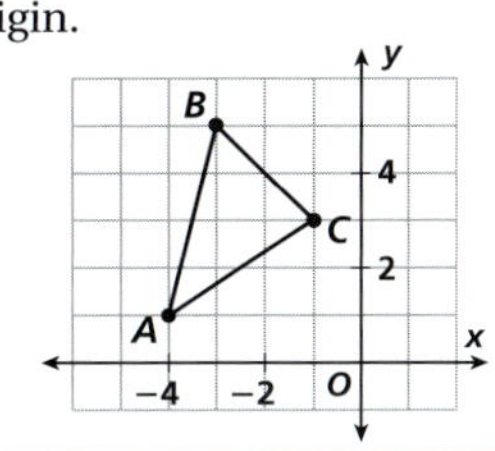

Standardized Test Prep

Standardized Test Prep

Chapter 12

Purpose: *To provide review and practice for Chapters 1–12 and standardized tests*

Assessment Resources

Cumulative Tests (Levels A, B, C)
Assessment Resources. . . . pp. 259–270

State-Specific Test Practice Online
KEYWORD: MR4 TestPrep

Test Prep Doctor

For item 9, remind students to use the divisibility rules rather than doing the actual division. Eliminate **A** because it is divisible by 2, 3, 4, 6, and 9. Eliminate **B** and **C** because they are not divisible by 4.

Answers

9. $V = \pi r^2 h$
$= (3.14)\left(\frac{6}{2}\right)^2(8)$
$= 226.08 \text{ in}^3$

10. The new *x* coordinates are the old y coordinates. The new *y* coordinates are the opposite of the old *x* coordinates.
$A(-4, 1) \rightarrow A'(1, 4)$
$B(-3, 5) \rightarrow B'(5, 3)$
$C(-1, 3) \rightarrow C'(3, 1)$

Student Handbook

Exponent
Base

Extra Practice

Extra Practice ■ Chapter 1

1A Whole Numbers and Exponents

LESSON 1-1

1. The area of Canada is 3,851,788 square miles. The area of the United States is 3,717,792 square miles. Which country has the greater area? Canada
2. In 2001, it was estimated that 14,902,000 students attended high school and 14,889,000 attended college. Were more students in high school or in college? high school

Write the numbers in order from least to greatest.

3. 783; 772; 1,702 772; 783; 1,702
4. 10,318; 1,308; 10,301 1,308; 10,301; 10,318
5. 34,903; 32,788; 32,679 32,679; 32,788; 34,903
6. 24,615; 24,829; 24,560 24,560; 24,615; 24,829
7. 1,345; 1,780; 1,356 1,345; 1,356; 1,780
8. 29,992; 22,929; 22,922 22,922; 22,929; 29,992

LESSON 1-2

Estimate each sum or difference to the place value indicated.

9. 7,685 + 8,230; thousands 16,000
10. 23,218 + 37,518; ten thousands 60,000
11. 52,087 − 35,210; ten thousands 10,000
12. 292,801 − 156,127; hundred thousands 100,000
13. 14,325 + 25,629; hundreds 39,900
14. 9,210 − 396; hundreds 8,800
15. Mr. Peterson needs topsoil for his garden. His rectangular garden is 78 in. long and 48 in. wide. A bag of topsoil covers an area of 500 square inches. How many bags should Mr. Peterson buy? 8 bags of topsoil
16. Natalie's family is having a picnic at an amusement park. The park is 153 miles from Natalie's house. If the family drives 55 mi/h, about how long will it take them to get to the park? about 3 hours

LESSON 1-3

Write each expression in exponential form.

17. $5 \times 5 \times 5 \times 5 \times 5 \times 5$ 5^6
18. $3 \times 3 \times 3 \times 3$ 3^4
19. $10 \times 10 \times 10 \times 10 \times 10$ 10^5
20. $2 \times 2 \times 2 \times 2$ 2^4
21. $7 \times 7 \times 7$ 7^3
22. 9×9 9^2

Find each value.

23. 8^2 64
24. 5^5 3,125
25. 6^3 216
26. 10^6 1,000,000
27. 9^1 9
28. 3^6 729
29. 4^3 64
30. 2^5 32
31. Patricia e-mailed a joke to 4 of her friends. Each of those friends e-mailed the joke to 4 other friends. If this pattern continues, how many people will receive the e-mail on the fifth round of e-mails? 4^5, or 1,024 people

Extra Practice

Extra Practice ■ Chapter 1

1B Using Whole Numbers

LESSON 1-4

Evaluate each expression.

1. $15 + 7 \times 3$ 36
2. $3 \times 3^2 + 13 - 5$ 35
3. $10 \div (3 + 2) \times 2^3 - 8$ 8
4. $4^2 - 12 \div 3 + (7 - 5)$ 14
5. $10 \times (25 - 11) \div 7 + 6$ 26
6. $(3 + 6) \times 18 \div 2 + 7$ 88

LESSON 1-5

Find each sum or product.

7. 15 + 7 + 23 + 5 50
8. 4 × 13 × 5 260
9. 34 + 16 + 22 + 18 90

Use the Distributive Property to find each product.

10. 5 × 54 270
11. 3 × 32 96
12. 7 × 26 182
13. 9 × 73 657

LESSON 1-6

14. The table shows the number of days it rained each month. How many days total did it rain in the year? 122 days

Month	Days of Rain	Month	Days of Rain
January	6	July	15
February	5	August	9
March	7	September	17
April	14	October	14
May	12	November	8
June	10	December	5

15. The coldest temperature in a city was 11°F. The warmest temperature that same year was 89°F. What is the difference between the highest and lowest temperatures? 78°F
16. Heather is a member of a dance company. She practices 14 hours a week. How many hours does she practice each year? (*Hint*: There are 52 weeks in a year.) 728 hours

LESSON 1-7 17–19. See p. A7.

Identify a pattern in each sequence. Use your pattern to name the next three terms.

17. 8, 16, 32, ■, ■, ■
18. 6, 11, 16, ■, ■, ■
19. 7, 21, 63, ■, ■, ■
20. 4, 20, 36, 52, ■, ■, ■ add 16; 68, 84, 100
21. 1, 3, 6, 10, 15, ■, ■, ■ add 2, 3, 4, etc.; 21, 28, 36
22. 100, 85, 70, ■, ■, ■ subtract 15; 55, 40, 25

Identify a pattern in each sequence. Use your pattern to name the missing terms.

23. 496, 248, 260, ■, 142, 71, ■ divide by 2, add 12; 130, 83
24. 1, 8, 4, 32, 16, ■, 64, 512, ■ multiply by 8, divide by 2; 128, 256

Extra Practice

Extra Practice ■ Chapter 2

2A Understanding Variables and Expressions

LESSON 2-1

Evaluate the expression to find the missing values in each table.

1.

y	23 + y
17	40
27	■
37	■

50, 60

2.

w	w × 3 + 10
4	22
5	■
6	■

25, 28

3.

x	x ÷ 8
40	5
48	■
56	■

6, 7

4.

a	2 × a − 1
5	■
10	■
15	■

9, 19, 29

5.

b	54 ÷ b
3	■
6	■
9	■

18, 9, 6

6.

c	18 + c ÷ 4
4	■
12	■
20	■

19, 21, 23

Find an expression for each table.

7.

t	■
7	35
8	40
9	45

$5 \times t$

8.

s	■
66	11
54	9
36	6

$s \div 6$

9.

z	■
52	26
60	30
68	34

$z \div 2$

LESSON 2-2

10. Earth has a diameter of 7,926 miles. Let d represent the diameter of the Moon, which is smaller than the diameter of Earth. Write an expression to show how much larger the diameter of Earth is than the diameter of the Moon. $7,926 - d$
11. Let p represent the number of players on a team. Write an expression to show how many players will be on 65 teams. $65p$
12. Marion scored 82 more points than Jody in a contest. Let j represent the number of points that Jody scored. Write an expression to show how many points Marion scored. $j + 82$

Write each phrase as a numerical or algebraic expression.

13. the sum of 322 and 18 $322 + 18$
14. the product of 7 and 12 7×12
15. the quotient of n and 8 $\frac{n}{8}$
16. 14 more than x $x + 14$

Write two word phrases for each expression. 17–22. See p. A7.

17. (23)(6)
18. $52 - p$
19. $y \div 4$
20. 8 + 4
21. $h - 96$
22. $13 \cdot m$

Extra Practice

Extra Practice ■ Chapter 2

2B Equations

LESSON 2-3

Determine whether the given value of each variable is a solution.

1. $a + 15 = 34$, when $a = 17$ no
2. $t - 9 = 14$, when $t = 23$ yes
3. Rachel says she is 5 feet tall. Her friend measured her height as 60 inches. Determine if the two measurements are equal. 5 feet is equal to 60 inches.

LESSON 2-4

Solve each equation. Check your answers.

4. $r + 13 = 36$ $r = 23$
5. $52 = 24 + n$ $n = 28$
6. Towns A, B, and C are located along Main Road, as shown on the map. Town A is 34 miles from town C. Town B is 12 miles from town C. Find the distance d between town A and town B. 22 miles

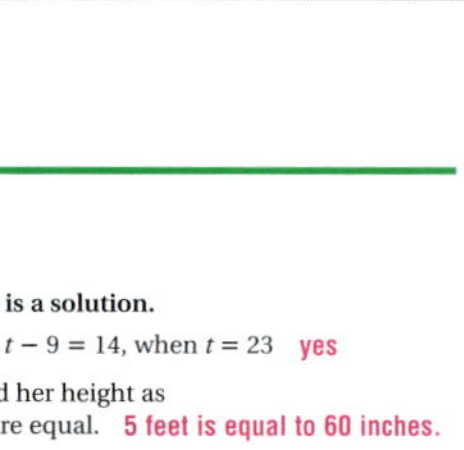

LESSON 2-5

7. Reggie withdrew $175 from his bank account to go shopping. After his withdrawal, there was $234 left in Reggie's account. How much money did Reggie have in his account before his withdrawal? $409

Solve each equation. Check your answers.

8. $z - 9 = 5$ $z = 14$
9. $v - 17 = 14$ $v = 31$
10. $24 = w - 6$ $w = 30$

LESSON 2-6

11. A squirrel can run 36 miles in 3 hours. Solve the equation $3m = 36$ to find the number of miles a squirrel can run in 1 hour. $m = 12$; a squirrel can run 12 miles in an hour.

Solve each equation. Check your answers.

12. $4y = 20$ $y = 5$
13. $21 = 3t$ $t = 7$
14. $72 = 9g$ $g = 8$
15. The area of a rectangle is 54 in^2. Its width is 6 in. What is its length? 9 inches

LESSON 2-7

Solve each equation. Check your answers.

16. $\frac{n}{4} = 6$ $n = 24$
17. $7 = \frac{t}{5}$ $t = 35$
18. $\frac{a}{8} = 12$ $a = 96$
19. Irene likes to run and ride a bike for exercise. Each day, she runs for one-third the time that she rides her bike. Yesterday, Irene ran for 15 minutes. How many minutes did she ride her bike? 45 minutes

Extra Practice

Extra Practice ■ Chapter 3

3A Understanding Decimals

LESSON 3-1

Write each in standard form, expanded form, and words.

1. 1.32 $1 + 0.3 + 0.02$, one and thirty-two hundredths
2. $0.6 + 0.003 + 0.0008$ 0.6038, six thousand thirty-eight ten-thousandths
3. five and three thousandths 5.003, $5 + 0.003$
4. Joshua ran 1.45 miles, and Jasmine ran 1.5 miles. Who ran farther? Jasmine ran farther.

Order the decimals from least to greatest.

5. 3.89, 3.08, 3.8 3.08, 3.8, 3.89
6. 20.65, 20.09, 20.7 20.09, 20.65, 20.7
7. 0.053, 0.43, 0.340 0.053, 0.340, 0.43

LESSON 3-2

8. The femur is the upper leg bone, and the tibia is one of the lower leg bones. The average length of the femur is 50.5 cm, and the average length of the tibia is 43.03 cm. Estimate the total length of the leg if the bones were placed end to end. about 93 cm

Estimate. Round to the indicated place value.

9. $5.856 - 1.3497$; hundredths 4.51
10. $4.7609 + 7.2471$; tenths 12.0

Estimate each product or quotient.

11. $20.84 \div 3.201$ about 7
12. 31.02×4.91 about 150
13. $39.76 \div 7.94$ about 5
14. $47.36 \div 7.66$ about 6
15. 9.518×11.1102 about 100
16. $70.61894 \div 1.879$ about 36

Estimate a range for the sum.

17. $8.38 + 24.92 + 4.8$ from 36 to 38.5
18. $38.27 + 2.99 + 15.32$ from 55 to 57

LESSON 3-3

Find each sum or difference.

19. $1.65 + 4.53 + 3.2$ 9.38
20. $2.2 + 6.8$ 9.0
21. $7 - 0.6$ 6.4

Evaluate $6.35 - s$ for each value of s.

22. $s = 3.2$ 3.15
23. $s = 2.108$ 4.242
24. $s = 5.0421$ 1.3079

LESSON 3-4

Multiply or divide.

25. $2{,}318 \times 1{,}000$ 2,318,000
26. $6{,}210 \div 100$ 62.1
27. 34.5×10^4 345,000

Convert each measure.

28. 3,450 g = ■ kg 3.45
29. 0.6 L = ■ mL 600
30. 23 cm = ■ m 0.23

Extra Practice ■ Chapter 3

3B Multiplying and Dividing Decimals

LESSON 3-5

Write each number in scientific notation.

1. 60,000 6×10^4
2. 423,800 4.238×10^5
3. 8,500,000 8.5×10^6

Write each number in standard form.

4. 5.632×10^5 563,200
5. 2.1×10^8 210,000,000
6. 1.425×10^4 14,250

LESSON 3-6

Find each product.

7. 0.5×0.7 0.35
8. 0.3×0.06 0.018
9. 6.12×5.9 36.108

Evaluate $4x$ for each value of x.

10. $x = 2.071$ 8.284
11. $x = 5.42$ 21.68
12. $x = 7.85$ 31.4

LESSON 3-7

Evaluate the expression $0.564 \div x$ for the given value of x.

13. $x = 4$ 0.141
14. $x = 12$ 0.047
15. $x = 2$ 0.282

LESSON 3-8

Find the quotient.

16. $4.5 \div 0.9$ 5
17. $59.7 \div 0.4$ 149.25
18. $8.32 \div 8$ 1.04

LESSON 3-9

19. Jocelyn has 3.5 yards of ribbon. She needs 0.6 yards of ribbon to make one bow. How many bows can Jocelyn make? 5 bows
20. Louie has a piece of wood that is 46.8 cm long. If he cuts the piece into 4 equal sections, how long will each section be? 11.7 cm

LESSON 3-10

Solve each equation. Check your answer.

21. $b - 5.2 = 2.6$ $b = 7.8$
22. $5t = 24.5$ $t = 4.9$
23. $\frac{p}{3} = 1.8$ $p = 5.4$
24. The area of a rectangle is 41 cm^2. Its length is 8.2 cm. What is its width? 5 cm
25. The area of Henry's kitchen is 168 ft^2. The cost of tile is $4.62 per square foot. What is the total cost to tile the kitchen? $776.16

Extra Practice ■ Chapter 4

4A Number Theory

LESSON 4-1

Tell whether each number is divisible by 2, 3, 4, 5, 6, 9, and 10.

1. 12,680 2, 4, 5, 10
2. 174 2, 3, 6
3. 1,638 2, 3, 6, 9
4. 735 3, 5

Tell whether each number is prime or composite.

5. 97 prime
6. 9 composite
7. 111 composite
8. 101 prime
9. 373 prime
10. 256 composite
11. 37 prime
12. 153 composite

LESSON 4-2

List all the factors of each number.

13. 28 1, 2, 4, 7, 14, 28
14. 51 1, 3, 17, 51
15. 70 1, 2, 5, 7, 10, 14, 35, 70
16. 15 1, 3, 5, 15
17. 24 1, 2, 3, 4, 6, 8, 12, 24
18. 38 1, 2, 19, 38

Write the prime factorization for each number.

19. 48 $2^4 \cdot 3$
20. 72 $2^3 \cdot 3^2$
21. 81 3^4
22. 90 $2 \cdot 3^2 \cdot 5$
23. 150 $2 \cdot 3 \cdot 5^2$
24. 99 $3^2 \cdot 11$

LESSON 4-3

Find the GCF of each set of numbers.

25. 15 and 35 5
26. 16 and 40 8
27. 22 and 68 2
28. 6, 36, and 60 6
29. 27, 36, and 54 9
30. 14, 28, and 63 7
31. 26, 65, and 78 13
32. 8, 20, and 32 4
33. 12, 42, and 72 6
34. Alice has 42 red beads and 24 white beads. What is the greatest number of bracelets Alice can make if each bracelet has the same number of red beads and the same number of white beads and if every bead is used? 6 bracelets
35. The flower shop has 16 red roses and 28 white carnations. What is the greatest number of arrangements that can be made if each arrangement has the same number of roses and the same number of carnations and if every flower is used? 4 arrangements
36. Lisa has 15 bars of strawberry-scented soap, 30 bars of almond-scented soap, and 25 bars of orange-scented soap. She wants to make gift baskets. Each basket must have the same number of scented soap. What is the greatest number of gift baskets Lisa can make if every bar of soap is used? 5 gift baskets

Extra Practice ■ Chapter 4

4B Understanding Fractions

LESSON 4-4

Write each decimal as a fraction or mixed number.

1. 0.31 $\frac{31}{100}$
2. 1.9 $1\frac{9}{10}$
3. 2.53 $2\frac{53}{100}$
4. 0.07 $\frac{7}{100}$

Write each fraction or mixed number as a decimal.

5. $1\frac{7}{8}$ 1.875
6. $\frac{5}{9}$ $0.\overline{5}$
7. $6\frac{3}{5}$ 6.6
8. $\frac{5}{6}$ $0.8\overline{3}$

Order the fractions and decimals from least to greatest.

9. 0.3, $\frac{3}{5}$, 0.53 0.3, 0.53, $\frac{3}{5}$
10. 0.8, 0.67, $\frac{7}{8}$ 0.67, 0.8, $\frac{7}{8}$
11. 0.68, $\frac{2}{3}$, $\frac{3}{4}$ $\frac{2}{3}$, 0.68, $\frac{3}{4}$

LESSON 4-5

Find two equivalent fractions for the given fraction. Possible answers given.

12. $\frac{3}{12}$ $\frac{1}{4}$, $\frac{6}{24}$
13. $\frac{3}{15}$ $\frac{1}{5}$, $\frac{9}{45}$
14. $\frac{6}{16}$ $\frac{3}{8}$, $\frac{12}{32}$
15. $\frac{4}{10}$ $\frac{2}{5}$, $\frac{8}{20}$

Find the missing number that makes the fractions equivalent.

16. $\frac{4}{5} = \frac{■}{20}$ 16
17. $\frac{8}{12} = \frac{2}{■}$ 3
18. $\frac{6}{7} = \frac{■}{28}$ 24
19. $\frac{24}{3} = \frac{■}{1}$ 8

Write each fraction in simplest form.

20. $\frac{6}{10}$ $\frac{3}{5}$
21. $\frac{7}{9}$ $\frac{7}{9}$
22. $\frac{4}{16}$ $\frac{1}{4}$
23. $\frac{2}{6}$ $\frac{1}{3}$

LESSON 4-6

Compare. Write <, >, or =.

24. $\frac{2}{5}$ ■ $\frac{4}{5}$ $<$
25. $\frac{5}{6}$ ■ $\frac{7}{8}$ $<$
26. $\frac{1}{3}$ ■ $\frac{9}{27}$ $=$
27. $\frac{9}{15}$ ■ $\frac{2}{5}$ $>$
28. Natalie lives $\frac{1}{6}$ mile from school. Peter lives $\frac{3}{10}$ mile from school. Who lives closer to the school? Natalie

Order the fractions from least to greatest.

29. $\frac{3}{5}, \frac{5}{9}, \frac{4}{5}$ $\frac{5}{9}, \frac{3}{5}, \frac{4}{5}$
30. $\frac{1}{6}, \frac{3}{7}, \frac{1}{3}$ $\frac{1}{6}, \frac{1}{3}, \frac{3}{7}$
31. $\frac{1}{2}, \frac{5}{8}, \frac{7}{12}$ $\frac{1}{2}, \frac{7}{12}, \frac{5}{8}$

LESSON 4-7

Write each mixed number as an improper fraction.

32. $3\frac{1}{4}$ $\frac{13}{4}$
33. $6\frac{5}{7}$ $\frac{47}{7}$
34. $1\frac{2}{9}$ $\frac{11}{9}$
35. $2\frac{7}{10}$ $\frac{27}{10}$
36. Brett's favorite soup recipe calls for $\frac{14}{4}$ cups of chicken broth. Write $\frac{14}{4}$ as a mixed number. $3\frac{1}{2}$

Extra Practice ■ Chapter 4

4C Introduction to Fraction Operations

LESSON 4-8

1. A teaspoon is equivalent to $\frac{1}{6}$ fluid ounce. How many fluid ounces are 2 teaspoons equivalent to? $\frac{1}{3}$
2. Gerry walked $\frac{5}{8}$ of a mile and then ran $\frac{3}{8}$ of a mile. What is the total distance Gerry covered? 1 mile

Subtract. Write your answer in simplest form.

3. $1 - \frac{7}{9}$ $\frac{2}{9}$
4. $2\frac{5}{6} - 1\frac{1}{6}$ $1\frac{2}{3}$
5. $5\frac{7}{10} - 3\frac{3}{10}$ $2\frac{2}{5}$
6. $2 - \frac{3}{4}$ $1\frac{1}{4}$
7. $2\frac{5}{8} - 2\frac{3}{8}$ $\frac{1}{4}$
8. $6\frac{11}{12} - 3\frac{5}{12}$ $3\frac{1}{2}$
9. $6\frac{2}{3} - 5\frac{1}{3}$ $1\frac{1}{3}$
10. $7\frac{3}{4} - 4\frac{3}{4}$ 3
11. $10\frac{4}{5} - 2\frac{2}{5}$ $8\frac{2}{5}$

Evaluate $\frac{11}{12} - x$ for each value of x. Write your answer in simplest form.

12. $x = \frac{7}{12}$ $\frac{1}{3}$
13. $x = \frac{1}{12}$ $\frac{5}{6}$
14. $x = \frac{2}{12}$ $\frac{3}{4}$
15. $x = \frac{10}{12}$ $\frac{1}{12}$
16. $x = \frac{5}{12}$ $\frac{1}{2}$
17. $x = \frac{3}{12}$ $\frac{2}{3}$

LESSON 4-9

Multiply. Write your answer in simplest form.

18. $2 \cdot \frac{1}{5}$ $\frac{2}{5}$
19. $3 \cdot \frac{1}{6}$ $\frac{1}{2}$
20. $2 \cdot \frac{2}{11}$ $\frac{4}{11}$
21. $4 \cdot \frac{3}{7}$ $\frac{12}{7}$ or $1\frac{5}{7}$
22. $6 \cdot \frac{2}{3}$ 4
23. $8 \cdot \frac{3}{10}$ $\frac{12}{5}$ or $2\frac{2}{5}$
24. $5 \cdot \frac{7}{8}$ $\frac{35}{8}$ or $4\frac{3}{8}$
25. $7 \cdot \frac{6}{11}$ $\frac{42}{11}$ or $3\frac{9}{11}$
26. $3 \cdot \frac{11}{12}$ $\frac{11}{4}$ or $2\frac{3}{4}$

Evaluate $5x$ for each value of x. Write your answer in simplest form.

27. $x = \frac{1}{6}$ $\frac{5}{6}$
28. $x = \frac{3}{5}$ 3
29. $x = \frac{3}{8}$ $\frac{15}{8}$ or $1\frac{7}{8}$
30. $x = \frac{4}{5}$ 4
31. $x = \frac{2}{7}$ $\frac{10}{7}$ or $1\frac{3}{7}$
32. $x = \frac{2}{10}$ 1
33. $x = \frac{2}{9}$ $\frac{10}{9}$ or $1\frac{1}{9}$
34. $x = \frac{5}{12}$ $\frac{25}{12}$ or $2\frac{1}{12}$
35. $x = \frac{13}{15}$ $\frac{13}{3}$ or $4\frac{1}{3}$
36. There are 16 players on the baseball team. Of these players, $\frac{1}{4}$ are girls. How many girls play on the baseball team? 4 girls
37. Sue made a beaded necklace. Of the 54 beads she used, $\frac{1}{6}$ were white. How many white beads did Sue use? 9 white beads
38. Connie is using 16 flowers to make an arrangement. Of the 16 flowers, $\frac{3}{8}$ are roses. How many roses are in her arrangement? 6 roses

644 *Extra Practice*

Extra Practice ■ Chapter 5

5A Multiplying and Dividing Fractions

LESSON 5-1

Multiply. Write each answer in simplest form.

1. $\frac{1}{10} \cdot \frac{5}{6}$ $\frac{1}{12}$
2. $\frac{8}{9} \cdot \frac{3}{4}$ $\frac{2}{3}$
3. $\frac{5}{7} \cdot \frac{3}{10}$ $\frac{3}{14}$
4. $\frac{7}{10} \cdot \frac{3}{7}$ $\frac{3}{10}$
5. $\frac{5}{12} \cdot \frac{6}{7}$ $\frac{5}{14}$
6. $\frac{2}{3} \cdot \frac{3}{8}$ $\frac{1}{4}$

Evaluate the expression $a \cdot \frac{1}{10}$ for each value of a. Write the answer in simplest form.

7. $a = \frac{4}{5}$ $\frac{2}{25}$
8. $a = \frac{2}{3}$ $\frac{1}{15}$
9. $a = \frac{5}{9}$ $\frac{1}{18}$

LESSON 5-2

Multiply. Write each answer in simplest form.

10. $\frac{1}{4} \cdot 1\frac{2}{3}$ $\frac{5}{12}$
11. $2\frac{3}{5} \cdot \frac{1}{3}$ $\frac{13}{15}$
12. $\frac{7}{8} \cdot 1\frac{1}{3}$ $1\frac{1}{6}$

Find each product. Write the answer in simplest form.

13. $1\frac{1}{3} \cdot 1\frac{3}{5}$ $2\frac{2}{15}$
14. $4 \cdot 2\frac{6}{7}$ $11\frac{3}{7}$
15. $2\frac{1}{4} \cdot 2\frac{1}{2}$ $5\frac{5}{8}$
16. $\frac{2}{3}$ of $1\frac{2}{3}$ $1\frac{1}{9}$
17. $\frac{1}{6}$ of 5 $\frac{5}{6}$
18. $\frac{2}{5}$ of $4\frac{1}{2}$ $1\frac{4}{5}$

LESSON 5-3

Find the reciprocal.

19. $\frac{7}{9}$ $\frac{9}{7}$
20. $\frac{2}{13}$ $\frac{13}{2}$
21. $\frac{1}{12}$ $\frac{12}{1}$ or 12
22. $\frac{8}{5}$ $\frac{5}{8}$

Divide. Write each answer in simplest form.

23. $\frac{1}{6} \div 3$ $\frac{1}{18}$
24. $\frac{4}{7} \div 2$ $\frac{2}{7}$
25. $\frac{8}{9} \div 6$ $\frac{4}{27}$
26. $1\frac{4}{5} \div 1\frac{1}{4}$ $1\frac{11}{25}$
27. $2\frac{1}{2} \div 1\frac{3}{4}$ $1\frac{3}{7}$
28. $3\frac{1}{5} \div 1\frac{3}{10}$ $2\frac{6}{13}$

LESSON 5-4

Solve each equation. Write the answer in simplest form.

29. $\frac{3}{5}a = 12$ $a = 20$
30. $6b = \frac{3}{7}$ $b = \frac{1}{14}$
31. $\frac{3}{8}x = 5$ $x = 13\frac{1}{3}$
32. $3s = \frac{7}{9}$ $s = \frac{7}{27}$
33. $\frac{5}{12}m = 3$ $m = 7\frac{1}{5}$
34. $\frac{9}{10}t = 6$ $t = 6\frac{2}{3}$
35. Joanie used $\frac{2}{3}$ of a box of invitations to invite friends to her birthday party. If she sent out 12 invitations, how many total invitations were in the box? 18 invitations
36. The coach gave $\frac{3}{4}$ of the football team extra laps to run. If he gave 33 players extra laps, how many players are on his team? 44 players

Extra Practice 645

Extra Practice ■ Chapter 5

5B Adding and Subtracting Fractions

LESSON 5-5

1. There are 18 girls on the dance team. Barrettes are sold in packs of 6. Ponytail holders are sold in packs of 2. What is the least number of packs they could buy so that each girl has a barrette and a ponytail holder and none are left over? 3 packs of barrettes and 9 packs of ponytail holders

Find the least common multiple (LCM).

2. 9 and 15 45
3. 12 and 16 48
4. 10 and 12 60

LESSON 5-6

Estimate each sum or difference by rounding to 0, $\frac{1}{2}$, or 1.

5. $\frac{7}{8} + \frac{7}{15}$ about $1\frac{1}{2}$
6. $\frac{5}{6} + \frac{1}{11}$ about 1
7. $\frac{7}{12} - \frac{4}{9}$ about 0

Use the table for Exercises 8 and 9.

8. The table shows the number of hours each day that Michael worked. About how many hours did Michael work on Monday and Tuesday? about 10 hours
9. About how many more hours did he work on Thursday than on Friday? about $1\frac{1}{2}$

Michael's Work Schedule	
Day	**Hours Worked**
Monday	$4\frac{5}{6}$
Tuesday	$5\frac{1}{4}$
Thursday	$6\frac{1}{10}$
Friday	$4\frac{5}{12}$

LESSON 5-7

Add or subtract. Write each answer in simplest form.

10. $\frac{3}{5} + \frac{2}{3}$ $1\frac{4}{15}$
11. $\frac{7}{8} - \frac{1}{6}$ $\frac{17}{24}$
12. $\frac{3}{7} + \frac{1}{2}$ $\frac{13}{14}$

LESSON 5-8

Find each sum or difference. Write the answer in simplest form.

13. $18\frac{1}{3} + 16\frac{1}{6}$ $34\frac{1}{2}$
14. $5\frac{3}{4} + 3\frac{5}{12}$ $9\frac{1}{6}$
15. $12\frac{1}{2} - 8\frac{2}{5}$ $4\frac{1}{10}$

LESSON 5-9

Subtract. Write each answer in simplest form.

16. $4\frac{2}{5} - 2\frac{9}{10}$ $1\frac{1}{2}$
17. $9\frac{1}{6} - 5\frac{5}{6}$ $3\frac{1}{3}$
18. $6 - 1\frac{7}{12}$ $4\frac{5}{12}$

LESSON 5-10

Solve each equation. Write the solution in simplest form.

19. $a + 5\frac{3}{10} = 9$ $a = 3\frac{7}{10}$
20. $1\frac{3}{8} = x - 2\frac{1}{4}$ $x = 3\frac{5}{8}$
21. $6\frac{5}{6} = t + 1\frac{2}{3}$ $t = 5\frac{1}{6}$

646 *Extra Practice*

Extra Practice ■ Chapter 6

6A Organizing Data

LESSON 6-1 1–2. Complete answers on p. A7.

1. Each year a community holds a 5 km race. In 1998, 1,345 people participated in the race. In 1999, 1,415 people participated. In 2000, 1,532 people participated. In 2001, 1,607 people participated, and in 2002, 1,781 people participated. Use the data to make a table. Then use your table to describe how participation changed over time. The participation increased from 1998 to 2002.
2. Make a table using the basketball data below. Then use your table to tell which player had the most points, rebounds, and assists.

   In 1,560 games, Kareem Abdul-Jabbar scored 38,387 points, grabbed 17,440 rebounds, and made 5,660 assists. In 897 games, Larry Bird scored 21,791 points, grabbed 8,974 rebounds, and made 5,695 assists. In 963 games, Bill Russell scored 14,522 points, grabbed 21,620 rebounds, and made 4,100 assists. points: Kareem Abdul-Jabbar, rebounds: Bill Russell, assists: Larry Bird

LESSON 6-2

Find the range, mean, median, and mode of each data set.

3.

Number of Books Read						
8	4	6	8	8	5	3

range: 5, mean: 6, median: 6, mode: 8

4.

Distance Biked (mi)					
22	25	33	22	24	30

range: 11, mean: 26, median: 24.5, mode: 22

5.

Points Scored				
16	18	23	13	15

range: 10, mean: 17, median: 16, no mode

6.

Hours Worked							
37	42	43	38	39	40	45	40

range: 8, mean: 40.5, median: 40, mode: 40

LESSON 6-3

7. a. The table shows a student's test scores. Find the mean, median, and mode of the test scores. mean: 85.5, median: 84.5, mode: no mode

   b. On the next test the student scored a 92. Find the mean, median, and mode with the new test score. mean: 86.8, median: 87, mode: no mode

Test Scores			
78	82	87	95

8. The table shows the number of hours a student spent on homework during each of 10 weeks.

Number of Hours Spent on Homework										
Week	1	2	3	4	5	6	7	8	9	10
Hours	8	9	9	9	7	9	8	0	7	2

   a. Which numbers are outliers? 0 and 2

   b. Find the mean, median, and mode with and without the outliers. with: mean = 6.8, median = 8, mode = 9; without: mean = 8.25, median = 8.5, mode = 9

   c. How does including the outliers affect the mean? the median? the mode? The mean decreases by 1.45, the median decreases by 0.5, and the mode is unaffected.

9. The daily temperatures for the first eight days of April were 52°F, 63°F, 61°F, 54°F, 52°F, 55°F, 68°F, and 75°. What are the mean, median, and mode of this data set? Which one best describes the data set? 9. See p. A7.

Extra Practice 647

Extra Practice ■ Chapter 6

6B Displaying and Interpreting Data

LESSON 6-4

Use the bar graph to answer each question.

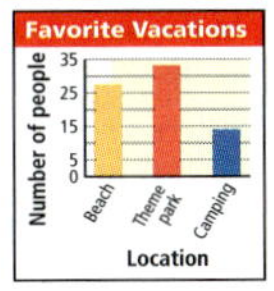

1. Which type of vacation received the most votes? theme park
2. Which types of vacations received more than 20 votes? beach, theme park

LESSON 6-5 3. Complete answer on p. A7.

3. Each student in Mr. Ander's class recorded the sport he or she plays. Make a tally table to organize the data. Which sport do most students play? baseball

baseball	basketball	soccer
tennis	baseball	football
baseball	football	soccer
basketball	basketball	baseball
baseball	basketball	baseball

4. Use your tally table from Exercise 3 to make a cumulative frequency table.
   4–5. See p. A7.
5. Use the data of students' heights to make a frequency table with intervals. Then use your frequency table to make a histogram.

Heights of Students (in.)							
63	58	48	60	60	65	56	57
56	62	61	58	59	55	64	50

LESSON 6-6

Name the ordered pair for each location on the grid.

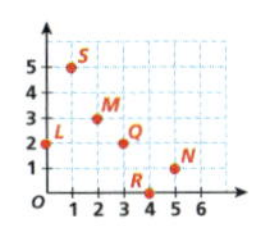

6. *L* (0, 2)
7. *M* (2, 3)
8. *R* (4, 0)

Graph and label each point on a coordinate grid. 9–11. See p. A8.

9. *A*(0, 3)
10. *B*(5, 3)
11. *C*(2, 1)

LESSON 6-7 12. Complete answer on p. A8.

12. Use the data in the table to make a double-line graph. Did toy sales increase or decrease for store A? increase

Toy Sales	January	March	May	July
Store A	$460	$580	$950	$1200
Store B	$520	$450	$880	$1250

LESSON 6-8 13. See p. A8.

13. Explain why this bar graph is misleading.

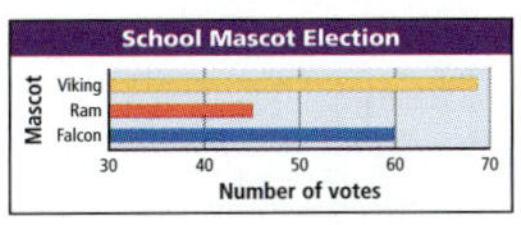

LESSON 6-9

14. Make a stem-and-leaf plot of the data about time spent on homework. Then use your stem-and-leaf plot to find the mean, median, and mode of the data. mean: 35, median: 35, mode: 35
    14. Complete answer on p. A8.

Time Spent Doing Homework (min)				
15	35	60	65	15
10	35	60	20	35

Extra Practice ■ Chapter 7

7A Lines and Angles

LESSON 7-1

Use the diagram to name each geometric figure.

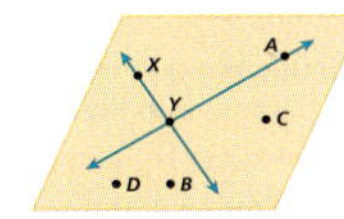

1. three points *X, Y, B*
2. two lines $\overleftrightarrow{XY}$; $\overleftrightarrow{AY}$
3. a point shared by two lines *Y*
4. a plane *BCD*

Use the diagram to give a possible name to each figure.

5. three different line segments $\overline{XY}$; $\overline{XZ}$; $\overline{YZ}$
6. three ways to name the line $\overleftrightarrow{XY}$; $\overleftrightarrow{XZ}$; $\overleftrightarrow{YZ}$
7. six different rays $\overrightarrow{XY}$; $\overrightarrow{XZ}$; $\overrightarrow{YX}$; $\overrightarrow{YZ}$; $\overrightarrow{ZY}$; $\overrightarrow{ZX}$
8. another name for ray *XY* ray *XZ*

LESSON 7-2

Use a protractor to measure each angle. Tell what type of angle it is.

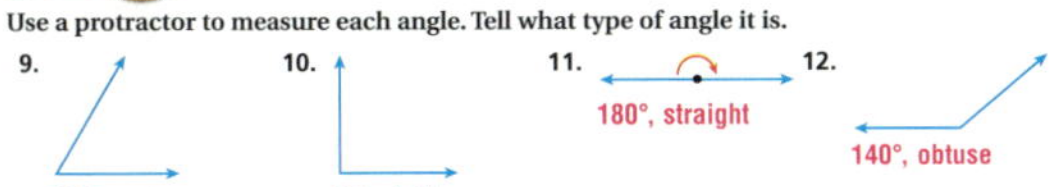

9. 60°, acute
10. 90°, right
11. 180°, straight
12. 140°, obtuse

LESSON 7-3

Find each unknown angle measure.

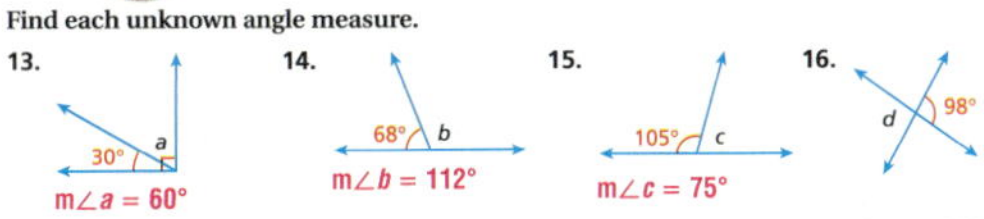

13. $m\angle a = 60°$
14. $m\angle b = 112°$
15. $m\angle c = 75°$
16. $m\angle d = 98°$

LESSON 7-4

Classify each pair of lines.

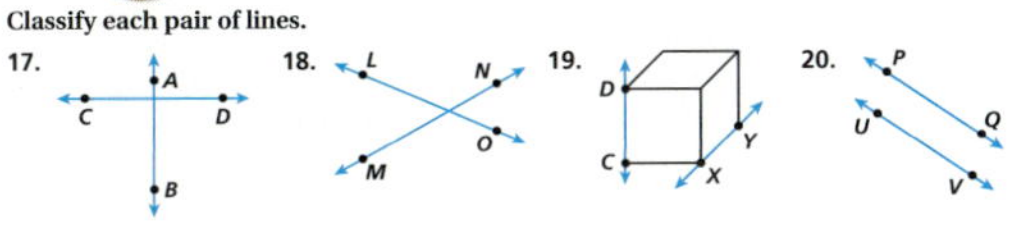

17. perpendicular
18. intersecting
19. skew
20. parallel

Extra Practice ■ Chapter 7

7B Polygons

LESSON 7-5

Use the diagram to find the measure of each indicated angle.

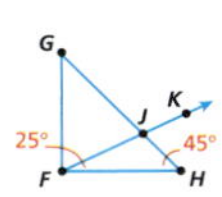

1. $\angle FJH$ 110°
2. $\angle FJG$ 70°

Classify each triangle.

3. The sum of the lengths of the sides is 14 in. isosceles
4. The sum of the lengths of the sides is 45 ft. equilateral
5. The sum of the lengths of the sides is 20 ft. scalene

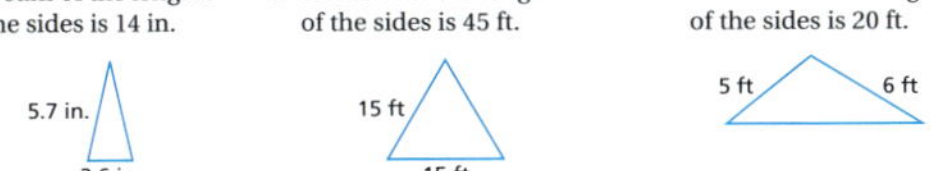

LESSON 7-6

Give the most descriptive name for each figure.

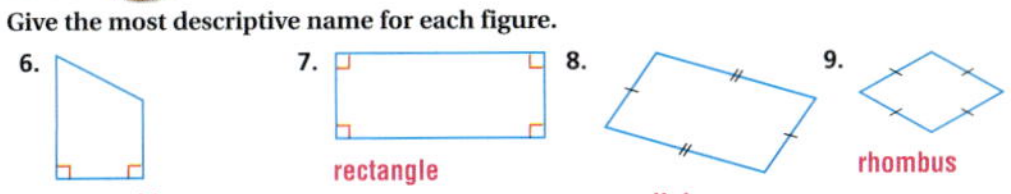

6. trapezoid
7. rectangle
8. parallelogram
9. rhombus

Complete each statement.

10. A parallelogram with four right angles can be a ___?___ or a ___?___. rectangle, square
11. A quadrilateral with exactly two parallel sides is a ___?___. trapezoid

LESSON 7-7

Name each polygon, and tell whether it appears to be regular or not regular.

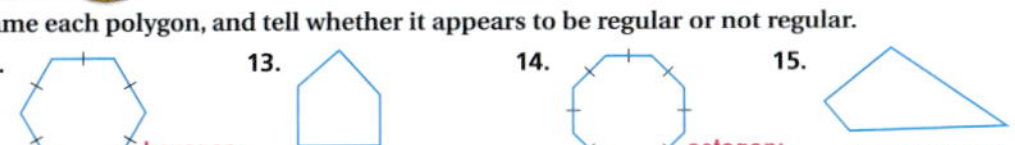

12. hexagon; regular
13. pentagon; not regular
14. octagon; regular
15. quadrilateral; not regular

LESSON 7-8 16–17. See p. A8.

Identify a possible pattern. Use the pattern to draw the missing figure.

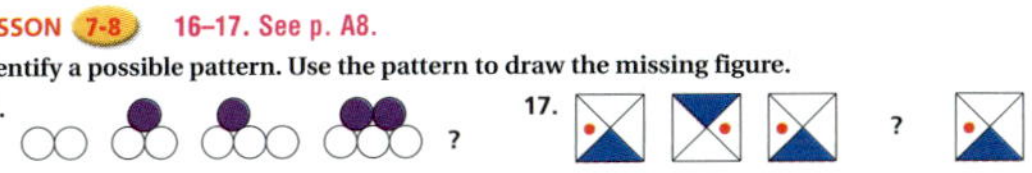

16. ?
17. ?

Extra Practice ■ Chapter 7

7C Polygon Relationships

LESSON 7-9

Decide whether the figures in each pair are congruent. If not, explain.

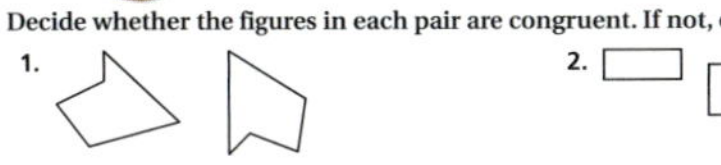

1. The figures are congruent.
2. The figures are not congruent. Both are rectangles, but they have different sizes.

LESSON 7-10

Tell whether each is a translation, rotation, or reflection.

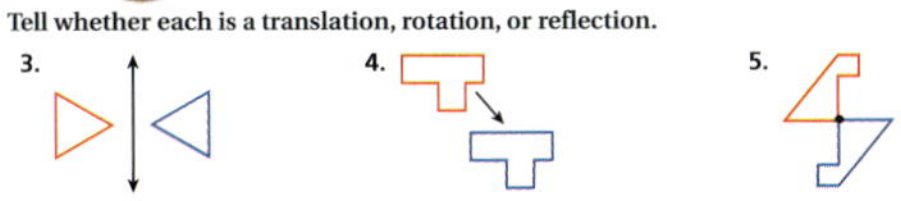

3. reflection
4. translation
5. rotation

Draw each transformation.

6. Draw a translation 2 cm to the right.
7. Draw a 90° clockwise rotation about the point.
8. Draw a vertical reflection across the line.

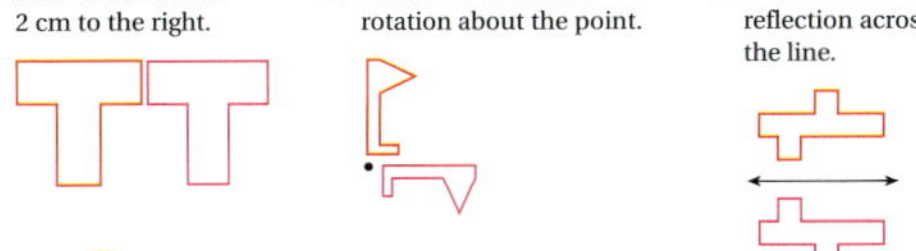

LESSON 7-11

Determine whether each dashed line appears to be a line of symmetry.

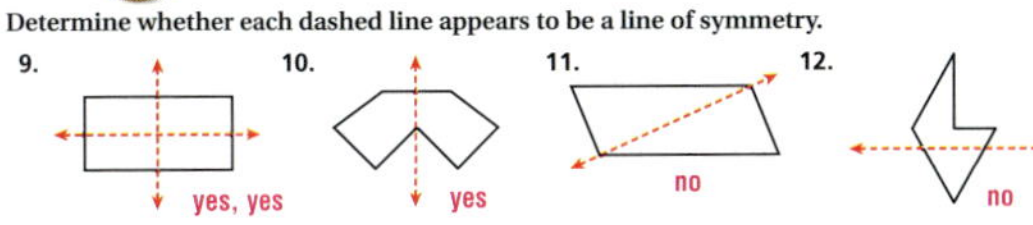

9. yes, yes
10. yes
11. no
12. no

LESSON 7-12 13–15. See p. A8.

Identify whether each shape can tessellate the plane. Make a drawing to show your answer.

13. yes
14. yes
15. no

Extra Practice ■ Chapter 8

8A Ratios and Proportions

LESSON 8-1

Use the table to write each ratio.

Types of Books in Doug's Collection			
Reference	10	Comic	7
Mystery	8	Poetry	5
Biography	3	Cooking	4

1. cooking books to poetry books 4:5
2. biography books to total books 3:37
3. Write three equivalent ratios to compare the number of biography books with the number of cooking books. 3:4, 6:8, 9:12
4. A pack of 12 pens costs \$5.52. A pack of 8 pens costs \$3.92. Which is the better deal? the pack of 12 pens

LESSON 8-2

Find the value of n in each proportion.

5. $\frac{5}{4} = \frac{n}{12}$ $n = 15$ 6. $\frac{2}{9} = \frac{4}{n}$ $n = 18$ 7. $\frac{6}{10} = \frac{n}{5}$ $n = 3$ 8. $\frac{7}{8} = \frac{21}{n}$ $n = 24$
9. To make 2 quarts of punch, Jenny adds 16 grams of juice mix to 2 quarts of water. How much mix does Jenny need to make 3 quarts of punch? 24 grams

LESSON 8-3

10. The highest island peak is Puncak Jaya, in Indonesia, which is 16,503 feet tall. How many yards tall is Puncak Jaya? 5,501 yards

LESSON 8-4

11. The two triangles are similar. Find the missing length y and the measure of $\angle B$. $y = 4$ in.; $m\angle B = 27°$

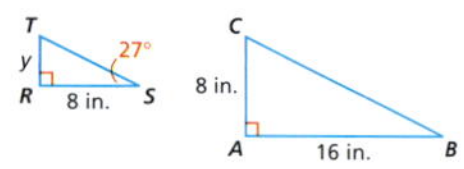

LESSON 8-5

12. A telephone pole casts a shadow that is 32 yd long. At the same time, a yardstick casts a shadow that is 4 yd long. How tall is the telephone pole? 8 yards

LESSON 8-6

Use the map to answer each question.

13. On the map, the distance from State College to Belmont is 2 cm. What is the actual distance between the two locations? 10 miles
14. Henderson City is 83 miles from State College. How many centimeters apart should the two locations be placed on the map? 16.6 cm

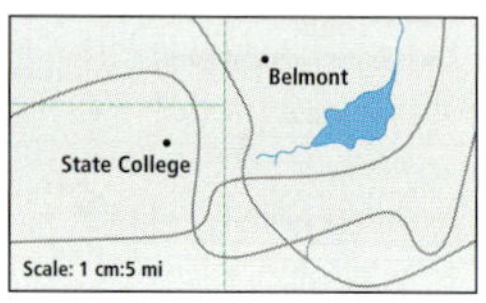

Extra Practice ■ Chapter 8

8B Percents

LESSON 8-7

Write each percent as a fraction in simplest form.

1. 50% $\frac{1}{2}$ 2. 34% $\frac{17}{50}$ 3. 8% $\frac{2}{25}$ 4. 12% $\frac{3}{25}$
5. Michael's baseball team won 85% of its games. Write 85% as a fraction in simplest form. $\frac{17}{20}$

Write each percent as a decimal.

6. 13% 0.13 7. 76% 0.76 8. 5% 0.05 9. 70% 0.7
10. At the toy store, sales increased by 26%. Write 26% as a decimal. 0.26

LESSON 8-8

Write each decimal as a percent.

11. 0.56 56% 12. 0.092 9.2% 13. 0.4 40% 14. 0.735 73.5%

Write each fraction as a percent.

15. $\frac{2}{5}$ 40% 16. $\frac{4}{25}$ 16% 17. $\frac{7}{16}$ 43.75% 18. $\frac{7}{8}$ 87.5%
19. In Mrs. Piper's class, $\frac{17}{20}$ of the students have a pet. What percent of the students in the class have pets? 85%

LESSON 8-9

20. A theater sold a total of 570 tickets for a new movie. Of those tickets, 30% were children's tickets. How many children's tickets were sold? 171 tickets
21. Kathy has listened to 80% of the music on a CD. If 26 minutes have passed, how many more minutes of music are left on the CD? 6.5 minutes
22. Find 30% of 98. 29.4 23. Find 15% of 220. 33 24. Find 5% of 72. 3.6

LESSON 8-10

25. Ashley wants to buy a sweater regularly priced at \$19.95. It is on sale for 25% off the regular price. About how much will she pay for the sweater after the discount? \$15.00
26. Margo and her three friends went to dinner. The bill was \$34.62. They left a tip that was 15% of the bill. About how much was the tip? \$5.25
27. Patricia is buying new roller skates that cost \$59.99. The sales tax rate is 7%. About how much will the total cost of the roller skates be? \$64.20

Extra Practice ■ Chapter 9

9A Integers

LESSON 9-1

Name a positive or negative number to represent each situation.

1. 120 feet below sea level −120 2. saving \$22 +22 3. a decrease of 5° −5

Graph each integer and its opposite on a number line. 4–7. See p. A8.

4. +1 5. −5 6. −3 7. +2

Find the absolute value of each integer.

8. −15 15 9. 11 11 10. −2 2 11. 25 25
12. Death Valley, California, has an elevation of −282 feet. The city of Long Beach, also in California, has an elevation of 170 feet. Which location is farther from sea level? Use absolute value to explain your answer. Death Valley is farther from sea level because | −282 | > | 170 |.

LESSON 9-2

Compare each pair of integers. Write < or >.

13. 15 ▒ −19 > 14. −7 ▒ −10 > 15. −3 ▒ 7 < 16. −8 ▒ 2 <

Order the integers in each set from least to greatest.

17. −6, 5, −2 −6, −2, 5 18. 12, −25, 10 −25, 10, 12 19. −1, −3, 4, 0 −3, −1, 0, 4
20. On Monday, the temperature was 3°C. On Tuesday, the temperature was −4°C. On Wednesday, the temperature was −1°C. On which day was the temperature the coldest? Tuesday

LESSON 9-3

Name the quadrant where each point is located.

21. A Quadrant IV 22. R y-axis, no quadrant 23. C Quadrant I 24. T Quadrant III

Give the coordinates of each point.

25. B (−3, 2) 26. S (−4, 0) 27. D (4, −2) 28. U (−3, −2)

Graph each point on a coordinate plane. 29–32. See p. A8.

29. $M(2, -1)$ 30. $W(-4, -2)$
31. $A(2, 3)$ 32. $H(-1, 3)$

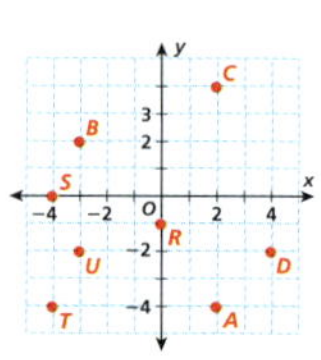

33. Graph and connect the points (5, 5), (5, −3), and (−2, −3) on a coordinate plane. What kind of figure do you have? triangle
34. Graph the points (2, 3), (2, −3), and (−4, 3). What ordered pair gives the location of a fourth point that will make the figure a square? (−4, −3)

Extra Practice ■ Chapter 9

9B Integer Operations

LESSON 9-4

Find each sum.

1. 3 + (−4) −1 2. −5 + 8 3 3. 6 + (−2) 4 4. −9 + 4 −5

Evaluate $y + 2$ for each value of y.

5. $y = -5$ −3 6. $y = -1$ 1 7. $y = 3$ 5 8. $y = -8$ −6
9. In the morning, the temperature was −3°C. By the afternoon, the temperature had risen 7°C. What was the temperature in the afternoon? 4°C

LESSON 9-5

Find each difference.

10. 10 − 6 4 11. −5 − (−3) −2 12. 12 − (−4) 16 13. −3 − 2 −5

Evaluate $a - (-5)$ for each value of a.

14. $a = -6$ −1 15. $a = 2$ 7 16. $a = 1$ 6 17. $a = -5$ 0

LESSON 9-6

Find each product.

18. 5 · (−2) −10 19. −3 · (−7) 21 20. −4 · 4 −16 21. 8 · (−9) −72

Evaluate $3x$ for each value of x.

22. $x = -5$ −15 23. $x = 8$ 24 24. $x = -9$ −27 25. $x = 0$ 0

LESSON 9-7

Find each quotient.

26. 20 ÷ (−4) −5 27. −48 ÷ (−6) 8 28. −24 ÷ 8 −3 29. −18 ÷ (−2) 9

Evaluate $\frac{n}{4}$ for each value of n.

30. $n = -36$ −9 31. $n = 44$ 11 32. $n = -12$ −3 33. $n = -60$ −15

LESSON 9-8

Solve each equation. Check your answers.

34. $5 + y = 1$ $y = -4$ 35. $b - 8 = -6$ $b = 2$ 36. $-6 + m = -2$ $m = 4$

Solve each equation. Check your answers.

37. $-6g = 30$ $g = -5$ 38. $-3c = -9$ $c = 3$ 39. $7r = -42$ $r = -6$
40. $\frac{n}{-8} = 7$ $n = -56$ 41. $\frac{x}{-2} = -6$ $x = 12$ 42. $\frac{s}{6} = -9$ $s = -54$

Extra Practice ■ Chapter 10

10A Perimeter, Area, and Circumference

LESSON 10-1

Find the perimeter P of each rectangle.

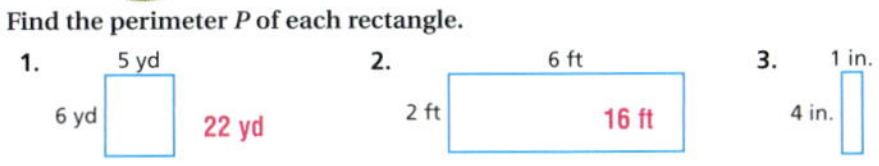

1. 22 yd
2. 16 ft
3. 10 in.

Find each unknown measure.

4. What is the value of b if the perimeter equals 82 cm? $b = 20$ cm

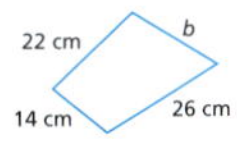

5. What is the perimeter of the polygon? 36 cm

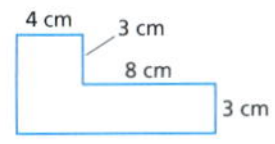

LESSON 10-2

Find the area of each figure.

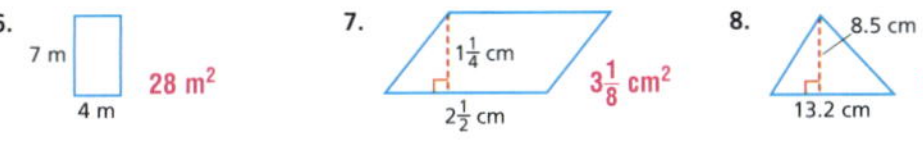

6. 28 m²
7. $3\frac{1}{8}$ cm²
8. 56.1 cm²

LESSON 10-3

Find the area of each polygon.

9. 45 in²
10. 204 ft²

LESSON 10-4

Find how the perimeter and area of each figure change when the dimensions change.

11. Divide each measurement by 3. The perimeter decreases by a factor of 3, and the area decreases by a factor of 3^2, or 9.
12. Multiply each measurement by 2. The perimeter increases by a factor of 2, and the area increases by a factor of 2^2, or 4.

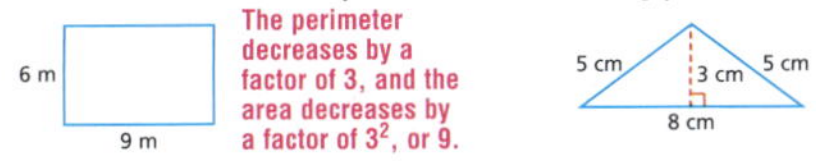

LESSON 10-5

Find each missing value to the nearest hundredth. Use 3.14 for π.

13. $d = 6$ ft, $C = ?$ 18.84 ft
14. $r = 12$ cm, $C = ?$ 75.36 cm
15. $C = 17.27$ in., $d = ?$ 5.5 in.

Extra Practice ■ Chapter 10

10B Volume and Surface Area

LESSON 10-6

Identify the number of faces, edges, and vertices on each solid figure.

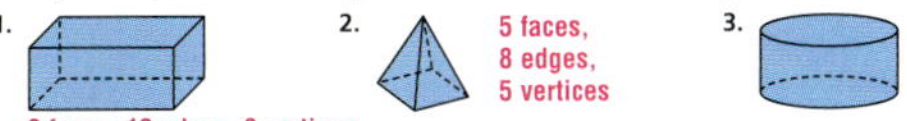

1. 6 faces, 12 edges, 8 vertices
2. 5 faces, 8 edges, 5 vertices
3. 2 faces, 2 edges, 0 vertices

LESSON 10-7

Find the surface area S of each solid figure.

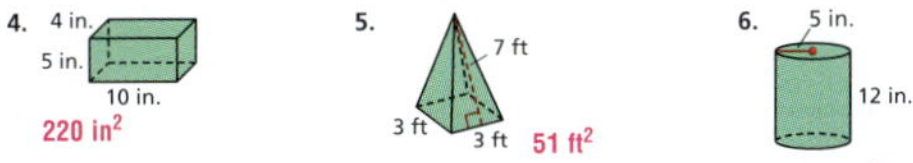

4. 220 in²
5. 51 ft²
6. 533.8 in²

LESSON 10-8

Find the volume of each prism.

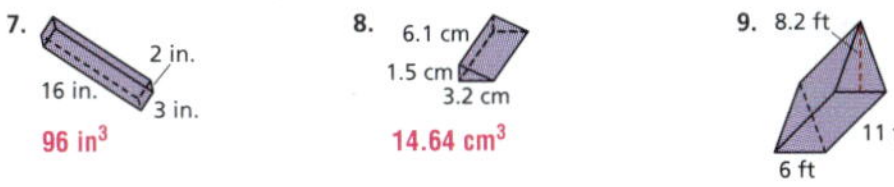

7. 96 in³
8. 14.64 cm³
9. 270.6 ft³

LESSON 10-9

Find the volume V of each cylinder to the nearest cubic unit.

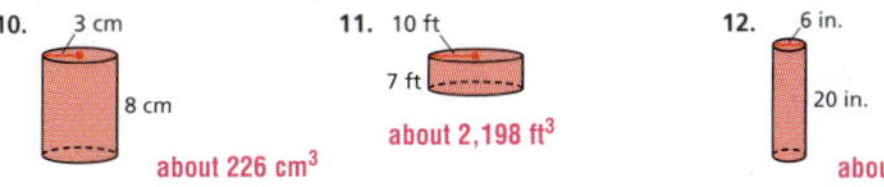

10. about 226 cm³
11. about 2,198 ft³
12. about 565 in³

13. Which cylinder has the greater volume? cylinder A; 3,014.4 cm³ > 2,486.88 cm³

Extra Practice ■ Chapter 11

11A Understanding Probability

LESSON 11-1

Write *impossible, unlikely, as likely as not, likely,* or *certain* to describe each event.

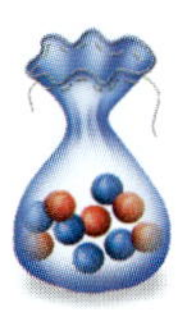

1. picking a green marble from this bag of marbles impossible
2. picking a red marble from this bag of marbles as likely as not
3. The chance of winning a sweepstakes is 3%. Write this probability as a decimal and as a fraction. 0.03, $\frac{3}{100}$
4. A particular brand of cereal is offering a prize in each box. There is a 34% chance the toy will be a rubber ball, a 50% chance it will be a small figurine, and a 16% chance it will be a game. Is it more likely that the prize will be a rubber ball or a game? a rubber ball

LESSON 11-2

For each experiment, identify the outcome shown and the sample space.

5.  blue; {yellow, blue, red, green}
6. 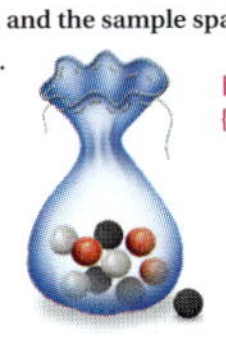 black; {black, red, white}

For one month, Maggie recorded the weather. She organized her results in a frequency table.

Weather	Sunny	Cloudy	Rainy
Frequency	17	6	7

7. Find the experimental probability of cloudy weather. $\frac{1}{5}$
8. Based on Maggie's findings, is it more likely that the weather will be cloudy or rainy? rainy

LESSON 11-3

9. What is the probability of rolling an even number on a fair number cube? $\frac{1}{2}$
10. What is the probability of randomly choosing the letter *T* from the letters *M, A, T, H, E, M, A, T, I, C, S*? $\frac{2}{11}$
11. The weather report stated that there is a 42% chance of snow today. What is the probability that it will **not** snow? 58%
12. During its grand opening, a store is giving away prizes. The chance of winning a prize is 0.16. Find the probability of **not** winning a prize. 0.84

Extra Practice ■ Chapter 11

11B Using Probability

LESSON 11-4

1. Miguel is buying a new car. He has three choices for the exterior color: black, silver, or blue. He has two choices for the interior color: black or brown. How many different color combinations can Miguel choose from? 6
2. For breakfast, Brianna can have oatmeal, cold cereal, or eggs and then a banana, an apple, or an orange. How many different breakfast combinations can Brianna choose from? 9
3. At summer camp, the campers participate in 3 different activities each day: hiking, swimming, and arts and crafts. How many different ways can these 3 activities be arranged? 6

LESSON 11-5

4. Find the probability of spinning red on the spinner and choosing a red marble from the bag. $\frac{1}{5}$
5. Find the probability of spinning yellow and choosing a marble that is not yellow. $\frac{3}{20}$
6. Find the probability of spinning a color that is not blue and choosing a marble that is not blue. $\frac{9}{20}$
7. You toss two fair coins and roll a fair number cube. What is the probability that both coins will land heads up and the cube will show a number greater than 4? $\frac{1}{12}$

LESSON 11-6

8. A local survey stated that 26% of the population has a pet dog. Out of 600 people, how many people can you predict will have a pet dog? 156 people
9. You roll a fair number cube 54 times. How many times do you predict that you will roll a number less than 3? 18 times
10. A promotion team is selling tickets for unreserved seats to a concert. The promotion team estimates that 75% of the people who purchase a ticket will actually attend the concert. If the stadium seats 15,000 people and the promotion team wants to have all of the seats full at the concert, how many concert tickets should they sell? 20,000 tickets

Extra Practice ■ Chapter 12

12A Introduction to Functions

LESSON 12-1

Write an equation for a function that gives the values in each table. Use the equation to find the value of y for the indicated value of x.

1.

x	1	2	3	4	5	10
y	7	9	11	13	15	■

$y = 2x + 5$; 25

2.

x	3	5	7	9	11	13
y	5	11	17	23	29	■

$y = 3x - 4$; 35

3.

x	12	10	8	6	4	2
y	18	■	22	24	26	28

$y = 30 - x$; 20

4.

x	0	4	8	12	16	36
y	2	3	4	5	6	■

$y = \frac{1}{4}x + 2$; 11

Write an equation for the function. Tell what each variable you use represents.

5. The length of a rectangle is 4 cm less than 3 times its width. $\ell = 3w - 4$; w = width of rectangle, ℓ = length of rectangle
6. Darren's age is 5 more than 2 times Nicole's age. $d = 2n + 5$; d = Darren's age, n = Nicole's age
7. Monica is a hair stylist. She kept track of the number of appointments she had for the week. On Thursday, she had 4 appointments for a total of \$112. On Friday, she had 7 appointments for a total of \$196. On Saturday, she had 6 appointments for a total of \$168. Write an equation to represent the function. $m = \$28a$; m = money earned, a = number of appointments

LESSON 12-2

Use the given x-values to write solutions of the equation as ordered pairs.

8. $y = 6x + 2$ for $x = 1, 2, 3, 4$ (1, 8), (2, 14), (3, 20), (4, 26)
9. $y = 5x - 9$ for $x = 2, 3, 4, 5$ (2, 1), (3, 6), (4, 11), (5, 16)

Determine whether the ordered pair is a solution to the given equation.

10. (2, 3); $y = x + 1$ yes
11. (9, 7); $y = 3x - 12$ no
12. (6, 12); $y = 4x - 10$ no
13. (4, 6); $y = \frac{1}{2}x + 4$ yes

Use the graph of the linear function to find the value of y for the given value of x.

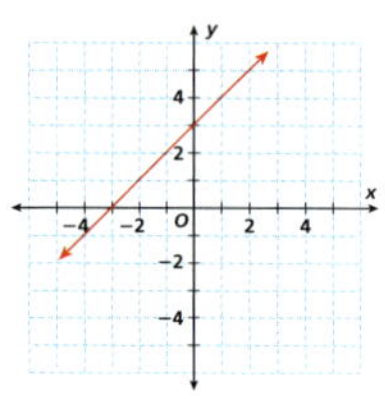

14. $x = 0$ $y = 3$
15. $x = -2$ $y = 1$
16. $x = 2$ $y = 5$

Graph the function described by the equation on a separate sheet of paper.

17. $y = 4x - 3$
18. $y = x + 1$

17–18. See p. A8.

Extra Practice ■ Chapter 12

12B Coordinate Geometry

LESSON 12-3

Give the coordinates of the vertices of the figure after the given translations.

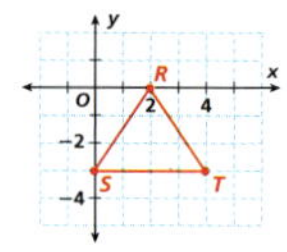

1. Translate triangle RST 3 units right and 4 units up. $R(5, 4)$; $S(3, 1)$; $T(7, 1)$
2. Translate triangle RST 5 units left and 2 units down. $R(-3, -2)$; $S(-5, -5)$; $T(-1, -5)$

LESSON 12-4

Give the coordinates of the vertices of each figure after the given reflection.

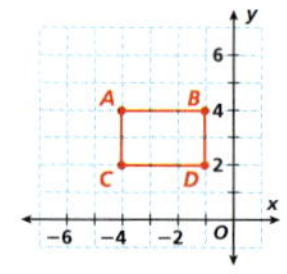

3. Reflect rectangle $ABCD$ across the x-axis. $A(-4, -4)$; $B(-1, -4)$; $C(-4, -2)$; $D(-1, -2)$
4. Reflect rectangle $ABCD$ across the y-axis. $A(4, 4)$; $B(1, 4)$; $C(4, 2)$; $D(1, 2)$

LESSON 12-5

Give the coordinates of the vertices of each figure after the given rotation.

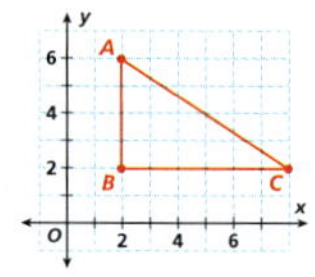

5. Rotate triangle ABC clockwise 90° about the origin. $A(6, -2)$; $B(2, -2)$; $C(2, -8)$
6. Rotate triangle ABC counterclockwise 180° about the origin. $A(-2, -6)$; $B(-2, -2)$; $C(-8, -2)$

LESSON 12-6

7–8. original horizontal dimension: 4 units; original vertical dimension: 2 units

Write the dimensions of the figure. Stretch or shrink the figure as stated, and give the new dimensions.

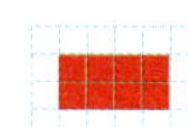

7. Increase the vertical dimension by a factor of 3. new dimensions: horizontal, 4 units; vertical, 6 units
8. Decrease the horizontal dimension by multiplying by $\frac{1}{4}$. new dimensions: horizontal, 1 unit; vertical, 2 units

Skills Bank

Skills Bank → Review Skills

Place Value—Hundreds Through Hundred-thousandths

You can use a place-value chart to read and write numbers.

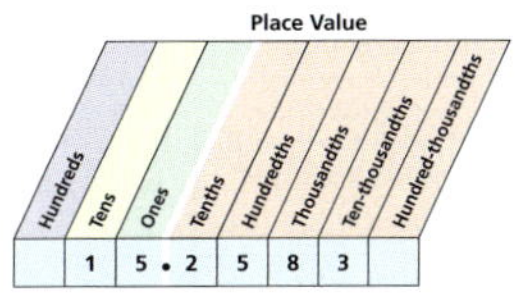

EXAMPLE

What is the place value of the digit 3 in 15.2583?

The digit 3 is in the ten-thousandths place.

PRACTICE

Write the place value of the underlined digit.

1. 0.4562<u>9</u> hundred-thousandths
2. 3<u>4</u>.071 ones
3. 6,<u>1</u>90.05 hundreds
4. 0.208<u>1</u>9 thousandths
5. 1<u>0</u>3.526 tens
6. 3.72<u>1</u>1 hundredths
7. 2.160<u>8</u> ten-thousandths
8. <u>9</u>72.8562 hundreds

Compare and Order Whole Numbers

As you read a number line from left to right, the numbers are ordered from least to greatest.

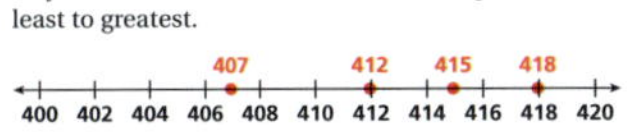

You can use a number line and place value to compare whole numbers. Use the symbols > (is greater than) and < (is less than).

EXAMPLE

Compare. Write <, >, or =.

A 412 ■ 418
418 is to the right of 412 on a number line.
412 < 418

B 415 ■ 407
1 ten is greater than 0 tens.
415 > 407

PRACTICE

Compare. Write <, >, or =.

1. 419 ■ 410 >
2. 9,161 ■ 8,957 >
3. 5,036 ■ 5,402 <
4. 617 ■ 681 <
5. 700 ■ 698 >
6. 1,611 ■ 1,489 >

Round Whole Numbers

You can use a number line or rounding rules to round whole numbers to the nearest 10, 100, 1,000, or 10,000.

EXAMPLE 1

Round 547 to the nearest 10.

Look at the number line.

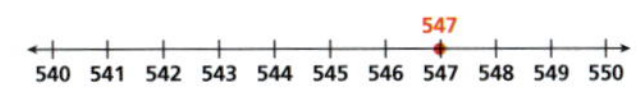

547 is closer to 550 than to 540. So 547 rounded to the nearest 10 is 550.

ROUNDING RULES

If the digit to the right is 5 or greater, increase the digit in the rounding place by 1.

If the digit to the right is less than 5, keep the digit in the rounding place the same.

EXAMPLE 2

Round 12,573 to the nearest 1,000.

12,573 — *Find the digit in the thousands place.*

↑ *Digit is 5 or greater. Add 1.* — *Look at the digit to its right.*

12,573 rounded to the nearest 1,000 is 13,000.

PRACTICE

Round each number to the given place value.

1. 15,638; nearest 100 15,600
2. 37,519; nearest 1,000 38,000
3. 9,298; nearest 10 9,300
4. 69,504; nearest 10,000 70,000
5. 852; nearest 1,000 1,000
6. 33,449; nearest 100 33,400

Round Decimals

You can use rounding rules to round decimals to the nearest whole number, tenth, hundredth, or thousandth.

EXAMPLE

Round each decimal to the given place value.

A 5.16; whole number
1 < 5 So 5.16 rounds to 5.

B 13.4056; hundredth
5 ≥ 5 So 13.4056 rounds to 13.41.

PRACTICE

Round each decimal to the given place value.

1. 3.982; tenth 4.0
2. 6.3174; hundredth 6.32
3. 1.471; whole number 1
4. 48.1526; hundredth 48.15
5. 5.03654; thousandth 5.037
6. 0.083; tenth 0.1

Place Value Patterns

You can use basic facts and place value to solve math problems mentally.

EXAMPLE

Solve mentally.

A 300 + 200
Basic fact: 3 + 2 = 5 — *Think: 3 hundreds + 2 hundreds*
300 + 200 = 500

B 200 × 600
Basic fact: 2 × 6 = 12 — *Think: There are four zeros in the factors, so place four zeros in the product.*
200 × 600 = 120,000

PRACTICE

Solve mentally.

1. 500 + 400 900
2. 80 − 50 30
3. 700 × 30 21,000
4. 2,500 ÷ 50 50
5. 1,200 + 600 1,800
6. 20 × 9,000 180,000
7. 650 − 300 350
8. 320 ÷ 8 40

Roman Numerals

Instead of using place value, as with the decimal system, combinations of letters are used to represent numbers in the Roman numeral system.

I = 1	V = 5	X = 10
L = 50	C = 100	D = 500
M = 1,000		

No letter can be written more than three times in a row. If a letter is written before a letter that represents a larger value, then subtract the first letter's value from the second letter's value.

EXAMPLE

Write each decimal number as a Roman numeral and each Roman numeral as a decimal number.

A 3
3 = I + I + I = III

B 9
9 = X − I = IX

C CLV
CLV = 100 + 50 + 5 = 155

D XC
XC = 100 − 10 = 90

PRACTICE

Write each decimal number as a Roman numeral and each Roman numeral as a decimal number.

1. 12 XII
2. 25 XXV
3. 209 CCIX
4. 54 LIV
5. VIII 8
6. LXXII 72
7. XIX 19
8. MMIV 2004

Addition

Addition is used to find the total of two or more quantities. The answer to an addition problem is called the *sum*.

EXAMPLE

4,617 + 5,682

Step 1: Add the ones.	**Step 2:** Add the tens.	**Step 3:** Add the hundreds. Regroup.	**Step 4:** Add the thousands.
4,617 + 5,682 9	4,617 + 5,682 99	1 4,617 + 5,682 299	1 4,617 + 5,682 10,299

The sum is 10,299.

PRACTICE

Find the sum.

1. 711 + 591 1,302
2. 2,580 + 2,345 4,925
3. 21,470 + 13,329 34,799
4. \$165 + \$304 \$469
5. 6,905 + 872 7,777
6. 47,231 + 3,254 50,485

Subtraction

Subtraction is used to take away one quantity from another quantity or to compare two quantities. The answer to a subtraction problem is called the *difference*. The difference tells how much greater or smaller one number is than the other.

EXAMPLE

780 − 468

Step 1: Subtract the ones. Regroup.	**Step 2:** Subtract the tens.	**Step 3:** Subtract the hundreds.
7 10 7 8 0 − 4 6 8 2	7 10 7 8 0 − 4 6 8 1 2	7 10 7 8 0 − 4 6 8 3 1 2

The difference is 312.

PRACTICE

Find the difference.

1. 6,785 − 2,426 4,359
2. 3,000 − 1,930 1,070
3. 932 − 868 64
4. 41,003 − 22,500 18,503
5. \$1,075 − \$918 \$157
6. 12,035 − 640 11,395

Multiply Whole Numbers

Multiplication is used to combine groups of equal amounts. The answer to a multiplication problem is called the *product*.

EXAMPLE

105 × 214

Step 1: Think of 214 as 2 hundreds, 1 ten, and 4 ones. Multiply by 4 ones.	**Step 2:** Multiply by 1 ten, or 10.	**Step 3:** Multiply by 2 hundreds, or 200.	**Step 4:** Add the partial products.
105 × 214 420 ← 4 × 105	105 × 214 420 1050 ← 10 × 105	105 × 214 420 1050 21000 ← 200 × 105	105 × 214 420 1050 +21000 22,470

The product is 22,470.

PRACTICE

Find the product.

1. 350 × 112 **39,200**
2. 3,218 × 231 **743,358**
3. 187 × 136 **25,432**
4. 5,028 × 225 **1,131,300**
5. 642 × 428 **274,776**
6. 2,039 × 570 **1,162,230**

Multiply by Powers of Ten

You can use mental math to multiply by powers of ten.

EXAMPLE

4,000 × 100

Step 1: Look for a basic fact using the nonzero part of the factors.	**Step 2:** Add the number of zeros in the factors. Place that number of zeros in the product.
4 × 1 = 4	4,000 × 100 = 400,000

The product is 400,000.

PRACTICE

Multiply.

1. 600 × 100 **60,000**
2. 90 × 1,000 **90,000**
3. 2,000 × 10 **20,000**
4. 400 × 10 **4,000**
5. 10,000 × 1,000 **10,000,000**
6. 7,100 × 1,000 **7,100,000**

Divide Whole Numbers

Division is used to separate a quantity into equal groups. The answer to a division problem is known as the *quotient*.

EXAMPLE

672 ÷ 16

Step 1: Write the first number inside the long division symbol and the second number to the left. Place the first digit of the quotient.	**Step 2:** Multiply 4 by 16, and place the product under 67.	**Step 3:** Bring down the next digit of the dividend.
16)672 *16 cannot go into 6, so try 67.*	4 16)672 *Subtract 64 from 67.* − 64 3	42 16)672 *Divide 32 by 16.* − 64↓ 32 −32 0

The quotient is 42.

PRACTICE

Find the quotient.

1. 578 ÷ 34 **17**
2. 736 ÷ 8 **92**
3. 826 ÷ 118 **7**
4. 945 ÷ 45 **21**
5. 6,312 ÷ 263 **24**
6. 5,989 ÷ 53 **113**

Divide with Zeros in the Quotient

Sometimes when dividing, you need to use zeros in the quotient as placeholders.

EXAMPLE

3,648 ÷ 12

Step 1: Divide 36 by 12 because 12 > 3.	**Step 2:** Place a zero in the quotient because 12 > 4.	**Step 3:** Bring down the 8.
3 12)3,648	30 12)3,648 −36↓ 04	304 12)3,648 −36↓ 048 −48 0

The quotient is 304.

PRACTICE

Find the quotient.

1. 424 ÷ 4 **106**
2. 5,796 ÷ 28 **207**
3. 540 ÷ 18 **30**
4. 7,380 ÷ 123 **60**
5. 12,045 ÷ 3 **4,015**
6. 10,626 ÷ 21 **506**

Compatible Numbers

Compatible numbers are numbers that are easy to compute mentally. They are often based on groups of 10 or on basic facts.

EXAMPLE 1

A 7 + 6 + 3 + 4

(7 + 3) + (6 + 4) *Make groups of 10.*

10 + 10

20

B 2 × 32 × 5

(2 × 5) × 32 *Make a group of 10.*

10 × 32

320

EXAMPLE 2

Estimate 358 ÷ 9.

Basic fact: 36 ÷ 9 = 4 *360 is compatible with 9. 360 ÷ 9 = 40*

358 ÷ 9 ≈ 40

PRACTICE

Use compatible numbers to solve.

1. 15 + 42 + 38 + 25 **120**
2. 4 × 3 × 25 **300**
3. 17 + 51 + 23 + 19 **110**
4. 6 × 15 × 4 **360**
5. 11 + 123 + 57 + 9 **200**
6. 2 × 7 × 20 × 5 **1,400**

Estimate by rounding to find compatible numbers.

7. 473 ÷ 80 **6**
8. 118 ÷ 4 **30**
9. 57 ÷ 11 **5**

Mental Math

You can use the Distributive Property to find products mentally.

EXAMPLE

6 × 32

Step 1: Write 32 as the sum of a multiple of 10 and a one-digit number.	**Step 2:** Use the Distributive Property.	**Step 3:** Use mental math to multiply and then to add.
6 × 32 6 × (30 + 2)	6 × (30 + 2) (6 × 30) + (6 × 2)	(6 × 30) + (6 × 2) 180 + 12 = 192

PRACTICE

Use the Distributive Property to find each product.

1. 5 × 66 **330**
2. 3 × 42 **126**
3. 8 × 21 **168**
4. 7 × 84 **588**
5. 5 × 93 **465**
6. 4 × 75 **300**

Properties

Addition and multiplication follow some properties, or laws. Knowing the addition and multiplication properties can help you evaluate expressions.

Addition Properties		
Commutative	You can add numbers in any order.	5 + 1 = 1 + 5
Associative	When you are only adding, you can group any of the numbers together.	(9 + 3) + 2 = 9 + (3 + 2)
Identity Property of Zero	The sum of any number and zero is equal to the number.	9 + 0 = 9

Multiplication Properties		
Commutative	You can multiply numbers in any order.	5 × 8 = 8 × 5
Associative	When you are only multiplying, you can group any of the numbers together.	(4 × 9) × 7 = 4 × (9 × 7)
Identity Property of One	The product of any number and one is equal to the number.	6 × 1 = 6
Property of Zero	The product of any number and zero is zero.	5 × 0 = 0
Distributive	When you multiply a number times a sum, you can find the sum first and then multiply, or multiply each number in the sum and then add.	6 × (4 + 5) = 6 × 4 + 6 × 5

EXAMPLE

Tell which property is shown in the equation (3 + 4) + 7 = 3 + (4 + 7).

The Associative Property of Addition is shown.

PRACTICE

Tell which property is shown. **1–9.** **See p. A8.**

1. 6 × (3 × 2) = (6 × 3) × 2
2. 12 × 9 = 9 × 12
3. 0 + *d* = *d*
4. *k* × 1 = *k*
5. 8 + 5 = 5 + 8
6. 2 × (3 + 10) = (2 × 3) + (2 × 10)
7. 2 + (3 + 4) = (2 + 3) + 4
8. 99 × 0 = 0
9. *y*(3 + 10) = 3*y* + 10*y*

Fractional Part of a Region

You can use fractions to name parts of a whole. The denominator tells how many equal parts are in the whole. The numerator tells how many of those parts are being considered.

EXAMPLE

Tell what fraction of each region is shaded.

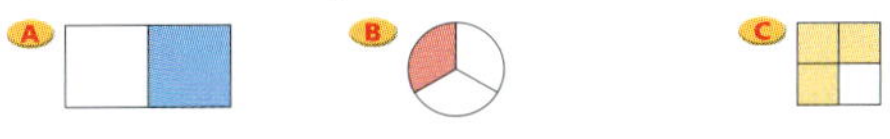

A $\frac{1}{2}$ B $\frac{1}{3}$ C $\frac{3}{4}$

PRACTICE

Tell what fraction of each region is shaded.

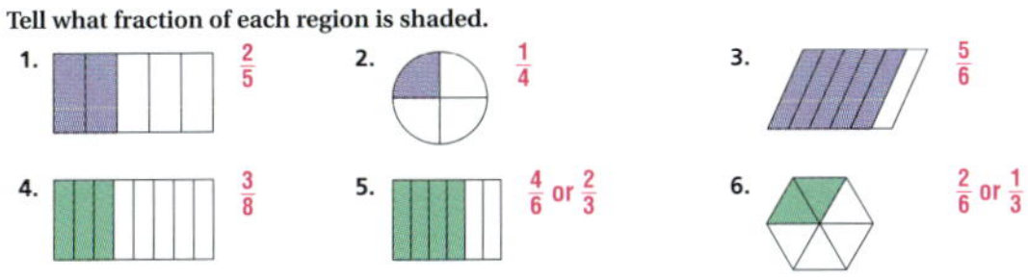

1. $\frac{2}{5}$
2. $\frac{1}{4}$
3. $\frac{5}{6}$
4. $\frac{3}{8}$
5. $\frac{4}{6}$ or $\frac{2}{3}$
6. $\frac{2}{6}$ or $\frac{1}{3}$

Fractional Part of a Set

You can use fractions to name part of a set. The denominator tells how many items are in the set. The numerator tells how many of those items are being used.

EXAMPLE

Tell what fraction of each set are stars.

A ■☆■☆●☆●■■■

3 out of 10 shapes are stars.

$\frac{3}{10}$ of the shapes are stars.

B ☆●☆☆●☆☆

5 out of 7 shapes are stars.

$\frac{5}{7}$ of the shapes are stars.

PRACTICE

Tell what fraction of each set is shaded.

1. ☆☆☆☆☆☆ $\frac{3}{6}$, or $\frac{1}{2}$
2. ■■■■□ $\frac{4}{5}$
3. ●☆○○☆○ $\frac{2}{6}$, or $\frac{1}{3}$
4. ■■■□□ $\frac{3}{5}$
5. ●○□□ $\frac{1}{4}$
6. ☆●■○□△♡ $\frac{3}{7}$

Pictographs

Pictographs are graphs that use pictures to display data. Pictographs include a key to tell what each picture represents.

EXAMPLE

How many students chose red as their favorite color?

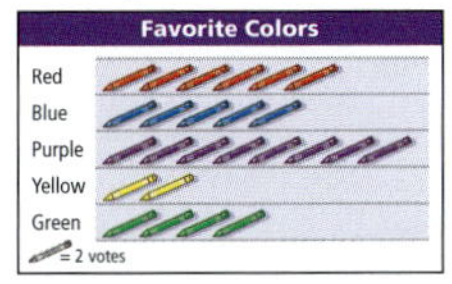

Each ✏ stands for 2 students.

There are 6 ✏ in the row for red.

$6 \times 2 = 12$

So 12 students chose red as their favorite color.

PRACTICE

Use the pictograph for Exercises 1–4.

1. How many tickets did theater A sell? 120 tickets
2. Which theater sold the most tickets? theater B
3. How many more tickets did theater C sell than theater D? 20 more tickets
4. Theater E sold 180 tickets. How would this be shown on the pictograph? 9 ticket symbols

Use the pictograph for Exercises 5–7.

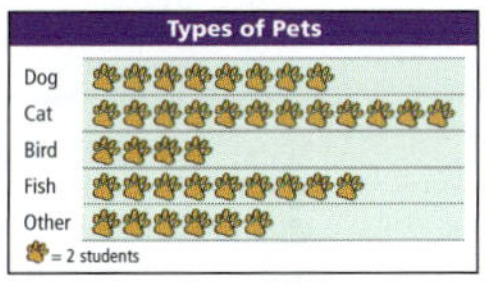

Mr. Carr took a survey of sixth-graders in his school. He asked them which type of pet they have. He recorded the data in a table.

5. How many students have pet birds? 8
6. How many more students have pet cats than pet fish? 6
7. How many students were surveyed? 78
8. Elizabeth took a survey of her neighbors. She recorded the number of children in each family in a table. Use the data to make a pictograph. 8. See p. A8.

Children	Families
0	1
1	6
2	4
3 or more	2

Line Plots

You can display data on a line plot. Line plots use a number line and *x*'s or another symbol to show frequency of values. By looking at a line plot, you can quickly see the range of the data, mode, and outliers.

EXAMPLE 1

Use the line plot to answer the following questions.

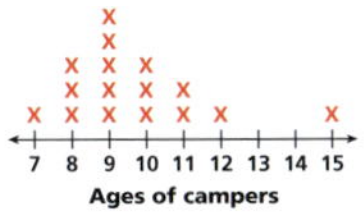

A **What is the range of the data?**
The line plot has data plotted from 7 to 15. The range is 7 to 15. $15 - 7 = 8$

B **How many campers are age 10?**
On the line plot, there are 3 *x*'s above the 10 mark. There are 3 campers who are age 10.

C **What is the mode of the data?**
The mode is the number that occurs most often. Age 9 has the most *x*'s. The mode is 9.

EXAMPLE 2

Students in Mr. Gordon's class ran several miles a week. The number of miles run by the students is recorded in the table. Organize the data in a line plot.

Number of Miles Run	3	4	5	6	7	8	9	10
Number of Students	5	0	6	4	3	7	0	2

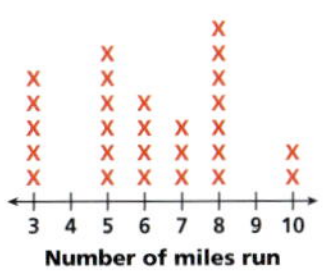

PRACTICE

Mark's baseball coach kept track of the number of hits by each player on the team. He organized the data in a line plot. Use the line plot for Exercises 1–3.

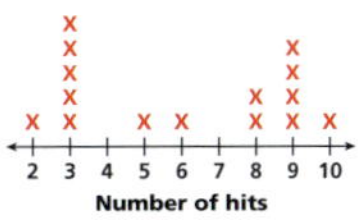

1. How many players had 4 hits? 0 players
2. How many players had more than 6 hits? 7 players
3. What is the mode of the data? 3 hits
4. Betty participates in a summer reading program. The frequency table shows the number of books read by the students in the program. Organize the data in a line plot. 4. See p. A8.

Summer Reading Program						
No. of Books Read	5	8	10	11	12	15
No. of Students	4	9	5	10	6	3

Measure Length to the Nearest $\frac{1}{16}$ Inch 6.5.1, 6.7.6

Each inch on this ruler is separated into 16 equal parts. Each mark is $\frac{1}{16}$ inch.

EXAMPLE

What is the length of the pencil?

Count the number of $\frac{1}{16}$ marks after the 5-inch mark. There are 3 marks.

The pencil is $5\frac{3}{16}$ inches long.

PRACTICE

Use a ruler to find the length of each object to the nearest $\frac{1}{16}$ inch.

1.  $3\frac{15}{16}$ in.
2.  $2\frac{1}{16}$ in.
3.  $1\frac{11}{16}$ in.

Read Scales

A *scale* is similar to a number line with numbers or marks placed at fixed intervals. You can find scales on graphs and on measuring instruments, such as rulers and thermometers.

EXAMPLE

What temperature is shown on the thermometer?

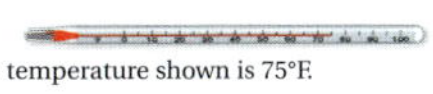

The scale goes from 0°F to 100°F in intervals of 5°F. The temperature shown is 75°F.

PRACTICE

Read each scale.

1. 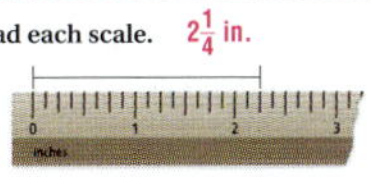 $2\frac{1}{4}$ in.
2.  36°C
3.  $2\frac{1}{2}$ cups

Time

Seconds, minutes, hours, days, weeks, months, and years are units you can use to measure time.

EXAMPLE

Which instrument would you use to measure how long it takes to read a page in a book?

A digital clock shows hours and minutes.

An analog clock shows hours, minutes, and seconds.

A calendar shows days, weeks, months, and years.

Since it would take less than a day to read a page in a book, you could use a digital clock or an analog clock.

PRACTICE

Name the appropriate instrument and unit to measure time for each event.

1. completing 6th grade calendar; months
2. running a mile analog clock; minutes or seconds
3. eating lunch analog clock, digital clock; minutes
4. Earth revolving around the Sun calendar; year

Right Triangle Trigonometry

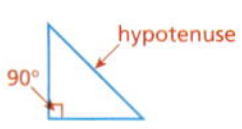

A right triangle has one right angle.
The side opposite the right angle is called the *hypotenuse*.
The hypotenuse is the longest side of a right triangle.
The other sides of a right triangle are called *legs*.

EXAMPLE

Determine if the triangle is a right triangle. If so, identify the hypotenuse.

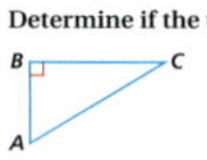

$\triangle ABC$ has a 90° angle.
$\triangle ABC$ is a right triangle.
Line segment CA is the hypotenuse.

PRACTICE

Determine if each triangle is a right triangle. If so, identify the hypotenuse.

1. 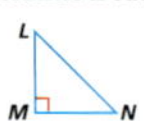 yes; $\overline{LN}$
2. 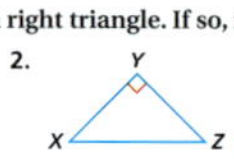 yes; $\overline{XZ}$
3. 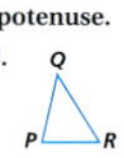 no

Skills Bank Preview Skills

Graph Cumulative Frequency

You have seen how to make a cumulative frequency table for a data set. You can also graph the cumulative frequencies for a data set.

EXAMPLE

The midterm test scores for Mr. Andrews's math class are given in the table at right. Then make a cumulative frequency table. Then make a histogram of the cumulative frequencies.

Midterm Test Scores					
70	86	70	74	77	95
82	62	69	79	7	80
87	68	72	72	91	87
98	73	64	81	77	73
99	76	68	95	85	80

Divide the data into equally sized intervals.

Midterm Score	Frequency	Cumulative Frequency
60–64	2	2
65–69	3	5
70–74	8	13
75–79	4	17
80–84	4	21
85–89	4	25
90–94	1	26
95–99	4	30

The cumulative frequency column shows a running total of all frequencies.

The frequency tells the number of times an event, category, or group occurs.

To make a histogram of the cumulative frequencies, draw a bar for the cumulative frequency for each interval.

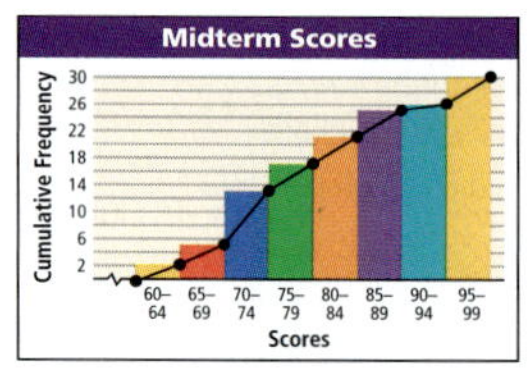

To make a line graph of the cumulative frequencies, place points in the lower left corner of the first bar and upper right corner of every bar. Then connect those points with line segments, as shown.

PRACTICE

1. Make a cumulative frequency histogram and line graph for the data set. See p. A8.

Students' Heights (cm)					
160	130	142	153	164	160
161	162	132	155	140	130
150	145	140	138	166	155
154	155	160	160	155	158

Relative Frequency and Relative Frequency Distributions

In a data set, the relative frequency of a data value is that value's frequency divided by the total number of data values.

$$\text{relative frequency} = \frac{\text{frequency}}{\text{total number of data values}}$$

Relative frequencies can be shown in tables or displayed in histograms.

EXAMPLE

The average class size in 20 schools is given in the table. Make a relative frequency table and a relative frequency histogram of the data.

Average Class Size				
22	25	20	28	31
37	24	19	29	32
38	35	19	32	34
38	25	38	26	33

Divide the data into equally sized intervals.

Class Size	Frequency	Relative Frequency
19–23	4	$\frac{4}{20} = \frac{1}{5}$
24–28	5	$\frac{5}{20} = \frac{1}{4}$
29–33	5	$\frac{5}{20} = \frac{1}{4}$
34–38	6	$\frac{6}{20} = \frac{3}{10}$

There are 20 data points. Divide each frequency by 20 to find the relative frequency.

To make a histogram, draw a bar for each relative frequency.

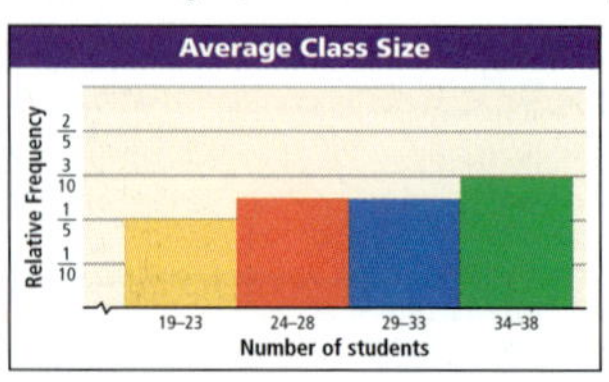

PRACTICE

1. Survey the students in your class and record the number of books read by each student in the past year. Make a relative frequency histogram to display the data. Check students' answers.

Skills Bank · Science Skills

Convert Units 6.5.2, 6.5.9

When you know how different units relate, you can use formulas to convert from one unit to another.

Length	Weight	Capacity
1 foot (ft) = 12 inches (in.)	1 pound (lb) = 16 ounces (oz)	1 pint (pt) = 2 cups (c)
1 yard (yd) = 3 feet	1 ton (T) = 2,000 pounds	1 quart (qt) = 2 pints
1 mile (mi) = 1,760 yards		1 gallon (gal) = 4 quarts

EXAMPLE 1

Complete.

A **24 feet = ▒ yards**

Think: 3 feet = 1 yard

24 feet ÷ 4 feet per yard = 6 yards

24 feet = 6 yards

B **3 gallons = ▒ quarts**

Think: 1 gallon = 4 quarts

3 gallons × 4 quarts per gallon = 12 quarts

3 gallons = 12 quarts

Temperature can be measured in degrees Fahrenheit (°F) or in degrees Celsius (°C). To change from one temperature scale to the other, you can use the following formulas:

$$F = \left(\frac{9}{5} \times C\right) + 32 \qquad C = \frac{5}{9} \times (F - 32)$$

EXAMPLE 2

Convert between temperature scales.

A **50°C = ▒ °F**

$F = \left(\frac{9}{5} \times 50\right) + 32$

$F = 122$

B **68°F = ▒ °C**

$C = \frac{5}{9} \times (68 - 32)$

$C = 20$

PRACTICE

Complete.

1. 72 in. = ▒ ft 6
2. 5 lb = ▒ oz 80
3. 2 mi = ▒ yd 3,520
4. 500 lb = ▒ T 0.25
5. 8 pt = ▒ c 16
6. 5 pt = ▒ qt 2.5
7. 8 yd = ▒ in. 288
8. 2 T = ▒ oz 64,000
9. 3 gal = ▒ c 48

Convert between temperature scales.

10. 32°F = ▒ °C 0
11. 30°C = ▒ °F 86
12. 50°F = ▒ °C 10
13. 100°C = ▒ °F 212
14. 77°F = ▒ °C 25
15. 41°F = ▒ °C 5

Compute Measurements of Combined Units

Sometimes a measurement is given in a combination of units. For example, a piece of wood may measure 3 feet 4 inches. You can add or subtract measurements that are a combination of units.

EXAMPLE 1

4 ft 8 in. + 5 ft 6 in.

Step 1: Line up the units.	Step 2: Add the inches.	Step 3: Add the feet.	Step 4: Rewrite the answer in simplest form.
4 ft 8 in. + 5 ft 6 in.	4 ft 8 in. + 5 ft 6 in. 14 in.	4 ft 8 in. + 5 ft 6 in. 9 ft 14 in.	*Think: 12 in. = 1 ft* 9 ft 14 in. = 10 ft 2 in.

The sum is 10 ft 2 in.

EXAMPLE 2

3 hr 20 min − 1 hr 50 min

Step 1: Line up the units.	Step 2: Regroup if needed.	Step 3: Subtract the minutes.	Step 4: Subtract the hours.
3 hr 20 min − 1 hr 50 min	2 hr 80 min − 1 hr 50 min	2 hr 80 min − 1 hr 50 min 30 min	2 hr 80 min − 1 hr 50 min 1 hr 30 min

The difference is 1 hr 30 min.

PRACTICE

Add.

1. 7 ft 2 in. + 6 ft 8 in. 13 ft 10 in.
2. 8 lb 6 oz + 4 lb 12 oz 13 lb 2 oz
3. 2 gal 1 qt + 4 gal 1 qt 6 gal 2 qt
4. 12 ft 11 in. + 3 ft 4 in. 16 ft 3 in.
5. 4 hr 12 min + 3 hr 42 min 7 hr 54 min
6. 152 yd 2 ft + 75 yd 6 in. 227 yd 2 ft 6 in.
7. 5 yd 2 ft 3 in. + 8 yd 1 ft 8 in. 14 yd 11 in.
8. 2 hr 36 min 45 s + 5 hr 42 min 20 s 8 hr 19 min 5 s

Subtract.

9. 20 ft 8 in. − 7 ft 6 in. 13 ft 2 in.
10. 10 yd 1 ft − 5 yd 2 ft 4 yd 2 ft
11. 6 lb 5 oz − 2 lb 8 oz 3 lb 13 oz
12. 12 h 13 min − 6 h 25 min 5 hr 48 min
13. 5 min 15 s − 4 min 55 s 20 s
14. 3 mi 550 yd − 1 mi 760 yd 1 mi 1,550 yd
15. 4 gal 1 c − 3 qt 1 pt 3 gal 1 pt 1 c
16. 1 day − 8 hr 36 min 15 hr 24 min

Compare Units 6.5.3

When converting area from one unit to another, you must remember that area is measured in square units.

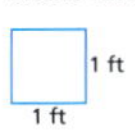

1 square foot = 1 foot × 1 foot
= 12 inches × 12 inches
= 144 square inches

Customary Units for Area	
1 square foot (ft^2) = 144 square inches (in^2)	1 acre (a) = 4,850 square yards (yd^2)
1 square yard (yd^2) = 9 square feet (ft^2)	1 acre (a) = 43,560 square feet (ft^2)
1 square yard (yd^2) = 1,296 square inches (in^2)	1 square mile (mi^2) = 640 acres (a)

Multiply to convert from larger units to smaller units.

Divide to convert from smaller units to larger units.

EXAMPLE 1

Find the area of the rectangle in square feet and in square inches.

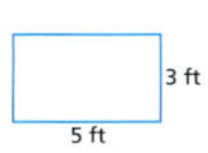

3 ft × 5 ft = 15 ft^2 *Think: 1 ft^2 = 144 in^2*

15 ft^2 = 15 × 144 in^2 = 2,160 in^2

EXAMPLE 2

Which is the greater area, 3 yd^2 or 25 ft^2?

3 yd^2 = 3 × 9 ft^2 = 27 ft^2 *Think: 1 yd^2 = 9 ft^2*

27 ft^2 > 25 ft^2

3 yd^2 > 25 ft^2

PRACTICE

1. Find the area of the rectangle in square yards and square feet. 60 yd^2; 540 ft^2
2. A plot of land is 1.5 miles long and 1 mile wide. What is the area of the land in square miles and in acres? 1.5 mi^2; 960 acres

Compare. Write <, >, or =.

3. 12,500 yd^2 ▒ 3 acres <
4. 6 yd^2 ▒ 42 ft^2 >
5. 4 ft^2 ▒ 576 in^2 =
6. 5 yd^2 ▒ 6,500 in^2 <
7. 2.3 mi^2 ▒ 1,430 acres >
8. 0.5 acre ▒ 21,700 ft^2 >

Surface Area to Volume Ratio

Surface area is the sum of the areas of all the faces or surfaces of a solid figure. *Volume* is the amount of space within the solid figure. Area is a measurement of two dimensions, length and width. Volume is a measure of three dimensions, length, width, and height. A surface area to volume ratio compares the surface area and volume of a solid.

EXAMPLE 1

Find the surface area and volume of the rectangular prism.

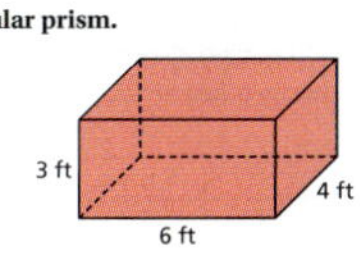

$S = 2wh + 2\ell w + 2\ell h$
$= (2 \times 4 \times 3) + (2 \times 6 \times 4) + (2 \times 6 \times 3)$
$= 24 + 48 + 36$
$= 108 \text{ ft}^2$

$V = \ell \times w \times h$
$= 6 \times 4 \times 3$
$= 72 \text{ ft}^3$

EXAMPLE 2

What is the surface area to volume ratio for the cube?

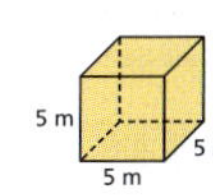

$S = 6s^2$
$= 6 \times 5 \times 5$
$= 150 \text{ m}^2$

$V = \ell \times w \times h$
$= 5 \times 5 \times 5$
$= 125 \text{ m}^3$

The ratio of surface area to volume for the cube is 150 m^2:125 m^3 or 6 m^2:5 m^3.

PRACTICE

Find the surface area and volume of each rectangular prism.

1. 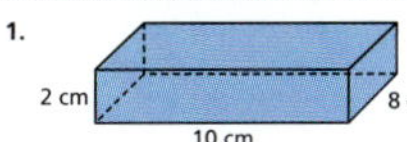

$S = 232 \text{ cm}^2$; $V = 160 \text{ cm}^3$

2. 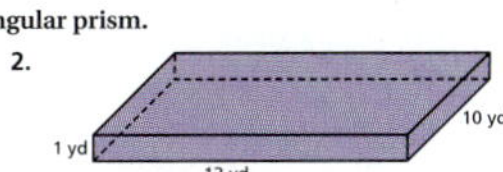

$S = 284 \text{ yd}^2$; $V = 120 \text{ yd}^3$

3. a rectangular prism with $\ell = 13$ km, $w = 10$ km, and $h = 3$ km $S = 398 \text{ km}^2$; $V = 390 \text{ km}^3$
4. a cube with sides of length 2.5 ft $S = 37.5 \text{ ft}^2$; $V = 15.625 \text{ ft}^3$

Write the surface area to volume ratio for each solid.

5. 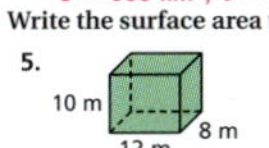

592 m^2:960 m^3, or 37 m^2:60 m^3

6. 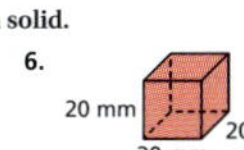

2,400 mm^2:8,000 mm^3, or 3 mm^2:10 mm^3

7. a rectangular prism with $\ell = 5$ ft, $w = 4$ ft, and $h = 11$ ft 238 ft^2:220 ft^3, or 119 ft^2:110 ft^3
8. a rectangular prism with $\ell = 8$ dm, $w = 8$ dm, and $h = 4$ dm 256 dm^2:256 dm^3, or 1 dm^2:1 dm^3

Solve Literal Formulas

Formulas are equations that show a relationship between two or more quantities. Formulas can be used to find missing information or to calculate a quantity. For example, the formula $A = \ell w$ is used to find the area of a rectangle. We can solve the formula $A = \ell w$ for w using the same rules used to solve equations.

EXAMPLE

A **Solve $A = \ell w$ for w.**

$A = \ell w$

$\frac{A}{\ell} = \frac{\ell w}{\ell}$ *Divide both sides by ℓ.*

$\frac{A}{\ell} = w$

B **The formula $V = \ell wh$ is used to find the volume of a rectangular prism. Solve $V = \ell wh$ for h.**

$V = \ell wh$

$\frac{V}{\ell} = \frac{\ell wh}{\ell}$ *Divide both sides by ℓ.*

$\frac{V}{\ell} = wh$

$\frac{V}{\ell w} = \frac{wh}{w}$ *Divide both sides by w.*

$\frac{V}{\ell w} = h$

PRACTICE

Solve.

1. The formula $d = rt$ is used to find distance.
   Solve $d = rt$ for r. $r = \frac{d}{t}$
2. The formula $P = 2\ell + 2w$ is used to find the perimeter of a rectangle.
   Solve $P = 2\ell + 2w$ for ℓ. $\ell = \frac{P - 2w}{2}$
3. The formula $V = \pi r^2 h$ is used to find the volume of a cylinder.
   Solve $V = \pi r^2 h$ for h. $h = \frac{V}{\pi r^2}$
4. The formula $C = \frac{5}{9}(F - 32)$ is used to convert from degrees Fahrenheit to degrees Celsius.
   Solve $C = \frac{5}{9}(F - 32)$ for F. $F = \left(\frac{9}{5} \times C\right) + 32$
5. The formula $A = \frac{1}{2}bh$ is used to find the area of a triangle.
   Solve $A = \frac{1}{2}bh$ for b. $b = \frac{2A}{h}$
6. The formula $I = Prt$ is used to find simple interest.
   Solve $I = Prt$ for P. $P = \frac{I}{rt}$

Exponential Function Behavior

Data that changes exponentially increases or decreases by a common factor.

The Richter scale is used to express the magnitude of earthquakes. Each counting number represents a magnitude that is 10 times stronger than the one before it.

Magnitude	Relative Strength
0	1
1	10^1
2	10^2
3	10^3
4	10^4
5	10^5
6	10^6
7	10^7
8	10^8

EXAMPLE

How much stronger is an earthquake of magnitude 4 than one of magnitude 2?

An earthquake of magnitude 4 has a relative strength of 10^4. An earthquake of magnitude 2 has a relative strength of 10^2.

An earthquake of magnitude 4 is 10^2, or 100, times stronger than an earthquake of magnitude 2.

PRACTICE

1. In 1976, an earthquake in China registered 8 on the Richter scale. In 1999, an earthquake in Colombia registered 6 on the Richter scale. Which earthquake was weaker and by what factor?
   the earthquake in Colombia; 100 times weaker
2. Earthquake A registered 3 on the Richter scale. Earthquake B was 10,000 times stronger than earthquake A. What was the magnitude of earthquake B? 7

You can see exponential population growth by observing bacteria in an environment with unlimited resources. Use the graph of bacteria growth for Exercises 3–5.

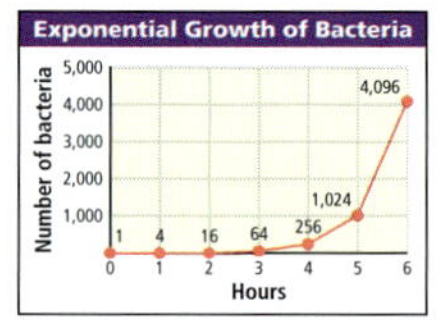

3. By what factor does the graph show the bacteria population increasing each hour? 4
4. If the bacteria continue to grow at this rate, how many bacteria will there be in 8 hours? Write the answer in both exponential and standard form.
   4^8; 65,536 bacteria
5. How many more bacteria will there be in 10 hours than in 8 hours? 4^2, or 16, times more bacteria

Half-life

Half-life is the time that it takes for half of a certain amount of radioactive material to decay. You can use information about the half-life of an element to determine how much of a sample will remain after a given time or to find the age of a sample.

EXAMPLE 1

The half-life of sodium-24 is 15 hours. If you have a 6 g sample of sodium-24, how much will remain after 45 hours?

Every 15 hours, one-half of the sample decays.

Time	0 hours	15 hours	30 hours	45 hours
Amount of Sample	6 g	3 g	1.5 g	0.75 g

After 45 hours, 0.75 g of sodium-24 will remain.

EXAMPLE 2

The half-life of bismuth-212 is 60.5 minutes. If you have a 4 g sample of bismuth-212 from a sample that was originally 16 g, how old is the sample?

Every 60.5 minutes, one-half of the sample decays.

Time	0 min	60.5 min	121 min
Amount of Sample	16 g	8 g	4 g

The sample is 121 minutes old.

PRACTICE

Solve.

1. Radium-226 has a half-life of 1,600 years. How many years will it take for an 8 g sample to decay to 0.5 g? 6,400 years
2. Cobalt-60 has a half-life of 5.26 years. A 10 g sample of cobalt-60 has decayed to 1.25 g. How old is the sample? 15.78 years
3. Iodine-131 has a half-life of 8.07 days. How much of a 4.4 g sample will there be after 40.35 days? 0.1375 g
4. A sample of phosphorus-24 decayed from 12 g to 1.5 g in 42.9 days. What is the half-life of phosphorus-24? 14.3 days
5. You have a 0.6 g sample of sodium-24. The half-life of sodium-24 is 15 hours. The original sample size was 9.6 g. How old is the sample? 60 hours

Selected Answers

Selected Answers

Chapter 1

1-1 Exercises

1. Mount Aconcagua **2.** Mediterranean Sea **3.** 349, 642, 726 **4.** 103, 513, 915 **5.** 497, 809, 1,264 **7.** Mississippi River **9.** 279, 367, 597 **11.** 705, 810, 946 **13.** 111, 1,523, 2,913 **15.** < **17.** > **19.** < **21.** 924, 591, 341 **23.** 911, 747, 439, 291 **25.** 5,480, 5,389, 5,349 **27.** Montana, California, Texas **33.** 300,000 **35.** 6,000 **37.** twenty-four thousand, four hundred ninety-eight **39.** four million, six hundred five thousand, nine hundred twenty-six **41.** 15,903,108

1-2 Exercises

1. 7,000 **2.** 10,000 **3.** 1,500 bottles of water **4.** 300 bottles **5.** 11,000 **7.** 5,000 **9.** 150 softballs **11.** 500 **13.** 3,000 **15.** 40,000 **17.** 900,000 **19.** 10 mi^2 **21.** 40,000 mi^2 **27.** 24,058 **29.** 3,568 **31.** 41 **33.** 10,521 **35.** 70,007

1-3 Exercises

1. 8^3 **2.** 7^2 **3.** 4^4 **4.** 5^5 **5.** 16 **6.** 27 **7.** 625 **8.** 64 **9.** 1,024 people **11.** 9^4 **13.** 6^5 **15.** 3^2 **17.** 243 **19.** 81 **21.** 1 **23.** 100,000,000 **25.** $16 \times 16 \times 16$ **27.** $31 \times 31 \times 31 \times 31 \times 31 \times 31$ **29.** 4 **31.** $17 \times 17 \times 17 \times 17 \times 17 \times 17$ **33.** 1,000,000 **35.** 6,561 **37.** 361 **39.** 57 **41.** > **43.** < **45.** > **47.** 1,024 cells **49.** 8; 2^8 or 256 **51.** 3; 2^3 or 8 **55.** 1,000 + 300 + 50 + 4 **57.** 400,000 + 10,000 + 6,000 + 700 + 90 + 8 **59.** 800 **61.** 20,000 **63.** 2,000,000

1-4 Exercises

1. 33 **2.** 15 **3.** 51 **4.** 50 **5.** 4 **6.** 58 **7.** $138 **9.** 10 **11.** 25 **13.** 47 **15.** 19 **17.** 24 **19.** 1,250 pages **21.** 18 **23.** 62 **25.** 22 **27.** 40 **29.** $(7 + 2) \times 6 - (4 - 3) = 53$ **31.** $5^2 - 10 + (5 + 4^2) = 36$ **33.** $9^2 - 2 \times (15 + 16) - 8 = 11$ **35.** 60 m^3 **39.** 8,245; 8,452; 8,732 **41.** 11,901; 12,681; 12,751 **43.** 50,000 **45.** 121 **47.** 9 **49.** C

1-5 Exercises

1. 40 **2.** 80 **3.** 280 **4.** 320 **5.** 120 **6.** 416 **7.** 156 **8.** 84 **9.** 99 **10.** 156 **11.** 108 **12.** 174 **13.** 40 **15.** 250 **17.** 108 **19.** 426 **21.** 125 **23.** 147 **25.** 40 **27.** 480 **29.** 111 **31.** 153 **33.** 198 **35.** 56 **37.** 275 **39.** 340 **41.** 192 **43.** $175 **45.** 70 plants **47.** $208 **53.** 31,000 **55.** 81 **57.** 64 **59.** B

1-6 Exercises

1. 364 astronauts **2.** 832 medals **3.** 64,890 golf balls **5.** 462 **7.** 111; pencil and paper **9.** 515,844; calculator **11.** 210; mental math **13.** 11,286; calculator **15.** 446,121; calculator **17.** 11,822,400 mi **21.** 4^4 **23.** 10^3 **25.** 14 **27.** C

1-7 Exercises

1. 60, 72, 84 **2.** 30, 15, 0 **3.** 36, 34, 45 **4.** 34, 38, 32 **5.** 18, 486 **6.** 4, 10 **7.** 18, 27 **8.** 64, 4 **9.** 69, 84, 100 **11.** 49, 41, 46 **13.** 24, 720 **15.** 200, 25 **17.** 1, 3, 9, 27, 81 **19.** 100, 93, 86, 79, 72 **21.** 57°F **25.** 3,000 **27.** 14,000 **29.** 125 **31.** 256

Chapter 1 Extension

1. 5 **3.** 7 **5.** 11 **7.** 2 **9.** $(1 \times 16) + (1 \times 8) + (0 \times 4) + (1 \times 2) + (0 \times 1)$ **11.** $(1 \times 16) + (1 \times 8) + (1 \times 4) + (1 \times 2) + (0 \times 1)$ **13.** $(1 \times 16) + (0 \times 8) + (1 \times 4) + (0 \times 2) + (0 \times 1)$ **15.** $(1 \times 16) + (0 \times 8) + (0 \times 4) + (1 \times 2) + (0 \times 1)$ **17.** 1100 **19.** 10010 **21.** 1 **23.** 10110 **25.** > **27.** <

Chapter 1 Study Guide and Review

1. sequence, term **2.** base, exponent **3.** order of operations **4.** evaluate **5.** 8,731; 8,735; 8,737; 8,740 **6.** 53,337; 53,341; 53,452; 53,456 **7.** 8,791; 81,790; 87,091; 87,901 **8.** 2,651; 22,561; 25,615; 26,551 **9.** 91,363; 93,613; 96,361; 96,631 **10.** 10,101; 10,110; 11,010; 11,110 **11.** 1,000 **12.** 6,000 **13.** 20,000 **14.** 800 **15.** 5^3 **16.** 3^4 **17.** 7^5 **18.** 8^2 **19.** 4^4 **20.** 1^3 **21.** 256 **22.** 16 **23.** 27 **24.** 1 **25.** 125 **26.** 100 **27.** 59 **28.** 11 **29.** 26 **30.** 17 **31.** 5 **32.** 45 **33.** 9 **34.** 3 **35.** 30 **36.** 520 **37.** 80 **38.** 1,080 **39.** 40 **40.** 320 **41.** 100 **42.** 130 **43.** 168 **44.** 135 **45.** 204 **46.** 152 **47.** 216 **48.** 165 **49.** 52 **50.** 423 **51.** 62° **52.** 186 bars **53.** 17 and 28 **54.** 13 and 16 **55.** 40 and 120 **56.** 32 and 64

Chapter 2

2-1 Exercises

1. 56, 65 **2.** 108, 120 **3.** $x - 5$ **4.** $w + 9$ **5.** 400, 600 **7.** $x \div 8$ **9.** $x - 2$ **11.** 32 **13.** 77 **15.** 56 **17.** 32 zlotys **23.** 876; 972; 1,298 **25.** 40,000 **27.** 3^3 **29.** 10^4

2-2 Exercises

1. $4{,}028 - m$ **2.** $279 - 125$ **3.** $15x$ **4–7.** Possible answers given. **4.** the sum of r and 87; r plus 87 **5.** the product of 345 and 196; 345 times 196 **6.** the quotient of 476 and 28; 476 divided by 28 **7.** the difference of d and 5; five less than d **9.** $5x$ **11.** $325 \div 25$ **13.** $137 + 675$ **15.** $j - 14$ **17–23.** Possible answers given. **17.** take away 19 from 243; 243 minus 19 **19.** 75 multiplied by 342; the product of 342 and 75 **21.** the product of 45 and 23; 45 times 23 **23.** the difference of 228 and b; b less than 228 **25.** $15 \div d$ **27.** $67m$ **29.** $678 - 319$ **31.** $d \div 4$ **37.** = **39.** 32 **41.** 1,331

2-3 Exercises

1. no **2.** yes **3.** yes **4.** no **5.** yes **6.** no **7.** 53 feet is equal to 636 inches. **9.** no **11.** yes **13.** no **15.** yes **17.** no **19.** yes **21.** no **23.** yes **25.** yes **27.** no **29.** yes **31.** yes **33.** no **35.** 3 **37.** 12 **39.** 13 **41.** no **47.** 81 **49.** 216 **51.** 6 **53.** 75

2-4 Exercises

1. $x = 36$ **2.** $y = 37$ **3.** $n = 19$ **4.** $t = 69$ **5.** $p = 18$ **6.** $c = 16$ **7.** 6 blocks **9.** $r = 7$ **11.** $b = 25$ **13.** $z = 9$ **15.** $g = 16$ **17.** 6 m **19.** $n = 7$ **21.** $y = 19$ **23.** $h = 78$ **25.** $b = 69$ **27.** $t = 26$ **29.** $m = 22$ **31.** 37°C **37.** 20,000 **39.** 100,000 **41.** 36 **43.** C

2-5 Exercises

1. $p = 17$ **2.** $x = 19$ **3.** $a = 31$ **4.** $y = 22$ **5.** $n = 33$ **6.** $d = 41$ **7.** $y = 25$ **9.** $a = 38$ **11.** $a = 97$ **13.** $p = 33$ **15.** $s = 31$ **17.** $x = 36$ **19.** $a = 21$ **21.** $f = 14$ **23.** $r = 154$ **25.** $g = 143$ **27.** $m = 18$ **29.** $k = 64$ **31.** $r - 14{,}162 = 248$; $r = 14{,}410$ ft **37.** 81 **39.** 144 **41.** equation **43.** D

2-6 Exercises

1. $x = 3$ **2.** $w = 9$ **3.** $a = 9$ **4.** $b = 8$ **5.** $c = 11$ **6.** $n = 6$ **7.** 45 ft **9.** $a = 4$ **11.** $x = 4$ **13.** $t = 7$ **15.** $m = 11$ **17.** $387w = 104{,}247$; about 250 to 350 miles **19.** $y = 9$ **21.** $y = 8$ **23.** $y = 20$ **25.** $z = 40$ **27.** $y = 23$ **29.** $y = 18$ **31.** $y = 8$ **33.** $a = 14$ **35.** $x = 3$ **37.** 188 segments **39.** 2 times more **43.** < **45.** 17 **47.** 73

2-7 Exercises

1. $y = 12$ **2.** $z = 28$ **3.** $r = 63$ **4.** 90 min **5.** $d = 36$ **7.** $m = 49$ **9.** $c = 96$ **11.** $c = 165$ **13.** $c = 48$ **15.** $c = 180$ **17.** $c = 432$ **19.** 15 seconds **23.** 162, 486 **25.** 14 **27.** 123

Chapter 2 Extension

1., **3.**, **5.**, **7.**, **9.**

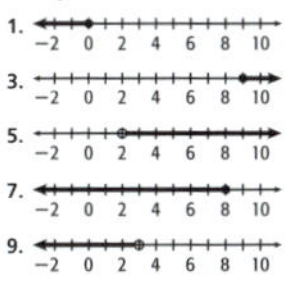

11. $y \geq 5$ **13.** $c > 1$ **15.** $r \leq 14$ **17.** $r \leq 3$ **19.** $k < 4$ **21.** $p > 11$;

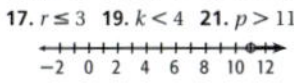

23. $a < 20{,}320$

Chapter 2 Study Guide and Review

1. algebraic expression **2.** equation **3.** variable **4.** constant **5.** 7, 6 **6.** 6, 10 **7.** $p \times 6$ **8.** $s \div 2$ **9.** $15 + b$ **10.** 6×5 **11.** $9t$ **12.** $g \div 9$ **13–19.** Possible answers given. **13.** the product of 4 and z; 4 times z **14.** 54 divided by 6; the quotient of 54 and 6 **15.** 3 minus y; the difference of 3 and y **16.** y minus 3; the difference of y and 3 **17.** 15 plus x; the sum of 15 and x **18.** m divided by 20; the quotient of m and 20 **19.** the sum of 5,100 and 64; 64 added to 5,100 **20.** yes **21.** no **22.** yes **23.** yes **24.** no **25.** no **26.** $x = 6$ **27.** $n = 14$ **28.** $c = 29$ **29.** $y = 6$ **30.** $p = 27$ **31.** $w = 9$ **32.** $b = 11$ **33.** $n = 44$ **34.** $p = 16$ **35.** $d = 57$ **36.** $k = 45$ **37.** $d = 9$ **38.** $p = 63$ **39.** $n = 67$ **40.** $r = 14$ **41.** $w = 144$ **42.** $h = 60$ **43.** $p = 167$ **44.** $v = 8$ **45.** $y = 9$ **46.** $c = 7$ **47.** $n = 2$ **48.** $s = 8$ **49.** $t = 10$ **50.** $a = 8$ **51.** $y = 8$ **52.** $r = 42$ **53.** $t = 15$ **54.** $y = 18$ **55.** $n = 72$ **56.** $z = 52$ **57.** $b = 100$ **58.** $n = 77$ **59.** $p = 90$

Chapter 3

3-1 Exercises

1. 1 + 0.9 + 0.08; one and ninety-eight hundredths **2.** 10.041; 10 + 0.04 + 0.001 **3.** 0.0765; seven hundred sixty-five ten-thousandths **4.** 0.04 + 0.007 + 0.0002; four hundred seventy-two ten-thousandths **5.** osmium **6.** 9.35, 9.5, 9.65 **7.** 4.09; 4.1; 4.18 **8.** 12.09, 12.39, 12.92 **9.** 7 + 0.08 + 0.009 + 0.0003; seven and eight hundred ninety-three ten-thousandths **11.** 7.15; 7 + 0.1 + 0.05 **13.** the Chupaderos meteorite **15.** 1.5, 1.56, 1.62 **17.** nine and seven thousandths **19.** ten and twenty-two thousandths **21.** > **23.** = **25.** three hundredths **27.** one tenth **29.** 4.034, 1.43, 1.424, 1.043, 0.34 **31.** Ross 154 **33.** Alpha Centauri, Proxima Centauri **37.** > **39.** 3^5 **41.** 13^3

3-2 Exercises

1. about 12 miles **2.** 1.8 **3.** 12 **4.** 16.20 **5.** 5.5 **6.** 10 **7.** 120 **8.** 7 **9.** from 44 to 46.5 **10.** from 40 to 42 **11.** about 450 miles **13.** 3.4 **15.** 5.157 **17.** 20 **19.** 6 **21.** from 14 to 17 **23.** 48 **25.** 17 **27.** $0.22, $0.10, $0.08, $0.04 **29.** $(12 \times 8) - (18 \times 4) = 24$; about 24 cents **37.** 10,000 **39.** 24 **41.** C

3-3 Exercises

1. 20.2 mi **2. a.** 5.95 mi **b.** 12.65 mi **3.** 3 **4.** 5.6 **5.** 8 **6.** 4.9 **7.** 2.98 **8.** 3.55 **9.** 0.5888 **10.** 4.948 **11.** 36.115 **13.** 9 **15.** 4 **17.** 6.4 **19.** 25 **21.** 8.15 **23.** 2.46 **25.** 9.81 **27.** 25.839 **29.** 4.4308 **31.** 52.836 **33.** 29.376 **35.** 84.966 **37.** $72.42 **39.** 0.196 **43.** 343 **45.** 100,000 **47.** 108 **49.** $4m + 2$

3-4 Exercises

1. 593,700 **2.** 0.71925 **3.** 609,120 **4.** cm **5.** L **6.** g **7.** mL **8.** 7,000 **9.** 5 **10.** 500 **11.** 0.18 **13.** 15,090 **15.** 741,000 **17.** 4.2516 **19.** m **21.** mL **23.** 0.25 **25.** 18 **27.** 10,000 **29.** 0.06087 **31.** 11.18 **33.** 0.06 **35.** 7,540 **37.** table B; 91.6 cm longer **43.** $a = 34$ **45.** $w = 75$ **47.** $p = 9$ **49.** $t = 80{,}000$

3-5 Exercises

1. 6.2×10^4 **2.** 5.0×10^5 **3.** 6.913×10^6 **4.** 1.3×10^5 **5.** 7.015×10^6 **6.** 2.0×10^4 **7.** 6,793,000 **8.** 14,000 **9.** 382,000 **10.** 94,010,000 **11.** 3,300 **12.** 18,850 **13.** 9.0×10^4 **15.** 1.607×10^6 **17.** 6.0×10^6 **19.** 1.8×10^3 **21.** 5.04×10^7 **23.** 1,630,000 **25.** 21,400 **27.** 811,640,000 **29.** 91,060,000 **31.** 7,210 **33.** 720 **35.** 6,954 **37.** 1.1205×10^5 **39.** 4.562×10^3 **41.** 6.5342×10^4 **43.** 4.0×10^3 **45.** 1.8×10^4 **47.** 1.95×10^4 **49.** 3×10^5 km/s; 1.125×10^3 ft/s **51.** $150{,}000 = 1.5 \times 10^5$ **53.** 579,000,000; 5.79×10^8 **59.** Multiply by three, then add 5; 276, 281 **61.** $y = 52$ **63.** D

3-6 Exercises

1. $1.68 **2.** $74.32 **3.** 0.24 **4.** 0.0040 **5.** 0.21 **6.** 0.072 **7.** 16.52 **8.** 22.90 **9.** 35.63 **11.** $1.96 **13.** 2.25 **15.** 0.128 **17.** 0.12 **19.** 0.000015 **21.** 17.227 **23.** 1.148 **25.** 2.5914 **27.** 0.294 **29.** 26.46 **31.** 1.6632 **33.** 12.2122 **35.** 15.662 **37.** 73.5 **39.** Mercury and Mars **41.** 3.42 lb **45.** multiply by 7, subtract 4 **47.** 163 + 24 **49.** $y \div 8$ or $\frac{y}{8}$

3-7 Exercises

1. 0.23 **2.** 0.12 **3.** 0.35 **4.** 0.18 **5.** 0.078 **6.** 0.052 **7.** 0.104 **8.** 0.026 **9.** $8.82 **11.** 0.22 **13.** 0.27 **15.** 0.171 **17.** 0.076 **19.** 0.9 **21.** 2.1432 **23.** 0.0989 **25.** 0.126 **27.** 14.371 **29.** $13.25 **35.** 225,971; 2,004,801; 298,500,004 **37.** < **39.** C

3-8 Exercises

1. 5 **2.** 34.5 **3.** 17 **4.** 6 **5.** 6 **6.** 264.125 **7.** 54.6 mi/h **8.** 9 lb **9.** 6 **11.** 8 **13.** $213.\overline{3}$ **15.** $11.3\overline{1}$ **17.** 5 **19.** 11.6 gal **21.** 6.3 **23.** 191.1 **25.** 184.74 **27.** 1,270 **29.** 201,000 **31.** 12.2 **33.** 12.2 **35.** 9.44 **37.** 232 bills; $4,640 **39.** 63.5 mi/h **41.** 78.38 mi **45.** 360 **47.** 360 **49.** $y = 12$ **51.** 8.304, 8.05, 8.009 **53.** 30.75, 30.709, 30.211

3-9 Exercises

1. 10 belts **2.** 6 packs **3.** 2.25 m **5.** 8 bunches **7.** 3 packs **9.** 4 floors **13.** 45; 54; 63 **15.** 1,366 **17.** 6

3-10 Exercises

1. $a = 7.1$ **2.** $n = 1.4$ **3.** $c = 12.8$ **4.** $x = 6.01$ **5.** $d = 3.488$ **6.** $m = 0.4$ **7.** 60.375 m^2 **8.** 16.2 cm **9.** $b = 9.3$ **11.** $r = 20.8$ **13.** $a = 10.7$ **15.** $f = 6.56$ **17.** $z = 4$ **19.** $3.00/kg **21.** $t = 51.9$ **23.** $m = 8.1$ **25.** $m = 4.367$ **27.** $w = 78.034$ **29.** $c = 36.14$ **31.** $a = 4.6$ **33. a.** 19.5 units, 21 units **b.** 50.5 units **35. a.** 1,900,000 kg **b.** 1.9×10^6 kg **37.** 9 capsules **43.** 30 **45.** $n = 18$ **47.** A

Chapter 3 Extension

1. 4 **3.** 1 **5.** 5 **7.** 14.3 **9.** 17.72 **11.** 35.61

Chapter 3 Study Guide and Review

1. front-end estimation **2.** Scientific notation **3.** Clustering **4.** 5 + 0.6 + 0.08; five and sixty-eight hundredths **5.** 1 + 0.007 + 0.0006; one and seventy-six ten-thousandths **6.** 1 + 0.2 + 0.003; one and two hundred three thousandths **7.** 20 + 3 + 0.005; twenty-three and five thousandths **8.** 1.12, 1.2, 1.3 **9.** 11.07, 11.17, 11.7 **10.** 0.033, 0.3, 0.303 **11.** 5.009, 5.5, 5.950 **12.** 11.32 **13.** 2.3 **14.** 80 **15.** 9 **16.** 24.85 **17.** 5.3 **18.** 2.58 **19.** 2.8718 **20.** 126,000 **21.** 0.546 **22.** 6,700,000 **23.** 1.806 **24.** 8,900 **25.** 0.18 **26.** 5.5×10^5 **27.** 7.23×10^3 **28.** 1.3×10^6 **29.** 1.48×10^1 **30.** 30,200 **31.** 429,300 **32.** 1,700,000 **33.** 5,390 **34.** 9.44 **35.** 0.865 **36.** 0.0072 **37.** 24.416 **38.** 1.03 **39.** 0.72 **40.** 3.85 **41.** 2.59 **42.** $3.64 **43.** 8.1 **44.** $6.1\overline{6}$ **45.** $3.87\overline{6}$ **46.** 52.275 **47.** 0.75 m **48.** 14 containers **49.** 9 cars **50.** $a = 13.38$ **51.** $y = 2.62$ **52.** $n = 2.29$ **53.** $p = 60.2$ **54.** $5.00

Chapter 4

4-1 Exercises

1. 2, 4 **2.** 2, 3, 4, 6, 9 **3.** none **4.** 3, 9 **5.** composite **6.** prime **7.** composite **8.** composite **9.** 3 **11.** 3, 5, 9 **13.** 2, 4 **15.** 2 **17.** composite **19.** prime **21.** composite **23.** prime **25.** composite **27.** prime **29.** no, no, no, no **31.** yes, no, yes, no, no, no, no **39.** 53, 59, 61, 67, 71, 73, 79, 83, 89, and 97 **41.** 2, 3, 6 **43.** 90 **49.** < **51.** > **53.** $16n$ **55.** $b = 27$ **57.** $y = 29$

4-2 Exercises

1. 1, 2, 3, 4, 6, 12 **2.** 1, 3, 7, 21 **3.** 1, 2, 4, 13, 26, 52 **4.** 1, 3, 5, 15, 25, 75 **5.** $2^4 \cdot 3$ **6.** $2^2 \cdot 5$ **7.** $2 \cdot 3 \cdot 11$ **8.** $2 \cdot 17$ **9.** 1, 2, 3, 4, 6, 8, 12, 24 **11.** 1, 2, 3, 6, 7, 14, 21, 42 **13.** 1, 67 **15.** 1, 5, 17, 85 **17.** 7^2 **19.** $2^2 \cdot 19$ **21.** 3^4 **23.** $2^2 \cdot 5 \cdot 7$ **33.** $3^2 \cdot 11$ **35.** $2^2 \cdot 71$ **37.** $2^3 \cdot 3 \cdot 5 \cdot 7$ **39.** $2^2 \cdot 5 \cdot 37$ **41.** $2^2 \cdot 5^2$ **43.** 7^3 **45.** 340; $2^2 \cdot 5 \cdot 17$ **47.** 142; $2 \cdot 71$ **51.** 7,000 **53.** 40,000 **55.** 625 **57.** 81 **59.** $p = 8$ **61.** $b = 77$

4-3 Exercises

1. 9 **2.** 8 **3.** 7 **4.** 15 **5.** 6 **6.** 9 **7.** 4 arrangements **9.** 14 **11.** 2 **13.** 4 **15.** 12 **17.** 3 teams **19.** 12 **21.** 5 **23.** 2 **25.** 75 **27.** 4 **29.** 5 **31.** 4 **33.** 3 **35.** 6 rows **37.** 12 **41.** 51 **43.** 23 **45.** $n = 3$ **47.** $a = 13$ **49.** C

4-4 Exercises

1. $\frac{3}{20}$ **2.** $1\frac{1}{4}$ **3.** $\frac{43}{100}$ **4.** $2\frac{3}{5}$ **5.** 0.4 **6.** 2.875 **7.** 0.125 **8.** 4.1 **9.** 0.21, $\frac{2}{3}$, 0.78 **10.** $\frac{1}{6}$, $\frac{5}{16}$, 0.67 **11.** $\frac{1}{9}$, 0.3, 0.52 **13.** $5\frac{71}{100}$ **15.** $3\frac{23}{100}$ **17.** $2\frac{7}{10}$ **19.** $6\frac{3}{10}$ **21.** 1.6 **23.** 3.275 **25.** 0.375 **27.** 0.625 **29.** $\frac{1}{9}$, 0.29, $\frac{3}{8}$ **31.** $\frac{1}{10}$, 0.11, 0.13 **33.** 0.31, $\frac{3}{7}$, 0.76 **35.** $92\frac{3}{10}$ **37.** $107\frac{17}{100}$ **39.** $0.1\overline{6}$; repeats **41.** $0.41\overline{6}$; repeats **43.** > **45.** $4\frac{1}{2}$, 4.48, 3.92 **47.** 125.25, 125.205, $125\frac{1}{5}$ **49.** Jill **55.** distributive **57.** 2; 4.32 **59.** 4; 16.7552 **61.** C

4-5 Exercises

1–4. Possible answers given. **1.** $\frac{2}{3}$, $\frac{8}{12}$ **2.** $\frac{1}{4}$, $\frac{2}{8}$ **3.** $\frac{1}{2}$, $\frac{5}{10}$ **4.** $\frac{3}{8}$, $\frac{9}{24}$ **5.** 25 **6.** 3 **7.** 21 **8.** $\frac{1}{5}$ **9.** $\frac{1}{3}$ **10.** $\frac{1}{4}$ **11.** $\frac{3}{5}$ **13–19.** Possible answers given. **13.** $\frac{1}{5}$, $\frac{5}{25}$ **15.** $\frac{1}{6}$, $\frac{2}{12}$ **17.** $\frac{2}{5}$, $\frac{8}{20}$ **19.** $\frac{3}{5}$, $\frac{9}{15}$ **21.** 8 **23.** 6 **25.** 140 **27.** $\frac{2}{3}$ **29.** $\frac{3}{7}$ **31.** $\frac{1}{7}$ **33.** $\frac{2}{7}$ **35.** $\frac{1}{4} = \frac{2}{8}$ **37.** $\frac{2}{4} = \frac{1}{2}$ **39.** $\frac{8}{10} = \frac{4}{5}$ **41.** Use two of the $\frac{1}{4}$ tsp measuring spoons; use four of the $\frac{1}{8}$ tsp measuring spoons. **43.** 12 bracelets **45.** 644, 640, 271, 204 **47.** < **49.** < **51.** < **53.** C

4-6 Exercises

1. > **2.** < **3.** = **4.** yes **5.** $\frac{1}{5}$, $\frac{3}{8}$, $\frac{2}{3}$ **6.** $\frac{1}{4}$, $\frac{1}{3}$, $\frac{2}{5}$ **7.** $\frac{1}{8}$, $\frac{2}{7}$, $\frac{5}{9}$ **9.** < **11.** < **13.** = **15.** $\frac{3}{7}$, $\frac{1}{2}$, $\frac{3}{5}$ **17.** $\frac{1}{3}$, $\frac{3}{8}$, $\frac{4}{9}$ **19.** $\frac{5}{9}$, $\frac{13}{18}$, $\frac{5}{6}$ **21.** < **23.** > **25.** > **27.** $\frac{3}{10}$, $\frac{2}{5}$, $\frac{1}{2}$ **29.** $\frac{1}{5}$, $\frac{7}{15}$, $\frac{2}{3}$ **31.** $\frac{1}{2}$, $\frac{7}{12}$, $\frac{5}{8}$ **33.** advertising **35.** United States, China, Brazil **39.** 6.8 **41.** 4.9 **43.** 3.7 **45.** 3.1 **47.** 4.5×10^1 **49.** 3.19×10^2 **51.** 4.05×10^5 **53.** D

4-7 Exercises

1. $2\frac{2}{5}$ **2.** $\frac{5}{4}$ **3.** $\frac{8}{3}$ **4.** $\frac{9}{7}$ **5.** $\frac{12}{5}$ **7.** $8\frac{3}{5}$ **9.** $\frac{20}{9}$ **11.** $\frac{13}{3}$ **13.** $\frac{25}{6}$ **15.** $\frac{19}{5}$ **17.** 4; whole number **19.** $8\frac{3}{5}$; mixed number **21.** $8\frac{7}{10}$; mixed number **23.** 15; whole number **25.** $\frac{53}{11}$ **27.** $\frac{93}{5}$ **29.** ■ = 7; ● = 11 **31.** ■ = 7; ● = 7 **33.** ■ = 18; ● = 3 **35.** ■ = 2; ● = 15 **37.** ■ = 3; ● = 3 **39.** ■ = 1; ● = 17 **41.** $\frac{141}{5}$; $\frac{73}{2}$ **43.** femur, tibia, fibula, humerus, ulna **49.** 81 **51.** 256 **53.** $y = 42$ **55.** B

4-8 Exercises

1. $\frac{1}{2}$ ft **2.** $1\frac{2}{5}$ **3.** $7\frac{1}{7}$ **4.** $3\frac{1}{3}$ **5.** $5\frac{1}{6}$ **6.** $\frac{3}{5}$ **7.** $\frac{2}{5}$ **8.** $\frac{6}{5}$, or $1\frac{1}{5}$ **9.** $\frac{1}{5}$ **11.** $\frac{2}{7}$ **13.** $1\frac{3}{5}$ **15.** $\frac{6}{5}$, or $1\frac{1}{5}$ **17.** $\frac{1}{10}$ **19.** $\frac{5}{8}$ **21.** $\frac{14}{33}$ **23.** $\frac{2}{3}$ **25.** $13\frac{1}{3}$ **27.** $\frac{17}{24}$ **29.** $\frac{2}{13}$ **31.** $\frac{11}{17}$ **33.** $15\frac{1}{4}$ **35.** $1\frac{3}{4}$ hr **37.** 1 ft **43.** 1 **45.** Add 7; 31 **47.** Add 3, then subtract 2; 6

4-9 Exercises

1. $\frac{8}{9}$ **2.** $\frac{2}{5}$ **3.** 3 **4.** $3\frac{1}{9}$ **5.** $\frac{3}{7}$ **6.** $\frac{8}{11}$ **7.** 6 **8.** 6 **9.** 8 **10.** 6 **11.** 9 **12.** 10 **13.** 27 boys **15.** $\frac{3}{4}$ **17.** $\frac{4}{5}$ **19.** $\frac{6}{11}$ **21.** 10 **23.** 6 **25.** 2 **27.** 7 **29.** 15 **31.** 5 **33.** $\frac{48}{5}$, or $9\frac{3}{5}$ **35.** 45 **37.** > **39.** = **41.** < **43.** > **45.** 253, $225\frac{1}{2}$, 231 **51.** base = 4, exponent = 8 **53.** base = 12, exponent = 1 **55.** 38 **57.** 128

Chapter 4 Extension

1. A: 0, 2, 4, 6, … B: 1, 3, 5, 7, …

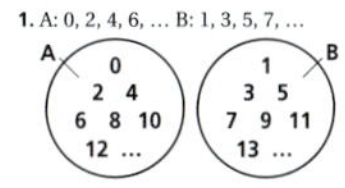

intersection: empty
union: all whole numbers

3. A: 1, 2, 3, 4, 6, 8, 9, 12, 18, 24, 36, 72 B: 1, 2, 3, 4, 6, 9, 12, 18, 36

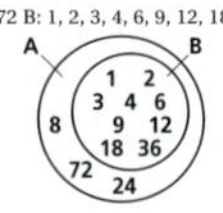

intersection: 1, 2, 3, 4, 6, 9, 12, 18, 36
union: 1, 2, 3, 4, 6, 8, 9, 12, 18, 24, 36, 72
5. yes **7.** no

Chapter 4 Study Guide and Review

1. improper fraction; mixed number **2.** repeating decimal; terminating decimal **3.** prime number; composite number
4. 2 **5.** 2, 3, 5, 6, 9, 10 **6.** 2, 3, 6, 9
7. 2, 4 **8.** 2, 5, 10 **9.** 3
10. composite **11.** composite
12. prime **13.** composite
14. prime **15.** composite
16. composite **17.** prime
18. composite **19.** prime
20. 1, 2, 3, 4, 5, 6, 10, 12, 15, 20, 30, 60 **21.** 1, 2, 3, 4, 6, 8, 9, 12, 18, 24, 36, 72 **22.** 1, 29 **23.** 1, 2, 4, 7, 8, 14, 28, 56 **24.** 1, 5, 17, 85
25. 1, 71 **26.** $5 \cdot 13$ **27.** $2 \cdot 47$
28. $2 \cdot 5 \cdot 11$ **29.** 3^4 **30.** $3^2 \cdot 11$
31. $2^2 \cdot 19$ **32.** 97 **33.** $5 \cdot 11$
34. $2 \cdot 23$ **35.** 12 **36.** 25 **37.** 9
38. $\frac{37}{100}$ **39.** $1\frac{4}{5}$ **40.** $\frac{2}{5}$ **41.** 0.875
42. 0.4 **43.** $0.\overline{7}$
44–46. Possible answers given.
44. $\frac{2}{3}, \frac{8}{12}$ **45.** $\frac{8}{10}, \frac{16}{20}$ **46.** $\frac{1}{4}, \frac{2}{8}$
47. $\frac{7}{8}$ **48.** $\frac{3}{10}$ **49.** $\frac{7}{10}$ **50.** > **51.** >
52. $\frac{3}{8}, \frac{2}{3}, \frac{7}{8}$ **53.** $\frac{3}{12}, \frac{1}{3}, \frac{4}{6}$ **54.** $\frac{34}{9}$
55. $\frac{29}{12}$ **56.** $\frac{37}{7}$ **57.** $3\frac{5}{6}$ **58.** $3\frac{2}{5}$
59. $5\frac{1}{8}$ **60.** 1 **61.** $\frac{3}{4}$ **62.** $\frac{3}{5}$ **63.** $6\frac{5}{7}$
64. $\frac{5}{7}$ **65.** $\frac{3}{4}$ **66.** $2\frac{4}{7}$ **67.** $\frac{8}{9}$

Chapter 5

5-1 Exercises

1. $\frac{1}{6}$ **2.** $\frac{1}{10}$ **3.** $\frac{3}{7}$ **4.** $\frac{1}{2}$ **5.** $\frac{1}{6}$ **6.** $\frac{4}{33}$
7. $\frac{2}{15}$ **9.** $\frac{3}{25}$ **11.** $\frac{1}{15}$ **13.** $\frac{2}{7}$ **15.** $\frac{7}{15}$
17. $\frac{9}{20}$ **19.** $\frac{2}{15}$ **21.** $\frac{1}{8}$ **23.** $\frac{3}{20}$
25. $\frac{4}{15}$ **27.** $\frac{1}{14}$ **29.** = **31.** $\frac{1}{4}$ cup
33. a. Multiply by $\frac{1}{4}$. **b.** $\frac{1}{12}$
35. $\frac{3}{8}$ **39.** 21 **41.** > **43.** < **45.** J

5-2 Exercises

1. $\frac{5}{6}$ **2.** $\frac{2}{3}$ **3.** $\frac{11}{14}$ **4.** $1\frac{1}{7}$ **5.** $\frac{13}{15}$ **6.** $\frac{8}{11}$
7. $2\frac{1}{16}$ **8.** $2\frac{3}{5}$ **9.** $21\frac{5}{7}$ **11.** $\frac{5}{7}$ **13.** $\frac{13}{14}$
15. $1\frac{1}{3}$ **17.** $1\frac{1}{3}$ **19.** $2\frac{2}{7}$ **21.** $15\frac{1}{2}$
23. $23\frac{1}{2}$ **25.** $3\frac{3}{5}$ **27.** $2\frac{3}{4}$ **29.** $2\frac{1}{3}$
31. $\frac{1}{6}$ **33.** $13\frac{3}{4}$ **35.** $16\frac{1}{10}$ **37.** $2\frac{1}{5}$
39. 2 **41.** $10\frac{2}{9}$ **43.** $2\frac{2}{15}$ **45.** 240 people **47.** 90 people **53.** $k = 45$
55. D

5-3 Exercises

1. $\frac{7}{2}$ **2.** $\frac{9}{5}$ **3.** $\frac{9}{1}$, or 9 **4.** $\frac{11}{3}$ **5.** $\frac{5}{18}$
6. $1\frac{5}{7}$ **7.** $\frac{1}{12}$ **8.** 4 **9.** $\frac{9}{50}$ **10.** $\frac{1}{2}$
11. $\frac{8}{7}$ **13.** $\frac{8}{3}$ **15.** $\frac{11}{8}$ **17.** $\frac{7}{6}$ **19.** $\frac{7}{32}$
21. $\frac{2}{27}$ **23.** $\frac{3}{10}$ **25.** $1\frac{1}{4}$ **27.** $\frac{2}{9}$
29. $1\frac{17}{25}$ **31.** $4\frac{2}{3}$ **33.** $3\frac{6}{7}$ **35.** $2\frac{2}{11}$
37. $\frac{33}{50}$ **39.** $\frac{4}{5}$ **41.** $1\frac{8}{27}$ **43.** yes
47. yes **49.** yes **51.** greater than
53. 25 in. **55.** 16 bags **61.** add 3; 16 **63.** B

5-4 Exercises

1. $z = 16$ **2.** $n = \frac{3}{20}$ **3.** $x = 7\frac{1}{2}$
4. 24 **5.** $t = \frac{2}{21}$ **7.** $r = 15$
9. $y = 20$ **11.** 4 cans **13.** $m = 1\frac{1}{2}$
15. $z = \frac{7}{40}$ **17.** $b = 14$ **19.** $w = 2$
21. $d = \frac{1}{32}$ **23.** $n \div 4 = \frac{1}{2}, n = 2$
25. a. $\frac{3}{8}$ cup **b.** $1\frac{1}{2}$ cups
27. 30 pages **29. a.** 200 people **b.** 10 people **33.** 11,000
35. 15,000 **37.** 11,000 **39.** 718,000
41. 4.2034 **43.** 503 **45.** C

5-5 Exercises

1. 3 packs of pencils and 4 packs of erasers **2.** 15 **3.** 36 **4.** 6 **5.** 20
6. 12 **7.** 48 **8.** 24 **9.** 40 **10.** 30
11. 63 **12.** 45 **13.** 150 **15.** 8
17. 20 **19.** 18 **21.** 12 **23.** 24
25. 66 **27.** 60 **29.** 140 **31.** 12
33. a–b. Possible answers given.
a. 16, 20, 28 **b.** 18, 30, 42 **c.** 12
d. 120, 144, 168, and 192 **39.** 679; 879; 978 **41.** yes **43.** no **45.** 20.8
47. 710,000

5-6 Exercises

1. about 1 **2.** about $\frac{1}{2}$ **3.** about $\frac{1}{2}$
4. about 0 **5.** 16 mi **6.** 1 mi
7. about $\frac{1}{2}$ **9.** about 0 **11.** 4 tons
13. $3\frac{1}{2}$ tons **15.** > **17.** < **19.** >
21. $1\frac{1}{2}$ **23.** $6\frac{1}{2}$ **25.** $30\frac{1}{2}$ **27.** $1\frac{1}{4}$ in.
29. 1 in. **33.** 8^5 **35.** 7^6
37. $a \times 2 + 4$ **39.** 16.06 **41.** 3.12

5-7 Exercises

1. $\frac{5}{12}$ ton **2.** $\frac{4}{9}$ **3.** $\frac{3}{10}$ **4.** $\frac{4}{15}$ **5.** $\frac{13}{14}$
7. $\frac{1}{6}$ cup **9.** $\frac{7}{12}$ **11.** $\frac{9}{20}$ **13.** $1\frac{2}{15}$
15. $1\frac{1}{8}$ **17.** $\frac{7}{15}$ **19.** $\frac{7}{18}$ **21.** $\frac{1}{3}$
23. $\frac{28}{33}$ **25.** $\frac{7}{18}$ **27.** $\frac{11}{14}$ **29.** 0 **31.** $\frac{1}{2}$
33. $\frac{3}{4}$ **35.** $\frac{13}{18}$ **37.** $\frac{2}{3}$ lb **39.** $\frac{9}{40}$ lb
45. 16 **47.** 343 **49.** G

5-8 Exercises

1. $10\frac{5}{12}$ **2.** $4\frac{13}{24}$ **3.** $6\frac{1}{12}$ **4.** $5\frac{5}{14}$
5. $4\frac{1}{4}$ **7.** $6\frac{7}{12}$ **9.** $6\frac{1}{4}$ **11.** $9\frac{7}{10}$
13. $8\frac{1}{3}$ **15.** $34\frac{1}{2}$ **17.** $3\frac{51}{90}$ **19.** $20\frac{13}{36}$
21. $12\frac{5}{24}$ **23.** $\frac{7}{10}$ **25.** $2\frac{3}{5}$ **27.** $13\frac{1}{4}$
29. 5 **31.** $1\frac{1}{12}$ **33.** $34\frac{2}{3}$
35. a. $26\frac{3}{5}$ lb **b.** $2\frac{1}{10}$ lb **c.** $11\frac{1}{10}$ lb
37. $\frac{1}{2}$ mi **39.** $7\frac{1}{2}$ yd **41.** $5\frac{1}{6}$ yd
45. 146,500 **47.** 209,467,000
49. $\frac{2}{15}$ **51.** $\frac{1}{12}$ **53.** A

5-9 Exercises

1. $\frac{3}{4}$ **2.** $5\frac{4}{9}$ **3.** $1\frac{2}{3}$ **4.** $2\frac{1}{3}$ **5.** $2\frac{3}{5}$
7. $3\frac{4}{5}$ **9.** $7\frac{7}{8}$ **11.** $4\frac{13}{18}$ **13.** $2\frac{1}{5}$
15. $1\frac{3}{7}$ **17.** $4\frac{1}{4}$ **19.** $2\frac{13}{20}$ **21.** $3\frac{7}{8}$
23. $2\frac{7}{8}$ **25.** $3\frac{8}{11}$ **27.** $4\frac{11}{12}$ **29.** $1\frac{5}{6}$
31. $\frac{1}{12}$ **33.** $13\frac{5}{12}$ **35.** $\$4\frac{1}{5}$
37. $1\frac{1}{12}$ yd^2 **39.** $1\frac{11}{12}$ yd^2
43. 16 **45.** $\frac{3}{28}$ **47.** $\frac{3}{14}$ **49.** J

5-10 Exercises

1. $4\frac{1}{2}$ **2.** $8\frac{4}{9}$ **3.** $5\frac{5}{8}$ **4.** $4\frac{1}{10}$
5. $57\frac{3}{4}$ in. **7.** $3\frac{5}{8}$ **9.** $4\frac{5}{6}$
11. $1\frac{1}{2}$ **13.** $1\frac{1}{6}$ **15.** $9\frac{7}{12}$
17. $1\frac{1}{8}$ **19.** $7\frac{7}{18}$ **21.** $4\frac{23}{33}$
23. $9\frac{17}{24}$ **25.** $9\frac{9}{10}$ **27.** $12\frac{3}{8}$
29. $4\frac{3}{4}$ ft **31.** $53\frac{9}{10}$
33. a. $15\frac{3}{4}$ min **b.** yes **39.** 9,198; 10,462; 11,320 **41.** A

Chapter 5 Study Guide and Review

1. reciprocals **2.** least common denominator **3.** $\frac{1}{3}$ **4.** $\frac{15}{28}$ **5.** $\frac{1}{10}$
6. $\frac{7}{25}$ **7.** $\frac{5}{81}$ **8.** $\frac{3}{14}$ **9.** $\frac{9}{10}$ **10.** $1\frac{1}{4}$
11. 2 **12.** $\frac{4}{21}$ **13.** $\frac{3}{20}$ **14.** $\frac{5}{9}$
15. $a = \frac{1}{8}$ **16.** $b = 2$ **17.** $m = 17\frac{1}{2}$
18. $g = \frac{2}{15}$ **19.** $r = 10\frac{4}{5}$ **20.** $s = 50$
21. 30 **22.** 48 **23.** 27 **24.** 60
25. 225 **26.** 660 **27.** about 1
28. about $\frac{1}{2}$ **29.** about 11
30. about $2\frac{1}{2}$ **31.** $\frac{33}{40}$ **32.** $\frac{3}{4}$
33. $\frac{1}{15}$ **34.** $\frac{5}{24}$ **35.** $4\frac{7}{10}$ **36.** $3\frac{1}{18}$
37. $11\frac{43}{60}$ **38.** $4\frac{1}{12}$ **39.** $1\frac{13}{20}$
40. $2\frac{3}{8}$ **41.** $\frac{11}{30}$ gal **42.** $3\frac{2}{3}$ **43.** $1\frac{1}{2}$
44. $5\frac{2}{3}$ **45.** $2\frac{5}{8}$ **46.** $6\frac{13}{14}$ **47.** $1\frac{1}{8}$
48. $4\frac{3}{4}$ ft **49.** $30\frac{3}{20}$ **50.** $14\frac{11}{12}$
51. $5\frac{5}{12}$ **52.** $3\frac{4}{9}$ **53.** $5\frac{7}{15}$ **54.** 7 oz

Chapter 6

6-1 Exercises

1.

Day	High Temperature (°F)
Mon	72
Tue	75
Wed	68
Thu	62
Fri	55

2. Possible answer: The daily high peaked on Tuesday and then dropped for the remainder of the week. The temperature will continue to drop over the weekend.

3.

Test	Grade
1st	70
2nd	75
3rd	80
4th	85
5th	90

5.

Date	Thickness (in.)
December 3	1
December 18	2
January 3	5
January 18	11
February 3	17

Possible answer: around January 10
9. 5.234×10^6 **11.** 1.2078×10
13. 0.08 **15.** D

6-2 Exercises

1. range = 19, mean = 54, median = 51, mode = 48
3. range = 23, mean = 57.2, median = 55, no mode
5.

State	Mean Score
Connecticut	509
Maine	500
Massachusetts	513
New Hampshire	519
Rhode Island	500
Vermont	508

range = 19, mean = 508.2, median = 508.5, mode = 500
9. $\frac{4}{11}, \frac{5}{8}, \frac{2}{3}$ **11.** $\frac{18}{35}$ **13.** $\frac{7}{12}$ **15.** B

6-3 Exercises

1. a. mean = 4.75, median = 5, no mode **b.** mean = 10, median = 7, no mode **2.** with: mean = 45.4, median = 42, no mode; without: mean = 40.2, median = 40, no mode **3.** mean = 225, median = 187.5, mode = 240; median
5. with: mean = 710.4, median = 788, no mode; without: mean = 877.75, median = 868, no mode
7. mean ≈ 118.29, median = 128, no mode **13.** 1, 2, 3, 4, 6, 9, 12, 18, 36 **15.** 3 **17.** 4 **19.** B

6-4 Exercises

1. green **2.** black, white, red
3.

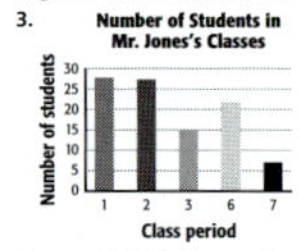

4.

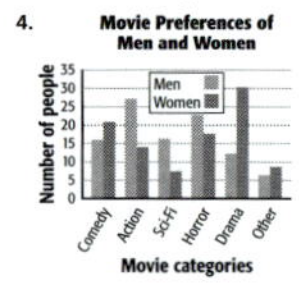

5. orange
7.

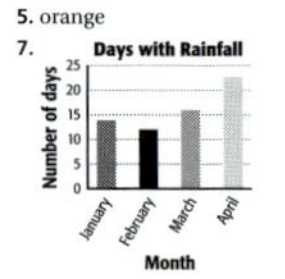

9. 14 million mi^2
11. about 8.14 million mi^2
17. $7z$ **19.** $\frac{12}{5}$ **21.** $\frac{29}{4}$ **23.** A

6-5 Exercises

1.

Type of Instrument	
Trumpet	卌
Drums	II
Tuba	I
Trombone	III
French horn	IIII

2.

Type of Instrument		
Age	Frequency	Cumulative Frequency
Trumpet	5	5
Drums	2	7
Tuba	1	8
Trombone	3	11
French horn	4	15

3.

Number of Years of Each Presidential Term			
Number (Intervals)	0–4	5–8	9–12
Frequency	26	15	1

4.

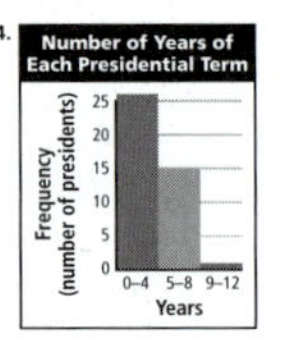

5.

Pets	
Dog	卌 I
Cat	卌
Bird	IIII
Fish	III
Hamster	II

7.

Final Medal Standing at the Summer Olympic Games for the Top 25 Countries					
Number (intervals)	0–20	21–40	41–60	61–80	81–100
Frequency	14	8	3	0	2

9a.

Favorite Sport	
Basketball	II
Football	IIII
Track and field	III
Soccer	II
Hockey	II
Tennis	I
Baseball	I

b.

Favorite Sport		
Age	Frequency	Cumulative Frequency
Basketball	2	2
Football	4	6
Track & field	3	9
Soccer	2	11
Hockey	2	13
Tennis	1	14
Baseball	1	15

11.

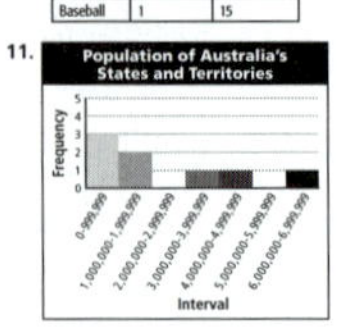

15. $1 + 0.2 + 0.03$; one and twenty-three hundredths
17. $20 + 6 + 0.07$; twenty-six and seven hundredths **19.** $\frac{1}{6}$
21. $\frac{9}{2}$ **23.** C

6-6 Exercises

1. (2, 3) **2.** (0, 7) **3.** (7, 6) **4.** (9, 1)
5. (4, 5) **6.** (11, 4)
7–10.

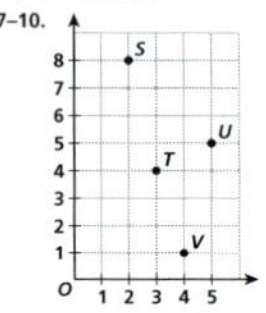

11. (3, 0) **13.** (1, 4) **15.** (11, 7)
17–22.

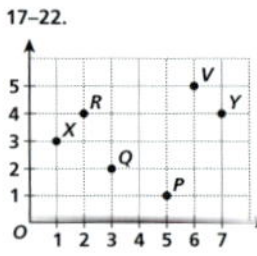

23. A **25.** C **27.** P **29.** (9, 8)
31. (1, 5) **33.** (9, 0) **39.** $2 \cdot 3^2$
41. $3 \cdot 11$ **43.** $1\frac{1}{2}$ min

6-7 Exercises

1.

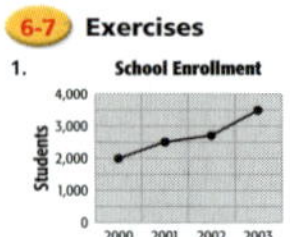

2. 2000 **3.** decrease
4.

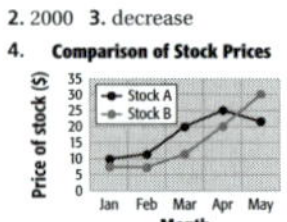

5.

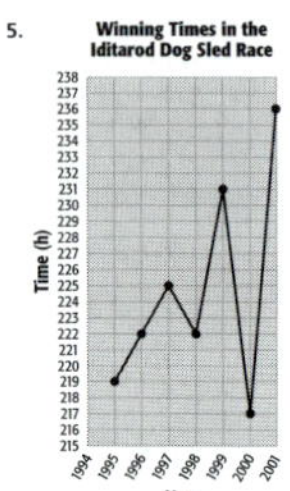

7. 1999 **9.** about 17 lb
11.

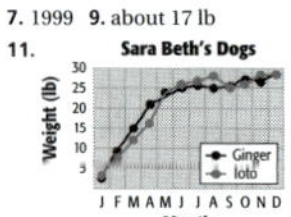

15. Distributive Property
17. \$5.00

6-8 Exercises

1. The top bar represents 40 years, but the other bars represent only 10 years. **2.** that the community center had more volunteers in the past than now **3.** Kerry does not begin biking from home.
4. It appears that Kerry biked farther in the 30-minute period when she actually did not. **5.** The vertical axis begins at 430 rather than at zero. **7.** The yearly increments change. **9.** Possible answer: line graph, because you can easily see the changes from month to month **15.** 60 **17.** 120
19. $\frac{3}{28}$ **21.** $\frac{3}{14}$

6-9 Exercises

1.

Stems	Leaves
3	7 9
4	0 5 8
5	1 6

Key: 3|7 means 37

2. 10 **3.** 44 **4.** 27.8 **5.** 32
6. no mode **7.** 34 **9.** 41
11. 52 **13.** 42 **15.** B
17.

Stems	Leaves
8	0 1 2 3 4 7 8 9
9	2 4 4 5 9
10	0 1 3 9
11	
12	4 5

Key: 8|0 means 80

21. 21.47 **23.** 23.45 **25.** D

Chapter 6 Extension

1. 2 **3.** 13 **5.** 15
7.

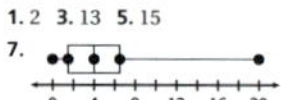

9. Class 1 **11.** 18

Chapter 6 Study Guide and Review

1. histogram, bar graph
2. ordered pair **3.** mode
4.

Snake Lengths (ft)	
Anaconda	35 ft
Diamond python	21 ft
King cobra	19 ft
Boa constrictor	16 ft

5. range = 7; mean = 37; median = 38; mode = 39
6. with outlier: mean ≈ 14.29, median = 11, mode = 12; without outlier: mean ≈ 10.33, median = 10.5, mode = 12
7. with outlier: mean = 31, median = 32, mode = 32; without outlier: mean = 35.75, median = 33, mode = 32
8. with outlier: mean ≈ 19.67; median = 14, mode = none; without outlier: mean = 13.2; median = 13, mode = none
9. 8th grade
10.

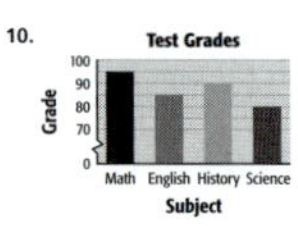

11.

Points Scored			
Points (Intervals)	3–5	6–8	9–11
Frequency	3	2	1

12.

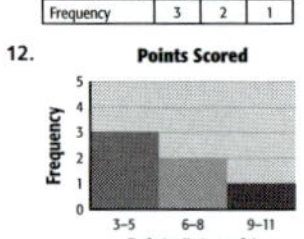

13. (4, 1)
14. (3, 2)
15.

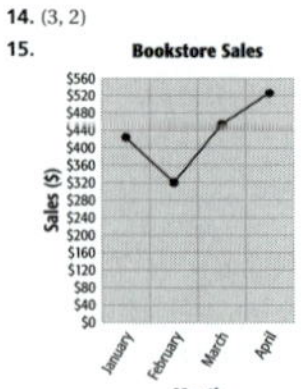

16. April
17. Sales decreased from January to February and then increased from February to April.
18. The scale starts out in increments of 1 mile and then changes to 5 miles.
19. Basketball Scores

Stems	Leaves
2	0 2 6 8
3	4
4	0 4 6

Key: 2|0 means 20

20. least value = 20, greatest value = 46, mean = 32.5, median = 31, no mode, range = 26

Chapter 7

7-1 Exercises

1–2. Possible answers given.
1. M and N **2.** $\overleftrightarrow{KN}$ **3.** K
4–6. Possible answers given.
4. JKL **5.** $\overrightarrow{AC}$ and $\overrightarrow{AB}$
6. $\overrightarrow{AC}, \overrightarrow{BC}, \overrightarrow{BA}$, and $\overrightarrow{CA}$ **7.** $\overleftrightarrow{AB}$
9–13. Possible answers given.
9. $\overrightarrow{DF}$; $\overrightarrow{ED}$ **11.** FGH **13.** $\overline{WX}, \overline{XY}, \overline{YZ}, \overline{ZY}, \overline{YW}$, and $\overline{ZX}$ **15.** C
17–19. Possible answers given.
17. $\overrightarrow{CA}$ and $\overrightarrow{CB}$ **19.** $\overrightarrow{BC}$ and $\overrightarrow{CB}$
21. 2 mi **25.** $7p$ **27.** 51.85
29. 17

7-2 Exercises

1. 90°, right **2.** 135°, obtuse
3. 60°, acute
4.

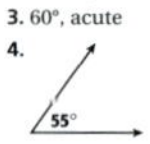

5.

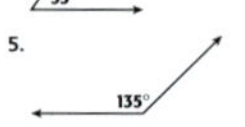

6.

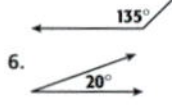

7.

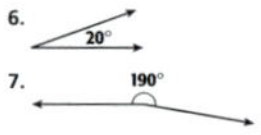

8. actual measure: 140°
9. actual measure: 55°
10. actual measure: 30°
11. 40°, acute **13.** 180°, straight
15. 20°, acute
17.

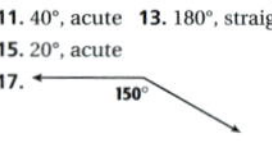

19.

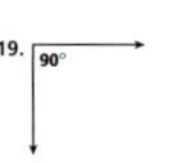

21.

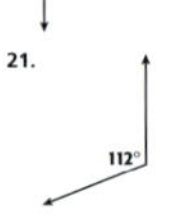

23. 

25. actual measure: 53°
27. actual measure: 125°
29–31. Possible answers given.

29. 31. 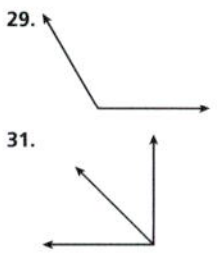

33. obtuse 35. acute angle
41. 64 43. 1 45. 2.8 47. 2.4

7-3 Exercises

1. adjacent 2. adjacent
3. m∠a = 9° 4. m∠b = 30°
5. adjacent 7. m∠c = 78°
9. angles 1, 5, 6, 7, and 8
11. 108°, 108°, 72° 13. 39°
15. 49.91° 17. 43° 19. 35°
21. 105° 23. 44° 25. 45°
27. a. 47°, 137°
b. 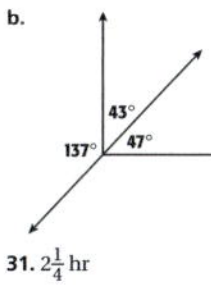

31. $2\frac{1}{4}$ hr

7-4 Exercises

1. intersecting 2. perpendicular
3. perpendicular 5. skew
7. intersecting 9. perpendicular
11. $\overleftrightarrow{BC}$, $\overleftrightarrow{FG}$, and $\overleftrightarrow{EH}$ 13. Possible answer: $\overleftrightarrow{AD}$ and $\overleftrightarrow{GH}$ 15. never
17. always 19. never 25. 7 27. $\frac{5}{6}$
29. $8\frac{5}{7}$ 31. $6\frac{3}{8}$

7-5 Exercises

1. obtuse triangle 2. 98° 3. 82°
4. equilateral 5. isosceles 7. 60°
9. scalene 11. yes; right 13. no
15. 1.9 cm; isosceles 17. $1\frac{1}{6}$ ft; equilateral
19–21. Possible answers given.

19. 21. 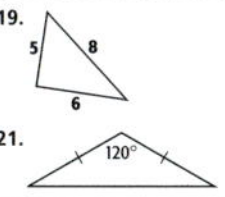

23. right triangle 25. 27
27. 360° 29. 3 31. 1

32–35. 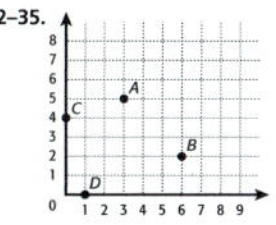

7-6 Exercises

1. rectangle 2. trapezoid
3. square 4. quadrilateral
5. squares 6. right 7. quadrilateral
9. parallelogram 11. trapezoid
13. quadrilateral, parallelogram, rectangle, rhombus, square; square 15. quadrilateral, parallelogram, rhombus; rhombus 17. never 19. always
21. sometimes 23. sometimes
25. not possible 27. not possible
29. a. If the frame is 10 in. by 13. in., the total length of the sides is 46 in., not 38 in.
b. Possible answer: The dimensions, 10 in. by 13 in., were too long, so try shorter lengths. c. 8 in. by 11 in. 35. 147°

7-7 Exercises

1. polygon, hexagon, regular
2. polygon, quadrilateral, not regular 3. polygon, triangle, regular 4. 40 m 5. not a polygon
7. not a polygon 9. not formed by line segments 11. not formed by line segments 13. hexagon
15. always 17. never 19. a. 2
b. 5 23. 10, 13, 16, 19, 22 25. $\frac{9}{10}$
27. $\frac{1}{5}$ 29. C

7-8 Exercises

1. Rotate figure 90° clockwise;

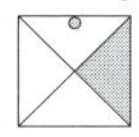

2. Move dot and triangle 1 position counterclockwise;

3–9. Possible answers given.
3. purple, purple, red, yellow, green, yellow; next five beads: red, yellow, green, yellow, purple
5. The number of objects doubles each time.

7. 

9. 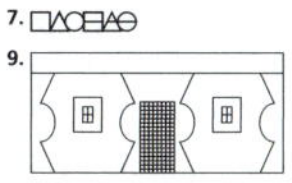

13. $7^2 \times 4^2$ 15. C

7-9 Exercises

1. not congruent 2. congruent
3. figure A 5. The figures are irregular hexagons that are not congruent.
7. Possible answer: 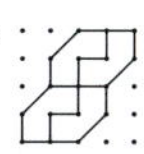

11. $t = 9$ 13. $s = 48$ 15. 3 17. C

7-10 Exercises

1. reflection 2. rotation
3. translation

4. 

5. 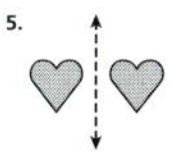

7. translation
9. 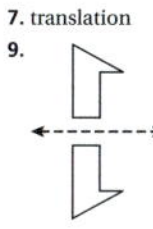

11. 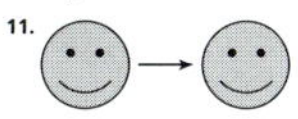

13. 

15. A
17. 

21. 12 23. 72 25. 36 27. D

7-11 Exercises

1. yes 2. no 3. yes 4. 4 lines of symmetry 5. 3 lines of symmetry
6. 5 lines of symmetry 7. 4 lines of symmetry 8. 3 lines of symmetry 9. no 11. no
13. 8 lines of symmetry 15. 1 line of symmetry 17. a. yes b. no
21. one hundred one and twenty-five hundredths
23. twelve billion, thirty million, nine hundred twenty-one thousand 25. C

7-12 Exercises

1. yes 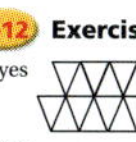

2. yes 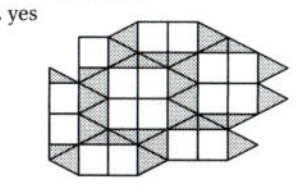

3. yes 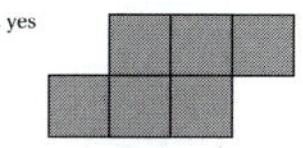

4. no 

5. no 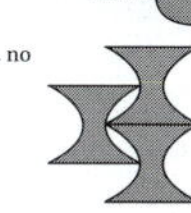

6. yes 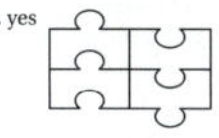

7. yes 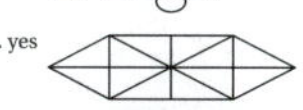

9. yes 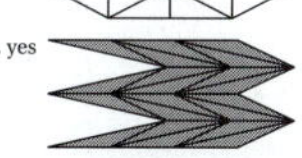

11. no 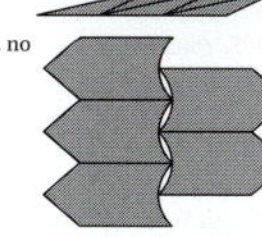

13. B 17–19. Possible answers given. 17. b plus 13, 13 more than b 19. 26 minus c, c less than 26 21. D

Chapter 7 Extension

1. 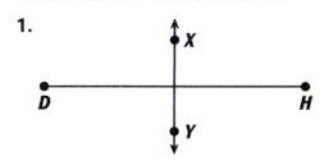

3. 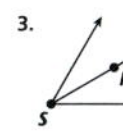

5. 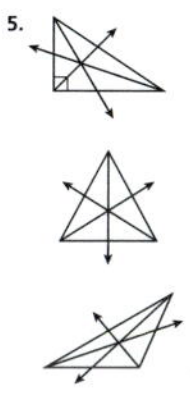

The point of intersection for all three triangles is within the triangles.

Chapter 7 Study Guide and Review

1. trapezoid 2. polygon
3. Possible answer: $\overrightarrow{ED}$; $\overleftrightarrow{DA}$
4. acute 5. obtuse 6. acute
7. $b = 27°$ 8. $d = 98°$
9. perpendicular 10. skew
11. obtuse scalene
12. parallelogram 13. triangle; regular 14. rectangle; not regular
15. 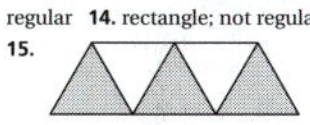

16. not congruent 17. congruent
18. translation 19. yes 20. yes

Chapter 8

8-1 Exercises

1. 3:10 2. 10:41 3. 41:16
4. Possible answer: $\frac{2}{9}, \frac{4}{18}, \frac{6}{27}$
5. the 8-ounce bag 7. 19:3
9. Possible answers: 6:9, 2:3, 12:18
11. 10 to 7, 10:7, $\frac{10}{7}$ 13. four to thirty, 4:30, $\frac{4}{30}$ 15. 6:5; Possible answer: 12:10, 18:15 17. 8:8; Possible answer: 4:4, 1:1 19. 4 × 6 and 24 × 36 21. 6 million ft^3
25. $c = 1.1$ 27. $t = 51.54$ 29. $7\frac{7}{9}$
31. $12\frac{1}{2}$

8-2 Exercises

1. Possible answer: $\frac{6}{3} = \frac{2}{1}$ 2. $n = 4$
3. $t = 7$ 4. $c = 2$ 5. $26.25
7. $d = 16$ 9. $x = 3$ 11. $p = 2$
13. $p = 3$ 15. $p = 4$ 17. $p = 75$
19. $p = 1.5$ 21. 113 euros, 160 Canadian dollars, 828 renminbi, 440 shekels, and 910 pesos 27. mode 29. mean

8-3 Exercises

1. 18 in. 2. 24 cups 3. 250,000 lb
5. 434 yd 7. 144 9. 31 11. 14
13. < 15. = 17. 4 pints
19. a. 9 yd b. 324 in. c. about 810 cm 23. 16 25. 35 27. B

8-4 Exercises

1. x = 4 cm, m∠G = 37°
2. 7.5 in. 3. n = 3 in., m∠M = 110°
5. sides: $\overline{AC}$ and $\overline{XY}$, $\overline{XW}$ and $\overline{AB}$, $\overline{BC}$ and $\overline{WY}$; angles: X and A, W and B, Y and C 7. m∠E = 78°, m∠L = 78°, m∠M = 51°; the length of $\overline{ML}$ is 21 in. 9. yes
11. 50 ft 15. 0.4 17. 10.007 19. D

8-5 Exercises

1. 15 ft 2. 19.5 ft 3. 18 ft 5. 104 in.
7. 120 m 11. 30 13. 92 15. B

8-6 Exercises

1. 300 ft 2. 3 m 3. no 5. 2.5 in.
7. 3 in. 9. 1.25 in. 11. 0.875 in.
13. 357 km 15. 297.5 km
17. a. about 4 cm b. which direction Wichita Falls is from San Antonio 21. $2.71 23. $1\frac{1}{5}$ 25. $\frac{15}{28}$
27. H

8-7 Exercises

1. 2. 3. 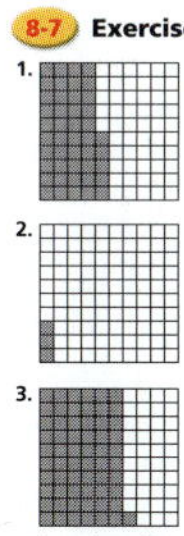

4. $\frac{1}{4}$ 5. $\frac{4}{5}$ 6. $\frac{27}{50}$ 7. $\frac{23}{25}$ 8. 0.72
9. 0.04 10. 0.9 11. 0.64
13. 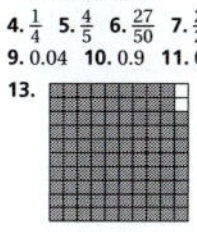

15. $\frac{1}{5}$ 17. $\frac{11}{100}$ 19. $\frac{16}{25}$ 21. $\frac{4}{5}$
23. 0.13 25. 0.6 27. 0.07
29. $\frac{23}{100}$, 0.23 31. $\frac{49}{100}$, 0.49
33. $\frac{37}{100}$, 0.37 35. $\frac{2}{25}$, 0.08
37. $\frac{47}{50}$, 0.94 39. 0, 0
41. $\frac{15}{100} = \frac{3}{20}$ 43. oldies 45. 1
49. 660 cm 51. range = 71, mean = 33.2, median = 21, no mode 53. D

8-8 Exercises

1. 39% 2. 12.5% 3. 80%
4. 11.2% 5. 44% 6. 87.5%
7. 70% 8. 50% 9. 75% 11. 55%
13. 30.8% 15. 1% 17. 2% 19. 30%
21. 45% 23. 68.75% 25. 40%
27. 4%, $\frac{1}{25}$ 29. 45%, $\frac{9}{20}$
31. 81%, $\frac{81}{100}$ 33. 39%, $\frac{39}{100}$
35. 80%, 0.8 37. 83.33%, 0.83
39. 6.67%, 0.07 41. 72.73%, 0.73
43. < 45. = 47. > 49. $\frac{21}{50}$, 0.43, 45% 51. $\frac{1}{4}$, 26%, 0.7 53. 0.125, 14%, $\frac{9}{20}$ 55. a. about 67% b. 0.99
c. 38% 61. 19 63. 5
65. 67. 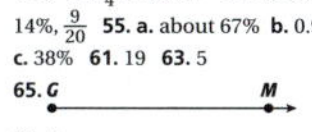

8-9 Exercises

1. 44 T-shirts 2. 15 hr 3. 6.72
4. 156 5. 97.75 6. 37.8 7. 6 dolls
9. 30 min 11. 28.6 13. 4.5
15. 2.28 17. 65.6 19. 40.56
21. 5 23. a. 9 feet b. 108 ft^2
25. 12 atoms of hydrogen, 6 atoms of carbon, and 6 atoms of oxygen 29. 7 31. 726 33. 2 · 19
35. $2^3 \cdot 3 \cdot 5$ 37. B

8-10 Exercises

1. about $7.65 2. about $2.40
3. about $151.20 5. about $18.75
7. about $11.00 9. about $55.65
11. yes 13. a. Kentucky; about $0.38 b. New York; $0.04 less than in North Carolina 17. 1, 2, 4, 5, 10, 20 19. 1, 59 21. mean = 24, median = 23, mode = 23
23. 25. 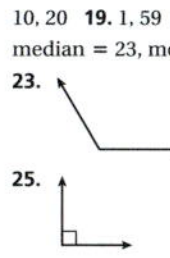

Chapter 8 Extension

1. $148.75 3. $32 5. $250
7. $367.20 9. $37,500

Chapter 8 Study Guide and Review

1. discount 2. equivalent ratios
3. percent 4. corresponding angles 5. Possible answers: 2:4, 3:6, 6:12 6. 12 oz for $2.64
7. $n = 9$ 8. $n = 3$ 9. $n = 14$
10. $n = 2$ 11. 48 cups

12. $2\frac{1}{2}$ or 2.5 13. n = 11 in.; m∠A = 90° 14. 94 ft 15. 43.75 mi
16. 3 in. 17. $\frac{3}{4}$ 18. $\frac{3}{50}$ 19. $\frac{3}{10}$
20. 0.08 21. 0.65 22. 0.2
23. 89.6% 24. 70% 25. 5.7%
26. 12% 27. 70% 28. 25%
29. 87.5% 30. 80% 31. 6.25%
32. 12 33. 5.94 34. 117 tickets
35. about $19.00 36. about $4.35
37. about $1.08

Chapter 9

9-1 Exercises

1. +5 2. −15
3.–6. 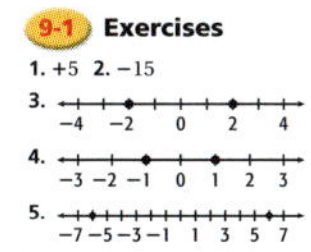

7. 3 8. 4 9. 1 10. 2 11. +50
13. +7
15. 17. 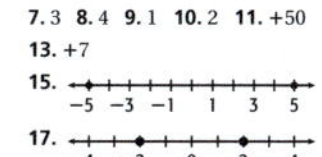

19. 4 21. 3 27. +92 29. +25
31. 419 33. 723 35. 35
37. 295 39. −10,924 41. +2; −2
43. C 47. 17,395 < 17,465 < 17,498 < 17,509 49. 24

9-2 Exercises

1. > 2. < 3. < 4. −2, 0, 9
5. −5, −4, 3, 7 6. −6, −1, 8, 10
7. 3:30 A.M. 9. > 11. −6, −3, 11
13. −12, 0, 1, 5 15. −10, −6, 4, 7, 24
17. Pacific 19. > 21. <
23. < 25. −31, −18, −9
27. −25, −4, 15, 31 29. −50, −27, 38, 42, 69 31. 68 > −7 or −7 < 68
37. 1 39. 48 41. 0 43. 1.1
45. 0.125 47. J

9-3 Exercises

1. III 2. no quadrant 3. (−2, 0)
4. (1, 2)
5–6. 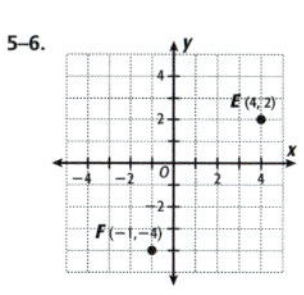

7. no quadrant 9. I 11. (3, −4)
13. (4, 4)
15–18. 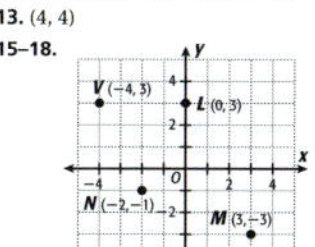

19. IV 21. II
22–27. 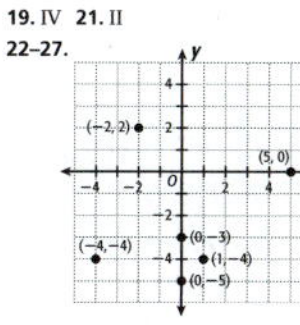

31. Atlantic Ocean 35. $\frac{3}{40}$
37. parallelogram 39. trapezoid

9-4 Exercises

1. 3 + 2 = 5 2. 4 3. −5 4. 1
5. −12 6. 0 7. −4 8. −600 ft
9. 6 + (−2) = 4 11. 11 13. 0
15. 6 17. −16 19. −8
21. −15 23. 29 ft
25. 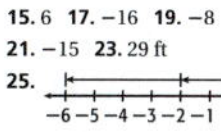

27. 7 29. −11 31. 3 33. 4
35. −1 37. −17 39. −6°F 45. 12
47. 9 49. 60 51. 50

9-5 Exercises

1. 6 − 5 = 1 2. 3 3. −3 4. 14
5. 2 6. 8 7. −9 9. −4 11. 11
13. 10 15. 7 17. −12 19. 10
21. 21 23. 5 25. 3 27. 19 m
29. 9°F 33. 27 35. 125 37. A

9-6 Exercises

1. 24 2. −10 3. −21 4. 9 5. 0
6. 16 7. 9 9. 33 11. −36 13. 45
15. −32 17. 48 19. −24 21. −12
23. −28 25. −6 27. 36 29. −24
31. −42 33. 14 35. 54 37. −90
39. −90 41. −42 45. $y = 9$
47. $n = 48$ 49–51. Possible answers given: 49. $\frac{2}{4}, \frac{4}{8}$ 51. $\frac{4}{6}, \frac{6}{9}$
53. C

9-7 Exercises

1. 8 2. −5 3. 3 4. −2 5. 10
6. −15 7. 5 9. −8 11. −5 13. 1
15. 3 17. −23 19. −6 21. 1
23. −5 25. 3 27. 8 29. 9
31. a. −3° F b. −2° F 33. 110; an increase of 110 seals each year from 1971 to 1975 37. $n = 2.5$
39. $w = 2.4$ 41. $2\frac{1}{3}$ or $\frac{7}{3}$ 43. C

9-8 Exercises

1. $m = 12$ 2. $a = -5$ 3. $z = 9$
4. $b = -8$ 5. $w = 54$ 6. $c = -7$
7. $g = 4$ 9. $t = -14$ 11. $y = -19$
13. $j = 8$ 15. $a = -52$
17. $k = -20$ 19. $x = 17$
21. $k = -4$ 23. $a = 19$
25. $g = -3$ 27. $f = -8$
29. $c = -9$ 31. $b = 32$
33. $a = 8$ 35. $j = -45$
37. $n = -4$ 39. $j = -3$ 41. −60 ft
43. $x = -18{,}500$; sponge
45. −24 ft 49. $2^2 \times 19$ 51. 2^4
53. 3 × 7 55. $2^2 \times 3 \times 13$
57. $1\frac{1}{4}$, or $\frac{5}{4}$ 59. $\frac{9}{10}$ 61. $\frac{19}{28}$ 63. B

Chapter 9 Extension

1. 1, $\frac{1}{3}, \frac{1}{9}$ 3. 1, $\frac{1}{6}, \frac{1}{36}$ 5. 7^{-1} 7. 8^{-2}
9. 3^{-1} 11. 7^{-2} 13. 4^{-2} 15. 3^{-4}
17. $\frac{1}{27}$ 19. $\frac{1}{729}$ 21. 256 23. $\frac{1}{64}$
25. $\frac{1}{125}$ 27. 1 29. 1

Chapter 9 Study Guide and Review

1. opposites 2. x-axis, y-axis
3. coordinate plane, quadrants

4. positive numbers, negative numbers 5. +10 6. –50

7. 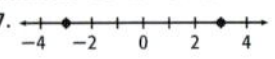

8. 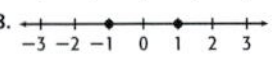

9. 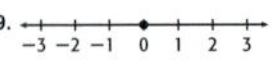

10. 2 11. 1 12. 0 13. < 14. < 15. < 16. > 17. > 18. < 19. –1, 2, 4 20. –3, 0, 4 21. –3, –2, 0, 1 22. –8, –6, 0 23. –7, –4, 7 24. –5, –1, 3, 7 25. (–2, –3) 26. (1, 0) 27. III 28. II

29–31.

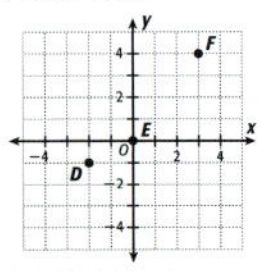

32. –2 33. 0 34. 1 35. –5 36. –17 37. 8 38. –8 39. 9 40. 10 41. 0 42. 13 43. –6 44. –10 45. 6 46. 6 47. –8 48. –16 49. 45 50. –3 51. 3 52. 2 53. –2 54. –12 55. –9 56. $w = 4$ 57. $a = -12$ 58. $q = -7$ 59. $x = -5$

Chapter 10

10-1 Exercises

1. 2 in. 2. 28 cm 3. 40 m 4. 22.6 in. 5. 7 yd 7. 96 in. 9. 7 cm 11. 42 m 13. 6 in. 15. 42 in. 17. 2 km 19. 8.5 cm 23. 24.68 25. 74.35 27. 3 29. 42

10-2 Exercises

1. about 8.5 square units 2. about 9 square units 3. 98 mm^2 4. 100.1 in^2 5. 48 ft^2 6. 21 cm^2 7. 3 yd^2 8. 33 mm^2 9. about 6 square units 11. 125 mi^2 13. 260 ft^2 15. 37 m^2 17. B 21. 1 min 23. $\frac{7}{8}$ 25. $5\frac{13}{15}$

10-3 Exercises

1. 32 in^2 2. 18.84 cm^2 3. 54 cm^2 5. 640 yd^2 7. a. Draw a 1,100 km by 600 km rectangle and a 1,100 km by 750 km by 1,300 km right triangle. b. 1,072,500 km^2 c. about $\frac{1}{7}$ 11. $1\frac{11}{16}$ 13. $\frac{1}{21}$

10-4 Exercises

1. The perimeter is divided by 3, and the area is divided by 9, or 3^2. 2. The perimeter is multiplied by 2, and the area is multiplied by 4, or 2^2. 3. The perimeter is multiplied by 4, and the area is multiplied by 16, or 4^2. 5. a. 4,800 ft^2 b. 280 ft, 320 ft 7. 18 × 24 11. 19 × 3 13. 5 + 9

10-5 Exercises

1. circle G, diameter $\overline{EF}$, and radii $\overline{GF}$, $\overline{GE}$, and $\overline{GD}$ 2. 31.4 mm 3. 12.56 in. 4. 154 ft^2 5. 616 cm^2 7. 4.71 m 9. 0.5 in. 11. 9.63 cm^2 13. 14 m, 43.96 m, 153.86 m^2 15. 2.25 cm, 1.13 cm, 3.97 cm^2 17. a.

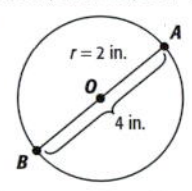

b. 12.56 in. c. 12.56 in^2 19. a. about 85 in^2 b. no 23. 5.1 25. 16.08 27. $\frac{5}{8}, \frac{1}{2}, \frac{3}{8}$ 29. $\frac{7}{10}, \frac{3}{5}, \frac{3}{10}$

10-6 Exercises

1. 5 faces, 8 edges, 5 vertices 2. 7 faces, 15 edges, 10 vertices 3. 5 faces, 8 edges, 5 vertices 4. hexagonal prism 5. square pyramid 6. cube 7. 5 faces, 9 edges, 6 vertices 9. 6 faces, 12 edges, 8 vertices 11. rectangular prism 13. square pyramid, yes 15. cone, no 17. B, C, and D 19. B 21. true 23. true 25. 8; octagonal pyramid 29. 108, 91, 89, 75, 24, 5 31. 19, 18, 17, 15, 13 33. 0.03 35. 488.41

10-7 Exercises

1. 94 in^2 2. 112 in^2 3. 132 ft^2 4. 2,640 cm^2 5. 326.56 ft^2 6. 747.32 in^2 7. 108 cm^2 9. 14,650.75 in^2 11. 381.48 ft^2 13. 1,274.84 in^2 15. 126.39 in^2 17. 188.4 m^2 19. $13\frac{1}{2}$ km^2 21. 345.4 in^2 23. a. $\ell = 10$ m; $h = 5$ m; 700 m^2 b. $\ell = 9$ in.; $h = 3$ in.; 342 in^2 27. 14.2 ft 29. < 31. <

10-8 Exercises

1. 162 cm^3 2. 64 in^3 3. 10 ft^3 4. 351 m^3 5. 320 ft^3 6. 2,500 dm^3 7. 1 × 1 × 10 and 2 × 5 × 1 9. 79.36 in^3 11. 54 m^3 13. 71.72 ft^3 15. 480 in^3 17. 474.375 km^3 19. 20 ft 21. 8.96 g/cm^3, 19.32 g/cm^3, 5.02 g/cm^3, 0.4 g/cm^3, 10.5 g/cm^3 23. Check to see if the egg floats in water; if it does, then the egg is spoiled. 27. 2 29. 40 31. 3 33. D

10-9 Exercises

1. 754 m^3 2. 157 cm^3 3. 3,140 in^3 4. 31 in^3 5. cylinder B 7. 314 ft^3 9. 31 cm^3 and 283 cm^3 11. 138 in^3 13. 4 m^3 15. 100.48 in^3 17. 659.67 yd^3 19. 176 ft^3 21. no 23. 12 mm 27. no 29–31. Possible answers given. 29. $\frac{4}{6}$ 31. $\frac{3}{4}$

Chapter 10 Study Guide and Review

1. polyhedron 2. volume 3. perimeter, circumference 4. diameter 5. 33.9 in. 6. 7 ft 7. 12 in^2 8. 154 in^2 9. 135 cm^2 10. 175.5 ft^2 11. The perimeter is multiplied by 2, and the area is

multiplied by 4, or 2^2. 12. 31.4 ft 13. 50.24 cm 14. 9 m 15. 11 ft 16. 154 cm^2 17. 616 cm^2 18. 5 faces, 8 edges, 5 vertices, square pyramid 19. 6 faces, 12 edges, 8 vertices, rectangular prism 20. 125 m^2 21. 102 cm^2 22. 384 cm^3 23. 6,300 in^3 24. 353 m^3 25. 2,308 ft^3

Chapter 11

11-1 Exercises

1. certain 2. unlikely 3. 0.4, $\frac{2}{5}$ 4. boy and girl 5. likely 7. unlikely 9. 0.3, 30% 11. likely 13. as likely as not 15. unlikely 17. AB negative 21. $\frac{3}{8}$ 23. II 25. III

11-2 Exercises

1. 6; {2, 4, 6, 8} 2. $\frac{8}{15}$ 3. no hits; 1 hit 5. HTH, {HHH, HHT, HTH, THH, HTT, THT, TTH, TTT} 7. $\frac{13}{25}$ 9. {yellow T-shirt, red T-shirt, green T-shirt} 11. {Anna & Joel, Roseann & Anna, Joel & Roseann} 13. {2, 4, 6} 17. 9, 11, 13 19. 25, 36, 49 21. G

11-3 Exercises

1. $\frac{1}{2}$ 2. $\frac{2}{5}$ 3. 74% 4. 0.3 5. $\frac{1}{6}$ 7. $\frac{4}{9}$ 9. 0.96 11. $\frac{5}{6}$ 13. 1 15. $\frac{1}{6}$ 17. 0 19. < 21. = 23. < 25. 0.1 27. $\frac{3}{8}$ 29. 0% 31. The probability of rolling each number is $\frac{1}{8}$. 33. $\frac{2}{7}$ 37. 0.2 m 39. 9,000,000 m

41.

Month	Number of Days
January	31
February	28 or 29
March	31
April	30
May	31
June	30
July	31
August	31
September	30
October	31
November	30
December	31

11-4 Exercises

1. 6 2. 20 3. 9 5. 27 7. 36 9. a. 12 b. 18 15. 169 17. 32 19. $\frac{4}{9}$ 21. $\frac{13}{30}$

11-5 Exercises

1. $\frac{1}{4}$ 2. $\frac{1}{4}$ 3. $\frac{1}{9}$ 5. $\frac{1}{3}$ 7. $\frac{1}{2}$ 9. $\frac{1}{3}$ 11. 0 13. 1 15. = 17. < 19. $\frac{1}{25}$ 25. $m = \frac{1}{14}$ 27. 142°

11-6 Exercises

1. 600 2. 10 3. 200 tickets 5. 32 7. 129 donors 9. about 20 11. about 20 15. $y = -120$ 17. $j = 45$ 19. 6 21. 1 22. D

Chapter 11 Extension

1. 1 to 29 3. 29 to 1 5. 1 to 5; 5 to 1 7. 2 to 4; 4 to 2 9. 1 to 1; 1 to 1 11. 7 to 3 13. 94 to 6

Chapter 11 Study Guide and Review

1. equally likely 2. experiment; outcome 3. probability 4. theoretical probability 5. sample space 6. certain 7. 0.75; $\frac{3}{4}$ 8. white 9. $\frac{4}{15}$ 10. $\frac{1}{4}$ 11. $\frac{1}{2}$ 12. 75% 13. 18 combinations 14. $\frac{1}{6}$ 15. $\frac{1}{8}$ 16. about 100 items 17. about 25 times 18. about 1,575 teenagers 19. about 100 students

Chapter 12

12-1 Exercises

1. $y = x - 6$; 3 2. $y = 5x + 1$; 51 3. $j = b - 6$ 4. $c = \$0.50b$ 5. $y = 4x$; 28 7. $c = 12s - 2$ 9. $p = 150m$ 11. $y = x + 3.4$; 2.4, 8.4 13. $f = \$2.50 + \$0.90m$ 15. $58.50 = 6.5h$; 9 hours 21. –20 23. 30

12-2 Exercises

1. (1, 8); (2, 14); (3, 20); (4, 26) 2. (1, –2); (2, –4); (3, –6); (4, –8) 3. no 4. yes 5. 2 6. 1

7.

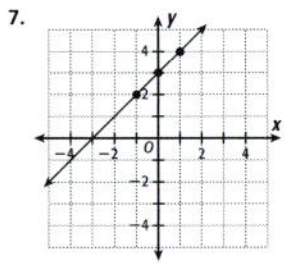

8.

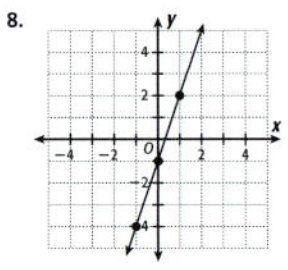

9. (1, –3); (2, –7); (3, –11); (4, –15) 11. yes 13. 1 15. 3

17.

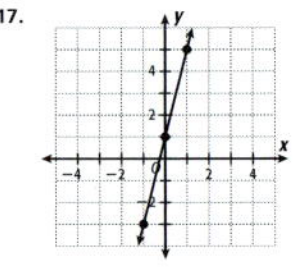

19. –3, –2, –1, 0;

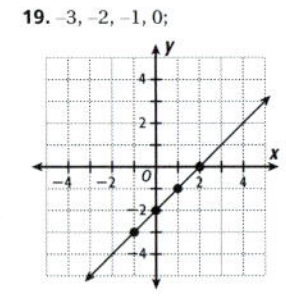

21. (1, 14)

23.

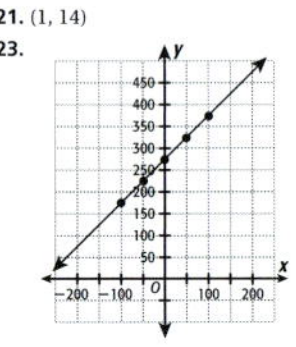

25. –196°C 29. $\frac{1}{4}$ 31. $\frac{2}{5}$ 33. IV 35. II 37. A

12-3 Exercises

1. $R'(2, 5)$, $S'(3, 2)$, $T'(-1, 2)$ 2. $M'(0, -1)$, $N'(1, 0)$, $O'(2, -1)$, $P'(2, -3)$, $Q'(0, -3)$ 3. $R'(0, 7)$, $S'(1, 4)$, $T'(-3, 4)$ 5. $A'(-9, -3)$, $B'(-4, -3)$, $C'(-4, -5)$, $D'(-9, -5)$ 7. $P'(1, -2)$, $Q'(5, -1)$, $R'(5, -5)$, $S'(1, -4)$ 11. –12 13. –5

12-4 Exercises

1. $P'(3, -1)$, $Q'(1, -1)$, $R'(1, -4)$ 2. $A'(-2, -2)$, $B'(0, -2)$, $C'(1, 0)$, $D'(-3, 0)$ 3. $N'(0, -4)$, $Q'(0, -1)$, $P'(2, 0)$ 5. $A'(-1, 3)$, $B'(-3, 3)$, $C'(-3, 1)$, $D'(-1, 1)$ 7. WOW 11. 1×10^2 13. A

12-5 Exercises

1. $Q'(-1, 1)$, $R'(-4, 1)$, $S'(-4, 2)$, $T'(-1, 2)$ 2. $Q'(1, 1)$, $R'(1, 4)$, $S'(2, 4)$, $T'(2, 1)$ 3. $A'(0, 0)$, $B'(3, -4)$, $C'(0, -4)$ 5. $P'(1, -1)$, $N'(1, -4)$, $M'(2, -4)$, $L'(2, -1)$ 7. $A'(3, 0)$, $B'(3, -3)$, $C'(0, -3)$, $D'(0, 0)$ 9. $B'(3, -3)$, $C'(3, 0)$, $E'(4, 0)$ 11. math 13. no 17. 0.445, 0.9, 1.06, 1.2 19. 7.09, 7.9, 7.99, 7.999 21. J

12-6 Exercises

1. vertical: 3, horizontal: 5; vertical: 3, horizontal: 15
2. vertical: 3, horizontal: 5; vertical: 30, horizontal: 5
3. vertical: 3, horizontal: 5; vertical: 3, horizontal: 1
4. vertical: 3, horizontal: 5; vertical: 1, horizontal: 15
5. Original dimensions:
top = 1 × 8, center = 3 × 2,
bottom = 1 × 8
New dimensions:
top = 1 × 40, center = 3 × 10,
bottom = 1 × 40
7. Original dimensions:
top = 1 × 8, center = 3 × 2,
bottom = 1 × 8
New dimensions:
top = 1 × 4, center = 3 × 1,
bottom = 1 × 4
9. a.

b. 12 units; 48 units 15. 2 17. 5

Chapter 12 Study Guide and Review

1. output, input 2. linear equation 3. function 4. $y = 2x + 2$; $y = 18$ when $x = 8$ 5. $y = x \div 2 + 1$; $y = 4$ when $x = 6$ 6. $y = 5x - 2$; $y = 53$ when $x = 11$ 7. ℓ = length; w = width; $\ell = 4w$ 8. (1, –3); (2, –1); (3, 1); (4, 3) 9. (1, 8); (2, 9); (3, 10); (4, 11) 10. yes 11. no 12. $y = 2$

13.

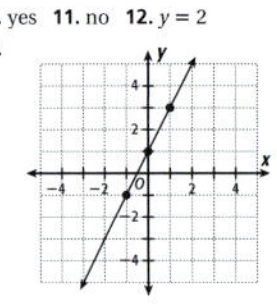

14. $A'(0, 1)$, $B'(2, 1)$, $C'(0, -1)$, $D'(2, -1)$ 15. $F'(-6, 1)$, $G'(-2, 3)$, $H'(-1, 1)$ 16. $F'(-3, -4)$, $G'(1, -2)$, $H'(2, -4)$ 17. $A'(-1, -2)$, $B'(-2, -1)$, $C'(-5, -3)$, $D'(-4, -4)$ 18. $A'(1, -1)$, $B'(3, -2)$, $C'(3, -5)$, $D'(1, -4)$ 19. original dimensions: 6 × 9
new dimensions: 12 × 9

Additional Answers

Chapter 1

Lesson 1-6

For Exercises 7–16, the calculation method depends on the individual's skill level. Possible methods are given.

7. 111; pencil and paper

8. 55; pencil and paper

9. 515,844; calculator

10. 575; pencil and paper

11. 210; mental math

12. 298; mental math

13. 11,286; calculator

14. 350; mental math

15. 446,121; calculator

16. 213; mental math

19. Possible answer: Use paper and pencil if the numbers are small but not easy to work with. Use mental math if the numbers are small and easy to work with. Use a calculator if there are several large numbers.

Lesson 1-7

1. Add 12 to each term to get the next term; 60, 72, 84.

2. Subtract 15 from each term to get the next term; 30, 15, 0.

3. Add 11 to one term and subtract 2 from the next; 36, 34, 45.

4. Subtract 6 from one term and add 4 to the next; 34, 38, 32.

5. Multiply each term by 3 to get the next term; 18, 486.

6. Divide one term by 10 and multiply the next term by 5; 4, 10.

7. Multiply one term by 6 and divide the next term by 2; 18, 27.

8. Divide each term by 4 to get the next term; 64, 4.

9. Add 10 to the first term, 11 to the next, 12 to the next, etc.; 69, 84, 100.

10. Subtract 1 from the first term, 2 from the next, 3 from the next, etc.; 85, 80, 74.

11. Add 5 to one term and subtract 8 from the next; 49, 41, 46.

12. Subtract 2 from one term and add 3 to the next; 5, 8, 6.

13. Multiply the first term by 2, the next by 3, the next by 4, etc.; 24, 720.

14. Divide one term by 2 and divide the next term by 5; 60, 3.

15. Divide one term by 4 and then multiply the next term by 2; 200, 25.

16. Divide one term by 2 and then multiply the next term by 3; 90, 135.

Performance Assessment

1. 10 rounds

2. Possible answer: $2^1, 2^3, 2^5, \ldots$; the sequence was created by adding 2 to the exponent of each term to get the next term. To extend the pattern, continue adding 2 to the previous term's exponent. The next term in the sequence is 2^7.

3. Possible answer: $4^2 \cdot (4 \div 2 + 1) - 41$; using PEMDAS, do the operations in the parentheses first from left to right, starting with the division, to get 3. Then simplify 4^2 to give you 16. Do any multiplication and division from left to right next, so $16 \cdot 3 = 48$. Then subtract $48 - 41$ to get 7 for your answer.

Chapter 2

Lesson 2-5

33. The value of n is greater than or equal to 15. Any number less than 15 will result in a negative number.

34. The student subtracted 17 from 54 instead of doing the inverse operation, addition.

Performance Assessment

1. $n \div 6$; 8

2. $60h$

3. 13 is not a solution; $x = 12$

Chapter 3

Hands-On Lab 3B

Activity 2

Try This

1. 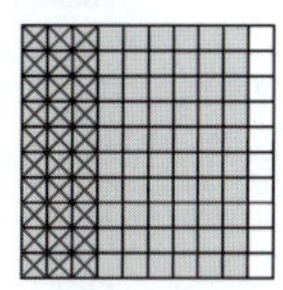 0.6

2. 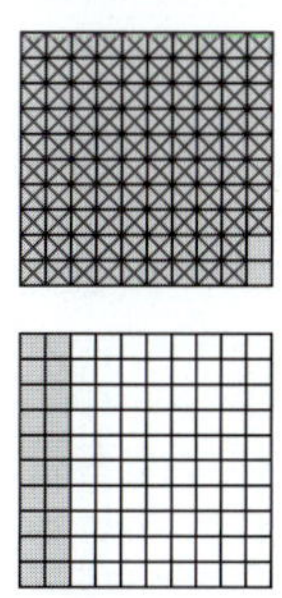 0.22

3. 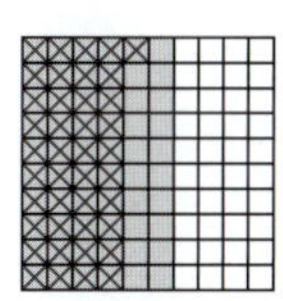 0.19

4. 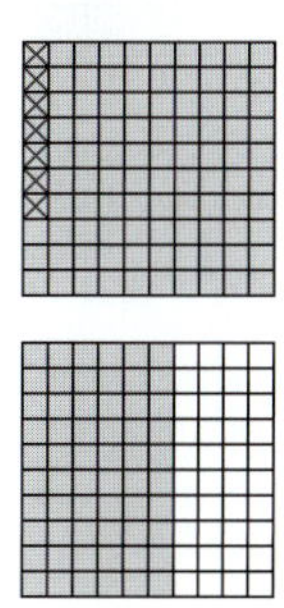 1.53

5. 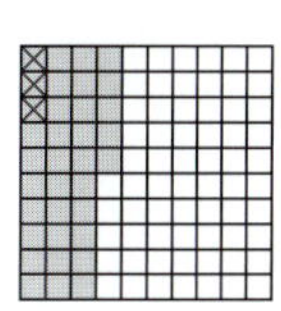 0.32

6. 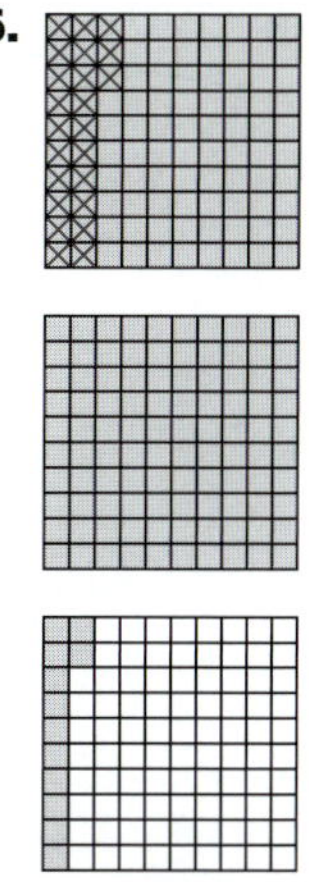 1.89

7. 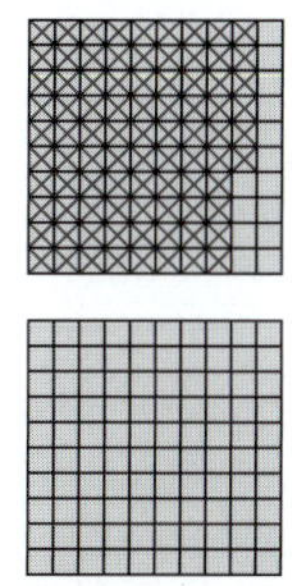 1.14

8. 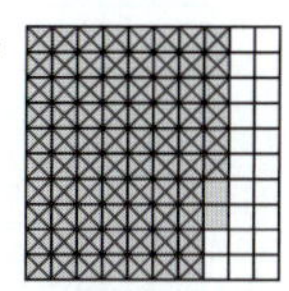 0.02

9. 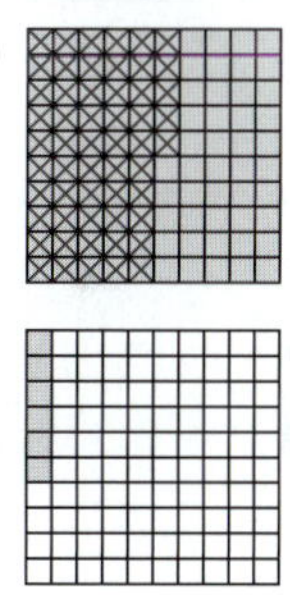 0.51

Hands-On Lab 3D

Activity 2

Try This

1. 1.01

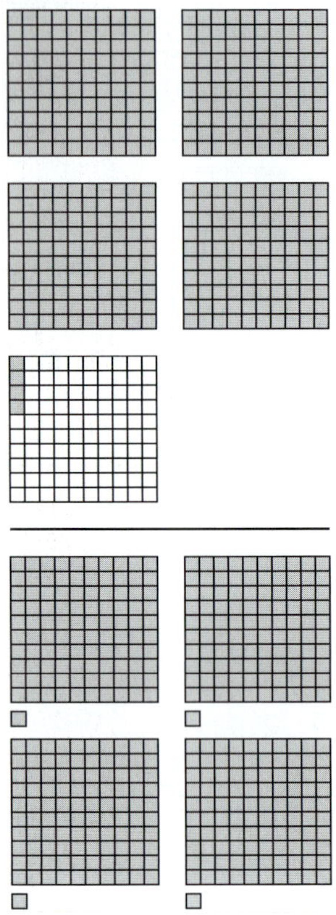

2. 0.65

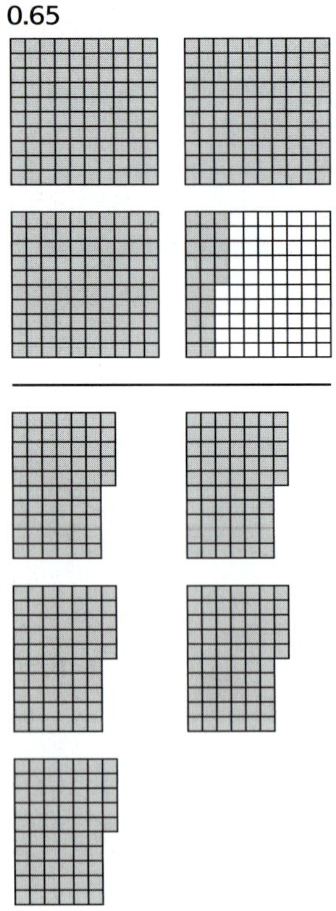

3. 6

4. 7

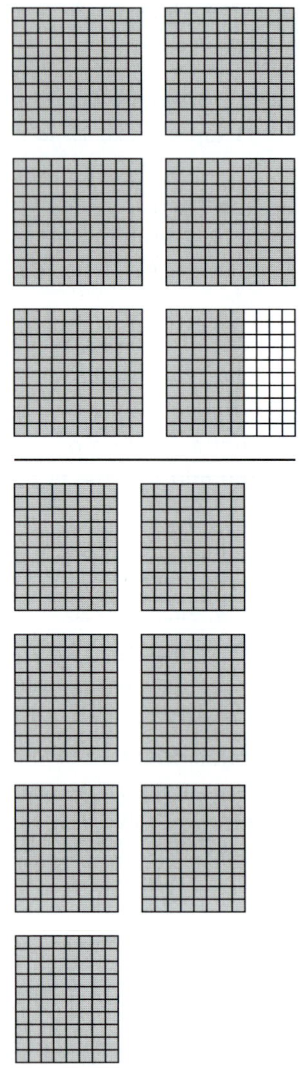

Performance Assessment

1. \$54.96; \$18.32

2. The price per movie on Friday is \$3.75. The price per movie on Saturday is \$3.25. Emily should rent on Saturday.

3. $12 \div 1.25 = 9.6$; Michael can make 9 complete costumes.

Chapter 4

Extension

1. A: 0, 2, 4, 6, …

B: 1, 3, 5, 7, …

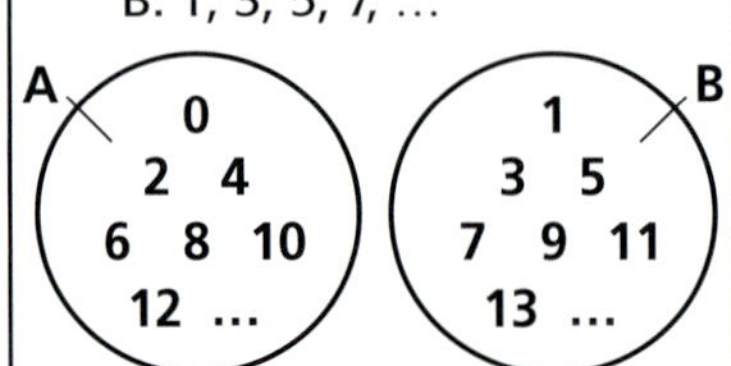

intersection: empty

union: all whole numbers

2. A: 1, 2, 3, 6, 9, 18

B: 1, 2, 4, 5, 8, 10, 20, 40

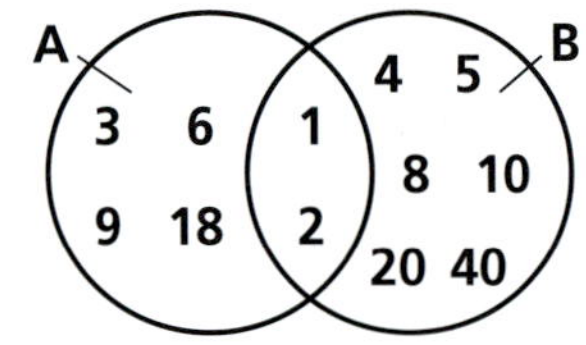

intersection: 1, 2

union: 1, 2, 3, 4, 5, 6, 8, 9, 10, 18, 20, 40

3. A: 1, 2, 3, 4, 6, 9, 12, 18, 24, 36, 72

B: 1, 2, 3, 4, 6, 9, 12, 18, 36

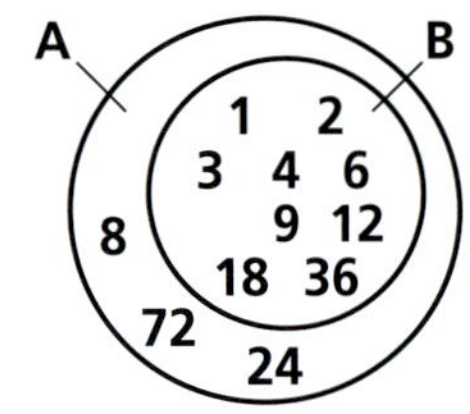

intersection: 1, 2, 3, 4, 6, 9, 12, 18, 36

union: 1, 2, 3, 4, 6, 8, 9, 12, 18, 24, 36, 72

4. A: 2, 4, 6, …

B: 4, 6, 8, 9 …

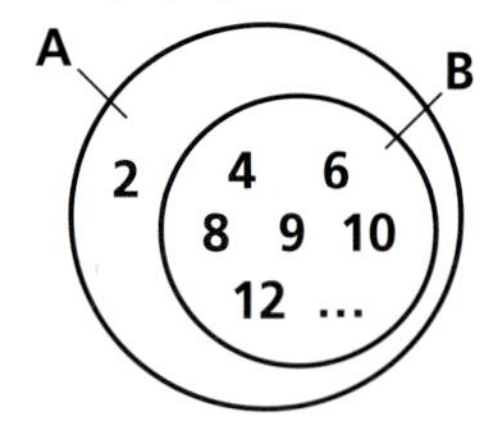

intersection: even numbers except 2

union: 2 and all composite numbers

9. Possible answer: Draw a Venn diagram with set A as the factors of one number and set B as the factors of the other number. The greatest number in the intersection is the GCF. Possible example:

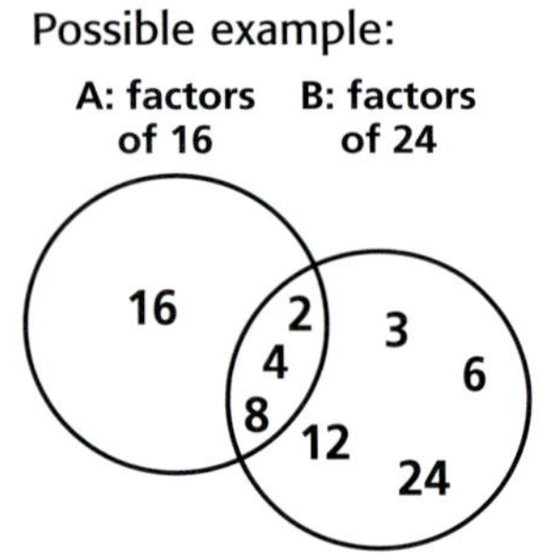

10. Possible answer: Draw 3 circles, one for the factors of each number. Common factors will appear in the area where all three circles overlap. The GCF is the greatest number in this section. Possible example:

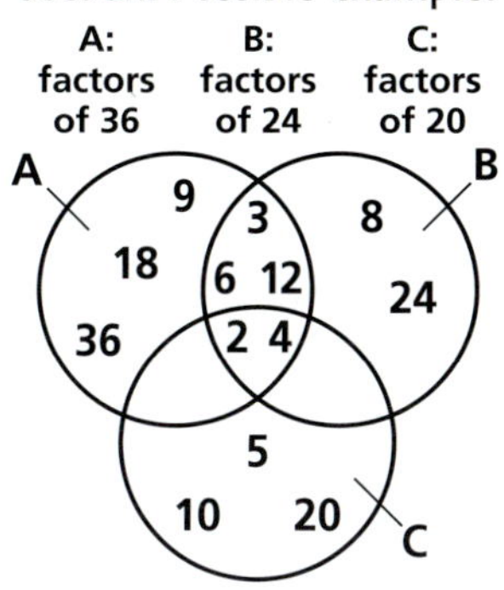

Performance Assessment

1. 3 groups; find the GCF of 9 and 15
2. $1 - \left(\frac{1}{8} + \frac{3}{8} + \frac{2}{8}\right) = \frac{2}{8}$
3. $4\frac{2}{5}$; $\frac{22}{5}$; 4.4

Chapter 5

Performance Assessment

1. 420
2. 60 of each; 4 packages of cereal samples and 3 packages of pamphlets; LCM of 15 and 20 is $2 \times 2 \times 3 \times 5$
3. 2; Round both addends up to 1, resulting in an overestimate.

Chapter 6

Lesson 6-1

1.

Day	High Temperature (°F)
Mon	72
Tue	75
Wed	68
Thu	62
Fri	55

3.

Test	Grade
1st	70
2nd	75
3rd	80
4th	85
5th	90

4. Possible answer: Joe's grades improved steadily over the five exams. Joe will do well on his sixth exam.

5.

Date	Thickness (in.)
December 3	1
December 18	2
January 3	5
January 18	11
February 3	17

Possible answer: It became safe to ice-skate around January 10.

Lesson 6-2

5.

State	Mean Score
Connecticut	509
Maine	500
Massachusetts	513
New Hampshire	519
Rhode Island	500
Vermont	508

range = 19, mean = 508.2, median = 508.5, mode = 500

Lesson 6-4

4. **Movie Preferences of Men and Women**

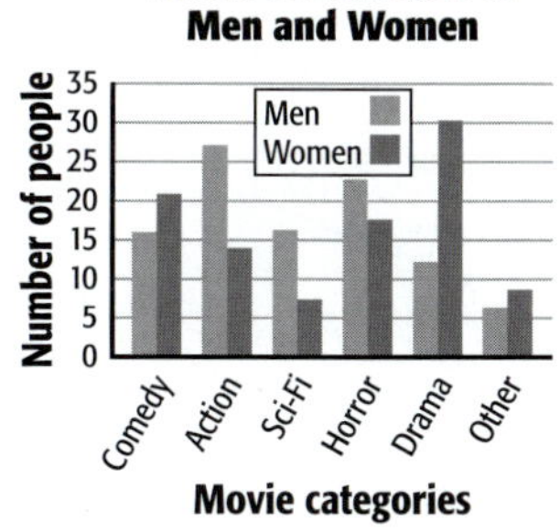

7. **Days with Rainfall**

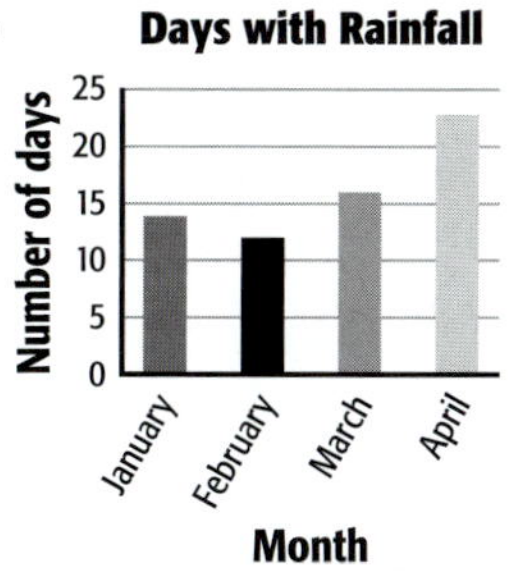

8. **Heart Rates (Beats per Minute) Before and After Exercise**

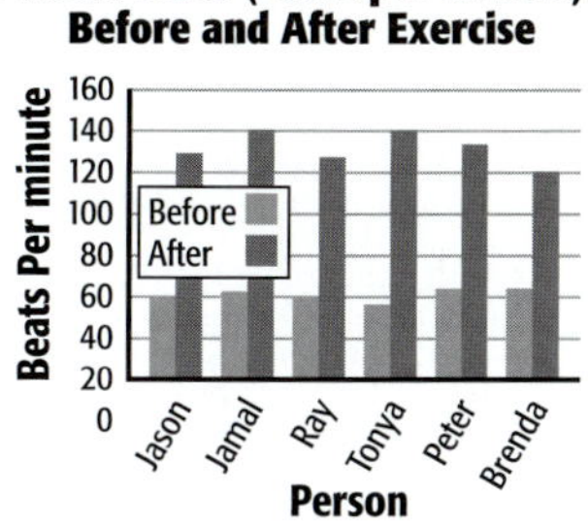

Lesson 6-5

1.

Type of Instrument	
Trumpet	卌
Drums	II
Tuba	I
Trombone	III
French horn	IIII

Tuba

2.

Type of Instrument		
Age	Frequency	Cumulative Frequency
Trumpet	5	5
Drums	2	7
Tuba	1	8
Trombone	3	11
French horn	4	15

3.

Number of Years of Each Presidential Term			
Number (Intervals)	0–4	5–8	9–12
Frequency	26	15	1

4. **Number of Years of Each Presidential Term**

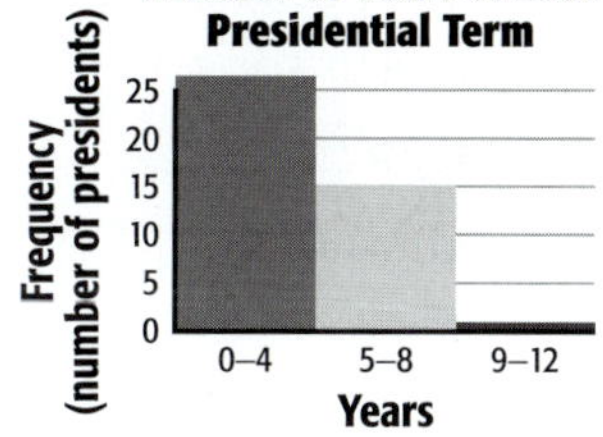

5.

Pets	
Dog	卌 I
Cat	卌
Bird	IIII
Fish	III
Hamster	II

Dog

6.

Pets Owned by Students		
Pet	Frequency	Cumulative Frequency
Dog	6	6
Cat	5	11
Bird	4	15
Fish	3	18
Hamster	2	20

7.

Final Medal Standing at the Summer Olympic Games for the Top 25 Countries					
Number (intervals)	0–20	21–40	41–60	61–80	81–100
Frequency	14	8	3	0	2

8. **Number of Olympic Medals Won**

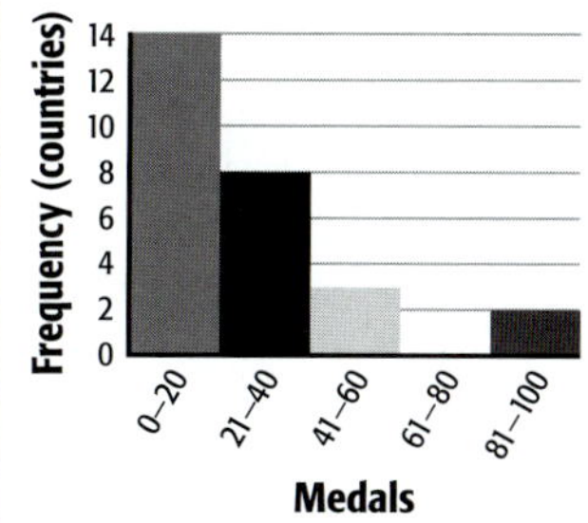

9. a.

Favorite Sport	
Basketball	II
Football	IIII
Track and field	III
Soccer	II
Hockey	II
Tennis	I
Baseball	I

b.

Favorite Sport		
Age	Frequency	Cumulative Frequency
Basketball	2	2
Football	4	6
Track & field	3	9
Soccer	2	11
Hockey	2	13
Tennis	1	14
Baseball	1	15

10.

Populations of Australia's States and Territories	
Census	Frequency
0–999,999	3
1,000,000–1,999,999	2
2,000,000–2,999,999	0
3,000,000–3,999,999	1
4,000,000–4,999,999	1
5,000,000–5,999,999	0
6,000,000–6,999,999	1

11. **Populations of Australia's States and Territories**

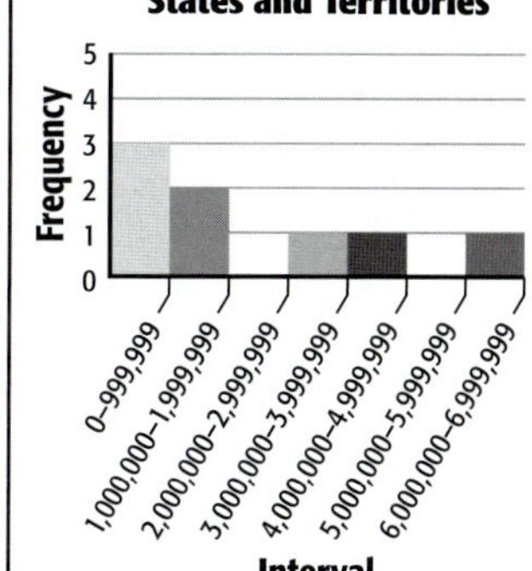

Lesson 6-6

7–10.

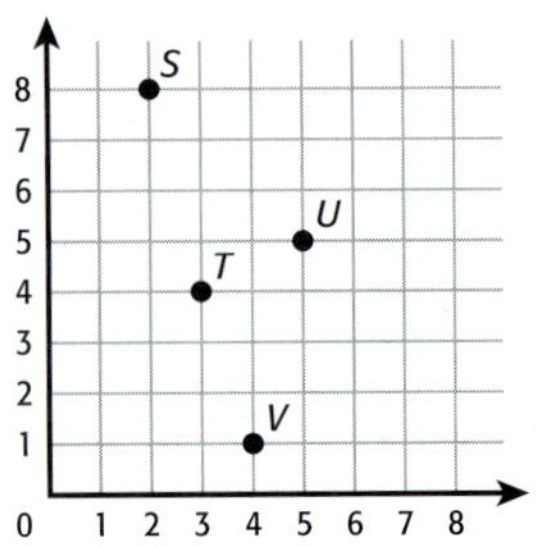

17–22.

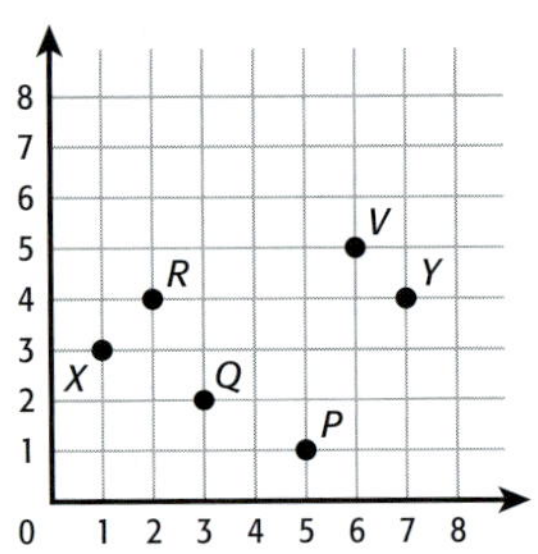

36. The point located at (3, 2) will be one unit farther to the right of (2, 3) and one unit lower.

Lesson 6-7

5.

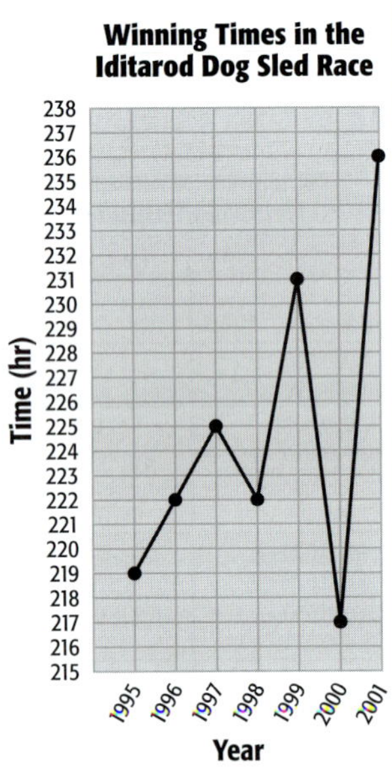

11.

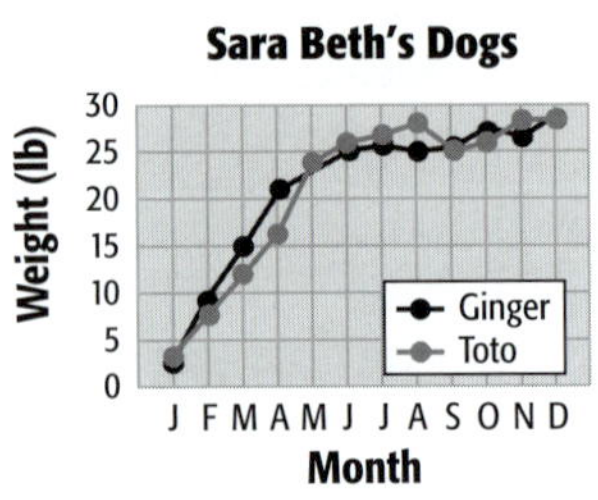

12.

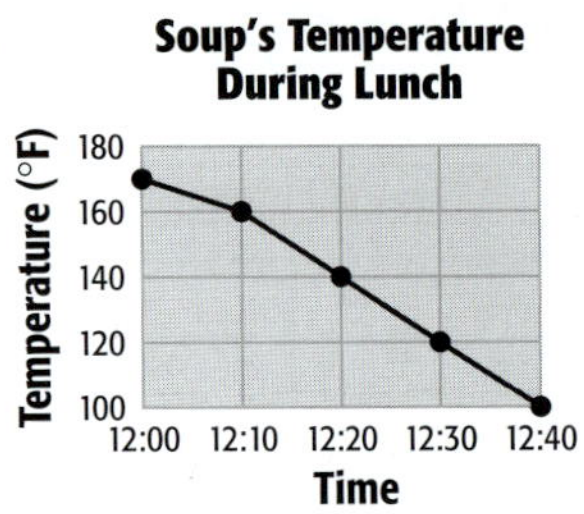

Possible answer: The soup cools as time passes.

Lesson 6-8

14. Possible answer: They could gather cholesterol data on patients not taking the medication and then make a double-line graph to compare the two sets of data.

Lesson 6-9

1. **Daily High Temperatures (°F)**

Stems	Leaves
3	7 9
4	0 5 8
5	1 6

Key: 3|7 means 37

8. **Heights of Plants (cm)**

Stems	Leaves
1	2 5
2	0 7 8 8
3	0 7
4	0 7

Key: 1|2 means 12

17. **Number of Cars with One Passenger**

Stems	Leaves
8	0 1 2 3 7 8 9
9	2 4 4 5 9
10	0 1 3 9
11	
12	4 5

Key: 8|0 means 80

19. **Ages in Josh's Family**

Stems	Leaves
1	3 5 7 9
2	
3	
4	5
5	
6	9

Key: 1|3 means 13

Performance Assessment

1.

Daily High Temperatures	
Day	**Temperature (°F)**
Mon	54
Tues	62
Wed	65
Thu	60
Fri	62

range: 11°F, mean: 60.6°F, median: 62°F, mode: 62°F

2. The highest possible average that Emily can have is 65.7.

3. Team B

Chapter 7

Lesson 7-2

19.

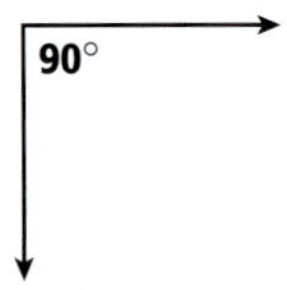

20.

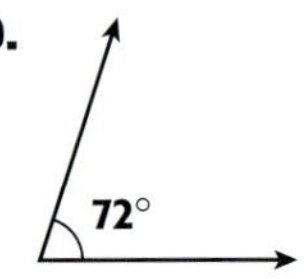

21.

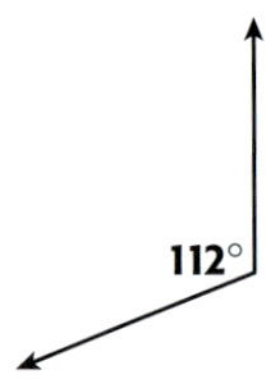

22.

23.

24.

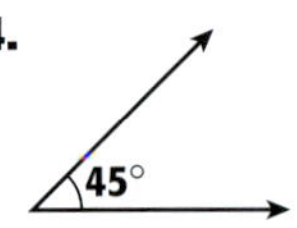

Lesson 7-3

24.

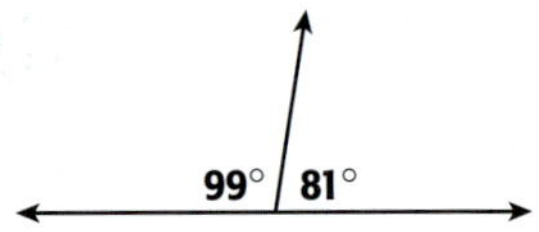

27. a. 47°, 137°

b.

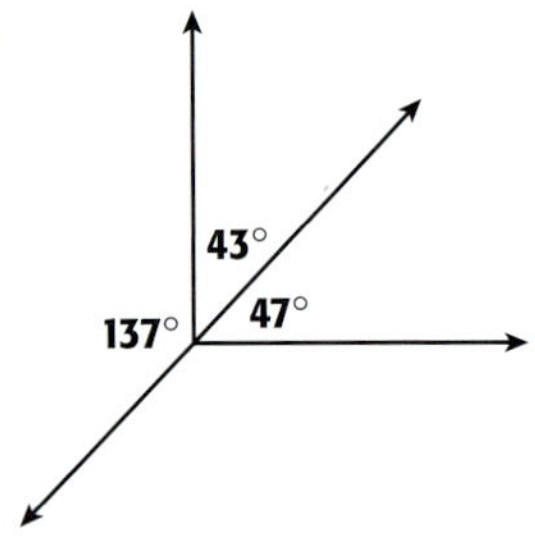

28. Possible answer:

If angle x is 25°, what is the measure of angle y?

Lesson 7-4

20. a. Possible answer:

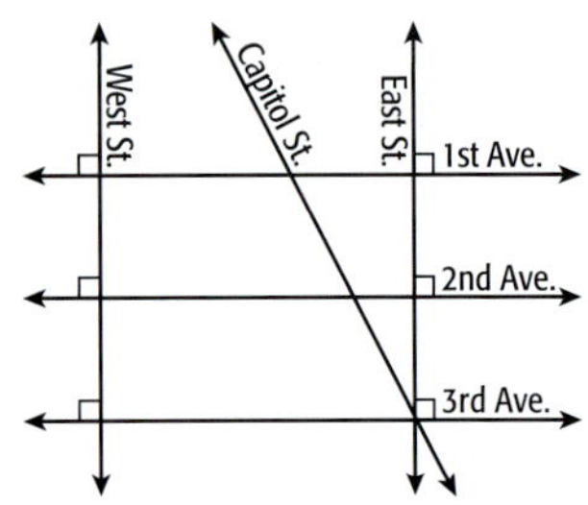

b. Possible answer:

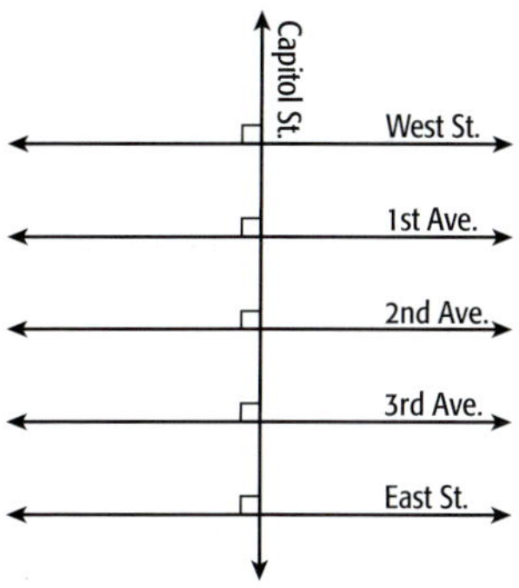

Lesson 7-6

31. A rectangle is a parallelogram and must have two pairs of parallel sides. The figure could be a trapezoid.

32. A square is a rectangle because it contains four right angles; it is a rhombus because all four of its sides are congruent.

Lesson 7-9

7. 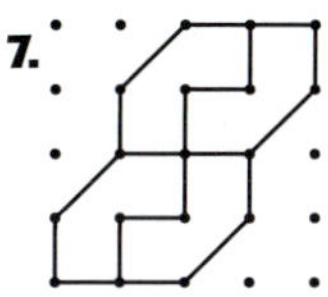

8. 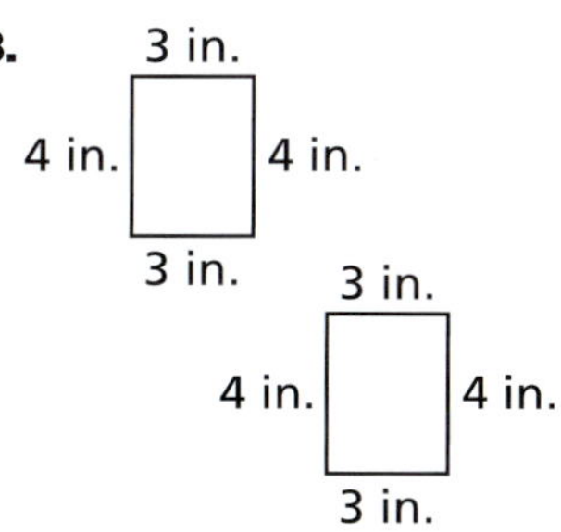

Check students' answers. Side lengths measure between 2 in. and 6 in. Corresponding side lengths of the two rectangles should be congruent.

Lesson 7-12

1. yes

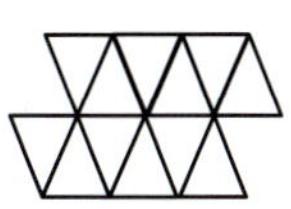

2. yes

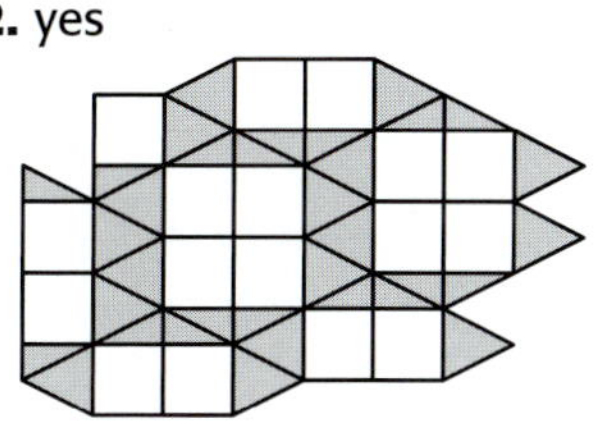

3. yes

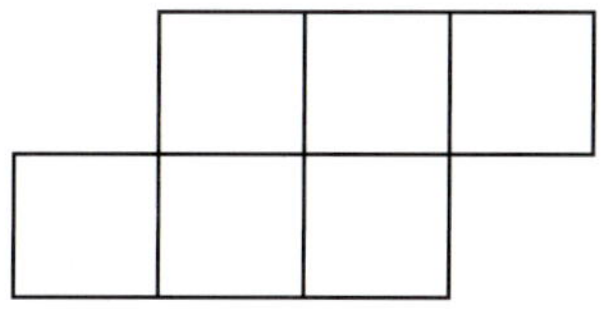

4. no

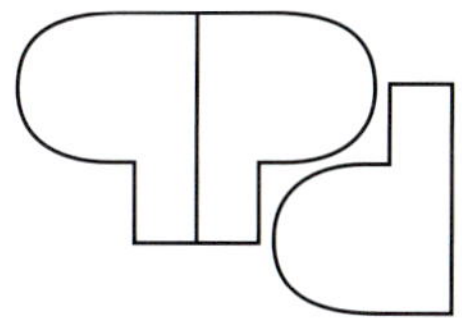

5. no

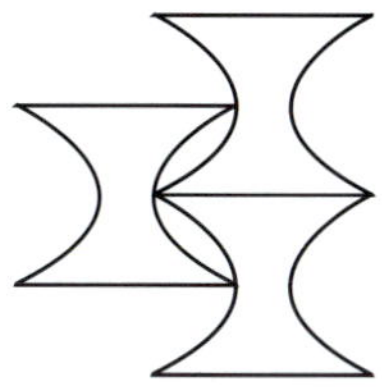

6. yes

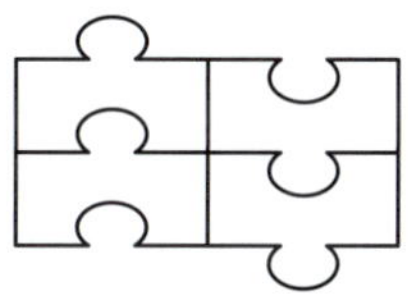

7. yes

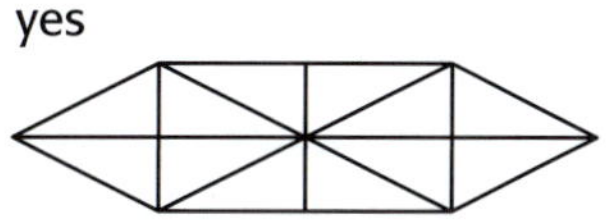

8. no

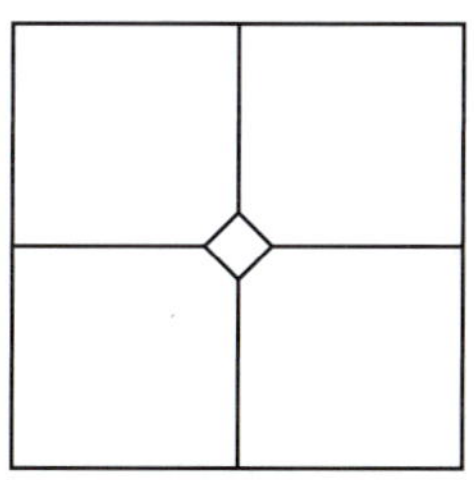

9. yes

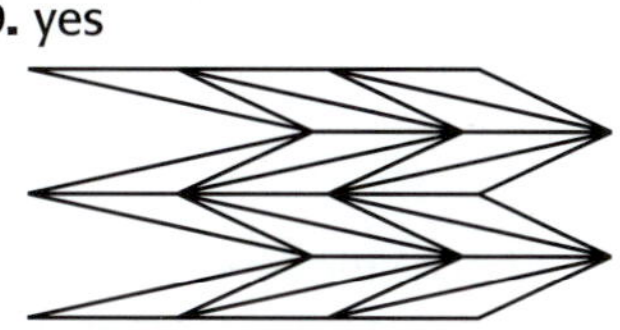

10. yes

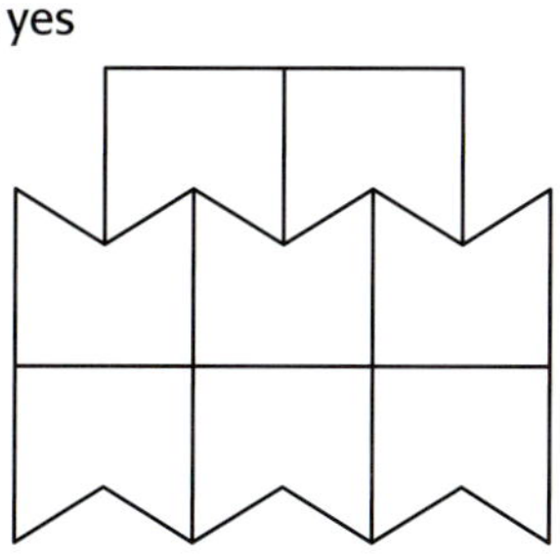

11. no

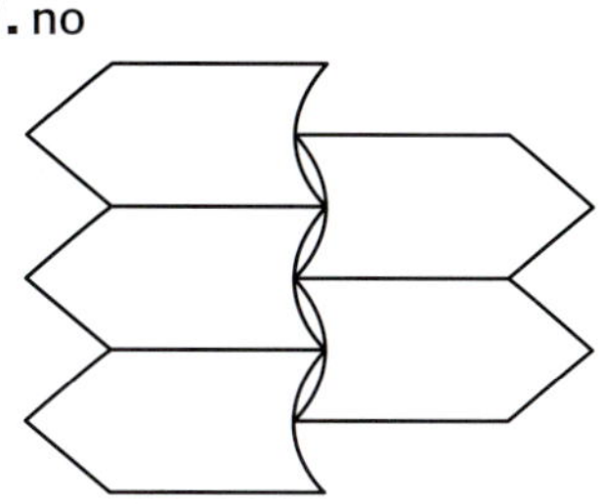

12. yes

15. A figure or figures can tessellate a plane if they can be arranged repeatedly to cover the plane without any gaps.

16. 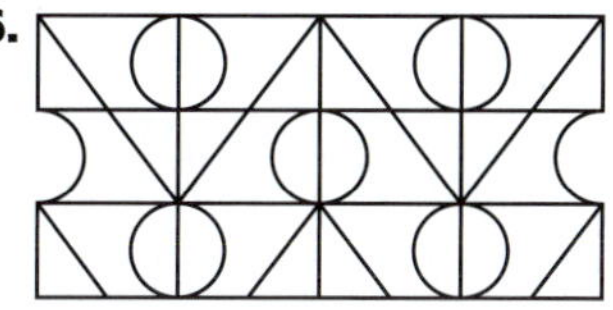

Extension

4. 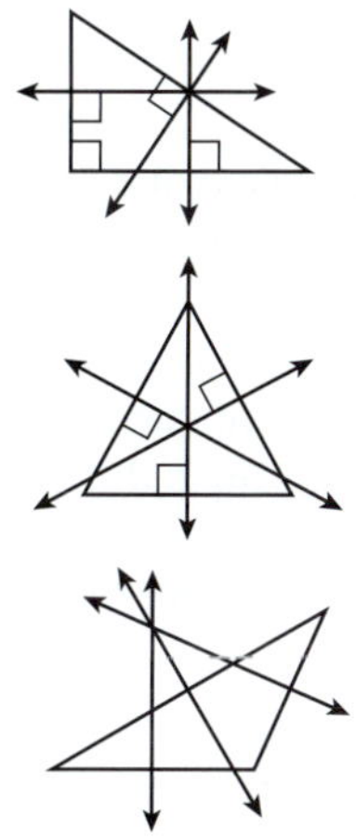

The point of intersection for the obtuse triangle is outside the triangle. For the right triangle, the point of intersection is on a side of the triangle. For the acute triangle, the point of intersection is inside the triangle.

5. 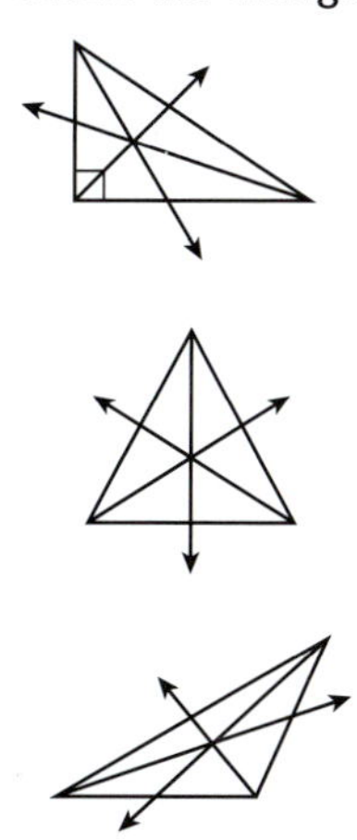

For all three triangles, each point of intersection is within the triangle.

Performance Assessment

1. 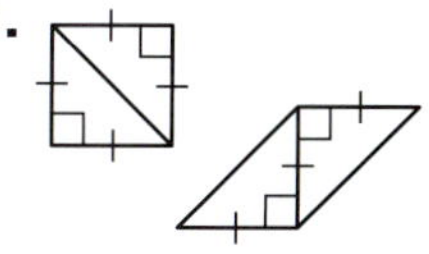

a square, a parallelogram with the height congruent to the base and the acute angles equal to 45°

2. The lines are perpendicular and intersect to form right angles. Since vertical angles are always congruent and two adjacent angles are congruent in this drawing, all four angles are congruent. Each angle measures 90°.

3. $90 + 3x = 360$; $x = 90$; rectangle

Chapter 8

Lesson 8-5

14.

Stems	Leaves
8	5 6 9
9	1 2 7
10	0 2 3 4

Key: 8|5 means 85

Lesson 8-7

1. 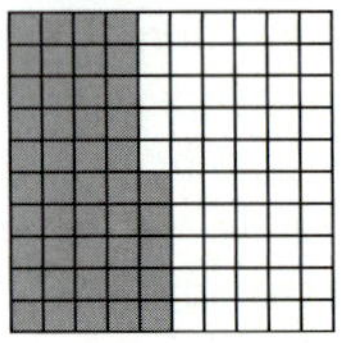

2. 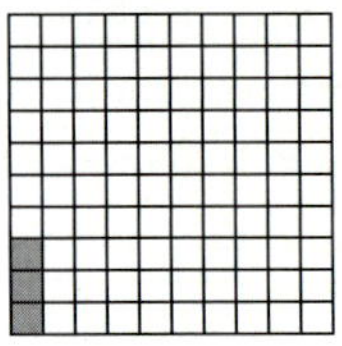

3. 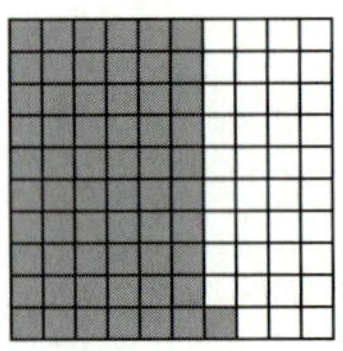

12. 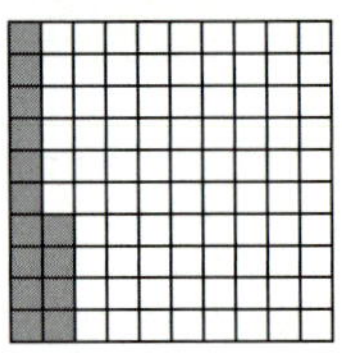

13. 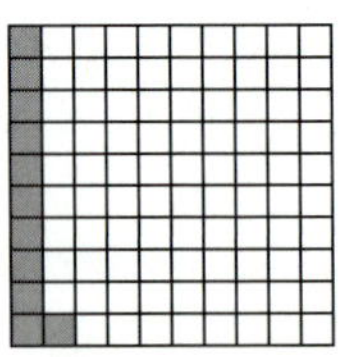

14. 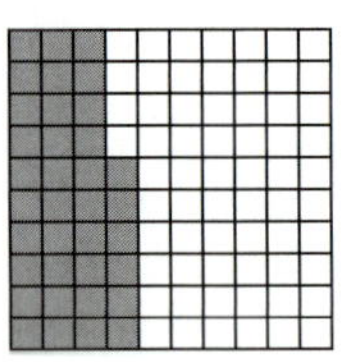

42. 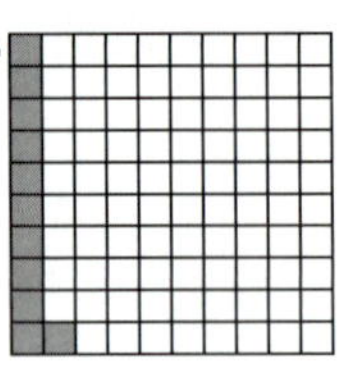

11% = 0.11

Lesson 8-8

64. 

65. 

66. 

67. 

Lesson 8-10

14. \$19.46 is 40% of the initial price. This amount should be subtracted from the initial price to find the sale price of \$29.19.

15. A discount amount is subtracted from the original cost. The sales tax and tip amount are added to the total cost. A discount could be used at a store where something is 25% off. Sales tax is added to the price of almost every item you buy, such as a television. Tip is the amount you add to the bill for the waiter when you are in a restaurant.

16. No; the price is more if you take off 50% and then 20% than if you take off 70%. \$10.00 × 50% = \$5.00; \$5.00 × 20% = \$1.00 and \$5.00 − \$1.00 = \$4.00; \$10.00 × 70% = \$7.00 and \$10.00 − \$7.00 = \$3.00

Performance Assessment

1. \$2.60; \$2.65; \$2.50; \$2.48; 8 for \$19.84 is the best deal

2. 30%; (28 − 19.6) ÷ 28 = 8.4 ÷ 28 = 0.3

Chapter 9

Performance Assessment

1. 1°F

2. 2

3. $x = 20{,}320 - 20{,}602$; $x = -282$ feet

Chapter 10

Lesson 10-3

7. a. Check students' drawings; they should draw a 1,100 km by 600 km rectangle and a 1,100 km by 750 km by 1,300 km right triangle.

b. 660,000 km^2 + 412,500 km^2 = 1,072,500 km^2

c. South Australia is about $\frac{1}{8}$ of the total area.

8. Possible answer:

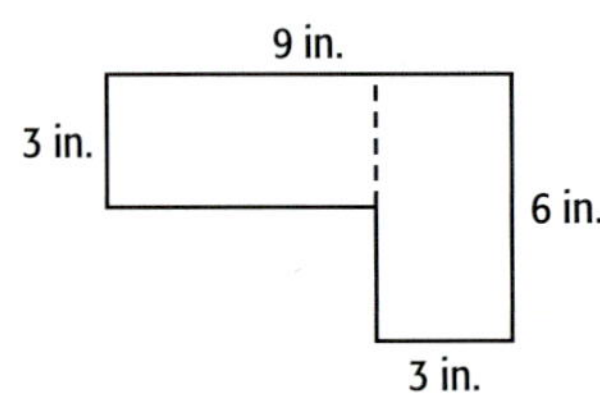

Draw a line dividing the figure into the two parts. Find the area of each part and add the areas.
$A = (3 \times 6) + (3 \times 6) = 36$ in^2

Lesson 10-4

2. Check students' drawings for a 2 cm by 4 cm rectangle. The perimeter is multiplied by 2, and the area is multiplied by 4, or 2^2.

3. When the dimensions of the triangle are multiplied by 4, the perimeter is multiplied by 4, and the area is multiplied by 16, or 4^2.

4. Check students' drawings for triangle with side lengths 1.25 cm, 1.25 cm, and 1.5 cm. The perimeter is divided by 2 and the area is divided by 4, or 2^2.

Performance Assessment

1. a. 24 in., 27.6 in^2

b. 24 in., 36 in^2

c. 24 in., 24 in^2
The three figures have the same perimeter; figure B has the greatest area.

2. 8 feet × 8 feet

Chapter 11

Performance Assessment

1. 18

2. Spinner 1: $P(r) = \frac{1}{3}$, $P(g) = \frac{1}{3}$, $P(y) = \frac{1}{3}$,
Spinner 2: $P(r) = \frac{1}{5}$, $P(g) = \frac{2}{5}$, $P(y) = \frac{2}{5}$,
Spinner 3: $P(r) = \frac{1}{4}$, $P(g) = \frac{1}{2}$, $P(y) = \frac{1}{4}$;
Spinner 1 is fair because each outcome is equally likely; spinner 3

3. $\frac{5}{7}$; use

$$\frac{\text{marbles that are not green}}{\text{total marbles}}$$

or $$1 - \frac{\text{marbles that are green}}{\text{total marbles}}$$

Chapter 12

Lesson 12-2

8.

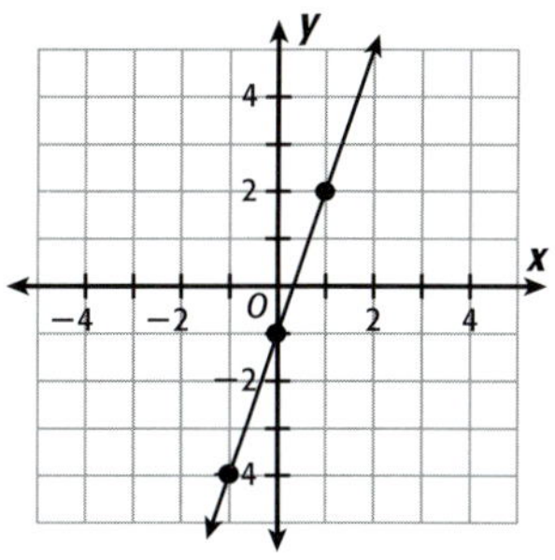

17.

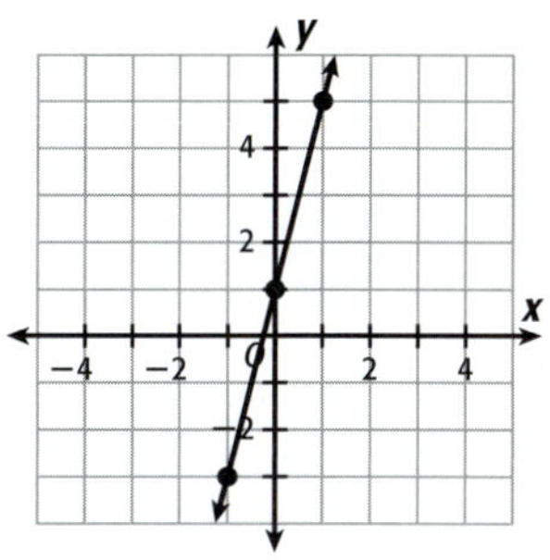

18.

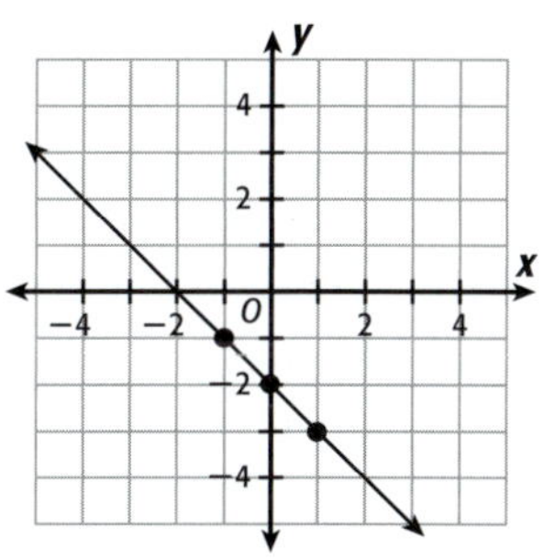

19. −3, −2, −1, 0

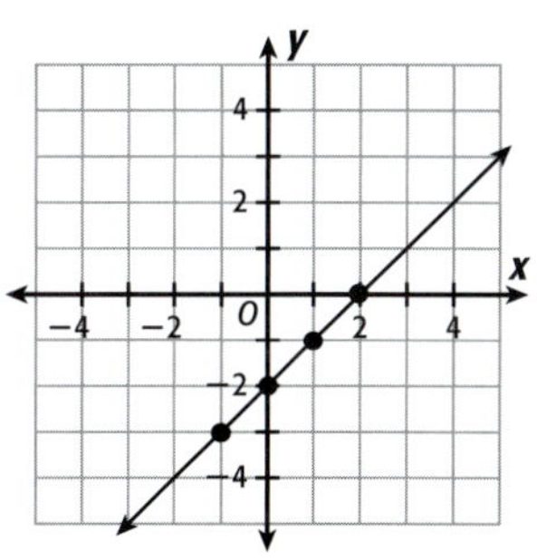

20. −6, −4, −2, 0

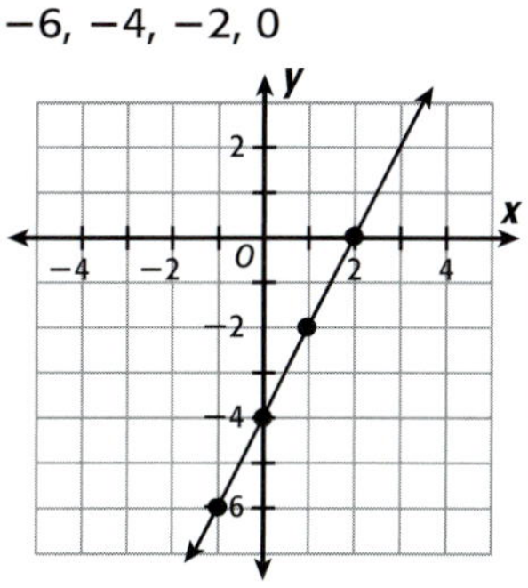

Lesson 12-6

5.

Top Rectangle
Vertical: 1, no change
Horizontal: 8 to 40
Center Rectangle
Vertical: 3, no change
Horizontal: 2 to 10
Bottom Rectangle
Vertical: 1, no change
Horizontal: 8 to 40

6.

Top Rectangle
Vertical: 1 to $\frac{1}{3}$
Horizontal: 8, no change
Center Rectangle
Vertical: 3 to 1
Horizontal: 2, no change
Bottom Rectangle
Vertical: 1 to $\frac{1}{3}$
Horizontal: 8, no change

7.

Top Rectangle
Vertical: 1, no change
Horizontal: 8 to 4
Center Rectangle
Vertical: 3, no change
Horizontal: 2 to 1
Bottom Rectangle
Vertical: 1, no change
Horizontal: 8 to 4

9. a.

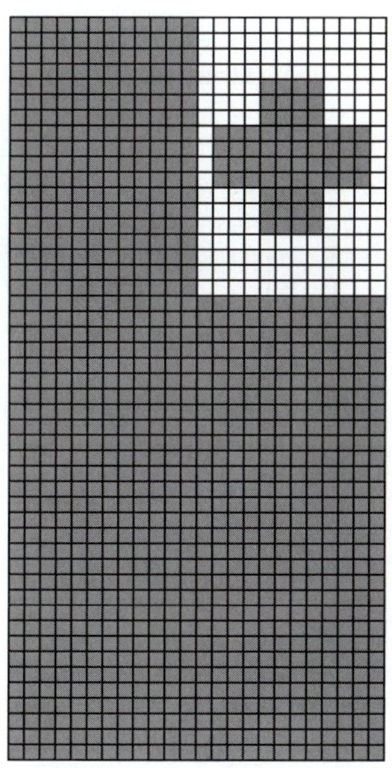

b. 12 units; 48 units

Performance Assessment

1. 12 square units; count the units on the graph or multiply the length by the width
2. E' (1, −2), F' (2, −5), G' (4, −4), H' (3, −1)

Extra Practice

1B Using Whole Numbers

17. Multiply by 2; 64, 128, 256.

18. Add 5; 21, 26, 31.

19. Multiply by 3; 189, 567, 1,701.

2A Understanding Variables and Expressions

17. 23 times 6; the product of 23 and 6

18. Take away p from 52; 52 minus p.

19. y divided by 4; the quotient of y and 4

20. the sum of 8 and 4; 8 plus 4

21. the difference of h and 96; 96 less than h

22. 13 groups of m; the product of 13 and m

6A Organizing Data

1. The participation increased from 1998 to 2002.

5K Race	
Year	**Number of Participants**
1998	1,345
1999	1,415
2000	1,532
2001	1,607
2002	1,781

2. points: Kareem Abdul-Jabbar, rebounds: Bill Russell, assists: Larry Bird

Basketball Data			
Player	**Points**	**Rebounds**	**Assists**
Jabbar	38,387	17,440	5,660
Bird	21,791	8,974	5,695
Russell	14,522	21,620	4,100

9. mean: 60, median: 58, mode: 52; median

6B Displaying and Interpreting Data

3.

Sport Played	
Sport	**Number of Students**
Baseball	𝍸 I
Basketball	IIII
Tennis	I
Soccer	II
Football	II

4.

Sport Played		
Sport	**Frequency**	**Cumulative Frequency**
Baseball	6	6
Basketball	4	10
Tennis	1	11
Soccer	2	13
Football	2	15

5.

Heights of Students (in.)				
Number (intervals)	46–50	51–55	56–60	61–65
Frequency	2	1	8	5

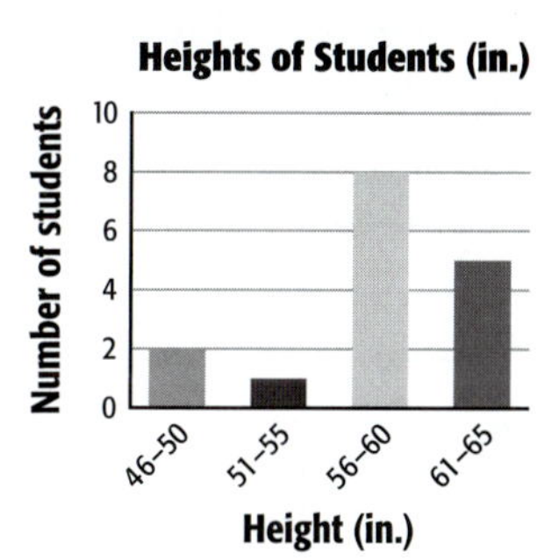

9–11.

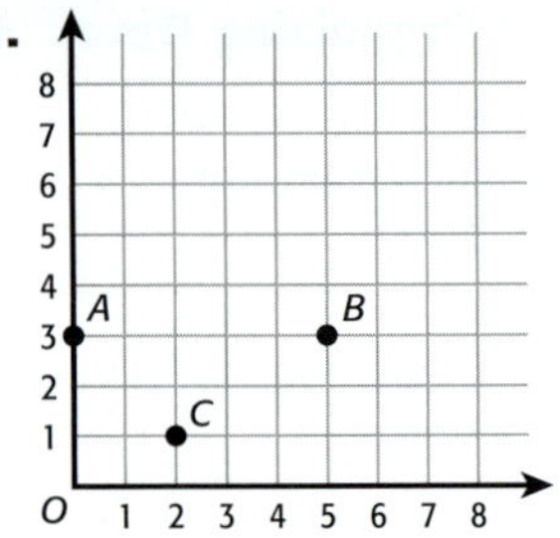

12. increase

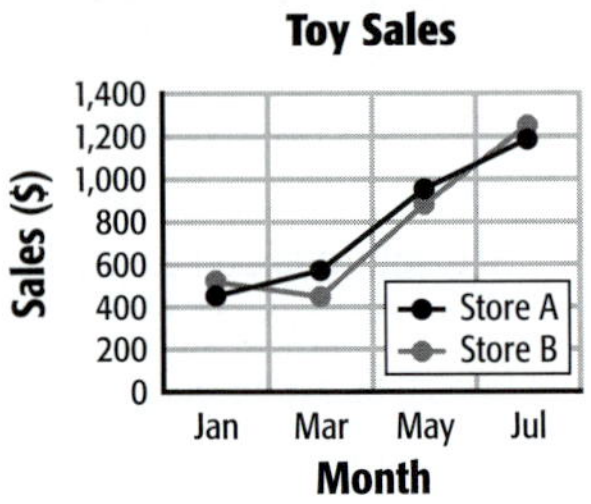

13. The scale starts at 30. It looks as if twice as many people voted for Viking, but in reality Viking received only about 20 more votes than Ram.

14. mean: 35, median: 35, mode: 35

Time Spent Doing Homework (min)

Stems	Leaves
1	0 5 5
2	0
3	5 5 5
4	
5	
6	0 0 5

Key: 1|0 means 10

7B Polygons

16.

17.

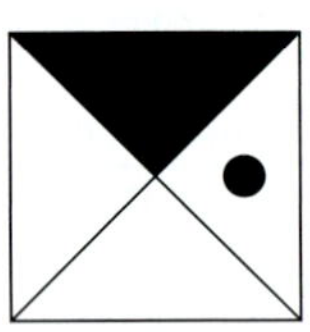

7C Polygon Relationships

13.

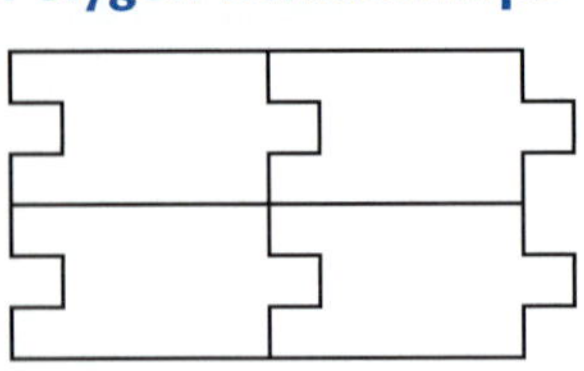

14.

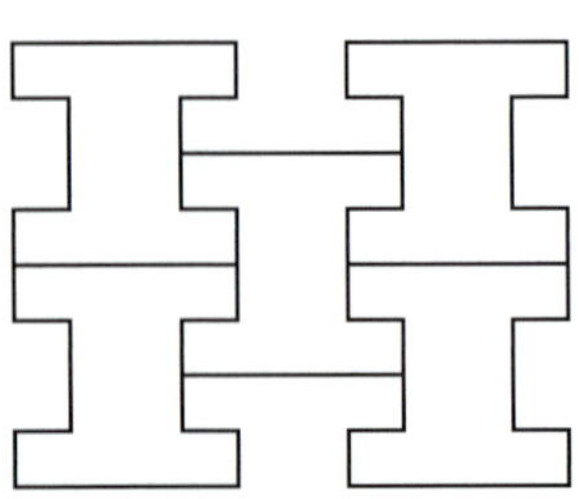

15.

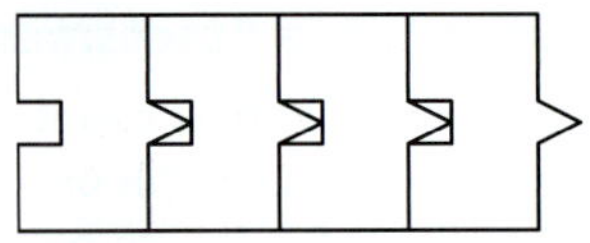

9A Integers

4.

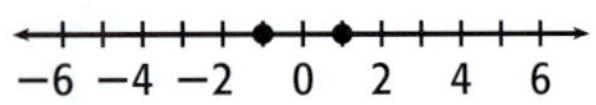

5.

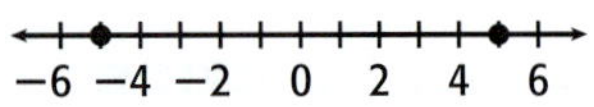

6.

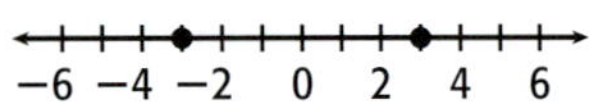

7.

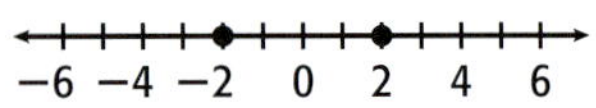

29–32.

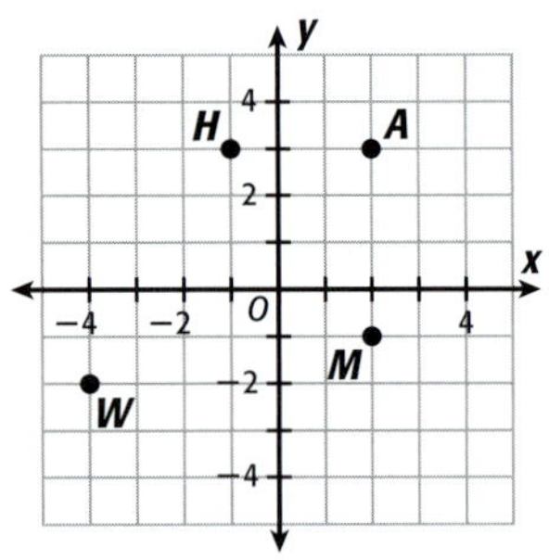

12A Introduction to Functions

17.

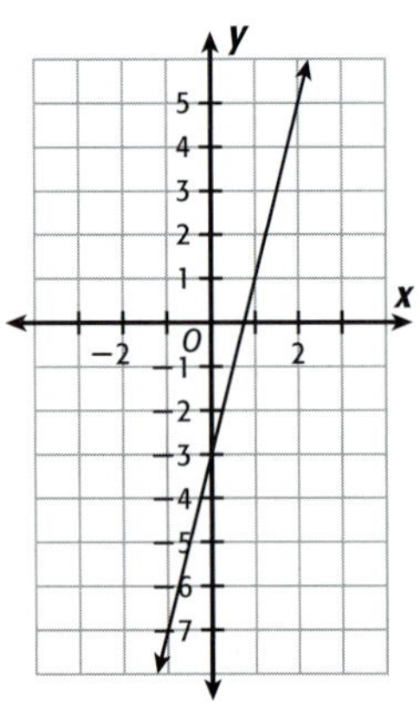

18.

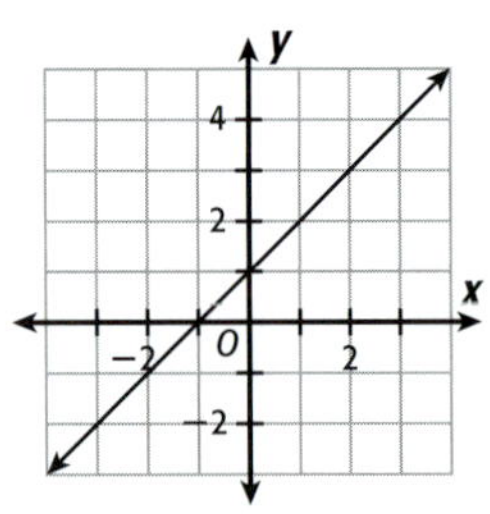

Skills Bank

Properties

1. Associative Property of Multiplication
2. Commutative Property of Multiplication
3. Identity Property of Zero
4. Identity Property of One
5. Commutative Property of Addition
6. Distributive Property
7. Associative Property of Addition
8. Multiplicative Property of Zero
9. Distributive Property

Pictographs

8.

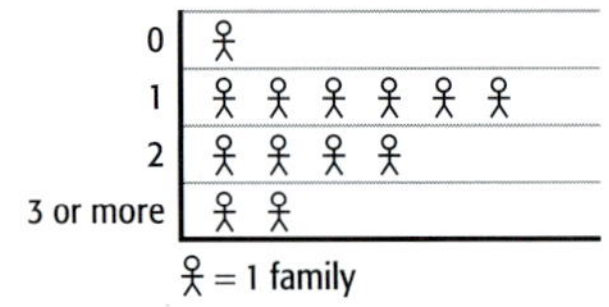

Line Plots

4.

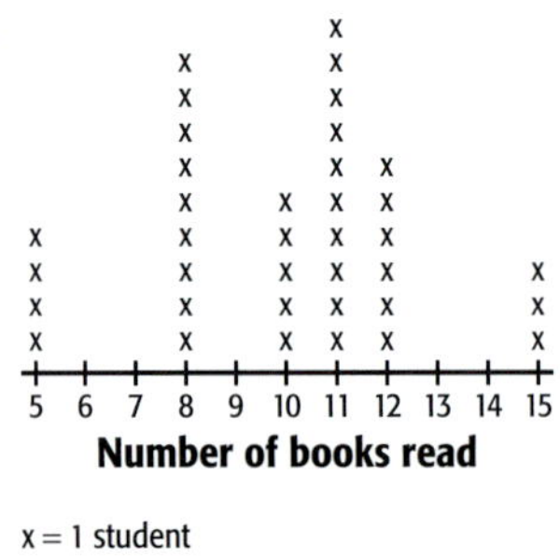

Graph Cumulative Frequency

1. Possible answer:

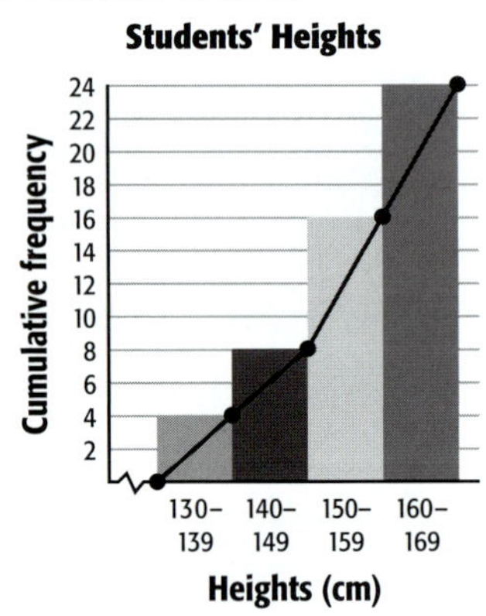

Lesson Quizzes

Lesson 6-1

1.

Age	bpm
newborn	135
2 yr	110
6 yr	95
10 yr	87
20 yr	71
40 yr	72
60 yr	74

Lesson 6-4

3.

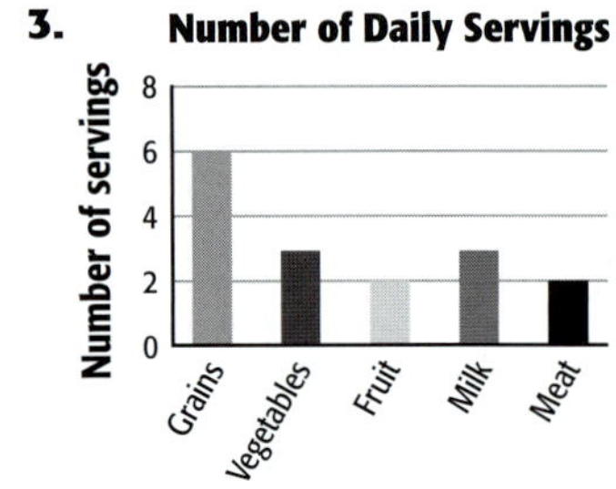

Lesson 6-5

1.

Number of Vacation Days	
1–5	卌
6–10	卌 I
11–15	IIII
16–20	III

2.

Frequency of Number of Vacation Days		
Number	Frequency	Cumulative Frequency
1–5	5	5
6–10	6	11
11–15	4	15
16–20	3	18

Lesson 6-7

1.

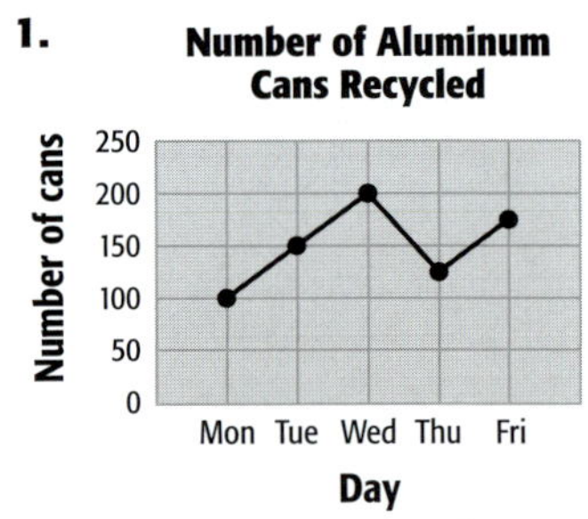

Lesson 6-9

1.

Stems	Leaves
2	1 2 7 8 9
3	0 1 4 4 5 6
4	0 2 6 9
5	2

Key: 2|1 means 21

Credits

Photo

Abbreviations used: (t) top, (c) center, (b) bottom, (l) left, (r) right, (bkgd) background

Cover (all), Pronk & Associates.

Title page (all), Pronk & Associates.

Master icons—teens (All): Sam Dudgeon/HRW.

Author photos by Sam Dudgeon/HRW

Problem Solving Handbook: xx (c), Pictor/Alamy Photos; xxi (c), Victoria Smith/HRW; xxii (tr), Charles Gupton/CORBIS; xxiii (tr), Victoria Smith/HRW; xxiv (tr), Victoria Smith/HRW; xxvi (tr), Cleo Freelance Photography/Painet Inc.; xxvii (tr), Sam Dudgeon/HRW; xxvii (cl), Victoria Smith/HRW; xxvii (bl), Victoria Smith/HRW; xxi (b), Sam Dudgeon/HRW. **Chapter 1:** 2–3 (bkgd), Getty Images/FPG International; 2 (br), Jeff Greenberg/MR/Photo Researchers, Inc.; 4 (tr), Steve Ewert Photography; 7 (tl), Image Source/elektraVision/PictureQuest; 9 (cr), Iowa State Fair; 11 (tr), National Geographic Image Collection/James Amos; 12 (tr), CORBIS/Bettmann; 15 (tr), CORBIS/Lester V. Bergman; 17 (br), Michael Dunning/Getty Images/FPG International; 21 (b), Sam Dudgeon/HRW; 23 (tr), National Geographic Image Collection/Kenneth Garrett; 23 (tc), National Geographic Image Collection/Kenneth Garrett; 24 (cr), DINODIA/Art Directors & TRIP Photo Library; 24 (tr), PhotoDisc-Digital Image copyright 2004 PhotoDisc; 27 (c), Sam Dudgeon/HRW; 27 (r), PhotoDisc - Digital Image copyright 2004 PhotoDisc; 27 (tl), C.K. Lorenz/Photo Researchers; 28 (tr), NASA; 36 (b), Layne Kennedy/CORBIS; 37 (tr), Greg Ryan/Sally Beyer; 37 (br), Greg Ryan/Sally Beyer; 38 (br), Jenny Thomas/HRW; 44 (cr), Kobal Collection/Lucasfilm/20th Century Fox. **Chapter 2:** 46–47 (bkgd), Christian Michaels/Getty Images/FPG International; 46 (br), Peter Yang/HRW Photo; 51 (tl), Sam Dudgeon/HRW; 55 (tr), AP Photo/NASA; 55 (all patches), NASA; 57 (bl), David A. Northcott/CORBIS; 59 (br), CORBIS/Brandon D. Cole; 62 (tr), Franklin Jay Viola/Viola's Photo Visions; 65 (tl), Peter Yang/HRW; 66 (c - Lincoln), Library of Congress; 66 (tr - Kennedy), AP Photo; 66 (bkgd - flag), Corbis Images; 69 (tr), National Geographic Image Collection/Bianca Lavies; 72 (tr), Darwin Dale/Photo Researchers, Inc.; 73 (tr), Takeshi Takahara/Photo Researchers, Inc.; 73 (c - oyster), Eric Kamp/Index Stock Imagery/PictureQuest; 78 (r), H. Mark Weidman; 79 (tl, tr), H. Mark Weidman. **Chapter 3:** 88–89 (bkgd), AP/Wide World Photos; 88 (br), Peter Yang/HRW Photo; 93 (cr), Jerry Schad/Photo Researchers, Inc.; 96 (tr), Getty Images/Stone; 99 (tr), PhotoDisc - Digital Image copyright 2004 PhotoDisc; 102 (tr), Steven E. Sutton/Duomo Photography; 105 (tl), Getty Images/Stone; 110 (c), Peter Van Steen/HRW; 113 (b), Peter Van Steen/HRW; 117 (tl), George Hall/Check Six; 120 (tr), NASA/Photo Researchers, Inc.; 123 (tr), CORBIS; 123 (br), Bettmann/CORBIS; 124 (tr), Peter Van Steen/HRW; 126 (br), Victoria Smith/HRW; 127 (tr), Getty Images/Stone; 130 (tl), Larry Stevens/Nawrocki Stock Photo; 131 (tr), Peter Yang/HRW; 134 (tr), CORBIS/Richard Hamilton Smith; 137 (tr), SuperStock; 138 (tc), Michelle Bridwell/HRW Photo; 140 (bc), Artville - Digital Image copyright 2004 Artville; 140 (br), Layne Kennedy/CORBIS; 141 (tr), John L. Gilkey; 142 (cr), Jenny Thomas/HRW. **Chapter 4:** 150–151 (bkgd), Peter Van Steen/HRW; 150 (br), Digital Image copyright 2004 PhotoDisc; 152 (tr), Darren Carroll/HRW; 155 (tl), Mike Norton/Animals Animals/Earth Scenes; 159 (tr, t), Frans Lanting/Minden Pictures; 161 (tr), PhotoEdit; 163 (tl), Frans Lanting/Minden Pictures; 165 (b), Digital Image copyright 2004 PhotoDisc; 167 (tr), Bettman/Corbis; 167 (tc), ball - Randy Faris/Corbis; signature - Courtesy of the National Baseball Hall of Fame; 170 (cr), Pat Lanza/FIELD/Bruce Coleman, Inc.; 175 (tr), Bob Krist/CORBIS; 175 (br), Wendell Metzen/Bruce Coleman, Inc.; 176 (t), Peter Van Steen/HRW; 177 (egg, shamrock, acorn, shell, rock), PhotoDisc - Digital Image copyright 2004 PhotoDisc; 177 (key, penny), EyeWire - Digital Image copyright 2004 EyeWire; 177(cicada), Artville - Digital Image copyright 2004 Artville; 178 (tr), Steve Cohen/FoodPix; 181 (tl), Peter Van Steen/HRW; 182 (tr), Peter French/Bruce Coleman, Inc.; 185 (tl), Science Photo Library/Photo Researchers, Inc.; 185 (cr), Sam Dudgeon/HRW Photo; 187 (bl), Tom Brakefield/CORBIS; 188 (tr), Mike Norton/Animals Animals/Earth Scenes; 188 (tc), Peter Van Steen/HRW; 191 (tl), Gary Meszaros/Bruce Coleman, Inc.; 195 (tr), David Ryan/Photo 20-20/PictureQuest; 195 (cr), Bob Rowan/Progressive Image/CORBIS; 198 (cr), Michael Meissner, Courtesy Cataloochee Ski Area; 198 (br), Trail Map, Courtesy Cataloochee Ski Area; 199 (r), Richard T. Nowitz/CORBIS; 200 (br), Randall Hyman/HRW; 206 (br), Peter Van Steen/HRW. **Chapter 5:** 208–209 (bkgd), Peter Van Steen/HRW; 208 (br), Digital Image copyright 2004 PhotoDisc ; 215 (tl), Merlin D. Tuttle/Bat Conservation International; 216 (tr), Ken Karp/HRW; 217 (bkgd), Beverly Barrett/HRW; 219 (tr), Corbis Images; 222 (tr), Jenny Thomas Photography/HRW; 225 (tr), Allen Blake Sheldon/Animals Animals/Earth Scenes; 226 (tr), Lori Grinker/Contact Press Images/PictureQuest; 227 (cl), Beverly Barrett/HRW; 227 (tr), Beverly Barrett/HRW; 229 (tl), Frans Lanting/Minden Pictures; 231 (br), Maximilian Stock Ltd./FoodPix; 232 (tr), Pictor International/PictureQuest; 232 (c), Beverly Barrett/HRW; 232 (b), Beverly Barrett/HRW; 236 (tr), James Martin/Getty Images/Stone; 237 (cr), SuperStock; 239 (tl), SuperStock; 239 (tr), Raymond A. Mendez/Animals Animals/Earth Scenes; 239 (tc), Mark Moffett/Minden Pictures; 245 (tr), SuperStock; 245 (br), Gerry Ellis/Minden Pictures; 245 (cr), National Geographic Image Collection/Paul Chesley; 245 (bl), Eric Hosking/CORBIS; 246 (tr), Frans Lanting/Minden Pictures; 247 (c), Frans Lanting/Minden Pictures; 249 (tl), Gerry Ellis/Minden Pictures; 256 (tr), Private Collection/Edmond Von Hoorick/SuperStock; 257 (cr), Alan Pitcairn/Grant Heilman Photography; 259 (t), Victoria Smith/HRW; 260 (cr), National Geographic Image Collection/Medford Taylor; 260 (tr), Digital Image copyright 2004 EyeWire; 261 (tr), Digital Image copyright 2004 PhotoDisc; 261 (b), National Geographic Image Collection/James P. Blair; 262 (br), Ken Karp/HRW; 269 (br), 1998 Image Farm Inc.

Chapter 6: 270–271 (bkgd), Charles W. Campbell/CORBIS; 270 (br), CORBIS; 272 (tc), Reuters NewMedia Inc./CORBIS; 272 (bl), Carl and Ann Purcell/Index Stock Imagery, Inc.; 275 (tr), Jenny Thomas Photography/HRW; 278 (tr), Trent Nelson/The Salt Lake Tribune/CORBIS Sygma; 278 (cl), AFP PHOTO/George FREY/Corbis; 281 (tr), NASA/Science Photo Library/Photo Researchers, Inc.; 283 (piano), Victoria & Albert Museum/Art Resource, NY; 283 (phonograph), U.S. Department of the Interior, National Park Service, Edison National Historic Site; 283 (tape recorder), Index Stock/Alamy Photos; 283 (CD player), Pintail Pictures/Alamy Photos; 284 (cl), Sharon Smith/Bruce Coleman, Inc.; 284 (cr), Tim Davis/Photo Researchers, Inc.; 284 (tl), SuperStock; 284 (tr), Dr. Eckart Pott/Bruce Coleman, Inc.; 290 (t - whorl), Leonard Lessin/Peter Arnold; 290 (t - arch), Federal Bureau of Investigation; 290 (t - loop), Archive Photos; 291 (tl), Rob Crandall/Stock Connection/ PictureQuest; 297 (tr), Bettmann/CORBIS; 301 (tc), Corbis Images/PictureQuest; 301 (tr), David Madison/Bruce Coleman, Inc.; 301 (br), Shane Young/AP/Wide World Photos; 304 (tr, c, cr), John Bavosi/Science Photo Library/Photo Researchers, Inc.; 305 (tr), Bryan Berg; 310 (bl), Courtesy Northwest Michigan Maritime Museum; 311 (t), William Hebert/Courtesy of the Frederick Meijer Gardens; 311 (cr), Courtesy of the Frederick Meijer Gardens; 312 (br), Randall Hyman/HRW. **Chapter 7:** 320–321 (bkgd), Art by Jane Dixon/HRW; 320 (br), Zhi Xiong China Tourism Press/Getty Images/The Image Bank; 326 (tr), Peter Van Steen/HRW; 332 (c), Michael Kelley/Getty Images/Stone; 332 (bc), (TempSport/CORBIS; 334 (tl, tr), Peter Van Steen/HRW; 334 (cl), Werner Forman Archive/Piers Morris Collection/Art Resource, NY; 334 (cr), P. W. Grace/Photo Researchers, Inc.; 334 (tl), Peter Van Steen/HRW; 336 (t), Walter Bibikow/Index Stock Imagery/PictureQuest; 337 (tl), Emmanuel Faure/SuperStock; 337 (tr), Peter Van Steen/HRW; 337 (cl), Pictor/Alamy Photos; 337 (cr), Peter Van Steen/HRW; 344 (bc), Bob Krist/CORBIS; 344 (t, c), Peter Van Steen/HRW; 347 (tl), Peter Newark's Western Americana; 351 (tl), David Forbert/SuperStock; 352 (c), PhotoDisc - Digital Image copyright 2004 PhotoDisc; 352 (bc), PhotoDisc - Digital Image copyright 2004 PhotoDisc; 353 (tr), PhotoDisc - Digital Image copyright 2004 PhotoDisc; 354 (tl), Beverly Barrett/HRW; 354 (tc, tr), Peter Van Steen/HRW; 354 (cl), Eric Grave/Science Source/Photo Researchers, Inc.; 354 (c), M. Abbey/Photo Researchers, Inc.; 354 (cr), Kim Taylor/Bruce Coleman/PictureQuest; 356 (tr), Lowe Art Museum, The University of Miami/SuperStock; 357 (cr, bl), Peter Van Steen/HRW; 358 (c), Peter Van Steen/HRW; 359 (tr), Steve Vidler/SuperStock; 359 (br), Nicholas DeVore/Getty Images/Stone; 361 (b), Michael Boys/CORBIS; 362 (tr), Ken Karp/HRW Photo; 364 (cl), Original Artwork "Sunflowers" by Mary Backer/Sam Dudgeon/HRW Photo; 364 (c), "Sunflowers" by Mary Backer/Sam Dudgeon/HRW Photo; 364 (cr), "Sunflowers" by Mary Backer/Sam Dudgeon/HRW Photo; 368 (cl), Sam Dudgeon/ HRW; 369 (tr), John Greim/Index Stock Imagery/PictureQuest; 372 (cr), corbis images.com; 372 (tr), Anna Clopet/CORBIS; 372 (br), Digital Image copyright 2004 PhotoDisc; 373 (tr), puzzle: http://www.tessellations.com; photo by: Sam Dudgeon/HRW; 373 (c), Ralph A. Clevenger/CORBIS; 375 (tr, cl, c, cr), John Warden/ SuperStock; 375 (br), Victoria Smith/HRW; 380 (cr), David Muench/CORBIS; 381 (tr), Dave G. Houser/CORBIS; 388 (br), Sam Dudgeon/HRW. **Chapter 8:** 390–391 (bkgd), Ralph A. Clevenger/CORBIS; 390 (br), Beth Davidow; 392 (tr), Reuters NewMedia Inc./CORBIS; 401 (tl), Ronald Zak/AP/Wide World Photos; 401 (cl), Imapress/Jean Claude N'Diaye/The Image Works; 401 (tr), AFP/CORBIS; 401 (cr), Jon Bower/Alamy Photos; 402 (tr), Corbis Images; 404 (l), Pierre Gleizes/Still Pictures; 406 (tr), Mary Cassatt, The Boating Party, Chester Dale Collection, Photograph (c) 2002 Board of

Trustees, National Gallery of Art, Washington. 1893/1894, oil on canvas; 407 (tr), Jenny Thomas/HRW; 409 (tr), Sharon McNeill/Bethel Area Chamber of Commerce; 412 (tl), National Geographic Image Collection; 412 (cr), Getty Images/Stone; 412 (tr), W. Perry Conway/CORBIS; 414 (cr), E. R. Degginger/Bruce Coleman, Inc.; 418 (tr), Jenny Thomas/HRW Photo; 421 (tr), Michelle Bridwell/PhotoEdit/PictureQuest; 425 (l), Bettman/CORBIS; 426 (tr), StockFood/Grand; 427 (tr), Corbis Images; 429 (tr), Lucy Schaly/Beaver County Times/AP/Wide World Photos; 432 (t), Peter Van Steen/HRW; 432 (cr), J. & P. Wegner/Animals Animals/Earth Scenes; 435 (cl), Sam Dudgeon/HRW; 438 (b), SuperStock; 438 (tr), Bettmann/CORBIS; 439 (br), Andy Levin/Photo Researchers, Inc.; 439 (t), Joseph Sohm; ChromoSohm Inc./CORBIS; 440 (br), Ken Karp/HRW; 440 (cl), Mimmo Jodice/CORBIS; 440 (c), Massimo Listri/CORBIS; 440 (cr), Gianni Dagli Orti/CORBIS; 446 (b), Digital Vision. **Chapter 9:** 448–489 (bkgd), Christopher A. Klein/National Geographic Artist/ National Geographic Image Collection; 448 (br), Peter Yang/HRW; 450 (tr - Death Valley), CORBIS/Bill Ross; 450 (tr - snow), Patrick J. Endres/Alaska Photographics.com; 453 (tl), National Geographic Image Collection/Emory Kristof; 454 (tr), CORBIS/AFP; 455 (cl), Chris Trotman/Duomo Photography; 463 (br), AFP/CORBIS; 465 (tr), CORBIS/Douglas Peebles; 468 (tl), Robert Emmett Bright/Photo Researchers; 473 (tr), Dennis Degnan/CORBIS; 476 (tr), AP/Wide World Photos; 479 (cr), Francisco Marquez; 485 (tl), Robert Frerck/Odyssey/Chicago; 488-489 (c), James R. Fisher/ Photo Researchers, Inc.; 489 (tr), Ian Adams Photography; 489 (b), Richard Stockton/ Alamy Photos; 490 (br), Jenny Thomas/HRW. **Chapter 10:** 498–499 (bkgd), Sam Dudgeon/HRW; 498 (br), Getty Images/FPG International; 500 (tr), Courtesy of Sigma Tau Pharmaceuticals, Inc.; 500 (b), Victoria Smith/HRW; 502 (tr), Barry L. Runk/Grant Heilman Photography; 502 (br), PhotoDisc - Digital Image copyright 2004 PhotoDisc; 504 (tr), Larry Lefever/Grant Heilman Photography; 507 (tr), William Hamilton/SuperStock; 507 (c), NASA/Science Source/Photo Researchers, Inc.; 507 (b), Sam Dudgeon/HRW; 511 (tr), Kwame Zikomo/SuperStock; 512 (tl, tr), Getty Images/The Image Bank; 516 (tr), Sam Dudgeon/HRW; 516 (cr), Corbis Images; 519 (cr), William Sallez/Duomo/CORBIS; 519 (tl), Mark McKeown/The Paris News/AP/Wide World Photos; 521 (bl), Omni Photo Communications Inc./Index Stock Imagery/PictureQuest; 524 (tr), Michael S. Yamashita/CORBIS; 525 (c), Sam Dudgeon/HRW; 525 (b), Peter Van Steen/HRW; 526 (cl), Barry L. Runk/Grant Heilman Photography; 526 (c, cr), Charles D. Winters/Photo Researchers, Inc.; 526 (bl, bc, br), Peter Van Steen/HRW; 527 (tl), Richard Cummins/CORBIS; 527 (tc), Joseph Sohm; ChromoSohm Inc./CORBIS; 527 (tr), Hisham Ibrahim/CORBIS; 533 (tl), Mark C. Burnett/Photo Researchers, Inc.; 535 (all), Peter Van Steen/HRW; 535 (c, t), Sam Dudgeon/HRW; 537 (tr), TEK Image/Science Photo Library/Photo Researchers, Inc.; 537 (c), Runk/Schoenberger/Grant Heilman Photography; 537 (b), Charles D. Winters/Photo Researchers, Inc.; 538 (cr), U.S. Department of the Interior, National Park Service, Edison National Historic Site; 538 (t), LOC/Science Source/Photo Researchers, Inc.; 542 (cr), Lee Snider/CORBIS; 542 (br), Colin Anderson/CORBIS; 543 (b), John G. Shedd Aquarium/Edward B. Lines, Jr.; 543 (tr), Colin Anderson/CORBIS; 543 (cr), Brandon D. Cole/CORBIS; 544 (c), Tim Davis/HRW Photo; 550 (b), Gianni Dagli Orti/CORBIS. **Chapter 11:** 552–553 (bkgd), SuperStock; 552 (br), Bill Varie/CORBIS; 557 (tr), TEK Image/Science Photo Library/Photo Researchers, Inc.; 558 (penny, nickel), EyeWire - Digital Image copyright 2004 EyeWire; 560 (quarter, dime), EyeWire - Digital Image copyright 2004 EyeWire; 560 (dollar coin), Sam Dudgeon/HRW; 560 (l), Sam Dudgeon/HRW; 565 (l), Peter Van Steen/HRW; 567 (l), CORBIS; 569 (b), Getty Images/Stone; 571 (l), Prof. K. Seddon & Dr. T. Evans, Queen's University, Belfast/Science Photo Library/Photo Researchers, Inc.; 574 (tr), IT Stock International/Index Stock Imagery/PictureQuest; 575 (tl), Digital Image copyright 2004 EyeWire; 577 (r), Pat Doyle/CORBIS; 580 (tr), The Old Farmer's Almanac* and Design Mark* are used with permission of Yankee Publishing Inc., Dublin, New Hampshire; 583 (hockey team), Reuters NewMedia Inc./CORBIS; 583 (tr), Jeff Curtes/CORBIS; 583 (two girls), Yvette Cardozo/Index Stock Imagery/PictureQuest; 583 (bkgd), Fotopic/Index Stock Imagery/PictureQuest; 583 (tc), Richard T. Nowitz/CORBIS; 586 (br), Aneal Vohra/Unicorn Stock Photos; 587 (b), Larry Miller/Photo Researchers, Inc.; 588 (br), Ken Karp/HRW; 592 (penny, nickel), Digital Image copyright (c) 2004 EyeWire; 593 (b), (c)Stockbyte - four food images; frame - Victoria Smith/HRW; 594 (spinner), Sam Dudgeon/HRW; 594 (penny), EyeWire - Digital Image copyright 2004 EyeWire; 594 (cards), Scott Van Osdol/HRW. **Chapter 12:** 596–597 (bkgd), Reuters NewMedia Inc./CORBIS; 596 (br), Maximilian Stock Ltd./Science Photo Library/Photo Researchers, Inc.; 601 (l), Catherine Karnow/CORBIS; 604 (tr), Ken Karp/HRW; 607 (tr), Simon Fraser/MRC Unit, Newcastle General Hospital/Science Photo Library/Photo Researchers, Inc.; 610 (t), Philip Gould/CORBIS; 613 (tc), Owen Franken/CORBIS; 613 (tr), Sam Dudgeon/HRW; 616 (tr), AFP/CORBIS; 619 (tl), A Fish Story by Dutch artist Gustave Verbeek. From his comic strip series Upside-Down, which appeared in the New York Herald from 1903 to 1905./Photo by Sam Dudgeon/HRW; 620 (tr), Peter Van Steen/HRW; 623 (tr), Maxine Cass; 624 (br), Carol Kitman; 625 (t), Joseph Sohm, ChromoSohm Inc./CORBIS; 625 (br), Walt Marz; 626 (br), Tim Davis/HRW Photo; 632 (br), Sam Dudgeon/HRW.

Back matter: 634(tl), Sam Dugeon/HRW Photo; 635 (br), Sam Dudgeon/HRW Photo.

Art

Illustrations: All work, unless otherwise noted, contributed by Holt, Rinehart & Winston.

Problem Solving Handbook: Page xxv (tr), Argosy; xxviii (all), Jeffrey Oh; 1 (tr), Argosy. **Chapter 1:** Page 4 (cr, bl), Argosy; 5 (inset), Ortelius Design; 7 (tr), Ortelius Design; 7 (c), Argosy; 10-11 (charts), Argosy; 12 (tc), Greg Geisler; 15 (cr), Argosy; 20 (all), Argosy; 26 (br), Argosy; 36 (cr), Ortelius Design; 38 (tr), Micah Harman; 44-45 (br), Argosy. **Chapter 2:** Page 48 (tr), Jeffrey Oh; 51 (cr), Argosy; 52 (tr), Don Dixon; 53 (all), Greg Geisler; 55 (tl), Argosy; 61 (tl), Argosy; 63 (cr), Ortelius Design; 64 (cr), Dan Stuckenschnieder; 64 (bc), Jeffrey Oh; 68 (all), Argosy; 71 (tr), Dan Stuckenschneider/Uhl Studios; 71 (cr), Ortelius Design; 72 (cr), Argosy; 78 (tc, cl), Ortelius Design; 79 (br), Nenad Jakesevic; 80 (tr), Ted Williams; 86 (cr), Ortelius Design; 87 (br), Argosy. **Chapter 3:** Page 92 (c), Argosy; 95 (tr), Cameron Eagle; 109 (c), Ortelius Design; 111 (tl, cl), Jeffrey Oh; 117 (c), Argosy; 128 (tr), Cindy Jeftovic; 130 (cr), Ortelius Design; 133 (cr), Kim Malek; 140 (c), Ortelius Design; 141 (b), Nenad Jakesevic; 142 (all), Greg Geisler; 148 (br), Mark Heine. **Chapter 4:** Page 156 (tc), Greg Geisler; 163 (tr), Jeffrey Oh; 170 (tc), Lance Lekander; 172 (tr), Stephen Durke/Washington Artists; 176 (all), Argosy; 181 (cr), Jeffrey Oh; 182 (tr), Argosy; 183 (cr), Greg Geisler; 191 (cr), Jeffrey Oh; 198 (tc), Ortelius Design; 199-200 (all), Darlene Olivia McElroy. **Chapter 5:** Page 212 (tr), Lori Bilter; 215 (tr), Jeffrey Oh; 215 (cr), Brent Spraggins; 225 (t), Argosy; 229 (cr), Argosy (Chart)/John Etheridge(Bagel); 235 (tr), Cindy Jeftovic; 236 (c), Stephen Durke/Washington Artists; 237 (cr), Argosy; 249 (cr), Brent Spraggins; 252 (tr), Ted Williams; 255 (tr), Ted Williams; 257 (cl), Ortelius Design; 259 (cr), Argosy; 260 (tc), Ortelius Design; 261 (cr), Argosy; 262 (cr), John Etheridge.
Chapter 6: Page 277 (tc), Stephen Durke/Washington Artists; 279 (cr), Stephen Durke/Washington Artists; 280 (tr), Argosy; 280 (br), Ortelius Design; 293 (tr), Ortelius Design; 294 (c), Dan Stuckenschneider/Uhl Studios; 296 (cr), Argosy; 300 (cr), Jeffrey Oh; 301 (tr, c/chart), Argosy; 303 (cr), Argosy; 310 (tr), Ortelius Design; 311 (br), Argosy; 318 (br), Jeffrey Oh. **Chapter 7:** Page 325 (r), John Etheridge; 329 (all), Cindy Jeftovic; 337 (br), Stephen Durke/Washington Artists; 338 (tr), Jane Sanders; 338 (cr), Nenad Jakesevic; 343 (br), Argosy; 352 (tr), Jane Sanders; 354 (cr), Eric Larsen; 355 (br), Jeffrey Oh; 359 (r/map), Ortelius Design; 365-366 (all), Lance Lekander; 368 (all), Argosy; 371-372 (all), Argosy; 380 (tc, b), Ortelius Design; 381 (c), Nenad Jakesevic. **Chapter 8:** Page 393–394 (all), Greg Geisler; 398 (tr), Fian Arroyo; 400 (tc), Argosy; 400 (cr), John Etheridge; 401 (cr), Argosy; 408 (cr), Jeffrey Oh; 409 (c), James Hindermeier; 410-411 (all), Argosy; 413 (t), Don Dixon; 414 (c), John Hartwell; 414 (tr), Argosy; 415 (tr), Eric Larsen; 416 (br), Brent Spraggins; 417 (br), Cindy Jeftovic; 421 (cr), Jane Sanders; 429 (tr), Argosy; 435 (tr), Argosy; 436 (c), Greg Geisler; 438 (c), Ortelius Design; 440 (r), Eric Larsen; 443 (cr), Argosy; 447 (br), Ortelius Design. **Chapter 9:** Page 450 (tr), Stephen Durke/Washington Artists; 453 (cr), Argosy; 457 (c), Argosy; 461 (tr), Ortelius; 466 (br), Argosy; 472 (br), Argosy; 478 (br), Argosy; 479 (tr), Andrea Tachiera; 482 (tr), Nenad Jakesevic; 482 (c), Greg Giesler; 485 (tr), Argosy; 488 (tc), Ortelius Design; 489 (cr), Argosy; 490 (tr), John Hartwell; 496 (b), Argosy. **Chapter 10:** Page 503 (cr), Argosy; 503 (cr), Ortelius Design; 509 (tr), Jeffrey Oh; 510 (cr), Ortelius Design; 512 (all), Argosy; 530 (tr), John Etheridge; 541 (c), Jeffrey Oh; 542 (tr), Ortelius Design; 544 (tr), Cindy Jeftovic. **Chapter 11:** Page 554 (tr), John Etheridge; 554 (bl), Eric Larsen; 557 (cr), Argosy; 558 (tc), Eric Larsen; 568 (cl), Eric Larsen; 570 (bc), John Etheridge; 573 (tr), Kim Malek; 575 (tr), Jeffrey Oh; 586 (bl), Ortelius Design; 586 (cr), Argosy; 586 (tr), Ortelius Design; 588 (r), Darlene Olivia McElroy; 592 (cr), Jeffrey Oh; 593 (cr), Argosy; 593 (bc), Jeffrey Oh. **Chapter 12:** Page 598 (c), Jeffrey Oh; 615 (br), Argosy; 619 (tr), Argosy; 621 (c), Argosy; 623 (cr), Ortelius Design; 624 (tr, c), Ortelius Design; 626 (r), John Etheridge. **Back matter:** Page 639 (cr), Argosy; 652 (br), Argosy; 658 (tr), Jeffrey Oh; 658-659 (all), Jeffrey Oh; 662 (t), Argosy; 671 (all), Argosy; 673-674 (all), Argosy.

Glossary

internet connect

Glossary Online: *go.hrw.com*
Keyword: MR4 Glossary

A

absolute value The distance of a number from zero on a number line; shown by $| \ |$. (p. 451)

Example: $|-5| = 5$

acute angle An angle that measures less than 90°. (p. 326)

acute triangle A triangle with all angles measuring less than 90°. (p. 344)

addend A number added to one or more other numbers to form a sum. For example, in the expression $4 + 6 + 7$, 4, 6, and 7 are addends.

Addition Property of Opposites The property that states that the sum of a number and its opposite equals zero.

Example: $12 + (-12) = 0$

adjacent angles Angles in the same plane that have a common vertex and a common side; in the diagram, $\angle a$ and $\angle b$ are adjacent angles. (p. 332)

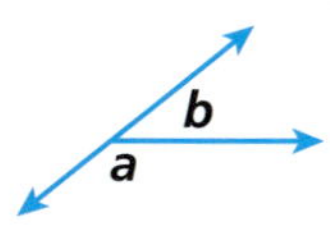

algebraic expression An expression that contains at least one variable. (p. 48)

Example: $x + 8$, $4(m - b)$

algebraic inequality An inequality that contains at least one variable.

Example: $x + 3 > 10$, $5a > b + 3$

alternate exterior angles A pair of angles formed by two lines intersected by a third line; in the diagram, the pairs of alternate exterior angles are $\angle a$ and $\angle d$ and $\angle b$ and $\angle c$. (p. 341)

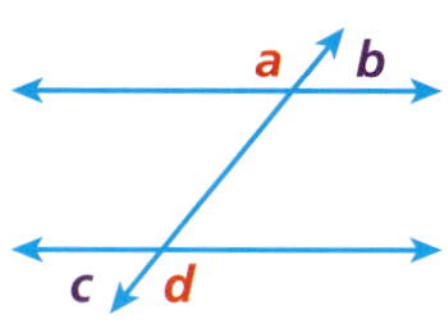

alternate interior angles A pair of angles formed by two lines intersected by a third line; in the diagram, the pairs of alternate interior angles are $\angle r$ and $\angle v$ and $\angle s$ and $\angle t$. (p. 341)

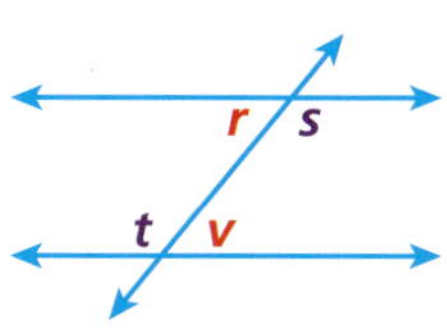

angle A figure formed by two rays with a common endpoint called the vertex. (p. 326)

angle bisector A line, segment, or ray that divides an angle into two congruent angles; in the diagram, $\overrightarrow{MP}$ is an angle bisector of $\angle NMO$.

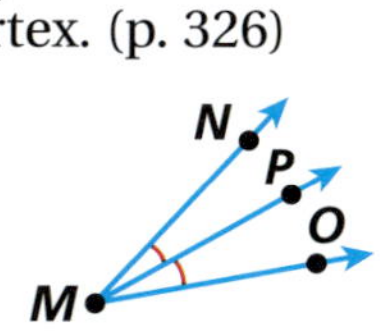

area The number of square units needed to cover a given surface. (p. 504)

Associative Property

Addition: The property that states that for three or more numbers, their sum is always the same, regardless of their grouping. (p. 24)

Example: $2 + 3 + 8 = (2 + 3) + 8 = 2 + (3 + 8)$

Multiplication: The property that states that for three or more numbers, their product is always the same, regardless of their grouping. (p. 24)

Example: $2 \cdot 3 \cdot 8 = (2 \cdot 3) \cdot 8 = 2 \cdot (3 \cdot 8)$

asymmetrical Not identical on either side of a central line; not symmetrical.

average The sum of the items in a set of data divided by the number of items in the set; also called *mean.* (p. 275)

axes The two perpendicular lines of a coordinate plane that intersect at the origin. (p. 458)

B

bar graph A graph that uses vertical or horizontal bars to display data. (p. 284)

base (in numeration) When a number is raised to a power, the number that is used as a factor is the base. (p. 12)

Example: $3^5 = 3 \cdot 3 \cdot 3 \cdot 3 \cdot 3$; 3 is the base.

base (of a polygon or three-dimensional figure) A side of a polygon; a face of a three-dimensional figure by which the figure is measured or classified. (p. 524)

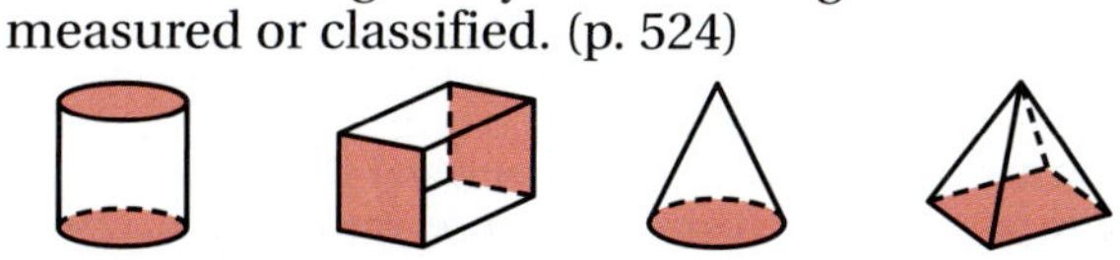

Bases of a cylinder | Bases of a prism | Base of a cone | Base of a pyramid

base-10 system A number system in which all numbers are expressed using the digits 0–9. (p. 34)

binary number system A number system in which all numbers are expressed using only two digits, 0 and 1. (p. 34)

bisect To divide into two congruent parts. (p. 378)

box-and-whisker plot A graph that displays the highest and lowest quarters of data as whiskers, the middle two quarters of the data as a box, and the median. (p. 308)

break (graph) A zigzag on a horizontal or vertical scale of a graph that indicates that some of the numbers on the scale have been omitted. (p. 285)

C

capacity The amount a container can hold when filled.

Celsius A metric scale for measuring temperature in which 0°C is the freezing point of water and 100°C is the boiling point of water; also called *centigrade*.

center (of a circle) The point inside a circle that is the same distance from all the points on the circle. (p. 516)

center (of rotation) The point about which a figure is rotated.

certain (probability) Sure to happen; having a probability of 1. (p. 554)

circle The set of all points in a plane that are the same distance from a given point called the center. (p. 516)

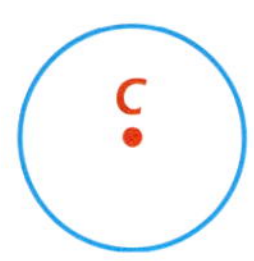

circle graph A graph that uses sections of a circle to compare parts to the whole and parts to other parts. (p. 430)

circumference The distance around a circle. (p. 516)

clockwise A circular movement in the direction shown.

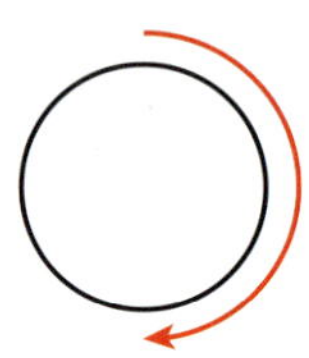

clustering A method used to estimate a sum when all addends are close to the same value. (p. 96)

Example: 27, 29, 24, and 26 all cluster around 25.

combination An arrangement of items or events in which order does not matter. (p. 579)

common denominator A denominator that is the same in two or more fractions. (p. 179)

Example: The common denominator of $\frac{5}{8}$ and $\frac{2}{8}$ is 8.

common factor A number that is a factor of two or more numbers.

Example: 8 is a common factor of 16 and 40.

common multiple A number that is a multiple of each of two or more numbers.

Example: 15 is a common multiple of 3 and 5.

Commutative Property

Addition: The property that states that two or more numbers can be added in any order without changing the sum. (p. 24)

Example: 8 + 20 = 20 + 8

Multiplication: The property that states that two or more numbers can be multiplied in any order without changing the product. (p. 24)

Example: $6 \cdot 12 = 12 \cdot 6$

compatible numbers Numbers that are close to the given numbers that make estimation or mental calculation easier. (p. 8)

compensation When a number in a problem is close to another number that is easier to calculate with, the easier number is used to find the answer. Then the answer is adjusted by adding to it or subtracting from it. (p. 28)

complementary angles Two angles whose measures add to 90°. (p. 333)

composite number A number greater than 1 that has more than two whole-number factors. (p. 153)

compound event An event made up of two or more simple events. (p. 574)

cone A three-dimensional figure with one vertex and one circular base. (p. 525)

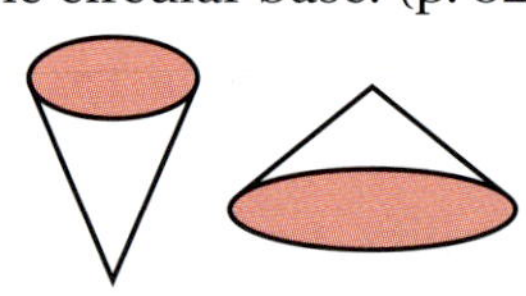

congruent Having the same size and shape. (p. 332)

congruent angles Angles that have the same measure. (p. 331)

congruent segments Segments that have the same length. (p. 330)

constant A value that does not change. (p. 48)

coordinate One of the numbers of an ordered pair that locate a point on a coordinate graph. (p. 458)

coordinate plane (coordinate grid) A plane formed by the intersection of a horizontal number line called the x-axis and a vertical number line called the y-axis. (pp. 294, 458)

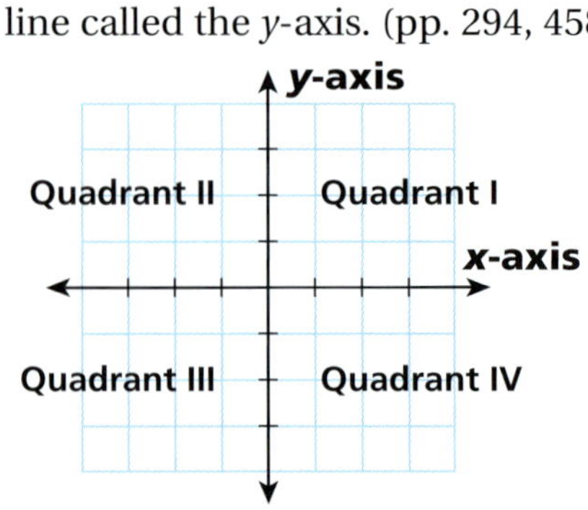

correspondence The relationship between two or more objects that are matched.

corresponding angles (for lines) A pair of angles formed by two lines intersected by a third line; in the diagram, the pairs of corresponding angles are $\angle m$ and $\angle q$, $\angle n$ and $\angle r$, $\angle o$ and $\angle s$, and $\angle p$ and $\angle t$. (p. 341)

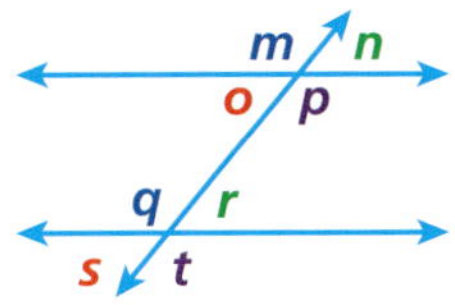

corresponding angles (in polygons) Matching angles of two or more polygons. (p. 405)

corresponding sides Matching sides of two or more polygons. (p. 405)

counterclockwise A circular movement in the direction shown.

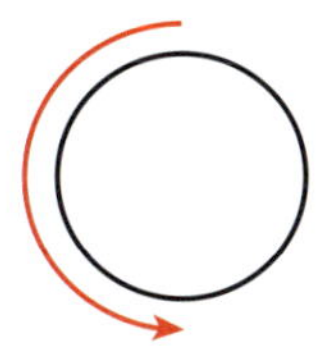

cross product The product of numbers on the diagonal when comparing two ratios. (p. 399)

Example: $\frac{2}{3} \times \frac{4}{6}$ $2 \cdot 6 = 12$

$3 \cdot 4 = 12$

cube (geometric figure) A rectangular prism with six congruent square faces.

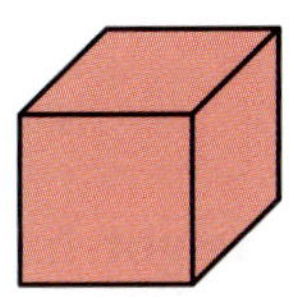

cube (in numeration) A number raised to the third power. (p. 12)

cumulative frequency The sum of successive data items. (p. 290)

customary system of measurement The measurement system often used in the United States.

Example: inches, feet, miles, ounces, pounds, tons, cups, quarts, gallons

cylinder A three-dimensional figure with two parallel, congruent circular bases connected by a curved lateral surface. (p. 524)

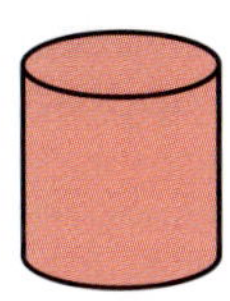

D

decagon A polygon with ten sides.

degree The unit of measure for angles or temperature.

denominator The bottom number of a fraction that tells how many equal parts are in the whole.

Example: $\frac{3}{4}$ ⟵ denominator

diagonal A line segment that connects two non-adjacent vertices of a polygon.

diameter A line segment that passes through the center of a circle and has endpoints on the circle, or the length of that segment. (p. 516)

difference The result when one number is subtracted from another.

dimension The length, width, or height of a figure.

discount The amount by which the original price is reduced. (p. 432)

Distributive Property The property that states if you multiply a sum by a number, you will get the same result if you multiply each addend by that number and then add the products. (p. 25)

Example: $5(20 + 1) = (5 \cdot 20) + (5 \cdot 1)$

dividend The number to be divided in a division problem.

Example: In $8 \div 4 = 2$, 8 is the dividend.

divisible Can be divided by a number without leaving a remainder. (p. 152)

divisor The number you are dividing by in a division problem.

Example: In $8 \div 4 = 2$, 4 is the divisor.

dodecahedron A polyhedron with 12 faces.

domain The set of all possible input values of a function.

double-bar graph A bar graph that compares two related sets of data. (p. 285)

double-line graph A graph that shows how two related sets of data change over time. (p. 298)

E

edge The line segment along which two faces of a polyhedron intersect. (p. 524)

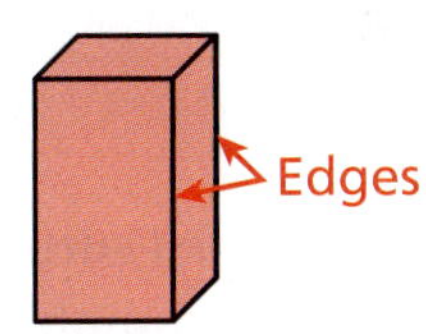

elements The words, numbers, or objects in a set. (p. 196)

empty set A set that has no elements. (p. 196)

endpoint A point at the end of a line segment or ray.

equally likely outcomes Outcomes that have the same probability. (p. 564)

equation A mathematical sentence that shows that two expressions are equivalent. (p. 58)

equilateral triangle A triangle with three congruent sides. (p. 345)

equivalent Having the same value.

equivalent fractions Fractions that name the same amount or part. (p. 172)

equivalent ratios Ratios that name the same comparison. (p. 392)

estimate (n) An answer that is close to the exact answer and is found by rounding or other methods.

estimate (v) To find an answer close to the exact answer by rounding or other methods.

evaluate To find the value of a numerical or algebraic expression. (p. 20)

even number A whole number that is divisible by two.

event An outcome or set of outcomes of an experiment or situation.

expanded form A number written as the sum of the values of its digits.

Example: 236,536 written in expanded form is 200,000 + 30,000 + 6,000 + 500 + 30 + 6.

experiment In probability, any activity based on chance, such as tossing a coin. (p. 558)

experimental probability The ratio of the number of times an event occurs to the total number of trials, or times that the activity is performed. (p. 558)

exponent The number that indicates how many times the base is used as a factor. (p. 12)

Example: $2^3 = 2 \times 2 \times 2 = 8$; 3 is the exponent.

exponential form A number is in exponential form when it is written with a base and an exponent. (p. 12)

Example: 4^2 is the exponential form for $4 \cdot 4$.

expression A mathematical phrase that contains operations, numbers, and/or variables.

F

face A flat surface of a polyhedron. (p. 524)

factor A number that is multiplied by another number to get a product. (p. 156)

factor tree A diagram showing how a whole number breaks down into its prime factors. (p. 157)

Fahrenheit A temperature scale in which 32°F is the freezing point of water and 212°F is the boiling point of water.

fair When all outcomes of an experiment are equally likely, the experiment is said to be fair. (p. 564)

first quartile The median of the lower half of a set of data; also called *lower quartile.* (p. 308)

formula A rule showing relationships among quantities.

Example: $A = \ell w$ is the formula for the area of a rectangle.

fraction A number in the form $\frac{a}{b}$, where $b \neq 0$.

frequency table A table that lists items together according to the number of times, or frequency, that the items occur. (p. 290)

front-end estimation An estimating technique in which the front digits of the addends are added and then the sum is adjusted for a closer estimate. (p. 97)

function An input-output relationship that has exactly one output for each input. (p. 598)

function table A table of ordered pairs that represent solutions of a function. (p. 598)

G

graph of an equation A graph of the set of ordered pairs that are solutions of the equation.

greatest common factor (GCF) The largest common factor of two or more given numbers. (p. 160)

H

height In a triangle or quadrilateral, the perpendicular distance from the base to the opposite vertex or side. (p. 505)

In a prism or cylinder, the perpendicular distance between the bases. (pp. 531, 534)

heptagon A seven-sided polygon.

hexagon A six-sided polygon.

histogram A bar graph that shows the frequency of data within equal intervals. (p. 291)

hypotenuse In a right triangle, the side opposite the right angle.

I

Identity Property of One The property that states that the product of 1 and any number is that number.

Identity Property of Zero The property that states the sum of zero and any number is that number.

image A figure resulting from a transformation.

impossible (probability) Can never happen; having a probability of 0. (p. 554)

improper fraction A fraction in which the numerator is greater than or equal to the denominator. (p. 182)

Example: $\frac{5}{5}, \frac{5}{3}$

indirect measurement The technique of using similar figures and proportions to find a measure. (p. 409)

inequality A mathematical sentence that shows the relationship between quantities that are not equal. (p. 76)

Example: $5 < 8$, $5x + 2 \geq 12$

input The value substituted into an expression or function. (p. 598)

integers The set of whole numbers and their opposites. (p. 450)

interest The amount of money charged for borrowing or using money, or the amount of money earned by saving money. (p. 436)

interior angles Angles on the inner sides of two lines intersected by a third line. In the diagram, $\angle c$, $\angle d$, $\angle e$, and $\angle f$ are interior angles. (p. 341)

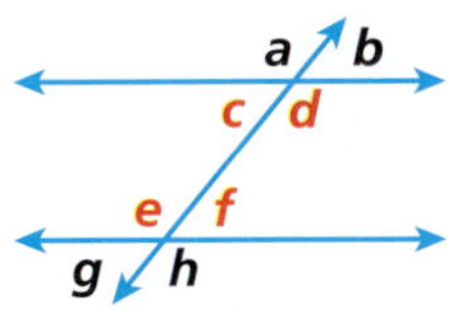

interquartile range The difference between the upper and lower quartiles of a data set. (p. 309)

intersecting lines Lines that cross at exactly one point. (p. 336)

intersection (sets) The set of elements common to two or more sets. (p. 196)

interval The space between marked values on a number line or the scale of a graph.

inverse operations Operations that undo each other: addition and subtraction, or multiplication and division.

isosceles triangle A triangle with at least two congruent sides. (p. 345)

L

lateral surface In a cylinder, the curved surface connecting the circular bases; in a cone, the curved surface that is not a base.

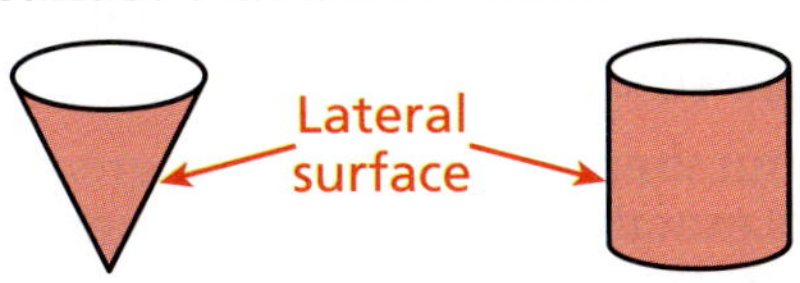

least common denominator (LCD) The least common multiple of two or more denominators. (p. 242)

least common multiple (LCM) The smallest number, other than zero, that is a multiple of two or more given numbers. (p. 232)

like fractions Fractions that have the same denominator. (p. 178)

Example: $\frac{5}{12}$ and $\frac{3}{12}$ are like fractions.

line A straight path that extends without end in opposite directions. (p. 322)

line graph A graph that uses line segments to show how data changes. (p. 297)

line plot A number line with marks or dots that show frequency.

line of reflection A line that a figure is flipped across to create a mirror image of the original figure. (p. 365)

line of symmetry The imaginary "mirror" in line symmetry. (p. 369)

line segment A part of a line between two endpoints. (p. 323)

line symmetry A figure has line symmetry if one half is a mirror-image of the other half. (p. 369)

linear equation An equation whose solutions form a straight line on a coordinate plane. (p. 605)

lower extreme The least number in a set of data. (p. 308)

lower quartile The median of the lower half of a set of data; also called *first quartile.* (p. 308)

M

mean The sum of the items in a set of data divided by the number of items in the set; also called *average.* (p. 275)

median The middle number or the mean (average) of the two middle numbers in an ordered set of data. (p. 275)

metric system of measurement A decimal system of weights and measures that is used universally in science and commonly throughout the world.

Example: centimeters, meters, kilometers, gram, kilograms, milliliters, liters

midpoint The point that divides a line segment into two congruent line segments.

mixed number A number made up of a whole number that is not zero and a fraction. (p. 167)

mode The number or numbers that occur most frequently in a set of data; when all numbers occur with the same frequency, we say there is no mode. (p. 275)

multiple The product of any number and a whole number is a multiple of that number.

Multiplication Property of Zero The property that states that the product of any number and 0 is 0.

N

negative number A number less than zero. (p. 450)

net An arrangement of two-dimensional figures that can be folded to form a polyhedron. (p. 530)

numerator The top number of a fraction that tells how many parts of a whole are being considered.

Example: $\frac{4}{5}$ ⟵ numerator

numerical expression An expression that contains only numbers and operations. (p. 20)

O

obtuse angle An angle whose measure is greater than 90° but less than 180°. (p. 326)

obtuse triangle A triangle containing one obtuse angle. (p. 344)

octagon An eight-sided polygon.

odd number A whole number that is not divisible by two.

odds A comparison of favorable outcomes and unfavorable outcomes. (p. 584)

opposites Two numbers that are an equal distance from zero on a number line. (p. 450)

order of operations A rule for evaluating expressions: first perform the operations in parentheses, then compute powers and roots, then perform all multiplication and division from left to right, and then perform all addition and subtraction from left to right. (p. 20)

ordered pair A pair of numbers that can be used to locate a point on a coordinate plane. (p. 294)

origin The point where the x-axis and y-axis intersect on the coordinate plane; (0, 0). (p. 458)

outcome A possible result of a probability experiment. (p. 558)

outlier A value much greater or much less than the others in a data set. (p. 278)

output The value that results from the substitution of a given input into an expression or function. (p. 598)

overestimate An estimate that is greater than the exact answer. (p. 8)

P

parallel lines Lines in a plane that do not intersect. (p. 336)

parallelogram A quadrilateral with two pairs of parallel sides. (p. 348)

pentagon A five-sided polygon.

percent A ratio comparing a number to 100. (p. 418)

Example: $45\% = \frac{45}{100}$

perfect square A square of a whole number. (p. 31)

Example: $5 \cdot 5 = 25$, and $7^2 = 49$; 25 and 49 are perfect squares.

perimeter The distance around a polygon. (p. 500)

permutation An arrangement of items or events in which order is important. (p. 578)

perpendicular bisector A line that intersects a segment at its midpoint and is perpendicular to the segment. (p. 378)

perpendicular lines Lines that intersect to form right angles. (p. 336)

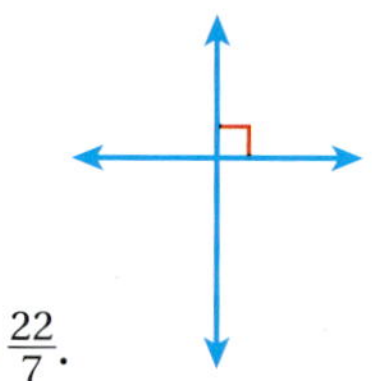

pi (π) The ratio of the circumference of a circle to the length of its diameter; $\pi \approx 3.14$ or $\frac{22}{7}$. (p. 516)

plane A flat surface that extends forever. (p. 322)

point An exact location in space. (p. 322)

polygon A closed plane figure formed by three or more line segments that intersect only at their endpoints. (p. 352)

polyhedron A three-dimensional figure in which all the surfaces or faces are polygons. (p. 524)

positive number A number greater than zero. (p. 450)

power A number produced by raising a base to an exponent.

Example: $2^3 = 8$, so 2 to the 3rd power is 8.

prediction A guess about something that will happen in the future. (p. 580)

prime factorization A number written as the product of its prime factors. (p. 156)

Example: $10 = 2 \cdot 5$, $24 = 2^3 \cdot 3$

prime number A whole number greater than 1 that has exactly two factors, itself and 1. (p. 153)

principal The initial amount of money borrowed or saved. (p. 436)

prism A polyhedron that has two congruent, polygon-shaped bases and other faces that are all rectangles. (p. 524)

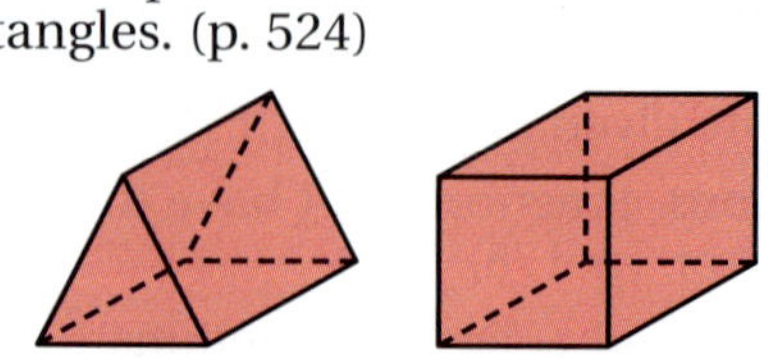

probability A number from 0 to 1 (or 0% to 100%) that describes how likely an event is to occur. (p. 554)

product The result when two or more numbers are multiplied.

proper fraction A fraction in which the numerator is less than the denominator. (p. 182)

Example: $\frac{3}{4}, \frac{1}{12}, \frac{7}{8}$

proportion An equation that states that two ratios are equivalent. (p. 398)

protractor A tool for measuring angles.

pyramid A polyhedron with a polygon base and triangular sides that all meet at a common vertex. (p. 525)

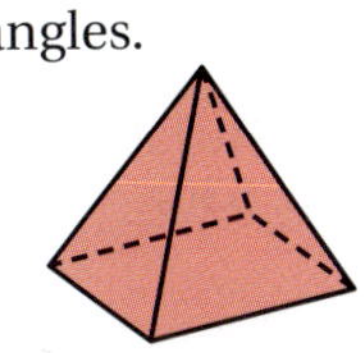

Q

quadrant The x- and y-axes divide the coordinate plane into four regions. Each region is called a quadrant. (p. 458)

quadrilateral A four-sided polygon. (p. 348)

quartile Three values, one of which is the median, that divide a data set into fourths. See also *first quartile, third quartile.* (p. 308)

quotient The result when one number is divided by another.

R

radius A line segment with one endpoint at the center of a circle and the other endpoint on the circle, or the length of that segment. (p. 516)

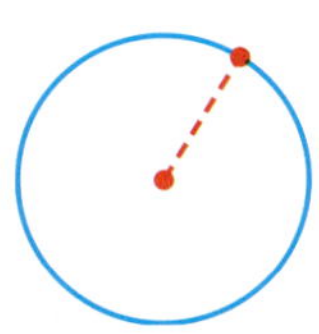

random numbers In a set of random numbers, each number has an equal chance of being selected. (p. 589)

range (in statistics) The difference between the greatest and least values in a data set. (p. 275)

rate A ratio that compares two quantities measured in different units. (p. 393)

Example: The speed limit is 55 miles per hour, or 55 mi/h.

rate of interest The percent charged or earned on an amount of money; see *simple interest.* (p. 436)

ratio A comparison of two quantities by division. (p. 392)

Example: 12 to 25, 12:25, $\frac{12}{25}$

ray A part of a line that starts at one endpoint and extends forever. (p. 323)

reciprocal One of two numbers whose product is 1. (p. 222)

Example: The reciprocal of $\frac{2}{3}$ is $\frac{3}{2}$.

rectangle A parallelogram with four right angles. (p. 348)

rectangular prism A polyhedron whose bases are rectangles and whose other faces are rectangles.

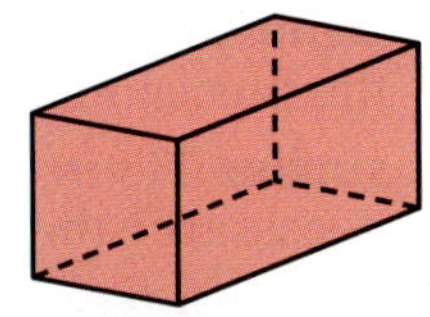

reflection A transformation of a figure that flips the figure across a line. (p. 365)

regular polygon A polygon with congruent sides and angles. (p. 352)

repeating decimal A decimal in which one or more digits repeat infinitely. (p. 168)

Example: $0.757575... = 0.\overline{75}$

rhombus A parallelogram with all sides congruent. (p. 348)

right angle An angle that measures 90°. (p. 326)

right triangle A triangle containing a right angle. (p. 344)

rotation A transformation in which a figure is turned around a point. (p. 365)

rounding Replacing a number with an estimate of that number to a given place value.

Example: 2,354 rounded to the nearest thousand is 2,000; 2,354 rounded to the nearest 100 is 2,400.

S

sales tax A percent of the cost of an item, which is charged by governments to raise money. (p. 432)

sample space All possible outcomes of an experiment. (p. 558)

scale The ratio between two sets of measurements. (p. 412)

scale drawing A drawing that uses a scale to make an object proportionally smaller than or larger than the real object. (p. 412)

scale model A proportional model of a three-dimensional object.

scalene triangle A triangle with no congruent sides. (p. 345)

scientific notation A method of writing very large or very small numbers by using powers of 10. (p. 114)

second quartile The median of a set of data. (p. 308)

segment A part of a line between two endpoints. (p. 323)

sequence An ordered list of numbers. (p. 31)

set A group of items. (p. 196)

side A line bounding a geometric figure; one of the faces forming the outside of an object.

significant figures The figures used to express the precision of a measurement. (p. 138)

similar Figures with the same shape but not necessarily the same size are similar. (p. 405)

simple interest A fixed percent of the principal. It is found using the formula $I = Prt$, where P represents the principal, r the rate of interest, and t the time. (p. 436)

simplest form (of a fraction) A fraction is in simplest form when the numerator and denominator have no common factors other than 1. (p. 173)

simplify To write a fraction or expression in simplest form.

simulation A model of an experiment, often one that would be too difficult or too time-consuming to actually perform. (p. 562)

skew lines Lines that lie in different planes that are neither parallel nor intersecting. (p. 336)

solid figure A three-dimensional figure.

solution of an equation A value or values that make an equation true. (p. 58)

solution of an inequality A value or values that make an inequality true. (p. 76)

solve To find an answer or a solution.

square (geometry) A rectangle with four congruent sides. (p. 348)

square (numeration) A number raised to the second power. (p. 12)

Example: In 5^2, the number 5 is squared.

square number The product of a number and itself.

Example: 25 is a square number. $5 \cdot 5 = 25$

square root One of the two equal factors of a number. (p. 603)

Example: $16 = 4 \cdot 4$, or $16 = -4 \cdot -4$, so 4 and -4 are square roots of 16.

standard form (in numeration) A way to write numbers by using digits.

Example: Five thousand, two hundred ten in standard form is 5,210.

stem-and-leaf plot A graph used to organize and display data so that the frequencies can be compared. (p. 305)

straight angle An angle that measures 180°. (p. 326)

subset A set contained within another set. (p. 197)

substitute To replace a variable with a number or another expression in an algebraic expression.

sum The result when two or more numbers are added.

supplementary angles Two angles whose measures have a sum of 180°. (p. 333)

surface area The sum of the areas of the faces, or surfaces, of a three-dimensional figure. (p. 530)

T

term (in a sequence) An element or number in a sequence. (p. 31)

terminating decimal A decimal number that ends or terminates. (p. 168)

Example: 6.75

tessellation A repeating pattern of plane figures that completely cover a plane with no gaps or overlaps. (p. 373)

theoretical probability The ratio of the number of equally likely outcomes in an event to the total number of possible outcomes. (p. 564)

third quartile The median of the upper half of a set of data; also called *upper quartile.* (p. 308)

tip The amount of money added to a bill for service; usually a percent of the bill. (p. 432)

transformation A change in the size or position of a figure. (p. 365)

translation A movement (slide) of a figure along a straight line. (p. 365)

trapezoid A quadrilateral with exactly one pair of parallel sides. (p. 348)

tree diagram A branching diagram that shows all possible combinations or outcomes of an event. (p. 570)

triangle A three-sided polygon.

Triangle Sum Theorem The theorem that states that the measures of the angles in a triangle add to 180°.

triangular prism A polyhedron whose bases are triangles and whose other faces are rectangles.

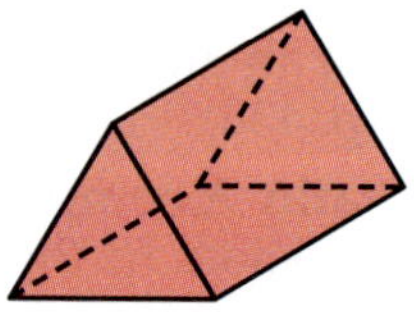

U

underestimate An estimate that is less than the exact answer. (p. 8)

union The set of all elements that belong to two or more sets. (p. 196)

unit conversion The process of changing one unit of measure to another.

unit rate A rate in which the second quantity in the comparison is one unit. (p. 393)

Example: 10 cm per minute

unlike fractions Fractions with different denominators. (p. 179)

Example: $\frac{3}{4}$ and $\frac{1}{2}$ are unlike fractions.

upper extreme The greatest number in a set of data. (p. 308)

upper quartile The median of the upper half of a set of data; also called *third quartile.* (p. 308)

V

variable A symbol used to represent a quantity that can change. (p. 48)

Venn diagram A diagram that is used to show relationships between sets. (p. 196)

vertex On an angle or polygon, the point where two sides intersect; on a polyhedron, the intersection of three or more faces; on a cone or pyramid, the top point. (pp. 326, 524)

vertical angles A pair of opposite congruent angles formed by intersecting lines; in the diagram, $\angle a$ and $\angle c$ are vertical angles and $\angle b$ and $\angle d$ are vertical angles. (p. 332)

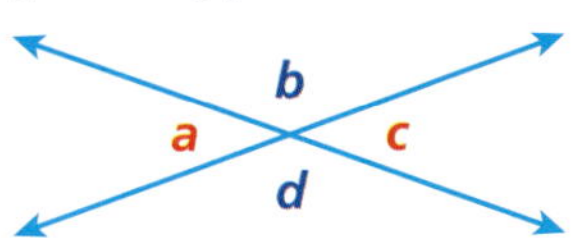

volume The number of cubic units needed to fill a given space. (p. 534)

X

***x*-axis** The horizontal axis on a coordinate plane. (p. 458)

***x*-coordinate** The first number in an ordered pair; it tells the distance to move right or left from the origin, (0, 0). (p. 458)

Example: 5 is the x-coordinate in (5, 3).

Y

***y*-axis** The vertical axis on a coordinate plane. (p. 458)

***y*-coordinate** The second number in an ordered pair; it tells the distance to move up or down from the origin, (0, 0). (p. 458)

Example: 3 is the y-coordinate in (5, 3).

Z

zero pair A number and its opposite, which add to 0.

Example: 18 and −18

Index

A

B

C

H

O

P

S

T

U

V

W

X

Y

Z

2005 NAEP Mathematics Framework Grade 8 Assessment

I. Number Properties

1. Number sense

- **a.** Use place value to model and describe integers and decimals.
- **b.** Model or describe rational numbers or numerical relationships using number lines and diagrams.
- **d.** Write or rename rational numbers.
- **e.** Recognize, translate between, or apply multiple representations of rational numbers (fractions, decimals, and percents) in meaningful contexts.
- **f.** Express or interpret numbers using scientific notation from real life contexts.
- **g.** Find or model absolute value or apply to problem situations.
- **i.** Order or compare rational numbers (fractions, decimals, percents, or integers) using various models and representations (e.g., number line).
- **j.** Order or compare rational numbers including very large and small integers, and decimals and fractions close to zero.

2. Estimation

- a. Establish or apply benchmarks for rational numbers and common irrational numbers (e.g., π) in contexts.
- **b.** Make estimates appropriate to a given situation by:
  - identifying when estimation is appropriate,
  - determining the level of accuracy needed,
  - selecting the appropriate method of estimation, or
  - analyzing the effect of an estimation method on the accuracy of results.
- **c.** Verify solutions or determine the reasonableness of results in a variety of situations including calculator and computer results.
- **d.** Estimate square or cube roots of numbers less than 1,000 between two whole numbers.

3. Number operations

- **a.** Perform computations with rational numbers.
- **d.** Describe the effect of multiplying and dividing by numbers, including the effect of multiplying or dividing a rational number by:
  - zero, or
  - a number less than zero, or
  - a number between zero and one, or
  - one, or
  - a number greater than one.
- **e.** Provide a mathematical argument to explain operations with two or more fractions.
- **f.** Interpret rational number operations and the relationship between them.
- **g.** Solve application problems involving rational numbers and operations using exact answers or estimates as appropriate.

4. Ratios and proportional reasoning

- **a.** Use ratios to describe problem situations.
- **b.** Use fractions to represent and express ratios and proportions.
- **c.** Use proportional reasoning to model and solve problems (including rates, scaling, and similarity).
- **d.** Solve problems involving percentages (including percent increase and decrease, interest rates, tax, discount, tips, or part/whole relationships).

5. Properties of number and operations

- **a.** Describe odd and even integers and how they behave under different operations.
- **b.** Recognize, find, or use factors, multiples, or prime factorization.
- **c.** Recognize or use prime and composite numbers to solve problems.
- **d.** Use divisibility or remainders in problem settings.
- **e.** Apply basic properties of operations.
- **f.** Explain or justify a mathematical concept or relationship (e.g., explain why 17 is prime).

2005 NAEP Mathematics Framework Grade 8 Assessment

II. Measurement

1. Measuring physical attributes

b. Compare objects with respect to length, area, volume, angle measurement, weight, or mass.

c. Estimate the size of an object with respect to a given measurement attribute (e.g., area).

g. Select or use appropriate measurement instrument to determine or create a given length, area, volume, angle, weight, or mass.

h. Solve mathematical or real-world problems involving perimeter or area of plane figures such as triangles, rectangles, circles, or composite figures.

j. Solve problems involving volume or surface area of rectangular solids, cylinders, prisms, or composite shapes.

k. Solve problems involving indirect measurement such as finding the height of a building by comparing its shadow with the height and shadow of a known object.

l. Solve problems involving rates such as speed or population density.

2. Systems of measurement

a. Select or use appropriate type of unit for the attribute being measured such as length, area, angle, time, or volume.

b. Solve problems involving conversions within the same measurement system such as conversions involving square inches and square feet.

c. Estimate the measure of an object in one system given the measure of that object in another system and the approximate conversion factor. For example:

- Distance Conversion: 1 kilometer is approximately $\frac{5}{8}$ of a mile.
- Money Conversion: US dollar is approximately 1.5 Canadian dollars.
- Temperature Conversion: Fahrenheit to Celsius

d. Determine appropriate size of unit of measurement in problem situation involving such attributes as length, area, or volume.

e. Determine appropriate accuracy of measurement in problem situations (e.g., the accuracy of each of several lengths needed to obtain a specified accuracy of total length) and find the measure to that degree of accuracy.

f. Construct or solve problems (e.g., floor area of a room) involving scale drawings.

2005 NAEP Mathematics Framework Grade 8 Assessment

III. Geometry

1. Dimension and shape

a. Draw or describe a path of shortest length between points to solve problems in context.

b. Identify a geometric object given written description of its properties.

c. Identify, define, or describe geometric shapes in the plane and in 3-dimensional space given a visual representation.

d. Draw or sketch from a written description polygons, circles, or semicircles.

e. Represent or describe a three-dimensional situation in a two-dimensional drawing using perspective.

f. Demonstrate an understanding about the two- and three-dimensional shapes in our world through identifying, drawing, modeling, building, or taking apart.

2. Transformation of shapes and preservation of properties

a. Identify lines of symmetry in plane figures or recognize and classify types of symmetries of plane figures.

c. Recognize or informally describe the effect of a transformation on two-dimensional geometric shapes (reflections across lines of symmetry, rotations, translations, magnifications, and contractions).

d. Predict results of combining, subdividing, and changing shapes of plane figures and solids (e.g., paper folding, tiling, and cutting up and rearranging pieces).

e. Justify relationships of congruence and similarity, and apply these relationships using scaling and proportional reasoning.

f. For similar figures, identify and use the relationships of conservation of angle and of proportionality of side length and perimeter.

3. Relationships between geometric figures

b. Apply geometric properties and relationships in solving simple problems in two- and three-dimensions.

c. Represent problem situations with simple geometric models to solve mathematical or real-world problems.

d. Use the Pythagorean Theorem to solve problems.

f. Describe or analyze simple properties of, or relationships between, triangles, quadrilaterals, and other polygonal plane figures.

g. Describe or analyze properties and relationships of parallel or intersecting lines.

4. Position and direction

a. Describe relative positions of points and lines using the geometric ideas of midpoint, points on common line through a common point, parallelism, or perpendicularity.

b. Describe the intersection of two or more geometric figures in the plane (e.g., intersection of a circle and a line).

c. Visualize or describe the cross-section of a solid.

d. Represent geometric figures using rectangular coordinates on a plane.

5. Mathematical reasoning

a. Make and test a geometric conjecture about regular polygons.

2005 NAEP Mathematics Framework Grade 8 Assessment

IV. Data Analysis and Probability

1. Data representation

- **a.** Read or interpret data, including interpolating or extrapolating from data.
- **b.** Given a set of data, complete a graph and then solve a problem using the data in the graph (circle graphs, histograms, bar graphs, line graphs, scatter plots).
- **c.** Solve problems by estimating and computing with data from a single set or across sets of data.
- **d.** Given a graph or a set of data, determine whether information is represented effectively and appropriately (circle graphs, histograms, bar graphs, line graphs, scatter plots).
- **e.** Compare and contrast the effectiveness of different representations of the same data.

2. Characteristics of data sets

- **a.** Calculate, use, or interpret mean, median, mode, or range.
- **b.** Describe how mean, median, mode, range, or interquartile ranges relate to the shape of the distribution.
- **c.** Identify outliers and determine their effect on mean, median, mode, or range.
- **d.** Using appropriate statistical measures, compare two or more data sets describing the same characteristic for two different populations or subsets of the same population.
- **e.** Visually choose the line that best fits given a scatter plot and informally explain the meaning of the line. Use the line to make predictions.

3. Experiments and samples

- **a.** Given a sample, identify possible sources of bias in sampling.
- **b.** Distinguish between a random and non-random sample.
- **d.** Evaluate the design of an experiment.

4. Probability

- **a.** Analyze a situation that involves probability of an independent event.
- **b.** Determine the theoretical probability of simple and compound events in familiar contexts.
- **c.** Estimate the probability of simple and compound events through experimentation or simulation.
- **d.** Distinguish between experimental and theoretical probability.
- **e.** Determine the sample space for a given situation.
- **f.** Use a sample space to determine the probability of the possible outcomes of an event.
- **g.** Represent probability using fractions, decimals, and percents.
- **h.** Determine the probability of independent and dependent events. (Dependent events should be limited to linear functions with a small sample size.)
- **j.** Interpret probabilities within a given context.

2005 NAEP Mathematics Framework Grade 8 Assessment

V. Algebra

1. Patterns, relations, and functions

a. Recognize, describe, or extend numerical and geometric patterns using tables, graphs, words, or symbols.

b. Generalize a pattern appearing in a numerical sequence or table or graph using words or symbols.

c. Analyze or create patterns, sequences, or linear functions given a rule.

e. Identify functions as linear or non-linear or contrast distinguishing properties of functions from tables, graphs, or equations.

f. Interpret the meaning of slope or intercepts in linear functions.

2. Algebraic representations

a. Translate between different representations of linear expressions using symbols, graphs, tables, diagrams, or written descriptions.

b. Analyze or interpret linear relationships expressed in symbols, graphs, tables, diagrams, or written descriptions.

c. Graph or interpret points that are represented by ordered pairs of numbers on a rectangular coordinate system.

d. Solve problems involving coordinate pairs on the rectangular coordinate system.

e. Make, validate, and justify conclusions and generalizations about linear relationships.

g. Identify or represent functional relationships in meaningful contexts including proportional, linear, and common non-linear (e.g., compound interest, bacterial growth) in tables, graphs, words, or symbols.

3. Variables, expressions, and operations

a. Write algebraic expressions, equations, or inequalities to represent a situation.

b. Perform basic operations, using appropriate tools, on linear algebraic expressions (including grouping and order of multiple operations involving basic operations, exponents, roots, simplifying, and expanding).

4. Equations and inequalities

a. Solve linear equations or inequalities (e.g., $ax + b = c$ or $ax + b = cx + d$ or $ax + b > c$).

b. Interpret "=" as an equivalence between two expressions and use this interpretation to solve problems.

c. Analyze situations or solve problems using linear equations and inequalities with rational coefficients symbolically or graphically (e.g., $ax + b = c$ or $ax + b = cx + d$).

d. Interpret relationships between symbolic linear expressions and graphs of lines by identifying and computing slope and intercepts (e.g., know in $y = ax + b$, that a is the rate of change and b is the vertical intercept of the graph).

e. Use and evaluate common formulas [e.g., relationship between a circle's circumference and diameter ($C = \pi d$), distance and time under constant speed].

Table of Measures

METRIC	CUSTOMARY
Length	
1,000 millimeters (mm) = 1 meter (m)	1 foot (ft) = 12 inches (in.)
100 centimeters (cm) = 1 meter	1 yard (yd) = 36 inches
10 decimeters (dm) = 1 meter	1 yard = 3 feet
1 kilometer (km) = 1,000 meters	1 mile (mi) = 5,280 feet
	1 mile = 1,760 yards
Capacity	
1,000 milliliters (mL) = 1 liter (L)	1 cup (c) = 8 fluid ounces (fl oz)
100 centiliters (cL) = 1 liter	1 pint (pt) = 2 cups
10 deciliters (dL) = 1 liter	1 quart (qt) = 2 pints
1 kiloliter (kL) = 1,000 liters	1 quart = 4 cups
	1 gallon (gal) = 4 quarts
	1 gallon = 16 cups
Mass/Weight	
1,000 milligrams (mg) = 1 gram (g)	1 pound (lb) = 16 ounces (oz)
100 centigrams (cg) = 1 gram	1 ton (T) = 2,000 pounds
10 decigrams (dg) = 1 gram	
1 kilogram (kg) = 1,000 grams	
1 tonne (t) = 1,000 kilograms	

TIME

1 minute (min) = 60 seconds (s)	1 year (yr) = 12 months (mo)
1 hour (hr) = 60 minutes	1 year = 365 days
1 day = 24 hours	1 leap year = 366 days
1 week (wk) = 7 days	